ENVIRONMENTAL ORGANIC CHEMISTRY

Second Edition

ENVIRONMENTAL ORGANIC CHEMISTRY

Second Edition

Rene P. Schwarzenbach
Philip M. Gschwend
Dieter M. Imboden

WILEY-INTERSCIENCE

A JOHN WILEY & SONS, INC., PUBLICATION

Library of Congress Cataloging-in-Publication Data Is Available

ISBN 0-471-35750-2

Printed in the United States of America.

10 9 8 7 6 5 4

Contents

Preface   xi

PART I
Introduction   1

    1 General Topic and Overview   3

        1.1   Introduction / 4

        1.2   Assessing Organic Chemicals in the Environment:
               The Needs and the Tasks / 6

        1.3   What Is This Book All About? / 8

    2 An Introduction to Environmental Organic Chemicals   13

        2.1   Introduction / 14

        2.2   The Makeup of Organic Compounds / 14

        2.3   Classification, Nomenclature, and Examples of
               Environmental Organic Chemicals / 31

        2.4   Questions and Problems / 51

PART II
Equilibrium Partitioning Between Gaseous, Liquid,
and Solid Phases   55

    3 Partitioning: Molecular Interactions and Thermodynamics   57

        3.1   Introduction / 59

        3.2   Molecular Interactions Determining the Partitioning
               of Organic Compounds Between Different Phases / 59

        3.3   Using Thermodynamic Functions to Quantify
               Molecular Energies / 73

        3.4   Using Thermodynaimc Functions to Quantify Equilibrium
               Partitioning / 84

        3.5   Using Partition Constants/Coefficients to Assess the
               Equilibrium Distribution of Neutral Organic Compounds in
               Multiphase Systems / 93

        3.6   Questions and Problems / 95

vi

4 Vapor Pressure   **97**

4.1 Introduction / 98

4.2 Theoretical Background / 98

4.3 Molecular Interactions Governing Vapor Pressure / 110

4.4 Availability of Experimental Vapor Pressure Data and Estimation Methods / 118

4.5 Questions and Problems / 126

5 Activity Coefficient and Solubility in Water   **133**

5.1 Introduction / 135

5.2 Thermodynamic Considerations / 135

5.3 Molecular Interpretation of the Excess Free Energy of Organic Compounds in Aqueous Solutions / 142

5.4 Effect of Temperature and Solution Composition on Aqueous Solubility and Activity Coefficient / 154

5.5 Availability of Experimental Data; Methods for Estimation of Aqueous Activity Coefficient and Aqueous Solubility / 172

5.6 Questions and Problems / 175

6 Air–Organic Solvent and Air–Water Partitioning   **181**

6.1 Introduction / 182

6.2 Thermodynamic Considerations / 183

6.3 Air–Organic Solvent Partitioning / 185

6.4 Air–Water Partitioning / 197

6.5 Questions and Problems / 208

7 Organic Liquid–Water Partitioning   **213**

7.1 Introduction / 214

7.2 Thermodynamic Considerations / 214

7.3 Comparison of Different Organic Solvent–Water Systems / 216

7.4 The *n*-Octanol–Water Partition Constant / 223

7.5 Dissolution of Organic Compounds in Water from Organic Liquid Mixtures—Equilibrium Considerations (Advanced Topic) / 235

7.6 Questions and Problems / 239

8 Organic Acids and Bases: Acidity Constant and Partitioning Behavior   **245**

8.1 Introduction / 246

8.2 Thermodynamic Considerations / 246

8.3 Chemical Structure and Acidity Constant / 256

8.4   Availability of Experimental Data; Methods for
Estimation of Acidity Constants / 261

8.5   Aqueous Solubility and Partitioning Behavior of Organic
Acids and Bases / 268

8.6   Questions and Problems / 272

**9   Sorption I: General Introduction and Sorption Processes
Involving Organic Matter   275**

9.1   Introduction / 277

9.2   Sorption Isotherms, Solid–Water Distribution
Coefficients (K_{id}), and the Fraction Dissolved (f_{iw}) / 280

9.3   Sorption of Neutral Organic Compounds from Water
to Solid-Phase Organic Matter (POM) / 291

9.4   Sorption of Neutral Compounds to "Dissolved"
Organic Matter (DOM) / 314

9.5   Sorption of Organic Acids and Bases to Natural
Organic Matter (NOM) / 321

9.6   Questions and Problems / 326

**10   Sorption II: Partitioning to Living Media – Bioaccumulation
and Baseline Toxicity   331**

10.1   Introduction / 333

10.2   Partitioning to Defined Biomedia / 335

10.3   Bioaccumulation in Aquatic Systems / 349

10.4   Bioaccumulation in Terrestrial Systems / 361

10.5   Biomagnification / 366

10.6   Baseline Toxicity (Narcosis) / 374

10.7   Questions and Problems / 381

**11   Sorption III: Sorption Processes Involving
Inorganic Surfaces   387**

11.1   Introduction / 389

11.2   Adsorption of Nonionic Organic Compounds to
Inorganic Surfaces from Air / 391

11.3   Sorption of Nonionic Organic Compounds to Inorganic
Surfaces in Water / 408

11.4   Adsorption of Ionized Organic Compounds from
Aqueous Solutions to Charged Mineral Surfaces / 417

11.5   Surface Reactions of Organic Compounds *(Advanced Topic)* / 441

11.6   Questions and Problems / 448

Part III
Transformation Processes   459

12 Thermodynamics and Kinetics of Transformation Reactions   461

12.1 Introduction / 462

12.2 Thermodynamics of Transformation Reactions / 463

12.3 Kinetic Aspects of Transformation Reactions / 468

12.4 Well-Mixed Reactor or One-Box Model / 482

12.5 Questions and Problems / 486

**13 Chemical Transformations I: Hydrolysis and Reactions
Involving Other Nucleophilic Species   489**

13.1 Introduction, Overview / 491

13.2 Nucleophilic Substitution and Elimination
at Saturated Carbon Atoms / 495

13.3 Hydrolytic Reactions of Carboxylic and
Carbonic Acid Derivatives / 513

13.4 Hydrolytic Reactions of Phosphoric and Thiophosphoric
Acid Esters / 536

13.5 Effects of Dissolved Metal Species and of Mineral Oxide
Surfaces on Hydrolytic Reactions (*Advanced Topic*) / 540

13.6 Questions and Problems / 546

14 Chemical Transformations II: Redox Reactions   555

14.1 Introduction, Overview / 556

14.2 Thermodynamic Considerations of Redox Reactions / 559

14.3 Reaction Pathways and Kinetics of Redox Reactions / 580

14.4 Questions and Problems / 602

15 Direct Photolysis   611

15.1 Introduction / 613

15.2 Some Basic Principles of Photochemistry / 614

15.3 Light Absorption by Organic Chemicals in Natural Waters / 627

15.4 Quantum Yield and Rate of Direct Photolysis / 641

15.5 Effects of Solid Sorbents (Particles, Soil Surfaces) on
Direct Photolysis / 649

15.6 Questions and Problems / 650

**16 Indirect Photolysis: Reactions with Photooxidants
in Natural Waters and in the Atmosphere   655**

16.1 Introduction / 656

16.2 Indirect Photolysis in Surface Waters / 658

16.3 Indirect Photolysis in the Atmosphere (Troposphere)—
Reactions with Hydroxyl Radical (HO$^{\bullet}$) / 672

16.4 Questions and Problems / 683

17 Biological Transformations   **687**

17.1 Introduction / 689

17.2 Some Important Concepts about Microorganisms / 694

17.3 Biochemical Strategies of Microbial Organic Chemists / 702

17.4 Rates of Biotransformations: Uptake / 734

17.5 Rates of Biotransformations: Microbial Growth / 739

17.6 Rates of Biotransformations: Enzymes / 750

17.7 Questions and Problems / 767

PART IV
Modeling Tools: Transport and Reaction   **775**

18 Transport by Random Motion   **777**

18.1 Introduction: Advection and Diffusion / 779

18.2 Random Motion / 780

18.3 Random Motion at the Molecular Level:
Molecular Diffusion Coefficients / 798

18.4 Diffusion in Porous Media / 815

18.5 Other Random Transport Processes in the Environment / 825

18.6 Questions and Problems / 828

19 Transport Through Boundaries   **833**

19.1 The Role of Boundaries in the Environment / 835

19.2 Bottleneck Boundaries / 839

19.3 Wall Boundaries / 848

19.4 Diffusive Boundaries / 866

19.5 Spherical Boundaries *(Advanced Topic)* / 871

19.6 Questions and Problems / 883

20 Air–Water Exchange   **887**

20.1 Introduction / 889

20.2 Measurement of Air–Water Transfer Velocities / 896

20.3 Air–Water Exchange Models / 906

20.4 Air–Water Exchange in Flowing Waters / 921

20.5 Influence of Surface Films and Chemical Reactions
on Air–Water Exchange *(Advanced Topic)* / 929

20.6 Questions and Problems / 939

21 Box Models   **945**

   21.1 Principles of Modeling / 947

   21.2 One-Box Models / 955

   21.3 Two-Box Models / 982

   21.4 Dynamic Properties of Linear Multidimensional Models
(Advanced Topic) / 991

   21.5 Questions and Problems / 1000

22 Models in Space and Time   **1005**

   22.1 One-Dimensional Diffusion/Advection/Reaction Models / 1006

   22.2 Turbulent Diffusion / 1019

   22.3 Horizontal Diffusion: Two–Dimensional Mixing / 1030

   22.4 Dispersion (*Advanced Topic*) / 1038

   22.5 Questions and Problems / 1044

**Part V
Environmental Systems and Case Studies**   **1049**

23 Ponds, Lakes, and Oceans   **1051**

   23.1 Linear One-Box Models of Lakes, Ponds, and Oceans / 1054

   23.2 The Role of Particles and the Sediment–Water Interface / 1059

   23.3 Two-Box Models of Lakes / 1075

   23.4 One-Dimensional Continuous Lake Models (*Advanced Topic*) / 1082

   23.5 Questions and Problems / 1093

24 Rivers   **1101**

   24.1 Transport and Reaction in Rivers / 1102

   24.2 Turbulent Mixing and Dispersion in Rivers / 1120

   24.3 A Linear Transport/Reaction Model for Rivers / 1130

   24.4 Questions and Problems / 1141

25 Groundwater   **1147**

   25.1 Groundwater Hydraulics / 1148

   25.2 Time-Dependent Input into an Aquifer (*Advanced Topic*) / 1160

   25.3 Sorption and Transformations / 1170

   25.4 Questions and Problems / 1179

Appendix   **1185**

Bibliography   **1213**

Index (Subject Index, Compound Index, List of Illustrative Examples)   **1255**

PREFACE

"Don't worry, we will never do it again!"
This is the promise we sincerely made almost 10 years ago to our families, friends, and colleagues after having survived together the writing of the first edition of our textbook *Environmental Organic Chemistry*, and made once more after finishing the companion *Problems Book* two years later. But keeping such promises and keeping up with this rapidly expanding, exciting field of environmental sciences seem to be two things that are mutually exclusive. Hence, with fading memories of what it was really like, and flattered by the success of the first edition of our textbook, we decided to take on the challenge again; maybe at first not realizing that we have grown older and, as a consequence, that our professional lives have become much more diverse and busy than they used to be. Furthermore, what began as a minor revision and updating of the first edition soon developed its own dynamics, completely overturned old chapters and created new ones. During this process it became clear to us that the integration of the *Problems Book* with its two additional system chapters on rivers and groundwater into the main book would shift the gravity of the new edition toward the system approach, however, not at the expense of the fundamental chemical principles, but by adding more physics and mathematical modeling. This is now the product of four years of struggling with an immense amount of recent literature, as well as of continuously suffering from being on the horns of a dilemma; that is, the attempt to provide a fundamental text combining background theory, illustrative examples, and questions and problems, and, at the same time, to give a state-of-the-art account of a rather broad and interdisciplinary field. However, it would be completely wrong to view the writing of this second edition solely as an ordeal; on the contrary, particularly the many exciting discussions with numerous students and colleagues have been very rewarding and most enjoyable. We hope that some of this joy will also be felt by our readers.

What is this book all about? Everything you ever needed to know for assessing the environmental behavior of organic chemicals and more? Not quite, but we hope a great deal of it, and certainly more than in the first edition. As in the first edition, our major goal is to provide an understanding of how molecular interactions and macroscopic transport phenomena determine the distribution in space and time of organic compounds released into natural and engineered environments. We hope to do this by teaching the reader to utilize the *structure* of a given chemical to deduce that chemical's intrinsic *physical properties* and *reactivities*. Emphasis is placed on *quantification* of phase transfer, transformation, and transport processes at each level. By first considering each of the processes that act on organic chemicals one at a time, we try to build bits of knowledge and understandings that later in the book are combined in mathematical models to assess organic compound behavior in the environment.

Who should read and use this book, or at least keep it on their bookshelf? From our experience with the first edition, and maybe still with a little bit of wishful thinking, we are inclined to answer this question with "Everybody who has to deal with organic pollutants in the environment". More specifically, we believe that the theoretical explanations and mathematical relationships discussed are very useful for *chemistry* professors and students who want both fundamental explanations and concrete applications that the students can use to remember those chemical principles. Likewise, we suggest that *environmental and earth science* professors and their students can utilize the chemical property information and quantitative descriptions of chemical cycling to think about how humans are playing an increasingly important role in changing the Earth system and how we may use specific chemicals as tracers of environmental processes. Further, we believe that *civil and environmental engineering* professors and students will benefit from detailed understanding of the fundamental phenomena supporting existing mitigation and remedial designs, and they should gain insights that allow them to invent the engineering approaches of the future. *Environmental policy and management* professors and students should also benefit by seeing our capabilities (and limitations) in estimating chemical exposures that result from our society's use of chemicals. Finally, *chemists and chemical engineers in industry* should be able to use this book's information to help make "green chemistry" decisions, and *governmental regulators and environmental consultants* should use the book to be better able to analyze the problem sites they must assess and manage.

To meet the needs of this very diverse audience, we have tried, wherever possible, to divide the various chapters or topics into more elementary and more advanced parts, hoping to make this book useful for beginners as well as for people with more expertise. At many points, we have tried to explain concepts from the very beginning level (e.g., chemical potential) so that individuals who do not recall (or never had) their basic chemistry can still develop insights into and understand the origin and limits of modeling calculations and correlation equations. We have also incorporated numerous references throughout the text to help people who want to follow particular topics further. Finally, by including many illustrative examples, we have attempted to show environmental practitioners how to arrive at quantitative results for particular cases of interest to them. Hence, this book should serve as a text for introductory courses in environmental organic chemistry, as well as a source of information for hazard and risk assessment of organic chemicals in the environment. We hope that with this textbook, we can make a contribution to the education of environmental scientists and engineers and, thus, to a better protection of our environment.

Acknowledgments. Those who have ever written textbooks know that the authors are not the only ones who play an important role in the realization of the final product. Without the help of many of our co-workers, colleagues and students, it would have taken another millennium to finish this book. We thank all of them, but above all Béatrice Schwertfeger who, together with Lilo Schwarz and Cécile Haussener produced the entire camera-ready manuscript. Furthermore, we acknowledge Toni Bernet for his professional help with the final layout, and Sabine Koch for helping in assembling the reference list. We are especially grateful to Dieter Diem for reviewing the whole manuscript and, particularly, for producing the compound and subject indices. Another key role was played by Werner Angst who drew most of the more complicated structures and reaction schemes, and who helped with the compound index and with thousands of small details. Furthermore, we are particularly indepted to Kai-Uwe Goss, whose significant input into Part II of the book and whose review of several other chapters are especially acknowledged. Many important comments and criticisms were made by other colleagues and students including Andreas Kappler and Torsten Schmidt who reviewed Parts I to III, Mike McLachlan (Part II), Stefan Haderlein (Chapters 9 to 11), Beate Escher and Zach Schreiber (Chapter 10), Lynn Roberts (Chapters 13 and 14), Martin Elsner, Luc Zwank and Paul Tratnyek (Chapter 14), Andrea Ciani and Silvio Canonica (Chapters 15 and 16), Werner Angst, Colleen Cavanaugh, Hans Peter Kohler, Rainer Meckenstock and Alexander Zehnder (Chapter 17), and Frank Peeters (Part IV). We are also deeply indepted to the Swiss Federal Institute of Environmental Sciences and Technology (EAWAG) and the Swiss Federal Institute of Technology in Zurich (ETHZ) for significant financial support, which made it possible to produce a low-cost textbook. Finally with no further promises but with some guilty feelings, we thank our families for their patience and support, particularly our wives Theres Schwarzenbach, Colleen Cavanaugh, and Sibyl Imboden who will hopefully still recognize their husbands after what again must have seemed an endless preoccupation with THE NEW BOOK!

René P. Schwarzenbach
Dübendorf and Zürich, Switzerland

Philip M. Gschwend
Cambridge, Massachussetts, USA

Dieter M. Imboden
Zürich, Switzerland

Part I

Introduction

Chapter 1

GENERAL TOPIC AND OVERVIEW

1.1   **Introduction**

1.2   **Assessing Organic Chemicals in the Environment: The Needs and the Tasks**

1.3   **What Is This Book All About?**
The (Impossible?) Goals of the Book
A Short Guide to the Book
The "Zoo" of Symbols, Subscripts, and Superscripts:
     Some Remarks on Notation

1.1   Introduction

For many decades now, human society has purposefully released numerous synthetic organic chemicals to our environment in an effort to control unwanted organisms such as weeds, insect pests, rodents, and pathogens. For example, DDT was sprayed more than 50 years ago to control mosquitoes (Storer, 1946). This insecticide application was so successful in reducing the incidence of malaria that DDT was once called one of "life's great necessities" (Scientific American, 1951). Similarly, other biocides like pentachlorophenol (PCP, a molluscicide for control schistosomiasis and an industrial disinfectant; Weinbach, 1957) or tributyl tin (used to inhibit fouling on boat hulls) have proven extremely effective remedies for problems of humankind caused by other organisms on earth. In all of these cases, we intentionally introduce organic chemicals into our environment.

Meanwhile, many other chemicals have enabled our society to accomplish great technical advances. For example, we have learned to recover fossil hydrocarbons from the earth and use these for heating, for transportation fuels, and for synthetic starting materials. Likewise, synthetic compounds like tetraethyllead, chlorinated solvents, freons, methyl t-butyl ether (MTBE), polychlorinated biphenyls (PCBs), and many others (see Chapter 2) have enabled us to develop products and perform industrial processes with greater efficiencies and safety. However, it has become quite apparent that even such "contained" applications always result in a certain level of discharge of these compounds to the environment.

In retrospect, it is not surprising to see that quite a large portion of these synthetic chemicals have caused many problems. It was found that the biocides that were aimed at particular target organisms also harmed nontarget organisms. For example, very soon after its initial use, DDT was found to affect fish such as lake trout (Surber, 1946; Burdick et al., 1964). Moreover, chemicals used in one geographic locale were seen to disperse widely (see example given in Fig. 1.1) and appear in the tissues of various plants and animals (e.g., PCBs in British wildlife, Holmes et al., 1967; PCBs in Dutch fish, shellfish, and birds, Koeman et al., 1969). Perhaps still more startling was the recognition that some of those persistent nonbiocide chemicals also caused ill health in some organisms. For example, bio-uptake of PCBs to toxic levels has now been recognized for many years (Risebrough et al., 1968; Gustafson, 1970).

Considering these early lessons, and considering present society's continuously expanding utilization of materials, energy, and space accompanied by an increasing use of anthropogenic organic chemicals, it seems obvious that the contamination of water, soil, and air with such compounds will continue to be a major issue in environmental protection. Note that we term chemicals *anthropogenic* if they are introduced into the environment primarily or exclusively as a consequence of human activity.

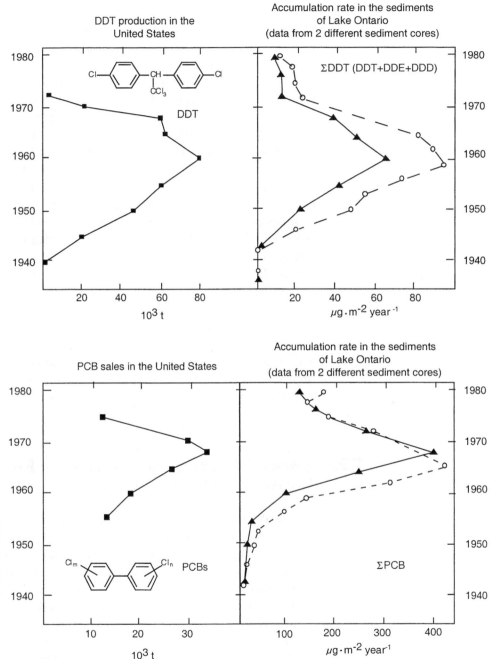

Figure 1.1 Historical records of the sales/production volumes of (*a*) DDT and (*b*) PCBs, and the similarity of these time-varying trends to the accumulation rates of these chemicals in the sediments of Lake Ontario (from Eisenreich et al., 1989).

When addressing the issue of anthropogenic organic chemicals in the environment, one often tends to emphasize the consequences of spectacular accidents or the problems connected with hazardous waste management (e.g., waste water treatment, waste incineration, and dump sites). These are certainly very significant problems. But of at least equivalent importance is the chronic contamination of the environment due to the use of chemicals. According to the Organization of Economic Cooperation and Development (OECD) the global annual industrial production of synthetic chemicals and materials exceeds 300 million tons. Furthermore, there are presently over 100,000 (mostly organic) synthetic chemicals in daily use, and this number increases continuously.

Although some of these everyday chemicals are not of direct environmental concern, numerous compounds are continuously introduced into the environment in large quantities (e.g., solvents, components of detergents, dyes and varnishes, additives in plastics and textiles, chemicals used for construction, antifouling agents, herbicides, insecticides, and fungicides). Furthermore, there are many biologically active compounds such as hormones and antibiotics used in human and veterinary applications, that may already raise concern when introduced into a given ecosystem at comparably low quantities (see Chapter 2). Hence, in addition to problems related to accidents and waste management, a major present and future task encompasses identification and possibly replacement of those widely used synthetic chemicals that may present unexpected hazard to us. Furthermore, new chemicals must be environmentally compatible—we must ensure that these compounds do not upset important processes and cycles of ecosystems. All of these tasks require knowledge of (1) the processes that govern the transport and transformations of anthropogenic chemicals in the environment and (2) the effects of such chemicals on organisms (including humans), organism communities, and whole ecosystems. The first topic is the theme of this book. Our focus is on anthropogenic *organic* chemicals, and we discuss these primarily from the perspective of *aquatic* environments: groundwaters, streams and rivers, ponds and lakes, and estuaries and oceans. We note, however, that the ubiquity of water on Earth and its interactions with soils, sediment beds, organisms, and the atmosphere implies that understanding chemical fates in aquatic realms closely corresponds to the delineation of their fates in the environment as a whole.

1.2   Assessing Organic Chemicals in the Environment: The Needs and the Tasks

Organic chemicals that are introduced into the environment are subjected to various physical, chemical, and biological processes. These processes act in an interconnected way in environmental systems to determine the overall fate of the compound (e.g., in a lake, Fig. 1.2). They can be divided into two major categories: processes that leave the structure of a chemical (i.e., its "identity") unchanged, and those that transform the chemical into one or several products of different environmental behavior and effect(s). The first category of processes includes transport and mixing phenomena within a given environmental compartment (e.g., in a water body) as well as transfer processes between different phases and/or compartments (e.g., water–air exchange, sorption and sedimentation, sediment–water exchange, uptake by organisms). The second type of processes leads to alterations of the structure of a compound. It includes chemical, photochemical, and/or biological (above all microbial) transformation reactions. It is important to recognize that, in a given environmental system, all of these processes may occur simultaneously, and, therefore, different processes may strongly influence each other.

When confronted with any practical question concerning the environmental behavior of an organic chemical, one obviously needs to be able to quantify each of the individual processes occurring in the system considered. Quite often it may be relatively easy to

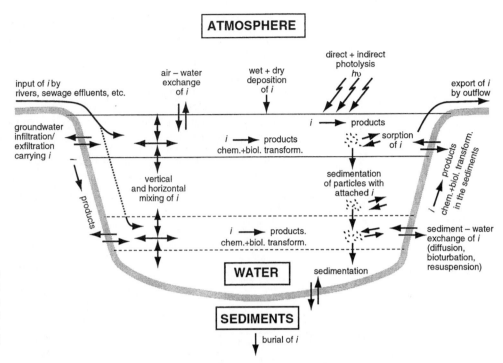

ATMOSPHERE

input of *i* by
rivers, sewage effluents, etc.

air – water
exchange
of *i*

wet + dry
deposition
of *i*

direct + indirect
photolysis
hv

export of *i*
by outflow

i ⟶ products

groundwater
infiltration/
exfiltration
carrying *i*

i ⟶ products
chem.+biol. transform.

sorption
of *i*

products
chem.+biol. transform.
in the sediments

products

vertical
and horizontal
mixing of *i*

sedimentation
of particles with
attached *i*

i ⟶ products.
chem.+biol. transform.

sediment – water
exchange of *i*
(diffusion,
bioturbation,
resuspension)

WATER

sedimentation

SEDIMENTS

burial of *i*

Figure 1.2 Processes that determine the distribution, residence time, and sinks of an organic chemical *i* in a lake. This example illustrates the various physical, chemical, and biological processes that a compound is subjected to in the environment.

identify those processes that are not relevant in a given situation. For example, by inspecting the physicochemical properties of the compound, we may immediately conclude that sorption to particles and sedimentation in a lake is not important. Or from looking at the structure of the chemical, we may disregard hydrolysis (i.e., reaction with water) as a relevant transformation reaction. In any case, whether a process turns out to be important or not, we have to be able to quantify all relevant compound-specific and system-specific parameters that are required to describe this process (Fig. 1.3). To this end, we need to develop a feeling of how chemical structures cause the molecular interactions that govern the various transfer and reaction processes. We should stress that without an understanding of the molecular level, a sound assessment of the environmental behavior of organic compounds is not possible. On the other hand, this knowledge of basic chemistry is far from enough to cope with the complexity of environmental systems. Hence, we also need to learn how to quantify all relevant environmental factors (Fig. 1.3) required to describe a particular process. Finally, in order to be able to understand and evaluate the dynamic behavior of a given compound in the environment, we have to acquaint ourselves with the basis principles of transport and mixing phenomena, and we have to learn how to use models of appropriate complexity to evaluate and describe the interplay between all the physical, chemical, and biological processes occurring in a given system.

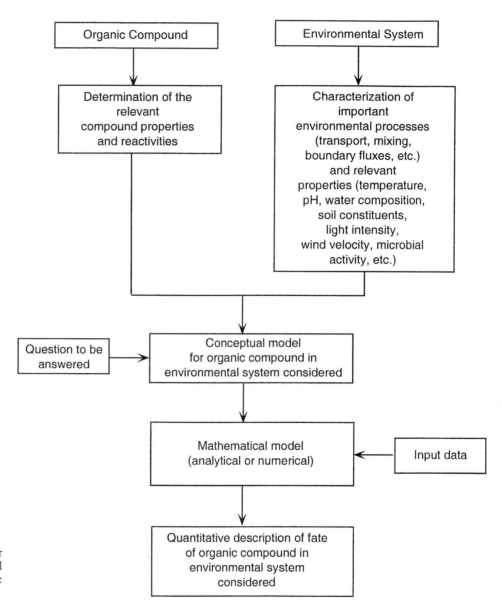

Figure 1.3 General scheme for evaluation of the environmental behavior of anthropogenic organic compounds.

What Is This Book All About?

The (Impossible?) Goals of This book

Considering the needs and tasks discussed above, the reader may wonder whether it is reasonable, or even at all possible, to cover such a broad area including environmental chemistry, environmental physics, and (mathematical) modeling in a single textbook. Further, besides being intended as a textbook for students in environmental sciences and engineering, earth sciences, chemistry, and physics, this book should serve as a reference for practitioners who need to solve "real world problems". And, last but not least, it should characterize of the state of the art of the field of environmental organic chemistry at the beginning of this century.

This all sounds overambitious, and, in a sense, of course, it is. Nevertheless, we strongly feel that an integral view of the whole field, from the microscopic scale of molecular interactions up to the macroscopic scale of whole environmental system dynamics, is necessary for a sound assessment of organic chemicals in the environment. Therefore, this book should be considered as an attempt to introduce and integrate the most important aspects of all relevant topics, but not an exhaustive treatment of particular subjects. For readers who want to pursue certain topics in greater detail, numerous literature citations have been included (although this is unusual for a textbook). Furthermore, basic principles are emphasized and simplified pictures are sometimes used to help the less experienced reader to enhance her or his intuitive perception of a given process. In order to help bring together theory and practice, we have also introduced numerous illustrative examples. These show the reader the step-by-step elements of some common types of calculations used in this field. Also, each chapter ends with a set of questions to point to the most important aspects of a chapter and to inspire qualitative discussions, and problems that allow teachers to explore the depth of understanding of their students. Finally, we have chosen to write the text in a somewhat colloquial style to enhance the palatability of the fundamental discussions; we hope that the professionals among our readers will make an allowance for this effort to teach.

A Short Guide to the Book

The book is divided into five parts. We progressively work our way from primarily compound-related aspects (intrinsic compound properties and reactivities) up to whole environmental system considerations. In the second introductory chapter (Chapter 2), we turn our attention to the main actors of this book: *anthropogenic organic chemicals*. We review some terminology and basic concepts used in organic chemistry, and we take a glimpse at the structures of several different important classes of environmental organic chemicals.

Part II is devoted to *equilibrium* considerations of the partitioning of organic chemicals between gaseous, liquid, and solid phases. In Chapter 3 we address the molecular interactions that determine the partitioning behavior of organic chemicals, and we review the most important thermodynamic concepts used to describe partitioning. An important goal of Chapter 3 is to set the stage for building a conceptual framework that allows us to treat partitioning processes in a holistic way. In the consecutive chapters, we then try to further develop this framework step by step. We start out in Chapter 4 with a discussion of *vapor pressure*, which is a direct measure of the forces between the compound's molecules when present in the pure condensed form of the chemical. In Chapter 5, we tackle a quite complicated subject; that is, we try to understand what structural features determine how much an organic chemical likes (or dislikes) to be dissolved in water. We see how the *aqueous activity coefficient* of a compound and its *water solubility* are interrelated, and how they are influenced by the presence of other water constituents including dissolved salts and organic cosolvents. In Chapters 6 and 7 we then evaluate the *air–liquid* and *organic liquid–water partitioning* of organic compounds. Both chapters build directly upon the concepts derived in Chapters 3 to 5. The discussion of basic aspects of partitioning processes is concluded in Chapter 8, which deals with compounds that may undergo proton transfer reactions in aqueous solution: *organic acids and bases*. Because ionic organic molecules show a different partitioning behavior than their neutral counterparts, the ability to deduce the fraction of a chemical present as an ionized species is necessary to predict phase equilibria of organic acids and bases.

Armed with the basic knowledge and insights acquired in Chapters 3 through 8, we then deal with partitioning processes involving more complex and environmentally more relevant condensed phases including *natural organic matter* (Chapter 9), *organisms* (Chapter 10), and *inorganic solids* (Chapter 11). As we will learn in these chapters, associations with these natural "sorbents" are pivotal to the transport, distribution, and fate, as well as for the effects of organic chemicals in the environment.

Part III is devoted to abiotic and biological *transformation processes*. For our discussions we divide these processes into three major categories, i.e., *chemical* (Chapters 13 and 14), *photochemical* (Chapters 15 and 16), and *biologically mediated* (Chapter 17) transformation and/or degradation reactions. In Chapter 12, we review some of the basic thermodynamic and kinetic concepts needed for Chapters 13 through 17, and we introduce the simplest mass balance model: the well-mixed reactor or one-box model. Chapter 13 deals primarily with reactions of organic chemicals with *nucleophiles*, in particular, with water (*hydrolysis*). In Chapter 14, we discuss *redox reactions* of organic chemicals and we address some of the most important biogeochemical processes that determine the redox conditions in a given environment (e.g., in soils, aquifers, sediments, landfills, and hazardous waste sites).

In Chapter 15 we address the consequences of the *direct interaction* of organic compounds *with sunlight*. This also forces us to evaluate the light regime in natural systems, in particular, in surface waters. Chapter 16 then deals with reactions of organic chemicals with photochemically produced reactive species (*photooxidants*) in surface waters and in the atmosphere. Note that in Chapters 15 and 16, the focus is on quantification of these processes rather than on a discussion of reaction pathways.

We conclude Part III with Chapter 17, in which some aspects of microbial transformations of anthropogenic compounds in the environment are addressed. The aim of this chapter is to provide some insights into the strategies applied by microorganisms to break down xenobiotic ("foreign to organisms") organic chemicals and to demonstrate concepts useful in quantifying such microbial transformation reactions.

Tired of so much chemistry , some readers may now turn with pleasure to Part IV, in which transport and mixing phenomena are explained. Furthermore, Part IV provides the conceptual and mathematical framework for building models for the quantitative description of the dynamic behavior of organic chemicals in environmental systems.

Part IV is organized in the following way: Chapter 18 gives an overview of transport phenomena in the environment by grouping them into just two categories: *directed transport* and *random processes*. While directed transport (advection, transport under the influence of gravitation, etc.) will be treated in detail in Chapter 22, the discussion in Chapter 18 focuses on transport by randomness. We start with different kinds of diffusion phenomena, discuss Fick's laws, introduce the concept of molecular diffusion, and end with a first glance at turbulent transport. Chapter 19 deals with transport at boundaries. Here we choose a novel approach by classifying the enormous variety of boundaries in natural systems as just one of three types: *bottleneck, wall, or diffusive boundaries*. In case the reader considers this systematic treatment of boundary processes laborious, we hope that she or he will later enjoy the ease with which such diverse problems as air–water exchange, the dilution of a pollutant patch by dispersion, or the dynamics of sorption on particles in water can be understood based on the more fundamental things learned earlier. Chapter 20 is devoted to one very important boundary in the environment, the *air–water interface*.

The remaining two chapters of Part IV set the basis for the more advanced environmental models discussed in Part V. Chapter 21 starts with the simple one-box model already discussed at the end of Chapter 12. One- and two-box models are combined with the different boundary processes discussed before. Special emphasis is put on linear models, since they can be solved analytically. Conceptually, there is only a small step from multibox models to the models that describe the spatial dimensions as continuous variables, although the step mathematically is expensive as the model equations become partial differential equations, which, unfortunately, are more complex than the simple differential equations used for the box models. Here we will not move very far, but just open a window into this fascinating world.

Finally, in **Part V** we present some *case studies* aimed at illustrating how to combine all the theories and concepts developed throughout the book. The environmental systems that we have chosen to do this include lakes (Chapter 23), rivers (chapter 24), and aquifers (chapter 25). These cases will also demonstrate how far one can go with simple models that do not need a large computer but just rely on the mathematical understanding of the user and perhaps on a simple pocket calculator to get quantitative results.

The "Zoo" of Symbols, Subscripts, and Superscripts: Some Remarks on Notation

This book combines information from a wide spectrum of disciplinary cultures, each having its own nomenclature and rules for how to express certain physical or chemical quantities. For example, in basic chemistry, particularly in physical chemistry, the rules are tough and everything is strictly regulated, whereas in physics, freedom of choice is rather large. Hence, one dilemma that we have to cope with in this book is to satisfy all these different worlds.

When considering the numerous compound- and system-specific properties, parameters and factors used for the assessment of organic compounds in the environment, it is virtually impossible to use a consistent, throughout-the-book, always-to-the-last-detail unambiguous symbolism. Of course, we could come close to perfectionism by introducing a large set of different symbols carrying numerous subscripts and superscripts. Probably only the perfectionists among the readers would appreciate such an undertaking. Thus, in order to find a compromise, we have chosen a somewhat pragmatic way in that we try to be strictly consistent only within parts of the book that are closely interrelated. Thus, for example, in Part II, where we talk about the partitioning of a given compound between various different phases, we use a subscript *i* for identifying those quantities that are compound-specific. The goal is to familiarize the reader with these quantities in order to help him or her to learn to distinguish them from system properties. Later in the book, particularly in Parts IV and V, where numerous mathematical expressions are used, we generally abandon this convention unless it is required for clarification. The reason is not to overload the mathematical expressions with unnecessary subscripts.

Another problem that the reader has to live with is that certain symbols (e.g., α, λ, f) are used to denote several different quantities. The reason is that in the different disciplines, these symbols are commonly used for a given purpose, and we have sought to maintain the same nomenclature as much as possible.

In summary, this textbook has been designed to acquaint the reader with the basic principles of organic compound behavior in the environment and to provide conceptual tools and pertinent information necessary to evaluate and describe quantitatively the dynamics of anthropogenic organic chemicals in natural systems. Special emphasis is placed on the *interrelationship between chemical structure and environmental behavior* of organic compounds. The information contained in this book has been collected from many areas of basic and applied science and engineering, including chemistry, physics, biology, geology, limnology, oceanography, pharmacology, agricultural sciences, atmospheric sciences, and hydrogeology, as well as chemical, civil, and environmental engineering. This reflects the multidisciplinary approach that must be taken when studying the dynamics of organic compounds in the environment. However, there are still numerous gaps in our knowledge, which will become apparent in the text. It is, therefore, our hope that this book will motivate students to become active in this important field of research.

Chapter 2

An Introduction to Environmental Organic Chemicals

2.1   Introduction

2.2   The Makeup of Organic Compounds
Elemental Composition, Molecular Formula, and Molar Mass
Electron Shells of Elements Present in Organic Compounds
Covalent Bonding
Bond Energies (Enthalpies) and Bond Lengths. The Concept of Electro-
        negativity
Oxidation State of the Atoms in an Organic Molecule
Illustrative Example 2.1: *Determining the Oxidation States of the Carbon
        Atoms Present in Organic Molecules*
The Spatial Arrangement of the Atoms in Organic Molecules
Delocalized Electrons, Resonance, and Aromaticity

**2.3   Classification, Nomenclature, and Examples of Environmental
        Organic Chemicals**
The Carbon Skeleton of Organic Compounds: Saturated,
        Unsaturated, and Aromatic Hydrocarbons
Organohalogens
Oxygen-Containing Functional Groups
Nitrogen-Containing Functional Groups
Sulfur-Containing Functional Groups
Phosphorus-Containing Functional Groups
Some Additional Examples of Compounds Exhibiting More Complex
        Structures

2.4   Questions and Problems

2.1 Introduction

When confronted with the plethora of natural and man-made organic chemicals released into the environment (Blumer, 1975; Stumm et al., 1983), many of us may feel overwhelmed. How can we ever hope to assess all of the things that happen to each of the substances in this menagerie, encompassing so many compound names, formulas, properties, and reactivities? It is the premise of our discussions in this book that each chemical's *structure*, which dictates that compound's "personality," provides a systematic basis with which to understand and predict chemical behavior in the environment. Thus in order to quantify the dynamics of organic compounds in the macroscopic world (Parts IV and V), we will need to learn to visualize organic molecules in the microscopic environments in which they exist. However, before we do that in Parts II and III, we first need to refresh our memories with some of the terminology and basic chemical concepts of organic chemistry used throughout this book (Section 2.2). For readers with little background in organic chemistry, it may be useful to consult the introductory chapters of an organic chemistry textbook in addition to this section. On the other hand, professional chemists might want to continue directly with Section 2.3, where we will try to give an overview of some important groups of environmentally relevant organic chemicals.

2.2 The Makeup of Organic Compounds

To understand the nature and reactivity of organic molecules, we first consider the pieces of which organic molecules are made. This involves both the various atoms and the chemical bonds linking them. First, we note that most of the millions of known natural and synthetic (man-made) organic compounds are combinations of only a few elements, namely carbon (C), hydrogen (H), oxygen (O), nitrogen (N), sulfur (S), phosphorus (P), and the halogens fluorine (F), chlorine (Cl), bromine (Br), and iodine (I). The chief reason for the almost unlimited number of stable organic molecules that can be built from these few elements is the ability of carbon to form up to four stable carbon–carbon bonds. This permits all kinds of three-dimensional *carbon skeletons* to be made, even when the carbon atoms are also bound to *heteroatoms* (i.e., to elements other than carbon and hydrogen). Fortunately, despite the extremely large number of existing organic chemicals, knowledge of a few governing rules about the nature of the elements and chemical bonds present in organic molecules will already enable us to understand important relationships between the structure of a given compound and its properties and reactivities. These properties and reactivities determine the compound's behavior in the environment.

Elemental Composition, Molecular Formula, and Molar Mass

When describing a compound, we have to specify which elements it contains. This information is given by the *elemental composition* of the compound. For example, a chlorinated hydrocarbon, as the name implies, consists of chlorine, hydrogen, and carbon. The next question we then have to address is how many atoms of each of these elements are present in one molecule. The answer to that question is given by the *molecular formula*, for example, four carbon atoms, nine hydrogen atoms, and one chlorine atom: C_4H_9Cl. The molecular formula allows us to calculate the *molecular mass* of the compound, which is the sum of the masses of all atoms present in the molecule. The atomic masses of the elements of interest to us are given in Table 2.1. Note that for many of the elements, there exist naturally occurring stable *isotopes*. Isotopes are atoms that have the same number of protons and electrons (which determines their chemical nature), but different numbers of neutrons in the nucleus, thus giving rise to different atomic masses. Examples of elements exhibiting isotopes that have a significant natural abundance are carbon ($^{13}C:^{12}C = 0.011:1$), sulfur ($^{34}S:^{32}S = 0.044:1$), chlorine ($^{37}Cl:^{35}Cl = 0.32:1$), and bromine ($^{81}Br:^{79}Br = 0.98:1$). Consequently, the atomic masses given in Table 2.1 represent averaged values of the naturally occurring isotopes of a given element (e.g., average carbon is 1.1% at 13 u + 98.9% at 12 u = 12.011 u). Note that 1 u (unified atomic mass unit) is approximately equal to 1.6605×10^{-27} kg. Using these atomic mass values we obtain a molecular mass (or molecular weight) of 92.57 u for one single molecule with the molecular formula C_4H_9Cl. If we take the amount of 1 mole of pure substance (that is, the Avogadro's number $N_A = 6.02 \times 10^{23}$ identical units, here molecules), this amount weighs 92.57 grams and is named the *molar mass* of the (pure) substance. So, 1 mole (abbreviation 1 mol) of any (pure) substance always contains the same amount of molecules and its mass in grams is what each molecule's mass is in u.

Given the molecular formula, we now have to describe how the different atoms are connected to each other. The description of the exact connection of the various atoms is commonly referred to as the *structure* of the compound. Depending on the number and types of atoms, there may be many different ways to interconnect a given set of atoms which yield different structures. Such related compounds are referred to as *isomers*. Furthermore, as we will discuss later, there may be several compounds whose atoms are connected in exactly the same order (i.e., they exhibit the same structure), but their spatial arrangement differs. Such compounds are then called *stereoisomers*. It should be pointed out, however, that, quite often, and particularly in German-speaking areas, the term *structure* is also used to denote both the connectivity (i.e., the way the atoms are connected to each other) as well as the spatial arrangement of the atoms. The term *constitution* of a compound is then sometimes introduced to describe solely the connectivity.

Electron Shells of Elements Present in Organic Compounds

Before we can examine how many different structures exist with a given molecular formula (e.g., C_4H_9Cl), we have to recall some of the rules concerning the number and nature of bonds that each of the various elements present in organic molecules may form. To this end, we first examine the electronic characteristics of the atoms involved.

Both theory and experiment indicate that the electronic structures of the noble gases [helium (He), neon (Ne), argon (Ar), krypton (Kr), xenon (Xe), and radon (Rn)] are especially nonreactive; these atoms are said to contain "filled shells" (Table 2.1). Much of the chemistry of the elements present in organic molecules is understandable in terms of a simple model describing the tendencies of the atoms to attain such filled-shell conditions by gaining, losing, or, most importantly, sharing electrons.

Table 2.1 Atomic Mass, Electronic Configuration, and Typical Number of Covalent Bonds of the Most Important Elements Present in Organic Molecules

Element[a]			Mass[b] (u)	Number of Electrons in Shell					Net Charge of Kernel	Number of Covalent Bonds Commonly Occurring in Organic Molecules
Name	Symbol	Number		K	L	M	N	O		
Hydrogen	H	1	1.008	1					1+	1
Helium	He	2		2					0	
Carbon	C	6	12.011	2	4				4+	4
Nitrogen	N	7	14.007	2	5				5+	3,(4)[c]
Oxygen	O	8	15.999	2	6				6+	2,(1)[d]
Fluorine	F	9	18.998	2	7				7+	1
Neon	Ne	10		2	8				0	
Phosphorus	P	15	30.974	2	8	5			5+	3,5
Sulfur	S	16	32.06	2	8	6			6+	2,4,6(1)[d]
Chlorine	Cl	17	35.453	2	8	7			7+	1
Argon	Ar	18		2	8	8			0	
Bromine	Br	35	79.904	2	8	18	7		7+	1
Krypton	Kr	36		2	8	18	8		0	
Iodine	I	53	126.905	2	8	18	18	7	7+	1
Xenon	Xe	54		2	8	18	18	8	0	

[a] The underlined elements are the noble gases. [b] Based on the assigned atomic mass constant of u = atomic mass of $^{12}C/12$; abundance-averaged values of the naturally occurring isotopes. [c] Positively charged atom. [d] Negatively charged atom.

16

The first shell (K-shell) holds only two electrons [helium structure; first row of the periodic system (p.s.) of the elements] ; the second (L-shell; second row of the p.s.) holds eight; the third (M-shell; third row of the p.s.) can ultimately hold 18, but a stable configuration is reached when the shell is filled with eight electrons (argon structure). Thus, among the elements present in organic molecules, hydrogen requires two electrons to fill its outer shell (one it supplies, the other it must get somewhere else), while the other important atoms of organic chemistry require eight, that is, an octet configuration (see Table 2.1). It is important to realize that the number of electrons supplied by a particular atom in its outer shell (the so-called valence electrons; note that the remainder of the atom is referred to as *kernel*) chiefly determines the chemical nature of an element, although some significant differences between elements exhibiting the same number of outer-shell electrons do exist. The latter is due in large part to the different energetic status of the electrons in the various shells, reflecting the distance of the electrons from the positively charged nucleus. We address the differences between such elements (e.g., nitrogen and phosphorus, oxygen and sulfur) at various stages during our discussions.

Covalent Bonding

The means by which the atoms in organic molecules customarily complete their outer-shell or *valence-shell* octet is by sharing electrons with other atoms, thus forming so-called *covalent bonds*. Each single covalent bond is composed of a pair of electrons, in most cases one electron contributed by each of the two bonded atoms. The covalent bond may thus be characterized as a mutual deception in which each atom, though contributing only one electron to the bond, "feels" it has both electrons in its effort to fill its outer shell. Thus we visualize the bonds in an organic compound structure as electron pairs localized between two positive atomic nuclei; the electrostatic attraction of these nuclei to these electrons holds the atoms together. The simple physical law of the attraction of unlike charges and the repulsion of like charges is the most basic force in chemistry, and it will help us to explain many chemical phenomena.

Using the simple concept of electron sharing to complete the valence-shell octet, we can now easily deduce from Table 2.1 that H, F, Cl, Br, and I should form one bond (monovalent atoms), O and S two (bivalent atoms), N and P three (trivalent atoms), and C four (tetravalent atom) bonds in a neutral organic molecule. With a few exceptions (particularly for S and P), which we will address later in Section 2.3, this concept is valid for the majority of cases that are of interest to us. For our compound with the molecular formula of C_4H_9Cl, we are now ready to draw all the possible *structural isomers* by simply applying these valency rules. Fig. 2.1 shows that there are four different possibilities. With this example we also take the opportunity to get acquainted with some of the common conventions used to symbolize molecular structures. Since it is clearer to separate shared and unshared electron pairs, the former (the actual covalent bond) are written as straight lines connecting two atomic symbols, while the unshared valence electrons are represented by pairs of dots (line 1 in Fig. 2.1). This representation clearly shows the nuclei and all of the electrons we must visualize. To simplify the drawing, all lines indicating bonds to hydrogen as well as the dots for unshared (nonbonding) electrons are frequently not shown (line 2). For further convenience we may, in many cases, eliminate all the bond lines without

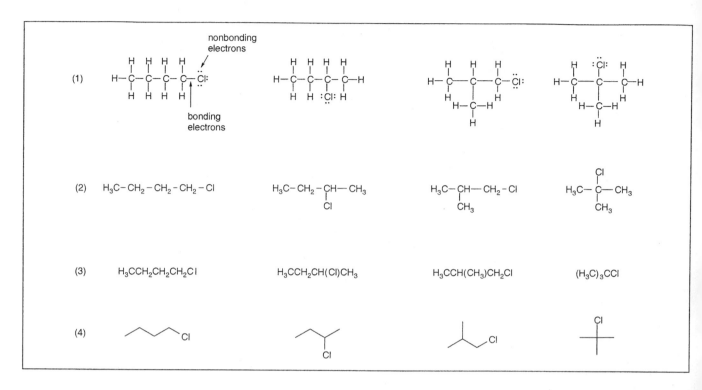

Figure 2.1 Conventions for symbolizing the molecular structures of the four butyl chloride (or chlorobutane) isomers.

loss of clarity, as illustrated in line 3. Note that branching is indicated by using parentheses in this case. Finally, especially when dealing with compounds exhibiting a large number of carbon atoms, it is very convenient to just sketch the carbon skeleton as depicted in line 4. Each line is thus a skeletal bond and is assumed to have carbons at each end unless another element is shown. Furthermore, no carbon–hydrogen bonds are indicated, but are assumed present as required to make up full bonding (four bonds) at each carbon atom. To distinguish the various carbon–carbon bonds, bond lines are placed (whenever possible), at about 120°, roughly resembling the true physical bond angle (see below).

At this point in our discussion about chemical bonds and structural formulas, we should stress that structural isomers may exhibit very different properties and reactivities. For example, the rates of hydrolysis (reaction with water, see Chapter 13) of the four butyl chlorides shown in Fig. 2.1 are quite different. While the hydrolytic half-life (time required for the concentration to drop by a factor of 2) of the first and third compound is about 1 year at 25°C, it is approximately 1 month for the second compound, and only 30 seconds for the fourth compound. When we compare the two possible structural isomers with the molecular formula C_2H_6O, we can again find distinct differences in that the well-known ethanol (CH_3CH_2OH) is a liquid at ambient conditions while dimethylether (CH_3OCH_3) is a gas. These examples should remind us that differences in the arrangement of a single collection of atoms may mean very different environmental behavior; thus we must learn what it is about compound structure that dictates such differences.

So far we have dealt only with single bonding between two atoms. There are, however, many cases in which atoms with more than one "missing" electron in their outer shell form *double bonds* or, sometimes, even *triple bonds;* that is, two atoms

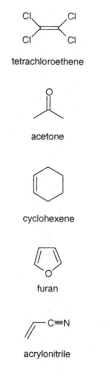

tetrachloroethene

acetone

cyclohexene

furan

acrylonitrile

Figure 2.2 Some simple molecules exhibiting double and/or triple bonds. Note that we use notation type 4 in Fig. 2.1.

share either two or even three pairs of electrons to complete their valence shells. A few examples of compounds exhibiting double or triple bonds are given in Fig. 2.2. We note that a double bond is indicated by a double line and, logically, a triple bond by three parallel lines between the corresponding atoms. We also note from Fig. 2.2 that there are compounds with ring structures (that may or may not contain double bonds). Rings are usually composed predominantly of carbon atoms, but they may also contain heteroatoms (i.e., elements other than carbon or hydrogen such as O, N) in the ring (see Section 2.3).

Bond Energies (Enthalpies) and Bond Lengths. The Concept of Electronegativity

An important aspect of chemical bonding that we need to address is the strength of a chemical bond in organic molecules; that is, we should have a general idea of the energy involved in holding atoms together in a covalent bond. The most convenient measure of bond energy is indicated by the bond *dissociation enthalpy*, ΔH^0_{AB}. For a diatomic molecule, this is defined as the *heat* change of the *gas phase* reaction:

$$A-B \rightarrow A^{\bullet} + B^{\bullet}\; ; \; \Delta H^0_{AB} \qquad (2\text{-}1)$$

at constant pressure and temperature [e.g., 1.013 bar (= 1 atm) and 25°C]. Here ΔH^0_{AB} also contains the differences in translational, rotational (only AB), and vibrational (only AB) energies between *educt* (A–B) and *products* (A$^{\bullet}$, B$^{\bullet}$). Unfortunately, it is not possible to directly measure bond dissociation (or formation) enthalpies for each of the different bonds present in a molecule containing more than one bond; they have to be determined indirectly, commonly through thermochemical studies of evolved heat (calorimetric measurements) in reactions such as combustion. These studies yield only enthalpies of overall reactions, where several bonds are broken and formed, respectively. The individual bond dissociation (or formation) enthalpies have then to be deduced from this data in various ways. The results are commonly shown in tables as average strengths for a particular type of bond, valid for gas phase reactions at 25°C and 1.013 bar. Table 2.2 summarizes average bond enthalpies (and bond lengths) of some important covalent bonds. From these data, some general conclusions about covalent bonds can be drawn, and a very useful concept can be derived to evaluate the uneven distribution of the electrons in a chemical bond: the concept of *electronegativity*.

Electronegativity. When visualizing a chemical bond, it is appropriate to imagine that the "electron cloud" or averaged electron position located between the two nuclei is, in general, distorted toward the atom that has the higher attraction for the electrons, that is, the atom that is more *electronegative*. This results in the accumulation of negative charge at one end of the bond (denoted as δ −) and a corresponding deficiency at the other end (denoted as δ +):

$$-\underset{|}{\overset{|}{C}}\!\!\overset{\delta+}{-}\!X^{\delta-}$$

Among the elements present in organic molecules, we intuitively (and correctly) predict that the smaller the atom (hence allowing a closer approach of the bonding

Table 2.2 Average Bond Lengths (Å) and Average Bond Enthalpies ($kJ \cdot mol^{-1}$) of Some Important Covalent Bonds[a]

Bond	Length/Enthalpy	Bond	Length/Enthalpy	Bond	Length/Enthalpy
		Diatomic Molecules			
H–H	0.74/436	F-F	1.42/155	O=O	1.21/498
H–F	0.92/566	Cl–Cl	1.99/243	N≡N	1.10/946
H–Cl	1.27/432	Br–Br	2.28/193		
H–Br	1.41/367	I–I	2.67/152		
H–I	1.60/298				
		Covalent Bonds in Organic Molecules			
Single bonds[b]					
H–C	1.11/415	C–C	1.54/348	C–F	1.38/486
H–N	1.00/390	C–N	1.47/306	C–Cl	1.78/339
H–O	0.96/465	C–O	1.41/360	C–Br	1.94/281
H–S	1.33/348	C–S	1.81/275	C–I	2.14/216
Double and triple bonds					
C=C	1.34/612	C=O[d]	1.20/737	C≡C	1.16/838
C=N	1.28/608	C=O[e]	1.20/750	C≡N	1.16/888
C=S[c]	1.56/536	C=O[f]	1.16/804		

[a] Bond length/bond enthalpy. Note that 1 Å equals 0.1 nm. [b] Bond lengths are given for bonds in which none of the partner atoms is involved in a double or triple bond. In such cases bond lengths are somewhat shorter. [c] In carbon disulfide. [d] In aldehydes. [e] In ketones. [f] In carbon dioxide.

electrons to the positively charged nucleus) and the higher the net charge of the kernel (nucleus plus the electrons of the inner, filled shells; see Table 2.1), the greater will be that atom's tendency to attract additional electrons. Hence, as indicated in Table 2.3, within a row in the Periodic Table (e.g., from C to F), electronegativity increases with increasing kernel charge, and within a column (e.g., from F to I), electronegativity decreases with increasing kernel size. The most commonly used quantitative scale to express *electronegativity* (Table 2.3) has been devised by Pauling (1960). On this scale, a value of 4.0 is arbitrarily assigned to the most electronegative atom, fluorine, and a value of 1.0 to lithium. The difference in electronegativity between two atoms A and B is calculated from the extra bond energy in A–B versus the mean bond energies of A–A and B–B in which the electrons should be equally shared. The reason for deriving relative electronegativities based on bond energies is that we interpret the extra bond strength in such a polarized bond to be due to the attraction of the partial positive and negative charges.

Let us discuss the importance of charge separation in bonds involving atoms of different electronegativity, for example, C and N, O, or Cl. The extent of partial ionic character in such *polar* covalent bonds is a key factor in determining a compound's

Table 2.3 Electronegativities of Atoms According to the Scale Devised by Pauling (1960)

Charge of Kernel:	+ 1	+ 4	+ 5	+ 6	+ 7	
	H 2.2					Increasing Size of Kernel
		C 2.5	N 3.0	O 3.5	F 4.0	
			P 2.2	S 2.5	Cl 3.0	
					Br 2.8	
					I 2.5	

behavior and reactivity in the environment. The polarization in bonds is important in directing the course of chemical reactions in which either these bonds themselves or other bonds in the vicinity are broken. Furthermore, the partial charge separation makes each bond between dissimilar atoms a *dipole*. The (vector) sum of all bond dipoles in a structure yields the total dipole moment of the molecule, an entity that can be measured. However, it is the dipole moments of individual bonds that are most important with respect to the interactions of a given compound with its molecular surroundings.

From Table 2.3 it can be seen that according to Pauling's scale, carbon is slightly more electron-attracting than hydrogen. It should be noted, however, that the electron-attracting power of an atom in isolation differs from that attached to electron-attracting or electron-donating substituents in an organic molecule. For example, many experimental observations indicate that carbon in $-CH_3$ is significantly less electron-attracting than hydrogen. We may rationalize this by recognizing that each additional hydrogen contributes some electron density to the carbon and successively reduces that central atom's electronegativity. In conclusion, we should be aware that the electronegativity values in Table 2.3 represent only a rough scale of the relative electron-attracting power of the elements. Hence in bonds between atoms of similar electronegativity, the direction and extent of polarization will also depend on the type of substitution at the two atoms.

Hydrogen Bonding. One special result of the polarization of bonds to hydrogen which we should highlight at this point is the so-called *hydrogen bonding*. As indicated in Table 2.1, hydrogen does not possess any inner electrons isolating its nucleus (consisting of just one proton) from the bonding electrons. Thus, in bonds of hydrogen with highly electronegative atoms, the bonding electrons are drawn strongly to the electronegative atom, leaving the proton exposed at the outer end of the covalent bond. This relatively bare proton can now attract another electron-rich center, especially heteroatoms with nonbonding electrons, and form a *hydrogen bond* as schematically indicated below by the dotted line:

$$- X^{\delta-} - H^{\delta+} \ldots : Y^{\delta-} - \qquad X, Y = N, O, \ldots$$

In organic molecules this is primarily the case if X and Y represent nitrogen or oxygen.

If the electron-rich center forms part of the same molecule, one speaks of an *intra*molecular hydrogen bond; if the association involves two different molecules, it is referred to as an *inter*molecular hydrogen bond. Although, compared to covalent bonds, such hydrogen bonds are relatively weak (15 to 20 $kJ \cdot mol^{-1}$), they are of enormous importance with respect to the spatial arrangements and interactions of molecules.

We now return to Table 2.2 to note a few simple generalities about bond lengths and bond strengths in organic molecules. As one can see, bond lengths of first-row elements (C, N, O, F) with hydrogen are all around 1 Å (0.1 nm). Bonds involving larger atoms (S, P, Cl, Br, I) are longer and weaker. Finally, double and triple bonds are shorter and stronger than the corresponding single bonds; and we notice that the bond enthalpies of double and triple bonds are often somewhat less than twice and three times, respectively, the values of the single atom bonds (important exception: C=O bonds).

To get an appreciation of the magnitude of bond energies, it is illustrative to compare bond enthalpies to the energy of molecular motion (translational, vibrational, and rotational), which, at room temperature, is typically on the order of a few tens of kilojoules per mole. As can be seen from Table 2.2, most bond energies in organic molecules are much larger than this, and, therefore, organic compounds are, in general, stable to thermal disruption at ambient temperatures. At high temperatures, however, the energy of intramolecular motion increases and can then exceed certain bond energies. This leads to a thermally induced disruption of bonds, a process that is commonly referred to as pyrolysis (heat splitting).

It is important to realize that the persistence of organic compounds in the environment is due to the relatively high energy (of activation) needed to break bonds and not because the atoms in a given molecule are present in their lowest possible energetic state (and, therefore, would not react with other chemical species). Hence, many organic compounds are nonreactive for kinetic, not thermodynamic, reasons. We will discuss the energetics and kinetics of chemical reactions in detail later in this book. Here a simple example helps to illustrate this point. From daily experience we know that heat can be gained from burning natural gas, gasoline, fuel oil, or wood. As we also know, all these fuels are virtually inert under environmental conditions until we light a match, and then provide the necessary initial activation energy to break bonds. Once the reaction has started, enough heat is liberated to keep it going. The amount of heat liberated can be estimated from the bond enthalpies given in Table 2.2. For example, when burning methane gas in a stove, the process that occurs is the reaction of the hydrocarbon, methane, with oxygen to yield CO_2 and H_2O:

$$CH_4(g) + 2O_2(g) \rightarrow CO_2(g) + 2H_2O(g) \qquad (2\text{-}2)$$

In this gas phase reaction we break four C–H and two O=O "double" bonds and we make two C=O and four O–H bonds. Hence, we have to invest $(4 \times 415) + (2 \times 498)$ $= +2656$ kJ·mol^{-1}, and we gain $(2 \times 804) + (4 \times 465) = -3468$ kJ·mol^{-1}. The estimated heat of reaction at 25°C for reaction 2-2 is therefore, -812 kJ·mol^{-1} (the experimental value is -802 kJ·mol^{-1}), a quite impressive amount of energy. We recall from basic chemistry that, by convention, we use a minus sign to indicate that the reaction is *exothermic*; that is, heat is given off to the outside. A positive sign is assigned to the heat of reaction if the reaction consumes heat by taking energy into the product structures; such reactions are then called *endothermic*.

The example illustrates that enthalpy can be gained when nonpolar bonds, as commonly encountered in organic molecules, are broken and polar bonds, such as those in carbon dioxide and water, are formed. Reactions which involve the transfer of electrons between different chemical species are generally referred to as *redox reactions*. Such reactions form the basis for the energy production of all organisms. From this point of view we can consider organic compounds as energy sources.

Oxidation State of the Atoms in an Organic Molecule

When dealing with transformation reactions, it is important to know whether electrons have been transferred between the reactants. For evaluating the number of electrons transferred, it is convenient to examine the (formal) oxidation states of all atoms involved in the reaction. Of particular interest to us will be the oxidation state of carbon, nitrogen, and sulfur in a given organic molecule, since these are the elements most frequently involved in organic redox reactions.

The terms *oxidation* and *reduction* refer, respectively, to the loss and gain of electrons at an atom or ion. An oxidation state of zero is assigned to the uncharged element; a loss of Z electrons is then an oxidation to an oxidation state of + Z. Similarly, a gain of electrons leads to an oxidation state lower by an amount equal to the number of gained electrons. A simple example is the oxidation of sodium by chlorine, resulting in the formation of sodium chloride:

$$\overset{\cdot}{Na}{}^{0} \xrightarrow{\text{oxidation}} Na^{+1} + e^{-}$$

$$:\overset{\cdot\cdot}{\underset{\cdot\cdot}{Cl}}\cdot{}^{0} + e^{-} \xrightarrow{\text{reduction}} :\overset{\cdot\cdot}{\underset{\cdot\cdot}{Cl}}:{}^{-1}$$

To bridge the gap between full electron transfer in ionic redox reactions (as shown in the example above) and the situation with the shared electrons encountered in covalent bonds, one can formally assign the possession of the electron pair in a covalent bond to the more electronegative atom of the two bonded atoms. By doing so, one can then count the electrons on each atom as one would with simple inorganic ions. For any atom in an organic molecule, the oxidation state may be computed by adding 0 for each bond to an identical atom; – 1 for each bond to a less electronegative atom or for each negative charge on the atom; and + 1 for each bond to a more electronegative atom or for each positive charge. We note that in C–S, C–I, and even C–P bonds the electrons are attributed to the heteroatom, although the electronegativities of these heteroatoms are very similar to that of carbon. Finally, we should also point out that Roman instead of Arabic numbers are frequently used to express the oxidation state of a covalently bound atom.

Some examples demonstrating how to determine the oxidation states of the carbon atoms present in organic molecules are given in Illustrative Example 2.1. More examples covering the other elements that may assume several different oxidation states (i.e., N, S, P) will be given in Section 2.3 in our discussion of functional groups.

Illustrative Example 2.1

Determining the Oxidation States of the Carbon Atoms Present in Organic Molecules

Problem

Determine the oxidation state of each carbon present in (a) iso-pentane, (b) acetic acid, (c) trichloroethene, and (d) 4-methylphenol (*p*-cresol). Note that in organic molecules hydrogen always assumes an oxidation state of +I, chloride of –I, and oxygen (in most cases, see Section 2.3) of –II.

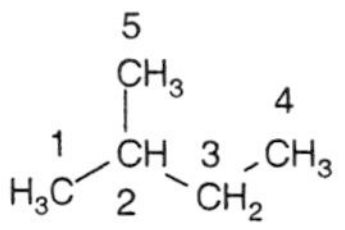

2-methyl-butane
(iso-pentane)

Answer (a)

The carbons of the methyl groups (C_1, C_4, C_5) are bound to three hydrogens and one carbon; hence their oxydation state is $3(-I) + (0) = -III$. The methylene group (C_3) is bound to two hydrogens and two carbons, which yields $2(-I) + 2(0) = -II$. Finally, the methene group (C_2) exhibits an oxidation state of $(-I) + 3(0) = -I$.

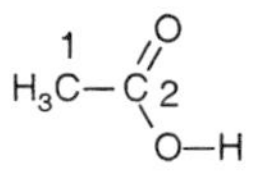

acetic acid

Answer (b)

As in example (a) the oxidation state of the carbon of the methyl group (C_1) is –III, while the oxidation state of the carboxylic carbon (C_2) is $(+II) + (+I) + (0) = +III$. Hence the "average oxidation state" of carbon in acetic acid is 0.

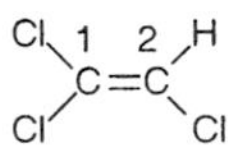

trichloroethene

Answer (c)

In trichloroethene the oxidation states of the two carbons are $2(+I) + 0 = +II$ for (C_1) and $(-I) + (+I) + 0 = 0$ for (C_2).

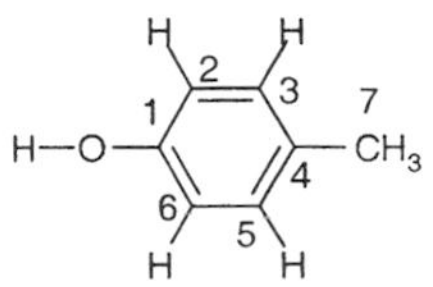

4-methyl-phenol
(*p*-cresol)

Answer (d)

The carbons present in the benzene ring exhibit oxidation states of $(+I) + 2(0) = +I$ for (C_1), $(-I) + 2(0) = -I$ for (C_2, C_3, C_5, C_6), and $3(0) = 0$ for (C_4); respectively. The oxidation state of the methyl carbon, is again –III.

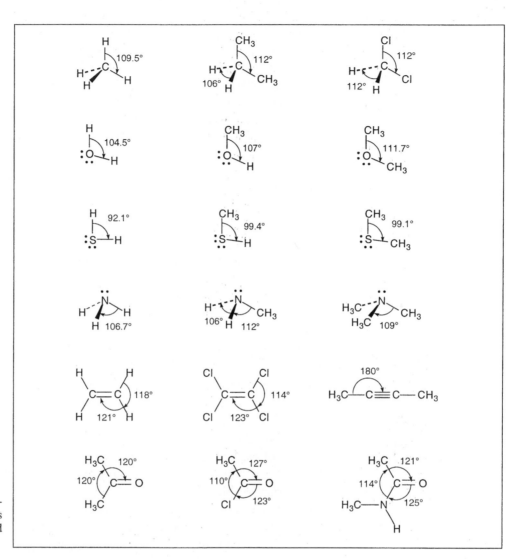

Figure 2.3 Examples of bond angles in some simple molecules (from Hendrickson et al., 1970, and March 1992).

The Spatial Arrangement of the Atoms in Organic Molecules

To describe the steric arrangement of the atoms in a molecule, in addition to bond lengths we need to know something about the angles between the bonds, the sizes of the atoms, and their freedom to move within the molecule, (e.g., rotations about bonds).

Bond Angles. A simple but very effective rule that we may apply when considering *bond angles* in molecules is that the electrons accept the closeness to one another because of pairing, but that each pair of electrons, shared (i.e., involved in a chemical bond) or unshared, wants to stay as far as possible from the other pairs of electrons [for details see valence shell electron pair repulsion (VSEPR) theory; Pfennig and Frock, 1999]. This means that in the case of a carbon atom with four single bonds, the bonds will generally point toward the corners of a *tetrahedron*. In the symmetrical case, that is, when a carbon is bound to four identical *substituents* [i.e., atoms or groups of atoms as –H in CH_4, or –Cl in CCl_4, or –CH_3 in $C(CH_3)_4$], the bond angles are 109.5°. In most cases, however, each carbon atom is bound to different substituents, which leads to minor variations in the bond angles, as illustrated by some examples given in Fig. 2.3. For *saturated* carbon atoms, that is, carbon atoms not involved in a double or triple bond, the C–C–C bond angles are typically about 112°, except for ring systems containing less than six ring atoms, where bond angles may be considerably smaller. With respect to the heteroatoms N, O, P, and S, we see from the examples given in Fig. 2.3 that the nonbonded electron pairs behave as if they point to imaginary substituents, thus giving rise to bent or pyramidal geometry (provided that the heteroatoms are also only single-bonded to other atoms).

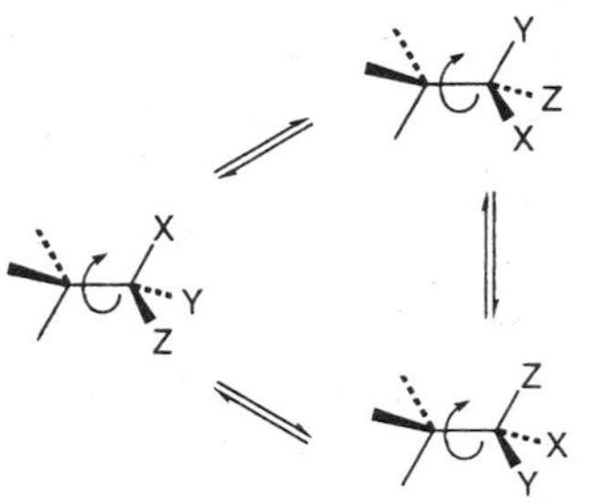

Stereoisomerism. We should note that the association of electrons in a single, or sigma (σ), bond allows rotation about the axis of the linkage (Fig. 2.4). Such rotation does not disrupt the bonding electron pair (i.e., it does not break the bond), and therefore under ambient temperatures the substituents attached to two carbons bonded by a sigma bond are usually not "frozen" in position with respect to one another. Thus the spatial arrangement of groups of atoms connected by such a single bond may change from time to time owing to such rotation, but such geometric distributions of the atoms in the structure are usually not separable from one another since interconversions occur during separation. However, as discussed below, even if fast rotations about a single bond occur, *stereoisomerism* is possible. *Stereoisomers are compounds made up of the same atoms bonded by the same sequence of bonds, but having different three-dimensional structures that are not superimposable.*

Figure 2.4 Rotation about a σ-bond leading to various spatial arrangements of the atoms in a molecule.

When considering stereoisomerism, one commonly distinguishes two different cases. First, there are molecules that are alike in every respect but one: they are *mirror images* of each other that are not superimposable. We refer to such molecules as being *chiral*. Note that, in general, any object for which the image and mirror image are distinguishable (e.g., our left and right hands), is denoted to be chiral. For example, if in a molecule a carbon atom is bound to four different substituents (as is the case in the herbicide mecoprop; Fig. 2.5), two structural isomers are possible. In this context one sometimes refers to such a carbon atom as a center of chirality. Mirror-image isomers are called *enantiomers* or *optical isomers* (because they rotate the plane of polarized light in opposite directions). In general, we may say that enantiomers have identical properties in a symmetrical molecular environment, but that their behavior may differ quite significantly in a chiral environment. Most importantly, they may react at very different rates with other chiral species. This is the reason why many compounds are biologically active, while their enantiomers are not. For example, the "R-form" of mecoprop (see Fig. 2.5) is an active herbicide, whereas the "S-form" is biologically rather inactive (Bosshardt, 1988).

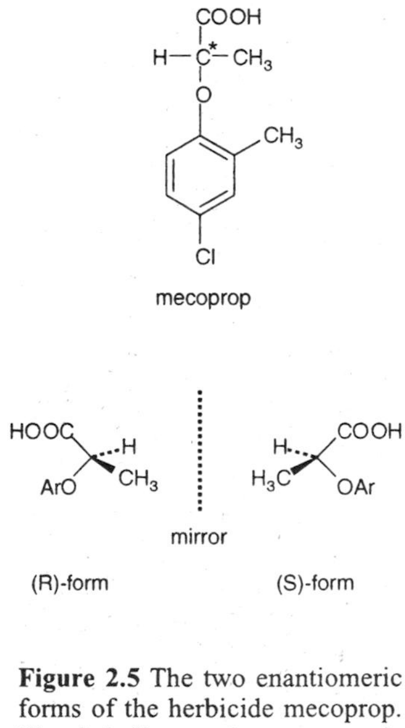

Figure 2.5 The two enantiomeric forms of the herbicide mecoprop. The * indicates the asymmetric carbon center. Ar denotes the aromatic substituent.

The second type of stereoisomerism encompasses all other cases in which the three-dimensional structures of two isomers exhibiting the same connectivity among the atoms are not superimposable. Such stereoisomers are referred to as *diastereomers*. Diastereomers may arise due to different structural factors. One possibility is the presence of more than one chiral moiety. For example, many natural products contain 2 to 10 asymmetric centers per molecule, and molecules of compound classes such as polysaccharides and proteins contain hundreds. Thus, organisms may build large molecules that exhibit highly stereoselective sites, which are important for many biochemical reactions including the transformation of organic pollutants.

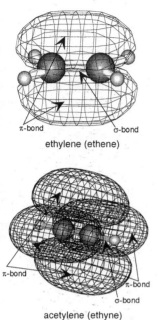

π-bond   σ-bond

ethylene (ethene)

π-bond   π-bond

σ-bond

acetylene (ethyne)

Figure 2.6 Simplified picture of a double (ethylene) and triple (acetylene) bond, respectively.

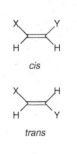

cis

trans

maleic acid
(*cis*)

fumaric acid
(*trans*)

Figure 2.7 *Cis/trans* isomerism at double bonds exhibiting two substituents.

Another important form of diastereoisomerism results from restricted rotation around bonds such as are encountered with double bonds and/or ring structures. When considering the geometry of a *double bond*, we imagine a combination of two different types of bonds between two atoms. One of the bonds would be equivalent to a single bond, that is, a bond in which the pair of electrons occupies the region around the axis between the doubly bonded atoms. We can picture the second bond, which is called a π-bond (e.g., carbon–carbon, carbon–oxygen, carbon–nitrogen, carbon–sulfur, nitrogen–oxygen), by imagining the two bonding *π-electrons* to be present in an "electron cloud" located above and below a plane in which the axes of all other bonds (real or imaginary in the case of nonbonded electron pairs) lay, as in the case of ethylene (Fig. 2.6). The atoms closest to a carbon–carbon double bond are in a plane with bond angles of about 120° (see examples given in Fig. 2.3). Rotation about the axis would mean that we would have to break this bond. In triple-bond compounds, as in the case for acetylene (ethyne), there are two π-bond electron clouds which are orthogonal to each other, thus leading to a linear (bond angles = 180°) configuration (Fig. 2.6).

Let us now consider a compound XHC=CHY in which X,Y ≠ H. In this case, there are two isomers (sometimes also called *geometric* isomers), which are distinct (and, in principle, separable) because we can no longer rotate about the C–C bond (Fig. 2.7). To distinguish between the two isomers, one commonly uses the terms *cis* and *trans* to describe the relative position of two *substituents* (atoms or groups other than hydrogen). The term *cis* is used if the two substituents are on the same side of the double bond, the term *trans* if they are across from one another. As in the other cases of isomerism, we note that such closely related compounds may exhibit quite different properties. For example, the boiling points of *cis*- and *trans*-1,2-dichloroethene (X=Y=Cl), are 60 and 48°C, respectively. More pronounced differences in properties between *cis/trans* isomers are observed when interactions between two substituents (e.g., intramolecular hydrogen bonding) occur in the *cis* but not in the *trans* form, as is encountered with maleic and fumaric acid (see margin). These two compounds are so different that they have been given different names. For example, their melting points differ by more than 150°C and their aqueous solubilities by more than a factor of 100.

The organization of atoms into a ring containing less than 10 carbons also prevents free rotation. Consequently, *cis* and *trans* isomers are also possible in such ring systems; the *cis* isomer is the one with two substituents on the same side of the ring (i.e., above or below); the *trans* isomer exhibits a substituent on either side (Fig. 2.8).

In ring systems with more than two substituted carbons, more isomers are possible. An example well known to environmental scientists and engineers is 1,2,3,4,5,6-hexachlorocyclohexane (HCH). Three of the possible 8 isomers (the so-called α-, β-, γ-isomers) are particularly important from an environmental point of view:

α-isomer   β-isomer   γ-isomer

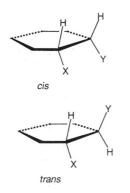

cis

trans

Figure 2.8 *Cis/trans* isomerism in ring systems, such as in cyclohexane.

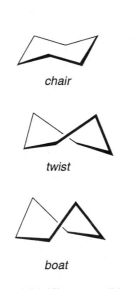

chair

twist

boat

Figure 2.9 Different possible conformations of a six-membered ring (e.g., cyclohexane).

At this point we should reiterate that the relative positions of atoms in many structures are continuously changing. The term *different conformations* of a molecule is used if two different three-dimensional arrangements of the atoms in a molecule are rapidly interconvertible, as is the case if free rotations about sigma bonds are possible. If rotation is not possible, we speak of different *configurations*, which represent isomers that can be separated. Obviously, the conformation(s) with the lowest energy [the most stable form(s)] is(are) the one(s) in which a molecule will preferentially exist. In the case of six-membered rings such as cyclohexane, three stable conformations (i.e., the chair, twist, and boat form) exist (Fig. 2.9).

Depending on the type of substituents, usually one of the forms is the most stable one. In the case of the HCH isomers, this is the chair form. If we have a closer look at the chair form, we see that six of the bonds linking substituents to the ring are directed differently from the other six:

The six *a*xial bonds are directed upward or downward from the "plane" of the ring, while the other six *e*quatorial bonds are more within the "plane." Conversion of one chair form into another converts all axial bonds into equatorial bonds and *vice versa*. In monosubstituted cyclohexanes, for electronic reasons, the more stable form is usually the one with the substituent in the equatorial position. If there is more than one substituent, the situation is more complicated since we have to consider more combinations of substituents which may interact. Often the more stable form is the one with more substituents in the equatorial positions. For example, in α-1,2,3,4,5,6-hexachlorocyclohexane (see above) four chlorines are equatorial (aaeeee), and in the β-isomer all substituents are equatorial. The structural arrangement of the β-isomer also greatly inhibits degradation reactions [the steric arrangement of the chlorine atoms is unfavorable for dehydrochlorination (see Chapter 13) or reductive dechlorination; see Bachmann et al. 1988].

Delocalized Electrons, Resonance, and Aromaticity

Having gained some insights into the spatial orientation of bonds in chemicals and the consequences for the steric arrangement of the atoms in an organic molecule, we can now proceed to discuss special situations in which electrons move throughout a region covering more than two atoms. The resulting bonds are often referred to as "delocalized chemical bonds." From an energetic point of view, this diminished constraint on the positions of these electrons in the bonds results in their having lower energy and, as a consequence, the molecule exhibits greater stability. For us the most important case of delocalization is encountered in molecules exhibiting multiple π-*bonds* spaced so they can interact with one another. We refer to such a series of π-bonds as *conjugated*. To effectively interact, π-bonds must be adjacent to each other and the σ-bonds of all atoms involved must *lie in one plane*. In such a conjugated system, we can qualitatively visualize the π-electrons to be smeared over the whole region, as is illustrated for propenal (also known as *acrolein*) in Fig. 2.10. If we try to consider acrolein's structure by indicating the extreme possible positions of these conjugated electrons, we may write

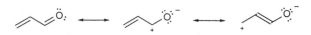

acrolein

Figure 2.10 Schematic picture of π-electron delocalization in acrolein (CH$_2$=CH–CHO).

The back-and-fourth arrows are not intended to suggest the three structures are interconvertible, but rather that the location of the four electrons is best thought of as something in between situation of such extreme possibilities. This freedom in electron positions results in what we call their delocalization, and the visualization of a given molecule by a set of localized structures is called the *resonance* method for representing a structure. The relative contributions of the extremes to the overall resonance structure is determined by their relative stabilities. The stabilizing effect of delocalization is most pronounced in so-called *aromatic systems*. The best-known aromatic system is that of benzene, where we have three conjugated double bonds in a six-membered ring:

Note again that each of the "static structures" alone does not represent the molecule, but that the molecule is a hybrid of these structures. Thus, as shown in parentheses, the electrons in the conjugated π-bonds of benzene are sometimes denoted with a circle.

In *substituted* benzenes (i.e., benzenes in which hydrogen is substituted by another atom or group of atoms), depending on the type and position of the substituents, the different resonance forms may exhibit somewhat different stabilities. Therefore, their contributions are different.

A quantitative estimate of the stabilization or *resonance energy* of benzene (which cannot be directly measured) may be obtained by determining the heat evolved when hydrogen is added to benzene and to cyclohexene to yield cyclohexane:

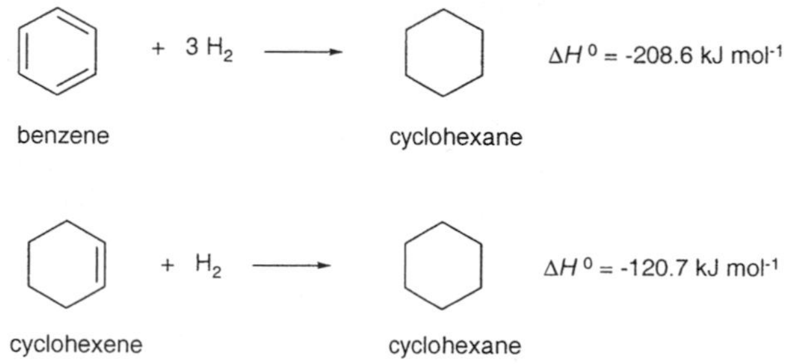

+ 3 H$_2$ $\longrightarrow$ ΔH^0 = -208.6 kJ mol^{-1}

benzene   cyclohexane

+ H$_2$ $\longrightarrow$ ΔH^0 = -120.7 kJ mol^{-1}

cyclohexene   cyclohexane

If each of the double bonds in benzene were identical to the one in cyclohexane, the heat of hydrogenation of benzene would be three times the heat evolved during hydrogenation of cyclohexene. The values given above show that there is a large discrepancy between the "expected" ($-120.7 \times 3 = -362$ kJ $\cdot$ mol^{-1}) and the measured

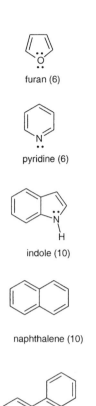

furan (6)

pyridine (6)

indole (10)

naphthalene (10)

phenanthrene (14)

Figure 2.11 Some additional examples of organic compounds that are aromatic (in parentheses, number of π-electrons).

(-208.6 kJ·mol^{-1}) ΔH^0 value. Hence, benzene is about 150 kJ·mol^{-1} more stable than would be expected if there were no resonance interactions among the π-electrons. Large stabilization energies are observed not only in components containing a benzene ring but, in general, in *cyclic π-bond* systems with $4n + 2$ (i.e., 6, 10, 14...) electrons. In the early days of organic chemistry, it was recognized that the benzene ring is particularly unreactive compared to acyclic (noncyclic) compounds containing conjugated double bonds. The quality that renders such ring systems especially stable was and still is referred to as *aromaticity*. Some additional examples of aromatic ring systems are given in Fig. 2.11. We note that there are aromatic compounds containing heteroatoms that contribute either one (e.g., pyridine) or two electrons (e.g., furan, indole) to the conjugated π-electron systems. Also note that some polycyclic compounds are referred to as polycyclic *aromatic* compounds, although they are not aromatic throughout their structure in a strict sense (e.g., pyrene with 16 electrons in its π-bond system, see below; Fig. 2.13).

As already indicated for acrolein and for the five-membered *heteroaromatic* rings (e.g. furan; Fig. 2.11), resonance may also be important between nonbonded electrons on a single atom and a π-bond system. For example, an unshared electron pair of oxygen greatly contributes to the stabilization of the carboxylate anion:

Similarly, the two unshared electrons of the nitrogen in aniline are in resonance with the aromatic π-electron system:

aniline

As we will see in Chapter 8, the delocalization of the unshared electron pair in aniline has an important impact on the acid/base properties of anilines as compared to aliphatic amino compounds.

In summary, delocalization of electrons enhances stability, and we can visualize delocalized bonding by using the resonance method. In later chapters we will learn more about the effects of resonance on chemical equilibrium and on the kinetics of chemical reactions of organic compounds.

Classification, Nomenclature, and Examples of Environmental Organic Chemicals

When grouping or classifying environmental organic chemicals, instead of taking a strictly structural approach, it is common practice to use compound categories based on some physical chemical properties, and/or based on the source or the use of the chemicals. Terms such as "volatile organic compounds" (VOCs), "hydrophobic compounds," "surfactants," "solvents," plasticizers," "pesticides" (herbicides, insecticides, fungicides, etc.), organic dyes and pigments, or mineral-oil products are very common in the literature. Using such compound categories undoubtedly has its practical value; however, one has to be aware that each of these groups of chemicals usually encompasses compounds of very different structures. Thus the environmental behavior of chemicals within any such group can vary widely.

If we look at organic compounds from a structural point of view, we have already seen that organic molecules are composed of a skeleton of carbon atoms. This skeleton is sheathed in hydrogens, with *heteroatoms* (i.e., halogens, O, N, S, P) or groups of heteroatoms inserted in or attached to that skeleton. Such "sites" are called *functional groups*, since they are commonly the site of reactivity or function. Hence, the functional groups deserve our special attention if we want to understand or assess a compound's behavior in the environment. For classifying organic chemicals according to structural features, we may then have to base the classification on the type of the carbon skeleton, the type of functional group(s) present, or a combination of both. Consequently, particularly when dealing with a compound exhibiting several functionalities, its assignment to a given compound class may be somewhat arbitrary.

lindane

There exists a systematic nomenclature for naming individual organic compounds. For a more detailed treatment of this topic we refer, however, to any organic chemistry textbook. It should be noted, however, that especially in environmental organic chemistry, one frequently uses so-called "common" or "trivial" names instead of a compound's systematic name. Furthermore, quite frequently, the systematic nomenclature is applied incorrectly. For a given compound, one may therefore find a whole series of different names in the literature (see, for example, synonym listings in the Merck Index), which, unfortunately, may sometimes lead to a certain confusion. For example, one of the 1,2,3,4,5,6-hexachlorocyclohexane isomers, lindane (see margin), also goes by the following diverse aliases: γ-HCH, γ-benzene hexachloride, gamma hexachlor, ENT 7796, Aparasin, Aphtiria, γ-BHC, Gammalin, Gamene, Gamiso, Gammexane, Gexane, Jacutin, Kwell, Lindafor, Lindatox, Lorexane, Quellada, Streunex, Tri-6, and Viton. In this book, however, we shall consider problems encountered with nomenclature to be only of secondary importance as long as we are always aware of the exact structure of the compound with which we are dealing.

In the following discussion we will try to pursue three goals at the same time. Our first goal is to provide an overview of the most important functional groups present in environmentally relevant organic chemicals. The second goal is to take a first qualitative look at how such functional groups may influence the chemistry of given

CH$_3$–

methyl

CH$_3$– CH$_2$–

ethyl

CH$_3$– CH$_2$– CH$_2$–

n-propyl

CH$_3$– CH$_2$– CH$_2$– CH$_2$–

n-butyl

R— CH$_2$ —

primary

R$_1$
  \
   CH—
  /
R$_2$

secondary

R$_1$
 |
R$_2$ — C—
 |
R$_3$

tertiary

ortho or 1,2-

meta or 1,3-

para or 1,4-

organic compounds. Finally, by introducing the nature and origins of specific compounds and compound classes, we will try to familiarize ourselves with the most important "chemical actors" that will be with us throughout the book.

The Carbon Skeleton of Organic Compounds: Saturated, Unsaturated, and Aromatic Hydrocarbons

Let us start out by a few comments about the terms used to describe carbon skeletons encountered in organic molecules. When considering a *hydrocarbon* (i.e., a compound consisting of only C and H) or a *hydrocarbon group* (i.e., a hydrocarbon *substituent*) in a molecule, the only possible "functionalities" are carbon–carbon double and triple bonds. A carbon skeleton is said to be *saturated* if it has no double or triple bond, and *unsaturated* if there is at least one such bond present. Hence, in a hydrocarbon, the term "saturated" indicates that the carbon skeleton contains the maximum number of hydrogen atoms compatible with the requirement that carbon always forms four bonds and hydrogen one. A saturated carbon atom is one that is singly bound to four other separate atoms.

Carbon skeletons exhibiting no ring structures (i.e., only unbranched or branched chains of carbon atoms) are named *aliphatic*, those containing one or several rings are *alicyclic*, or in the presence of an aromatic ring system, *aromatic*. Of course, a compound may exhibit an aliphatic as well as an alicyclic and/or an aromatic entity. The hierarchy of assignment of the compound to one of these three subclasses is commonly aromatic over alicyclic over aliphatic. A saturated aliphatic hydrocarbon is called an *alkane* or a paraffin. If considered as a substituent, it is referred to as an *alkyl* group. Alkyl groups represent the most ubiquitous hydrocarbon substituents present in environmental organic chemicals. Their general formula is –C$_n$H$_{2n+1}$. Examples of common alkyl groups are methyl, ethyl, propyl, and butyl groups (see margin). The prefix *n*, which stands for *normal*, is used to denote an unbranched alkyl chain, *iso* means that there are two methyl groups at the end of an otherwise straight chain, and *neo* is used to denote three methyl groups at the end of the chain. Alkyl groups are further classified according to whether they are primary, secondary, or tertiary. Hence, an alkyl group is referred to as *primary* if the carbon at the point of attachment is bonded to only one other carbon, as *secondary* (*s*-) if bonded to two other carbons, and as *tertiary* (*t*-) if bonded to three other carbons (see margin). Note that R stands for a carbon-centered substituent.

With respect to *unsaturated* hydrocarbons we should note that compounds exhibiting one or several double bonds are often called *alkenes* or *olefins*. Finally, we need to add a brief note concerning the nomenclature in aromatic systems, particularly, in six-numbered rings such as benzene. Here the terms *ortho-*, *meta-*, and *para-*substitution are often used to express the relative position of two substituents in a given ring system. Identically, we could refer to those isomers as 1,2-(*ortho*), 1,3-(*meta*), 1,4-(*para*) disubstituted compounds (see margin).

Hydrocarbons are ubiquitously in the environment. Natural hydrocarbons range widely in size from methane to *β*-carotene (Fig. 2.12); many other branched, ole-finic, cyclic, and aromatic hydrocarbons are found in fossil fuels, or tend to derive

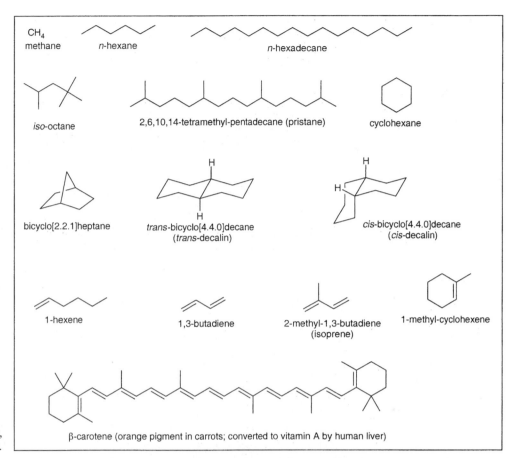

CH₄
methane   *n*-hexane   *n*-hexadecane

iso-octane   2,6,10,14-tetramethyl-pentadecane (pristane)   cyclohexane

bicyclo[2.2.1]heptane   *trans*-bicyclo[4.4.0]decane
(*trans*-decalin)   *cis*-bicyclo[4.4.0]decane
(*cis*-decalin)

1-hexene   1,3-butadiene   2-methyl-1,3-butadiene
(isoprene)   1-methyl-cyclohexene

β-carotene (orange pigment in carrots; converted to vitamin A by human liver)

Figure 2.12 Examples of aliphatic, alicyclic, and olefinic hydrocarbons.

from synthetic processing of fossil fuels (Figs. 2.12 and 2.13). When we recognize that the global annual production of liquid petroleum products (e.g., gasoline, kerosene, heating oils) is about 3 billion metric tons, it should be no surprise that producing, transporting, processing, storing, using, and disposing of these hydrocarbons poses major problems. Such organic chemicals are released to the atmosphere when we pump gasoline into the tanks of our cars, or are introduced to the surfaces of our streets when these same cars leak crankcase oil. Thus it is not only the spectacular instances such as the wreck of the *Exxon Valdez* or blowout of the IXTOC-II off-shore oil well through which petroleum hydrocarbons pollute the environment. Furthermore, although they share a common source, the various hydrocarbons in the exceedingly complex mixture that is oil certainly do not behave the same in the environment. Some constituents are noted for their tendency to vaporize while others clearly prefer to bind to solids; some oil hydrocarbons are extremely unreactive ("methane is the billiard ball of organic chemistry") while others interact beautifully with light; some are quite nontoxic while others are renowned for their carcinogenicity. Thus petroleum hydrocarbons have motivated a great deal of the research in environmental organic chemistry, and the individual components are a good example of why chemical structures must be visualized to predict fate. We should note at this point that hydrocarbons are not capable of forming hydrogen bonds to polar species such as water. As a consequence, many of these compounds are *hydrophobic* ("water-hating").

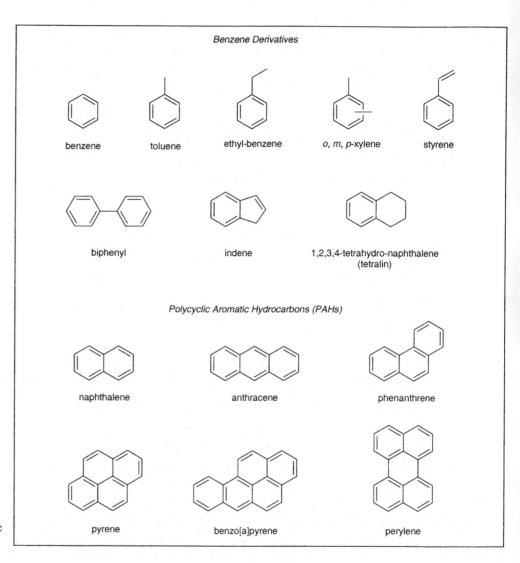

Figure 2.13 Examples of aromatic hydrocarbons.

Among the very large number of hydrocarbons present in the environment, we want to specifically mention two groups of compounds that are of particular interest to many environmental chemists and engineers: the so-called BTEX compounds (BTEX stands for *b*enzene, *t*oluene, *e*thylbenzene, and the three *x*ylene isomers; Fig. 2.13), and the *p*olycyclic *a*romatic *h*ydrocarbons (PAHs; Fig. 2.13). The BTEX components are important gasoline constitutents, and they are also widely used as solvents. They are very common soil and groundwater pollutants, and, particularly because of the high toxicity of benzene, they have triggered much research in the field of soil and groundwater remediation. The major sources of PAHs in the environment include the combustion of fossil fuels (gasoline, oil, coal), forest fires, (accidental) direct inputs of mineral oils, and the use of creosotes as wood preservatives. We should note that we produce PAHs when barbecuing steaks or other meat. From a human health perspective, PAHs have drawn considerable interest primarily because some of them (e.g., benzo(a)pyrene, Fig. 2.13) are very potent carcinogens. This is the main reason why PAHs are considered to be among the most important air pollutants. Furthermore, because of their high tendency to bioaccumulate (Chapter 10),

PAHs are also of great ecotoxicological concern. It is, therefore, not surprising that these compounds are among the most intensively investigated organic pollutants. Hence, they will serve us as important model compounds throughout this book.

Organohalogens

Many of the organic chemicals that are of particular environmental concern contain one or several halogen atoms, especially chlorine (Cl), and, to a lesser extent, fluorine (F) and/or bromine (Br). Although the number of known naturally produced organo-halogens (e.g., produced by marine algae) has increased drastically over the last two decades, the problems connected with the ubiquitous occurrence of halogenated compounds in the environment are primarily due to anthropogenic inputs. There are several reasons for the still-vast industrial production of halogen-containing chemicals. First, because of their high electronegativity (Table 2.3), particularly fluorine and chlorine form rather strong bonds with carbon (Table 2.2). Hence, in many cases, substitution of carbon-bound hydrogens by halogens enhances the inertness of a molecule (and thus also its persistence in the environment). For example, many agrochemicals capable of enzyme inhibition are fluorine-stabilized analogues of the natural enzyme substrate (e.g., they exhibit a CF_3- instead of a CH_3-group; see Key et al., 1997). Furthermore, we should note that halogens (including fluorine) present in organic compounds, particularly when bound to aromatic carbon atoms, have a very weak tendency to be engaged in hydrogen bonds with H-donors such as water. Therefore, as we will learn in later chapters, the presence of the larger halogens (i.e., chlorine and bromine) renders a compound more *hydrophobic*, which increases its tendency to partition into organic phases including organisms. This is one reason why numerous pesticides contain one or several halogens.

From a quantitative point of view, the major halogenated compounds used and re-leased into the environment are the *polyhalogenated hydrocarbons*, that is, com-pounds that are made up only of carbon, hydrogen, and halogens. By choosing the appropriate type and number of halogens, inert, nonflammable gases, liquids, and solids exhibiting specific physical-chemical properties that make them suitable for specific purposes can be designed. Prominent examples include the polyhalogenated C_1- and C_2-compounds that are used in enormous quantities (i.e., in tens to hundreds of thousands of tons per year) as aerosol propellants, refrigerants, blowing agents for plastic forms, anaesthetics, and, last but not least, as solvents for various purposes (see examples given in Table 2.4). Due to their high volatility and persistence in the troposphere, particularly the *fluoro-* and *chlorofluorocarbons* (FCs and CFCs, also called *freons*) are of great concern because of their significant stratospheric ozone-depletion and global-warming potential. This is the reason why these compounds are being replaced by *hydrofluoro-* and *hydrochlorofluorocarbons* (HFCs and HCFCs, respectively), which exhibit significantly shorter atmospheric lifetimes (Wallington et al., 1994). However, they are still not the ultimate solution to the problem.

The *chlorinated solvents,* including dichloromethane, tri- and tetrachloroethene, and 1,1,1-trichloroethane (see Table 2.4), are still among the top groundwater pollutants. Under oxic conditions these compounds are quite persistent and, because they are also quite mobile in the subsurface, they can lead to the contamination of large groundwater areas. As we will discuss in detail in Chapters 14 and 17, in anoxic

Table 2.4 Examples of Important Industrially Produced C_1- and C_2-Halocarbons

Compound Name(s)	Formula	Major Use
Dichlorodifluoromethane (CFC-12)	CCl_2F_2	Aerosol propellants, refrigerants (domestic, automobile air conditioning), flowing agents for plastic foams, etc.
Trichlorofluoromethane (CFC-11)	CCl_3F	
Chlorodifluoromethane (HCFC-22)	$CHClF_2$	
1,1,1,2-Tetrafluoroethane (HCFC-134a)	$CF_3\text{-}CHF$	
1,1-Dichloro-1-fluoroethane (HCFC-141b)	$CCl_2F\text{-}CH_3$	
Dichloromethane (Methylene chloride)	CH_2Cl_2	solvent
Trichloroethene (Trichloroethylene, TRI)	$CHCl=CCl_2$	solvent
Tetrachloroethene (Tetrachloroethylene, Perchloroethylene, PER)	$CCl_2=CCl_2$	solvent
1,1,1-Trichloroethane	$CCl_3\text{-}CH_3$	solvent

environments (i.e., in the absence of oxygen), such polyhalogenated hydrocarbons may be reduced to less halogenated compounds by a process called *reductive dehalogenation*. We should note that substitution of a carbon-bound hydrogen (oxidation state +I) by a halogen (−I) increases the oxidation state of the corresponding carbon atom by +II. Hence, for example, in tetrachloromethane (CCl_4) the carbon atom is fully oxidized (i.e., its oxidation state is +IV). Therefore, it should not be surprising that some of the polyhalogenated hydroarbons are very effective electron acceptors in abiotic and biological redox reactions.

The C_1- and C_2-halocarbons are, of course, not the only group of persistent polyhalogenated hydrocarbons that are of great environmental concern. Because of their hydrophobic character and tendency to bioaccumulate (see Chapter 10), a variety of *polychlorinated aromatic hydrocarbons* have gained a very bad reputation. Recently, some of these classical environmental contaminants have again been subject of considerable debates, because some of the compounds are suspected to exhibit endocrine-disrupting properties (i.e., they may disturb the hormonal balance in organisms; see Keith, 1997). Examples of such compounds include the *polychlorinated biphenyls* (PCBs) and the *polychlorinated terphenyls* (PCTs) (see Fig. 2.14). These chemicals are commonly used as complex technical mixtures of different *congeners;* that is isomers and compounds exhibiting different numbers of chlorine atoms but having the same source (i.e., chlorination of a biphenyl skeleton). Note that there are 209 possible PCB congeners (Ballschmiter et al., 1992), and as many

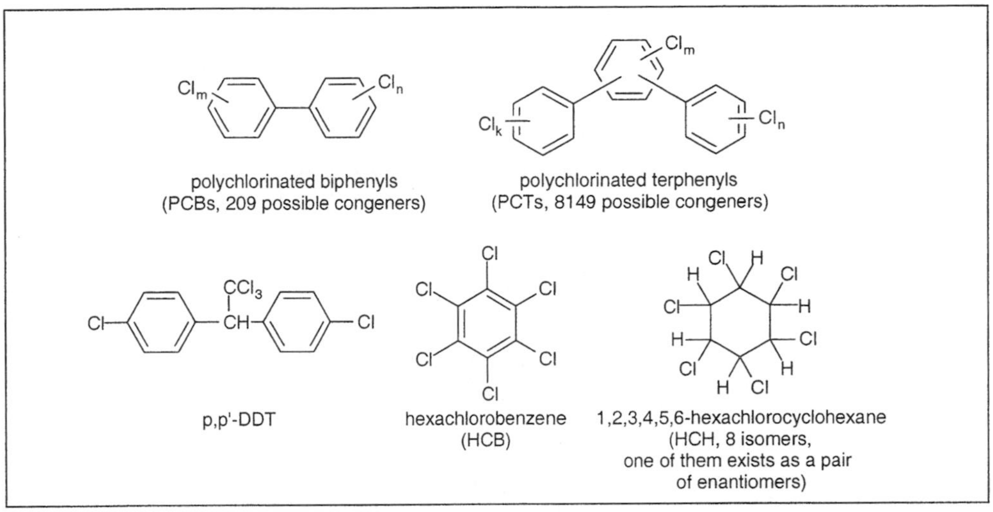

Figure 2.14 Some examples of classical polychlorinated hydrocarbons that are globally distributed.

as 8149 possible PCT congeners (Wester et al., 1996). To date, more than 1 million metric tons of PCBs and PCTs have been produced and have been used in waxes, printing inks, paints and lacquers and as capacitor dielectric fluids, transformer coolants, hydraulic fluids, heat-transfer fluids, lubricants, plasticizers, and fire retardants. They are lost to the environment during production and storage, and particularly from disposal sites. Although PCBs and PCTs have been banned or at least severely restricted in numerous countries, they are still ubiquitous in the environment. Together with other polychlorinated hydrocarbons (e.g., the classical organochlorine pesticides, p,p'-DDT, HCB, and HCH; see Fig. 2.14), they can be detected everywhere in the world, at the bottom of the ocean as well as in arctic snow, indicating the powerful environmental transport mechanisms acting on such chemicals (Vallack et al., 1998; Kallenborn et al., 1998). A group of compounds that is structurally related to the PCBs, and that cause similar environmental problems, are the *polybrominated biphenyls* (PBBs) that are used in large quantities as flame retardants (de Boer et al., 2000).

Finally, we should note that large amounts of known and unknown chlorinated compounds are formed and released to the environment due to the use of chlorine in waste and drinking water disinfection, and in bleaching processes in the pulp and paper industry. Important in this group are, for example, the trihalomethanes (THMs; $CHCl_3$, $CHBrCl_2$, $CHBr_2Cl$, and $CHBr_3$).

Oxygen-Containing Functional Groups

Among the heteroatoms present in natural and anthropogenic organic compounds, oxygen plays a unique role because it is part of a large number of important functional groups. Due to its high electronegativity (Table 2.3), oxygen forms polar bonds with hydrogen, carbon, nitrogen, phosphorus, and sulfur. As a consequence, these functionalities have a significant impact on the physical-chemical properties as well as on the reactivity of a given compound. In the following we will discuss some functional groups involving solely oxygen. Additional oxygen-containing functionalities will be addressed below when discussing functional groups exhibiting also other heteroatoms (i.e., N, S, P). Finally, we should note again that, in general,

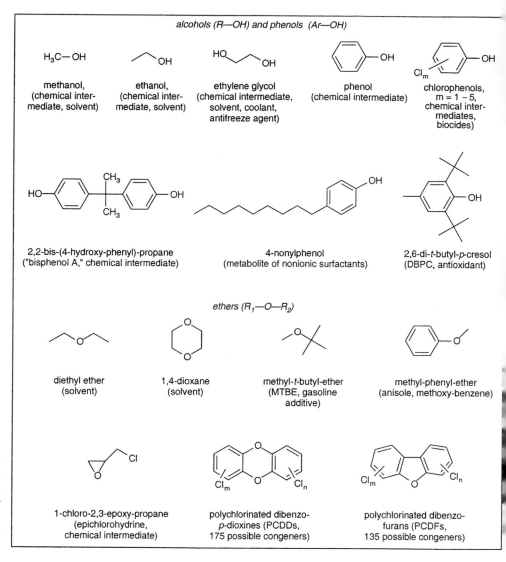

Figure 2.15 Examples of important industrially and/or environmentally relevant chemicals exhibiting either alcohol or ether functionalities. Some of the common uses and/or sources of the chemicals are given in parentheses.

oxygen forms two bonds and that its oxidation state in organic molecules is commonly –II, except for peroxides (e.g., R–O–O–R), where its oxidation state is –I.

Alcohol and Ether Functions. Let us start by looking at the simplest oxygen functions, that is, an oxygen atom that forms single bonds to a carbon and a hydrogen atom, or to two carbon atoms. The former group is called a *hydroxyl* group or *alcohol* function (R–OH), while in the latter case we speak of an *ether* function (R$_1$–O–R$_2$). Some illustrative examples of industrially and/or environmentally relevant chemicals exhibiting such functional groups are given in Fig. 2.15.

As we will see in Part II, both alcohol and ether functions have a great influence on the physical-chemical properties and the partitioning behavior of an organic chemical, because of the ability of the oxygen atom to participate in hydrogen bonds. We should, however, note that this ability is significantly greater in the case of the hydroxyl group, because R–OH may act as both H-donor and H-acceptor, while R$_1$–O–R$_2$ is only an H-acceptor. Furthermore, depending on the nature of R, the

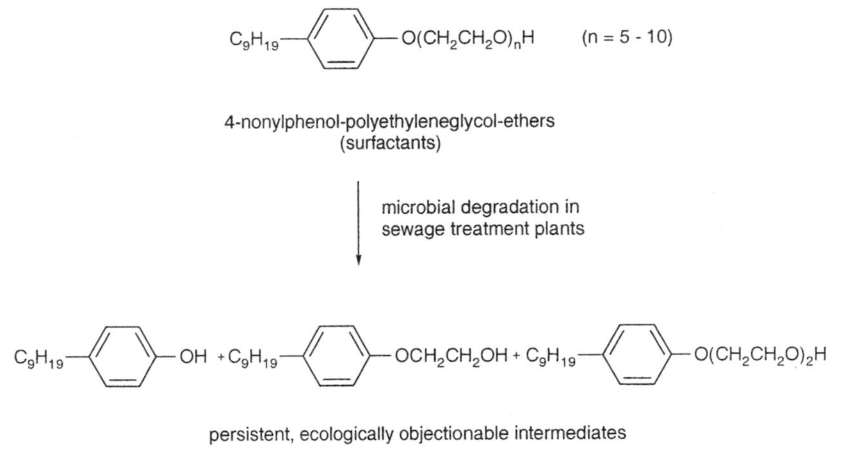

Figure 2.16 One example of a group of widely used chemicals that, as a consequence of biological waste-water treatment, are converted into persistent objectionable degradation intermediates.

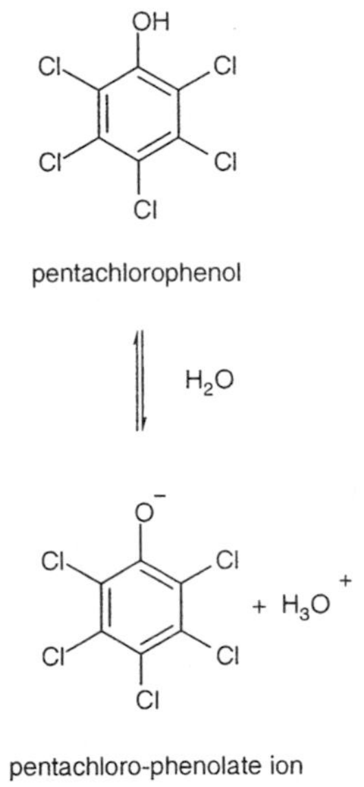

R–OH group may dissociate in aqueous solution; that is, it may act as a weak acid. This is primarily the case for aromatic alcohols, including *phenols*, that are substituted with electron-withdrawing substituents. A prominent example is the biocide pentachlorophenol (see margin), which in aqueous solution at pH 7.0, is present primarily (> 99%) as pentachloro-phenolate anion. As will be discussed in detail in Chapter 8 and in Part III, such charged species have very different properties and reactivities as compared to their neutral counterparts. Finally, we should note that phenols, particularly those that are substituted with electron-donating substituents (e.g., alkyl groups), may be oxidized in the environment, leading to a variety of products (Chapters 14 and 16). Some phenols that are especially easily oxidized and that form stable radicals are even used as antioxidants (e.g., DBPC, Fig. 2.15) in petroleum products, rubber, plastics, food packages, animal feeds, and so on (Kirk-Othmer, 1992).

Before we turn to other functional groups a few remarks about some of the compounds listed in Fig. 2.15 are necessary. First, we should note that with an annual global production of between 15 to 20 million metric tons (!), methanol and methyl-*t*-butyl ether (MTBE) are among the top high-volume industrial chemicals. While methanol does not have to be considered as a major environmental problem, MBTE, which is used as oxygenate in gasoline to improve the combustion process, has become a widespread pollutant in the aquatic environment, particularly in groundwater (Squillace et al., 1997). The reason is the significant water solubility of MBTE and its rather high persistence toward biodegradation. In Chapter 7 we will address the issue of the dissolution of chemicals from complex mixtures such as gasoline into water.

Phenols. Phenolic compounds are used in very large quantities for a variety of industrial purposes. They may also be formed in the environment by abiotic (e.g., hydroxylation in the atmosphere; see Chapter 16) or biological processes (Chapter 17). A prominent example for the latter case is the formation of 4-nonylphenol from the microbial

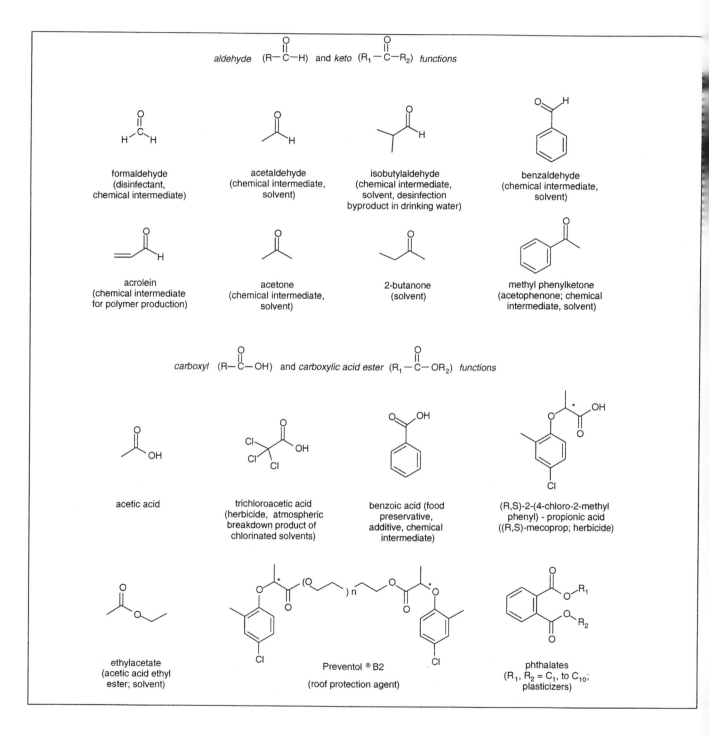

Figure 2.17 Examples of important industrially and/or environmentally relevant chemicals exhibiting carbon–oxygen functions with a doubly bound oxygen. Some of the common uses and/or sources of the chemicals are given in parentheses.

degradation of 4-nonylphenol polyethyleneglycol ethers (Fig. 2.16) that are used as nonionic surfactants (Giger et al., 1984; Ahel et al., 1996). We will make some general comments on surfactants later when discussing some other types of commercially important surface active chemicals. The case of the 4-nonylphenol polyethyleneglycol ethers is an illustrative example demonstrating that under certain circumstances, microbial transformation processes may lead to compounds that are of significantly greater concern than the parent compounds. Particularly 4-nonylphenol has been shown to exhibit some endocrine-disrupting activity. This is the major reason why in many countries the use of 4-nonylphenol polyethyleneglycol ethers has been significantly restricted.

Polychlorinated Dibenzo-(p)-Dioxins and Dibenzo-Furans. Another group of compounds that we need to specifically address are the polychlorinated *dibenzo-p-dioxins* (PCDDs) and *dibenzo-furans* (PCDFs) (Fig. 2.15). The PCDDs and PCDFs are not intentionally produced but are released into the environment from various combustion processes and as a result of their occurrence as unwanted byproducts in various chlorinated chemical formulations (e.g., chlorinated phenols, chlorinated phenoxy herbicides; see Alcock and Jones, 1996). Because some of the PCDD and PCDF congeners are very toxic (e.g., 2,3,7,8-tetrachloro dibenzo-*p*-dioxin, see margin), there have been and still are considerable efforts to assess their sources, distribution, and fate in the environment. Similarly to the PCBs or DDT (see above), the PCDDs and PCDFs are highly hydrophobic and very persistent in the environment. It is therefore not surprising that they have also been detected everywhere on earth (Brzuzy and Hites, 1996; Lohmann and Jones, 1998; Vallack et al., 1998). Finally, we should note that polybrominated diphenylethers (PBDEs, see margin) that, like the PBBs (see above), are used as flame retardants, are of increasing environmental concern (de Boer et al., 2000).

Aldehyde and Keto Functions. There are a number of other carbon–oxygen functions in which one oxygen forms a *double bond* with the carbon atom. If there is only one oxygen involved, depending on whether the carbon is bound to a hydrogen or not, one refers to the function as an *aldehyde* (R–CHO) or *keto* (R_1–CO–R_2) group, respectively (Fig. 2.17). As in the case of the ether oxygens, the aldehyde and keto oxygens are H-acceptors, which makes some simple aldehydes and ketones suitable solvents for many purposes (e.g., acetaldehyde, acetone, 2-butanone; see Fig. 2.17). Furthermore, because these functional groups are quite reactive, many important chemical intermediates used in industry exhibit an aldehyde or keto group. Finally, we should note that a variety of simple aldehydes such as isobutyraldehyde (Fig. 2.17) may be formed during water treatment. Since some of these compounds have an extremely unpleasant odor, their formation (e.g. as disinfection byproducts in drinking water) may cause considerable problems (Froese et al., 1999).

Carboxyl Groups. If we oxidize an aldehyde group, that is, if we replace the hydrogen by hydroxyl, we obtain an acidic function that is referred to as *carboxylic acid* (R–COOH) group (Fig. 2.17). As the name implies, carboxylic acids may dissociate in aqueous solution. Depending on R, their pK_a-values (see Chapter 8) are in the range between 0 and 6. Therefore, compounds exhibiting one or several carboxylic groups are present in natural waters primarily in dissociated (anionic) forms. Furthermore, we should point out that carboxylic acid functions are both strong H-donors and H-acceptors. Hence, the presence of a carboxylic function increases the water solubility of a compound significantly. The presence of such compounds in the environment may be due to direct input (e.g., as herbicides such as mecoprop, Fig. 2.17) or due to abiotic or biological transformation of other compounds (see Part III). For example, the halogenated acetic acids, including mono-, di-, and trichloroacetic acid, as well as trifluoroacetic acid, have been detected in high concentrations ($> 1 \ \mu g \cdot L^{-1}$) in rainwater. These acids are considered to be, at least in

2,3,7,8-tetrachloro-
dibenzo-*p*-dioxin

polybrominated diphenylethers
(PBDEs)

part, atmospheric oxidation products of some of the chlorinated and fluorinated ethanes and ethenes discussed above (see, e.g., Berg et al., 2000). Carboxylic acids in the environment are also formed from *hydrolysis* reactions of carboxylic acid *derivatives* such as *carboxylic acid esters* and *amides* (see below). We will discuss such reactions in detail in Chapter 13.

Carboxylic Acid Ester Functions. As illustrated by Fig. 2.17, we generally talk of an ester function if we replace the OH of an acid by an OR group. Hence, because of the lack of an OH, an ester function (R_1–$COOR_2$) can only act as a H-acceptor. Therefore, it has a smaller impact on a compound's water solubility as compared to the corresponding acid group. Ester functions can be found in many natural (e.g., oils and fats) and anthropogenic chemicals. One important group of man-made chemicals exhibiting carboxylic ester functions are the *phthalates* (Fig. 2.17), which are diesters of phthalic acid. The R_1 and R_2 in Fig. 2.17 denote hydrocarbon groups, in most cases alkyl groups consisting of between 1 and 10 carbon atoms. The annual global production of phthalates, which are used primarily as plasticizers, exceeds 1 million metric tons. Phthalates are ubiquitous in the environment, and they are among the most notorious laboratory contaminants encountered when analyzing environmental samples. Carboxylic ester functions are also frequently encountered in compounds used as pesticides, or, more generally, as biocides. In some cases, the biologically active compound is actually the "free" carboxylic acid. The ester, which can be designed to have very different physical-chemical properties, is then used to provide the active compound continuously at the target site by (slow) hydrolysis of the ester function. An example is Preventol® B2 (Fig. 2.17), which is applied to prevent root penetration in bituminous construction materials such as roofing felts, sealants, insulations, and asphalt mixtures. When hydrolyzed, Preventol® B2 releases both (R,S)-mecoprop enantiomers (see Fig. 2.5 and Section 2.1, discussion on chirality). In a study on the occurrence and behavior of pesticides during artificial groundwater infiltration of roof runoff (Bucheli et al., 1998a), both mecoprop enantiomers were detected in roof runoff in much higher concentrations (up to 500 μg $\cdot$ L^{-1}) than in the corresponding rainwater. It could be shown (Bucheli et al., 1998b) that Preventol® B2 present in bituminous sealing sheets used on flat roofs was the source of these compounds. A comparison of the (R,S) mecoprop loads from flat roofs and from agricultural applications into surface waters in Switzerland revealed that these loads were in the same order of magnitude!

We conclude this brief discussion of carbon–oxygen functions by noting that, particularly among the pesticides, there are numerous compounds that are *carbonic acid* derivatives including *carbonic acid esters* ("carbonates," R_1OCOOR_2). In this functional group, the carbon atom is fully oxidized, and therefore, hydrolysis yields CO_2 and the corresponding alcohols (see Chapter 13). We will encounter other carboxylic and carbonic acid derivatives in the following when introducing additional heteroatoms.

Nitrogen-Containing Functional Groups

Among the heteroatoms present in organic molecules, nitrogen and sulfur (which will be discussed later) are somewhat special cases in that, like carbon, they may assume many different oxidation states. Therefore, there are many nitrogen- and

Table 2.5 Some Important Nitrogen-Containing Functional Groups Present in Anthropogenic Organic Compounds

Group	Name (oxidation state of nitrogen)	Group	Name (oxidation state of nitrogen)
$R_1-\overset{\overset{R_2}{\mid}}{\underset{\underset{R_4}{\mid}}{N}}{}^+-R_3$	ammonium (-III)	$R_1-NH-NH-R_2$	hydrazo (-II)
$R_1-\overset{R_2}{\underset{R_3}{N}}$	amino[a] (-III) (amine)	$R_1\overset{R_2}{\underset{}{N=N}}$	azo (-I)
$R_1-\overset{O}{C}-\overset{R_2}{\underset{R_3}{N}}$	carboxylic acid amide[a] (-III)	$R-N\overset{OH}{\underset{H}{}}$	hydroxyl-amine (-I)
$R-C\equiv N$	cyano, nitrilo (-III)	$R-N\overset{O}{}$	nitroso (+I)
$R_1\overset{O}{\underset{R_2}{N}}-\overset{}{\underset{R_4}{N}}R_3$	urea (-III)	$R-\overset{+}{N}\overset{O^-}{\underset{O}{}}$	nitro (+III)
$R_1\overset{O}{\underset{R_2}{N}}-O-R_3$	carbamate (-III)	$R-O-\overset{+}{N}\overset{O^-}{\underset{O}{}}$	nitrato (+V) (nitrate)

[a]Primary if $R_2 = R_3 = H$; secondary if $R_2 = H$ and $R_3 \neq H$; tertiary if $R_2 \neq H$ and $R_3 \neq H$.

sulfur-containing functional groups exhibiting very different properties and reactivities. Table 2.5 gives an overview of the most important functional groups containing nitrogen, carbon and/or oxygen, but no other heteroatoms. As is evident from this table, nitrogen forms, in general, three, and in some special cases, four bonds, and it may assume oxidation states between –III and +V. In the following we will confine our discussions to a few of the functional groups listed in Table 2.5. Some of the others will be addressed in later chapters (e.g., the carboxylic and carbonic acid derivatives including amides, nitriles, ureas, and carbamates will be discussed in the context of hydrolysis reactions in Chapter 13).

Amino Group. Nitrogen-containing groups often have a significant impact on the physical-chemical properties and reactivities of environmentally relevant chemicals in which they occur (Fig. 2.18). A very important group is the *amino group* that is present in numerous natural (e.g., amino acids, amino sugars) and anthropogenic chemicals. For example, aromatic amino compounds, particularly *aminobenzenes* (*anilines;* Fig. 2.18) are important intermediates for the synthesis of numerous different chemicals including dyes, pharmaceuticals, pesticides, antioxidants, and many more (Fishbein, 1984). Furthermore, benzene rings carrying amino groups are present in a variety of other, more complex molecules. In addition, as we will see in Chapters 13 and 14, hydrolysis of various acid derivatives including the ones mentioned above, as well as reductive transformations of aromatic azo- and nitro-

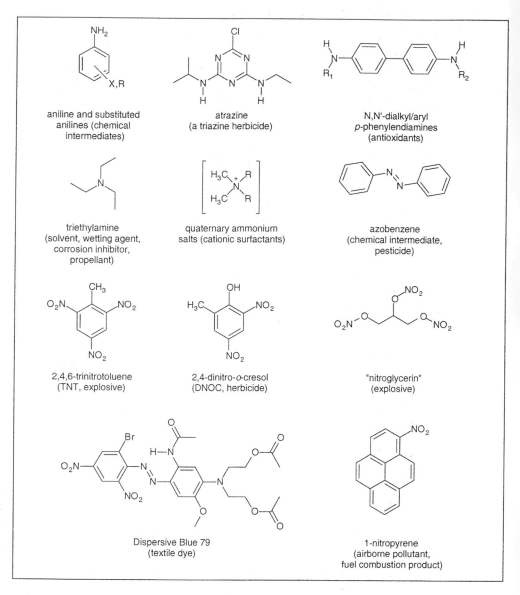

Figure 2.18 Examples of industrially and/or environmentally important chemicals exhibiting primarily nitrogen-containing functional groups. Some of the common uses and/or sources of the chemicals are given in parentheses.

aniline and substituted anilines (chemical intermediates)

atrazine (a triazine herbicide)

N,N'-dialkyl/aryl p-phenylendiamines (antioxidants)

triethylamine (solvent, wetting agent, corrosion inhibitor, propellant)

quaternary ammonium salts (cationic surfactants)

azobenzene (chemical intermediate, pesticide)

2,4,6-trinitrotoluene (TNT, explosive)

2,4-dinitro-o-cresol (DNOC, herbicide)

"nitroglycerin" (explosive)

Dispersive Blue 79 (textile dye)

1-nitropyrene (airborne pollutant, fuel combustion product)

compounds (see below), may lead to the formation of aromatic amino compounds in the environment.

From a chemical point of view, amino groups have several effects on the properties and on the reactivity of a given compound. Amino groups may engage in hydrogen bonding, both as H-acceptors and (to a somewhat lesser extent) as H-donors (only primary and secondary amines; see footnote in Table 2.5). However, in contrast to the weakly acidic OH, amino groups are *weak bases*. Hence, in aqueous solution they may acquire a proton, thus forming a cationic *ammonium* species (see Chapter 8). This is particularly important for aliphatic amino groups (e.g., triethylamine, Fig. 2.18) that are significantly stronger bases as compared to aromatic amino groups. Note that for use as surfactants, some amines are alkylated to *quaternary ammonium compounds,* thus forming stable cations (see example given in Fig. 2.18). We will address some of the special properties of surfactants below when discussing another functional group that is widely used in this type of chemicals, the sulfonate group.

A second important feature of amino groups, particularly when bound to aromatic systems, is their capability to act as a π-electron donor (see Section 2.1). An interesting example is the azo dye, Disperse Blue 79 (Fig. 2.18), where the extreme shift in light absorption (blue color) is achieved by introducing a strong electron-donating substituent (i.e., a dialkylamino group) in *para*-position to the azo group at one of the benzene rings, and two strong electron-withdrawing substituents (i.e., two nitro groups, see below) in *ortho*- and *para*-position at the other benzene ring. We will discuss the effect of such substituents on the light absorption of organic compounds, and we will explain the reason why aromatic azo compounds absorb light quite strongly in the visible wavelength range (hence their use as dyes) in Chapter 15 when treating direct photolysis reactions.

Like some phenolic groups, amino groups bound to an aromatic system (e.g., benzene) may be oxidized in the environment. In some cases such transformations may lead to the formation of products that are of considerable concern (Chapter 14). Finally, some aromatic amines that form stable radicals are actually also used as antioxidants (see example given in Fig. 2.18 and Kirk-Othmer, 1992).

Nitro Group. The second nitrogen-containing functional group to discuss here is the nitro group. Numerous synthetic high-volume chemicals contain one or several nitro groups that are bound to an aromatic system, in particular to a benzene ring. Aromatic compounds exhibiting nitro groups are pivotal intermediates in various branches of the chemical industry, and they are also present in many end products including explosives (e.g., TNT, Fig. 2.18), agrochemicals (e.g., DNOC, Fig. 2.18), and dyes (e.g., Dispersive Blue 79, Fig. 2.18). In addition, a large number of nitroaromatic compounds are formed in the atmosphere by photochemically induced reactions of aromatic hydrocarbons involving hydroxyl radicals and nitrous oxides (Atkinson, 2000). Furthermore, such compounds have also been shown to be produced during fuel combustion, for example, in car engines (Tremp et al., 1993). Hence, such nitration processes may lead to a significant input of quite toxic compounds in the environment (e.g., nitropyrene, Fig. 2.18; see also Fiedler and Mücke, 1991).

In order to understand how a nitro group affects the properties and reactivity of a compound, we recall that the nitro group has a strong *electron-withdrawing* character and that it may delocalize π-electrons (see Chapter 8). Thus, particularly aromatic nitro groups have a big influence on the electron distribution in a molecule (or in parts of the molecule). This affects properties such as the acidity constant of an acid or base group bound to an aromatic system (Chapter 8), the specific interaction of aromatic compounds with electron donors such as oxygen atoms present at clay mineral surfaces (Chapter 11), or the light absorption properties of the compound (Chapter 15). Furthermore, because in a nitro group the nitrogen exhibits an oxidation state of +III (Table 2.5), this group may act as an oxidant (i.e., as an electron acceptor). This is most evident in explosives (e.g., TNT, Fig. 2.18) where the nitro group functions as built-in oxidant, thus ensuring a very fast oxidation of the molecule leading to the liberation of a large amount of energy in a very short time period. This effect can also

Table 2.6 Some Important Sulfur-Containing Functional groups Present in Anthropogenic Organic Compounds

Group	Name (oxidation state of sulfur)	Group	Name (oxidation state of sulfur)
R—SH	thiol, mercaptan (-II)	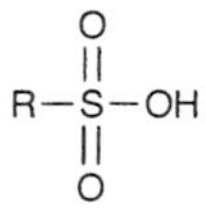 O‖R—S—OH‖O	sulfonic acid (+IV)
R₁—S—R₂	thioether, sulfide (-II)	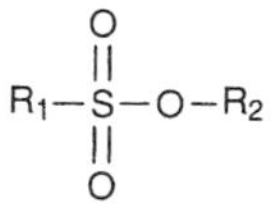 O‖R₁—S—O—R₂‖O	sulfonic acid ester (+IV)
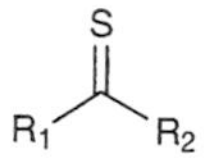 S‖R₁⋀R₂	thiocarbonyl (-II)	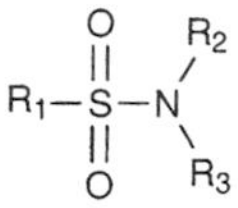 O R₂‖ /R₁—S—N‖ \O R₃	sulfonic acid amide, sulfonamide (+IV)
R₁—S—S—R₂	disulfide (-I)	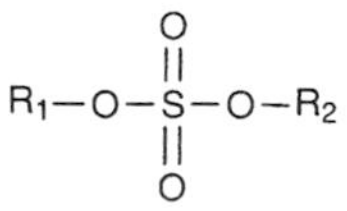 O‖R₁—O—S—O—R₂‖O	sulfuric acid ester, sulfate (+VI)
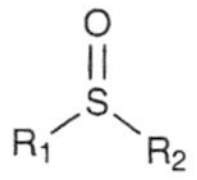 O‖S R₁ R₂	sulfoxide (0)		
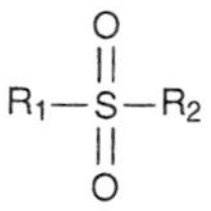 O‖R₁—S—R₂‖O	sulfone (+II)		

be achieved with nitrate groups (e.g., nitroglycerin, Fig. 2.18). Finally, in the environment, under reducing conditions, nitro groups may be reductively transformed into nitroso-, hydroxylamine-, and ultimately, amino groups (see Table 2.5), yielding products that may be of similar or even greater concern than the parent nitro compound (Chapter 14). This is particularly important when dealing with the assessment of ammunition waste sites, which, in many countries, cause severe problems with respect to soil and groundwater contaminations (Haderlein et al., 2000).

Sulfur-Containing Functional Groups

Sulfur in organic molecules may assume various different oxidation states (i.e., –II to +VI, see examples given in Table 2.6). Because sulfur is a third-row element with d-orbitals, it is capable of *valence-shell expansion* to accommodate more electrons than are allowed by the simple octet rule. Hence, as can be seen from Table 2.6, sulfur may be surrounded by 10 (e.g., sulfoxide) or even 12 (e.g., sulfone, sulfonic acid and its derivatives) electrons, and this capacity allows for (redox) reactions that are inaccessible to the corresponding oxygen analogues. Other important differences from oxygen include its significantly smaller electronegativity (Table 2.3) and its much weaker tendency to be engaged in H-bonding. Furthermore, as compared to oxygen, sulfur forms weaker bonds with carbon and hydrogen (Table 2.3), causing mercapto groups to be more acidic (Chapter 8) and more nucleophilic

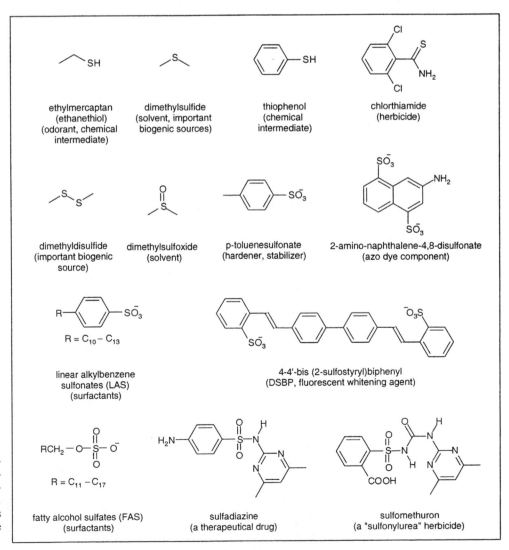

R₁, R₂ sulfoxides

sulfone

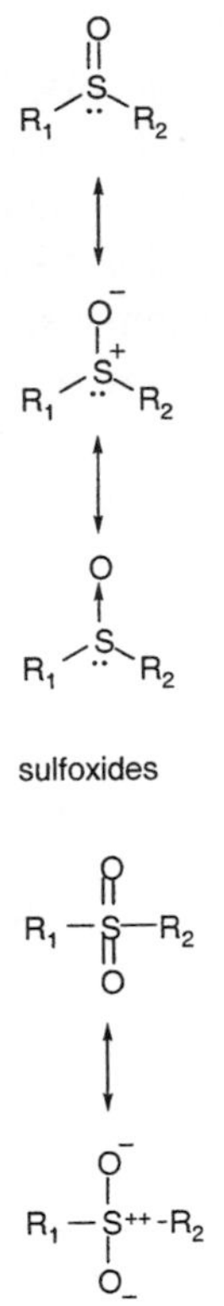

Figure 2.19 Examples of "special" double bonds between sulfur and oxygen.

(Chapters 13 and 14) as compared to their oxygen analogues. Finally, we should point out that sulfur may not only form regular π-double bonds (as in a thiocarbonyl, Table 2.6), but may also undergo a special type of double bond in which the sulfur's d-orbitals participate (indicated by an arrow between S and O in Fig. 2.19). Such double bonds exist in functional groups in which sulfur is engaged in more than two bonds. In such cases we may represent the functional group by two resonance structures, but the bond is nevertheless localized. We should note that this type of double bond does not significantly change the geometry at the atoms involved (in contrast to the regular π-bond). Hence, the geometry at the sulfur atom is pyramidal. In fact, unsymmetrical sulfoxides ($R_1 \neq R_2$ in Fig. 2.19) are chiral (like an asymmetrically substituted carbon atom, see Section 2.2) in that the oxygen and the lone electron pair play the role of the other two substituents.

Figure 2.20 gives some examples of chemicals in which sulfur-containing functional groups govern the compound's properties and/or reactivity. Among the compounds exhibiting *reduced* sulfur atoms (–I, –II), the low-molecular-weight mercaptans,

Figure 2.20 Examples of industrially and/or environmentally important chemicals exhibiting a sulfur-containing functional group. Some (not all) of the common uses and/or sources of the chemicals are given in parentheses.

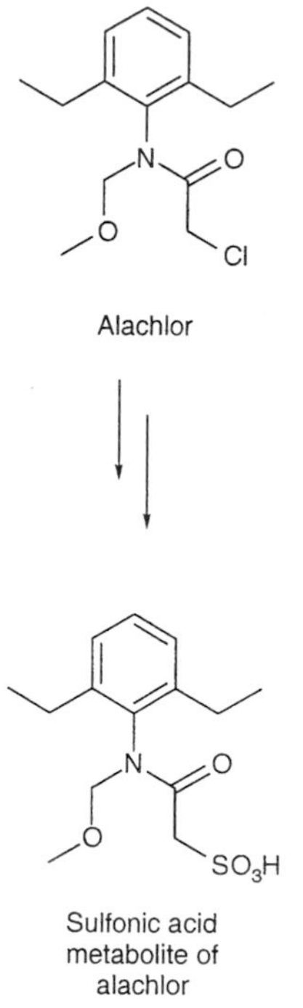

Alachlor

Sulfonic acid
metabolite of
alachlor

Figure 2.21 Gluthathione S-transferase enzyme-mediated reaction ultimately yielding a sulfonated metabolite (from Field and Thurman, 1996).

dialkylsulfides, and dialkyldisulfides have anthropogenic and very important biogenic sources. For example, dimethyl sulfide is one of the major volatile sulfur compounds in the global sulfur cycle. Furthermore, thioether (R_1–S–R_2) or thioester (R_1–CO–SR_2) groups can be found in many pesticides. Finally, again in the world of pesticides, a C=S (e.g., chlorthiamide, Fig. 2.20) and, particularly, a P=S function (see below) is sometimes introduced instead of a C=O or P=O function, respectively. This change is made primarily with the goal of lowering the general toxicity of the compound. In some cases, the sulfur is then enzymatically replaced by an oxygen in the target organism(s), thus increasing the toxic effect of the compound at the target site.

With respect to the groups exhibiting oxidized sulfur atoms, aromatic sulfonic acid groups, in particular, benzene- and naphthalene sulfonic acids, are the most important ones (see in Fig. 2.20). Because sulfonic acid groups have very low pK_a values (< 1, Chapter 8), they dissociate completely in aqueous solution, thus forming the corresponding negatively charged sulfonates. Hence, such groups significantly increase the *hydrophilicity* (and thus the water solubility) of a compound. This is the major reason why aromatic sulfonate groups are present in a variety of commercially important chemicals including surfactants (e.g., linear alkylbenzene sulfonates, LAS, Fig. 2.20), anionic azo dyes, fluorescent whitening agents (Fig. 2.20), construction chemicals, and many more. In addition, introduction of a sulfonic acid group with the goal of forming a highly water-soluble metabolite is a detoxification strategy of a variety of organisms (see example given in Fig. 2.21). It is therefore not too surprising that numerous compounds exhibiting aromatic sulfonate groups have been detected in waste waters and leachates, as well as in natural waters (Altenbach and Giger, 1995; Field and Thurman, 1996; Stoll and Giger, 1998; Suter et al., 1999).

Among the compounds containing sulfonate groups the linear alkylbenzene sulfonates (LAS) have drawn particular interest, because they are produced and used in very large quantities (i.e., > 2 million metric tons per year; Ainsworth, 1996). They have been widely detected in aquatic (Matthis et al., 1999) as well as in terrestrial environments (where they are introduced primarily as a result of sewage sludge amendment; Jensen, 1999). We should note that LAS as well as other commercially available surfactants [e.g., alkylphenol polyethylene glycol ethers (Fig. 2.16), quarternary ammonium salts (Fig. 2.18), or fatty alcohol sulfates (Fig. 2.20)] are not single substances, but are mixtures of compounds of different carbon chain lengths. Owing to their *amphiphilic* character (partly hydrophilic and partly hydrophobic), the surfactants have special properties that render them unique among environmental chemicals. In aqueous solutions they distribute in such a manner that their concentration at the interfaces of water with gases or solids is higher than in the inner regions of the solution. This results in a change of system properties, for example, a lowering of the interfacial tension between water and an adjacent nonaqueous phase, and in a change of wetting properties. Furthermore, inside the solution, on exceeding certain concentrations surfactants form aggregates, called *micelles*. Hence, surfactants may keep otherwise-insoluble compounds in the aqueous phase, and they form an important part of any kind of detergent. Surfactants are also widely used as wetting agents, dispersing agents, and emulsifiers in all kinds of consumer products and industrial applications (Piorr, 1987).

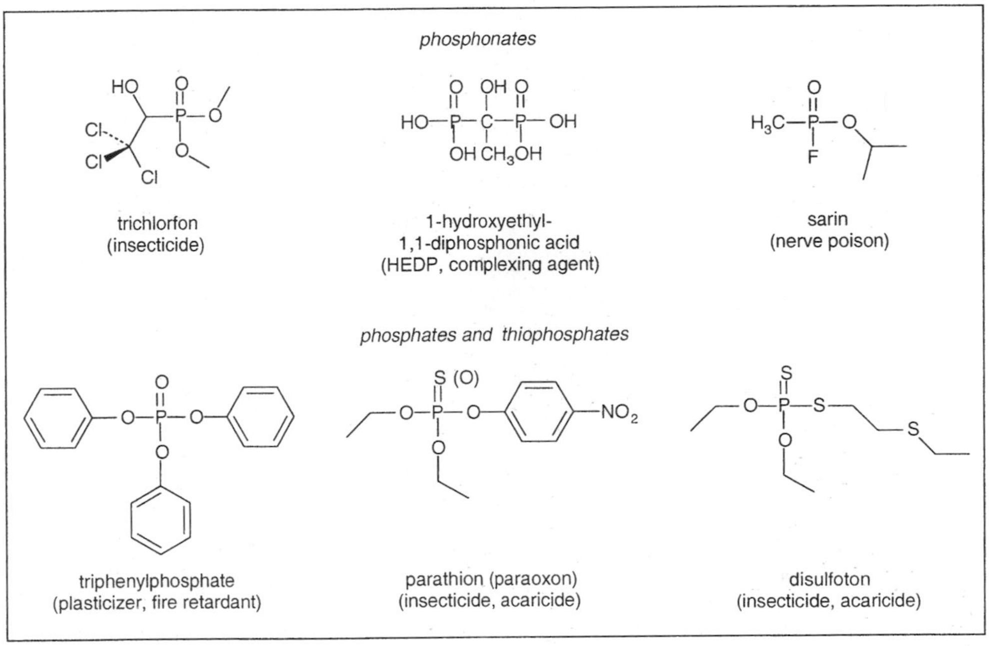

Figure 2.22 Examples of important anthropogenic chemicals exhibiting a phosphorus-containing functional group. Some of the common uses of the chemicals are given in parentheses.

We conclude this section by noting that, as in the case of other acids, there are various sulfonic and sulfuric acid derivatives including esters and amides. Fig. 2.20 shows two examples of sulfonic acid amides that are representatives of important groups of drugs (sulfodiazine) and herbicides (sulfometuron), respectively.

Phosphorus-Containing Functional Groups

Although in principle, phosphorus may, like nitrogen, assume a variety of oxidation states (i.e., –III to +V), in environmentally relevant compounds, it is present primarily in the more oxidized forms [i.e., +III (e.g., phosphonic acid derivatives) and, +V (e.g., phosphoric and thiophosphoric acid derivatives); see Fig. 2.22]. In these oxidation states, P forms in most cases three single bonds and one double bond, the latter commonly with either an oxygen or a sulfur atom. As in the case of the other third-row element sulfur (Fig. 2.19), these double bonds are "special" double bonds (Fig. 2.23) that do not change the geometry at the atoms involved. From the examples given in Fig. 2.22 it can be seen that particularly esters and thioesters of phosphonic-, phosphoric-, and thiophosphoric acid are used for a variety of purposes including as plasticizers and/or flame retardants (Carlsson et al., 1997), and as pesticides, mostly as insecticides and acaricides (Tomlin, 1994; Hornsby et al., 1996). We will address these types of compounds in more detail in Chapter 13 when discussing hydrolysis reactions. Finally, we note that phosphonates (exhibiting one P–C bond) that contain several phosphonic acid groups are increasingly used as chelating agents (Nowack, 1998). Some phosphonic acid derivatives are extremely toxic nerve gases and, therefore, used as chemical weapons (e.g., sarin, Fig. 2. 22).

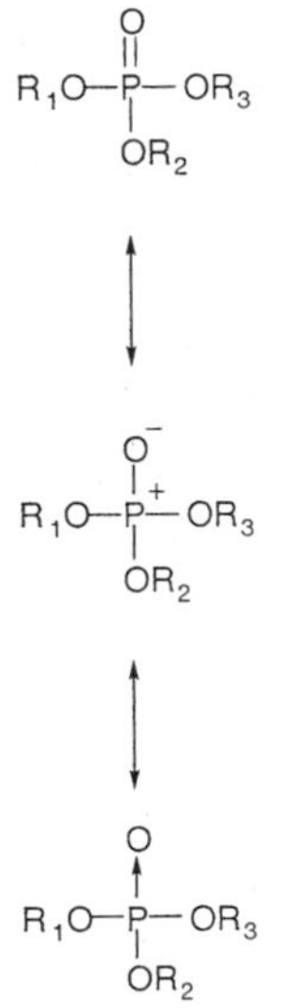

Some Additional Examples of Compounds Exhibiting More Complex Structures

To round out our tour through the jungle of environmental organic chemicals, we shall have a brief look at some additional examples of compounds that have more

Figure 2.23 Examples of a "special" double bond between phosphorus and oxygen indicated by the arrow between O and P.

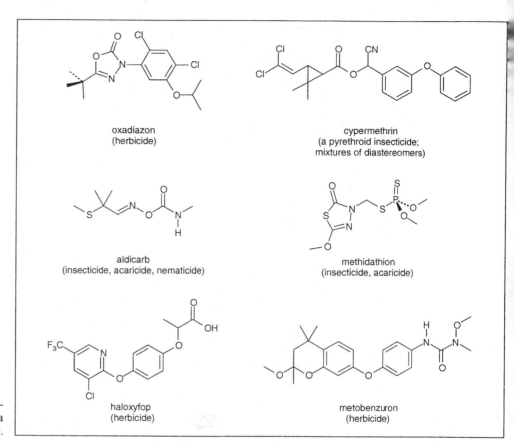

Figure 2.24 Some additional examples of pesticides exhibiting a somewhat more complex structure.

complex structures, both with respect to the carbon skeleton as well as with respect to the type and number of functional groups (Figs. 2.24 and 2.25). Obviously, assessing the environmental behavior of such compounds poses a particular challenge, since the mutual electronic and steric interactions of functional groups may influence the properties and reactivities of a compound in a way that cannot be *a priori* predicted from looking at the various functionalities in isolation. Nevertheless, with the general knowledge that we will acquire in Parts II and III on how structural moieties influence the environmental behavior of organic compounds, we should be able to put ourselves into the position to also tackle the assessment of such "complex" chemicals in the environment.

Some examples of compounds with structures that are somewhat more complex than the structures of the majority of the chemicals that we will encounter throughout the book include pesticides and pharmaceuticals. Such compounds are designed to be biologically highly active. Pesticides and pharmaceuticals are used by human society to exert a desired biological effect at a given target site. Obviously, from an environmental point of view, such compounds are of particular concern when they are present at locations where they are not supposed to be. For pesticides, this has been recognized for a long time, and therefore, there is quite a lot of information available on the environmental behavior of such compounds (e.g., Tomkin, 1994; Hornsby et al., 1996; Montgomery, 1997). In contrast, for pharmaceutical chemicals, and in particular, for drugs and hormones used in human and veterinary applications,

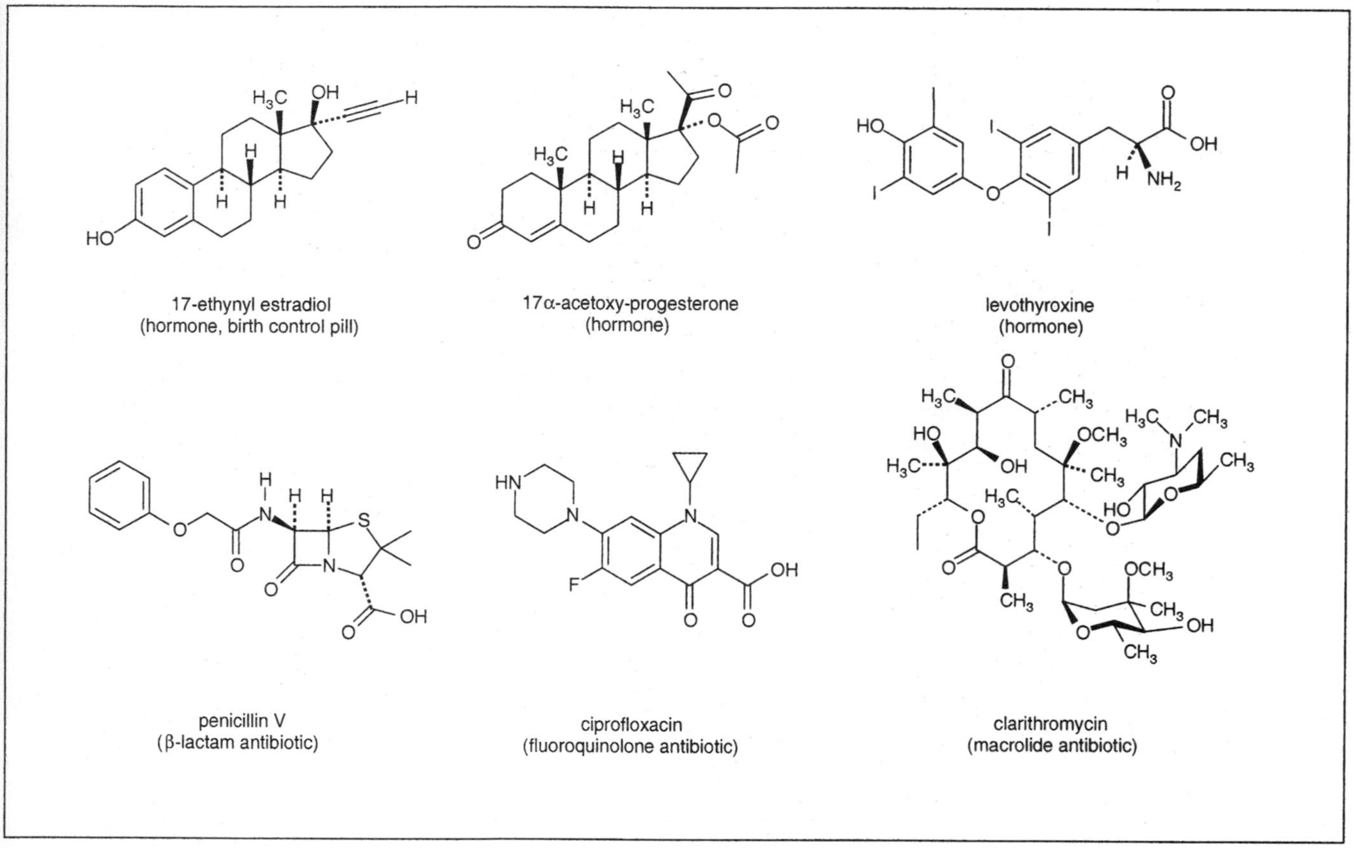

Figure 2.25 Some examples of pharmaceutical chemicals exhibiting a somewhat more complex structure. Note that such compounds have been detected in waste waters and in surface waters (Ternes and Wilken, 1999).

environmental aspects are only an emerging issue (Halling-Sørensen et al., 1998; Ternes and Wilken, 1999; Sedlak et al., 2000; Kolpin et al., 2002).

2.4   Questions and Problems

Questions

Q 2.1

Which are the most common elements encountered in organic chemicals? What is a *heteroatom* in an organic molecule?

Q 2.2

Explain the simplest model used to describe the tendency of the various elements present in organic chemicals to undergo covalent bonding. For which elements is this simple model not strictly applicable?

Q 2.3

What does the *structure* of a given compound describe? What are *structural isomers*?

Q 2.4

What types of covalent bonds exist between the atoms present in organic molecules? What factors determine the strength of covalent bonds? Give some examples of very strong and very weak bonds.

Q 2.5

Which atoms present in organic molecules may exhibit different *oxidation states*? Explain in words how you assign the oxidation states to the different atoms present in a given molecule.

Q 2.6

What is *stereoisomerism*? What type of stereoismerism exists? Give some examples of different types of stereoisomers.

Q 2.7

In what context do you use the terms *cis* and *trans*? In what context do you use *ortho*, *meta*, and *para*?

Q 2.8

Explain the terms *delocalized electrons, resonance,* and *aromaticity*. Give examples of compounds for which these terms apply.

Q 2.9

What are the major sources of the BTEX compounds in the environment? Why are these compounds considered to be a problem?

Q 2.10

What are the major sources of PAHs in the environment? What are the major problems connected with this class of compounds?

Q 2.11

Why do so many high-volume production chemicals contain several halogen atoms? Why are polyhalogenated hydrocarbons rather persistent in the environment?

Q 2.12

Give at least five examples of polyhalogenated hydrocarbons and explain what kind of environmental problems they cause.

Q 2.13

How may an OH-group influence the behavior of an organic compound in aqueous solution?

Q 2.14

What is an *ester* function? Give some examples of environmentally relevant chemicals exhibiting *carboxylic acid ester* functions.

Q 2.15

How do amino groups and nitro groups affect the properties of (a) aliphatic, and (b) aromatic compounds?

Q 2.16

Name some important differences in the chemical nature of oxygen and sulfur. How is this reflected in sulfur-containing functional groups?

Q 2.17

For which major purpose(s) are sulfonic acid groups introduced into man-made organic chemicals?

Q 2.18

What are the characteristics and the major uses of surfactants? Give some examples of important commercially available surfactants.

Q 2.19

What is special about double bonds between oxygen and sulfur or phosphorus?

Q 2.20

What are the most common phosphorus-containing functional groups encountered in environmentally relevant compounds? What are the major uses of such compounds?

Problems

P2.1 *Determining the Oxidation State of the Atoms in an Organic Molecule*

Determine the oxidation states of the numbered atoms in the following organic molecules:

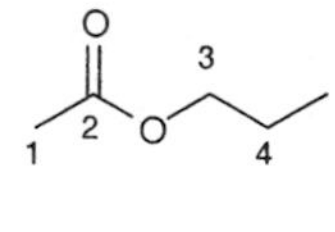

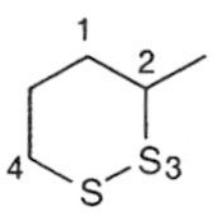

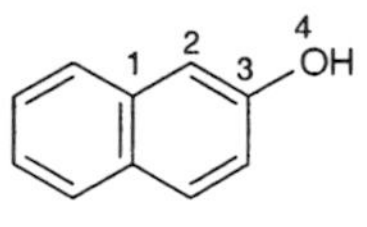

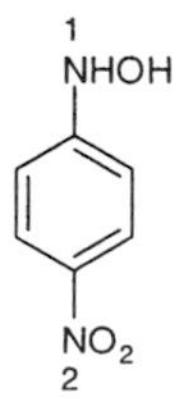

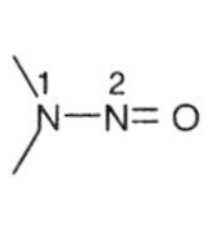

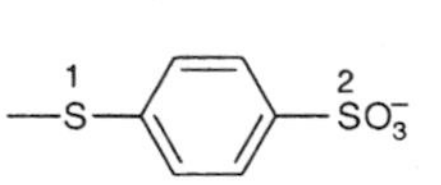

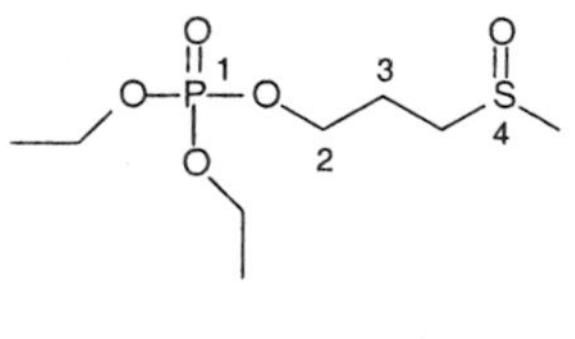

P2.2 *Assessing the Number of Stereiosomers*

Write down all possible stereoisomers of 1,2,3,5- and 1,2,4,5-tetrachlorocyclo-hexane. Which of them are chiral, that is, which ones exist as pair of enantiomers?

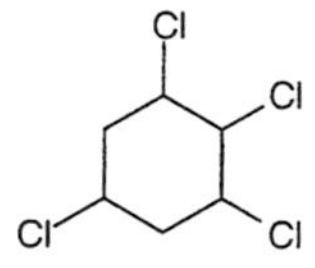

1,2,3,5-tetrachlorocyclohexane 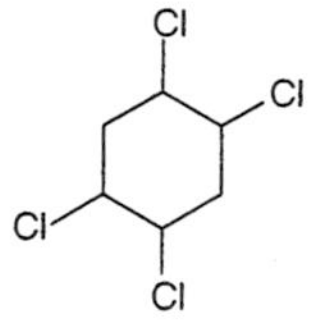   1,2,4,5-tetrachlorocyclohexane

PART II

Equilibrium Partitioning Between Gaseous, Liquid, and Solid Phases

To assess the behavior of an organic chemical in natural or engineered systems, one must determine how the molecules of that compound distribute themselves among the different environmental compartments or phases present. Such phases may include air, water, organic matter, mineral solids, and even organisms. The chapters of Part II (Chapters 3 through 11) examine the pertinent compound properties and environmental factors that are needed for quantifying such *partitioning*.

In these chapters, we focus on *equilibrium* situations and the associated problem of calculating the distribution of a compound between the different phases, when no net exchange occurs anymore. There are many situations in which it is correct to assume that phase transfer processes are fast compared to the other processes (e.g., transformations) determining a compound's fate. In such cases, it is appropriate to describe phase interchanges with an equilibrium approach. One example would be partitioning of compounds between a parcel of air and the aerosols suspended in it. Another case might be partitioning between the pore water and solids in sediment beds.

Of course, in the real world there are also many situations where partitioning equilibrium is not reached. However, even in such cases, parameters are extremely useful which characterize what the equilibrium distributions would be given sufficient time. These parameters can be used to ascertain a compound's tendency to accumulate in particular environmental compartments. For example, knowing the relevant partitioning parameters we would deduce that a freon like CF_2Cl_2 would tend to move to the atmosphere or that a pesticide like DDT would concentrate in biota. Moreover, we can use quantitative descriptions of partitioning equilibria to identify the direction of chemical transport when environmental compartments are exchanging chemicals. Hence, some time after an underground gasoline spill, we can evaluate whether gasoline components are still dissolving into the adjacent groundwater and volatilizing into the overlying soil gases. Finally, the equilibrium partition constant is needed for calculating the rate of transfer of a compound across interfaces.

As we have already pointed out in Chapter 1, one of our primary goals is to gain insight into how a compound's structure determines its fate in the environment. When dealing with phase partitioning, we need to develop an understanding of the *intermolecular interactions* between a compound's molecules and a given molecular environment in which it occurs. Thus, we need to see how structural moieties are related to measurable, macroscopic quantities. Throughout the chapters in Part II we will, therefore, put considerable effort into visualizing and quantifying (1) the interaction energies arising from contacts between the molecules (*enthalpies*), and (2) the freedom of motion of the molecules in a given phase (*entropies*). This will enable us to build a conceptual framework that allows us to treat partitioning processes in a holistic way. We will try to do this step by step. First, we will evaluate some simple, well-defined systems. These will include the partitioning of organic compounds between the gas phase and the pure compound (Chapter 4), between the pure compound and water (Chapter 5), between the gas phase and bulk liquid phases including organic solvents and water (Chapter 6), and finally between organic solvents and water (Chapter 7). For simplicity, the discussions in Chapters 4 through 7 will be confined to nonionic organic species only. The partitioning behavior of ionizable organic acids and bases will be treated separately in Chapter 8 together with a discussion of acidity constants.

Armed with the basic knowledge and insights acquired in Chapters 3 through 8, we will then be able to tackle partitioning processes involving more complex environmental phases (or matrices) including natural organic matter (NOM; Chapter 9), biota (Chapter 10), and environmentally relevant solids (Chapter 11). In these cases, the major challenge will be to identify and quantify environmental factors that determine the partitioning behavior of a given type of organic compound. Hence, in Chapters 9 through 11 the molecular nature of NOM, of biological systems, and of solid surfaces will be important issues with respect to partitioning processes of organic chemicals.

In conclusion, it should be noted that Part II is organized in a way that each chapter builds, at least in part, on the knowledge acquired in the preceeding chapters. Thus, particularly for the inexperienced reader, it is advisable to follow the chapters in the given sequence. Furthermore, even for the experienced reader, it might be useful to read Chapter 3. In this chapter, all the background theory and terminology used in the following chapters is summarized and discussed.

Chapter 3

PARTITIONING: MOLECULAR INTERACTIONS AND THERMODYNAMICS

3.1 Introduction

3.2 Molecular Interactions Determining the Partitioning of Organic Compounds Between Different Phases
Partitioning "Reactions"
Origins of Intermolecular Attractions
Box 3.1: *Classification of Organic Compounds According to Their Ability to Undergo Particular Molecular Interactions*
Relative Strengths of Dispersive Energies Between Partitioning Partners
A First Glance at Equilibrium Partition Constants
Examples of Absorption from the Gas Phase
Example 1: Vapor Pressure and Molecular Interactions in the Pure Liquid Compound
Example 2: Air–Solvent Partitioning
Examples of Adsorption from the Gas Phase
Example 3: Air–Solid Surface Partitioning

3.3 Using Thermodynamic Functions to Quantify Molecular Energies
Chemical Potential
Fugacity
Pressure and Fugacities of a Compound in the Gas State
Reference States and Standard States
Fugacities of Liquids and Solids
Activity Coefficient and Chemical Potential
Excess Free Energy, Excess Enthalpy, and Excess Entropy

3.4   **Using Thermodynaimc Functions to Quantify Equilibrium Partitioning**
Equilibrium Partition Constants and Standard Free Energy of Transfer
Effect of Temperature on Equilibrium Partitioning
Using Linear Free Energy Relationships (LFERs) to Predict and/or to
   Evaluate Partition Constants and/or Partition Coefficients
Box 3.2: *Partition Constants, Partition Coefficients, and Distribution Ratios
   – A Few Comments on Nomenclature*
Concluding Remarks

3.5   **Using Partition Constants/Coefficients to Assess the Equilibrium
Distribution of Neutral Organic Compounds in Multiphase Systems**
Illustrative Example 3.1: *The "Soup Bowl" Problem*

3.6   **Questions and Problems**

3.1   Introduction

The goal of this chapter is to summarize the concepts needed in the following chapters for a qualitative and quantitative treatment of environmental equilibrium partitioning processes. We start by developing some general understanding of the intermolecular forces that govern the partitioning of organic compounds between different phases. This is done by visualizing the interactions between molecules using examples that will be treated in more detail in later chapters. Then, in Sections 3.3 and 3.4 we discuss how molecular interactions are characterized via thermodynamic functions that enable us to quantitatively describe the distribution of molecules participating in a specific partitioning process. Here, we will focus on those thermodynamic entities that are relevant to our problem: Gibbs free energy (G), enthalpy (H), entropy (S), chemical potential (μ), fugacity (f), activity (a), and activity coefficient (γ). In Section 3.4 we also discuss some common "extrathermodynamic" approaches with which we may estimate data for new chemicals of interest from experience with old ones. Finally, in Section 3.5 we address the issue of how we may generally calculate the fraction of a substance's occurrence in any environmental phase of interest at equilibrium with other environmental media.

3.2   Molecular Interactions Determining the Partitioning of Organic Compounds Between Different Phases

Partitioning "Reactions"

The partitioning of an organic compound, i, between two phases, 1 and 2, may be thought of like a chemical reaction in which "bonds" are broken and formed. In this case, however, the "bonds" involve intermolecular attraction energies, which are, however, much weaker than covalent bonds. For example, if the process of interest involves moving i from within phase 1 (i.e., *desorption* from phase 1) to within a different phase 2 (i.e., *absorption* into phase 2), or vice versa, we may write the partitioning "reaction":

$$1{:}i{:}1 + 2{:}2 \rightleftharpoons 1{:}1 + 2{:}i{:}2 \qquad (3\text{-}1)$$

where the colons indicate intermolecular "attractions" which are broken and formed during the exchange. Here we show the compound, i, to be inside phase 1 ($1{:}i{:}1$) and phase 2 ($2{:}i{:}2$) by putting it between two numbers.

We distinguish this *absorptive* exchange (Eq. 3-1) from one in which i partitions to an interface. In this new case, the process should be viewed as an *adsorption* of i to the surface of phase 2:

$$1{:}i{:}1 + 1{:}2 \rightleftharpoons 1{:}1 + 1{:}i{:}2 \qquad (3\text{-}2)$$

Here, the reaction shown as Eq. 3-2 indicates the presence of an interface between phases 1 and 2. Unlike the case of *absorption* where attractions between 2 and 2 had to be broken and ones between 1 and 1 were made, now in this *adsorption* case the intermolecular bonds between 1 and 2 must be broken as such bonds between 1 and 1 are made.

This "reaction" point of view enables us to organize our thinking about partitioning. First, we must identify the combinations of materials that are juxtaposed before and after the partitioning process. Second, we must ascertain what kinds of chemical structure elements (e.g., $-CH_2-$, $-OH$) are present in the partitioning molecules (i.e., in i) and the material of which each participating phase is made (i.e., 1 or 2 above). This allows us to identify the kinds of intermolecular interactions that control the strengths of the "bonds" that are broken and formed. Finally, we need to consider the numbers of interactions, or areas of contact, which are changed in the process.

To understand the extent of such partitioning processes, we have to evaluate how various parts of i are attracted to structures of phases 1 and 2. It will be the summation of all these attractions that are broken and formed that will dictate the relative affinity of i for the two competing phases with which it could associate. Since these attractive forces stem from uneven electron distributions, we need to discuss where in the structures of organic chemicals and in condensed phases there are electron enrichments and deficiencies. Subsequently, we can examine the importance of these uneven electron distributions with respect to attracting molecules to other materials.

Origins of Intermolecular Attractions

The attractive forces between *uncharged molecules* generally result from the electron-deficient regions in a molecule attracting electron-rich counterparts in neighboring molecules or the atoms making up surfaces. The total affinity of molecules for one another comes from the summation of all attractions. The resulting interactions (Fig. 3.1) can be divided into two categories:

(1) "*Nonspecific*" interactions that exist between any kinds of molecules, no matter what chemical structure these molecules may have. These nonspecific interactions are generally referred to as *van der Waals (vdW)* interactions. They are a superposition of the following components:

(i) Attractions between *time-varying, uneven electron distributions* in adjacent molecules are the origin of *London dispersive energies*. The intensity of such unevenness in a particular molecule or material is related to its polarizability. As a result, the strength of intermolecular attraction energies arising from these time-varying dipoles is proportional to the product of the polarizabilities of each of the interacting sets of atoms.

(ii) Dipole-induced dipole interactions are the source of *Debye energies*. Dipoles exist within chemical structures because of the juxtaposition of atoms with different electronegativities (e.g., an oxygen bonded to a carbon atom). When such a permanent dipole moment in one chemical is juxtaposed to material with a time-averaged even electron distribution, then the first molecule causes an uneven electron distribution to form in the second material. The strength of the resultant intermolecular attraction is proportional to the product of the dipole moment of the first molecule and the

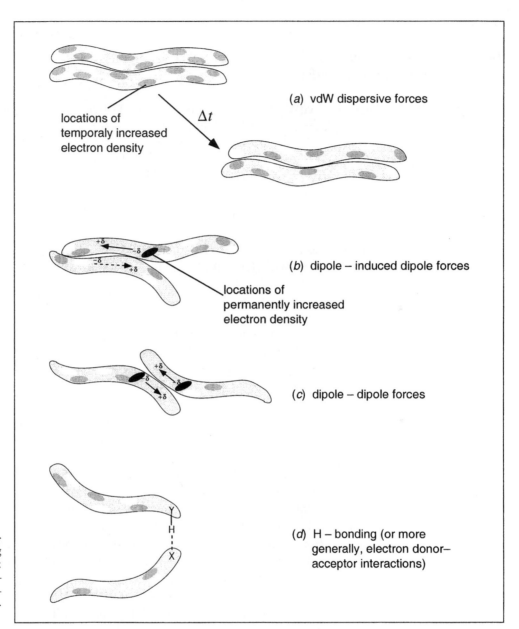

locations of
temporaly increased
electron density

Δt

(*a*) vdW dispersive forces

(*b*) dipole – induced dipole forces

locations of
permanently increased
electron density

(*c*) dipole – dipole forces

(*d*) H – bonding (or more
generally, electron donor–
acceptor interactions)

Figure 3.1 Illustration of the various molecular interactions arising from uneven electron distributions: (*a*) dispersive forces, (*b*) dipole–induced dipole forces, (*c*) dipole–dipole forces, (*d*) electron acceptor–electron donor forces.

polarizability of the second molecule.

(iii)   Dipole–dipole interactions are the cause of *Keesom energies*. In this case, permanent dipoles in each substance cause the molecules to orient so that the two dipoles face each other in a head-to-tail fashion. The strengths of these attractions are proportional to the product of the dipole moments of the two interacting molecules and depend on the orientation of the interacting partners.

(2)   *Specific* interactions (Fig. 3.1*d*) that result from particular molecular structures that enable relatively strong local attractions between permanently electron-poor parts of a chemical structure (e.g., the hydrogen attached to an oxygen) and corresponding permanently electron-rich sites of another molecule (e.g., the nonbonded electrons

Box 3.1   Classification of Organic Compounds According to Their Ability to Undergo Particular Molecular Interactions

Compounds that undergo only vdW interactions (London plus Debye plus Keesom interactions) are commonly referred to as *apolar*. Examples include alkanes, chlorinated benzenes, and PCBs.

If a chemical exhibits a functionality that has either donor or acceptor character but not both, we call such a compound *monopolar*. Examples include structures with an ether function, $-C-O-C-$ (an electron donor or H-acceptor), a keto group, $>C=O$ (an electron donor or H-acceptor), or an aromatic ring carrying electron withdrawing substituents (an electron acceptor).

Some molecules contain moieties like amino ($-NH_2$), hydroxyl ($-OH$), and carboxyl groups ($-COOH$) that exhibit both donor and acceptor properties. We refer to these compounds as *bipolar*.

For large, complex compounds, it is often not obvious how the whole compound should be classified. Such compounds may exhibit functional groups that participate in locally strong polar interactions. However, due to the large size of the molecule, the overall behavior of the compound is dominated by vdW-interactions. It has, therefore, become common practice to divide the world of chemicals into only two categories, namely, *polar* and *nonpolar* compounds. The nonpolar chemicals include all those chemicals whose molecular interactions are dominated by vdW forces.

of atoms like oxygen and nitrogen). These specific interactions, which we will refer to as *polar* interactions are, of course, only possible between molecules that exhibit complementary structural moieties, that is, if one moiety acts as an *electron acceptor* (often also referred to as *H-donor*) and the other one as an *electron donor* (or *H-acceptor*). Hence, polar interactions can be classified as *electron donor–acceptor* (EDA) or *hydrogen donor–acceptor* (HDA) interactions. Note that both terms are widely used in the literature.

As indicated in Box 3.1, the ability (or inability) of a given compound to undergo specific interactions can be used to divide organic chemicals into different categories. This classification will ultimately be useful when we want to determine whether we should include various factors for quantifying the contributions of these forces in our estimates of the energies controlling specific absorption or adsorption associations in which we are interested.

In the absence of electron donor-acceptor interactions, the London dispersive energy is the dominant contributor to the overall attractions of many molecules to their surroundings. Hence, understanding this type of intermolecular interaction and its dependency on chemical structure allows us to establish a baseline for chemical attractions. If molecules exhibit stronger attractions than expected from these interactions, then this implies the importance of other intermolecular forces. To see the superposition of these additional interactions and their effect on various partitioning phenomena below, we have to examine the role of dispersive forces in more detail, because these forces generally dominate the vdW interactions (Fowkes, 1991; Good and Chaudhury, 1991). The goal of the following discussion is to derive a quantitative measure for these forces. In Chapters 4 and 5 we will then address approaches to quantify polar interactions.

Relative Strengths of Dispersive Energies Between Partitioning Partners

As noted above, London dispersive interactions occur even between molecules of apolar compounds like alkanes, that *on average over time* exhibit a rather smooth distribution of electrons throughout the whole molecular structure. This interaction occurs in all chemicals because there are momentary (order of femtosecond timescales) displacements of the electrons within the structure such that short-lived electron-rich and electron-poor regions temporarily develop. This continuous movement of electrons implies the continuous presence of short-lived dipoles in the structure. This fleeting dipole is felt by neighboring molecules whose electrons respond in a complementary fashion. Consequently, there is an intermolecular attraction between these molecular regions. In the next moment, these attractive interactions shift elsewhere in the molecule.

To think about the strength of dispersive attractions, we consider a situation in which a molecule, i, is moved from a gas phase and mixed into (i.e., absorbed by) a liquid made of the substance, 1:

$$i(g) \ + \ 1{:}1 \ \rightarrow \ 1{:}i{:}1 \qquad\qquad (3\text{-}3)$$

where the parenthetic g indicates that i is coming from a gas phase. In this particular partitioning process, we assume ideal behavior in the gas phase, that is, we ignore intermolecular attractions in this phase. Hence, we can focus on the forming interactions between the molecules of i and the liquid 1 medium. Even if the structure of i does not give rise to permanently uneven electron distributions, there will at least be a dispersive interaction energy with 1.

Considering one molecule of i next to one molecule of 1, we have a dispersive attraction energy, $\Delta_{\text{disp}}g$, given by (Israelachvili, 1992):

$$\Delta_{\text{disp}}g \ / \text{J per interaction} = - \ (\ 3/2 \ ) \ (\ I \ / \ \sigma^6 \ )(\ \alpha_i \, \alpha_1 \ ) \ / \ (4\pi\varepsilon_0)^2 \qquad (3\text{-}4)$$

where I is equal to $(I_i \ I_1/I_i{+}I_1)$ and I_i and I_1 are the first ionization energies of chemicals i and 1, respectively,

> σ is the distance of separation between the temporary dipoles,
> α_i is the polarizability of i,
> α_1 is the polarizability of 1, and
> ε_0 is the permittivity of vacuum.

Generally, molecules exhibit I values between 8 and 12 eV (i.e., 1.3 and 2×10^{-18} J), and the separations between molecules must be related to the molecular sizes. The polarizability, α, of a molecule is related to its ability to develop uneven electron distributions in response to imposed electric fields on femtosecond timescales. Since visible light corresponds to electromagnetic radiation with frequencies around

10^{15} Hz, a material's ability to respond to light, as indicated by a property like the refractive indices, n_{Di}, is related to the material's polarizability. This relationship is known as the Lorenz-Lorentz equation (Israelachvili, 1992):

$$\alpha_i / (4\pi\varepsilon_0) = [(n_{Di}^2 - 1)/(n_{Di}^2 + 2)](3M_i / 4\pi \rho N_A) \tag{3-5}$$

where M_i is the molar mass,
        ρ is the density, and
        N_A is Avogadro constant.

It is instructive to look at the refractive indices for a variety of chemical structures (Table 3.1.) What one quickly sees is that polar compounds *are not the same as polarizable* compounds. Indeed, polarizability is more related to chemical structure features like overall size (higher homologs within a compound family have greater polarizabilities), and presence of conjugated electron systems (benzene is more polarizable than hexane; polarizability increases in the order: benzene < naphthalene < pyrene). Finally, molecules with large atoms containing nonbonded electrons far from the nucleus (e.g., bromine, iodine) are generally more polarizable. After this brief diversion, now we continue to use refractive indices to estimate polarizabilities.

We may modify the Lorenz-Lorentz expression, if we note that $M_i / \rho N_A$ is an estimate of an individual molecule's volume. Assuming the molecule is spherical, we may deduce that:

$$M_i / \rho N_A = (4\pi / 3)(\sigma / 2)^3 \tag{3-6}$$

Note that here we assume that the distance separating the temporary dipoles in adjacent molecules is, on average, the same as the sum of the radii of those molecules. Therefore, we find:

$$(3M_i / 4\pi \rho N_A) = (\sigma)^3 / 8 \tag{3-7}$$

Substituting this result into the Lorenz-Lorentz equation, and then using that result in Eq. 3-4, we find:

$$\Delta_{disp}g / \text{J per interaction} = -(3I / 256)\left[\frac{n_{Di}^2 - 1}{n_{Di}^2 + 2}\right]\left[\frac{n_{D1}^2 - 1}{n_{D1}^2 + 2}\right] \tag{3-8}$$

Finally, the molecule i does not interact with one solvent molecule, 1, but rather is surrounded by a number of molecules. This "stoichiometry" (i.e., ratio of i to 1) is given by the ratio of the total surface area, TSA_i (m²), of i and the contact area, CA (m²), of i with each solvent molecule. Hence, the integrated intermolecular interaction may be:

$$\Delta_{disp}G / \text{J} \cdot \text{mol}^{-1} = N_A(TSA_i / CA)\Delta_{disp}g$$

$$= -N_A (TSA_i / CA)(3 I / 256)\left[\frac{n_{Di}^2 - 1}{n_{Di}^2 + 2}\right]\left[\frac{n_{D1}^2 - 1}{n_{D1}^2 + 2}\right] \tag{3-9}$$

Since parameters like the solvent contact area, the first ionization potential, and the

Table 3.1 Some Examples of Refractive Indices, n_{Di}, of Organic Compounds. Note that Larger Values of n_{Di} Imply Greater Molecular Polarizability and Dispersive Interactions with a Molecule's Surroundings[a]

Compound	Structure	Refractive Index[a]
Methanol	CH_3-OH	1.326
Acetone		1.342
Ethanol	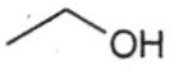 OH	1.359
Acetic acid	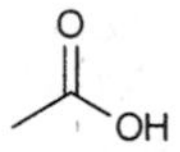 OH	1.370
Hexane		1.372
Octanol	 OH	1.427
Ethylene glycol	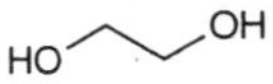 HO OH	1.429
Trichloromethane	$CHCl_3$	1.444
Benzene		1.498
Chlorobenzene	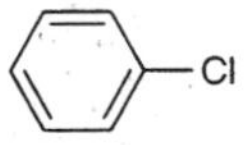 Cl	1.523
Nitrobenzene	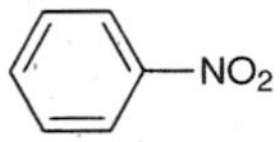 NO_2	1.550
Naphthalene	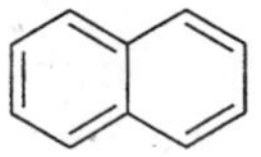	1.590
Tribromomethane	$CHBr_3$	1.601
Pyrene	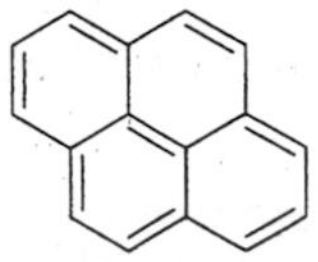	1.770

[a] Data from Lide (1995).

distance of separation between neighboring molecules are fairly invariant, one may expect the dispersive energies to vary between various molecules as:

$$\Delta_{disp}G \ / \ J \ mol^{-1} \approx -constant \ (TSA_i) \left[\frac{n_{Di}^2-1}{n_{Di}^2+2}\right]\left[\frac{n_{D1}^2-1}{n_{D1}^2+2}\right] \qquad (3\text{-}10)$$

This result suggests that we can look at the partitioning of various compounds (i.e., vary i) from the gas phase and expect that their relative tendencies to go into or onto differing media (i.e., vary the chemical nature of medium 1) will depend, at least in part, on predictable dispersive force attractions. Partitioning that is in excess of what we expect from this baseline attractive energy for any chemical must indicate the presence of functional groups in that chemical's structure and/or in the interacting medium that allows additional attractive intermolecular forces [Debye, Keesom, or EDA (HDA) interactions]. Hence, in the next section, we examine partitioning of various chemicals into and onto different defined media to see the roles of chemical structure.

A First Glance at Equilibrium Partition Constants

To explore how molecular structures give rise to intermolecular attractions, and these in turn dictate phase partitioning of those molecules, we need to introduce a parameter that quantifies the relative abundance of the molecules of a given organic compound i in each phase *at equilibrium*. First, we note that we consider the reversible partitioning of a compound i between a phase 1 and a phase 2. As we have done before, but now only mentioning the chemical which is partitioning, we can write this process as a "reaction" by arbitrarily choosing phase 2 as "reactant" and phase 1 as product:

$$i \text{ in phase 2} \rightleftharpoons i \text{ in phase 1}$$

$$\text{"reactant"} \quad \text{"product"}$$

(3-11)

The equilibrium situation can thus be described by an *equilibrium partition constant*, K_{i12}, which we define as:

$$K_{i12} = \frac{\text{concentration of } i \text{ in phase 1}}{\text{concentration of } i \text{ in phase 2}}$$

(3-12)

Note that we have chosen i in phase 2 as the "reactant" in order to have the abundance of i in phase 1 in the numerator of Eq. 3-12. Furthermore, for practical purposes, we define a constant expressed as a ratio of concentrations rather than activities (see Section 3.4). Finally, we consider only situations in which the compound is present as a solute, that is, at low concentrations at which it does not significantly affect the properties of the bulk phase.

K_{i12} is related to a (Gibbs) free energy term, $\Delta_{12}G_i$ by a Boltzmann-type expression (e.g., Atkins, 1998):

$$K_{i12} = \text{constant} \cdot e^{-\Delta_{12}G_i / RT}$$

(3-13)

$$\ln K_{i12} = -\frac{\Delta_{12}G_i}{RT} + \ln \text{(constant)}$$

where we will refer to $\Delta_{12}G_i$ as the *free energy of transfer* of i from phase 2 to phase 1, R is the universal gas constant (8.31 $J \cdot mol^{-1}K^{-1}$), and T is the absolute temperature in Kelvin. The constant in Eq. 3-13 depends on how we express the abundance of the compounds in the two phases (e.g., as partial pressure, mole fraction, or molar concentration). We will address this issue as well as the derivation of $\Delta_{12}G_i$ in Sections 3.3 and 3.4 when we discuss some important thermodynamic functions. At this point, we note that $\Delta_{12}G_i$ expresses the free energy change (per mole i) for the

ideal gas (g) (no interactions)

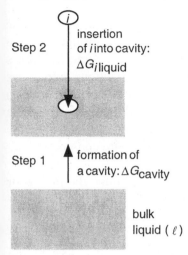

Step 2 | insertion of i into cavity: $\Delta G_{i\,\text{liquid}}$

Step 1 | formation of a cavity: ΔG_{cavity}

bulk liquid (ℓ)

$\Delta_{g\ell}G_i = \Delta G_{\text{cavity}} + \Delta G_{i\,\text{liquid}}$

Figure 3.2 Absorption of a compound i from an ideal gas phase *into* a bulk liquid.

process of taking molecules of the compound i out of phase 2 and putting them into phase 1 under standard conditions. Hence, $\Delta_{12}G_i$ sums both the *enthalpic* and *entropic* effects (see Section 3.3) that result from the changes occurring in the intermolecular interactions in both phases when removing or adding molecules of compound i. These changes are caused by differences in the interactions of i with the molecules forming the bulk phases between which the compound is partitioned (e.g., i:1 interactions), and also by changes in the interactions among the bulk phase molecules themselves (e.g., 1:1 interactions). As we will soon see, these latter contributions to $\Delta_{12}G_i$ are particularly important for partitioning processes involving aqueous phases (i.e., 1 = H_2O) . If the overall $\Delta_{12}G_i$ is negative, the compound prefers to be in phase 1 as compared to phase 2. Thus, at equilibrium, its abundance will be higher in phase 1 (and vice versa, if $\Delta_{12}G_i$ is positive). By examining the relative sizes of $\ln K_{i12}$ as a function of chemical structure, we will be able to see when particular types of intermolecular interactions become important.

Examples of Absorption from the Gas Phase

Let us now look at some partitioning data. We will discuss the two general cases: (i) the partitioning of organic compounds between the (ideal) gas phase and a bulk liquid (*absorption*, Fig. 3.2), and (ii) the partitioning between the gas phase and a solid surface (*adsorption*, Fig. 3.3). In order to emphasize the influence of single pairs of molecule:molecule interactions, we confine our discussion to the partitioning of neutral organic compounds from an *ideal gas phase* (e.g., air). This is another way of saying that there are no intermolecular attractions to break when i leaves the gas phase. First, we consider partitioning "reactions" of the form:

$$i(g) + 1{:}1 \rightleftharpoons 1{:}i{:}1 \tag{3-14}$$

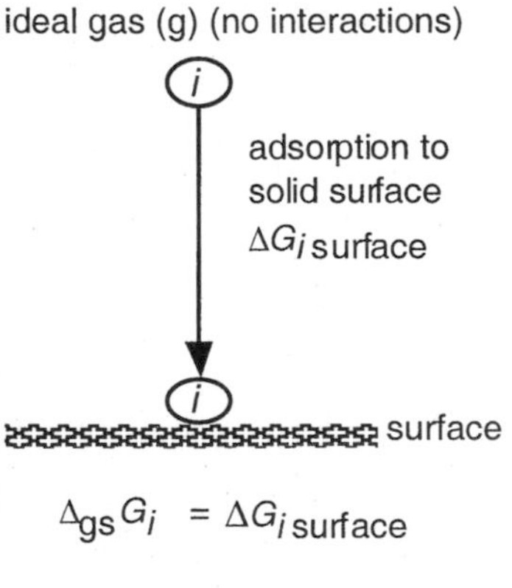

ideal gas (g) (no interactions)

i

adsorption to solid surface $\Delta G_{i\,\text{surface}}$

i surface

$\Delta_{gs}G_i = \Delta G_{i\,\text{surface}}$

Figure 3.3 Adsorption of a compound i from an ideal gas phase *to* a surface.

As is illustrated in Fig. 3.2, when considering the *ab*sorption of a gaseous molecule into a bulk liquid, the free energy term can be broken up into a term describing the energy that has to be spent to create a cavity in the liquid (ΔG_{cavity} = breaking 1:1 interactions in Eq. 3-3) and a term describing the free energy change caused upon insertion of the compound into the cavity ($\Delta G_{i\text{liquid}}$ = making i:1 interactions in Eq. 3-3). The former energy involves disruption of solvent:solvent interactions as we discussed earlier. The latter free energy involves the formation of solute:solvent attractions. It is easy to conceive that the overall free energy of transfer, and thus the partition constant of a compound i (see Eq 3-12), will depend strongly on the type of compound as well as on the type of bulk liquid considered. Therefore, in our following examples, we will inspect gas-liquid equilibrium partition constants of members of a variety of different compound classes for bulk liquids exhibiting very different properties: (1) the pure organic liquid compound itself, (2) one apolar organic liquid, hexadecane, and (3) one polar solvent, water.

Example 1: Vapor Pressure and Molecular Interactions in the Pure Liquid Compound

We start out by evaluating the intermolecular interactions among the molecules of a given compound in its own pure liquid state by considering the equilibrium partitioning with the gas phase:

$$i(g) + i:i \rightleftharpoons i:i:i \qquad (3\text{-}15)$$

In this case, the free energy of transfer of i from the pure liquid to an ideal gas phase (i.e., air) $\Delta_{aL}G_i$, and thus the corresponding gas (air)–liquid equilibrium partition constant, K_{iaL} (see Eq. 3-16 below), are direct measures of the attractive forces between like molecules in the liquid (recall we assume no interactions among gas phase molecules). Note that for the following discussion we use a subscript "a" (air) to denote the ideal gas phase. Furthermore a capital "L" is used to describe the pure liquid in order to distinguish from other liquid phases (subscript ℓ). Finally, the superscript $*$ indicates that we are dealing with a property of a pure compound.

Commonly, gas-liquid partitioning is expressed by the saturated liquid vapor pressure, p_{iL}^*, of the compound i. This important chemical property will be discussed in detail in Chapter 4. Briefly, p_{iL}^* is the pressure exerted by the compound's molecules in the gas phase above the pure liquid at equilibrium. Since this pressure generally involves only part of the total pressure, we often refer to it as a partial pressure due to the chemical of interest. In this case, when there is no more build up of vapor molecules in a closed system, we say that the gas phase is "saturated" with the compound. Note that because p_{iL}^* is strongly temperature dependent, when comparing vapor pressures of different compounds to see the influence of chemical structure, we have to use p_{iL}^* values measured at the same temperature (which also holds for all other equilibrium constants discussed later; see Section 3.4).

For comparison of chemical's partition constants between air and different bulk liquids, it is useful to express p_{iL}^* as a constant, K_{iaL}, that describes the relative amount of the compound in the two phases in molar concentrations (i.e., $\text{mol}\cdot\text{L}^{-1}$):

$$K_{iaL} = \frac{C_{ia}^{sat}}{C_{iL}}\left(\frac{\text{mol}\cdot\text{L}^{-1}\ \text{air}}{\text{mol}\cdot\text{L}^{-1}\ \text{pure liquid compound}}\right) \qquad (3\text{-}16)$$

Using the ideal gas law ($pV = nRT$), the saturation concentration in the air, $C_{ia}^{sat}\,(= n\,/\,V)$, can be calculated from p_{iL}^* by:

$$C_{ia}^{sat} = \frac{p_{iL}^*}{RT} \qquad (3\text{-}17)$$

C_{iL} can be derived from the density, ρ_{iL}, of the liquid compound and from its molar mass, M_i:

$$C_{iL} = \frac{\rho_{iL}}{M_i}\left(\frac{\text{g}\cdot\text{L}^{-1}}{\text{g}\cdot\text{mol}^{-1}}\right) \qquad (3\text{-}18)$$

Note that C_{iL} is the inverse of the molar volume of the liquid compound, which we denote as $\overline{V}_{iL}$. Substitution of Eqs. 3-17 and 3-18 into Eq. 3-16 yields:

$$K_{iaL} = \frac{M_i}{\rho_{iL}\cdot RT}\,p_{iL}^* \qquad (3\text{-}19)$$

Armed with such partition constants, we can calculate the free energies involved in exchanging substances between a gas and their own pure liquids.

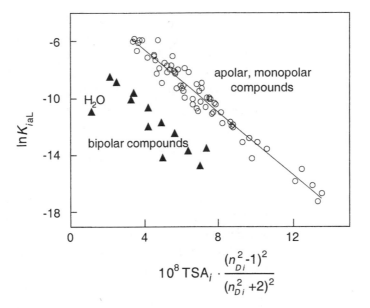

Figure 3.4 Plot of ln K_{iaL} (25°C) (Eq. 3-19) versus the dispersive vdW-parameter defined by Eq. 3-10 with 1 = i. Note that TSA$_i$ is in cm^2 mol^{-1}.

Now we are prepared to observe how chemical structures control this particular case of partitioning: the distribution of molecules between their own pure liquid phase and an equilibrated gas phase. To indicate the importance of the dispersion forces among the molecules in their pure liquid phase, we examine K_{iaL} values determined for a wide variety of liquid compounds as a function of their $\Delta_{disp}G$ defined above by Eq. 3-10 (Fig. 3.4). The compounds chosen are composed of less than 10 carbon atoms and exhibit no more than one functional group. Hence, the functional group can be considered to contribute significantly to the overall capability of the molecules to interact with neighboring molecules. Compound classes are *n*-alkanes, alkylated benzenes, polychlorinated methanes and ethanes (include permanent dipoles), aliphatic ketones (include permanent dipoles and can act as electron donors), aldehydes (include permanent dipoles and can act as electron donors), carboxylic acid esters (include permanent dipoles and can act as electron donors), alcohols (include permanent dipoles, can act as electron acceptors and donors), and carboxylic acids (include permanent dipoles and can act as electron acceptors and donors). Note that these functional groups are discussed in Section 2.3 (Figs. 2-15 and 2-17).

As demonstrated for all *apolar* and *monopolar* compounds (see Box 3.1 for definitions), which cannot undergo electron donor-acceptor interactions with like molecules in their pure liquid, a good inverse linear correlation is found between ln K_{iaL} and our metric of $\Delta_{disp}G$. Some of the variation in ln K_{iaL} is due to the polarizability contribution (ranges over about a factor of 1.6), while some is also due to changes in TSA$_i$ (factor of about 7). This factor of about 10 in $\Delta_{disp}G$ is consistent with the ln K_{iaL} changing by about 10 ln units (Fig. 3.4.) Note that some of the scatter is due to the compounds classified as "monopolar" having additional dipole–induced dipole, dipole–dipole interactions, and possibly also having some weak bipolar character (e.g., the ketones, aldehydes, esters). Furthermore, the TSA$_i$ value used is only a very crude approximation of the actual contact area among the molecules. Nevertheless, Fig. 3.4 nicely shows the dominating role of dispersive vdW interactions in determining the air/pure liquid partitioning (i.e., the vapor pressure) of apolar and monopolar organic compounds.

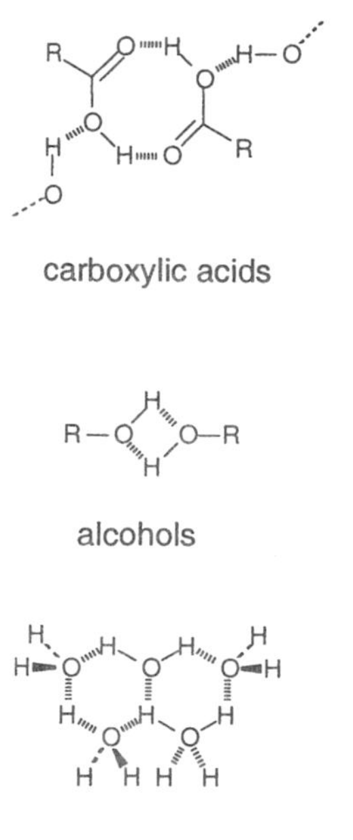

carboxylic acids

alcohols

water

Figure 3.5 H-bonding in various pure liquids.

The truly bipolar compounds that may form rather strong hydrogen bonds in the pure liquids (Fig. 3.5) have, however, distinctly lower K_{iaL} values than expected solely from their vdW interactions. The most extreme case is water, which has a K_{iaL} value that is almost four orders of magnitude smaller than the value one would expect for a nonpolar compound with similar size and dispersive vdW interactions. From this finding, we see the great importance of superimposing hydrogen-bonding interactions on the dispersive interactions of molecules like water.

Example 2: Air–Solvent Partitioning

In our next example we compare the partitioning of the same set of compounds between air (gas phase) and two very different solvents, hexadecane and water. These two liquids are chosen to represent two extreme cases, both with respect to the free energy costs of changing solvent:solvent interactions as well as with respect to the type of interactions the solvent molecules may have with the organic solute. In the case of hexadecane, all compounds, irrespective of their polarity, can undergo only vdW interactions with these hydrocarbon solvent molecules. Furthermore, the free energy cost for cavity formation reflects breaking only vdW interactions among the hexadecane molecules. Thus, as is nicely illustrated by Fig. 3.6a, for *all compounds* (apolar, monopolar, and bipolar), an inverse relationship is found between $\ln K_{iah}$ and the dispersive vdW parameter of the compound expressed as product of the solute's total surface area and refractive index estimator of polarizability (Eq. 3-10).

The situation is completely different in the air/water partitioning system (Fig. 3.6b). As is evident, very large differences in K_{iaw} are observed between members of different compound classes. For example, the air/water partition constants of the *n*-alkanes are more than two orders of magnitude larger than those of the corresponding ethers, and even five orders of magnitude larger than those of the alcohols exhibiting a similar dispersive vdW parameter. These differences reflect the different abilities of the compound molecules to undergo polar interactions with the water molecules, interactions that help to counterbalance the rather large free energy costs for creating a cavity in the bulk water. Thus, in contrast to the partitioning from air to a solvent like hexadecane, apolar compounds such as the *n*-alkanes are "expelled" from the bulk water phase. This is not because they do not have attractive vdW interactions with the water molecules, but rather because of the high costs of cavity formation (breaking $H_2O:H_2O$ interactions). This effect is also seen within each series of compounds that differ only by entities that add vdW interactions (i.e., CH_2-groups). K_{iaw} increases with increasing size of the molecule (Fig. 3.6b), which is in contrast to the situation found in the air/hexadecane system (Fig. 3.6a).

Examples of Adsorption from the Gas Phase

Now we shift to cases which allow us to gain insights into the intermolecular forces between organic molecules and a given surface (Fig. 3.3). By inspecting gas/solid adsorption constants of a variety of compounds interacting with two surfaces exhibiting very different properties (i.e., quartz versus teflon), we will learn a few somewhat surprising facts. For instance, we will see that in this case, a nonpolar hydrocarbon interacts more strongly with a polar surface than with a nonpolar surface. Intuitively, this might not have been expected.

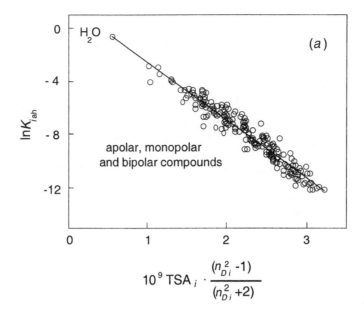

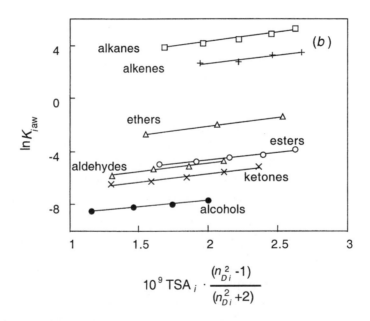

Figure 3.6 Plot of the natural logarithms of the partition constants at 25°C of a series of apolar, monopolar, and bipolar organic compounds between air and (*a*) *n*-hexadecane (*n*-$C_{16}H_{34}$) and (*b*) water versus the dispersive vdW- parameter of the compounds defined by Eq. 3-10. Note that from Eq. 3-10 only the compound part is used because the solvent part (1) is the same for all compounds, and that TSA_i is in cm^2 mol^{-1}.

Example 3: Air–Solid Surface Partitioning

In our final example, we consider the partitioning of a small set of organic compounds between air and two different solid surfaces, teflon and quartz (Fig. 3.7). The two surfaces differ distinctly in their properties in that the teflon surface is made up of atoms that cannot participate in EDA interactions, while the quartz surface (which exhibits OH-groups), has a strongly bipolar character (much like a water surface). In the case of air/surface partitioning, the partition constants reflect the interactions of a given organic compound with the aggregate of atoms making up the surface. In contrast to air/bulk liquid partitioning, for these surface interactions no cavity as in the solvent has to be formed. Hence, in this case (Fig. 3.3), the free energy change on

Figure 3.7 Plots of the natural logarithms of the air/surface partition constants, K_{ias}, of a series of apolar and monopolar compounds for two different surfaces (i.e., teflon and quartz) versus the dispersive vdW-parameter of the compounds defined by Eq. 3-10. Note that from Eq. 3-10 only the compound part is used because the solvent part (1) is the same for all compounds, and that TSA$_i$ is in cm^2 mol^{-1}. (Data at 25°C from Goss and Schwarzenbach, 1998.)

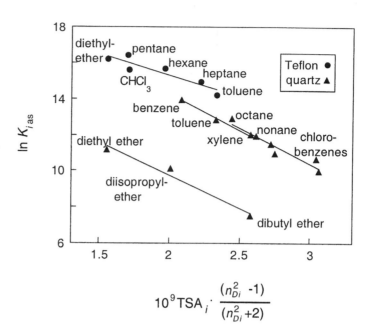

exchange does not include a term like ΔG_{cavity}. Had the solid been immersed in a liquid so that there were liquid molecules:solid surface interactions, insertion of our partitioning substance i at the surface would have required us to consider the free energy of making room for the adsorbate (e.g., breaking 1:2 interactions in partitioning "reaction" shown in Eq. 3-2).

In the case of the teflon surface where only vdW interactions are possible for any adsorbing molecule, a plot of lnK_{ias} versus the sorbate's dispersive vdW parameter yields a straight line for apolar and polar compounds (Fig. 3.7). In contrast, for the bipolar quartz surface, apolar and monopolar compounds are separated into several groups according to their ability (or inability) to undergo polar interactions. One interesting additional detail that can be seen from Fig. 3.7 is that nonpolar or weakly monopolar compounds such as alkanes and alkylbenzenes are slightly more attracted to the polar quartz surface as compared to the nonpolar teflon surface. This finding may be somewhat surprising. Intuitively, we still might have the idea that nonpolar compounds are attracted more strongly by nonpolar counterparts as compared to polar counterparts. This expectation is not generally true. The reason in this particular case is that the ability of a quartz surface to undergo vdW interactions is larger than that of a teflon surface (see Section 11.2). In this context it is important to realize that compounds which we denote as being *hydrophobic* (i.e., disliking water) are actually attracted to water *surfaces*. One example illustrating this point is the thin gasoline films that all of us have seen on the surface of polluted lakes or rivers. Obviously, in this case, the attractive (vdW) forces between the *hydrophobic* hydrocarbons and the water molecules at the water surface overcome the (vdW) forces among the hydrocarbon molecules themselves that would favor the formation of oil droplets. Hence, the term *hydrophobicity* of a compound should only be used in connection with a compound's tendency to be *dissolved in a bulk water phase*. In such cases, the balance between the free energy costs for cavity formation *and* the free energy gains due to the interactions of the compound with the water molecules is important. Moreover, as has become evident from our above discussion, the hydrophobicity of organic compounds will increase with increasing size of the molecules. For a given size, hydrophobicity will be maximal for compounds that can only undergo vdW interactions with water. We will come back to this point in Chapter 5, when discussing water solubilities and aqueous activity coefficients of organic compounds.

With these first insights into the molecular interactions that govern the partitioning of organic compounds between different phases in the environment, we are now prepared to tackle some thermodynamic formalisms. We will need these parameters and their interrelationships for quantitative treatments of the various phase transfer processes discussed in the following chapters.

3.3   Using Thermodynamic Functions to Quantify Molecular Energies

Chemical Potential

When considering the relative energy status of the molecules of a particular compound in a given environmental system (e.g., benzene in aqueous solution), we can envision the molecules to embody different forms of energies. Some energies are those associated with the molecule's chemical bonds and bond vibrations, flexations, and rotations. Other energies include those due to whole-molecule translations, reorientations, and interactions of the molecules with their surroundings. The whole energy content is the *internal* energy and is dependent on the temperature, pressure, and chemical composition of the system. When we talk about the "energy content" of a given substance, we are usually not concerned with the energy status of a single molecule at any given time, but rather with an average energy status of the entire population of one type of organic molecule (e.g., benzene) in the system. To describe the (average) "energy status" of a compound i mixed in a *milieu* of substances, Gibbs (1873, 1876) introduced an entity referred to as *total free energy*, G, of this system, which could be expressed as the sum of the contributions from all of the different components present:

$$G(p,T,n_1,n_2,....n_i,..n_N) = \sum_{i=1}^{N} n_i \mu_i$$

(3-20)

where n_i is the amount of compound i (in moles) in the system containing N compounds. The entity μ_i, which is referred to as *chemical potential* of the compound i, is then given by:

$$\mu_i (\text{J} \cdot \text{mol}^{-1}) \equiv \left[\frac{\partial G(\text{J})}{\partial n_i(\text{mol})} \right]_{T,p,nj \neq i}$$

(3-21)

Hence, μ_i expresses the Gibbs free energy (which we denote just as free energy) added to the system at *constant T, P, and composition* with each added increment of compound i. Let us now try to evaluate this important function μ_i. When adding an *incremental number* of molecules of i, free energy is introduced in the form of internal energies of substance i as well as by the interaction of i with other molecules in the system. As more i is added, the composition of the mixture changes and, consequently, μ_i changes as a function of the amount of i. We should note that μ_i is sometimes also referred to as the *partial molar free energy*, G_i, of a compound. Finally, we recall that $G_i (\text{J} \cdot \text{mol}^{-1})$ is related to the *partial molar enthalpy*, $H_i (\text{J} \cdot \text{mol}^{-1})$, and

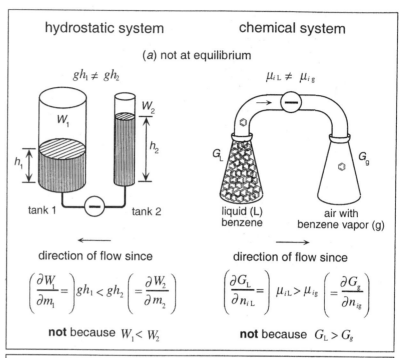

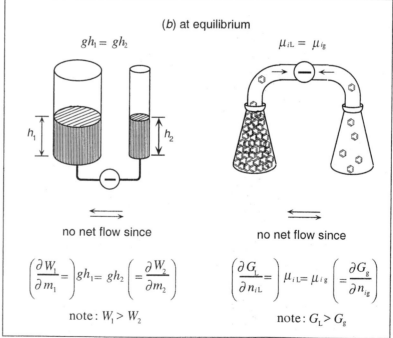

Figure 3.8 Conceptualization of the potential functions in a hydrostatic system and in a simple chemical system. (*a*) In the unequilibrated hydrostatic system, water will flow from reservoir 2 of higher hydrostatic potential (=gh_2, where g is the acceleration due to gravity and h_2 is the observable height of water in the tank) to reservoir 1 of lower hydrostatic potential; total water volumes (i.e., total potential energies W_1 and W_2) do not dictate flow. Similarly, benzene molecules move from liquid benzene to the head space in the nonequilibrated chemical system, not because there are more molecules in the flask containing the liquid, but because the molecules initially exhibit a higher chemical potential in the liquid than in the gas. (*b*) At equilibrium, the hydrostatic system is characterized by equal hydrostatic potentials in both reservoirs (not equal water volumes) and the chemical system reflects equal chemical potentials in both flasks (not equal benzene concentrations).

partial molar entropy, S_i (J.mol^{-1} K^{-1}), by the well-known general relationship:

$$\mu_i \equiv G_i = H_i - T \cdot S_i \qquad (3\text{-}22)$$

Gibbs (1876) recognized that the chemical potential could be used to assess the tendency of component i to be transferred from one system to another or to be transformed within a system. This is analogous to the use of hydrostatic head potential for identifying the direction of flow between water reservoirs (Fig. 3.8a). We know that equilibrium (no net flow in either direction) is reached, when the hydrostatic head potentials of the two reservoirs are equal (Fig. 3.8b). Similarly, chemical equilibrium is characterized by equal chemical potentials for each of the constituents. As with hydrostatic head potential, chemical potential is an *intensive* entity, meaning it is independent of the size of the system (in contrast to the total free energy G, which is an *extensive* function).

Fugacity

Unfortunately, unlike hydraulic head potentials, there is no way of directly observing chemical potentials. Consequently, the concept of *fugacity* was born. Lewis (1901) reasoned that rather than look into a system and try to quantify all of the chemical potential energies carried by the various components of interest, it would be more practical to assess a molecule's *urge to escape or flee* that system (hence "fugacity" from Latin *fugere*, to flee). If one could quantify *the relative tendencies* of molecules to flee various situations, one could simultaneously recognize the *relative* chemical potentials of the compounds of interest in those situations. Based on the differences in their chemical potentials, one could quantify the direction (higher μ_i to lower μ_i) and extent to which a transfer process would occur.

Pressure and Fugacities of a Compound in the Gas State

Let us quantify first the "fleeing tendency" or fugacity of molecules in a gas (just about the simplest molecular system) in a way we can observe or measure. Imagine a certain number of moles (n_i) of a pure gaseous compound i confined to a volume, V, say in a closed beaker, at a specific temperature, T. The molecules of the gaseous compound exert a pressure p_i on the walls of the beaker (a quantity we can feel and measure) as they press upon it seeking to pass (Fig. 3.9a). It is not difficult to imagine that if the gas molecules wish to escape more "insistently" (i.e., a higher chemical potential as a result, for example, of the addition of more i molecules to the gas phase in the beaker), their impact on the walls will increase. Consequently, we will measure a higher gas pressure. For an *ideal* gas, the pressure is perfectly proportional to the amount of gaseous compound. Stating this quantitatively, we see that *at constant T*, the incremental change in chemical potential of the gaseous compound i may be related to a corresponding change in pressure (deduced from the Gibbs-Duhem equation; see, e.g., Prausnitz, 1969, p. 17):

$$(d\mu_{ig})_T = \frac{V}{n_{ig}} dp_i \tag{3-23}$$

In this case we can substitute V/n_{ig} with RT/p_i:

$$(d\mu_{ig})_T = \frac{RT}{p_i} dp_i \tag{3-24}$$

As mentioned above, the absolute value of the chemical potential cannot be measured but we can measure the absolute value of pressure or the amount of substance in the gas phase. Hence, we may define a standard value of the chemical potential of the gaseous compound i, μ_{ig}^0, by defining a standard amount and standard pressure in the form of one variable, the standard pressure $p_i = p_i^0$ (commonly 1 bar). We do this by integration of Eq. 3-24:

$$\int_{\mu_{ig}^0}^{\mu_{ig}} (d\mu_{ig})_T = RT \int_{p_i^0}^{p_i} \frac{1}{p_i} \, dp_i \qquad (3\text{-}25)$$

which yields:

$$\mu_{ig} = \mu_{ig}^0 + RT \ln\left[\frac{p_i}{p_i^0}\right] \qquad (3\text{-}26)$$

Let us now look at the situation in which we deal with *real* gases, that is, with a situation in which intermolecular forces between the molecules cannot be neglected (as will be even more the case for liquids and solids, see below). These forces influence the (partial) pressure of the gas molecules, but not the amount of the gaseous compound(s). This real pressure is called fugacity.

In contrast to the pressure of an ideal gas, the fugacity is not only a function of the amount of substance and temperature, but also of the composition (types and amounts of gaseous compounds present) of the gaseous system and of the total pressure. The fugacitiy of a gaseous compound is, however, closely related to its partial pressure. To account for the nonideality of the gas, one can relate these terms by using a fugacity coefficient, θ_{ig}:

$$f_{ig} = \theta_{ig} p_i \qquad (3\text{-}27)$$

It is now easy to see that the correct expression for the chemical potential of a gaseous compound i is not based on pressure but on fugacity:

$$\mu_{ig} = \mu_{ig}^0 + RT \ln\left[\frac{f_i}{p_i^0}\right] \qquad (3\text{-}28)$$

Note that for the standard state one defines ideal gas behavior, that is, $f_i^0 = p_i^0$ (commonly 1 bar).

Under typical environmental conditions with atmospheric pressure, gas densities are very low so that we set $\theta_i = 1$. In other words, for all our following discussion, we will assume that any compound will exhibit ideal gas behavior (i.e., we will use Eq. 3-26 instead of Eq. 3-28).

In a mixture of gaseous compounds having a total pressure p, p_i is the partial pressure of compound i, which may be expressed as:

$$p_i = x_{ig} p \qquad (3\text{-}29)$$

where x_{ig} is the mole fraction of i:

$$x_{ig} = \frac{n_{ig}}{\sum_j n_{jg}} \qquad (3\text{-}30)$$

and $\Sigma_j n_{jg}$ is the total number of moles present in the gas, and p is the total pressure. Thus, the fugacity of a gas i in a mixture is given by:

$$f_{ig} = \theta_{ig} x_{ig} p$$
$$\cong p_i$$

(3-31)

Reference States and Standard States

Before we discuss the fugacities of compounds in liquid and solid phases, a few remarks on the choice of reference states are necessary. As we have seen in the development of Eq. 3-26, we face one obvious difficulty. Since we cannot compute an absolute value for the chemical potential, we must be content with computing *changes* in the chemical potential as caused by changes in the independent variables of temperature, pressure, and composition. This difficulty, however, is apparent rather than fundamental; it is really no more than an inconvenience. It results because the relationships between the chemical potential and the physically measurable quantities are in the form of differential equations, which upon integration yield only differences. With the choice of an appropriate reference state, it is usually possible to express the energetics of a given process in rather simple terms. This is similar to concepts used in everyday life, where we choose reference states to express the magnitude of entities, for example, the altitude of a mountain relative to sea level.

When we consider a change in the "energy status" of a compound of interest [e.g., the transfer of organic molecules from the pure liquid phase to the overlying gas phase (vaporization), as discussed in Section 3.2], we try to do our energy-change bookkeeping in such a way that we concern ourselves with only those energetic properties of the molecules that undergo change. During the vaporization of liquid benzene, for example, we will not worry about the internal energy content of the benzene molecules themselves, since these molecules maintain the same bonds, and practically the same bond motions, in both the gaseous and liquid states. Rather, we will focus on the energy change associated with having benzene molecules in new surroundings. Benzene molecules in gas or liquid phases will, therefore, feel different attractions to their neighboring molecules and will contain different orientational and translational energies since in a liquid the molecules are packed fairly tightly together, while in the gas they are almost isolated. This focus on only the changing aspects is the guiding consideration in our choice of *reference states*. For each chemical species of interest, we want to pick a form (a reference state of the material) that is closely related to the situation at hand. For instance, it would be silly (though feasible) to consider the energy status of elemental carbon and hydrogen of which the benzene molecule is composed as the reference point with which evaporating benzene should be compared. Instead, we shall be clever and, in this case, choose the "energy status" of pure liquid benzene as a reference state, since liquid benzene includes all of the internal bonding energies common to both the gaseous and liquid forms of the compound.

In the field of environmental organic chemistry, the most common *reference states* used include: (1) the *pure liquid state*, when we are concerned with phase transfer processes; (2) the *infinite dilution state*, when we are dealing with reactions of organic chemicals in solution (e.g., proton transfer reactions in water, see Chapter 8); and (3) the elements in their naturally occurring forms (e.g., C, H_2, O_2, N_2, and Cl_2), when we are interested in reactions in which many bonds are broken and/or formed. Certainly, other reference states may be chosen as convenience dictates, the guiding principle being that one can clearly see how the chemical species considered in a given system is related to this state. Once we have chosen an appropriate reference state, we also must specify the conditions of our reference state; that is, the pressure and the concentration. These conditions are referred to as *standard conditions* and, together with the reference state, form the *standard state* of a chemical species. We then refer to μ_i^0 in Eqs. 3-26 and 3-27 as the *standard chemical potential*, a value that quantifies the "energy status" under these specific conditions. Since we are most often concerned with the behavior of chemicals in the earth's near-surface ecosystems, 1 bar (10^5 Pa or 0.987 atm) is usually chosen as standard pressure. Furthermore, we have to indicate the temperature at which we consider the chemical potential. If not otherwise indicated we will commonly assume a temperature of 298 K (25°C). In summary, as long as we are unambiguous in our choice of reference state, standard conditions, and temperatures, hopefully chosen so that both the starting and final states of a molecular change may be clearly related to these choices, our energy bookkeeping should be fairly straightforward.

Fugacities of Liquids and Solids

Let us now continue with our discussion of how to relate the chemical potential to measurable quantities. We have already seen that the chemical potential of a gaseous compound can be related to pressure. Since substances in both the liquid and solid phases also exert vapor pressures, Lewis reasoned that these pressures likewise reflected the escaping tendencies of these materials from their condensed phases (Fig. 3.9). He thereby extended this logic by defining the fugacities of pure liquids (including subcooled and superheated liquids, hence the subscript "L") and solids (subscript "s") as a function of their vapor pressures, p_{iL}^* :

$$f_{iL} = \gamma_{iL} \cdot p_{iL}^*$$
$$f_{is} = \gamma_{is} \cdot p_{is}^* \tag{3-32}$$

where γ_i now accounts for nonideal behavior resulting from molecule–molecule interactions. These activity coefficients are commonly set equal to 1 when we decide to take as the reference state the pure compound in the phase it naturally assumes under the conditions of interest. The molecules are viewed, therefore, as "dissolved" in like molecules, and this condition is defined to have "ideal" mixing behavior.

If we consider, for example, compound i in a liquid mixture, e.g., in organic or in aqueous solution; (subscript "ℓ" see Fig. 3.9d), we can now relate its fugacity in the mixture to the fugacity of the pure liquid compound by [note that for convenience, we have chosen the *pure liquid compound* (superscript *) as our *reference state*]:

$$f_{i\ell} = \gamma_{i\ell} \cdot x_{i\ell} \cdot f_{iL}^*$$
$$= \gamma_{i\ell} \cdot x_{i\ell} \cdot p_{iL}^* \tag{3-33}$$

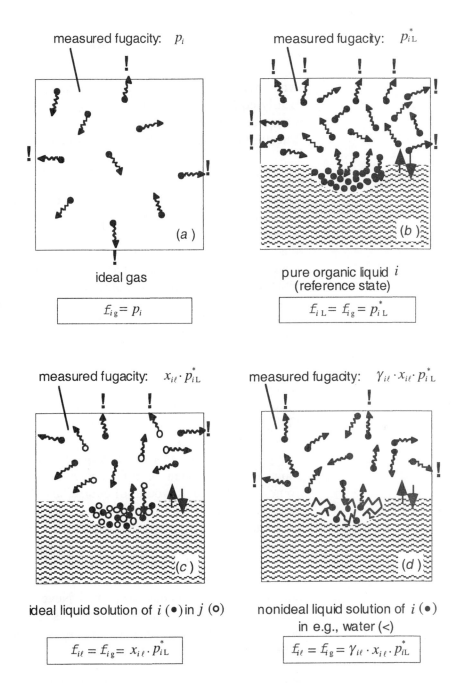

measured fugacity: p_i

measured fugacity: p_{iL}^*

ideal gas

$$f_{ig} = p_i$$

pure organic liquid i
(reference state)

$$f_{iL} = f_{ig} = p_{iL}^*$$

measured fugacity: $x_{i\ell} \cdot p_{iL}^*$

measured fugacity: $\gamma_{i\ell} \cdot x_{i\ell} \cdot p_{iL}^*$

Figure 3.9 Conceptualization of the fugacity of a compound i (a) in an *ideal gas*; (b) in a *pure* liquid compound i; (c) in an *ideal* liquid mixture; and (d) in a *nonideal* liquid mixture (e.g., in aqueous solution). Note that in (b), (c), and (d), the gas and liquid phases are in equilibrium with one another.

ideal liquid solution of i (●) in j (○)

$$f_{i\ell} = f_{ig} = x_{i\ell} \cdot p_{iL}^*$$

nonideal liquid solution of i (●)
in e.g., water (<)

$$f_{i\ell} = f_{ig} = \gamma_{i\ell} \cdot x_{i\ell} \cdot p_{iL}^*$$

where $x_{i\ell}$ is the mole fraction of i (Eq. 3-30) in the mixture (or solution). If the compounds form an *ideal* mixture (Fig. 3.9c), implying that no nonideal behavior results from interactions among unlike molecules, $\gamma_{i\ell}$ is equal to 1 and Eq. 3-33 represents the well-known *Raoult's Law*.

Activity Coefficient and Chemical Potential

Using the concept of fugacity we can now, in analogy to the gaseous phase (Eq. 3-28), express the chemical potential of a compound i in a liquid solution by:

$$\mu_{i\ell} = \mu_{iL}^* + RT \ln \left[\frac{f_{i\ell}}{p_{iL}^*} \right] \qquad (3\text{-}34)$$

where we have chosen the pure liquid compound as reference state. Note that μ_{iL}^* is nearly equal to the standard free energy of formation $\Delta_f G_i^{\circ}(L)$ of the pure liquid compound, which is commonly defined at 1 bar and not at p_{iL}^*. Hence $\mu_{iL}^* \cong \mu_{iL}^0$. Substitution of Eq. 3-33 into Eq. 3-34 then yields:

$$\mu_{i\ell} = \mu_{iL}^* + RT \ln \gamma_{i\ell} \cdot x_{i\ell} \qquad (3\text{-}35)$$

Generally, the expression $f_i / f_{ref} = \gamma_i \cdot x_i = a_i$ is referred to as the *activity* of the compound. That is, a_i is a measure of how active a compound is in a given state (e.g., in aqueous solution) compared to its standard state (e.g., the pure organic liquid at the same T and p). Since γ_i relates a_i, the "apparent concentration" of i, to the real concentration x_i, it is only logical that one refers to γ_i as the *activity coefficient*. It must be emphasized here *that the activity of a given compound in a given phase is a relative measure and is, therefore, keyed to the reference state*. The numerical value of γ_i will therefore depend on the choice of reference state, since, as we have seen in Section 3.2, molecules of i in different reference states (i.e., liquid solutions) interact differently with their surroundings.

It is instructive to examine the activity coefficients of a few organic compounds in different solvents using the pure organic liquid compound as the reference state (Table 3.2). These $\gamma_{i\ell}$ values express the escaping tendency of a given compound from the respective solution *relative to the compound's escaping tendency from its own pure liquid*. Note that with this choice, for each compound i, we define a different molecular environment as reference state. At first glance, this might seem to be somewhat strange. It would, of course, also be feasible and, if the necessary data were available, even meaningful, to choose the same reference state for all compounds (e.g., infinitely dilute in hexadecane). By doing so, activity coefficients for other molecular environments (e.g., water) could be more easily compared, because they would all be related to a situation where all compounds exhibit solely vdW interactions. Another option would be to use the ideal gas phase as the reference state where no interactions occur at all. Nevertheless, as we will see in the following chapters, the choice of the pure liquid as reference state has significant practical advantages, in that the pertinent properties of the pure liquid are known or can be estimated for many organic compounds.

But let us now inspect the $\gamma_{i\ell}$ values for the various chemicals given in Table 3.2. As we would probably have expected intuitively from our discussions in Section 3.2, $\gamma_{i\ell}$ values close to 1 are found in those cases in which molecular interactions in the solution are nearly the same as in the pure liquid compound. For example, when the intermolecular interactions in a pure liquid are dominated by vdW interactions, and when solutions also exhibit only vdW interactions between the solute and solvent and between the solvent molecules themselves, we have $\gamma_{i\ell}$ values close to 1. Examples include solutions of nonpolar and monopolar compounds in an apolar solvent (e.g., n-hexane, benzene, and diethylether in hexadecane), as well as solutions of nonpolar solutes in monopolar solvents (e.g., n-hexane in chloroform). In contrast, if we consider situations in which strong polar interactions are involved between the solute

Table 3.2 Activity Coefficients [a], $\gamma_{i\ell}$, of Hexane (apolar), Benzene (monopolar), Diethylether (monopolar), and Ethanol (bipolar) in Different Solvents at Infinite Dilution at 25°C. Reference: Pure Liquid Organic Compound.

Solvent	Solute			
(*Polarity*)	Hexane (*apolar*)	Benzene (*monopolar, H-acceptor*)	Diethylether (*monopolar, H-acceptor*)	Ethanol (*bipolar*)
n-Hexadecane (n-$C_{16}H_{34}$) (*apolar*)	~ 1	~ 1	~ 1	35
Trichloromethane ($CHCl_3$) (*monopolar, H-donor*)	1.8	0.8	0.3	4.5
Ethanol (C_2H_5OH) (*bipolar*)	12	5.4	n.a.[b]	1
Water (*bipolar*)	460000	2500	130	3.6

[a]Data from Gmehling et al. (1994).   [b]n.a. = not available.

molecules in their pure liquid and/or among the solvent molecules, but not in the solute:solvent interactions, activity coefficients of much greater than 1 may be found. For example, the activity coefficient of ethanol in hexadecane is 35 (Table 3.2). This significant deviation from 1 is due to the polar interactions of the ethanol molecules in their pure liquid, which cannot be counterbalanced by the vdW interactions in solution. The most striking example in Table 3.2 is, however, the solution of n-hexane in water (γ_{iw} = 460000). As pointed out in Section 3.2, this extremely high escaping tendency of nonpolar organic compounds from bulk water is not due to a lack of attractive interactions between the organic molecules and the water molecules, but to the very high free energy costs for cavity formation.

Finally, inspection of Table 3.2 shows also that there are cases in which $\gamma_{i\ell}$ can be even smaller than 1. An example is a solution of diethylether in chloroform. Here, the solute is an electron donor (H-acceptor), while the chloroform solvent is an electron acceptor (H-donor). In this case, the solute and solvent both acquire additional intermolecular interactions that were unavailable to them in their pure liquid forms. The monopolar diethylether (only vdW interactions in its pure liquid) can add polar interactions to its vdW attractions with the molecules of the monopolar chloroform solvent exhibiting a complementary electron acceptor property.

Excess Free Energy, Excess Enthalpy, and Excess Entropy

Before we go on and apply Eq. 3-35 to describe the partitioning of a compound i between two phases, a few comments are necessary on the various partial molar free energy terms included in Eq. 3-35. First, we rewrite Eq. 3-35 by splitting the term that expresses the difference in partial molar free energy of a compound i between its actual situation in a given solution and its situation in the reference state:

$$\mu_{i\ell} = \mu_{iL}^* \quad + \quad RT \ln x_{i\ell} \quad + \quad RT \ln \gamma_{i\ell} \qquad (3\text{-}36)$$

$$T \cdot S_{imix}^{ideal} \qquad\qquad \text{partial molar}$$
$$\textit{excess} \text{ free}$$
$$\text{energy } ``G_{i\ell}^E\text{''}$$

As already pointed out, $\gamma_{i\ell}$ is 1 if a compound forms an ideal solution. In this rather rare case, the term $RT \ln \gamma_{i\ell}$, which we denote as *partial molar excess free energy* of compound i in solution ℓ, $G_{i\ell}^E$, is 0. This means that the difference between the chemical potential of the compound in solution and its chemical potential in the reference state is only due to the different concentration of the compound i in the two states. The term $R \ln x_{i\ell} = S_{imix}^{ideal}$ expresses the partial molar entropy of *ideal* mixing (a purely statistical term) when diluting the compound from its pure liquid ($x_{iL} = 1$) into a solvent that consists of otherwise like molecules.

Let us now have a closer look at this excess free energy term. Note that, for simplicity, in the following and throughout the book we will drop the term "partial molar" and just talk about the excess free energy of a given compound in a given molecular environment. To evaluate the excess free energy term, it is useful to first make some general comments on the various enthalpic and entropic contributions (Eq. 3-22) to the free energy of a given compound in a specific molecular environment. We will do this in a somewhat simplistic way. In brief, the *enthalpy* term represents all attractions or attachments of a compound's molecule to its surroundings. These include *inter*-molecular interactions as discussed in Section 3.2 as well as the internal attraction or bonds (*intra*molecular forces, e.g., bond energies, see Section 2.2). Thus, the enthalpic contributions may be thought of as the "glue" holding the parts of a molecule to its surroundings. As we have already pointed out several times, when we are only interested in the partitioning of organic compounds we choose a reference state in a way that we have only to deal with changes in *inter*molecular interactions when comparing the energy of a compound in various molecular environments.

The *entropy* term is best imagined as involving the "freedom" or latitude of orientation, configuration, and translation of the molecules involved. When molecules are forced to be organized or confined, work must be done. As a consequence, energy must be spent in the process. Conversely, the more ways the molecule can twist and turn, the more freedom the bonding electrons have in moving around in the molecular structure, then the more "randomness" exists. As a result, the entropy terms are larger. This leads to a more negative free energy term (see Eq. 3-22). By analogy to Eq. 3-22, we can express the excess free energy term in Eq. 3-36 as:

$$G_{i\ell}^E = RT \ln \gamma_{i\ell} = H_{i\ell}^E - T \cdot S_{i\ell}^E \qquad (3\text{-}37)$$

where $H_{i\ell}^E$ and $S_{i\ell}^E$ are the (partial molar) excess enthalpy and excess entropy, respectively, of the compound i in phase ℓ. Let us now inspect the enthalpic and entropic contributions to $G_{i\ell}^E$ for four simple compounds in hexadecane and in water (Table 3.3). Also shown in Table 3.3 are the corresponding values for the ideal gas phase (i.e., G_{ig}^E, H_{ig}^E, S_{ig}^E), which are, of course, identical with the free energy, the enthalpy, and the entropy of vaporization of the pure liquid compounds, respectively,

Table 3.3 Excess Free Energies, Enthalpies, and Entropies of Hexane (apolar), Benzene (monopolar), Diethylether (monopolar), and Ethanol (bipolar) in the Ideal Gas Phase, in Hexadecane, and in Water at Infinite Dilution.[a] All Data at 25°C. Reference: Pure Liquid Organic Compound.

Phase Compound (i)	$G^E_{i\,\text{phase}}$ (kJ·mol⁻¹)	=	$H^E_{i\,\text{phase}}$ (kJ·mol⁻¹)	−	$T\,S^E_{i\,\text{phase}}$ (kJ·mol⁻¹)	$S^E_{i\,\text{phase}}$ (J·mol⁻¹K⁻¹)
Gas Phase						
Hexane	4.0	=	31.6	−	27.6	92.6
Benzene	5.3	=	33.9	−	28.6	96.0
Diethylether	0.8	=	27.1	−	26.3	88.2
Ethanol	6.3	=	42.6	−	36.3	122.0
Hexadecane						
Hexane	-0.2	=	0.6	−	0.8	2.7
Benzene	0.4	=	3.5	−	3.1	9.7
Diethylether	0.0	=	1.9	−	1.9	6.4
Ethanol	8.8	=	26.3	−	17.5	58.7
Water						
Hexane	32.3	=	−0.4	+	32.7	−109.7
Benzene	19.4	=	2.2	+	17.2	−58.4
Diethylether	12.0	=	−19.7	+	31.7	−106.3
Ethanol	3.2	=	−10.0	+	13.2	−44.3

[a] Data from Abraham et al. (1990) and Lide (1995).

which we will discuss in detail in Chapter 4. Here, we just note from the examples given in Table 3.3 that when considering a compound in the ideal gas state relative to the pure liquid, both enthalpy costs as well as entropy gains are important in determining the overall excess free energy. The rather high excess enthalpy and excess entropy values observed for ethanol can be fully rationalized by the ability of this bipolar compound to undergo quite strong polar interactions within the pure liquid (which is not the case for the other compounds). This results in a stronger "glue" among the molecules and, therefore, in a higher (positive) H^E_{ig}. For the same reasons, the ethanol molecules have less freedom to move around in their own liquid phase, which leads to a larger entropy gain when transferred to the (ideal) gas phase (where freedom is maximal!).

A very different picture is found for the compounds in hexadecane. Here, the apolar and monopolar compounds show almost ideal behavior (i.e., $G^E_{i\ell} \approx 0$) because in their own liquids, as well as in hexadecane, they can undergo only vdW interactions. In the case of ethanol, again, a significant enthalpy cost and entropy gain is found, which can be explained with the same arguments used above for the gas phase. The absolute $H^E_{i\ell}$ and $T \cdot S^E_{i\ell}$ values are, however, smaller as compared to the gas phase, because ethanol undergoes vdW interactions with the hexadecane–solvent molecules, and because the freedom to move around in hexadecane is smaller than in the gas phase.

Finally, the most interesting, and maybe somewhat puzzling, case is the aqueous phase. We might perhaps not have guessed that the excess enthalpies in water are close to zero or even negative for all four compounds, even for the apolar hexane. The very high fugacity of hexane in water (as compared to its pure liquid) is therefore not due to enthalpic reasons. Rather this is caused by a very large negative entropy contribution. This effect is also significant for the other three compounds (see Table 3.3). This significant loss in entropy when transferring an organic molecule from one liquid phase (the pure compound) to another liquid phase (water) is, at first glance, rather surprising (compare water with hexadecane). Hence, solutions in water involve some special intermolecular interactions which we need to unravel when we want to understand the environmental partitioning of organic compounds. We will learn more about these secrets in Chapter 5. Our next step here is to demonstrate how partition constants between two different phases are linked to the corresponding excess free energy terms.

3.4   Using Thermodynamic Functions to Quantify Equilibrium Partitioning

The next task in our general discussion of equilibrium partitioning of organic compounds between two different phases is to visualize how equilibrium partition constants as defined by Eq. 3-12 are related to the various energy terms discussed in Section 3.3. Then, we will be interested in the free energy term $\Delta_{12}G_i$ introduced in Eq. 3-13, and we will briefly address the effect of *temperature* on equilibrium partitioning. Finally, we will make a few comments on some simple *linear free energy relationships (LFERs)*, which, when applied with appropriate caution, are extremely powerful extrathermodynamic tools (i.e., empirical approaches that cannot be derived strictly from thermodynamic theory) used to predict and/or evaluate partition constants of organic compounds in different two-phase systems.

Equilibrium Partition Constants and Standard Free Energy of Transfer

Let us consider a system in which two bulk phases, 1 and 2 (e.g., air and water, an organic phase and water), are in contact with each other at a given temperature and pressure. We assume that the two phases are in equilibrium with each other with respect to the amounts of all chemical species present in each. We now introduce a *very small amount* of a given organic compound i into phase 2 (i.e., the properties of both bulk phases are not significantly influenced by the introduction of the compound). After a short time, some molecules of compound i will have been transferred from phase 2 (reactant) to phase 1 (product) as portrayed in Eq. 3-11. At this point we write down the chemical potentials of i in the two phases according to Eq. 3-36:

$$\mu_{i1} = \mu_{iL}^* + RT \ln x_{i1} + RT \ln \gamma_{i1}$$
$$\mu_{i2} = \mu_{iL}^* + RT \ln x_{i2} + RT \ln \gamma_{i2}$$

$$(3\text{-}38)$$

The difference between the two chemical potentials (which corresponds to the free energy of a reaction, see Chapter 12), is then given as:

$$\mu_{i1} - \mu_{i2} = RT \ln \frac{x_{i1}}{x_{i2}} + RT \ln \frac{\gamma_{i1}}{\gamma_{i2}} \qquad (3\text{-}39)$$

It is easy to see that at the very beginning of our experiment, μ_{i1} will be smaller than μ_{i2} $(x_{i1} < x_{i2})$ and, hence, the difference will be negative. Consequently, a net transfer of compound i from phase 2 to phase 1 will occur until equilibrium (i.e., $\mu_{i1} = \mu_{i2}$) is reached. Then, at equilibrium, we obtain after some rearrangement:

$$\ln K'_{i12} = \ln \frac{x_{i1}}{x_{i2}} = -\frac{(RT \ln \gamma_{i1} - RT \ln \gamma_{i2})}{RT} \qquad (3\text{-}40)$$

which is equivalent to:

$$K'_{i12} = e^{-(RT \ln \gamma_{i1} - RT \ln \gamma_{i2}) / RT}$$

or:

$$K'_{i12} = e^{-\Delta_{12} G_i / RT} \qquad (3\text{-}41)$$

where K'_{i12} is the partition constant on a mole fraction basis. We distinguish this mole fraction basis by using a superscript prime. Comparison of Eq. 3-41 with Eq. 3-13 reveals that, when expressing the abundance of the compound i in mole fractions, the constant in Eq. 3-13 is equal to 1. Furthermore, and more importantly, we can see now that the free energy of transfer $\Delta_{12} G_i$ equals the difference between the (partial molar) excess free energies of i in the two phases under specified conditions:

$$\Delta_{12} G_i = G_{i1}^E - G_{i2}^E \qquad (3\text{-}42)$$

At this point let us address the problem of expressing abundance of compounds in a bulk phase. In environmental chemistry, the most common way to express concentrations is not by mole fraction, but by the number of molecules per unit volume, for example, as moles per liter of solution ($mol \cdot L^{-1}$, M). This *molar* concentration scale is sometimes not optimal (volumes are, for example, dependent on T and p, whereas masses are not; hence, the use of concentration data normalized per kilogram of seawater is often seen in the oceanographic literature). However, the molar scale is widely used. We can convert mole fractions to molar concentrations by:

$$C_{i\ell} = \frac{x_{i\ell} \, (\text{mol (total mol)}^{-1})}{\overline{V}_\ell (\text{L (total mol)}^{-1})} \qquad (3\text{-}43)$$

where $C_{i\ell}$ is the concentration (moles per liter) of i in phase ℓ and $\overline{V}_\ell$ is the molar volume of the mixture or solution. When we deal with a mixture of several components (e.g., organic solvent/water mixtures in Chapter 5) we will generally apply *Amagat's Law* as a first approximation. That is, we assume that the components of the liquid phase mix with no change in volume due to intermolecular interactions:

$$\overline{V}_\ell = \sum_j x_j \overline{V}_j \qquad (3\text{-}44)$$

where x_j and $\overline{V}_j$ are the mole fractions and molar volumes, respectively, of the pure component j. For *aqueous solutions* of moderately or only sparingly soluble compounds, we can usually neglect the contribution of the organic solute to the molar volume of the mixture. This means that we set $\overline{V}_\ell$ equal to $\overline{V}_w$, the molar volume of water ($\overline{V}_w = 0.018$ L·mol^{-1} at 25°C).

Substitution of x_i by $C_{i\ell} \cdot \overline{V}_\ell$ in Eq. 3-40 then yields the partition constant, K_{i12}, expressed in molar concentrations (note that we now omit the prime superscript):

$$\ln K_{i12} = \ln \frac{C_{i1}}{C_{i2}} = -\ln \frac{\overline{V}_1}{\overline{V}_2} - \frac{\left(RT \ln \gamma_{i1} - RT \ln \gamma_{i2}\right)}{RT} \tag{3-45}$$

This is equivalent to:

or:

$$K_{i12} = \frac{\overline{V}_2}{\overline{V}_1} \cdot e^{-\left(RT \ln \gamma_{i1} - RT \ln \gamma_{i2}\right)/RT} \tag{3-46}$$

$$K_{i12} = \frac{\overline{V}_2}{\overline{V}_1} \cdot e^{-\Delta_{12} G_i /RT}$$

Comparison of Eq. 3-46 with Eq. 3-13 shows that, when expressing the abundance of the compound i in molar concentrations, the constant in Eq. 3-13 corresponds to ($\overline{V}_2 / \overline{V}_1$). Note again that this is strictly true only when we consider infinitely dilute solutions of i in the two phases. Note also that one could debate whether this constant term should be incorporated into the $\Delta_{12}G_i$ term, that is, $\Delta_{12}G_i^c = \Delta_{12}G_i + RT \ln(\overline{V}_1 / \overline{V}_2)$, where $\Delta_{12}G_i^c$ is the free energy of transfer expressed on a molar concentration base (Vitha and Carr, 2000). In our following discussions we will, however, use Eq. 3-46 for relating partition constants with free energies of transfer.

Using the excess free energy, enthalpy, and entropy values given for our four model compounds in Table 3.3, we can now easily calculate how these compounds partition between the various phases (i.e., between air and hexadecane, air and water, and hexadecane and water, respectively) at equilibrium. Table 3.4 summarizes the results of these calculations. These results reflect, of course, what we have already discussed above when inspecting the excess energy terms of the compounds in the various phases. In Chapters 6 and 7 we will address in detail the partitioning of organic compounds between air and liquids (including water), and organic phases and water, respectively. Here we just note again the very important entropy contributions to the overall excess free energy of transfer of a compound i if water is one of the phases involved.

Effect of Temperature on Equilibrium Partitioning

So far, we have considered the equilibrium partitioning of an organic compound at a given temperature and pressure. Since partition constants are commonly reported for only one particular temperature (e.g., 25°C, as is the case for the data summarized in Appendix C), we need to be able to extrapolate these values to other conditions of temperature.

We should note that in most cases in environmental organic chemistry, we can neglect the effect of pressure changes on equilibrium partitioning. Exceptions might

Table 3.4 Air–Hexadecane, Air–Water, and Hexadecane–Water Equilibrium Partitioning of Hexane, Benzene, Diethylether, and Ethanol: Free Energies, Enthalpies, and Entropies of Transfer, as well as Partition Constants Expressed on a Molar Base (i.e., mol ·L^{-1}phase 1/mol · L^{-1}phase 2)

Phase 1/Phase 2 Compound (i)	$\Delta_{12}G_i$ (kJ·mol^{-1})		$\Delta_{12}H_i$ (kJ·mol^{-1})		$T\Delta_{12}S_i$ (kJ·mol^{-1})	$\Delta_{12}S_i$ (kJ·mol^{-1})	$K_{i12}{}^a$
Air/Hexadecane							
Hexane	4.2	=	31.0	–	26.8	89.9	2.2×10^{-3}
Benzene	4.9	=	30.4	–	25.5	85.6	1.7×10^{-3}
Diethylether	0.8	=	25.2	–	24.4	81.9	8.7×10^{-3}
Ethanol	−2.5	=	16.3	–	18.8	73.3	3.3×10^{-2}
Air/Water							
Hexane	−28.3	=	32.0	–	60.3	202.3	6.5×10^{1}
Benzene	−14.1	=	29.7	–	43.8	147.0	2.1×10^{-1}
Diethylether	−11.2	=	46.8	–	58.8	194.6	6.6×10^{-2}
Ethanol	3.1	=	52.6	–	49.5	166.3	2.0×10^{-4}
Hexadecane/Water							
Hexane	−32.5	=	1.0	–	33.5	112.4	3.0×10^{4}
Benzene	−19.0	=	1.3	–	20.3	68.1	1.3×10^{2}
Diethylether	−12.0	=	21.6	–	33.6	112.8	7.7×10^{0}
Ethanol	5.6	=	36.3	–	30.7	103.0	6.4×10^{-3}

^{a} Eq. 3-13 with const. = $\overline{V}_1 / \overline{V}_2$; molar volumes at 25°C and 1 bar: $\overline{V}_{\text{ideal gas}} = 24.73 \text{ L} \cdot \text{mol}^{-1}$, $\overline{V}_{\text{hexadecane}} = 0.293 \text{ L} \cdot \text{mol}^{-1}$, $\overline{V}_{\text{water}} = 0.018 \text{ L} \cdot \text{mol}^{-1}$

include cases of very high pressure, as for example, in the deep sea (>200 bar) or in deep groundwater. For these particular applications we refer to the corresponding literature (e.g., Prausnitz, 1969; Atkins, 1998).

Here, we confine our discussion to the *temperature dependence* of partitioning. As a starting point we consider the differentiation of $\ln K_{i12}$ (Eq. 3-46) with respect to temperature:

$$\frac{d \ln K_{i12}}{dT} = \frac{d \ln \text{ constant}}{dT} - \frac{1}{R} \cdot \frac{d \left(\Delta_{12}G_i / T \right)}{dT} \tag{3-47}$$

Let us first look at the temperature dependence of the constant. Using the mole fraction basis, this constant is equal to 1 and therefore temperature independent if mole fractions or partial pressures, respectively, are used to express the abundance of i in a given liquid or in the gas phase, respectively. In contrast, when using molar concentrations, the constant is given by the ratio of the molar volumes of the two phases. These are, of course, influenced by temperature. However, as a first approximation we may neglect this relatively small effect (< 10% in the temperature range between 0°C and 30°C), and rewrite Eq. 3-47 as:

$$\frac{d \ln K_{i12}}{dT} = -\frac{1}{R} \cdot \frac{d \left(\Delta_{12}G_i / T \right)}{dT} \tag{3-48}$$

Applying the well-known *Gibbs-Helmholtz equation* (Atkins, 1998), we may express the right-hand term of Eq. 3-48 as:

$$-\frac{1}{R} \cdot \frac{d\left(\Delta_{12}G_i \, / \, T\right)}{dT} = \frac{1}{R} \cdot \frac{\Delta_{12}H_i}{T^2} \tag{3-49}$$

which leads to another well-known equation, the *van't Hoff equation*:

$$\frac{d \ln K_{i12}}{dT} = \frac{\Delta_{12}H_i}{RT^2} \tag{3-50}$$

Note that Eqs. 3-49 and 3-50 are very general equations which also apply, for example, to describing temperature dependencies of reaction equilibrium constants, as will be discussed in Chapters 8 and 12 (of course, with the appropriate reaction free energy and enthalpy terms).

If we assume that $\Delta_{12}H_i$ is *constant* over a small temperature range (say between T_1 and T_2), Eq. 3-50 can be integrated. The result of this integration is:

$$\ln \frac{K_{i12}\left(T_2\right)}{K_{i12}\left(T_1\right)} = -\frac{\Delta_{12}H_i}{R}\left(\frac{1}{T_2} - \frac{1}{T_1}\right)$$

or:

$$K_{i12}(T_2) = K_{i12}(T_1) \cdot e^{-\frac{\Delta_{12}H_i}{R}\left(\frac{1}{T_2}-\frac{1}{T_1}\right)} \tag{3-51}$$

Note that by measuring K_{i12} values at various temperatures, $\Delta_{12}H_i$ can be obtained from a linear regression (i.e., a least square fit) of $\ln K_{i12}$ versus $1/T$:

$$\ln K_{i12} = -\frac{A}{T} + B \tag{3-52}$$

Since the slope A (in K) of the regression line is given by $A = \Delta_{12}H_i \, / \, R$, the $\Delta_{12}H_i$ value can be obtained by:

$$\Delta_{12}H_i = R \cdot \text{slope } A \tag{3-53}$$

It should be pointed out that if one of the phases considered is the gas phase, and if K_{i12} is expressed in molar concentrations (including the gas phase), $\Delta_{12}H_i$ in Eqs. 3-51 and 3-53 has to be replaced by $\Delta_{12}H_i + RT_{av}$, where T_{av} is the average temperature (in K) of the temperature range considered (for details see Atkinson and Curthoys, 1978). Finally, we should note that the temperature dependence of K_{i12} (and other equilibrium constants) over *large* temperature ranges can be approximated by a function of the type:

$$\ln K_{i12} = a_1 + \frac{a_2}{T} + a_3 \ln T + a_4 T \tag{3-54}$$

The parameters a_1, a_2, a_3, and a_4 are obtained similarly as A and B in Eq. 3-53 by fitting experimental K_{i12} data obtained at different temperatures.

Table 3.5 Effect of Temperature on Equilibrium Partition Constants as a function of $\Delta_{12}H_i$

$\Delta_{12}H_i$ (kJ · mol^{-1})	Factor[a]
−20	0.75
−10	0.86
0	1.00
10	1.16
20	1.33
30	1.53
40	1.77
50	2.04
60	2.36
70	2.72

[a] Average "increase" (factor) of K_{i12} per 10°C increase in temperature.

Table 3.5 gives the average change in K_{i12} per 10°C increase/decrease in temperature for various $\Delta_{12}H_i$ values. A much more comprehensive table which is extremely useful for assessing the temperature dependence of equilibrium constants as well as of reaction rate constants is Table D1 in Appendix D.

Using the numbers given in Table 3.5 we can now inspect Table 3.4 in order to get some feeling of the temperature dependency of partition constants. Except for the hexadecane/water partitioning of hexane and benzene, there is a significant effect of temperature on the partition constants, particularly if one of the phases is the gas phase. For example, the air/water partition constant of diethylether is about 4 times larger at 25°C as compared to 5°C ($\Delta_{12}H_i = 46.8$ kJ·mol^{-1}). As we will see later in various other chapters, in cases in which equilibrium is not established, temperature may have an important effect on the direction of fluxes of compounds between environmental compartments.

Using Linear Free Energy Relationships (LFERs) to Predict and/or to Evaluate Partition Constants and/or Partition Coefficients

We conclude this section by a few general remarks about *extrathermodynamic* approaches. These quantitative methods involve empirical approaches that cannot be derived strictly from thermodynamic theory. They are widely used to predict and/or to evaluate partition constants and/or partition coefficients (see Box 3.2 for nomenclature) of organic compounds. There are many situations in which some of the data required to assess the partitioning behavior of a compound in the environment are not available, and, therefore, have to be estimated. For example, we may need to know the water solubility of a given compound, its partition coefficient between natural organic matter and water, or its adsorption constant from air to a natural surface. In all these, and in many more cases, we have to find means to predict these unknown entities from one or several known quantities.

The basic idea behind the most common approaches used for predicting partition constants (including vapor pressure and water solubility or partition coefficients) is to express the (unknown) free energy of transfer, $\Delta_{12}G_i$, of a given compound in the two-phase system of interest by one or several other (known) free energy terms chosen in a way that these terms can be linearly related to $\Delta_{12}G_i$. We will encounter and discuss such *linear free energy relationships (LFERs)* in various other chapters of this book. Here we will confine ourselves to some general remarks sketching the basic idea. It should be pointed out that, in practice, such LFERs are sometimes used without the necessary caution. Our considerations of molecular interactions and our discussion on the excess energy terms of organic compounds in various phases will help us throughout the book to develop a more critical attitude toward such LFERs, which is necessary for a proper application of these powerful tools.

To illustrate, we first consider a simple one-parameter LFER approach that is very widely used and, unfortunately, often also abused in environmental organic chemistry. In this approach, a linear relationship is assumed between the free energies of transfer of a series of compounds in two different two-phase systems:

$$\Delta_{12}G_i = a \cdot \Delta_{34}G_i + \text{constant} \tag{3-55}$$

Box. 3.2   Partition Constants, Partition Coefficients, and Distribution Ratios, A Few Comments on Nomenclature

In the literature there is sometimes a certain confusion about the proper use of the terms "partition constant," "partition coefficient," and "distribution ratio." Throughout this book, we will use these terms in the following way. We will talk of a partition constant or a partition coefficient when we consider only *one chemical species* in each phase. Thereby, we will reserve the term partition *constant* for those cases where we deal with the equilibrium partitioning between two *well-defined* phases, where we can be sure that the proportionality factor between the concentrations in the two phases is actually a concentration-independent constant at given conditions. Examples include the air–pure surface partition constant (Chapter 4), the air–pure water partition constant (Chapter 6), and the *n*-octanol–water partition constant (Chapter 7). In all other cases where this proportionality factor may vary somewhat with different related phases, we will talk about a partition *coefficient*. A prominent example is the natural organic matter–water partition coefficient that we will discuss in Chapter 9. Furthermore, we will use the very general term "distribution *ratio*" when we deal with situations where we just want to express the ratio of total concentrations of a given chemical in two phases. Examples include the equilibrium distribution ratio of organic acids or bases in air-water, organic solvent–water, or natural organic matter–water systems (where these compounds may be present both as neutral and charged species (Chapters 8 and 9), and the natural solid-water distribution ratio of a chemical where *various* different sorption mechanisms may be responsible for the presence of the compound in the solid phase (Chapter 11). Finally, we should note that several other terms including "distribution constants," "distribution coefficients," and "accumulation factors" are often used in the literature to describe partitioning. We will generally not use these terms except for our discussion on bioaccumulation, where we will adopt the commonly used "bioaccumulation factor" (Chapter 10).

where very often one of the phases is the same in the two systems (e.g., 2 = 4). In terms of partition constants/coefficients, Eq. 3-55 can be written as (see Eq. 3-46; note that decadic instead of natural logarithms are commonly used):

$$\log K_{i12} = a \cdot \log K_{i34} + \text{constant}' \tag{3-56}$$

Table 3.6 gives some prominent examples of such LFERs. Note that in all cases indicated in Table 3.6, the two systems related by the LFER have one phase in common (i.e., air, water). As should be evident from our basic considerations of molecular interactions, such LFERs can work properly only if various criteria are fulfilled. Important aspects that have to be taken into account include the type of molecular interactions that the compounds considered may undergo in the various phases, as well as the factors that determine the free energy costs of cavity formation, if bulk liquid phases are involved. Hence, for example, it should not come as a big surprise that very poor correlations are generally found when trying to relate partition constants/coefficients of a series of compounds of different polarities between two systems that contain phases exhibiting very different properties (e.g., air/hexadecane and air/water partitioning). On the other hand, rather good correlations can be expected when considering two similar systems (e.g., two organic phase/water systems), particularly when choosing groups of compounds that undergo the same type

Table 3.6 Examples of Simple One-Parameter Linear Free Energy Relationships (LFERs) for Relating Partition Constants and/or Partition Coefficients in Different Two-Phase Systems (Including the Pure Compound as Phase)

Partition Constants/Coefficients Correlated	LFER	Discussed in Chapter
Octanol–water partition constant and aqueous solubility of the pure liquid compound	$\log K_{iow} = -a \cdot \log C_{iw}^{sat} + b$	7
Natural organic carbon–water partition coefficient and octanol–water partition constant	$\log K_{ioc} = a \cdot \log K_{iow} + b$	9
Lipid–water partition coefficient and octanol–water partition constant	$\log K_{ilipw} = a \cdot \log K_{iow} + b$	10
Air–solid surface partition constant and vapor pressure of the pure liquid compound	$\log K_{ias} = a \cdot \log p_{iL}^{*} + b$	11
Air–particle partition coefficient and air–octanol partition constant	$\log K_{iap} = a \cdot \log K_{iao} + b$	11

of interactions in a given phase. In these cases, an LFER established from a set of compounds with known partition constants /coefficients in both systems (from which the slope a and the constant term in Eq. 3-56 can be determined by a linear regression analysis) can be used with good success to predict partition constants/ coefficients of compounds for which the partition constants/coefficients are known only in one of the systems considered (e.g., prediction of natural organic matter– water partition coefficients from octanol–water partition constants, see Chapter 9).

However, LFERs of the type of Eq. 3-56 may not only be used as predictive tools; they may also serve other purposes. For example, they may be very helpful to check reported experimental data for consistency (i.e., to detect experimental errors). They may also enable us to discover unexpected partitioning behavior of a given compound, for example, if a compound is an outlier, but, based on its structure, is expected to fit the LFER. Finally, as will be discussed in various other chapters, if for a given set of model compounds such LFERs have been established for various two-phase systems, where one of the phases is not very well characterized (e.g., various natural organic matter–water systems, different atmospheric particle–air systems), the slopes a of the respective LFERs may yield some important information on the nature of the phases considered (e.g., to detect differences or similarities among the phases).

A second, very different general approach to predict the partition constant of a compound in a given two-phase system assumes that the free energy of transfer term for the whole molecule can be expressed by a linear combination of terms that describe the free energy of transfer of parts of the molecules (at the extreme of the atoms of which the molecule is made up):

$$\Delta_{12}G_i = \sum_{parts} \Delta_{12}G_{part\,of\,i} + \text{special interaction terms} \tag{3-57}$$

Stated in terms of partition constants, this becomes:

$$\log K_{i12} = \sum_{parts} \Delta \log K_{part\,of\,i12} + \text{special interaction terms} \tag{3-58}$$

The special interaction terms are necessary to describe *intra*molecular interactions between different parts of the molecule that cannot be accounted for when considering the transfer of the isolated parts. Obviously, this type of approach has a big advantage in that it allows one to estimate a partition constant based solely on the compound's structure. Good results can be anticipated particularly in those cases where the partition constant of a structurally closely related compound is known, and thus only the contributions of the parts that are different between the two compounds have to be added and/or subtracted, respectively. The most advanced and most widely used method that is based on this concept is the structural group contribution method for estimating octanol–water partition constants. We will discuss this method in Chapter 7.

Finally, we should note that there are a series of more sophisticated methods available that may be used for estimating partition constants. We will discuss the most promising approaches that are based on a direct quantification of molecular interactions later in the following chapters.

Concluding Remarks

The goal of this chapter was to take a first glimpse at the molecular interactions that govern the partitioning behavior of organic compounds between gaseous, liquid, and solid phases, and to recall how simple thermodynamic concepts, in particular, chemical potential, can be used to quantify equilibrium partitioning. In the following chapters, we will discuss important measurable quantities that we need to know when assessing environmental partitioning of organic chemicals. We will continue our effort to visualize the molecule:molecule interactions as well as the freedom of motion of the molecules in a given phase, in order to understand the enthalpic and entropic contributions to the free energy status of the molecules of a given compound in a given molecular environment relative to the pure liquid compound (which we have chosen to be our reference state). By doing so we will hopefully improve our ability to rationalize how pertinent compound properties are related to the compound's structure. It is very important to realize that developing some skills in structure-property considerations is essential for a critical evaluation of experimental data, and, particularly, for a proper use of predictive tools (e.g., LFERs) used to estimate such properties when experimental data are not available.

3.5 **Using Partition Constants/Coefficients to Assess the Equilibrium Distribution of Neutral Organic Compounds in Multiphase Systems**

Our final task in this chapter is to demonstrate how partition constants/coefficients can be used to calculate the equilibrium distribution of a compound i in a given multiphase system. As already pointed out earlier, for simplicity, we consider only neutral species. As we will see in Chapter 8, the equilibrium partitioning of ionogenic compounds (i.e., compounds that are or may also be present as charged species, as, for example, acids or bases) is somewhat more complicated to describe. However, the general approach discussed here is the same.

We start out by considering a very simple example, the partitioning of a compound i between two bulk phases 1 and 2 exhibiting the volumes V_1 and V_2. As discussed in the previous section, at equilibrium the molar concentrations C_{i1} and C_{i2} of i in the two phases are related by the corresponding equilibrium partition constant/coefficients:

$$K_{i12} = \frac{C_{i1}}{C_{i2}} \tag{3-59}$$

It is now easy to see that we may calculate the fraction of the total amount of i present at equilibrium in phase 1, f_{i1}, simply by:

$$f_{i1} = \frac{\text{mass of } i \text{ in phase 1}}{\text{total mass of } i} = \frac{C_{i1} \cdot V_1}{C_{i1} \cdot V_1 + C_{i2} \cdot V_2} \tag{3-60}$$

Dividing the numerator and denominator of the right-hand side of Eq. 3-60 by $(C_{i1} \cdot V_1)$ yields:

$$f_{i1} = \frac{1}{1 + \dfrac{C_{i2}}{C_{i1}} \cdot \dfrac{V_2}{V_1}} \tag{3-61}$$

By substituting Eq. 3-59 into Eq. 3-61 and by defining the (volume) ratio of the two phases $r_{12} = V_1/V_2$, one obtains:

$$f_{i1} = \frac{1}{1 + \dfrac{1}{K_{i12} \cdot r_{12}}} \tag{3-62}$$

and analogously for the fraction of i in phase 2:

$$f_{i2} = \frac{1}{1 + K_{i12} \cdot r_{12}} \tag{3-63}$$

Of course, in a two-phase system, $f_{i1} + f_{i2}$ must be equal to 1 (which can be easily checked). Note that Eqs. 3-62 and 3-63 are also valid if one of the phases is a solid (e.g., solid–water partitioning in a lake or in an aquifer, or solid–air partitioning in the atmosphere). In such cases, K_{i12} is often expressed by the ratio of mole of i per mass of solid concentration and mole of i per volume concentration, and therefore, r_{12} is then given by the ratio of the mass of solid and the volume of the bulk liquid or gas phase present in the system considered.

The equations derived for calculating the fractions of total i present in each phase at equilibrium in a two-phase system (Eqs. 3-62 and 3-63) can be easily extended to a multiphase system containing n phases (e.g., to a "unit world"). If we pick one phase (denoted as phase 1) as the reference phase and if we use the partition constants of i between this phase and all other phases present in the system:

$$K_{i1n(n\neq1)} = \frac{C_{i1}}{C_{in(n\neq1)}} \qquad (3\text{-}64)$$

then, the fraction of i in phase 1 is given by:

$$f_{i1} = \frac{1}{1 + \sum\limits_{n=2}^{n} \dfrac{1}{K_{i1n}} \cdot \dfrac{1}{r_{1n}}} \qquad (3\text{-}65)$$

Note that the partition constant/coefficient of i for any other two phases in the system can be calculated from the $K_{i1n(n\neq1)}$ values. Thus, for example, K_{i23} is given by:

$$K_{i23} = \frac{C_{i2}}{C_{i3}} = \frac{C_{i1}\,/\,C_{i3}}{C_{i1}\,/\,C_{i2}} = \frac{K_{i13}}{K_{i12}} \qquad (3\text{-}66)$$

Obviously, as is demonstrated by Illustrative Example 3.1, since any of the phases can be chosen as phase 1, Eq. 3-65 can be used to calculate the fraction of total i at equilibrium in each of the phases present in the system.

Illustrative Example 3.1

The "Soup Bowl" Problem

Problem

A covered soup bowl contains 1 L of a very diluted, cold soup (25 °C), 1 L of air, and a floating blob of fat of a volume of 1 mL. The system also contains 1 mg of naphthalene. Estimate the amount of naphthalene you would ingest if you were to eat only the fat blob. Assume that equilibrium is established.

In the Appendix C you find the air–water partition constant (K_{iaw}) of naphthalene and its octanol–water partition coefficient (K_{iow}) that you use as surrogate for the fat–water partition coefficient, K_{ifw}). Note that these entities are given as ratios of molar concentrations. Use the fat (octanol) as phase 1 and calculate the fat–air (octanol–air) partition constant, K_{ifa}:

i=naphthalene

$K_{iaw} = 10^{-1.76}$

$K_{ifw} \approx K_{iow} = 10^{3.36}$

$r_{fw} = 10^{-3}$

$r_{fa} = 10^{-3}$

$$K_{ifa} = \frac{K_{ifw}}{K_{iaw}} = \frac{10^{3.36}}{10^{-1.76}} = 10^{5.12}$$

Insertion of K_{ifw}, K_{ifa}, r_{fw}, r_{fa} into Eq. 3-65 yields the fraction in the fat blob:

$$f_{if} = \frac{1}{1 + \dfrac{1}{10^{3.36}} \cdot \dfrac{1}{10^{-3}} + \dfrac{1}{10^{5.12}} \cdot \dfrac{1}{10^{-3}}} \cong 0.7$$

Hence, you would take up *0.7* mg of the 1mg total naphthalene if only eating the fat blob, or you would take up only *0.3* mg when leaving the fat blob, and just eating the soup (the part in the air can more or less be neglected).

3.6   Questions and Problems

Questions

Q 3.1

Give at least 3 reasons why, in environmental organic chemistry, it is so important to understand the *equilibrium* partitioning behavior of a given organic compound between gaseous, liquid, and solid phases.

Q 3.2

How is the equilibrium partition constant defined? To which thermodynamic function(s) is the partition constant related, and which molecular factors determine its magnitude, in the case of

(a) Partitioning between the gas phase and a bulk liquid?
(b) Partitioning between the gas phase and the surface of a condensed phase?
(c) Partitioning between two bulk liquid phases?
(d) Partitioning between a bulk liquid and a solid surface?

Q 3.3

Give at least three examples of environmentally relevant classes of (a) apolar, (b) monopolar, and (c) bipolar compounds. In the case of the monopolar compounds, indicate whether they are electron donors (H-acceptors) or electron acceptors (H-donors).

Q 3.4

Fig. 3.4 shows that when plotting the air–pure liquid compound partition constants of a large number of chemicals versus their dispersive vdW parameters, the apolar and monopolar compounds fall more or less on one line, while the bipolar compounds do not show this behavior. Explain these findings. For which kind of bulk liquids (give examples) would you expect that in a similar plot, *all* compounds (including the bipolar ones) should fit one line?

Q 3.5

The apolar compound *n*-hexane is considered to be quite *hydrophobic* ("water-hating"). Does this mean that there are repulsive forces between hexane and water molecules?

Q 3.6

One of your friends has difficulty understanding what the *chemical potential* of a given compound in a given system expresses. Try to explain it in words to him or her. What do the quantities *fugacity* and *activity* describe? How are they related to the *activity coefficient*?

Q 3.7

Somebody claims that the activity coefficient of *n*-hexane in water is close to 1. Table 3.2 indicates, however, that this is not at all true, but that γ_{iw} of hexane is 460000! Why could this person also be right?

Q 3.8

What are the advantages and disadvantages of choosing the pure liquid compound as reference state?

Q 3.9

2-butanone (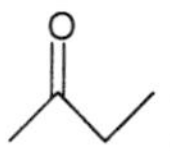) is an important chemical intermediate. When using the pure liquid compound as reference state, in which solvents (give examples) would you expect that this compound has an activity coefficient of (a) close to 1, (b) smaller than one, and (c) larger than one? (Table 3.2 might be helpful.)

Q 3.10

Which thermodynamic function needs to be known for assessing the temperature dependence of equilibrium partitioning? How can this function be derived from experimental data? What caution is advised when extrapolating partition constants from one temperature to another temperature?

Q 3.11

Explain in words the basic idea behind simple one-parameter LFERs for evaluation and/or prediction of equilibrium partition constants. What are the most common approaches? What are the dangers when using such LFERs?

Q 3.12

In Table 3.6 some simple LFERs relating partition constants/coefficients are given. These include vapor pressure and water solubility. Why are these properties, in principle, also partition constants? What is the difference to other partition constants?

Problems

P 3.1 *How Much of the Benzene Initially Present in a Water Sample Has Partitioned into the Headspace of the Sampling Flask?*

You are the boss of a commercial analytical laboratory and your job is to check all results before they are sent to the customers. One day you look at the numbers from the analysis of benzene in BTEX (see Chapter 2) contaminated groundwater samples. For a given sample, your laboratory reports a benzene concentration in water of $100 \ \mu g \cdot L^{-1}$.

Knowing the problems associated with the analysis of volatile organic compounds, you inquire about the handling of the samples. Here we go! The samples (100 mL) were put into 1 L flasks, which were then sealed and stored at 5°C for several days. Then, in the cooling room, an aliquot of the water was withdrawn and analyzed for benzene. What was the original concentration of benzene in the water sample? Assume that equilibrium is established between the gas phase and the water and neglect adsorption of benzene to the glass walls of the bottle. The data required to answer this question can be found in Table 3.4.

Chapter 4

VAPOR PRESSURE

4.1   Introduction

4.2   Theoretical Background
Aggregate State and Phase Diagram: Normal Melting Point (T_m),
    Normal Boiling Point (T_b), and Critical Points (T_c, p_{ic}^*)
Thermodynamic Description of the Vapor Pressure–Temperature
    Relationship
Illustrative Example 4.1: *Basic Vapor Pressure Calculations*

4.3   Molecular Interactions Governing Vapor Pressure
Enthalpy and Entropy Contributions to the Free Energy
    of Vaporization
Trouton's Rule of Constant Entropy of Vaporization
    at the Boiling Point
Quantifications of van der Waals and of Polar Interactions
    Determining Vapor Pressure of Pure Liquids

**4.4   Availability of Experimental Vapor Pressure Data and
Estimation Methods**
Experimental Data
Vapor Pressure Estimation Methods for Liquids
Entropy of Fusion and the Vapor Pressure of Solids
Box 4.1: *Parameters Used to Estimate Entropies of Phase Change
    Processes*

4.5   Questions and Problems

4.1 Introduction

Transport and transformation processes in the atmosphere are among the key processes that govern the distribution and fate of organic chemicals in the environment. In addition, other gaseous phases, such as air pockets in unsaturated soils or bubbles in biological water treatment facilities, may significantly influence the behavior of organic compounds in natural or engineered systems. Hence, one important aspect in our treatment of the partitioning of organic compounds in the environment is the quantitative description of how much a compound likes or dislikes being in the gas phase as compared to other relevant (condensed) phases. In this chapter, we will focus primarily on the equilibrium partitioning of an organic compound between the gas phase and the pure compound itself. That is, we will treat the (*saturation*) *vapor pressure* of organic compounds.

The vapor pressure of a compound is not only a measure of the maximum possible concentration of a compound in the gas phase at a given temperature, but it also provides important quantitative information on the attractive forces among the compound's molecules in the condensed phase. As we will see below, vapor pressure data may also be very useful for predicting equilibrium constants for the partitioning of organic compounds between the gas phase and other liquid or solid phases. Finally, we should note that knowledge of the vapor pressure is required not only to describe equilibrium partitioning between the gas phase and a condensed phase, but also for quantification of the *rate of evaporation* of a compound from its pure phase or when present in a mixture.

In the following sections, we will first look at some thermodynamic aspects concerning the vapor pressure of organic compounds (Section 4.2). This theoretical background will not only enable us to assess vapor pressure data at any given temperature, it will also allow us to deepen our insights into the molecular interactions between organic compounds that we started to discuss in Chapter 3. Note that in Section 4.3, we will introduce a simple model for quantification of molecular interactions that we will continue to use in the following chapters.

4.2 Theoretical Background

To begin, it is instructive to visualize what the molecules of a substance do to establish an equilibrium vapor pressure. We can do this by using a kinetic-molecular description, where we consider the case in which the rate of evaporation balances the rate of condensation. Let us consider a condensed pure compound (either liquid or solid) in equilibrium with its vapor phase (see Fig. 3.9*b*). At a given temperature, a certain number of molecules thermally jostling about in the condensed phase will continuously acquire sufficient energy to overcome the forces of attraction to their neighboring molecules and escape from the condensed phase. Meanwhile in the vapor phase, there will be continuous collisions of some vapor molecules with the surface of the condensed phase. A fraction of the colliding molecules will have so little kinetic energy, or will dissipate their energy upon collision with the condensed surface, that rather than bounce back into the vapor phase, they will be combined into the condensed phase. At a given temperature, these opposing processes of evaporation and condensation reach an equilibrium state that is controlled primarily by molecule–molecule attractions in the condensed phase and is characterized by the amount of molecules in the vapor above the condensed phase. This gas phase amount is expressed as the equilibrium vapor pressure, p_i^*, of the compound i. Recall that we use the superscript $*$ to denote that we look at the (partial) pressure exerted by the compound's molecules at saturation. Furthermore, note that when we speak of the gas phase, for simplicity, we are assuming that all compounds behave like an *ideal gas*. This means that we do not consider the composition of the gas phase. Thus, it does not matter in the following whether the gas is air (mostly N_2 and O_2), an inert gas such as helium or argon, or the saturated vapor of the compound itself. In all these cases, we assume that the various species present in the gas phase do not "feel" each other. This is not appropriate for situations under "high pressure" (>10 bar), as would be seen for gas phases in the ocean or deep groundwater (e.g., greater than 100 m below the water surface) or in pressurized reactors. In such cases, the deviations from ideality begin to exceed about 5% and molecule:molecule interactions must be considered (see Prausnitz, 1969).

From daily life, we know that at ambient conditions of temperature (e.g., 25°C) and pressure (e.g., 1 bar), some organic chemicals are gases, some are liquids, and others are solids when present in their *pure form*. It may perhaps be somewhat trivial, but we should recall that when we talk about a pure chemical, we mean that only molecules of that particular compound are present in the phase considered. Hence, in a pure gas, the partial pressure of the compound is equal to the total pressure. As already addressed to a certain extent in Chapter 3, a pure compound will be a liquid or a solid at ambient conditions, if the forces between the molecules in the condensed phase are strong enough to overcome the tendency of the molecules to fly apart. In other words, if the enthalpy terms (which reflect the "glue" among the molecules in the liquid) outweigh the entropy terms (which is a measure of "freedom" gained when going from the liquid phase to the gas phase), then one has a positive free energy term and the material will exist as a liquid or solid. Conversely, if this free energy term is negative, then the compound is a gas at given conditions (e.g., 25°C and 1 bar). This is illustrated by the series of n-alkanes, where the C_1–C_4 compounds are gases, the C_5–C_{17} compounds are liquids, and the compounds with more than 18 carbon atoms are solids at 25°C and 1 bar total pressure (Fig. 4.1.) This family of hydrocarbons exhibits a vapor pressure range of more than 15 orders of magnitude ranging from 40.7 bar or 4.07×10^6 Pa (C_2H_6) down to about 10^{-14} bar or 10^{-9} Pa (n–$C_{30}H_{62}$). Note that there is no vapor pressure defined for methane at 25°C because methane cannot exist in a defined condensed form at this temperature, even at a very high pressure (see below). In the following, we will use these n-alkanes to illustrate some important general points.

Aggregate State and Phase Diagram: Normal Melting Point (T_m), Normal Boiling Point (T_b), and Critical Points (T_c, p_{ic}^*)

According to the Gibbs phase rule (number of degrees of freedom = number of components − number of phases + 2; see Atkins, 1998), for a system containing a single chemical distributed between two phases at equilibrium, there is only one

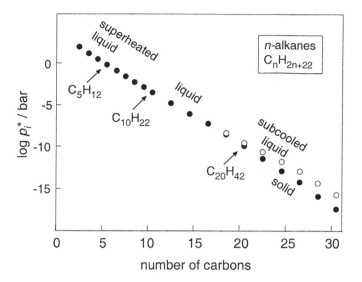

Figure 4.1 Vapor pressure at 25°C of *n*-alkanes as a function of chain length. The subcooled liquid vapor pressures have been calculated by extrapolation of p_{iL}^* values determined above the melting point (Eq. 4-8). Data from Daubert (1997) and Lide (1995).

degree of freedom. Therefore, by choosing a temperature of interest (i.e., using the one degree of freedom), everything else is fixed. Here, the vapor pressure of the compound in the gas phase is fixed. This dependence of vapor pressure on temperature can be conveniently diagramed in a pressure–temperature plot (Fig. 4.2). Such a "phase diagram" also identifies some important single temperature/pressure points. The diagram also allows us to assess the aggregate state (i.e., solid, liquid, gas, supercritical fluid) of the compound under various conditions of temperature and pressure. Let us look at this phase diagram more closely by using four *n*-alkanes (Table 4.1) as illustrative examples.

First we inspect the *normal melting points* (T_m) of the compounds. Note that because T_m, T_b and T_c already have a subscript denoting that they are compound specific parameters, we omit the subscript *i*. T_m is the temperature at which the solid and the liquid phase are in equilibrium at 1.013 bar (= 1 atm) total external pressure. At 1 bar total pressure, we would refer to T_m as *standard* melting point. As a first approximation, we assume that small changes in pressure do not have a significant impact on the melting point. Extending this, we also assume that T_m is equal to the *triple point temperature* (T_t). This triple point temperature occurs at only one set of pressure/temperature conditions under which the solid, liquid, and gas phase of a pure substance all simultaneously coexist in equilibrium.

Among our model compounds (Table 4.1), only *n*-eicosane has a T_m value that is above 25°C; that is, it is the only alkane in this group that is a solid at room temperature. The three other compounds have much lower melting points, which means that, in these cases, we would have to lower the temperature at least to −29.7, −138.4, and −182.5°C in order to "freeze" *n*-decane, *n*-butane, and methane, respectively.

Let us now perform a little experiment with *n*-eicosane. We place pure (solid) *n*-eicosane at 25°C in an open vessel (vessel 1, Fig. 4.3*a*) and in a closed vessel (vessel 2, Fig. 4.3*b*). In the open vessel we have an ambient total pressure of 1 atm or

Table 4.1 Normal Melting Points (T_m), Normal Boiling Points (T_b), and Critical Points (T_c, p^*_{ic}) of some n-Alkanes. Note that temperatures are given in °C and not in K[a]

Compound	T_m (°C)	T_b (°C)	T_c (°C)	p^*_{ic} (bar)	Location of Ambient Temperature (i.e., 25°C in Fig. 4.2 ($T_1...T_4$)	Aggregate State at 25°C
Methane (CH_4)	−182.5	−164.0	−82.6	46.04	T_4	gas
n-Butane (C_4H_{10})	−138.4	−0.5	152.0	37.84	T_3	gas
n-Decane ($C_{10}H_{22}$)	−29.7	174.1	344.5	21.04	T_2	liquid
n-Eicosane ($C_{20}H_{42}$)	36.8	343.0	496.0	11.60	T_1	solid

[a] All data from Lide (1995).

Figure 4.2 Simplified phase diagram of a pure organic chemical. Note that the boundary between the solid and liquid phase has been drawn assuming the chemical's melting point (T_m) equals its triple point (T_t), the temperature–pressure condition where all three phases coexist.) In reality, T_m is a little higher than T_t for some compounds and a little lower for others.

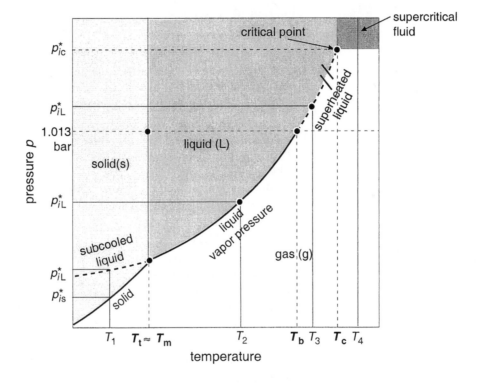

1.013 bar (exerted mostly by the nitrogen and oxygen molecules in the air). In contrast, in the closed vessel, we start out with a vacuum, that is, we allow no molecules other than eicosane in this vessel. Now we wait until equilibrium between the solid and vapor phase is reached, and then we measure the partial pressure of n-eicosane in the gas phase in each vessel. In the closed vessel, the total pressure will be equal to the vapor pressure, p^*_{is}, of solid eicosane. At 25°C this is 10^{-8} bar or 10^{-3} Pa. In our phase diagram in Fig. 4.2, this pressure/temperature point is represented by the point on the bold line at T_1. Now the question is: What is the partial pressure of eicosane in the gas phase in equilibrium with the solid phase in the open vessel 1? Is it also equal to p^*_{is}? The answer is yes because, particularly in the case of a solid compound, for pressures less than about 10 bar the total system pressure has a small influence on p^*_{is}. In general, at pressures near 1 bar we can assume that the difference in the partial pressures between the situations depicted by Figs. 4.3a and b will be less than 0.5% for most organic compounds (Atkins, 1998).

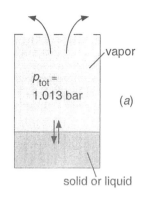

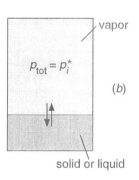

Figure 4.3 Open (*a*) and closed (*b*) vessel containing a pure condensed phase and a vapor phase. In case (*a*) the total pressure (1 bar) is exerted by the compound molecules, and by other gaseous species (e.g., O_2, N_2) that do not significantly alter the composition of the condensed phase. In case (*b*) the total pressure is equal to the partial pressure of the compound molecules; that is, there are no other gaseous species present.

Note, however, that in the open vessel the compound would vanish because molecules could continuously leave the vessel and thus would have to be replenished from the condensed phase to keep a constant saturation vapor pressure.

Returning to our experiment, if we now increase the temperature, then we observe that p_{is}^* of *n*-eicosane increases. In our phase diagram, we move on the solid line from T_1 toward T_m. At T_m, the compound melts and becomes a liquid. Above T_m, a further increase of temperature leads, of course, to a further increase of the vapor pressure which we now denote as p_{iL}^*, indicating that we are now dealing with the vapor pressure of a *liquid* (L) compound (e.g., pressure/temperature point at T_2 in Fig. 4.2). We continue to raise the temperature until p_{iL}^* reaches 1 atm (1.013 bar), which equals the total external pressure in vessel 1. Now we have a very different situation in vessel 1 as compared to vessel 2. In the open vessel 1, the compound *boils*, while in vessel 2, boiling cannot occur (there is no escape for the molecules). The temperature, T_b, at which p_{iL}^* is 1 atm is referred to as the *normal boiling point* temperature (or *standard* boiling point temperature, if p_{iL}^* is 1 bar). Note that, historically, the standard pressure has been taken to be 1 atm (1.013 bar) and that, therefore, most T_b values are still reported as *normal* boiling points, which are somewhat higher than the *standard* boiling points. However, for practical purposes, we will neglect these small differences and just refer to *the* boiling point, T_b. The boiling point of *n*-eicosane is 343°C (Table 4.1). We should recall that boiling means that in an open system, vaporization can occur throughout the bulk of the liquid and the vapor can expand freely into the surroundings. Hence, in contrast to the melting point, the boiling point of a compound depends strongly on the external pressure. A well-known example illustrating this fact is the boiling point of pure water. This is 100°C at 1.013 bar; but at lower pressures such as would apply on the summit of Mt. Everest (0.357 bar external pressure), pure water boils at lower temperatures (about 73°C on Mt. Everest, which renders cooking rather tedious!).

At temperatures above the boiling point (e.g., T_3), at a given external pressure (e.g., 1.013 bar) a compound exists only in the gas phase. For ambient temperatures, this is the case only for a limited number of organic chemicals. Examples are *n*-butane and methane, which have boiling points of –0.5 and –164°C, respectively (Table 4.1). Other examples include some of the halogenated methanes such as the pesticide methyl bromide (CH_3Br), or some of the freons (e.g., CCl_2F_2, $CClF_3$; see Appendix C).

In contrast to the open system, in our closed system (Fig. 4.3*b*) we can increase the temperature above the boiling point and create a situation where we have a vapor pressure, p_{iL}^*, of greater than 1.013 bar. We take advantage of this fact, for example, in pressure cookers or autoclaves, which allow us to cook food or kill bacteria at elevated water temperatures. In such a case, we still have both a liquid and a gas phase (e.g., pressure/temperature point on bold broken line and T_3 in Fig. 4.2). We then commonly refer to the liquid phase as being a *superheated* liquid. For gases at ambient temperature, this means that, in order to be able to store them as liquids (for example, in a pressure bottle) we have to increase their partial pressure in the gas phase until we reach the vapor pressure of the superheated liquid. For *n*-butane (which we use, for example, as fuel for barbeque) this pressure is 2.5 bar and for CCl_2F_2 (a freon that has been widely used as a propellant and foaming agent) the corresponding p_{iL}^* value is 5.6 bar at 25°C (see Appendix C). In the case of methane, however, we would encounter some serious difficulties if we tried to condense this compound to a liquid at 25°C. Before we try to understand this problem with methane, we first continue our experiment with *n*-eicosane in the closed vessel. If we continue to raise the temperature, we build up more and more molecules in the gas phase (increasing the gas density) at the same time as we continuously decrease the density of the liquid. Finally, we reach a point where the density of the vapor is equal to that of the remaining liquid, meaning that we do not have two distinguishable phases anymore. This pressure/temperature point is called the critical point of the compound (T_c, p_{ic}^*, see Fig. 4.2). For *n*-eicosane, the corresponding T_c and p_{ic}^* values are 496°C and 11.6 bar, respectively. Above these values the compound exists only as one phase, which is commonly referred to as *supercritical fluid*. Methane has a critical temperature of –82.6°C (Table 4.1). Hence, liquid methane will exist only below this temperature. In our phase diagram this means that methane belongs to those rather few chemicals for which the ambient temperature is above T_c (e.g., T_4 in Fig. 4.2). Other prominent examples of such chemicals are O_2 (T_c = –118.6°C) and N_2 (T_c = –147°C.)

Before we turn to a quantitative description of the vapor pressure–temperature boundaries, we need to define one important additional vapor pressure value: the *subcooled liquid vapor pressure* of a compound. Imagine what is happening if we cool liquid eicosane from an elevated temperature (e.g., T_2 in Fig. 4.2) to a temperature below its melting (or freezing) point (e.g., T_1 in Fig. 4.2). Above the melting point (T_m = 36.8°C) we observe a decrease in p_{iL}^* according to the solid line describing the liquid–gas boundary. Below the melting point we follow another solid line now describing the solid–gas boundary until we reach $p_{is}^*(T_1)$. We note that, below the melting point, the decrease in vapor pressure with decreasing temperature is steeper than in the region above the melting point, where the compound is a liquid. This finding can be rationalized by recognizing that the energy required to transfer molecules from the solid to the gas phase is higher than transferring them from the corresponding liquid to the gas phase. Hence, below the melting point, if we continued to move along the liquid–gas boundary (dashed line in Fig. 4.2) at T_1, we would have reached another vapor pressure value, p_{iL}^*, which is larger than the corresponding p_{is}^* of the solid compound (examples in Fig. 4.1). This p_{iL}^* value, which is referred to as the vapor pressure of the *subcooled liquid*, is an important entity, because it tells us something about the molecular interactions of the compound in its pure liquid at a temperature where the compound is actually a solid! At this point it might be somewhat unclear why this is so important to know. Knowledge of the properties of the subcooled liquid compound are necessary for understanding and quantifying the molecular situations in environments in which molecules exist in a liquid state (e.g., dissolved in water), although they would be solids if pure. This is a major reason why we have chosen the pure liquid compound as reference state for describing partitioning processes.

Thermodynamic Description of the Vapor Pressure–Temperature Relationship

Liquid–Vapor Equilibrium. In order to quantify the vapor pressure–temperature relationship (bold line in Fig. 4.2) we start out by considering the *liquid–vapor equilibrium.* To this end, we first write down the chemical potentials of a given compound i in the gas phase and in its pure liquid, respectively (see Eq. 3-36; note that $x_{ig} = p_i / p^0$):

$$\mu_{ig} = \mu_{iL}^* + RT \ln (p_i / p^0) + RT \ln \gamma_{ig} \qquad (4\text{-}1)$$

$$\mu_{iL} = \mu_{iL}^* + RT \ln x_{iL} + RT \ln \gamma_{iL} \qquad (4\text{-}2)$$

where p^0 is the standard pressure (1 bar), and $RT \ln \gamma_{ig}$ is the excess free energy of the compound, G_{ig}^E, in the gas phase (see Chapter 3).

Note that for the pure liquid (Eq. 4-2) we explicitly show both the ideal mixing entropy term ($RT \ln x_{iL}$) and the excess free energy term ($RT \ln \gamma_{iL}$), although these two terms are both equal to zero, when we choose the pure liquid to be our reference state. We do this to show that we are dealing here with a partitioning process, and we must consider both forms of the chemical of interest relative to the same reference. The amount of the compound in the gas phase is described by its partial pressure, p_i. Note also that, without writing it down explicitly, we always have to divide any concentration terms by the corresponding standard concentration in the reference state. For the gas phase, we have chosen a standard pressure, p_i^0, of 1 bar since that is close to the pressure we usually have on the surface of the earth. At *liquid–vapor equilibrium* (i.e., $\mu_{ig} = \mu_{iL}$), at a given temperature we then obtain:

$$\ln \frac{p_i / p^0}{x_{iL}} = -\frac{\left(RT \ln \gamma_{ig} - RT \ln \gamma_{iL} \right)}{RT} \qquad (4\text{-}3)$$

Substituting p_i / x_{iL} by p_{iL}^* (the saturation vapor pressure of the pure liquid compound, since $x_{iL}=1$) and by realizing that in this case, $\Delta_{12}G_i$ (see Eq. 3-46) is simply given by G_{ig}^E (the excess free energy of the compound in the gas phase; see examples given in Table 3.2) we may rewrite Eq. 4-3 as:

$$G_{ig}^E = -RT \ln(p_{iL}^* / p^0) \qquad (4\text{-}4)$$

By denoting G_{ig}^E as $\Delta_{vap}G_i$, the free energy of vaporization of the liquid compound, and by omitting to write down every time that we have to divide p_{iL}^* by p^0 (which is commonly 1 bar), we get:

$$\Delta_{vap}G_i = -RT \ln p_{iL}^* \qquad (4\text{-}5)$$

From Eq. 4-5 we can see that $\Delta_{vap}G_i$ will be positive at temperatures at which the vapor pressure is smaller than the standard pressure (i.e., 1 bar), which is, of course, the case at temperatures below the boiling point. At T_b, $p_{iL}^* = p^0$, and therefore:

$$-RT \ln 1 = 0 = \Delta_{vap}G_i(T_b) = \Delta_{vap}H_i(T_b) - T_b \cdot \Delta_{vap}S_i(T_b)$$

or: $\qquad\qquad\qquad\qquad\qquad\qquad\qquad\qquad\qquad\qquad\qquad\qquad (4\text{-}6)$

$$T_b \cdot \Delta_{vap}S_i(T_b) = \Delta_{vap}H_i(T_b)$$

Hence, at the boiling point, the compound molecules in the liquid state can "fly apart" because their gain in entropy on vaporizing now matches the enthalpic attractions that are trying to hold them together. Above the boiling point, $\Delta_{vap}G_i$ will be negative (because $T\Delta_{vap}S_i > \Delta_{vap}H_i$). That is, we have to apply compound partial pressures greater than 1 bar to be able to keep a liquid phase present.

We have also seen that we can treat the vapor pressure like an equilibrium constant K_{i12}. Hence, the temperature dependence of p_{iL}^* can be described by the *van't Hoff* equation (Eq. 3-50):

$$\frac{d \ln p_{iL}^*}{dT} = \frac{\Delta_{vap}H_i(T)}{RT^2} \qquad (4\text{-}7)$$

In this case, this equation is commonly referred to as *Clausius-Clapeyron* equation (e.g., Atkins, 1998). We can integrate Eq. 4-7 if we assume that $\Delta_{vap}H_i$ is *constant* over a given temperature range. We note that $\Delta_{vap}H_i$ is zero at the critical point, T_c, it rises rapidly at temperatures approaching the boiling point, and then it rises more slowly at lower temperatures (Reid et al., 1977). Hence, over a narrow temperature range (e.g., the ambient temperature range from 0°C to 30°C) we can express the temperature dependence of p_{iL}^* by (see Eq. 3-51):

$$\ln p_{iL}^* = -\frac{A}{T} + B \qquad (4\text{-}8)$$

where $A = \Delta_{vap}H_i / R$.

For liquids, plotting the observed log p_{iL}^* $(= \ln p_{iL}^* / 2.303)$ versus inverse T (K) over the ambient temperature range (Fig. 4.4) yields practically linear relations, as expected from Eq. 4-8. Therefore, over narrow temperature ranges in which there are some vapor pressure data available, Eq. 4-8 can be used to calculate vapor pressures at any other temperature *provided that the aggregate state of the compound does not change within the temperature range considered, i.e., that the compound does not become a solid*. If the temperature range is enlarged, the fit of experimental data may be improved by modifying Eq. 4-8 to reflect the temperature dependence of ΔH_{vap}. This is done by the introduction of a third parameter C:

$$\ln p_{iL}^* = -\frac{A}{T+C} + B \qquad (4\text{-}9)$$

Eq. 4-9 is known as the Antoine equation. It has been widely used to regress experimental data. Values for A, B, and C can be found for many compounds in the literature (e.g., Lide, 1995, Daubert, 1997). Note, however, that when using Eqs. 4-8 and 4-9 to extrapolate vapor pressure data below the melting point, one gets an estimate of the vapor pressure of the subcooled liquid compound at that temperature (e.g., naphthalene in Fig. 4.4).

Solid–Vapor Equilibrium. In a very similar way as for the liquid–vapor equilibrium, we can derive a relationship for the temperature dependence of the vapor pressure of the solid compound. By analogy to Eq. 4-5, we write:

$$\ln p_{is}^* = -\frac{\Delta_{sub}G_i}{RT} \qquad (4\text{-}10)$$

where we have replaced the free energy of vaporization by the free energy of

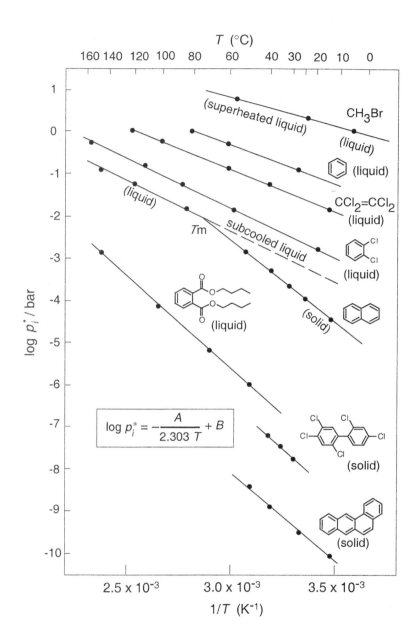

Figure 4.4 Temperature dependence of vapor pressure for some representative compounds. Note that the decadic logarithm is used ($\ln p_i^* = 2.303 \log p_i^*$).

sublimation (transfer from a solid to the gas phase), $\Delta_{sub}G_i$. Note that $\Delta_{sub}G_i$ is given by the difference between the excess free energy of the compound in the gas phase, G_{ig}^E, and its excess free energy in the solid phase, G_{is}^E. Because the excess free energy in the solid phase is negative (the fugacity in the solid phase is smaller than in the liquid phase due to lattice formation), $\Delta_{sub}G_i$ is larger than $\Delta_{vap}G_i$ of the subcooled liquid compound by a term that is commonly referred to as the *free energy of fusion*, $\Delta_{fus}G_i \ (= \Delta_{fus}H_i - T\Delta_{fus}S_i)$:

$$\Delta_{sub}G_i = \Delta_{fus}G_i + \Delta_{vap}G_i \qquad (4\text{-}11)$$

In terms of enthalpy and entropy, this means:

$$\Delta_{sub}H_i = \Delta_{fus}H_i + \Delta_{vap}H_i \qquad (4\text{-}12)$$

and:

$$\Delta_{sub}S_i = \Delta_{fus}S_i + \Delta_{vap}S_i \qquad (4\text{-}13)$$

The first thermodynamic expression above states that the intermolecular attraction forces we must overcome to sublime the molecules of a substance are equal to the sum of the forces required to first melt it and then vaporize it. Likewise, the increased randomness obtained as molecules sublime is the same as the sum of entropies associated with the sequence of melting and vaporizing. Consequently, if we can predict such thermodynamic terms for vaporization or melting, we already know the corresponding parameters for sublimation.

Now, $\Delta_{fus}G_i$ is equal to the negative excess free energy of the compound in the solid state G_{is}^{E}, since we have chosen the liquid state as our reference state. This free energy change is given by:

$$\Delta_{fus}G_i = \Delta_{sub}G_i - \Delta_{vap}G_i = RT \ln \frac{p_{iL}^*}{p_{is}^*} \qquad (4\text{-}14)$$

or:

$$p_{iL}^* = p_{is}^* \cdot e^{+\Delta_{fus}G_i / RT} \qquad (4\text{-}15)$$

In other words, $\Delta_{fus}G_i$ is the free energy required to convert the compound's molecules from the pure solid state to the pure liquid state. Knowledge of $\Delta_{fus}G_i$ at a given temperature is extremely important for estimating other properties of the subcooled liquid compound. As can be qualitatively seen from Fig. 4.2, $\Delta_{fus}G_i$ decreases with increasing temperature (the solid and broken bold vapor pressure lines approach each other when moving toward the melting point). At the melting point, T_m, $\Delta_{fus}G_i$ becomes zero, and, by analogy to the situation at the boiling point (Eq. 4-6), we can write:

$$T_m \cdot \Delta_{fus}S_i(T_m) = \Delta_{fus}H_i(T_m) \qquad (4\text{-}16)$$

Also by analogy to the case for the liquid compound, we can describe the temperature dependence of p_{is}^* by (see Eq. 4-8):

$$\ln p_{is}^* = -\frac{A}{T} + B \qquad (4\text{-}17)$$

where $A = \Delta_{sub}H_i / R$. One may also add a third parameter (like C in Eq. 4-9) to correct for the temperature dependence of $\Delta_{sub}H_i$. Illustrative Example 4.1 shows how to derive and apply Eqs. 4-8 and 4-17. It also demonstrates how to extract free energies, enthalpies, and entropies of vaporization and fusion from experimental vapor pressure data.

Illustrative Example 4.1   Basic Vapor Pressure Calculations

Consider the chemical 1,2,4,5-tetramethylbenzene (abbreviated TeMB and also called durene). In an old *CRC Handbook of Chemistry and Physics* you find vapor pressure data that are given in mm Hg (torr; see left margin).

Problem

Estimate the vapor pressure p_i^*, of TeMB (*in bar and Pa*) at 20°C and 150°C using the experimental vapor pressure data given in the margin. Also express the result in *molar concentration* (mol·L⁻¹) and in *mass concentration* (g·L⁻¹) of TeMB in the gas phase.

Answer

Convert temperatures in °C to K (add 273.2) and calculate $1/T$ values. Also, take the natural logarithms of the p_i^* values. Note that at 45.0 and 74.6°C, TeMB is a solid. Hence, group the data according to the aggregate state of the compound:

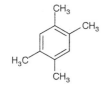

1,2,4,5-tetramethylbenzene
(TeMB)

M_i = 134.2 g mol⁻¹
T_m = 79.5°C
T_b = 195.9°C

T (°C)	p_i^* (mm Hg)
45.0 s[a]	1
74.6 s[a]	10
104.2	40
128.1	100
172.1	400
195.9	760

[a] Means that TeMB is a solid at these temperatures

Solid Compound ($T < T_m$)		
$1/T$ (K⁻¹)	0.003143	0.002875
ln p_{is}^* / mm Hg	0	2.303

Liquid Compound ($T > T_m$)				
$1/T$ (K⁻¹)	0.002650	0.002492	0.002246	0.002132
ln p_{iL}^* / mm Hg	3.689	4.605	5.991	6.633

Perform least squares fits of ln p_i^* versus $1/T$ (see Fig. 1). The results are:

Solid compound: $$\ln p_{is}^* / \text{mm Hg} = -\frac{8609 \text{ K}}{T} + 27.1 \tag{1}$$

Liquid compound: $$\ln p_{iL}^* / \text{mm Hg} = -\frac{5676 \text{ K}}{T} + 18.7 \tag{2}$$

Note that if we had converted mm Hg to bar (1 mm Hg = 0.00133 bar), the intercepts of Eqs. 1 and 2 would be 20.5 and 12.1, respectively.

Insert $T = 293.2$ K (= 20°C) into Eq. 1, calculate ln p_{is}^*, and get p_{is}^*:

$$p_{is}^* (20°C) = 0.10 \text{ mm Hg} = 0.000133 \text{ bar} = 13.3 \text{ Pa}$$

For calculating p_{iL}^* at 150°C, set $T = 423.2$ K in Eq. 2. The resulting p_{iL}^* value is:

$$p_{iL}^* (150°C) = 198 \text{ mm Hg} = 0.264 \text{ bar} = 26400 \text{ Pa}$$

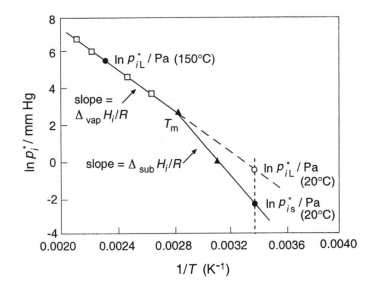

Figure 1 Temperature dependence of the vapor pressure of TeMB: plot of $\ln p_i^*$ / mm Hg versus $1/T$.

Hence, the vapor pressure of TeMB is more than three orders of magnitude greater at 150°C as compared to 20°C, which illustrates the strong temperature dependence of this compound property.

For calculating the molar and mass concentrations in the gas phase, assume that TeMB behaves like an ideal gas ($pV = nRT$). Then the gas phase concentration, C_{ig}, is given by:

$$C_{ig} = \frac{n_{ig}}{V_g} = \frac{p_i^*}{RT}$$

With $R = 0.0831$ L·bar·mol^{-1}·K^{-1} and $T = 293.2$ K or 423.2 K, respectively, the calculated concentrations are (note that 1 mol TeMB corresponds to 134.2 g):

$$C_{ig}(20\,°C) = \frac{(0.000133)}{(0.0831)(293.2)} = 5.5 \times 10^{-6} \text{ mol·L}^{-1} = 7.3 \times 10^{-4} \text{ g·L}^{-1}$$

and

$$C_{ig}(150\,°C) = \frac{(0.264)}{(0.0831)(423.2)} = 7.5 \times 10^{-3} \text{ mol·L}^{-1} = 1.01 \text{ g·L}^{-1}$$

Problem

Estimate the free energy ($\Delta_{fus}G_i$, in kJ·mol^{-1}), the enthalpy ($\Delta_{fus}H_i$, kJ·mol^{-1}), and the entropy ($\Delta_{fus}S_i$, kJ·mol^{-1}K^{-1}) of fusion of TeMB at 20°C using the vapor pressure data given above.

Answer

Insert $T = 293.2$ K into Eq. 2 to estimate the vapor pressure of subcooled TeMB at 20°C:

$$p_{iL}^*(20°C) = 0.52 \text{ mm Hg} = 0.00069 \text{ bar} = 69 \text{ Pa}$$

Hence, at 20°C, p_{iL}^* is about five times larger as compared to p_{is}^*. Note, however, that you have extrapolated this value over quite a large temperature range. Use Eq. 4-14 to calculate $\Delta_{fus}G_i$:

$$\Delta_{fus}G_i \,(20\,°C) = RT\,\ln\frac{p_{iL}^*}{p_{is}^*} = (2.44\,\text{kJ}\cdot\text{mol}^{-1})\,\ln\frac{(0.00069)}{(0.000133)} = 4.0\,\text{kJ}\cdot\text{mol}^{-1}$$

Estimate $\Delta_{fus}H_i$ from the average $\Delta_{vap}H_i$ and $\Delta_{sub}H_i$ that you can derive from the slopes of the regression lines Eqs. 1 and 2:

$$\Delta_{fus}H_i = \Delta_{sub}H_i - \Delta_{vap}H_i = R\cdot\text{slope 1} - R\cdot\text{slope 2}$$

$$= (8.3145\ \text{J}\cdot\text{mol}^{-1}\,\text{K}^{-1})\,(8609\ \text{K}) - (8.3145\ \text{J}\cdot\text{mol}^{-1}\,\text{K}^{-1})\,(5676\ \text{K})$$

$$= 71.6\ \text{kJ}\cdot\text{mol}^{-1} - 47.2\ \text{kJ}\cdot\text{mol}^{-1} = 24.4\ \text{kJ}\cdot\text{mol}^{-1}$$

Since $\Delta_{fus}S_i = (\Delta_{fus}H_i - \Delta_{fus}G_i)\,/\,T$, you get:

$$\Delta_{fus}S_i = (24.4\ \text{kJ}\cdot\text{mol}^{-1} - 4.0\ \text{kJ}\cdot\text{mol}^{-1})\,/\,293.2\ \text{K} = 69.6\ \text{J}\cdot\text{mol}^{-1}\,\text{K}^{-1}$$

Note again that in these calculations, all $\Delta_{12}H_i$ values have been assumed to be constant over the temperature range considered. Therefore, all changes in $\Delta_{fus}G_i$ (which is zero at T_m) are attributed to a change in $\Delta_{fus}S_i$ with changing temperature. This is, of course, not exactly correct.

4.3   Molecular Interactions Governing Vapor Pressure

Enthalpy and Entropy Contributions to the Free Energy of Vaporization

Now we can see how a chemical's structure causes it to have a particular vapor pressure. This is possible because, as a first approximation, the free energy of vaporization, $\Delta_{vap}G_i$, mostly differs from compound to compound due to differences in those substances' enthalpies of vaporization, $\Delta_{vap}H_i$. These enthalpies reflect the sum of intermolecular attractions that act to hold those liquid molecules together. Thus, we can expect that substances that exhibit high vapor pressures have structures that do not enable the molecules to have strong intermolecular attractions. Conversely, molecules with low vapor pressures must have structures that cause the molecules to be substantially attracted to one another.

Moreover, this relation between chemical structure and vapor pressure also holds because enthalpies and entropies of vaporization are directly related, in general. Recall that the entropy of vaporization reflects the difference of a molecule's freedom in the gas phase versus the liquid phase ($\Delta_{vap}S_i = S_{ig} - S_{iL}$). At ambient pressures, we may assume that differences in $\Delta_{vap}S_i$ between different compounds are primarily due to differences in molecular freedom in the liquid phase. (The freedom of the molecules in the gas phase is not that different between compounds). Hence, not surprisingly, molecules that exhibit stronger intermolecular attractions

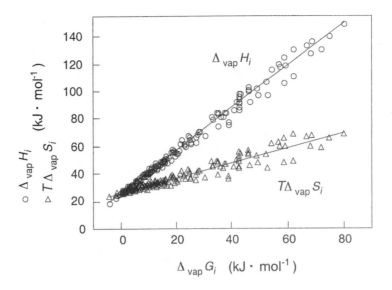

Figure 4.5: Plot of $\Delta_{vap}H_i$ and $T\Delta_{vap}S_i$ versus $\Delta_{vap}G_i$ for a wide variety of organic compounds at 25°C. At the intercept (i.e., for $\Delta_{vap}G_i = 0$) the value for $\Delta_{vap}H_i$ ($= T\Delta_{vap}S_i$) obtained from a linear regression analysis is 25.8 kJ·mol⁻¹.

(and hence greater $\Delta_{vap}H_i$ values) have lower values of S_{iL}, causing higher values of $\Delta_{vap}S_i$. Since the total free energy of vaporization is given by:

$$\Delta_{vap}G_i(T) = \Delta_{vap}H_i(T) - T\Delta_{vap}S_i(T) \qquad (4\text{-}18)$$

correlated differences from compound to compound in $\Delta_{vap}H_i$ and $\Delta_{vap}S_i$ result in changes in $\Delta_{vap}G_i$ which are proportional to either $\Delta_{vap}H_i$ or $\Delta_{vap}S_i$! We can see this quantitatively if we look at the enthalpic and entropic contributions to the total free energy of vaporization for a large number of compounds (Fig. 4.5).

In general, we see that the enthalpic contribution is larger than the entropic one, but also that these contributions co-vary. This is true for a very diverse group of compounds at a given temperature (25°C), including apolar, monopolar, and bipolar compounds. Hence, if we view the forces between the molecules (the "glue") to be reflected primarily in the enthalpy term, then p_{iL}^* is a direct measure of these forces in the pure liquid.

Trouton's Rule of Constant Entropy of Vaporization at the Boiling Point

An interesting point in Fig. 4.5 is the $\Delta_{vap}H_i$ intercept where $\Delta_{vap}G_i = 0$. At this point, $\Delta_{vap}H_i$ is 25.8 kJ · mol⁻¹. This point represents a compound with a boiling point of 25°C. Hence, for this compound the entropy of vaporization at T_b can be calculated by (Eq. 4-6):

$$\Delta_{vap}S_i(T_b) = \frac{\Delta_{vap}H_i(T_b)}{T_b} = \frac{25.8 \text{ kJ mol}^{-1}}{298 \text{ K}} = 86.6 \text{ J} \cdot \text{mol}^{-1} \text{ K}^{-1}$$

This $\Delta_{vap}S_i(T_b)$ value is typical for many other organic compounds that boil at very different temperatures (Table 4.2). In fact, long ago, Trouton (1884) recognized that the entropy of vaporization at the boiling point for many apolar and monopolar substances is more or less constant: between *85 and 90* J · mol⁻¹ K⁻¹. Note that the "constancy" of $\Delta_{vap}S_i(T_b)$ implies that there must be a close relationship between $\Delta_{vap}H_i(T_b)$ and T_b.

Table 4.2 Variations in Normal Boiling Points, Liquid Vapor Pressure at 25°C, Observed Enthalpies and Entropies, and Predicted Entropies of Vaporization at the Boiling Point of Substituted Benzenes and Some Other Compounds [a]

Compound	Substituent(s) or Formula	T_b (°C)	$\log p_{iL}^*$/Pa at 25 °C	Observed $\Delta_{vap}H_i(T_b)$ (kJ·mol⁻¹)	Observed $\Delta_{vap}S_i(T_b)$ (J·mol⁻¹ K⁻¹)	Predicted $\Delta_{vap}S_i(T_b)$ Eq. 4-20 (J·mol⁻¹ K⁻¹)	Predicted $\Delta_{vap}S_i(T_b)$ Eq. 4-21 (J·mol⁻¹ K⁻¹)
Substituted Benzenes							
Benzene	–H	80.1	4.10	30.7	86.9	85.3	86.0
Methylbenzene (Toluene)	–CH₃	110.6	3.57	33.2	86.5	86.0	86.0
Ethylbenzene	–CH₂CH₃	136.2	3.09	35.6	87.0	86.6	86.0
n-Propylbenzene	–(CH₂)₂CH₂	159.2	2.65	38.2	88.4	87.0	86.4
n-Butylbenzene	–(CH₂)₃CH₂	183.3	2.15	39.9	87.4	87.5	86.8
Fluorobenzene	–F	84.7	4.00	31.2	87.2	85.5	86.0
Chlorobenzene	–Cl	131.7	3.20	35.2	86.9	86.5	86.0
1,2-Dichlorobenzene	2x–Cl	180.0	2.26	39.7	87.6	87.4	86.0
1,4-Dichlorobenzene	2x–Cl	174.0	2.37	38.8	86.8	87.3	86.0
Nitrobenzene	–NO₂	210.8	1.48	40.8	84.3	88.0	86.0
Aminobenzene (Aniline)	–NH₂	169.1	1.95	42.4	95.9	96.0	91.0
Hydroxybenzene (Phenol)	–OH	181.8	1.74	45.7	100.5!	100.6	101.0
Benzylalcohol	–CH₂OH	205.3	1.18	50.5	105.6!	114.2	101.0
Other Compounds							
n-Hexane	CH₃(CH₂)₄CH₃	68.7	4.30	28.9	84.6	85.1	87.2
n-Decane	CH₃(CH₂)₈CH₃	174.1	2.24	38.8	86.8	87.3	88.8
n-Hexadecane	CH₃(CH₂)₁₄CH₃	287.0	–0.73	53.9	96.2	89.2	91.2
Ethanol	CH₃CH₂OH	78.3	3.90	38.6	115.8!	110.9	116.9
Naphthalene	see Fig. 2.13	218.0	1.75	43.7	89.0	88.1	86.0
Anthracene	see Fig. 2.13	341.0	–1.15	54.8	89.0	90.0	86.0
Phenanthrene	see Fig. 2.13	339.0	–1.14	53.0	86.6	90.0	86.0

[a] Data from Lide (1995), and Delle Site (1997).

Kistiakowsky (1923) utilized the Clapeyron equation and the ideal gas law to derive an expression to estimate each individual compound's $\Delta_{vap}S_i(T_b)$ in which the chemical's boiling temperature is used:

$$\Delta_{vap}S_i(T_b) = (36.6 + 8.31 \ln T_b) \quad J \cdot mol^{-1} K^{-1} \qquad (4\text{-}19)$$

This expression reflects a weak relationship between the apolar or monopolar compound boiling temperature and entropy of vaporization, but substantially verifies Trouton's empirical observation.

Examination of the $\Delta_{vap}S_i(T_b)$ for various apolar and monopolar compounds reveals some small differences which are understandable in the light of intermolecular forces operating in the liquid phase. For example, elongate molecules such as *n*-hexadecane show higher $\Delta_{vap}S_i(T_b)$ than their corresponding shorter-chain homologues (e.g., *n*-hexane, see Table 4.2). This makes sense since the longer molecules have more contact area for each molecule and thus have a greater tendency to organize in parallel, maximizing the vdW attractions. This decrease in S_{iL} translates into a larger $\Delta_{vap}S_i \ (= S_{ig} - S_{iL})$.

For bipolar organic liquids, especially for hydrogen-bonding liquids such as alcohols and amines, the tendency to orient in the liquid phase, due to these highly directional intermolecular attractions, is greatly increased by this intermolecular interaction. We can see the effect of this in the significantly larger entropies of vaporization of bipolar chemicals, like aniline, phenol, benzyl alcohol, or ethanol (Table 4.2).

Fishtine (1963) provided a set of empirical factors, K_F, which correct the Kistiakowsky estimation of $\Delta_{vap}S_i(T_b)$ for such polar interactions:

$$\Delta_{vap}S_i(T_b) = K_F \ (36.6 + 8.31 \ln T_b) \quad J \cdot mol^{-1} K^{-1} \qquad (4\text{-}20)$$

K_F values are equal to 1.0 for apolar and many monopolar compounds. For compounds exhibiting weakly bipolar character (e.g., esters, ketones, nitriles), a modest correction with a K_F of about 1.04 can be made. Significant corrections are necessary for primary amines (K_F = 1.10), phenols (K_F = 1.15), and aliphatic alcohols (K_F = 1.30). For a more comprehensive compilation of K_F values, we refer to the literature (e.g., Grain, 1982a).

By considering the important structural features of molecules, Myrdal et al. (1996) have developed an alternative way for estimating $\Delta_{vap}S_i(T_b)$. In their approach, which is also based on Trouton's rule, both the flexibility of the molecule (i.e., the presence of single-bonded atoms in long chains) and the inclusion of moieties able to participate in polar interactions are taken into account:

$$\Delta_{vap}S_i(T_b) = (86.0 + 0.4 \ \tau + 1421 \ HBN) \quad J \cdot mol^{-1} K^{-1} \qquad (4\text{-}21)$$

where

$\tau = \Sigma \ (SP3 + 0.5 \ SP2 + 0.5 \ RING) - 1$ is the effective number of torsional bonds,

SP3 is the number of nonring, nonterminal atoms bound to four other atoms (where the nonbonded electron pairs in NH, N, O, and S should be counted as a "bond"),

SP2 is the number of nonring, nonterminal atoms singly bound to two other atoms and doubly bound to a third partner atom,

RING indicates the number of independent ring systems in the compound, and

τ is set equal to zero if its value is negative (for more explanations of parameter τ see Box 4.1).

HBN is the hydrogen bond number and is defined by the following equation:

$$HBN = \frac{\sqrt{OH + COOH} + 0.33\sqrt{NH_2}}{M_i} \qquad (4\text{-}22)$$

where OH, COOH, and NH_2 represent the number of hydroxy, carboxylic acid, and amino groups, respectively, and M_i is the molar mass of the compound.

As is indicated by the examples given in Table 4.2, both methods (Eqs. 4-20 and 4-21) provide reasonable estimates of $\Delta_{vap}S_i(T_b)$. Such equations, along with the generally applicable integrated Clapeyron expression, establish a highly flexible means of estimating compound vapor pressures as a function of temperature (see Section 4.4 for examples).

Quantifications of Van der Waals and of Polar Interactions Determining Vapor Pressure of Pure Liquids

By looking at vapor pressures as a function of chemical structures, we can conclude that the vapor pressure of a liquid or a subcooled liquid depends on the size of the molecule, on its specific ability to undergo vdW interactions, and on its specific ability to be engaged in polar interactions. For example, each addition of a $-CH_2-$ group in the series of the n-alkanes (Fig. 4.1) or in the series of the n-alkyl-benzenes (Table 4.2) leads to a *decrease* in p_{iL}^* at 25°C by about a factor of 3 (or an *increase* in T_b by between 20 and 25 degrees). Similarly, when increasing the number of rings in aromatic hydrocarbons (e.g., benzene $\rightarrow$ naphthalene $\rightarrow$ anthracene, phenanthrene), T_b increases and p_{iL}^* decreases significantly (Table 4.2).

The very significant effect of the presence of a bipolar group on T_b and p_{iL}^* can be nicely seen when comparing hydroxybenzene (phenol) and toluene (Table 4.2). In this case, the difference in T_b (~ 70°C) and in p_{iL}^* (factor of 40 at 25°C) can be attributed primarily to the polar interactions among the phenol molecules because both compounds have similar sizes and a similar specific ability to be engaged in vdW interactions.

Let us now try to derive a model that allows us to express quantitatively the molecular interactions that govern the liquid vapor pressure. We do this, not primarily with the goal of developing a predictive tool for estimating p_{iL}^*, but to introduce a conceptual approach that we will extend and apply later when discussing other partitioning processes.

The basic idea is that we assume that we can *separate the free energy contributions of the vdW and polar interactions and that these contributions are additive:*

$$\Delta_{vap}G_i = \Delta_{vap}G_i^{vdW} + \Delta_{vap}G_i^{polar} \tag{4-23}$$

Note that $\Delta_{vap}G_i^{vdW}$ encompasses dispersive (i.e., London), dipole-induced dipole (i.e., Debye), and dipole–dipole (i.e., Keesom) contributions (Section 3.2). However, in most organic liquids, dipole interactions are generally of secondary importance. Hence, as a first approximation, we consider only the dispersive interactions. Then we can use the approach described in Section 3.2 to quantify the vdW term (Eq. 3-10, Fig. 3.4). Since $\ln p_{iL}^* = -\Delta_{vap}G_i / RT$, we may express Eq. 4-23 as:

$$\ln p_{iL}^* = a(TSA_i)\left(\frac{n_{Di}^2 - 1}{n_{Di}^2 + 2}\right)^2 + b(HD_i)(HA_i) + c \tag{4-24}$$

where we have also introduced a compound-specific H-donor (HD_i) and a compound-specific H-acceptor (HA_i) descriptor for quantification of $\Delta_{vap}G_i^{polar}$. Note that a, b, and c are proportionality and scaling coefficients that also contain the term $(RT)^{-1}$.

There are a variety of methods for estimating the total surface area, TSA_i, of a given compound on a molar base; here we use a simple one; that is, we estimate TSA_i from the molar volume, $\overline{V}_i$, and by assuming that the molecules are perfect spheres:

$$TSA_i \cong 4\pi N_A \left(\overline{V}_{iL}\frac{3}{4\pi N_A}\right)^{\frac{2}{3}} \tag{4-25}$$

where N_A is the Avogadro's number, and the molar volume of a pure liquid compound can be calculated from its molar mass and its density $\overline{V}_{iL} = M_i / \rho_i$.

The *H-bond descriptors,* HD_i (donor property) and HA_i (acceptor property), of a compound depend on the type and number of functional groups in a molecule. Using spectroscopic and chromatographic measurements on a larger number of chemicals, Abraham and coworkers (Abraham et al., 1994a and b) have derived an empirical parameter set of α_i and β_i values that can be used as a quantitative measure of the H-donor and H-acceptor properties of a compound on a molar base (Table 4.3). The functional groups exhibiting the strongest *bipolar* character are alcohols and carboxyl acids; that is, these compound classes have values of α_i and β_i which are well above zero. Interestingly, most of the *monopolar* compounds exhibit predominantly H-acceptor characteristics. Furthermore, there are significant differences in the α_i- and β_i-values between different polyhalogenated alkanes. These compounds also possess H-donor characteristics, due to the electron-withdrawing nature of the halogens. Finally, water has a strong tendency to undergo interactions with both H-acceptors and H-donors. When considering the small size of the water molecules, this is what makes water such a special solvent.

Table 4.3 α_i- and β_i-Values for Some Selected Compounds[a]

Compound (Class), Functional Group		α_i (H-Donor)	β_i (H-Acceptor)
Alkanes	(C_nH_{2n+2})	0	0
1-Alkenes	$(1\text{-}C_nH_{2n})$	0	0.07
Aliphatic ethers	(ROR')	0	0.45
Aliphatic aldehydes	(RCHO)	0^b	0.45
Carboxylic acid esters	(RCOOR')	0^b	0.45
Aliphatic ketones	(RCOR')	0^b	0.51
Aliphatic amines	$(R\text{–}NH_2)$	0.16	0.61
Aliphatic alcohols	(R–OH)	0.37	0.48
Carboxylic acids	(R–COOH)	0.60	0.45
Benzene		0	0.14
Methylbenzene		0	0.14
Ethylbenzene		0	0.15
Dimethylbenzene		0	0.16
Trimethylbenzene		0	0.19
Chlorobenzene		0	0.07
1,2-Dichlorobenzene		0	0.04
1,3-Dichlorobenzene		0	0.02
Chlorobenzene	$(Cl_n, n > 2)$	0	0
Aniline		0.26	0.41
Benzaldehyde		0	0.39
Phenol		0.60	0.31
Pyridine		0	0.52
Naphthalene		0	0.20
Indane		0	0.17
Acenaphthene		0	0.20
Fluorene		0	0.20
Phenanthrene		0	0.26
Anthracene		0	0.26
Fluoranthene		0	0.20
Benzo(a)fluorene		0	0.20
Pyrene		0	0.29
Benzo(a)anthracene		0	0.33
Chrysene		0	0.33
Perylene		0	0.40
Benzo(a)pyrene		0	0.44
Benzo(ghi)perylene		0	0.46
Dichloromethane		0.10	0.05
Trichloromethane		0.15	0.02
Tetrachloromethane		0	0
1,1,1-Trichloroethane		0	0.09
1,1,2,2-Tetrachloroethane		0.16	0.12
Trichloroethene		0.08	0.03
Tetrachloroethene		0	0
Water		0.82	0.35

[a] Data from Abraham et al. (1994a and b). [b] Some other sources (Reichardt, 1988; Fowkes et al., 1990) indicate that aldehydes, esters, and ketones also exhibit a weak H-donor property.

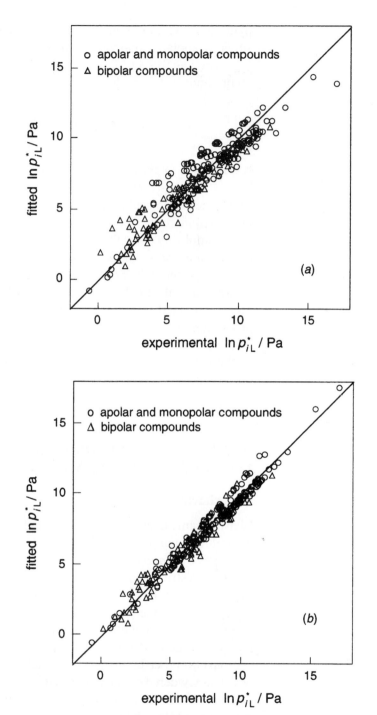

Figure 4.6 Fitted versus experimental liquid vapor pressures of a series of organic compounds, including apolar and monopolar (o) as well as bipolar (Δ) representatives. The set of compounds is the same as in Figs. 3.4 and 3.6a. (a) Data fitted by using Eq. 4.27. (b) Data fitted by the equation ln p_{iL}^* = 1.02 lnK_{iah} − 13.8 (α_i) (β_i) + 15.8, where K_{iah} is the air-*n*-hexadecane partition constant.

Substitution of α_i and β_i and of TSA$_i$ from Eq. 4-25 into Eq. 4-24 yields:

$$\ln p_{iL}^* = a\left[\left(\overline{V}_{iL}\right)^{2/3}\left(\frac{n_{Di}^2-1}{n_{Di}^2+2}\right)^2\right] + b\,(\alpha_i)\,(\beta_i) + c \tag{4-26}$$

Note that the constant term $4\pi N_A\,(3/4\pi N_A)^{2/3}$ has been included into the coefficient *a*. Considering the same set of compounds included in Fig. 3.4, we see that adding the hydrogen-bonding interaction term allows us to reasonably predict the vapor pressures of apolar, monopolar, *and* bipolar compounds (Fig. 4.6*a*). Using a multiple regression analysis of Eq. 4-26 against the experimental p_{iL}^* values at 25°C of the compounds of the model set yields:

$$\ln p_{iL}^* / \text{Pa} = -4.49 \left[\left(\overline{V}_{iL} \right)^{2/3} \left(\frac{n_{Di}^2 - 1}{n_{Di}^2 + 2} \right)^2 \right] - 15.1 \, (\alpha_i) \, (\beta_i) + 14.5 \qquad (4\text{-}27)$$

where $\overline{V}_{iL}$ is in cm^{3}mol^{-1}.

The rather large scatter in the data points shown in Fig. 4.6*a* could be reduced by using a more refined approach for quantifying the nonspecific interaction parameter. For example, by using the air–hexadecane partition constant K_{iah} (see Section 3.2) as a more appropriate measure of the vdW interactions, the predicted vapor pressures are even closer to their corresponding observed values. Furthermore, in the literature (Abraham et al., 1994a), an additional polarity/polarizability parameter (π_i) is commonly included in this type of models. This parameter improves the quantitative prediction of the aqueous activity coefficients (see Section 5.3). That is, it seems to be of some importance when polar water molecules surround an organic compound. In the case of vapor pressure, however, introduction of this additional parameter does not significantly improve the result.

4.4 Availability of Experimental Vapor Pressure Data and Estimation Methods

Experimental Data

Many organic chemicals of environmental concern have rather low vapor pressures at ambient temperatures (Appendix C). Since simple measurements of vapor pressures by *manometric methods* or by determining *boiling points at reduced pressures* are restricted to relatively volatile compounds ($p_i^* > 1$ to 10 Pa), more sophisticated methods have to be applied for compounds of low volatility ($p_i^* < 1$ Pa). The methods most widely used are *gas saturation* and *effusion* [see Delle Site (1997) for a review of these and other methods]. In the case of *gas saturation*, a saturated vapor phase is produced by passing an inert gas, air, nitrogen, or oxygen (when a combustion procedure is used for analysis) through a thermostated column packed with the powdered compound or with an analyte-coated inert support. The saturation pressure of the substance is represented by its partial vapor pressure. Usually, the vapor is collected on liquid or solid traps and the substance is determined by suitable means. The *effusion methods* determine the vapor pressure at constant temperature from the measurement of the weight loss through a small orifice into a vacuum.

An attractive alternative to the direct measurement of vapor pressure is the use of *gas chromatographic retention* to estimate p_i^* (e.g., Hinckley et al., 1990). This method is based on the evaluation of the partitioning behavior of a given compound between the gas phase (i.e., the mobile phase) and a bulk organic phase (i.e., the stationary phase) at different temperatures. The method hinges on the selection of an appropriate reference compound for which accurate vapor pressure data is available, as well as on the choice of an appropriate stationary phase, in which both compound and reference exhibit similar activity coefficients. Note that for solid compounds, since the molecules are *dissolved* in the stationary phase, the gas chromatographic method yields the vapor pressure of the subcooled liquid (p_{iL}^*).

Inspection of the literature shows that vapor pressure data are readily available for many high-to-medium-volatility compounds (i.e., compounds with T_b < 400°C). These data can be found in data compilations (e.g., Daubert, 1997; Mackay et al., 1992–1997; Lide, 1995). For compounds exhibiting very low vapor pressures, the data are more scattered throughout the literature with the exception of agrochemicals (e.g., Montgomery, 1997). Furthermore, for such compounds, p_i^* values obtained by different methods and/or different laboratories may vary by as much as a factor of 2 to 3, in some cases, by more than an order of magnitude. In addition, in many cases, vapor pressure data have been determined at elevated temperatures, and ambient values must be extrapolated. Such data should, therefore, be treated with the necessary caution. One way of deciding which vapor pressure should be selected is to compare the experimental data with values predicted using other compound properties (see below).

Finally, very often vapor pressures are reported only for one particular temperature (e.g., 20°C or 25°C, as in Appendix C). Since vapor pressure is strongly dependent on temperature, it is necessary to be able to extrapolate such values over the ambient temperature range. Hence, it is necessary to know the enthalpy of vaporization or sublimation at ambient temperature. As we have seen in Section 4.3, for *liquid* compounds, a proportionality between $\Delta_{vap}H_i$ and $T\Delta_{vap}S_i$ is observed (Fig. 4.5). This means that $\Delta_{vap}G_i$ is proportional to $\Delta_{vap}H_i$. This can be used to derive an extremely useful empirical relationship between $\Delta_{vap}H_i$ and ln p_{iL}^* (or log p_{iL}^*) *for a given temperature T_1* (Goss and Schwarzenbach, 1999a):

$$\Delta_{vap}H_i(T_1) = -a \log p_{iL}^*(T_1) + b \tag{4-28}$$

At 25°C (298 K), the linear regression derived for the data set shown in Fig. 4.7 is:

$$\Delta_{vap}H_i(298\,K)/(kJ\cdot mol^{-1}) = -8.80(\pm 0.07)\log p_{iL}^*(298\,K)/Pa + 70.0(\pm 0.2) \tag{4-29}$$

Note that in contrast to Fig. 4.7, we use the decadic logarithm in Eq. 4-29 and that this relationship holds over a very large vapor pressure range (> 15 orders of magnitude).

Assuming that this $\Delta_{vap}H_i$ value is constant over the ambient temperature range, it can be used to estimate p_{iL}^* at other temperatures (see also Eq. 3-51):

$$p_{iL}^*(T) = p_{iL}^*(298\,K)\cdot e^{-\frac{\Delta_{vap}H_i(298\,K)}{R}\left[\frac{1}{T}-\frac{1}{298\,K}\right]} \tag{4-30}$$

It should be pointed out again that Eq. 4-29 applies to the vapor pressure of the *liquid* compound. For solids, the difference between p_{is}^* and p_{iL}^* can be estimated using the melting point temperature of the compound, see below (Eq. 4-40).

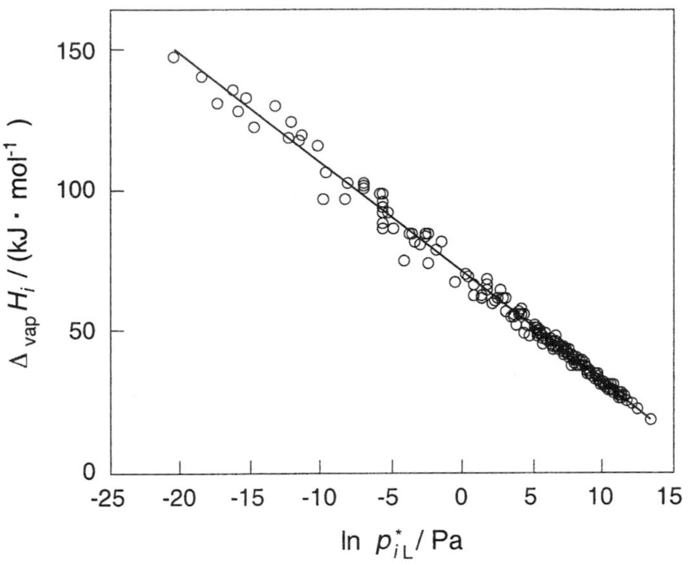

Figure 4.7 Plot of $\Delta_{vap}H_i$ versus $\ln p_{iL}^*$ for a large number of apolar, monopolar, and bipolar compounds. Note that some bipolar outliers are not included. (For details see Goss and Schwarzenbach, 1999a.)

Vapor Pressure Estimation Methods for Liquids

One strategy for estimating the vapor pressure of (subcooled) liquid compounds is to derive multiple parameter regression equations that relate the free energy of vaporization (and thus $\ln p_{iL}^*$) to other properties and/or structural descriptors of the compound. The goal of all these approaches is to express the molecular interactions that determine $\Delta_{vap}G_i$ by readily accessible entities. Examples of such parameters include constitutional descriptors (e.g., partial charges), shape descriptors (e.g., topological indices), geometrical descriptors (e.g., surface area, molar volume), and quantum-chemical descriptors (e.g., dipole moment, quadrupole moment, polarizability). For an overview of these methods, we refer to the literature (Delle Site, 1997; Liang and Gallagher, 1998).

Here, we confine our discussion to an approach that can be easily handled because it requires only knowledge of the chemical's structure, its normal boiling point, and, if the compound is a solid, its melting point. Note that if T_b and T_m are not available, they can also be estimated (for details see Boethling and Mackay, 2000). Various equations using this approach have been proposed (Delle Site, 1997; Myrdal and Yalkowski, 1997), but they are all based on the same general idea. To predict the *liquid vapor pressure curve* below the boiling point [see solid and broken (below T_m) bold line in Fig. 4.2], we use the Clausius-Clapeyron equation:

$$\frac{d \ln p_{iL}^*}{dT} = -\frac{\Delta_{vap}H_i(T)}{RT^2} \tag{4-7}$$

and properties of the compound at the boiling point, T_b. As we recall (Section 4.2), at the boiling point the enthalpy of vaporization can be related to the entropy of vaporization:

$$\Delta_{vap}H_i(T_b) = T_b \cdot \Delta_{vap}S_i(T_b) \tag{4-6}$$

This entropy change can be estimated with reasonable accuracy (Section 4.3).

Hence, for temperatures very close to the boiling point, we integrate Eq. 4-7 by assuming that $\Delta_{vap}H_i(T) = \Delta_{vap}H_i(T_b)$ = constant (see Section 4.2). However, in most cases, one would like to estimate the vapor pressure at temperatures (e.g., 25°C) that are well below the boiling point of the compound. Therefore, one has to account for the temperature dependence of $\Delta_{vap}H_i$ below the boiling point. A first approximation is to assume a linear temperature dependence of $\Delta_{vap}H_i$ over the temperature range considered, that is, to assume a constant heat capacity of vaporization, $\Delta_{vap}C_{pi}$ (the difference between the vapor and liquid heat capacities). Thus, if the heat capacity of vaporization, $\Delta_{vap}C_{pi}(T_b)$, at the boiling point is known, $\Delta_{vap}H_i(T)$ can be expressed by (e.g., Atkins, 1998):

$$\Delta_{vap}H_i(T) \cong \Delta_{vap}H_i(T_b) + \Delta_{vap}C_{pi}(T_b) \cdot (T - T_b) \tag{4-31}$$

Substitution of Eqs. 4-31 and 4-6 into Eq. 4-7 and integration from 1 bar to p_{iL}^* and from T_b to T then yields:

$$\ln p_{iL}^* / \text{bar} \cong -\frac{\Delta_{vap}S_i(T_b)}{R}\left[\frac{T_b}{T} - 1\right] + \frac{\Delta_{vap}C_{pi}(T_b)}{R}\left[\left(\frac{T_b}{T} - 1\right) - \ln\frac{T_b}{T}\right] \tag{4-32}$$

In the literature, various suggestions have been made of how to estimate $\Delta_{vap}S_i(T_b)$ and $\Delta_{vap}C_{pi}(T_b)$. One approach that works well primarily for prediction of vapor pressures of relatively low boiling compounds (i.e., $T_b < \sim 300°C$) was proposed by Mackay et al. (1982). In this approach, the Kistiakowsky-Fishtine expression (Eq. 4-20) is used to estimate $\Delta_{vap}S_i(T_b)$, and it is assumed that, particularly for smaller molecules, the ratio of $\Delta_{vap}C_{pi}(T_b) / \Delta_{vap}S_i(T_b)$ has an average value of 0.8 ($\pm$ 0.2). Inserting Eq. 4-20 and substituting $\Delta_{vap}C_{pi}(T_b)$ by 0.8 $\Delta_{vap}S_i(T_b)$ into Eq. 4-32 thus yields:

$$\ln p_{iL}^* / \text{bar} \cong -K_F(4.4 + \ln T_b)\left[1.8\left(\frac{T_b}{T} - 1\right) - 0.8\ln\frac{T_b}{T}\right] \tag{4-33}$$

Another approach has been put forward by Myrdal and Yalkowsky (1997), which the authors contend is superior for high boiling compounds. Using Eq. 4-21 to estimate $\Delta_{vap}S_i(T_b)$ and an additional empirical equation for quantification of $\Delta_{vap}C_{pi}(T_b)$:

$$\Delta_{vap}C_{pi}(T_b) \cong -(90 + 2.1\tau) \quad \text{J} \cdot \text{mol}^{-1}\,\text{K}^{-1} \tag{4-34}$$

they propose the following equation for estimating vapor pressures of organic compounds:

$$\ln p_{iL}^* / \text{bar} = -(21.2 + 0.3\tau + 177\,\text{HBN})\left(\frac{T_b}{T} - 1\right) + (10.8 + 0.25\,\tau) \cdot \ln\frac{T_b}{T} \tag{4-35}$$

As discussed in Section 4.3, the two parameters τ and HBN, which describe the overall flexibility and the hydrogen-bonding capacity, respectively, of the molecules, can be easily derived from the structure of the compound.

Table 4.4 shows that these relatively simple approaches work quite well for compounds with boiling points not exceeding 300°C. Larger discrepancies to experimental values up to a factor of 10 have to be expected for very high boiling

Table 4.4 Comparison of Predicted and Experimental Vapor Pressures at 25°C for Selected (Subcooled) Liquid Organic Compounds[a]

Compound	T_b/°C	T_m/°C	K_F	Predicted (Eq. 4-33) p_{iL}^*/Pa	τ	HBN	Predicted (Eq. 4-35) p_{iL}^*/Pa	Experimental p_{iL}^*/Pa
Benzene	80.1	5.5	1.00	1.3×10^4	0	0	1.3×10^4	1.3×10^4
1,2,4,5-Tetramethylbenzene	195.9	79.5	1.00	8.6×10^1	0	0	7.2×10^1	$6.9 \times 10^{1\ b}$
Naphthalene	218.0	80.2	1.00	3.2×10^1	0	0	2.4×10^1	$3.0 \times 10^{1\ c}$
Anthracene	341.0	217.5	1.00	5.6×10^{-2}	0	0	4.3×10^{-2}	$1.0 \times 10^{-1\ c}$
Pyrene	403.0	156.0	1.00	2.0×10^{-3}	0	0	1.5×10^{-3}	$1.3 \times 10^{-2\ c}$
Tetrachlorethene	121.4	−22.4	1.01	2.4×10^3	0	0	2.2×10^3	2.5×10^3
Hexane	69.0	−95.0	1.00	2.1×10^4	3	0	1.7×10^4	2.0×10^4
Hexadecane	287.0	18.2	1.00	9.6×10^{-1}	13	0	2.3×10^{-2}	1.9×10^{-1}
Aniline	184.0	−6.3	1.10	1.7×10^2	0	0.0035	7.9×10^1	1.3×10^2
Benzylalcohol	205.3	−15.2	1.30	5.7×10^0	0	0.0092	1.7×10^1	1.5×10^1

[a] Data from Appendix C, if not otherwise indicated. [b] See Illustrative Example 4.1. [c] Average value determined by gas chromatography and estimated from Eq. 4-35 using experimental $\Delta_{fus}S_i$ (T_m) values. Data from Hinckley et al. (1990).

compounds. For such compounds, however, the experimental data are often not very accurate. Note again that any approach using solely boiling point data can predict only the (subcooled) liquid vapor pressure. Hence, for compounds that are solids at the temperature of interest, one has to estimate additionally the contribution of fusion; that is, we have to predict the solid–vapor boundary below the melting point (solid bold line below T_m in Fig. 4.2).

Entropy of Fusion and the Vapor Pressure of Solids

In a very similar way as discussed above for estimating p_{iL}^* from boiling point data, one can treat the vapor pressure curve *below the melting point*. Again we use the Clausius-Clapeyron equation:

$$\frac{d \ln p_{is}^*}{dT} = \frac{\Delta_{sub}H_i(T)}{RT^2} \qquad (4\text{-}36)$$

Since we are only interested in the ratio of p_{is}^*/p_{iL}^* at a given temperature (i.e., in the contribution of melting), we can subtract Eq. 4-7 from Eq. 4-36 to get:

$$\frac{d \ln p_{is}^*/p_{iL}^*}{dT} = \frac{\Delta_{sub}H_i(T) - \Delta_{vap}H_i(T)}{RT^2} = \frac{\Delta_{fus}H_i(T)}{RT^2} \qquad (4\text{-}37)$$

If, as a first approximation, we assume that $\Delta_{fus}H_i$ is constant over the temperature range below the melting point, and if we substitute Eq. 4-16 into Eq. 4-37, we can integrate Eq. 4-37 from 1 ($p_{is}^* = p_{iL}^*$ at T_m!) to p_{is}^*/p_{iL}^* and from T_m to T, respectively. We then obtain for $T \leq T_m$:

$$\ln \frac{p_{is}^*}{p_{iL}^*} = -\frac{\Delta_{fus}S_i(T_m)}{R}\left[\frac{T_m}{T} - 1\right] \qquad (4\text{-}38)$$

Hence, now we are left with the problem of estimating the entropy of fusion at the melting point. Unfortunately, $\Delta_{fus}S_i(T_m)$ (Table 4.5) is much more variable than $\Delta_{vap}S_i(T_b)$ (Table 4.2). This might be expected since $\Delta_{fus}S_i(T_m)$ is equal to $S_{iL}(T_m) - S_{is}(T_m)$ and both of these entropies can vary differently with compound structure. One reason is that molecular symmetry is an important determinant of the properties of a solid substance in contrast to a liquid, where the orientation of a molecule is not that important (Dannenfelser et al., 1993). Nevertheless, as demonstrated by Myrdal and Yalkowski (1997), a reasonable estimate of $\Delta_{fus}S_i(T_m)$ can be obtained by the empirical relationship (Table 4.5):

$$\Delta_{fus}S_i(T_m) \cong (56.5 + 9.2\,\tau - 19.2 \log \sigma) \quad \text{J} \cdot \text{mol}^{-1}\,\text{K}^{-1} \qquad (4\text{-}39)$$

where

τ is the effective number of torsional bonds (see Box 4.1), and

σ is the rotational symmetry number that describes the indistinguishable orientations in which a compound may be positioned in space (Box 4.1).

Table 4.5 Comparison of Experimental and Predicted (Eq. 4-39) Entropies of Fusion at the Normal Melting Point[a]

| Compound | T_m (°C) | Experimental | | Predicted (Eq. 4-39) | | |
		$\Delta_{fus}H_i\,(T_m)$ (kJ·mol⁻¹)	$\Delta_{fus}S_i\,(T_m)$ (J·mol⁻¹K⁻¹)	τ	σ	$\Delta_{fus}S_i\,(T_m)$
Benzene	5.5	10.0	35.7	0	12	35.8
n-Butylbenzene	−88.0	11.2	60.5	2	2	69.1
1,4-Dichlorobenzene	52.7	17.2	52.8	0	4	45.0
Naphthalene	80.2	18.6	52.7	0	4	45.0
Phenanthrene	101.0	18.1	48.6	0	2	50.7
Fluoranthene	107.8	18.9	49.6	0	2	50.7
Pyrene	151.2	17.1	40.3	0	4	45.0
Decane	−29.7	28.8	118.3	7	2	115.1
Eicosane	36.8	69.9	225.6	17	2	207.1
Benzoic acid	122.4	18.1	45.8	0	2	50.7
2,2',4,5,5'-Pentachlorobiphenyl	77.0	18.8	53.6	0	1	56.5
p,p'-DDT	109.0	27.4	71.6	1	1	65.7

[a] Data from Hinckley et al. (1990) and Lide (1995)

Obviously, for compounds exhibiting no rotational symmetry axis, σ is equal to 1 (which is the case for many of the more complex environmental chemicals). For benzene, on the other hand, $\sigma = 12$ (there are six twofold rotational axes), while for 1,4-dichlorobenzene $\sigma = 4$ (only two twofold rotational axes). Some examples of the application of Eq. 4-36 are given in Table 4.5. For a detailed discussion of the symmetry aspects (i.e., the derivation of σ) we refer to the articles by Dannenfelser et al. (1993) and Dannenfelser and Yalkowsky (1996). Finally, we should note that Eq. 4-39 does not work well for small spherical molecules or for polar compounds for which H-bonding has a significant impact on $\Delta_{fus}S_i\,(T_m)$. Hence, there is certainly room for improvement of this empirical relationship.

Substitution of Eq. 4-39 into Eq. 4-38 then gives ($R = 8.31$ J·mol⁻¹K⁻¹):

$$\ln \frac{p_{is}^*}{p_{iL}^*} = -(6.80 + 1.1\tau - 2.3 \log \sigma)\left[\frac{T_m}{T} - 1\right] \tag{4-40}$$

which can be used to estimate p_{is}^* from the subcooled liquid vapor pressure p_{iL}^*, and vice versa. Note that insertion of Eq. 4-40 into Eq. 4-14 yields an estimate of the free energy of fusion:

$$\Delta_{fus}G_i = +(56.5 + 9.2\,\tau - 19.2 \log \sigma)\,[T_m - T] \quad \text{J·mol}^{-1} \tag{4-41}$$

an entity that will be important for estimating other properties of the subcooled liquid such as water solubility.

Box 4.1   Parameters Used to Estimate Entropies of Phase Change Processes

In phase change processes, the overall entropy changes, $\Delta_{12}S_i$, can be understood by considering the degrees of freedom lost when molecules in one condition (e.g., as a liquid) are packed into a less free, new condition (e.g., as a solid). Such transformations have been viewed as involving three contributions to the change in molecular freedom: (1) positional, (2) conformational, and (3) rotational (Yalkowsky, 1979; Dannenfelser and Yalkowsky, 1996):

$$\Delta_{12}S_i = \Delta_{12}S_{i\,\text{positional}} + \Delta_{12}S_{i\,\text{conformational}} + \Delta_{12}S_{i\,\text{rotational}}$$

For the process of condensation (i.e., opposite direction to vaporization), the positional freedom loss involves about -86 J·mol^{-1} K^{-1}, while for the process of freezing (i.e., opposite direction to fusion), the positional freedom loss involves -50 to -60 J·mol^{-1} K^{-1}.

When a substance is packed into a liquid from a gas or into a solid from a liquid, the molecules also have a reduced ability to assume the various conformations. This loss of freedom is reflected in $\Delta_{12}S_{i\,\text{conformational}}$. Different conformations arise from the ability of structures to rotate around single bonds. For example, consider 1-bromo-2-chloro-ethane. Viewing the two carbons and the chlorine substituent as co-existing in a plane, we recognize that the bromine atom can occur in the same plane opposite the chlorine atom, or above the plane or behind the plane:

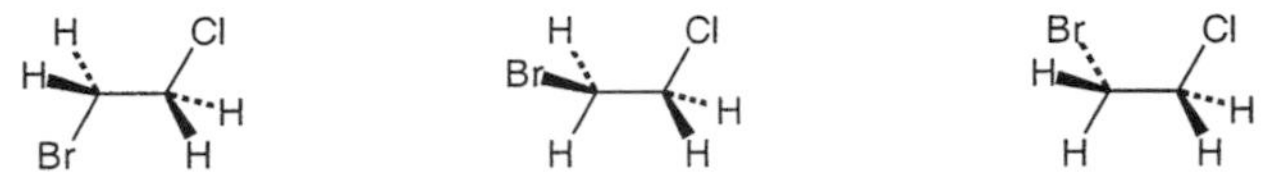

This amounts to rotating around the single bond connecting the two carbons. Every bond capable of such rotations offers three distinguishable orientations. Hence, if we increased the chain length by one $-CH_2$ unit, there would be $3 \times 3 = 9$ distinguishable conformations. Note that three atoms in such a chain do not enable conformation variation since three points always determine a single plane. Hence, $\Delta_{12}S_{i\,\text{conformational}}$ increases as the number of bonds capable of rotation minus two (equivalent to number of nonterminable sp^3 atoms in chain; note hydrogens are not sp^3 atoms).

Atoms in chain that include doubly bonded moieties do not offer as much conformational variety. Consider methyl ethyl ketone; rotation around the bond between the carbonyl carbon and the C$_3$ allow two (not three) distinguishable conformers:

Hence, such atoms need to be discounted in their contribution to $\Delta_{12}S_{i\,\text{conformational}}$ and this is done by applying a factor of 0.5 times the number of such sp^2 members of a chain. This discounting also applies to ring systems. Hence, we can estimate a parameter, τ:

$$\tau = (\text{number of nonterminable sp}^3) + 0.5 \times (\text{number of nonterminate sp}^2) + 0.5 \times (\text{ring systems}) - 1$$

and the number of distinguishable conformers is approximately 3^τ. Empirically the observed data for the entropy of fusion at T_m are best fit using 2.85^τ. With this estimate, one finds $\Delta_{12}S_{i\,\text{conformational}}$ is approximately R ln (number of distinguishable conformers) $\approx R \ln (2.85^\tau) = 9.2 \; \tau$ (see Eq. 4-39). For the case of 1-bromo-2-chloro-ethane with τ of 1, $R \ln (2.85^\tau) = 9 \; J \cdot mol^{-1} \; K^{-1}$. As chains get longer, the magnitude of this contribution grows quickly (see Table 4.5).

Changes in rotational freedom and $\Delta_{12}S_{i \, rotational}$ can be understood by considering the "symmetry" of a molecule. This entropy contribution may be quantified by a parameter, σ, quantifying the number of indistinguishable ways a given molecule can exist in space. The more indistinguishable orientations there are, the easier it is to convert the molecules to a more packed phase (hence making the absolute value of $\Delta_{12}S_{i \, rotational}$ smaller). One may begin by assessing whether a three dimensional view of a given molecule looks the same from above and below (i.e., is there a plane of symmetry in the plane of paper on which a molecule can be drawn?) A molecule like vinyl chloride does not look the same ($\sigma=1$), while DDE does ($\sigma=2$). Next, one may ask is there a way to rotate a molecule around an axis perpendicular to any plane of symmetry (e.g., perpendicular to the paper on which the molecule is drawn) and have orientations that look the same. In this sense, vinyl chloride and DDE have only one orientation that look the same, but 1,4-dichlorobenzene looks the same from above and below as well as if it is rotated 180° ($\sigma=2\times2$) and benzene looks the same from above and below and every time it is rotated 60° ($\sigma=2\times6$). The product of these numbers of indistinguishable orientations yields the symmetry number, σ. The higher a molecule's symmetry number, the less freedom change there is associated with packing or unpacking molecules. In the case of the entropy of fusion, $\Delta_{12}S_{i \, rotational} = R \ln \sigma = 19.2 \log \sigma$. When σ is 1, $\Delta_{12}S_{i \, rotational}$ is zero; and when σ is 12, the absolute value is about 20 $J \cdot mol^{-1} \; K^{-1}$. Note that the sign depends on whether one considers unpacking (more freedom so $\Delta_{12}S_{i \, rotational}$ has positive sign) or the packing (e.g., freezing or condensation) direction of phase change.

vinyl chloride (chloroethene)   DDE   benzene   1,4-dichlorobenzene

---

4.5   Questions and Problems

Questions

Q 4.1

Give at least five examples of environmentally relevant organic chemicals that are (a) solids, (b) liquids, and (c) gases at 25°C.

Q 4.2

Why are certain chemicals gases at ambient conditions?

Q 4.3

Propane ($T_b = -42.1°C$, $T_c = 101.2°C$) is a gas at 25°C. How can you "produce" liquid propane (give two options)?

Q 4.4

What is the difference between the *normal* and the *standard* boiling point?

Q 4.5

Explain in words the terms *subcooled liquid*, *superheated liquid*, and *supercritical fluid*.

Q 4.6

Why is the excess free energy of a solid, G_{is}^E, negative? How is G_{is}^E related to the free energy of fusion, $\Delta_{fus}G_i$? How does G_{is}^E change with temperature? At which temperature is G_{is}^E equal to zero?

Q 4.7

How are the (subcooled) liquid and solid vapor pressures of a given compound at a given temperature related to each other?

Q 4.8

The two isomeric polycyclic aromatic hydrocarbons phenanthrene and anthracene are solids at 25°C. Although these compounds have almost the same boiling point (see below), their vapor pressures at 25°C differ by more than one order of magnitude (see Appendix C). Explain these findings. What differences would you expect for the subcooled liquid vapor pressures of the two compounds at 25°C?

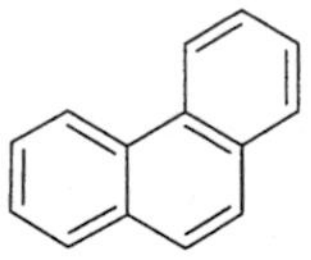

phenanthrene
$T_m = 101.0\ °C$
$T_b = 339.0\ °C$

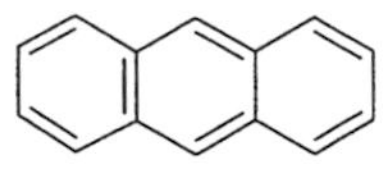

anthracene
$T_m = 217.5\ °C$
$T_b = 341.0\ °C$

Q 4.9

Which thermodynamic function needs to be known for assessing the temperature dependence of the vapor pressure of a given compound? How can this function be derived from experimental data? What caution is advised when extrapolating vapor pressure data from one temperature to another temperature?

Problems

P 4.1 *A Solvent Spill*

You teach environmental organic chemistry and for a demonstration of partitioning processes of organic compounds you bring a glass bottle containing 10 L of the common solvent tetrachloroethene (perchloroethene, PCE) into your class room. After closing the door you stumble and drop the bottle. The bottle breaks and the solvent is spilled on the floor. Soon you can smell the solvent vapor in the air. (The odor threshold of PCE is between 8 and 30 mg·m⁻³). Answer the following questions:

(a)   What is the maximum PCE concentration that you can expect in the air in the room ($T = 20°C$)? How much of the solvent has evaporated if you assume that the air volume is 50 m³? (Neglect any adsorption of PCE on the walls and on the furniture).

(b)   If the same accident happened in your sauna (volume 15 m³, $T = 80°C$), what maximum PCE concentration would you and your friends be exposed to there?

In the *CRC Handbook of Chemistry and Physics* (Lide, 1995) you find the following vapor pressure data for PCE:

$T/°C$	25	50	75	100
p_i^*/kPa	2.42	8.27	22.9	54.2

All other necessary data can be found in Appendix C.

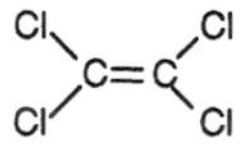

tetrachloroethene
(PCE)

P 4.2 *How Much Freon Is Left in the Old Pressure Bottle?*

In a dump site, you find an old 3-liter pressure bottle with a pressure gauge that indicates a pressure of 2.7 bar. The temperature is 10°C. From the label you can see that the bottle contains Freon 12 (i.e., dichlorodifluoromethane, CCl_2F_2). You wonder how much Freon 12 is still left in bottle. Try to answer this question. In the *CRC Handbook of Chemistry and Physics* (Lide, 1995) you find the following data on CCl_2F_2:

$T/°C$	−25	0	25	50	75
p_i^*/kPa	123	308	651	1216	2076

Using these data, estimate the free energy ($\Delta_{cond}G_i$), the enthalpy ($\Delta_{cond}H_i$), and the entropy ($\Delta_{cond}S_i$) of condensation of Freon 12 at 25°C. Note that condensation is the opposite of vaporization (watch out for the signs of the three quantities).

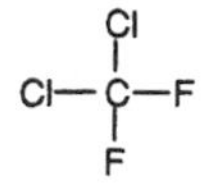

dichlorodifluoromethane
(Freon 12)

P 4.3 *What Are the Differences Between Freon 12 and Its Replacement HFC-134a? (From Roberts, 1995)*

Hydrofluorocarbon 134a (1,1,1,2-tetrafluoroethane) is used as a replacement for Freon 12 (see Problem 4.2) for refrigeration applications. (Why is such a replacement necessary and what is the advantage of HFC-134a from an environmental protection point of view?) Some vapor pressure data for Freon 12 is given in Problem 4.2. The vapor pressure data of HFC-134a have been determined very carefully and are as follows:

$T/°C$	−40.0	−30.0	−20.0	−10.0	0	+10.0
p_i^*/kPa	51.6	84.7	132.9	200.7	292.9	414.8

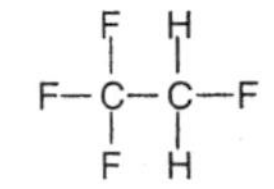

1,1,1,2-tetrafluoroethane
(HFC-134a)

(a)   Determine the normal boiling points (in°C) of these compounds from the data provided.

(b)   At what temperature (in°C) will they have an equal vapor pressure?

(c)   Compare the (average) enthalpies ($\Delta_{vap}H_i$) and entropies ($\Delta_{vap}S_i$) of vaporization of the two compounds at the temperatures calculated under (b). Can you rationalize any differences you observe between the two compounds?

(d)   Automobile air conditioners commonly operate at temperatures between 30 and 50°C. Are the vapor pressures of the two compounds significantly (i.e., greater than 10%) different in this temperature region?

P 4.4   *A Public Toilet Problem*

Pure 1,4-dichlorobenzene (1,4-DCB) is still used as a disinfectant and airfreshener in some public toilets. As an employee of the health department of a large city you are asked to evaluate whether the 1,4-DCB present in the air in such toilets may pose a health problem to the toilet personnel who are exposed to this compound for several hours every day. In this context you are interested in the maximum possible 1,4-DCB concentration in the toilet air at 20°C. Calculate this concentration in g per m^3 air assuming that

(a)   You go to the library and get the vapor pressure data given below from an old edition of the *CRC Handbook of Chemistry and Physics*.

(b)   You have no time to look for vapor pressure data, but you know the boiling point ($T_b = 174.0°C$) and the melting point ($T_m = 53.1°C$) of 1,4-DCB.

Compare the two results. What would be the maximum 1,4-DCB concentration in the air of a public toilet located in Death Valley (temperature 60°C)? Any comments?

$T/°C$	29.1s	44.4s	54.8	84.8	108.4	150.2
$p_i^*/mm\ Hg$	1	4	10	40	100	400

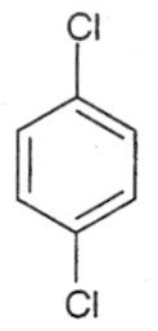

1,4-dichlorobenzene
(1,4-DCB)

P 4.5 *True or False?*

Somebody bets you that at 60°C, the vapor pressure of 1,2-dichlorobenzene (1,2-DCB) is smaller than that of 1,4-dichlorobenzene (1,4-DCB), but that at 20°C, the opposite is true; that is, p_i^* (1,2-DCB, 20°C) > p_i^* (1,4-DCB, 20°C). Is this person right? If yes, at what temperature do both compounds exhibit the same vapor pressure? Try to answer these questions by using only the T_m and T_b values given in Appendix C.

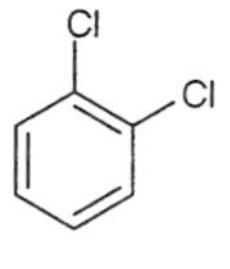

1,2-dichlorobenzene
(1,2-DCB)

P 4.6 *Estimating Vapor Pressure Data*

Since you live in a cold area, you are more interested in the vapor pressure of organic compounds at 0°C as compared to 25°C. Estimate the vapor pressures at 0°C from (i) the p_i^* values given in Appendix C for 25°C, and (ii) only using the T_m and T_b values (also given in Appendix C) for the following compounds:

(a) Methacrylate:

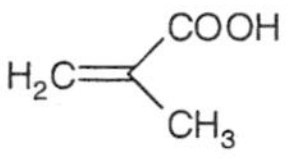

(b) Dimethyl phthalate:

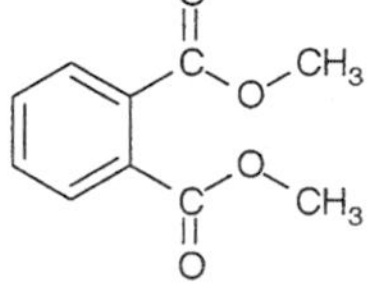

(c) 2,3,7,8-Tetrachlorodibenzo-p-dioxin:

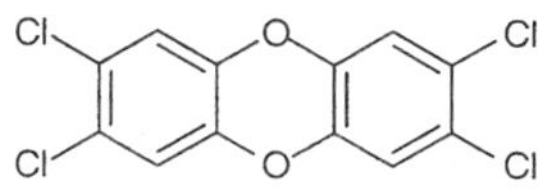

Compare and discuss the results.

P 4.7 *Evaluating Experimental Vapor Pressure Data of the Widely Used Pesticide Lindane*

Using the *Knudsen effusion technique* and *highly purified* samples of lindane [(γ-HCH), one of the most widely used and most frequently detected organochlorine pesticides; see Willet et al. (1998)], Boehncke et al. (1996) determined the vapor pressure of this compound in the temperature range between 20 and 50°C. For this temperature range they derived the following relationship (note that the melting point of lindane is 112.5°C; its boiling point is 323.4°C):

$$\ln p_i^* / Pa = -\frac{11754 \ K}{T} + 34.53 \tag{1}$$

Wania et al. (1994) used *commercial* lindane and a *gas saturation method*, and they obtained for the temperature range between −30° and +30°C:

$$\ln p_i^* / Pa = -\frac{12816 \ K}{T} + 39.12 \tag{2}$$

Finally, Hinckley et al. (1990), using the *gas chromatographic technique*, reported:

$$\ln p_i^* / Pa = -\frac{8478 \ K}{T} + 25.67 \tag{3}$$

for the temperature range between 40 and 85°C.

(a)   Calculate the vapor pressure and the enthalpy of sublimation of lindane at 25°C from each of these three equations, and compare the different values. Why does Eq. 3 yield such a different result as compared to Eqs. 1 and 2? Try to explain the differences between Eq. 1 and Eq. 2. Which equation would you recommend for estimating the vapor pressure of lindane in the ambient temperature range?

(b)   Estimate the free energy of fusion ($\Delta_{fus}G_i$) of lindane at 25°C, (i) from the data given above (Eqs. 1–3), and (ii) using only the normal melting point temperature. Any comments?

(c)   Estimate the vapor pressure of lindane at 25°C from its boiling and melting point temperatures given above. Use both equations given in Section 4.4 (Eqs. 4-33 and 4-35) to estimate p_{iL}^*, and Eq. 4-40 to get p_{is}^*. Compare the results with the p_i^* values derived from the experimental data.

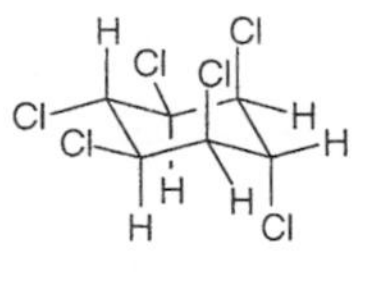

1,2,3,4,5,6-Hexachlorocyclohexane
(γ-HCH, Lindane)

Chapter 5

ACTIVITY COEFFICIENT AND SOLUBILITY IN WATER

5.1   Introduction

5.2   Thermodynamic Considerations
        Solubilities and Aqueous Activity Coefficients of Organic Liquids
        Solubilities and Aqueous Activity Coefficients of Organic Solids
        Solubilities and Aqueous Activity Coefficients of Organic Gases
        Illustrative Example 5.1: *Deriving Liquid Aqueous Solubilities, Aqueous
            Activity Coefficients, and Excess Free Energies in Aqueous Solution from
            Experimental Solubility Data*
        Concentration Dependence of the Aqueous Activity Coefficient

5.3   Molecular Interpretation of the Excess Free Energy of Organic
        Compounds in Aqueous Solutions
        Enthalpic and Entropic Contributions to the Excess Free Energy
        Molecular Picture of the Dissolution Process
        Model for Description of the Aqueous Activity Coefficient
        Box 5.1: *Estimating Molar Volumes from Structure*
        Illustrative Example 5.2: *Evaluating the Factors that Govern the Aqueous
            Activity Coefficient of a Given Compound*

5.4   Effect of Temperature and Solution Composition on Aqueous Solubility
        and Activity Coefficient
        Temperature
        Illustrative Example 5.3: *Evaluating the Effect of Temperature on Aqueous
            Solubilities and Aqueous Activity Coefficients*
        Dissolved Inorganic Salts
        Illustrative Example 5.4: *Quantifying the Effect of Inorganic Salts on
            Aqueous Solubility and Aqueous Activity Coefficients*

Organic Cosolvents *(Advanced Topic)*
Illustrative Example 5.5: *Estimating the Solubilities and Activity*
    Coefficients of Organic Pollutants in Organic Solvent–Water Mixtures

5.5   Availability of Experimental Data; Methods for Estimation of Aqueous Activity Coefficient and Aqueous Solubility
Experimental Data
Prediction of Aqueous Solubilities and/or Aqueous Activity Coefficients

5.6   Questions and Problems

5.1   Introduction

Whether an organic compound "likes" or "dislikes" being surrounded by liquid water, or alternatively whether water "likes" or "dislikes" to accommodate a given organic solute, is of utmost importance to the environmental behavior and impact of that compound. Due to its small size and hydrogen-bonding characteristics, water is a rather exceptional solvent. Indeed, environmentally relevant compounds have aqueous solubilities ranging over more than ten orders of magnitude — from completely soluble compounds (i.e., miscible) to levels of saturation that are so low that the concentration can be measured only with very sophisticated methods (Appendix C). In this chapter, we will discuss and try to visualize the molecular factors that cause this immense range of results associated with transferring an organic compound from a nonaqueous phase to an aqueous solution (or vice versa).

We will start our discussion by considering a special case, that is, the situation in which the molecules of a pure compound (gas, liquid, or solid) are partitioned so that its concentration reflects equilibrium between the pure material and aqueous solution. In this case, we refer to the equilibrium concentration (or the *saturation* concentration) in the aqueous phase as the *water solubility* or the *aqueous solubility* of the compound. This concentration will be denoted as C_{iw}^{sat}. This compound property, which has been determined experimentally for many compounds, tells us the maximum concentration of a given chemical that can be dissolved in pure water at a given temperature. In Section 5.2, we will discuss how the aqueous activity coefficient *at saturation*, γ_{iw}^{sat}, is related to aqueous solubility. We will also examine when we can use γ_{iw}^{sat} as the activity coefficient of a compound in *diluted* aqueous solution, γ_{iw}^{∞} (which represents a more relevant situation in the environment).

In the next step in Section 5.3, we will explore how chemical structures of the solutes govern their aqueous activity coefficients. This will be done by inspecting how the chemical structures of the solutes correspond to different enthalpic and entropic contributions to the excess free energy of putting those substances in aqueous solution. Using these insights we will extend the molecular interaction model that we introduced and applied in Chapter 4 to quantitatively describe activity coefficients in pure water. In Section 5.4, we will then deal with the effects of temperature and of certain dissolved water constituents that may be present in the environment (i.e., inorganic ions, organic cosolutes and cosolvents) on the solubilities and the aqueous activity coefficients of organic compounds. Finally, in Section 5.5 we will comment on experimental methods and on predictive tools used to estimate aqueous solubilities and aqueous activity coefficients of organic compounds.

5.2   Thermodynamic Considerations

Solubilities and Aqueous Activity Coefficients of Organic Liquids

Let us first imagine an experiment in which we bring a pure, water-immiscible organic *liquid* into contact with pure water at a given temperature and ask what will happen. Intuitively, we know that some organic molecules will leave the organic phase and dissolve into water, while some water molecules will enter the organic liquid. After some time, so many organic molecules will have entered the water that some will begin to return to the organic phase. When the fluxes of molecules into and out of the organic phase are balanced, the system has reached a state of equilibrium. At this point, the amount of organic molecules in the water is the water solubility of that liquid organic compound. Similarly, the amount of water molecules in the organic phase reflects the solubility of water in that organic liquid.

To describe this process thermodynamically, at any instant in time during our experiment, we can express the chemical potentials of the organic compound i in each of the two phases (Chapter 3). For the compound in the *organic liquid phase,* we have:

$$\mu_{iL} = \mu_{iL}^* + RT \ln \gamma_{iL} \cdot x_{iL} \tag{5-1}$$

where we use the subscript L to indicate the pure liquid organic phase, although it contains some water molecules. For the compound in the aqueous phase, the corresponding expression of its chemical potential is:

$$\mu_{iw} = \mu_{iL}^* + RT \ln \gamma_{iw} \cdot x_{iw} \tag{5-2}$$

where we use the subscript w to refer to parameters of the compound i in the *water.* Note that both expressions relate chemical potential to the same reference potential, μ_{iL}^*. Hence at any given time, the difference in chemical potentials of the "product" (solutes in aqueous solution) minus the "reactant" (i in its pure liquid) molecules is given by:

$$\mu_{iw} - \mu_{iL} = RT \ln \gamma_{iw} \cdot x_{iw} - RT \ln \gamma_{iL} \cdot x_{iL} \tag{5-3}$$

In the beginning of our experiment, μ_{iL} is much larger than μ_{iw} (x_{iw} is near zero). Therefore, a net flux of organic molecules from the organic phase (higher chemical potential) to the aqueous phase (lower chemical potential) occurs. This process continues and x_{iw} increases until the chemical potentials (or the fugacities) become equal in both phases. At this point, equilibrium is reached and we may say: $\gamma_{iw} x_{iw} = \gamma_{iL} x_{iL}$ and $f_{iw} = f_{iL}$! Once at equilibrium, we obtain:

$$\ln \frac{x_{iw}^{sat}}{x_{iL}} = \frac{RT \ln \gamma_{iL} - RT \ln \gamma_{iw}^{sat}}{RT} \tag{5-4}$$

where now we use the superscript "sat" to indicate that we are dealing with a saturated aqueous solution of the compound. In Eq. 5-4 we also retain the product of the gas constant and system temperature, RT, to indicate that the ratio of concentrations in the two phases is related to a difference in free energies (i.e., each term, $RT \ln \gamma$, is a free energy term for one mole of molecules in a particular state).

For the majority of the compounds of interest to us, we can now make two important simplifying assumptions. First, in the organic liquid, the mole fraction of water is small compared with the mole fraction of the compound itself; that is, x_{iL} remains nearly 1 (see Table 5.1). Also, we may assume that the compound shows ideal behavior in its water-saturated liquid phase; that is, we set $\gamma_{iL} = 1$. With these assumptions, after some rearrangement, Eq. 5-4 simplifies to:

Table 5.1 Mole Fraction of Some Common Organic Liquids Saturated
with Water [a]

Organic Liquid i	x_{iL}	Organic Liquid i	x_{iL}
n-Pentane	0.9995	Chlorobenzene	0.9981
n-Hexane	0.9995	Nitrobenzene	0.9860
n-Heptane	0.9993	Aminobenzene	0.9787
n-Octane	0.9994		
n-Decane	0.9994	Diethylether	0.9501
n-Hexadecane	0.9994	Methoxybenzene	0.9924
		Ethyl acetate	0.8620
Trichloromethane	0.9948	Butyl acetate	0.9000
Tetrachloromethane	0.9993	2-Butanone	0.6580
Trichloroethene	0.9977	2-Pentanone	0.8600
Tetrachloroethene	0.9993	2-Hexanone	0.8930
Benzene	0.9975	1-Butanol	0.4980
Toluene	0.9976	1-Pentanol	0.6580
1,3-Dimethylbenzene	0.9978	1-Hexanol	0.7100
1,3,5-Trimethylbenzene	0.9978	1-Octanol	0.8060
n-Propylbenzene	0.9958		

[a] Data from a compilation presented by Demond and Lindner (1993).

$$\ln x_{iw}^{sat} = -\frac{RT \ln \gamma_{iw}^{sat}}{RT} = -\frac{G_{iw}^{E,sat}}{RT} \qquad (5\text{-}5)$$

where $G_{iw}^{E,sat}$ is the *excess free energy* of the compound in *saturated* aqueous solution
(see Chapter 3).

Now we can see a key result. The aqueous mole fraction solubility of an organic
liquid is simply given by the inverse aqueous activity coefficient:

$$x_{iw}^{sat} = \frac{1}{\gamma_{iw}^{sat}} \quad \text{for liquids}$$

or in the more usual molar units (Eq. 3-43): $\qquad (5\text{-}6)$

$$C_{iw}^{sat} = \frac{1}{\overline{V}_w \cdot \gamma_{iw}^{sat}} \quad \text{for liquids}$$

where $\overline{V}_w$ is the molar volume of water (0.018 L/mol).

Obviously, we can also say that for a liquid compound, the aqueous activity coefficient
at saturation is given by the inverse of its mole fraction solubility:

$$\gamma_{iw}^{sat} = \frac{1}{x_{iw}^{sat}}$$

or: $\qquad (5\text{-}7)$

$$\gamma_{iw}^{sat} = \frac{1}{\overline{V}_w \cdot C_{iw}^{sat}}$$

Solubilities and Aqueous Activity Coefficients of Organic Solids

When considering the solubility of a solid organic compound in water, conceptually we can imagine first converting it to the liquid state and then proceeding as above for a liquid compound. The free energy cost involved in the solid-to-liquid conversion is referred to as the free energy of fusion, $\Delta_{fus}G_i$ (Chapter 4). This entity can be derived from experimental vapor pressure data (Eq. 4-14):

$$\Delta_{fus}G_i = RT \ln \frac{p_{iL}^*}{p_{is}^*} \tag{5-8}$$

It can also be estimated from the melting point of the compound (Eq. 4-41).

Now, we can express the difference in chemical potential as:

$$\begin{aligned} \mu_{iw} - \mu_{is} &= \mu_{iw} - (\mu_{iL} - \Delta_{fus}G_i) \\ &= RT \ln \gamma_{iw} \cdot x_{iw} - (RT \ln \gamma_{iL} \cdot x_{iL} - \Delta_{fus}G_i) \end{aligned} \tag{5-9}$$

By setting x_{iL} and γ_{iL} equal to 1, and by proceeding as above for liquids, we then obtain at equilibrium ($\mu_{iw} - \mu_{is} = 0$):

$$x_{iw}^{sat}(s) = \frac{1}{\gamma_{iw}^{sat}} \cdot e^{-\Delta_{fus}G_i/RT} \quad \textit{for solids}$$

or in molar units: $\tag{5-10}$

$$C_{iw}^{sat}(s) = \frac{1}{\overline{V}_w \cdot \gamma_{iw}^{sat}} \cdot e^{-\Delta_{fus}G_i/RT} \quad \textit{for solids}$$

Now it is clear that the solubilities of organic solids in water are dependent on both the incompatibility of the chemicals with the water *and* the ease with which the solids are converted to liquids.

One may also see how the aqueous activity coefficient is related to solubility for organic substances that are solids:

$$\gamma_{iw}^{sat} = \frac{1}{x_{iw}^{sat}(s)} \cdot e^{-\Delta_{fus}G_i/RT}$$

or: $\tag{5-11}$

$$\gamma_{iw}^{sat} = \frac{1}{\overline{V}_w \cdot C_{iw}^{sat}(s)} \cdot e^{-\Delta_{fus}G_i/RT}$$

Recalling the concept of a *subcooled* liquid compound as one that has cooled below its freezing temperature without becoming solid (Chapter 4.2), we may evaluate the solubility of such a hypothetical liquid, $C_{iw}^{sat}(L)$, from Eq. 5-11 as:

$$\gamma_{iw}^{sat} = \frac{1}{\overline{V}_w \cdot C_{iw}^{sat}(L)} \tag{5-12}$$

where the liquid compound solubility is related to the actual experimental solubility of the solid compound by:

$$C_{iw}^{sat}(L) = C_{iw}^{sat}(s) \cdot e^{+\Delta_{fus}G_i / RT} \tag{5-13}$$

Solubilities and Aqueous Activity Coefficients of Organic Gases

The aqueous solubility of a gaseous compound is commonly reported for 1 bar (or 1 atm = 1.013 bar) partial pressure of the pure compound. One of the few exceptions is the solubility of O_2 which is generally given for equilibrium with the gas at 0.21 bar, since this value is appropriate for the earth's atmosphere at sea level. As discussed in Chapter 3, the partial pressure of a compound in the gas phase (ideal gas) at equilibrium above a liquid solution is identical to the fugacity of the compound in the solution (see Fig. 3.9d). Therefore equating fugacity expressions for a compound in both the gas phase and an equilibrated aqueous solution phase, we have:

$$p_i = \gamma_{iw} \cdot x_{iw} \cdot p_{iL}^* \tag{5-14}$$

Now we can see how to express the mole fraction solubility of a gaseous organic substance as a function of the partial pressure p_i:

$$x_{iw}^{p_i} = \frac{1}{\gamma_{iw}^{p_i}} \cdot \frac{p_i}{p_{iL}^*} \quad \text{for gases}$$

or in molar units: $\tag{5-15}$

$$C_{iw}^{p_i} = \frac{1}{\overline{V}_w \cdot \gamma_{iw}^{p_i}} \cdot \frac{p_i}{p_{iL}^*} \quad \text{for gases}$$

It thus follows that the aqueous activity coefficient of a gaseous pure compound is related to the solubility by:

$$\gamma_{iw}^{p_i} = \frac{1}{x_{iw}^{p_i}} \cdot \frac{p_i}{p_{iL}^*}$$

or: $\tag{5-16}$

$$\gamma_{iw}^{p_i} = \frac{1}{\overline{V}_w \cdot C_{iw}^{p_i}} \cdot \frac{p_i}{p_{iL}^*}$$

Note that $\gamma_{iw}^{p_i}$ is not necessarily constant with varying p_i. In fact, evaluation of the air–water equilibrium distribution ratio as a function of p_i is one of the methods that can be used to assess the concentration dependence of γ_{iw} of an organic compound, regardless whether the compound is a gas, liquid, or solid at the temperature considered (see below).

If, for sparingly soluble gases, we assume that $\gamma_{iw}^{p_i}$ is independent of concentration (even at saturation, i.e., at $p_i = p_{iL}^*$, where the compound is also present as a liquid), then we can calculate the solubility of the *superheated liquid compound*, $C_{iw}^{sat}(L)$, from the actual solubility determined at p_i (e.g., at 1 bar) by:

$$C_{iw}^{sat}(L) = C_{iw}^{p_i} \cdot \frac{p_{iL}^*}{p_i} \tag{5-17}$$

Some example calculations demonstrating how to derive γ_{iw} and G_{iw}^E values from experimental solubility data are given in Illustrative Example 5.1.

Illustrative Example 5.1

Deriving Liquid Aqueous Solubilities, Aqueous Activity Coefficients, and Excess Free Energies in Aqueous Solution from Experimental Solubility Data

Problem

Calculate the C_{iw}^{sat} (L), γ_{iw}^{sat} and G_{iw}^{E} of (a) di-n-butyl phthalate, (b) γ-1,2,3,4,5,6-hexachlorocyclohexane (γ-HCH, lindane), and (c) chloroethene (vinyl chloride) at 25°C using the data provided in Appendix C.

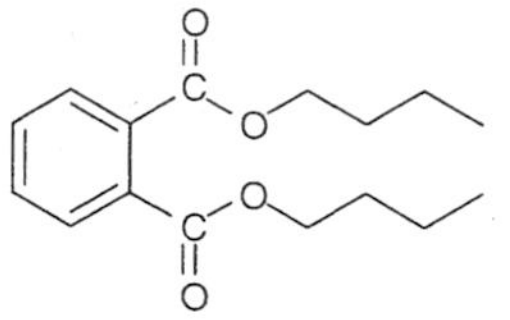

di-n-butyl phthalate

C_{iw}^{sat} (25°C) = 3.4 × 10^{-5} mol·L^{-1}
T_m < 25°C

Answer (a)

Di-n-butylphthalate is a liquid at 25°C. Hence, $C_{iw}^{sat} = C_{iw}^{sat}$ (L) = 3.4×10^{-5} mol·L^{-1}, and (Eq. 5-7):

$$\gamma_{iw}^{sat} \cong \frac{1}{(0.018\ L\cdot mol^{-1})(3.4 \times 10^{-5}\ mol\cdot L^{-1})} = 1.6 \times 10^6$$

which yields an excess free energy of:

$$G_{iw}^{E,sat} = RT \ln \gamma_{iw}^{sat} = (8.314\ J\cdot mol^{-1}\ K^{-1})\ (298.1\ K)\ (14.3) = 35.4\ kJ\cdot mol^{-1}$$

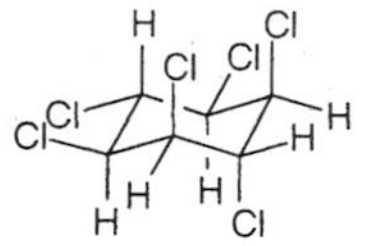

1,2,3,4,5,6-hexachlorocyclohexane
(γ-HCH)

C_{iw}^{sat} (25°C) = 2.5 × 10^{-5} mol·L^{-1}
T_m = 113°C

Answer (b)

γ-HCH is a solid at 25°C. To calculate the solubility of liquid γ-HCH, estimate first the free energy of fusion from the normal melting point temperature (Eq. 4-41, see also Problem 4.7):

$$\Delta_{fus}G_i \cong (56.5)\ [386 - 298] \cong 5.0\ kJ\cdot mol^{-1}$$

Insert the values for C_{iw}^{sat} and $\Delta_{fus}G_i$ into Eq. 5-13 to get C_{iw}^{sat} (L):

$$C_{iw}^{sat}\ (L) = (2.5 \times 10^{-5})\cdot e^{+(5.0/2.48)} = 1.9 \times 10^{-4}\ mol\cdot L^{-1}$$

Insertion of C_{iw}^{sat} (L) into Eq. 5-12 yields:

$$\gamma_{iw}^{sat} \cong 2.9 \times 10^5, \text{ and } G_{iw}^E = 31.2\ kJ\cdot mol^{-1}$$

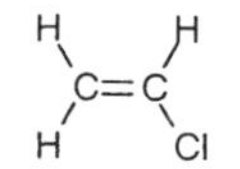

Chloroethene (vinyl chloride)

C_{iw}^{1bar} (25°C) = 4.4 × 10^{-2} mol·L^{-1}
p_{iL}^{*} (25°C) = 3.55 bar

Answer (c)
Chloroethene (Vinylchloride) is a gas at 25°C. Calculate first the solubility of superheated liquid vinylchloride (Eq. 5-17):

$$C_{iw}^{sat}\ (L) = (4.4 \times 10^{-2})\left(\frac{3.55}{1}\right) = 1.6 \times 10^{-1}\ mol\cdot L^{-1}$$

This yields

$$\gamma_{iw}^{sat} \cong 3.5 \times 10^2, \text{ and } G_{iw}^E = 14.5\ kJ\cdot mol^{-1}$$

Table 5.2 Comparison of Activity Coefficients and Corresponding Excess Free Energies of a Series of Organic Compounds in Dilute and Saturated Aqueous Solution at 25°C (recall that $G_{iw}^{E} = RT \ln \gamma_{iw}$)

Compound	γ_{iw}^{sat}	$G_{iw}^{E,sat\ a}$ (kJ·mol^{-1})	γ_{iw}^{∞}	$G_{iw}^{E,\infty\ b}$ (kJ·mol^{-1})
Methanol	miscible	miscible	1.6	1.2
Ethanol	miscible	miscible	3.7	3.2
Acetone	miscible	miscible	7.0	4.8
1-Butanol	7.0×10^1	10.5	5.0×10^1	9.7
Phenol	6.3×10^1	10.3	5.7×10^1	10.0
Aniline	1.4×10^2	12.3	1.3×10^2	12.1
3-Methylphenol	2.5×10^2	13.7	2.3×10^1	13.5
1-Hexanol	9.0×10^2	16.9	8.0×10^2	16.5
Trichloromethane	7.9×10^2	16.5	8.2×10^2	16.6
Benzene	2.5×10^3	19.4	2.5×10^3	19.4
Chlorobenzene	1.4×10^4	23.7	1.3×10^4	23.5
Tetrachloroethene	7.5×10^4	27.8	5.0×10^4	26.8
Naphthalene	6.7×10^4	27.5	6.9×10^4	27.6
1,2-Dichlorobenzene	6.2×10^4	27.3	6.8×10^4	27.6
1,3,5-Trimethylbenzene	1.3×10^5	29.2	1.2×10^5	29.0
Phenanthrene	2.0×10^6	35.9	1.7×10^6	35.5
Anthracene	2.5×10^6	36.5	2.7×10^6	36.7
Hexachlorobenzene	4.3×10^7	43.6	3.5×10^7	43.0
2,4,4'-Trichlorobiphenyl	5.6×10^7	44.2	4.7×10^7	43.8
2,2',5,5'-Tetrachlorobiphenyl	7.0×10^8	46.5	7.5×10^7	44.9
Benzo(a)pyrene	3.2×10^8	48.5	2.7×10^8	48.1

[a] Data from Appendix C using enthalpy and entropy of fusion values given by Hinckley et al. (1990), and Lide (1995). [b] Data from Sherman et al. (1996), Staudinger and Roberts (1996), Mitchell and Jurs (1998).

Concentration Dependence of the Aqueous Activity Coefficient

From an environmental point of view, it is often of most interest to know the activity coefficient of an organic compound in *dilute* aqueous solution. This activity coefficient is commonly denoted as γ_{iw}^{∞}, and is referred to as *limiting* activity coefficient or *infinite dilution* activity coefficient.

As we have shown above, activity coefficients can be deduced from the aqueous solubilities (together with vapor pressure or melting data, as necessary). In this case, the activity coefficient reflects the compatibility of the organic solute with water solutions that may have been significantly modified by the presence of the solute itself. It is important to know when such values of γ_{iw}^{sat} will be the same as the corresponding γ_{iw}^{∞} values. Table 5.2 shows a comparison of γ_{iw}^{sat} values obtained from solubility measurements (Eqs. 5-6 and 5-10) with γ_{iw}^{∞} values determined by various methods (that will be addressed in Section 5.5) for a series of compounds covering a very large range in activity coefficients. As is evident, even for compounds exhibiting a substantial aqueous solubility (e.g., 1-butanol, phenol), the differences between the activity coefficients in dilute solution and in saturated solution are not larger than about 30%. In fact, particularly for the more sparingly soluble compounds, the differences are well within the range of error of the experimental data. Hence, for compounds exhibiting activity coefficients larger than about 100 (which represents the majority of the chemicals of interest to us), we will assume that γ_{iw} is independent of the concentration of the compound (and, therefore, we will typically omit any superscript). By making this assumption, we imply that the organic solutes do not "feel" each other in the aqueous solution even under saturation conditions. Or to put it more scientifically, we assume that the solvation of a given organic molecule by water molecules is not influenced by the other organic molecules present. But, as we will see in the following, this assumption is not always true!

5.3 Molecular Interpretation of the Excess Free Energy of Organic Compounds in Aqueous Solutions

Enthalpic and Entropic Contributions to the Excess Free Energy

Water is a very unique solvent that has two outstanding characteristics: (1) the small size of its molecules, and (2) the strong hydrogen bonding between these molecules. Hence, when we consider the molecular factors that govern the free energy of the transfer of an organic compound from its pure liquid into a pure aqueous phase, we have to be aware that it takes quite a number of water molecules to surround each organic molecule. Also, the water molecules adjacent to the organic solute are in a special situation with respect to forming hydrogen bonds as compared to the other bulk water molecules.

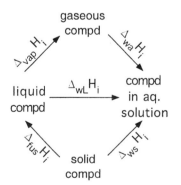

Figure 5.1 Enthalpies of the various phase transfers that can be used to derive the excess enthalpy, H_{iw}^E, of an organic compound in saturated and in dilute aqueous solution.

Before we deal with these molecular aspects in detail, it is instructive to inspect the enthalpic (H_{iw}^E) and entropic ($-T \cdot S_{iw}^E$) contributions to the excess free energies of various organic compounds in aqueous solution (Table 5.3). Values representative of *saturated* aqueous solutions of the compounds have been derived from measurements of the enthalpies of dissolution of the liquids (i.e., $H_{iw}^E = \Delta_{wL}H_i$, Fig. 5.1) or solids ($H_{iw}^E = \Delta_{ws}H_i - \Delta_{fus}H_i$, Fig. 5.1). Data suited to *dilute* conditions have been obtained from enthalpies of air–water partitioning (i.e., $H_{iw}^E = \Delta_{vap}H_i + \Delta_{wa}H_i$, Fig. 5.1). Since in both the saturated and dilute solution, the excess free energies are indistinguishable [data in Table 5.2 gives G_{iw}^E (dilute) = 0.989 $G_{iw}^{E,sat}$ (saturated) − 0.038, R^2 = 0.99], the entropy contributions have been calculated using one (average) G_{iw}^E value. Note that the experimental data reported in the literature show considerable scatter, particularly when comparing H_{iw}^E values determined for saturated conditions with those determined for dilute solutions. Therefore, the numbers given in Table 5.3 should be treated with some caution. Nevertheless, these data allow us to draw some important general conclusions.

The first and most important feature that can be seen from the data (Table 5.3) is that the excess enthalpies of the smaller-sized compounds are close to zero (i.e., between −10 and +10 kJ · mol^{-1}). This is even true for apolar compounds such as tetrachloroethene or hexane. Hence in these cases, the intermolecular interactions that must be disrupted to remove a small organic molecule from its pure liquid (i.e., the enthalpy

Table 5.3 Enthalpic (H_{iw}^E) and Entropic (S_{iw}^E) Contributions to the Excess Free Energy of a Series of Organic Compounds in Saturated ("Sat") and Dilute ("Dil") Aqueous Solution at 20 to 25°C. The Compounds are Ordered by Increasing Size Expressed by Their Molar Volume

Compound	Molar Volume[a] (cm^3·mol^{-1})	G_{iw}^E (kJ·mol^{-1})	H_{iw}^E (kJ·mol^{-1}) Sat[b]/Dil[c,d]	$T \cdot S_{iw}^E$ (kJ·mol^{-1}) Sat/Dil
Trichloromethane	81	17	–2/3	–19/–20
Benzene	89	19	2/4	–21/–23
2-Butanone	90	8	–7/–5	–15/–13
Trichloroethene	90	22	–4/2	–26/–20
Phenol	90	10	1/8	–9/–2
Aniline	91	12	2	–10
Tetrachloromethane	97	23	–4/–2	–27/–25
Tetrachloroethene	102	27	–5/3	–32/–24
Benzaldehyde	102	19	4/10	–15/–9
4-Methylphenol	103	13	2/11	–11/–2
Diethylether	104	18	–20/–14	–31/–25
Benzylalcohol	104	12	–7	–19
Methylbenzene	106	23	2/6	–21/–17
2-Pentanone	106	11	–7	–19
Diethylsulfide	108	18	–1/–1	–19/–19
1-Pentanol	109	13	–8	–21
n-Pentane	116	29	–2	–31
1,4-Dimethylbenzene	123	26	3/9	–23/–16
Naphthalene	130	28	9/12	–19/–16
n-Hexane	132	32	~0	–32
1,3,5-Trimethylbenzene	139	29	8	–21
n-Propylbenzene	139	29	2	–27
1-Octanol	158	23	–3	–26
n-Octane	164	40	6	–34
Hexachlorobenzene	167	43	11/27	–32/–16
Phenanthrene	171	36	17/46	–19/+10!
Anthracene	171	37	20/43	–17/+6!
Benzo(a)pyrene	223	48	25/61	–23/+13!

[a] Calculated from density and molar mass. [b] Data from Whitehouse (1984), Abraham et al. (1990), and Shiu et al. (1997). [c] Data from Dewulf et al. (1995), Dohnal and Fenclová (1995), Staudinger and Roberts (1996), and Alaee et al. (1996). [d] Enthalpies of vaporization from Hinckley et al. (1990), and Lide (1995).

of vaporization) are more or less replaced by intermolecular interactions of equal strength in the water.

Only for larger apolar and weakly monopolar compounds (e.g., PAHs, PCBs) are significantly more positive H_{iw}^E values found. Indeed, if we examine the H_{iw}^E values within single compound classes, we can see that this parameter becomes more positive as the sizes of the structures increase (e.g., benzene, naphthalene, anthracene, benzo(a)pyrene).

Thus, for small organic compounds (molar volumes $< 150 \text{ cm}^3 \text{ mol}^{-1}$), it is the unfavorable entropy term that dominates the excess free energy of solution. Since these chemicals were historically studied first, this is probably the origin of the "sense" that entropic effects determine the "hydrophobicity" of organic compounds. However, since larger organic compounds have increasingly disfavorable enthalpic contributions, when we are interested in these substances both enthalpy and entropy must be considered. At this point it should be noted that for these compounds (e.g., hexachlorobenzene, phenanthrene, anthracene) the H_{iw}^E values derived for saturated and dilute conditions show considerable differences (Table 5.3). In all these cases the H_{iw}^E values are significantly larger for dilute conditions. This difference in excess enthalpy is obviously compensated by an increase in excess entropy, since G_{iw}^E is more or less independent of concentration (see above). To date, however, there are not enough experimental data available to assess whether this is a real phenomenon, or whether these findings are due to experimental artifacts.

Molecular Picture of the Dissolution Process

Let us now try to visualize the various molecular changes that determine the enthalpies and entropies of transfering an organic molecule from its pure liquid into water. As already pointed out, one of the key concerns in this process is how the water molecules surrounding the organic compound arrange themselves to optimize their own interactions from an energetic point of view. Since water is an "associated" liquid, meaning that its molecules are hydrogen-bonded so extensively that they act as "packets" of several H_2O molecules tied together, one must also consider the organic solute's influence on water molecules that are not in direct contact with the organic solute.

In the classical model view, it is thought that the water molecules form an ice-like structure around the organic molecule (Frank and Evans, 1945; Shinoda, 1977). This results from the need of water molecules to maximize their hydrogen bonding. Since the apolar portions of organic solutes cannot participate in this type of intermolecular interaction, the water molecules lining the "solute cavity" were believed to orient so as to maximize their hydrogen bonding to the waters away from the solute. Such orientation would limit the directions these cavity-lining water molecules could face, thereby having the effect of "freezing" them in space. This freezing effect would give rise to an enthalpy gain and an entropy loss, which would be in accordance with the experimental solubility data.

However, the results from numerous, more recent experimental and theoretical studies support an alternative picture (Blokzijl and Engberts, 1993; Meng and Kollman, 1996). In this scenario, the water surrounding a nonpolar solute maintains, but does not enhance, its H-bonding network. One can imagine that, at ambient temperatures, the packets of water molecules adjacent to an apolar organic molecule lose only a very small proportion of their total hydrogen bonds (i.e., the packet:packet interactions). By doing so, they are able to host an apolar solute of limited size without losing a significant number of their H-bonds (Blokzijl and Engberts, 1993). Hence, the introduction of a relatively small apolar or weakly polar organic solute that undergoes primarily vdW interactions should not provoke a significant loss in enthalpy due to the breaking of H-bonds among the water molecules. For such solutes it is, therefore, not surprising that the enthalpy that has to be spent to isolate the compounds from their pure liquid (i.e., the enthalpy of vaporization) is about equal to the enthalpy gained from the vdW interactions of the organic molecules with the water molecules in the aqueous solution. Examples of such compounds include benzene, tetrachloromethane, tetrachloroethene, methylbenzene (toluene), n-pentane, 1,4-dimethylbenzene, and n-hexane (Table 5.3).

The factors that determine the large unfavorable entropy terms for these compounds are somewhat more difficult to rationalize. First, there is a diminishing effect of the favorable entropy of dissolving (or mixing) a (large) organic compound in a solvent consisting of very small molecules, which is, of course, particularly true for water. This excess entropy term can be as big as $-8 \text{ kJ} \cdot \text{mol}^{-1}$, depending on the size of the organic compound. Note that a difference of about $6 \text{ kJ} \cdot \text{mol}^{-1}$ (i.e., $RT \ln 10$) means a factor of 10 difference in the activity coefficient. However, as can be seen from Table 5.3, the actual negative entropy contributions found for the apolar compounds mentioned above are much larger (i.e., $20 - 30 \text{ kJ} \cdot \text{mol}^{-1}$). Hence, there must be other factors that contribute significantly to this large negative entropy. It is conceivable that the water molecules forming the hydration shell lose some of their freedom of motion as compared to the bulk water molecules when accommodating an (apolar) organic compound. Alternatively, the organic compound itself could experience such a loss of freedom when being transferred from its pure liquid into an environment that is more "rigid," because it is now surrounded by many solvent molecules that are interconnected by hydrogen bonds. Moving from a liquid to a more solid-like environment (thus losing translational, rotational, and flexing freedom) could explain the quite substantial differences in excess entropy found between rigid aromatic (e.g., benzene, methylbenzene, naphthalene) and aliphatic compounds (e.g., pentane, hexane) of similar size (Table 5.3). Indeed, we have already noticed these differences when discussing entropies of fusion in Section 4.4 (Table 4.5) and the involved magnitudes are similar.

Let us now examine what happens to the enthalpy and entropy of solution in water if we introduce a polar group on a small nonpolar organic structure. Generally, the presence of a monopolar or bipolar group leads to a decrease in the enthalpy term and an increase in the entropy term. For example, we can see such changes if we contrast data for 2-pentanone with that for pentane (Table 5.3). Both of these thermodynamic parameters imply that the polar moiety promotes the new compound's solubility over the unsubstituted structure. Note that in the case of bipolar compounds (e.g., alcohols), the effect might not seem as dramatic as may be expected (e.g., compare pentane, 2-pentanone, and 1-pentanol in Table 5.3). But one has to keep in mind that for bipolar compounds (in contrast to the monopolar compounds), polar attractions in the pure organic liquid have to be overcome as part of the total energy of transferring the compound to water.

To rationalize the effect of polar groups on H_{iw}^{E} and S_{iw}^{E}, we can imagine that polar interactions with the water molecules around the solute cavity replace some of the hydrogen bonds between the water molecules. As indicated by the experimental data, this loss of water:water interaction enthalpy seems to be compensated by the enthalpy gained from the organic solute:water polar interactions. At this point it should also be mentioned that additional polarization effects could enhance the interaction between the organic solute and the water molecules in the hydration shell

(Blokzijl and Engberts, 1993). To explain the entropy gain, we can imagine that a (partial) "loosening up" of the waters surrounding an organic solute will increase the freedom of motion of both the water molecules and the organic solute involved.

So far, we have considered rather small-sized organic molecules. Larger molecules such as the PAHs or the PCBs exhibit large positive excess enthalpies (Table 5.3). Apparently, with increasing apolar solute size, water is not able to maintain a maximum of hydrogen bonds among the water molecules involved. Hence, for these types of compounds the excess enthalpy term may become dominant (Table 5.3).

In summary, we can conclude that the excess free energy of an organic compound in aqueous solution, and thus its activity coefficient, depends especially on (1) the size and the shape of the molecule, and (2) its H-donor and/or H-acceptor properties.

Model for Description of the Aqueous Activity Coefficient

Let us now extend our molecular descriptor model introduced in Chapter 4 (Eqs. 4-26 and 4-27) to the aqueous activity coefficient. We should point out it is not our principal goal to derive an optimized tool for prediction of γ_{iw}, but to develop further our understanding of how certain structural features determine a compound's partitioning behavior between aqueous and nonaqueous phases. Therefore, we will try to keep our model as simple as possible. For a more comprehensive treatment of this topic [i.e., of so-called linear solvation energy relationships (LSERs)] we refer to the literature (e.g., Kamlet et al., 1983; Abraham et al., 1990; Abraham, 1993; Abraham et al., 1994a and b; Sherman et al., 1996).

First, we consider how a compound's size may influence its activity coefficient, which is related to its liquid aqueous solubilities (Section 5.2). Generally, within any one compound class, we have already seen that the excess free energy of solution in water becomes more positive as we consider larger and larger members of each compound class. In each case, we are increasing the size of the molecules in the compound class by adding apolar portions to the overall structure (e.g., $-CH_2-$ groups). Consequently, the integral interactions with the solvent water molecules become increasingly unfavorable.

In light of such empirical trends, and as is illustrated by Fig. 5.2, we should not be surprised to see that relationships of the following forms can be found for individual compound classes:

$$\ln \gamma_{iw} = a \cdot (size_i) + b$$

or:

$$\ln C_{iw}^{sat}(L) = -c \cdot (size_i) + d$$

(5-18)

The size parameter in such correlations can come from molecular weights, molar volumes, or other related parameters. One such parameter is the estimate of compound size based on the incremental contributions of the atoms involved. Such an approach is the basis for methods like those of McGowan (see Box 5.1 below).

Having means to estimate relative solute sizes, we recognize that we can now estimate a new compound's aqueous activity coefficient and/or liquid solubility from

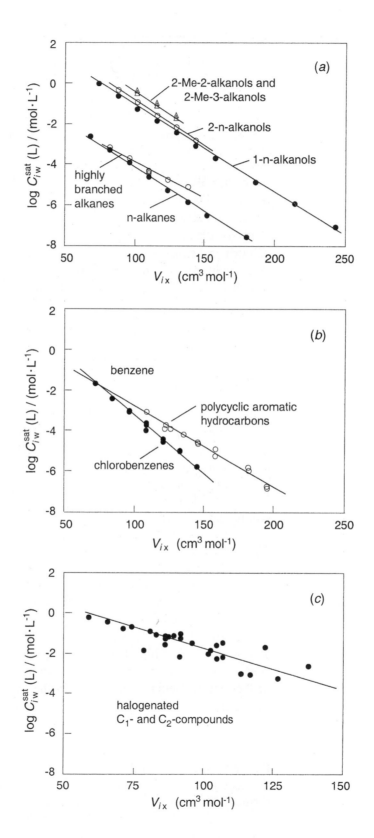

Figure 5.2 Aqueous solubility of the (subcooled) liquid compound at 25°C as a function of the estimated molar volume (V_{ix}, see Box 5.1) of the molecule for various compound classes. The linear regression equations and correlation coefficients (R^2) for the various sets of compounds are given in Table 5.4. Note that for practical reasons, decadic instead of natural logarithms are used.
(a) n-alkanes (C_4–C_{16}), highly branched alkanes (C_5–C_9), 1-3-methyl-3-alkanols (C_6–C_8); *(b)* chlorobenzenes (Cl_1–Cl_6), polycyclic aromatic hydrocarbons (benzene-benzo(a)pyrene); *(c)* polyhalogenated methanes, ethanes, and ethenes. Data from Appendix C and from data compilations reported by Ruelle and Kesselring (1997a and b).

Table 5.4 Linear Relationships Between log C_{iw}^{sat} (L) and V_{ix} [a] for the Various Sets of Compounds Shown in Fig. 5.2 (all data for 25°C).

Set of Compounds	n [c]	c	d	R^2
		log C_{iw}^{sat} (L) $/(mol \cdot L^{-1})$ $= -c \cdot V_{ix} + d$ [b]		
n-Alkanes	8	0.0442	0.34	0.99
Branched alkanes	7	0.0349	−0.38	0.97
Primary alkanols	10	0.0416	3.01	0.99
Secondary alkanols	5	0.0435	3.52	0.99
Tertiary alkanols	6	0.0438	4.01	0.99
Chlorinated benzenes	13	0.0556	2.27	0.99
Polycyclic aromatic hydrocarbons	13	0.0399	1.90	0.99
Polyhalogenated C_1- and C_2-compounds	27	0.0404	1.85	0.86

[a] Molar volume in $cm^3 \cdot mol^{-1}$ estimated by the method discussed in Box. 5.1. [b] Eq. 5-18; note that decadic instead of natural logarithms are used. [c] Number of compounds.

knowledge of the liquid solubilities of other chemicals in its compound class (see examples given in Table 5.4).

While the relations of chemical size and solubility are gratifying to recognize, we still notice that each compound class exhibits its own behavior (Fig. 5.2). Hence, we may wonder if there is any means to account for variations from compound class to compound class. Based on our visualizations of organic solute intermolecular interactions, it is not surprising to learn that parameters that quantify the importance of interactions like hydrogen bonding can be used to adjust for differences between compound classes.

Thinking in analogy to our discussions of the influence of molecular structure on vapor pressures (Eqs. 4-24 to 4-27), we can try to express $\ln\gamma_{iw}$ by a series of terms describing the various molecular interactions and freedoms of motions when transferring a compound from its pure liquid to water. Unlike the cases discussed in Chapter 4, where one of the phases was the gas phase, now we need to account for both the molecular interactions between the compound and the water and the interactions in the pure liquid. This latter group of interactions, however, can simply be characterized by using the vapor pressure of the compound as a quantitative measure of the intermolecular interactions in the pure liquid. Our problem then reduces to describing the transfer of an organic compound from the gas phase to water:

$$\ln \gamma_{iw} = -\ln p_{iL}^* + \text{terms describing the gas–water transfer} \qquad (5-19)$$

It is easy to see that for describing the solvation of an organic solute in water we need to account not only for the size of the molecule (or of the cavity that needs to be formed), but also for the vdW and hydrogen-bonding interactions of the solute with the water molecules. By assuming that the average vdW, H-donor, and H-acceptor properties of the water forming the hydration shell do not vary much with the type of organic solute that they surround, we can include these properties in a correlation equation with appropriate scaling coefficients:

$$\ln \gamma_{iw} = -\ln p_{iL}^* + s\left[(V_i)^{2/3}\left(\frac{n_{Di}^2 - 1}{n_{Di}^2 + 2}\right)\right] + a\,(\alpha_i) + b(\beta_i) + vV_i + \text{constant} \quad (5\text{-}20)$$

vdW (dispersive)   H-donor H-acceptor   size

Note that our *multiparameter LFER* Eq. 5-20 includes two terms that contain a volume term (a quantitative measure of the volume of one mole molecules) as a size parameter ("vdW," "size"-terms). This V_i value can be the molar volume, $\overline{V}_i$, of the compound (derived from the molar mass and the density of the compound, see Chapters 3 and 4), or it can be an estimated entity (see Box 5.1). Therefore, we denote this term as V_i and not $\overline{V}_i$. We will, however, use the term "molar volume" even if we refer to estimated V_i values.

A question that one might ask is whether it is necessary to include two volume terms in Eq. 5-20, because one could imagine that these two terms are strongly correlated

---

Box 5.1 Estimating Molar Volumes from Structure

A very common way of expressing the bulk size of the molecules of a given compound (or more precisely of 1 mole of the compound) is to use the "molar volume," V_i, of the compound. As we have already discussed in Chapters 3 and 4, we can derive V_i from the molar mass and from the liquid density (i.e., $V_i = \overline{V}_i = M_i / \rho_{iL}$ a given temperature. This way of defining V_i has, however, certain disadvantages when we want to express the bulk size of a given compound molecule in equations such as Eq. 5-20. First, because ρ_{iL} is a bulk property, for polar compounds (e.g., alcohols) that have a network-like hydrogen-bond structure, the calculated V_i value represents a molar volume that reflects not only the *intrinsic molecular volume* but also the bulk structure. Second, adjustments have to be made when dealing with compounds that are solids. Therefore, various methods for estimating V_i values from the structure of the compound have been developed (for an overview see Chapter 18 and Yalkowski and Banerjee, 1992; Mackay et al., 1992–1997). Although each of these methods yields different absolute V_i values, the various data sets correlate reasonably well with each other (Mackay et al., 1992–1997). A simple method that seems to work almost as well as the more sophisticated approaches has been proposed by McGowan and coworkers (McGowan and Mellors, 1986; Abraham and McGowan, 1987). In this method, each element is assigned a characteristic atomic volume (see table below) and the total volume, which is denoted as V_{ix}, is calculated by just summing up all atomic volumes and by subtracting 6.56 cm³ mol⁻¹ for each bond, no matter whether single, double, or triple. Thus, V_{ix} for benzene is calculated as $V_{ix} = (6)(16.35) + (6)(8.71) - (12)(6.56) = 71.6$ cm³ mol⁻¹, an example that illustrates how trivial the calculation is. Of course, by this method, identical V_{ix} values will be obtained for structural isomers, which is, however, a reasonable first approximation for many applications. Note again that for each bond between two atoms, 6.56 cm³ mol⁻¹ is to be subtracted. Some example calculations are included in some of the illustrative examples.

Characteristic Atomic Volumes in cm³ mol⁻¹ (From Abraham and McGowan, 1987)

C	16.35	H	8.71	O	12.43	N	14.39	P	24.87	F	10.48	Cl	20.95
Br	26.21	I	34.53	S	22.91	Si	26.83						

to each other. In fact, when applying Eq. 5-20 to nonpolar organic solvents (see Chapter 6), it is sufficient to use only the vdW term (which decreases γ_{iw} because s is negative; see below). We can, however, easily see that in the special case of water as a solvent, we need to include an additional size term in order to address the large entropy costs when inserting an organic solute into the bulk water. Figure 5.3a shows that with this equation, the aqueous activity coefficients of over 250 compounds covering a wide variety of compound classes can be collapsed onto one line reasonably well.

This result is accomplished without considering the dipolarity/polarizability characteristics that one can expect to play a role in a polar solvent such as water. Consequently, it can be expected that the still-large scatter observed in the data shown in Fig. 5.3a can be further reduced, if one adds another parameter that takes into account these aspects. One widely used additional parameter (in addition to α_i and β_i) that is thought to express the dipolarity/polarizability of an organic compound is a parameter commonly denoted as π_i. Note that several sets of π_i values have been derived that may be somewhat different in absolute numbers (e.g., π_i values reported by Li et al., 1993). Table 5.5 summarizes π_i values for some representative compounds. Inspection of Table 5.5 shows that π_i ranges between 0 for the apolar alkanes up to almost 2 for aromatic compounds exhibiting several polar groups (e.g., 4-nitrophenol). For more details, particularly with respect to the derivation of this not-so-easy-to-interpret parameter, we refer to the literature (e.g., Abraham et al., 1991 and 1994a, and references cited therein). Inclusion of π_i into Eq. 5-20 then yields:

$$\ln \gamma_{iw} = -\ln p_{iL}^* + s\left[(V_i)^{2/3}\left(\frac{n_{Di}^2 - 1}{n_{Di}^2 + 2}\right)\right] + p(\pi_i) + a(\alpha_i) + \qquad (5\text{-}21)$$

$$\text{vdW (dispersive)} \quad \text{dipolarity/} \quad \text{H-donor}$$
$$\text{polarizability}$$

$$+b(\beta_i) + vV_i + \text{constant}$$
$$\text{H-acceptor} \quad \text{size}$$

As is illustrated by Fig. 5.3b, with this extended equation, the fit of the experimental data can be improved significantly. The best fit equation obtained from the experimental data set is:

$$\ln \gamma_{iw} = -\ln p_{iL}^* / \text{bar} - 0.572\left[(V_{ix})^{2/3}\left(\frac{n_{Di}^2 - 1}{n_{Di}^2 + 2}\right)\right] - 5.78\,(\pi_i) - 8.77\,(\alpha_i)$$

$$\qquad (5\text{-}22)$$

$$-11.1\,(\beta_i) + 0.0472\,V_{ix} + 9.49$$

Note that for the derivation of Eq. 5-22 we have adopted a very simple characteristic atomic volume contribution method estimating V_i (see Box 5.1), which we denote as V_{ix}. Since the various methods commonly used to assess "molar volumes" yield quite different absolute values (see e.g., Mackay et al., 1992–1997), V_{ix} values in cm^3 mol^{-1} calculated by this method should be used when applying Eq. 5-22. Hence, if, in addition to V_{ix}, p_{iL}^*, n_{Di}, p_i, a_i, and b_i are known or can be estimated for a given

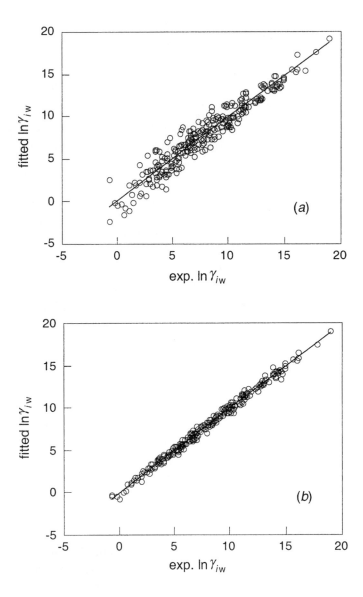

compound (e.g., Platts et al., 2000), its activity coefficient and therefore also its liquid aqueous solubility (Eqs. 5-7 and 5-12) can be predicted from Eq. 5-22 within a factor of 2 to 3.

It should be noted that when replacing the London dispersive interactions term by other properties such as, for example, the air–hexadecane partition constant, by expressing the surface area in a more sophisticated way, and/or by including additional terms, the predictive capability could still be somewhat improved. From our earlier discussions, we should recall that we do not yet exactly understand all the molecular factors that govern the solvation of organic compounds in water, particularly with respect to the entropic contributions. It is important to realize that for many of the various molecular descriptors that are presently used in the literature to model γ_{iw} or related properties (see Section 5.5), it is not known exactly how they contribute to the excess free energy of the compound in aqueous solution. Therefore, when also considering that some of the descriptors used are correlated to each other (a fact that

Table 5.5 Some Representative π_i Values

Compound or Group of Compounds	$\pi_i{}^a$	Compound or Group of Compounds	$\pi_i{}^a$
Alkanes	0.00	1,2,3-Trimethylbenzene	0.61
Cycloalkanes	0.10	1,3,5-Trimethylbenzene	0.52
1-Alkenes	0.08	Naphthalene	0.92
1-Alkines	0.23	Acenaphthene	1.04
Aliphatic ethers (ROR')	0.25	Chlorobenzene	0.65
Aliphatic aldehydes (RCHO)	0.65	1,2-Dichlorobenzene	0.76
Aliphatic ketones (RCOR')	0.68	1,4-Dichlorobenzene	0.75
Aliphatic carboxylic acid esters (RCOOR')	0.55-0.60	1,2,3-Trichlorobenzene	0.86
Aliphatic amines (RNH$_2$)	0.35	1,3,5-Trichlorobenzene	0.73
Primary aliphatic alcohols (R-CH$_2$OH)	0.42	1,2,3,4-Tetrachlorobenzene	0.92
Secondary aliphatic alcohols (RR'CHOH)	0.36	1,2,4,5-Tetrachlorobenzene	0.86
Aliphatic carboxylic acids (RCOOH)	0.63	Benzaldehyde	1.00
Trichloromethane	0.49	Benzonitrile	1.11
Tetrachloromethane	0.38	Nitrobenzene	1.11
1,1,2,2-Tetrachloroethane	0.76	Phenol	0.89
Tetrachloroethene	0.42	Alkylphenol	0.80-0.90
Tribromomethane	0.68	2-Chlorophenol	0.88
Benzene	0.52	4-Chlorophenol	1.08
Toluene	0.52	2-Nitrophenol	1.05
1,2-Dimethylbenzene	0.56	4-Nitrophenol	1.72
1,4-Dimethylbenzene	0.52		

^{a} Data from Abraham et al. (1994a).

is often not recognized in the literature!), our policy should be to use as few and as *clearly defined* parameters as possible. There is certainly still room for further improvements in this area of research. Nevertheless, as is demonstrated by the examples discussed in Illustrative Example 5.2, Eq. 5-22 is very useful to assess which molecular factors primarily determine the aqueous activity coefficient (or the excess free energy in water) of a given compound.

A very important conclusion that we can draw from our effort to use insights on inter-molecular interactions to develop a means to estimate γ_{iw} is that this important compound property is very sensitive to changes in the structure of a compound. Hence, as we will also notice in the following chapters, in any simple structure–property or property–property relationship involving γ_{iw} (or $C_{iw}^{sat}(L)$), we have to be careful to confine a given equation to a set of compounds for which structural differences either are not reflected, or are proportionally reflected in the type of molecular descriptors used in Eq. 5-22. Otherwise, we are in danger of mixing apples with oranges (and grapes!). For example, as already addressed above, it is common practice to try to correlate the aqueous activity coefficient (or the liquid aqueous solubility as in Fig. 5.2) with the size (molar volume, total surface area) of the organic molecule. As is illustrated by Fig. 5.2, good correlations can be expected only for sets of compounds that fulfill the above-mentioned criteria. Fig. 5.2a shows, for example, that even sets of quite closely related compounds such as n-alkanes and highly branched alkanes, or primary, secondary, and tertiary aliphatic alcohols, exhibit different linear relationships between liquid aqueous solubility and molar volume. In the case of the apolar alkanes (i.e., $\pi_i = \alpha_i = \beta_i = 0$), the differences must be due to the different shapes of the n-alkanes as compared to the highly branched ones. In the case of the aliphatic alcohols, the differences between the three sets of compounds can be found primarily in the polar interaction terms of the alcohol moieties. Within each series, however, very good correlations are obtained. Two other examples where quite satisfying correlations are obtained, are shown in Fig. 5.2b. The rather good correlation found for the apolar, rigid chlorinated benzenes (i.e., $\alpha_i = \beta_i = 0$) does not come as a surprise, because these compounds exhibit also very similar π_i values (Table 5.5). In the case of the PAHs, however, the correlation does hold only because the polar parameters (i.e., π_i and β_i) increase both with increasing size. Finally, Fig. 5.2c shows a group of compounds, the polyhalogenated C_1- and C_2-compounds, for which, intuitively, we might have expected a much better result. A closer inspection of the polar parameters of the various compounds shows, however, that the rather large scatter could have been anticipated. For example, the π_i, α_i and β_i values of the similarly sized 1,1,2,2-tetrachloroethane and tetrachloroethene differ substantially (0.76, 0.16, 0.12 versus 0.42, 0.0, 0.0, respectively), which is reflected in the 20-times-higher liquid aqueous solubility of 1,1,2,2-tetrachloroethane as compared to tetrachloroethene. This example should remind us again that such simple one-parameter correlations work, in general, only for limited sets of "structurally closely related" compounds for which they may, however, be very powerful predictive tools. Obviously, as shown by the examples in Fig. 5.2, it may not always be clear whether two compounds are structurally closely related with respect to the factors that govern their aqueous activity coefficients. In such cases inspection of the type of parameters used in Eq. 5-22 may be very helpful for selecting appropriate reference compounds.

Illustrative Example 5.2

Evaluating the Factors that Govern the Aqueous Activity Coefficient of a Given Compound

Problem

Calculate the activity coefficients as well as the excess free energies of n-octane (Oct), 1-methylnaphthalene (1-MeNa), and 4-t-butylphenol (4-BuPh) in aqueous solution at 25°C using Eq. 5-22. Compare and discuss the contributions of the various terms in Eq. 5-22.

Answer

Get the p_{iL}^* values from the data given in Appendix C. Note that 4-BuPh ($T_m = 99°C$) is a solid at 25°C (use Eq. 4-40 to estimate p_{iL}^* from p_{is}^*). Calculate V_{ix} using the method described in Box 5.1. Get the n_{Di} values of the compounds from Lide (1995). Use the α_i, and β_i, and π_i values given in Tables 4.3 and 5.5. The resulting data sets for the three compounds are given in the margin. Recall that $G_{iw}^E = RT \ln \gamma_{iw}$. Insertion of the respective values into Eq. 5-22 yields the following result:

n-octane (Oct)

$p_{iL}^* = 1826 \text{ Pa}$

$V_{ix} = 123.6 \text{ cm}^3 \text{ mol}^{-1}$

$n_{Di} = 1.397$

$\pi_i = \alpha_i = \beta_i = 0$

1-methyl-naphthalene (1-MeNa)

$p_{iL}^* = 8.33$ Pa
$V_{ix} = 122.6$ cm^3 mol^{-1}
$n_{Di} = 1.617$
$\pi_i = 0.90$
$\alpha_i = 0$
$\beta_i = 0.20$

4-t-butyl-phenol

$p_{iL}^* = 6.75$ Pa
$V_{ix} = 133.9$ cm^3 mol^{-1}
$n_{Di} = 1.517$
$\pi_i = 0.89$
$\alpha_i = 0.56$
$\beta_i = 0.39$

Term	Oct		1-MeNa		4-BuPh	
	$\Delta \ln \gamma_{iw}$ —	(G_{iw}^E) (kJ·mol^{-1})	$\Delta \ln \gamma_{iw}$ —	(G_{iw}^E) (kJ·mol^{-1})	$\Delta \ln \gamma_{iw}$ —	(G_{iw}^E) (kJ·mol^{-1})
$-\ln p_{iL}^*$	+4.00	(+9.9)	+9.40	(+23.3)	+9.61	(+23.8)
$-$ vdW^a	−3.42	(−8.5)	−4.94	(−12.2)	−4.53	(−11.2)
$-5.78\,\pi_i$	0		−5.20	(−12.9)	−5.14	(−12.7)
$-8.77\,\alpha_i$	0		0		−4.91	(−12.2)
$-11.12\,\beta_i$	0		−2.22	(−5.5)	−4.33	(−10.7)
$+0.0472\,V_{ix}$	+5.83	(+14.4)	+5.7	(+14.3)	+6.32	(+15.7)
+ constant	+9.49	(+23.5)	+9.49	(+23.5)	+9.49	(+23.5)
$\ln \gamma_{iw}\,(G_{iw}^E)$	15.9	(39.3)	12.2	(30.5)	6.51	(16.2)
exp. value	16.0		12.5		7.15	

^{a} dispersive vdW $= 0.572 \left[(V_{ix})^{2/3} \left(\dfrac{n_{Di}^2 - 1}{n_{Di}^2 + 2} \right) \right]$

First, you note that, although the three compounds are of comparable size, there are significant differences in their γ_{iw} (i.e., G_{iw}^E) values.

As is evident, the lack of any polar interactions with the water molecules is the major cause for the large hydrophobicity of Oct, although this compound exhibits the highest vapor pressure (which facilitates the transfer of Oct from the pure liquid into another phase as compared to the other two compounds). Comparison of 1-MeNa with Oct reveals that the lower activity coefficients (i.e., the higher liquid water solubilities) of aromatic compounds as compared to aliphatic compounds of similar size are primarily due to the relatively large polarizability term (π_i) of aromatic structures. Finally, from comparing 4-BuPh with 1-MeNa it can be seen that H-bond interactions (α_i, β_i terms) may decrease γ_{iw} by several orders of magnitude (note that for these two compounds, all other terms contribute similarly to the overall γ_{iw}).

5.4   Effect of Temperature and Solution Composition on Aqueous Solubility and Activity Coefficient

So far, we have focused on how differences in molecular structure affect the solubilities and activity coefficients of organic compounds in pure water at 25°C. The next step is to evaluate the influence of some important environmental factors on these properties. In the following we consider three such factors: temperature, ionic strength (i.e., dissolved salts), and organic cosolutes. The influence of pH of the aqueous solution, which is most important for acids and bases, will be discussed in Chapter 8.

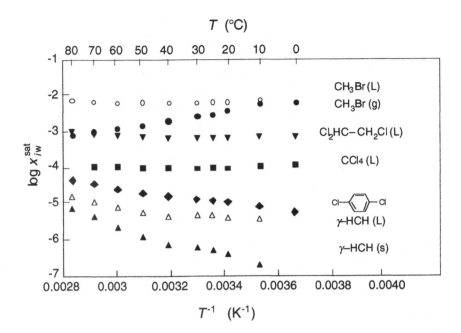

Figure 5.4 Effect of temperature on the mole fraction solubility in water of some halogenated hydrocarbons. γ-HCH is γ-1,2,3,4,5,6-hexachlorocyclohexane (lindane; for structure see Illustrative Example 5.1). Data from Horvath (1982).

Temperature

Let us consider the temperature dependence of the mole fraction solubility of organic *liquids*. Since $x_{iw}^{sat} / x_{iL} \cong x_{iw}^{sat}$ ($x_{iL} \cong 1$) represents the partitioning constant between the aqueous phase and the pure liquid, for a narrow temperature range, its temperature dependence is given by (Section 3.4):

$$\ln x_{iw}^{sat}(L) = -\frac{H_{iw}^{E}}{R} \cdot \frac{1}{T} + \text{constant} \qquad (5\text{-}23)$$

When expressing aqueous solubility in molar units we may write Eq. 5-23 as:

$$\ln C_{iw}^{sat}(L) = -\frac{H_{iw}^{E}}{R} \cdot \frac{1}{T} + \text{constant}' \qquad (5\text{-}24)$$

Now constant' $= $ constant $-\log \overline{V}_w$ and we assume a temperature-independent molar volume ($\overline{V}_w$) of the aqueous solution (see Section 3.4).

For the majority of the (subcooled) liquid compounds, the excess enthalpy, H_{iw}^{E}, is quite small and may even be negative at 25°C (Table 5.3). Thus, for a temperature range between 0 and 80°C, the change in the liquid solubility with increasing temperature is therefore rather small (Fig. 5.4). For some compounds like $CH_3Br(L)$, $CHCl_2$–CH_2Cl, and CCl_4, a solubility minimum is found at ambient temperatures. This occurs because, at low temperatures, H_{iw}^{E} is negative and, in general, H_{iw}^{E} becomes more positive with increasing temperature [in contrast to $\Delta_{vap}H_i$, which decreases with increasing temperature (see Chapter 4)]. This observation can be explained by the fact that at elevated temperatures, some of the hydrogen bonds among the water molecules forming the hydration shell are broken, which leads to a more positive excess enthalpy. Thus, when applying Eqs. 5-23 or 5-24, we know that H_{iw}^{E}

is not constant over the whole ambient temperature range and we can see some curvature in the plots of log x_{iw}^{sat} versus $1/T$ (Fig. 5.4). This is, however, not too much of a problem since the temperature effect is small anyway. For most com-pounds x_{iw}^{sat} (L) (or C_{iw}^{sat} (L)) will vary less than a factor of 2 between 0 and 30°C. Only for the larger rigid, apolar compounds such as PAHs, PCBs, and polychlorinated dibenzodioxines (PCDDs), is the effect of temperature on the liquid aqueous solubility significant [see Illustrative Example 5.3, case (b)].

When we are interested in the actual solubilities of *solids or gases*, however, the effect of temperature becomes much more important (e.g., CH$_3$Br(g) and γ-HCH(s) in Fig. 5.4). Now we must consider the total enthalpy change when transferring a molecule from the solid or gas phase, respectively, to water. This total enthalpy change includes the sum of the enthalpy of the phase change (i.e., conversion of a solid into a subcooled liquid or a gas into a superheated liquid at the temperature of interest) and the excess enthalpy of solution. Hence, for solids the temperature dependence of solubility over a narrow temperature range is given by:

$$\ln C_{iw}^{sat}(s) = -\frac{\Delta_{fus}H_i + H_{iw}^E}{R} \cdot \frac{1}{T} + \text{constant} \qquad (5\text{-}25)$$

and for gases:

$$\ln C_{iw}^{1bar}(g) = -\frac{-\Delta_{vap}H_i + H_{iw}^E}{R} \cdot \frac{1}{T} + \text{constant} \qquad (5\text{-}26)$$

Note that, in general, the resulting enthalpy change will be positive in the case of solids (due to the large positive $\Delta_{fus}H_i$) and negative (dominating positive $\Delta_{vap}H_i$) in the case of gases. Consequently, the solubility of solids increases with increasing temperature, since the "cost" of melting decreases with increasing temperature (and becomes zero at the melting point). Conversely, the difficulty in condensing gaseous organic compounds increases with increasing temperature; thus, heating an aqueous solution tends to diminish solubilities of (organic) gases through this term. Some applications of Eqs. 5-23 to 5-26 are given in the Illustrative Example 5.3.

Illustrative Example 5.3   **Evaluating the Effect of Temperature on Aqueous Solubilities and Aqueous Activity Coefficients**

Problem

Estimate the solubilities, C_{iw}^{sat}, the activity coefficients, γ_{iw}^{sat}, and the excess enthalpies, H_{iw}^E, in water of (a) trichloromethane at 5°C, (b) dibenzofuran at 10°C, and (c) chloroethene (vinyl chloride) at 40°C.

Answer (a)

Since trichloromethane is a liquid at ambient temperatures, the magnitude of change in its solubility with solution temperature is dictated by its excess enthalpy, H_{iw}^E (Eq. 5-23). Generally, for low-molecular-weight compounds, you can expect that H_{iw}^E will not be too different from zero (± 10 kJ·mol^{-1}; see Table 5.3). Hence, at 5°C both

trichoromethane
(chloroform)

$T_m = -63°C$

$T / (°C)$	$x_{iw}^{sat\ a}$
0	0.001514
10	0.001366
20	0.001249
25	0.001203
30	0.001168

^{a} Data from Horvath (1982).

C_{iw}^{sat} and γ_{iw}^{sat} should not differ too much from the corresponding values at 25°C. In fact, inspection of the experimental data reported by Horvath (1982) shows that between 0 and 30°C the mole fraction solubility of trichloromethane *decreases*, but only about 20% (see margin). Since H_{iw}^{E} increases with increasing temperature, use only the two x_{iw}^{sat} values given at 0 and 10°C to estimate x_{iw}^{sat} at 5°C (Eq. 5-23):

$$\ln x_{iw}^{sat} = \frac{799}{T} - 9.418 \qquad (1)$$

Note again that the excess enthalpy of solution of trichloromethane between 0 and 10°C is slightly negative, i.e., $H_{iw}^{E} = -(799\ K)(8.31) = -6.6\ kJ \cdot mol^{-1}$.

Insertion of $T = 278.2$ K into Eq. 1 yields $x_{iw}^{sat}(5°C) = 0.001436$ or, in molar concentrations (Eq. 3-43):

$$C_{iw}^{sat}(5°C) = x_{iw}^{sat} / \overline{V}_w = (0.001436) / (0.018) = 8.0 \times 10^{-2}\ mol \cdot L^{-1}$$

The aqueous activity coefficient is given by $1 / x_{iw}^{sat}$ (trichloromethane is a liquid; see Eq. 5.6):

$$\gamma_{iw}^{sat}(5°C) = 1 / (0.001436) = 7.0 \times 10^{2}$$

dibenzofuran

$M_i = 168.2\ g \cdot mol^{-1}$

$T_m = 87°C$

$T / (°C)$	$C_{iw}^{sat\ a} / (g \cdot m^{-3})$
5	1.92
15	3.04
25	4.75
35	7.56
45	11.7

^{a} Data from Shiu et al. (1997).

Answer (b)

Dibenzofuran is a solid at ambient temperatures. Hence, the enthalpy of solution ($\Delta_{ws}H_i$) is given by the sum of the enthalpy of fusion ($\Delta_{fus}H_i$) and the excess enthalpy in aqueous solution (H_{iw}^{E}) (Fig. 5.1 and Eq. 5-25). In a paper by Shiu et al. (1997) you find aqueous solubility data expressed in $g \cdot m^{-3}$ for dibenzofuran at various temperatures (see margin). For simplicity, assume that $\overline{V}_w$ is temperature independent.

Calculate $1/T$ in K^{-1} and $\log C_{iw}^{sat}$:

$1 / T\ (K^{-1})$	0.003597	0.003472	0.003356	0.003247	0.003145
$\log C_{iw}^{sat} / (g \cdot m^{-3})$	0.282	0.483	0.677	0.879	1.069

and perform a least square fit of $\log C_{iw}^{sat}$ versus $1 / T$:

$$\log C_{iw}^{sat} / (g \cdot m^{-3}) = -\frac{1742}{T} + 6.536 \qquad (2)$$

From the slope one obtains an average $\Delta_{ws}H_i$ [= (1742)(2.303)(8.31)] value of 33.4 $kJ \cdot mol^{-1}$. Note that because we use decadic logarithms, the slope in Eq. 2 is equal to $\Delta_{ws}H_i /(2.303)\ R$. Hence, the aqueous solubility increases by about a factor of 1.6 per 10 degrees increase in temperature (Table 3.5). Insertion of $T = 283.2$ K into Eq. 2 yields:

$$\log C_{iw}^{sat}(10°C) = 0.385 \text{ or } C_{iw}^{sat}(10°C) = 2.43\ g \cdot m^{-3}$$

or, in molar concentrations:

$$C_{iw}^{sat}(10°C) = \frac{(2.43 \times 10^{-3})}{(168.2)} = 1.44 \times 10^{-5} \ mol \cdot L^{-1}$$

To get the activity coefficient, estimate first the aqueous solubility of subcooled liquid dibenzofuran at 283 K (Eq. 5-12). To the end, estimate $\Delta_{fus}G_i$ at 10°C from T_m using Eq. 4-41:

$$\Delta_{fus}G_i(10°C) = (56.5 + 0 - 19.2 \log 2)(77) = 3.9 \ kJ \cdot mol^{-1}$$

Insertion into Eq. 5-13 yields

$$C_{iw}^{sat}(L, 10°C) = (1.44 \times 10^{-5})(5.3) = 7.6 \times 10^{-5} \ mol \cdot L^{-1}$$

and thus (Eq. 5-12)

$$\gamma_{iw}^{sat}(10°C) = 1 / (7.6 \times 10^{-5})(0.018) = 7.3 \times 10^{5}$$

To estimate H_{iw}^{E}, assume a constant $\Delta_{fus}H_i$ below the melting point:

$$\Delta_{fus}H_i \cong \Delta_{fus}H_i(T_m) = T_m \cdot \Delta_{fus}S_i(T_m)$$

Use Eq. 4-39 to estimate $\Delta_{fus}S_i(T_m)$:

$$\Delta_{fus}S_i(T_m) = (56.5 + 0 - 19.2 \log 2) = 50.8 \ J \cdot mol^{-1} \ K^{-1}$$

This yields

$$\Delta_{fus}H_i = (50.8)(360) = 18.3 \ kJ \cdot mol^{-1}$$

and

$$H_{iw}^{E} = \Delta_{ws}H_i - \Delta_{fus}H_i = 33.4 - 18.3 = 15.1 \ kJ \cdot mol^{-1}$$

Note that this H_{iw}^{E} value represents an average value for the ambient temperature range.

chloroethene
(vinyl chloride)

$T_b = -13.4°C$

$T / (°C)$	$x_{iw}^{sat\,a}$	$p_{iL}^{*} / bar^{\,a}$
0	0.00158	1.70
25	0.000798	3.86
50	0.000410	7.69

^{a} Data from Horvath (1982).

Answer (c)

Chloroethene (vinyl chloride) is a gas at the temperature considered. Hence, the enthalpy of solution ($\Delta_{wa}H_i$) is given by the sum of the enthalpy of condensation ($\Delta_{cond}H_i$, which is equal to the negative enthalpy of vaporization) and the excess enthalpy in aqueous solution (H_{iw}^{E}) (Fig. 5.1 and Eq. 5-26). Horvath et al. (1982) gives the solubilities of chloroethene at 0°C, 25°C, and 50°C and 1 bar partial pressure. Also given are the vapor pressures of the superheated liquid at these three temperatures.

After conversion of °C to K, perform a least square fit of ln x_{iw}^{1bar} versus $1 / T$:

$$\ln x_{iw}^{1bar} = +\frac{2375}{T} - 15.134 \qquad (3)$$

From the slope you obtain a $\Delta_{wa}H_i$ [= − (2375) (8.31)] value of −19.7 kJ·mol^{-1}, meaning that the solubility of chloroethene decreases by about a factor of 1.3 per 10 degrees increase in temperature (Table 3.5). Insertion of T = 313.2 K into Eq. 3 yields x_{iw}^{1bar} (40°C) = 0.000526 or in molar concentration (Eq. 3-43):

$$C_{iw}^{1bar} (40°C) = 2.9 \times 10^{-2} \text{ mol·L}^{-1}$$

To get the activity coefficient of chloroethene (Eq. 5-15) calculate its vapor pressure at 40°C using the least square fit of ln p_{iL}^* versus 1 / T:

$$\ln p_{iL}^* / \text{bar} = - \frac{2662}{T} + 10.283 \tag{4}$$

Insertion of T = 313.2 K into Eq. 4 yields a p_{iL}^* value of 5.95 bar, which yields a γ_{iw}^{sat} value of (Eq. 5-16):

$$\gamma_{iw}^{1bar} (40°C) = \frac{1}{(0.000526)} \cdot \frac{1 \text{ bar}}{5.95 \text{ bar}} = 3.2 \times 10^2$$

From the slope in Eq. 4 you can obtain $\Delta_{vap}H_i$ [= (2662) (8.31)] = 22.1 kJ·mol^{-1}. Thus, one obtains an average H_{iw}^E value of:

$$H_{iw}^E = \Delta_{wa}H_i + \Delta_{vap}H_i = -19.7 + 22.1 = + 2.4 \text{ kJ·mol}^{-1}$$

which means that the activity coefficient of chloroethene is more or less constant over the ambient temperature range.

Dissolved Inorganic Salts

When considering saline environments (e.g., seawater, salt lakes, subsurface brines), we have to consider the effects of dissolved inorganic salt(s) on aqueous solubilities and on activity coefficients of organic compounds. Although the number of studies that have been devoted to this topic is rather limited, a few important conclusions can be drawn. Qualitatively, it has been observed that the presence of the predominant inorganic ionic species found in natural waters (i.e., Na^+, K^+, Mg^{2+}, Ca^{2+}, Cl^-, HCO_3^-, SO_4^{2-}) generally decrease the aqueous solubility (or increase the aqueous activity coefficient) of nonpolar or weakly polar organic compounds. Furthermore, it has been found that the magnitude of this effect, which is commonly referred to as *salting-out*, depends on the compound and on the type of ions present.

Long ago, Setschenow (1889) established an *empirical* formula relating organic compound solubilities in saline aqueous solutions ($C_{iw,salt}^{sat}$) to those in pure water (C_{iw}^{sat}):

$$\log\left(\frac{C_{iw}^{sat}}{C_{iw,salt}^{sat}}\right) = K_i^s [\text{salt}]_{tot}$$

or: $\tag{5-27}$

$$C_{iw,salt}^{sat} = C_{iw}^{sat} \cdot 10^{-K_i^s [\text{salt}]_{tot}}$$

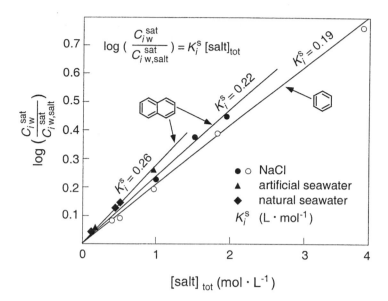

Figure 5.5 Effect of salt concentrations on the aqueous solubility of benzene (McDevit and Long, 1952), and naphthalene (Gordon and Thorne, 1967a).

where $[\text{salt}]_{\text{tot}}$ is the total *molar* salt concentration and K_i^s is the *Setschenow* or *salting constant* (unit M^{-1}). This salting constant relates the effectiveness of a particular salt or combination of salts to the change in solubility of a given compound i. For a particular salt (e.g., NaCl) or salt mixture (e.g., seawater; for composition see Table 5.6), Eq. 5-27 is valid over a wide range of salt concentrations (Fig. 5.5). Note that the "salting-out" effect increases exponentially with increasing salt concentration. K_i^s-values for a given organic solute and salt composition can be determined experimentally by linear regression of experimental solubilities measured at various salt concentrations (i.e., plots of $\log C_{iw}^{\text{sat}}$ versus $[\text{salt}]_{\text{tot}}$). We should point out that at very high salt concentrations, the effect of the dissolved salts on the molar volume of the solution has to be taken into account. However, as a first approximation, in many cases (e.g., seawater) we may neglect the effect. Written in terms of activity coefficients, Eq. 5-27 is:

$$\gamma_{iw,\text{salt}} = \gamma_{iw} \cdot 10^{+K_i^s [\text{salt}]_{\text{tot}}} \tag{5-28}$$

Hence $\gamma_{iw,\text{salt}}$ increases exponentially with increasing salt concentration.

Note that if K_i^s has been determined from solubility measurements, $\gamma_{iw,\text{salt}}$ is strictly valid only for saturated conditions. For dilute solutions $\gamma_{iw,\text{salt}}$ can be determined from measurements of air–water or organic solvent–water partition constants at different salt concentrations. From the few compounds for which $\gamma_{iw,\text{salt}}$ has been determined by both solubility and air–water or solvent–water partitioning experiments, because of the large scatter in the data, it is not clear whether K_i^s varies with organic solute concentration. It can, however, be concluded that, if there is an effect, it is not very large.

Before we inspect K_i^s values of a variety of organic compounds for seawater (the most important natural saline environment), we first take a look at the salting-out efficiencies of various ion combinations. Since it is very difficult to quantify the contribution of individual ions, salting constants are available only for combined salts. Nevertheless, the data in Table 5.6 illustrate that smaller ions that form

Table 5.6 Salt Composition of Seawater and Salting Constants for Benzene, Naphthalene, and 1-Naphthol at 25°C for Some Important Salts

Salt	Weight (g ·mol^{-1})	Mole fraction in seawater [b] x_{salt}	Salting constant [a] K_i^s (benzene) (L ·mol^{-1})	K_i^s (naphthalene) (L ·mol^{-1})	K_i^s (1–naphthol) (L ·mol^{-1})
NaCl	58.5	0.799	0.19	0.22	0.21
MgCl$_2$	95.3	0.104		0.30	0.33
Na$_2$SO$_4$	142.0	0.055	0.53	0.72	
CaCl$_2$	110.0	0.020		0.32	0.35
KCl	74.5	0.017	0.16	0.19	0.18
NaHCO$_3$	84.0	0.005		0.32	
KBr				0.13	0.13
CsBr				0.01	
(CH$_3$)$_4$NCl					–0.36
(CH$_3$)$_4$NBr			–0.15		

[a] Data from McDevit and Long (1952), Gordon and Thorne (1967a,b), Almeida et al. (1983), and Sanemasa et al, (1984). [b] Gordon and Thorne (1967a,b).

hydration shells with more water molecules (e.g., Na$^+$, Mg^{2+}, Ca^{2+}, Cl$^-$) have a bigger effect than larger ions that tend to bind water molecules only very weakly (e.g., Cs$^+$, N(CH$_3$)$_4^+$, Br$^-$). In fact, larger organic ions such as tetramethyl-ammonium (N(CH$_3$)$_4^+$) can even have an opposite effect; that is, they promote solubility (or decrease the activity coefficient). Note that such *salting-in* effects can also be observed for very polar compounds that may strongly interact with certain ions (Almeida et al., 1984). In a simple way, we can rationalize the salting-out of nonpolar and weakly polar compounds by imagining that the dissolved ions compete (successfully) with the organic compound for solvent molecules. Many of the environmentally relevant ions bind water molecules quite tightly in aqueous solution, which can be seen even macroscopically in that the volume of the aqueous solution is reduced. As a consequence, the freedom of some water molecules to solvate an organic molecule is disrupted, and depending on the type of salt and/or compound present, may lead to a loss or gain in solubility (Leberman and Soper, 1995). Furthermore, the solvation of an organic compound, particularly when it is large and nonpolar, requires a large number of water molecules. Hence, we may intuitively anticipate that larger nonpolar organic compounds will exhibit higher K_i^s values as compared to smaller and/or more polar compounds.

Let us now inspect the K_i^s values of some organic compounds in seawater. Using the data given in Table 5.6 we can make our own artificial seawater (at least with respect to the major ion composition) by dissolving an appropriate amount of the corresponding salts in water. The weight of 1 mole of "seawater salt" is given by (0.799) (58.5) + (0.104) (95.3) + (0.055) (142) + (0.02) (110) + (0.017) (74.5) + (0.005) (84) = 68.35 g. Hence, if we dissolve 34.17 g of seawater-salt in 1 L, we obtain a seawater with a salinity of 34.2‰, which corresponds to a total molar salt concentration ($[\text{salt}]_{\text{tot}}$ in Eq. 5-27) of 0.5 M. As has been demonstrated by various studies, the differences between K_i^s values determined in artificial and real seawater are usually only marginal. Furthermore, since seawater is dominated by one salt, that is, NaCl (Table 5.6), as a first approximation K_i^s values determined for sodium chloride can be used as a surrogate. Let us illustrate this by calculating $K_{i,\text{seawater}}^s$ for naphthalene using the data given in Table 5.6. If we assume that naphthalene does not specifically interact with any of the inorganic ions present, we may estimate $K_{i,\text{seawater}}^s$ by summing up the contributions of the various salts present (Gordon and Thorne, 1967a,b):

$$K_{i,\text{seawater}}^s \cong \sum_k K_{i,\text{salt }k}^s \cdot x_k \qquad (5\text{-}29)$$

where x_k is the mole fraction and $K_{i,\text{salt }k}^s$ is the salting constant of salt k in the mixture. For naphthalene we then obtain (Table 5.6):

$$K_{i,\text{seawater}}^s = (0.799)\,(0.22\ \text{M}^{-1}) + (0.104)\,(0.30\ \text{M}^{-1}) + (0.055)\,(0.72\ \text{M}^{-1}) + (0.02)$$
$$(0.32\ \text{M}^{-1}) + (0.01)\,(0.19\ \text{M}^{-1}) + (0.005)\,(0.32\ \text{M}^{-1}) = 0.26\ \text{M}^{-1}$$

which compares very favorably with the experimental value for seawater (average value 0.27 M^{-1}). The K_i^s value of naphthalene for NaCl is 0.22 M^{-1}. Hence, the contribution of the other salts is only 0.04. With insertion of the two K_i^s values into Eq. 5-28 and assuming a $[\text{salt}]_{\text{tot}} = 0.5$ M (typical seawater), we obtain $\gamma_{iw,\text{salt}} / \gamma_{iw}$ ratios of 1.66 for $K_i^s = 0.22$ and 1.84 for $K_i^s = 0.265$, respectively. In general, the error introduced when using $K_{i,\text{NaCl}}^s$ instead of $K_{i,\text{seawater}}^s$ is only in the order of 10%, which is often well within the experimental error of K_i^s measurements. Therefore, in the data set given in Table 5.7, some K_i^s values determined for NaCl have been included. Some more data can be found in the review by Xie et al. (1997).

A few general comments on the data given in Table 5.7 are necessary. First, where available, average K_i^s values taken from different studies are reported. The ranges indicated for these values show that in general, one has to expect rather large uncertainties (i.e., up to ± 20%) in the reported K_i^s values. Furthermore, it should also be noted that Table 5.7 contains values determined from solubility as well as from partitioning (i.e., air–water, organic solvent–water) experiments. Finally, the results of the few studies in which the effect of temperature on salting-out has been investigated (Whitehouse, 1984; Zhou and Mopper, 1990; Dewulf et al., 1995; Alaee et al., 1996) suggest that K_i^s increases somewhat with decreasing temperature. Unfortunately, due to the relatively large scatter in the data, no quantitative relationship can be derived. As a first approximation, the data given in Table 5.7 should, however, also be applicable at temperatures other than 25°C.

Inspection of Table 5.7 shows that our conclusion drawn above from our simple picture of the salting effect, which is that smaller and/or polar compounds should exhibit smaller K_i^s values as compared to larger, nonpolar compounds, is more or less confirmed by the experimental data. When considering the rather limited experimental data set, and the relatively large uncertainty in the data, it is, however, presently not feasible to derive any reliable quantitative relationship using molecular descriptors that would allow prediction of K_i^s values of other compounds. One

Table 5.7 Salting Constants for Some Organic Compounds for Seawater

Compound	K_i^s (L·mol^{-1})	Compound	K_i^s (L·mol^{-1})
Halogenated C$_1$- and C$_2$-Compounds [a,b,c,d]		*Substituted Benzenes and Phenols* [b,d,e,f,h]	
Trichloromethane	0.2	Benzene	0.20 (±0.02)
Tetrachloromethane	0.2	Toluene	0.24 (±0.03)
Methylbromide	0.15	Ethylbenzene	0.29 (±0.02)
Dichlorodifluoromethane	0.29	1,2-Dimethylbenzene	0.30
Trichlorofluoromethane	0.30	1,3-Dimethylbenzene	0.29
1,1-Dichloroethane	0.2	1,4-Dimethylbenzene	0.30
1,2-Dichloroethane	0.2	*n*-Propylbenzene (NaCl)	0.28
1,1,1-Trichloroethane	0.25	Chlorobenzene (NaCl)	0.23
Trichloroethene	0.21 (±0.01)	1,4-Dichlorobenzene (NaCl)	0.27
Tetrachloroethene	0.24 (±0.02)	Benzaldehyde	0.20 (±0.04)
		Phenol	0.13 (±0.02)
Miscellaneous Aliphatic Compounds [e,f]		2-Nitrophenol	0.13 (±0.01)
Pentane (NaCl)	0.22	3-Nitrophenol	0.15
Hexane (NaCl)	0.28	4-Nitrophenol	0.17
1-Butanal	0.3	4-Nitrotoluene	0.16
1-Pentanal	0.3	4-Aminotoluene	0.17
1-Hexanal	0.4		
1-Heptanal	0.5	*Polycyclic Aromatic Compounds* [e,h,i,j,k,l]	
1-Octanal	0.6	Naphthalene	0.28 (±0.04)
1-Nonanal	~ 1.0	Fluorene (NaCl)	0.27
1-Decanal	~ 1.0	Phenanthrene	0.30 (±0.03)
Dimethylsulfide	0.17	Anthracene	0.30 (±0.02)
2-Butanone	0.20	Fluoranthene (NaCl)	0.34
		Pyrene	0.30 (±0.02)
PCBs [e,g]		Chrysene (NaCl)	0.34
Biphenyl	0.32 (±0.05)	Benzo[a]pyrene	0.34
Various PCBs (dichloro to hexachloro)	0.3–0.4	Benzo[a]anthracene (NaCl)	0.35
		1-Naphthol (NaCl)	0.23

[a] Warner and Weiss (1995). [b] Dewulf et al. (1995). [c] DeBruyn and Saltzman (1997). [d] Peng and Wang (1998). [e] Sanemasa et al. (1984). [f] Zhou and Mopper (1990). [g] Brownawell (1986). [h] Hashimoto et al. (1984). [i] Eganhouse and Calder (1975). [j] Whitehouse (1984). [k] Gordon and Thorne (1967b). [l] Almeida et al. (1983).

class of compounds that does not quite fit the qualitative picture is the *n*-alkanals (Table 5.7). One possible cause for the unexpectedly high salting constants of these compounds is their tendency to form diols in aqueous solution (Bell and McDougall, 1960). For example, acetaldehyde (R = CH$_3$, see margin) forms about 50% diol in pure water. If, in saltwater, the aldehyde/diol ratio is changed in favor of the aldehyde, one would expect a stronger salting-out effect, because it can be assumed that the diol form is more easily accommodated in water as compared to the aldehyde form. An additional reason for the large K_i^s values of the larger-chain aldehydes could be the fact that the effect of salt on the activity coefficients of flexible molecules is larger than the effect on the more rigid compounds. However, there are presently no reliable data available to verify this hypothesis.

In summary, we can conclude that at moderate salt concentrations typical for seawater (~ 0.5 M), salinity will affect aqueous solubility (or the aqueous activity coefficient) by a factor of between less than 1.5 (small and/or polar compounds) and about 3 (large, nonpolar compounds, n-alkanals). Hence, in marine environments for many compounds, salting-out will not be a major factor in determining their partitioning behavior. Note, however, that in environments exhibiting much higher salt concentrations [e.g., in the Dead Sea (5 M) or in subsurface brines near oil fields], because of the exponential relationship (Eq. 5-28), salting-out will be substantial (see also Illustrative Example 5.4).

Illustrative Example 5.4

Quantifying the Effect of Inorganic Salts on Aqueous Solubility and Aqueous Activity Coefficients

Problem

Estimate the solubility and the activity coefficient of phenanthrene in (a) seawater at 25°C and 30‰ salinity, and (b) a salt solution containing 117 g NaCl per liter water.

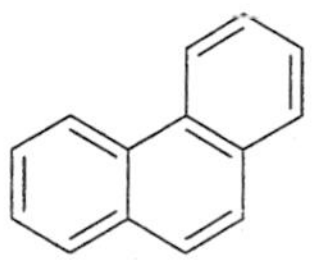

i = phenanthrene

C_{iw}^{sat} (25°C) $= 6.3 \times 10^{-6}$ mol·L^{-1}
$T_m = 101°C$
γ_{iw}^{sat} (25°C) $= 2.0 \times 10^{6}$
(see Table 5.2)

Answer (a)

At 25°C phenanthrene is a solid. Because the free energy contributions of phase change (i.e., melting, or condensation in the case of a gas) to the overall free energy of solution are not affected by salts in the solution, it is the aqueous activity coefficient that is increased as salt concentration increases (Eq. 5-28). Hence, the actual solubility C_{iw}^{sat} decreases by the same factor (Eq. 5-27). The K_i^s value of phenanthrene is 0.30 M^{-1} (Table 5.7). Since 34.2‰ salinity corresponds to a total salt concentration of 0.5 M (see text), [salt]$_{tot}$ for 30‰ is equal to 0.44 M. Insertion of these values into Eq. 5-28 yields:

$$\gamma_{iw}^{sat} \cdot 10^{+(0.30)(0.44)} = (1.34)\,\gamma_{iw}^{sat} = 2.7 \times 10^{6}$$

The aqueous solubility in 30‰ seawater is then given by:

$$C_{iw,salt}^{sat} = C_{iw}^{sat} / (1.34) = 4.7 \times 10^{-6} \text{ mol·L}^{-1}$$

Hence, in 30‰ seawater γ_{iw}^{sat} increases (C_{iw}^{sat} decreases) by about 30% as compared to pure water.

Answer (b)

Use the K_i^s value given for seawater as a surrogate for the NaCl solution. 117 g NaCl per liter correspond to a molar concentration of 2 M. Thus:

$$\gamma_{iw,NaCl}^{sat} = \gamma_{iw}^{sat} \cdot 10^{+(0.30)(2.0)} = (4.0)\,\gamma_{iw}^{sat} = 8.0 \times 10^{6}$$

and

$$C_{iw,salt}^{sat} = 1.6 \times 10^{-6} \text{ mol·L}^{-1}$$

Problem

At oil exploitation facilities it is common practice to add salt to the wastewater in order to decrease the solubility of the oil components, although in the wastewater treatment one then has to cope with a salt problem. Calculate how much NaCl you have to add to 1 m^3 of water in order to increase the activity coefficient of *n*-hexane by a factor of ten. How much Na$_2$SO$_4$ would be required to do the same job?

$i = n$ - hexane

Answer

In order to increase the activity coefficient of a given compound by a factor of ten, the exponent in Eq. 5-28 has to be equal to 1:

$$K_i^s \, [\text{salt}]_{\text{tot}} = 1$$

The K_i^s value for hexane for NaCl is 0.28 M^{-1} (Table 5.7). Then a total salt concentration [salt]$_{\text{tot}}$ = 1 / 0.28 M^{-1} = *3.57* M is needed, which corresponds to an amount of *208.8* kg·m^{-3}.

For estimating the amount of Na$_2$SO$_4$ required, assume a similar relative K_i^s value (relative to NaCl) as determined for benzene (i.e., 0.53 M^{-1} for Na$_2$SO$_4$ versus 0.19 M^{-1} for NaCl, see Table 5.6):

$$K_i^s \, (\text{hexane, Na}_2\text{SO}_4) = (0.28)(0.53) / (0.19) = 0.78 \text{ M}^{-1}$$

Thus in the case of Na$_2$SO$_4$, the required [salt]$_{\text{tot}}$ is 1 / 0.78 M^{-1} = *1.28* M or *181.8* kg·m^{-3}, which is about the same amount as the NaCl needed although, on a molar base, Na$_2$SO$_4$ is much more potent as a salting-out agent.

Advanced Topic

Organic Cosolvents

So far we have considered only situations in which a given organic compound was present as the sole *organic* solute in an aqueous solution. Of course, in reality, in any environmentally relevant aquatic system there will be numerous other natural and/or anthropogenic organic chemicals present that may or may not affect the solubility or, even more important, the activity coefficient of the compound of interest to us. We will treat this issue of organic cosolutes in Chapter 7 when discussing the organic phase–water partitioning of organic compounds present in complex mixtures (e.g., gasoline, oil, PCBs). In this section we will focus on the effect of highly water-soluble organic compounds (i.e., *organic cosolvents*) that *may completely change the solvation properties* of an "aqueous" phase. We may encounter such situations in industrial waste waters or at waste disposal sites where, because of careless dumping procedures, leachates may contain a high portion of organic solvent(s). Furthermore, one of the remediation techniques for contaminated soils is to "wash" the soil with mixtures of water and water-miscible cosolvents (Li et al., 1996). Finally, from an analytical point of view, knowledge of how cosolvents influence the activity coefficient of a given organic compound in organic solvent–water mixtures is pivotal for choosing appropriate mobile phases in reversed-phase liquid chromatography.

Let us start with some comments on the experimental data available on effects of cosolvents on the aqueous solubility and aqueous activity coefficient of organic pollutants. First we should point out that the majority of the systematic studies on this topic have focused on the effects of *completely water-miscible organic solvents* (CMOSs, e.g., methanol, ethanol, propanol, acetone, dioxane, acetonitrile, dimethylsulfoxide, dimethylformamide, glycerol, and many more) and on the *solubility* of sparingly soluble organic *solids*. A large portion of the available data has been collected for drugs and has been published in the pharmaceutical literature. With respect to environmentally more relevant compounds, most investigations have been confined to PAHs (Morris et al., 1988; Dickhut et al., 1989; Li et al., 1996; Fan and Jafvert, 1997) and PCBs (Li and Andren, 1994). Few studies have investigated the impact of CMOSs on the solubility (Groves, 1988) or on the activity coefficient in dilute solution (Munz and Roberts, 1986; Jayasinghe et al., 1992) of *liquid* organic compounds. Note that solubility experiments involving organic liquids are more difficult to interpret because of the partitioning of the cosolvent(s) into the liquid organic phase, which may lead to significant changes in its composition (Groves, 1988). In certain cases, the composition of the liquid phase may even affect the crystal structure of a solid compound, thus complicating the interpretation of solubility data (Khossravi and Connors, 1992). Finally, only very limited data are available on the effect of *partially miscible organic solvents* (PMOSs, e.g., *n*-alcohols ($n > 3$), ethers, halogenated C_1- and C_2-compounds, substituted benzenes) on the aqueous solubility or aqueous activity coefficient of organic pollutants in the presence (Pinal et al., 1990 and 1991) or in the absence (Li and Andren, 1994; Coyle et al., 1997) of a CMOS. Thus, our following discussion will be devoted primarily to water–CMOS systems.

Let us first look at some qualitative aspects of how CMOSs affect the activity coefficient, and thus the solubility and partitioning behavior, of a given organic compound when present in a water/CMOS mixture. The following general conclusions are illustrated by the examples given in Figs. 5.6 to 5.8 and in Table 5.8.

First, we point out that, in general, the activity coefficient of an organic solute, $\gamma_{i\ell}$, decreases (i.e., solubility increases) in an *exponential way* with increasing volume fraction of CMOS. (Note that we use the subscript ℓ to denote that we are dealing with a liquid solution, and, in the following, we do not distinguish between $\gamma_{i\ell}$ values at saturated and dilute solutions.) Second, a significant effect (i.e., > factor 2) is observed only at cosolvent volume fractions greater than 5 to 10% (depending on the solvent). Below 1%, the effect can more or less be neglected (see below). Hence, when conducting experiments, we do not have to worry about significant changes in the activities of organic solutes in an aqueous phase when adding a small amount of a CMOS, as is, for example, common practice when spiking an aqueous solution with a sparingly soluble organic compound dissolved in an organic solvent. Third, the magnitude of the cosolvent effect, as well as its dependence on the amount of cosolvent present, is a function of both the type of cosolvent (Fig. 5.7, Table 5.8) and the type of organic solute (Figs. 5.6 and 5.8) considered. For example, the activity coefficient (or the mole fraction solubility) of naphthalene decreases (increases) by a factor of about 15 when going from pure water to a 40% methanol/60% water mixture, while the effect is about 3 times smaller or 20 times larger when glycerol or acetone, respectively, are the cosolvents (Table 5.8). Furthermore, as can be seen from Fig. 5.8, in 20% methanol/80% water (volume

Figure 5.6 Illustration of the effect of a completely water-miscible solvent (CMOS, i.e., methanol) on the activity coefficient of organic compounds in water–organic solvent mixtures: decadic logarithm of the activity coefficient as a function of the volume fraction of methanol. Note that the data for naphthalene (Dickhut et al., 1989; Fan and Jafvert, 1997) and for the two PCBs (Li and Andren, 1994) have been derived from solubility measurements; whereas for the anilins (Jayasinghe et al., 1992), air–water partition constants determined under dilute conditions have been used to calculate $\gamma_{i\ell}$.

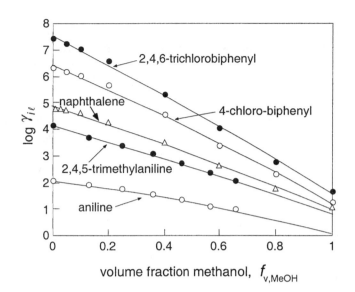

Figure 5.7 Effect of three different CMOSs (i.e., methanol, ethanol, propanol) on the activity coefficient of 2,4,6-trichlorobiphenyl. Data from Li and Andren (1994).

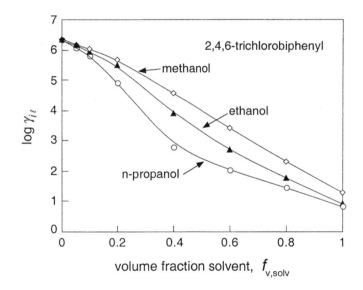

Figure 5.8 Ratio of the activity coefficient in water (γ_{iw}) and in methanol/water [20% (v : v) and 40% (v : v) methanol] as a function of the molar volume (V_{ix}, see Box 5.1) of the solute: $\log(\gamma_{iw}^{sat}/\gamma_{i\ell}^{sat}) = a \cdot V_{ix} + b$. The three compound classes include the following compounds:

Anilines: aniline, 4-methyl-aniline, 3,4-dimethyl-aniline, 2,4,5-trimethyl-aniline; $f_{v,MeOH} = 0.2$: $a = 0.00700$, $b = -0.309$; $f_{v,MeOH} = 0.4$: $a = 0.0128$, $b = -0.432$.

PAHs: naphthalene, anthracene, phenanthrene, pyrene, perylene; $f_{v,MeOH} = 0.2$: $a = 0.0104$, $b = -0.668$; $f_{v,MeOH} = 0.4$: $a = 0.0147$, $b = -0.469$.

PCBs: 4-chlorobiphenyl, 2,4,6-trichlorobiphenyl, 2,2',4,4',6,6'-hexachlorobipheny; $f_{v,MeOH} = 0.2$: $a = 0.0955$, $b = -0.704$; $f_{v,MeOH} = 0.4$: $a = 0.0180$, $b = -0.848$.

Data from Morris et al. (1988), Jayasinghe et al. (1992), Li and Andren (1994), Fan and Jafvert (1997).

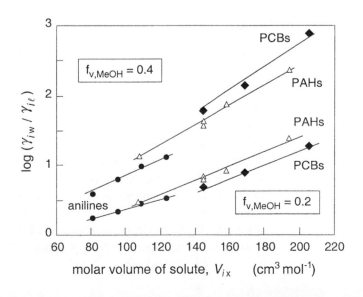

Table 5.8 Effect of Various CMOSs on the Activity Coefficient or Mole Fraction Solubility, Respectively, of Naphthalene at Two Different Solvent/Water Ratios ($f_{v,solv} = 0.2$ and 0.4)

Cosolvent	Structure	Solubility[b] Parameter $(M Pa)^{1/2}$	Naphthalene[a] $\gamma_{iw}^{sat} / \gamma_{i\ell}^{sat} = x_{i\ell}^{sat} / x_{iw}^{sat}$	
			$f_{v,solv} = 0.2$ $(\sigma_i^c)^c$	$f_{v,solv} = 0.4$
Glycerol	OH | HOCH$_2$–CH–CH$_2$OH	36.2	2.5 (2.0)	5.5
Ethyleneglycol	HOCH$_2$—CH$_2$OH	34.9	3 (2.4)	9
Methanol	CH$_3$OH	29.7	3.5 (2.7)	14
Dimethylsulfoxide (DMSO)	H$_3$C $\!\!\diagdown\!\!$ S=O H$_3$C $\!\!\diagup\!\!$	26.7	5.5 (3.7)	3.6
Ethanol	H$_3$CCH$_2$OH	26.1	7 (4.2)	48
Propanol	H$_3$CCH$_2$CH$_2$OH	24.9	17 (6.2)	180
Acetonitrile	H$_3$C—C≡N	24.8	14 (5.7)	140
Dimethylformamide	O ‖ H—C—N(CH$_3$)$_2$	24.8	15 (5.9)	130
1,4-Dioxane	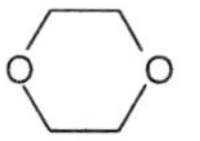 (dioxane ring)	20.7	14 (5.7)	180
Acetone	O ‖ H$_3$C—C—CH$_3$	19.7	20 (6.5)	270

[a] Data from Dickhut et al. (1989), Li et al. (1996), and Fan and Jafvert (1997). [b] Hildebrand solubility parameter taken from Barton (1991). The parameter is defined as the square root of the ratio of the enthalpy of vaporization and the molar volume of the liquid. [c] Cosolvency power for the range $0 < f_{v,solv} < 0.2$; see Eq. 5-30.

fractional methanol, $f_{v,MeOH} = 0.2$) the activity coefficients of aniline ($V_{ix} = 82$ cm^3 mol^{-1}) and 2,2',4,4',6,6'-hexachlorobiphenyl ($V_{ix} = 206$ cm^3 mol^{-1}) are smaller by a factor of about 2 and 20, respectively, as compared to pure water, while at 40% methanol/60% water ($f_{v,MeOH} = 0.4$) the corresponding factors are 4 and 750, respectively. In general, we may assume that for a given cosolvent–water system, the effect of the cosolvent on $\gamma_{i\ell}$ will be larger for large, nonpolar solutes that are only sparingly soluble in water (e.g., PCBs, PAHs) as compared to more polar, small molecules (e.g., aniline) exhibiting higher water solubilities. Finally, with respect to the "cosolvency-power" of CMOSs, we can see from Table 5.8 that, qualitatively, the more "water-like" solvents such as glycerol, ethylene glycol, or methanol, have a much smaller impact on the activity coefficient of an organic solute as compared to organic solvents for which hydrogen bonding is important, but not the overall dominating factor. Note that the (Hildebrand) solubility parameter given in Table 5.8 (see footnote for explanation) is a measure of the cohesive forces among the molecules in the pure solvent. As can be seen, qualitatively, there is a trend that solvents exhibiting higher cohesive forces in their pure liquid tend to have a smaller cosolvent effect when mixed with water.

Before we address cosolvency in a more quantitative way, it is useful to try to picture how a cosolvent affects the solvation of an organic solute on a molecular level. From the examples given in Table 5.8 we can see that CMOSs are relatively small molecules with strong H-acceptor and/or H-donor properties. When mixed with water, they are, therefore, able to break up some of the hydrogen bonds between the water molecules and thus form a new H-bonded "mixed solvent" that will change its property in a rather complex way as a function of the nature and of the relative amount of the cosolvent. As we have seen in Section 5.3, for smaller and/or more polar organic compounds, the excess free energy in pure aqueous solution is dominated by the entropic contribution. Only for larger, particularly for nonpolar solutes, is the enthalpy term also significant (Table 5.3). Hence, it is reasonable to assume and is supported by experimental data (e.g., Bustamante et al., 1998) that in water-rich mixtures (i.e., $f_{v, solv} < 0.5$), the observed decrease in excess free energy (increase in solubility) of organic compounds with increasing CMOS/water ratio is primarily due to a substantial increase in the excess entropy, which may even compensate possible increases in excess enthalpy. Since these changes in excess enthalpy and entropy with changing cosolvent–water composition are, in general, not linearly correlated with each other, a nonlinear relationship between excess free energy (or log $\gamma_{i\ell}$) and $f_{v,solv}$ can be expected and, as illustrated by Fig. 5.7, is actually found for many organic solute/CMOS/water systems.

Considering the rather complex factors that determine the excess free energy of an organic solute in a CMOS/water mixture, it is not too surprising that any quantitative models developed for describing cosolvent effects have only rather limited predictive capabilities. The models are, however, quite well suited for fitting experimental data, and for estimating activity coefficients of structurally closely related compounds in a given CMOS/water system for which experimental data are available.

For a discussion of the various approaches taken to quantify cosolvent effects we refer to the literature (e.g., Li and Andren, 1995; Li et al., 1996; Fan and Jafvert, 1997). For our purpose here, we adopt the most simple empirical approach where we assume a *log-linear* relationship between activity coefficient (or mole fraction solubility) of a given compound and volume fraction of the CMOS over a narrow $f_{v,solv}$-range (i.e., $\Delta f_{v,solv} = 0.2$) confined by $f_{v,solv}^1$ and $f_{v,solv}^2$. Hence, for a given organic solute i and a given CMOS/water system, we get (note that we omit the subscript "solv" to indicate the CMOS):

$$\log \gamma_{i\ell}(f_v) = \log \gamma_{i\ell}(f_v^1) - \sigma_i^c \cdot (f_v - f_v^1) \qquad (5\text{-}30)$$

or:

$$\log x_{i\ell}^{sat}(f_v) = \log x_{i\ell}^{sat}(f_v^1) + \sigma_i^c \cdot (f_v - f_v^1) \qquad (5\text{-}31)$$

where the slope σ_i^c, which is dependent on both the solute and the cosolvent, is given by $[\log \gamma_{i\ell}(f_v^1) - \log \gamma_{i\ell}(f_v^2)] / (f_v^2 - f_v^1)$, and $f_v^1 \leq f_v \leq f_v^2$. σ_i^c is commonly referred to as the cosolvency power of the solvent for the solute i. Note, however, that σ_i^c is not a constant, but changes with increasing f_v. If $f_v^1 = 0$, that is, if we consider only the range between pure water and a given cosolvent fraction f_v^2 (e.g., 0.2), Eq. 5-30 simplifies to:

$$\log \gamma_{i\ell}(f_v) = \log \gamma_{iw} - \sigma_i^c \cdot f_v \qquad (5\text{-}32)$$

with $\sigma_i^c = [\log \gamma_{iw} - \log \gamma_{i\ell}(f_v^2)] / f_v^2$. Eq. 5-32 can also be written as:

$$\gamma_{i\ell}(f_v) = \gamma_{iw} \cdot 10^{-\sigma_i^c \cdot f_v} \qquad (5\text{-}33)$$

Hence, this approach is very similar to the one used for describing the effect of salt on aqueous solubility and aqueous activity coefficient (Eqs. 5-27 and 5-28). Some example calculations using Eq. 5-30 or 5-31, respectively, are given in Illustrative Example 5.5. Finally, we should note that the mole fractions of two solvents in a binary mixture are related to the volume fractions by:

$$x_1 = \frac{1}{1 + \dfrac{(1 - f_{v1})}{f_{v1}} \cdot \dfrac{\overline{V}_1}{\overline{V}_2}} \quad \text{and} \quad x_2 = 1 - x_1 \qquad (5\text{-}34)$$

We conclude this section with some brief comments on the cosolvent effects of partially miscible organic solvents (PMOSs). These solvents include very polar liquids such as n-butanol, n-butanone, n-pentanol, or o-cresol, but also nonpolar organic compounds such as benzene, toluene, or halogenated methanes, ethanes, and ethenes. For the polar PMOS, a similar effect as for the CMOS can be observed; that is, these solvents decrease the activity coefficient of an organic solute when added to pure water or to a CMOS/water mixture (Pinal et al., 1990; Pinal et al., 1991; Li and Andren, 1994). For the less polar PMOS there is not enough data available to draw any general conclusions.

Illustrative Example 5.5

Estimating the Solubilities and the Activity Coefficients of Organic Pollutants in Organic Solvent – Water Mixtures

Problem

Estimate the solubility and the activity coefficient of (a) naphthalene, and (b) benzo(a)pyrene in a 30% methanol/70% water (v : v) mixture at 25°C.

i = naphthalene

C_{iw}^{sat} (25°C) = 2.5×10^{-4} mol·L^{-1}
T_m = 80.2°C
γ_{iw}^{sat} (25°C) = 6.7×10^4
(see Table 5.2)

Answer (a)

As in the case of inorganic salts (Illustrative Example 5.4) the free energy contributions of phase change to the overall free energy of solution are not affected by CMOS (with some exceptions in which the solvent changes the crystal structure of a solid, see text). Hence, you need only estimate the effect of the CMOS on the activity coefficient. Use the $\gamma_{iw}^{sat} / \gamma_{i\ell}^{sat}$ ratios given for naphthalene in Table 5.8 for $f_{v,MeOH} = 0.2$ and $f_{v,MeOH} = 0.4$ to estimate $\gamma_{i\ell}^{sat}$ for $f_{v,MeOH} = 0.3$ by interpolation, using the log-linear relationship Eq. 5-30. Calculate first the log $\gamma_{i\ell}^{sat}$ values for $f_{v,MeOH} = 0.2$ and 0.4, respectively:

$$\log \gamma_{i\ell}^{\text{sat}} (f_{\text{v,MeOH}} = 0.2) = \log (\gamma_{iw}^{\text{sat}}) / 3.5) = 4.28$$

$$\log \gamma_{i\ell}^{\text{sat}} (f_{\text{v,MeOH}} = 0.4) = \log (\gamma_{iw}^{\text{sat}}) / 14) = 3.68$$

The slope σ_i^c in Eq. 5-30 is then obtained by:

$$\sigma_i^c = (4.28 - 3.68) / (0.2) = 3.0$$

which yields (Eq. 5-30):

$$\log \gamma_{i\ell}^{\text{sat}} (f_{\text{v,MeOH}} = 0.3) = 4.28 - (3.0)(0.1) = 3.98$$

or:

$$\gamma_{i\ell}^{\text{sat}} (f_{\text{v,MeOH}} = 0.3) = 9.5 \times 10^3$$

which is about a factor of 7 smaller than γ_{iw}^{sat}. This also means that the mol fraction solubility of naphthalene will be about a factor of 7 larger (Eq. 5-31), that is:

$$x_{i\ell}^{\text{sat}} = (7)(2.5 \times 10^{-4})(0.018) = 3.2 \times 10^{-5}.$$

The mole fraction of methanol in a 30% methanol/70% water mixture is 0.16 (Eq. 5-34). The molar volume of methanol is 0.0406 L·mol⁻¹. Hence, the molar volume of the mixture can be calculated as (Eq. 3-43, note that we assume additivity):

$$V_{i\ell} = (0.16)(0.0406) + (0.84)(0.018) = 0.022 \text{ L·mol}^{-1}$$

Hence, the molar solubility of naphthalene in the mixture is:

$$C_{i\ell}^{\text{sat}} (f_{\text{v,MeOH}} = 0.3) = (3.2 \times 10^{-5}) / (0.022) = 1.45 \times 10^{-3} \text{ mol·L}^{-1}$$

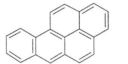

$i = $ benzo(a)pyrene

$C_{iw}^{\text{sat}} (25°C) = 7.2 \times 10^{-9} \text{ mol·L}^{-1}$
$T_m = 176.5°C$
$\gamma_{iw}^{\text{sat}} (25°C) = 3.2 \times 10^8$
(see Table 5.2)

Answer (b)

For benzo(a)pyrene there are no experimental data available. Estimate $\gamma_{iw}^{\text{sat}} / \gamma_{i\ell}^{\text{sat}}$ for $f_{\text{v,MeOH}} = 0.2$ and 0.4, respectively using the data given for other PAHs in Fig. 5.8. From the original data sets reported in the literature, derive the corresponding linear relationships between $\log \gamma_{iw}^{\text{sat}} / \gamma_{i\ell}^{\text{sat}}$ and the molar volume, V_{ix} (in cm³ mol⁻¹), calculated by the method described in Box 5.1 (see caption Fig. 5.8):

$$f_{\text{v,MeOH}} = 0.2: \quad \log (\gamma_{iw}^{\text{sat}} / \gamma_{i\ell}^{\text{sat}}) = (0.0104) V_{ix} - 0.668 \qquad (1)$$

$$f_{\text{v,MeOH}} = 0.4: \quad \log (\gamma_{iw}^{\text{sat}} / \gamma_{i\ell}^{\text{sat}}) = (0.0147) V_{ix} - 0.469 \qquad (2)$$

Insertion of $\log \gamma_{iw}^{\text{sat}}$ and V_{ix} (195.0 cm³ mol⁻¹) of benzo(a)pyrene into Eqs. 1 and 2 yields:

$$\log \gamma_{i\ell}^{\text{sat}} (f_{\text{v,MeOH}} = 0.2) = 7.15$$

$$\log \gamma_{i\ell}^{\text{sat}} (f_{\text{v,MeOH}} = 0.4) = 6.10$$

which yields a slope $\sigma_i^c = (7.15 - 6.10) / (0.2) = 5.25$ demonstrating that the effect of the cosolvent methanol is more pronounced for the more hydrophobic benzo(a)pyrene as compared to the less hydrophobic naphthalene ($\sigma_i^c = 3.0$, see above). Insertion of the according values into Eq. 5-28 then yields:

$$\log \gamma_{i\ell}^{\text{sat}} (f_{\text{v,MeOH}} = 0.3) = 7.15 - (5.25)(0.1) = 6.63$$

or

$$\gamma_{i\ell}^{\text{sat}} (f_{\text{v,MeOH}} = 0.3) = 4.2 \times 10^6$$

which is about 75 times smaller than γ_{iw}^{sat}. Hence, the corresponding mole fraction solubility is about a factor of 75 larger (Eq. 5-31), that is:

$$x_{i\ell}^{\text{sat}} = (75)(7.2 \times 10^{-9})(0.018) = 9.7 \times 10^{-9}$$

and by analogy to case (a):

$$C_{i\ell}^{\text{sat}} (f_{\text{v,MeOH}}) = (9.7 \times 10^{-9}) / (0.022) = 4.4 \times 10^{-7} \text{ mol} \cdot \text{L}^{-1}$$

5.5 Availability of Experimental Data; Methods for Estimation of Aqueous Activity Coefficient and Aqueous Solubility

Experimental Data

We have already seen from our above discussions that organic chemicals cover a very wide range of aqueous solubilities, that is, from completely miscible down to 10^{-10} mol $\cdot$ L^{-1} and below (see Appendix C). Because of these low solubilities and owing to the analytical limitations in the past, many organic substances have acquired the reputation of "being insoluble in water" (e.g., Lide, 1995). From an environmental point of view such a statement is, of course, not correct at all because water is one of the major transport and reaction media for organic compounds in natural systems. Hence, for assessing the behavior and the effects of organic pollutants in the environment, accurate data on aqueous solubilities and aqueous activity coefficients are of utmost importance.

The conventional method of determining *aqueous solubility* is to equilibrate an excess amount of the pure chemical of interest with water in a batch reactor. Equilibrium is achieved by gently shaking or by slowly stirring with a magnetic stirrer. The aim is to prevent formation of emulsions or suspensions and thus avoid extra experimental procedures such a filtration or centrifugation which may be required to ensure that a "true" solution is obtained. This method, which is commonly referred to as *shake flask method* works quite well for more soluble compounds. For more sparingly soluble chemicals such as higher alkanes, PAHs, PCBs, polychlorinated dioxins, and dibenzo-furans, however, experimental difficulties can still occur because of the formation of emulsion or microcrystal suspensions, and because of adsorption phenomenon if filtration is necessary. For such compounds, the *generator column method* has been found to produce much more accurate solubility data. In this method, a solid support (e.g., glass bead) is coated with the chemical, packed in an open tubular column, and water is run through at a precise flow rate to achieve equilibrium. Subsequently, the aqueous effluent is assessed for the organic solute by using an appropriate analytical technique. For a more detailed discussion of the various aspects of the experimental determination of aqueous solubilities, we refer to Yalkowsky and Banerjee (1992) and to the handbooks published by Mackay et al. (1992–1997). These latter handbooks also contain large compilations of water solubilities of a variety of important compound classes. Additional experimental data may be found in smaller compilations reported by Ruelle and Kesselring (1997a,b), Montgomery (1997), and Mitchell and Jurs (1998). As with our comments on experimental vapor pressure data (Section 4.4), we should point out that, particularly for sparingly soluble compounds, aqueous solubilities determined by different methods and/or different laboratories may vary by as much as a factor of 2 to 3, and in some cases even by more than an order of magnitude. Such data should, therefore, be treated with the necessary caution. Again, one way of deciding which solubility value should be selected is to compare the experimental data with predicted values using other compound properties or solubility data from structurally related compounds.

Several experimental methods are available to determine activity coefficients of organic compounds in dilute aqueous solution. A critical review of the various direct and indirect measurement methods can be found in the article by Sherman et al. (1996). For compounds exhibiting small γ_{iw} values (i.e., high water solubilities), differential ebulliometry or dewpoint techniques are frequently used. Here, the effect of the solute on the boiling point of the solvent (i.e., water), or on the dewpoint of the solvent vapor, respectively, is determined under constant pressure. The measured changes then allow us to derive the activity coefficient of the compound in the solvent. For compounds that exhibit larger γ_{iw} values, particularly for the more volatile compounds, the techniques of head space gas chromatography and gas stripping seem to provide reliable activity coefficients in dilute aqueous solution. Among the indirect approaches, derivation of γ_{iw} values from experimental evaluation of the partitioning of a compound between an organic solvent (e.g., hexadecane, n-octanol) and water is the most widely used method. We will come back to this issue in Chapter 7.

In summary, we can conclude that there is quite a large experimental database on aqueous solubilities and/or aqueous activity coefficients of organic compounds available in the literature. In this context, we should recall that for compounds exhibiting γ_{iw} values greater than about 100, we may assume that γ_{iw} is concentration independent (Section 5.2). Thus, if only γ_{iw} is known for a given compound (either experimentally determined or predicted; see below), we can estimate its aqueous solubility by using Eqs. 5-6, 5-10, or 5-15, respectively (see also Illustrative Example 5.2). If neither the aqueous solubility nor the aqueous activity coefficient is known for a given compound (e.g., for a new chemical), we may use one of the various methods that have been developed for estimating these important compound properties.

Prediction of Aqueous Solubilities and/or Aqueous Activity Coefficients

Any general approach (i.e., any approach that is not restricted to a confined set of structurally related compounds) for prediction of aqueous solubilities and/or aqueous activity coefficients has to cope with the intrinsic difficulty of describing precisely what is happening when introducing an organic compound in water. We have discussed this problem in detail in Section 5.3, where we have also seen that by using molecular descriptors such as the solvatochromic parameters we are able to model the aqueous activity coefficient of any kind of organic compound with reasonable success. The problem with these and similar approaches (Yalkowsky and Banerjee, 1992) is only that often not all required descriptors are readily available for a given compound.

Therefore, similar to the attempts made to estimate vapor pressure (Section 4.4) there have been a series of quite promising approaches to derive topological, geometric, and electronic molecular descriptors for prediction of aqueous activity coefficients from chemical structure (e.g., Mitchell and Jurs, 1998; Huibers and Katritzky, 1998). The advantage of such quantitative structure property relationships (QSPRs) is, of course, that they can be applied to any compound for which the structure is known. The disadvantages are that these methods require sophisticated computer software, and that they are not very transparent for the user. Furthermore, at the present stage, it remains to be seen how good the actual predictive capabilities of these QSPRs are.

A completely different method that has been shown to be particularly useful for estimating activity coefficients in *nonaqueous* solutions is based on a group contribution approach. The best known and most widely used version of this approach is the UNIFAC method (Hansen et al., 1991; Kan and Tomson, 1996). A similar approach that is, however, focused on aqueous solutions is referred to as AQUAFAC (Myrdal et al., 1993). The basic idea of this type of approach is to express enthalpic and entropic contributions to the excess free energy by summing up interactive terms of parts of the solute and solved molecules, particularly of their functional groups. A large number of such interaction parameters have been derived from a statistical analysis of experimental data on vapor–liquid partitioning. As already mentioned above, UNIFAC works best for nonaqueous mixtures or mixtures that contain only a very limited amount of water. When dealing with solutions exhibiting significant amounts of water, the present limitations of UNIFAC, but also of AQUAFAC (Sherman et al., 1996; Fan and Jafvert, 1997), are probably primarily due to the difficulties in properly expressing the entropic contributions resulting from the unique properties of the solvent water.

We conclude our short discussion of estimation methods for aqueous solubilities and aqueous activity coefficients by restating that simple one-parameter relationships [e.g., relationships between liquid aqueous solubility and molar volume (Table 5.4) or octanol–water partition constant (Section 7.4)] may also be quite powerful predictive tools, provided that we confine a given equation to a set of *structurally closely related* compounds. In this context, we should emphasize again (and again!) that only with a sufficient understanding of the molecular interactions that determine the excess free energy of a given compound in a given molecular environment (here in aqueous solution) will we be able to define which compounds are structurally related with respect to a given partitioning process. This will become even more evident in the following chapters.

5.6 Questions and Problems

Questions

Q 5.1

What is meant by the term *water solubility* or *aqueous solubility* of a given compound? What is the range of aqueous solubilities encountered when dealing with environmentally relevant compounds?

Q 5.2

How is the aqueous activity coefficient of a compound related to the aqueous solubility, if the compound is (a) a liquid, (b) a solid, and (c) a gas under the prevailing conditions? Comment on any assumptions that you make when answering this question.

Q 5.3

The excess enthalypy in aqueous solution (H_{iw}^{E}) of 2-butanone is smaller than that of the similarly sized phenol (Table 5.3), although one can assume that phenol has stronger polar interactions with the water molecules. Try to explain these findings.

2-butanone   phenol

Q 5.4

The excess entropy in aqueous solution (S_{iw}^{E}) of *n*-hexane is significantly more negative as compared to the similarly sized naphthalene (Table 5.3). Try to explain this difference.

n-hexane   naphthalene

Q 5.5

Figure 5.2 and Table 5.4 show that, for a given class of strucurally closely related compounds, a linear relationship exists between liquid aqueous solubility and size of the molecule (Eq. 5-18). Note that in both Fig. 5.2 and Table 5.4 decadic instead of natural logarithms are used:

$$\log C_{iw}^{sat}(L) = -c \cdot V_{ix} + d$$

Try to answer the following questions:

(a) Why are the slopes c of the regression lines quite similar for the *n*-alkanes and the alkanols (Fig. 5.2*a*), and why do these two groups of compounds exhibit such different intercepts d? Why are there any significant differences in the intercepts between primary, secondary, and tertiary alkanols?

(b) Why do the chlorobenzenes and PAHs (Fig. 5.2*b*) exhibit different slopes?

(c) Why is there such a large scatter in the data of the halogenated C_1- and C_2-compounds (Fig. 5.2*c*)?

Q 5.6

Explain qualitatively how the aqueous solubility of a (a) liquid, (b) solid, and (c) gaseous compound changes with temperature. Which thermodynamic quantity(ies) do you need to know for quantifying this temperature dependence?

Q 5.7

Explain in words how environmentally relevant inorganic salts affect the aqueous solubility of a (a) liquid, (b) solid, and (c) gaseous compound. Is it true that the effect is linearly related to the concentration of a given salt? What is the magnitude of the effect of salt on the aqueous activity coefficient of organic compounds in typical seawater?

Q 5.8

Explain in words how organic cosolvents affect the activity coefficients in water-solvent mixtures? Which organic solvents are most effective? Is it true that the effect of an organic cosolvent is linearly related to its volume fraction in the solvent–water mixture? Below which volume fraction can the effect of an organic cosolvent be neglected?

Q 5.9

Derive Eq. 5-34 by realizing that the number of moles of a given compound present in a given volume, V_L, of the pure liquid of that compound is given by $V_L / \overline{V}_L$.

Problems

P 5.1 *Calculating Aqueous Activity Coefficients and Excess Free Energies in Aqueous Solution from Experimental Solubility Data*

Calculate the aqueous activity coefficients, γ_{iw}^{sat}, and the excess free energies in aqueous solution, G_{iw}^E (in kJ·mol^{-1}), of (a) *n*-decane (n-$C_{10}H_{22}$), (b) 2,3,7,8-tetra-chlorodibenzo-p-dioxin, and (c) bromomethane (CH_3Br) at 25°C using the data provided in Appendix C.

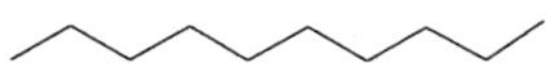

n-decane

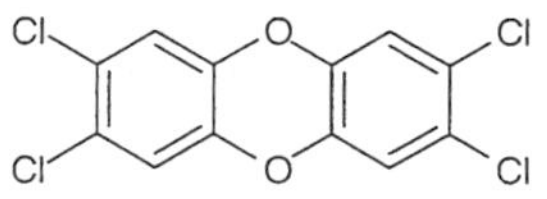

2,3,7,8-tetrachloro-
dibenzo-*p*-dioxin

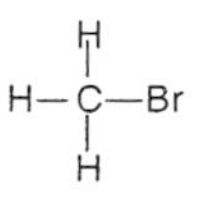

bromomethane

P 5.2 *A Tricky Stock Solution*

You work in an analytical laboratory and you are asked to prepare 250 mL of a 0.5 M stock solution of anthracene in toluene (ρ^{20} (toluene) = 0.87 g·cm^{-3}) as solvent. You look up the molar mass of anthracene, go to the balance, weigh out 22.3 g of this compound, put it into a 250 mL volumetric flask, and then fill the flask with toluene. To your surprise, even after several hours of intensive shaking, there is still a substantial portion of undissolved anthracene present in the flask, although your intuition tells you that these two aromatic compounds should form a near-ideal liquid mixture.

(a) What is the problem?

(b) Give an estimate of how much anthracene has actually been dissolved (in grams).

(c) What is anthracene's concentration (in molar units) in the stock solution (at 20°C)? The necessary data can be found in Appendix C.

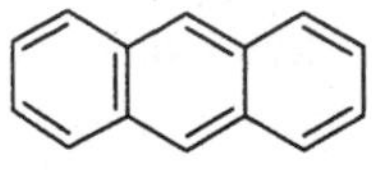

anthracene
$T_m = 217.5\ °C$
$T_b = 341.0\ °C$

P 5.3 Explaining the Differences in Aqueous Solubility Between n-Hexane, di-n-Propylether, and n-Hexanol

As can be seen from the data in Appendix C, the aqueous solubilities of n-hexanol ($C_{iw}^{sat} = 6.2 \times 10^{-2}\ mol\cdot L^{-1}$) and di-$n$-propylether ($C_{iw}^{sat} = 3.0 \times 10^{-2}\ mol\cdot L^{-1}$) exceed that of n-hexane ($C_{iw}^{sat} = 1.5 \times 10^{-4}\ mol\cdot L^{-1}$) by more than two orders of magnitude.

(a) Try to explain the differences in C_{iw}^{sat} between the three compounds based on their abilities to undergo intermolecular interactions.

(b) Use Eq. 5-22 to evaluate the various factors that determine the aqueous solubilities of the three compounds. You can find all necessary data in Tables 4.3 and 5.5 and in Appendix C.

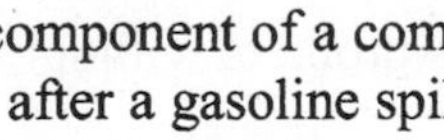

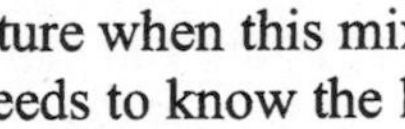

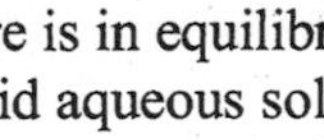

| $n_{Di} =$ | n-hexanol 1.418 | di-n-proylether 1.381 | n-hexane 1.375 |

P 5.4 Estimating Aqueous Solubilities Using Solubility Data of Structurally Related Compounds (adapted from Roberts, 1995).

As will be discussed in Chapter 7, for estimating the aqueous saturation concentration of a given component of a complex mixture when this mixture is in equilibrium with water (e.g., after a gasoline spill), one needs to know the liquid aqueous solubility of the pure compound of interest. Shown below are the aqueous solubilities of some hydrocarbons present in gasoline that are all liquids at 25°C. Estimate the aqueous solubilities (in molar units) of the two gasoline compounds isoctane (2,2,4-trimethyl-pentane) and 1-heptene using the experimental data reported below and/or using one of the equations given in Table 5.4. Comment on the selection of the set of reference compounds that you use for your estimates.

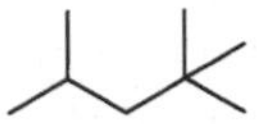

isoctane    1-heptene

Compound	M_i (g·mol^{-1})	T_b (°C)	C_{iw}^{sat} (25°C) (mg·L^{-1})
1-pentene	70.1	30.0	148
2-methyl-1-pentene	84.2	60.7	78
1-hexene	84.2	63.4	50
4-methyl-1-pentene	84.2	53.9	48
2,2-dimethylbutane	86.2	49.7	12.8
2,2-dimethylpentane	100.2	79.2	4.4
2,2,3-trimethylbutane	100.2	80.9	4.4
3-methylhexane	100.2	92.0	3.3
1-octene	112.2	121.3	2.7
2-methylheptane	114.2	117.6	0.85
1-nonene	126.3	146.9	1.12
3-methyloctane	128.3	143.0	1.42
2,2,5-trimethylhexane	128.3	124.0	1.15

P 5.5 *Evaluating the Effect of Temperature on the Aqueous Solubility and on the Aqueous Activity Coefficient of a Solid Compound*

Living in a cold area, you want to know the aqueous solubility and the aqueous activity coefficient of organic compounds at 1°C rather than at 25°C.

(a) Estimate C_{iw}^{sat} (in molar units) and γ_{iw}^{sat} of 1,2,3,7-tetrachlorodibenzo-*p*-dioxin at 1°C using aqueous solubilities of this compound determined at more elevated temperatures by Friesen and Webster (1990):

(b) Also calculate the average excess enthalpy (H_{iw}^E) of the compound in water for the temperature range considered (in kJ·mol^{-1}). Why are you interested in this quantity? Comment on any assumption that you make.

T / °C	7.0	11.5	17.0	21.0	26.0
C_{iw}^{sat} / mol·L^{-1}	7.56×10^{-10}	8.12×10^{-10}	12.5×10^{-10}	14.9×10^{-10}	22.6×10^{-10}

Hint: You can solve this problem without any lengthy calculations!

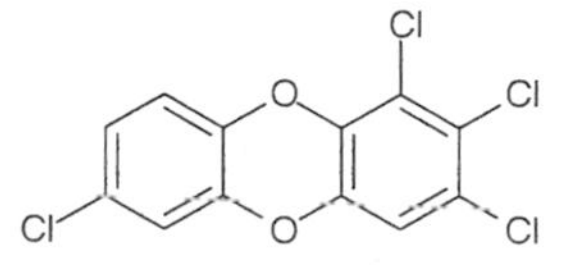

1,2,3,7-tetrachloro-*p*-dibenzodioxin
$M_i = 322.0$ g·mol^{-1}
$T_m = 175.0$ °C

P 5.6 *Evaluating the Effect of Temperature on the Solubility and/or the Activity Coefficient of a Gaseous Compound (Freon 12) in Freshwater and in Seawater*

For an assessment of the global distribution of persistent volatile halogenated hydrocarbons, the solubility and activity coefficients of such compounds in natural waters need to be known. Warner and Weiss (1985) have determined the solubilities of dichlorodifluoromethane (Freon 12) at 1 bar partial pressure at various temperatures in freshwater and in seawater (35.8‰ salinity):

Freshwater:

$T\,/\,°C$	0.9	9.6	19.9	29.9	40.7
$C_{iw}^{1bar}\,/\,mol \cdot L^{-1}$	9.0×10^{-3}	5.6×10^{-3}	3.5×10^{-3}	2.5×10^{-3}	1.8×10^{-3}

Seawater:

$T\,/\,°C$	4.8	9.2	20.4	29.6	39.9
$C_{iw}^{1bar}\,/\,mol \cdot L^{-1}$	4.9×10^{-3}	4.0×10^{-3}	2.4×10^{-3}	1.7×10^{-3}	1.3×10^{-3}

(a) Estimate the solubilities (in molar units) of Freon 12 in freshwater and in seawater at 1 bar partial pressure at 5 and 25°C.

(b) Calculate the activity coefficients of Freon 12 in freshwater and seawater at these temperatures by using the vapor pressure data given in Problem 4.2.

(c) Derive the average excess enthalpy (H_{iw}^E in kJ·mol) of Freon 12 in freshwater and seawater for the ambient temperature range (i.e., 0 – 40°C).

(d) Comment on any differences found between freshwater and seawater.

P 5.7 *A Small Bet with an Oceanographer*

A colleague of yours who works in oceanography bets you that both the solubility as well as the activity coefficient of naphthalene are larger in seawater (35‰ salinity) at 25°C than in distilled water at 5°C. Is this not a contradiction? How much money do you bet? Estimate C_{iw}^{sat} and γ_{iw}^{sat} for naphthalene in seawater at 25°C and in distilled water at 5°C. Discuss the result. Assume that the average enthalpy of solution ($\Delta_{ws}H_i$, Fig. 5.1) of naphthalene is about 30 kJ·mol^{-1} over the ambient temperature range. All other data can be found in Tables 5.3 and 5.7 and in Appendix C.

P 5.8 *Evaluating the Effect of Different Cosolvents on the Retention Time of Organic Compounds in Reversed-Phase Liquid Chromatography*

The retention time of an organic compound in reversed-phase liquid chromatography is heavily influenced by the activity coefficient of the compound in the mobile phase, which commonly consists of a CMOS/water mixture.

(a) Estimate by what factor the activity coefficients of naphthalene and anthracene change when switching the mobile phase from 30% methanol / 70% water (v : v) to 30% acetonitrile / 70% water.

(b) What is the effect on the absolute and relative retention times of the two compounds when leaving all other parameters invariant?

In the literature (Pinal et al., 1991) you find data showing that the activity coefficient of anthracene is 400 times smaller in 40% acetonitrile / 60% water (v : v) as compared to pure water. All other necessary data can be found in Table 5.8, Illustrative Example 5.5, and Appendix C.

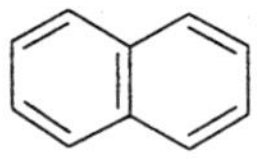

naphthalene 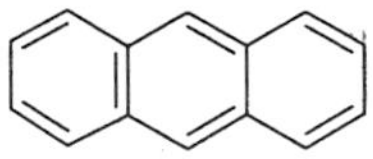   anthracene

Chapter 6

AIR–ORGANIC SOLVENT AND AIR–WATER PARTITIONING

6.1   Introduction

6.2   Thermodynamic Considerations
Raoult's Law
Henry's Law and the Henry's Law Constant
Effect of Temperature on Air–Liquid Partitioning

6.3   Air–Organic Solvent Partitioning
Air–Organic Solvent and Other Partition Constants
Comparison of Different Organic Solvents
LFERs Relating Partition Constants in Different Air–Solvent Systems
Model for Description of Air–Solvent Partitioning
Temperature Dependence of Air–Organic Solvent Partition Constants
Applications
Illustrative Example 6.1: *Assessing the Contamination of Organic Liquids by Air Pollutants*

6.4   Air–Water Partitioning
"The" Henry's Law Constant
Effect of Temperature on Air–Water Partitioning
Effect of Solution Composition on Air–Water Partitioning
Illustrative Example 6.2: *Evaluating the Direction of Air–Water Gas Exchange at Different Temperatures*
Illustrative Example 6.3: *Assessing the Effect of Solution Composition on Air–Aqueous Phase Partitioning*
Availability of Experimental Data
Estimation of Air–Water Partition Constants
Illustrative Example 6.4: *Estimating Air–Water Partition Constants by the Bond Contribution Method*

6.5   Questions and Problems

6.1 Introduction

The transfer of chemicals between air and aqueous phases is one of the key processes affecting the fates of many organic compounds in the environment. Examples include exchanges of volatile and semivolatile compounds between air and rain or fog droplets, between the atmosphere and rivers, lakes, or the oceans, and between residual soil water and soil gases below ground. Additionally, we are sometimes interested in the transfers of organic chemicals from other bulk liquids to neighboring gas phases. An example involves the evaporation of benzene from gasoline. In all of these instances, we need to know the equilibrium distribution constant or coefficient of the substance partitioning between the liquid and gas phases of interest.

In light of our discussions of the molecular factors determining excess free energies of organic compounds in gas phases (Chapter 4) and in liquid phases (particularly in aqueous phases, Chapter 5), we are now in a good position to understand equilibrium partitioning of organic compounds between air and bulk liquids. In this chapter, we will focus on air–water partitioning (Section 6.4). But before we do this, we will first examine the equilibrium partitioning of organic compounds between air and various organic liquids (Section 6.3). In addition to having instances where such partitioning is important, we can use these partitioning data to further illustrate how chemical structure controls chemical behavior. By comparing air/liquid partition constants of model compounds interacting with solvents ranging from apolar (e.g., hexane, hexadecane) to bipolar (e.g., methanol, ethylene glycol), we can deepen our understanding of the molecular factors that govern partitioning processes involving bulk liquid organic phases. This will be of importance later in our discussions of partitioning processes involving natural organic phases, including natural organic matter (Chapter 9) as well as organic phases present in living organisms (Chapter 10).

n-octanol

At this point, we note that one particular organic solvent, *n-octanol*, is still widely used as a surrogate for many natural organic phases. The air–octanol partition constant (see below) and the octanol–water partition constant (Chapter 7) of a compound have been extremely popular parameters for relating partition coefficients involving natural organic phases by applying simple one-parameter LFERs (see examples given in Table 3.5). Hence, to put us in the position to critically analyze such LFERs, it is necessary that we learn more about the properties of this "famous" solvent.

We begin, however, our discussion of air/liquid phase partitioning by reiterating some general thermodynamic considerations that we will need throughout this chapter.

6.2   Thermodynamic Considerations

Raoult's Law

Assuming ideal gas behavior, the equilibrium partial pressure, p_i, of a compound above a liquid solution or liquid mixture is a direct measure of the fugacity, $f_{i\ell}$, of that compound in the liquid phase (see Fig. 3.9 and Eq. 3-33).

Note that we make a distinction between a solution and a mixture. When we talk of a solution, we imply that the organic solute is not a major component of the bulk liquid. Therefore, that presence of a dissolved organic compound does not have a significant impact on the properties of the bulk liquid. In contrast, in a mixture we recognize that the major components contribute substantially to the overall nature of the medium. This is reflected in macroscopic properties like air–liquid surface tensions and in molecule-scale phenomena like solubilities of trace constitutents.

In any case, we may write the equilibrium condition (Eq. 3-33):

$$f_{i\ell} = p_i = \gamma_{i\ell} \cdot x_{i\ell} \cdot p_{iL}^* \tag{6-1}$$

Let us now consider two special cases. In the first case, we assume that the compound of interest forms an *ideal* solution or mixture with the solvent or the liquid mixture, respectively. In assuming this, we are asserting that the chemical enjoys the same set of intermolecular interactions and freedoms that it has when it was "dissolved in a liquid of itself" (reference state). This means that $\gamma_{i\ell}$ is equal to 1, and, therefore, for any solution or mixture composition, the fugacity (or the partial pressure of the compound i above the liquid) is simply given by:

$$f_{i\ell} = p_i = x_{i\ell} \cdot p_{iL}^* \tag{6-2}$$

Eq. 6-2 is known as *Raoult's Law*.

In some cases of organic mixtures, we can apply Eq. 6-2 without too much inaccuracy; however, assuming that $\gamma_{i\ell}$ is equal to 1 can be quite inappropriate in many other cases (see Section 7.5).

Henry's Law and the Henry's Law Constant

In this chapter we will focus on another special case, that is, the case in which we assume that $\gamma_{i\ell}$ is different from 1 *but is constant over the concentration range considered*. This situation is primarily met when we are dealing with dilute solutions. As we have seen for the solvent water (Table 5.2), for many organic compounds of interest to us, γ_{iw} does not vary much with concentration, even up to saturated solutions. Hence, for our treatment of air–water partitioning, as well as for our examples of air/organic solvent partitioning at dilute conditions, we will assume that $\gamma_{i\ell}$ is constant. This allows us to modify Eq. 6-1 to a form known as *Henry's Law*:

$$f_{i\ell} = p_i = \gamma_{i\ell} \cdot p_{iL}^* \cdot x_{i\ell} = K_{iH}'(\ell) \cdot x_{i\ell} \tag{6-3}$$

$$K_{iH}'(\ell) = \frac{p_i}{x_{i\ell}} = \gamma_{i\ell} \cdot p_{iL}^* = \text{constant} \tag{6-4}$$

where $K'_{iH}(\ell)$ is the *Henry's Law Constant* of solute i for the solvent ℓ. Note that we use the superscript prime to indicate that the equilibrium partition constant is expressed on a partial pressure and mole fraction basis, as was originally done by Henry (Atkins, 1998). Hence when we give a numerical value for $K'_{iH}(\ell)$, we have to express it as fraction of the standard pressure (which is 1 bar).

In the environmental chemistry literature, it is common to refer to "the" Henry's Law constant of a given compound when the solvent in question is water. In the following, we will adopt this nomenclature and denote the air–water partition constant as defined by Eq. 6-4 simply as K'_{iH} (i.e., we omit to indicate the solvent)

Two other common ways of expressing air–liquid equilibrium partitioning are to use molar concentrations for i (i.e., $\mathrm{mol \cdot L^{-1}}$), either only in the liquid or in both the liquid and the gas phase. In the first case, we simply have to convert mole fractions to molar concentrations (Eq. 3-43):

$$K_{iH}(\ell) = \frac{p_i}{C_{i\ell}} = K'_{iH}(\ell) \cdot \overline{V_\ell} = \gamma_{i\ell} \cdot p^*_{iL} \cdot \overline{V_\ell} \tag{6-5}$$

where $\overline{V_\ell}$ is the molar volume of the bulk liquid (e.g., in $\mathrm{L \cdot mol^{-1}}$), and $K_{iH}(\ell)$ (no prime superscript) now has units like $(\mathrm{bar \cdot L \cdot mol^{-1}})$ or $(\mathrm{Pa \cdot L \cdot mol^{-1}})$. Note again that when pure water is the solvent, we denote $K_{iH}(\ell)$ simply as K_{iH}. Furthermore, we should point out that, particularly in the engineering literature but also in many handbooks, K_{iH} values are often given in units of $(\mathrm{Pa \cdot m^3 mol^{-1}})$. In this case, the liquid phase concentration is in units of $(\mathrm{mol \cdot m^{-3}})$. No matter in what units we express this parameter, it always reflects the same relative concentrations of the partitioning chemical in the gaseous and liquid phases.

For practical applications, such as for assessing the equilibrium distribution of a given compound in a multiphase system, it is most convenient to use a "dimensionless" air–solvent partition constant. This form uses molar concentrations in both phases. In this case, we denote the air–liquid partition constant as $K_{ia\ell}$. Since $C_{ia} = p_i / RT$ (Section 3.2), we then obtain:

$$K_{ia\ell} = \frac{C_{ia}}{C_{i\ell}} = K_{iH}(\ell) / RT = \gamma_{i\ell} \cdot \overline{V_\ell} \cdot p^*_{iL} / RT \tag{6-6}$$

If $K_{ia\ell}$ and p^*_{iL} are known for a given compound, that chemical's activity coefficient (and thus its excess free energy via Eq. 3-37) in the liquid phase can be calculated:

$$\gamma_{i\ell} = K_{ia\ell} \cdot \frac{RT}{\overline{V_\ell} \cdot p^*_{iL}} \tag{6-7}$$

Note that many of the activity coefficients of organic compounds in dilute aqueous solution, γ^∞_{iw}, that we used in Chapter 5 were derived from experimental air–water partition constants (K_{iaw}) using Eq. 6-7. Finally, we should point out that in the literature, similar to air–solid surface partitioning (Section 11.1), partition constants are quite often reported as the reciprocal quantity of the *air–solvent* partition constants as defined above, that is, as *solvent–air* partition constants. However, it does not really matter in what form such constants are given, as long as we pay careful attention to how they are defined, the temperature at which they are given and the units of concentration that are used.

Effect of Temperature on Air–Liquid Partitioning

Temperature influences air–bulk liquid partitioning of a compound i chiefly in two ways: (1) by its effect on the activity coefficient of the compound in the liquid phase and (2) by its effect on the compound's liquid vapor pressure. In the cases where Henry's law applies (Eq. 6-4), for a narrow temperature range, we may write the familiar relationship (Section 3.4):

$$\ln K'_{i\mathrm{H}}(\ell) = -\frac{\Delta_{a\ell} H_i}{R} \cdot \frac{1}{T} + \text{constant} \qquad (6\text{-}8)$$

where $\Delta_{a\ell} H_i$ is the standard enthalpy of transfer of i from the liquid to the gas phase. This enthalpy change is given by the difference between the excess enthalpy of the compound in the gas phase ($H_{ig}^{\mathrm{E}} \equiv \Delta_{vap} H_i$) and in the liquid phase ($H_{i\ell}^{\mathrm{E}}$):

$$\Delta_{a\ell} H_i = \Delta_{vap} H_i - H_{i\ell}^{\mathrm{E}} \qquad (6\text{-}9)$$

Hence, if no experimental value for $\Delta_{a\ell} H_i$ is available (i.e., from measurements of $K'_{i\mathrm{H}}(\ell)$ at different temperatures), it can be obtained from experimental (or estimated) $\Delta_{vap} H_i$ and $H_{i\ell}^{\mathrm{E}}$ values. Finally, we should note that Eq. 6-8 applies in a strict sense only if we express the amount of the compound in the gas and liquid phase as partial pressure and mole fraction, respectively. However, if we assume that the molar volume of the liquid, $\overline{V}_\ell$, is not significantly affected by temperature changes, we may also apply Eq. 6-8 to describe the temperature dependence of $K_{i\mathrm{H}}(\ell)$ (Eq. 6-5) with a constant term that is given by "constant $+ \ln \overline{V}_\ell$." Furthermore, if we express the amount of the compound in the gas phase in molar concentrations (Eq. 6-6), then we have to add the term RT_{av} to $\Delta_{a\ell} H_i$ where T_{av} (in K) is the average temperature of the temperature range considered (see Section 3.4):

$$\ln K_{ia\ell} = -\frac{\Delta_{a\ell} H_i + RT_{av}}{R} \cdot \frac{1}{T} + \text{constant} \qquad (6\text{-}10)$$

6.3 Air–Organic Solvent Partitioning

Air–Organic Solvent and Other Partition Constants

Now we turn our attention to the equilibrium constants, $K_{ia\ell}$, that quantify the partitioning of organic compounds between air and various liquids of very diverse solvent characteristics. As discussed in Chapter 3, the air–liquid partitioning behaviors of organic compounds differ greatly when comparing an apolar organic solvent like n-hexadecane with the polar solvent water (Fig. 3.6). This stems from the differences in *solute:solvent and solvent:solvent* molecular interactions, and we have introduced a mathematical model to sum the effects of these interaction energies (Eqs. 4-26 and 5-21).

Since thermodynamic properties are independent of the reaction pathway and only depend on the starting and ending conditions, we know that the partitioning of organic chemicals between air and liquids can be related to other equilibrium processes (Fig. 6.1). For example, there is a direct relationship between a chemical's air–organic solvent partition constant ($K_{ia\ell}$), its partitioning between air and water (K_{iaw}), and its organic solvent–water partition constant ($K_{i\ell w}$), provided that we consider only the water-saturated (i.e. "wet") organic phase:

$$K_{ia\ell}\left(=\frac{C_{ia}}{C_{i\ell}}\right) = \frac{K_{iaw}(=C_{ia}/C_{iw})}{K_{i\ell w}(=C_{i\ell}/C_{iw})} \qquad (6\text{-}11)$$

This brings up an important question. To what extent is the $K_{ia\ell}$ value that has been determined experimentally using the "dry" organic solvents different from the value calculated from the air–water and organic solvent–water partition constants (Eq. 6-11)? Several methods yield such equilibrium constants, including use of head space analysis (Park et al., 1987), chromatographic techniques (Gruber et al., 1997), generator columns (Harner and Mackay, 1995; Harner and Bidleman, 1998a), or a fugacity meter (Kömp and McLachlan, 1997a). In order to answer this question, we have to evaluate how the cross "contaminations" of the organic solvent with water and of the water with organic solvent affect the activity coefficients of the compound of interest in both the organic and aqueous phases. Also, these solvent modifications may affect the molar volumes of the liquids. For pure nonpolar organic solvents that are only sparingly soluble in water and contain only very little water at saturation (e.g., $x_L > 0.99$, Table 5.1), we may justifiably neglect such molar volume effects. However, for more polar solvents the situation is not so clear. For example, if we use n-octanol as the organic solvent, there will be roughly one water molecule for every four octanol molecules in the organic phase at equilibrium with water ($x_L \cong 0.8$, Table 5.1). This means that the molar volume of "dry" octanol ($\overline{V}_\ell = 0.16$ L·mol^{-1}) is about 20% larger than that of "wet" octanol ($\overline{V}_\ell \cong 0.13$ L·mol^{-1}). In contrast, there will be only about one octanol molecule for every 10,000 water molecules in octanol-saturated water. This has no significant impact on the molar volume of the aqueous phase.

Furthermore, for most compounds of interest to us, the octanol molecules present as cosolutes in the aqueous phase will have only a minor effect on the other organic compounds' activity coefficients. Also, the activity coefficients of a series of apolar, monopolar, and bipolar compounds in wet versus dry octanol shows that, in most cases, $\gamma_{i\ell}$ values changes by less than a factor of 2 to 3 when water is present in wet octanol (Dallas and Carr, 1992; Sherman et al., 1996; Kömp and McLachlan, 1997a). Hence, as a first approximation, for nonpolar solvents, for n-octanol, and possibly for other solvents exhibiting polar groups, we may use Eq. 6-11 as a first approximation to estimate air–"dry" organic solvent partition constants for organic compounds as illustrated in Fig. 6.2. Conversely, experimental $K_{ia\ell}$ data may be used to estimate K_{iaw} or $K_{i\ell w}$, if one or the other of these two constants is known.

Comparison of Different Organic Solvents

Let us now evaluate how the $K_{ia\ell}$ values of different compounds are affected by the chemical nature of the organic solvent. To this end, we consider a set of five model compounds (Fig. 6.3) exhibiting very different structures that enable them to

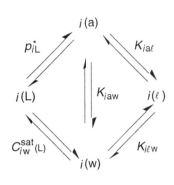

Figure 6.1 Partitioning processes for a chemical, i, considered in Chapters 3 to 7. a = air (gas phase), ℓ = water-immiscible organic solvent, L = pure liquid organic compound, and w = aqueous phase.

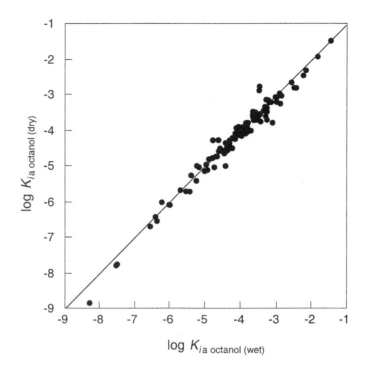

Figure 6.2 Experimentally determined air–"dry octanol" partition constants versus calculated (Eq. 6-11) air–"wet octanol" partition constants. Data from Harner and Mackay (1995), Gruber et al. (1997), Harner and Bidleman (1998a,b), Abraham et al. (2001).

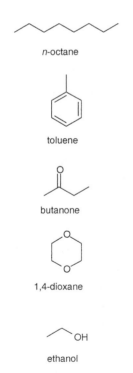

Figure 6.3 Structures of the model compounds used for evaluating partitioning in various air–solvent systems.

participate to varying degrees in dispersive, dipole:dipole, dipole:induced dipole, H-donor, and H-acceptor interactions (Tables 4.3 and 5.4). For these compounds, the $K_{ia\ell}$ values have been determined for six organic solvents that differ quite substantially in "polarity" (Table 6.1; Park et al., 1987). Note that the parameters π_i, α_i, and β_i used to characterize *solutes* cannot be directly used to describe *solvent* properties. Nevertheless, they provide a qualitative measure of the interactions that molecules of a given organic solvent may undergo. Thus, the solvent hexadecane and toluene can be classified as apolar and weakly monopolar, respectively. Dichloromethane represents a solvent with some H-donor character, in addition to participating in dispersive and polar/polarizable interactions. And *n*-octanol, methanol, and ethylene glycol are H-donor as well as H-acceptor solvents of increasing polarity. Also note that these solvents exhibit quite different molar volumes (e.g., $\overline{V}_{\text{hexadecane}} = 0.293$ L·mol^{-1}, $\overline{V}_{\text{methanol}} = 0.040$ L·mol^{-1}). This has a significant influence on the absolute values of the corresponding $K_{ia\ell}$'s of a given compound in the various solvents (Eq. 6-6). Nevertheless, for visualizing the effects of solvation of the compounds in the various solvents, we can use these air–solvent partition constants instead of the corresponding $K_{iH}(\ell)$ values (Eq. 6-4), which would be directly related to the $\Delta_{a\ell}G_i$ values.

In the air–organic solvent combinations considered in Table 6.1, all of these compounds partition favorably into the organic phase (i.e., $K_{ia\ell} \ll 1$). This is even true for the partitioning of the bipolar chemical, ethanol, into the apolar solvent, hexadecane. Even more importantly, it is true for the apolar solute, *n*-octane, dissolving into the highly bipolar solvent, ethylene glycol. Note that this glycol derivative represents one of the most "water-like" organic solvents (Table 5.8). The latter finding illustrates again the unique properties of the solvent water, since in water, the activity coefficient of *n*-octane is about 10^7 (Chapter 5) as compared to

Table 6.1 Measured Air–Solvent Partition Constants ($K_{ia\ell}$) and Calculated Activity Coefficients ($\gamma_{i\ell}$, Eq. 6-7) at 25°C of Five Model Compounds Exhibiting Different H-Donor, H-Acceptor, and Polarity/Polarizability Properties for Some Organic Solvents

| Compound i[b] | log p_{iL}^*/Pa | Solvent ℓ: log $K_{ia\ell}$ ($\gamma_{i\ell}$ in parentheses)[a] | | | | | |
		Hexadecane (apolar)	Toluene (monopolar, H–A)	Dichloromethane (bipolar, primarily H-D)	n-Octanol (bipolar)	Methanol (bipolar)	Ethyleneglycol (bipolar)
n-Octane (apolar)	3.27	−3.7 (0.92)	−3.9 (1.7)	−3.7 (4.1)	−3.4 (3.2)	−2.9 (46)	−1.1 (1800)
Toluene (monopolar, H–A)	3.58	−3.4 (0.96)	−3.8 (1.0)	−4.0 (0.96)	−3.3 (2.0)	−3.2 (10)	−2.4 (52)
Butanone (monopolar, H–A)	4.10	−2.3 (3.2)	−3.1 (1.4)	−3.9 (0.44)	−2.8 (2.1)	−3.3 (2.5)	−2.6 (8.4)
1,4-Dioxane (monopolar, H–A)	3.69	−2.9 (2.4)	−3.6 (1.2)	−4.3 (0.42)	−3.2 (2.1)	−3.6 (3.4)	−3.3 (4.9)
Ethanol (bipolar)	3.90	−1.5 (35)	−2.3 (15)	−2.7 (9.2)	−3.2 (1.1)	−3.9 (1.01)	−3.5 (1.9)

[a] Data from Park et al. (1987). [b] The structures are given in Fig. 6.3.

only 1800 in ethylene glycol. In general, with few exceptions (e.g., ethanol), we may assume that the activity coefficient of most organic compounds in an organic solvent will be much smaller than their γ_{iw} values. In many cases, $\gamma_{i\ell}$ is less than 100 or even less than 10. Consequently, compounds with very small liquid–vapor pressures will also exhibit very small $K_{ia\ell}$ values. This is, of course, not a surprising result because p_{iL}^* is itself an air–organic solvent partition constant. Hence, like vapor pressure, the air–organic solvent partition constants may vary by many orders of magnitude within a compound class. This contrasts the air–water partition constants for the same sets of compounds. Such chemically related groups of compounds commonly have K_{iaw} values that span a much more narrow range (Section 6.4). Finally, we should note that these findings also indicate that, in the environment, we may anticipate that most of the chemicals of interest to us will partition favorably from the air into condensed natural organic phases (see Chapters 10 and 11).

LFERs Relating Partition Constants in Different Air–Solvent Systems

Another important lesson that we can learn from the data presented in Table 6.1 is that the activity coefficient of an organic compound in an organic solvent depends strongly on the prospective involvements of both the partitioning compound *and* the solvent for dispersive, dipolar, H-donor, and H-acceptor intermolecular interactions. This implies that we may need to represent the properties of both the solute and the solvent when we seek to correlate air–liquid partition constants of structurally diverse substances. Thus, if the types of intermolecular interactions of a variety of solutes interacting with two chemically distinct solvents 1 and 2 are very different, a one-parameter LFER for all compounds, *i,* of the form:

$$\log K_{ia1} = a \cdot \log K_{ia2} + b \qquad (6\text{-}12)$$

is inadequate to correlate partition constants (Fig. 6.4). For example, hexadecane interacts only via vdW forces with all partitioning substances. Thus, solute interactions with this hydrocarbon and the corresponding $K_{ia\ell}$ values will reflect only these energies. Another solvent like octanol may, however, participate in various combinations of dispersive, polar, H-acceptor, and H-donor interactions with solutes of diverse structures. Thus the $K_{ia\ell}$ values for octanol may involve a sum of effects. These sums of intermolecular attractions may not correlate with the vdW-alone interactions that hexadecane can offer.

Recalling our earlier discussions of one-parameter LFERs (Section 5.3), we should be able to predict when we can anticipate that Eq. 6-12 is applicable to a given set of compounds and solvents. Obviously, if we consider two apolar solvents (e.g., cyclohexane and hexadecane) where chiefly dispersive interactions predominate between these solvents and all solute molecules, then we can expect to find an LFER encompassing apolar, monopolar, and bipolar compounds (e.g., Fig. 6.5). Further-more, we can also anticipate success developing an LFER when combining different types of compounds partitioning into two closely related polar solvents (e.g., methanol and ethanol). In this case, we can assume that the contributions to the excess free energies of solution in both solvents are due to very similar polar interactions in the two solvents (e.g., Fig. 6.6). Finally, if two solvents are considered that exhibit rather different abilities to interact through polar mechanisms,

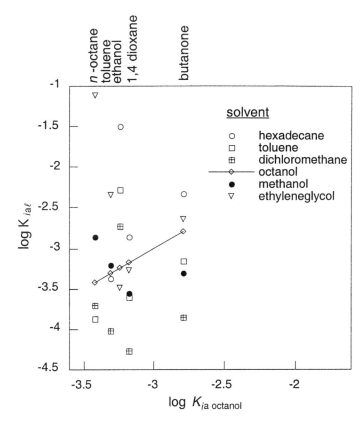

Figure 6.4 Plots of log $K_{ia\ell}$ versus log $K_{ia\,octanol}$ for the model compounds and solvents listed in Table 6.1. Data for 25°C. Note the different scales on the x- and y-axes.

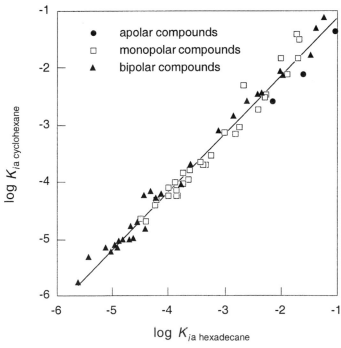

Figure 6.5 Plot of the decadic logarithms of the air–cyclohexane versus the air–hexadecane partition constants of a series of apolar, monopolar, and bipolar compounds. Data from Abraham et al. (1994b).

we can expect LFERs to hold only for strictly apolar compounds or for closely related sets of polar compounds. Let us, for example, consider air–olive oil partitioning versus air–octanol partitioning of a range of compounds. The polar groups in olive

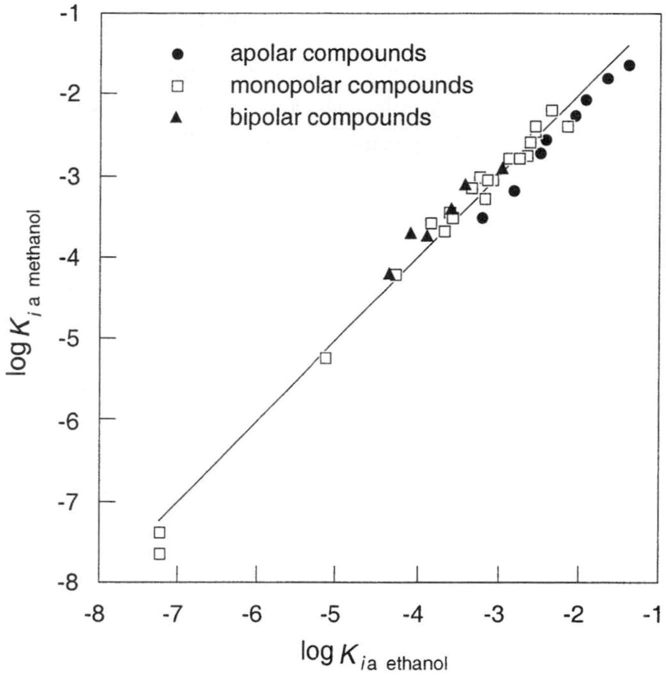

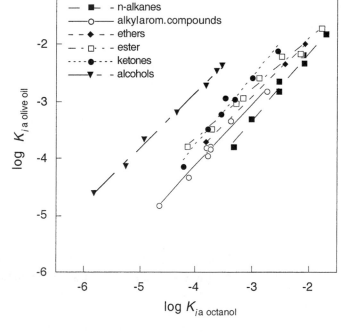

Figure 6.6 Plots of the decadic logarithms of the air–methanol versus the air–ethanol partition constants of a series of apolar, monopolar, and bipolar com-pounds. Data from Tiegs et al. (1986) and Abraham et al. (1998 and 1999).

Figure 6.7 Plot of the decadic logarithms of the air–olive oil partition coefficients versus the air–octanol partition constants for various sets of structurally related apolar, monopolar, and bipolar compounds. Note that olive oil is a mixture of compounds that may vary in composition. Therefore, we refer to $K_{ia\,olive\,oil}$ as the air–olive oil partition *coefficient* (and not constant, see Box 3.2). Adapted from Goss and Schwarzenbach (2001). The a and b values for the LFERs (Eq. 6-12) are: alkanes ($a = 1.15$, $b = 0.16$), alkyl aromatic compounds ($a = 1.08$, $b = 0.22$), ethers ($a = 0.97$, $b = 0.01$), esters ($a = 0.88$, $b = -0.14$), ketones ($a = 1.21$, $b = 1.06$), alcohols ($a = 0.98$, $b = 1.07$).

CH₂—O—COR₁
CH₂—O—COR₂
CH₂—O—COR₃

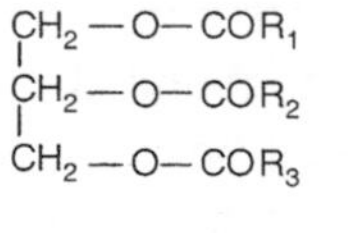

General structure of olive oil, R_1, R_2, R_3 = C_{14}, C_{16}, C_{18} saturated or unsaturated (for details see Hui, 1996).

oil are monopolar caboxylic acid esters (see margin), while *n*-octanol is a bipolar solvent. As demonstrated in Fig. 6.7, good LFERs are found for sets of compounds involving homologues (i.e., compounds differing only by the number of –CH₂– units) or families of compounds for which the polar properties change proportionally with size (e.g., PAHs; see also Section 5.3). Of course, we may be able to combine various sets of compounds that are not too different in polarity into one LFER (e.g., the ethers and esters in Fig. 6.7) with only limited loss in precision.

Model for Description of Air–Solvent Partitioning

We previously used our insights regarding the solute:water and water:water intermolecular interactions to assemble a mathematical model for estimating a compound's aqueous activity coefficient (Section 5.3, Eqs. 5-19 to 5-22). Now we can easily modify this model for the prediction of air–organic solvent partitioning. First, since $K_{ia\ell}$ is proportional to the product, $\gamma_{i\ell} \cdot p_{iL}^*$ (Eq. 6-6), we can remove the $-\ln p_{iL}^*$ term in Eq. 5-21 (which reflects the free energy of transfer from the pure liquid to the gas phase). Next, we do not need to include a specific volume term. This was previously included to account for the large entropy costs associated with inserting an organic solute into bulk water (i.e., forming the solute cavity). In organic solvents the free energy costs for creating a cavity are much smaller than in water, and they are not a dominating contribution to the overall $\Delta_{a\ell}G_i$. Furthermore, the cavity term is proportional to the size of the molecule and, therefore, correlates with the dispersive energy term. Hence, for organic solvents, by analogy to Eq. 5-21, we may express $\log K_{ia\ell}$ as:

$$\ln K_{ia\ell} = s\left[V_{ix}^{2/3}\left(\frac{n_{Di}^2 - 1}{n_{Di}^2 + 2}\right)\right] + p(\pi_i) + a(\alpha_i) + b(\beta_i) + \text{constant} \qquad (6\text{-}13)$$

Note that V_{ix} is in cm³ mol⁻¹.

As is illustrated in Fig. 6.8 for the air–olive oil system, this multiparameter LFER Eq. 6-13 is able to fit the experimental $K_{ia\ell}$ data quite satisfactorily.

The set of coefficients (s, p, a, b, constant) obtained from fitting experimental $K_{ia\ell}$ values for olive oil, as well as for some other organic solvents, are summarized in Table 6.2. These constants clearly quantify the importance of the individual intermolecular interactions for each solvent. For example, n-hexadecane has nonzero s and p coefficients, representing this solvent's ability to interact via dispersive and polarizability mechanisms. But the a and b coefficients are zero, consistent with our expectation from hexadecane's structure that hydrogen bonding is impossible for this hydrocarbon. At the other extreme in "polarity," methanol has nonzero coefficients for all of the terms, demonstrating this solvent's capability to interact via all mechanisms.

Indeed, we can use the coefficient values to directly see how chemical structures enable specific kinds of intermolecular interactions. For example, focusing on the a values in Table 6.2, we can contrast the relative importance of H-bonding accepting for these different liquids. We are probably not surprised to see that the two alcohols, octanol and methanol, are the most effective (as indicated by a values between -8 an -9) at donating their oxygen atom's nonbonded electrons to an H-donor partner. We may also expect that olive oil (contains $-C(=O)O-$ as part of structure) and acetonitrile (CH_3CN) may be able to donate nonbonded electrons from their oxygen and nitrogen atoms, respectively, to hydrogen-bonding partners. Hence, we anticipate these liquids will have nonzero a coefficients, and the "best-fit" values show this is true but that they hydrogen-bond less effectively than the two alcohols. Note that benzene and trichloromethane have nonzero a coefficients. Although the a values are small compared to those of the alcohols, their significant

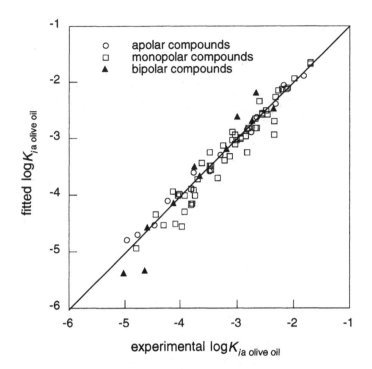

Figure 6.8 Fitted (Eq. 6-13) versus experimental air–olive oil partition coefficients for a series of compounds including those in Fig. 6.7. Note that some of the relatively large scatter in the data may be due to the fact that olive oils from different origins may differ in composition.

difference from zero in the fitting of $K_{ia\ell}$ values means that the π-electrons of the benzene ring can be donated somewhat to a hydrogen donor compound (or other electron-deficient positions of substances). Likewise, the nonbonded electrons of trichloromethane's chlorine atoms must be somewhat available to share with electron-deficient moieties, although, when inspecting the b coefficient, this compound is a much stronger electron acceptor (H-donor). Similar "structure-activity" interpretations can be made for all the other LFER parameters in Table 6.2.

In summary, multiparameter LFERs such as the ones given for some organic solvents in Table 6.2 are very useful in many respects. First, they allow one to get an estimate of the $K_{ia\ell}$ value of a given compound for a given solvent, provided that the compound's π_i, α_i, and β_i values are known. Second, such LFERs characterize a given solvent with respect to its ability to host different apolar, monopolar, and/or bipolar solutes (see above). This may help us anticipate where organic chemicals will accumulate. Next, we can use such multiparameter LFER information to rationalize when a simple one-parameter LFER (Eq. 6-12) should be appropriate. For example, we can see now that an LFER between air–olive oil partition coefficients and air–n-octanol partition constants can be expected to hold only for confined sets of compounds, and not for the universe of chemicals (Fig. 6.7). Another result that we could have anticipated from such data sets is the existence of single LFERs for the solvent system methanol/ethanol (Fig. 6.6). Finally, comparison of the a and b values permits us to attach quantitative reasoning to our (sometimes incorrect) intuitive reasoning regarding the interactions of chemicals with one another. For example, by comparing the relevant a and b coefficients of the alcohols and water, we now know that water is a much stronger H-acceptor, but that all these solvents are similar in their ability to act as H-donors toward organic solutes.

Table 6.2 Air–Organic Solvent Partitioning: Multiparameter LFERs (Eq. 6-13) for Some Organic Solvents at 25°C[a]

Solvent	Coefficient for Parameter					$\ln K_{ia\ell}$ range[b]	R^2	n[c]
	vdW s	(π_i) p	(α_i) a	(β_i) b	constant			
n-Hexadecane (apolar)	-1.67	-1.49	0	0	2.01	1.8 to 12.6	0.98	302
Benzene (monopolar)	-1.38	-3.47	-1.45	0	0.26	4.4 to 13.6	0.96	36
Olive oil (monopolar)	-1.74	-2.83	-4.47	0	2.86	-3.9 to 11.6	0.94	79
Trichloromethane (bipolar)[d]	-1.47	-3.41	-0.96	-4.21	1.27	-3.5 to 18.8	0.95	108
Acetonitrile (bipolar)	-0.49	-2.14	-2.30	-0.82	0.46	-1.3 to 5.7	0.98	42
n-Octanol (bipolar)	-1.47	-2.60	-8.32	-3.71	2.42	2.6 to 19.6	0.97	254
Methanol (bipolar)	-1.26	-3.41	-8.92	-3.90	1.57	1.7 to 14.4	0.97	65

[a] Data from Maher and Smith (1979), Ocampo and Klinger (1983), Srivastava et al. (1986), Tiegs et al. (1986), Abraham et al. (1987), Abraham et al. (1994a and b), Abraham et al. (1998), Abraham et al. (1999). [b] Range of experimental $K_{ia\ell}$ values that were used to establish the LFER. [c] Number of compounds. [d] Primarily H-donor (electron acceptor).

Temperature Dependence of Air–Organic Solvent Partition Constants

As indicated by Eqs. 6-8 to 6-10, the temperature dependence of $K_{ia\ell}$ is determined by the corresponding $\Delta_{a\ell}H_i$. This enthalpy is given by the difference between the enthalpy of vaporization ($\Delta_{vap}H_i$) and the excess enthalpy of the compound in the organic phase ($H_{i\ell}^E$). For most organic solvents and compounds, we may assume that $H_{i\ell}^E$ is much smaller than $\Delta_{vap}H_i$. For example, $H_{i\ell}^E$ is less than a tenth of $\Delta_{vap}H_i$ in the case of hexadecane (Abraham et al., 1990) and n-octanol as solvents (Harner and Mackay, 1995; Harner and Bidleman, 1996; Gruber et al., 1997). Hence, as a first approximation, we may use $\Delta_{vap}H_i$ to assess the effect of temperature on $K_{ia\ell}$ (i.e., $\Delta_{a\ell}H_i \cong \Delta_{vap}H_i$). This means that, like vapor pressure (Chapter 4), $K_{ia\ell}$ values are strongly temperature dependent. Finally, we should recall from Section 4.4 (Eq. 4-29) that we may estimate $\Delta_{vap}H_i$ from the liquid–vapor pressure of the compound.

Applications

We conclude this section with a few comments on the practical importance of considering air (or gas)–organic solvent partitioning. First, knowledge of the respective $K_{ia\ell}$ value(s) is, of course, required to assess how much a given organic liquid (e.g., olive oil) will tend to become "contaminated" by organic chemicals present in the air around it. This problem might be of interest in our private and/or professional lives (see Illustrative Example 6.1). Second, when analyzing organic compounds by gas chromatography, it is of great importance to know how specific compounds partition between the gaseous mobile phase (i.e., H_2, He, N_2) and the stationary phase. This latter phase is commonly a liquid organic coating at the inner surface of a glass or silica capillary column. In fact, for choosing the appropriate stationary phase (e.g., polar versus nonpolar) for the separation of a given group of compounds, it is necessary to understand the molecular factors that determine the activity coefficients of the compounds in various stationary phases. This information can be gained from analyzing $K_{ia\ell}$ values of the compounds for different solvents.

Furthermore, air–organic solvent partition constants, in particular the air–octanol partition constant, are widely used to evaluate and/or predict the partitioning of organic compounds between air and natural organic phases. Such organic phases are present, for example, in aerosols or soils (Chapters 9 and 11) or as part of biological systems (Chapter 10).

Finally, the relationships between the air–organic solvent, the air–water, and the organic solvent–water partition constants of a given compound (Eq. 6-11) will make it very easy to understand organic solvent–water partitioning, which we will treat in Chapter 7.

Illustrative Example 6.1

Assessing the Contamination of Organic Liquids by Air Pollutants

Problem

You live in a town where air pollution caused primarily by traffic is quite substantial. From a recent article in the local newspaper you have learned that the benzene concentration in the air in your area may reach up to 10 parts per billion on a volume base (i.e., 10 ppbv). You wonder to what extent the olive oil that you use for your salad, and that you have left in an open bottle on the table on your balcony, is contaminated with this rather toxic compound. Calculate the maximum concentration of benzene in the olive oil assuming an average temperature of 25°C and a total pressure of 1 bar. Use the ideal gas law to convert ppbv to molar concentrations.

i = benzene

Answer

With 10 ppbv the partial pressure of benzene in the air is $p_i = 10 \times 10^{-9}$ bar $= 10^{-8}$ bar, which corresponds to a concentration of:

$$C_{ia} = \frac{p_i}{RT} = \frac{10^{-8}}{(0.0831)(298)} = 4.0 \times 10^{-10} \text{ mol} \cdot \text{L}^{-1} = 0.03 \,\mu g \cdot \text{L}^{-1}$$

For estimating the air–olive oil partition coefficient, calculate first the air–octanol partition constant from the air–water (K_{iaw}) and octanol–water (K_{iow}) partition constants given in Appendix C (Eq. 6-11):

$$K_{iao} = \frac{K_{iaw}}{K_{iow}} = \frac{10^{-0.65}}{10^{+2.17}} = 10^{-2.82}$$

Use the LFER shown in Fig. 6-7 for alkyl aromatic compounds ($\log K_{ia \text{ olive oil}} = 1.08 \log K_{iao} + 0.22$) to estimate the air–olive oil partition coefficient:

$$\log K_{ia \text{ olive oil}} = (1.08)(-2.82) + 0.22 = -2.83$$

An alternative way of estimating the air–olive oil partition coefficient is to apply the LFER Eq. 6-13 using the constants given in Table 6.2 for the air–olive oil system:

$$\ln K_{ia \text{ olive oil}} = -1.74\left[V_{ix}^{2/3}\left[\frac{n_{D_i}^2 - 1}{n_{D_i}^2 + 2}\right]\right] - 2.83(\pi_i) - 4.47(\alpha_i) + 2.86$$

The corresponding parameters for benzene are: $V_{ix} = 71.6$ cm^3 mol^{-1} (Box 5.2), $n_{Di} = 1.50$ (Table 3.1), $\pi_i = 0.52$ (Table 5.4), $\alpha_i = 0$ (Table 4.3). Insertion of these values in the above equation yields:

$$\ln K_{ia \text{ olive oil}} = (-1.74)(5.07) - (2.83)(0.52) - (4.47)(0) + 2.86 = -7.44$$

or:

$$\log K_{ia \text{ olive oil}} = -3.23$$

Hence, both estimates yield a $K_{ia \text{ olive oil}}$ value for benzene of about 10^{-3}, and thus a maximum benzene concentration in the olive oil of $10^3 \, C_{ia} = 30 \,\mu g \cdot \text{L}^{-1}$. Considering that the drinking water standard for benzene is 5 $\mu g \cdot L^{-1}$ this concentration should, therefore, not create a serious problem for your health assuming that you do not consume tremendous amounts of olive oil each day.

Problem

In your laboratory refrigerator (5°C) you store pure cyclohexane that you use for extracting organic trace contaminants from water samples for subsequent analysis by gas chromatography. Among the compounds of interest is tetrachloroethene (also called perchloroethene or PCE). One day you realize that somebody is using tetrachloroethene in the laboratory. In fact, you can even smell the compound in the air (odor threshold values: $0.03 - 0.1$ mg$\cdot L^{-1}$). You are worried that your cyclohexane is "contaminated," particularly, because you have realized that the bottle was not well sealed in the refrigerator. Calculate the concentration of PCE in the air that, at 5°C, would be sufficient to "produce" an equilibrium PCE concentration in the cyclohexane of 1 $\mu g \cdot \mu L^{-1}$, which you would consider to be a problem for your analysis.

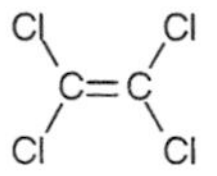

i = tetrachloroethene
(PCE)

Answer

Use the air–n-hexadecane partition constant of PCE ($K_{ia\,\text{hexadecane}} = 2.5 \times 10^{-4}$ at 25°C; Abraham et al., 1994a) as surrogate for the air–cyclohexane partition constant of PCE (Fig. 6.5). Furthermore, for determining the temperature dependence of $K_{ia\,\text{hexadecane}}$ assume that $\Delta_{a\,\text{hexadecane}} H + RT_{av} \cong \Delta_{\text{vap}} H_i + RT_{av}$ (Section 3.4). For PCE, this value is about 40 kJ $\cdot$ mol^{-1} (Lide, 1995). Hence, at 5°C, the $K_{ia\,\text{hexadecane}}$ value is about 0.3 times the value at 25°C (Table 3.5); that is, $K_{ia\,\text{hexadecane}} \cong 7.5 \times 10^{-5}$. This means that the PCE concentration in the air required to produce a concentration in the cyclohexane of 1 $\mu g \cdot mL^{-1}$ or 1 mg$\cdot L^{-1}$ is:

$$C_{ia} = (7.5 \times 10^{-5})\,(1)\ \text{mg} \cdot L^{-1} = 0.000075\ \text{mg} \cdot L^{-1}$$

which is about 400 times lower than the odor threshold. Thus, your cyclohexane is in great danger of getting contaminated by the PCE in the air.

6.4   Air–Water Partitioning

"The" Henry's Law Constant

For our discussion of air–water partitioning, we start by rewriting Eq. 6-5 for water as the solvent (Eq. 6-6):

$$K_{iH} = \frac{p_i}{C_{iw}} = \gamma_{iw} \cdot p_{iL}^* \cdot \overline{V}_w \tag{6-14}$$

Recall that K_{iH} is commonly referred to in environmental literature as "the" Henry's law constant. The "dimensionless" Henry constant is denoted K_{iaw} and is related to K_{iH} by (Eq. 6-6):

$$K_{iaw} = \frac{K_{iH}}{RT} \tag{6-15}$$

Inspection of Eq. 6-14 reveals that we do not need to learn anything new to understand air–water equilibrium partitioning of neutral organic compounds. All we have to do is to recall how chemical structures (controlling intermolecular interactions) and environmental factors (e.g., temperature, presence of salts or organic cosolvents in the aqueous phase) affect the vapor pressure and the aqueous activity coefficient of a given compound. Hence, our discussion of air–water partitioning can be quite brief.

First, consider how structural moieties affect the Henry's Law constant. We can see that within a class of apolar or weakly polar compounds (e.g., n-alkanes, chlorinated benzenes, alkylbenzenes, PCBs, PAHs), the K_{iaw} values vary by less than one order of magnitude (see data in Appendix C.) This is also true for sets of compounds that differ only by apolar moieties (e.g., polyalkyl- or polychlorophenols). This is in contrast to vapor pressure and aqueous solubility data for the same families of compounds. These latter properties vary by five or more orders of magnitude within any one group of those compounds. We can rationalize these findings by recalling that an increase in size of the compound leads to an increase in γ_{iw} (or a decrease in water solubility), as well as to a decrease in p_{iL}^*. Hence, the effect of molecular size is canceled out to a large degree when multiplying γ_{iw} with p_{iL}^* (Eq. 6-14).

toluene

$K_{iaw}(25°C) = 2.5 \times 10^{-1}$

OH

phenol

$K_{iaw}(25°C) = 2.5 \times 10^{-5}$

However, as is illustrated by the two substituted benzenes, toluene and phenol (see margin), the presence of a polar group has a tremendous effect on K_{iaw}. Replacing an apolar moiety with a bipolar hydrogen-bonding one leads to a decrease in both γ_{iw} (it increases C_{iw}^{sat}) and p_{iL}^*. Thus, K_{iaw} values differ widely between apolar and bipolar derivatives.

We may also recall that for most compounds of interest, we can assume that γ_{iw} is more or less independent of concentration (Section 5.2). Hence, in Eq. 6-14, we may substitute γ_{iw} by γ_{iw}^{sat}. This activity coefficient, in turn, can be expressed by the liquid aqueous solubility of the compound (i.e., $\gamma_{iw}^{sat} = 1/(C_{iw}^{sat}(L) \cdot \overline{V}_w)$; Eq. 5-12). Using this relation, we then obtain:

$$K_{iH} \cong \gamma_{iw}^{sat} \cdot p_{iL}^* \cdot \overline{V}_w \cong \frac{p_{iL}^*}{C_{iw}^{sat}(L)} \tag{6-16}$$

For solid compounds we may also write:

$$K_{iH} \cong \frac{p_{is}^*}{C_{iw}^{sat}(s)} \tag{6-17}$$

because the free energy term relating the liquid and solid vapor pressure and the liquid and solid aqueous solubility, respectively (Eqs. 4-15 and 5-13), cancels when dividing the two entities. From a practical point of view, Eqs. 6-16 and 6-17 are very interesting, because they tell us that we may estimate the Henry's law constant of a compound directly from its vapor pressure and its aqueous solubility. In fact, many of the K_{iH} or K_{iaw} values listed in data compilations (including the data given in Appendix C) have been derived in this way. Comparison of calculated with experimental K_{iaw} values (compare values given in parentheses in Appendix C, or see article by Brennan et al., 1998) shows that, in most cases, Eqs. 6-16 and 6-17 yield very satisfactory estimates (less than a factor of 2 deviation).

Effect of Temperature on Air–Water Partitioning

As indicated by Eq. 6-9, the standard enthalpy of transfer of a compound i from water to air is given by:

$$\Delta_{aw}H_i = \Delta_{vap}H_i - H_{iw}^E \qquad (6\text{-}18)$$

Typically, for many smaller organic molecules, H_{iw}^E is rather small (i.e., $|H_{iw}^E| <$ 10 kJ $\cdot$mol^{-1}; Table 5.3). As a result, for such small compounds, similar to the situation encountered in air–organic solvent partitioning, $\Delta_{aw}H_i$ will not be very different from the enthalpy of vaporization of the compound (Table 6.3), and therefore, the effect of temperature on air–water partitioning, will, in general, be significant (see Illustrative Example 6.2).

There are, however, also many cases in which $\Delta_{aw}H_i$ differs quite substantially from $\Delta_{vap}H_i$. Due to their relatively high positive H_{iw}^E values (see Table 5.3), large, apolar compounds exhibit a significantly smaller $\Delta_{aw}H_i$ as compared to $\Delta_{vap}H_i$ (see examples given in Table 6.3). Nevertheless, even in these cases, $\Delta_{aw}H_i$ is still quite large, so that the effect of temperature on K_{iaw} cannot be neglected.

For monopolar compounds (e.g., ethers, ketones, aldehydes), $\Delta_{aw}H_i$ may even be larger than $\Delta_{vap}H_i$. This happens because of the additional polar interactions in the aqueous phase, leading to negative H_{iw}^E values (Table 5.3).

At this point we should note that it is not a trivial task to measure accurately $\Delta_{aw}H_i$ values. This is particularly true for very hydrophobic compounds. Therefore, it is also not too surprising that experimentally determined $\Delta_{aw}H_i$ values reported by different authors may differ substantially (see examples given in Table 6.3). Furthermore, particularly for many very hydrophobic compounds, there seems to be a discrepancy between $\Delta_{aw}H_i$ values derived from measurements of K_{iaw} at different temperatures (Eq. 6-10) under *dilute* conditions, and $\Delta_{aw}H_i$ values calculated from the enthalpy of vaporization and the enthalpy of solution ($\Delta_{wL}H_i = H_{iw}^E$; see Fig. 5.1; note that $\Delta_{wa}H_i = -\Delta_{aw}H_i$). Note that this latter approach reflects *saturated* conditions. Nevertheless, before using an experimentally determined $\Delta_{aw}H_i$ value, it is advisable to check this value for consistency with that calculated from $\Delta_{aw}H_i$ and H_{iw}^E.

Effect of Solution Composition on Air–Water Partitioning

To evaluate the effects of salts or organic cosolvents on air–water (or more correctly, air–aqueous phase or air–organic solvent/water mixture) partitioning, we may simply apply the approaches discussed in Section 5.4 (Eqs. 5-27 and 5-29). Thus, knowing how salt affects a compound's aqueous solubility, while having no effect on its saturation vapor pressure, we deduce that the impact of salt on K_{iaw} may be expressed by:

$$K_{iaw,saltwater} = K_{iaw} \cdot 10^{+K_i^s [\text{salt}]_{tot}} \qquad (6\text{-}19)$$

Table 6.3 Experimental Standard Enthalpies of Vaporization ($\Delta_{vap}H_i$) and Standard Enthalpies of Transfer from Water to Air ($\Delta_{aw}H_i$) of Selected Organic Compounds at 25°C[a]

Compound Name	$\Delta_{vap}H_i$ (kJ·mol^{-1})	$\Delta_{aw}H_i$ (kJ·mol^{-1})	Compound Name	$\Delta_{vap}H_i$ (kJ·mol^{-1})	$\Delta_{aw}H_i$ (kJ·mol^{-1})
n-Hexane	32	32	2,5-Dichlorobiphenyl	74	47
n-Heptane	37	34	2,4,4'-Trichlorobiphenyl	78	50
n-Octane	41	36	2,2',5,5'-Tetrachlorobiphenyl	81	52
Cyclohexane	33	33	γ-Hexachlorocyclohexane (γ-HCH)	70	46
Benzene	34	30	Diethylether	27	42, 47
Methylbenzene	38	32	Methyl-t-butylether (MTBE)	30	61
Ethylbenzene	42	39	Methanol	37	45
1,2-Dimethylbenzene	43	33	1-Hexanol	62	68
1,4-Dimethylbenzene	42	33	Cyclohexanol	62	70
Naphthalene	56	45, 47	1-Octanol	71	74
Anthracene	70	27, 47, 48	Phenol	58	48, 57
Phenanthrene	74	30, 47, 60	2-Methylphenol	62	46, 59
Pyrene	87	43, 56	4-Methylphenol	62	49, 60
Dichlorodifluoromethane	21	27	Aminobenzene (Aniline)	52	54
Trichlorofluoromethane	27	23	Butanone	35	42
Dichloromethane	29	30	2-Hexanone	43	49
Trichloromethane	31	35	Hexanol	53	62
Trichloroethene	35	33, 38, 39	Benzaldehyde	49	45, 56
Tetrachloroethene	40	37	Butylacetate	44	52
Chlorobenzene	41	30, 43	Dimethylsulfide	28	30
Hexachlorobenzene	76	49	Diethylsulfide	36	37

[a] Data from Abraham et al. (1990), Zhou and Mopper (1990), Kucklick et al. (1991), Ten Hulscher et al. (1992), Dewulf et al. (1995), Dohnal and Fenclovà (1995), Lide (1995), Alaee et al. (1996), Staudinger and Robert (1996), Allen et al. (1998), Goss and Schwarzenbach (1999a).

Illustrative Example 6.2

Evaluating the Direction of Air–Water Gas Exchange at Different Temperatures

Problem

What is the direction (into water? or out of water?) of the air–water exchange of benzene for a well-mixed shallow pond located in the center of a big city in each of the following seasons: (a) a typical summer situation ($T = 25°C$), and (b) a typical winter situation ($T = 5°C$)? In both cases, the concentrations detected in air and water are $C_{ia} = 0.05$ mg·m^{-3} and $C_{iw} = 0.4$ mg·m^{-3}. Assume that the temperature of the water and of the air is the same.

i = benzene

Answer (a)

The air–water partition constant, K_{iaw}, of benzene is 0.22 at 25°C (Appendix C), The quotient of the concentrations of benzene in the air and in water is:

$$\frac{C_{ia}}{C_{iw}} = \frac{0.05}{0.4} = 0.125$$

Hence, at 25°C, $C_{ia}/C_{iw} < K_{iaw}$, and therefore, there is a net flux from the water to the air (the system wants to move toward equilibrium).

Answer (b)

The $\Delta_{aw}H_i$ the value of benzene is 30 kJ·mol (Table 6.3). With $\Delta_{aw}H_i + RT_{av}$ ($T_{av} = 288$ K) $= 30 + 2.4 = 32.4$ kJ·mol^{-1}, you get a K_{iaw} value at 5°C of (Table 3.5):

$$K_{iaw}\ (5°C) = 0.4\ K_{iaw}\ (25°C) = 0.05$$

Thus, at 5°C the ratio $C_{ia}/C_{iw} > K_{iaw}$; therefore, this time there is a net flux from the air to the water.

This example shows that the direction of gas exchange may be strongly influenced by temperature.

Note that in Eq. 6-19 we neglect the effect of the dissolved salt on the molar volume of the aqueous phase. This is a reasonable first approximation if we deal with salt solutions that are not too concentrated (e.g., seawater; see Illustrative Example 6.3).

For assessing $K_{ia\ell}$ values for organic solvent/water mixtures, we can estimate the activity coefficient of the compound of interest in the liquid phase using Eq. 5-30. Inserting this value, together with p_{iL}^*, and the appropriate molar volume of the solvent mixture into Eq. 6-6 (see Illustrative Example 6.3), then yields the corresponding $K_{ia\ell}$.

Illustrative Example 6.3

Assessing the Effect of Solution Composition on Air–Aqueous/Phase Partitioning

Problem

Recall Problem 3.1. You are the boss of an analytical laboratory and, this time, you check the numbers from the analysis of chlorobenzene in water samples of very different origins, namely (a) moderately contaminated groundwater, (b) seawater ([salt]$_{tot}$ ≈ 0.5 M), (c) water from a brine ([salt]$_{tot}$ ≈ 5.0 M), and (d) leachate of a hazardous-waste site containing 40% (v : v) methanol. For all samples, your laboratory reports the same chlorobenzene concentration of 10 $\mu g \cdot L^{-1}$. Again the sample flasks were unfortunately not completely filled. This time, the 1 L flasks were filled with 400 mL liquid, and stored at 25°C before analysis. What were the original concentrations (in $\mu g \cdot L^{-1}$) of chlorobenzene in the four samples?

i = chlorobenzene

p^*_{iL} (25°C) = 0.016 bar

γ^{sat}_{iw} = 14000 (Table 5.2)

K_{iaw} (25°C) = 0.16

K^s_i = 0.23 M^{-1} (Table 5.7)

Answer

For calculating the original concentration of a compound i in a two-phase system that contains an air volume V_a and a liquid volume V_ℓ, divide the total mass of i present by the volume of the liquid phase:

$$C^{orig}_{i\ell} = \frac{C_{i\ell} \cdot V_\ell + C_{ia} \cdot V_a}{V_\ell} \qquad (1)$$

Substitute C_{ia} by $K_{ia\ell} \cdot C_i$ into Eq. (1) and rearrange the equation to get

$$C^{orig}_{i\ell} = C_{i\ell}\left(1 + K_{ia\ell}\frac{V_a}{V_\ell}\right) \qquad (2)$$

Case (a) (ℓ = water)

Insert $C_{i\ell}$ = 10 mg·L^{-1}, K_{iaw} = 0.16, and V_a / V_ℓ = 1.5 into Eq. (2) to get an original concentration of *12.4* mg·L^{-1}.

Case (b) (ℓ = seawater) and (c) (ℓ = brine)

In this case use Eq. 6-19 to calculate $K_{ia\ell}$:

$$K_{ia\ell} = K_{iaw} \cdot 10^{+K^s_i [salt]_{tot}} \qquad (6-19)$$

Insertion of K_{iaw}, K^s_i and [salt]$_{tot}$ into Eq. 6-19 yields for case (b):

$$K_{ia\ell} = (0.16)(1.30) = 0.21 \text{ and, therefore, } C^{orig}_{i\ell} = 13.2 \text{ mg·L}^{-1}$$

for *case (c):*

$$K_{ia\ell} = (0.16)(14.1) = 2.26 \text{ and, therefore, } C^{orig}_{i\ell} = 43.9 \text{ mg·L}^{-1}$$

Case (d) (ℓ = 40% methanol / 60% water)

Use the linear relationship between log ($\gamma^{sat}_{iw} / \gamma^{sat}_{i\ell}$) and the molar volume, V_{ix} (in

$cm^3\ mol^{-1}$) shown in Fig. 5.8 for PCBs and $f_{v,MeOH} = 0.4$ to estimate $\gamma_{i\ell}^{sat}$ for chlorobenzene:

$$\log (\gamma_{iw}^{sat} / \gamma_{i\ell}^{sat}) = (0.0180)\ V_{ix} - 0.850$$

Insertion of γ_{iw}^{sat} and V_{ix} (83.8 $cm^3\ mol^{-1}$) of chlorobenzene yields a $\gamma_{i\ell}^{sat}$ value of 3070. The molar volume of methanol is 40 $cm^3\ mol^{-1}$. Hence, when assuming that, as a first approximation, Amagat's law (Eq. 3-44) is valid (which is not exactly true in this case), the molar volume of the 40% methanol / 60% water mixture is (Eq. 5-34): $\cong (0.23)\ (40) + (0.77)\ (18) = 23.3\ cm^3\ mol^{-1} = 0.0231\ L \cdot mol^{-1}$. Inserting this value together with $\gamma_{i\ell}^{sat}$ and p_{iL}^* into Eq. 6-6 yields:

$$K_{ia\ell} = (3070)\ (0.0231)\ (0.016) / (0.0831)\ (298) = 0.046$$

and therefore $C_{i\ell}^{orig} = 10.07\ mg \cdot L^{-1}$.

Availability of Experimental Data

The experimental determination of air–water partition constants is not an easy task to perform, particularly when dealing with compounds exhibiting very small K_{iaw} values. Although the available values of experimental K_{iaw} are steadily growing in number, compared to vapor pressures, aqueous solubilities, or n-octanol–water partition constants, the data are still quite limited. Compilations of experimental air–water partition constants can be found in handbooks such as the one published by Mackay et al. (1992–1997), or in review articles including those by Staudinger and Roberts (1996), or Brennan et al. (1998). Note that in some cases, considerable differences (i.e., up to an order of magnitude) may exist between experimental K_{iaw} values reported by different authors. Therefore, it is advisable to "check" such values by comparison with estimated ones. For example, one may see if experimental results appear reasonable by using the ratio of vapor pressure and aqueous solubility (Eqs. 6-16 and 6-17) or an LFER such as the one given below (Eq. 6-22).

There are two general experimental approaches commonly used for determining air–water partition constants, the *static* and the *dynamic* equilibration approach. A detailed description of the different existing variations of the two methods can be found in the review by Staudinger and Roberts (1996) and in the literature cited therein. Here we will confine ourselves to a few remarks on the general concepts of these experimental approaches.

The *static* equilibrium approach is, in principle, straightforward. In this method, the air–water partition constant is directly determined by measuring concentrations of a compound at a given temperature in the air and/or water in closed systems (e.g., in a gas-tight syringe, or in sealed bottles). If chemical concentrations are measured only in one phase, the concentration in the other is assessed as difference to the total amount of i in the system. In this approach, the error in determining K_{iaw} can be reduced either by equilibrating a given volume of an aqueous solution of a compound subsequently with several given volumes of solute-free air (e.g., in a syringe; see Problem 6.5), or by using multiple containers having different headspace-to-liquid volume ratios. The main experimental challenges of the static methods are to ensure that equilibrium is reached and also maintained during sampling, and to minimize sampling errors. Since with the static approach, it is possible to use neither very large nor very small air-to-water volume ratios, these methods are primarily suited for compounds with no extreme preference for one of the phases. In more extreme cases, dynamic methods may provide much better results.

The most widely applied *dynamic* method is the batch air or gas stripping technique. By using a stripping apparatus, bubbles of air or another inert gas are produced near the bottom of a vessel and then rise to the surface of the solution, the exit gas achieving equilibrium with the water. Hence, this experimental design requires that the velocity of the rising bubbles is sufficiently small and the height of the well-mixed water column is sufficiently great to establish air–water equilibrium. Furthermore, the bubbles need to be large enough so that *ad*sorption at the air–water interface can be neglected (this interface may be important for very hydrophobic compounds; see Section 11.2). If all this is achieved, the air–water partition constant can be determined by measuring the decrease in water concentration, C_{iw}, as a function of time (Mackay et al., 1979):

$$C_{iw}(t) = C_{iw}(0) \cdot e^{-\frac{K_{iaw} \cdot G}{V_w} \cdot t} \qquad (6\text{-}20)$$

where G is the gas volume flow per unit time and V_w is the volume of the aqueous solution. Hence, if G and V_w are known, K_{iaw} can be deduced from the slope of the linear regression of $\ln C_{iw}(t)$ versus t:

$$\ln C_{iw}(t) = -\,\text{slope} \cdot t + \text{constant} \qquad (6\text{-}21)$$

where $K_{iaw} = (\text{slope})(V_w/G)$. Note that, conversely, if K_{iaw} is known, Eq. 6-20 allows one to estimate the time required to purge a given compound from a water sample (e.g., the time required to lower its concentration to 1% of the initial concentration) for a given gas flow rate. This issue may be important when dealing with the behavior of organic pollutants in water treatment plants. It also pertains to problems in analytical chemistry, where the purge-and-trap method is widely used to enrich volatile compounds from water samples (Standard Methods for Examination of Water and Wastewater, 1995; see Problem 6.6).

An alternative dynamic approach to gas stripping is the concurrent flow technique, which is based on the use of a wetted wall column apparatus. Compound-laden water is introduced continuously at the top of a wetted wall column where it comes into contact with a compound-free gas stream flowing concurrently down the column. As with gas stripping, the major challenge is to allow sufficient contact time to ensure phase equilibrium is reached by the time the two streams reach the bottom of the column. The two streams are separated at the bottom of the column, and either solvent extracted or trapped on solid-phase sorbents for subsequent analysis. To determine the K_{iaw} value of a given compound, the system is run for a set amount of time. Given knowledge of the flow rates employed along with the compound masses present in the separated phase streams, K_{iaw} can be calculated. With this method, a rigorous mass balance can be conducted.

Estimation of Air–Water Partition Constants

As already discussed, the K_{iaw} value of a given compound may also be approximated by the ratio of its vapor pressure and its aqueous solubility (Eqs. 6-16 and 6-17). When using this approximation one has to be aware that, particularly for compounds exhibiting very small p_{iL}^* and/or C_{iw}^{sat} values, rather large errors may be introduced due to the uncertainties in the experimental vapor pressure and solubility data (see Sections 4.4 and 5.5).

Another possibility to predict K_{iaw} is to use our multiparameter LFER approach. As we introduced in Chapter 5, we may consider the intermolecular interactions between solute molecules and a solvent like water to estimate values of γ_{iw} (Eq. 5-22). Based on such a predictor of γ_{iw}, we may expect a similar equation can be found to estimate K_{iaw} values, similar to that we have already applied to air–organic solvent partitioning in Section 6.3 (Table 6.2). Considering a database of over 300 compounds, a best-fit equation for K_{iaw} values which reflects the influence of various intermolecular interactions on air–water partitioning is:

$$
\ln K_{iaw} = -0.540 \left[(V_{ix})^{2/3} \left(\frac{n_{Di}^2 - 1}{n_{Di}^2 + 2} \right) \right] - 5.71(\pi_i) - 8.74(\alpha_i) - 11.2(\beta_i)
$$
$$
+ 0.0459 V_{ix} + 2.25 \qquad (R^2 = 0.99)
$$

(6-22)

Note that the values of the coefficients s, p, a, b and v in Eq. 6-22 are slightly different from those in Eq. 5-22, because a larger set of compounds has been used for their derivation. Nevertheless, Eq. 6-22 is, of course, identical to the part in Eq. 5-22 that describes the transfer of a compound from the gas phase to the aqueous phase. Hence, we do not need to repeat our discussion of the various terms describing this process. Furthermore, our comments made on the various other methods developed for estimating aqueous activity coefficients, including QSPRs or group contribution methods such as UNIFAC or AQUAFAC (see Section 5.5), also apply directly for the methods suggested to predict K_{iH} values (Staudinger and Roberts, 1996; Brennan et al., 1998). In all cases, the key problem is the same; we need a quantitative description of the solubilization of an organic compound in the complex solvent water.

We conclude this section by addressing a simple LFER approach to estimate K_{iaw} values based solely on chemical structure. The underlying idea of this LFER of the type Eq. 3-57 (Section 3.4) was introduced by Hine and Mookerjee (1975) and expanded by Meyland and Howard (1991). In this method, each bond type (e.g., a C–H bond) is taken to have a substantially constant effect on $\Delta_{aw}G_i$, regardless of the substance in which it occurs. This assumption is reasonably valid for simple molecules in which no significant interactions between functional groups take place. Hence, the method is interesting primarily from a didactic point of view, in that we can see how certain substructural units affect air–water partitioning.

Table 6.4 Bond Contributions for Estimation of log K_{iaw} at 25°C [a]

Bond [b]	Contribution	Bond [b]	Contribution
C – H	+0.1197	C_{ar} – OH	–0.5967 [c]
C – C	–0.1163	C_{ar} – O	–0.3473 [c]
C – C_{ar}	–0.1619	C_{ar} – N_{ar}	–1.6282
C – C_d	–0.0635	C_{ar} – S_{ar}	–0.3739
C – C_t	–0.5375	C_{ar} – O_{ar}	–0.2419
C – CO	–1.7057	C_{ar} – S	–0.6345
C – N	–1.3001	C_{ar} – N	–0.7304
C – O	–1.0855	C_{ar} – I	–0.4806
C – S	–1.1056	C_{ar} – F	+0.2214
C – Cl	–0.3335	C_{ar} – C_d	–0.4391
C – Br	–0.8187	C_{ar} – CN	–1.8606
C – F	+0.4184	C_{ar} – CO	–1.2387
C – I	–1.0074	C_{ar} – Br	–0.2454
C – NO_2	–3.1231	C_{ar} – NO_2	–2.2496
C – CN	–3.2624	CO – H	–1.2102
C – P	–0.7786	CO – O	–0.0714
C = S	+0.0460	CO – N	–2.4261
C_d – H	+0.1005	CO – CO	–2.4000
C_d = C_d	–0.0000 [d]	O – H	–3.2318
C_d – C_d	–0.0997	O – P	–0.3930
C_d – CO	–1.9260	O – O	+0.4036
C_d – Cl	–0.0426	O = P	–1.6334
C_d – CN	–2.5514	N – H	–1.2835
C_d – O	–0.2051	N – N	–1.0956 [e]
C_d – F	+0.3824	N = O	–1.0956 [e]
C_t – H	–0.0040	N = N	–0.1374
C_t ≡ C_t	–0.0000 [d]	S – H	–0.2247
C_{ar} – H	+0.1543	S – S	+0.1891
C_{ar} – C_{ar}	–0.2638 [f]	S – P	–0.6334
C_{ar} – C_{ar}	–0.1490 [g]	S = P	+1.0317
C_{ar} – Cl	+0.0241		

[a] Data from Meylan and Howard (1991). [b] C: single-bonded aliphatic carbon; C_d: olefinic carbon; C_t: triple-bonded carbon; C_{ar}: aromatic carbon; N_{ar}: aromatic nitrogen; S_{ar}: aromatic sulfur; O_{ar}: aromatic oxygen; CO: carbonyl (C = O); CN: cyano (C ≡ N). *Note:* The carbonyl, cyano, and nitrofunctions are treated as single atoms. [c] Two separate types of aromatic carbon-to-oxygen bonds have been derived: (a) the oxygen is part of an –OH function, and (b) the oxygen is not connected to hydrogen. [d] The C = C and C ≡ C bonds are assigned a value of zero by definition (Hine and Mookerjee, 1975). [e] Value is specific for nitrosamines. [f] Intraring aromatic carbon to aromatic carbon. [g] External aromatic carbon to aromatic carbon (e.g., biphenyl).

Table 6.4 summarizes bond contribution values derived by Meyland and Howard (1991) from a large data set for a temperature of 25°C. These values can be used to calculate log K_{iaw} by simple addition of these bond contributions:

$$\log K_{iaw}(25°C) = \sum_k (\text{number of bonds type } k)\,(\text{contribution of bond type } k) \quad (6\text{-}23)$$

Most of the symbols in Table 6.4 are self-explanatory. For example, C–H is a singly bonded carbon–hydrogen subunit; C_{ar}–Cl is a chlorine bound to an aromatic carbon; and C–C_d is a carbon bound to an doubly bonded (olefinic) carbon. Some groups, such as the carbonyl group (C=O), are treated as a single "atom." Just looking at the signs and values of the bond contribution, we readily see that units such as C–H bonds tend to encourage molecules to partition into the air, while other units like O–H groups strongly induce molecules to remain associated with the water. These tendencies correspond to expected behaviors deduced qualitatively from our earlier considerations of intermolecular interactions of organic molecules with water (Chapter 5). Some sample calculations are performed in Illustrative Example 6.4. This simple bond contribution approach is usually accurate to within a factor of 2 or 3. One major drawback, however, is that it does not account for special inter-molecular or intramolecular interactions that may be unique to the molecule in which a particular bond type occurs. Therefore, additional correction factors may have to be applied (Meylan and Howard, 1991). Furthermore, the limited applicability of this simple approach for prediction of K_{iaw} values of more complex molecules has to be stressed.

Illustrative Example 6.4

Estimating Air–Water Partition Constants by the Bond Contribution Method

Problem

Estimate the K_{iaw} values at 25°C of (a) *n*-hexane, (b) benzene, (c) diethylether, and (d) ethanol using the bond contribution values given in Table 6.4. Compare these values with the experimental air–water partition constants given in Table 3.4. Note that for a linear or branched alkane (i.e., hexane) a correction factor of +0.75 log units has to be added (Meylan and Howard, 1991).

i = n-hexane

Answer (a)

log K_{iaw} (*n*-hexane) = 14 (C–H) + 5 (C–C) + 0.75 = *1.84*.

(The experimental value is 1.81)

i = benzene

Answer (b)

log K_{iaw} (benzene) = 6 (C_{ar}–H) + 6 (C_{ar}–C_{ar}) = *–0.66*.

(The experimental value is –0.68)

i = diethylether

Answer (c)

log K_{iaw} (diethylether) = 10 (C–H + 2 (C–C) + 2 (C–O) = *–1.21*.

(The experimental value is –1.18)

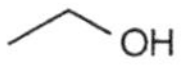

i = ethanol

Answer (d)

$\log K_{iaw}$ (ethanol) = 5 (C–H) + 1 (C–C) + 1 (C–O) + 1 (O–H) = -3.84.

(The experimental value is -3.70)

6.5   Questions and Problems

Questions

Q 6.1

Give examples of situations in which you need to know the equilibrium partition constant of an organic pollutant between (a) air and an organic liquid phase, and (b) air and water.

Q 6.2

How is the Henry's law constant defined? For which conditions is it valid?

Q 6.3

How do organic chemicals generally partition between a gas phase (i.e., air) and an *organic* liquid phase? Which molecular factors determine the magnitude of $K_{ia\ell}$?

Q 6.4

Why was *n*-octanol chosen as a surrogate for natural organic phases? Why not another solvent such as *n*-hexane, methylbenzene, trichloromethane, or diethylether? Why is the use of any organic solvent as general surrogate of a natural organic phase somewhat questionable?

Q 6.5

Table 6.1 shows that *n*-octane partitions much more favorably from air into *n*-octanol than into ethyleneglycol. In contrast, for dioxane (see structure in Fig. 6.3), the corresponding $K_{ia\ell}$ values are more or less identical. Try to rationalize these findings.

Q 6.6

Has temperature a significant effect on the partitioning of organic compounds between air and a bulk liquid phase? How does $K_{ia\ell}$ change with increasing temperature?

Q 6.7

Describe in general terms in which cases you would expect that the enthalpy of transfer of an organic compound from a bulk liquid phase (including water) to air $(\Delta_{a\ell}H_i)$ is (a) larger, (b) about equal, and (c) smaller than the enthalpy of vaporization ($\Delta_{vap}H_i$) of the compound. Give some specific examples for each of these cases.

Q 6.8

Within a given class of apolar or weakly polar compounds (e.g., alkanes, chlorobenzenes, alkylbenzenes, PCBs), the variation in the air–octanol partition constants (K_{iao}) is much larger than the variation in the air–water partition constants (K_{iaw}). For example, the K_{iao} values of the chlorinated benzenes vary between $10^{-3.5}$ (chlorobenzene) and 10^{-7} (hexachlorobenzene, see Harner and Mackay, 1995), whereas their K_{iaw} values are all within the same order of magnitude (Appendix C). Try to explain these findings.

Q 6.9

What is the effect of dissolved salt on air–water partitioning? How is this effect related to the total salt concentration?

Problems

P 6.1 *A Small Ranking Exercise*

Rank the four compounds (I–IV) indicated below in the order of increasing tendency to distribute from (a) air into hexadecane (mimicking an apolar environment), (b) air to olive oil, and (c) air to water. Use the α_i, β_i, and V_{ix} values given in Table 4.3 and calculated by the method given in Box 5.1. Assume, that the four compounds have about the same n_{Di} value. Do not perform unnecessary calculations. Comment on your choices. Finally, check your result (c) by applying the bond contributions given in Table 6.4.

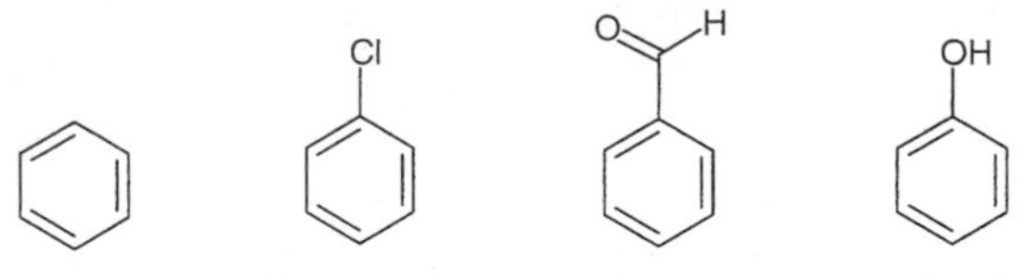

| benzene | chlorobenzene | benzaldehyde | phenol |
| I | II | III | IV |

P 6.2 *Raining Out*

Because of the increasing contamination of the atmosphere by organic pollutants, there is also a growing concern about the quality of rainwater. In this context, it is interesting to know how well a given compound is scavenged from the atmosphere by rainfall. Although for a quantitative description of this process, more sophisticated models are required, some simple equilibrium calculations are quite helpful.

Assume that PCE, MTBE, and phenol (see below) are present in the atmosphere at low concentrations. Consider now a drop of water (volume ~ 0.1 mL, pH = 6.0) in a volume of 100 L of air [corresponds about to the air–water ratio of a cloud (Seinfeld, 1986)]. Calculate the fraction of the total amount of each compound present in the water drop at 25°C and at 5°C assuming equilibrium between the two phases. Use the data given in Appendix C and in Table 6.3, and comment on any assumption that you make.

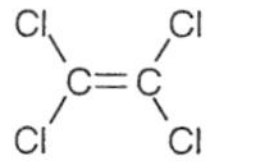

tetrachloroethene
(PCE)

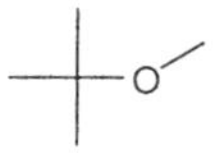

methyl-*t*-butylether
(MTBE)

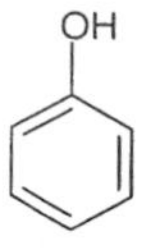

phenol

P 6.3 *Evaluating the Direction of Air–Water Gas Exchange in the Arctic Sea*

C_1- and C_2-halocarbons of natural and anthropogenic origin are omnipresent in the atmosphere and in seawater. For example, for 1,1,1-trichloroethane (also called methyl chloroform, MCF), typical concentrations in the northern hemisphere air and in Arctic surface waters are $C_{ia} = 0.9$ mg·m^{-3} air and $C_{iw} = 2.5$ mg·m^{-3} seawater (Fogelqvist, 1985). Using these concentrations, evaluate whether there is a net flux of MCF between the air and the surface waters of the Arctic Ocean assuming a temperature of (a) 0°C, and (b) 10°C. If there is a net flux, indicate its direction (i.e., sea to air or air to sea). Assume that the salinity of the seawater is 35‰. You can find all the necessary data in Appendix C, and in Tables 5.7 and 6.3.

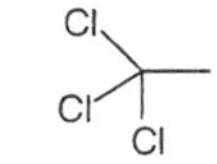

i = 1,1,1,-trichloroethane
(methyl chloroform, MCF)

P 6.4 *Getting the "Right" Air–Water Partition Constant for Benzyl Chloride*

In Chapter 24 the rate of elimination by gas exchange of benzyl chloride (BC) in a river will be calculated. To this end the K_{iaw} value of BC must be known. In the literature (Mackay and Shiu, 1981), you can find only vapor pressure and water solubility data for BC (see below). Because BC hydrolyzes in water with a half-life of 15 hours at 25°C (see Chapter 13), you wonder whether you can trust the aqueous solubility data. Approximate the K_{iaw}-value of BC by vapor pressure and aqueous solubility, and compare it to the value obtained by applying the bond contributions given in Table 6.4. (Use the K_{iaw}-value of toluene that you can find in the Appendix C as a starting value.) Which value do you trust more?

Hint: Use also other compound properties that are available or that can be estimated to perform simple plausibility tests on the experimental vapor pressure and aqueous solubility data of BC at 25°C.

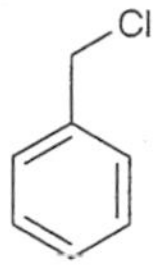

i = benzyl chloride (BC)

$T_m = -39°C$

$T_b = 179.3°C$

$p_{iL}^* (25°C) = 1.7 \times 10^{-3}$ bar

$C_{iw}^{sat} (25°C) = 3.5 \times 10^{-3}$ mol·L^{-1}

P 6.5 *Experimental Determination of the Air–Water Partition Constant of CF₃I (From Roberts, 1995)*

Not all stratospheric ozone destruction is caused by freons: up to 25% of the Antarctic "ozone hole" has been attributed to halons, compounds frequently used as fire extinguishers. A halon is a bromofluorocarbon; examples include CF_3Br, CF_2BrCl, and $BrF_2C–CF_2Br$. Because of their potential for damage to the environment, *production* of halons was banned as of Jan. 1, 1994 as part of an international agreement, although *use* of fire extinguishers containing halons is still allowed. Nevertheless, the chemical industry is still anxiously searching for alternatives to halons. One such promising alternative that has emerged is CF_3I, a gas with a boiling point of –22.5°C.

You are trying to measure the air–water partition constant of CF_3I. The method you are using is one of multiple equilibration. Essentially, a glass syringe containing 17 mL of water (but no headspace) is initially saturated with CF_3I. A very small sample (1 μL) of the aqueous phase is removed and is injected into a gas chromatograph, and the peak area is recorded ("initial peak area"). Next, 2 mL of air is drawn up into the syringe, which is closed off and shaken for 15 minutes to equilibrate the air and water phases. The air phase is dispelled from the syringe, 1 μL of the aqueous phase is injected into the GC, and the new peak area is recorded ("first equilibration"). The process of adding 2 mL of air, shaking the syringe, dispelling the gas phase, and reanalyzing the aqueous phase is repeated several times:

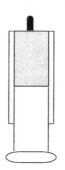 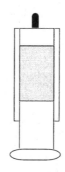 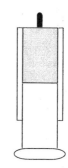

1. Fill with saturated CF_3I solution; analyze

2. Add 2 mL air to 17 mL water; equilibrate

3. Dispel air and reanalyze

4. Repeat steps 2 and 3 as required.

Shown below are data for CF_3I that you have obtained using the technique just described. Derive a mathematical relationship between peak area and the number of equilibration steps, and use this relationship to determine the K_{iaw} value for this compound from the appropriate regression of the experimental data provided. Assume that the peak area is linearly proportional to the concentration of CF_3I in the aqueous phase.

Experimental data (note all equilibrations conducted at room temperature):

Equilibration Number	Peak Area (mV·s)	Equilibration Number	Peak Area (mV·s)
0	583.850	6	21.370
0	532.089	7	13.726
1	287.789	7	11.134
1	291.891	7	10.581
2	152.832	8	7.285
2	158.352	8	5.282
3	95.606	9	4.435
3	105.630	9	3.173
4	61.371	10	1.726
4	56.450	10	2.606
5	41.332	11	1.440
5	36.071	11	1.754

Hint: Make yourself clear that the peak area remaining after the headspace is dispelled after an equilibration (i.e., the nth one) is related to the peak area determined in the previous measurement (i.e., the $(n$-1$)$th equilibrium) by:

$$(\text{Area})_{\text{measured at equil. } n} = f_{\text{w}} \cdot (\text{Area})_{\text{measured at equil. } n-1}$$

where f_{w} is the fraction of the total mass present in the water at equilibrium. Note that f_{w} is constant because the air-to-water volume ratio is always the same, and because it can be assumed that K_{iaw} is independent of concentration.

P 6.6   *Purge and Trap: How Long Do You Need to Purge to Get 90% of a Given Compound Out of the Water?*

The purge-and-trap method (see Section 6.4) is a common method to enrich volatile organic compounds from water samples. In your apparatus, you purge a 1 L water sample with a gas (air) volume flow of 1.5 L gas per minute at a temperature of 25°C. The compounds that you are interested in include tetrachloroethene, chlorobenzene and methyl-*t*-butylether (MTBE). Calculate the time required to purge 90% of each compound from the water. Any comments? How much time would you save if you would increase the temperature from 25°C to 35°C? What could be a problem when raising the temperature too much? You can find all necessary data in Appendix C and in Table 6.3.

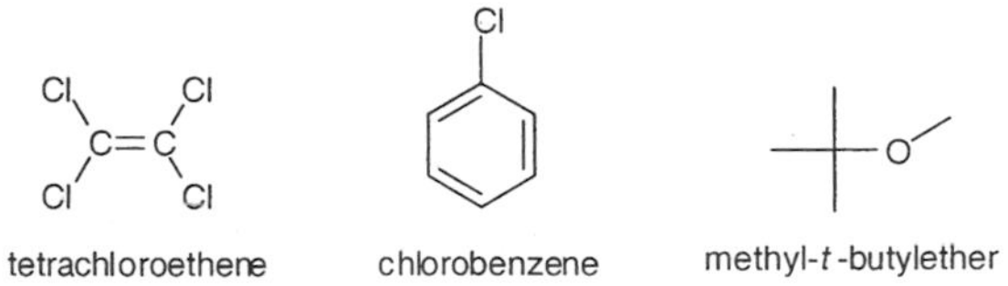

tetrachloroethene   chlorobenzene   methyl-*t*-butylether

Chapter 7

Organic Liquid–Water Partitioning

7.1 Introduction

7.2 Thermodynamic Considerations
The Organic Solvent–Water Partition Constant
Effect of Temperature and Salt on Organic Solvent–Water Partitioning

7.3 Comparison of Different Organic Solvent–Water Systems
General Comments
LFERs Relating Partition Constants in Different Solvent–Water Systems
Model for Description of Organic Solvent–Water Partitioning
Illustrative Example 7.1: *Evaluating the Factors that Govern the Organic Solvent–Water Partitioning of a Compound*

7.4 The *n*-Octanol–Water Partition Constant
General Comments
Availability of Experimental Data
One-Parameter LFERs for Estimation of Octanol–Water Partition Constants
Polyparameter LFERs for Estimation of the Octanol–Water Partition Constant
The Atom/Fragment Contribution Method for Estimation of the Octanol–Water Partition Constant
Illustrative Example 7.2: *Estimating Octanol–Water Partition Constants from Structure Using the Atom/Fragment Contribution Method*
Illustrative Example 7.3: *Estimating Octanol–Water Partition Constants Based on Experimental K_{iow}'s of Structurally Related Compounds*

7.5 Dissolution of Organic Compounds in Water from Organic Liquid Mixtures—Equilibrium Considerations (Advanced Topic)
Illustrative Example 7.4: *Estimating the Concentrations of Individual PCB Congeners in Water that Is in Equilibrium with an Aroclor and an Aroclor/Hydraulic Oil Mixture*

7.6 Questions and Problems

7.1 Introduction

In this chapter we will focus on the equilibrium partitioning of neutral organic compounds between aqueous solutions and *water-immiscible, well-defined* organic liquids. Our focus will be on situations in which the organic compound is present at a low enough concentration that it does not have a significant impact on the properties of either bulk liquid.

As will be discussed in Chapters 9 to 11, the distribution of neutral organic compounds between water and natural solids (e.g., soils, sediments, and suspended particles) and organisms can in many cases be viewed as a partitioning process between the aqueous phase and organic phases present in those solids. This conceptualization even applies somewhat to "solids" that are alive! As early as 1900, investigators studying the uptake of nonpolar drugs by organisms discovered that they could use water-immiscible organic solvents like olive oil or *n*-octanol as a reasonable surrogate for organisms insofar as accumulation of pharmaceutically important organic molecules from the water was concerned (Meyer, 1899; Overton, 1899). Although the extent of uptake from water into these solvents was not identical to that into organisms, it was proportional. That is, *within a series of compounds*, higher accumulation into an organism corresponded to more favorable partitioning into the organic solvent. More recently, environmental chemists have found similar correlations with soil humus and other naturally occurring organic phases (Chapter 9).

Furthermore, knowledge of the molecular factors that determine the partitioning of an organic compound between a liquid organic phase and water is of great interest in environmental analytical chemistry. This is particularly important when dealing with enrichment (i.e., extraction from water samples) or separation steps (i.e., reversed-phase liquid chromatography). Finally, understanding pure solvent–water partitioning will also be applicable to the problem of dissolving organic compounds in water when those organic substances are present in complex mixtures. In practice, we need such knowledge when dealing with contamination of the environment by mixtures such as gasoline, petroleum, or PCBs (Section 7.5).

We start, however, with some general thermodynamic considerations (Section 7.2). Then, using our insights gained in Chapter 6, we compare solvent–water partition constants of a series of model compounds for different organic solvents of different polarity (Section 7.3). Finally, because *n*-octanol is such a widely used organic solvent in environmental chemistry, we will discuss the octanol–water partition constant in somewhat more detail (Section 7.4).

7.2 Thermodynamic Considerations

The Organic Solvent–Water Partition Constant

In Section 3.4, we derived the equilibrium partition constant of a compound between two bulk liquid phases (Eq. 3-40). Denoting the organic phase with a subscript ℓ, we express the organic solvent–water partition constant on a mole fraction basis (superscript prime) as:

$$K'_{i\ell w} = \frac{x_{i\ell}}{x_{iw}} = \frac{\gamma_{iw}}{\gamma_{i\ell}} \qquad (7\text{-}1)$$

Hence, $K'_{i\ell w}$ is simply given by the ratio of the activity coefficients of the compound in water and in the organic phase. Note that this result applies whether the partitioning compound is a gas, liquid, or solid as a pure substance under the conditions of interest because the dissolved molecules exist in a liquid-like form in both phases. For many of the compounds of interest to us, we know that γ_{iw} can be quite large (e.g., 10^2 to $>10^8$, see Table 5.2). In contrast, in most organic solvents, organic compounds exhibit rather small activity coefficients (e.g., < 1 to 10^2, see Tables 3.2 and 6.1). Consequently, we can expect that in many cases, the magnitudes of organic solvent–water partition constants will be dominated by γ_{iw}. As a result, within a series of structurally related compounds, we may *generally find increasing organic solvent–water partition constants with decreasing (liquid) water solubilities* [recall that $\gamma_{iw} \cong \gamma_{iw}^{sat}$ is given by $(\overline{V}_w \cdot C_{iw}^{sat}(\text{L}))^{-1}$; Section 5.2].

A more common way of expressing organic solvent–water partition constants is to use molar concentrations in both phases (Eq. 3-45):

$$K_{i\ell w} = \frac{C_{i\ell}}{C_{iw}} = \frac{\overline{V}_w}{\overline{V}_\ell} \cdot \frac{\gamma_{iw}}{\gamma_{i\ell}} \qquad (7\text{-}2)$$

where $\overline{V}_w$ and $\overline{V}_\ell$ are the molar volumes of water and the organic solvent, respectively. Note that in Eq. 7-2 we have to use the molar volumes of the mutually saturated liquid phases (e.g., water which contains as much octanol as it can hold and water-saturated octanol). Considering the rather limited water solubility of most water-immiscible organic solvents, we can assume that we can often justifiably use the molar volume of pure water (i.e., 0.018 L·mol⁻¹ at 25°C). Similarly for apolar and weakly polar organic solvents, we may use the molar volume of the water-free solvent. Only for some polar organic solvents, may we have to correct for the presence of water in the organic phase (e.g., water-wet *n*-octanol has a $\overline{V}_\ell$ value of 0.13 L·mol⁻¹ as compared to 0.16 L·mol⁻¹ for "dry" octanol).

If we may assume that the mutual saturation of the two liquid phases has little effect on γ_{iw} and $\gamma_{i\ell}$, we may relate $K_{i\ell w}$ to the respective air–solvent and air–water partition constants (see Eq. 6-11):

$$K_{i\ell w} = \frac{K_{iaw}(= C_{ia} / C_{iw})}{K_{ia\ell}(= C_{ia} / C_{i\ell})} \qquad (7\text{-}3)$$

Effect of Temperature and Salt on Organic Solvent–Water Partitioning

As for any partition constant, over a temperature range narrow enough that the enthalpy of transfer may be assumed nearly constant, we may express the temperature dependence of $K_{i\ell w}$ by:

$$\ln K_{i\ell w} = -\frac{\Delta_{\ell w} H_i}{R} \cdot \frac{1}{T} + \text{constant} \qquad (7\text{-}4)$$

where $\Delta_{\ell w} H_i$ is the enthalpy of transfer of i from water to the organic solvent. This enthalpy difference is given by the difference between the excess enthalpies of the compound in the two phases:

$$\Delta_{\ell w} H_i = H_{i\ell}^E - H_{iw}^E \qquad (7\text{-}5)$$

The magnitude of the excess enthalpy of a given compound in the organic phase depends, of course, on the natures of both the solvent and the solute. For many compounds H_{iw}^E has a fairly small absolute value (e.g., Table 5.3). Substantial deviation from zero (i.e., $\left| H_{iw}^E \right| > 10$ kJ·mol^{-1}) occurs for some small monopolar compounds (e.g., diethylether, $H_{iw}^E = -20$ kJ·mol^{-1}) and for large apolar or weakly monopolar compounds (e.g., PCBs, PAHs which exhibit positive H_{iw}^E values, Table 5.3). Typically, $H_{i\ell}^E$ for organic solutes and organic solvents does not exceed ± 10 kJ·mol^{-1} (Section 6.3). Exceptions include small bipolar compounds in apolar solvents (e.g., the excess enthalpy of solution for ethanol in hexadecane is $+26$ kJ·mol^{-1}, see Table 3.3). Since, at the same time, such compounds tend to have negative H_{iw}^E values, the $\Delta_{\ell w} H_i$ value may become substantial (e.g., $+36$ kJ·mol^{-1} for hexadecane–water partitioning of ethanol, Table 3.4). However, for the majority of cases we are interested in, we can assume that organic solvent–water partitioning is only weakly dependent on temperature.

Using a similar approach, one may deduce how other factors should influence organic liquid–water partitioning. For example, we know that the addition of common salts (e.g., NaCl) to water containing organic solutes causes the aqueous activity coefficients of those organic solutes to increase. Since ionic substances are not compatible with nonpolar media like apolar organic solvents, one would not expect salt to dissolve in significant amounts in organic solvents. Consequently, the influence of salt on activity coefficients of organic solutes in organic solvents would likely be minimal. Combining these insights via Eq. 7.2, we can now calculate that the influence of salt on organic liquid–aqueous solution partitioning of organic compounds will entirely correspond to the impact of this factor on the aqueous activity coefficient, and hence (see Eq. 5-28):

$$K_{i\ell w, salt} = K_{i\ell w} \cdot 10^{+K_i^s [salt]_{tot}} \qquad (7\text{-}6)$$

7.3   Comparison of Different Organic Solvent–Water Systems

General Comments

Since the organic solvent–water partition constant of a given compound is determined by the ratio of its activity coefficients in the two phases (Eq. 7-2), we can rationalize how different compounds partition in different organic solvent–water systems. Consider the values of log $K_{i\ell w}$ for a series of compounds i partitioning into five organic solvents ℓ exhibiting different polarities (Table 7.1). First, focus on the partitioning behavior of the apolar and weakly monopolar compounds (octane, chlorobenzene, methylbenzene). These undergo primarily vdW interactions (i.e., n-octane, chlorobenzene, methylbenzene for which α_i and β_i are small or even zero). In general, such compounds partition very favorably from water into organic

Table 7.1 Organic Solvent–Water Partition Constants of a Series of Compounds for Various Organic Solvents at 25°C [a]

Compound i (Solute) (α_i/β_i) [b]	Structure	n-Hexane (0.00/0.00) [b,c] $\log K_{ihw}$	Methylbenzene (Toluene) (0.00/0.14) [b,c] $\log K_{itw}$	Diethylether (0.00/0.45) [b,c] $\log K_{idw}$	Trichloromethane (Chloroform) (0.15/0.02) [b,c] $\log K_{icw}$	n-Octanol (0.37/0.48) [b,c] $\log K_{iow}$
n-Octane (0.00/0.00)		6.08	5.98	6.03	6.01	5.53
Chlorobenzene (0.00/0.07)	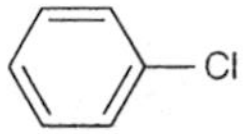	2.91			3.40	2.78
Methylbenzene (0.00/0.14)	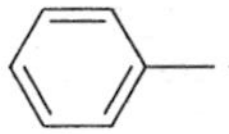	2.83	3.14	3.07	3.43	2.66
Pyridine (0.00/0.52)	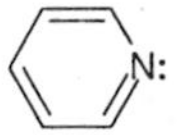	−0.21	0.29	0.08	1.43	0.65
Acetone (0.04/0.51)	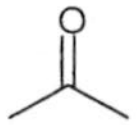	−0.92	−0.31	−0.21	0.72	−0.24
Aniline (0.26/0.41)	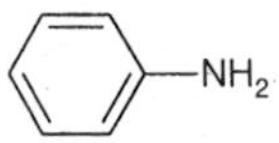	0.01	0.78	0.85	1.23	0.90
1-Hexanol (0.037/0.48)	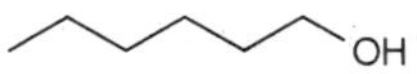	0.45	1.29	1.80	1.69	2.03
Phenol (0.60/0.31)	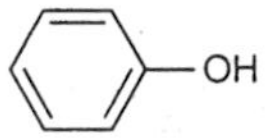	−0.89	0.12	1.58	0.37	1.49
Hexanoic Acid (0.60/0.45)	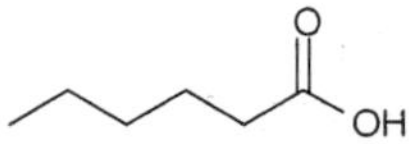	−0.14	0.48	1.78	0.71	1.95

[a] Data from Hansch and Leo (1979). [b] (α_i/β_i); Abraham et al. (1994a). [c] The α_i and β_i values correspond to the values of the solvents when present as solute! They are not necessarily identical when the compounds act as solvent. However, they give at least a good qualitative idea of the polar properties of the solvent.

solvents. This is not too surprising since these compounds have rather large γ_{iw}-values (Chapter 5). Furthermore, their log $K_{i\ell w}$, values do not vary much among the different organic solvents. For example, n-octane's partition coefficients vary only by about a factor of 4 for the five solvents shown in Table 7.1. For the strictly apolar solutes, lower values of log $K_{i\ell w}$, can be expected in bipolar solvents such as n-octanol. In the case of such a bipolar solvent, some solvent:solvent polar interactions have to be overcome when forming the solute cavity.

In contrast, partitioning from water into organic solvents may be somewhat enhanced if the solvents exhibit complementary polarity to monopolar solutes. One example is the partitioning of methylbenzene (toluene) between water and trichloromethane (Table 7.1). Each additional polar effect may become very substantial if the solute is strongly monopolar. This is illustrated by the trichloromethane-water partition constants of pyridine and acetone. Both of these solutes are quite strong H-acceptors or electron donors (i.e., $\beta_i \cong 0.5$). This causes these solutes to be strongly attracted to trichloromethane's hydrogen and results in significantly higher log K_{icw} values of these two compounds than for the other solvent–water systems. Note that the electron-accepting properties of trichloromethane (and of other polyhalogenated methanes and ethanes, e.g., dichloromethane, see Table 6.1) make such solvents well suited for the extraction of electron-donating solutes from water or other environmentally relevant matrices including soil or sediment samples.

When considering bipolar solutes (e.g., aniline, 1-hexanol, phenol, hexanoic acid), we can see that depending on the relative magnitudes of the solvent's α_i and β_i values, solute:solvent interactions may become quite attractive. For example, for aniline, for which $\alpha_i < \beta_i$, trichloromethane is still the most favorable solvent, whereas for phenol ($\alpha_i > \beta_i$), diethylether wins over the others. Finally, due to the lack of polar interactions in hexane, bipolar solutes partition rather poorly from water into such apolar solvents (Table 7.1).

LFERs Relating Partition Constants in Different Solvent–Water Systems

Often we may want to quantitatively extrapolate our experience with one organic solvent–water partitioning system to know what to expect for new systems. This is typically done using a linear free energy relationship of the form:

$$\log K_{i1w} = a \cdot \log K_{i2w} + b \tag{7-7}$$

where partitioning of solute, i, between some organic liquid, 1, and water is related to the partitioning of the same solute between another organic liquid, 2, and water. However, we should recall from our qualitative discussion of the molecular factors that govern organic solvent–water partitioning that such simple LFERs as shown in Eq. 7-7 will not always serve to correlate $K_{i\ell w}$ values of a large variety of compounds for structurally diverse solvent–water systems. Nonetheless, there are numerous special cases of groups of compounds and/or pairs of organic solvents for which such LFERs may be applied with good success. Obvious special cases include all those in which the molecular interactions of a given group of compounds are similar in nature in both organic phases. This is illustrated in Fig. 7.1 for the two solvents hexadecane and octanol (subscripts h and o, respectively). In this case, a

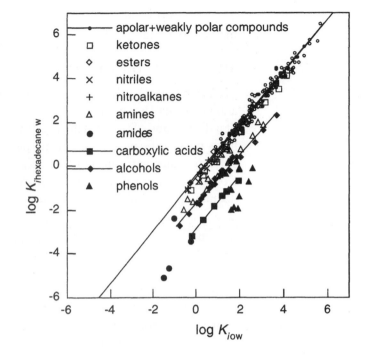

Figure 7.1 Plot of the decadic logarithms of the hexadecane–water partition constants versus the octanol–water constants for a variety of apolar, monopolar, and bipolar compounds. Data from Abraham (1994b). The a and b values for some LFERs (Eq. 7-7) are: apolar and weakly monopolar compounds ($a = 1.21$, $b = 0.43$; Eq. 7-8), aliphatic carboxylic acids ($a = 1.21$, $b = -2.88$), and aliphatic alcohols ($a = 1.12$, $b = -1.74$).

good correlation is found for all *apolar* and *weakly polar* compounds, for which the vdW interactions are the dominating forces in both organic solvents:

$$\log K_{ihw} = 1.21 \, (\pm \, 0.02) \cdot \log K_{iow} - 0.43 \, (\pm \, 0.06)$$

$$(N = 89, \, R^2 = 0.97)$$

(7-8)

The slope of greater than 1 in Eq. 7-8 indicates that structural differences in the solutes have a somewhat greater impact on their partitioning behaviors in the hexadecane–water, as compared to the octanol–water system. This can be rationalized as arising from the different free-energy costs related to the cavity formation in the two solvents, which is larger in the bipolar octanol (see discussion in Chapter 6).

A second important feature shown in Fig. 7.1 is that, for the two organic solvents considered, the more polar compounds do not fit well in the LFER expressed in Eq. 7-8. This is particularly true for bipolar solutes. Here, LFERs may be found only for structurally related compounds. For example, good correlations exist for a homologous series of compounds such as the aliphatic carboxylic acids or the aliphatic alcohols. In these cases, within the series of compounds, the polar contribution is constant; that is, the compounds differ only in their ability to undergo dispersive vdW interactions. This example shows that we have to be careful when applying one-parameter LFERs to describe systems in which more than one intermolecular interaction is varying. Such is the case when we are dealing with diverse groups of partitioning chemicals and/or with structurally complex organic phases including natural organic matter (Chapter 9) or parts of organisms (Chapter 10). If we are, however, aware of the pertinent molecular interactions that govern the partitioning of a given set of organic compounds in the organic phase–water

systems considered, appropriately applied one-parameter LFERs of the type Eq. 7-
may be extremely useful predictive tools.

Model for Description of Organic Solvent–Water Partitioning

Multiparameter LFERs for description of air–organic solvent (Eq. 6-13, Table 6.2
and air–water (Eq. 6-21) partition constants have been developed. If we can assume
that dissolution of water in the organic solvent and of the organic solvent in the water
have no significant effects on the partitioning of a given compound, the organic
solvent–water partition constant, $K_{i\ell w}$, is equal to K_{iaw} divided by $K_{ia\ell}$ (Eq. 7-3).
Consequently, we can develop a multiparameter equation for $K_{i\ell w}$ and immediately
deduce the coefficients from these earlier LFERs:

$$\ln K_{i\ell w} = s\left[V_{ix}^{2/3}\left(\frac{n_{D_i}^2 - 1}{n_{D_i}^2 + 2} \right) \right] + p(\pi_i) + a(\alpha_i) + b(\beta_i) + v(V_{ix}) + \text{constant} \qquad (7\text{-}9)$$

In this case, the coefficients s, p, a, b, v, and constant in Eq. 7-9 reflect the
differences of the solvent interaction parameters (i.e., dispersive, polar, H-donor
H-acceptor properties, and cavity formation) for water and organic solvent
considered. As for the other multiparameter LFERs discussed in earlier chapters, for
a given solvent–water system, these coefficients can be obtained by fitting an
appropriate set of experimental $K_{i\ell w}$ values using the chemical property parameters
V_{ix}, n_{Di}, π_i, α_i, and β_i. If such experimental data are not available, but if a multiparameter LFER has been established for the corresponding air–organic solvent
system (Eq. 6-13, Table 6.2), Eq. 7-9 can be derived by simply subtracting Eq. 6-13
from Eq. 6-21, provided that we are dealing with water-immiscible organic solvents.
Conversely, a multiparameter LFER for air–solvent partitioning can be obtained by
subtracting Eq. 7-9 from Eq. 6-21. When doing so, one has to be careful to use
equations that have been established with the same molecular parameter sets (e.g.,
the same calculated molar volumes (see Box 5.1), as well as the same compilations
of published π_i, α_i, and β_i values. Furthermore, the equations that are combined
should preferably cover a similar range of compounds used for their derivation.
Finally, we should note again that we are assuming that dissolution of water in the
organic solvent and of the organic solvent in the water have no significant effect on
the partitioning of a given compound (Section 6.3).

Such multiparameter LFERs have been developed for a few organic solvent–water
systems (Table 7.2.) The magnitudes of the fitted coefficients, when combined with
an individual solute's V_{ix}, n_{Di}, π_i, α_i, β_i values, reveal the importance of each intermolecular interaction to the overall partitioning process for that chemical. To
interpret the various terms, we note that these coefficients reflect the differences of
the corresponding terms used to describe the partitioning of the compounds from air
to water and from air to organic solvent, respectively (see Chapter 6). Some applications of Eq. 7-9 are discussed in Illustrative Example 7.1.

Table 7.2 Organic Solvent–Water Partitioning: Multiparameter LFERs Eq. 7-8 for Some Selected Organic Solvents at 25°C [a]

Solvent	Coefficient for Parameter							
	vdW s	(π_i) p	(α_i) a	(β_i) b	V_{ix} v	constant	$\ln K_{i\ell w}$ range [b]	R^2	n [c]
n-Hexadecane (apolar)	0.75	–3.61	–8.06	–11.41	0.069	–0.16	–10.0 to 6.11	0.99	302
Trichloromethane	–0.01	0.01	–3.45	–3.38	0.042	0.30	–2.1 to 3.7	1.00	40
Diethylether	–0.11	–0.12	–0.49	–10.54	0.100	–0.10	–0.5 to 5.6	0.97	35
n-Octanol	0.62	–2.53	–0.35	–7.88	0.063	–0.25	–3.2 to 13.0	0.98	260

[a] Data from Abraham et al. (1994a,b). [b] Range of $K_{i\ell w}$ values that were used to establish the LFER. [c] Number of compounds.

Illustrative Example 7.1

Evaluating the Factors that Govern the Organic Solvent–Water Partitioning of a Compound

Problem

Calculate the n-hexadecane–water ($\ln K_{ihw}$) and the n-octanol–water ($\ln K_{iow}$) partition constants at 25°C of n-octane (Oct), 1-methylnaphthalene (1-MeNa) and 4-t-butylphenol (4-BuPh) using the polyparameter LFER, Eq. 7-9, with the coefficients given in Table 7.2. Compare and discuss the contributions of the various terms in Eq. 7-9 for the three compounds in the two solvent–water systems. Note that the three compounds have already been used in Illustrative Example 5.2 to evaluate the polyparameter LFER describing the aqueous activity coefficient.

Answer

Get the n_{Di} values of the compounds from Lide (1995). Use the α_i, β_i and π_i values given in Tables 4.3 and 5.5. The resulting data sets for the three compounds are given in the margin. Insertion of the respective values into Eq. 7-9 with the appropriate coefficients (Table 7.2) yields the following results:

n-octane (Oct)

$p_{iL}^* = 1826$ Pa
$V_{ix} = 123.6$ cm^3 mol^{-1}
$n_{Di} = 1.397$
$\pi_i = 0$
$\alpha_i = 0$
$\beta_i = 0$

1-methylnaphthalene (1-MeNa)

$p_{iL}^* = 8.33$ Pa
$V_{ix} = 122.6$ cm^3 mol^{-1}
$n_{Di} = 1.617$
$\pi_i = 0.90$
$\alpha_i = 0$
$\beta_i = 0.20$

4-t-butylphenol (4-BuPh)

$p_{iL}^* = 6.75$ Pa
$V_{ix} = 133.9$ cm^3 mol^{-1}
$n_{Di} = 1.517$
$\pi_i = 0.89$
$\alpha_i = 0.56$
$\beta_i = 0.39$

Term	Oct		1-MeNa		4-BuPh	
	$\Delta \ln K_{ihw}$	$\Delta \ln K_{iow}$	$\Delta \ln K_{ihw}$	$\Delta \ln K_{iow}$	$\Delta \ln K_{ihw}$	$\Delta \ln K_{iow}$
$s \cdot$ disp. vdW [a]	+4.47	+3.70	+6.47	+5.35	+5.94	+4.91
$+ p \cdot (\pi_i)$	0	0	−3.25	−2.28	−3.21	−2.25
$+ a \cdot (\alpha_i)$	0	0	0	0	−4.51	−0.20
$+ b \cdot (\beta_i)$	0	0	−2.28	−1.58	−4.45	−3.07
$+ v \cdot (V_{ix})$	+8.52	+7.78	+8.46	+7.72	+9.24	+8.44
$+$ constant	−0.16	−0.25	−0.16	−0.25	−0.16	−0.25
$\ln K_{i\ell w}$	12.8	11.2	9.24	8.96	2.85	7.58
observed	13.3	11.9	9.21	9.19	2.20	7.23

[a] dispersive vdW attractions $= \left[(V_{ix})^{2/3} \left(\dfrac{n_{Di}^2 - 1}{n_{Di}^2 + 2} \right) \right]$

First, note that the three compounds are of similar size. Hence, the two terms that reflect primarily the differences in the energy costs for cavity formation and the differences in the dispersive interactions of the solute (i.e., $v \cdot V_{ix}$ and $s \cdot$ disp. vdW) in water and in the organic solvent are of comparable magnitudes for the three compounds. Note that the values in the table reflect variations on a natural logarithm scale. So, for example, the effect of the product, $v \cdot V_{ix}$, is to vary K_{iow} by a factor of 5 between these compounds and the product, $s \cdot$ disp vdW, also contributes a factor of 5 variation to these compounds' K_{iow} values. Because of the higher costs of cavity formation in the water as compared to n-hexadecane and n-octanol, both terms promote partitioning into the organic phase (i.e., they have positive values). This promoting effect is somewhat larger in the *n*-hexadecane–water system than in the *n*-octanol–water system because of the somewhat higher costs of forming a cavity in the bipolar solvent, *n*-octanol.

Significant differences in the partition constants of the three compounds, in particular for the *n*-hexadecane–water system, are also due to the polar interactions, also including the dipolarity/polarizability parameter, π_i. For the two organic solvent–water systems considered, due to the strong polar interactions of mono- and bipolar compounds in the water as compared to the organic phase, all these terms are negative. Therefore, these polar intermolecular interactions decrease the $K_{i\ell w}$ value. These polar effects are more pronounced in the *n*-hexadecane–water system (e.g., 1-MeNa partitioning reduced by a factor of 26) as compared to the *n*-octanol–water system (e.g., 1-MeNa partitioning reduced by a factor of 10).

Finally, with respect to the H-acceptor properties of the solvents (*a*-term), water and *n*-octanol are quite similar. Therefore, for a hydrogen-bonding solute like 4-BuPh, the corresponding product, $a \cdot (\alpha_i)$, is close to zero. This is not the case for the hexadecane–water system where loss of hydrogen bonding in this alkane solvent causes both the H-acceptor and H-donor terms to contribute factors of about 100 to 4-BuPh's value of K_{ihw}.

7.4   The *n*-Octanol–Water Partition Constant

General Comments

Because *n*-octanol is still the most widely used organic solvent for predicting partitioning of organic compounds between natural organic phases and water, we need to discuss the octanol–water partition constant, K_{iow}, in more detail. Note that in the literature, K_{iow} is often also denoted as P or P_{ow} (for *partitioning*). From the preceding discussions, we recall that *n*-octanol has an *amphiphilic* character. That is, it has a substantial apolar part as well as a bipolar functional group. Thus, in contrast to smaller bipolar solvents (e.g., methanol, ethylene glycol), where more hydrogen bonds have to be disturbed when creating a cavity of a given size, the free-energy costs for cavity formation in *n*-octanol are not that high. Also, the presence of the bipolar alcohol group ensures favorable interactions with bipolar and monopolar solutes. Hence, *n*-octanol is a solvent that is capable of accommodating any kind of solute. As a result, the activity coefficients in octanol (Fig. 7.2) of a large number of very diverse organic compounds are between 0.1 (bipolar small compounds) and 10 (apolar or weakly polar medium-sized compounds). Values of γ_{io} exceeding 10 can be expected only for larger hydrophobic compounds, including highly chlorinated biphenyls and dibenzodioxins, certain PAHs, and some hydrophobic dyes (Sijm et al., 1999). Therefore, the K_{iow} values of the more hydrophobic compounds (i.e., $\gamma_{iw} \gg 10^3$) are primarily determined by the activity coefficients in the aqueous phase.

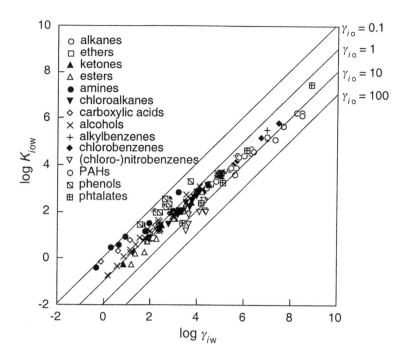

Figure 7.2 Plot of the decadic logarithms of the octanol–water partition constants versus the aqueous activity coefficients for a variety of apolar, monopolar, and bipolar compounds. The diagonal lines show the location of compounds with activity coefficients in octanol (calculated using Eq. 7-2) of 0.1, 1, 10, and 100, respectively.

For sets of compounds with the same functional group and variations in their apolar structural portion, we can also see that γ_{io} is either constant or varies proportionally to γ_{iw} (Fig. 7.2). Thus, for such groups of compounds, we find one-parameter LFERs of the type:

$$\log K_{iow} = a \cdot \log \gamma_{iw} + b \tag{7-10}$$

Since γ_{iw} is more or less equal to γ_{iw}^{sat} for many low solubility compounds ($\gamma_{iw} > ca.$ 50), we have $\gamma_{iw} = (\overline{V}_w \cdot C_{iw}^{sat}(L))^{-1}$. Considering such sets of compounds, we can rewrite Eq. 7-10 as:

$$\log K_{iow} = -a \cdot \log C_{iw}^{sat}(L) + b' \tag{7-11}$$

where $b' = b - a \cdot \log \overline{V}_w = b + 1.74\, a$ (at 25°C). Note that in Eq. 7-11, $C_{iw}^{sat}(L)$ is expressed in $mol \cdot L^{-1}$.

Such correlation equations have been derived for many classes of compounds (Table 7.3). These examples illustrate that very good relationships are found when only members of a specific compound class are included in the LFER. One can also reasonably combine compound classes into a single LFER if only compounds that exhibit similar intermolecular interaction characteristics are used (e.g., alkyl and chlorobenzenes; aliphatic ethers and ketones; polychlorinated biphenyls and polychlorinated dibenzodioxins).

When properly applied, LFERs of these types may be quite useful for estimating K_{iow} from γ_{iw} or $C_{iw}^{sat}(L)$. Additionally, these relationships can be used to check new K_{iow} and/or $C_{iw}^{sat}(L)$ values for consistency.

Table 7.3 LFERs Between Octanol–Water Partition Constants and Aqueous Activity Coefficients or Liquid Aqueous Solubilities at 25°C for Various Sets of Compounds: Slopes and Intercepts of Eqs. 7-10 and 7-11

Set of Compounds	$a^{a,b}$	b^a	b'^b	$\log K_{iow}$ range^c	R^2	n^d
Alkanes	0.85	−0.87	0.62	3.0 to 6.3	0.98	112
Alkylbenzenes	0.94	−1.04	0.60	2.1 to 5.5	0.99	15
Polycyclic aromatic hydrocarbons	0.75	−0.13	1.17	3.3 to 6.3	0.98	11
Chlorobenzenes	0.90	−0.95	0.62	2.9 to 5.8	0.99	10
Polychlorinated biphenyls	0.85	−0.70	0.78	4.0 to 8.0	0.92	14
Polychlorinated dibenzodioxins	0.84	−0.79	0.67	4.3 to 8.0	0.98	13
Phthalates	1.09	−2.16	−0.26	1.5 to 7.5	1.00	5
Aliphatic esters (RCOOR')	0.99	−1.27	(0.45)^e	−0.3 to 2.8	0.98	15
Aliphatic ethers (R–O–R')	0.91	−0.90	(0.68)^e	0.9 to 3.2	0.96	4
Aliphatic ketones (RCOR')	0.90	−0.89	(0.68)^e	−0.2 to 3.1	0.99	10
Aliphatic amines (R–NH$_2$, R–NHR')	0.88	+0.03	(1.56)^e	−0.4 to 2.8	0.96	12
Aliphatic alcohols (R–OH)	0.94	−0.76	(0.88)^e	−0.7 to 3.7	0.98	20
Aliphatic carboxylic acids (R–COOH)	0.69	−0.10	(1.10)^e	−0.2 to 1.9	0.99	5

^{a} Eq. 7-10. ^{b} Eq. 7-11. ^{c} Range of experimental values for which the LFER has been established. ^{d} Number of compounds used for LFER. ^{e} Only for compounds for which $\log K_{iow} > \sim 1$.

Availability of Experimental Data

The most common experimental approaches for determination of octanol–water partition constants are quite similar to those for water solubility. These employ shake flask or generator column techniques (Mackay et al., 1992-1997). The "shake flask method," in which the compound is partitioned in a closed vessel between given volumes of octanol and water, is restricted to compounds with K_{iow} values of less than about 10^5. The reason is that for more hydrophobic compounds the concentration in the aqueous phase becomes too low to be accurately measured, even when using very small octanol-to-water volume ratios. Hence, for more hydrophobic compounds "generator columns," coupled with solid sorbent cartridges, are commonly used. Briefly, large volumes of octanol-saturated water (up to 10 L) are passed through small columns, packed with beads of inert support material that are coated with octanol solutions (typically 10 mL) of the compound of interest. As the water passes through the column, an equilibrium distribution of the compound is established between the immobile octanol solutions and the slowly flowing water. By collecting and concentrating the chemical of interest with a solid sorbent cartridge from large volumes of the effluent water leaving the column, enough material may be accumulated to allow accurate quantification of the trace level water load. This result, along with knowledge of the volume of water extracted and the concentration of the compound in the octanol, ultimately provides the K_{iow} value.

As for vapor pressure and aqueous solubility, there is quite a large experimental database on octanol–water partition constants available in the literature (see, e.g., Mackay et al., 1992–1997; Hansch et al., 1995). Up to K_{iow} values of about 10^6, the experimental data for neutral species are commonly quite accurate. For more hydrophobic compounds, accurate measurements require meticulous techniques. Hence, it is not surprising to find differences of more than an order of magnitude in the K_{iow} values reported by different authors for a given highly hydrophobic compound. Such data should, therefore, be treated with the necessary caution. Again, as with other compound properties, one way of deciding which value should be selected is to compare the experimental data with predicted values using other compound properties or K_{iow} data from structurally related compounds.

One-parameter LFERs for Estimation of Octanol–Water Partition Constants

There are also various methods for *estimating* the K_{iow} of a given compound. This can be done from other experimentally determined properties and/or by using molecular descriptors derived from the structure of the compound. We have already discussed some of the approaches (and their limitations) when evaluating the one-parameter LFERs correlating K_{iow} with aqueous solubility (Eq. 7-11, Table 7.3) or with other organic solvent–water partition constants (Eqs. 7-7 and 7-8). A related method that is quite frequently applied is based on the retention behavior of a given compound in a liquid-chromatographic system [high-performance liquid chromatography (HPLC) or thin-layer chromatography (TLC)]. Here, the organic solute is transported in a polar phase (e.g., water or a water/methanol mixture) through a porous stationary phase which commonly consists of an organic phase that is bound to a silica support (e.g., C_2–C_{18} alkyl chains covalently bound to silica beads). As the compounds of interest move through the system, they partition between the organic phase and the polar mobile phase.

Hence, in analogy to organic solvent–water systems, particularly for sets of structurally related apolar or weakly polar compounds for which solute hydrophobicity primarily determines the partitioning behavior, good correlations between K_{iow} and the *s*tationary-phase/*m*obile-phase partition constant, K_{ism}, of a given compound may be obtained. Since, in a given chromatographic system, the travel time or *retention time*, t_i, of a solute i is directly proportional to K_{ism}, an LFER of the following form is obtained:

$$\log K_{iow} = a \cdot \log t_i + b \qquad (7\text{-}12)$$

To compare different chromatographic systems, however, it is more useful to use the *relative retention time* (also called the *capacity factor,* k_i'). This parameter is defined as the retention of the compound relative to a nonretained chemical species, such as a very polar organic compound or an inorganic species such as nitrate:

$$k_i' = [(\, t_i - t_0\,) / t_0] \qquad (7\text{-}13)$$

where t_0 is the travel time of the nonretained species in the system. Eq. 7-12 is then written as:

$$\log K_{iow} = a \cdot \log\!\left(\frac{t_i - t_0}{t_0}\right) + b'$$

or: $\qquad\qquad\qquad\qquad\qquad\qquad\qquad\qquad\qquad\qquad (7\text{-}14)$

$$\log K_{iow} = a \cdot \log k_i' + b'$$

It should be pointed out that the coefficients a and b or b' in Eqs. 7-12 and 7-14 must be determined using appropriate reference compounds for *each* chromatographic system. With respect to the choice of the organic stationary phase and reference compounds (type, range of hydrophobicity) and the goodness of the LFER, in principle the same conclusions as drawn earlier for organic solvent–water systems are valid. For a given set of structurally related compounds, reasonably good correlations may be obtained. Finally, we should note that when using an organic solvent–water mixture as mobile phase, the (rather complex) effect of the organic cosolvent on the activity coefficient of an organic compound in the mobile phase (Section 5.4) has to be taken into account when establishing LFERs of the type Eqs. 7-12 and 7-14.

In summary, appropriate use of chromatographic systems for evaluating the partitioning behavior of organic compounds between nonaqueous phases and water (e.g., octanol–water) offers several advantages. Once a chromatographic system is set up and calibrated, many compounds may be investigated at once. The measurements are fast. Also, accurate compound quantification (which is a prerequisite when using solvent–water systems) is not required. For more details and additional references see Lambert (1993) and Herbert and Dorsey (1995).

Polyparameter LFERs for Estimation of the Octanol–Water Partition Constant

It is also possible to estimate K_{iow} via polyparameter LFERs such as Eq. 7-9 (with the coefficients in Table 7.2), provided that all the necessary parameters are known for the compound of interest. Note that such polyparameter LFERs are also used to characterize stationary phases in chromatographic systems such as the ones described above (Abraham et al., 1997). Such information provides the necessary knowledge about the molecular interactions between a given set of compounds and a given stationary phase. This understanding is very helpful for establishing logical one-parameter LFERs (Eqs. 7-12 and 7-14) for prediction of K_{iow} values.

The Atom/Fragment Contribution Method for Estimation of the Octanol–Water Partition Constant

Finally, the fragment or group contribution approach is widely used for predicting K_{iow} values solely from the structure of a given compound. We have already introduced this approach in very general terms in Section 3.4 (Eqs. 3-57 and 3-58), and we have discussed one application in Section 6.4 when dealing with the prediction of the air–water partition constants (Eq. 6-22, Table 6.4). We have also pointed out that any approach of this type suffers from the difficulty of quantifying electronic and steric effects between functional groups present within the same molecule. Therefore, in addition to simply adding up the individual contributions associated with the various structural pieces of which a compound is composed, numerous correction factors have to be used to account for such *intramolecular* interactions. Nevertheless, because of the very large number of experimental octanol–water partition constants available, the various versions of fragment or group contribution methods proposed in the literature for estimating K_{iow} (e.g., Hansch et al., 1995; Meylan and Howard, 1995) are much more sophisticated than the methods available to predict other partition constants, including K_{iaw}.

The classical and most widely used fragment or group contribution method for estimating K_{iow} is the one introduced originally by Rekker and co-workers (Rekker, 1977) and Hansch and Leo (Hansch and Leo, 1979; Hansch and Leo, 1995; Hansch et al., 1995). The computerized version of this method (known as the CLOGP program; note again the P is often used to denote K_{iow}) has been initially established by Chou and Jurs (1979) and has since been modified and extended (Hansch and Leo, 1995). The method uses primarily single-atom "fundamental" fragments consisting of isolated types of carbons, hydrogen, and various heteroatoms, plus some multiple-atom "fundamental" fragments (e.g., –OH, –COOH, –CN, –NO$_2$). These fundamental fragments were derived from a limited number of simple molecules. Therefore, the method also uses a large number of correction factors including unsaturation and conjugation, branching, multiple halogenation, proximity of polar groups, and many more (for more details see Hansch et al., 1995).

In the following, the atom/fragment contribution method (AFC method) developed by Meylan and Howard (1995) is used to illustrate the approach. This method is similar to the CLOGP method, but it is easier to see its application without using a computer program. Here, we confine ourselves to a few selected examples of fragment coefficients and correction factors. This will reveal how the method is applied and how certain important substructural units quantitatively affect the *n*-octanol–water partitioning of a given compound. For a more detailed treatment of this method including a discussion of its performance, we refer to the literature (Meylan and Howard, 1995).

Using a large database of K_{iow} values, fragment coefficients and correction factors were derived by multiple linear regression (Tables 7.4 and 7.5 give selected values of fragment coefficients and of some correction factors reported by Meylan and Howard, 1995). For estimating the log K_{iow} value of a given compound at 25°C, one simply adds up all fragment constants, f_k, and correction factors, c_j, according to the equation

$$\log K_{iow} = \sum_k n_k \cdot f_k + \sum_j n_j \cdot c_j + 0.23 \qquad (7\text{-}15)$$

where n_k and n_j are the frequency of each type of fragment and specific interaction, respectively, occurring in the compound of interest.

The magnitudes of the individual atom/fragment coefficients give us a feeling for the contribution of each type of substructural unit (e.g., a functional group) to the overall K_{iow} of a compound. Recall that in most cases, the effect of a given subunit on K_{iow} is primarily due to its effect on the aqueous activity coefficient of the compound, and to a lesser extent on γ_{io}. First, we note that any aliphatic, olefinic, or aromatic carbon atom has a positive fragment coefficient and therefore increases log K_{iow}. For aliphatic carbons, the coefficient decreases with increased branching. This can be rationalized by the smaller size of a branched versus nonbranched compound resulting in reduced cavity "costs." Furthermore, because of the higher polarizability of π-electrons, olefinic and aromatic carbon atoms have a somewhat smaller coefficient as compared to the corresponding aliphatic carbon. Except for aliphatically bound fluorine, all halogens increase K_{iow} significantly. This *hydrophobic* effect of the halogens increases, as expected, with the size of the halogens (i.e., I > Br > Cl > F), and it is more pronounced for halogens bound to aromatic carbon as compared to halogens on aliphatic carbon. The latter fact can be explained by the interactions of the nonbonded electrons of the halogens with the π-electron system, causing a decrease in the polarity of the corresponding carbon–halogen bond.

With respect to the functional groups containing oxygen, nitrogen, sulfur and phosphorus (see also Chapter 2), in most cases, such polar groups decrease log K_{iow} primarily due to hydrogen bonding. This *hydrophilic* effect is, in general, more pronounced if the polar group is aliphatically bound. Again, interactions of nonbonded or π-electrons of the functional group with the aromatic π-electron system (i.e., by resonance, see Chapter 2) are the major explanation for these findings. Note that in the case of isolated double bonds, this resonance effect is smaller. It is only one-third to one-half of the effect of an aromatic system.

As illustrated by the examples in Table 7.5, application of correction factors is necessary in those cases in which electronic and/or steric interactions of functional groups within a molecule influence the solvation of the compound. A positive correction factor is required if the interaction decreases the overall H-donor and/or

Table 7.4 Selected Atom/Fragment Coefficients, f_k for log K_{iow} Estimation at 25°C (Eqs. 7-15 and 7-16) [a]

Atom/Fragment	f_k	Atom/Fragment	f_k
Carbon		*Carbonyls*	
–CH$_3$	0.55	al–CHO	–0.94
–CH$_2$–	0.49	ar–CHO	–0.28
–CH<	0.36	al–CO–al	–1.56
>C<	0.27	ol–CO–al	–1.27
=CH$_2$	0.52	ar–CO–al	–0.87
=CH– or =C<	0.38	ar–CO–ar	–0.20
C$_{ar}$	0.29	al–COO– (ester)	–0.95
		ar–COO– (ester)	–0.71
Halogens		al–CON< (amide)	–0.52
al–F	0.00	ar–CON< (amide)	0.16
ar–F	0.20	>N–COO– (carbamate)	0.13
al–Cl	0.31	>N–CO–N< (urea)	1.05
ol–Cl	0.49	al–COOH	–0.69
ar–Cl	0.64	ar–COOH	–0.12
al–Br	0.40		
ar–Br	0.89	*Nitrogen-Containing Groups*	
al–I	0.81	al–NH$_2$	–1.41
ar–I	1.17	al–NH–	–1.50
		al–N<	–1.83
		ar–NH$_2$, ar –NH–, ar–N<	–0.92
Aliphatic Oxygen		al–NO$_2$	–0.81
al–O–al	–1.26	ar–NO$_2$	–0.18
al–O–ar	–0.47	ar–N=N–ar	0.35
ar–O–ar	0.29	al–C≡N	–0.92
al–OH	–1.41	ar–C≡N	–0.45
ol–OH	–0.89		
ar–OH	–0.48		
al–O–(P)	–0.02	*Sulfur-Containing Groups*	
ar–O–(P)	0.53	al–SH	
		ar–SH	
Heteroatoms in Aromatic		al–S–al	–0.40
Systems		ar–S–al	0.05
Oxygen	–0.04	al–SO–al	–2.55
Nitrogen in five-member ring	–0.53	ar–SO–al	–2.11
Nitrogen in six-member ring	–0.73	al–SO$_2$–al	–2.43
Nitrogen at fused ring location	0.00	ar–SO$_2$–al	–1.98
Sulfur	0.41	al–SO$_2$N<	–0.44
		ar–SO$_2$N<	–0.21
		ar–SO$_3$H	–3.16
Phosphorus			
⪢P=O	–2.42		
⪢P=S	–0.66		

[a] Data from Meylan and Howard (1995); total number of fragment constants derived: 130; al = aliphatic attachment, ol = olefinic attachment; ar = aromatic attachment.

Table 7.5 Examples of Correction Factors, c_j, for log K_{iow} Estimation at 25°C (Eqs. 7-15 and 7-16) [a]

Description	c_j	Description	c_j
Factors Involving Aromatic Ring Substituent Positions [b]			
o–OH/–COOH	1.19	*o*–N< /two arom. N	1.28
o–OH/–COO–(ester)	1.26	*o*–CH$_3$/–CON< (amide)	–0.74
o–N /–CON< (amide)	0.62	2 × *o* –CH$_3$/–CON< (amide)	–1.13
o–OR/arom. N	0.45	*p*–N /–OH	–0.35
o–OR/two arom. N	0.90	*o,m,p*–NO$_2$/–OH or –N<	0.58
o–N< /arom. N	0.64	*p*–OH/COO–(ester)	0.65
Miscellaneous Factors			
More than one aliph. –COOH	–0.59	Symmetric triazine ring	0.89
More than one aliph. –OH	0.41	Fused aliphatic ring	
α-Amino acid	–2.02	connection [c]	–0.34

[a] Data from Meylan and Howard (1995); total number of correction factors derived: 235.
[b] *o* = *ortho*, *m* = *meta*, *p* = *para* substitution. [c] See Illustrative Example 7.2.

H-acceptor capability of the compound. The factor is negative, if the opposite is true. Examples of the former case are *ortho*-substitutions in aromatic systems leading to *intramolecular* H-bonding (e.g., –COOH/–OH; –COOR/–OH; –OH/–OH), or substituents in any position that decrease the electron density at a polar group (e.g., –OH or –N< with –NO$_2$). Examples in which negative correction factors have to be applied include *ortho*-substitutions in aromatic systems that cause a disturbance of the resonance of a polar group with the aromatic system (the attachment has a more aliphatic character, e.g., –CH$_3$/–CONH$_2$), or the presence of several polar groups leading to a higher overall polarity of the molecule. For a more comprehensive collection of all the 235 correction factors derived for this method, see the paper by Meylan and Howard (1995). Some examples of the use of Eq. 7-15 are given in Illustrative Example 7.2.

Finally, one should recognize that if the log K_{iow} value of a structurally related compound (rel.compd) is known, the estimation expression is simplified and the accuracy of the result is improved using:

$$\log K_{iow} = \log K_{iow}(\text{rel.compd}) - \sum_k n_k \cdot f_k + \sum_k n_k \cdot f_k - \sum_j n_j \cdot c_j + \sum_j n_j \cdot c_j \quad (7\text{-}16)$$

fragments corrections
removed added removed added

Some applications of Eq. 7-16 are given in Illustrative Example 7.3.

Illustrative Example 7.2

Estimating Octanol–Water Partition Constants from Structure Using the Atom/Fragment Contribution Method

Problem

Estimate the K_{iow} values at 25°C of (a) ethylacetate, (b) 2,3,7,8-tetrachloro-dibenzodioxin, (c) the herbicide 2-s-butyl-4,6-dinitrophenol (Dinoseb), (d) the insecticide parathion, and (e) the hormone testosterone using solely the fragment coefficients and correction factors given in Tables 7.4 and 7.5 (Eq. 7-15).

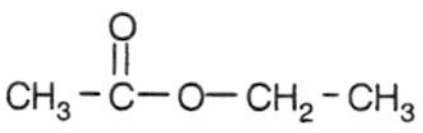

ethyl acetate

Answer (a)

Fragment	f_k	$\times$	n_k	$=$	Value
$-CH_3$	0.55		2		1.10
$-CH_2-$	0.49		1		0.49
al–COO–(ester)	–0.95		1		–0.95
					+0.23

$$\log K_{iow} \text{ (est.)} \quad 0.87$$
$$\text{(exp.)} \quad 0.73$$

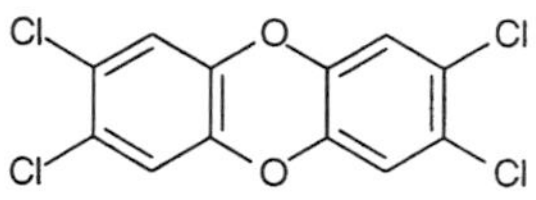

2,3,7,8-tetrachlorodibenzodioxin

Answer (b)

Fragment	f_k	$\times$	n_k	$=$	Value
C_{ar}	0.29		12		3.48
ar–Cl	0.64		4		2.56
ar–O–ar	0.29		2		0.58
					+0.23

$$\log K_{iow} \text{ (est.)} \quad 6.85$$
$$\text{(exp.)} \quad 6.53$$

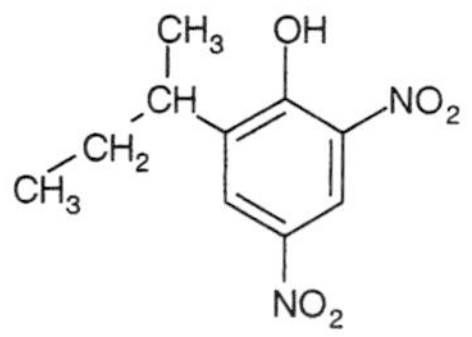

2-s-butyl-4,6-dinitrophenol
(Dinoseb)

Answer (c)

Fragment	f_k	$\times$	n_k	$=$	Value
$-CH_3$	0.55		2		1.10
$-CH_2-$	0.49		1		0.49
$-CH<$	0.36		1		0.36
C_{ar}	0.29		6		1.65
ar–OH	–0.48		1		–0.48
ar–NO$_2$	–0.18		2		–0.36

Corr. Factor	c_j	$\times$	n_j	$=$	Value
o,m,p–NO$_2$/OH	0.58		1		0.58
					+0.23

$$\log K_{iow} \text{ (est.)} \quad 3.57$$
$$\text{(exp.)} \quad 3.56$$

Note: Since it is not clear whether in the case of two nitro substituents the correction factor has to be applied twice, both scenarios have been calculated. Comparison with the experimental value suggests that the correction factor has to be applicated only once.

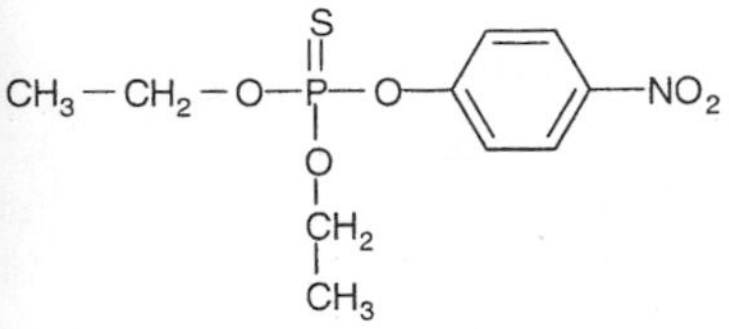

Answer (d)

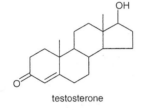

parathion

Fragment	f_k	$\times$	n_k	$=$	Value
–CH$_3$	0.55		2		1.10
–CH$_2$–	0.49		2		0.98
al–O–P	–0.02		2		–0.04
P=S	–0.66		1		–0.66
ar–O–P	+0.53		1		0.53
C$_{ar}$	0.29		6		1.74
ar–NO$_2$	–0.18		1		–0.18
					+0.23

$$\log K_{iow} \text{ (est.)} \quad 3.70$$
$$\text{(exp.)} \quad 3.83$$

Answer (e)

testosterone

Fragment	f_k	$\times$	n_k	$=$	Value
–CH$_3$	0.55		2		1.10
–CH$_2$–	0.49		8		3.92
–CH<	0.36		4		1.44
>C<	0.27		2		0.54
=CH– or =C<	0.38		2		0.76
al–OH	–1.41		1		–1.41
ol–CO–al	–1.27		1		–1.27

Corr. Factor	c_j	$\times$	n_j	$=$	Value
fused aliph ring.corr.	–0.34		6		–2.04
					+0.23

$$\log K_{iow} \text{ (est.)} \quad 3.27$$
$$\text{(exp.)} \quad 3.32$$

Illustrative Example 7.3

Estimating Octanol–Water Partition Constants Based on Experimental K_{iow}'s of Structurally Related Compounds

Problem

Estimate the K_{iow} values at 25°C of the following compounds based on the experimental K_{iow} values of the indicated structurally related compounds: (a) benzoic acid dimethylaminoethyl ester from benzoic acid ethyl ester (log K_{iow} = 2.64), (b) the insecticide methoxychlor from DDT (log K_{iow} = 6.20), (c) the insecticide fenthion from parathion [log K_{iow} = 3.83, see Ill. Ex. 7.2, Answer (d)], and (d) the hormone estradiol from testosterone [log K_{iow} = 3.32, see Ill. Ex. 7.2, Answer (e)].

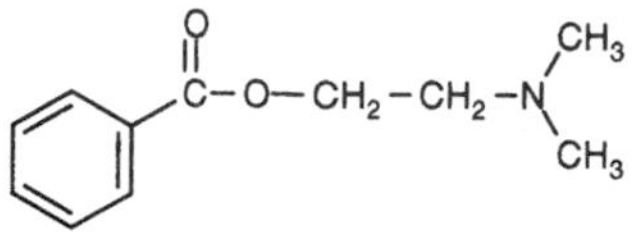

benzoic acid dimethylaminoethylester

Answer (a)

Fragment		f_k	×	n_k	=	Value
Starting	K_{iow}					2.64
Add	$-CH_2-$	0.55		1		0.49
	$-CH_3$	0.55		1		0.55
	al–N<	−1.83		1		−1.83
				$\log K_{iow}$ (est.)		*1.85*
					(exp.)	2.06

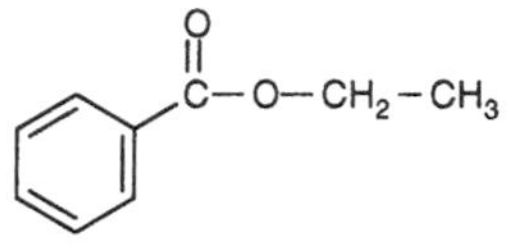

benzoic acid ethylester

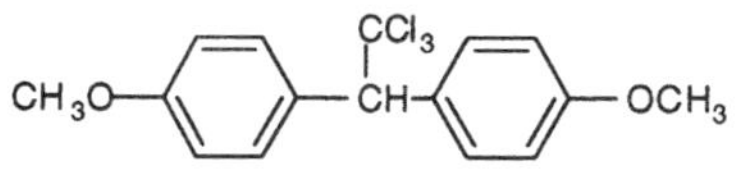

Answer (b)

methoxychlor

Fragment		f_k	×	n_k	=	Value
Starting	K_{iow}					6.20
Remove	ar–Cl	0.64		2		−1.28
Add	$-CH_3$	0.55		2		1.10
	al–O–ar	−0.47		2		−0.94
				$\log K_{iow}$ (est.)		*5.08*
					(exp.)	5.08

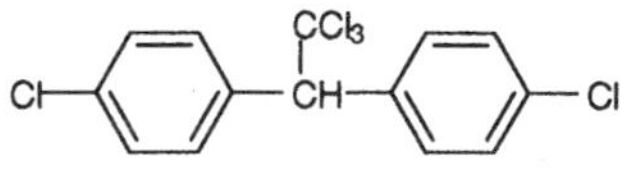

DDT

Hence, by substitution of 2 chlorine atoms by two methoxy groups the K_{iow} value is lowered by one order of magnitude. Such insights are used by chemical manufacturers to adjust chemical properties to suit specific purposes.

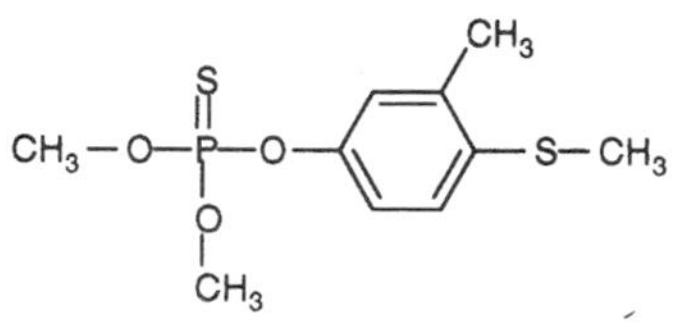

Answer (c)

fenthion

Fragment		f_k	×	n_k	=	Value
Starting	K_{iow}					3.83
Remove	$-CH_2-$	0.49		2		−0.98
	ar–NO$_2$	−0.18		1		+0.18
Add	$-CH_3$	0.55		2		+1.10
	ar–S–al	0.05		1		+0.05
				$\log K_{iow}$ (est.)		*4.18*
					(exp.)	4.10

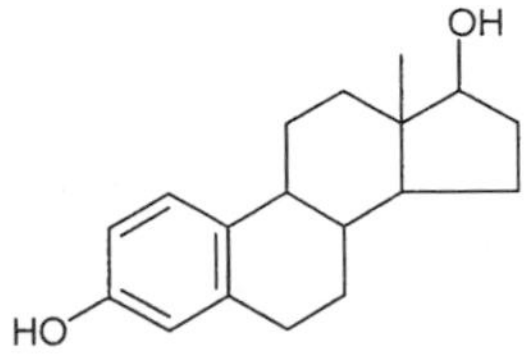

estradiol

Answer (d)

Fragment/ Corr. Factor		f_k/c_j	$\times$	n_k/n_j	$=$	Value
Starting	K_{iow}					3.32
Remove	$-CH_3$	0.55		1		-0.55
	$-CH_2-$	0.49		2		-0.98
	$>C<$	0.27		1		-0.27
	$=CH-$ or $=C<$	0.38		2		-0.76
	ol–CO–al	-1.27		1		$+1.27$
	fused aliph. ring.corr.	-0.34		2		$+0.68$?
Add	C_{ar}	0.29		6		$+1.74$
	ar–OH	0.48		1		$\underline{-0.48}$
				$\log K_{iow}$ (est.)		*3.97*
					or	*3.29*

Advanced Topic   **7.5**

Dissolution of Organic Compounds in Water from Organic Liquid Mixtures—Equilibrium Considerations

There are numerous cases of environmental contaminations in which we need to know how organic compounds present in *liquid organic mixtures* partition into an aqueous phase. Such cases include the dissolution of compounds from the mixtures into water from so-called *light non aqueous phase liquids* (LNAPLs, e.g., gasoline, diesel fuel, heating oil) or *dense non aqueous phase liquids* (DNAPLs, e.g., coal tars, creosotes, chlorinated solvent mixtures, PCBs, hydraulic oils containing PCBs). The density distinction is made because if the liquid mixture density is greater than that of water, then the mixture tends to "fall" through water bodies and reside at loci like bedrock underlying aquifers or at river bottoms. In contrast, LNAPLs float on water tables or at the air–water interface.

In this section, we evaluate the factors that determine the concentration of a given component of an LNAPL or DNAPL in an adjacent aqueous phase that is *in equilibrium* with the organic mixture. Hence, we consider a snapshot of the situation where we assume a constant composition of the liquid organic mixture. Of course, in reality, when exposed continuously to "clean" water, the composition of an LNAPL or DNAPL may change significantly with time, because a given mixture will become depleted in the more water–soluble compounds. Furthermore, depending on the contact time and contact area between the organic phase and the water, equilibrium may not be established. Therefore, a mass transfer approach has to be taken to describe the dissolution process. However, even for modeling the dissolution kinetics, the equilibrium partitioning of a given compound needs to be known to quantify the mass transfer gradient (see Part IV).

As a starting point for describing the equilibrium partitioning of a given compound *i* between a liquid organic mixture (subscript "mix") and an aqueous phase (subscript "w"), we rewrite Eq. 7-1 as:

$$x_{iw} = x_{imix} \cdot \gamma_{imix} \cdot \gamma_{iw}^{-1} \qquad (7\text{-}17)$$

or in terms of molar concentrations (Eq. 7-2):

$$C_{iw} = C_{imix} \cdot \overline{V}_{mix} \cdot \gamma_{imix} \cdot \left(\gamma_{iw} \cdot \overline{V}_w \right)^{-1} \qquad (7\text{-}18)$$

In order to calculate the aqueous concentration of compound i at equilibrium, one needs to know its mole fraction, x_{imix}, in the mixture (or its molar concentration, C_{imix}, and the molar volume, $\overline{V}_{mix}$, of the mixture), as well as its activity coefficients in the organic (γ_{imix}) and the aqueous (γ_{iw}) phases. Very often, when dealing with complex mixtures, $\overline{V}_{mix}$ is not known and has to be estimated. At a first approximation, this can be done from the density, ρ_{mix}, of the liquid mixture, and by assuming an average molar mass, $\overline{M}_{mix}$, of the mixture components:

$$\overline{V}_{mix} \cong \frac{\overline{M}_{mix}}{\rho_{mix}} \qquad (7\text{-}19)$$

For example, the molar mass of gasoline has been estimated as near $110 \ \mathrm{g \cdot mol^{-1}}$ and that of coal tar as near $150 \ \mathrm{g \cdot mol^{-1}}$ (Masters, 1998; Picel et al., 1988).

Let us now consider the various factors that may influence the equilibrium partitioning of an organic compound between an organic mixture and water. First, there are some cases in which the organic mixture originally contains a significant amount of a highly water–soluble compound that, at equilibrium with a water phase, may have dissolved to a great extent into the water leading to a cosolvent effect as discussed in Section 5.4. Prominent examples include the presence of oxygenated compounds such as methyl-t-butyl ether (MBTE), methanol, or ethanol in gasoline (Cline et al., 1991; Poulsen et al., 1992; Heermann and Powers, 1998). We recall from Section 5.4 that we can neglect the cosolvent effect in the water if the volume fraction of the organic solvent does not exceed 0.01 (1%). However, in some countries, by law, such polar compounds may make up 10 to 20% of the gasoline. In these cases, cosolvent effects in the "aqueous" phase may be significant (for more details see, e.g., Heermann and Powers, 1998). To illustrate, the activity coefficient of naphthalene in a 20% ethanol/80% water mixture is 7 times smaller than in pure water (Table 5.8). In the following, we will focus on those cases for which we may assume that the effect of other dissolved mixture constituents on the *aqueous* activity coefficient of a given compound is minimal. Furthermore, we also neglect the effect of salts on γ_{iw} (which has the opposite effect of a cosolvent, see Section 5.4, Eq. 5-28), which we would need to consider when dealing with the pollution of the marine environment or groundwater brines. Thus, for compounds for which $\gamma_{iw} \cong \gamma_{iw}^{sat}$ (Section 5.2), we may substitute the term $(\gamma_{iw} \overline{V}_w)^{-1}$ in Eq. 7-18 by the *liquid* aqueous solubility, $C_{iw}^{sat}(L)$, of the compound:

$$C_{iw} = C_{imix} \cdot \overline{V}_{mix} \cdot \gamma_{imix} \cdot C_{iw}^{sat}(L) \qquad (7\text{-}20)$$

or with $x_{imix} = C_{imix} \cdot \overline{V}_{mix}$:

$$C_{iw} = x_{imix} \cdot \gamma_{imix} \cdot C_{iw}^{sat}(L) \qquad (7\text{-}21)$$

By rearranging Eq. 7-20, we may then also express the organic mixture–water partition coefficient, $K_{i\text{mix w}}$, as:

$$K_{i\text{mix w}} = \frac{C_{i\text{mix}}}{C_{iw}} = \frac{1}{\gamma_{i\text{mix}} \cdot \overline{V}_{\text{mix}} \cdot C_{iw}^{\text{sat}}(\text{L})} \qquad (7\text{-}22)$$

Let us now evaluate in which cases we may, as a first approximation, assume that *Raoult's law* is valid. Said another way, in what cases may we set $\gamma_{i\text{mix}}$ in Eqs. 7-20 to 7-22 equal to 1? From our earlier discussions on the molecular factors that determine the magnitude of the activity coefficient of an organic compound in an organic liquid, we would expect more or less ideal behavior, that is, $\gamma_{i\text{mix}}$ values not too different from 1 for (i) apolar compounds in mixtures in which the major components undergo primarily vdW interactions, and (ii) monopolar compounds in the same situation, but with the restriction that in the mixture there are no major constituents exhibiting a significant complementary polarity. As confirmed by experimental data and by model calculations using approaches such as the UNIFAC method (e.g., Peters et al., 1999b), examples meeting the above criteria include aliphatic hydrocarbons and BTEX compounds (Fig. 2.13) present in most gasolines and in other fuels (Cline et al., 1991; Hemptinne et al., 1998; Heermann and Powers, 1998; Garg and Rixey, 1999), the components of mixtures of chlorinated solvents (Broholm and Feenstra, 1995), PAHs present in diesel fuels, coal tars, and creosotes (Lane and Loehr, 1992; Lee et al., 1992a,b), and PCB congeners present in pure PCB mixtures (e.g., Aroclor 1242) or in hydraulic oils consisting of other types of compounds [e.g., trialkylphenylphosphates (Luthy et al., 1997a); see also Illustrative Example 7.4]. In all these cases, the $\gamma_{i\text{mix}}$ values determined were found to meet the Raoult's law criteria within less than a factor of 2 to 3, and, therefore, for practical purposes, $\gamma_{i\text{mix}}$ can be approximated as 1.

We should note that, particularly for bipolar compounds such as, for example, certain additives in gasoline (e.g., phenolic compounds, aromatic amines, see Chapter 2), larger deviations from ideal behavior have to be expected (Schmidt et al., 2002). In addition, it should be pointed out that in mixtures containing major quantities of polar compounds, the activity coefficients of the various mixture compounds may change with time if these polar constituents are depleted during the dissolution process. Furthermore, when using organic mixture–water partition coefficients as defined by Eq. 7-22, changes in the molar volume of the mixture as a consequence of the preferential dissolution of the more water-soluble components may have to be considered. Finally, we should be aware that the preferential dissolution of more soluble compounds in a mixture leads to a higher concentration of the less soluble compounds and thus to increasing concentrations in the aqueous phase. This has to be taken into account when evaluating the long-term dynamics of complex organic mixtures in the environment (e.g., Mackay et al., 1996; Peters et al., 1999a).

Illustrative Example 7.4

Fyrquel 220

Chemical structure of Fyrquel 220 hydraulic oil (R is primarily *n*-butyl)

Estimating the Concentrations of Individual PCB Congeners in Water that Is in Equilibrium with an Aroclor and an Aroclor/Hydraulic Oil Mixture

Aroclor 1242 is a commercial PCB mixture with an average chlorine content of 42%, an average molar mass $\overline{M}_{\text{Aroclor}}$ of about 265 g·mol^{-1}, and a density ρ_{Aroclor} of 1.39 g·cm^{-3} at 25°C. Luthy et al. (1997a) have determined the composition of a pure Aroclor 1242 mixture, and they have measured the aqueous concentrations of some individual congeners established at 25°C in equilibrium with (a) the real Aroclor 1242 mixture and (b) a mixture of 5% (v/v) Aroclor 1242 in a hydraulic oil (Fyrquel 220) consisting of trialkyl-phenylphosphates (see margin), with an average molar mass $\overline{M}_{\text{Fyrquel}}$ of about 380 g·mol^{-1} at a density r_{Fyrquel} of 1.14 g·cm^{-3} at 25°C.

Problem

Among the congeners investigated was 2,2',5,5'-tetrachlorobiphenyl (TeClBP), which was determined to be present in the Aroclor 1242 mixture at about 3.2 mass percent (i.e., mass fraction $m_{i\,\text{Aroclor}} = 0.032$ g$_i$·g$_{\text{Aroclor}}^{-1}$). The measured aqueous concentrations for this compound were 1.11 μg·L^{-1} (case a) and 0.10 μg·L^{-1} (case b), respectively. Are these concentrations reasonable? What aqueous TeClBP concentrations would you have predicted from the above information, when assuming that Raoult's law is valid in both cases?

i - 2,2'5,5'-tetrachlorobiphenyl
(TeClBP)

$M_i = 292.0$ g·cm^{-3}
$T_m = 86.5$°C
$C_{iw}^{\text{sat}} = 10^{-7}$ mol·L^{-1}
(see Appendix C)

Answer (a)

Convert the mass fraction ($m_{i\,\text{Aroclor}} = 0.032$) of TeClBP in the Aroclor 1242 mixture into the mole fraction by using the average molar mass, $\overline{M}_{\text{Arcolor}}$, of 265 g·mol^{-1}:

$$x_{i\,\text{Aroclor}} = m_{i\,\text{Aroclor}} \cdot \frac{\overline{M}_{\text{Aroclor}}}{M_i} = (0.032)\frac{(265)}{(292)} = 0.029$$

Estimate the liquid aqueous solubility of TeClBP from its aqueous solubility using T_m (Eqs. 5-13 and 4-41). The resulting C_{iw}^{sat} (L) value is 3.5×10^{-7} mol·L^{-1}. Insert this value together with the above calculated $x_{i\,\text{Aroclor}}$ value and $\gamma_{i\,\text{Aroclor}} = 1$ into Eq. 7-19 to get the estimated aqueous concentration of TeClBP:

$$C_{iw} = (0.029)(1)(3.5 \times 10^{-7}) = 1.0 \times 10^{-8}\,\text{mol·L}^{-1}$$

or about *3 μg·L^{-1}*. This value is three times higher than the measured one, but it is well within the same order of magnitude. Since it is rather unlikely that the apolar TeClBP has a $\gamma_{i\,\text{Aroclor}}$ value significantly smaller than 1, the discrepancy is more likely to be due to uncertainties in the measured mole fraction in the Aroclor 1242 mixture, in the aqueous concentration, and/or in the subcooled liquid solubility of TeClBP.

Answer (b)

Calculate the molar volumes of Aroclor 1242 and Fyrquel from the corresponding average molar masses and densities:

$$\overline{V}_{\text{Aroclor}} = \frac{265}{1.39} = 191 \text{ cm}^3 \text{ mol}^{-1} = 0.191 \text{ L} \cdot \text{mol}^{-1}$$

$$\overline{V}_{\text{Fyrquel}} = \frac{380}{1.14} = 333 \text{ cm}^3 \text{ mol}^{-1} = 0.333 \text{ L} \cdot \text{mol}^{-1}$$

Consider now one liter of the 5% (v/v) Aroclor/Fyrquel mixture. In this liter there are $(0.05)/(0.191) = 0.262$ moles total PCBs and $(0.95)/(0.333) = 2.85$ moles Fyrquel compounds, which yield a total number of 3.11 moles. The mole fraction of TeClBP in this mixture is then given by [recall that $x_{i\,\text{Aroclor}} = 0.029$, see Answer (a)]:

$$x_{i\,\text{mix}} = \frac{(0.029)(0.262)}{(3.11)} = 0.0024$$

Insertion of this value together with $C_{iw}^{\text{sat}} (\text{L}) = 3.5 \times 10^{-7} \text{ mol} \cdot \text{L}^{-1}$ and $\gamma_{i\,\text{mix}} = 1$ into Eq. 7-19 yields

$$C_{iw} = (0.0024)\,(1)\,(3.5 \times 10^{-7} \text{ mol} \cdot \text{L}^{-1}) = 8 \times 10^{-9} \text{ mol} \cdot \text{L}^{-1}$$

or about *0.25* μg·L^{-1}. Again, this value is about 3 times higher than the measured one, which is very consistent with the result obtained for the pure Aroclor 1242 mixture. This suggests that the activity coefficients of TeClBP in Aroclor 1242 and in the Aroclor 1242/Fyrquel mixture are quite similar. From a molecular interaction point of view, this conclusion seems also reasonable, since the phosphate esters making up the Fyrquel mixtures are monopolar and exhibit about the same π_i values as the apolar PCBs (Abraham et al., 1994b).

7.6   Questions and Problems

Questions

Q 7.1

Give several reasons why it is important to know something about the partitioning behavior of a given compound between organic solvents and water.

Q 7.2

Which is(are) the dominating factor(s) determining the organic solvent–water partitioning of the majority of organic compounds of environmental concern?

Q 7.3

Why is the effect of temperature on organic solvent–water partitioning of organic compounds in many cases not very significant? What maximum $\left| \Delta_{\ell w} H_i \right|$ values would you expect? Give examples of solutes and organic solvents for which you would expect (a) a substantial positive (i.e., > 10 kJ·mol^{-1}) and (b) a substantial negative (i.e., < -10 kJ·mol^{-1}) $\Delta_{\ell w} H_i$ value.

Q 7.4

When comparing the $K_{i\ell w}$ values of the stimulant amphetamine for the solvents trichloromethane (chloroform, log K_{icw} = 2.20), n-octanol (log K_{iow} = 1.80), and n-heptane (log K_{ihw} = 0.40), one can see that they differ quite substantially. Try to explain these differences.

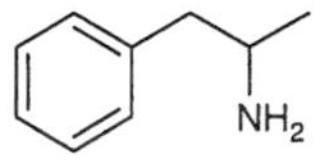

i = amphetamine
(2-aminopropylbenzene)

Q 7.5

Imagine a compound for which $\Delta_{\ell w}G_i$ is equal to zero ($G_{i\ell}^E = G_{iw}^E$) in each of the solvent–water systems trichloromethane (chloroform)–water, n-octanol–water, and n-hexadecane–water. A friend of yours claims that the $K_{i\ell w}$ (= $C_{i\ell} / C_{iw}$) values of such a compound are 0.22, 0.11, and 0.06, respectively, for the three solvent–water systems. Another friend disagrees and claims that the $K_{i\ell w}$ values are all equal to 1. Who is right and why?

Q 7.6

What are the prerequisites for a successful estimation of $K_{i\ell w}$ (e.g., K_{iow}) by liquid chromatography?

Q 7.7

What are the major difficulties of any atom/fragment contribution method for estimation of solvent–water partition constants from structure?

Q 7.8

What are the major factors determining the aqueous concentration of a constituent of a liquid organic mixture (LNAPL, DNAPL) that is in equilibrium with an aqueous phase? Explain Raoult's law and give some practical examples of (a) cases in which you can apply it to estimate the concentration of a given LNAPL or DNAPL constituent in water that is in equilibrium with the organic liquid, and (b) cases in which Raoult's law does not hold.

Q 7.9

When flushing a gasoline-contaminated soil in a laboratory column with clean water, Mackay et al. (1996) observed that after 5 pore volumes (i.e., after 5 times replacing the water in the column), the benzene concentration in the effluent decreased from initially 370 to about 75 $\mu g \cdot L^{-1}$, while the 1,2-dimethylbenzene concentration increased from 1200 to 1400 $\mu g \cdot L^{-1}$. Try to explain these findings.

Problems

P 7.1 *Estimating Activity Coefficients of Organic Compounds in Organic Solvents*

Calculate the activity coefficients of (a) *n*-octane, and (b) aniline in water-saturated (see Table 5.1) *n*-hexane (γ_{ih}), toluene (γ_{it}), diethylether (γ_{id}), chloroform (γ_{ic}), *n*-octanol (γ_{io}), and in water γ_{iw} from the $K_{i\ell w}$ values given in Table 7.1. The aqueous solubilities of the two compounds are given in Appendix C. Compare and discuss the results.

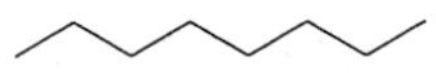

n-octane 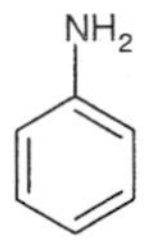   aniline

P 7.2 *Some Additional K_{iow} Estimation Exercises Using the Atom/Fragment Contribution Method*

Estimate the K_{iow} values of the four compounds indicated below (a) by using only fragment constants and correction factors (Eq. 7-15), and (b) by starting with the K_{iow} value of a structurally related compound (Eq. 7-16) that you choose from Appendix C. Discuss the results by comparing them with the indicated experimental K_{iow} values.

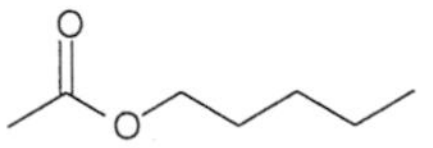

n-pentyl acetate
(exp. log K_{iow} = 2.23)

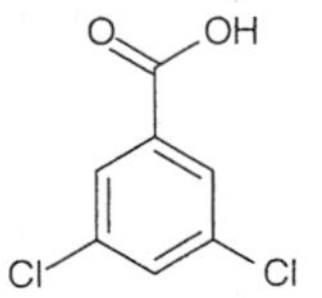

3,5-dichloro benzoic acid
(exp. log K_{iow} = 3.00)

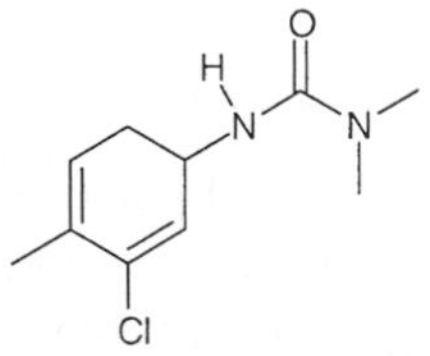

chlortoluron
(exp. log K_{iow} = 2.41)

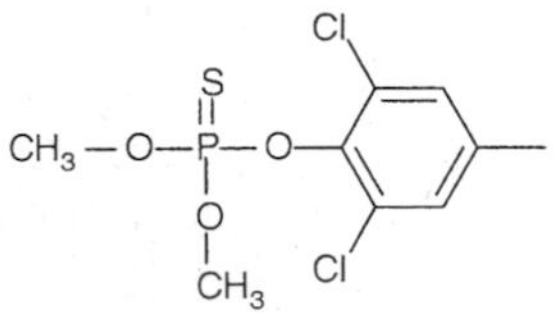

tolclofos-methyl
(exp. log K_{iow} = 4.56)

P 7.3 *Extraction of Organic Pollutants from Water Samples*

For analyzing organic pollutants in water, the compounds are commonly preconcentrated by adsorption, stripping (see Problem 6.6), or extraction with an organic solvent. You have the job to determine the concentration of 1-naphthol in a contaminated groundwater by using gas chromatography. You decide to extract 20 mL water samples with a convenient solvent. In the literature (Hansch and Leo, 1979) you find the following $K_{i\ell w}$ value for a series of solvents:

Solvent ℓ	$\log K_{i\ell w}$
n-hexane	0.52
benzene	1.89
trichloromethane (chloroform)	1.82
ethyl acetate (acetic acid ethyl ester)	2.60
n-octanol	2.90

Are you surprised to find such big differences in the $K_{i\ell w}$ values of 1-naphtol for the various solvents? If not, try to explain these differences. You choose ethyl acetate as solvent for the extraction. Why not *n*-octanol?

Now you wonder how much ethyl acetate you should use. Calculate the volume of ethyl acetate that you need at minimum if you want to extract at least 99% of the total 1-naphthol present in the water sample. Are you happy with this precon-centration step? Somebody tells you that it would be much wiser to extract the sample twice with the goal to get each time 90% of the total compound out of the water (which would also amount to 99%), and then pool the two extracts. How much total ethyl acetate would you need in this case? Finally, another colleague suggests to add 3.56 g NaCl to the 20 mL sample in order to improve the extraction efficiency. How much less ethyl acetate would be required in the presence of the salt? (*Hint:* Consult Table 5.7.) Is there any other effect that the addition of NaCl would have on the extraction, and is this effect favorable for the analytical procedure chosen?

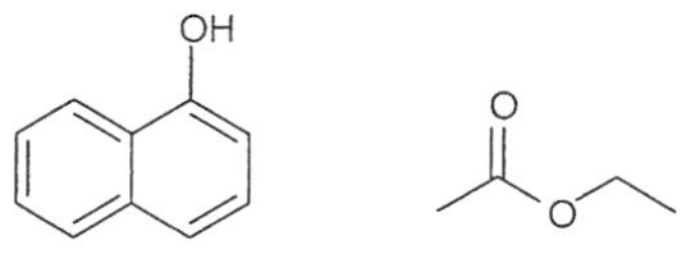

i = 1-naphthol   ethyl acetate

P 7.4   *A Small Accident in Your Kitchen*

In your kitchen ($T = 25°C$) you drop a small bottle with 20 mL of the solvent 1,1,1-trichloroethane (methyl chloroform, MCF) that you use for cleaning purposes. The bottle breaks and the solvent starts to evaporate. The doors and the windows are closed. On your stove there is an open pan containing 2 L of cold olive oil. Furthermore, on the floor there is a large bucket that is filled with 50 L of water. The air volume of the kitchen is 30 m³. Calculate the concentration of MCF in the air, in the water in the bucket, and in the olive oil at equilibrium by assuming that the adsorption of MCF to any other phases/surfaces present in the kitchen can be neglected. Consider MCF as an apolar compound. You can find some important physical–chemical data in Appendix C and in Fig. 6.7. Comment on any assumption that you make.

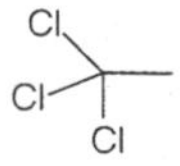

1,1,1-trichloroethane
(methyl chloroform, MCF)

P 7.5 *Evaluating Partition Constants of Chlorinated Phenols in Two Different Organic Solvent-Water Systems*

Kishino and Kobayashi (1994) determined the *n*-octane–water (K_{ioctw}) and *n*-octanol–water (K_{iow}) partition constants of a series of chlorinated phenols (see Table below). Plot the log K_{ioctw} values versus the log K_{iow} values of the 13 compounds. Inspect the data and derive meaningfull LFERs of the type Eq. 7-7 for subsets of compounds. Discuss your findings in terms of the molecular interactions that govern the partitioning of the chlorinated phenols in the two different solvent–water systems.

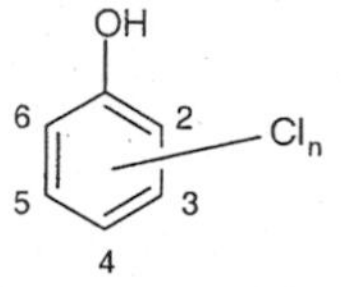

chlorinated phenols

	compound	log K_{ioctw}	log K_{iow} [a]
1	Phenol	−0.99	1.57
2	2-Chlorophenol	0.74	2.29
3	3-Chlorophenol	−0.31	2.64
4	4-Chlorophenol	−0.41	2.53
5	2,3-Dichlorophenol	1.27	3.26
6	2,4-Dichlorophenol	1.21	3.20
7	2,5-Dichloropehol	1.31	3.36
8	2,6-Dichlorophenol	1.48	2.92
9	3,5-Dichlorophenol	0.41	3.60
10	2,4,5-Trichlorophenol	1.76	4.02
11	2,4,6-Trichlorophenol	2.05	3.67
12	2,3,4,6-Tetrachlorophenol	2.58	4.24
13	Pentachlorophenol	3.18	5.02

[a] Values given in Appendix C may differ somewhat from the ones determined by Kishino and Kobayashi (1994).

P 7.6 *Assessing the Dissolution Behavior of Gasoline Components*

Gasoline is a mixture of primarily aliphatic (>50%) and aromatic (~ 30%) hydrocarbons with an average molar mass, $\overline{M}_{gas}$, of about 105 g·mol^{-1} and a density of about 0.75 g·m^{-3} (Cline et al., 1991). In addition it contains a variety of additives including, for example, oxygenates (see Section 7.5), antioxidants, corrosion inhibitors, detergents, antifreezing agents, dyes, and many more (Owen, 1989).

You are confronted with a gasoline spill underneath a gas station. Among the compounds that are of great concern with respect to groundwater pollution are benzene and 3,4-dimethylaniline (DMA). You know that the spilled gasoline contains 2 volume percent benzene and 10 mg·L^{-1} DMA. Furthermore, in the literature you find experimental gasoline–water partition coefficients (Eq. 7-22) of 300 L$_w$·L$_{gasoline}^{-1}$ for benzene (Cline et al., 1991) and 30 L$_w$·L$_{gasoline}^{-1}$ for DMA [determined by Schmidt et al. (2002) at pH8 where DMA is present primarily as neutral species (see Chapter 8)]. Note that these coefficients have been determined for other brands of gasoline. Answer now the following questions.

(a) Using the gasoline–water partition coefficients reported in the literature (see above), calculate the activity coefficients of benzene and DMA in the gasoline mixture of interest. Which of the two values do you trust more?

(b) What benzene and DMA concentration would you expect in groundwater that is in equilibrium with a large pool of the spilled gasoline at 25°C (i.e., assume that the gasoline composition is not altered significantly by the dissolution of the components in the aqueous phase).

(c) In the aqueous phase that is in equilibrium with the spilled gasoline, you measure a naphthalene concentration of 1 mg·L^{-1}. How much naphthalene does the gasoline contain? Comment on any assumption that you make.

You find all necessary data in Appendix C.

Hint: To estimate the mole fraction of a given gasoline component from its volume fraction, use Eq. 5-34 by assuming a binary mixture of the component with a solvent that has the average molar volume of the whole gasoline mixture.

benzene 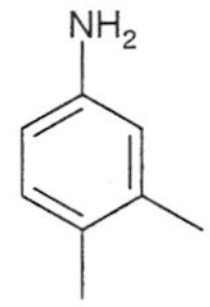   3,4-dimethylaniline   naphthalene
                      (3,4-DMA)

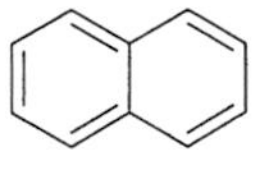

Chapter 8

ORGANIC ACIDS AND BASES: ACIDITY CONSTANT AND PARTITIONING BEHAVIOR

8.1   **Introduction**

8.2   **Thermodynamic Considerations**
Organic Acids, Acidity Constant
Organic Bases
Effect of Temperature on the Acidity Constant
Speciation in Natural Waters
Illustrative Example 8.1: *Assessing the Speciation of Organic Acids and Bases in Natural Waters*

8.3   **Chemical Structure and Acidity Constant**
Overview of Acid and Base Functional Groups
Inductive Effects
Delocalization Effects
Proximity Effects

8.4   **Availability of Experimental Data; Methods for Estimation of Acidity Constants**
Experimental Data
Estimation of Acidity Constants: The Hammett Correlation
Illustrative Example 8.2: *Estimating Acidity Constants of Aromatic Acids and Bases Using the Hammett Equation*

8.5   **Aqueous Solubility and Partitioning Behavior of Organic Acids and Bases**
Aqueous Solubility
Air–Water Partitioning
Illustrative Example 8.3: *Assessing the Air–Water Distribution of Organic Acids and Bases in a Cloud*
Organic Solvent–Water Partitioning

8.6   **Questions and Problems**

8.1 Introduction

Until now, we have confined our discussions about compound properties to *neutral* chemical species. Some important environmental organic chemicals may, however, undergo proton transfer reactions resulting in the formation of *charged* species (e.g., anions or cations: see examples given in Tables 8.1 and 8.2). These charged species have *very different properties and reactivities as compared to their neutral counterparts*; thus, it is important to know whether and to what extent the molecules of an organic compound may form ions in a given environmental system.

A proton transfer can occur only if an *acid* (HA), that is, a proton *donor* (Brønsted and Pedersen, 1924), reacts with a *base* (B), that is, a *proton acceptor*, since isolated protons are quite unstable species:

$$\begin{array}{rcl} HA & \rightleftharpoons & A^- + H^+ \\ H^+ + B & \rightleftharpoons & BH^+ \\ \hline HA + B & \rightleftharpoons & BH^+ + A^- \end{array} \qquad (8\text{-}1)$$

Note that A^- is called the *conjugate* base of HA and BH^+ the conjugate acid of B. Proton transfer reactions as described by Eq. 8-1 are usually *very fast and reversible*. It makes sense then that we treat such reactions as *equilibrium processes*, and that we are interested in the equilibrium distribution of the species involved in the reaction. In this chapter we confine our discussion to proton transfer reactions in *aqueous solution*, although in some cases, such reactions may also be important in nonaqueous media. Our major concern will be the speciation of an organic acid or base (neutral versus ionic species) in water under given conditions. Before we get to that, however, we have to recall some basic thermodynamic aspects that we need to describe acid–base reactions in aqueous solution.

8.2 Thermodynamic Considerations

Organic Acids, Acidity Constant

Let us first consider the reaction of an organic acid (HA) with water which, in reaction Eq. 8-1, plays the role of the base:

$$HA + H_2O \rightleftharpoons H_3O^+ + A^- \qquad (8\text{-}2)$$

So far, we have used the pure liquid compound as reference state for describing the thermodynamics of transfer processes between different media (Chapter 3). When treating reactions of several different chemical species in one medium (e.g., water) it is, however, much more convenient to use the infinite dilution state in that medium as the reference state for the solutes. Hence, for acid–base reactions in aqueous solutions, in analogy to Eq. 3-34, we may express the chemical potential of the solute i as:

$$\mu_i = \mu_i^{0'} + RT \ln \frac{\gamma_i'[i]}{(\gamma_i^{0'} = 1)[i]^0} \qquad (8\text{-}3)$$

In this case, the standard chemical potential corresponds to the standard free energy of formation of the species i in *aqueous solution*, that is:

$$\mu_i^{0'} \equiv \Delta_f G_i^0 (aq) \tag{8-4}$$

at given p^0 and T (1 bar, e.g. 25°C), γ_i' is the activity coefficient, $[i]$ is the actual concentration of the species i, and $[i]^0$ is its concentration in the *standard* state. For our treatment of acid–base reactions we will set $[i]^0 = 1$ M.

Note that we use the prime superscript to denote the infinite dilution reference state (as opposed to the pure liquid state), and that we omit any subscript to indicate that we are dealing with aqueous solutions. Also note that because we have chosen the aqueous solution as reference state, γ_i' will in many cases not be substantially different from 1. Exceptions are the charged species at high ionic strength as, for example, encountered in seawater (see below).

Before we go on to define the acidity constant of a given organic acid in water, we need to introduce a thermodynamic convention for scaling such constants. We do this relative to H_3O^+ in that we define the dissociation of H_3O^+ in water to have a standard free-energy change $\Delta_r G^0 = 0$, which means that the equilibrium constant of this reaction is equal to 1:

$$H_3O^+ \rightleftharpoons H_2O + H^+ \ ; \Delta_r G^0 = 0 \ J \cdot mol^{-1} \ or \ K = 1 \tag{8-5}$$

Note that this is equivalent to the definition that the free energy of formation of the proton in aqueous solution is equal to zero [i.e., $\Delta_f G_{H^+}^0 (aq) = 0$ at any temperature (for more details see, e.g., Stumm and Morgan, 1996)]. Hence, we can rewrite Eq. 8-2 as:

$$HA \rightleftharpoons H^+ + A^- \tag{8-6}$$

By setting $[i]^0 = 1$ M and $\mu_{H^+}^{0'} \equiv \Delta_f G_{H^+}^0 (aq) = 0$ J·mol^{-1} we can now express the chemical potentials of the three species in reaction Eq. 8-6 by:

$$\mu_{HA} = \mu_{HA}^{0'} + RT \ln \gamma_{HA}' [HA]$$

$$\mu_{H^+} = RT \ln \gamma_{H^+}' [H^+] \tag{8-7}$$

$$\mu_{A^-} = \mu_{A^-}^{0'} + RT \ln \gamma_{A^-}' [A^-]$$

Let us now first consider the situation in which equilibrium of reaction Eq. 8-6 is not yet reached. From Eqs. 3-20 and 3-21 (Chapter 3) it is easy to see that if the reaction proceeds by the increment dn_r, the change in the total free energy of the system considered is given by:

$$dG = -\mu_{HA} \, dn_r + \mu_{H^+} \, dn_r + \mu_{A^-} \, dn_r \tag{8-8}$$

The quantity dG/dn_r is referred to as the *free energy of reaction*, which we denote as $\Delta_r G$. Hence:

$$\Delta_r G = -\mu_{HA} + \mu_{H^+} + \mu_{A^-} \tag{8-9}$$

We will discuss this quantity in more detail in Chapter 12 where we introduce reactions for which we may not assume that they are in equilibrium. Here, we are interested only in the equilibrium situation; that is, the situation in which $\Delta_r G = 0$. By inserting Eq. 8-7 into Eq. 8-9 and setting $\Delta_r G = 0$ we obtain after some rearrangement:

$$0 = -\mu_{HA}^{0'} + \mu_{A^-}^{0'} + RT \ln \frac{\left(\gamma_{H^+}' [H^+]\right)\left(\gamma_{A^-}' [A^-]\right)}{\left(\gamma_{HA}' [HA]\right)} \tag{8-10}$$

The algebraic sum of the *standard chemical potentials* is called the *standard free energy of reaction* and is denoted as $\Delta_r G^0$. In our case:

$$\Delta_r G^0 = -\mu_{HA}^{0'} + \mu_{A^-}^{0'} = -\Delta_f G_{HA}^0(aq) + \Delta_f G_{A^-}^0(aq) \tag{8-11}$$

Hence, the equilibrium constant, K_{ia}, which is commonly referred to as *acidity constant* or *acid dissociation constant* of the acid $i = HA$, is given by:

$$K_{ia} = \frac{\left(\gamma_{H^+}' [H^+]\right)\left(\gamma_{A^-}' [A^-]\right)}{\left(\gamma_{HA}' [HA]\right)} = e^{-\Delta_r G^0 / RT} \tag{8-12}$$

When determining K_{ia} values of organic acids, one generally uses techniques by which the hydrogen ion activity [pH = $-\log (\gamma_{H^+}' [H^+])$] is measured, while HA and A$^-$ are determined as molar concentrations. Thus, many acidity constants reported in the literature are so-called "mixed acidity constants" which are operationally defined for a given aqueous medium (e.g., 0.05 – 0.1 M salt solution):

$$K_{ia}^* \equiv (\gamma_{H^+}' [H^+]) \frac{[A^-]}{[HA]} = K_{ia} \frac{\gamma_{HA}'}{\gamma_{A^-}'} \tag{8-13}$$

In some cases, the "true" thermodynamic K_{ia} values are extrapolated by measurements of K_{ia}^* values at different ionic strengths, and by estimating the activity coefficients, particularly γ_{A^-}', using, for example, the Davies equation (for details see Stumm and Morgan, 1996). Considering the uncertainties in the measurements, it is, however, reasonable to assume that at the moderate to low ionic strength at which K_{ia}^* values are determined, we can set $K_{ia}^* \cong K_{ia}$. For our following discussions, we will also assume that γ_{HA}' and γ_{A^-}' are approximately 1. Note that when we are dealing with saline waters (e.g., seawater), the K_{ia}^* values may differ more substantially from K_{ia} [e.g., a factor of 2 has been found for some phenols (Demianov et al. 1995)].

When assuming that $\gamma_{HA}' = \gamma_{A^-}' = 1$, and by using the common chemical shorthand of

$pX = -\log X$, we can rewrite Eq. 8-13 as:

$$\log\frac{[A^-]}{[HA]} = \log K_{ia} - \log\left(\gamma'_{H^+}[H^+]\right) = pH - pK_{ia} \qquad (8\text{-}14)$$

Eq. 8-14 allows us now to visualize the meaning of the *acidity constant* for a given organic compound. We can see that the pK_{ia} is a measure of the *strength of an organic acid* relative to the acid–base pair H_3O^+ / H_2O. For example, the pK_{ia} identifies at which hydrogen ion activity (expressed by the pH) our organic acid is present in equal parts in the dissociated (A^-) and nondissociated (HA) forms:

$$[A^-] = [HA] \quad \text{at} \quad pH = pK_{ia} \qquad (8\text{-}15)$$

If the pK_{ia} of an organic acid is very low (i.e., $pK_{ia} \cong 0$ to 3), we speak of a *strong organic acid*. A strong acid has a high tendency to deprotonate even in an aqueous solution of high H^+ activity (low pH). Examples of strong organic acids are trifluoroacetic acid, 2,4-dinitrobenzoic acid, and 2,4,6-trinitrophenol (see Table 8.1). Consequently, at ambient pH values (i.e., pH = 4 to 9), such acids will be present in natural waters predominantly in their dissociated form, that is, as anions. The other examples given in Table 8.1 show that organic acids of environmental concern cover a broad pK_{ia} range. Logically, *weaker acids* are those with higher pK_{ia} values. Hence, very weak acids (i.e., $pK_{ia} \geq 10$) will be present in natural waters primarily in their nondissociated form. Many important organic acids, however, have pK_{ia} values between 4 and 10. In these cases, exact knowledge of the pK_{ia} value is necessary since, as already pointed out, the environmental behavior of the dissociated form of the molecule is very different from that of the nondissociated form.

Organic Bases

By analogy with the acids, we can define a basicity constant for the reaction of an organic base ($i = B$) with water:

$$B + H_2O \rightleftharpoons OH^- + BH^+$$

$$(8\text{-}16)$$

$$K_{ib} = \frac{\left(\gamma'_{OH^-}[OH^-]\right)\left(\gamma'_{BH^+}[BH^+]\right)}{\left(\gamma'_B[B]\right)}$$

Here the reaction of a neutral base with water results in the formation of a cation. To compare acids and bases on a uniform scale, it is convenient to use the acidity constant of the conjugate acid ($i = BH^+$) as a measure of the base strength:

$$BH^+ \rightleftharpoons H^+ + B$$

$$K_{ia} = \frac{\left(\gamma'_{H^+}[H^+]\right)\left(\gamma'_B[B]\right)}{\left(\gamma'_{BH^+}[BH^+]\right)} \qquad (8\text{-}17)$$

K_{ib} and K_{ia} are quantitatively interrelated by the *ionization constant of water* (*ion product of water*), K_w:

Table 8.1 Examples of Neutral Organic Acids

Acid i (HA)		pK_{ia} [a] (25°C)	Fraction in Neutral (Acid) Form at pH 7 α_{ia} [b]
Name	Structure		

Carboxylic Acids $(R\text{-}COOH \rightleftharpoons R\text{-}COO^- + H^+)$

Trifluoroacetic acid (TFA)	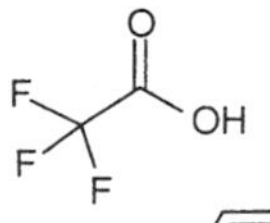	0.40	< 0.001
2,6-Dinitrobenzoic acid	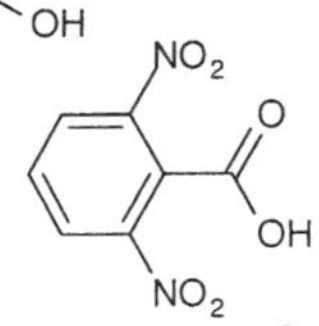	1.14	< 0.001
4-Nitrobenzoic acid	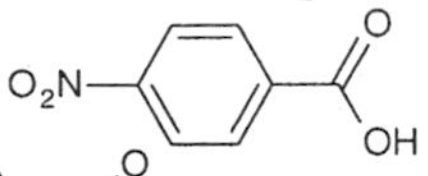	3.44	<0.001
Benzoic acid	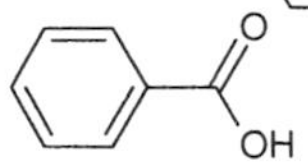	4.19	0.002
Acetic acid	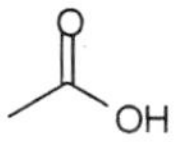	4.75	0.006
Hexanoic acid	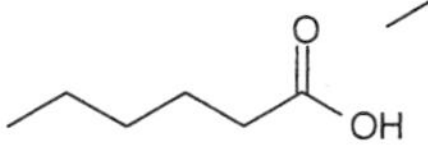	4.89	0.008

Phenolic Groups $(Ar\text{-}OH \rightleftharpoons Ar\text{-}O^- + H^+)$

2,4,6-Trinitrophenol	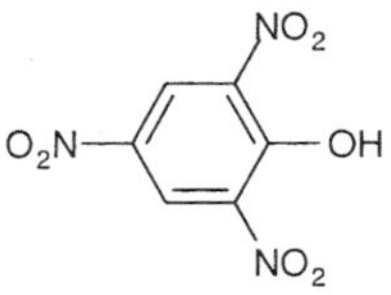	0.38	< 0.001
Pentachlorophenol	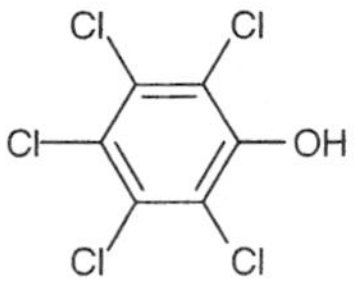	4.75	0.006
2-Nitrophenol	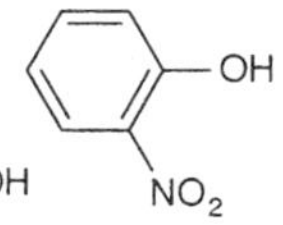	7.20	0.613
2-Naphthol	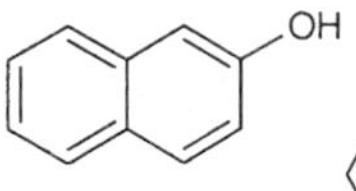	9.51	0.997
Phenol	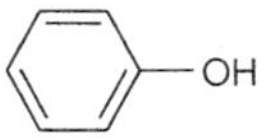	9.90	0.998
2,4,6-Trimethylphenol	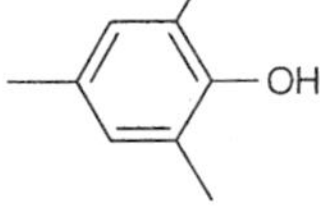	10.90	> 0.999

Miscellaneous Groups $(AH \rightleftharpoons A^- + H^+)$

1-Naphthalene-sulfonic acid	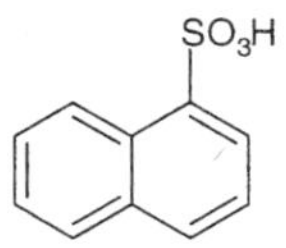	0.57	< 0.001
p-Toluenesulfonic acid	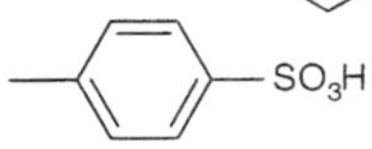	0.70	< 0.001
Thioacetic acid	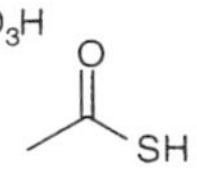	3.33	< 0.001
Thiophenol	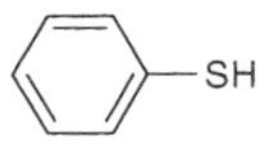	6.50	0.240
Ethanethiol	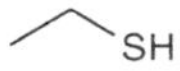	10.61	>0.999
Aliphatic alcohols	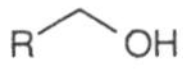	> 14	>> 0.999

[a] From Dean (1985) and Lide (1995). [b] See Eq. 8-21.

Table 8.2 Examples of Neutral Organic Bases

Base i (B)		$pK_{ia}{}^{a}$ $(= pK_{BH^+})$ $(25°C)$	Fraction in Neutral (Base) Form at pH 7 $(1-\alpha_{ia})^{b}$
Name	Structure		

Aliphatic and Aromatic Aminogroups $(Ar\text{-}$ or $R\text{-}\overset{|}{\underset{|}{N}}{}^{\pm}H \rightleftharpoons Ar\text{-}$ or $R\text{-}\overset{|}{\underset{|}{N}}{:} + H^+)$

4-Nitroaniline	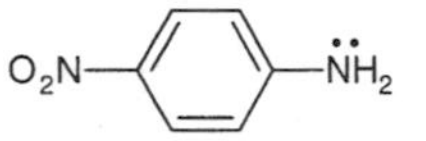 $O_2N\text{—}\langle\ \rangle\text{—}\ddot{N}H_2$	1.01	< 0.999
1-Naphthylamine	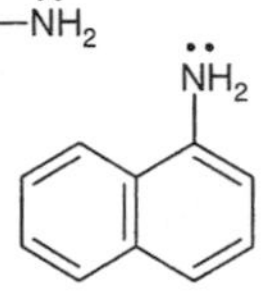	3.92	0.999
4-Chloroaniline	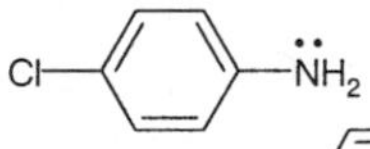 $Cl\text{—}\langle\ \rangle\text{—}\ddot{N}H_2$	3.99	0.999
Aniline	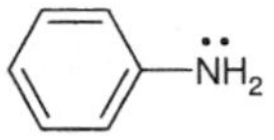 $\langle\ \rangle\text{—}\ddot{N}H_2$	4.63	0.996
N,N-Dimethylaniline	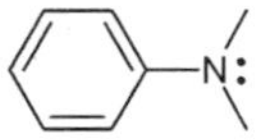	5.12	0.987
Trimethylamine	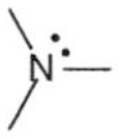	9.81	0.002
n-Hexylamine	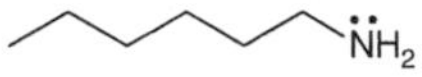 $\sim\!\!\sim\!\!\sim\ddot{N}H_2$	10.64	< 0.001
Piperidine	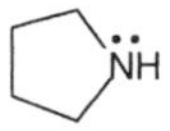	11.12	< 0.001

Heterocyclic Nitrogen $(\overset{\|}{N}{}^+\!\!-H \rightleftharpoons \overset{\|}{N}{:} + H^+)$

4-Nitropyridine	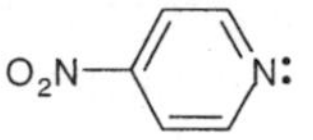 $O_2N\text{—}\langle\ \rangle\text{N}{:}$	1.23	> 0.999
4-Chloropyridine	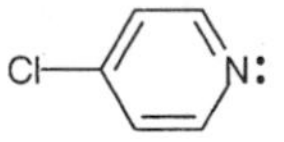 $Cl\text{—}\langle\ \rangle\text{N}{:}$	3.83	> 0.999
Pyridine	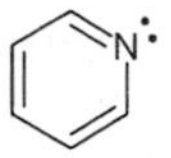	5.25	0.983
Isoquinoline	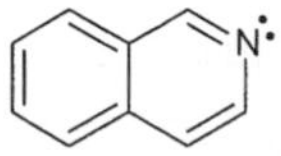	5.40	0.975
Benzimidazole	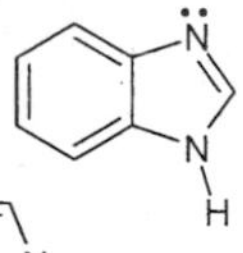	5.53	0.967
Imidazole	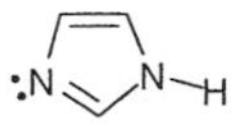	7.00	0.500
Benzotriazole	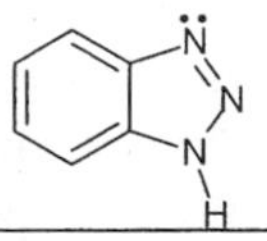	8.50	0.031

[a] From Dean (1985) and Lide (1995). [b] See Eq. 8-21.

Table 8.3 Acidity Constants (pK_{ia}) of Some Organic Acids and of H_2O at Different Temperatures

Acid (i) (HA, BH$^+$) [a]	pK_{ia} [b]				
	0°C	10°C	20°C	30°C	40°C
4-Nitrobenzoic acid		3.45	3.44	3.44	3.45
Acetic acid	4.78	4.76	4.76	4.76	4.77
2-Nitrophenol	7.45	7.35	7.24	7.15	
Imidazole	7.58	7.33	7.10	6.89	6.78
4-Aminopyridine	9.87	9.55	9.25	8.98	8.72
Piperidine	11.96	11.61	11.28	10.97	10.67
H_2O	14.94	14.53	14.16	13.84	13.54

[a] For structures see Tables 8.1 and 8.2. [b] From Dean (1985) and Schwarzenbach et al. (1988).

$$K_w = K_{ia} \cdot K_{ib} = \left(\gamma'_{H^+}[H^+]\right)\left(\gamma'_{OH^-}[OH^-]\right) = 1.01 \times 10^{-14} \qquad (8\text{-}18)$$

at 25°C for pure water. Note that K_w is strongly temperature dependent (see also Table D2, Appendix D). Using our pX nomenclature:

$$pK_{ia} = pK_w - pK_{ib} \qquad (8\text{-}19)$$

From Eq. 8-19 it follows that the stronger the acid (low pK_{ia}), the weaker is the basicity of its conjugate base (high pK_{ib}), while the stronger the base (low pK_b), the weaker its conjugate acid (high pK_a). Thus, a neutral base with a pK_{ib} value < 3 (i.e., the pK_{ia} of the conjugate acid > 11!) will be present in water predominantly as a cation at ambient pH values. Some examples of important organic bases are shown in Table 8.2.

Effect of Temperature on Acidity Constants

In analogy to the temperature dependence of equilibrium partition constants (Eqs. 3-47 to 3-54, Section 3.4), the effect of temperature on K_{ia} over a small temperature range can be described by:

$$K_{ia}(T_2) = K_{ia}(T_1) \cdot e^{-\frac{\Delta_r H^0}{R}\left(\frac{1}{T_2} - \frac{1}{T_1}\right)} \qquad (8\text{-}20)$$

where $\Delta_r H^0$ is the *standard enthalpy of reaction* of the reactions Eqs. 8-6 and 8-17, respectively. In general, $\Delta_r H^0$ is very small for strong acids and increases with increasing pK_{ia} value. Hence, for stronger acids we may neglect the effect of temperature on K_{ia}, whereas for very weak acids this effect is very substantial. For example, the ionization constants of piperidine and water (K_w) change by about one order of magnitude between 0 and 30°C, whereas for 4-nitrobenzoic acid or acetic acid almost no temperature dependence is observed (Table 8.3).

Speciation in Natural Waters

Given the pK_{ia} of an organic acid or base, we can now ask to what extent this compound is ionized in a natural water; that is, what are the relative abundances of the neutral versus the charged species? The pH of a natural water is primarily determined by various *inorganic* acids and bases (e.g., H_2CO_3, HCO_3^-, CO_3^{2-}) which are usually present at much higher concentrations than the compounds that interest us (Stumm and Morgan, 1996, Chapters 3 and 9). These acids and bases act as hydrogen ion buffers (pH buffers), meaning that the addition of a very small quantity of acid or base will lead to a much smaller change in pH as compared to a nonbuffered solution. We can easily visualize this buffering effect by the following simple example. Let us assume that a hypothetical acid–base pair has a pK_{ia} = 7.00, so its undissociated and its dissociated forms are present at equal concentrations in one liter of water, say 10^{-3} mol·L^{-1}. According to Eq. 8.14, the pH of this aqueous solution will then be:

$$pH = 7.00 + \log \frac{10^{-3} \, mol \cdot L^{-1}}{10^{-3} \, mol \cdot L^{-1}} = 7.00$$

If we now add 10^{-5} moles of a *strong* organic acid (i.e., we add 10^{-5} moles H^+), for example, 2,4,6-trinitrophenol (which would correspond to a total concentration of this compound of 10^{-5} mol·L^{-1} or 2 mg·L^{-1}), the pH would change by less than 0.01 units:

$$pH = 7.00 + \log \frac{0.99 \times 10^{-3} \, mol \cdot L^{-1}}{1.01 \times 10^{-3} \, mol \cdot L^{-1}} = 6.991$$

As a first approximation then, we may assume that adding a "trace" organic acid or base (where trace < 0.1 mM) to a natural water will, in most cases, not significantly affect the pH of the water.

For a given pH, we may now express the fraction of our organic acid (denoted as HA, the same holds for BH^+) present in the acid form in the water, α_{ia}, by:

$$\alpha_{ia} = \frac{[HA]}{[HA]+[A^-]} = \frac{1}{1+\dfrac{[A^-]}{[HA]}}$$

$$= \frac{1}{1+10^{(pH-pK_{ia})}}$$

(8-21)

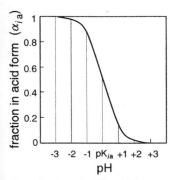

Figure 8.1 Fraction in acid form as function of pH. Note that at pH = pK_{ia}, the acid and base forms are present at equal concentrations, i.e., [HA] = [A$^-$]; [BH$^+$] = [B].

Tables 8.1 and 8.2 give calculated α_{ia} and ($1-\alpha_{ia}$) values, respectively, for various acids and bases in water at pH 7. Fig. 8.1 shows schematically the speciation of a given acid (or base) as a function of pH. Some example calculations are given in Illustrative Example 8.1. It should be reemphasized that the neutral and ionic "forms" of a given neutral acid (base) behave very differently in the environment. Depending on the process considered, either the neutral or ionic species may be the dominant factor in the compound's "reactivity," even if the relative amount of that

Illustrative Example 8.1

Assessing the Speciation of Organic Acids and Bases in Natural Waters

Problem

Calculate the fraction of (a) pentachlorophenol (PCP), (b) 3,4-dimethylaniline (DMA), and (c) *ortho*-phthalic acid (*o*-PA) present at 25°C as neutral species in a raindrop (pH = 4.0) and in lake water (pH = 8.0). For *o*-PA calculate also the fractions of the other two species present.

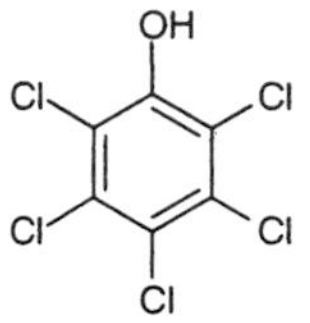

i = pentachlorophenol
(PCP)
$pK_{ia} = 4.75$

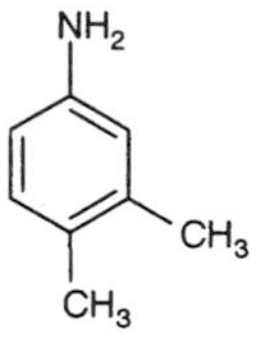

i = 3,4-dimethylaniline
(DMA)
$pK_{ia} = 5.28$

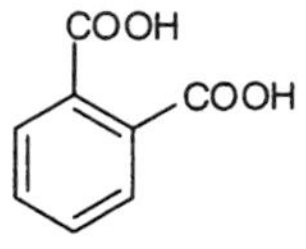

i = *ortho*-phthalic acid
(*o*-PA)
$pK_{ia1} = 2.89$
$pK_{ia2} = 5.51$

Answer (a)

For PCP, the acid form is the neutral species (i.e., fraction = α_{ia}). Insertion of pK_{ia} and the appropriate pH into Eq. 8-21 yields:

$$\alpha_{ia} \text{ at pH } 4.0 = 0.85 \quad \text{and} \quad \alpha_{ia} \text{ at pH } 8.0 = 0.00056$$

Answer (b)

For DMA, the base form is the neutral species (i.e., fraction = $1 - \alpha_{ia}$). Hence, $1 - \alpha_{ia}$ is given by (use Eq. 8-21):

$$(1 - \alpha_{ia}) = \frac{10^{(pH-pK_{ia})}}{1+10^{(pH-pK_{ia})}} = \frac{1}{1+10^{(pK_{ia}-pH)}} \tag{1}$$

Insertion of pK_{ia} and pH into Eq. 1 yields:

$$(1 - \alpha_{ia}) \text{ at pH } 4.0 = 0.050 \quad \text{and} \quad (1 - \alpha_{ia}) \text{ at pH } 8.0 = 0.998$$

Answer (c)

o-PA is a diprotonic acid (i.e., H_2A) with the two acidity constants (denote $\gamma'_{H^+}[H^+]$ as $\{H^+\}$):

$$K_{ia1} = \frac{\{H^+\}[HA^-]}{[H_2A]} \quad \text{and} \quad K_{ia2} = \frac{\{H^+\}[A^{2-}]}{[HA^-]}$$

Hence, the fraction of the three species at a given pH can be expressed as:

$$\alpha_{H_2Aa} = \frac{[H_2A]}{[H_2A]+[HA^-]+[A^{2-}]} = \frac{1}{1+\dfrac{[HA^-]}{[H_2A]}+\dfrac{[A^{2-}]}{[H_2A]}}$$

With $\dfrac{[HA^-]}{[H_2A]} = \dfrac{K_{ia1}}{\{H^+\}}$ and $\dfrac{[A^{2-}]}{[H_2A]} = \dfrac{K_{ia1}\cdot K_{ia2}}{\{H^+\}^2}$ one gets:

$$\alpha_{H_2Aa} = \frac{1}{1+10^{(pH-pK_{ia1})}+10^{(2pH-pK_{ia1}-pK_{ia2})}} \tag{2}$$

In analogy, you can derive the equation for the other two species. The result is:

$$\alpha_{HA\bar{a}} = \frac{1}{1+10^{(pK_{ia1}-pH)}+10^{(pH-pK_{ia2})}} \tag{3}$$

and:

$$\alpha_{A^{2-}} = \frac{1}{1+10^{(pK_{ia1}+pK_{ia2}-2pH)} + 10^{(pK_{ia2}-pH)}} \tag{4}$$

Insertion of pK_{ia1} and pK_{ia2} and of the appropriate pH value into Eqs. 2 to 4 then yields

at pH 4.0: $\alpha_{H_2Aa} = 0.07$; $\alpha_{HA_a^-} = 0.90$; $\alpha_{A^{2-}} = 0.03$

at pH 8.0: $\alpha_{H_2Aa} = <10^{-7}$; $\alpha_{HA_a^-} = 0.003$; $\alpha_{A^{2-}} = 0.997$

---

species is very low.

So far, we have dealt with organic acids and bases that possess only one acid or base group in the pK_a range of interest. There are, however, compounds with more than one acid or base function. An example of a "two-protic acid" is given in Fig. 8.2a. In such cases, it is possible that a molecule is present in aqueous solution as a doubly charged anion. Similarly, as illustrated by 1,2-diaminopropane in Fig. 8.2b, a "two-protic" base may form doubly charged cations. A very interesting case involves those compounds that have both acidic and basic functions, such as amino acids and the hydroxy-isoquinoline shown in Fig. 8.2c. Here it is not always possible to unambiguously specify a proton transfer reaction in terms of the actual chemical species involved. In the case of a simple amino acid, for example, proton transfer may occur by two different pathways:

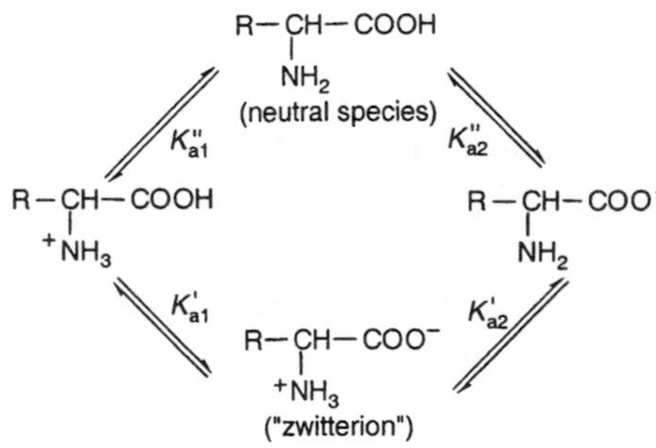

Although four microscopic acidity constants $(K'_{a1}, K''_{a1}, K'_{a2}, K''_{a2})$ may be defined, only two apparent (macroscopic) acidity constants K_{a1}, and K_{a2}, may be determined experimentally (Fleck, 1966, Chapter 5):

$$K_{ia1} = K'_{a1} + K''_{a1} \; ; \; K_{ia2} = \frac{K'_{a2} K''_{a2}}{K'_{a2} + K''_{a2}} \tag{8-22}$$

By comparing the magnitude of K_{a1} and K_{a2} with the K_a values of structurally comparable acid functions, one can, however, conclude whether *zwitterion* formation is important. In the case of the amino *acids* with $pK_{ia1} = 2$ to 3 and $pK_{ia2} = 9$ to 11, zwitterion formation is very likely, since pK_{ia1} is very similar to that of a carboxylic acid carrying an electron-withdrawing α-substituent (compare with chloroacetic acid in Table 8.1), and the pK_{ia2} corresponds to that of an aliphatic amine (see examples in Table 8.2). In contrast, for 7-hydroxy-isoquinoline (Fig. 8.2c), zwitterion formation is very unlikely, since the pK_{ia1} corresponds to that exhibited by the nitrogen in

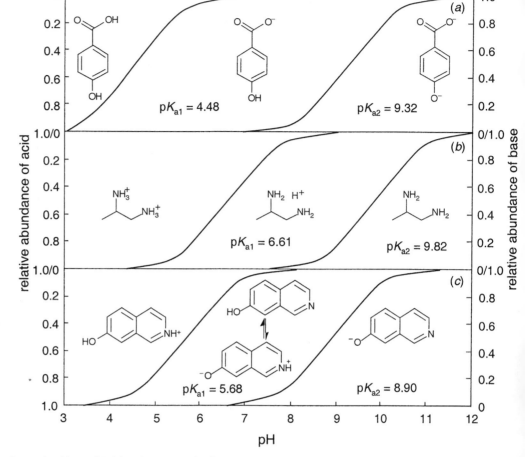

Figure 8.2 Relative amounts of the conjugate acid–base species as a function of pH for some compounds exhibiting more than one acid or base moiety:
(*a*) 4-hydroxy benzoic acid,
(*b*) 1,2-diaminopropane, and
(*c*) 7-hydroxy-isoquinoline.

isoquinoline (Table 8.2), and the pK_{ia1} is more typical of monosubstituted 2-naphthols (compare with 2-naphthol, Table 8.1). Finally, note that for compounds such as amino acids and hydroxy-isoquinolines, a pH value exists at which the *average net charge* of all species present is zero. This pH value is called the *isoelectric pH* and is given by:

$$pH_{isoelectric} = \frac{1}{2}(pK_{ia1} + pK_{ia2})$$

(8-23)

8.3   Chemical Structure and Acidity Constant

Overview of Acid and Base Functional Groups

Tables 8.1 and 8.2 give the range of pK_{ia} values for some important functional groups that have either proton-donor or proton-acceptor properties. As already pointed out, we are primarily interested in compounds having pK_{ia} values in the range of 3 to 11; therefore, the most important functional groups we have to consider include aliphatic and aromatic carboxyl groups, aromatic hydroxyl groups (e.g., phenolic compounds), aliphatic and aromatic amino groups, nitrogen atoms incorporated in aromatic compounds, and aliphatic or aromatic thiols. The range in pK_{ia} values for a given functional group may vary by many units because of the structural

Table 8.4 Inductive and Resonance Effects of Some Common Substituents [a]

Effect [b]	Substituents
	Inductive
+ I	O^-, NH^-, alkyl
– I	SO_2R, NH_3^+, NO_2, CN, F, Cl, Br, COOR, I, COR, OH, OR, SR, phenyl, NR_2
	Resonance
+ R	F, Cl, Br, I, OH, OR, NH_2, NR_2, NHCOR, O^-, NH^-
– R	NO_2, CN, CO_2R, $CONH_2$, phenyl, COR, SO_2R

[a] From Clark and Perrin (1964). [b] A plus sign means that the effect increases the pK_a; a minus sign means that the effect decreases the pK_a.

characteristics of the remainder of the molecule. Depending on the type and number of substituent groups on the aromatic ring, for example, the pK_{ia} values for substituted phenols may differ by almost 10 units (Table 8.1). It is necessary, therefore, that we make an effort to understand the effects of various structural entities on the acid or base properties of a given functional group. To this end, we recall that the standard free energy, $\Delta_r G^0$, for the proton dissociation reaction is given by the difference in the standard free energies of formation of the acid and conjugate base in aqueous solution (Eq. 8-11). Therefore, when comparing acidity constants of compounds exhibiting a specific acid or base functional group, the question is simply how much the rest of the molecule favors (decreases the free energy of formation) or disfavors (increases the free energy of formation) the ionic versus the neutral form of the compound in aqueous solution. Hence, we have to evaluate electronic and steric effects of substituents on the relative stability of the acid–conjugate base couple considered.

Inductive Effects

Let us first consider a simple example, the influence of a chloro-substituent on the pK_a of butyric acid:

$i =$	$CH_3CH_2CH_2COOH$	$\underset{\overset{|}{Cl}}{CH_2}CH_2CH_2COOH$	$CH_3\underset{\overset{|}{Cl}}{CH}CH_2COOH$	$CH_3CH_2\underset{\overset{|}{Cl}}{CH}COOH$
pK_{ia}	4.81	4.52	4.05	2.86

In this example, we see that if we substitute a hydrogen atom by chlorine, which is much more electronegative than hydrogen (see Chapter 2), the pK_a of the carboxyl group decreases. Furthermore, the closer the electron-withdrawing chlorine substituent is to the carboxyl group, the stronger is its effect in decreasing the pK_a. We can intuitively explain these findings by realizing that any group that will have an electron-withdrawing effect on the carboxyl group (or any other acid function) will help to accommodate a negative charge and increase the stability of the ionized form. In the case of an organic base, an electron-withdrawing substituent will, of course, destabilize the acidic form (the cation) and, therefore, also lower the pK_a. This effect is called a *negative inductive effect* (–I). Table 8.4 shows that most functional groups with which we are concerned have inductive electron-withdrawing (–I) effects, and only a few have electron-donating (+I) effects such as, for example, alkyl groups:

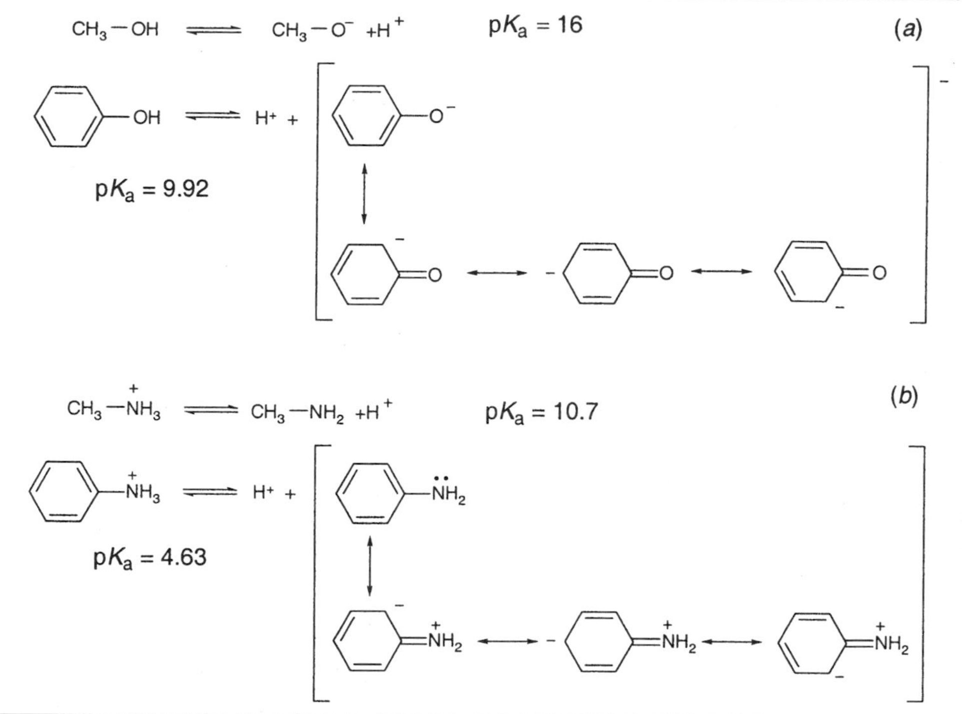

$$CH_3-OH \rightleftharpoons CH_3-O^- + H^+ \qquad pK_a = 16 \qquad\qquad (a)$$

$pK_a = 9.92$

$$CH_3-\overset{+}{N}H_3 \rightleftharpoons CH_3-NH_2 + H^+ \qquad pK_a = 10.7 \qquad (b)$$

$pK_a = 4.63$

Figure 8.3 Effect of delocalization on the pK_a of $-OH$ and $-NH_3^+$.

$i =$	CH_3COOH	CH_3CH_2COOH
pK_{ia}	4.75	4.87

As illustrated by the chlorobutyric acids discussed above, in *saturated* molecules inductive effects usually fall off quite rapidly with distance.

Delocalization Effects

In unsaturated chemicals, such as aromatic or olefinic compounds (i.e., compounds with "mobile" π-electrons; see Chapter 2), the inductive effect of a substituent may be felt over larger distances (i.e., more bonds). In such systems, however, another effect, the *delocalization of electrons*, may be of even greater importance. In Chapter 2, we learned that the delocalization of electrons (i.e., the "smearing" of π-electrons over several bonds) may significantly increase the stability of an organic species. In the case of an organic acid, delocalization of the negative charge may, therefore, lead to a considerable decrease in the pK_a of a given functional group, as one can see from comparing the pK_a of an aliphatic alcohol with that of phenol (Fig. 8.3a). Analogously, by stabilizing the neutral species, the delocalization of the free electrons of an amino group has a very significant effect on the pK_a of the conjugated ammonium ion (see Fig. 8.3b).

In the next step, we introduce a substituent on the aromatic ring which, through the aromatic π-electron system, may develop shared electrons (i.e., through "resonance" or "conjugation") with the acid or base function (e.g., the $-OH$ or $-NH_2$ group). For example, the much lower pK_a value of *para*-nitrophenol as compared with *meta*-nitrophenol may be attributed to additional resonance stabilization of the anionic species by the *para*-positioned nitro group (see Fig. 8.4). In the *meta* position, only the electron-withdrawing negative inductive effect of the nitro group is felt by the –OH group. Other substituents that increase acidity (i.e., that lower the pK_a, "–R" effect) are listed in Table 8.4. All of these substituents can help to accommodate electrons. On the other hand, substituents with heteroatoms having nonbonding electrons that may be in resonance with the *p*-electron system, have an *electron-donating resonance effect* (+R, see examples given in Table 8.4), and will therefore decrease acidity (i.e., increase pK_a). Note that many groups that have a negative inductive effect (–I) at the same time have a positive resonance effect (+R). The overall impact of such substituents depends critically on their location in the molecule. In monoaromatic molecules, for example, resonance in the *meta* position is negligible, but will be significant in both the *ortho* and *para* positions.

Proximity Effects

Another important group of effects are *proximity effects*; that is, effects arising from the influence of substituents that are physically close to the acid or base function under consideration. Here, two interactions are important: *intra*molecular (within

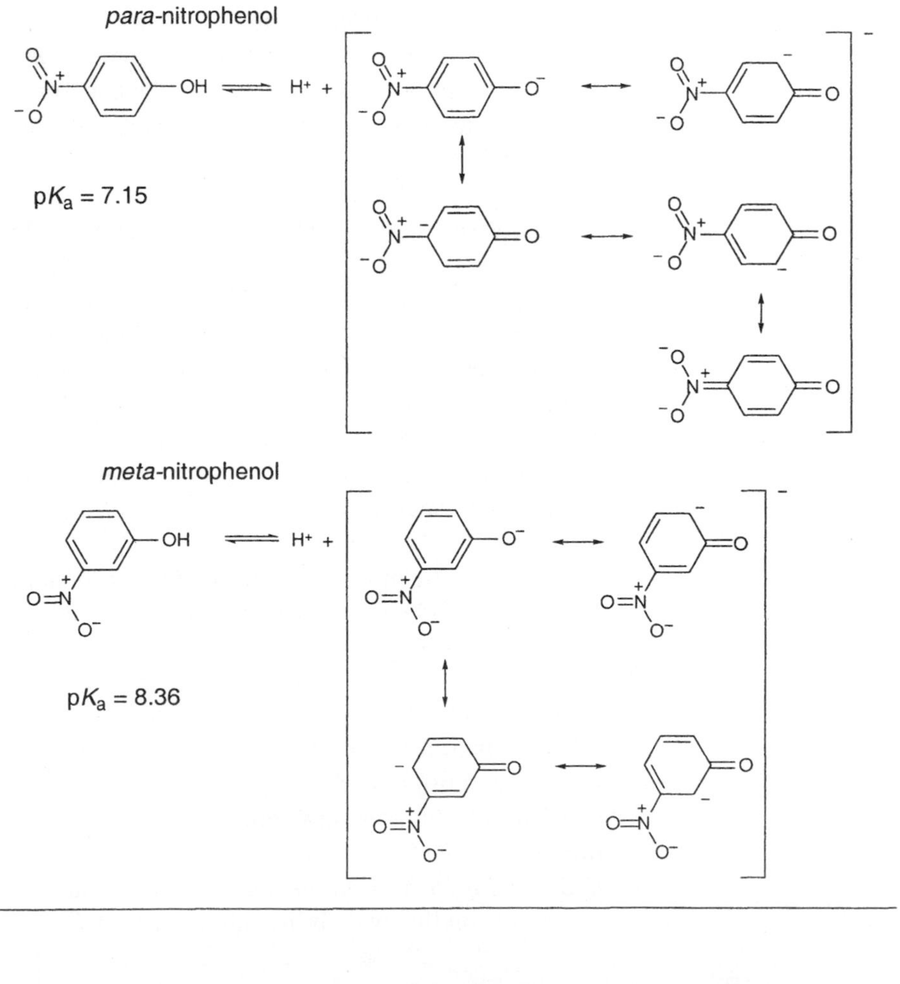

Figure 8.4 Influence of the position of a nitro substituent on the pK_a of a phenolic hydrogen.

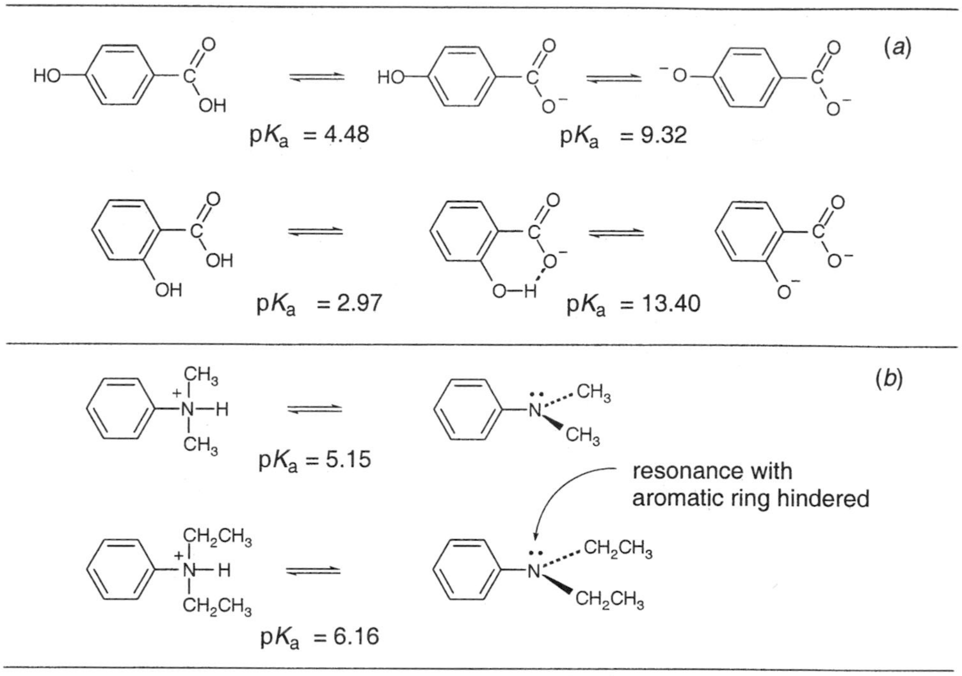

Figure 8.5 Examples of proximity effects on acidity constants: (a) hydrogen bonding and (b) steric interactions.

the same molecule) *hydrogen bonding and steric effects*. An example of the effect of intramolecular hydrogen bonding is given in Fig. 8.5a. The stabilization of the carboxylate anion by the hydroxyl hydrogen in *ortho*-hydroxy-benzoic acid (salicylic acid) leads to a much lower pK_{a1} value and to a much higher pK_{a2} value compared with *para*-hydroxobenzoic acid, in which no intramolelcular hydrogen bonding is possible.

In some cases, steric effects may have a measurable impact on the pK_a of a given acid or base function. This involves steric constraints that inhibit optimum solvation of the ionic species by the water molecules (and thus increase the pK_a), or hinder the resonance of the electrons of a given acid or base group with other parts of the molecule by causing these groups to twist with respect to one another and to avoid coplanarity. For example, the large difference found between the pK_a of N,N-dimethylaniline and of *N,N*-diethylaniline (Fig. 8.5b) is partially due to the larger ethyl substituents that limit free rotation and, thus, the orientation of the free electrons of the nitrogen atom and, thus, their resonance with the π-electrons of the aromatic ring.

In summary, the most important factors influencing the pK_a of a given acid or base function are inductive, resonance, and steric effects. The impact of a substituent on the pK_a depends critically on where the substituent is located in the molecule relative to the acid or base group. In one place, a given substituent may have only one of the mentioned effects, while in another location, all effects may play a role. It is quite difficult, therefore, to establish simple general rules for quantifying the effect(s) of structural entities on the pK_a of an acid or base function. Nevertheless, in certain restricted cases, a quantification of the effects of substituents on the pK_a value is possible by using LFERs. In the next section, we discuss one example of such an approach, the *Hammett correlation* for aromatic compounds. First, however, a few comments on the availability of experimental pK_{ia} values are necessary.

Availability of Experimental Data; Methods for Estimation of Acidity Constants

Experimental Data

Acidity has long been recognized as a very important property of some organic compounds. Experimental methods for determining acidity constants are well established, and there is quite a large database of pK_{ia} values of organic acids and bases (e.g., Kortüm et al., 1961; Perrin, 1972; Serjeant and Dempsey, 1979; Dean, 1985; Lide, 1995). The most common procedures discussed by Kortüm et al. (1961) include titration, determination of the concentration ratio of acid–base pairs at various pH values using conductance methods, electrochemical methods, and spectrophotometric methods. It should be again noted that pK_{ia} values reported in the literature are often "mixed acidity constants" (see Section 8.2), that are commonly measured at 20 or 25°C, and at a given ionic strength (e.g., 0.05 – 0.1 M salt solution). Depending on the type of measurement and the conditions chosen, therefore, reported pK_{ia} may vary by as much as 0.3 pK_a units. Also, as discussed above, primarily depending on the strength of an acid or base, the effect of temperature may be more or less pronounced (Table 8.3).

Estimation of Acidity Constants: The Hammett Correlation

In Chapters 6 and 7 we used LFERs to quantify the effects of structural entities on the partitioning behavior of organic compounds. In an analogous way, LFERs can be used to *quantitatively* evaluate the influence of structural moieties on the pK_a of a given acid or base function, particularly if only electronic effects are important.

Long ago, Hammett (1940) recognized that for *substituted benzoic acids* (see Fig. 8.6) the effect of substituents in either the *meta* or *para* position on the standard free energy change of dissociation of the carboxyl group could be expressed as the sum of the free energy change of the dissociation of the unsubstituted compound, $\Delta_r G_H^0$, and the contributions of the various substituents; $\Delta_r G_j^0$:

$$\Delta_r G^0 = \Delta_r G_H^0 + \sum_j \Delta_r G_j^0 \tag{8-24}$$

To express the effect of substituent j on the pK_a, Hammett introduced a constant σ_j, that is defined as:

$$\sigma_j = \frac{-\Delta_r G_j^0}{2.303\ RT} \tag{8-25}$$

Since σ_j differs for *meta* and *para* substitutions, there are two sets of σ_j values, $\sigma_{j\text{meta}}$ and $\sigma_{j\text{para}}$. *Ortho* substitution is excluded since, as we have already seen, proximity effects, which are difficult to separate from electronic factors, may play an important role. Since $\Delta_r G^0 = -2.303\ RT \log K_a$, we may write Eq. 8-24 in terms of acidity constants (note that in the following we omit the subscript i to denote the acid function):

$$\log \frac{K_a}{K_{aH}} = \sum_j \sigma_j \quad \text{or} \quad pK_a = pK_{aH} - \sum_j \sigma_j \tag{8-26}$$

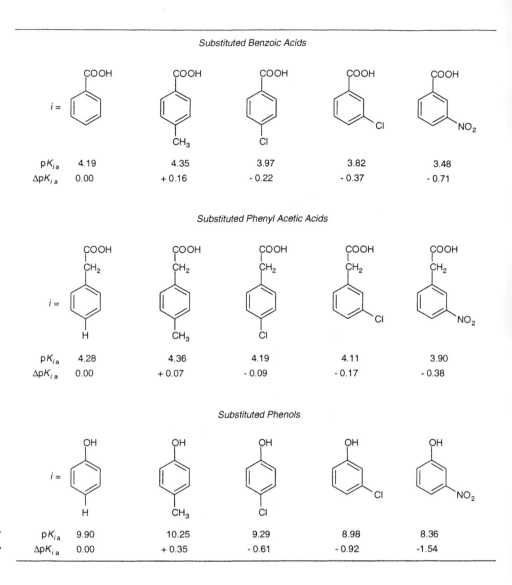

Figure 8.6 Effect of ring substituents on the pK_a of benzoic acid, phenyl acetic acid, and phenol.

Table 8.5 lists $\sigma_{j\text{meta}}$ and $\sigma_{j\text{para}}$ values for some common substituent groups. Note that these σ values are a quantitative measurement of the effect of a given substituent on the pK_a of *benzoic acid*. As we would expect from our previous discussion, the sign of the σ value reflects the net electron-withdrawing (positive sign) or electron-donating (negative sign) character of a given substituent in either the *meta* or *para* position. For example, we see that –NO_2 and –C≡N are strongly electron-withdrawing in both positions, while the electron-providing groups, –NH_2 or –$N(CH_3)_2$, are strongly electron-donating in the *para* position, but show a much weaker effect in the *meta* position. The differences between $\sigma_{j\text{meta}}$ and $\sigma_{j\text{para}}$ of a given substituent are due to the difference in importance between the inductive and resonance effects which, as we mentioned earlier, may have opposite signs (see Table 8.4).

Let us now examine the effects of the *same* substituents on the pK_a of another group of acids, the substituted phenyl acetic acids. As we might have anticipated, Fig. 8.6 shows that each of the various substituents exerts the same relative effect as in their benzoic counterparts; however, in the case of phenyl acetic acid, the greater separation between

Table 8.5 Hammett Constants for Some Common Substituents [a]

Substituent j	$\sigma_{j\text{meta}}$	$\sigma_{j\text{para}}$	Substituent j	$\sigma_{j\text{meta}}$	$\sigma_{j\text{para}}$	$\sigma^-_{j\text{para}}$
$-H$	0.00	0.00	$-OCH_3$	0.11	−0.24	−0.12
$-CH_3$	−0.06	−0.16	$-OCOCH_3$	0.36	0.31	
$-CH_2CH_3$	−0.06	−0.15	$-CHO$	0.36	0.22	1.03
$-CH_2CH_2CH_2CH_3$	−0.07	−0.16	$-COCH_3$	0.38	0.50	0.82
$-C(CH_3)_3$	−0.10	−0.20	$-COOCH_3$	0.33	0.45	0.66
$-CH=CH_2$	0.08	−0.08	$-CN$	0.62	0.67	0.89
$-Ph$ [b]	0.06	0.01	$-NH_2$	−0.04	−0.66	
$-CH_2OH$	0.07	0.08	$-NHCH_3$	−0.25	−0.84	
$-CH_2Cl$	0.12	0.18	$-N(CH_3)_2$	−0.15	−0.83	
$-CCl_3$	0.40	0.46	$-NO_2$	0.73	0.78	1.25
$-CF_3$	0.44	0.57	$-SH$	0.25	0.15	
$-F$	0.34	0.05	$-SCH_3$	0.13	0.01	
$-Cl$	0.37	0.22	$-SOCH_3$	0.50	0.49	
$-Br$	0.40	0.23	$-SO_2CH_3$	0.68	0.72	
$-I$	0.35	0.18	$-SO_3^-$	0.05	0.09	
$-OH$	0.10	−0.36				

[a] Values taken from Dean (1985) and Shorter (1994 and 1997). [b] Phenyl.

substituent and reaction site makes the impact less pronounced than in the benzoic acid. Plotting pK_{aH}–pK_a values for meta- and para-substituted phenyl acetic acids versus $\sum_j \sigma_j$ values results in a straight line with a slope ρ of less than 1 (Fig. 8.7). In this case, introduction of a substituent on the aromatic ring has only about half the effect on the pK_a as compared with the effect of the same substituent on the pK_a of benzoic acid. Thus, ρ is a measure of how sensitive the dissociation reaction is to substitution as compared with substituted benzoic acid and is commonly referred to as the *susceptibility factor* that relates one set of reactions to another. If we consider another group of acids, the substituted β-phenyl propionic acids, where the substituents are located at even greater distances from the carboxyl group, yet smaller ρ values are expected and found ($\rho = 0.21$, Fig. 8.7).

If we express these findings in energetic terms, we obtain:

$$\Delta_r G^0 = \Delta_r G_H^0 - \rho \, 2.303 \, RT \sum_j \sigma_j \tag{8-27}$$

The classical form of the *Hammett Equation* is Eq. 8.27, expressed in terms of equilibrium constants (i.e., acidity constants):

$$\log K_a = \log K_{aH} + \rho \sum_j \sigma_j$$

or:
$$pK_a = pK_{aH} - \rho \sum_j \sigma_j \tag{8-28}$$

Examples of pK_{aH} and ρ values for quantification of aromatic substituent effects on the pK_a values of various types of acids are given in Table 8.6.

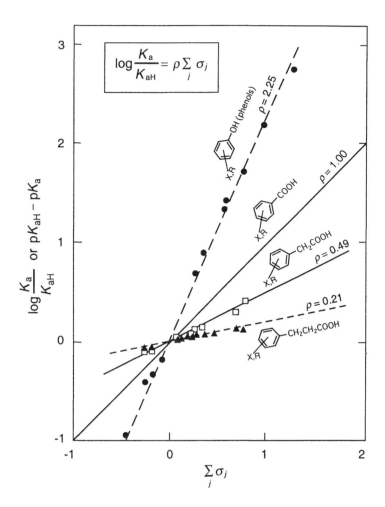

Figure 8.7 Hammett plots for *meta*- and *para*-substituted phenols, phenylacetic acids, and 3-phenylpropionic acids (data from Serjeant and Dempsey, 1979).

Note that for compound classes such as phenols, anilines, and pyridines where the acid (base) function is in resonance with the aromatic ring, the ρ values obtained are significantly greater than 1 (Table 8.6); that is, the electronic effect of the substituents is greater than in the case of benzoic acid.

In many cases, the simple approach of using $\sigma_{j\text{meta}}$ and $\sigma_{j\text{para}}$ values is applied with reasonable success. One should be aware, however, that good correlations are not always obtained, simply implying that in those cases one has not incorporated all of the molecular interactions into the LFER that play a role in that particular system. This is usually encountered when substituents exhibit a more complex interaction with the reaction center or when substituents interact with one another.

A simple case where the general σ constants in Table 8.5 do not succeed in correlating acidity constants is when the acid or base function is in *direct resonance* with the substituent. This may occur in cases such as substituted phenols, anilines, and pyridines. For example, owing to resonance (see Fig. 8.4), a *para* nitro group decreases the pK_a of phenol much more than would be predicted from the $\sigma_{j\text{para}}$ constant obtained from the dissociation of *p*-nitrobenzoic acid. In such "resonance" cases (another example would be the anilines), a special set of σ values (denoted as $\sigma^-_{j\text{para}}$) has been derived (Table 8.5) to try to account for both inductive and resonance

Table 8.6 Hammett Relationships for Quantifications of Aromatic Substituent Effects on the Toxicity of Various Acids [a]

Acid	pK_{aH} (pK$_a$ of unsubstituted compound)	ρ
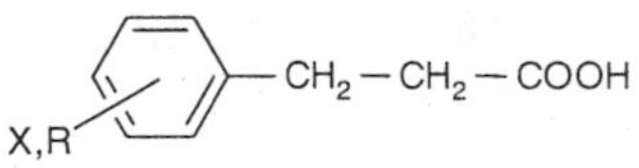 X,R—⟨benzene⟩—CH_2-CH_2-COOH	4.55	0.21
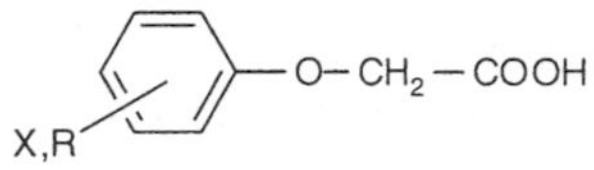 X,R—⟨benzene⟩—$O-CH_2-COOH$	3.17	0.30
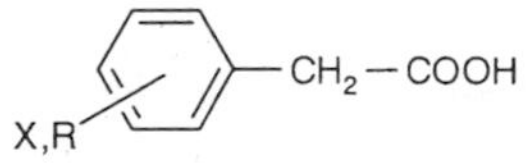 X,R—⟨benzene⟩—CH_2-COOH	4.30	0.49
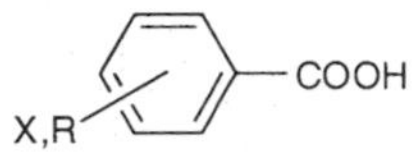 X,R—⟨benzene⟩—$COOH$	4.19	1.00 (by definition)
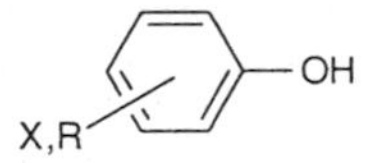 X,R—⟨benzene⟩—OH	9.90	2.25 [b]
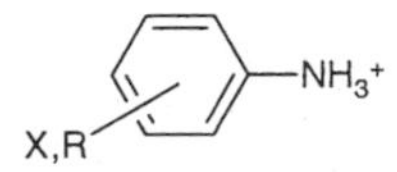 X,R—⟨benzene⟩—NH_3^+	4.63	2.90 [b]
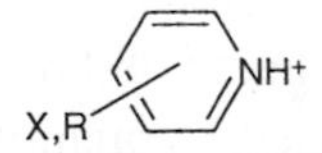 X,R—⟨pyridinium⟩NH^+	5.25	5.90 [b]

[a] Eq. 8-28; Data from Williams (1984). [b] Use σ_{jpara}^- instead of σ_{jpara} for substituents that are in direct resonance will the acid function (Table 8.5).

effects. If these values are employed, good correlations are obtained, as shown for *meta-* and *para*-substituted phenols in Fig. 8.7.

We have stated earlier that because of proximity effects, no generally applicable σ_j values may be derived for *ortho* substitution. Nevertheless, one can determine a set of *apparent* σ_{jortho} values for a specific type of reaction, as for example, for the dissociation of substituted phenols. Table 8.7 gives such apparent σ_{jortho} constants for estimating pK$_a$ values of substituted phenols and anilines. Of course, in cases of multiple substitution, substituents may interact with one another, thereby resulting in larger deviations of experimental from predicted pK$_a$ values. Some example calculations using the Hammett equation are given in Illustrative Example 8.2.

In our discussion of the Hammett correlation, we have confined ourselves mostly to benzene derivatives. Of course, a similar approach can be taken for other aromatic systems, such as for the derivatives of polycyclic aromatic hydrocarbons and heterocyclic aromatic compounds. For a discussion of such applications, we refer to

Table 8.7 Examples of Apparent Hammett Constants for *ortho*-Substitution in Phenols and in Anilines [a]

Substituent j	$\sigma_{j\,ortho}^{phenols}$	$\sigma_{j\,ortho}^{anilines}$	Substituent j	$\sigma_{j\,ortho}^{phenols}$	$\sigma_{j\,ortho}^{anilines}$
– CH$_3$	–0.13	0.10	– OH		–0.09
– CH$_2$CH$_2$CH$_2$CH$_3$	–0.18		–OCH$_3$	0.00	0.02
– CH$_2$OH	0.04		–CHO	0.75	
– F	0.54	0.47	–NH$_2$		0.00
– Cl	0.68	0.67	–NO$_2$	1.24	1.72
– Br	0.70	0.71			
– I	0.63	0.70			

[a] Data from Clark and Perrin (1964) and Barlin and Perrin (1966).

papers by Clark and Perrin (1964), Barlin and Perrin (1966), and Perrin (1980). Using the Hammett equation as a starting point, a variety of refinements using more sophisticated sets of constants have also been suggested. The interested reader can find a treatment of these approaches, as well as compilations of substituent constants, in various textbooks (e.g., Lowry and Schueller-Richardson, 1981; Williams, 1984; Exner, 1988) and in data collections (e.g., Harris and Hayes, 1982; Dean, 1985; Hansch et al., 1991, 1995). These references also give an overview of parallel approaches, such as the *Taft correlation* developed to predict pK_a values in *aliphatic and alicyclic* systems.

In summary, in this section we have discussed the electronic and steric effects of structural moieties on the pK_a value of acid and base functions in organic molecules. We have seen how LFERs can be used to quantitatively describe these electronic effects. At this point, it is important to realize that we have used such LFERs to evaluate the relative stability and, hence, the relative energy status of organic species in aqueous solution (e.g., anionic vs. neutral species). It should come as no surprise then that we will find similar relationships when dealing with chemical reactions other than proton transfer processes in Chapter 13.

Illustrative Example 8.2

Estimating Acidity Constants of Aromatic Acids and Bases Using the Hammett Equation

Problem

Estimate the pK_a values at 25°C of (a) 3.4,5-trichlorophenol (3,4,5-TCP), (b) pentachlorophenol (PCP), (c) 4-nitrophenol (4-NP), (d) 3,4-dimethylaniline (3,4-DMA, pK_a of conjugate acid), and (e) 2,4,5-trichlorophenoxy acetic acid (2,4,5-T)

Use the Hammett relationship Eq. 8-26:

$$pK_a = pK_{aH} - \rho \sum_j \sigma_j$$

to estimate the pK_a values of compounds (a)–(e). Get the necessary σ_j, pK_{aH}, and ρ values from Tables 8.5, 8.6, and 8.7.

Answer (a)

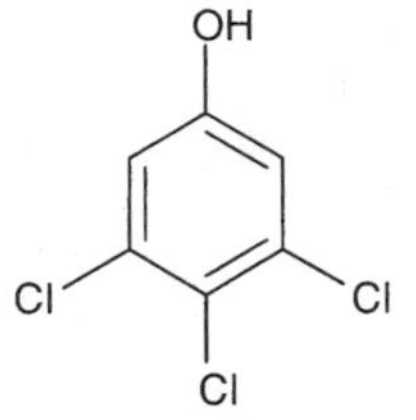

pK_{aH} (phenol)	9.90
ρ	2.25
σ_{meta} (Cl)	0.37
σ_{para} (Cl)	0.22

3,4,5-TCP

$$pK_a = 9.90 - (2.25)\,[2\,(0.37) + 0.22] = 7.74$$

The reported experimental value is 7.73 (Schellenberg et al., 1984).

Answer (b)

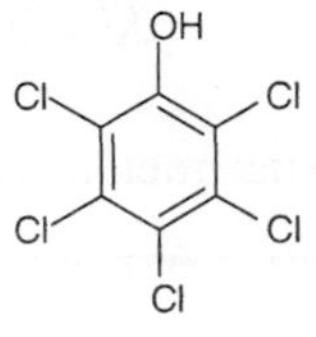

pK_{aH} (phenol)	9.90
ρ	2.25
$\sigma_{ortho}^{phenols}$ (Cl)	0.68
σ_{meta} (Cl)	0.37
σ_{para} (Cl)	0.22

TCP

$$pK_a = 9.90 - (2.25)\,[2\,(0.68) + 2\,(0.37) + 0.22] = 4.68$$

The reported experimental values are 4.75 (Schellenberg et al., 1984), and 4.83 (Jafvert et al., 1990).

Answer (c)

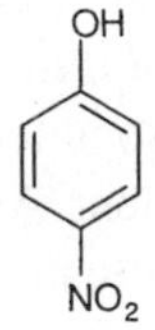

pK_{aH} (phenol)	9.90
ρ	2.25
σ_{para}^- (NO$_2$)	1.25

4-NP

$$pK_a = 9.90 - (2.25)\,(1.25) = 7.09$$

Note that because the nitro group is in resonance with the OH-group, the σ_{para}^- and not the σ_{para} value has to be used. The reported experimental values are 7.08, 7.15, and 7.18 (see Schwarzenbach et al., 1988, and refs. cited therein).

Answer (d)

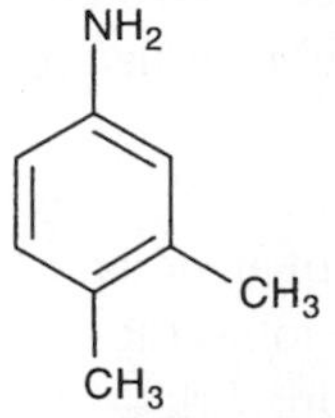

pK_{aH} (aniline)	4.63
ρ	2.89
σ_{meta} (CH$_3$)	- 0.06
σ_{para} (CH$_3$)	- 0.16

3,4-DMA

$$pK_a = 4.63 - (2.90)\,(-0.06 - 0.16) = 5.27$$

The reported experimental value is 5.28 (Johnson and Westall, 1990).

Answer (e)

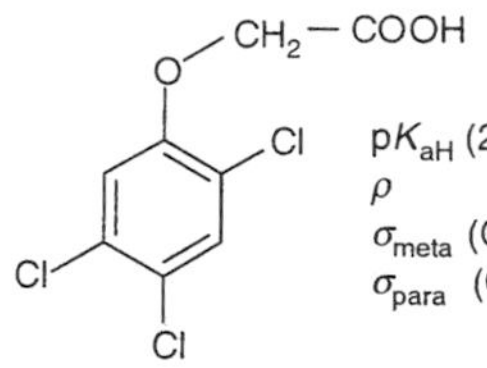

$$pK_{aH} \text{ (2-CPAA)} \quad 3.05$$
$$\rho \quad 0.30$$
$$\sigma_{meta} \text{ (Cl)} \quad 0.37$$
$$\sigma_{para} \text{ (Cl)} \quad 0.22$$

2,4,5-T

Since there are no σ_{ortho} values available, use 2-chlorophenoxy acetic acid (2-CPAA, $pK_a = 3.05$, Lide, 1995) as the starting value.

$$pK_a = 3.05 - (0.30)\,(0.37 + 0.22) = 2.87$$

The reported experimental values are 2.80 and 2.85 (Jafvert et al., 1990).

8.5   Aqueous Solubility and Partitioning Behavior of Organic Acids and Bases

Aqueous Solubility

The water solubility of the ionic form (salt) of an organic acid or base is generally several orders of magnitude higher than the solubility of the neutral species [which we denote as the *solubility* (C_{iw}^{sat}) of the compound]. The *total* concentration of the compound (nondissociated and dissociated forms) at saturation, $C_{iw,tot}^{sat}$, is, therefore, strongly pH-dependent. As has been demonstrated for pentachlorophenol (Arcand et al., 1995) and as is illustrated schematically for an organic acid in Fig. 8.8 (line *a*) at low pH, the saturation concentration is given by the solubility of the neutral compound. At higher pH values, $C_{iw,tot}^{sat}$ is determined by the fraction in neutral (acidic) form, α_{ia} (Eq. 8-21):

$$C_{iw,tot}^{sat} = \frac{C_{iw}^{sat}}{\alpha_{ia}} \qquad \text{for an organic acid} \tag{8-29}$$

Eq. 8-29 is valid, of course, only up to the solubility product of the salt of the ionized organic species (which is dependent on the type of counterion(s) present). Unfortunately, solubility data of organic salts are scattered and not systematically understood.

In the case of an organic base, the situation is symmetrical to the one shown in Fig. 8.8 in that the ionic (acid) form dominates at low pH. Hence, $C_{iw,tot}^{sat}$ is then given by [Fig. 8.8 (line *b*)]:

$$C_{iw,tot}^{sat} = \frac{C_{iw}^{sat}}{1 - \alpha_{ia}} \qquad \text{for an organic base} \tag{8-30}$$

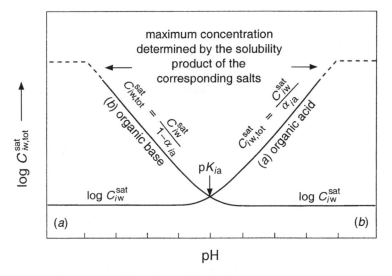

Figure 8.8 Schematic representation of the total aqueous solubility of (*a*) an organic acid, and (*b*) an organic base as a function of pH. Note that for simplicity the same pK_{ia} values and maximum solubilities of the neutral and charged (salt) species have been assumed.

Air–Water Partitioning

When considering the air–water equilibrium partitioning of an organic acid or base, we may, in general, assume that the ionized species will not be present in the gas phase. The air–water distribution ratio of an organic acid, D_{iaw} (note that we speak of a ratio and not of a partition constant since we are dealing with more than one species), is then given by:

$$D_{iaw} = \frac{[HA]_a}{[HA]_w + [A^-]_w} \qquad (8\text{-}31)$$

Multiplication of Eq. 8-31 with $[HA]_w / [HA]_w$ (= 1) and rearrangement shows that D_{iaw} is simply given by the product of the fraction in nondissociated form (α_{ia}) and the air–water partition constant of the neutral compound (K_{iaw}):

$$D_{iaw} = \frac{[HA]_w}{[HA]_w + [A^-]_w} \cdot \frac{[HA]_a}{[HA]_w} = \alpha_{ia} \cdot K_{iaw} \qquad \text{for an organic acid} \qquad (8\text{-}32)$$

By analogy we obtain:

$$D_{iaw} = (1 - \alpha_{ia}) K_{iaw} \qquad \text{for an organic base} \qquad (8\text{-}33)$$

An application of Eqs. 8-32 and 8-33 is given in Illustrative Example 8.3.

Illustrative Example 8.3

Assessing the Air–Water Distribution of Organic Acids and Bases in a Cloud

Problem

The air–water volume ratio (V_a / V_w) in a cloud is about 10^6 (Seinfeld, 1986). Consider now a given cloud volume that contains a certain total amount of (a) 2-4-dinitro-6-methyl phenol (DNOC), and (b) 4-chloroaniline (4-CA). Calculate the fraction of total DNOC and 4-CA, respectively, present in the water phase at equilibrium at 10°C for pH 2, 4, and 6. Neglect the effect of temperature on the acidity constant.

Answer (a)

The fraction of a given acid i present in the cloud water, f_{icw}, is given by (see Section 3.5 and Eq. 8-32):

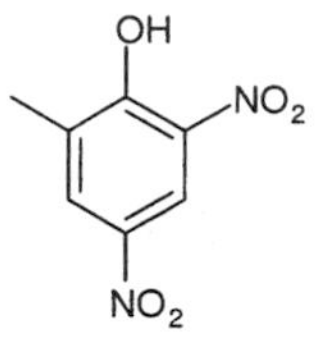

OH
NO$_2$
NO$_2$

i = 2,4-dinitro-6-methylphenol
(DNOC)

pK_{ia} (25°C) = 4.31
K_{iaw} (25°C) = 3.0 x 10^{-5}
$\Delta_{aw}H_i \cong$ 70 kJ · mol^{-1}

$$f_{icw} = \frac{1}{1 + D_{iaw}\dfrac{V_a}{V_w}} = \frac{1}{1 + \alpha_{ia} \cdot K_{iaw}\dfrac{V_a}{V_w}} \tag{1}$$

With $\Delta_{aw}H_i + RT_{av} = 72.4$ kJ · mol^{-1}, you get a K_{iaw} value at 10°C of (Eq. 3-50):

$$K_{iaw}(283\ \text{K}) = K_{iaw}(298\text{K}) \cdot e^{-\frac{72400}{8.31}\left[\frac{1}{283} - \frac{1}{298}\right]} = 0.21\, K_{iaw}(298\text{K}) = 6.4 \times 10^{-6}$$

Insertion of this value together with $V_a/V_w = 10^6$ and $\alpha_{ia} = (1 + 10^{pH-4.31})^{-1}$ into Eq. 1 yields:

$$f_{icw} = \frac{1}{1 + \dfrac{6.4}{1 + 10^{(pH-4.31)}}}$$

The resulting f_{icw} values are *0.14* (pH 2), *0.19* (pH 4), and *0.89* (pH 6). Hence, in contrast to apolar and weakly polar compounds (see Problem 6.2), DNOC partitions very favorably from the gas phase into an aqueous phase. It is, therefore, not surprising that this compound as well as other nitrophenols have been found in rather high concentrations (> 1 μg · L^{-1}) in rainwater (Tremp et al., 1993).

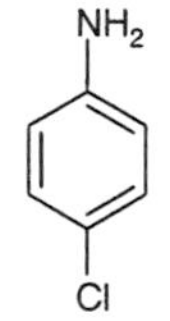

Answer (b)
Since 4-CA is a base, the fraction in the cloud water is given by (Eq. 8-33):

NH$_2$
Cl

i = 4-chloroaniline
(4-CA)

pK_{ia} (25°C) = 4.00
K_{iaw} (25°C) = 4.4 x 10^{-5}
$\Delta_{aw}H_i \cong$ 50 kJ · mol^{-1}

$$f_{icw} = \frac{1}{1 + (1 - \alpha_{ia}) \cdot K_{iaw}\dfrac{V_a}{V_w}} \tag{2}$$

With $\Delta_{aw}H_i + RT_{av} = 52.4$ kJ · mol^{-1}, you get a K_{iaw} value at 10°C of:

$$K_{iaw}(283\ \text{K}) = 0.33\, K_{iaw}(298\ \text{K}) = 1.4 \times 10^{-5}$$

Insertion of this value together with $V_a / V_w = 10^6$ and $(1 - \alpha_{ia}) = (1 + 10^{(4.00-pH)})^{-1}$ into Eq. (2) yields:

$$f_{icw} = \frac{1}{1 + \dfrac{14}{1 + 10^{(4.00-pH)}}}$$

The resulting f_{icw} values are *0.88* (pH 2), *0.13* (pH 4), and *0.067* (pH 6). This result shows that, like the phenols, aniline can be expected to be washed out quite efficiently from the atmosphere.

Organic Solvent–Water Partitioning

In contrast to air–water partitioning, the situation may be a little more complicated when dealing with organic solvent–water partitioning of organic acids and bases. As an example, Fig. 8.9 shows the pH dependence of the n-octanol–water distribution ratios, D_{iow} (HA, A$^-$), of four pesticides exhibiting an acid function:

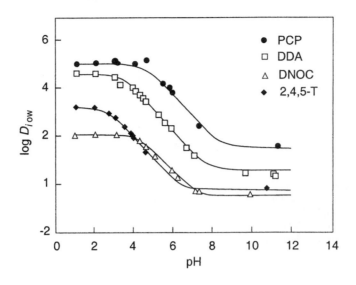

Figure 8.9 The pH dependence of the *n*-octanol–water distribution ratio of pentachlorophenol (PCP, pK_{ia} = 4.75), 4-chloro-α-(4-chlorophenyl) benzene acetic acid (DDA, pK_{ia} = 3.66), 2-methyl-4,6-dinitrophenol (DNOC, pK_{ia} = 4.46), and 2,4,5-trichlorophenoxy acetic acid (2,4,5-T, pK_{ia} = 2.83). (from Jafvert et al., 1990).

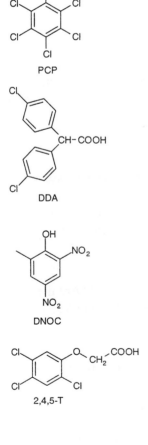

PCP

DDA

DNOC

2,4,5-T

$$D_{iow} = \frac{[HA]_{o,tot}}{[HA]_w + [A^-]_w} \tag{8-34}$$

where $[HA]_{o,tot}$ is the total concentration of HA in octanol. Since in octanol, not only the nondissociated acid but also ion pairs (with inorganic counterions) as well as ionic organic species may be present (Jafvert et al., 1990; Strathmann and Jafvert, 1998), D_{iow} of an acid may have a significant value even at high pH, particularly when dealing with *hydrophobic* acids. For pentachlorophenol (PCP, pK_{ia} = 4.75), for example, at pH 12 (virtually all PCP present as phenolate in the aqueous phase) and 0.1 M KCl, a D_{ow} (A⁻) value of about 100 has been determined (Fig. 8.9). Note that the partitioning of the ionic species depends on the type and concentration of the counterions present in the aqueous phase (Fig. 8.10). Hence, for calculating the organic phase–water equilibrium distribution ratio of an organic acid or base, a variety of species in both phases have to be considered (for details, see Jafvert et al., 1990; Strathmann and Jafvert, 1998).

From Fig. 8.9 it can be seen that for organic acids (and similarly for organic bases, Johnson and Westall, 1990), in the case of *n*-octanol, the partition constant of the neutral species is more than two orders of magnitude larger than the distribution ratio of the ionic species. Note that for less polar solvents, particularly, for apolar and weakly monopolar solvents, we can anticipate an even larger difference (Kishino and Kobayashi, 1994). Hence, at pH < pK_{ia} +2 for acids and pH > pK_{ia} −2 for bases, the neutral species is the dominant species in determining the organic solvent–water ratio, $D_{i\ell w}$, of the compound. At these pH values, by analogy to the air–water distribution ratio (Eqs. 8-32 and 8-33) we may express $D_{i\ell w}$ by:

$$D_{i\ell w} \cong \alpha_{ia} \cdot K_{i\ell w} \qquad \text{for organic acids} \tag{8-35}$$

and

$$D_{i\ell w} \cong (1 - \alpha_{ia}) \cdot K_{i\ell w} \qquad \text{for organic bases} \tag{8-36}$$

We should note, however, that when we are dealing with natural (organic) phases which may exhibit charged functionalities, this simple approach is no longer applicable. We will come back to this issue in Chapters 9,10, and 11.

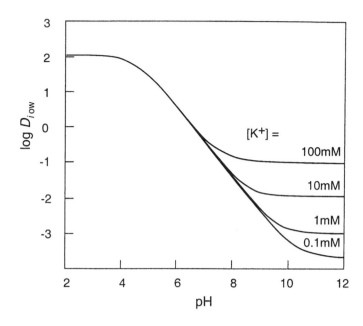

Figure 8.10 Calculated octanol–water distribution ratio of 2,4-dinitro-6-methylphenol (DNOC, pK_a = 4.46) as a function of pH and K^+ concentration (adapted from Jafvert et al. 1990).

8.6   Questions and Problems

Questions

Q 8.1

Name at least five different acid and/or base functions present in environmental organic chemicals. Which factors determine the pK_a of a given acid or base function? Indicate the pK_a ranges of the various functions.

Q 8.2

Explain the terms *inductive* and *resonance effect* of substituents. What makes a substituent exhibit a *negative* resonance effect? Which types of substituents have a positive resonance effect? Can a given substituent exhibit at the same time a negative inductive and a positive resonance effect? If yes, give some examples of such substituents.

Q 8.3

How are the Hammet $\sigma_{j\text{meta}}$ and $\sigma_{j\text{para}}$ substituent constants defined? Are there cases in which the $\sigma_{j\text{para}}$ values are not applicable? If yes, give some examples.

Q 8.4

For –OH and –OCH$_3$, the $\sigma_{j\text{meta}}$ values are positive, whereas $\sigma_{j\text{para}}$ is negative (Table 8.5). Try to explain these findings.

Q 8.5

As indicated below, 1-naphthylamine and quinoline exhibit very different susceptibility factors ρ (2.81 versus 5.90) in the corresponding Hammett equations. Try to explain this fact.

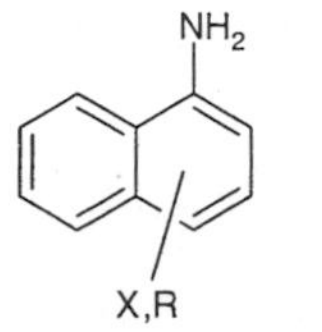

1-naphthylamine

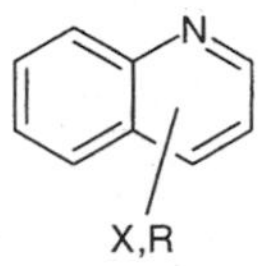

quinoline

$pK_a = 3.85 - 2.81 \sum_j \sigma_j$ $pK_a = 4.88 - 5.90 \sum_j \sigma_j$

(from Dean, 1985)

Q 8.6

The two isomers 2,4,6-trichlorophenol and 3,4,5-trichlorophenol have quite different pK_{ia} values. What are the reasons for this big difference?

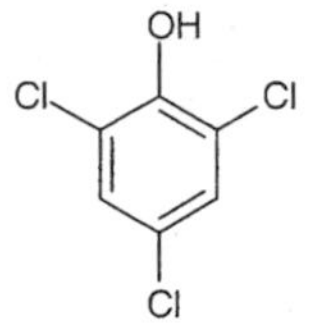

i = 2,4,6-trichlorophenol
(pK_{ia} = 6.15)

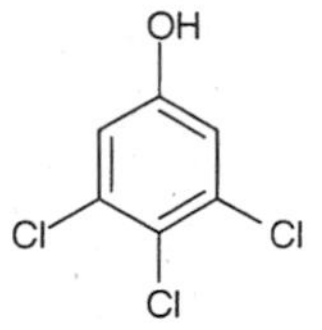

3,4,5-trichlorophenol
(pK_{ia} = 7.73)

Q 8.7

The pK_{ia} of the herbicide sulcotrion is 3.13 (Tomlin, 1997). Would you have expected that this compound is such a strong acid? Write down the structure of the conjugate base of sulcotrion and try to explain the rather strong acidity of this herbicide.

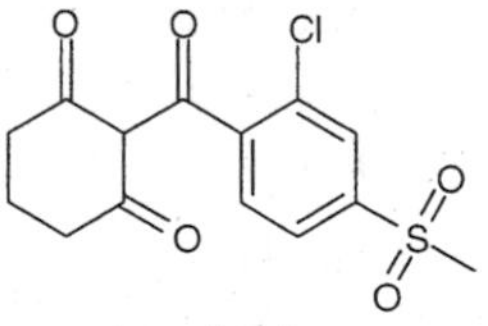

i = sulcotrion

Q 8.8

Give examples of compounds for which the aqueous solubility (a) increases, and (b) decreases significantly, when changing the pH 4 from 4 to 7.

Q 8.9

Consider the organic solvent–water partitioning of organic acids and bases. In which cases and/or under which conditions can you neglect the partitioning of the charged species into the organic phase?

Problems

P 8.1 *Estimation of Acidity Constants and Speciation in Water of Aromatic Organic Acids and Bases*

Represent graphically (as shown in Fig. 8.1) the speciation of (a) 4-methyl-2,5-dinitrophenol, (b) 3,4,5-trimethylaniline, and (c) 3,4-dihydroxybenzoic acid as a function of pH (pH-range 2 to 12) at 25°C. Estimate, if necessary, the pK_{ia} values of the compounds.

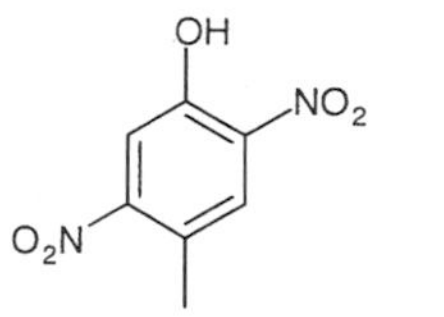

i = 2,5-dinitro-4-methyl phenol	3,4,5-trimethyl-aniline	3,4-dihydroxy-benzoic acid
		pK_{ia1} = 4.48
		pK_{ia2} = 8.83
		pK_{ia3} = 12.60

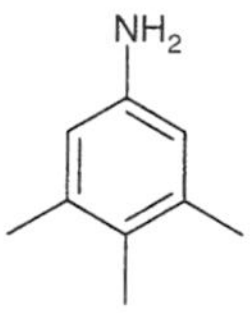

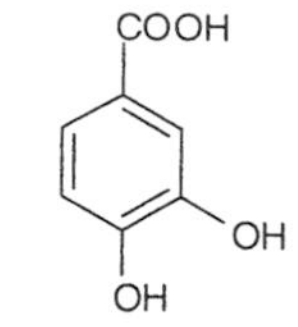

P 8.2 Air–Water Equilibrium Distribution of Organic Acids and Bases in Fog

Represent graphically the approximate fraction of (a) total 2,3,4,6-tetrachlorophenol and (b) total aniline present in the water phase of a dense fog (air–water volume ratio $\cong 10^5$) as a function of pH (pH-range 2 to 7) at 5 and 25°C. Neglect any adsorption to the surface of the fog droplet. Assume a $\Delta_{aw}H_i$ value of about 70 kJ·mol^{-1} for 2,3,4,6-tetrachlorophenol, and 50 kJ·mol^{-1} for aniline. All other data can be found in Appendix C.

P 8.3 Extracting Organic Acids and Bases from Water Samples

You have the job to determine the concentrations of 2,4,6-trichlorophenol (2,4,6-TCP) and 4-ethyl-2,6-dimethylpyridine (EDMP) in wastewater samples from an industrial site. You decide to extract the compounds first into an organic solvent, and then analyze them by liquid chromatography. From the $K_{i\ell w}$ values reported for the two compounds for various solvent–water systems, you conclude that there seems to be no single solvent that is optimally suited to extract the two compounds simultaneously. Would this be wise anyway? If there were such a solvent, at what pH would you carry out the extraction? What would be the problem? Anyway, you decide to extract first 2,4,6-TCP with butylacetate (subscript b) and then EDMP with trichloromethane (chloroform, subscript c). Give the pH-conditions at which you perform the extractions and calculate how much solvent you need at minimum in each case if you want to extract at least 98% of the compounds present in a 100 mL water sample.

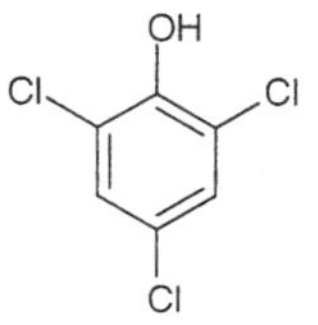

i = 2,4,6-trichlorophenol (2,4,6-TCP)	i = 4-ethyl-2,6-dimethyl pyridine (EDMP)
log K_{ibw} = 3.60	log K_{icw} = 3.70
pK_{ia} = 6.15	pK_{ia} = 7.43

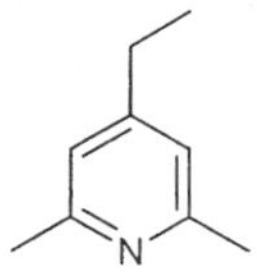

Chapter 9

SORPTION I: GENERAL INTRODUCTION AND SORPTION PROCESSES INVOLVING ORGANIC MATTER

9.1   **Introduction**

9.2   **Sorption Isotherms, Solid–Water Distribution Coefficients (K_{id}), and the Fraction Dissolved (f_{iw})**
Qualitative Considerations
Quantitative Description of Sorption Isotherms
The Solid–Water Distribution Coefficient K_{id}
Illustrative Example 9.1: *Determining K_{id} Values from Experimental Data*
Dissolved and Sorbed Fractions of a Compound in a System
The Complex Nature of K_{id}

9.3   **Sorption of Neutral Organic Compounds from Water to Solid-Phase Organic Matter (POM)**
Overview
Structural Characteristics of POM Relevant to Sorption
Determination of K_{ioc} Values and Availability of Experimental Data
Estimation of K_{ioc} Values
K_{ioc} as a Function of Sorbate Concentration
Illustrative Example 9.2: *Evaluating the Concentration Dependence of Sorption of Phenanthrene to Soil and Sediment POM*
Illustrative Example 9.3: *Estimating Pore Water Concentrations in a Polluted Sediment*
Effect of Temperature and Solution Composition on K_{ioc}
Illustrative Example 9.4: *How Much Does the Presence of 20% Methanol in the "Aqueous" Phase Affect the Retardation of Phenanthrene in an Aquifer?*

9.4 **Sorption of Neutral Compounds to "Dissolved" Organic Matter (DOM)**
Qualitative Description of DOM–Solute Associations
Determination of K_{iDOC} Values and Availability of Experimental Data
DOM Properties Governing the Magnitude of K_{iDOC}
Effect of pH, Ionic Strength, and Temperature on K_{iDOC}
LFERs Relating K_{iDOC} Values to K_{iow} Values
Illustrative Example 9.5: *Evaluating the Effect of DOM on the Bioavailability of Benzo(a)pyrene*

9.5 **Sorption of Organic Acids and Bases to Natural Organic Matter (NOM)**
Effect of Charged Moieties on Sorption: General Considerations
Sorption of Compounds Forming Anionic Species (Organic Acids)
Sorption of Compounds Forming Cationic Species (Organic Bases)

9.6 **Questions and Problems**

9.1   Introduction

The process in which chemicals become associated with solid phases is generally referred to as *sorption*. It is *ad*sorption if the molecules attach to a two-dimensional surface, while it is *ab*sorption if the molecules penetrate into a three-dimensional matrix. This phase transfer process may involve vapor molecules or dissolved molecules associating with solid phases.

Sorption is extremely important because it may dramatically affect the fate and impacts of chemicals in the environment. Such importance is readily understood if we recognize that structurally identical molecules behave very differently if they are: (a) in the gas phase or (b) surrounded by water molecules and ions as opposed to (c) clinging onto the exterior of solids or (d) buried within a solid matrix (Fig. 9.1). Clearly, the environmental transport of waterborne molecules must differ from the movements of the same kind of molecules attached to particles that settle. Also, transport of a given compound in porous media such as soils, sediments, and aquifers is strongly influenced by the compound's tendency to sorb to the various components of the solid matrix. Additionally, only dissolved molecules are available to collide with the interfaces leading to other environmental compartments such as the atmosphere; thus phase transfers are controlled by the dissolved species of a chemical (Chapters 19 and 20). Similarly, since molecular transfer is a prerequisite for the uptake of organic pollutants by organisms, the *bioavailability* of a given compound and thus its rate of biotransformation or its toxic effect(s) are affected by sorption processes (Chapters 10 and 17). Furthermore, some sorbed molecules are substantially shaded from incident light; therefore, these molecules may not experience direct photolysis processes. Moreover, when present inside solid matrices, they may never come in contact with short-lived, solution-phase photooxidants like OH-radicals (Chapters 15 and 16). Finally, since the chemical natures of aqueous solutions and solid environments differ greatly (e.g., pH, redox conditions), various chemical reactions including hydrolysis or redox reactions may occur at very different rates in the sorbed and dissolved states (Chapters 13 and 14). Hence, we must understand solid–solution and solid–gas phase exchange phenomena before we can quantify virtually any other process affecting the fates of organic chemicals in the environment.

Unfortunately, when we are dealing with natural environments, sorption is very often not an exchange between one homogeneous solution/vapor phase and a single solid medium. Rather, in a given system some combination of interactions may govern the association of a particular chemical (called the *sorbate*) with any particular solid or mixture of solids (called the *sorbent(s)*). Consider the case of 3,4-dimethylaniline (3,4-dimethyl aminobenzene, Fig. 9.2). This compound is a weak base with $pK_{ia} = 5.28$ (see Illustrative Example 8.2); hence, it reacts in aqueous solution to form some 3,4-dimethyl ammonium cations. For the fraction of molecules that remain uncharged, this organic compound may escape the water by penetrating the natural organic matter present in the system. Additionally, such a nonionic molecule may displace water molecules from the region near a mineral surface to some extent and be held there by London dispersive and polar interactions. These two types of sorption mechanisms are general and will operate

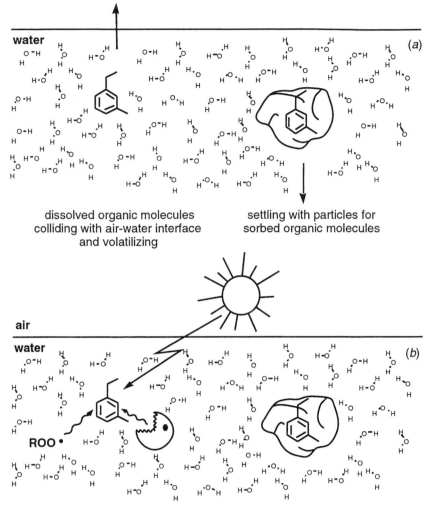

(a)

dissolved organic molecules
colliding with air-water interface
and volatilizing

settling with particles for
sorbed organic molecules

(b)

dissolved organic molecules are more accessible to light, to other dissolved
chemicals, and to microorganisms than sorbed molecules

Figure 9.1 Illustration of some processes in which sorbed species behave differently from dissolved molecules of the same substance. (*a*) Dissolved species may participate directly in air–water exchange while sorbed species may settle with solids. (*b*) Dissolved species may react at different rates as compared with their sorbed counterparts due to differential access of other dissolved and solid-phase "reactants."

for any organic chemical and any natural solid. Additionally, since the sorbate is ionizable in the aqueous solution, then electrostatic attraction to specific surface sites exhibiting the opposite charge will promote sorption of the ionic species. Finally, should the sorbate and the sorbent exhibit mutually reactive moieties (e.g., in Fig. 9.2 a carbonyl group on the sorbent and an amino group on the sorbate), some portion of a chemical may actually become bonded to the solid. All of these interaction mechanisms operate simultaneously, and the combination that dominates the overall solution–solid distribution will depend on the structural properties of the organic sorbate and the solid sorbent of interest.

In this and the following two chapters, we will focus on *solid–aqueous solution* and *solid-air exchange* involving natural sorbents. We will try to visualize the sets of molecular interactions involved in each of the above-mentioned sorption processes. With such pictures in our minds, we will seek to rationalize what makes various sorption mechanisms important under various circumstances. Establishing the critical compound properties and solid characteristics will enable us to understand

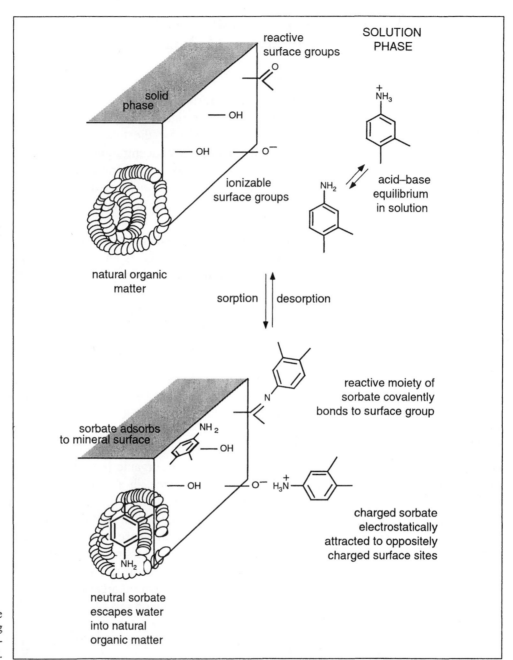

Figure 9.2 Some sorbent–sorbate interactions possibly controlling the association of a chemical, (3,4-dimethylaniline), with natural solids.

when and to what extent predictive approaches for quantification of sorption may be applied. Ultimately, we should gain some feeling for what structural features of a chemical and what characteristics of solids (and solutions) are important to sorptive interactions. In this chapter, we will focus on sorption processes involving natural organic matter. In Chapter 10 we will address sorption to living media, that is, we will treat bioaccumulation. Finally, sorption from water to mineral surfaces and gas–solid phase transfers will be discussed in Chapter 11. To start out, we will first address some general aspects needed to quantify *sorption equilibrium* in a given system. Kinetic aspects of sorption processes will be treated in Chapter 19.

9.2 Sorption Isotherms, Solid–Water Distribution Coefficients (K_{id}), and the Fraction Dissolved (f_{iw})

Qualitative Considerations

When we are interested in the *equilibrium* distribution of a chemical between the solids and solution present in any particular volume of an aquatic environment, we begin by considering how the total sorbate concentration, C_{is} (e.g., mol·kg^{-1}) depends on chemical's concentration in the solution, C_{iw} (e.g., mol·L^{-1}). The relationship of these two concentrations is commonly referred to as a *sorption isotherm*. The name *isotherm* is used to indicate that this sorption relationship applies only at a constant temperature.

Experimentally determined sorption isotherms exhibit a variety of shapes for diverse combinations of sorbates and sorbents (Fig. 9.3). The simplest case (Fig. 9.3*a*) is the one in which the affinity of the sorbate for the sorbent remains the same over the observed concentration range. This is the so-called *linear* isotherm case. It applies to situations where partitioning *into* a homogeneous organic phase is dominating the overall sorption, and/or at low concentrations where the strongest adsorption sites are far from being saturated. The second types of behavior (Figs. 9.3*b* and *c*) reflect those situations in which at higher and higher sorbate concentrations it becomes more and more difficult to sorb additional molecules. This occurs in cases where the binding sites become filled and/or remaining sites are less attractive to the sorbate molecules. In the extreme case (Fig. 9.3*c*), above some maximum C_{is} value, all sites are "saturated" and no more additional sorption is possible. Isotherms of the type shown in Figs. 9.3*b* and *c* are encountered in studies of *ad*sorption processes to organic (e.g., activated carbon) or inorganic (e.g., clay mineral) surfaces. Of course, in a soil or sediment, there may be more than one important sorbent present. Therefore, the overall sorption isotherm may reflect the superposition of several individual isotherms that are characteristic for each specific type of sorbent. When such a case involves an adsorbent (e.g., soot, clay mineral) exhibiting a limited number of sites with a high affinity for the sorbate (type (c) isotherm) that dominates the overall sorption at low concentrations, plus a partitioning process (e.g., into natural organic matter; type (a) isotherm) predominating at higher concentrations, then a mixed isotherm is seen (Fig. 9.3*b* or *d*). Likewise, superimposition of multiple adsorption isotherms results in a mixed isotherm looking like an isotherm of type (*b*) (Weber et al., 1992).

Another case that is less frequently encountered involves the situation in which previously sorbed molecules lead to a modification of the sorbent which favors further sorption (Fig. 9.3*e*). Such effects have been seen in studies involving anionic or cationic surfactants as sorbates. In some of these cases, a sigmoidal isotherm shape (Fig. 9.3*f*) has been observed, indicating that the sorption-promoting effect starts only after a certain loading of the sorbent.

In summary, depending on the composition of a natural bulk sorbent and on the chemical nature of the sorbate, multiple sorption mechanisms can act simultaneously and the resulting isotherms may have a variety of different shapes. We should note that it is not possible to prove a particular sorption mechanism

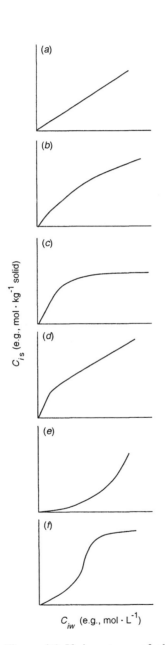

Figure 9.3 Various types of observed relationships between concentrations of a chemical in the sorbed state, C_{is}, and the dissolved state, C_{iw}. Note that similar relationships apply to the sorption of gaseous compounds to solid sorbents.

applies from the shape of the isotherm. Nevertheless, the isotherm type and its degree of nonlinearity must be consistent with the sorption mechanism(s) prevailing in a given situation.

Quantitative Description of Sorption Isotherms

A very common mathematical approach for fitting experimentally determined sorption data using a minimum of adjustable parameters employs an *empirical* relationship known as the *Freundlich* isotherm:

$$C_{is} = K_{iF} \cdot C_{iw}^{n_i} \qquad (9\text{-}1)$$

where K_{iF} is the Freundlich constant or capacity factor [(e.g., Eq. 9-1 in $(\text{mol} \cdot \text{kg}^{-1})$ $(\text{mol} \cdot \text{L}^{-1})^{-n_i})$]; and n_i is the Freundlich exponent. Note that for a correct thermodynamic treatment of Eq. 9-1 we would always have to use dimensionless activities of compound i in both the sorbed and aqueous phase in order to obtain a dimensionless K_{iF}. However, in practice C_{is} and C_{iw} are expressed in a variety of concentration units. Therefore, K_{iF} is commonly reported in the corresponding units, which also means that for $n_i \neq 1$, K_{iF} depends nonlinearly on the units in which C_{iw} is expressed (see Illustrative Example 9.1 and Problem 9.5).

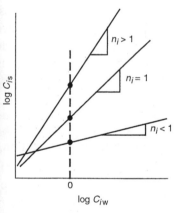

The relationship Eq. 9-1 assumes there are multiple types of sorption sites acting in parallel, with each site type exhibiting a different sorption free energy and total site abundance. The exponent is an index of the diversity of free energies associated with the sorption of the solute by multiple components of a heterogeneous sorbent (Weber and Digiano, 1996). When $n_i = 1$, the isotherm is linear and we infer constant sorption free energies at all sorbate concentrations (Fig. 9.3a); when $n_i < 1$, the isotherm is concave downward and one infers that added sorbates are bound with weaker and weaker free energies (Fig. 9.3b); finally when $n_i > 1$, the isotherm is convex upward and we infer that more sorbate presence in the sorbent enhances the free energies of further sorption (Fig. 9.3e). K_{iF} and n_i can be deduced from experimental data by linear regression of the logarithmic form of Eq. 9-1 (Fig. 9.4; see also Illustrative Example 9.1):

Figure 9.4 Graphic representation of the Freundlich isotherm Eq. 9-2 for the three cases $n_i > 1$, $n_i = 1$, and $n_i < 1$. Note that n_i and $\log K_{iF}$ are obtained from the slope (n_i) and intercept ($\log K_{iF}$ indicated by the points at $\log C_{iw} = 0$) of the regression line.

$$\log C_{is} = n_i \log C_{iw} + \log K_{iF} \qquad (9\text{-}2)$$

If a given isotherm cannot be described by Eq. 9-2, then some assumptions behind the Freundlich multi-site conceptualization are not valid. For example, if there are limited total sorption sites that become saturated (case shown in Fig. 9.3c), then C_{is} cannot increase indefinitely with increasing C_{iw}. In this case, the *Langmuir* isotherm may be a more appropriate model:

$$C_{is} = \frac{\Gamma_{\max} \cdot K_{iL} \cdot C_{iw}}{1 + K_{iL} \cdot C_{iw}} \qquad (9\text{-}3)$$

where $\Gamma_{\max}$ represents the total number of surface sites per mass of sorbent. In the ideal case, $\Gamma_{\max}$ would be equal for all sorbates. However, in reality, $\Gamma_{\max}$ may vary somewhat between different compounds (e.g., because of differences in sorbate size). Therefore, it usually represents the maximum achievable surface concentration of a given compound i (i.e., $\Gamma_{\max} = C_{is,\max}$). The constant K_{iL}, which is

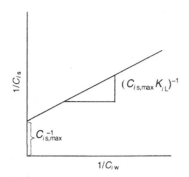

Figure 9.5 Graphic representation of the Langmuir isotherm Eq. 9-4. Note that $C_{is,\max}$ and K_{iL} can be derived from the slope and intercept of the regression line (see also Illustrative Example 9.1).

commonly referred to as the Langmuir constant, is defined as the equilibrium constant of the sorption reaction:

$$\text{surface site} + \text{sorbate in aqueous solution} \rightleftharpoons \text{sorbed sorbate}$$

Note that in this approach, since K_{iL} is constant, this implies a constant sorbate affinity for all surface sites. To derive K_{iL} and $C_{is,\max}$ from experimental data, one may fit $1/C_{iw}$ versus $1/C_{is}$:

$$\frac{1}{C_{is}} = \left(\frac{1}{C_{is,\max} \cdot K_{iL}} \right) \frac{1}{C_{iw}} + \frac{1}{C_{is,\max}} \tag{9-4}$$

and use the slope and intercept to extract estimates of the isotherm constants (Fig. 9.5).

There are many cases in which the relationship between sorbed concentrations and dissolved concentrations covering a large concentration range cannot be described solely by a linear, a Langmuir, or even a Freundlich equation (e.g., cases d and f in Fig. 9.3). In these cases, combinations of linear-, Langmuir-, and/or Freundlich-type equations may need to be applied (e.g., Weber et al., 1992; Xing and Pignatello, 1997; Xia and Ball, 1999). Among these *distributed reactivity models* (Weber et al., 1992), the simplest case involves a pair of sorption mechanisms involving absorption (e.g., linear isotherm with partition coefficient, K_{ip}) and site-limited adsorption (e.g., Langmuir isotherm), and the resultant combined equation is:

$$C_{is} = K_{ip} \cdot C_{iw} + \frac{C_{is,\max} \cdot K_{iL} \cdot C_{iw}}{1 + K_{iL} \cdot C_{iw}} \tag{9-5}$$

Another form that fits data from sediments known to contain black carbon (e.g., soot) uses a combination of a linear isotherm and a Freundlich isotherm (Accardi-Dey and Gschwend, 2002):

$$C_{is} = K_{ip} C_{iw} + K_{iF} C_{iw}^{n_i} \tag{9-6}$$

These *dual-mode* models have been found to be quite good in fitting experimental data for natural sorbents that contain components exhibiting a limited number of more highly active adsorption sites as well as components into which organic compounds may absorb (Huang et al., 1997; Xing and Pignatello, 1997; Xia and Ball, 1999). At low concentrations, the Langmuir or the Freundlich term may dominate the overall isotherm, while at high concentrations (e.g., $K_{iL} \cdot C_{iw} \gg 1$), the absorption term dominates (see Section 9.3).

The Solid–Water Distribution Coefficient, K_{id}

To assess the extent to which a compound is associated with solid phases in a given system at equilibrium (see below), we need to know the ratio of the compound's total equilibrium concentrations in the solids and in the aqueous solution. We denote this solid-water distribution coefficient as K_{id} (e.g., in $L \cdot kg^{-1}$ solid):

$$K_{id} = \frac{C_{is}}{C_{iw}} \tag{9-7}$$

(When writing natural solid–water distribution or partition coefficients, we will use a somewhat different subscript terminology than used for air–water or organic solvent–water partitioning; that is, we will not indicate the involvement of a water phase by using a subscript "w".) When dealing with nonlinear isotherms, the value of this ratio may apply only at the given solute concentration (i.e., if n_i in Eq. 9-1 is substantially different from 1). Inserting Eq. 9-1 into Eq. 9-7, we can see how K_{id} varies with sorbate concentration:

$$K_{id} = K_{iF} \cdot C_{iw}^{n_i-1} \tag{9-8}$$

For practical applications, one often assumes that K_{id} is constant over some concentration range. We can examine the reasonableness of such a simplification by differentiating K_{id} with respect to C_{iw} in Eq. 9-8 and rearranging the result to find:

$$\frac{dK_{id}}{K_{id}} = (n_i - 1)\frac{dC_{iw}}{C_{iw}} \tag{9-9}$$

So the assumption about the constancy of K_{id} is equivalent to presuming either: (a) the overall process is described by a linear isotherm ($n_i - 1 = 0$), or (b) the relative concentration variation, (dC_{iw}/C_{iw}), is sufficiently small that when multiplied by ($n_i - 1$) the relative K_{id} variation, (dK_{id}/K_{id}), is also small. For example, if the sorbate concentration range is less than a factor of 10, when multiplied by ($n_i - 1$) with an n_i value of 0.7, then the solid–water distribution coefficient would vary by less than a factor of 3.

Illustrative Example 9.1

NO₂ structure image

1,4-dinitrobenzene (1,4-DNB)

C_{iw} (μmol·L^{-1})	C_{is} (μmol·kg^{-1})
0.06	97
0.17	241
0.24	363
0.34	483
0.51	633
0.85	915
1.8	1640
2.8	2160
3.6	2850
7.6	4240
19.5	6100
26.5	7060

Determining K_{id} Values from Experimental Data

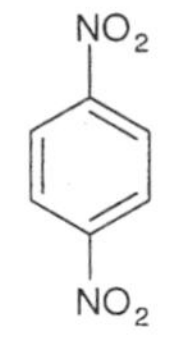

A common way to determine K_{id} values is to measure sorption isotherms in *batch* experiments. To this end, the equilibrium concentrations of a given compound in the solid phase (C_{is}) and in the aqueous phase (C_{iw}) are determined at various compound concentrations and/or solid–water ratios. Consider now the sorption of 1,4-dinitrobenzene (1,4-DNB) to the homoionic clay mineral, K$^+$-illite, at pH 7.0 and 20°C. 1,4-DNB forms electron donor-acceptor (EDA) complexes with clay minerals (see Chapter 11). In a series of batch experiments, Haderlein et al. (1996) measured the data at 20°C given in the margin.

Problem

Using this data, estimate the K_{id}-values for 1,4-DNB in a K$^+$-illite-water suspension (pH 7.0 at 20°C) for equilibrium concentrations of 1,4-DNB in the aqueous phase of 0.20 μM and of 15 μM, respectively.

Answer

Plot C_{is} versus C_{iw} to see the shape of the sorption isotherm (Fig. 1):

For K_{id} at $C_{iw} = 0.20$ μM, assume a linear isotherm for the concentration range 0–0.5 μM. Perform a least squares fit of C_{is} versus C_{iw} using only the first four data

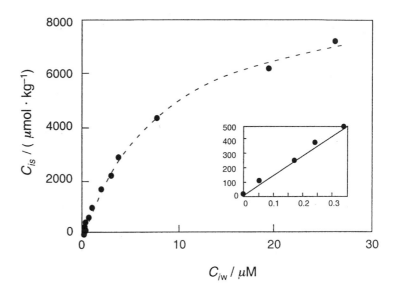

Figure 1 Plot of C_{is} versus C_{iw}. The dotted line represents the fitted Langmuir equation (see below).

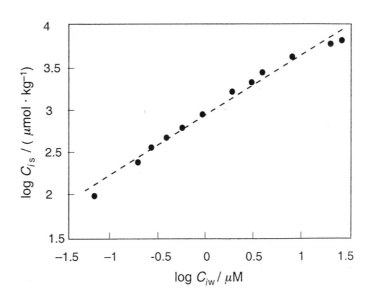

Figure 2 Plot of log C_{is} versus log C_{iw} using the whole data set.

points and the origin (see insert in Fig. 1). The resulting regression equation is:

$$C_{is} = 1425\ C_{iw}\quad (R^2 = 1.0)$$

Hence, you get a K_{id} value (slope) of *1425* L·kg^{-1} that is valid for the whole concentration range considered (i.e., $C_{iw} \leq 5\ \mu M$).

For deriving K_{id} at $C_{iw} = 15\ \mu M$, fit the experimental data with the Freundlich equation (Eq. 9.1). To determine the K_{iF} and n_i values use Eq. 9.2 (i.e., perform a least squares fit of log C_{is} versus log C_{iw} using all data points).

The resulting regression here is:

$$\log C_{is} = 0.70 \log C_{iw} + 2.97\quad (R^2 = 0.98)$$

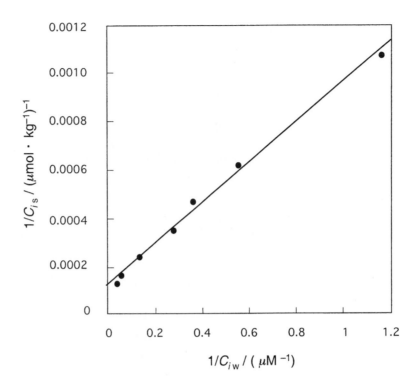

Figure 3 Plot of $1/C_{is}$ versus $1/C_{iw}$ for the data points with $C_{iw} > 0.5$ μM.

Hence, $K_{iF} = 10^{2.97} \cong 1000$ (μmol·kg^{-1} μM$^{-0.70}$) [see comment on units of K_{iF} below Eq. 9-1] and $n_i = 0.70$; therefore (Eq. 9-8):

$$K_{id} = 1000 \cdot C_{iw}^{-0.3}$$

Insertion of $C_{iw} = 15$ μM yields a K_{id} value of about *450* L·kg^{-1}.

Note that this K_{id} value is significantly smaller than the K_{id} obtained in the linear part of the isotherm (i.e., at low 1,4-DNB concentrations). Furthermore, as can be seen from Fig. 2, the Freundlich equation overestimates C_{is} (and thus K_{id}) at both the low and the high end of the concentration range considered. In fact, inspection of Fig. 2 reveals that at very high concentrations, the K$^+$-illite surface seems to become saturated with 1,4-DNB, which is not surprising considering that only limited adsorption sites are available. In such a case, the sorption isotherm can also be approximated by a Langmuir equation (Eq. 9-3).

To get the corresponding K_{iL} and C_{imax} values, use Eq. 9-4 (i.e., perform a least squares fit of $1/C_{is}$ versus $1/C_{iw}$). Use only the data with $C_{iw} > 0.5$ μM to get a reasonable weighting of data points in the low and high concentration range.

The resulting regression equation is:

$$\frac{1}{C_{is}} = 0.000753 \frac{1}{C_{iw}} + 0.000152 \quad (R^2 = 0.99)$$

yielding a C_{imax} of 6600 μmol·kg^{-1} and a K_{iL} value of 0.201 L·μmol^{-1}. At very low concentrations (i.e., $K_{iL} \cdot C_{iw} \ll 1$), which includes $C_{iw} = 0.20 \ \mu M$, K_{id} is given by the linear relationship:

$$K_{id} = K_{iL} \cdot C_{imax} = (0.201) \ (6600) = 1320 \ \text{L} \cdot \text{kg}^{-1}$$

which is somewhat smaller than the K_{id} value determined from the linear regression analysis using only the first four data points (i.e., $K_{id} = 1425 \ \text{L} \cdot \text{kg}^{-1}$, see above). This is not too surprising when considering that the Langmuir model assumes that all surface sites exhibit the same affinities for the sorbate. This is not necessarily the case, as it is likely that sites with higher affinities are occupied first. Therefore, a linear fit of data points determined at low concentrations can be expected to yield a higher apparent sorption coefficient as compared to the coefficient calculated from nonlinear extrapolation of data covering a wide concentration rate.

Inserting $C_{iw} = 15 \ \mu M$ into Eq. 9-3 with the above derived K_{iL} and C_{imax} values yields a C_{is} value of $(6600)(0.201)(15)/[1+(0.201)(15)] = 4950 \ \mu\text{mol} \cdot \text{kg}^{-1}$, and thus a K_{id} of $4950 \ / \ 15 = 330 \ \text{L} \cdot \text{kg}^{-1}$. This value is somewhat smaller than the one derived from the Freundlich equation ($450 \ \text{L} \cdot \text{kg}^{-1}$; see above). These calculations show that when estimating K_{id} values from experimental data, depending on the concentration range of interest, one has to make an optimal choice with respect to the selection of the experimental data points as well as with respect to the type of isotherm used to fit the data.

Dissolved and Sorbed Fractions of a Compound in a System

Armed with a K_{id} for a case of interest, we may evaluate what fraction of the compound is dissolved in the water, f_{iw}, for any environmental volume containing both solids and water, but only these phases:

$$f_{iw} = \frac{C_{iw} \cdot V_w}{C_{iw} V_w + C_{is} M_s} \tag{9-10}$$

where V_w is the volume of water (e.g., L) in the total volume V_{tot}, and M_s is the mass of solids (e.g., kg) present in that same total volume. Now if we substitute the product $K_{id} \cdot C_{iw}$ from Eq. 9-7 for C_{is} in Eq. 9-10, we have:

$$f_{iw} = \frac{C_{iw} V_w}{C_{iw} V_w + K_{id} C_{iw} M_s}$$

$$= \frac{V_w}{V_w + K_{id} M_s} \tag{9-11}$$

Finally, noting that we refer to the quotient, M_s/V_w, as the solid–water phase ratio, r_{sw} (e.g., $\text{kg} \cdot \text{L}^{-1}$) in the environmental compartment of interest, we may describe the fraction of chemical in solution as a function of K_{id} and this ratio:

$$f_{iw} = \frac{1}{1+(M_s/V_w)K_{id}}$$

$$= \frac{1}{1+r_{sw}\cdot K_{id}} \tag{9-12}$$

Such an expression clearly indicates that for substances exhibiting a great affinity for solids (hence a large value of K_{id}) or in situations having large amounts of solids per volume of water (large value of r_{sw}), we predict that correspondingly small fractions of the chemical remain dissolved in the water. Note the fraction associated with solids, f_{is}, must be given by $(1-f_{iw})$ since we assume that no other phases are present (e.g., air, other immiscible liquids).

The fraction of the total volume, V_{tot}, that is not occupied by solids, the *porosity*, ϕ, is often used instead of r_{sw} to characterize the solid–water phase ratio in some environmental systems like sediment beds or aquifers. In the absence of any gas phase, ϕ is related to parameters discussed above by:

$$\phi = \frac{V_w}{V_{tot}} = \frac{V_w}{V_w+V_s} \tag{9-13}$$

where, V_s, the volume occupied by particles, can be calculated from M_s/ρ_s (where ρ_s is the density of the solids and is typically near 2.5 kg L^{-1} for many natural minerals.) Thus, we find the porosity is also given by:

$$\phi = \frac{V_w}{V_w+M_s/\rho_s} = \frac{1}{1+r_{sw}/\rho_s} \tag{9-14}$$

and solving for r_{sw} yields the corresponding relation:

$$r_{sw} = \rho_s\frac{1-\phi}{\phi} \tag{9-15}$$

Finally, in the soil and groundwater literature, it is also common to use still a third parameter called *bulk density*, ρ_b. Bulk density reflects the ratio, M_s/V_{tot}, so we see it is simply given by $\rho_s(1-\phi)$. Thus, knowing bulk density we have r_{sw} is equal to ρ_b/ϕ. It is a matter of convenience whether r_{sw}, ϕ, or ρ_b is used.

The application of such solution- versus solid-associated speciation information may be illustrated by considering an organic chemical, say 1,4-dimethylbenzene (DMB), in a lake and in flowing groundwater. In lakes, the solid–water ratio is given by the suspended solids concentration (since $V_w \approx V_{tot}$), which is typically near 10^{-6} kg$\cdot$L^{-1}. From experience we may know that the K_{id} value for DMB in this case happens to be 1 L$\cdot$kg^{-1}; therefore we can see that virtually all of this compound is in the dissolved form in the lake:

$$f_{iw} = \frac{1}{1+10^{-6}\cdot 1} \cong 1$$

In contrast, now consider the groundwater situation; ρ_s for aquifer solids is about 2.5 kg$\cdot$L^{-1} (e.g., quartz density is 2.65 kg$\cdot$L^{-1}); ϕ is often between 0.2 and 0.4. If in our

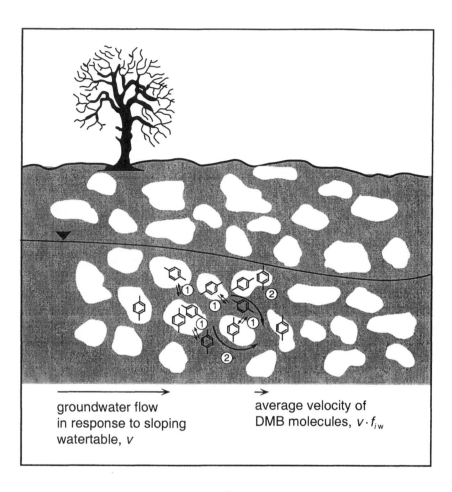

Figure 9.6 Illustration of the retardation of 1,4-dimethylbenzene (DMB) transport in groundwater due to: (1) reversible sorptive exchange between water and solids, and (2) limiting transport of DMB to that fraction remaining in the flowing water. As dissolved molecules move ahead, they become sorbed and stopped, while molecules sorbed at the rear return to the water and catch up. Thus, overall transport of DMB is slower than that of the water itself.

groundwater flow in response to sloping watertable, v

average velocity of DMB molecules, $v \cdot f_{iw}$

particular groundwater situation $\phi = 0.2$, and $r_{sw} = 10 \ \text{kg} \cdot \text{L}^{-1}$, we predict that the fraction of DMB in solution, again assuming K_{id} of 1 L·kg^{-1}, is drastically lower than in the lake:

$$f_{iw} = \frac{1}{1 + 10 \cdot 1} \cong 0.09$$

So we deduce that only one DMB molecule out of 11 will be in the moving groundwater at any instant (Fig. 9.6). This result has implications for the fate of the DMB in that subsurface environment. If DMB sorptive exchange between the aquifer solids and the water is fast relative to the groundwater flow and if sorption is reversible, we can conclude that the whole population of DMB molecules moves at one-eleventh the rate of the water. The phenomenon of diminished chemical transport speed relative to the water seepage velocity is referred to as *retardation*. It is commonly discussed using the *retardation factor, R_{fi}*, which is simply equal to the reciprocal of the fraction of molecules capable of moving with the flow at any instant, f_{iw}^{-1} (see Chapter 25).

Many situations require us to know something about the distribution of a chemical between a solution and solids. Our task then is to see how we can get K_{id} values suited for the cases that concern us. As we already pointed out above, these K_{id} values are determined by the structures of the sorbates as well as the composition of the aqueous phase and the sorbents.

The Complex Nature of K_{id}

The prediction of K_{id} for any particular combination of organic chemical and solids in the environment can be difficult, but fortunately many situations appear reducible to fairly simple limiting cases. We begin by emphasizing that the way we defined K_{id} means that we may have lumped together many chemical species in each phase. For example, referring again to Fig. 9.2, we recognize that the total concentration of the dimethylaniline in the sorbed phase combines the contributions of molecules in many different sorbed forms. Even the solution in this case contains both a neutral and a charged species of this chemical. Thus, in a conceptual way, the distribution ratio for this case would have to be written as:

$$K_{id} = \frac{C_{ioc} \cdot f_{oc} + C_{imin} \cdot A_{surf} + C_{iex} \cdot \sigma_{surf\,ex} \cdot A_{surf} + C_{irxn} \cdot \sigma_{surf\,rxn} \cdot A_{surf}}{C_{iw,neut} + C_{iw,ion}} \quad (9\text{-}16)$$

where C_{ioc} is the concentration of sorbate i associated with the natural organic matter (expressed as organic carbon) present ($mol \cdot kg^{-1}$ oc)

   f_{oc} is the weight fraction of solid which is natural organic matter (expressed as organic carbon, i.e., kg oc $\cdot$ kg^{-1} solid)

   C_{imin} is the concentration of sorbate i associated with the mineral surface ($mol \cdot m^{-2}$)

   A_{surf} is the specific surface area of the relevant solid

   C_{iex} is the concentration of ionized sorbate drawn towards positions of opposite charge on the solid surface ($mol \cdot mol^{-1}$ surface charges)

   $\sigma_{surf\,ex}$ is the net concentration of suitably charged sites on the solid surface (mol surface charges $\cdot m^{-2}$) for ion exchange

   C_{irxn} is the concentration of sorbate i bonded in a reversible reaction to the solid ($mol \cdot mol^{-1}$ reaction sites)

   $\sigma_{surf\,rxn}$ is the concentration of reactive sites on the solid surface (mol reaction sites $\cdot m^{-2}$)

   $C_{iw,neut}$ is the concentration of uncharged chemical i in solution ($mol \cdot L^{-1}$)

   $C_{iw,ion}$ is the concentration of the charged chemical i in solution ($mol \cdot L^{-1}$)

All terms in Eq. 9-16 may also deserve further subdivision. For example, $C_{ioc} \cdot f_{oc}$ may reflect the sum of adsorption and absorption mechanisms acting to associate the chemical to a variety of different forms of organic matter (e.g., living biomass of microorganisms, partially degraded organic matter from plants, plastic debris from humans, etc.). Similarly, $C_{imin} \cdot A_{surf}$ may reflect a linear combination of the interactions of several mineral surfaces present in a particular soil or sediment with a single sorbate. Thus, a soil consisting of montmorillonite, kaolinite, iron oxide, and quartz mineral components may actually have $C_{imin} \cdot A_{surf} = C_{imont} \cdot a \cdot A_{surf} + C_{ikao} \cdot b \cdot A_{surf} + C_{iiron\ ox} \cdot c \cdot A_{surf} + C_{iquartz} \cdot d \cdot A_{surf}$ where the parameters a, b, c, and d are the area fractions exhibited by each mineral type. Similarly, $C_{irxn} \cdot \sigma_{rxn} \cdot A_{rxn}$ may reflect bonding to several different kinds of surface moieties, each with its own reactivity with the sorbate (e.g., 3,4-dimethylaniline). For now, we will work from the simplified expression which is Eq. 9-16, primarily because there are few data available allowing rational subdivisions of soil or sediment differentially sorbing organic chemicals beyond that reflected in this equation.

It is very important to realize that only particular combinations of species in the numerator and denominator of complex K_{id} expressions like that of Eq. 9-16 are involved in any one exchange process. For example, in the case of dimethylaniline (DMA) (Fig. 9.2), exchanges between the solution and the solid-phase organic matter:

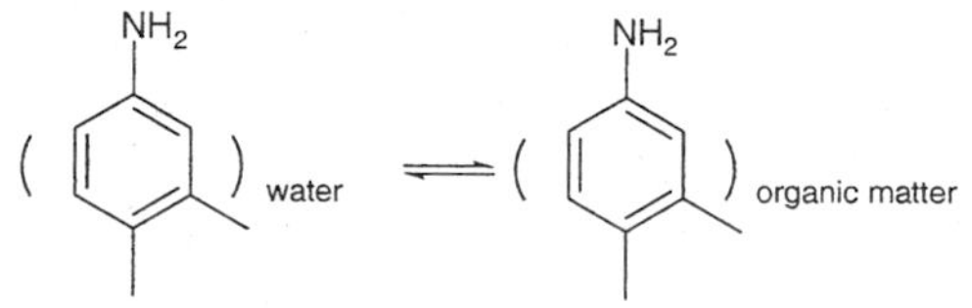

$$(9\text{-}17)$$

reflect establishing the same chemical potential of the uncharged DMA species in the water and in the particulate natural organic phase. As a result, a single free energy change and associated equilibrium constant applies to the sorption reaction depicted by Eq. 9-17. Similarly, the combination:

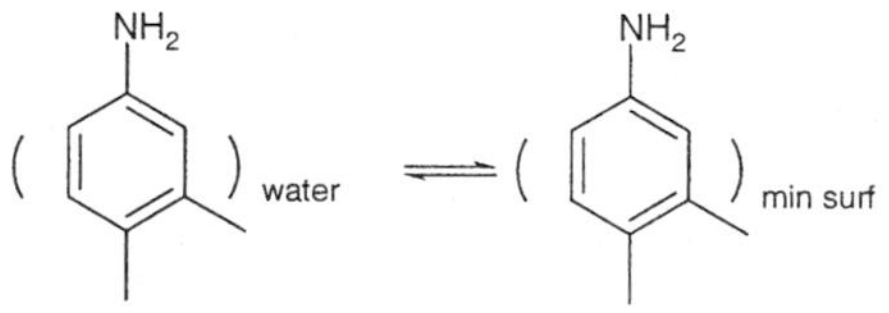

$$(9\text{-}18)$$

would indicate a simultaneously occurring exchange of uncharged aniline molecules from aqueous solution to the available mineral surfaces. Again, this exchange is characterized by a unique free energy difference reflecting the equilibria shown in Eq. 9-18. Likewise, the exchange of:

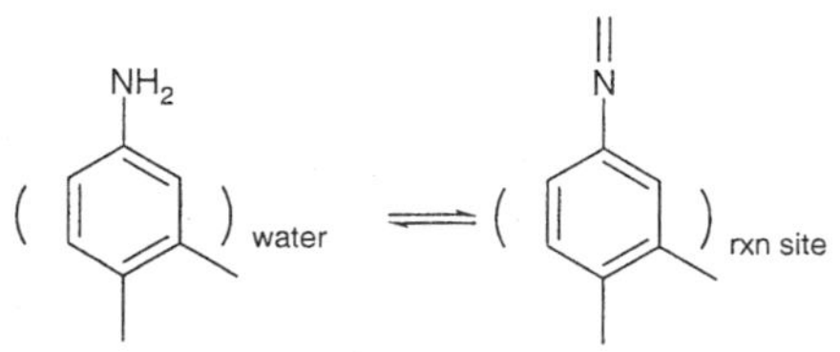

$$(9\text{-}19)$$

should be considered if it is the neutral sorbate which can react with components of the solid. Note that such specific binding to a particular solid phase moiety may prevent rapid desorption, and therefore such sorbate–solid associations may cause part or all of the sorption process to appear irreversible on some time scale of interest.

So far we have considered sorptive interactions in which only the DMA species was directly involved. In contrast, it is the charged DMA species (i.e., anilinium ions) that is important in the ion exchange process:

$$(\quad)_{\text{water}} \; \rightleftharpoons \; (\quad)_{\text{ion exchange site}} \qquad (9\text{-}20)$$

Of course, the anilinium ion in solution is quantitatively related to the neutral aniline species via an acid–base reaction having its own equilibrium constant (see Chapter 8). But we also emphasize that the solution–solid exchange shown in Eq. 19-20 has to be described using the appropriate equilibrium expression relating corresponding species in each phase. The influence of each sorption mechanism is ultimately reflected by all these equilibria in the overall K_{id} expression, and each is weighted by the availability of the respective sorbent properties in the heterogeneous solid (i.e., f_{oc}, σ_{ex}, σ_{rxn} or the various A_{surf} values). By combining information on the individual equilibria (e.g., Eqs. 9-17 through 9-20) with these sorbent properties, we can develop versions of the complex K_{id} expression (Eq. 9-16) which take into account the structure of the chemical we are considering. In the following, we discuss these individual equilibrium relationships.

9.3 Sorption of Neutral Organic Compounds from Water to Solid-Phase Organic Matter (POM)

Overview

Among the sorbents present in the environment, organic matter plays an important role in the overall sorption of many organic chemicals. This is true even for compounds that may undergo specific interactions with inorganic sorbent components (see Chapter 11). We can rationalize this importance by recognizing that most surfaces of inorganic sorbents are polar and expose a combination of hydroxy- and oxy-moieties to their exterior. These polar surfaces are especially attractive to substances like water that form hydrogen bonds. Hence, in contrast to air–solid surface partitioning (Section 11.2), the adsorption of a nonionic organic molecule from water to an inorganic surface requires displacing the water molecules at such a surface. This is quite unfavorable from an energetic point of view. However, absorption of organic chemicals into natural organic matter or adsorption to a hydrophobic organic surface does not require displacement of tightly bound water molecules. Hence, nonionic organic sorbates successfully compete for associations with solid-phase organic matter.

Therefore, we may not be too surprised to find that nonionic chemicals show increasing solid–water distribution ratios for soils and sediments with increasing amounts of natural organic matter. This is illustrated for tetrachloromethane (carbon tetrachloride, CT) and 1,2-dichlorobenzene (DCB) when these two sorbates were examined for their solid–water distribution coefficients using a large number of soils and sediments (Fig. 9.7, Kile et al., 1995.)

Note that the common analytical methods for determining the total organic material present in a sorbent often involve combusting the sample and measuring evolved

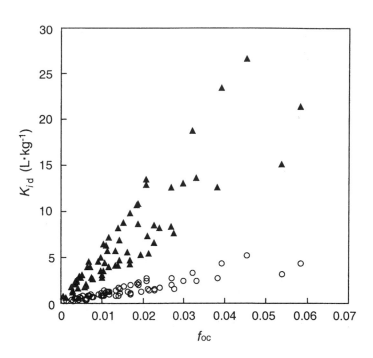

Figure 9.7 Observed increase in solid–water distribution ratios for the apolar compounds, tetrachloromethane (o) and 1,2-dichlorobenzene (▲) with increasing organic matter content of the solids (measured as organic carbon, f_{oc}, see Eq. 9-21) for 32 soils and 36 sediments. Data from Kile et al. (1995).

tetrachloromethane

1,2-dichlorobenzene

CO_2. Therefore, the abundance of organic material present is often expressed by the weight fraction that consisted of reduced carbon:

$$f_{oc} = \frac{\text{mass of organic carbon}}{\text{total mass of sorbent}} \quad (\text{kg oc} \cdot \text{kg}^{-1}\text{solid}) \qquad (9\text{-}21)$$

Obviously, it is actually the total organic mass consisting of carbon, hydrogen, oxygen, nitrogen, etc. within the solid phase that acts to sorb the chemical of interest (i.e., the f_{om} in kg om·kg^{-1} solid). Natural organic matter is typically made up of about half carbon (40 to 60% carbon); hence, f_{om} approximately equals $2 \cdot f_{oc}$ and these two metrics are reasonably correlated.

Returning to the sorption observations (Fig. 9.7), as the mass fraction of organic carbon, f_{oc}, present in the solids approaches zero, the K_{id} values for both compounds become very small. Even at very low f_{oc} values (i.e., $f_{oc} \cong 0.001$ kg oc·kg^{-1} solid), sorption to the organic components of a natural sorbent may still be the dominant mechanism (see Chapter 11).

In order to evaluate the ability of natural organic materials to sorb organic pollutants, it is useful to define an organic carbon normalized sorption coefficient:

$$K_{ioc} = \frac{K_{id}}{f_{oc}} = \frac{C_{ioc}}{C_{iw}} \qquad (9\text{-}22)$$

where C_{ioc} is the concentration of the total sorbate concentration associated with the natural organic carbon (i.e., mol·kg^{-1} oc). Note that in this case, it is assumed that organic matter is the dominant sorbent; that is, C_{is} is given by $C_{ioc} \cdot f_{oc}$, the first term in the numerator of Eq. 9-16. Clearly the value of K_{ioc} differs for tetrachloromethane and 1,2-dichlorobenzene (the slopes differ in Fig. 9.7), and it is generally true that

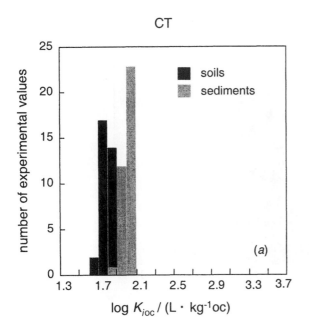
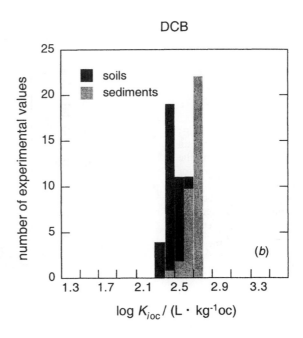

Figure 9.8 Frequency diagrams showing the variability in the log K_{ioc} values of (a) tetrachloromethane (CT) and (b) 1,2-dichlorobenzene (DCB) for 32 soils (dark bars) and 36 sediments (light bars). The range of f_{oc} values of the soils and sediments investigated is indicated in Fig. 9.7. Data from Kile et al. (1995).

each chemical has its own "organic carbon normalized" solid–water partition coefficient, K_{ioc}.

The K_{id} value of a given compound shows some variation between different soils and sediments exhibiting the same organic carbon content (Fig. 9.7). This indicates that *not only the quantity, but also the quality* of the organic material present has an influence on K_{id}. Normalizing to the organic carbon contents of each soil and sediment, we can examine this variability for both tetrachloromethane and 1,2-dichlorobenzene sorbing to a variety of soils and sediments of very different origins (Fig. 9.8.) All the K_{ioc} values lie within a factor of about 2 (i.e., $\pm 2\sigma \sim \pm 0.3$ log units). We should emphasize that these data include only K_{ioc} values determined in the linear range of the isotherms by a single research group. The data show that for these two apolar compounds, soil organic matter seems on average to be a somewhat poorer "solvent" as compared to sediment organic matter (Fig. 9.8). In fact, the average K_{CToc} values are 60 ± 7 L·kg^{-1} oc for the 32 soils and 100 ± 11 L·kg^{-1} oc for the 36 sediments investigated; similarly the average K_{DCBoc} values are 290 ± 42 L·kg^{-1} oc and 500 ± 66 L·kg^{-1} oc, for the soils and sediments, respectively. Apparently, the sources of organic matter in terrestrial settings leave residues that are somewhat more polar than the corresponding residues derived chiefly in water bodies. Thus, variations in K_{ioc} may primarily reflect differences in the chemical nature of the organic matter. Using data from numerous research groups, Gerstl (1990) also examined the variability of log K_{ioc} values for 13 other nonionic compounds. He found the K_{ioc} observations to be log normally distributed and to exhibit relative standard deviations for log K_{ioc} values of about $\pm 1\sigma \sim \pm 0.3$ log units. An example is the herbicide atrazine, for which more than 200 observations were compiled (Fig. 9.9). DDT and lindane, two apolar compounds, exhibited similar variability in their log K_{ioc} values as did atrazine. The variations can be attributed to the different methods applied by different groups and the variability in the

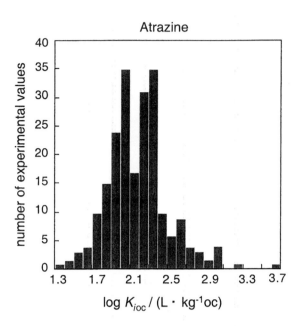

Figure 9.9 Frequency diagram illustrating the variability in the log K_{ioc} values determined for atrazine for 217 different soil and sediment samples. The numbers on the X-axis indicate the center of a log K_{ioc} range in which a certain number of experimental K_{ioc} values fall. Data compiled by Gerstl (1990).

atrazine

qualitative nature of the organic matter in the wide range of soils and sediments used. In sum, careful determinations of nonionic organic compound absorption into natural organic matter appear to yield log K_{ioc} values to about $\pm\,0.3$ log units ($\pm\,1\sigma$) precision.

Structural Characteristics of POM Relevant to Sorption

Let us now consider what the organic materials in soil and sediment sorbents are. As has become evident from numerous studies (see e.g., Thurman, 1985; Schulten and Schnitzer, 1997; Hayes, 1998), the natural organic matter present in soils, sediments, groundwaters, surface waters, atmospheric aerosols, and in wastewaters may include recognizable biochemicals like proteins, nucleic acids, lipids, cellulose, and lignin. But also, these environmental media contain a menagerie of macromolecular residues due to *diagenesis* (the reactions of partial degradation, rearrangement, and recombination of the original molecules formed in *biogenesis*). Naturally, the structure of such altered materials will depend on the ingredients supplied by the particular organisms living in or near the environment of interest. Moreover, the residues will tend to be structurally randomized. For example, soil scientists have deduced that the recalcitrant remains of woody terrestrial plants make up a major portion of the natural organic matter in soils (e.g., Schulten and Schnitzer, 1997). Such materials also make up an important fraction of organic matter suspended in freshwaters and deposited in associated sediments. Similarly, marine chemists have found that the natural organic matter, suspended in the oceans at sites far from land, consists of altered biomolecules such as polysaccharides and lipids that derived from the plankton and were subsequently modified in the environment (Aluwihare et al., 1997; Aluwihare and Repeta, 1999). At intermediate locales, such as large lakes and estuaries, the natural organic material in sediments and suspended in water appears to derive from a variable mixture of terrestrial organism and aquatic organism remains. An often-studied subset of these altered complex organic substances are commonly referred to as *humic substances* if they are soluble or extractable in aqueous base (and insoluble in organic solvents), and *humin* or *kerogen* if they are not. The humic substances are further subdivided into *fulvic acids* if they are soluble in both acidic and basic solutions and *humic acids* if they are not soluble at pH 2. For a detailed overview of the present knowledge of humic materials, we refer to the literature (e.g., Hayes and Wilson, 1997; Davies and Gabbour, 1998; Huang et al., 1998; Piccolo and Conte, 2000). Here, we address only the most important structural features that are relevant to sorption of organic pollutants.

First, we note that natural organic matter that potentially acts as a sorbent occurs in a very broad spectrum of molecular sizes from the small proteins and fulvic acids of about 1 kDa to the huge complexes of solid wood and kerogen ($>>$ 1000 kDa). Furthermore, natural organic matter is somewhat polar in that it contains numerous oxygen-containing functional groups including carboxy-, phenoxy-, hydroxy-, and carbonyl-substituents (Fig. 9.10). Depending on the type of organic material considered, the number of such polar groups may vary quite significantly. For example, highly polar fulvic acids may have oxygen-to-carbon mole ratios (O/C ratios) of near 0.5 (Table 9.1). More mature organic matter (i.e., organic matter that has been exposed for longer time to higher pressures and temperatures in buried sediments) have O/C ratios around 0.2 to 0.3, and these evolve toward coal values below 0.1 (Brownlow, 1979). These polar groups may become involved in H-bonding, which may significantly affect the three-dimensional arrangements and water content of these macromolecular media. Since many of the polar groups are acidic (e.g., carboxylic acid groups, phenolic groups) and because they undergo complexation with metal ions (e.g., Ca^{2+}, Fe^{3+}, Al^{3+}), pH and ionic strength have some impact on the tendency for the natural organic matter to be physically extended (when charged groups repulse one another) or coiled and forming domains that are not exposed to outside aqueous solutions. This may be particularly important in the case of "dissolved" organic matter (see Section 9.4).

In summary, we can visualize the natural organic matter as a complex mixture of macromolecules derived from the remains of organisms and modified after their release to the environment through the processes of diagenesis. This organic matter exhibits hydrophobic and hydrophilic domains. There is some evidence that the aggregate state of the organic matter may include portions with both fluid and rigid character. Borrowing terms commonly used in polymer chemistry, the inferred fluid domains have been referred to as "rubbery," and the more rigid ones as "glassy" domains (Leboeuf and Weber, 1997; Xing and Pignatello, 1997). Other nomenclature uses the terms *soft* and *hard* carbon, respectively (Weber et al., 1992; Luthy et al., 1997b). The glassy domain may contain nanopores (i.e., microvoids of a few nanometers size) that are accessible only by (slow) diffusion through the solid phase (Xing and Pignatello, 1997; Aochi and Farmer, 1997; Xia and Ball, 1999; Cornelissen et al., 2000). This would result in slow sorption kinetics (Pignatello and Xing, 1996). Thus, the natural organic matter may include a diverse array of compositions, resulting in both hydrophobic and hydrophilic domains, and formed into both flexible and rigid subvolumes. This picture suggests nonionic organic compounds may both *ab*sorb into flexible organic matter and any voids of rigid portions, as well as *ad*sorb onto any rigid organic surfaces.

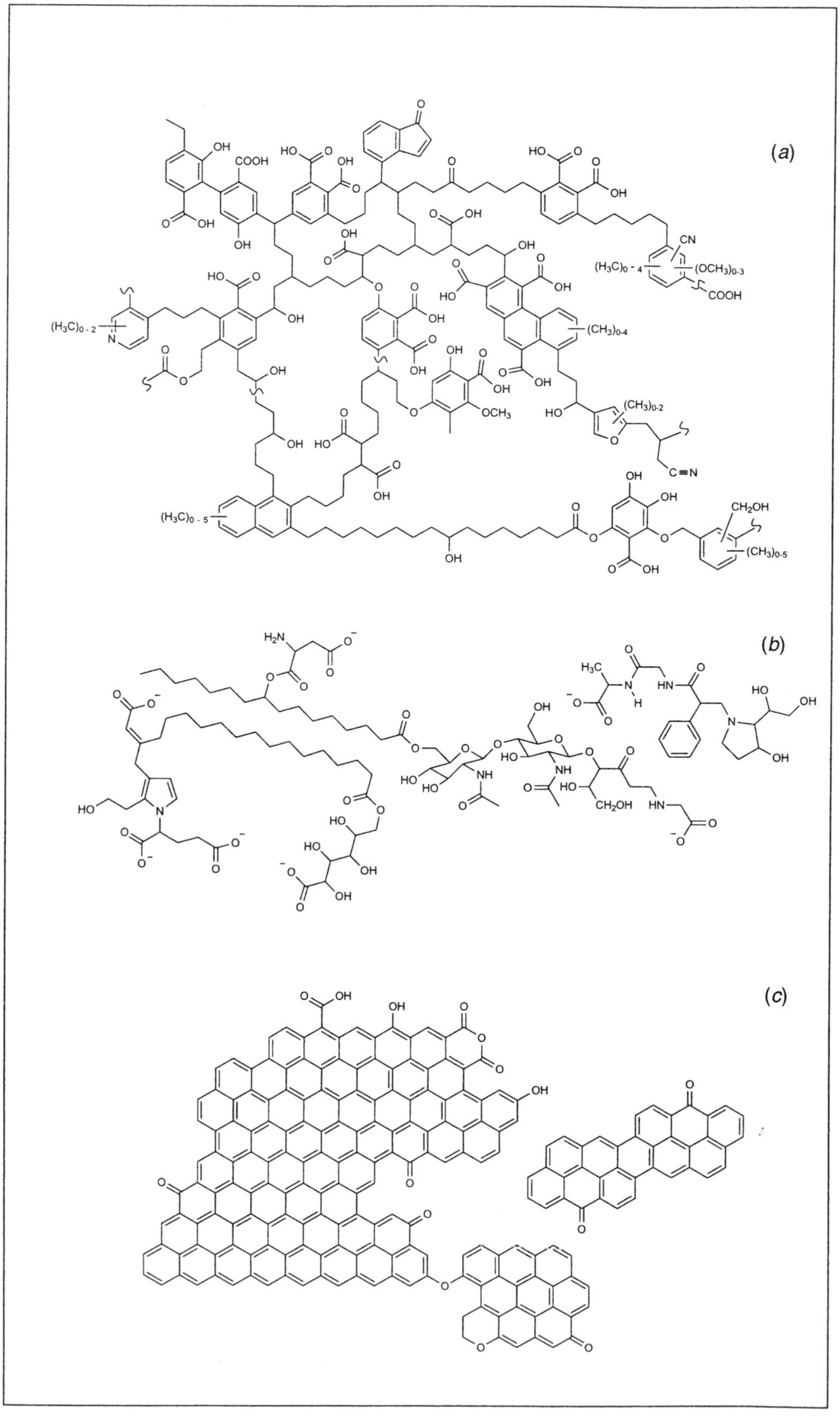

Figure 9.10 (*a*) Schematic soil humic acid structure proposed by Schulten and Schnitzer (1997). Note that the "~" symbols stand for a linkages in the macromolecules to more of the same types of structure. (*b*) Schematic seawater humic substances structure proposed by Zafiriou et al. (1984). (*c*) Schematic black carbon structure proposed by Sergides et al. (1987).

Table 9.1 Properties of Organic Components that May Act as Sorbents of Organic Compounds in the Environment

Component	Mole Ratio				Molecular Mass average (u)[a]	% Aromaticity	Reference[b]
	C	H	N	O			
Biogenic Molecules							
Proteins	1.0	1.6	0.4	0.2		<10	1
Collagen (protein)	1.0	1.7	0.19	0.31		<10	2
Cellulose (polysaccharide)	1.0	1.7	<0.01	0.84	$ca.\ 10^6$ (cotton)	<10	2
Chitin (polysaccharide)	1.0	1.8	0.13	0.64		<10	2
Lignin (alkaline extract)	1.0	1.1	<0.01	0.40		28	2
Lignin (org. solvent extract)	1.0	0.98	<0.01	0.33		34	2
Diagenetic Materials							
Fulvic acids							
soil leachate	1.0	1.04	<0.1	0.53		36	3
brown lake water	1.0	0.88	<0.1	0.55		35	3
river water	1.0	1.62	<0.1	1.09		29	4
groundwater	1.0	1.04	<0.1	0.51		24	3
Suwannee River fulvic acid	1.0	0.87	0.1	0.53	2000	25	7
Humic acids							
brown lake water	1.0	0.80	<0.1	0.54		40	3
river water	1.0	1.48	<0.1	0.91		38	4
"average" soil	1.0	1.15	0.07	0.50	7800		5
Aldrich	1.0	0.78	0.01	0.44	9000	41	6, 7
Suwannee River	1.0	0.94	0.02	0.61	3200	42	6, 7
Humin	1.0	1.9	0.5	1.1			8
Kerogen	1.0	0.4 to 1		0.05 to 0.3			9
Combustion-Derived Materials							
NIST diesel soot	1.0	0.1	0.016			~100	10
BC from Boston Harbor sediment	1.0	1.0	0.07				10

[a] Mass average. [b] References: 1. Oser (1965); 2. Xing et al. (1994); 3. Haitzer et al. (1999); 4. Zhou et al. (1995); 5. Schulten and Schnitzer (1997); 6. Arnold et al. (1998); 7. Chin et al. (1994); 8. Garbarini and Lion (1986); 9. Brownlow (1979); 10. Accardi-Dey and Gschwend (2002). [c] BC = black carbon.

In addition to the natural organic matter present due to biogenesis and diagenesis, other identifiable organic sorbents, mostly derived from human activities, can be present (and would be included in an f_{oc} measurement). Examples include combustion byproducts (soots and fly ash), plastics and rubbers, wood, and non-aqueous-phase liquids. The most potent among these other sorbents are various forms of black carbon (BC). Black carbon involves the residues from incomplete combustion processes (Goldberg, 1985). The myriad existing descriptors of these materials (soot, smoke, black carbon, carbon black, charcoal, spheroidal carbonaceous particles, elemental carbon, graphitic carbon, charred particles, high-surface-area carbonaceous material) reflect either the formation processes or the operational techniques employed for their characterization. BC particles are ubiquitous in sediments and soils, often contributing 1 to 10% of the f_{oc} (Gustafsson and Gschwend, 1998). Such particles can be quite porous and have a rather apolar and aromatic surface (Table 9.1). Consequently, they exhibit a high affinity for many organic pollutants, particularly for planar aromatic compounds. Therefore significantly higher apparent K_{ioc} values may be observed in the field as compared to values that would be predicted from simple partitioning models (Gustafsson et al., 1997; Naes et al., 1998; Kleineidam et al. 1999; Karapanagioti et al., 2000).

Another example involves wood chips or sawdust used as fills at industrial sites. Wood is also a significant component of solid waste, accounting for up to 25 wt% of materials at landfills that accept demolition wastes (Niessen, 1977). Wood is composed primarily of three polymeric components: lignin (25–30% of softwood mass), cellulose (40–45% of softwood mass), and hemicellulose (remaining mass) (Thomson, 1996). As has been shown by Severton and Banerjee (1996) and Mackay and Gschwend (2000), sorption of hydrophobic organic compounds by wood is primarily controlled by sorption to the lignin. This is not too surprising when considering the rather polar character of cellulose and hemicellulose as compared to lignin (compare O/C and H/C ratios in Table 9.1). Also synthetic polymers such as polyethylene (Barrer and Fergusson, 1958; Rogers et al., 1960; Flynn, 1982; Doong and Ho, 1992; Aminabhavi and Naik, 1998), PVC (Xiao et al., 1997), and rubber (Barrer and Fergusson, 1958, Kim et al., 1997) and many others are well known to absorb nonionic organic compounds. If such materials are present in a soil, sediment, or waste of interest, then they will serve as part of the organic sorbent mix. Finally, a special organic sorbent that may be of importance, particularly, when dealing with contamination in the subsurface, is nonaqueous phase liquids (NAPLs, Hunt et al., 1988; Mackay and Cherry, 1989). These liquids may be immobilized in porous media and serve as absorbents for passing nonionic organic compounds (Mackay et al., 1996). In such cases we may apply partition coefficients as discussed in Section 7.5 (Eq. 7-22) to describe sorption equilibrium, but we have to keep in mind that the chemical composition of the absorbing NAPL will evolve with time.

In conclusion, sorption of neutral organic chemicals to the organic matter present in a given environmental system may involve partitioning into, as well as adsorption onto, a variety of different organic phases. Thus, in general, we cannot expect linear isotherms over the whole concentration range, and we should be aware that predictions of overall K_{ioc} values may have rather large errors if some of the important organic materials present are not recognized (Kleineidam et al., 1999).

Conversely, with appropriate site-specific information, reasonable estimates of the magnitude of sorption coefficients can be made (see below).

Determination of K_{ioc} Values and Availability of Experimental Data

K_{ioc} values are available for a large number of chemicals in the literature. The vast majority of these K_{ioc}'s have been determined in batch experiments in which a defined volume of water is mixed with a given amount of sorbent, the resultant slurry is spiked with a given amount of sorbing compound(s), and then the system is equilibrated with shaking or stirring. After equilibrium is established, the solid and aqueous phases are mostly separated by centrifugation or filtration. In most studies, only the aqueous phase is then analyzed for the partitioning substance, and its concentration in the solid phase is calculated by the difference between the total mass added and the measured mass in the water. Direct determinations of solid phase concentrations are usually only performed to verify that other loss mechanisms did not remove the compound from the aqueous phase (e.g., due to volatilization, adsorption to the vessel, and/or degradation). K_{ioc} is then calculated by dividing the experimentally determined K_{id} ($= C_{is} / C_{iw}$) value by the fraction of organic carbon, f_{oc}, of the sorbent investigated (Eq. 9-22).

Of course, a meaningful K_{ioc} value is obtained only if sorption to the natural organic material is the dominant process. This may be particularly problematic for sorbents exhibiting very low organic carbon contents. Also, solid–water contact times are sometimes too short to allow sorbates to reach all the sorption sites that are accessible only by slow diffusion (Xing and Pignatello, 1997); thus, assuming sorptive equilibrium may not be appropriate. This kinetic problem can be especially problematic for equilibrations that employ sorbate solutions flowing through columns containing the solids. Finally, errors may be introduced due to incomplete phase separations causing the presence of water (containing dissolved compound) with the solid phase, as well as colloids (containing sorbed compound) in the aqueous phase. Hence, the experimentally determined apparent solid–water distribution coefficient, K_{id}^{app}, is not equal to the "true" K_{id} but is given by:

$$K_{id}^{app} = \frac{C_{is} + C_{iw} V_{ws}}{C_{iw} + C_{iDOC}[DOC]} \qquad (9\text{-}23)$$

where C_{is} is the compound concentration on the separated particles (mol·kg⁻¹ solid)

C_{iw} is the compound concentration in the water (mol·L⁻¹)

V_{ws} is the volume of water left with the separated particles (L·kg⁻¹ solid)

C_{iDOC} is the compound concentration associated with colloids (mol·kg⁻¹ oc)

[DOC]   is the concentration of organic matter in the colloids (expressed as C) remaining with the bulk water (kg oc·L⁻¹)

By dividing the numerator and denominator of Eq. 9-23 by C_{iw}, and then substituting C_{is}/C_{iw} by K_{id} and C_{iDOC}/C_{iw} by K_{iDOC}, we may rewrite Eq. 9-23 as:

$$K_{id}^{\text{app}} = \frac{K_{id} + V_{ws}}{1 + K_{iDOC} \cdot [\text{DOC}]} \qquad (9\text{-}24)$$

The expression indicates that the apparent solid–water distribution coefficient will equal the "true" one only if $V_{ws} \ll K_{id}$ and if $K_{iDOC} \cdot [\text{DOC}] \ll 1$. For weakly sorbing compounds (low K_{id}), this equation suggests that the experimental K_{id}^{app}, and thus K_{ioc}^{app}, may be erroneously high. For compounds that do tend to sorb (high K_{iDOC}) and in situations where organic colloids are substantial (high $[\text{DOC}]$), batch observations of solid–water partitioning produce lower distribution coefficients than K_{id}. Note that these phase separation difficulties are probably one of the major explanations for the so-called "solids concentration effect" in which K_{id} appears to decrease with greater and greater loads of total solids and thus DOM colloids in batch sorption systems (Gschwend and Wu, 1985). Note also that these problems may also be important for other colloid-containing systems such as where sorption to clay minerals plays a major role (see Chapter 11). Finally, particularly in older studies, radiolabeled chemicals of poor purity were used, and this can also have an influence on the result (Gu et al., 1995).

Considering all these experimental problems, as well as the natural variability of natural organic sorbents, it should not be surprising that K_{ioc} values reported in the literature for a given chemical may vary by up to an order of magnitude or even more. This is particularly true for polar compounds for which uncontrolled solution conditions like pH and ionic strength may also play an important role. Thus, when selecting a K_{ioc} value from the literature, one should be cautious. In this context, it should be noted that K_{ioc} values are log-normally distributed (normal distribution of the corresponding free energy values), and therefore log K_{ioc} values, not K_{ioc} values, should be averaged when several different K_{ioc}'s have been determined (Gerstl, 1990).

For the following discussions, we will primarily use K_{ioc} values from compilations published by Sabljic et al. (1995) and Poole and Poole (1999). According to these authors, the values should be representative for POM–water *absorption* (i.e., they have been derived from the linear part of the isotherms). Furthermore, many of the reported K_{ioc}'s are average values derived from data reported by different authors. Distinction between different sources of sorbents (e.g., soils, aquifer materials, freshwater, or marine sediments) has not been made. Nevertheless, at least for the apolar and weakly monopolar compounds, these values should be reasonably representative for partitioning to soil and sediment organic matter.

Estimation of K_{ioc} values

Any attempt to estimate a K_{ioc} value for a compound of interest (with its particular abilities to participate in different intermolecular interactions) should take into account the structural properties of the POM present in the system considered. To this end, the use of multiparameter LFERs such as the one that we have applied for description of organic solvent–water partitioning (Eq. 7-9) would be highly desirable (Poole and Poole, 1999). Unfortunately, the available data do not allow such analyses, largely due to the very diverse solid phase sources from which reported K_{ioc} values have been derived.

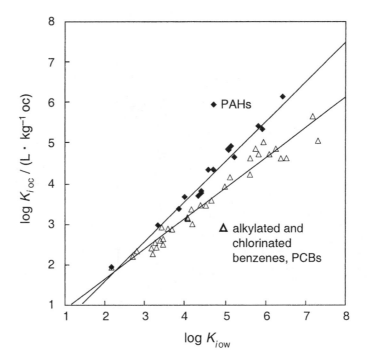

Figure 9.11 Plot of log K_{ioc} versus log K_{iow} for PAHs (◆) and for a series of alkylated and chlorinated benzenes and biphenyls (PCBs) (Δ). The slopes and intercepts of the linear regression lines are given in Table 9.2.

Therefore, for estimates of K_{ioc}'s it is more feasible to use compound class-specific LFERs. These include correlations of log K_{ioc} with molecular connectivity indices (or topological indices; for an overview see Gawlik et al., 1997), with log C_{iw}^{sat} (L) (analogous to Eq. 7-11), and with log K_{iow}. Although molecular connectivity indices or topological indices have the advantage that they can be derived directly from the structure of a chemical, they are more complicated to use and do not really yield much better results than simpler one-parameter LFERs using C_{iw}^{sat} (L) or K_{iow} as compound descriptors.

Since C_{iw}^{sat} (L) and K_{iow} can be related to each other (Eq. 7-11, Table 7.3), here we will confine ourselves to log K_{ioc} – log K_{iow} relationships. Table 9.2 summarizes the slopes a and intercepts b derived for some sets of organic compounds by fitting:

$$\log K_{ioc} = a \cdot \log K_{iow} + b \qquad (9\text{-}25)$$

Note that the K_{ioc} values used tend to represent mostly absorption into soil and sediment POM. Therefore, any estimates using equations such as the ones given in Table 9.2 should be considered to be within a factor of 2 to 3. Furthermore, such LFERs should be applied very cautiously outside the log K_{ioc} – log K_{iow} range for which they have been established. This is particularly critical for LFERs that have been derived only for a relatively narrow log K_{iow} range (e.g., the phenyl ureas).

Let us make some general comments on this type of LFER. First, reasonable correlations are found for sets of compounds that undergo primarily London dispersive interactions (Fig. 9.11; alkylated and chlorinated benzenes, chlorinated biphenyls). Good correlations are also found for sets of compounds in which polar interactions change proportionally with size (PAHs) or remain approximately

Figure 9.12 Plot of log K_{ioc} versus log K_{iow} for a alkylated and halogenated (R_1 = alkyl, halogen) phenylureas (R_2 = R_3 = H; $\triangle$, halogen, see margin below), phenyl-methylureas (R_2 = CH_3, R_3 = H, $\square$), and phenyl-dimethylureas (R_2 = R_3 = CH_3, $\bullet$). The slope and intercept of the linear regression using all the data is given in Table 9.2 (Eq. 9-26i); each subset of ureas would yield a tighter correlation if considered alone (e.g., Eq. 9-26j).

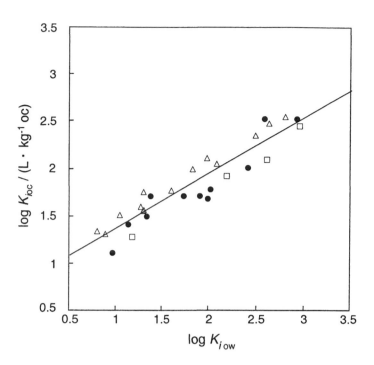

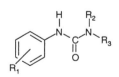

constant (chlorinated phenols). These results are reasonable based on our previous discussions of organic solvent–water partitioning (Chapter 7).

It should also be not too surprising that poorer results are obtained when trying to correlate sets of compounds with members exhibiting significantly different H-acceptor and/or H-donor properties. This is the case for the halogenated C_1-, C_2-, and C_3-compounds. Combining the entire set leads to an R^2 of only 0.68 (Eq. 9-26d). Focusing on the chloroalkenes, the correlation is much stronger (R^2 of 0.97 although N is only 4); while for the polyhalogenated alkanes with and without bromine correlations are much more variable. This can be understood if we recall that compounds like CH_2Cl_2 exhibit H-donor and H-acceptor capabilities (e.g., for $CHCl_2$ α_i = 0.10 and β_i = 0.05) while Cl_3CCH_3 has only H-acceptor ability (α_i = 0, β_i = 0.09) and CCl_4 has neither (Table 4.3). Hence, lumping such sets of compounds in a single-parameter LFER should yield variability as is seen. Such H-bonding variability also occurs within the large set of phenyl ureas that are used primarily as herbicides (some with –NH_2, others with –NH–CH_3, and finally some with –N(CH_3)$_2$). In the case of the phenyl ureas, a significantly better correlation can be obtained for any subset of these compounds exhibiting consistent H-bonding on the terminal amino group (Fig. 9.12). Consequently, more highly correlated relationships with a single parameter like log K_{iow} are also found for these subsets (Table 9.2). These examples demonstrate that care has to be taken when selecting a set of compounds for the establishment of one-parameter LFERs. Hence, published LFERs relating log K_{ioc} values to log K_{iow} or related parameters (liquid aqueous solubilities or chromatographic retention times; see Gawlik et al., 1997, for review) should be checked to see that the "training set" of sorbing compounds have chemical structures that ensure that they participate in the same intermolecular interactions into the two partitioning media.

Table 9.2 LFERs Relating Particulate Organic Matter–Water Partition Coefficients and Octanol–Water Partition Constants at 20 to 25°C for Some Sets of Neutral Organic Compounds: Slopes and Intercepts of Eq. 9-25 [a]

Equation	Set of Compounds	$\log K_{ioc}$ [b] $= a \cdot \log K_{iow} + b$		$\log K_{iow}$ range [c]	R^2	N [d]
		a	b			
9-26a	Alkylated and chlorinated benzenes, PCBs (± apolar)	0.74	0.15	2.2 to 7.3	0.96	32
9-26b	PAHs (monopolar)	0.98	−0.32	2.2 to 6.4	0.98	14
9-26c	Chlorinated phenols (neutral species; bipolar)	0.89	−0.15	2.2 to 5.3	0.97	10
9-26d	C_1- and C_2-halocarbons (apolar, monopolar, and bipolar)	0.57	0.66	1.4 to 2.9	0.68!	19
9-26e	Only chloroalkanes (mix)	0.42	0.93		0.59!	9
9-26f	Only chloroalkenes (± apolar)	0.96	−0.23		0.97	4
9-26g	Only compounds including bromine (mix)	0.50	0.81		0.49!	6
9-26h	All phenylureas (bipolar)	0.49	1.05	0.5 to 4.2	0.62!	52
9-26i	Only alkylated and halogenated phenylureas, phenyl-methylureas, and phenyl-dimethylurea (bipolar) [e]	0.59	0.78	0.8 to 2.9	0.87	27
9-26j	Only alkylated and halogenated phenylureas (bipolar) [f]	0.62	0.84	0.8 to 2.8	0.98	13

[a] Data from Sabljic et al. (1995), Chiou et al. (1998), and Poole and Poole (1999). The data for chlorinated phenols have been taken in part from Schellenberg et al. (1984). [b] K_{ioc} in $L \cdot kg^{-1}$ oc. [c] Range of experimental values for which LFER has been established. [d] Number of compounds. [e] Number of compounds used for LFER. See Fig. 9.12, all compounds. [f] See Fig. 9.12, only Δ.

K_{ioc} as a Function of Sorbate Concentration

Let us now come back to the issue of linearity of the isotherm and dependency of K_{id} on the sorbate concentration. In numerous field studies in which both particle-associated and dissolved concentrations of PAHs are measured, apparent K_{ioc} values are up to two orders of magnitude higher than one would have predicted from a simple absorption model (Gustafsson and Gschwend, 1999). If a natural soil or sediment matrix includes impenetrable hydrophobic solids on which the chemical of interest may *ad*sorb, the overall K_{ioc} value must reflect both *ab*sorption into recent natural organic matter and *ad*sorption onto these surfaces.

We start out by considering the effect of such adsorption sites on the isotherms of *apolar and weakly monopolar* compounds. For these types of sorbates, hydrophobic organic surfaces and/or nanopores of carbonaceous materials are the most likely sites of adsorption. Such hydrophobic surfaces may be present due to the inclusion of particles like coal dust, soots, or highly metamorphosed organic matter (e.g., kerogen). Because of the highly planar aromatic surfaces of these particular materials, it is reasonable to assume that planar hydrophobic sorbates that can maximize the molecular contact with these surfaces should exhibit higher affinities, as compared to other nonplanar compounds of similar hydrophobicity.

Let us evaluate some experimental data. To this end, we use a dual-mode model (Eq. 9-6). This model is a combination of a linear absorption (to represent the sorbate's mixing into natural organic matter) and a Freundlich equation (as seen for adsorption to hydrophobic surfaces or pores of solids like activated carbons):

$$C_{is} = K_{ip}C_{iw} + K_{iF}C_{iw}^{n_i} \tag{9-6}$$

The value of the partition coefficient, K_{ip}, is given by the product, $f_{oc}K_{ioc}$, where f_{oc} and K_{ioc} apply only to the natural organic matter into which the sorbate can penetrate. The value of K_{iF} is less well understood, but recent observations suggest it should be related to the quantity of adsorbent present (e.g., the fraction of "black carbon" in a solid matrix, f_{bc}) and the particular compound's black-carbon-normalized adsorption coefficient (e.g., K_{ibc}). Typical values of the Freundlich exponent are near 0.7. Hence, in a first approximation the data should fit:

$$C_{is} = f_{oc}K_{ioc}C_{iw} + f_{bc}K_{ibc}C_{iw}^{0.7} \tag{9-27}$$

Observations certainly fit this type of dual-mechanism model. For example, Xia and Ball (1999) recently examined the sorption of several organic compounds to an aquifer sediment. They measured that sediment's f_{oc} to be 0.015. Using a literature value of the $K_{pyrene\ oc}$ of $10^{4.7}$ (Gawlik et al., 1997), it is clear that the pyrene sorption they observed greatly exceeded expectations based on only f_{oc} times $K_{pyrene\ oc}$ (Figure 9-13a). Subtracting this absorption contribution to the total $K_{pyrene\ d}$ and using a recently reported value for $K_{pyrene\ bc}$ of $10^{6.5}$ (Bucheli and Gustafsson, 2000), the data indicate f_{bc} in this aquifer sediment contributed about 0.6% of the solid mass (a large fraction of that Miocene sediment's remaining reduced carbon content). Using this f_{bc}, the entire pyrene sorption isotherm was well fit using Eq. 9-27 (Figure 9-13a). Moreover, fixing f_{bc} at 0.006, the isotherms for the other sorbates tested by Xia and

Figure 9.13 (*a*) Freundlich isotherm (closed diamonds) observed for pyrene sorption to an aquifer sediment by Xia and Ball (1999.) Lower line represents expectations of $K_{id} = f_{oc} K_{ioc}$ with $f_{oc} = 0.015$ (measured) and $K_{ioc} = 10^{4.7}$ (Gawlik et al., 1997), while upper line shows predictions assuming $K_{id} = f_{oc} K_{ioc} + f_{bc} K_{ibc} C_{iw}^{-0.3}$ with $f_{bc} = 0.006$ (fitted) and using $K_{ibc} = 10^{6.5}$ from Bucheli and Gustafsson (2000) and Accardi-Dey and Gschwend (2001).

(*b*) Holding f_{bc} at 0.006, values of K_{ibc} can be estimated for all the other sorbates (all planar) tested on this aquifer sediment: benzene, chlorobenzene, 1,2-dichlorobenzene, naphthalene, 1,2,4-trichlorobenzene, fluorene, 1,2,4,5-tetrachlorobenzene, phenanthrene, and pyrene.

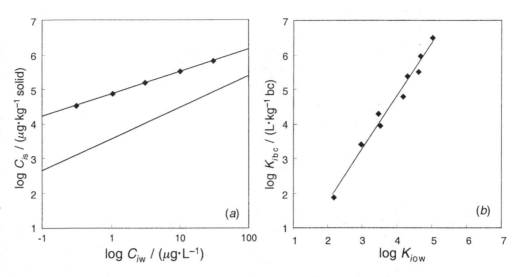

Ball can be used to extract those compounds' K_{ibc} values in $[(\mu g \cdot kg^{-1} bc)(\mu g \cdot L^{-1})^{-n_i}]$. These data suggest that for planar sorbates a value of K_{ibc} can be estimated via:

$$\log K_{ibc} \cong 1.6 \log K_{iow} - 1.4 \qquad (N = 9, R^2 = 0.98) \qquad (9\text{-}28)$$

Consistent with experience with adsorbents like activated carbons, the fitted K_{ibc} values are greater for sorbates with larger hydrophobicities (Fig. 9.13*b*). Note that when using Freundlich isotherms, the K_{iF} value depends nonlinearly on the units in which the concentration in the aqueous phase is expressed (see Problem 9.5).

Neglecting the contribution of adsorption, especially for planar compounds and at low concentrations, may cause substantial underestimation of K_{id}. This is shown in Illustrative Example 9.2 for phenanthrene sorption to various soils and sediments (Huang et al., 1997).

Illustrative Example 9.2

Evaluating the Concentration Dependence of Sorption of Phenanthrene to Soil and Sediment POM

Huang et al. (1997) measured sorption isotherms for phenanthrene on 21 soils and sediments. All isotherms were nonlinear with Freundlich exponents n_i (Eq. 9-1) between 0.65 and 0.9. For example, for a topsoil (Chelsea I) and for a lake sediment (EPA-23), interpolating the isotherm data yields the following "observed" sorbed concentrations, C_{is}, in equilibrium with dissolved concentrations, C_{iw}, of 1 $\mu g \cdot L^{-1}$ and 100 $\mu g \cdot L^{-1}$, respectively:

C_{iw} ($\mu g \cdot L^{-1}$)	C_{is} ($\mu g \cdot kg^{-1}$ solid)	
	Chelsea-I	EPA-23
1	3,200	1,700
100	91,000	51,000

Problem

Using Eq. 9-27, estimate the equilibrium solid phase concentrations, C_{is}, of phenanthrene for this topsoil and this sediment for aqueous concentrations, C_{iw}, of 1 and 100 $\mu g \cdot L^{-1}$. Compare these values with the concentrations obtained from interpolation of the sorption isotherms (see above):

$$C_{is} = f_{oc} K_{ioc} C_{iw} + f_{bc} K_{ibc} C_{iw}^{0.7} \qquad (9-27)$$

i - phenanthrene

$\log K_{iow} = 4.57$
$\log K_{ioc} = 4.3$

Answer

For this nonionic, planar compound, you have to take into account both absorption to POM and adsorption to a high-affinity sorbent (e.g., black carbon). For Chelsea I soil, f_{oc} was measured as 0.056 kg oc$\cdot$kg^{-1} solid. The f_{bc} was not measured, but in sediment samples it is typically between 1 and 10% of the f_{oc} (Gustafsson and Gschwend, 1998). Use the full range of 1 to 10% to see the possible impact of adsorption to black carbon (i.e., f_{bc} = 0.00056 to 0.0056 kg bc$\cdot$kg^{-1} solid). Assume n_i = 0.7 and use Eq. 9-28 to estimate K_{ibc}:

$$\log K_{ibc} = 1.6 \log K_{iow} - 1.4 = (1.6)(4.57)) - 1.4 = 5.9$$

Insertion of f_{oc}, K_{ioc}, f_{bc}, and K_{ibc} into Eq. 9-27 yields C_{is} values for C_{iw}=1 and 100 $\mu g \cdot L^{-1}$, respectively:

for C_{iw} =   1 $\mu g \cdot L^{-1}$:

$$C_{is} = (0.056)(10^{4.3})(1) + (0.00056 \text{ to } 0.0056)(10^{5.9})(1)^{0.7}$$
$$= 1100 + (440 \text{ to } 4400) = \textit{1540 to 5500 } \mu g \cdot kg^{-1} \text{ solid}$$
$$(\text{observed } 3200 \ \mu g \cdot kg^{-1} \text{ solid})$$

Note that a calculation based only on the product, $f_{oc}K_{ioc}$ would underestimate the oberved values by about a factor of three.

for C_{iw} = 100 $\mu g \cdot L^{-1}$:

$$C_{is} = (0.056)(10^{4.3})(100) + (0.00056 \text{ to } 0.0056)(10^{5.9})(100)^{0.7}$$
$$= 110,000 + (11,000 \text{ to } 110,000) = \textit{121,000 to 220,000 } \mu g \cdot kg^{-1} \text{ solid}$$
$$(\text{observed } 91,000 \ \mu g \cdot kg^{-1} \text{ solid})$$

In this case, the estimate based only on $f_{oc}K_{ioc}$ is very close to the experimental result, indicating that for high substrate concentrations partitioning into POM is the dominant sorption mechanism.

For EPA-23 lake sediment, f_{oc} was measured as 0.026 kg oc$\cdot$kg^{-1} solid. Again assuming the same K_{ioc} and K_{ibc} values for phenanthrene and the same f_{bc} range, one obtains:

for C_{iw} = 1 $\mu g \cdot L^{-1}$:

$$C_{is} = 520 + (210 \text{ to } 2100) = \textit{730 to 2600 } \mu g \cdot kg^{-1} \text{ solid}$$
$$(\text{observed } 1700 \ \mu g \cdot kg^{-1} \text{ solid})$$

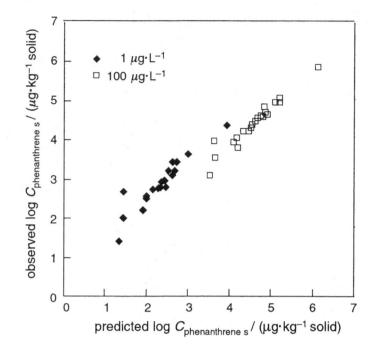

Figure 1 Predicted versus experimental sorbed phenanthrene concentrations for a series of soils and sediments. The diamonds indicate solids equilibrated with 1 $\mu g \cdot L^{-1}$; squares are for solids equilibrated with 100 $\mu g \cdot L^{-1}$.

and:

for $C_{iw} = 100 \ \mu g \cdot L^{-1}$:

$C_{is} = 52{,}000 + (5200 + 52{,}000) = \mathit{57{,}000 \ \text{to} \ 104{,}000} \ \mu g \cdot kg^{-1}$ solid
(observed 51,000 $\mu g \cdot kg^{-1}$ solid)

Hence, as for the Chelsea I topsoil, $f_{oc} \cdot K_{ioc}$ underestimates the observed sorption by about a factor of 3 at $C_{iw} = 1 \ \mu g \cdot L^{-1}$, whereas at 100 $\mu g \cdot L^{-1}$, sorption is dominated by POM–water partitioning.

Predictions of sorbed phenanthrene concentrations equilibrated with 1 or 100 $\mu g \cdot L^{-1}$ dissolved concentrations for *all* the soils and sediments (assuming $f_{bc} = 0.05 \ f_{oc}$) examined by Huang et al. (1997) are shown in Fig. 1. Values for 100 $\mu g \cdot L^{-1}$ fall somewhat below the extrapolation from 1 $\mu g \cdot L^{-1}$ observations, indicating the shift from *ad*sorption to *ab*sorption as the dominating mechanism.

For compounds other than PAHs, unfortunately there are not enough data available that would allow a more general analysis of the concentration dependence of K_{ioc} values. Nevertheless, a few additional observations may give us some better feeling of the magnitude of this dependence. For example, for sorption of smaller apolar and weakly monopolar compounds (e.g., benzene, chlorobenzene, 1,2-dichlorobenzene, tetrachloroethene, dibromoethane) to soil (Chiou and Kile, 1998) or aquifer materials (Xia and Ball, 1999 and 2000), not more than a factor of 2 difference in K_{ioc} was found between low and high sorbate concentrations. A somewhat more pronounced effect (i.e., factor 2 to 3) was observed for sorption of the more polar herbicides atrazine (Sposito et al., 1996; for structure see Fig. 9.9) and diuron (Spurlock and Biggar, 1994a; for structure, see margin) as well as for 3,5-dichlorophenol to soil organic carbon (Chiou and Kile, 1998). Note again that in the case of polar compounds, other high-affinity sites present in POM may cause the nonlinear sorption behavior.

In summary, we should realize that, when applying K_{ioc} values determined at high concentrations or derived from LFERs such as Eq. 9-26 (Table 9.2), we may underestimate equilibrium sorption at low concentrations (i.e., below 1% of the compound's solubility) by a factor of 2 or more. Due to competition with other sorbates present in a natural system, the effect of specific adsorption could, however, be significantly attenuated (Chiou et al., 2000). Furthermore, the abundance of specific adsorption sites may be rather low in certain environments. Hence, in cases in which the effect can be expected to be moderate and in which we need only to get an order of magnitude estimate of K_{ioc}, we may decide to neglect the nonlinearity of the isotherm. In other situations, however, such as the PAH case discussed in Illustrative Example 9.3, adsorption to carbonaceous materials or other high-affinity sorbents present in significant abundances has to be taken into account.

Illustrative Example 9.3   **Estimating Pore Water Concentrations in a Polluted Sediment**

Commonly, we are interested in estimating the concentration of a chemical in the pore water of a sediment bed. For example, we may be concerned that a pollutant is diffusing out of a contaminated sediment bed into the overlying waters; the rate of this exchange will be proportional to the difference in dissolved concentrations in the pore water and the overlying water (see Chapter 19). Or we may be interested in the biouptake of the pollutants by invertebrates living in the bed (see Chapter 10). Such uptake may be proportional to the pore water concentrations, which themselves are proportional to the concentrations in the solids.

Problem

In a sediment core from Boston Harbor (off Spectacle Island, Massachussetts, USA), McGroddy et al. (1996) measured sediment and pore water concentrations of pyrene and 2,2',4,5,5'-pentachlorobiphenyl (PCB101). For a set of depths, she obtained the following values given below. Estimate the "mean" pore water concentrations of these two chemicals. Assume that f_{bc} is 10% of f_{oc} in this area (Gustafsson and Gschwend, 1998). Neglect sorption of PCB101 to black carbon since this is not a planar molecule.

Answer

Rearranging Eq. 9-27 and solving for $C_{i\text{pore water}}$ ($= C_{iw}$) yields:

$$C_{i\text{pore water}} = C_{is} / (f_{oc} K_{ioc} + f_{bc} K_{ibc} C_{i\text{pore water}}^{-0.3}) \tag{1}$$

Note that because of the nonlinearity of the Freundlich term, you may not isolate

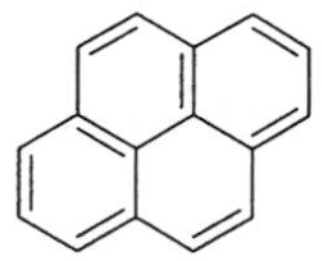

pyrene

log K_{iow} = 5.13
log K_{ibc} = 6.50
K_{ibc} in $(\mu g \cdot kg^{-1}bc)(\mu g \cdot L^{-1})^{-0.7}$
(Bucheli and Gustafsson, 2000)

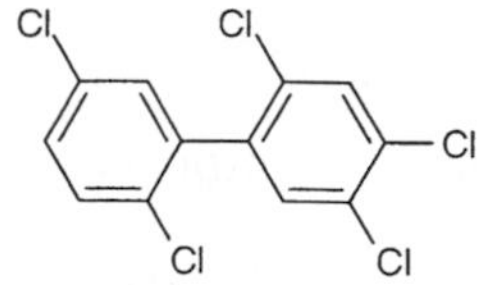

2,2',4,5,5'-pentachlorobiphenyl
(PCB101)
log K_{iow} = 6.36

Depth interval (cm below surface)	f_{oc} (kg oc·kg^{-1} solid)	Concentration in sediment C_{is}/(μg·kg^{-1} solid)	
		Pyrene	PCB101
6-8	0.035	1900	15
8-10	0.034	1700	15
10-12	0.030	2600,1600	22,30
means	0.032	2000	20

$C_{i\,pore\ water}$. Therefore, to solve Eq. 1, you will have to guess the solution and revise your guess until you satisfy the equation.

Estimate the K_{ioc} values of the compounds from K_{iow} (Eqs. 9-26b and 9-26a, respectively):

pyrene: $\log K_{ioc} = (0.98)(5.13)-0.32 = 4.7$

PCB101: $\log K_{ioc} = (0.74)(6.36)+0.15 = 4.9$

Insertion of C_{is} = 2000 μg·kg^{-1} solid together with f_{oc} = 0.032 and f_{bc} = 0.1 f_{oc} = 0.0032 and the appropriate K_{ioc} and K_{ibc} values into Eq. 1 allows an estimate for pyrene via:

$$C_{i\,pore\ water}/(\mu g \cdot L^{-1}) = 2000/[(0.032)(10^{4.7}) + 0.003)(10^{6.5})(C_{i\,pore\ water})^{-0.3}]$$

By trial and error, one may find $C_{i\,pore\,water} \cong 0.1\,\mu g \cdot L^{-1}$, not too far from the measured value of 0.04 μg·L^{-1} (McGroddy et al., 1996). Neglecting adsorption to black carbon would have yielded a much higher estimate:

$$C_{i\,pore\ water} = 2000/[(0.032)(10^{4.7})] = 1.2\ \mu g \cdot L^{-1}$$

For PCB101, the pore water concentration at equilibrium is estimated to be:

$$C_{i\,pore\ water} = 20/[0.032)(10^{4.9})] = 0.008\ \mu g \cdot L^{-1}$$

which is, again, quite similar to the measured value of 0.004 μg·L^{-1} (McGroddy et al., 1996).

Effect of Temperature and Solution Composition on K_{ioc}

Temperature. Like organic solvent–water systems (Eq. 7-4), over a narrow temperature range, we may express the temperature dependence of K_{ioc} by:

$$\ln K_{ioc} = -\frac{\Delta_{POMw}H_i}{R} \cdot \frac{1}{T} + \text{const.} \qquad (9\text{-}29)$$

with:

$$\Delta_{POMw}H_i = H_{iPOM}^E - H_{iw}^E \qquad (9\text{-}30)$$

where H_{iPOM}^E and H_{iw}^E are the excess enthalpies of i in the sorbed phase (an average value) and in the aqueous phase, respectively. Previously (Table 5.3) we saw that for many small organic compounds, H_{iw}^E is relatively small. Significant negative deviations from zero can be expected primarily for small monopolar compounds (e.g., –20 to –30 kJ·mol^{-1}) and large apolar compounds (e.g., +20 to 30 kJ·mol^{-1}). Note that H_{iPOM}^E represents the average excess enthalpy of the compound for the various sorption sites and may, therefore, also depend on the concentration range considered. Hence, at low concentrations where sorption to specific sites may dominate the overall K_{ioc} (see above), H_{iPOM}^E may be different as compared to higher concentrations where partitioning is the major process. For sorption of apolar and weakly polar compounds to hydrophobic sites, including surfaces of carbonaceous materials, we may still assume that H_{iw}^E is primarily determining $\Delta_{POMw}H_i$ (Bucheli and Gustafsson, 2000). For polar compounds that may undergo H-bond interactions with specific sites present in POM, however, H_{iPOM}^E may change significantly with concentration. For example, for sorption of diuron and other phenylurea herbicides to soils, H_{iPOM}^E values were found to vary between about –40 to –50 kJ·mol^{-1} at very low concentrations to ±10 kJ·mol^{-1} at concentrations where nonspecific partitioning can be assumed to dominate the overall sorption (Spurlock, 1995). Since for these compounds, H_{iw}^E is about +10 kJ·mol^{-1} (Spurlock, 1995), one obtains $\Delta_{POMw}H_i$ values of between –50 and –60 kJ·mol^{-1} at very low concentrations, and between –20 and 0 kJ·mol^{-1} for the high concentration range. Hence, as in many other cases, partition coefficients are rather insensitive to temperature, whereas adsorption coefficients increase by about a factor of 2 with a 10 degree decrease in temperature. Consequently, in such situations, one can expect that the nonlinearity of the isotherm increases with decreasing temperature.

pH and Ionic Strength. For sorption of apolar and weakly polar compounds to POM we may neglect the effect of pH and ionic strength on the sorption properties of the organic phase. However, particularly for higher salt concentrations, we may have to take into account the effect of salt on the activity coefficient of the compound in the aqueous phase. As discussed in Section 5.4 and applied to air–water partitioning in Chapter 6 (Eq. 6-19), we may quantify this effect simply by:

$$K_{ioc,\text{saltwater}} = K_{ioc} \cdot 10^{+K_i^s[\text{salt}]_{\text{tot}}} \qquad (9\text{-}31)$$

K_i^s values for seawater are given in Table 5.7 for a series of compounds. Using a typical value of 0.3 to 0.35 for PAHs and $[\text{salt}]_{\text{tot}} = 0.5$ M for seawater, we can see that the presence of the salt will increase K_{ioc} by a factor of about 1.5.

Organic Cosolvents. Sometimes we have to deal with the behavior of organic chemicals present in aqueous solutions containing an appreciable amount of an organic cosolvent (see Section 5.4). Let us examine the effect of such organic cosolvents on the K_{ioc} value of a given compound by looking at sorption to POM simply as a partitioning process between water and a complex organic solvent exhibiting a molar volume $\overline{V}_{oc}$ and a density ρ_{oc}. Hence, by analogy to Eq. 7-2 we express K_{ioc} as:

$$K_{ioc} = \frac{\gamma_{iw} \cdot \overline{V}_w}{\gamma_{ioc} \cdot \overline{V}_{oc} \cdot \rho_{oc}} \qquad (9\text{-}32)$$

We can now evaluate how a given organic cosolvent will affect the various parameters in Eq. 9-32. In Section 5.4 we discussed the dependence of the activity coefficient of a compound in a solvent–water mixture on the fraction of the cosolvent. We have seen that, depending on solute and cosolvent considered, this dependence may be quite complex (Figs. 5.6 and 5.7; Table 5.8). In the following discussion, we confine ourselves to rather small cosolvent concentrations (i.e., $f_v < 0.2$ to 0.3) for which we may assume a log-linear relationship (Eq. 5-32). We may then express the activity coefficient, $\gamma_{i\ell}$, of compound i in the solvent–water mixture as:

$$\gamma_{i\ell}(f_v) = \gamma_{iw} \cdot 10^{-\sigma_i^c \cdot f_v} \qquad (5\text{-}33)$$

Inspection of Table 5.8 shows that for naphthalene, for a variety of organic cosolvents, σ_i^c varies between 2.0 (for glycerol) and 6.5 (for acetone). Recall that σ_i^c is referred to as cosolvency power. Hence, in a 20% cosolvent–80% water mixture (i.e., $f_v = 0.2$), $\gamma_{i\ell}$ will be lower by a factor of between 2.5 (glycerol) and 20 (acetone) as compared to γ_{iw}. Considering this rather large effect, as a first approximation, we may neglect the change in the molar volume of the solvent–water mixture, particularly when dealing with cosolvents with relatively small $\overline{V}_\ell$ values. For example, the molar volume of a 20% methanol ($\overline{V}_\ell = 0.0406$ L·mol^{-1}) – 80% water mixture is 0.02 L·mol^{-1} as compared to 0.018 L·mol^{-1} for pure water (see Illustrative Example 5.5). Thus, if the cosolvent has no effect on the properties of the POM sorbent, the effect of the cosolvent on K_{ioc} (or K_{id}), could be described simply by:

$$K_{ioc,solv/w} = K_{ioc} \cdot 10^{-\sigma_i^c \cdot f_v} \qquad (9\text{-}33)$$

Note again that σ_i^c depends both on the solute i and on the type of the cosolvent. Furthermore, σ_i^c is not a constant and may vary depending on the f_v-range for which it has been determined. In general, σ_i^c will increase with increasing f_v (e.g., Figs. 5.6 and 5.7).

Depending on the type of sorbate, sorbent, and cosolvent considered, we should not necessarily assume that the effect of the cosolvent on the sorption properties of the sorbent may be completely neglected. In order to account for this effect, the empirical model (Eq. 9-33) can be extended by introducing a second parameter α (Rao et al., 1985):

$$K_{ioc,solv/w} = K_{ioc} \cdot 10^{-\alpha \cdot \sigma_i^c \cdot f_v} \qquad (9\text{-}34)$$

Note that α quantifies how the organic cosolvent changes the nature of the sorbent with respect to its quantity (e.g., by swelling and including cosolvent in the sorbent!)

and quality. Thus α will depend not only on the type of sorbent and cosolvent, but also on the type of compound (hence, we could also use a subscript i). Note that in the literature α is often assumed to be independent of the solute considered, which is reasonable when dealing with apolar compounds or structurally related compounds. Furthermore, because it is likely that the cosolvent has a different impact on the various different sorbents that may be present in a given environment (e.g., carbonaceous material versus humic components), α may vary with f_v and even with solute concentrations. Nevertheless, at least for a "water-like" cosolvent such as methanol, the simple model Eq. 9-34 has been applied with reasonable success to fit experimental data, particularly for POM–water partitioning of nonpolar or weakly polar compounds. In most of the studies involving *methanol* as cosolvent, α has been found to be close to 1. That is, this cosolvent did not have a significant impact on the organic sorbent (Rao et al., 1990; Spurlock and Biggar, 1994b; Bouchard, 1998). An example calculation is given in Illustrative Example 9.4. For more details we refer to the literature (e.g., Nkedi-Kizza et al., 1985; Fu and Luthy, 1986; Rao et al., 1990; Wood et al., 1990; Lee et al., 1993; Kimble and Chin, 1994; Nzengung et al., 1996; Lee and Rao, 1996; Bouchard, 1998).

Illustrative Example 9.4

How Much Does the Presence of 20% Methanol in the "Aqueous" Phase Affect the Retardation of Phenanthrene in an Aquifer?

Problem

Consider the transport of phenanthrene in an aquifer exhibiting a porosity, ϕ, of 0.2, and an average density, ρ_s, of the aquifer material of 2.5 kg solid·L^{-1}. Furthermore, assume that the average organic carbon content of the aquifer material is 0.5% (i.e., $f_{oc} = 0.005$ kg oc·kg^{-1} solid). Calculate the retardation factor R_{fi} (i.e., f_{iw}^{-1}. see Section 9.2) of phenanthrene in this aquifer if the groundwater consists (a) of pure water, (b) of a 20% methanol/80% water volume mixture, and (c) of a 30% methanol/70% water volume mixture.

i - phenanthrene

Answer

The retardation factor R_{fi} is given by (Eq. 9-12):

$$R_{fi} = f_{iw}^{-1} = 1 + r_{sw} \cdot K_{id}$$

or with $r_{sw} = \rho_s \dfrac{1-\phi}{\phi}$:

$$R_{fi} = 1 + \rho_s \frac{1-\phi}{\phi} K_{id}$$

Inserting the values for ρ_s and ϕ yields:

$$R_{fi} = 1 + 10 \text{ (kg solid·L}^{-1}) \, K_{id} \text{ (L·kg}^{-1} \text{ solid)} \tag{1}$$

Case (a)

Assume that partitioning to the POM is the major sorption mechanism. Use an average K_{ioc} value of 2×10^4 L·kg^{-1} oc to estimate K_{id} (Eq. 9-22):

$$K_{id} = f_{oc} \cdot K_{ioc} = (0.005)(2 \times 10^4) = 100 \text{ L} \cdot \text{kg}^{-1} \text{ solid}$$

Insertion of this value into Eq. 1 above yields an R_{fi} value of about *1000* if the groundwater consists of pure water.

Case (b)

To estimate the effect of the methanol on K_{id}, assume that α is about 1 in Eq. 9-34. Get the σ_i^c value of phenanthrene from the PAH data summarized in Fig. 5.8 using (see figure caption):

$$\log\left(\gamma_{iw}^{\text{sat}} / \gamma_{i\ell}^{\text{sat}}\right) = (0.0104) V_{ix} - 0.668 \quad \text{for } f_{v,\text{MeOH}} = 0.2$$

(see also Illustrative Example 5.5, Answer (b)).

In this case, σ_i^c is given by (see also Section 5.4; Eqs. 5-30 and 5-31):

$$\sigma_i^c = \log\left(\gamma_{iw}^{\text{sat}} / \gamma_{i\ell}^{\text{sat}}\right) / f_{v,\text{MeOH}} = \left[(0.0104) V_{ix} - 0.668\right] / (0.2)$$

The V_{ix} of phenanthrene calculated by the method described in Box 5.1 is 145.4 cm^3 mol^{-1}, and, therefore:

$$\sigma_i^c = 0.84 / 0.2 = 4.2$$

Insertion of this value into Eq. 9-35 with $\alpha = 1$ yields:

$$K_{ioc,20\%\text{MeOH}} = 0.14 \, K_{ioc}$$

Hence, the 20% methanol lowers the apparent K_{ioc} by a factor of about 7; that is, $R_{fi} \cong 140$ instead of 1000 in the case of pure groundwater.

Case (c)

Use the σ_i^c value derived above for the 0 to 20% methanol range. Note that in reality, σ_i^c may be somewhat larger at this higher methanol fraction. Insertion of $\sigma_i^c = 4.2$ into Eq. 9-35 with $\alpha = 1$ and $f_v = 0.3$ yields:

$$K_{ioc,30\% \text{ MeOH}} = 0.055 \, K_{ioc}$$

which yields an R_{fi} value of *55*.

9.4   Sorption of Neutral Compounds to "Dissolved" Organic Matter (DOM)

"Dissolved" organic matter (DOM) is operationally defined as the organic material that passes a filter, commonly having pore size near 1 μm. DOM is usually quantified as "dissolved" organic carbon (DOC, mg C·L^{-1}) since the organic matter is burned and the evolved carbon dioxide is actually quantified. In natural waters, DOM includes a wide range of constituents from truly dissolved, small molecules like acetate, to macromolecules such as humic acids, to filter-passing submicro-meter-sized particles such as viruses (Gustafsson and Gschwend, 1997).

Qualitative Description of DOM–Solute Associations

First let us consider some of the effects of DOM on the behavior of a given compound that can be observed in an aqueous solution. Very much as in the case of an organic cosolvent (Section 5.4), one can find that DOM enhances an organic compound's apparent solubility (e.g., Chiou et al., 1986; Chin et al., 1997), and diminishes its air–water distribution ratio (Mackay et al., 1979; Brownawell, 1986). These effects are observed at DOC concentrations that are far below the concentration of any cosolvent required to produce a similar effect (e.g., 10 mg C·L^{-1}). Furthermore, it has been demonstrated in various studies that the bioavailability of a given chemical in an aqueous solution, and thus its uptake by organisms (i.e., bioaccumulation, toxicity), may significantly decrease when DOM is added to the solution (e.g., Leversee et al., 1983; Traina et al., 1996; Haitzer et al., 1999; see also Chapter 10). Finally, it has been shown that due to interactions with DOM constituents, the light-processing characteristics of chemicals may be altered significantly. For example, it has been found that the presence of DOM diminishes the ability of polycyclic aromatic hydrocarbons (PAHs) to fluoresce (Gauthier et al., 1986 and 1987; Backhus and Gschwend, 1990). In fact, this effect has even been used in various studies for quantifying the association of PAHs with DOM.

Considering all these findings, it is reasonable to envision the interaction of organic pollutants with DOM primarily as a sorption process. That is, we consider at least some of the DOM constituents as a distinct nonaqueous organic phase. This is not surprising since filter-passing DOM certainly contains colloidal components (i.e., DOM that does not settle, but does form a "particle-like" medium). This picture is also consistent with the results of measurements of the mobility of a fluorinated probe molecule in DOM solutions using ^{19}F-NMR (Hinedi et al. 1997). Hence, we can imagine that the most effective "DOM microsorbents" will be those that are able to accommodate a neutral organic chemical in a way that allows it to escape completely from the aqueous phase. We can look at such microsorbents as organic *colloids*, that is, as microparticles or macromolecules that are small enough to move primarily by Brownian motion, as opposed to gravitational settling. This means that the size of such colloids may range from about a nanometer to a micrometer in diameter. Based on this perception, the relevant DOM microsorbents in natural waters would include organic polymers as well as organic coatings on very small inorganic particles (e.g., aluminosilicates, iron oxides).

The K_{ioc} values for sorption of a given compound to DOM, which we will denote as K_{iDOC}, will depend on the fraction of the DOM which is large enough to act as a sorbent (related to the molecular weight distribution of the DOM), as well as on the organic matter's chemical structure (e.g., degree of aromaticity and presence of polar functionalities).

Determination of K_{iDOC} Values and Availability of Experimental Data

First, we need to comment on the available database and on the methods by which K_{iDOC} values are determined. Compared to soil–water or sediment–water systems, the determination of K_{iDOC} values is somewhat more difficult because the organic and the aqueous phase cannot be easily separated. One separation approach uses dialysis techniques, where a solution containing the DOM is separated by a membrane from a solution containing the dissolved organic pollutant (Carter and Suffet, 1982; McCarthy and Jimenez, 1985). The membrane allows the organic compound of interest to diffuse through, but it must be impermeable for the (larger) DOM constituents. After equilibration, the partitioning substance is measured on both sides of the membrane and any "extra" concentration on the DOM-containing side is attributed to sorption to that DOM. The major drawback of this classic method of K_{iDOC} determination is the rather long time required to reach equilibrium between the two solutions, and the fact that some smaller DOM constituents may diffuse through the membrane.

Therefore, various other techniques that do not require any separation of the DOM from the aqueous phase have been applied. These methods include solubility enhancement measurements (Chiou et al., 1986; Chin et al., 1997; Hunchack-Kariouk et al., 1997), measurements of the impact of DOM on the gas exchange behavior of the solute (i.e., "gas purge techniques"; Hassett and Milicic, 1985; Lüers and ten Hulscher, 1996), fluorescence-quenching techniques (Gauthier et al., 1986; Backhus and Gschwend, 1990; Schlautman and Morgan, 1993; Kumke et al., 1994; Traina et al., 1996), and direct solid-phase microextraction (SPME) of the dissolved organic species (Pörschmann et al., 1997; Doll et al., 1999). All of these methods have advantages and disadvantages. The solubility enhancement method is, in principle, applicable to any kind of compound, but it yields only one point of the sorption isotherm, that is, the K_{iDOC} value at saturation. Furthermore, the formation of microcrystals and/or emulsions may create artifacts. With the gas purge method, a large concentration range may be covered, but the method is restricted to volatile and semivolatile compounds; more importantly, due to the continuous removal of the solute, equilibrium may not always be maintained. The advantages of the fluorescence-quenching and the SPME technique are that they can cover a large concentration range. Moreover, with care they do not disturb the system significantly during the measurement. The disadvantages of the fluorescence-quenching technique are that it is restricted to fluorescent compounds and that artifacts may be produced by dynamic quenching mechanisms. Finally, the drawback of the SPME method is that it yields optimal results only for compounds that can be extracted sufficiently into the polymer film on the fiber used to concentrate the solutes from the aqueous solution.

To date, the majority of studies on the sorption of organic compounds to DOM have been conducted with PAHs. Therefore, in contrast to the situation found for soil, aquifer, or sediment particulate organic matter, it is not possible to provide a database of K_{iDOC} values for members of other compound classes. Nevertheless, the available data allow us to draw some important general conclusions concerning the major factors that govern sorption of apolar and weakly polar compounds to DOM. We should note that for these compounds, linear isotherms have been generally observed over the entire concentration range.

DOM Properties Governing the Magnitude of K_{iDOC}

Depending on the DOM considered, the K_{iDOC} values of a given sorbate (e.g., PAHs like phenanthrene and pyrene) have been reported to vary by up to two orders of magnitude (Chin et al., 1997; Georgi, 1998; Laor et al., 1998; Haitzer et al., 1999; Kopinke et al., 1999). The size of the colloids or macromolecules, their conformation (coiled or extended), and their composition (e.g., aromaticity, type and number of polar groups) should influence sorption of neutral organic pollutants. Hence, in order to quantitatively describe the sorption characteristics of a given DOM, some quantitative measure of each of these *bulk* properties is required.

phenanthrene

pyrene

Presently the most commonly used parameters include: (1) weight-averaged or number-averaged molecular weight, (2) molar UV-light absorptivities, $\varepsilon_i(\lambda)$, at a given wavelength (e.g., $\lambda = 254$ nm or 280 nm), (3) degree of aromaticity determined by ^{1}H- and/or ^{13}C-NMR, and (4) stoichiometric ratios of the various elements (i.e., H/C, H/O, O/C, (N+O)/C) used to express the degree of saturation and the overall polarity of the DOM. Obviously, some of these parameters correlate with each other (Chin et al., 1994 and 1997; Haitzer et al., 1999). For example, the molar absorptivities at 254 or 280 nm also reflect the degree of aromaticity of a given DOM (see also Chapters 15 and 16). Furthermore, aromaticity and H/C ratio are (inversely) related. Finally, H/O, O/C, and (N+O)/C all reflect the polarity of a given DOM, although they do not give any information on its H-donor and H-acceptor properties. Hence, when trying to relate K_{iDOC} values to DOM properties, one has to be aware of possible cross-correlations. One should also be aware that these parameters represent only a crude characterization of the DOM and are, therefore, not comparable to the parameters that we have used to describe the vdW and H-bonding interactions in the polyparameter LSER models of well-defined organic solvents:water partitioning systems (Chapter 7).

For now, we focus on bulk DOM parameters. Using pyrene as a common sorbate, 17 different DOMs were tested for their ability to sorb this PAH at 25°C. Correlations of the resultant K_{iDOC} values with the bulk properties of the DOMs were then sought (Chin et al., 1997; Georgi, 1998). For pyrene sorption, the molar absorptivities at 280 nm (reflecting aromaticity) and the O/C ratio (reflecting overall polarity) were found to yield a significant correlation (Fig. 9.14):

$$\log K_{\text{pyrene,DOC}} / (\text{L} \cdot \text{kg}^{-1} \text{ oc}) =$$

$$1.45 \log \varepsilon_i(280 \text{ nm}) / (\text{L} \cdot \text{mol}^{-1} \text{oc} \cdot \text{cm}^{-1}) - 1.70 \ (\text{O/C}) + 1.14 \qquad (9\text{-}35)$$

$$(N = 17, R^2 = 0.89)$$

It is interesting to note that the DOMs used for the derivation of Eq. 9-35 included a

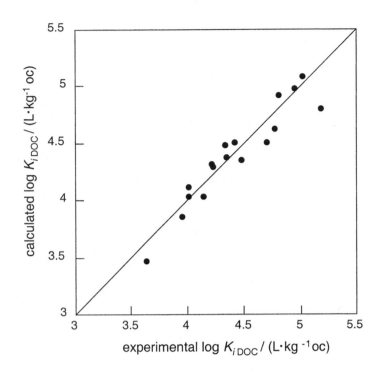

Figure 9.14 Plot of calculated (Eq. 9-35) versus experimental K_{iDOC} values of pyrene (pH 7, 25°C) for 17 different DOMs covering a wide range of molecular sizes, aromaticities, and polarities. Data from Chin et al. (1997) and Georgi (1998).

wide variety of fulvic and humic materials from creeks, rivers, swamps, lakes, soils, and dump sites, as well as commercial humic acids. In addition, 6 of the 17 $K_{pyrene,DOC}$ values were determined by solubility enhancement (Chin et al., 1997), whereas the rest were measured by SPME (Georgi, 1998).

Effect of pH, Ionic Strength, and Temperature on K_{iDOC}

pH and Ionic Strength. Because DOM constituents exhibit a rather large number of acidic (e.g., carboxylic acid, phenolic) and other polar groups that may dissociate and/or form complexes with metal ions (e.g., Na^+, K^+, Mg^+, Ca^{2+}, Al^{3+}), pH and major ion composition of the solution affect the sorption characteristics of a given DOM. When carboxyl groups are ionized, their electrostatic repulsions cause the DOM to spread out in solution. When divalent cations like calcium are bound to such functional groups, they enable bridging of like-charged groups and therefore cause the macromolecules to coil. The results of the limited number of studies on this subject demonstrate that these effects may be quite complex (Schlautmann and Morgan, 1993; Ragle et al., 1997; Georgi, 1998). Hence, depending on the nature of the DOM constituents, changes in pH and ion composition may or may not have a significant impact on the number of hydrophobic and hydrophilic domains that are relevant for sorption of a given neutral organic compound. Consequently, based on the very limited available data for PAHs, it is not possible to quantitatively predict the effect of solution composition on the K_{iDOC} value of a given compound. Nevertheless, a few general comments on trends and magnitudes of such effects can be made.

First, at low to moderate ionic strengths (i.e., $I < 0.1$ M), we may, as a first approximation, neglect the effect of dissolved salts on K_{iDOC} for apolar and weakly polar compounds. Since there are very little data available for higher ionic strength, it is not possible to draw any general conclusion for seawater. We recall, however, that at high salt concentrations, salting-out phenomena as described in Chapter 5 also need to be taken into account.

When considering the effect of pH, there is a general trend that the K_{iDOC} values of apolar and weakly polar compounds decrease with increasing pH. This effect is more pronounced for larger compounds as compared to smaller compounds. Furthermore, the biggest effects are found for DOMs that consist primarily of smaller sized, highly polar constituents such as fulvic acids. In these cases, differences in K_{iDOC} of large PAHs of up to a factor of 2 to 3 or even more have been measured between pH 4 and 10 (Schlautman and Morgan, 1993; Georgi, 1998). We can rationalize these findings by envisioning an increase in the number of negatively charged functional groups with increasing pH; these may lead to the destruction of hydrophobic DOM domains (e.g., by uncoiling of macromolecules and/or by disaggregation of DOM components). For the larger and less polar humic acids, however, these effects are not very substantial. Therefore, particularly within the pH-range typically encountered in natural waters (i.e., pH 6 to 9), in many cases we neglect the effect of pH on K_{iDOC}.

Temperature. There are also very few experimental data available on the effect of temperature on DOM–water partitioning. As always, we examine the influence of temperature by considering the magnitude of the involved excess enthalpy terms in the context of an equation:

$$\ln K_{iDOC} = -\frac{\Delta_{DOMw} H_i}{R} \cdot \frac{1}{T} + \text{const.} \tag{9-36}$$

where $\Delta_{DOMw} H_i$ is the (average) standard enthalpy of transfer of i from water to the various DOM constituents and is given by:

$$\Delta_{DOMw} H_i = H_{iDOM}^E - H_{iw}^E \tag{9-37}$$

$\Delta_{DOMw} H_i$ values of between -20 and -40 kJ $\cdot$ mol^{-1} have been reported for six PAHs including fluoranthene and benzo(a)pyrene (Lüers and ten Hulscher, 1996). For such compounds $\Delta_{DOMw} H_i$ is likely dominated by the H_{iw}^E term. That is, we assume that H_{iDOM}^E is smaller since organic sorbate:organic sorbent interactions are more nearly ideal than organic solute:water interactions. The observed $\Delta_{DOMw} H_i$ values are only slightly less negative than the corresponding negative H_{iw}^E values. Hence, in most cases, K_{iDOC} will be only weakly to moderately temperature dependent, that is, less than a factor of 2 with a 10 degree change in temperature. Note that a negative $\Delta_{DOMw} H_i$ value means that K_{iDOC} decreases with increasing temperature.

LFERs Relating K_{iDOC} Values to K_{iow} Values

Analogous to the use of simple one-parameter LFERs for relating POM–water partition coefficients (Eq. 9-25, Table 9.2) *for a given DOM*, we may correlate log K_{iDOC} values with log K_{iow} values:

$$\log K_{iDOC} = a \cdot \log K_{iow} + b \tag{9-38}$$

Figure 9.15 Equilibrium sorption to a commercial humic acid (Roth-HA): Correlations (Eq. 9-38) between log K_{iDOC} and log K_{iow} for series of compounds including PAHs (open circles, log K_{iDOC} = 0.91 log K_{iow} + 0.16; R^2 = 0.98), some arenes including biphenyls and *trans*-stilbene (closed triangles, log K_{iDOC} = 0.94 log K_{iow} − 0.29, R^2 = 0.99), and alkanes (C_7 – C_{11}; open squares, log K_{iDOC} = 0.80 K_{iow} + 0.24, R^2 = 0.98). Note that the units of K_{iDOC} are L·kg⁻¹ oc. Data are from Georgi (1998).

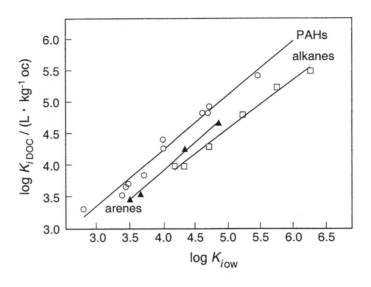

Figure 9.16 LFERs (Eq. 9-38) relating log K_{iDOC} and log K_{iow} of a set of PAHs for seven different humic or fulvic acids. Note that the units of K_{iDOC} are L·kg⁻¹ oc. Data from Georgi (1998). The slopes and intercepts of the various LFERs are (1) 0.91/0.16, (2) 0.84/0.37, (3) 0.86/0.26, (4) 0.86/0.19, (5) 0.87/−0.12, (6) 0.79/0.06, (7) 0.72/−0.03.

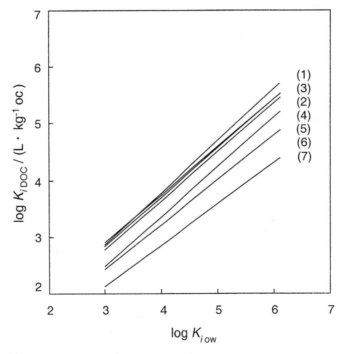

Considering the complex structural features of DOM, we may expect to find good correlations only for confined sets of compounds exhibiting similar molecular interaction properties. This has been seen for the sorption of a series of PAHs, arenes, and alkanes to a commercial humic acid (Fig. 9.15). Note that linear isotherms were observed over the whole concentration range. The significant differences in K_{iDOC} between PAHs and alkanes exhibiting the same K_{iow} value may be explained by the stronger vdW interactions of the PAHs with the aromatic structures of the humic acid and/or polar interactions of the PAHs with H-donor groups.

When applying LFERs such as Eq. 9-38, the variability in compound properties and the structural variability of filter-passing natural organic matter are important. The resultant K_{iDOC} of a given compound may vary by more than one order of magnitude depending on the DOM considered. Hence, the slopes and intercepts of LFERs established for a given set of structurally related compounds (e.g., PAHs) for various DOMs, may be similar in some cases but may differ significantly in other cases (see examples given in Fig. 9.16). Nevertheless, because of the lack of more sophisticated approaches, in practice, simple LFERs such as Eq. 9-38 are presently the only means to make any reasonable predictions of K_{iDOC} values. But dependable predictions (i.e., within a factor of 2 to 3) can be expected only if an LFER is used that was established for a set of compounds *and* natural organic matter that is representative for the case considered (see Illustrative Example 9.5).

Illustrative Example 9.5

Evaluating the Effect of DOM on the Bioavailability of Benzo(a)pyrene (BP)

In a variety of studies, it has been shown that organic pollutants associated with DOM are not bioavailable to aquatic organisms (e.g., Leversee et al., 1993; Traina et al., 1996; Haitzer et al., 1999). Hence, the *bioconcentration factor*, BCF_{iDOC}, of a given compound i in the presence of DOM can be expressed by:

$$BCF_{iDOC} = f_{iw} \cdot BCF_i$$

where BCF_i is the bioconcentration factor determined in pure water (for definitions see Chapter 10) and f_{iw} is the fraction that is truly dissolved in the water (see Eqs. 9-10 to 9-12).

Problem

In order to estimate how the bioavailability of benzo(a)pyrene (BP) is affected by DOM, you want to assess the speciation of this compound as a function of DOM quantity and quality. To this end, calculate the f_{iw} value of BP for aqueous solutions (pH 7, 25°C) containing (a) 10 mg DOC·L⁻¹ and 100 mg DOC·L⁻¹, respectively, and (b) assuming DOM qualities as reflected by the LFERs #1 and #7 in Fig. 9.16 (see figure caption for slopes and intercepts). Note that DOM #1 represents a humic acid that exhibits a high affinity for PAHs, whereas DOM #7 is a fulvic acid with a low affinity. Hence, the two DOMs may represent extreme cases with respect to sorption of apolar and weakly polar compounds in natural waters.

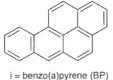

i = benzo(a)pyrene (BP)

log K_{iow} = 6.13
(see Appendix C)

Answer

When setting $r_{sw} = [DOC]$ and $K_{id} = K_{iDOC}$ in Eq. 9-12, the fraction in dissolved form can be calculated by:

$$f_{iw} = \frac{1}{1 + [DOC] \cdot K_{iDOC}} \tag{1}$$

Calculate the maximum and minimum K_{iDOC} for BP by inserting its K_{iow} value into the LFERs #1 and #7, respectively:

max. value: $\log K_{iDOC} = (0.91)(6.13) + 0.16 = 5.74$

min value: $\log K_{iDOC} = (0.72)(6.13) - 0.03 = 4.38$

Hence, K_{iDOC} may vary between 2.4×10^4 and 5.5×10^5 L·kg^{-1} oc. Insertion of these values into Eq. 1 (above) yields f_{iw} values:

for $[DOC] = 10^{-5}$ kg oc·L^{-1}:   between about *0.8* and *0.15*

for $[DOC] = 10^{-4}$ kg oc·L^{-1}:   between about *0.3* and *0.02*.

This means that depending on the quantity and "quality" of the DOM present, between 2% (100 mg oc·L^{-1}, high affinity DOM) and 80% (10 mg oc·L^{-1}, low affinity DOM) of the total BP present may be bioavailable.

More generally, this is the fraction of BP that is available for processes that only act on the dissolved species (e.g., air–water exchange, see Chapter 20). Hence, for a strongly sorbing compound such as BP, sorption to DOM may significantly influence this PAH's environmental behavior. In such cases, one obviously needs to get a more accurate K_{iDOC} value for the system considered.

9.5   Sorption of Organic Acids and Bases to Natural Organic Matter (NOM)

Effect of Charged Moieties on Sorption: General Considerations

Depending on their pK_{ia} value(s), organic acids and bases may be partially or even fully ionized at ambient pH. When considering the sorption of ionized species, in addition to vdW and hydrogen bonding interactions, we also have to take into account electrostatic interactions with charged species present in natural sorbents (Fig. 9.2). These interactions are attractive if the charges of the sorbent and solute have opposite signs, and they are repulsive if both exhibit the same sign. Furthermore, some acid and base functions may also form chemical bonds with certain sorbent moieties (see example given in Fig. 9.2 and Eq. 9-19). In these cases, quantification of sorption may become even more complicated, because the abundances and reactivities of such sorptive sites need to be quantified. We will discuss a few examples of such cases later in this section and in Chapter 11. Here we focus on the sorption of organic acids and bases to natural organic matter (NOM). In the following discussion we do not distinguish between sorption to DOM or POM. We quantify the overall sorption as a *distribution ratio*, D_{ioc}, since more than one dissolved and sorbed species may exist.

At ambient pH values, NOM is negatively charged, primarily due to the presence of carboxylic acid groups. Therefore, bulk NOM primarily acts as a cation exchanger. This means we may expect that negatively charged organic species will sorb more weakly to NOM than their neutral counterparts. Nevertheless, in situations in which

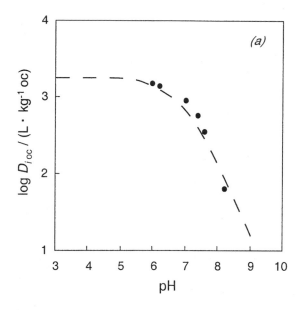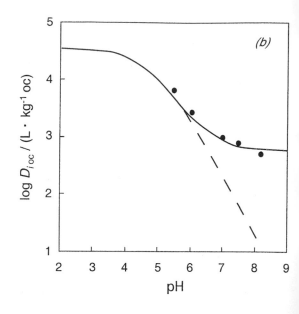

Figure 9.17 Sorption of chlorinated phenols to sediment organic matter: D_{ioc} values determined for (*a*) 2,4,5-trichlorophenol and (*b*) pentachlorophenol as a function of pH. The solid line represents Eq. 9-41, the dashed line Eq. 9-42. Data from Schellenberg et al. (1984).

the anionic species predominates, its sorption to NOM may not always be neglected (see below). For positively charged species, on the other hand, we may expect stronger association with NOM constituents as compared to the neutral compound.

Sorption of Compounds Forming Anionic Species (Organic Acids)

When dealing with weak acids (e.g., phenolic compounds) that exhibit only one acidic group, we may generally derive D_{ioc} using (Schellenberg et al., 1984; Lee et al., 1990; Jafvert, 1990; Severtson and Banerjee, 1996; Gundersen et al, 1997):

$$D_{ioc} = \frac{[HA]_{oc} + [A^-]_{oc}}{[HA]_w + [A^-]_w} \tag{9-39}$$

Since the acid and its conjugate base in solution are related via (Chapter 8):

$$K_{ia} = \frac{\{H^+\}[A^-]_w}{[HA]_w} \tag{9-40}$$

we find:

$$D_{ioc} = \alpha_{ia} \cdot K_{ioc}^{HA} + (1 - \alpha_{ia}) \cdot K_{ioc}^{A^-} \tag{9-41}$$

where K_{ioc}^{HA} and $K_{ioc}^{A^-}$ are the sorption coefficients for the neutral and the charged species, respectively, and α_{ia} is the fraction of the compound in acidic (i.e., nondissociated) form (see Chapter 8):

$$\alpha_{ia} = \frac{1}{1 + K_{ia}/\{H^+\}} = \frac{1}{1 + 10^{pH - pK_{ia}}} \tag{8-21}$$

As pointed out above, we may assume that $K_{ioc}^{A^-}$ will, in general, be significantly smaller than K_{ioc}^{HA}. This is clearly seen for sorption of 2,4,5-trichlorophenol (pK_{ia} = 6.94) to sediment organic matter (Fig. 9.17*a*). For this type of compound, sorption of the neutral species dominates D_{ioc} up to a pH of about 2 units above the pK_{ia} of the

2,4,5-trichlorophenol
(TCP)

pentachlorophenol
(PCP)

$R = C_{10} - C_{13}$
linear akylbenzene sulfonates
(LAS)

compound. Hence, up to this pH ($pK_{ia} + 2$), we may neglect sorption of the anion. The overall sorption to NOM simplifies to:

$$D_{ioc} = \alpha_{ia} \cdot K_{ioc}^{HA} \qquad (9\text{-}42)$$

Note that when applying Eqs. 9-41 and 9-42 we assume that K_{ioc}^{HA} and $K_{ioc}^{A^-}$ does not significantly depend on the pH of the solution.

At pH > ~ pK_{ia} + 2, particularly when dealing with hydrophobic acids such as pentachlorophenol (PCP, pK_{ia} = 4.75), the sorption of the anionic species may be important (Fig. 9.17b). In the case of PCP, the $K_{ioc}^{A^-}$ value is about 500 L·kg^{-1} oc. This is somewhat less than 2 orders of magnitude smaller than the corresponding K_{ioc}^{HA} value. In this case, Eq. 9-41 fits the experimental data well (Fig. 9.17b).

We should point out that when describing the sorption of an organic anion over a large pH-range, the pH dependence of $K_{ioc}^{A^-}$ has to be taken into account (Jafvert, 1990). This is particularly important for highly acidic, amphiphilic compounds that are present exclusively as anions at ambient pH. A prominent example involves the linear alkylbenzene sulfonates (LASs) that are widely used detergents (see also Section 2.3). For these compounds, it was found that sorption to sediments (primarily to the organic matter present) decreased by almost an order of magnitude between pH 5 and 10 (Westall et al., 1999). Furthermore, sorption isotherms were nonlinear (n_i < 1 in Eq. 9-1) and were also dependent on the major cation composition of the aqueous phase. This result is not too surprising when considering that the NOM constituents (as well as other surface sites, see Chapter 11) become increasingly negatively charged with increasing pH, thus making it more difficult to accommodate a negatively charged sorbate. Similarly, we can rationalize the increasing sorption found with increasing concentration of positively charged cations (e.g., Ca^{2+}). Besides the possibility of ion pair formation with the anionic sorbate forming a neutral or even cationic species more inclined to sorb, such cations may also complex with the anionic moieties of the NOM, reducing the negative charge and the electrostatic repulsion of the anionic sorbate (Westall et al., 1999). Furthermore, nonlinear sorption may be explained by both a limited number of sites for accommodation of negatively charged species as well as increasing electrostatic repulsion with increasing sorbate concentration. Hence, all of these factors need to be considered when dealing with the sorption of anionic species. For a more detailed treatment of this topic, we refer to the literature (e.g., Jafvert, 1990; Stapleton et al., 1994; Westall et al., 1999). Finally, we should note that when dealing with sorption of surfactants such as the LASs, at higher concentrations, we also have to take into account the formation of micelles in solution as well as on surfaces (see Chapter 11).

Sorption of Compounds Forming Cationic Species (Organic Bases)

As mentioned above, sorption of the cationic form of an organic base to negatively charged sites in natural sorbents may dominate the overall sorption of the compound, at least over a certain pH-range. Examples of such cases include aliphatic and aromatic amines (Davis 1993; Lee et al., 1997; Fabrega et al., 1998), N-heterocyclic compounds (Zachara et al., 1986; Brownawell et al., 1990), and triorganotin compounds (Weidenhaupt et al., 1997; Arnold et al., 1998; Berg et al.,

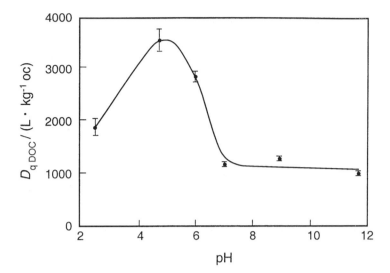

Figure 9.18 Sorption of quinoline to Aldrich humic acid (AHA): Influence of pH on the overall D_{qDOC} value (adapted from Nielsen et al., 1997).

2001; see also below). Because of the limited number of cation exchange sites present in a natural sorbent, sorption isotherms of the cationic species will be nonlinear and competition with other cations present (e.g., Na^+, K^+, Mg^{2+}, Ca^{2+}, Al^{3+}) may occur. Furthermore, except for the permanently negatively charged siloxane surfaces of clay minerals (see Chapter 11), the number of negatively charged sites present in a given natural sorbent is strongly pH-dependent. Hence, the quantitative description of sorption of organic cations to natural sorbents is a rather complex task. We will discuss some examples in more detail in Chapter 11.

Here we confine ourselves to some observations of the sorption of organic bases to NOM. Consider the pH-dependence (Fig. 9.18) of the sorption of quinoline (subscript q, for structure, see margin) to Aldrich humic acid (AHA). In this case, the D_{qDOC} value shows a maximum at about pH 5. This corresponds to the pK_{ia} of the compound. At high pHs (i.e., pH > 7) when virtually all of the quinoline is in its nonionic form, the overall sorption is primarily determined by partitioning of this neutral species (Q) to AHA:

$$D_{qDOC} \cong \frac{[Q]_{oc}}{[Q]_w} = K_{qDOC}^Q \qquad \text{at high pHs} \qquad (9\text{-}43)$$

With decreasing pH, the fraction of the cationic form of quinoline (QH^+) increases and the sorbed cations increase too. However, at the same time the number of negatively charged AHA moieties decreases. This leads to the maximum observed at pH 5. Now the partitioning reflects:

$$D_{qDOC} \cong \frac{[Q]_{oc} + [QH^+]_{oc}}{[Q]_w + [QH^+]_w} \qquad \text{pHs below 7} \qquad (9\text{-}44)$$

In this case, due to electrostatic interactions, the maximum D_{qDOC} is about a factor of 4 larger than the K_{qDOC}^Q (see Eq. 9-43 for partitioning of the neutral species).

An even more pronounced case involves the sorption of the two biocides, tributyltin (TBT) and triphenyltin (TPT) to AHA. Because of their very high toxicity toward

pK_{ia} = 4.90

quinoline

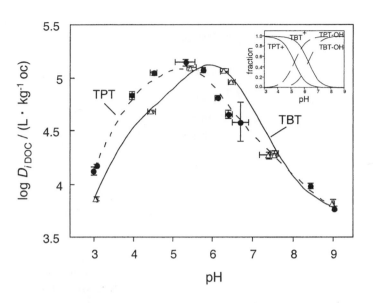

Figure 9.19 Aldrich humic acid (AHA)–water distribution ratio (D_{iDOC}) of TBT ($\triangle$) and TPT ($\bullet$) as a function of pH. Each point was determined from a sorption isotherm. Error bars are the standard deviation of the slope of the linear isotherm and of pH. The lines were calculated using the model described by Arnold et al. (1998). The insert shows the speciation of TBT and TPT as a function of pH. Adapted from Arnold et al. (1997 and 1998).

R = n-C$_4$H$_9$:TBT
pK_{ia} = 6.3
R = C$_6$H$_5$: TPT
pK_{ia} = 5.2

aquatic organisms, TBT and TPT are of considerable environmental concern (Fent, 1996). Again, sorption varies strongly with pH (Fig. 9.19). In these cases, the D_{iDOC} at the pH corresponding to these compounds' pK_{ia} values is enhanced by more than a factor of 10 over the partitioning of the neutral species (TBTOH, TPTOH). In fact, even at pH 8, where the abundance of TBT$^+$ or TPT$^+$ is very small, sorption of the cation was still found to dominate the overall sorption (Arnold et al., 1998).

These findings can be rationalized by postulating an inner sphere complex formation (i.e., by ligand exchange of a water molecule) between the tin atom of the charged species and negatively charged ligands (i.e., carboxylate, phenolate groups) present in the humic acid. The observed pH dependence of the overall AHA–water distribution ratio of the two compounds could be described successfully with a semiempirical, discrete log K_j spectrum model using four discrete complexation sites in AHA exhibiting fixed pK_{aj} values of 4, 6, 8, and 10 (Fig. 9.19). Note that K_j is the complexation constant of TBT$^+$ or TPT$^+$, respectively, with the ligand type j [i.e., a carboxyl (pK_{aj} = 4,6) or phenolate (pK_{aj} = 8,10) group]. For more details, refer to Arnold et al. (1998).

We conclude this section by noting that sorption of charged species to NOM is generally fast and reversible, provided that no real chemical reactions take place that lead to the formation of covalent bonds (i.e., to "bound residues"; see chapter 14). This conclusion is based on experimental data and on the assumption that in aqueous solution the more polar NOM sites are more easily accessible as compared to the more hydrophobic domains. For charged species, we may, therefore, assume that equilibrium is established within relatively short time periods. Hence, for example, in the case of TBT and TPT, contaminated sediments may represent an important source for these highly toxic compounds in the overlying water column (Berg et al., 2001).

9.6   ## Questions and Problems

Questions

Q 9.1

Give five reasons why it is important to know to what extent a given chemical is present in the sorbed form in a natural or engineered system.

Q 9.2

What are the most important natural sorbents and sorption mechanisms for (a) apolar compounds, (b) polar compounds, and (c) ionized compounds?

Q 9.3

What is a sorption isotherm? Which types of sorption isotherms may be encountered when dealing with sorption of organic compounds to natural sorbents? Does the shape of a sorption isotherm tell you anything about the sorption mechanism(s)? If yes, what? If no, why not?

Q 9.4

Write down the most common mathematical expressions used to describe sorption isotherms. Discuss the meaning of the various parameters and describe how they can be derived from experimental data.

Q 9.5

Why is natural organic matter (NOM) such an important sorbent for all organic compounds? What types of organic phases may be present in a given system? What are the most important properties of NOM with respect to the sorption of organic compounds?

Q 9.6

How is the K_{ioc} (K_{iDOC}) value of a given compound defined? How large is the variability of K_{ioc} (K_{iDOC}) for (a) different particulate organic phases (POM), and (b) different "dissolved" organic phases (DOM)? Which are the major structural factors of POM or DOM that cause this variability?

Q 9.7

As noted in Section 9.3 (Fig. 9.8), the average K_{ioc} values of 1,2-dichlorobenzene determined by Kile et al. (1995) for uncontaminated soil–water and sediment–water partitioning are about 300 and 500 L·kg^{-1} oc, respectively. However, for heavily contaminated soils and sediments, these authors found significantly higher K_{ioc} values (700 − 3000 L·kg^{-1} oc), although isotherms were linear over a wide concentration range. Try to explain these findings.

Q 9.8

How do (a) pH, (b) ionic strength, and (c) temperature affect the sorption of *neutral* organic compounds to dissolved and particulate organic matter? Give examples of compound–organic phase combinations in which you expect (i) a minimum, and (ii) a maximum effect.

Q 9.9

How does the presence of a completely miscible organic cosolvent (CMOS) affect the speciation of an organic compound in a given environment (e.g., in an aquifer)? What are the most important parameters determining the effect of an organic cosolvent? How can this effect be quantified?

Q 9.10

What is the major difference between the sorption of neutral and the sorption of charged organic species to NOM? Qualitatively describe the pH dependence of the NOM–water partitioning of (a) an organic acid, and (b) an organic base.

Problems

P 9.1 *What Fraction of Atrazine Is Present in Dissolved Form?*

Atrazine is still one of the most widely used herbicides. Estimate the fraction of total atrazine present in truly dissolved form (a) in lake water exhibiting 2 mg $POC \cdot L^{-1}$, (b) in marsh water containing 100 mg $solids \cdot L^{-1}$, if the solid's organic carbon content is 20%, and (c) in an aquifer exhibiting a porosity of 0.2 by volume, a density of the minerals present of 2.5 $kg \cdot L^{-1}$, and an organic carbon content of 0.5%. Assume that partitioning to POM is the major sorption mechanism. You can find K_{ioc} values for atrazine in Fig. 9.9. Comment on which value(s) you select for your calculations.

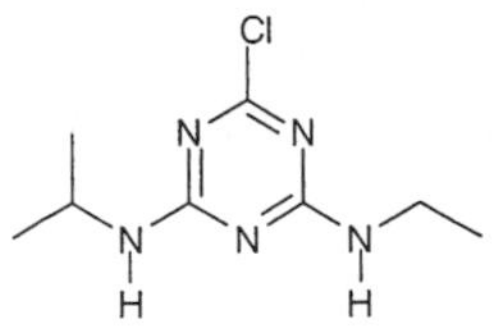

atrazine

P 9.2 *Estimating the K_{ioc} Value of Isoproturon from K_{ioc}'s of Structurally Related Compounds*

Urea-based herbicides are widely used despite the concern that they may contaminate groundwater beneath agricultural regions. You have been asked to evaluate the sorption behavior of the herbicide isoproturon in soils.

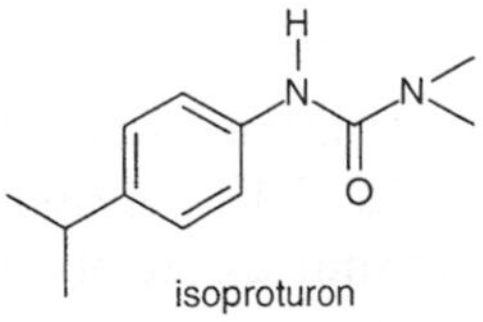

isoproturon

Unable to find information on this specific compound, you collect data on some structurally related compounds:

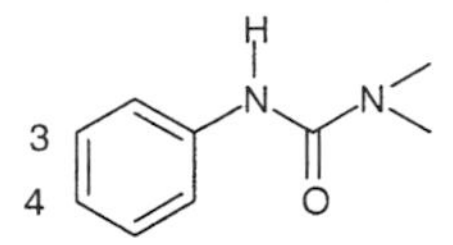 compound	$\log K_{iow}$	$\log K_{ioc}$ (L·kg^{-1}oc)
4-methyl	1.33	1.51
3,5-dimethyl	1.90	1.73
4-chloro	1.94	1.95
3,4-dichloro	2.60	2.40
3-fluoro	1.37	1.73
4-methoxy	0.83	1.40

What K_{ioc} do you estimate for isoproturon? Do you use all compounds for deriving an LFER?

P 9.3 *Evaluating the Transport of 1,2-Dichloropropane in Groundwater*

A group of investigators from the USGS recently discovered a large plume of the soil fumigant 1,2-dichloropropane (DCP) in the groundwater flowing away from an airfield. The aquifer through which the DCP plume is passing has been found to have a porosity of 0.3. The aquifer solids consist of 95% quartz (density 2.65 g·mL^{-1}; surface area 0.1 m^2·g^{-1}), 4% kaolinite (density 2.6 g·mL^{-1}; surface area 10 m^2·g^{-1}), 1% iron oxides (density 3.5 g·mL^{-1}; surface area 50 m^2·g^{-1}), and organic carbon content of 0.2%. What retardation factor [R_{fi} ($= f_{iw}^{-1}$; see Eq. 9-12)] do you expect at minimum (assumption that only POM is responsible for sorption) for DCP transport in the plume assuming that sorptive exchanges are always at equilibrium?

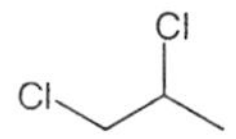

1,2-dichloropropane
(log K_{iow}=2.28; Montgomery, 1997)

P 9.4 *Estimating the Retardation of Organic Compounds in an Aquifer from Breakthrough Data of Tracer Compounds*

Using tritiated water as conservative tracer, an average retardation factor, R_{fi} (f_{iw}^{-1}; see Eq. 9.12) of about 10 was determined for chlorobenzene in an aquifer. (a) Assuming this retardation factor reflects absorption only to the aquifer solids' POM, what is the average organic carbon content (f_{oc}) of the aquifer material if its minerals have density 2.5 kg·L^{-1} and if the porosity is 0.33? (b) Estimate the R_{fi} values of 1,3,5-trichlorobenzene (1,3,5-TCB) and 2,4,6-trichlorophenol (2,4,6-TCP) in this aquifer (pH = 7.5, T = 10°C) by assuming that absorption into the POM present is the major sorption mechanism. Why can you expect to make a better prediction of R_{fi} for 1,3,5-TCB as compared to 2,4,6-TCP?

You can find all necessary information in Table 9.2 and in Appendix C. Comment on all assumptions that you make.

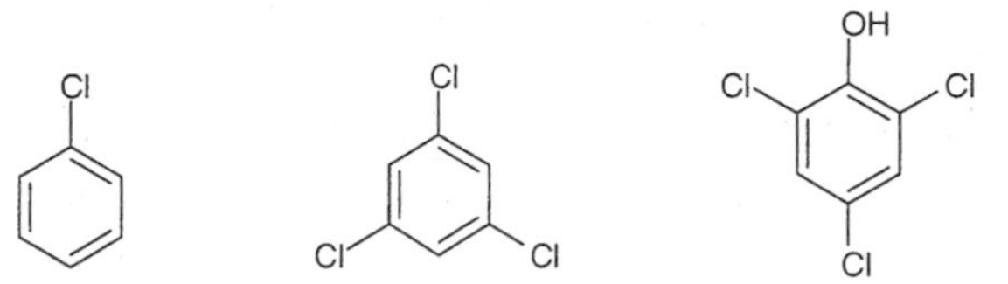

chlorobenzene   1,3,5-trichlorobenzene   2,4,6-trichlorophenol
                            (1,3,5-TCB)   (2,4,6-TCP)

P 9.5   *Evaluating the Concentration Dependence of Equilibrium Sorption of 1,2,4,5-Tetrachlorobenzene (TeCB) to an Aquitard Material*

Xia and Ball (1999) measured sorption isotherms for a series of chlorinated benzenes and PAHs for an aquitard material ($f_{oc} = 0.015$ kg oc $\cdot$ kg^{-1} solid) from a formation believed to date to the middle to late Miocene. Hence, compared to soils or recent sediment POM, the organic matter present in this aquitard material can be assumed to be fairly mature and/or contain char particles from prehistoric fires. A nonlinear isotherm was found for TeCB (fitting Eq. 9-2) and the following Freundlich parameters were reported: $K_{TeCB\,F} = 128\,(\text{mg}\cdot\text{g}^{-1})(\text{mg}\cdot\text{mL}^{-1})^{-n_{TeCB}}$ and $n_{TeCB} = 0.80$. For partitioning of TeCB to this material (linear part of the isotherm at higher concentrations), the authors found a K_{ioc} value of 4.2×10^4 L $\cdot$ kg^{-1}oc.

(a) Calculate the apparent K_{ioc} values of TeCB for the aquitard material for aqueous TeCB concentrations of C_{iw}=1, 10, and 100 μg $\cdot$ L^{-1} using the Freundlich isotherm given above. Compare these values to the K_{ioc} values given above for POM–water partitioning. Comment on the result.

(b) At what aqueous TeCB concentration (μg $\cdot$ L^{-1}) would the contribution of *ad*sorption to the overall K_{ioc} be only half of the contribution of *ab*sorption, (partitioning)?

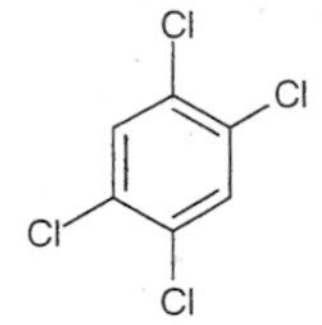

1,2,4,5-tetrachlorobenzene
(TeCB)

Note: When using Freundlich isotherms, be aware that the numerical value of K_{iF} depends nonlinearly on the unit in which the concentration in the aqueous phase is expressed. Hence for solving this problem, you may first convert μg $\cdot$ L^{-1} to mg $\cdot$ mL^{-1} or you may express the Freundlich equation using, for example, μg $\cdot$ kg^{-1} and μg $\cdot$ L^{-1}, respectively:

$$K_{\text{TeCB F}} = 128(10^6 \,\mu g \cdot kg^{-1})(10^6 \,\mu g \cdot L^{-1})^{-0.8}$$
$$= 128 \cdot 10^6 \cdot 10^{-4.8}(\mu g \cdot kg^{-1})(\mu g \cdot L^{-1})^{-0.8}$$
$$= 2030(\mu g \cdot kg^{-1})(\mu g \cdot L^{-1})^{-0.8}$$

P 9.6 *Is Sorption to Dissolved Organic Matter Important for the Environmental Behavior of Naphthalene?*

Somebody claims that for naphthalene, sorption to DOM is generally unimportant in the environment. Is this statement correct? Consult Illustrative Example 9.5 to answer this question.

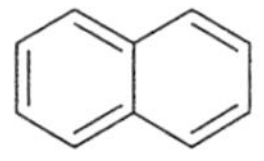

naphthalene

P 9.7 *Assessing the Speciation of a PCB-Congener in a Sediment–Pore Water System*

Consider a surface sediment exhibiting a porosity $\phi = 0.8$, solids with average density $\rho_s = 2.0 \, kg \cdot L^{-1}$ solid, a particulate organic carbon content of 5%, and a DOC concentration in the pore water of 20 mg $DOC \cdot L^{-1}$. Estimate the fractions of the total 2,2',4,4'-tetrachlorobiphenyl (PCB47) present in truly dissolved form in the porewater and associated with the pore water DOM. Assume that absorption into the organic material is the major sorption mechanism and that $K_{i\text{DOC}} \cong 1/3 \, K_{ioc}$. Estimate K_{ioc} using Eq. 9-26a with the K_{iow} value of PCB47 given in Appendix C.

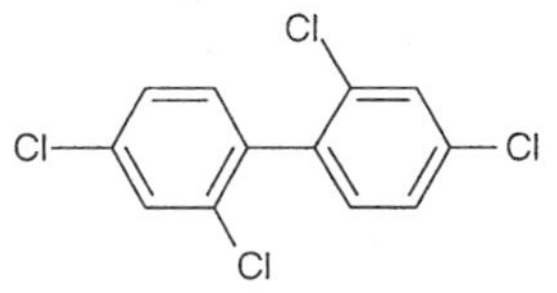

2,2',4,4'-tetrachlorobiphenyl
(PCB47)

Chapter 10

Sorption II: Partitioning to Living Media – Bioaccumulation and Baseline Toxicity

10.1 Introduction

10.2 Partitioning to Defined Biomedia
The Composition of Living Media
Equilibrium Partitioning to Specific Types of Organic Phases Found in
    Organisms
A Model to Estimate Equilibrium Partitioning to Whole Organisms
Parameters Used to Describe Experimental Bioaccumulation Data
Illustrative Example 10.1: *Evaluating Bioaccumulation from a Colloid-
    Containing Aqueous Solution*
Illustrative Example 10.2: *Estimating Equilibrium Bioaccumulation Factors
    from Water*
Illustrative Example 10.3: *Estimating Equilibrium Bioaccumulation Factors
    from Air*

10.3 Bioaccumulation in Aquatic Systems
Bioaccumulation as a Dynamic Process
Evaluating Bioaccumulation Disequilibrium – Example: Biota–Sediment
    Accumulation Factors
Using Fugacities or Chemical Activities for Evaluation of Bioaccumulation
    Disequilibrium
Illustrative Example 10.4: *Calculating Fugacities or Chemical Activities to
    Evaluate Bioaccumulation*

10.4 Bioaccumulation in Terrestrial Systems
Transfer of Organic Pollutants from Air to Terrestrial Biota
Air–Plant Equilibrium Partitioning
Illustrative Example 10.5: *Evaluating Air–Pasture Partitioning of PCBs*
Uptake of Organic Pollutants from Soil

10.5 Biomagnification
Defining Biomagnification
Biomagnification Along Aquatic Food Chains and Food Webs
Biomagnification Along Terrestrial Food Chains

10.6 Baseline Toxicity (Narcosis)
Quantitative Structure–Activity Relationships (QSARs) for Baseline
    Toxicity
Critical and Lethal Body Burdens
Illustrative Example 10.6: *Evaluating Lethal Body Burdens of Chlorinated
    Benzenes in Fish*

10.7 Questions and Problems

10.1   Introduction

The discovery in the 1960s and early 1970s that some organic chemicals such as DDT and PCBs were reconcentrated from the environment into organisms like birds and fish inspired many people's concern for our environment. Since such *bioaccumulation* of chemicals might eventually cause them to be transferred from the environment through food webs to higher organisms, including humans (Fig. 10.1), it became very important to understand how a chemical's properties affected these transfers.

Now we know that these accumulation processes may involve (1) direct partitioning between air and water and living media (e.g., grass, trees, phytoplankton, zooplankton), and/or (2) a more complicated sequence of transfer processes in that compounds are taken up with food and then transported internally to various parts of the organism. In many cases, phase partitioning equilibrium may not be established between certain compartments within an organism (e.g., liver, storage fats) and the environmental media in which the organism lives. This is particularly true for compounds that are metabolized by the organism. It is also true in situations in which the exchange with the environment is very slow. For example, chemical exchanges between the tissues of mammals or fish with the media that they use to breath (i.e., air and water, respectively) can be quite prolonged. As a consequence of the latter, persistent compounds may be present at significantly higher concentrations in certain tissues of higher organisms (e.g., in lipid phases) than one would predict by using a simple partitioning model between this tissue and the media surrounding the organism (e.g., water, air). In such situations, one often speaks of *biomagnification* of a given compound along a food chain.

We begin our discussion by first considering equilibrium partitioning of organic chemicals between defined biological materials and water or air (Section 10.2). This will enable us to recognize in which part(s) of a given organism a given chemical will tend to accumulate. Furthermore, such equilibrium considerations are very useful for assessing the potential of a given compound to bioaccumulate, an insight that is useful when we need to judge the wisdom of using particular chemicals for purposes that ultimately result in their release to the environment. Such equilibrium considerations are also important for evaluating the chemical gradients driving chemical transfers in real field situations where concentration data have been determined (Sections 10.3 and 10.4). This insight would allow us to identify environmental compartments such as contaminated sediments that are most needing cleanup. Then we will examine the process of biomagnification and how we might understand the changes in a chemical's concentration along a food chain (Section 10.5). Finally, in Section 10.6 we will learn how equilibrium partitioning considerations can be used to assess a compound's effectiveness for inducing *narcotic* effects in a given organism. This type of toxicity, which is also referred to as *nonspecific toxicity*, is caused primarily by partitioning of the compound into biological membranes, and is commonly also referred to as *baseline toxicity*. It tells us something about the minimum toxicity of a given compound toward a given organism.

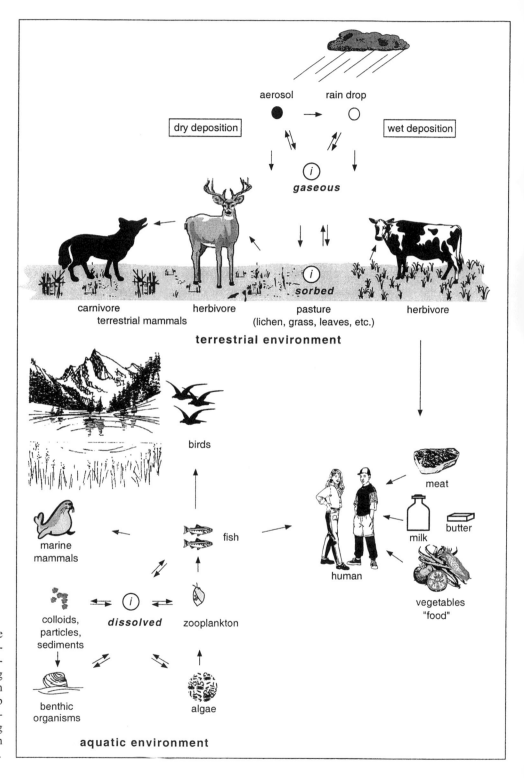

Figure 10.1 Examples of the transfer of a compound *i* from various media within the environment to organisms including humans by partitioning between contacting media and by food web transfers. These examples illustrate the complexity of anticipating the extent of bioaccumulation in aquatic and terrestrial food chains.

In summary, the major goal of this chapter is to enhance our understanding of the various factors that determine where and to what extent organic chemicals accumulate in living media. We should note that knowledge of the locally differing (internal) concentrations of a given organic chemical in organisms (e.g., at the site of enzyme inhibition or site in a tissue of hormone binding) is pivotal for any sound assessment of the chemical's (eco)toxicity (Sijm and Hermens, 2000). Of course, for this purpose it would be most advantageous to know the exact concentration of the compound at the site of toxic action, but presently this is not possible in most cases. Nonetheless, knowledge of average concentrations in the whole organism or in the major tissues into which chemicals tend to accumulate may be sufficient to answer questions about the likelihood of adverse effects resulting from accumulation of chemicals through food webs or for estimating the influence of large masses of biota (e.g., forests; Wania and McLachlan, 2001) on the overall fluxes of organic compounds in the environment.

10.2   Partitioning to Defined Biomedia

The Composition of Living Media

To anticipate the accumulation of xenobiotic organic chemicals in the tissues of organisms, we start by developing an awareness of the "chemical nature" of those living materials. By doing this, we hope to envision the intermolecular interactions that attract organic chemicals into organisms, much as we could see the interactions that control a specific compound's affinity for solvents of various structures (recall Tables 6.1 and 7.1). Recall that the compounds of interest to us are only about 1 nm in size; hence, as in the case of natural organic matter (Chapter 9), here we are interested in organic portions of organisms that are much larger than this (e.g., proteins or lipids) and not low-molecular-weight components like acetate or glucose that contribute only a few percent to organism biomass.

In addition to water and inorganic solids (salts dissolved in cell fluids, shells, and bones), organisms consist of a mix of organic substances. Some of these are macromolecules (e.g., globular proteins, cellulose). Some combine to form subcellular and tissue "structures" built with combinations of lipids, proteins, carbohydrates, and some specialized polymers like cutin or lignin (Fig. 10.2). These diverse organic materials cause organisms to have diverse macromolecular, cellular, and tissue portions that may be apolar, monopolar, and/or bipolar.

For animals, it is generally the protein fraction that predominates on the whole-organism basis, followed by carbohydrate components, and then a variable lipid content (Table 10.1). Since lipids serve both as ubiquitous structural components (e.g., phospholipids in membranes) and as energy reserves (especially triacylglycerides), the contributions of the total lipids may vary widely from organism to organism and tissue to tissue in the same organism. For example, on a *dry weight basis*, the lipid contents of phytoplankton typically range between about 10 and 30%, but this fraction may go as low as 1% (Shifrin and Chisholm, 1981; Stange and Swackhamer, 1994; Berglund et al., 2000). Similar ranges can be found in fish (Henderson and Tocker, 1987; Ewald, 1996; Berglund et al., 2000), zooplankton (Berglund et al., 2000), and in benthic invertebrates (Morrison et al., 1996; Cavaletto and Gardner, 1998). Note that in the case of benthic invertebrates (e.g., amphipods, shrimp), the lipid content may even exceed 40%, and may vary within one genus by up to a factor of five depending on the physiological condition

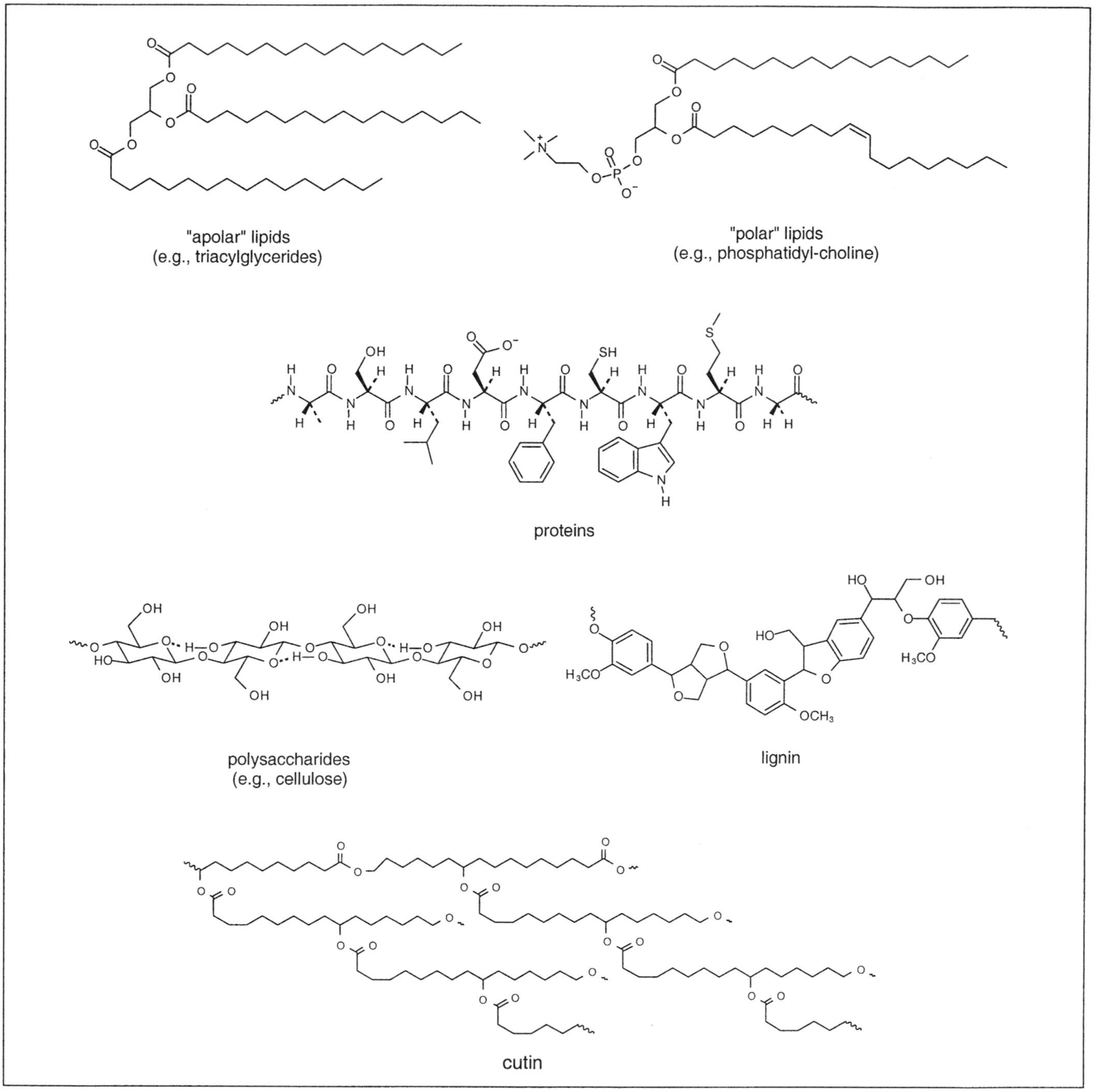

Figure 10.2 Examples of natural polymers relevant for sorption of organic pollutants in living media. Note that we consider triacylglycerides as primarily apolar although they contain monopolar (ester) groups.

(Cavaletto and Gardner, 1998). Likewise, the lipid contents of a given phytoplankton species may vary by a factor of two to three depending on its growth phase and/or environmental conditions (Shifrin and Chisholm, 1981; Stange and Swackhamer, 1994). Within a single organism, the composition can also vary widely as exhibited by lipids in caribou: (1) muscle has only 1 to 2% lipid, (2) liver has 4−13% lipid, and (3) fatty tissues have almost 80% lipid content. From our daily experience, we know that mammals including humans may exhibit quite different lipid contents, and that within one individual this lipid content may vary

Table 10.1 Chemical Composition of Some Organisms (dry-weight, ash-free basis)

Organism (reference) [a]	%Lipid	%Protein	%Carbohydrate	Other
bacterium (1)				
Escherichia coli	10	60	5	25% DNA/RNA
phytoplankton (2, 3)	20 ± 10	50 ± 15	30	
lichen (4)				
Cladonia spp.	2	3	94	
vascular land plants				
"grasses" (5, 6, 7)	0.5 – 2	15 – 25		1 – 5% cutin
deciduous leaves (oak, maple) (8, 9)	3	15	42	26% lignin
pine needles (8)	28	8	47	17% lignin
"wood" (10)	4	1	66	29% lignin
apples (11)	5	0	95	
almonds (11)	56	21	22	
spinach (11)	0	50	50	
aquatic invertebrates				
zooplankton (2)	15 – 35	60 – 70	10	
copepod (8)	10	65	25	
amphipod (*Pontoporeia hoyi*) (12)	9 – 46			
shrimp (*Mysis relicta*) (12)	10 – 41			
oyster (8)	12	55	33	
zebra mussel (13)	8 – 12	50 – 60	30 – 40	
polychaete (*Abarenicola pacifica*) (14)	12			
chironomid larvae (15)	6 – 13	65 – 71	21 – 23	
terrestrial invertebrates				
earthworm (*Lumbricus rubellus*) (16)	5			
aquatic vertebrates				
<u>fatty fish</u>				
trout filet (11)	30	70	<3	
lake trout (17)	6 – 18			
salmon (11, 18)	11			
<u>lean fish</u>				
cod filet (18)	0.7			
pike filet (18)	0.7			
terrestrial vertebrates				
deer meat (11)	10	90	<1	
beef (roast) (11)	20	80	<3	
caribou muscle (19)	5 – 12			

[a] (1) Neidhardt et al. (1990); (2) Parsons and Takahashi (1973); (3) Shifrin and Chisholm (1981); (4) Ahti and Hepburn (1967); (5) Huston et al. (2001); (6) Böhme et al. (1999); (7) Tolls and McLachland (1994); (8) Hunt (1979); (9) Aber and Martin (1999); (10) Thompson (1996); (11) Rombauer et al. (1997); (12) Cavaletto and Gardner (1998); (13) Napela et al. (1993); (14) Weston et al. (2000); (15) Beattie (1978); (16) Ma et al. (1998); (17) Thomann and Connolly (1984); (18) Ewald (1996); (19) Kelly and Gobas (2001).

substantially with time. Finally, we should note that terrestrial grasses and tree leaves apparently have a relatively low lipid content of less than 5% (Aber and Martin, 1999; Böhme et al., 1999).

Lipids are, of course, not the only variable component in living media. In animals, proteins make up the largest portion of the total body mass, but they may be secondary contributors in certain tissues (Table 10.1). In plants, protein is more variable. On a dry-weight basis in phytoplankton, protein contents may range between 30% and 60% (Shifrin and Chisholm, 1981), whereas they occur at about 20% of grasses and leaves (10 – 25%, Huston, 1992; Aber and Martin, 1999), are highly variable contributors in fruits and nuts (Rombauer et al., 1997), and are virtually absent in wood. On the other hand, the wood of vascular land plants consists primarily of lignin (~ 1/3) and carbohydrates, in particular, cellulose and hemicellulose (~ 2/3; Thompson, 1996). Finally, as realized by many people who watch their diet carefully, fruits like apples are rich in carbohydrates, nuts like almonds have a lot of fat, and vegetables like spinach are virtually fat-free.

When considering that organic pollutants tend to partition very differently into the various organic materials of which living media are made up, it is, therefore, not surprising that concentrations of compounds expressed per total weight or volume may vary significantly for "organisms" exposed to the same environment. Hence, in order to understand how and to what extent organic chemicals accumulate in living media, we first need to inspect how well the various biological "solvents" may accumulate such solutes.

Equilibrium Partitioning to Specific Types of Organic Phases Found in Organisms

Now we have seen that the materials that may accommodate organic pollutants in organisms include a mix of proteins, polysaccharides, lipids, lignin, and other polymers like cutin. In order to assess how much each of these organic phases contributes to the uptake of a given compound by a particular organism, one would like to know how the compound partitions between environmental media and these various defined types of organic phases. Unfortunately, such experimental partitioning data are rather scarce. Furthermore, the available data have been determined for surrogate organic phases that may not accurately mimic the actual organic materials present in the living media. For example, liposomes, the microscopic vesicles constructed by aggregating polar lipids in the laboratory and used to investigate the lipid–water partitioning of organic pollutants, often exist in a somewhat different aggregate state as compared to their presence in biological membranes. Likewise the use of a single triacylglyceride, triolein, or a single protein, bovine serum albumin, to represent the behavior of structurally diverse fats and proteins is surely oversimplified. Also, we note that for all phases except for the lipids, the available data have been collected for relatively small and/or polar compounds. Thus, prediction of the partitioning behavior of large, hydrophobic compounds such as PCBs, polychlorinated dibenzodioxins, PAHs, and so forth between these phases and water or air is difficult. Nevertheless, from the available data we can get at least a semiquantitative indication as to which phases have to be considered in a given case.

Table 10.2 Organic Phase–Water Partition Coefficients of Some Organic Compounds for Various Organic Phases: Octanol (o), Triacylglycerides (tag), Liposomes (lips), Proteins (prot), Cutin (cut), and Lignin (lig). (All values except K_{iow} in $L \cdot kg^{-1}$ org. phase for 20–25°C if not otherwise stated. Note that the densities of the organic phases are close to 1 kg org. phase $\cdot L^{-1}$ org. phase.)

Compound	log K_{iow}	log K_{itagw} [a]	log K_{ilipsw} [b]	log K_{iprotw}	log K_{icutw} [c]	log K_{iligw} [d]
Apolar and Weakly Monopolar Compounds						
Toluene	2.69	2.77			2.60	1.80
1,2-Dimethylbenzene	3.16	3.25	2.98		2.81	2.08
Chlorobenzene	2.78	2.97	2.90		2.70	
1,3,5-Trichlorobenzene	4.19	4.36	4.32	2.14 [e]		
1,2,3,4-Tetrachlorobenzene	4.64	4.68		2.32 [e]		
Pentachlorobenzene	5.18	5.27	5.26	2.64 [e]		
Naphthalene	3.33			3.21 [f], 2.08 [g]		
Perylene	6.25			5.23 [f]		
Monopolar and Bipolar Compounds						
4-Chlorophenol	2.42		2.94 [h]	2.30 [i], 1.60 [g]		
2,4-Dichlorophenol	3.09		3.55 [h]			2.30
Aniline	0.90	0.91	1.63	1.30 [f]		
3-Nitroaniline	1.37		2.17	0.70 [g]		
Nitrobenzene	1.85	2.15	2.01	1.88 [f]		
Chlorothion [k]	3.65		3.83 [l]	2.92 [e]		

[a] Triolein–water system (Chiou, 1985). [b] L-α-dimyristoylphosphatidylcholine liposome–water system a 35°C (Vaes et al., 1997). [c] Welke et al. (1998). [d] Mackay and Gschwend (2000). [e] Protein fraction of a cytosol isolate from a pond snail (*Symnae stagnalis*) (Legierse, 1998). [f] Bovine serum albumin (BSA) (Helmer et al., 1965; Backhus and Gschwend, 1990). [g] Bovine hemoglobin (BH) (Kiehs et al., 1966). [h] L-α-dioleylphosphatidylcholine liposome–water system (Escher and Schwarzenbach, 1996). [i] Vaes et al. (1996). [k] 3-chloro-4-nitrophenyldimethyl-phosphorothionate. [l] Legierse (1998).

We begin by examining some organic phase–water partition coefficients of a series of compounds for some defined organic phases mimicking living media constituents: octanol, triolein, liposomes, proteins, cutin, and lignin (Table 10.2). The main structural features of these absorbents (Fig. 10.2) allow us to anticipate whether they provide a primarily apolar molecular environment (e.g., triolein), a monopolar system (e.g., cutin), or a bipolar medium (proteins, lignin). Note that the quantitatively prominent carbohydrates (e.g., cellulose, chitin) have a very low tendency to accumulate organic chemicals from water (data not shown; Garbarini and Lion, 1986; Xing et al., 1996). This is not too surprising considering the presence of hydroxyl moieties throughout the structures of these polymers (Fig. 10.2). This structural feature causes polysaccharides to interact very well with water all through their structure. Thus apolar and monopolar compounds cannot outcompete with the solvent for associations with these thoroughly bipolar polymers. In addition, some abundant carbohydrates such as cellulose and chitin tend to be crystalline in organisms; this solid state condition also discourages uptake of partitioning compounds.

Hence, we consider carbohydrates to be of minor importance for the accumulation of organic chemicals.

Compared to carbohydrates, lignin and cutin, important polymers in terrestrial plants, show a higher affinity for many organic pollutants. Small apolar and weakly monopolar molecules (e.g., chlorobenzene, toluene, 1,2-dimethylbenzene) have been shown to exhibit only a slightly smaller affinity to cuticular polymer matrices than to a solvent like octanol (Table 10.2; Welke et al., 1998). Thus, the cuticular coatings on leaves are considered to be an important absorber for apolar and weakly polar hydrophobic compounds. This can be understood by noting the apolar/monopolar nature (i.e., long alkyl chains and ester groups) of cutin (Fig. 10.2). In the case of lignin, partition coefficients for such apolar substances are somewhat smaller (i.e., up to an order of magnitude; Table 10.2). Although the data are very limited, lignin–water partition coefficients appear to be more comparable to K_{ioc} values, with K_{iligw} values in the order of $K_{iow}^{0.7}$ for the small aromatic hydrocarbons and chlorinated phenols examined to date (Severtson and Banerjee, 1996; MacKay and Gschwend, 2000.) This is actually not too surprising since soil organic matter is thought to derive in large part from preservation of the lignin portions of plants and it has been observed that it exhibits many lignin-like structures (Fig. 9.10).

A somewhat more difficult task is to assess the affinity of proteins with respect to accumulation of chemicals in organisms. As noted above, proteins are ubiquitous and quantitatively prominent macromolecules in organisms. Although they contain a rather large number of polar groups, certain proteins may also exhibit substantial hydrophobic character. This is due to their inclusion of "hydrophobic amino acids" in the polymeric chains. While all the amino acids contribute to chains with repeating units (–NH–CHR–C(O)O–), some of the "R groups" include apolar moieties like –CH₃, –CH(CH₃)₂, –CH₂–CH(CH₃)₂, and –CH₂–(C₆H₅) (Fig. 10.2). Moreover, many protein chains fold back on themselves, enabling the hydrophobic R groups to be arranged so as to occur on the inside of the resultant coiled macromolecules. This results in the formation of nonaqueous "particles"; for example, the globular protein (see sketch in the margin), serum albumin (molecular mass of 65,000 u), that occurs in the blood plasma forms an elliptical particle whose long axis is about 13 nm and whose short axis is about 3 nm in its native state.

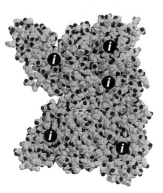

Human serum albumine

Due to differences in amino acid composition (i.e., R groups) and macromolecular configuration, proteins vary considerably in their ability to accommodate organic solutes. For example, the protein–water partition coefficient of the bipolar compound, 4-chlorophenol, quantifying this compound's association with bovine serum albumin (BSA) is a factor of 5 greater than its corresponding coefficient for bovine hemoglobin (BH). For the weakly monopolar compound, naphthalene, the difference is about a factor of 10 (Table 10.2), and a still smaller protein, myoglobin (ca. 16,000 u), does not sorb any measurable amount of naphthalene (Kiehs et al., 1966.) Thus, the formation of hydrophobic domains within these protein biopolymers appears to be a key feature allowing them to serve as sorbents of apolar and monopolar organic compounds. This would imply that so-called beta proteins which occur naturally in extended configurations (e.g., protein in hair, collagen) would not absorb apolar and monopolar compounds well from water. Therefore, it is not

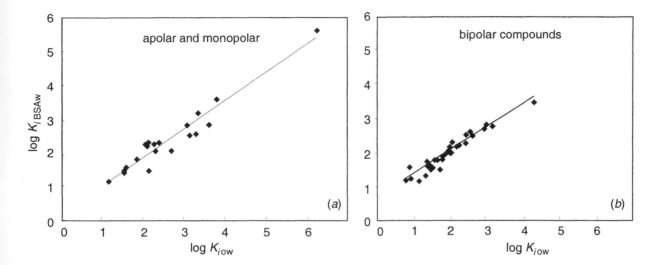

Figure 10.3 Correlations of the K_{iBSAw} values with the corresponding K_{iow} values for (*a*) apolar and monopolar compounds and (*b*) bipolar compounds (data from Helmer et al., 1965; Backhus and Gschwend, 1990; Vaes et al., 1996; and Yuan et al., 1999).

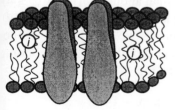

possible to establish a single partition coefficient for each compound of interest interacting with the diverse mixture of proteins found in organisms.

With these caveats in mind, we briefly examine the K_{iprotw} values available for the globular protein, BSA. Separating the compounds that have been tested into apolar/monopolar and bipolar subsets, we find that each group sorbs to BSA as a function of the chemical's hydrophobicity as measured by the corresponding K_{iow}'s (Figure 10.3). Numerous other observations of protein-related phenomena (e.g., enzyme inhibition) have also been found to exhibit correlations between the solute concentrations in the medium required to disrupt a particular target protein's actions and the hydrophocities of the added chemicals as quantified by their K_{iow} values (Hansch et al., 1995). These data also suggest that solutes partition between many diverse globular proteins and water in a manner that is proportional to the solute's K_{iow}. The LFERs relating the logarithms of protein–water partition coefficients with the corresponding octanol–water partition constants typically exhibit slopes of about 0.7 (Hansch and Leo, 1995):

$$\log K_{iprotw} = (\sim 0.7) \log K_{iow} \qquad (10\text{-}1)$$

Since such protein–water partition coefficients are less than the corresponding lipid–water coefficients (see below), proteins will play a significant role in the overall accumulation of the compounds only in organisms and tissues with high protein and low lipid contents.

Now we consider the partitioning of organic compounds to lipids. Lipids also encompass a diverse group including apolar and polar materials. Examples of lipids that could form a primarily apolar molecular environment are the triacylglycerides (Fig. 10.2). Such chemical structures are common in fat storage in organisms and hence their quantitative presence in organisms is quite variable. The triacylglycerides, triolein and tricaprylin, have been used as laboratory surrogates for mimicking the "fat" in organisms (Chiou, 1985, Bahadur et al., 1999). Examples of polar lipids are the phosphatidylcholines (Fig. 10.2). These are found primarily in biological membranes (see sketch in the margin); hence this component of the lipid

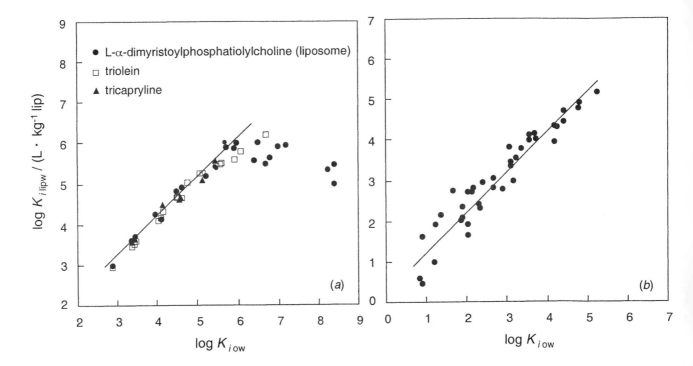

Figure 10.4 (*a*) Lipid–water partition coefficients of a series of chlorinated and brominated benzenes, biphenyls, naphthalenes, and dibenzodioxins plotted versus their octanol–water partition constants. The lipid phases include L–α–dimyristoylphosphatidyl-choline (•, 2 × C_{14} fatty acid; Gobas et al., 1988), triolein (□, 3 × C_{18} fatty acid; Chiou, 1985), and tricapry-line (▲, 3 × C_8 fatty acid; Bahadur et al., 1999). The correlation line is from fit of compounds with log K_{iow} ≤ 6 (see Eq. 10-2). (*b*) Lipid–water partition coefficients of a diverse group of monopolar and bipolar compounds plotted versus their octanol–water partition constants. The lipid phases are vesicles of L–α-dimyristoyl-phosphatidylcholine (2 × C_{14} fatty acid; Vaes et al., 1997) and L–α–dioleyl-phosphatidylcholine (2 × C_{18} fatty acid; Escher and Schwarzenbach, 2000).

pool always exists in organisms at a relatively consistent level. Vesicles constructed of such polar lipids (liposomes) have been used as surrogates for investigating membrane–water partitioning (Gobas et al., 1988; Escher and Schwarzenbach 1996; van Wezel et al., 1996; Rowe et al., 1997; Vaes et al., 1997; Escher et al. 2000). Since all lipids include hydrocarbon portions in their structures, it is not surprising that hydrophobic chemicals partition very favorably from water into such phases, much as they do into apolar organic solvents such as hexane or hexadecane (see Chapter 7). For apolar compounds that may undergo only vdW interactions, and that exhibit octanol–water partition constants smaller than about 10^6 (log K_{iow} ≤ 6) very similar K_{ilipw} values are found for both nonpolar and polar lipids (Fig. 10.4*a*) Thus, as we did for natural organic matter (Chapter 9), we may derive a one-parameter LFER relating lipid–water partition coefficients with the corresponding K_{iow} values for apolar compounds with log K_{iow} ≤ 6 (see Fig. 10.4*a*):

$$\log K_{ilipw} = 0.91 \log K_{iow} + 0.50 \ (R^2 = 0.98; N = 36)$$

or:

$$K_{ilipw} = 3.2 \times K_{iow}^{0.91}$$

(10-2)

Note that since the specific weight of lipids is close to 1 kg lipid·L^{-1} lipid, the estimated K_{ilipw} value can be expressed either in units of L·kg^{-1} lipid or "dimensionless" as L·L^{-1} lipid.

For apolar compounds with larger K_{iow} values and particularly in the case of the liposomes, significantly smaller K_{ilipw} values than predicted from Eq. 10-2 have been reported (Fig. 10.4*a*). This change in the observed trend may be due to a difficulty in accommodating large molecules in the structured and size-limited lipid bilayers of membranes (Opperhuizen et al., 1985). However, there may also be artifacts in the measurements. Such artifacts may involve slow uptake kinetics (i.e., observed distributions may not reflect equilibrium). Or the experiments may be complicated by colloidal/complexation associations of such highly hydrophobic compounds in the aqueous solution (see Chapter 9), causing higher aqueous phase abundances in the experiments than the real concentration of the freely dissolved molecules. (We will take this up again in Illustrative Example 10.1.)

Recalling our discussions of organic solvent–water partitioning (Chapter 7), we should not be too surprised that polar compounds partition differently into polar lipids than they do into nonpolar lipids. For example, aniline partitions about a factor of five times more favorably into the dimyristoylphosphatidylcholine vesicle as compared to the primarily apolar triolein phase; in contrast, the apolar compound, chlorobenzene, exhibits about the same partition coefficients (Table 10.2). Furthermore, the log $K_{i\text{lipw}}$ values of various monopolar and bipolar compounds observed using two different polar lipids, when examined versus their corresponding log $K_{i\text{ow}}$ values, exhibit a significant scatter (Fig. 10-4b). For any particular log $K_{i\text{ow}}$ value, this data indicate we could estimate log $K_{i\text{lipw}}$ values to within about a factor of 3. This scatter simply indicates that there are variable polar interactions of the compounds in the lipid phases as compared to octanol. As already addressed in Chapters 7 and 9 for organic solvents and natural organic matter, in the case of polar compounds we may expect to find LFERs of the type Eq. 10-2 only for confined sets of structurally closely related compounds. We will get back to this issue in Section 10.6 where we will discuss nonspecific membrane toxicity.

A Model to Estimate Equilibrium Partitioning to Whole Organisms

Accumulation of organic chemicals in organisms is a dynamic process that may involve several uptake, elimination, and depuration routes acting simultaneously. For many xenobiotic compounds, the physical uptake and depuration processes can occur "passively," that is without an organism's explicit effort to transport these substances. Often the biological membranes, designed to keep the fluids and their contents on the inside and the outside separate from one another, prove to be incapable of excluding nonionic organic chemicals. To evaluate the resultant unintended presence of xenobiotic chemicals in organisms, it is useful then to estimate the concentration that would be established in the organism if that organism achieved partitioning equilibrium with its surroundings (e.g., water, air, soil, sediment).

Recognizing that organisms are heterogeneous mixtures of diverse organic phases, here we calculate this equilibrium condition by assuming that (1) each organic phase acts independently and comes to an equilibrium with all other biological phases, and (2) the total organism accumulation of the chemical can be estimated as the sum of uptakes into each part. Hence, as we have done for estimation of the overall sorption of organic compounds by natural solids (Eq. 9-16, Section 9.2), we consider a whole organism–environmental medium mixed partition coefficient, $K_{i\text{bio}}$:

$$K_{i\text{bio}} = \frac{f_{\text{lip}} \cdot C^*_{i\text{lip}} + f_{\text{prot}} \cdot C^*_{i\text{prot}} + f_{\text{lig}} \cdot C^*_{i\text{lig}} + f_{\text{cut}} \cdot C^*_{i\text{cut}} + \ldots}{C_{i\text{med}}} \quad (10\text{-}3)$$

where $f_{\text{org.phase}}$ is the fraction of organism made up of the corresponding phase [lipid (lip), protein (prot), lignin (lig), cutin (cut), etc.], and is expressed, for example, as (kg org. phase·kg^{-1} dry organism)

$C_{\text{org.phase}}^{*}$ is the concentration of the compound in a given organic phase *at equilibrium with the concentration in the environmental medium* (therefore the superscript *), and is expressed, for example, as (mol·kg^{-1} dry phase)

$C_{i\text{med}}$ is the concentration of the compound in the environmental medium (e.g., water, air, soil) considered to be in equilibrium with the organism, and is expressed, for example, as (mol·L^{-1} med or mol·kg^{-1} med)

Hence, the units of $K_{i\text{bio}}$ are, for example, (L med·kg^{-1} dry organism) if med is air or water. Whenever possible, one should express the concentration in the organism on a dry weight basis since this will assist us in using our chemical intuition; unfortunately we need to be aware that throughout the literature one finds experimentally reported concentrations on a wet-weight (w.w.) or fresh weight (f.w.) basis.

Noting that the equilibrium concentration in each particular organic phase represented in the numerator of Eq. 10-3 can be divided by the concentration in the environmental medium, we can modify the equation using the respective partition coefficient between that organic phase and the environmental medium, $K_{i\text{org.phase med}}$:

$$K_{i\text{bio}} = f_{\text{lip}} \cdot K_{i\text{lipmed}} + f_{\text{prot}} \cdot K_{i\text{protmed}} + f_{\text{lig}} \cdot K_{i\text{ligmed}} + f_{\text{cut}} \cdot K_{i\text{cutmed}} + \dots \quad (10\text{-}4)$$

Hence, if the composition of the organism (i.e., f_{phase} values) and if the various partition coefficients in Eq. 10-4 can be estimated, we can calculate an approximate value for the $K_{i\text{bio}}$. As we saw in discussing organism compositions (Table 10.1) and defined media–water partition coefficients (Table 10.2), the uncertainties in the inputs to Eq. 10-4 imply that such calculations can be expected to provide only an order of magnitude estimate.

In the literature, the estimated chemical's equilibrium concentration, $C_{i\text{bio}}^{*}$, in the organism (i.e., the numerator of Eq. 10-3) is called the *theoretical bioaccumulation potential*, TBP_i, and is given by:

$$C_{i\text{bio}}^{*} = TBP_i \text{ (e.g., in mol·kg}^{-1}\text{ dry organism)} = K_{i\text{bio}} \cdot C_{i\text{med}} \quad (10\text{-}5)$$

Often we are interested in apolar or weakly monopolar pollutants (PCBs, chlorinated solvents, PAHs) and in organisms/tissues that contain significant lipid content (>5% on a dry weight basis). Also, investigators sometimes specially sample the fatty tissues (e.g., in some animals or humans). In such cases, the terms $f_{\text{lip}} C_{\text{lip}}^{*}$ or $f_{\text{lip}} K_{i\text{lipmed}}$ dominate in Eqs. 10-3 and 10-4 (see Illustrative Examples 10.1 to 10.3). Consequently it is reasonable to assume that the measured concentrations chiefly reflect the compounds present in the lipid phase and the concentrations can be normalized to the lipid content of the organism. Thus, the lipid normalized, $K_{i\text{bio,lip}}$, is simply given by:

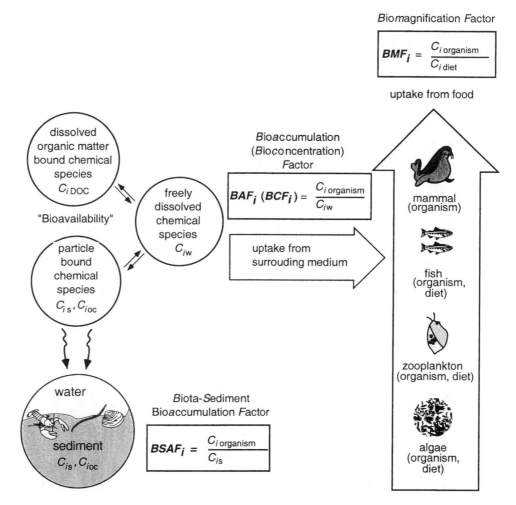

Figure 10.5 Terms and parameters frequently used to describe accumulation of chemicals in aquatic organisms. Note that the term "bioaccumulation" (BAF_i) is used to describe the total accumulation by all possible routes (e.g., passive uptake, intake by food and digestion, etc.). The term bioconcentration (BCF_i) is sometimes used if uptake occurs only from the dissolved phase. A very similar approach is used in terrestrial systems just by exchanging the environmental media, water and sediment, with air and soil respectively.

$$K_{i\text{bio,lip}} = K_{i\text{lipmed}} \qquad (10\text{-}6)$$

and correspondingly:

$$C^*_{i\text{bio,lip}} = TBP_{i\text{lip}} \text{ (e.g., in mol} \cdot \text{kg}^{-1} \text{ lip)} = K_{i\text{lipmed}} \cdot C_{i\text{med}} \qquad (10\text{-}7)$$

Parameters Used to Describe Experimental Bioaccumulation Data

The various parameters (e.g., BCF_i, BMF_i, $BSAF_i$) introduced in Fig. 10.5 for the aquatic environment represent various ways of relating the *actual* concentration determined for a given compound in an organism or organism compartment (e.g., in the lipids) to the compound's concentration in the medium from which the compound is thought to be primarily taken up. Of course, a given compound can be taken up simultaneously by various routes. For example, animals living at the sediment–water interface can experience passive uptake of dissolved compound from water (bioconcentration) as well as uptake of the sorbed molecules on sediment particles or present in their diet (biomagnification). Thus to describe the net result of all uptake and elimination processes taking place, one simplifies and defines a bioaccumulation factor, BAF_i, which relates the *actual* concentration in the organism to the *actual* concentration in one medium of the environment in which the organism primarily lives:

$$BAF_i = \frac{C_{i\text{organism}}}{C_{i\text{med}}} \tag{10-8}$$

In aquatic systems, $C_{i\text{med}}$ (i.e., C_{iw}) denotes the truly dissolved concentration (e.g., not including the fraction sorbed to DOM); in the terrestrial environment this may be the truly gaseous concentration (i.e., C_{ia}, not including any fraction in aerosols). However, the total filter-passing concentrations are widely used in the literature and these may not collect only the truly dissolved or truly vapor-phase species. This is particularly a problem for very hydrophobic and/or nonvolatile compounds that have a strong tendency to associate with particles, since some of these can pass filters. Failure to use the right value of $C_{i\text{med}}$ causes the experimentally determined BAF_i to be erroneously low (see Illustrative Example 10.1). For organisms living at the sediment–water interface, $C_{i\text{med}}$ is often set equal to C_{is}, the concentration in the particulate phase, that is, $BAF_i = BSAF_i$ (see Fig. 10.5). This is, of course, somewhat arbitrary since these organisms live both in the sediment and in the water column, and these two compartments are usually not in equilibrium with each other.

In the following we will use these concepts to evaluate some experimental data and to make some estimates of the tendency of a given chemical to accumulate in a given organism. We start out with examples from the aquatic environment, and then consider the terrestrial system.

Illustrative Example 10.1   **Evaluating Bioaccumulation from a Colloid-Containing Aqueous Solution**

Problem

Stange and Swackhamer (1994) studied the accumulation of a large number of PCB congeners in three phytoplankton species including *Anabaena sp.* in laboratory cultures. The apparent BAF_i (or BCF_i) values determined after 12 days of exposure for *Anabaena sp.* were 2.5×10^4 L·kg^{-1} d.w. for PCB52 (2,2',5,5'-tetrachlorobiphenyl) and 7.3×10^4 L·kg^{-1} d.w. for PCB180 (2,2', 3,4,4',5,5'-heptachlorobiphenyl). The lipid content of *Anabaena sp.* after the exposure period was 5.3%, that is, $f_{\text{lip}} = 0.053$ kg lip·kg^{-1} d.w. The concentration of dissolved organic carbon (DOC) in the culture was 74 mg C·L^{-1}.

Estimate the $K_{i\text{bio}}$ values (Eq. 10-4) for PCB52 and PCB180 for *Anabaena sp.* and compare them to the apparent BAF_i values given above. Assume that with a rather high lipid content of 5.3%, the role of proteins as sorbents in *Anabaena sp.* can be neglected for apolar compounds such as PCBs (see Illustrative Example 10.2, pentachlorobenzene). Could any discrepancy found between $K_{i\text{bio}}$ and BAF_i be explained by the fact that the compounds associated with the organic colloids present in the culture medium?

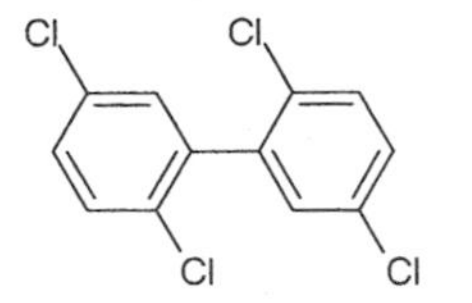

2,2',5,5'-tetrachlorobiphenyl
(PCB52)

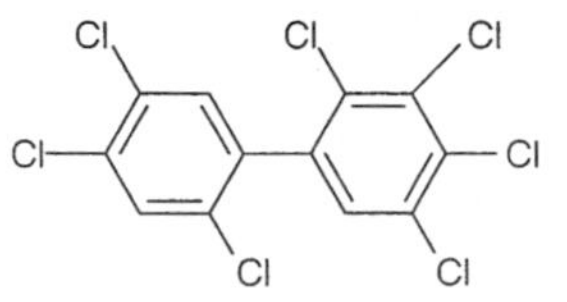

2,2',3,4,4',5,5'-heptachlorobiphenyl
(PCB 180)

Answer (a)

Estimate the K_{ilipw} values of PCB52 and PCB180 from their K_{iow} values (Appendix C) using Eq. 10-2:

PCB52: $K_{ilipw} = 1.2 \times 10^6$ L·kg^{-1} lip (see Illustrative Example 10.3)

PCB180: $K_{ilipw} = 1.6 \times 10^7$ L· kg^{-1} lip

Calculate the corresponding K_{ibio} values by multiplying K_{ilipw} by f_{lip} (Eq. 10-4):

PCB52: $K_{ibio} = (1.2 \times 10^6)(0.053) = 6.4 \times 10^4$ L·kg^{-1} d.w.

PCB180: $K_{ibio} = (1.6 \times 10^7)(0.053) = 8.5 \times 10^5$ L·kg^{-1} d.w.

Hence, the estimated K_{ibio} values are about 2.6 and 12 times higher than the actual measured bioaccumulation factors. The result for PCB180 appears substantially outside expected estimation errors. If all of the discrepancies were due to the speciation of the compounds in the culture media [and not because of kinetic effects or due to the large size of the molecules (see Fig. 10.4a)], this would mean that the fraction in truly dissolved form f_{iw} (see Chapter 9, Illustrative Example 9.5) would have to be 0.39 (1/2.6) for PCB52 and 0.086 (1/12) for PCB180. From these values and the DOC concentration given in the problem statement, calculate the K_{iDOC} values that would be required for the two compounds (see Illustrative Example 9.5, Eq. 1):

$$K_{iDOC} = \frac{(1 - f_{iw})}{f_{iw}[\text{DOC}]} \tag{1}$$

Insertion of the appropriate values into Eq. 1 yields:

PCB52: $K_{iDOC} = (0.61)/[(0.39)(7.4 \times 10^{-5}$ kg DOC·L$^{-1})]$ $= 2.1 \times 10^4$ L·kg^{-1} DOC

PCB180: $K_{iDOC} = (0.914)/[(0.086)(7.4 \times 10^{-5}$ kg DOC·L$^{-1})] = 1.5 \times 10^5$ L·kg^{-1} DOC

These values are similar to the average K_{ioc} values predicted from Eq. 9-26a in Table 9.2 for soil and sediment organic matter:

PCB52: $K_{ioc} = 4.7 \times 10^4$ L·kg^{-1} oc

PCB180: $K_{ioc} = 3.9 \times 10^5$ L·kg^{-1} oc

Hence, in principle, we could explain the discrepancies between predicted and observed BAF_i values by the reduction of the bioavailability of the compounds caused by sorption to colloidal organic matter present in the culture media. We should note, however, that depending on the nature of the "dissolved" organic matter (i.e., molecular size distribution, aromaticity, polarity, etc.), the K_{iDOC} value of a compound may vary considerably (see Section 9.4).

Illustrative Example 10.2   **Estimating Equilibrium Bioaccumulation Factors from Water**

Problem

Legierse (1998) investigated the uptake of chlorothion (for structure see margin) and of a series of chlorinated benzenes from water into the pond snail *Lymnea stagnalis*. On a wet-weight (w.w.) basis, the pond snail contained 0.9% total lipids (0.4% polar, 0.5% apolar), 2.8% proteins, and 96% polar components including primarily water. Estimate the equilibrium bioaccumulation factors (K_{ibio}) at 20°C for chlorothion and pentachlorobenzene and compare them to the experimental BAF_i values given in the margin. Note that in this case, we would also refer to the observed concentration ratio as a bioconcentration factor (BCF_i), because we consider uptake only from water and not from diet or ingestion of particles (see Fig. 10.5).

Answer

Assume that only the lipid and protein fractions are important for the bioaccumulation of the two compounds in the snail. Further assume that the lipid–water partition coefficients of the compounds are similar for polar and nonpolar lipids, i.e., $K_{itagw} \cong K_{ilipsw}$ (see Table 10.2). Note that this assumption is reasonable for pentachlorobenzene, but that for the polar chlorothion you may somewhat overestimate the overall lipid–water partition coefficient when using K_{ilipsw}. The K_{iprotw} values observed by Legierse (1998) were 830 and 450 $L \cdot kg^{-1}$ protein, while very rough estimates based on 0.7 $\log K_{iow}$ would yield values of 360 and 4000 $L \cdot kg^{-1}$ protein, respectively. Using these values of K_{ilipsw} and K_{iprotw} for the two compounds together with $f_{lip} = 0.009$ kg lip $\cdot kg^{-1}$ w.w. and $f_{prot} = 0.028$ kg prot $\cdot kg^{-1}$ w.w. in Eq. 10-4 yields:

chlorothion:

$$K_{ibio} = (0.009)(6800) + (0.028)(360 \text{ to } 830)$$

$$= 61 + (10 \text{ to } 23) \, L \cdot kg^{-1} \text{ w.w.}$$

$$= 70 \text{ to } 80 \, L \cdot kg^{-1} \text{ w.w.}$$

pentachlorobenzene:

$$K_{ibio} = (0.009)(182'000) + (0.028)(450 \text{ to } 4000)$$

$$= 1638 + (12 \text{ to } 110) \, L \cdot kg^{-1} \text{ w.w.}$$

$$= 1650 \text{ to } 1750 \, L \cdot kg \text{ w.w.}$$

Hence, the estimated K_{ibio} values are within a factor of three (chlorothion) and two (pentachlorobenzene), respectively, of the experimental BAF_i values. But recall that when dealing with living media, due to the rather large uncertainties in parameter estimation, any predicted K_{ibio} values have to be considered good to within factors of 2 to 3. Finally, note that in the case of the polar chlorothion, the contribution of the protein fraction to the overall accumulation may be significant (about 20%), whereas for the apolar pentachlorobenzene, this contribution can be neglected.

chlorothion

$BAF_{i \text{ exp.}} = 30 \, L \cdot kg^{-1}$ w.w

pentachlorobenzene

$BAF_{i \text{ exp.}} = 900 \, L \cdot kg^{-1}$ w.w

Illustrative Example 10.3

Estimating Equilibrium Bioaccumulation Factors from Air

Problem

Kömp and McLachlan (1997b) determined plant–air equilibrium partition coefficients, K_{ipa} of a series of PCB congeners for several plants, including ryegrass. From their data, for PCB52 (2,2',5,5'-tetrachlorobiphenyl) a ryegrass–air partition coefficient of 3×10^6 L·kg^{-1} d.w. at 25°C can be derived. Note that air–plant partitioning is highly temperature dependent. We will address that issue in Section 10.4. Assuming ryegrass has 2% lipid, 20% protein, and 4% cutin (Tolls and McLachlan, 1994), estimate K_{ipa} for PCB52 for this grass at 25°C.

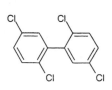

2,2',5,5'-tetrachlorobiphenyl
(PCB52)

$\log K_{iow} = 6.11$
$\log K_{iaw} = -1.70$

Answer

First, estimate the pure phase–*water* partition coefficients:

lipid (Eq. 10-2): $\log K_{ilipw} = 0.91 (6.11) + 0.5 = 6.1$

protein (Eq. 10-1): $\log K_{iprotw} \sim 0.7 (6.11) = 4.3$

cutin: $\log K_{icutw} \sim \log K_{iow} = 6.1$

Now utilize Eq. 10-4:

$$K_{ibio} = (0.02)(10^{6.1}) + (0.2)(10^{4.3}) + (0.04)(10^{6.1})$$
$$= 25000 + 4000 + 50000$$
$$= 79000 \text{ L water·kg}^{-1} \text{ d.w.}$$

To relate this result to partitioning from air, recall that:

$$K_{ipa} = \frac{K_{ipw}}{K_{iaw}}$$

Using the K_{iaw} value given, you find:

$$K_{ipa} = \frac{79000}{0.02} = 4.0 \times 10^6 \text{ L air·kg}^{-1} \text{ d.w.}$$

This matches the reported value quite well.

10.3 Bioaccumulation in Aquatic Systems

Bioaccumulation as a Dynamic Process

In the real world, equilibrium partitioning between an organism and its surroundings may not be achieved, even if a compound is not metabolized by the organism. Therefore, the observed BAF_i value (Eq. 10-8) may differ from the theoretical equilibrium expectation, K_{ibio} (Eq. 10-4). In fact, because the accumulation of a given chemical may depend on several different processes occurring at the same time (see example "fish" in Fig. 10.6), BAF_i may change continuously with time for

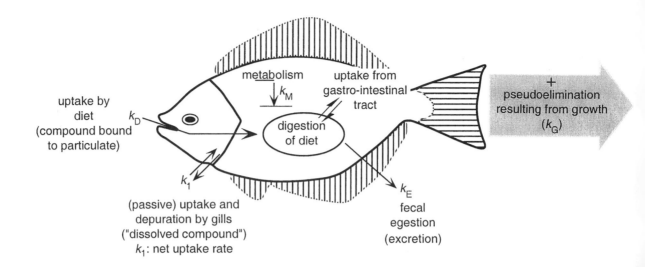

Figure 10.6 Illustration of some processes determining bioaccumulation of a chemical in a fish. The various k values can be formulated as first-order rate constants for description of the kinetics dependent on the physiology and behavior of the fish (for more advanced models, see Gobas and Morrison, 2000).

a given organism. Hence, in a given natural system (e.g., a lake), even two individuals of the same species (e.g., two trouts) may exhibit different BAF_i values.

It is useful to briefly consider a simple conceptional model that considers simultaneous inputs and outputs of a compound in an organism, the one-box approach (see Section 12.4 for a general discussion of one-box models). In this approach we assume that the organism (i.e., the fish) is a well-mixed reactor (which, of course, it is not), and we define all processes as first-order reactions. The temporal change in concentration of a given compound i in the fish, $C_{i\text{fish}}$, can then be described simply by:

$$dC_{i\text{fish}}/dt =$$

$$\underbrace{k_1(K_{i\text{fishw}}C_{iw} - C_{i\text{fish}})}_{\text{gill exchange}} + \underbrace{k_D(K_{i\text{fish diet}}C_{\text{diet}})}_{\text{uptake with food}} - \underbrace{k_E(K_{i\text{fish excreta}}C_{i\text{excreta}})}_{\text{excretion}} - \underbrace{k_M C_{i\text{fish}}}_{\text{metabolism}} - \underbrace{k_G C_{i\text{fish}}}_{\text{growth}} \quad (10\text{-}9)$$

where k_1, k_D, k_E, k_M, and k_G are the first-order rate constants (dim$\cdot$T^{-1}) for the various processes indicated in Fig. 10.6. $K_{i\text{fishw}}$, $K_{i\text{fish diet}}$, and $K_{i\text{fish excreta}}$ are the equilibrium constants for partitioning of chemical i between the various combinations of fish, water, diet, and excreta. This equation indicates that the chemical's concentration in the fish can evolve over time as the various mechanisms act simultaneously. (We will see how to get the time-evolving solution of this expression in Chapter 20.)

For now, we consider one special situation called the *steady-state* case. In this case, the concentration in the organism (i.e., in the fish) does not change with time (i.e., $dC_{i\text{fish}}/dt = 0$). This will occur if the total rate of chemical elimination equals the total rate of its uptake. By setting $dC_{i\text{fish}}/dt$ equal to zero in Eq. 10-9, we can solve for the steady-state concentration in the fish (indicated by the superscript ∞):

$$C_{i\text{fish}}^{\infty} = (k_1 K_{i\text{fishw}}C_{iw} + k_D K_{i\text{fish diet}}C_{\text{diet}} - k_E K_{i\text{fish excreta}}C_{i\text{excreta}})/(k_1 + k_M + k_G) \quad (10\text{-}10)$$

or dividing this result by the water concentration, C_{iw}, we have:

$$BAF_{ifish}^{\infty} \;=\; C_{ifish}^{\infty}/\,C_{iw}^{\infty} \tag{10-11}$$

$$= (k_1 K_{ifishw} + k_D K_{ifish\,diet} BAF_{idiet} - k_E K_{ifish\,excreta} C_{iexcreta}/C_{iw})/(k_1 + k_M + k_G)$$

Note that $BAF_{idiet} = C_{idiet}/C_{iw}$. We should point out that when using Eqs. 10-10 or 10-11, we assume that all parameters including BAF_{idiet} are constant. This is, of course, not generally true. For example, the lipid content of a fish can vary widely through its life (e.g., Henderson and Tocher, 1987), thus the value of K_{ifishw} would vary correspondingly. Likewise, the specific growth rate of organisms does not remain constant over their entire life, so the "dilution" of the chemical does not continuously happen at the same rate. Nevertheless, this simple mathematical budget may help us to qualitatively understand some of the observations made in field measurements.

As a first example, let us consider the case in which uptake and elimination is dominated by exchanges at the gills. In this case, the BAF_{ifish}^{∞} would correspond to:

$$BAF_{ifish}^{\infty} \;=\; (k_1 K_{ifishw})/(k_1) = K_{ifishw} = C_{ifish}^{\infty}/\,C_{iw}^{\infty} \tag{10-12}$$

The result is the equilibrium coefficient, K_{ifishw}, and this result could be predicted by the approach described in Section 10.2 (Eq. 10-4).

In contrast, if uptake is via exchange at the gills but losses occur primarily by metabolism ($k_M > k_1$ and k_G), the BAF_{ifish}^{∞} will be smaller than K_{ifishw}:

$$BAF_{ifish}^{\infty} \;=\; (k_1 K_{ifishw})/(k_M) \;<\; [(k_1 K_{ifishw})/(k_1) = K_{ifishw}] \tag{10-13}$$

Hence, depending on the rates of the various exchange and transformation process-es, "phase equilibrium" for the chemical i may not be achieved under real-world conditions.

Evaluating Bioaccumulation Disequlibrium – Example: Biota–Sediment Accumulation Factors

Despite the likelihood of disequilibrium problems, it remains important to try to predict the extent of bioaccumulation as a step to evaluating the hazard posed in situations of interest. One likely cause of disequilibrium involves cases in which an organism is exposed to more than one environmental medium. For example, this is the case for organisms living at the sediment–water interface (e.g., clams, polychaetes, amphipods, insect larvae). Sediment beds often exhibit much higher chemical contamination of hydrophobic compounds than the overlying water column. Hence, organisms may take up contaminants from sediments and simultaneously release them to the water column (Fig. 10-7). In order to eat, such organisms often ingest particles from the sediment bed and thereby incur high exposures. In order to "breath," these organisms must also pump water containing oxygen in and out, thus coming into intense contact with the less contaminated medium. Consequently, we may hypothesize that in such organisms, concentrations of persistent chemicals will be established that lie in between those concentrations that we would anticipate if the organism was in equilibrium with either the water column or the sediment. In other words, we would expect $C_{iorganism}$ and BAF_i values

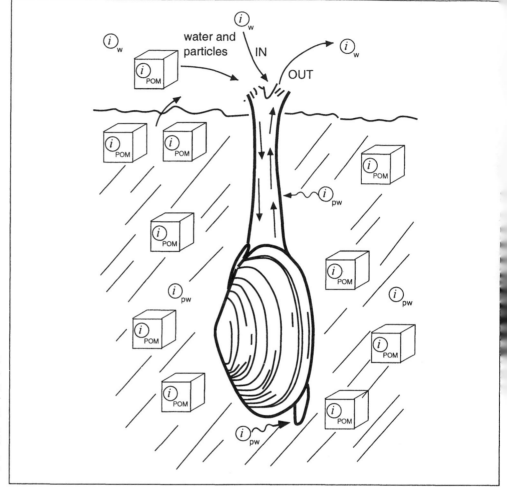

Figure 10.7 Schematic illustration of uptake and depuration of chemicals by a benthic organism, a clam dwelling near the sediment–water interface. i_w is the truly dis-solved compound in the water column, i_{POM} is the compound asso-ciated with the sediment organic matter, i_{pw} is the compound dis-solved in the sediment porewater.

that are greater than predicted from equilibrium considerations involving the water, but $C_{i\,organism}$ and $BSAF_i$ values that are smaller than the ones predicted from equilibrium with the sediment.

Let us check this hypothesis by inspecting some field data on the accumulation of PCBs by benthic organisms. Before we look at the observational data, we need to estimate what we would expect if equilibrium were established between the organism and the water column or sediment, respectively. In our discussion we will consider primarily animals with significant lipid levels (>5% by dry weight implying lipid-to-protein ratios of greater than 0.1), and PCBs whose affinities for lipids are much greater than 10 times their affinities for proteins (e.g., for PCB52, the $K_{i\,lipw}$ is about 10^6 and its $K_{i\,protw}$ is on the order of 10^4). Thus, we will assume that the PCBs are present primarily in the lipid phases of the animals and we will solve for equilibrium lipid normalized concentrations. Likewise for the sediment bed, we will assume that the PCBs are chiefly absorbed into the natural organic matter of those solids. Finally, by focusing on persistent PCB congeners, we are probably justified in neglecting any effect of metabolism on the observed animal concentrations.

For the water column, we can use Eqs. 10-6 and 10-7 to get an estimate of TBP_{ilip} (e.g., in $mol \cdot kg^{-1}$ lip). For accumulation from water, we get:

$$TBP_{ilip} \text{ (from water)} = K_{ilipw} \cdot C_{iw} \tag{10-14}$$

where C_{iw} is the truly dissolved concentration in the water (e.g. in $mol \cdot L^{-1}$). For PCBs, we may use Eq. 10-2 to estimate K_{ilipw} from K_{iow}:

$$K_{ilipw} / (L \cdot kg^{-1} \text{ lip}) = 3.2 \times K_{iow}^{0.91} \tag{10-15}$$

In order to describe the equilibrium situation with respect to the sediment, we assume that the PCBs are sorbed primarily by the natural organic matter. That is, we express the sediment pore water concentration (subscript pw) as a function of the corresponding measured sediment concentration by using $K_{id} = f_{oc} K_{ioc}$ (see Chapter 9):

$$C_{ipw} = \frac{C_{ised}}{f_{oc} K_{ioc}} = \frac{C_{ioc}}{K_{ioc}} \tag{10-16}$$

where C_{ioc} ($= C_{ised} / f_{oc}$) is the organic carbon normalized concentration of the compound in the sediment (e.g., in $mol \cdot kg^{-1}$ oc) and C_{ipw} (e.g., in $mol \cdot L^{-1}$) is the truly dissolved concentration in the pore water. Now for the sediment bed, we estimate:

$$TBP_{ilip} \text{ (from sediment)} = K_{ilipw} \cdot C_{ipw} \tag{10-17}$$

or by expressing C_{ipw} in terms of K_{ioc} and C_{ioc} (Eq. 10-16):

$$TBP_{ilip} \text{(from sediment)} = \frac{K_{ilipw}}{K_{ioc}} \cdot C_{ioc} \tag{10-18}$$

As discussed in Chapter 9 for PCBs, we may estimate K_{ioc} from K_{iow} using Eq. 9-26a in Table 9.2:

$$K_{ioc} / (L \cdot kg^{-1} \text{ oc}) = 1.4 \times K_{iow}^{0.74} \tag{10-19}$$

By analogy to the case in the water column, we can define an equilibrium constant reflecting PCB partitioning between the organism lipid pool and the sediment organic carbon pool:

$$K_{ilipoc} = \frac{C_{ilip}}{C_{ioc}} \tag{10-20}$$

And this ratio, in turn, can be estimated as:

$$K_{ilipoc} = \frac{K_{ilipw}}{K_{ioc}} \tag{10-21}$$

For PCBs, insertion of Eqs. 10-15 and 10-19 into Eq. 10-20 yields:

$$K_{ilipoc} / (kg \text{ oc} \cdot kg^{-1} \text{ lip}) = 2.3 \times K_{iow}^{0.17} \qquad \text{for PCBs} \tag{10-22}$$

Note that when we consider the situation at equilibrium, the exact pathways involved in PCB uptake and depuration are not important to the end result (e.g., whether the chemical transport into the organism occurred via the dissolved phase or by direct ingestion of sediment particles and/or diet organisms).

Let us now use these expressions to compare predicted versus measured bioaccumulation and biota–sediment bioaccumulation for PCBs in some benthic organisms. Morrison et al. (1996) reported concentrations of individual PCB congeners for the sediment beds and the overlying water column in Western Lake Erie. From these data, they could demonstrate that the overlying waters were highly undersaturated with each of the individual PCBs with respect to equilibrium with the sediments (more than a factor of 10; compare C_{ioc}/C_{iw} values with K_{ioc} values in Table 10.3). They also acquired PCB concentration data for phytoplankton, as well as in several bottom-dwelling animals including zebra mussels (*Dreissena polymorpha*), caddisfly larvae (*Hydropsyche alterans*), an amphipod (*Gammarus fasciatus*), and a crayfish (*Orconectes propinquus*). Furthermore, they examined the lipid contents of the organisms and the organic carbon contents of the sediments.

From these data, experimental BAF_i and $BSAF_i$ values can be derived:

$$BAF_{ilip} = \frac{C_{ilip}}{C_{iw}} \tag{10-23}$$

$$BSAF_{ilipoc} = \frac{C_{ilip}}{C_{ioc}} \tag{10-24}$$

and can be compared to estimated values assuming phase equilibrations (i.e., K_{ilipw} and K_{ilipoc}). Note that the estimated equilibrium values are calculated with data for 25°C; that is, we neglect the (small) temperature dependence of K_{ilipw} and K_{ilipoc}. Furthermore, the experimental data, from which the BAF_{ilip} and $BSAF_{ilipoc}$ values have been derived, are mean values with relative standard deviations between 20 and 50%. Despite these qualifications, some important conclusions can be drawn (Table 10.3).

First, for a given compound, the observed BAF_{ilip} and $BSAF_{ilipoc}$ values were very similar (± a factor of 2) for all sediment-dwelling organisms considered. Second, as we have anticipated, the BAF_{ilip} values were significantly larger than the corresponding K_{ilipw} values, whereas the $BSAF_{ilipoc}$ values were significantly smaller than the respective K_{ilipoc} values (see ratios in italics in Table 10.3). Finally, with increasing hydrophobicity of the PCB congener, deviation from equilibrium with the water phase increases (i.e., factor 2 to 3 for PCB52 to 5 to 9 for PCB153), and it decreases with respect to the sediment (i.e., factor 15 to 30 for PCB52 to 5 to 10 for PCB153). A possible explanation for these findings is that the depuration rate decreases with increasing hydrophobicity of the compound.

It is interesting to compare these observed $BSAF_{ilipoc}$ values (Table 10.3) with values reported for other sediment-dwelling organisms and/or locations. Tracey and Hansen (1996) have compiled numerous $BSAF_{ilipoc}$ data for PCBs and some data for polycyclic aromatic hydrocarbons (PAHs). The reported $BSAF_{ilipoc}$ values for PCBs

Table 10.3 Estimated Equilibrium (K_{ilipw}, K_{ilipoc}) versus Measured Lipid-Normalized Bioaccumulation Factors (BAF_{ilip}) and Biota–Sediment Bioaccumulation Factors ($BSAF_{ilipoc}$) for Three PCB Congeners in Phytoplankton and Various Sediment-Dwelling Organisms in Western Lake Erie [a]

	PCB52 (2,2',5,5')	PCB101 (2,2',4,5,5')	PCB153 (2,2',4,4',5,5')
Measured Concentrations			
C_{iw}/(ng·L^{-1})	0.028	0.018	0.006
C_{ised}/(ng·kg^{-1} d.w.)	4000	5000	5800
Properties			
K_{iow}	1.3×10^6	2.5×10^6	1.6×10^7
K_{ioc}/(L·kg^{-1} oc) [b]	4.7×10^4	6.6×10^4	3.0×10^5
C_{ioc}/C_{iw} [c]	1.9×10^6	3.8×10^6	1.3×10^7
K_{ilipw}/(L·kg^{-1} lip) [d]	1.2×10^6	2.1×10^6	1.1×10^7
K_{ilipoc}/(kg oc·kg^{-1} lip) [e]	25	28	39
Relative to the Water Column	BAF_{ilip} [f]/(L·kg^{-1} lip); BAF_{ilip}/K_{ilipw} [g]		
phytoplankton	2.5×10^6; *2.1*	7.8×10^6; *3.7*	3.2×10^7; *3.0*
caddisfly larvae	3.6×10^6; *3.0*	1.3×10^7; *6.2*	1.0×10^8; *9.9*
amphipod	3.5×10^6; *3.0*	7.9×10^6; *3.8*	6.8×10^7; *6.0*
zebra mussel	4.0×10^6; *3.3*	1.2×10^7; *5.7*	9.0×10^7; *8.0*
crayfish	2.4×10^6; *2.0*	6.2×10^6 *3.0*	6.1×10^7; *5.5*
Relative to the Sediment Bed ($f_{oc} = 0.074$)	$BSAF_{ilipoc}$ [h]/(kg oc·kg^{-1} lip); $BSAF_{ilipoc}$/K_{ilipoc} [g]		
phytoplankton	1.3; *0.05*	2.1; *0.08*	2.5; *0.06*
caddisfly larvae	1.8; *0.07*	3.4; *0.12*	7.8; *0.20*
amphipod	1.8; *0.07*	2.1; *0.08*	5.1; *0.13*
zebra mussel	2.1; *0.08*	3.2; *0.11*	6.9; *0.18*
crayfish	0.7; *0.03*	1.7; *0.06*	4.6; *0.12*

[a] Calculated from data reported by Morrison et al. (1996). [b] Eq. 10-19. [c] Ratio of experimentally determined concentrations. [d] Eq. 10-15. [e] Eq. 10-22. [f] Eq. 10-23. [g] Italic numbers. [h] Eq. 10-24.

are between 0.5 and about 6 kg oc·kg^{-1} lip. Furthermore, mean $BSAF_{ilipoc}$ values were found to increase with increasing K_{iow} for $K_{iow} < 10^7$. For more hydrophobic compounds, the $BSAF_{ilipoc}$ values tend to decrease again, which could reflect slower uptake and/or steric hindrance as discussed in Section 10.2.

Compared to PCBs, the $BSAF_{ilipoc}$ values for PAHs reported in the literature are almost one order of magnitude smaller (i.e., < 0.1 to 0.5 kg oc·kg^{-1} lip; Tracey and Hansen, 1996; Clarke and McFarland, 2000). This has been found not only for sediment dwelling organisms, but also for accumulation of chemicals in earthworms, where, in analogy to the biota–sediment accumulation factor, one can define a biota–soil accumulation factor (Ma et al., 1998; Krauss et al., 2000). These findings can be partially rationalized by the fact that PAHs tend to have higher K_{ioc} values (or additionally they experience stronger sorption to soot carbon) than PCBs exhibiting the same K_{iow} values (see Fig. 9.11). For estimating K_{ioc} values of PAHs we can use Eq. 6-29b in Table 9.2:

$$K_{ioc} / (L \cdot kg^{-1} \text{ oc}) = 0.5 \times K_{iow}^{0.98} \tag{10-25}$$

If we now assume that, at least for the apolar lipids, Eq. 10-15 is also valid for estimation of K_{ilipw} of PAHs, by analogy to Eq. 10-22, we would then predict the K_{ilipoc} of PAHs to be:

$$K_{ilipoc} / (kg \text{ oc} \cdot kg^{-1} \text{ lip}) = 6.4 \times K_{iow}^{-0.08} \qquad \text{for PAHs} \tag{10-26}$$

Hence, for a PAH with $K_{iow} = 10^6$ we would obtain a K_{ilipoc} of about 2 kg oc$\cdot$kg^{-1} lip as compared to 24 kg oc$\cdot$kg^{-1} lip for a PCB with the same K_{iow}. This would explain the differences found in the field. We should, however, stress again that K_{ioc} values as predicted from LFERs such as Eqs. 10-19 and 10-25 reflect sediment organic matter–water partitioning and not sorption to highly active sorbents (e.g., soot) that may be present in sediments. Thus, very low $BSAF_{ilipoc}$ values found in the field not only may reflect disequilibrium but also may be due to the presence of such sorbents, which are particularly important for sorption of PAHs (see Illustrative Example 9.3).

Using Fugacities or Chemical Activities for Evaluation of Bioaccumulation Disequilibrium

Now we have seen how bioaccumulation disequilibrium can be evaluated using a comparison of observed bioaccumulation factors (BAF_i and $BSAF_i$) with expectations from phase equilibrium considerations (K_{ilipw} and K_{ilipoc}). Using the latter factors, we can also calculate the concentrations of a given compound that we would expect in an organism if it were in equilibrium with either the water or the sediment, respectively, and then compare these concentrations with the actual measured ones.

Another way of evaluating disequilibrium, particularly when considering the partitioning of a given compound between several environmental compartments, is to compare the fugacities or chemical activities of the compound in the various compartments. We recall from Chapter 3 that it is the difference in activity or fugacity, and not concentration, that determines in which direction a net flux of compound will occur. Such transfers will continue until the fugacities or activities are equal in the interacting phases.

Evaluating a chemical's activity, a_i(phase), in any phase of interest corresponds to contrasting its concentration to whatever concentration would be expected at equilibrium with the reference state. Choosing the pure liquid organic compound as the reference state implies that the chemical's activity is equal to 1 when it occurs at its *liquid* solubility in water or its pure *liquid* vapor pressure in air. Since we have learned many approaches for relating concentrations in nonaqueous phases (NOM, lipids, air) to their corresponding equilibrium concentrations in water ($K_{iphasew}$ values), one convenient approach for calculating a chemical's activity in a particular medium is simply (1) to apply such partitioning constants to calculate the corresponding aqueous concentrations in equilibrium with that phase and then (2) to normalize the result to the compound's liquid solubility:

$$a_i(\text{phase}) = \frac{C_{iphase}}{C_{iphase}^{sat}} = \frac{C_{iphase}}{K_{iphasew} C_{iw}^{sat}(\text{L})} \qquad (10\text{-}27)$$

For chemicals in condensed phases (e.g., NOM, lipids), such an activity calculation is sensitive to the system temperature as described by (Eq. 3-51):

$$a_i(\text{phase}, T_1) = a_i(\text{phase}, T_{ref}) \exp(-\Delta_{phasew} H_i / R\,(1/T_1 - 1/T_{ref})) \qquad (10\text{-}28)$$

Of course, we can also use the fugacity of a given compound in a given molecular environment (e.g., a solvent) since this is a measure of the fleeing tendency of the compound from that environment (Fig. 3.9). A quantitative measure of fugacity is the partial pressure, p_i, that the compound molecules would exert in the gas phase, if the gas phase were in equilibrium with the phase under consideration, and behaved like an ideal gas. Hence, for calculating the fugacity, f_i (phase), of a compound in a given condensed phase we need to know the compound's gas-condensed phase partition coefficient K_{iH}(phase) (see Section 6.2):

$$f_i(\text{phase}) = K_{iH}(\text{phase}) \cdot C_{iphase} \qquad (10\text{-}29)$$

where K_{iH}(phase) is defined as the equilibrium ratio of the partial pressure of the compound and its molar concentration in the phase considered. Hence, the units of K_{iH}(phase) are, for example, $\text{Pa} \cdot (\text{mol} \cdot \text{L}^{-1})^{-1}$ or $\text{Pa} \cdot (\text{mol} \cdot \text{kg}^{-1})^{-1}$. Note that in the commonly used fugacity models, K_{iH}(phase) is referred to as *fugacity capacity* of the compound in the given phase (Mackay, 1979; Mackay et al., 1992 – 1997).

When calculating the fugacities of a given compound in a given phase, one faces a few difficulties. First, except for *the* Henry's law constant K_{iH} (recall that we refer to the air–water partition constant as *the* Henry's law constant, Section 6.4), there are very few experimental data available for partitioning of organic compounds between the gas phase (i.e., air) and important environmental phases including biological media. The relative abundance of corresponding phase–water partition coefficients is one practical advantage of using the chemical activity approach described above. However, K_{iH}(phase) values can generally be estimated from the corresponding condensed phase–water partition coefficient and the air–water partition constant, K_{iaw}, of the compound:

$$K_{iH}(\text{phase}) = \frac{K_{iaw} \cdot RT}{K_{iphasew}} \qquad (10\text{-}30)$$

Second, in contrast to partitioning between water and natural condensed phases, partitioning involving the gas phase is strongly temperature dependent (see Chapters 4 and 6, and Section 10.4). Consequently, when using fugacities to assess compound

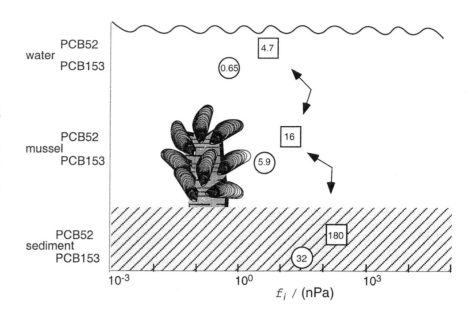

Figure 10.8 Calculated fugacities (see Illustrative Example 10.4) of PCB52 (data in squares) and PCB153 (data in circles) in the water and sediment, and in the lipids of zebra mussels in Western Lake Erie at 25°C, indicating that the mussels exhibit concentrations that are intermediate between equilibrium with the water column and the sediment bed. The corresponding activities are given in Illustrative Example 10.4.

fluxes between environmental compartments, one has to be careful to apply the appropriate K_{iH}(phase, T) values.

Let us now come back to our example of PCB bioaccumulation in sediment-dwelling organisms in Lake Erie (Table 10.3). To assess that situation, this time we will compare the chemical activities (see Illustrative Example 10.4) and fugacities (Fig. 10.8) of the PCB congeners, PCB52 and PCB153, in the lake water, zebra mussel tissues, and the bed sediment. The calculations of these values are performed in Illustrative Example 10.4. For simplicity, a temperature of 25°C has been assumed. Note that, if we assume that $\Delta_{12}H_i$ for air–lipid and air–natural organic matter partitioning are similar to the $\Delta_{vap}H_i$ values of the compounds (81 kJ·mol^{-1} for PCB52 and 91 kJ·mol^{-1} for PCB153), and that $\Delta_{aw}H_i$ (air–water partitioning) is in the order of 50 and 70 kJ·mol^{-1}, respectively, for the two compounds, then per 10 degree decrease in temperature, the fugacities shown in Fig. 10.8 would decrease by a factor of 2 and 2.7 in the aqueous phase, and by factors of 3 and 3.5 in the organic phases, respectively (see Tabel D 1, Appendix D).

Neglecting this concern for inaccuracy due to temperature considerations, the PCB activities or fugacities (which of course yield the same relative results if the partition coefficient inputs are all accurate) prove to be highest in the bed sediments (by factors of about 10 and 5 over the mussels for PCB52 and PCB153, respectively) and lowest in the water (by factors of 0.3 and 0.1 less than the mussels for PCB52 and PCB153, respectively). Hence, the PCB52 is closer to equilibrium with the water, whereas PCB153 is closer to equilibrium with the sediment (see discussion above). It appears that the mussels are acquiring these PCBs from their diet, which undoubtedly included resuspended sediments, and they are releasing these contaminants back to the relatively clean water. One can also immediately see that both compounds are highly underconcentrated in the aqueous phase as compared to the sediment (about a factor of 40).

Although these activity and fugacity results exactly correspond to the results deduced using BAF_{ilip}, K_{ilipw}, $BSAF_{ilipoc}$, and K_{ilipoc} values (Table 10.3), the use of activities or fugacities may provide a somewhat more transparent picture of a given situation. Note that the ratios of the activities or fugacities between lipid and water and lipid and sediment correspond to the BAF_{ilip}/K_{ilipw} and $BSAF_{ilipoc}/K_{ilipoc}$ ratios, respectively (Table 10.3). In the following we will use fugacity considerations for evaluating the transfer of organic compounds from air or soil to biological media in terrestrial systems.

Illustrative Example 10.4

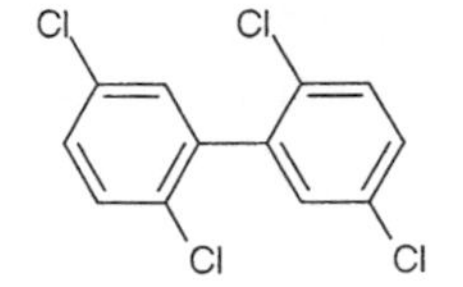

2,2',4,4'-tetrachlorobiphenyl
(PCB52)

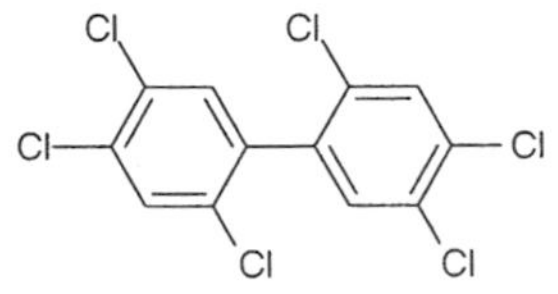

2,2',4,4',5,5'-hexachlorobiphenyl
(PCB 153)

Calculating Fugacities or Chemical Activities to Evaluate Bioaccumulation

Morrison et al. (1996) reported the concentrations of PCB52 and PCB153 in the water column, the sediments, and zebra mussels from Lake Erie. The observed concentrations are shown below. They also found that the sediment had an f_{oc} of 0.074 kg oc·kg^{-1} sediment and the mussels had an f_{lip} of 0.013 kg lip·kg^{-1} w.w.

	PCB52	PCB153
C_{iw} / (ng·L^{-1})	0.028	0.006
C_{ised} / (ng·kg^{-1} d.w.)	4000	5800
$C_{imussel}$ / (ng·kg^{-1} w.w.)	1500	7100

Focusing on the water and mussel concentrations in the table, one might infer that both PCBs were substantially biomagnified since the concentration ("parts per trillion" basis) in the mussel appears to be 10^5 and 10^6 times greater than in the water.

Problem

In order to examine the roles of both the water column *and* the sediment bed in PCB bioaccumulation by the mussels, estimate the fugacities (f_i) and chemical activities (a_i, relative to the pure liquid chemical reference state) at 25°C for both PCB congeners in the water, in the sediment, and in the mussels.

Some useful information is given in the following table (data for 25°C):

	PCB52	PCB153
M_i / (g·mol^{-1})	292	361
C_{iw}^{sat} (L) / M	$10^{-6.1}$	$10^{-7.0}$
K_{iow}	$10^{6.18}$	$10^{7.15}$
K_{iaw}	$10^{-1.7}$	$10^{-1.8}$
K_{ioc} (= $1.4 \times K_{ow}^{0.74}$, Eq. 10-19)	$10^{4.7}$	$10^{5.4}$
K_{ilipw} (= $3.2 \times K_{ow}^{0.91}$, Eq. 10-2)	$10^{6.1}$	$10^{7.0}$

Answer

To estimate either the fugacities or the chemical activities, we first calculate the concentrations in the water, C_{iw}^*, equilibrated with each observed concentration:

in water, $C_{iw}^* = C_{iw}$

in sediment, $C_{iw}^* = \dfrac{C_{ised}}{f_{oc} K_{ioc}}$

in mussel, $C_{iw}^* = \dfrac{C_{imussel}}{f_{lip} K_{ilipw}}$

Multiplying this result by RTK_{iaw} (note: $RT = 2.48 \times 10^6$ Pa L·mol^{-1}), we find the corresponding fugacities (Eq. 10-29). Dividing this fugacity result by the vapor pressures over the pure liquid or, more directly, dividing the equilibrated water concentrations by the liquid solubilities yields the chemical activities:

	PCB52	PCB153	Ratio to water	
Fugacities				
in water $(= C_{iw} \cdot K_{iaw} \cdot RT)$	4.7 nPa	0.65 nPa	1	1
in sed $\left(= \dfrac{C_{ised}}{f_{oc} K_{ioc}} \cdot K_{iaw} \cdot RT \right)$	180 nPa	32 nPa	40	50
in mussel $\left(= \dfrac{C_{imussel}}{f_{lip} K_{ilipw}} \cdot K_{iaw} \cdot RT \right)$	16 nPa	5.9 nPa	3	9
Chemical activities				
in water $\left(= \dfrac{C_{iw}}{C_{iw}^{sat}(L)} \right)$	1.2×10^{-7}	1.7×10^{-7}	1	1
in sed $\left(\dfrac{C_{ised}}{f_{oc} K_{ioc} \cdot C_{iw}^{sat}(L)} \right)$	4.5×10^{-6}	8.0×10^{-6}	40	50
in mussel $\left(= \dfrac{C_{imussel}}{f_{lip} K_{ilipw} \cdot C_{iw}^{sat}(L)} \right)$	3.8×10^{-7}	1.5×10^{-6}	3	9

Whether we consider the fugacities or the chemical activities, we find the same result. Both PCBs exhibit chemical potentials that are between what we find for the water and the sediment (see Fig. 10.8).

10.4   Bioaccumulation in Terrestrial Systems

Compared to aquatic systems, it is currently more difficult to assess bioaccumulation of organic chemicals in terrestrial systems. First, particularly when dealing with terrestrial plant materials, we may not *a priori* assume that only the lipid components predominate as the material into which the compounds partition; polymers like lignin and cutin can also be important. Second, the composition of plant materials and terrestrial animals, in which bioaccumulation measurements have been made, is often not reported. So, estimates of equilibrium partition coefficients and assessments of chemical activities or fugacities are not possible. Furthermore, partitioning from the gas phase (i.e., air) is strongly temperature dependent. As a consequence, seasonal fluctuations in air–plant partitioning processes are generally much more pronounced than fluctuations in partitioning coefficients involving water. Therefore, our present ability to predict bioaccumulation in terrestrial systems is still rather limited. Nevertheless, a few important general aspects can be addressed in a quantitative way.

Transfer of Organic Pollutants from Air to Terrestrial Biota

We start by considering the bioaccumulation of *S*emivolatile *O*rganic *C*hemicals (SOCs) between the atmosphere and plants. The term SOCs is commonly used in discussions involving persistent, apolar or weakly polar compounds exhibiting low vapor pressures. SOCs include important classes of environmental pollutants like polychlorinated aromatic hydrocarbons (e.g., DDT, DDE, DDD, HCB, PCBs), and polychlorinated dibenzo-*p*-dioxins (PCDDs) and -dibenzofurans (PCDFs), as well as polycyclic aromatic hydrocarbons (PAHs). Such compounds partition very favorably from air into organic phases (see Chapter 6). Furthermore, because of their high hydrophobicities, from an overall mass flux point of view, uptake from soil water via roots will, in general, be much less important for such compounds as compared to more water-soluble chemicals (Chiou et al., 2001). Therefore, atmosphere–terrestrial plant partitioning is an important route for such compounds to enter agricultural food chains (e.g., as depicted in Fig. 10.1: pasture → cattle → cows' milk, butter, meat → humans; McLachlan, 1996; Thomas et al., 1998a, 1999). Such air-to-biota transfers are also important in terrestrial wildlife food chains (e.g., pasture → caribou → wolf; Kelly and Gobas, 2001). Furthermore, due to the extensive partitioning, terrestrial vegetation plays an important role in the cycling of SOCs in the environment (Simonich and Hites, 1995; Wania and McLachlan, 2001; Trapp et al., 2001). Finally, certain more long-lived terrestrial plant media including lichens, moss, pine needles, and tree bark have been and are being used to monitor atmospheric pollution with SOCs (Calamari et al., 1994 and 1995; Tremolada et al., 1996; Simonich and Hites, 1997). In the following we will discuss some examples of the transfer of airborne SOCs to terrestrial systems. However, in order to be able to calculate fugacities in plant materials, we need to consider first some aspects of air–plant tissue equilibrium partitioning.

Air–Plant Equilibrium Partitioning

Experimentally determined air–plant equilibrium partition coefficients of SOCs as well as of other organic compounds are rather scarce. Note that, in the following,

Table 10.4 LFERs Relating Air–Plant Partition Coefficients, K_{iap}, and Air–Octanol Partition Constants, K_{iao}, of PCBs for Various Herbs and Grasses at 25°C.(K_{iap} values are in kg dry weight·L^{-1}.) [a]

Plant Species	$\text{Log } K_{iap}/(\text{kg d.w.} \cdot \text{L}^{-1}) = a \cdot \log K_{iao} + b$			
	a	b	Log K_{iao} Range	R^2
ryegrass (*Lolium multiflorum*)	1.15	+3.07	−9.9 to −7.2	0.98
clover (*Trifolium repens*)	0.70	−0.83	−8.8 to −7.4	0.86
plantain (*Plantago lanceolata*)	0.87	+0.64	−8.8 to −7.4	0.98
hawks beard (*Crepis biennis*)	0.74	−0.55	−9.4 to −7.4	0.97
yarrow (*Achillea millefolium*)	0.57	−2.33	−9.5 to −7.4	0.93

[a] Derived from partition coefficients reported by Kömp and McLachlan (1997b) as volume air/volume plant by assuming a density of 1 kg·L^{-1} and by using the dry-weight percentage given by Böhme et al. (1999).

similar to vapor pressure and to air–solvent partitioning (Chapter 6), we define K_{iap} as the ratio of the equilibrium concentrations in the air and in the plant:

$$K_{iap} = \frac{C_{ia}}{C_{ip}} \qquad (10\text{-}31)$$

and not vice versa as we did when discussing bioaccumulation from air in the Illustrative Example 10.3, and as is often done in the literature. Thus, a very small K_{iap} value means that the compound partitions very favorably into the plant material.

Kömp and McLachlan (1997b) determined air–plant partition coefficients of a series of PCB congeners at different temperatures for different grass and herb species. From this laboratory data, two important general conclusions can be drawn. First, significant differences in the partitioning behavior between the plants were observed. Partition coefficients varied by up to a factor of 20 for different plant species accumulating a single PCB congener. This was particularly true for the more volatile compounds. Kömp and McLachlan (1997b) quantitatively illustrated this variability by showing that quite different regression coefficients were obtained when air–plant partition coefficients were correlated with the corresponding air–octanol partition coefficients, K_{iao}, of the compounds (see a and b values in Table 10.4):

$$\log K_{iap} = a \cdot \log K_{iao} + b \qquad (10\text{-}32)$$

Since the composition of the various plants tested was not determined (i.e., lipid, cutin, lignin), the differences cannot be examined in light of variable contributions. Nevertheless, the substantially different slopes observed indicate that not only the quantity but also the quality of the plant biomass was important in determining air–plant partitioning. Note that a very similar interspecies variability has been observed for the same plants in a field study (Böhme et al., 1999).

The second important result is the strong temperature dependence of K_{iap}. For example, for PCB52 and PCB153, the model compounds that we focused on to illustrate bioaccumulation in aquatic systems (see above), $\Delta_{ap}H_i$ values are about 90 kJ·mol^{-1} and 110 kJ·mol^{-1}, respectively. Hence, in regions with large temperature fluctuations, K_{iap} values, and thus concentrations found in plants, may differ by more than one order of magnitude between warm and cold seasons, provided that the kinetics of air–plant exchange are fast enough. For example, Kelly and Gobas (2001) found that concentrations of PCBs in lichens collected from a tundra system in the arctic region of Canada in the spring exceeded those in lichens collected in the summer by factors of 10 or even more. Since, the concentrations of PCBs in the arctic air are significantly smaller in the winter than in the summer (Hung et al., 2001), these differences can be attributed primarily to the strong effect of temperature on air–plant partitioning.

In summary, we note that air–plant partition coefficients of a given compound may vary significantly between different plant species and this partitioning is strongly temperature dependent. However, for practical purposes, it may be sufficient to use an average K_{iap} value and an average $\Delta_{ap}H_i$ value for calculating fugacities in plant materials. This would allow an initial assessment of air–plant equilibration in a given field situation or prediction of maximum concentrations in plant materials that one would anticipate for a given average air concentration (see Illustrative Example 10.5).

Illustrative Example 10.5   **Evaluating Air–Pasture Partitioning of PCBs**

Problem

Thomas et al. (1998b) conducted a field study of the air-to-pasture transfer of a series of PCB congeners at a rural site in northwest England. The pasture consisted of a mixture of grasses and herbs. The average concentrations of the three congeners PCB52, PCB153, and PCB180 determined in 12 samples collected between the end of April and mid-October 1996 are given in the table below together with the air–octanol partition constants and the average $\Delta_{ap}H_i$ values of the compounds reported by Kömp and McLachlan (1997b). Also included are the average concentrations of the three congeners in the air above that region. The mean temperature during the sampling period was about 10°C. Calculate the fugacities (nPa) of the three PCB congeners in the air and the pasture, as well as the concentrations of the compounds in the pasture (ng·kg^{-1} d.w.) that would be established at equilibrium with the average air concentrations.

Answer

Fugacity in Air

Convert concentration to partial pressure by first converting it to mol·L^{-1} (division by the molar mass, M_i, see Appendix C) and subsequent multiplication by RT (2.35×10^6 Pa·L·mol^{-1} at 10°C):

Air–Octanol Partition Constants (K_{iao}), Average Enthalpies of Air–Pasture Transfer ($\Delta_{ap}H_i$), and Average Concentrations in Pasture (C_{ip}) and Air (C_{ia}) of Three PCB Congeners

	PCB52 (2,2',5,5')	PCB153 (2,2',4,4',5,5')	PCB180 (2,2',3,4,4',5,5')
$\log K_{iao}$ [a]	−8.0	−9.1	−9.8
$\Delta_{ap}H_i$ / (kJ·mol⁻¹) [b]	90	110	130
C_{ip} / (ng·kg⁻¹ d.w.) [c]	40	65	18
C_{ia} / (pg·m⁻³) [c]	6.8	3.2	0.7

[a] Average between calculated (i.e., $\log K_{iao} = \log K_{iaw} - \log K_{iow}$, Eq. 6-11) and experimental value (Kömp and McLachland, 1997a). [b] Kömp and McLachlan (1997b). [c] Thomas et al. (1998b).

PCB52: $f_i(\text{air}) = (6.8 \times 10^{-3} \text{ pg·L}^{-1}/292 \times 10^{12} \text{ pg·mol}^{-1}) (2.35 \times 10^6 \text{ Pa·L·mol}^{-1})$

$$= 5.5 \times 10^{-11} \text{ Pa}$$

$$= 0.055 \text{ nPa}$$

and, accordingly,

PCB153: $f_i(\text{air}) = 0.021$ nPa

PCB180: $f_i(\text{air}) = 0.0042$ nPa

Fugacity in Pasture

Calculate an average K_{iap} value by inserting the K_{iao} value into the five equations given in Table 10.4:

$$\overline{K}_{iap} = \frac{1}{5}\Sigma K_{iap} \text{ (estimated)}$$

The resulting average K_{iap} values at 25°C are:

PCB52: $K_{iap} = -\frac{1}{5} (10^{-6.1} + 10^{-6.4} + 10^{-6.3} + 10^{-6.5} + 10^{-6.9}) = 10^{-6.4}$

and, accordingly,

PCB153: $K_{iap} = 10^{-7.3}$

PCB180: $K_{iap} = 10^{-7.9}$

Use Eq. 3-51 (or the factors given in Table D1 in Appendix D) to convert the K_{iap} values from 25°C to 10°C [neglect the contribution of RT_{av} (Eq. 6-10); $\Delta_{ap}H_i$ values are given in the above table].

PCB52: $K_{iap}(10°C) = 0.15 \, K_{iap}(25°C) = 6.0 \times 10^{-8}$ kg d.w.·L⁻¹

PCB153: $K_{iap}(10°C) = 0.095 \, K_{iap}(25°C) = 4.8 \times 10^{-9}$ kg d.w.·L⁻¹

PCB180: $K_{iap}(10°C) = 0.062 \, K_{iap}(25°C) = 7.8 \times 10^{-10}$ kg d.w.·L⁻¹

Calculate now the concentrations in the air that would be established *in equilibrium* with the actual concentrations in the pasture:

PCB52: $C_{ia}^{eq} = K_{iap} \cdot C_{ip} = (6.0 \times 10^{-8}$ kg d.w.$\cdot$L^{-1}) (40 ng$\cdot$kg^{-1} d.w.)

$\qquad\qquad = 2.4 \times 10^{-15}$ g$\cdot$L^{-1}

$\qquad\qquad = 2.4$ pg$\cdot$m^{-3}

and accordingly,

PCB153: $C_{ia}^{eq} = 3.1 \times 10^{-16}$ g$\cdot$L^{-1} = 0.31 pg$\cdot$m^{-3}

PCB180: $C_{ia}^{eq} = 1.4 \times 10^{-17}$ g$\cdot$L^{-1} = 0.014 pg$\cdot$m^{-3}

Convert these concentrations to partial pressures as done above for the actual air concentrations. The resulting fugacities are:

			f_i(air,obs.) (April – October)
PCB52:	f_i(pasture)	= *0.019* nPa	0.055 nPa
PCB153:	f_i(pasture)	= *0.0020* nPa	0.021 nPa
PCB180:	f_i(pasture)	= *0.000084* nPa	0.0042 nPa

Comparison of these values with the fugacities of the compounds in air (see above) shows that PCB52 is rather close to equilibrium (less than a factor of 3), whereas the other two congeners exhibit fugacities in the pasture that are a factor of 10 and 50, respectively, lower than the fugacities in the air. Therefore, equilibrium concentrations in the pasture would be a factor of about 3, 10, and 50 times higher for PCB52, PCB153, and PCB 180, respectively, as compared to the actual measured ones (i.e., *120, 650,* and *900* ng$\cdot$kg^{-1} d.w., respectively).

This finding that deviation from equilibrium increases with decreasing volatility of the compounds (i.e., decreasing K_{iap} value) has also been observed in a field study by Böhme et al. (1999), who determined the concentrations of a large number of SOCs including PCBs, PCDDs, PCDFs, and PAHs in a variety of plants. A plausible explanation for these observations is that gas exchange becomes increasingly kinetically limited with increasing size of the molecule. Note that for SOCs of very low volatility, that are primarily bound to particles in the atmosphere, particle deposition may become the dominant process for the overall transfer of the compound from the atmosphere to plants (for more details see McLachlan, 1999, and Böhme et al., 1999).

Uptake of Organic Pollutants from Soil

Another important uptake route for chemicals by terrestrial plants and organisms living in soils (e.g., earthworms) is uptake from soil interstitial water and/or by ingestion of soil particles. In particular, uptake of contaminants from soils by certain plant species is of great interest, because this process, which is referred to as phytoremediation, is used to clean up contaminated sites. The idea of phyto-remediation is that compounds partition from contaminated soils into the plant roots; subsequently the compounds may be metabolized in the plant tissues, transported to woody tissues for storage, or transported to the leaves where the chemical can be transferred to the atmosphere (e.g., Thompson et al. 1998). Uptake of contaminants from soil water to plant is particularly important for more water-soluble compounds including numerous pesticides and solvents. Note that for more polar compounds including various pesticides, plant components other than lipids (i.e., proteins, carbohydrates, and even plant water) may also become relevant reservoirs for the overall accumulation of the chemical in the plant, particularly if the plant (or crop) considered has a very low lipid content (Chiou et al., 2001).

When evaluating the accumulation of chemicals in plants and soil organisms, we have a similar situation as encountered when dealing with organisms living at the sediment–water interface. That is, we are confronted with the fact that there may be a simultaneous exchange with more than one environmental medium (e.g., the atmosphere and the soil). Hence, to assess the direction in which chemicals are transferred, we need to determine the fugacities in all relevant compartments. However, this task is not too difficult, since we have already discussed all the relevant partition processes between biological media, air, water, and soil organic matter, respectively.

For example, when we are interested in the accumulation of SOCs in earthworms, we may adopt a similar approach as we used for sediment-dwelling organisms. Since earthworms have a significant lipid content (ca. 5%, Table 10.1) and we are interested in relatively hydrophobic substances, we use lipid- and organic carbon-normalized biota–soil accumulation partition coefficients (K_{ilipoc}) and bioaccu-mulation factors ($BSAF_{ilipoc}$). These correspond exactly to the biota–sediment accumulation partition coefficients and factors defined by Eqs. 10-21 and 10-24 (for an application see Problem P 10.2).

10.5   Biomagnification

Defining Biomagnification

In the environment, one frequently observes the concentration of a given compound in organisms increasing as one examines successive organisms along a given food chain (see example given in Fig. 10.9). The term *biomagnification* has been introduced to describe this phenomenon. The associated biomagnification factor, BMF_i, is therefore defined as the ratio of the chemical concentration in the organism of the higher trophic level divided by the same chemical's concentration in the organism contributing a major part of the diet (Fig. 10.5):

$$BMF_i = \frac{C_{iorganism}}{C_{idiet}} \qquad (10\text{-}33)$$

Eq. 10-33 can also be written in terms of the observed bioaccumulation factors (Eq.

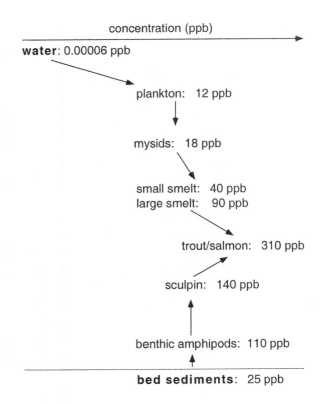

concentration (ppb)

water: 0.00006 ppb

plankton: 12 ppb

mysids: 18 ppb

small smelt: 40 ppb
large smelt: 90 ppb

trout/salmon: 310 ppb

sculpin: 140 ppb

benthic amphipods: 110 ppb

bed sediments: 25 ppb

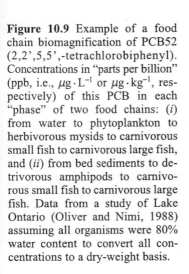

Figure 10.9 Example of a food chain biomagnification of PCB52 (2,2',5,5',-tetrachlorobiphenyl). Concentrations in "parts per billion" (ppb, i.e., $\mu g \cdot L^{-1}$ or $\mu g \cdot kg^{-1}$, respectively) of this PCB in each "phase" of two food chains: (*i*) from water to phytoplankton to herbivorous mysids to carnivorous small fish to carnivorous large fish, and (*ii*) from bed sediments to detrivorous amphipods to carnivorous small fish to carnivorous large fish. Data from a study of Lake Ontario (Oliver and Nimi, 1988) assuming all organisms were 80% water content to convert all concentrations to a dry-weight basis.

10-8) of the organism considered and of its diet if both have been referenced to the same environmental medium ($C_{i\text{med}}$):

$$BMF_i = \frac{BAF_{i\text{organism}}}{BAF_{i\text{diet}}} \qquad (10\text{-}34)$$

We should point out that although the name of this bio*magnification* factor suggests a buildup in organisms of higher trophic level, BMF_i values may be smaller than 1. Such is the case, for example, for compounds that are efficiently metabolized in the organism but not in its diet; that is, $BAF_{i\text{organism}} < BAF_{i\text{diet}}$. Furthermore, if the compound concentrations are expressed on a total organism weight basis, a $BMF_i \neq 1$ does not necessarily mean that the concentration of the compound in a particularly important location in the organism (e.g., in the membrane lipids where a toxic effect may occur, see Section 10.6) is different in the organism and its diet. A $BMF_i \neq 1$ may simply reflect differences in composition of the organism and its diet. For example, if, compared to its diet, an organism exhibits a much higher fraction of organic phases that are favorable for the compound to partition into (in particular, lipids), a BMF_i value > 1 is likely to be found (see e.g., Leblanc, 1995; Kucklick and Baker, 1998). On the other hand, a higher lipid content of the diet will tend to decrease the whole body weight-normalized BMF_i.

However, biomagnification factors significantly different from 1 may also reflect real processes building up the effective concentrations in a given organism's tissues. Said more accurately, the chemical activity or fugacity of the compound may prove to be greater as one moves up the food chain:

"real biomagnification" when:

$$a_{i\,organism} > a_{i\,diet} \tag{10-35}$$

or equivalently:

$$f_{i\,organism} > f_{i\,diet} \tag{10-36}$$

Let us see why such "real" biomagnification may occur. First, in many organisms, the organic mass making up the diet is substantially degraded while in the animal gut. For example, herbivorous fish typically excrete about 30 to 40% of their ingested food while carnivorous fish excrete about 20% (Brett and Groves, 1979). Moreover, particular biochemical fractions of the diet may be especially utilized; for example, Gobas et al. (1999) observed that the gut contents of a rock bass contained only about one-third the lipid content of the crayfish species that formed the predominent prey. Hence, biomagnification would be encouraged in the herbivore's or carnivore's gastrointestinal tract, since it can be assumed that large parts of the lipids and proteins present in the diet, but not a biochemically recalcitrant xenobiotic compound, are degraded there. This digestion of the sorbent leads, of course, to an increase in fugacity of the compound in the gastrointestinal tract of an organism as compared to the original diet. Because elimination by passive depuration through the gills or lungs is slow in higher organisms, concentration levels well above predicted equilibrium values may thus be established in such organisms for long time periods (i.e., even for the lifetime of the organism).

Biomagnification Along Aquatic Food Chains and Food Webs

Thus, there are various compound- and organism-specific factors that determine whether a "real" biomagnification occurs in a given food chain or food web. Armed with our improved understanding of these factors, let us examine some reported examples of specific organochlorine compound concentrations in organisms forming simple food chains or food webs (Fig. 10-10). Knowing that this type of apolar compound tends to accumulate predominantly in the lipids of organisms, we normalize all the observed concentrations by the lipid contents of these organisms yielding values of $C_{i\,lip}$ ($\mu g \cdot kg^{-1}$ lip) for each organism or organism tissue. To the extent that this lipid-dominance is correct, the resultant values are then linearly proportional to the individual compound's activities or fugacities [Eqs. 10-27 and 10-29 (phase = lip)].

As is evident in planktonic food webs (e.g., Fig. 10.10a, Harding et al., 1997; Paterson et al., 1998), individual persistent hydrophobic compounds such as PCB congeners do not tend to show large biomagnification. Often ranges of $C_{i\,lip}$ values observed at one trophic level overlap the ranges of observations made at higher levels. For example, in this case the plankton samples showed lipid-normalized concentrations of PCB153 ranging from 25 to 170 $\mu g \cdot kg^{-1}$ lip, while the juvenile fish at a higher trophic level exhibited 90 to 500 $\mu g \cdot kg^{-1}$ lip. Using mean values, $C_{i\,lip}$ values of the zooplankton are indistinguishable from the phytoplankton they eat, while the fish are only a factor of 3 "magnified."

Apparently, in smaller organisms like the plankton, depuration and/or excretion is fast enough that a significant disequilibrium between organism and diet may not be

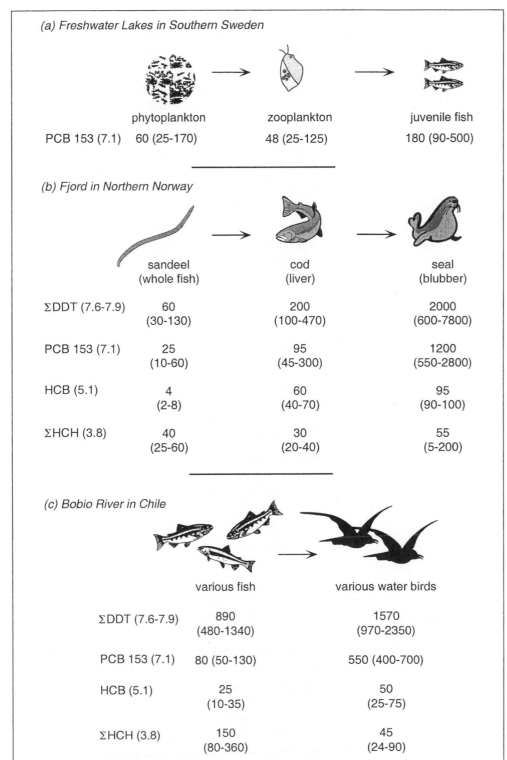

Figure 10.10 Average values of lipid-normalized concentrations (ranges in parentheses) of some organochlorine compounds: PCB153, ΣDDT = o,p-DDT + p,p-DDT + o,p-DDE + p,p-DDE, ΣHCHs = α- + β- + δ-hexachlorocyclohexane, and HCB = hexachlorobenzene (for structures see Fig. 2.14) in organisms belonging to some food chains ($\log K_{iow}$ values are given in parentheses after the compound names). All concentrations are expressed in $\mu g \cdot kg^{-1}$ lip. (*a*) Planktonic food webs in 19 lakes in Southern Sweden (Berglund et al., 2000). The average lipid contents were 5.4, 8.8, and 6.6% for the phytoplankton, zooplankton, and fish, respectively. (*b*) Local marine food chain in a fjord in Northern Norway (Ruus et al., 1999). (*c*) Fish and fish-eating water birds from the Santa Barbara location, Bobio River, Chile (Focardi et al., 1996).

established, even for highly hydrophobic compounds. However, in higher organisms, such disequilibrium with the environment may occur. Thus, if an organism's diet serves as the main source of a contaminant, and in digesting the diet the compound's fugacity in the gastrointestinal (GI) tract is increased by factors between 1 and 10, then the larger animal will exhibit a body fugacity between that of its GI tract and the environmental medium in which it respires.

Empirically, it is found that this disequilibrium and thus the BMF_i value tend to increase with increasing hydrophobicity of the compound (compare concentrations given in Figs. 10.10b and c; note that log K_{iow} values are given in parentheses). This is probably due mostly to slower depuration processes. Thus, for example, when considering biomagnification of PCBs, one usually observes a change in the mixture's composition, that is, a shift toward higher chlorinated congeners along a given food chain (see e.g., Russel et al., 1995; Feldman and Titus, 2001; Jackson et al., 2001). In contrast, a faster depuration process may explain the diminished biomagnification of the much less hydrophobic HCHs (log K_{iow} about 3.8) as compared to the other more hydrophobic organochlorine compounds. In addition, the BMF_i value of 0.3 for ΣHCH observed for the water birds (Fig. 10.10c) can be attributed to an increased metabolic transformation of these compounds in the birds (Focardi et al., 1996). Finally, we should point out that biomagnification in aquatic food chains or food webs may have a significant impact on humans, primarily on those for which aquatic organisms form an important diet. For example, in a survey conducted between 1990 and 1995 in Canada (Greizerstein et al., 1999), it was found that the lipid-normalized PCB concentrations in human milk of Inuits (who consume a lot of fish) were about one order of magnitude higher (i.e., ΣPCB = 400 $\mu g \cdot kg^{-1}$ lip) as compared to the other population in the same region.

Biomagnification Along Terrestrial Food Chains

Influence of Temperature Differences Along the Food Chain. Let us now evaluate some field data regarding biomagnification of semivolatile organic compounds (SOCs) in some terrestrial food chains. First, we consider an agricultural food chain involving air → pasture plants → (cow) → cow's milk → human milk (Fig. 10.11). Again, here we focus on three specific PCB congeners (i.e., PCB52, PCB153, PCB180). Since these PCBs have been seen to be persistent in the environment, it may be reasonable to initially assume that they are not rapidly degraded in any compartment of this food chain (although we will revisit this at the end). Their high hydrophobicities allow us to assume they primarily exist in the lipid fractions of the biological elements in the food chain.

McLachlan (1996) reported concentration data for these PCBs in the food chain from a study conducted in the region of Bayreuth, Germany (Table 10.5). Comparing cow's milk and pasture grasses, the biomagnification factors, $BMF_i = C_{ilip,cow}/C_{ip}$, are greater than 1 for PCB153 and PCB180 (i.e., 15 and 24 kg d.w.$\cdot kg^{-1}$ lip, respectively; Table 10.5), although PCB52 showed a value of only about 0.3. At first glance, this suggests that there is biomagnification of these two congeners from pasture to cow's milk. However, if the plants are assumed to contain about 5% lipid-plus-cutin (Table 10.1), then these lipid-normalized biomagnification factors, $BMF_{ilip} = C_{ilip,cow}/C_{ilip,p}$, of the two larger PCBs would have values close to 1.

However, to do these comparisons correctly, we really need to contrast the fugacities (or chemical activities) of the compounds in the different media. But unlike aquatic

Table 10.5 Average Concentrations of PCB52, PCB153, and PCB180 in Air (C_{ia}), Pasture (C_{ip}), Cow's milk (lipid normalized, $C_{ilip,cow}$), and Human Milk (lipid normalized, $C_{ilip,human}$) in the Region of Bayreuth, Germany (1989 – 91) (Data from McLachlan, 1996. Also given are the estimated air–lipid partition coefficients and the used $\Delta_{alip}H_i$ values.)

	PCB52 (2,2',5,5')	PCB153 (2,2',4,4',5,5')	PCB180 (2,2',3,4,4',5,5')
C_{ia} / (pg·m^{-3})	27	20	4.2
C_{ip} / (ng·kg^{-1} d.w.) ^{a}	80	400	80
$C_{ilip,cow}$ / (ng·kg^{-1} lip)	27	5900	1900
$C_{ilip,human}$ / (ng·kg^{-1} lip)	1500	175000	71300
log K_{ialip} / (L·kg^{-1} lip) ^{b}	–7.8	–8.8	–9.5
$\Delta_{alip}H_i$ / (kJ·mol^{-1}) ^{c}	81	91	97

^{a} Calculated from wet weight values by assuming 25% dry weight. ^{b} Calculated for 25°C from K_{ilipw} and K_{iaw}: $K_{ialip} = K_{iaw}/K_{ilipw}$; K_{ilipw} estimated using Eq. 10-2. ^{c} Values reported by Kömp and McLachlan (1997a) for vaporization, i.e., $\Delta_{alip}H_i \cong \Delta_{vap}H_i$.

systems, terrestrial food chains often contain interacting elements that exist at very different temperatures. For example, to calculate the fugacities of compounds in mammals, it is most appropriate to use a temperature of 37°C. We might even be interested in the fugacities of the PCBs in the grasses after they have been eaten by the cow, and then we would want those values at 37°C also. So in addition to the pertinent air–lipid partition coefficients for each congener at 25°C (Table 10.4), we need to adjust these partition coefficients for the temperature of interest using information on the enthalpies of air–lipid transfers ($\Delta_{alip}H_i$, Table 10.5).

Now we can calculate each congener's fugacity in each medium (Fig. 10.11). (The fugacities have been derived in the same way as described in the Illustrative Examples 10.4 and 10.5.) Due to seasonal variations, the resulting values represent averages and therefore should be viewed as order-of-magnitude values. Furthermore, the fugacities given in Fig. 10.11 differ somewhat from those calculated by McLachlan (1996) due to slightly different air–pasture and air–lipid partition coefficients used for their derivation. We should also point out that for PCB180, we have extrapolated the K_{ilipw} from K_{iow} values using Eq. 10-2. This means that the real K_{ilipw} value could be significantly smaller. This would translate into a larger K_{ialip} value, and therefore higher calculated fugacities of this compound in the lipids. Clearly one needs to be careful in interpreting such calculated fugacities, particularly in situations where we are close to equilibrium.

Nevertheless, some interesting general conclusions can be drawn from this case (Fig. 10.11). First, we note that the deviation from air–pasture equilibrium increases with decreasing volatility of the compound. This may imply the slower uptake of more hydrophobic compounds by the plants (McLachlan, 1999; Böhme et al., 1999). Second, using the temperature considerations, the ratio of fugacities, f_i, in the cow's

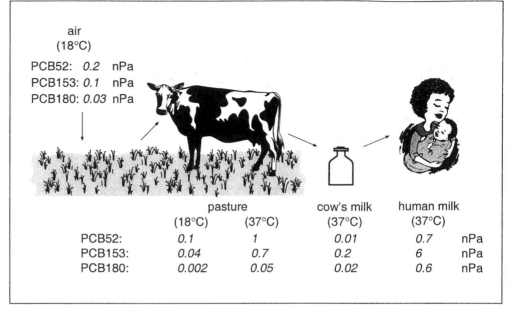

Figure 10.11 Calculated average fugacities of PCB52, PCB153, and PCB180 in air, pasture, cow's milk, and human milk in the region of Bayreuth, Germany in 1989/90. The average concentrations from which the fugacities have been calculated are given in Table 10.5 together with the partition coefficients and $\Delta_{12}H_i$ values for air–lipid partitioning. The average partition coefficients, K_{iap}, and corresponding $\Delta_{ap}H_i$ values are given in Illustrative Example 10.5.

milk versus the grass becomes 0.01, 0.3 and 0.4 for PCB52, PCB153, and PCB180, respectively.

Now we see the interesting result that at 37°C the fugacities of the compounds in the milk are lower than in the plants forming the diet of the cow. This result derives from the increased fugacities of the "diet" after it is taken in and its temperature is raised in the gut of the animal. On the other hand, with the digestion of the grasses causing the loss of sorbent in the GI tract and thereby increasing the fugacities on undigested components, one would expect higher chemical fugacities at the higher trophic level.

Hence, in this particular case we could speak of a "biodilution" rather than of a biomagnification. The especially low observed fugacity of PCB52 in cow's milk could be partially due to metabolic transformation of this compound in the cow. Finally, we note that fugacities in human milk fat are between one and two orders of magnitude higher than in cow's milk fat. These findings suggest that humans were not acquiring these PCBs chiefly from cow's milk, but from other sources, for example, from aquatic food sources (fish, seafood, etc.) that contain much higher levels of such compounds (see Section 10.3), and/or that humans are very efficient in biomagnifying compounds from their food sources.

Let us conclude this example by comparing these data with data collected in another, similar study (Thomas et al., 1998a; Table 10.6). As can be seen in the data, comparable BMF_i values were found for cow's milk as well as for cow's meat. Hence it seems that such compounds accumulate in a very similar way in the milk and the muscle of the animals. For those among us who like to eat liver, we should note that PCB levels of various congeners were generally found to be a factor of two to four times higher in cow's liver as compared to other body fat (Thomas et al., 1999).

Table 10.6 Biomagnification Factors (i.e., Animal Fat-Feed Concentration Ratios), BMF_i for PCB52, PCB153, and PCB180 for Transfer from Pasture to Animal Fat

Animal Compartment	BMF_i / (kg d.w.·kg^{-1} lip)		
	PCB52 (2,2',5,5')	PCB153 (2,2',4,4',5,5')	PCB180 (2,2',3,4,4',5,5')
Cow's milk [a]	0.34	15	24
Cow's milk [b]	n.d. [c]	20	19
Cow's meat [b]	0.40	13	20

[a] Germany 1989/90; data from McLachlan (1996); see Table 10.5. [b] England 1996/97; data from Thomas et al. (1998a). [c] n.d. = not determined.

Influence of Seasonal Temperature Differences. In our second example, we consider a wildlife food chain involving the transfer of our PCB model compounds from lichen/willow leaves to caribou to wolves in an arctic environment (Kelly and Gobas, 2001). Using observations of chemical concentrations, we calculate the fugacities of a few PCBs as we have done before but always at 37°C (Fig. 10.12). The sampling for the analyses of the PCBs in lichens and in willow leaves was performed a few years later than for the animal tissues (i.e., 1994/97 instead of 1992). However, when considering the very similar seasonal concentration pattern of the compounds in arctic air during these years (Hung et al., 2001), the calculated fugacities should be reasonably representative.

We can see some interesting features. First, we notice the much higher fugacities (i.e., concentrations) of the compounds in the pasture in the spring as compared to the summer, although air concentrations are generally higher in the summer as compared to the winter (Hung et al., 2001). This is because the air–plant partition

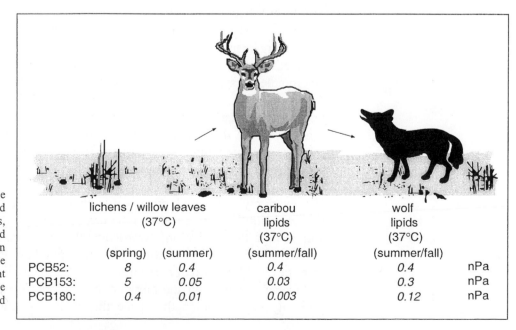

Figure 10.12 Calculated average fugacities of PCB52, PCB153, and PCB180 in pasture (i.e., lichens, willow leaves), caribou lipids, and wolf lipids in an arctic food chain in Canada (Bathurst Region). The caribou and wolf data represent averages of male and female animals. Data from Kelly and Gobas (2001).

	lichens / willow leaves (37°C)		caribou lipids (37°C)	wolf lipids (37°C)	
	(spring)	(summer)	(summer/fall)	(summer/fall)	
PCB52:	8	0.4	0.4	0.4	nPa
PCB153:	5	0.05	0.03	0.3	nPa
PCB180:	0.4	0.01	0.003	0.12	nPa

coefficients are highly temperature dependent. For example, the K_{iap} value of PCB153 ($\Delta_{ap}H_i = 110$ kJ·mol^{-1}, see Illustrative Example 10.5) changes by almost a factor of 5 per 10°C change in temperature. Hence, the large seasonal concentration variations found for PCBs in arctic plant tissues are likely due primarily to the large seasonal temperature fluctuations in such regions. Since the body concentrations of contaminants such as PCBs in animals like caribou and wolves represent a time-averaging over months and years, such time-varying exposures make it difficult to determine any biomagnification.

Another important result is that, very similar to the agricultural food chain discussed above (Fig. 10.11), there is no real biomagnification from pasture to consumers (in this case the caribou). In fact, the caribou consistently exhibited fugacities that were much less than their springtime diet.

Finally, for the higher chlorinated congeners (PCB153, PCB180), biomagnification is observed between animal prey (i.e., the caribou) and predator (i.e., the wolf). This may be explained if the wolf digests much of its caribou diet and thereby substantially increases the PCB fugacities in its GI tract. If excretion and respiration only slowly return the PCBs to the arctic air from where they came, then the wolf may exhibit a continuous fugacity excess relative to its prey.

Finally, it is interesting to note that in the fall, female animals exhibited somewhat lower PCB and other organochlorine compound concentrations than male animals. This may be explained in part by lactation throughout the nursing period that occurs during the summer (for more details see Kelly and Gobas, 2001).

10.6   Baseline Toxicity (Narcosis)

We conclude this chapter by briefly addressing toxicological issues that primarily result from accumulation of chemicals in biological membranes. Organic chemicals can exert a variety of toxic effects in organisms. Depending on both the chemical and the organism, these may include unintended interactions of the contaminant or its reactive metabolites with critical functional components like enzymes or genetic macromolecules (DNA, RNA). Further, even without covalent bond formation, some nonpolar compounds can physically associate with enzymes or genetic materials and disrupt their normal three-dimensional structure. This may harm the functioning of the macromolecule (e.g., association of dioxins with DNA, or estrogen disruptors binding to hormone recognition sites). Such effects require quite specific chemical interactions with components of the organism.

However, any compound, even if it is chemically inert, if present at high enough concentrations in biological membranes can change those membranes' properties and disrupt their functions. Consequently, membrane-associated processes like photosynthesis, energy transduction, transport in or out of the cell, enzyme activities, transmission of nerve impulses, and so on may deteriorate (see van Wezel and Opperhuizen, 1995 and literature cited therein). Since these effects seem to be primarily dependent on the space that contaminating molecules occupy in the membrane, and not on the chemical structure of the compound, one commonly refers to this type of toxicity as *nonspecific toxicity*. Since such nonspecific interactions include disrupting nerve functions, they are also often referred to as *narcosis*. Furthermore, because any additional specific toxic effect would increase the overall toxicity of a compound, one also uses terms like *minimum toxicity* or *baseline toxicity* to describe this mode of toxic action.

Quantitative Structure–Activity Relationships (QSARs) for Baseline Toxicity

Let us now evaluate how we can assess the baseline toxicity of organic chemicals in a quantitative way. We have already mentioned that certain membrane functions may be disrupted if a chemical occupies a certain volume fraction of that membrane. This means that for two compounds of the same size, we would anticipate that when they are present at equal concentrations in the membrane they would exert the same effect. Furthermore, since the majority of chemicals of interest to us do not differ in size by more than a factor of 3 to 4 (compare molar volumes in Chapter 5, e.g., Fig. 5.2), the membrane concentration required for any compound to cause a narcotic effect will be in the same order of magnitude. Therefore, we may expect that the concentration of a compound required in an environmental medium (e.g., water, air) to cause a narcotic effect in an organism should be inversely proportional to the tendency of the compound to accumulate from that medium into biological membranes.

In fact, for chemicals that exhibit only baseline toxicity, linear free energy relationships (LFERs) relating the environmental concentration required to exert a certain effect on (i.e., the effective concentration, EC_i), or even kill (i.e., the lethal concentration, LC_i), an organism have been found to correlate with parameters used to describe organic phase–water partitioning (e.g., Lipnick, 1995). The simplest and most widely used descriptor is the octanol–water partition constant. Hence, in the literature, one can find numerous LFERs [also called *quantitative structure–activity relationships (QSARs)*] relating LC_{i50} values (concentration in $mol \cdot L^{-1}$ required to kill 50% of a given population after a certain time period, i.e., 24 h, 48 h, 96 h) with K_{iow}s:

$$\log \frac{1}{LC_{i50}} = a \cdot \log K_{iow} + b \qquad (10\text{-}37)$$

For example, such a relationship can be seen to apply to the data of Hutchinson et al. (1978), who reported the concentrations of numerous apolar and slightly monopolar compounds to inhibit microalgal photosynthesis by a factor of 2. When the effective concentrations (in $mol \cdot L^{-1}$) are examined versus each compound's octanol–water partition coefficients, one finds the following relations:

$$\log \frac{1}{EC_i} = 0.93 \log K_{iow} + 5.17 \text{ (for } Chlamydomonas\ angulosa\text{)} \qquad (10\text{-}38)$$

and:

$$\log \frac{1}{EC_i} = 0.82 \log K_{iow} + 4.93 \text{ (for } Chlorella\ vulgaris\text{)} \qquad (10\text{-}39)$$

These trends clearly reveal that chemicals with greater tendencies to partition into membranes require lower concentrations in the culture media to cause

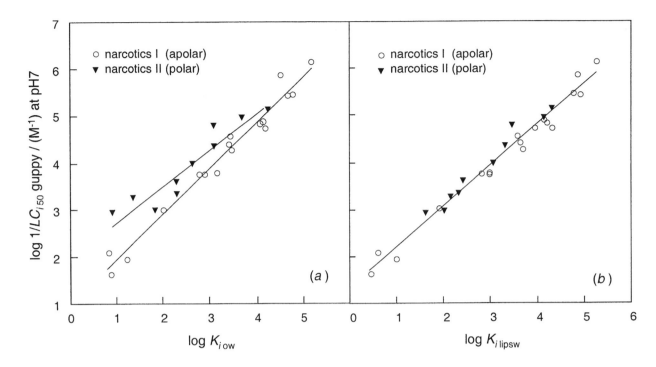

Figure 10.13 Plot of log $1/LC_{i50}$ for guppies (small tropical fish, *Poecilia reticulata*) versus (*a*) log octanol–water partition constant (log K_{iow}), and (*b*) log L-α-dimyristoylphosphatidylcholine liposome–water coefficient (log K_{ilipsw}) for a series of neutral apolar and weakly polar (o) and polar (▼) compounds. Data from Gobas et al. (1988), Vaes et al. (1998), and Gunatilleka and Poole (1999).

photosynthetic inhibition. Such an effect could be caused by mixing enough of each compound into the chloroplast membranes to change their properties in a way that their function is compromised.

However, as we have already pointed out above (Fig. 10.4), K_{iow} is not always an optimal choice for mimicking membrane–water partitioning, particularly when one is dealing with a diverse group of chemicals including apolar, monopolar, and bipolar compounds. This can be seen in an analysis of the log $1/LC_{i50}$ values for a series of compounds of different polarities tested for their ability to kill guppies (Fig. 10.13*a*). Plotting the data against each compound's log K_{iow} yields different regression lines for some monopolar and bipolar compounds as compared to some apolar and weakly monopolar compounds. Such findings have led to the suggestion that we must divide chemicals exhibiting only baseline toxicity into two categories: polar and nonpolar narcotics (Verhaar et al., 1992). However, when we use a more appropriate membrane mimic like liposomes to study the contaminant's buildup in the biological membranes, then we find that the resulting liposome–water partition coefficients, K_{ilipsw}, can be used to develop a single LFER for all the compounds (Fig. 10.13*b*):

$$\log \frac{1}{LC_{i50}} = 0.87 \log K_{ilipsw} + 1.3 \ (R^2 = 0.99; \ N = 27) \tag{10-40}$$

Note that other promising approaches for relating toxicity to molecular properties include the use of polyparameter LFERs, as we have discussed in Chapters 6 and 7 for describing air–organic phase and organic phase–water partitioning (Gunatilleka and Poole, 1999).

The problem of using octanol as a surrogate for membranes is even more pronounced when dealing with membrane–water partitioning of weak organic acids

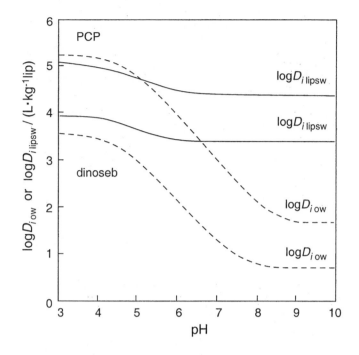

Figure 10.14 Log octanol–water distribution ratios (log D_{iow}, broken lines) and log L-α-dioleylphosphatidylcholine liposome–water distribution ratios (log D_{ilipsw}, solid lines) of pentachlorophenol (PCP) and 2-*sec*-butyl-4,6-dinitrophenol (dinoseb) as a function of pH at 25°C and 100 mM ionic strength. Data from Escher et al. (2000).

pentachlorobenzene
(PCP)

2-*sec*-butyl-4,6-dinitrophenol
(dinoseb)

or bases. Consider the octanol–water and liposome–water partitioning behavior of the two weak acids pentachlorophenol (PCP, $pK_a = 4.75$) and 2-*sec*-butyl-4,6-dinitrophenol (dinoseb, $pK_a = 4.62$). We evaluate how these compounds' organic phase–water distribution ratios, $D_{iorg.phase\,w}$, vary as a function of pH (Fig. 10.14):

$$D_{iorg.phase\,w} = \frac{[HA]_{org.phase,tot}}{[HA]_w + [A^-]_w} \tag{10-41}$$

where $[HA]_{org.phase,tot}$ is the total concentration of the acid in the organic phase (neutral + anionic species) and $[HA]_w$ and $[A^-]_w$ are the concentrations of the neutral and deprotonated species in the aqueous phase (see also Section 8.5). For both compounds, the distribution ratios are clearly pH-dependent. However, the pH-dependence is much more pronounced in the octanol–water system. The reason is that phenolates and, in general, charged organic species partition much more favorably from water into polar lipids (such as present in membranes) as compared to octanol (for more details see Escher and Schwarzenbach, 1996; Escher et al., 2000). These amphiphilic compounds are able to position themselves so that their polar ends may interact with the polar moieties of the lipid bilayer while the nonpolar part is accommodated in the nonpolar regions of the liposomes; such accommodation is not favored in octanol as a bulk medium.

Recognizing the complexity of quantifying both apolar and bipolar compound partitioning to membranes with a single parameter, let us now inspect the acute toxicity (expressed by the LC_{i50}) of dinoseb and of a series of chlorinated phenols (including PCP) toward guppies at pH 7. We compare these ionizable compounds' toxicities to the toxicities of a series of chlorinated benzenes that act only as narcotics. For most of the phenols, but particularly dinoseb and PCP, a much higher toxicity is seen than one would predict from either their D_{iow} (Fig. 10.15a) or D_{ilipsw}

(Fig. 10.15b) values and the trend derived from the chlorinated benzenes alone (i.e., the data points for these compounds lie well above the regression line indicating baseline toxicity). This enhanced toxicity of the phenols may not be too surprising since these compounds are known to exert specific effects, including interference with the energy transduction of cells by destroying the electrochemical proton gradient (i.e., uncoupling; for more details, e.g., Escher et al., 2001). We should note that comparison of the actually measured toxicity of a compound with its baseline toxicity predicted from QSARs such as Eq. 10-40 allows one to recognize whether the compound also exerts a specific mode of toxic action. To this end, a toxic ratio, TR_i, can be defined to quantify this excess toxicity. For example, if death of the organism is the measured toxicity end point, TR_i is given by (Verhaar et al., 1992):

$$TR_i = \frac{LC_{i50} \text{ (baseline, predicted from a QSAR)}}{LC_{i50} \text{ (actually measured)}} \qquad (10\text{-}42)$$

As can be seen from Fig. 10.15a, if we would use D_{iow} for prediction of baseline toxicity, we would get a TR_i value of almost 10^4 for dinoseb. But this result would be at least partially due to our underestimating the partitioning of this compound from water to the membranes. A more correct TR_i value is the one taking into account the significant partitioning of the anion to the membrane by using the D_{ilipw} value; that is, a more realistic TR_i value for dinoseb is only in the order of 10^2 (Fig. 10.15b). Note that in this case, we assume that the anionic species exhibits the same narcotic effect as the neutral species. This example shows again that we have to be very careful in choosing appropriate molecular parameters for describing partitioning of organic compounds to biological media.

Critical and Lethal Body Burdens

Finally, we can use our insights into bioaccumulation phenomena to understand concepts such as critical body residues (CBR_i, McCarty et al., 1992), critical body burdens, CBB_i, and lethal body burdens, LBB_i (e.g., Sijm and Hermens, 2000; Barron et al., 2001). These parameters reflect attempts to set values quantifying generally applicable concentrations of a chemical in an organism that are found to elicit a particular toxic effect. For example, it has been proposed that a small range of chemical concentrations between 2 and 8 mmol kg^{-1} w.w. of nonionic compounds always induces a narcotic effect for all organisms; this "threshold" concentration in organisms has been called the critical body residue level (CBR_i) (McCarty et al., 1992). Similarly, the critical body burdens (CBB_i, if the observed effect is not death) and the lethal body burdens (LBB_i, if the organism is killed) have been quantified for various organism/compound combinations to identify relatively sensitive organisms for particular contaminants (e.g., Sijm and Hermens, 2000; Barron et al., 2001.)

However, having seen that organisms are variable mixtures of organic materials (and they have variable water contents), we should now recognize that whole body- and wet weight-normalized concentrations may not be the most useful parameter. As discussed in this chapter, compounds accumulate very differently into the various compartments of an organism. Thus, for example, an organism exhibiting a large portion of storage lipids (where no toxic effects occur) may have a significantly higher CBB_i as compared to an organism with a very low lipid content. Such a

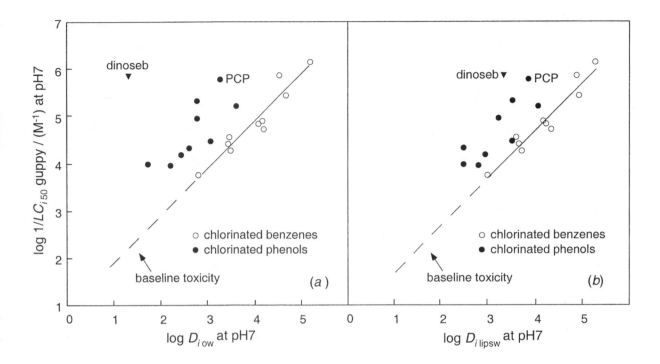

Figure 10.15 Plot of log $1/LC_{i50}$ for guppies versus (a) log octanol–water distribution ratio (log D_{iow}, Eq. 10-41), and (b) log liposome–water distribution ratio (log D_{ilipsw}, Eq. 10-41) at pH 7 for a series of chlorinated benzenes (○) and chlorinated phenols (●); as well as for the herbicide 2-sec-butyl-4,6-dinitrophenol (dinoseb) (▼). The liposomes used were L-α-dimyristoyl-phosphatidylcholine (chlorinated benzenes) and L-α-dioleyl-phosphatidylcholine (chlorinated phenols and dinoseb). The pH dependence of D_{iow} and D_{ilipsw} of pentachlorophenol (PCP) and dinoseb is shown in Fig. 10.14. Data from Saarikoski and Viluskela (1992), Gobas et al. (1988), Escher and Schwarzenbach (1996), and Gunatilleka and Poole (1999).

species may thus be classified (erroneously) as less sensitive to the compound. Consequently, for many hydrophobic compounds a lipid-normalized CBB_{ilip} or LBB_{ilip} would already be a more reasonable parameter, although only the concentration in the membrane lipids may be of toxicological relevance. For apolar or weakly polar compounds that partition more or less equally well into the various lipid compartments, this would not be too much of a problem, and CBB_{ilip} or LBB_{ilip} would be a reasonable measure. For polar compounds, and in particular weak organic acids and bases, we would expect organisms with a high storage lipid content to be more sensitive, because polar compounds tend to partition more favorably into polar lipids (as present in membranes) as compared to apolar storage lipids. Hence, these should have a smaller apparent CBB_{ilip} or LBB_{ilip} value.

Finally, we should point out that in natural systems, organisms may be exposed to a variety of chemicals at the same time (e.g., after an oil spill, in PCB-contaminated sediments, etc.). Since in the case of narcotic effects, the chemical nature of the compound is not so important, it is easy to imagine that it is the sum of all compounds accumulating in an organism's membrane that will lead to an adverse effect. Thus, the CBB_i and LBB_i values measured for a single mixture component may be significantly lower than the value that would be measured if a clean organism were exposed to the compound alone. Similarly, an already contaminated organism may exhibit a much lower EC_i or LC_i value when exposed to a single chemical as compared to an organism that, prior to exposure, had lived in a clean environment. Hence, although for narcotic effects the CBB_i or LBB_i values are quite high for a single compound (see Illustrative Example 10.6), because of its concentration-additive nature, this mode of action may be of ecotoxicological significance in contaminated environments (van Wezel et al., 1996b).

Illustrative Example 10.6   **Evaluating Lethal Body Burdens of Chlorinated Benzenes in Fish**

Problem

Sijm et al. (1993) determined LBB_i and 96h- LC_{i50} values of a series of chlorinated benzenes in guppies (*Poecilia reticulata*). The fish had an average weight of 0.18 g w.w. and a total lipid content of about 5% w.w. For 1,2,3-trichlorobenzene (TrCB) the LBB_i values found were 2.7, 2.0, and 2.4 mmol·kg^{-1} w.w. for exposure to 5.6, 3.8, and 1.9 μmol TrCB·L^{-1}, respectively. The death of the fish occurred after 2.4, 24, and 96 hours. Note that similar LBB_i values (2 – 8 mmol·kg^{-1} w.w.) were found for the other compounds. The 96 h LC_{i50} value of TrCB was determined to be 1.9 μmol·L^{-1}.

(a) Estimate the volume fraction of TrCB in the membrane lipids of the guppies for a LBB_i value of 2.4 mmol·kg^{-1} w.w.

(b) Calculate the theoretical bioaccumulation potential TBP_i (Eq. 10-5) of TrCB at an aqueous concentration corresponding to the 96 h LC_{i50} value and compare this value with the LBB_i value given above.

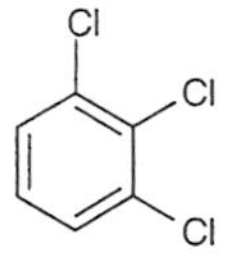

i = 1,2,3-trichlorobenzene
(TrCB)
$\log K_{iow}$ = 4.14

Answer (a)

Assume that TrCB accumulates primarily in the lipid phases of the fish, and that accumulation into the polar membrane lipids (about 25% of the total lipids; van Wezel et al., 1995) and storage lipids is about equal. Hence, the lipid normalized concentration of TrCB in guppy is:

$$C_{ilip} = LBB_i / f_{lip} = (2.4 \text{ mmol·kg}^{-1} \text{ w.w.}) / (0.05 \text{ kg lip·kg}^{-1} \text{ w.w.})$$

$$= 48 \text{ mmol·kg}^{-1} \text{ lip}$$

Estimate the molar volume, V_i, of TrCB from its density and molar mass given in Appendix C (see also Eq. 3-18):

$$V_i = M_i / \rho_i = (181.5 \text{ g·mol}^{-1}) / (1.49 \text{ g·cm}^{-3}) = 121 \text{ cm}^3·\text{mol}^{-1}$$

The total volume of TrCB in one kg of lipid is then given by:

$$V_{tot,TrCB} = (48 \times 10^{-3} \text{ mol}) (0.121 \text{ L·mol}^{-1}) = 0.0058 \text{ L}$$

Assuming that the density of the lipids is close to 1 kg lip·L^{-1}, this means that the lethal volume fraction of apolar narcotics in membranes is on the order of *1%* (fraction $\cong 0.01$ L compound·L^{-1} lipid), which is quite considerable.

Answer (b)

Use Eq. 10-7 to estimate the TBP_{ilip} value:

$$TBP_{ilip} = K_{ilipw} \cdot LC_{i50} \tag{1}$$

Estimate K_{ilipw} from K_{iow} using the LFER Eq. 10-2. The resulting K_{ilipw} value is 1.9×10^4 L·kg^{-1} lip. Insertion of this value together with the corresponding LC_{i50} into Eq. 1 yields:

$$TBP_{ilip} = (1.9 \times 10^4 \text{ L·kg}^{-1} \text{ lip}) (1.9 \times 10^{-6} \text{ mol·L}^{-1}) = 36 \text{ mmol·kg}^{-1} \text{ lip}$$

which corresponds to a LBB_i value of:

$$LBB_i = TBP_{ilip} \cdot f_{lip} = (36 \text{ mmol·kg}^{-1} \text{ lip}) (0.05 \text{ kg lip·kg}^{-1} \text{ w.w.})$$
$$= 1.8 \text{ mmol·kg}^{-1} \text{ w.w.}$$

This is in very good agreement with the LBB_i values determined experimentally. Note, however, that for the same compound and type of fish, significantly higher (factor of 5) 96 h LC_{i50} values have been reported (Gunatilleka and Poole, 1999), indicating that depending on the test conditions (including the conditions of the organisms) one has to expect some variability in the toxicity data for a given compound and organism.

10.7   Questions and Problems

Questions

Q 10.1

Why can bioaccumulation, in general, not be treated as an equilibrium process?

Q 10.2

What are the most important organic phases (polymers) present in living media into which organic pollutants may partition? Characterize these phases with respect to their ability to "dissolve" organic solutes via various intermolecular interaction mechanisms. In which cases are which phases important?

Q 10.3

Describe in words, the parameters: (1) bioconcentration factor, (2) biomagnification factor, (3) biota-sediment- and biota-soil- accumulation factor, and (4) bioaccumulation factor.

Q 10.4

Through what mechanisms can bioaccumulation lead to a "real" biomagnification?

Q 10.5

Why do biomagnification and biota-sediment-accumulation factors of persistent chemicals in aquatic systems tend to increase with increasing hydrophobicity of the compounds?

Q 10.6

Why is it useful to compare the activities or fugacities of a given compound in various environmental compartments?

Q 10.7

Explain in words how you calculate the activity or fugacity of a given compound in a given environmental compartment (e.g., water, sediment, lipid components of an organism, etc.). What are the major problems encountered when calculating (or estimating) fugacity values?

Q 10.8

What is meant by the terms "baseline toxicity" or "narcosis"? Why is this type of toxicity relevant in the environment?

Q 10.9

When is the *n*-octanol–water partition constant an acceptable parameter for the quantification of nonspecific membrane toxicity? In what situations does it fail?

Problems

P 10.1 *Evaluating the Accumulation of Chlorobenzenes from Water to Tissues of the Soybean Plant*

Tam et al. (1996) investigated the uptake from water of a series of chlorinated benzenes by various tissues (leaves, petals, stems, roots) of the soybean plant. For two of the seven compounds investigated, they obtained the following apparent equilibrium leaf-water (BAF_{ileafw}) and root-water (BAF_{irootw}) bioaccumulation factors (or bioconcentration factors since uptake is only from water, see Fig.10.5):

	1,2-Dichlorobenzene (DCB)	1,2,3,4-Tetrachlorobenzene (TeCB)
BAF_{ileafw}/(L·kg^{-1} w.w.)	7.9×10^1	1.0×10^3
BAF_{irootw}/(L·kg^{-1} w.w.)	2.2×10^1	2.9×10^2

The lipid and water contents of the fresh leaves were 1.9% and 82%, respectively, those of the roots, 0.38% and 92%.

Estimate the K_{ibio} values (Eq. 10-4) of the two compounds partitioning to the soybean leaves and roots and compare these values with the experimentally determined BAF_i values given above. How important are materials contributing to the plant biomass other than the lipids (i.e., proteins, polysaccharides, water) with respect to the accumulation of the two compounds in soybean? Use Table 10.2 and Appendix C for solving this problem.

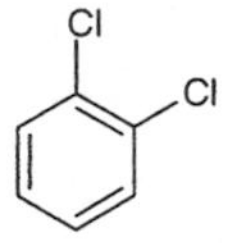

1,2-dichlorobenzene
(DCB)

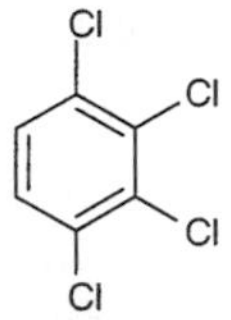

1,2,3,4-tetrachlorobenzene
(TeCB)

P 10.2 *Evaluating the Bioaccumulation of PAHs by Earthworms in Contaminated Soils*

In a field study, Ma et al. (1998) determined the concentration of PAHs in earthworms (*Limbricus rubellus*) present in various contaminated soils. The concentrations measured for phenanthrene and benzo[a]pyrene in the worms, C_{iworm}, collected from two of the soils investigated are given below together with the respective soil concentrations of the two compounds. The f_{oc} values of the soils were 0.038 (soil 1) and 0.015 (soil 2) kg oc $\cdot$ kg^{-1} solid. The average lipid content of the worms was 0.012 ± 0.06 kg lip $\cdot$ kg^{-1} f.w. (f.w. = fresh weight).

Calculate the lipid- and organic carbon–normalized biota-soil accumulation factors ($BASF_{ilipoc}$) for the two compounds and soils. Compare these values with estimated equilibrium accumulation factors (K_{ilipoc}). All necessary data can be found in Appendix C. Discuss why the resulting $BASF_i$ and K_i values differ.

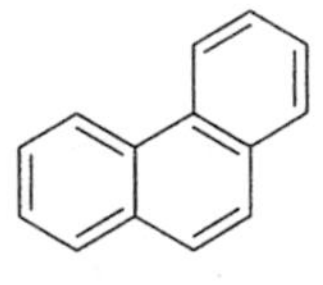

phenanthrene

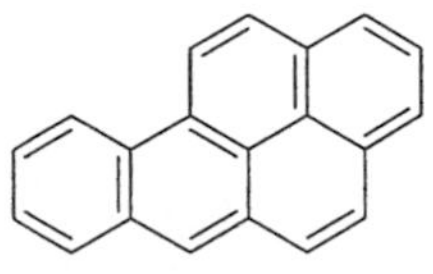

benzo[a]pyrene

	Phenanthrene	Benzo[a]pyrene
Soil 1:		
$C_{iworm}/(\mu g \cdot kg^{-1}$ f.w.$)$	17	28
$C_{is}/(\mu g \cdot kg^{-1}$ solid$)$	840	990
Soil 2:		
$C_{iworm}/(\mu g \cdot kg^{-1}$ f.w.$)$	4	6
$C_{is}/(\mu g \cdot kg^{-1}$ solid$)$	22	27

P 10.3 *Evaluating Phenanthrene and Pyrene Concentrations in Agricultural Plants*

Böhme et al. (1999) conducted a field study in which the concentrations of numerous SOCs were determined in a variety of agricultural plants in the region of Bayreuth, Germany. For phenanthrene and pyrene, they measured the following concentrations in ryegrass (C_{irye}) and yarrow (C_{iyar}). They also reported average air concentrations (C_{ia}) of the two compounds:

	Phenanthrene	Pyrene
$C_{irye}/(\mu g \cdot kg^{-1} \text{ d.w.})$	18	11
$C_{iyar}/(\mu g \cdot kg^{-1} \text{ d.w.})$	160	33
$C_{ia}/(ng \cdot m^{-3})$ [a]	2.7	0.18
$\Delta_{ap}H_i/(kJ \cdot mol^{-1})$ [b]	90	100

[a] Concentration in gaseous form. [b] Estimated value.

Estimate the concentrations of the two compounds in the plants ($\mu g \cdot kg^{-1}$ d.w.) that would be established in equilibrium with the measured concentrations in the air (a) by using the LFERs given in Table 10.4 (note that these equations were derived for PCBs), and (b) by assuming that the compounds are primarily accumulated in the lipids ($f_{lip} = 0.02$ kg lip·kg^{-1} d.w. for ryegrass and 0.015 kg lip·kg^{-1} d.w. for yarrow) of the plants, and that Eq. 10-2 can be also applied to PAHs. Also assume an average temperature of 15°C, and use the $\Delta_{ap}H_i$ values given above to quantify the effect of temperature in both air–plant and air–lipid partitioning. Do the values you estimate match the observed results? Discuss any discrepancies.

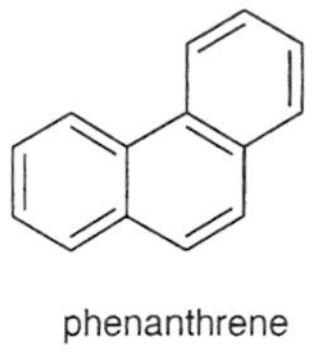

phenanthrene 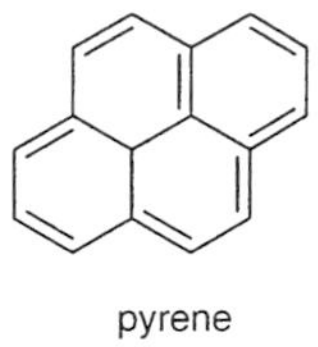   pyrene

P 10.4 *Assessing the Fugacities of Hexachlorobenzene (HCB) and 2,3,7,8-Tetra-chlorodibenzodioxin (2,3,7,8-TCDD) in an Agricultural Food Chain*

McLachlan (1996) measured the concentrations of a series of SOCs in an agricultural food chain in Bayreuth, Germany. For HCB and 2,3,7,8-TCDD he reported the following concentrations:

		HCB	2,3,7,8-TCDD
Air [a]	C_{ia} / (pg·m^{-3})	460	0.0027
Soil [b]	C_{is} / (ng·kg^{-1} d.w.)	360	0.05
Grass [c]	C_{ip} / (ng·kg^{-1} d.w.)	85	0.006
Cow's milk	$C_{ilip,cow}$ / (ng·kg^{-1} lip)	9000	0.19
Human milk	$C_{ilip,hum}$ / (ng·kg^{-1} lip)	230000	3.6
$\Delta_{vap} H_i$ / (kJ·mol^{-1})		90	105

[a] Gaseous concentration. [b] $f_{oc} = 0.02$ kg oc·kg^{-1} solid. [c] Various grasses, use average K_{iap} value (Table 10.4).

Calculate the fugacities of the compounds at 18°C (average temperature during the sampling period) for air, soil, and grass, and at 37°C for grass, cow's milk and human milk. Critically comment on the results. Assume that HCB and 2,3,7,8-TCDD can be treated like PCBs, and use $\Delta_{vap} H_i$ to account for the temperature dependence of the partitioning between the air and the various condensed phases. Use the data given in Appendix C.

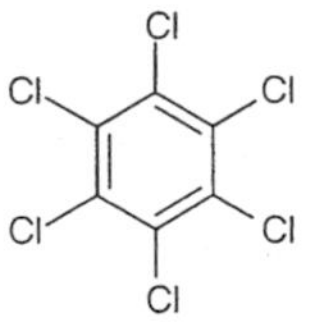

hexachlorobenzene
(HCB)

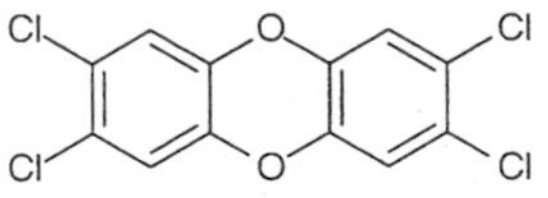

2,3,7,8-tetrachlorodibenzodioxin
(2,3,7,8-TCDD)

P 10.5 *Estimating Lethal Body Burdens in Fish Using LC$_{i50}$ values*

Van Wetzel et al. (1995) determined LBB_i values for some chlorinated phenols in fathead minnows (*Pimepholes promela*) at about 20°C in the laboratory. The fish had an average weight of 0.68 g w.w. and a lipid content of 4.4% of the wet weight. The LBB_i value found for 2,4-dichlorophenol (DCP) at pH 6.2 in short-term experiments (1 hour) at elevated concentrations was about 2 mmol·kg^{-1} w.w. Compare this value with the theoretical bioaccumulation potential (TBP_i) of this compound at an aqueous concentration corresponding to the LC_{i50} value. Note that the experimental LBB_i value is given on a wet weight basis. Assume that the LBB_i values for fathead minnows are similar to those for guppies, that is, use Eq. 10-40 to estimate LC_{i50} from K_{ilipsw} given in Table 10.2. For pK_{ia} see Appendix C.

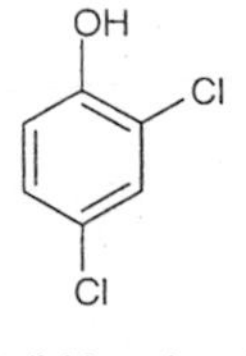

2,4-dichlorophenol
(DCP)

P 10.6 *Estimating the Lethal Body Burden (LBB$_i$) for 4-Nonylphenol for the Marine Amphipod, Ampelisca abdita, and Setting a Corresponding Sediment Quality Criterion*

The marine amphipod *Ampelisca abdita* is a suspension and deposit feeder and it often occurs as a numerically dominant member of benthic communities (Fay et al., 2000). It can also serve as an important food source for bottom-feeding fish. 4-nonylphenol is a degradation product from detergents (see Fig. 2.16). Its relevant physical-chemical properties are given in Appendix C.

You are interested in the well-being of *Ampelisca abdita*, living in a harbor whose sediments are contaminated with 4-nonylphenol. You remember that the lethal volume fraction of narcotic chemicals in membranes is about 0.01 L compound $\cdot$ L^{-1} lipid. If the sediment contains 2% organic carbon by weight, and the amphipod is assumed to accumulate body burdens up to equilibrium with the sediments on which it lives, what sediment concentration of 4-nonylphenol should be deemed acceptable with respect to baseline toxicity? Assume a log $K_{i\text{lipsw}}$ = 5.5 for 4-nonylphenol. Use Eq. 9-26c (alkylated and chlorinated benzenes!) for estimating $K_{i\text{oc}}$. Compare your result with the findings of Fay et al. (2000), who observed a die-off of half the amphipods when they were exposed to about 0.16 g 4-nonylphenol $\cdot$ kg^{-1} sediment.

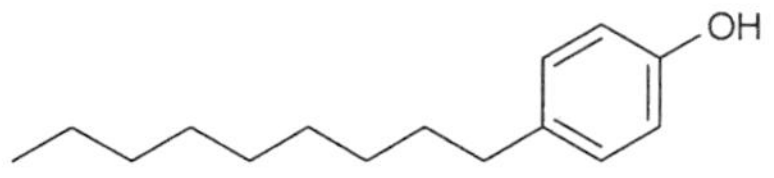

4-nonylphenol

Chapter 11

SORPTION III: SORPTION PROCESSES INVOLVING INORGANIC SURFACES

11.1 Introduction

11.2 Adsorption of Nonionic Organic Compounds to Inorganic Surfaces from Air
Characterization of Mineral Surfaces
Model for Energetics Controlling Air–Surface Adsorption of Apolar and Monopolar Compounds
Van der Waals and Polar Surface Parameters of Some Selected Liquid and Solid Surfaces
Estimating Air–Surface Adsorption Coefficients of Apolar and Monopolar Compounds; Applications
Illustrative Example 11.1: *Estimating the Fraction of Phenanthrene Present in the Gas Phase and Sorbed to the Walls of a Vessel*
Illustrative Example 11.2: *Sorption of Tetrachloroethene from Air to Moist and Dry Soil*
Illustrative Example 11.3: *Gas–Particle Partitioning of Organic Compounds to Environmental Tobacco Smoke: Adsorption or Absorption?*

11.3 Sorption of Nonionic Organic Compounds to Inorganic Surfaces in Water
Partitioning of Apolar and Weakly Monopolar Compounds into the Region Near Mineral Surfaces
Surface Adsorption Due to Electron Donor-Acceptor Interactions
Illustrative Example 11.4: *Estimating the Retardation of Trinitrotoluene Transport in Groundwater*

11.4 **Adsorption of Ionized Organic Compounds from Aqueous Solutions to Charged Mineral Surfaces**

Charging of Mineral Surfaces in Water

Box 11.1: *Estimating the Surface Charge $\sigma_{surf\,ex}$ of Oxides when H^+ and OH^- are the Potential-Determining Ions*

Conceptualization of the Ion Exchange Sorptive "Reactions," Free Energies, and Equilibrium Constants

Ion Exchange of Organic Cations

Box 11.2: *General Derivation of Ion Exchange Isotherms for Cationic Organic Compound, Competing with Monovalent, Inorganic Cations, from Solution Containing Monovalent Co-Ions*

Illustrative Example 11.5: *Transport of Di-Isopropanol-Amine (DIPA) in Groundwater from a Sour Gas Processing Plant*

Effects of Hydrophobicity of an Organic Sorbate's R Group

Illustrative Example 11.6: *Estimating Dodecyl Sulfonate Sorption to Alumina at Different pHs*

"Sorption" Due to Formation of Additional "Solid" Phases

11.5 **Surface Reactions of Organic Compounds** *(Advanced Topic)*

Organic Sorbate–Natural Organic Matter Reactions

Organic Sorbate–Inorganic Solid Surface Reactions

Illustrative Example 11.7: *Estimating the Adsorption of Benzoic Acid to Goethite*

11.6 **Questions and Problems**

11.1 **Introduction**

*Ad*sorption of organic molecules to polar *inorganic* surfaces can be important in a variety of environmental situations. For example, organic sorbates of all polarities are strongly attracted to "dry" inorganic surfaces from air (Fig. 11.1*a*). Thus organic chemical sorption to sun-dried surface soils can be very important, especially in arid regions. Adsorption of organic compounds from aqueous solutions to polar inorganic surfaces is also significant if those sorbates' structures motivate surface association. Such sorption-driving forces include sorbate hydrophobicity that discourages the sorbate from remaining in the bulk aqueous solution (Fig. 11.1*b*), specific sorbate-surface attractions arising from electron donor-acceptor interactions (Fig. 11.1*c*), and specific sorbate-surface attractions arising from the presence of complementary charges on the two partners (Fig.11.1*d*). These sorption mechanisms are important in a variety of environmental situations. For example, (weak) sorption to minerals due to sorbate hydrophobicity is seen for natural solids like those in sand and gravel aquifers that exhibit very little organic content. In the case of electron donor-acceptor effects, these are seen for certain aluminosilicates (e.g., clay minerals) interacting with aromatic compounds like the explosive, TNT. Electrostatic interactions must be considered for all charged organic compounds like cationic and anionic surfactants since natural inorganic solids are generally charged themselves when they are submerged in water. All of these sorption mechanisms are forms of *physisorption* since the sorbate-surface associations do not involve the formation of covalent bonds. This contrasts the cases where the organic sorbate participates in a bond-breaking and bond-making surface reaction that we refer to as *chemisorption* (Fig. 11.1*e*). Such surface bonding interactions are also described as the formation of *inner sphere* complexes since the surface and the organic sorbate approach one another close enough to enable overlap of the orbitals responsible for bonding. This bonding process is important when the inorganic surface exposes metals like iron or aluminum that utilize the organic sorbate as one of their complement of ligands.

In all of these cases, the structure of the organic sorbate, the composition of the surface, and the conditions of the vapor or solution exchanging with the solid must be considered. However, it is important to note that with some experience in thinking about the organic chemicals and environmental situation involved, we can usually anticipate which one or two sorption mechanisms will predominate. For example, in Chapter 9 we wrote an expression reflecting several simultaneously active sorption mechanisms, each with their own equilibrium descriptor, to estimate an overall solid–water distribution coefficient for cases of interest (Eq. 9-16):

$$K_{id} = \frac{C_{ioc} \cdot f_{oc} + C_{imin} \cdot A_{surf} + C_{iex} \cdot \sigma_{surf\,ex} \cdot A_{surf} + C_{irxn} \cdot \sigma_{surf\,rxn} \cdot A_{surf}}{C_{iw,neut} + C_{iw,ion}} \quad (9\text{-}16)$$

where in light of Fig. 11.1 we recognize C_{imin} is a sorbed species drawn to a mineral surface due to hydrophobicity or due to electron donor-acceptor interactions (Figs. 11.1*b* and *c*), C_{iex} is a sorbed species associated with the surface due to electrostatic attractions, and C_{irxn} is a sorbed species bonded to the surface. As discussed in Chapter 8, a particular compound's structure (i.e., pK_{ia}) and the solution pH dictates whether it is charged. In the absence of an ionizable species, we quickly disregard

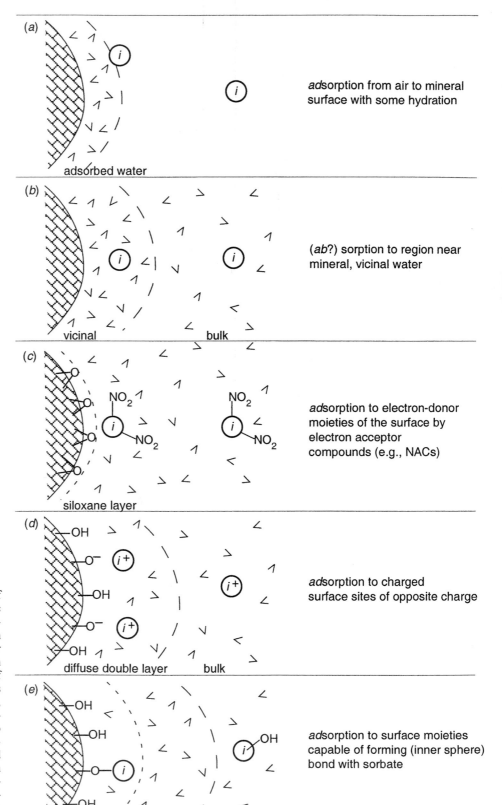

Figure 11.1. Schematic views of various ways in which an organic chemical, *i*, may sorb to natural *inorganic* solids: (*a*) adsorption from air to surfaces with limited water presence, (*b*) partitioning from aqueous solutions to the layer of "vicinal water" adjacent to surfaces that serves as an absorbent liquid, (*c*) adsorption from aqueous solution to specific surface sites due to electron donor-acceptor interactions, (*d*) adsorption of charged molecules from aqueous solution to complementarily charged surfaces due to electrostatic attractions, and (*e*) chemisorption due to surface bonding or inner sphere complex formation.

(*a*) *ad*sorption from air to mineral surface with some hydration

(*ab*?) sorption to region near mineral, vicinal water

(*c*) *ad*sorption to electron-donor moieties of the surface by electron acceptor compounds (e.g., NACs)

(*d*) *ad*sorption to charged surface sites of opposite charge

(*e*) *ad*sorption to surface moieties capable of forming (inner sphere) bond with sorbate

the terms involving $C_{iw,ion}$ and C_{iex} in Eq. 9-16. Likewise, the absence of detectable natural organic matter in or on a solid of interest ($f_{oc} \approx 0$) would cause us to disregard the term involving C_{ioc}. No matter which terms one imagines to be important, we need to discuss how to evaluate the relevant equilibria between *a single gaseous/ dissolved species and the single corresponding sorbed species*. We have already done this in Chapter 9 for the partitioning of nonionic organic compounds between water solutions and natural organic matter.

In this chapter, we examine mechanisms causing organic chemical sorption to polar inorganic surfaces from air and from aqueous solutions. We will begin by considering the adsorption of nonionic organic substances to (polar) inorganic surfaces from air (Section 11.2) and from water (Section 11.3). Then we will evaluate ionized compounds in solution having electrostatic attractions to charged mineral surfaces (Section 11.4). We will conclude with surface interactions involving sorbate–sorbent bonding (Section 11.5). As we proceed through this progression, we will identify solid properties like surface area (A_{surf}), surface charge ($\sigma_{surf\,ex}$), and reactive surface sites ($\sigma_{surf\,rxn}$) that are the key sorbent factors used to "tune" the intensity of each interaction for new cases of interest. In this manner, we should continue to systematically develop our understanding of how the structures of sorbates dictate their sorption behavior in solid-containing environments of interest.

11.2 Adsorption of Nonionic Organic Compounds to Inorganic Surfaces from Air

Transfer processes between the gas phase (i.e., air) and natural surfaces are frequently neglected in the environmental assessment of organic chemicals. Commonly, it is assumed that nonionic organic compounds partition primarily between the bulk phases air and water or air and organic matter, respectively. However, as we will see in our following discussions, there are situations in which *ad*sorption, particularly to inorganic surfaces, may be equally or even more important than *ab*sorption into a bulk phase. Such situations include sorption of gaseous chemicals to snow or ice, to the surfaces of small water droplets (< 10μm) as they occur in fog, to inorganic aerosols as encountered above the sea (i.e., salts) or above deserts (i.e., mineral dust), and, last but not least, to abundant soil minerals such as quartz and clays at low humidity. Furthermore, knowledge of the adsorption behavior of organic chemicals from the gas phase onto surfaces of condensed phases is important for indoor air quality assessments, and for the design of engineered systems, including air purification systems (e.g., filters). Unfortunately, to date, studies are still rather scarce in which exchanges of organic chemicals between the gas phase and environmentally important inorganic surfaces have been systemati-cally investigated. Nevertheless, since adsorption from the gas phase to a *surface* of a condensed phase is somewhat easier to treat than sorption processes involving bulk phases (where cavity formation has to be taken into account, see Figs. 3.2 and 3.3), some important general insights into this process can be gained with the limited experimental data.

Characterization of Mineral Surfaces

Besides liquid and solid (snow, ice) water surfaces, mineral surfaces play the most important role with respect to adsorption processes involving gaseous organic compounds in the environment. Thus, a brief description of the chemical nature of mineral surfaces is helpful to understand our subsequent discussion of such sorption phenomena.

Many common minerals expose a surface to the exterior which consists of hydroxyl groups protruding into the medium from a "checkerboard" plane of electron-deficient metals (e.g., Si, Al, Fe) and electron-rich ligands (e.g., hydroxyl, carbo-nate) (Fig. 11.2a). Like water molecules, these surface groups typically include a combination of hydrogen donors (e.g., $-OH$, $-OC(O)OH$) and acceptors ($-OH$, $-OC(O)OH$, and $-O-$). Hence, such bipolar surfaces interact with molecules adjacent to the mineral surface via vdW, H-donor, and H-acceptor forces. We can use data on pure liquid:silica solid attractions (Fowkes, 1964) to understand the relative contributions of such molecule:surface interactions (Fig. 11.2b). While all sorbates are attracted to all surfaces by vdW forces, stronger attractions per unit surface area of silica are observed as complementary functional groups are included on the surface and the sorbate that are capable of H-bonding (note that the free energies of adsorption, $\Delta_{asurf}G_i$, are in $mJ \cdot m^{-2}$). This surface attraction energy is very strong for an H-donor and H-acceptor sorbate like water ($460 \ mJ \cdot m^{-2}$), implying that such surfaces strongly prefer to bind water over small nonionic organic compounds. Thus, the overall energy change resulting from adsorption of organic chemicals directly to such solids would have to reflect the high "cost" of desorption of water from the same surface.

As minerals almost never exist in the environment without some exposure to water vapor, one may reasonably expect that water molecules are always present on these natural polar surfaces. The extent of this coverage is proportional to the activity of water in the atmosphere (i.e., the ratio of water's partial pressure to its saturation pressure at the particular temperature). Commonly such water activity is quantified in terms of "relative humidity" (RH) where 100% RH implies the air is equilibrated with pure water liquid (i.e., the reference state for water). $p_{H_2O}^*$ at ambient temperatures corresponds to about 1 mol of water per m^3 of air. Under this 100% RH condition, the activity of the water molecules on the mineral surface exposed to air can be assumed to be similar to that at bulk water surfaces. Hence, we can view the mineral as being fully covered with a liquid water film. This film is likely to have surface properties like those of pure water. But one has the first monolayer of water present on the surface (i.e., a one-water-molecule thick layer everywhere) at RH conditions much less than 100%. For example, many pure mineral oxides exhibit monolayer coverage at about 30% RH (Goss and Schwarzenbach, 1999b). Compositionally hetcrogeneous soils have been reported to have monolayer quantities of water (i.e., about 1% water content by weight) at less than 20% RH (Chiou and Shoup, 1985). At greater relative humidities, this layer of water grows thicker and thicker (Fig. 11.3). For example, Chiou and Shoup (1985) observed that adsorbed water on a soil equaled about three times monolayer coverage at 80% RH. The variably water-covered surfaces exhibit great differences in their affinities for organic compounds sorbing from the gas phase (Fig. 11.4: compare adsorption of 1,3-dichlorobenzene to dry

(*a*)

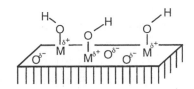

(*b*)   sorbate interactions with polar inorganic surfaces:

van der Waals

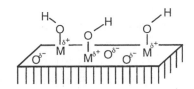

heptane on silica

$\Delta_{asurf}G_{heptane} \cong 100 \text{ mJ·m}^{-2}$

(28 kJ·mol⁻¹ heptane)

van der Waals
weak H-bond acceptor

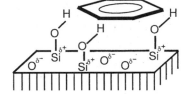

benzene on silica

$\Delta_{asurf}G_{benzene} \cong 140 \text{ mJ·m}^{-2}$

(28 kJ·mol⁻¹ benzene)

van der Waals
H-bond acceptor

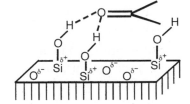

acetone on silica

$\Delta_{asurf}G_{acetone} \cong 160 \text{ mJ·m}^{-2}$

(28 kJ·mol⁻¹ acetone)

van der Waals
H-bond acceptor
and donor

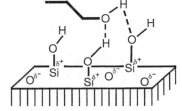

propanol on silica

$\Delta_{asurf}G_{propanol} \cong 180 \text{ mJ·m}^{-2}$

(33 kJ·mol⁻¹ propanol)

van der Waals
H-bond acceptor
and donor

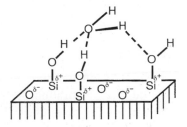

water on silica

$\Delta_{asurf}G_{water} \cong 460 \text{ mJ·m}^{-2}$

(30 kJ·mol⁻¹ water)

Figure 11.2 (*a*) A schematic view of a mineral surface exhibiting loci of partial positive charges where metal atoms occur (indicated by M's in the picture) and partial negative charges where linking anions occur (shown as oxygen atoms) and hydroxyls extending to the exterior. (*b*) Interactions of organic chemicals wetting silica surfaces and estimates of the $\Delta_{asurf}G_i$ values derived from surface tension data (Fowkes, 1964). Recall that $\Delta_{asurf}G_i$ represents the free energy for moving i from the surface to air; therefore, these values are positive.

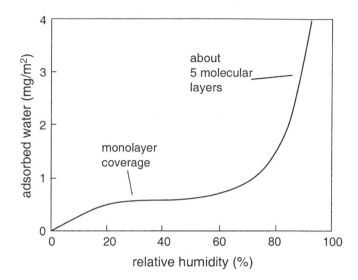

Figure 11.3 Typical adsorption isotherm for water on a mineral oxide surface.

soil, the same soil equilibrated with 50% RH, and the same soil equilibrated with 90% RH). Hence, because organic compounds are not able to replace water molecules at the surface, it is these partially water-wet mineral surfaces that we must consider as adsorbents of organic chemicals from air.

Model for Energetics Controlling Air–Surface Adsorption of Apolar and Monopolar Compounds

Now we are ready to discuss the free energy of surface adsorption of nonionic compounds on the surface of a condensed liquid or solid phase from air. We picture this transfer:

$$i(\text{air}) + \text{surface} \rightleftharpoons i(\text{surface}) \tag{11-1}$$

and emphasize as discussed above and in Chapter 3 that this process does not require any displacement of competing molecules, nor the formation of a cavity.

Since adsorption to a surface is directly proportional to the surface area of the condensed phase, it is most useful to define the equilibrium partition coefficient, $K_{i\text{asurf}}$, as the concentration of the compound in the gas phase (i.e., air, subscript "a") divided by the concentration per *unit surface area* (therefore subscript "surf" of the condensed phase:

$$K_{i\text{asurf}} = \frac{C_{ia}(\text{e.g.,}\,\text{mol}\cdot\text{m}^{-3})}{C_{i\text{surf}}(\text{e.g.,}\,\text{mol}\cdot\text{m}^{-2})} \tag{11-2}$$

Hence, $K_{i\text{asurf}}$ is expressed in m^{-1}. Furthermore, we assume that $K_{i\text{asurf}}$ is not dependent on the concentration of the compound at the surface; that is, we assume that we are far from saturating the surface with compound i. This then corresponds to a "linear adsorption isotherm" for a homogeneous mineral surface, since the sorbate molecules do not feel each other in the gas phase or at the solid surface. Finally, we should point out that in the literature, contrary to the notation used here, gas/solid partition coefficients are often expressed in a reciprocal way; that is, the reported

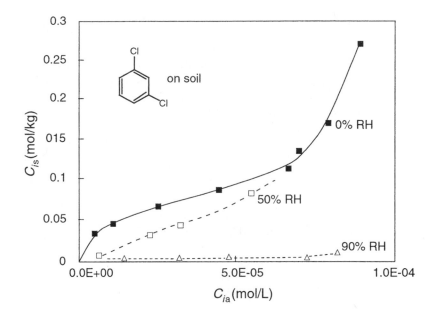

Figure 11.4 Vapor sorption isotherms of 1,3-dichlorobenzene on Woodburn soil at 20°C as a function of relative humidity (RH). Adapted from Chiou and Shoup (1985).

partition coefficient is given as the concentration in the condensed phase over the concentration in the gas phase.

In analogy to our molecular interaction model used to describe partitioning between bulk phases (Chapters 4 to 7), we can approximate the air–surface partition coefficient, K_{iasurf}, by summing the contributing interaction energies we make at the surface and assuming no significant interactions in the gas phase (Goss 1997a and b):

$$\ln K_{iasurf} = a(CA_i)(vdW_i)(vdW_{surf}) + b(HD_i)(HA_{surf}) + c(HA_i)(HD_{surf}) + d \quad (11\text{-}3)$$

where CA_i is the actual contact area of the molecule with the surface, vdW_i and vdW_{surf} are measures of the specific vdW interactions of the compound i and of the surface, $surf$, respectively, and HA_i, HD_i, HA_{surf}, and HD_{surf} are the corresponding parameters for describing the hydrogen-bonding acceptor and donor interactions between the compound and the surface. The factors a, b, c, and d are empirical proportionality coefficients.

Note that in Eq. 11-3, in contrast to the equation used for describing bulk phase partitioning, CA_i is not the total surface area (TSA_i) of the organic molecule. Rather it reflects the contact area between the molecule and the surface. But as a first approximation, we will assume that CA_i is proportional to TSA_i. By doing so, we neglect any special steric aspects that may affect the ability of a sorbate to closely approach the surface moieties with which it interacts.

Let us first consider the various compound-specific parameters in Eq. 11-3. In Chapters 4 to 7, we used the refractive index to quantify the dispersive vdW properties of a given compound [i.e., $vdW_i \propto (n_{Di}^2 - 1)/(n_{Di}^2 + 2)$, see Section 3.2]. We also noted that other parameters such as the air–hexadecane partition constant, K_{iah}, would have been more appropriate to describe vdW_i; however, due to lack of experimental K_{iah} values, we have chosen not to do so here. Instead, we use the liquid vapor pressure, p_{iL}^*, of the compound to describe vdW_i, although, because of polar interactions of bipolar compounds in their pure liquids, we will thus have to restrict our discussion to *apolar and monopolar* compounds. We do this primarily because in current practice, simple LFERs relating experimental air–solid (not surface, therefore, subscript "s") partition coefficients, K_{ias}, to vapor pressure data, are still widely used:

$$\log K_{ias} = a \cdot \log p_{iL}^* + b \tag{11-4}$$

Hence our following discussion will also allow us to critically evaluate such LFERs.

Using Eq. 4-24 and realizing that we consider only compounds for which the product $(\text{HD}_i)(\text{HA}_i) = 0$, we have:

$$\ln p_{iL}^* = a'(\text{TSA}_i)(\text{vdW}_i)^2 + d' \tag{11-5}$$

Since vdW_i varies by only about a factor of 1.5 among different compounds, there is an almost linear relationship between vdW_i and vdW_i^2 ($R^2 = 0.99$). Hence, we can write:

$$\ln p_{iL}^* = a''(\text{TSA}_i)(\text{vdW}_i) + d'' \tag{11-6}$$

Rearrangement of Eq. 11-6 and insertion into Eq. 11-3 together with the Abraham H-bond descriptors (Table 4.3), $\alpha_i(= \text{HD}_i)$ and β_i ($= \text{HA}_i$), yields the polyparameter LFER:

$$\ln K_{iasurf} = a'''(\text{vdW}_{surf}) \ln p_{iL}^* + b(\alpha_i)(\text{HA}_{surf}) + c(\beta_i)(\text{HD}_{surf}) \\ + d'''(\text{vdW}_{surf}) + \text{constant} \tag{11-7}$$

Note that the ratio of CA_i to TSA_i has been included in a''' and d'''. The next step is to find parameters for quantifying the van der Waals and hydrogen bonding terms for the surfaces. Goss (1997a and b) demonstrated that the square root of the van der Waals component of the surface free energy (e.g., derived from surface tensions in the case of liquids) can be used with good success. By using appropriate reference compounds, vdW_i values can be determined by contact angles for solids and interfacial tension measurements for liquids, or by inverse gas chromatography (for details and a literature review see Goss 1997a and b). As discussed below, quantification of HA_{surf} and HD_{surf} is best done relative to a bipolar reference surface such as the water surface.

The surface parameters for some important condensed phases are given in Table 11.1 (see below). However, in order to apply these parameters to estimate K_{iasurf} values, we first need to know the fitting parameters a''', b, c, d''', and the constant in Eq. 11-7. These parameters can be determined by multiple regression if, for a large enough set of apolar and monopolar compounds, experimental K_{iasurf} values for adsorption to a well-characterized bipolar surface are available. As demonstrated by Goss (1997b) and Roth et al. (2002), a well-suited surface for this purpose is the bulk water surface. Using the air–bulk water surface partition constants reported by Roth et al. (2002) for 43 organic compounds exhibiting a wide range of polarities, an equation for estimation of K_{iasurf} *for any apolar or monopolar compound to any surface* can be derived for 15°C (288K):

$$\ln K_{iasurf}(288K)\,/\,m^{-1} = \begin{aligned} & 0.135(vdW_{surf})\ln p_{iL}^*(288K)\,/\,Pa - 2.06(vdW_{surf}) \\ & -8.44(\alpha_i)(HA_{surf}) \\ & -11.1(\beta_i)(HD_{surf}) \\ & +19.5 \qquad\qquad (N=43,\ R^2=0.93) \end{aligned}$$

$$(11\text{-}8)$$

Note that for the water surface, HA_{surf} and HD_{surf} have been set to 1.0; that is, the H-donor and H-acceptor properties of the water are implicitly included in the b- and c-terms. Therefore, the HA_{surf} and HD_{surf} values of bipolar surfaces such as mineral oxide or salt surfaces exhibit values not too different from 1 (Table 11.1).

When applying Eq. 11-8, one has to be aware that the predicted K_{iasurf} values represent only approximations, particularly if steric factors are important. Furthermore, Eq. 11-8 is valid for 15°C. For extrapolation to other temperatures, by analogy to estimating $\Delta_{vap}H_i$ from p_{iL}^* (Eq. 4-29), we may use the empirical relationship adapted from Goss and Schwarzenbach (1999a) to estimate $\Delta_{asurf}H_i$ from K_{iasurf} at 15°C (288K):

$$\Delta_{asurf}H_i(288K)\,/\,(kJ\cdot mol^{-1}) = 9.83(\pm 0.28)\log K_{iasurf}(288K)\,/\,m^{-1} + 90.5(\pm 1.4)$$

$$(11\text{-}9)$$

Hence, when assuming $\Delta_{asurf}H_i$ to be constant over the ambient temperature range, we get:

$$K_{iasurf}(T) = K_{iasurf}(288K)\cdot e^{-\frac{\Delta_{asurf}H_i(288K)+RT_{av}\left(\frac{1}{T}-\frac{1}{288}\right)}{R}}$$

$$(11\text{-}10)$$

where T_{av} is $[0.5\,(T+288)]$ K.

Van der Waals and Polar Surface Parameters of Some Selected Liquid and Solid Surfaces

Before we discuss some applications of the polyparameter-LFER Eq. 11-8, it is useful to have a short look at the surface parameters given for some important condensed phases in Table 11.1. Since the large majority of monopolar compounds are H-acceptors (i.e., $\alpha_i = 0$, $\beta_i \neq 0$), we focus our discussion primarily on the vdW_{surf} and HD_{surf} values.

We should note that, in general, solid surfaces must be expected to exhibit chemical and morphological heterogeneities. In this case different surface sites would have to be described by different surface parameter values. However, the adsorbed water film that is always found at ambient conditions on hydrophilic surfaces levels out these heterogeneities. Hence, the minerals and salt surfaces in Table 11.1 can be characterized by single values just like homogeneous surfaces.

Let us start out by looking at the vdW parameter of some important surfaces. Inspection of Table 11.1 shows that vdW_{surf} may vary significantly between about

Table 11.1 Van der Waals (vdW_{surf}), H-Acceptor (Electron Donor) (HA_{surf}), and H-Donor (Electron Acceptor) (HD_{surf}) Values for Some Condensed Phases at 15°C (If Not Otherwise Stated) [a]

Surface	Relative Humidity RH/(%) [c]	vdW_{surf} [b] $((mJ)^{1/2} m^{-1})$	HA_{surf} [d]	HD_{surf} [d]
Inorganic Surfaces				
Water	100	4.7	1.0	1.0
Water (0°C)	100	4.7	n.a.	n.a.
Ice (0°C)	100	5.4	n.a.	n.a.
Quartz (SiO_2)	45	6.8	0.89	1.06
	90	5.3	0.88	0.85
Ca-kaolinite	30	7.0	n.a.	1.12
	90	4.7	n.a.	0.75
Hematite (Fe_2O_3)	30	6.5	n.a.	0.92
	90	4.8	n.a.	0.76
Limestone ($CaCO_3$)	40	5.4	1.20	0.96
	90	4.9	1.01	0.91
Corundum (Al_2O_3)	40	5.4	1.13	1.00
	90	4.8	1.05	0.89
KNO_3	35	7.1	0.99	0.75
	60	6.8	1.00	0.70
$(NH_4)_2SO_4$	35	6.6	1.08	0.73
	70	6.3	1.34	0.53
NH_4Cl	35	6.8	1.19	0.75
	60	6.3	1.21	0.69
NaCl	35	6.2	1.11	0.79
	60	6.0	1.06	0.77
Organic Surfaces				
Paraffin wax ($H-(CH_2)_n-H$)	n.r.	5.0	0	0
Teflon ($F-(CF_2)_n-F$)	n.r.	4.2	0	0
Nylon 6,6	?	6.0		
Activated carbon	n.r.	~ 11		
Graphite	?	10.7–11.5		

[a] Data from Goss (1997b), Goss and Schwarzenbach (1999b and 2002). [b] Square root of the van der Waals component of the surface free energy. [c] n.r. = not relevant, ? = not reported. [d] Relative to the H-acceptor and H-donor properties of water; n.a. = not available.

4 (teflon) and 12 (e.g., graphite) $[(mJ)^{1/2} m^{-1}]$. This means, for example, that an apolar compound ($\alpha_i = \beta_i = 0$) with a p_{iL}^* value of 1 Pa ($\ln p_{iL}^* = 0$ in Eq 11-8) will adsorb more than seven orders of magnitude more strongly to graphite than to teflon (i.e., K_{iasurf} decreases from about 10^5 to less than 10^{-2} m^{-1}).

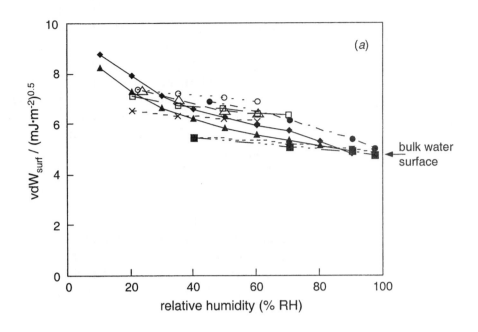

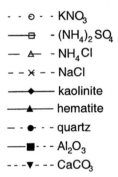

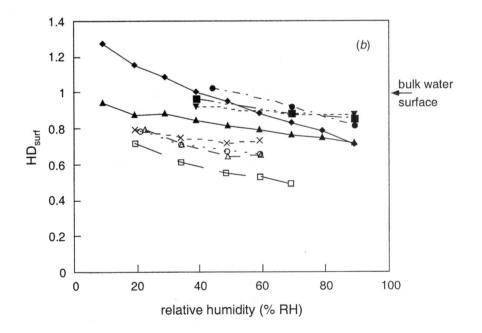

Figure 11.5 (*a*) Van der Waals parameter vdW$_{surf}$ and (*b*) H-donor parameter HD$_{surf}$ of various mineral surfaces as a function of relative humidity (RH) at 15 to 20°C. On the righthand side, the corresponding values of the bulk water surface are indicated. Note that HD$_{surf}$ of water is set equal to 1.0, and all other HD$_{surf}$ values are determined relative to the water surface. Adapted from Goss and Schwarzenbach (1999b and 2002).

An interesting observation can be made when comparing the vdW$_{surf}$ value of the bulk water surface (vdW$_{surf}$ = 4.7) with those of a series of representative *mineral oxide surfaces* (i.e., quartz, kaolinite, hematite, limestone, corundum). As is evident from Fig. 11.5*a*, in all cases, vdW$_{surf}$ decreases significantly with increasing relative humidity (RH) and approaches the bulk water surface value when approaching water vapor saturation. This can be explained as follows: most organic molecules cannot compete with water for the sorption sites at the mineral surface. Hence, in the presence of water, organic molecules can only adsorb on top of the adsorbed water film (as pictured in Fig. 11.1*a*). With increasing number of water layers, the vdW interactions at the surface of the adsorbed water films become more and more independent of the type of surface that is underneath the water molecules. Thus, at 90% RH, which corresponds to an average of about 5 to 9 molecular layers of adsorbed water (Fig. 11.3), an apolar compound cannot tell the difference between a quartz (SiO_2) and a corrundum (Al_2O_3) surface, while at 20% RH the differences are significant. Note that, as indicated by Fig. 11.5a, between 30 and 90% RH the values can be estimated by linear interpolation. Some values for low (30 – 45%) and high (90%) RH are given in Table 11.1.

Before we look at some other surfaces, we should briefly address the H-donor (electron acceptor) properties, HD_{surf}, of the mineral oxides discussed so far. As can be seen from Fig. 11.5b (data for mineral oxides) and Table 11.1, HD_{surf} values decrease with increasing RH and become more similar with increasing RH. Furthermore, between 30 and 90% RH the HD_{surf} values can also be estimated by linear interpolation. However, in contrast to the vdW parameter, at 90% RH this value is smaller than that of the bulk water surface. This may have to do with the orientation of the water molecules caused by the nearby solid surface, but an unambiguous explanation is still missing. Between 90 and 100% RH, when the thickness of the adsorbed water layer rapidly grows, one can anticipate that this difference disappears.

Salt surfaces are particularly interesting when dealing with aerosols in the marine environment. Compared to mineral surfaces, the salt surface parameters exhibit a considerably weaker dependence on relative humidity (Fig. 11.5 and Table 11.1). This can be explained by the fact that all four salts considered are hygroscopic and, therefore, adsorb water in significant amounts at low RH. However, the type of underlying salt still has a strong influence on this water layer because one can assume that this layer represents a saturated solution of the salt. Hence, it is not surprising that the vdW_{surf} and HD_{surf} values do not become equal for different salts at high RH, and that they do not match the values of the pure bulk water surface. Note that no data above 60 or 70% RH are available, but that it can be assumed that these surface parameters do not change at higher humidities (Goss and Schwarzenbach, 1999b). Finally, for the *hydrophobic* surfaces (e.g., paraffin wax, polyethylene, polyvinylchloride, polystyrene, teflon) to which water molecules only weakly adsorb, the effect of humidity can probably be neglected.

Estimating Air–Surface Adsorption Coefficients of Apolar and Monopolar Compounds; Applications

Les us first consider the cases in which Eq. 11-8 reduces to a simple one-parameter LFER relating K_{iasurf} to vapor pressure:

$$\ln K_{iasurf}(288K) / m^{-1} = m \cdot \ln p_{iL}^*(288K) / Pa + constant \qquad (11\text{-}11)$$

These cases include the following situations:

(a) The H-bonding interactions between the compounds and a given surface are zero (i.e., only vdW interactions are important). This is the case for all apolar surfaces (i.e., $HA_{surf} = HD_{surf} = 0$) or for any surface if only apolar compounds are considered (i.e., $\alpha_i = \beta_i = 0$). In both cases, both the slope m and the constant in the LFER Eq. 11-11 are dependent only on the vdW_{surf} value of the surface:

$$\ln K_{i\text{asurf}}(288\text{K}) / \text{m}^{-1} = 0.135(\text{vdW}_{\text{surf}})\ln p_{iL}^{*}(288\text{K}) / \text{Pa} + [19.5 - 2.06(\text{vdW}_{\text{surf}})]$$

$$(11\text{-}12)$$

This means that for an apolar surface, all apolar or monopolar compounds should fit the same straight line. We have already demonstrated this for a teflon surface in Fig. 3.7 (note that the vdW parameter used in this graph is proportional to p_{iL}^{*}, Eqs. 11-5 and 11-6).

(b) Within a series of monopolar compounds, the H-bonding interaction term is constant (i.e., α_i or β_i, respectively, is a constant). From the α_i and β_i values given in Table 4.3, we can see that virtually all monopolar compounds are H-acceptors (or electron donors). Therefore, we need only to consider the term $(\beta_i) \cdot \text{HD}_{\text{surf}}$ in Eq. 11-8. Hence, we would expect to be able to derive an LFER of the type of Eq. 11-11 for a series of monopolar compounds exhibiting the same β_i values:

$$\ln K_{i\text{asurf}}(288\text{K}) / \text{m}^{-1} = 0.135(\text{vdW}_{\text{surf}})\ln p_{iL}^{*}(288\text{K}) / \text{Pa}$$
$$+ [19.5 - 2.06(\text{vdW}_{\text{surf}}) - 11.1(\beta_i)(\text{HD}_{\text{surf}})]$$

$$(11\text{-}13)$$

with β_i = constant for the series of compounds considered.

When comparing groups of compounds with different (constant) β_i values, we would expect them to fit parallel straight lines with different constant terms. Again, this is nicely illustrated in Fig. 3.7 for a quartz surface for some alkanes and chlorinated benzenes (both $\beta_i \cong 0$), some methylbenzenes ($\beta_i \cong 0.15$), and some aliphatic ethers ($\beta_i \cong 0.45$).

(c) Within a series of compounds the polar interaction term changes proportionally to the $\ln p_{iL}^{*}$ term. For example, for polycyclic aromatic hydrocarbons (PAHs), the term 11.1 (β_i) HD_{surf} rises proportionally with increasing number of rings and thus with decreasing $\ln p_{iL}^{*}$. Hence, for the adsorption of PAHs to a hydrogen-donating (electron-accepting) surface (i.e., $\text{HD}_{\text{surf}} > 0$), a plot of $\ln K_{i\text{asurf}}$ vs. $\ln p_{iL}^{*}$ is expected to yield a straight line but with a slope that is steeper than in the cases discussed above. This is confirmed by experimental data (Storey et al., 1995).

This analysis of Eq. 11-8 illustrates in a very general way that we have to be careful when applying one-parameter LFERs to air–surface partitioning processes. We have seen that depending on the type of sorbent (here a liquid or solid surface), it may be possible to correlate partitioning data of a wide variety of different chemicals with one single LFER, while in other cases, different LFERs will apply for different compound classes. For example, we would expect one single LFER describing the adsorption of PCBs ($\beta_i \cong 0$) and PAHs (β_i proportional to $\ln p_{iL}^{*}$) to an apolar surface (e.g., teflon), but two very distinct LFERs for the two compound classes for adsorption to an H-donating surface (e.g., quartz).

As demonstrated by Illustrative Example 11.1, the model Eq. 11-8 enables us to make some interesting calculations concerning the partitioning of organic pollutants between the gas phase (e.g., air) and solid surfaces. However, we should point out again that there are various difficulties in applying this model to real-world problems. For example, we need to know the types and particularly the areas of the dominating (accessible) surfaces present in a given system. To date, experimental data are reported in most cases on a per-mass and not on a per-surface-area basis because surface areas are not very well known. Furthermore, surface areas and surface properties may change with changing conditions (e.g., with changing humidity). In addition, the overall partitioning process may also be strongly dominated by absorption of a compound into a bulk phase (e.g., water, natural organic material), so that adsorption to a surface is not important anymore. Nevertheless, as is demonstrated by Illustrative Examples 11.2 and 11.3, Eq. 11-8 can be very helpful for evaluating experimental vapor/condensed phase partitioning data to gain insight on whether adsorption or absorption is the dominant process in a given case, or to assess the relative importance of various sorption mechanisms in a given system.

Illustrative Example 11.1

Estimating the Fraction of Phenanthrene Present in the Gas Phase and Sorbed to the Walls of a Vessel

Problem

Consider two closed air-sampling vessels made out of (a) teflon and (b) glass (assume like quartz) with an air volume $V_a = 10^{-3} m^3$ (1 L) and an inner surface area of $A_{surf} = 6 \times 10^{-2} m^2$. In these vessels you capture air samples that you want to analyze for phenanthrene. Calculate the fraction of the total phenanthrene present in the air in the two vessels after adsorption equilibrium between the gas phase and the walls of the vessel has been established at 15°C (288 K) and 50% relative humidity. Assume that only adsorption at the surface of the walls is important. (Note that in the case of teflon, absorption could also be important.)

Answer (a)

Teflon is an apolar sorbent that undergoes only vdW interactions. Insert the vdW$_{surf}$ parameter of teflon (4.2, independent of RH, see Table 11.1) into Eq. 11-12 to obtain:

$$\ln K_{ia\,teflon}(288K)/m^{-1} = 0.567 \ln p_{iL}^*(288K)/Pa + 10.85 \tag{1}$$

Note that this equation is valid for adsorption of any apolar or monopolar organic compound from air to teflon. In Appendix C you find the vapor pressure of *solid* phenanthrene at 25°C ($T_m = 101°C$; log $p_{is}^*/Pa = -1.66$ or ln $p_{is}^*/Pa = -3.82$). Estimate first its liquid vapor pressure using Eq. 4-40 ($\tau = 0$, $\sigma = 2$, see Table 4.5):

$$\ln p_{iL}^*(298\,K)/Pa = -3.82 + [6.80 + 0 - (2.3)(0.3)]\left(\frac{374}{298} - 1\right) = -2.26$$

or log $p_{iL}^* = -0.98$. Insert this value into Eq. 4-29 to get $\Delta_{vap}H_i$:

$$\Delta_{vap}H_i(298K)/(kJ \cdot mol^{-1}) = (-8.80)(-0.98) + 70.0 = 78.6\ kJ \cdot mol^{-1}$$

i = phenanthrene

Therefore (Eq. 4-30):

$$\ln p^*_{iL}(288\,\text{K})/\text{Pa} = -2.26 - \frac{78600}{8.31}\left(\frac{1}{288} - \frac{1}{298}\right) = -3.36$$

Insertion of this value into Eq. 1 then yields:

$$\ln K_{ia\,\text{teflon}}(288\text{K})/\text{m}^{-1} = (0.567)(-3.36) + 10.85 = 8.94$$

or:

$$K_{ia\,\text{teflon}}(288\text{ K}) = 7.7 \times 10^3 \text{ m}^{-1}$$

The fraction of phenanthrene present at equilibrium in the gas phase (air) is given by:

$$f_{ia} = \frac{C_{ia}V_a}{C_{ia}V_a + C_{i\text{teflon}}A_{\text{teflon}}} = \frac{1}{1 + \dfrac{C_{i\text{teflon}}}{C_{ia}}\dfrac{A_{\text{teflon}}}{V_a}} = \frac{1}{1 + \dfrac{1}{K_{ia\,\text{teflon}}}\cdot\dfrac{A_{\text{teflon}}}{V_a}} \quad (2)$$

where $C_{i\text{teflon}}$ is the concentration of phenanthrene per unit surface area. Insert the given values of A_{teflon} and V_a together with the estimated $K_{ia\,\text{teflon}}$ value into Eq. 2 to get:

$$f_{ia} = \frac{1}{1 + (0.00013\,\text{m})(60\,\text{m}^{-1})} = 0.992$$

Thus, virtually all phenanthrene is still present in the gas phase.

Answer (b)

Quartz is a bipolar sorbent that exhibits quite strong H-donor and H-acceptor properties. Phenanthrene is monopolar with $\alpha_i = 0.0$ and $\beta_i = 0.26$ (Table 4.3). Hence, in this case (Eq. 11-8):

$$\ln K_{ia\text{surf}}(288\text{K})/\text{m}^{-1} = 0.135(\text{vdW}_{\text{surf}})\ln p^*_{iL}(288\text{K})/\text{Pa} - 2.06(\text{vdW}_{\text{surf}}) \\ -11.1(0.26)(\text{HD}_{\text{surf}}) + 19.5 \quad (3)$$

Interpolate linearly the vdW_{surf} and HD_{surf} values given in Table 11.1 for quartz to obtain the appropriate values for 50% RH:

$$\text{vdW}_{\text{quartz}}(50\%) = 6.8 - (5/45)(1.5) = 6.6$$

$$\text{HD}_{\text{quartz}}(50\%) = 1.06 - (5/45)(0.21) = 1.04$$

Insertion of these values together with $\ln p^*_{iL}(288\text{ K})/\text{Pa}$ calculated above (answer a) into Eq. 3 yields:

$$\ln K_{i\text{quartz}}(288\text{ K})/\text{m}^{-1} = (0.135)(6.6)(-3.36) - 2.06(6.6) - (11.1)(0.26)(1.04) + 19.5 \\ = -0.09$$

or:

$$K_{i\text{aquartz}}(288\text{ K}) = 0.91 \text{ m}^{-1}$$

The fraction of phenanthrene in the air is then in this case (see Eq. 2):

$$f_{ia} = \frac{1}{1+(1.1\,\text{m})(60\,\text{m}^{-1})} = 0.015$$

which means that in the glass vessel, 98.5%(!) of the compound would be "lost" to the surfaces. Would you have guessed this large difference between the two vessels?

Illustrative Example 11.2   **Sorption of Tetrachloroethene from Air to Moist and Dry Soil**

Consider the prospects for tetrachloroethene (PCE) to be transported in the gas phase from a contaminated soil out to the atmosphere. For this to occur [either through diffusion (see Chapter 18) or in response to venting during remediation], this PCE must substantially exist in the soil gas.

Problem

Estimate the fraction of PCE present in soil gas (air) at equilibrium for the following two cases:

Case 1: Soil air at 90% RH and a temperature of 15°C

Case 2: Soil air at 40% RH and a temperature of 15°C

assuming the following soil compositions and properties:

93% quartz (specific surface area $A_q = 10^2\ \text{m}^2\ \text{kg}^{-1}$)
6% kaolinite ($A_k = 10^4\ \text{m}^2\ \text{kg}^{-1}$)
1% organic matter ($f_{oc} = 0.005$)
density of soil material; $\rho_s = 2.6\ \text{kg} \cdot \text{L}^{-1}$
total soil porosity, $\phi = 0.35$

Further assume that at these relative humidities, the water present in the soil does not act as a bulk phase, and that the sorption isotherm is linear.

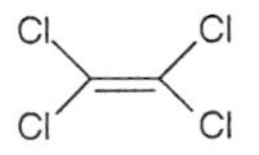

tetrachloroethene
(PCE)

Answer

Neglecting the water as a distinct bulk phase, the fraction of PCE in the soil air is given by (Eq. 3-65):

$$f_{ia} = \frac{1}{1 + K_{iasoil}^{-1} r_{as}^{-1}} \tag{1}$$

where $r_{as} = V_a/M_s$ is the air volume–soil mass ratio (e.g., in L air·kg^{-1} solid), and K_{iasoil} is the bulk air–soil partition coefficient on a weight base (i.e., in L air·kg^{-1} solid). Assuming that the void space is filled only with soil air, in analogy to Eq. 9-15, in both cases, r_{as}^{-1} is given by:

$$r_{as}^{-1} = \rho_s \frac{1-\phi}{\phi} = (2.6)\frac{(0.65)}{(0.35)} \cong 5\ \text{kg solid} \cdot \text{L}^{-1}\ \text{air}$$

The bulk air–soil partition coefficient, K_{iasoil}, encompasses *ab*sorption into soil organic matter as well as *ad*sorption to the (water-covered) quartz (subscript q) and kaolinite (subscript k) surfaces, respectively. K_{iasoil} can be expressed by:

$$K_{iasoil} / (\text{kg solid} \cdot \text{L}^{-1} \text{ air}) = \frac{C_{ia}}{f_{oc} C_{ioc} + f_q A_q C_{iq} + f_k A_k C_{ik}}$$

$$= \frac{1}{f_{oc} K_{iaoc}^{-1} + 10^3 f_q A_q K_{iaq}^{-1} + 10^3 f_k A_k K_{iak}^{-1}} \qquad (2)$$

where f_{oc}, f_q, and f_k are the fractions of organic carbon (0.005 kg oc·kg^{-1} solid), of quartz (0.93 kg quartz·kg^{-1} solid), and of kaolinite (0.06 kg kaolinite·kg^{-1} solid), respectively. Note that since the specific surface areas and the K_{iasurf} values are expressed in m^2·kg^{-1} and m^{-1}, respectively, a factor 10^3 has to be introduced for conversion of m^3 to L.

K_{iaoc} can be approximated by the natural organic matter–water partition coefficient, K_{ioc}, and by the air–water partition constant, K_{iaw} (Eq. 6-11):

$$K_{iaoc} = \frac{K_{iaw}}{K_{ioc}} \qquad (3)$$

Note that we assume that the relative humidity has no significant effect on K_{iaoc} (which, by Eq. 3 is estimated for 100% RH).

The K_{iaw} value of PCE is 1.2 at 25°C (Appendix C). Use Eq. 9.26*f* to estimate K_{ioc} from K_{iow} (Appendix C):

$$\log K_{ioc} = 0.96 \log K_{iow} - 0.23 = (0.96)(2.88) - 0.23 = 2.53$$

yielding a K_{iaoc} value (Eq. 3) of $(1.2)/10^{2.53} = 3.5 \times 10^{-3}$ L air·kg^{-1} oc at 25°C. Assuming that $\Delta_{aoc} H_i + RT_{av} \cong \Delta_{vap} H_i + RT_{av} \cong 40$ kJ·mol^{-1} (Table 6.3), you obtain a K_{iaoc} value at 15°C of about 2.0×10^{-3} L air·kg^{-1} oc (Eq. 6-10).

Since PCE is an apolar compound, use Eq. 11-12 to estimate K_{iaq} and K_{iak}. The vapor pressure of PCE at 25°C is $p_{iL}^* = 2.5 \times 10^3$ Pa (Appendix C). At 15°C ($\Delta_{vap} H_i = 40$ kJ·mol^{-1}, see above), p_{iL}^* is estimated to be 1.4×10^3 Pa or $\ln p_{iL}^* = 7.25$.

Case 1: Insert $\ln p_{iL}^*$ together with the vdW$_{surf}$ parameter given for quartz and kaolinite at 90% RH into Eq. 11-12 to obtain:

$$\ln K_{iaq} (15°\text{C}, 90\% \text{ RH})/\text{m}^{-1} = (0.135)(5.3)(7.25) + [19.5 - (2.06)(5.3)] = 13.8$$

or:

$$K_{iaq} \cong 10^6 \text{ m}^{-1}$$

and:

$$\ln K_{iak} (15°\text{C}, 90\% \text{ RH})/\text{m}^{-1} = (0.135)(4.7)(7.25) + [19.5 - (2.06)(4.7)] = 14.4$$

or:

$$K_{iak} \cong 2 \times 10^6 \text{ m}^{-1}$$

Insertion of K_{iaq}, K_{iak}, K_{iaoc} (see above) together with the corresponding f and A values into Eq. 2 yields:

$$K_{iasoil}\,(15°C,\,90\%\,RH) = 1/[(0.005)(2 \times 10^{-3})^{-1} + 10^3(0.93)10^2\,10^{-6}$$
$$+ 10^3(0.06)10^4(0.5)10^{-6}]$$
$$= 1/[2.5 + 0.093 + 0.3] \cong 0.35 \text{ kg solid} \cdot L^{-1} \text{ air}$$

Note that, in this case, partitioning to NOM is the major sorption mechanism. Insert this value together with the r_{as} value calculated above to obtain:

$$f_{ia}\,(15°C, 90\%\,RH) = \frac{1}{1+(0.35)^{-1}(5)} = 0.065$$

Case 2: Estimate the vdW_{surf} values for quartz and kaolinite at 40% RH by extrapolation and interpolation, respectively, from the data given in Table 11.1:

$$vdW_q\,(40\%\,RH) = 6.8 + (1/9)(1.5) \cong 7.0$$
$$vdW_k\,(40\%\,RH) = 7.0 - (1/6)(2.3) \cong 6.6$$

Insertion of these values into Eq. 11-12 yields:

$$K_{iaq}\,(15°C, 40\%\,RH) = 1.5 \times 10^5 \text{ m}^{-1}$$
$$K_{iak}\,(15°C, 40\%\,RH) = 2.3 \times 10^5 \text{ m}^{-1}$$

Hence, at 40% RH, PCE adsorbs by about a factor of 10 more strongly to the mineral surfaces as compared to 90% RH. Insertion of these values together with all other parameters discussed above into Eq. 1 then yields:

$$K_{iasoil}\,(15°C, 40\%\,RH) = 1/[2.5 + 0.6 + 2.6] \cong 0.18 \text{ kg solid} \cdot L^{-1} \text{ air}$$

Hence, at 40% RH adsorption to the mineral surfaces, particularly to kaolinite, is of equal importance as absorption into NOM. Insertion of this value into Eq. 1 gives:

$$f_{ia}\,(15°C, 40\%\,RH) = \frac{1}{1+(0.18)^{-1}(5)} = 0.035$$

This example shows that with decreasing humidity, the fraction of PCE present in the gas phase decreases due to increasing adsorption to the mineral surface.

Illustrative Example 11.3   **Gas–Particle Partitioning of Organic Compounds to Environmental Tobacco Smoke: Adsorption or Absorption?**

The behavior and health effects of environmental tobacco smoke (ETS) depends upon how certain compounds (e.g., PAHs, nicotine, carbazole) are distributed between the gas and the particulate phases. Using a desorption technique, Liang and Pankow (1996) have determined gas/ETS particle partition coefficients for a series of alkanes, PAHs, and for some nitrogen-containing compounds including nicotine at 20°C. They report K_{ip} values that are defined as:

$$K_{ip} = \frac{\text{mass of compound } i / \mu\text{g particle}}{\text{mass of compound } i / \text{m}^3 \text{ air}}$$

Hence, the K_{ip} values are given in m³ air μg^{-1} particle. For the n-alkanes investigated (i.e., $C_{16} - C_{22}$), they report the following one-parameter LFER:

$$\log K_{ip} (293 \text{ K, } 60\% \text{ RH})/(\text{m}^3 \mu\text{g}^{-1}) = -0.89 \log p^*_{iL}/\text{Torr} - 7.44 \ (R^2 = 0.99) \quad (1)$$

With respect to the nature of the particles, they found that the ETS particles were essentially fluid in nature and that they consisted primarily of organic material with a water content of about 15% (by weight). The authors postulate that the gas/particle partitioning of organic compounds in environmental tobacco smoke is governed primarily by *ab*sorption of the compounds into the condensed phase and not by adsorption at the surface.

Problem

Assuming that the reported K_{ip} values represent equilibrium coefficients, try to find an argument that supports the authors' hypothesis that gas/ETS particle partitioning is governed by *ab*sorption rather than by *ad*sorption. (*Hint:* Calculate the specific particle surface area, A_p (e.g., in m² g⁻¹ particle), that would be necessary if *ad*sorption to the surface would be the key process).

Answer

Note that K_{ip} in Eq. 1 is defined as the reciprocal of K_{iasurf} in Eq. 11-2. Multiply Eq. 1 by −1 to obtain a relationship between $K'_{ip} = K_{ip}^{-1}$ and p^*_{iL}, and express p^*_{iL} in Pa (1 Torr = 133.3 Pa):

$$\log K'_{ip} /(\mu\text{g particle} \cdot \text{m}^{-3} \text{ air}) = 0.89 \log p^*_{iL} / \text{Pa} + 5.32 \quad (2)$$

Hence, the "experimental" K_{ip} value of an n-alkane with, for example, a p^*_{iL} value of 1 Pa ($\log p^*_{iL} = 0$) at 20°C is:

$$K'_{ip} = 10^{5.32} \ \mu\text{g particle} \cdot \text{m}^{-3} \text{ air} = 2.1 \times 10^5 \ \mu\text{g particle} \cdot \text{m}^{-3} \text{ air}$$

Apply Eq. 11-12 to estimate the K_{iasurf} value of an *apolar* compound with $p^*_{iL} = 1$ Pa at 15°C by using an apparent vdW$_{surf}$ value that you can calculate from the slope of the LFER Eq. 2 (slope = 0.135 vdW$_{surf}$; note that the slope is independent of the units in which the partition constant is expressed):

$$\text{vdW}_{surf} = \frac{0.89}{0.135} = 6.6$$

and, therefore:

$$\ln K_{iasurf} (288\text{K}) / \text{m}^{-1} = 0 + [19.5 - 2.06(6.6)] = 5.90$$

or:

$$K_{iasurf} (288\text{K}) = 3.7 \times 10^2 \text{ m}^{-1}$$

at 15°C. Insertion of this value into Eq. 11-9 yields a $\Delta_{\text{asurf}}H_i$ of 116 kJ·mol^{-1}, which means that at 20°C (Eq. 11-10):

$$K_{i\text{asurf}} (293\text{K}) = 8.6 \times 10^2 \text{ m}^{-1}$$

The relationship between K'_{ip} and $K_{i\text{asurf}}$ is given by:

$$K_{i\text{asurf}} (\text{m}^{-1}) = K'_{ip}(\mu\text{g particle}\cdot\text{m}^{-3}\text{ air})\cdot A_p \text{ (m}^2\text{ g}^{-1}\text{ particle}) \ 10^{-6} \text{ (g particle}\cdot\mu\text{g}^{-1}\text{particle})$$

Thus, the specific surface area required if *ad*sorption governed the partitioning is:

$$A_p = \frac{10^6 \ K_{i\text{asurf}}}{K'_{ip}} = \frac{(10^6)(8.6\times10^2)}{(2.1\times10^5)} \cong 4100 \text{ m}^2\text{g}^{-1} \text{ particle}$$

This requisite surface area is more than three orders of magnitude higher than one would expect for organic aerosols (e.g., 2 m^2 g^{-1} particle for organic-rich aerosols from urban atmospheres (Liang et al., 1997)). This result strongly suggests that *ab*sorption is the key process.

11.3   Sorption of Nonionic Organic Compounds to Inorganic Surfaces in Water

When inorganic solids are fully submerged in water, nonionic organic chemical sorption is still observed (Fig. 11.6). Such sorption may prove to be significant under certain conditions. For example, solids in some environmental systems do not include enough natural organic matter to dominate sorption. Consequently, association of organic solutes with mineral surfaces may be the only sorption process, even if this process adds negligibly whenever natural organic matter is present. This situation exists in subsurface environments, such as aquifer solids derived from sand-and-gravel beach deposits or aluminosilicate clay-rich subsoil horizons, where there are very small organic contents (Schwarzenbach and Westall, 1981; Banerjee et al., 1985; Piwoni and Banerjee, 1989; Hundal et al., 2001). Additionally, engineered systems such as clay liners are often used to isolate organic wastes buried below ground; and while we may be interested in the impact of this low-permeability material on the subsurface hydraulics, we also need to understand the ability of those aluminosilicate minerals to bind organic pollutants and inhibit offsite transport (Boyd et al., 1988). Also, sorption to certain mineral surfaces, even if insignificant from a mass balance point of view, may be critical to quantify when dealing with surface-catalyzed transformations (Ulrich and Stone, 1989). Finally, laboratory glass surfaces (e.g., Carmo et al., 2000) and metal-sampling vessels may sorb nonionic compounds from aqueous solutions, confusing subsequent data interpretation.

Empirically, we sometimes find that nonionic organic compound sorption on inorganic solids is best described with linear isotherms. Such is the case for observations of pyrene sorption on kaolinite suspensions (Fig. 11.6a; Backhus, 1990). Generally,

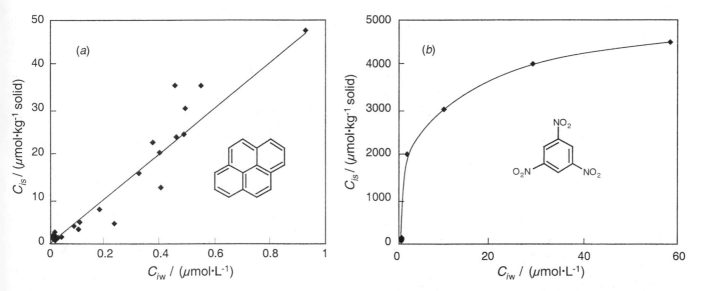

Figure 11.6 Sorption isotherms for two kinds of nonionic organic compounds from aqueous solutions to suspended kaolinite: (a) slightly monopolar compound, pyrene, showing a linear isotherm up to its solubility (Backhus, 1990), and (b) monopolar compound, 1,3,5-trinitrobenzene, showing a hyperbolic isotherm (Haderlein et al., 1996).

such cases involve organic sorbates whose structures do not enable them to have strong specific interactions with polar surfaces. In contrast, other compounds, such as certain nitroaromatic compounds, exhibit Langmuir isotherms for their sorption on clay minerals (Fig. 11.6b; Haderlein et al., 1996). These compounds can specifically interact with sites on the mineral surface, and they show stronger binding as evidenced by enthalpies of sorption. Let us examine these cases one at a time.

Partitioning of Apolar and Weakly Monopolar Compounds into the Region Near Mineral Surfaces

Numerous investigators have reported sorption of apolar and weakly monopolar organic compounds to mineral surfaces (Yaron et al., 1967; Bailey et al., 1968; Mills and Biggar, 1969a; Boucher and Lee, 1972; van Bladel and Moreale, 1974; Schwarzenbach and Westall, 1981; Estes et al., 1988; Szecsody and Bales, 1989; Zhang et al., 1990; Backhus 1990; Schlautman and Morgan, 1994; Mader et al., 1997; Hundal et al., 2001). Often the observations are made with natural solids, and so it is extremely difficult to eliminate all the natural organic matter that may be present. For example, Hundal et al. (2001) used natural samples of montmorillonites to assess the sorption of phenanthrene to such high-surface-area fines, and they carefully quantified the f_{oc} values of those clays to all be near 0.001. Given phenanthrene's K_{ioc} of $10^{4.2}$ L·kg^{-1}oc, their observed K_{id} values between 10 and 40 L·kg^{-1} solid were close to the product, $f_{oc} K_{ioc} \approx 20$ L·kg^{-1} solid. Nonetheless, investigators like Mader et al. (1997) have observed significantly retarded transport of several aromatic compounds by water flowing through laboratory columns packed with synthetic (and hence organic-carbon-free) hematite (α-Fe$_2$O$_3$) or corundum (α-Al$_2$O$_3$). K_{id} values for many substances on a variety of inorganic solids have magnitudes between 0.1 and 100 L·kg^{-1} solid, with larger values found for less-water-soluble compounds and higher-surface-area solids. Clearly some mechanism acts to affiliate, even if only in a limited way, such organic chemicals with inorganic media under environmental conditions.

*Ad*sorption of nonionic organic compounds to hydrophilic inorganic surfaces submerged in water would require these organic sorbates to displace water molecules already adhering to the polar surface. Since apolar and weakly monopolar organic compounds do not hydrogen bond substantially (α_i and β_i values for such compounds are low or even zero), these organic sorbates cannot interact with polar surfaces as effectively as water can. Hence, surface *ad*sorption from water to fully water-wet hydrophilic solids probably does not explain the sorption of apolar and monopolar organic compounds to minerals (recall Fig. 11.1*b*).

Two alternative explanations have been suggested which are both quite speculative. First, portions of mineral surfaces of intermediate polarity (e.g., siloxane regions, –Si–O–Si–) may permit some exchange of polar water and nonpolar organic sorbates (Hundal et al., 2001). Such surfaces occur in minerals like the faces of aluminosilicates. However, amorphous solids like silica (–Si–OH) and alumina (–Al–OH) have very hydrophilic exteriors when these inorganic materials are suspended in water. Yet these amorphous materials still clearly show sorption of apolar substances (e.g., Mills and Biggar, 1969a; Schwarzenbach and Westall, 1981; Estes et al., 1988; Szecsody and Bales, 1989; Farrell et al., 1999).

Another possibility is that organic sorbates partition from the bulk aqueous solution into the "special" water immediately adjacent to solid surfaces or filling the nanometer-sized pores of these solids. Due to interactions with the solid as described in Section 11.2, water molecules near inorganic surfaces are more organized than corresponding molecules located in the bulk solution. Such surface-ordered water films, called *vicinal water*, may extend for nanometers away from the solid surface. The volume of this vicinal water per mass of sorbent is related to the intraparticle porosity and surface area. Consequently, the amount of such special near-surface water per mass of solid is greater for fine, porous silica (~ 0.5 mL$\cdot$g^{-1}) than for quartzite sand (< 0.001 mL$\cdot$g^{-1}) and greater for expandable montmorillonite (~ 0.5 mL$\cdot$g^{-1}) than for the two-layer clay kaolinite (< 0.02 mL$\cdot$g^{-1}). Such volumes may approach a milliliter per gram in highly porous solids (Mikhail et al., 1968a, b; Ogram et al., 1985). In view of this concept of vicinal water, apolar and mono-polar compound sorption to inorganic solids may be best viewed as partitioning between relatively disorganized bulk water and this special volume of ordered water near the solid's surface:

$$(i\,)_{\text{bulk water}} + n(\text{H}_2\text{O})_{\text{vicinal}} \rightleftharpoons (i\,)_{\text{vicinal water}} + n(\text{H}_2\text{O})_{\text{bulk}} \qquad (11\text{-}14)$$

Note that such a conceptualization still requires a corresponding "desorption" of some number, n, of water molecules from the vicinal region to make a cavity near the solid surface in which i must fit. This solvent desorption process contributes an additional free energy term to the overall organic compound sorption process due to breaking of water:water interactions in this layer.

While the exact sorption mechanism may not be clear, the available literature suggests the following generalizations. First, coarser particles (e.g., silica sand) exhibit less binding per mass of solid than corresponding finer particles made of the same material (e.g., porous silica). This is presumably due to the influence of increased solid *surface area* per mass of sorbent. Thus, values of sorption coefficients for minerals ($K_{i\text{min}} = C_{i\text{min}}/C_{i\text{water}}$) are more useful if they are normalized to the solid's surface area rather than its mass. The second tendency we see is that for any

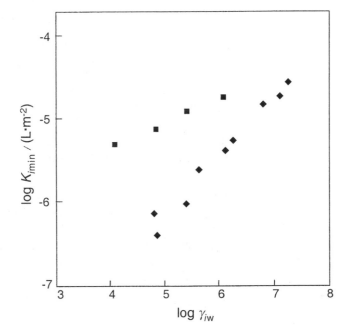

Figure 11.7 Apolar and slightly monopolar organic compound sorption coefficients, K_{imin} (L · m^{-2}), observed for exchange between aluminum oxides and aqueous solutions plotted as a function of sorbate hydrophobicity (as measured by γ_{iw}): apolar series of chlorobenzenes (squares, Schwarzenbach and Westall, 1981) and aromatic and chloroaromatic compounds (diamonds, Mader et al., 1997).

one sorbent, binding increases within a series of sorbates as a function of their hydrophobicities (e.g., as reflected by aqueous activity coefficients). This can be seen in the data for various aromatic compounds binding to aluminum oxides (Fig. 11.7).

Another clue as to the nature of nonionic compound binding to mineral surfaces comes from studies on the effect of temperature on this process. Increasing the system temperature resulted in diminished sorption (Mills and Biggar, 1969b; Boucher and Lee, 1972; van Bladel and Moreale, 1974; Mader et al., 1997). This result shows that the overall process in these cases was exothermic. Since the dissolution of such nonpolar compounds in water is generally an endothermic process (recall Table 5.3), we may reasonably anticipate that some of this energy yield on mineral sorption came from the removal of those chemicals from aqueous solution. Indeed, in the studies noted above, after accounting for solution enthalpies, the remaining steps in mineral binding of nonpolar sorbates proved to be energetically neutral or even slightly endothermic. From these results it appears that strong molecule:surface interactions are not involved.

In light of these observations, it appears that nonionic organic compounds of low polarity exhibit a weak tendency to partition from bulk aqueous solution into the region near mineral surfaces. Although uncertain, the mechanism may involve partitioning between two aqueous solvents: the bulk medium and the water film near the mineral surface (i.e., vicinal water), which may be energetically more favorable. If the tendency of sorbates to escape aqueous solutions is an important factor, as it appears from the very limited available data, we expect the free energy of sorption of neutral organic compounds to minerals to be inversely related to the free energy of aqueous dissolution of those same (liquid) chemicals.

Using the limited empirical observations of the magnitudes of K_{imin} values, one may examine the question of when nonpolar chemical sorption to mineral surfaces from aqueous solution dominates association with natural organic matter (i.e., $A_{surf} K_{imin} > f_{oc} K_{ioc}$ in Eq. 9-16). Mineral surface sorption of nonpolar compounds starts to become important with respect to mass balance considerations when f_{oc} is below 0.001 (Schwarzenbach and Westall, 1981: Banerjee et al., 1985). However, this "threshold" depends on the sorbate's hydrophobicity (Karickhoff, 1984; Banerjee et al., 1985) and the ratio of available inorganic surface area to organic content (Hassett et al., 1980; Means et al., 1980; Karickhoff, 1984). Competition between sorption to natural organic matter and to mineral surfaces is probably also a function of other soil or sediment properties. While we may not be able to predict such sorption yet, it is clear that the hydrophobic portions of organic chemicals prefer to escape bulk solution for the region near particle surfaces. This factor will also be important when we examine the sorption of organic compounds participating in ion exchanges.

In sum, we do not yet know how to predict sorption of apolar and monopolar organic compounds to mineral surfaces submerged in water. If empirical results are available from structurally related compounds, parameters quantifying sorbate hydropho-bicity may help us anticipate the intensity of surface associations for new com-pounds in the same compound classes (i.e., interpolating data such as that shown in Fig. 11.7).

Surface Adsorption Due to Electron Donor-Acceptor Interactions

Some nonionic organic compounds exhibit much stronger mineral surface affinities than we see for apolar and weakly monopolar compounds like chlorobenzenes and PAHs. In these cases, the organic sorbates are able to displace water from the mineral surface and participate in fairly strong sorbate:sorbent intermolecular interactions. Example compounds include *nitroaromatic compounds* (NACs) such as the explosive, trinitrotoluene (TNT), or the herbicide, 2,4-dinitro-6-methyl-phenol, also called dinitro-*o*-cresol (DNOC).

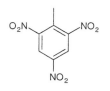

trinitrotoluene
(TNT)

2,4-dinitro-6-methylphenol
(DNOC)

Such NACs do not exhibit linear isotherms when they sorb under certain conditions with aluminosilicate clays (Fig. 11.6*b*; Haderlein and Schwarzenbach, 1993; Haderlein et al., 1996). Rather they show saturation behavior indicating an association with specific sites on the solid surfaces. This specific site interaction is also indicated by the observations of competitive effects among different NACs in sorption experiments. Further, the sorption enthalpies have been found to be much greater than excess enthalpies of aqueous solution of these sorbates (e.g., 4-methyl-2-nitro-phenol exhibits a sorption enthalpy of -41.7 kJ·mol^{-1}). These data all indicate that there is a strong specific interaction of NACs with the aluminosilicate clay surfaces.

Another important feature of this type of sorption is that the nature of the cations serving as the counterions on the faces of clays affects the intensity of NAC sorption. These minerals bear a net negative charge in their interior due to iso-morphic substitutions (e.g., inclusion of a Al^{3+} in place of a Si^{4+} in the crystal lattice). As a result, the clay's exterior is covered by cations like potassium and sodium. NAC sorption is much greater when the clay has adsorbed potassium rather than calcium or sodium. This result has been interpreted to mean that the large hydrated ions of sodium and calcium can block NAC access to sites on a clay's surface, while the much less hydrated potassium ions can serve as counterions without blocking NAC

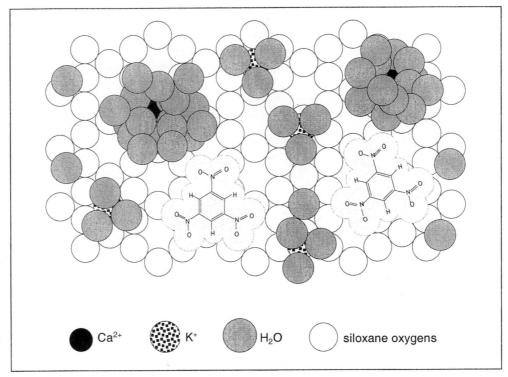

Figure 11.8 Schematic representation of the basal siloxane layer of a clay mineral in the presence of hydrated exchangeable cations and adsorbed NACs. Adsorbed NACs (here 1,3,5-trinitrobenzene) form co-planar EDA complexes with weakly hydrated siloxane oxygens (adapted from Weissmahr et al. 1997).

access to the surface (Fig. 11.8).

Due to the strong electron-withdrawing nature of nitro substituents on aromatic rings (e.g., recall Hammett σ_j values, Table 8.5; note that other electron-withdrawing substituents such as $-C\equiv N$ have a similar effect, Weissmahr et al., 1999), many NACs have a strong ability to attract electron-donating (ED) moieties toward the π-clouds of these aromatic compounds (i.e., the NACs are electron acceptors). Suitably electron-rich moieties exist at the siloxanes of aluminosilicate minerals. As long as these siloxane oxygens are not blocked by highly hydrated cations, NACs can form an electron donor–electron acceptor complex with the siloxane oxygens:

$$NAC + (Si) > O:H_2O \rightleftharpoons (Si) > O:NAC + H_2O \qquad (11\text{-}15)$$

and we define a thermodynamic equilibrium constant for such exchange:

$$K_{NAC,EDA} \,/\, (L \cdot mol^{-1}\, sites) = \frac{[(Si) > O:NAC]}{[(Si) > O:H_2O][NAC]} \qquad (11\text{-}16)$$

where this constant has the subscript EDA to remind us that it reflects *electron donor-acceptor* surface interactions and it has the units that result from the ratio, (mol $NAC \cdot kg^{-1}$ solid) (mol sites $\cdot kg^{-1}$ solid)$^{-1}$ (mol $NAC \cdot L^{-1}$)$^{-1}$.

Values of $K_{NAC,EDA}$ have been measured for a large number of NACs (Table 11.2; Haderlein et al. 1996). Generally, the values increase as more electron-withdrawing nitro substituents occur on a sorbate (e.g., compare 4-nitrotoluene at 820 vs. 2,4-dinitrotoluene at 120,000 vs. 2,4,6-trinitrotoluene at 300,000). Also ring substituents

Table 11.2 Adsorption of Nonionic Nitroaromatic Compounds (NACs) to Aluminosilicate Clays: (a) Surface Area Factors, f_{saf}, for Different Clays Expressing Maximum Sorption Sites Relative to Kaolinite, and (b) $K_{NAC,EDA}$ Values (L·mol^{-1}sites) Measured for Several NACs on K$^+$-Kaolinite Allowing Estimates of K_{NACd} Values Due to Electron Donor-Acceptor Interactions (Eq. 11-20)

(a)	Aluminosilicate Clay	Surface Area Factor (f_{saf})
	Kaolinite	1
	Illite	6
	Montmorillonite	12

(b)	Compound	$K_{NAC, EDA}$ (L·mol^{-1} sites) [a]
	Nitrobenzene	100
	1,2-Dinitrobenzene	70
	1,4-Dinitrobenzene	31,000
	2-Nitrotoluene	50
	3-Nitrotoluene	420
	4-Nitrotoluene	820
	2,4-Dinitrotoluene	120,000
	2,6-Dinitrotoluene	1,700
	2,4,6-Trinitrotoluene (TNT)	300,000
	2-Amine-4,6-dinitrotoluene	50,000
	4-Amino-2,6-dinitrotoluene	1,800
	2,6-Diamino-4-nitrotoluene	180
	2,4-Dinitro-6-methyl-phenol (DNOC)	450,000
	2,4-Dinitro-6-sec-butyl-phenol (Dinoseb)	1,100

[a] Data from Haderlein et al. (1996).

that prevent close approach of the ring system to the siloxane face (e.g., *sec*-butyl in Dinoseb) strongly lower the value of the $K_{NAC,EDA}$. The extent of complexation with any particular clay mineral also depends on the abundance of siloxane sites per mass of clay. This surface area factor is greatest for an expandable clay like montmorillonite (about 12 times more than kaolinite), intermediate for a nonexpandable three-layer clay like illite (about 6 times more than kaolinite), and least for a two-layer clay like kaolinite.

Recognizing that [total sites] = [(Si) > O:NAC] + [(Si) > O:H$_2$O)], we can eliminate the term [(Si) > O:H$_2$O] in Eq. 11-16, and after rearranging find:

$$[(Si) > O:NAC] = \frac{[\text{total sites}] \cdot K_{NAC,EDA} \cdot [NAC]}{1 + K_{NAC,EDA}[NAC]} \tag{11-17}$$

or:

$$K_{NACd} = \frac{[(Si) > O:NAC]}{[NAC]} = \frac{[\text{total sites}]K_{NAC,EDA}}{1 + K_{NAC,EDA}[NAC]} \tag{11-18}$$

In most aquatic systems where Na^+, K^+, Mg^{2+}, and Ca^{2+} serve as the predominant cations, it is only the fraction of siloxane surface covered by potassium counterions that proves to be accessible to NACs. Hence, for natural solids in the real world, the total available sites for such EDA interactions can be approximated by the product:

$$[\text{total sites}] = f_{\text{clay}} \cdot f_{K^+\text{clay}} \cdot f_{\text{saf}} \cdot (6 \times 10^{-3} \text{ mol sites} \cdot \text{kg}^{-1} \text{ K}^+\text{-kaolinite}) \quad (11\text{-}19)$$

where f_{clay} is the clay mineral (not clay size) content of the solids (kg clay·kg^{-1} solid)

$f_{K^+\text{clay}}$ is the fraction of cationic counterion charges contributed by weakly hydrated cations like potassium (kg K$^+$-clay·kg^{-1} clay)

f_{saf} is the average surface area factor reflecting the ratio of siloxane surface availability of the clay minerals present versus kaolinite

6×10^{-3} mol sites·kg^{-1} K$^+$-kaolinite is the typical value for the maximum site density on kaolinite

Finally, combining the expression for the total sites with Eq. 11-18, we have:

$$K_{\text{NACd}} / (\text{L} \cdot \text{kg}^{-1} \text{ solid}) = \frac{f_{\text{clay}} \cdot f_{K^+\text{clay}} \cdot f_{\text{saf}} (6 \times 10^{-3}) K_{\text{NAC,EDA}}}{1 + K_{\text{NAC,EDA}}[\text{NAC}]} \quad (11\text{-}20)$$

Note that for simplicity we assume that $f_{K^+\text{clay}}$ and $K_{\text{NAC,d}}$ are linearly related, which is not necessarily the case (see Weissmahr et al., 1999). Eq. 11-20 indicates that K_{NACd} is constant at low concentrations (i.e., $[\text{NAC}] \ll 1/K_{\text{NAC,EDA}}$) and declines at higher levels. Now one may use insights on the clay mineralogy of natural solids and the cationic composition of aqueous solutions in which they are bathed to estimate the sorption of NACs (see Illustrative Example 11.4). Often this sorption mechanism is even more important than absorption to NOM for NACs with fairly large $K_{\text{NAC,EDA}}$ values (Weissmahr et al., 1999). Finally, we should note that when present in mixtures, competition for sites between different NACs may strongly influence the transport of these contaminants (Fesch et al., 1998).

Illustrative Example 11.4   **Estimating the Retardation of Trinitrotoluene Transport in Groundwater**

Problem

Due to past munitions production and use, the explosive NAC, trinitrotoluene (TNT), occurs in a ground water at 0.1 μM. You need to evaluate this compound's mobility in this oxic aquifer knowing it has the characteristics shown below.

Mineralogy: 75% quartz, 20% feldspar, 5% illite, 0.2% organic matter (NOM)

Density of aquifer material: $\rho_s = 2.6$ kg · L^{-1}

Aquifer porosity: $\phi = 0.31$

Groundwater composition: pH = 7.1, $[Na^+] = 0.5$ mM, $[K^+] = 0.04$ mM, $[Ca^2] = 1.3$ mM, $[Cl^-] = 1.5$ mM, $[HCO_3^-] = 1.6$ mM

What retardation factor do you estimate?

Answer

The retardation of subsurface transport of TNT arises from this compound's absorption into NOM and adsorption onto mineral siloxane surfaces covered with weakly hydrated cations like potassium (but not sodium and calcium). While components of feldspars exhibit some siloxane surfaces, here we anticipate that most of the siloxanes occur in the aluminosilicate clay minerals (e.g., illite) because these particles have such high specific surface areas (Table 11.3). Hence, the total K_{TNTd} for TNT may be found at this site:

$$K_{TNTd} = \frac{f_{oc}[TNT]_{oc} + f_{illite}[TNT]_{illite}}{[TNT]_{water}} = f_{oc}K_{TNToc} + K_{TNTd,illite}$$

The K_{iow} of TNT is $10^{1.86}$ (Haderlein et al., 1996). Looking at Table 9.2, you see that nitroaromatic compounds are not included in any of the compound classes listed. But you also note that the log K_{iow} – log K_{ioc} relationships of substituted aromatic compounds have slopes between 0.7 and 1.0 and intercepts near 0. Hence, you estimate the log K_{ioc} of TNT:

$$\log K_{TNToc} \sim (0.7 \text{ to } 1.0) \log K_{TNTow} = 1.3 \text{ to } 1.9$$

and with f_{oc} of 0.001, this implies $f_{oc}K_{TNToc}$ is only 0.02 to 0.08 L · kg^{-1}.

Adsorption to the K^+-covered siloxane surfaces of the clay, illite, can be estimated using Eq. 11-20. $K_{TNT,EDA}$ is 300,000 L · mol^{-1} and the surface area factor, f_{saf}, for illite is 6 (Table 11.2). Since the ground water contains so much calcium relative to potassium (30:1), only a very small fraction of the cation exchange sites on the illite are covered with weakly hydrated potassium ions; you assume f_{K^+clay} is about 0.01. Thus, you estimate:

$$K_{TNTd,illite} \cong (0.05)(0.01)(6)(6 \times 10^{-3})(300,000)/[1 + (300,000)(1 \times 10^{-7} \text{ M})] \cong 5 \text{ L} \cdot \text{kg}^{-1}$$

This result is much larger than the value of $f_{oc} \cdot K_{TNToc}$, so you realize that the EDA interaction with the aquifer clay dominates sorption of TNT in this case. Finally, to estimate TNT's retardation factor, R_{fTNT} ($= f_{TNTw}^{-1}$; see Eq. 9-12):

$$R_{fTNT} = 1 + \rho_s \frac{1-\phi}{\phi} K_{TNTd} = 1 + (2.6)\frac{(1-0.31)}{(0.31)} \cdot 5 \cong 30$$

This result implies that the TNT will move through the subsurface at a rate that is 1/30 the rate of the groundwater velocity. You also note that if the TNT concentrations anywhere in the plume are above 1×10^{-6} M, then the $K_{TNTd,illite}$ would be smaller (second term in the denominator of Eq. 11-20 won't be negligible) and the retardation factor will correspondingly decrease.

O_2N — NO_2

NO_2

trinitrotoluene
(TNT)

Adsorption of Ionized Organic Compounds from Aqueous Solutions to Charged Mineral Surfaces

Now we consider organic compounds that exhibit at least one ionic group in their structure (e.g., $-COO^-, -NH_3^+, -SO_3^-$). These charged organic compounds can occur in two regions: dissolved in the water layer immediately adjacent to the surface (Fig. 11.1d) or actually bonded to the surface (Fig. 11.1e). Much of the work on this topic has been performed by investigators interested in surfactants, since the inclusion of a charged moiety on an otherwise nonpolar chemical skeleton renders the resultant compound *amphiphilic* (one part liking water, the other part liking apolar media) and capable of participating in many interesting interfacial phenomena. Also, much progress in this topic has been made by researchers studying inorganic surfaces (e.g., metal oxides) and how these minerals are affected by organic *ligands* (compounds that bind metals; from the Latin *ligare*: to bind).

Experiments often, but not always, show that the sorption isotherms for charged organic sorbates interacting with natural solids are nonlinear (Fig. 11.9). Said another way, the solid–water distribution ratio may change markedly as a function of the sorbate's own dissolved concentration. The extent of solid association of charged organic compounds also varies as a function of factors like solution pH, since this property governs both the presence of charges on mineral surfaces and the fraction of sorbate in an ionized form (Section 9.5). Solution ionic strength and ionic composition also affect the sorption of charged organic chemicals, especially if inorganic ions compete with organic ones for binding sites. The mineral composition of the sorbent is also a key factor since the charge density on a given solid is dependent on that material's particular responses to the surrounding conditions. Due to the combined effects of all of these factors, *a priori* estimation of charged organic chemical sorption to natural soils and sediments is more difficult than for neutral organic compounds where everything is often reducible to a few key factors (e.g., K_{ioc} and f_{oc}). In the following sections, we examine in more detail the nature of the interactions of charged molecules with charged surfaces and discuss how we estimate the extent of such (ad)sorption. But first, we consider the mechanisms that control the presence of surface charges, $\sigma_{surf\,ex}$, on solids in water.

Charging of Mineral Surfaces in Water

Almost all mineral particles in natural waters are charged. Owing to ionizable surface groups, virtually every solid presents a charged surface to the aqueous solution. If this surface charge is of opposite sign to that exhibited by an organic functional group, then there will be an electrostatic attraction between the organic sorbate in the bulk solution and the particle surface. This is the same interaction force drawing inorganic counterions like Na^+ and Ca^{2+} near a negatively charged surface in water. And so the organic ions will accumulate in the thin film of water surrounding the particle as part of the population of charges in solution balancing the charges on the solid surface (Fig. 11.10). Conversely, organic molecules with charges of like sign at the surface will be repulsed from the near-surface water. These electrostatic effects act similarly for all charged sorbates.

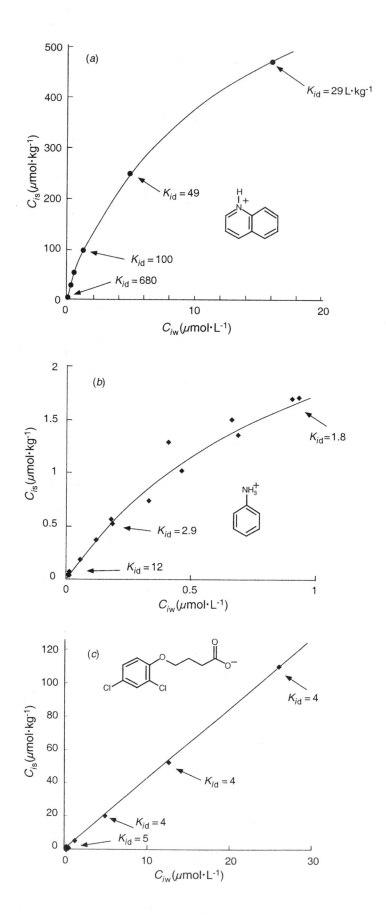

Figure 11.9 Sorption isotherms for some charged organic compounds interacting with natural solids: (*a*) quinolinium cation on a subsoil of $f_{oc} = 0.024$ and cation exchange capacity of 84 mmol/kg (Zachara et al., 1986), (*b*) anilinium cation on a surface soil with $f_{oc} = 0.013$ and cation exchange capacity of 112 mmol/kg (Lee et al., 1997), and (*c*) sorption of 4-(2,4-dichloro-phenoxy)-butyrate anion on a sediment with $f_{oc} = 0.015$ and unknown anion exchange capacity (Jafvert, 1990).

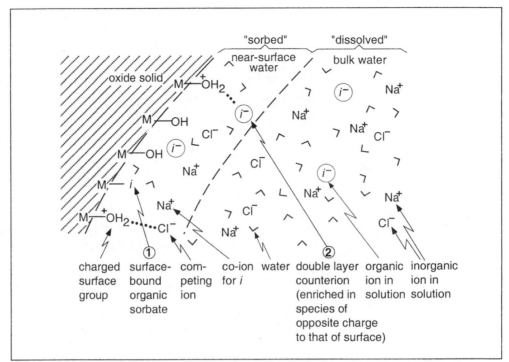

Figure 11.10 A positively charged oxide particle in water attracts anionic species including organic ones (e.g., i^-) to the near-surface water. Some of these anionic species may also react with the surface, displacing other ligands (e.g., H_2O or OH^-), to form surface-bound sorbate. M in the solid refers to atoms like Si, Al, or Fe.

To evaluate the importance of charge-derived interactions, we need to know how many charges are on the surfaces of solids (i.e., $\sigma_{surf\,ex}$ in mol charges m^{-2}). The adjacent surficial layer of water must contain an excess of ions called *counterions* (e.g., Cl^- and i^- in Fig. 11.10) that carry charge equal in magnitude and opposite in sign to that exhibited by the particle surface. The thickness of this ion-rich water layer, which is sometimes called the diffuse double layer, varies inversely with the ionic strength of the solution. The e^{-1}-characteristic thickness is given by $0.28 \times I^{-0.5}$ nm, where I is the solution's ionic strength in molar units (Morel, 1983). For typical ionic compositions of natural waters ($\sim 10^{-3}$ to 0.5 M), this means that most [i.e., $(1-e^{-1})\,100\% = 63\%$] of the counterions are packed into a layer of water between 0.3 and 10 nm thick, and nearly all [i.e., $(1-e^{-3})\,100\% = 95\%$] are within 1–30 nm of the surface. It is worth noting that this range is very similar to the 1–10 nm range postulated to reflect ordered vicinal water (Drost-Hansen, 1969); thus, this is indeed a very special microscopic water environment.

The amount of charges on particle surfaces depends on the mineralogy of the solid and the nature of the aqueous solution in which it occurs. Several important kinds of surfaces are common in the environment (Table 11.3). Here we especially consider: (1) oxides or oxyhydroxides, (2) alumino-silicates or clay minerals, and (3) natural organic matter and other solids like carbonates.

Oxides/Oxyhydroxides. For natural solids that are oxides or oxyhydroxides (e.g., quartz, SiO_2; goethite, α–FeOOH; gibbsite, $Al(OH)_3$), their water-wet surface is covered by hydroxyl groups (recall Fig. 11.2). These hydroxyl moieties can undergo proton-exchange reactions with the aqueous solution much like dissolved acids:

$$\equiv\text{M–OH}_2^+ \;\rightleftharpoons\; \equiv\text{M–OH} + \text{H}^+ \tag{11-21}$$

$$\equiv\text{M–OH} \;\rightleftharpoons\; \equiv\text{M–O}^- + \text{H}^+ \tag{11-22}$$

where $\equiv$M refers to an atom like Si, Fe, or Al in the particle surface. We may define acid–base equilibrium constants for those reactions (neglecting activity coefficients here):

$$K_{a1} = [\equiv\text{M–OH}][\text{H}^+] \,/\, [\equiv\text{M–OH}_2^+] \tag{11-23}$$

$$K_{a2} = [\equiv\text{M–O}^-][\text{H}^+] \,/\, [\equiv\text{M–OH}] \tag{11-24}$$

These surface acid equilibrium constants differ from their solution counterparts in that they reflect both an intrinsic reactivity of the particular O-H bond and an electrostatic free energy of moving H^+ to and from a charged surface:

$$K_{a1} = K_{a1}^{\text{int}} \cdot e^{zF\Psi/RT} \tag{11-25}$$

and:

$$K_{a2} = K_{a2}^{\text{int}} \cdot e^{zF\Psi/RT} \tag{11-26}$$

where $z = +1$ for the exchanging ion, H^+ in this case

F is the Faraday constant ($96{,}485 \; \text{C} \cdot \text{mol}^{-1}$)

Ψ is the surface potential relative to the bulk solution (V or $\text{J} \cdot \text{C}^{-1}$)

R is the gas constant ($8.31 \; \text{J} \cdot \text{mol}^{-1} \, \text{K}^{-1}$)

T is the absolute temperature (K)

At higher and higher pH's, Ψ becomes less and less positive as reaction 11-21 proceeds to the right and more and more negative as reaction 11-22 continues to the right. This variation in surface charge buildup makes it increasingly more difficult to move H^+ away from an increasingly negatively charged oxide surface as solution pH is increased. The magnitude of this effect is calculated with the exponential terms in Eqs. 11-25 and 11-26.

It is also possible for some reactive inorganic species (e.g., Fe^{3+} or PO_4^{3-}) to strongly bind to the surface. In such a case, these inorganic ions along with H^+ and OH^- are responsible for establishing the extent of charging on the solid surface. The combination of ions responsible for this charge formation are called "potential (Ψ) determining." For now, we neglect specific adsorption of inorganic ions and their effects on surface charge (e.g., see Dzombak and Morel, 1990, for examples of specific sorption and its associated impact on surface charge for ferrihydrite, $\text{Fe(OH)}_3(\text{s})$).

For the case at hand, it is easy to see that the amount of $\equiv\text{MOH}_2^+$ and $\equiv\text{MO}^-$ species on the solid surface control the surface's charge. The concentration of this charge, $\sigma_{\text{surf ex}}$ (mol charges $\cdot \text{m}^{-2}$) can be estimated (neglecting other specifically sorbed species) by the difference between positive and negative site concentrations:

$$\sigma_{\text{surf ex}} = [\equiv M\text{--}OH_2^+] - [\equiv M\text{--}O^-] \tag{11-27}$$

where the surface species concentrations are given in units of $mol \cdot m^{-2}$ of exposed surface. When these two surface species are present in equal concentrations, the surface exhibits zero net charge (also $\Psi = 0$); we call the solution pH that establishes this condition the pH of zero point of charge, or pH_{zpc}. This pH_{zpc} can be calculated if we know the intrinsic acidities of $\equiv MOH_2^+$ and $\equiv MOH$:

$$[\equiv MOH_2^+] = [\equiv M\text{--}O^-] \qquad \text{at } pH_{zpc} \tag{11-28}$$

Substituting from Eqs. 11-23 to 11-26 and recalling $\Psi_{zpc} = 0$, we have:

$$[\equiv M\text{--}OH][H^+]_{zpc} / K_{a1}^{int} = K_{a2}^{int} [\equiv M\text{--}OH] / [H^+]_{zpc} \tag{11-29}$$

Simplifying Eq. 11-29 allows us to relate pH_{zpc} and the intrinsic acidities of the surface:

$$[H^+]_{zpc}^2 = K_{a1}^{int} K_{a2}^{int} \tag{11-30}$$

$$pH_{zpc} = 0.5 \, (pK_{a1}^{int} + pK_{a2}^{int}) \tag{11-31}$$

Equation 11-31 shows that an oxyhydroxide's pH_{zpc} is midway between the intrinsic pK_a's of its surface groups. Now, when the aqueous solution pH is below the pH_{zpc}, we have the condition $[\equiv MOH_2^+] > [\equiv MO^-]$, and the solid exhibits a net positive surface charge. Conversely, when we are above the solid's pH_{zpc}, then $[\equiv MO^-] > [\equiv MOH_2^+]$, the surface is negatively charged, and it becomes increasingly so at higher pH. Note that at neutral pH values most surfaces present in natural systems exhibit a net negative charge due to such surface reactions plus adsorption of NOM.

Our task now is to estimate the concentration of surface charge for solid oxides that interest us as a function of solid and solution properties. For cases in which H^+ and OH^- are potential-determining, such estimates can be done by solving for the abundances of the important surface species using two sets of information: (1) knowledge of the intrinsic acidities for the oxide of interest, and (2) the feedback relationship of surface potential on surface charge density. For a salt with cations and anions carrying the same charge (e.g., Na^+ and Cl^- or Ca^{2+} and SO_4^{2-}), this charge density, $\sigma_{\text{surf ex}} = [\equiv MOH_2^+] - [\equiv MO^-]$, can be calculated (Stumm and Morgan, 1996, and as shown in Box 11.1). Figure 11.11 illustrates the results of such calculations on charge density at pH's below and above an oxide's pH_{zpc} for aqueous solutions of $I = 0.001$ and 0.5 M.

These results may be understood with a specific example. If we were interested in amorphous iron oxide with a pK_{a1}^{int} of 7 and a pK_{a2}^{int} of 9 (Table 11.3) and a surface hydroxyl concentration, $[\equiv MOH]$, of $2 \times 10^{-6} \, mol \cdot m^{-2}$, we could estimate that this

---

Box 11.1   Estimating the Surface Charge $\sigma_{surf\,ex}$ of Oxides when H^+ and OH^- are the Potential-Determining Ions (after Stumm and Morgan, 1996)

$$\sigma_{surf\,ex}(\text{mol charges} \cdot m^{-2}) = [\equiv MOH_2^+] - [\equiv MO^-] \qquad (11\text{-}27)$$

where

[$\equiv MOH_2^+$] is the concentration of protonated surface sites ($mol \cdot m^{-2}$)

[$\equiv MO^-$] is the concentration of deprotonated surface sites ($mol \cdot m^{-2}$)

Using acidity relationships (Eqs. 11-23 through 11-26) to substitute for the charged surface species:

$$\sigma_{surf\,ex} = [\equiv MOH][H^+](K_{a1}^{int})^{-1}e^{-F\psi/RT} - [\equiv MOH][H^+]^{-1}K_{a2}^{int} \cdot e^{F\psi/RT} \qquad (11\text{-}32)$$

where

[$\equiv MOH$] is the concentration of hydroxyl groups on the solid surface ($mol \cdot m^{-2}$),

K_{a1}^{int} and K_{a2}^{int} are the intrinsic acidity constants of the $\equiv MOH_2^+$ and $\equiv MOH$ sites, respectively

ψ is the surface potential ($V = J \cdot C^{-1}$)

R is the gas constant ($8.31\ J \cdot K^{-1}\ mol^{-1}$)

T is the absolute temperature (K)

F is the Faraday constant ($96{,}485\ C \cdot mol^{-1}$)

The surface charge and surface potential are related:

$$\psi = \frac{2RT}{zT}\sinh^{-1}\left[\left[\frac{\pi F^2 10^{-3}}{2\varepsilon RTI}\right]^{0.5}\sigma_{surf\,ex}\right] \qquad (11\text{-}33)$$

where

z is the valence of ions in the background electrolyte (e.g., NaCl, $z = 1$)

ε is the dielectric constant of water ($7.2 \times 10^{-10}\ C \cdot V^{-1}\ m^{-1}$ at 25°C)

I is the solution ionic strength ($mol \cdot L^{-1}$; 10^{-3} is needed to convert L to m^3)

Substituting Eq. 11-33 into Eq. 11-32 yields:

$$\sigma_{surf\,ex} = [\equiv MOH][H^+](K_{a1}^{int})^{-1}\exp\left(-\frac{2}{z}\sinh^{-1}\left[\left[\frac{\pi F^2 10^{-3}}{2\varepsilon RTI}\right]^{0.5}\sigma_{surf\,ex}\right]\right)$$

$$-[\equiv MOH][H^+]^{-1}(K_{a2}^{int})\exp\left(+\frac{2}{z}\sinh^{-1}\left[\left[\frac{\pi F^2 10^{-3}}{2\varepsilon RTI}\right]^{0.5}\sigma_{surf\,ex}\right]\right) \qquad (11\text{-}34)$$

For any oxide with particular K_{a1}^{int}, K_{a2}^{int}, and [$\equiv MOH$] and any solution with specific pH, temperature, and ionic strength, Eq. 11-34 can be solved for $\sigma_{surf\,ex}$ by trial and error (e.g., using a spreadsheet program).

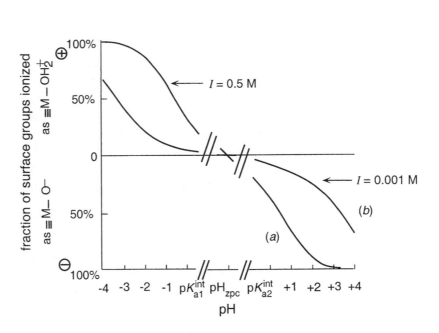

Figure 11.11 Variation of surface charge on a solid oxide (25°C) due to protonation and deprotonation of surface hydroxyls (at 2×10^{-6} mol·m^{-2}) as a function of solution pH for (a) $I = 0.5$ M of a dissolved salt with both the cation and the anion exhibiting one charge (1:1 electrolyte) and (b) $I = 0.001$ M 1:1 electrolyte. Note the breaks in the curves between unspecified values of pK_{a1}^{int} and pK_{a2}^{int}.

solid would have about 5×10^{-8} mol·m^{-2} of positive charges on its surface at pH 6 in freshwater of $I = 10^{-3}$ M (i.e., <10% as $\equiv MOH_2^+$ in Fig. 11.11). In salty water of $I = 0.5$ M and pH = 6, the same solid would have about 6×10^{-7} mol·m^{-2} of positive charges (i.e., ~30% $\equiv MOH_2^+$ in Fig. 11.11). If the solution pH was 7 instead of 6, the surface charge density would decrease by almost a factor of 3. It would not be until pH was increased to above 8 [pH$_{zpc}$ = 0.5(7 + 9) = 8] that this particular iron oxide solid would start to show a net negative surface charge.

Typically, surface charge densities in the range of 10^{-6} to 10^{-8} mol·m^{-2} are seen for oxides (Table 11.3) at circumneutral pHs. This implies that 10^{-6} to 10^{-8} moles of counterions, including some charged organic molecules, will accumulate opposite each meter squared of surface area due to electrostatic attractions. This property of surfaces is often referred to as the solid's cation exchange capacity (CEC) or anion exchange capacity (AEC), depending whether the solid exhibits net negative or positive charging, respectively. Note again, this treatment neglects the influence of specifically sorbed ions which would neutralize some of this surface charge [e.g., Fe(OH)$^{2+}$ or HPO$_4^{2-}$ bound to the surface].

Aluminosilicate Clays. Clay minerals present a different case with regard to assessing their surface charge. These mixed aluminum oxides and silicon oxides (thus aluminosilicates) expose two kinds of surface to the external media, and therefore the same particles may exhibit both a CEC and an AEC at the same time (Table 11.3). First, the edges of these flake-shaped minerals are somewhat like aluminum oxides in their behavior and respond to pH changes in solution much like pure aluminum oxides (e.g., pH$_{zpc}$ of kaolinite edge ~7; Williams and Williams, 1978). The consequent anion exchange capacity observed empirically for clays is up to 0.1 mol·kg^{-1} for a wide variety of clays (Grim, 1968), but this value changes with solution pH and ionic strength. In contrast, as we have discussed before, the faces of these platey particles have a "siloxane" structure (–Si–O–Si–) which does not leave free hydroxyl groups (–Si–OH) to participate in proton exchange reactions with the bulk solution. Instead, the faces exhibit a charge due to cation substitutions for the aluminum or silicon atoms within the internal structure. These "isomorphic" substitutions involve cations of lower total positive charge (e.g., Al^{3+} for Si^{4+} or Mg^{2+} for Al^{3+}). The result is a fixed and permanent charge deficiency that looks like a negative surface charge to the surrounding solution. Empirical measures of this negative surface charge or CEC are made by assessing the maximum concentrations of weakly bound cations such as ammonium, NH_4^+, that can be sorbed. Table 11.3 shows the results of such cation exchange capacity tests on three common clays, montmorillonite, illite, and kaolinite. Expandable three-layer clays like montmorillonite exhibit the highest CEC's near $1 \; mol \cdot kg^{-1}$ or 1.4×10^{-6} moles of charged sites per meter squared (assuming a specific surface area of $700 \; m^2 \cdot g^{-1}$; Grim, 1968). On the other extreme, two-layer kaolinite clays exhibit the lowest CECs of about $0.1 \; mol \cdot kg^{-1}$ (Grim, 1968). This is chiefly due to their greatly reduced specific surface areas compared to the expandable three-layer clays, since per unit area these kaolinites actually have greater charge density, $\sim 10^{-5} \; mol \cdot m^{-2}$.

Particulate Natural Organic Matter and Other Solids. Particulate natural organic matter may also contribute to the assemblage of charged sites of solids in water. This is mostly due to ionization reactions of carboxyl groups (–COOH), and at higher pH values, phenolic groups (aromatic ring –OH). Such acidic moieties have been found at about 1 to 10 mmol per gram of natural organic matter. Depending on the surrounding molecular environment, the carboxyl moieties exhibit pK_a's ranging from about 3 to 6. Consequently, the extent of charge buildup in the organic portion of natural particles will vary as a function of pH.

Still other solid phases like carbonates are common in nature, and these materials also exhibit surface charging due to excess M^{2+} or CO_3^{2-} on the solid's surface forming surface species like $\equiv MOC(O)O^-$ (Table 11.3). Realizing there will almost always be charges on particle surfaces submerged in water, we can now examine their impact with regard to sorbing ionized organic chemicals from solution.

Conceptualization of the Ion Exchange Sorptive "Reactions", Free Energies, and Equilibrium Constants

Due to surface charging, an ion *exchange* "reaction" mechanism can occur and enable the accumulation of a mixture of ions in the near-surface water (e.g., exchanging i^- for Cl^- in Fig. 11.10). Since this water layer remains tightly associated with the solid, any organic ions that it includes appear "sorbed." Further, it is sometimes possible for such ions to displace other ligands and bond to the solid surface. This *chemisorption forms a second sorbed species that is distinct from like-structured organic ions dissolved in the near-surface water* (e.g., $M–i \neq M–OH_2^+$ plus i^- nearby). Depending on the type of surface involved, the "R–" group and the charged moiety(ies) of the sorbate, and the ionic composition of the solution, one or both of these sorbed species may coexist in significant proportions (Stone et al., 1993). In this section, we will treat the ionic organic compound sorption when there is no bonding to the surface. Later (Section 11.5), we will pick up the topic of surface reactions.

When organic chemicals include structural components that are ionized, new interactions become important insofar as the forces attracting (or repulsing) these sor-

Table 11.3 Sorbent Properties of "Pure Solids" Commonly Present in Aquatic Environments

Sorbent	Compositions	Specific Surface Area (m²·g⁻¹)	CEC (mol·m⁻²)	AEC (mol·m⁻²)	pK_{a1}^{int}	pK_{a2}^{int}	pH_{zpc}	Ref.[b]
Oxides								
Quartz	SiO_2	0.14	8×10^{-8} [a]		(−3)	7	2	1, 2
Amorphous silica	SiO_2	500	8×10^{-8} [a]		(−3)	7	2	1, 2
Goethite	α-FeOOH	46		2×10^{-8} [a]	6	9	7.5	2
Amorphous iron oxide	$Fe(OH)_3$	600		5×10^{-8} [a]	7	9	8	2, 4, 5
Alumina	Al_2O_3	15		7×10^{-8} [a]	7	10	8.5	2
Gibbsite	$Al(OH)_3$	120	2×10^{-8} [a]		5	8	6.5	2, 6
Aluminosilicates								
Na-montmorillonite	$Na_3Al_7Si_{11}O_{30}(OH)_6$	600–800	0.9 to 2×10^{-6}	3 to 4×10^{-7}			2.5	6, 7
Kaolinite	$Al_2Si_2O_5(OH)_2$	12	0.2 to 1×10^{-5}	0.6 to 2×10^{-5}			4.6	6, 7
Illite	$KAl_3Si_3O_{10}(OH)_2$	65–100	1 to 6×10^{-6}	3×10^{-7}				
Organic								
Humus	$C_{10}H_{12}N_{0.4}O_6$	1	1 to 10×10^{-3}					8, 9
Carbonate								
Calcite	$CaCO_3$	1		9×10^{-6}			8–9.5	10, 11

[a] Calculated CEC and AEC values assume solution pH = 7, ionic strength of 10^{-2} M, T = 293 K, solid-site density of 2×10^{-6} mol·m⁻², and use of Eq. 11-32 or 11-34. Negative logarithms of intrinsic acidity constants (pK^{int}) are rounded off to the nearest unit. [b] References: 1. Parks (1965); 2. Schindler and Stumm (1987); 3. Mikhail et al. (1968a,b); 4. Tipping (1981); 5. Dzombak and Morel (1990); 6. Davis (1982); 7. Grim (1968); 8. Chiou et al. (1990); 9. Khan (1980); 10. Zullig and Morse (1988); 11. Somasundaran and Agar (1967).

bates to solid surfaces are concerned (Fig. 11.10). First, since most mineral surfaces in water are charged, there is an electrostatic interaction between any charged molecule and such water-wet solids. We will call this interaction energy $\Delta_{elect}G_i$. Note that in the following discussion, *i denotes a charged organic species* (e.g., a deprotonated acid, A⁻, or a protonated base, BH⁺). When the organic sorbate *i* (with charge z_i) and sorbent (with surface potential Ψ) are oppositely charged, the electrostatic attraction strongly promotes adsorption and has the magnitude:

$$\Delta_{elect}G_i = z_i F \Psi \tag{11-35}$$

where F is the Faraday constant and Ψ (in volts) is the electric potential difference between the bulk solution (assumed to be at zero potential) and the particle surface. Additionally, as we saw for nonionic organic compounds (Section 11.3), the hydrophobic part of a sorbate's structure encourages its transfer into the near-surface region. This free energy contribution will be termed $\Delta_{hydrophobic}G_i$. Together, these interactions promote an ion exchange "reaction" to occur (Fig. 11.10 shows exchange of two anions):

$$i + \text{comp.ion:surf} \rightleftharpoons i\text{:surf} + \text{comp.ion} \tag{11-36}$$

where *i* is the organic ion participating in the exchange, "surf" represents the presence of charged sites on the solid in water, "comp. ion" is the inorganic ion (in this case of the same charge) with which *i competes*, and the colons indicate surface association without bonding.

The accumulation of sorbed organic ions relative to ion concentrations in the bulk solution is due to a free energy increment:

$$\Delta_{\text{surf water}}G_i = z_i F \Psi + \Delta_{hydrophobic}G_i \tag{11-37}$$

$$= -RT \ln \left([i\text{:surf}] / [i]\right) \tag{11-38}$$

$$= -RT \ln K_{id\ ex} \tag{11-39}$$

where $K_{id\ ex}$ (L·kg⁻¹) is the sorbed-to-dissolved *distribution ratio* of the ionic organic compound accumulated due to ion *exchange*, and we note that we have substituted chemical concentrations for chemical activities.

Since the displacement of the competing ion results in the "recovery" of its own $\Delta_{elect}G_{comp.ion}$, the overall *exchange* process involves a free energy change (indicated by subscript, ex):

$$\Delta_{\text{surf water,ex}}G_i = (z_i F \Psi + \Delta_{hydrophobic}G_i) - z_{comp.ion}F\Psi \tag{11-40}$$

$$\cong \Delta_{hydrophobic}G_i \tag{11-41}$$

$$= -RT \ln \left([i\text{:surf}][\text{comp.ion}]/[i][\text{comp.ion:surf}]\right) \tag{11-42}$$

$$= -RT \ln K_{iex} \tag{11-43}$$

assuming the two competing ions approach the charged surface to the same extent on average (surface potential varies with distance from the solid surface). Note that K_{iex} (dimensionless) $\neq K_{id\ ex}$ (L·kg⁻¹), but rather:

$$K_{i\text{ex}} = K_{id\,\text{ex}} / K_{\text{comp.ion d ex}} \qquad (11\text{-}44)$$

Due to the hydrophobic free energy advantage of the organic sorbate, more i can accumulate in the near-surface water than the competing inorganic ions it displaces! However, electroneutrality requires no net charge buildup in this region. So this additional sorption energy requires the cotransfer of a "co-ion" of *opposite* charge to i (e.g., Na$^+$ combined with i^- in Fig. 11.10). This situation is expressed with a *charge balance equation for the near-surface region*:

$$[\text{comp.ion:surf}] + [i\text{:surf}] = \text{surface exchange capacity} + [\text{co-ion:surf}] \qquad (11\text{-}45)$$

where   [comp.ion:surf] is the concentration of competing counterions (mol·kg^{-1})

"surface exchange capacity" (i.e., CEC or AEC) is the concentration of charges on the surface of the solid (mol sites·kg^{-1})

[co-ion:surf] represents the co-ions of opposite charge to i that are present near the surface

Since the transfer of these oppositely charged co-ions against the electrostatic potential requires $-z_i F \Psi$ (note that $z_{\text{co-ion}} = -z_i$), accumulation in the near surface water is given (focusing on the monovalent case here):

$$[\text{co-ion:surf}]^* = [\text{co-ion}]\exp(-z_i FY/RT) = [\text{co-ion}]\exp(\Delta_{\text{hydrophobic}}G_i/RT) \qquad (11\text{-}46)$$

where [co-ion:surf]* is the concentration normalized per volume of near-surface water and taking the co-ion to be a monovalent, positively charged ion here. To convert this to the concentration per mass of solid sorbent, we need to multiply this result by the volume of near-surface water per mass of solid, V_{vic} (L·kg^{-1}):

$$[\text{co-ion:surf}] = V_{\text{vic}}[\text{co-ion:surf}]^* = V_{\text{vic}}[\text{co-ion}]\exp(\Delta_{\text{hydrophobic}}G_i/RT)$$

$$\qquad (11\text{-}47)$$

$$= K_{i\text{ex}} V_{\text{vic}} [\text{co-ion}]$$

The size of V_{vic} may be estimated using knowledge of a solid's specific surface area, A_{surf}, and the e^{-1} thickness of the diffuse double layer (recall thickness estimated using $0.28 \times I^{-0.5}$ nm). Assuming this thickness ranges from 0.3 to 10 nm and the specific surface areas of natural solids range from 1 (reported for some aquifer solids, Schwarzenbach and Westall, 1981) to 700 m^2 g^{-1} for a finely dispersed clay (Table 11.3), we see that V_{vic} may range from less than 0.001 L·kg^{-1} for sandy materials in a salty environment to about 1 L·kg^{-1} when ionic strength is low and particles have high specific surface areas.

Ion Exchange of Organic Cations

Now we can completely evaluate the extent of organic ion accumulation as dissolved ions in the near-surface water adjacent to a charged particle's surface (i.e., [i:surf]) using a combination of equations like Eqs. 11-36, 45, and 47. We will illustrate this by considering ion exchange of organic cations ($i = $ BH$^+$). Specific expressions for [i:surf] or the corresponding distribution ratio, $K_{id\,\text{ex}}$, quantifying this surface concentration relative to its corresponding bulk solution concentration are derived in Box 11.2 for the cases of protonated organic bases (BH$^+$) competing with

---

Box 11.2   General Derivation of Ion Exchange Isotherms for Cationic Organic Compound ($i = BH^+$). Competing with Monovalent, M^+ (e.g., Na^+ or K^+) Inorganic Cations, for Cation Exchange Capacity (CEC), from Solution Containing Monovalent Co-Ion (e.g., Cl^-).

1. "Sorption reactions" and associated equilibrium:

monovalent ion, M^+, exchange $i + M{:}surf \rightleftharpoons M^+ + i{:}surf$ (11-48)

equilibrium constant: $[i{:}surf][M^+]/[i][M{:}surf] = \exp(-\Delta_{hydrophobic}G_i / RT) = K_{iex}$ (11-49)

where concentrations are used in place of activities.

coexchange of pair into volume of water near surface, $i + co\text{-}ion \rightleftharpoons i{:}surf + co\text{-}ion{:}surf$ (11-50)

Anion accumulation against electrostatic repulsion must balance hydrophobic forces attracting "extra" i to the surface water volume (here called V_{vic} in $L \cdot kg^{-1}$):

$$([i{:}surf]/[i])_{excess} = [co\text{-}ion{:}surf]/[co\text{-}ion] = V_{vic} \exp(-\Delta_{hydrophobic}G_i / RT) = V_{vic} K_{iex}$$ (11-51)

so:

$$[co\text{-}ion{:}surf] = V_{vic} K_{iex} [co\text{-}ion]$$ (11-52)

2. Sum of cations must equal cation exchange capacity (CEC) plus any anions near surface:

$$[i{:}surf] + [M{:}surf] = CEC + [co\text{-}ion{:}surf]$$ (11-53)

substituting for $[M{:}surf]$ in Eq. 11-49 and using Eqs. 11-51 and 11-53:

$$K_{iex} = \frac{[i{:}surf][M^+]}{[i](CEC + V_{vic}K_{iex}[co\text{-}ion] - [i{:}surf])}$$ (11-54)

and rearranging:

$$[i{:}surf] = \frac{(CEC + V_{vic}K_{iex}[co\text{-}ion])K_{iex}}{[M^+] + K_{iex}[i]}[i]$$ (11-55)

This result indicates a hyperbolic dependency of $[i{:}surf]$ on $[i]$ when $CEC \gg V_{vic} K_{iex} [co\text{-}ion]$ (i.e., a Langmuir type isotherm, Section 9.2).

At low concentrations of $[i]$, this implies:

$$K_{id\,ex} = \frac{[i{:}surf]}{[i]} \cong \frac{(CEC + V_{vic}K_{iex}[co\text{-}ion])K_{iex}}{[M^+]}$$ (11-56)

---

monovalent inorganic cations (M^+).

Now we can estimate the concentrations of organic ions in the near-surface water as we change the concentration of the dissolved species (Fig. 11.12a). At low organic cation concentrations (i.e., $K_{iex}[i] \ll [M^+]$, the bound-to-dissolved ratio is constant:

$$K_{id\,ex} = ((CEC + V_{vic}[\text{co-ion}]K_{iex})\,K_{iex})\,/\,[\text{comp.ion}]$$

$$(11\text{-}57)$$

$$= \text{constant} \qquad \text{for low } [i]$$

At "high" levels of i (i.e., $[M^+] \ll K_{iex}[i]$), Eq. 11-55 indicates that the organic counterion concentrations must asymptotically approach a constant value set by the total surface charge density (the cation exchange capacity of the clay) as long as CEC $\gg$ $V_{vic}[\text{co-ion}]K_{iex}$:

$$[i:\text{surf}] = (CEC) \qquad \text{for high } [i] \qquad (11\text{-}58)$$

However, if the value of the product, $V_{vic}[\text{co-ion}]K_{iex}$, is large (see below) such as occurs when the R group of the organic ion causes K_{iex} to be large, then the sorbed ion concentration does not level off at this CEC threshold!

Let us complete this theoretical treatment by considering some observations of a small organic cation, ethyl ammonium (i = EA, $pK_{ia} \sim 10$), associating with a single type of solid surface, Na-montmorillonite, suspended in 10^{-2} M NaCl aqueous solution (Cowan and White, 1958). These workers measured the CEC (= $\sigma_{surf\,ex}\,A_{surf}$) of their clay sorbent to be nearly 1 mol·kg^{-1} (recall that this surface charge is not sensitive to solution ionic strength like the oxides, since it arises principally from isomorphic substitutions). The ionic strength implies that the double layer thickness is about 3 nm; and so combined with the large specific surface area for montmorillonite near 7×10^5 m^2 kg^{-1} (Table 11.3), we expect V_{vic} to be about 1 L·kg^{-1} in this case. We do not need to consider bond formation of the EA ion with the clay surface because this substance cannot react with montmorillonite surface moieties. Nor must we be concerned with neutral amine absorption into particulate organic matter in this case, since we have chosen solids that lack this phase.

$$CH_3 - CH_2 - NH_3^+$$
ethyl ammonium
(EA)

Before any EA was added to the montmorillonite suspension, the negative surface charges on the surface of the clay were balanced by an excess of hydrated sodium cations relative to hydrated chloride ions accumulated in the thin film of water surrounding the particles. When a small quantity of EA was added to the suspension (insufficient to change the ionic strength), an *ion exchange reaction* occurred, resulting in some of the EA ions exchanging with Na$^+$ counterions near the solid's surface. This ion exchange can be expressed (Eq. 11-36; note that we denote the positively charged species as EA):

$$\text{EA} + \text{Na:surf} \rightleftharpoons \text{EA:surf} + \text{Na}^+ \qquad (11\text{-}59)$$

where again the colons in the bound reactant and product shown indicate association without bond formation. At pHs $\ll$ 10 we can neglect the neutral species, CH$_3$CH$_2$NH$_2$, because the positively charged species is much more abundant than its neutral conjugate base. Assuming that the aqueous activity coefficients cancel ($\gamma'_{EA} \cong \gamma'_{Na^+}$) and assuming that $\gamma_{EA:surf} \cong \gamma_{Na:surf}$, we use concentrations and define an overall ion exchange coefficient:

$$K_{EAex} = [\text{EA:surf}][\text{Na}^+]\,/\,[\text{EA}][\text{Na:surf}] \qquad (11\text{-}60)$$

and an ethyl ammonium ion solid–water distribution ratio:

$$K_{EAd\,ex} = [EA:surface] / [EA] \tag{11-61}$$

Using the derivation in Box 11.2 with $i = EA$, $M^+ = Na^+$ and co-ion $= Cl^-$, we find that the sorbed EA concentration depends on the dissolved EA:

$$[EA:surf] = \{(((CEC + V_{vic}[Cl^-]K_{EAex}) \, K_{EAex}) / ([Na^+] + K_{EAex} [EA])\} \, [EA] \tag{11-62}$$

where [EA:surf] represents the sorbed concentration $(mol \cdot kg^{-1})$

       [EA] is the dissolved concentration $(mol \cdot L^{-1})$

       CEC $= \sigma_{surf\,ex} A_{surf}$ is the cation exchange capacity of the clay $(mol \cdot kg^{-1})$

       $[Cl^-]$ is the concentration of the co-ion that may partition into the near-surface water to balance excess EA there (i.e., $>$ CEC) $(mol \cdot L^{-1})$

       K_{EAex} is the equilibrium constant for this ion exchange between monovalent organic and inorganic ions $(-)$

       $[Na^+]$ is the concentration of the competing monovalent cation, $(mol \cdot L^{-1})$

In this expression, we see the important factors dictating the extent of this accumulation of organic ions near the charged particle surface. First, as the intensity of particle charging is increased (i.e., greater CEC), then the extent of sorption grows. Further, assuming K_{EAex} is near 1 for EA since its R group is not very hydrophobic, in this case CEC ($= 960$ mmol kg^{-1}) $>> V_{vic}[Cl^-] K_{EA\,ex}$ ($\approx (1$ L $\cdot$ kg$^{-1})(10$ mmol $\cdot$ L$^{-1})(1)$ $= 10$ mmol $\cdot$ kg^{-1}). Consequently, we expect EA's isotherm to asymptotically approach the clays' CEC and not exceed this sorption limit.

These theoretical expectations correspond nicely to the data. The observed isotherm can be fit with $K_{EAex} = 2$ (Fig. 11.12b). Since $K_{EAex} = 1$ would imply no preference between the sodium and the ethyl ammonium ions, this fit value of K_{EAex} indicates only a little selection of the organic cation over the sodium ion, presumably because of a little hydrophobicity of the ethyl substituent. As expected, at elevated organic sorbate levels, the bound-versus-dissolved distribution ratio (K_{EAdex}) actually declines, and the isotherm is hyperbolic (Fig. 11.12b). We also deduce that for EA concentrations less than about 10^{-2} M (i.e., less than the Na$^+$ concentration), we have a constant K_{EAdex} of about 200 $(mol \cdot kg^{-1})/(mol \cdot L^{-1})$ based on expression 11-62 with CEC of 0.96 mol $\cdot$ kg^{-1}, K_{EAex} of 2, and $[Na^+]$ of 0.01 mol $\cdot$ L^{-1}. With the ionic strength of 10^{-2} M, the characteristic length of the diffuse double layer is about 3 nm; together with an estimate of this montmorillonite's surface area (~ 700 m$^2 \cdot$ g), we can calculate that this distribution ratio corresponds to about 60 $(mol \cdot L^{-1}$ near surface water) $(mol \cdot L^{-1}$ bulk water$)^{-1}$. Clearly, the electrostatic attraction of the negatively charged clay face for the organic cation is concentrating ethyl ammonium ions in the water near the particles. Such near-surface accumulation amounts to "sorption" because the water of the diffuse double layer/vicinal water layer does not move relative to the solid.

The remaining issue involves assessing the impact of the hydrophobicity of the R

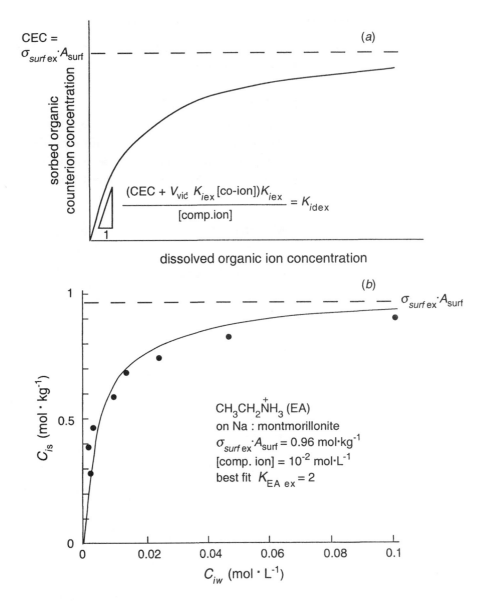

Figure 11.12 (*a*) Schematized Langmuir isotherm showing the variation of sorbed organic counterions (i.e., those within the near-surface water layer) vs. dissolved (or bulk solution) organic ion concentration changes: CEC = $\sigma_{surf\,ex}\,A_{surf}$ is the total surface charge density, V_{vic} is the volume of near-surface water, [co-ion] is the concentration of ion of opposite charge balancing excess organic sorbate in the near-surface region, $K_{i\,ex}$ is the exchange reaction equilibrium constant, and [comp.ion] is the concentration of competing counterions. (*b*) For the specific case of ethyl ammonium (EA) sorption to montmorillonite (CEC measured to be 0.96 mol·kg^{-1}) from a 10 mM aqueous NaCl solution; a Langmuir isotherm with best fit K_{EAex} = 2 matches the experimental data well (data from Cowan and White, 1958).

group and its effect on the magnitude of $K_{i\,ex}$. But for organic ions whose R groups are not very hydrophobic, we can already anticipate the impacts of such ion exchange (e.g., see Illustrative Example 11.5).

Effects of Hydrophobicity of an Organic Sorbate's R Group

Thus the remaining problem involves the question of how various hydrophobic portions of charged organic sorbates influence their sorption (i.e., the magnitude of K_{iex}). Presumably the sorbate's chemical structure determines the preference of the sorbate for the near-particle water region versus the bulk solution. Cowan and White (1958) also investigated the sorption of a series of alkyl ammonium ions to the same Na-montmorillonite (Fig. 11.13). A very interesting pattern emerged: the longer the alkyl chain, the steeper was the initial isotherm slope. Exactly parallel results have been seen for sorption of other amphiphiles, such as negatively charged n-alkyl benzene sulfonates binding to positively charged alumina particles (Somasundaran et

Illustrative Example 11.5   **Transport of Di-Isopropanol-Amine (DIPA) in Groundwater from a Sour Gas Processing Plant**

Problem

Di-isopropanol-amine (DIPA) is used to remove hydrogen sulfide from natural gas supplies (Goar, 1971). Unfortunately, this compound has been found as a groundwater contaminant at a total concentration of about 1 mM levels near such a sour gas processing plant. Consider an aquifer with the following characteristics:

Mineralogy: 70% quartz, 5% calcite, 25% montmorillonite, 0.2% organic matter (NOM)

Cation exchange capacity: CEC = 90 mmol $\cdot$ kg^{-1}

Density of aquifer material: $\rho_s = 2.6$ kg $\cdot$ L^{-1}

Total aquifer porosity: $\phi = 0.40$

Groundwater composition: pH = 8.0, [Na$^+$] = 20 mM; [Ca^{2+}] = 1 mM; [Cl$^-$] = 20 mM; [HCO$_3^-$] = 1 mM.

Estimate the retardation factor for DIPA in this aquifer.

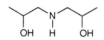

di-isopropanol-amine
(DIPA)

p$K_{\text{DIPAH}^+\, a} = 8.88$

Answer

At pH 8, DIPA mostly exists as a cation (DIPAH$^+$):

$$\alpha_{\text{DIPAH}^+a} = (1 + 10^{(\text{pH}-\text{p}K\text{DIPAH}+a)})^{-1} = (1 + 10^{-0.88})^{-1} = 0.88$$

Since the neutral compound is very polar, you do not expect it to be extensively absorbed into natural organic matter. Hence, you consider cation exchange to be the major sorption mechanism. Therefore:

$$K_{\text{DIPAd}} \cong \alpha_{\text{DIPAH}^+a} \cdot K_{\text{DIPAH}^+d\text{ ex}} \tag{1}$$

Considering the groundwater composition, you assume that DIPAH$^+$ is mostly competing with monovalent Na$^+$. In this case, use Eq. 11-55 (Box 11.2) to solve for $K_{\text{DIPAH}^+d\text{ ex}}$:

$$K_{\text{DIPAH}^+d\text{ex}} = \frac{[\text{DIPAH}^+:\text{surf}]}{[\text{DIPAH}^+]} = \frac{(\text{CEC} + V_{\text{vic}}K_{\text{DIPAH}^+\text{ex}}[\text{Cl}^-])K_{\text{DIPAH}^+\text{ex}}}{[\text{Na}^+] + K_{\text{DIPAH}^+\text{ex}}[\text{DIPAH}^+]} \tag{2}$$

Assuming that there is no preference for DIPAH$^+$ versus Na$^+$ since the compound is quite hydrophilic, then $K_{\text{DIPAH}^+\text{ex}}$ is about 1. Also taking V_{vic} to be about 0.01 L $\cdot$ kg^{-1}, then CEC = 90 mmol $\cdot$ kg^{-1} is much larger than $V_{\text{vic}} K_{\text{DIPAH}^+\text{ex}}$ [Cl$^-$] $\cong 0.2$ mmol $\cdot$ kg^{-1}. Simplifying, you estimate:

$$K_{\text{DIPAH}^+d\text{ex}} = \frac{(90 \text{ mmol} \cdot \text{kg}^{-1})(1)}{(20 \text{ mmol} \cdot \text{L}^{-1}) + (1)(1 \text{ mmol} \cdot \text{L}^{-1})} = 4.3 \text{ L} \cdot \text{kg}^{-1}$$

Insertion of this value together with $a_{DIPAH}+$ into Eq. 1 yields:

$$K_{DIPAd} = (0.88)(4.3) = 3.8 \ L \cdot kg^{-1}$$

(Note that laboratory observations by Luther et al. (1998) find a K_{DIPAd} value of about $3 \ L \cdot kg^{-1}$ for such a case.)

The retardation factor is then calculated as (see Illustrative Example 11.4):

$$R_{fDIPA} = 1 + (2.6)\frac{(1 - 0.4)}{(0.4)}(3.8) \cong 16$$

---

al., 1984). Further, above about eight carbons in the chain, the extent of binding *could far exceed the clay's cation exchange capacity and the isotherms no longer conform to the Langmuir model.* These observations imply another sorption mechanism must be occurring simultaneously with the simple exchange of one charged ion for another.

Such effects are very likely due to the increasing hydrophobicity of the R groups involved. By favoring chemical partitioning to the near surface from the bulk solution, hydrophobic effects augment the electrostatic forces and thereby enhance the tendency of the sorbates to collect near the particle surface (Somasundaran et al., 1984). Presumably this "extra" transfer occurs because the hydrophobic portion of the organic ion prefers to escape the bulk water and move into the near-surface water more than the co-ion (Cl^- in the case of the alkyl ammonium ions) is electrostatically inhibited from entering this layer. Such partitioning of co-ions from aqueous solution into organic solvents has also been observed (Jafvert et al., 1990). Thus, we anticipate little differences in sorption for organic chemicals due to moieties of like charge (e.g., $-COO^-$ vs. $-SO_3^-$) if they do not react with the surface since the electrostatic attraction to a surface is fairly nonselective; but we do expect substantial variations between sorbates if they differ in the hydrophobicity of their nonpolar parts.

Recognizing the need to maintain electroneutrality near the solid's surface, a second exchange process has been postulated (Brownawell et al., 1990). The excessive accumulation (relative to the surface CEC) of large R–substituted ions implies a mechanism in which organic ions, together with oppositely charged inorganic co-ions necessary to maintain electroneutrality, partition into the medium adjacent to charged particles (Fig. 11.10). This means, in the case of the sorption of the alkyl ammonium ions to Na-montmorillonite shown in Fig. 11.13, the charge balance equation in the near-surface region must be (after Eq. 11-45 or 11-53):

$$RNH_3{:}surf + Na{:}surf = CEC + Cl{:}surf \tag{11-63}$$

This model explains why the total sorbed concentrations of organic ions can exceed the solid's CEC (Fig. 11.13) and why the isotherms of all the alkyl amines are not well fit with a series of simple Langmuir isotherms.

If the nonpolar portion of the organic cation is not very hydrophobic, then the Lang-

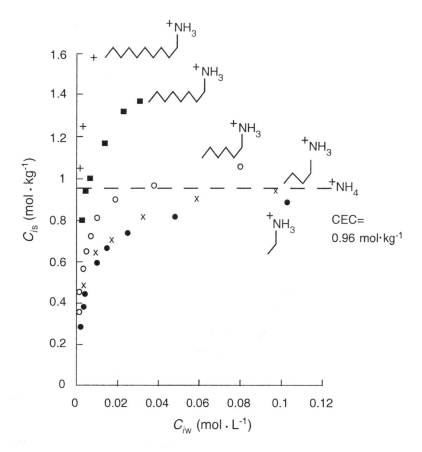

Figure 11.13 Adsorption isotherms for a series of alkyl ammonium compounds on sodium montmorillonite (adapted from Cowan and White. 1958). The horizontal dashed line indicates the cation exchange capacity of the clay.

muir ion exchange process will predominate and the combined result will appear near-hyperbolic (e.g., butyl ammonium in Fig 11.13). However, if the extra exchange process predominates because the hydrophobicity of R is great, the isotherm will appear almost linear (e.g., decyl ammonium in Fig. 11.13). Finally, if both processes are important, a much more complex isotherm is possible (recall Eq. 11-55). Obviously, the shape of the isotherm that an experimentalist would see depends on the range of dissolved concentrations utilized and the combination of parameters that apply in each case.

To provide an estimate of constants, K_{iex}, suitable for use in Eq. 11-55, let us try to isolate the contribution of the sorbate's hydrophobicity using some available data. For the alkyl ammonium ions exchanging with sodium cations in the data of Cowan and White (1958), we have (see Eqs. 11-37 to 11-44):

$$K_{iex} = K_{id\,ex} / K_{Na^+ d\,ex} = \exp\left[(-\Delta_{\text{hydrophobic}} G_i)/RT\right] \qquad (11\text{-}64)$$

where $i = RNH_3^+$, and:

$$K_{id} = \exp\left[(-z_i F \Psi - \Delta_{\text{hydrophobic}} G_i)/RT\right] \qquad (11\text{-}65)$$

implying:

$$-RT \ln K_{id} = -RT \ln K_{iex} + z_i F \Psi \qquad (11\text{-}66)$$

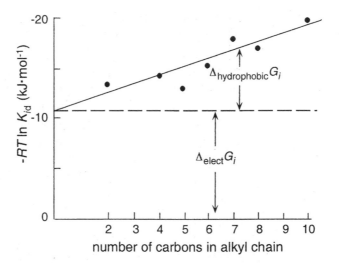

Figure 11.14 Variation in observed ion exchange free energy change ($-RT \ln K_{id}$) for a series of alkyl ammonium ions associating with a sodium montmorillonite (see also Fig. 11.13). All K_{id}'s taken at low organic ion concentrations. The least-squares fit line yields an intercept of -10.9 kJ·mol^{-1}.

Thus, if we examine the variation in $-RT \ln K_{id}$ for a series of alkyl amines participating in ion exchange, we should see how R groups affect the value of $-RT \ln K_{iex}$ while the product, $z_i FY$, remains constant. Since this hydrophobic effect appears to regularly increase with the size of the nonpolar part of the chemical structure (Cowan and White, 1958; Somasundaran et al., 1984), we may reasonably propose this energy term is composed of "excess free energy of solution in water" contributions from each of the nonpolar parts of the structure. Consequently, we expect for the alkyl ammonium ions studied by Cowan and White (1958):

$$-RT \ln K_{iex} = \Delta_{\text{hydrophobic}} G_i \cong m \, \Delta_{\text{hydrophobic}} G_{CH2} \tag{11-67}$$

and together with Eq. 11-66, have:

$$-RT \ln K_{id} = -(m \, \Delta_{\text{hydrophobic}} G_{CH2}) + z_i FY \tag{11-68}$$

where m is the number of methylene ($-CH_2-$) groups in each sorbate's alkyl chain, and $\Delta_{\text{hydrophobic}} G_{CH2}$ is the hydrophobic contribution made by each methylene driving these sorbates into the diffuse double layer-vicinal water layer.

Figure 11.14 shows the variation of $-RT \ln K_{id}$ for the alkyl ammonium ions (when these organic sorbates are present at levels much less than Na$^+$) as a function of the number of methylenes in the alkyl chains. The least-squares correlation line through the data yields:

$$-RT \ln K_{id} = -10.9 - m(0.75 \text{ kJ·mol}^{-1}) \tag{11-69}$$

This intercept in this fitted result implies that the alkyl ammonium ions experienced an electrostatic attraction to the clay surface corresponding to:

$$z_i F Y \cong -10.9 \text{ kJ·mol}^{-1} \tag{11-70}$$

corresponding to $Y \cong -0.11$ V, a typical surface potential. Also we see $\Delta_{\text{hydrophobic}} G_{CH2} = -0.75$ kJ·mol^{-1}. Examination of the variation in aqueous solubilities for compound classes like alkanes or alcohols (Chapter 5) as a function of additional methylene groups reveals that G_{iw}^E changes by almost 4 kJ·mol^{-1} for each methylene increase in chain length. Thus, the $\Delta_{\text{hydrophobic}} G_{\text{CH}_2}$ contributing to $\Delta_{\text{surf water}} G_i$ in Eq. 11-37 corresponds to "a relief" of about 20% of the excess free energy of aqueous solution per methylene group. Somasundaran et al. (1984) noted that inclusion of the phenyl group in alkyl aryl sulfonates increases the ion exchange sorption tendency of these amphiphiles to a degree corresponding to lengthening the alkyl chain by 3–4 methylene groups. This is consistent with increasing the nonpolar structure's hydrophobicity to the same extent [i.e., $\Delta \log K_{\text{ow}}(\text{phenyl}) \sim 1.68$ and $\Delta \log K_{\text{ow}}(3\text{–}4$ methylenes$] \sim 1.59$ to 2.12). Thus, we may be justified in estimating the hydrophobic contribution to be about 20% of the excess free energy of aqueous solution in the corresponding hydrocarbon (Section 5.2):

$$K_{iex} \cong \exp\left[(-\Delta_{\text{hydrophobic}} G_i)/RT\right] \tag{11-64}$$

$$\cong \exp\left(+0.2\, G_{iw}^{\text{E}}/RT\right) \tag{11-71}$$

$$\cong \exp\left(+0.2\,[RT \ln 55.3 / C_{iw}^{\text{sat}}(\ell,\text{L})]/RT\right) \tag{11-72}$$

$$\cong 2\,[C_{iw}^{\text{sat}}(\ell,\text{L})]^{-0.2} \tag{11-73}$$

The data of Cowan and White (1958) yield the empirical result:

$$K_{iex}(\text{RNH}_3^+) \sim 1.1\,[C_{iw}^{\text{sat}}(\text{L})]^{-0.19} \tag{11-74}$$

using the solubilities of the corresponding alkanes. Such expressions predict K_{iex} of decyl amine to be between 20 (Eq. 11-74) and 50 (Eq. 11-73), since $10^{-6.57}$ M is the liquid solubility of decane. The data of Cowan and White (1958) imply a $K_{iex}(\text{decyl}$ amine$)$ of 36.

There is little doubt that hydrophobic phenomena are playing a role in determining the extent of amphiphilic sorption; however, a great deal more work is necessary before these approaches are proven to be robust. Illustrative Example 11.6 shows how one might estimate an anionic surfactant's adsorption.

Finally, we must note that work performed using the mix of solids that occur in real-world soils and sediments suggests the heterogeneity of the natural sorbents is also very important to charged organic species (Brownawell et al., 1990). It appears that, in addition to complex sorption mechanisms acting for a given kind of charged solid, one sees the influence of more than one solid surface type at the same time. Thus, estimating sorption to such real-world solids may require fits summing several isotherms (see example in Fig. 11.15 for two Langmuir isotherms superimposed to fit roughly the experimental data), but presumably each reflecting the involvement of different solid materials that make up the complex medium we simply call a soil, subsoil, or sediment.

"Sorption" Due to Formation of Additional "Solid" Phases

To conclude, we should also point out that condensed "phases" containing the organic compound of interest can be formed at the particle surfaces. Such phenomena occur in cases involving ionic organic compounds that form micelles/hemimicelles

Illustrative Example 11.6   **Estimating Dodecyl Sulfonate Sorption to Alumina at Different pHs**

Fuerstenau and Wakamatsu (1975) examined alumina as a sorbent for dodecyl sulfonate (DS) ions. Alkyl sulfonates, $R-SO_3^-$, do not participate in substantial ligand exchange reactions with alumina.

Problem

Estimate the alumina–water distribution coefficients of DS, K_{DSd}, from 2 mM NaCl: for (a) 50 μM DS solution and pH 7.2 and (b) 5 μM DS solution and pH 5.2. Assume the following properties of the alumina:

$$pH_{zpc} = 9, \quad pK_{a1}^{int} = 7.5, \text{ and } pK_{a2}^{int} = 10.5; \quad A_{surf} = 1.5 \times 10^4 \text{ m}^2 \text{ kg}^{-1};$$

$$5 \times 10^{-6} \text{ mol} \equiv \text{Al-OH sites} \cdot \text{m}^{-2} \text{ alumina}; \quad V_{vic} = 0.05 \text{ L} \cdot \text{kg}^{-1}$$

$$CH_3-(CH_2)_{10}-CH_2-\overset{\overset{O}{\|}}{\underset{\underset{O}{\|}}{S}}-O^-$$

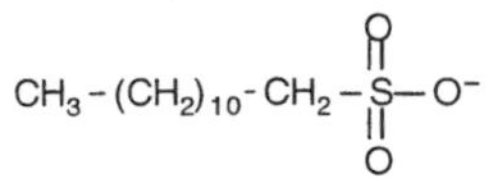

dodecyl sulfonate

Answer

DS is attracted to alumina at pHs below this solid's pH_{zpc} due to both electrostatic attraction to this solid's surface charge concentration (or anion exchange capacity, AEC) and the hydrophobicity of the 12-carbon-long chain. For sorption competition with the monovalent ion, Cl⁻, we use Eq. 11-55 (Box 11.2), recognizing that in this case we are interested in an organic anion attracted to the positively charged suspended alumina solid:

$$K_{DSd} = \{(AEC + V_{vic} [Na^+] K_{DS\ ex}) K_{DS\ ex}\} / \{[Cl^-] + K_{DS\ ex} [DS^-]\}$$

First we need the alumina's AEC. With the alumina's pK_{a1}^{int} of 7.5, pK_{a2}^{int} of 10.5 (hence the pH_{zpc} of 9), [≡Al–OH] of 5×10^{-6} mol · m⁻², and I of 2×10^{-3} mol · L⁻¹, we use Eq. 11-34 (Box 11.1) and solve by trial and error to find:

$$\sigma_{surf\ ex} \text{ (pH 7.2)} = 5.6 \times 10^{-8} \text{ mol (+ sites)} \cdot \text{m}^{-2}$$

and:

$$\sigma_{surf\ ex} \text{ (pH 5.2)} = 3.8 \times 10^{-7} \text{ mol (+ sites)} \cdot \text{m}^{-2}$$

These results indicate that the alumina has about 7 times more positively charged sites at pH 5.2 as compared to pH 7.2. Since the specific surface area of the alumina is 1.5×10^4 m² kg⁻¹, these results imply AEC values:

$$AEC \text{ (pH 7.2)} = A_{surf} \times \sigma_{surf\ ex} \text{ (pH 7.2)} = 8.4 \times 10^{-4} \text{ mol} \cdot \text{kg}^{-1}$$

and:

$$AEC \text{ (pH 5.2)} = A_{surf} \times \sigma_{surf\ ex} \text{ (pH 5.2)} = 5.7 \times 10^{-3} \text{ mol} \cdot \text{kg}^{-1}$$

Next, evaluate the preference of the surface region for DS over chloride using Eqs. 11-73 and 11-74:

$$K_{DS\ ex} \approx (1 \text{ to } 2) \times (C_{iw}^{sat} (1,L))^{-0.2}$$

From Appendix C, we find $C_{dodecane\ w}^{sat} (L) = 10^{-7.52}$ M, so this implies $K_{DS\ ex} \approx 30$ to 60.

Now you are ready to estimate the DS sorption for each case:

$$K_{DSd} \approx \{(\text{AEC} + V_{vic}\,[\text{Na}^+]\,K_{DS\,ex})\,K_{DS\,ex}\} \,/\, \{[\text{Cl}^-] + K_{DS\,ex}\,[\text{DS}^-]\}$$

at pH 7.2 and 50 mM DS:

$$K_{DSd} \approx \frac{\{(8.4\times10^{-4}\,\text{mol}\cdot\text{kg}^{-1}) + (0.05\,\text{L}\cdot\text{kg}^{-1})(2\times10^{-3}\,\text{mol}\cdot\text{L}^{-1})(30\text{ to }60)\}(30\text{ to }60)}{\{(2\times10^{-3}\,\text{mol}\cdot\text{L}^{-1}) + (30\text{ to }60)(5\times10^{-5}\,\text{mol}\cdot\text{L}^{-1})\}}$$

$$= \frac{\{(8.4\times10^{-4}\,\text{mol}\cdot\text{kg}^{-1}) + (30\text{ to }60\times10^{-4}\,\text{mol}\cdot\text{kg}^{-1})\}(30\text{ to }60)}{\{(2\times10^{-3}\,\text{mol}\cdot\text{L}^{-1}) + (1.5\text{ to }3\times10^{-3}\,\text{mol}\cdot\text{L}^{-1})\}}$$

$$= 30 \text{ to } 80 \, \text{L}\cdot\text{kg}^{-1}.$$

Comparing the values of the two terms in the sum in the numerator indicates that DS is mainly partitioning into near-surface water due to its hydrophobicity; the sum in the denominator implies that the dissolved DS concentration is already large enough to cause isotherm nonlinearity.

At pH 5.2 and 5 mM DS:

$$K_{DSd} \approx \frac{\{(5.7\times10^{-3}\,\text{mol}\cdot\text{kg}^{-1}) + (0.05\,\text{L}\cdot\text{kg}^{-1})(2\times10^{-3}\,\text{mol}\cdot\text{L}^{-1})(30\text{ to }60)\}(30\text{ to }60)}{\{(2\times10^{-3}\,\text{mol}\cdot\text{L}^{-1}) + (30\text{ to }60)(5\times10^{-6}\,\text{mol}\cdot\text{L}^{-1})\}}$$

$$= \frac{\{(57\times10^{-4}\,\text{mol}\cdot\text{kg}^{-1}) + (30\text{ to }60\times10^{-4}\,\text{mol}\cdot\text{kg}^{-1})\}\,(30\text{ to }60)}{\{(2\times10^{-3}\,\text{mol}\cdot\text{L}^{-1}) + (0.15\text{ to }0.3\times10^{-3}\,\text{mol}\cdot\text{L}^{-1})\}}$$

$$= 120 \text{ to } 300 \, \text{L}\cdot\text{kg}^{-1}.$$

Now the sum in the numerator indicates that both ion exchange and hydrophobic partitioning with a co-ion (here Na^+) are important. The sum in the denominator shows that at this dissolved DS concentration the K_{DSd} is still on the linear portion of the isotherm.

For comparison, Fuerstenau and Wakamatsu (1975) observed K_{DSd} (pH 7.2) of about 15 and K_{DSd} (pH 5.2) of about 180. Within the limits of our knowledge concerning parameters like V_{vic} (maybe known to $\pm$ factor of 3) and $K_{DS\,ex}$ (maybe known to $\pm$ factor of 2), the correspondence between the model estimates and observations is as good as can be expected.

Note: these data also indicated a preference for DS over the competing chloride concentration of between 30 and 60. This implies the dodecylsulfonate was accumulated in the diffuse double layer surrounding the alumina relative to its bulk solution concentration more than an order of magnitude more preferentially than the inorganic chloride adsorbate. As a consequence, it is perhaps not surprising that Fuerstenau and Wakamatsu (1975) observed the accumulation of hemimicelles on the alumina at only about 400 μM (pH 7.2) and about 7 μM (pH 5.2) bulk DS-concentrations (see discussion of hemimicelles).

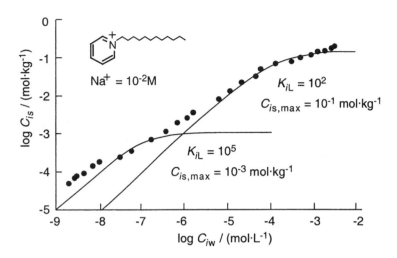

Figure 11.15 Observed sorption of dodecylpyridinium on a soil (EPA-12) exhibiting an overall cation exchange capacity of 0.135 mol·kg⁻¹. Two Langmuir isotherms (defined with particular values of $C_{is,max}$ and K_{iL}, recall Eq. 9-5) are placed on the data to illustrate how different portions of the observed isotherm may reflect the influence of different materials in the complex soil sorbent or possibly different mechanisms (data from Brownawell et al., 1990).

(Fuerstenau, 1956; Somasundaran et al., 1964; Chandar et al., 1983, 1987) and inorganic salt precipitates (Jafvert and Heath, 1991). These phases sometimes become very important at higher sorbate concentrations in an isotherm. If these phases remain attached to the solids, the organic compound appears to be sorbed and isotherms appear to be very steep! In these instances, we are no longer discussing "sorption" but rather phase equilibria. Nonetheless, we should note that enhanced concentration of organic and inorganic ions near the solid surface arising due to sorptive mechanisms may promote the formation of such new phases. In this context, we consider the case of hemimicelles below.

Hemimicelles. Amphiphilic compounds sometimes "sorb" via a special phenomenon called hemimicelle formation (Fuerstenau, 1956; Somasundaran et al., 1964: Chandar et al., 1983, 1987). Hemimicelle formation plays a critical role in amphiphile "sorption" to minerals when the organic ions are present at relatively high dissolved concentrations, about 0.001–0.01 of their *c*ritical *m*icelle *c*oncentrations (CMC, i.e., the level at which they self-associate in the bulk solution). When the organic sorbate levels are low, the sorption mechanism is like the ion exchange mechanism we discussed above (Fig. 11.16, part I). At some point in a titration of sorbents by micelle-forming compounds, presumably due to both electrostatic and hydrophobic effects enhancing the near-surface concentrations, amphiphile concentrations build up in the near-particle region to a point where self-aggregation of the molecules occurs in that thin water layer (Fig. 11.16, part IIa). This in turn allows rapid coagulation of the resultant micelle with the oppositely charged particle surface. Such aggregation smothers that subarea of the particle's surface charge with what have been called hemimicelles (Fig. 11.16, part IIb). Electrophoretic mobility measurements clearly demonstrate the neutralization of the particle's charges in this steep portion of the isotherm, even going so far as to reverse the surface charge (e.g., Chandar et al., 1987). The onset of this particle coating by hemimicelles occurs at different dissolved concentrations for various amphiphiles, but is near millimolar levels ($\geq$100 mg/L) for decyl-substituted amphiphiles and is near micromolar levels ($\geq$100 μg/L) for octadecyl derivatives. For the case of sorption of dodecyl sulfonate to alumina at pH 5.2 discussed in Illustrative Example 11.6, setting [DS:surf] equal to the CMC of

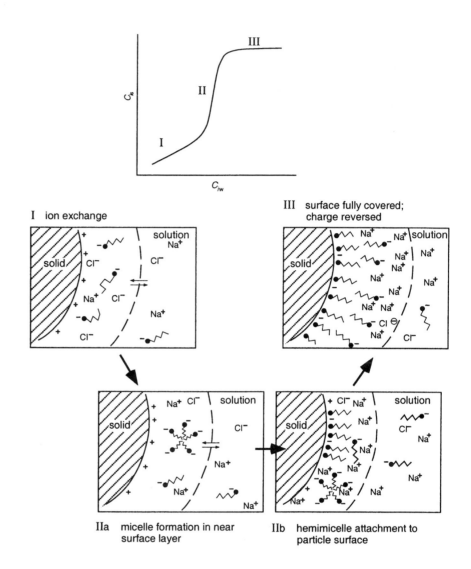

Figure 11.16 Relationship between sorbed and dissolved amphiphile concentrations (upper isotherm plot). These different parts of the isotherm reflect changes in the solid surface as sorption proceeds, possibly explainable by the following: in portion (I) with low dissolved concentrations, sorption occurs via ion exchange and related mechanisms. At some point, sufficient near-surface concentration enhancement occurs that micelles form there (IIa) and rapid coagulation between oppositely charged micelles and the surface follows (IIb). When the surface becomes fully coated with such micelles, additional sorption is stopped (III). In portion III, the solid surface charge is converted from one sign to the other, implying sorbates must become physically associated with the particle surface, as opposed to simply being present in the diffuse double layer or the vicinal water layer.

this compound leads one to expect hemimicelle formation at about 1 μM. Fuerstenau and Wakamatsu (1975) saw hemimicelles in this case at 6 μM. In any case, the bulk solution concentration is much less than the CMC. It appears that the near-surface concentrations elevated by factors of 100 or more, derived from accumulation of these amphiphiles in the thin film of water near the particle surface, are enriching just enough to achieve critical micelle concentrations in this near-surface water layer. Continued increase in amphiphile concentration results in the particle surface becoming increasingly coated by hemimicelles, apparently while the near-surface water maintains its concentration near that of the CMC. Finally, the entire particle surface is covered with a bilayer of amphiphile molecules; the particle's surface charge is now that of the surfactant; and the addition of more amphiphile to the solution does not yield any higher sorbed loads (Fig. 11.16, part III). This especially extensive degree of sorption may be the cause of macroscopic phenomena such as dispersion of coagulated colloids and particle flotation.

Surface Reactions of Organic Compounds

Until this point, we have focused on cases in which we could neglect chemical bond formation between the sorbate and materials in the solid phase. However, at least two kinds of surface reactions are known to be important for sorption of some chemicals (referred to as *chemisorption*). Simply, some organic substances can form covalent bonds with the NOM in a sediment or soil (see Fig. 9.2); other organic sorbates are able to serve as ligands of metals on the surfaces of inorganic solids (Fig. 11.1e). We discuss these processes below.

Organic Sorbate–Natural Organic Matter Reactions

First, some organic sorbates can react with organic moieties contained within the natural organic matter of a particulate phase. Especially prominent in this regard are organic bases like substituted anilines (Hsu and Bartha, 1974, 1976; Fabrega-Duque et al., 2000; Li et al., 2000; Weber et al., 2001). Due to their low pK_{ia}s (~ 5), the aromatic amine functionality is mostly not protonated at natural water pHs, and the nonbonded electrons can therefore attack carbonyl moieties in the NOM:

$$\text{(11-75)}$$

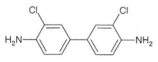

3,3'-dichlorobenzidine

When compounds like 3,3'-dichlorobenzidine or aniline are mixed with sediment, they become irretrievable using organic solvents that should remove them from absorbed positions within natural organic matter or using salt solutions that should displace them from ion exchange sites (Appleton et al., 1980; Weber et al., 2001). Conditions that promote hydrolysis (see Chapter 13) do release much of these added amino derivatives. Thus, it appears that reactions between the basic amine and carbonyl functionalities in the natural organic matter explain the strong sorption seen (Stevenson, 1976).

Such reactions often proceed slowly over hours, days, and even years, so the extent of this sorption due to organic chemical:organic chemical reactions is difficult to predict. Furthermore, such bond-forming sorption is sometimes irreversible on the timescales of interest, and we might not wish to include these effects in a K_{id} expression reflecting sorption equilibrium. Nonetheless, this condensation-type sorption is very important to reducing the mobility and bioavailability of such compounds (Li et al., 2000; Weber et al., 2001).

Organic Sorbate–Inorganic Solid Surface Reactions

A second type of surface reaction involves bonding of the organic compounds with atoms (e.g., metals) exposed on the surface of the solid (Table 11.4). In these cases a water or a hydroxyl bound to a metal on the solid is displaced by the organic sorbate:

$$\equiv M - OH_2^+ + i \rightleftharpoons \ \equiv M - i + H_2O \tag{11-76}$$

or:

$$\equiv M - OH + i \rightleftharpoons \ \equiv M - i + OH^- \tag{11-77}$$

Table 11.4 Examples of Organic Sorbates Reacting with Mineral Surfaces [a]

Rxn number		Ref. [b]
(1)	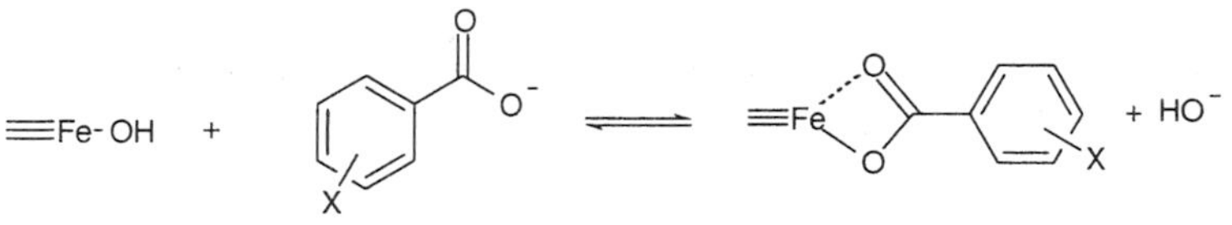 $\equiv$Fe–OH + [substituted benzoates] $\rightleftharpoons$ $\equiv$Fe–[benzoate complex] + HO$^-$	1,2
(2)	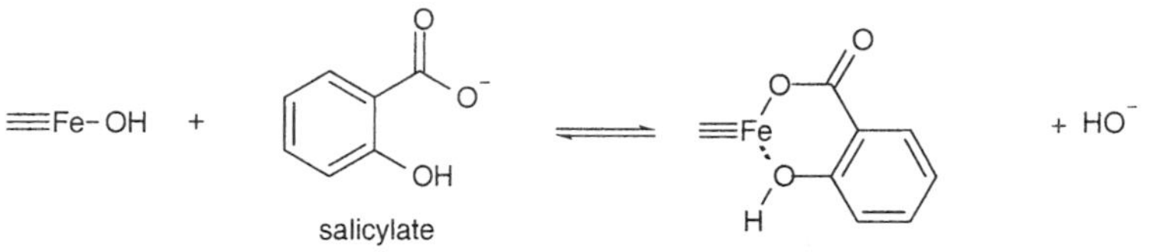 $\equiv$Fe–OH + [salicylate] $\rightleftharpoons$ $\equiv$Fe–[salicylate complex] + HO$^-$	2,3,4
(3)	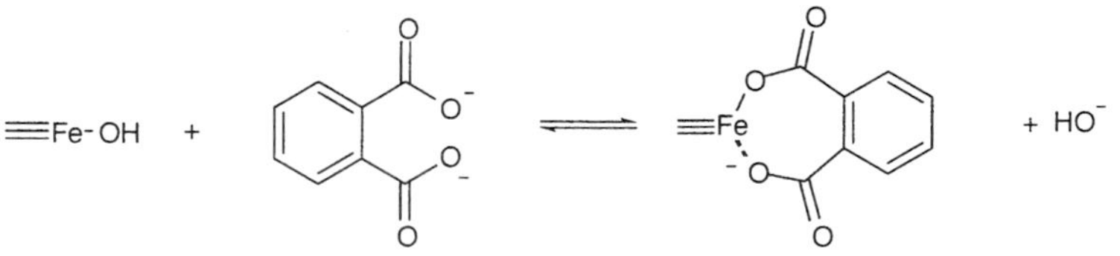 $\equiv$Fe–OH + [o-phthalate] $\rightleftharpoons$ $\equiv$Fe–[o-phthalate complex] + HO$^-$	2,3,5,6
(4)	 $\equiv$Fe–OH + [oxalate] $\rightleftharpoons$ $\equiv$Fe–[oxalate complex] + HO$^-$	3,7
(5)	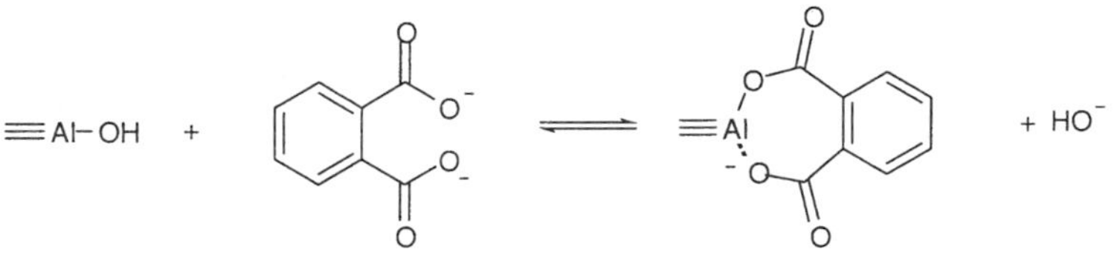 $\equiv$Al–OH + [o-phthalate] $\rightleftharpoons$ $\equiv$Al–[o-phthalate complex] + HO$^-$	8
(6)	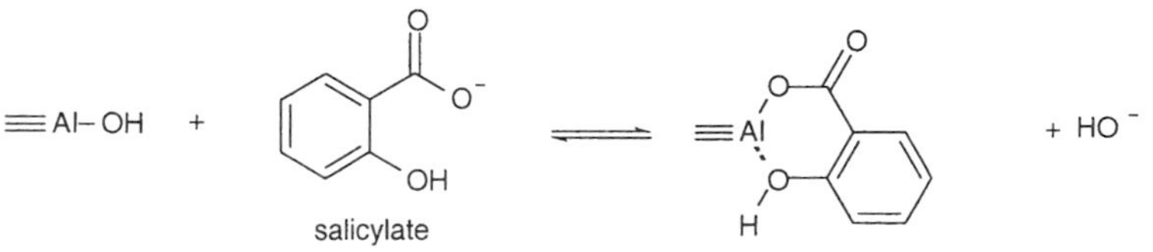 $\equiv$Al–OH + [salicylate] $\rightleftharpoons$ $\equiv$Al–[salicylate complex] + HO$^-$	8
(7)	 $\equiv$Ti–OH + [substituted o-catechols] $\rightleftharpoons$ $\equiv$Ti–[catechol complex] + H$_2$O	9

[a] Only limited information is available regarding the bonding of species to water-wet surfaces; thus the bonding of the sorbates shown here is conjecture. [b] 1. Kung and McBride (1989); 2. Evanko and Dzombak (1999); 3. Balistrieri and Murray (1987); 4. Yost et al. (1990); 5. Lövgren (1991); 6. Ali and Dzombak (1996a); 7. Mesuere and Fish (1992); 8. Stumm et al. (1980); 9. Vasudevan and Stone (1996).

Such reactions can occur on oxides of Fe, Al, and Ti. Carboxylic acids and phenols are common reactive moieties of the sorbates i. Often the resultant bound species is not really known. Given this additional sorption mechanism, the situation becomes more complicated, and multiple equilibria must be modeled (Stone et al. 1993) to account for all species:

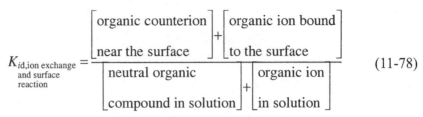

$$K_{\substack{id,\text{ion exchange} \\ \text{and surface} \\ \text{reaction}}} = \dfrac{\left[\begin{array}{c}\text{organic counterion} \\ \text{near the surface}\end{array}\right] + \left[\begin{array}{c}\text{organic ion bound} \\ \text{to the surface}\end{array}\right]}{\left[\begin{array}{c}\text{neutral organic} \\ \text{compound in solution}\end{array}\right] + \left[\begin{array}{c}\text{organic ion} \\ \text{in solution}\end{array}\right]} \qquad (11\text{-}78)$$

In the following we neglect the neutral organic species, though this may not always be appropriate (e.g., catechols, see reaction 7 in Table 11.4). We can then separate Eq. 11-78 into parts:

$$K_{\substack{id,\text{ion exchange} \\ \text{and surface} \\ \text{reaction}}} = \dfrac{[\text{organic counterion}]}{[\text{organic ion in solution}]} + \dfrac{[\text{organic ion bound to surface}]}{[\text{organic ion in solution}]} \qquad (11\text{-}79)$$

Note that $K_{id,\text{ ion exchange and surface reaction}}$ expresses the sorption equilibrium of the charged organic species. Hence, for calculating the overall sorption of the compound it has to be multiplied by the fraction of the ionized organic compound present in aqueous solution. Using a previous result (e.g., Eq. 11-55, assuming monovalent ions and an anionic organic sorbate) we get:

$$K_{\substack{id,\text{ion exchange} \\ \text{and surface} \\ \text{reaction}}} = \dfrac{(\text{AEC} + V_{\text{vic}}K_{iex}[\text{co-ion}])K_{iex}}{[\text{comp.ion}] + K_{iex}[\text{organic ion in solution}]}$$
$$+ \dfrac{[\text{organic ion bound to surface}]}{[\text{organic ion in solution}]} \qquad (11\text{-}80)$$

Now our task is to develop an expression to predict the last term. To do this, we begin by writing the reaction involved:

$$i\text{:surface} + \text{L–M}\equiv \;\rightleftharpoons\; i\text{–M}\equiv + \text{L:surface} \qquad (11\text{-}81)$$

where i–M≡ and L–M≡ are an organic compound and an inorganic ligand like –OH bonded to the solid as indicated by the hyphen. The ions, i:surface and L:surface, are present at "ion exchange" concentrations in the immediate vicinity of the reaction sites that differ from their bulk solution concentration. Note that the ligand, L, is probably not the same as the competing ion in the ion exchange "reaction." Such a reaction reflects a free energy change that we will refer to as $\Delta_{\text{rxn}}G_i$ and a corresponding equilibrium expression:

$$K_{irxn} = \dfrac{[i-\text{M}\equiv][\text{L:surface}]}{[i\text{:surface}][\text{L}-\text{M}\equiv]} \qquad (11\text{-}82)$$

If we can assume that there are a finite number of key reactive sites on the solid, $\sigma_{\text{surf rxn}}$ $(\text{mol}\cdot\text{m}^{-2})$, then we have:

$$A_{\text{surf}} \cdot \sigma_{\text{surf rxn}} = [i\text{–M}\equiv] + [\text{L–M}\equiv] \tag{11-83}$$

with A_{surf} equal to the specific particle surface area (m^2 kg^{-1}). Therefore, we can rewrite Eq. 11-82:

$$K_{irxn} = \frac{[i - \text{M} \equiv][\text{L}:\text{surface}]}{[i:\text{surface}](A_{\text{surf}} \cdot \sigma_{\text{surf rxn}} - [i - \text{M} \equiv])} \tag{11-84}$$

We also recall that the concentrations of ions in the layer of water next to the particle surface can be related to the corresponding species in the bulk solution:

$$[\text{L}:\text{surface}] = [\text{L}^-]_{\text{bulk}} \cdot e^{-\Delta_{\text{elect}}G_i/RT} \tag{11-85}$$

and:

$$[i:\text{surface}] = [i]_{\text{bulk}} \cdot e^{-\Delta_{\text{elect}}G_i/RT} \cdot e^{-\Delta_{\text{hydrophobic}}G_i/RT} \tag{11-86}$$

Using these relations in Eq. 11-84, we have:

$$K_{irxn} = \frac{[i - \text{M} \equiv][\text{L}^-]_{\text{bulk}} \cdot e^{-\Delta_{\text{elect}}G_i/RT}}{(A_{\text{surf}} \cdot \sigma_{\text{surf ex}} - [i - \text{M} \equiv])[i]_{\text{bulk}} \cdot e^{-\Delta_{\text{elect}}G_i/RT} \cdot e^{-\Delta_{\text{hydrophobic}}G_i/RT}}$$

$$= \frac{[i - \text{M} \equiv][\text{L}^-]_{\text{bulk}}}{(A_{\text{surf}} \cdot \sigma_{\text{surf ex}} - [i - \text{M} \equiv])[i]_{\text{bulk}} K_{iex}} \tag{11-87}$$

where K_{iex} is equal to exp $(-\Delta_{\text{hydrophobic}}G_i/RT)$ as shown in Eqs. 11-40 to 11-43. Simplifying and rearranging, we then find:

$$[i - \text{M} \equiv] = \frac{\sigma_{\text{surf rxn}} \cdot A_{\text{surf}} \cdot K_{iex} \cdot K_{irxn} \cdot [i]_{\text{bulk}}}{[\text{L}^-]_{\text{bulk}} + K_{iex} \cdot K_{irxn} \cdot [i]_{\text{bulk}}} \tag{11-88}$$

Thus another Langmuir isotherm is expected with the maximum bound concentrations equal to $\sigma_{\text{surf rxn}} \cdot A_{\text{surf}}$ and the K_{iL} given by $K_{irxn} K_{iex} [\text{L}^-]_{\text{bulk}}^{-1}$. Returning to our overall K_{id} expression (Eq. 11-80), we can now write:

$$K_{\substack{id,\text{ion exchange} \\ \text{and surface} \\ \text{reaction}}} = \frac{(\sigma_{\text{surf ex}} \cdot A_{\text{surf}} + V_{\text{vic}} \cdot K_{iex}[\text{co-ion}])K_{iex}}{[\text{comp.ion}] + K_{iex} + [i]}$$

$$+ \frac{\sigma_{\text{surf rxn}} \cdot A_{\text{surf}} \cdot K_{iex} \cdot K_{irxn}}{[\text{comp.ligand}] + K_{iex} \cdot K_{irxn} \cdot [i]} \tag{11-89}$$

As for nonreacting organic ions, we need information on the ion exchange tendency of the chemical of interest (K_{iex} or $\Delta_{\text{hydrophobic}}G_i$); we also need a means to assess K_{irxn}.

Various investigators have utilized surface complexation modeling along with reasonable hypotheses concerning the surface species formed (and hence the adsorption reaction stoichiometry) to extract values of the product, $K_{iex} K_{irxn}$, for cases of interest (e.g., Mesuere and Fish, 1992; Ali and Dzombak, 1996a and b; Vasudevan and Stone, 1996; Evanko and Dzombak, 1998, 1999). For example, Evanko and Dzombak (1999) fitted data for four carboxylic acids (benzoic, 1-naphthoic, 3,5-dihydroxybenzoic, and 6-phenylhexanoic acid) sorbing to goethite from 10 mM NaCl solutions. They considered a sorption reaction of the form:

$$\equiv Fe-OH + i + H^+ \;\rightleftharpoons\; \equiv Fe-i + H_2O \qquad (11\text{-}90)$$

and fitted an "intrinsic" equilibrium constant, K_1^{int}, after accounting for electrostatic contributions (recall Eqs. 11-25 and 11-26):

$$K_1^{int} = [\equiv Fe-i] \,/\, \gamma_i'[i] \; \gamma_{H^+}'[H^+] \, [\equiv Fe-OH] \qquad (11\text{-}91)$$

Adding the reaction, $H_2O = H^+ + OH^-$ with $K_w = 10^{-14}$, this reaction is equivalent to:

$$\equiv Fe-OH + i \;\rightleftharpoons\; Fe-i + OH^- \qquad (11\text{-}92)$$

where we now imply that hydroxide ion is the surface ligand that is replaced by the organic acid. For this reaction 11-92, the equilibrium constant is the product, $K_{iex} K_{irxn}$. This product is related to the previous "intrinsic" constant:

$$K_{iex} K_{irxn} = 10^{-14} \; K_1^{int} \qquad (11\text{-}93)$$

For the four mono carboxylic acids investigated by Evanko and Dzombak (1998, 1999), a value of K_1^{int} near 10^8 was always found. The value was a little higher for the acids with larger R groups (ranging from benzoic at $10^{7.89}$ to phenyl hexanoic at $10^{8.68}$). This range (factor of 6) is consistent with our expectations from the K_{iex} contribution, since use of Eqs. 11-73 and 74 would cause K_{iex}(benzoate) to be about 2 to 4 and K_{iex}(phenyl hexanoate) to be about 10 to 20.

Investigators have also noted other dependencies of K_1^{int} values on the structure of the organic sorbates. First, K_1^{int} values increase with the addition of moieties like carboxyl groups or phenolic hydroxyls in positions (e.g., *ortho* to one another on an aromatic ring) that allow them to multiply bind to surface metals (e.g. Evanko and Dzombak, 1998). Also the values of K_1^{int} increase for ligands with greater pK_as (Vasudevan and Stone, 1996). This may be interpreted as the greater the tendency to "hold" a proton (i.e., greater pK_a), the greater will be the affinity for bonding to a metal on an oxide surface.

Returning to our effort to anticipate the overall sorption of organic compounds that may act both as counterions and as surface ligands, we can recognize all the terms in the second half of Eq. 11-89:

$$\sigma_{surf\,rxn} A_{surf} = [\equiv Fe-OH] \qquad (11\text{-}94)$$

and:

$$[\text{comp.ligand}] = [OH^-] \qquad (11\text{-}95)$$

With the empirical measures of K_1^{int} reported in the growing literature and understandings of the "stoichiometries" of both the ion exchange and ligand exchange processes, we can now estimate the solid–water distribution ratios of such ionic organic sorbates (see Illustrative Example 11.7).

We should point out that many organic sorbates, and especially bidentate ones like phthalate and salicylate, can apparently form more than one bound surface species.

Illustrative Example 11.7   **Estimating the Adsorption of Benzoic Acid to Goethite**

Problem

Estimate the goethite–water distribution coefficients from 10 mM NaCl aqueous solutions of benzoic acid (pK_{ia} = 4.1) at 50 μM at pHs 4, 5, and 6 to a synthetic iron oxyhydroxide, goethite, with the following properties (from Evanko and Dzombak, 1998):

surface area, A_{surf}: 79×10^3 m^2 kg^{-1}

total [$\equiv$FeOH]: 2.3×10^{-6} mol sites $\cdot$ m^{-2}

pK_{a1}^{int} = 7.68; pK_{a2}^{int} = 8.32

and assume V_{vic} = 0.1 L $\cdot$ kg^{-1}

Answer

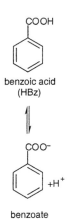

benzoic acid
(HBz)

benzoate
(Bz$^-$)

To find the overall goethite–water distribution coefficient for benzoic acid (HBz), assume that only adsorption of the deprotonated species (benzoate, Bz$^-$) is important:

$$K_{HBzd} = (1 - \alpha_{HBza})\, K_{Bz^-d}, \text{ ion exchange and surface reaction} \tag{1}$$

where $(1 - \alpha_{HBza})$ is the fraction of total benzoic acid present as benzoate (see Section 8.2):

$$(1 - \alpha_{HBza}) = 10^{pH - pK_{HBza}} / (1 + 10^{pH - pK_{HBza}}) \tag{2}$$

Hence, the fraction of benzoate present at the different pH values is:

at pH 4: $(1 - \alpha_{HBza}) = 0.44$

at pH 5: $(1 - \alpha_{HBza}) = 0.89$

at pH 6: $(1 - \alpha_{HBza}) = 0.99$

Use Eq. 11-89 to estimate K_{Bz^-d} (omit the remainder of the subscript; also note that i = Bz$^-$ in the following equations):

$$K_{Bz^-d} = \frac{(\sigma_{surfex} \cdot A_{surf} + V_{vic} \cdot K_{iex}[co-ion])K_{iex}}{[comp.ion] + K_{iex}[i]}$$

$$+ \frac{\sigma_{surfrxn} \cdot A_{surf} \cdot K_{iex} \cdot K_{irxn}}{[comp.ligand] + K_{iex} \cdot K_{irxn}[i]} \tag{3}$$

Using Eq. 11-34 (Box 11.1) and the goethite surface site density, its pK_{a1}^{int} and pK_{a2}^{int} values, and the ionic strength of the solution, find the intensity of surface charging of the goethite at the three pH values of interest (note that $\sigma_{surfex} A_{surf}$= AEC):

σ_{surfex} (pH 4) = 13 $\times 10^{-7}$ mol $\cdot$ m^{-2} $\rightarrow$ $\sigma_{surfex} \cdot A_{surf}$ = 0.10 mol $\cdot$ kg^{-1}

σ_{surfex} (pH 5) = 4.9 $\times 10^{-7}$ mol $\cdot$ m^{-2} $\rightarrow$ $\sigma_{surfex} \cdot A_{surf}$ = 0.039 mol $\cdot$ kg^{-1}

σ_{surfex} (pH 6) = 1.7 $\times 10^{-7}$ mol $\cdot$ m^{-2} $\rightarrow$ $\sigma_{surfex} \cdot A_{surf}$ = 0.013 mol $\cdot$ kg^{-1}

At pH 4, about 60% of the surface sites are positively charged, while this proportion drops to about 20% at pH 5 and less than 10% at pH 6.

Now estimate the preference for benzoate over chloride as a counterion using information on the aqueous solubility of benzene (Eqs. 11-73 and 74):

$$K_{iex} \sim (1 \text{ to } 2) \, (C_w^{sat}(\text{benzene}))^{-0.2} = (1 \text{ to } 2) \, (10^{-1.64})^{-0.2} = 2 \text{ to } 4$$

For the ion exchange sorption process, note that [co-ion] = [Na$^+$] = 10^{-2} mol $\cdot$ L^{-1} and [comp.ion] = [Cl$^-$] = 10^{-2} mol L^{-1}.

Next, consider the factors determining benzoate bonding to the goethite surface. From Evanko and Dzombak (1999), you have $K_1^{int} = 10^{7.89}$ for the adsorption reaction:

$$[\equiv\!\text{FeOH}] + [\text{H}^+] + [\text{Bz}^-] \leftrightarrow [\equiv\!\text{FeBz}] + \text{H}_2\text{O}$$

which you can convert to the equivalent ligand exchange reaction:

$$[\equiv\!\text{FeOH}] + [\text{Bz}^-] \rightleftharpoons [\equiv\!\text{FeBz}] + \text{OH}^-$$

with $K_{iex} \cdot K_{irxn} = K_1^{int} \cdot K_w = 10^{7.89} 10^{-14} = 10^{-6.1}$.

Also note that [comp.ligand] = [OH$^-$] = $10^{-14}/10^{-\text{pH}}$. Finally, estimate $\sigma_{surf\,rxn}$ as $[\equiv\!\text{FeOH}] = 2.3 \times 10^{-6}$ mol $\cdot$ m^{-2} $\times$ 79 $\times$ 10^3 m^2 kg^{-1} = 0.18 mol $\cdot$ kg^{-1}. Note that this value probably should be reduced at each pH by the concentration of $[\equiv\!\text{FeOH}_2^+]$ that has been formed; for example, at pH 4, $[\equiv\!\text{FeOH}] \cong 0.18 - 0.10 = 0.08$ mol $\cdot$ kg^{-1}. This should also cause us to suspect the increasing importance of a new ligand exchange process in this case:

$$[\equiv\!\text{FeOH}_2^+] + [\text{Bz}^-] \rightleftharpoons [\equiv\!\text{FeBz}] + \text{H}_2\text{O}$$

which would have a different (?larger) intrinsic equilibrium constant.

Finally, insert all the values in Eq. 3, to obtain (recognizing that $[i]$ = [Bz$^-$] must be less than the total 50 mM added due to some fraction sorbed, i.e., $< [(1-\alpha_{HBza}) \times 5 \times 10^{-5}]$) at pH 4:

$$K_{Bz^-d} = \frac{(0.10 + 0.1 \times (2 \text{ to } 4)[0.01])(2 \text{ to } 4)}{[0.01] + (2 \text{ to } 4)[< 0.44 \times 5 \times 10^{-5}]} + \frac{(0.18)(10^{-6.1})}{[10^{-14}/10^{-4}] + (10^{-6.1})[< 0.44 \times 5 \times 10^{-5}]}$$

$$\cong \frac{(0.10)(2 \text{ to } 4)}{[0.01]} + \frac{(1 \times 10^{-7})}{[10^{-10}]}$$

$$\cong \{(20 \text{ to } 40) + (10^3)\}$$

$$\cong 1000 \, \text{L} \cdot \text{kg}^{-1}$$

which yields a K_{HBzd} value (Eq. 1) of about *400* to *500* L·kg^{-1}. Evanko and Dzombak observed K_{HBzd} (pH 4) = 120 L $\cdot$ kg^{-1}.

Likewise K_{HBzd} (pH 5) ≈ *120* L · kg^{-1} and K_{HBzd} (pH 6) ≈ *14* L · kg^{-1}. Evanko and Dzombak (1998) observed K_{HBzd} (pH 5) ~ 50 L · kg^{-1} and K_{HBzd} (pH 6) < 6 L · kg^{-1}.

In all these estimates, it appears that the bound benzoate predominates over the benzoate present as counterions in the diffuse double layer.

The relative importance of these surface species varies greatly as a function of pH. Hence, accurate predictions of the sorption of such organic ligands on mineral oxides requires applying more than one empirical surface reaction equilibrium constant to calculate the contributions of each bound species (see Evanko and Dzombak, 1999 for examples).

Finally, we can also evaluate K_{irxn} recognizing that the tendency to form chemical linkages to solid surface atoms correlates with the likelihood of forming comparable complexes in solution (Stumm et al., 1980; Schindler and Stumm, 1987; Dzombak and Morel, 1990). That is, the free energy change associated with the exchange shown by Eq. 11-81 appears energetically similar to that for a process occurring between two dissolved components:

$$M–L^{z+} + R^- \rightleftharpoons M–R^{z+} + L^- \qquad (11\text{-}96)$$

Thus, it may be feasible to estimate K_{irxn} from the solution-phase exchange reaction, characterized by its equilibrium constant:

$$K_{irxn} \approx K_{i\,\text{ligand exchange in solution}} = \frac{[M-R^{z+}][L^-]}{[M-L^{z+}][R^-]} \qquad (11\text{-}97)$$

A substantial database is available to quantify such solution equilibria (e.g., Martell and Smith, 1977; Morel, 1983).

11.6   Questions and Problems

Questions

Q 11.1

Give five examples of environmentally relevant situations in which adsorption of organic vapors on inorganic surfaces is important.

Q 11.2

For what kind of compounds and in which environmentally relevant cases is adsorption of organic chemicals to inorganic surfaces in water important? Give five examples.

Q 11.3

What intermolecular interactions and corresponding free energy contributions ($\Delta ?G_i$) would you suspect to be important for the following sorbate:sorbent:solution combinations:

(a) 1,4-dichlorobenzene partitioning between air and quartz sand?
(b) phenol partitioning between air and Teflon?
(c) phenol partitioning between air and quartz sand?
(d) benzophenone partitioning between air and quartz sand?
(e) 1,4-dichlorobenzene partitioning between water and quartz sand?
(g) benzophenone partitioning between water and quartz sand?
(g) phenol partitioning between water and quartz sand?
(h) trinitrotoluene partitioning between water and quartz sand?
(i) trinitrotoluene partitioning between water and K$^+$-kaolinite?
(j) benzyl ammonium between water and quartz sand?
(h) *ortho*-phthalic acid between water and quartz sand?

Indicate in each case the intermolecular interaction forces, the key structural features of the sorbate, the site type(s) of the sorbent involved, and the environmental parameters influencing sorption.

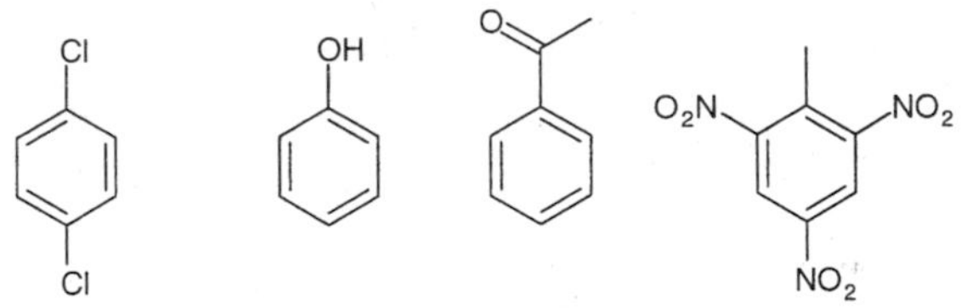

| 1,4-dichlorobenzene | phenol | benzophenone | trinitrotoluene |

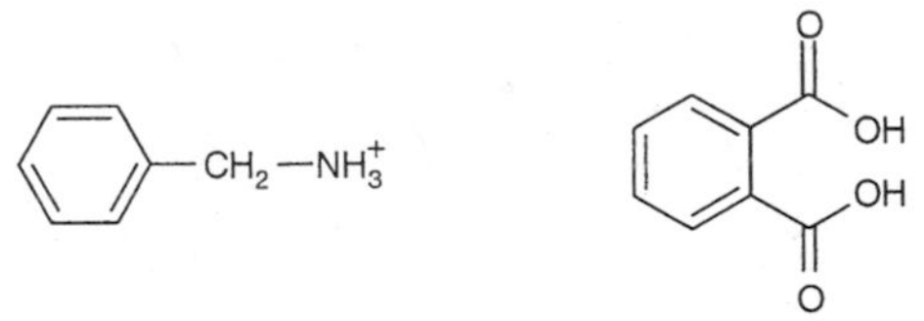

benzyl ammonium
$pK_{ia} = 9.33$

ortho-phthalic acid
$pK_{ia1} = 2.89$
$pK_{ia2} = 5.51$

Q 11.4

Why does the sorption of organic vapors to polar inorganic surfaces generally decrease with increasing humidity? Why does the relative humidity have a negligible influence on sorption of organic vapors to apolar surfaces?

Q 11.5

Storey et al. (1995) reported K_{iasurf} values for the adsorption of *n*-alkanes and PAHs from air to quartz at 25 – 30% RH and 70 – 75% RH, respectively. When plotting $\ln K_{iasurf}$ versus $\ln p_{iL}^*$ (Eq. 11-11) for the various data sets, the following slopes m are obtained:

	n-alkanes	PAHs
$m(25 - 30\%$ RH)	1.04	1.22
$m(70 - 75\%$ RH)	0.96	1.18

Would you have expected to find steeper slopes for the PAHs as compared to the *n*-alkanes? If yes, why? Why are the slopes at low RH somewhat steeper than the ones corresponding at high relative humidities?

Q 11.6

Consider two apolar compounds exhibiting a factor of 10 difference in liquid vapor pressures. What differences do you expect for the two compounds in their (a) air–Teflon, and (b) air–graphite adsorption coefficients?

Q 11.7

Explain why the following sorbate pairs exhibit the relative K_{id}'s indicated for sorption to a siloxane surface:

(a) K_{id} (2,4-dinitrotoluene) $\gg K_{id}$ (2-nitrotoluene)
(b) K_{id} (1,4-dinitrobenzene) $\gg K_{id}$ (1,2-dinitrobenzene)
(c) K_{id} (DNOC) $\gg K_{id}$ (Dinoseb)

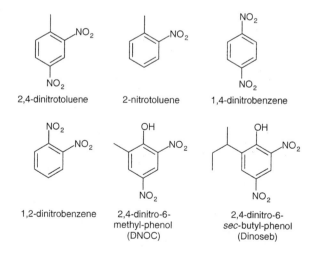

2,4-dinitrotoluene   2-nitrotoluene   1,4-dinitrobenzene

1,2-dinitrobenzene   2,4-dinitro-6-methyl-phenol (DNOC)   2,4-dinitro-6-sec-butyl-phenol (Dinoseb)

Q 11.8

Why do minerals have charges when they are submerged in water?

Q 11.9

Indicate whether the following solids are positively charged, neutral, or negatively charged when they occur in water at pH 7 (neglect specific adsorbates like phosphate or ferric iron species):

(a) quartz (SiO_2)
(b) natural organic matter

(c) goethite (FeOOH)

(d) gibbsite(Al(OH)$_3$)

(e) kaolinite

Q 11.10

Which of the two compounds do you think would sorb more to kaolinite in water at pH 6?

(a) pyrene or pyrene sulfonate?

(b) butyl ammonium or butyrate?

(c) propyl ammonium or octyl ammonium?

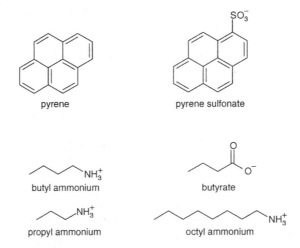

pyrene   pyrene sulfonate

butyl ammonium   butyrate

propyl ammonium   octyl ammonium

Q 11.11

How can organic ions accumulate in excess of the ion exchange capacity near a pure mineral solid submerged in water?

Q 11.12

If organic ions are not bonded to a mineral's surface, why do they still not migrate past the minerals in a groundwater flow?

Problems

P 11.1 *How Much Benzo[a]pyrene Is Adsorbed to a Glass Fiber Filter?*

You want to sample air using a glass fiber filter to determine the concentration of particulate benzo[a]pyrene (BaP). Since you are working in a region and during a season with low relative humidity (assume 50% RH, temperature 15°C), you are worried that the filter may adsorb significant quantities of gaseous BaP.

(a) If the filter has a surface area of 1 m^2 for the SiO$_2$ fibers, what total mass of BaP (in nanograms) would you expect to be adsorbed to the filter if it reached equilibrium with a gaseous BaP concentration of 0.1 ng BaP · m^{-3}?

(b) Assuming the air also had a particulate BaP concentration of 0.1 ng BaP · m^{-3} (i.e., BaP happens to be half gaseous and half sorbed in the air), how many cubic meters of air should you send through the filter to be sure to have 10 times as much particulate BaP on the filter as compared to adsorbed BaP? (Assume the filter is 100% efficient at capturing particulate BaP.)

(c) How would your sampling volume change (increase or decrease needed volume by what factor) if the weather were characterized by 90% RH and 25°C, assuming the same gaseous and particulate BaP concentrations?

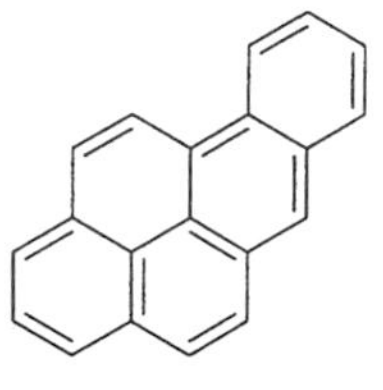

i = benzo[a]pyrene (BaP)

M_i = 252.3 g·mol^{-1}

p_{iL}^{*} (25°C) = 2 × 10^{-5} Pa

$\Delta_{vap} H_i$ (25°C) = 110 kJ·mol^{-1}

α_i = 0 ; β_i = 0.44 (Table 4.3)

P 11.2 *Designing a Sorption Treatment to Remove 1,1,2,2-Tetrachloroethane from a Waste Gas Stream*

A process in your company generates waste gases that need to be vented to the outside at a rate of 1 m^3 per hour. In particular, you must be sure that the 1,1,2,2-tetrachloroethane present at 100 ppmv (i.e., 100 × 10^{-6} m^3 of 1,1,2,2-tetrachloroethane vapor per m^3 of total gas) will be removed from the gas-stream before discharge. Someone suggests you construct an adsorbent column filled with alumina (Al$_2$O$_3$) and run the gas through that column to capture the 1,1,2,2-tetrachloroethane.

(a) If the waste gas stream is quite dry (i.e., 60% RH) and warm (30°C), how many hours of waste gas can you treat with a 10 m^3 tank of alumina (packed bed porosity 0.3, density 4 g · mL^{-1}, specific surface area of 10 m^2 g^{-1}, and assumed surface properties like those measured for corundum), assuming the 1,1,2,2-tetrachloroethane "breaks through" at a volume equal to the tank's void volume (i.e., number of cubic meters in tank that are filled with gas) divided by the venting gas flow rate and by the equilibrium fraction of 1,1,2,2-tetrachloroethane in the gas phase?

(b) If you could construct a tank with the same void volume and surface area of silica, would it be more effective? What about activated carbon? Explain your reasoning.

$$Cl_2HC–CHCl_2$$

1,1,2,2-tetrachloroethane
(see Appendix C)

P 11.3 *Air–Particle Partitioning in the Atmosphere: Evaluation of Experimental Data*

In a series of smog chamber experiments, Liang et al. (1997) have studied the air/particle partitioning behavior of a series of *n*-alkanes (C_{16}–C_{14}) and of a group of polycyclic aromatic hydrocarbons (PAHs). The model aerosol materials investigated included solid ammonium sulfate ($(NH_4)_2SO_4$), liquid dioctyl phthalate, and secondary organic aerosol generated from the photooxidation of whole gasoline vapor. Partition coefficients were also measured for ambient *n*-alkanes sorbing to urban particulate material (UPM) during summer smog episodes in the Los Angeles metropolitan area. The authors report K_{ip} values (in $m^3 \, \mu g^{-1}$) that are defined as (see also Illustrative Example 11.3):

$$K_{ip} = \frac{\text{mass of compound } i \,/\, \mu g \text{ particle}}{\text{mass of compound } i \,/\, m^3 \text{air}}$$

For the partitioning of the *n*-alkanes with $(NH_4)_2SO_4$ and with UPM they obtained the following two one-parameter LFERs:

$(NH_4)_2SO_4$ (32°C, 10% RH): $\log K_{ip} / m^3 \, \mu g^{-1} = -0.96 \log p_{iL}^* \,/\, \text{torr} - 7.66$

UPM (37°C, 42% RH): $\log K_{ip} / m^3 \, \mu g^{-1} = -1.03 \log p_{iL}^* \,/\, \text{torr} - 6.68$

The estimated specific surface areas, a_s (which are always subject of debate!), are $17.5 \, m^2 \, g^{-1}$ for $(NH_4)_2SO_4$ and $2 \, m^2 \, g^{-1}$ for UPM.

Estimate the K_{iasurf} values of *n*-octadecane (Eq. 11-12) for $(NH_4)_2SO_4$ and UPM at the conditions of temperature and RH indicated above. Use an (extrapolated) vdW_{surf} value of 7.2 for $(NH_4)_2SO_4$ at 10% RH (Goss and Schwarzenbach, 1999b). For UPM use the slope of the LFER (i.e., −1.03) to derive vdW_{surf}. What assumptions do you make? Convert the K_{iasurf} values to K_{ip} values (or vice versa, see Illustrative Example 11.3). Compare the estimated values with the experimental values obtained by inserting the appropriate p_{iL}^* value into the above LFERs. Try to find explanations for possible discrepancies.

$$CH_3–(CH_2)_{16}–CH_3$$

n-octadecane
(see Appendix C)

P 11.4 *Where Do Organic Compounds Sit in a Fog Droplet? Inside or at the Surface?*

Several studies have shown that the concentrations of many organic pollutants in fog water are much higher than would be expected from the compound's equilibrium air/water partition constant (see Chapter 6), K_{iaw} (= gaseous concentration of compound *i* in the air/dissolved concentration of compound *i* in pure bulk water). In order to describe the observed enrichment of compounds in fog water, an enrichment factor EF can be defined (see Goss, 1994 and references cited therein):

$$\text{EF} = \frac{K_{iaw}}{D_{iaw}}$$

where D_{iaw} = *total* concentration of i in the gas phase/total concentration of i in the fog droplet ($D_{iaw} = C_{ia}/C_{iw\,tot}$). One possible cause for an enrichment (i.e., $D_{iaw} < K_{iaw}$) is the presence of colloidal organic material in the fog droplet, with which the organic compounds may associate (see Chapter 9). Another possibility suggested by several authors (e.g., Perona, 1992; Valsaraj et al., 1993; Goss, 1994) is enrichment by adsorption at the air/water interface, that is, at the surface of the fog droplet. Is this a reasonable assumption for any organic compound? Estimate the enrichment factor due to surface adsorption at equilibrium for a fog droplet (consisting of pure water) of 8 mm diameter with a surface area (A_d) to volume (V_d) ratio, r_{sv}, of 7500 cm^2 cm^{-3} for (a) tetrachloroethene, (b) phenanthrene, and (c) benzo[a]pyrene at 15°C. Neglect the fact that the surface is curved. *(Hint:* Express the total concentration, $C_{iw\,tot}$, in the fog droplet by $(A_d \cdot C_{isurf} + V_d \cdot C_{iw}) / V_d$ where $A_d / V_d = r_{sv}$ and $C_{isurf} = C_a / K_{iasurf}$. C_{isurf}, C_{iw}, and C_{ia} are the surface concentration, the bulk water concentration, and the bulk air concentration of i, respectively. K_{iasurf} can be estimated by Eq. 11-8).

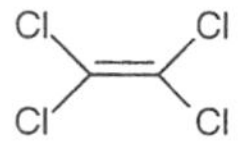

i = tetrachloroethene	i = phenanthrene	i = benzo[a]pyrene
p_{iL}^* (288K) = 1400 Pa	p_{iL}^* (288K) = 0.043 Pa	p_{iL}^* (288K) = 4.2 × 10^{-6} Pa
α_i = 0 ; β_i = 0	α_i = 0 ; β_i = 0.26	α_i = 0 ; β_i = 0.44 (Table 4.3)

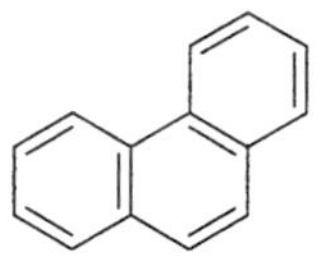

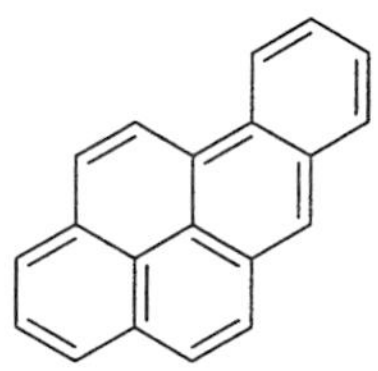

P 11.5   *What Fraction of the TNT Is Dissolved in the River?*

2,4,6-Trinitrotoluene (TNT) is discovered being carried down a turbid river at concentrations of 10 ppb (i.e., 10 μg of TNT per liter water). Colleagues tell you that the suspended solids consist of 15 mg montmorillonite L^{-1} and 1 mg oc $\cdot$ L^{-1}.

Estimate the fraction (%) of TNT dissolved in the river water with the peak concentration TNT and assuming:

(a) the montmorillonite is fully available to sorb TNT, or
(b) none of the montmorillonite siloxanes are available to sorb TNT.

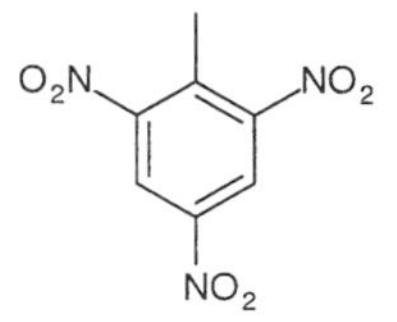

i = 2,4,6-trinitrotoluene (TNT)
M_i = 227.1 g·mol^{-1}
K_{ioc} = 20 L·kg^{-1}oc

P 11.6 *Estimating the Arrival of 2,6-Dinitrotoluene at a Water Supply Well*

2,6-Dinitrotoluene (DNT) is found in a groundwater sample at 10 ppm. Estimate the time of arrival ($t \approx$ distance/(velocity x fraction dissolved)) at a water supply well located 500 meters down gradient assuming: "plug flow" (i.e., no dispersion) at 0.3 m day^{-1}, through the mostly sandy quartz aquifer of porosity 0.3; the aquifer solids contain natural organic matter at 0.1% by weight and illite at 6% by weight; and the groundwater has pH 5, dissolved oxygen at 1 ppm (hence no reduction of DNT; see Chapter 14), Na$^+$ of 1 mM, K$^+$ of 0.1 mM, and Cl$^-$ at 1.1 mM.

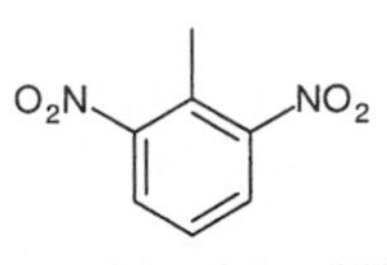

2,6-dinitrotoluene (DNT)
M_i = 168.1 g·mol^{-1}
K_{ioc} = 25 L·kg^{-1}oc

P 11.7 *Developing a Landfill Liner to Retain Organic Wastes*

You have been charged to evaluate competing options for lining a landfill in which the fairly soluble organic solvent, nitrobenzene (NB), will be buried. One vendor tells you that one can use kaolinite, a common clay found in your geographical region, to provide a liner that is 5 cm thick, has 35% porosity, and a density of 2.65 g · cm^{-3} solid, and whose hydraulic conductivity only lets water in the landfill flow through the liner at a seepage velocity of 1 cm per year.

(a) If the kaolinite liner does not crack, how long (in years) would you estimate it would take before any NB would "break through" the liner assuming the percolating water has pH 6.5, dissolved oxygen of 200 μM, nitrobenzene concentrations up to 1/10th its solubility, and that an analysis of the kaolinite reveals the mole ratio of bound cations to be: Na$^+$:K$^+$:Ca^{2+} is 4:1:10?

(b) Assume you can require augmenting the wastes with K$_2$CO$_3$, which causes the kaolinite to always have a ratio of bound ions: Na$^+$:K$^+$:Ca^{2+} at 0.1:9:1. How long (in years) would you estimate it would take before any NB would "break through" the liner in this case?

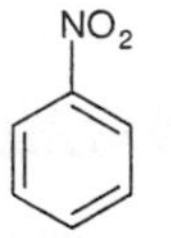

nitrobenzene
(see Appendix C)

P 11.8 *Evaluating the Sorption of an Organic Anion, 2,4-Dichlorophenoxy-Butyrate, to Negatively Charged Natural Solids*

In Fig. 11.9c, some sorption data are shown indicating that an organic anion, 2,4-dichlorophenoxy-butyrate (DB$^-$), sorbed to a sediment from water of pH 7.9 despite the sediment's overall negative charge (as evidenced by its CEC of about 140 mmol · kg^{-1}; Jafvert, 1990).

(a) The sediment also contained 1.5% organic carbon (f_{oc} = 0.015). Given a pK$_{ia}$ of

4.95 for this acid, can you account for the observed K_{id} values near 4 to 5 $L \cdot kg^{-1}$ assuming the neutral DB species partitioned into the NOM of the sediment? What K_{id} value ($L \cdot kg^{-1}$) do you expect from such absorption?

(b) You suspect the hydrophobicity of this organic anion also causes it to accumulate near the mineral surface *against the electrostatic repulsion* it feels via an exchange reaction:

$$DB^- + Na^+ \rightleftharpoons DB{:}surf + Na{:}surf \qquad\qquad K_{DB\ ex}\ (L^2 \cdot kg^{-2})$$

(i) Write a charge balance equation for the near-surface water assuming a solution composition of 20 mM NaCl. *(Note:* The solution actually contained about 2.5 mM Ca^{++}.)

(ii) Write an equilibrium equation (i.e., relating $K_{DB\ ex}$ to chemical concentrations) for the exchange reaction shown (i.e., here neglect the term, [Cl:surf], expecting it will be small relative to [DB:surf]).

(iii) Derive an isotherm equation describing [DB:surf] as a function of [DB$^-$] by combining the equilibrium quotient relation with your near-surface charge balance equation. *(Hint:* $(1 + \varepsilon)^{0.5} \cong 1 + \varepsilon/2$ when ε is a small number.)

(iv) What value of $K_{DB\ ex}$ would be necessary to account for the observed $K_{DB\ d}$ values (i.e., [DB:surf]/[DB$^-$]) if the solution composition was 20 mM NaCl?

(v) Would such a $K_{DB\ ex}$ be "reasonable"? Explain your reasoning (in light of electrostatic and hydrophobic energies required).

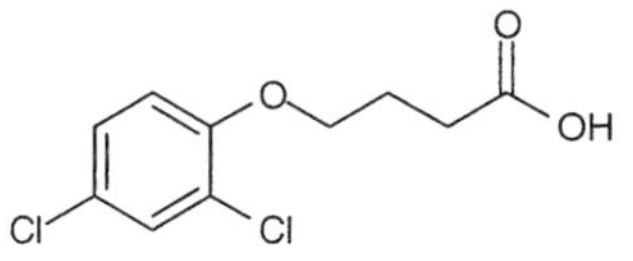

i = 2,4-dichlorophenoxy butyric acid (DB)

M_i = 249 $g \cdot mol^{-1}$
pK_{ia} = 4.95
K_{iow} = 2.5 × 10^3

P 11.9 *Designing a Reactor to Remove Aniline from a Wastewater*

You have been charged with removing the aniline present at 100 ppm in the water of a 100 m^3 tank.

(a) One colleague suggests you add alum ($Al_2(SO_4)_3$) and NaOH to make a 100 mg $\cdot L^{-1}$ suspension of negatively charged amorphous aluminum hydroxide ($Al(OH)_3$) particles at pH 10 and 10 mM Na_2SO_4. Assume an $\equiv$AlOH surface density of 6 × 10^{-6} mol $\cdot m^{-2}$, a specific surface area of 800 m$^2 \cdot g^{-1}$, and $pK_{a1}^{int} = 7$ and $pK_{a2}^{int} = 9$.

Will the aniline sorb to these negatively charged aluminum hydroxide particles and be carried to the bottom of the tank? Calculate the fraction of aniline sorbed to the particles before settling.

(b)   Another colleague suggests you add Na:montmorillonite clay and HCl to make a 100 mg L^{-1} suspension of clay particles at pH 3 and 10 mM NaCl. Assume a CEC of 1×10^{-6} mol $\cdot$ m^{-2} and AEC of 5×10^{-7} mol $\cdot$ m^{-2} at pH 3, a specific surface area of 7×10^5 m^2 kg^{-1}, and a pH$_{zpc}$ = 2.5.

Will the aniline sorb to these mixed charged clay particles and be carried to the bottom of the tank?   Calculate the fraction of aniline sorbed to the particles before settling.

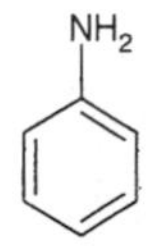

aniline
(see Appendix C)

P 11.10 *What Mechanism Accounts for the Benzidine Sorption in Sediments and Soils?*

Zierath et al. (1980) measured sorption isotherm data for benzidine on sediments and soils. Using Missouri River sediment with f_{oc} = 0.023 kg oc $\cdot$ kg^{-1} solid, CEC = 190 mmol $\cdot$ kg^{-1}, and a specific surface area A_{surf} = 131 m^2 g^{-1}, they obtained the following sorption data:

C_{iw} / (μmol $\cdot$ L^{-1})	C_{is} / (μmol $\cdot$ L^{-1})
20	1500
30	3000
120	5300
200	7600
340	9300

You are interested in discerning what mechanism or mechanisms were responsible for the benzidene sorption observed with Missouri River sediment. To examine this question, you assume the experimental 1 g:10 mL suspensions had a pH of 6 and a salt content of 1 mM NaCl. Given these assumptions, what sorption mechanism would predominate? Justify your answer using estimates of K_{id} assuming (i) first assuming absorption into organic matter predominates and (ii) then assuming adsorption to ion exchange sites predominates.

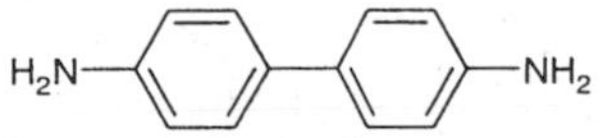

benzidine
(see Appendix C)

P 11.11 *Impact of Diquat Sorption on Its Biodegradation*

The presence of montmorillonite in microbial cultures has been seen to reduce the rate of diquat (D) biodegradation (Weber and Coble, 1968). It has been hypothesized that this is due to significant diquat adsorption to the clay. Neglecting the coexchange of diquat with chloride into the near-surface water since the organic R group is not very hydrophobic, what fraction (%) of the diquat in 1 mM diquat solutions would be adsorbed to a 10 mg $\cdot L^{-1}$ montmorillonite (assume CEC = 1 mol $\cdot kg^{-1}$) suspension at pH 6 and 10^{-2} M NaCl and assuming $K_{D\,ex}$ is:

$$K_{D\,ex} = [D\text{:surf}_2][Na^+] / [D^{2+}][Na\text{:surf}]^2 = 3 \text{ kg} \cdot L^{-1}$$

(Hint: $(1-\varepsilon)^{0.5} \cong 1 - \varepsilon/2$ *for small values of* ε.*)*

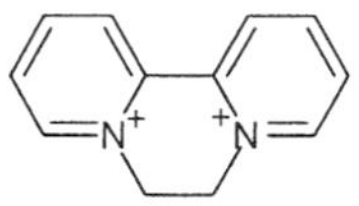

diquat

P 11.12 *Adsorption of Organic Ions to Iron Oxides from Seawater*

Balistrieri and Murray (1987) evaluated the sorption of organic acids on positively charged goethite (FeOOH) particles suspended at 6.6 g $\cdot L^{-1}$ in 0.53 M NaCl solutions to mimic seawater salt. They observed the following trend for *ortho*-phthalic acid added at 200 μM:

pH	3	4	5	6	7	8
% adsorbed	60	65	50	25	5	5

Why is the extent of adsorption largest near pH 4?

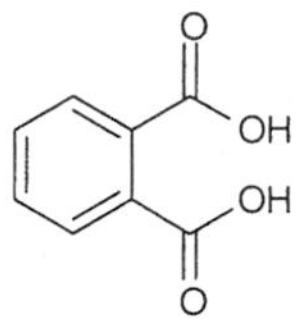

ortho-phthalic acid
$pK_{i\,a1} = 2.89$
$pK_{i\,a2} = 5.51$

PART III

TRANSFORMATION PROCESSES

(a)

2   3,4-dichloroaniline

↓ microbial oxidation

3,3',4,4'-tetrachloro-azobenzene

(b)

tetrachloroethene

↓ microbial or chemical reduction

chloroethene
(vinyl chloride)

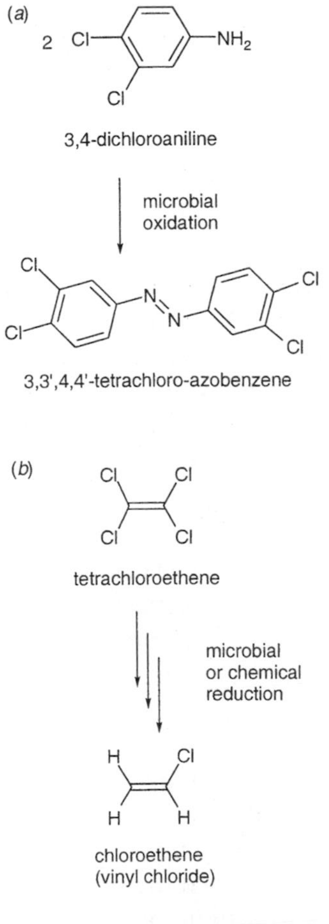

Two examples illustrating the formation of toxic products observed in *(a)* aerobic soils and *(b)* anaerobic landfills and aquifers (see Chapter 14).

So far, we have been concerned exclusively with the partitioning of organic compounds between different environments, that is, with processes that leave the molecular structure of a compound unaltered. In Part III (Chapters 12–17), we now turn our attention to processes by which a compound is converted to one or several products. Hence, we talk about processes (*reactions*) in which chemical bonds are broken and new bonds are formed. In many cases, such transformation reactions lead to products that are less harmful as compared to the parent compounds. In the ideal case, a xenobiotic compound is *mineralized* (i.e., it is converted to CO_2, H_2O, NO_3^-, NH_4^+, Cl^-, Br^-, etc.), in a given environmental compartment. However, there are also numerous examples demonstrating that transformation products may accumulate that are of equal or even greater environmental concern than the parent compounds. In such cases, it is, of course, necessary to worry also about the environmental behavior of the products, which very often exhibit quite different properties, reactivities, and toxicities. Two prominent examples of the conversion of two common xenobiotic organic chemicals (i.e., 3,4-dichloroaniline, tetrachloroethene; see Chapter 2) to very hazardous products are shown in the margin. We will encounter other examples in the following chapters.

For our discussions of the various transformation reactions that organic chemicals undergo in the environment, it is convenient and common use to divide these processes into three major categories: *chemical, photochemical,* and *biologically mediated* transformation reactions. The former two types of reactions are commonly referred to as *abiotic* transformation processes. *Chemical* reactions encompass all reactions that occur in the dark and without mediation of organisms. They are the topic of Chapters 13 and 14, and of parts of Chapter 16. We will subdivide these reactions into those where there is no net electron transfer occurring between the organic compound and a reactant in the environment (Chapter 13), and into *redox reactions* (Chapters 14 and 16), where electrons are either transferred from (oxidation) or to (reduction) the organic chemicals. In Chapter 13, our emphasis will be put on reactions of organic compounds with *nucleophiles* including water (i.e., *hydrolysis),* and on reactions involving bases (e.g., *elimination* reactions). Chapter 14 will be devoted to a general introduction to redox processes, and to a discussion of redox reactions of organic chemicals taking place primarily in landfills, soils, aquifers, and sediments. Here our focus will be on reduction reactions occurring under anoxic (i.e., in the absence of molecular oxygen) conditions. In Chapter 16, finally, we will discuss oxidation reactions involving very reactive oxidants (e.g., hydroxyl radical, singlet oxygen) that act as electrophiles and that are important in water treatment facilities, in surface waters, and in the atmosphere. Since some of these oxidants may be formed by both chemical and photochemical processes, we will treat this topic after addressing some general basic photochemical aspects in Chapter 15. Note that the term *indirect photolysis* is commonly used to denote reactions of organic compounds with reactive species that are formed as a result of the incidence of (sun)light on a water body or in the atmosphere. In contrast, one speaks of *direct photolysis* if a compound undergoes transformation as a consequence of direct absorption of light. We will discuss this process in detail in Chapter 15.

Our last and most difficult topic in the assessment of transformations of organic chemicals in the environment is biologically mediated reactions, which we will address in Chapter 17. Although organic compounds may be transformed by many different organisms (including humans), the most important living actors involved in biotransformations of anthropogenic organic chemicals in the environment are the microorganisms. Hence, our discussions will emphasize *microbial* transformation reactions. Note that biological transformations are usually the only process by which a xenobiotic compound may be completely mineralized in the environment. Hence, when assessing the environmental impact of a given compound, its *biodegradability* is one of the key issues. Unfortunately, as we will see later, because of the complexity of the factors that govern microbial transformation reactions, it is, in many cases, very difficult to make any sound prediction of the rates of such processes in a given natural system. Hence, the determination of biotransformation rates of organic chemicals in the environment generally hinges on the availability of appropriate field data that can be analyzed by using quantitative models (Part V). Consequently, our discussion of biotransformations will necessarily be somewhat more qualitative in nature as compared to the discussions of all other processes addressed in this book. Nevertheless, some general knowledge of the factors that determine the abundance and composition of microbial communities in the environment, and some basic insights into the types of reactions that microorganisms may catalyze, are a prerequisite for anticipating, or at least understanding after the fact, biologically mediated transformations in environmental systems.

Finally, it should be pointed out that Part III will add a new important element that we need to describe organic compounds in natural systems, that is, *time*. So far, we have dealt only with equilibrium concepts (e.g., with the partitioning of organic compounds between different phases), but we have not addressed the question of how fast such equilibria are reached. Thus, in Chapter 12 we will introduce the time axis, that is, we will describe the temporal evolution of a compound concentration due to the influence from various transformation and transport processes. In Part IV we will go one step further and also add space into our considerations.

Chapter 12

THERMODYNAMICS AND KINETICS OF TRANSFORMATION REACTIONS

12.1 Introduction

12.2 Thermodynamics of Transformation Reactions
Illustrative Example 12.1: *Energetics of Syntrophic Cooperation in Methanogenic Degradation*
Illustrative Example 12.2: *Transformation of Methyl Bromide to Methyl Chloride and Vice Versa*

12.3 Kinetic Aspects of Transformation Reactions
Phenomenological Description of Reaction Kinetics
First-Order Kinetics
Box 12.1: *Some Important Background Mathematics: The First-Order Linear Inhomogeneous Differential Equation (FOLIDE)*
First-Order Reaction Including Back Reaction
Reaction of Higher Order
Catalyzed Reactions
Box 12.2: *An Enzyme-Catalyzed Reaction (Michaelis-Menten Enzyme Kinetics)*
Arrhenius Equation and Transition State Theory
Linear Free-Energy Relationships
Impact of Solution Composition on Reaction Rates

12.4 Well-Mixed Reactor or One-Box Model
Illustrative Example 12.3: *A Benzyl Chloride Spill Into a Pond*

12.5 Questions and Problems

12.1   Introduction

There are several general questions that have to be addressed when dealing with transformation reactions of organic compounds in the environment:

1. Is there only one or are there several different reactions by which a given compound may be transformed under given environmental conditions?

2. What are the reaction products?

3. What are the *kinetics* of the different reactions, and what is the resultant overall rate by which the compound is "eliminated" from the system by these reactions?

4. What is the influence of important environmental variables such as temperature, pH, light intensity, redox condition, ionic strength, presence of certain solutes, concentration and type of solids, microbial activity, and so forth, on the transformation behavior of a given compound?

We will address all of these questions for each type of reaction considered in the following chapters. In this chapter, we will summarize the most important theoretical concepts needed for a quantitative treatment of transformation reactions of organic chemicals. We start out by extending our earlier discussions on the application of thermodynamic theory to assess a compound's behavior in the environment (see Chapters 3 and 8), this time with the goal of answering the question whether a given reaction is energetically favorable under the conditions prevailing in a given system. Furthermore, as already addressed in Chapter 8, we want to know to what extent a (reversible) reaction has proceeded when equilibrium is reached. When considering abiotic reactions, we will be interested in these questions to see whether a reaction may occur spontaneously. With respect to microbially mediated reactions, these questions are important for assessing whether a microorganism may gain energy from a given reaction, that is, whether there is a benefit for the microorganism to catalyze the reaction.

In Chapter 8, we addressed proton transfer reactions, which we have assumed to occur at much higher rates as compared to all other processes. So in this case we always considered equilibrium to be established instantaneously. For the reactions discussed in the following chapters, however, this assumption does not generally hold, since we are dealing with reactions that occur at much slower rates. Hence, our major focus will not be on thermodynamic, but rather on kinetic aspects of transformation reactions of organic chemicals. In Section 12.3 we will therefore discuss the mathematical framework that we need to describe zero-, first- and second-order reactions. We will also show how to solve somewhat more complicated problems such as enzyme kinetics.

Finally, in the last section of this chapter, we will introduce the simplest approach for modeling the dynamic behavior of organic compounds in laboratory and field systems: the *one-box model* or *well-mixed reactor*. In this model we assume that all system properties and species concentrations are the same throughout a given volume of interest. This first encounter with dynamic modeling will serve several purposes. First, besides batch reactors, well-mixed flow-through reactors are often used to evaluate the reaction kinetics for a given transformation process, particularly when one deals with heterogeneous systems (e.g., surface-catalyzed hydrolysis, microbial transformation). Hence, it is useful to get acquainted with the mathematical description of such reactors. Second, using such simple models will help us to see how we have to treat the various transformation processes in order to be able to combine them later with transfer and transport processes (Chapters 18 to 22) in models of different complexities (Chapters 23 to 25). Finally, the use of simple one-box models will allow us to make some interesting calculations on the residence times of reactive organic chemicals in well-mixed environmental systems such as, for example, the epilimnion of a lake, a small pond, or part of the atmosphere.

12.2 Thermodynamics of Transformation Reactions

Since most of the reactions discussed in the following chapters take place in aqueous media, we confine our thermodynamic considerations to reactions occurring in dilute aqueous solutions. For the gas-phase reactions of organic compounds with highly reactive oxidants (i.e., reactions in the atmosphere; Chapter 16), we will assume that these reactions are always energetically favorable and, thus, proceed spontaneously.

In Chapter 8, where we treated acid–base equilibria, we have seen that when dealing with reactions in dilute aqueous solutions, the appropriate choice of reference state for solutes is the *infinite dilution state in water*. The chemical potential of a compound i can then be expressed as:

$$\mu_i = \mu_i^{0'} + RT \ln \frac{\gamma_i' [i]}{(\gamma_i^{0'} = 1)[i]^0} \tag{8-3}$$

where γ_i' is the activity coefficient, $[i]$ is the actual concentration of the species i, $[i]^0$ is its concentration in the *standard* state, and the prime superscript is used to denote the infinite dilution reference state (in distinction to the pure organic liquid state). Recall that $\mu_i^{0'}$ corresponds to the standard free energy of formation of the species i in aqueous solution; that is:

$$\mu_i^{0'} \equiv \Delta_f G_i^0 (aq) \tag{8-4}$$

at given p^0 and T (e.g., 1 bar, 25°C). Also note that because we have chosen the infinitely dilute aqueous solution as reference, γ_i' will in many cases not be substantially different from 1.

Let us now consider a general reversible chemical reaction:

$$aA + bB + \ldots \rightleftharpoons pP + qQ + \ldots \tag{12-1}$$

where $a, b, \ldots p, q, \ldots$ are the *stoichiometric coefficients* of the reaction; that is, the coefficients that describe the relative number of moles of each reactant consumed or produced by a given reaction. At a given composition [A], [B],..., [P], [Q],... of the system (we are not interested in other system components that are not involved in the reaction), it is easy to see from Eqs. 3-19 and 3-20 (Chapter 3) that, if the reaction proceeds by the increment dn_r so that $dn_A = adn_r$, $dn_B = bdn_r$ ..., to $dn_P = pdn_r$, $dn_C = qdn_r$ ..., we cause a change in the total free energy of the system which is given by:

$$dG = -a\mu_A dn_r - b\mu_B dn_r - ... + p\mu_P dn_r + q\mu_Q dn_r$$

$$dG = (-a\mu_A - b\mu_B - ... + p\mu_P + q\mu_Q + ...)dn_r \tag{12-2}$$

The quantity dG/dn_r, which is a measure of the free energy change in the system as the reaction progresses, is referred to as the *free energy of reaction*, which we denote as $\Delta_r G$. We use the subscript "r" to distinguish the free energy of reaction from the free energy of transfer that we used in Part II. Hence,

$$\Delta_r G = -a\mu_A - b\mu_B - ... + p\mu_P + q\mu_Q + ... \tag{12-3}$$

Inserting Eq. 8-3 for each species taking part in the reaction into Eq. 12-3, and assessing a standard concentration $[i]^0 = 1$ M for all species considered [note that, of course, 1 M is a hypothetical concentration for most of the organic compounds of interest, because their water solubilities are usually much lower (see Chapter 5)], we then obtain after some rearrangements:

$$\Delta_r G = -a\mu_A^{0'} - b\mu_B^{0'} - ... + p\mu_P^{0'} + q\mu_Q^{0'} + ...$$

$$+RT\ln\frac{\left(\gamma_P'[P]\right)^p\left(\gamma_Q'[Q]\right)^q...}{\left(\gamma_A'[A]\right)^a\left(\gamma_B'[B]\right)^b...} \tag{12-4}$$

The algebraic sum of the *standard chemical potentials* of the products and reactants is called the *standard free energy of reaction*, and is denoted as $\Delta_r G^0$:

$$\Delta_r G^0 = -a\mu_A^{0'} - b\mu_B^{0'} - ... + p\mu_P^{0'} + q\mu_Q^{0'} + ...$$

$$= -a\Delta_f G_A^0(aq) - b\Delta_f G_B^0(aq) - ... + p\Delta_f G_P^0(aq) + q\Delta_f G_Q^0(aq) + ... \tag{12-5}$$

Hence, $\Delta_r G^0$ is a measure of the free energy of the reaction when all species are present in their standard state (where concentrations are assumed to be 1 M, and activity coefficients are set to 1). We should recall that a negative $\Delta_r G^0$ would mean that, *under standard conditions*, the reaction Eq. 12-1 would proceed spontaneously from left to right; if $\Delta_r G^0$ is positive, the reaction would proceed spontaneously in the opposite direction.

By substituting Eq. 12-5 into Eq. 12-4, and by using the notation $\{i\}$ for expressing the *activity* of a given species (i.e., $\{i\} = \gamma_i'[i]$), we can rewrite Eq. 12-4 as:

$$\Delta_r G = \Delta_r G^0 + RT\ln\frac{\{P\}^p\{Q\}^q...}{\{A\}^a\{B\}^b...} \tag{12-6}$$

By defining a reaction quotient, Q_r, as:

$$Q_r = \frac{\{P\}^p \{Q\}^q \cdots}{\{A\}^a \{B\}^b \cdots} \tag{12-7}$$

we can rewrite Eq. 12-6 as

$$\Delta_r G = \Delta_r G^0 + RT \ln Q_r \tag{12-8}$$

It is important to realize that $\Delta_r G$ can be heavily influenced by the Q_r term. Hence, even if $\Delta_r G^0$ is positive (i.e., the standard reaction is *endergonic* and does not, therefore, occur spontaneously under standard conditions), a reaction may still proceed in a given system (i.e., it is *exergonic* because of a very small Q_r value). A small Q_r could be due to a very small activity (or concentration, because $\gamma_i^{0'}$ values are commonly not very different from 1, see below) of one or several product(s), and/or a very high activity (concentration) of one or several reactant(s) present in the system at a given time. Conversely, a large Q_r value (e.g., accumulation of products, depletion of reactant(s)) may lead to a positive $\Delta_r G$, although $\Delta_r G^0$ may be negative. Illustrative Example 12.1 helps to make this point clear.

When we are interested in the *equilibrium* composition in a given system (i.e., in the situation where $\Delta_r G = 0$), the Q_r value is not a variable anymore but is given by the equilibrium reaction constant, K_r. This is related to $\Delta_r G^0$ by the equation (see also Chapter 8):

$$\ln K_r \equiv \ln \frac{\{P\}^p \{Q\}^q \cdots}{\{A\}^a \{B\}^b \cdots} = -\frac{\Delta_r G^0}{RT} \tag{12-9}$$

Hence, if $\Delta_r G^0$ is known or can be calculated from $\Delta_f G_i^0(aq)$ values, the extent to which a given reaction proceeds until reaching equilibrium can be assessed (see Illustrative Example 12.2).

Finally, we should recall from Chapter 8 that when assuming a constant *standard enthalpy of reaction*, $\Delta_r H^0$, over a small temperature range, the relationship between K_r values at two different temperatures is given by

$$K_r(T_2) = K_r(T_1) \cdot e^{-\frac{\Delta_r H^0}{R}\left(\frac{1}{T_2}-\frac{1}{T_1}\right)} \tag{see 8-20}$$

Illustrative Example 12.1   **Energetics of Syntrophic Cooperation in Methanogenic Degradation**

Introduction

The small amount of energy available in methanogenic processes (see Chapter 14) forces the microorganisms involved into a very efficient cooperation. In many cases, neither partner can operate without the other; that is, together they may exhibit a metabolic activity that neither one could accomplish on its own. Such cooperations are called *syntrophic relationships* (for details see Schink, 1997). A classical example is the *Methanobacillus meliansky* culture, which is a co-culture of two partner organisms, strain S and strain M.o.H. The two strains cooperate in the conversion of ethanol to acetate and methane by interspecies hydrogen transfer, as follows:

$$\text{Strain S:} \qquad CH_3CH_2OH + H_2O \rightarrow CH_3COO^- + H^+ + 2H_2 \qquad (1)$$

ethanol   acetate

$$\text{Strain M.o.H.:} \qquad 4H_2 + CO_2 \rightarrow CH_4 + 2H_2O \qquad (2)$$

methane

The critical point in this syntrophic cooperation is that the reaction that Strain S is catalyzing (Eq. 1) is only exergonic (negative $\Delta_r G$) if the H_2 partial pressure (or the corresponding H_2 concentration) in the aqueous phase is low enough. Hence, strain M.o.H. has to remove hydrogen efficiently from the solution.

Problem

Calculate the maximum p_{H2} at which strain S can just gain some energy from catalyzing reaction Eq. 1 at pH 7 and 25°C, by assuming that both ethanol and acetate are present at a concentration of 10^{-6} M.

Answer

$\Delta_f G^0_{CH_3CH_2OH}(aq) = -181.8 \, kJ \cdot mol^{-1}$
$\Delta_f G^0_{CH_3COO^-}(aq) = -369.4 \, kJ \cdot mol^{-1}$
$\Delta_f G^0_{H_2O}(\ell) = -237.2 \, kJ \cdot mol^{-1}$
$\Delta_f G^0_{H_2}(g) = 0 \, kJ \cdot mol^{-1}$
$\Delta_f G^0_{H^+}(aq) = 0 \, kJ \cdot mol^{-1}$

In the literature (e.g., Thauer et al. 1977; Hanselmann, 1991) you find the $\Delta_f G^0(aq)$ values at 25°C for all the species involved in reaction Eq. 1. Note that by convention, the free energies of formation of the elements in their naturally occurring most stable form, as well as of the proton in aqueous solution, are set to zero. From these values, calculate the standard free energy of reaction Eq. 1:

$$\Delta_r G^0 = -(-181.8) - (-237.2) + (-369.4) + (0) + (0) = +49.6 \, kJ \cdot mol^{-1}$$

Hence, under standard conditions ($\{CH_3CH_2OH\} = \{CH_3COO^-\} = \{H^+\} = \{H_2O\} = 1$, $p_{H2} = 1$ bar), this reaction is endergonic. In order to get a negative free energy of reaction, $\Delta_r G$, the term $RT\ln Q$ has to be smaller than $-\Delta_r G^0$ (Eq. 12-8, note that $\{H_2O\} = 1$):

$$RT \ln Q < -49.6 \, kJ \cdot mol \text{ or } Q < 2 \times 10^{-9}$$

Therefore:

$$Q = \frac{\{CH_3COO^-\}\{H^+\} \, p_{H_2}^2}{\{CH_3CH_2OH\}} = \frac{(10^{-6})(10^{-7}) \, p_{H_2}^2}{(10^{-6})} = 10^{-7} p_{H_2}^2$$

is less than 2×10^{-9} if $p_{H2} < 0.14$ bar.

Of course, in order to sustain growth, a significantly more negative $\Delta_r G$ is required, which is achieved by a hydrogen partial pressure of $< 10^{-3}$ bar (Schink, 1997).

Illustrative Example 12.2

Transformation of Methyl Bromide to Methyl Chloride and Vice Versa

Problem

Consider the reversible transformation of the soil fumigant methyl bromide (CH_3Br) to methyl chloride (CH_3Cl) in aqueous solution (a nucleophilic substitution reaction, see Chapter 13):

$$CH_3Br + Cl^- \rightleftharpoons CH_3Cl + Br^-$$

In which direction does this reaction occur at 25°C in a contaminated groundwater containing 50 mM Cl^-, 1 mM Br^-, and (a) 10 times more CH_3Cl than CH_3Br.? (b) How does your answer change if there is 1000 times more CH_3Cl than CH_3Br? What is the relative abundance of CH_3Br and CH_3Cl at equilibrium assuming constant Cl^- and Br^- concentrations? Also assume that the activity coefficients of all species (including the charged species) are ~ 1, that is, $\{i\} \cong [i]$.

Answer

$\Delta_f G^0_{Cl^-} (aq) = -131.3\,kJ \cdot mol^{-1}$
$\Delta_f G^0_{Br^-} (aq) = -104.0\,kJ \cdot mol^{-1}$
$\Delta_f G^0_{CH_3Br} (g) = -28.2\,kJ \cdot mol^{-1}$
$\Delta_f G^0_{CH_3Cl} (g) = -58.0\,kJ \cdot mol^{-1}$

In the literature you find $\Delta_f G^0_i (aq)$ values for chloride and bromide. However, for CH_3Br and CH_3Cl, only values for the free energy of formation in the gas phase, $\Delta_f G^0_i (g)$, are given. This is, of course, not a serious problem, because the difference between $\Delta_f G^0_i (aq)$ and $\Delta_f G^0_i (g)$ is the free energy of transfer between the gas phase and water, which is directly related to the Henry's law constant (Section 6.4). Note that we have to take into account that we use molar concentrations (and not mole fractions) in the aqueous phases. Hence, as discussed in Section 3.4 (Eqs. 3-41 to 3-46), the free energy of transfer from the gas phase to the aqueous phase is given by $-[\Delta_{aw} G_i + RT \ln \overline{V}_w / (L\,mol^{-1})]$, and, therefore:

$$\Delta_f G^0_i (aq) = \Delta_f G^0_i (g) - \left[\Delta_{aw} G_i + RT \ln \overline{V}_w / (L \cdot mol^{-1})\right] \tag{1}$$

or

$$\Delta_f G^0_i (aq) = \Delta_f G^0_i (g) + RT \ln K_{iH} \tag{2}$$

where K_{iH} is the Henry's law constant expressed in $bar \cdot L \cdot mol^{-1}$.

$C^{sat}_{CH_3Br\,w} = 10^{-0.78}\,mol \cdot L^{-1}$
$C^{sat}_{CH_3Cl\,w} = 10^{-0.98}\,mol \cdot L^{-1}$

In Appendix C you find the water solubilities, C^{sat}_{iw} of the two gases at 25°C and 1 bar partial pressure, which allows you to estimate K_{iH}:

$$K_{iH} \cong \frac{1}{C^{sat}_{iw}}$$

The approximated K_{iH} values are 6.0 $bar \cdot L \cdot mol^{-1}$ for CH_3Br, and 9.6 $bar \cdot L \cdot mol^{-1}$ for CH_3Cl. Insertion of these values into Eq. (2) together with the $\Delta_f G^0_i (g)$ values found for the two compounds in the literature (see above) yields:

$$\Delta_f G^0_{CH_3Br} (aq) = -23.7\,kJ \cdot mol^{-1}, \text{ and } \Delta_f G^0_{CH_3Cl} (aq) = -52.4\,kJ \cdot mol^{-1}$$

Using these values, the standard free energy of reaction in aqueous solution can now be calculated:

$$\Delta_r G^0 = -(-23.7) - (-131.3) + (-52.4) + (-104.0) = -1.4 \text{ kJ} \cdot \text{mol}^{-1}$$

In order to get the free energy of reaction under the given conditions (a) and (b) calculate the respective Q_r values (Eq. 12-8; set $\{i\} = [i]$):

$$\text{(a)} \quad Q_r = \frac{(10)[\text{Br}^-]}{(1)[\text{Cl}^-]} = \frac{(10)(10^{-3})}{(1)(5 \times 10^{-2})} = 0.2$$

$$\text{(b)} \quad Q_r = \frac{(10^3)[\text{Br}^-]}{(1)[\text{Cl}^-]} = \frac{(10^3)(10^{-3})}{(1)(5 \times 10^{-2})} = 20$$

Insertion of $\Delta_r G^0$ and Q_r into Eq. 12-8 then yields:

$$\text{(a)} \quad \Delta_r G = -1.4 + 2.48 \ln 0.2 = -5.4 \text{ kJ} \cdot \text{mol}^{-1}$$

$$\text{(b)} \quad \Delta_r G = -1.4 + 2.48 \ln 20 = +6.0 \text{ kJ} \cdot \text{mol}^{-1}$$

Hence, at the solution composition prevailing in case (a) the reaction occurs spontaneously from left to right; that is, CH_3Br is converted to CH_3Cl. However, in case (b), the reaction proceeds in the opposite direction; that is, CH_3Cl is converted to CH_3Br! This result indicates that at equilibrium, the $[CH_3Cl]$ to $[CH_3Br]$ ratio has to lie between 10 and 1000.

To answer the question about the $[CH_3Cl]$ to $[CH_3Br]$ ratio at equilibrium, calculate the equilibrium reaction constant, K_r (Eq. 12-9):

$$\ln K_r = -\frac{\Delta_r G^0}{RT} = -\frac{(-1.4 \text{ kJ} \cdot \text{mol}^{-1})}{(2.48 \text{ kJ} \cdot \text{mol}^{-1})} = 0.56; \quad K_r = 1.76$$

Since

$$K_r = \frac{[CH_3Cl][\text{Br}^-]}{[CH_3Br][\text{Cl}^-]}$$

the ratio $[CH_3Cl]$ to $[CH_3Br]$ *at equilibrium* for fixed $[\text{Cl}^-]$ and $[\text{Br}^-]$ is given by:

$$\frac{[CH_3Cl]}{[CH_3Br]} = K_r \cdot \frac{[\text{Cl}^-]}{[\text{Br}^-]} = \frac{(1.76)(5 \times 10^{-2})}{(10^{-3})} = 88$$

12.3 Kinetic Aspects of Transformation Reactions

Phenomenological Description of Reaction Kinetics

From experimental data or by analogy to the reactivity of compounds of related structure, we can often derive an empirical rate law for the transformation of a given compound. The *rate law* is a mathematical function, specifically a differential equation, describing the turnover rate of the compound of interest as a function of the concentrations of the various species participating in the reaction. Hence, the rate law reflects the overall reaction on a *macroscopic* level. In a very general way, we can write the macroscopic rate law for the transformation rate (i.e., for the disappearance rate) of an organic compound (*i*) as:

$$\frac{d[\text{org}]}{dt} = -k[i]^i[\text{B}]^b[\text{C}]^c \dots \tag{12-10}$$

where the exponents *i*, *b*, *c*, ... indicate the *order of the reaction* with respect to the corresponding species: org, B, C, ... Note that when dealing with more complex reactions, particularly in heterogeneous systems, the (phenomenological) reaction orders do not necessarily have to be integer numbers and they may also change with concentration. The *total order n* of the reaction is given by the sum of the exponents $n = i + b + c + \dots$, and *k* is called the *nth-order reaction rate constant*, and has the dimension $[(\text{ML}^{-3})^{1-n}\,\text{T}^{-1}]$.

Before we take a look at some typical rate laws encountered with chemical reactions in the environment, some additional comments are necessary. It is important to realize that the empirical rate law Eq. 12-10 for the transformation of an organic compound does not reveal the *mechanism* of the reaction considered. As we will see, even a very-simple-looking reaction may proceed by several distinct reaction steps (*elementary molecular changes*) in which chemical bonds are broken and new bonds are formed to convert the compound to the observed product. Each of these steps, including back reactions, may be important in determining the overall reaction rate. Therefore, the reaction rate constant, *k*, may be a composite of reaction rate constants of several elementary reaction steps.

It is particularly important to be aware of this point when one wants to derive quantitative structure-reactivity relationships for a specific reaction of a series of structurally related compounds (i.e., if one wants to relate the rate constants to certain properties of the compounds). In these cases, one may find that such relationships do not hold for some of the compounds considered, the reason being that for different compounds, different elementary reaction steps may be rate determining. Examples will be discussed in the following chapters.

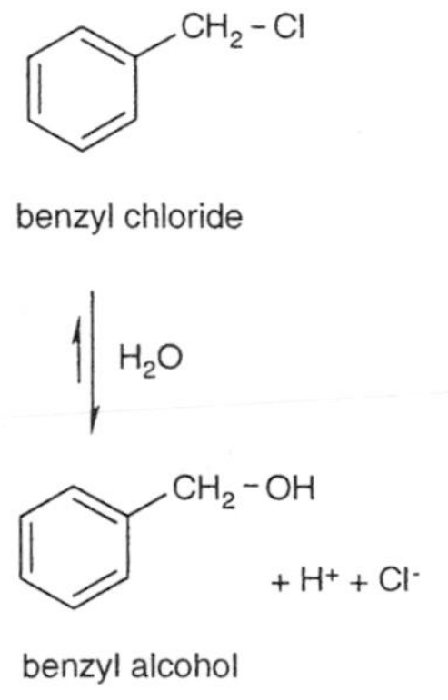

First-Order Kinetics

We start our discussion of specific reaction rate laws by examining the results of a simple experiment in which we observe how the concentration of benzyl chloride (Fig. 12.1) changes as a function of time in aqueous solutions of pH 3, 6, and 9 at 25°C (Fig. 12.2). When plotting the concentration of benzyl chloride (denoted as [A]) as a function of time, we find that we get an *exponential* decrease in concentration independent of pH (Fig. 12.2*a*). Hence, we find that the turnover rate of benzyl chloride is always proportional to its current concentration. This can be expressed mathematically by a *first-order rate* law:

$$\frac{d[\text{A}]}{dt} = -k[\text{A}] \tag{12-11}$$

benzyl chloride

H₂O

benzyl alcohol

+ H⁺ + Cl⁻

Figure 12.1 Reaction of benzyl chloride with water (a nucleophilic substitution reaction, see Chapter 13).

Figure 12.2 Decrease in benzyl chloride concentration (*a*) plotted directly and (*b*) plotted logarithmically as a function of time in aqueous solution at three different pHs. The right graph reflects the fact that Eq. 12-12 can be transformed into the form $\ln([A]_t/[A]_0) = -kt$.

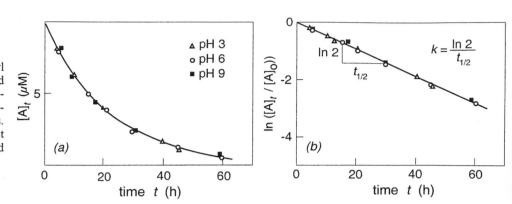

where k is referred to as the *first-order rate constant* and has the dimension $[T^{-1}]$. Since Eq. 12-11 is a linear differential equation, the first-order reaction is often called a linear reaction and k the linear rate constant. Integration of Eq. 12-11 from $[A] = [A]_0$ at $t = 0$ to $[A] = [A]_t$ at time t yields the mathematical description of the curve in Fig. 12.2*a* (see Box 12.1, Case a):

$$[A]_t = [A]_0 \cdot e^{-kt} \tag{12-12}$$

Eq. 12-12 implies that the logarithm of the ratio $[A]_t/[A]_0$ yields a straight line through the origin with slope $-k$. Thus, if data from kinetic experiments are plotted as in Fig. 12.2, we can both check whether the reaction is first order in $[A]$ and determine the rate constant k using a linear regression analysis. We note that in the case of first-order kinetics, the *half-life*, $t_{1/2}$, of the compound (i.e., the time in which its concentration drops by a factor of 2) is *independent of concentration* and equal to:

$$t_{1/2} = \frac{\ln 2}{k} = \frac{0.693}{k} \tag{12-13}$$

From Fig. 12.1 we notice that the transformation of benzyl chloride to benzyl alcohol involves a water molecule which was not included in the rate law (Eq. 12-11). If the water molecule were involved in the slowest step of the reaction, that is, in the *rate-determining* step, the reaction should be described by a *second-order rate* law:

$$\frac{d[A]}{dt} = -k'[A][B] \tag{12-14}$$

where we have denoted H_2O as B, and k' is now referred to as *second-order rate constant* with dimension $[M^{-1}L^3T^{-1}]$. Yet, since water is present in a large excess, its concentration (~ 55.3 M) is not altered significantly during the course of the reaction: $[B]_t \cong [B]_0 \sim 55.3$ M. Hence, by setting $k = k'[B]_0$, we again obtain a first-order rate law (Eq. 12-11). In this and all other cases in which we simplify the rate law by assuming the concentrations of certain species to be constant, we use the prefix *pseudo-* to indicate that, on a molecular level, more species take part in the reaction than are incorporated in the rate law. Without knowing more about the reaction mechanism, we cannot say whether for the case of benzyl chloride the rate law is *pseudo-first-order* or truly first order. In the former case, we refer to k as the *pseudo-first-order rate constant*.

Box 12.1   Some Important Background Mathematics: The First-Order Linear Inhomogeneous Differential Equation (FOLIDE)

In our modeling efforts it will often be our goal to express, or at least to approximate, all dynamic processes by (pseudo-)first-order rate laws. This will enable us to perform "back-of-the-envelope" calculations for a quick assessment of the relative importance of the various transport and transformation processes that govern the behavior of a given organic compound in a given system, and to get a first idea of the temporal variation in concentration of the compound in the system. With this approach, we will always have to deal with the same type of differential equation, the *first-order linear inhomogeneous differential equation (FOLIDE)*. It has the general form:

$$\frac{dy}{dt} = J - ky \tag{1}$$

Here y denotes the time-dependent or *dynamic variable* (usually a concentration), J is the so-called *inhomogeneous* or *input term,* and k is the overall *first-order rate constant,* which can be a sum of several first-order rate constants each describing a different process. If y has the dimension of a concentration $[ML^{-3}]$, then J has $[ML^{-3}T^{-1}]$. Note that J as well as k can be time-dependent. We discuss the solution of Eq. 1 by starting with the simplest case and then move to the more complex ones.

Case a. $J = 0$, k = constant

The corresponding equation:

$$\frac{dy}{dt} = -ky \tag{2}$$

is called the first-order linear *homogeneous* differential equation. It has the solution (Fig. *a*):

$$y(t) = y_0 e^{-kt} \tag{3}$$

with y_0 the initial value $y(t = 0)$. For $k > 0$, the so-called *steady state* $y_\infty = y(t \to \infty)$ is zero. Strictly speaking it takes an infinite amount of time until y becomes zero. In practice the exponential expression becomes extremely small once (kt) is much larger than 1. For instance, $e^{-3} = 0.05$, thus at:

$$t_{5\%} = \frac{3}{k} \tag{4}$$

$y(t)$ has dropped to 5% of its initial value. The half-life, $t_{1/2}$, (i.e., the time when $y(t)$ has dropped to one half of y_0), is:

$$t_{1/2} = \frac{\ln 2}{k} = \frac{0.693}{k} \tag{5}$$

If k is negative, $y(t \to \infty)$ grows infinitely.

Case b. J, k = constant

For $k > 0$, the solution is (Fig. *b*):

$$y(t) = y_0 e^{-kt} + y_\infty (1 - e^{-kt}) = y_\infty + (y_0 - y_\infty)e^{-kt} \qquad (6)$$

with the steady-state:

$$y_\infty = \frac{J}{k} \qquad (7)$$

Note that Eq. 4 can still be looked upon as a measure for the time needed to approach steady-state.

Case c. Variable input $J(t)$, k = constant

For $k > 0$, the solution is:

$$y(t) = y_0 e^{-kt} + \int_0^t e^{-k(t-t')} J(t')dt' \qquad (8)$$

The first term, identical to the first term in Eq. 6, describes the exponential decay of the initial value y_0. The integration time t' in the second term runs from $t' = 0$, the initial time, to $t' = t$, the time for which y is evaluated (the "present"). Because of the term $e^{-k(t-t')}$ the integral represents a weighted sum of the input J during the time interval between 0 and t. Inputs that occurred far back in time $[(t - t')$ large] have little or no influence on the actual value $y(t)$. In fact, for $(t - t') > 3/k$ the weight of J has dropped to less than 5% (see Eq. 4). An example is shown in Fig. c.

Case d. Variable input $J(t)$, variable $k(t)$

For completeness the most general solution of Eq.1 is given below:

$$y(t) = y_0 e^{-\Phi(t)} + e^{-\Phi(t)} \int_0^t e^{\Phi(t')} J(t')dt' \qquad (9)$$

$$\Phi(t) = \int_0^t k(t')dt'$$

Note that even if the coefficients are time dependent, the differential equation is still linear. However, if J or k explicitly depend on the variable y, the equation would be nonlinear.

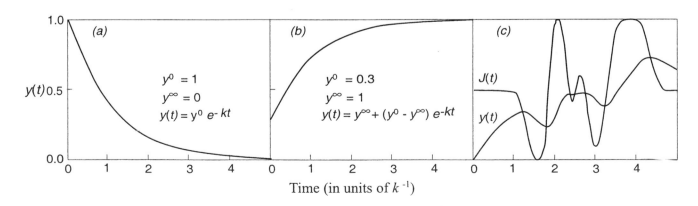

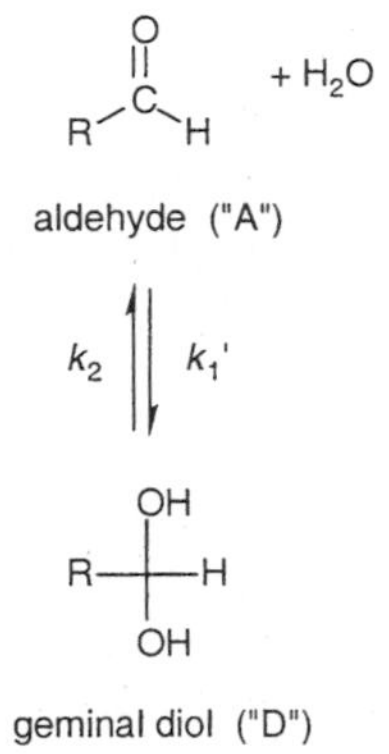

First-Order Reaction Including Back Reaction

aldehyde ("A")

geminal diol ("D")

Figure 12.3 Many aldehydes react substantially with water in a reversible fashion to yield a diol (see also Bell and McDougall, 1960).

When treating the overall transformation kinetics of an organic compound as we have done for the hydrolysis of benzyl chloride (Eq. 12-11), we assume that the *reverse* reaction (i.e., the formation of benzyl chloride from benzyl alcohol) can be neglected. For many of the reactions discussed in the following chapters we will make this assumption either because the reverse reaction has an extremely small rate constant (i.e., the reaction is practically *irreversible*), or because the concentration(s) of the reactant(s) are very large as compared to the concentration(s) of the product(s). There are, however, situations in which the reverse reaction has to be taken into account. We have already encountered such a reaction in Illustrative Example 12.1. To demonstrate how to handle the reaction kinetics in such a case, we use the hydration of an aldehyde to yield a diol (Fig. 12.3). This example will also illustrate how the equilibrium reaction constant, K_r, is related to the kinetic rate constants, k_1 and k_2, of the forward and reverse reaction.

The hydration of an aldehyde RCHO can be expressed with two (pseudo-first-order) rate laws consisting of an elimination (minus sign) and a production (plus sign) term:

$$\frac{d}{dt}[RCHO] = -k_1'[H_2O][RCHO] + k_2[RCH(OH)_2]$$
$$= -k_1[RCHO] + k_2[RCH(OH)_2] \tag{12-15}$$

For example, for formaldehyde (R = H) at neutral pH, the pseudo-first-order rate constant for the hydration reaction (forward reaction), $k_1 = k_1'$ [H_2O], is about 10 s^{-1} and the first-order rate constant for dehydration, k_2, is about 5×10^{-3} s^{-1}. In Chapter 20 we will use this example to show that the reactivity of compounds can influence the kinetics of air/water exchange if both processes (reaction and exchange) occur on similar time scales.

Substituting A for the aldehyde and D for the diol, we can rewrite Eq. 12-15 for A and a corresponding equation for D:

$$\frac{d[A]}{dt} = -k_1[A] + k_2[D] \tag{12-16a}$$

$$\frac{d[D]}{dt} = +k_1[A] - k_2[D] \tag{12-16b}$$

The concentrations for which $d[A]/dt = 0$ and $d[B]/dt = 0$ are called the equilibrium or steady-state concentrations, $[A]_\infty$ and $[B]_\infty$. From Eq. 12-16:

$$\frac{[D]_\infty}{[A]_\infty} = \frac{k_1}{k_2} = K_r \tag{12-17}$$

Here the equilibrium reaction constant, K_r, which in Eq. 12-9 has been derived from thermodynamic considerations, is given by the ratio of the rate constants of the forward and the reverse reaction. Inserting the rate constants for formaldehyde into Eq. 12-17 yields:

$$K_r = \frac{k_1}{k_2} = \frac{10 \text{ s}^{-1}}{5 \times 10^{-3} \text{ s}^{-1}} = 2 \times 10^3 \tag{12-18}$$

Hence, at equilibrium and neutral pH, formaldehyde dissolved in water occurs 99.8% as the diol. In the case of acetaldehyde ($R = CH_3$), this fraction is about 50%.

Although the solution of a set of differential equations usually involves some more sophisticated mathematical tools (see Chapter 21), for our problem we can simplify the procedure. Addition of the two equations, Eqs. 12-16a and b, yields:

$$\frac{d[A]}{dt} + \frac{d[D]}{dt} = \frac{d}{dt}([A] + [D]) = 0 \tag{12-19}$$

This equation states that the total concentration, $[A]_{tot} = [A]_0 + [D]_0$, is constant. Thus, in Eq. 12-16a we can make the substitution $[D] = [A]_{tot} - [A]$, which yields a kinetic expression in the form of a FOLIDE:

$$\frac{d[A]}{dt} = -k_1[A] + k_2([A]_{tot} - [A]) \tag{12-20}$$

$$= k_2[A]_{tot} - (k_1 + k_2)[A]$$

According to Box 12.1 (Case b, Eq. 6), this has the solution:

$$[A]_t = [A]_0 \, e^{-(k_1+k_2)t} + [A]_\infty \, (1 - e^{-(k_1+k_2)t}) \tag{12-21}$$

where

$$[A]_\infty = \frac{k_2}{k_1 + k_2}[A]_{tot} \tag{12-22}$$

We can also use this procedure to track the diol:

$$[D]_t = [A]_{tot} - [A]_t \tag{12-23}$$

As an interesting fact, we can learn from Eq. 12-21 that the time to steady-state (or time to equilibrium) depends on the *sum* of the forward and reverse reaction rate constants. Thus, even if one rate constant is very small, time to equilibrium can be small, provided that the other rate constant is large. By using Eq. 4 in Box 12.1 (95% of equilibrium reached) we obtain:

$$t_{5\%} = \frac{3}{(10 + 5 \times 10^{-3})\text{s}^{-1}} \cong 0.3 \text{s} \tag{12-24}$$

Reactions of Higher Order

Although most reactions with which we are concerned are not truly first order, it is convenient for modeling purposes to make assumptions that allow us to reduce the order of the reaction law, ideally to pseudo-first order. For example, when considering reactions of organic chemicals with environmental reactants for which we can assume that their concentrations do not change significantly during the reaction (e.g., OH^-, Cl^-, Br^-, HS^-, O_2), we can lower the reaction order by setting the concentrations of these species constant (as we have done with water in the examples discussed above). In the case of second-order reactions, we can thus often achieve a pseudo-first-order rate law.

There are however, cases where higher-order rate laws have to be applied to describe adequately the reaction kinetics of a xenobiotic compound. The mathematics involved in solving these more complex situations may be found in textbooks on chemical kinetics (e.g., Frost and Pearson, 1961; Laidler, 1965; Brezonik, 1994).

Catalyzed Reactions

There are quite a few situations in which rates of transformation reactions of organic compounds are accelerated by reactive species that do not appear in the overall reaction equation. Such species, generally referred to as *catalysts*, are continuously regenerated; that is, they are not consumed during the reaction. Examples of catalysts that we will discuss in the following chapters include reactive surface sites (Chapter 13), electron transfer mediators (Chapter 14), and, particularly enzymes, in the case of microbial transformations (Chapter 17). Consequently, in these cases the reaction cannot be characterized by a simple reaction order, that is, by a simple power law as used for the reactions discussed so far. Often in such situations, reaction kinetics are found to exhibit a gradual transition from first-order behavior at low compound concentration (the compound "sees" a constant steady-state concentration of the catalyst) to zero-order (i.e., constant term) behavior at high compound concentration (all reactive species are "saturated"):

$$\frac{d[A]}{dt} = -k[A], \text{ for low } [A]$$

$$(12\text{-}25)$$

$$\frac{d[A]}{dt} = -J, \text{ for high } [A]$$

where k and J have the dimensions $[T^{-1}]$ and $[ML^{-3}\,T^{-1}]$, respectively.

There are a variety of mathematical expressions with which we could fit the required dependence of the transformation rate on the compound concentration.

In Box 12.2, a simple model for a special kind of catalyzed reaction, the Michaelis-Menten enzyme kinetics, is presented, which leads to the following kinetic expression:

$$\frac{d[A]}{dt} = -J\,\frac{[A]}{[A] + (J/k)} \qquad (12\text{-}26)$$

As shown in Fig. 12.4, this equation exhibits the behavior described by Eq. 12-25. In fact, it became so popular that often it is mistakenly given a deeper theoretical meaning even in cases where it simply serves as a fitting curve. Therefore, the reader should not forget that there are other curves built from two parameters (J, k) which

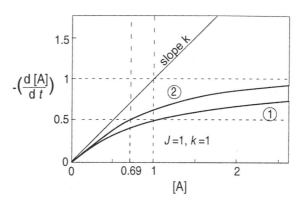

Figure 12.4 Comparison of the reaction rate functions Eqs. 12-27 (curve 1) and 12-28 (curve 2). The straight line through the origin indicates the initial slope of size k. Half-saturation concentrations are at $\ln 2\ (J/k)$ and (J/k), respectively. Parameter values $J = 1$, $k = 1$.

exhibit the same behavior as Eq. 12-25 for very small or very large [A] values but are different in between. For instance,

$$\frac{d[A]}{dt} = -J(1 - e^{-(k/J)[A]}) \tag{12-27}$$

has the same properties as Eq. 12-25, but its *half-saturation concentration*, the concentration of A where the rate of change is $J/2$, is smaller than for Eq. 12-26 (see Fig. 12.4).

To summarize, we should make a clear distinction between the application of Eq. 12-26 as a model for enzyme kinetics (Box 12.2) or as a fitting curve. In the latter case, we have to remember that other functions, such as Eq. 12-27, may fulfill this purpose as well.

Arrhenius Equation and Transition State Theory

If we want to understand and describe the influence of environmental factors, especially temperature, on chemical reaction rates, and if we want to see how transformation rates vary as a function of the chemical structure of a compound, we need to take a closer look at these reactions on a molecular level. As mentioned already, a chemical reaction often proceeds in several sequential elementary steps. Frequently, one step in the reaction sequence occurs at a much slower rate than all the others.

Box 12.2   An Enzyme-Catalyzed Reaction (Michaelis-Menten Enzyme Kinetics)

We consider a catalyzed reaction in which an organic compound, A, is transformed to a product, P, via a complex formed by A with a catalyst E. Although the letter E for catalyst reminds us that frequently the catalyst is an enzyme (see Chapter 17), other species such as reactive surface sites can also take over this role. A typical catalyzed reaction can be divided into three steps:

Step 1: Substrate A reacts with E to form a complex AE. This reaction is assumed to be reversible and fast as compared to Step 2. It can be expressed by an equilibrium constant, K_E:

$$A + E \rightleftharpoons AE \quad ; \quad K_E = \frac{k_f}{k_b} = \frac{[AE]}{[A][E]} \tag{1}$$

Step 2: AE is transformed by a *first-order reaction* (rate constant k_E) into the product-catalyst complex PE:

$$AE \xrightarrow{k_E} PE \quad : \quad \frac{d[AE]}{dt} = -k_E[AE] \tag{2}$$

Step 3: PE decays into P and E (hence, E is regenerated) in a reaction that is fast as compared to reaction Eq. 2.

In order to derive the overall rate law for the transformation of A to P we assume that the total catalyst concentration $[E]_{tot}$ is constant and that the concentration of PE is very small (Step 3 is assumed to be fast). Hence, $[E]_{tot} = [E] + [AE]$. Using Eq. 1 to substitute for $[E]$:

$$[E]_{tot} = [E] + [AE] = [AE]\left(1 + \frac{1}{K_E[A]}\right) \tag{3}$$

Since in Step 1 the rate of disappearance of A is equal to the rate of formation of AE we write

$$-\left(\frac{d[A]}{dt}\right)_{Step\ 1} = +\left(\frac{d[AE]}{dt}\right)_{Step\ 1} \tag{4}$$

With Eqs. 2 and 4 we can now write down the total mass balance of $[AE]$:

$$\frac{d[AE]}{dt} = \left(\frac{d[AE]}{dt}\right)_{step\ 1} - \left(\frac{d[AE]}{dt}\right)_{step\ 2} = \frac{d[A]}{dt} - k_E[AE] \tag{5}$$

It can be assumed that, once the reaction is in progress, concentrations of intermediates including AE are at *steady-state*, that is, $d[AE]/dt = 0$. Thus, Eq. 5 can be transformed into:

$$\frac{d[A]}{dt} = -k_E[AE] \tag{6}$$

Solving Eq. 3 for $[AE]$ and substituting into Eq. 6 yields:

$$\frac{d[A]}{dt} = -k_E \frac{[E]_{tot}}{1 + \frac{1}{K_E[A]}} = -k_E[E]_{tot}\frac{[A]}{[A] + \frac{1}{K_E}} \tag{7}$$

In the notation of Eq. 12-31 we find that the maximum reaction rate $\left([A] \gg K_E^{-1}\right)$ is $J = k_E[E_{tot}]$ while the first-order reaction rate at small concentrations $\left([A] \ll K_E^{-1}\right)$ is $k = k_E[E]_{tot}K_E$.

To summarize, the described model of enzyme catalysis leads to the kinetic expression:

$$\frac{d[A]}{dt} = -J\frac{[A]}{[A] + (J/k)} \tag{12-26}$$

with $J = k_E[E]_{tot}$, $k = k_E[E]_{tot}K_E$.

This step acts as a "bottleneck" in the overall process and determines the overall rate of the reaction. Consequently, this slowest step is generally the one in which we are most interested.

Let us begin by taking a look at the effect of temperature on the rate of a chemical reaction. Experimentally, we commonly find that the reaction rate constant varies as an exponential function of temperature. This can be mathematically expressed by the so-called *Arrhenius equation:*

$$k = A \cdot e^{-E_a/RT} \tag{12-28}$$

where A is called the *preexponential factor* or *frequency factor,* and E_a is referred to as the *activation energy*. The unit of A corresponds to the unit of the rate constant k. For a given reaction A and E_a can be derived by linear regression of a plot of $\ln k$ versus $1/T$:

$$\ln k = \ln A - E_a / RT \tag{12-29}$$

Hence, very similar to the equilibrium partition constants (e.g., Eq. 3-51) or equilibrium constants (e.g., Eq. 8-20), rate constants at two different temperatures T_1 and T_2 are related by:

$$\ln \frac{k(T_1)}{k(T_2)} = \frac{E_a}{R} \left(\frac{1}{T_2} - \frac{1}{T_1} \right) \tag{12-30}$$

Therefore, knowledge of E_a allows us to use Eq. 12-30 to assess the effect of temperature on reaction rate constants. Note that when deriving and using Eqs. 12-29 and 12-30 to calculate rate constants at different temperatures, one assumes that A and E_a are temperature independent. This is a reasonable first approximation if the temperature range considered is not too large, and *if we are dealing with only one reaction that causes the compound to disappear*. Table D1 in Appendix D shows that, depending on the magnitude of E_a (E_a values typically range between 40 and 130 kJ·mol⁻¹), an increase (decrease) of 10°C may accelerate (slow down) a reaction by a factor of between 2 and 6.

For a qualitative interpretation of the Arrhenius equation we consider a simple elementary bimolecular reaction in aqueous solution:

$$B + C \rightarrow D + E \tag{12-31}$$

Let us now deduce the factors that control the rate of conversion of B and C to D and E by imagining the transformation process is portrayed well by what is known as a collision rate model. (Strictly speaking, the collision rate model applies to gas phase reactions; here we use it to describe interactions in solution where we are not specifying the roles played by the solvent molecules.) First, in order to be able to react, the molecules B and C have to encounter each other and collide. Hence, the rate of reaction depends on the frequency of encounters of B and C, which is proportional to the product of their concentrations. The rate is also related to how fast B and C move in the aqueous solution. Next, the rate is proportional to the probability that B and C meet with the "right orientation" to be able to react, which we may refer to as the "orientation probability". Third, only a fraction of collisions have a sufficient amount of energy (greater then or equal to E_a) to break the relevant bonds in B and C that make the reactants change to products D and E. The fraction of species exhibiting an energy greater than E_a is given by $e^{-E_a/RT}$, which corresponds exactly to the exponential term in the Arrhenius equation. Consequently, the collision frequency and the orientation probability factors are included in the preexponential factor A as well as in the concentration dependence in the rate law:

$$\text{rate} = -A \cdot e^{-E_a/RT}[B][C] \tag{12-32}$$

This simple model allows us to qualitatively rationalize the Arrhenius equation.

For a more quantitative approach to chemical kinetics, in particular for understanding and applying linear free-energy relationships to estimate transformation rates of organic chemicals from structure (see below), we need to acquaint ourselves with a more sophisticated theoretical framework, the so-called *activated complex* or *transition state theory*. We can view this theory as a thermodynamic approach to chemical kinetics. To explain, we consider the highest energy state (i.e., the transition state) that the reactants of an elementary reaction go through on their way to form products. We then propose a "structure" of this high-energy species, commonly referred to as an *activated complex*, denoted as $BC^{\ddagger}$. Specific examples of activated complexes will be discussed in the following chapters. Since the transition state represents the point of bond changing highest in energy, and therefore the point most difficult to reach, the rate of formation of the activated complex determines the overall rate of the elementary reaction, and of the overall reaction if we consider the rate-limiting step. The quantitative treatment of this approach hinges on the postulate *that the reactants establish an equilibrium with the activated complex*. This assumption is somewhat unusual in that the activated complex has only a transitory existence since it lies at an energy maximum (rather than a minimum). The second assumption, which is derived from statistical mechanics, is that *all activated complexes proceed onto products with a fixed first-order rate constant, which is given by* kT/h, where k and h are the Boltzmann (1.38×10^{-23} J·K^{-1}) and Planck (6.63×10^{-34} J·s) constants, respectively (for details see Atkins, 1998). Hence, we can express our reaction as consisting of two steps:

$$B + C \rightleftharpoons BC^{\ddagger} \rightleftharpoons D + E \tag{12-33}$$

Based on these two postulates, the rate law of a reaction is then simply given as the product of the universal rate constant times the concentration of the activated complex:

$$\text{rate} = \left(\frac{kT}{h}\right)[BC]^{\ddagger} \tag{12-34}$$

By assuming activity coefficients of 1 for all species involved, $[BC^{\ddagger}]$ can be expressed in terms of the reactant concentrations and the equilibrium constant $K^{\ddagger}$:

$$K^{\ddagger} = \frac{[BC^{\ddagger}]}{[B][C]} \tag{12-35}$$

and

$$\text{rate} = \left(\frac{kT}{h}\right)K^{\ddagger}[B][C] \tag{12-36}$$

From Section 12-2 (Eq. 12-9) we also recall that $K_r = e^{-\Delta_r G^0 / RT}$. Hence, we get:

$$\text{rate} = \left(\frac{kT}{h}\right) \cdot e^{-\Delta^{\ddagger} G^0 / RT} [B][C] \tag{12-37}$$

where $\Delta^{\ddagger} G^0$ is referred to as the standard free energy of activation. Since $\Delta^{\ddagger} G^0 = \Delta^{\ddagger} H^0 - T\Delta^{\ddagger} S^0$, we can also write:

$$\text{rate} = \left(\frac{kT}{h}\right) e^{\Delta^{\ddagger} S^0 / R} \cdot e^{-\Delta^{\ddagger} H^0 / RT} [B][C] \tag{12-38}$$

where $\Delta^{\ddagger} S^0$ and $\Delta^{\ddagger} H^0$ are the standard entropy and enthalpy of activation, respectively. The rate constant is then given by:

$$\text{rate constant} = \left(\frac{kT}{h}\right) e^{\Delta^{\ddagger} S^0 / R} \cdot e^{-\Delta^{\ddagger} H^0 / RT} \tag{12-39}$$

which resembles the Arrhenius rate law (Eq. 12-28). Note that E_a represents a *potential* energy of activation: the difference between $\Delta^{\ddagger} H^0$ and E_a is given by the *kinetic* energy of activation, which is usually small as compared to E_a. For our bimolecular reaction, this difference is given by $1RT$ (e.g., Atkins, 1998), and we can rewrite Eq. 12-39 as:

$$\text{rate constant} = \frac{kT}{h} e^{\Delta^{\ddagger} S^0 / R} \cdot e^{-(E_a - 1RT)/RT}$$

$$\tag{12-40}$$

$$= \frac{kT}{h} e^{\Delta^{\ddagger} S^0 / R} \cdot e^{+1} \cdot e^{-E_a / RT}$$

Thus, based on transition state theory, the preexponential factor A in the Arrhenius equation is interpreted to encompass the universal rate constant and the entropy of activation, the latter factor corresponding to what we have called "orientation probability" in our simple collision rate model. From Eq. 12-40 we also recognize that A is linearly dependent on temperature, but it is easy to see how that is usually masked by the overwhelming exponential temperature dependence of the activation energy term. Finally, we note that A values may range over several orders of magnitude, depending on the reaction considered. Unimolecular dissociation or elimination reactions usually exhibit large preexponential factors (A between 10^{12} and 10^{16} s^{-1}) because the entropy of activation is only slightly negative or even positive. Bimolecular reactions, on the other hand, commonly exhibit A values between about 10^7 and 10^{12} M^{-1} s^{-1} (Harris and Wamser, 1976; Mabey and Mill, 1978).

Linear Free-Energy Relationships

In earlier chapters we have already applied linear free-energy relationships (LFERs) to evaluate and/or to predict *equilibrium partition constants* (Chapters 4 to 11) as well as *equilibrium reaction constants* (i.e., using the Hammett relationship; Chapter 8). The basic idea behind all the approaches that we have discussed so far was to use empirical relationships to quantify the free energy of transfer ($\Delta_{12} G_i$), or the standard free energy of reaction ($\Delta_r G^0$), respectively, for a given compound and process from corresponding free-energy terms determined for reference compounds and processes.

The goal was always to find simple *linear* relationships between the respective free-energy terms (or the logarithms of the corresponding equilibrium constants).

As discussed above, when using the transition state theory to describe reaction kinetics, we postulate an equilibrium between the reactants and the activated complex for the rate-limiting step of a given reaction; that is, we define a standard free-energy term that we have referred to as the standard free energy of activation, $\Delta^{\ddagger}G^0$, of the reaction. Hence, the transition state theory provides us with a very handy framework to evaluate and/or to predict reaction rates by using LFERs in a similar way as we have done for equilibrium constants. The only difference and the major difficulty is, however, that this time we need a quantitative measure of $\Delta^{\ddagger}G^0$, instead of $\Delta_r G^0$. This means that we must define the "structure" of the activated complex, and we must be able to assess how structural differences in the reactants influence the (relative) stability of this activated complex. As one can easily imagine, this task is rather difficult because it involves the understanding of not only the *electronic* [inductive (polar), resonance] and *steric* effects of structural moieties of the reactants on $\Delta^{\ddagger}G^0$, but also the effects of changes in solvation, which, in the case of water, may be rather complex. Nevertheless, as will be illustrated by several examples in the following chapters, LFERs relating rate data are not only very useful for predictions of reaction rate constants but also for the evaluation of kinetic data, since an LFER (or the lack of an LFER) might give useful insights to reaction mechanisms. Finally, for successful applications of LFERs to kinetic data, two important points have to be observed. First, when considering a given type of reaction, an LFER relating rate constants of a series of structurally closely related compounds to any parameters chosen to describe electronic and steric effects of structural moieties on $\Delta^{\ddagger}G^0$ will have a chance to be successful only if, for the compounds considered, the same elementary reaction step(s) is (are) rate limiting. Second, one has to be very careful to choose appropriate descriptors for quantifying electronic and steric effects on $\Delta^{\ddagger}G^0$ of a given type of reaction. We will come back to these important points in the following chapters. For a more comprehensive treatment of this topic we refer to the literature (e.g., Exner, 1988: Hansch and Leo, 1995)

Impact of Solution Composition on Reaction Rates

As a final point we need to make a few comments on the impact of solution composition on reaction rates. Unfortunately, because many organic compounds are only sparingly soluble in water, a large portion of the kinetic data available in the literature on reactions that are of interest to us have been determined in organic solvents or in organic solvent–water mixtures instead of pure water. Since the intermolecular interactions of reactants and of activated complexes (particularly if they are charged) with solvent molecules (called solvation) may involve significant energies, making or breaking these interactions can have an important effect on $\Delta^{\ddagger}G^0$. Thus, significantly different rate data are obtained for different solvent systems, in particular for reactions involving ionic species, when we compare polar solvents like water with nonpolar ones. As illustrated by the reaction rate constants given in Table 12.1 for the base-catalyzed hydrolysis of benzoic acid ethyl ester (ethyl benzoate, see margin) in various organic solvent–water systems, differences of more than one order of magnitude may be found. Hence, we have to be cautious when applying kinetic data obtained in organic solvents and/or organic solvent–water systems to purely aqueous

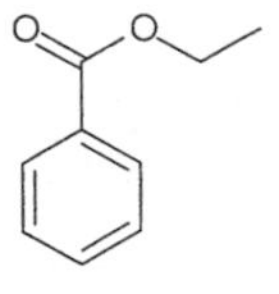

ethyl benzoate

Table 12.1 Reaction Rate Constants for the Base-Catalyzed Hydrolysis of Benzoic Acid Ethyl Ester (Ethyl Benzoate) in Various Organic Solvent–Water Mixtures [a]

Mixture	k_B $(M^{-1} s^{-1})$
Water	3.0×10^{-2}
60% Ethanol / 40% water	1.2×10^{-3}
60% Acetone / 40% water	2.8×10^{-3}
40% Dioxane / 60% water	7.0×10^{-3}
70% Dioxane / 30% water	3.3×10^{-3}

[a] Data from Mabey and Mill (1978)

solution, particularly for reactions involving polar reactants. In every case, the feasibility of extrapolating rate constants from nonaqueous to aqueous solutions has to be carefully checked (for further discussion of solvent effects see Shorter, 1973).

Furthermore, in aqueous solutions, the influence of dissolved organic and inorganic species (e.g., buffer solutions used in laboratory experiments, the major ions and dissolved organic matter present in natural waters, trace metals, mineral oxide surfaces) on transformation rates has to be evaluated in each case. As we will see in the following chapters, such species may act as reactants or catalysts, or they may influence the reaction rate indirectly.

12.4   Well-Mixed Reactor or One-Box Model

We conclude this chapter by introducing a simple tool with which we will be able to put the reactivities of organic compounds into an environmental context: the *well-mixed reactor* or *one-box model*.

In Chapter 21 we will show that the construction of any environmental model first consists of the appropriate choice of a boundary between the system and the outside world (see Fig. 21.1). Here we choose the simplest possible system [i.e., a homogeneous (completely mixed) box that is connected to the outside world by an input I and an output O, Fig. 12.5]. We consider one single chemical compound that shall be described either by the total amount in the system, $\mathcal{M}$, or by its mean concentration, $C = \mathcal{M}/V$, where V is the volume of the system. Note that for simplicity we omit the subscript i in the following derivation. Several reaction processes, R_j, act on the compound in the interior of the system; we characterize them by the total rate R_{tot}:

$$R_{tot} = \sum_j R_j \quad [MT^{-1}] \tag{12-41}$$

Then the mass balance for the compound in the well-mixed box is simply:

$$\frac{d\mathcal{M}}{dt} = I - O - R_{tot} \quad [MT^{-1}] \tag{12-42}$$

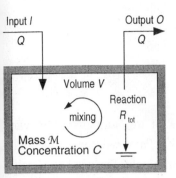

Input I

Q

Output O

Q

Volume V

mixing

Reaction

R_{tot}

Mass $\mathcal{M}$
Concentration C

Figure 12.5 Schematic representation of a simple one-box model including constant input (I), output (O), and transformation reactions in the bulk phase. $\mathcal{M}$ is the total mass of the compound in the constant volume V, C is the total concentration, and Q is the flow rate.

All quantities on the right-hand side have the dimension mass per time!

If the volume V is constant, we can rewrite this equation using $\mathcal{M} = VC$ and dividing both sides by V:

$$\frac{dC}{dt} = \frac{I}{V} - \frac{O}{V} - \frac{R_{tot}}{V} \quad [\text{ML}^{-3}\,\text{T}^{-1}] \tag{12-43}$$

Expressed in words, this equation says that the temporal change of the concentration in the reactor is equal to the input per unit volume and time minus the output and total reaction per unit volume and time.

According to the definitions given in Fig. 12.5, the input I is not influenced by what happens in the system; it is a so-called external variable. This is not true for the two elimination processes, O and R_{tot}. Thus, in order to solve Eqs. 12-42 or 12-43 we have to determine how these terms depend on the "system state." The only quantities which characterize the "system state" are $\mathcal{M}$ or C. In fact, there is only one independent state variable since $\mathcal{M}$ and C are proportional to each other.

The simplest assumptions we can make are the following:

(1) The output O shall be due to the outflow of water (or another fluid) leaving the system with flow rate Q (dimension L^3T^{-1}):

$$O = Q \cdot C = \frac{Q}{V}\mathcal{M} = k_w\mathcal{M} \tag{12-44}$$

where

$$k_w = \frac{Q}{V} \quad [\text{T}^{-1}] \tag{12-45}$$

is the flushing or dilution rate of the reactor.

Note that since the reactor is assumed to be completely mixed, the outflowing water (or solvent) must carry the concentration C.

(2) All reaction processes are first-order, that is:

$$R_{tot} = \sum_j R_j = \sum_j k_j\mathcal{M} = k_{tot}\mathcal{M} \tag{12-46}$$

with $k_{tot} = \sum_j k_j$.

Inserting these expressions into Eq. 12-42 yields:

$$\frac{d\mathcal{M}}{dt} = I - k_w\mathcal{M} - k_{tot}\mathcal{M} = I - (k_w + k_{tot})\mathcal{M} \tag{12-47}$$

$$\frac{dC}{dt} = \frac{I}{V} - \frac{O}{V} - \frac{R_{tot}}{V} \quad [\text{ML}^{-3}\,\text{T}^{-1}] \tag{12-48}$$

In Eqs. 12-47/48 we recognize the first-order linear inhomogeneous differential equation (FOLIDE, Box 12.1). Depending on whether the input I and the different rate constants (k_w, k_j) are constant with time, their solutions are given in Eqs. 6,8, or 9 of Box 12.1.

For constant coefficients, a steady-state can be calculated from Eq. 7 in Box 12.1:

and thus:

$$M_\infty = \frac{I}{k_w + k_{tot}} \tag{12-49a}$$

$$C_\infty = \frac{I/V}{k_w + k_{tot}} = \frac{M_\infty}{V} \tag{12-49b}$$

Note that the denominator of these equations consists of the sum of all relevant (transport and reaction) rate constants, $k_w + k_{tot} = k_w + \Sigma k_j$. Thus, we can immediately assess the relative influence of these processes on C_∞^j or M_∞ by looking at the relative size of all k-values. This is also important when we are interested in time-to-steady-state (Eq. 7 in Box 12.2):

$$t_{5\%} = \frac{3}{k_w + \sum_j k_j} \tag{12-50}$$

One simple fast reaction or a large k_w value can make M_∞, C_∞, and $t_{5\%}$ small.

In Eqs. 12-47/48 the input term is an exterior quantity and not determined by the model itself. If the substance is added to the system through the inflowing water (or other fluid), we can express I by the input concentration: $I = Q \cdot C_{in}$. Inserting this into Eq. 12-49b yields:

$$C_\infty = \frac{(Q/V)C_{in}}{k_w + k_{tot}} = \frac{k_w}{k_w + k_{tot}} C_{in} \tag{12-51}$$

This equation can be very helpful for assessing the behavior of a substance in a natural system. If input and system concentration at steady-state are equal, k_{tot} must be zero. In turn, for any reactive substance C_∞ must be smaller than C_{in}. A simple application of the one-box model is given in Illustrative Example 12.3.

Illustrative Example 12.3   A Benzyl Chloride Spill Into a Pond

Problem

Due to an accident right before Christmas, an unknown amount of benzyl chloride (BzC) (see Fig. 12.1) is introduced into a small, well-mixed pond that is used as a drinking-water reservoir. Working for the State Water Authority, you are asked to estimate how much BzC has entered the pond, and more importantly, how long it will take until the concentration will have dropped below 1 $\mu g \cdot L^{-1}$. Because of the Christmas holiday, it takes you 5 days until you are ready to make the first measurement. At this time (i.e., 5 days after the spill) the measured BzC concentration in the pond water is 50 $\mu g \cdot L^{-1}$. A second measurement 5 days later shows that the concentration has dropped to 23.6 $\mu g \cdot L^{-1}$. Besides answering the above questions, you also want to identify the major removal mechanism(s) for BzC in the pond. Assume that all relevant systems and compound parameters are constant during the relevant time period.

Pond Characteristics

Volume $V = 10^5$ m^3
Surface $A = 2 \times 10^4$ m^2
Water through-flow $Q = 10^3$ m^3 d^{-1}
Average water temperature 5°C

Benzyl Chloride (BzC)

Hydrolysis half-life at 25°C = 15 h
Activation Energy E_a = 80 kJ·mol^{-1}

Answer

From the two measured BzC concentrations at day 5 and day 10 you can calculate the total (pseudo-)first-order rate constant k_{tot}, using the simplest case of the FOLIDE (Box 12.1, Case a, Eq. 3):

$$\ln \frac{[\text{BzC}](10\text{d})}{[\text{BzC}](5\text{d})} = -k_{tot} \cdot \Delta t \quad (\Delta t = 5\text{d})$$

and thus:

$$k_{tot} = \frac{\ln \dfrac{(50 \ \mu g \cdot L^{-1})}{(23.6 \ \mu g \cdot L^{-1})}}{5 \text{ d}} = 0.15 \text{ d}^{-1}$$

which corresponds to a half-life (Eq. 12-13) of 4.6 d. The initial concentration right after the spill (day 0) can then be derived from:

$$[\text{BzC}]_0 = \frac{[\text{BzC}]_{5d}}{e^{-kt}} = \frac{50 \ \mu g \cdot L^{-1}}{e^{-0.75}} = 106 \ \mu g \cdot L^{-1}$$

Multiplication with the volume of the pond (see pond characteristics) yields the estimated total input:

$$\text{Total input} = (106 \text{ mg} \cdot \text{m}^{-3}) \, (10^5 \text{ m}^3) = 10.6 \text{ kg}$$

For calculating the time required for the BzC concentration to drop to 1 $\mu g \cdot L^{-1}$, you can, for example, start out from the measurement made 10 days after the spill:

$$t = \frac{\ln \dfrac{(23.6 \ \mu g \cdot L^{-1})}{(1 \ \mu g \cdot L^{-1})}}{k_{tot} \, (= 0.15 \text{ d}^{-1})} = 21 \text{ d}$$

Hence, from the time of the spill it takes (10+21) d or about *one month* to reach 1 $mg \cdot L^{-1}$.

For evaluating the major removal mechanism(s), calculate the hydrolysis rate of BzC at 5°C using the half-life at 25°C (i.e., 15 h) and an activation energy, E_a of 80 kJ·mol^{-1}. Insertion of E_a into Eq. 12-30 with $T_1 = 298$ K and $T_2 = 278$ K shows that the hydrolysis rate constant of BzC at 5°C is about ten times smaller (or the half-life is ten times larger) as compared to 25°C (see also Table D1 of Appendix D). Thus, the elimination of BzC by hydrolysis occurs with a rate constant, k_h, of

$$k_h = 0.1 \cdot \frac{\ln 2}{15 \text{ h}} = 0.0462 \text{ h}^{-1} = 0.11 \text{ d}^{-1}$$

Comparison of k_h with the calculated $k_{tot} = 0.15$ d^{-1} shows that *abiotic hydrolysis* is the most important removal mechanism for BzC in the pond (~*75%*); thus, you have to worry about the transformation product benzyl alcohol (Fig. 12.1). About *7%* is removed by *flushing* ($k_w = V/Q = 0.01$ d^{-1}), and the rest by other processes. Considering the properties of benzyl chloride (e.g., K_{iow}, K_{iaw}, see Appendix C), the most likely additional elimination processes are gas exchange and biotransformation (see later chapters).

12.5   Questions and Problems

Questions

Q 12.1

Give some examples of important transformation reactions that organic pollutants may undergo in the environment.

Q 12.2

What are the most important questions that have to be addressed when dealing with transformation reactions in the environment?

Q 12.3

Does a positive standard free energy of reaction, $\Delta_r G^0$, mean that a given reaction does not occur spontaneously under any condition?

Q 12.4

Consider a reversible reaction with a $\Delta_r G^0$ value of -20 kJ·mol^{-1}. In which direction will this reaction occur at the point at which the reaction quotient, Q_r, is 1000? What is the equilibrium reaction constant, K_r, of this reaction?

Q 12.5

What is meant by the term *rate law* of a transformation reaction of a given organic compound? What does it describe?

Q 12.6

What is the *total order n* of a given reaction? What are the most common reaction orders in environmental organic chemistry? Does the total order of a given reaction tell you anything about the reaction mechanism? What is a *pseudo-nth-order rate constant?*

Q 12.7

What are the dimensions of a (i) zero-order, (ii) first-order, and (iii) second-order reaction rate constant? What is the *half-life* of a given compound with respect to a given reaction? In which case(s) is the half-life independent of the concentration of the compound?

Q 12.8

What is a *catalyzed* reaction? How does the rate of disappearance of a given compound in an enzyme- or surface-catalyzed reaction typically depend on the concentration of the compound?

Q 12.9

What are the assumptions made when describing a catalyzed reaction by a Michaelis-Menten type rate law? Write down the Michaelis-Menten rate law and discuss the various terms by using a graphical representation.

Q 12.10

What are the assumptions made when using the Arrhenius equation to describe the effect of temperature on the transformation rate of a given compound? What is the range of activation energies typically observed for transformation reactions of organic compounds in the environment?

Q 12.11

Consider a compound that is transformed by two different reactions exhibiting very different activation energies: E_a (reaction 1) = 50 kJ·mol^{-1} and E_a (reaction 2) = 100 kJ·mol^{-1}. Assume that both reactions can be described by a pseudo-first-order rate law, and that at 25°C, they are equally important; that is, equal amounts of the two different transformation products are formed. What is the product distribution that you would expect at (i) 5°C and (ii) 40°C?

Q 12.12

What is a linear differential equation? What special properties does it have?

Q 12.13

What does "time to steady-state" mean? Why is there no unique definition of time to steady-state for a linear first-order differential equation?

Q 12.14

Convince yourself that the solution of the FOLIDE with variable input $J(t)$ (Eq. 8 of Box 12.1) includes, as a special case, the solution for constant J (Eq. 6 of Box 12.1).

Q 12.15

Consider the linear one-box model. What can we learn from a comparison of C_{in} and C_{∞} regarding the behavior of the chemical in the system?

Q 12.16

In a linear well-mixed reactor model the flushing rate is k_w = 0.5 h^{-1}, the total reaction rate constant of a specific chemical k_{tot} = 1.5 h^{-1}. What is the "retention factor" of the reactor for the considered chemical, that is, what percentage of the chemical is reacting in the reactor? How does this percentage change when the input of the chemical is doubled?

Problems

P 12.1 *Evaluating the Transformation of Hexachloroethane to Tetrachloroethene in an Aqueous Ferrous/Ferric Iron Solution*

Consider the transformation of hexachloroethane (HCA) to tetrachloroethene (PCE) in an acidic (why acidic?) aqueous solution at 25°C containing 0.5 mM Fe^{2+}(aq), 5 mM Fe^{3+}(aq), 20 mM Cl^-, and 1 μM HCA:

$$Cl_3C - CCl_3 \; + 2\ Fe^{2+}(aq) \rightleftharpoons Cl_2C = CCl_2 + 2\ Fe^{3+}(aq) + 2\ Cl^-$$

$$\text{HCA} \qquad\qquad\qquad\qquad\qquad \text{PCE}$$

What type of reaction is this? To what extent is HCA transformed to PCE? Calculate the [PCE] / [HCA] ratio at equilibrium. Assume that all activity coefficients are 1. In the literature you find the following data:

$$\Delta_f G^0_{Fe^{2+}} (aq) = -78.9 \text{ kJ} \cdot \text{mol}^{-1}$$

$$\Delta_f G^0_{Fe^{3+}} (aq) = -4.6 \text{ kJ} \cdot \text{mol}^{-1}$$

$$\Delta_f G^0_{Cl^-} (aq) = -131.3 \text{ kJ} \cdot \text{mol}^{-1}$$

$$\Delta_f G^0_{HCA} (g) = -54.9 \text{ kJ} \cdot \text{mol}^{-1}; \quad K_{HCAH} (25°C) = 3.96 \text{ L} \cdot \text{bar} \cdot \text{mol}^{-1}$$

$$\Delta_f G^0_{PCE} (g) = +20.5 \text{ kJ} \cdot \text{mol}^{-1}; \quad K_{PCEH} (25°C) = 27.9 \text{ L} \cdot \text{bar} \cdot \text{mol}^{-1}$$

P 12.2 *Investigating the Elimination Process of a Chemical in a Well-Mixed Reactor*

The behavior of a chemical is investigated in a well-mixed reactor (volume V, flow rate Q) by measuring the outflow concentration C_{out} at steady-state for different input concentrations C_{in}. The results are given in the table below. (a) Determine the order of the elimination process and formulate the differential equation which describes the chemical in the reactor. (b) How long does it take for the outflow concentration to drop from 40 mmol·L^{-1} to 2 mmol·L^{-1}, if at time t_1, C_{in} drops to zero instantaneously?

C_{in} /(mmol·L^{-1})	C_{out} /(mmol·L^{-1})
10	8
25	20
40	32
65	52

P 12.3 *Nonlinear Biodegradation in a Reactor*

A chemical is continuously introduced into a well-mixed reactor (input concentration C_{in}, discharge rate Q, reactor volume V). The chemical is biodegraded and also flushed into the outlet. The degradation rate can be approximated by the function:

$$\left(\frac{dC}{dt} \right)_{biodeg} = \begin{cases} -k_r C \left(1 - \dfrac{C}{C_{crit}} \right) & \text{if } 0 \leq C \leq C_{crit} \\ 0 & \text{if } C > C_{crit} \end{cases}$$

(a) Formulate the system's equation and calculate the steady-state concentration(s) of the system. (b) If there are several steady-states, to which steady-state does the system move? *Hint*: Make a qualitative analysis by just looking at the size of dC/dt.

Numbers (the parameters are constant):

$$V = 500 \text{ L}$$
$$Q = 100 \text{ L} \cdot \text{h}^{-1}$$
$$k_r = 0.5 \text{ h}^{-1}$$
$$C_{crit} = 100 \text{ mg} \cdot \text{L}^{-1}$$
$$C_{in} = 110 \text{ mg} \cdot \text{L}^{-1}$$

Further question: What happens with the system when the input concentration C_{in} is either increased to 150 mg·L^{-1} or lowered to 50 mg·L^{-1}?

Chapter 13

CHEMICAL TRANSFORMATIONS I: HYDROLYSIS AND REACTIONS INVOLVING OTHER NUCLEOPHILIC SPECIES

13.1 Introduction, Overview
Illustrative Example 13.1: *Evaluating the Thermodynamics of Hydrolysis Reactions*

13.2 Nucleophilic Substitution and Elimination at Saturated Carbon Atoms
Nucleophilic Displacement of Halogens at Saturated Carbon Atoms
Box 13.1: *The Concept of Hard and Soft Lewis Acids and Bases (HSAB)*
Illustrative Example 13.2: *Some More Reactions Involving Methyl Bromide*
Illustrative Example 13.3: *1,2-Dibromoethane in the Hypolimnion of the Lower Mystic Lake, Massachusetts*
Polyhalogenated Alkanes — Elimination Mechanisms

13.3 Hydrolytic Reactions of Carboxylic and Carbonic Acid Derivatives
Carboxylic Acid Esters
Illustrative Example 13.4: *Deriving Kinetic Parameters for Hydrolysis Reactions from Experimental Data*
Illustrative Example 13.5: *Calculating Hydrolysis Reaction Times as a Function of Temperature and pH*
Carboxylic Acid Amides
Carbamates
Quantitative Structure-Reactivity Considerations
Hammett Relationship
Illustrative Example 13.6: *Estimating Hydrolysis Rate Constants Using the Hammett Relationship*
Brønsted Relationship

13.4 Hydrolytic Reactions of Phosphoric and Thiophosphoric Acid Esters

13.5 Effects of Dissolved Metal Species and of Mineral Oxide Surfaces on Hydrolytic Reactions (*Advanced Topic*)
Effects of Dissolved Metal Species
Effects of Mineral Oxide Surfaces

13.6 Questions and Problems

13.1 Introduction, Overview

In Chapter 2 we noted that covalent bonds between two atoms of different electronegativity (e.g., carbon and halogens, carbon and oxygen, phosphorus and oxygen) are polar; that is, one of the atoms carries a partial positive charge (e.g., carbon, phosphorus), whereas the other one exhibits a partial negative charge (e.g., halogen, oxygen). In organic molecules, such a polar bond may become the site of a chemical reaction in that either a *nucleophilic* species (nucleus-liking and, hence, an electron-rich species) is attracted by the electron-deficient atom of the bond, or an *electrophilic* species (electron-liking and, hence, an electron-poor species) is attracted by the partial negative charge. In the environment, the majority of the chemical species that may chemically react with organic compounds are inorganic *nucleophiles* (see examples given in Table 13.1). Because of the large abundance of such nucleophiles in the environment (note that water itself is a nucleophile), reactive *electrophiles* are very short-lived. Therefore, reactions of organic compounds with electrophiles occur usually only in light-induced or biologically mediated processes (see Chapters 16 and 17), or in engineered systems (e.g., water treatment), where such species are added.

As can be derived from Table 13.1, nucleophilic species possess a partial or full negative charge and/or have nonbonded valence electrons. As a consequence of an encounter with an organic molecule exhibiting a polar bond, the electron-rich atom of the nucleophile may form a bond with the electron-deficient atom in the organic molecule, thus causing a modification of the organic compound. Since a new bond is formed by this process, another bond has to be broken at the atom at which the reaction occurs. This usually (but not always) means that a group (or atom) is split off from the organic compound. Such a group (or atom) is commonly referred to as a *leaving group*. As is illustrated by the examples given in Table 13.2 common leaving groups in organic chemicals include halides (reactions 1, 2, 4), alcohol moieties (reactions 5, 6, 7), and some more complex groups such as phosphates (reaction 3). We will address the factors that determine whether a particular part of a molecule is a good leaving group in detail later when discussing various types of reactions. At this point we just notice that, in general, a good leaving group is an entity that forms a stable species in aqueous solution. For example, if the leaving group is an anion (e.g., Cl^-, Br^-, ArO^-, ArS^-, RO^-), we can usually relate the ease with which it dissociates from the molecule with the ease with which its conjugate acid (e.g., HCl, HBr, HOR, HSR) dissociates in aqueous solution. This latter capability is expressed by the pK_{ia} of the acid. We recall from Chapter 8 that the pK_{ia} of an acid is a measure of the relative stability of its nondissociated versus dissociated form in water. Hence, for example, as we will see when discussing ester or carbamate hydrolysis (reactions 5 to 7 in Table 13.2), we can expect the conjugate bases of alcohols with low pK_{ia} values (e.g., *p*-nitrophenol, $pK_{ia} = 7.06$, reaction 6) to be better leaving groups than those whose conjugated acids have high pK_{ia}'s (e.g., ethanol, $pK_{ia} = 16$, reaction 5).

But let us go back to our considerations of environmentally relevant nucleophiles. Because of its great abundance, water plays a pivotal role among the nucleophiles present in the environment. A reaction in which a water molecule (or hydroxide ion) substitutes for another atom or group of atoms present in an organic molecule is commonly called a *hydrolysis* reaction. We note that in a hydrolysis reaction, the

Table 13.1 Examples of Important Environmenal Nucleophiles

increasing nucleophilicity for reaction at a saturated carbon

ClO_4^-
H_2O
NO_3^-
F^-
SO_4^{2-}, CH_3COO^-
Cl^-
HCO_3^-, HPO_3^{2-}
NO_2^-
PhO^{-a}, Br^-, OH^-
I^-, CN^-
HS^-, R_2NH^b
$S_2O_3^{2-}$, SO_3^{2-}, PhS^-

[a] $Ph = C_6H_5$ (phenyl)
[b] $R = CH_3$, C_2H_5

Table 13.2 Examples of Environmentally Relevant Chemical Reactions Involving Nucleophiles and/or Bases

Reactants	Products	Reaction Number

Nucleophilic Substitutions at Saturated Carbon Atoms

$CH_3Br + H_2O \longrightarrow CH_3OH + H^+ + Br^-$ (1)

Methyl bromide   Methanol

$CH_3Cl + HS^- \longrightarrow CH_3SH + Cl^-$ (2)

Methyl chloride   Methane thiol (Methyl mercaptan)

$CH_3O-\overset{\overset{O}{\|}}{P}(OCH_3)_2 + H_2O \longrightarrow CH_3OH + {}^-O-\overset{\overset{O}{\|}}{P}(OCH_3)_2 + H^+$ (3)

Trimethylphosphate   Methanol   Dimethylphosphate

β-Elimination

$Cl_2HC-CHCl_2 + HO^- \longrightarrow Cl_2C{=}CHCl + Cl^- + H_2O$ (4)

1,1,2,2-Tetrachloroethane   Trichloroethene

Ester Hydrolysis

$H_3C-\overset{\overset{O}{\|}}{C}-OCH_2CH_3 + H_2O \longrightarrow H_3C-\overset{\overset{O}{\|}}{C}-O^- + HO-CH_2CH_3 + H^+$ (5)

Ethyl acetate (Acetic acid ethylester)   Acetate   Ethanol

$(C_2H_5O)_2\overset{\overset{S}{\|}}{P}-O-\langle\rangle-NO_2 + HO^- \longrightarrow (C_2H_5O)_2\overset{\overset{S}{\|}}{P}-O^- + HO-\langle\rangle-NO_2$ (6)

Parathion   O,O-Diethyl-thiophosphoric acid   4-Nitrophenol

Carbamate Hydrolysis

$H_3CNH-\overset{\overset{O}{\|}}{C}-O-\langle\rangle + H_2O \longrightarrow CH_3NH_2 + CO_2 + HO-\langle\rangle$ (7)

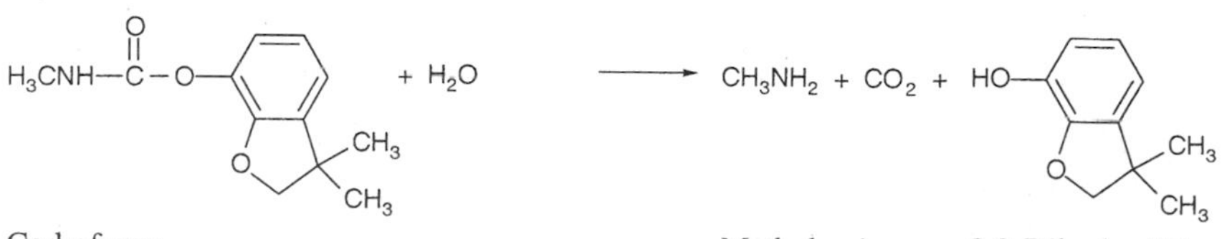

Carbofuran   Methylamine   2,3-Dihydro-3,3-dimethyl-7-benzo-furanol

compound is transformed into more polar products that have quite different properties. Therefore, the products have different environmental behaviors than the starting chemical. We also note that the products of hydrolysis are often of somewhat less environmental concern as compared to the parent compound. This is, however, not necessarily true for the products of reactions involving nucleophiles other than water or hydroxide ion. Examples of such nucleophiles include cyanide (CN^-, e.g., in hazardous waste sites) and, particularly, inorganic and organic reduced sulfur species (e.g., HS^-, S_n^{2-}, (n = 2 – 4), RS^-, ArS^-), which, as we will see later, are very potent nucleophiles that may be present at significant concentrations in anaerobic environments (see, e.g., Roberts et al. 1992; Miller et al., 1998).

In this chapter, we will address primarily mechanistic and kinetic aspects of reactions involving nucleophiles and/or bases (in the case of elimination reactions). We should, however, recall that, under certain conditions, for thermodynamic reasons, a reaction may not proceed spontaneously (see, e.g., Illustrative Example 13.1). For most *hydrolysis reactions* we may usually assume that *under ambient conditions of pH, reactant and product concentrations*, the reaction proceeds spontaneously and to an extent that, for practical purposes, we may consider it to be *irreversible*. This is shown by the calculations in Illustrative Example 13.1. The result of the first calculation (reaction 1) needs, however, some comments.

Illustrative Example 13.1   **Evaluating the Thermodynamics of Hydrolysis Reactions**

Problem

Consider the hydrolyses of methyl bromide (Reaction 1 in Table 13.2) and ethyl acetate (reaction 5 in Table 13.2) at 25°C and pH 7.0 in a contaminated groundwater containing 100 mM Cl^- and 1 mM Br^-. Assume that neither the pH nor the inorganic species concentrations change significantly during the reaction. Also assume that the activity coefficients of all species are 1. To what extent do the two compounds hydrolyze under these conditions?

Reaction 1: $CH_3Br + H_2O \rightarrow CH_3OH + H^+ + Br^-$

Reaction 5: $CH_3COOC_2H_5 + H_2O \rightarrow CH_3COO^- + HOCH_2CH_3 + H^+$

Answer

In the literature you find all the necessary $\Delta_f G^0(aq)$ values for the species involved in the two reactions. See Illustrative Example 12.1 for a detailed description of how to make the necessary calculations. The resulting standard free energies of reaction are:

Reaction 1: $\Delta_r G^0 = -28.4 \ kJ \cdot mol^{-1}$

Reaction 5: $\Delta_r G^0 = +19.0 \ kJ \cdot mol^{-1}$

Thus, at standard conditions (pH 0, all species at 1 M concentration), reaction 5 occurs in the opposite direction. In fact, carboxylic acid esters such as ethyl acetate may be synthesized under acidic conditions in water/alcohol mixtures. However,

Species (all aqueous)	$\Delta_f G_i^0(aq)$ ($kJ \cdot mol^{-1}$)
CH_3Br	– 13.8
CH_3OH	– 175.4
$CH_3COOC_2H_5$	– 332.5
CH_3COO^-	– 369.1
C_2H_5OH	– 181.6
$H_2O (\ell)$	– 237.2
Br^-	– 104.0
H^+	0

under the conditions prevailing in the groundwater, both reactions exhibit negative $\Delta_r G$ values (very small Q_r, see Eq. 12-8) and thus proceed spontaneously to the right. The resulting product-reactant ratios *at equilibrium* are:

Reaction 1: $\ln \dfrac{[CH_3OH]}{[CH_3Br]} = 34.5$ or $\dfrac{[CH_3OH]}{[CH_3Br]} = 9.6 \times 10^{14}$

Reaction 5: $\ln \dfrac{[CH_3COO^-][C_2H_5OH]}{[CH_3COOC_2H_5]} = 8.4$ or $\dfrac{[CH_3COO^-][C_2H_5OH]}{[CH_3COOC_2H_5]} = 4.7 \times 10^3$

When comparing the hydrolysis of methyl bromide with its reaction with Cl⁻ under the same conditions (i.e., [Cl⁻] = 100 mM, see Illustrative Example 13.2), we see that from a thermodynamic point of view, the hydrolysis reaction is heavily favored (compare $\Delta_r G^0$ values). This does not mean that the methyl bromide present is primarily transformed into methanol instead of methyl chloride (which it would be, if the reaction were to be *thermodynamically* controlled). In fact, in this and all other cases discussed in this chapter, we will assume that the reactions considered will be *kinetically* controlled; that is, the relative importance of the various transformation pathways of a given compound will be determined by the relative reaction rates and not by the respective $\Delta_r G^0$ values. Thus, in our example, because Cl⁻ is about a 10^3 times better nucleophile as compared to water (see Section 13.2) and because its concentration is about 10^3 times smaller than that of water (0.05 M versus 55.3 M), the two reactions would be of about equal importance under the conditions prevailing in this groundwater. Note that the product methyl chloride would subsequently also hydrolyze to yield methanol, though at a much slower rate. We will come back to this problem in Section 13.2 (Illustrative Example 13.2).

In the following, we will choose sets of compounds exhibiting various types of structures to introduce and discuss some general fundamental reaction mechanisms with which we need to become familiar in order to be able to assess chemical reactivities of organic compounds in the environment. We will also use these examples to discuss some approaches taken to derive quantitative structure-reactivity relationships for a given type of reaction. We start out with nucleophilic substitutions and -elimination reactions using simple halogenated hydrocarbons as model compounds (Section 13.2). We then turn our attention to the hydrolysis of carboxylic acid esters which will help us to get acquainted with the structural features that control the reactivities of a variety of other carboxylic and carbonic acid derivatives (Section 13.3). In the next step, using the combined knowledge of Sections 13.2 and 13.3 we consider reactions of phosphoric and thiophosphoric acid derivatives (Section 13.4). In all our discussions in this chapter, our main focus will be on reactions in *homogeneous aqueous solution* (called *homogeneous reactions* because we assume the medium is the same throughout its volume). We will, however, also address some heterogeneous reactions (e.g., at solid surfaces), although there is still rather little data available to derive general rules for describing such processes (Section 13.5).

13.2 Nucleophilic Substitution and Elimination at Saturated Carbon Atoms

Nucleophilic Displacement of Halogens at Saturated Carbon Atoms

With our first example of chemical reactions, we want to get acquainted with a very important type of reaction in organic chemistry, that is, with *nucleophilic substitution at a saturated* carbon atom. Since halogens are very common constituents of man-made organic chemicals, we consider their displacement by environmentally relevant nucleophiles. In these cases the halogen plays the role of the leaving group.

To describe aliphatic nucleophilic substitution reactions, it is useful to consider two different reaction mechanisms representing two extreme cases (Figs. 13.1 and 13.2). In the first case shown in Fig. 13.1, we picture the reaction to occur because a nucleophile (e.g., Nu$^-$) "attacks" the carbon atom from the side opposite to the leaving group, L$^-$ (e.g., halide). In the transition state, which is postulated to exhibit a trigonal bipyramidal geometry, the nucleophile is then thought to be partly bound to the carbon atom, and the leaving group is postulated to be partly dissociated. Hence, in such a simple picture, we consider the nucleophile to push the leaving group out of the molecule.

S_N2 *Mechanism.* In this first case, the standard free energy of activation $\Delta^\ddagger G^0$, and thus the rate of the reaction, depends strongly on both the capability of the nucleophile to initiate a substitution reaction and the willingness of the organic molecule to undergo that reaction. The former factor may be expressed by the relative nucleophilicity of the nucleophile, an entity that can be quantified (see discussion below). The latter contribution to $\Delta^\ddagger G^0$, however, is more difficult to quantify since it incorpo-

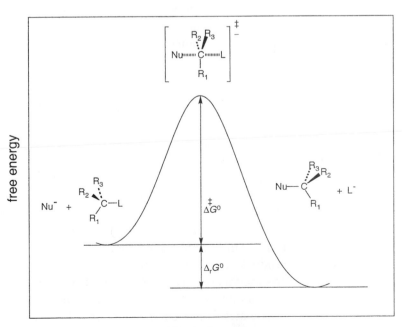

Figure 13.1 Two-dimensional portrayal of relative free energies exhibited by the reactants, activated complex, and products of an S_N2 reaction.

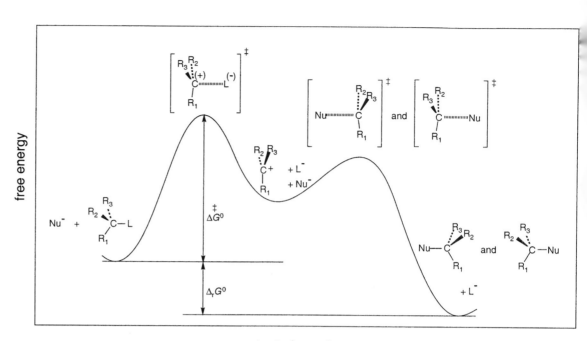

Figure 13.2 Two-dimensional portrayal of the relative free energies exhibited by reactants, activated complex, an intermediate, and product of an S_N1 reaction.

rates various electronic and steric factors that are strongly determined by the structure of the organic molecule. As we can imagine, $\Delta^\ddagger G^0$ depends upon the facility with which the nucleophile can get to the site of reaction (i.e., how much steric hindrance there is), upon the charge distribution at the reaction center, and upon how easily the leaving group will split from the molecule.

If a reaction occurs by this first mechanism, it is commonly termed an S_N2 *reaction (i.e., substitution, nucleophilic, bimolecular)*. It represents an example of a simple elementary bimolecular reaction, as we discussed in Section 12.3, and it therefore follows a second-order kinetic rate law:

$$\frac{d[R_1R_2R_3C-L]}{dt} = -k[Nu^-][R_1R_2R_3C-L] \qquad (13\text{-}1)$$

where k is a second-order rate constant (e.g., $M^{-1}\,s^{-1}$).

S_N1 Mechanism. A second mechanism, differing substantially from the first, is one in which we postulate that the substitution reaction occurs in two steps. As illustrated by Fig. 13.2, in the first (rate-determining) step, the leaving group is completely dissociated from the organic compound. Because the leaving group takes both electrons with it, a (planar) carbocation is formed as an *intermediate* (i.e., a temporary product existing in an energy minimum along the reaction path). In the second, faster step, this reactive carbocation then combines with a nucleophile to form a product. In this case, the reaction rate depends solely on how easily the leaving group dissociates from the molecule. Since the "structure" of the activated complex can be assumed to resemble the structure of the intermediate (see Fig. 13.2), an important factor determining $\Delta^\ddagger G^0$ is the stability of the carbocation formed. Hence this mecha-

nism is favored in cases where the carbocation is stabilized, for example, by resonance.

If a reaction occurs exclusively by this second mechanism, the observed rate law is first-order:

$$\frac{d[R_1R_2R_3C-L]}{dt} = -k[R_1R_2R_3C-L] \tag{13-2}$$

where k is now a first-order rate constant (e.g., s^{-1}). The reaction is then said to occur by an S_N1 *(i.e., substitution, nucleophilic, unimolecular) mechanism.* Note that in aqueous solution, the S_N1 mechanism will, in general, strongly favor the formation of the hydrolysis product (i.e., substitution of $-L$ by $-OH$) because the nucleophiles are not involved in the rate-limiting step, and water molecules are present in such overwhelming abundance that one of them will have the greatest likelihood of colliding with the reactive carbocation.

The Relative Effectiveness of Nucleophiles to Displace Leaving Groups. Let us first look at some reactions that occur predominantly by an S_N2 mechanism. This will allow us to evaluate the relative nucleophilicities of some important environmental nucleophiles. We consider nucleophilic substitution in aqueous solution of methyl halides (CH$_3$L, L = F, Cl, Br, I), which are important volatile compounds in the marine and freshwater environments (Zafiriou, 1975; Pearson, 1982a). In Fig. 13.3, the second-order rate constants at 25°C for the reactions of the methyl halides with various nucleophiles are plotted for each L. From these data, we may derive two important, quite generally applicable, conclusions. First, we recognize that a given methyl halide shows the same relative reactivity toward the various nucleophiles as the other methyl halides. These findings were first quantified in a linear free energy relationship by Swain and Scott (1953):

$$\log\left(\frac{k_{Nu}}{k_{H_2O}}\right) = s \cdot n_{Nu,CH_3Br} \tag{13-3}$$

where k_{Nu} is the *second-order* rate constant for a nucleophilic displacement by a nucleophile of interest, k_{H_2O} is the *second-order* rate constant for nucleophilic attack by water (the standard nucleophile), n is a measure of the attacking aptitude or nucleophilicity of the nucleophile of interest, and s reflects the sensitivity of the organic molecule to nucleophilic attack. The n values of some important environmental nucleophiles determined for the reaction with *methyl bromide* (CH$_3$Br) in aqueous solution are given in Table 13.3. It should be pointed out that currently, particularly in pharmaceutical, toxicological, and basic chemical applications (Hansch and Leo, 1995), the relative nucleophilicity of inorganic and organic nucleophiles is quantified using another reference reaction, substitution of *methyl iodide* (CH$_3$I) in *methanol*:

$$\log\left(\frac{k_{Nu}}{k_{CH_3OH}}\right) = s' \cdot n_{Nu,CH_3I} \tag{13-4}$$

Some $n_{Nu,CH3I}$ values are given in Table 13.4. Note that in this case, $n_{Nu,CH3I}$ is by definition set to zero for methanol (and not for water as in Eq. 13-3). Eq. 13-4 has the

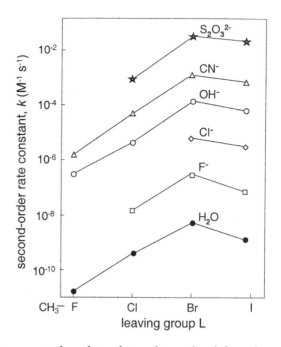

Figure 13.3 Rate constants for reactions of methyl halides with various nucleophiles (data from Hughes, 1971; Mabey and Mill, 1978).

Table 13.3 Relative Nucleophilicities of Some Important Environmental Nucleophiles: n-Values Determined from the Reaction with Methyl Bromide or n-Hexyl Bromide in Water (Eq. 13-3, $s = 1$)

Nucleophile	$n_{\mathrm{Nu,CH_3Br}}$ [a]
ClO_4^-	<0
H_2O	0
NO_3^-	1.0
F^-	2.0
SO_4^{2-}	2.5
CH_3COO^-	2.7
Cl^-	3.0
HCO_3^-, HPO_4^{2-}	3.8
Br^-	3.9
OH^-	4.2
I^-	5.0
CN^-, HS^-	5.1
$S_2O_3^{2-}$	6.1 [b]
PhS^-	6.8 [b]
S_4^{2-}	7.2 [b]

[a] Data from Hine (1962). [b] Data from Haag and Mill (1988a).

advantage that $n_{\mathrm{Nu,CH_3I}}$ values have been determined for a large number of *organic* nucleophiles; it has the disadvantage that the special properties of the solvent water are not taken into account and that the n-values are not scaled to water acting as nucleophile. Therefore, for quantification of S_N2 reaction rates in aqueous media, we prefer to use Eq. 13-3. Nevertheless, as is evident from Fig. 13.4, since the n-values in the two systems more or less parallel one another, $n_{\mathrm{Nu,CH_3I}}$ values can be used to get a rough estimate of $n_{\mathrm{Nu,CH_3Br}}$ values of nucleophiles for which such values are not available. As a crude approximation, $n_{\mathrm{Nu,CH_3Br}}$ corresponds to about 2/3 of the value of $n_{\mathrm{Nu,CH_3I}}$. Thus, from some of the $n_{\mathrm{Nu,CH_3I}}$ values given in Table 13.4, we can conclude that certain organic nucleophiles (pyridine, $PhNH_2$, PhO^-, $(C_2H_5)_2NH$) that are representatives for constituents of natural organic matter (NOM, see Chapter 9) will have $n_{\mathrm{Nu,CH_3Br}}$ values between 3.5 and 5. This means, for example, that certain amino groups present in NOM are up to 10^5 times stronger nucleophiles than water when considering the reaction with methyl bromide. In fact, Gan et al. (1994) have postulated that S_N2 reactions of methyl bromide with such NOM constituents are important sinks for this widely used fumigant in soils. In addition, Table 13.4 shows that azide (N_3^-), which is very often used to inhibit microbial activity in environmental samples or experimental systems, is about as strong a nucleophile as Br^-. Hence, when using this poison in large concentrations, one has to be cautious not to induce unwanted S_N2 reactions in a given sample or system. Similar considerations should be made when using concentrated phosphate buffers in experimental systems. Finally, we note again that reduced inorganic and organic sulfur species are the most potent nucleophiles present in the environment.

The major factors influencing the nucleophilicity of a chemical species (and thus the magnitude of its n value) are the ease with which the nucleophile can leave the solution to get to the reaction center and the ability of the bonding atom to donate its electrons to form the transition state. Hence, nucleophilicity increases with decreasing solvation energy of the nucleophile. Since the valence electrons of larger atoms (e.g.,

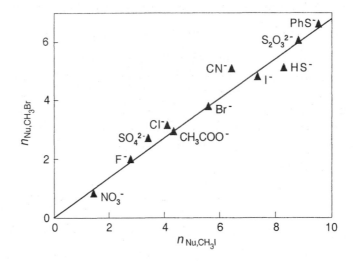

Figure 13.4 Plot of n_{Nu,CH_3Br} versus n_{Nu,CH_3I}, values for a series of nucleophiles (see Tables 13.3 and 13.4). A linear regression calculation yields the relationship: $n_{Nu,CH_3Br} = 0.68\, n_{Nu,CH_3I}$ ($R^2 = 0.98$).

Table 13.4 Relative Nucleophilicities of Some Important Environmental Nucleophiles: n-Values Determined From the Reaction with Methyl Iodide in Methanol (Eq. 13-4, $s' = 1$)

Nucleophile	n_{Nu,CH_3I} [a]
CH_3OH	0
NO_3^-	~1.5
F^-	~2.7
SO_4^{2-}	3.5
HCO_3^{2-}, HPO_4^{2-}	3.8
CH_3COO^-	4.4
Cl^-	4.4
Pyridine	5.2
$PhNH_2$	5.7
PhO^-	5.8
Br^-, N_3^-	5.8
CN^-	6.7
$(C_2H_5)_2NH$	~7.0
I^-	7.4
HS^-	~8
$S_2O_3^{2-}$	8.9
PhS^-	9.9

[a] Data from Pearson et al. (1968).

S, I) are more polarizable (they are further away from the nucleus), and since larger "soft" species have, in general, lower solvation energies, they are better nucleophiles as compared to smaller "hard" species. (For definition of the term "hard" and "soft" nucleophiles, see Box 13.1.) Thus, we can qualitatively understand why, for example, nucleophilicity increases from F^- to Cl^- to Br^- to I^- (Table 13.3), and why HS^- is a stronger nucleophile than OH^-.

As a standard for the sensitivity values, s in Eq. 13-3 is set equal to 1.0 for S_N2 reactions of methyl bromide. Hence, a compound for which the reaction rate of nucleophilic substitution is more dependent than methyl bromide on the nucleophilicity of the attacking group will have an s value greater than 1, and one that is less dependent will have a smaller value. It is important to be aware that, when we use Eq. 13-3 to evaluate and/or predict rates of S_N2 reactions involving different organic substrates, we assume that the relative nucleophilicities of the nucleophiles (expressed by n_{Nu,CH_3Br}) remain the same. Hence, we lump all the differences which are due to the types of leaving groups and other structural characteristics influencing the reactions into the sensitivity value s (which might not always be correct). Furthermore, in principle, the n_{Nu,CH_3Br} values (as well as s) are valid only for a given temperature (i.e., 25°C). However, when considering that the activation energies of various S_N2 reactions do not vary that much for an order of magnitude assessment of reaction rates (for which the Swain-Scott model can be used), we may assume n_{Nu,CH_3Br} as well as s to be temperature independent.

From the few examples available, it is difficult to derive general rules for a clear assessment of the structural factors that determine the s values for a given compound. As a general trend it can be observed that, at least in "simple" molecules, leaving groups exhibiting a "hard" heteroatom (e.g., O, Cl) yield s-values that are somewhat smaller than 1 [e.g., trimethylphosphate ($s = 0.90$); dimethylsulfate ($s = 0.83$); ethylene oxide ($s = 0.96$); benzylchloride (Fig. 12.1, $s = 0.86$); all values from Hansch and Leo, 1995]. On the other hand, in the case of softer leaving groups s values greater than 1 have been found [e.g., methyl iodide ($s = 1.2$)]. A similar trend has been observed by Roberts et al. (1992), who found increasing ratios of the second-order

Box 13.1 The Concept of Hard and Soft Lewis Acids and Bases (HSAB)

Pearson (1963) proposed dividing Lewis acids (i.e., electron acceptors, "electrophiles") and Lewis bases (i.e., electron donors, "nucleophiles") into different categories referred to as "hard" and "soft." *Hard* acids and bases are relatively small, exhibit a high electronegativity, and have a low polarizability. *Soft* acids and bases encompass those species that are relatively large, and of low electronegativity and high polarizability. Consequently, "hardness" can be associated with a relatively large amount of ionic character and "softness" with a large amount of covalent character in a bond or activated complex. Obviously, in such a scheme, "hard" and "soft" are not absolute, but gradually varying qualities when comparing different chemical species. Nevertheless, for a qualitative assessment of the relative importance of reactions of Lewis acids and Lewis bases, the so-called "HSAB" (hard and soft acid–base) rules suggested by Pearson (1963) may be quite useful.

Rule 1: *Equilibrium*. Hard acids prefer to associate with hard bases and soft acids associate with soft bases.

Rule 2: *Kinetics*. Hard acids react readily with hard bases and soft acids with soft bases.

Environmentally relevant nucleophiles (Lewis bases) may be classified according to Pearson's HSAB principle, as "hard," "soft," or borderline (possessing intermediate hard/soft character) as follows (Larson and Weber, 1994):

Hard: $OH^-, H_2PO_4^-, HCO_3^-, NO_3^-, SO_4^{2-}, F^-, Cl^-, NH_3, CH_3OO^-$ ...

Borderline: $H_2O, SO_3^{2-}, Br^-, C_6H_5NH_2$ ...

Soft: $HS^-, S_n^{2-}, RS^-, PhS^-, S_2O_3^{2-}, I^-, CN^-$ ...

Similarly, relevant electrophiles (Lewis acids) including A-type metal cations (hard), bivalent transition metal ions (borderline), and B-type metal ions (soft) can be categorized (see Stumm and Morgan 1996). Note that in organic molecules, the atom where a nucleophile attacks (i.e., the electrophilic site) may possess harder (e.g., C=O, P=O) or softer (e.g., CH_3-X) character.

---

rate constants, k_{HS^-}/k_{OH^-}, for the reaction of CH_2Cl_2, CH_2BrCl, and CH_2Br_2 with HS^- and OH^-, respectively. Note that in this sequence the halogen substituents become "softer."

The type of leaving group is, of course, not the only factor determining s. For example, if other groups are located near the reaction site and may strongly interact (electronically and/or sterically) with a nucleophile in the transition state, then s values very different from 1 may be obtained. Furthermore, the Swain-Scott model (Eq. 13-3) may only be poorly applicable to a broad range of nucleophiles. Such a case has been reported by Lippa (Lippa, 2002), who have investigated the reactions of some chloroacetamide herbicides (i.e., alachlor and propachlor, see margin for structure) with a series of strong nucleophiles including Br^-, SCN^-, OH^-, N_3^-, I^-, HS^-, $S_2O_3^{2-}$, PhS^-, and S_n^{2-}. As has been demonstrated by Stamper et al. (1997) with HS^- as the nucleophile, the reaction takes place at the carbon carrying the chlorine atom. Within the range of nucleophilicities studied by Lippa and ((i.e., ~ 4 < $n_{Nu,CH3Br}$ < ~ 7, see Table 13.3), s

H₃C

alachlor

propachlor

values of about 1.6 were obtained for both compounds. Finally, s values very different from 1 and/or poor correlation between log k_{Nu} and $n_{Nu,CH3Br}$ may be found for compounds that do not react primarily by an S_N2 mechanism.

Let us now estimate at what approximate concentration a given nucleophile must be present in a natural water in order to compete with H_2O in an S_N2 reaction with a simple alkyl halide (i.e., CH_3L, L = Cl, Br, I). For simplicity we set $s = 1$. For the nucleophiles listed in Table 13.3, the calculated concentrations, $[Nu]_{50\%}$, at which the two reactions are equally important ($k_{Nu} [Nu]_{50\%} = k_{H2O} [H_2O]$) are:

$$[Nu]_{50\%} = 55.3 \times 10^{-n_{Nu,CH3Br}} \tag{13-5}$$

Depending on the relative nucleophilicities, $[Nu]_{50\%}$ ranges from micromolar to molar concentrations (Table 13.5). Although these values represent only order-of-magnitude estimates, they allow some important conclusions. First, in uncontaminated freshwaters (where bicarbonate typically occurs at about 10^{-3} M, chloride and sulfate occur at about 10^{-4} M, and hydroxide is micromolar or less, Stumm and Morgan, 1996), the concentrations of nucleophiles are usually too small to compete successfully with water in S_N2 reactions involving aliphatic halides. Hence the major reaction will be the displacement of the halide by water molecules. In salty or contaminated waters, however, nucleophilic substitution reactions other than hydrolysis may occur. Zafiriou (1975), for example, has demonstrated that in seawater ($[Cl^-] \approx 0.5$ M) an important sink for methyl iodide is transformation to methyl chloride:

$$CH_3I + Cl^- \rightarrow CH_3Cl + I^- \tag{13-6}$$

The half-life with respect to chemical transformation of CH_3I in seawater at 20°C was determined to be 20 days, as compared to about 200 days in freshwater (reaction with H_2O yielding CH_3OH). In a case of a groundwater contamination with several alkyl bromides, Schwarzenbach et al. (1985) reported the formation of dialkyl sulfides under sulfate-reducing conditions in an aquifer. They postulated that in an initial reaction, primary alkyl bromides reacted with HS^- by an S_N2 mechanism to yield the corresponding mercaptans (thiols):

$$RCH_2Br + HS^- \rightarrow RCH_2SH + Br^- \tag{13-7}$$

These mercaptans then reacted further to yield rather hazardous products. We return to this case later. Some additional applications of the Swain-Scott model are given in Illustrative Examples 13.2 and 13.3.

A further conclusion that we may draw from Table 13.5 is that the S_N2 reactions of aliphatic halides with OH^- should be unimportant at pH values below about 10. Since the hydrolysis of a carbon–halogen bond is commonly not catalyzed by acids, one can assume that in most cases, the hydrolysis rate of aliphatic halides will be *independent of pH* at typical ambient conditions. Hence, regardless of whether hydrolysis occurs by an S_N1 or S_N2 mechanism (or a mixture of both, see below), the reaction may be described by a first-order rate law. The first-order rate constant is then commonly denoted as k_N ($= k_{H2O} \cdot [H_2O]$) to express neutral hydrolysis. Note that if the

Table 13.5 Calculated Concentration of Nucleophile Required to Compete with Water in an S_N2 Reaction with Alkyl Halides Assuming an s Value of 1

Nucleophile	$[Nu]_{50\%}{}^a$ (M)
NO_3^-	~6
F^-	~6 × 10^{-1}
SO_4^{2-}	~2 × 10^{-1}
Cl^-	~6 × 10^{-2}
HCO_3^-	~9 × 10^{-3}
HPO_4^-	~9 × 10^{-3}
Br^-	~7 × 10^{-3}
OH^-	~4 × 10^{-3}
I^-	~6 × 10^{-4}
HS^-	~4 × 10^{-4}
CN^-	~4 × 10^{-4}
$S_2O_3^{2-}$	~4 × 10^{-5}
S_4^{2-}	~4 × 10^{-6}

^{a} Eq. 13-5 using the $n_{Nu,CH3Br}$ values given in Table 13.3.

Illustrative Example 13.2   **Some More Reactions Involving Methyl Bromide**

Problem

Estimate the half-life in days (with respect to chemical transformation) of methyl bromide (CH_3Br) present at low concentration (i.e., < 1 mM) in a homogeneous aqueous solution (pH = 7.0, T = 25°C) containing 100 mM Cl^-, 2 mM NO_3^-, 1 mM HCO_3^-, and 0.1 mM CN^-. In pure water at pH 7.0 and 25°C, the half-life of CH_3Br is about 20 days.

Answer

Since all nucleophiles are present in excess concentrations (i.e., $\gg [CH_3Br]_0$), the reaction of CH_3Br can be expressed by a pseudo-first-order law with a pseudo-first-order rate constant, k_{obs}, that is given by:

$$k_{obs} = k_{H_2O}[H_2O] + \sum_j k_{Nu_j}[Nu_j] \tag{1}$$

Inspection of Table 13.5 shows that the reactions with NO_3^- and OH^- can be neglected. For estimation of the rate constants for the reactions with the other nucleophiles, use the rearranged form of Eq. 13.3 with $s = 1$:

$$k_{Nu} = k_{H_2O} \cdot 10^{n_{Nu,CH_3Br}} \tag{2}$$

Insert $n_{Nu,CH3Br}$ values from Table 13.3 into Eq. 2, and substitute k_{Nu} for each nucleophile j into Eq. 1:

$$k_{obs} = k_{H_2O}\left\{[H_2O] + 10^3[Cl^-] + 10^{3.8}[HCO_3^-] + 10^{5.1}[CN^-]\right\} \tag{3}$$

Insertion of the concentrations of the various nucleophiles into Eq. 3 then yields:

$$k_{obs} = k_{H_2O}(55.5 + 100 + 6.3 + 12.6) = k_{H_2O}(174.4)$$

This calculation shows that the reaction with chloride is about twice as important as the neutral hydrolysis, while the reactions with the other two nucleophiles only make up about 10% of the overall transformation rate of CH_3Br. Note that, in some cases, a minor reaction might still be important because a more persistent toxic product may be formed (in this case acetonitrile CH_3CN). Since in pure water:

$$t_{1/2} = \frac{\ln 2}{k_N} = \frac{\ln 2}{k_{H_2O}[H_2O]}$$

you may use the reported hydrolysis half-life (20d) to find:

$$k_{H_2O} = \frac{0.693}{(55.3 \text{ M})(20\text{d})} = 6.3 \times 10^{-4} \text{M}^{-1}\text{d}^{-1}$$

Therefore, the half-life of CH_3Br in the aqueous solution is:

$$t_{1/2} = \frac{\ln 2}{k_{obs}} = \frac{0.693}{(6.3 \times 10^{-4} \text{M}^{-1}\text{d}^{-1})(174.4 \text{ M})} = \sim 6 \text{ d}$$

Illustrative Example 13.3

1,2-Dibromoethane in the Hypolimnion of the Lower Mystic Lake, Massachusetts

Problem

Various studies suggest that in pure water, the major transformation reaction of the widely used pesticide, 1,2-dibromoethane (1,2-DBE), is neutral hydrolysis to yield the final product ethylene glycol (Roberts et al., 1993). Based on measurements at high temperatures, Jeffers and Wolfe (1996) have estimated a hydrolysis half-life of 6.4 years for 1,2-DBE at 25°C, corresponding to a k_N value of 3.5×10^{-9} s^{-1}. The reported Arrhenius activation energy for this reaction is: $E_a =$ 108 kJ $\cdot$ mol^{-1}. Estimate how large the concentration of $\Sigma[S_n^{2-}](= [^-S-S^-] + [^-S-S-S^-] + [^-S-S-S-S^-])$ species expressed as $[S_n^{2-}]$ would have to be in the anoxic hypolimnion of Mystic Lake, Massachusetts at 10°C (see water composition given in margin) in order to lower the half-life of 1,2-DBE by a factor of 100 as compared to the half-life determined by hydrolysis alone. Compare this calculated concentration with the actual measured concentration of S_n^{2-} given below. Assume that the initial reaction with the reduced sulfur species (HS^-, S_n^{2-}) present is an S_N2 reaction at one of the carbon atoms and not reductive debromination (a process that we will discuss in Chapter 14). What products would you expect from the reaction of 1,2-DBE with the polysulfide species?

$$BrCH_2 - CH_2Br$$

1,2-Dibromoethane
(1,2-DBE)

Lake Water Chemistry for Anoxic Hypolimnion of the Lower Mystic Lake (Miller et al., 1998):

pH = 6.8
$[Cl^-] = 0.4$ M
$[HS^-] = 3 \times 10^{-3}$ M
$[S_n^{2-}] = 9 \times 10^{-5}$ M

Answer

With $E_a = 108$ kJ $\cdot$ mol^{-1}, k_N at 10°C will be about 10% of that at 25°C (see Table D1 Appendix D). Hence, $k_N = 3.5 \times 10^{-10}$ s^{-1}, corresponding to a hydrolysis half-life of 64 years. To calculate the required $[S_n^{2-}]$ concentration to reach a half-life of 0.64 years or a k_{obs} of 3.5×10^{-8} s^{-1}, write an equation analogous to Eq. 3 in Illustrative Example 13.2, by using the n_{Nu,CH_3Br} value of S_4^{2-} for $[S_n^{2-}]$:

$$k_{obs} = 3.5 \times 10^{-8} \text{ s}^{-1} = k_{H_2O}\{[H_2O] + 10^3[Cl^-] + 10^{5.1}[HS^-] + 10^{7.2}[S_n^{2-}]\} \quad (1)$$

Division of both sides of Eq. 1 by k_{H_2O} ($= 3.5 \times 10^{-10}$ s^{-1}/55.3 M) and insertion of the concentrations reported for Cl$^-$ and HS$^-$ in the hypolimnion of Mystic Lake yields:

$$5550 \text{ M} = 55.3 \text{ M} + 400 \text{ M} + 378 \text{ M} + 10^{7.2} [S_n^{2-}] \text{ M}$$

which yields a required S_n^{2-} concentration of:

$$[S_n^{2-}] = 3 \times 10^{-4} \text{ M}$$

This is only a factor 3 to 4 higher than the concentration of such species calculated by Miller et al. (1998) for the hypolimnion of the lower Mystic Lake. In fact, for certain environments (e.g., in salt marsh pore water) substantially higher concentrations of polysulfides have been reported (Lippa, 2002). Hence, such a calculation shows the importance of such species in sulfur-rich environments.

As has been found by Schwarzenbach et al. (1985) in a contaminated aquifer under sulfate-reducing conditions, reactions of alkyl dihalides where the halides are not bound to the same carbon atom may lead to cyclic polysulfides. The most probable mechanism is an initial substitution of one of the halides by S_n^{2-}, followed by an *intramolecular substitution* of the second halide (a so-called S_Ni reaction). Thus, for 1,2-DBE one can formulate the reaction as follows:

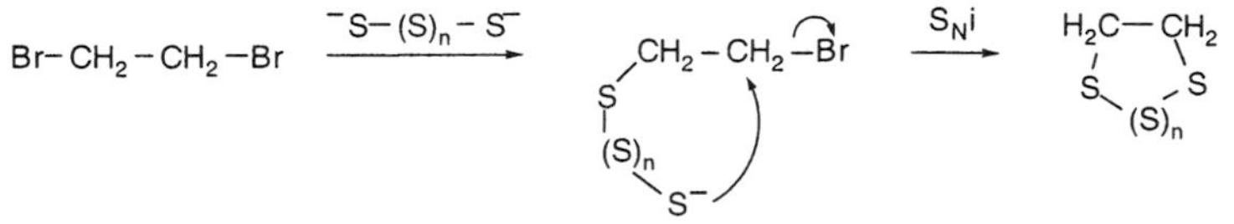

Hence, the most likely products are ethylene, di-, tri-, and tetrasulfide:

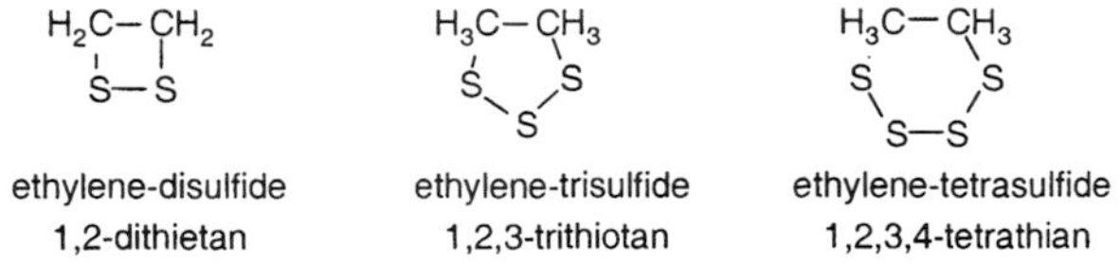

| ethylene-disulfide | ethylene-trisulfide | ethylene-tetrasulfide |
| 1,2-dithietan | 1,2,3-trithiotan | 1,2,3,4-tetrathian |

rate of transformation of a given halogenated compound is found to be pH-dependent in the ambient pH-range (i.e., pH 5–9), this is an indication that the compound also reacts by one or several other mechanism(s) (e.g., β-elimination, see below).

Leaving Groups. We now return to Fig. 13.3 to learn something about the various halogens as leaving groups. It is tempting to assume that the weaker a nucleophile (i.e., the smaller its $n_{\mathrm{Nu,CH3Br}}$ value, see Table 13.3), the better leaving group it should be. Hence we would expect the reactivities of the methyl halides to decrease in the order $CH_3F > CH_3Cl > CH_3Br > CH_3I$. However, what is experimentally found (Fig. 13.3) is almost the opposite, namely, the reaction rate decreases in the order $CH_3Br \sim CH_3I > CH_3Cl > CH_3F$. The major reason for these findings is the increasing strength of the C–X bond (that has to be broken) when going from C–I to C–F (Table 2.2). This bond-strength factor proves to be dominant in determining the much slower reaction rates of C–Cl and, in particular, C–F bonds as compared to C–Br and C–I.

Let us now look at some examples to illustrate what we have discussed so far to get a feeling of how structural moieties influence the mechanisms, and to see some rates of nucleophilic substitution reactions of halogenated hydrocarbons in the environment. Table 13.6 summarizes the (neutral) hydrolysis half-lives of various mono-halogenated compounds at 25°C. We can see that, as anticipated, for a given type of compound, the carbon–bromine and carbon–iodine bonds hydrolyze fastest, about 1–2 orders of magnitude faster than the carbon–chlorine bond. Furthermore, we note that for the compounds of interest to us, S_N1 or S_N2 hydrolysis of carbon–fluorine bonds is likely to be too slow to be of great environmental significance.

When comparing the hydrolysis half-lives of the alkyl halides in Table 13.6, we notice that the reaction rates increase dramatically when going from primary to secondary to tertiary carbon–halogen bonds. In this series, increasing the stabilization

Table 13.6 Hydrolysis Half-Lives and Postulated Reaction Mechanisms at 25°C of Some Monohalogenated Hydrocarbons at Neutral pH [a]

Compound	Type of Carbon to Which L is Attached	$t_{1/2}$(Hydrolysis)				Dominant Mechanism(s) in Nucleophilic Substitution Reactions
		L = F	Cl	Br	I	
R—CH$_2$—L	primary	≈30 yr [b]	340 d [b]	20–40 d [c]	50–110 d [d]	S_N2
H$_3$C$\diagdown$CH—L / H$_3$C	secondary		38 d	2 d	3 d	$S_N2 \ldots S_N1$
CH$_3$ / H$_3$C—$\vert$—L / CH$_3$	tertiary	50 d	23 s			S_N1
CH$_2$=CH—CH$_2$—L	allyl		69 d	0.5 d	2 d	$S_N2 \ldots S_N1$
⬡—CH$_2$—L	benzyl		15 h	0.4 h		$S_N2 \ldots S_N1$

[a] Data taken from Robertson (1969) and Mabey and Mill (1978). [b] R = H. [c] R = H, C$_1$ to C$_5$-n-alkyl. [d] R = H, CH$_3$.

of the carbocation by the electron-donating methyl groups decreases the activation energy needed to form this intermediate, thereby shifting the reaction to an increasingly S_N1-like mechanism. Similarly, faster hydrolysis rates and increasing S_N1 character can be expected if stabilization is possible by resonance with a double bond or an aromatic ring. As indicated by the denotation $S_N2 \ldots S_N1$ in Table 13.6, it is in some cases not possible (nor feasible) to assign a strict S_N2 or S_N1 character to a given nucleophilic substitution reaction. We recall that we refer to an S_N2 mechanism if the nucleophile plays the most important role it can play in the nucleophilic substitution reaction. In the other extreme, in the S_N1 case, the nucleophile is not relevant at all for determining the reaction rate.

It is now easy to imagine that depending on the nucleophile and on various steric (e.g., steric hindrance) and electronic (e.g., stabilization by conjugation) factors, the relative importance of the nucleophile may well lie somewhere in between these two extremes. We may, therefore, simply look at such cases as exhibiting properties intermediate between S_N1 and S_N2 mechanisms.

With respect to possible product formation, we have seen that other nucleophiles may compete with water only if they are present at appreciable concentrations (see Table 13.5) and if the reaction occurs by an S_N2-like mechanism. An interesting example illustrating the above-mentioned intermediate situation is the previously mentioned case study of a groundwater contamination by primary and secondary alkyl bromides. In this case, among other compounds, a series of short-chain alkyl bromides (Fig. 13.5) were introduced continuously into the ground by wastewater also containing high concentrations of sulfate (SO_4^{2-}). Due to the activity of sulfate-reducing bacteria, hydrogen sulfide (H_2S/HS^-) was formed. This sulfide, in turn,

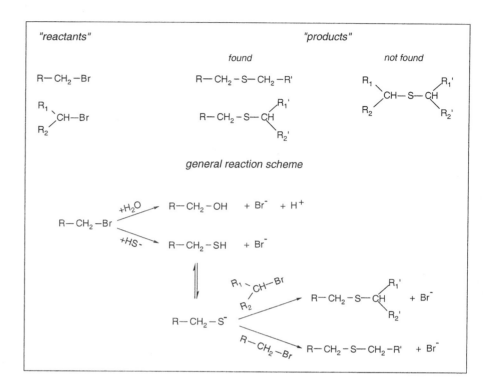

Figure 13.5 Alkyl bromides ("reactants") leaked into groundwater and dialkyl sulfides found several years later; the reaction scheme shown can account for the products seen (for details, see Schwarzenbach et al., 1985).

reacted with the alkyl bromides to yield alkyl mercaptans (or thiols; Fig. 13.5). The mercaptans (RSH/RS⁻), which are even better nucleophiles than H_2S/HS^-, then reacted further with other alkyl bromide molecules, resulting in the formation of a whole series of dialkyl sulfides and other hazardous products (for more details see Schwarzenbach et al., 1985). Of interest to us here is the fact that all possible dialkyl sulfides exhibiting at least one primary alkyl group were found, but that no compounds with two secondary alkyl groups could be detected. These results suggest that the secondary alkyl bromides were reacting chiefly via an S_N1 mechanism, thereby yielding secondary alcohols. It was not until the primary alkyl mercaptans, which are particularly strong nucleophiles, appeared that the secondary bromides also became involved in a more S_N2-like reaction.

Polyhalogenated Alkanes—Elimination Mechanisms

So far, we have considered only monohalogenated compounds. However, there are a variety of polyhalogenated alkanes that are of great environmental concern. Table 13.7 summarizes some of the kinetic data available on the reactivity of such compounds in aqueous solution. Additional kinetic data can be found in Roberts et al. (1993) and Jeffers and Wolfe (1996). Some important conclusions can be drawn from these data. First, we notice that polyhalogenated methanes hydrolyze extremely slowly under environmental conditions. This result is mostly due to steric hindrance and to back-bonding by the relatively electron-rich bulky halogens (Hughes, 1971). Hence, nucleophilic substitution reactions at the carbon atoms of such compounds are typically of minor environmental significance. However, as we will see later, the polyhalogenated methanes as well as other polyhalogenated compounds may, under certain environmental conditions, react by another reaction pathway, namely, reductive dehalogenation (see Chapter 14).

From the reaction products of the polyhalogenated ethanes and propanes shown in Table 13.7 we deduce that such halogenated compounds may react in aqueous solution by yet another type of reaction, so-called *β-elimination*. In this reaction, in addition to the leaving group (L$^-$), a proton is lost from an adjacent carbon atom (hence the prefix *β*-) and a double bond is formed:

$$\begin{array}{c} | \quad | \\ -\text{C}-\text{C}- \\ | \quad | \\ \text{H} \quad \text{L} \end{array} \xrightarrow{\textit{β-elimination}} \quad \diagdown\text{C}=\text{C}\diagup \quad + \text{H}^+ + \text{L}^- \tag{13-8}$$

If L = halogen, this type of reaction is referred to as *dehydrohalogenation*. Thus, when assessing the fate of halogenated compounds in natural waters, this process has to be considered in addition to nucleophilic substitution. The question then is what structural features and environmental conditions determine whether only one or both of these two competing types of reactions will be important.

Generally, *β*-elimination is important in molecules in which nucleophilic substitution is sterically hindered and/or in which relatively acidic protons are present at carbon atoms adjacent to the carbon carrying the leaving group. These criteria are optimally met in 1,1,2,2-tetrachloroethane and in pentachloroethane. In these compounds four or five electron-withdrawing chlorine atoms render the hydrogen(s) more acidic and, simultaneously, these relatively large halide substituents hinder nucleophilic attack. In water 1,1,2,2-tetrachloroethane is converted more or less quantitatively to trichloroethene and pentachloroethane to tetrachloroethene, respectively, by a so-called *E2 (elimination, bimolecular) elimination*. That is, the elimination takes place in a "concerted" fashion in which a base (e.g., OH$^-$) and the polyhalogenated compound interact and form only one transition state (see Fig. 13.6). The reaction, therefore, follows a second-order kinetic rate law:

$$\text{rate} = -k[\text{OH}^-]\left[\begin{array}{c} | \quad | \\ -\text{C}-\text{C}- \\ | \quad | \\ \text{H} \quad \text{L} \end{array}\right] \tag{13-9}$$

In this, as in many other cases in aqueous solution, OH$^-$ plays the role of the base. Note that for compounds such as 1,1,2,2-tetrachloroethane and pentachloroethane, the base catalyzed reaction is important at quite low pH values (I_{NB} = 4.5, i.e., pH at which the neutral and base catalyzed reaction are equally important, see Table 13.7 and Section 13.3). In fact, for polyhalogenated alkanes a small I_{NB} value (e.g., <7) is indicative of an E2 reaction, or, in special cases, of an E1$_{CB}$ reaction; see below. Some other examples of compounds reacting by an E2-mechanism include 1,1,2-trichloroethane, 1,1,2-tribromoethane, and 1,2-dibromo-3-chloroethane (see Table 13.7). A high I_{NB} value (e.g., >10) does not, however, necessarily exclude *β*-elimination, because this reaction may also occur with water as base, or by an alternative to the S$_N$1 mechanism (i.e., an E1 mechanism, see below).

For 1,1,2,2-tetrachloroethane, we can picture an E2 reaction in a very similar way as we have done with the S$_N$2 reaction (Fig. 13.6). Here, however, one species (usually OH$^-$) plays the role of a base that induces the breaking of a C-H bond at a *β*-carbon by leaving both electrons to the carbon atom. The resulting activated complex then contains a carbon which is partially (or, in the most extreme case, fully) negatively

Table 13.7 Kinetic Data on Nucleophilic Substitution and Nonreductive Elimination (Dehydrohalogenation) Reactions of Some Polyhalogenated Hydrocarbons in Aqueous Solution at 25°C [a]

Compound Name (Structure)	Major Product(s)	k_N (s⁻¹)	$E_{a,N}$ (kJ·mol⁻¹)	k_B (M⁻¹·s⁻¹)	$E_{a,B}$ (kJ·mol⁻¹)	I_{NB} [b]	$t_{1/2}$ at pH 7
Dichloromethane [c] (CH_2Cl_2)	(CH_2O)	3×10^{-11}		2×10^{-8}		~11	~700 yr
Trichloromethane [d] ($CHCl_3$)	($HCOOH$)	3×10^{-12}	123	9×10^{-5}	105	6.5	~2×10^3 yr
Tetrachloromethane [e] (CCl_4)	not reported	5×10^{-10}	113			>10	~40 yr
Tribromomethane [c] ($CHBr_3$)	($HCOOH$)			3×10^{-4}		<7	~700 yr
1,2-Dichloroethane [f] ($ClCH_2$–CH_2Cl)	$HOCH_2$–CH_2OH	3×10^{-10}	104	2×10^{-6}	98	~10	~70 yr
1,2-Dibromoethane [f] ($BrCH_2$–CH_2Br)	$HOCH_2$–CH_2OH (>75%) CH_2=$CHBr$	3.5×10^{-9}	108	4×10^{-5}	95	~10	~6.5 yr
1,1,1-Trichloroethane [d] (CCl_3–CH_3)	CH_3–$COOH$ (~80%) CH_2=CCl_2 (~20%)	2×10^{-8}	116			>10	~1 yr
1,1-Dichloro-1-fluoroethane [f] (CCl_2F–CH_3)	not reported	~1×10^{-9}	109			>10	~20 yr
1-Chloro-1,1-difluoroethane [f] ($CClF_2$–CH_3)	not reported	~1×10^{-12}	113			>10	~2×10^4 yr
1,1,2–Trichloroethane [d,g] ($CHCl_2$–CH_2Cl)	CH_2=CCl_2	$<10^{-12}$		1.5×10^{-3}	88	<5	~140 yr

Compound	Product						$t_{1/2}$
1,1,2-Tribromoethane[f] (CHBr₂CH₂Br)	not reported	1×10^{-10}	113	2×10^{-1}	81	~5	~1 yr
1,1,1,2-Tetrachloroethane[f] (CCl₃–CH₂Cl)	not reported	4×10^{-10}	95	3.5×10^{-4}	100	~8	~45 yr
1,1,2,2-Tetrachloroethane (CHCl₂–CHCl₂)	CHCl=CCl₂	$<1 \times 10^{-10}$	93	5×10^{-1}	78	<4.5	160 d
Pentachloroethane[d,g] (CCl₃–CHCl₂)	CCl₂=CCl₂	8×10^{-10}	95	2.7×10^{1}	80	~4.5	3 d
1,1,2,2-Tetrachloro-1-fluoroethane[f] (CCl₂F–CHCl₂)	not reported			$< 10^{-9}$		>10	$> 10^{5}$ yr
1,2-Dibromo-3-chloropropane (DBCP)[h] (CH₂Br–CHBr–CH₂Cl)	CH₂=CBr–CH₂OH (>95% at 85°C)	2.5×10^{-10}	93	6×10^{-3}	93	~6.5	~25 yr
γ-Hexachlorocyclohexane (lindane, HCH)[g]	not reported	7×10^{-10}		2.5×10^{-2}		~6.5	~7 yr
DDT[g]	DDE	1.5×10^{-9}		1×10^{-2}		~7	~7 yr

[a] Most data are extrapolated from experimental data obtained at elevated temperatures. [b] I_{NB} = pH at which neutral and base-catalyzed reaction are equally important, i.e., $k_N = k_B [OH^-]$, see also Section 13.3. [c] Mabey and Mill (1978). [d] Jeffers et al. (1989). [e] Jeffers et al. (1996). [f] Jeffers and Wolfe (1996). [g] Roberts et al. (1993). [h] Burlinson et al. (1982).

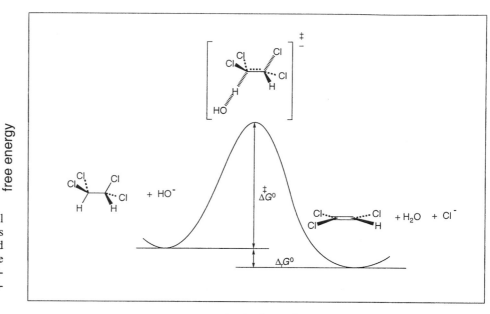

Figure 13.6 Two-dimensional portrayal of relative free energies exhibited by the reactants, activated complex, and products of the β-elimination reactions of 1,1,2,2-tetrachloroethane by an E2 mechanism.

extent of reaction
(reaction coordinate)

β-hexachlorocyclohexane
(no antiplanar H–C–C–Cl)

γ-hexachlorocyclohexane
(three antiplanar H–C–C–Cl)

charged. Hence, any group that stabilizes the negative charge at this carbon atom by induction or resonance will enhance the reaction rate. Note that this is equivalent to our earlier statement that the reaction occurs faster the more acidic the proton(s) is (are) at the β-carbon(s). The electrons of the breaking (or broken) C-H bond now play the role of a nucleophile by attacking the leaving group from the backside (as the electrons of the nucleophile in an S_N2 reaction), thus causing the breaking of the C-L bond and the formation of a double bond. The steric requirements for optimal E2 elimination are, therefore, an antiplanar configuration of the atoms involved in the reaction as depicted in Fig. 13.6. Consequently, in ring systems, elimination might in some cases be hindered owing to such steric factors (the inability of the β-protons to be antiplanar to leaving groups as in β-hexachlorocyclohexane, which, in contrast to the γ-isomer, does not show any measurable reactivity).

The role of the leaving group in elimination reactions can, in general, be looked at in a very similar way as in S_N reactions. As illustrated by the relative amounts of elimination products formed by the base-catalyzed reactions of the pesticide 1,2-dibromo-3-chloropropane (DBCP, Eq. 13-10), bromide is a better leaving group than chloride (Burlinson et al., 1982):

(13-10)

We note that in this case the elimination products, that is, 2-bromo-3-chloropropene (BCP) and 2,3-dibromopropene (DBP), are allylic halides. Consequently, these both hydrolyze in relatively fast steps most likely via S_N1 reactions (see the example given in Table 13.6) to form 2-bromoallyl alcohol (BAA).

As mentioned above, in some special cases, a compound may react by a so-called $E1_{CB}$ mechanism. This process shares many features with E2 reactions. Both are initiated by attack of a base on a labile hydrogen. They differ, however, in that, in a $E1_{CB}$ reaction, degradation begins as a normal reversible acid–base reaction, and not by a concerted action as depicted in Fig. 13.6 for E2 reactions. Such a mechanism has been observed primarily for alkyl substrates that have highly acidic hydrogens and that also possess features capable of providing significant carbanion stabilization, such as aryl substituents as is the case of DDT, or for substrates that lack good leaving groups in β-position as in the case of 1,1-dichloro-2,2,2-trifluoroethane (for more examples and references see Roberts et al., 1993).

As indicated in Table 13.7, 1,2-dibromoethane ($BrCH_2$–CH_2Br) and 1,1,1-trichloro-ethane (CH_3–CCl_3) are examples in which both hydrolysis and elimination are important. If in such cases the reactions occur by S_N2 and E2 mechanisms, respectively, the ratio of the hydrolysis versus elimination products should vary with varying pH and temperature, since the two competing reactions likely exhibit different pH and temperature dependencies. On the other hand, if the reaction mechanisms were more S_N1- and E1-like, a much less pronounced effect of temperature or pH on product formation would be expected, since the rate-determining step in aqueous solution may be considered to be identical for both reactions:

$$\text{(13-11)}$$

We note that in Eq. 13-11 we have introduced the E1 (elimination, unimolecular) reaction, which commonly competes with the S_N1 reaction provided that an adjacent carbon atom carries one or several hydrogen atoms that may dissociate. We also note that similar to what we have stated earlier for nucleophilic substitution reactions, elimination reactions may occur by mechanisms between the E2 and E1 extremes.

From the experimental data available for the reactivities of 1,2-dibromoethane (EDB) and 1,1,1-trichloroethane (TrCE), it is not possible to draw sound conclusions as to the mechanisms and the pH and temperature dependence of product formation of the reactions of these compounds in water. It is, however, interesting to note that the overall reaction rate of TrCE was found to be pH independent below pH 11, and that temperature had no significant influence on the product formation in the temperature range between 25 and 80°C (Haag and Mill, 1988). These findings indicate that this compound undergoes S_N1- and E1-type reactions in aqueous solution. It should also be pointed out that the primary hydrolysis products of both EDB (i.e., $BrCH_2$-CH_2OH) and TrCE (i.e., CH_3-CCl_2OH) subsequently hydrolyze again in relatively fast reactions to yield the final products, ethylene glycol and acetic acid, respectively (see Table 13.7).

With these examples we conclude our discussion of nucleophilic substitution and β-elimination reactions involving saturated carbon–halogen bonds in environmental chemicals. For more extensive treatment of this topic, including the use of polar substituent constants to derive quantitative structure-reactivity relationships for E2 reactions of polyhalogenated alkanes, we refer to the review by Roberts et al., (1993). Before we go on discussing another group of reactions, we need, however, to make some final remarks about S_N and E reactions of halogenated compounds. First, we note that the activation energies of the reactions in which halogens are removed from saturated carbons in organic molecules by an S_N or E mechanism are between 80 and 120 $kJ \cdot mol^{-1}$. Hence, these reactions are quite sensitive to temperature; that is, a difference in 10°C means a difference in reaction rate of a factor of 3-5 (Eq. 12-30, Table D1, Appendix D). Second, we have seen that a compound may react by several competing reactions. In these cases, the general rate law will be a composite of the rate laws of the individual reactions:

$$\text{rate} = -\left\{ k_N + k_{EN} + \left(k_B + k_{EB} \right)[OH^-] + \sum_j k_{Nu_j}[Nu_j] \right\} C_{iw} \qquad (13\text{-}12)$$

where C_{iw} is the concentration of the dissolved halogenated compound i in water, k_N and k_{EN} are the (pseudo)first-order rate constant for the neutral, and k_B and k_{EB} are the second-order rate constants for the base-catalyzed hydrolysis and elimination reactions, respectively, and k_{Nuj} is the second-order rate constant of the S_N2 reaction with any other particular nucleophile j. Note that k_{Nuj} may be estimated using the Swain-Scott relationship (Eq. 13-6). We recall that by assuming constant pH and constant nucleophile concentration(s), Eq. 13-12 can be reduced to a pseudo-first-order rate law with a pseudo-first-order rate constant k_{obs} that is given by:

$$k_{obs} = k_N + k_{EN} + (k_B + k_{EB})[OH^-] + \sum_j k_{Nu_j}[Nu_j] \qquad (13\text{-}13)$$

We should point out, however, that depending on the relative importance of the various reactions, k_{obs} may not be a simple function of pH and temperature, and that product formation may strongly depend on these two variables. Furthermore, we note that many environmentally important organic compounds exhibit halogen atoms bound to a carbon–carbon double bond, be it an olefinic (e.g., chlorinated ethenes) or an aromatic (e.g., chlorinated benzenes, PCBs) system. In many cases, under environmental conditions, these carbon–halogen bonds undergo S_N or E reactions at extremely slow rates, and we therefore may consider these reactions to be unimportant.

13.3 Hydrolytic Reactions of Carboxylic and Carbonic Acid Derivatives

In this section we consider a second important type of reaction in which a nucleophile attacks a carbon atom, but this time a carbon that is doubly bound to a heteroatom and singly bound to at least one other heteroatom. The major difference from the cases discussed in the previous section is that we are now considering nucleophilic reactions involving an *unsaturated* carbon atom exhibiting *multiple* bonds to other more electronegative atoms. Also, structural parts connected by singly bound heteroatoms may serve as leaving groups. Since in the environment such functional groups react predominantly with the nucleophiles, H_2O and OH^-, we confine ourselves to hydrolytic reactions. As an illustration we may consider the reaction of a carboxylic *acid derivative* with OH^-, a reaction that, in many cases, occurs by the general reaction mechanism:

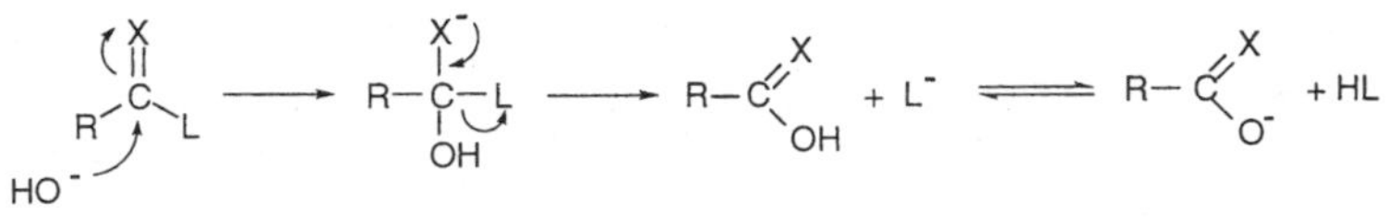

$$\text{(13-14)}$$

where X may be O, S, or NR. The most common leaving groups, L^-, include RO^-, RS^-, and $R_1R_2N^-$ (Fig. 13.7). We note that if hydrolysis of such a functionality occurs by a mechanism similar to Eq. 13-14, the reaction products include the acid (under basic conditions usually present as the conjugate base) and the leaving group, which, in most cases of interest to us, is an alcohol, thiol, or an amine.

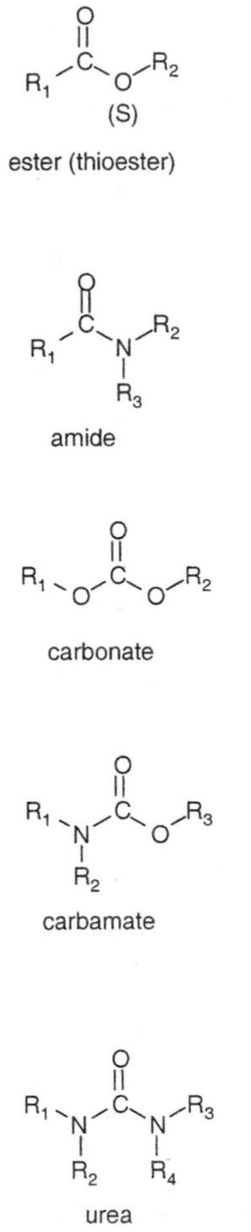

In the following discussions we first look at hydrolytic reactions of ester functions. As already mentioned in Chapter 2, ester functions are among the most common acid derivatives present in natural as well as man-made chemicals (e.g., lipids, plasticizers, pesticides). In a general way, an ester bond is defined as:

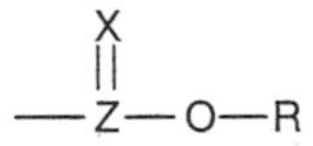

$$\begin{matrix} X \\ \| \\ -Z-O-R \end{matrix}$$

where Z = C, P, S; X = O, S; and R is a carbon-centered substituent. Hence, hydrolysis of an ester bond yields the corresponding acid and the alcohol. If O is replaced by S, the functional group is referred to as thioester. Such thioesters are quite common in phosphoric acid and thiophosphoric acid derivatives (see Section 13.4) that are used as pesticides. We first consider, however, the hydrolysis of a more familiar group of esters, the *carboxylic acid esters* (see Fig. 13.7). We use this type of functionality to discuss some general mechanistic and structural aspects of hydrolysis that are valid for esters and other carboxylic and carbonic acid derivatives.

Figure 13.7 Examples of carboxylic and carbonic acid derivatives. R_1, R_2, R_3, R_4 denote carbon-centered substituents.

Carboxylic Acid Esters

Hydrolysis half-lives of carboxylic acid esters, defined as:

$$t_{1/2(\text{hydrolysis})} = \frac{\ln 2}{k_h} \qquad \text{(13-15)}$$

where k_h is the pseudo-first-order hydrolysis rate constant, typically vary widely as a

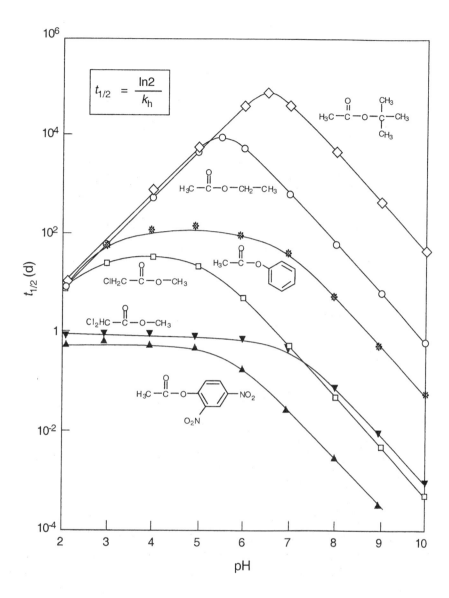

Figure 13.8 Variation of hydrolysis half-life at 25°C for several carboxylic acid esters as a function of solution pH due to changing contributions of the acid-catalyzed, neutral, and base-catalyzed mechanisms.

function of pH (Fig. 13-8). This dependency arises because three separate reactions, one catalyzed by H⁺, a second depending on OH⁻, and a third resulting from attack by H_2O, occur simultaneously. Recognizing that the curve sections that decrease with a slope of −1 as a function of pH reflect reactions mediated by OH⁻, we notice that for all compounds, reaction with OH⁻ ("base catalysis") is important even at pH values below pH 7, and that acid catalysis (curve portions with slope of +1) is relevant only at relatively low pHs and only for compounds showing rather slow hydrolysis kinetics. By taking into account the acid-catalyzed (k_A, e.g., M⁻¹ s⁻¹), neutral (k_{H_2O}, e.g., M⁻¹ s⁻¹), and base-catalyzed (k_B, e.g., M⁻¹ s⁻¹) reactions, we can express the observed (pseudo-first-order) hydrolysis rate constant, k_h (e.g., s⁻¹), at constant pH as:

$$k_h = k_A[H^+] + k_{H_2O}[H_2O] + k_B[OH^-] \qquad (13\text{-}16)$$

and since $[H_2O]$ generally remains constant, we can simplify to:

$$k_h = k_A[H^+] + k_N + k_B[OH^-] \qquad (13\text{-}17)$$

Figure 13.9 Schematic representation of the relative contribution of the acid-catalyzed, neutral, and base-catalyzed reactions to the overall hydrolysis rate as a function of solution pH: (a) neutral reaction rate is significant over some pH range; (b) the contributions of the neutral reaction can always be neglected.

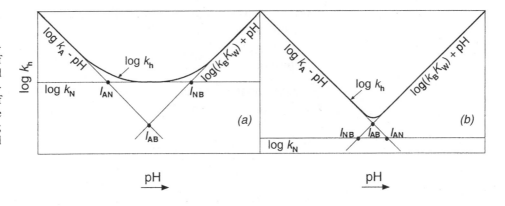

where: $$k_N = k_{H_2O} \cdot [H_2O] \qquad (13\text{-}18)$$

If k_A, k_N, and k_B are known for a given compound, we can calculate the pH values at which two reactions are equally important. As is schematically shown in Fig. 13.9a and b, these pH values are given by the intersections, I, of the lines representing the contributions of each reaction to the overall reaction rate as a function of pH. Note that Fig. 13.9 is drawn on a logarithmic scale. Hence, for example, I_{AB} is the pH at which $k_A[H^+] = k_B[OH^-]$. If we set pH $= -\log[H^+]$ and $[OH^-] = K_w/[H^+]$, then we obtain $I_{AB} = 0.5 \log(k_A/k_B K_w)$. If the neutral reaction (pH independent reaction with H_2O) is dominant over a wider pH range (extreme case shown in Fig. 13.9a), then I_{AB} is only of theoretical value since both acid- and base-catalyzed reactions are unimportant at this pH. Similarly, if the neutral reaction is never important (extreme case shown in Fig. 13.9b), then neither I_{AN} nor I_{NB} have much practical meaning. Before we discuss hydrolysis reactions in more detail, we should again stress that the neutral, acid-catalyzed, and base-catalyzed reactions are three very different reactions exhibiting different reaction mechanisms and, hence, different kinetic parameters. Thus, for example, because of different activation energies, the relative importance of each reaction (as is expressed by the I values) depends on temperature. Furthermore, as we will see when discussing quantitative structure-reactivity relationships (i.e., LFERs, see below), substituents usually have quite different effects on the neutral, acid-catalyzed, and base-catalyzed hydrolyses. Consequently, one has to be very careful to apply LFERs only to the kinetics of each individual reaction and not to the overall reaction (unless, of course, the overall reaction reflects only one dominant mechanism). Illustrative Examples 13.4 and 13.5 demonstrate how to derive all necessary kinetic parameters for assessing the hydrolysis behavior of a carboxylic acid ester from experimental data, and how to apply these parameters. Note that the general procedure outlined in these examples is also applicable to many other hydrolysis reactions. But let us now look at the various hydrolytic mechanisms of carboxylic acid esters more closely.

Illustrative Example 13.4   **Deriving Kinetic Parameters for Hydrolysis Reactions from Experimental Data**

Consider the hydrolysis of 2,4-dinitrophenyl acetate (DNPA), a compound for which the acid-catalyzed reaction is unimportant at pH > 2 (see Fig. 13.8). In a laboratory class, the time course of the change in concentration of DNPA in homogeneous aqueous solution has been followed at various conditions of pH and temperature using an HPLC method (for details see Klausen et al., 1997).

Problem

Determine the (pseudo-)first-order reaction rate constants, k_h, for this reaction at pH 5.0 and pH 8.5 at 22.5°C using the data sets given below:

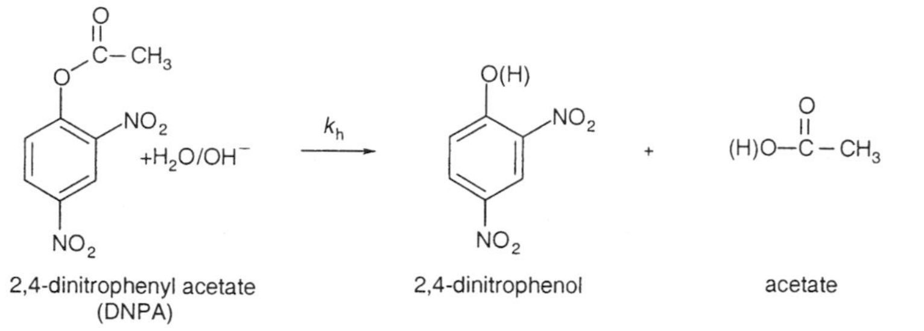

| 2,4-dinitrophenyl acetate (DNPA) | 2,4-dinitrophenol | acetate |

pH 5.0[a], $T = 22.5$°C		pH 8.5, $T = 22.5$°C	
Time (min)	[DNPA (μM)]	Time (min)	[DNPA (μM)]
0	100.0	0	100.0
11.0	97.1	4.9	88.1
21.5	95.2	10.1	74.3
33.1	90.6	15.4	63.6
42.6	90.1	25.2	47.7
51.4	88.5	30.2	41.2
60.4	85.0	35.1	33.8
68.9	83.6	44.0	26.6
75.5	81.5	57.6	17.3

[a] Note that very similar results were also found at pH 4.0 and 22.5°C.

Answer

Assuming a (pseudo-)first-order rate law, k_h can be determined from a least squares fit of $\ln([DNPA]_t / [DNPA]_0)$ versus time (see also Figure below):

$$\ln([DNPA]_t / [DNPA]_0) = -k_h \cdot t \qquad (1)$$

The resulting k_h values are:

$$k_h(\text{pH } 5.0, 22.5°C) = 2.6 \times 10^{-3} \text{ min}^{-1} = 4.4 \times 10^{-5} \text{ s}^{-1}$$

$$k_h(\text{pH } 8.5, 22.5°C) = 3.1 \times 10^{-2} \text{ min}^{-1} = 5.1 \times 10^{-4} \text{ s}^{-1}$$

Note that k_h increases with increasing pH, indicating that the base-catalyzed reaction is important, at least at higher pH values.

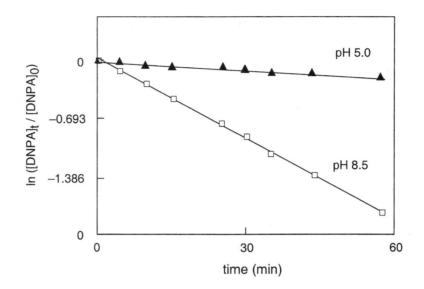

Problem

Using the data given above, derive the rate constants for the neutral (k_N) and base-catalyzed (k_B) hydrolysis of DNPA at 22.5°C. At what pH are the two reactions equally important?

Answer

When assuming that the acid-catalyzed reaction is not important in the pH-range considered, Eq. 3-17 simplifies to:

$$k_h = k_N + k_B \cdot [OH^-] \tag{2}$$

The fact that very similar k_h values have been found at pH 4.0 and pH 5.0 indicates that up to pH 5.0, the base-catalyzed reaction can be neglected, and therefore:

$$k_N \,(22.5°C) = k_h \,(\text{pH } 5.0, 22.5°C) = 4.4 \times 10^{-5} \text{ s}^{-1}$$

Using this k_N-value, k_B can be determined by rearranging Eq. 2:

$$k_B(22.5°C) = \frac{k_h(\text{pH } 8.5, 22.5°C) - k_N(22.5°C)}{[OH^-]}$$

with the hydroxide concentration given by (see Eq. 8-18) :

$$[OH^-] = \frac{K_w}{[H^+]}$$

Note that the ionization constant of water, K_w, is strongly temperature dependent. At 22.5°C, $K_w = 10^{-14.08}$ (Table D2 in Appendix D). Hence, at pH 8.5 (i.e., $[H^+] = 10^{-8.5}$), $[OH^-] = 10^{-5.58}$ and:

$$k_B(22.5°C) = \frac{4.7 \times 10^{-4}}{10^{-5.58}} = 180 \text{ M}^{-1} \text{ s}^{-1}$$

The pH value, I_{NB}, at which the neutral and the base-catalyzed reactions are of equal importance is deduced by (see Fig. 13.9):

$$I_{NB} = \log \frac{k_N}{k_B \cdot K_w} = \log \frac{4.4 \times 10^{-5}}{180 \cdot 10^{-14.08}} = 7.5$$

Thus, at pH 8.5, the hydrolysis of DNPA is dominated by the base-catalyzed reaction.

Problem

Derive the Arrhenius activation energy, E_a, for the neutral hydrolysis of DNPA using the data given in the margin.

Answer

According to Eq. 12-29, the temperature dependence of a rate constant can be described by:

$$\ln k = -\frac{E_a}{R} \cdot \frac{1}{T} + \text{const.}$$

Note that for the temperature range considered, E_a is assumed to be constant. Convert temperatures in °C to K and calculate $1/T$ values. Also take the natural logarithms of the k_N values (see margin).

T (°C)	k_N / s^{-1}
17.7	3.1×10^{-5}
22.5	4.4×10^{-5}
25.0	5.2×10^{-5}
30.0	7.5×10^{-5}

$1/T$ / K^{-1}	$\ln k_N$ / s^{-1}
0.00344	-10.38
0.00338	-10.03
0.00335	-9.86
0.00330	-9.50

Perform a least squares fit of $\ln k_N$ versus $1/T$. The resulting slope is:

$$\text{slope} = -\frac{E_a}{R} = -6318 \text{ K}$$

and therefore:

$$E_a = -R \cdot \text{slope} = 8.31 \cdot (6318) = 52.5 \text{ kJ} \cdot \text{mol}^{-1}$$

The E_a value determined for the base-catalyzed reaction is 60.0 kJ $\cdot$ mol^{-1} (data not shown).

Illustrative Example 13.5   **Calculating Hydrolysis Reaction times as a Function of Temperature and pH**

Problem

Calculate the time required to decrease the concentration of DNPA (see Illustrative Example 13.4) by hydrolysis to 50% (half-life) and to 5% of its initial concentration (a) in the epilimnion of a lake ($T = 22.5$°C, pH = 8.5), and (b) in the hypolimnion of the same lake ($T = 5$°C, pH = 7.5).

Answer

The hydrolysis half-life is calculated by:

$$t_{1/2} = \frac{\ln 2}{k_h} = \frac{0.693}{k_h} \tag{12-13}$$

By analogy, the time required to reduce the concentration to 5% (i.e., $[DNPA]_t$ / $[DNPA]_0 = 0.05$) is given by (see Eq. 1, Illustrative Example 13.4):

$$t_{0.05} = \frac{\ln(1/0.05)}{k_h} = \frac{3}{k_h} \tag{1}$$

(a) Calculate k_h (Eq. 2, Illustrative Example 13.4) for 22.5°C and pH 8.5 using the above derived k_N and k_B values and $[OH^-] = 10^{-5.58}$ M:

$$k_h (22.5°C) = (180) (10^{-5.58}) + 4.4 \times 10^{-5} = 5.1 \times 10^{-4} \text{ s}^{-1}$$

Note that at pH 8.5 and 22.5°C, hydrolysis is dominated by the base-catalyzed reaction. Insertion of k_h into Eqs. 12-13 and 1 then yields:

$$t_{1/2}(22.5°C) = \frac{0.693}{5.1 \times 10^{-4} \text{ s}^{-1}} = 1360 \text{ s} = 22.7 \text{ min}$$

$$t_{0.05} (22.5°C) = \frac{3}{5.1 \times 10^{-4} \text{ s}^{-1}} = 5880 \text{ s} = 1.63 \text{ h}$$

(b) Calculate the k_N and k_B values for 5°C (278.2 K) from the corresponding rate constants derived above for 22.5°C (295.7 K) using (see Eq. 12-30):

$$k(T_1) = k(T_2) \cdot e^{(E_a/R)(1/T_2 - 1/T_1)}$$

where $T_2 = 295.7$ K and $T_1 = 278.2$ K, and E_a is the activation energy given in Illustrative Example 13.4. The results obtained are:

$$k_N (5°C) = 1.1 \times 10^{-5} \text{ s}^{-1} \text{ and } k_B (5°C) = 38.6 \text{ M}^{-1} \text{ s}^{-1}.$$

Since $K_w = 10^{-14.73}$ at 5°C (Table D2, Appendix D), the OH^- concentration at pH 7.5 is $10^{-7.23}$ M, resulting in a k_h-value of:

$$k_h (5°C) = (38.6) (10^{-7.23}) + 1.1 \times 10^{-5} = 1.3 \times 10^{-5} \text{ s}^{-1}$$

Note that in contrast to the epilimnion, in the hypolimnion the hydrolysis of DNAP is dominated by the neutral reaction. The corresponding reaction times are:

$$t_{1/2}(5°C) = \frac{0.693}{1.3 \times 10^{-5} \text{ s}^{-1}} = 53300 \text{ s} = 14.8 \text{ h}$$

$$t_{0.05} (5°C) = \frac{3}{1.3 \times 10^{-5} \text{ s}^{-1}} = 230000 \text{ s} = 62.9 \text{ h}$$

Hence, under the assumed conditions, DNPA hydrolyzes about 40 times faster in the epilimnion of the lake as compared to the hypolimnion.

Table 13.8 Rate Constants k_A, k_N, and k_B, Half-Lives at pH 7, and I Values for Hydrolysis of Some Carboxylic Acid Esters at 25°C [a]

Compound

$$R_1-\overset{\overset{\displaystyle O}{\|}}{C}-O-R_2$$

R_1	R_2	k_A (M^{-1} s^{-1})	k_N (s^{-1})	k_B (M^{-1} s^{-1})	$t_{1/2}$ (pH 7)	I_{AN} [b,c,e]	I_{AB} [c,e]	I_{NB} [d,e]
CH$_3$–	–CH$_2$CH$_3$	1.1×10^{-4}	1.5×10^{-10}	1.1×10^{-1}	2 yr	(5.9)	5.5	(5.1)
CH$_3$–	–C(CH$_3$)$_3$	1.3×10^{-4}		1.5×10^{-3}	140 yr		6.5	
H–	–C(CH$_3$)$_3$	2.7×10^{-3}	1.0×10^{-6}	1.7×10^{0}	7 d	2.6	5.6	7.8
CH$_3$–	–CH=CH$_2$	1.4×10^{-4}	1.1×10^{-7}	1.0×10^{1}	7 d	3.1	(4.6)	6.0
CH$_3$–	–⟨phenyl⟩	7.8×10^{-5}	6.6×10^{-8}	1.4×10^{0}	38 d	3.1	(4.8)	6.7
CH$_3$–	–⟨phenyl with NO$_2$, O$_2$N⟩		1.1×10^{-5}	9.4×10^{1}	10 h			7.1
CH$_2$Cl–	–CH$_3$	8.5×10^{-5}	2.1×10^{-7}	1.4×10^{2}	14 h	2.6	(3.9)	5.2
CHCl$_2$–	–CH$_3$	2.3×10^{-4}	1.5×10^{-5}	2.8×10^{3}	40 min	1.2	(3.5)	5.7
CHCl$_2$–	–⟨phenyl⟩		1.8×10^{-3}	1.3×10^{4}	4 min			7.1

[a] Data from Mabey and Mill (1978) except for *tert*-butyl formate (R$_1$ = H, R$_2$ = C(CH$_3$)$_3$; Church et al., 1999). [b] $I_{AN} = \log (k_A/k_N)$. [c] $I_{AB} = 1/2 \log (k_A/k_B K_w)$. [d] $I_{NB} = \log (k_N/k_B K_w)$. [e] Parentheses indicate that one or both of the processes is too slow to contribute significantly to the overall rate.

In Table 13.8 the hydrolysis rate constants and I values at 25°C are given for some carboxylic acid esters including the compounds shown in Fig. 13.8. Note that the activation energies of ester hydrolysis reactions (data not shown) span quite a wide range between about 40 and 80 kJ · mol^{-1} (Kirby, 1972; Mabey and Mill, 1978). Hence, depending on structure and reaction mechanisms reaction rates will change by a factor of between 2 and 3 for a 10-degree change in temperature (see Section 12.3 and Table D1 in Appendix D). The data in Table 13.8 illustrate some general findings about the influence of structural moieties on the rate of the different hydrolytic reactions. First, we see that between the various compounds, relatively small differences are observed in the magnitude of k_A, which is in contrast to the large differences found for the k_N and k_B values, respectively. We also see that structural differences in the leaving group (i.e., the alcohol) seem not to have a big influence on k_A, suggesting that dissociation of the leaving group is not rate determining. Let us try to rationalize these findings by looking at the reaction mechanisms of *acid-catalyzed hydrolysis*. We consider the mechanism believed to reflect the situation for most carboxylic acid esters; that is, the one in which the reaction proceeds through a

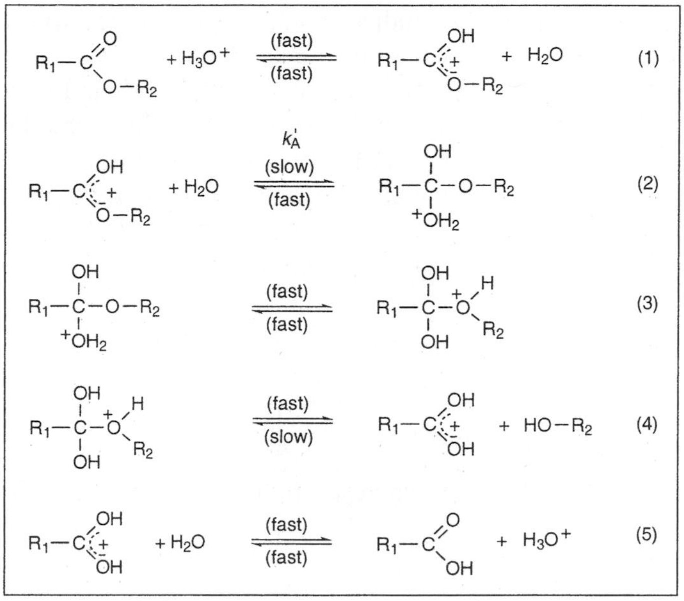

Figure 13.10 Reaction scheme for the acid-catalyzed hydrolysis of carboxylic acid esters.

tetrahedral intermediate. Fig. 13.10 shows the postulated reaction scheme for this reaction. We recall that each of the elementary reaction steps is, in principle, reversible, and that the overall reaction rate of any chemical reaction is determined by the rate(s) of the slowest step(s).

Acid-Catalyzed Hydrolysis. In acid-catalyzed ester hydrolysis the species that undergoes the rate-determining step is the protonated ester (Fig. 13.10). When the molecule is in this protonated form, the enhanced depletion of electrons near the central carbon promotes the approach of an electron-rich oxygen of a water molecule. Hence, the hydrolysis rate depends on the fraction of compound molecules that are protonated. This fraction, in turn, depends on how strong a base the ester function is. If we define an acidity constant (see Chapter 8) for the protonated species

$$K_a = \frac{\left[R_1 - C \underset{O-R_2}{\overset{O}{}} \right] \cdot \left[H^+ \right]}{\left[R_1 - C \underset{\overset{+}{O}-R_2}{\overset{OH}{}} \right]} \tag{13-19}$$

then we can express the concentration of the protonated ester molecule as:

$$\left[R_1 - C \underset{\overset{+}{O}-R_2}{\overset{OH}{}} \right] = \frac{1}{K_a} \cdot \left[R_1 - C \underset{O-R_2}{\overset{O}{}} \right] \cdot \left[H^+ \right] \tag{13-20}$$

As indicated in Fig. 13.10, the slowest, and therefore rate-determining, reaction step is then the nucleophilic attack of a water molecule at the carbonyl carbon of the protonated species. This carbonyl is much more susceptible to nucleophilic attack than in the neutral ester. Since the dissociation of the (protonated) leaving group (HO–R_2) is fast (forward portion of reaction 4 in Fig. 13.10), the rate of ester disappearance through acid-catalyzed hydrolysis is given by:

$$\frac{d\left[R_1 - C{\overset{O}{\underset{O-R_2}{\Big\langle}}} \right]}{dt} = - k'_A \cdot \left[R_1 - C{\overset{OH}{\underset{\overset{+}{O}-R_2}{\Big\langle}}} \right] \cdot \left[H_2O \right] \qquad (13\text{-}21)$$

or, when substituting Eq. 13-20 into Eq. 13.21:

$$\frac{d\left[R_1 - C{\overset{O}{\underset{O-R_2}{\Big\langle}}} \right]}{dt} = - \frac{k'_A}{K_a} \cdot \left[R_1 - C{\overset{O}{\underset{O-R_2}{\Big\langle}}} \right] \cdot \left[H^+ \right] \cdot \left[H_2O \right] \qquad (13\text{-}22)$$

$$= - k_A \cdot \left[R_1 - C{\overset{O}{\underset{O-R_2}{\Big\langle}}} \right] \cdot \left[H^+ \right]$$

Hence, the second-order rate constant k_A is given by a combination of other constants:

$$k_A = \frac{k'_A [H_2O]}{K_a} \qquad (13\text{-}23)$$

Now we are in a better position to understand, at least qualitatively, why acid-catalyzed ester hydrolysis is relatively insensitive to electronic substituent effects. When considering the influence of an electron-withdrawing substituent on k_A, we can easily see that this substituent has two effects that work against each other. On the one hand, the substituent will decrease the $\Delta^{\ddagger}G^0$ of the rate-limiting step (i.e., increase k_A', see Reaction 2 in Fig. 13.10), while on the other hand, it will render the ester group more acidic thereby increasing the K_a of the protonated ester. As a result, any electron-withdrawing substituents make the neutral and base-catalyzed reactions more effective than the associated acid-catalyzed mechanism at near-neutral pH conditions (see discussion below). Said another way, acid-catalyzed hydrolysis will primarily be important for esters exhibiting neither electron-withdrawing substituents nor good leaving groups (i.e., also not electron-withdrawing in nature), as is the case, for example, for alkyl esters of aliphatic carboxylic acids.

Before we turn to discussing neutral and base-catalyzed hydrolysis of ester functions, we need to reflect on what structural features determine how good a leaving

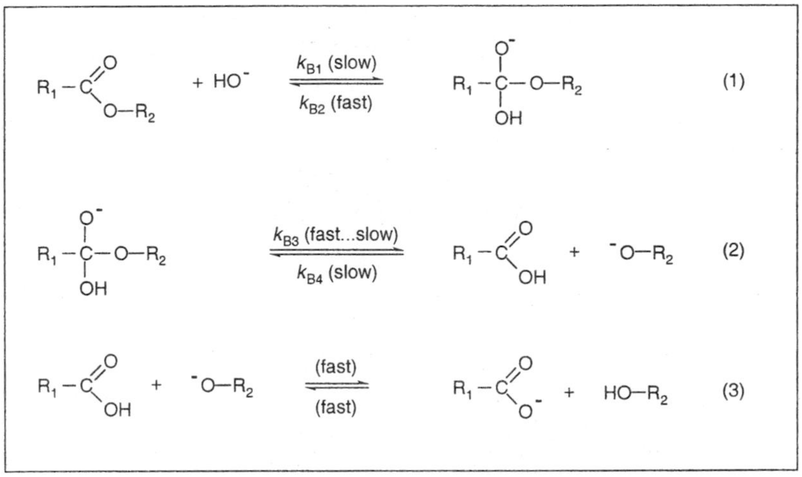

Figure 13.11 Reaction scheme for the base-catalyzed hydrolysis of carboxylic acid esters.

group a given alcohol moiety is. We have postulated that under acidic conditions, the alcohol dissociates as a neutral molecule, and that the dissociation step is not rate determining. However, in some cases under neutral, and always under basic, conditions, the alcohol moiety leaves as an anionic species (i.e., RO⁻). In these cases, the rate of dissociation of the alcohol moiety may influence the overall reaction rate. As a rule of thumb, we can relate the ease with which the RO-group dissociates with the ease with which the corresponding alcohol dissociates in aqueous solution, expressed by its pK_a value. Note that we use here again a thermodynamic argument to describe a kinetic phenomenon.

Base-Catalyzed Hydrolysis. Let us now look at the reaction of a carboxylic ester with OH⁻, that is, the *base-catalyzed hydrolysis.* The reaction scheme for the most common reaction mechanism is given in Fig. 13.11. As indicated in reaction step 2, in contrast to the acid-catalyzed reaction (Fig. 13.10), the breakdown of the tetrahedral intermediate, I, may be kinetically important. Thus we write for the overall reaction rate:

$$\frac{d\left[R_1-C\underset{O-R_2}{\overset{O}{\diagup}} \right]}{dt} = -k_{B3} \cdot [I] \tag{13-24}$$

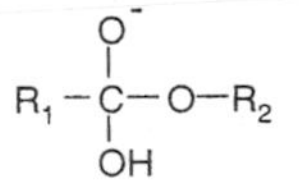

If the chain of events "backs up" at the tetrahedral intermediate (I), then this species quickly reaches an unchanging or steady-state concentration and we may write:

intermediate I

$$\frac{d[I]}{dt} = 0 = +k_{B1} \cdot \left[R_1-C\underset{O-R_2}{\overset{O}{\diagdown}} \right] \cdot \left[HO^- \right] - k_{B2} \cdot [I] - k_{B3} \cdot [I] + k_{B4} \cdot \left[R_1-C\underset{OH}{\overset{O}{\diagup}} \right] \cdot \left[R_2-O^- \right]$$

$$\tag{13-25}$$

Recognizing that R_1COOH and R_2O^- are very quickly removed by deprotonation and protonation, respectively, we may neglect the fourth term on the right-hand side of Eq. 13-25. Thus, we solve for the concentration of the intermediate I at steady state:

$$[I] = \frac{k_{B1} \cdot \left[R_1 - C \underset{O-R_2}{\overset{O}{\diagdown}} \right] \cdot \left[HO^- \right]}{k_{B2} + k_{B3}} \tag{13-26}$$

and substituting in the overall rate expression:

$$\frac{d \left[R_1 - C \underset{O-R_2}{\overset{O}{\diagdown}} \right]}{dt} = -\frac{k_{B1} \cdot k_{B3}}{k_{B2} + k_{B3}} \cdot \left[R_1 - C \underset{O-R_2}{\overset{O}{\diagdown}} \right] \cdot \left[HO^- \right] \tag{13-27}$$

we derive a rate law in terms of starting compounds and:

$$k_B = \frac{k_{B1} k_{B3}}{k_{B2} + k_{B3}} \tag{13-28}$$

For good leaving groups ($k_{B3} \gg k_{B2}$), k_B is equal to k_{B1}, meaning that solely the formation of the tetrahedral intermediate is rate determining. This is usually the case for esters exhibiting an *aromatic* alcohol moiety (e.g., for phenyl esters, where pK_a of the phenol is < 10). In the hydrolysis of alkyl esters (pK_a > 15), however, k_{B3} may be even smaller than k_{B2}, reflecting the loss of an alkoxide versus hydroxide ion from the tetrahedral intermediate. In these cases, k_B is not equal to k_{B1}, but is given by Eq. 13-28. We will come back to this issue later when discussing quantitative structure-reactivity relationships (e.g., LFERs, see below).

Neutral Hydrolysis. We finish our discussion of the major hydrolysis mechanisms of carboxylic acid esters by looking at the neutral (pH independent) reaction at the carbonyl carbon. From the reaction scheme given in Fig. 13.12, we see that, very similar to what we have postulated for the base-catalyzed reaction, the dissociation

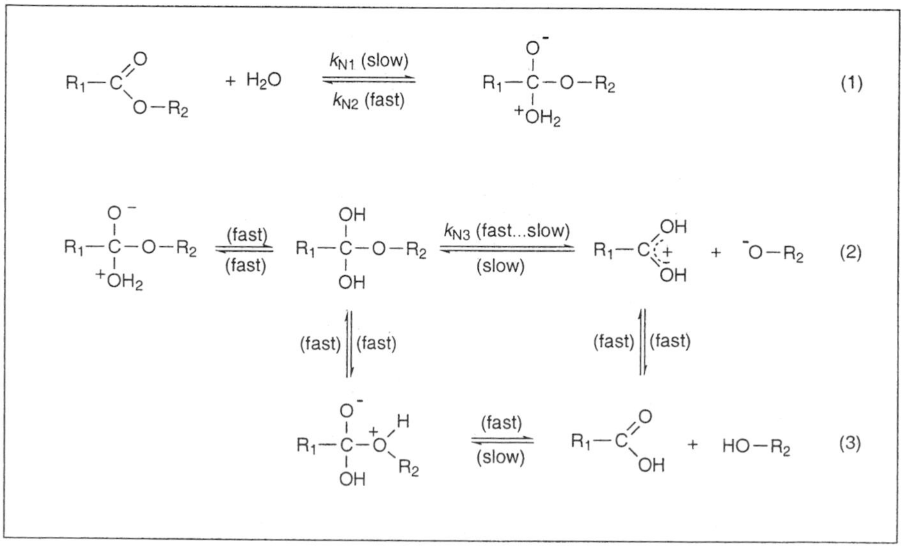

Figure 13.12 Reaction scheme for the neutral hydrolysis of carboxylic acid esters.

Table 13.9 Comparison of k_N and k_B Values of Some Carboxylic Acid Esters at 25°C and Influence of Leaving Group and Polar Substituents on k_N and k_B [a]

Compound	pK_a of ROH	Relative Value		k_B/k_N (M^{-1})
		k_N	k_B	
$H_3C-\overset{\text{O}}{\overset{\|}{C}}-O-CH_2CH_3$	≈ 16	1	1	7.3×10^8
$H_3C-\overset{\text{O}}{\overset{\|}{C}}-O-\text{(phenyl)}$	9.98	440	13	2.1×10^7
$H_3C-\overset{\text{O}}{\overset{\|}{C}}-O-\text{(2-O}_2\text{N, 4-NO}_2 \text{ phenyl)}$	3.96	73000	850	8.5×10^6
$H_2CCl-\overset{\text{O}}{\overset{\|}{C}}-OCH_3$	≈ 15	1	1	6.6×10^8
$HCCl_2-\overset{\text{O}}{\overset{\|}{C}}-OCH_3$	≈ 15	71	20	1.9×10^8
$HCCl_2-\overset{\text{O}}{\overset{\|}{C}}-O-\text{(phenyl)}$	9.98	8600	93	6.3×10^6

[a] Data derived from Table 13.8.

of the leaving group (expressed by k_{N3}) may be rate determining. In the neutral case, however, the situation is somewhat more complicated since, particularly for poor leaving groups (i.e., alkoxy groups), the alcohol moiety may leave as a neutral species and not as an anion (reaction path 3 in Fig. 13.12). This might have to be taken into account when deriving or applying LFERs to k_N values.

Let us now compare the relative importance of the neutral versus base-catalyzed hydrolysis of carboxylic acid esters. Inspection of Table 13.8 and Fig. 13.8 reveals that the relative importance of these two processes (i.e., the magnitude of the I_{NB} value) depends on both the goodness of the leaving group and on substitution in the acid part of the molecule. From the examples given in Table 13.9 we can see that structural changes, particularly with respect to the leaving group, but also to a certain extent with respect to substitution in the acid part, have a greater impact on k_N than on k_B. We can intuitively rationalize these findings by imagining that structural differences influencing the $\Delta^{\ddagger}G^0$ of the reaction will be more strongly felt by the weak nucleophile H_2O as compared to the much stronger (more electron-rich) nucleophile OH$^-$. Consequently, carboxylic acid esters exhibiting good leaving groups and/or electron-withdrawing substituents in the acid part of the molecule will have relatively high I_{NB} values (I_{NB} values of up to $7 - 7.5$). In these cases, neutral hydrolysis has to be considered at ambient pH values, and, as we recall from our earlier discussion, the acid-catalyzed reaction can be neglected (see examples given in Fig. 13.8 and Table 13.8). On the other hand, rate-decreasing substituents (i.e., alkyl groups) will decrease I_{NB} and increase I_{AN}, leading to a situation as depicted in Fig. 13.9b.

Additional Reaction Mechanisms. So far we have confined our discussion to the most common case of ester hydrolysis, that is, the case in which the reaction takes place at the carbonyl carbon. In some cases, however, an ester may also react in water by an S_N-type or E-type mechanism (see Section 13.2) with the acid moiety (i.e., $^-OOC - R_1$) being the leaving group. The S_N-type reactions occur primarily with esters exhibiting a tertiary alcohol group. The products of this reaction are the same as the products of the common hydrolysis reaction. In the case of elimination, however, products are different since the ester is converted to the olefin and the corresponding conjugate base of the acid:

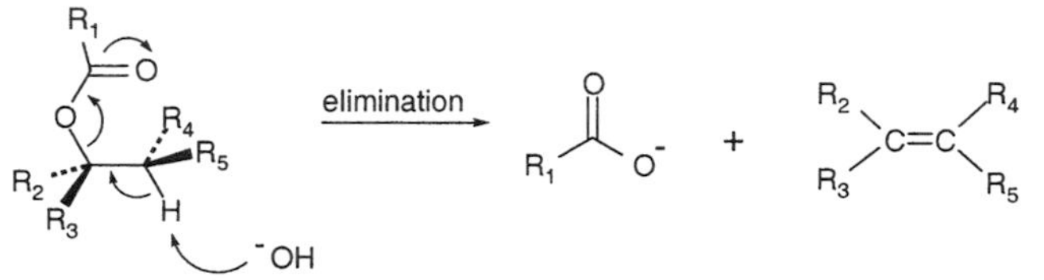

$$(13\text{-}29)$$

Elimination according to Eq. 13-29 will be important for compounds exhibiting acidic protons, in β-position to the alcoholic carbon forming the ester bond.

Finally, if the α-carbon of the acid moiety (i.e., the carbon bound to the carbonyl carbon) is substituted by an electron-withdrawing group that renders the α-hydrogens more acidic, the ester may hydrolyze by an elimination mechanism involving a ketene intermediate:

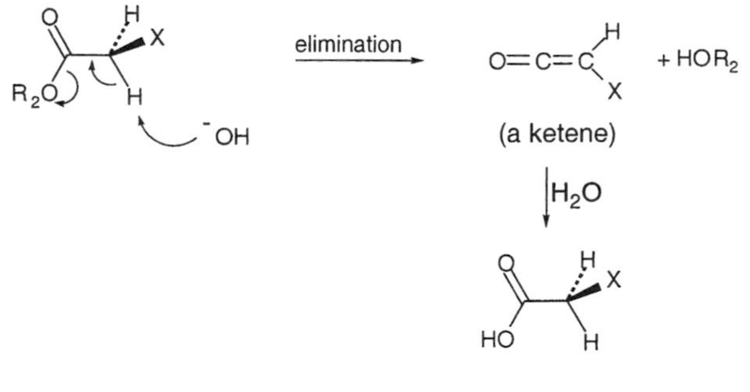

$$(13\text{-}30)$$

where the second step of reaction 13-30 is an addition of H_2O to the carbon–carbon double bond (for more details see, e.g., March, 1992). We will encounter analogous mechanisms when discussing the hydrolysis of carbamates (see below).

Carboxylic Acid Amides

Amide functions are very important linkages in natural compounds (e.g., in proteins) and some simple amides are used in industry. Furthermore, numerous herbicides contain amide groups (Montgomery, 1997). Generally an amide bond is defined as:

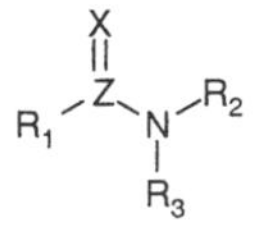

where R_2 and R_3 are hydrogens or carbon-centered substituents.

Table 13.10 Rate Constants k_A and k_B, Half-Lives at pH 7, and I_{AB} Values for Hydrolysis of Some Amides at 25°C [a]

Compound

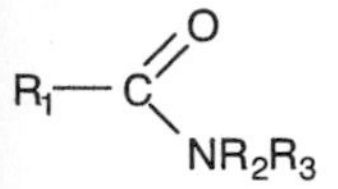

R_1	R_2	R_3	k_A $(M^{-1} s^{-1})$	k_B $(M^{-1} s^{-1})$	$t_{1/2}$ (pH 7)	I_{AB}
CH_3-	$-H$	$-H$	8.4×10^{-6}	4.7×10^{-5}	4000 yr	6.6
$i\text{-}C_3H_9-$	$-H$	$-H$	4.6×10^{-6}	2.4×10^{-5}	7700 yr	6.6
CH_2Cl-	$-H$	$-H$	1.1×10^{-5}	1.5×10^{-1}	1.5 yr	4.9
CH_3-	$-CH_3$	$-H$	3.2×10^{-7}	5.5×10^{-6}	40,000 yr	6.4
CH_3-	$-CH_3$	$-CH_3$	5.2×10^{-7}	1.1×10^{-5}	20,000 yr	6.3

[a] Data from Mabey and Mill (1978).

The hydrolysis of carboxylic acid amides (i.e., Z = C and X = O) can be treated much like the hydrolysis of carboxylic acid esters; that is, a similar structure-reactivity pattern is found (Talbot, 1972). Compared to ester functions, however, amide functions are in general much less reactive since the $-NR_2R_3$ group is less electronegative than the $-OR_2$ group. Even more important, the $-NR_2R_3$ group is a much poorer leaving group [the pK_a's of amines ($R_1R_2NH \rightleftharpoons R_1R_2N^-$) are much larger than those of alcohols ($ROH \rightleftharpoons RO^-$)]. Due to these factors, and because amide groups are quite basic, neutral hydrolysis is usually unimportant relative to the acid- or base-catalyzed reaction (case *b* in Fig. 13.9). Furthermore, because the amide group is more basic than the ester group:

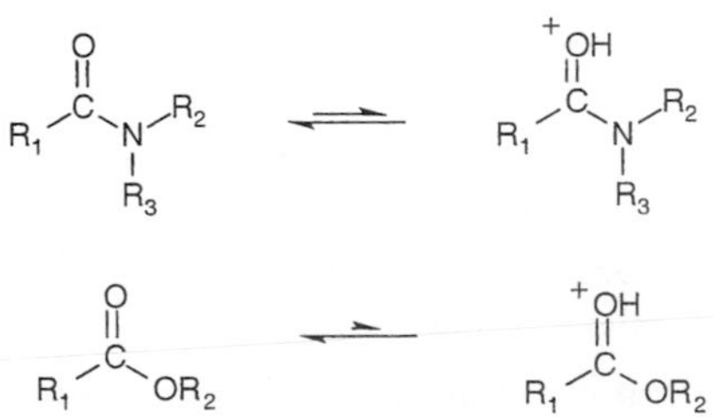

the I_{AB} values of amides are commonly higher than those of ester functions. In many cases, k_A and k_B are of similar magnitude, unless the amide function is substituted with electron-withdrawing groups or atoms (see examples given in Table 13.10; e.g., $R_1 = CH_2Cl$). Note that because the hydrolysis half-lives of most of the compounds shown in Table 13.10 are very large under ambient conditions, many of the values given are only order-of-magnitude estimates that have been extrapolated from measurements conducted at elevated temperatures and extreme pH values. Finally, we note that activation energies are typically between 80 and 90 kJ·mol^{-1} for the acid-catalyzed hydrolysis of amide functions, and between 50 and 80 kJ·mol^{-1} for the base-catalyzed reaction (Mabey and Mill, 1978).

Carbamates

The next group of compounds that we want to look at are derivatives of carbamic acid ($HO-CO-NH_2$), that is, the carbamates. Carbamates are widely used as herbicides and insecticides. The carbamate function exhibits both an ester and an amide-type functionality (Montgomery, 1997):

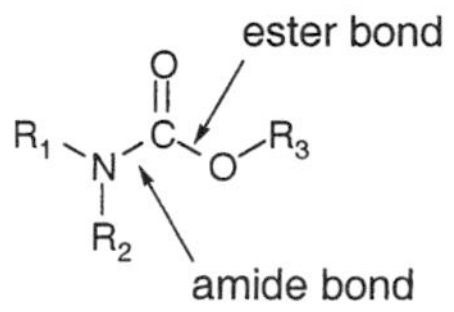

where R_1 and R_2 are hydrogens or carbon-centered substituents, and R_3 is a carbon-centered substituent. Hence, a carbamate has two potential leaving groups: an alcohol and an amine. Since, in most cases, the alcohol moiety will be the better leaving group, the initial hydrolysis reaction commonly occurs by cleavage of the ester bond. Initial breaking of the amide bond may, for example, occur if R_3 is an alkyl group, and R_1 and R_2 are aromatic rings that are substituted with electron-withdrawing substituents (and thus stabilize the $R_1R_2N^-$ anion in aqueous solution). However, regardless of the reaction mechanism, the hydrolysis of carbamates eventually yields the alcohol (R_3OH), the amine (R_1R_2NH), and CO_2 (see below).

Since base catalysis plays an important role in the hydrolytic breakdown of carbamates, this process has been investigated quite extensively. However, very few data are available on the neutral reaction, which, in some cases, may be significant at ambient pH values. The acid-catalyzed reaction, on the other hand, can generally be neglected. This is presumably because so many electron-withdrawing atoms surround the central carbon that protonation of the carboxyl oxygen insignificantly enhances its susceptibility to nucleophilic attack.

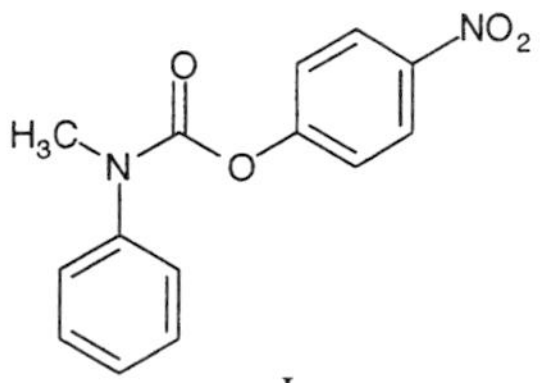

I

$t_{1/2}$ (pH7) = 275 yr

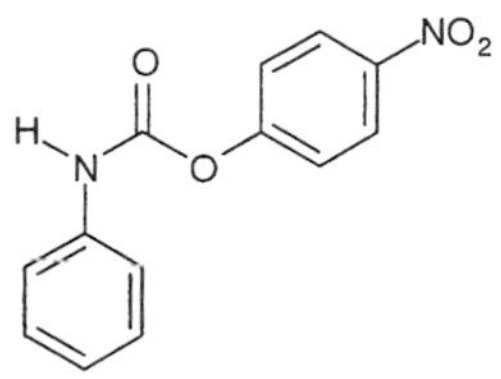

II

$t_{1/2}$ (pH7) = 25 s

When considering the base-catalyzed hydrolysis of carbamate functions, the critical question is whether one of the groups bound to the nitrogen (R_1, R_2) is a hydrogen atom. This becomes obvious when we compare the $t_{1/2}$ values of compounds I and II (see margin). First, we realize that although the *p*-nitrophenol group is a good leaving group, the base-catalyzed hydrolysis of 4-nitrophenyl N-methyl-N-phenyl carbamate (I) is very slow. In this case, by analogy to what we have postulated for most ester and amide functions, the rate-determining step is the formation of a tetrahedral intermediate (see reaction step 1 in Fig. 13.13). We note that the hydrolysis of the ester bond is generally followed by a fast decarboxylation reaction yielding the corresponding amine (reaction step 4 in Fig. 13.13). An analogous reaction occurs if the amine is the leaving group, yielding an unstable carbonic acid monoester that also hydrolyzes rapidly. As is illustrated by the k_B values of some other N,N-disubstituted carbamates (see Table 13.11), we can conclude that such compounds are generally quite resistant to base-catalyzed hydrolysis. From the data in Table 13.11 we can also see that the base-catalyzed hydrolysis of N,N-disubstituted carbamates is somewhat insensitive to the nature of the alcohol moiety, indicating that dissociation of the leaving group is not rate determining.

In contrast, for the series of the N-monosubstituted carbamates (R_1 = H, Table 13.11),

Table 13.11 Rate Constants k_B (and k_N), Half-Lives at pH 7, and I_{NB} Values for Hydrolysis of Some Simple Carbamates at 25°C [a,b]

Compound

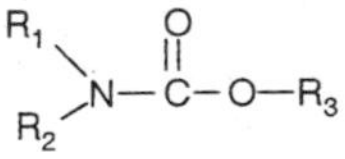

R_1	R_2	R_3	k_N (s^{-1})	k_B $(M^{-1} s^{-1})$	$t_{1/2}$ [c] (pH 7)
CH_3-	CH_3-	$-CH_2CH_3$	NA	4.5×10^{-6}	50,000 yr
CH_3-	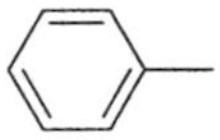 (phenyl)	$-CH_2CH_3$	NA	4.0×10^{-6}	55,000 yr
CH_3-	CH_3-	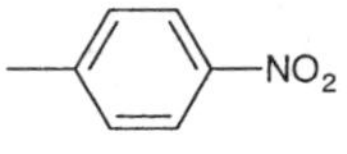 (aryl)$-NO_2$	NA	4.0×10^{-4}	550 yr
CH_3-	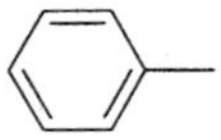 (phenyl)	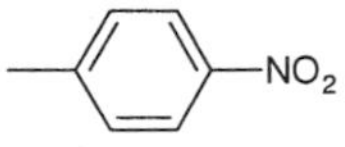 (aryl)$-NO_2$	NA	8.0×10^{-4}	275 yr
$H-$	CH_3-	$-CH_2CH_3$	NA	5.5×10^{-6}	40,000 yr
$H-$	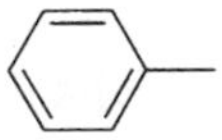 (phenyl)	$-CH_2CH_3$	NA	3.2×10^{-5}	7,000 yr
$H-$	CH_3-	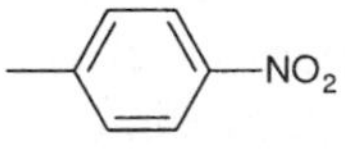 (aryl)$-NO_2$	NA	6.0×10^2	3 h
$H-$	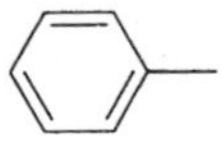 (phenyl)	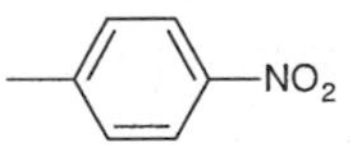 (aryl)$-NO_2$	NA	2.7×10^5	25 s
$H-$	CH_3-	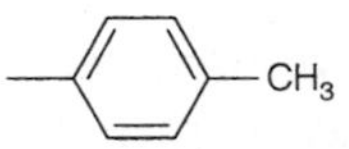 (aryl)$-CH_3$	6.0×10^{-8}	5.6×10^{-1}	70 d [d,e]
$H-$	CH_3-	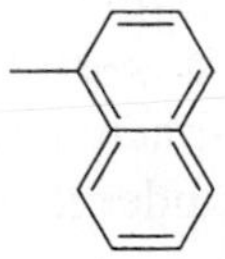 (naphthyl)	9.0×10^{-7}	5.0×10^1	33 h [d,f]

[a] Data from Dittert and Higuchi (1963), Williams (1972, 1973), Vontor et al. (1972), and El-Amamy and Mill (1984). [b] NA = not available. [c] Half-life for base-catalyzed reaction; actual half-life may be shorter. [d] Half-life for neutral and base-catalyzed reaction. [e] $I_{NB} = 7.01$. [f] $I_{NB} = 6.25$.

the reaction occurs much faster in most cases. It is also very sensitive to the type of alcohol moiety. This indicates that, in this case, dissociation of the leaving group is rate limiting, and that the reaction proceeds by a different mechanism. The generally accepted mechanism (e.g., Bender and Homer, 1965; Williams, 1972) involves deprotonation of the amide function (similar to esters with acidic protons at the carbon

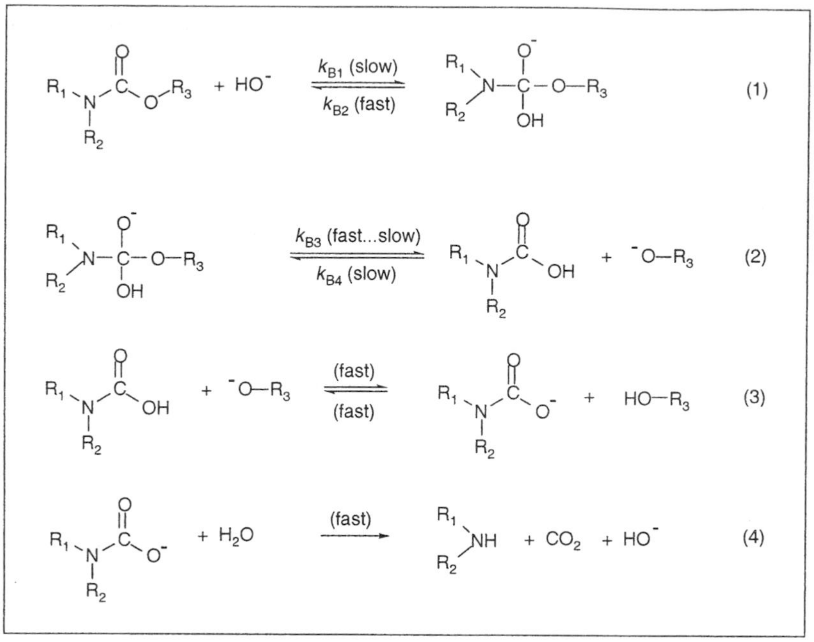

Figure 13.13 Reaction scheme for the base-catalyzed hydrolysis of carbamates when the mechanism involves a tetrahedral intermediate.

adjacent to the ester group, see Eq. 13-30), with subsequent elimination of the alkoxy group (Fig. 13-14). The resulting isocyanate ($R_1N = C = O$) is then rapidly converted to the amine and CO_2 by addition of water and subsequent decarboxylation.

Comparison of the k_B values of N,N-disubstituted versus N-monosubstituted carbamates (Table 13.11) shows that if the alcohol moiety is a good leaving group (i.e., aromatic ring-carrying electron-withdrawing substituents), differences in half-lives for the base-catalyzed reaction of up to 10 orders of magnitude may be found. Only if the leaving group is very poor (e.g., R_3 = alkyl) can the reaction occurring via a tetrahedral intermediate (Fig. 13.13) compete with the reaction involving an elimination step (Fig. 13.14).

From Table 13.11 we also see that the N-methyl carbamates often have significantly smaller k_B values as compared to the corresponding N-phenyl carbamates. We can rationalize these findings by the anion-stabilizing effect of the phenyl group, an effect also reflected by the N–H proton being more acidic than that of the N-methyl compound. Thus, the N-phenyl carbamates have a greater fraction in the deprotonated species available to undergo elimination.

As is indicated in Table 13.11, hydrolysis half-lives of the N,N-disubstituted carbamates (and the monosubstituted carbamates exhibiting poor leaving groups) are very large at ambient pH values. Although there are virtually no rate data available on the neutral reaction of such slowly hydrolyzing compounds, one could speculate that, as found for the very slowly hydrolyzing esters (Fig. 13.8), the neutral reaction should not be too important. Hence, k_N will be too small to be of environmental significance. For the more reactive N-monoalkyl carbamates, however, the neutral reaction might have to be considered. From the few compounds for which k_N values have been reported (Table 13.11), and by using our chemical intuition, we may con-

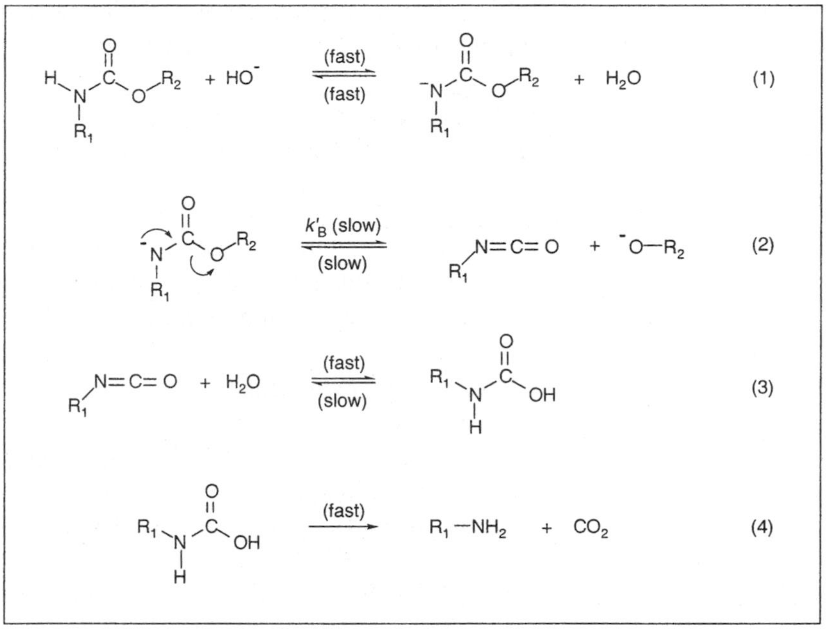

Figure 13.14 Reaction scheme for the base-catalyzed hydrolysis of carbamates when the mechanism involves an elimination step.

clude that the relative importance of the neutral reaction increases (i.e., the I_{NB} value increases) with increasing reactivity. Finally, we note that activation energies for the base-catalyzed hydrolysis of carbamates span a wide range of between 50 and 100 kJ·mol^{-1} (Christenson, 1964).

Quantitative Structure-Reactivity Considerations

In order to understand available kinetic data better and potentially allow estimation of reaction rates of new compounds within a compound class, we often want to develop quantitative relationships between the structures of individual compounds and their reactivities. Such relationships usually involve situations in which the standard free energies of activation, $\Delta^{\ddagger}G^0$, vary systematically with chemical structure changes. In these cases, we may try to apply linear free-energy relationships (LFERs) in a manner similar to our approach for evaluating or estimating equilibrium constants (e.g., acidity constants in Chapter 8). The use of such LFERs to relate kinetic data for a given reaction of a series of structurally related compounds hinges on the ability to express quantitatively the *electronic* and *steric* effects of structural moieties of the reactants on $\Delta^{\ddagger}G^0$. When dealing with hydrolytic reactions of carboxylic and carbonic acid derivatives, in general, both types of effects have to be taken into account. In this context, it has to be pointed out that quantification of steric effects is much more difficult as compared to electronic effects (see, e.g., Exner, 1988). In our following discussion we will, therefore, confine ourselves to cases in which electronic effects predominate. For approaches in which steric effects are included, particularly the approach introduced by Taft and co-workers, we refer to the literature (Taft 1956; Pavelich and Taft, 1957; MacPhee et al., 1978, Williams, 1984).

Hammett Relationship

In our first example we evaluate the influence of *meta* and *para* ring substituents on the base-catalyzed hydrolysis of substituted benzoic acid esters:

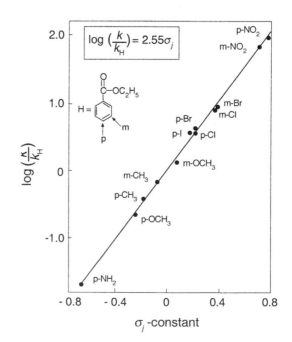

In this case, structural changes are made at points in the structure that are well away from the reaction site. As discussed above (Fig. 13.11), we can assume a tetrahedral activated complex exhibiting a negative charge that is not significantly delocalized into the ring system. Thus, intuitively, we expect that through an *inductive* effect, electron-withdrawing ring substituents (–X) will stabilize the negatively charged activated complex relative to the uncharged ground state; that is, they will decrease $\Delta^{\ddagger}G^0$ as compared to the unsubstituted compound and, therefore, increase k relative to k_H (unsubstituted compound). Conversely, an electron-donating substituent exerts the opposite effect. As we have seen in Chapter 8, the inductive effect of aromatic substituents in meta or para positions may be quantitatively expressed by the *Hammett* $\sigma_{j\text{meta}}$ and $\sigma_{j\text{para}}$ substituent constants (see Table 8.5). It comes as no surprise that, in this case, where we deal primarily with inductive effects, the rate constants of the base-catalyzed hydrolysis of meta and para monosubstituted benzoic acid ethyl esters (R =C$_2$H$_5$) can be related successfully by the *Hammett* equation (Fig. 13.15):

$$\log\left(\frac{k}{k_H}\right) = \rho\sigma_{jm,p} \tag{13-31}$$

From the linear fit shown in Fig. 13.15, a ρ value of 2.55 (Exner, 1988) is obtained at 25°C, indicating a substantial influence of the substituents on the reaction rate. For

Figure 13.15 Effects of substituents on the base-catalyzed hydrolysis of benzoic acid ethyl esters in ethanol:water (85:15) at 25°C. Relative reaction rates are correlated with Hammet σ_j constants (data from Tinsley, 1979).

example, a nitro group in the meta or para position increases the rate of hydrolysis by a factor of 100. It should be pointed out that the data shown in Fig. 13.15 were not obtained in pure water, but in a mixture of ethanol and water (85:15), for which the most complete data set is available.

A quite similar ρ value ($\rho = 2.47$) is found for the same reaction in acetone/water (3:2, Exner, 1988), but a significantly smaller value ($\rho = 1.77$) is derived in pure water (Drossman et al., 1988). These findings are consistent with the results obtained for the base-catalyzed hydrolysis of meta and para substituted benzoic acid methyl esters (R = CH_3, instead of CH_2CH_3), for which at 25°C, the corresponding ρ values are 2.38 in acetone/water (3:2, Exner, 1988) and 1.67 in pure water (Smith and Menger, 1969; Steinberg and Lena, 1995), respectively. The fact that ρ is much smaller in water as compared to organic solvent/water mixtures can be rationalized by the better ability of the solvent, water, to polarize the carboxyl bond, which leads to a reduction of the substituent effect. Hence, as already pointed out in Chapter 12 (Table 12.1), one has to be very careful when trying to extrapolate hydrolysis rate data from nonaqueous to aqueous solutions. Furthermore, we should note that ρ is also a function of temperature since, as implied by Eqs. 12-44 and 12-45, the effect of temperature on the rate constant is different for compounds exhibiting different $\Delta^{\ddagger}H^0$ (or E_a) values, the relative size of which is determined by the type of substituent(s) present. For example, in the case of the hydrolysis of the benzoic acid ethyl esters in ethanol/water (85:15), ρ decreases from 2.55 at 25°C to 2.13 at 50°C (Exner, 1988).

III

IV

Since in the case of the benzoic acid methyl and ethyl esters we have left the alcohol moiety (–O–R) invariant, ρ primarily reflects structural effect on the k_{B1} term in Eq. 13-28. Although the dissociation of the leaving group may also determine the overall k_B term it is the same within the series considered. For substituted phenyl benzoates (III) and phenyl acetates (IV), the situation is somewhat different, since the substituents exhibit effects on k_{B1} as well as on k_{B3}. These effects occur parallel to each other. For example, an electron-withdrawing substituent increases k_{B1} by an inductive effect and, at the same time, it renders the alcohol moiety a better leaving group (it decreases the pK_a of the alcohol). If the dissociation of the leaving group (i.e., the phenolate species) is rate determining, we expect a ρ value similar to or even greater than that found for the benzoic acid esters discussed above (although there is an oxygen between the phenyl group and the carbonyl carbon which renders the electronic effect of substituents on k_{B1} smaller as compared to the benzoic acid ethyl esters). However, the observed ρ value derived from k_B values of some substituted phenyl acetic acid esters in aqueous solution is very small, being on the order of 1 (Drossman et al., 1988). This result implies that the effect of electron-withdrawing substituents in enhancing the combined k_B expression (Eq. 13-28) is even less than the impact of the same substituents on k_{B1} rates for benzoic acid methyl and ethyl esters. This suggests that for the phenyl esters, the rate of dissociation of the leaving group is not significant in determining the overall reaction rate. This example nicely demonstrates how such LFERs may be useful for prediction of rate constants and may also give valuable hints regarding rate-determining steps. An additional example which also demonstrates how to derive a Hammett equation from kinetic data is given in Illustrative Example 13.6.

Illustrative Example 13.6 **Estimating Hydrolysis Rate Constants Using the Hammett Relationship**

Consider the base catalyzed hydrolysis of 3,4,5-trichlorophenyl-N-phenyl carbamate:

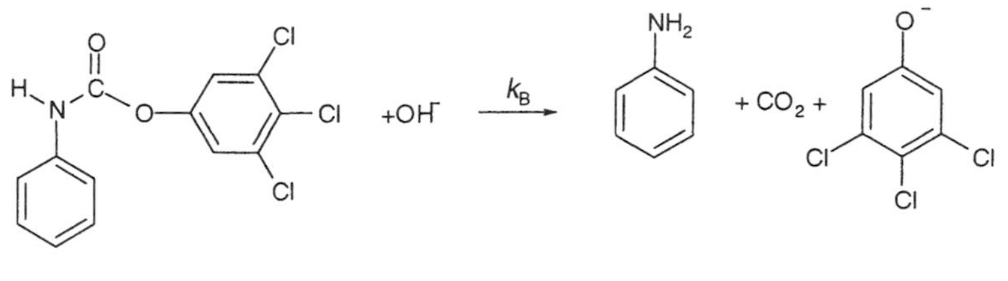

Problem

Estimate the second-order rate constant, k_B, at 25°C for reaction Eq. 1 using k_B values reported below for other substituted phenyl N-phenyl carbamates.

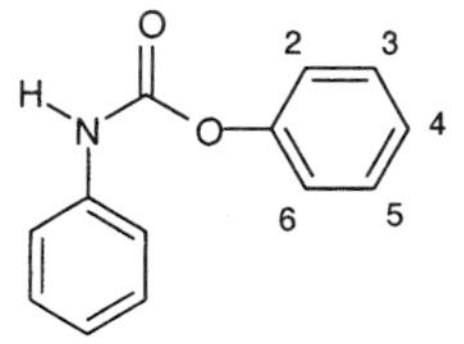

Substituent	σ_j [a]	k_B (M^{-1} s^{-1}) [b]
4–OCH$_3$	–0.24	2.5×10^1
4–CH$_3$	–0.16	3.0×10^1
4–Cl	0.22	4.2×10^2
3–Cl	0.37	1.8×10^3
3–NO$_2$	0.73	1.3×10^4
4–NO$_2$	0.78 (1.25) [c]	2.7×10^5

[a] See Table 8.5. [b] Data from references given in Table 13.11. [c] σ^-_{jpara}

Answer

Use the Hammett equation to relate the k_B values of the six compounds for which data are available:

$$\log k_B = \rho \cdot \sum_j \sigma_j + c$$

where ρ is susceptibility factor and c is a constant corresponding to the log k_B of the unsubstituted compound (i.e., of phenyl N-phenyl carbamate, see margin).

Determine ρ and c from a least squares fit of log k_B versus σ_j (see Table 8.5) for the monosubstituted compounds found in the literature (see also Fig. below):

$$\log k_B \text{ (in M}^{-1}\text{ s}^{-1}) = 2.82 \sum_j \sigma_j + 2.02 \ (R^2 = 0.99) \tag{2}$$

As is evident from the plot of log k_B versus σ_j for the nitro substituent in 4-position, the σ^-_{jpara} (i.e., 1.25) and not the σ_{jpara} (i.e., 0.78) has to be used, indicating that resonance with the phenolic group is important. Together with the rather high ρ value of 2.82, this suggests that the dissociation of the leaving group (i.e., the phenolate) is the rate-determining step (in contrast to the phenyl esters). This is consistent with the elimination mechanism proposed in Fig. 13.14.

Insertion of $\sum_j \sigma_j = 2 \cdot (0.37) + (0.22) = 0.96$ for 3,4,5-trichlorophenyl-N-phenyl carbamate into Eq. 2 then yields:

$$\log k_B \text{ (in M}^{-1}\text{ s}^{-1}) = 4.73 \text{ or } k_B = 5.3 \times 10^4 \text{ M}^{-1}\text{ s}^{-1} \text{ (at 25°C)}$$

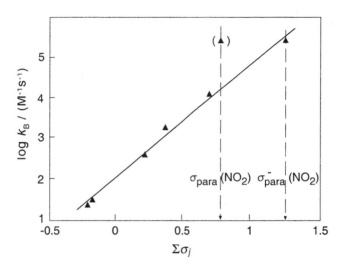

Brønsted Relationship

As we have seen, the Hammett equation can, in principle, be applied to both equilibria and rate data. This implies that in certain cases, it is feasible to relate rate constants to equilibrium constants when both reflect the effects of a given structural moiety. In a general form a rate-equilibrium relationship can be written in terms of the corresponding changes in free energies of activation and of equilibration:

$$\Delta\Delta^{\ddagger}G^{0} = \beta\Delta\Delta_{r}G^{0} \tag{13-32}$$

where the first Δ indicates the incremental differences between the $\Delta^{\ddagger}G^{0}$ and $\Delta_{r}G^{0}$ values, respectively, of a series of structurally related compounds. In terms of rate and equilibrium constants, Eq. 13-32 can be expressed as:

$$\log\frac{k}{k_{H}} = \beta \cdot \log\frac{K}{K_{H}} \tag{13-33}$$

where the subscript H denotes a reference compound (e.g., the unsubstituted compound). A very common application of Eq. 13-33 is the use of acidity constants (K_{a}) of a (sub)structural (sub)unit (e.g., the K_{a} of a leaving group) to relate rate constants for hydrolytic reactions of a series of compounds. In this case, Eq. 13-33 is commonly referred to as a *Brønsted relationship*, and can be written as:

$$\log k = -\beta \cdot pK_{a} + C \tag{13-34}$$

where C is the logarithm of the rate constant of the compound for which the corresponding pK_{a} value is zero.

An example in which rate constants are related to equilibrium constants involves the base-catalyzed hydrolysis of N-phenyl carbamates (Fig. 13.16). As discussed above, these compounds hydrolyze with the dissociation of the alcohol moiety being the rate-determining step. Hence, by using the pK_{a}'s of the leaving groups (phenols and aliphatic alcohols), we find a nice correlation to the rates of these reactions.

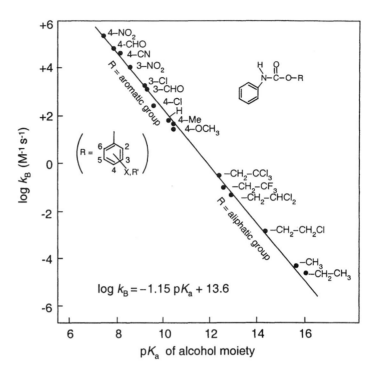

Figure 13.16 Correlation (Brønsted plot) of base-catalyzed hydrolysis rates (log k_B) of carbamates as a function of the pK_a of the alcohol moiety for a series of N-phenyl carbamates. Data from references given in Table 13.11.

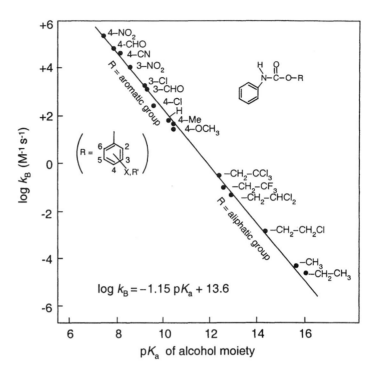

(thio)phosphoric acid ester

(thio)phosphoric acid thioester

Figure 13.17 General structure of phosphoric and thiophosphoric acid (thio)esters. Note that often $R_1 = R_2 = CH_3$ or C_2H_5.

13.4 Hydrolytic Reactions of Phosphoric and Thiophosphoric Acid Esters

The last group of compounds whose hydrolyses we will discuss are the phosphoric and thiophosphoric acid (thio)esters (Fig. 13.17). We use these examples to gain some insight into the reactivity of compounds exhibiting a pentavalent phosphorus atom. Also this treatment will show that the concepts we have discussed so far in this chapter can be used to understand hydrolysis reactions of other kinds of compounds. Because of their significant biological activity (cholinesterase inhibition), esters and thioesters of phosphoric acid and thiophosphoric acid are widely applied as insecticides (Khan, 1980; Gianessi and Anderson, 1995; Montgomery, 1997). Note that the thionate (P = S) esters exhibit a lower mammalian toxicity than corresponding oxonate (P = O) esters and that, for this reason, they are more widely used. The thionate esters are, however, converted to the oxonates by oxidation both inside and outside of organisms (Eto, 1979; Hassal, 1990). In addition to the use of the phosphoric acid esters as biocides, some trialkyl and triaryl phosphates are used in very large quantities in fire-resistant hydraulic fluids and as fire-retardant plasticizers (see Chapter 2). Consequently, such compounds are of great environmental significance and concern. Interestingly, despite the widespread use of phosphate and thiophosphate esters and thioesters, there are still few data available in the open literature on the reactivity of such compounds in aqueous solution. In addition, rate constants reported for a given compound often differ by more than an order of magnitude between different authors. Nevertheless, it is possible to draw some important general conclusions from the available data concerning the hydrolytic decomposition of this group of compounds. Furthermore, the following examples shall give us an additional opportunity to deepen our knowledge of organic reactions involving nucleophilic species.

Table 13.12 Rate Constants k_A, k_N, and k_B, Half-Lives, $t_{1/2}$, at pH 7, and I_{NB} Values for Hydrolysis of Some Phosphoric and Thiophosphoric Acid Triesters at 25°C [a]

Compound Name	Structural Formula	k_A [b]	k_N (s^{-1})	k_B $(M^{-1} s^{-1})$	$t_{1/2}$ (pH 7)	I_{NB}
Trimethylphosphate	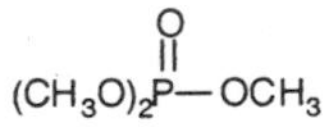	NI	1.8×10^{-8}	1.6×10^{-4}	1.2 yr	10.0
Triethylphosphate	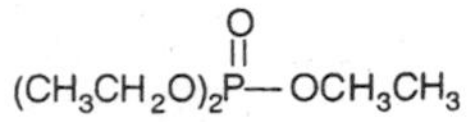	NI	$\approx 4 \times 10^{-9}$	8.2×10^{-6}	≈ 5.5 yr	10.7
Triphenylphosphate	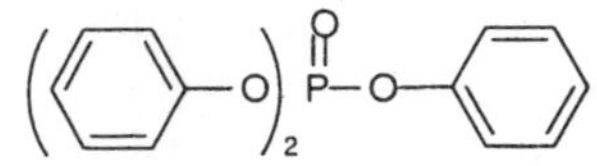	NI	$< 3 \times 10^{-9}$	2.5×10^{-1}	320 d	< 6
Paraoxon	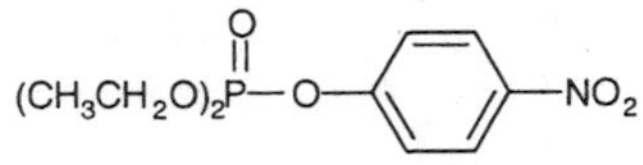	NI	7.3×10^{-8}	3.9×10^{-1}	72 d	7.3
Parathion	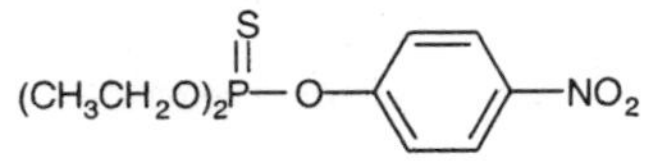	NI	8.3×10^{-8}	5.7×10^{-2}	89 d	8.2
Methylparathion	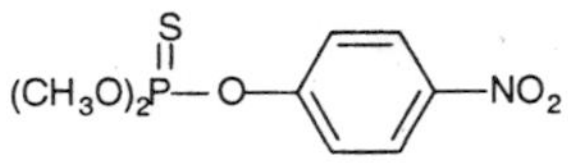	NI	1.2×10^{-7}	1.1×10^{-2}	67 d	9.0
Thiometon [c]	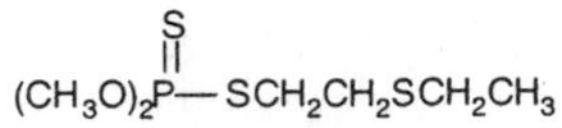	NI	1.1×10^{-7}	6.4×10^{-3}	73 d	9.4
Disulfoton [c]	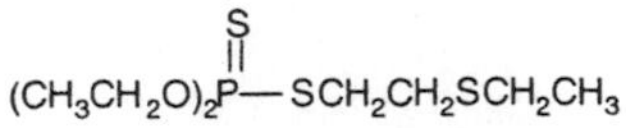	NI	1.4×10^{-7}	2.0×10^{-3}	57 d	10.0
Diazoxon [c]	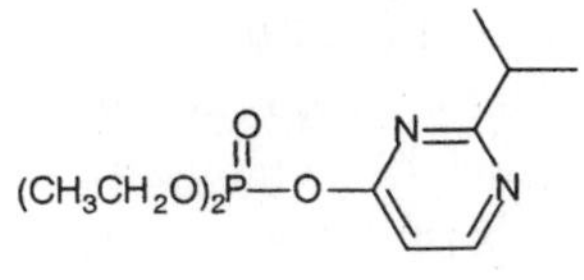	6.5×10^{-1}	2.8×10^{-7}	7.6×10^{-2}	23 d	8.6 [d]
Diazinon [c]	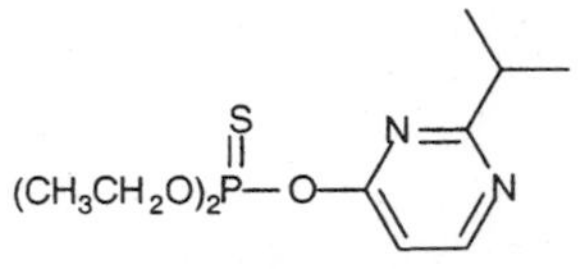	2.1×10^{-2}	4.3×10^{-8}	5.3×10^{-3}	178 d	8.9 [e]

[a] Data from Faust and Gomaa (1972), Mabey and Mill (1978), and Wanner et al. (1989). [b] NI = not important. [c] At 20°C. [d] $I_{AN} = 6.4$. [e] $I_{AN} = 5.7$.

As for other esters, hydrolysis of phosphoric and thiophosphoric acid triester occurs via acid-catalyzed, base-catalyzed, and neutral mechanisms (Table 13.12). We note that in the following discussion we are concerned primarily with acid triesters, although the hydrolysis products of these compounds, that is, the di- and monoesters, are also of environmental concern inasmuch as they usually seem to react at slower rates as compared to the triesters (Mabey and Mill, 1978; Wolfe, 1980).

When trying to understand the reactivity of phosphate and thiophosphate esters, it is important to realize that such compounds may react like alkyl halides by nucleophilic displacement (S_N2) both at the phosphorus atom (with an alcohol moiety being the leaving group) and at the carbon bound to the oxygen of an alcohol moiety (with the diester being the leaving group):

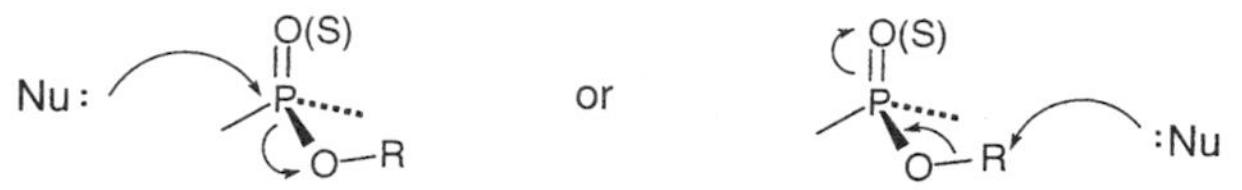

$$(13-35)$$

Note that the reaction at the phosphorus atom is postulated to occur by an S_N2 (no intermediate formed) rather than by an addition mechanism such as we encountered with carboxylic acid derivatives (Kirby and Warren, 1967). As we learned in Section 13.2, for attack at a saturated carbon atom, OH^- is a better nucleophile than H_2O by about a factor of 10^4 (Table 13.2). Toward phosphorus, which is a "harder" electrophilic center (see Box 13.1), however, the relative nucleophilicity increases dramatically. For triphenyl phosphate, for example, OH^- is about 10^8 times stronger than H_2O as a nucleophile (Barnard et al., 1961). Note that in the case of triphenyl phosphate, no substitution may occur at the carbon bound to the oxygen of the alcohol moiety, and therefore, neutral hydrolysis is much less important as compared to the other cases (see I_{NB} values in Table 13.12). Consequently, the *base-catalyzed reaction* generally occurs at the phosphorus atom leading to the dissociation of the alcohol moiety that is the best leaving group (P–O cleavage), as is illustrated by the reaction of parathion with OH^-:

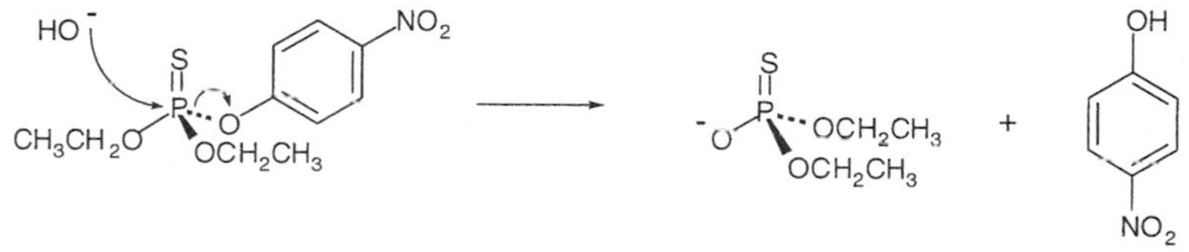

$$(13-36)$$

Depending on the alcohol moieties present (i.e., quality of leaving group(s), presence of an aliphatic alcohol moiety), the *neutral reaction* as well as reactions with soft nucleophiles (e.g. HS^-, CN^-, see Box 13.1) may also proceed by nucleophilic substitution at a carbon atom (C–O cleavage). This is the case for trialkyl phosphates such as trimethyl and triethyl phosphate:

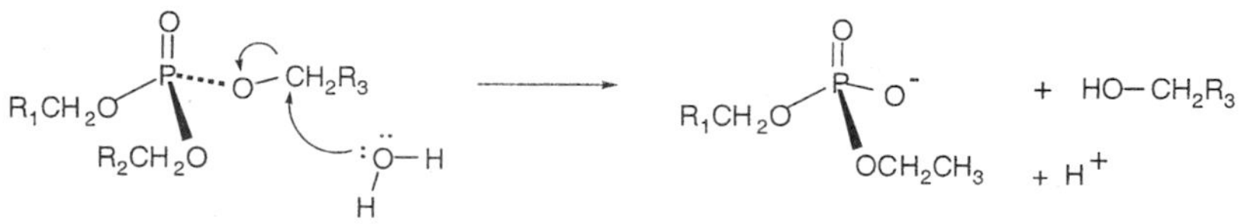

$$(13-37)$$

Note that if the reaction occurs by mechanism Eq 13-37 in analogy to what we have encountered with S_N2 reactions of primary alkyl halides, methyl esters will react faster than the corresponding ethyl or other primary alkyl esters. Of course, if a good leaving group is present, the neutral reaction may proceed by both reaction mechanisms, that is, C–O as well as P–O cleavage. For example, Weber (1976) found that at 70°C and pH 5.9, parathion reacted 90% by C–O cleavage. At lower temperatures, a higher proportion of the neutral reaction occurred by P–O cleavage. This observation can be explained by the higher activation energy of the reaction involving C–O cleavage as compared to P–O cleavage. This simple example shows us that when dealing with phosphoric acid and thiophosphoric acid derivatives, we have to be aware that under different conditions, different hydrolysis mechanisms may predominate.

In most cases, hydrolysis of phosphoric and thiophosporic acids is quite insensitive to acid catalysis unless there is a base function present in one of the alcohol moieties. If such a base is protonated, the reactivity is enhanced. Examples are the two insecticides diazoxon and diazinon (Table 13.12), where protonation of one of the nitrogens of a pyrimidine ring renders the alcohol moiety a much better leaving group. Furthermore, comparison of the relative reactivities of phosphoric and thiophosphoric esters indicates that, in many cases, the thionates hydrolyze somewhat more slowly than the corresponding oxonate esters. This can be rationalized by the higher electronegativity of oxygen as compared to sulfur. The presence of oxygens makes the phosphorus atoms somewhat more electrophilic (enhancement of the reaction at the P atom) as well as the diester a better leaving group (C–O cleavage, see Eq. 13-35). However, due to the large scatter in experimental data, these differences are not always obvious.

There are quite a few phosphoric and thiophosphoric acid derivatives exhibiting one thioester and two (often identical) ester groups ($R_1 = R_2$ = methyl or ethyl, see Fig. 13.17). In these cases, the situation is now even more complicated. Depending on R_1, R_2, and R_3, the compound may react by P–O, P–S, C–O, and C–S cleavage, giving rise to a variety of possible products. If R_1 and R_2 are methyl or ethyl, the base-catalyzed reaction generally occurs by P–S cleavage with ^-S–R_3 being the leaving group. The neutral reaction, however, may proceed by cleavage of P–S, C–O, or C–S, each alone or in combination. The C–S cleavage may preferably occur if the R_3 moiety contains a nucleophilic group, which, by internal nucleophilic attack (S_Ni) may favor this reaction pathway. Such internal attacks accelerate the overall disappearance rate of the compound (Eq. 13-38). Examples are the systox-type compounds that contain a nucleophilic sulfide group:

$$(13\text{-}38)$$

$$(R = -CH_3, -CH_2CH_3)$$

$$HOCH_2CH_2SCH_2CH_3 \quad + \; H^+$$

Such intramolecular reactions cause the rate of neutral hydrolysis of, for example, demeton S, to be faster than that of the corresponding sulfoxide and sulfone (see $t_{1/2}$ values at low pH, Fig. 13.18). This occurs even though in the latter two cases the ^-S–R_3 moiety should be a better leaving group when considering P–S cleavage. However, both the $-SO-$ and $-SO_2-$ groups are much weaker nucleophiles than $-S-$, and will not, therefore, favor C–S cleavage by an S_Ni mechanism. Note, however, that the k_B values of both the sulfoxide and particularly the sulfone are much larger than that of demeton S. This may result from two factors, the above-mentioned differences in ^-S–R_3 as leaving groups, and, perhaps more important, the effect of the $-SO-$ and $-SO_2-$ groups on the acidity of the protons at the adjacent carbon atoms. This factor may allow yet another reaction mechanism to become important, that is, β-elimination, similar to the case we discussed earlier for carboxylic acid esters (Eq. 13-29):

Figure 13.18 Variation of hydrolysis half-life of three thiophosphoric acid esters with solution pH indicating the relative insensitivity exhibited by demeton S due to the importance of an S_Ni mechanism for that compound (data from Muhlmann and Schrader, 1957).

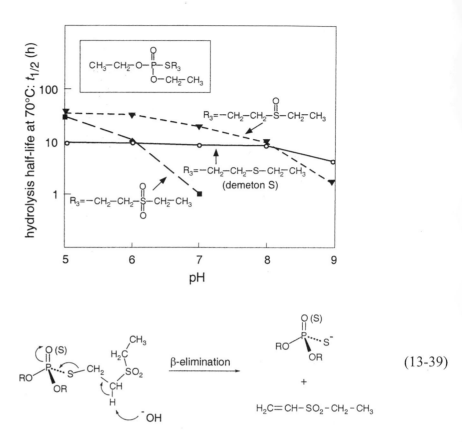

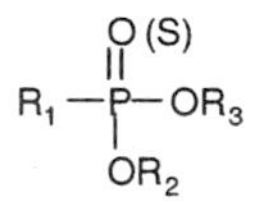

Figure 13.19 Examples of other phosphorus-containing hydrolyzable functionalities. In these compounds, carbons are directly bonded to phosphorus. Note that if $-O-R_3$ is replaced by $-S-R_3$ one obtains the corresponding thioesters.

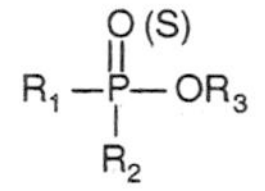

(13-39)

As is unfortunately true for many investigations, the studies reflected by the data shown in Fig. 13.18 did not include analysis for transformation products. Therefore, we may only speculate regarding reaction mechanisms. Nevertheless, we may conclude that when dealing with the hydrolysis of phosphoric and thiophosphoric acid derivatives, as well as with other phosphorus-containing hydrolyzable functionalities (see Fig. 13.19), one has to be aware that various reaction mechanisms may apply. Consequently, depending on the environmental conditions prevailing, product distribution, at least with respect to intermediates formed, may vary considerably.

Advanced Topic 13.5

Effects of Dissolved Metal Species and of Metal Oxide Surfaces on Hydrolytic Reactions

We conclude this chapter by briefly looking at the impact of metal species on the abiotic hydrolysis of organic chemicals. We begin by looking at the effects of *dissolved* metal species. We should note in many natural systems, the fraction in dissolved form of these metal species that are capable of exerting a significant effect on hydrolysis reactions of organic chemicals (e.g., Al(III), Co(II), Cu(II), Fe(III), Ni(II), Pb(II), and Zn(II)) may be rather small (see, e.g., Smolen and Stone, 1997a). Nevertheless, in certain natural or engineered systems, such processes may be important [note, for example, that Cu(II) is added in significant amounts to some agricultural soils as a fungicide (Hassall, 1990)]. Furthermore, in many enzyme-catalyzed hydrolysis reactions, interactions between metal centers and the organic substrate play a pivotal role (see Chapter 17). Finally, such interactions are also responsible for certain surface-catalyzed hydrolytic transformations of organic pollutants.

Effects of Dissolved Metal Species

Dissolved metal species (i.e., aquo complexes or other inorganic and organic metal complexes) may influence hydrolytic reactions of organic compounds in various ways (Chin, 1991; Suh, 1992). Here, we focus on the direct effects that metal species may have on the addition (or the attack) of the nucleophile and/or on the dissociation of the leaving group. First, similar to the case of proton catalysis, by coordinating a hydrolyzable function in a molecule, metal species may decrease the electron density at a given atom (e.g., carbon, phosphorus) thus facilitating the attack of a nucleophile (e.g., H_2O, OH^-). Second, again by analogy to protonation, dissociation of a leaving group may be promoted by interaction of the leaving group with a metal center. Third, nucleophiles bound to a metal center (e.g., OH^-, nucleophilic part of an organic ligand, e.g., Huang, 1997) may be involved in a given reaction. As illustrated by the following examples, depending on the metal ion and organic compound considered, only one or several of these possible mechanisms may be important. Finally, as also illustrated below, we should note that in certain cases, interaction with a metal species may actually inhibit the hydrolytic transformation of an organic chemical (Huang and Stone, 1999).

Let us start with the simplest case, in which a given metal species does not complex the organic molecule but, via one of its ligands, acts directly as a nucleophile. As demonstrated by various authors (e.g. Buckingham and Clark, 1982; Plastourgou and Hoffmann, 1984), metal hydroxo species that are formed by deprotonation of a coordinated water may exhibit a significant nucleophilicity. By the reaction of a series of metal hydroxo complexes, MOH^{v+}, with 2,4-dinitrophenyl acetate (DNPA), Buckingham and Clark (1982) found a Brønsted relationship (Eq. 13-34) between the second-order rate constant, $k_{MOH^{v+}}$, and the pK_a of the corresponding metal aquo complex, $MOH_2^{(v+1)+}$, with a rather small β value of 0.33. Note that depending on the type of other ligands present, the charge of such aquo complexes may also be negative (i.e., v is a negative number). For this type of reaction the authors postulated a mechanism analogous to the neutral or base-catalyzed hydrolysis, respectively (Figs. 13.11 and 13.12):

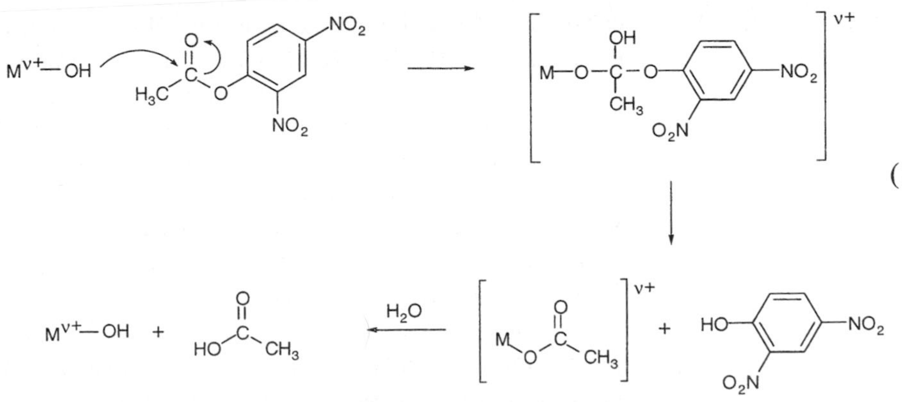

$$(13\text{-}40)$$

For example, for the cobalt complex, *trans*-Co(NH$_3$)$_4$NO$_2$OH$^+$ (pK_a of *trans*-Co(NH$_3$)$_4$NO$_2$OH = 7.16), they determined a $k_{MeOH^{v+}}$ value of about 1×10^{-1} M^{-1} s^{-1} at 25°C. Hence, from their data set one would predict a similar $k_{MeOH^{v+}}$ value for the reaction of DNPA with CuOH$^+$ (pK_a of Cu(aq)$^{2+}$ = 7.5), one of the species that might be important in the metal catalyzed hydrolysis of organic pollutants in the environment (Smolen and Stone, 1997a). Comparison of this $k_{MeOH^{v+}}$ value with the k_N ($\sim 5 \times 10^{-5}$ s^{-1}) and k_B ($\sim 2 \times 10^{2}$ M^{-1} s^{-1}) values that we have derived for DNPA in Illustrative Example 13.4 shows that at pH 7.4 (I_{NB} for this reaction) a CuOH$^+$ concentration of 5×10^{-4} M would be required to compete with the neutral and base-catalyzed reaction. Of course, at pH 7.4 and depending on the water composition, other copper species may be present in natural waters that may or may not be important nucleophiles. It should also be pointed out that with increasing pH, the importance of dissolved copper species and of dissolved metal species in natural waters in general, decreases again because of solubility limitations and because of adsorption of the metal species to solid surfaces (Stumm and Morgan, 1996).

But let us now turn to the other mechanisms by which dissolved metal species may affect hydrolysis reactions of organic chemicals. One mechanism involves coordination of the hydrolyzable moiety in the molecule. Since compared to most other environmentally relevant metals, dissolved Cu(II)-species possess properties most suitable for this type of mechanism (Smolen and Stone, 1997a), we consider some examples of Cu(II) catalysis. In a study on the divalent metal ion-catalyzed hydrolysis of various phosphorothionate and phosphorooxonate ester pesticides, Smolen and Stone (1997a) found that, in aqueous buffer solutions at pH values between 5 and 7, the hydrolysis of several thionate esters including methylchlorpyrifos (V) and ronnel (VI) was accelerated by more than two orders of magnitude in the presence of 1 mM total Cu(II). The effect of other metals was found to be much smaller (Pb(II)), or even insignificant (Ni(II)). Furthermore, more significant Cu(II) catalysis was observed for the thionate esters as compared to the corresponding oxonates [e.g., methylchlorpyrifos oxon (VII)]. In this latter case, Cu(II) and Pb(II) showed a similar effect while catalysis by Ni(II), Co(II), and Zn(II) was negligible. These findings are consistent with the assumption that in the case of the oxonates, the observed catalytic effect was probably primarily due to the reaction of the compounds with Cu(II) and Pb(II) hydroxo species (as discussed above, see Eq. 13-40), since Pb(II) can be assumed to form much weaker complexes with N- and O-ligands as compared to Cu(II). Note that these metal hydroxo species can be considered to be harder nucleophiles as compared to water, and therefore they will react primarily at the P-atoms (like OH$^-$, see Section 13.4). For the thionate esters, however, one has to postulate coordination of the S atom to the metal center (i.e., Cu(II)), which leads to facilitated attack of H$_2$O at the phosphorus center. Thereby one could even imagine that a coordinated water molecule acts as nucleophile in an intramolecular reaction. In the case of methylchlorpyrifos (V), the Cu(II) could even form a bidentate complex also involving the nitrogen atom in the aromatic ring. From the fact that a very similar catalytic effect was observed for ronnel (VI), Smolen and Stone (1997a) concluded, however, that this nitrogen atom is not significantly coordinated. This seems a reasonable assumption when considering the rather low basicity of this aromatic nitrogen [the pK_a of the protonated form is < 5; for more details see Smolen and Stone (1997a)]. With the same argument, one can probably also exclude the

V

VI

VII

coordination of one of the oxygen atoms of the OCH$_3$ groups, although this has been postulated by various authors (see, e.g., Larson and Weber, 1994).

Finally, with respect to the products formed by the metal-catalyzed hydrolysis of these phosporothionate and phosphorooxonate ester pesticides, it should be pointed out again that both discussed mechanisms favor P–O cleavage over C–O cleavage. Hence, the product distribution is very different from that obtained in the neutral hydrolysis (primarily C–O cleavage, see Section 13.4; Smolen and Stone, 1997a).

In addition to the phosphorus and thiophosphorus acid derivatives, there are many other hydrolyzable groups that may be coordinated by dissolved metal ions, and thus may undergo a metal-catalyzed hydrolysis. There are quite a number of cases where bidentate complexes involving oxygen and/or nitrogen atoms as ligands may be formed. Classical examples are the metal-catalyzed hydrolysis of α-amino esters of peptides (e.g., Hay and Morris, 1976; Sutton and Buckingham, 1987):

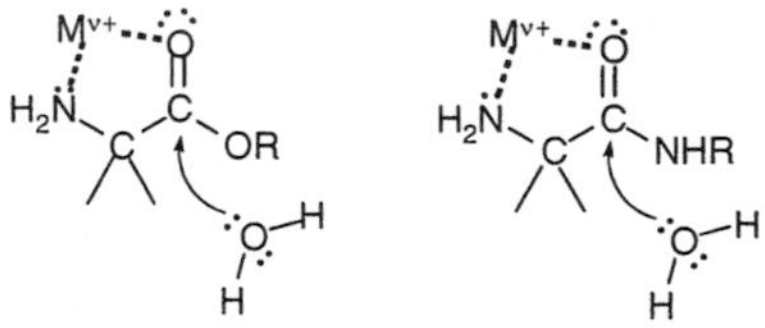

Other compounds that may undergo similar processes include a variety of amide, carbonate, hydrazide, and sulfonylurea agrochemicals (Huang, 1997).

One quite prominent example is the Cu^{2+}-catalyzed decomposition of aldicarb, a widely used systematic pesticide (Bank and Tyrell, 1985). In this case the most likely reaction mechanism is not a facilitated hydrolysis but a β-elimination (see also Huang, 1997):

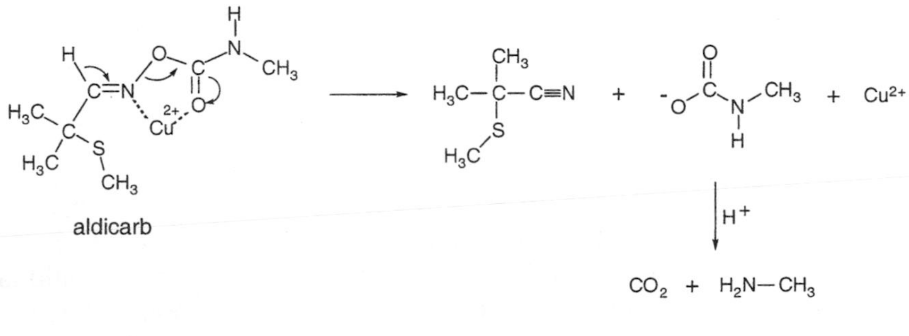

(13-45)

With our last example, we should reiterate that complexation of a compound by a metal ion may not necessarily mean that a transformation reaction (e.g., hydrolysis, β-elimination) is always accelerated. In contrast, in certain cases, the reaction may even be inhibited. For example, Huang and Stone (1999) demonstrated that secondary amides such as the herbicide naptalam may, due to deprotonation of the amide nitrogen, form a bidentate metal complex that is much less susceptible toward hydrolysis as compared to the noncomplexed compound:

less reactive
towards hydrolysis

In contrast, tertiary amides are subject to metal catalyzed hydrolysis.

In summary, we note that dissolved metal ions may affect both the rates as well as the mechanism(s) of hydrolytic transformations of organic chemicals. In many cases, metal ions enhance transformation rates. However, inhibitory effects may also occur. It is important to realize that the effect of a given metal depends significantly on its speciation in aqueous solution, and that different metals exhibit very different abilities to promote or inhibit hydrolytic reactions. Using the environmentally relevant metals that may be present in dissolved form in natural systems at appreciable concentrations, Cu(II) seems to play the most important role in the chemical catalysis of hydrolytic transformations of organic pollutants. For a more detailed discussion of this topic, including approaches for quantification of the rate of dissolved metal-catalyzed hydrolysis reaction (which is a rather difficult task), we refer to the literature (e.g., Smolen and Stone, 1997a; Huang, 1997).

Effects of Metal Oxide Surfaces

Although there are still only a rather limited number of studies available on the effects of metal oxide surfaces on hydrolytic transformations of organic compounds, a few important general conclusions can be drawn. First, analogous to the effects of dissolved metals, (lattice-)bound metal atoms present at the surface may coordinate a hydrolyzable moiety, thus catalyzing (or in some cases inhibiting) a given reaction (Torrents and Stone, 1991, 1994; Smolen and Stone, 1997b). A very instructive example is the catalysis of the hydrolysis of phenyl picolinate (Fig. 13.20a) by various oxide surfaces (Torrents and Stone, 1991). As can be seen from Fig. 13.21, the hydrolysis of this compound is significantly accelerated in the presence of iron and, particularly, titanium oxide surfaces. No effects were observed for Al_2O_3 and SiO_2. Torrents and Stone proposed that the observed rate enhancement is most probably due to formation of a five-membered bidentate complex involving a surface bound Al(III) or Ti(IV), as well as the carboxyl oxygen and the pyridinal nitrogen atom, respectively (Fig. 13.20a). Such a complex would facilitate the attack of a nucleophile (e.g., H_2O). This hypothesis is supported by the observation that hydrolysis of the isomeric phenyl isonicotinate (Fig. 13.20b), which can form only weak monodentate complexes, was not catalyzed by any of the mineral oxides investigated.

In addition, when the nitrogen ligand is replaced by an oxygen ligand, as is the case for phenyl salicylate (VIII, see margin), catalysis is also observed for Al_2O_3 but not for SiO_2 (Torrents and Stone, 1994). The observation that TiO_2 exceeds FeOOH and Al_2O_3 in its ability to catalyze the hydrolysis of organic compounds capable of forming complexes with the corresponding surface-bound metal atoms has been made not only for carboxylic acid esters, but also for a series of phosphorooxonate and

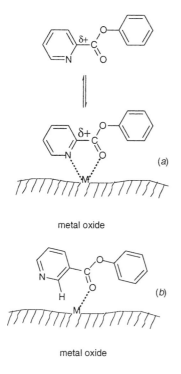

metal oxide

(a)

(b)

metal oxide

Figure 13.20 (a) Phenyl picolinate may coordinate to a surface bound metal by forming a five-membered bidentate complex. (b) The isomeric phenyl isonicotinate may form only a (weak) monodentate complex.

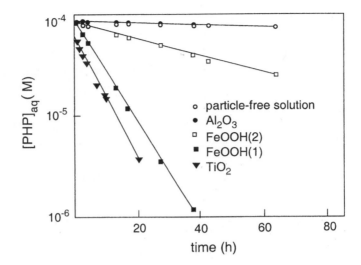

Figure 13.21 Effect of various mineral oxides on loss of phenyl picolinate (PHP) from solution via hydrolysis. The solutions contained 10 g L^{-1} oxide, 1×10^{-3} M acetate buffer (pH 5), and 5×10^{-2} M NaCl. From Torrents and Stone (1991).

VIII

phenyl salicylate

phosphorothionate pesticides (Smolen and Stone, 1998). These findings can be rationalized by the considerably higher Lewis acidity (i.e., electrophilicity) of Ti(IV) as compared to Al(III) or Fe(III). Furthermore, in contrast to many of the dissolved metals addressed above, these metals generally form stronger complexes with O-ligands than with N- or S-ligands (Stumm and Morgan, 1996). Consequently, surface catalysis of the hydrolysis of compounds exhibiting oxygen-donor groups (e.g., C = O, P = O, –OH) can be expected to be more pronounced. The results of the available studies so far support this expectation. Thus, in contrast to Cu(II) in solution, the hydrolysis of some phosphorooxonate esters was found to be much more strongly catalyzed by TiO_2, Al_2O_3 and FeOOH (Smolen and Stone, 1997b) as compared to phosphorothionate esters, for which very small effects were observed (Smolen and Stone, 1997b; Dannenberg and Pekkonen, 1998). Almost negligible mineral oxide surface catalysis was also found by Huang (1997) for some carbonate, hydrazide, and sulfonylurea agrochemicals.

In addition to rate-enhancing effects caused by the complexation of a given compound at a metal oxide surface, one can also imagine that hydroxo groups bound to the surface may act as nucleophiles. Thus, such surface sorption may cause the overall hydrolysis rate of a given compound to increase (Hoffmann, 1990). Furthermore, as postulated by Stone (1989), electrostatic forces and other forces operative in the interfacial region between surface and bulk solution can cause accumulation of reactants (organic compound and nucleophile), thus facilitating the reaction. In any case, whatever the mechanism(s) of a mineral oxide surface-catalyzed reaction may be, it is easy to see that quantification of such processes, and thus the assessment of their relative importance in a given natural system (e.g., soil, aquifer), is rather difficult for several reasons. On one hand, as illustrated by Fig. 13.21, different surfaces (even if the same metal is involved, e.g., Fe(III)) may exhibit very different reactivities that will depend in a complex way on the surface structure and surface heterogeneity. It will also depend on important environmental variables such as pH and ionic strength. On the other hand, in the environment, there will be a variety of inorganic and organic (i.e., natural organic matter constituents) species that may compete with a given organic chemical for the reactive surface sites (Torrents and Stone, 1993a and b). Hence, at the present time, we have to content ourselves with a rather qualitative treatment of this topic. This also holds for hydrolytic reactions catalyzed by other solids, particularly by clay minerals, where we can assume that the same type of mechanism(s) may be effective. Here, depending on the cations adsorbed and depending on the degree of water saturation, significant differences in catalytic activity may be found for the types of compounds considered above (e.g., Larson and Weber, 1994).

With these remarks on heterogeneous processes we conclude our discussion of hydrolytic reactions and other reactions involving nucleophilic species. We should point out that we have taken a close look at only a few representative structural moieties that may undergo these types of reactions in the environment. Nevertheless, the general knowledge that we have acquired in this chapter should put us in a much better position to evaluate the importance of such reactions for other functional groups that form part of environmental organic chemicals.

13.6   Questions and Problems

Questions

Q 13.1

Explain in words what a nucleophilic substitution reaction is. At what kind of atoms do such reactions primarily occur? What is(are) the mechanism(s) and the corresponding rate law(s) of such reactions?

Q 13.2

What are the major factors determining the rates of nucleophilic substitution reactions?

Q 13.3

Explain the terms *hard* and *soft* Lewis acids and bases.

Q 13.4

Rank the following nucleophiles in order of increasing nucleophilicity with respect to nucleophilic substitution reactions at a saturated carbon atom. Comment on your choice.

$$Br^-, OH^-, NO_3^-, CN^-, ClO_4^-, S_2O_3^{2-}, H_2O$$

Q 13.5

Explain in words what the Swain-Scott relationship describes and discuss in which cases it may be applied.

Q 13.6

Give at least four examples of good leaving groups.

Q 13.7

Explain in words what a β-elimination reaction is. What is the difference between an E1 and an E2 mechanism?

Q 13.8

Which structural and environmental factors favor an elimination mechanism over a nucleophilic substitution mechanism?

Q 13.9

When comparing the hydrolysis rate constants of a series of carboxylic acid esters (Table 13.8), it can be seen that the values for the acid-catalyzed reactions are all of the same magnitude, whereas the rate constants for the base-catalyzed reactions vary by several orders of magnitude. Explain these findings.

Q 13.10

What do the terms I_{AN}, I_{AB}, and I_{NB} express?

Q 13.11

Inspection of Table 13.12 shows that the I_{NB} values for hydrolysis of trimethyl- and triethylphosphate are ≥ 10, whereas I_{NB} of triphenylphosphate is < 6. Try to explain these findings.

Q 13.12

As can also be seen from Table 13.12, acid-catalyzed hydrolysis is unimportant for many phosphoric and thiophosphoric acid triesters. Among the exceptions are diaxonon and diazinon ($I_{AN} = 6.4$ and 5.7, respectively). Try to explain why.

Q 13.13

Rank the carbamates I–VI in order of increasing reactivity with respect to base-catalyzed hydrolysis. Comment on your choice.

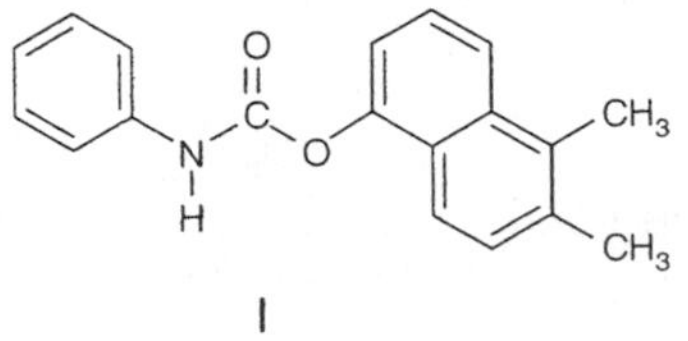

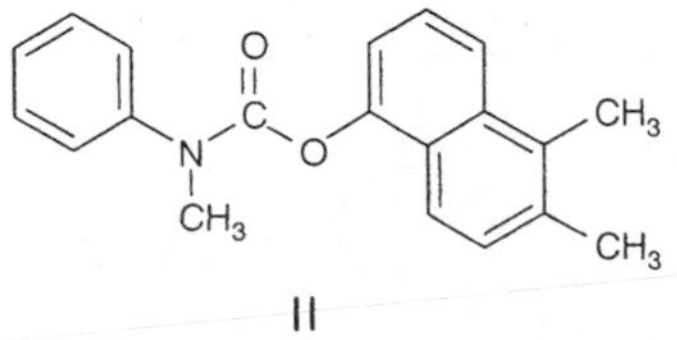

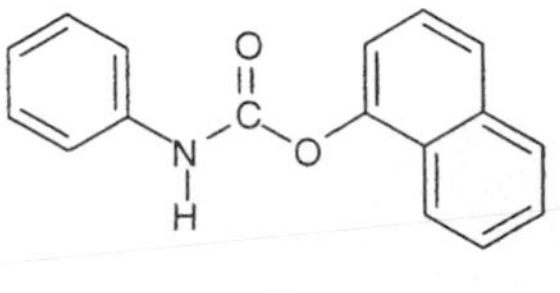

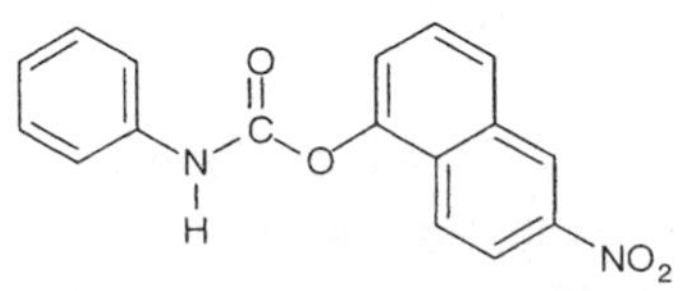

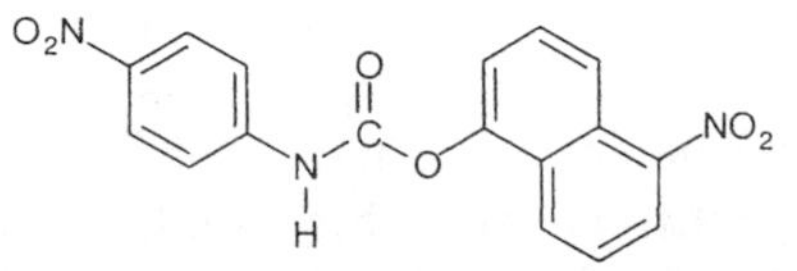

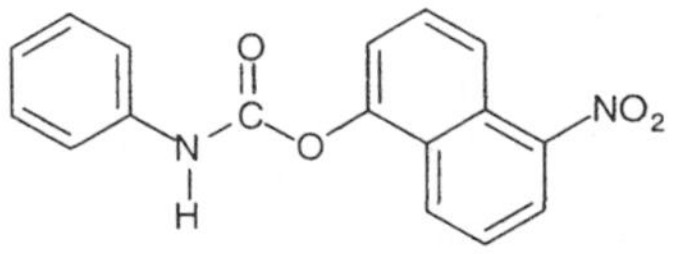

Problems

P 13.1 *What Happens to Trimethylphosphate in Seawater and in a Leachate?*

The hydrolysis half-life of trimethylphosphate ($(CH_3O)_3PO$, TMP) in pure water is 1.2 yr at 25°C and pH 7.0 (Table 13.13). A colleague in oceanography claims that in sterile seawater, he observed a half-life for TMP of only about 80 days at 25°C and pH 7. Is this result reasonable? What are the major products of the abiotic transformation of TMP in seawater?

Because you are more interested in groundwater contamination, you wonder how fast TMP would be transformed by chemical reactions at 10°C and pH 8.0 in a leachate from a waste disposal site containing 0.25 M Cl^-, 0.05 M Br^-, and 10^{-4} M CN^-. Calculate the approximate half-life of TMP under these conditions by trusting your colleague's measurements and by assuming that all relevant reactions exhibit about the same activation energy of 95 kJ·mol^{-1}. Also assume an *s*-value of 0.9 in the Swain-Scott relationship (Eq. 13-3).

P 13.2 *Chemical Transformation of Polychlorinated Ethanes in a Lake*

Assume that the three polychlorinated ethanes, 1,1,2,2-tetrachloroethane, 1,1,1,2-tetrachloroethane, and pentachloroethane are introduced into a lake by an accident. Calculate the half-life for chemical transformation of each of the three compounds in (a) the epilimnion of the lake ($T = 25$°C, pH 8.5) and (b) the hypolimnion of the lake ($T = 5$°C, pH 7.5). Furthermore, indicate for each compound the pH (for the epilimnion and for the hypolimnion) at which the neutral and the base-catalyzed reaction would be equally important. What is(are) the transformation product(s) of these compounds? Explain the different reactivities of the three compounds. You can find all necessary data in Table 13.7.

P 13.3 *Why Do DDT and Methoxychlor Exhibit Such Different pH-Hydrolytic Half-Life Profiles? What Is(Are) the Major Transformation Products?*

Wolfe et al. (1977) have determined the rate constants for the neutral and base-catalyzed transformation of the two classical insecticides DDT and methoxychlor in water at 27°C:

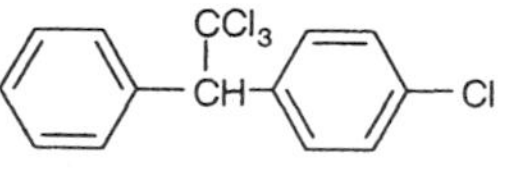

DDT	methoxychlor
$k_N = 1.9 \times 10^{-9}$ s^{-1}	$k_N = 1.2 \times 10^{-8}$ s^{-1}
$k_B = 9.9 \times 10^{-3}$ M^{-1} s^{-1}	$k_B = 3.8 \times 10^{-4}$ M^{-1} s^{-1}

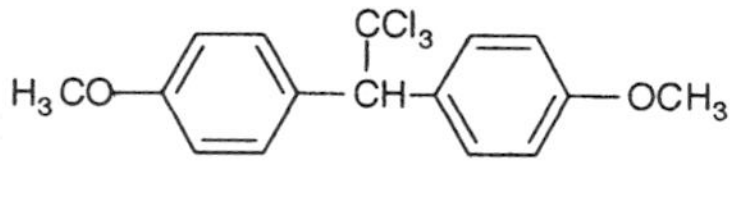

Inspection of the k_N and k_B values shows that DDT has an almost 10 times smaller k_N value than methoxychlor, while the opposite is true for the k_B values. Try to explain these differences. Which reaction (neutral or base-catalyzed) dominates the transformation of the two compounds at pH 7 and at pH 9? What are the corresponding half-lives at the two pH values at 27°C, and what major product(s) do you expect to be formed? Depict the corresponding reaction pathways.

Hint: For the neutral reaction of methoxychlor, Wolfe et al. (1977) also proposed an alternative mechanism, that is, an S_N1 mechanism at the CCl_3-group followed by a 1,2-migration of one of the phenyl moieties to yield 1,2-*bis* (*p*-methoxyphenyl)-2-hydroxy-1,1-dichloroethane:

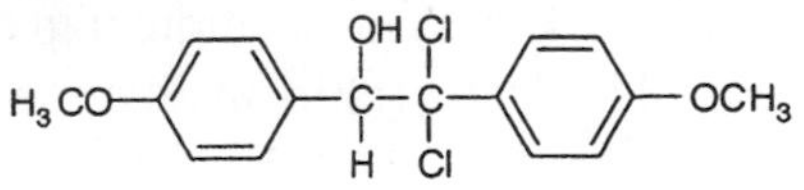

Is this product stable in aqueous solution?

P 13.4 *Hydrolysis of Mono- and Dihaloalkanes in Aqueous Solution: Mechanistic Evaluation of Reaction Rates and Activation Parameters*

In a review of kinetic studies of the solvolysis of alkyl halides in water, Robertson (1967) has reported rate data as well as enthalpies and entropies of activation for a series of alkyl halides. Using this data set given below, try to rationalize the factors that determine primarily the rate of (neutral) hydrolysis of such compounds in aqueous solution. Can you make any suggestions about the mechanism by which the various compounds hydrolyze? Try to classify the compounds very roughly according to the S_N1 versus S_N2 character of the reaction. Note, for example, that CH_3Cl and $(CH_3)_3CCl$ exhibit very similar $\Delta^{\ddagger}H^0$ values, and yet their half-lives differ by more than 6 orders of magnitude!

Half-lives ($t_{1/2}$), Enthalpies ($\Delta^{\ddagger}H^0$) and Entropies ($\Delta^{\ddagger}S^0$) of Activation for the (Neutral) Hydrolysis of a Series of Alkyl Halides in Water at 25°C and pH 7.0 [a]

Compound	$t_{1/2}$ (pH 7.0) (h)	$\Delta^{\ddagger}H^0$ (kJ·mol⁻¹)	$\Delta^{\ddagger}S^0$ (J·mol⁻¹ K⁻¹)
CH_3Cl	8056	100.3	− 51.4
CH_3Br	473	95.7	− 42.2
$CH_3CH_2CH_2Br$	688	92.4	− 56.8
$(CH_3)_2CHCl$	907	100.2	− 33.9
$(CH_3)_2CHBr$	50.5	101.8	− 5.9
$CH_3CCl_2CH_3$	21.2	108.1	+ 22.2
$CH_3CBrClCH_3$	1.1	102.5	+ 28.0
$CH_3CBr_2CH_3$	4.2	107.6	+ 33.4
$(CH_3)_3CCl$	0.0065	99.5	+ 60.2

[a] Data from Robertson (1967).

P 13.5 *Assessing the Hydrolysis Half-Life of tert-Butyl Formate*

Various studies on the fate of the gasoline additive methyl-*t*-butyl ether (MTBE) have shown that it can be oxidized to *t*-butyl-formate (TBF); which happens particularly in the atmosphere:

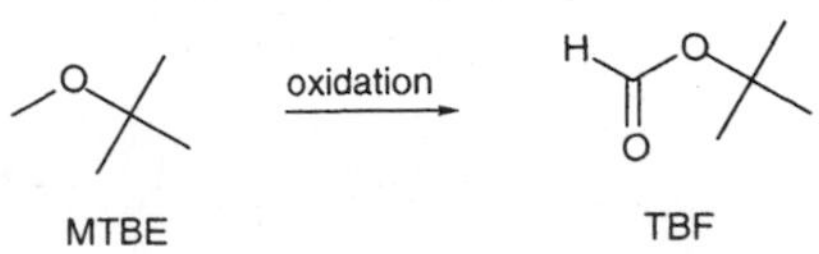

MTBE   TBF

Church et al. (1999) have investigated the hydrolysis of TBF as a function of pH and temperature. The rate constants for the acid-catalyzed, neutral, and base-catalyzed reactions are given in Table 13.8 (R_1 = H, R_2 = C(CH$_3$)$_3$). The corresponding E_a values are 60, 80, and 90 kJ·mol^{-1}, respectively. Calculate the hydrolysis half-lives of TBF (a) in an acidic rain drop (pH = 2.5) at 5°C, (b) in surface water (pH = 8.0) at 15°C, and (c) in an alkaline solution at pH 12 and 25°C.

P 13.6 *Synthesizing the "Right" Carbamates*

You work in the chemical industry and you are asked to synthesize two different carbamates of either Type I or Type II (see below). One carbamate should have a hydrolysis half-life of approximately 1 month at 25°C and pH 8.0, while the hydrolysis half-life of the other one should be about 10 months at 25°C and pH 9.0. You assume that only the base-catalyzed reaction is important at the pHs of interest, and you search the literature for k_B values for these types of compounds. For some Type I compounds k_B values are given in Illustrative Example 13.6, and for some Type II compounds you find the data given below. What are the structures of the molecules that you are going to synthesize in order to get the desired half-lives?

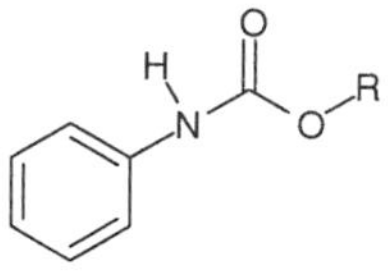

Type I 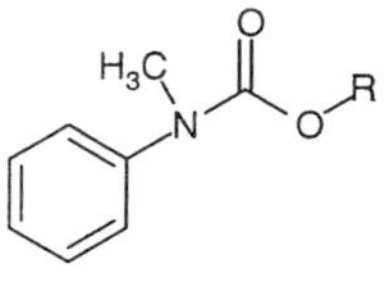   Type II

Second-Order Rate Constants k_B at 25°C for the Hydrolysis of Some Substituted Phenyl N-Methyl-N-Phenyl Carbamates (Type II) [a]

R	k_B / (M^{-1} s^{-1})	R	k_B / (M^{-1} s^{-1})
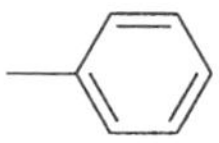 (phenyl)	7.5×10^{-5}	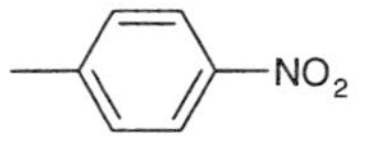 (phenyl)—NO$_2$	3.9×10^{-4}
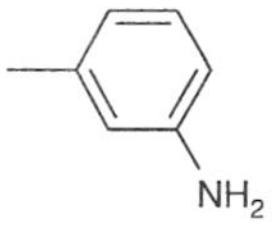 (phenyl, NH$_2$)	2.8×10^{-5}	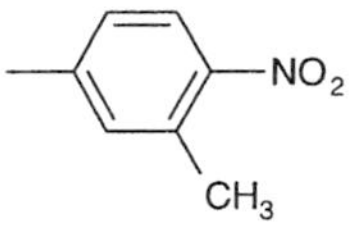 (phenyl, NO$_2$, CH$_3$)	3.3×10^{-4}
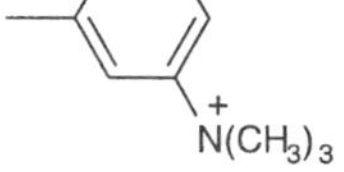 (phenyl, N(CH$_3$)$_3^+$)	2.5×10^{-4} [b]		

[a] Data from references given in Table 13.11. [b] The σ_{meta} value for the ±N(CH$_3$)$_3^+$ group is +0.88.

P 13.7 *Multiple Structure-Reactivity Correlations: Evaluating and Predicting Alkaline Hydrolysis Rates of Acyl- and Aryl-Substituted Phenyl Benzoates*

Kirsch et al. (1968) investigated the base-catalyzed hydrolysis of 24 *meta*- and *para*-disubstituted benzoic acid phenyl esters in acetonitrile/water (1 : 2) at 25°C:

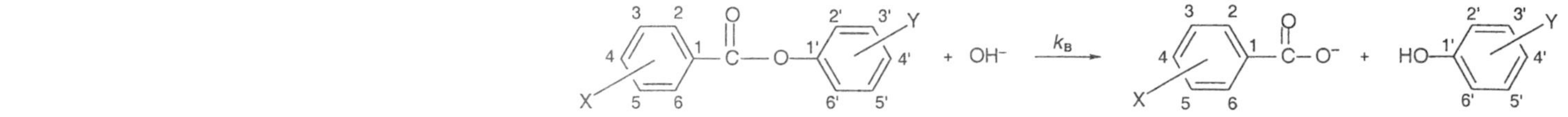

For these compounds they derived the following Hammett equation:

$$\log k_B \text{ (in } M^{-1} s^{-1}) = 2.01 \ \sigma_x + 0.95 \ \sigma_y - 1.23 \ (R^2 = 0.99) \tag{1}$$

where both σ_x and σ_y ranged between about -0.8 and $+0.8$. What does Eq. 1 tell you about the rate-determining step of this reaction? Why can you not use Eq. 1 to predict k_B values of substituted benzoic acid phenyl esters in *aqueous solution*? In the literature (Drossman et al., 1988) you find k_B values of a few monosubstituted phenyl benzoates ($X = H$) that were determined at 25°C in aqueous solution. From these values you derive the Hammett relationship:

$$\log k_B \text{ (in } M^{-1} s^{-1}) = 0.75 \ \sigma_y - 0.22 \ (R^2 = 0.98) \tag{2}$$

Explain the difference in the ρ_y (0.95 versus 0.75) and $\log k_{B,H}$ ($X = Y = H$; -1.23 versus -0.22) values obtained for the two solvent systems. Make a guess of the magnitude of ρ_x in aqueous solution and estimate the hydrolysis   half-life of 4-nitro benzoic acid 4-nitrophenyl ester ($X = Y = 4\text{-}NO_2$) in water at pH 8.0 and 25°C by assuming that only the base-catalyzed reaction is important.

P 13.8 *Hydrolysis of an Insecticide in a River*

After a fire in a chemical storehouse at Schweizerhalle, Switzerland, in November 1986, several tons of various pesticides, solvents, dyes, and other raw and intermediate chemicals were flushed into the Rhine River (Capel et al., 1988; Wanner et al., 1989). Among these chemicals was the insecticide disulfoton, of which 3500 kg were introduced into the river water (11°C, pH 7.5). During the 8 days "travel time" from Schweizerhalle to the Dutch border, 2500 kg of this compound were "eliminated" from the river water. Somebody wants to know how much of this elimination was due to abiotic hydrolysis. Since in the literature you do not find any good kinetic data for the hydrolysis of disulfoton, you make your own measurements in the laboratory. Under all selected experimental conditions, you observe (pseudo)first-order kinetics, and you get the results given below.

Determine the k_{obs}-value for the conditions in the river (11°C, pH 7.5), and calculate how much disulfoton was transformed by hydrolysis over the 8 days. What are the most likely hydrolysis products?

$$(C_2H_5O)_2\overset{\overset{\displaystyle S}{\|}}{P}\!-\!S\!-\!CH_2\!-\!CH_2\!-\!S\!-\!CH_2CH_3$$

disulfoton

Temperature	k_{obs}/s^{-1} [a]		
°C	pH 6.0	pH 11.98	pH 11.72
20		1.3×10^{-5}	
30	4.0×10^{-7} [b]		3.6×10^{-5}
40	9.6×10^{-7}		
45	1.5×10^{-6}		
50	2.9×10^{-6}		

[a] $k_{obs} = k_n$. [b] A similar k_{obs}-value was obtained at pH 4.0 and 30°C.

P 13.9 Base-Catalyzed Hydrolysis of Diethyl Phenylphosphates: Mechanistic Considerations Using LFERs

It is commonly assumed that the base-catalyzed hydrolysis of substituted dialkyl (i.e., dimethyl or diethyl) phenyl phosphates occurs by nucleophilic attack of OH^- at the phosphorus with the phenolate being the leaving group (see also Section 13.4):

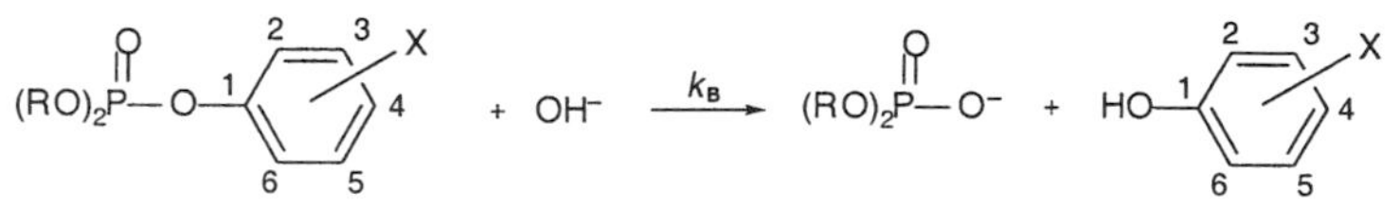

Furthermore, it has been postulated that, when considering a series of such compounds, the relative reactivity (i.e., the relative magnitude of the k_B values) is determined primarily by the relative electrophilicity of the phosphorus atom and not by the relative "goodness" of the leaving group (i.e., the phenolates). Is this hypothesis correct? Try to answer this question by evaluating the Hammett (Eq. 13-31) and Brønsted (Eq. 13-34) relationship that you can derive from the k_B values reported by van Hooidonk and Ginjaar (1967) for a series of *meta*- and *para*-substituted diethyl phenyl phosphates (see below). Do you include all compounds in the Brønsted relationship? If not, which ones do you exclude and why? How is the ρ-value derived for

Second-Order Rates Constants, k_B, and pK_{ia} Values of the Phenol Moieties for a Series of Monosubstituted Diethyl Phenyl Phosphates at 25°C

Substituent X (and position)	$\log k_B$ [a] (k_B in M^{-1} s^{-1})	pK_{ia} [b] (0.1 M KCl)	Substituent X (and position)	$\log k_B$ [a] (k_B in M^{-1} s^{-1})	pK_{ia} [b] (0.1 M KCl)
4–OCH$_3$	− 3.55	10.12	3–Cl	− 2.81	9.13
4–C$_2$H$_5$	− 3.49	10.18	3–Br	− 2.78	9.06
3–CH$_3$	− 3.45	10.09	4–COCH$_3$	− 2.49	8.01
H	− 3.33	9.89	3–NO$_2$	− 2.19	8.39
3–OCH$_3$	− 3.21	9.65	4–CN	− 2.19	7.85
4–Cl	− 2.94	9.35	4–NO$_2$	− 1.96	6.99
4–Br	− 2.90	9.27			

[a] Data from van Hooidonk and Ginjaar (1967). [b] Experimental values in 0.1 M KCl determined by van Hooidonk and Ginjaar (1967). Note that these values differ somewhat from the values given for some of the compounds in the Appendix C.

the Hammett correlation related to the β-value obtained from the Brønsted equation?

P 13.10 Estimating the Hydrolysis Half-Life of Methyl-3,4-Dichlorobenzene-Sulfonate in Homogeneous Aqueous Solution

A colleague of yours who investigates the fate of benzene sulfonates and benzene sulfonate esters in natural waters is interested in the stability of methyl-3,4-dichlorobenzene sulfonate (MDCBS) in aqueous solution. Because he has not read Chapter 13 of *Environmental Organic Chemistry* he asks you to help him to estimate the hydrolysis half-life of this compound in water at 25°C and at 5°C. In the literature you find rate constants for the neutral hydrolysis of some substituted methyl benzene sulfonates at

25°C, and you learn that the activation energies of these reactions are in the order of 85 kJ · mol^{-1} (Robertson, 1967). Using the data given below, estimate the neutral hydrolysis half-life of MDCBS at 25°C and 5°C. Postulate the most likely reaction mechanism for the hydrolysis of MDCBS. Do you expect that the reaction will be pH dependent in natural waters?

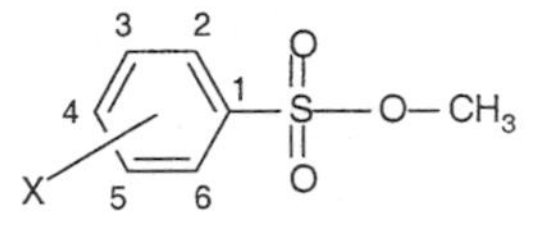

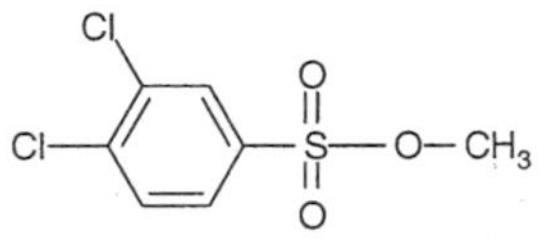

First-Order Rate Constants for the Neutral Hydrolysis of Some Substituted Methyl Benzenesulfonates in Aqueous Solution at 25°C [a]

Substituent(s)	$k_N \times 10^6$ (s^{-1})	Substituent(s)	$k_N \times 10^6$ (s^{-1})
4–CH$_3$	8.0	4–Br	9.4
4–OCH$_3$	6.0	3–NO$_2$	52.7
3–CH$_3$–4–CH$_3$	6.6	4–NO$_2$	62.5

[a] Data from Robertson (1967).

P 13.11 *Assessing the Hydrolysis Half-Life of a Highly Strained Hydrocarbon*

Highly strained hydrocarbons such as quadricyclane (structure see below) may serve as high-performance aviation fuels (Hill et al., 1997). It is, therefore, important to know the environmental behavior of such compounds, particularly with respect to spills. In this context, Hill et al. (1997) have studied the hydrolysis of quadricyclane in aqueous solution at pH values between 3 and 4 as well as in soil slurries exhibiting pH values between 4.6 and 6.4. They found that in homogeneous aqueous solution at a given pH, the disappearance of quadricyclane followed pseudo-first-order kinetics, and that two major products (i.e., nortricyclyl alcohol and exo-5-norbornen-2-ol) were formed at a ratio of about 15 : 1:

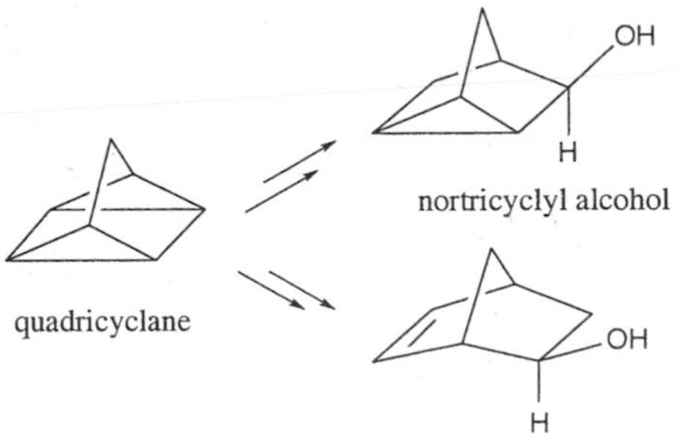

(1)

Furthermore, they observed that the most important factor affecting the reactivity of quadricyclane in the soils was pH.

Estimate the hydrolysis half life of quadricyclane in aqueous solution at pH 4.6 and 6.4 using the experimental data given below (note that there is considerable scatter in the data!). Propose possible reaction mechanisms for the conversion of quadricyclane to the above-mentioned products (Eq. 1).

Pseudo First Order Rate Constant, k_{obs}, for the Disappearance of Quadricyclane at Various pH Values in Aqueous Solution (Data from Hill et al., 1997).

pH	k_{obs} (min^{-1})	pH	k_{obs}(min^{-1})
3.00	2.7×10^{-2}	3.58	7.5×10^{-3}
3.12	2.0×10^{-2}	3.87	2.6×10^{-3}
3.18	1.45×10^{-2}	4.00	1.0×10^{-3}
3.50	6.0×10^{-3}		

Chapter 14

CHEMICAL TRANSFORMATIONS II: REDOX REACTIONS

14.1 **Introduction, Overview**

14.2 **Thermodynamic Considerations of Redox Reactions**
Half Reactions and (Standard) Reduction Potentials
Illustrative Example 14.1: *Calculating Standard Reduction Potentials from Free Energies of Formation*
One-Electron Reduction Potentials
Processes Determining the Redox Conditions in the Environment
Illustrative Example 14.2: *Establishing Mass Balances for Oxygen and Nitrate in a Given System*
Evaluating the Thermodynamics of Redox Reactions under Environmental Conditions
Illustrative Example 14.3: *Calculating the Reduction Potential of an Aqueous Hydrogen Sulfide (H_2S) Solution as a Function of pH and Total H_2S Concentration*
Illustrative Example 14.4: *Calculating Free Energies of Reaction from Half Reaction Reduction Potentials*

14.3 **Reaction Pathways and Kinetics of Redox Reactions**
Factors Determining the Rate of Redox Reactions
Reduction of Nitroaromatic Compounds (NACs)
Illustrative Example 14.5: *Estimating Rates of Reduction of Nitroaromatic Compounds by DOM Components in the Presence of Hydrogen Sulfide*
Reductive Dehalogenation Reactions of Polyhalogenated C_1- and C_2-Compounds
Oxidation Reactions

14.4 **Questions and Problems**

14.1 Introduction, Overview

In Chapter 13 we have confined ourselves to transformation reactions in which no net electron transfer occurred from (i.e., oxidation of) or to (i.e., reduction of) the organic compound of interest. Many important pathways by which organic chemicals are transformed in the environment involve oxidative and reductive steps, especially when we consider photochemical and biologically mediated transformation processes. Oxidation and reduction reactions may, however, also occur abiotically in the dark. We should note that some of the reactions we discuss may be catalyzed by biological molecules (e.g., iron porphyrins, quinoid compounds) released from organisms (e.g., after cell lysis). This has led to a certain confusion with respect of the use of the term "abiotic" for such reactions. For the following discussion, we adopt the definition of Macalady et al. (1986), who suggest that a reaction is abiotic if it does not directly involve the participation of metabolically active organisms. Of course, this does not imply that "abiotic" redox reactions are not heavily influenced by biological (particularly microbial) activity, since the availability of suitable reactants for electron transfer reactions is determined largely by biological processes.

At this point, we should first ask ourselves how we can recognize whether an organic compound has been oxidized or reduced during a reaction. The easiest way to do that is to check whether there has been a net change in the oxidation state(s) of atoms like C, N, or S (see Chapter 2) involved in the reaction. For example, if a chlorine atom in an organic molecule is replaced by a hydrogen atom, as is observed in the transformation of DDT to DDD:

$$+ H^+ + 2e^- \longrightarrow \qquad\qquad + Cl^- \qquad (14\text{-}1)$$

the oxidation state of the carbon atom at which the reaction occurs changes from +III to +I. The oxidation states of all other atoms remain the same. Hence, conversion of DDT to DDD requires a total of two electrons to be transferred from an electron donor to DDT. This type of reaction is termed a *reductive dechlorination*. Note that the species that donates the electrons is oxidized during this process. Thus, in any electron transfer reaction, one of the reactants is oxidized while the other one is reduced. Hence, we term such reactions *redox* reactions. Since our focus is on the organic pollutant, we speak of an oxidation reaction if the pollutant is oxidized or of a reduction reaction if the pollutant is reduced. Let us now compare the reaction discussed above (Eq. 14-1) with another reaction that we discussed in Section 13.2, dehydrochlorination. Here, as is illustrated by the transformation of DDT to DDE:

$$+ HO^- \longrightarrow \qquad\qquad + H_2O + Cl^- \qquad (14\text{-}2)$$

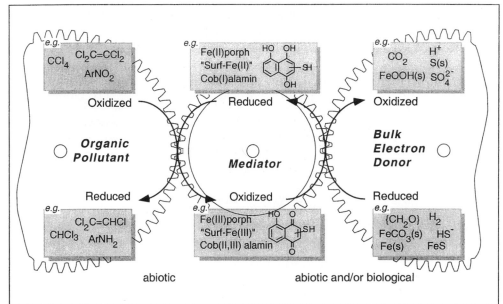

Figure 14.1 Schematic representation depicting the importance of electron transfer mediators as well as the concurrence of microbial and abiotic processes for reductive transformations of organic pollutants. Adapted from Schwarzenbach et al. (1997).

the change in oxidation state of one of the carbon atoms involved in the reaction is compensated by the change in oxidation state of the adjacent carbon atom. Hence, dehydrochlorination requires no net electron transfer from or to the compound, and therefore we shall not consider this reaction to be a redox reaction. Another elimination reaction, the *dihalo-elimination*, is, however, again a redox reaction. If we consider, for example, the dihalo-elimination of hexachloroethane (HCA) to tetrachloroethene (perchloroethylene, PCE), a reaction that has been observed to occur in groundwater systems (e.g., Criddle et al., 1986):

$$
\begin{array}{c}
\text{HCA} \;+\; 2e^- \;\longrightarrow\; \text{PCE} \;+\; 2Cl^-
\end{array}
\tag{14-3}
$$

we realize that during this reaction, the oxidation states of both carbon atoms are altered by -1. Hence, as in our first example (Eq. 14-1), the reduction of HCA to PCE requires two electrons to be transformed from an electron donor to HCA.

Let us now take a brief look at some important redox reactions of organic pollutants that may occur abiotically in the environment. We first note that only a few functional groups are oxidized or reduced abiotically. This contrasts with biologically mediated redox processes by which organic pollutants may be completely mineralized to CO_2, H_2O and so on. Table 14.1 gives some examples of functional groups that may be involved in chemical redox reactions. We discuss some of these reactions in detail later. In Table 14.1 only overall reactions are indicated, and the species that act as a sink or source of the electrons (i.e., the oxidants or reductants, respectively) are not specified. Hence, Table 14.1 gives no information about the actual reaction mechanism that may consist of several reaction steps.

Table 14.1 Examples of Some Simple Redox Reactions That May Occur Chemically in the Environment [a]

Oxidized Species	Reduction ⇌ Oxidation	Reduced Species	Equation Number
Change in Oxidation State of Carbon Atom(s)			
$R-COOH + 2H^+ + 2e^-$	⟵	$R-CHO + H_2O$	(14-4)
$O=\!\!\langle\bigcirc\rangle\!\!=O + 2H^+ + 2e^-$	⇌	$HO-\langle\bigcirc\rangle-OH$	(14-5)
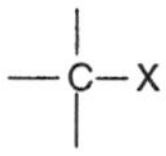 $-\overset{\|}{\underset{\|}{C}}-X \;(X=Cl,Br,I) + H^+ + 2e^-$	⟶	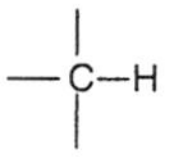 $-\overset{\|}{\underset{\|}{C}}-H + X^-$	(14-6)
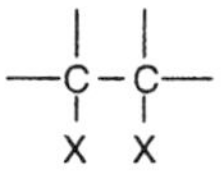 $-\overset{\|}{\underset{X}{C}}-\overset{\|}{\underset{X}{C}}- \;(X=Cl,Br,I) + 2e^-$	⟶	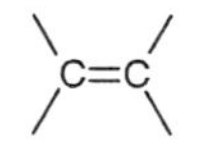 $\text{C=C} + 2X^-$	(14-7)
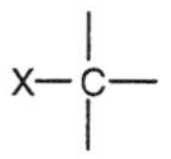 $2\,X-\overset{\|}{\underset{\|}{C}}- \;(X=Cl,Br,I) + 2e^-$	⟶	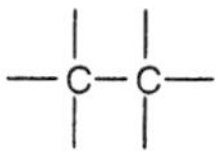 $-\overset{\|}{\underset{\|}{C}}-\overset{\|}{\underset{\|}{C}}- + 2X^-$	(14-8)
Change in Oxidation State of Nitrogen Atom(s) [b]			
$\text{(Ar)}-NO_2 + 6H^+ + 6e^-$	⇌⋯	$\text{(Ar)}-NH_2 + 2H_2O$	(14-9)
$\text{(Ar)}-N=N-\text{(Ar)} + 2H^+ + 2e^-$	⇌	$\text{(Ar)}-NH-NH-\text{(Ar)}$	(14-10)
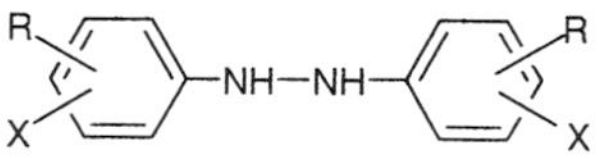 $\text{(Ar)}-NH-NH-\text{(Ar)} + 2H^+ + 2e^-$	⇌	$2\;\text{(Ar)}-NH_2$ 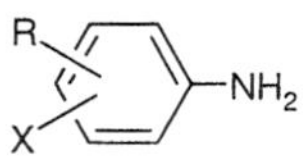	(14-11)
Change in Oxidation State of Sulfur Atom(s) [c]			
$R-S-S-R + 2H^+ + 2e^-$	⇌	$2\,R-SH$	(14-12)
$R-\overset{O}{\overset{\|\|}{S}}-R' + 2H^+ + 2e^-$	⇌⋯	$R-S-R' + H_2O$	(14-13)

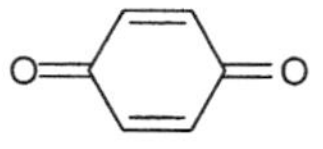

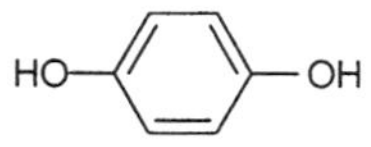

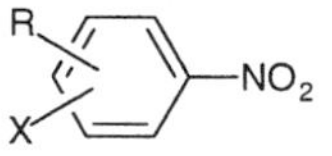

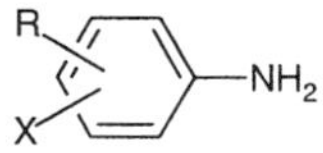

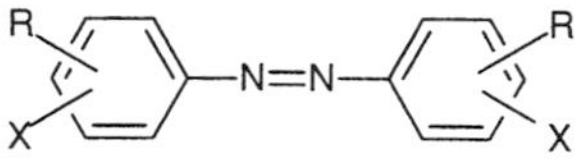

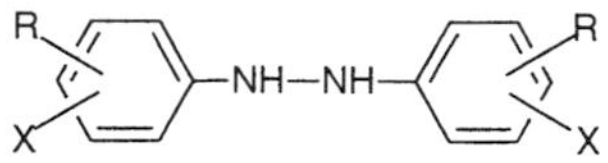

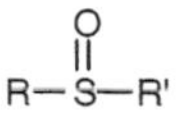

[a] Note that some reactions are reversible (indicated by ⇌), whereas others are irreversible under environmental conditions. The dotted arrow indicates that, in principle, a reaction is possible, but no clear evidence exists showing that the reaction proceeds abiotically in the dark. [b] For oxidation states of nitrogen in various functional groups see Table 2.5. [c] For oxidation states of sulfur in various functional groups see Table 2.6.

Furthermore, we should point out that in the environment, it is not known in many cases which species act as electron donors or acceptors in an observed redox reaction of a given organic pollutant. Such reactions may be catalyzed by electron transfer mediators that are present only at low concentrations, but that are continuously regenerated by chemical and/or biological processes involving the actual bulk electron donors present in the system (Fig. 14.1). Hence, in contrast to the reactions discussed in Chapter 13, we are in a much more difficult position with respect to quantification of reaction rates. Consequently, with our present knowledge of redox reactions of organic pollutants in the environment, we frequently have to content ourselves with a rather qualitative description of such processes. This may include an assessment of the environmental (redox) conditions that must prevail to allow a reaction to occur spontaneously, and an assessment of the relative reactivities of a series of related compounds in a given system.

14.2   Thermodynamic Considerations of Redox Reactions

For most of the abiotic reactions discussed so far (e.g., hydrolysis) the free energy change, $\Delta_r G$, of the reaction considered is negative under typical environmental conditions. These reactions occur spontaneously. Therefore, we did not discuss the thermodynamics of such reactions extensively. When looking at redox reactions of organic pollutants, the situation is quite different. Here, depending on the redox conditions prevailing in a given (micro)environment, an electron transfer to or from an organic compound may or may not be thermodynamically feasible. Depending on the redox conditions (which are predominantly determined by microbially mediated processes), electron acceptors (oxidants) or donors (reductants) that may react abiotically in a thermodynamically favorable reaction with a given organic chemical may or may not be present in sufficient abundance. Furthermore, as we have seen when discussing hydrolysis reactions, a reaction may also not occur at a significant rate for kinetic reasons. Nevertheless, thermodynamic considerations are very helpful as a first step in evaluating the redox conditions under which a given organic compound might undergo an oxidation or reduction reaction. Furthermore, since most of the redox reactions in the environment are biologically mediated, the evaluation of how much energy an organism may derive from a given reaction (e.g., see Illustrative Example 12.1) may provide very useful insights to the sequences in which important biological redox reactions occur in the environment, and the kinds of organisms expected under given conditions (see Thauer et al., 1977; Hanselmann, 1991; Schink, 1997). Hence, the following remarks on thermodynamic aspects of redox reactions also form an important base for our discussions of biological transformation processes in Chapter 17.

Half Reactions and (Standard) Reduction Potentials

We start with a simple reversible redox reaction for which we can directly measure the free energy of reaction, $\Delta_r G$, with a galvanic cell. This example helps us introduce the concept of using (standard) reduction potentials for evaluating the energetics (i.e., the free energies) of redox processes. Let us consider the reversible interconversion of 1,4-benzoquinone (BQ) and hydroquinone (HQ) (reaction 14-5 in Table 14.1). We perform this reaction at the surface of an inert electrode (e.g.,

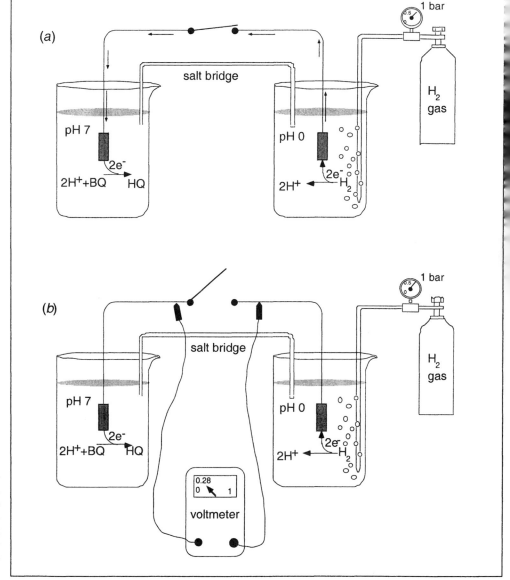

Figure 14.2 Illustration of electrochemical processes occurring in solutions with benzoquinone-hydroquinone and proton–hydrogen couples: (a) processes cycling electrons through connected system, and (b) voltage measured between separated half reactions.

platinum, graphite) that is immersed in an aqueous solution buffered at pH 7 (i.e. $\{H^+\} = 10^{-7}$) and containing BQ and HQ (see Fig. 14.2). The electron transfer occurs through a wire that connects the electrode with another inert electrode (e.g., platinum) that is immersed in an aqueous solution maintained at pH 0 (i.e. $\{H^+\} = 1$) and bubbled with molecular hydrogen (partial pressure $p_{H_2} = 1$ bar). The latter electrode is referred to as the standard hydrogen electrode (SHE). At this electrode, hydrogen is oxidized or H^+ is reduced depending upon the direction of electron flow:

$$2\,H^+ + 2\,e^- \rightleftharpoons H_2(g) \tag{14-14}$$

Note that by convention we always write a *half reaction* such as Eq. 14-14 as a *reduction*; that is, the oxidized species appears on the left side of the equation. Also, we omit the denotion (aq) for the dissolved species. In our experimental setup the reaction occurring at the other electrode is:

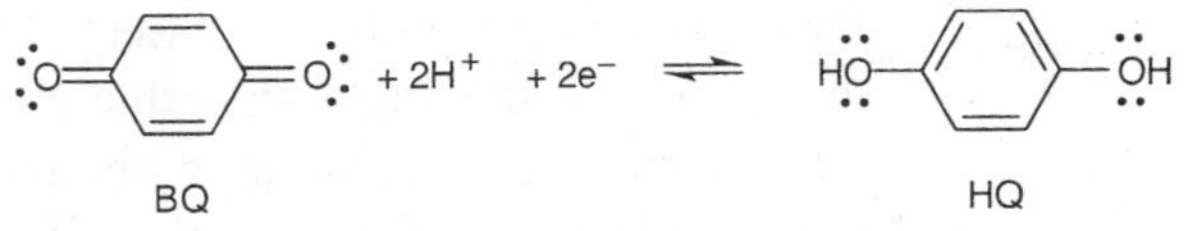

$$\text{BQ} \qquad\qquad\qquad\qquad\qquad \text{HQ}$$

(14-5)

The platinum electrodes serve to transport electrons from H_2 on one side to BQ on the other. Hence, the reaction we are actually considering is given by:

$$\text{BQ} + \text{H}_2(\text{g})\ (1\ \text{bar}) + 2\ \text{H}^+\ (\text{pH 7}) \rightleftharpoons \text{HQ} + 2\ \text{H}^+\ (\text{pH 0})$$

(14-15)

As discussed in Chapter 12 (Eq. 12-4), the $\Delta_r G$ value of reaction Eq. 14-15 is given by (note that $p_{H_2} = 1$ bar):

$$\Delta_r G = \Delta_r G^0 + RT \ln \frac{\{HQ\}(1)^2}{\{BQ\}(10^{-7})^2 \cdot 1}$$

(14-16)

or for any other proton activity in the second cell:

$$\Delta_r G = \Delta_r G^0 + RT \ln \frac{\{HQ\}}{\{BQ\}\{H^+\}^2}$$

(14-17)

With a potentiometer we can now measure the potential difference, ΔE, between the two electrodes. During our potentiometric measurement, no current is flowing between the two electrodes. If we assume electrochemical equilibrium at the electrode surface (which implies that we have a fast reversible half reaction at each electrode), then the potential difference, ΔE, is directly related to the $\Delta_r G$ of the reaction (e.g., Atkins, 1998) by:

$$\Delta_r G = -nF\Delta E$$

(14-18)

where n is the number of electrons transferred ($n = 2$ in our example) and F is the Faraday constant (= electric charge of 1 mole of electrons = 96485 Coulomb (C)·mol^{-1}). Note that we assign a positive sign to ΔE if the reaction as it is written in Eq. 14-15 proceeds spontaneously from left to right, that is, if the oxidized species (i.e., BQ) is spontaneously reduced by $H_2(g)$.

Since in experiments such as the one we have just discussed, it is only possible to determine potential differences between two electrodes (and not the absolute potential of each half cell), it is now useful to choose a reference system to which all measured potential differences may be related. In accord with the IUPAC 1953 Stockholm convention, the standard hydrogen electrode (SHE) is commonly selected as the reference electrode to which we arbitrarily assign a zero value of electrical potential. This is equivalent to assigning (arbitrarily) a standard free energy change, $\Delta_r G^0$, of zero at all temperatures to the half reaction:

$$\text{H}^+ + e^- \rightleftharpoons \frac{1}{2}\ \text{H}_2(\text{g}) \qquad\qquad K_r = 1,\ \Delta_r G^0 = 0\ \text{kJ} \cdot \text{mol}^{-1}$$

(14-19)

Note that this is also equivalent to setting the standard free energies of formation, $\Delta_f G^0$, of the proton and the electron in aqueous solution equal to zero.

By using this convention, we may now assign the measured ΔE value completely to the reaction occurring at the other electrode, in our example, to the half reaction Eq. 14-5. Instead of ΔE, we then use the term E_H, the subscript H indicating that the potential is given relative to the SHE. Hence, we can rewrite Eq. 14-18 as:

$$\Delta_r G = - nFE_H \tag{14-20}$$

Substitution of Eq. 14-20 into Eq. 14-17 and *conversion to decadic logarithms* yields for our half reaction 14-5:

$$E_H = E_H^0 - \frac{2.303\,RT}{nF} \log \frac{\{HQ\}}{\{BQ\}\{H^+\}^2} \tag{14-21}$$

where $E_H^0 = -\Delta_r G^0 / nF$. This type of equation is commonly referred to as the *Nernst equation* of an electrode reaction. Hence, the Nernst equation *is equivalent to expressing the free energy of a reaction (Eq. 12-8) in terms of potentials rather than in terms of free energies:*

$$E_H = E_H^0 - \frac{2.303\,RT}{nF} \log Q_r \tag{14-22}$$

where Q_r is the reaction quotient (Eq. 12-7). E_H^0 is called the *standard redox potential* or *standard reduction potential* since we always write the half reaction as a reduction. E_H^0 is the potential we would measure against the SHE if all species involved in the (reversible) reaction were in their standard states of unit activity (recall that we use the "infinite dilution state" as the reference state). In our example, E_H would be equal to E_H^0 if $\{BQ\} = \{HQ\}$ and $\{H^+\} = 1$. The E_H^0 value of reaction Eq. 14-5 is +0.70 V at 25°C. Hence, at standard conditions, the value for $\Delta_r G^0$ ($= -nF E_H^0$) of this half reaction (actually of the reaction of aqueous BQ with gaseous hydrogen under standard conditions) is –135 kJ·mol^{-1} corresponding to an equilibrium constant $K_r = 10^{+23.7}$. This value indicates that at pH 0 we would thermodynamically be able to reduce BQ almost completely to HQ using molecular hydrogen at 1 bar pressure.

Since we are dealing with redox reactions occurring in the environment we should be more interested in standard redox potential values (or $\Delta_r G^0$ values) that are more representative of typical natural conditions. We can calculate such values easily by assigning a typical concentration (or, more precisely, activity) value to the major water constituents that are involved in a given redox reaction. For example, we can define a $E_H^0(W)$ value (the W indicating conditions typical for natural waters) by setting the pH equal to 7, the concentration of chloride to 10^{-3} M (if we consider a dechlorination reaction, see Table 14.1), of bromide to 10^{-5}M, and so on, but by leaving organic oxidant and reductant at unit activity. As we can see from Eq. 14-21, in our example, only the hydrogen ion activity is relevant. At 25°C the term 2.303 RT/F has a value of 0.059 V. Hence, the $E_H^0(W)$ value for the half reaction (Eq. 14-5) is $E_H^0(W) = E_H^0 - (0.059/2) \times 14 = 0.28$ V. This corresponds to a $\Delta_r G^0(W)$ value

Table 14.2 Standard Reduction Potentials and Average Standard Free Energies of Reaction (per Electron Transferred) at 25 °C of Some Redox Couples that Are Important in Natural Redox Processes (The reactions are ordered in decreasing $E_H^0(W)$ values.) [a]

	Halfreaction		E_H^0 (V)	$E_H^0(W)$ (V)	$\Delta_r G^0(W)/n$ [c] (kJ·mol⁻¹)
	Oxidized Species	Reduced Species			
(1a)	$O_2(g) + 4 H^+ + 4 e^- = 2 H_2O$		+1.23	+0.81	−78.3
(1b)	$O_2(aq) + 4 H^+ + 4 e^- = 2 H_2O$		+1.19	+0.77	−74.3
(2)	$2 NO_3^- + 12 H^+ + 10 e^- = N_2(g) + 6 H_2O$		+1.24	+0.74	−72.1
(3)	$MnO_2(s) + HCO_3^- (10^{-3}) + 3 H^+ + 2 e^- = MnCO_3(s) + 2 H_2O$			+0.53 [b]	−50.7 [b]
(4)	$NO_3^- + 2 H^+ + 2 e^- = NO_2^- + H_2O$		+0.85	+0.43	−41.6
(5)	$NO_3^- + 10 H^+ + 8 e^- = NH_4^+ + 3 H_2O$		+0.88	+0.36	−35.0
(6)	$FeOOH(s) + HCO_3^- (10^{-3} M) + 2 H^+ + e^- = FeCO_3(s) + 2 H_2O$			−0.05 [b]	+ 4.8 [b]
(7)	CH_3COCOO^- (pyruvate) $+ 2 H^+ + 2 e^- = CH_3CHOHCOO^-$ (lactate)			−0.19	+17.8
(8a)	$HCO_3^- + 9 H^+ + 8 e^- = CH_4(aq) + 3 H_2O$		+0.21	−0.20	+19.3
(8b)	$CO_2(g) + 8 H^+ + 8 e^- = CH_4(g) + 2 H_2O$		+0.17	−0.24	+23.6
(9)	$SO_4^{2-} + 9 H^+ + 8 e^- = HS^- + 4 H_2O$		+0.25	−0.22	+20.9
(10)	$S(s) + 2 H^+ + 2 e^- = H_2S(aq)$		+0.14	−0.27	+26.0
(11a)	$2 H^+ + 2 e^- = H_2(aq)$		+0.08	−0.33	+31.8
(11b)	$2 H^+ + 2 e^- = H_2(g)$		0.00	−0.41	+40.0
(12)	$6 CO_2(g) + 24 H^+ + 24 e^- = C_6H_{12}O_6(glucose) + 6 H_2O$		−0.01	−0.43	+41.0

[a] Note that most of the electron transfer reactions involving these redox couples are biologically mediated. Data from Thauer et al. (1977) and Stumm and Morgan (1995). [b] Note that these values correspond to [HCO_3^-] = 10^{-3} M. [c] n = number of electrons transferred.

of $−54.0$ kJ·mol⁻¹, and an equilibrium constant $K_r(W) = 10^{+9.5}$. In the following, we primarily use $E_H^0(W)$ values for evaluating the energetics of redox reactions under natural conditions.

The standard reduction potentials of some environmentally important redox couples and of some organic redox couples are given in Tables 14.2 and 14.3. We should point out that many of the half reactions that we consider do not occur reversibly at an electrode surface, so that we would not be able to measure the corresponding E_H values using a galvanic cell. Nevertheless it is very convenient to express the free energy change of a half reaction by assigning the appropriate standard reduction potentials, that is, $E_H^0 = -\Delta_r G^0/nF$. One possibility is to calculate such reduction potentials from thermodynamic data, such as (estimated) standard free energies of formation ($\Delta_f G^0(aq)$, Eq. 12-5) of the various species involved in the half reaction (see Illustrative Example 14.1).

Table 14.3 Standard Reduction Potentials and Average Standard Free Energies of Reaction (per Electron Transferred) at 25°C of Some Organic Redox Couples in Aqueous Solution (The reactions are ordered in decreasing $E_H(W)$ values.) [a]

	Oxidized Species		Reduced Species	E_H^0 (V)	$E_H^0(W)$ [b] (V)	$\Delta_r G^0(W)/n$ [c] (kJ·mol^{-1})
(1)	$CCl_3-CCl_3 + 2e^-$	=	$Cl_2C=CCl_2 + 2Cl^-$	+0.95	+1.13	-109.0
(2)	$CBr_4 + H^+ + 2e^-$	=	$CHBr_3 + Br^-$	+0.89	+0.83	-80.1
(3)	$CCl_4 + H^+ + 2e^-$	=	$CHCl_3 + Cl^-$	+0.79	+0.67	-64.7
(4)	$CHBr_3 + H^+ + 2e^-$	=	$CH_2Br_2 + Br^-$	+0.67	+0.61	-58.9
(5)	$Cl_2C=CCl_2 + H^+ + 2e^-$	=	$Cl_2C=CHCl + Cl^-$	+0.70	+0.58	-56.0
(6)	$CHCl_3 + H^+ + 2e^-$	=	$CH_2Cl_2 + Cl^-$	+0.68	+0.56	-54.0
(7)	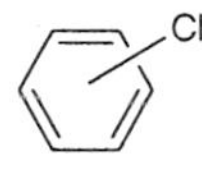 (C$_6$Cl$_6$ ring) $+ H^+ + 2e^-$	=	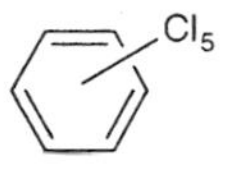 (C$_6$HCl$_5$ ring) $+ Cl^-$	+0.68	+0.56	-54.0
(8)	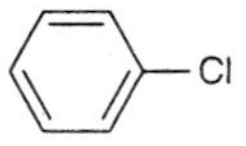 (chlorobenzene) $-Cl + H^+ + 2e^-$	=	 (benzene) $+ Cl^-$	+0.54	+0.42	-40.5
(9)	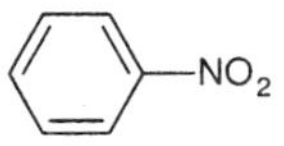 (phenyl)$-NO_2 + 6H^+ + 6e^-$	=	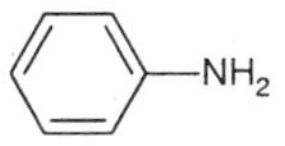 (phenyl)$-NH_2 + 2H_2O$	+0.83	+0.42	-40.5
(10)	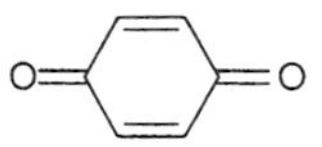 $O=$(cyclohexadiene)$=O + 2H^+ + 2e^-$	=	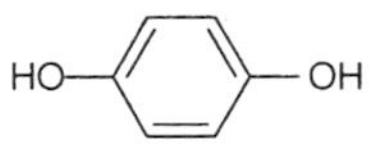 $HO-$(benzene)$-OH$	+0.70	+0.28	-27.0
(11)	$H_3C-\overset{O}{\overset{\|}{S}}-CH_3 + 2H^+ + 2e^-$	=	$H_3C-S-CH_3 + H_2O$	+0.57	+0.16	-15.4
(12)	(phenyl)$-N=N-$(phenyl) $+ 4H^+ + 4e^-$	=	2 (phenyl)$-NH_2$ 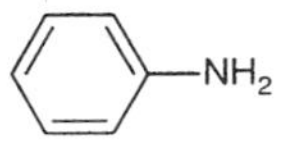	+0.31	-0.10	+9.7
(13)	$CH_3-\overset{O}{\underset{O}{\overset{\|}{\underset{\|}{S}}}}-CH_3 + 2H^+ + 2e^-$	=	$H_3C-\overset{O}{\overset{\|}{S}}-CH_3 + H_2O$	+0.17	-0.24	+23.2
(14)	$R-S-S-R + 2H^+ + 2e^-$ (cystine)	=	$2R-SH$ (cysteine)	+0.02	-0.39	+37.6

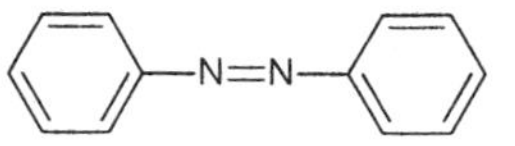

[a] Estimated from thermodynamic data Dean (1985); Vogel et al. (1987); Krop et al. (1994); Roberts et al. (1996); Totten and Roberts (2001). [b] $[H^+] = 10^{-7}$, $\{Cl^-\} = 10^{-3}$, $\{Br^-\} = 10^{-5}$. [c] n = number of electrons transferred.

Illustrative Example 14.1   **Calculating Standard Reduction Potentials from Free Energies of Formation**

Problem

Consider the half reaction in aqueous solution:

$$2\ NO_3^- + 12\ H^+ + 10\ e^- \rightleftharpoons N_2(g) + 6\ H_2O \tag{1}$$

which is catalyzed by microorganisms and is commonly referred to as denitrification. Calculate the E_H^0 and $E_H^0(W)$ values of this reaction at 25°C using the $\Delta_f G^0$ values given for the various species in the left margin. What are the E_H^0 and $E_H^0(W)$ values of the half reaction:

$$\frac{1}{5}NO_3^- + \frac{6}{5}H^+ + e^- \rightleftharpoons \frac{1}{10}N_2(g) + \frac{3}{5}H_2O \tag{2}$$

$\Delta_f G_{NO_3^-}^0\ (aq) = -111.3\ \text{kJ} \cdot \text{mol}^{-1}$

$\Delta_f G_{H^+}^0\ (aq) = \quad 0\ \text{kJ} \cdot \text{mol}^{-1}$

$\Delta_f G_{e^-}^0\ (aq) = \quad 0\ \text{kJ} \cdot \text{mol}^{-1}$

$\Delta_f G_{N_2}^0\ (g) = \quad 0\ \text{kJ} \cdot \text{mol}^{-1}$

$\Delta_f G_{H_2O}^0\ (\ell) = -237.2\ \text{kJ} \cdot \text{mol}^{-1}$

Answer

For reaction 1, $\Delta_r G^0$ is given by:

$$\Delta_r G^0 = -2(-111.3) - 0 - 0 + 0 + 6(-237.2) = -1200.6\ \text{kJ} \cdot \text{mol}^{-1}$$

Recall that $\Delta_r G^0$ of the half reaction 1 is actually the $\Delta_r G^0$ of the reaction of NO_3^- with H_2 under standard conditions. Since $E_H^0 = -\Delta_r G^0/nF$, you obtain (note that $F = 96485\ \text{C} \cdot \text{mol}^{-1} = 96485\ \text{J} \cdot \text{V}^{-1}\ \text{mol}^{-1}$):

$$E_H^0 = \frac{(1200.6\ \text{kJ} \cdot \text{mol}^{-1})}{(10)(96.5\ \text{kJ} \cdot \text{mol}^{-1}\ \text{V}^{-1})} = +1.24\ V$$

To calculate the potential at pH 7, use the Nernst equation (Eq. 14-22) for reaction 1. At 25°C (where $2.303\ RT/F = 0.059\ V$) this is:

$$E_H = E_H^0 - \frac{0.059\ V}{10} \log \frac{\{H_2O\}^6\ p_{N_2}}{\{NO_3^-\}^2 \{H^+\}^{12}} \tag{3}$$

With all species except H^+ [$\{H^+\} = 10^{-7}$] at standard conditions, you obtain:

$$E_H^0(W) = E_H^0 - \frac{0.059\ V}{10} \log(10^{-7})^{-12}$$

$$= 1.24 - 0.50 = +0.74\ V$$

Note that when using this $E_H^0(W)$ value as standard potential, you have to write Eq. 3 as:

$$E_H = E_H^0(W) - \frac{0.059\ V}{10} \log \frac{\{H_2O\}^6\ p_{N_2}}{\{NO_3^-\}^2 \left(\{H^+\}/10^{-7}\right)^{12}} \tag{4}$$

Finally, the E_H^0 and $E_H^0(W)$ values calculated for half reaction 2 are identical to these obtained for half reaction 1. Eq. 2 just expresses the same half reaction for the transfer of one electron. Hence, $\Delta_r G^0$ of reaction 2 is 10 times smaller as compared to $\Delta_r G^0$ of reaction 1, but at the same time we divide only by $n = 1$ instead of $n = 10$, which yields the same values for the E_H^0 and for the $E_H^0(W)$. The corresponding

Nernst equation Eq. 4 thus becomes:

$$E_H = E_H^0(W) - 0.059 \text{ V} \log \frac{\{H_2O\}^{3/5} p_{N_2}^{1/10}}{\{NO_3\}^{1/5}(\{H^+\}/10^{-7})^{6/5}} \tag{5}$$

Problem

Consider the reduction of hexachloroethane (C_2Cl_6) to tetrachloroethene (C_2Cl_4):

$$C_2Cl_6 + 2 \text{ e}^- \rightleftharpoons C_2Cl_4 + 2 \text{ Cl}^- \tag{6}$$

Calculate the E_H^0 and $E_H^0(W)$ values of this reaction at 25°C using $\Delta_r G^0$ values that you can find in the literature.

$\Delta_f G_{Cl^-}^0(aq) = -131.3 \text{ kJ} \cdot \text{mol}^{-1}$

$\Delta_f G_{C_2Cl_6}^0(g) = -54.9 \text{ kJ} \cdot \text{mol}^{-1}$

$\Delta_f G_{C_2Cl_4}^0(g) = +20.5 \text{ kJ} \cdot \text{mol}^{-1}$

Answer

In the literature you find the $\Delta_f G_i^0(aq)$ value for Cl^-; but for C_2Cl_6 and C_2Cl_4 only values for the free energy of formation in the gas phase are available. As shown in Illustrative Example 12.2 (Eq. 2), the free energy of the two compounds in the aqueous phase can be calculated from the gas-phase data and the Henry's law constants (expressed in bar·L·mol⁻¹):

$$\Delta_f G_i^0(aq) = \Delta_f G_i^0(g) + RT \ln K_{iH} \tag{7}$$

The K_{iH} values of C_2Cl_6 and C_2Cl_4 are 3.95 bar·L·mol⁻¹ and 27.5 bar·L·mol⁻¹, respectively. Insertion of these values into Eq. 7 together with the $\Delta_f G_i^0(g)$ values found for the two compounds in the literature yields:

$$\Delta_f G_{C_2Cl_6}^0(aq) = -51.5 \text{ kJ} \cdot \text{mol}^{-1} \text{ and } \Delta_f G_{C_2Cl_4}^0(aq) = +28.7 \text{ kJ} \cdot \text{mol}^{-1}$$

Hence, the standard free energy of reaction Eq. 6 can now be calculated (note that we omit $\Delta_f G_{e^-}^0(aq)$ which is zero):

$$\Delta_r G^0 = -(-51.5) + (+28.7) + 2(-131.3) = -182.4 \text{ kJ} \cdot \text{mol}^{-1}$$

The corresponding E_H^0 value is given by:

$$E_H^0 = \frac{-(-182.4 \text{ kJ} \cdot \text{mol}^{-1})}{(2)(96.5 \text{ kJ} \cdot \text{mol}^{-1} \text{ V}^{-1})} = 0.95 \text{ V}$$

Hence, the Nernst equation can be written as:

$$E_H = 0.95 \text{ V} - \frac{0.059 \text{ V}}{2} \log \frac{\{C_2Cl_4\}\{Cl^-\}^2}{\{C_2Cl_6\}} \tag{8}$$

Insertion of $\{Cl^-\} = 10^{-3}$ and setting $\{C_2Cl_4\} = \{C_2Cl_6\} = 1$ yields the $E_H^0(W)$ value:

$$E_H^0(W) = 0.95 \text{ V} + 3(0.059 \text{ V}) = 1.13 \text{ V}$$

Note that when using $E_H^0(W)$ instead of E_H^0, the Nernst equation Eq. 8 becomes:

$$E_H = 1.13 \text{ V} - \frac{0.059 \text{ V}}{2} \log \frac{\{C_2Cl_4\}(\{Cl^-\}/10^{-3})^2}{\{C_2Cl_6\}} \tag{9}$$

Problem

Consider the reduction of nitrobenzene (NB) to aniline (An; $Ar = C_6H_5$):

$$ArNO_2 + 6\ e^- + 6\ H^+ \rightleftharpoons ArNH_2 + 2\ H_2O \qquad (10)$$

$$\text{NB} \qquad\qquad\qquad \text{An}$$

Calculate the E_H^0 and $E_H^0(W)$ values of this reaction at 25°C using the $\Delta_f G^0$ values that you can find in the literature.

$\Delta_f G_{NB}^0(\ell) = +146.2\ \text{kJ} \cdot \text{mol}^{-1}$

$\Delta_f G_{An}^0(\ell) = +149.1\ \text{kJ} \cdot \text{mol}^{-1}$

$\Delta_f G_{H_2O}^0(\ell) = -237.2\ \text{kJ} \cdot \text{mol}^{-1}$

$C_{NBw}^{sat}(\ell) = 0.017\ \text{mol} \cdot \text{L}^{-1}$

$C_{Anw}^{sat}(\ell) = 0.39\ \text{mol} \cdot \text{L}^{-1}$

Answer

In this case, $\Delta_f G_i^0$ values are available only for the pure liquid compounds (Dean, 1985). Also known are the aqueous solubilities of the two compounds. Since for the solvent H_2O the reference state is the pure liquid, you may directly use $\Delta_f G_{H_2O}^0(\ell)$. For NB and An, however, you need to calculate $\Delta_f G_i^0(aq)$, that is, the standard free energy of formation in aqueous solution at a concentration of 1 M. From Chapters 3 and 5 you recall that transferring a compound from its pure liquid to water is given by the term $RT \ln x_{iw}\gamma_{iw}$. In this case, you want x_{iw} at 1 M. Therefore, you obtain:

$$\Delta_f G_i^0(aq) = \Delta_f G_i^0(\ell) + RT \ln x_{iw}(1\ M) + RT \ln \gamma_{iw} \qquad (11)$$

Now you can express the $\Delta_r G^0$ of reaction 10 as:

$$\Delta_r G^0 = -\Delta_f G_{NB}^0(\ell) + \Delta_f G_{An}^0(\ell) + 2\,\Delta_f G_{H_2O}^0(\ell)$$

$$-RT \ln x_{NBw}(1\ M) + RT \ln x_{Anw}(1\ M) \qquad (12)$$

$$-RT \ln \gamma_{NBw} + RT \ln \gamma_{Anw}$$

Note that $x_{NBw}(1\ M) \cong x_{Anw}(1\ M)$. Furthermore, as a first approximation, we may assume that the aqueous activity coefficients of both compounds are independent of concentration, Thus, $\gamma_{iw} \cong \gamma_{iw}^{sat}$, and since $\gamma_{iw}^{sat} \cong (C_{iw}^{sat}(\ell) \cdot \overline{V}_w)^{-1}$ (Eq. 5.7), we obtain:

$$\Delta_r G_H^0 = -\Delta_f G_{NB}^0(\ell) + \Delta_f G_{An}^0(\ell) + 2\Delta_f G_{H_2O}^0(\ell)$$

$$+RT \ln\left[C_{NBw}^{sat}(\ell) / C_{Anw}^{sat}(\ell) \right] \qquad (13)$$

Insertion of the $\Delta_f G_i^0(\ell)$ and C_{iw}^{sat} values given above into Eq. 13 yields:

$$\Delta_r G^0 = -(146.2) + (149.1) + (-474.4) + (2.48)(-3.13) = -479.3\ \text{kJ} \cdot \text{mol}^{-1}$$

which corresponds to an E_H^0 value of *+0.83* V.

Hence, the Nernst equation can be written as:

$$E_H = 0.83\ \text{V} - \frac{0.059\ \text{V}}{6} \log \frac{\{ArNH_2\}}{\{ArNO_2\}\{H^+\}^6} \qquad (14)$$

Insertion of $\{H^+\} = 10^{-7}$ and $\{ArNH_2\} = \{ArNO_2\}$ yields the $E_H^0(W)$ value:

$$E_H^0(W) = 0.83 \text{ V} - \frac{(0.059)(42)}{(6)} = +0.42 \text{ V}$$

Note that when using $E_H^0(W)$ instead of E_H^0, the Nernst equation Eq. 14 becomes:

$$E_H = 0.42 \text{ V} - \frac{0.059 \text{ V}}{(6)} \log \frac{\{ArNH_2\}}{\{ArNO_2\}(\{H^+\}/10^{-7})^6}$$

One-Electron Reduction Potentials

So far, except for the iron(III)/iron(II) couple [reaction (6) in Table 14.2], we have considered reduction potentials of half reactions with an overall transfer of an even number of electrons (i.e., 2, 4, 6, etc.). However, in many *abiotic* multielectron redox processes, particularly if organic compounds are involved, the actual electron transfer frequently occurs by a sequence of one-electron transfer steps (Eberson, 1987). The resulting intermediates formed are often very reactive, and they are not stable under environmental conditions. In our benzoquinone example, BQ is first reduced to the corresponding semiquinone (SQ), which is then reduced to HQ:

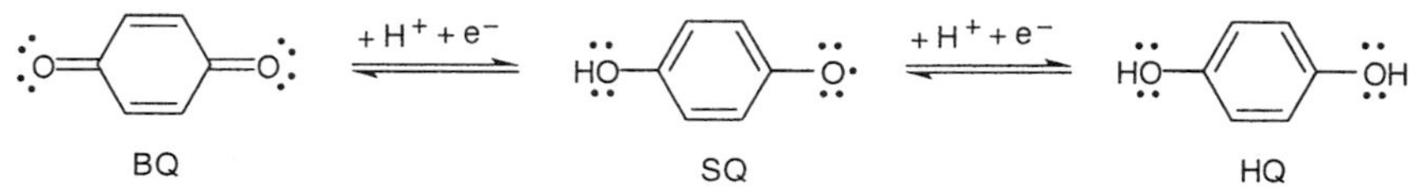

(14-23)

Each of these subsequent one-electron steps has its own $E_H^0(W)$ value (Neta, 1981). We denote the reduction potential for the transfer of the first electron by $E_H^1(W)$, and for the transfer of the second electron by $E_H^2(W)$:

$$BQ + H^+ + e^- \rightleftharpoons SQ; \qquad E_H^1(W) = +0.10 \text{ V} \qquad (14\text{-}24)$$

$$SQ + H^+ + e^- \rightleftharpoons HQ; \qquad E_H^2(W) = +0.46 \text{ V} \qquad (14\text{-}25)$$

[Note that in the literature one often finds the notation E_{m7}^1 and E_{m7}^2 for $E_H^1(W)$ and $E_H^2(W)$, respectively].

From these values we see that the free energy change is much less negative [smaller $E_H^0(W)$ value] for the transfer of the first electron to BQ as compared to the transfer of the second electron to SQ. Conversely, there is more energy required to oxidize HQ to SQ as compared to the oxidation of SQ to BQ. In general, we can assume that the formation of an organic radical is much less favorable from an energetic point of view, as compared to the formation of an organic species exhibiting an even number of electrons. From this we may conclude that the first one-electron transfer between an organic chemical and an electron donor or acceptor is frequently the rate-limiting step. Thus, when we are interested in relating thermodynamic and kinetic data (e.g., through LFERs), we need to consider primarily the E_H values of this rate-limiting step, that is, the E_H value of the first one-electron transfer (see Section 14.3). We should be aware that if this first step is endergonic (i.e., positive $\Delta_r G$ value for 1 e$^-$-transfer), the overall reaction may still be exergonic (i.e., negative $\Delta_r G$ value for 2 e$^-$-transfer), and the whole reaction may proceed spontaneously (Eberson, 1987).

Therefore, in evaluating whether or not a given redox reaction is possible under given conditions, we need to consider the E_H values of the overall reaction.

Finally, we should note that the E_H^0 value of a multielectron transfer half reaction is given by the *average* of the respective standard one-electron reduction potentials. This is easy to rationalize when recalling that the overall standard free energy of reaction of a sequence of reaction steps is given by the sum of the $\Delta_r G^0$ values of each step. Hence, we may write:

$$\Delta_r G^0 = \Delta_r G^1 + \Delta_r G^2 + \ldots + \Delta_r G^n = \sum_{k=1}^{n} \Delta_r G^k \tag{14-26}$$

Substitution of $\Delta_r G^0$ by $-nFE_H^0$ and $\Delta_r G^k$ by $-FE_H^k$ into Eq. 14-26 and rearrangement yields:

$$E_H^0 = \frac{1}{n} \sum_{k=1}^{n} E_H^k \tag{14-27}$$

and, similarly:

$$E_H^0(W) = \frac{1}{n} \sum_{k=1}^{n} E_H^k(W) \tag{14-28}$$

Thus, the $E_H^0(W)$ value of the overall reaction Eq. 14-23 ($BQ + 2\,H^+ + 2\,e^- = HQ$) is $(0.10\ V + 0.46\ V)/2 = 0.28\ V$.

Processes Determining the Redox Conditions in the Environment

Before we proceed to evaluate the thermodynamics of redox reactions at environmental conditions, we need to make a few remarks on microbial processes that determine the redox conditions in the environment.

We can get a general idea about the maximum free energies that microorganisms may gain from catalyzing various redox reactions from the data in Table 14.2. On earth, photosynthetic harvesting of solar energy is the main cause for nonequilibrium redox conditions. In the process of photosynthesis, organic compounds exhibiting reduced states of carbon, nitrogen, and sulfur are synthesized, and at the same time oxidized species including O_2 (oxic photosynthesis) or oxidized sulfur species (anoxic photosynthesis) are produced. Using glucose as a model organic compound, we can express oxic photosynthesis by combining Eqs. (12) and (1) in Table 14.2. Note that we have to take the reversed form of Eq. (1). Since we are looking at the overall process, it is convenient to write the reaction with a stoichiometry corresponding to the transfer of one electron:

$$\frac{1}{4}CO_2(g) + \frac{1}{4}H_2O \rightleftharpoons \frac{1}{24}C_6H_{12}O_6 + \frac{1}{4}O_2(g) \tag{14-29}$$

The standard free energy change per electron transferred, $\Delta_r G^0(W)/n$, of reaction Eq. 14-29 can now be simply derived from Table 14.2 by adding the $\Delta_r G^0(W)$ value of reaction (12) ($+41.0\ kJ \cdot mol^{-1}$) and reversed reaction (1) ($+78.3\ kJ \cdot mol^{-1}$): $\Delta_r G^0(W)/n = +119.3\ kJ \cdot mol^{-1}$. Thus, on a "per-electron basis," under standard conditions (pH 7), we have to invest $119.3\ kJ \cdot mol^{-1}$ to (photo)synthesize glucose from CO_2 and H_2O. In our standard redox potential picture using $E_H^0(W)$ values, this is equivalent

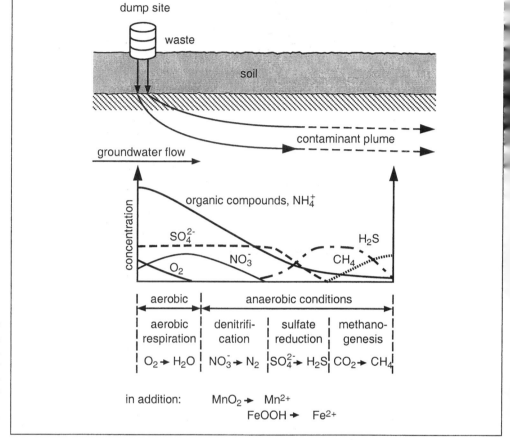

Figure 14.3 Variation in concentrations of important dissolved redox species along the flowpath of a contaminant plume in groundwater. This sequence results in several zones of characteristic microbial metabolism and corresponding redox conditions (adapted from Bouwer et al., 1984).

to promoting one mole of electrons from a potential of +0.81 to –0.43 V (see Table 14.2).

The chemical energy stored in reduced chemical species (including organic pollutants) can now be utilized by organisms that are capable of catalyzing energy-yielding redox reactions. For example, from Table 14.2 we can deduce that in the oxidation of glucose [reversed reaction (12)], oxygen is the most favorable oxidant (i.e., electron acceptor) from an energetic point of view, at least if O_2 is reduced all the way to H_2O (which is commonly the case in biologically mediated processes). The $\Delta_r G(W)/n$ value for the reaction of glucose with O_2 (reversed reaction Eq. 14-29) is, of course, –119.3 kJ·mol^{-1}. The next "best" electron acceptors would be NO_3^- (if converted to N_2), then $MnO_2(s)$, and so on going down the list in Table 14.2.

Interestingly, the chemical reaction sequence given in Table 14.2 (that is based on standard free energy considerations) is, in essence, paralleled by a spatial and/or temporal succession of different microorganisms in the environment. In other words, in a given (micro)environment, the organisms that tend to be dominant are those capable of utilizing the "best" electron acceptor(s) available, where the "best" electron acceptor is the one exhibiting the highest reduction potential. These microorganisms then in turn determine the redox conditions in that (micro)environment. This sequential utilization of electron acceptors can be seen if we look at the dynamics of some dissolved redox species along the flowpath of a confined contaminant plume in the ground (Fig. 14.3). For simplicity, we assume a situation where we have a constant input of reduced (e.g., organic compounds, NH_4^+) and oxidized species (e.g., O_2, NO_3^-, SO_4^{2-}). As is shown in Figure 14.3, natural or synthetic organic compounds (the major electron donors) are degraded over the whole length of the plume. As long as there is molecular oxygen present, *aerobic respiration* takes place. This includes the oxidation of organic compounds and NH_4^+ (to NO_3^-) and the consumption of O_2. We should point out that in aerobic respiration, oxygen not only plays the role of a terminal electron acceptor, but it is also a cosubstrate in many important biologically catalyzed reactions. This is the reason why we usually make such a sharp distinction between oxic, suboxic, and anoxic conditions (see also Chapter 17).

Once the oxygen is consumed, *denitrification* (see Illustrative Example 14.2) is observed until nitrate is virtually absent. In the region where denitrification occurs, one often observes the reductive dissolution of oxidized manganese phases [e.g., $MnO_2(s)$, $MnOOH(s)$], which may or may not be biologically catalyzed. Under those conditions iron is still present in oxidized forms [e.g., $FeOOH(s)$]. Then, a marked decrease in redox potential occurs when only electron acceptors are left in significant abundance that exhibit low reduction potentials (see Table 14.2). This redox sequence has led to a somewhat different terminology in that one speaks of the *oxic* (aerobic), *suboxic* (denitrification, manganese reduction), and *anoxic* conditions (low redox potential). Processes involving electron acceptors exhibiting a low redox potential include, in sequence: *iron reduction, sulfate reduction, fermentation,* and *methanogenesis.*

Illustrative Example 14.2   **Establishing Mass Balances for Oxygen and Nitrate in a Given System**

Problem

Consider a situation in which bank filtrate of a polluted river is used for drinking-water supply. Among other water constituents, dissolved organic material (measured as dissolved organic carbon, DOC), ammonia, oxygen, and nitrate are continuously monitored in the river and in a well located at a distance of 10 meters from the river bank. The average values obtained for the four parameters are given below. Inspection of the field data shows that a significant portion of the organic material and virtually all NH_4^+ and O_2 are eliminated by microbial processes during infiltration, but that the infiltrated water still exhibits 75% of the nitrate concentration observed in the river. Are these findings reasonable when assuming that no additional (water) input occurs during infiltration, and that the organic material is oxidized to CO_2, NH_4^+ is oxidized to NO_3^-, and NO_3^- is reduced to N_2?

Species (measured parameter)	Concentration in the River	Concentration in the Well
Organic material (CH_2O) ([DOC])	4.2 mg $C \cdot L^{-1}$	1.2 mg $C \cdot L^{-1}$
Ammonia (NH_4^+) ([NH_4^+–N])	2.1 mg $N \cdot L^{-1}$	< 0.1 mg $N \cdot L^{-1}$
Dissolved oxygen (O_2) ([O_2])	9.6 mg $O_2 \cdot L^{-1}$	< 0.1 mg $O_2 \cdot L^{-1}$
Nitrate (NO_3^-) ([NO_3^-–N])	2.1 mg $N \cdot L^{-1}$	1.6 mg $N \cdot L^{-1}$

Answer

Use the half reactions given in Table 14.2 to establish electron balances for the various processes:

(i) Oxidation of organic material:

$$CH_2O + H_2O = CO_2 + 4 H^+ + 4 e^-$$

(ii) Oxidation of ammonia (nitrification):

$$NH_4^+ + 3 H_2O = NO_3^- + 10 H^+ + 8 e^-$$

(iii) Reduction of oxygen:

$$O_2 + 4 H^+ + 4 e^- = 2 H_2O$$

(iv) Reduction of nitrate (denitrification):

$$NO_3^- + 6 H^+ + 5 e^- = \frac{1}{2} N_2 + 3 H_2O$$

Calculate how many electrons are produced and consumed, respectively, by the various processes during infiltration:

(i) $\Delta[CH_2O] = -3.0$ mg $C \cdot L^{-1}$	$= -0.25$ mM $CH_2O \cdot L^{-1}$	$= +1.0$ mM e^-
(ii) $\Delta[NH_4^+] = -2.1$ mg $N \cdot L^{-1}$	$= -0.15$ mM $NH_4^+ \cdot L^{-1}$	$= +1.2$ mM e^-
(iii) $\Delta[O_2] = -9.6$ mg $O_2 \cdot L^{-1}$	$= -0.3$ mM $O_2 \cdot L^{-1}$	$= -1.2$ mM e^-

Total electrons without considering denitrification	$= +1.0$ mM e^-

Thus, in order to balance the electrons, 1 mM e^- have to be consumed by denitrification. Hence, the calculated consumption of nitrate is 1 mM $e^- = 0.2$ mM $NO_3^- = 2.8$ mg $N \cdot L^{-1}$, which is more than is present in the river water. Note, however, that reaction (ii) produces 0.15 mM $= 2.1$ mg $N \cdot L^{-1}$ nitrate. Thus, one would expect a net decrease in nitrate of only 0.7 mg $N \cdot L^{-1}$, which compares well with the observed 0.5 mg $N \cdot L^{-1}$ decrease. The measured concentration changes of the four water constituents are, therefore, reasonable.

Problem

For remediation of an aquifer that has been contaminated with toluene, ground-water is pumped through the contaminated zone, then pumped back to the surface, saturated with air or supplied with nitrate, and finally introduced into the ground again. The idea of this quite widely applied procedure is to stimulate those indigenous microorganisms that are capable of mineralizing a given substrate, in this case toluene, by supplying the necessary oxidants (Hunkeler et al., 1995). Calculate how much water is at least required to supply sufficient O_2 or NO_3^-, respectively, for degradation of 1 kg of toluene, when assuming (i) that toluene is not mobilized by this procedure, (ii) that it is completely mineralized to CO_2 and H_2O, and (iii) that the water contains either 10 mg $O_2 \cdot L^{-1}$ or 100 mg $NO_3^- \cdot L^{-1}$, respectively. Note that much more NO_3^- could, in principle, be dissolved in the water, but that the maximum allowed concentration is commonly limited by the water authorities.

toluene
$M_i = 92.1$ g $\cdot$ mol^{-1}

Answer

To calculate how many electrons have to be transferred to O_2 or NO_3^- respectively, when oxidizing 1 mole of toluene (C_7H_8) to CO_2, determine first the average oxidation state of the carbon atoms present in toluene (see also examples discussed in Illustrative Example 2.1). Since this compound is made up only of carbon and hydrogen atoms (the oxidation state of H is +I), you can just consider the hydrogen/carbon ratio, which yields an average carbon oxidation state of $-8/7$. Considering that the oxidation state of carbon in CO_2 is +IV, it is easy to see that a total of $(4 - (-8/7)) \times 7 = 36$ moles of electrons have to be transferred. The overall reactions are, therefore:

$$C_7H_8 + 9\,O_2 = 7\,CO_2 + 4\,H_2O$$

if O_2 is the oxidant (4 electrons per O_2, see Table 14.2), and, with NO_3^- as the electron acceptor (5 electrons per NO_3^-, Table 14.2):

$$C_7H_8 + \frac{36}{5}NO_3^- + \frac{36}{5}H^+ = 7\,CO_2 + \frac{18}{5}N_2 + \frac{38}{5}H_2O$$

Consequently, 9 moles of O_2 or 7.2 moles of NO_3^- are required to mineralize 1 mole of toluene. Since 1 kg of toluene corresponds to 10.86 moles, this means that at least 97.72 moles O_2 or 78.12 moles NO_3^- have to be provided by the water that is pumped through the contaminated zone. Thus, in the case of O_2 (10 g $O_2 \cdot m^{-3}$), the total water volume required is:

$$V = \frac{97.72 \text{ mol}}{(10 \text{ g} \cdot m^{-3})/(32 \text{ g} \cdot mol^{-1})} = 312.7 \text{ m}^3$$

If NO_3^- (100 g $NO_3^- \cdot m^{-3}$) is used, the calculated water volume is only:

$$V = \frac{78.1 \text{ mol}}{(100 \text{ g} \cdot m^{-3})/(62 \text{ g} \cdot mol^{-1})} = 48.5 \text{ m}^3$$

The temporal and/or spatial succession of redox processes illustrated in Fig. 14.3 for a groundwater case is also observed in other environments in which access to oxygen and other electron acceptors is limited. Examples include sediment beds and poorly mixed lakes and ocean basins. Finally, we should note that, in certain cases, the apparent redox sequence may be reversed when following, for example, a plume because the availability of stronger oxidants may increase with increasing distance from a landfill or hazardous waste site. For a more detailed discussion of the biogeochemical processes that determine the redox conditions in natural systems, we refer to the literature (Drever, 1988; Morel and Hering, 1993; Appelo and Postma, 1993; Stumm and Morgan, 1996; Christensen et al., 2000 and 2001).

Evaluating the Thermodynamics of Redox Reactions under Environmental Conditions

Let us now come back to the question of how to assess whether a given organic compound may, in principle, undergo a redox reaction in a given environmental system. For such an assessment we need to know the standard reduction potentials of the half reaction involving the compound of interest and its oxidized or reduced transformation product, and of the environmental oxidant/reductant couple involved. Since we often do not know the oxidant or reductant, we need to assign an E_H value to the environmental system we are considering. Unfortunately, unlike the situation with proton transfer reactions where we may use pH as a master variable, it is usually not possible to assign an unequivocal E_H value to a given natural water (Stumm and Morgan, 1996). Many environmentally significant redox processes are slow, and therefore we cannot assume equilibrium between all redox couples present. That also means that measurements of redox potentials of natural waters using an inert electrode and a reference electrode are often difficult to interpret, inasmuch as many important redox pairs do not show reversible electrochemical behavior at the electrode surface. This is particularly true for more oxidizing environments (aerobic conditions, denitrifying conditions) since the electrode does not respond to redox couples involving oxygen or inorganic nitrogen species. Under more reducing conditions, E_H measurements may be of some value, since there are often certain redox couples present to which the electrode does respond. Such couples include manganese species (Mn^{III}, Mn^{IV}/Mn^{II}), iron species (Fe^{III}/Fe^{II}), and certain organic compounds (e.g., quinones/hydroquinones). When measuring redox potentials in the field as well as in the laboratory, the SHE is often not used as a reference electrode for practical reasons. The most common reference electrodes are the saturated calomel electrode (SCE, $E_H^0 = +0.24$ V at 25°C) and the silver–silver chloride electrode ($E_H^0 = +0.22$ V at 25°C). The measured potentials are, however, easily converted to the hydrogen scale by adding the appropriate E_H^0 value of the reference electrode (e.g., +0.24 V in the case of SCE) to the measured value.

Owing to the difficulties in assigning a meaningful E_H value to a given natural system, it is helpful to use the $E_H^0(W)$ values of the most important biogeochemical redox processes (Table 14.2 and Fig. 14.4) as a framework for evaluating under which general redox conditions a given organic compound might undergo a certain redox reaction.

Let us illustrate this point with a few examples. By inspecting Fig. 14.4 we can see

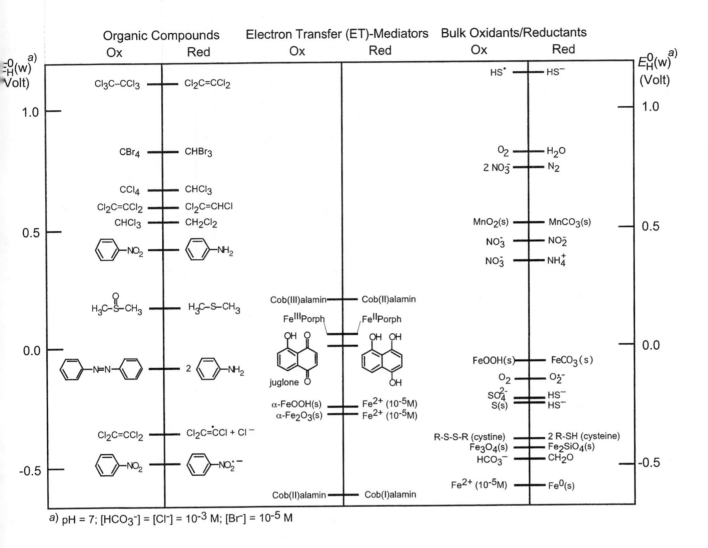

Organic Compounds   Electron Transfer (ET)-Mediators   Bulk Oxidants/Reductants

a) pH = 7; $[HCO_3^-]$ = $[Cl^-]$ = 10^{-3} M; $[Br^-]$ = 10^{-5} M

Figure 14.4 Selection of environmentally relevant redox couples including organic pollutants such as nitroaromatic and halogenated compounds, as well as examples of electron transfer mediators and important bulk reductants. The values given represent reduction potentials at pH 7 at equal (except otherwise indicated) concentrations of the redox partners but at environmental con-centrations of the major anions involved: $[HCO_3^-]$ = $[Cl^-]$ = 10^{-3} M; $[Br^-]$ = 10^{-5} M; Porph = porphyrin.

that hexachloroethane may be reduced to tetrachloroethene [reaction 1 in Table 14.3] under any environmental redox conditions. The reduction of nitrobenzene to aniline [reaction 9 in Table 14.3] is possible only at redox conditions typical for environments in which iron reduction, sulfate reduction, or fermentation occurs. Aniline may be oxidized to azobenzene [reverse reaction 12 in Table 14.3] under aerobic, denitrifying, and manganese reducing conditions. Hence, in the subsurface where pollutants may be transported through various redox zones, nitrobenzenes may first get reduced to the corresponding anilines, which then may be converted to azobenzenes upon reaching more oxidizing environments.

At this point, we might wonder how reasonable it is to use $E_H^0(W)$ values as given in Tables 14.2 and 14.3 for assessing whether a reaction will occur spontaneously in a given natural system. The species involved will not, of course, be present at standard concentrations. To evaluate this problem, let us compare the E_H value of a 10^{-4} M aqueous hydrogen sulfide (H_2S) solution at pH 8 with the $E_H^0(W)$ value of reaction 10 in Table 14.2 (–0.27 V). The calculated value (see Illustrative Example 14.3) is –0.18 V, which is still in the same ballpark. Of course, if we want to evaluate the free energy of reaction of a redox reaction involving redox couples that exhibit very similar $E_H^0(W)$ values, we need to take into account the actual concentrations (activities) of the species involved. Some examples demonstrating how to calculate the free energy of reaction, $\Delta_r G$, of a redox reaction from the corresponding half reaction reduction potentials are given in Illustrative Example 14.4.

Illustrative Example 14.3   **Calculating the Reduction Potential of an Aqueous Hydrogen Sulfide (H_2S) Solution as a Function of pH and Total H_2S Concentration**

Problem

Derive the general Nernst equation for expressing the E_H value of a hydrogen sulfide solution as a function of pH and $\{H_2S\}_{tot}$ by assuming that H_2S is oxidized to elemental sulfur [reaction 10 in Table 14.2]. Calculate the E_H for a 10^{-4} M H_2S_{tot} solution at pH 8 and 25°C.

Answer

The Nernst equation for reaction 10 in Table 14.2 is:

$$E_H = E_H^0 - \frac{0.059 \text{ V}}{2} \log \frac{\{H_2S\}}{\{H^+\}^2} \tag{1}$$

where $E_H^0 = 0.14$ V. Since H_2S dissociates in aqueous solution:

$$H_2S \rightleftharpoons HS^- + H^+ \quad ; \quad pK_a = 7.0 \text{ at } 25°C \tag{2}$$

the actual H_2S concentration at a given pH is (see Eq. 8-21):

$$\{H_2S\} = \frac{1}{1 + K_a / \{H^+\}} \{H_2S\}_{tot} \tag{3}$$

Substitution of $E_H^0 = 0.14$ V and Eq. 3 into Eq. 1 yields the desired Nernst equation:

$$E_H(\{H_2S\}_{tot}, pH, 25°C) = +0.14 \text{ V} - \frac{0.059 \text{ V}}{2} \log \frac{\{H_2S\}_{tot}}{\{H^+\}[\{H^+\} + K_a]} \tag{4}$$

By setting $\{H_2S\}_{tot} = 10^{-4}$, $\{H^+\} = 10^{-8}$, and $K_a = 10^{-7}$, one obtains an E_H value of -0.18 V.

Note that for describing the E_H value of a hydrogen sulfide solution, instead of reaction 10 in Table 14.2, we could also use the redox couple involving $S(s)$ and HS^- (instead of H_2S):

$$S(s) + H^+ + 2 e^- = HS^- \quad ; \quad E_H^0 = -0.06 \text{ V} \tag{5}$$

The corresponding Nernst equation is then:

$$E_H(\{H_2S\}_{tot}, pH, 25 °C) = -0.06 \text{ V} - \frac{0.059 \text{ V}}{2} \log \frac{K_a \{H_2S\}_{tot}}{\{H^+\}[\{H^+\} + K_a]} \tag{6}$$

The results from Eq. 4 and Eq. 6 must be identical if H_2S, HS^-, and $S(s)$ are all at equilibrium with one another. We can see that this is true by noting that $[(-0.059\ V)/2]\log K_a = +0.20\ V$. Using this in Eq. 6, we find:

$$E_H\left(\{H_2S\}_{tot}, pH, 25°C\right) = (-0.06 + 0.20)\ V - \frac{0.059\ V}{2}\log\frac{\{H_2S\}_{tot}}{\{H^+\}\left[\{H^+\} + K_a\right]}$$

$$= +0.14\ V - \frac{0.059\ V}{2}\log\frac{\{H_2S\}_{tot}}{\{H^+\}\left[\{H^+\} + K_a\right]}$$

Illustrative Example 14.4

Calculating Free Energies of Reaction from Half Reaction Reduction Potentials

Problem

Determine which of the following reactions may occur spontaneously in aqueous media. Calculate the corresponding $\Delta_r G$ value.

(a) The reduction of azobenzene (AzB) to aniline (An) [reaction 12 in Table 14.3] by H_2S assuming that $S(s)$ is formed under:

(i) standard environmental conditions ("W" conditions), or
(ii) at pH 9, $\{H_2S\}_{tot} = 10^{-4}$, and with $\{AzB\} = 10^{-8}$ and $\{An\} = 10^{-6}$.

What would be the $\{An\}/\{AzB\}$ ratio at equilibrium at pH 9 and 10^{-4} M H_2S assuming an initial azobenzene concentration of $5 \cdot 10^{-7}$ M?

(b) The oxidation of dimethyl sulfide (DMS) to dimethyl sulfoxide (DMSO) [reverse reaction (11) in Table 14.3] by $FeOOH(s)$ assuming that $FeCO_3(s)$ is formed under

(i) standard environmental conditions ("W" conditions), or
(ii) at pH 9, $\{HCO_3^-\} = 10^{-2}$, and $\{DMS\}/\{DMSO\} = 10^4$.

Answer

The $\Delta_r G$ of a reaction is related to the difference, ΔE_H, of the reduction potentials of the corresponding half reactions by Eq. 14-18:

$$\Delta_r G = -nF\ \Delta E_H \tag{1}$$

where $\Delta E_H = E_H(\text{oxidant}) - E_H(\text{reductant})$. Thus, if ΔE_H is positive, then $\Delta_r G$ is negative, and the reaction may occur spontaneously.

Answer (a)

Using the $E_H^0(W)$ values, the Nernst equations for the two half reactions are (see also Illustrative Example 14.2):

$$E_H(AzB\ /\ An) = -0.10 - \frac{0.059\ V}{4}\log\frac{\{An\}^2}{\{AzB\}\left(\{H^+\}/10^{-7}\right)^4} \tag{2}$$

$$E_H(\text{S(s)} / \text{H}_2\text{S}) = -0.27 - \frac{0.059 \text{ V}}{2} \log \frac{\{\text{H}_2\text{S}\}_{\text{tot}}}{(\{\text{H}^+\} / 10^{-7})\left[(\{\text{H}^+\} / 10^{-7}) + K_a / 10^{-7}\right]} \qquad (3)$$

Note that setting the activity of H^+ in the standard state to 10^{-7}, K_a also has to be divided by 10^{-7} in Eq. 3. [Note that $K_a/10^{-7} = (\{\text{H}^+\}/10^{-7})\{\text{HS}^-\}/\{\text{H}_2\text{S}\}$.]

(i) At standard environmental conditions, ΔE_H is given by the difference of the E_H^0 (W) values; that is:

$$\Delta E_H(\text{W}) = E_H^0(\text{W}) (\text{AzB/An}) - E_H^0(\text{W}) (\text{S(s)/H}_2\text{S}) = (-0.10) - (-0.27) = +0.17 \text{ V}$$

Hence, the reaction:

$$\text{AzB} + 2\,\text{H}_2\text{S} \rightleftharpoons 2\,\text{An} + 2\,\text{S(s)} \qquad (4)$$

occurs spontaneously from left to right at standard environmental conditions. The $\Delta_r G^0(\text{W})$ value at these conditions is (note that 4 electrons are transferred):

$$\Delta_r G^0(\text{W}) = -(0.17 \text{ V}) (4) (96.5 \text{ kJ} \cdot \text{mol}^{-1} \text{ V}^{-1}) = -65.6 \text{ kJmol}^{-1}$$

(ii) Insertion of the corresponding activities of the various species into Eqs. 2 and 3 yields the E_H values for the conditions specified above:

$$E_H(\text{AzB} / \text{An}) = -0.10 - \frac{0.059 \text{ V}}{4} \log \frac{(10^{-6})^2}{(10^{-8})(10^{-9} / 10^{-7})^4} = -0.16 \text{ V}$$

$$E_H(\text{S(s)}/\text{H}_2\text{S}) = -0.27 - \frac{0.059 \text{ V}}{2} \log \frac{(10^{-4})}{(10^{-9}/10^{-7})(10^{-9}/10^{-7} + 10^{-7}/10^{-7})}$$
$$= -0.21 \text{ V}$$

In this case, $\Delta E_H = E_H(\text{AzB/An}) - E_H(\text{S(s)/H}_2\text{S})$, and we find:

$$\Delta E_H = (-0.16) - (-0.21) = +0.05 \text{ V}$$

Therefore:

$$\Delta_r G = (0.05 \text{ V}) (4) (96.5 \text{ kJ} \cdot \text{mol}^{-1} \text{ V}^{-1}) = -19.3 \text{ kJ} \cdot \text{mol}^{-1}$$

Hence, reaction Eq. 4 still occurs from left to right, although it is much closer to equilibrium as compared to standard environmental conditions.

Since the H_2S concentration is much higher than the AzB concentration, it remains more or less constant during the reaction. The E_H value of the system is therefore determined by $E_H(\text{S(s)/H}_2\text{S}) = -0.21$ V (see above). At equilibrium, $E_H(\text{AzB/An})$ has to be equal to -0.21 V (i.e., $\Delta E_H = 0$). Insertion of this value and $\{\text{H}^+\} = 10^{-9}$ into Eq. 2 above yields after some rearrangement:

$$\log \frac{\{\text{An}\}^2}{\{\text{AzB}\}} \cong 0, \quad \text{i.e.,} \quad \frac{\{\text{An}\}^2}{\{\text{AzB}\}} \cong 1$$

By denoting the equilibrium concentration of An as x, and with an initial AzB concentration of 5×10^{-7} M, we may write (note that one AzB produces two An):

$$\frac{x^2}{(5\times10^{-7})-\frac{1}{2}x}=1 \quad \text{or} \quad x^2+\frac{1}{2}x-5\cdot10^{-7}=0$$

Solving this equation yields about 10^{-6} for {An} and about 10^{-12} for {AzB}. Hence virtually all AzB is reduced to An.

Answer (b)

Using the $E_H^0(W)$ values, the two Nernst equations are:

$$E_H(DMSO/DMS) \;=\; +0.16\ V - \frac{0.059\ V}{2}\log\frac{\{DMS\}}{\{DMSO\}\big(\{H^+\}/10^{-7}\big)^2} \tag{5}$$

$$E_H(FeOOH(s)/FeCO_3(s)) =$$

$$-0.05\ V - \frac{0.059\ V}{1}\log\frac{1}{(\{HCO_3^-\}/10^{-3})(\{H^+\}/10^{-7})^2} \tag{6}$$

The ΔE_H value for the reaction:

$$DMS + 2\ FeOOH(s) + 2\ HCO_3^- + 2\ H^+ \rightleftharpoons DMSO + 2\ FeCO_3(s) + 3\ H_2O \tag{7}$$

is given by:

$$\Delta E_H = -E_H\,(DMSO/DMS) + E_H\,(FeOOH(s)/FeCO_3(s))$$

(i) At standard environmental conditions:

$$\Delta E_H \;=\; -E_H^0(W)\,(DMSO/DMS) + E_H^0(W)\,(FeOOH(s)/FeCO_3(s))$$
$$=\; -0.16\ V - 0.05\ V = -0.21\ V,$$

and, therefore, $\Delta_r G^0(W) = -(-0.21\ V)\,(2)\,(96.5\ kJ\cdot mol\cdot V^{-1}) = +40.5\ kJ\cdot mol^{-1}$.

Hence, reaction Eq. 7 does not occur spontaneously from left to right. In fact, the result shows that DMSO could be reduced by $FeCO_3(s)$ [also seen from Fig. 14.4, since $FeCO_3(s)$ lies below DMSO].

(ii) Insertion of the corresponding activities of the species involved into Eqs. (5) and 6 yields:

$$E_H(DMSO/DMS) \;=\; 0.16\ V \;-\; \frac{0.059\ V}{2}\log\frac{10^4}{(10^{-9}/10^{-7})^2} = -0.08\ V$$

$$E_H(FeOOH(s)/FeCO_3(s)) =$$

$$-0.05\ V - 0.059\ V\log\frac{1}{(10^{-2}/10^{-3})(10^{-9}/10^{-7})^2} = -0.22\ V$$

which yields a ΔE_H of $-(-0.08\ V) - 0.22\ V = -0.14\ V$.

In this case, $\Delta_r G = -(-0.14\ V)\,(2)\,(96.5\ kJ\cdot mol\cdot V^{-1}) = +27\ kJ\cdot mol^{-1}$, which is a similar situation as found for standard environmental conditions.

14.3 Reaction Pathways and Kinetics of Redox Reactions

Factors Determining the Rate of Redox Reactions

Having considered the reduction potentials of "overall" half reactions, we can decide whether from a thermodynamic point of view a given compound may undergo oxidation or reduction to yield a specific product in a given environment. We now have to tackle the more difficult part, the kinetics of such reactions. As pointed out earlier, a compound may react with several different electron acceptors (oxidants) or electron donors (reductants), and the relative importance of such species present in a given system may be strongly influenced by complex biogeochemical processes. Furthermore, depending on the type of compound(s) and the oxidant(s) or reductant(s) involved, various reaction steps – sorption/desorption to/from unreactive sorbents (e.g., NOM), adsorption to a reactive surface, actual electron transfer, or regeneration of oxidant(s) or reductant(s) – may determine the overall transformation rate. Thus, in different systems, not only the absolute rates but also the relative rates of oxidation or reduction of a series of compounds may be quite different, even if the compounds are structurally closely related. This can be seen in the reduction of four substituted nitrobenzenes (Eq. 14-9, Table 14.1) in three different systems (Fig. 14.5). In the DOM/H_2S system, the range of reactivity of the four compounds spans four orders of magnitude. In contrast, only two orders of magnitude variability is found for the reaction with an iron(II)porphyrin, and in the ferrogenic (i.e., iron reducing) aquifer columns, all compounds are reduced at the same rate. Obviously, different rate-limiting processes are responsible for the observed overall transformation rates in the three systems (see below). This example shows that, in contrast to the reactions discussed in Chapter 13, prediction of rates of redox reactions in natural or technical systems will be rather difficult. Nevertheless, as we will see in the following, knowledge from studies in well-defined model systems may help us to develop a framework for assessing pathways and rates of redox reactions of organic chemicals in more complex systems. Before we illustrate this approach by some examples, we first need to make some general remarks on the factors that determine the kinetics of redox reactions.

As discussed in Section 14.2, the oxidative or reductive transformation of an organic compound commonly requires two electrons (or, more generally, an even number of electrons) to be transferred to yield a stable product. In many cases, however, the two electrons are transferred in sequential steps (Eberson, 1987). With the transfer of the first electron, a radical species is formed which, in general, is much more reactive than the parent compound. Hence, the overall transformation rate will often be determined by the rate of transfer of the first electron from or to the organic compound. Therefore, we should be particularly interested in those compound-specific properties that are relevant for this first one-electron reaction.

In a very simple way, we may picture a one-electron transfer reaction (e.g., the transfer of the first electron from a reductant R to an organic compound (e.g., an organic pollutant P) schematically as:

$$P + R \rightleftharpoons (PR) \rightleftharpoons [PR \leftrightarrow P^{\cdot-} R^{\cdot+}]^{\ddagger} \rightleftharpoons (P^{\cdot-} R^{\cdot+}) \rightleftharpoons P^{\cdot-} + R^{\cdot+} \quad (14\text{-}30)$$

$$\text{educts} \quad \text{precursor} \quad \text{transition} \quad \text{successor} \quad \text{products}$$
$$\text{complex} \quad \text{state} \quad \text{complex}$$

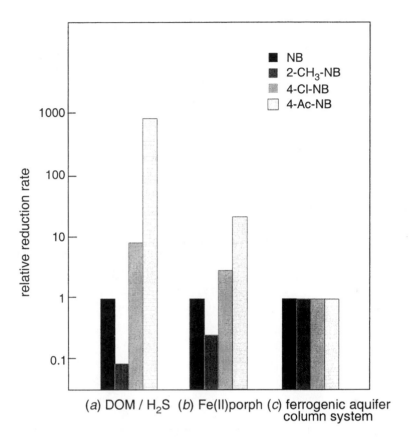

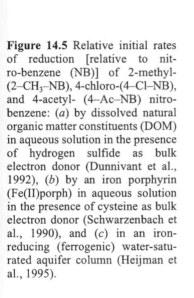

Figure 14.5 Relative initial rates of reduction [relative to nitro-benzene (NB)] of 2-methyl-(2–CH₃–NB), 4-chloro-(4–Cl–NB), and 4-acetyl- (4–Ac–NB) nitro-benzene: (*a*) by dissolved natural organic matter constituents (DOM) in aqueous solution in the presence of hydrogen sulfide as bulk electron donor (Dunnivant et al., 1992), (*b*) by an iron porphyrin (Fe(II)porph) in aqueous solution in the presence of cysteine as bulk electron donor (Schwarzenbach et al., 1990), and (*c*) in an iron-reducing (ferrogenic) water-saturated aquifer column (Heijman et al., 1995).

Note that in this context, one often speaks of an *inner-sphere* mechanism if there is a strong electronic coupling between R and P in the transition state, and conversely, of an *outer-sphere* mechanism, if the interaction is weak (Eberson, 1987).

From Eq. 14-30 we see that we may divide a one-electron transfer into various steps (maybe somewhat artificially). First, a precursor complex (PR) has to be formed; that is, the reactants have to meet and interact. Hence, electronic as well as steric factors determine the rate and extent at which this precursor complex formation occurs. Furthermore, in many cases, redox reactions take place at surfaces, and therefore, the sorption behavior of the compound may also be important for determining the rate of transformation. In the next step, the actual electron transfer between P and R occurs. The activation energy required to allow this electron transfer to happen depends strongly on the "willingness" of the two reactants to lose and gain, respectively, an electron. Finally, in the last steps of reaction sequence Eq. 14-30, a successor complex may be postulated which decays into the products.

In the following we will try to illustrate these general points by discussing two specific types of redox reactions: the reduction of aromatic nitro groups (Eq. 14-9) and the reductive dehalogenation of polyhalogenated C_1- and C_2-compounds (Eqs. 14-6 to 14-8). These two cases represent two very different types of reactions. In the first case, the transfer of the first electron is reversible, whereas in the second case, it is typically irreversible and involves the breaking of a bond. In the latter case, therefore, one speaks of a *dissociative* electron transfer. Furthermore, compounds

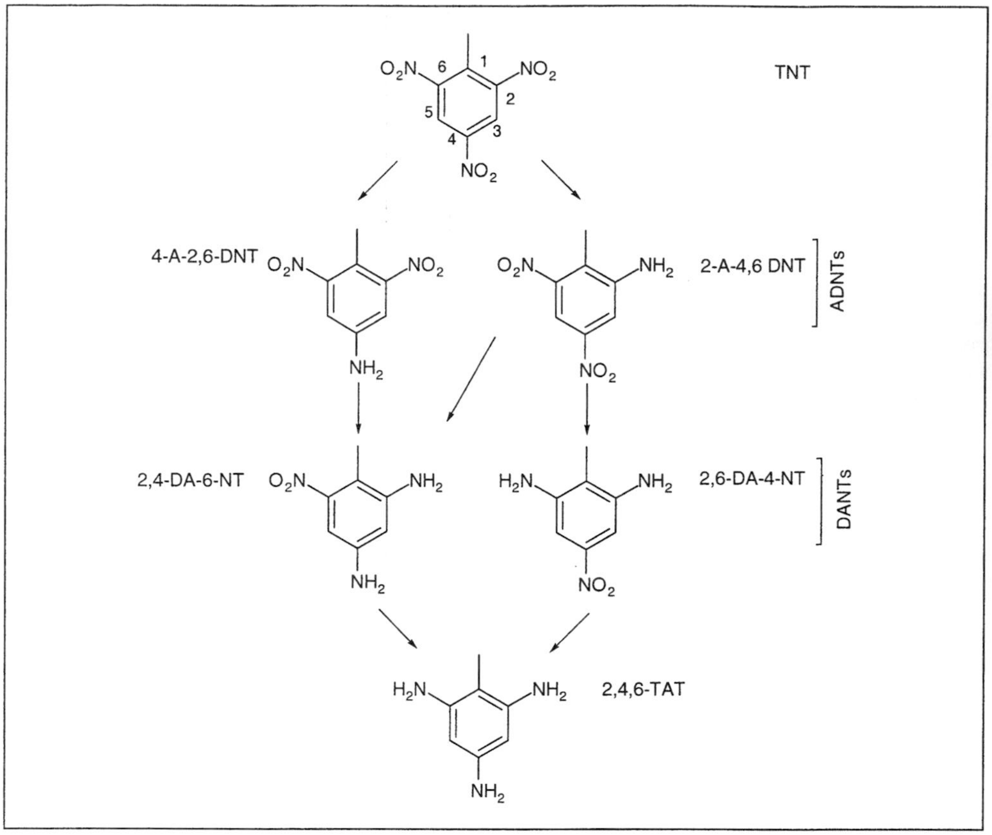

Figure 14.6 Reduction pathway of 2,4,6-trinitrotoluene to aminodi-nitrotoluenes (ADNTs), diamino-nitrotoluenes (DANTs), and tri-aminotoluene (TAT).

undergoing reductive dehalogenation may also react by a two-electron transfer mechanism, which may yield different products as compared to the one-electron transfer reaction. Note that we have chosen two examples representing *reductive* transformations of organic pollutants. The reason is that chemical *oxidations* involving oxidants other than reactive species formed by photochemical processes (Chapter 16) are somewhat less important in natural systems (in contrast to "engineered" systems, e.g., water treatment). We will add a few comments on chemical oxidations of organic pollutants at the end of this section.

Reduction of Nitroaromatic Compounds (NACs)

Aromatic nitro groups ($ArNO_2$) are present in many environmentally relevant chemicals including pesticides, dyes, and explosives (Chapter 2). As is illustrated by nitrobenzene in Fig. 14.4, reduction to the corresponding amino compounds is thermodynamically feasible under redox conditions below about +0.4 V. Hence, it is not surprising that reduction of NACs has been observed in many anaerobic soils and sediments (Spain et al., 2000). In most cases, the corresponding amino compounds (IV in Eq. 14-31) were found as the major reduction products, although stable intermediates (i.e., the nitroso(II) and the hydroxylamine(III) compound) are formed during reduction of an aromatic nitro group:

$$ArNO_2 \xrightarrow[-H_2O]{+2e^- + 2H^+} ArNO \xrightarrow{+2e^- + 2H^+} ArNHOH \xrightarrow[-H_2O]{+2e^- + 2H^+} ArNH_2 \qquad (14\text{-}31)$$
$$\text{I} \qquad\qquad\qquad \text{II} \qquad\qquad\qquad \text{III} \qquad\qquad\qquad \text{IV}$$

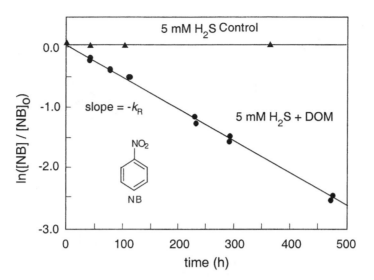

Figure 14.7 Reduction of nitro-benzene (NB) in 5 mM aqueous hydrogen sulfide solution in the absence (▲) and presence (●) of DOM (Hyde County, 66 mg DOC/L) at pH 7.2 and 25°C: Plot of ln ([NB]/[NB]$_0$) versus time. [NB]$_0$ and [NB] are the concentrations at time zero and t, respectively. Adapted from Dunnivant et al. (1992).

In laboratory model systems using reduced DOM constituents (Dunnivant et al., 1992), Fe(II) adsorbed to iron (hydr)oxides (Klausen et al., 1995), or zero-valent iron metal (Agrawal and Tratnyek, 1996) as reductants, the nitroso- and particularly the hydroxylamino compounds have been observed as reaction intermediates, but were ultimately also converted to the corresponding amino-compounds.

From a practical point of view, reduction of NACs is of great interest for two reasons. First, the amino compounds formed may exhibit a considerable (eco)toxicity, and therefore may be of even greater concern as compared to the parent compounds. Additionally, the reduced products may react further with natural matrices, in particular with natural organic matter, thus leading to "bound residues" (see sections on oxidations below). One prominent example involves the reduction products of the explosive, 2,4,6-trinitrotoluene (TNT; see Fig. 14.6), particularly the two isomeric diaminonitrotoluenes (2,4-DA-6-NT and 2,6-DA-4-NT) and the completely reduced triaminotoluene (TAT). These have been found to bind irreversibly to organic matter constituents present in soils (Achtnich et al., 2000) and sediments (Elovitz and Weber, 1999). This process offers interesting perspectives for the treatment of NAC contaminated sites. In fact, a dual step anaerobic/aerobic soil slurry treatment process has been developed for remediation of TNT contaminated soils (Lenke et al., 2000).

Let us now turn to some kinetic considerations of NAC reduction. As an example, consider the time courses of nitrobenzene (NB) concentration in 5 mM aqueous hydrogen sulfide (H$_2$S) solution in the absence and presence of natural organic matter (Fig. 14.7). As is evident, although reduction of NB by H$_2$S to nitrosobenzene and further to aniline (Eq. 14-31) is very favorable from a thermodynamic point of view (see Fig. 14.4), it seems to be an extremely slow process. However, when DOM is added to the solution, reduction occurs at an appreciable rate (Fig. 14.7). In order to understand these findings, some general kinetic aspects of redox reactions involving NACs should be recognized.

First, the transfer of the first electron is in many cases the rate-limiting process in the

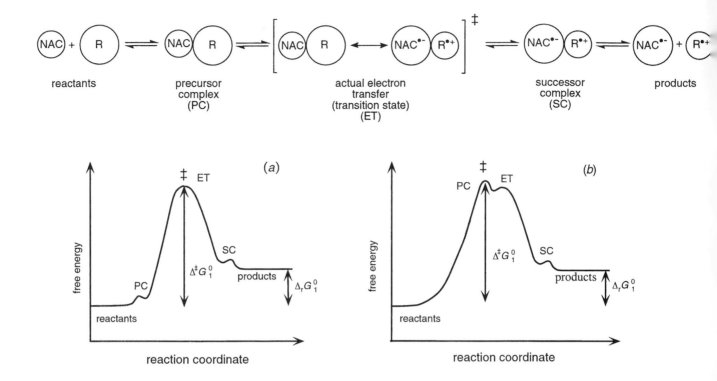

Figure 14.8 Simplified scheme for the transfer of the first electron from a reductant R to a NAC (adapted from Eberson, 1987). Panels (*a*) and (*b*) show free energy profiles of reactions where the actual electron transfer (*a*) or other steps such as precursor formation (*b*) are rate determining. Note that the subscript 1 is used to denote transfer of one electron to the NAC.

overall reduction or oxidation of an organic pollutant. In the case of NACs at ambient pH (i.e., pH 6 – 9), the transfer of the first electron yields a nitroaromatic radical anion $ArNO_2^{\bullet-}$ (the pK_a values of $ArNO_2^{\bullet-}$ radicals are well below 5; Neta and Meisel, 1976):

$$ArNO_2 + e^- \rightleftharpoons ArNO_2^{\bullet-}; \quad E_H^1(ArNO_2) \qquad (14\text{-}32)$$

where $E_H^1(ArNO_2)$ denotes the one-electron standard reduction potential of the halfreaction Eq. 14-32 at pH $\geq$ 6. Since for NACs, the formation of the radical anion is reversible, $E_H^1(ArNO_2)$ values can be measured, for example, by pulse radiolysis, and are available for a variety of such compounds (see examples given in Table 14.4).

We can envision a simple reaction scheme of the various steps that may determine the overall rate of a one-electron transfer reaction between a reductant R and a NAC (Fig. 14.8). Depending on the reductant(s) involved in the reaction, the actual transfer of the electron and/or other steps such as precursor complex formation or successor complex dissociation may be rate determining for the formation of the nitroaromatic radical anion, $ArNO_2^{\bullet-}$. Fig. 14.8*a* depicts the situation in which the actual electron transfer is the rate-determining step. The transition state in such a reaction is energetically closer to the radical products than to the precursor complex. The standard free energy of activation, $\Delta^{\ddagger}G_1^0$, can then be assumed to be proportional to the standard free energy change, $\Delta_r G_1^0$, of the reaction (note that we use a subscript 1 to denote transfer of the first electron to the NAC):

$$\Delta^{\ddagger}G_1^0 = a' \Delta_r G_1^0 + \text{constant}' \qquad (14\text{-}33)$$

Table 14.4 Names, Abbreviations, and One-Electron Reduction Potentials [$E_H^1(ArNO_2)$; Eq. 14-32] of a Series of Substituted Nitrobenzenes

Compound	Abbreviation	$E_H^1(ArNO_2)$ [a] (mV)
2,4,6-Trinitrotoluene	TNT	−280 [b]
2-Amino-4,6-dinitrotoluene	2-A-4,6-DNT	−400 [b]
4-Amino-2,6-dinitrotoluene	4-A-2,6-DNT	−440 [b]
2,4-Diamino-6-nitrotoluene	2,4-DA-6-NT	−505 [b]
2,6-Diamino-4-nitrotoluene	2,6-DA-4-NT	−495 [b]
2,4-Dinitrotoluene	2,4-DNT	−380
2,6-Dinitrotoluene	2,6-DNT	−400
Nitrobenzene	NB	−485
2-Methylnitrobenzene	2-CH₃-NB	−590
3-Methylnitrobenzene	3-CH₃-NB	−475
4-Methylnitrobenzene	4-CH₃-NB	−500
2-Chloronitrobenzene	2-Cl-NB	−485
3-Chloronitrobenzene	3-Cl-NB	−405
4-Chloronitrobenzene	4-Cl-NB	−450
2-Acetylnitrobenzene	2-Ac-NB	−470
3-Acetylnitrobenzene	3-Ac-NB	−505
4-Acetylnitrobenzene	4-Ac-NB	−360
1,2-Dinitrobenzene	1,2-DNB	−290
1,3-Dinitrobenzene	1,3-DNB	−345
1,4-Dinitrobenzene	1,4-DNB	−260
3-Aminonitrobenzene	3-NH₂-NB	−500
4-Aminonitrobenzene	4-NH₂-NB	< −560

[a] Values from references cited in Hofstetter et al. (1999). [b] Values from Hofstetter et al. (1999) and Riefler and Smets (2000).

Since log k_R is proportional to $\Delta^{\ddagger} G_1^0/2.3\,RT$ (Chapter 12), where k_R denotes the reaction rate constant, we may also write this linear free energy relationship as:

$$\log k_R = a \frac{\Delta_r G_1^0}{2.3\,RT} + \text{constant} \qquad (14\text{-}34)$$

From Section 14.2 we recall that:

$$\Delta_r G_1^0 = -F[E_H^1(ArNO_2) - E_H^1(R^{\cdot-})] \qquad (14\text{-}35)$$

where $n = 1$ and $E_H^1(R^{\cdot-})$ is the one-electron standard reduction potential of the half reaction $R^{\cdot-} + e^- = R$. Insertion of Eq. 14-35 into Eq. 14-34 then yields:

$$\log k_R = -a \frac{[E_H^1(ArNO_2) - E_H^1(R^{\cdot-})]}{2.3\,RT\,/\,F} + \text{constant} \qquad (14\text{-}36)$$

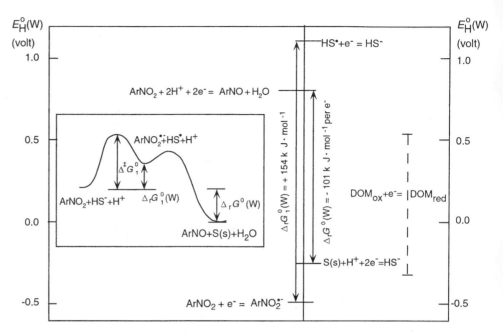

Figure 14.9 Energetic considerations for the reduction of nitrobenzene to nitrosobenzene ($Ar = C_6H_5$) with HS^- as electron donor at environmental ("W") standard conditions. Note that $\Delta_r G^0(W) = -nF \cdot \Delta E_H^0(W)$. $\Delta_r G_1^0(W)$ is the standard free energy of reaction for the transfer of the first electron; $\Delta_r G^0$ is the overall standard free energy of reaction for the transfer of both electrons.

We should emphasize that we expect Eq. 14-36 to hold only if the actual electron transfer is rate limiting. If other steps in the reaction sequence are partially or fully rate limiting (e.g., precursor formation, Fig. 14.8b), other factors have to be taken into account for evaluating and/or interpreting reduction rates (see below).

Now we are in the position to rationalize the observations made in Fig. 14.7 by looking at the energetics of the reduction of nitrobenzene to nitrosobenzene by hydrogen sulfide in homogeneous aqueous solution (Fig. 14.9; $Ar = C_6H_5$):

$$ArNO_2 + HS^- + H^+ \rightleftharpoons ArNO + S(s) + H_2O \qquad (14\text{-}37)$$

In this case we assume that the actual electron transfer is rate limiting (Dunnivant et al., 1992). Hence, as is evident from Fig. 14.9, although the overall reaction is strongly exergonic (i.e., $\Delta_r G^0 = -101 \ kJ \cdot mol^{-1}$), the transfer of the first electron is highly endergonic (i.e., $\Delta_r G_1^0 = +154 \ kJ \cdot mol$), suggesting a large $\Delta^\ddagger G_1^0$ value. Consequently, nitrobenzene, as well as other nitroaromatic compounds (Dunnivant et al., 1992), reacts only very slowly with hydrogen sulfide under these conditions. Upon addition of natural organic matter to an H_2S solution, reduced DOM constituents may be formed that exhibit more negative reduction potentials than $HS^\bullet$ (see range of $E_H^0(W)$ values of $DOM_{ox/red}$ couples, dashed line in Fig. 14.9). Such DOM species (e.g., hydroquinone and mercaptohydroquinone moieties; Dunnivant et al., 1992; Perlinger et al., 1996) may reduce NACs at a much faster rate (Fig. 14.7). Because such species may be re-reduced continuously and at higher rates by the bulk reductant H_2S, they may act as electron transfer mediators (Fig. 14.1). The rapid regeneration of these species yields a steady-state concentration of reduced DOM constituents, which explains the pseudo-first order kinetics observed for the disppearance of nitrobenzene (and of other NACs) shown in Fig. 14.7.

For evaluating the reduction kinetics of NACs in a given natural system, the relative reaction rates of a series of NACs with known $E_H^1(ArNO_2)$ values can be used to

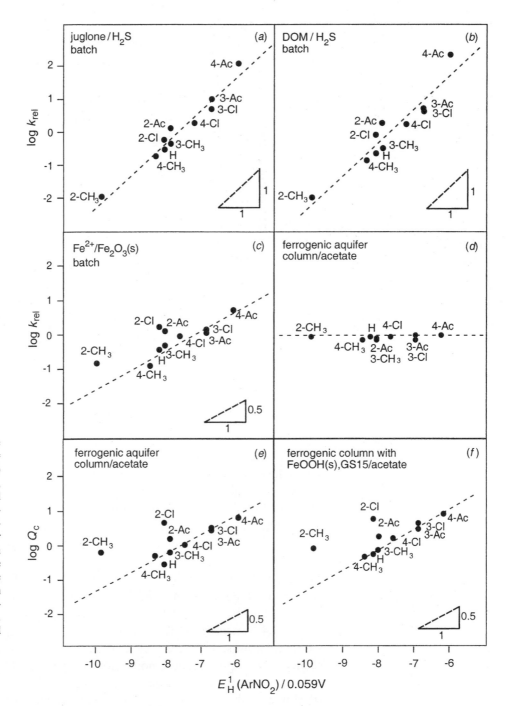

Figure 14.10 Plots of the relative $(a–d)$ reaction rate constants [k_{rel} = k_R(NAC)/ k_R(4-Cl-NB)] and (e, f) competition coefficients (log Q_c, Eq. 14-39) of 10 monosubstituted nitrobenzenes versus their one-electron reduction potentials (divided by 0.059 V, see Eq. 14-36) for some laboratory batch and column systems. For abbreviations see Table 14.4; note that only the substituent is indicated. Data from Schwarzenbach et al. (1990); Dunnivant et al. (1992); Heijman et al. (1995); Klausen et al. (1995); and Hofstetter et al. (1999).

probe whether the actual electron transfer is limiting. To this end, the rate constants observed for the reduction of the NACs can be analyzed using the LFER Eq. 14-36 with $E_H^1(R^{\cdot-})$ = constant and 2.3 RT/F = 0.059 V at 25°C:

$$\log k_R = a \frac{E_H^1(ArNO_2)}{0.059 \text{ V}} + b \qquad (14\text{-}38)$$

If in a given case, a significant correlation is found between log k_R and $E_H^1(ArNO_2)$, 0.059V with a slope of close to 1.0, it can be concluded that, for the series of NACs

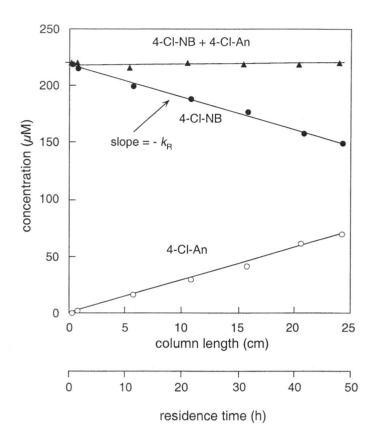

Figure 14.11 Reduction of NACs in ferrogenic aquifer solid columns. Concentrations of 4-chloronitrobenzene (4-Cl-NB, •) and of its transformation product 4-chloroaniline (4-Cl-An, o) vary with position (and time) in the column. Also indicated is the excellent mass balance (▲). Adapted from Heijman et al. (1995).

reduced mercaptojuglone

(or any other compound class) considered, the actual transfer of the electron from the reductant to the compounds is rate determining (Eberson, 1987). If a much weaker dependency of log k_R on $E_H^1(ArNO_2)$ (i.e., a slope of $a \ll 1$), or no correlation at all is found, then other reaction steps and/or other processes are important, including, for example, precursor complex formation, the (slow) regeneration of reactive sites, and/or just plain mass transport (Scherer et al., 2001).

The examples given in Fig. 14.10 illustrate these different cases. Figures 14.10a–d show plots of the logarithms of the relative reaction rate constants (relative to k_R of 4-chloro-nitrobenzene) of 10 monosubstituted nitrobenzenes (for abbreviations see Table 14.4) versus the $E_H^1(ArNO_2)/0.059$ V values (Eq. 14-36) of the compounds for some environmentally relevant systems. As can be seen, for the reaction with reduced mercaptojuglone (model for reduced DOM constituents under sulfate-reducing conditions, see structure in margin) formed by addition of H_2S to juglone (Perlinger et al., 1996), as well as in the DOM/H_2S system, a very strong correlation with a slope of close to 1.0 is found (Figs. 14.10a,b). This indicates that in both systems, the transfer of the electron is rate determining. Note that in these cases, k_{rel} represents the ratio of two pseudo-first order rate constants, and that very similar correlations were obtained for other NACs including TNT, ADNTs, and DANTs (Fig. 14.6) as well as dinitrobenzene isomers (Dunnivant et al., 1992; Hofstetter et al., 1999).

A completely contrasting situation is shown in Fig. 14.10d. Here, k_{rel} represents the ratio of apparent zero-order rate constants determined for the reduction of the model

Figure 14.12 Reduction of 4-chloronitrobenzene (4-Cl-NB) in aqueous solution in the presence of 17 $m^2 L^{-1}$ magnetite and an initial concentration of 2.3 mM Fe(II) at pH 7 and 25°C: plot of ln ([4-Cl-NB]/[4-Cl-NB]$_0$) versus time (■). [4-Cl-NB]$_0$ and [4-Cl-NB] are the concentrations at time zero and t, respectively. Adapted from Klausen et al. (1995). Note that experimental points deviate from pseudo-first-order behavior for long observation times. 4-Cl-NB was not reduced in suspensions of magnetite without Fe(II) ($\triangledown$), or solutions of Fe(II) without magnetite ($\triangle$).

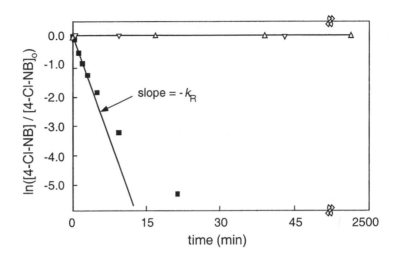

NACs in a laboratory column system containing aquifer material from the banks of a river-groundwater infiltration site (Fig. 14.11). The columns were run under ferrogenic conditions. Note that zero-order kinetics suggests that the reactive sites were always saturated such as encountered in enzyme kinetics at saturation (Box 12.2). In this system, all model compounds as well as other NACs including again TNT, ADNTs, and DANTs (data not shown, see Hofstetter et al., 1999) reacted at virtually the same rate. However, when present in mixtures, the compounds showed competition for the reactive sites. A competition quotient, Q_c (competition with the reference compound 4-Cl-NB present at about equal concentrations) was determined for all model compounds:

$$Q_c = \frac{k_R(NAC)}{k_R(4\text{-}Cl\text{-}NB)} \qquad (14\text{-}39)$$

Hence, these Q_c values are a quantitative measure for the relative affinities of the various NACs to the reactive sites. Figs. 14.10e and f show plots of log Q_c versus $E_H^1(ArNO_2)/0.059$ V of the 10 monosubstituted benzenes. A virtually identical picture was obtained for the log Q_c values derived from an aquifer solid column and from a column containing FeOOH-coated sand and a culture of the iron-reducing bacterium, *Geobacter metallireducens* (GS15). Furthermore, a similar pattern (Fig. 14.10c) was found when correlating relative initial pseudo-first-order rate constants determined for NAC reduction by Fe(II) species adsorbed to iron oxide surfaces (Fig. 14.12) or pseudo-first-order reaction constants for reaction with an iron porphyrin (data not shown; see Schwarzenbach et al., 1990). Fig. 14.12 shows that Fe(II) species adsorbed to iron oxide surfaces are very potent reductants, at least for NACs ($t_{1/2}$ of a few minutes in the experimental system considered).

From all these observations and relative behaviors it can be concluded that formation of a precursor complex or regeneration of reactive sites is important in determining the overall rate of NAC reduction by surface-bound iron(II) species. Therefore, in this reaction scenario, a much weaker correlation between log k_{rel} and $E_H^1(ArNO_2)/0.059$ V can be expected and is actually obtained (e.g., Fig. 14.10d). In fact, the apparent correlation of the 3- and 4-substituted nitrobenzenes (slope = ~0.5) may be due to a co-correlation between $E_H^1(ArNO_2)$ and the tendency of the

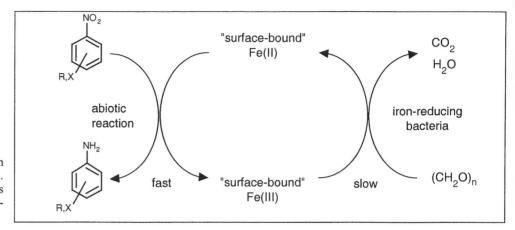

Figure 14.13 General reaction scheme proposed by Heijman et al. (1995) for the reduction of NACs in aquifer columns under ferrogenic conditions.

compounds to interact with an iron(II) center. This can be rationalized by postulating an electron donor-acceptor (EDA) complex between the iron center and the NAC, similarly to the EDA complex thought to form between NACs and the siloxane surfaces of clay minerals (see Chapter 11). Considering the high reactivities of the NACs with iron oxide "surface-bound" Fe(II) species (Fig. 14.12), the much slower and uniform reduction rates of the NACs in the columns can then be attributed to a (rate-determining) "slow" regeneration of "surface-bound" Fe(II) by iron-reducing bacteria (Fig. 14.13). This biological step is, of course, the same for all compounds.

These examples of NAC reduction clearly demonstrate that, in general, any *a priori* predictions of rates of redox reactions involving organic pollutants will be very difficult, particularly when we are dealing with heterogeneous systems (i.e., soils, sediments). However, the examples also show that by using a series of structurally closely related compounds as "reactive probes" for which the pertinent properties (e.g., the E_H^1-values) are known, important information about the type and reactivity of the relevant reductants (or oxidants, see below and Chapter 16) present in a given system can be obtained. In fact, based on the relative reactivities of a series of NACs, reduced iron species were identified as predominant reductants in an anaerobic leachate-contaminated aquifer (Rügge et al., 1998). Finally, as illustrated by Illustrative Example 14.4, if a system can be "calibrated" by an LFER of the type Eq. 14-36, predictions of relative and even absolute reaction rates of structurally related compounds may be feasible.

Illustrative Example 14.5

Estimating Rates of Reduction of Nitroaromatic Compounds by DOM Components in the Presence of Hydrogen Sulfide

Problem

Simulating sulfate-reducing conditions, Dunnivant et al. (1992) investigated the reduction of a series of monosubstituted nitrobenzenes in aqueous hydrogen sulfide solutions containing DOM from various sources. They found that for a given H_2S concentration and pH, the pseudo-first-order rate constants, k_R (Fig. 14.7), were linearly related to the total dissolved organic carbon (DOC) concentrations. Thus, for a given DOM, second-order rate constants:

$$k_{DOM} = k_R / [DOC] \tag{1}$$

could be derived for the various compounds. Interestingly, for a variety of DOMs from various sources (natural waters, dump sites), the k_{DOM} values of a given compound were all in the same order of magnitude. Therefore, from these data, an LFER relating average k_{DOM} values to the corresponding $E_H^1(ArNO_2)$ values can be derived for given conditions (e.g., 5 mM H_2S, pH 7, 25°C):

$$\log k_{DOM} / [h^{-1}(mgC/L)^{-1}] = 1.0 \frac{E_H^1(ArNO_2)}{0.059\ V} + 4.0 \tag{2}$$

Estimate the relative reactivity of TNT and 2,4-DA-6-NT (see Table 14.4 and Fig. 14.6) under these conditions, as well as the half-lives ($t_{1/2}$) of the two compounds at a DOC concentration of 10 mg C·L^{-1}.

Answer

The relative reactivity of the two compounds is independent of the DOC concentration and is given by Eq. 2:

$$k_{rel} = \frac{k_{DOM}(TNT)}{k_{DOM}(2,4\text{-}DA\text{-}6\text{-}NT)} = 10^{1.0\left[E_H^1(TNT)-E_H^1(2,4\text{-}DA\text{-}6\text{-}NT)\right]/0.059\ V} \tag{3}$$

Note that in this system a difference of 59 mV in the $E_H^1(ArNO_2)$ values of two compounds means that their relative reactivity differs by one order of magnitude. For the two compounds considered, the difference in their $E_H^1(ArNO_2)$ values is (Table 14.4):

$$E_H^1(TNT) - E_H^1(2,4\text{-}DA\text{-}6\text{-}NT) = (-280) - (-505) = +225\ mV$$

Insertion of this value into Eq. 3 yields a k_{rel} value of:

$$k_{rel} = 10^{(1.0)(225)/59} = 6.5 \times 10^3$$

Hence, TNT reacts about 4 orders of magnitude faster than its reduction product 2,4-DA-6-NT.

For calculating $t_{1/2}$, insert $E_H^1(TNT)$ into Eq. 2. This yields a second-order rate constant:

$$\log k_{DOM}(TNT) = 1.0 \frac{-0.280}{0.059} + 4.0 = -0.75$$

or $k_{DOM}(TNT) = 0.18\ h^{-1}$ (mg C/L)$^{-1}$. The $t_{1/2}$ value of TNT is then given by:

$$t_{1/2}(TNT) = \frac{\ln 2}{k_{DOM}(TNT) \cdot [DOC]} = \frac{(0.69)}{(0.18)(10)} h = 0.39\ h$$

Hence, TNT reacts extremely rapidly in this system. In contrast, the half-life of 2,4-DA-6-NT would be *106* days.

Reductive Dehalogenation Reactions of Polyhalogenated C_1- and C_2-Compounds

Halogenated organic compounds, and in particular polychlorinated hydrocarbons, are among the most ubiquitous environmental pollutants. Under oxic conditions many of these compounds are quite persistent. This is especially true for highly halogenated compounds such as the polyhalogenated C_1- and C_2-compounds (Table 2.4), and a variety of polyhalogenated aromatic compounds (e.g., polychlorinated benzenes, PCBs, PCDDs, see Figs. 2.14 and 2.15). However, under reducing conditions these compounds can undergo reductive dehalogenation; that is, one or even two halogens are lost from the molecule as a consequence of an electron transfer to the compound (see examples Eqs. 14-1 and 14-3). This type of reaction is of great interest from an ecotoxicological point of view as the products often exhibit very different toxicities. Such reactions are also important in environmental engineering, due to their potential applicability in the treatment of wastes as well as in remediation approaches to removing such compounds from contaminated soils and aquifers.

From a thermodynamic point of view, reductive dehalogenations are feasible with most reductants present in anaerobic environments (Fig. 14.4, Table 14.3). In fact, some of the half reactions (e.g., reaction 1 in Table 14.3) have even more positive $E_H^0(W)$ values than the half reactions involving oxygen. Hence, it is not too surprising that microorganisms have been found that grow on halogenated hydrocarbons as sole terminal electron acceptors (McCarty, 1997; Fetzner, 1998). In our discussion of abiotic dehalogenation reactions we will confine ourselves to polyhalogenated *aliphatic* and *olefinic* compounds for the following reasons. First, although thermodynamically feasible, abiotic reduction of polyhalogenated *aromatic* hydrocarbons is rather slow in natural systems. Most of the observed reductions of such compounds including chlorinated benzenes and PCBs can be attributed primarily to microbial processes (Chapter 17). Furthermore, in contrast to polyhalogenated aliphatic and olefinic hydrocarbons, halogenated aromatic compounds react very slowly with most zero-valent metals (e.g., iron, zinc) that are widely applied in reactive walls or barriers for remediation of contaminated groundwater (Tratnyek, 1996; Scherer et al., 2000). The reason is that dehalogenation of an aromatic system occurs most easily by initial hydrogenation (and not by direct electron transfer). In general, such hydrogenations can be achived only by enzymatic reactions (Chapter 17) or, in the case of engineered systems, by using hydrogen and an appropriate catalyst such as nickel or palladium (Schüth and Reinhard, 1998). Therefore, when considering only abiotic processes, reduction of polyhalogenated aromatic compounds is less important.

Let us now turn to the reduction of polyhalogenated C_1- and C_2-compounds. First we note that many of these compounds react by several reaction pathways that may yield different intermediate and/or final products. Furthermore, the relative importance of the various pathways will, in general, depend in a rather complex way on a variety of environmental factors including the nature of the reductant, tem-perature, pH, and presence of dissolved or particulate chemical species. The reason is that the reactive intermediates formed by transfer of one or two electrons to a polyhalogenated C_1- or C_2-compound may undergo a variety of subsequent

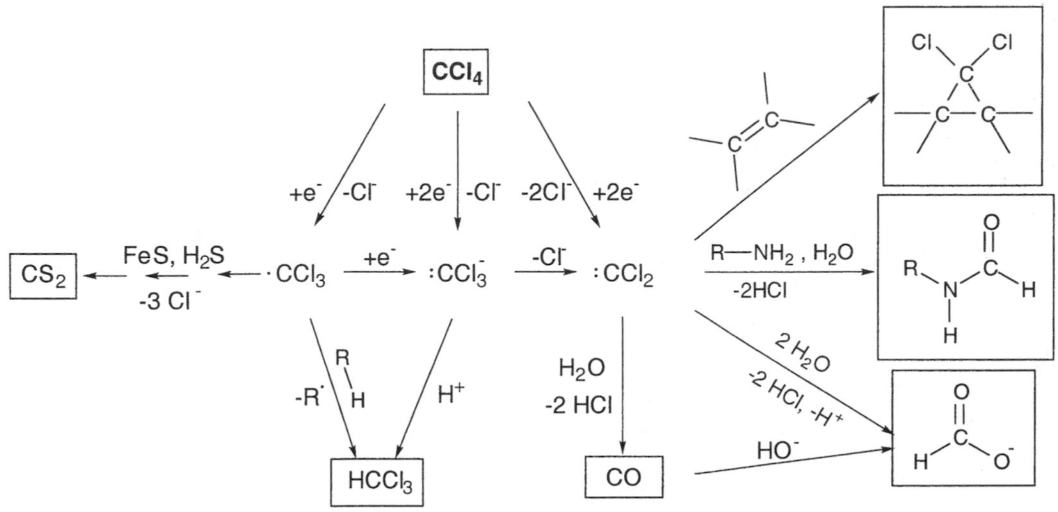

Figure 14.14 Postulated reaction pathways for the reduction of CCl$_4$ and products detected in various different systems.

reactions. This is the reason why mass balances are often rather poor in studies involving such compounds. Consequently, predictions not only of reaction rates but also of the types and relative amounts of products formed in a given system are very difficult. The following two examples illustrate this point.

Consider the reduction of polyhalogenated methanes. For example, several major pathways have been postulated for the reduction of tetrachloromethane in various systems (Fig. 14.14). As can be seen, there are three possible reactive intermediates [i.e., the trichloromethyl radical ($\cdot CCl_3$), the trichloromethyl anion ($:CCl_3^-$), and dichlorocarbene ($:CCl_2$)], which may further react to yield a variety of products. In systems that do not contain appreciable amounts of organic constituents that could react with dichlorocarbene (e.g., amino groups or electron-rich double bonds; see Buschmann et al., 1999), or that do not exhibit reduced sulfur species leading to the formation of CS$_2$ (Kriegmann-King and Reinhard, 1992; Lewis et al., 1996; Devlin and Müller, 1999), the major products commonly found are CHCl$_3$, CO, and/or HCOO$^-$. The relative amount of CHCl$_3$ found depends strongly on the type of reductant (two versus one electron transfer) and/or on other factors including pH and the presence of organic materials from which a hydrogen atom can be abstracted by the trichloromethane radical. Thus, yields of CHCl$_3$ of smaller than 10% to over 90% may be found (for details see Kriegmann-King and Reinhard, 1994; Amonette et al., 2000; Butler and Hayes, 2000; Pecher et al., 2002). Note that CHCl$_3$ may be reduced further, though generally at slower rates than CCl$_4$ (see below). This example demonstrates the difficulty of predicting the product distribution from the reduction of polyhalogenated methanes.

In our second example we consider the reduction of chlorinated ethenes including the prominent solvents tetrachloroethene (perchloroethylene, PCE) and trichloroethene (TCE). An overview of the hypothesized reaction sequence for reduction of these compounds by zero-valent iron (Fe(0)) has been constructed (Fig. 14.15; Arnold and Roberts, 2000). Identical or very similar reaction schemes have been

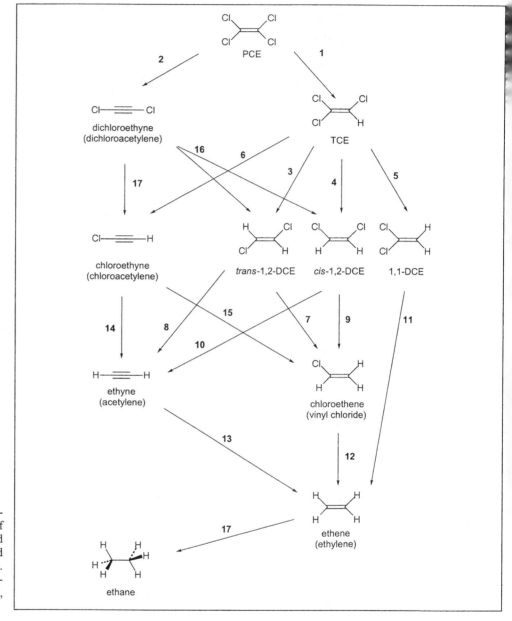

Figure 14.15 Hypothesized reaction sequence for reduction of chlorinated ethenes and related compounds by Fe(0). Adapted from Arnold and Roberts (2000). Abbreviations are PCE (tetrachloroethene), TCE (trichloroethene), and DCE (dichloroethene).

postulated for reduction of these compounds by other reductants including zerovalent zinc (Zn(0); Arnold and Roberts, 1998), and iron sulfide (FeS; Butler and Hayes, 1999), and by cob(I)alamin, which is involved in the enzymatic reduction of halogenated ethenes by a variety of anaerobic bacteria (Glod et al., 1997a and b). However, as we can imagine, depending on the relative rates of the various reactions (reactions 1 to 17 in Fig. 14.15) different reaction products may accumulate in different systems as reaction intermediates. For example, the reduction of TCE by Zn(0) (Arnold and Roberts, 1998) yielded about 50% *trans*-1,2-DCE (reaction 3), 20% *cis*-1,2-DCE (reaction 4), and about 30% chloroacetylene (reaction 6). This last product was further reduced very rapidly to acetylene (about 90%, reaction 14) and to vinyl chloride (about 10%, reaction 15). Note that the formation of vinyl chloride is of particular concern because of the rather high toxicity of this compound. In contrast to the reaction with Zn(0), the reduction of TCE by cob(I)alamin led primarily to the formation of *cis*-1,2-DCE (> 70%) and only about 5% to *trans*-1,2-DCE and about 20 to 25% to the other products (Glod et al., 1997a). Finally, using a clean Fe(0) surface, Arnold and Roberts (2000) found that the major pathway (~ 90%) for PCE and TCE reduction was β-elimination (reactions 2 and 6 in Fig. 14.15). These examples again demonstrate that prediction of the relative importance of various reaction pathways and thus prediction of product distribution of reductive dehalogenation reactions is a rather difficult task.

When considering the *kinetics* of reductive dehalogenations, we should also point out that, compared to the already-complex situation encountered with NAC reduction (see above), we have to cope with additional difficulties. First, a given halogenated compound may react by different pathways that may be initiated by the transfer of one or two electrons. Second, since the overall reaction involves the breaking of one or even two carbon-halogen bonds, the reaction is irreversible, and therefore we cannot *a priori* assume that the free energy of reaction ($\Delta_r G^o$) is proportional to the free energy of the transition state ($\Delta^{\ddagger} G^o$). Thus, in many cases, one- or two-electron reduction potentials may not be the appropriate measures for derivation of LFERs. Similar problems hold for other molecular descriptors that try to quantify the tendency of a compound to acquire an electron and/or to describe the strength of the bond that has to be broken. Such descriptors include the energy of the *l*owest *u*noccupied *m*olecular *o*rbital (LUMO), the *v*ertical (electron) *a*ttachment *e*nergy (VAE), and the bond dissociation enthalpy (Scherer et al., 1998; Perlinger et al., 2000, Burrow et al., 2000; Liu et al., 2000). An additional difficulty for polyhalogenated compounds is that the accuracy with which all these descriptors, and in particular bond dissociation enthalpies, can be measured or estimated is not that great. Finally, we should be aware that the bond dissociation enthalpy describes the strength of a given bond in the parent molecule and not in the anionic radical species that may be formed upon addition of an electron. Therefore, this parameter does not necessarily provide the pertinent information in all cases considered.

In summary, the overall rate of reductive dehalogenation of a given compound in a given system may be determined by various rather complex steps, and may, therefore, be influenced by several compound properties. Furthermore, even within a series of structurally related compounds, the relative importance of the various steps may differ, thus rendering any quantitative structure reactivity relationships (QSARs) rather difficult. This also means that calibration of a given system with a small set of model compounds for estimating absolute reaction rates will be even more difficult as compared to the situation with NAC reduction (see above).

Consequently, with the present knowledge of reductive dehalogenation reactions of C_1- and C_2-compounds, and considering the "quality" of the various molecular descriptors, only qualitative, or at most semiquantitative, predictions of the relative reactivities of a confined set of structurally related compounds in a given system are possible. Nevertheless, evaluation of such relative reactivities in different systems may provide important insights into such reactions, which will be demonstrated by the following two examples. These two examples will, however, also illustrate the

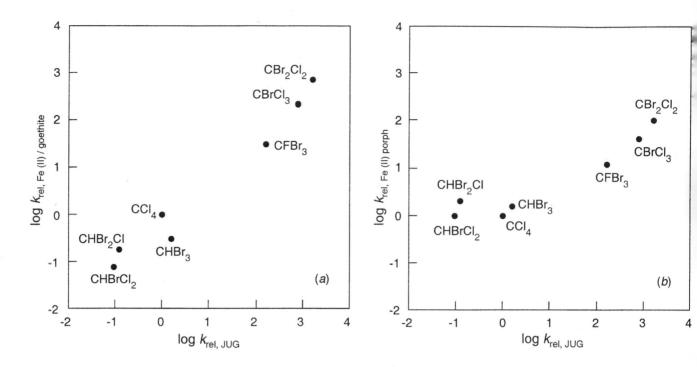

Figure 14.16 Comparison of the relative rates (relative to CCl_4) of reduction of a series of polyhalogenated methanes by reduced iron species versus rates with mercaptojuglone (JUG, for structure see above). Plots of the logarithms of the relative reaction rates for (a) Fe(II)/goethite and (b) Fe(II)porph versus JUG. Data from Perlinger et al. (1998) and Pecher et al. (2002).

above-discussed intrinsic difficulties with which one has to cope with when trying to derive QSARs for reductive dehalogenation reactions, or, more generally, for any redox reaction in natural systems.

In our first example we look at the relative reduction rates of polyhalogenated *methanes* in various systems. The decadic logarithms of the relative rates of reduction of a series of polyhalogenated methanes by Fe(II) associated with goethite (Fig. 14.16a) and by an iron(II)porphyrin in aqueous solution (Fig. 14.16b) are plotted versus the relative rates of reduction of the compounds by reduced mercaptojuglone (mimicking reduced DOM constituents in the presence of H_2S; for structure see above). Qualitatively, we can see from these data that the more highly halogenated compounds tend to react faster than do the corresponding less halogenated compounds. For example, the trihalomethanes generally react more slowly than the corresponding tetrahalomethanes. Note that in the three systems considered in Fig. 14.16, CBr_4 reacted much too fast to be measured, whereas the reaction of $CHCl_3$ was much too slow (Perlinger et al., 1998; Pecher et al., 2002). Similar observations have been made in other systems. For example, with zero-valent iron, CCl_4 reacts about two orders of magnitude faster than $CHCl_3$ (Scherer et al., 1998). A second important qualitative general conclusion that can be drawn is that brominated compounds react significantly faster than their chlorinated analogues (compare $CBrCl_3$ with CCl_4). This can be rationalized as due to the lower bond strength of the C–Br versus C–Cl bond (Table 2.2). Similarly, we may expect that a carbon-iodine (C–I) bond will, in general, be cleaved more easily than a C–Br bond, but a C–F bond will be much harder to break.

Despite the scatter in the data, we can see that the relative rates for the series of halogenated methanes considered are quite similar in the Fe(II)/goethite and the mercaptojuglone system. Both span a range of about four orders of magnitude. In

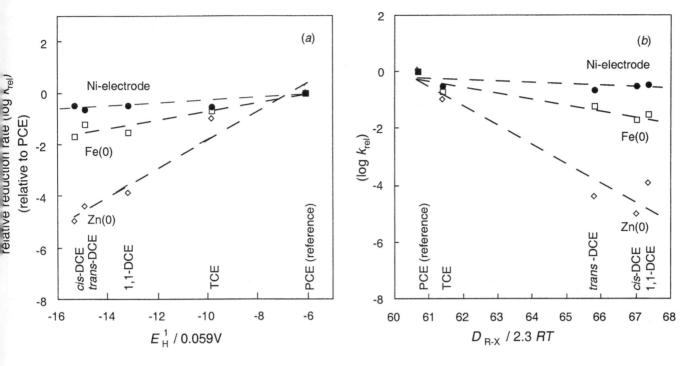

Figure 14.17 Reduction of chlorinated ethenes (for structures see Fig. 14.15) at a nickel electrode and by two zero-valent metals [Fe(0), Zn(0)]. Decadic logarithms of the relative overall reduction rates plotted (*a*) against $E_H^1/0.059$ V (analogous to Eq. 14-38; E_H^1 values from Arnold and Roberts, 1998), and (*b*) against the C–Cl bond energy (D_{R-X}) divided by 2.3 RT (D_{R-X} values from Perlinger et al., 2000). The absolute surface-normalized second-order rate constants for PCE are 3×10^{-3} L·m^{-2} s^{-1} (Ni-electrode at –1.0 V; Liu et al., 2000), 6×10^{-7} L·m^{-2} s^{-1} (Fe(0); average value reported by Scherer et al., 1998), and 8×10^{-5} L·m^{-2} s^{-1} (Zn(0); Arnold and Roberts, 1998).

contrast, in the Fe(II) porphyrin system, the range in reactivity is only two orders of magnitude (Fig. 14.16*b*). This suggests that different reaction steps are rate limiting and/or that the compounds react by a different mechanism in the case of the reaction with the iron porphyrin as compared to the other two reductants. In fact, an (outer-sphere) one-electron mechanism has been proposed for the iron porphyrin system (Perlinger et al., 1998), whereas both initial one- and two-electron transfers have been postulated for the mercaptojuglone (Perlinger et al., 1998) and for the Fe(II)/ goethite systems (Amonette et al., 2000; Pecher et al., 2002).

In our second example we look at the reduction of chlorinated ethenes at a nickel electrode and at the surfaces of two zero-valent metals [Fe(0), Zn(0)]. To gain insight into the rate-limiting process(es) in these cases, we consider how the relative overall reduction rates (relative to PCE) of PCE, TCE, and the three DCE isomers (see Fig. 14.15 for structures) vary as a function of two common descriptors used in QSARs, the one-electron reduction potential (E_H^1; Fig. 14.17*a*) and the bond dissociation energy (D_{R-X}; Fig. 14.17*b*). In all these systems, the reduction rates were found to be significantly slower than diffusion of the compounds to the respective surfaces. Therefore, the large differences in the relative reactivities of the compounds between the systems reflect differences in the actual reaction at the metal surface.

Inspection of Fig. 14.17 reveals that at the nickel electrode, all compounds react at about the same rate, although they exhibit very different F_H^1 and D_{R-X} values. Recall that 0.059 V or 5.7 kJ·mol^{-1} (2.3 RT) difference means one order of magnitude difference in reactivity if the slope of the corresponding LFER (see Eqs. 14-33 to 14-38) is 1.0. Similarly, a rather narrow range in reactivity is found for the reaction with Fe(0), which is reflected in the very small slopes of +0.17 and –0.20 of the lines fitted to the points (Figs. 14.17*a* and *b*). Hence, in these two systems, the actual dissociative electron transfer does not seem to be the most important step in determining the overall reaction rate. As discussed above, other steps such as precursor complex formation and/or regeneration of reactive sites could also be rate limiting and would explain such a behavior. In fact, Klausen et al. (1995) postulated that precursor complex formation and regeneration of reactive sites were determinant factors in the reduction of NACs in various iron(II)/iron oxide systems. This is also a likely explanation of the very narrow range in reactivity found for the reduction of a series of polychlorinated ethanes and ethenes by iron sulfide (slopes of +0.14 and −0.20 in the "LFERs" using E_H^1 and D_{R-X} as descriptors; for details see Butler and Hayes, 2000).

Yet another possibility for the poor correlations of the reaction rates with E_H^1 or D_{R-X} in the iron(0) and nickel(0) systems, is the presence of alternative reaction mechanisms. Using a clean Fe(0) surface, Arnold and Roberts (2000) found that for dihaloelimination of PCE (reaction 2 in Fig. 14.15), TCE (reaction 6), and the three DCEs (reactions 8, 10, and 11), the reaction rates increased significantly with decreasing number of halogens, (i.e., PCE $\ll$ TCE $<$ DCEs). They postulated the formation of a di-σ-bonded intermediate at the iron surface (followed by a hydride transfer and/or abstraction of a sorbed hydrogen atom), which would favor the less halogenated ethenes. Hence, depending on the relative importance of this reaction pathway, the observed overall reaction rate at a given metal(0) surface may be quite insensitive to differences in E_H^1 or D_{R-X}.

In contrast to the iron and nickel systems, greater slopes (i.e., +0.56 and −0.67, respectively) are obtained for the corresponding LFERs for the reaction with Zn(0) (Fig. 14.17). This dependency indicates that the actual electron transfer is important in determining the overall rate of reduction. Similar results were also found for the reaction of polyhalogenated methanes and ethanes by the iron(II) porphyrin and by the mercaptojuglone discussed above. Although the uncertainty in the D_{R-X} values used is probably rather large, the slopes of −0.46 (Fe(0)porph) and −0.82 (JUG) reported by Perlinger et al. (2000) for the respective LFERs are consistent with the conclusions drawn above (Fig. 14.16) for these systems.

Oxidation Reactions

When we think about oxidation of organic pollutants in the environment, we immediately wonder about the importance of dioxygen in such reactions. From our daily experience, we know that most organic compounds (fortunately) do not react spontaneously at significant rates with molecular oxygen, although the overall reactions would, in general, be exergonic. Hence, the reason for the inertness of many organic pollutants with respect to molecular oxygen (i.e., not activated by photolytic or biological processes) must be a kinetic one. Indeed, the standard reduction potential for transferring one electron to molecular oxygen yielding superoxide [$pK_a(O_2H^{\bullet})$ = 4.88; Ilan et al., 1987]:

$$O_2(1 \text{ M}) + e^- \rightleftharpoons O_2^{\bullet -}(1 \text{ M}); \quad E_H^0(W) = -0.16 \text{ V} \qquad (14\text{-}40)$$

shows that in aqueous solutions at pH 7, molecular oxygen is only a very weak oxidant (much as H_2S is a weak reductant with respect to the transfer of the first

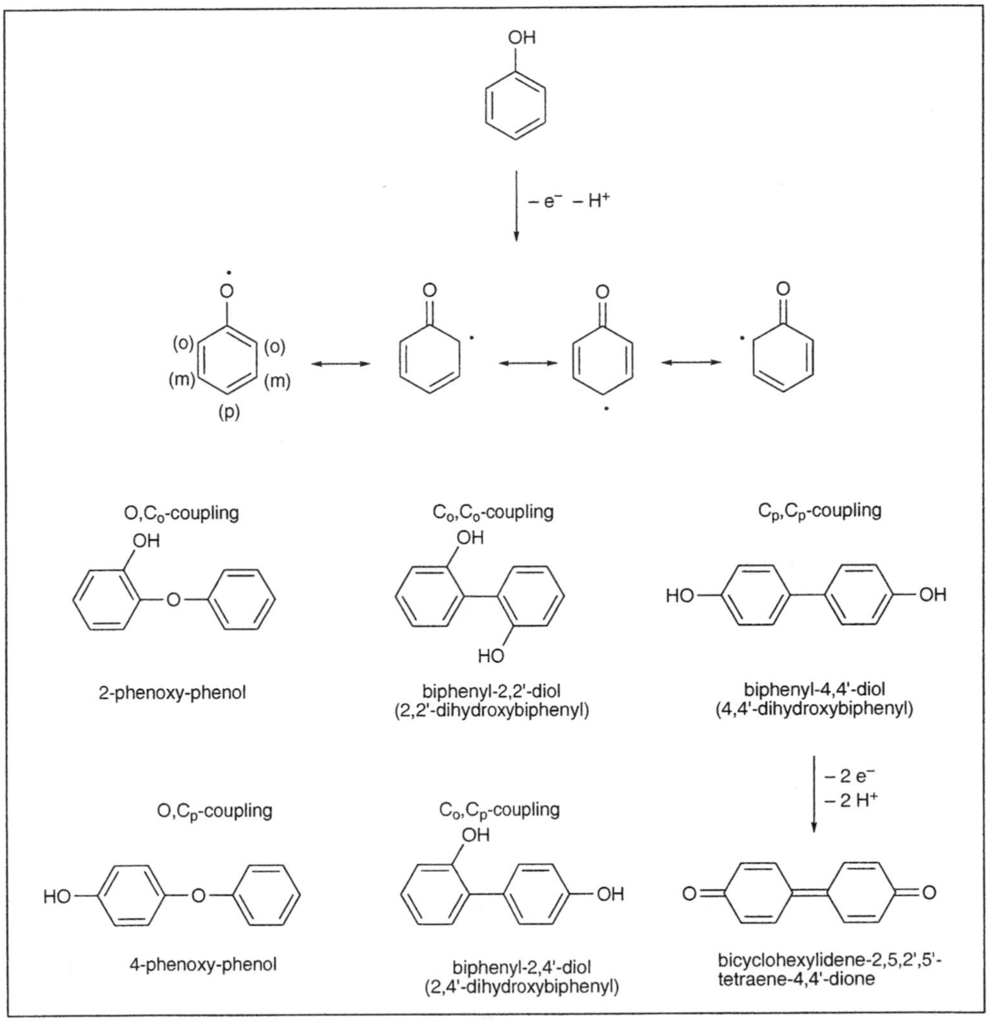

Figure 14.18 Some simple products and postulated mechanisms for oxidative coupling of phenol. The subscripts "o" and "p" are used to denote *ortho*- and *para*-position, respectively. Note that many more products can be formed particularly from substituted phenols (see, e.g., Dec and Bollag, 1994; Yu et al., 1994).

electron; see Fig. 14.4). Consequently, only compounds that easily "lose" an electron will react with molecular oxygen at significant rates. Example of such compounds include certain phenols (ArOH) and anilines (ArNH$_2$), especially those that are substituted with electron-donating groups (e.g., antioxidants; see Figs. 2.15 and 2.18), and mercaptans (R–SH, Ar–SH; see Fig. 2.20). The same type of compounds may also react with manganese (III/IV) oxides (Stone, 1987; Ulrich and Stone, 1989; Laha and Luthy, 1990; Klausen et al. 1997), and possibly with iron(III)(hydr)oxides, which are the most abundant solid oxidants present in the environment. Furthermore, chelating agents including, for example, phosphonates may be oxidized by molecular oxygen in the presence of manganese (Nowack and Stone, 2000). Finally, at hazardous waste sites, industrially used oxidizing agents may be involved in the oxidation of organic compounds. An important example is Cr(VI), which can occur as highly soluble and highly toxic chromate anion (HCrO$_4^-$ or CrO$_4^{2-}$, $pK_{a2} = 6.49$). Chromate has been shown to oxidize alkyl- and alkoxy-substituted phenols at appreciable rates, particularly at low pH (Elovitz and Fish, 1994 and 1995).

Let us first have a short look at some reaction pathways, and then make a few comments on the kinetics of such oxidation processes. As mentioned above, phenols and

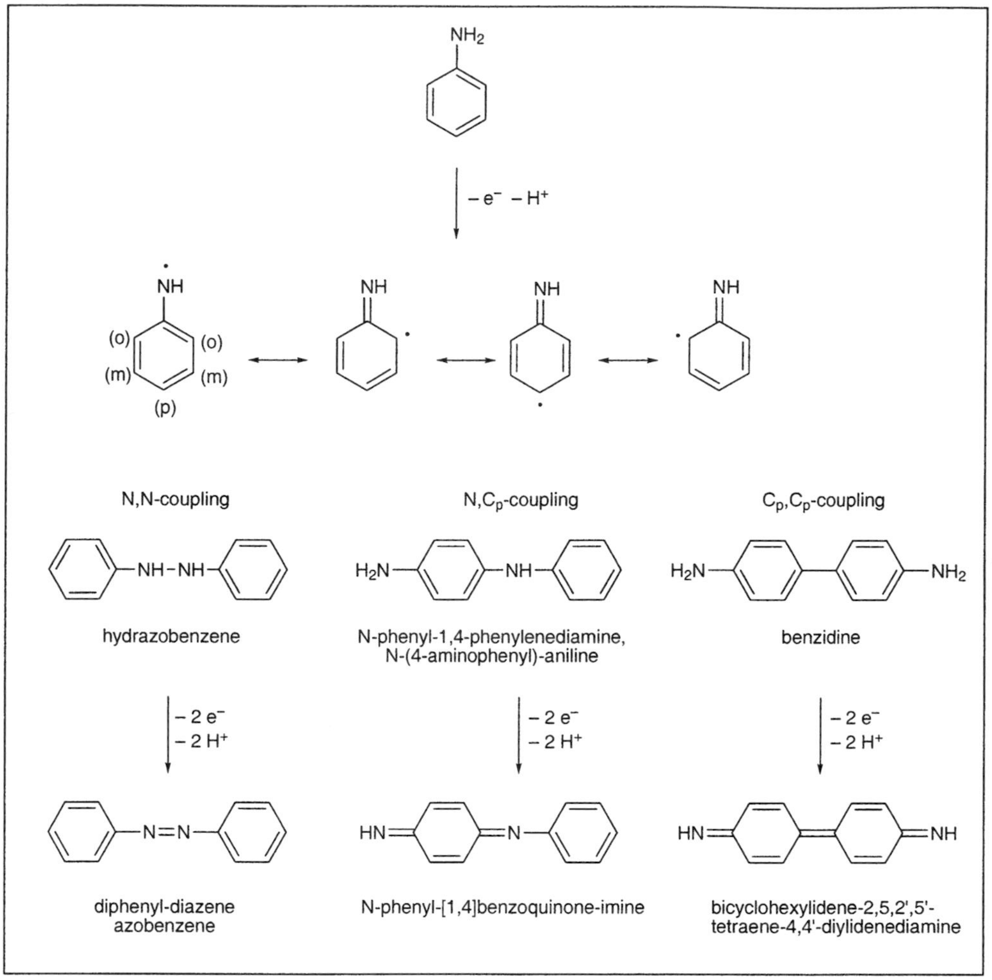

Figure 14.19 Some simple products and postulated mechanisms for oxidative coupling of aniline (adapted from Laha and Luthy, 1990). The subscripts "o" and "p" are used to denote *ortho-* and *para-* position, respectively. Note that many more products can be formed particularly from substituted anilines (see, e.g., Dec and Bollag, 1995).

aromatic amines (e.g., anilines) exhibiting electron-donating substituents including alkyl- and alkoxy-groups are particularly susceptible to chemical oxidation. In both cases the initial reaction leads to the formation of a radical that may be stabilized by delocalization in the ring. These radicals may then undergo a whole suite of reactions leading to a variety of different products (Figs. 14.18 and 14.19). In addition, such radicals may react with natural organic matter components and may polymerize, thus forming so-called "bound residues." Note that such processes, often referred to as "oxidative coupling" reactions, may be initiated by abiotic oxidants such as MnO_2 and by (extracellular) oxidoreductive enzymes (e.g., Hatcher et al., 1993; Dec and Bollag, 1994 and 1995; Burgos et al., 1996). We should, however, also note that covalent binding of aromatic amines to natural organic matter may also occur by other types of reactions including nucleophilic addition to carbonyl moieties present (Weber et al., 1996; Thorn et al, 1996).

With respect to the kinetics of oxidation reactions, the same comments as made in Section 14.2 are, of course, valid. To illustrate, we consider the oxidation of substituted phenols and anilines by MnO_2 and of substituted phenols by $HCrO_4^-$. By analogy to the type of LFER used to evaluate NAC reduction (Eq. 14-38), we can relate oxidation reaction rates to the one-electron standard oxidation potentials of

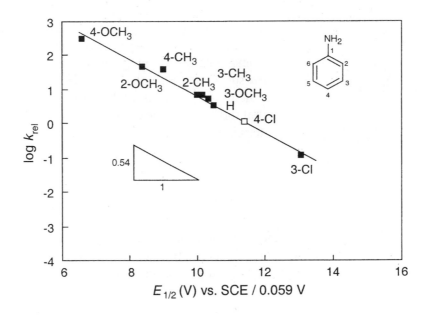

Figure 14.20 Plot of the decadic logarithms of the relative initial pseudo-first-order rate constants (relative to 4-Cl-aniline) versus $E_{1/2}$ (ArX$^{\bullet}$)/0.059 V for the oxidation of a series of mono-substituted anilines by MnO_2 at pH 6.4. Data from Klausen et al. (1997).

the compounds, $E^1_{H,ox}$ (ArXH; X = O, NH):

$$\log k_R = a' \frac{E^1_{H,ox}(ArXH)}{0.059 \ V} + b' \qquad (14\text{-}41)$$

Note that $E^1_{H,ox}$ (ArXH) is equivalent to the negative value of the standard reduction potential of the half reaction:

$$ArX^{\bullet} + e^- + H^+ \rightleftharpoons ArXH \quad ; \ E^1_H(ArX^{\bullet}) \qquad (14\text{-}42)$$

That is, $E^1_{H,ox}$ (ArXH) $= - E^1_H$ (ArX$^{\bullet}$). For this type of reaction, E^1_H (ArX$^{\bullet}$) is positive. Hence, the more positive this value, the more difficult it is to oxidize the compound. For many phenols and anilines, polarographic half-wave potentials, $E_{1/2}$(ArX$^{\bullet}$), determined at pH values where the compound is present in its neutral form, are available. These values should reasonably parallel the oxidation potentials of the compounds, and therefore can also be used to relate oxidation rate constants:

$$\log k_R = -a \frac{E_{1/2}(ArX^{\bullet})}{0.059 \ V} + b \qquad (14\text{-}43)$$

As an example, we consider the oxidation of a series of monosubstituted anilines by MnO_2 in batch systems. In this case, quite a good correlation between log k_R (expressed relative to k_R of 4-chloroaniline) and $E_{1/2}$(ArX$^{\bullet}$) is obtained (Fig. 14.20). The slope of −0.54 indicates that, similar to what we have postulated for the reduction of NACs by surface-bound Fe(II) (see Fig. 14.10d), the overall reaction rate is determined not solely by the actual electron transfer but also by other steps such as precursor complex formation. Comparable results (slopes of between −0.5 and −0.6) were obtained for the reaction of MnO_2 at pH 4 with a series of substituted anilines (Laha and Luthy, 1990), and with a series of substituted phenols at pH 4.4 (Stone, 1987). In all these cases, only initial pseudo-first-order rate constants determined with "clean" MnO_2 were considered. In the presence of solutes such as Mn^{2+} that may adsorb to the oxide surface, much slower reaction rates and much

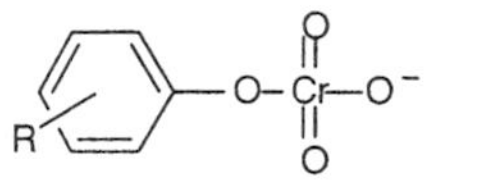

substituted phenyl-
chromate

smaller (less negative) slopes are observed (Klausen et al., 1997). This indicates that regeneration of reactive sites (as in the Fe(II)/iron oxide system, see Section 14.2) becomes rate determining. In contrast, for the oxidation of a series of *meta*- and *para*-substituted phenols by $HCrO_4^-$ at pH 2, a slope of close to –1 is found (see Problem 14.6), indicating that in this case the actual electron transfer and not the initial formation of the corresponding phenylchromate (see margin) is primarily rate determining (for mechanistic details see Elovitz and Fish, 1995).

In conclusion we should stress that quantification of rates of redox reactions in natural systems is difficult. Numerous compound- and system-specific factors may influence the overall reaction rate. Evaluation of the relative reactivities of a series of structurally related compounds that are likely to react by the same reaction mechanism(s), may, however, provide important insight into the processes determining a given reaction in a given system. Such information may allow at least order-of-magnitude estimates of how fast a given compound will undergo oxidation or reduction in that system.

14.4   Questions and Problems

Questions

Q 14.1

How do you determine whether a given transformation reaction of an organic compound is a redox reaction? Which elements in an organic molecule are primarily involved in redox reactions?

Q 14.2

Which of the following transformation reactions (from left to right) is an oxidation? a reduction? not a redox reaction? In the cases of redox reactions, write the complete half reactions.

(i) $R - CHO \rightarrow R - CH_2OH$

(ii) $CH_2Br - CHBr - CH_2Cl \rightarrow CH_2 = CBr - CH_2Cl$

(iii) $CH_2Br - CHBr - CH_2Cl \rightarrow CH_2 = CH - CH_2Cl$

(iv) $CBr_4 \rightarrow CH_2Br_2$

(v) $CBr_4 \rightarrow HCOO^-$

(vi)

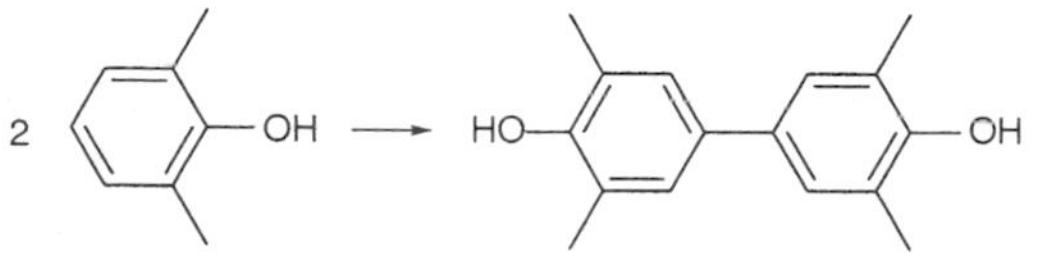

(vii)

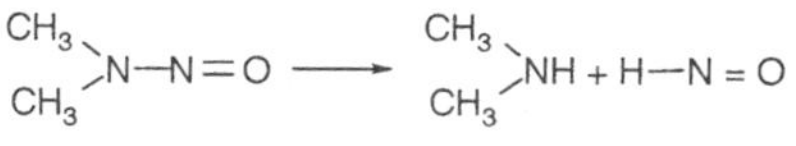

(viii) —NHOH ⟶ —NO

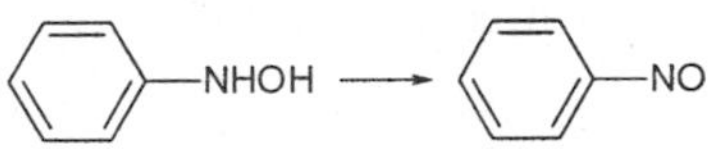

(ix) 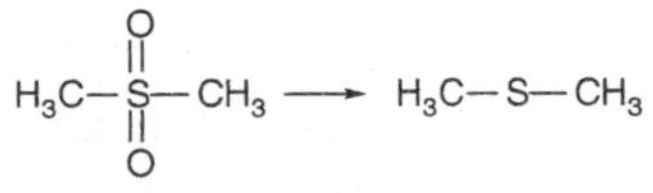 $H_3C-\overset{\overset{O}{\|}}{\underset{\underset{O}{\|}}{S}}-CH_3 \longrightarrow H_3C-S-CH_3$

(x) $2\ CH_3 - SH \rightarrow CH_3 - S - S - CH_3$

Q 14.3

Fumigants (volatile pesticides) are used in large quantities in warm regions to control soil-borne pests. Wang et al. (2000) studied the transformation of various fumigants including propargyl bromide (PBr; $CH \equiv C - CH_2Br$) and chloropicrin (PC, for structure see below) in aqueous solution containing the fertilizer ammonium thiosulfate (ATS). At 1 mM ATS and 20°C the half-lives of PBr and PC were 34 h and 114 h, respectively, as compared to 3100 h and 2000 h in pure water. A mass balance showed that in the case of PBr, one thiosulfate was consumed and one Br^- was produced per PBr transformed, which is consistent with an S_N2 reaction (see Section 13.2). In the case of PC, however, four thiosulfates were consumed and two Cl^- were liberated per PC transformed:

$$CCl_3-NO_2 + 4\ S_2O_3^{2-} + ? \rightarrow ? + 2\ Cl^- + ?$$

chloropicrin   thiosulfate
(PC)

$$^-:S-\overset{\overset{O}{\|}}{\underset{\underset{O}{\|}}{S}}-O^-$$

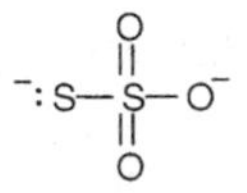

thiosulfate

Since no product analyses were carried out, the authors did not speculate on the type of reaction occurring. Can you help them? Try to complete the above reaction equation. *Hint:* See reactions (iv) and (x) in Q 14.2.

Q 14.4

Explain in words how the standard reduction potential of a half reaction is defined. What are the most common standard conditions used for defining reduction potentials in environmental chemistry?

Q 14.5

What does the Nernst equation describe?

Q 14.6

What is the difference between the standard reduction potentials of the following two half reactions (i) and (ii)?

(i) $ArNO_2 + 6\,e^- + 6\,H^+ \rightleftharpoons ArNH_2 + 2\,H_2O$

(ii) $1/6\,ArNO_2 + e^- + H^+ \rightleftharpoons 1/6\,ArNH_2 + 1/3\,H_2O$

Q 14.7

Calculate the standard reduction potential under environmental ("W") conditions of reaction (ix) in Q 14.2 ($CH_3 - SO_2 - CH_3 / CH_3 - S - CH_3$) using the data given in Table 14.3.

Q 14.8

Which of the following reactions is thermodynamically feasible under "environmental" standard conditions ("W" conditions)? Do not make any calculations!

(i) The reduction of nitrobenzene to aniline by hydrogen sulfide assuming that elemental sulfur ($S(s)$) is formed.

(ii) The oxidation of dimethylsulfide (CH_3-S-CH_3) to dimethylsulfoxide ($CH_3-SO-CH_3$) by goethite ($FeOOH(s)$) assuming that siderite ($FeCO_3(s)$) is formed.

(iii) The oxidation of aniline to azobenzene (reverse reaction 12 in Table 14.3) by manganese oxide ($MnO_2(s)$) assuming that manganese carbonate ($MnCO_3(s)$) is formed.

Q 14.9

What factors determine the overall rate of oxidation or reduction of a given organic compound in a given environmental system?

Q 14.10

What is an electron transfer mediator? Give some examples of environmentally relevant species that may act as such mediators.

Q 14.11

What are common molecular descriptors used to correlate rates of oxidative or reductive transformations of structurally related compounds?

Q 14.12

Why is it so difficult to derive generally applicable quantitative structure-reactivity relationships (QSARs) for redox reactions involving organic compounds? What is particularly problematic when dealing with reductive dehalogenation reactions?

Q 14.13

Agrawal and Tratnyek (1996) have investigated the reduction of a series of NACs spanning a wide range of $E_H^1(ArNO_2)$ values by zero-valent iron metal grains in batch reactors. The observed reduction kinetics were first order, and virtually identical first-order rate constants were obtained for all NACs investigated. Furthermore, the reduction rates increased with increasing mixing rates. Try to give an explanation for these findings.

Problems

P 14.1 *Calculating Standard Reduction Potentials* E_H^0 *and* $E_H^0(W)$ *From Free Energies of Formation*

Calculate the E_H^0 and $E_H^0(W)$ values for the half reactions of the following redox couples in aqueous solution at 25°C using the $\Delta_f G^0$ values given. Find any required additional information in Appendix C.

(a) SO_4^{2-} / HS^-

$\Delta_f G^0_{SO_4^{2-}}(aq) = -744.6 \text{ kJ} \cdot \text{mol}^{-1}$; $\Delta_f G^0_{HS^-}(aq) = +12.05 \text{ kJ} \cdot \text{mol}^{-1}$

$\Delta_f G^0_{H_2O}(\ell) = -237.2 \text{ kJ} \cdot \text{mol}^{-1}$

(b)

DCB   CB

$\Delta_f G^0_{DCB}(g) = +82.7 \text{ kJ} \cdot \text{mol}^{-1}$; $\Delta_f G^0_{CB}(g) = +99.2 \text{ kJ} \cdot \text{mol}^{-1}$

$\Delta_f G^0_{Cl^-}(aq) = -131.3 \text{ kJ} \cdot \text{mol}^{-1}$

Compare the result with the values given for dechlorination of other chlorobenzenes in Table 14.3. Comments?

(c)

TCE   *cis* - DCE

$\Delta_f G^0_{TCE}(\ell) = +14.1 \text{ kJ} \cdot \text{mol}^{-1}$; $\Delta_f G^0_{DCE}(\ell) = +21.44 \text{ kJ} \cdot \text{mol}^{-1}$

$\Delta_f G^0_{Cl^-}(aq) = -131.3 \text{ kJ} \cdot \text{mol}^{-1}$

Compare the result with the E_H^0 values of the PCE/TCE couple in Table 14.3. Comments?

P 14.2 *Some Additional Questions Concerning the Bioremediation of Contaminated Aquifers*

You are involved in the remediation of an aquifer that has been contaminated with 2-methylnaphthalene. Similar to the toluene case discussed in Illustrative Example 14.2, the aquifer is flushed with air-saturated water that is pumped into the ground at one place and withdrawn nearby. Calculate how much water is at least required to supply sufficient oxygen for the microbial mineralization of 1 kg of 2-methylnaphthalene assuming that the water contains 10 mg $O_2 \cdot L^{-1}$.

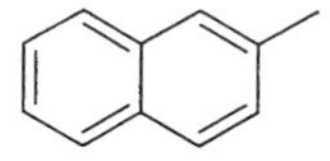

2-methylnaphthalene

$M_i = 142.2$ g $\cdot$ mol^{-1}

P 14.3 *What Redox Zones Can Be Expected in This Laboratory Aquifer Column?*

You work in a research laboratory and your job is to investigate the microbial degradation of organic pollutants in laboratory aquifer column systems. You supply a column continuously with a synthetic groundwater containing 0.3 mM O_2, 0.5 mM NO_3^-, 0.5 mM SO_4^{2-}, and 1 mM HCO_3^-, as well as 0.1 mM benzoic acid butyl ester, which is easily mineralized to CO_2 and H_2O. The temperature is 20°C and the pH is 7.3 (well buffered). Would you expect sulfate reduction or even methanogenesis to occur in this column? Establish an electron balance to answer this question.

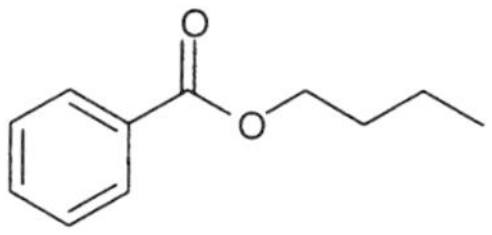

benzoic acid butyl ester
(butyl benzoate)
$M_i = 178.2$ g $\cdot$ mol^{-1}

P 14.4 *Calculating Reduction Potentials of Half Reactions at Various Conditions of pH and Solution Composition*

Calculate the half reaction reduction potentials of the following redox couples in aqueous solution at 25°C under the conditions indicated using (i) E_H^0 and/or (ii) $E_H^0(W)$ as starting point (see Tables 14.2 and 14.3). Compare the calculated E_H values with the corresponding $E_H^0(W)$ values.

(a) $MnO_2(s)$ / $MnCO_3(s)$, pH 8.5, $\{HCO_3^-\} = 10^{-4}$

(b) Nitrobenzene ($ArNO_2$) / aniline ($ArNH_2$), pH 9.0, $\{ArNH_2\}$ /$\{ArNO_2\} = 10^5$

P 14.5 *Are These Two Redox Reactions Thermodynamically Favorable?*

Somebody claims that the two redox reactions (a) and (b) involving organic compounds are thermodynamically favorable at 25°C in aqueous solution under the indicated conditions. Is this correct? Calculate the $\Delta_r G$ values of the reactions. Use the information below and summarized in Tables 14.2 and 14.3 to answer this question.

(a) The oxidation of hydroquinone (HQ) to benzoquinone (BQ) [reverse of reaction 10 in Table 14.3] by $Fe^{3+}(aq)$ in the presence of $Fe^{2+}(aq)$ under the following conditions:

$\{Fe^{3+}(aq)\} = \{Fe^{2+}(aq)\} = 10^{-3}$; $E_H^0(Fe^{3+}(aq)/Fe^{2+}(aq)) = 0.77$ V,

pH 2 [assume that at this pH, Fe^{3+} and Fe^{2+} are present solely as aquo ions (aq).]

$\{HQ\} = 10^{-7}$, $\{BQ\} = 10^{-5}$

What would be the $\{HQ\} / \{BQ\}$ ratio *at equilibrium*? Comment on any assumptions you made.

(b) The reduction of dimethylsulfone (DMSF) to dimethylsulfoxide (DMSO) (reaction 13 in Table 14.3) in a 10 mM H_2S_{tot} solution at pH 8, $\{DMSF\} = 10^{-6}$, and $\{DMSO\} = 10^{-7}$. Assume that H_2S is oxidized to elemental sulfur.

P 14.6 *Evaluating the Effect of Substituents on the One-Electron Reduction Potentials of Nitroaromatic Compounds*

Inspection of Table 14.4 reveals that the type and position of substituents have a significant impact on the one-electron reduction potential, $E_H^1(ArNO_2)$, of NACs. Try to answer the following questions by considering electronic and/or steric effects (see also Chapter 8). NB = nitrobenzene.

(a) Why does $E_H^1(ArNO_2)$ increase (become less negative) in the sequence 4-NH_2-NB < 4-CH_3-NB < NB < 4-Cl-NB < 4 Ac-NB < 4-NO_2-NB?

(b) Why has 3-Cl-NB a less negative $E_H^1(ArNO_2)$ value than 4-Cl-NB, whereas the opposite is true for 3-Ac-NB versus 4-Ac-NB and 3-NO_2-NB (1,3-DNB) versus 4-NO_2-NB (1,4-DNB)?

(c) In many cases, *ortho*-substituted NACs have a more negative $E_H^1(ArNO_2)$ value as compared to the para-substituted isomers (e.g., 2-CH_3-NB < 4-CH_3-NB; 2-Cl-NB < 4-Cl-NB, 2-Ac-NB < 4-Ac-NB, 1,2-DNB < 1,4-DNB). What could be the major reason for these findings?

(d) Rank the three substituted nitronaphthalenes shown below (I–III) in the order of increasing $E_H^1(ArNO_2)$ values. Comment on your choice.

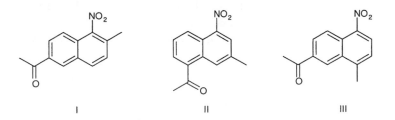

I   II   III

P 14.7 *Evaluating Relative Reduction Rates in an Anaerobic Sediment*

Jafvert and Wolfe (1987) studied the rate of disappearance of a series of halogenated ethanes in an anaerobic sediment-water slurry. They found the following initial pseudo-first-order rate constants, k_{obs}, for the various compounds:

Compound Name	Structure	k_{obs} [a] (s^{-1})
1,2-Dichloroethane	$CH_2Cl–CH_2Cl$	$\ll 2 \times 10^{-7}$
1,2-Dibromoethane	$CH_2Br–CH_2Br$	3.5×10^{-6}
1,2-Diiodoethane	$CH_2I–CH_2I$	4.8×10^{-4}
1,1,2,2-Tetrachloroethane	$CHCl_2–CHCl_2$	1.2×10^{-6}
Hexachloroethane	$CCl_3–CCl_3$	3.2×10^{-4}

[a] Sediment-to-water ratio = 0.075 (r_{sw}), pH 6.5, apparent E_H value = −0.14 V.

Try to explain qualitatively the observed differences in reactivity. Are there compounds in this table for which other reactions than reductive dehalogenation may be important under these conditions? If yes, which ones, and what kind of reaction do they undergo?

P 14.8 *What Are the Pathways of the Reduction of 1,1,1-Trichloroethane by Zero-Valent Metals and Bimetallic Reductants?*

Information concerning the pathways and products of reactions of polyhalogenated solvents with zero-valent metals may be critical to the success of *in situ* treatment techniques. Fennelly and Roberts (1998) have investigated the reduction of 1,1,1-trichloroethane (1,1,1-TCA) by Fe(O) and Zn(O) and by two bimetallic (nickel/iron and copper/iron) reductants. The following products were detected at measurable concentrations as intermediates and/or final products:

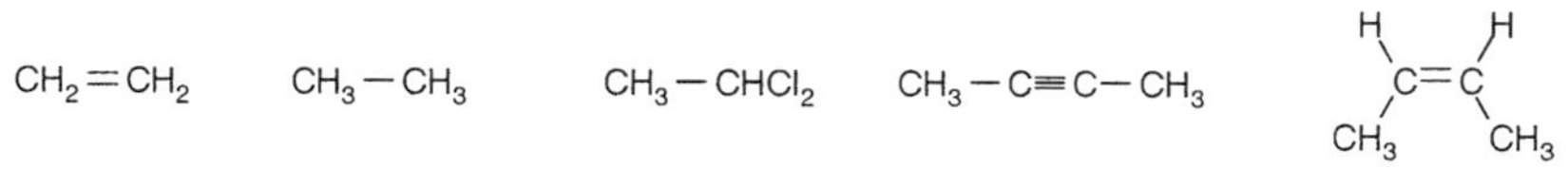

ethene | ethane | 1,1-dichloroethane (1,1-DCA) | 2-butyne | *cis*-2-butene

Not observed were chloroethane (CH_3-CH_2Cl) and vinyl chloride (CH_2=$CHCl$). An interesting finding was that 1,1-DCA reacted much too slowly to represent an intermediate in the formation of ethane. The authors postulated a scheme involving successive one- or two-electron reduction steps to form radicals and carbenes to explain the absence of other observable intermediates, as well as the formation of products originating from radical or possibly from carbene coupling. Try to construct such a hypothesized reaction scheme yourself.

P 14.9 *Evaluating and Estimating the Rates of Oxidation of Phenolic Compounds by Chromium(VI) in Homogeneous Aqueous Solution*

Many chromate-contaminated sites have high concentrations of Cr(VI) (up to 0.2 M!) and low pH localized within one or more plume areas. Elovitz and Fish (1994 and 1995) have investigated the oxidation kinetics of a series of substituted phenols (ArOH) by Cr(VI) present primarily as $HCrO_4^-$ at the conditions prevailing in these experiments. At a fixed pH, the reaction rate could be described by an empirical pseudo-second-order rate law:

$$-\frac{d[ArOH]}{d(t)} = k_{ArOH}[HCrO_4^-][ArOH]$$

Note that k_{ArOH} is a rather complex function of pH, and decreases by several orders of magnitude between pH 2 and 7 (for details see Elovitz and Fish, 1995). For a series of substituted phenols, the k_{ArNO2} values determined at pH 2 are given in the table below together with the $E_{1/2}$ values of the compounds. Note that the $E_{1/2}$ values are expressed relative to the standard calomel electrode (SCE); that is, they are lower by –0.24 V values reported relative to the SHE. Also given is a plot of log k_{ArOH} versus $E_{1/2}/0.059$ V, which shows that, when including all compounds into one LFER of the type Eq. 14-43, there is scatter in the data.

(a) Try to find an explanation for the scatter observed. Are there subsets of compounds that should yield a much better LFER? Which ones?

(b) Estimate the k_{ArNO2} values of 3,5-dimethoxyphenol ($E_{1/2}$ = 0.60 V) and 3-chlorophenol ($E_{1/2}$ = 0.74 V) using (i) the LFER established for all compounds (see figure legend), and (ii) the LFER that you have derived from an "intelligently" chosen subset of the compounds. Compare and discuss the results.

(c) Calculate the half-life of 3,5-dimethoxyphenol in a 1 mM chromate solution at pH 2. What chromate concentration would be required to oxidize 4-nitrophenol ($E_{1/2}$ = 0.92 V) in aqueous solution at pH 2 with a half-life of less than one month? Comment on the result.

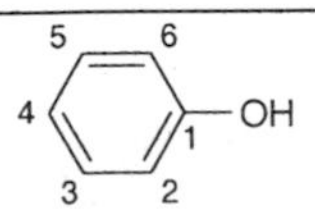

Compound	k_{ArOH} [a] $(M^{-1} s^{-1})$	$E_{1/2}$ [b] (V vs SCE)
1 H (phenol)	2.6×10^{-5}	0.63
2 4-methyl	1.2×10^{-3}	0.54
3 2,4-dimethyl	9.4×10^{-3}	0.46
4 3,4-dimethyl	2.8×10^{-3}	0.51
5 2,6-dimethyl	1.2×10^{-3}	0.43
6 2,4,6-trimethyl	5.3×10^{-3}	0.39
7 4-methoxy	2.2×10^{-1}	0.41
8 2,6-dimethoxy	4.4×10^{-1}	0.32
9 3,4-dimethoxy	3.7×10^{0}	0.35
10 2-methoxy-4-aldehyde	1.1×10^{-4}	0.60
11 2-methoxy-4-methyl	3.8×10^{-1}	0.37
12 4-chloro	2.3×10^{-5}	0.65

[a] Data from Elovitz and Fish (1994). [b] Data from Suatoni et al., 1961.

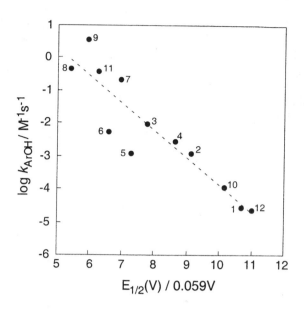

Plot of log k_{ArOH} versus $E_{1/2}$ / 0.059 V for the compounds listed in the table. The linear regression (Eq. 14-43) including all data points (dotted line) yields: log k_{ArOH} = -0.83 ($E_{1/2}$ / 0.059V) + 4.4, (R^2 = 0.92).

Chapter 15

DIRECT PHOTOLYSIS

15.1 Introduction

15.2 Some Basic Principles of Photochemistry
Light Absorption by Chemical Species: Molar Extinction Coefficients
Illustrative Example 15.1: *Determining Decadic Molar Extinction Coefficients of Organic Pollutants*
Chemical Structure and Light Absorption
The Fate of Excited Chemical Species: Quantum Yields

15.3 Light Absorption by Organic Chemicals in Natural Waters
Light and Light Attenuation in Natural Water Bodies
Specific Rate of Light Absorption by an Organic Pollutant
Near-Surface Specific Rate of Light Absorption of an Organic Pollutant
Example Illustrating the Calculation of the Near-Surface Specific Rate of Light Absorption of an Organic Pollutant
Specific Rate of Light Absorption of an Organic Pollutant in a Well-Mixed Water Body
Light-Screening Factors
Illustrative Example 15.2: *Using the Screening Factor S(λ_m) to Estimate the Total Specific Light Absorption Rate of PNAP in the Epilimnion of a Lake*

15.4 Quantum Yield and Rate of Direct Photolysis
First-Order Rate Constant for Quantification of Direct Photolysis
Illustrative Example 15.3: *Estimating the Photolysis Half-Life of a Weak Organic Acid in the Well-Mixed Epilimnion of a Lake*
Determination of Quantum Yields and Chemical Actinometry (*Advanced Topic*)

15.5 Effects of Solid Sorbents (Particles, Soil Surfaces) on Direct Photolysis
Effects of Particles in Water
Direct Photolysis on Soil Surfaces

15.6 Questions and Problems

15.1 Introduction

So far, we discussed chemical reactions in which the reacting molecules were in their so-called electronic *ground state*. We termed these reactions chemical reactions, mainly because temperature had a significant effect on reaction rates. We might recall that heat has primarily an impact on the translational, rotational, and vibrational energy of an organic molecule, but that a (small) change in temperature leaves the electronic ground state of the molecule essentially unchanged. In Chapter 2 we classified in a very simplified manner the electronic ground state of a molecule by assigning the valence electrons of the various atoms to three categories, that is, part of a σ-bond, a π-bond (or a delocalized π-bond system), or a nonbonding electron localized on an atom, usually a heteroatom. In the ground state, one commonly assigns the electrons that are engaged in chemical bonds to *bonding orbitals* (i.e., σ- or π-orbitals), whereas the electrons localized on an atom are said to occupy so-called *nonbonding orbitals* (i.e., *n*-orbitals). If a molecule is now exposed to ultraviolet (uv) or visible (vis) light (of interest to us is the wavelength range of solar radiation that may promote phototransformations of organic pollutants at the earth surface, i.e., 290–600 nm, see Section 15.3), electrons may get promoted from bonding or nonbonding orbitals to so-called *antibonding orbitals* (i.e., σ^*- or π^*-orbitals). The molecule is then said to be in an *excited state,* that is, it has become a much more reactive species as compared to the reactivity it exhibits in the electronic ground state. As we will see in Section 15.2, an excited species may undergo a variety of processes.

In this chapter we consider the case in which a given organic pollutant absorbs light and, as a consequence of that light absorption, undergoes transformation. This process is commonly referred to as *direct photolysis*. In Chapter 16, we will address those cases in which organic chemicals are transformed by energy transfer from another excited species (e.g., components of natural organic matter), or by reaction with very reactive, short-lived species formed in the presence of light (e.g., hydroxyl radicals, singlet oxygen, ozone, peroxy radicals, etc.). These processes are usually summarized by the term *indirect photolysis*. In both direct and indirect photolysis our discussion will present the concepts used to quantify these processes, rather than trying to evaluate reaction pathways and transformation products. Furthermore, our treatment of direct photolysis will focus primarily on reactions in aqueous solution because in the gas phase, this process presently cannot be quantified due to the lack of the pertinent parameters. The discussion of indirect photolysis will, however, encompass both water and air, because, particularly the reaction with hydroxyl radicals is strongly determining the residence time of many organic pollutants in the troposphere. Before we can begin to treat direct and indirect photolysis, we need to review a few basic principles of photochemistry.

15.2 Some Basic Principles of Photochemistry

Light Absorption by Chemical Species: Molar Extinction Coefficients

To picture the process of light absorption by a chemical species, we recall that we may look upon light as having both wave- and particle-like properties (see, e.g., Turro, 1978; Finlayson-Pitts and Pitts, 1986). As a wave, we consider light to be a combination of oscillating electric and magnetic fields perpendicular to each other and to the direction of propagation of the wave. The distance between two consecutive maxima is the wavelength λ, which is inversely proportional to the frequency v commonly expressed by the number of cycles passing a fixed point in 1 sec:

$$\lambda = \frac{c}{v} \tag{15-1}$$

where c is the speed of light in a vacuum, 3.0×10^8 m·s^{-1}.

A more particle-oriented consideration of light shows that light is quantized and is emitted, transmitted, and absorbed in discrete units, so-called *photons* or *quanta*. The energy E of a photon or quantum (the unit of light on a molecular level) is given by:

$$E = hv = h\frac{c}{\lambda} \tag{15-2}$$

where h *is* the Planck constant, 6.63×10^{-34} J·s. Note that the energy of a photon is dependent on its wavelength. On a molar basis the unit of light is commonly called an *einstein,* although the IUPAC has decided to discard this term. Hence, 1 einstein is the equivalent to 6.02×10^{23} (= 1 mol) photons or quanta. The energy of light of wavelength λ (nm) is:

$$E = 6.02 \times 10^{23} \cdot h\frac{c}{\lambda} = \frac{1.196 \times 10^5}{\lambda} \text{kJ} \cdot \text{einstein}^{-1} \tag{15-3}$$

It is instructive to compare the light energies at different wavelengths with bond energies typically encountered in organic molecules. Table 15.1 shows that the energy of uv and visible light is of the same order of magnitude as that of covalent bonds. Thus, in principle, such bonds could be cleaved as a consequence of light absorption. Whether reactions take place depends on the probability with which a given compound absorbs light of a given wavelength, and on the probability that the excited species undergoes a particular reaction.

When a photon passes close to a molecule, there is an interaction between the electromagnetic field associated with the molecule and that associated with the radiation. If, and only if, the radiation is absorbed by the molecule as a result of this interaction, can the radiation be effective in producing photochemical changes (Grotthus-Draper law, see, e.g., Finlayson-Pitts and Pitts, 1986). Therefore, the first thing we need to be concerned about is the probability with which a given compound absorbs uv and visible light. This information is contained in the compounds *uv/vis absorption spectrum,* which is often readily available or can be easily measured with a spectrophotometer.

Table 15.1 Typical Energies for Some Single Bonds and the Approximate Wavelengths of Light Corresponding to This Energy [a]

Bond	Bond Energy E [b] (kJ·mol^{-1})	Wavelength λ (nm)
O–H	465	257
H–H	436	274
C–H	415	288
N–H	390	307
C–O	360	332
C–C	348	344
C–Cl	339	353
Cl–Cl	243	492
Br–Br	193	620
O–O	146	820

[a] Compare Eq. 15-3. [b] Values from Table 2.2.

Let us consider the absorption of light by a solution of a given chemical in particle-free water contained in a transparent vessel, for example, in a quartz cuvette. The quantitative description of light absorption by such a system is based on two empirical laws. The first law, Lambert's law, states that the *fraction* of radiation absorbed by the system is independent of the intensity of that radiation. Note that this law is not valid when very high intensities of radiation are employed (e.g., when using lasers). The second law, Beer's law, states that the *amount* of radiation absorbed by the system is proportional to the number of molecules absorbing the radiation. Beer's law is valid as long as there are no significant interactions (e.g., associations) between the molecules. From these two laws, the well-known *Beer-Lambert law* is obtained that relates the light intensity, $I(\lambda)$, emerging from the solution to the incident light intensity, $I_0(\lambda)$ (e. g., in einstein·cm^{-2} s^{-1}):

$$I(\lambda) = I_0(\lambda) \times 10^{-[\alpha(\lambda) + \varepsilon_i(\lambda)C_i]l} \tag{15-4}$$

or:

$$\text{absorbance} \quad A(\lambda) \equiv \log \frac{I_0(\lambda)}{I(\lambda)} = [\alpha(\lambda) + \varepsilon_i(\lambda)C_i]l \tag{15-5}$$

where C_i is the concentration of the compound i of interest in moles per liter (M), l is the path length of the light in the solution commonly expressed in centimeters, $\alpha(\lambda)$ is the decadic absorption or attenuation coefficient of the medium in cm^{-1} (i.e., of the water that may or may not contain other light-absorbing species), and $\varepsilon_i(\lambda)$ is the *decadic molar absorption coefficient* of the compound i at wavelength λ in M^{-1} cm^{-1}. Here $\varepsilon_i(\lambda)$ is a measure of the probability that the compound i absorbs light at a particular wavelength.

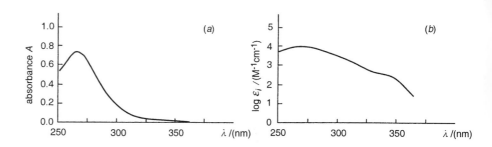

Figure 15.1 Electronic absorption spectrum of nitrobenzene in aqueous solution at 0.1 mM. (*a*) Absorbance *A* as a function of wavelength. (*b*) Log ε_i as a function of wavelength.

Using a spectrophotometer and an appropriate solvent and reference solution (i.e., the same liquid phase as the one containing the compound so that absorption effects other than by the chemical cancel), the absorbance *A* of a solution of the compound:

$$A(\lambda) = \varepsilon_i(\lambda) \cdot C_i \cdot l \qquad (15\text{-}6)$$

can be measured as a function of wavelength in a cuvette exhibiting a specific width (e.g., 1, 5, or 10 cm). In the spectrophotometer, it can be assumed that the path length of the light within the cell is more or less identical with the cell width (provided that there is no light scattering occurring within the cell, for example, due to the presence of particles). In this case, $\alpha(\lambda)$, in Eq. 15-4 and 15-5 is commonly referred to as *beam attenuation coefficient*.

Figure 15.1*a* gives an example of an electronic absorption spectrum, that is, a *uv/vis spectrum* of an organic compound. From this spectrum, ε_i may be calculated for each wavelength (see Illustrative Example 15.1) and plotted as a function of wavelength as shown in Fig. 15.1*b*. Other examples are given in Figs. 15.2–15.5 and discussed in the next section. Note that in these figures, $\varepsilon_i(\lambda)$ is expressed on a logarithmic scale since it may range over several orders of magnitude.

Illustrative Example 15.1   **Determining Decadic Molar Extinction Coefficients of Organic Pollutants**

nitrobenzene

Somebody measures the electronic absorption spectrum of an 0.1 mM solution of nitrobenzene in pure water. For wavelengths below 310 nm, the spectrum is recorded using a 1 cm cuvette; for higher wavelengths, a 5 cm cuvette is used. The following absorbances are recorded:

Wavelength λ/nm	Absorbance (1 cm) A	Wavelength λ/nm	Absorbance (5 cm) A
250	0.54	310	0.70
265 (λ_{max})	0.76	320	0.40
280	0.60	330	0.29
290	0.43	340	0.28
300	0.25	350	0.14
		360	0.07
		370	0.01

Problem

Calculate the decadic molar extinction coefficients of nitrobenzene for the wavelengths indicated above.

Answer (a)

Rearrange Eq. 15-6 to calculate $\varepsilon_i(l)$:

$$\varepsilon_i(\lambda) = \frac{A(\lambda)}{C_i \cdot l}$$

where $C_i = 10^{-4}$ M and $l = 1$ cm or 5 cm, respectively. The results are (see also Fig. 15.1b):

λ/nm	ε_i/M^{-1} cm^{-1}	λ/nm	ε_i/M^{-1} cm^{-1}
250	5400	320	800
265 [a]	7600	330	580
280	6000	340	560
290	4300	350	280
300	2500	360	140
310	1400	370	20

[a] λ_{max}

As we can see from the spectra shown in Figs. 15.2–15.5, organic compounds may absorb light over a wide wavelength range exhibiting one or several absorption maxima. Each absorption maximum can be assigned to a specific electron transition, for example, a $\pi \rightarrow \pi^*$ or $n \rightarrow \pi^*$ transition. Note that particularly in the excited state, members of a population of molecules are distributed among various vibrational and rotational states, and that, therefore, broad absorption bands (resulting from numerous unresolved narrow bands) rather than sharp absorption lines are observed in the uv/vis spectrum.

Often in the literature, only the wavelengths (λ_{max}) and corresponding ε values (ε_{imax}) of the absorption maxima are reported. These values can be used for a preliminary assessment of whether a compound might absorb ambient light. For quantification of photochemical processes, however, the whole spectrum must be known.

There is in fact a large body of uv/vis spectra of organic compounds available in the literature, although most of these data were obtained in organic solvents (e.g., Pretsch et al., 2000). It should be noted that absorption spectra are susceptible to solvent effects, especially if the solute undergoes hydrogen bonding with the solvent molecules. Nevertheless, particularly if a spectrum was recorded in a polar organic solvent (e.g., methanol, acetonitrile) or an organic solvent-water mixture, it can be

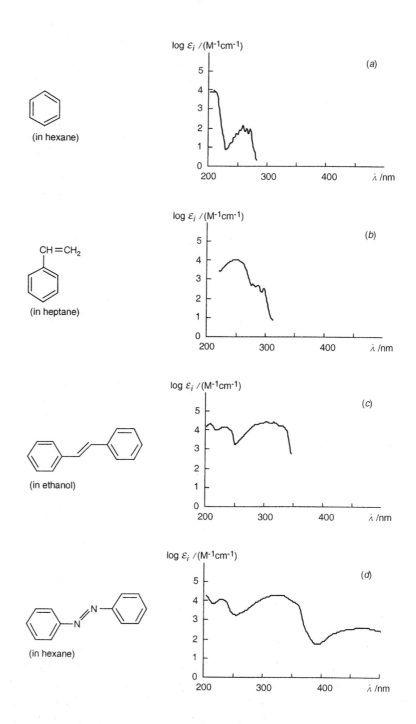

Figure 15.2 Electronic absorption spectra of (*a*) benzene, (*b*) styrene, (*c*) *trans*-stilbene, and (*d*) azobenzene (data from Pretsch et al., 2000).

used to approximate the rate of light absorption of a given chemical in aqueous solution when exposed to sunlight. This is, of course, only feasible if the speciation of the chemical is the same in the organic solvent and in water, which may not be the case if the compound exhibits acid or base functionalities.

Chemical Structure and Light Absorption

As indicated by the examples given in Figs 15.1–15.5, light absorption of organic compounds in the wavelength range of interest to us (i.e., 290-600 nm) is in most

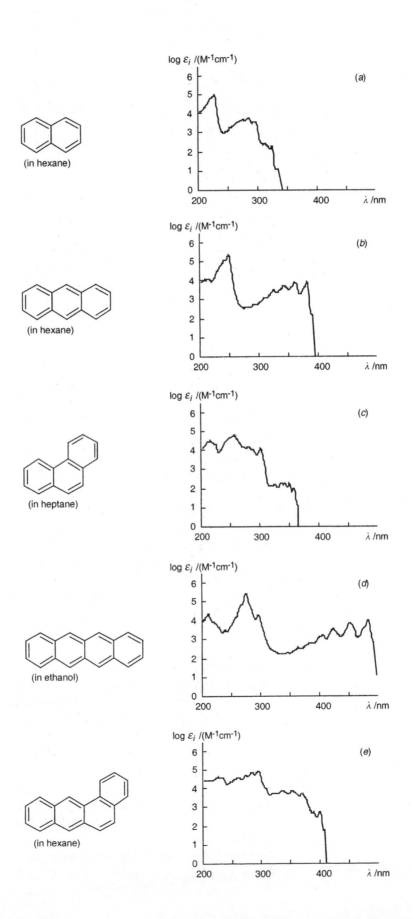

Figure 15.3 Electronic absorption spectra of (*a*) naphthalene, (*b*) anthracene, (*c*) phenanthrene, (*d*) 2,3-benzanthracene (naphthacene), and (*e*) 1,2-benzanthracene (benz(a)-anthracene) (data from Pretsch et al., 2000).

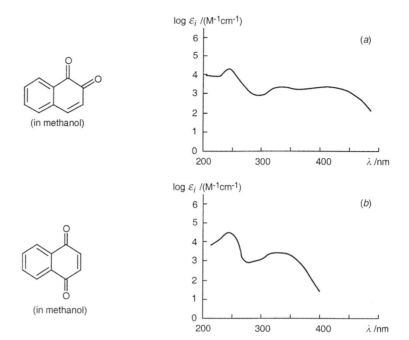

Figure 15.4 Electronic absorption spectra of (*a*) 1,2-naphthoquinone and (*b*) 1,4-naphthoquinone (data from Pretsch et al., 2000).

cases associated with the presence of a delocalized π-electron system. Hence, aromatic rings and conjugated double bonds in particular may form a so-called *chromophore*, a structural moiety that exhibits a characteristic uv/vis absorption spectrum. In such systems, the most probable electron transitions are promotions of π-electrons from bonding to antibonding π-orbitals. Such transitions are commonly referred to as $\pi \rightarrow \pi^*$ transitions, and generally they give rise to the most intense absorption bands in the spectrum. If a π-electron system contains atoms with non-bonding electrons (i.e., hetero-atoms), so-called $n \rightarrow \pi^*$ transitions may also be observed. These transitions commonly occur at longer wavelengths (i.e., lower energy) as compared to the $\pi \rightarrow \pi^*$ transitions, and they usually exhibit a significantly smaller molar extinction coefficient. For example, in the spectrum of nitrobenzene (Fig. 15.1), the absorption band with a maximum at 267 nm ($\varepsilon_{imax} \cong 7500 \ M^{-1} \ cm^{-1}$) may be assigned to a $\pi \rightarrow \pi^*$ transition, while the much less intense band at 340 nm ($\varepsilon_{imax} \cong 150 \ M^{-1} \ cm^{-1}$) is due to an $n \rightarrow \pi^*$ transition.

Let us now consider some aspects concerning the relationship between chemical structure and light absorption of organic compounds. We will confine ourselves to a few general remarks. For a more detailed discussion of this topic, we refer to the literature (Silverstein et al., 1991).

We start out by looking at light absorption by chromophores that consist of a series of conjugated double bonds. Such chromophores are not very frequently encountered in xenobiotic organic compounds, but they play an important role in natural materials like pigments present in photosynthetic cells (e.g., carotenoid pigments, porphyrins). In straight-chain polyenes, each additional conjugated double bond shifts the absorption maximum of the lowest energy (i.e., highest wavelength) $\pi \rightarrow \pi^*$ transition by about 30 nm to higher wavelengths (a so-called bathochromic shift). This is a general phenomenon, and it may be stated that, in general, the more conjugation in a

Table 15.2 Correlation Between Wavelength of Absorbed Radiation by a Given Object (e.g., an Aqueous Solution) and Observed Color of the Object When Exposed to White Light [a]

Absorbed Light		Observed Color
Wavelength (nm)	Corresponding Color	
400	Violet	Yellow-green
425	Indigo blue	Yellow
450	Blue	Orange
490	Blue-green	Red
510	Green	Purple
530	Yellow-green	Violet
550	Yellow	Indigo blue
590	Orange	Blue
640	Red	Blue-green
730	Purple	Green

[a] From Pretsch et al. (1983).

molecule, the more the absorption is displaced toward higher wavelengths (and therefore reflecting lower required light energies). This may also be seen when comparing the absorption spectra of benzene (Fig. 15.2a), styrene (Fig. 15.2b), and stilbene (Fig. 15.2c), or the spectra of a series of polycyclic aromatic hydrocarbons exhibiting different numbers of rings (Fig. 15.3). Note that in the case of polycyclic aromatic compounds, not only the number of rings but also the way in which the rings are fused together determines the absorption spectrum. For example, large differences exist between the spectra of anthracene (Fig. 15.3b) and phenanthrene (Fig. 15.3c), or between 2,3-benzanthracene (Fig. 15.3d) and 1,2-benzanthracene (Fig. 15.3e).

Comparison of the absorption spectra of stilbene (Fig. 15.2c) and azobenzene (Fig. 15.2d) shows another interesting feature. The replacement of the two double-bonded carbon atoms by two nitrogen atoms leads to additional, intensive $n \rightarrow \pi^*$ transitions. In this case, the additional absorptions lay in the visible wavelength range (i.e., above 400 nm, see Table 15.2). Substituted azobenzenes are widely used dyes, and it has been recognized for quite some time that large amounts of such compounds enter the environment (Weber and Wolfe, 1987).

Quinoid-type is another important group of chromophores that absorb light over a wide wavelength range that includes visible light (Fig. 15.4). Quinoid-type chromophores are important constituents of naturally occurring organic material (e.g., humic and fulvic acids), and they are partly responsible for the yellow color of natural waters containing high concentrations of dissolved organic matter. In Chapter 14 we discussed the role of quinoid-type compounds in abiotic reduction processes of organic pollutants. From a photochemical point of view, quinoid compounds are

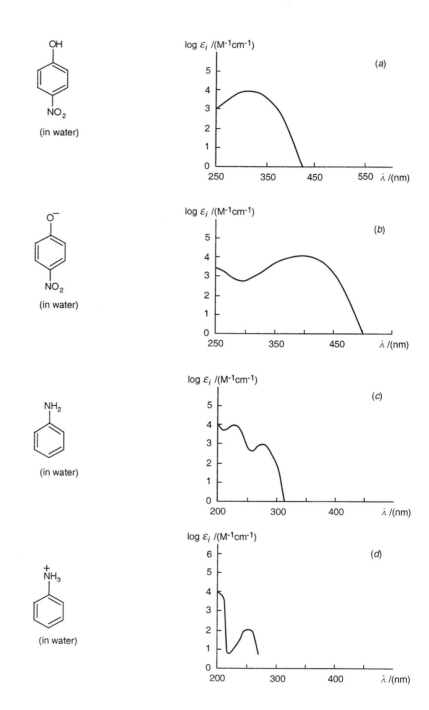

Figure 15.5 Electronic absorption spectra of (a) 4-nitrophenol, (b) 4-nitrophenolate, (c) aniline, and (d) anilinium ion (data from Pretsch et al., 2000).

interesting because they may act as sensitizers for indirect photolytic processes (see Chapter 16).

We conclude our short discussion of relationships between chemical structure and light absorbance by considering some cases in which an acid or base function forms part of a chromophore. Important examples of compounds exhibiting such chromophores are phenols and anilines. As is evident from the spectra shown in Fig. 15.5, deprotonation of a phenolic group results in a substantial bathochromic shift

(shift to longer wavelengths), which is due mostly to delocalization of the negative charge (see Chapter 8). Consequently, depending on the pK_a of a given phenol, light absorption by the phenolic species may vary significantly over the ambient pH range. In the case of aromatic amines (Fig. 15.5c,d), protonation of the amino group results in a so-called hypsochromic shift (shift to shorter wavelengths), because, as a consequence of protonation, the chromophore is altered in that the nitrogen atoms no longer possess nonbonded electrons that may delocalize into the aromatic system. Since protonation of aromatic amines occurs only at relatively low pH values (pH < ~ 5, see Chapter 8), this effect is only important in acidic waters (e.g., in an acidic rain droplet).

In conclusion, from an environmental photochemistry point of view, the most important chromophores present in organic compounds consist of conjugated π- electron systems that may or may not interact with the nonbonded electrons of heteroatoms. In addition, there are certain cases in which xenobiotic organic compounds that themselves do not absorb light above 300 nm undergo *charge transfer transitions* when complexed to a transition metal. A prominent example is the iron(III)-EDTA complex that absorbs light above 300 nm. As a consequence, EDTA, a widely used complexing agent that absorbs no light above 250 nm and that is very resistant to microbial and chemical degradation, may undergo direct photolytic transformation in surface waters, provided that enough iron(III) is available (Frank and Rau, 1990; Kari et al., 1995; Kari and Giger, 1995). Finally, compounds that have two or more noninteracting chromophores exhibit an absorption spectrum corresponding to the superposition of the spectra of the individual chromophores.

The Fate of Excited Chemical Species: Quantum Yields

When a chemical species has been promoted to an excited state, it does not remain there for long. There are various physical or chemical processes that the excited species may undergo. Fig. 15.6 summarizes the most important reaction pathways. As indicated, there are several *physical processes* by which an excited species may return to the ground state; that is, it is not structurally altered by these processes. For example, a species in the first excited state (the state it is commonly promoted to as a consequence of light absorption) may convert to a high vibrational level of the ground state, and then cascade down through the vibrational levels of the ground state by giving off its energy in small increments of heat to the environment. This process is referred to as *internal conversion*. Alternatively, an excited molecule may directly, or after undergoing some change to another excited state (by so-called *intersystem crossing),* drop to some low vibrational level of the ground state all at once by giving off the energy in the form of light. These luminescent processes are called fluorescence and phosphorescence, respectively. Finally, an excited species may transfer its excess energy to another molecule in the environment in a process called *photosensitization*. The excited species thus drops to its ground state while the other molecule becomes excited. Compounds that, after light absorption, efficiently transfer their energy to other chemical species are referred to as *photosensitizers*. The chemical species that efficiently accept the electronic energy are called *acceptors* or *quenchers*. We will come back to photosensitized processes later when discussing indirect photolysis of organic pollutants (Chapter 16). A more detailed treatment of the various physical processes of excited species is given by Roof (1982) and by

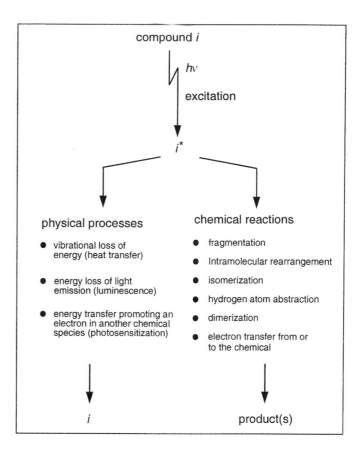

Figure 15.6 Physical processes and chemical reactions of a photochemically excited organic species.

March (1992). An extensive discussion may be found in an appropriate textbook (e.g., Calvert and Pitts, 1967; Turro, 1991).

In addition to the physical processes mentioned above, there are a variety of *chemical reactions* that an excited species may undergo (Fig. 15.6). These reactions are of interest when considering direct photolysis of organic pollutants, because only chemical reactions lead to a transformation and thus to a removal of the compound from a given system. Note that the chemical processes indicated in Fig. 15.6 represent primary steps in the photolytic transformation of a given compound, and that the products of these primary steps may further react by either photochemical, chemical, or biological processes. Consequently, it can be very difficult to identify and quantify all photochemical transformation products, particularly in natural waters or in soils where a variety of possible reactants are present. Some examples of photochemical transformations are given in Fig. 15.7. For more examples and detailed discussions we refer to the literature (e.g., Mill and Mabey, 1985; March, 1992; Larson and Weber, 1994; Boule, 1999). It should be noted that the pathway(s) and the rate(s) of photochemical transformations of excited species in solution commonly depend strongly on the solvent, and in many cases also on the solution composition (e.g., pH, oxygen concentration, ionic strength; Mill and Mabey, 1985). Thus, it is advisable to use data from experiments conducted in solutions with water as the major (> 90%) or even sole solvent and with a solution composition representative of the natural system considered in assessing the photochemical transformation of a

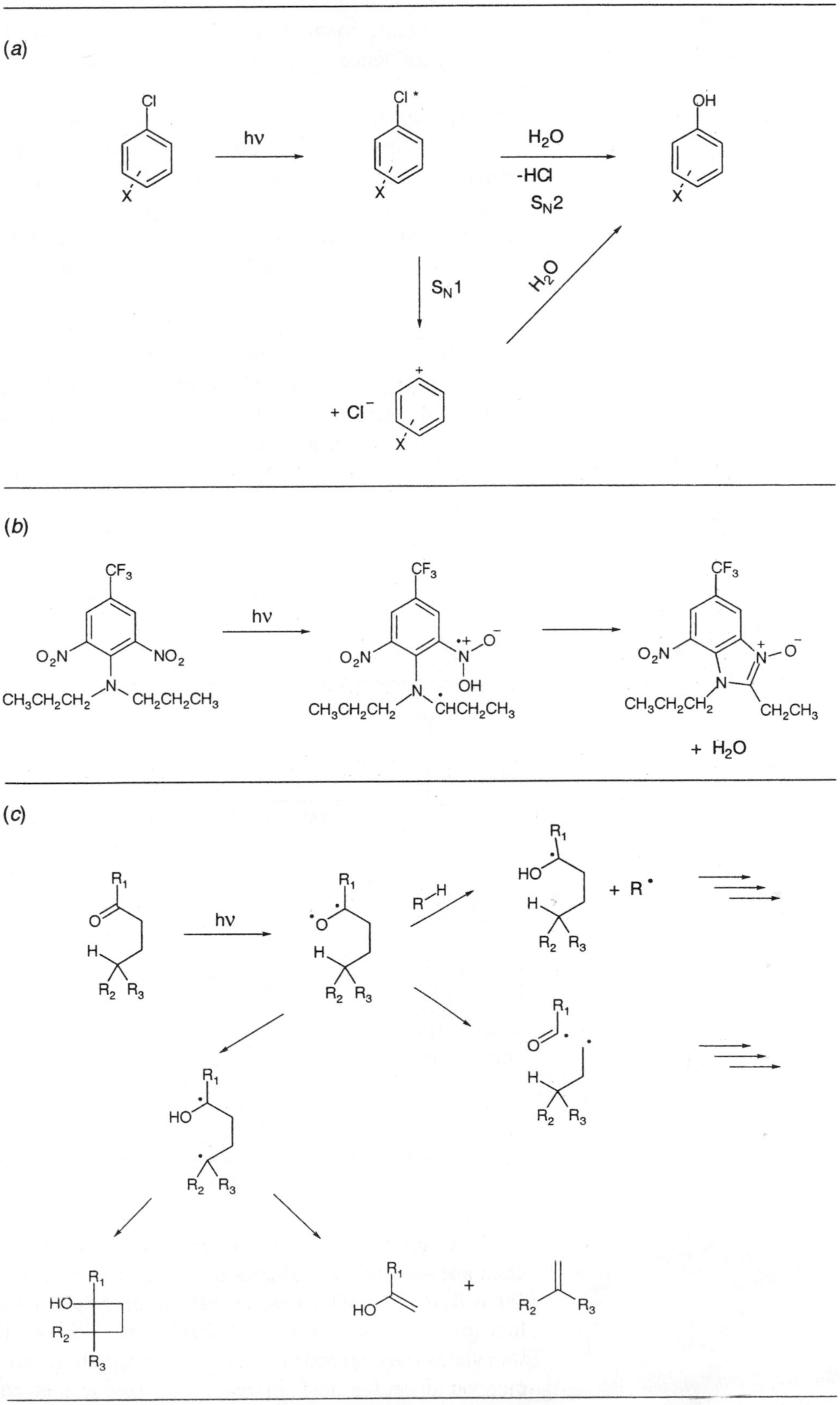

(a)

(b)

(c)

Figure 15.7 Examples of direct photochemical reaction pathways: (a) substituted chlorobenzenes, (b) trifluralin, and (c) a ketone (from Mill and Mabey, 1985).

given compound in the environment. In this context it is necessary to point out that certain organic cosolvents (e.g., acetone) are good sensitizers and may, therefore, strongly influence the photolytic half-life of the compound.

Finally, compared to the chemical reactions discussed in the previous chapter, photochemical transformations of organic compounds usually exhibit a much weaker temperature dependence. Reactions of excited species in aqueous solutions have activation energies of between 10 and 30 kJ·mol^{-1} (Mill and Mabey, 1985). Hence, a 10°C increase (decrease) in temperature accelerates (slows down) a reaction only by a factor of between 1.15 and 1.5 (see Table 3.5).

As we have seen, an excited organic molecule may undergo several physical and chemical processes. The relative importance of the various processes depends, of course, on the structure of the compound and on its environment (e.g., type of solvent, presence of solutes). For each individual process j, we may, for a given environment, define a *quantum yield* $\Phi_{ij}(\lambda)$ which denotes the fraction of the excited molecules of a given compound i that react by that particular physical or chemical pathway:

$$\Phi_{ij}(\lambda) = \frac{\text{number of molecules } i \text{ reacting by pathway } j}{\substack{\text{total number of molecules } i \text{ excited by absorption} \\ \text{of radiation of wavelength } \lambda}} \qquad (15\text{-}7)$$

Since the absorption of light by an organic molecule is, in general, a one-quantum process, we may also write Eq. 15-7 as:

$$\Phi_{ij}(\lambda) = \frac{\text{number of molecules } i \text{ reacting by pathway } j}{\substack{\text{total number of photons (of wavelength } \lambda \text{) absorbed} \\ \text{by the system owing to the presence of the compound } i}} \qquad (15\text{-}8)$$

From an environmental chemist's point of view, it is often not necessary to determine all the individual quantum yields for each reaction pathway (which is, in general, a very difficult and time-consuming task). Rather we derive a lumped quantum yield which encompasses all reactions that alter the structure of the component. This lumped parameter is commonly referred to as *reaction quantum yield* and is denoted as $\Phi_{ir}(\lambda)$:

$$\Phi_{ir}(\lambda) = \frac{\text{total number (i.e., moles) of molecules } i \text{ transformed}}{\substack{\text{total number (i.e., moles) of photons (of wavelength } \lambda \text{) absorbed} \\ \text{by the system due to the presence of the compound } i}} \qquad (15\text{-}9)$$

Unfortunately, there are no simple rules to predict reaction quantum yields from chemical structure, and, therefore, $\Phi_{ir}(\lambda)$ values have to be determined experimentally. We will address such experimental approaches in Section 15.4, and confine ourselves here to a few general remarks. First, we should note that, in principle, reaction quantum yields may exceed unity in cases in which the absorption of a photon by a given compound causes a chain reaction to occur that consumes additional compound molecules. Such cases are, however, very unlikely to happen with organic pollutants in natural waters, mainly because of the rather low pollutant concentrations, and because of the presence of other water constituents that may inhibit chain reactions (Roof, 1982). Consequently, in the discussions that follow we always assume maximum reaction quantum yields of 1.

A second aspect that needs to be addressed is the wavelength dependence of Φ_{ir}. Although vapor-phase reaction quantum yields differ considerably between different wavelengths, they are in many cases approximately wavelength independent (at least over the wavelength range of a given absorption band, corresponding to one mode of excitation) for reactions of organic pollutants in aqueous solutions (Zepp, 1982). Hence, reaction quantum yields determined at a given wavelength (preferably at a wavelength at or near the maximum specific light absorption rate of the compound, see below) may be used to estimate the overall transformation rate of a given compound. Note, however, that if a compound absorbs light over a broad wavelength range exhibiting several maxima of light absorption (e.g., azo dyes), quantum yields may have to be determined for various wavelengths (e.g., Haag and Mill 1987).

15.3   Light Absorption by Organic Compounds in Natural Waters

Light and Light Attenuation in Natural Water Bodies

When dealing with the exposure of a natural water body to sunlight, unlike the situation encountered in a spectrophotometer, one cannot consider the radiation to enter the water perpendicular to the surface as a collimated beam. Sunlight at the surface of the earth consists of direct and scattered light (the latter is commonly referred to as sky radiation) entering a water body at various angles (Fig. 15.8). The solar spectrum at a given point at the surface of the earth depends on many factors including the geographic location (latitude, altitude), season, time of day, weather conditions, air pollution above the region considered, and so on (Finlayson-Pitts and Pitts, 1986). In our discussion we address some of the approaches taken for either calculating or measuring light intensities at the surface or in the water column of natural waters. For a more detailed treatment of this topic we refer, however, to the literature (e.g., Smith and Tyler, 1976; Zepp and Cline, 1977; Zepp, 1980; Baker and Smith, 1982; Leifer, 1988).

Let us consider a well-mixed water body of volume V (cm^3) and (horizontal) surface area A (cm^2) that is exposed to sunlight. Recall that we speak of a well-mixed water body when mixing is fast compared to all other processes. This means that the system is homogeneous with respect to all water constituents and properties, including optical properties such as the light attenuation coefficient of the medium. This is true in many cases such as in shallow water bodies or when we are interested in only the surface layer of a given natural water. If we denote the incident light intensity at a given wavelength (λ), which is commonly referred to as *spectral photon fluence rate* (e.g., einstein $\cdot$ cm^{-2} s^{-1}) as $W(\lambda)$, we can express the light intensity at $z_{mix} = V/A$ (the mean depth of the mixed water body in cm) by applying the Lambert-Beer law (see Section 15.2):

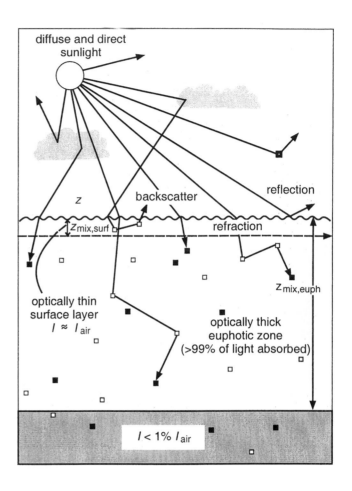

Figure 15.8 Fate of photons in a natural water body (adapted from Zafiriou, 1983). Open squares indicate reflective particles; filled squares are absorptive particles or molecules.

$$W(z_{mix}, \lambda) = W(\lambda) \cdot 10^{-\alpha_D(\lambda) z_{mix}} \qquad (15\text{-}10)$$

where $\alpha_D(\lambda)$ (in cm^{-1}) is commonly referred to as the apparent or *diffuse attenuation coefficient*. The diffuse attenuation coefficient [which often is also denoted as $K_T(\lambda)$] can be determined *in situ* by measuring the light intensities at the surface and at the depth z_{mix} (e.g., Baker and Smith, 1982; Winterle et al., 1987):

$$\alpha_D(\lambda) = \frac{1}{z_{mix}} \log \frac{W(\lambda)}{W(z_{mix}, \lambda)} \qquad (15\text{-}11)$$

Hence, $\alpha_D(\lambda)$ is a measure of how much radiation is absorbed by the mixed water layer *over a vertical distance* z_{mix}. As schematically depicted in Fig. 15.8, light arrives at the surface of the water at various angles and is then refracted at the air–water interface. Less than 10% of the incident light is usually lost due to backscatter and reflection (Zafiriou, 1983). Within the water column, it is (1) scattered by suspended particles, and (2) absorbed by particles and dissolved species, especially natural organic matter. From Fig. 15.8 it can be easily deduced that the average path length of light at any wavelength λ will be larger than z_{mix}. For a given well-mixed water body, we can define a distribution function, $D(\lambda)$, as the ratio of the average light path length $l(\lambda)$ and z_{mix}:

$$D(\lambda) = \frac{l(\lambda)}{z_{mix}} \qquad (15\text{-}12)$$

Recall that the beam attenuation coefficient $\alpha(\lambda)$ (Eq. 15-4) of a given solution is a measure of the attenuation of a collimated beam entering the solution perpendicular to the surface (e.g., to the surface of a cuvette in a spectrophotometer). In such a case, the path length of the light, $l(\lambda)$, is equal to the inside width of the cuvette (which we could also denote as z_{mix}), provided that no significant scattering occurs and thus $D(\lambda)$ is equal to 1. Scattering is mainly due to particles present in the water. For our discussion and calculations, we initially assume a situation in which particles play a minor role, such as encountered in nonturbid waters. Later we address the impact of particles on light attenuation in a natural water body, as well as the effect of particles on photolytic transformation rates of organic pollutants. For a nonturbid water, we may determine $\alpha(\lambda)$ in a spectrophotometer, and use this entity to estimate the diffuse attenuation coefficient $\alpha_D(\lambda)$ by multiplication with $D(\lambda)$:

$$\alpha_D(\lambda) = D(\lambda)\alpha(\lambda) \qquad (15\text{-}13)$$

Hence, $\alpha(\lambda)$ represents the attenuation coefficient of the water per unit pathlength. Application of Eq. 15-13 requires, however, that we can obtain a good estimate for $D(\lambda)$, which is not easily done, especially when dealing with deeper water columns and for very turbid waters. For shallow depths (e.g., the top 50 cm of a nonturbid natural water body), $D(\lambda)$ is primarily determined by the ratio of direct and sky radiation, and by the angle of refraction of direct radiation. For these cases, $D(\lambda)$ may be calculated by computer programs (see Zepp, 1980 and references cited therein). Estimated values of $D(\lambda)$ for near-surface uv and blue light (450 nm) range between 1.05 and 1.30, depending on the solar zenith angle (Zepp, 1980). In very turbid waters where light scattering by particles is also significant, $D(\lambda)$ values of up to 2.0 have been determined (Miller and Zepp, 1979a).

Let us now return to our well-mixed water body and ask how much light *of a given wavelength* λ is absorbed by the water column per unit surface area and time. We may calculate this rate of light absorption by simply calculating the difference between the incident light intensity and the light intensity at depth z_{mix} (Eq. 15-10):

$$\begin{aligned}\text{Rate of light absorption by} \\ \text{the water body } \textit{per unit surface area} &= W(\lambda) - W(z_{mix}, \lambda) \\ &= W(\lambda)[1 - 10^{-\alpha_D(\lambda)z_{mix}}]\end{aligned} \qquad (15\text{-}14)$$

The average rate of light absorption *per unit volume* is then obtained by multiplying Eq. 15-14 by the total irradiated surface area (yielding the total number of photons absorbed per unit time by the whole water body), and dividing by the total volume:

$$\begin{aligned}\text{Rate of light absorption by} \\ \text{the water body } \textit{per unit volume} &= W(\lambda)[1 - 10^{-\alpha_D(\lambda)z_{mix}}]\frac{A}{V} \\ &= \frac{W(\lambda)[1 - 10^{-\alpha_D(\lambda)z_{mix}}]}{z_{mix}}\end{aligned} \qquad (15\text{-}15)$$

Specific Rate of Light Absorption by an Organic Pollutant

If we now add a pollutant exhibiting a molar extinction coefficient $\varepsilon_i(\lambda)$ to a given water body, the attenuation coefficient $\alpha(\lambda)$ [not $\alpha_D(\lambda)$!] is altered to $\alpha(\lambda) + \varepsilon_i(\lambda)C_i$ where C_i is the aqueous concentration of the pollutant in dissolved form in moles per liter. In most cases, however, the pollutant concentration in a natural water will be low, and light absorption by the pollutant will be small as compared to the light absorption by all other chromophores present. Consequently, the rate of sunlight absorption by the water body (Eq. 15-15) will essentially be unchanged. The (small) fraction of light $F_i(\lambda)$ absorbed by the pollutant i is given by:

$$F_i(\lambda) = \frac{\varepsilon_i(\lambda)C_i}{\alpha(\lambda) + \varepsilon_i(\lambda)C_i} \qquad (15\text{-}16)$$

or, since $\varepsilon_i(\lambda)\,C_i << \alpha(\lambda)$:

$$F_i(\lambda) \cong \frac{\varepsilon_i(\lambda)}{\alpha(\lambda)} C_i \qquad (15\text{-}17)$$

Multiplication of the rate of light absorption by the water body per unit volume (Eq. 15-15) with $F_i(\lambda)$ (Eq. 15-17) then yields the entity that we are most interested in, that is, the rate of light absorption by the compound per unit volume denoted as $I_a(\lambda)$:

$$\begin{aligned}
I_a(\lambda) &= \frac{W(\lambda) \cdot \varepsilon_i(\lambda) \cdot [1 - 10^{-\alpha_D(\lambda)z_{mix}}]}{z_{mix} \cdot \alpha(\lambda)} C_i \\
&= k_a(\lambda) \cdot C_i
\end{aligned} \qquad (15\text{-}18)$$

where $k_a(\lambda)$ is commonly referred to as the *specific rate of light absorption* (not a first-order rate constant!) of a given compound in a given system. Hence, $k_a(\lambda)$ expresses the amount (moles) of photons of wavelength λ that are absorbed per unit time per mole of compound i present in the system considered (which in our case is a well-mixed water body of mean depth z_{mix}). Using the units introduced above, that is, $W(\lambda)$ in einstein $\cdot$ cm^{-2} s^{-1} nm^{-1}, $\varepsilon_i(\lambda)$ in M^{-1} cm^{-1}, z_{mix} in cm, and $\alpha(\lambda)$ and $\alpha_D(\lambda)$ in cm^{-1}, $k_a(\lambda)$ has the units:

$$\begin{aligned}
[k_a(\lambda)] &= \frac{\text{einstein cm}^{-2}\,\text{s}^{-1}\,\text{nm}^{-1}\text{cm}^{-1}\,(\text{mol compound } i)^{-1}\text{L}}{\text{cm} \cdot \text{cm}^{-1}} \\
&= \text{einstein cm}^{-3}\,\text{L}\,(\text{mol compound } i)^{-1}\,\text{s}^{-1}\,\text{nm}^{-1} \\
&= 10^3\,\text{einstein}\,(\text{mol compound } i)^{-1}\,\text{s}^{-1}\,\text{nm}^{-1}
\end{aligned} \qquad (15\text{-}19)$$

Thus, we have to express $W(\lambda)$ in *millieinstein* per square centimeter per second to obtain the (desired) units of einstein per mole compound per second for $k_a(\lambda)$. The unit of $I_a(\lambda)$ is then einstein per liter per second.

Near-Surface Specific Rate of Light Absorption of an Organic Pollutant

There are two extreme cases for which Eq. 15-18 can be simplified. The first case applies to the situation where very little light (e.g., less than 5%) is absorbed by the system, that is, the situation in which $\alpha_D(\lambda)z_{mix} < 0.02$. This is true for a very shallow mixed water body (i.e., the top few centimeters of a natural water body, laboratory tubes), or a water body exhibiting a very low $\alpha_D(\lambda)$ value (e.g., distilled water, open ocean water). If $\alpha_D(\lambda)\, z_{mix} < 0.02$, we can make the following approximation (with $10^{-exp} \cong 1 - 2.3 \cdot exp$):

$$1 - 10^{-\alpha_D(\lambda)z_{mix}} \cong 2.3 \cdot \alpha_D(\lambda)z_{mix} \tag{15-20}$$

and:

$$k_a^0(\lambda) = \frac{2.3 \cdot W(\lambda)\alpha_D^0(\lambda)\, \varepsilon_i(\lambda)}{\alpha(\lambda)} \tag{15-21}$$

where $k_a^0(\lambda)$ is the *near-surface* (superscript 0) specific rate of light absorption at wavelength λ.

Since $\alpha_D^0(\lambda) = D^0(\lambda) \cdot \alpha(\lambda)$ (Eq. 15-13), we obtain:

$$k_a^0(\lambda) = 2.3 \cdot W(\lambda)D^0(\lambda)\varepsilon_i(\lambda) \tag{15-22}$$

or $k_a^0(\lambda) = 2.3 \cdot Z(\lambda)\varepsilon_i(\lambda)$ where $Z(\lambda) = W(\lambda)D^0(\lambda)$.

As pointed out earlier, for shallow depths $D^0(\lambda)$ can be approximated by computer calculations. Also, with the same computer programs (SOLAR, or GCSOLAR, which is an updated version of SOLAR; see Zepp and Cline, 1977; Leifer, 1988), $W(\lambda)$ values may be estimated for a given geographic location, season, and time of day. The programs also allow one to take into account the effects of overcast skies. Tables 15.3 and 15.4 give calculated $W(\lambda)$ as well as $Z(\lambda)$ values for midday (noon) at sea level at latitude 40°N for a midseason clear summer and winter day, respectively. In addition, the tables contain 24-hour averaged $Z(\lambda)$ values that take into account diurnal fluctuations in sunlight intensity. This data set is important for comparison of photolysis rates with the rates of other processes that determine the fate of a given compound in a water body. Note again that the $Z(\lambda)$ values given in Tables 15.3 and 15.4 are applicable only to shallow waters (i.e., $z_{mix} \leq \sim 50$ cm).

To allow simple back-of-the-envelope calculations, all $W(\lambda)$ and $Z(\lambda)$ values given in Tables 15.3 and 15.4 are *integrated values* over a specified wavelength range. The indicated wavelength represents the center of a given wavelength range. For example, $W(\text{noon}, 310 \text{ nm})$ is the total number of photons (expressed in millieinsteins) per square centimeter of surface and per second at midday *integrated* over the wavelength range between 308.75 and 311.25 nm. The midday average light intensity within this wavelength range is then given by $W(\text{noon}, 310 \text{ nm})/\Delta\lambda$ with $\Delta\lambda = 2.5$ nm, and, therefore, expressed in millieinstein $\cdot$ cm^{-2} s^{-1} nm^{-1}.

Example Illustrating the Calculation of the Near-Surface Specific Rate of Light Absorption of an Organic Pollutant

Let us use a practical example to get acquainted with some aspects of solar irradiance in natural waters, and to illustrate how to use light data as presented in Tables 15.3 and 15.4 for estimating specific light absorption rates of organic pollutants. We first want to calculate the near-surface specific light absorption rate k_a^0 of *para*-nitro-acetophenone (PNAP) at 40°N latitude (sea level) at noon on a clear midsummer day. Note that for this

para-nitro-acetophenone
(PNAP)

Table 15.3 $W(\lambda)$ and $Z(\lambda)$ Values for a Midsummer Day at 40°N Latitude (Sea Level) Under Clear Skies [a]

λ (Center) (nm)	λ Range ($\Delta\lambda$) (nm)	$W(\text{noon},\lambda)$ [b,d]	$Z(\text{noon},\lambda)$ [b,d]	$Z(24\,\text{h},\lambda)$ [c,d]
		(millieinstein·cm^{-2} s^{-1})		(millieinstein·cm^{-2} d^{-1})
297.5	2.5	1.08(–9)	1.19(–9)	2.68(–5)
300.0	2.5	3.64(–9)	3.99(–9)	1.17(–4)
302.5	2.5	1.10(–8)	1.21(–8)	3.60(–4)
305.0	2.5	2.71(–8)	3.01(–8)	8.47(–4)
307.5	2.5	4.55(–8)	5.06(–8)	1.62(–3)
310.0	2.5	7.38(–8)	8.23(–8)	2.68(–3)
312.5	2.5	1.07(–7)	1.19(–7)	3.94(–3)
315.0	2.5	1.43(–7)	1.60(–7)	5.30(–3)
317.5	2.5	1.71(–7)	1.91(–7)	6.73(–3)
320.0	2.5	2.01(–7)	2.24(–7)	8.12(–3)
323.1	3.75	3.75(–7)	4.18(–7)	1.45(–2)
330.0	10	1.27(–6)	1.41(–6)	5.03(–2)
340.0	10	1.45(–6)	1.60(–6)	6.34(–2)
350.0	10	1.56(–6)	1.71(–6)	7.03(–2)
360.0	10	1.66(–6)	1.83(–6)	7.77(–2)
370.0	10	1.86(–6)	2.03(–6)	8.29(–2)
380.0	10	2.06(–6)	2.24(–6)	8.86(–2)
390.0	10	2.46(–6)	2.68(–6)	8.38(–2)
400.0	10	3.52(–6)	3.84(–6)	1.20(–1)
420.0	30	1.40(–5)	1.51(–5)	4.77(–1)
450.0	30	1.77(–5)	1.90(–5)	6.04(–1)
480.0	30	1.91(–5)	2.04(–5)	6.52(–1)
510.0	30	1.99(–5)	2.12(–5)	6.82(–1)
540.0	30	2.10(–5)	2.22(–5)	7.09(–1)
570.0	30	2.13(–5)	2.25(–5)	7.14(–1)
600.0	30	2.13(–5)	2.24(–5)	7.19(–1)
640.0	50	3.54(–5)	3.72(–5)	1.22

[a] $Z(\lambda) = W(\lambda) \cdot D^0(\lambda)$. Midsummer refers to a solar declination of +20° (late July). [b] Values derived from data of Zepp and Cline (1977). [c] Values derived from Leifer (1988); note: $Z(24\,\text{h},\lambda)=L(\lambda)/2.303$ in this reference. [d] Numbers in parentheses are powers of 10.

geographic location, the result of our calculation will represent the maximum value for PNAP to be expected in a natural water exposed to sunlight. Fig. 15.9 is a graphical representation of Eq. 15-22 derived from the $Z(\lambda)$ [$= W(\lambda)D^0(\lambda)$] and $\varepsilon_i(\lambda)$ values for PNAP given in Table 15.5. As mentioned above, all $W(\lambda)$ and $Z(\lambda)$ values are integrated values over a given wavelength range $\Delta\lambda$, and $\varepsilon_i(\lambda)$ is the average molar extinction coefficient of the compound within this range. Hence, since:

$$k_a^0(\lambda) = 2.3 \cdot Z(\lambda)\varepsilon_i(\lambda) \qquad (15\text{-}23)$$

Table 15.4 $W(\lambda)$ and $Z(\lambda)$ Values for a Midwinter Day at 40°N Latitude (Sea Level) Under Clear Skies [a]

λ (Center) (nm)	λ Range ($\Delta\lambda$) (nm)	$W(noon,\lambda)$ [b,d]	$Z(noon,\lambda)$ [b,d]	$Z(24\ h,\lambda)$ [c,d]
		(millieinstein·cm^{-2} s^{-1})		(millieinstein·cm^{-2} d^{-1})
297.5	2.5	0.00(0)	0.00(0)	0.00(0)
300.0	2.5	1.00(−10)	1.22(−10)	2.22(−6)
302.5	2.5	4.98(−10)	6.11(−10)	1.31(−5)
305.0	2.5	2.31(−9)	2.83(−9)	5.17/ (−5)
307.5	2.5	6.12(−9)	7.46(−9)	1.47(−4)
310.0	2.5	1.16(−8)	1.42(−8)	3.26(−4)
312.5	2.5	2.41(−8)	2.94(−8)	6.04(−4)
315.0	2.5	3.69(−8)	4.50(−8)	9.64(−4)
317.5	2.5	4.92(−8)	6.02(−8)	1.39(−3)
320.0	2.5	6.78(−8)	8.28(−8)	1.84(−3)
323.1	3.75	1.23(−7)	1.51(−7)	3.58(−3)
330.0	10	4.63(−7)	5.68(−7)	1.37(−2)
340.0	10	5.66(−7)	6.97(−7)	1.87(−2)
350.0	10	6.03(−7)	7.46(−7)	2.16(−2)
360.0	10	6.36(−7)	7.95(−7)	2.47(−2)
370.0	10	6.94(−7)	8.63(−7)	2.70(−2)
380.0	10	7.48(−7)	9.34(−7)	2.94(−2)
390.0	10	1.07(−6)	1.33(−6)	2.75(−2)
400.0	10	1.55(−6)	1.93(−6)	3.95(−2)
420.0	30	6.19(−6)	7.75(−6)	1.58(−1)
450.0	30	7.92(−6)	1.00(−5)	2.03(−1)
480.0	30	8.59(−6)	1.09(−5)	2.21(−1)
510.0	30	9.02(−6)	1.15(−5)	2.31(−1)
540.0	30	9.37(−6)	1.20(−5)	2.40(−1)
570.0	30	9.47(−6)	1.21(−5)	2.40(−1)
600.0	30	9.57(−6)	1.23(−5)	2.44(−1)
640.0	50	1.65(−5)	2.12(−5)	4.25(−1)

[a] $Z(\lambda) = W(\lambda) \cdot D^0(\lambda)$. Midwinter refers to a solar declination of −20° (late January). [b] Values derived from data of Zepp and Cline (1977). [c] Values derived from Leifer (1988); note: $Z(24\ h,\lambda) = L(\lambda)/2.303$ in this reference. [d] Numbers in parentheses are powers of 10.

the $k_a^0(\lambda)$ values calculated in Table 15.5 are also integrated values over the indicated $\Delta\lambda$ range. The curves drawn in Fig. 15.9 have been constructed by using the average values within a given $\Delta\lambda$ range, that is, by using $Z(\lambda)/\Delta\lambda$ and $k_a(\lambda)/\Delta\lambda$ values, respectively.

As can be seen from Fig. 15.9 and from Tables 15.3 and 15.4, the solar irradiance at the surface of the earth shows a sharp decrease in the uv-B region (uv-B: 280–320 nm) with virtually no intensity below 290 nm. Hence, only compounds absorbing light above 290 nm undergo direct photolysis. Owing to the sharp decrease in light

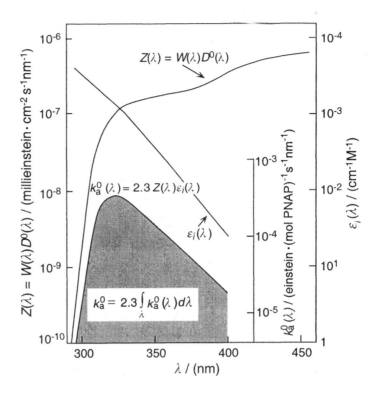

Figure 15.9 Graphical representation of the calculation of the near-surface specific light absorption rate, k_a^0, for *para*-nitro-acetophenone (PNAP) for a clear-sky midday, midsummer at 40°N latitude. The shaded area corresponds to the total rate. Note that the y axes are on logarithmic scales.

intensity in the uv-B region, compounds absorbing light primarily in the uv-B and lower uv-A regions (uv-A: 320–400 nm) show a maximum in $k_a^0(\lambda)$ even if they do not exhibit a maximum in $\varepsilon_i(\lambda)$ in that wavelength region. For example, as Fig. 15.9 and Table 15.5 show, PNAP absorbs sunlight primarily in the wavelength range between 305 and 370 nm, with a maximum $k_a^0(\lambda)$ value between 320 and 330 nm. Thus, when evaluating the direct photolytic transformation of PNAP within the water column of a natural water body, we may need to know $\alpha_D(\lambda)$ values only for a relatively narrow wavelength range (see discussion below).

Integration of Eq. 15-23 over the wavelength range over which the chemical absorbs light (i.e., 295 – 420 nm for PNAP) yields the near-surface specific rate of light absorption by the compound (see hatched area in Fig. 15.9):

$$k_a^0 = \int_\lambda k_a^0(\lambda)d\lambda = 2.3\int_\lambda Z(\lambda)\varepsilon_i(\lambda)d\lambda \qquad (15\text{-}24)$$

When using integrated $Z(\lambda)$ values as given in Tables 15.3 and 15.4, the integral in Eq. 15-24 is approximated by a sum:

$$k_a^0 \approx \sum k_a^0(\lambda)\Delta\lambda = 2.3 \sum Z(\lambda)\cdot\varepsilon_i(\lambda) \qquad (15\text{-}25)$$

For PNAP, the calculation of k_a^0(noon) for a midsummer day at 40°N latitude is shown in Table 15.5. The result k_a^0(noon) $=14.5 \times 10^{-3}$ einstein (mol PNAP)$^{-1}$ s^{-1} indicates that, near the surface of a natural water body, a total of 14.5 millieinstein are absorbed per second per mole of PNAP present in dilute solution. This means that each PNAP molecule is excited once every 70 seconds. The corresponding calculated near-surface specific rate of light absorption averaged over one day is

Table 15.5 Calculation of the Near-Surface Total Specific Light Absorption Rate k_a^0 of p-Nitroacetophenone (PNAP) at 40°N Latitude at Noon on a Clear Midsummer Day

λ (Center) (nm)	λ Range ($\Delta \lambda$) (nm)	Solar Irradiance		PNAP
		Z(noon,λ) [a] (millieinstein· cm^{-2} s^{-1})	$\varepsilon_i(\lambda)$ [b] (cm^{-1} M^{-1})	$k_a^0(\lambda) = 2.3\, Z(\lambda)\varepsilon_i(\lambda)$ [einstein (mol PNAP)$^{-1}$ s^{-1}] $10^3\, k_a^0(\lambda)$
297.5	2.5	1.19(−9)	3790	0.01
300.0	2.5	3.99(−9)	3380	0.03
302.5	2.5	1.21(−8)	3070	0.09
305.0	2.5	3.01(−8)	2810	0.20
307.5	2.5	5.06(−8)	2590	0.30
310.0	2.5	8.23(−8)	2380	0.45
312.5	2.5	1.19(−7)	2180	0.60
315.0	2.5	1.60(−7)	1980	0.73
317.5	2.5	1.91(−7)	1790	0.79
320.0	2.5	2.24(−7)	1610	0.83
323.1	3.75	4.18(−7)	1380	1.33
330.0	10	1.41(−6)	959	3.12
340.0	10	1.60(−6)	561	2.06
350.0	10	1.71(−6)	357	1.42
360.0	10	1.83(−6)	230	0.97
370.0	10	2.03(−6)	140	0.66
380.0	10	2.24(−6)	81	0.41
390.0	10	2.68(−6)	45	0.28
400.0	10	3.84(−6)	23	0.22
420.0	30	1.51(−5)	0	0
450.0	30	1.90(−5)	0	0

$$k_a^0 = \Sigma\, k_a^0(\lambda) = 14.5 \cdot 10^{-3}$$
$$\text{einstein (mol·PNAP)}^{-1}\,\text{s}^{-1}$$

[a] Values from Table 15.3; numbers in parentheses are powers of 10. [b] Values are taken from Leifer (1988).

k_a^0 (24 h) = 532 einstein (mol PNAP)$^{-1}$ d^{-1} (calculation not shown), which is about 40% of the midday value extrapolated to 24 h [i.e., k_a^0 (noon) = 1250 einstein (mol PNAP)$^{-1}$ d^{-1}].

Figure 15.10 shows the variation in the k_a^0 (24 h) values of PNAP as a function of season and decadic latitude in the northern hemisphere. As can be seen, few differences are found in the summer between different latitudes. During the other seasons, however, a significant decrease in k_a^0 is observed with increasing latitude. For example, at 30°N, k_a^0 (24 h) of PNAP is about twice as large in the summer as compared to the winter, while at 60°N, the difference is more than a factor of 20. It should be pointed out that temporal and geographical variations in light intensity are most pronounced in the uv-B and low uv-A regions. Consequently, for compounds such as PNAP that absorb light mostly in

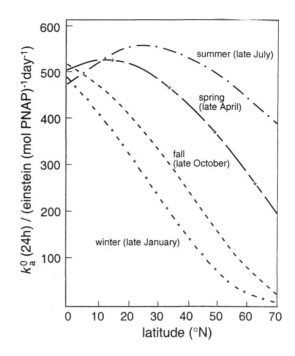

Figure 15.10 Calculated 24 h averaged near-surface specific light absorption rates, $k_a^0(24\,h)$, for PNAP as a function of season and latitude in the northern hemisphere (data from Leifer, 1988).

that wavelength range, the rate of light absorption is very sensitive to diurnal, seasonal, and geographic changes (Zepp and Cline, 1977).

Specific Rate of Light Absorption of an Organic Pollutant in a Well-Mixed Water Body

So far we have considered the rate of light absorption by a pollutant present at low concentration near the surface of a water body, that is, within a zone in which only very little light is absorbed [i.e., $\alpha_D(\lambda)z_{mix} < 0.02$]. The opposite extreme is the situation in which nearly all of the light is absorbed in the mixed water body considered [i.e. $\alpha_D(\lambda)z_{mix} > 2$]. In this case, $1 - 10^{-\alpha_D(\lambda)z_{mix}} \cong 1$ and Eq. 15-18 simplifies to:

$$I_a^t(\lambda) = \frac{W(\lambda)\varepsilon_i(\lambda)}{z_{mix}\alpha(\lambda)} \cdot C_i \tag{15-26}$$

and:

$$k_a^t(\lambda) = \frac{W(\lambda)\varepsilon_i(\lambda)}{z_{mix}\alpha(\lambda)} \tag{15-27}$$

where the superscript t denotes *total* light absorption rate. Note that Eqs. 15-26 and 15-27 are valid only if $\varepsilon(\lambda)C_i << \alpha(\lambda)$ (i.e., for a dilute solution of the pollutant). In this case, $k_a^t(\lambda)$ is dependent on $\alpha(\lambda)$ and not on $\alpha_D(\lambda)$ [the pathlength of the light has no effect on $k_a(\lambda)$, since all light is absorbed within the mixed water body considered].

In most natural waters, light and particularly uv-light (which is of most importance for photolytic transformations of organic pollutants) is absorbed primarily by organic

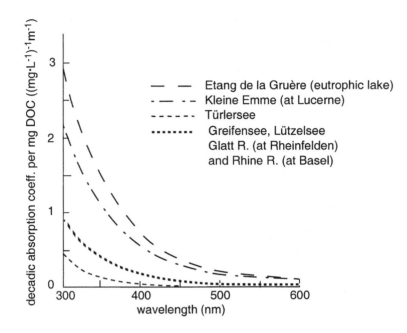

— — Etang de la Gruère (eutrophic lake)
·— ·· Kleine Emme (at Lucerne)
- - - - - Türlersee
········ Greifensee, Lützelsee
Glatt R. (at Rheinfelden)
and Rhine R. (at Basel)

Figure 15.11 Decadic beam attenuation coefficients calculated per milligram of DOC as a function of wavelength for a series of water samples from various Swiss lakes and rivers (data from Haag and Hoigné, 1986).

constituents, especially dissolved organic matter. In Fig. 15.11, α values (i.e., beam attenuation coefficients) determined per milligram per liter of dissolved organic carbon (DOC) for filtered water samples derived from various Swiss lakes and rivers are plotted as a function of wavelength. These examples illustrate that natural organic materials interact with solar light primarily between 300 and 600 nm, that $\alpha(\lambda)$ generally decreases with increasing wavelength (particularly in the uv-B and uv-A region), but that significant differences exist between the specific light attenuation coefficients ($\alpha(\lambda)/[DOC]$) of waters from different origins. These latter findings are not too surprising since, for example, humic and fulvic acids derived from different sources have been postulated to exhibit different types and relative abundance of chromophores (Chapter 9 and Zafiriou et al., 1984).

We may calculate the k_a^t value for a well-mixed water body in which virtually all light is absorbed by:

$$k_a^t = \Sigma k_a^t(\lambda) = \frac{1}{z_{mix}} \Sigma \frac{W(\lambda)\varepsilon_i(\lambda)}{\alpha(\lambda)} \qquad (15\text{-}28)$$

In Table 15.6, k_a^t(24 h) is calculated for PNAP for the well-mixed epilimnion of a small eutrophic lake (our example is Greifensee in Switzerland; $z_{mix} = 5$ m, [DOC] = 4 mg C·L^{-1}; $\alpha(\lambda)$ values are given in Table 15.6) on a clear midsummer day at 47.5°N latitude. The result is k_a^t(24 h) = 22.5 einstein (mol PNAP)$^{-1}$ day^{-1}. This absorption rate implies that each PNAP molecule is excited only about once an hour in this "opaque water case," much less than once a minute in the transparent extreme (see above). Before we can be sure that our assumption that mixing (typical vertical mixing rates in the epilimnion of Swiss lakes are, for example, between 1 and 10 day^{-1}) is fast as compared to the photolytic transformation of PNAP in the epilimnion of this lake is correct, we must discuss quantum yields.

Table 15.6 Calculation of the 24 h Averaged Total Specific Light Absorption Rate k_a^t of p-Nitroacetophenone (PNAP) at ~47.5°N Latitude on a Clear Midsummer Day in the Well-Mixed Epilimnion of Greifensee in Switzerland

λ(Center) (nm)	λ Range ($\Delta\lambda$) (nm)	Intensity of Solar Radiation at the Surface $W(24\,h,\lambda)$ [a] (millieinstein cm^{-2} d^{-1})	Average Beam Attenuation Coefficient $\alpha(\lambda)$ [b] (cm^{-1})	PNAP [c] $\varepsilon_i(\lambda)$ (cm^{-1} M^{-1})	$k_a^t(\lambda) = W(24\,h,\lambda)\cdot\varepsilon_i(\lambda)/z_{mix}\cdot\alpha(\lambda)$ (cm^{-1} M^{-1}) [einstein (mol PNAP)$^{-1}$ d^{-1}]
297.5	2.5	1.12(−5)	0.0430	3790	0.00
300.0	2.5	5.92(−5)	0.0415	3380	0.01
302.5	2.5	2.10(−4)	0.0395	3070	0.03
305.0	2.5	5.49(−4)	0.0375	2810	0.08
307.5	2.5	1.14(−3)	0.0355	2590	0.17
310.0	2.5	1.99(−3)	0.0335	2380	0.28
312.5	2.5	3.06(−3)	0.0320	2180	0.42
315.0	2.5	4.26(−3)	0.0305	1980	0.55
317.5	2.5	5.25(−3)	0.0290	1790	0.68
320.0	2.5	6.75(−3)	0.0275	1610	0.79
323.1	3.75	1.23(−2)	0.0260	1380	1.31
330.0	10	4.34(−2)	0.0220	959	3.86
340	10	5.53(−2)	0.0185	561	3.35
350.0	10	6.20(−2)	0.0150	357	2.95
360.0	10	6.87(−2)	0.0125	230	2.53
370.0	10	7.34(−2)	0.0100	140	2.05
380.0	10	7.86(−2)	0.0083	81	1.53
390.0	10	7.38(−2)	0.0069	45	0.96
400.0	10	1.06(−1)	0.0055	23	0.89
420.0	30	4.23(−1)	0.0042	0	0.00
450.0	30	5.37(−1)	0.0028	0	0.00
458.0	30	5.80(−1)	0.0019	0	0.00
Σ495.5–600.0	114.5	1.85(0)	0.0010	0	0.00
		$\Sigma W(\lambda) = 3.99(0)$ millieinstein·cm^{-2} d^{-1}			$k_a^t = \Sigma\, k_a^t(\lambda) = 22.5$ einstein (mol PNAP)$^{-1}$ d^{-1}

[a] $W(\lambda)$ values estimated from Leifer (1988) for 50°N latidude; values in parentheses indicate powers of 10. [b] Average value for the indicated λ-range; see Fig. 15.9. [c] Well-mixed epilimnion with mean depth $z_{mix} = 500$ cm; $\varepsilon_i(\lambda)$ values taken from Leifer (1988).

Light-Screening Factors

Before we turn to discussing reaction quantum yields, we want to introduce an approximation for calculating $k_a(\lambda)$ values of a given compound in a well-mixed water body in which neither very little light [situation described by the near-surface value $k_a^0(\lambda)$] nor all light [described by $k_a^l(\lambda)$] is absorbed. For this purpose, we introduce a *light-screening factor*, $S(\lambda)$, which is defined as the ratio of the $k_a(\lambda)$ value for a mixed water body of depth z_{mix} (Eq. 15-18) and the near-surface specific rate of light absorption $k_a^0(\lambda)$ (Eq. 15-22); therefore:

$$S(\lambda) = \frac{W(\lambda)\varepsilon_i(\lambda)[1-10^{-\alpha_D(\lambda)z_{mix}}]}{z_{mix}\alpha(\lambda)\cdot 2.3\cdot W(\lambda)D^0(\lambda)\varepsilon_i(\lambda)}$$

(15-29)

$$= \frac{[1-10^{-\alpha_D(\lambda)z_{mix}}]}{2.3\cdot z_{mix}\alpha_D^0(\lambda)}$$

Note that $S(\lambda)$ is a function both of $\alpha_D^0(\lambda)$ (the diffuse attenuation coefficient near the surface) and of $\alpha_D(\lambda)$ (the average diffuse attenuation coefficient for the whole water column of depth z_{mix}). Using $S(\lambda)$, we may express $k_a(\lambda)$ as:

$$k_a(\lambda) = k_a^0(\lambda)\cdot S(\lambda)$$

(15-30)

and approximate k_a by:

$$k_a = \Sigma k_a^0(\lambda)\cdot S(\lambda)$$

(15-31)

For crude estimates of $k_a(\lambda)$, we may make two assumptions that simplify the calculation of $S(\lambda)$. First, we may set $\alpha_D(\lambda) \cong \alpha_D^0(\lambda)$ if the water body considered is not too deep (e.g., only a few meters). Second, for nonturbid waters, we may assume an average $D(\lambda)$ value of $D \cong 1.2$ (see Zepp and Cline, 1977). Then, Eq. 15-29 simplifies to:

$$S(\lambda) = \frac{[1-10^{-(1.2)\alpha(\lambda)z_{mix}}]}{(2.3)(1.2)z_{mix}\alpha(\lambda)}$$

(15-32)

where $\alpha(\lambda)$ is the beam attenuation coefficient that we may easily determine with a spectrophotometer. Finally, if the compound of interest absorbs light only over a relatively narrow wavelength range (e.g., PNAP, Fig. 15.9), one may use a single (average) $\alpha(\lambda)$ [or $\alpha_D(\lambda)$] value for this wavelength range to calculate the effect of light attenuation on the specific rate of light absorption of the compound. For example, if we choose the α value at the wavelength λ_m of the *maximum specific rate of light absorption* [not the maximum $\varepsilon_i(\lambda)$!] of the compound (i.e., 323 nm in the case of PNAP, see Fig. 15.9), k_a may be approximated by:

$$k_a \approx S(\lambda_m)\Sigma k_a^0(\lambda)$$
$$\approx S(\lambda_m)k_a^0$$

(15-33)

For practical applications, Eq. 15-33 is extremely useful, since in many cases experimental data are available only for the near-surface specific rate of light absorption or, even more frequently, for the total near-surface direct photolytic transformation rate of a given pollutant. An example demonstrating the use of screening factors is given in Illustrative Example 15.2.

Illustrative Example 15.2   **Using the Screening Factor $S(\lambda_m)$ to Estimate the Total Specific Light Absorption Rate of PNAP in the Epilimnion of a Lake**

Problem

In Table 15.6, the 24 h total specific light absorption rate, k_a^t (24 h), of PNAP is calculated (estimated) for a clear midsummer day in the well-mixed epilimnion of Greifensee in Switzerland (47°N latitude). For a latitude of 40°N the near-surface total specific light absorption rate k_a^0 (noon) of PNAP is calculated for a clear midsummer day in Table 15.5. Can you use this k_a^0 (noon) value to estimate k_a^t (24 h) of PNAP in Greifensee? If yes, perform the calculation and compare the resulting k_a^t(24 h) value with the k_a^t(24 h) value obtained in Table 15.5. Assume that on a daily basis, k_a^0 (noon) is about 2.5 times larger than k_a^t(24 h) [compare k_a^t (24 h) given in Fig. 15.10 with k_a^0 (noon)·(86400 s·d^{-1})].

Answer

Figure 15.10 shows that for the summer, k_a^0 values are not very different between 40°N and 50°N latitudes. Hence, the k_a^0 values of 1.45×10^{-2} einstein·(mol PNAP)$^{-1}$ s^{-1} given in Table 15.5 can be used to estimate k_a in the epilimnion of Greifensee. Inspection of Table 15.5 also shows that the maximum light absorption rate of PNAP is at about 330 nm. At this wavelength the average beam attenuation coefficient of Greifensee water is α(330 nm) = 0.022 cm^{-1} (Table 15.6). Insertion of this value together with z_{mix} = 500 cm into Eq. 15-32 yields:

$$S(330\text{nm}) = \frac{1}{(2.3)(1.2)(500 \text{ cm})(0.022 \text{ cm}^{-1})} = 0.033$$

Hence, k_a(noon) of PNAP in the epilimnion of Greifensee is estimated to be (Eq. 15-33):

$$k_a(\text{noon}) = S(330\text{nm}) \cdot k_a^0(\text{noon}) = (0.033)(1.45 \times 10^{-2} \text{ einstein (mol PNAP)}^{-1}\text{s}^{-1})$$
$$= 4.79 \times 10^{-4} \text{ einstein (mol PNAP)}^{-1}\text{s}^{-1}$$
$$= 41.4 \text{ einstein (mol PNAP)}^{-1}\text{d}^{-1}$$

When assuming that k_a(24 h) = 0.4 k_a(noon), one obtains a k_a(24 h) value of *16.6* einstein (mol PNAP)$^{-1}$ d^{-1}, which, considering all approximations made, is in reasonable agreement with the value estimated in Table 15.6 [i.e., 22.5 einstein (mol PNAP)$^{-1}$ d^{-1}'].

15.4 Quantum Yield and Rate of Direct Photolysis

First-Order Rate Constant for Quantification of Direct Photolysis

In Section 15.2 we defined a reaction quantum yield, $\Phi_{ir}(\lambda)$, describing the total number of compound molecules (e.g., moles compound) transformed by a chemical reaction per total number of photons (e.g., einsteins) absorbed by a given system resulting from the presence of the compound (Eq. 15-9). In Eq. 15-18, we denoted the rate of light absorption of wavelength λ by the pollutant per unit volume (e.g., einstein per liter per second) as $I_a(\lambda)$. It is now easy to see that the product of these two entities describes the number of compound molecules transformed per unit volume per time [e.g., (mol compound i) per liter per second]. This is also equal to the concentration change per unit time in a given system, or the rate of transformation of the pollutant:

$$
\begin{aligned}
Rate \text{ of direct } photolysis & \\
\text{(subscript "p") at wavelength } \lambda = -\left(\frac{dC_i}{dt}\right)_{p,\lambda} & \\
= \Phi_{ir}(\lambda)I_a(\lambda) & \\
= \Phi_{ir}(\lambda)k_a(\lambda)C_i & \\
= k_p(\lambda)C_i &
\end{aligned}
\tag{15-34}
$$

where $k_p(\lambda) = \Phi_{ir}(\lambda)k_a(\lambda)$ is the direct photolysis first-order rate constant at wavelength λ. It has the unit of time^{-1} (e.g., per second or per day). We recall that we may express direct photolysis as a first-order process only if $\varepsilon_i(\lambda)C_i \ll \alpha(\lambda)$. The (total) rate of direct photolytic transformation of a given pollutant in a well-mixed water body is then given by:

$$
\text{(total) } rate = -\left(\frac{dC_i}{dt}\right)_p = \left[\sum k_p(\lambda)\right]C_i
\tag{15-35}
$$

If we assume *that the quantum yield is independent of wavelength*, we simply multiply the total specific light absorption rate of the compound in a well-mixed water body by Φ_{ir} to obtain k_p:

$$
k_p = \Phi_{ir} \cdot k_a
\tag{15-36}
$$

Similarly, the near-surface first-order rate constant k_p^0 for direct photolysis is given by:

$$
k_p^0 = \Phi_{ir} \cdot k_a^0
\tag{15-37}
$$

Finally, using Eq. 15-33, k_p and k_p^0 may be related by:

$$
k_p = S(\lambda_m) k_p^0
\tag{15-38}
$$

As indicated by Eq. 15-36, to estimate the rate of direct photolysis of a pollutant in a given system, one needs to know the k_a value as well as the reaction quantum yield for the compound considered. As we have extensively discussed, k_a values may be estimated with the help of spreadsheet calculations or computer programs. However,

Table 15.7 Direct Photolysis Reaction Quantum Yields of Some Selected Organic Pollutants in Aqueous Solution

Compound	Wavelength [a] (nm)	Solvent Other Than Water [b] (pH)	Reaction Quantum Yield (Φ_{ir}) [c]	Ref [d]
Naphthalene	313		1.5×10^{-2}	1
1-Methylnaphthalene	313		1.8×10^{-2}	1
2-Methylnaphthalene	313		5.3×10^{-3}	1
Phenanthrene	313		1.0×10^{-2}	1
Anthracene	313		3.0×10^{-3}	1
Pyrene	313,366		2.1×10^{-3}	1
1,2-Benzanthracene	313,366,sun	1 % AN	3.2×10^{-3}	2
Benzo(a)pyrene	313,366,sun	1–20 % AN	8.9×10^{-4}	2
3,4-Dichloroaniline	313,polychr.	pH 7–10	4.4×10^{-2}	3
3,5-Dichloroaniline	313,sun	pH 4–10	5.2×10^{-2}	3
Pentachlorophenolate	314,polychr.	pH 8–10	1.3×10^{-2}	3
4-Nitrophenol	313,polychr.	pH 2–4	1.1×10^{-4}	3
4-Nitrophenolate	365,polychr.	pH 9–10	8.1×10^{-6}	3
Nitrobenzene	313		2.9×10^{-5}	4
4-Nitrotoluene	366		5.2×10^{-3}	4
2,4-Dinitrotoluene	313		2.0×10^{-3}	4
2,4,6-Trinitrotoluene	313,366,sun		2.1×10^{-3}	5

[a] Wavelength or wavelength range at which Φ_{ir} has been determined, sun = sunlight, polychr. = polychromatic artificial light (> 290 nm). [b] AN = acetonitrile. [c] Moles of compound converted per moles of photons absorbed. [d] (1) Zepp and Schlotzhauer (1979); (2) Mill et al. (1981); (3) Lemaire et al. (1985); (4) Simmons and Zepp (pers. comm.); (5) Mabey et al. (1983).

there are presently no rules for predicting Φ_{ir} values from chemical structure. Thus, quantum yields have to be determined experimentally. Below we shall briefly outline the most widely used procedures, but refer to the literature for a more detailed discussion (Zepp, 1978; Roof, 1982; Zepp, 1982; Mill and Mabey, 1985; Leifer, 1988).

In Table 15.7 the reaction quantum yields are given for some selected organic pollutants. As can be seen, reaction quantum yields vary over many orders of magnitude, with some compounds exhibiting very small Φ_{ir} values. However, since the reaction rate is dependent on both k_a and Φ_{ir} (Eq. 15-34), a low reaction quantum yield does not necessarily mean that direct photolysis is not important for that compound. For example, the near-surface direct photolytic half-life of 4-nitrophenolate ($\Phi_{ir} = 8.1 \times 10^{-6}$) at 40°N latitude is estimated to be in the order of only a few hours, similar to the half-life of the neutral 4-nitrophenol, which exhibits a Φ_{ir} more than 10 times larger (Lemaire et al., 1985). The reason for the similar half-lives is the much higher rate of light absorption of 4-nitrophenolate as compared to the neutral species, 4-nitrophenol (compare uv/vis spectra in Fig. 15.5 and Illustrative Example 15.3). As a second example, comparison of the near-surface photolytic half-lives (summer, 40°N latitude) of the two isomers phenanthrene ($t_{1/2} \sim 60$ days) and anthracene ($t_{1/2} \sim 5$ days, Zepp and Schlotzhauer, 1979) shows that the smaller Φ_{ir} value of anthracene (as compared to phenanthrene) is by far outweighed by its much higher k_a value (compare uv/vis spectra in Figs. 15.3b and c). These two examples illustrate again that both k_a and Φ_{ir} are important factors in determining the rate of direct photolysis in natural waters. For more examples of quantum yields as well as a discussion of product formation of direct photolysis of some important compound classes we refer to the literature [Kramer et al., 1996 (fluorescent whitening agents); Pagni and Sigman, 1999 (PAHs and PCBs); Richard and Grabner, 1999 (phenols); Méallier, 1999 (some pesticides)].

Illustrative Example 15.3

Estimating the Photolysis Half-Life of a Weak Organic Acid in the Well-Mixed Epilimnion of a Lake

Problem

Estimate the 24 h averaged direct photolysis half-life of 4-nitrophenol (4NP) (a) near the surface, and (b) in the well-mixed epilimnion of Greifensee (pH = 7.5; $z_{mix} = 500$ cm; $\alpha(\lambda)$ values given in Table 15.6) on a clear midsummer day. Furthermore, a friendly colleague has already calculated the 24 h averaged near-surface total specific light absorption rates of the nondissociated (HA) and dissociated (A$^-$) species:

$$k_a^0(24\,\text{h}, \text{HA}) = 4.5 \times 10^3 \text{ einstein} \cdot (\text{mol HA})^{-1} \text{d}^{-1} \quad (\lambda_m \cong 330 \text{ nm})$$
$$k_a^0(24\,\text{h}, \text{A}^-) = 3.2 \times 10^4 \text{ einstein} \cdot (\text{mol A}^-)^{-1} \text{d}^{-1} \quad (\lambda_m \cong 400 \text{ nm})$$

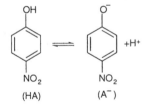

(HA) (A$^-$)

4-nitrophenol

Answer (a)

4NP is a weak acid with a pK_{ia} of 7.11 (see Chapter 8 and Appendix C). The near-surface rate of photolysis of 4NP (HA + A$^-$) is given by:

$$-\frac{d[\text{4NP}]_{tot}}{dt} = k_p^0(24\,\text{h, 4NP}) \cdot [\text{4NP}]_{tot} \quad (1)$$

$k_p^0(24\,\text{h, 4NP})$ is a "lumped" rate constant that can be expressed as the sum of the k_p^0 values for the nondissociated and dissociated species, respectively:

$$k_p^0(24\,\text{h, 4NP}) = (\alpha_{ia}) \cdot k_p^0(24\,\text{h, HA}) + (1 - \alpha_{ia}) \cdot k_p^0(24\,\text{h, A}^-) \quad (2)$$

where $\alpha_{ia} = (1 + 10^{(\text{pH} - pK_{ia})})^{-1}$ (Eq. 8-22). Insert the quantum yields given in Table 15.7 for 4-nitrophenol (1.1×10^{-4}) and 4-nitrophenolate (8.1×10^{-6}) together with the k_a^0 values given above into Eq. 15-37 to get:

$$k_p^0(24\,\text{h, HA}) = (1.1 \times 10^{-4} \text{ (mol HA) einstein}^{-1})(4.5 \times 10^3 \text{ einstein (mol HA)}^{-1}\text{d}^{-1}$$
$$= 0.5 \text{ d}^{-1}$$
$$k_p^0(24\,\text{h, A}^-) = (8.1 \times 10^{-6})(3.2 \times 10^4) = 0.26 \text{ d}^{-1}$$

Insertion of these values into Eq. 2 together with $\alpha_{ia} = 0.29$ at pH 7.5 yields:

$$k_p^0 (24\,h, 4NP) = (0.29)(0.5\ d^{-1}) + (0.71)(0.26\ d^{-1}) = 0.33\ d^{-1}$$

and:

$$t_{1/2}(24\,h, 4NP) = (\ln 2)/0.33\ d^{-1}) \cong 2\ d$$

Note that at pH 7.5, the direct photolytic transformations of HA and A$^-$, respectively, contribute about equally to the overall near-surface direct photolysis rate of 4NP.

Answer (b)

For the well-mixed epilimnion, in analogy to case (a), the overall rate of direct photolysis of 4-NP can be expressed as:

$$-\frac{d[4NP]_{tot}}{dt} = k_p (24\,h, 4NP) \cdot [4NP]_{tot} \tag{3}$$

where $k_p(24\,h, 4NP)$ is now the depth-averaged first-order photolysis rate constant, and is given by:

$$k_p (24\,h, 4NP) = \alpha_{ia} \cdot k_p (24\,h, HA) + (1 - \alpha_{ia}) k_p (24\,h, A^-) \tag{4}$$

Use the screening factors at maximum light absorption $S(\lambda_m)$ (Eq. 15-38) to estimate the corresponding $k_p(24\,h, HA)$ and $k_p(24\,h, A^-)$ values:

$$k_p (24\,h, HA) = S(330\ nm) \cdot k_p^0 (24\,h, HA) \tag{5}$$

$$k_p (24\,h, A^-) = S(400\ nm) \cdot k_p^0 (24\,h, A^-) \tag{6}$$

Insert the $\alpha(\lambda_m)$ values given in Table 15.6 together with $z_{mix} = 500$ cm into Eq. 15-32 to get:

$$S(330\ nm) = \frac{1}{(2.3)(1.2) \cdot (500\ cm)(0.0185\ cm^{-1})} = 0.039$$

$$S(400\ nm) = \frac{1}{(2.3)(1.2) \cdot (500\ cm)(0.0055\ cm^{-1})} = 0.013$$

Hence, in the epilimnion of Greifensee, for light with $\lambda = 330$ nm and 400 nm, the average photon fluence rate is only about 4 and 13%, respectively, of the near-surface photon fluence rate. Insertion of these values into Eqs. (5) and (6) together with the k_p^0 values calculated above (case a) yields:

$$k_p (24\,h, HA) = (0.039)(0.5\ d^{-1}) = 0.020\ d^{-1}$$

$$k_p (24\,h, A^-) = (0.13)(0.26\ d^{-1}) = 0.034\ d^{-1}$$

and with $\alpha_{ia} = 0.29$ at pH 7.5 (Eq. 4):

$$k_p(24 \text{ h, 4NP}) = (0.29)(0.0185 \text{ d}^{-1}) + (0.71)(0.0338 \text{ d}^{-1})$$

$$= (0.0054 \text{ d}^{-1} + 0.0240 \text{ d}^{-1}) = 0.029 \text{ d}^{-1}$$

and:

$$t_{1/2}(24 \text{ h, 4NP}) = (\ln 2)/0.029 \text{ d}^{-1} = 24 \text{ d}$$

Hence, when considering the whole epilimnion, the direct photolysis half-life of 4-NP is about 10 times longer as compared to the half-life at the surface. Note that in contrast to the near-surface situation, because of the very different screening factors, the reaction of the dissociated species is about four times more important in determining the overall direct photolysis rate of 4NP in the well-mixed epilimnion.

Advanced Topic

Determination of Quantum Yields and Chemical Actinometry

In the most common procedures used to determine reaction quantum yields, an (oxygenated) dilute solution of the compound (preferably in distilled water or distilled water containing a low amount of a polar organic solvent) is irradiated by constant intensity monochromatic radiation in a photochemical apparatus (e.g., optical bench, merry-go-round reactor). In the laboratory, various light sources are available to investigate photolytic processes and to determine quantum yields. The most common lamps include low-, medium-, and high-pressure mercury lamps, xenon lamps, and lasers. These lamps are used in connection with various filter systems to obtain the desired monochromatic or polychromatic light (Calvert and Pitts, 1966; Zepp, 1982; Mill and Mabey, 1985). For determination of quantum yields in the uv-B and uv-A region, two-filter systems (a short description is given by Mill and Mabey, 1985) are widely used in connection with medium- and high-pressure mercury lamps to isolate the 313- and 366-nm bands. Because many important environmental pollutants absorb light primarily in the uv-region, a large number of quantum yields reported in the literature have been determined at 313 and/or 366 nm.

Distilled rather than natural water is often used as the solvent for determination of quantum yields for two major reasons. First, the total absorbance of the solution at the wavelength of irradiation should not exceed 0.02. Second, and more important, the presence of natural water constituents (e.g., humic material, nitrate) could enhance the total photolytic transformation rate by indirect photolytic processes as described in Chapter 16. Zepp and Baughman (1978) have argued that for many chemicals Φ_{ir} obtained in distilled water is nearly the same as that observed in natural waters (at least in uncontaminated freshwaters), because concentrations of natural water constituents that could undergo reactions with or quench photolysis of excited pollutants are generally very low. Furthermore, the effects of molecular oxygen, which may act as a quencher, can also be studied in distilled water.

From measurements of the concentration C_i of the compound i as a function of exposure time, the first-order photolysis rate constant, $k_p(\lambda)$, is then determined by calculating the slope of a plot of $\ln C_i/C_{i0}$ versus time (see Section 12.3). Since the absorbance of the solution is less than 0.02, $k_p(\lambda)$ is given as [Zepp, 1978; see also analogy to Eq. 15-22 with $D(\lambda)=(A/V)\cdot l(\lambda)$]:

$$k_p(\lambda) = 2.3 \cdot W(\lambda)\frac{A}{V} \cdot \varepsilon_i(\lambda) \cdot l(\lambda) \cdot \Phi_r(\lambda) \qquad (15\text{-}39)$$

where $W(\lambda)(A/V)$ is the incident light intensity per unit volume of the cell (e.g., a quartz vessel with total surface A and volume V), and $l(\lambda)$ is the cell pathlength that can be determined experimentally for the selected λ value (Zepp, 1978). Hence, $\Phi_{ir}(\lambda)$ can be calculated by:

$$\Phi_{ir}(\lambda) = \frac{k_p(\lambda)}{2.3 \cdot W(\lambda)(A/V) \cdot \varepsilon_i(\lambda) \cdot l(\lambda)} \qquad (15\text{-}40)$$

provided that the light intensity term $W(\lambda)(A/V)$ is known. This light intensity term may be determined by exposing a *chemical actinometer* to the light source in the same way and at the same time that the compound of interest is exposed. A chemical actinometer is a solution of a photoreactive reference compound (subscript R) that reacts with a well-known reaction quantum yield, $\Phi_{r,R}(\lambda)$, preferably with an approximately similar half-life as the compound for which $\Phi_{ir}(\lambda)$ is to be determined. There are two types of chemical actinometers: (1) concentrated solutions of some chemicals that absorb virtually all of the incident light, and (2) chemical actinometers that only weakly absorb the monochromatic radiation (i.e., for which the absorbance is less than 0.02). In the first case the reaction proceeds by zero-order kinetics and the reaction rate is given by [see Eq. 15-26 with $W(\lambda)/z_{mix} = W(\lambda)\cdot(A/V)$, and $\alpha(\lambda) = \varepsilon_R(\lambda)C_R$]:

$$rate_R = W(\lambda)\frac{A}{V} \cdot \Phi_{r,R}(\lambda) \qquad (15\text{-}41)$$

and so the rate of light-input is given by:

$$W(\lambda)\frac{A}{V} = \frac{rate_R}{\Phi_{r,R}(\lambda)} \qquad (15\text{-}42)$$

This result may be substituted into Eq. 15-40 to calculate $\Phi_{ir}(\lambda)$. Classical actinometers that are used in this way include the potassium ferrioxalate actinometer that can be employed both in the uv and visible spectral region the Reinecke's salt actinometer (visible region), and the *ortho*-nitrobenzaldehyde actinometer (uv region). For further description of these actinometers we refer to the literature (e.g., Leifer, 1988, pp. 148–151).

For the dilute solution actinometer (i.e., absorbance < 0.02), Eq. 15-39 applies also for the description of $k_{p,R}(\lambda)$. From this equation, $W(\lambda)(A/V)\cdot l(\lambda)$ may be determined by:

$$W(\lambda)\frac{A}{V} \cdot l(\lambda) = \frac{k_{p,R}(\lambda)}{2.3\varepsilon_R(\lambda) \cdot \Phi_{r,R}(\lambda)} \qquad (15\text{-}43)$$

Substitution of Eq. 15-43 into Eq. 15-40 then yields the reaction quantum yield of the compound of interest at wavelength λ:

$$\Phi_{ir}(\lambda) = \frac{k_p(\lambda) \cdot \varepsilon_R(\lambda)}{k_{p,R}(\lambda) \cdot \varepsilon_i(\lambda)} \cdot \Phi_{r,R}(\lambda) \qquad (15\text{-}44)$$

To obtain an environmental quantum yield or quantum efficiency, Dulin and Mill (1982) suggested exposing dilute solutions of both the pollutant and the chemical actinometer to sunlight. The quantum efficiency in this way represents an averaged value over the wavelength range over which the compound absorbs sunlight. It can be estimated from the measured first-order rate constants k_p and $k_{p,R}$, and from the ratio of the specific sunlight absorption rates of pollutant and reference compound *calculated* (Eq. 15-25) for the time and locations of the experiments:

$$\Phi_{ir}(\text{sunlight}) = \frac{k_p k_{a,R}^{calc}}{k_{p,R} k_a^{calc}} \cdot \Phi_{r,R}(\text{sunlight}) \qquad (15\text{-}45)$$

Eq. 15-45 assumes that the ratio of the total light absorbed by the pollutant and the chemical actinometer is constant over changes in seasons, latitudes, and sky conditions. The validity of this approach depends, of course, on the reliability of the simulated solar spectral irradiances. Since variations in sunlight intensities as a consequence of weather, diurnal, and/or seasonal changes are most pronounced in the uv-B and low uv-A region, the largest errors arising with this approach can be expected for chemicals that have a maximum specific light absorption rate in this wavelength region (i.e., 290 – 350 nm). Nevertheless this outdoor approach to determine Φ_{ir} of a given compound may be very useful, particularly in cases in which the quantum yield is wavelength-dependent.

In principle, any organic compound [with known $\Phi_{ir}(\lambda)$ or $\Phi_{ir}(\text{sunlight})$] that absorbs light in the appropriate wavelength range, and that exhibits a photolytic half-life similar to that of the compound of interest, could be used as a dilute solution actinometer (Zepp, 1978, 1982). In practice, however, such compounds are often difficult to find. Dulin and Mill (1982) discussed the criteria that need to be fulfilled by a good chemical actinometer, particularly when used for sunlight experiments. They described a binary chemical actinometer approach that is applicable primarily to measure radiation intensities in the uv region. The major advantage of this type of actinometer is that the quantum yield and thus the half-life of the actinometer chemical is adjustable, thus ensuring that both actinometer and pollutant are exposed to the same levels of light. This is particularly important in cases where the compounds are exposed to sunlight over a longer period of time (e.g., several hours to days).

The basic principle of a binary actinometer lies in the bimolecular photoreaction of a photosensitive species (the reference compound R) with a photoinsensitive reactant R_j:

$$R + R_j \xrightarrow{\ hv\ } \text{defined products} \qquad (15\text{-}46)$$

The rate of the reaction, that is, the rate of conversion of R (which is measured) is then given by:

$$rate(\lambda) = -\left(\frac{d[R]}{dt}\right)_{\lambda} = k_{a,R}(\lambda) \cdot \Phi_{r,R}(\lambda)[R]$$

(15-47)

$$\text{with } \Phi_{r,R}(\lambda) = \Phi_{r,R}^{0}(\lambda) + k_j[R_j]$$

$\Phi_{r,R}^{0}(\lambda)$ is the extrapolated quantum yield in the absence of R_j, and k_j is a measure for the yield of the reaction of the excited R with R_j, and has the units of (mol R converted) einstein^{-1} M^{-1}. For practical purposes, R_j should be present in excess concentration (i.e., $[R_j] >> [R]$). Thus, if over a reasonable concentration range of R_j, $\Phi_{r,R}^{0}(\lambda) << k_j[R_j]$, $\Phi_{r,R}(\lambda)$ (and thus the photolytic half-life of R at a given light intensity) can be varied linearly with $[R_j]$. Two useful examples of such binary actinometers are p-nitroanisole(PNA)/pyridine and p-nitroacetophenone (PNAP)/pyridine. In their excited states, both PNA and PNAP undergo a nucleophilic displacement reaction with pyridine:

R$_1$ = –OCH$_3$ (PNA)
R$_1$ = –COCH$_3$ (PNAP)

(15-48)

The reaction follows the kinetics described by Eq. 15-47. Both PNA and PNAP absorb light in the uv region (< 400 nm) and show a constant $\Phi_{ir}(\lambda)$ over this wavelength range.

The PNA/pyridine actinometer is useful for very fast reactions. In sunlight, it can be adjusted to half-lives between a few minutes and about 12 h. The upper time limit is determined by the (very high) k_a^0 value of PNA [~ 5000 einstein (mol PNA)$^{-1}$ d^{-1} for a midsummer day at 40°N latitude], and the (rather small) $\Phi_{r,R}^{0}$ value of 3×10^{-4} (Dulin and Mill, 1982; Leifer, 1988). In comparison, as we discussed earlier, for the same geographic latitude and time, PNAP exhibits a k_a^0 value that is about 10 times smaller [i.e., 532 einstein (mol PNAP)$^{-1}$ day^{-1}], and it has an even smaller quantum yield as compared to PNA ($\Phi_{r,R}^{0} < 10^{-5}$, Dulin and Mill, 1982). Hence, although PNAP absorbs sunlight at an appreciable rate, its (direct) photolytic half-life would be very large in a natural water. In the presence of pyridine in a test vessel, however, the photolytic half-life can be adjusted to range between a few hours and several months (for details see Dulin and Mill, 1982, or Leifer, 1988).

At this point we should note that when extrapolating photolysis rates determined in test vessels to natural water bodies, the geometry of the test vessel must be taken into account. For example, sunlight photolysis rates measured in cylindrical test tubes have been found to be greater by a factor of 1.5–2.2 as compared to the rates determined in flat dishes (Dulin and Mill, 1982; Haag and Hoigné, 1986). The major reason for these findings are lens effects of the curved glass and the fact that the tubes are exposed to light from all sides.

15.5 Effects of Solid Sorbents (Particles, Soil Surfaces) on Direct Photolysis

Effect of Particles in Water

We conclude this chapter by briefly addressing the effects of solid sorbents on direct photolytic transformations of organic pollutants in natural waters. We start out by considering the effect of particles on direct photolysis in surface waters. As we have already mentioned, particles may contribute to the light attenuation in a water body, both by light absorption and light scattering. Depending on which effect is predominant, the rate of direct photolysis of a dissolved species in a given system may be decreased or enhanced. In most cases, a decrease in direct photolysis rates of dissolved organic pollutants is observed with increasing particle concentrations, indicating that light absorption is the more important factor (Miller and Zepp, 1979b). In some clay suspensions, however, Miller and Zepp (1979b) observed an increase in the photolysis rate of a ketone which was attributed to increases in the mean light path length caused by scattering.

A more complicated issue is the photolytic transformation of sorbed compounds. Predictions of direct photolysis rates of organic pollutants sorbed to a solid surface are impeded by the fact that a compound may be shielded from the light (see below). In addition, the uv/vis absorption spectrum of a given compound may be significantly different in the sorbed state as compared to the dissolved state (e.g., Parlar, 1980). Similar effects may be observed when a (hydrophobic) pollutant is associated with dissolved or colloidal natural organic matter. Furthermore, owing to the different molecular environment, sorbed species may exhibit very different reaction quantum yields and photoreaction distributions. For example, kinetic and product studies of photoreactions of some highly hydrophobic nonionic organic chemicals indicated that the compounds were in a microenvironment that was less polar than water and that was a considerably better hydrogen donor (Miller and Zepp, 1979b). This is consistent with our picture that hydrophobic compounds preferably sorb in organic components of natural solids (see Chapter 9).

Direct Photolysis on Soil Surfaces

Photolysis on soil surfaces is an important degradation pathway for a variety of organic compounds including agrochemicals or chemicals introduced to soils by sewage sludge applications. From the results of numerous studies it can be concluded that both direct and indirect photolytic transformation reactions of a given chemical may be quite different on soils compared to homogeneous or heterogeneous aqueous systems (see references cited in Balmer et al., 2000). Despite this fact, and although the evaluation of photodegradation on soil is required for numerous chemicals by registration authorities, systematic investigations of this process are still lacking. Large difficulties are encountered when designing experiments that allow the various factors that determine photochemical degradation on soils to be evaluated.

Presently the most common experimental approach to study photodegradation of organic compounds on soils is to expose a series of thin, spiked soil layers (thickness typically between 0.25 and 2 mm) to a light source. The overall disappearance rate coefficient of the compound, which is generally reported as photodegradation rate coefficient, is then determined by measuring the total loss of compound from the soil layers as a function of time. However, these reported rates are of rather limited value, because they always depend on the layer thickness of the soil and in most cases also on transport kinetics, which should in fact be treated separately. Because light penetration into soils is very limited (i.e., 0.1 to 0.5 mm; Herbert and Miller, 1990) and wavelength dependent, the fraction of total compound actually exposed to light depends on the type of soil, on the thickness of the soil layer, and on the light absorption spectrum of the compound. Thus, the rate of transport (i.e., retarded diffusion, see Chapter 18) of the compound from dark locations to irradiated zones within the soil layer will heavily influence the observed overall elimination rate. Since transport depends on the gas/solid partitioning behavior of the compound, and since sorption is strongly influenced by temperature and humidity (Chapter 11), these parameters also need to be controlled in experiments. For future studies, experimental approaches are needed that allow determination of the actual photolysis rate constants, and that are independent of layer thickness and transport velocity of a compound. Only then can the influence of the surfaces on light absorption rate and quantum yield be evaluated and quantified. A first promising step in this direction has been reported by Balmer et al. (2000).

15.6   Questions and Problems

Questions

Q 15.1

What happens if an organic compound absorbs light in the uv/vis range?

Q 15.2

What is the difference between direct and indirect photolysis?

Q 15.3

Explain in words the Beer-Lambert law.

Q 15.4

How is the decadic molar extinction coefficient, $\varepsilon_i(\lambda)$, of a given compound defined? What is described by the decadic light absorption or light attenuation coefficient, $\alpha(\lambda)$?

Q 15.5

What structural features are required in order to allow an organic compound to absorb sunlight at a significant rate? Give examples of compounds that absorb sunlight (a) only in the uv-range, (b) in both the uv and visible range. What color is a concentrated solution of azobenzene (Fig. 15.2d and Table 15.2)?

Q 15.6

Give examples of the physical and chemical processes that a photochemically excited organic compound may undergo.

Q 15.7

What does the reaction quantum yield, $\Phi_{ir}(\lambda)$, exactly describe? In aqueous solution one usually assumes that Φ_{ir} is independent of λ. Is this assumption always correct? Can you give an example where reaction quantum yields may have to be determined at different wavelengths?

Q 15.8

Why can one not just use the decadic light absorption coefficient, $\alpha(\lambda)$, to describe light attenuation in a given water body? How is the diffuse attenuation coefficient, $\alpha_D(\lambda)$, related to $\alpha(\lambda)$?

Q 15.9

Which water constituents primarily determine the magnitude of the diffuse attenuation coefficient $\alpha_D(\lambda)$?

Q 15.10

Which light penetrates deeper into a natural water body, uv or visible light? Why?

Q 15.11

What does the light-screening factor, $S(\lambda)$, exactly describe? For what can it be used?

Q 15.12

How is the first-order direct photolysis rate constant of a given compound defined? What is(are) the prerequisite(s) that direct photolysis can be expressed by a first-order rate law?

Q 15.13

Enumerate *all* factors that determine the direct photolysis rate of a given compound in surface waters. Explain how you can determine and estimate the various parameters required to calculate the direct photolysis rate constant for a given situation.

Problems

P 15.1 *Estimating Light Penetration into a Natural Water Body*

Consider a well-mixed, nonturbid water body with a dissolved organic carbon concentration (DOC) of 4 mg $C \cdot L^{-1}$. The decadic beam attenuation coefficients, $\alpha(\lambda)$, determined for a water sample at five wavelengths are the following (see other examples given in Fig. 15.11):

λ/nm	300	350	400	450	500
$\alpha(\lambda)$/cm^{-1} [a]	0.042	0.015	0.006	0.003	0.002

[a] Values taken from Table 15.6 for epilimnic water from Greifensee, Switzerland.

Calculate for each wavelength indicated above the thickness of the water layer required to attenuate sunlight by a factor of 2. At what depth is 99% of the incoming light of a given wavelength absorbed by the water body (euphotic zone, see Fig. 15.8)? Assume an average value for the distribution function, $D(\lambda)$, in Eq. 15-12, of 1.2. Comment on the results.

P 15.2 *Is Everything Okay with These Light Penetration Measurements?*

During the summer, somebody measures how much light coming into a small mesotrophic lake is absorbed in the water column as a function of depth. He gets the following results:

Depth (m)	T (°C)	% of Incoming Light Intensity of Wavelength λ		
		300 nm	400 nm	500 nm
0	20.3	100	100	100
1	20.2	10	60	90
2.5	20.3	0.3	30	75
5	16.5	< 0.1	3	40
7.5	10.2		0.3	24
10	8.4		< 0.1	18

Calculate the diffuse light attenuation coefficients, $\alpha_D(\lambda)$, as a function of depth for the three indicated wavelengths. Do the results of these light penetration measurements make sense?

P 15.3 *Estimating the Near-Surface Total Specific Light Absorption Rates of Nitrobenzene*

Estimate the k_a^0 value of nitrobenzene present at low concentration in a natural water body on a clear midsummer day as well as on a clear midwinter day (both at noon and averaged over 24 h) at 40°N latitude and sea level (see Table 15.3 and 15.4). For nitrobenzene you find the following $\varepsilon_i(\lambda)$ values in the literature:

λ/nm	300	310	320	330	340	350	360	370
$\varepsilon_i(\lambda)/(\text{M}^{-1}\text{cm}^{-1})$	2500	1400	800	580	560	280	140	20

Compare the noon values on a daily base with the 24 h averaged values. Comment on the difference between summer and winter. Perform your calculations by choosing appropriate wavelength ranges from the data given in Tables 15.3 and 15.4.

P 15.4 *Photolysis or Hydrolysis? Which Process Is More Important for the Elimination of the Insecticide Carbaryl from a Shallow Water Body?*

Consider a well-mixed oligotrophic shallow water body (z_{mix}) = 1 m; α(300 nm) = 0.2 m^{-1}, α(320 nm) = 0.15 m^{-1}; α(350 nm) = 0.1 m^{-1}; pH = 7.0; T = 15°C) exposed to sunlight on a clear summer day at 40°N latitude. Due to spraying of the insecticide carbaryl in the surroundings, there is a significant input of this compound into the water body. As an employee of the company that manufactures this insecticide, you are asked how persistent this compound is in this water body. Calculate the half-life of carbaryl in the water under the given conditions by assuming that photolysis and abiotic hydrolysis are the dominant elimination mechanisms. In the literature you find the following data for carbaryl (Roof, 1982; Table 13.11):

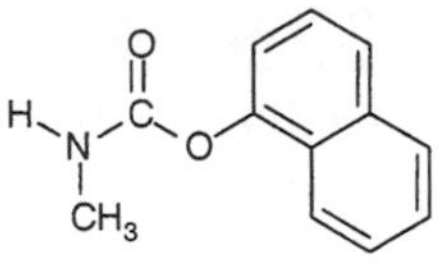

$k_N(25°C)$ $= 9.0 \times 10^{-7} s^{-1}$
$k_B(25°C)$ $= 5.0 \times 10^1 M^{-1} s^{-1}$
$\Phi_r(313 \text{ nm}) = 0.006$ (mol carbaryl) einstein^{-1}

carbaryl

and the decadic molar extinction coefficients:

λ/nm	ε_i /(M^{-1} cm^{-1})	λ/nm	ε_i /(M^{-1} cm^{-1})
297.5	1480	315	261
300	918	317.5	235
302.5	741	320	101
305	532	323	45
307.5	427	330	11
310	356	340	< 1
312.5	288		

P 15.5 *Estimating the Direct Photolysis Half-Life of Pentachlorophenol in Chesapeake Bay (adapted from Roberts, 1995)*

Shown below are some data concerning light absorption in the uv/vis range by pentachlorophenol (PCP). Absorption values for the neutral species are given in ethanol, those for the dissociated species at pH 10 in aqueous solution.

Absorbance (Eq. 15-5, $l = 1$ cm) in Ethanol:

Concentration: 1.8×10^{-4} M

λ/nm	300	305	310	315
A	0.452	0.513	0.305	0.115

Concentration: 1.8×10^{-3} M

λ/nm	320	330
A	0.350	0.040

Absorbance (Eq. 15.5, $l = 1$cm) in Water at pH 10:

Concentration: 1.0×10^{-4} M

λ/nm	300	305	310	315	320
A	0.270	0.325	0.390	0.465	0.510
λ/nm	325	330	335	340	
A	0.440	0.294	0.170	0.084	345
λ/nm	350	360			0.043
A	0.021	0.005			

(a) Calculate the decadic molar extinction coefficients of pentachlorophenol and pentachlorophenolate as a function of λ. Which form do you expect to be more susceptible to direct photolysis, when assuming that both species exhibit the same quantum yield $\Phi_{ir} = 0.013$?

(b) Estimate the half-life (24 h average) with expect to direct photolysis of total pentachlorophenol spilled in the well-mixed Chesapeake Bay in December (pH 6.2, water temperature 15°C, ionic strength 0.03 M, average depth 5m), with $\alpha(\lambda)$ values:

λ/nm	300	310	320	330
$\alpha(\lambda)$/cm^{-1}	0.052	0.042	0.034	0.028
λ/nm	340	350	360	370
$\alpha(\lambda)$/cm^{-1}	0.023	0.019	0.016	0.013

What portion of the uv/vis spectrum is responsible for most of the direct photolysis? Use the data given in Table 15.4 for your estimate.

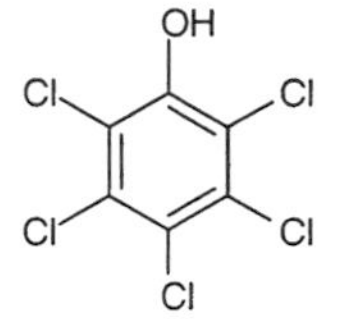

pentachlorophenol
$pK_{ia} = 4.75$

P 15.6   Does Direct Photolysis Affect the Phenanthrene-to-Anthracene Ratio in Aerosol Droplets?

Gschwend and Hites (1981) observed that the two closely related polycyclic aromatic hydrocarbons, phenanthrene and anthracene, occur in a ratio of about 3-to-1 in urban air. In contrast, sedimentary deposits obtained from remote locations (e.g., Adirondack mountain ponds) exhibited phenanthrene-to-anthracene ratios of 15-to-1. You hypothesize that these chemicals are co-carried in aerosol droplets from Midwestern U.S. urban environments via easterly winds to remote locations (like the Adirondacks) where the aerosol particles fall out of the atmosphere and rapidly accumulate in the ponds' sediment beds without any further compositional change (i.e., the phenanthrene-to-anthracene ratio stops changing after the aerosols leave the air). If summertime direct photolysis was responsible for the change in phenanthrene-to-anthracene ratio, estimate how long the aerosols would have to have been in the air. Comment on the assumptions that you make. What are your conclusions?

Chapter 16

INDIRECT PHOTOLYSIS: REACTIONS WITH PHOTOOXIDANTS IN NATURAL WATERS AND IN THE ATMOSPHERE

16.1 **Introduction**

16.2 **Indirect Photolysis in Surface Waters**
Overview
Kinetic Approach for Reactions with Well-Defined Photooxidants
Illustrative Example 16.1: *Estimating Near-Surface Hydroxyl Radical Steady-State Concentrations in Sunlit Natural Waters*
Reactions with Hydroxyl Radical (HO$^\bullet$)
Illustrative Example 16.2: *Estimating the Indirect Photolysis Half-Life of Atrazine in a Shallow Pond*
Reactions with Singlet Oxygen ($^1O_2$)
Reactions with Reactive DOM Constituents ($^3DOM^*$, ROO$^\bullet$, RO$^\bullet$, etc.)

16.3 **Indirect Photolysis in the Atmosphere (Troposphere)—Reactions with Hydroxyl Radical (HO$^\bullet$)**
Sources and Typical Concentrations of HO$^\bullet$ in the Troposphere
Rate Constants and Tropospheric Half-Lives for Reactions with HO$^\bullet$
Estimation of Gas-Phase HO$^\bullet$ Reaction Rate Constants
Illustrative Example 16.3: *Estimating Tropospheric Half-Lives of Organic Pollutants*

16.4 **Questions and Problems**

16.1 Introduction

In Chapter 15 we dealt with reactions of organic pollutants occurring as a consequence of direct light absorption by the pollutant. In the environment, particularly in the atmosphere, in surface waters, and on soil or plant surfaces, there are, however, other light-induced processes that may lead to the transformation of a given organic compound. Such processes are initiated through light absorption by other chemicals in the system. They are, therefore, commonly referred to as *indirect* or *sensitized* photolysis. (The term "sensitized" is sometimes reserved for indirect photoreactions involving energy transfer; see below.)

There are several important reactive species that are generated as a consequence of light absorption in the aquatic environment including atmospheric waters (Table 16.1; Faust, 1999). Some of these species are also important reactants in the gas phase in the atmosphere (e.g., HO$^\bullet$, O$_3$, NO$_3^\bullet$), and in water treatment plants where they are chemically produced (e.g., HO$^\bullet$, O$_3$, HO$_2^\bullet$, R$^\bullet$, ROO$^\bullet$, CO$_3^{\bullet-}$). From the one-electron reduction potentials given in Table 16.1, we can see that all these species are rather strong oxidants compared to molecular oxygen (^{3}O$_2$). Although some reduction reactions may occur in natural waters (e.g., polyhalogenated compounds involving photochemically produced hydrated electrons; Zepp and Ritmiller, 1995), in general, organic pollutants are *oxidized* by such species. Hence, one commonly refers to these species as *transient photooxidants*. The term "transient" is used to indicate that these photooxidants are rather short-lived, because they are rapidly removed (i.e., *scavenged*) by physical (i.e., energy transfer) or chemical processes. Note that those species in which oxygen atoms play an important role are often also summarized as *reactive oxygen species* (ROS, Blough and Zepp, 1995).

Table 16.1 Standard One-Electron Reaction Potentials, E_H^1, in Aqueous Solution at 25°C of Some Environmentally and Technically Important (Photo)Oxidants [a]

Oxidant	Reaction in Water	E_H^1 / V
HO$^\bullet$	HO$^\bullet$ + e$^-$ = HO$^-$	1.9
O$_3$	O$_3^\bullet$ + e$^-$ = O$_3^-$	1.0
^{1}O$_2$	^{1}O$_2$ + e$^-$ = O$_2^{\bullet-}$	0.83
HO$_2^\bullet$ / O$_2^{\bullet-}$	HO$_2^\bullet$ + e$^-$ = HO$_2^-$	0.75
^{3}O$_2$	^{3}O$_2$ + e$^-$ = O$_2^{\bullet-}$	–0.16
ArO$^\bullet$ [b]	ArO$^\bullet$ + e$^-$ = ArO$^-$	0.79
R,X–ArO$^\bullet$ [b]	R,X–ArO$^\bullet$ + e$^-$ = R,X–ArO$^-$	0.2 – 1.2
RO$^\bullet$ [c]	RO$^\bullet$ + e$^-$ = RO$^-$	1.2
ROO$^\bullet$ [c]	ROO$^\bullet$ + e$^-$ = ROO$^-$	0.77
CO$_3^{\bullet-}$	CO$_3^{\bullet-}$ + e$^-$ = CO$_3^{2-}$	1.6
NO$_3^\bullet$	NO$_3^\bullet$ + e$^-$ = NO$_3^-$	2.3

[a] Data from Sulzberger et al. (1997) and Faust (1999). E_H^1 values of additional species can be found in Faust (1999). [b] Ar = phenyl. [c] R = alkyl.

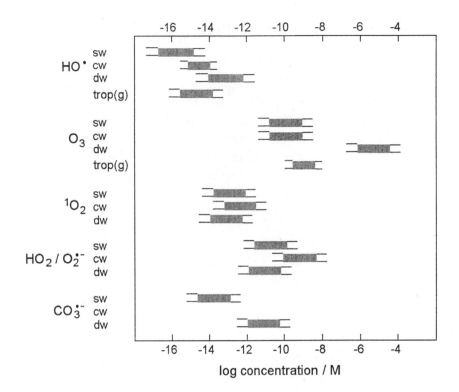

Figure 16.1 Ranges of steady-state concentrations of reactive oxygen species in sunlit surface waters (sw), sunlit cloud waters (cw), drinking-water treatment (dw), and the troposphere (trop(g)). Data from Sulzberger et al. (1997) and Atkinson et al. (1999).

Due to the high reactivities of the various photooxidants, they are commonly present in the environment at very low concentrations (see examples given in Fig. 16.1). Therefore, although the second-order rate constants (Eq. 12-14) for their reactions with organic pollutants are very large (see below), the overall transformation rates may not be significant in certain environments. We should also note that it is often difficult to directly measure the concentrations of these photooxidants. Thus, *steady-state* concentrations are frequently determined indirectly by measuring transformation rates of appropriate probe molecules that react selectively with a given photooxidant, or by estimates of production and scavenging rates (see below).

In this chapter, we will confine ourselves to illustrating indirect photolysis by a few important reactions of organic pollutants with photooxidants in surface waters [i.e., reactions with $HO^•$, $^1O_2$, and $^3UC^*$ (unknown chromophore); Section 16.2] as well as in the gas phase in the troposphere ($HO^•$; Section 16.3). We will focus our discussion on the assessment of transformation rates of organic pollutants in these systems by assuming typical steady-state concentrations of the various reactive species. For a more detailed discussion of the rather complex processes that determine the steady-state concentrations of these and other photooxidants, as well as the breakdown products of the organic pollutants in environmental systems, we refer to the literature (e.g., aqueous media: Blough and Zepp, 1995; Hoigné, 1997; Boule, 1999; Canonica and Freiburghaus, 2001; troposphere: Atkinson, 1989, 1994, 2000).

Indirect Photolysis in Surface Waters

Overview

The most important light absorbers that may induce indirect photolytic transformations of organic pollutants in natural waters are the chromophores present in dissolved organic material (DOM). A little calculation illustrates that each of these chromophores is excited numerous times during one day. Let us look at the well-mixed epilimnion of Greifensee (small eutrophic lake in Switzerland at 47.5°N, latitude), our model lake that we use throughout the remainder of this chapter. From the $\alpha(\lambda)$ values given in Table 15.6 we can deduce that virtually all light between 290 and 600 nm is absorbed within the epilimnion ($z_{mix} = 5$ m) of this lake. As indicated by Table 15.6, on a clear summer day the total number of photons absorbed in the wavelength range that is important for indirect photolysis (290–600 nm) is about 4 millieinstein cm^{-2} d^{-1}. Hence, per liter of epilimnion water, 8×10^{-3} moles of photons are absorbed per day. The question now is how many chromophores are present to absorb these photons. We may make an upper estimate of this number by assuming that each chromophore contains at least 10 carbon atoms. With a dissolved organic carbon content (DOC) of 4 mg C·L^{-1} we obtain a maximum chromophore concentration of about 30 μM. This means that, on the average, each chromophore present in the epilimnion of Greifensee would be excited 270 times per day or more than 10 times per hour. As one can easily imagine (Fig. 16.2), reactions of these excited chromophores may lead to alterations within the dissolved organic matter, to direct transformation of an organic pollutant i, as well as to the formation of a variety of reactive species (e.g., $^1O_2, RO^{\bullet}, HO_2^{\bullet}, O_2^{\bullet-}, HO^{\bullet}$).

As we have discussed in Section 15.2, an excited chemical species may undergo various physical and chemical processes (Fig. 15.6). In the case of natural organic matter constituents, by far the most important acceptor (*quencher*) of excited unknown chromophores (UCs) is molecular oxygen in its ground-state (triplet oxygen, $^3O_2$). Since promotion of $^3O_2$ to its first excited state (singlet oxygen $^1O_2$) requires only 94 kJ·mole^{-1}, almost all chromophores absorbing light in the uv- and visible-wavelength range may (after intersystem crossing to a triplet state) transfer their absorbed light energy to an oxygen molecule. The resulting $^1O_2$ may then react by chemical reactions with certain organic pollutants (see below).

As indicated in Fig. 16.2, in addition to energy transfer, chemical reactions of excited UCs ($^1UC^*, ^3UC^*$) may lead to the formation of other reactive oxygen species (ROS) that may react with organic pollutants. Such ROS include DOM-derived oxyl- and peroxyl radicals (RO$^{\bullet}$, ROO$^{\bullet}$), superoxide radical anions ($O_2^{\bullet-}$) that may be further reduced to H_2O_2, and hydroxyl radicals (HO$^{\bullet}$). In the case of HO$^{\bullet}$, however, DOM is a net sink rather than a source. Finally, some of the $^3UC^*$ may react directly with certain more easily oxidizable pollutants (see below).

In addition to DOM, there are other water constituents that upon absorption of light may yield transient photooxidants. The most prominent examples are nitrate (NO_3^-), nitrite (NO_2^-), and various Fe(II)- and Fe(III) complexes. In many freshwaters, photolysis of NO_3^- and NO_2^- appears to be the major source for HO$^{\bullet}$ (Blough and Zepp, 1995):

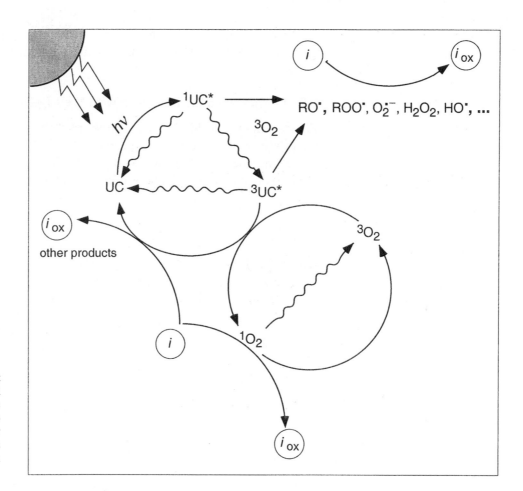

Figure 16.2 Pathways for indirect photolysis of an organic compound *i* involving excited natural organic matter constituents. UC refers to unknown chromophores. Wavy arrows symbolize radiationless transition (adapted from Zafiriou et al., 1984).

$$NO_3^- \xrightarrow{h\upsilon} NO_3^{-*} \longrightarrow NO_2 + O^{\cdot -} \xrightarrow{H_2O} HO^{\cdot} + OH^-$$

$$(16\text{-}1)$$

$$NO_2^- \xrightarrow{h\upsilon} NO_2^{-*} \longrightarrow NO + O^{\cdot -} \xrightarrow{H_2O} HO^{\cdot} + OH^-$$

The production of $HO^{\cdot}$ involving Fe(II) and Fe(III) species as well as the involvment of reactive iron species in pollutant transformation is primarily important in iron-rich waters exhibiting a low pH. Such is the case in surface waters contaminated with acidic mine drainage (Allen et al., 1996). Finally, we note that reactions of $HO^{\cdot}$ with some inorganic water constituents may yield more long-lived radicals which, although less reactive than $HO^{\cdot}$, may become significant for pollutant transformation under certain conditions. Examples are the carbonate radical and the bromine radical (seawater):

$$HO^{\cdot} + HCO_3^- / CO_3^{2-} \rightarrow CO_3^{\cdot -} + H_2O / OH^-$$

$$HO^{\cdot} + Br^- \qquad \rightarrow Br^{\cdot} + OH^-$$

$$(16\text{-}2)$$

For sunlit surface waters and drinking water treated by ozonation, carbonate radical steady-state concentrations were estimated to be typically two orders of magnitude higher than $HO^{\cdot}$ concentrations (see Fig. 16.1). Thus, this process may become important for compounds that react with $CO_3^{\cdot -}$ by less than a factor of 100 to 1000 more slowly as compared to $HO^{\cdot}$. Such compounds include the more easily oxidizable compound classes such as electron-rich anilines and phenols (Canonica and Tratnyek, 2002), and chemicals containing reduced sulfur (Huang and Mabury, 2000).

Kinetic Approach for Reactions with Well-Defined Photooxidants

For describing the kinetics of indirect photolysis of organic pollutants involving well-defined photooxidants (e.g., HO$^\bullet$, $^1O_2$, $CO_3^{\bullet-}$), we adapt the approach suggested by Hoigné et al. (1989) and Mill (1989). The rate of formation, $r_{f,ox}(\lambda)$, of a given photooxidant (Ox) by radiation of wavelength λ may be described by:

$$r_{f,ox}(\lambda) = \left(\frac{d[Ox]}{dt}\right)_\lambda = k_{a,A}(\lambda)\cdot\Phi_{r,A}(\lambda)\cdot[A] \qquad (16\text{-}3)$$

where $k_{a,A}(\lambda)$ [einstein (mol A)$^{-1}$ s^{-1}] is the specific light absorption rate of the bulk chromophore(s) A involved (e.g., DOM, NO_3^-), leading to the production of Ox, [A] is the (bulk) concentration of the responsible chemicals (e.g., [DOM] or [NO_3^-]), and $\Phi_{r,A}(\lambda)$ is the overall quantum efficiency for the production of Ox. Hence, $\Phi_{r,A}(\lambda)$ is a lumped parameter taking into account the reactions of the excited chromophores with other chemical species including $^3O_2$. Consequently, $\Phi_{r,A}(\lambda)$ is only a constant if all relevant parameters (e.g., the concentration of $^3O_2$) are kept constant in the system considered. The total rate of production of Ox is then given by integration of Eq. 16-3 over the wavelength range that is significant for the formation of Ox (i.e., range over which A absorbs light of sufficient energy for production of Ox):

$$r_{f,ox} = \frac{d[Ox]}{dt} = \int_\lambda k_{a,A}(\lambda)\cdot\Phi_{r,A}(\lambda)\cdot d\lambda\,[A]$$
$$\cong \left(\sum_\lambda k_{a,A}(\lambda)\cdot\Phi_{r,A}(\lambda)\Delta\lambda\right)[A] \qquad (16\text{-}4)$$

This wavelength range may be very narrow as, for example, for the production of HO$^\bullet$ from NO_3^- ($\lambda = 290 - 340$ nm), or it may be rather broad as, for example, for the production of $^1O_2$ from DOM ($\lambda = 290$ to ~600 nm).

Since photooxidants are quite reactive species, they are also "consumed" by various processes. These include physical quenching such as generation of heat in H_2O in the case of $^1O_2$, or chemical reactions with various water constituents (e.g., with DOM, HCO_3^-/CO_3^{2-}). If we assign to each of these Ox-consuming processes, j, a pseudo-first-order rate constant, $k_{ox,j}$ (hence, we also keep the concentrations of all consuming species, j, constant), we may describe the rate of consumption, $r_{c,ox}$ of Ox by:

$$r_{c,ox} = -\frac{d[Ox]}{dt} = \sum_j (k_{ox,j})[Ox] \qquad (16\text{-}5)$$

where $k_{ox,j} = k'_{ox,j}[j]$. Note that the Ox-consuming processes are chemical processes and that, therefore, $r_{c,ox}$ is light-independent.

Let us now consider a shallow water body that is exposed to noon sunlight. If we keep everything in the system constant, we will reach a steady state in which

$r_{f,ox}^0 = r_{c,ox}^0$, that is, the photooxidant will be present at a steady state concentration $[Ox]_{ss}^0$ of:

$$[Ox]_{ss}^0 = \frac{\sum_\lambda k_{a,A}^0(\lambda)\Phi_{r,A}(\lambda)[A]}{\sum_j k_{ox,j}} \qquad (16\text{-}6)$$

where we have introduced the superscript "0" to indicate near-surface light conditions. An example calculation is given in Illustrative Example 16.1.

Now we have an easy way of describing the indirect photolysis of a pollutant by a pseudo-first-order rate law, provided that the compound considered does not significantly affect $[Ox]_{ss}^0$, and that we are able to measure or estimate $[Ox]_{ss}^0$. The near-surface rate of loss of the pollutant is then given by:

$$\text{near-surface rate of loss of pollutant} = -\left(\frac{dC_i}{dt}\right)_{ox}^0 \qquad (16\text{-}7)$$

$$= k_{p,ox}'[Ox]_{ss}^0 C_i = k_{p,ox}^0 C_i$$

where $k_{p,ox}'$ and $k_{p,ox}^0$ are the second-order and near-surface pseudo-first-order reaction rate constants, respectively, for reaction of the pollutant i with Ox.

Unfortunately, it is not possible in most cases to quantify a given photooxidant by a direct measurement. By analogy to chemical actinometry (Section 15.4), however, one may use a probe or reference compound (P, subscript ref) with known $k_{p,ox,ref}'$ to determine $[Ox]_{ss}^0$ in a given natural water. This involves adding the chemical at a known concentration to the water, illuminating, and measuring the compound's disappearance. Since the probe compound disappearance kinetics also obeys Eq. 16-7, $[Ox]_{ss}^0$ can then be calculated from the slope of a correlation of ln [P] versus time:

$$\ln \frac{[P]}{[P_0]} = -k_{p,ox,ref}'[Ox]_{ss}^0 \cdot t \qquad (16\text{-}8)$$

A requirement for such a measurement is, of course, that the probe compound does not react by any other pathway. We discuss some of these probe systems later when discussing some specific photoreactants. At this point we recall, however, that if $[Ox]_{ss}^0$ values are determined in a cuvette or other photochemical vessel, the geometry of the vessel has to be taken into account when extrapolating the values to natural water bodies (see Section 15.4).

The average steady state photooxidant concentrations for longer periods of time (e.g., one to several days) may be roughly estimated from the measured value by multiplication with the ratio of the (computed or measured) integrated average light intensities (integrated over the wavelength range of maximum production of Ox) prevailing during the two time periods:

$$[Ox]_{ss}^0\left(\begin{array}{c}\text{estimation}\\\text{period}\end{array}\right) = [Ox]_{ss}^0(\text{noon})\left(\begin{array}{c}\text{experimental}\\\text{period}\end{array}\right)\frac{\sum Z(\text{est.period})}{\sum Z(\text{exp.period})} \qquad (16\text{-}9)$$

We also recall that when considering near-surface light conditions, we have to apply $Z(\lambda)$ [and not $W(\lambda)$ values]. For example, for a summer day at 40°N latitude, we may use the $Z(\lambda)$ values given in Table 15.3 to estimate the 24 h average Ox steady state concentration from the concentration measured at noon by:

$$[\text{Ox}]_{ss}^0(24\text{ h}) = [\text{Ox}]_{ss}^0(\text{noon})\frac{\sum Z(24\text{ h},\lambda)}{86400\sum Z(\text{noon},\lambda)} \qquad (16\text{-}10)$$

Note that a conversion factor of 86400 has to be introduced to make the two sets of $Z(\lambda)$ values compatible with respect to their conventional units (s^{-1} and d^{-1}, respectively; see also Illustrative Example 16.1).

In principle, by analogy to the direct photolytic processes, measurements of near-surface steady-state concentrations of photooxidants may be used to estimate average Ox concentrations in a well-mixed water body by applying an (average) light-screening factor (see Eqs. 15-29 to 15-33) to the near-surface rate of Ox production (and thus to $[\text{Ox}]_{ss}^0$; see Eq. 16-6):

$$[\text{Ox}]_{ss} = [\text{Ox}]_{ss}^0 S(\lambda_{\max}) \qquad (16\text{-}11)$$

However, to apply Eq. 16-11, an appropriate $\lambda_{\max}$ has to be selected. That is, one has to know in which wavelength region maximum Ox production takes place.

Illustrative Example 16.1

Estimating Near-Surface Hydroxyl Radical Steady-State Concentrations in Sunlit Natural Waters

Problem

Estimate the near-surface hydroxyl radical steady-state concentration at noon ($[\text{HO}^\bullet]_{ss}^0(\text{noon})$) and averaged over a day ($[\text{HO}^\bullet]_{ss}^0(24\text{ h})$) in Greifensee (47.5°N) on a clear summer day. Assume that photolysis of nitrate (NO_3^-) and nitrite (NO_2^-) are the major sources, and that DOM, HCO_3^-, and CO_3^{2-} are the major sinks for $HO^\bullet$ in Greifensee. The concentrations of the various species are given in the margin.

$[NO_3^-]$	=	150 μM
$[NO_2^-]$	=	1.5 μM
$[DOC]$	=	4 mg oc $\cdot$ L^{-1}
$[HCO_3^-]$	=	1.2 mM
$[CO_3^{2-}]$	=	0.014 mM

Answer

Assuming average wavelength-independent quantum yields for the photolysis of nitrate and nitrite, $[\text{HO}^\bullet]_{ss}^0(\text{noon})$ is given by (Eqs. 16-4, 16-5, and 16-6; see also Eqs. 15-25 and 15-37):

$$[\text{HO}^\bullet]_{ss}^0(\text{noon}) = \frac{\Phi_{r,NO_3^-} \cdot k_{a,NO_3^-}^0(\text{noon}) \cdot [NO_3^-] + \Phi_{r,NO_2^-} \cdot k_{a,NO_2^-}^0(\text{noon}) \cdot [NO_2^-]}{k'_{\text{HO}^\bullet,\text{DOC}}[DOC] + k'_{\text{HO}^\bullet,HCO_3^-}[HCO_3^-] + k'_{\text{HO}^\bullet,CO_3^{2-}}[CO_3^{2-}]} \qquad (1)$$

In the literature you find the decadic molar extinction coefficients, $\varepsilon_i(\lambda)$, for nitrate (Gaffney et al., 1992) and nitrite (Fischer and Warneck, 1996). Using these $\varepsilon_i(\lambda)$ values, $k_{a,NO_3^-}^0(\text{noon})$ and $k_{a,NO_2^-}^0(\text{noon})$ may be calculated as we have calculated the k_a^0 value of PNAP in Table 13.5. For Greifensee (47.5°N, in the summer not so different from 40°N; see Fig. 15.10) one obtains $k_{a,NO_3^-}^0$ (noon) $= 2.0 \times 10^{-5}$ einstein (mol $NO_3^-)^{-1}$ s^{-1} and $k_{a,NO_2^-}^0$ (noon) $= 6.0 \times 10^{-4}$ einstein (mol $NO_2^-)^{-1}$ s^{-1}. NO_3^- absorbs light between 290 and 340 nm with a maximum light absorption rate at 320 nm (λ_{max}). For NO_2^- the range is much wider, 290–400 nm with $\lambda_{max} = 360$ nm.

Comparison of the two k_a^0 values shows that NO_2^- absorbs about 30 times more light as compared to NO_3^-. Furthermore, the quantum yield of NO_2^- at 360 nm [$\Phi_{r,NO_2^-} = 0.028$ (mol $HO^\bullet$) einstein^{-1}] is 4 times larger than that of NO_3^- at 320 nm [$\Phi_{r,NO_3^-} = 0.007$ (mol $HO^\bullet$) einstein^{-1}] (Jankowski et al., 1999). Consequently, on a per-mole basis, NO_2^- produces about two orders of magnitude more $HO^\bullet$ as compared to NO_3^-:

$$r_{f,HO^\bullet}^0 (\text{noon})/(M \cdot s^{-1}) = (1.4 \times 10^{-7})[NO_3^-] + (1.7 \times 10^{-5})[NO_2^-] \qquad (2)$$

Note that in Greifensee, NO_3^- is 100 times more abundant than NO_2^- (see concentrations given in margin). Hence, both NO_3^- and NO_2^- contribute about equally to the near-surface production of $HO^\bullet$ in this lake.

With respect to the consumption of $HO^\bullet$, you can also find the corresponding second-order rate constants $k_{HO^\bullet,j}'$ in the literature. For reaction of $HO^\bullet$ with DOM, an average rate constant $k_{HO^\bullet,DOC}' \cong 2.5 \times 10^4$ L $\cdot$ (mg oc)$^{-1}$ s^{-1} can be used (Larson and Zepp, 1988; Brezonik and Fulkerson-Brekken, 1998). For reactions with HCO_3^- and CO_3^{2-}, the rate constants are $k_{HO^\bullet,HCO_3^-}' = 1.0 \times 10^7$ M^{-1} s^{-1} and $k_{HO^\bullet,CO_3^{2-}}' = 4.0 \times 10^8$ M^{-1} s^{-1} (Larson and Zepp, 1988). Insertion of all these rate constants together with Eq. 2 into Eq. 1 yields:

$$[HO^\bullet]_{ss}^0 (\text{noon})/M = \frac{(1.4 \times 10^{-7})[NO_3^-] + (1.7 \times 10^{-5})[NO_2^-]}{(2.5 \times 10^4)[DOC] + (1.0 \times 10^7)[HCO_3^-] + (4.0 \times 10^8)[CO_3^{2-}]}$$

where [DOC] has to be expressed in mg oc $\cdot$ L^{-1} and all other concentrations in mol $\cdot$ L^{-1}. With the above given concentrations of the various species involved, one then obtains a $HO^\bullet$ steady-state concentration of:

$$[HO^\bullet]_{ss}^0 (\text{noon}) = \frac{(1.4 \times 10^{-7})(1.5 \times 10^{-4}) + (1.7 \times 10^{-5})(1.5 \times 10^{-6})}{(2.5 \times 10^4)(4) + (1.0 \times 10^7)(1.2 \times 10^{-3}) + (4.0 \times 10^8)(1.4 \times 10^{-5})}$$

$$\cong 4 \times 10^{-16} \text{ M}$$

For estimation of $[HO^\bullet]_{ss}^0$ (24 h), insert the Z(noon, λ) and Z(24 h, λ) values given in Table 15.3 into Eq. 16-10 for $\lambda < 400$ nm. The result is:

$$[HO^\bullet]_{ss}^0 (24 \text{ h}) = \frac{(56.0 \times 10^{-2})}{86400(1.48 \times 10^{-5})} [HO^\bullet]_{ss}^0 (\text{noon}) = 0.44 \, [HO^\bullet]_{ss}^0 (\text{noon})$$

$$= (0.44)(4 \times 10^{-16}) = 1.8 \times 10^{-16} \text{ M}$$

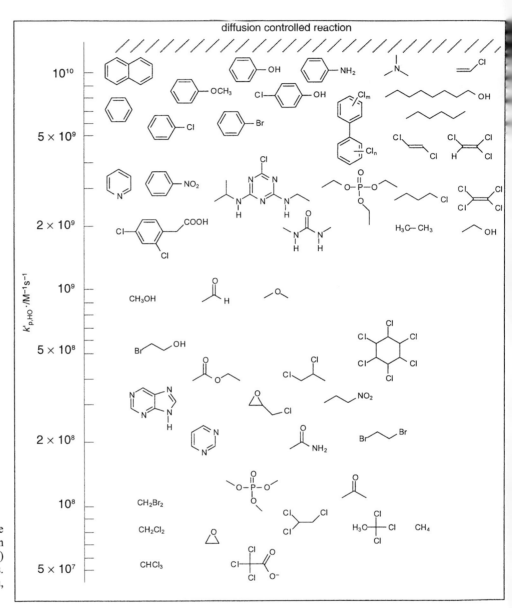

Figure 16.3 Second-order rate constants for reaction with HO• in aqueous solution ($k'_{p,HO•}$; Eq. 16-7) for a series of organic compounds. Data from http://allen.rad.nd.edu, and Haag and Yao (1992).

Reactions with Hydroxyl Radical (HO•)

Because of its high reactivity, direct observation of hydroxyl radicals is very difficult in natural waters. Most of the evidence for the existence of HO• derives from product analysis studies and from studies of relative photolytic reactivities of a series of compounds. Because HO• reacts with many organic compounds at nearly diffusion-controlled rates (Fig. 16.3), various organic substrates that do not undergo other photolytic transformations may be used as probe molecules (for more details see Hoigné, 1997; Vaugham and Blough, 1998).

HO• reacts with organic pollutants primarily in two different ways: (i) by electrophilic addition to a double bond or an aromatic system (Eq. 16-12), and (ii) by abstraction of a hydrogen atom from a carbon atom (Eq. 16-13):

$$(16\text{-}12)$$

$$R - H \xrightarrow{\ \text{HO}^\bullet\ } R^\bullet + H_2O \qquad (16\text{-}13)$$

As can be seen from Fig. 16.3, nearly diffusion-controlled reaction rates (i.e., $k'_{\text{p,HO}^\bullet} = 5 \cdot 10^9 - 10^{10} \ \text{M}^{-1} \text{s}^{-1}$) are observed for compounds exhibiting (i) aromatic rings and/or carbon–carbon double bonds with electron-donating substituents, and/or (ii) aliphatic groups from which an H-atom can be easily abstracted. However, even in the presence of electron-withdrawing substituents, addition reactions to aromatic systems or double bonds still occur at appreciable rates (e.g., nitrobenzene, tetrachloroethene, see Fig. 16.3). Hence, HO$^\bullet$ is not a very selective (photo)oxidant. This is of particular interest for the removal of pollutants during oxidative drinking-water treatment (Haag and Yao, 1992). Furthermore, many pesticides exhibit $k'_{\text{p,HO}^\bullet}$ values $> 10^9 \ \text{M}^{-1} \text{s}^{-1}$. Therefore, reaction with HO$^\bullet$ may be an important removal process for such compounds in nitrate-rich surface waters in agricultural areas (see Kolpin and Kolkhoff, 1993, and Illustrative Example 16.2).

Significantly slower rates are found only for compounds that do not exhibit any aromatic ring or carbon–carbon double bond, and for aliphatic compounds with no easily abstractable H-atoms. Such H-atoms include those that are bound to carbon atoms carrying one or several electronegative heteroatoms or groups. (Note that the stabilization of a carbon radical (R$^\bullet$) is similar to that of a carbocation.) We will come back to such structure-reactivity considerations in Section 16.3, when discussing reaction of HO$^\bullet$ with organic pollutants in the gas phase (i.e., in the atmosphere).

Illustrative Example 16.2

Estimating the Indirect Photolysis Half-Life of Atrazine in a Shallow Pond

Problem

Consider a shallow, well-mixed pond (average depth = 2 m) in an agricultural area at 40°N latitude. The following concentrations have been determined in the pond water: $[\text{DOC}] = 4 \ \text{mg oc} \cdot \text{L}^{-1}$, $[\text{HCO}_3^-] = 1 \ \text{mM}$, $[\text{CO}_3^{2-}] = 0.01 \ \text{mM}$, $[\text{NO}_3^-] = 0.5$ mM, $[\text{NO}_2^-] = 0.003$ mM. Calculate the 24 h averaged half-life of atrazine in this pond for clear summer day conditions by using the beam-attenuation coefficients, $\alpha(\lambda)$, given for Greifensee in Table 15.6. Assume that reaction with HO$^\bullet$ is the only important indirect photolysis mechanism for atrazine.

atrazine

$k'_{\text{p,HO}^\bullet} = 3 \times 10^9 \ \text{M}^{-1} \text{s}^{-1}$

Answer

Use Eq. 2 in Illustrative Example 16.1 to estimate the near-surface production rate of HO$^\bullet$ in the pond. To account for light attenuation in the water column, apply light-screening factors for the wavelength of maximum light absorption of NO$_3^-$ ($\lambda_{\text{max}} = 320$ nm) and NO$_2^-$ ($\lambda_{\text{max}} = 360$ nm):

$$r_{\text{f,HO}^\bullet}(\text{noon})/(\text{M} \cdot \text{s}^{-1}) = (1.4 \times 10^{-7})[\text{NO}_3^-] \cdot S(320\,\text{nm}) + (1.7 \times 10^{-5})[\text{NO}_2^-] \cdot S(360\,\text{nm})$$

With $\alpha(320 \text{ nm}) = 0.0275 \text{ cm}^{-1}$ and $\alpha(360 \text{ nm}) = 0.0125 \text{ cm}^{-1}$ (Table 15.6), the corresponding $S(\lambda_{max})$ values are (Eq. 15-32):

$$S(320 \text{ nm}) = [(2.3)(1.2)(200)(0.0275)]^{-1} = 0.066$$

$$S(360 \text{ nm}) = [(2.3)(1.2)(200)(0.0125)]^{-1} = 0.14$$

and therefore:

$$r_{f,HO^\bullet}(\text{noon})/(\text{M} \cdot \text{s}^{-1}) = (9.2 \times 10^{-9})[\text{NO}_3^-] + (2.4 \times 10^{-6})[\text{NO}_2^-] \tag{1}$$

The HO$^\bullet$ steady-state concentration at noon is then given by (see Eq. 3 in Illustrative Example 16.1):

$$[\text{HO}^\bullet]_{ss}(\text{noon}) = \frac{(9.2 \times 10^{-9})[\text{NO}_3^-] + (2.4 \times 10^{-6})[\text{NO}_2^-]}{(2.5 \times 10^4)[\text{DOC}] + (1.0 \times 10^7)[\text{HCO}_3^-] + (4.0 \times 10^8)[\text{CO}_3^{2-}]} \tag{2}$$

Insertion of all concentrations into Eq. 2 yields:

$$[\text{HO}^\bullet]_{ss}(\text{noon}) = \frac{(9.2 \times 10^{-9})(5.0 \times 10^{-4}) + (2.4 \times 10^{-6})(3.0 \times 10^{-6})}{(2.5 \times 10^4)(4.0) + (1.0 \times 10^7)(10^{-3}) + (4.0 \times 10^8)(10^{-5})}$$

$$= 1.0 \times 10^{-16} \text{M}$$

which on a 24 h average corresponds to (see Illustrative Example 16.1):

$$[\text{HO}^\bullet]_{ss}(24 \text{ h}) = 0.44 [\text{HO}^\bullet]_{ss}(\text{noon}) = 4.4 \times 10^{-17} \text{ M}$$

The indirect photolysis half-life of atrazine (assuming that reaction with HO$^\bullet$ is the dominant process) is then given by (Eq. 16-7):

$$t_{1/2} = \frac{\ln 2}{k'_{p,HO^\bullet}[\text{HO}^\bullet]_{ss}(24 \text{ h})} = \frac{0.69}{(3 \times 10^9)(4.4 \times 10^{-17})} = 5.3 \times 10^6 \text{ s} \cong 60 \text{ d}$$

Note that the near-surface concentration of HO$^\bullet$ in this pond is:

$$[\text{HO}^\bullet]_{ss}^0(24 \text{ h}) = (0.44)\frac{(1.4 \times 10^{-7})(5 \times 10^{-4}) + (1.7 \times 10^{-5})(3.0 \times 10^{-6})}{(2.5 \times 10^4)(4.0) + (1.0 \times 10^7)(10^{-3}) + (4 \times 10^8)(10^{-5})}$$

$$= 1.1 \times 10^{-15} \text{ M}$$

which is about 25 times higher than the [HO$^\bullet$] averaged over the whole water column. Hence, the half-life of atrazine at the surface in the pond is only about 2.5 days.

Reactions with Singlet Oxygen ($^1\text{O}_2$)

As indicated in Fig. 16.2, $^1\text{O}_2$ is formed primarily by energy transfer from $^3\text{UC}^*$ to $^3\text{O}_2$. The most important consumption mechanism for $^1\text{O}_2$ is physical quenching by water. At DOM concentrations typical for most surface waters (DOC $< 20 \text{ mg C} \cdot \text{L}^{-1}$),

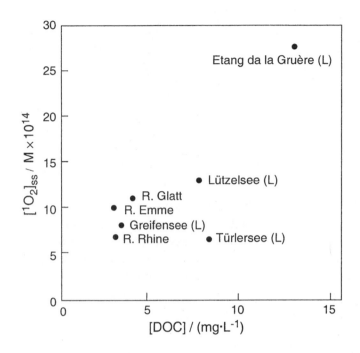

Figure 16.4 Observed $[^1O_2]_{ss}^0$ in water samples from some Swiss rivers (R) and lakes (L) as a function of the dissolved organic carbon (DOC) concentrations of these waters. The results apply for noontime light intensity on a clear summer day at 47.5°N (data from Haag and Hoigné, 1986).

quenching of $^1O_2$ by DOM can be neglected (Haag and Hoigné, 1986). Hence, the near-surface steady-state concentration of $^1O_2$ in a natural water is directly proportional to the DOM concentration. Note, however, that different types of aquatic DOM may exhibit quite different overall quantum yields for $^1O_2$ production. This is illustrated by Fig. 16.4, in which $[^1O_2]_{ss}^0$ values for various lake and river waters are plotted against the DOC concentration. As can be seen from this plot, in waters exhibiting DOC values between 3 and 4 mg oc$\cdot$L^{-1}, maximum $^1O_2$ steady-state concentrations in the order of 7 to 11×10^{-14} M are detected on a summer day at 47.5°N latitude. Since the variation in $[^1O_2]_{ss}^0$ is broader than the DOC concentration range, and since $^1O_2$ consumption rates (via quenching by water) are the same for all waters, this variation in $[^1O_2]_{ss}^0$ must be due to differences in the light absorbance by the UCs present and/or in the quantum yields for production of $^1O_2$. In fact, good correlations between natural water uv absorbance or fluorescence and singlet oxygen steady-state concentrations have been found (Shao et al., 1994).

The $[^1O_2]_{ss}^0$ values shown in Fig. 16.4 were determined by using furfuryl alcohol (FFA, I) as a probe compound (Haag et al., 1984a). Another frequently used "trapping agent" for $^1O_2$ determination is 2,5-dimethylfuran (2,5-DMF, II) (Zepp et al., 1977):

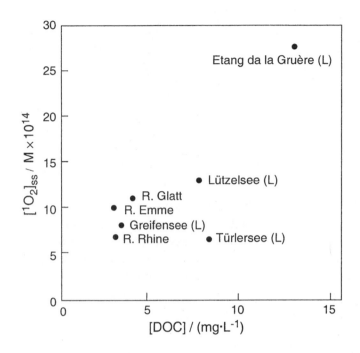

With these compounds (as with other dienes), $^1O_2$ undergoes a so-called *Diels-Alder-reaction* (March 1992), forming an endoperoxide intermediate that further reacts to yield various products. For example, the reaction of FFA with $^1O_2$ is (Haag et al., 1984a):

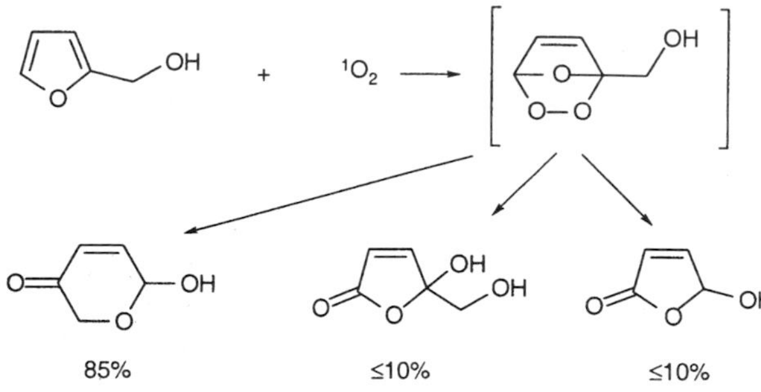

$$(16\text{-}14)$$

Besides its properties as a reactant in addition reactions, $^1O_2$ is also a significantly better electron acceptor (oxidant) than $^3O_2$ (Table 16.1). However, because of its low steady-state concentration in natural waters, it is an important photooxidant for only a few very reactive types of organic compounds. Such compounds include those exhibiting structural moieties that may undergo Diels-Alder reactions, those containing electron-rich double bonds (i.e., double bonds that are substituted with electron-donating groups), or compounds exhibiting functional groups that are easily oxidized, including reduced sulfur groups (e.g., sulfides), anilines, and phenols (see also Section 14.3). Fig. 16.5 summarizes some of the kinetic data available for reactions of organic compounds with $^1O_2$ in water. As can be seen, for phenolic compounds the transformation rate is pH dependent, since the phenolate species (A$^-$) is much more reactive toward oxidation by $^1O_2$ as compared to the neutral phenol (HA) [i.e., $k'_{p,{}^1O_2,HA} \ll k'_{p,{}^1O_2,A^-}$].

Since $^1O_2$ behaves as an *electrophile*, one can assume that electron-donating substituents on an organic compound will, in general, increase its reactivity, while electron-withdrawing substituents will have the opposite effect. In the case of phenolic compounds, the effect(s) of the substituent(s) on the pK_a (see Chapter 8), and thus on the concentration of the reactive phenolate species present at a given pH, may be more important than the effect(s) of the substituent(s) on $k_{p,{}^1O_2}$. In this case, the overall transformation rate is dominated by the rate of transformation of the anionic species (A$^-$):

$$\text{rate} = -\left(\frac{dC_{it}}{dt}\right)_{{}^1O_2} \cong (1-\alpha_{ia}) \cdot k'_{p,{}^1O_2,A^-} [{}^1O_2]_{ss} C_{it} \qquad (16\text{-}15)$$

where C_{it} is the total phenol concentration ([A$^-$] + [HA]) and $(1-\alpha_{ia}) = [1 + 10^{(pKia-pH)}]^{-1}$ is the fraction in the dissociated (anionic) form (see Eq. 8-21). This effect is illustrated in Fig. 16.5, where for the phenolic compounds, $(1-\alpha_{ia}) \cdot k'_{p,{}^1O_2}$ is plotted as a function of pH. Note that at lower pH-values the contribution of the nondissociated phenol may become important (e.g., for 4-methylphenol, see Fig. 16.5). In these cases, the overall rate of transformation is given by:

$$\text{rate} = -\left(\frac{dC_{it}}{dt}\right)_{{}^1O_2} = \left[\alpha_{ia} \cdot k'_{p,{}^1O_2,HA} + (1-\alpha_{ia}) \cdot k'_{p,{}^1O_2,A^-}\right] \cdot [{}^1O_2]_{ss} C_{it} \qquad (16\text{-}16)$$

The right-hand scale in Fig. 16.5 gives the calculated half-lives for indirect photolysis involving $^1O_2$ for the various compounds in the well-mixed epilimnion of

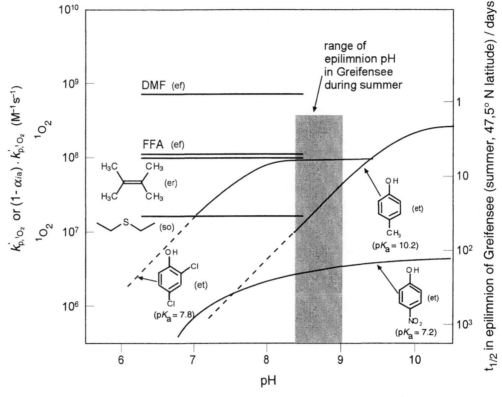

Figure 16.5 Second-order rate constants [multiplied by $(1 - \alpha_{ia})$ for phenols] for reactions of several compounds with $^1O_2$ (left scale) as a function of pH. The abbreviations in parentheses indicate the reaction type: ef = endoperoxide formation (Eq. 16-14); er = ene reaction; et = electron transfer; so = sulfur oxidation. The scale on the right indicates the half-lives of the compounds in the epilimnion of Greifensee on a clear summer day (data from Scully and Hoigné, 1987).

Greifensee (z_{mix} = 5 m) on a clear summer day. The half-lives are based on a measured $[^1O_2]_{ss}^0$ (noon) value of 8×10^{-14} M, corresponding to a $[^1O_2]_{ss}^0$ (24 h) value of about 3.5×10^{-14} M (factor 0.44; see Illustrative Example 16.1). When assuming that virtually all light is absorbed within the epilimnion {i.e., $S(\lambda_m) = [2.3 \cdot (1.2) \cdot \alpha(\lambda_m) \cdot z_{mix}]^{-1}$}, and taking the α value at 410 nm [$\alpha(410$ nm) = 0.005 cm^{-1}; see Table 15.6], an average $^1O_2$ concentration $[^1O_2]_{ss}$ (24 h) = 4×10^{-15} M is calculated (Eq. 16-11). The choice of 410 nm is based on the findings by Haag et al. (1984b) that some humic and fulvic materials exhibit a maximum in $^1O_2$ production around this wavelength.

With the $[^1O_2]_{ss}^0$ (24 h) value calculated above, the half-life of a phenol with respect to photooxidation by $^1O_2$ in the epilimnion of Greifensee is then given by:

$$t_{1/2} = \frac{\ln 2}{(1-\alpha_a) \cdot k'_{p,^1O_2,A^-}(4 \times 10^{-15})} \sec = \frac{2 \times 10^9}{(1-\alpha_{ia}) \cdot k'_{p,^1O_2,A^-}} \text{days} \qquad (16\text{-}17)$$

From the half-lives indicated in Fig. 16.5 it can be seen that for most pollutants, the assumption of a well-mixed epilimnion (typical mixing rates 1 – 10 d^{-1}) with respect to indirect photolysis with $^1O_2$ is a reasonable assumption. Furthermore, for compounds exhibiting $k'_{p,^1O_2}$ values [or $(1 - \alpha_{ia})$ $k'_{p,^1O_2}$ values for phenolate species] greater than 10^7 M^{-1} s^{-1}, during the summer, photooxidation by $^1O_2$ is equal to, or more important than, depletion of the concentration by dilution with inflowing water [$t_{1/2}$(dilution) in the epilimnion of Greifensee on the order of 70 days]. We should recall, however, that only a few compound classes exhibit such large $k'_{p,^1O_2}$ values, and that, therefore, $^1O_2$ must be considered to be a rather selective photooxidant.

Table 16.2 Chemical Structure, Hammett Constants, and Pseudo-First-Order Rate Constants for Suwanee River Fulvic Acid (SRFA) Sensitized Photolysis of a Series of Phenyl Urea Herbicides (PUHs)

Compound	No.[c]	–R$_1$	–R$_2$	–R$_3$	σ^+_{jpara}	σ^+_{jmeta}	$\sigma^+_{jpara}+\sigma^+_{jmeta}$	$k^0_{p,sens}$ [h^{-1}]
Metoxuron	1	–OCH$_3$	–Cl	–CH$_3$	–0.78	0.37	–0.41	0.63
CGA 24482	2	3,4-tetramethylene		–CH$_3$	–0.30[d]	–0.07[d]	–0.37	0.52
GGA 16519	3	–CH$_2$CH$_3$	–H	–CH$_3$	–0.30	0	–0.30	0.47
IPU	4	–CH(CH$_3$)$_2$	–H	–CH$_3$	–0.28	0	–0.28	0.44
CGA 17767	5	–CH(CH$_3$)$_2$	–H	–CH$_2$CH$_3$	–0.28	0	–0.28	0.32
CGA 17092	6	–C(CH$_3$)$_3$	–H	–CH$_3$	–0.26	0	–0.26	0.31
Fenuron	7	–H	–H	–CH$_3$	0	0	0	0.18
Chlorotoluron	8	–CH$_3$	–Cl	–CH$_3$	–0.31	0.37	0.06	0.22
GCA 18414	9	–CH(CH$_3$)$_2$	–Cl	–CH$_2$CH$_3$	–0.28	0.37	0.09	0.16
Fluometuron	10	–H	–CF$_3$	–CH$_3$	0	0.43	0.43	0.051
Diuron	11	–Cl	–Cl	–CH$_3$	0.11	0.37	0.48	0.055

Column group headers: Substituents (–R$_1$, –R$_2$, –R$_3$); Hammett Constants[a] (σ^+_{jpara}, σ^+_{jmeta}, $\sigma^+_{jpara}+\sigma^+_{jmeta}$); $k^0_{p,sens}$[b].

[a] Hansch et al. (1991). [b] SRFA: 2.5 mg oc · L^{-1}, $\lambda > 320$ nm; data from Gerecke et al. (2001). [c] Numbers in Fig. 16.6. [d] Hammett constant for 3,4-tetramethylene assumed to correspond to –R$_1$ = – R$_2$ = –CH$_2$CH$_3$.

Reactions with Reactive DOM Constituents (^{3}DOM*, ROO$^{\bullet}$, RO$^{\bullet}$, etc.)

We conclude our discussion of indirect photolysis in water by briefly addressing an example in which the "photoreactant" is not well defined as in the case of HO$^{\bullet}$ or ^{1}O$_2$. In this case, prediction of absolute reaction rates is more difficult. As has been shown in various studies, reactive DOM species include short-lived excited triplet states of DOM constituents (e.g., aromatic ketones), which we will denote as ^{3}DOM*, and, possibly, more long-lived radicals including ROO$^{\bullet}$ and RO$^{\bullet}$ species (Faust and Hoigné, 1987; Canonica et al., 1995; Canonica et al., 2000; Canonica and Freiburghaus, 2001). In the following, we will focus on reactions with excited triplet states of DOM constituents (^{3}DOM*). We should note, however, that these reactions cannot be completely separated from reactions with ROO$^{\bullet}$ or RO$^{\bullet}$, which are, however, thought to be of minor importance (Canonica et al., 1995).

Direct reactions of ^{3}DOM* with an organic chemical, which are often referred to as *photosensitized reactions*, may be classified as energy, electron- or hydrogen-transfer reactions. Energy transfer may cause, for example, *cis-trans*-isomerization of double bonds (Zepp et al., 1985). The other two mechanisms lead to an oxidation of the organic pollutant. Again, reaction rates are fastest with easily oxidizable compounds such as electron-rich phenols (Canonica et al., 1995; Canonica and Freiburghaus, 2001). Such reactions may, however, also be important for less reactive compounds such as phenylurea herbicides, for which DOM-mediated phototransformation may be the only relevant elimination mechanism in surface waters

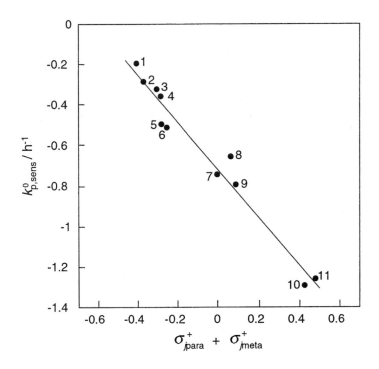

Figure 16.6 Hammett plot for the oxidation of a series of substituted phenylurea herbicides by $^3DOM^*$. The linear regression line is given in Eq. 16-19. The compound names and structures as well as the corresponding $\sigma^+_{j\mathrm{para}}$, $\sigma^+_{j\mathrm{meta}}(=\sigma_{j\mathrm{meta}})$, and $k^0_{\mathrm{p,sens}}$ values are given in Table 16.2. Adapted from Gerecke et al. (2001).

(Gerecke et al., 2001). As is shown by Fig. 16.6, the DOM-sensitized photooxidation of a series of substituted phenylureas (see Table 16.2):

$$(16\text{-}18)$$

can be described reasonably well by a Hammett relationship:

$$\log k^0_{\mathrm{p,sens}} \,/\, \mathrm{h}^{-1} = -1.16\,\Sigma\sigma^+_j - 0.72 \qquad (16\text{-}19)$$

where $k^0_{\mathrm{p,sens}}$ is the observed pseudo-first-order rate constant determined for the photosensitized reaction of a substituted phenyl urea in a solution of 2.5 mg oc·L^{-1} Suwanee River fulvic acid (SRFA) at pH 8 exposed to broadband irradiation ($\lambda >$ 320 nm) in a photoreactor (for details see Gerecke et al., 2001). Note that, in analogy to using $\sigma^-_{j\mathrm{para}}$ for delocalization of a negative charge (Chapter 8), $\sigma^+_{j\mathrm{para}}$ instead of $\sigma_{j\mathrm{para}}$ values have to be used in the Hammett equation in cases where a positive charge or a radical is delocalized, as is, for example, the case for radical formation upon the oxidation of anilines, phenols, and phenylureas (for more details and a comprehensive compilation of σ^+_j constants see Hansch et al., 1991, and Hansch and Leo, 1995; some σ^+_j values can be found in Table 16.6).

An interesting result found by Canonica and Freiburghaus (2001) for the DOM-sensitized oxidation of electron-rich phenols is that $k^0_{\mathrm{p,sens}}$ was more or less proportional to the concentration of the dissolved organic carbon, independent of the source of the DOM. In addition, for some phenylureas, Gerecke et al. (2001)

obtained very similar carbon-normalized $k^0_{p,sens}$ values for SRFA and Greifensee water-dissolved organic matter. Although generalization of these results is not yet possible, Gerecke et al. (2001) showed that for a given lake (Greifensee), rate data determined in the laboratory could be successfully used to model the DOM-mediated phototransformation of the two phenylurea herbicides isoproturon and diuron (Table 16.2) in the epilimnion of the lake. The critical parameter that has to be estimated (or determined) is the (wavelength-dependent) quantum yield coefficient that describes the efficiency by which light absorption by DOM leads to the transformation of the compound of interest. For more details on this topic we refer to the papers of Canonica et al. (1995), Canonica and Freiburghaus (2001), and Gerecke et al. (2001).

16.3   Indirect Photolysis in the Atmosphere (Troposphere)— Reactions with Hydroxyl Radical (HO•)

Long-range transport of an organic pollutant in the environment will occur when the compound has a sufficiently long tropospheric lifetime. This lifetime is partially determined by wet or dry deposition and/or by vapor transfers into surface waters. In earlier chapters, we have already addressed important issues with respect to such phase exchange processes (Chapters 6 and 11). Additionally, the tropospheric lifetime of a pollutant strongly depends on its reactivity with photooxidants and, to a lesser extent, on direct photolysis. Since light absorption rates and reaction quantum yields are very difficult to quantify for direct photolysis reactions of organic chemicals in the atmosphere (Atkinson et al., 1999), we will confine our discussion of photolytic transformations to reactions with photooxidants, in particular, HO•.

As we have already noticed in Section 16.2 when discussing reactions of organic chemicals with ROS in the aqueous phase, HO• is a very reactive, rather nonselective oxidant. Other important photooxidants present in the troposphere include ozone (O_3) and the nitrate radical ($NO_3^•$). Although these species are generally present in significantly higher concentrations as compared to HO• (see Fig. 16.1), they are rather selective oxidants, and are, therefore, only important reactants for chemicals exhibiting specific functionalities. O_3 reacts primarily with compounds containing one or several electron-rich carbon–carbon double bonds such as alkenes (Atkinson, 1994; Atkinson et al., 1995; Grosjean et al., 1996a and b). The nitrate radical, which is particularly important at night when HO• radicals are less abundant (Atkinson et al., 1999), also reacts with compounds exhibiting electron-rich carbon–carbon double bonds. In addition, $NO_3^•$ undergoes reactions with polycyclic aromatic hydrocarbons (PAHs) and with compounds exhibiting reduced sulfur and/or nitrogen functionalities (Atkinson, 1994). In the case of PAHs, such reactions yield rather toxic nitroaromatic compounds, such as 1- and 2-nitronaphthalene from the reaction of naphthalene with $NO_3^•$ (Sasaki et al., 1997). Hence, reactions beside that with HO• should be considered in such cases.

Sources and Typical Concentrations of HO• in the Troposphere

The presence of relatively low levels of O_3 in the troposphere is important because photolysis of O_3 in the troposphere occurs in the wavelength region of 290–335 nm to form the excited oxygen, $O(^1D)$, atom. $O(^1D)$ atoms are either deactivated to ground-state oxygen, $O(^3P)$, atoms, or react with water vapor to generate $HO^\bullet$ radicals:

$$O_3 + h\nu \qquad\qquad \rightarrow O_2 + O(^1D)(290 > \lambda \le 335 \text{ nm})$$

$$O(^1D) + N_2, O_2 \quad \rightarrow O(^3P) + N_2, O_2 \qquad\qquad (16\text{-}20)$$

$$O(^1D) + H_2O \qquad \rightarrow 2 \, HO^\bullet$$

At 298 K and atmospheric pressure with 50% relative humidity, about 0.2 $HO^\bullet$ are produced per $O(^1D)$ atom formed. Photolysis of O_3 in the presence of water vapor is the major tropospheric source of $HO^\bullet$, particularly in the lower troposphere where water vapor mixing ratios are high (for an explanation of the term "mixing ratio" see below). Other sources of $HO^\bullet$ in the troposphere include the photolysis of nitrous acid (HONO), the photolysis of formaldehyde and other carbonyls in the presence of NO, and the dark reactions of O_3 with alkanes. Note that all these processes involve quite complicated reaction schemes. For a discussion of these reaction schemes we refer to the literature (e.g., Atkinson, 2000).

At this point, we need to make a few comments on how gaseous concentrations of chemical species in the atmosphere are commonly expressed. A widely used approach is to give the fraction of the total volume that is occupied by the gaseous species considered. This is referred to as the (*volume*) *mixing ratio*. Mixing ratios are frequently expressed as ppmv ($= 10^{-6}$), ppbv ($= 10^{-9}$), or pptv (10^{-12}). When assuming ideal gas behavior, for given p and T, mixing ratios are proportional to partial pressures and mole fractions. Thus, they can be easily converted to molar concentrations by (see also Chapter 4):

$$\text{concentration in mol} \cdot L^{-1} = \frac{\text{mixing ratio}}{RT} \cdot p \qquad (16\text{-}21)$$

where p is the total pressure. For atmospheric gas-phase reactions one commonly expresses concentrations of reactive species not in $mol \cdot L^{-1}$ but in $molecule \cdot cm^{-3}$. Thus, at 298 K and $p = 1$ bar, concentration and mixing ratio are related by:

$$\text{concentration in molecule} \cdot cm^{-3} = \frac{\text{mixing ratio} \, (6.022 \times 10^{23})(1)}{(0.0831)(10^3)(298)} \qquad (16\text{-}22)$$

$$= (\text{mixing ratio})(2.43 \times 10^{19})$$

Direct spectroscopic measurements of $HO^\bullet$ close to ground level show peak daytime $HO^\bullet$ concentrations in the range of 2 to 10×10^6 $molecule \cdot cm^{-3}$ for mid-latitude northern hemisphere sites in the summer (Atkinson et al., 1999). These measurements show a distinct diurnal profile, with a maximum $HO^\bullet$ concentration around solar noon. Model calculations suggest that, in addition to exhibiting a diurnal profile, the $HO^\bullet$ concentration depends on season and latitude. Thus, for example, the mean monthly surface $HO^\bullet$ concentrations (24 h averages) at 35°N, latitude are estimated to be in the order of 2×10^5 $molecule \cdot cm^{-3}$ in January and 2×10^6 $molecule \cdot cm^{-3}$ in July, compared to about 1.2×10^6 $molecule \cdot cm^{-3}$ during the whole year at the Equator. The summer/winter $HO^\bullet$ concentration ratio increases with increasing latitude because of the increasing light differences (for more details see

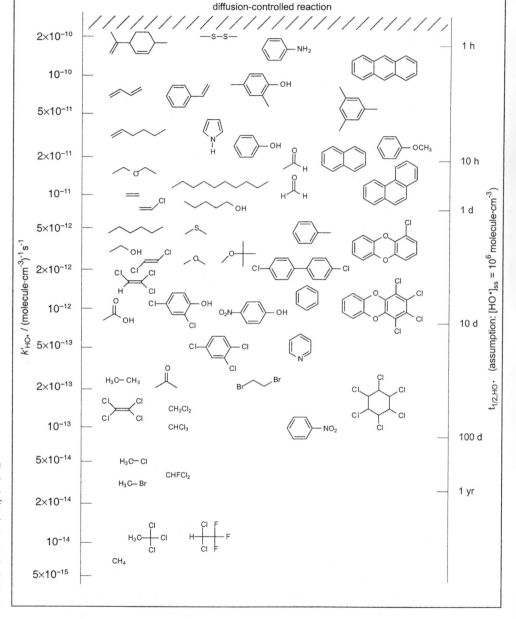

Figure 16.7 Second-order rate constants and half-lives for reaction of HO˙ radicals in the troposphere at 298 K for a series of organic compounds. For calculation of the half-lives a HO˙ steady-state concentration of 10^{-6} molecule·cm^{-3} has been assumed. Data from Atkinson (1989), Atkinson (1994), Anderson and Hites (1996), Brubaker and Hites (1997).

references cited in Atkinson et al., 1999). For practical purposes (e.g, for use in environmental fate models), it is reasonable to assume a diurnally, seasonally, and annually averaged global tropospheric HO˙ concentration of 1×10^6 molecule·cm^{-3}.

Rate Constants and Tropospheric Half-Lives for Reactions with HO˙

Second-order rate constants, $k'_{HO˙}$, allow us to estimate tropospheric half-lives $t_{1/2,HO˙}$ (Eq. 16-23) for environmentally relevant chemicals (see Fig. 16.7 for some examples):

$$t_{1/2,HO˙} = \frac{\ln 2}{k'_{HO˙}[HO˙]_{ss}}$$

(16-23)

Note that the units of $k'_{HO\cdot}$ are $(molecule \cdot cm^{-3})^{-1} s^{-1}$ and that $t_{1/2,HO\cdot}$ has been calculated for an average global tropospheric $HO\cdot$ concentration of 1×10^6 molecule $\cdot cm^{-3}$ at $25°C$. We should also note that, as a first approximation, we neglect the effect of temperature on $k'_{HO\cdot}$, because over the temperature range of the lower troposphere $(-40°$ to $30°C)$, this effect is less than a factor of 2 to 3 for many compounds of interest to us (Atkinson 1989 and 1994).

Comparison of Figs. 16.7 and 16.3 shows that relative reaction rates of organic chemicals with $HO\cdot$ follow more or less the same general pattern in air and in water; that is, compounds containing electron-rich double bonds or aromatic systems and/or easily abstractable H-atoms react faster as compared to compounds exhibiting no such functionalities. However, the differences in absolute rates are much more pronounced in the gas phase (about a range of 10^4 for the compounds considered) as compared to the solution phase (about a range of 10^2 for the same scope of compounds). The major reason is that in the aqueous phase, particularly for addition reactions of $HO\cdot$ to double bonds or aromatic systems, there is a rapid release of energy from the intermediate adduct to the solvent molecules. This stabilizes the intermediate as compared to the gas phase, where the reaction is much more reversible. Hence, in the gas phase, such reactions are more selective, and this is reflected in larger differences in reactivity between compounds. As a consequence only for compounds for which H-abstraction is the major reaction mechanism, correlation between reaction rate constants in the aqueous phase ($k'_{p,HO\cdot}$) and in the gas phase $k'_{HO\cdot}$ can be expected. So far, a reasonable correlation was found only for alkanes by Haag and Yao (1992):

$$\log k'_{p,HO\cdot} = 0.8 \log k_{HO\cdot} + 1.7 \,(R^2 = 0.76) \tag{16-24}$$

where $k_{HO\cdot}$ was converted from $(molecule \cdot cm^{-3})^{-1} s^{-1}$ to the same units as $k'_{p,HO\cdot}$, i.e., $M^{-1} s^{-1}$. The lack of such correlations is, in fact, somewhat unfortunate, because, as we will see in the following, $k_{HO\cdot}$ values can be estimated with quite good success from the structure of the compound.

Estimation of Gas-Phase HO• Reaction Rate Constants

Various methods for the estimation of gas-phase $HO\cdot$ reaction rate constants for organic compounds have been proposed, ranging from estimation methods for single classes to generalized estimation methods for the complete range of organic compounds. Many of these methods utilize molecular properties of the chemical including ionization energy, NMR chemical shifts, bond-dissociation energies, and infrared absorption frequencies, or they involve transition-state calculations (see references given in Kwok and Atkinson, 1995). However, most of these estimation methods are restricted in their use because of the limited database concerning the pertinent molecular properties. Therefore, we discuss here a structure-reactivity approach that was originally proposed by Atkinson (1986) and has been extended by Kwok and Atkinson (1995). This approach hinges on the assumption that the total rate constant, $k'_{HO\cdot}$, can be expressed by the sum of four rate constants, each of which describes one of the four different basic reaction mechanisms: (a) H-atom abstraction from C–H or O–H bonds [$k'_{HO\cdot}$ (H-abstr.)]; (b) $HO\cdot$ addition to >C=C< or

Table 16.3 Group Rate Constants (k_{prim}, k_{sec}, k_{tert}, k_{OH}) and Group Substituent Factors, $F(X)$, at 298 K for H Abstraction at C–H or O–H bonds [a]

Group Rate Constants (10^{12} cm^3 molecule^{-1} s^{-1}) $k_{prim} = 0.136$; $k_{sec} = 0.934$; $k_{tert} = 1.94$; $k_{OH} = 0.14$					
Group Substituent X	$F(X)$	Group Substituent X	$F(X)$	Group Substituent X	$F(X)$
–CH$_3$	1.00	–CF$_3$	0.071	–OCF$_3$ ⎫	
–CH$_2$– ⎫		–CHF$_2$	0.13	–OCF$_2$– ⎪	0.17
> CH – ⎬ "C"	1.23	–CH$_2$F	0.61	–OCHF$_2$ ⎬	
> C < ⎭		–CF$_2$–	0.018	–OCH$_2$F ⎭	
–C$_6$H$_5$ ⎫		–CHO ⎫	0.75	–OCH$_2$CF$_3$	0.44
> C = C < ⎬	1.0	> CO ⎭			
–C ≡ C – ⎭		–CH$_2$ – CO(–) ⎫		–NH$_2$ ⎫	
–F	0.094	> CH – CO(–) ⎬	3.9	> NH ⎬	9.3
–Cl	0.38	≡ C – CO(–) ⎭		> N – ⎭	
–Br	0.28	–COOR	0.74	–SH ⎫	
–I	0.53	–COCF$_3$	0.11	–S– ⎬	7.8
		–CN	0.19	–SS– ⎭	
		–CH$_2$CN	0.12	–OPO(OR)$_2$ ⎫	20.5
–CH$_2$Cl ⎫		–NO$_2$	0.0	–SPO(OR)$_2$ ⎭	
–CHCl$_2$ ⎪		–CH$_2$NO$_2$	0.14	3-member-ring	0.02
–CHCl– ⎬	0.36	–OH	3.5	4-member-ring	0.28
> CCl– ⎭		–OR	8.4	5-member-ring	0.64
				≥6-member-ring	1.0
–CCl$_3$	0.069				
–CCl$_2$F	0.044				
–CClF$_2$	0.031				

[a] Data from Kwok and Atkinson (1995); R = alkyl group.

–C≡C– bonds [$k'_{HO\bullet}$(DB)]; (c) HO$^{\bullet}$ addition to aromatic rings [$k'_{HO\bullet}$(Ar)]; and (d) interaction with N-, P-, and S-containing functional groups [$k'_{HO\bullet}$(NPS)]:

$$k'_{HO\bullet} = k'_{HO\bullet}(\text{H-abstr.}) + k'_{HO\bullet}(\text{DB}) + k'_{HO\bullet}(\text{Ar}) + k'_{HO\bullet}(\text{NPS}) \qquad (16\text{-}25)$$

Each of these four second-order rate constants can be estimated from the structure of the compound of interest using group rate constants and substituent factors that have been derived from a large set of experimental data. This method has proven to be quite successful for prediction of $k'_{HO\bullet}$ for compounds that are well represented in this database (i.e., predictions within a factor of 2 to 3). Larger deviations are found, for example, for certain ethers and, particularly, for polyhalogenated compounds exhibiting several fluorine atoms (see Table 3 in Kwok and Atkinson, 1995). Finally, for multifunctional molecules, the calculation may exceed the rate for diffusion-controlled reactions. In these cases, a "maximum" rate constant of $\sim 2 \times 10^{-10}$ (molecule $\cdot$ cm^{-3})$^{-1}$ s^{-1} should be used. In the following, we will briefly sketch this method and give some example calculations. For a more detailed discussion we refer to the literature (Kwok and Atkinson, 1995).

H-Atom Abstraction from C–H and O–H Bonds. The estimation of $k'_{HO\cdot}$ (H-abstr.) is based on group rate constants (see Table 16.3) for –CH$_3$ (k_{prim}), –CH$_2$– (k_{sec}), >CH– (k_{tert}), and –OH (k_{OH}) that reflect the presence of the "standard" substituent –CH$_3$ [i.e., k_{tert} is the second-order rate constant for the abstraction of the tertiary C–H in (CH$_3$)$_3$C–H]. These standard values are then modified by multiplication with factors $F(X)$ for substituents connected to the group considered:

$$
\begin{aligned}
k'_{HO\cdot} \text{ (H–abstr.)} = \ & k_{prim}F(X) + k_{prim}F(X^I) + \ldots && \text{for each –CH}_3 \\
& + k_{sec}F(X)F(X^I) + k_{sec}F(X^{II})F(X^{III}) + \ldots && \text{for each –CH}_2\text{–} \\
& + k_{tert}F(X)F(X^I)F(X^{II}) + k_{tert}F(X^{III})F(X^{IV})F(X^V) + \ldots && \text{for each >CH–} \\
& + k_{OH}F(X) + k_{OH}F(X^I) + \ldots && \text{for each –OH}
\end{aligned}
\tag{16-26}
$$

where X, X^I, X^{II}, etc. denote the various substituents. $F(X)$ values for some common substituents are given in Table 16.3. More values can be found in the literature (Kwock and Atkinson, 1995). Some example calculations are performed in Illustrative Example 16.3.

HO$^\bullet$ Addition to >C=C< and –C≡C-Bonds. For HO$^\bullet$ addition to double or triple bonds, an analogous approach is taken as for H abstraction. For a given double bond or system of two conjugated double bonds, a group rate constant is defined. These group rate constants (see examples given in Table 16.4) reflect the $k'_{HO\cdot}$ (DB) value if the double or triple bond(s) is(are) substituted by alkyl substituent groups. This means that for alkyl substituents, the group substituent factors, $C(X)$ (see examples given in Table 16.5) that are used to modify the group rate constant are set equal to 1.0. Hence, for example, the contribution of a double bond that is carrying two substituents X and X$'$ in *trans*-position is given by:

$$
k'_{HO\cdot} \text{ (X–CH=CH–X}', trans) = k(trans\text{–CH=CH–}) \, C(X) \, C(X^I)
\tag{16-27}
$$

A simple illustration is the estimation of $k'_{HO\cdot}$ of *trans*-dichloroethene, which is calculated by [note that $k'_{HO\cdot} = k'_{HO\cdot}$(DB), Eq. 16-25]:

$$
k'_{HO\cdot} \text{ (Cl–CH=CH–Cl)} = k(trans\text{–CH=CH–}) \, C(Cl) \, C(Cl)
\tag{16-28}
$$

Inserting the corresponding values from Tables 16.4 and 16.5, respectively, into Eq. 16-28 yields a $k'_{HO\cdot}$ value for *trans*-dichloroethene of $k'_{HO\cdot} = (64.0)(0.21)(0.21) \times 10^{-12} = 2.8 \times 10^{-12}$ cm^3 molecule^{-1} s^{-1}, which corresponds very well with the experimental value of 2.5×10^{12} cm^3 molecule^{-1} s^{-1} (Atkinson, 1994). Another example calculation is given in Illustrative Example 16.3.

Table 16.4 Group Rate Constants for HO˙ Addition to Double and Triple Bonds at 298 K [a]

Structural Unit	$10^{12} \cdot k$(unit) (cm^{3}molecule^{-1}s^{-1})	Structural Unit	$10^{12} \cdot k$(unit) (cm^{3}molecule^{-1}s^{-1})
CH$_2$=CH–	26.3	>C=C–C=C<	
CH$_2$=C<	51.4		
cis–CH=CH–	56.4		
trans–CH=CH–	64.0	5 H, 1 substituent	105
–CH=C<	86.9	4 H, 2 substituents	142
>C=C<	110	3 H, 3 substituents	190
HC≡C	7.0	2 H, 4 substituents	260
–C≡C–	27		

[a] Data from Kwok and Atkinson (1995).

Table 16.5 Group Substituent Factors, $C(X)$, at 298 K for HO˙ Addition to Carbon–Carbon Double and Triple Bonds [a]

Substituent Group X	$C(X)$	Substituent Group X	$C(X)$
–F	0.21	–COCH$_3$	0.90
–Cl	0.21	–COOR	0.25
–Br	0.26	–OR	1.3
–CH$_2$Cl	0.76	–CN	0.16
–CHO	0.34	–CH$_2$OH	1.6

[a] Data from Kwok and Atkinson (1995); R = alkyl group.

HO˙-Addition to Monocyclic Aromatic Compounds and Biphenyls. The rate constants for HO˙ addition to aromatic rings are simply calculated by using the rate constant for the unsubstituted compound and the Hammett-σ_j^+ constants. Thus, $k'_{HO˙}$(Ar) for addition to a substituted benzene ring can be estimated by:

$$\log k'_{HO˙}(Ar)/(cm^3\,molecule^{-1}\,s^{-1}) = -11.7 - 1.34 \sum \sigma_j^+ \qquad (16\text{-}29)$$

where σ_j^+ are the substituent constants that we have already encountered in Section 16.2, and where it is assumed that HO˙ adds to the substituent-free position yielding the least positive or most negative value of $\sum \sigma_j^+$ (i.e., the carbon with the highest electron density). Furthermore, steric hindrance is neglected, and therefore, σ_{jortho}^+ is set equal to σ_{jpara}^+. Finally, we should note that σ_{jmeta}^+ is equal to σ_{jmeta} (Table 8.5). Some σ_j^+ values are listed in Table 16.6. More data can be found in Hansch et al. (1991). Example calculations are given in Illustrative Example 16.3.

HO˙ Interaction with N-, P-, and S-Containing Groups. A rather limited data set is available for quantification of the interaction of HO˙ with nitrogen-, phosphorus- and sulfur-containing functional groups. The group rate constants for some func-

Table 16.6 Electrophilic Aromatic Substituent Constants [a]

Substituent j	$\sigma^+_{j\text{para}} = \sigma^+_{j\text{ortho}}$	$\sigma^+_{j\text{meta}}$	Substituent j	$\sigma^+_{j\text{para}} = \sigma^+_{j\text{ortho}}$	$\sigma^+_{j\text{meta}}$ [b]
–H	0.00	0.00	–OH	–0.92	0.10
–CH$_3$	–0.31	–0.06	–OCH$_3$	–0.78	0.11
–CH$_2$CH$_3$	–0.30	–0.06	–OC$_6$H$_5$ (phenyl)	–0.50	0.25
–CH(CH$_3$)$_2$	–0.28	–0.06	–OCOCH$_3$	–0.19	0.36
–CH$_2$CH$_2$CH$_2$CH$_3$	–0.29	–0.07	–CHO	0.73	0.36
–C(CH$_3$)$_3$	–0.26	–0.10	–COOCH$_3$	0.49	0.33
–CH=CH$_2$	–0.16	–0.08	–COOH	0.42	0.37
–C$_6$H$_5$ (phenyl)	–0.18	0.06	–COO$^-$	–0.02	–0.03
–CH$_2$OH	–0.04	0.07	–CN	0.66	0.56
–CH$_2$Cl	–0.01	0.12	–NH$_2$	–1.30	–0.16
–CF$_3$	–0.61	0.44	–NHCH$_3$	–1.81	–0.25
–F	–0.07	0.34	–N(CH$_3$)$_2$	1.70	–0.16
–Cl	0.11	0.37	–NHCOCH$_3$	–0.60	0.21
–Br	0.14	0.40	–NO$_2$	0.79	0.73
–I	0.14	0.35	–SCH$_3$	–0.60	0.13

[a] $\sigma^+_{j\text{para}}$ values from Hansch et al. (1991) and Hansch and Leo (1995); $\sigma^+_{j\text{meta}} = \sigma_{j\text{meta}}$ (see Table 8.5).

Table 16.7 Group Rate Constants for HO$^\bullet$ Interaction with N-, P-, and S-Containing Groups [a]

Structural Unit	$10^{12} k$(unit) (cm^3 molecule^{-1} s^{-1})	Structural Unit	$10^{12} k$(unit) (cm^3 molecule^{-1} s^{-1})
RNH$_2$	21	RSH	32.5
RR'NH	63	RSR	1.7
RR'R''N	66	RSSR	225
RR'N–NO	0	$\equiv$P=O	0
RR'N–NO$_2$	1.3	$\equiv$P=S	53

[a] Data from Kwock and Atkinson (1995); R,R',R'' = alkyl groups.

tional groups are summarized in Table 16.7. Example calculations are given in Illustrative Example 16.3.

In summary, when applied to the type of compounds from which the group rate constants and substituent factors given in Tables 16.3 to 16.7 have been derived, the estimation of $k'_{\text{HO}^\bullet}$ from structure yields quite satisfactory results. As is demonstrated by the examples given in Illustrative Example 16.3, in many cases only one or two structural moieties dominate the overall $k'_{\text{HO}^\bullet}$ value. Therefore, calculation of $k'_{\text{HO}^\bullet}$ can often be simplified by taking into account only a few terms of the overall estimation equation (Eq. 16-25).

Illustrative Example 16.3   **Estimating Tropospheric Half-Lives of Organic Pollutants**

Problem

Estimate the $t_{1/2,HO\cdot}$ values (Eq. 16-23) at 25°C of the following compounds: (a) isooctane (2,2,4-trimethylpentane), (b) 1,1,2-trichloroethane, (c) tetrahydrofuran, (d) 1-heptene, (e) *cis*-1,3-dichloropropene, (f) 3-nitrotoluene (1-methyl-3-nitrobenzene), and (g) dibenzo-*p*-dioxin. Assume an average HO· steady-state concentration of 1×10^6 molecule·cm^{-3} air.

isooctane

Answer (a)

For this molecule, only H abstraction is important, Hence, $k'_{HO\cdot} = k'_{HO\cdot}$(H-abstr.), which is given by (Table 16.3):

$$
\begin{aligned}
k'_{HO\cdot}\text{(H-abstr.)} &= 5\, k_{prim}F(\text{"C"}) + k_{sec}F(\text{"C"}) + k_{tert}F(\text{"C"}) \\
&= 5\,(k_{prim} + k_{sec} + k_{tert})\,F(\text{"C"}) \\
&= [5(0.136) + (0.934) + (1.94)]\,(1.23) \times 10^{-12}\ \text{cm}^3\,\text{molecule}^{-1}\,\text{s}^{-1} \\
&= 4.4 \times 10^{-12}\ \text{cm}^3\,\text{molecule}^{-3}\,\text{s}^{-1}
\end{aligned}
$$

[The experimental value is 3.9×10^{-12} cm^3 molecule^{-3} s^{-1}; Atkinson (1989)].

Insertion of this value into Eq. 16-23 together with $[HO\cdot]_{ss} = 10^6$ molecule·cm^{-3} yields a half-life of:

$$
t_{1/2,HO\cdot} = \frac{0.69}{4.4 \times 10^{-6}}\,s \cong 2\ d
$$

1,1,2-trichloroethane

Answer (b)

As for isooctane [see (a)], only H abstraction has to be taken into account (Table 16.3):

$$
\begin{aligned}
k'_{HO\cdot} &= k'_{HO\cdot}\text{(H-abstr.)} = k_{sec}\,F(\text{Cl})\,F(\text{CHCl}_2) + k_{tert}\,F(\text{Cl})\,F(\text{Cl})\,F(\text{CH}_2\text{Cl}) \\
&= [(0.934)(0.38)(0.36) + (1.94)(0.38)(0.38)(0.36)] \times 10^{-12}\ \text{cm}^3\,\text{molecule}^{-1}\,\text{s}^{-1} \\
&= 2.3 \times 10^{-13}\ \text{cm}^3\,\text{molecule}^{-1}\,\text{s}^{-1}
\end{aligned}
$$

[The experimental value is 3.2×10^{-13} cm^3 molecule^{-3} s^{-1}; Atkinson (1989).]

This corresponds to an estimated $t_{1/2,HO\cdot} \cong 35\ d$.

tetrahydrofuran

Answer (c)

Again as in cases (a) and (b) H abstraction is important:

$$
k'_{HO\cdot} = k'_{HO\cdot}\text{(H-abstr.)} = 2\,k_{sec}\,F(\text{"C"})\,F(\text{OR}) + 2\,k_{sec}\,F(\text{"C"})\,F(\text{CH}_2\text{OR})
$$

No $F(\text{CH}_2\text{OR})$ value is available. However, it can be assumed that the overall $k'_{HO\cdot}$ is dominated by the first term since $F(\text{OR}) = 8.4$ (see Table 16.3). Thus:

$$k'_{HO\bullet} = 2(0.934)(1.23)(8.4) \times 10^{-12} \text{ cm}^3 \text{ molecule}^{-1} \text{ s}^{-1}$$

$$= 1.9 \times 10^{-11} \text{ cm}^3 \text{ molecule}^{-1} \text{ s}^{-1}$$

[The experimental value is 1.6×10^{-11} cm^{3}molecule^{-1} s^{-1}; Atkinson (1989).]

This corresponds to an estimated $t_{1/2,HO\bullet} \cong 10$ h.

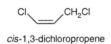

1-heptene

Answer (d)

This reaction of 1-heptene is dominated by addition of HO$^\bullet$ to the double bond (Table 16.4), but abstraction of H-atoms cannot be completely neglected (Table 16.3):

$$k'_{HO\bullet} = k'_{HO\bullet}(DB) + k'_{HO\bullet}(\text{H-abstr.})$$

$$= (CH_2 = CH-) + 3\,k_{sec}\,F("C")^2 + k_{sec}\,F("C")\,F(CH_3) + k_{prim}\,F("C")$$

$$= [26.3 + (3)(0.934)(1.23)^2 + (0.934)(1.23)(1) + (0.136)(1.23)] \times 10^{-12} \text{ cm}^3 \text{ molecule}^{-1} \text{ s}^{-1}$$

$$= 3.2 \times 10^{-11} \text{ cm}^3 \text{ molecule}^{-1} \text{ s}^{-1}$$

[The experimental value is 3.6×10^{-11} cm^3 molecule^{-3} s^{-1}; Atkinson (1989).]

Note that for the double bond we have chosen a factor for a saturated carbon, $F("C")$, which is not really appropriate. However, because of the dominating role of HO$^\bullet$ addition to the double bond, an $F(C=C)$ value is difficult to derive. Nevertheless, the result shows that estimated and experimental value are in very good agreement. Using the estimated $k'_{HO\bullet}$ value, a $t_{1/2,HO\bullet} \cong 6$ h is obtained.

Answer (e)

cis-1,3-dichloropropene

The reaction of *cis*-1,3-dichloropropene is dominated by addition of HO$^\bullet$ to the double bond (Tables 16.4 and 16.5):

$$k'_{HO\bullet} = k'_{HO\bullet} = (cis-CH-C<)\,C(Cl)\,C(CH_2Cl)$$

$$= (56.4)(0.21)(0.76) \times 10^{-12} \text{ cm}^3 \text{ molecule}^{-1} \text{ s}^{-1}$$

$$= 9.0 \times 10^{-12} \text{ cm}^3 \text{ molecule}^{-1} \text{s}^{-1}$$

[The experimental value is 8.4×10^{-12} cm^3 molecule^{-3} s^{-1}; Atkinson (1989).]

Thus, the estimated half-life is $t_{1/2,HO\bullet} \cong 21$ h.

Answer (f)

3-nitrotoluene

For 3-nitrotoluene, addition of HO$^\bullet$ to the aromatic ring is dominating. Hence, $k'_{HO\bullet} \cong k_{HO\bullet}(Ar)$. Use Eq. 16-29 to estimate $k'_{HO\bullet}(Ar)$:

$$\log k'_{HO\bullet}(Ar) = -11.7 - 1.34 \sum \sigma_j^+ \qquad (1)$$

Check which of the positions 2, 4, 5, or 6 has the most negative (least positive). $\sum \sigma_j^+$ value. Note that positions 2, 4, and 6 have the same $\sum \sigma_j^+ = -0.31 + 0.79 = 0.48$.

$\Sigma\sigma_j^+$ for position 5 is $-0.06 + 0.73 = 0.67$. Insert the value of 0.48 into Eq. 1 to obtain:

$$\log k_{HO\bullet}'(Ar) = -11.7 - 0.64 = -12.34$$

Because there are three equivalent positions to which $HO\bullet$ can add, the rate constant can be assumed to be three times the rate constant estimated above:

$$k_{HO\bullet}' = 3\, k_{HO\bullet}'(Ar) = 1.4 \times 10^{-12}\ cm^3\ molecule^{-3}\ s^{-1}$$

[The experimental value is $1.1 \times 10^{-12}\ cm^3\ molecule^{-3}\ s^{-1}$; Atkinson (1989).]

This corresponds to an estimated $t_{1/2,HO\bullet}$ value of about *17* d.

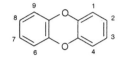

dibenzo[1.4]dioxin

Answer (g)

The only possible reaction is addition of $HO\bullet$ to the aromatic rings. Note that all positions have the same $\Sigma\sigma_j^+$ value:

$$\Sigma\sigma_j^+ = \Sigma\sigma_{OC_6H_5,ortho}^+ + \Sigma\sigma_{OC_6H_5,meta}^+ = -0.50 + 0.25 = -0.25$$

Note that we neglect the second aromatic ring; that is, we consider addition of $HO\bullet$ to one of the benzene rings as being equivalent to addition to:

$$\underset{I}{\text{(structure: benzene ring with O—C}_6\text{H}_5 \text{ and O—C}_6\text{H}_5 \text{ substituents)}}$$

Insertion of $\Sigma\sigma_j^+ = -0.25$ into Eq. 16-29 yields:

$$\log k_{HO\bullet}' = \log 4\, k_{HO\bullet}'(Ar) = -11.7 + 0.34 = -10.76$$

or:

$$k_{HO\bullet}'(Ar) = 1.7 \times 10^{-11}\ cm^3\ molecule^{-1}\ s^{-1}$$

Because there are two benzene rings to which $HO\bullet$ can add, the rate constant for dibenzo-*p*-dioxin can be assumed to be twice the rate constant estimated for compound I above:

$$k_{HO\bullet}' = 3.4 \times 10^{-11}\ cm^3\ molecule^{-1}\ s^{-1}$$

[The experimental value is $1.5 \times 10^{-11}\ cm^3\ molecule^{-3}\ s^{-1}$; Kwok et al. (1994).]

The estimated half-life is: $t_{1/2,HO\bullet} = 22$ h.

16.4 Questions and Problems

Questions

Q 16.1

What is meant by the terms *indirect photolysis* and *sensitized photolysis*? What types of reactions does indirect photolysis of organic pollutants include in natural waters? in the atmosphere?

Q 16.2

Why does one speak of *transient photooxidants*? Give examples of such species. How can their concentrations be measured? What are their major sources and sinks in natural waters? in the atmosphere?

Q 16.3

Describe the general kinetic approach that can be used to quantify indirect photolysis involving well-defined photooxidants in: (a) the water column of surface waters and (b) the atmosphere.

Q 16.4

Why is the hydroxyl radical (HO$\cdot$) a more important photooxidant in the atmosphere as compared to surface waters? What kind of reactions do organic pollutants undergo with HO$\cdot$? How structure specific are these reactions?

Q 16.5

Somebody claims that the volume-averaged steady-state concentration of singlet oxygen ($[^1O_2]_{ss}$) in a well-mixed water body of 1 m depth is more or less independent of the DOM concentration, whereas at the surface it is proportional to [DOM], at least for [DOC] < 20 mg oc$\cdot$L^{-1}. Could this person be correct, and if yes, why?

Q 16.6

Why is $^1O_2$ a more selective photooxidant than HO$\cdot$? With what type of organic pollutants does $^1O_2$ primarily react?

Q 16.7

Why is it difficult to quantify DOM-mediated photolysis of organic pollutants? What type of reactions do $^3DOM^*$ species undergo with organic chemicals?

Problems

P 16.1 *How Important Are Reactions of Phenols with HO$\cdot$ and $^1O_2$ in a Shallow Pond?*

Consider the shallow well-mixed pond (average depth = 2 m, T = 25°C; pH = 8.5; $\alpha(\lambda)$ see Table 15.6; mean residence time of the water: 35d) introduced in Illustrative Example 16.2. In this pond, a midday, near-surface steady-state concentration of $^1O_2$ ($[^1O_2]_{ss}^0$(noon)) of 8×10^{-14} M has been determined using FFA as probe molecule (see Eq. 16-12). Recall that maximum $^1O_2$ production occurs at

410 nm. The steady-state HO$^{\bullet}$ concentrations in this pond have already been estimated in Illustrative Example 16.2. Calculate the 24 h averaged photolysis half-life of 4-methoxy- and 2-chlorophenol in this pond by assuming that only the reactions with HO$^{\bullet}$ and $^{1}O_2$ are important (note that for 4-methoxyphenol reactions with ^{3}NOM$^{\bullet}$ may also be relevant). What is the relative importance of the two processes? How important are they compared to flushing? (see Section 12.4). In the literature you find the following data (Tratnyek et al., 1991):

4-methoxyphenol
$pK_a = 10.2$

2-chlorophenol
$pK_a = 8.5$

$k_{p,^{1}O_2}^{'}(ArO^{-}) = 7 \times 10^{8}\,M^{-1}\,s^{-1}$ $k_{p,^{1}O_2}^{'}(ArO^{-}) = 2 \times 10^{8}\,M^{-1}\,s^{-1}$

$k_{p,^{1}O_2}^{'}(ArOH) = 3 \times 10^{7}\,M^{-1}\,s^{-1}$ $k_{p,^{1}O_2}^{'}(ArOH) = 5 \times 10^{6}\,M^{-1}\,s^{-1}$

For the reaction with HO$^{\bullet}$ you can assume that both compounds react with a nearly diffusion controlled rate constant (see Fig. 16.3).

P 16.2 *Evaluating Reactions of Organic Pollutants with the Carbonate Radical (CO$_3^{\bullet-}$)*

As pointed out in Section 16.2 (Eq. 16-2), the major source for $CO_3^{\bullet-}$ radicals in advanced oxidation processes, and possibly also in surface water, is reaction of HCO_3^{-} and CO_3^{2-} with hydroxyl radical (HO$^{\bullet}$). The major sink for $CO_3^{\bullet-}$ is regarded to be DOM. The rate constant for reaction of $CO_3^{\bullet-}$ with DOM is in the order of $4 \times 10^{1}\,L\cdot(mg\cdot oc)^{-1}\,s^{-1}$ (Larson and Zepp, 1988), which is about 600 times smaller than for the reaction of HO$^{\bullet}$ with DOM constituents (see Illustrative Example 16.1).

(a) Estimate the near-surface carbonate radical steady-state concentration at noon ($[CO_3^{\bullet-}]_{ss}^{0}$(noon)) and averaged over a day $[CO_3^{\bullet-}]_{ss}^{0}$ (24 h)) in Greifensee (47.5°N) on a clear summer day. The concentrations of the relevant water constituents are given in Illustrative Example 16.1. *(Hint:* Note that about 15% of the HO$^{\bullet}$ produced is consumed by HCO_3^{-}/CO_3^{2-}). Compare the carbonate radical concentrations with the $[HO^{\bullet}]_{ss}^{0}$ concentrations calculated in Illustrative Example 16.1.

(b) For reactions of *meta*- and *para*-substituted anilines, Canonica and Tratnyek (2002) reported the following QSAR:

$$\log k_{p,CO_3^{\bullet-}}^{'}(ArNH_2)/(M^{-1}\,s^{-1}) = -0.98\sum \sigma_j^{+} + 8.71$$

Calculate the 24 h averaged near-surface half-lives of 3,4-dimethylaniline (3,4-DMA) and 4-nitroaniline (4-NA) with respect to reaction with $CO_3^{\bullet-}$. How important are these reactions relative to the reactions of the two compounds with HO$^{\bullet}$? Note that in analogy to using σ_{jpara}^{-} for delocalization of a negative charge (Chapter 8), σ_{jpara}^{+} values have to be used in cases where a positive charge or a radical is delocalized (σ_j^{+} values from Hansch et al., 1991; see also Table 16.6).

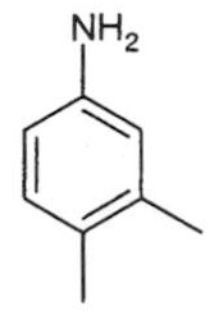

3,4-dimethylaniline

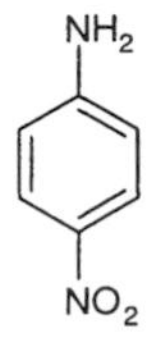

4-nitroaniline

$$\sigma^+_{CH_3 para} = -0.37$$

$$\sigma^+_{CH_3 meta} = -0.06$$

$$\sigma^+_{NO_2 para} = 0.79$$

P 16.3 *Estimating Atmospheric Half-Lives of Organic Pollutants with Respect to Reaction with HO*·

Estimate the $k'_{HO^·}$ and $t_{1/2,HO^·}$ (for $[HO^·]_{ss} = 10^6$ molecule·cm^{-3}) values at 25°C of the following compounds using the group rate constants and substituent factors and constants given in Section 16.3 and in Tables 16.3 to 16.6. Compare the estimated values of $k'_{HO^·}$ with the experimental values given in parentheses. Try to explain any larger discrepancies (i.e., > factor 3).

(a)

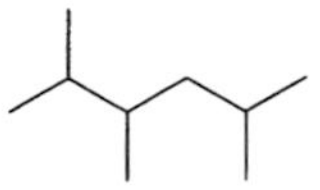

2,3,5-trimethylhexane (exp: 8 x 10^{-12} cm^3 molecule^{-1}s^{-1})

(b)

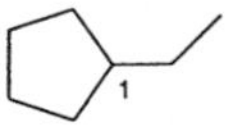

ethylcyclopentane (exp: 9 x 10^{-12} cm^3 molecule^{-1}s^{-1})

[Note that for the carbon denoted as 1, the 5-number-ring factor (Table 16.3) has to be applied.]

(c)

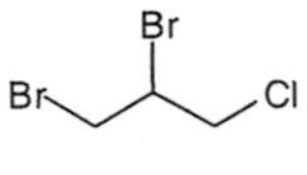

1,2-dibromo-3-chloropropane (exp: 4 x 10^{-13} cm^3 molecule^{-1}s^{-1})

(d)

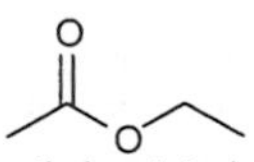

ethylacetate (exp: 1.8 x 10^{-12} cm^3 molecule^{-1}s^{-1})

(e)

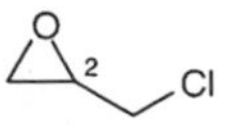

1-chloro-2,3-epoxypropane (exp: 4 x 10^{-13} cm^3 molecule^{-1}s^{-1})

[Note that for the carbon denoted as 2, the 3-number-ring factor (Table 16.3) has to be applied.]

(f)

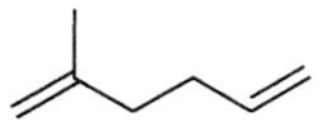

2-methyl-1,5-hexadiene (exp: 9.6 x 10^{-11} cm^3 molecule^{-1}s^{-1})

(g)

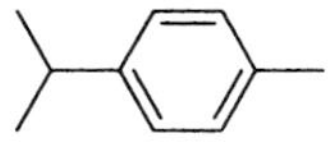

α-terpinene (exp: 3.6 x 10^{-10} cm^3 molecule^{-1}s^{-1})

(h)

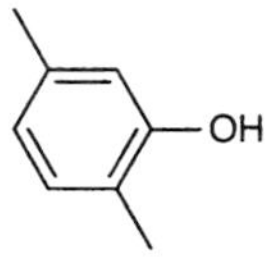

2,5-dimethylphenol (exp: 4 x 10^{-11} cm^3 molecule^{-1}s^{-1})

(i)

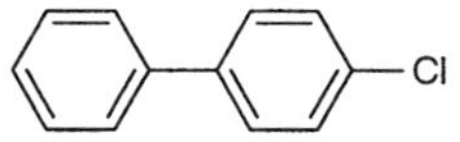

4-chlorobiphenyl (exp: 4 x 10^{-12} cm^3 molecule^{-1}s^{-1})

(j)

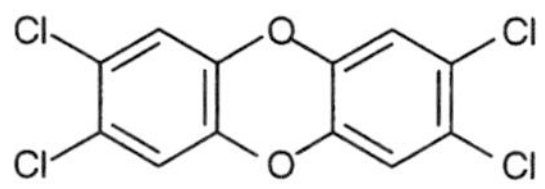

2,3,7,8-tetrachloro-dibenzo[1.4]dioxin (exp: 4 x 10^{-12} cm^3 molecule^{-1}s^{-1})

Chapter 17

BIOLOGICAL TRANSFORMATIONS

17.1 Introduction
Illustrative Example 17.1: *Is a Proposed Microbial Transformation Thermodynamically Feasible?*

17.2 Some Important Concepts about Microorganisms
Microbial Ecology and Interactions
Enzymology

17.3 Biochemical Strategies of Microbial Organic Chemists
Chemical Structure-Biodegradability Studies
Hydrolyses
Illustrative Example 17.2: *What Products Do You Expect from the Microbial Degradation of EDB?*
Illustrative Example 17.3: *What Products Do You Expect from the Microbial Degradation of Linuron?*
Oxidations Involving Electrophilic Oxygen-Containing Bioreactants
Illustrative Example 17.4: *What Products Do You Expect from the Microbial Degradation of Vinyl Chloride in an Oxic Environment?*
Reductions Involving Nucleophilic Electron-Rich Bioreactants
Illustrative Example 17.5: *What Products Do You Expect from the Microbial Degradation of DDT in a Reducing Environment?*
Additions of Carbon-Containing Moieties or Water

17.4 Rates of Biotransformations: Uptake
Substrate Bioavailability and Uptake Kinetics

17.5   Rates of Biotransformations: Microbial Growth
Monod Population Growth Kinetics
Box 17.1: *Monod-Limiting-Substrate Models of Microbial Population Growth*
Illustrative Example 17.6: *Evaluating the Biodegradation of Glycerol by Microorganisms Growing on that Substrate in a Well-Mixed Tank*
Illustrative Example 17.7 *Estimating the Time to Degrade a Spilled Chemical*

17.6   Rates of Biotransformations: Enzymes
Michaelis-Menten Enzyme Kinetics
An Example of Enzyme Kinetics: Hydrolases
Box 17.2: Kinetic Expressions for Enzymatic Hydrolysis of R–L Under Simplifying Assumptions
Estimating Biotransformation Rates for Michaelis-Menten Cases
Illustrative Example 17.8: *Evaluating the Co-Metabolic Biodegradation of Trichloroethene by Microorganisms Growing on Methane in a Well-Mixed Tank*
Illustrative Example 17.9: *Estimating Biotransformation Rates of an Organic Pollutant in a Natural System*

17.7   Questions and Problems

17.1 Introduction

Reactions mediated by organisms constitute a very important set of transformations affecting the fates of almost all organic compounds in both natural and engineered environments. As for abiotic chemical and photochemical reactions, these biochemical processes change the structure of the organic chemical of interest thereby removing that particular compound from an environmental system. One major goal of this chapter is to characterize the rates of such losses allowing biotransformations to be included in mass balance considerations for compounds of interest. The products that result from biotransformations, of course, exhibit their own partitioning properties, reactivities, overall fates, and effects. Hence, the appearance of these compounds may inspire new mass balance and toxicity considerations (e.g., the appearance of vinyl chloride from biotransformations of other chlorinated solvents).

When we speak of biologically mediated transformations of organic compounds, we do not mean that these compounds are fully mineralized. *Mineralization* may be defined as the complete conversion of an organic chemical to stable inorganic forms of C, H, N, P, and so on. Consequently, mineralization generally entails several successive biological transformations. As a result, experimental observations relying solely on the appearance of CO_2 to quantify the loss rate of a specific organic chemical place a lower limit on the initial transformation rate of this organic compound (i.e., $d[CO_2]/dt \leq -d[i]/dt$).

Biochemical transformations of organic compounds are especially important because many reactions, although thermodynamically feasible, occur extremely slowly due to kinetic limitations. For example, we might be interested in the question of whether benzene can be biodegraded under naturally occurring methanogenic conditions (see Illustrative Example 17.1). Such *natural attenuation* of this toxic aromatic substance may be thermodynamically allowed under the perceived conditions. But these conditions may not be accurate (e.g., the benzene and methane chemical activities in the system). Also other environmental factors may cause the rate to be unobservably slow. One possibility is that the relevant microorganisms are simply not active in the environment of interest.

In this chapter we focus on transformations of xenobiotic organic compounds catalyzed by microorganisms. This immensely diverse group includes the most ubiquitous agents of biochemical processes determining the fates of organic chemicals in the environment. We first describe a few general principles pertaining to the ecology and enzymatic capabilities of microorganisms, especially insofar as these characteristics influence what we can expect these organisms to do (Section 17.2). Next, we focus on the major strategies microorganisms use to initiate xenobiotic compound transformations: hydrolysis, oxidation, reduction, and additions. In Section 17.3, the goal is to develop a better understanding of what structural features of organic chemicals are susceptible to microbial attack by knowing the mechanisms of these biochemical reactions. These relatively detailed descriptions also provide some basis for mathematical formulations with which rates of such processes may be described. Finally, in Sections 17.4 through 17.6, we discuss

Figure 17.1 Sequence of events in the overall process of biotransformations: (1) bacterial cell containing enzymes takes up organic chemical, *i*, (2) *i* binds to suitable enzyme, (3) enzyme:*i* complex reacts, producing the transformation product(s) of *i*, and (4) the product(s) is(are) released from the enzyme. Several additional processes may influence the overall rate such as: (5) transport of *i* from forms that are unavailable (e.g., sorbed) to the microorganisms, (6) production of new or additional enzyme capacity [e.g., due to turning on genes (induction), due to removing materials which prevent enzyme operation (activation), or due to acquisition of new genetic capabilities via mutation or plasmid transfer], and (7) growth of the total microbial population carrying out the biotransformation of *i*.

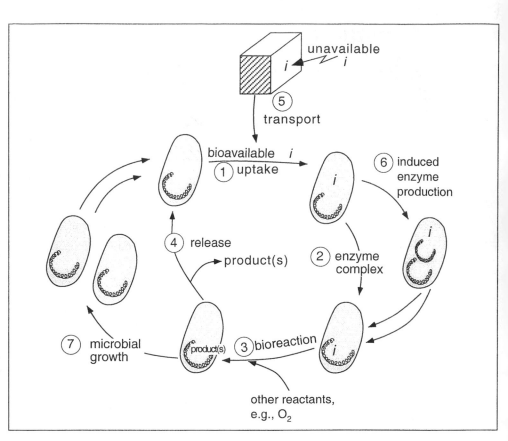

how we might approach predicting the overall rate of any particular biodegradation. Several very different steps in the process may end up controlling the overall rate (Fig. 17.1). For example, processes that limit the rate of delivery of substrate molecules to micro-organisms (*i* or O_2 as examples in Fig. 17.1) may determine the biotransformation rate. This effect is often cited as the reason why very hydrophobic aromatic hydrocarbons are oxidized more slowly in the environment than their less hydrophobic congeners. Mass transport has also been recognized as important for situations in which microorganisms form associations where some cells are buried beneath many others as in biofilms and flocs (Rittmann and McCarty, 2001). In these cases, the diminished rate of molecular diffusion of the chemical or of necessary cosubstrates like O_2 across intervening media (e.g., through polymeric microbial secretions, see Stewart, 1998 for review) may dictate the overall rate of xenobiotic chemical biotransformation. Obviously, one must know the average distribution of cells and other media, the tendencies of chemicals to partition between phases (e.g., solid–water partitioning), and the abilities of these molecules to move in each phase (e.g., by diffusion) to describe these transport processes mathematically. Probably, this is most critical whenever we are dealing with situations where most of a chemical of concern is sorbed, such as encountered in sediments, soils, and aquifers for very hydrophobic compounds.

Other potential limitations may influence biological transformations of organic compounds. First, organic compound uptake by the isolated cells may be rate limiting. Since many of the chemicals of interest to us are foreign to the micro-organisms, in general the cells may not have systems associated with their exterior membranes to pick up the substance *actively* and carry it rapidly to the interior of the cell. However, for nonpolar substances, the lipid-rich cell membranes, designed to separate charged metabolites within the cell from the ionic aqueous medium on the outside, allow hydrophobic chemicals to dissolve in these cellular boundaries (recall our discussions of biota–water partitioning, Chapter 10). Thus we see *passive* uptake as nonpolar species diffuse through the membranes to the interior of microbial cells. This uptake rate may limit the overall biotransformations of interest. Castro et al. (1985) interpreted the relative rates found for reduction of a series of organohalides by cultures of *Pseudomonas putida* as reflecting permeability limitations. In such a case, once the chemical arrives at the cellular site of enzymatic processing, it is rapidly transformed, so the overall chemical depletion rate is governed by the rate of substrate permeation into the cells. If permeation is the bottleneck, organic compound structural features affecting this uptake rate will be most important.

Once the chemical and enzymes coexist, however, other processes still remain that may govern the overall biodegradation (Fig. 17.1). First, the presence of the chemical may cause the organism to produce more enzyme for the degradation of this substance. Such an increase in enzymes may be due to: (1) *induction* causing genes to be turned on for enzyme expression or (2) *derepression* causing existing enzymes to be activated for their catalytic roles. The requisite enzymatic apparatus may be *constitutive* (i.e., it is always in place), but its abundance may still be enhanced should conditions encourage that. It is also possible that genetic *mutation* results in an enzyme that operates more effectively. Obviously, if greater quantities or more effective enzymes result, the overall internal processing of the chemical will speed up. Although conditions may cause more or less of constitutive enzymes to be present, once these catalysts are available, the biotransformation rate may depend on the specific interactions of the compound with the enzyme: both insofar as they associate with one another and then as they become involved in bond breaking and making. These interactions are best quantified using *Michaelis-Menten enzyme kinetics*, which we discuss in Section 17.6.

Finally, if the metabolism of the chemical results in substantial energy yield and/or cell-building materials, then the microorganism may increase in cell numbers in response. Then, the overall rate of biodegradation will be dictated by the rate of microbial population increase. In these cases, *microbial population dynamics,* an approach developed by *Monod* (1949), must be included in our analysis of the chemical of interest. This too will be expanded upon in Section 17.5.

In sum, biotransformations may be limited by: (1) delivery of the chemical to the organisms' metabolic apparatus capable of transforming the chemical, (2) the enzyme's ability to mediate the initial transformation of the chemical, or (3) the growth of a population of microorganisms in response to the presence of a new substrate. Depending on what limits the rate of biotransformation, different mathematical frameworks are required to describe the kinetics of the process both with respect to the nature of the equations and the parameters they require.

In this chapter, we treat biotransformations from an organic chemist's point of view, seeking to reveal how the structure of the chemical of interest dictates its initial biotransformation(s). We do not attempt to review the immense literature on biodegradation observations, nor deal with the complete metabolic pathways that are the focus of general biochemistry texts and are now found on the web (e.g., the University of Minnesota site at http://umbbd.ahc.umn.edu/index.html). We also do not describe much about microbial biochemistry and ecology, since excellent texts on these topics are available (e.g., Gottschalk, 1986; Madigan et al., 2000). Although biotransformation rates are probably the least well understood inputs to overall chemical fate modeling, we will find generalities to better anticipate, or at least interpret after the fact, biologically mediated reactions in environmental systems.

Illustrative Example 17.1

Is a Proposed Microbiological Transformation Thermodynamically Feasible?

Problem

A colleague suggests that benzene (C_6H_6) can be "naturally attenuated" by microorganisms in ground water under conditions of methanogenesis producing bicarbonate and methane. Check the thermodynamic feasibility of this transformation at 25°C assuming: the solution pH is 7, the bicarbonate concentration is 1 mM, the methane concentration is 100 μM, and the benzene concentration is 1 μM. In the literature (Hanselmann, 1991) you find the free energies of formation given at 25°C in the margin.

benzene
(C_6H_6)

$\Delta_f G^0_{C_6H_6} (\ell)$ = +123.0 kJ · mol^{-1}

$\Delta_f G^0_{H_2O} (\ell)$ = −237.2 kJ · mol^{-1}

$\Delta_f G^0_{HCO_3^-} (aq)$ = −586.9 kJ · mol^{-1}

$\Delta_f G^0_{CH_4} (g)$ = −50.8 kJ · mol^{-1}

$\Delta_f G^0_{H^+} (aq)$ = 0 kJ · mol^{-1}

Answer

First you need to balance a reaction in which some of the carbons in benzene (redox state −I) are oxidized to bicarbonate carbons (redox state +IV) and the rest are reduced to methane (redox state −IV):

$$C_6H_6 \rightarrow x \ HCO_3^- + y \ CH_4 \tag{1}$$

This situation requires $x + y = 6$. Further, the total number of electrons provided by the oxidation of benzene to HCO_3^- (5x electrons) must end up in the reduced product, methane (i.e., 3y electrons). Thus, from the two equations, $5x = 3y$ and $x + y = 6$ you get $x = 2.25$ and $y = 3.75$. To obtain the 2.25 times 3 = 6.75 moles of oxygen in the bicarbonate product, you need to add 6.75 moles of water to the benzene:

$$C_6H_6 + 6.75 \ H_2O \rightarrow 2.25 \ HCO_3^- + 3.75 \ CH_4 \tag{2}$$

Finally, to balance the reaction stoichiometry (and charge), we see that the 19.5 hydrogens of the reactants end up as 17.25 hydrogens on the bicarbonate and methane products, requiring 2.25 more protons to be adding to the products:

$$C_6H_6 + 6.75 \ H_2O \rightarrow 2.25 \ HCO_3^- + 3.75 \ CH_4 + 2.25 \ H^+ \tag{3}$$

Note that you could also express Eq. (3) as:

$$C_6H_6 + 4.5 \ H_2O \rightarrow 2.25 \ CO_2 + 3.75 \ CH_4 \tag{4}$$

Next, you need to calculate the $\Delta_r G^0$ for reaction 3 (see Section 12.2):

$$\Delta_r G^0 = -\Delta_f G^0_{C_6H_6}(aq) - 6.75\, \Delta_f G^0_{H_2O}(\ell) \\ + 2.25\, \Delta_f G^0_{HCO_3^-}(aq) + 3.75\, \Delta_f G^0_{CH_4}(aq) + 2.25\Delta_f G^0_{H^+}(aq) \tag{5}$$

To obtain $\Delta_f G^0_{C_6H_6}(aq)$, use $\Delta_f G^0(\ell)$ and adjust it with this compound's solution phase activity when benzene occurs at "1 M" standard concentration (see also Illustrative Example 14.1, nitrobenzene problem):

$$\Delta_f G^0_{C_6H_6}(aq) = \Delta_f G^0(\ell) + RT \ln x_{C_6H_6w}(1M) + RT \ln \gamma_{C_6H_6w} \tag{6}$$

Since x_{C6H6w} (1 M) is about 0.019 (note that the molar volume of benzene is 88.6 cm^3 mol^{-1}, and, therefore, there are about 50 moles H$_2$O in a 1 M aqueous benzene solution), and since $RT\ln\gamma_{C6H6w} = G^E_{C_6H_6} = 19.4$ kJ·mol^{-1} (Table 5.2), you obtain:

$$\Delta_f G^0_{C_6H_6}(aq) = 123.0 - 9.8 + 19.4 = +132.6 \text{ kJ·mol}^{-1}$$

To obtain $\Delta_f G^0_{CH_4}(aq)$, use $\Delta_f G^0_{CH_4}(g)$ and adjust it with this compound's K_H value given in Appendix C (= $10^{2.82}$ bar·mol·L^{-1}; see also Illustrative Example 12.2):

$$\Delta_f G^0_{CH_4}(aq) = \Delta_f G^0_{CH_4}(g) + RT \ln K_{CH_4H}$$

$$= -50.8 + 16.1 = -34.7 \text{ kJ·mol}^{-1}$$

Inserting all the $\Delta_f G^0(aq)$ values into Eq. 5 then yields:

$$\Delta_r G^0 = (-132.6) - (6.75)(-237.2) + (2.25)(-586.9) + (3.75)(-34.7) + 0$$
$$= +17.9 \text{ kJ·mol}^{-1}$$

Hence, under standard conditions (i.e., all reactants and products at 1 M, and H$_2$O as pure liquid) the reaction is unfavorable. At the conditions prevailing, the $\Delta_r G$ is given by (see Section 12.2, Eqs. 12-7 and 12-8, and Eq. 3 above):

$$\Delta_r G = \Delta_r G^0 + RT \ln Q_r$$
$$= \Delta_r G^0 + RT \ln \frac{\{HCO_3^-\}^{2.25}\{CH_4\}^{3.75}\{H^+\}^{2.25}}{\{C_6H_6\}\{1\}^{6.75}} \tag{7}$$

Assuming an activity coefficient, of close to 1 for HCO$_3^-$, insertion of the respective concentrations into Eq. 7 then yields:

$$\Delta_r G = +17.9 + 2.48 \ln \frac{(10^{-3})^{2.25}(10^{-4})^{3.57}(10^{-7})^{2.25}}{(10^{-6})(1)}$$

$$= +17.9 - (2.48)\ln(10^{-31.5}) = -162 \text{ kJ·mol}$$

Thus, under the conditions prevailing in the aquifer, the reaction is thermodynamically feasible.

17.2   Some Important Concepts About Microorganisms

Microbial Ecology and Interactions

Although plants and animals can transform many xenobiotic organic chemicals, from an environmental system mass balance point of view, microorganisms often play the most important role in degrading organic compounds. Microorganisms include diverse bacteria, protists, and fungi. Representatives are present virtually everywhere in nature, even under extreme conditions of temperature, pressure, pH, salinity, oxygen, nutrients, and low water content.

Although widely present, only a fraction of the microorganisms are metabolically active in a particular environment, depending on the conditions. Dramatic differences such as oxic (O_2 present) versus anoxic (O_2 absent) conditions (see also Chapter 14), and less visual ones like high versus low nutrient or trace metal activities, affect the particular mixture of species that are active at any one place and time. Fortunately, with respect to biotransformations of organic compounds, subtle effects may not be too important since many organisms exhibit similar biochemical pathways and capabilities. However, environmental factors such as whether oxygen is present or not can have an extremely important influence on the metabolic capabilities of the microorganisms living at a site. As an example, the biodegradation of many hydrocarbons in oxic settings is well known; however, in oxygen's absence, these compounds are often much more persistent (Fig. 17.2). Another example would be the commonly observed transformation of some chlorinated solvents (e.g., CCl_4) under highly reducing conditions, while in oxic situations these compounds are persistent (Bouwer and McCarty, 1983b; Wilson and Wilson, 1985; Fogel et al., 1986). Since factors like temperature, pH, and oxygen concentration affect the composition, growth rate, and enzymatic contents of the microbial community, perhaps it is not surprising that such environmental conditions not only influence the rates of biologically mediated transformations, but sometimes they dictate whether these processes operate at all.

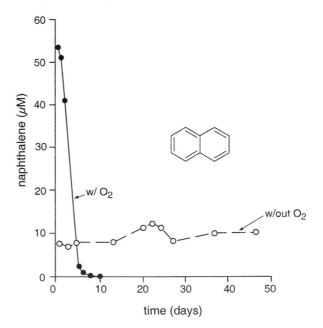

Figure 17.2 Variation in time courses of naphthalene degradation by microorganisms in laboratory soil–water incubations with O_2 present (●) or no O_2 present (o). Data from Mihelcic and Luthy, 1988.

Another important aspect to consider when dealing with biochemical transformations is how different microorganisms can form associations that benefit one another. It has long been recognized that "cooperating" species, termed *consortia*, may be required to execute a particular sequence of transformations (Gray and Thornton, 1928). For example, 3-chlorobenzoate is degraded by a consortium of bacteria (Dolfing and Tiedje, 1986). The first species removes the chlorine from the aromatic ring:

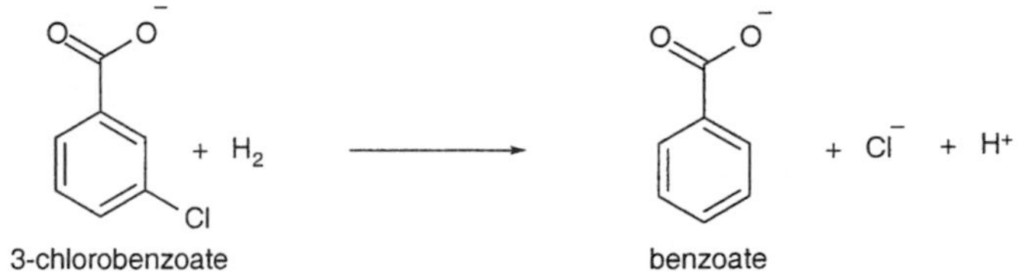

$$\text{3-chlorobenzoate} + H_2 \longrightarrow \text{benzoate} + Cl^- + H^+ \qquad (17\text{-}1)$$

The degradation is continued by a second bacterial species:

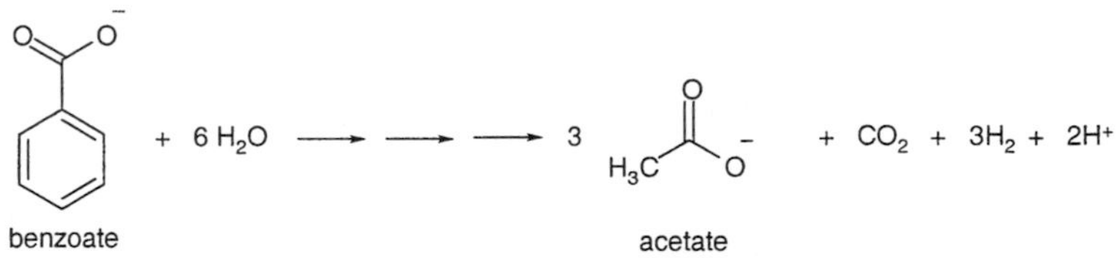

$$\text{benzoate} + 6\,H_2O \longrightarrow \longrightarrow \longrightarrow 3\ \text{acetate} + CO_2 + 3H_2 + 2H^+ \qquad (17\text{-}2)$$

The H_2 produced by this second bacterium is subsequently used by the first microorganism, as well as by organisms that use it to make methane from CO_2. Finally, another species utilizes the acetate and releases methane:

$$\text{acetate} + H^+ \longrightarrow \longrightarrow \longrightarrow CH_4 + CO_2 \qquad (17\text{-}3)$$

Since the build up of products (benzoate, H_2, and acetate) is prevented, the catalytic removal of 3-chlorobenzoate by the first microorganism can continue. This example of *syntrophy* illustrates how cooperation of different bacteria, each exhibiting specific enzymatic capabilities, may enable the degradation of a compound that would not have been performed by a single organism. In another case, seven different microorganisms were needed to process the herbicide, Dalapon (2,2-dichloropropionic acid; Senior et al., 1976). Such findings suggest that if certain microbial species are absent or inactive so that the products of biocatalysis are not removed, overall biochemical pathways can be shut down.

Microorganisms also interact via exchange of genetic material between species (Madigan et al., 2000). For example, *plasmids*, which are relatively short lengths of DNA, occur in many, but not all, bacteria and they have been found to code for a variety of degradative enzymes. Unlike the "normal" process in which DNA is passed "vertically" from mother to daughter during cell multiplication, plasmid DNA can be exchanged "horizontally" between non-progeny organisms. Thus, such exchange allows additional metabolic tools, or combinations of tools, to be acquired and utilized in a particular prokaryotic species. If the resulting metabolic capability proves beneficial, for example by providing a new energy or carbon source or by

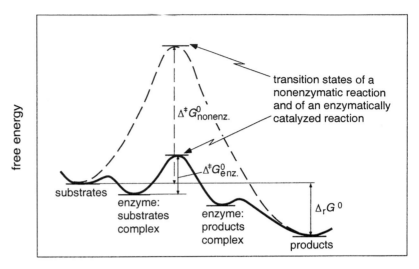

Figure 17.3 Schematic representation of the change in activation energy barrier for an enzymatically mediated reaction as compared to the analogous non-catalyzed chemical reaction.

eliminating a critical toxicant, the newly enhanced microorganism will maintain the genetic code and the derived enzymatic facility. However, should the extra tools prove not very worthwhile (i.e., not helping confer competitive advantage), then the newly acquired approaches will probably be lost. Through an amazing balance of metabolic and genetic cooperation, the microbial communities in the environment maximize their ability to live there. Since the introduction of xenobiotic organic chemicals could present new nutritional opportunities (source of C, N, S...) or toxic threats, it is reasonable to expect microorganisms and their community structure to change in response to unfamiliar compounds in their milieu.

Enzymology

Organisms facilitate reactions via two important approaches. The first approach involves the use of special proteins, called enzymes, that serve as catalysts (Nelson and Cox, 2000). These reusable "biological tools" promote substrate interactions and thereby lower the free energy of activation that determines the transformation rate (Fig. 17.3). Enzymes can lower the activation energy of reactions by several tens of kJ per mole, thereby speeding the transformations by factors of 10^9 or more. Some of this reaction rate enhancement arises because enzymes form complexes with substrates; by holding the reacting compounds in an advantageous orientation with respect to one another, the enzymes facilitate substrate interaction. Additionally, enzymes include polar and charged structural components that can alter the electron densities of the bound reactants (e.g., see the use of zinc by hydrolases in Section 17.3). This lowers the barriers to breaking existing bonds and encourages new bonds to be made.

Second, organisms may invest metabolic energy to synthesize reactive species. For example, before it is used to oxidize hydrocarbons, O_2 is converted to a much more reactive oxidant by complexation and reducing it with a compound the organisms had to spend energy to make (see Section 17.3). This scheme is similar to one previously discussed in photochemical transformations where, by absorption of light, activated species are formed that are much more reactive (Chapter 16).

In light of these insights, some principles controlling the types of enzymatic tools that microorganisms maintain can be inferred. First, since organisms evolved to deal with chemicals like amino acids, sugars, and fatty acids, and the corresponding proteins, carbohydrates, and lipids, we should expect much of their enzymatic apparatus is suited to performing the metabolic job of synthesizing (anabolism) and degrading (catabolism) such compounds. The corollary of this point is that recently invented chemicals with structures that differ from those usually processed by organisms are sometimes not met with a ready and abundant arsenal of suitable enzymatic tools when microorganisms first encounter them. Humans too have experienced this phenomenon: our own catabolic capabilities are specialized for handling L-amino acids (see margin); and thus synthetic D-amino acids made of the same atoms linked together in the same way but in the mirror image form can be used as food additives to stimulate our taste buds but not add to our caloric intake! Our expectation then is that structurally unusual chemicals will be somewhat refractory in the environment with respect to microbial transformations.

A second key principle is that enzymes do not have perfect substrate specificity. Enzymes, although "designed" for binding and catalyzing reactions of particular chemicals, may also have some ability to bind and induce reactions in structurally similar compounds. This imperfect enzyme specificity suggests that chemicals, exhibiting some structural part very like those of common substrates, may undergo biotransformation. For example, 5-phenyl pentanoic acid may become involved with the enzymatic apparatus designed to handle fatty acids (Fig. 17.4, Nelson and Cox, 2000). Thus, a certain level of biodegradation may occur for xenobiotic chemicals with structural similarities to naturally occurring organic compounds. This is part of the basis for the phenomenon of *co-metabolism* (Horvath, 1972; Alexander, 1981), in which nontarget compounds are found to be degraded by enzyme systems intended for transforming other substrates.

This imperfect binding specificity principle also helps us to understand why chemicals called *competitive inhibitors* may block the active sites of enzymes. These inhibitors are structurally like the enzyme's appropriate substrate, enabling them to bind. But these compounds may be somewhat, or even completely, unreactive. Such enzyme inhibition appears to explain the limited microbial dehalogenation of 3-chlorobenzoate in the presence of 3,5-dichlorobenzoate (Suflita et al., 1983). In this case, 3,5-dichlorobenzoate is initially transformed to 3-chlorobenzoate:

$$2e^- + H^+ \qquad\qquad + Cl^- \qquad 2e^- + H^+ \qquad\qquad + Cl^- \qquad (17\text{-}4)$$

3,5-dichlorobenzoate   3-chlorobenzoate   benzoate

Subsequent degradation of the 3-chlorobenzoate does not proceed until most of the 3,5-dichlorobenzoate is transformed. The explanation for this finding is that the dichloro aromatic substrate competes with the monochloro compound for the same enzyme active site. As a result, 3,5-dichlorobenzoate acts as a competitive inhibitor of the biochemical removal of 3-chlorobenzoate. Once nearly all the dichloro

D-amino-acid   L-amino-acid

mirror plane

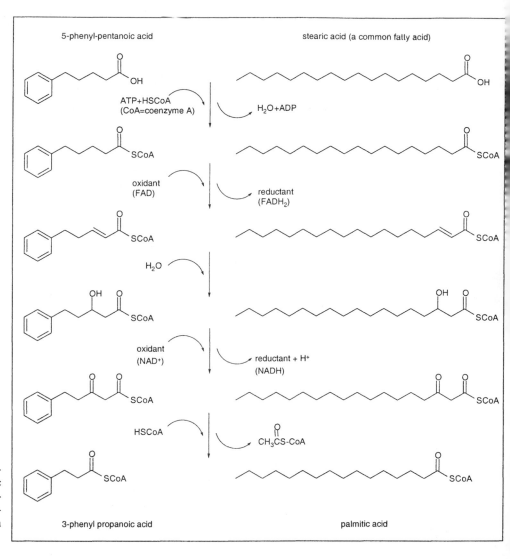

Figure 17.4 Parallel "beta oxidation" pathways for a xenobiotic substituted benzene, 5-phenyl pentanoic acid, and a naturally occurring fatty acid, stearic acid (Nelson and Cox, 2000).

compound is degraded, 3-chlorobenzoate molecules can increasingly access the enzyme site and transformation to benzoate proceeds.

This type of substrate–substrate interaction may be especially important for contaminants introduced as mixtures in the environment. One example is the mix of polycyclic aromatic hydrocarbons co-introduced in oil spills or in coal tar wastes. Laboratory observations of the biodegradation of such hydrocarbons show that some of these components can inhibit the removal of other compounds in the mix (Guha et al., 1999). Such results indicate that the rate of a particular chemical's biotransformation may be a function of factors such as the presence of competing substrates interacting with the same enzyme systems.

A third generality regarding the metabolic aptitude of organisms is that they always seem to have some relatively nonspecific enzymes available just for the purpose of attacking and utilizing unexpected or unwanted compounds (Schlichting et al., 2000). This may be analogous to our carrying a Swiss Army knife readily available to pry, hammer, pick at, slice, uncork, punch, and tweeze as the occasion presents itself. Organisms have long been bombarded with chemicals made by other species, and consequently, they have always needed to eliminate some of this chemical noise. The strategy for many bacteria often entails an initial oxidative step, converting the insulting chemical signal into something more polar. The resultant product may now fit into the common metabolic pathways; or since it is more water soluble, it may be returned to the environment. The lack of substrate specificity designed into this enzymatic capability concomitantly causes such proteins not to be especially abundant in most organisms. From the organism's point of view, it would simply be energetically too expensive to maintain a high concentration of these enzymes (like carrying *ten* Swiss Army knives instead of one!) Nonetheless, this principle implies that sometimes organic chemicals, that are unusual to the relevant organisms, may be slowly degradable via the use of such relatively nonspecific enzymes.

Whatever the origin of a capability for initial attack, the goal of these reactions is to transform the compound into one or several products structurally more similar to chemicals that microorganisms are used to metabolizing. As a result, after one or just a few transformations, the resulting chemical product(s) can be included in the common metabolic pathways and be fully degraded. A good example of this is the bacterial degradation of substituted benzenes. Initially these aromatic hydrocarbons are oxidized to catechol (*ortho*-dihydroxybenzene) or its derivatives by an oxygenase (using O_2 and NADH, see Section 17.3) and a dehydrogenase:

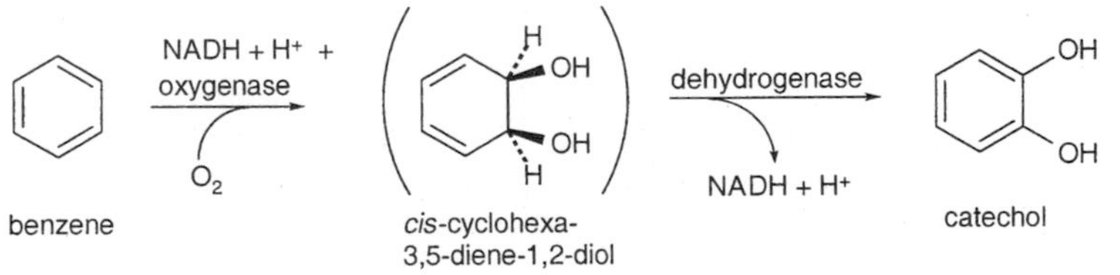

$$(17-5)$$

benzene   *cis*-cyclohexa-3,5-diene-1,2-diol   catechol

Catechol and its derivatives are also produced in the metabolism of numerous natural aromatic compounds, like salicylate or vanillate (Suzuki et al., 1991; Brunel and Davison, 1988):

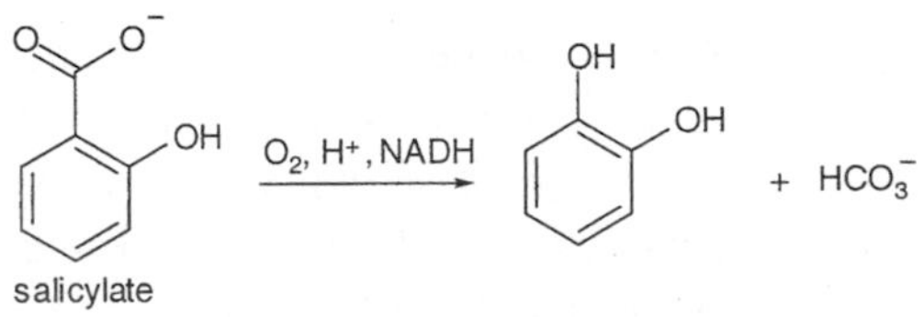

$$(17-6)$$

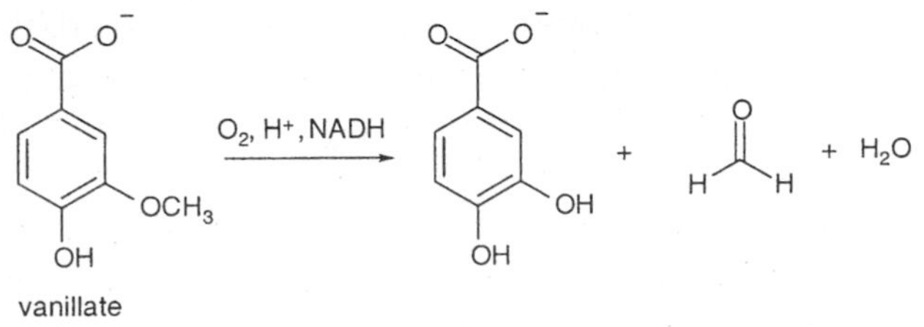

salicylate

$$(17-7)$$

vanillate

As a result, pathways are available in many microorganisms for processing di-hydroxybenzene derivatives. Such continued breakdown proceeds via enzymatic pathways which open the ring between the two hydroxyl substituents (ortho cleavage) or adjacent to them (meta cleavage). The resulting chain compounds are converted to small, useful metabolites. Thus for nonsubstituted aromatic compounds, the initial oxidation of the aromatic ring is the key to delivering the xenobiotic compound into pre-existing catabolic pathways.

Exceptions exist to this tendency for ready incorporation of the initial transformation products of xenobiotic compounds into a common pathway. First, occasionally a product is formed which is unreactive in subsequent steps in a particular microorganism. Such partially degraded compounds have been referred to as *dead-end metabolites* (Knackmuss, 1981). An example of this is the 5-chloro-2-hydroxymuconic acid semialdehyde produced by the meta cleavage of 4-chlorocatechol by a particular pseudomonad species:

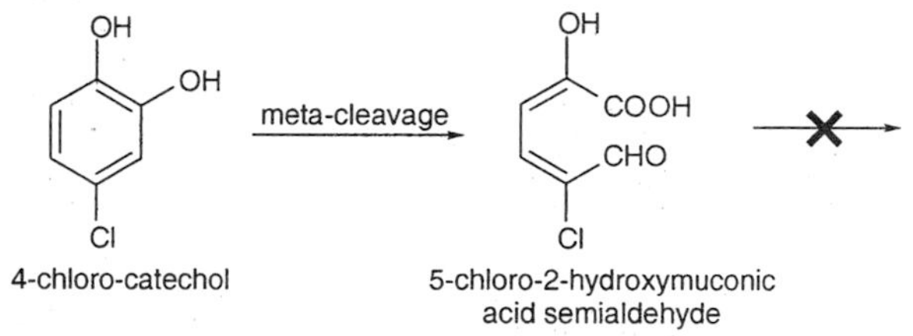

4-chloro-catechol   5-chloro-2-hydroxymuconic acid semialdehyde   (17-8)

Apparently, the presence of the chloro substituent blocks the next reaction, which normally operates on 2-hydroxymuconic acid semialdehyde to produce 2-oxo-pent-4-enoic acid:

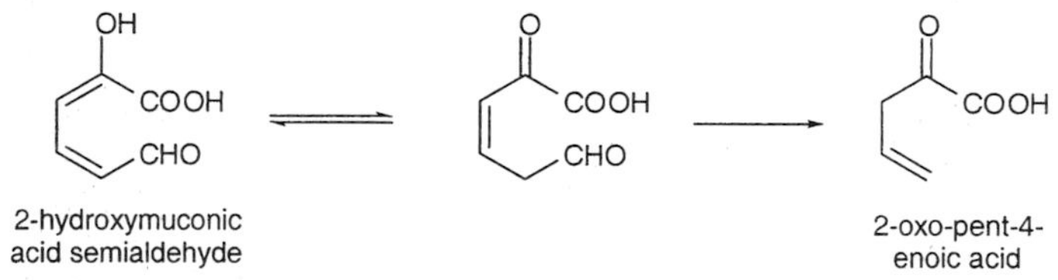

2-hydroxymuconic acid semialdehyde   2-oxo-pent-4-enoic acid   (17-9)

If a meta-cleavage pathway is the only one available to a particular microorganism, then the 5-chloro-2-hydroxymuconic acid semialdehyde (Eq.17-8) will accumulate unless another "initial" biotransformation is performed on it.

Another exception to the tendency for initial biotransformation products to be readily directed into subsequent steps in common metabolic pathways involves the formation of so-called *suicide metabolites* (Knackmuss, 1981). These problematic products result when the biological transformations yield a compound which subsequently attacks the enzymes involved. If this attack debilitates one of these enzymes, the successful operation of the relevant metabolic pathway is stopped. An example is the production of acyl halides from 3-halocatechol in certain microorganisms (e.g., Bartels et al., 1984):

3-halo-catechol   an acyl halide   (17-10)

Such acyl halides react rapidly with nucleophiles, and consequently these compounds may bind to nucleophilic moieties (e.g., –SH) of the enzyme near the site of their initial production:

$$\text{Enz-Nu:} \quad + \quad \overset{\displaystyle O}{\underset{\displaystyle R}{\overset{\|}{C}}}\hspace{-0.5em}\diagdown_X \quad \longrightarrow \quad \text{Enz-Nu-}\overset{+}{C}\hspace{-0.3em}\diagup^{\displaystyle O}_{\hspace{-0.5em}R} \quad + \quad X^- \qquad (17\text{-}11)$$

The resulting change in the enzyme's structure may prevent its continued operation, such as is seen for the *meta*-dioxygenase which formed an acyl halide from 3-halo-catechol. Thus, if biodegradation of the original compound (e.g., 3-halo-catechol) is to proceed, either a rapid hydrolysis of the acyl halide must occur before it can do damage (Mars et al., 1997; Kaschabek et al., 1998) or another metabolic pathway such as ring opening "distal" to the halogen (i.e., between the 1 and 6 carbons, Riegert et al., 1999) must be utilized. By avoiding these potential metabolic pathway problems, many xenobiotic compounds are successfully biotransformed after a suitable initial enzymatic attack is made.

Finally, we should recognize that not all of the enzymes which organisms are genetically capable of producing are always present (i.e., they are not *constitutive*). In response to a new stimulus, such as the introduction of an organic compound, organisms can turn on the production of appropriate enzymes. Those enzymes are referred to as *inducible*. This process of gearing up for a particular metabolic activity may make the description of the rate of a pollutant's biotransformation somewhat complicated because a time of no apparent activity, or a *lag period,* would be seen. There are other possible reasons for lag periods including:

1. enzymes that are already available are "repressed" (made to be ineffective) and some time must pass or some condition must change before they are altered so as to become active,

2. time may be required for a few bacteria to multiply to significant numbers (see Section 17.5),

3. an interval may be necessary for mutations to enable development of enzymes able to perform new or more efficient transformations,

4. plasmid/DNA transfers may be required to allow existing microorganisms to develop or combine suitable enzymatic tools, or

5. it may simply be that particular species of microorganisms must "immigrate" to the environmental region of interest or that cysts or spores that are present must germinate.

These phenomena exhibit very different timescales, ranging from minutes for enzyme induction and derepression to days for sufficient population growth to undetermined lengths of time for "successful" mutations. Thus one or more of these factors may make it difficult to predict how fast a chemical in a particular environmental setting will undergo even the initial step of biodegradation.

17.3   Biochemical Strategies of Microbial Organic Chemists

Our goal now is to learn to anticipate if an organic compound is biodegradable. In this effort, we will think like an organic chemist. Looking at the structure of a particular compound, we first direct our attention to the functional groups present. Next, we imagine the conversions needed to change the substrate's structure so it becomes a more desirable product (e.g., more easily degradable, less toxic). Finally, we consider the catalysts and reagents available for use. Two steps will help us see how microbial chemists do this. First, we discuss what occurs when various chemical functionalities are exposed to microorganisms. This approach reveals that there are only a modest number of reaction types that these microbial organic chemists use to *initiate the breakdown* of xenobiotic compounds. These include (Table 17.1):

1. *hydrolysis* through enzymatically mediated nucleophilic attacks (rxns 17-12, 13, 14),

2. *oxidation* using an electrophilic form of oxygen (rxns 17-15, 16, 17),

3. *reduction* by a nucleophilic form of hydride (rxn 17-18) or reduced metals like cobalt (rxn 17-22),

4. *additions* (e.g., using electrophilic carboxyl or nucleophilic hydroxyl) or free radical H-abstraction and addition of fumarate (rxns 17-19, 20, 21).

Recognizing these types of microbial reactions, we need to discuss what biochemical tools are available. The pertinent tools are the *enzymes* (catalysts) and *coenzymes* (reagents); so the chemical nature of these biomolecules and the mechanisms employed to change substrate structures will be examined in the second part of this section. As a result, we can anticipate which mechanisms will be used as a means of initial attack (Table 17.1) given the structure of the organic compound of interest (i.e., its functional groups). Further, we can see what *cosubstrates* (e.g., O_2,) are needed and that may be limited in the environment in which the microorganism lives. Additionally, this will provide a basis for understanding some of the mathematical expressions describing the rates of bioreactions when enzyme-associated processes limit the overall biotransformation rate (i.e., steps 2, 3, or 4 shown in Fig. 17.1).

Chemical Structure–Biodegradability Studies

In order to identify structural features of organic compounds that enhance or inhibit their biodegradability, recent studies have sought structure–biodegradability relationships (e.g., Boethling et al., 1994; Gamberger et al., 1996; Rorije et al., 1999; Loonen et al., 1999; Tunkel et al., 2000). In these efforts, data are assembled quantifying a compound's ease of biodegradation on a scale from "labile" to "recalcitrant". For example, Japanese investigators examined the ability of micro-organisms in aqueous suspensions to degrade about 900 organic chemicals exhibiting a diverse range of functional groups (the MITI data set; Takatsuki et al., 1995). A readily degradable substance was defined as one that experimentally showed consumption of dissolved O_2 exceeding more than 60% of the theoretical oxygen demand needed for complete mineralization of the organic chemical

Table 17.1 Important Metabolic Approaches and Tools of Microorganisms for Initiating the Breakdown of Organic Pollutants

Chemical / Environmental Factors	Metabolic Approach and "Tools"

Uneven Electron Density Due to Heteroatoms; Oxic or Anoxic Environment

1. If leaving group on sterically accessible position

then **nucleophilic attack**, e.g., via $-CH_2-S^-$

$$(17\text{-}12)$$

2. If ester-like structure (Z = C, P, or S; X = O, S, or NR)

then **hydrolysis** with
a) base catalysis (e.g., serine$-O^-$) or
b) acid catalysis (Zn^{2+} and $-COO^-$)

$$(17\text{-}13)$$

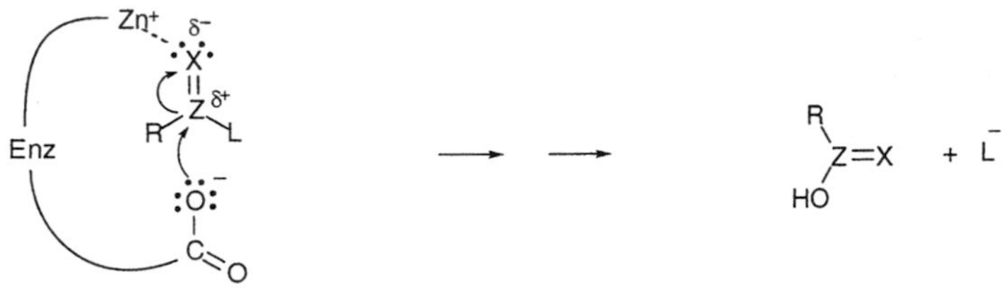

$$(17\text{-}14)$$

Generally Even Electron Density; Oxic Environment

3. If π, n electrons (N or S), alkyl σ-electrons

then **oxidation** via electrophilic oxygen (e.g. $Fe^{(IV)}O\bullet$)

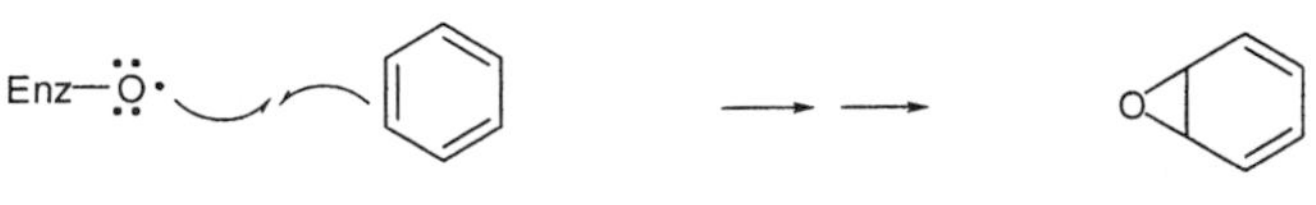

$$(17\text{-}15)$$

$$(17\text{-}16)$$

$$(17\text{-}17)$$

4. If aliphatic $-C(=O)-$, $-OH$ have partially oxidized / reduced carbon atoms

then **reduction / oxidation** via nucleophilic H^- transfers (e.g., from NAD(P)H) and Lewis acid catalysis (e.g., Zn^{2+})

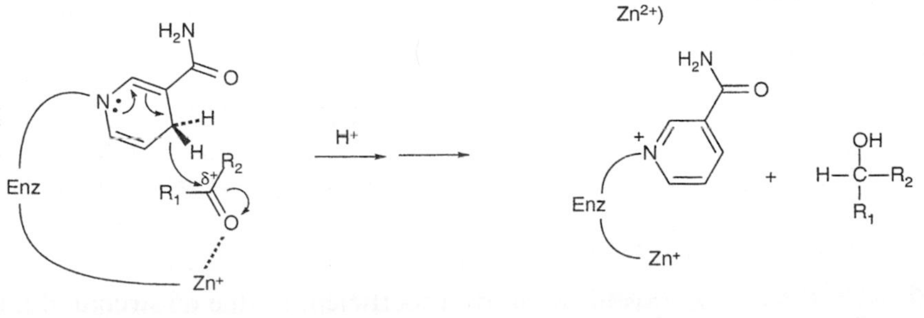

$$(17\text{-}18)$$

Table 17.1 (cont.)

Chemical / Environmental Factors	Metabolic Approach and "Tools"

Generally Even Electron Density; Anoxic Environment

5. If benzyl, allyl, alkyl σ-electrons

then **H-abstraction** via –S• radical
and addition to fumarate

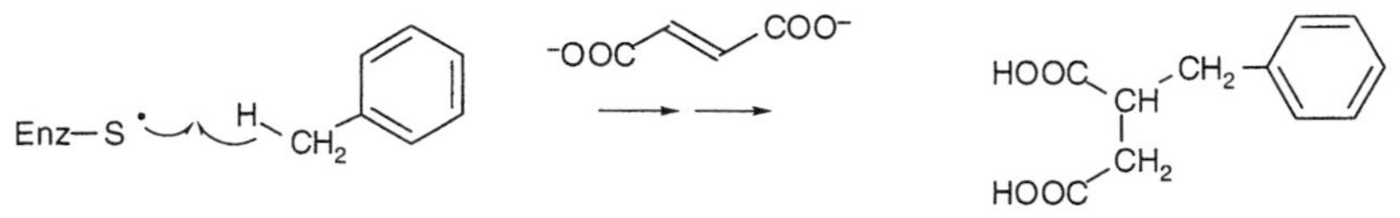

(17-19)

6. If alkene

then **addition of water**

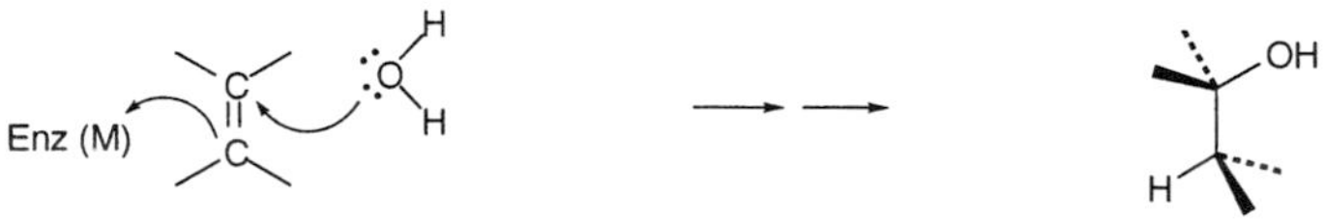

(17-20)

7. If enol (including phenol)

then **addition of carboxyl group** via –B–COOH

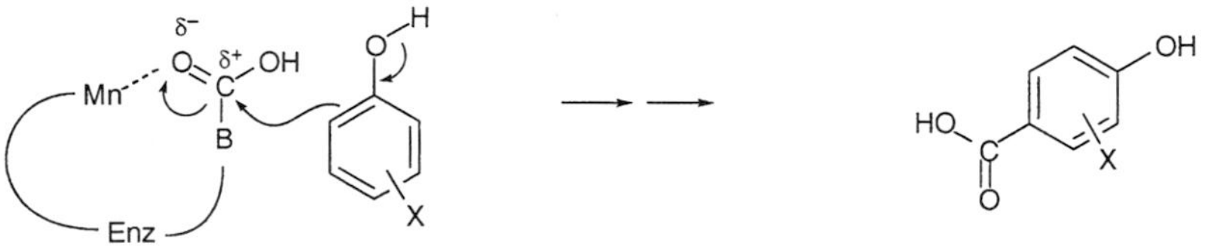

(17-21)

Sterically Limited Access Due to Substitution; Anoxic (Micro)Environment

8. If polyhalogenated

then **electron transfer** from metals (Co$^{(I)}$, Fe$^{(II)}$)
to halogen and addition of H$^+$ to the compound

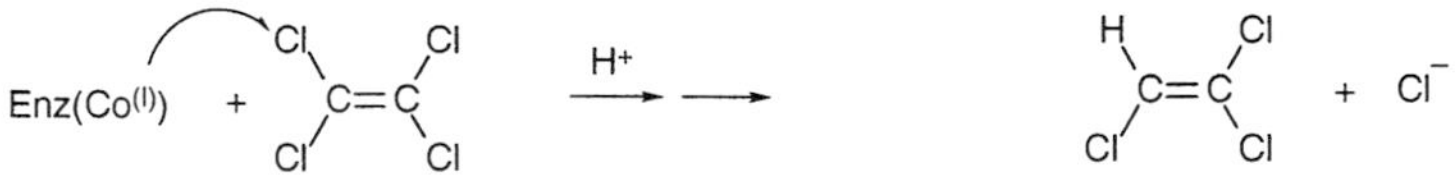

(17-22)

within 28 days. The databases were then statistically analyzed looking for structural features of the substrates that widely correlate with the empirical biodegradability observations using an optimization of equations like (Tunkel et al., 2000):

$$Y_i = a_0 + a_1 f_1 + a_2 f_2 + \ldots + a_n f_n + a_{mw}MW + e_i \qquad (17\text{-}23)$$

where Y_i is the probability of chemical's "ready biodegradation"

a_0 is the model intercept

a_n is the optimized coefficient for the nth structural fragment

f_n is the number of occurrences of structural fragment n in the chemical

a_{mw} is the optimized coefficient for the compound's molecular mass

MW is the compound's molecular mass

e_i is the equation's error term with mean value of zero

Table 17.2 Examples of Fragment Coefficients (Eq. 17-23) Indicating How a Given Structural Moiety Influences the Aerobic Biodegradability of a Compound (from Tunkel et al., 2000).

fragment n	coefficient a_n
labile	
–CHO	+0.34
–C(O)OR	+0.28
–C(O)NR$_2$	+0.22
–C(O)OH	ca. +0.2
–OH	ca. +0.1
recalcitrant	
mw	–0.001
tert–N	–0.03
quat–C	–0.04
aliphatic–Cl	–0.03
arom.–Cl	–0.1
arom.– NO$_2$	–0.24

For *oxic* conditions, data such as the MITI results have been analyzed using Eq. 17-23 or more complex mathematical forms. Several structure–biodegradability tendencies are found. First, it is widely seen that compounds including hydrolyzable groups like carboxylic acid esters (C(O)OR), amides (C(O)NR$_2$), and anhydrides or phosphorus acid esters are readily degraded (Boethling et al., 1994; Rorije et al., 1999; Loonen et al., 1999). This tendency is reflected in the positive coefficients in the optimized linear model (Table 17.2). Perhaps this result is not surprising since such structural features are very common in proteins, polysaccharides, and lipids and therefore in all organisms on earth. Thus all organisms have hydrolytic enzymes that process such functional groups (see Section 17.6). Next, the presence of hydroxy (–OH), formyl (–CHO), and carboxy (–COOH) moieties usually indicate a compound's biodegradation will be facile (Tunkel et al., 2000). Again, such oxygen-containing moieties are very common in the primary metabolites of organisms (e.g., tricarboxylic acid cycle metabolites). Consequently, xenobiotic compounds that contain such functional groups are generally degraded with available microbial enzymes under aerobic conditions.

In contrast, some structural features consistently diminish the ease of bio-degradability (see fragments with minus sign in Table 17.2). For example, chloro and nitro groups, particularly on aromatic rings, are seen to correlate with chemical recalcitrance (Boethling et al., 1994; Klopman and Tu, 1997; Rorije et al., 1999; Loonen et al., 1999; Tunkel et al., 2000). Although such moieties do occur in natural products, they are part of the so-called secondary metabolic pathways, implying that lesser transformation activities typically occur in most organisms. Further, certain structural features like quaternary carbons (CR$_1$R$_2$R$_3$R$_4$ with no R = H) and tertiary nitrogens (NR$_1$R$_2$R$_3$ with no R = H) appear to discourage biodegradability. Again, we do not usually find these features in primary metabolites. Hence the enzymatic tools needed for transforming the naturally occurring analogs of xenobiotic substrates tend to be present at lower activities or even not at all. Examples would be oxidases like cytochrome P450 or dehaloreductases; such enzymes are inducible (i.e., their synthesis is turned on when suitable substrates are present), but they may not be maintained at high activities.

Such structure–activity relationships (e.g., the fitted Eq. 17-23) of aerobic biodegradation prove capable of predicting qualitatively a compound's lability in 80 to 90% of the cases. As one might expect, structural influences on biodegradability exhibit fragment–fragment "interactions" that complicate the simple models that treat each functional group in isolation (Loonen et al., 1999). For example, the presence of a carboxylic acid moiety improves the degradability of an otherwise recalcitrant aromatic substrate; conversely, the addition of a halogen to a chlorinated compound disproportionately diminishes the biodegradability of that substance under aerobic conditions.

Another approach to capturing such interactions involves using "artificial intelligence rules" to anticipate biodegradability (Gamberger et al., 1996). In this approach, a specific compound is expected to be readily biodegradable if it fulfills a set of conditions. For example, Gamberger et al. (1996) developed a rule like: A compound is readily degradable if (1) it has one C-O bond, AND (2) it has no quaternary carbons, AND (3) it has no rings. Using the MITI database, these workers were able to correctly classify most compounds with only six such rules.

One should note that specific chemical groups may influence biodegradability differently between oxic and anoxic conditions (Rorije et al., 1998). Fragments like hydroxyls, carboxyls, and esters encourage biodegradability under both conditions, implying the biological reagents are insensitive to O_2. However, a prominent distinction involves increased halogenation. This structural feature tends to facilitate biodegradation under anoxic conditions but retards transformation in oxic systems. This observation indicates that the biological reagents of anaerobes (e.g., reductive dehalogenases) operate in the absence of O_2 (see below). Unfortunately, the databases necessary to explore structure–activity trends in anoxic environments are far less extensive than the corresponding sets under oxic conditions.

These structure–activity considerations give us many useful insights. For example, one may use them to improve chemical designs so that the compounds can accomplish the technical tasks for which they are prepared (e.g., acting as a herbicide), but not cause subsequent environmental damage when their use is completed. Further, one can potentially foresee treatment procedures that take advantage of alternating oxic and anoxic conditions to complete the degradation of mixtures of diverse organic compounds. However, these data do not allow quantitative estimates of biodegradation rates, parameters that we would need to anticipate the fates of specific compounds in environmental and engineered systems. For those parameters, we need to examine the specific processes that limit the overall biodegradation of xenobiotic compounds of interest to us (see Sections 17.4, 17.5, and 17.6). Nonetheless, studies of structure-biodegradability allow us to focus on a modest number of reaction types that microorganisms use widely to *initiate the breakdown* of xenobiotic compounds (Table 17.1).

The major focus in the discussions below is on the chemical nature of the enzymatic catalysts and coenzymes used in the initial transformation step. We will also pay some attention to the details of these enzymatic mechanisms. This will provide a basis for understanding how mathematical expressions describing the associated transformation rates can be derived when enzyme-catalyzed reactions limit the overall biotransformation rate (i.e., steps 2, 3 or 4 shown in Fig. 17.1).

Hydrolyses

The first important metabolic approach used by microorganisms to initiate transformations of xenobiotic compounds involves hydrolyses. This set of reactions can occur under all environmental conditions. Also the enzymes that catalyze these degradation reactions are typically constitutive (i.e., always present), although their activity levels can be regulated (e.g., hydrolytic dehalogenases, Janssen et al., 2001).

Table 17.3 Some Microbially Mediated Hydrolysis Reactions

Substrate	Product(s)	Reference

alkyl halides

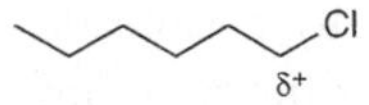

$+ H_2O \longrightarrow$ ... $+ Cl^- + H^+$ *a*

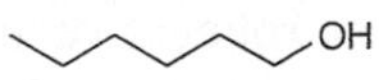

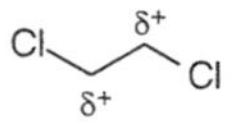

$+ H_2O \longrightarrow$ ... $+ Cl^- + H^+$ *b*

$\overset{\delta+}{CH_2Cl_2} \qquad + H_2O \longrightarrow H_2CO + 2Cl^- + 2H^+$ *c*

esters

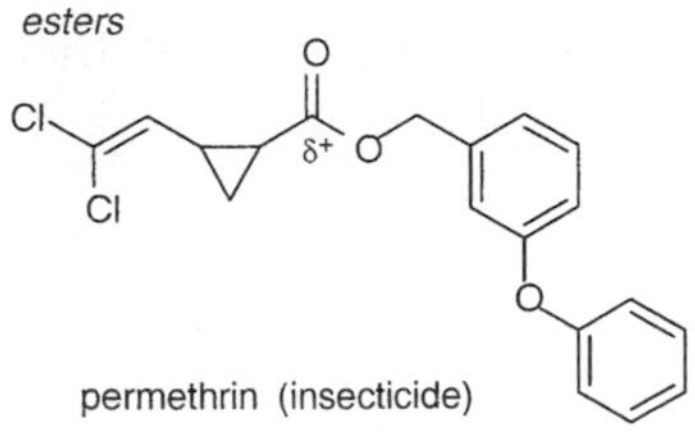

permethrin (insecticide)   *d*

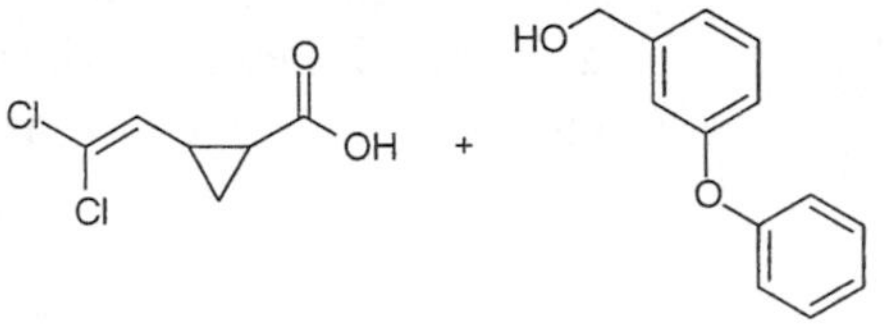

amides

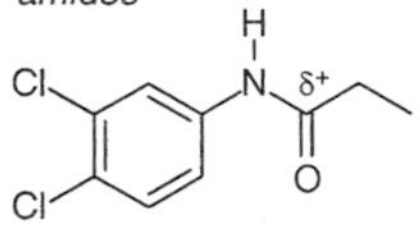

propanil (herbicide)   *e*

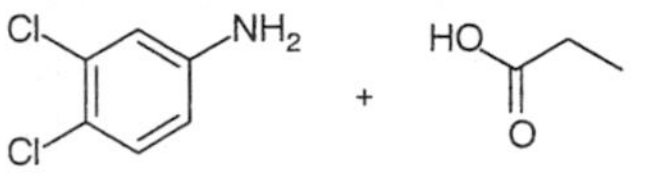

carbamates

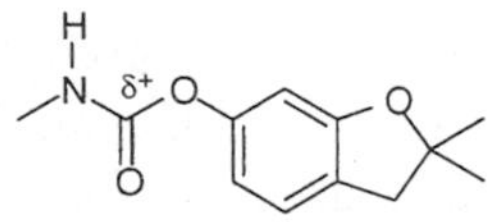

carbofuran (insecticide) $CH_3NH_2 + CO_2 +$ ...   *f*

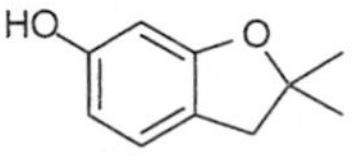

ureas

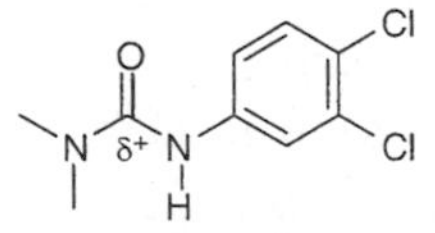

diuron (herbicide) $(CH_3)_2NH + CO_2 +$ ...   *g*

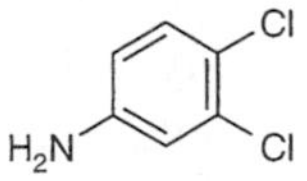

phosphates

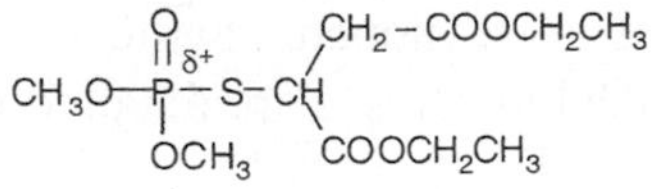

malathion (insecticide)   *h*

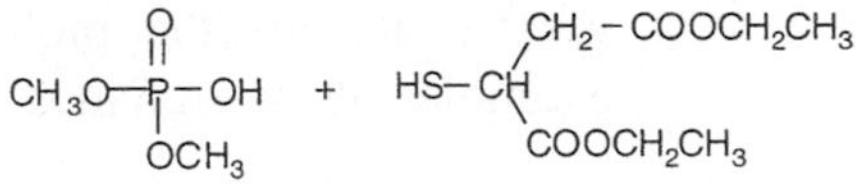

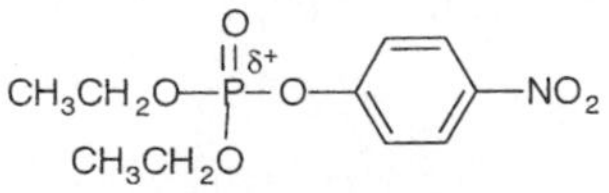

paraoxon (insecticide)   *i*

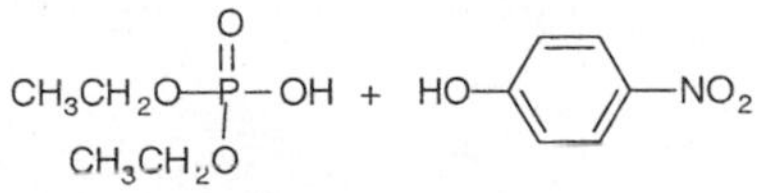

a Scholtz et al., 1987a,b; *b* Keuning et al., 1985; *c* Kohler-Staub and Leisinger, 1985; *d* Maloney et al., 1988; *e* Chiska and Kearney, 1970; Bartha, 1971; *f* Chaudry and Ali, 1988; Ramanand et al., 1988; *g* Englehardt et al., 1973; *h* Rosenberg and Alexander, 1979; *i* Munnecke, 1976

By examining the reactants and products of microbial hydrolysis transformations (Table 17.3), we see that xenobiotic organic compounds react at the same structural positions as expected for reactions with HO$^-$ or H$_2$O (see Chapter 13). That is, the target compound always exhibits a central atom that is electron deficient (indicated by δ^+) and thus is susceptible to nucleophilic attack. For example, when a compound has an electronegative leaving group, L, bound to a saturated carbon allowing backside access (see also Fig. 13.1), then the following biologically mediated reaction occurs:

$$
\begin{array}{ccccc}
H & & & & \\
\backslash & & \overset{\delta^+\ \ \delta^-}{C-L} \xrightarrow{\ S_N2\ } & HO-C & + H^+ + L^- \\
O: & \to & & & \\
/ & & & & \\
H & & & & \\
\end{array}
\tag{17-24}
$$

As for comparable abiotic hydrolyses, these biochemical transformations yield alcohols. Likewise, when we have carboxylic acid esters or related compounds (Section 13.3), we have net bioreactions:

$$
H_2O \ + \ \underset{R}{\overset{\overset{X^{\delta-}}{\|}}{Z}}{}^{\delta+}_{\ \ L^{\delta-}} \ \longrightarrow \ \underset{R}{\overset{\overset{X}{\|}}{Z}}{}_{OH} \ + \ H^+ + L^-
\tag{17-25}
$$

where Z is C, P, or S and X is O, NR, or S. Again as for abiotic hydrolyses of esters and amides, these biochemical transformations yield acids and corresponding alcohols or amines. Such transformations facilitate release of the xenobiotic compound from the organism back into its aquatic environment and/or enable the continued catabolic breakdown of the substance.

When mediated by microorganisms, hydrolyses generally proceed faster than the comparable abiotic reactions (e.g., Munnecke, 1976; Wolfe et al., 1980c). Since the cytoplasmic pH of organisms is not particularly different from the range of pH values seen in natural waters, these organisms must be using enzymatic methods to facilitate these reactions. First, organisms use strong nucleophiles (e.g., –S$^-$) in an enzymatic context to make the initial attack. Moreover, they can bind the hydrolyzable moiety with an electron-withdrawing substituent to enhance the rate of attack in a manner analogous to what we have seen in acid- or metal-catalyzed hydrolyses (Chapter 13.3 and 13.5). Also, the enzymes hold the reacting species in the right positions with respect to one another, thereby enhancing the rate of their interaction by reducing any unfavorable entropy change of reaction (i.e., making a negative $\Delta_r S$ of the catalytic step more positive). The combination of such influences greatly hastens the biological process over nonenzymatic mechanisms. Some examples of the enzymes that accomplish hydrolyses are detailed in the following.

Biological Hydrolysis of Alkyl Halides with Glutathione. One widespread hydrolysis approach used by organisms involves the use of the tripeptide, glutathione (GSH):

This tripeptide (γ-glutamic acid-cysteine-glycine) contains a thiol moiety which is an excellent nucleophile. In other enzymatically catalyzed hydrolyses, another amino acid nucleophile, the carboxylate anion of aspartate, is used (Li et al., 1998; Janssen et al., 2001). In both cases, the leaving group is displaced in an S_N2 reaction process (see Chapter 13.2).

Focusing here on GSH, compounds susceptible to nucleophilic attack like halides and epoxides have been found to be transformed using GSH in the first step (Vuilleumier, 1997). For example, this is the case for dichloromethane (Stucki et al., 1981; Kohler-Staub and Leisinger, 1985):

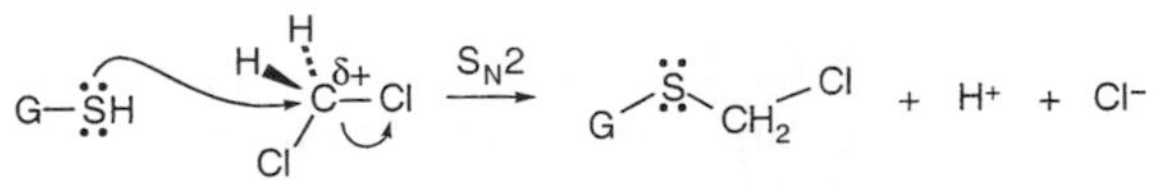

$$\text{G-SH} \quad \xrightarrow{\quad S_N2 \quad} \quad \text{G-S-CH}_2\text{-Cl} + \text{H}^+ + \text{Cl}^- \qquad (17\text{-}26)$$

This reaction is mediated by an enzyme called a glutathione transferase, which facilitates the encounter of GSH and the compound it is attacking (Mannervik, 1985; Vuilleumier, 1997). Formation of the GSH adduct (i.e., the compound formed when the two reactants are attached to one another) permits the attack of water on the previously chlorinated carbon; and since the resulting intermediate is not particularly stable in this case, it decomposes, releasing formaldehyde and regenerating glutathione in a reaction much like the dehydration of *gem*-diols:

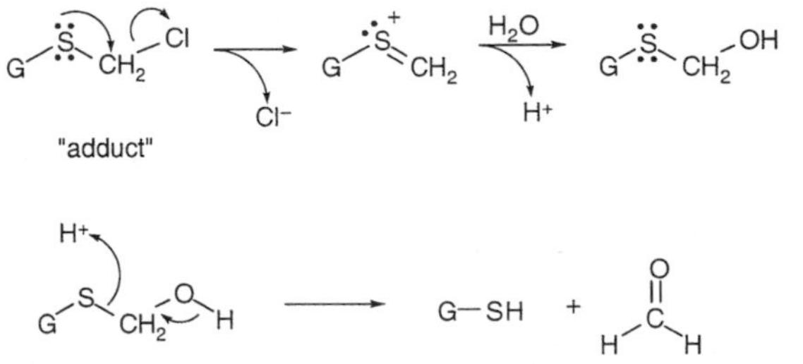

$$\text{G-S-CH}_2\text{-Cl} \xrightarrow{\quad} \text{G-S}^+\text{-CH}_2 \xrightarrow[\text{H}^+]{\text{H}_2\text{O}} \text{G-S-CH}_2\text{-OH}$$

"adduct"

$$\text{H}^+ \quad \text{G-S-CH}_2\text{-O-H} \longrightarrow \text{G-SH} + \underset{\text{H}\quad\text{H}}{\overset{\text{O}}{\text{C}}} \qquad (17\text{-}27)$$

The overall result is equivalent to hydrolysis of both of the original carbon–chlorine bonds:

$$\text{CH}_2\text{Cl}_2 + \text{H}_2\text{O} \longrightarrow \text{CH}_2\text{O} + 2\,\text{H}^+ + 2\,\text{Cl}^- \qquad (17\text{-}28)$$

In sum, the excellent bionucleophile, GSH, is used to get the reaction started, and subsequent hydrolysis steps cause it to go to completion. The process works well enough that some bacterial species (e.g., a *Hyphomicrobium* isolate) can grow on methylene chloride as its sole source of carbon (Stucki et al., 1981). This microorganism is also capable of degrading other dihalomethanes, CH_2BrCl, CH_2Br_2, and CH_2I_2 (Kohler-Staub and Leisinger, 1985). As mentioned above, other dehalogenases rely on moieties of amino acids like cysteine (Keuning et al., 1985; Scholtz et al., 1987a, b) or aspartic acid (Li et al., 1998: Janssen et al. 2001) contained within the protein structure. Like GSH-transferases, these enzymes first form the enzyme adduct:

$$\text{Enz-Nu}: \quad + \quad \text{R-X} \longrightarrow \text{Enz-Nu-R} + \text{X}^- \qquad (17\text{-}29)$$

"adduct"

Illustrative Example 17.2

What Products Do You Expect from Microbial Degradation of 1,2-Dibromoethane (EDB?)

Problem

Ethylene dibromide (EDB, also 1,2-dibromoethane) is a xenobiotic compound used as a gasoline additive and a soil fumigant. What initial biodegradation product would you expect from this compound?

Answer

Br–CH₂CH₂–Br

1,2-dibromoethane
(EDB)

From the structure of this organic compound, we note the presence of a good leaving group, Br⁻, attached to a saturated carbon accessible from the back side. Thus, we anticipate microorganisms would begin to degrade this compound using a nucleophilic attack on the carbon to which the bromide is attached, with subsequent hydrolysis of the enzyme adduct:

$$Enz–Nu^- + Br–CH_2CH_2–Br \rightarrow Enz–Nu–CH_2CH_2–Br + Br^-$$

$$\rightarrow Enz–Nu^- + HO–CH_2CH_2–Br$$

Thus the expected initial product is 2-bromoethanol. This is what Poelarends et al. (1999) found when they examined a *Mycobacterium* that could grow on EDB. Perhaps not surprisingly, they also found the second bromide was subsequently hydrolyzed too:

$$Enz–Nu^- + Br–CH_2CH_2–OH \rightarrow Enz–Nu–CH_2CH_2–OH + Br^-$$

$$\rightarrow Enz–Nu^- + \underset{CH_2—CH_2}{\overset{O}{\triangle}}$$

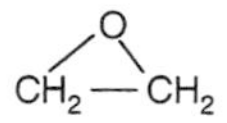

making the product, ethylene oxide. This product was ultimately mineralized to CO_2 and H_2O.

In a subsequent step, the enzyme adduct then detaches the alkyl group as an alcohol product. Sometimes, monohalogenated compounds act as growth inhibitors because the adduct formed is difficult to hydrolyze and thereby incapacitates the enzyme (recall suicide metabolites). However, assuming the adduct can be hydrolyzed, then the enzyme is prepared to serve again.

Enzymatic Hydrolysis Reactions of Esters. Xenobiotic compounds containing esters or other acid derivatives in their structures (e.g., amides, carbamates, ureas, etc., see Table 17.3) are often readily hydrolyzed by microorganisms. To understand how enzymatic steps can be used to transform these substances, it is instructive to consider the hydrolases (i.e., enzymes that catalyze hydrolysis reactions) used by organisms to split naturally occurring analogs (e.g., fatty acid esters in lipids or amides in proteins). The same chemical processes, and possibly even some of the same enzymes themselves, are involved in the hydrolysis of xenobiotic substrates.

We begin by considering the enzymes using the hydroxyl group of serine (Bruice et al., 1962; Fersht, 1985) or the thiol group of cysteine to initiate the hydrolytic attack:

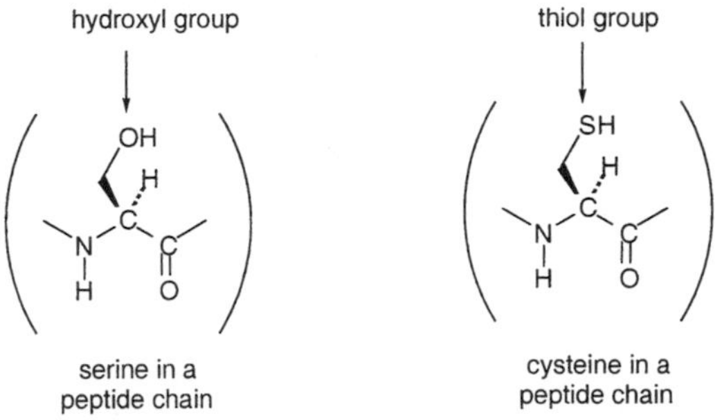

These amino acid functional groups play the role of nucleophiles and attack electron-deficient central carbons or other central atoms (e.g., P, S) in esters or other acid derivatives. Hence, the enzymatic mechanism in this case operates similarly to the base-catalyzed process discussed in Chapter 13.3. The general process proceeds by the following steps (Fig. 17.5). First, the substrate associates with the free enzyme (I) in a position suited for nucleophilic attack. To improve the ability of the nucleophile to form a bond with the electron-deficient atom, other amino acids (e.g., histidine and aspartic acid) may assist in this and subsequent steps by proton transfers (this group of amino acids is often referred to as the "charge relay system"). This amounts to converting the serine or the cysteine to the more nucleophilic conjugate bases, RO^- or RS^-, respectively (see II in Fig. 17.5). Subsequently, the nucleophile attacks, forming a tetrahedral intermediate (III) much as we saw previously in the chemical hydrolysis process (Section 13.3). Decomposition of this intermediate leads to the release of an alcohol, an amine, or another leaving group from the original compound. Continued processing involves enzyme-assisted attack of water on the enzyme adduct (IV, again with the help of the charge relay system) and release of the acid compound.

Experience with various carboxylic acid esters and amides shows that either initial attack (the "acylation" step shown from II → IV) or release of the acid compound (the "deacylation" step shown from IV → I) can be rate limiting, depending on the kind of compound hydrolyzed. For amides, the initial attack and release of an amine is often the slow step (Fersht, 1985), since these nitrogen compounds are very strong bases and thus poor leaving groups. In contrast, esters often have the deacylation steps as the bottleneck to reaction completion (Fersht, 1985). Since ionizable groups of various amino acids play critical roles in these hydrolysis reactions, such hydrolases exhibit some sensitivity to the medium pH. Generally, these enzymes operate best at near-neutral pHs, since more acidic conditions protonate the histidine and thereby negate its involvement.

Another approach for hydrolysis of acid derivatives utilizes metal-containing hydrolases (see I in Fig. 17.6). In these enzymes, a metal, for example a zinc atom, is included in the active site of the enzyme (Fersht, 1985). The process begins with the association of the carbonyl oxygen (or its equivalent in other compounds) with a ligand position on this metal (II). As we found for the acid-catalyzed hydrolysis mechanism discussed in Chapter 13.3, this association with an electropositive metal

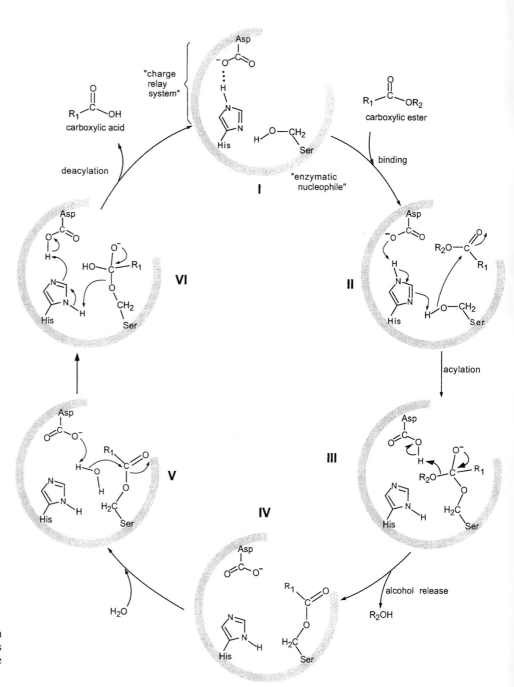

Figure 17.5 Schematized reaction sequence showing the hydrolysis of an ester by a serine hydrolase (Fersht, 1985).

atom causes the central atom (e.g., a carbonyl carbon in an ester) to be even more electron deficient. The result is a very susceptible position for attack by nucleophilic moieties, such as the carboxylate of a nearby glutamate. An anhydride-type tetrahedral intermediate is formed (III), resulting in the release of the alcohol portion of the original ester (IV). The new enzyme-acid complex is more suited to attack by water than the original compound. Thus attack by water on (V) ultimately leads to the severance of the acid's covalent linkage to the glutamate (VI). Ligand exchange at the zinc permits the release of the acid product. Thus, this overall biological hydrolysis operates analogously to an acid-catalyzed abiotic reaction.

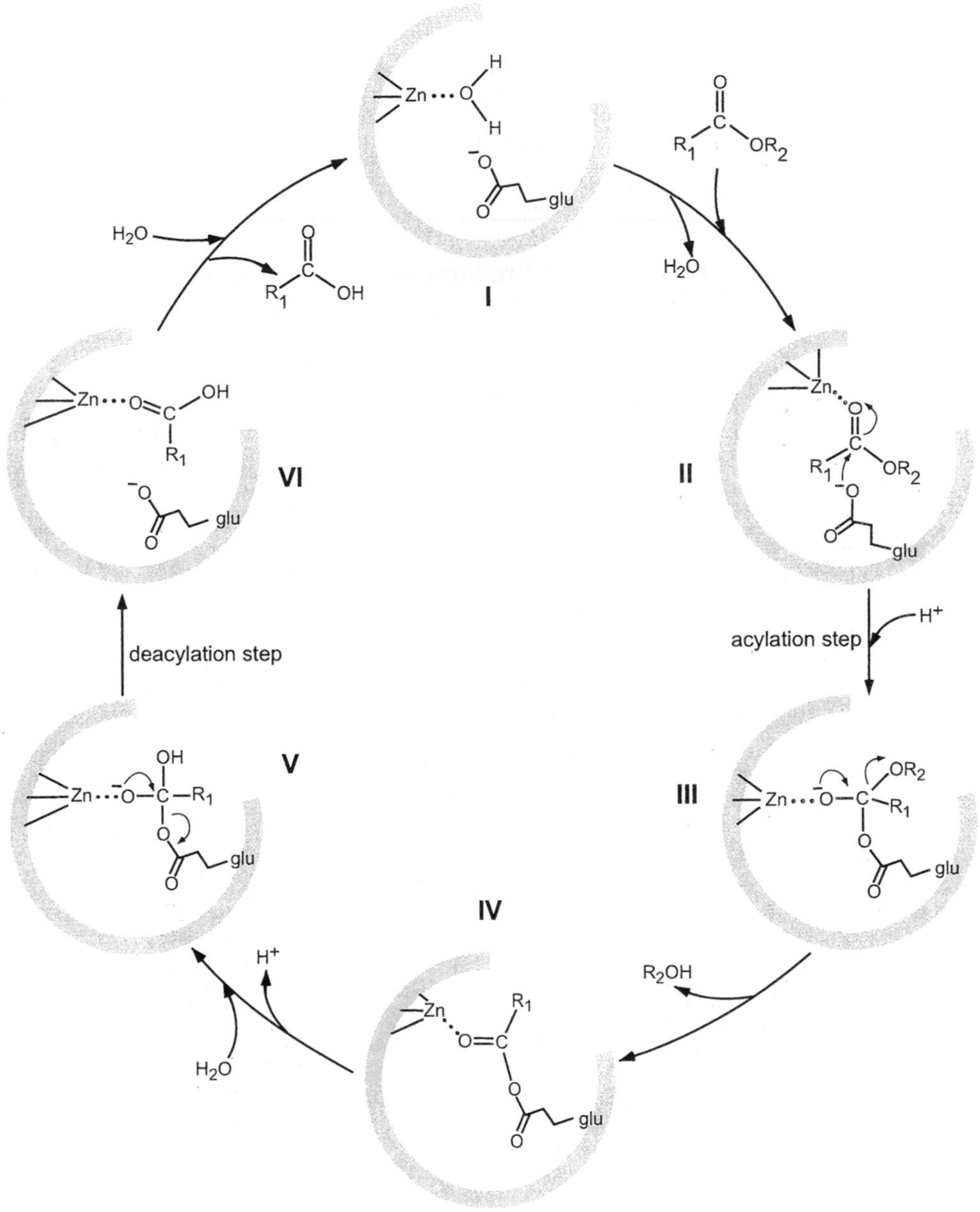

Figure 17.6 Reaction sequence for hydrolysis of an ester by a zinc-containing hydrolase (Fersht, 1985).

Now recognizing that hydrolases use mechanisms parallel to abiotic "base" and "acid" catalysis mechanisms, we can anticipate how the rates of transformations of related xenobiotic compounds will vary. This may help us understand and/or develop predictive relationships between the rates of biological and comparable chemical hydrolyses as long as these processes are rate limited by similar mechanisms. Empirically, it is found that enzymes that catalyze hydrolysis are not always very selective within a structurally related group of xenobiotic compounds. This was noted for enzymes induced by various kinds of substrates including halides (Keuning et al., 1985; Scholtz et al., 1987a b; Li et al., 1998), phenyl amides and ureas (Englehardt et al., 1973), and phosphoric and thiophosphoric esters (Munnecke, 1976; Rosenberg and Alexander, 1979). Also, hydrolysis enzymes suited for attack of xenobiotic compounds are commonly constitutive. Thus Wanner et al. (1989) were not surprised to see biodegradation of some hydrolyzable insecticides with no lag period, after these compounds were spilled into the Rhine River. To sum, if a (xenobiotic) organic compound contains a hydrolyzable functionality, one should anticipate that microorganisms will attack at that point of the structure.

Illustrative Example 17.3   **What Products Do You Expect from Microbial Degradation of Linuron?**

Problem

Linuron (see structure below) is a herbicide used by potato, rice, and wheat farmers. What initial biodegradation product would you expect from this compound?

Answer

From the structure of this compound, you see two good leaving groups in the chlorides, Cl⁻, but you note that they are attached to electron-rich carbons (note π electrons) that cannot be approached from the back side. Hence, we do not expect microorganisms to use a nucleophilic attack displacing those chlorides.

You also note the presence of an "amide" group (more accurately, an urea); in this case, R–Z(=X)–L is R–C(=O)–NHR. Thus, microbial attack likely starts with hydrolysis of this part of the structure:

$$\text{Enz–Nu}^- + \text{R–C(=O)–NHR} \rightarrow \text{Enz–Nu–C(=O)–R} + \text{NH}_2\text{R}$$

$$\rightarrow \text{Enz–Nu}^- + \text{R–C(=O)–OH}$$

Thus, we anticipate the initial products are 3,4-dichloroaniline and N-methyl, N-methoxy-carbamic acid.

This is what was found by Englehardt et al. (1973) for this compound using a hydrolyzing enzyme purified from a *Bacillus* sp.:

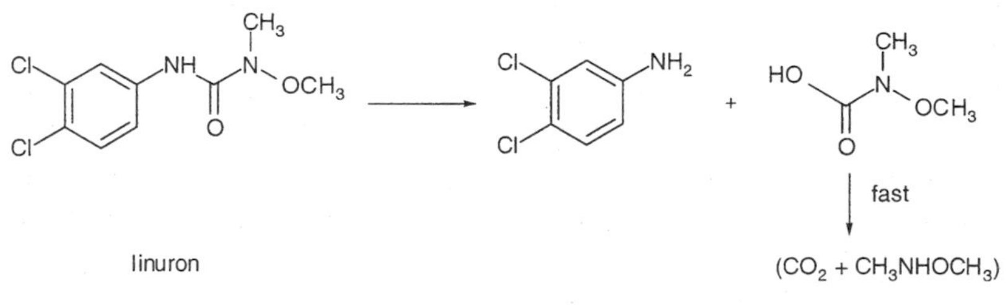

linuron   (CO₂ + CH₃NHOCH₃)

In their studies the enzyme activity decreased when reagents were added that complex with thiols; hence it was concluded that cysteine was probably the key enzymatic nucleophile. The production of this enzyme could be induced by exposure of the bacteria to various phenylamide- or phenylurea-containing herbicides and fungicides. The enzyme was also capable of hydrolyzing a variety of other phenylamides and phenylureas, albeit at somewhat different rates.

Oxidations Involving Electrophilic Oxygen-Containing Bioreactants

The second approach used by microorganisms to initiate the degradation of xenobiotic organic compounds involves the use of electrophilic forms of oxygen to oxidize organic substrates. Since O_2 in the environment is not very reactive with most organic compounds (see Eq. 14-40 and associated text), organisms must invest metabolic energy to convert this oxygen into a more effective oxidant. Generally, this involves enzymes with metals (e.g., oxygenases) and coenzymes like NAD(P)H (see below). If both atoms of oxygen from O_2 are transferred, the enzymes are called *dioxygenases*. In contrast, *monooxygenases* deliver only a single oxygen atom. Since O_2 is a co-substrate, this approach is generally only possible in oxic environments (Table 17.1 for reactions 17-15, 17-16, and 17-17).

The bio-oxidants that are formed are highly suited to attacking organic molecules by attracting the most readily available electrons in the substrates' structures (Table 17.4). Often these are π-electrons of aromatic rings and carbon–carbon double bonds:

$$\text{dioxygenase (O}_2\text{)} \ + \ \overset{\diagdown}{\underset{\diagup}{C}}\!\underset{\diagdown}{\overset{\parallel}{C}} \ \longrightarrow \ \text{dioxygenase} \ + \ \begin{matrix} O-C \\ | \quad \diagdown \\ O-C \end{matrix} \qquad (17\text{-}30)$$

$$\text{monooxygenase (O}_2\text{)} \ + \ \overset{\diagdown}{\underset{\diagup}{C}}\!\underset{\diagdown}{\overset{\parallel}{C}} \ \longrightarrow \ \text{monooxygenase} \ + \ \underset{\text{an epoxide}}{\overset{O}{C-C}} \ + \ H_2O \qquad (17\text{-}31)$$

In other structures, these are the nonbonded electrons of sulfur or nitrogen:

$$\text{monooxygenase (O}_2\text{)} \ + \ R_1\!\!\overset{\displaystyle ..}{\underset{\displaystyle ..}{S}}\!\!R_2 \ \longrightarrow \ \text{monooxygenase} \ + \ R_1\!\overset{\overset{O}{\parallel}}{\underset{}{S}}\!R_2 \ + \ H_2O \qquad (17\text{-}32)$$

In the absence of such π- and *n*-electrons, oxidation can involve the σ-electrons of carbon-hydrogen bonds (Ullrich, 1972; White and Coon, 1980, Guengerich and MacDonald, 1984):

$$\text{monooxygenase (O}_2\text{)} \ + \ \underset{H}{\overset{R_2}{\underset{\diagup}{R_1\!-\!\!C\!-\!R_3}}} \ \longrightarrow \ \text{monooxygenase} \ + \ \underset{\underset{H}{\diagup O}}{\overset{R_2}{R_1\!-\!\!C\!-\!R_3}} \ + \ H_2O \qquad (17\text{-}33)$$

Whether one oxygen atom or two oxygen atoms are transferred to the organic chemical, the key that permits these reactions to occur is the attraction of the biologically produced oxidant to the electrons of the organic chemical. Thus it is clear that organisms can accomplish this by using an enzymatically prepared form of electrophilic oxygen.

Recognition of these "electrophilic attack" approaches of microorganisms for executing oxygenation reactions enables us to make important predictions regarding such biotransformations. For example, if we consider attack of phenol by the electrophilic oxygen of a monooxygenase, we might expect that the compound reacts like an enol:

$$\text{(17-34)}$$

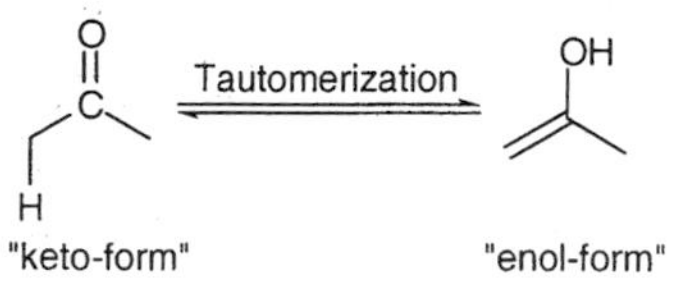

phenol

"keto-form"   "enol-form"

and the π-electrons adjacent to the hydroxyl moiety are most nucleophilic. Hence, we are not surprised to see the addition *ortho* (but not *meta*) to the hydroxyl moiety as seen for phenol monooxygenases (Suske et al., 1999):

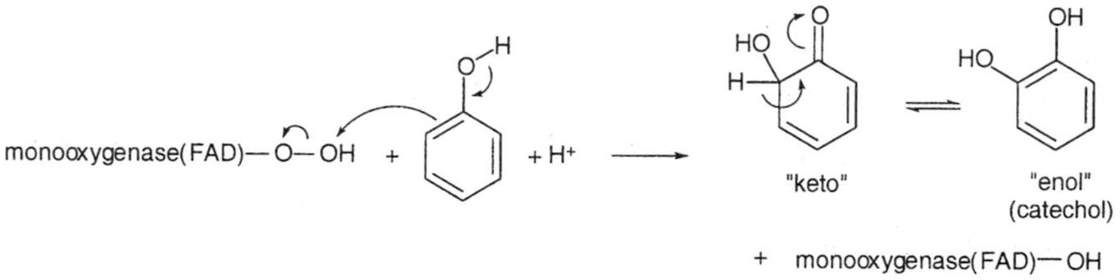

$$\text{(17-35)}$$

For alkenes, the presence of π-electrons encourages the reaction to occur at the double bond rather than at saturated positions of the molecule:

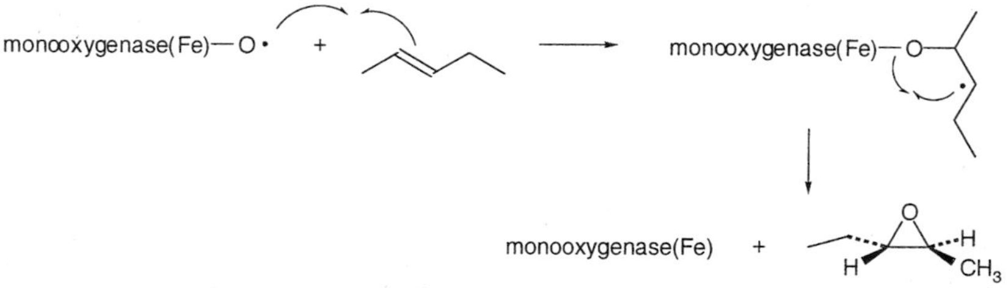

$$\text{(17-36)}$$

Furthermore, one might expect that electron-withdrawing substituents on aromatic systems (e.g., positive σ values) make the π-electrons of the ring system less nucleophilic and thus the rate of attack of the electrophilic oxygen on that ring slower. This is consistent with observations on a series of substituted benzoates where steric factors do not dominate (Knackmuss, 1981):

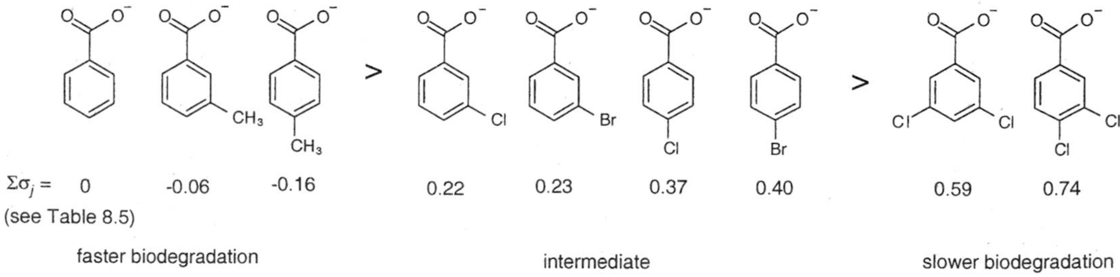

| $\Sigma\sigma_j =$ | 0 | -0.06 | -0.16 | 0.22 | 0.23 | 0.37 | 0.40 | 0.59 | 0.74 |

(see Table 8.5)

faster biodegradation   intermediate   slower biodegradation

This trend in relative biodegradability is consistent with a situation in which the rate-limiting step involves an electrophilic attack on the π-electrons of the ring. Such data also illustrate how variations in chemical structure may affect the rates of biotransformations.

Table 17.4 Some Microbially Mediated Oxidations

Substrate	Product	Reference

π-electron attack

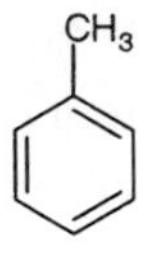

toluene

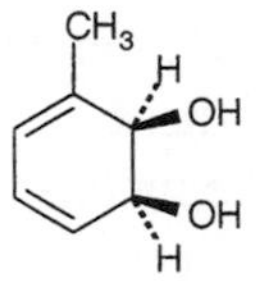

cis-1,2-dihydroxy-3-methyl-
cyclohexa-3,5-diene

a

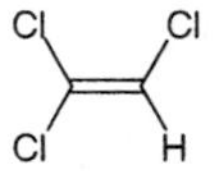

trichloroethene

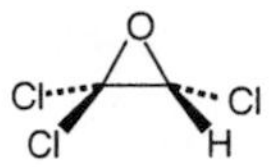

trichloro-oxirane
(trichloro ethylene oxide)

b

n-electron attack

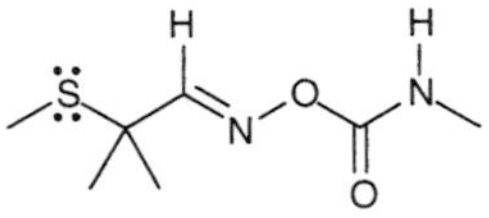

aldicarb

c

σ-electron attack

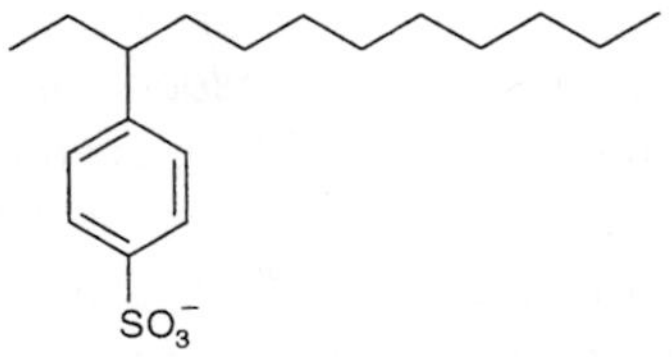

4-(3-dodecyl)-benzenesulfonate

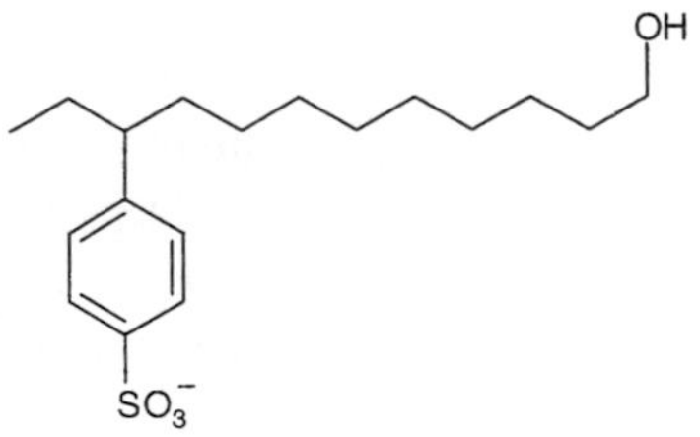

d

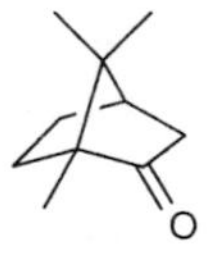

camphor

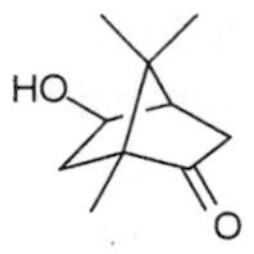

e

R—CH₃ ⟶ R—CH₂OH ⟶ R—CHO ⟶ R—COOH

$$R\text{—}CH_3 \longrightarrow R\text{—}CH_2OH \longrightarrow R\text{—}CHO \longrightarrow R\text{—}COOH$$

alkane   alcohol   aldehyde   carboxylic acid

f

[a] Gibson et al., 1970; Yeh et al., 1977; [b] Little et al., 1988; [c] Jones, 1976; [d] Swisher, 1987; [e] Hedegaard and Gunsalus, 1965; Conrad et al., 1965; [f] Ratledge, 1984.

An Example of a Biological Oxygenation: Cytochrome P450 Monooxygenases. Organisms have developed a very interesting approach for preparing and using electrophilic oxygen. We illustrate this methodology with the case of cytochrome P450 monooxygenases, a widespread and well-studied bio-oxidation system (Ullrich, 1972; White and Coon, 1980; Guengerich and MacDonald, 1984; Groves and Han, 1995; Peterson and Graham-Lorence, 1995; Schlichting et al., 2000). The active site of these enzymes is an iron held within an organic ring system called a *porphyrin*. The metal-ring system combination is called a *heme* and is carried within a protein environment:

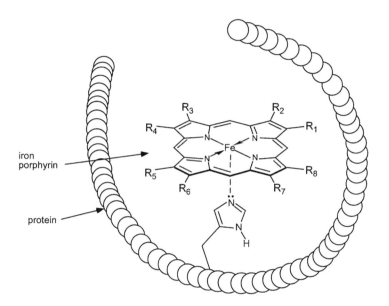

The entire iron-porphyrin-protein complex is called a *cytochrome* and such proteins are important electron-transfer components of cells. Generally, access to the macromolecular region in which the oxidation reactions occur is via a hydrophobic "channel" through the protein (Mueller et al., 1995). As a result, organic substrates are transferred from aqueous solution into the enzyme's active site primarily due to their hydrophobicity and are limited by their size. This important feature seems very appropriate: hydrophobic molecules are "selected" to associate with this enzyme, and these are precisely the ones that are most difficult for organisms to avoid accumulating from a surrounding aquatic environment.

After substrate:enzyme complex formation, (I → II in Fig. 17.7), the enzyme begins to be prepared for reaction. The iron atom is converted from Fe^{III} to Fe^{II} by an enzyme called a reductase and a reduced nicotinamide adenine dinucleotide, NAD(P)H, generated by a separate energy-yielding metabolism (III). Then molecular oxygen is bound by the iron (IV). Next a second electron is transferred to the iron-oxygen complex, again ultimately from NAD(P)H. The resulting anionic oxygen quickly protonates, and we now have a biologically produced analog to hydrogen peroxide (V like HO-OH). Subsequent steps form an electrophilic oxygen (illustrated by $[Fe^{IV}-O]^{+\cdot}$ in Fig. 17.7 intermediate VI). This highly reactive species can now attract electrons from a neighboring source. Not coincidentally, since the

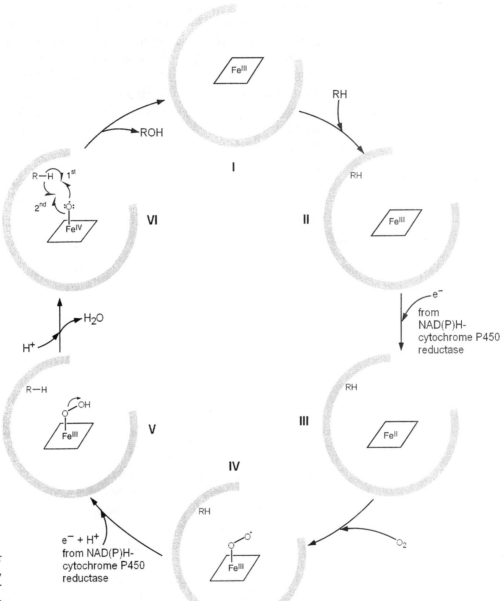

Figure 17.7 Reaction sequence for oxidation of an organic substrate, RH, by a cytochrome-P450 mono-oxygenase (after Ullrich, 1972, Schlichting et al., 2000).

enzyme has bound an organic chemical earlier in the sequence, the electrophilic oxygen sees this chemical as its first opportunity to acquire electrons. The ultimate result is the addition of an oxygen atom to the substrate. Often the product itself is reactive (e.g., an epoxide), and it may subsequently add H_2O. The product is now more polar than the original substrate, and it will tend to escape the nonpolar protein to the surrounding aqueous cytoplasm if it is not further degraded.

Certain very specialized organisms can survive on their unique abilities to execute difficult oxidations. For example, bacteria called *methanotrophs* use a monooxygenase to degrade methane, and *white rot fungi* use a peroxidase to oxidize lignin, the refractory polymer in woody tissues. These organisms must exhibit quite high activities of such enzymes to allow them to survive. Interestingly, methane monooxygenase and lignin peroxidase have been found to be quite capable of oxidizing many organic pollutants too (e.g., Wilson and Wilson, 1985; Fogel et al., 1986; Bumpus et al., 1985; Hammel et al., 1986; Chang and Cohen, 1995, 1996; Chang and Criddle, 1997; Alvarez-Cohen and Speitel, 2001). Most other organisms do not maintain high levels of nonspecific monooxygenases, but such metabolic tools are ubiquitous in aerobic organisms. Thus, we can anticipate that in the presence of molecular oxygen, biooxidation of xenobiotic organic compounds will occur, albeit sometimes slowly relative to biological processing of more typical metabolites. The products of such oxidation reactions may be more effectively returned to the aquatic medium or catabolized in subsequent enzymatic steps.

Illustrative Example 17.4

What Products Do You Expect from Microbial Degradation of Vinyl Chloride in an Oxic Environment?

Problem

Vinyl chloride (VC) is the monomer from which the common polymer, polyvinyl chloride, is produced. Hence, VC is produced on a very large scale. Not surprisingly, occasional releases of the rather toxic VC to the environment occur. What initial biodegradation product would you expect from VC if it were released into an oxic environment?

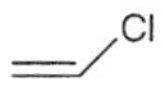

chloroethene
(vinyl chloride, VC)

Answer

From the structure of this compound, you see a good leaving group in the chloride, Cl^-. But since the chloride is attached to a carbon that is enveloped by a π-cloud of electrons (see Fig. 2.6a), it cannot be approached by an electron-rich nucleophile as used in microbial hydrolases. Hence, we do not expect microorganisms to use a nucleophilic attack and a hydrolysis approach.

But noting the presence of the π-cloud of electrons, an initial metabolic attack using an electrophilic "bioreagent" might be expected. Thus, microbial attack in oxic environments likely starts with use of an oxidase on this part of the structure:

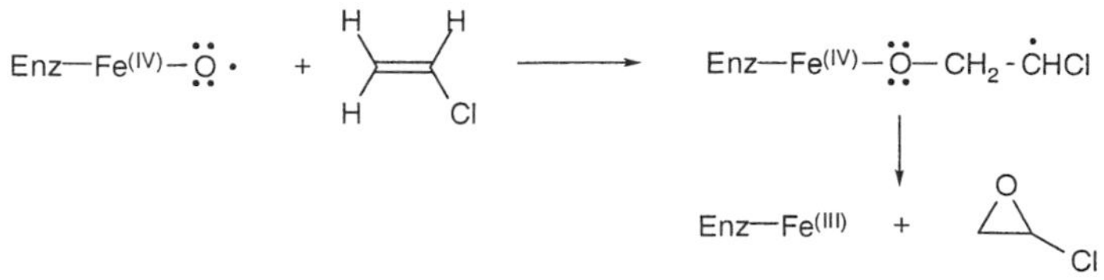

Thus, we anticipate the initial product is the corresponding epoxide.

This is indeed what was found by Hartmans and de Bont (1992) for VC degraded using a monooxygenase from a *Mycobacterium aurum*, a bacterium found to be capable of growth on VC as its carbon and energy sources.

Reductions Involving Nucleophilic Electron-Rich Bioreactants

A third prominent set of biologically mediated reactions used for the initial transformations of xenobiotic compounds are reductions. As discussed in Chapter 14, reduction reactions entail transferring electrons to the organic compound of interest. Microbially mediated reductive transformations involve the same structural moieties that are susceptible to abiotic reductions (Table 17.5). The common characteristic for the structures at the point of reduction is that electron-withdrawing

Table 17.5 Some Microbially Mediated Reductions

Substrate	Product(s)	Reference

carbonyl group

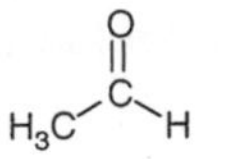

acetaldehyde → H_3C-CH_2-OH ethanol   a

nitro group

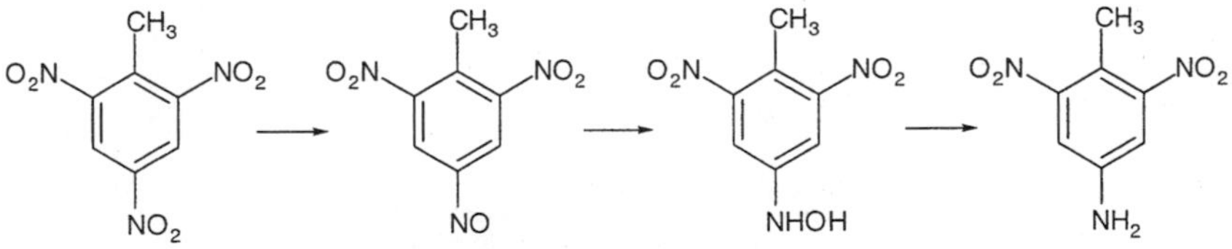

2,4,6-trinitro-toluene (TNT) → 2,6-dinitro-4-nitroso-toluene → N-(4-methyl-3,5-di-nitro-phenyl)-hydroxylamine → 4-amino-2,6-dinitro-toluene   b

sulfoxide and sulfone groups

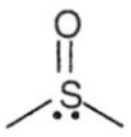

dimethyl sulfoxide (DMSO) → dimethyl sulfide   c

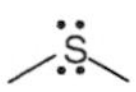

halides (reductive dehalogenation)

d

CCl_4 carbon tetrachloride → $CHCl_3$ chloroform

$ClCH_2CH_2Cl$ 1,2-dichloroethane (ethylene chloride) → $H_2C=CH_2$ ethene (ethylene)   e

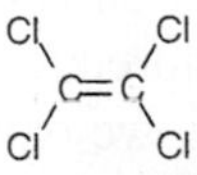

tetrachloroethene → trichloroethene   f

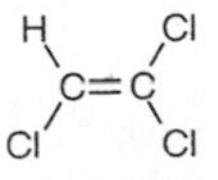

[a] Gottschalk, 1986; [b] Sommerville et al., 1995; Vorbeck et al., 1998; Pak et al., 2000; [c] Weiner et al., 1988; Zinder and Brock, 1978; Griebler and Slezak, 2001; [d] Castro et al., 1985; [e] Holliger et al., 1992; [f] Gantzer and Wackett, 1991; Glod et al., 1997a; Wohlfarth and Diekert, 1999.

heteroatoms are causing the formation of oxidized central atoms (recall Tables 2.5 and 2.6):

$$
\underset{\text{C}(+\text{II})}{R_1\!\!\overset{\displaystyle\overset{O}{\|}}{\underset{}{C}}\!\!R_2}
\qquad
\underset{\text{C}(+\text{III})}{R\!-\!\overset{\displaystyle Cl}{\underset{\displaystyle Cl}{C}}\!-\!Cl}
\qquad
\underset{\text{N}(+\text{III})}{R\!-\!\overset{+}{N}\!\!\overset{\nearrow O}{\underset{\searrow O^-}{}}}
\qquad
\underset{\text{S}(+\text{II})}{R_1\!-\!\overset{\displaystyle\overset{O}{\|}}{\underset{\displaystyle\underset{O}{\|}}{S}}\!-\!R_2}
$$

Due to the electron-withdrawing nature of the attached heteroatoms, the oxidized C, N, and S atoms prove to be quite amenable to nucleophilic attacks in which electrons are added to the central atom. This is an important strategy employed by microorganisms for beginning the degradation of such compounds.

Biological Reduction by Alcohol Dehydrogenases. Bioreductions of atoms doubly bound to oxygen are consistent with a mechanism for attack that would involve a nucleophilic form of hydrogen. Indeed such bioreductants occur in organisms. The reactive portion of one such reductant, the nicotinamide ring of NADH or NADPH (referred to as NAD(P)H), is:

$$
\text{NAD(P)H} \rightleftharpoons \text{NAD(P)}^+ + (H\!:^-) \qquad (17\text{-}37)
$$

This molecule has the marvelous property of holding a nucleophilic form of hydrogen on the ring at the position opposite to the nitrogen atom. This hydrogen is encouraged to be "thrown off" as hydride ($H\!:^-$) since a stabilized aromatic π-electron system can be established using the nonbonded electrons of the N atom. The hydride is added to the electron-deficient part of an organic substrate being reduced. NAD(P)$^+$ can later be reduced back to NAD(P)H as a result of other metabolic processes (e.g., tricarboxylic acid cycle).

To illustrate the role of NAD(P)H in reductive reactions, we consider a class of enzymes called alcohol dehydrogenases (Fersht, 1985). This group of enzymes is ubiquitous, and they have been studied extensively in a variety of organisms (e.g., bacteria, yeast, mammals). These enzymes catalyze the conversion of alcohols ($R_1R_2CH\text{--}OH$) to carbonyl compounds ($R_1R_2C{=}O$) and vice versa (e.g., first reaction in Table 17.5) by facilitating the interactions of the substrate of interest with NAD(P)$^+$/NAD(P)H. Note that although such enzymes are called dehydrogenases indicating they facilitate the oxidation of the substrate by H_2 removal (alcohol $\rightarrow$ carbonyl + H_2); they can also catalyze reductions in the opposite direction. Hence these enzymes are members of the oxidoreductases (enzymes that catalyze oxidations and reductions). In the following, we examine this interaction in some detail to gain insights into the factors governing the overall rates.

Alcohol dehydrogenases enable NAD(P)H to reduce carbonyls as shown in Figure 17.8. First, the enzyme (I) binds the reductant (i.e., NAD(P)H) at a position

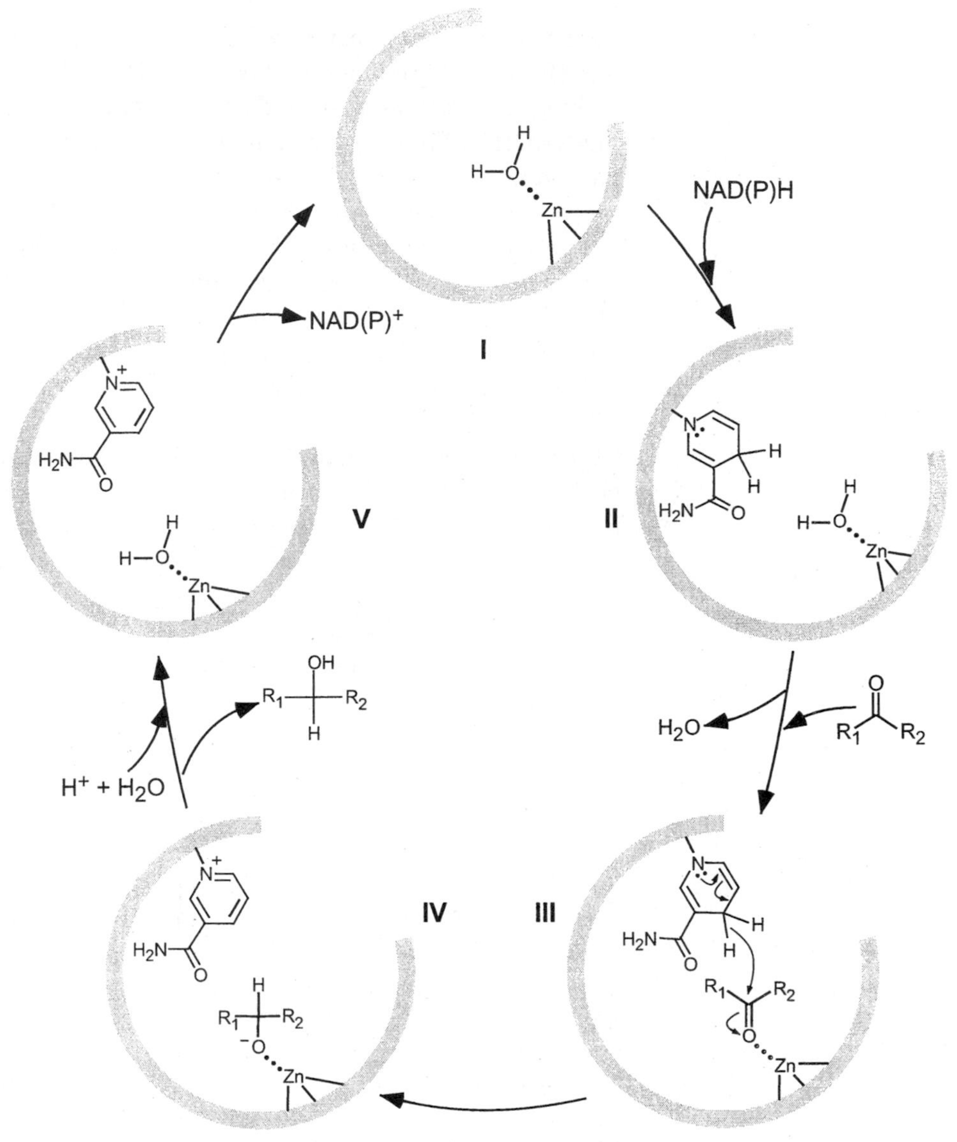

Figure 17.8 Reaction sequence for reduction of a carbonyl compound by a dehydrogenase (Fersht, 1985).

advantageous for interaction with the compound to be reduced (II). Next, the carbonyl oxygen of the organic substrate associates with an electron-accepting moiety (III); in the figure this involves the carbonyl associating as a ligand with a zinc atom in the enzyme (displacing a water molecule that previously occupied the position in II). The complexation of the oxygen of the organic substrate with the zinc serves to draw more electron density away from the carbonyl carbon, as well as to orient the organic substrate suitably for interaction with the previously bound NAD(P)H. Now the hydride carried by NAD(P)H can attack the carbonyl carbon, reducing this functional group to an alcoholic moiety in a two-electron transfer step (III → IV). Acquisition of a proton (ultimately from the aqueous medium, but initially from acidic groups near the reaction site) completes the structural changes and results in the release of the alcohol product from the enzyme (IV → V). Finally, the oxidized NAD(P)$^+$ must be discharged from the enzyme; a new reductant must be bound, and only then can the reaction proceed again.

Several of the above-mentioned steps could limit the overall rate of this transformation (Fersht, 1985): binding of NAD(P)H, binding of the carbonyl compound, bond making and breaking in the reduction reaction, release of the alcohol, and release of the NAD(P)$^+$. For the reduction of aromatic aldehydes (Ar–CHO) with dehydrogenase, release of the product alcohol controls the overall rate of the process (IV $\rightarrow$ V); but for acetaldehyde (CH$_3$CHO) reduction by this same enzyme, it is hydride attack that is rate limiting (III $\rightarrow$ IV). These differences should remind us that particular enzymes exhibit some specificity with respect to the chemicals they process. Overall, the transformation proceeds because organisms utilize reactive reductants (NAD(P)H) and because the enzyme lowers the free energy of activation for the reaction to proceed (recall Fig. 17.3).

NAD(P)H is required in the bioreductions of many xenobiotic compounds. However, NADH or NADPH are not necessarily the entities that directly react with the organic substrate. Rather, the prosthetic groups of other enzymes are themselves reduced by NAD(P)H, and the resulting reduced enzyme components are the actual reactants involved in bond making and breaking of the xenobiotic substance. A prominent set of examples involves the flavin-dependent oxidoreductases. The key reactive portion in these *flavoproteins* is the three-ring flavin (FAD) which is reduced by NAD(P)H to FADH$_2$:

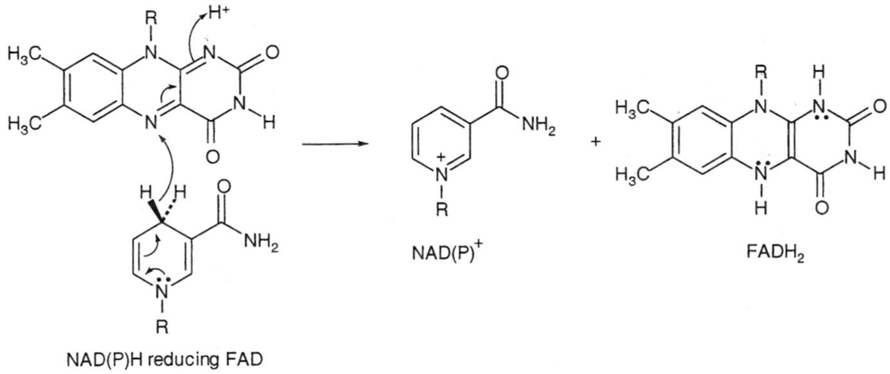

NAD(P)H reducing FAD

(17-38)

The resulting flavin can now participate in either hydride transfer or a pair of 1-electron transfers since it can exist in three stable oxidation states:

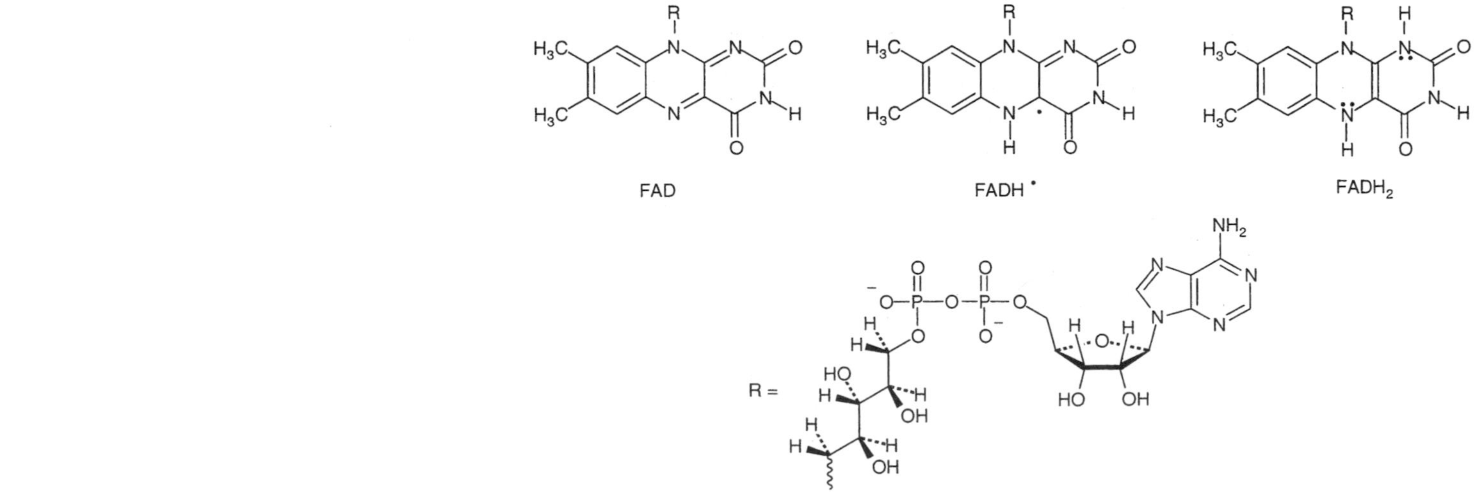

Due to the variety of enzymatic environments in which FAD/ FADH$^{\bullet}$/FADH$_2$ can be held, enzymes that contain them exhibit redox potentials ranging from -0.45 to $+0.15$ V at pH 7 (Bugg, 1997). Xenobiotic compounds whose structures suggest they can be reduced via free radical mechanisms or via hydride transfers may be reduced by such flavoproteins.

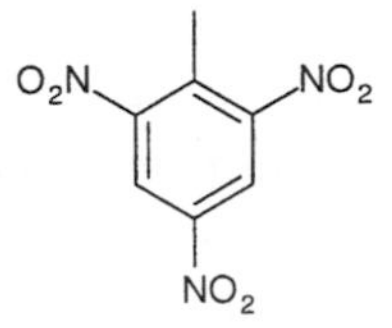

An example of a xenobiotic compound reduced through the actions of such flavoproteins is TNT (trinitrotoluene). This compound is transformed by oxygen-insensitive nitroreductases (referred to as type I) that convert aromatic nitro groups sequentially to nitroso- and then hydroxylamino- moieties (Somerville et al., 1995; Vorbeck et al., 1998; Pak et al., 2000):

TNT

$$\text{ArNO}_2 \xrightarrow[\text{H}_2\text{O}]{+2e^- + 2H^+} \text{ArNO} \xrightarrow{+2e^- + 2H^+} \text{ArNHOH} \qquad (17\text{-}39)$$

nitro-   nitroso-   hydroxylamino-

Most nitroreductases found in bacteria to date fall into this type I category. Type I nitroreductive transformations may be limited by the first of two electron transfers in a tight sequence of one-electron transfers since the enzymatic rates correlate with the corresponding $E_H^1(\text{ArNO}_2)$ (see Eq. 14-32) values (Riefler and Smets, 2000). However, it has also been noted that the free energies of the one-electron and two-electron reductions correlate with one another, and therefore this thermodynamic data may not distinguish between the one- vs. two-electron possibilties (Nivinskas et al., 2001).

A second type of nitroreductase, referred to as Type II, is also known for anaerobic organisms (Somerville et al. 1995; Kobori et al., 2001). These enzymes are susceptible to disruption by O$_2$ and are believed to proceed stepwise through one-electron transfers. Such enzymes are capable of catalyzing the degradation of nitroesters like nitroglycerin (Blehert et al., 1999):

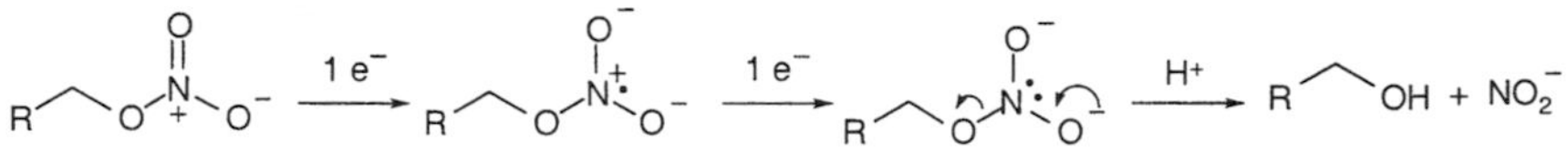

$$(17\text{-}40)$$

A distinguishing feature is that nitrite (NO$_2^-$) is released from the organic compound. These reductases are also flavoproteins and serve to demonstrate the reactive versatility of the flavin moiety.

Biological Reduction With Reduced Metals in Enzymes. Some reductions occur on organic substrates whose structures are not suited to "hydride" attack at the carbon. For example, in the case of bromotrichloromethane, the bulky chloride moieties sterically block a nucleophile's approach to the highly substituted carbon:

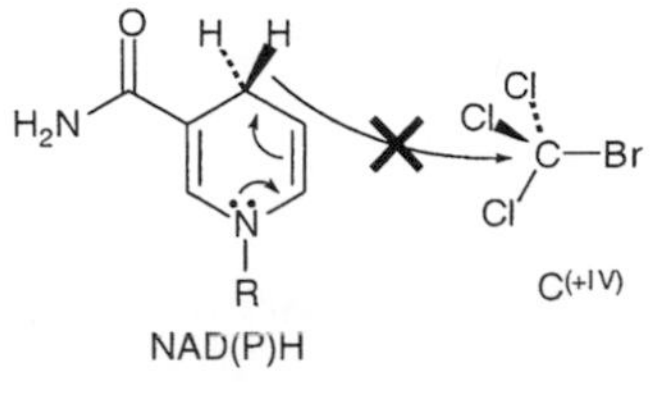

$$(17\text{-}41)$$

NAD(P)H

tetrachloroethene
a vinyl halide
(chlorines on unsaturated carbon)

2,3,4,5,6-pentachlorobiphenyl
an aryl halide
(chlorines on unsaturated carbon)

In the case of vinyl and aryl halides, the π-electron clouds deter approach of a hydride to the carbons.

Yet, microbially mediated reductions of polyhalogenated compounds have long been shown to occur (Hill and McCarty, 1967; Parr and Smith, 1976; Bouwer and McCarty, 1983a,b; Vogel et al., 1987; Egli et al., 1987; Ganzer and Wackett, 1991; Mohn and Tiedje, 1992). Recent studies have even reported reductions of aryl (Woods et al., 1999) and vinyl halides (Magnuson et al., 1998; Wohlfarth and Diekert, 1999; Magnuson et al., 2000). In these cases, transition-metal coenzymes (e.g., with reduced Fe, Co, or Ni) have been found to be capable of reductively dehalogenating these chemicals (e.g., Wood et al., 1968; Zoro et al., 1974; Krone et al., 1989a, b; Gantzer and Wackett, 1991; Holliger et al., 1992).

As a first example, Castro and co-workers (Wade and Castro, 1973ab; Castro and Bartnicki, 1975; Bartnicki et al., 1978; and Castro et al., 1985) extensively documented the reactivities of one such set of bioreductants, the heme proteins (see cytochrome P450 discussed above), with various alkyl halides. During respiration, electrons are transferred to the iron atoms in cytochromes, causing those irons to be in their Fe(II) state. In this state, the cytochromes have low reduction potentials (i.e., $E_H^0(W)$ values between 0 and 0.4 V), thermodynamically sufficient to reduce many halogenated compounds (see Table 14.3 and Fig. 14.4). These hemes have also been shown capable of reducing other functional groups such as nitro substituents ($-NO_2$, Ong and Castro, 1977). Based on studies of hemes outside their protein carriers, the reaction sequence is thought to begin when the halogenated compound is complexed with the reduced iron atom in an axial position (Fig. 17.9). Lysis of the carbon–halogen bond occurs after an inner sphere (i.e., direct) single-electron transfer from

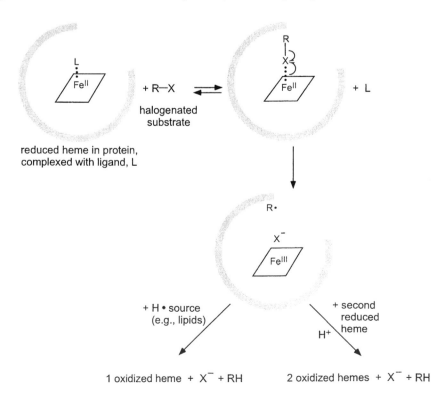

Figure 17.9 Reduction of an akyl halide by a heme-protein such as cytochrome P450 (after Hanzlik, 1981; Castro et al., 1985).

the iron to the accessible halogen (not the central carbon). For isolated hemes in solution, this is the rate-limiting step. The result is a very stable halide product (X^-) and a highly reactive carbon radical ($R^\bullet$). Since free radicals are formed, dimers can occur as products:

$$CCl_4 + e^- \longrightarrow {}^\bullet CCl_3 + Cl^- \tag{17-42}$$

$$2\ {}^\bullet CCl_3 \longrightarrow Cl_3C\!-\!CCl_3 \tag{17-43}$$

When $R^\bullet$ escapes and collides with a good $H^\bullet$ donor (e.g., unsaturated lipids), a hydrogen is abstracted, yielding RH and "damaging" the structure of the $H^\bullet$-donor (Hanzlik, 1981):

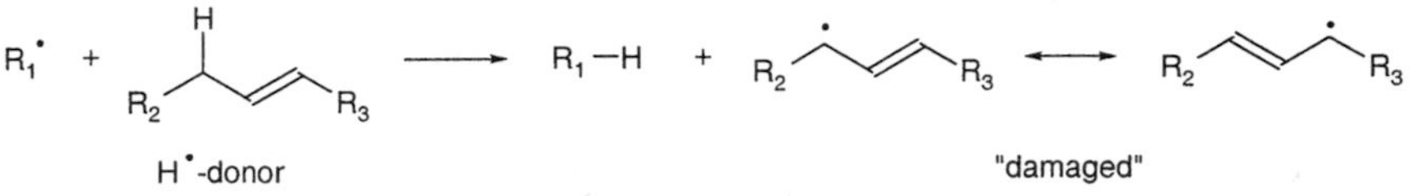

$$R_1^\bullet + \underset{\text{H}^\bullet\text{-donor}}{\cdots} \longrightarrow R_1\!-\!H + \underset{\text{"damaged"}}{\cdots} \tag{17-44}$$

Finally, should a second reduced heme provide a second electron to $R^\bullet$, a carbanion would be formed which would immediately be protonated on entering any surrounding aqueous cytoplasm or solution:

$$R^\bullet \xrightarrow[\text{Fe(II)heme}]{\text{Fe(III)heme}} R\!:^- \xrightarrow[\text{H}_2\text{O}]{\text{HO}^-} RH \tag{17-45}$$

Alternatively, if a suitable leaving group is present at a vicinal position (e.g., in hexachloroethane, HCA), elimination may result forming the corresponding unsaturated compound (e.g., tetrachloroethene, PER):

$$\underset{\text{HCA}}{Cl_3C\!-\!CCl_3} \xrightarrow[\text{Fe(II)heme}]{\text{Fe(III)heme}} Cl_3C\!-\!{}^\bullet CCl_2 + Cl^- \tag{17-46}$$

$$Cl_3C\!-\!{}^\bullet CCl_2 \xrightarrow[\text{Fe(II)heme}]{\text{Fe(III)heme}} \left[Cl_3C\!-\!{:}CCl_2 \right]^- \longrightarrow \underset{\text{PER}}{Cl_2C\!=\!CCl_2} + Cl^- \tag{17-47}$$

In the case of polychlorinated ethenes, cobalt-containing enzymes have been found to be capable of using reduction reactions to dechlorinate these compounds (Neumann et al., 1996; Schumacher et al., 1997; Magnuson et al., 1998, 2000):

$$Cl_2C\!=\!CXCl + Enz\!-\!Co(I) + H^+ \rightarrow \rightarrow HClC\!=\!CXCl + Cl^- + Enz\!-\!Co(III) \tag{17-48}$$

where X = H, F, or Cl. With only one exception, all the enzymes studied to date that catalyze this type of bioreduction on a vinyl carbon have been found to contain a corrinoid cofactor, that is, a ring system containing a cobalt much like an iron porphyrin (I in Fig. 17.10, Wohlfarth and Diekert, 1999). Studies using a particular corrinoid, cobalamin or vitamin B12, indicate that this reduction process is feasible when the cobalt is reduced to its +I state and exhibits a redox potential of about

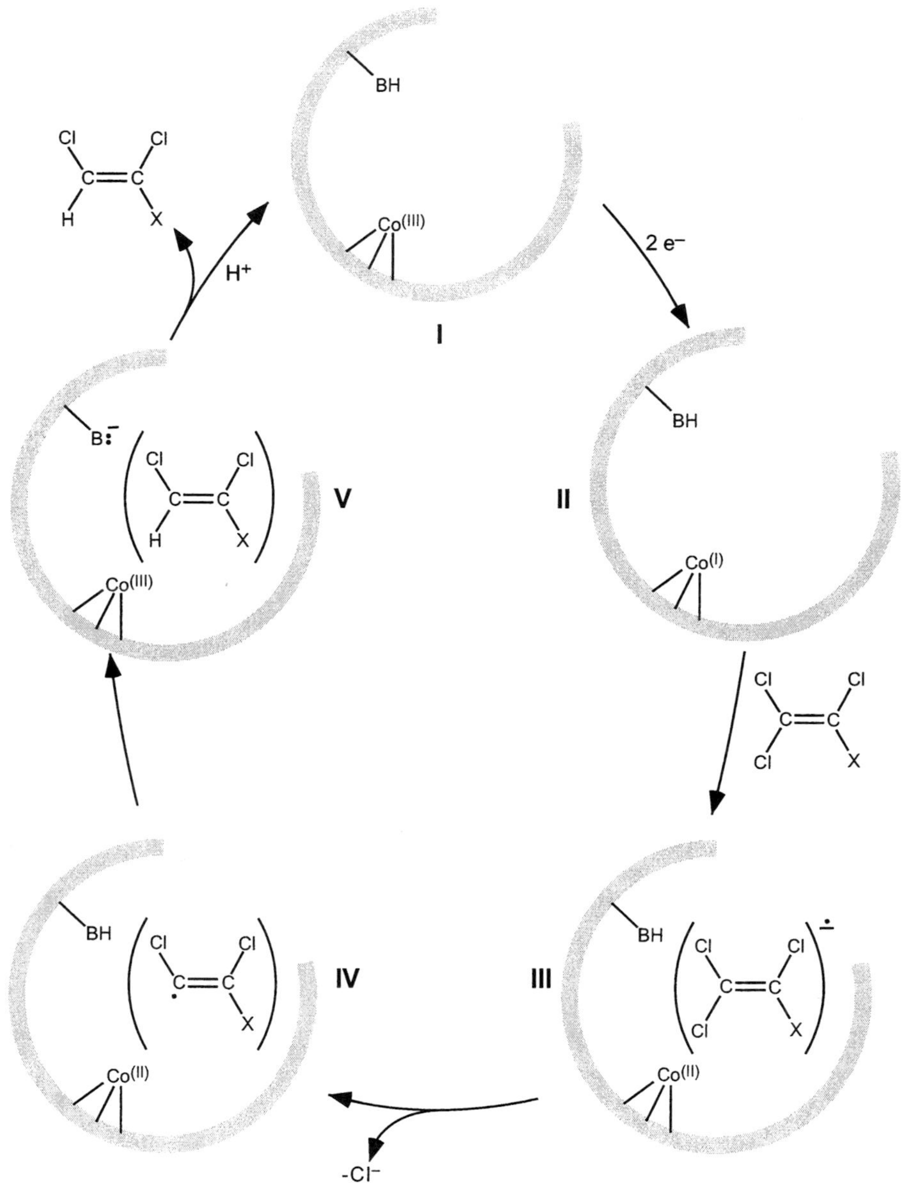

Figure 17.10 Reduction of tetrachloroethene by a cobalt-containing reductase (after Glod et al.1997a; Wohlfarth and Diekert, 1999).

–350 mV (Gantzer and Wackett, 1991; Glod et al., 1997a; Shey and van der Donk, 2000). The resulting "super-reduced corrinoid" is both a very strong nucleophile, and it is also capable of transferring a single electron to the halogens of substrates like tetra- and trichloroethene (see II → III in Fig. 17.10). The radical anion product releases chloride (Cl^-) (III → IV in Fig. 17.10), apparently leaving a very short-lived vinyl radical! This radical is very quickly reduced, possibly by the nearby Co(II) atom (IV → V in Fig. 17.10). Upon protonation, the substrate has had one of its previously bound halogens replaced by a hydrogen. The enzyme has returned to its Co(III) base state (I in Fig. 17.10), awaiting reduction to the Co(I) condition before the cycle can be repeated. The key products of this sequence acting on trichlorethene prove to be *cis*- and *trans*-dichloroethene (as opposed to 1,1-dichloroethene), since the initial electron reductions are favored at the carbon with the most chlorines attached.

This mechanism apparently shifts to an addition reaction across the double bond for less chlorinated compounds like the dichloroethenes and vinyl chloride (Glod et al., 1997b):

$$\text{Enz—}\overset{..}{\text{Co}}{}^{(I)} \ + \ \text{HXC} \!=\! \text{CHCl} + \text{H}^+ \longrightarrow \text{Enz—Co}^{(III)} \!-\! \text{CXH} \!-\! \text{CH}_2\text{Cl} \qquad (17\text{-}49)$$

The addition product is then reduced and breaks down, yielding chloride and the Co(II) form of the enzyme:

$$\text{e}^- \ + \ \text{Enz—Co}^{(III)} \!-\! \text{CXH} \!-\! \text{CH}_2 \!-\! \text{Cl} \longrightarrow \text{Enz—Co}^{(II)} \ + \ \text{XHC} \!=\! \text{CH}_2 \ + \ \text{Cl}^- \qquad (17\text{-}50)$$

The *cis*- and *trans*-dichloroethenes reacting by this addition mechanism are transformed more slowly than the tetra- and trichloroethenes that could form them and more slowly than the vinyl chloride that their reactions would form. Thus, these particular dichlorocompounds accumulate when microorganisms reduce tetra- and trichloroethene in anoxic environments contaminated by these solvents (see Glod et al., 1997b and references therein). Reduction reactions of 1,1-dichloroethene with the super-reduced corrinoids are more likely to form the very toxic product, vinyl chloride.

Amazingly, some microorganisms are capable of using halogenated compounds as terminal electron acceptors (McCarty, 1997; Schumacher et al., 1997; Fetzner, 1998; Wohlfarth and Diekert, 1999). This amounts to these bacteria using compounds like tetrachloroethene instead of O_2, NO_3^-, or SO_4^{2-} to "breathe"!

Several factors may limit the overall rate of enzymatic reductive reactions. First, the electron transfer to the reactive metal (e.g., Co, Fe, or Ni) may be limiting. It is also possible that access of the organic substrates to the reduced metals contained within enzyme microenvironments may be limited. Mass transfer limitation is even more important in intact bacterial cells. For example, Castro et al. (1985) found that rates of heme-catalyzed reductive dehalogenations were independent of the heme content of the cells.

In summary, organisms can use biological reductants such as NAD(P)H, capable of hydride transfer (two electron transfer), and reduced flavoproteins and metallo-proteins, capable of single electron donation. Although not necessarily intended to interact with xenobiotic organic compounds, when such organic chemicals come in contact with suitably reactive bioreductants in vivo, reductions can occur.

Illustrative Example 17.5

What Products Do You Expect from Microbial Degradation of DDT in a Reducing Environment?

Problem

1,1,1-Trichloro-2,2-di(4-chlorophenyl)ethane (DDT) is an insecticide used widely in the past against mosquitos. What initial biodegradation product would you expect from DDT if it accumulated in a reducing environment?

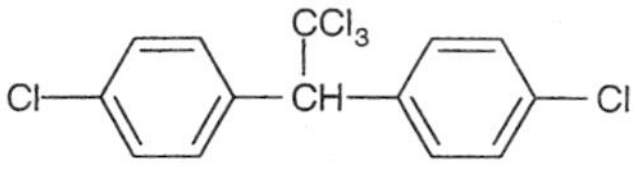

DDT

Answer

From the structure of this compound, you see a good leaving group in the chloride Cl$^-$, on several parts of this compound's structure. But since the chlorines are attached to a carbon that is also substituted with three other large groups, or they are on aryl carbons, they cannot be approached readily from the back side by an enzymatic nucleophile as used in microbial hydrolases. Hence, we do not expect microorganisms to use a nucleophilic attack and a hydrolysis approach.

Since the environment is reducing, you infer that O_2 is absent. Thus, an initial metabolic attack using an electrophilic form of oxygen would not be expected.

Rather a nucleophilic reduction of the trichloromethyl group appears most likely since the three chlorines in close proximity may attract electrons from a reduced enzymatic prosthetic group (e.g., a reduced metal, Me^{red}):

$$\text{Enz--Me}^{red} + \text{Cl--CCl}_2\text{--R} + H^+ \rightarrow \text{Enz--Me}^{ox} + \text{Cl}^- + \text{H--CCl}_2\text{--R} \qquad (1)$$

analogous to the reduction of carbon tetrachloride (CCl_4) to trichloromethane ($CHCl_3$) shown in Table 17.5.

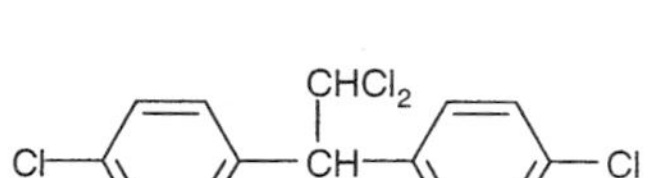

DDD

The predicted product, 1,1-dichloro-2,2-di(4-chlorophenyl)ethane (DDD), is indeed found to be formed from DDT in a variety of anoxic systems like sewage sludges and sediments (Montgomery, 1997).

Additions of Carbon-Containing Moieties or Water

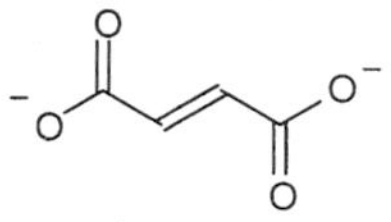

A final strategy used by microorganisms to initiate the degradation of xenobiotic organic compounds involves adding carbon-containing groups or water to the substrate. Two such carbon-containing groups are fumarate and bicarbonate (see margin). These transformations chiefly occur when the xenobiotic compound structure does not contain any functional group susceptible to attack under the conditions that apply. For example, such additions are seen in anoxic environments where use of electrophilic forms of oxygen held by oxygenases is precluded. By attaching carbon-containing moieties, the resulting addition compounds prove to be much more susceptible to further degradation, and so "investment" of reactants like fumarate proves to be worthwhile to the microorganism. The chemistries of these additions are described below so that we can anticipate where they are likely to apply.

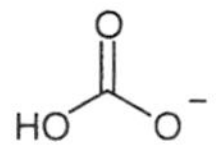

fumarate

bicarbonate

Fumarate Additions. One strategy used by some microorganisms to initiate the transformations of some organic compounds involves the addition of fumarate to the substrate:

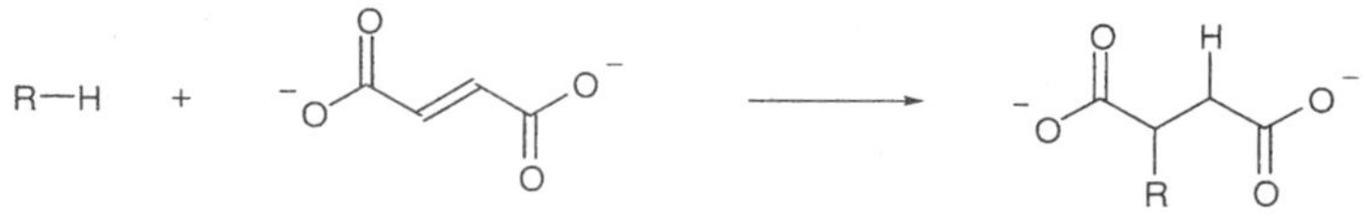

$$\text{R--H} + \text{fumarate} \longrightarrow \text{product} \qquad (17\text{-}51)$$

To accomplish this, the enzymatic catalyst uses a site containing a free radical (Eklund and Fontecave, 1999). For example, in the enzyme, benzylsuccinate synthase, a free radical is formed from a glycine in the protein chain (see margin; Krieger et al., 2001). This glycyl radical can abstract a hydrogen atom from a nearby cysteine residue and begin a catalytic cycle (I → II in Fig. 17.11; Spormann and Widdel, 2000). The cysteine-S-yl radical can subsequently abstract a hydrogen atom from suitable sites on the target organic substrate (II → IV). Benzyl and allyl hydrogens are especially reactive since the free radicals formed are resonance stabilized; hence compounds like toluene, xylene, cresols, and methylnaphthalene have been found to be transformed in this way (Beller and Spormann, 1997; Müller et al., 1999; Arnweiler et al., 2000; Müller et al., 2001). The substrate radical can now add to the double bond of fumarate, held in the active site of the enzyme (V). Returning the previously abstracted hydrogen from the sulfanyl moiety to the addition product yields the original organic substrate with a succinate ($^-OOC-CH_2-CH_2-COO^-$) added to the position of easiest hydrogen abstraction (VI). This addition product can be converted to the corresponding coenzyme A derivative which is now more readily degradable. For example, toluene is converted to benzoyl-CoA, a central intermediate in the anaerobic degradation pathway of many anaerobic substrates (Harwood and Gibson, 1997; Heider and Fuchs, 1997):

$$(17\text{-}52)$$

Due to the involvement of free radicals, this biological strategy is "poisoned" by the presence of the diradical, O_2.

Recently, evidence for this microbial transformation initiation strategy has also been seen for alkanes incubated in anaerobic cultures (Kropp et al., 2000; Rabus et al., 2001). In one case, a denitrifying bacterium was involved, while in the other case, the alkane was metabolized by a sulfate-reducing organism. Consistent with the hydrogen abstraction–fumarate addition mechanism (Fig. 17.11), succinate was found attached to the subterminal carbons of alkanes like hexane and dodecane:

$$H_3C-CH_2-(CH_2)_{\overline{n}}CH_3 \longrightarrow H_3C-\overset{\bullet}{C}H-(CH_2)_{\overline{n}}CH_3 \longrightarrow \begin{array}{c} HOOC-CH-CH_2-COOH \\ | \\ H_3C-CH-(CH_2)_{\overline{n}}CH_3 \end{array} \quad (17\text{-}53)$$

This secondary carbon position offers the best combination of "free radical stability" and ability to approach the enzyme's reactive site. This addition product could be further transformed to yield 2-carboxy-substituted compounds (So and Young, 1999), derivatives that are subsequently used in pathways involving fatty acids.

Carboxylations. Another approach used by microorganisms to initiate the degradation of some xenobiotic organic compounds in anoxic settings involves converting the substrate to a carboxylated derivative:

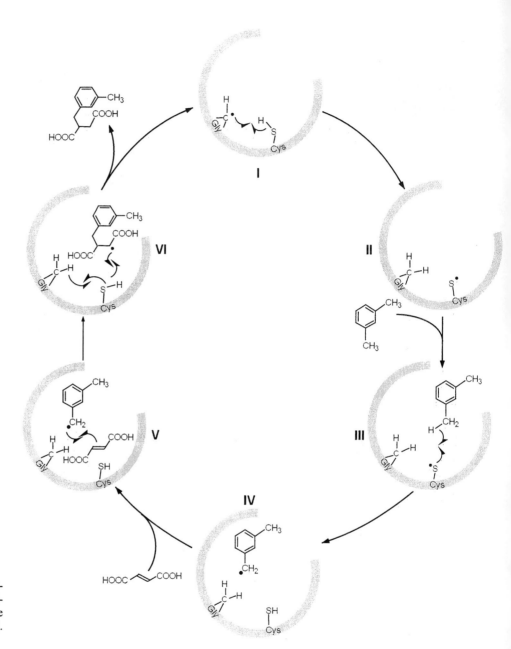

Figure 17.11 Sequence of reactions showing the action of benzylsuccinate synthase on *m*-xylene (after Spormann and Widdel, 2000).

$$RH + HCO_3^- \rightarrow R - COO^- + H_2O \qquad (17\text{-}54)$$

Such transformations have long been recognized for organic compounds capable of enol/keto interconversions:

$$(17\text{-}55)$$

enol form   keto form

Examples include phenols, cresols, and chlorophenols, and these compounds have been found to be degraded after an initial carboxylation (Tschech and Fuchs, 1989;

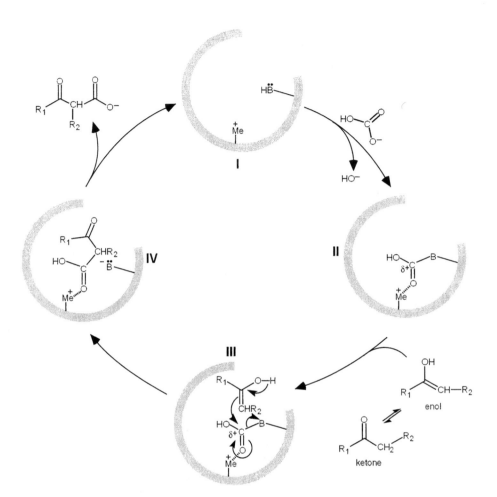

Figure 17.12 Enzymatic carboxylation reaction in which CO_2 from bicarbonate is held by a base (e.g., biotin) and magnesium, so as to present an electrophilic carbon to an enol-containing substrate (after Dugas, 1996).

Bisaillon et al., 1991; Rudolphi et al., 1991; Heider and Fuchs, 1997; Huang et al., 1999). Such aromatic compounds exhibit some nucleophilicity in the π-electrons at their *ortho* and *para* positions, and they may attack electrophiles held enzymatically. Under anaerobic conditions, bicarbonate can be bound by enzymes containing metals like magnesium and cofactors like biotin to form an electrophile (see I → II in Fig. 17.12). When an enol attacks the carbon derived from the bicarbonate (III), a carboxylated derivative of the original substrate is formed (IV). Tschech and Fuchs (1989) have observed carbon isotope exchange in the carboxyl position of 4-hydroxy-benzoate, implying the reversibility of this addition reaction. After an initial aromatic ring carboxylation, further degradation can proceed via the benzoyl-CoA pathway (Heider and Fuchs, 1997).

An important extension of this carboxylation strategy has recently been suggested by Zhang and Young (1997) who studied the degradation of two polycyclic aromatic hydrocarbons (PAHs), naphthalene and phenanthrene, in a sulfate-reducing sediment. The naphthalene was found to be converted to 2-naphthoic acid and the phenanthrene to phenanthrene carboxylic acid (see below). Further degradation to CO_2 was rapid after the presumptive initial carboxylation. Possibly carbon dioxide derivatives can be made so electrophilic by enzymatic interactions (shown as interactions with a metal, Me^+, and a base, B:, see Fig. 17.12) that even the π-electrons of PAHs are able to attack and add a carboxyl group.

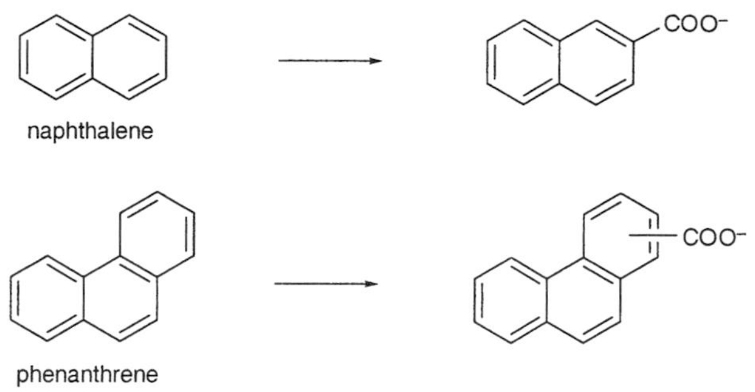

naphthalene

phenanthrene

Hydrations of Double Bonds. Finally, a few scattered observations indicate that anaerobic microorganisms can use water to begin the degradation of isolated double bonds. Schink (1985a) reported that 1-hexadecene was degraded in the absence of O_2 to CH_4 and CO_2. He found that 1-hexadecanol (i.e., the product of 1-hexadecene hydration) was also degraded by the same microorganisms without lag phase, consistent with his hypothesis that this alcohol was an initial metabolite of the alkene. Likewise, he also found that acetylene ($H–C\equiv C–H$) was hydrated to form acetaldehyde (H_3CCHO) in the initial anaerobic transformation of that alkyne (Schink, 1985b). Similar reactions are known for natural compounds like the hydration of oleic acid to 10-hydroxystearic acid (Yang et al., 1993).

However, the enzymology of alkene and alkyne hydration is not well known. Recently, Meckenstock et al. (1999) discovered that the enzyme responsible for anaerobic hydration of acetylene contains a tungsten atom and an [Fe–S] cluster. This may hint that the enzyme uses the tungsten as a Lewis acid to activate the double bond. Possibly, the [Fe–S] cluster then serves to deliver a hydroxide as known in many common metabolite hydrations (Flint and Allen, 1996). Having introduced an oxygen moiety in an initial hydration, anaerobic bacteria may now be able to continue the biodegradation of such compounds.

17.4   Rates of Biotransformations: Uptake

Having seen the major types of biologically mediated *initial* transformations of xenobiotic compounds, we now consider the processes that determine the biotransformation *rates*. These processes include (Fig. 17.1): transfers from unavailable to bioavailable chemical species, substrate uptake, enzymatic catalysis of substrate transformation, and microbial population growth. Recognizing the rate-limiting process for situations of interest is a critical step toward including biotransformations in mass balance models aimed at characterizing the fates of specific compounds in environmental and engineered systems of interest.

Substrate Bioavailability and Uptake Kinetics

The first process that can limit initial biotransformation rates involves delivery of the organic substrates from the environment to the microorganism's exterior. This corresponds to a case in which step 5 in Fig. 17.1 is rate-limiting. This situation is commonly referred to as a matter of *bioavailability*.

Often, organic compounds are distributed among more than one interchanging gaseous, liquid, or solid phase in environmental systems. A classic example involves the occurrence of an organic pollutant in sediment beds as both a "dissolved-in-the-pore-water" species and as a "sorbed-to-the-particles" species (see Chapter 9). If substrate molecules are "buried" in media that do not permit direct transfer into the microorganisms, then the overall degradation sequence may be limited by the chemical's rate of transfer from such unavailable forms to a bioavailable one. For the case of unavailable sorbed molecules needing to move into an aqueous solution in which microorganisms occur, this would involve the rate of desorption:

$$(d[i]_{total} / dt)_{bio} = -k_{desorb} (M_s/V_t) ([i]_{sorbed} - K_{id}[i]_w)$$

$$\approx -k_{desorb} (M_s/V_t) [i]_{sorbed}$$

$$\approx -k_{desorb} (M_s/V_t) (1-f_{iw}) [i]_{total} \qquad (17\text{-}56)$$

where k_{desorb} is the rate constant reflecting mass transfer of an unavailable i species to become biologically available i [T^{-1}]

  (M_s/V_t) is the mass of solids per total volume (M solid $\cdot$ L^{-3})

  $[i]_{sorbed}$ is the concentration of the chemical species (e.g., sorbed) that is not directly available for biological interactions [M M^{-1} solid]

  K_{id} is the equilibrium solid-water distribution constant [L^3 M^{-1} solid] (see Chapter 9)

  $[i]_w$ is the concentration [M L^{-3}] of the bioavailable chemical species (e.g., dissolved if these solute molecules are taken up by the microorganisms)

  f_{iw} is the fraction of the compound present in the bioavailable form (e.g., fraction dissolved as discussed in Chapter 9)

When k_{desorb} is very slow (or even zero as when the compound is encapsulated in an authigenic mineral), $[i]_{sorbed} \gg K_{id}[i]_w$ so we can neglect the second term in the gradient driving transfer. In this case, we refer to the compound as experiencing *sequestration*. The parameter, $(1-f_{iw})$, quantifies the extent of a compound's sequestration in a particular case of interest when we are justified to assume that the dissolved fraction is equal to the bioavailable fraction. Quantitative evaluation of k_{desorb} is taken up in Section 19.5.

When transfers of molecules from media where the microorganisms do not occur (e.g., solids, NAPLs, gases) to phases where they are present (e.g., in water) are not rate-limiting, then it is possible that uptake by the microorganisms can be the slowest step in the sequence. Such a situation implies that the chemical fugacity of the chemical of interest in that system is not equal to the chemical fugacity inside the cells where the relevant enzymatic apparatus occurs. Now we have a case in which the rate of biodegradation is expressed:

$$(d[i]_{\text{total}} / dt)_{\text{bio}} = - k_{\text{uptake}} (V_w/V_t) ([i]_w - [i]_{\text{bio}}/K_{i\text{bio}})$$

$$\approx - k_{\text{uptake}} (V_w/V_t) [i]_w$$

$$\approx - k_{\text{uptake}} (V_w/V_t) (f_{iw}) [i]_{\text{total}} \qquad (17\text{-}57)$$

where k_{uptake} is the rate constant reflecting mass transfer of a bioavailable i species outside the organisms to i inside the organisms [T^{-1}]

(V_w/V_t) is the volumetric fraction occupied by water (–)

$[i]_w$ reflects the concentration of chemical species that is directly available for biological uptake [M L^{-3}]

$[i]_{\text{bio}}$ is the substrate concentration within the organisms [M M$_{\text{bio}}^{-1}$]

$K_{i\text{bio}}$ is the equilibrium biota-water distribution constant [L^3 M$_{\text{bio}}^{-1}$] (See Chapter 10)

This does not mean that only dissolved molecules can be directly taken up by the organisms! Rather, it means that uptake is proportional to the difference in chemical activities/fugacities between the outside and the inside of the organism. When uptake is limiting, then $[i]_w \gg [i]_{\text{bio}}/K_{i\text{bio}}$, so this latter quotient can be neglected. One consequence of the presence of multiple chemical species is that the *bioavailability* of the compound of interest is reduced by a factor that relates the concentration "driving" chemical uptake (e.g., dissolved) to the total concentration (e.g., dissolved plus sorbed). Hence, we see that a compound's bioavailability in this case is quantified by the fraction which is pushing chemical uptake (i.e., f_{iw} in Eq. 17-57 when we assume the dissolved species plays this role).

The parameter, k_{uptake}, can have different physical meanings depending on the situation. For example, if the relevant microorganisms occur embedded in a *floc* (a suspended aggregate of microorganims) or a *biofilm* (a natural polymeric coating containing cells on a solid surface), then this mass transfer coefficient must reflect the substrate's diffusivity in the floc or biofilm and the average distance into the film that the compound must diffuse [i.e., k_{uptake} is proportional to $D_{i\text{biofilm}} \cdot$ (film thickness$/2$)$^{-2}$]. Rittmann and McCarty (2001) have developed a comprehensive treatment of such biofilm kinetics. Stewart (1998) has reviewed information on substrate diffusivities in microbially generated polymeric media.

Alternatively, k_{uptake}, may represent the rate constant associated with transfers from immediately outside the cell to the interior spaces where the relevant enzymes occur. In this case, we need to focus on how long it takes for microorganisms to be equilibrated with medium in which they occur (step 1 in Fig. 17.1). This requires an evaluation of the time for chemicals to be transported from the outside of the microorganisms across their cell envelope to the relevant enzymes. The cell envelope of bacteria differs from species to species (Sikkema et al., 1995) and can even be changed by a single microbial strain in ways that affect cross-membrane transport in response to environmental conditions (e.g., Pinkart et al., 1996; Ramos et al., 1997). However, a general description is useful in the context of molecular

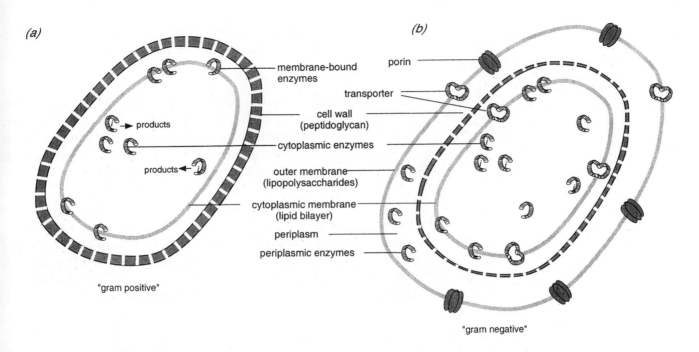

Figure 17.13 Barriers separating periplasmic enzymes (gram-negative bacteria) and cytoplasmic enzymes (gram-negative and gram-positive bacteria) from substrates in the surrounding environment (Madigan et al., 2000).

uptake. First, bacterial envelopes exhibit one of the following series of chemical barriers (Fig. 17.13, Madigan et al., 2000). In so-called "gram-positive bacteria" (named because they stain with a particular dye), the outside of the cell is separated from the inside by two barriers: a porous cell wall made from a mix of peptides and sugars and a lipid bilayer. This contrasts with the "gram-negative bacteria" (which do not stain with the same dye) that use three barriers: an outer lipopolysaccharide membrane, a thin peptide-sugar cell wall, and an inner cytoplasmic lipid membrane. These barriers separate the cytoplasmic and periplasmic enzymes from organic substrates that occur in the exterior environment.

The lipid membranes are a few nanometers thick. They contain proteins whose role is to actively transport particular target chemicals across these nonpolar barriers. The outer membranes of gram negative bacteria also have protein channels called porins that allow passage of small polar and charged substrates.

For the most part, nonpolar chemicals move across the layers of the cell envelope by passive diffusion (Nikaido, 1979; Konings et al., 1981; Sikkema et al., 1995; Bugg et al., 2000). Experimental observations using nonmetabolizable compounds like trichlorobenzene (Ramos et al., 1997; Ramos et al., 1998) and metabolically incompetent mutants with phenanthrene (Bugg et al., 2000) show that passive transfer nonpolar compounds occurs quickly, taking only seconds to minutes.

Since passive permeation into cells occurs on time scales of hours or less, nonpolar compounds that are somewhat recalcitrant (i.e., lasting for days or more in the environment) are probably not limited by their ability to enter microorganisms from surrounding aqueous media. This expectation is supported by the observations that the relative rates of dehalogenation of a series of α,ω-dichlorinated alkanes by resting cells of *Rhodococcus erythropolis Y2* paralleled the results using a purified dehalogenase enzyme (Damborsky et al., 1996). Clearly the transport of these hydrophobic substrates to the dehalogenase enzyme in the cells did not control the overall degradation.

Recently, it has been shown that a mechanism, referred to as an efflux pump, exists to actively transport nonpolar substance *out* of some bacteria (Ramos et al., 1998; Bugg et al., 2000). This partially explains why some microorganisms are capable of maintaining the integrity of their membranes and surviving in situations where high concentrations of hydrophobic substances occur.

We should also note that microorganisms have "water-filled pores" (porins) through their exterior membranes (Fig. 17.13) which permit the passive entrance of small hydrophilic substances (Madigan et al., 2000). In studies of enteric bacteria, passive glucose uptake exhibited transmembrane uptake time scales of less than a millisecond (Nikaido, 1979). Thus, the rate of passive uptake of small, hydrophilic molecules (< 500 molecular mass units) via membrane pores of bacteria is not likely to cause them to avoid biodegradation for prolonged times.

Porins are somewhat selective. Larger hydrated chemical species diffuse in more slowly. Negatively charged chemicals experience electrostatic repulsions that limit their ability to enter. This may explain the situation for highly charged organic compounds like ethylene diamine tetraacetate (EDTA) which are thought to be recalcitrant because of slow or nonexistent passive uptake by most microorganisms. Only cation-complexed (e.g., with Ca^{++}) forms whose charge is neutralized appear suited for uptake (see Egli, 2000).

$$^-OOC\,CH_2 \diagdown \quad \diagup CH_2-COO^-$$
$$N-CH_2-CH_2-N$$
$$^-OOC\,CH_2 \diagup \quad \diagdown CH_2-COO^-$$

EDTA

However, active uptake mechanisms have now been found in some bacteria for various xenobiotic organic anions. These include 4-chlorobenzoate (Groenewegen et al., 1990), 4-toluene sulfonate (Locher et al., 1993), 2,4-D (Leveau et al., 1998), mecoprop and dichlorprop (Zipper et al., 1998), and even aminopolycarboxylates (Egli, 2001). Such active uptake appears to be driven by the proton motive force (i.e., accumulation of protons in bacterial cytoplasm). These transport mechanisms exhibit saturation kinetics (e.g., Zipper et al., 1998), and so their quantitative treatment is the same as other enzyme-limited metabolic processes (discussed below as Michaelis-Menten cases).

Finally, assuming bioavailability and biouptake do not limit the rate of biodegradation, then we expect the biodegradation kinetics to reflect the rate of growth due to utilization of a substrate or the rate of enzyme processing of that compound (both discussed in more details below). In these cases, the rate of biotransformation, k_{bio}, has been studied as a function of the substrate's concentration in the aqueous media in which the microorganisms or enzymes occur. Hence, for an environmental system, we can write:

$$(d[i]_{total} / dt)_{bio} = -V_w\,V_t^{-1}\,(d[i]_w/dt)_{bio} = -V_w\,V_t^{-1}\,k_{bio}\,[i]_w$$

$$= -V_w\,V_t^{-1}\,k_{bio}\,f_{iw}\,[i]_{total} \tag{17-58}$$

where k_{bio} is a pseudo first order biodegradation rate coefficient (T^{-1}).

The parameter, f_{iw}, again serves to quantify the bioavailability of the chemical relative to its total concentration in the system and the coefficient, k_{bio}, represents the rate of processing a fully available chemical species. Now we need to examine the processes that likely dictate the magnitude of k_{bio}, and we consider two situations in the following. In the first case, the rate of growth of a population of microorganisms may be controlled by the utilization of the organic chemical of interest, and when this is so, the rate of cell number multiplication controls the overall rate of chemical loss (step 7 in Fig. 17.1). Since long ago microbial population growth was mathematically modeled as a function of a limiting substrate by Monod (1949), this type of situation will be referred to here as a *Monod-type* case. In contrast, when a specific enzyme-mediated reaction limits the overall rate of removal of a chemical of interest (steps 2, 3 and 4 in Fig. 17.1), we shall refer to this as a *Michaelis-Menten-type* case. Almost 100 years ago these investigators developed a mathematical model reflecting the dependency of the rates of enzymatic processes on substrate parameters (Nelson and Cox, 2000).

17.5 Rates of Biotransformations: Microbial Growth

Monod Population Growth Kinetics

In the following, we consider situations in which the concentration of the chemical of interest to us limits the growth of the degrading microorganisms (step 7 in Fig. 17.1). Such situations may occur when there is a large new input of a labile substrate into the environment. To illustrate, let's examine the removal of *p*-cresol by a laboratory culture (Fig. 17.14; Smith et al., 1978). In this case, microorganisms that could grow on *p*-cresol were enriched from pond water. When a suspension of these microorganisms at 10^7 cells$\cdot$L^{-1} (corresponding to about 1 to 10 μg biomass per liter) was exposed to a 40 to 50 μM solution of *p*-cresol (ca. 5000 μg substrate$\cdot$L^{-1}), it initially appeared that the chemical was not metabolized because no change in *p*-cresol concentration could be detected. In this case, since the microorganisms were selected to be metabolically competent to degrade *p*-cresol, it can be assumed that the apparent absence of degradation was not due to their enzymatic deficiencies. Rather, initially the microbial population was too small to have any *discernible* impact on the *p*-cresol mass. Obviously, the cells multiplied very quickly in the period from 2 to 16 hours (Fig. 17.14), and when they finally reached abundances greater than about 10^9 cells$\cdot$L^{-1}, enough bacteria were present to cause significant substrate depletion. Thus, to describe the time course of this chemical removal in such cases, the microbial population dynamics in the system has to be quantified. This can be done using the microbial population modeling approach developed by Monod (1949).

We begin by considering the relationship of cell numbers to time for a growing population limited by a substrate like *p*-cresol. In response to a new growth opportunity, the cell numbers increase exponentially, and this period of so-called *exponential growth* can be described using:

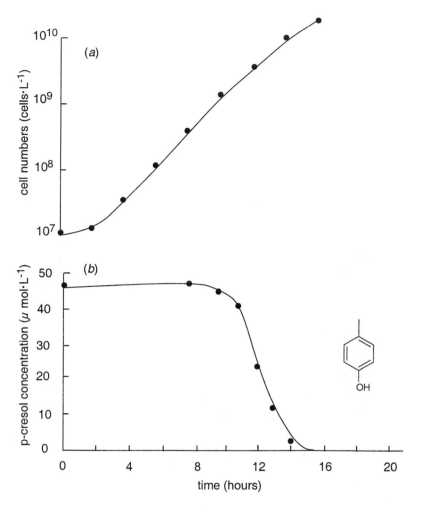

Figure 17.14 Time courses for cell numbers and *p*-cresol concentrations in a laboratory experiment (Smith et al., 1978) immediately after this substrate and bacteria capable of degrading *p*-cresol were mixed together.

$$\frac{d[\text{B}]}{dt} = \mu \cdot [\text{B}] \quad \text{or} \quad [\text{B}]_t = [\text{B}]_0 \cdot e^{\mu \cdot t} \tag{17-59}$$

where [B] is the cell abundance (cells·L^{-1}), and μ is the specific growth rate (h^{-1}). Hence, ln [B] will change in direct proportion to t:

$$\ln[\text{B}]_{t_2} = \ln[\text{B}]_{t_1} + \mu \cdot (t_2 - t_1) \tag{17-60}$$

Put another way, during exponential growth the microbial population will double in number for every time interval, $t = (\ln 2)/\mu$.

Monod (1949) recognized that the growth rate of a microbial species was related to the concentration of a critical substance (or substances) sustaining its growth. More "food" means faster growth, at least up to a certain point when the *maximum growth rate*, $\mu_{\max}$, is achieved. At this point some other factor becomes limiting. Monod mathematically related this population growth response to the concentration of the substance limiting growth with the expression:

$$\mu = \frac{\mu_{\max}[i]}{K_{i\text{M}} + [i]} \tag{17-61}$$

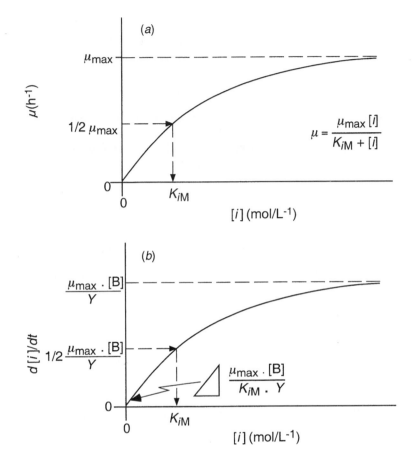

Figure 17.15 Relationships of: (*a*) microbial population specific growth rate, μ, versus substrate concentration after Monod (1949), and (*b*) consequent substrate disappearance rate, $d[i]/dt$, versus substrate concentration.

where μ_{max} is the fastest possible growth rate (e.g., h^{-1}) corresponding to the situation when the limiting chemical is present in excess

$[i]$ is the concentration ($mol \cdot L^{-1}$) of the growth-limiting chemical

K_{iM} is the Monod constant ($mol \cdot L^{-1}$) equivalent to the chemical concentration of i at which population growth is half maximal

K_{iM} is often denoted with K_s in the microbiology literature, but here we use the subscript iM to emphasize Monod growth on i is rate-limiting. This formulation yields a hyperbolic dependency of μ on $[i]$ (Fig. 17.15 and case 1 in Box 17.1). That is, when i is present at low levels ($[i] \ll K_{iM}$), its concentration limits the rate of increase in cell numbers. However, when there is a surplus level of the "food" chemical, other factors limit the rate of population increase.

The conceptual model expressed by Eq. 17-61 implies that no other substance is simultaneously limiting microbial population growth. This assumption may be invalid; for example, an electron acceptor like O_2 may be simultaneously needed for the degradation of the organic chemical of interest. Such dual-limiting substrate cases require modifying Eq. 17-61 to reflect the impacts of both chemicals (see Case 2 in Box 17.1). An interesting remediation situation where such co-substrates are important involves the combination of the electron donor, H_2, and unusual electron acceptors like tetrachloroethene (Haston and McCarty, 1999). Half saturation constants for H_2 used to dechlorinate compounds like tetrachloroethene are near 10 to 100 nM (Smatlak et al., 1996; Ballapragada et al., 1997). Since this key electron donor is commonly at low nanomolar concentrations in subsurface environments of interest, the dual-substrate limiting model may be most appropriate (Haston and McCarty, 1999). Whenever environmental settings exhibit concentrations of the

---

Box 17.1   Monod Limiting-Substrate Models of Microbial Population Growth

Case 1: Single limiting substrate, i (Monod, 1949):

$$\mu = \frac{\mu_{max} \cdot [i]}{K_{iM} + [i]} \tag{1}$$

where μ is the specific growth rate of the microbial population $[T^{-1}]$

   μ_{max} is the maximum growth rate $[T^{-1}]$ of the microbial population in the presence of excess i,

   $[i]$ is the concentration of substrate i $[ML^{-3}]$

   K_{iM} is the concentration of i when growth is half-maximal $[ML^{-3}]$

Case 2: Limitation by two substrates, i and j, whose simultaneous bioreaction limits population growth (Bae and Rittmann, 1996):

$$\mu = \mu_{max} \cdot \frac{[i]}{K_{iM} + [i]} \cdot \frac{[j]}{K_{jM} + [j]} \tag{2}$$

Examples:

i may be an organic compound like toluene and j would be an inorganic species like O_2, or i may be an inorganic species like H_2 and j could be an organic compound like tetrachloroethene.

When either $[i] \gg K_{iM}$ or $[j] \gg K_{jM}$, Eq. 2 simplifies to Eq. 1.

Case 3: Limitation by n growth substrates processed by the same single bottleneck metabolic step (Guha et al., 1999):

$$\mu_{tot} = \sum_{i=1}^{n} \mu_i \tag{3}$$

and:

$$\mu_i = \frac{\mu_{imax} \cdot [i]}{K_{iM} + \sum_{j=1}^{n} \frac{K_{iM}}{K_{jM}} [j]} \tag{4}$$

where μ_{imax} is the maximum specific growth rate when i is limiting

   K_{iM} is the half-saturation coefficient for substrate, i

   K_{jM} is the half-saturation coefficient for each of the competing substrates, j, including i

Example:

Substrates are a mixture of structurally similar hydrocarbons all oxidized for growth by a common rate-limiting enzymatic process.

limiting co-substrates that are near or below the applicable half saturation constants, additional terms will be needed to accurately quantify microbial growth kinetics (e.g., Case 2 in Box 17.1). However, we may often be justified to use the one-limiting-substrate model (i.e., Case 1 in Box 17.1).

It is also possible that more than one organic substrate contributes carbon and energy to a single microbial species. If such substrates are all processed by the same enzymatic approach, and therefore compete with one another at a single metabolic transformation step, then overall population growth will reflect the contributions of numerous substrates (see Case 3 in Box 17.1). This may occur when structurally related mixtures of compounds (e.g., a mix of polycyclic aromatic hydrocarbons) that commonly co-occur in the environment are simultaneously degraded (Guha et al., 1999). In this case, the overall growth of the microbial population reflects the use of all the substrates (see Eq. 3 in Box 17.1). Since all the substrates are competing for the same enzyme, the chemical concentration in the denominator of each compound's Monod expression is modified by a factor ($\sum (K_{iM} / K_{jM}) \cdot [j]$ from $j = 1$ to n in Eq. 4 of Box 17.1) that quantifies the relative usage of the competing chemicals, j, at the bottleneck step. The degradation for any one organic chemical can be enhanced or diminished relative to the case where it serves as the only growth-limiting substrate! It is possible that the overall microbial growth rate will be enhanced by the use of multiple compounds and thereby increase the rate of removal of a single substrate of interest. It is also possible that competitive effects involving access to a key enzyme cause particular substrates to be "ignored" and their degradation decreased (Guha et al., 1999). Each case will reflect a balance of these effects.

For now, we focus on cases in which microbial growth is limited by the substrate of interest to us (i.e., Case 1 in Box 17.1). In order to evaluate changes in the limiting chemical's concentration, we need to relate microbial growth to changes in compound concentration. This can be done by recognizing that degrading a certain amount of chemical mass enables a proportional enhancement in microbial biomass:

$$\frac{d[B] / dt \, (\text{cell} \cdot L^{-1} h^{-1})}{d[i] / dt \, (\text{mol} \cdot L^{-1} h^{-1})} = Y \left(\frac{\text{cells grown}}{\text{moles of substrate } i \text{ used}} \right) \quad (17\text{-}62)$$

This proportionality is called the *yield* of the particular biological process, and it is commonly denoted as Y. For carbon-limiting substrates oxidized by aerobes, biomass yields are usually near 0.5 g biomass $\cdot g^{-1}$ carbon (Neidhardt et al., 1990). Using yield information relevant to a particular compound/microbial species combination, we can now relate the production rate of new cells to the disappearance rate of the chemical of concern:

$$\mu \cdot [B] = Y \cdot \left(-\frac{d[i]}{dt} \right) \quad (17\text{-}63)$$

And, upon rearranging:

$$\frac{d[i]}{dt} = -\frac{\mu \cdot [B]}{Y} \quad (17\text{-}64)$$

Substituting Monod's mathematical description of microbial population growth, μ, as a function of substrate concentration, $[i]$, (Eq. 17-61) into Eq. 17-64, we have:

$$\frac{d[i]}{dt} = -\frac{\mu_{max} \cdot [B]}{Y(K_{iM} + [i])} \cdot [i]$$

(17-65)

This variation of chemical removal rate, $d[i]/dt$, with its concentration, $[i]$, has a hyperbolic form (Fig. 17.15).

The relationship of microbial growth and chemical removal (Eqs. 17-64 and 17-65) implies that when the chemical is present at low levels ($[i] \ll K_{iM}$), its instantaneous rate of degradation (i.e., at a particular time t, and microbial abundance, $[B]_t$) is linearly proportional to its concentration and the "concentration" of the microorganisms "reacting" with it:

$$\frac{d[i]}{dt} \approx -\frac{\mu_{max}}{K_{iM} \cdot Y}[B]_t[i]$$

(17-66)

$$= -k'_{bio}[B]_t[i]$$

(17-67)

where k'_{bio} ($L^3 cell^{-1}T^{-1}$) is effectively a second order rate coefficient equal to μ_{max} $K_{iM}^{-1} Y_i^{-1}$ when $[i] \ll K_{iM}$. In contrast, when the chemical is present in large amounts relative to the microbial community needs ($[i] \gg K_{iM}$), then its rate of removal becomes independent of its concentration:

$$\frac{d[i]}{dt} \approx -\frac{\mu_{max}}{Y} \cdot [B]_t$$

(17-68)

$$= -k'_{bio}[B]_t$$

(17-69)

Consequently, k'_{bio} ($ML^{-3}T^{-1}$) is equal to $\mu_{max} Y_i^{-1}$ for $[i] \gg K_{iM}$. In both of these limiting cases, and obviously for the transitional conditions between, we need information on the factors quantifying microbial growth, μ_{max}, K_{iM}, and Y_i, and on the abundance of cells, $[B]_t$, to be able to predict transformation rates of the compound when it limits population growth (see Illustrative Example 17.6 on page 747).

Returning to the example of p-cresol degradation (Fig. 17.14), we can see how some of these biological parameters are deduced. First, we recognize that in the early part of the experiment, when p-cresol levels do not change (< 10 h), we have the condition $[p\text{-cresol}] \gg K_{p\text{-cresol M}}$, so the changing cell numbers reflect μ_{max}. From the upper portion of the figure, we see that cell numbers increase from about 10^7 cells$\cdot L^{-1}$ at 2 hours to a little more than 10^9 cells$\cdot L^{-1}$ at 10 hours. Using Eq. 17-60, we estimate:

$$\mu_{max} = \frac{\ln([B]_{t2}/[B]_{t1})}{(t_2 - t_1)} \approx \frac{\ln(100)}{8 \text{ h}} \cong 0.6 \text{ h}^{-1}$$

(17-70)

Examining the results between about 10 and 14 hours, we can estimate the yield factor:

$$Y = \frac{[B]_{14h} - [B]_{10h}}{[p\text{-}cresol]_{10h} - [p\text{-}cresol]_{14h}}$$

(17-71)

$$= \frac{(9.4 - 1.3) \times 10^9 \text{ cells} \cdot L^{-1}}{(44 - 1.3) \times 10^{-6} \text{ mol} \cdot L^{-1}} \cong 2 \times 10^{14} \text{ cells} \cdot \text{mol}^{-1}$$

Since bacterial cells weigh about 0.3 pg dry weight per cell (Madigan et al., 2000) and their dry mass is about 50% carbon, this yield appears reasonable (i.e., about 60 g of cells from 100 g of p-cresol).

Finally, to deduce K_{iM} we need to examine growth rates over a range of levels of the substrate. Inversion of Eq. 17-61 yields:

$$\frac{1}{\mu} = \frac{K_{iM}}{\mu_{max}} \cdot \frac{1}{[i]} + \frac{1}{\mu_{max}}$$

(17-72)

Hence a fit of $[i]^{-1}$ versus μ^{-1} yields an intercept $= \mu_{max}^{-1}$ and a slope divided by the intercept of K_{iM}. For example, using the data shown in Fig. 17.14, we may fit and find $\mu_{max} = 0.69 \text{ h}^{-1}$ and $K_{iM} = 6.1 \ \mu M$ using the whole time course.

Using such microbial population dynamics factors, we are now in a position to estimate biodegradation rates for compounds supporting growth like p-cresol. In the situation depicted in Fig. 17.14, we have for the early part of the experiment:

$$k'_{bio} = \frac{\mu_{max}}{Y} \quad (\text{since } [i] \gg K_{iM})$$

$$\approx \frac{0.69 \text{h}^{-1}}{2 \times 10^{14} \text{ cells} \cdot \text{mol}^{-1}} \approx 3 \times 10^{-15} \text{ mol} \cdot \text{cell}^{-1} \text{ h}^{-1}$$

(17-73)

Thus, early in the time course when $[i] \gg K_{iM}$, the rate of p-cresol removal was independent of p-cresol concentration and was continuously changing as the microorganism population increased, ranging from about 3×10^{-8} mol·L^{-1} h^{-1} at 2 h [when $[B]_{2h} \sim 10^7$ cells·L^{-1}] to about 3×10^{-6} mol·L^{-1} h^{-1} at 10 h [when $[B]_{10h} \sim 10^9$ cells·L^{-1}]. Subsequently, the rate of p-cresol removal became a function of the concentration of this substrate; so near the end of the incubation (say 14 hours with $[p$-cresol] at 3 μM) we have:

$$\frac{d[p\text{-}cresol]}{dt} = -\frac{\mu_{max}[B]_{14h}}{(K_{p\text{-}cresolM} + [p\text{-}cresol])Y}[p\text{-}cresol]$$

$$\cong \frac{(0.69 \text{ h}^{-1})(10^{10} \text{ cells} \cdot L^{-1})(3 \times 10^{-6} \text{ mol} \cdot L^{-1})}{(9 \times 10^{-6} \text{ mol} \cdot L^{-1})(2 \times 10^{14} \text{ cells} \cdot \text{mol}^{-1})}$$

(17-74)

$$\cong 1 \times 10^{-5} \text{ mol} \cdot L^{-1} \cdot h^{-1} \qquad \text{at 14 h}$$

Now, we can see how to estimate the biodegradation of chemicals that prove to be growth-limiting substrates of particular microbial species.

Table 17.6 Some Monod Biodegradation Parameters Obtained from Pure or Enrichment Cultures Grown on the Substrate Indicated (Yields Estimated Assuming Cell Mass of 0.3 pg per Cell)

Substrate i	Source of Microorganisms	μ_{max} (h^{-1})	K_{iM} (μM)	Y (cells·mol^{-1})
glucose [a]	pure cultures: *Vibrio, Aerobacter, E. coli, Achromobacter*	0.40–0.65	17–46	–
glycerol [b]	pure cultures: *Achromobacter, Aerobacter*	0.55, 1.2	11, 120	–
acetate [c]	*Pseudomonas* sp.	0.28	1600	7×10^{13}
p-cresol [d]	enrichments from pond water	0.69	6.4	2×10^{14}
quinoline [d]	enrichments from pond water	0.74	1.2	2×10^{14}
methyl parathion [d]	enrichments from creek water	0.61	10	2×10^{14}
malathion [e]	enrichments from river water	0.37	2.2	4×10^{10}
3-chlorobenzoate [c]	*Pseudomonas* sp.	0.13	50	2×10^{14}
methane [f]	enrichment from landfill	0.02	70	2×10^{13}
methane [g]	enrichment from aquifer solids	0.07	400	2×10^{13}
propane [f]	enrichment from landfill	0.02	20	1×10^{14}
toluene [f]	enrichments from aquifer solids	0.02	10	9×10^{13}
toluene [h]	enrichments from aquifer solids	0.03	10	2×10^{14}
phenol [h]	enrichments from aquifer	0.05	–	1×10^{14}
naphthalene [i]	enrichment from soil	0.23	200	2×10^{14}
phenanthrene [i]	enrichment from soil	0.037	4	3×10^{14}
pyrene [i]	enrichment from soil	0.0008	0.5	3×10^{14}
dichloromethane [j]	enrichments of *Pseudomonas* sp. mutants	0.11	–	10^{13} to 10^{14}
chloroethene (vinyl chloride) [k]	*Mycobacterium aurum* L1	0.04	3	5×10^{13}
tetrachloroethene [l]	various, reduction with H$_2$	0.04 to 0.3	0.1 to 200	*ca.* 10^{14}

[a] Jannasch, 1968. [b] Jannasch, 1967. [c] Tros et al., 1996. [d] Smith et al., 1978. [e] Paris et al., 1975. [f] Chang and Alvarez-Cohen, 1995. [g] Chang and Criddle, 1997. [h] Jenal-Wanner and McCarty, 1997. [i] Guha et al., 1999. [j] Brunner et al., 1980. [k] Hartmans and de Bont, 1992. [l] Rittmann and McCarty, 2001.

Various natural and xenobiotic compounds have been studied for their ability to be the sole support of growth for microorganisms (Table 17.6). Sometimes "enrichments" simply isolate a pre-existing subpopulation of bacteria from the mixture of organisms present in a natural sample; however, other times a mutation must occur which permits survival on the chemical provided (e.g., Brunner et al., 1980). From the data shown in Table 17.6, a few cautious generalizations can be made. First, maximum cell growth rates for acclimated cultures appear to correspond to doubling times of hours (recall

$t_{1/2} = \ln 2/\mu_{\max}$). This is not too different from cells grown on "normal" substrates like glucose. An exception may be for organisms growing on highly insoluble compounds like pyrene (Guha et al., 1999). Next, K_{iM} values reported for the few xenobiotic compounds studied in this manner are between μM and mM levels. Finally, the cell yields fall in the range such that 10 to 50% of xenobiotic compound carbon mass is translated into biomass carbon. An obvious exception to this is seen for malathion ($Y = 4 \times 10^{10}$ cells·mol^{-1} ≈ 10^{-4} g·cells·g^{-1} substrate). Malathion was found to be hydrolyzed to the monoacid and ethanol. The ethanol was used as a growth substrate, but the acid product was accumulated without being used further. Obviously, the gross yield in such cases would not be as large.

Given the typical values of $\mu_{\max}$, K_{iM}, and Y_i, under conditions where excess xenobiotic chemical is added to an environment containing some cells capable of living on it (i.e., $[i] >> K_{iM}$), we might expect k'_{bio} to be (Eqs. 17-68 and 17-69):

$$k'_{bio} \approx \frac{0.1 \text{ to } 1 \, h^{-1}}{10^{14} \text{ cell}/\text{mol}}$$

(17-75)

$$\approx 10^{-15} \text{ to } -10^{-14} \text{ mol} \cdot \text{cell}^{-1} \, h^{-1}$$

The factor controlling the time it would take for total compound degradation would be the abundance of the microbial subpopulation, $[B]_t$, capable of growth on the compound. This, then, brings us to the major weakness in trying to quantify degradation limited by microbial growth: how does one know what subset(s) of microorganisms are involved and what their abundance is for any particular occurrence of interest? As indicated in Illustrative Example 17.7, this may not be too much of a problem as long as other critical species such as electron acceptors like O_2 or nutrients such as nitrogen or phosphorus species are present in sufficient quantities to permit unchecked microbial growth. In some systems, especially ones we engineer, we may anticipate that the microbial population grows up to some steady-state condition reflecting a balance of growth versus microbial losses due to die off, wash out, or predation. In natural environments where microorganisms have to respond to dynamic conditions (e.g., inputs due to spills), some uncertainty will derive from our ignorance of the presence of suitable microbial species.

Illustrative Example 17.6

Evaluating the Biodegradation of Glycerol by Microorganisms Growing on that Substrate in a Well-Mixed Tank

Problem

You need to purify a waste water stream containing a readily biodegradable compound, glycerol, present in fairly high concentrations, 100 μM (9 ppm). If one can deliver this waste water into a well-mixed tank (often called a *continuously stirred tank reactor* or CSTR; see also Section 12.4) which is simultaneously receiving all other necessary supplies needed for microbial growth (e.g., O_2, nutrients), then one can build up a microbial population capable of degrading the glycerol to innocuous substances like CO_2 and H_2O before the water is discharged.

Calculate the output glycerol concentration (μM) after the microorganisms have increased their biomass to a steady-state level. Also calculate what the steady-state biomass level is (cells·m^{-3}). Assume you have a tank with $V = 10$ m^3, a waste water flow $Q = 50$ m^3 d^{-1}, and a microbial inoculum with growth properties like those shown in Table 17.6 for the *Aerobacter* sp., a die off coefficient b of 0.1 d^{-1}, and a glycerol-to-biomass yield of 10^{14} cells·mol^{-1}.

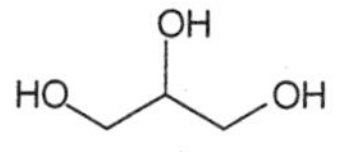

i = glycerol

Answer

The tank will reach steady-state conditions when the concentration of bacteria, [B], present has increased to a constant cell density dictated by a balance of their ability to grow on glycerol against their losses due to wash out and die off:

$$d[B]/dt \;\; = + \mu[B] - (Q/V)[B] - b[B] \tag{1}$$
$$= 0 \text{ at steady-state (ss)}$$

Therefore, the growth rate, μ, for this bacterial culture, after an initial phase of very rapid exponential growth, will ultimately settle down to a value fixed by the rates of wash out plus die off:

$$\mu_{ss} \;\; = (Q/V) + b \tag{2}$$
$$= (50 \text{ m}^3 \text{ d}^{-1}/10 \text{ m}^3) + (0.1 \text{ d}^{-1}) = 5.1 \text{ d}^{-1}$$

Note: Such a culture apparatus is widely used in the laboratory for the study of microorganism–substrate interactions and is referred to as a *chemostat*. The key idea is that at steady-state, the physical properties (volume and flow) determine the biological property (growth)!

Now you can solve for the resultant steady-state glycerol concentration using the Monod growth relation, Eq. 17-61, and values of $\mu_{max} = 28.8$ d^{-1} and $K_{iM} = 120$ μM from Table 17.6:

$$\mu_{ss} = (\mu_{max} \times [\text{glycerol}]_{ss}) / (K_{iM} + [\text{glycerol}]_{ss}) \tag{3}$$

implying:

$$[\text{glycerol}]_{ss} = (\mu_{ss} \cdot K_{iM}) / (\mu_{max} - \mu_{ss}) \tag{4}$$

so that:

$$[\text{glycerol}]_{ss} = (5.1 \text{ d}^{-1} \cdot 120 \ \mu\text{M}) / (28.8 \text{ d}^{-1} - 5.1 \text{ d}^{-1})$$
$$= 26 \ \mu\text{M}$$

Note that this result is independent of the concentration of glycerol in the input stream as long as the other necessary nutrients for growth are sufficient. This result is also independent of the original size of the inoculum. The bacteria simply increase in number until their use of glycerol in the tank ($100 \ \mu$M $- 26 \ \mu$M $= 74 \ \mu$M) corresponds to the bacterial biomass they form during a tank detention time. To obtain an estimate of this steady-state bacterial biomass, we use the glycerol mass balance equation:

$$\frac{d[\text{glycerol}]}{dt} = (Q_{in}/V)[\text{glycerol}]_{in} - (Q_{out}/V)[\text{glycerol}]_{ss} - \text{biodegradation}$$

where the glycerol biodegradation term is given by Eq. 17-64:

$$\frac{d[\text{glycerol}]}{dt} = (Q_{in}/V)[\text{glycerol}]_{in} - (Q_{out}/V)[\text{glycerol}]_{ss} - \mu_{ss}\,[\text{B}]\,/\,Y \qquad (5)$$

When the system is at steady state ($d[\text{glycerol}]/dt = 0$), then we may find:

$$[\text{B}]_{ss} = (Q/V)([\text{glycerol}]_{in} - [\text{glycerol}]_{ss})(Y)\,/\,\mu_{ss} \qquad (6)$$
$$= (50\ \text{m}^3\,\text{d}^{-1}/10\ \text{m}^3)(100\times10^{-3} - 26\times10^{-3}\,\text{mol}\cdot\text{m}^{-3})(10^{14}\,\text{cells}\cdot\text{mol}^{-1})/(5.1\ \text{d}^{-1})$$
$$= 7 \times 10^{12}\ \text{cells}\cdot\text{m}^{-3}$$

It is interesting to note that had the tank been inoculated with a 1-liter cell suspension at 10^9 cells$\cdot$L^{-1} (resulting in an initial cell density of only 10^5 cells$\cdot$L^{-1} in the tank), due to exponential growth at μ_{max} (28.8 d^{-1}), this steady-state cell density would be reached in only about 0.6 days.

Also note that the oxidation of glycerol completely to CO_2 requires 3.5 moles of O_2 for every mole of glycerol consumed. Thus, the degradation of 74 μmol glycerol$\cdot$L^{-1} can just be accomplished with the O_2 present at saturation (*ca.* 280 μM).

Illustrative Example 17.7   **Estimating the Time to Degrade a Spilled Chemical**

Problem

Imagine a case in which *p*-cresol (PC; for structure see Fig. 17.14) is spilled into a pond and dispersed to a initial concentration, $[\text{PC}]_0$, of 1 mM (*ca.* 100 ppm). How long (days) would you think it would be before this contaminant was mostly degraded by the indigenous microorganisms living in the pond water?

Answer

Some subpopulation of the microorganisms is likely capable of taking advantage of this opportunity and growing on this chemical. This implies that as they metabolize PC, they will rapidly increase in number. You can estimate the critical bacterial cell abundance, $[\text{B}]_{crit}$, by defining it as the cell density sufficient to consume all of the chemical during the next doubling interval:

$$[\text{B}]_{crit}\,/\,\text{cells}\cdot\text{L}^{-1} \approx [\text{PC}]_0\,/\,\text{mol}\cdot\text{L}^{-1}\cdot Y\,/\,\text{cells}\cdot\text{mol}^{-1} \qquad (1)$$

From Table 17.6, we find Y for a particular pond population was near 2×10^{14} cell$\cdot$mol^{-1}. Taken together with the initial PC concentrations:

$$[\text{B}]_{crit}\,/\,\text{cells}\cdot\text{L}^{-1} \approx 10^{-3}\ \text{mol}\ \text{L}^{-1}\cdot 2 \times 10^{14}\ \text{cells}\cdot\text{mol}^{-1} \qquad (2)$$
$$= 2 \times 10^{11}\ \text{cells}\cdot\text{L}^{-1}$$

If such cells were 1 μm in radius, each cell would occupy about 4×10^{-15} L; and if they were 10% carbon by weight, the $[B]_{crit}$ level would correspond to about 80 mg cell carbon per liter!

Since the population is increasing exponentially, we deduce the time period, t_{crit}, after the spill to reach $[B]_{crit}$ by calculating:

$$t_{crit} \approx \ln([B]_{crit} / [B]_0) / \mu_{max} \tag{3}$$

$$\approx \ln([PC]_0 \times Y / [B]_0) / \mu_{max} \tag{4}$$

From Table 17.6, you find μ_{max} is about 0.7 h^{-1} for some pond bacteria growing on PC. Unfortunately, you do not have any information on $[B]_0$, although the *total* bacterial cell numbers are typically near 10^9 cells·L^{-1}. Assuming 1% of this population participates in the use of PC as a growth substrate, we estimate:

$$t_{crit} \approx \ln[(10^{-3}\ M)\ (2 \times 10^{14}\ \text{cells·mol}^{-1}) / (10^7\ \text{cells·L}^{-1})] / 0.7\ \text{h}^{-1} \tag{5}$$

$$\approx \ln(2 \times 10^4) / 0.7\ \text{h}^{-1} = 14\ \text{h} \tag{6}$$

$[B]_0$ (cells·L^{-1})	t_{crit} (hours)
10^7	14
10^6	17
10^5	21
10^4	24
10^3	27

The table in the margin shows how t_{crit} estimates change if we vary the initial subpopulation cell concentration.

Clearly the result is not very sensitive to the guess of the initial cell abundance! Of course, all of this presumes that other factors do not limit the microbial population growth in this catastrophic incident-type case. Such factors would include difficulties of mass transport of pollutant molecules (e.g., in an oil slick or tar balls) to microorganisms in a water column (e.g., Uraizee et al., 1998). Also other critical substances like O_2 or the nutrients such as nitrogen or phosphorus species must be present in sufficient quantities to permit unchecked microbial growth.

17.6 Rates of Biotransformations: Enzymes

Now we consider situations in which transformation of the organic compound of interest does not cause growth of the microbial population. This may apply in many engineered laboratory and field situations (e.g., Semprini, 1997; Kim and Hao, 1999; Rittmann and McCarty, 2001). The rate of chemical removal in such cases may be controlled by the speed with which an enzyme catalyzes the chemical's structural change (e.g., steps 2, 3 and 4 in Fig. 17.1). This situation has been referred to as *co-metabolism*, when the relevant enzyme, intended to catalyze transformations of natural substances, also catalyzes the degradation of xenobiotic compounds due to its imperfect substrate specificity (Horvath, 1972; Alexander, 1981). Although the term, co-metabolism, may be used too broadly (Wackett, 1996), in this section we only consider instances in which enzyme-compound interactions limit the overall substrate's removal. Since enzyme-mediated kinetics were characterized long ago by Michaelis and Menten (Nelson and Cox, 2000), we will refer to such situations as *Michaelis-Menten cases*.

Michaelis-Menten Enzyme Kinetics

In general, Michaelis and Menten envisioned enzyme-mediated reactions as involving the following simple sequence:

binding step: i + Enzyme $\longleftrightarrow$ i:Enzyme   (17-76)

reaction step: i:Enzyme $\longrightarrow$ $product$:Enzyme   (17-77)

release step: $product$:Enzyme $\longrightarrow$ Enzyme + $product$ (17-78)

In this conceptualization, it is assumed that the product is rapidly released and removed so that back reactions do not occur. With this somewhat simplified view (i.e., compare this sequence with somewhat more detailed enzymatic processes examined in Section 17.3), a kinetics expression for the removal of i can be written assuming the reaction step is rate limiting (see Box 12.2 for derivation):

$$\frac{d[i]}{dt} = -\frac{k_E[\text{Enz}]_{tot}[i]}{K_{iMM} + [i]} \tag{17-79}$$

where k_E is the rate constant for the enzyme-catalyzed reaction step (mol compound $i \cdot \text{mol}^{-1}$ enzyme $\cdot \text{T}^{-1}$)

         $[\text{Enz}]_{tot}$ represents the total free and substrate-complexed enzyme concentration (mol enzyme $\cdot \text{L}^{-3}$)

         $[i]$ is the free substrate's concentration (mol compound $i \cdot \text{L}^{-3}$)

         K_{iMM} is the "half-saturation" constant reflecting concentration of i when the enzymatic rate is half maximal (mol compound $i \cdot \text{L}^{-3}$)

K_{iMM} is given the subscript, MM, to remind us that it reflects Michaelis-Menten enzyme kinetics as distinguished from K_{iM} used above to model microbial growth kinetics (see Monod cases above). Note, K_{iMM} is the same as K_E^{-1} in Box 12.2 when it's value represents the reciprocal of the equilibrium constant for the binding step.

For Michaelis-Menten cases we expect:

$$k_{bio} / \text{T}^{-1} = \frac{k_E[\text{Enz}]_{tot}}{K_{iMM} + [i]} \tag{17-80}$$

Commonly, $d[i]/dt$ is referred to as the "velocity" of the reaction and denoted v. This reaction velocity expression captures the dependency of $d[i]/dt$ on $[i]$ (see Fig. 17.16). Eq. 17-80 implies that the reaction velocity increases linearly with $[i]$ (slope of $k_E[\text{Enz}]_{tot} / K_{iMM}$) as long as $[i] \ll K_{iMM}$, and the reaction velocity is maximal (called $V_{max} = k_E[\text{Enz}]_{tot}$) when $[i]$ is much greater than K_{iMM}. Hence, Eq. 17-79 is commonly written:

$$v = \frac{V_{max}[i]}{K_{iMM} + [i]} \tag{17-81}$$

This kinetics expression indicates the rate's dependency on catalyst availability through the $[\text{Enz}]_{tot}$ term imbedded in V_{max} (see Illustrative Example 17.8 on page 765).

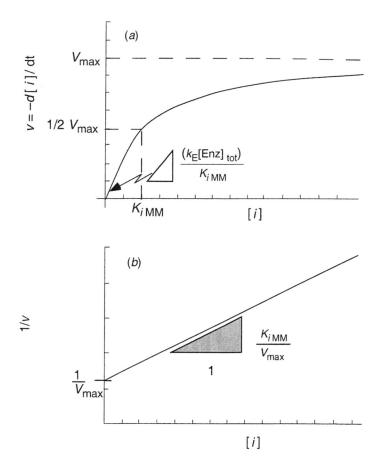

Figure 17.16 Relationships of biodegradation rate, v, to substrate concentration, $[i]$, when Michaelis-Menten enzyme kinetics is appropriate: (a) when plotted as hyperbolic relationship (Eq. 17-79 in text), or (b) when plotted as inverse equation, $1/v = (K_{iMM}/V_{max})(1/[i]) + (1/V_{max})$. Factors governing V_{max} and K_{iMM} differ depending on rate-limiting step in the enzymatic sequence.

The interpretations of the parameters, k_E and K_{iMM}, are not always as straightforward as implied by the simple reaction sequence shown above (Eqs. 17-76 to 17-78). Let's discuss *Michaelis-Menten cases* in more detail.

Organisms may not grow on particular substrates for various reasons. First, the available concentration of the substrate may be insufficient to support microbial multiplication. For example, Tros et al. (1996) showed that 3-chlorobenzoate was needed at concentrations above 10 μM to allow a *Pseudomonas* species to grow on it as a sole source of carbon and energy. Lower concentrations were metabolized if another growth substrate like acetate was provided. In such cases where the organisms have alternative and abundant primary substrates to use for growth (e.g., acetate), the transformations of the xenobiotic compounds are incidental processes occurring as secondary metabolism (i.e., involving enzymes that transform minor substances).

Alternatively, a population of organisms may be using a particular substrate for growth, and due to imperfect enzyme specificity, they also transform compounds with similar structures at the same time. These unintended substrates may be so recalcitrant that they alone cannot support microbial growth. An example of this involves a culture of bacteria that could grow on quinoline (Q); the microorganisms also degraded benzo[f]quinoline (BQ) if it was present (Fig. 17.17; Smith et al., 1978). However, transformation of BQ did not support cell growth (see upper panel of Fig. 17.17).

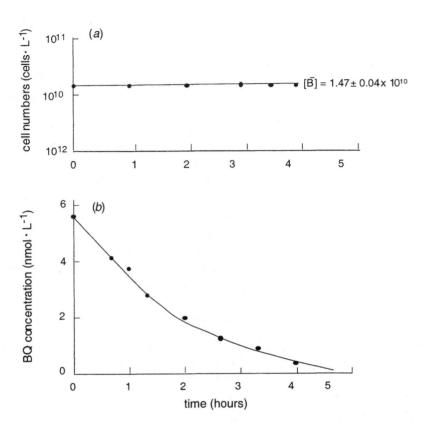

Figure 17.17 Time courses for cell numbers and benzo[f]quinoline (BQ) concentrations in a batch culture experiment (Smith et al. 1978).

quinoline (Q)

benzo[f]quinoline (BQ)

In such a case, the organisms have an enzyme capable of oxidizing Q. Only degradation of Q allows the bacteria to obtain carbon and energy sufficient to maintain and multiply the population. Due to imperfect substrate specificity, the same enzymatic reaction takes place with BQ too. Similar "co-metabolism" by enzymes intended for natural substrates has been reported for many xenobiotic compounds like chlorinated solvents (Semprini, 1997), chlorophenols (Kim and Hao, 1999), and chlorobiphenyls (Kohler et al., 1988).

Hence, sometimes phenomena associated with enzyme kinetics control the rate of biotransformations. If suitable enzymes are present in the microbial community, for example due to consumption of structurally related growth substrates, then we may see immediate degradation of compounds of interest like BQ when they are added to these metabolically competent microbial communities (Fig. 17.17). For such cases, if the abundance of the bacteria is varied, the rate of removal changes accordingly. Consequently, the removal of BQ could be described by a second-order rate law (Smith et al., 1978):

$$\frac{d[\text{BQ}]}{dt} = -k'_{\text{bio}} \cdot [\text{B}][\text{BQ}] \qquad (17\text{-}82)$$

where k'_{bio} is a second-order rate constant (e.g., $\text{L} \cdot \text{cell}^{-1} \, \text{h}^{-1}$), and $[\text{B}]$ is the concentration of bacterial cells degrading BQ ($\text{cell} \cdot \text{L}^{-1}$). k'_{bio} is marked with a prime to indicate that it is normalized to cell concentration. To be consistent with the Michaelis-Menten model (Eq. 17-79), $[\text{BQ}]$ must have been small compared to $K_{i\text{MM}}$

and the concentration of enzyme catalyzing BQ removal, $[Enz]_{tot}$, must have been proportional to the concentration of bacterial cells. This implies k'_{bio} was:

$$k'_{bio} = \frac{k_E [Enz]_{tot} / [B]}{K_{iMM}} \qquad (17\text{-}83)$$

Now we can see the types of biochemical factors that determine the rate constant, k'_{bio} for Michaelis-Menten cases: the ability of the enzyme to catalyze the transformation as reflected by the quotient, k_E / K_{iMM}, and the presence of enzyme in the microorganism population involved, as quantified by $[Enz]_{tot} / [B]$. In the following section, we develop some detailed kinetic expressions for one case of enzyme-mediated transformations. Examination of these results will help us to see how structural features of xenobiotic compounds may affect rates. Finally, we will improve our ability to understand the relative rates for structurally related chemicals that are transformed by the same mechanism and are limited at the same biodegradation step.

An Example of Enzyme Kinetics: Hydrolases

We begin by considering the degradation kinetics for organic compounds, R–L, whose biotransformation begins with an enzymatic hydrolysis reaction separating "L" from the rest of the molecule (e.g., Eqs. 17-12 to 14 in Table 17.1). Recalling the mechanisms discussed in Section 17.3 (Figs. 17.5 and 17.6), the steps of such a reaction sequence may be written:

Step 1: Enzyme:substrate association:

$$Enz-Nu \;+\; R-L \;\; \underset{k_{-1}}{\overset{k_1}{\rightleftharpoons}} \;\; \left(Enz-Nu{:}R-L\right) \qquad (17\text{-}84)$$

where the colon indicates the noncovalent enzyme-substrate association.

Step 2: Nucleophilic addition reaction, release of leaving group (alkylation or acylation step):

$$\left(Enz-Nu{:}R-L\right) \;\; \underset{k_{-2}}{\overset{k_2}{\rightleftharpoons}} \;\; \left(Enz-Nu-\overset{+}{R}\right) \;+\; L^- \qquad (17\text{-}85)$$

Step 3: Second nucleophilic reaction, release of remainder of substrate (dealkylation or deacylation step):

$$\left(Enz-Nu-\overset{+}{R}\right) \;+\; H_2O \;\; \underset{k_{-3}}{\overset{k_3}{\rightleftharpoons}} \;\; Enz-Nu \;+\; R-OH \;+\; H^+ \qquad (17\text{-}86)$$

Under special conditions, simplified kinetic expressions can be derived (Box 17.2). For example, when the initial nucleophilic reaction is the slowest step in the overall process (see Fig. 17.18a), the kinetic expression simplifies to:

$$\frac{d[R-L]}{dt} = -\frac{k_2 [Enz]_{tot} [R-L]}{(K_1)^{-1} + [R-L]} \qquad (17\text{-}87)$$

where $[Enz]_{tot}$ is the total hydrolase concentration (e.g., mol enzyme L^{-1})

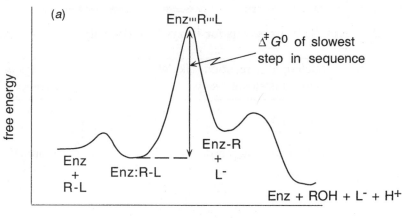

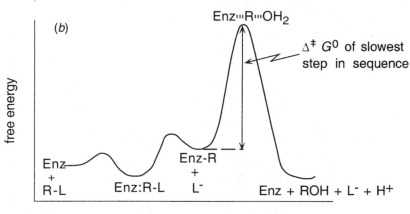

Figure 17.18 Schematic energy profiles for hydrolysis reactions carried out by enzymes when (a) the first nucleophilic reaction is rate limiting and (b) the second nucleophilic step is the slowest process in the sequence.

k_2 is the rate constant for the alkylation or acylation step

K_1 is the equilibrium constant quantifying enzyme:substrate association

Alternatively, if the second reaction (dealkylation or deacylation of the enzyme) is rate limiting (see Fig. 17.18b), we may deduce that:

$$\frac{d[R-L]}{dt} = -\frac{k_3[H_2O][Enz]_{tot}[R-L]}{(K_1)^{-1}(K_2)^{-1}[L^-] + (K_2)^{-1}[L^-][R-L] + [R-L]} \qquad (17\text{-}88)$$

where K_2 is the equilibrium constant quantifying the ratio of chemical species established in step 2 whenever step 3 proves to be the slowest in the sequence

   k_3 is the rate constant associated with the dealkylation of deacylation step in the enzyme-catalyzed hydrolysis

Both of these rate expressions (Eq. 17-87 and 17-88) are hyperbolic as a function of substrate concentration, [R–L]. That is, at low concentrations of [R–L], the rate linearly increases with [R–L]; but at high concentrations, the rate becomes independent of [R–L].

Box 17.2   Kinetic Expressions for Enzymatic Hydrolysis of R–L Under Simplifying Assumptions

Assume initial nucleophilic reaction (alkylation or acylation) is slowest step	Assume second nucleophilic reaction (dealkylation or deacylation)is slowest step

(1) $\dfrac{d[\text{R}-\text{L}]}{dt} = -k_2[\text{Enz}-\text{Nu:R}-\text{L}]$

Assume the complexation step is describable by an equilibrium constant:

(2) $K_1 = \dfrac{[\text{Enz}-\text{Nu:R}-\text{L}]}{[\text{Enz}-\text{Nu}][\text{R}-\text{L}]}$

(3) $[\text{Enz}]_{\text{tot}} = [\text{Enz}-\text{Nu}] + [\text{Enz}-\text{Nu:R}-\text{L}]$

Use Eqs. (2) and (3) to find:

(4) $[\text{Enz}-\text{Nu:R}-\text{L}] = \dfrac{[\text{Enz}]_{\text{tot}}[\text{R}-\text{L}]}{K_1^{-1} + [\text{R}-\text{L}]}$

Substitute (4) into (1)

(5) $\dfrac{d[\text{R}-\text{L}]}{dt} = -\dfrac{k_2[\text{Enz}]_{\text{tot}}[\text{R}-\text{L}]}{K_1^{-1} + [\text{R}-\text{L}]}$

giving

$V_{\text{max}} = k_2[\text{Enz}]_{\text{tot}}$

$K_{i\text{MM}} = K_i^{-1}$

---

(6) $\dfrac{d[\text{R}-\text{L}]}{dt} - k_3[\text{H}_2\text{O}][\text{Enz}-\text{Nu}-\text{R}]$

Assume the complexation step and the first nucleophilic reaction are describable by equilibrium constants:

(7) $K_1 = \dfrac{[\text{Enz}-\text{Nu:R}-\text{L}]}{[\text{Enz}-\text{Nu}][\text{R}-\text{L}]}$

(8) $K_2 = \dfrac{[\text{Enz}-\text{Nu}-\text{R}][\text{L}^-]}{[\text{Enz}-\text{Nu:R}-\text{L}]}$

(9) $[\text{Enz}]_{\text{tot}} = [\text{Enz}-\text{Nu}] + [\text{Enz}-\text{Nu:R}-\text{L}] + [\text{Enz}-\text{Nu}-\text{R}]$

Use Eqs. (7), (8), and (9) to find:

(10) $[\text{Enz}-\text{Nu}^+-\text{R}] = \dfrac{[\text{Enz}]_{\text{tot}}[\text{R}-\text{L}]}{[\text{L}^-]K_1^{-1}K_2^{-1} + [\text{L}^-][\text{R}-\text{L}]K_2^{-1} + [\text{R}-\text{L}]}$

Substitute (10) into (6)

(11) $\dfrac{d[\text{R}-\text{L}]}{dt} = -\dfrac{k_3[\text{H}_2\text{O}][\text{Enz}]_{\text{tot}}[\text{R}-\text{L}]}{[\text{L}^-]K_1^{-1}K_2^{-1} + [\text{L}^-][\text{R}-\text{L}]K_2^{-1} + [\text{R}-\text{L}]}$

giving

$V_{\text{max}} = k_3[\text{H}_2\text{O}][\text{Enz}]_{\text{tot}}$

$K_{i\text{MM}} = [\text{L}^-]K_1^{-1}K_2^{-1} + [\text{L}^-][\text{RL}]K_2^{-1}$

These kinetic expressions now enable us to understand what chemical and biological information is imbedded within the Michaelis-Menten type formulation typically used to describe such enzyme kinetics (Eq. 17-79 or 17-81; Fig. 17.16). First, we see that $K_{i\text{MM}}$ is given by:

$$K_{i\text{MM}} = \frac{1}{K_1} \tag{17-89}$$

when enzyme alkylation/acylation is rate limiting, or:

$$K_{i\text{MM}} = \frac{[\text{L}^-]}{K_1 K_2} + \frac{[\text{L}^-][\text{R}-\text{L}]}{K_2} \tag{17-90}$$

when enzyme dealkylation/deacylation is rate limiting.

This means that we can sometimes use physical chemical insights to enzyme-substrate complexation to anticipation K_{iMM} (i.e., when Eq. 17-89 applies), but we need to be aware that this parameter can be quite complicated in other cases (e.g., when Eq. 17-90 applies).

Next we see that V_{max} (the rate when [R–L] is large relative to K_{iMM}) is different for our two cases of hydrolysis limitation. This parameter always reflects the product of the rate constant of the slowest step and the concentration of total hydrolyzing enzyme present:

$$V_{max}(\text{1st nucleophilic attack slowest}) = -k_2[\text{Enz}]_{tot} \qquad (17\text{-}91)$$

or:

$$V_{max}(\text{2nd nucleophilic attack slowest}) = -k_3[\text{H}_2\text{O}][\text{Enz}]_{tot} \qquad (17\text{-}92)$$

If such enzymes occur at the same levels in relevant microbial populations, V_{max} may be directly related to other metrics of biomass presence such as cell numbers, biomass dry weight, or protein concentrations. In an attempt to enable extending results from one system to another (e.g., from laboratory observations to field situations), one often normalizes V_{max} by such biomass parameters. For example, in Table 17.7, the observed V_{max} values are normalized to the protein contents of the tested microbial populations or isolated enzymes, and the result is given as values V'_{max} (the prime is added to emphasize the normalization). To apply such information to new situations, one must multiply the normalized maximum velocities by a measure of the relevant enzyme concentration or biomass protein in the new system of interest (e.g., V'_{max} × microbial protein content in new case involving intact microorganisms). Of course, one is assuming that the ratio of enzyme to total protein is the same in the old and new situation.

We also note the importance of the rate constants k_2 and k_3 for these two rate-limiting situations. For conditions where hydrolyzable substrates are present in excess of their K_{iMM}'s, we might be able to understand the relative rates of hydrolysis of structurally related compounds by using our knowledge of how variations in chemical structure are known to influence the ease of nucleophilic attacks. In this context, an interesting case involves hydrolyses of a set of carboxylic acid esters with the same acid moiety by the enzyme, chymotrypsin (Fig. 17.19; Zerner et al., 1964). The hydrolyses of these compounds turn out to be limited at the second nucleophilic attack (deacylation) step. Since the same Enz-Nu-acid complex is cleaved for all members of this set of esters in this rate- limiting step (as in Eq. 17-86), we expect all of these esters to exhibit nearly the same overall rate of biotransformation. This is consistent with the rate data that vary by only ±15% (Fig. 17.19). Conversely, when esters of N-benzoyl glycine are hydrolyzed with the first nucleophilic attack (i.e., enzyme acylation) as the slowest step in the process (Epand and Wilson, 1963), the biotransformation rates are more variable (±40%) even among closely related compounds (Fig. 17.19). In these cases, as the alcohol moiety becomes larger (more steric hindrance) and more electron-donating, the rates of enzyme attack decrease (with the exception of the isobutyl compound). When an electron-withdrawing substituent, (pyridine-4-yl)-methyl, is present, the rate of enzyme acylation increases so much that the step is no longer rate limiting. This

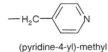

(pyridine-4-yl)-methyl

Table 17.7 Apparent Michaelis-Menten Parameters Reported for Microbial Degradation of Various Substrates

Substrate	K_{iMM}	V'_{max} (mol·kg^{-1} protein· s^{-1})
A. Natural populations [a]		
Toluene in seawater [c]	18 nM	6×10^{-10}
Biphenyl in seawater [c]	1.5 nM	4×10^{-8}
Toluene in stream [d]	2 μM	–
Benzene in lake water with methanotrophs [e]	2 μM	*ca.* 2×10^{-3}
m–Cresol in estuarine seawater [f]	6–17 nM	5–4000×10^{-9}
Chlorobenzene in estuarine seawater [f]	9–46 nM	2–4×10^{-8}
Trichlorobenzene in estuarine seawater [f]	25–38 nM	1–2×10^{-8}
Nitrilotriacetic acid in estuarine seawater [f]	290–580 nM	4–400×10^{-7}
B. Intact Microorganisms [b]		
Benzene by *Methylosinus trichosporium* [g]	–	2×10^{-3}
Toluene by *Cycloclasticus oligotrophus* [h]	100 nM	7×10^{-3}
Toluene by *Pseudomonas* sp. [h]	500 nM	8×10^{-4}
3-Chlorobenzoate by methanogenic consortium [i]	67 μM	1.1×10^{-4}
3,5-Dichlorobenzoate by methanogenic consortium [i]	47 μM	3.6×10^{-5}
Trichloroethene by methanotrophs (geom mean of N = 15) [j]	30 (1 to 200) μM	3×10^{-4} (8×10^{-6} to 1×10^{-2})
Trichloroethene by toluene or phenol degraders (N = 8) [j]	10 (1 to 80) μM	8×10^{-5} (3×10^{-5} to 3×10^{-4})
Trichloroethene by methanotroph mixed culture [k]	15 μM	3×10^{-5}
cis-Dichloroethene by methanotroph mixed culture [l]	30 μM	2×10^{-3}
Vinyl chloride by methanotroph mixed culture [l]	60 μM	2×10^{-3}
C. Cell-free extracts or isolated enzymes		
Naphthalene by sMMO from *Methylosinus sp.* [m]	40 μM	9×10^{-5}
4-Chlorobenzoic acid $\rightarrow$ 4-hydroxybenzoic acid [n]	30 μM	2×10^{-3}
Fluoroacetate $\rightarrow$ hydroxyacetate (glycolate) [o]	2.4 mM	–
Chloroacetate $\rightarrow$ hydroxyacetate [o]	20 mM	–
1-Chlorohexane $\rightarrow$ 1-hexanol [p]	40 μM	4.7×10^{-4}
1-Chloropropane $\rightarrow$ 1-propanol [q]	120 μM	1.7×10^{-4}
1-Bromopropane $\rightarrow$ 1-propanol [q]	20 μM	1.1×10^{-3}
1-Iodopropane $\rightarrow$ 1-propanol [q]	80 μM	1.3×10^{-3}
CH_2Cl_2 [r]	30 μM	1.7×10^{-2}
CH_2BrCl [r]	15 μM	1.5×10^{-2}
2-Hydroxybiphenyl $\rightarrow$ 2,3-dihydroxybiphenyl [s]	3 μM	6×10^{-2}
2-*sec*-Butylphenol $\rightarrow$ 2,3-dihydroxy-*sec*-butylphenol [s]	6 μM	4×10^{-2}
Linuron $\rightarrow$ (3,4-Dichlorophenyl)-1-methoxy-1-methylurea [t]	2 μM	2.5×10^{-3}
o-Nitrophenol $\rightarrow$ catechol [u]	8 μM	8×10^{-2}
Nitrobenzene $\rightarrow$ hydroxylaminobenzene [v]	5 μM	2×10^{0}

[a] Cell counts were converted to protein assuming 1×10^{-16} kg of protein per cell. [b] Assuming 0.3 pg dry mass per cell, and dry mass is 50% protein. [c] Reichardt et al., 1981. [d] Kim et al., 1995. [e] McNeill et al., unpub. [f] Bartholomew and Pfaender, 1983. [g] Burrows et al., 1984. [h] Button, 1998. [i] Suflita et al., 1983. [j] Alvarez-Cohen and Speitel, 2001. [k] Chang and Criddle, 1997. [l] Chang and Alvarez-Cohen, 1996. [m] Koh et al., 1993. [n] Marks et al., 1989. [o] Goldman, 1965. [p] Scholtz et al., 1987a. [q] Scholtz et al., 1987b. [r] Kohler-Staub and Leisinger, 1985. [s] Suske et al., 1997. [t] Englehardt et al., 1973. [u] Zeyer and Kocher, 1988. [v] Somerville et al., 1995.

latter result emphasizes the point that, if different steps are rate limiting within a compound family, we cannot expect the enzyme-catalyzed rate within the substrate set to exhibit consistent dependencies on substituent properties like polarity and steric nature.

The study of Scholtz et al. (1987a, b) is another instructive example of an enzymatic hydrolysis. These investigators found that in cell-free extracts (i.e., cells broken open and membranes and organelles removed before experimentation) from an *Arthrobacter* sp., alkyl halides hydrolyzed at similar rates as in pure cultures of the intact bacteria. Hence differential biouptake limitations can be neglected in this case. These investigators proposed an enzymatic sulfhydryl group was involved in the reaction leading ultimately to the corresponding alcohols. The V'_{max} values derived from their cell-free extract data for a series of alkyl halides are summarized in Table 17.8. If the first nucleophilic reaction step (enzyme alkylation) of the hydrolase was the slow step, we would expect the rates for alkyl chlorides, alkyl bromides, and alkyl iodides would differ since different C–X bonds were being broken in the rate-limiting step. Further, we might expect the longer carbon chain compounds to react somewhat more slowly due to steric hindrance of the initial nucleophilic attack. Generally the data in Table 17.8 indicate all of the alkyl halides shown reacted at about the same rate, which does not support either of these expectations. Thus, the initial nucleophilic attack is apparently not the rate-limiting step. Dealkylation (i.e., release of the alkyl group bound covalently to the enzyme after the initial attack) may indeed be slow; the inability to break up such adducts is frequently blamed for the toxic effects of *alkylating agents* (i.e., substrates that cause alkyl groups to attach to macromolecules so as to damage their function). Consequently we focus on the second nucleophilic attack since it appears to determine the rate of these alkyl halide hydrolyses. In this case, the leaving group (i.e., the enzyme itself) has become identical for all compounds involved, so we would not expect any rate variations because of this factor. Further, if owing to the action of the enzyme, the steric arrangement of the atoms involved in the second nucleophilic substitution reaction has already been optimized, then there might not be too much difference in steric accessibility between the various homologues. The

Table 17.8 Observed Maximal Rates (nmol·min^{-1} mg^{-1} protein) of Hydrolysis in Cell-Free Extracts of *Arthrobacter* sp. Grown with 1-Chlorobutane for Primary Halide Compounds Present at 5 mM (in Great Excess of their K_{iMM}) [a]

R—L + Enz—SH $\xrightarrow{\text{enzyme alkylation}}$ Enz—S—R $\xrightarrow[\text{enzyme dealkylation}]{H_2O}$ R—OH + Enz—SH , H$^+$ + L$^-$

"L"	R = Methyl-	Ethyl-	n-Propyl-	n-Butyl-
–Cl	–	–	–	40
–Br	–	70	70	40
–I	50	60	80	50

[a] Data from Scholtz et al. (1987a), corrected for fraction dissolved in the experimental media using K_{iaw} values and air–water ratios.

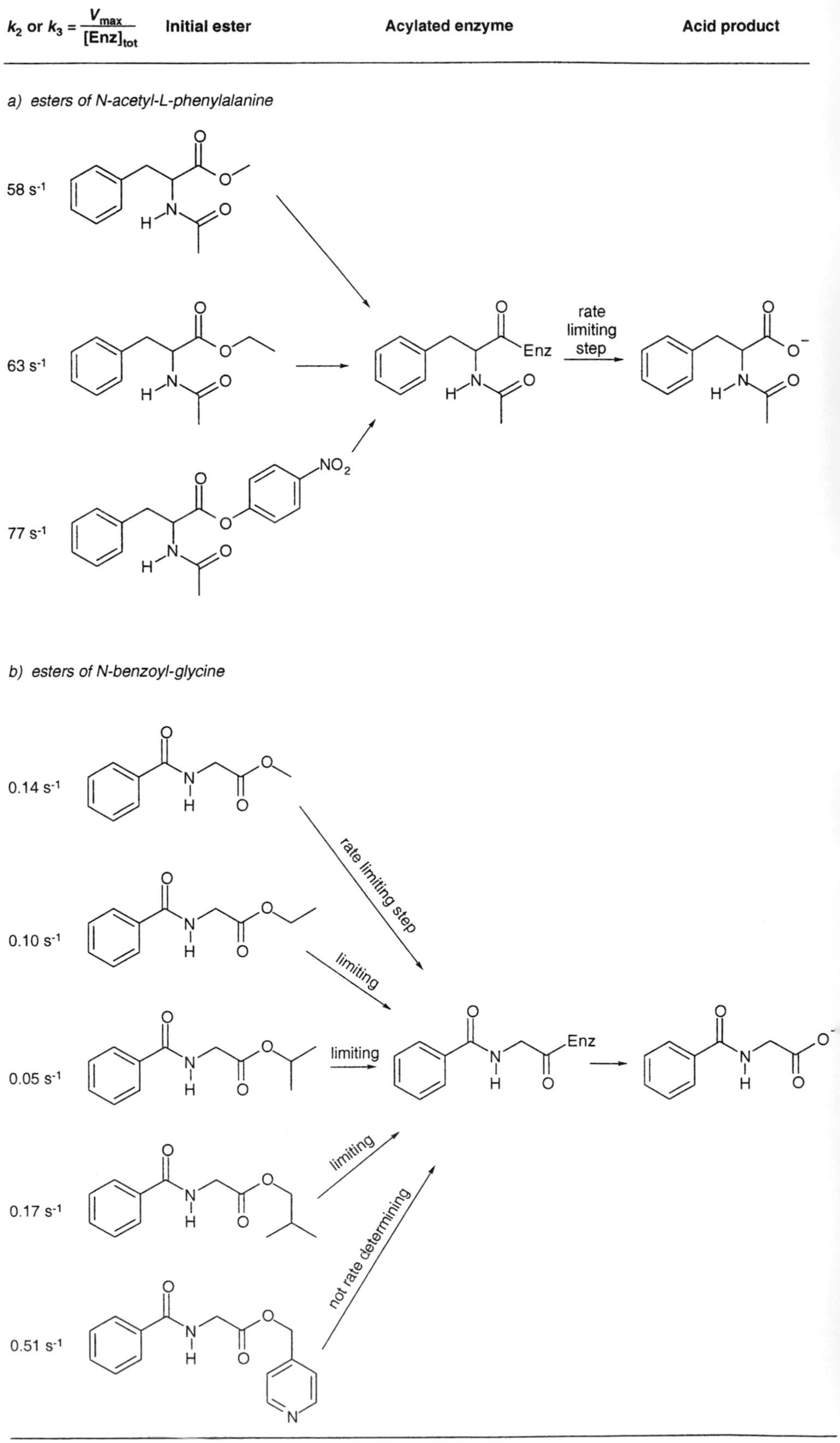

Figure 17.19 Rates of hydrolysis of two families of esters by a hydrolase, chymotrypsin. The esters of N-acetyl-L-phenylalanine exhibit very similar rates because the process in each case is limited by the same enzyme deacylation reaction (Zerner et al., 1964). The esters of N-benzoyl glycine exhibit rates varying by more than a factor of 3 because their hydrolyses are mostly limited by the initial enzyme acylation step (Epand and Wilson, 1963).

data shown in Table 17.8 support the conclusion that the release of the enzymatic sulfhydryl group is rate determining in this reaction. However, if another step in the dehalogenation of these alkyl halides is the same, such a step could be rate limiting. In either case, these data suggest it is possible to make reasonable estimates of V'_{max} values for other alkyl halides for this organism and these conditions. In sum, if one is correct in the assumption that the enzyme reaction mechanism is the same for all compounds in a family of interest, then it may be feasible to predict rates for new compounds in that family based on data from related data on other compounds.

With deeper understanding of the rate laws applicable to these hydrolases, now we need to deduce the parameters that combine to give corresponding k_{bio} values for Michaelis-Menten cases (Eq. 17-80). We may now see that the mathematical form we used earlier to describe the biodegradation of benzo[f]quinoline (Eq. 17-82) could apply in certain cases. Further we can rationalize the expressions used by others to model the hydrolysis of other pollutants when rates are normalized to cell numbers (e.g., Paris et al., 1981, for the butoxyethylester of 2,4-dichlorophenoxy acetic acid) or they are found to fall between zero and first order in substrate concentration (Wanner et al., 1989, for disulfoton and thiometon).

This understanding also helps us to anticipate how fast a set of related compounds should be hydrolyzed by a particular enzyme. First compounds with structures that encourage their binding to the hydrolase (i.e., increase K_1) will be processed more quickly. Interestingly, some hydrolases such as chymotrypsin (Berezin et al., 1970) and many oxidases like cytochrome P450 (Mueller et al., 1995) are known to bind their substrates in large measure owing to hydrophobic interactions. Put another way, compound binding in such instances should correlate with measures of compound hydrophobicity (e.g., γ^{sat}_{iw} or K_{iow}) although larger substrates may be less able to fit in the relevant enzyme's active site. Additionally, if we know something about the rate-limiting step, we can understand how structural variations within a compound class should influence enzymatic reaction coefficients like k_2 or k_3. Thus working with insights on susceptibility to comparable solution attacks (e.g., chemical hydrolysis rates by hydroxyl anion) and substrate hydrophobicity, one might find predictive relationships to describe relative biological hydrolysis rates of related sets of compounds.

Estimating Biotransformation Rates for Michaelis-Menten Cases

Now we are in a position to estimate biodegradation rates when the overall process is limited by enzyme kinetics. For cases of interest to us, we need information on: the enzymatic parameters, V'_{max} and K_{iMM}, for the compound, organisms, and conditions of interest. Often previous investigators have examined situations that are similar to new ones we are trying to model, and values of enzymatic parameters have been determined (Table 17.7). Half-saturation constants (K_{iMM}) for a wide variety of xenobiotic compounds are often in the micromolar range, although nanomolar and millimolar cases have also been measured. Perhaps not surprisingly, values of V'_{max} vary widely since active microbial species abundances also vary widely between natural and engineered systems. Nonetheless, for instances in which the environmental setting of interest is tested for the short-term degradation of the substrate as a function of that substrate's concentration, values of V'_{max} and K_{iMM} that allow reasonably accurate predictions can be found. For example, with information on a microbial species of interest, one can anticipate the degradation of a specific organic compound in an engineered reactor (see Illustrative Example 17.8; see below). Likewise, using laboratory information on an enzymatic process of interest, one may be able to predict the impact of biodegradation in a field setting (see Illustrative Example 17.9 on page 765).

To conclude, by recognizing cases of enzyme-limited biotransformation, these approaches allow us to estimate rates for a wide range of xenobiotic and trace natural organic compounds. We also need to emphasize several points of caution. In order to have confidence in techniques used to evaluate the reactivities of a compound, we must know several facts about the system. First, we must be sure that transport limitations are not controlling the overall rate; that is, desorption from nearby solids or transport into the cell must not be the slowest step in the process (steps 1 or 5 in Fig. 17.1). Second, we must know that the enzymes acting on the substrates of interest are essentially unchanging in their abundance (i.e., $[Enz]_{tot}$ is not increasing as indicated by steps 6 or 7 of Fig. 17.1). Finally, we must know K_{iMM} and V_{max} values. Since K_{iMM} values are frequently found to be in the micromolar or even nanomolar range (Table 17.7), we may expect millimolar and higher levels to exhibit saturation enzyme kinetics. Conversely, in Michaelis-Menten cases biodegradation rates for nanomolar and lower concentrations may be characterized with rate constants, k_{bio}, given by:

$$k_{bio} = V_{max}'' [B]/K_{iMM} \qquad (17\text{-}93)$$

where [B] reflects a measure of the biomass present in the environment of interest in the units to which V_{max}'' has been normalized. While K_{iMM} data may be applicable from case to case as long as the same key enzyme is involved, one should not expect the same to hold for V_{max}'' information since this parameter depends on the concentration of active enzyme, which in turn is certainly dependent on the abundance of the relevant microbial species. Despite such complications, understanding the nature and dependencies of enzyme-limited biotransformations has proven very helpful in problems like designing field remediations (e.g., Semprini, 1997; McCarty et al., 1998), and may eventually permit future clean-up designs using enzymes directly (e.g., Glod et al., 1997a).

Illustrative Example 17.8

Evaluating the Co-Metabolic Biodegradation of Trichloroethene by Microorganisms Growing on Methane in a Well-Mixed Tank (CSTR)

Problem

You need to purify a waste water stream containing a recalitrant compound, trichloroethene, present in fairly high concentrations, 100 μM (13 ppm). If one can deliver this waste water into a well mixed tank (i.e., a *continuously stirred tank reactor* or CSTR) virtually saturated with methane at 500 μM and oxygen at 1000 μM and all other nutrients needed for microbial *growth on methane*, then one can build up a methanotroph population capable of degrading the trichloroethene by cometabolism to innocuous substances like CO_2 and H_2O before the water is discharged.

Calculate the steady-state output trichloroethene concentration (μM) after the methanotrophs have increased their biomass to a steady state level (cell·m^{-3}) assuming a tank with a volume of either $V = 10$ m^3 or 50 m^3. Assume you have a waste water flow, Q, of 5 m^3 d^{-1}, a microbial inoculum with growth properties like those shown in Table 17.6 for the landfill-derived methane oxidizers, and a die-off coefficient b of 0.1 d^{-1}.

trichloroethene
(TCE)

Answer

The tank will reach steady-state conditions when the concentration of methane-oxidizing bacteria, [B], has increased to a constant cell density dictated by a balance of their ability to grow on methane against their losses due to wash out and die off:

$$d[B]/dt = +\mu[B] - (Q/V)[B] - b[B] \qquad (1)$$
$$= 0 \text{ at steady state}$$

Therefore, the growth rate, μ, for this bacterial culture, after an initial phase of very rapid exponential growth, will ultimately settle down to:

$$\mu_{\text{steady state}} = (Q/V) + b \qquad (2)$$

For the 10 m^3 tank, this will be: $= (5 \text{ m}^3 \text{ d}^{-1}/10 \text{ m}^3) + (0.1 \text{ d}^{-1})$ $= 0.6 \text{ d}^{-1}$

and for the 50 m^3 tank: $= (5 \text{ m}^3 \text{ d}^{-1}/50 \text{ m}^3) + (0.1 \text{ d}^{-1})$ $= 0.2 \text{ d}^{-1}$

Note that growth rate needed to maintain a steady-state methanotroph population in the 10 m^3 tank (0.6 d^{-1}) is greater than the μ_{max} for these bacteria (0.48 d^{-1})! This means that the physical flushing exceeds the ability of these organisms to multiply, and thus the methanotrophs will *wash out* leaving [B] = 0 in that case. Hence, we only consider the 50 m^3 tank below.

Now we can solve for the steady-state methane concentration assuming this substrate is limiting growth and using the Monod relation, Eq. 17-61 and values of $\mu_{\text{max}} = 0.48$ d^{-1} and $K_{i\text{M}} = 70$ μM from Table 17.6:

$$\mu = (\mu_{\text{max}} \times [\text{methane}]_{\text{steady state}}) / (K_{i\text{M}} + [\text{methane}]_{\text{steady state}}) \qquad (3)$$

implying:

$$[\text{methane}]_{\text{steady state}} = (\mu \cdot K_{i\text{M}})/(\mu_{\text{max}} - \mu) \qquad (4)$$

so that:

$$[\text{methane}]_{\text{steady state}} = (0.2 \text{ d}^{-1} \cdot 70 \text{ } \mu\text{M})/(0.48 \text{ d}^{-1} - 0.2 \text{ d}^{-1})$$
$$= 50 \text{ } \mu\text{M}$$

To obtain an estimate of this steady-state bacterial biomass, we use the methane mass balance equation:

$$d[\text{methane}]/dt = (Q_{\text{in}}/V)[\text{methane}]_{\text{in}} - (Q_{\text{out}}/V)[\text{methane}] - k_{\text{bio}} [\text{methane}]$$

where the methane biodegradation term is given by Eq. 17-64:

$$= (Q_{\text{in}}/V)[\text{methane}]_{\text{in}} - (Q_{\text{out}}/V)[\text{methane}] - \mu [B] / Y \qquad (5)$$

When the system is at steady state ($d[\text{methane}]/dt = 0$), then we may find:

$$[\text{B}]_{\text{steady state}} = (Q/V)([\text{methane}]_{\text{in}} - [\text{methane}]_{\text{steady state}})(Y) / \mu \tag{6}$$

$$= (5 \text{ m}^3 \text{ d}^{-1}/50 \text{ m}^3)(500 \times 10^{-3} - 50 \times 10^{-3} \text{ mol m}^{-3})(2 \times 10^{13} \text{ cell mol}^{-1})/(0.2 \text{ d}^{-1})$$

$$= 2.3 \times 10^{12} \text{ cell m}^{-3}$$

It is interesting to note that had the tank been inoculated with a 5 liter cell suspension at 10^9 cell$\cdot$L^{-1} (resulting in an initial cell density of only 2×10^4 cell$\cdot$L^{-1} in the tank), due to the slow exponential growth near μ_{max} (0.48 d^{-1}), this steady state cell density would only be reached after about 3 to 4 weeks!

We also note that the oxidation of methane completely to CO_2 requires 2 moles of O_2 for every mole of methane consumed. Thus, the degradation of 450 μmol methane L^{-1} can be accomplished with the O_2 present (*ca.* 1000 μM).

Once the methanotroph-containing CSR has been established, we expect it can cometabolically oxidize some of the TCE incidental to the use of the primary substrate, methane. The rate of TCE biotransformation is dictated by the effectiveness of the methane mono-oxygenase for interacting with TCE rather than methane. This enzymatic processing of TCE can be described using a Michaelis-Menten expression:

$$(d[\text{TCE}]/dt)_{\text{co-metab}} = V''_{\text{max}} [\text{B}] [\text{TCE}] / (K_{TCE\,\text{MM}} + [\text{TCE}]) \tag{7}$$

From Table 17.7, we find that methanotroph cultures exhibit V'_{max} values near 2.8×10^{-4} mol TCE kg^{-1} methanotroph protein s^{-1} and $K_{TCE\,\text{MM}}$ values near 30 μM. Assuming methanotrophs are 50% protein and each cell has 3×10^{-16} kg of biomass, then the V''_{max} is also equal to: 4.2×10^{-20} mol TCE cell^{-1} s^{-1} or 3.6×10^{-15} mol TCE cell^{-1} d^{-1}. Together with the steady-state cell density calculated above, this implies a TCE biodegradation rate "constant" near:

$$k_{\text{bio}} = V''_{\text{max}} [\text{B}] / (K_{TCE\,\text{MM}} + [\text{TCE}])$$

$$= 3.6 \times 10^{-15} \text{mol cell}^{-1} \text{ d}^{-1} \cdot 2.3 \times 10^{12} \text{ cell m}^{-3} / (30 \times 10^{-3} \text{ mol m}^{-3} + [\text{TCE}])$$

$$= 8.3 \times 10^{-3} \text{ d}^{-1} / (30 \times 10^{-3} \text{ mol m}^{-3} + [\text{TCE}])$$

We can find the steady-state concentration using the TCE mass balance equation:

$$d[\text{TCE}]/dt = (Q_{\text{in}}/V)[\text{TCE}]_{\text{in}} - (Q_{\text{out}}/V)[\text{TCE}] - k_{\text{bio}} [\text{TCE}]$$

$$= (Q_{\text{in}}/V)[\text{TCE}]_{\text{in}} - (Q_{\text{out}}/V)[\text{TCE}] - V''_{\text{max}} [\text{B}] [\text{TCE}] / (K_{TCE\,\text{MM}} + [\text{TCE}]) \tag{8}$$

which at steady state ($d[\text{TCE}]/dt = 0$) yields:

$$[\text{TCE}]_{\text{steady state}} = 48 \ \mu\text{M}$$

This implies only about half of the TCE is co-metabolically biodegraded by the methanotrophs in the 50 m^3 tank. Obviously, one could increase the size of the tank to improve this performance.

Illustrative Example 17.9

Estimating Biotransformation Rates of an Organic Pollutant in a Natural System

Problem

You are concerned about the fate of 2-nitrophenol (NP) found at a concentration of about 10 ppb (0.07 μM) in some groundwater. According to the literature (Zeyer and Kocher, 1988), this compound can be degraded aerobically by soil bacteria. The biodegradation pathway begins with an oxygenase that converts NP to catechol:

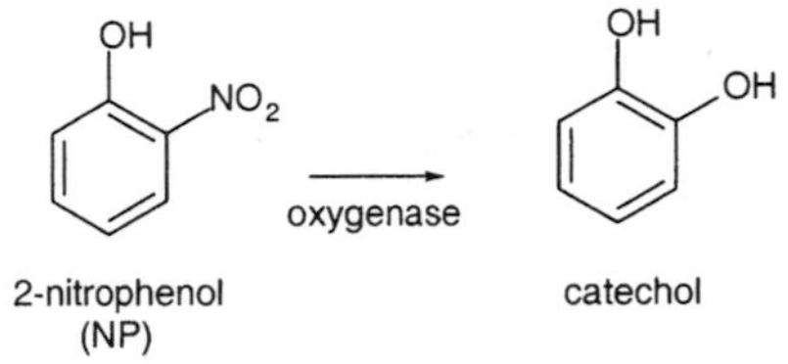

2-nitrophenol
(NP)
    catechol

In order to anticipate the biodegradation of NP, a soil pseudomonad is grown and enriched, and the protein fraction containing the oxygenase is isolated and purified by a factor of 40. With this protein isolate, you observed the rate of NP degradation as a function of that substrate's concentration (see data given below).

(1) Using these data, calculate the protein-normalized V'_{max} and $K_{NP\,MM}$ for this enzymatic reaction.

(2) Assuming the NP biotransformation is limited by reactions with such an oxygenase in groundwater microorganisms, what k_{bio} would you estimate for NP?

(3) Assuming the NP exhibits a K_d for the aquifer solids of 5 L·kg^{-1}, what overall half-life do you expect for NP in this groundwater assuming biodegradation is the chief removal mechanism?

[NP] (μM)	d[NP]/dt[Enz]$_{tot}$ rate of degradation (μmol·g^{-1} protein·min^{-1})
2	1040
3	1420
4	1920
5	2150
10	2480
20	2530

Answer

To deduce the enzyme parameters, fit the laboratory data using a variation of the Michaelis-Menten equation (Eq. 17-81) in which that hyperbolic equation is inverted to yield a linear form:

$$1/v = (K_{NP\,MM} / V_{max}) (1/[\text{substrate}]) + (1/V_{max}) \qquad (1)$$

Now we see that $1/v$ depends on $1/$[substrate] with a slope of $(K_{NP\,MM} / V_{max})$ and an intercept of $(1/V_{max})$. With the data from the table above, a least squares fit yields:

$$1/v = 0.00131 \ (1/[NP]) + 0.000263 \qquad (2)$$

From the intercept we have, $(1/V'_{max}) = 0.000263$. So the protein-normalized result is:

$$V'_{max} = 3800 \ mmol \cdot g^{-1} \ protein \ isolate \cdot min^{-1} \qquad (3)$$

Remembering this represents the result using a 40x purified fraction of the cell protein, to normalize to whole-cell protein, divide by this factor:

$$V''_{max} = 100 \ mmol \cdot g^{-1} \ protein \cdot min^{-1} \qquad (4)$$

To estimate the rate per g biomass (assume bacteria dry mass is 50% protein):

$$V'''_{max} = V''_{max} \ / \ 0.5 = 200 \ \mu mol \cdot g^{-1} \ dry \ biomass \cdot min^{-1} \qquad (5)$$

or per cell (assume a bacterial cell weighs between 0.1 and 1 picogram):

$$V''''_{max} = V'''_{max} = (0.1 \ to \ 1 \ pg \ cell) = 2 \times (10^{-17} \ to \ 10^{-16}) \ mol \cdot cell^{-1} \ min^{-1} \qquad (6)$$

Such "per cell" or "per biomass" maximum rates may be used with estimates of cell densities or bacterial biomass for systems of interest to approximate the V_{max} values for those cases.

Finally, recalling that the slope of the fit gave $(K_{NP\ MM} / V'_{max})$, we find:

$$K_{NP\ MM} = (K_{NP\ MM} / V'_{max})(V'_{max}) = (0.00131)(3800) = 5 \ \mu M \ (ca.\ 700 \ ppb) \qquad (7)$$

To estimate a rate applicable to the specific groundwater, one needs to "tune" for the abundance of microorganisms participating in NP removal there. Unfortunately, such information is typically unavailable. Here we use a plausible soil pseudomonad abundance of 10^6 cells $\cdot L^{-1}$. In this case, one estimates V_{max} to be:

$$V_{max} = 2 \times (10^{-17} \ to \ 10^{-16}) \ mol \cdot cells^{-1} \ min^{-1} \times 10^6 \ cells \cdot L^{-1}$$
$$= 2 \times (10^{-11} \ to \ 10^{-10}) \ mol \cdot L^{-1} \ min^{-1} \qquad (8)$$

Now we may find a value for k_{bio}:

$$k_{bio} = V_{max} \ / \ (K_{NP\ MM} + [NP]) \qquad (9)$$
$$= 2 \times (10^{-11} \ to \ 10^{-10}) \ mol \cdot L^{-1} \ min^{-1} \ / \ (5 \ \mu M + [NP]) \qquad (10)$$

Recalling that [NP] was present in the groundwater at 0.07 μM, we simplify:

$$k_{bio} \approx 2 \times (10^{-11} \ to \ 10^{-10}) \ mol \cdot L^{-1} \ min^{-1} \ / \ (5 \ \mu M) = 4 \times (10^{-6} \ to \ 10^{-5}) \ min^{-1} \qquad (11)$$

In light of the degradation rate estimated above, one may now estimate the overall biodegradation half-life of NP in the subsurface system. Assuming that the rates of desorption and uptake do not limit the overall biotransformation process, then we may expect (Eq. 17-58):

$$\frac{d[NP]_{tot}}{dt} = - V_w \ V_t^{-1} f_{NPw} \ k_{bio} \ [NP]_{tot} \qquad (12)$$

With K_{NPd} of 5 L·kg^{-1} and r_{sw} of 10 kg·L^{-1} (porosity of 0.2), we find:

$$f_{NPw} = (1 + r_{sw}K_{NPd})^{-1} = 0.02 \qquad (13)$$

The solution to Eq. 12 is:

$$[NP]_{tot,t} = [NP]_{tot,t=0} \exp(-V_w V_t^{-1} f_{NPw} k_{bio} t) \qquad (14)$$

So the half life is given:

$$t_{1/2} = \ln 2 / (f_{NPw} k_{bio}) \quad = 0.693 / (0.2 \times 0.02 \times 4 \times (10^{-6} \text{ to } 10^{-5}) \text{ min}^{-1})$$
$$= 4.3 \times (10^6 \text{ to } 10^7) \text{ minutes or 3000 to 30,000 days}$$

This result has a good deal of uncertainty, especially deriving from estimates of the abundance of NP-degrading bacteria.

17.7   Questions and Problems

Questions

Q 17.1

Why can the rate of biotransformation be greater than the rate of mineralization (i.e., conversion to CO_2, HNO_3, H_2SO_4, H_3PO_4, etc.)?

Q 17.2

List three processes that can limit the macroscopically observed rate of a chemical's biodegradation.

Q 17.3

Indicate the type of initial biotransformation and product you would expect for the following compounds if they occurred in an oxic stream. Note that you might expect no significant biological transformation in some cases.

(a) propane

(b) isopropanol (2-propanol)

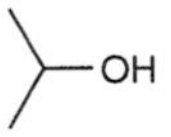

(c) n-propyl chloride

(d) chlorobenzene

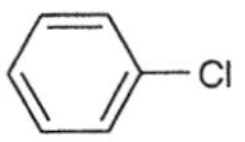

(e) n-propyl acetate

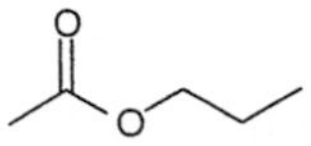

(f) di-n-propyl phosphate

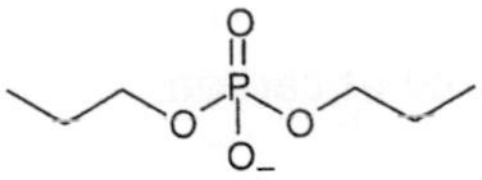

(g) 1,1,2-trichloropropene

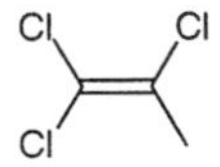

(h) 1,1,2,3-tetrachloropropene

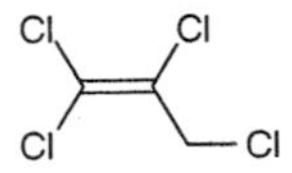

(i) acetone (2-propanone)

(j) propanoic acid

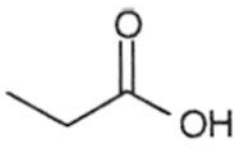

Q 17.4

Indicate the type of initial biotransformation and product you would expect for the compounds listed in Q 17.3 if they occurred in an anoxic groundwater. Note that you might expect no significant biological transformation in some cases.

Q 17.5

Give two reasons why an organic compound present in a system at "high" concentrations can be found to be biodegraded at a rate that is independent of that compound's concentration. That is, the loss can be modeled:

$$d[i]/dt = -k_{bio}[ML^{-3}T^{-1}]$$

Q 17.6

When a *labile compound* is catastrophically released to an environmental system, why does its removal typically show a lag period?

Q 17.7

What mathematical expression could be used to represent the growth rate of a microbial population limited by three substrates all used in the same utilization step? How many parameters do you need to know to apply such an expression to a new system of interest?

Q 17.8

Why are some compounds able to be biodegraded even though microorganisms cannot use them as a sole source of carbon and energy?

Q 17.9

What factor must be used to convert a V_{max} value determined using a pure culture of a particular microbial species and given in mol substrate $\cdot$ kg^{-1} protein $\cdot$ s^{-1} to a V_{max} value in units of mol substrate $\cdot$ L^{-1}s^{-1}? What does that factor represent?

Q 17.10

Given the enzymatic scenario depicted in Fig. 17.5, why would the enzymatic hydrolysis of an ester be sensitive to the environmental pH?

Q 17.11

Given the enzymatic oxidation scenario depicted in Fig. 17.7, why would the enzymatic oxidation of a hydrocarbon be enhanced by the addition of a substrate like formate?

Problems

P 17.1 *How Much Biomass Do You Expect?*

You want to grow a bacterial culture capable of using dichloromethane (CH$_2$Cl$_2$) as its sole source of carbon.

(a) Write a stochiometrically balanced reaction in which 50% of the carbon in CH_2Cl_2 is oxidized to CO_2 and the other 50% is used to make new biomass (CH_2O).

(b) If 50% of the carbon in this CH_2Cl_2 is converted to CO_2 (energy use) and the other 50% is used to synthesize new biomass (assume composition CH_2O), what yield in units of g cells·g^{-1} substrate do you expect?

(c) If a single cell weighs 0.3 pg (0.3×10^{-12} g), what yield do you expect in units of cells·mol^{-1} substrate?

(d) What values do you get if the substrate is toluene? Why are the results so different?

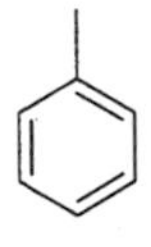

toluene

P 17.2 *Optimizing a Bioreactor*

Suppose you are interested in improving the degradation of glycerol by the 10 m^3 bioreactor discussed in Illustrative Example 17.6 (recall that glycerol was decreased by 74% in a 10 m^3 reactor in that case).

(a) One option is to vary the volume of the well-mixed reactor. What percent of incoming glycerol do you expect will be degraded if you double the tank volume to 20 m^3?

(b) Another suggestion is to feed the effluent from the first 10 m^3 tank into a second similar 10 m^3 tank (with additional O_2 and nutrients as necessary). What percentage of the original glycerol will be degraded in this second tank?

P 17.3 *How Effective Will This Low-O_2 Bioreactor Be at Removing Toluene?*

You have a 10 m^3 bioreactor containing a diverse mixture of bacteria. It is fed at 2 m^3 d^{-1} with a waste water containing 100 μM toluene (for structure see P 17.1). The waste water also contains a complex mixture of nontoxic organic chemicals. Due to the biodegradation of all the substances in the waste, the steady-state O_2 concentration in the reactor is only 3 μM. If the toluene oxidizers in the tank exhibit the properties shown below, what will the steady-state toluene concentration (μM) be exiting the tank? How would this result change if O_2 could be added to maintain a 30 μM steady-state concentration of O_2 in the reactor?

$\mu_{max} = 0.5 \ d^{-1}$ $Y = 2 \times 10^{14}$ cells·mol^{-1} die off at 0.1 d^{-1}

$K_{toluene \ M} = 10$ mM; $K_{oxygen \ M} = 1 \ \mu M$

P 17.4 *A Case of Oxygen-Limited Biodegradation*

Isopropanol (rubbing alcohol) is continuously discharged into a shallow (2 meters) pond. As a result, a bacterial species with μ_{max} of 0.3 hr^{-1}, $K_{isopropanol \ M}$ of 100 μM, and $Y_{isopropanol}$ of 2×10^{14} cells·mol^{-1} has increased in numbers throughout the pond.

After a time, the isopropanol concentration becomes constant at 30 μM (i.e., inputs are exactly balanced by biodegradation).

(a) If air-to-water exchange limits the input of oxygen to the pond to only 1×10^{-2} mol O_2 m^{-2} hr^{-1}, and this oxygen flux limits the degradation of isopropanol, what is the maximum rate of isopropanol degradation in the pond (mol·m^{-3} hr^{-1}) assuming it is completely mineralized to CO_2 and H_2O? Assume the pond is well-mixed vertically.

(b) For this isopropanol biodegradation rate, what bacterial cell density (cells·m^{-3}) is necessary? Assume the steady state $[O_2] = K_{O_2M}$.

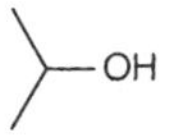

isopropanol
(2-propanol)

P 17.5 Can Phenanthrene Degradation Limit Bacterial Growth in an Oxic Sediment Bed?

Phenanthrene is present in an oxic sediment bed at 10 μmol·kg^{-1}. You know that some bacteria have been shown to use this polycyclic aromatic hydrocarbon as a growth substrate in the laboratory. Those bacteria exhibited a μ_{max} of 0.9 d^{-1}, a K_{iM} of 4 μmol·L^{-1}, and a yield of 3×10^{14} cells·mol^{-1} when their growth was limited by phenanthrene.

Assuming such bacteria in the oxic sediment bed are removed by predation at 0.2 d^{-1}, could they establish a steady-state population growing on phenanthrene dissolved in the porewater? Assume the sediment bed has a porosity of 0.6, $K_{phenanthrene\,d}$ is 200 L·kg^{-1}, and desorption is fast compared to biodegradation.

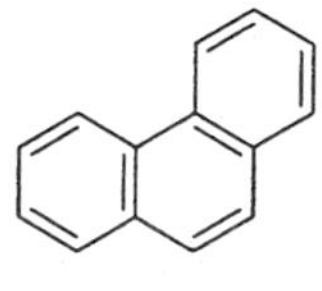

phenanthrene

P 17.6 Biodegradation of Nitrilotriacetic Acid (NTA) in a Lake

Ulrich et al. (1994) reported that the k_{bio} values necessary to explain the mass balance of nitrilotriacetic acid (NTA) present in a Swiss lake at 1 to 10 nM ranged from 0.02 to 0.05 d^{-1}. In another study, Bartholomew et al. (1983) tested an estuarine water for NTA biodegradation as a function NTA concentration, and they found V_{max} between 0.3 and 3 nmol·L^{-1} h^{-1} and K_{iMM} between 300 and 600 nM. Are the results of these two investigations consistent? Identify any assumptions you must make.

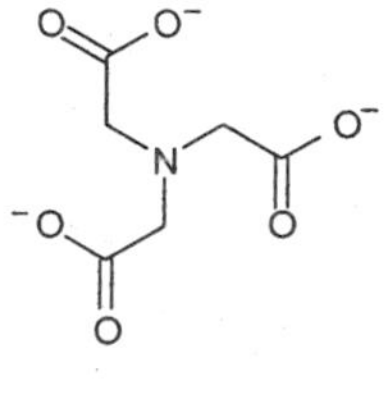

nitriloacetate
(NTA)

P 17.7 *Assessing the Rate of Toluene Biodegradation in a Shallow Stream*

Ground water leachate from a waste site causes a stream to have 4 μM toluene (for structure see P 17.1) at the point of exfiltration. This concentration decreases to about 0.01 μM some 400 meters downstream. Could this concentration change be due to biodegradation?

Assume the following apply: (a) the stream bottom is covered with a "film" of organisms 1200 g biomass$\cdot$m^{-2} of stream bottom; (b) using film scrapings in a laboratory suspension at 2 g biomass$\cdot$L^{-1}, you find that toluene added at 0.2 μM disappears at a rate of 1.2 hr^{-1}; and (c) using film scrapings in a laboratory suspension you find the rate of toluene degradation is half maximal at 2 μM. Also assume the stream is 0.1 m deep, flows at 200 m$\cdot$hr^{-1}, and does not incorporate significant additional exfiltrating ground water in the 400-meter-long reach in question. Finally, assume that toluene and O$_2$ mass transfers from the shallow water column into the biomass film are not rate limiting.

P 17.8 *Estimating the Biodegradability of Linuron in a River*

You are concerned about the longevity of the herbicide, linuron, leaching into a river from some neighboring farmland. Given the structure of this urea derivative, you expect it will be biodegraded via a hydrolysis mechanism. You recall a report of a hydrolase enzyme from a common bacterium that exhibits a half-saturation constant, K_{iMM}, for linuron of 2 μM and a maximum degradation rate, V'_{max}, for linuron of 2500 μmol$\cdot$kg^{-1} protein$\cdot$s^{-1}.

(a) Experimenting with some river water, you find that linuron added at 0.1 μM degrades with a half-life of 60 days. What would the hydrolase enzyme concentration (kg protein$\cdot$L^{-1}) in the river water have to be if biodegradation via such an enzyme accounted for all the linuron removal? Assuming a bacterial abundance in the river water of 10^9 cells$\cdot$L^{-1}, that they are 70% water and average 1 μm in diameter, and that their dry mass is half protein of specific density 1.5 g$\cdot$mL^{-1}, is the hydrolase concentration feasible?

(b) Would you expect linuron biodegradation in the river water to exhibit a lag phase?

(c) If linuron leached into the river at 0.01 μM, what biodegradation half life (days) would you expect it to have assuming the hydrolase concentration in the river water proved to be 3 $\times$ 10^{-10} kg protein$\cdot$L^{-1}?

linuron

P 17.9 *Biodegradation to Remove Trichloroethene from a Subsurface Site*

In order to degrade trichloroethene (TCE) contaminating some groundwater at 5 μM (660 ppb), you want to inject toluene (for structure see P 17.1) and O$_2$ below ground and grow a bacterial community capable of growing on the toluene and simultaneously co-metabolizing the TCE (after McCarty et al., 1998.)

$$
\begin{array}{c}
\text{Cl}\quad\quad\text{Cl}\\
\diagdown\quad\diagup\\
\text{C}=\text{C}\\
\diagup\quad\diagdown\\
\text{Cl}\quad\quad\text{H}
\end{array}
$$

trichloroethene
(TCE)

Using one liter of subsurface site material (containing 0.33 L of water and 0.67 L of solids; $K_{\text{toluene d}} = 0.1\ \text{L}\cdot\text{kg}^{-1}$) in an enclosed column in the laboratory, you flush it with water containing 100 μM toluene and O_2 (added as H_2O_2) in stoichiometric excess. You find the steady-state dry biomass is 10 mg biomass$\cdot$L^{-1} (i.e., 30 mg biomass$\cdot$L^{-1} of water). By varying the influent toluene concentration, you find the μ_{max} on this substrate is 1 d^{-1}, the die-off coefficient is 0.15 d^{-1}, the half-saturation constant with respect to dissolved toluene is 10 μM, and the dry biomass yield from toluene is 8×10^4 mg biomass$\cdot$mol^{-1} toluene.

(a) Assuming the laboratory parameters apply at the contaminated site, if toluene is injected at 100 μM, what half life (time for toluene concentration to decrease to half its initial concentration in days) would you expect for this growth substrate? Also assume toluene losses in the field should be modeled with:

$$(d[\text{toluene}]_{\text{total}}/dt)_{in\ situ} = -f_{\text{tolw}} \cdot k_{\text{bio}}\,[\text{toluene}]_{\text{total}}$$

where f_{tolw} is the fraction of toluene dissolved in the groundwater and you assume sorptive exchange is fast relative to biodegradation.

(b) Does the half life (time for toluene concentration to decrease to half its initial concentration in days) change if you inject toluene at 10 μM? Explain why it does or does not.

(c) You are interested in how closely treatment wells should be placed in the field. You reason that you want injected toluene to be present at significant concentrations everywhere in the treatment area. Using three times the toluene degradation half life (100 μM case) as a metric for the time in which toluene would be consumed below ground, how far (meters) into the aquifer would you expect the toluene-oxidizing bacteria to occur assuming the toluene/O_2 solution was injected at 40 L$\cdot$min^{-1} through a well with a 5-meter-long screen? Assume flow into the subsurface is cylindrical and recall that the porosity is 0.33.

(d) If the bacteria grown on toluene in the laboratory also co-metabolize TCE with $K_{\text{TCE MM}}$ of 10 μM and V'_{max} of 11 mol$\cdot$kg^{-1}protein$\cdot$d^{-1}, estimate the below-ground k_{bio} (d^{-1}) for trichloroethene assuming the bacteria establish a steady-state dry biomass equivalent to 10 mg dry biomass$\cdot$L^{-1} of intact aquifer and the bacteria are 50% protein. Also, assume $K_{\text{TCE d}} = 0.1\ \text{L}\cdot\text{kg}^{-1}$.

P 17.10 *Do Soil Methanotrophs Degrade Significant Amounts of Atmospheric Methane?*

Methane (CH_4) in the atmosphere is an important greenhouse gas. You wonder if this compound is degraded by bacteria that occur in the upper layers of oxic soils. You estimate that the diffusive flux of methane from the atmosphere (present at about

1500 ppbv or 6×10^{-8} mol$\cdot$L^{-1}) into the upper 10 centimeters of soil could be near 1×10^{-13} mol$\cdot$cm^{-2} s^{-1} assuming an average soil gas concentration in this layer of 500 ppbv.

For this problem, assume an air–water partition constant, $K_{CH_4aw} = 30$ and an organic carbon–water partition coefficient $K_{CH_4oc} = 10$ L$\cdot$kg^{-1}oc. Also assume the air-plus-water-filled soil porosity is 0.4, the soil solid density is 2.5 g$\cdot$mL^{-1}, the soil water content is 12% by weight, and the organic carbon content is 5% by weight. Finally, assume the methane biodegradation obeys: $d[\text{methane}]_{total}/dt = -f_{CH_4w} \cdot k_{bio}$ $[\text{methane}]_{total}$.

(a) What is the fraction in dissolved form for methane (f_{CH_4w}) in this surface soil?

(b) Now, consider the hypothesis that methanotrophs, bacteria that use methane as a growth substrate, consume the methane in the upper 10 cm soil layer. What value of K_{CH_4M} (mol$\cdot$L^{-1} water) would such a growing methanotroph population need to have if the methanotroph cell density in this upper soil layer was 10^6 cells$\cdot$g^{-1} soil and they exhibited a μ_{max} of 0.5 d^{-1} and a yield of 2×10^{13} cells$\cdot$mol^{-1}? Assume f_{CH_4w} is 0.1.

(c) Alternatively, the methane may be co-metabolized by bacteria growing on other substrates. What methanotroph cell density (cells$\cdot$g^{-1} soil) would be necessary to degrade the incoming methane if the co-metabolic parameters were: $K_{CH_4 MM} = 10$ μM and $V'_{max} = 1 \times 10^{-3}$ mol$\cdot$kg^{-1}protein$\cdot$s^{-1}? Is this feasible if the total bacteria are present at 10^9 cells$\cdot$g^{-1}? Assume methanotrophs are half protein in their dry mass and individually weigh 0.3 pg. Assume f_{CH_4w} is 0.1.

PART IV

Modeling Tools: Transport and Reaction

Working in natural systems, that is, conducting environmental organic chemistry, requires tools which we don't need in the laboratory. This includes combining different kinds of processes, for instance, transport and reaction processes. We do that by using mathematical equations to extrapolate knowledge acquired under well-controlled conditions (often in the laboratory) to natural (and often uncontrolled) conditions. Frequently, the mathematical concepts will be more complex than most of the equations which we have used so far. Many scientists would only call such sets of equations *models* and distinguish them from fundamental relations such as Henry's law. Yet, there are just gradual differences between them, although – as the reader will see – the mathematics may become rather complicated when we move forward from Henry's law to the modeling of turbulent transport of PCBs through the water column of Lake Superior.

In this and subsequent chapters we intend to guide the reader through the complex world of mathematical modeling. More specifically, we will discuss those kinds of environmental models which, as we believe, are central for tackling the task outlined in Chapter 1 that is, to evaluate the fate and behavior of anthropogenic organic compounds in the environment. The ultimate goal *would* be to describe the concentration distributions of selected compounds in space and time, $C_i(x,y,z,t)$, where x,y,z are the three space coordinates (Cartesian or others) and t is time. A model capable of fulfilling this task must combine characteristic properties related to the compounds with properties related to the environmental systems in which the compounds move (ocean, lakes, atmosphere, rivers, groundwater aquifers etc.). The former have been extensively treated in Parts II and III of this book; the latter are discussed in Part V.

Part IV, lying in between, prepares the ground for the construction of models of chemicals in real environmental systems. The path toward this goal can become extremely difficult and might turn away some readers (strong through their intentions may be), unless certain precautions are taken. First, we do not intend to climb the "$C_i(x,y,z,t)$-mountain" in one single tour de force. There will be steps and intermediate halts where climbers can rest or even turn back without loosing the insight gained so far. Those who continue the ascent will always be reminded that the very top of the mountain will never be reached – in fact, it cannot be reached by means of a model, since the top is the "real world," that is, real concentration distributions which could only be pictured by a "model of everything". Such a model would obviously violate one of the central goals of modeling, that is, simplicity.

Second, we should not mix up mathematical models with their representation as computer programs. Although the software market offers an increasing number of sophisticated, often rather user-friendly modeling programs, the uncritical application of such programs may be dangerous, especially if the result must serve as the basis for far-reaching and perhaps very costly measures. We strongly believe that every user of such programs should have some understanding of how models are built, how they work, and where they meet their limits. Often it is helpful to check the plausibility of modeling programs by comparing the results obtained for simple conditions with the analytical tools that will be derived in the following chapters. Obviously, this can be done only by a person who understands the different elements which are needed to describe a complex environmental system. Therefore, we put the emphasis on understanding the mathematical equations and their solutions that is, on the "reading skills" of the mathematical language. Only little will be said about computer programs, though we all know that, once the basics are understood, analytical solutions are now only seldom used for tackling actual problems. Again, we try to progress stepwise, from the familiar set of equations with one or more unknowns via simple linear differential equations to certain simple partial differential equations. The chapters are structured such that the reader can skip over certain more complicated parts at the end of each chapter without losing the ability to continue his or her ascent in the following chapter. Where suitable, mathematical tools will be presented in specially designed boxes to which we can refer whenever necessary.

Chapter 18

TRANSPORT BY RANDOM MOTION

18.1   **Introduction: Advection and Diffusion**
A Thought Experiment in a Train

18.2   **Random Motion**
Box 18.1: *Deterministic and Random Processes*
Bernoulli Coefficients
Normal Distributions
Box 18.2: *The Normal (Gaussian) Distribution*
Two Basic Descriptions of Transport by Random Motion: Mass Transfer
     Model and Gradient-Flux Law
Fick's First Law: Relating Spatial Changes in Concentration to Diffusive
     Fluxes
Mass Balance and Fick's Second Law
The Normal Distribution: An Important Solution of Fick's Second Law
     (*Advanced Topic*)
Box 18.3: *Mass Balance and Fick's Second Law in Three Dimensions*
Diffusion at Flat Boundaries
Symmetrical Diffusion at Flat Boundaries
Diffusion Into and From a Spherical Particle (*Advanced Topic*)

18.3   **Random Motion at the Molecular Level: Molecular Diffusion
        Coefficients**
Diffusivity and the Molecular Theory of the Ideal Gas
Diffusivities in Air
Illustrative Example 18.1: *Estimating Molar Volumes*
Illustrative Example 18.2: *Estimating Molecular Diffusivity in Air*
Diffusivities in Water
Box 18.4: *Temperature Dependence of Molecular Diffusivity in Fluids*
Illustrative Example 18.3: *Estimating Molecular Diffusivity in Water*

18.4   Diffusion in Porous Media
Diffusivity in Pores and Fick's Laws
Diffusion in Gas-Filled Pores: Knudsen Effect
Diffusion in Liquid-Filled Pores: Renkin Effect
Diffusion in the Unsaturated Zone of Soils
Diffusion of Sorbing Chemicals in Porous Media: Effective Diffusivity
Box 18.5: *Transport of Sorbing Chemicals in Porous Media and Break-through Time*
Illustrative Example 18.4: *Evaluating the Steady-State Flux of Benzene from Spilled Gasoline Through Soil to the Atmosphere*
Illustrative Example 18.5: *Interpreting Stratigraphic Profiles of Polychlorinated Naphthalenes in Lake Sediments*

18.5   Other Random Transport Processes in the Environment
A First Glimpse at Turbulent Diffusion
Diffusion Length Scales
Dispersion

18.6   Questions and Problems

18.1   Introduction: Advection and Diffusion

In natural systems there are two types of transport phenomena: (1) *transport by random motion*, and (2) *transport by directed motion*. Both types occur at a wide range of scales: from molecular to global distances, from microseconds to geological times. Well-known examples of these types are molecular diffusion (random transport) and advection in water currents (directed transport). There are many other manifestations such as *dispersion* as a random process (see Chapters 24 and 25) or settling of suspended particles due to gravitation as a directed transport. For simplicity we will subdivide such transport processes into those we will call *diffusive* for ones caused by random motions and those called *advective* for ones resulting from directed motions.

This chapter deals with the mathematical description of diffusion. Since it is easier to understand the nature of random transport by comparing it with directed transport, we briefly discuss advection as well. A more complete discussion of advective transport follows in Chapter 22.

A Thought Experiment in a Train

Imagine sitting in the dining car of that wonderful train which takes you through the steep mountains of eastern Switzerland to St. Moritz. While you travel uphill along the winding track, through loop-tunnels and narrow valleys, you order a cup of coffee. You add some milk and stir the coffee with your spoon. Then you lift the cup and take the first sip. Though in this wonderful environment it may sound frivolous, you suddenly ask yourself the question: How is the milk moving? You begin to analyze the situation. First, relative to the ground outside, the milk is traveling along the railway track toward St. Moritz. This motion is determined by the directed movement of the dining car. In a formal sense, the motion can be described by the three-dimensional velocity vector (i.e., by a set of three numbers which are the components of the velocity along the three axes of a Cartesian coordinate system: $\mathbf{v} = (v_x, v_y, v_z)$.) Note that all other objects in the dining car, the coffee cup, the table, yourself, etc. are transported along the same path. The transport is directed or advective.

Next you ask the question: How does the situation change when you lift the cup to your mouth? Obviously, this is a movement relative to the dining car; most objects in the car don't experience this movement of your arm. But the movement as such is still directed and shared by all the fluid elements in the cup and by the cup itself. Thus, this is still an advective motion, although on a smaller scale compared to the motion of the dining car. The combined effect on the milk molecules from the motions of car and cup can be expressed as the sum of two vectors: $\mathbf{v}_{tot} = \mathbf{v}_{car} + \mathbf{v}_{cup}$.

So far, we have not considered the effect of stirring the milk in the coffee. How should we characterize this motion? At first sight, the motion of the fluid produced by the spoon may look like advection as well, since it sets the coffee into a directed rotational motion. But this is not the only result. By stirring, the spoon has introduced enough mixing energy into the coffee such that the flow becomes turbulent. Although the movement of the spoon is not really random, it triggers the random process, that is, turbulent diffusion, which is necessary to achieve a uniformly mixed coffee and milk. If you could look even closer, you would see that the final job of mixing is done on the molecular level by random motions of the constituent molecules or molecular diffusion.

Before leaving our comfortable dining car, let us perform another experiment. For a moment you forget all your manners, take the salt shaker, put some salt into the sugar bowl, and mix both salt and sugar with your spoon. Not concerned that a later customer may not be very happy about your experiment, you consider how to classify your action as an advective or a diffusive process? The answer depends on the scale at which you are looking. At the scale of the sugar bowl, you have randomly mixed the individual sugar and salt grains so as to reach a homogeneous mixture of particles of both kinds. This is a diffusive process. Yet, at the level of the individual grains, the picture looks different: The molecules contained in an individual crystal were advectively, that is, jointly moved around. Thus, the distinction between advective and diffusive motion is context or scale dependent.

Real environmental transport processes contain all these phenomena that we encountered in the dining car. For instance, large ocean currents, such as the Gulf Stream, act as the dining car. Within these currents there are parcels of water called turbulent eddies which move relative to each other. In addition, small plants and animals carried in the current move relative to the surrounding water, take nutrients up, and mix them within their bodies.

In fact, wherever we look at a compartment of the environment, we usually find the simultaneous actions of advective and diffusive motion. In this chapter we focus on the latter.

18.2   Random Motion

Random motion is ubiquitous. At the molecular level, the thermal motions of atoms and molecules are random. Further, motions in macroscopic systems are often described by random processes. For example, the motion of stirred coffee is a turbulent flow that can be characterized by random velocity components. Randomness means that the movement of an individual portion of the medium (i.e., a molecule, a water parcel, etc.) cannot be described deterministically. However, if we analyze the *average effect* of many individual random motions, we often end up with a simple macroscopic law that depicts the mean motion of the random system (see Box 18.1).

We can analyze the connection between randomness at small scales and order at large scales by an infinite one-dimensional array of discrete boxes which are positioned along the x-axis (Csanady, 1973). The boxes are numbered by $m = 0, \pm 1, \pm 2, \ldots$ where the box $m = 0$ is situated at $x = 0$ (Fig. 18.1). Let us assume that at time $t = 0$ an object (molecule, particle, etc.) begins its random walk at box $m = 0$ (Fig. 18.1, top line). At fixed times $t = \Delta t, 2\Delta t, 3\Delta t \ldots$ the object jumps randomly either to the left or to the right. The path marked by A represents the path of an object which jumps twice to the left, then once to the right and once to the left again and finally twice to the right. At time $t = 6\Delta t$, the object happens to end up in the same box ($m = 0$) from which it started.

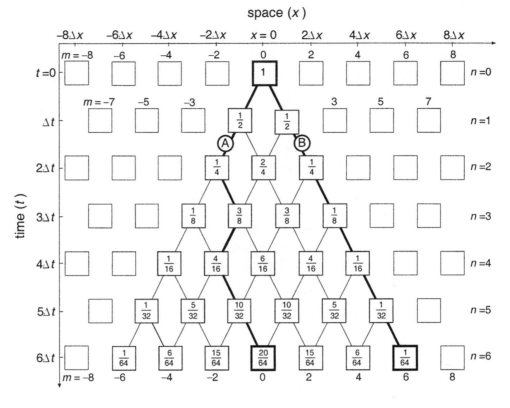

Figure 18.1 Random walk of an object through an infinite array of discrete boxes numbered by $m = 0$, $\pm 1, \pm 2, \ldots$. At time $t = 0$ the object is located in box $m = 0$ (probability 1) and then moves with equal probability to the two adjacent boxes $m = \pm 1$ (probabilities 1/2). The time steps are numbered by n. The resulting occupation probabilities, $p(n,m)$, of being in box m after time step n are the Bernoulli coefficients (Eq. 18-1). Curve A shows a typical individual path. Curve B represents the unlikely case in which the object jumps six times in the same direction.

Another path marked by B shows the extreme case of an object which jumps six times to the right. Its end position represents the largest distance which an object can travel within six time steps. Obviously, this distance grows with increasing number of time steps, n.

Bernoulli Coefficients

Although the path of an individual object cannot be predicted, we can conduct many experiments and then count the probability for an object to reach a given box m after n time steps, $p(n, m)$. You can easily convince yourself that these probabilities correspond to the numbers shown in Fig. 18.1, provided that at each box the chance is exactly 1/2 for the object to jump on either side. In Fig. 18.2 we have plotted $p(n, m)$ as a function of m for two times ($t = 4\,\Delta t$ and $t = 6\,\Delta t$). Note that m corresponds to the spatial coordinate, x, so the probability plots represent spatial distributions of objects which move at random from a given origin ($m = 0$ or $x = 0$) along the x-axis.

There is a simple recipe to construct the values of $p(n, m)$ for increasing values of n. When moving from one line of Fig. 18.1 (e.g., line $n = 2$) to the next ($n = 3$), 50% of the contents of box m is added to box $m-1$, the other 50% to box $m+1$. For instance, the value of the box at $n = 3$, $m = -1$ is derived from the two boxes at $\{n=2, m=-2\}$ and $\{n=2, m=0\}$ as follows: $1/2[(1/4)+(2/4)]=3/8$. The resulting numbers, known as the Bernoulli coefficients, can be expressed by the equation:

$$p(n,m) = \frac{n!}{2^n \left[\frac{1}{2}(n-m)\right]! \left[\frac{1}{2}(n+m)\right]!} \tag{18-1}$$

Box 18.1 Deterministic and Random Processes

Classical physics is based on the concept that things happen *deterministically*: *From A follows B and from B follows C and each of these successive outcomes can be described with some exact functional relationship.* Such are Newtons's laws and Maxwell's equations, to mention just two examples. Of course, scientists like Newton, Maxwell, and others were well aware of the seemingly irregular, unpredictable (random) nature of certain phenomena. The movement of the smoke from a chimney and the flow of the water in a river look rather irregular. Yet, these scientists were convinced that if only we were able to break down the description of the system to the smallest possible level (e.g., the molecules), the motion would turn out to be deterministic again.

Physicists and chemists invented an ingenious method to overcome the seemingly irregular nature of complex systems by the definition of macrovariables such as temperature (the average kinetic energy of molecules) or pressure (the average force of moving molecules exerted on a wall). They assumed that the average behavior of a large number of identical objects can be described by deterministic relationships (such as the law of the ideal gas), although the behavior of the individual objects cannot be known exactly.

At the beginning of the twentiest century, the French mathematician, Henri Poincaré, found that the solution of certain coupled nonlinear differential equations exhibits chaotic behavior although the underlying laws were fully deterministic. He pointed out that two systems starting with slightly different initial conditions would, after some time, move into very different directions. Since empirical observations are never exact in the mathematical sense but bear a finite error of measurement, the behavior of such systems could not be predicted beyond a certain point; these systems seem to be of random nature. A random process – in contrast to a deterministic process – is characterized by: *From A follows B with probability p_B, C with probability p_C etc.*

The idea of a fully deterministic world received its final blow from the modern physics of the twentiest century. Quantum mechanics abandoned the model of complete determinism. From a pragmatic point of view it is not relevant whether nature is inherently nondeterministic (as quantum theory states) or whether randomness is just a consequence of the complexity of natural systems. In fact, most descriptions (or models) of natural processes are made up of a mixture of deterministic and random elements.

---

where $n! = 1 \cdot 2 \cdot 3 \ldots (n-1) \cdot n$ is *n*-factorial. You may convince yourself that the coefficients are symmetric in m {i.e., $p(n,m) = p(n,-m)$} and that the sum over m for any n is 1. Note that the values of m from $-n$ to $+n$ run only through even or odd numbers depending on whether n is even or odd.

Normal Distributions

The analytical expression, Eq. 18-1, is not easy to evaluate for large values of n. Fortunately, the French mathematicians, DeMoivre and Laplace, found that with increasing n, the Bernoulli coefficients converge to the function:

$$p_n(m) = \left(\frac{2}{\pi n}\right)^{1/2} \exp\left\{-\frac{m^2}{2n}\right\} \tag{18-2}$$

In this expression n and m are no longer restricted to integer values. As shown in Fig. 18.2 for $n = 4$ and $n = 6$, the continuous representation of the Bernoulli coefficients is surprisingly good even for small n values; it becomes even better if n is large.

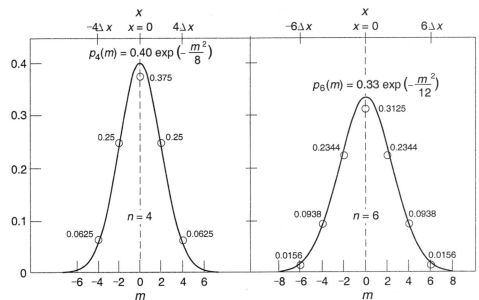

Figure 18.2 Bernoulli coefficients $p(n,m)$ for $n = 4$ and $n = 6$ (open circles with numbers) compared to the normal density approximation $p_n(m)$ by DeMoivre and Laplace. Note that in most mathematical handbooks, the Bernoulli coefficients are listed as $p(n,k) =$

$$\frac{1}{2^n} \frac{n!}{k!\,(n-k)!} \quad \text{where}$$

$$k = (1/2)(n + m).$$

A curve with the shape given by Eq. 18-2 is called a normal (or Gaussian) distribution. Usually it is denoted as $p_\sigma(x)$ where x is the spatial coordinate and σ is the standard deviation which characterizes the width of the distribution along the x-axis. The mathematical definition and properties of the normal distribution are presented in Box 18.2.

To show the relationship between $p_n(m)$ expressing the probabilities of numbers and $p_\sigma(x)$ describing a continuous spatial distribution of a quantity like concentration, we make use of the analogy between the integers n and m, which describe the simple random walk model shown in Fig. 18.1, and the time and space coordinates t and x, that is: $t = n\,\Delta t$ and $x = m\,\Delta x$. The incremental quantities, Δt and Δx, are characteristic for random motions; the latter is the *mean free path* which is commonly denoted as $\lambda = \Delta x$, the former is associated with the *mean velocity* $u_x = \Delta x/\Delta t = \lambda/\Delta t$. Thus, we get the following substitution rules:

$$m = x/\Delta x = x/\lambda \quad , \quad n = t/\Delta t = u_x\,t/\lambda$$

which we can insert into the exponential function of Eq. 18-2:

$$\exp\left\{-\frac{m^2}{2n}\right\} = \exp\left\{-\frac{(x/\lambda)^2}{2u_x t/\lambda}\right\} = \exp\left\{-\frac{x^2}{2(u_x t\lambda)}\right\}$$

To transform $p_n(m)$ into a normalized spatial distribution we still have to divide the nondimensional prefactor of Eq. 18-2 by the average distance between two adjacent occupied boxes. Since at any time only every second box is occupied (see Fig. 18.1), this distance is $2\Delta x = 2\lambda$. Finally we get:

$$p_\sigma(x) = \frac{p_n(m)}{2\lambda} = \left(-\frac{2}{\pi(u_x t/\lambda)}\right)^{1/2} \frac{1}{2\lambda} \exp\left\{-\frac{x^2}{2(u_x t\lambda)}\right\}$$

$$= \frac{1}{(2\pi)^{1/2}\sigma_x} \exp\left\{-\frac{x^2}{2\sigma_x^2}\right\}$$

(18-3a)

Box 18.2 The Normal (Gaussian) Distribution

The one-dimensional normal (or Gaussian) distribution along the x-axis is defined by

$$p_\sigma(x) = \frac{1}{(2\pi)^{1/2}\sigma} \exp\left(-\frac{x^2}{2\sigma^2}\right)$$

where σ is the standard deviation and σ^2 is the variance of the distribution.

$p_\sigma(x)$ has the following properties:

$$\text{Normalization:} \qquad \int_{-\infty}^{\infty} p_\sigma(x)\,dx = 1$$

$$\text{Mean value:} \qquad \bar{x} = \int_{-\infty}^{\infty} x\, p_\sigma(x)\,dx = 0$$

$$\text{Variance:} \qquad \overline{x^2} = \int_{-\infty}^{\infty} x^2\, p_\sigma(x)\,dx = \sigma^2$$

Integrals within finite boundaries have the following fixed values:

$$\int_{-\sigma}^{\sigma} p_\sigma(x)\,dx = 0.683$$

$$\int_{-2\sigma}^{2\sigma} p_\sigma(x)\,dx = 0.954$$

For other numerical values see Table A.1 in Appendix A.

---

where we have introduced the standard deviation (see Box 18.2):

$$\sigma_x = (u_x \lambda\, t)^{1/2} \qquad (18\text{-}3b)$$

This is an extremely important result. It says that the standard deviation of the one-dimensional spatial distribution of objects moving randomly along a space axis grows as the square root of time and that this growth is controlled by the product of the mean free path and the mean velocity, $(u_x\lambda)$. Note that this quantity has the dimension $[L^2 T^{-1}]$. Its meaning will become clear below.

Two Basic Descriptions of Transport by Random Motion: Mass Transfer Model and Gradient-Flux Law

Imagine that in Fig. 18.1 at some position x_o we put a virtual wall and calculate the net flux along the x-axis, F_x, of objects per unit time. F_x is the difference between the number of objects crossing the wall from left to right (i.e., in the positive x-direction) and the number passing from right to left. F_x can have either sign; a positive F_x means that there is a net flux of particles moving in the positive x-direction. It is easy to see that the net flux is proportional to the occupation difference between adjacent space elements. Furthermore, the flux is directed "against the gradient." That is, there is net transport from the box with a larger occupation number to the box with a smaller content.

There are two mathematical methods for formulating transport by random motion. The first, often called a *mass transfer model* (Cussler, 1984), relates the net flux to the difference in occupation numbers between two adjacent subsystems, A and B:

$$F_{A/B} = -\text{constant} \cdot \left[\text{occupation number B - occupation number A} \right]$$

In the framework of the model of Fig. 18.1, the occupation numbers would be the probabilities $p(n,m)$ of two adjacent boxes. For instance, the net flux in the time interval from $t = 4\,\Delta t$ to $5\,\Delta t$ between the boxes $m = 2$ and $m = 4$ is proportional to

$$-[p(4,4) - p(4,2)] = -[\frac{1}{16} - \frac{4}{16}] = \frac{3}{16}$$

In environmental systems fluxes are usually expressed as mass per unit area and per time (dimension $ML^{-2}T^{-1}$) and the occupation numbers as concentrations (dimension ML^{-3}). Then, the constant appearing in the flux expression must have the dimension of a velocity (LT^{-1}). Later we will use the term *transfer* or *exchange velocity* and designate it as $v_{A/B}$ or v_{ex} to discuss the speed with which mass is moved from subsystem A to subsystem B. To summarize, the *mass transfer model* takes the form:

$$F_{A/B} = -v_{A/B}(C_B - C_A) \qquad [ML^{-2}T^{-1}] \qquad (18\text{-}4)$$

The second model, the so-called *gradient-flux law*, is considered to be more fundamental, although it is based on a more restrictive physical picture. In contrast to the mass transfer model, in which no assumption is made regarding the spatial separation of subsystems A and B, in the gradient-flux law it is assumed that the subsystems and the distance between them, $\Delta x_{A/B}$, become infinitely small. For very small subsystems the term *occupation number* loses its meaning and must be replaced by *occupation density* or concentration. Obviously, the difference in occupation density tends toward zero, as well. Yet the ratio of the two differences, Δoccupation density : $\Delta x_{A/B}$, is equal to the spatial gradient of the occupation density and usually different from zero:

$$F_x = -\text{constant}\, \frac{d}{dx}\left(\text{occupation density}\right) \qquad (18\text{-}5)$$

where the minus sign indicates that the flux points against the gradient. Note that instead of the subscript A/B, which characterizes the flux in Eq. 18-4, we now use the subscript x to design the coordinate axes along which the flux occurs.

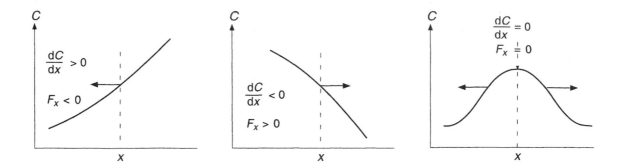

Figure 18.3 Relation between the variation of a concentration profile along x, $C(x)$, and the direction of the diffusive flux, F_x.

The relationship between the flux of a property and the spatial gradient of a related property called a *gradient-flux law* is typical for a whole class of physical processes. A list of processes that obey the gradient-flux law is given in Table 18.1. In all these processes, it is assumed that the flux is determined by the variation of a *local* property. This variation (such as the gradient of the occupation density) is the "driving force" for transport. Mathematically, the gradient is a local property of a function; nothing has to be known about the shape of the function elsewhere along the x-axis. It is this peculiarity which on one hand makes Eq. 18-5 so powerful, yet on the other hand restricts its validity. Later in Chapter 22, when we discuss turbulent diffusion, we will meet situations in which the "locality" of the flux is only approximately valid. Strictly speaking, the random walk model shown in Fig. 18.1 is such an example already, since the objects jump over the finite distance Δx.

Fick's First Law: Relating Spatial Changes in Concentration to Diffusive Fluxes

One well-known example of the gradient-flux law is *Fick's first law*, which relates the diffusive flux of a chemical to its concentration gradient and to the molecular diffusion coefficient:

$$F_x = -D \frac{dC}{dx} \qquad [\text{ML}^{-2}\text{T}^{-1}] \qquad (18\text{-}6)$$

where F_x is the mass flux per unit (cross-sectional) area and per unit time, D is molecular diffusivity, C is concentration, and dC/dx is the spatial gradient of C along the x-axis. The molecular diffusivity (or molecular diffusion coefficient) D has the dimension $[\text{L}^2\text{T}^{-1}]$. As shown in the following section, D depends on the diffusing chemical as well as on the medium through which it moves (e.g., water).

Note that in Eq. 18-6 the direction of the flux is indicated by the sign of F_x. If F_x is a positive number, the flux occurs in the positive x-direction. Likewise, the sign of the spatial gradient, dC/dx, is positive if C increases with x. Examples are given in Fig. 18.3.

Below we will show that diffusivity D can be interpreted in the framework of a random walk model (see Eqs. 18-16 and 18-17). Particularly, D is related to the random walk parameters, mean free path λ and mean velocity u_x, by the simple relation:

$$D = \frac{1}{2} \lambda u_x \qquad (18\text{-}7)$$

Table 18.1 Physical Processes Obeying the "Gradient-Flux Law"

Law	Equation	Variables and Conventional Units
Molecular diffusion (Fick)	$F_i = -D\dfrac{dC_i}{dx}$	F (mol m^{-2}s^{-1}) Mass flux C_i (mol m^{-3}) Concentration of chemical i D (m^{2}s^{-1}) Molecular diffusion coefficient
Conduction of heat (Fourier)	$F_{th} = -\kappa\dfrac{dT}{dx}$	F_{th} (W m^{-2}) Heat flux T (K) Temperature κ (W m^{-2}K^{-1}) Thermal conductivity
Transfer of momentum by viscous forces	$\tau = -\eta\dfrac{du}{dy}$	τ (kg m^{-1}s^{-2}) Shear stress η (kg m^{-1}s^{-1}) Dynamic viscosity du/dy (s^{-1}) Change of x-component of velocity (u) in y-direction
Flow of fluid through porous medium[a] (Darcy's Law, see Eq. 25-5)	$q = -K_q\dfrac{dh}{dx}$	q (m s^{-1}) Discharge per area h (m) Hydraulic head (pressure change along flowpath x) K_q (m s^{-1}) Hydraulic conductivity of medium
Electric conductivity[a,b] (Ohm's Law)	$j = +k\dfrac{dV}{dx}$	j (A m^{-2}) Electric current per area V (V) Voltage k (Ω^{-1}m^{-1}) Electric conductivity

[a] Manifestations of the gradient-flux law are not due to a random process but due the equilibrium between external force and internal friction. [b] The positive sign results from the special sign convention used for electric currents and fields.

If extended to three dimensions, the numerical factor on the right-hand side becomes 1/3 (Jeans, 1921):

$$D = \frac{1}{3} \lambda u \qquad (18\text{-}7a)$$

where u is the average molecular velocity in the three-dimensional space. These two equations are fundamental for a great variety of random walk situations and prove very useful. For instance, we can combine Eqs. 18-3b and 18-7 to calculate the average distance a population of objects has diffused in a one-dimensional case:

$$\sigma_x = (2Dt)^{1/2} \quad \text{measure of diffusion distance along one space axis} \qquad (18\text{-}8)$$

This law, independently found by Einstein (1905) and the Polish physicist Smoluchowski when they studied the Brownian motion of small particles, became an important clue for proving the real existence of atoms and molecules as discrete items of mass which move randomly.

Mass Balance and Fick's Second Law

In order to fully appreciate the consequences of the rather simple mathematical rules which describe the random walk, we move one step further and combine Fick's first law with the principle of mass balance which we used in Section 12.4 when deriving the one-box model. For simplicity, here we just consider diffusion along one spatial dimension (e.g., along the x-axis.)

We want to formulate the mass balance for a rectangular "test volume" ($V = \Delta x \Delta y \Delta z$) whose edges are parallel to the axes of a Cartesian coordinate system (Fig. 18.4). For the one-dimensional case, the flux is assumed to be parallel to the x-axis and independent of y and z, yet variable along the x-axis. It is expressed by $F_x(x)$ where the subscript x indicates the axis along which the flux occurs and the parenthetical x refers to the location on the x-axis (e.g., $x = 0$). If C is the average concentration in the test volume, then the change of total mass, VC, per unit time is:

$$\frac{d}{dt}(VC) = V\frac{dC}{dt} = a_x F_x(0) - a_x F_x(\Delta x) \quad [\text{MT}^{-1}] \qquad (18\text{-}9)$$
$$\text{flux in from left} - \text{flux out to right}$$

where $a_x = \Delta y \Delta z$ is the area of the test volume perpendicular to the x-axis. The signs on the right-hand side of Eq. 18-9 reflect the convention adopted for the mass flux in Eq. 18-6 (i.e., positive if pointed into the positive x-direction). Since V is constant, both sides can be divided by V:

$$\frac{dC}{dt} = \frac{\left[F_x(0) - F_x(\Delta x) \right]}{\Delta x} \qquad (18\text{-}10)$$

If the size of the test volume along the x-axis is made smaller and smaller (i.e., if $\Delta x \to 0$), both the numerator and the denominator on the right-hand side of Eq. 18-10 go to zero, but not the quotient. In fact, the derivative is defined as:

$$\lim_{\Delta x \to 0} \frac{\left[F_x(0) - F_x(\Delta x) \right]}{\Delta x} = -\frac{dF_x}{dx} \qquad (18\text{-}11)$$

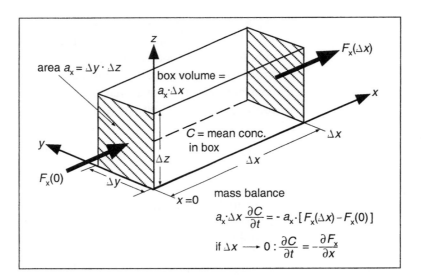

Figure 18.4 Rectangular test volume of size $\Delta x \cdot \Delta y \cdot \Delta z$ and mass flux along the x-axis.

Combining Eqs. 18-10 and 18-11 yields:

$$\frac{\partial C}{\partial t}\bigg|_{x=\text{constant}} = -\frac{\partial F_x}{\partial x}\bigg|_{t=\text{constant}} \quad [\text{ML}^{-3}\text{T}^{-1}] \tag{18-12}$$

Note that now we have two kinds of derivatives, one with respect to time, ($\partial C/\partial t$), and another with respect to space, ($\partial F_x/\partial x$). To indicate that each derivative is to be evaluated by keeping all *other* variables constant (mathematicians call it a *partial derivative*), we use the symbol ∂. Sometimes, as in Eq. 18-12, the variables to be kept constant are explicitly indicated as subscripts, but usually these reminders are omitted.

Eq. 18-12 and its two- and three-dimensional siblings which we will discuss in Chapter 22 are fundamental. In mathematics, such relations are said to obey *Gauss' theorem,* stating that the local concentration variation with time is equal to the negative spatial gradient of the flux along the x-axis. Note that at this point we have not made use of any explicit algebraic definition of F, so Eq. 18-12 is generally valid for all kinds of transport processes. For the special case of molecular diffusion, we get from Fick's first law, Eq. 18-6:

$$\frac{\partial C}{\partial t} = -\frac{\partial}{\partial x}\left(-D\frac{\partial C}{\partial x}\right) = D\frac{\partial^2 C}{\partial x^2} + \frac{\partial D}{\partial x}\frac{\partial C}{\partial x} \quad [\text{ML}^{-3}\text{T}^{-1}] \tag{18-13}$$

If diffusivity D is independent of x, which for molecular diffusion is usually the case, then $\partial D/\partial x$ is zero and the second term on the far right-hand side drops. We are left with *Fick's second law*:

$$\frac{\partial C}{\partial t} = D\frac{\partial^2 C}{\partial x^2} \quad [\text{ML}^{-3}\text{T}^{-1}] \tag{18-14}$$

This law states that the local concentration change with time due to a diffusive transport process is proportional to the second spatial derivative of the concentration.

As a special case, we consider a linear concentration profile along the x-axis $C(x) = a_o + a_1x$. Since the second derivative of $C(x)$ of such a profile is zero, diffusion leaves the concentrations along the x-axis unchanged. In other words, a linear profile is a steady-state solution of Eq. 18-14 ($\partial C/\partial t = 0$). Yet, the fact that C is constant does not mean that the flux is zero as well. In fact, inserting the linear profile into Fick's first law (Eq. 18-6) yields:

$$F_x = -D\frac{dC}{dx} = Da_1 = \text{constant}$$

Since the flux is independent of x, the substance which is transported along the x-axis can neither accumulate nor disappear, thus $\partial C/\partial t = 0$. This situation will be discussed in more detail in the next section dealing with transport through boundaries.

Advanced Topic

The Normal Distribution: An Important Solution of Fick's Second Law

Fick's second law (Eq. 18-14) is a second-order linear partial differential equation. Generally, its solutions are exponential functions or integrals of exponential functions such as the error function. They depend on the boundary conditions and on the initial conditions, that is, the concentration at a given time which is conveniently chosen as $t = 0$. The boundary conditions come in different forms. For instance, the concentration may be kept fixed at a "wall" located at x_o. Alternatively, the wall may be impermeable for the substance, thus the flux at x_o is zero. According to Eq. 18-6, this is equivalent to keeping $\partial C/\partial x = 0$ at x_o. Often it is assumed that the system is unbounded (i.e., that it extends from $x = -\infty$ to $+\infty$). For this case we have to make sure that the solution $C(x,t)$ remains finite when $x \to \pm\infty$. In many cases, solutions are found only by numerical approximations. For simple boundary conditions, the mathematical techniques for the solution of the diffusion equation (such as the Laplace transformation) are extensively discussed in Crank (1975) and Carslaw and Jaeger (1959).

As an illustration of unbounded diffusion along the x-axis, we consider the case of a chemical with total mass $\mathcal{M}^*$ per unit area (dimension ML^{-2}). At time $t = 0$, this compound is all concentrated at $x = 0$. A function which is infinite at $x = 0$ and zero otherwise but has a finite integral is called the *delta function*, defined by:

$$\delta(x) = \begin{cases} 0 & \text{for } x \neq 0 \\ \infty & \text{for } x = 0 \end{cases} \quad \text{with} \int_{-\infty}^{+\infty} \delta(x) = 1 \qquad (18\text{-}15)$$

A solution of Eq. 18-14 that obeys the required boundary conditions (i.e., it drops to zero for $x = \pm\infty$) is:

$$C(x,t) = \frac{\mathcal{M}^*}{2(\pi Dt)^{1/2}} \exp\left(-\frac{x^2}{4Dt}\right) \quad [ML^{-3}] \qquad (18\text{-}16)$$

This is a normal distribution (Box 18.2) with standard deviation:

$$\sigma_x = (2Dt)^{1/2} \qquad (18\text{-}17)$$

Box 18.3 Mass Balance and Fick's Second Law in Three Dimensions

In Q 18.6 (below) you are asked to derive the three-dimensional version of Gauss' theorem (Eq. 18-12). From there it is straightforward to show that, provided that diffusivity D is the same in all directions, the three-dimensional form of Fick's Second Law (Eq 18-14) has the form:

$$\frac{\partial C}{\partial t} = D\left(\frac{\partial^2 C}{\partial x^2} + \frac{\partial^2 C}{\partial y^2} + \frac{\partial^2 C}{\partial z^2}\right) \quad [ML^{-3}T^{-1}] \tag{1}$$

where x,y,z are the three Cartesian coordinates. Then the three-dimensional analogue to the special solution Eq. 18-16 is:

$$C(x,y,z,t) = \frac{M}{8(\pi Dt)^{3/2}} \exp\left(-\frac{x^2+y^2+z^2}{4Dt}\right) \quad [ML^{-3}] \tag{2}$$

Eq. 2 is a three-dimensional spherical normal distribution. Note that in contrast to M^* of Eq. 18-16, M now has the dimension of a mass. In fact, since $C(x,y,z,t)$ depends only on the spherical coordinate $r = (x^2+y^2+z^2)^{1/2}$, Eq. 2 is not really a three-dimensional spatial function but can be written as a function of r and t alone:

$$C(r,t) = \frac{M}{8(\pi Dt)^{3/2}} \exp\left(-\frac{r^2}{4Dt}\right) \quad [ML^{-3}] \tag{3}$$

and constant integral:

$$\int_{-\infty}^{\infty} C(x,t)dx = M^* \quad [ML^{-2}] \tag{18-18}$$

Now, everything falls into place: We set out to study the laws of random walk by using the simple model of Fig. 18.1 and found the Bernoulli coefficients. We then saw that for large n (which is equivalent to large times), the Bernoulli coefficients can be approximated by a normal distribution whose standard deviation, σ, grows in proportion to the square root of time, $t^{1/2}$ (Eq. 18-3). And now it turns out that the solution of the Fick's second law for unbounded diffusion is also a normal distribution. In fact, the analogy between Eqs. 18-3b and 18-17 gave the basis for the law by Einstein and Smoluchowski (Eq. 18-17) that we used earlier (Eq. 18-8). The expression $(2Dt)^{1/2}$ will also show up in other solutions of the diffusion equation.

In Box 18.3 the normal distribution solution of Fick's second law is extended to three dimensions.

Diffusion at Flat Boundaries

We now discuss two additional solutions of Fick's second law (Eq. 18-14) for particular boundary conditions. The first one deals with diffusion from a surface with fixed boundary concentration, C_o, into the semi-infinite space. The second one involves the disappearance ("erosion") of a concentration jump. Both cases will be important when dealing with the transport through boundaries (Chapter 19). No derivations will be given below. The interested reader is referred to Crank (1975) and Carslaw and Jaeger (1959) or to mathematical textbooks dealing with particular techniques for solving Eq. 18-14.

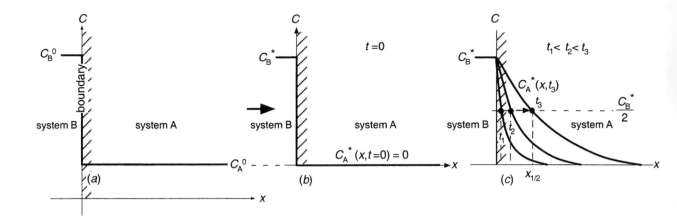

Figure 18.5 Diffusion at a concentration jump between area A and B. By redefining the concentrations in A and B, $C_A^* = C_A - C_A^o$; $C_B^* = C_B - C_A^o$, every situation can be reduced to the special case with $C_A^o = 0$. (a) to (c): concentration in B is kept constant at C_B^o. (d) and (e): In both media (A and B) transport is controlled by diffusion.

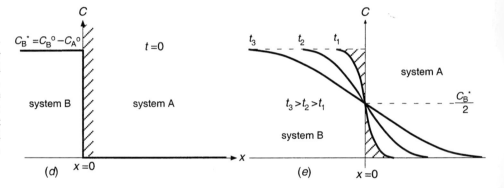

First, consider the diffusion of an organic compound across the boundary between two environmental systems, A and B. Imagine that at time $t = 0$, the surface of system A (e.g., an air bubble, a silt particle, etc.) is suddenly juxtaposed to a (very large) system B (e.g., the water of a lake, Fig. 18.5a). Mixing in system B is sufficient that the concentration of the selected compound at the boundary of the injected medium is kept at the constant value, C_B^o. This concentration is different from the initial concentration in A, C_A^o. In system A, transport occurs by diffusion only. We want to calculate the concentration in system A as it evolves in space and time, $C_A(x,t)$. For the time being, we will assume that the equilibrium distribution coefficient between A and B is 1. Hence, the concentration of A seeks to change to be equal to that of system B.

Since the governing differential equation (Eq. 18-14) is linear, we can focus on the difference in concentrations in the two systems by adjusting the concentrations such that the initial concentration in A is zero (Fig. 18.5b):

$$C_B^* = C_B^o - C_A^o = \text{constant for } x \leq 0$$

(18-19)

$$C_A^*(x,t) = C_A(x,t) - C_A^o \quad ; \quad C_A^*(x > 0, t = 0) = 0$$

Note that C_B^* is not necessarily a positive number. If initially $C_A^o > C_B^o$, C_B^* is negative. This case would describe the diffusion out of system A into B, yet the mathematical description is exactly the same.

We now observe how the concentration front penetrates into system A when time increases from t_1 to t_3 (Fig. 18.5c). In order to measure the penetration speed we can look, for instance, at the movement away from the boundary of the location where C_A^* reaches $C_B^*/2$, that is, half the concentration in system B. We call this point $x_{1/2}(t)$.

The solution of Eq. 18-14 with the boundary conditions Eq. 18-19 is given by:

$$C_A^*(x,t) = C_B^* \, \text{erfc}\left(\frac{x}{2(Dt)^{1/2}}\right) \tag{18-20}$$

where erfc(y) is the complement error function defined by (see Appendix A.2, Eq. A-8):

$$\text{erfc}(y) = \frac{2}{\sqrt{\pi}} \int_y^\infty e^{-z^2} dz \tag{18-21}$$

erfc(y) steadily decreases from erfc($-\infty$) = 2 to erfc(0) = 1 and erfc(∞) = 0, but most of the variation occurs between $y = -1$ and $y = 1$.

Expressed in the original concentration values we get from Eqs. 18-19 and 18-20:

$$C_A(x,t) = C_A^o + \left(C_B^o - C_A^o\right)\text{erfc}\left(\frac{x}{2(Dt)^{1/2}}\right) \tag{18-22}$$

To calculate the penetration of the front quantified by the distance of the *half-concentration* from the interface, $x_{1/2}$, we consult Table A.2 of Appendix A to look for that argument $y_{1/2}$ for which erfc($y_{1/2}$) = 0.5 and find by interpolation $y_{1/2}$ = 0.477. Inserting this value into Eq. 18-20 yields:

$$\frac{x_{1/2}}{2(Dt)^{1/2}} = y_{1/2} = 0.477$$

and

$$\tag{18-23}$$

$$x_{1/2} = 2y_{1/2}(Dt)^{1/2} = 0.954(Dt)^{1/2} \approx (Dt)^{1/2}$$

The concentration front penetrates the system A with a speed which is proportional to $t^{1/2}$. Except for the numerical factor, we again find the law by Einstein and Smoluchowski (Eq. 18-8). Note that if we had chosen another criterion to define the front, such as 0.9 C_B^*, $y_{1/2}$ would be replaced by $y_{0.9}$ for which erfc($y_{0.9}$) = 0.1. From Table A.2 of Appendix A we get $y_{0.9}$ = 0.086. Thus the prefactor in Eq. 18-23 would change (from 1 to 0.17), but not the law itself.

How much of the compound has crossed the interface at time t? To calculate the integrated mass flux we have to integrate $C_A^* = C_A - C_A^o$ from $x = 0$ to ∞. Thus from Eq. 18-20:

$$\mathcal{M}^*(t) = \int_0^\infty C_A^*(x,t)\,dx = (C_B^o - C_A^o)\int_0^\infty \text{erfc}\left(\frac{x}{2(Dt)^{1/2}}\right)dx \quad \left[\text{ML}^{-2}\right] \tag{18-24}$$

With the variable substitution $\xi = \dfrac{x}{2(Dt)^{1/2}}$ the integral becomes:

$$\mathcal{M}^*(t) = 2(Dt)^{1/2}\left(C_B^o - C_A^o\right)\int_0^\infty \text{erfc}(\xi)d\xi = 2\,\pi^{-1/2}(Dt)^{1/2}\left(C_B^o - C_A^o\right)$$
$$= 1.13(Dt)^{1/2}\left(C_B^o - C_A^o\right) \tag{18-25}$$

The integral mass flux increases linearly with the initial concentration difference, $C_B^o - C_A^o$, and with $(Dt)^{1/2}$.

The above expressions (Eqs. 18-20 to 18-25) remain valid if $C_B^o < C_A^o$. Also, note that these derivations assume the concentrations in the two phases are equal to one another at equilibrium (i.e., $K_{A/B} = 1$). We will see the result of changing this value of $K_{A/B}$ in Chapter 19.

Symmetrical Diffusion at Flat Boundaries

Next we consider a slightly different situation. Again two systems A and B are placed in contact via a flat boundary at $x = 0$, but now the concentration on either side of the boundary changes during the process. In other words, the assumption that mixing in system B is sufficient to maintain a constant B-side boundary concentration is dropped. This might be the case for components of a non-aqueous-phase liquid like gasoline after it is spilled on top of a groundwater table. Thus, we deal with a two-sided diffusive erosion process of an initial concentration jump (Fig. 18.5d). Again, assuming a value of $K_{A/B} = 1$, we redefine the concentrations such that the initial concentration in A is zero:

$$C_B^* = C_B - C_A^o \quad \text{and} \quad C_B^*(x, t = 0) = C_B^o - C_A^o$$
$$C_A^* = C_A - C_A^o \quad \text{and} \quad C_A^*(x, t = 0) = 0 \tag{18-26}$$

The solution to Fick's second law (Eq. 18-14), provided that the diffusion coefficients in both systems A and B are identical, is given by

$$C^*(x,t) = \frac{C_B^o - C_A^o}{2}\,\text{erfc}\left(\frac{x}{2(Dt)^{1/2}}\right) \tag{18-27}$$

Again we find the complementary error function, erfc(y), with the same argument as in Eq. 18-20. This time the solution applies to both sides of the interface: $x < 0$ for system B, $x > 0$ for system A. The interface is located at $x = 0$, where $C^*(0,t)$ is always equal to $(1/2)(C_B^o - C_A^o)$, since erfc(0) = 1. Note that the solution is symmetrical around $x = 0$ in the sense that the losses and gains are equal at equal distances from the boundary (Fig. 18.5e). The transformation back to the original concentrations (see Eq. 18-26) yields:

$$C(x,t) = C_A^o + \frac{C_B^o - C_A^o}{2}\,\text{erfc}\left(\frac{x}{2(Dt)^{1/2}}\right) \tag{18-28}$$

The concentration at the interface for $t > 0$ is time independent and equal to the arithmetic mean of C_A^o and C_B^o:

$$C(0,t) = \frac{C_B^o + C_A^o}{2} \tag{18-28a}$$

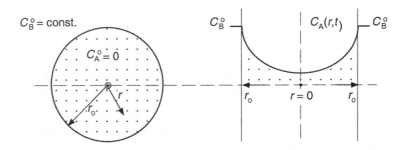

Figure 18.6 Diffusion from the environment (system B with constant concentration C_B^o) into a sphere with radius r_0 (system A).

As in Eq. 18-23 we can define a *penetration distance*. Now there is a *loss penetration* on one side of the boundary which is compensated for by the corresponding *gain penetration* on the other side of the boundary. Again, the penetration proceeds with a speed which is proportional to $(Dt)^{1/2}$. We can also ask how much of the compound has crossed the interface at time t. In Fig. 18.5e this quantity is represented for time t_1 by the hatched areas on either side of the boundary. The loss from system B corresponds to the gain of system A. By analogy to Eq. 18-25 the exchanged mass is given by:

$$\mathcal{M}_{ex}^*(t) = \int_0^x C^*(x,t)\mathrm{d}t = \pi^{-1/2}(Dt)^{1/2}(C_B^o - C_A^o)$$

(18-29)

$$= 0.564\,(Dt)^{1/2}(C_B^o - C_A^o)$$

Again, the total mass increases linearly with the initial concentration difference, $(C_B^o - C_A^o)$, with $t^{1/2}$, and with $D^{1/2}$.

Advanced Topic

Diffusion Into and From a Spherical Particle

The last case in this section deals with the sudden exposure of a spherical system A (particle, droplet, etc.) with initial concentration, C_A^o, to a constant exterior concentration, C_B^o (Fig. 18.6). Here again we assume that the equilibrium situation is represented by $K_{A/B} = 1$. At time $t = 0$, for the case in which $C_A^o < C_B^o$, the substance begins to move into the spherical system by diffusion (coefficient D). Now we are interested in the temporal evolution of the concentration inside the sphere and in the total exchanged mass at time t.

Diffusion into a sphere represents a three-dimensional situation; thus we have to use the three-dimensional version of Fick's second law (Box 18.3, Eq. 1). However, as mentioned before, by replacing the Cartesian coordinates x,y,z by spherical coordinates the situation becomes one-dimensional again. Eq. 3 of Box 18.3 represents one special solution to a spherically symmetric diffusion provided that the diffusion coefficient is constant and does not depend on the direction along which diffusion takes place (isotropic diffusion). *Note that diffusion into solids is not always isotropic, chiefly due to layering within the solid medium.* The boundary conditions of the problem posed in Fig. 18.6 requires that C is held constant on the surface of the sphere defined by the radius r_0.

As shown in mathematical textbooks, the Laplace operator (i.e., the right-hand differentials of Eq. 1 of Box 18.3), transformed into spherical coordinates, is:

$$\frac{\partial C}{\partial t} = D\left(\frac{\partial^2 C}{\partial r^2} + \frac{2}{r}\frac{\partial C}{\partial r}\right) \tag{18-30}$$

where r is the distance from the center of the sphere (Fig. 18.6). The solution of Eq. 18-30 is given by the infinite sum (Crank, 1975):

$$\frac{C(r,t) - C_A^o}{C_B^o - C_A^o} = 1 + \frac{2r_o}{\pi r}\sum_{k=1}^{\infty}\frac{(-1)^k}{k}\sin\frac{k\pi r}{r_o}\exp\left(-\frac{Dk^2\pi^2 t}{r_o^2}\right) \tag{18-31}$$

To generalize this result, we can express the equation by using dimensionless parameters. For example, if we normalize the spatial variable by the key spatial factor, r_0, we have:

$$r^* = \frac{r}{r_o} \quad (0 \le r^* \le 1 \text{ within the sphere}) \tag{18-32a}$$

Likewise, we can normalize t by the diffusion time, r_o^2 / D:

$$t^* = \frac{Dt}{r_o^2} \tag{18-32b}$$

Using these dimensionless lengths and times, we find the dimensionless concentration changes as:

$$\frac{C(r^*,t^*) - C_A^o}{C_B^o - C_A^o} = 1 + \frac{2}{\pi r^*}\sum_{k=1}^{\infty}\frac{(-1)^k}{k}\sin k\pi r^* \exp\left(-k^2\pi^2 t^*\right) \tag{18-33}$$

The concentration at the center of the sphere is given by the limit $r^* \to 0$:

$$\frac{C(0,t^*) - C_A^o}{C_B^o - C_A^o} = 1 + 2\sum_{k=1}^{\infty}(-1)^k \exp\left(-k^2\pi^2 t^*\right) \tag{18-34}$$

The total amount of diffusing substance which has entered or left the sphere at time t^*, $\mathcal{M}(t^*)$, is

$$\frac{\mathcal{M}(t^*)}{\mathcal{M}_\infty} = 1 - \frac{6}{\pi^2}\sum_{k=1}^{\infty}\frac{1}{k^2}\exp\left(-k^2\pi^2 t^*\right) \tag{18-35}$$

where $\mathcal{M}_\infty$ is the maximum mass exchange determined by the volume of the sphere and the initial concentration difference:

$$\mathcal{M}_\infty = \left(C_B^o - C_A^o\right)\frac{4\pi}{3}r_o^3 \quad [M] \tag{18-36}$$

Note that these results apply for the case in which the concentration in the environment (system B) is not changing.

The solutions (Eqs. 18-33 and 18-35) as a function of the dimensionless parameters are shown in Fig. 18.7. For instance, from Fig. 18.7b we can learn that half of the potential mass echange ($\mathcal{M}(t)/\mathcal{M}_\infty = 0.5$) has occurred at the nondimensional diffusion time $t_{1/2}^* \cong 0.03$, that is, in real time units (see Eq. 18-32b):

$$t_{1/2} \sim 0.03\frac{r_o^2}{D} \tag{18-37}$$

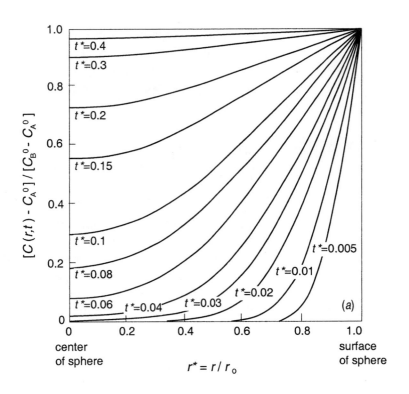

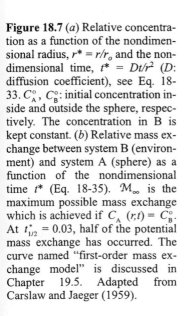

Figure 18.7 (*a*) Relative concentration as a function of the nondimensional radius, $r^* = r/r_o$ and the nondimensional time, $t^* = Dt/r^2$ (*D*: diffusion coefficient), see Eq. 18-33. C_A^o, C_B^o: initial concentration inside and outside the sphere, respectively. The concentration in B is kept constant. (*b*) Relative mass exchange between system B (environment) and system A (sphere) as a function of the nondimensional time t^* (Eq. 18-35). $\mathcal{M}_\infty$ is the maximum possible mass exchange which is achieved if C_A $(r,t) = C_B^o$. At $t_{1/2}^* = 0.03$, half of the potential mass exchange has occurred. The curve named "first-order mass exchange model" is discussed in Chapter 19.5. Adapted from Carslaw and Jaeger (1959).

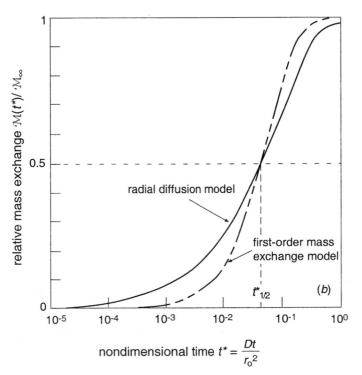

At this time, the concentration at the center of the sphere has barely changed from its initial value, C_A^o (see Fig. 18.7*a*).

These results will be used in Section 19.5 when we discuss the kinetics of sorptive exchange of a solute between particles and the water in which they are suspended.

We will then compare the dynamics of the radial diffusion model with a first-order exchange model which gives the same half-life as the radial model (Eq. 18-37). A preview of this comparison is given in Fig. 18.7*b*, which shows that the linear model underpredicts the exchange at short times and overpredicts it at long times.

18.3   Random Motion at the Molecular Level: Molecular Diffusion Coefficients

Diffusivity and the Molecular Theory of the Ideal Gas

The thermal movement of molecules often serves as a prototype of random motion. In fact, molecular diffusion is the result of the random walk of atoms and molecules through gaseous, liquid, solid, or mixed media. This section deals with molecular diffusion of organic substances in gases (particularly air) and in aqueous solutions. Diffusion in porous media (i.e., mixes of gases or liquids with solids) and in other media will be discussed in the following section.

Molecular diffusion deals with the relative motion of one kind of atom or molecule against a set of reference molecules. As explained in the introduction to this chapter (remember the trip in the dining car through the Swiss Alps), the reference system itself may move relative to some chosen coordinates. We called such directed motion advection. If one really looks very closely and wants to use crystal-clear definitions, it turns out that there is more than one way to choose the reference system. Each choice leads to a different separation between diffusion and advection, resulting in different diffusion coefficients.

If you want to better understand this subtle and somewhat confusing matter, then read Chapters 3 and 7 of the book by Cussler (1984) on diffusion, which gives a nice overview. Here we just mention one of his very didactic examples: Imagine two bulbs of equal volume connected by a capillary containing a stopcock. The left bulb contains 50 weight percent glycerol in water, the right bulb has pure water. When the stopcock is opened, glycerol diffuses from the left bulb toward the right, and water in the opposite direction, until both are evenly distributed in the bulbs. Against which reference system do we want to describe the diffusive flux? Of course, a quick (and convenient) answer would be "against the apparatus used for the experiment", but nature does not usually work with this kind of experimental crutch! In fact, volume changes less than 0.1% during the mixing process; that is, the center of volume hardly moves and thus would provide a reasonable reference system. In contrast, the mass average velocity is not zero, since the glycerol solution has a density of about 1.1 g cm^{-3}, as compared to pure water at 1.0 g cm^{-3}. Hence, initially the center of mass is located near the left bulb containing the glycerol, whereas at the end of the experiment it lies exactly in the middle between the bulbs. This means that diffusion against volume is different from diffusion against mass, and so are the corresponding diffusion coefficients. And to make things even more complicated, diffusion against the center of moles (that is, against the center of numbers of molecules) would yield a third coefficient. As it turns out the situation is different in gases and liquids, but in both phases diffusion against the center of volume is conceptually the best choice, although not all experimental setups yield this kind of diffusion coefficient.

There is an additional complication. Molecular diffusivity depends on both the diffusing chemical (designated by the subscript i) and the reference system (designed by a second subscript like w for water, a for air). Yet, the reference system could be any other gas or liquid (pure or mixture), like N_2, octanol, diesel fuel, etc. Chemical handbooks usually list diffusivities as *diluted two-component* or *binary* coefficients, for instance, the diffusion of tetrachloroethene relative to liquid water (D_{iw}), relative to pure gases (e.g., molecular nitrogen), or relative to air (D_{ia}). Strictly speaking, diffusion of a gaseous compound in air is not a two-component diffusion; yet it is usually treated as such, since the mixing ratio of the two major components of air, N_2 and O_2, is fairly constant. Thus the two-gas mix acts like a uniform reference system. If the diffusing species is diluted, that is, its concentration is much smaller than the one of the background substance, then the difference between the various diffusivities (mass, volume, moles) is not relevant. However, in real systems we may find the whole spectrum of mixtures, for instance, between liquid benzene and cyclohexane. Then, the coefficient of binary diffusion is a continuous function of the benzene-cyclohexane mixing ratio. At one end (pure benzene), we find the coefficient of *self-diffusion* of benzene, at the other end the coefficient of self-diffusion of cyclohexane. Coefficients of self-diffusion correspond to a special case of still another kind of diffusivity, that is, *tracer diffusion* at infinitely small tracer concentration. This coefficient is relevant, for instance, when we tag a small amount of benzene by radioactive carbon and watch how the tagged benzene is mixed into the untagged benzene.

Finally, the diffusion of a chemical may be influenced by another diffusing compound or by the solvent. The latter effect is known as solute-solvent interaction; it may become important when solute and solvent form an association that diffuses intact (e.g., by hydration). This may be less relevant for neutral organic compounds, but it plays a central role for diffusing ions. But even for noncharged particles the diffusivities of different chemicals may be coupled. The above example of the glycerol diffusing in water makes this evident: in order to keep the volume constant, the diffusive fluxes of water and glycerol must be coupled.

Unfortunately, environmental problems do not solely involve dilute binary systems. To mention just one example, the dissolution of a patch of spilled diesel fuel into the groundwater (see below; Illustrative Example 19.4) involves diffusion in multicomponent systems for which adequate data are extremely rare. In many cases diffusivities of diluted compounds in air or water must serve as best estimates for more complex systems. Yet, the above remarks should remind us that things may be more complicated.

In this section we discuss molecular diffusivities in the "standard environmental systems," air and water. Remember that the dimension of diffusivity is L^2T^{-1}. Thus, the standard metric unit is m^2s^{-1}. In most handbooks, diffusion coefficients are given in cm^2s^{-1}, the unit also adopted below. Note that $1\ cm^2s^{-1} = 10^{-4}\ m^2s^{-1}$.

Diffusivities in Air

According to the model of random walk in three dimensions, the diffusion coefficient of a molecule i, D_i, can be expressed as one-third of the product of its mean free path λ_i and its mean three-dimensional velocity u_i (Eq. 18-7a). In the framework of the molecular theory of gases, u_i is (e.g., Cussler, 1984):

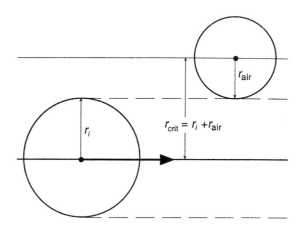

Figure 18.8 Scheme to calculate the mean free path λ_i of trace molecules in air.

$$u_i = \left(\frac{8RT}{\pi M_i}\right)^{1/2} \quad (\text{cm s}^{-1}) \tag{18-38}$$

where

R is the gas constant (8.31×10^7 g cm^{2}K^{-1}s^{-2}mol^{-1})

T is the absolute temperature (K)

M_i is the chemical's molar mass (g mol^{-1})

The mean free path λ_i of a molecule in air can be calculated from the sizes of the molecules involved. The most probable collision partners for a trace molecule (such as CFC–12) in air are molecular nitrogen (N_2) and oxygen (O_2). The trace molecule i is hit whenever its center gets closer to the center of an air molecule than the critical distance, $r_{crit} = r_i + r_{air}$ (Fig. 18.8). Picturing the molecules as spheres, the molecular radius r_i can be estimated from the collision cross-section A listed in chemical handbooks such as the *Tables of Physical and Chemical Constants* (Longman, London, 1973):

$$r_i = [A_i / \pi]^{1/2} \tag{18-39}$$

As an alternative r_i can be estimated from the molar volume $\overline{V}_i$. If we picture the molecules as spheres with radius r_i we get:

$$r_i = \left(\frac{3\overline{V}_i}{4\pi N_A}\right)^{1/3} \tag{18-40}$$

where $N_A = 6.02 \times 10^{23}$ mol^{-1} is Avogadro's number. Imagine that all molecules are fixed in space and that a selected trace molecule moves through space. As long as the center of the moving molecule keeps a distance larger than r_{crit}, the selected molecule does not collide with air molecules. Thus, if the trace molecule moves over the distance λ_i, it sweeps out the cylindrical volume $v^* = \pi r_{crit}^2 \lambda_i$.

How far can the CFC–12 molecule move on the average until the corresponding volume v^* contains an air molecule? The average volume occupied by an air molecule at total pressure p and temperature T is:

$$v_i = \frac{RT}{N_A p} = \frac{kT}{p} \qquad (18\text{-}41)$$

where $k = R/N$ is the Boltzmann constant. On the average, a collision between a trace and air molecule occurs when v^* and v_i are equal: $v_i = v^* = \pi r_{\text{crit}}^2 \lambda_i$. This expression can be solved for λ_i. Yet, it turns out that in order to account for the fact that all molecules are moving simultaneously, the result has to be reduced by the factor $2^{1/2}$, thus the final result is:

$$\lambda_i = \frac{RT}{N_A\, p}\, \frac{1}{\sqrt{2}\,\pi r_{\text{crit}}^2} = \frac{kT}{p}\, \frac{1}{\sqrt{2}\,\pi r_{\text{crit}}^2} \qquad (18\text{-}42)$$

Inserting Eqs. 18-38 and 18-42 into Eq. 18-7a yields:

$$D_{ia} = \frac{2}{3}\left(\frac{RT}{\pi M_i}\right)^{1/2} \frac{RT}{N_A\, p\, \pi r_{\text{crit}}^2} = \frac{2}{3 N_A}\left(\frac{R}{\pi}\right)^{3/2} \frac{T^{3/2}(1/M_i)^{1/2}}{p\, r_{\text{crit}}^2} \qquad (18\text{-}43)$$

The peculiar way in which the result on the far right-hand side of Eq. 18-43 is expressed makes it easier to understand the semiempirical relation proposed by Fuller et al. (1966) for the diffusivities of organic molecules in air:

$$D_{ia} = 10^{-3}\, \frac{T^{1.75}\left[(1/M_{\text{air}}) + (1/M_i)\right]^{1/2}}{p\left[\overline{V}_{\text{air}}^{1/3} + \overline{V}_i^{1/3}\right]^2} \quad (\text{cm}^2\text{s}^{-1}) \qquad (18\text{-}44)$$

where

T is the absolute temperature (K)

M_{air} is the average molar mass of air (28.97 g mol^{-1})

M_i is the chemical's molar mass (g mol^{-1})

p is the gas phase pressure (atm)

$\overline{V}_{\text{air}}$ is the average molar volume of the gases in air (≈ 20.1 cm^{3}mol^{-1})

$\overline{V}_i$ is the chemical's molar volume (cm^{3}mol^{-1})

Note that Eq. 18-44 is not dimensionally correct, and thus it is valid only if all the quantities are expressed in the units listed above. However, Eqs. 18-43 and 18-44 express essentially the same dependence on temperature, pressure, molecular size (volume or radius, see Eq. 18-40), and mass, although in the latter (Eq. 18-44) both molecular sizes and masses appear as a composite of the different molecules involved in the diffusion process.

The *Chapman-Enskog theory* (Chapman and Cowling, 1970) is a model which is positioned between the two approaches, the empirical relation by Fuller et al. (Eq. 18-44) and the theoretically stringent Equation 18-43. This theory improves the absolute size of the expression by taking into account the individual sizes and interactions of the diffusing molecules. However, the numerical values obtained with the model by Fuller et al. (Eq. 18-44) are still better than both the Chapman-Enskog theory and Eq. 18-43.

Whatever the best model, we expect that larger chemicals progress more slowly because their mean thermal velocity, u_i, is reduced and their larger cross-section

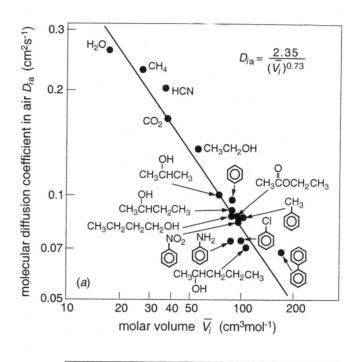

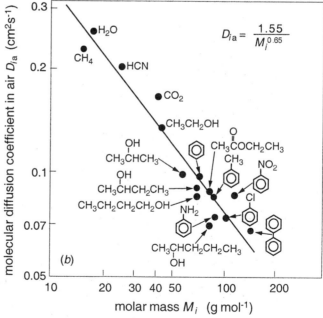

Figure 18.9 Molecular diffusion coefficients in air, D_{ia}, at 25°C for different molecules plotted as a function of: (a) their liquid molar volume, $\overline{V}_i$, (calculated as ratio of molar mass M_i to liquid density, ρ_{iL}), and (b) their molar mass, M_i. Data from references reviewed by Fuller et al. (1966) plotted on double-logarithmic scale.

diminishes their mean free path λ_i. As indicated in Fig. 18.9, diffusivities of molecules in air are usually on the order of 0.1 cm²s⁻¹. For small molecules like water, diffusivities are about 0.3 cm²s⁻¹ and drop to about 0.07 cm²s⁻¹ for organic molecules of molar mass M_i near 100 g mol⁻¹. Fig. 18.9a demonstrates that diffusion coefficients decrease approximately as the two-thirds power of molar volume $\overline{V}_i$, indicative of the importance of molecule cross-sectional area A_i. Furthermore, as shown in Fig. 18.9b, the chemical's molar mass may also serve as a useful measure of molecule size, and thus we see that molar masses are inversely correlated with gas-phase diffusivities.

We may now make use of our physicochemical understanding of diffusion to estimate chemical diffusivities in the gas phase and to explain the correlations found in Fig. 18.9. First, we need to estimate the molar volumes, $\overline{V}_i$, of the chemicals. One way is to estimate $\overline{V}_i$ by dividing the chemical's molar mass by its liquid density; this method was used for the preparation of Fig. 18.9a. Alternatively, we can sum the "size" of the atoms making up the chemical's structure. Table 18.2 shows "diffusion sizes" of various atoms deduced by regression of available diffusion data (Fuller et al., 1966). In Chapter 5, an alternative scheme by Abraham and McGowan (1987) was introduced to estimate molar volumes from structure (see Box 5.1), but these values were optimized to explain the strengths of intermolecular attractions. Thus, although the molar volumes estimated via the method of Abraham and McGowan (1987) correlate with those of Fuller et al. (1966), perhaps not surprisingly the scheme by Fuller et al. corresponds better to the results from the liquid density method. Values obtained using the different methods are compared in Table 18.3. Further examples are given in Illustrative Example 18.1.

As an example, let us estimate the diffusivity of benzene in air. According to Table 18.3, the molar volume of benzene calculated from liquid density (89 cm^{3}mol^{-1}) and from the component method by Fuller et al. (90.8 cm^{3}mol^{-1}) yield similar results. Incorporating these estimates into Eq. 18-44 yields:

$$D_{\text{benzene a}} (25°C, \ 1 \ \text{bar}) = \frac{(10^{-3})(298)^{1.75}\left(\dfrac{1}{29}+\dfrac{1}{78}\right)^{0.5}}{(1)\left[(20.1)^{1/3}+(90)^{1/3}\right]^2} \ \text{cm}^2\text{s}^{-1} = 0.090 \, \text{cm}^2\text{s}^{-1}$$

The experimental result is 0.096 cm^{2}s^{-1}. Fuller et al. (1966) found that diffusivities estimated from Eq. 18-44 match observations to within 10%.

An alternative technique involves adjusting diffusivities known for one chemical to approximate values for chemicals of related structure, recognizing that molecular velocities and thus molecular diffusivities vary roughly inversely with the square root of molecular mass (Eq. 18-38). We can use this readily available parameter to adjust the known diffusivity of a reference substance to the unknown diffusivity of substance i according to the following equation:

$$\frac{D_{i\,a}}{D_{\text{ref a}}} \approx \left[\frac{M_i}{M_{\text{ref}}}\right]^{-1/2} \tag{18-45}$$

Note that in this procedure the effect of molecular mean free path, that is, of molecular size is neglected. As an example we estimate diffusivity of toluene ($M_{\text{toluene}} = 92 \ \text{g mol}^{-1}$) from diffusivity of benzene ($M_{\text{benzene}} = 78 \ \text{g mol}^{-1}$) and get $D_{\text{toluene a}} \approx (0.096 \ \text{cm s}^{-1}) \ [92/78]^{-1/2} = 0.088$ cm s^{-1}. The experimental value is 0.086 cm s^{-1} (Gilliland, 1934).

We can also learn from Eq. 18-43 how the relevant environmental properties such as temperature T and pressure p influence $D_{i\,a}$. We rewrite the equation in the following way:

$$D_{i\,a} = \text{constant} \ T^{3/2} p^{-1} M_i^{-1/2} \text{A}_i^{-1} \tag{18-46}$$

Table 18.2 Estimation of Diffusion Volumes of Organic Molecules (Fuller et al., 1966)

Element	Volume Contribution (cm³mol⁻¹)
C	16.5
H	2.0
O	5.5
N	5.7
Cl	19.5
S	17.0
Rings	−20.2

Table 18.3 Comparison of Different Methods to Calculate the Molar Volume $\overline{V}_i$ of Organic Compounds. Values in cm³mol⁻¹

Compound	Composition	Volume from Structure		
		Box 5.1 [a]	Table 18.2 [b]	M_i / ρ_{iL} [c]
Benzene				
	6(C), 6(H) 12 (bonds), 1 (ring)	71.6	90.8	$\dfrac{78.1}{0.879} = 88.9$
Trichloroethene				
	2(C), 1(H), 3(Cl) 6 (bonds)	71.5	93.5	$\dfrac{131.4}{1.456} = 90.2$
Ethyl acetate				
	4(C), 2(O), 8(H) 13 (bonds)	74.7	93.0	$\dfrac{88.1}{0.901} = 97.8$
Trimethylamine				
	3(C), 1(N)), 9(H) 12 (bonds)	63.1	73.2	$\dfrac{59.1}{0.662} = 89.3$
Dimethylamine				
	2(C), 1(N), 7(H) 9 (bonds)	49.0	52.7	$\dfrac{45.1}{0.680} = 66.3$
Dimethyl sulfide				
	2(C), 1(S), 6(H) 8 (bonds)	55.4	62.0	$\dfrac{62.1}{0.846} = 73.4$

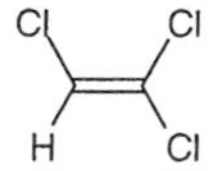

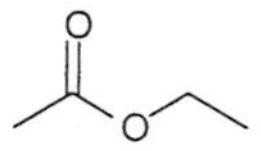

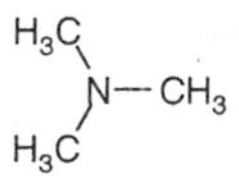

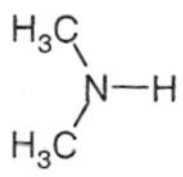

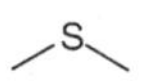

[a] From Abraham and McGowan (1987), see Box 5.1. [b] From Fuller et al. (1966), see Table 18.2.
[c] Method used in Figs. 18.8 and 18.10. M_i = molar mass (g mol⁻¹), ρ_{iL} = liquid density (g cm⁻³).

where A_i stands for the typical cross-sectional area of the molecules involved. This not only explains the correlation between D_{ia} and M_i or $\overline{V}_i$, respectively (note from Eq. 18-40 that A_i is proportional to $\overline{V}_i^{2/3}$), but it also predicts the pressure and temperature dependence of D_{ia}.

Experimental information on the temperature and pressure dependence of gaseous molecular diffusion coefficients is scarce. Most data refer to diffusivity of volatile

---

Illustrative Example 18.1 **Estimating Molar Volumes**

Problem

Estimate the molar volume $\overline{V}_i$ of dichlorodifluoromethane CCl_2F_2 (also called freon–12 or CFC–12) (a) from its liquid density and (b) with the element contribution method by Fuller (Table 18.2).

Answer (a)

By definition, molar volume $\overline{V}_i$, liquid density ρ_{iL}, and molar mass M_i are related by

$$\overline{V}_i = \frac{M_i}{\rho_{iL}} = \frac{120.9\,\text{g mol}^{-1}}{1.328\,\text{g cm}^{-3}} = 91.0\,\text{cm}^3\text{mol}^{-1}$$

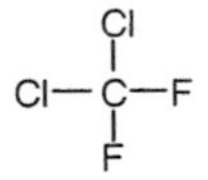

Answer (b)

Inspection of Table 18.2 shows that no contribution for F is available. However, by comparing the molar volumes of related compounds which contain different numbers of fluorine atoms we get the following differences in molar volume, $\Delta\overline{V}$, if one Cl is substituted by one F:

Compound	M_i (g mol^{-1})	ρ_{iL} (g cm^{-3})	$\overline{V}_i$ (cm^{3}mol^{-1})	$\Delta\overline{V}$ [a] (cm^{3}mol^{-1})
CCl_4	153.8	1.594	96.5	
CCl_3F	137.4	1.490	92.2	– 4.3
$CHCl_2F$	102.9	1.366	75.3	
$CHClF_2$	86.5	1.213	71.3	– 4.0

[a] Change of molar volume if one Cl is substituted by one F

These data indicate that the contribution of a fluorine atom to $\overline{V}_i$ is roughly 4 cm^{3}mol^{-1} less than the contribution of a chlorine atom. Since the latter is 19.5 cm^{3}mol^{-1} (Table 18.2), the contribution of F to $\overline{V}_i$ is estimated to be about 15.5 cm^{3}mol^{-1}. Thus:

$$\overline{V}(CCl_2F_2) = \overline{V}(C) + 2\overline{V}(Cl) + 2\overline{V}(F)$$

$$= (16.5 + 39.0 + 31.0)\,\text{cm}^3\text{mol}^{-1} = 86.5\,\text{cm}^3\text{mol}^{-1}$$

dichlorodifluoromethane
CCl_2F_2

$M_i = 120.9$ g mol^{-1}
$\rho_{iL} = 1.328$ g cm^{-3}

chemicals in air (see, e.g., Reid et al., 1977) or to binary systems, for instance N_2 and O_2 (Tokunaga et al., 1988). The inverse dependence of D_{ia} on pressure p seems to be fairly well confirmed. From a large set of binary diffusivities (that is the diffusion of one gas versus another), Marrero and Mason (1972) conclude that D_{ia} is proportional to $T^{1.724}$, yet the scattered data would also be compatible with the exponent in Eq. 18-46 of (3/2). Problem 18.5 deals with the T and p dependence of molecular diffusivity of dichlorodifluoromethane (CFC-12) in air. Different methods to calculate D_{ia} are compared in Illustrative Example 18.2.

Illustrative Example 18.2   **Estimating Molecular Diffusivity in Air**

Problem

Estimate the molecular diffusion coefficient in air, D_{ia}, of CFC-12 (see Illustrative Example 18.1) at 25°C: (a) from the mean molecular velocity and the mean free path, (b) from the molar mass, (c) from the molar volume, (d) from the combined molar mass and molar volume relationship by Fuller (Eq. 18-44), (e) from the molecular diffusivity of methane.

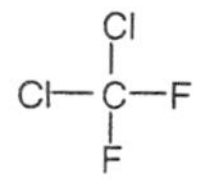

Answer (a)

To apply Eq. 18-43, we have to calculate the critical radius, r_{crit}, for the collision of a CFC-12 molecule with an air molecule (typically N_2 or O_2).

From the molar volume of CFC–12 (Illustrative Example 18.1a) we get:

i = CFC-12
T = 298.2 K (25°C)
p = 1 bar: gas phase pressure
M_{air} = 29 g mol^{-1}: molar mass of air
$\overline{V}_{air}$ = 20.1 cm^{3}mol^{-1}: molar volume of air
Collision cross-sections A for N_2 and O_2 (Tables of Physical and Chemical Constants, Longman, London, 1973):
    A (N_2) = 0.43 × 10^{-14} cm^2
    A (O_2) = 0.40 × 10^{-14} cm^2
M_i = 120.9 g mol^{-1}: molar mass of CFC-12
$\overline{V}_i$ = 91.0 cm^{3}mol^{-1}: molar volume of CFC-12
$D_{ref\,a}$ = 0.23 cm^{2}s^{-1}: molecular diffusivity of CH$_4$ in air at 25°C:

$$v_{i\,mol} = \frac{\overline{V}_i}{N_A} \sim \frac{91\,cm^3mol^{-1}}{6.0 \times 10^{23}\,mol^{-1}} = 1.5 \times 10^{-22}\,cm^3$$

If the CFC–12 is considered spherical, we get from Eq. 18-40:

$$r_i = \left(\frac{3v_{i\,mol}}{4\pi}\right)^{1/3} = 3.3 \times 10^{-8}\,cm$$

The most probable collision partners for a trace molecule in air are N_2 and O_2. Their collision cross sections are similar. Thus, we take one of the more abundant N_2 molecule. From Eq.18-39 we get:

$$r_{air} \approx r(N_2) = \left(\frac{A(N_2)}{\pi}\right)^{1/2} = \left(\frac{0.43 \times 10^{-14}\,cm^2}{3.14}\right)^{1/2} = 3.7 \times 10^{-8}\,cm$$

Thus:

$$r_{crit} = r_i + r(N_2) = 7.0 \times 10^{-8}\,cm$$

Inserting into Eq. 18-42 yields the mean free path of CFC-12 in air:

$$\lambda_i = \frac{8.31 \times 10^7 \times 298.2\,g\,cm^2s^{-2}\,mol^{-1}}{\sqrt{2} \times 3.142 \times 6.02 \times 10^{23}\,mol^{-1} \times 10^6\,g\,cm^{-1}s^{-2} \times (7.0 \times 10^{-8})^2\,cm^2} = 1.9 \times 10^{-6}\,cm$$

Note that in order to get a consistent set of units, the gas constant R is expressed as $R = 8.31 \times 10^7$ g cm^{2}s^{-2}mol^{-1}K^{-1}, and pressure p is 1 bar $= 10^6$ g cm^{-1}s^{-2}. According to Eq. 18-38, the mean three-dimensional molecular velocity is:

$$u_i = \left(\frac{8 \times 8.31 \times 10^7 \times 298.2 \text{ g cm}^2\text{s}^{-2}\text{ mol}^{-1}}{3.14 \times 120.9 \text{ g mol}^{-1}} \right)^{1/2} = 2.3 \times 10^4 \text{cm s}^{-1}$$

Thus, from Eq. 18-7a we finally get:

$$D_{ia} = \frac{1}{3} \times 2.3 \times 10^4 \text{cm s}^{-1} \times 1.9 \times 10^{-6} \text{cm} = \textit{1.5} \times \textit{10}^{-2} \textit{cm}^2\textit{s}^{-1}$$

A glance at Fig. 18.8 shows that typical D_{ia} values found for other molecules are of order 10^{-1}cm^{2}s^{-1}. The way λ_i was calculated explains at least partially why D_{ia} is rather small. It was implicitly assumed that whenever the centers of the CFC-12 and air molecules get closer than r_{crit}, the CFC-12 molecule collides in such a way as to completely "forget" its former direction and speed. In reality, slighter collisions only partially change the course of the molecules. The actual mean free path is the weighed average of distances between collisions of different strength.

Answer (b)

Calculate D_{ia} from the empirical relation to molar mass. From Fig. 18.9b follows:

$$D_{ia} [\text{cm}^{-2}\text{s}^{-1}] = \frac{1.55}{M_i^{0.65}} = \frac{1.55}{(120.9)^{0.65}} = \textit{6.9} \times \textit{10}^{-2} \text{ cm}^2\text{s}^{-1}$$

Answer (c)

Calculate D_{ia} from the empirical relation to (liquid) molar volume $\overline{V}_i$. From Fig. 18.9a follows:

$$D_{ia} [\text{cm}^2\text{s}^{-1}] = \frac{2.35}{(\overline{V}_i)^{0.73}} = \frac{2.35}{(91.0)^{0.73}} = \textit{8.7} \times \textit{10}^{-2} \text{cm}^2\text{s}^{-1}$$

Answer (d)

The semiempirical relationship by Fuller et al. (1966) yields (Eq. 18-44):

$$D_{ia} [\text{cm}^2\text{s}^{-1}] = 10^{-3} \frac{(298.2)^{1.75}[1/29 + 1/120.9]^{1/2}}{1 \times \left[(20.1)^{1/3} + (91.0)^{1/3} \right]^2} = \textit{8.5} \times \textit{10}^{-2} \text{ cm}^2\text{s}^{-1}$$

Answer (e)

Take Eq. 18-45 to calculate the diffusion coefficient of CFC-12 from the reference substance methane:

$$D_{ia} = D_{\text{ref a}} \left(\frac{M_i}{M_i} \right)^{-1/2} = 0.23 \text{cm}^2\text{s}^{-1} \times \left(\frac{120.9}{16} \right)^{-1/2} = \textit{8.4} \times \textit{10}^{-2} \text{cm}^2\text{s}^{-1}$$

The following table summarizes these results:

Diffusion Coefficient of CFC-12 in Air at 25°C, D_{ia}

Method	D_{ia} (cm^{2}s^{-1})
(a) Molecular theory	(1.5×10^{-2})
(b) From molar mass	6.9×10^{-2}
(c) From molar volume	8.7×10^{-2}
(d) From Fuller's combined expression	8.5×10^{-2}
(e) From diffusivity of CH_4	8.4×10^{-2}
Measured diffusivity in N_2 at 10°C [a]	9.1×10^{-2}

[a] From Monfort and Pellegatta (1991).

It is remarkable that the results from method c to e agree very well. Reasons were given above why the molecular theory (method a) underestimates the true value.

Diffusivities in Water

The diffusivities of organic solutes in water, D_{iw}, are also dependent on the diffusate's size. As illustrated in Fig. 18.10, size can still be characterized using molar volume or mass. Here we see diffusivities ranging from about 3×10^{-5} cm^{2}s^{-1} for small molecules to about 0.5×10^{-5} cm^{2}s^{-1} for those of molar mass near 300 g mol^{-1}. Diffusivities in water are about 10^4 times smaller than those in air. Based on what we have learned from the random walk model, most of the difference between D_{ia} and D_{iw} can be attributed to the density ratio between water and air (about 10^3), which leads to a much smaller mean free path in water. This is underlined by the inverse relationship between the molar volumes of the molecules and their diffusivities in both air and water (Figs. 18.9a and 18.10a). A similar relationship holds between molar mass and D_{iw} (Fig. 18.10b). Note, however, that a substance like radon-222 (Rn) deviates from this correlation. This is because the liquid density of radon is quite large (≈ 4.4 g cm^{-3}) compared to other substances (typically ≈ 1 g cm^{-3}). Thus, the molar mass of radon overestimates the relative size of this important geochemical radioactive noble gas. If diffusivity is related to molar volume instead, radon behaves as most other substances (Fig. 18.10a).

The random walk model is certainly less suitable for a liquid than for a gas. The rather large densities of fluids inhibit the Brownian motion of the molecules. In water, molecules move less in a "go-hit-go" mode but more by experiencing continuously varying forces acting upon them. From a macroscopic viewpoint, these forces are reflected in the viscosity of the liquid. Thus we expect to find a relationship between viscosity and diffusivity.

Let as look at a molecule moving relative to its neighbors with speed, u_i^*, where we have added the star to distinguish the *relative* molecular velocity from the absolute value as defined in Eq. 18-38. Viscous forces are usually assumed to be proportional to velocity differences, thus $\psi = f^* u_i^*$, where ψ is the force acting on the molecule and f^* is the friction factor. For the case of spherical particles that are much larger than the molecules of the liquid through which they are moving, f^* is given by Stokes' relation for the drag on a sphere (e.g., Lerman, 1979):

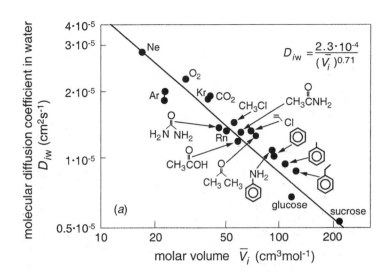

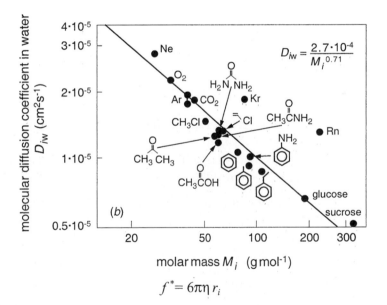

Figure 18.10 Molecular diffusion coefficients, D_{iw}, at 25°C of molecules in water plotted as a function of (a) their liquid molar volumes, V_i (calculated as ratio of molar mass to liquid density ρ_{iL}), and (b) their molar mass, M_i. Data from Hayduk and Laudie (1974) plotted on double-logarithmic scale.

$$f^* = 6\pi\eta\, r_i$$

where η is dynamic viscosity and r_i is the "effective" hydrodynamic radius of the molecule i. Thus:

$$\psi = f^* u_i^* = 6\pi\eta\, r_i u_i^* \tag{18-47}$$

The flux of molecules i per unit area and time, F_i, moving along a given direction at speed u_i^*, is the product of u_i^* and concentration C_i:

$$F_i = u_i^* C_i = \frac{\psi}{f^*} C_i = \frac{\psi}{6\pi\eta\, r_i} C_i \tag{18-48}$$

In order to evaluate this expression, we need to know the force ψ that is responsible for producing the molecular flux. It could be an external force such as an electric field acting on ions. Then evaluation of Eq. 18-48 would lead to the relationship between electric conductivity, viscosity, and diffusivity known as the *Nernst-Einstein relation*.

However, in the absence of an electric field and for uncharged particles diffusive motion is driven by an internal "force," the spatial gradient of the chemical potential, μ_i (see Chapters 3 and 8). Work is needed to move molecules from a location where their chemical potential is small to a location where μ_i is large. In turn, if the molecules move down the slope of μ_i, a "force" *per molecule* of size is driving them:

$$\psi = -\frac{1}{N_A}\frac{d\mu_i}{dx} \tag{18-49}$$

Since the chemical potential is usually formulated on a molar basis, division by the Avogadro's number N_A appears in the above expression. As the reader will realize, we were careful about using the word "force." In fact, ψ is an apparent, not a real force that pushes the molecules. It is the molecules' natural drift toward maximum entropy which drives them (Atkins, 1998). From Chapter 8 one gets:

$$\psi = -\frac{1}{N_A}\frac{d}{dx}\left\{\mu_i^{o'} + RT\ln a_i\right\} = -\frac{kT}{a_i}\frac{da_i}{dx} \tag{18-50}$$

where k is the Boltzmann constant, $\mu_i^{o'}$ is the chemical potential as pure liquid, and a_i the activity of the chemical in the aqueous solution. For small concentrations a_i and C_i are equal, thus:

$$\psi = -\frac{kT}{C_i}\frac{dC_i}{dx} \quad \text{for } a_i \approx C_i \tag{18-50a}$$

Inserting Eq. 18-50a into 18-48 yields:

$$F_i = -\frac{kT}{f^*}\frac{dC_i}{dx} = -\frac{kT}{6\pi\eta\, r_i}\frac{dC_i}{dx} \tag{18-51}$$

By comparing this result with Fick's first law (Eq. 18-6), we get the *Stokes-Einstein relation* between the diffusivity in aqueous solutions and the solution viscosity η:

$$D_{iw} = \frac{kT}{6\pi\eta\, r_i} \tag{18-52}$$

where

k = 1.381×10^{-23} kg m^{2}s^{-2}K^{-1} is the Boltzmann constant
η (kg m^{-1}s^{-1}) is dynamic viscosity
r_i (m) is the molecular radius

As a byproduct, we can learn from Eq. 18-50 that, in fact, it is not the gradient of concentration, C_i, but the chemical activity, a_i, that drives diffusion. Since at constant C_i, activity changes with temperature, ionic strength, and other parameters, a diffusive flux may actually occur even if the concentration gradient is zero.

As was the case for D_{ia}, the equations derived from the pure physical concepts are usually not the best numerical approximations of a given quantity, although they show which properties should enter into an empirical relationship. Othmer and Thakar (1953) derived the following expression with coefficients modified slightly by Hayduk and Laudie (1974):

$$D_{iw}(\text{cm}^2\text{s}^{-1}) = \frac{13.26 \times 10^{-5}}{\eta^{1.14}\,\overline{V}_i^{0.589}} \qquad (18\text{-}53)$$

where

η is the solution viscosity in centipoise (10^{-2} g cm^{-1}s^{-1}) at the temperature of interest

$\overline{V}_i$ is the molar volume of the chemical (cm^{3}mol^{-1})

Again one should be careful when applying this equation, since it is not correct dimensionally and thus can be used only when all the quantities are given in the prescribed units.

In this expression, the primary environmental determinant is the viscosity in the denominator. Note that the exponential is slightly larger than in the Stokes-Einstein relation (Eq. 18-52). Since viscosity decreases by about a factor of 2 between 0°C and 30°C, D_{iw} should increase by about the same factor over this temperature range. Furthermore, the influence of the molecule's size is also stronger in Eq. 18-53 than in 18-52 (note $r_i = $ constant $\overline{V}_i^{1/3}$). In Box 18.4 experimental information on the temperature dependence of D_{iw} is compared with the theoretical prediction from Eqs. 18-52 and 18-53.

Box 18.4 Temperature Dependence of Molecular Diffusivity in Fluids

(T is in Kelvin if not stated otherwise)

Due to the increasing kinetic energy of molecules with temperature we expect molecular diffusivities in gases and fluids to increase with T. For ideal gases we found that D_{ia} should be proportional to $T^{3/2}$ (Eq. 18-46). For fluids we can either use the *Stokes-Einstein relation* (Eq. 18-52), the empirical expression by Hayduk and Laudie (Eq. 18-53), or the so-called *activation theory* by Eyring (1936), which envisions molecular diffusion as an activation process that extracts the necessary energy out of the thermal energy pool. Here we will compare the three models with experimental data obtained for trichlorofluoromethane (CCl_3F, CFC-11) by Zheng et al. (1998).

A. Activation Theory (Eyring, 1936)

$$D_{iw}(T) = A_i \exp\left(-\frac{E_{ia}}{RT}\right) \qquad (1)$$

A_i and E_{ia} are compound specific parameters. Thus:

$$\frac{1}{D_{iw}}\frac{dD_{iw}}{dT} = \frac{E_{ia}}{R}\frac{1}{T^2} \qquad (2)$$

Fitting of experimental data for CFC-11 by Zheng et al. (1998) for the temperature range between 0°C and 30°C yields $A_i = 0.015$ cm^{2}s^{-1} and $E_{ia}/R = 2.18 \times 10^3$ K. The deviation between the fitted curve and the data is less than 3%. For comparison: Other values for E_{ia}/R are 2.42×10^3 K for CFC-12 (Zheng et al., 1998) and 2.29×10^3 K for CO_2 (Jähne et al., 1987b).

B. Stokes-Einstein Relation (Eq. 18-52)

If the temperature dependence of the molecular radius r_i is neglected we get:

$$\frac{1}{D_{iw}}\frac{dD_{iw}}{dT} = \frac{1}{T} - \frac{1}{\eta_w}\frac{d\eta_w}{dT} \qquad (3)$$

where η_w is dynamic viscosity of water (see table below). Note that according to Eq. 3 all aqueous diffusivities would have the same relative temperature dependence. Because of Eq. 2 this would also imply that the activation energy E_{ia} of all substances would be equal. As shown above for CFC-11, CFC-12, and CO_2, the latter is almost, but not strictly true. Hence, the temperature dependence of the molecular radius r_i, which in Eq. 3 is neglected, must be responsible for the additional variation among different compounds.

C. From Relation by Hayduk and Laudie (Eq. 18-53)

If the temperature dependence of the molar volume is neglected we get:

$$\frac{1}{D_{iw}}\frac{dD_{iw}}{dT} = -1.14 \frac{1}{\eta_w}\frac{d\eta_w}{dT} \qquad (4)$$

Again, the relative variation of D_{iw} with temperature is independent of the substance i.

In the following table the different models are applied to CFC-11. Note the excellent correspondence between the temperature variation calculated by the Stokes-Einstein relation (Eq. 3) and the expression by Hayduk and Laudie (Eq. 4), although both models overestimate the temperature effect compared to the activation model derived from the experimental data (Eq. 2).

Temperature Dependence of Molecular Diffusivity in Water of Trichlorofluoromethane (CFC-11)

T	η_w	$\frac{1}{\eta_w}\frac{d\eta_w}{dT}$	D_{iw} (10^{-6} cm^{2}s^{-1})		$\frac{1}{D_{iw}}\frac{dD_{iw}}{dT}$ (10^{-2}K^{-1})		
(°C)	(10^{-3} kg m^{-1}s^{-1})[a]	(10^{-2}K^{-1})	Measured[b] (Zheng et al., 1998)	Hayduk and Laudie (1974)[c]	Eq. 2 (measured)	Eq. 3	Eq. 4
0	1.787	−3.44	5.11	4.76	2.93	3.81	3.92
5	1.518	−3.16	5.89	5.74	2.82	3.52	3.60
10	1.307	−2.90	6.77	6.80	2.72	3.25	3.31
15	1.139	−2.68	7.74	7.96	2.63	3.03	3.06
20	1.002	−2.49	8.81	9.21	2.54	2.83	2.84
25	0.890	−2.30	9.98	10.54	2.45	2.64	2.62
30	0.797	−2.17	11.26	11.96	2.37	2.50	2.47

[a] Corresponds to centipoise (10^{-2}g cm^{-1}s^{-1}). [b] Measured data are fitted to the expression by Eyring with $A_i = 0.015$ cm^{2}s^{-1}, $E_{ia}/R = 2.18 \times 10^3$ K (Zheng et al., 1998). [c] With molar volume $\bar{V}_i = 92.2$ cm^{3}mol^{-1} (see Illustrative Example 18.1). Note that Zheng et al. (1998) – as do many other authors – use the wrong exponent for η_w (1.4 instead of 1.14). This error originates from a printing error in the abstract of Hayduk and Laudie's original publication. It explains why Zheng et al. get a larger discrepancy between this model and their data.

Let us evaluate Eqs. 18-52 and 18-53 for the case of trichloroethene (TCE) at 25°C. First, we estimate the molecule's size (i.e., its molar volume) using information given in Table 18.2:

$$\overline{V}_{TCE} = 2(C) + 1(H) + 3(Cl)$$
$$= 2\,(16.5) + 1\,(2.0) + 3\,(19.5)$$
$$= 93.5 \text{ cm}^3\text{mol}^{-1}$$

For comparison, in Table 18.3 we have also listed the molar volume of TCE calculated from its liquid density, $M_i/\rho_{iL} = 90.2$ cm^{3}mol^{-1}, and from the method proposed by Abraham and McGowan (1987) given in Box 5.1 (71.6 cm^{3}mol^{-1}). The molecular radius is approximated assuming the molecules are spherical (see Eq. 18-40):

$$r_{TCE} = \left[\frac{3\overline{V}_i}{4\pi N}\right]^{1/3} = \left[\frac{3 \times 93.5 \text{ cm}^3\text{mol}^{-1}}{4 \times 3.142 \times 6.02 \times 10^{23}\,\text{mol}^{-1}}\right]^{1/3} = 3.33 \times 10^{-8} \text{ cm} = 3.33 \times 10^{-10} \text{ m}$$

With $\eta = 0.89 \times 10^{-2}$ g cm^{-1}s^{-1} = 0.89 $\times$ 10^{-3} kg m^{-1}s^{-1}, the Stokes-Einstein relation (Eq. 18-52) yields:

$$D_{TCEw} = \frac{1.381 \times 10^{-23} \text{ kg m}^{-2}\text{s}^{-2}\text{K}^{-1} \times 298.2 \text{ K}}{6 \times 3.142 \times 0.89 \times 10^{-3} \text{ kg m}^{-1}\text{ s}^{-1} \times 3.33 \times 10^{-10} \text{ m}}$$
$$= 7.37 \times 10^{-10} \text{ m}^2\text{s}^{-1} = 7.37 \times 10^{-6} \text{ cm}^2\text{s}^{-1}$$

Likewise, Eq. 18-53 yields:

$$D_{TCEw} = \frac{13.26 \times 10^{-5}}{(0.89)^{1.14} \times (93.5)^{0.589}} \text{ cm}^2\text{s}^{-1} = 1.04 \times 10^{-5} \text{ cm}^2\text{s}^{-1}$$

Generally, Eq. 18-53 yields results that are correct to within 10%.

Due to the relationship by Othmer and Thaker (Eq. 18-53), one expects that liquid-phase diffusivities also can be estimated from values of related compounds:

$$\frac{D_{iw}}{D_{refw}} = \left[\frac{\overline{V}_i}{\overline{V}_{ref}}\right]^{-0.589} \tag{18-54}$$

This compares reasonably with the correlation found in Fig. 18.10*a*.

Again, molar mass is widely employed as relative index of molecular size, and a square-root functionality is often used for simplicity:

$$\frac{D_{iw}}{D_{refw}} \cong \left[\frac{M_i}{M_{ref}}\right]^{-1/2} \tag{18-55}$$

Fig. 18.10*b* indicates that within a class of molecules like benzene derivatives, the inverse relationship of diffusivity to molar mass is evident.

Illustrative Example 18.3   **Estimating Molecular Diffusivity in Water**

Problem

Estimate the molecular diffusion coefficient of dichlorodifluoromethan (CFC-12) in water, D_{iw}, at 25°C, by the following methods:
(a) From the Stokes-Einstein relation
(b) From the molar mass
(c) From the molar volume
(d) From the diffusion coefficient of methane (CH_4) in water
(e) From Hayduk and Laudie's semiempirical expression

See also Illustrative Examples 18.1 and 18.2.

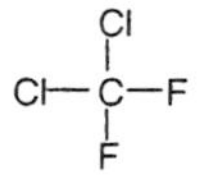

i = CFC-12
M_i = 120.9 g mol^{-1}
$\overline{V}_i$ = 91.0 cm^{3}mol^{-1}
r_i = 3.30 × 10^{-8} cm
Reference substance: methane in water at 25°C:
D_{refw} = 3.0 × 10^5 cm^{2}s^{-1}

Answer (a)

The regression line shown in Fig. 18.10b yields for CFC–12:

$$D_{iw}(cm^2s^{-1}) = \frac{2.7\times10^{-4}}{(M_i)^{0.71}} = \frac{2.7\times10^{-4}}{120.9^{0.71}} = 9.0\times10^{-6}\,cm^2s^{-1}$$

Answer (b)

From the regression line in Fig. 18.10a and $\overline{V}_i$ = 91.0 cm^{3}mol^{-1} follows:

$$D_{iw}(cm^2s^{-1}) = \frac{2.3\times10^{-4}}{\overline{V}_i^{0.71}} = \frac{2.3\times10^{-4}}{91.0^{0.71}} = 9.3\times10^{-6}\,cm^2s^{-1}$$

Answer (c)

From Eq. 18-55 follows:

$$D_{iw} = D_{refw}\left(\frac{M_i}{M_{ref}}\right)^{-1/2} = 3.0\times10^{-5}cm^2s^{-1}\times\left(\frac{120.9}{16}\right)^{-1/2} = 10.9\times10^{-6}cm^2s^{-1}$$

Answer (d)

In order to apply the Stokes-Einstein relation (Eq. 18-52) all variables have to be transformed into the correct units:

$$D_{iw} = \frac{1.381\times10^{-23}\,kg\,m^2s^{-2}K^{-1}\times\;298.2K}{6\times3.142\times0.89\times10^{-3}\,kg\,m^{-1}s^{-1}\times3.30\times10^{-10}\,m}$$

$$= 7.4\times10^{-10}\,m^2s^{-1} = 7.4\times10^{-6}\,cm^2s^{-1}$$

Note that this value is nearly equal to the one calculated earlier for trichloroethene, since the radii of the two molecules are approximately the same.

Answer (e)

The semiempirical relationship by Hayduk and Laudie (Eq. 18-53) yields:

$$D_{iw} = \frac{13.26 \times 10^{-5}}{(0.89)^{1.14} \times (91.0)^{0.589}} \text{ cm}^2\text{s}^{-1} = 10.6 \times 10^{-6} \text{ cm}^2\text{s}^{-1}$$

In summary:

As for Illustrative Example 18.2a (diffusivity of CFC-12 in air), these values agree fairly well with each other, except for the Stokes-Einstein relation, which was not meant to be a quantitative approximation but an expression to show qualitatively the relationship between diffusivity and other properties of both molecule and fluid.

Diffusion Coefficient of CFC-12 in Water at 25°C, D_{iw}

Method	D_{iw} (cm²s⁻¹)
(d) From Stokes-Einstein	7.4×10^{-6}
(a) From molar mass	9.0×10^{-6}
(b) From molar volume	9.3×10^{-6}
(c) From D_w of CH_4	10.9×10^{-6}
(e) From Hayduk and Laudie's relationship	10.6×10^{-6}

18.4   Diffusion in Porous Media

In this section, we consider a solute or vapor diffusing through fluid-filled pores of a porous medium (note that both liquids and gases are called fluids). There are several reasons why in this case the flux per unit bulk area (that is, per total area occupied by the medium) is different from the flux in a homogeneous fluid or gas system.

1. Usually diffusion through the solid matrix is negligible compared to diffusion through the fluids in the pores. Hence, only a reduced area is available for aqueous or gaseous diffusion. If the geometry of the pore space is random and isotropic, the corresponding flux reduction is given by the *porosity*, ϕ, where ϕ is defined as the volume fraction of the pores. (As an example of a nonisotropic porous media, consider the case where the pores are straight tubes aligned in the same direction; then diffusion along an axis perpendicular to the orientation of the tubes is zero, although ϕ is not.)

2. Since the pores are usually not straight, diffusion takes place over a longer distance than in a homogeneous medium. This effect is often described by *tortuosity, τ*, which measures the ratio of the mean path length connecting two arbitrary points and the length of the straight line between these points. Typically τ lies between 1.5 and 6. The porous diffusive flux is reduced by τ relative to the homogeneous flux.

3. Due to the small dimensions of the "channels" in porous media, viscous forces usually suppress turbulence. Hence, diffusion through the pore space occurs by molecular motions. If the size of the pores is small, molecular motions are reduced. In gas-filled pores, this is the case if the pore size is similar to or smaller than the

mean free path of the diffusing molecules (*Knudsen effect*). For the case of liquid-filled pores, it is the additional "drag-like" interactions with the pore wall which affects diffusion (*Renkin effect*).

The following is a short overview of how these effects are quantitatively described. Then we assess their consequences for the two Fickian laws.

Diffusivity in Pores and Fick's Laws

The diffusive flux equation per unit bulk area for chemical i can be written as:

$$F_i = -\phi \, D_{i\,\mathrm{pm}} \frac{\partial C_i}{\partial x} \qquad \left[\mathrm{ML_b^{-2}\,T^{-1}}\right] \qquad (18\text{-}56)$$

where ϕ is porosity $\left[\mathrm{L_f^3\,L_b^{-3}}\right]$

$D_{i\mathrm{pm}}$ is diffusivity of species i in the porous medium $\left[\mathrm{L_b^2\,T^{-1}}\right]$

C_i is concentration in the pore space $\left[\mathrm{ML_f^{-3}}\right]$

x is bulk distance along which F_i is measured $[\mathrm{L_b}]$

Note that for clarity, here we distinguish between the total (or bulk) spatial dimension $[\mathrm{L_b}]$ and the dimension of the fluid-filled pore space $[\mathrm{L_f}]$. Thus porosity, which otherwise would be nondimensional, has dimension $\mathrm{L_f^3\,L_b^{-3}}$.

The *in situ* concentration gradient along the real diffusion path is reduced by tortuosity τ. Thus the *in situ* flux is reduced by the same factor. This effect is incorporated in the porous medium diffusivity $D_{i\mathrm{pm}}$. If the pores are not too narrow, we get:

$$D_{i\mathrm{pm}} = \frac{D_i}{\tau} \qquad (18\text{-}57)$$

where D_i is the "open space" molecular diffusivity of species i in the medium which fills the pores (water, air, etc.).

When we repeat the procedure outlined in Section 18.2 (Eqs. 18-9 to 18-14) to derive Fick's second law, we must remember that the volume in which a concentration change is occurring due to the diffusive flux is reduced by porosity ϕ relative to the bulk volume. Thus, Eq. 18-12 carries an extra factor on the left-hand side:

$$\phi \frac{\partial C_i}{\partial t} = -\frac{\partial F_i}{\partial x} \qquad (18\text{-}58)$$

Inserting Eq. 18-56 yields

$$\phi \frac{\partial C_i}{\partial t} = \frac{\partial}{\partial x} (\phi \, D_{i\,\mathrm{pm}}) \frac{\partial C_i}{\partial x} \qquad (18\text{-}59)$$

If ϕ and $D_{i\mathrm{pm}}$ are constant along x, then after division by ϕ, Eq. 18-59 becomes:

$$\frac{\partial C_i}{\partial t} = D_{i\,\mathrm{pm}} \frac{\partial^2 C_i}{\partial x^2} \qquad (18\text{-}60)$$

This is the same result as in Eq. 18-14 with the only exception that diffusivity is replaced by D_{ipm}. The latter is usually smaller than the homogeneous diffusion coefficient, D_i. One effect is tortuosity (see Eq. 18-57), but sometimes additional influences are important. They are discussed next.

Diffusion in Gas-Filled Pores: Knudsen Effect

If the typical pore diameter, d_p, is of the same order of magnitude as the mean free path λ_i (Eq. 18-42) of the molecules migrating through gas-filled pores, then the molecules will frequently collide with the pore wall and the effective mean free path will be reduced. This effect is accounted for by the nondimensional *Knudsen number*, Kn:

$$Kn = \frac{\lambda_i}{d_p} \tag{18-61}$$

If Kn is of order 1 or larger, λ_i in Eq. 18-7a has to be replaced by the pore diameter, d_p. Hence, Eq. 18-43 is replaced by

$$D_{ipore} = \frac{1}{3} d_p u_i = \frac{1}{3} d_p \left(\frac{8RT}{\pi M_i} \right)^{1/2} \tag{18-62}$$

where u_i is the average three-dimensional molecular velocity (Eq. 18-38). According to Eq. 18-42 and Illustrative Example 18.2, λ_i is of order 10^{-6} cm. If R is inserted in the adequate units ($R = 8.315 \times 10^7$ g cm^{2}s^{-2}mol^{-1}K^{-1}), we get

$$D_{ipore}(cm^2 s^{-1}) = 4.80 \times 10^3 d_p \left(\frac{T}{M_i} \right)^{1/2} \tag{18-63}$$

Since Eq. 18-63 is dimensionally not correct, it is valid only if d_p is given in cm, T in Kelvin, and M_i in g mol^{-1}. Note that in contrast to Eq. 18-46, D_{ipore} is no longer proportional to p^{-1}, and its temperature dependence has changed to $T^{1/2}$ (instead of $T^{3/2}$). The above result is used in Illustrative Example 18.4 to calculate D_{ipore} of dichlorodifluoromethane (CFC-12) in different media.

Diffusion in Liquid-Filled Pores: Renkin Effect

In liquids, the mean free path is typically of the order of 10^{-10} m. Hence the Knudsen effect is not important (i.e., diffusing molecules collide with solvent molecules long before they typically arrive at a pore wall). However, diffusion is affected by a different mechanism, the viscous drag caused by the pore walls. This is known as the *Renkin effect* (Renkin, 1954). In essence, the ratio of pore diffusivity in the liquid-filled pore space and diffusivity in the free liquid, D_{ipore}/D_{ifree}, is a function of the nondimensional parameter

$$q_{iR} = \frac{d_i}{d_p} \tag{18-64}$$

where

      d_i is molecular diameter of the solute
      d_p is pore diameter

There are various empirical and theoretical equations of the form:

$$\frac{D_{ipore}}{D_{ifree}} = \text{function of } q_{iR} = f(q_{iR}) \tag{18-65}$$

(see, e.g., Renkin, 1954; Quinn et al., 1972). However, they all have:

$f(q_{iR}) \sim 1$ for $q_{iR} < 0.001$,

$f(q_{iR}) \sim 0.6$ for $q_{iR} = 0.1$, and

$f(q_{iR}) \to 0$ for $q_{iR} \to 1$.

This means that the Renkin effect sets in when d_p becomes smaller than about $1000\, d_i$ and suppresses diffusion completely when d_p approaches d_i. Diffusing molecules have a typical size of 10^{-8} to 10^{-7} cm. As an example, based on Renkin's original equation, for $q_{iR} = 0.4$ (that is, when the cross-section of the solute covers about $(0.4)^2 = 0.16$ or 16% of the pore opening cross-section), the diffusivity is reduced to 12% of diffusivity in the free liquid.

Diffusion in the Unsaturated Zone of Soils

In the unsaturated zone of soils, only part of the pores is filled with water while the rest is filled with air. Currie (1970) proposed the following expression for the diffusivity of a population of vapor molecules in unsaturated soil:

$$D_{iuz} = \frac{\theta_g^4}{\phi^{5/2}} D_{ia} = \frac{D_{ia}}{\tau_{uz}} \tag{18-66}$$

where

D_{iuz}	is diffusivity in the unsaturated zone
θ_g	is volumetric gas content of the soil; that is, that fraction of the total soil volume which is filled with air
ϕ	is total (air- and water-filled) porosity
D_{ia}	is molecular diffusivity in air
$\tau_{uz} = \phi^{5/2}/\theta_g^4$	is tortuosity of the unsaturated zone

Note that $0 \leq \theta_g \leq \phi \leq 1$. There are numerous other empirical expressions like Eq. 18-66 (see, e.g., Hillel, 1998). They all have in common that D_{iuz} becomes virtually zero if θ_g drops below 0.1. Remember that D_{iuz} of Eq. 18-66 is the diffusivity which appears in Fick's second law (Eq. 18-60). In contrast, D_{ipore}, which we introduced for the Knudsen and the Renkin effect, does not yet include the effect of tortuosity, but replaces D_i on the right-hand side of Eq. 18-57.

Equation 18-66 will be applied in Chapter 19 (Illustrative Example 19.2).

Diffusion of Sorbing Chemicals in Porous Media: Effective Diffusivity

Until now, we have tacitly assumed that the diffusing compound does not interact chemically with the solid matrix of the porous media. Yet, in a porous medium the solid-to-fluid ratio is several orders of magnitude larger than in the open water.

Thus, the sorption of chemicals on the surface of the solid matrix may become important even for substances with medium or even small solid-fluid equilibrium distribution coefficients. For the case of strongly sorbing chemicals only a tiny fraction of the chemical actually remains in the fluid. As diffusion on solids is so small that it usually can be neglected, only the chemical in the fluid phase is available for diffusive transport. Thus, the diffusivity of the total (fluid and sorbed) chemical, the *effective diffusivity* $D_{i\text{eff}}$, may be several orders of magnitude smaller than diffusivity of a nonsorbing chemical. We expect that the fraction which is not directly available for diffusion increases with the chemical's affinity to the sorbed phase. Therefore, the effective diffusivity must be inversely related to the solid–fluid distribution coefficient of the chemical and to the concentration of surface sites per fluid volume.

As shown in Box 18.5 (Eq. 11), $D_{i\text{eff}}$ is given by:

$$D_{i\,\text{eff}} = f_{if}\, D_{i\,\text{pm}} \tag{18-67}$$

The relative fraction of the fluid phase of chemical i, f_{if}, is defined in analogy to Eq. 9-12:

$$f_{if} = \frac{1}{1 + r_{sf} K_{is/f}} \tag{18-68}$$

where

r_{sf} is the solid-to-fluid phase ratio

$K_{is/f}$ is the solid–fluid equilibrium distribution of chemical i

As shown in Box 18.5, $D_{i\text{eff}}$ determines the speed at which a pollutant penetrates a porous media. The so-called *breakthrough time* indicates how long it takes until a pollutant has crossed a porous layer of a given thickness. In contrast, once both the fluid and sorbed concentrations have reached steady-state, the flux is solely controlled by $D_{i\text{pm}}$ and thus independent of sorption (independent of f_{if}). Illustrative Example 18.4 demonstrates the latter while the role of $D_{i\text{eff}}$ is shown in Illustrative Example 18.5.

(*Text continues on page 825*)

Box 18.5 Transport of Sorbing Chemicals in Porous Media and Breakthrough Time

We consider a sorbing chemical i which diffuses in the pore space of a porous media. The pores are filled with a fluid (liquid or gas). For the following discussion, it is helpful to distinguish between the "bulk" spatial dimension $[L_b]$, the spatial dimension of the fluid-filled pore space $[L_f]$, and spatial dimension of the solid phase $[L_s]$. Then, porosity ϕ has dimension $[L_f^3 L_b^{-3}]$, the concentration of i in the fluid, C_{if} $[M_i L_f^{-3}]$, diffusivity $[L_b^2 T^{-1}]$, and depth z $[L_b]$. For the solid phase mass we use the dimension $[M_s]$ in order to distinguish it from the mass of the chemical i $[M_i]$. Note that for the special case of a water-filled porous sediment, the subscript f (fluid) is replaced by w (water). Then the chosen notation is the same as in Chapter 9.

For the different concentrations the following notation is used:

$C_{it} = C_{id} + C_{ip}$ = Total (fluid and sorbed) concentration per bulk volume $[M_i L_b^{-3}]$

C_{id} = Dissolved concentration per bulk volume $[M_i L_b^{-3}]$

$$C_{ip} \qquad\qquad = \text{Sorbed concentration per bulk volume } [M_i L_b^{-3}]$$
$$C_{if} = C_{id}/\phi \qquad = \text{Dissolved concentration per water volume } [M_i L_f^{-3}]$$

$$C_{is} = \frac{C_{ip}}{(1-\phi)\rho_s} \qquad = \text{Sorbed concentration per solid phase mass } [M_i M_s^{-1}]$$

$$\phi \qquad\qquad = \text{Porosity of porous media } [L_f^3 L_b^{-3}]$$
$$\rho_s \qquad\qquad = \text{Density of solid sediment matrix } [M_p L_s^{-3}]$$

The solid matrix is assumed to be uniform such that the solid–fluid equilibrium can be formulated with a single distribution ratio:

$$K_{is/f} = C_{is}/C_{if} \qquad \left[L_f^3 M_s^{-1}\right] \tag{1}$$

The following equations hold:

$$C_{id} = f_{if}\, C_{it} \qquad , \qquad C_{ip} = (1-f_{if})\, C_{it} \tag{2}$$

$$f_{if} = \frac{1}{1 + r_{sf}\, K_{is/f}} \tag{3}$$

where

$$r_{sf} = \frac{\text{solid mass}}{\text{fluid volume}} = \rho_s \frac{1-\phi}{\phi} \qquad \left[M_s L_f^{-3}\right] \tag{4}$$

is the solid-to-fluid phase ratio (see Eq. 19-15 where f is replaced by w). In the sediment column, C_{id} and C_{ip} are described by the following expressions (see Eq. 18-60):

$$\frac{\partial C_{id}}{\partial t} = D_{ipm} \frac{\partial^2 C_{id}}{\partial z^2} - J_{sorption} \qquad \left[M_i L_b^{-3} T^{-1}\right] \tag{5}$$

$$\frac{\partial C_{ip}}{\partial t} = J_{sorption} \qquad \left[M_i L_b^{-3} T^{-1}\right] \tag{6}$$

$J_{sorption}$ describes the amount of chemical being sorbed (or desorbed, if $J_{sorption} < 0$) per unit bulk volume and time. From Eq. 2 (if f_{if} = constant):

$$J_{sorption} = \frac{\partial C_{ip}}{\partial t} = (1-f_{if})\frac{\partial C_{it}}{\partial t} = \frac{1-f_{if}}{f_{if}}\frac{\partial C_{id}}{\partial t} \tag{7}$$

Inserting into Eq. 5 and rearranging yields:

$$\frac{\partial C_{id}}{\partial t} + J_{sorption} = \left(1+\frac{1-f_{if}}{f_{if}}\right)\frac{\partial C_{id}}{\partial t} = \frac{1}{f_{if}}\frac{\partial C_{id}}{\partial t} = D_{ipm}\frac{\partial^2 C_{id}}{\partial z^2} \tag{8}$$

Thus:

$$\frac{\partial C_{id}}{\partial t} = f_{if}\, D_{ipm}\frac{\partial^2 C_{id}}{\partial z^2} \tag{9}$$

or because of Eq. 2:

$$\frac{\partial C_{it}}{\partial t} = f_{if}\, D_{ipm}\frac{\partial^2 C_{it}}{\partial z^2} = D_{ieff}\frac{\partial^2 C_{it}}{\partial z^2} \tag{10}$$

Thus, the effective diffusivity of the total concentration, C_{it}, of a sorbing chemical in a porous media, $D_{i\,eff}$, is given by:

$$D_{i\,eff} = f_f\, D_{i\,pm} \qquad (11)$$

where f_f is defined in Eq. 3 and D_{ipm} is diffusivity of the chemical in the fluid-filled pore space as defined in Eq. 18-57.

Breakthrough Time

The total flux of the chemical per unit bulk area across the porous media is approximated by the flux in the fluid phase. The latter follows from the first Fickian law (see Eq. 18-56):

$$F_i = -\phi\, D_{i\,pm}\, \frac{\partial C_{if}}{\partial x} \qquad \left[ML_b^{-2}\,T^{-1}\right] \qquad (12)$$

Yet, the *front* of a pollutant moves more slowly through the media since the fluid phase has to "drag along" the sorbed phase. In fact, the penetration time can be estimated with the familiar expression by Einstein and Smoluchowski (Eq. 18-8) by replacing D by the effective diffusivity, D_{ieff}. The time needed to cross a porous layer of thickness d, the so-called *breakthrough time*, is:

$$t_{\text{breakthrough}} = \frac{d^2}{2\,D_{i\,eff}} = \frac{d^2}{2 f_{if} D_{i\,pm}} \qquad (13)$$

Note that once the front has crossed the layer and both the fluid and sorbed concentrations have reached steady-state, the flux becomes independent of f_{if}. If the initial concentration in the porous media is different from zero, it takes less time to load the media to steady-state, and thus the breakthrough time is smaller.

Illustrative Example 18.4

Evaluating the Steady-State Flux of Benzene from Spilled Gasoline through Soil to the Atmosphere

i = benzene
M_i = 78.1 g mol^{-1}

Soil Properties:
Temperature: 15°C
40% porosity of which 75% is filled with air and 25% is filled with water.

Gasoline properties:
Total volume: 4000 L spread over an area of about 1000 m² and 0.01 m thick.
Density: 0.7 g cm^{-3}
Benzene content: 1% by mass
Average molecular mass of gasoline constituents: 100 g mol^{-1}.

Problem

An underground fuel storage tank has recently leaked gasoline below ground such that the hydrocarbon mixture (see Chapter 2) has formed a horizontally extensive layer sitting on a clay lens located 3 meters below the ground surface. When the benzene in the gasoline has had time to develop a steady-state profile in the soil gases from the spill to the ground surface, what will be the outward flux of benzene (g m^{-2}day^{-1}) from the soil surface located over the spill assuming only vertical diffusive transport through the soil air?

Answer

First note that benzene is primarily moving through the gas-filled pores. Diffusion through the water-filled pores is too slow to account for much of the total flux. To calculate the steady-state diffusive flux through the 3-meter-thick gas-filled pores, use Eq. 18-56 and replace D_{ipm} by diffusivity in the unsaturated zone, D_{iuz}, and ϕ by θ_g. The latter is the gas-filled void which amounts to 75% of the 40% total porosity. That is, $\theta_g = 0.30$.

$$F_i = -\theta_g\, D_{i\,uz}\, \frac{\partial C_i}{\partial z} \qquad \left[ML_b^{-2}\,T^{-1}\right] \qquad (1)$$

Next, you need the diffusivity of benzene in the unsaturated zone, $D_{i\ uz}$, which according to Eq. 18-66 is given by $D_{i\ air}/\tau_{uz}$. For the soil, estimate τ_{uz} from $(\phi)^{5/2}/(\theta_g)^4$ (Eq. 18-66), $\theta_g = 0.3$, and $\phi = 0.4$ Hence, τ_{uz} is 12.

To get the value of benzene's diffusivity in air at 15°C, we use this compound's experimentally reported value at 25°C of 0.096 cm²s⁻¹ and adjust this using the ratio: $(288.2K/298.2K)^{1.75}$ as indicated by the empirical relation of Fuller et al. (Eq. 18-44). This gives $D_{ia}(15°C)$ of 0.090 cm²s⁻¹. Combining with our previous deduction about the effect of tortuosity, we finally have $D_{iuz} = 0.0075$ cm²s⁻¹. This is equivalent to 7.5×10^{-7} m²s⁻¹.

For the concentration gradient at steady-state, the benzene concentration in the soil gas at 3 m depth has the vapor phase value that would be equilibrated with gasoline, and the concentration at the ground surface may be assumed to be zero (i.e., assuming the wind continuously carries away any benzene exiting the ground surface). The vapor-phase partial pressure of benzene at equilibrium with the gasoline is given by (Eq. 6-1):

$$p_i = \gamma_{igasoline}\ x_{igasoline}\ p_{iL}^*\ (15°C) \qquad (2)$$

Given the chemical similarity of benzene and the other hydrocarbons of which gasoline is made, we expect $\gamma_{igasoline}$ to be near 1. Indeed empirical evidence (see Section 7.5) suggests a value of 2 would be reasonable.

The benzene mole fraction in the gasoline is given by:

$$x_{igasoline} = \frac{1\ g_{benzene}/78\ g\ mol^{-1}}{100\ g_{gasoline}/100\ g\ mol^{-1}} = 0.013$$

Finally you need the vapor pressure of benzene at 15°C. Insert $p_{iL}^*(298\ K) = 0.13$ bar (Appendix C) and $\Delta_{vap}H_i = 34$ kJ mole (Table 6.3) into Eq. 4-30 to get:

$$p_{iL}^*(288\ K) = 0.13 \cdot e^{-\frac{34'000}{8.31}\left(\frac{1}{288}-\frac{1}{298}\right)} = 0.08\ bar$$

Inserting this vapor pressure together with the results above for activity coefficient and mole fraction into Eq. 2 yields the partial pressure of benzene expected in the soil gas adjacent to the gasoline:

$$p_i = (2)(0.013)(0.08)\ bar = 0.002\ bar$$

This corresponds to a vapor concentration of

$$C_i^0 = p_i/RT = 0.002/[(0.083)(288)] = 8.3 \times 10^{-5}\ mol\ L^{-1}$$

$$= 8.3 \times 10^{-2}\ mol\ m^{-3}$$

Now we have the steady-state concentration gradient (z is pointing upward):

$$\frac{dC_i}{dz} = \frac{0\ mol\ m^{-3} - 8.3 \times 10^{-2}\ mol\ m^{-3}}{3m} = -2.8 \times 10^{-2}\ mol\ m^{-4}$$

Finally, putting all the pieces together, we estimate the flux out of the ground by using Eq. 1:

$$F_i = -\theta_g D_{i\,uz} \frac{dC_i}{dz} = (-0.3 \times 7.5 \times 10^{-7} \, \text{m}^2\text{s}^{-1}) \times (-2.8 \times 10^{-2} \text{mol m}^{-4})$$

$$= 6.3 \times 10^{-9} \text{mol m}^{-2}\text{s}^{-1} = 5.4 \times 10^{-4} \text{mol m}^{-2}\text{d}^{-1} = 0.042 \text{ g m}^{-2}\text{d}^{-1}$$

Over a 1000 m^2 area this amounts to about 0.05 kg benzene per day. Since the initial spill involved about 28 kg of benzene (4000 L at 0.7 kg L^{-1} and containing 1% benzene by mass), such fluxes could be sustained for many weeks.

Note that benzene sorption to the soil solids and dissolution in the soil water do not affect the *steady-state flux*, but they influence the time needed until-steady state is reached (see Box 18.5).

Illustrative Example 18.5 | **Interpreting Stratigraphic Profiles of Polychlorinated Naphthalenes in Lake Sediments**

Problem

Gevao et al. (2000) recently reported the presence of polychlorinated naphthalenes in a lake sediment core taken in northwest England (see Table below for profile of pentachloronaphthalenes or PCNs). Like PCBs, these compounds were used by the electric industry as dielectric fluids in transformers and capacitors.

You are interested in the question of when pentachloronaphthalenes were first released in the area near the lake. Assume the lake sediments are always at 10°C, have 90% porosity, and their solids consist of 5% organic carbon.

Assuming a large pulse input in 1960 (i.e., causing a maximum concentration to appear in the 24–25 cm layer), is it reasonable to explain the observed concentrations in the 34–35 and 39–40 cm sections as due to diffusion from above?

i = pentachloronaphthalenes (PCNs)

M_i = 300.5 g·mol^{-1}
log K_{iow} = 6.2 (average value)

Answer

The simplest interpretation of the sedimentary profile is that the deepest layer with the first detectable pentachloronaphthalene corresponds to the first release. That puts the date at about 1936. However, you wonder if the first release could have come later (e.g., in 1960) and downward diffusion actually accounts for the appearance of detectable concentrations in the deeper sediments.

Using the formulation relating the diffusion distance and time (Eq. 18-8):

$$x_{\text{diffusion}} = (2Dt)^{1/2}$$

you see that you need a value of diffusivity for PCN in the sediment bed. It must reflect the effects of tortuosity and sorption. Combining Eqs. 18-57 and 18-67 yields:

$$D_{i\,\text{eff}} = f_{iw} D_{i\,\text{pm}} = f_{iw} D_{iw} / \tau \tag{1}$$

Depth (cm)	Date (year of deposition)	PCN Concentrations (ng kg^{-1} d.w.) [a]
0–1	2000	1800
4–5	1992	4300
9–10	1984	4900
14–15	1976	6200
19–20	1968	12000
24–25	1960	18000
29–30	1952	8000
34–35	1944	120
39–40	1936	120
44–45	1928	0.0
49–50	1920	0.0

[a] d.w. = dry weight.

where in the relative dissolved fraction, f_{iw}, the subscript f has been replaced by w to point out that water is the fluid. Since the bed is so porous, take τ to be 1.

Next, to deduce the fraction of PCN that is dissolved (i.e., f_{iw}), you need the K_{id} for these compounds. Assume that these apolar compounds sorb only to the particulate organic matter, and that sorption can be described by a linear isotherm (Section 9.3). Given the log K_{iow} = 6.2, use Eq. 9-26a (Table 9.2) to estimate K_{ioc}:

$$\log K_{ioc} = 0.74 \log K_{iow} + 0.15 = 4.7$$

Hence, the solid–water partition coefficient can be approximated by (Eq. 9-22):

$$K_{id} = f_{oc} K_{ioc} = 0.05 \times 10^{4.7} = 2500 \text{ mL g}^{-1} = 2500 \text{ cm}^3\text{g}^{-1}$$

Since the porosity of the sediment is 90%, you may also calculate the solid-to-water ratio of the sediment bed from Eq. 9-15:

$$r_{sw} = \rho_s \frac{1-\phi}{\phi} = 2.6 \text{ g cm}^{-3} \frac{0.1}{0.9} = 0.29 \text{ g cm}^{-3}$$

Combining these two results, we find the fraction dissolved:

$$f_{iw} = (1 + r_{sw} K_{id})^{-1} = (1 + 0.29 \text{ g cm}^{-3} \times 2500 \text{ cm}^3\text{g}^{-1})^{-1} = 0.0014$$

The second parameter that you need is D_{iw} (see Eq. 1). Start by estimating the size of PCNs using the approach of Fuller et al. (1966) (Table 18-2):

$$\overline{V_i} = 10 \text{ carbons} + 4 \text{ hydrogens} + 6 \text{ chlorines} - 2 \text{ rings}$$
$$= [(10)(16.5) + (3)(2.0) + (5)(19.5) - (2)(20.2)] \text{ cm}^3\text{mol}^{-1}$$
$$= 228 \text{ cm}^3\text{mol}^{-1}$$

Now using Eq. 18-53 from Hayduk and Laudie (1974), you obtain:

$$D_{iw}(\text{cm}^2\text{s}^{-1}) = \frac{13.26 \times 10^{-5}}{\eta^{1.14}\,\overline{V}_i^{0.589}} = \frac{13.26 \times 10^{-5}}{(1.307)^{1.14} \times 228^{0.589}} = 4.0 \times 10^{-6}\,\text{cm}^2\text{s}^{-1}$$

Insertion of this value together with the result for f_{iw} into Eq. 1 (with $\tau = 1$) then yields:

$$D_{i\text{eff}} = (0.0014)\,(4.0 \times 10^{-6}\,\text{cm}^2\text{s}^{-1}) = 5.6 \times 10^{-9}\,\text{cm}^2\text{s}^{-1}$$

Now you are ready to estimate the relevant diffusion distances. For the deepest depths in which PCNs appear, they are present at $120/18'000 = 0.0067$ of the peak concentration at 24–25 cm. As discussed with respect to Eq. 18-23, this implies that you are interested in the argument of the complementary error function where the erfc$(y_{0.0067}) = 0.0067$. In Appendix A, you find that $y_{0.0067}$ is about 1.9. Thus, you can solve:

$$x_{0.0067} = 1.9 \times 2(D_{i\text{eff}}\,t)^{0.5}$$

$$= 1.9 \times 2 \times (5.6 \times 10^{-9}\,\text{cm}^2\text{s}^{-1} \times 40\,\text{yr} \times 3.15 \times 10^7\,\text{s yr}^{-1})^{0.5} = 10\,\text{cm}$$

Thus, it appears reasonable to expect a small portion of the PCNs (about 1% of the peak concentration) to have diffusively migrated from a layer at about 24–25 cm down to one at about 35 cm. This is similar to the deepest layers in which PCNs are found, and given the uncertainties of the various estimated parameters, it is not too farfetched to argue that release to the lake occurred around 1960 (± 10 years).

18.5   Other Random Transport Processes in the Environment

The concept of diffusion as a process of random transport is not restricted to the case of molecular diffusion in liquids and gases. In fact, diffusion is such a powerful model for the description of transport that it can be applied to processes which extend over more than 20 orders of magnitude. On the very slow side, it can be applied to the migration of molecules in solids. This is especially important for the relative movement of atoms and molecules in minerals which, although being rather slow, becomes effective over geological time scales. At temperatures of several hundred degrees Celsius, as they prevail in the earth's mantle, diffusivities of atoms in minerals (e.g., in feldspar) are of the order 10^{-6} to 10^{-3} cm²s⁻¹. If these values are extrapolated to normal conditions (25°C, 1 bar), diffusivities become as small as 10^{-30} cm²s⁻¹ and less (Lerman, 1979). Diffusion in solids is usually highly anisotropic (i.e., its size depends on the direction relative to the orientation of the crystal lattice).

On the large side, the concept of diffusion also can be applied to macroscopic transport. This process is called *turbulent diffusion*. Turbulent diffusion is not based on thermal molecular motions, but on the mostly irregular (random) pattern of currents in water and air.

A First Glimpse at Turbulent Diffusion

Laminar flow is defined by a set of well-defined, distinct streamlines along which fluid elements flow without exchanging fluid with neighboring elements. Currents

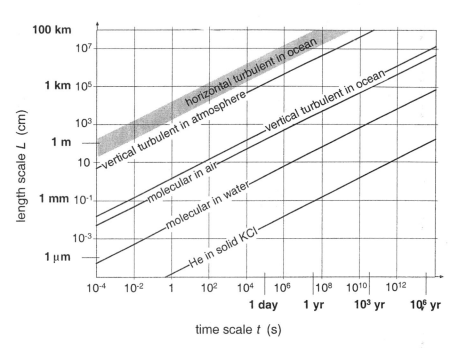

Figure 18.11 Diffusion distance, L, vs. diffusion time, t, for typical diffusivities calculated from the Einstein-Smoluchowski relation $L = (2Dt)^{1/2}$, Eq. 18-8. The following diffusivities, D, are used (values in cm^2s^{-1}): He in solid KCl at 25°C: 10^{-10}; molecular in water: 10^{-5}; molecular in air: 10^{-1}; vertical (turbulent) in ocean: 10^{0}; vertical (turbulent) in atmosphere: 10^{5}; horizontal (turbulent) in ocean: 10^{6} to 10^{8}. Values adapted from Lerman (1979).

of fluids (water, air, etc.) are rarely laminar. The Reynolds number, Re, a nondimensional quantity expressing the ratio between the forces of inertia and of viscosity, respectively, is defined by

$$Re = \frac{d\,v}{\eta_f / \rho_f} \qquad (18\text{-}69)$$

where d is the spatial dimension of the flow system or object around which the flow occurs (m)

  v is the typical flow velocity (m s^{-1})

  η_f is the dynamic viscosity of the fluid (kg m^{-1}s^{-1})

  ρ_f is the density of the fluid (kg m^{-3})

For laminar flow, Re has to be smaller than a critical value of about 0.1. As an example we take $v = 10^{-4}$ m s^{-1}, $(\eta_f/\rho_f) = 10^{-6}$ m^{2}s^{-1} (corresponding to a water temperature of 20°C) and ask how large d can be to keep Re below 0.1. We get $d < Re\,(\eta_f/\rho_f)/v = 0.1 \cdot 10^{-6}$ m^2 s$^{-1}/10^{-4}$ m s$^{-1} = 10^{-3}$ m. Thus, only in the submillimeter pore space of sediments or aquifers would the flow field be purely laminar. But it certainly would not be laminar in an open water body such as a lake or the ocean.

Turbulent flow means that, superimposed on the large-scale flow field (e.g., the Gulf Stream), we find random velocity components along the flow (longitudinal turbulence) as well as perpendicular to the flow (transversal turbulence). The effect of the turbulent velocity component on the transport of a dissolved substance can be described by an expression which has the same form as Fick's first law (Eq. 18-6), where the molecular diffusion coefficient is replaced by the so-called turbulent or eddy diffusion coefficient, E. For instance, for transport along the x-axis:

$$F_x = -E_x \frac{dC}{dx} \qquad [ML^{-2}T^{-1}] \qquad (18\text{-}70)$$

and similarly for the other components. As in Eq. 18-7, we can visualize E_x to result from the product of a typical velocity (the turbulent velocity v_{turb}) and a typical mean free path (λ_{turb}). Turbulent diffusivity is much larger than molecular diffusivity since the distances over which a molecule is transported by turbulence are much larger than the distances between molecular interactions, although the turbulent velocities are much smaller than molecular velocities. In fact, typical horizontal turbulent diffusion coefficients in lakes and oceans are between 10^2 and 10^8 cm^{2}s^{-1}. Compared to typical molecular diffusivities in water of 10^{-5} cm^{2}s^{-1}, turbulent diffusion in water can be up to 10^{13} times larger. In the atmosphere molecular diffusion is of order 10^{-1} cm^{2}s^{-1}, while horizontal turbulent diffusion is 10^8 cm^{2}s^{-1} or greater.

In natural systems (lakes, oceans, atmosphere) turbulent diffusion is usually anisotropic (i.e., much larger in the horizontal than vertical direction). There are two main reasons for that observation: (1) the extension of natural systems in the horizontal is usually much larger than in the vertical. Thus, the turbulent structures (often called eddies) that correspond to the mean free paths of random motions often look like pancakes; that is, they are flat along the vertical axis and mainly extended along the horizontal axes. (2) Often the atmosphere or the water body in a lake or ocean is density stratified (i.e., the density increases with depth). This compresses the eddies even further in the vertical. Gravitational forces keep the water parcels from moving too far away from the depth where they are neutrally buoyant, that is, where they have the same density as their environment. Thus, the anisotropic shape of the eddies results in turbulent diffusivities which differ in size along different spatial directions.

Diffusion Length Scales

The gradient-flux model to describe turbulent diffusion (Eq. 18-70) has the disadvantage that turbulent diffusivity, E_x, is scale dependent. As discussed in more detail in Chapter 22, in natural systems E_x increases with increasing horizontal scale of diffusion. This means that the speed with which two fluid parcels are separated by turbulence increases the further they are from each other. This is because turbulent structures (eddies) of increasing size become effective when the size of a diffusing patch becomes larger. Typical ranges of turbulent diffusivities in the environment are summarized in Table 18.4.

The relation between length and time scales of diffusion, calculated from the Einstein-Smoluchowski law (Eq. 18-8), are shown in Fig. 18.11 for diffusivities between 10^{-10} cm^{2}s^{-1} (helium in solid KCl) and 10^8 cm^{2}s^{-1} (horizontal turbulent diffusion in the atmosphere). Note that the relevant time scales extend from less than a millisecond to more than a million years while the spatial scales vary between 1 micrometer and a hundred kilometers. The fact that all these situations can be described by the same gradient-flux law (Eq. 18-6) demonstrates the great power of this concept.

Dispersion

In a unidirectional flow field (e.g., in a river or aquifer), there is an additional process of random transport called dispersion. Dispersion results from the fact that the velocities in adjacent streamlines are different. For instance, in a river the current velocities in the middle of the river bed are usually larger than on the sides. Due to lateral turbulence,

Table 18.4 Typical Molecular and Turbulent Diffusivities in the Environment

System	Diffusivity (cm^{2}s^{-1}) [a]
Molecular	
In water	$10^{-6} - 10^{-5}$
In air	10^{-1}
Turbulent in Ocean	
Vertical, mixed layer [b]	$0.1 - 10^4$
Vertical, deep sea	$1 - 10$
Horizontal [c]	$10^2 - 10^8$
Turbulent in Lakes	
Vertical mixed layer [b]	$0.1 - 10^4$
Vertical, deep water	$10^{-3} - 10^{-1}$
Horizontal [c]	$10^1 - 10^7$
Turbulent in Atmosphere	
Vertical	$10^4 - 10^{-5}$
Note: Horizontal transport mainly by advection (wind)	
Mixing in Rivers [d]	
Turbulent vertical	$1 - 10$
Turbulent lateral	$10 - 10^3$
Longitudinal by dispersion	$10^{-5} - 10^6$

[a] 1 cm^{2}s^{-1} = 8.64 m^{2}d^{-1}. [b] Maximum numbers for storm conditions. [c] Horizontal diffusivity is scale dependent, see Chapter 22. [d] See Chapter 23.

water parcels randomly switch between different streamlines. Thus, water parcels which spend more time on fast streamlines than on slow ones travel faster along the flow field, and *vice versa*. Therefore, a concentration cloud of a chemical dumped into a river within a short time period is transformed into an elongated cloud while it travels downstream.

The mathematical description of dispersion will be introduced in Section 22.4. Illustrative Examples follow in the chapters on rivers (Chapter 24) and on groundwater (Chapter 25).

18.6   Questions and Problems

Questions

Q 18.1

Explain the difference between the Bernoulli coefficient, $p(n,m)$, and the function $p_n(m)$ of Eq. 18-2.

Q 18.2

By which mathematical or physical principle are the Bernoulli coefficients and Fickian diffusion linked?

Q 18.3

Explain the relationship between Fick's first and second law.

Q 18.4

Determine the sign of $\partial C/\partial t$ at location x resulting from Fick's second law for the concentration profiles shown in Fig. 18.3. Can you develop an easy rule to assess the sign for an arbitrary curve at an arbitrary location?

Q 18.5

Derive analytical expressions for the flux F and the temporal concentration change $\partial C/\partial t$ for the following one-dimensional concentration distributions: (a) $C(x) = a + bx$; (b) $C(x) = a - bx - cx^2$; (c) $C(x) = C_o \exp(-ax)$; (d) $C(x) = a \sin(bx)$. The parameters a, b, c and diffusivity D are constant and positive.

Q18.6

Expand Fick's law and Gauss' theorem (Eq. 18-12) to three dimensions and derive Fick's second law for the general situation that the diffusivities D_x, D_y, and D_z are not equal (*anisotropic diffusion*) and vary in space. Show that the result can then be reduced to Eq. (1) of Box 18.3 provided that D is isotropic ($D_x = D_y = D_z$) and spatially constant.

Q 18.7

Make a qualitative sketch of the concentration profiles defined by Eqs. 18-22 and 18-28 and explain the physical reason that makes them different.

Q 18.8

Explain qualitatively the semiempirical relation of Fuller et al. (1966) for molecular diffusivity in air (Eq. 18-44). How is this expression related to the molecular theory of gases?

Q 18.9

Why can the diffusivity ratio of two chemicals in water be approximated by a power law of their molecular mass ratio? What is the exponent of the power law?

Q 18.10

Is it possible that the molecular diffusive flux in water along the x-axis is different from zero for a chemical that has constant concentration along x? Explain!

Q 18.11

What makes diffusion through a porous medium different? How do the relevant effects differ for diffusion in air and in water, respectively?

Q 18.12

Why is turbulent diffusion in oceans and lakes usually anisotropic? Explain the term *anisotropy* both in mathematical and normal language.

Q 18.13

Explain the difference and similarity between turbulent diffusion and dispersion.

Problems

P 18.1 Bernoulli Coefficients and Random Walk

(a) Calculate the Bernoulli coefficients for $n = 8$, $k = 0, \pm 2, \ldots$ and complete Fig. 18.1 to 8 time steps. Compare the coefficients with the corresponding approximative normal distribution function. (b) Show that the gradient flux law is valid at an imaginary wall at $m = -3$ for the time steps $n = 6$ to $n = 8$. Check signs carefully!

P 18.2 First and Second Derivatives

Draw concentration profiles, $C(x)$, with the following properties: (a) $C' = 0$; (b) $C' \neq 0$, $C'' = 0$; (c) all combinations of $\pm C' > 0$, $\pm C'' > 0$; (d) $C' = 0$, $C'' = 0$, $C'''\neq 0$. (Notation: $C' = \partial C/\partial x$, $C'' = \partial^2 C/\partial x^2$, $C''' = \partial^3 C/\partial x^3$)

P 18.3 Nonconstant Diffusivity

Calculate the mass flux F and the concentration change $\partial C/\partial t$ for the one-dimensional concentration distribution $C(x) = a + bx$, if diffusivity increases with x according to $D(x) = D_o + gx$. The parameters, a, b, g, and D_o are constant and positive.

Advanced Topic

trichloromethane (chloroform)

$M_i = 119.4$ g mol^{-1}
D_{iw} (CHCl$_3$) $= 5.5 \times 10^{-6}$ cm^{2}s^{-1}

P 18.4 Decrease of Chloroform Concentration by Diffusion

(a) In order to illustrate the slowness of molecular diffusion a colleague claims that one could place a droplet containing just 100 μg of chloroform into a stagnant water-filled pipe (diameter 2 cm) and it would take at least one month until the concentration along the pipe would nowhere exceed a concentration of 1 mg/L. How long does it really take? *Hint*: Assume that the initial distribution is "point-like" (Eq. 18-15) and then use Eq. 18-16.

(b) How long would it take if the chloroform were put into a three-dimensional water body and molecular diffusion were to occur along all three dimensions?

P 18.5 Temperature and Pressure Dependence of the Molecular Diffusion Coefficient in Air

Estimate the molecular diffusion coefficient of dichlorodifluoromethane (CCl$_2$F$_2$, CFC-12) in air, D_{ia}, at temperature $T = -10°$C and pressure $p = 0.5$ bar. In Illustrative Example 18.2 you calculated D_{ia} at $T = 25°$C, $p = 1$ bar to be about 8.5×10^{-2}cm^{2}s^{-1}.

P 18.6 Molecular Diffusion Coefficient of a PCB Congener in Water, D_{iw}

2',3,4,-trichlorobiphenyl
(TCBP)

(a) You are in desperate need for the molecular diffusion coefficient of 2',3,4-trichlorobiphenyl (TCBP) in water, D_{iw}, at 25°C but you cannot find any value in the literature. You are aware of several approximations. To make sure that they give reliable results you try three of them.

(b) As you know, the physicochemical properties of the various PCB congeners can be very different. Imagine you had to estimate D_{iw} for 2,4,4-trichlorobiphenyl instead. Do the estimated values differ? What is the problem?

(c) Estimate the relative change of D_{iw} of TCBP if the water temperature drops from 25°C to 0°C.

Hint: In case you need values for the dynamic viscosity of water; they are listed in Appendix B as a function of temperature T.

Advanced Topic

P 18.7 *Trifluoroacetic Acid in a Pond*

Trifluoroacetic acid is an atmospheric transformation product of freon substitutes. It enters surface waters by precipitation where it dissociates into its ionic form trifluoroacetate and then remains virtually inert.

We want to estimate how much of the trifluoroacetate (TFA) would diffuse into the pore space of a small pond (volume V, area A) within one year.

$CF_3-COOH \rightleftharpoons CF_3-COO^- + H^+$
trifluoroacetic acid   trifluoroacetate

Pond
$V = 5 \times 10^4 \, m^3$
$A = 2 \times 10^4 \, m^2$
no through-flow

Help: For simplicity assume that at a given time $t = 0$ the TFA concentration suddenly increases to some value, say $C_o = 1$ mg/L. Consider the diffusion of TFA into the sediment after 1 year by first assuming that the lake concentration C_o remains constant. Calculate the fraction of TFA found in the sediment pores, M_{sed}/M_o, where $M_o = VC_o$. If this fraction is not too large, your assumption ($C_o \sim$ constant) is justified. Use an effective molecular diffusivity of TFA in the sediment of $D_{eff} = 10^{-5}$ cm^{2}s^{-1}. Note that sediments mostly consist of water.

P 18.8 *Evaluating the Effectiveness of a Polyethylene Membrane for Retaining Organic Pollutants in a Relatively Dilute Wastewater*

You have been asked to comment on the likely effectiveness of a proposed 1-mm-thick linear low-density polyethylene (LLDPE) plastic sheet for retaining benzene present at ppm levels in a wastewater. To be considered effective, the plastic sheet must retain the benzene for at least 20 years.

Fortunately, you are aware of the study by Aminabhavi and Naik (1998). In that work, the investigators deduced the molecular diffusivities of several alkanes in plastics including LLDPE. Plotting their data as a function of chemical size (here, molar volumes), you have their results as shown in the figure below.

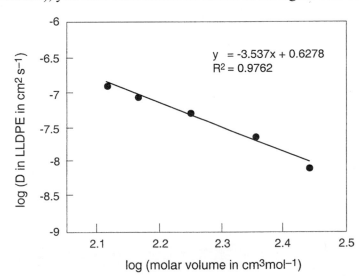

Not surprisingly, the diffusivities in the LLDPE are much less than corresponding diffusivities in water (e.g., $D_{hexane\ LLDPE} = 1.27 \times 10^{-7}$ cm^{2}s^{-1} vs. $D_{hexane\ water} = 0.80 \times 10^{-5}$ cm^{2}s^{-1}). As for other diffusivities, these data indicate that the diffusivities (in cm^{2}s^{-1}) of hydrocarbons in the LLDPE at 25°C are well correlated with the diffusates' molar volumes (in cm^{3}mol^{-1}):

$$\log D_{i\ LLDPE} = -3.5 \log \overline{V_i} + 0.63$$

If you define chemical breakthrough as the time it takes for the half-maximal concentrations to pass across the plastic membrane, will a 1 millimeter thick LLDPE sheet be effective at retaining benzene for 20 years?

P 18.9 *Daydreams During Environmental Organic Chemistry Lectures*

You are an excellent student taking a very exciting subject, Environmental Organic Chemistry. Arriving at the lecture hall early, you take a seat at the very front of the room. The lecture begins after all the students have taken their seats, and except for the wild machinations of your lecturer, the air in the room appears very still.

Suddenly, an attractive individual enters the back of the room and takes a seat there about 10 meters away from you. After only about 1 minute, you notice a pleasant scent, presumably associated with your newly arrived colleague.

Due to your amazing interest in the course topic, you wonder, "What must the horizontal eddy diffusity be in the lecture hall!"

Chapter 19

TRANSPORT THROUGH BOUNDARIES

19.1 The Role of Boundaries in the Environment

19.2 Bottleneck Boundaries
Simple Bottleneck Boundaries
A Simple Noninterface Bottleneck Boundary
Illustrative Example 19.1: *Vertical Exchange of Water in a Lake*
Two- and Multilayer Bottleneck Boundaries
Bottleneck Boundary Between Different Media
Illustrative Example 19.2: *Diffusion of a Volatile Compound from the Groundwater Through the Unsaturated Zone into the Atmosphere*

19.3 Wall Boundaries
Wall Boundary Between Identical Media
Wall Boundary Between Different Media
The Sediment–Water Interface as a Wall Boundary
Box 19.1: *Equilibrium of Sorbing Solutes at the Sediment–Water Interface*
Wall Boundary with Boundary Layer (Advanced Topic)
Illustrative Example 19.3: *Release of PCBs from the Historically Polluted Sediments of Boston Harbor*
Illustrative Example 19.4: *Dissolution of a Non-Aqueous-Phase Liquid (NAPL) into the Aqueous Phase*
Wall Boundary with Time-Variable Boundary Concentration

19.4 Diffusive Boundaries
Dispersion at the Edge of a Pollutant Front
Box 19.2: *Dilution of a Finite Pollutant Cloud Along One Dimension (Advanced Topic)*

19.5 Spherical Boundaries (Advanced Topic)
Bottleneck Boundary Around a Spherical Structure
Sorption Kinetics for Porous Particles Surrounded by Water
Box 19.3: *Spherical Wall Boundary with Boundary Layer*
Finite Bath Sorption
Illustrative Example 19.5: *Desorption Kinetics of an Organic Chemical from Contaminated Sediments*

19.6 Questions and Problems

19.1   The Role of Boundaries in the Environment

Many important processes in the environment *occur at boundaries*. Here we use the term *boundary* in a fairly general manner for surfaces at which properties of a system change extensively or, as in the case of *interfaces*, even discontinuously. *Interface boundaries* are characterized by a discontinuity of certain parameters such as density and chemical composition. Examples of interface boundaries are: the air–water interface of surface waters (ocean, lakes, rivers), the sediment–water interface in lakes and oceans, the surface of an oil droplet, the surface of an algal cell or a mineral particle suspended in water.

In the environment we frequently deal with a more vague kind of boundary. For instance, we have the boundary between the warm, less dense surface layer of a lake

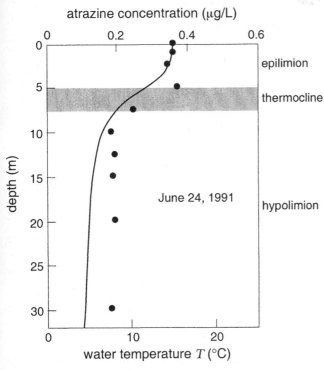

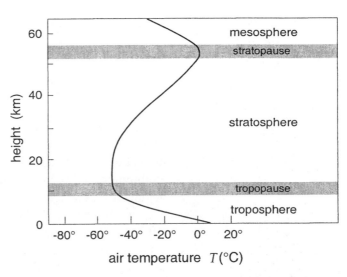

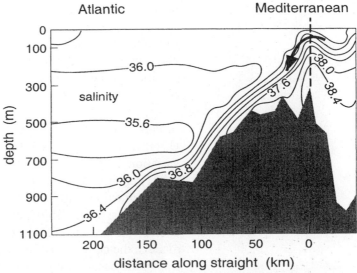

Figure 19.1 Examples of noninterface boundaries. (*a*) The thermocline between the epilimnion and the hypolimnion of Greifensee (Switzerland) characterized by a strong change of water temperature (line) and a corresponding distinct gradient of atrazine concentration (dots), a herbicide. From Ulrich et al., 1994. (*b*) The tropopause is the boundary between the troposphere and the stratosphere while the stratopause separates the stratosphere from the mesosphere. (*c*) The Straight of Gibraltar represents a boundary between the saline water of the Mediterranean and the less saline North Atlantic. The lines denote zones of constant salinity (standard salinity units). From Price et al., 1993.

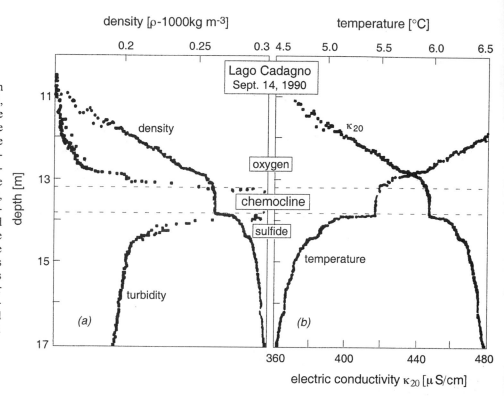

Figure 19.2 The chemocline in Lago di Cadagno (Switzerland), i.e., the boundary between the warm oxic surface water and the cold, anoxic deep water, is the preferential living space for sulfide-oxidizing bacteria (see maximum in the turbidity profile.) Note the abrupt change of temperature, electric conductivity (κ_{20}, reflecting the concentration of dissolved solids), and water density at the lower boundary of the chemocline as well as the rather homogeneous distribution of these parameters within the chemocline. The latter results from microturbulence produced by sinking bacteria and sharpens the gradient just below. From Wüest (1994).

called the *epilimnion* and the cold, denser deep-water layer called the *hypolimnion* (Fig. 19.1*a*). These boundaries are not interfaces in the strict sense but rather transition zones with a certain spatial extension. The boundary region between an epilimnion and a hypolimnion is called a *thermocline* (the zone in which water temperature changes substantially, though continuously). Other examples are the *tropopause*, that is, the boundary between the *troposphere* and the *stratosphere* (Fig. 19.1*b*), or the boundaries between adjacent water bodies which are connected by a straight, for instance, the straight of Gibraltar connecting the Mediterranean with the Atlantic Ocean where salinity and other properties exhibit spatial gradients which are distinctly greater than within the adjacent water bodies (Fig. 19.1*c*).

Boundaries are distinguished from the interior part of a system in two ways: (1) they play a pivotal role controlling the transport of energy and matter, and (2) they control chemical processes triggered by the contact of two systems with different chemical composition.

In Part V of this book we will discuss models of different environmental systems. Usually, such models will be extended in time as well as in space. To describe the variation in space we can either adopt the simpler scheme of box models (see Chapter 21) or introduce one or more continuously varying space coordinates (Chapter 22). Boundaries will be an essential part of both kinds of models. In the former, the boxes are separated by (interface or non-interface) boundaries; their appropriate choice can turn the construction of a model into a piece of art. In the latter kind of models, the continuous functions (such as vertical concentration profiles of chemicals in the ocean) are framed by so-called boundary conditions which can either be defined by a value (e.g., a concentration) or by a flux (e.g., a mass flux across the boundary). This chapter is devoted to the mathematical description of transport processes across different types of boundaries.

Boundaries are characterized by physical and chemical processes. For interface boundaries it is often sufficient to describe the *chemical processes* using equilibrium concepts; Chapters 4 to 7 and 9 deal with important examples. For instance, Henry's law reflects the equilibrium between two molecular fluxes occurring at the air–water interface, the flux of molecules from the gas phase into the water phase, and the inverse flux. Sometimes the transfer processes are combined with *quasi*-instantaneous chemical transformations such as acid/base reactions. Other transfer/reaction processes at boundaries are slower and require a kinetic description (in contrast to the equilibrium partitioning concept). An interesting example occurring at the so-called chemocline, that is, a noninterface boundary observed in Lago di Cadagno (Switzerland), is shown in Fig. 19.2. Here the thermocline is combined with strong chemical gradients produced by the contact between two chemical environments, the oxic surface water and the anoxic deep water which is rich in sulfide (HS^-). In the thin overlapping zone a special type of phototrophic sulfur bacterium (*Chromatium okenii*) that lives on the oxidation of HS^- to elementary sulfur finds ideal conditions of growth. Its concentration is reflected in the turbidity profile. Note the interesting vertical structure of temperature and salinity within the zone in which turbidity (a qualitative measure of biomass concentration) is large. It results from the micro-turbulence produced by the sinking bacteria (Wüest, 1994).

With respect to the *physical processes*, boundaries can be subdivided into just three classes. The distinction will be made according to the nature of the *resistance* to mass transfer across the boundary. We must recognize that this transfer is usually mediated by random motions. Thus, the resistance is like the inverse of a generalized diffusivity or transfer velocity, since both these quantities have the function of a *conductivity* (of mass, heat, momentum, etc.). For simplicity, the following discussion will be focused on the diffusion model (Eq. 18-6), although everything which will be said can also be adapted to the transfer model (Eq. 18-4).

The physical classification of the boundaries is made according to the shape of the generalized *diffusivity profile* across the boundary, $D(x)$, where x is a spatial coordinate perpendicular to the boundary (Fig. 19.3). Three types of boundaries are distinguished:

- bottleneck boundary

- wall boundary

- diffusive boundary

Remember that *diffusivity D* stands for a general parameter with dimension $[L^2T^{-1}]$ appearing in Fick's first law (Eq. 18-6) that relates the flux of a property (concentration, temperature, etc.) to the spatial gradient of the property. It can be a molecular diffusion coefficient, a coefficient of turbulent diffusion, or a dispersion coefficient.

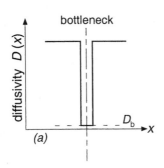

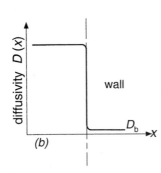

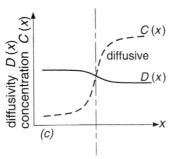

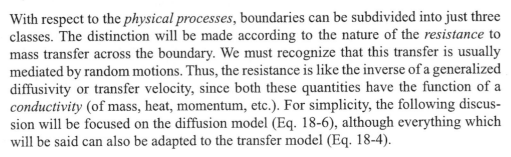

Figure 19.3 Three types of boundaries, each characterized by a different diffusivity profile, $D(x)$, along the spatial coordinate x perpendicular to the boundary. (*a*) bottleneck boundary, (*b*) wall boundary, and (*c*) diffusive boundary given by a distinct change of the water composition, here represented by $C(x)$. See text for further explanations.

The classification of all boundaries into just three types is made possible due to the enormous difference in magnitude of molecular and turbulent diffusivities (see Table 18.4). In Part V of this book, we will learn more about the conditions in which either molecular diffusion or turbulence are relevant. We will discover that as a general principle, transport by turbulence cannot cross interface boundaries (although turbulent kinetic energy can as demonstrated by the production of water waves by the wind). Mass crossing an interface boundary must "squeeze" itself through a zone in which transport occurs by molecular diffusion. If this zone separates two turbulent systems, it plays the role of a bottleneck which controls the overall mass flux. A *bottleneck boundary* (Fig. 19.3*a*) is like a two-lane bridge which separates two sections of a Los Angeles twelve-lane freeway. A typical bottleneck boundary is the water surface of rivers, lakes, and oceans. The atmosphere and the water are usually turbulent; but they are separated by a boundary zone in which transport occurs only by molecular processes.

A different situation is encountered at the bottom of a water body. The sediment–water interface is characterized by, on one side, a water column which is mostly turbulent (although usually less intensive than at the water surface), and, on the other side, by the pore space of the sediment column in which transport occurs by molecular diffusion. Thus, the turbulent water body meets a wall into which transport is slow, hence the term *wall boundary* (Fig. 19.3*b*). A wall boundary is like a one-sided bottleneck boundary, that is, like a freeway leading into a narrow winding road.

Finally at a *diffusive boundary* (Fig. 19.3*c*), diffusivity is of similar magnitudes on either side. Sometimes it is molecular, like at the contact between so-called mobile and immobile zones in groundwater; sometimes it is turbulent, as at the transition from the vigorously mixed surface waters to the thermocline of the ocean. In our picture we can visualize a diffusive boundary by the transition from a paved to an unpaved section of a road. A diffusive boundary may also mark the limit of a concentration patch floating in the current of a river. Here, the picture of a column of cars held back by a roadblock of some kind comes to mind. When the roadblock is removed and the cars begin to move along the road again, an observer from an airplane would see the head of the column (the boundary) becoming more and more spread out while the cars are speeding along the highway.

The distinction between these three types of boundaries will become clearer once we deal with examples. In the following sections the mathematical tools will be derived that are necessary to describe transport across these boundaries. They will then be applied to real environmental boundaries. We will also distinguish between different geometrical shapes of the boundary. Flat boundaries are easier to describe mathematically than spherical boundaries. The latter will be used to describe the exchange between suspended particles or droplets, and the surrounding fluid (algal cells in water, fog droplets in air, etc.). Furthermore, boundaries can be simple (one "layer") or have a multiple structure.

The air–water interface plays a key role among all natural boundaries. It controls the global distribution of many important natural and man-made chemicals (CO_2, CH_4,

CH$_3$SCH$_3$, CFCs, etc.), especially by mass exchange at the surface of the ocean. Therefore, Chapter 20 will be devoted just to this interface.

The approach pursued in this and the next chapter is focused on the common mathematical characteristics of boundary processes. Most of the necessary mathematics has been developed in Chapter 18. Yet, from a physical point of view, many different driving forces are responsible for the transfer of mass. For instance, air–water exchange (Chapter 20), described as either bottleneck or diffusive boundary, is controlled by the turbulent energy flux produced by wind and water currents. The nature of these and other phenomena will be discussed once the mathematical structure of the models has been developed.

19.2   Bottleneck Boundaries

Simple Bottleneck Boundaries

A *simple* bottleneck boundary is characterized by one single zone in which the transfer coefficient, or diffusivity, D, is significantly smaller than in the bulk portion on either side of the boundary. Fig. 19.3a gives a schematic view of a simple bottleneck boundary. In real systems the drop of D to the bottleneck value D_b is usually much smoother than shown in the figure. The term "bottleneck" indicates that the rate-determining step of the transfer across the boundary is controlled by the thin zone in which D is small and thus the resistance against this transport is large. As mentioned before, the most prominent example of a bottleneck boundary is the air–water exchange which determines the flux of chemicals from surface waters (oceans, lakes, rivers) to the atmosphere or vice versa. The physics of this boundary will be discussed in Chapter 20.

Before dealing with this and other examples, let us derive the mathematical tools which we need to describe the flux of a chemical across a simple bottleneck boundary. First, we recognize that for a *conservative* substance at *steady-state*, the flux, $F(x)$, along the boundary coordinate x orthogonal to the boundary must be constant. According to Fick's first law (Eq. 18-6) the flux is given by:

$$F(x) = -D(x)\frac{dC}{dx} = \text{constant} \qquad \left[ML^{-2}T^{-1}\right] \qquad (19\text{-}1)$$

where the notation $D(x)$ indicates that the diffusivity depends on x. In other words, the concentration gradient, dC/dx, must be inversely related to $D(x)$. Since within the bottleneck zone diffusivity, $D(x)$, is much smaller than in the two adjacent zones, the concentration gradient in the bottleneck zone must be much stronger than outside. Therefore, virtually all the concentration variation at the boundary is confined to the bottleneck. In Fig. 19.4 it is assumed that the concentration gradients in zones A and B are negligible compared to the gradient in the bottleneck, that is, the corresponding diffusivities in these zones are infinitely large compared to D_b in the bottleneck zone. Therefore the flux across the bottleneck can be calculated as if the concentrations in the adjacent boxes were held constant at C_A and C_B, respectively. If D_b within the bottleneck is constant,

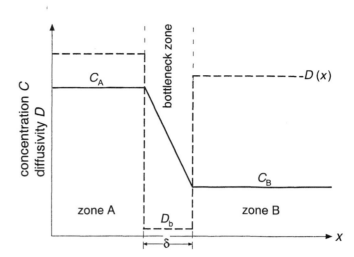

Figure 19.4 Transfer across a simple bottleneck boundary of thickness δ connecting two zones A and B. The solid line is concentration $C(x)$ and the dashed line is diffusivity $D(x)$.

then according to Eq. 19-1 the concentration gradient in the bottleneck must also be constant and equal to:

$$\left(\frac{dC}{dx}\right)_{\text{bottleneck}} = \frac{C_B - C_A}{\delta} \tag{19-2}$$

where δ is the thickness of the bottleneck. Combining Eqs. 19-1 and 19-2 yields:

$$F(x) = -\frac{D_b}{\delta}(C_B - C_A) = -v_b(C_B - C_A) \tag{19-3}$$

where:

$$v_b = \frac{D_b}{\delta} \qquad [\text{LT}^{-1}] \tag{19-4}$$

has the dimension of a velocity and is thus called the (boundary) *exchange or transfer velocity*. The sign of Eq. 19-3 is chosen such that a positive F indicates a flux in the positive x-direction. Since in the example shown in Fig. 19.4, C_B is smaller than C_A, F is positive. The flux (or rather the net flux, since random motions always cause back-and-forth exchange fluxes) is directed from the area of larger concentration (zone A) to the area of smaller concentration (zone B).

When comparing Eqs. 19-1 and 19-3, the reader may remember the discussion in Chapter 18 on the two models of random motion. In fact, these equations have their counterparts in Eqs. 18-6 and 18-4. If the exact nature of the physical processes acting at the bottleneck boundary is not known, the transfer model (Eqs. 18-4 or 19-3) which is characterized by a single parameter, that is, the transfer velocity v_b, is the more appropriate (or more 'honest') one. In contrast, the model which started from Fick's first law (Eq. 19-1) contains more information since Eq. 19-4 lets us conclude that the ratio of the exchange velocities of two different substances at the same boundary is equal to the ratio of the diffusivities in the bottleneck since both substances encounter the same thickness δ. Obviously, the bottleneck model will serve as one candidate for describing the air–water interface (see Chapter 20). However, it will turn out that observed transfer velocities are usually *not* proportional to molecular diffusivity. This demonstrates that sometimes the simpler and less ambitious model is more appropriate.

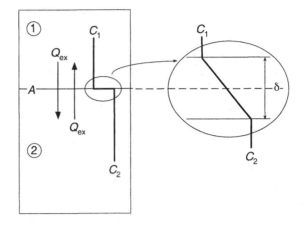

Figure 19.5 In multibox models, the exchange between two fairly homogeneous regions is expressed as the exchange flux of fluid (water, air etc.), Q_{ex}. Normalization by the contact area, A, yields the exchange velocity $v_{ex} = Q_{ex}/A$. This quotient can be interpreted as a bottleneck exchange velocity: $v_{ex} = D_b/\delta$.

A Simple Noninterface Bottleneck Boundary

Lakes and oceans are often vertically stratified. That is, two or more fairly homogeneous water layers are separated by zones of strong concentration and density gradients. In Chapter 21, two- and multibox models will be developed to describe the distribution of chemicals in such systems. In these models, volume fluxes, Q_{ex}, are introduced to describe the exchange of water and solutes between adjacent boxes (Fig. 19.5). Q_{ex} has the same dimension as, for instance, the discharge of a river, $[L^3T^{-1}]$. The net mass flux, ΣF_{net}, from box 1 into box 2 is given by:

$$\Sigma F_{net} = Q_{ex}(C_1 - C_2) \quad [MT^{-1}] \tag{19-5}$$

ΣF_{net} can be normalized by the area of the interface, A:

$$F_{net} = \frac{\Sigma F_{net}}{A} = \frac{Q_{ex}}{A}(C_1 - C_2) = v_{ex}(C_1 - C_2) \quad [ML^{-2}T^{-1}] \tag{19-6}$$

where $v_{ex} = Q_{ex}/A$ is the exchange or transfer velocity of Eq. 18-4. As in Eq. 19-4, v_{ex} can be interpreted as the quotient of the coefficient of diffusivity in the boundary zone, D_b, and the interface thickness δ (e.g., the thickness of the thermocline). This interpretation of Q_{ex} will be useful for the design of box models (see following chapters). A first application is given in Illustrative Example 19.1.

Illustrative Example 19.1   **Vertical Exchange of Water in a Lake**

Problem

Vertical temperature profiles in Greifensee, a lake near Zurich (Switzerland) with a surface area of 8.6 km^2 and a maximum depth of 32 m, show a distinct thermocline during summer and autumn (see figure). Imboden and Emerson (1978) determined the coefficient of vertical turbulent diffusion, E_z, to lie between 0.01 and 0.04 cm^{2}s^{-1} during this time of the year.

Estimate the bottleneck exchange velocity, v_{ex}, between the upper water volume (epilimnion) and the lower one (hypolimnion) of Greifensee and determine the mean water residence time of the water in the hypolimnion provided that vertical turbulent diffusion is the only exchange process between hypolimnion and epilimnion. Use an average thermocline thickness $\delta = 4$m.

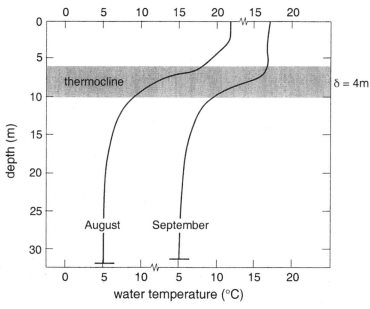

Orographic data of Greifensee (Switzerland) (taken from Table 21.1)

Volume:

Total, V_{tot} = 150×10^6 m³
Epilimnion, V_E = 50×10^6 m³
Hypolimnion, V_H = 100×10^6 m³

Area:

Surface, A_0 = 8.6×10^6 m²
At thermocline, A_{th} = 7.5×10^6 m²

Answer

From Eq. 19-4:

$$v_{ex} = \frac{E_z}{\delta} = \frac{(0.01 \text{ to } 0.04) \text{ cm}^2\text{s}^{-1}}{400 \text{ cm}} = (0.25 \text{ to } 1) \times 10^{-4} \text{ cm s}^{-1} = (2 \text{ to } 8) \times 10^{-2} \text{ m d}^{-1}$$

The vertical volumetric exchange flux of water between epilimnion and hypolimnion is:

$$Q_{ex} = A_{th} \cdot v_{ex} = 7.5 \times 10^6 \text{ m}^2 \times (0.02 \text{ to } 0.08) \text{ m d}^{-1}$$
$$= (1.5 \text{ to } 6) \times 10^5 \text{ m}^3\text{d}^{-1}$$

The mean water residence time is equal to the water volume divided by the volumetric flux:

$$\tau_H = \frac{V_H}{Q_{ex}} = \frac{100 \times 10^6 \text{ m}^3}{(1.5 \text{ to } 6) \times 10^5 \text{ m}^3\text{d}^{-1}} = 170 \text{ to } 670 \text{ d}$$

Note that τ_H is the inverse of the mean flushing rate constant $k_{w,H}$ defined according to Eq. 12-50. Yet, the upper limit of τ_H is hypothetical, since lakes like Greifensee are usually completely mixed during the winter.

Two- and Multilayer Bottleneck Boundaries

As a somewhat more complicated but also more realistic case, we consider the bottleneck to consist of two zones of (different) diffusivities, D_b^A and D_b^B, lying in between the two zones characterized by quasi-infinite D values and constant

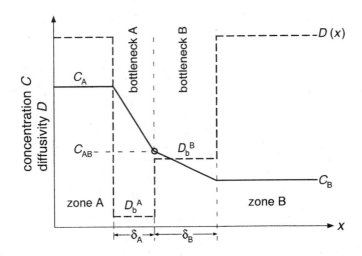

Figure 19.6 Transfer across a two-layer bottleneck boundary. Diffusivity in the bottleneck zone A is smaller than in zone B, but diffusivity in both zones is much smaller than in the adjacent bulk zones. The solid line is concentration $C(x)$, the dashed line is diffusivity $D(x)$. Note that here we assume that the phases in both zones are identical. In Fig. 19.7 the model is extended to a boundary with phase change.

concentrations, C_A and C_B (Fig. 19.6). We want to calculate the flux across this "sandwich" of bottlenecks. If we knew the concentration at the boundary between the two bottlenecks, C_{AB}, we could immediately write down the flux across either bottleneck. For instance, across bottleneck A:

$$F_{\text{bottleneck A}} = -D_b^A \frac{C_{AB} - C_A}{\delta_A} \qquad (19\text{-}7)$$

Thus, we ought to know how the total concentration change from zone A to B, $(C_A - C_B)$, is partitioned between the two bottlenecks. Remember that at steady-state the flux of a conservative substance along x must be constant (Eq. 19-1), thus:

$$F_{\text{bottleneck A}} = F_{\text{bottleneck B}}$$

or:

$$-\frac{D_b^A}{\delta_A}(C_{AB} - C_A) = -\frac{D_b^B}{\delta_B}(C_B - C_{AB}) \qquad (19\text{-}8)$$

Solving for C_{AB} and using the definitions:

$$v_b^A = \frac{D_b^A}{\delta_A} \qquad ; \qquad v_b^B = \frac{D_b^B}{\delta_B} \qquad (19\text{-}9)$$

yields after some algebraic manipulations:

$$C_{AB} = \frac{v_b^A C_A + v_b^B C_B}{v_b^A + v_b^B} \qquad (19\text{-}10)$$

Inserting Eq. 19-10 into one of the flux equations (e.g., Eq. 19.7) yields:

$$F = F_{\text{bottleneck A}} = F_{\text{bottleneck B}} = -\frac{v_b^A v_b^B}{v_b^A + v_b^B}(C_B - C_A) \qquad (19\text{-}11)$$

which can also be written in the form:

$$F = -v_{\text{tot}}(C_B - C_A) \qquad (19\text{-}12)$$

with:

$$\frac{1}{v_{tot}} = \frac{v_b^A + v_b^B}{v_b^A v_b^B} = \frac{1}{v_b^A} + \frac{1}{v_b^B} \tag{19-13}$$

Since the inverse of the transfer velocity, $(v_b)^{-1}$, can be interpreted as a *transfer resistance*, Eq. 19-13 expresses a simple, but very important rule: *the total resistance of two bottlenecks is equal to the sum of the resistances of the single bottlenecks.* This result can be easily generalized to three and more bottleneck zones (see Problem 19.3).

Bottleneck Boundary Between Different Media

Until now we have tacitly assumed that the boundary separates two identical media, for instance, hypolimnetic and epilimnetic water bodies, as in Illustrative Example 19.1. We can intuitively understand that (as stated by Eqs. 19-3 and 19-12) the net flux across the boundary is zero if the concentrations are equal on either side. Yet, how do we treat a boundary separating different phases, for instance, water and air? Obviously, the equations have to be modified since we cannot just subtract two concentrations, C_A and C_B, which refer to different phases, for instance, mole per m³ of water and mole per m³ of air. In such a situation the equilibrium (no flux) condition between the two phases is not given by $C_A = C_B$.

In order to extend the above theory, we have to make use of the equilibrium partition concepts described in Part II. Here we generally define the equilibrium partition function between any phase B and A by:

$$K_{B/A} = \left(\frac{C_B}{C_A}\right)_{equil} \tag{19-14}$$

For instance, if C_A is concentration in water in $mol\ m_{water}^{-3}$ and C_B concentration in air in $mol\ m_{air}^{-3}$, then $K_{B/A}$ is the (nondimensional) Henry's law constant, K_{iaw} (Eqs. 6-6 and 6-15).

At this point, we have to decide which system (A or B) is selected as the reference phase. Our choice determines the actual form of the overall transfer law and explains the asymmetry between the two phases which we meet, for instance, in the equations expressing air–water exchange (see Chapter 20, Eq. 20-3). Here we choose A as the reference system. Then:

$$C_A^{eq} = \frac{C_B}{K_{B/A}} \tag{19-15}$$

is the A-phase concentration in equilibrium with the B-phase concentration, C_B. In fact, C_A^{eq} is a property of system B, which is expressed in terms of the A-phase concentration scheme. For the example of air–water exchange, C_A^{eq} is the aqueous concentration which at equilibrium is "imprinted" by the atmospheric concentration (or partial pressure) of the substance under consideration.

Next we examine how the two-layer bottleneck exchange model (Eqs. 19-7 to 19-13) is modified by introducing a second phase. Fig. 19.7 sketches the concentration

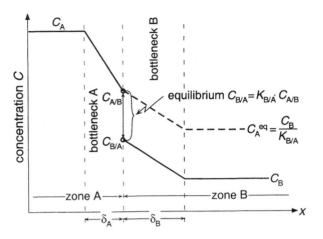

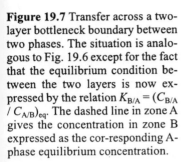

Figure 19.7 Transfer across a two-layer bottleneck boundary between two phases. The situation is analogous to Fig. 19.6 except for the fact that the equilibrium condition between the two layers is now expressed by the relation $K_{B/A} = (C_{B/A} / C_{A/B})_{eq}$. The dashed line in zone A gives the concentration in zone B expressed as the cor-responding A-phase equilibrium concentration.

profile across the two-layer bottleneck boundary. Compared to Fig. 19.6, the picture is modified only by the fact that at the interface separating the two phases the concentration jumps from $C_{A/B}$ on the A-phase side to $C_{B/A}$ on the B-phase side. The two boundary concentrations are assumed to be at equilibrium. Thus, according to Eq. 19-14:

$$C_{B/A} = K_{B/A}\, C_{A/B} \qquad (19\text{-}16)$$

The dashed line in Fig. 19.7 gives the concentration in zone B expressed as the corresponding A-phase equilibrium concentration. This modified representation is like an extrapolation of the A-phase concentration scheme into system B. In fact, it is the same as considering the variability of activity or fugacity of the chemical, rather than its concentration, through the adjacent media. Consequently, the concentration jump at the phase boundary disappears; the "concentration profile" (or more accurately the chemical activity profile) across the boundary looks like that shown in Fig. 19.6.

From here on the mathematical derivation is as before. First, we have to determine the contact concentration $C_{B/A}$ (or $C_{A/B}$) from the fact that the fluxes on either side of the interface are equal (see Eq. 19-8):

$$F = -\frac{D_b^A}{\delta_A}(C_{A/B} - C_A)$$

$$\qquad (19\text{-}17)$$

$$= -\frac{D_b^B}{\delta_B}(C_B - C_{B/A})$$

Substituting $C_{B/A}$ by $K_{B/A}C_{A/B}$ (Eq. 19-16) and solving for $C_{A/B}$ yields:

$$C_{A/B} = \frac{v_A C_A + v_B C_B}{v_A + v_B K_{B/A}} \qquad (19\text{-}18)$$

where the transfer velocities, v_A and v_B, are defined as in Eq. 19-9. Inserting into Eq. 19-17 and using Eq. 19-15 yields:

$$F = -v_{tot}\left(\frac{C_B}{K_{B/A}} - C_A\right) = -v_{tot}\left(C_A^{eq} - C_A\right) \qquad (19\text{-}19)$$

with:

$$\frac{1}{v_{tot}} = \frac{1}{v_A} + \frac{1}{v_B\, K_{B/A}} \qquad (19\text{-}20)$$

These expressions correspond to Eqs. 19-12 and 19-13, if the following substitutions are made:

$$C_B \to \frac{C_B}{K_{B/A}} \equiv C_A^{eq} \qquad \text{and} \qquad v_B \to v_B\, K_{B/A} \qquad (19\text{-}21)$$

Since system A was chosen as the reference, the variables of the nonreference phase B are modified according to the substitution rule of Eq. 19-21. This is the asymmetry between the two phases that was mentioned earlier.

Eqs. 19-19 and 19-20 represent a powerful tool for the description of multilayer bottleneck boundaries. In fact, the validity of the result extends beyond the special picture of a series of films across which transport occurs by molecular diffusion. Since the transfer velocities, v_A and $v_B K_{B/A}$, can be interpreted as inverse resistances, Eq. 19-20 states that the total resistance of a multilayer bottleneck boundary is equal to the sum of the individual resistances. Note that the resistance of the nonreference phase includes the additional factor $K_{B/A}$. In Problem 19.3, the above result shall be extended to three and more layers.

Two extreme situations of Eq. 19-20 will be discussed:

(1) $v_A \ll v_B\, K_{B/A}$: most of the resistance is located in layer A. Then:

$$v_{tot} \sim v_A \qquad \text{A - layer controlled transfer} \qquad (19\text{-}22)$$

(2) $v_A \gg v_B K_{B/A}$: most of the resistance is located in layer B. Then:

$$v_{tot} \sim v_B K_{B/A} \qquad \text{B - layer controlled transfer} \qquad (19\text{-}23)$$

For the latter case, the flux equation 19-19 becomes:

$$F = -v_B K_{B/A}\left(\frac{C_B}{K_{B/A}} - C_A\right) = -v_B\left(C_B - C_B^{eq}\right) \qquad (19\text{-}24)$$

where $C_B^{eq} = K_{B/A} C_A$ is the B-phase concentration at equilibrium with the A-phase concentration, C_A. In fact, Eq. 19-24 expresses the flux in a mathematical scheme in which now the *chemical in phase B* acts as the reference. Since for the case of Eq. 19-23 the A-phase boundary layer is not relevant for the overall rate of exchange, in this case it is reasonable to use phase B as the reference phase.

The above results will be useful for the two-film model of air–water exchange (Chapter 20). A very different bottleneck boundary, that is, the unsaturated zone of a soil, is discussed in Illustrative Example 19.2.

Illustrative Example 19.2   Diffusion of a Volatile Compound from the Groundwater Through the Unsaturated Zone into the Atmosphere

Problem

Mean aqueous concentrations of trichloroethene (TCE) in a contaminated aquifer were measured to be 25 µg/L. The water table is located 4 m below the soil surface. The saturated zone has a mean thickness of 50 m and an average porosity ϕ of 0.3. Water temperature in the aquifer is 10°C.

(a) Estimate the effective diffusivity in the gas phase of the unsaturated zone at 10°C. The moisture content in the unsaturated zone is 15%, the mean porosity 0.3.

(b) Calculate the vertical diffusive flux of TCE at steady-state from the aquifer through the unsaturated zone into the atmosphere.

(c) Estimate the time to steady-state of the diffusive flux. Assume as a rough estimation that due to sorption of TCE to the soil only 5% of the compound is present in the gas phase ($f_{ia} = 0.05$; see Illustrative Example 11.2).

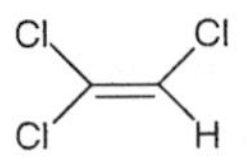

Answer (a)

From the empirical relation by Fuller et al. (Eq. 18-44) and the molar volume of TCE taken from Table 18.3 ($\bar{V} \sim 90$ cm³mol⁻¹) you get:

$$D_a(\text{TCE}, 10°C) = 0.077 \text{ cm}^2\text{s}^{-1}$$

The diffusivity through the unsaturated zone is (Eqs. 18-66):

$$D_{uz} = D_a(\text{TCE}, 10°C)/\tau_g$$

where τ_g is a function of the volumetric gas content, $\theta_g = 0.3 - 0.15 = 0.15$, and of porosity ϕ:

$$\tau_g^{-1} = \frac{\theta_g^4}{\phi^{5/2}} = \frac{(0.15)^4}{0.3^{5/2}} = 0.010$$

Thus:

$$D_{uz} = 7.7 \times 10^{-4} \text{ cm}^2\text{s}^{-1} = 6.7 \times 10^{-3} \text{ m}^2\text{d}^{-1}$$

trichloroethene

$M_i = 131.4$ g mol⁻¹

Aquifer:

Unsaturated zone

Depth, d	=	4 m
Porosity, ϕ	=	0.3
Moisture content	=	15 %
Temperature	=	10°C

Saturated zone

Porosity, ϕ	=	0.3
Water temperature	=	10°C
Aqueous TCE-conc.	=	25 µg/L

Answer (b)

The unsaturated zone can be modeled as a bottleneck boundary of thickness $\delta = 4$ m. The TCE concentration at the lower end of the boundary layer is given by the equilibrium with the aquifer and at the upper end by the atmospheric concentration of TCE, which is approximately zero. Thus, you need to calculate the nondimensional Henry coefficient of TCE at 10°C, $K_{\text{TCE a/w}}(10°C)$.

From Eq. 6-10:

$$\ln\frac{K_{\text{TCE a/w}}(10°C)}{K_{\text{TCE a/w}}(25°C)} = -\frac{\Delta_{aw}H_{\text{TCE}} + RT_{av}}{R}\left(\frac{1}{283} - \frac{1}{298}\right)$$

T_{av}: Average temperature (290 K)

From Appendix C: $K_{TCEa/w}(25°C) = 10^{-0.3} \cong 0.5$

From Table 6.3: $\Delta_{aw}H_{TCE} \cong 37 \text{ kJ mol}^{-1}$

Thus:

$$\ln\frac{K_{TCEa/w}(10°C)}{K_{TCEa/w}(25°C)} = -\frac{37,000 + 8.31 \times 290}{8.31}\left(\frac{1}{283} - \frac{1}{298}\right) = -0.843$$

$$K_{TCEa/w}(10°C) = K_{TCEa/w}(25°C)\,e^{-0.843} = 0.5 \times 0.43 = 0.22$$

Thus:

$$C_a^{eq} = C_w \cdot K_{TCEaw}(10°C) = 0.22 \times 25 \ \mu g L^{-1} = 5.5 \ \mu g L^{-1} = 5.5 \ mg \ m^{-3}$$

The diffusive flux of TCE through the unsaturated zone at steady-state is (see Illustrative Example 18.4, Eq. 1):

$$F = \theta_g D_{uz} \cdot \frac{C_a^{eq}}{\delta} = 0.15 \times 6.7 \times 10^{-3} \, m^2 d^{-1}\,\frac{5.5 \ mg \ m^{-3}}{4 \ m} = 1.4 \times 10^{-3} \ mg \ m^{-2} d^{-1}$$

Answer (c)

According to Box 18.5, Eq. 13, the breakthrough time depends on D_{ieff}, that is, on the mobile fraction of the chemical. Here this fraction is given by $f_{ia} = 0.05$. The layer thickness is equal to the depth of the unsaturated zone, $\delta = 4$ m.

$$t_{break\ through} = \frac{\delta^2}{2 f_{ia} D_{uz}} = \frac{(4\,m)^2}{2 \times 0.05 \times 6.7 \times 10^{-3}\,m^2 d^{-1}} = 24'000 \ d(!)$$

Note: The above example is based on the real case of a polluted aquifer in New Jersey (see Smith et al., 1996). According to detailed investigations on the site, which include the measurement of vertical concentration profiles in the unsaturated zone and flux chamber measurements, the authors conclude that vertical diffusion is not the only process causing vertical outgasing of TCE from the aquifer. Although they occasionally found in the unsaturated zone the linear concentration gradients which are indicative for a diffusive flux, often the profiles were not linear. They invoke the influence of vertical advection caused by air pressure fluctuations to explain these nonlinear profiles. The extremely large breakthrough time which was calculated in (c) is another indication that the flux cannot be purely diffusive. We will come back to this example in Chapter 22 (Illustrative Example 22.4).

19.3   Wall Boundaries

Wall boundaries are defined by an abrupt change of diffusivity $D(x)$ from a large value allowing virtually complete homogeneity to a value that is orders of magnitude smaller (Fig. 19.3b). Examples are the sediment–water interface in lakes and oceans, a spill of a nonaqueous-phase liquid (NAPL) exposed to air, or the surface of a natural particle suspended in water. In this section we deal with flat wall bound-

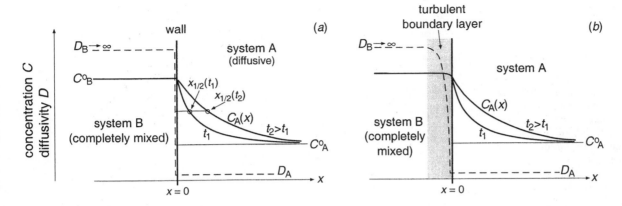

Figure 19.8 Diffusivity D and concentration C at wall boundary. (*a*) Schematic view of a wall boundary. Diffusivity drops abruptly from a very large value D_B, which guarantees complete mixing in system B, to the much smaller value D_A. The concentration penetrates into system A when time t grows. $x_{1/2}(t_i)$ is the "half-concentration depth" (Eq. 18-23) as a function of time. (*b*) In reality the change of D from the well-mixed system B into the diffusive system A is smooth (see text). Yet, the concentration profile in system A is well approximated by the idealized case shown in (*a*).

aries; spherical wall boundaries (the case of the suspended particle) are treated in Section 19.5.

The structure of turbulence in the transition zone from a fully turbulent fluid to a nonfluid medium (often called the *Prandtl* layer) has been studied intensively (see, for instance, Williams and Elder, 1989). Well-known examples are the structure of the turbulent wind field above the land surface (known as the planetary boundary layer) or the mixing regime above the sediments of lakes and oceans (benthic boundary layer). The vertical variation of $D(x)$ is schematically shown in Fig. 19.8*b*. Yet, in most cases it is sufficient to treat the boundary as if $D(x)$ had the shape shown in Fig. 19.8*a*.

Wall Boundary Between Identical Media

The mathematics of diffusion at flat wall boundaries has been derived in Section 18.2 (see Fig. 18.5*a–c*). Here, the well-mixed system with large diffusivity corresponds to system B of Fig. 18.5 in which the concentration is kept at the constant value C_B^o. The initial concentration in system A, C_A^o, is assumed to be smaller than C_B^o. Then the temporal evolution of the concentration profile in system A is given by Eq. 18-22. According to Eq. 18-23 the "half-concentration penetration depth", $x_{1/2}$, is approximatively equal to $(D_A t)^{1/2}$. The cumulative mass flux from system B into A at time t is equal to (Eq. 18-25):

$$\mathcal{M}(t) = \left(\frac{4}{\pi}\right)^{1/2} (D_A t)^{1/2} \left(C_B^o - C_A^o\right) \tag{19-25}$$

where D_A is diffusivity in system A. Note that the cumulative mass flux increases as the square root of the elapsed time t and is unlimited. That is, $\mathcal{M}(t)$ can theoretically increase to infinity. Equilibrium would be reached if the concentration in system A became everywhere C_B^o. Since both systems are assumed to be unbounded, complete equilibrium is never reached. Yet, the flux F into system A, that is, the time derivative of Eq. 19-25:

$$F(t) = \frac{d\mathcal{M}}{dt} = \left(\frac{D_A}{\pi t}\right)^{1/2} \left(C_B^o - C_A^o\right) \tag{19-26}$$

becomes zero for $t \rightarrow \infty$. Note that the equations are also valid if $C_B^o < C_A^o$. Then the flux $F(t)$ is negative; it describes the loss from system A to system B.

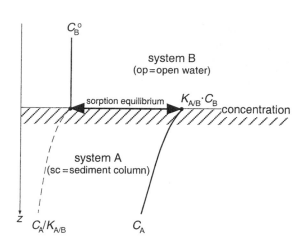

Figure 19.9 Schematic representation of the concentration profile of a compound across a wall boundary with phase change. For the case of the sediment–water interface, C_A is the *total* (dissolved and sorbed) concentration of a chemical in the sediment column (sc) whereas C_B^o represents the constant concentration in the overlying open water (op).

Imagine that system B is the water column of a lake and system A is the pore space of the lake sediments. In B, mixing is by turbulence and fairly intensive while in system A transport is by molecular diffusion. The above case corresponds to a situation in which at time t the concentration of a compound in the water suddenly rises to the value C_B^o. Then Eqs. 19-25 and 19-26 describe the cumulative and incremental mass flux of the compound into the infinitely deep sediment column.

Wall Boundary Between Different Media

The mathematics of the wall boundary model slightly changes if the media on either side of the interface are different. As an example, consider the volatilization of a dissolved chemical into the well-mixed atmosphere from a shallow puddle of water in which advective and turbulent motion is completely suppressed. Another example is the transport between a solid phase and a turbulent water body.

In Section 19.2 we treated the phase problem by choosing a reference system (for instance, water) to which the concentrations of the chemicals in other phases are related by equilibrium distribution coefficients such as the Henry's law constant. Here we employ the same approach. The following derivation is valid for an arbitrary wall boundary with phase change. The mixed system B is selected as the reference system. In order to exemplify the situation, Fig. 19.9 shows the case in which system A represents a sediment column and system B is the water overlying the sediments. This case will be explicitly discussed in Box 19.1.

As in Eq. 19-14, the equilibrium between the two concentrations on either side of the boundary shall be expressed by a general partition function:

$$\left(\frac{C_A}{C_B} \right)_{equilibrium} = K_{A/B} \tag{19-27}$$

System B is well mixed and its concentration kept constant at C_B^o. At $t = 0$, the compound starts to diffuse into system A. For simplicity, we assume that the initial concentration in A is zero, $C_A^o = 0$.

The cumulative and differential mass flux from B to A can be calculated from a slightly modified version of Eqs. 19-25 and 19-26:

$$\mathcal{M}(t) = \left(\frac{4}{\pi}\right)^{1/2} \left(D_A t\right)^{1/2} K_{A/B} \, C_B^o \tag{19-28}$$

$$F(t) = \frac{d\mathcal{M}}{dt} = \left(\frac{D_A}{\pi t}\right)^{1/2} K_{A/B} \, C_B^o \tag{19-29}$$

The additional factor $K_{A/B}$ reflects the fact that the A-side boundary concentration in equilibrium with C_B^o is given by Eq. 19-27. Note that D_A is the diffusivity of the chemical in the A phase, that is D_A is the coefficient which appears in Fick's second law (Eq. 18-14) formulated for the concentration C_A.

The Sediment–Water Interface as a Wall Boundary

For the special case of the sediment–water interface, D_A is determined by the aqueous diffusivity, the sediment structure (porosity, tortuosity, pore size), and the sorption property of the chemical. Let us demonstrate this by applying the theory of transport of sorbing chemicals in fluid-filled porous media, which we have derived in Chapter 18.4 and Box 18.5, to the special case of diffusion in the sediment column. Since for this particular situation the fluid in the pore space is water, the subscript f (fluid) is replaced by w (water) while the superscripts sc and op mean sediment column and open water.

Note that for the total (dissolved and particulate) concentration, C_t, the abrupt change of the solid-to-water-phase ratio, r_{sw} (Eq. 9-15), at the sediment surface acts like a phase change. The numerical example given in Table 19.1 demonstrates that the transition from the open water column of a lake or the ocean to the sediments involves an increase of r_{sw} by 5 to 6 orders of magnitude. Typically, in the open water, r_{sw}^{op} is of order 10^{-3} kg m^{-3} while in the sediment column r_{sw}^{sc} lies between 10^2 and 10^3 kg m^{-3}. Thus, at equilibrium the total (dissolved and sorbed) concentration per unit bulk volume on either side of the interface for compounds with small to moderate solid–water distribution ratios ($K_d < 10$ m^{3}kg^{-1}) is approximatively given by (see Box 19.1, Eq. 4):

$$K_{sc/op} \sim \frac{\phi^{sc}}{f_w^{sc}} = \phi^{sc}\left(1 + r_{sw}^{sc} K_d^{sc}\right) \quad , \quad \text{for} \quad K_d^{op} < 10 \text{ m}^3\text{kg}^{-1} \tag{19-30}$$

where ϕ^{sc} is porosity of the sediment column

$\qquad K_d^{sc}$ is the solid–pore water distribution ratio for a chemical (Eq. 9-7)

$\qquad f_w^{sc} = \left(1 + r_{sw}^{sc} K_d^{sc}\right)^{-1}$ is the dissolved fraction in the sediments (Eq. 9-12)

The equilibrium distribution $K_{sc/op}$ adopts the role of $K_{A/B}$ in the expression for the integrated and specific mass flux across the sediment–water interface (Eqs. 19-28 and 19-29). Since for a strongly sorbing chemical $K_{sc/op}$ can easily exceed 10^3 (see Table 19.1), at first sight Eqs. 19-28 and 19-29 seem to tell us that $\mathcal{M}(t)$ and $F(t)$ are orders of magnitude larger for sorbing than for nonsorbing species. Yet, this conclusion is premature. Remember that in these equations D_A is diffusivity in the A phase.

Table 19.1 Exchange of a Sorbing Chemical at the Sediment–Water Interface (Numerical example for penetration depth due to wall boundary flux).

The following assumptions are typical for lacustrine or marine sediments:

Solid-to-water-phase ratio r_{sw}

  Water column: $r_{sw}^{op} = 1 \text{ mg L}^{-1} = 10^{-3} \text{ kg m}^{-3}$
  (corresponds roughly to the concentration of suspended particles)

  Sediment column: porosity $\phi^{sc} = 0.8$, particle density $\rho_s = 2500 \text{ kg m}^{-3}$

  $r_{sw}^{sc} = \rho_s \dfrac{1 \pm \phi^{sc}}{\phi^{sc}} = 625 \text{ kg m}^{-3}$ (from Eq. 9-15)

Sorption partition coefficient: $K_d^{op} = K_d^{sc} = \begin{array}{ll} 1 \text{ m}^3\text{kg}^{-1} & \textit{Case A} \\ 10^3 \text{ m}^3\text{kg}^{-1} & \textit{Case B} \end{array}$

Effective diffusivity in the sediment: $D_w^{sc} = 10^{-6} \text{ cm}^2\text{s}^{-1}$

Dissolved fraction and equilibrium partition coefficient, $K_{sc/op}$, at sediment–water interface:

		Case A $K_d^{sc} = 1 \text{ m}^3\text{kg}^{-1}$	*Case B* $K_d^{sc} = 10^3 \text{ m}^3\text{kg}^{-1}$
Sediment column	$f_w^{sc} = (1 + r_{sw}^{sc} K_d)^{-1}$	1.6×10^{-3}	1.6×10^{-6}
Open water	$f_w^{op} = (1 + r_{sw}^{op} K_d)^{-1}$	0.999	0.5
$K_{sc/op}$ (Box 19.1, Eq.3)		500	2.5×10^5

Half-penetration depth (Eq. 19-34): $z_{1/2} = (f_w^{sc} D_w^{sc} t)^{1/2}$

t	nonsorbing	*Case A*	*Case B*
1 day	0.3 cm	0.01 cm	4×10^{-4} cm
1 year	6 cm	0.2 cm	7×10^{-3} cm
10^3 year	200 cm	7 cm	0.2 cm

Box 19.1   Equilibrium of Sorbing Solutes at the Sediment–Water Interface

We consider a flat sediment surface which is overlain by a completely mixed water column. A sorbing chemical is exchanged between the water and the sediment. Immediate sorption equilibrium at every local point in space is assumed.

The notation used was introduced in Chapter 9 and in Box. 18.5. Note that compared to the latter the subscript f (fluid) is replaced by w (for water). The superscripts sc and op mean "sediment column" and "open water".

$C_t = C_d + C_p$: Total (dissolved and sorbed) concentration per bulk volume $[\text{ML}_b^{-3}]$

$C_w = C_d / \phi^{sc}$: Dissolved concentration per pore water volume $[\text{ML}_w^{-3}]$

$C_s = \dfrac{C_p}{(1 - \phi^{sc})\rho_s}$: Sorbed (particulate) concentration per particle mass $[\text{MM}_p^{-1}]$

The sorption equilibrium is expressed by (K_d^{sc}: solid–water distribution ratio; Eq. 9-7):

$$K_d^{sc} = C_s / C_w \qquad \left[L_w^3 \, M_p^{-1} \right] \tag{9-7}$$

The boundary condition at the sediment–water interface which relates the diffusion equations on both sides of the boundary, is given by:

$$C_w^{sc} = C_w^{op} \tag{1}$$

For both phases (sc and op) the following definition holds:

$$C_w = \frac{C_d}{\phi} = \frac{f_w}{\phi} C_t \tag{2}$$

where f_w is the relative dissolved fraction of the total concentration C_t (Eq. 9-10). Thus the equilibrium condition for the *total* concentration at the sediment–water interface is (see also Eq. 3 of Box 18.5):

$$K_{sc/op} = \frac{C_t^{sc}}{C_t^{op}} = \frac{\left(\phi^{sc} / f_w^{sc} \right)}{\left(\phi^{op} / f_w^{op} \right)} = \frac{\phi^{sc}}{\phi^{op}} \cdot \frac{f_w^{op}}{f_w^{sc}} = \frac{\phi^{sc} \left(1 + r_{sw}^{sc} K_d^{sc} \right)}{\phi^{op} \left(1 + r_{sw}^{op} K_d^{op} \right)} \tag{3}$$

Since the open water column is nearly pure water ($\phi^{op} = 1$, $r_{sw}^{op} = 0$), for compounds with small to moderate solid–water distribution ratios ($K_d < 10 \text{ m}^3\text{kg}^{-1}$), the above *equilibrium partition coefficient* can be simplified to:

$$K_{sc/op} \approx K_{sc/pure\ water} = \frac{\phi^{sc}}{f_w^{sc}} = \phi^{sc} (1 + r_{sw}^{sc} K_d^{sc}) \tag{4}$$

That is, for the particular case of the sediment column, D_A is diffusivity of the total concentration, C_t, for which the second Fick's law is given by Eq. 10 of Box 18.5. From this equation we see that D_A adopts the form:

$$D_A \rightarrow f_w^{sc} D_w^{sc} = \frac{\phi_{sc}}{K_{sc/op}} D_w^{sc} \tag{19-31}$$

where D_w^{sc} stands for the porous media diffusivity D_{pm}. Remember that due to the effect of porosity diffusivity of the dissolved fraction in the pore space of the sediment column, D_w^{sc}, is usually smaller than diffusivity in the open water (see Eq. 18-57). In addition, in sediments with small pores diffusivity may be further reduced (Renkin effect, see Eq. 18-65).

We now insert Eq. 19-31 into Eqs. 19-28 and 19-29 and substitute $K_{sc/op}$ by Eq. 19-30:

$$\mathcal{M}_{sed}(t) = \left(\frac{4}{\pi} \right)^{1/2} \left(f_w^{sc} D_w^{sc} t \right)^{1/2} \frac{\phi^{sc}}{f_w^{sc}} C_w^{op} \tag{19-32}$$

$$= \left(\frac{4}{\pi} \right)^{1/2} \left(\frac{D_w^{sc} t}{f_w^{sc}} \right)^{1/2} \phi^{sc} C_w^{op}$$

$$F_{sed}(t) = \left(\frac{1}{\pi} \right)^{1/2} \left(\frac{D_w^{sc}}{f_w^{sc} t} \right)^{1/2} \phi^{sc} C_w^{op} \tag{19-33}$$

C_w^{op} is aqueous concentration in the open water column. For a strongly sorbing chemical ($K_{sc/op}$ of order 10^3), f_w^{sc} is of order 10^{-4}, so the apparent diffusivity of the chemical (D_A) is extremely small. This results mainly from the process of sorption (remember that a small f_w^{sc} means a large $K_{sc/op}$) while the effects of porosity and tortuosity on D_A are commonly much weaker. Intuitively we can understand this result. Since only the dissolved fraction of C_t is able to migrate by diffusion and since for a strongly sorbing species this fraction is extremely small, the dissolved fraction must drag the large fraction of the immobile sorbed chemical. Therefore, the diffusive migration of the *total* compound is much slower than that one of the *dissolved* species alone.

The "half-concentration penetration depth," $z_{1/2}$, for a sorbing species is approximately (see Eq. 18-23):

$$z_{1/2} = \left(f_w^{sc} D_w^{sc} t \right)^{1/2} \tag{19-34}$$

A numerical example is given in Table 19.1

To summarize for a sorbing chemical, mass exchange at the sediment–water interface can be treated like the exchange at a wall boundary with phase change. Sorption increases the specific and integrated mass exchange by the factor $(1/f_w^{sc})^{1/2}$; that is, it increases the capacity of the sediment to store the compound. At the same time, it slows down the speed at which the chemical penetrates the sediments (factor $(f_w^{sc})^{1/2}$). Note that the derived equations keep their validity if the chemical moves in the opposite direction, that is, if the total sediment concentration is larger than the concentration at equilibrium with the overlying water.

Advanced Topic

Wall Boundary with Boundary Layer

In the preceding section, the sediment surface was described as an intermedia wall boundary. Thereby we tacitly assumed that the "diffusion wall," that is, the location where diffusivity drops from D_B, to D_A coincides with the interface between the two media. As shown in Fig. 19.8b, the transition from a turbulent to a stagnant media includes a boundary layer in the former in which diffusivity drops in a characteristic manner.

As long as there is no phase change involved, the influence of the transition zone on mass transfer is negligible. The position of the boundary layer is slightly shifted, but the exchange flux is scarcely affected. This is no longer true if the boundary separates two different media, for instance, the water of a lake from the sediments. In this case the drop of diffusivity $D(x)$ and the increase of the partition ratio $K_{A/B}$ (Eq. 19-27) do not coincide (Fig. 19.10). Let us first develop the necessary mathematical tools to describe this new situation and then discuss an example for which the influence of the boundary layer may be relevant.

As before we consider the boundary between the completely mixed system B and the diffusive system A. The initial concentrations are C_A^o and C_B^o. On the B-side of the interface there is a stagnant boundary layer of thickness δ with constant diffusivity D_{bl}. At time $t = 0$, the two systems are brought into contact and mass exchange across the boundary is initiated. Let us assume that at time $t > 0$, the concentration on

Figure 19.10 Schematic view of concentration profile across a wall boundary between different media with a boundary layer of thickness δ on the B-side of the interface. Diffusivities are $D_B \rightarrow \infty$ in the completely mixed system B, $D_{bl} \ll D_B$ in the boundary layer, and D_A in system A. The intermedia equilibrium relationship at the interface is defined by $C_{A/B} / C_{B/A} = K_{A/B}$. C_A^o and C_B^o are initial concentrations (the latter is assumed to remain constant). $C_B^{eq} = C_A^o / K_{A/B}$ is the B-side concentration in equilibrium with C_A^o.

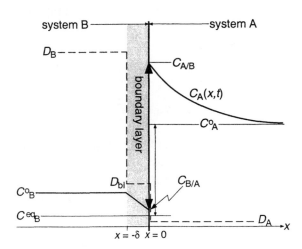

the B-side of the interface is $C_{B/A}$, which is at instantaneous equilibrium with the A-side boundary concentration $C_{A/B}$:

$$K_{A/B} = C_{A/B} / C_{B/A} \qquad (19\text{-}35)$$

where $K_{A/B}$ is the intermedia equilibrium distribution coefficient. If the chemical substance is conservative (nonreactive) on a time scale characterizing the migration of the molecules across the boundary, then the fluxes on either side of the boundary must be equal. On one hand, transport across the B-side boundary layer is (see Eq. 19-3):

$$F_{bl} = -\frac{D_{bl}}{\delta}\left(C_{B/A} - C_B^o\right) \qquad (19\text{-}36)$$

where the sign of F_{bl} has been chosen as usual; that is, such that a positive flux points into the positive x-direction. On the other hand, the flux into system A is (see Eq. 19-26):

$$F_A = \left(\frac{1}{\pi}\right)^{1/2}\left(\frac{D_A}{t}\right)^{1/2}\left(C_{A/B} - C_A^o\right) \qquad (19\text{-}37)$$

At any time the two fluxes must be equal: $F_{bl} = F_A$.

Replacing $C_{A/B}$ by $K_{A/B}C_{B/A}$ and solving for $C_{B/A}$ yields:

$$C_{B/A} = \frac{C_B^{eq} + \psi(t)C_B^o}{1 + \psi(t)} \qquad (19\text{-}38)$$

where $C_B^{eq} = C_A^o / K_{A/B}$ is the B-side concentration in equilibrium with the initial A-side concentration, and:

$$\psi(t) = \left(\frac{\pi t}{D_A}\right)^{1/2}\frac{D_{bl}}{\delta}\frac{1}{K_{A/B}} \qquad (19\text{-}39)$$

is a nondimensional, time-dependent function that makes the interface concentrations, $C_{B/A}$ and $C_{A/B}$, timedependent as well.

Now we insert Eq. 19-38 into one of the flux equations, for instance, into Eq. 19-36. After some rearrangement, we get:

$$F(t) = F_{bl}(t) = -\frac{D_{bl}}{\delta}\frac{C_B^{eq} - C_B^o}{1 + \psi(t)} = v_{bl}\frac{C_B^o - C_B^{eq}}{1 + \psi(t)} \qquad (19\text{-}40)$$

where $v_{bl} = D_{bl}/\delta$ is the boundary layer transfer velocity (see 19-4).

Note that the temporal evolution of both the boundary concentration, $C_{B/A}(t)$, and the flux across the boundary, $F(t)$, are controlled by the nondimensional function $\psi(t)$. This function defines a critical time scale for the switch between two regimes. We define t_{crit} as the time for which $\psi(t_{crit}) = 1$. From Eq. 19-39:

$$t_{crit} = \frac{D_A}{\pi}\left(\frac{\delta}{D_{bl}}\cdot K_{A/B}\right)^2 = \frac{D_A}{\pi}\left(\frac{K_{A/B}}{v_{bl}}\right)^2 \qquad (19\text{-}41)$$

We evaluate Eqs. 19-38 and 19-40 for two extreme times, for the initial situation ($t \ll t_{crit}$) and the long-term situation ($t \gg t_{crit}$):

$$C_{B/A}(t) \cong \begin{cases} C_B^{eq} = C_A^o / K_{A/B} & \text{for } t \ll t_{crit} \\ C_B^o & \text{for } t \gg t_{crit} \end{cases} \qquad (19\text{-}42)$$

$$F(t) \cong \begin{cases} v_{bl}\left(C_B^o - C_B^{eq}\right) & \text{for } t \ll t_{crit} \\ v_{bl}\cdot\dfrac{1}{\psi(t)}\left(C_B^o - C_B^{eq}\right) = \left(\dfrac{D_A}{\pi t}\right)^{1/2} K_{A/B}\left(C_B^o - C_B^{eq}\right) & \text{for } t \gg t_{crit} \end{cases} \qquad (19\text{-}43)$$

In these equations we recognize expressions which by now should have become familiar to us. During the initial phase of the exchange process ($t \ll t_{crit}$), boundary concentration and flux at the interface remind us of a (B-side controlled) bottleneck boundary with transfer velocity $v_{bl} = D_{bl}/\delta$ (see Eq. 19-19). The concentrations on either side are C_B^o and $C_B^{eq} = C_A^o/K_{A/B}$, where the latter is the B-side concentration in equilibrium with the initial A-side concentration C_A^o.

As time goes by, the interface looks more and more like a wall boundary. Eventually, the concentration gradient across the boundary layer becomes zero ($C_{B/A} \sim C_B^o$) and the flux takes the form of Eq. 19-26 with the extra factor $K_{A/B}$ expressing the partition equilibrium across the interface (see Eq. 19-29).

The transition from one regime to the other is shown in Fig. 19.11 as a function of the relative time:

$$\tau \equiv t/t_{crit} = [\psi(t)]^2 \qquad (19\text{-}44)$$

The concentration difference across the boundary layer and the boundary flux are normalized by their respective values attained for large times. It turns out that both are described by the same simple function of relative time τ:

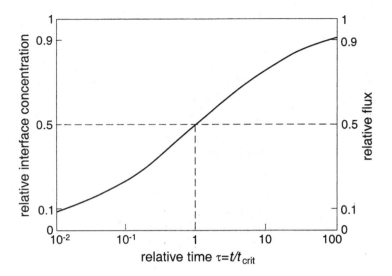

Figure 19.11 Wall boundary with boundary layer: Relative variation of concentration difference across boundary layer and relative boundary flux as a function of relative time $\tau = t / t_{crit}$ (Eq. 19-45).

$$\frac{C_{A/B}(t) - C_B^{eq}}{C_B^o - C_B^{eq}} = \frac{F(t)}{F(t \to \infty)} = \frac{\tau^{1/2}}{1 + \tau^{1/2}} \tag{19-45}$$

As shown in Fig. 19.11, most of the transition of $C_{A/B}(t)$ and $F(t)$ from one regime to the other occurs in the time interval between $\tau = 10^{-2}$ and 10^2.

The real time t is related to τ by $t = \tau \, t_{crit}$. Thus, in order to interpret Fig. 19.11 we should now get an idea of the size of t_{crit}. In Table 19.2 ranges of critical times t_{crit} are calculated for two types of wall boundaries for which diffusive boundary layers may be important. The first is the example of a gas–liquid interface. Here the gas phase is the mixed system (system B) which is connected to a liquid by a gaseous boundary layer. In the notation of Fig. 19.10, $K_{A/B}$ is then an inverse nondimensional Henry's law coefficient. According to Table 19.2 the influence of the boundary layer is important only for nonvolatile substances, provided that the transfer velocity v_{bl} is small. The second example depicts the boundary between an open water column (system B) and a porous medium (system A). Here $K_{A/B}$ measures the strength of sorption of a chemical in the porous media. Note that now diffusivity in system A, D_A, is inversely related to $K_{A/B}$. As it turns out, for strongly sorbing media, the boundary layer may control the exchange flux during a very long time period.

As shown in Illustrative Examples 19.3 and 19.4, often it is not immediately known whether an exchange process is controlled by transport across a boundary layer or by transport in the bulk phase. In Illustrative Example 19.3 we look at the case of resuspension of particles from the polluted sediments of Boston Harbor. We are interested in the question of what fraction of the pollutants sorbed to the particles (such as polychlorinated biphenyls) can diffuse into the open water column while the particles are resuspended due to turbulence produced by tidal currents in the bay. To answer this question we need to assess the possible role of the boundary layer around the particles.

In Illustrative Example 19.4 we look at the transfer between a non-aqueous-phase liquid (NAPL) such as diesel fuel and water. Although the example deals with an

Table 19.2 Influence of Boundary Layer on Mass Exchange at Wall Boundary

Definition of critical time (Eq. 19-41)

$$t_{crit} = \frac{D_A}{\pi} \left(\frac{\delta}{D_{bl}} \cdot K_{A/B} \right)^2 = \frac{D_A}{\pi} \left(\frac{K_{A/B}}{v_{bl}} \right)^2$$

A. Boundary Between Well-Mixed Gas (System B) and Stagnant Liquid (System A)

Transfer velocity across gaseous boundary layer: typically between 0.1 and 1 cm s^{-1} (up to 5 cm s^{-1}, see Fig. 20.2). $K_{A/B}$ is the nondimensional *liquid/gas* distribution coefficient (for air–water interface: inverse nondimensional Henry's law coefficient, i.e., $K_{ia/w}^{-1}$) with typical values between 10^{-3} and 10^3. D_A is the molecular gaseous diffusivity, typical size 0.1 cm^{2}s^{-1}.

Thus:

$$t_{crit} \sim 3 \times 10^{-2} \text{ cm}^2\text{s}^{-1} \left(\frac{K_{A/B}}{v_{bl}} \right)^2$$

Critical time t_{crit} (s)

$K_{A/B}$	$v_{bl} = 0.1$ cm s^{-1}	$v_{bl} = 1$ cm s^{-1}
10^{-3} (volatile)	3×10^{-6}	3×10^{-8}
1	3	3×10^{-2}
10^3 (nonvolatile)	3×10^6 (1 month)	3×10^4 (8 hours)

B. Boundary Between Well-Mixed Water (System B) and Porous Media (System A)

Typical transfer velocity across liquid layer: 10^{-3} cm s^{-1} (range 10^{-5} to 10^{-2} cm s^{-1}, see Section 20.2 and Illustrative Examples 19.3, 19.4). $K_{A/B}$ is the equilibrium partition coefficient with typical values between 1 and 10^4 (see Table 19.1). D_A is the aqueous molecular diffusivity in pore space (typical size 10^{-6} cm^{2}s^{-1}) divided by $K_{A/B}$.

Thus:

$$t_{crit} \sim 3 \times 10^{-7} \text{ cm}^2\text{s}^{-1} \frac{K_{A/B}}{v_{bl}^2}$$

Critical time t_{crit} (s)

$K_{A/B}$	$v_{bl} = 10^{-5}$ cm s^{-1}	$v_{bl} = 10^{-3}$ cm s^{-1}	$v_{bl} = 10^{-2}$ cm s^{-1}
1 (nonsorbing)	3×10^3 (1 hour)	0.3	3×10^{-3}
10^2	3×10^5 (3 days)	30	0.3
10^4 (strongly sorbing)	3×10^7 (1 year)	3×10^3 (1 hour)	30

artificial setup in the laboratory, there is a real story behind the experiments, that is, the pollution of groundwater by diesel fuel spilled into the aquifer (Schluep et al., 2001). It turns out that it is mainly the boundary layer on the water side of the NAPL–water interface that controls the solution of diesel fuel components into the water.

(Text continues on page 864)

Illustrative Example 19.3

Release of PCBs from the Historically Polluted Sediments of Boston Harbor

Problem

The bed sediments of Boston Harbor (Massachusetts, USA) have long accumulated organic contaminants like polychlorinated biphenyls (PCBs). As a result, investigators like McGroddy (1993) find surface sediment concentrations of compounds like 2,2',4,5,5'-pentachlorobiphenyl (PCB101) or 2,2'3,3',4,4'5-heptachlorobiphenyl (PCB170) near 30 ng per gram dry sediment material. Recently a major construction effort has moved the sewage discharges out of the harbor. Now the question arises, how long the existing legacy of polluted sediments will continue to release undesirable fluxes of these organic chemicals back to the water column.

Answer

To begin to answer this question, you are interested in deducing what process limits the bed-to-water-column releases of such highly sorptive chemicals. Intuitively, you expect diffusion out of the bed may be rate limiting. In order to assess the possible influence of a benthic boundary layer, calculate first from Eq. 19-41 the critical time t_{crit} for some representative compounds of concern. If this parameter is "large" for the chemicals of concern, then their release from the sediment bed will actually be controlled by diffusion through the thin layer of "stagnant" water lying just above the sediment–water interface.

2,2',4,5,5'-pentachlorobiphenyl
(PCB101)

2,2',3,3',4,4',5-heptachlorobiphenyl
(PCB170)

Property, Parameter	PCB101	PCB170
Molar volume, $\overline{V}_i$ (cm^{3}mol^{-1}) [a]	265	300
D_w at 15°C (cm^{2}s^{-1}) [b]	4.3×10^{-6}	4.0×10^{-6}
log K_{ow}	6.36	7.36
log K_{oc} (cm^3 g$_{oc}^{-1}$) [c]	4.86	5.60
log K_d^{sc} (cm^{3}g^{-1}solid) [d]	3.6	4.3
$K_{sc/op}$ [e]	2.0×10^3	1.0×10^4
v_{bl} (cm s^{-1}) [f]	2.9×10^{-4}	2.7×10^{-4}
t_{crit} (s) [g]	6.5×10^7	1.7×10^9
	(2 years)	(55 years)

[a] Using the diffusion volume contribution of Fuller et al. (1966); see Illustrative Example 18.1. [b] Using the expression of Hayduk and Laudie (1974); see Eq. 18-53. [c] Using log $K_{oc} = 0.74$ log K_{ow} + 0.15 (Eq. 9-29a). [d] Using $K_d = f_{oc}K_{oc}$; see Eq. 9-22. [e] Using Eq. 19-30 with $r_{sw}^{sc} = 0.63$ g cm^{-3} and $\phi^{sc} = 0.8$. [f] $v_{bl} = D_w/\delta$ assuming $\delta = 0.015$ cm. [g] Eq. 19-41; note that $K_{A/B} = K_{sc/op}$ (i.e., A = sediment column, B = open water above sediment).

Given the compound and environmental properties shown above, estimate the size of t_{crit} for PCB101 and PCB170.

These results indicate that the PCBs will be water–boundary layer limited in their release from the sediment bed for years after a step-function change in the overlying water column concentrations.

Note: In Chapter 23 we will further elaborate on the sediment–water exchange flux, especially in Box 23.2. and Table 23.6.

Illustrative Example 19.4

Dissolution of a Non-Aqueous-Phase Liquid (NAPL) into the Aqueous Phase

Problem

Mass exchange between a non-aqueous-phase liquid (NAPL) such as diesel fuel and water can be studied with the so-called slow stirring method (SSM). The SSM was designed to determine solubilities and octanol–water partition coefficients of organic compounds such as petroleum hydrocarbons. The kinetics of the exchange process also yield valuable information on the exchange between groundwater and a NAPL spilled into an aquifer.

The experimental setup of the SSM is shown in Fig. 19.12*a*. The temporal increase of the aqueous concentration of four diesel fuel components (benzene, *m*/*p*-xylene, naphthalene) is given in Figs. 19.12*b* to *d*. Relevant physicochemical properties are summarized in Table 19.3. Note that the two isomers *m*- and *p*-xylene exhibit virtually the same properties, and are, therefore, considered together.

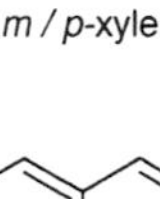

(a) In a first step the NAPL–water interface shall be described as a simple bottleneck boundary which separates two homogeneous mixed systems (NAPL, water). Convince yourself that an exchange velocity $v_{i\text{bl}} = 3.2 \times 10^{-4}$ cm s^{-1} for *m*/*p*-xylene explains the measured aqueous concentration change reasonably well. Calculate the corresponding water-side boundary layer thickness δ_{bl} and use the result to calculate $v_{i\text{bl}}$ for benzene and naphthalene.

(b) Why is it reasonable to assume that the exchange across the NAPL–water interface is indeed controlled by a boundary layer in the water and not in the NAPL?

(c) In the SSM setup the water is mixed while the NAPL is not. Modify the model developed in (a) by describing the interface as a wall boundary with a water-side boundary layer adjacent to a well-mixed water layer (system B in Fig. 19.10). Is the result very different from (a)? *Hint:* Calculate the critical time t_{crit} (Eq 19-41) and compare it to the time scale which is relevant for model (a).

(d) Closer inspection of the temporal change of the aqueous benzene concentration (Fig. 19.12*b*) shows that the steady-state concentration of benzene in the aqueous phase seems to lie about 10% below the equilibrium value with the NAPL phase. Explain the discrepancy.

i = benzene

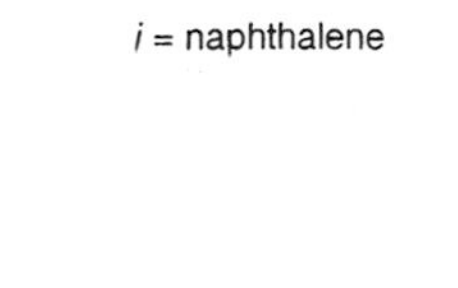

i = *m* / *p*-xylene

i = naphthalene

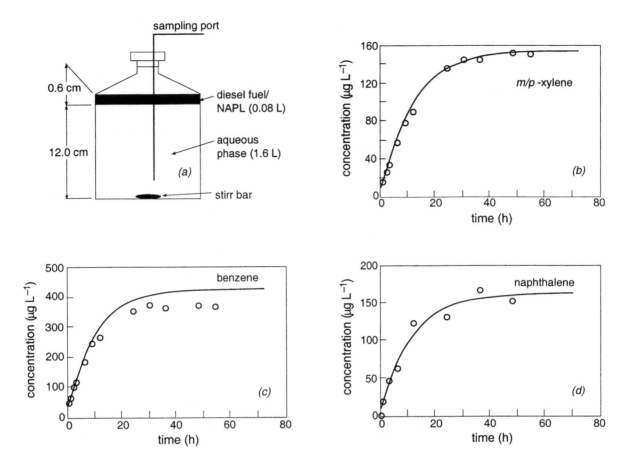

Figure 19.12 (a) Experimental setup to determine the exchange dynamics of a combined NAPL–water system using the slow stirring method (SSM). (b) – (d) Measured and calculated aqueous concentrations of benzene; m/p-xylene and naphthalene. The solid lines give the result of the linear bottleneck exchange model with an aqueous boundary layer thickness of $\delta_{bl} = 2.4 \times 10^{-2}$ cm = 240 μm (adapted from Schluep et al., 2000).

Answer (a)

The mass balance of the aqueous phase concentration, C_{iw}, yields:

$$V_w \frac{dC_{iw}}{dt} = A_{\text{interface}} F_{bl} = A_{\text{interface}} v_{ibl} \left(C_{iw/\text{NAPL}} - C_{iw} \right) \tag{1}$$

where V_w is the water volume, $A_{\text{interface}}$ the NAPL–water interface area, and $C_{iw/\text{NAPL}}$ the aqueous concentration in equilibrium with the concentration in the NAPL. The latter is (Eq. 7-21 with $\gamma_{i\text{NAPL}} = 1$):

$$C_{iw/\text{NAPL}} = x_{i\text{NAPL}} \cdot C_{iw}^{\text{sat}} (L) \tag{2}$$

where $C_{iw}^{\text{sat}}(L)$ is the aqueous solubility of the liquid compound i, and $x_{i\text{NAPL}}$ is its mole fraction in the NAPL. Note that we assume that the activity coefficients, $\gamma_{i\text{NAPL}}$, of these hydrocarbons in diesel fuel are all one. This is justified for compounds of diesel fuel which are structurally similar (see Section 7.5). Dividing the mass balance Eq. 1 by V_w yields:

$$\frac{dC_{iw}}{dt} = k_i \left(C_{iw/\text{NAPL}} - C_{iw} \right) \tag{3}$$

with $k_i = v_{ibl} A_{\text{interface}} / V_w = v_{ibl} / h_w$, where $h_w = 12$ cm is mean depth of the aqueous phase.

For constant $C_{iw/\text{NAPL}}$ and assuming that initially C_{iw} is zero, the differential equation has the solution (see Box 12.1):

$$C_{iw}(t) = C_{iw/NAPL}\left(1 - e^{-k_i t}\right) \qquad (4)$$

Fitting the measured m/p-xylene concentrations (Fig. 19.12b) with Eq. 4 yields $k_i = 2.7 \times 10^{-5} \, s^{-1}$. Thus:

$$v_{ibl} = k_i \, h_w = 2.7 \times 10^{-5} \, s^{-1} \times 12 \, cm = 3.2 \times 10^{-4} \, cm \, s^{-1}$$

The boundary layer thickness δ_{bl} can be calculated from v_{ibl} and D_{iw} of m/p-xylene (see Table 19.3) using the bottleneck model (Eq. 19-4):

$$\delta_{bl} = \frac{D_{iw}}{v_{ibl}} = \frac{7.8 \times 10^{-6} \, cm^2 s^{-1}}{3.2 \times 10^{-4} \, cm \, s^{-1}} = 2.4 \times 10^{-2} \, cm$$

From δ_{bl} we can calculate v_{ibl} and k_i for the other compounds (benzene, naphthalene, see table below). The corresponding model curves are shown in Figs. 19.12b–d.

	Benzene	m/p-Xylene	Naphthalene
$C_{iw/NAPL}{}^a$ ($\mu g \, L^{-1}$)	410	140	160
$K_{iNAPL/w}{}^b$	1.8×10^2	2.4×10^3	4.5×10^3
v_{ibl} ($s \, cm^{-1}$)	3.9×10^{-4}	3.2×10^{-4}	3.4×10^{-4}
k_i (s^{-1})	3.3×10^{-5}	2.7×10^{-5}	2.8×10^{-5}
$t_{i1/2} = \ln 2 / k_i$ (Box 12.1, Eq. 5) (h)	5.9	7.2	6.8

^{a} $C_{iw/NAPL} = x^o_{iNAPL} C^{sat}_{iw} (L)$ (see Eq. 2 and Table 19.3). Note that concentrations have been converted to $\mu g \, L^{-1}$. ^{b} $K_{iNAPL/w} = [\overline{V}_{NAPL} \cdot C^{sat}_{iw} (L)]^{-1}$; see Eq. 7-22 with $\gamma_{iNAPL} = 1$ and $\overline{V}_{NAPL} = \rho_{NAPL} \cdot M^{-1}_{NAPL}$ (see footnote in Table 19.3).

The result leads to the following conclusions:

1. The measured time to reach half the aqueous equilibrium concentration corresponds well with the calculated $t_{i1/2}$ values.

2. For m/p-xylene and naphthalene the aqueous equilibrium concentration (reached after about 2 days) agrees well with the calculated value, $C_{iw/NAPL}$. However, for benzene the measured value lies about 10% below the calculated equilibrium concentration (410 $\mu g \, L^{-1}$). This discrepancy will be explained below (Answer d).

Answer (b)

The overall exchange velocity of a two-layer bottleneck boundary is given by Eq. 19-20. If all concentrations are expressed in terms of the aqueous concentrations, the NAPL film exchange velocity carries the extra factor $K_{iNAPL/w}$. According to the above table, the partition coefficient $K_{iNAPL/w}$ lies between 1.8×10^2 for benzene and 4.5×10^3 for naphthalene. Thus, you expect that the NAPL-side exchange velocity is much larger than the aqueous exchange velocity and that therefore the latter controls the overall exchange.

Note: In Chapter 20 we will discuss the air–water exchange and find that indepen-

Table 19.3 Physicochemical Properties of Three Selected Diesel Fuel Compounds (Schluep et al., 2001) [a]

	Benzene	m/p-Xylene	Naphthalene
Molar mass M_i (g mol^{-1})	78.1	106.2	128.2
Initial concentration in diesel fuel, C_{iNAPL}^o (mol L^{-1})	9.7×10^{-4}	3.2×10^{-3}	5.5×10^{-3}
Initial mole fraction in NAPL, x_{iNAPL}^o (–)	2.4×10^{-4}	7.8×10^{-4}	1.4×10^{-3}
Liquid aqueous solubility at 20°C, C_{iw}^{sat} (L) (mol L^{-1})	2.2×10^{-2}	1.7×10^{-3}	9.0×10^{-4}
Liquid density at 20°C, ρ_{iL} (g cm^{-3})	0.877	0.873	1.150
Molecular diffusivity in water [b], D_{iw} (cm^{2}s^{-1})	9.4×10^{-6}	7.8×10^{-6}	8.2×10^{-6}
Molecular diffusivity in NAPL[b], D_{iNAPL} (cm^{2}s^{-1})	2.2×10^{-6}	1.8×10^{-6}	1.9×10^{-6}

[a] Relevant properties of diesel fuel (NAPL) used for the experiments: molar mass $M_{NAPL} = 202$ g mol^{-1}; liquid density $\rho_{NAPL} = 0.8207$ g cm^{-3}; dynamic viscosity at 20°C, η_{NAPL} (20°C) $= 3.64 \times 10^{-2}$ g cm^{-1}s^{-1}. [b] Approximated from relation by Hayduk and Laudie (1974), see Eq. 18-53; dynamic viscosity of water at 20°C, η_w (20°C) $= 1.002 \times 10^{-2}$ g cm^{-1}s^{-1}.

dently of the partition coefficient between air and water (the Henry's law constant), the air-side exchange velocity is about 10^3 times larger than the water-side value. The main reason is due to the difference of molecular diffusivities in air and water (see Figs. 18.8 and 18.10). Since diffusivity in water and diesel fuel is of similar magnitude (Table 19.3), it is reasonable to assume that the respective exchange velocities, v_{iw} and v_{iNAPL}, are of similar magnitude.

Answer (c)

As a refinement of the bottleneck model discussed in (a), we now treat the interface as a wall boundary between infinitely large reservoirs (water and NAPL) with a water-side boundary layer of thickness δ_{bl}. We calculate the critical time t_{icrit} from Eq. 19-41 using Table 19.3 and the parameters evaluated in (a). For benzene:

$$t_{icrit} = \frac{D_{iNAPL}}{\pi} \left(\frac{K_{iNAPL/w}}{v_{ibl}} \right)^2 = \frac{2.2 \times 10^{-6} \text{ cm}^2\text{s}^{-1}}{3.141} \left(\frac{180}{3.9 \times 10^{-4} \text{ cm s}^{-1}} \right)^2$$

$$= 1.48 \times 10^5 \text{ s} = 41 \text{ h (benzene)}$$

Correspondingly:

$$m/p\text{-xylene: } t_{icrit} = \frac{1.8 \times 10^{-6} \text{ cm}^2\text{s}^{-1}}{3.141} \left(\frac{2.4 \times 10^3}{3.2 \times 10^{-4} \text{ cm s}^{-1}} \right)^2 = 3.2 \times 10^7 \text{ s} = 370 \text{ d}$$

$$\text{naphthalene: } t_{icrit} = \frac{1.9 \times 10^{-6} \text{ cm}^2\text{s}^{-1}}{3.141} \left(\frac{4.5 \times 10^3}{3.4 \times 10^{-4} \text{ cm s}^{-1}} \right)^2 = 1.1 \times 10^8 \text{ s} = 3.4 \text{ yr}$$

Comparing these results with the half-equilibration time of the aqueous phase, $t_{i1/2}$ (see table above) we conclude that the aqueous concentration reaches its saturation value well before the exchange process switches from the boundary-layer-controlled to the NAPL-diffusion-controlled regime. Thus, diffusive transport of the diesel components from the interior of the NAPL to the boundary never controls the transfer process. Consequently, the simplex box model described in answer (a) is adequate.

Answer (d)

Until now we have tacitly assumed that the loss of certain chemical compounds from the NAPL to the aqueous phase does not significantly lower the concentration of these compounds in the remaining fuel. We can now check this assumption by looking at the relative partition of the component between the two phases when complete equilibrium is reached (concentration $C_{i\text{NAPL}}^{eq}$ and C_{iw}^{eq}):

$$V_{\text{NAPL}} C_{i\text{NAPL}}^{o} = V_{\text{NAPL}} C_{i\text{NAPL}}^{eq} + V_{w} C_{iw}^{eq} = C_{i\text{NAPL}}^{eq} \left(V_{\text{NAPL}} + \frac{V_{w}}{K_{i\text{NAPL}/w}} \right)$$

Thus:

$$\frac{C_{i\text{NAPL}}^{eq}}{C_{i\text{NAPL}}^{o}} = \frac{V_{\text{NAPL}}}{V_{\text{NAPL}} + \dfrac{V_{w}}{K_{i\text{NAPL}/w}}} = \frac{K_{i\text{NAPL}/w}}{K_{i\text{NAPL}/w} + \dfrac{V_{w}}{V_{\text{NAPL}}}}$$

According to Fig. 19.12a the water–NAPL volume ratio is 1.6 L / 0.08 L = 20. If the above concentration ratio is close to 1, the equilibration process between water and NAPL does not significantly deplete the NAPL. This situation is favored by large $K_{i\text{NAPL}/w}$ values. Benzene is the only case for which depletion may be important. Inserting the values for benzene yields:

$$\frac{C_{i\text{NAPL}}^{eq}}{C_{i\text{NAPL}}^{o}} = \frac{180}{180 + 20} = 0.90 \text{ (benzene)}$$

Thus when full equilibration is reached, 10% of the benzene has left the NAPL phase. This explains why the corresponding steady-state value in the aqueous phase (Fig. 19.12b) is 10% smaller than calculated in (a). For the other compounds the loss is much smaller.

Note: This illustrative example is based on a real investigation. The interested reader can find additional information and a more detailed treatment of the water–NAPL exchange problem in Schluep et al. (2001).

Wall Boundary with Time-Variable Boundary Concentration

Until now we have treated wall boundaries with constant concentration in the mixed system (system B in Fig. 19.8). Such situations are rare in nature. For instance, at the sediment–water interface of a lake the concentration of a chemical in the overlying water column is hardly constant during a period of several years. So we should find a method to extend the mathematical tools developed above. As it turns out, an analytical computation of the concentration in the diffusive system (the sediment column, for instance) as a function of depth and time, $C_t(x, t)$, may become rather cumbersome.

With a large choice of sophisticated computer tools at hand, it is now easy to calculate the solution of Eq. 10 of Box 18.5 for any boundary condition we want. Yet, computer programs do not necessarily provide a general understanding for how the system behaves. Let us therefore develop a semiquantitative picture by analyzing Fig. 18.5 and Fick's law (Eq. 18-14) for the case of a time-dependent boundary concentration, $C_B^o(t)$. As will be discussed in more detail in Chapter 22, the analysis of the time-dependent transport equation always involves a search for the adequate time scale. Here the relevant time scale, τ_B, is given by the rate of change of $C_B^o(t)$. For instance, if $C_B^o(t)$ changes exponentially, such that

$$C_B^o(t) = C_B^o(0)\, e^{\beta t} \tag{19-46}$$

then:

$$\tau_B = |\beta|^{-1} \tag{19-47}$$

Note that the rate β (dimension T^{-1}) can have either sign (increasing or decreasing boundary concentration). An alternative variation pattern of $C_B^o(t)$ that is common for natural systems is the case of a periodic variation:

$$C_B^o(t) = C_B^{mean}\left(1 + \gamma^{var}\sin\omega t\right) \ , \quad 0 \le \gamma^{var} \le 1 \tag{19-48}$$

where C_B^{mean} and γ^{var} are mean concentration and relative amplitude of variation. Here the time scale of variation, τ_B, is defined by the oscillation period which in turn is related to the angular frequency ω by:

$$\tau_B = \text{oscillation period} = \frac{2\pi}{\omega} \tag{19-49}$$

The time scale τ_B determines the penetration distance in system A beyond which the influence of the variation of C_B^0 can be disregarded. For instance, we can use the concept of "half-concentration depth" (Eq. 18-23) as a measure to assess the penetration distance of the time-dependent variation into system A. Thus:

$$x_A^{var} \sim \left(D_A \tau_B\right)^{1/2} \tag{19-50}$$

Table 19.4 Penetration Depth, x_A^{var}, of Time Variable Concentration Across a Wall Boundary (Eq. 19-50) for a Nonsorbing Chemical ($D_A = 10^{-5}$ cm^{2}s^{-1}) and a Sorbing Chemical ($D_A = 10^{-10}$ cm^{2}s^{-1})

Period ($2\pi/\omega$)	ω(s^{-1})	(Nonsorbing) x_A^{var} (cm)	(Sorbing) x_A^{var} (cm)
Day	7.3×10^{-5}	1	3×10^{-3}
Year	2.0×10^{-7}	20	6×10^{-2}
Centennial	2.0×10^{-9}	200	0.6

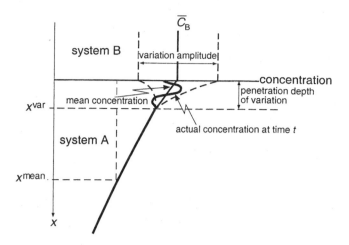

Figure 19.13 Different length scales of variation of concentration in system A due to a time-variable concentration in system B (wall boundary). $x^{\text{mean}} = (D_A t)^{1/2}$. x^{var} is defined in Eq. 19-51.

Then the influence of the changing boundary concentration is negligible at distances $x > x_A^{\text{var}}$. For the case of a periodic boundary concentration we can combine Eqs. 19-49 and 19-50:

$$x_A^{\text{var}} = \left(\frac{2\pi D_A}{\omega} \right)^{1/2} \tag{19-51}$$

In Table 19.4, numerical values are given for the penetration depth, x_A^{var}, of (1) a nonsorbing compound (diffusivity $D_A = 10^{-5}$ cm^{2}s^{-1}) and (2) and a sorbing compound ($D_A = 10^{-10}$ cm^{2}s^{-1}). The latter simulates the behavior of a chemical with $K_{\text{sc/op}} \sim 10^5$ (see Eqs. 19-30 and 19-31).

In Fig. 19.13, x_A^{var} is schematically shown together with the penetration of the mean concentration, x_A^{mean}, which increases as $t^{1/2}$ (see Eq. 18-23).

19.4   Diffusive Boundaries

A diffusive boundary connects two systems in which diffusivity is of similar size or equal (Fig. 19.3c). In contrast to a bottleneck boundary, which is characterized by one or several zones with significantly reduced diffusivity, or to a wall boundary, which exhibits an asymmetric drop in diffusivity, transport at a diffusive boundary is not very different from the inner part of the systems involved. What makes it a boundary is either a phase change (thus, the boundary is also an interface) or an abrupt change of one or several properties. By "property" we mean, for instance, the concentration of some chemical compound or of temperature.

In this section we will develop the mathematical tools to describe mass transfer at diffusive boundaries. Again, it is our intention to demonstrate that diffusive boundaries have common properties, although the physics controlling them may be different. We will then apply the mathematical tools to the process of dilution of a pollutant cloud in an aquatic system (ocean, lake, river). Here the boundary is produced by the localized (continuous or event-like) input of a chemical that first leads to a confined concentration patch. The patch is then mixed into its environment by diffusion or dispersion. Note that in this case the physical characteristics on both sides of the

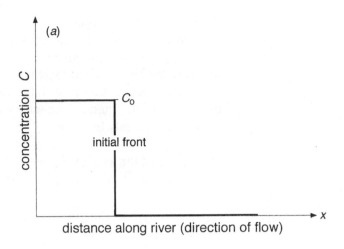

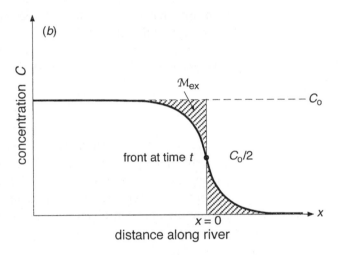

Figure 19.14 (*a*) A pollutant front in a river with an initial rectangular shape. This abrupt change is transformed into a smoother concentration profile while the front is moving downstream (*b*). The hatched area represents the integrated mass exchange $\mathcal{M}_{ex}$ across the front due to longitudinal dispersion.

boundary are alike; that is, we may find mixing due to molecular diffusion, turbulent diffusion, or dispersion.

In Chapter 20, the diffusive boundary scheme serves as one of several models to describe air–water exchange.

Dispersion at the Edge of a Pollutant Front

Let us assume that at time $t = 0$ a pollutant begins to be discharged continuously into a river through an outfall pipe. At some location downstream of the entry point, the pollutant will be completely mixed across the river. Provided that chemical or physical removal mechanisms can be disregarded, the average pollutant concentration in the river below this point is equal to $C_o = J/Q$, where J is the introduced pollutant mass per unit time and Q is the river discharge per unit time (assuming $Q \gg Q_{outfall}$).

Since the pollutant input is turned on suddenly, a pollutant front moves downstream. Initially, the shape of the front is rectangular (Fig. 19.14*a*). Due to longitudinal dispersion (see Chapters 18.5 and 22.4), the front gradually looses its rectangular shape while moving downstream (Fig. 19.14*b*). Longitudinal dispersion in rivers will be discussed in more detail in Chapter 24. At this point it is sufficient to remember that dispersion can be mathematically described as a diffusion-type process with a dispersion coefficient, E_{dis}, which has the usual dimension of L^2T^{-1}. In other words, the (moving) front is a diffusive boundary. The dispersion coefficients on both sides of the front usually have the same size. The front has no physical meaning (unless the pollutant concentration is so large that water density, viscosity, or other physical properties of the fluid are affected); it just results from the fact that the pollutant input was suddenly turned on.

Provided that the available river section on both sides of the front is quasi infinite, the slope of the front can be described by an expression like Eq. 18-27,

$$C(x,t) = \frac{C_o}{2}\,\mathrm{erfc}\left(\frac{x}{2(E_{dis}\,t)^{1/2}}\right) \qquad (19\text{-}52)$$

where x is the distance relative to the center of the (moving) front, erfc(...) is the complement error function (Appendix A), and $C_o = J/Q$ is the concentration jump across the boundary (ΔC in Eq. 18-27). The smoothing of the front corresponds to an exchange of pollutant mass across the front (represented by the hatched area in Fig. 19.14b). According to Eq. 18-29, the integrated transferred mass at time t is:

$$M_{ex}(t) = \left(\frac{1}{\pi}\right)(E_{dis}t)^{1/2}C_o = 0.564\,(E_{dis}t)^{1/2}C_o \qquad (19\text{-}53)$$

Strictly speaking, equations 19-52 and 19-53 are valid only if the pollutant cloud is infinitely long. A more realistic situation is treated in Box 19.2; here a pollutant patch of finite length L along the x-axis is eroded on both edges due to diffusion processes (turbulence, dispersion, etc.). Again, the boundary is of the diffusive type since the transport characteristics on both sides of the boundary are assumed to be identical.

Box 19.2   Dilution of a Finite Pollutant Cloud Along One Dimension (Advanced Topic)

We consider a pollutant cloud which at time $t = 0$ has the concentration C^o in the segment $x = \{-L/2, +L/2\}$ and zero outside. The cloud along the x-axis is eroded by diffusion (or dispersion) in the x-direction. The effective diffusivity is D. According to Crank (1975), the concentration distribution at time t is:

$$C(x,t) = \frac{C^o}{2}\left\{\mathrm{erf}\left(\frac{\frac{L}{2}-x}{2(Dt)^{1/2}}\right) + \mathrm{erf}\left(\frac{\frac{L}{2}+x}{2(Dt)^{1/2}}\right)\right\} \qquad (1)$$

where erf(...) is the error function (Appendix A).

We are interested in the temporal change of the maximum concentration. Due to the symmetry of the configuration, the maximum concentration is always located at $x = 0$. Thus, from Eq. 1:

$$C_{max} = C(x=0,t) = C^o\,\mathrm{erf}\left[\frac{L}{4(Dt)^{1/2}}\right] \qquad (2)$$

The figure gives a schematic view of $C(x,t)$ for increasing normalized time, $\tau = (Dt)^{1/2}/L$. Redrawn from Crank (1975).

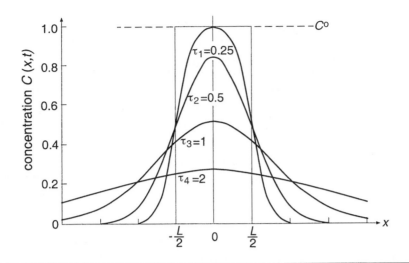

Diffusive Boundary Between Different Phases

Diffusive boundaries also exist between different phases. The best known example is the so-called surface renewal (or surface replacement) model of air–water exchange, an alternative to the stagnant two-film model. It will be discussed in Chapter 20.3.

Here we consider another example. Let us assume that in a short time a large quantity of diesel fuel is spilled on the surface of a river. At least for some time, the fuel forms a non-aqueous-phase liquid (NAPL) floating on the water surface. Due to the turbulence produced by the friction of the flowing water at the river bed and the wind blowing on top of the floating NAPL, fluid parcels from the interior of the respective fluids (water or diesel fuel) are brought to the fuel–water interface where they mutually exchange water and diesel components. This exchange lasts until another burst of turbulence separates them and brings fresh fluid parcels to the interface. Mass exchange across the water–NAPL interface during the contact time occurs by molecular diffusion; it can be described as a diffusive boundary exchange process similar to the one shown in Fig. 19.14. Yet, for the fuel components the equilibrium condition at the interface is not $C_{\text{left}} = C_{\text{right}}$, but is modified by some partitioning law as described in Illustrative Example 19.4. The modification introduced into the diffusive boundary model is the same as the one derived earlier for the non-equal-phase bottleneck boundary (Figs. 19.6. and 19.7).

The following mathematical derivation can be applied to any diffusive boundary. The situation is depicted in Fig. 19.15. As before we choose system A as the reference system. We assume that immediately after the formation of the new surface, the interface concentrations are at equilibrium. Thus, $C_{A/B}$ and $C_{B/A}$ are related by a partition coefficient (see Eq. 19-16):

$$C_{B/A} = K_{B/A}\, C_{A/B} \qquad (19\text{-}16)$$

After time t, a total mass $\mathcal{M}_A(t)$ has crossed the interface from system A to system B (or vice versa, if the situation is as shown in Fig. 19.15b). The corresponding gain in

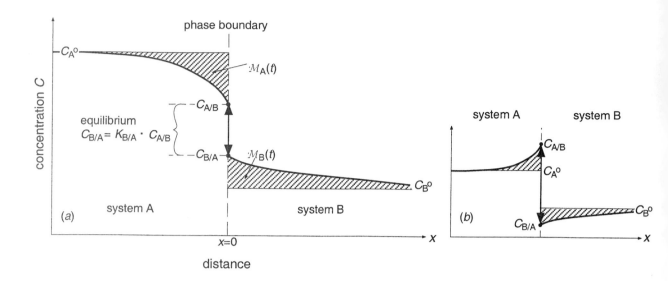

Figure 19.15 (*a*) Concentration profile at a diffusive boundary between two different phases. At the interface the instantaneous equilibrium between $C_{A/B}$ and $C_{B/A}$ is expressed by the partition coefficient $K_{B/A}$. The hatched areas show the integrated mass exchange after time t; $\mathcal{M}_A(t) = \mathcal{M}_B(t)$. (*b*) As before, but the size of $K_{B/A}$ causes a net mass flux in the opposite direction, that is from system B into system A.

system B is given by the hatched area $\mathcal{M}_B(t)$. According to Eq. 18-29, the two integrated fluxes are given by:

$$\mathcal{M}_A(t) = \left(\frac{1}{\pi}\right)^{1/2} (D_A t)^{1/2} (C_A^o - C_{A/B})$$

(19-54)

$$\mathcal{M}_B(t) = \left(\frac{1}{\pi}\right)^{1/2} (D_B t)^{1/2} (C_{B/A} - C_B^o)$$

The following mathematical manipulations have been used several times already. The boundary concentration $C_{A/B}$ is replaced by $C_{B/A}$. Since $\mathcal{M}_A$ and $\mathcal{M}_B$ have to be equal, we can solve Eq. 19-54 for $C_{B/A}$, insert the result into one of the above flux equations, and finally get:

$$\mathcal{M}_A(t) = \mathcal{M}_B(t) = \left(\frac{t}{\pi}\right)^{1/2} \frac{C_A^o - C_A^{eq}}{\dfrac{1}{D_A^{1/2}} + \dfrac{1}{K_{B/A} D_B^{1/2}}}$$

(19-55)

where:

$$C_A^{eq} = \frac{C_B^o}{K_{B/A}}$$

(19-15)

is the A-phase concentration that would be in equilibrium with C_B^o.

Remember that $\mathcal{M}_A$ and $\mathcal{M}_B$ represent the exchanged mass per unit area integrated over time t. Let us assume that the two phases remain in contact during the average exposure time, t_{exp}. In order to get the average mass flux per unit time, $\overline{F}$, we have to evaluate $\mathcal{M}_A$ (or $\mathcal{M}_B$) at $t = t_{exp}$ and divide it by t_{exp}:

$$\overline{F} = \frac{\mathcal{M}_A(t_{exp})}{t_{exp}} = \left(\frac{1}{\pi t_{exp}}\right)^{1/2} \frac{C_A^o - C_A^{eq}}{\dfrac{1}{D_A^{1/2}} + \dfrac{1}{K_{B/A} D_B^{1/2}}}$$

(19-56)

By analogy to Eqs. 19-19 and 19-20, $\overline{F}$ can be written as the product of the A-phase disequlibrium, $(C_A^o - C_A^{eq})$, and a transfer velocity v_{tot}:

$$\overline{F} = v_{tot}(C_A^o - C_A^{eq}) \tag{19-57}$$

where:

$$\frac{1}{v_{tot}} = \frac{1}{v_A} + \frac{1}{v_B K_{B/A}} \tag{19-58}$$

and:

$$v_A = \left(\frac{D_A}{\pi t_{exp}}\right)^{1/2}, \ v_B = \left(\frac{D_B}{\pi t_{exp}}\right)^{1/2} \tag{19-59}$$

Note that Eqs. 19-20 and 19-58 are identical. Thus, we can immediately identify the same limiting situations (see Eqs. 19-22 to 19-24):

1. $v_A \ll v_B K_{B/A} \Leftrightarrow D_A^{1/2} \ll K_{B/A} D_B^{1/2}$:

    A-layer controlled exchange: $v_{tot} = v_A$

2. $v_A \gg v_B K_{B/A} \Leftrightarrow D_A^{1/2} \gg K_{B/A} D_B^{1/2}$

    B-layer controlled exchange: $v_{tot} = K_{B/A} v_B$

The result for the diffusive boundary differs from the bottleneck exchange process in the following way:

1. v_A and v_B depend on the square root of the diffusivities, D_A and D_B.

2. The exposure time t_{exp} plays the role of the free parameter that in the bottleneck model is played by the film thickness δ.

In Chapter 20, the diffusive boundary model is employed as one of the models to describe air–water exchange.

Advanced Topic 19.5   Spherical Boundaries

In this section we deal with mass fluxes across spherical boundaries like the surface of gas bubbles in liquids, droplets in air, suspended particles or algal cells in water. It is true that suspended solids are rarely shaped like ideal spheres. Nonetheless, the following discussion can serve as a conceptual starting point from which more complex structures can be analyzed. Obviously, such situations require the application of numerical models, yet some of the principles, like the existence of characteristic length and time scales, will remain the same.

Mass exchange between finite solid particles and surrounding fluids or gases is of paramount importance for many environmental processes. Examples are the uptake of nutrients by cells, chemical transformation processes of vapor molecules in the atmosphere which occur in the liquid phase of cloud droplets, and the sorption of chemicals on natural particles.

In the following we proceed from the simple to the more complex models. First we describe the mass exchange by a fluid-side bottleneck model and assume that transport within the sphere can be disregarded either because it is fast or because the capacity of the particle surface is large enough to store all the substance taking part in the exchange process. Next, we look at the sorption of a solute into a porous spherical aggregate of smaller particles. Here the kinetics is controlled by diffusion in the pores of the aggregate. This means that the particle surface is described as a spherical wall boundary (for which the mathematical basis has been laid in Section 18.2). Finally the two models are combined. A critical time is determined that characterizes the transition from the bottleneck-controlled to the wall-boundary-controlled flux. Such consider-ations provide the basis for models of sorption kinetics of organic compounds to soil and sediment particles. If one combines the full complexity of sorption as discussed in Chapters 9 and 11 (nonlinearity, finite kinetics, etc.) with the complex shape of real particles, things become really difficult. Yet, understanding of the simple case is a prerequisite for dealing with the more complex one.

Bottleneck Boundary Around a Spherical Structure

Let us consider a spherical entity (particle, bubble, droplet) surrounded by a fluid medium (gas, liquid). The fluid and the sphere are separated by a fluid-side concentric boundary layer of thickness δ (Fig. 19.16). The concentration at the surface of the sphere is C_S; the fluid concentration outside the boundary layer is C_F. The equilibrium relation between S and F is defined by a partition coefficient $K_{S/F} = C_S/C_F^{eq}$ (Eq. 19-16).

Due to the spherical geometry of the surface, the concentration profile across the boundary layer is no longer a straight line as was the case for the flat bottleneck boundary (Fig. 19.4). We can calculate the steady-state profile by assuming that C_F and $C_F^{eq} = C_S/K_{S/F}$ are constant. Then, the integrated flux, ΣF, across all concentric shells with radius r inside the boundary layer ($r_o \leq r \leq r_o + \delta$) must be equal:

$$\Sigma F = -4\pi r^2 D_{bl} \frac{\partial C}{\partial r} = \text{constant} \qquad (19\text{-}60)$$

D_{bl} is diffusivity in the boundary layer and $(4\pi r^2)$ is the shell surface. Solving for the concentration gradient $\partial C/\partial r$ yields:

$$\frac{\partial C}{\partial r} = -\frac{\Sigma F}{4\pi D_{bl} r^2} \qquad (19\text{-}61)$$

If we integrate Eq. 19-61 from r_o to $r_o + \delta$, we get the concentration difference between the sphere and the fluid:

$$\Delta C \equiv C_F - C_F^{eq} = C_F - \frac{C_S}{K_{S/F}} \qquad (19\text{-}62)$$

$$= \int_{r_o}^{r_o+\delta} \frac{\partial C}{\partial r} dr = -\frac{\Sigma F}{4\pi D_{bl}} \int_{r_o}^{r_o+\delta} \frac{1}{r^2} dr = -\frac{\Sigma F}{4\pi D_{bl}} \left(\frac{1}{r_o} - \frac{1}{r_o+\delta} \right)$$

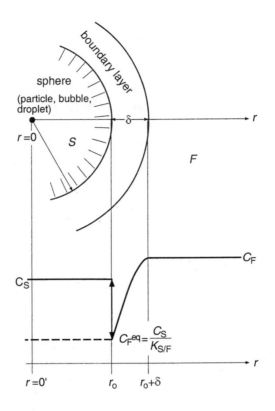

Figure 19.16 Spherical structure (particle, bubble, droplet) with radius r_o surrounded by a concentric fluid boundary layer with thickness δ. r is the spherical coordinate. The concentration inside the sphere is C_S. There is a phase change at the surface of the sphere with the equilibrium partition coefficient $K_{S/F} = C_S / C_F^{eq}$. C_F^{eq} is the fluid concentration in equilibrium with C_S.

Solving for ΣF yields the integrated flux across the spherical boundary:

$$\Sigma F = -\,4\pi D_{bl}\left(C_F - \frac{C_S}{K_{S/F}} \right)\left(\frac{1}{r_o} - \frac{1}{r_o + \delta} \right)^{-1} \quad [MT^{-1}] \qquad (19\text{-}63)$$

The sign of ΣF is such that ΣF is positive if the flux is directed from the sphere to the fluid (positive r-direction).

At first sight Eq. 19-63 does not resemble the type of equation that we found earlier for bottleneck boundaries (e.g., Eq. 19-19). That is not surprising since ΣF is an integrated flux (mass per unit time), while in the case of flat boundaries we have always dealt with specific fluxes (mass per unit area and time). If we divide ΣF by the surface of the sphere, $4\pi r_o^2$, after some algebraic rearrangements we get:

$$F = \frac{\Sigma F}{4\pi r_o^2} = -v_{eff}\left(C_F - \frac{C_S}{K_{S/F}} \right) \quad [ML^{-2}T^{-1}] \qquad (19\text{-}64)$$

where the effective exchange velocity is:

$$v_{eff} = \frac{D_{bl}}{\delta_{eff}} \quad \text{with} \quad \frac{1}{\delta_{eff}} = \left(\frac{1}{r_o} + \frac{1}{\delta} \right) \qquad (19\text{-}65)$$

This is an extremely useful result. The specific flux now looks like Eq. 19-3 where the boundary layer thickness δ is replaced by the effective thickness δ_{eff} given by Eq. 19-65. To understand the meaning of the latter, we discuss the following cases:

1. $\delta \ll r_0$: Then $\delta_{eff} \sim \delta$. The spherical boundary behaves like a flat bottleneck boundary. Compared to the boundary layer thickness d, the curvature of the sphere is too small to affect the flux.

2. $\delta \gg r_0$: Then $\delta_{eff} \sim r_0$. This is the case of a small sphere surrounded by a thick boundary zone. Then the flux across the surface of the sphere is solely controlled by the sphere radius r_0. As a special case, this includes the situation of a sphere suspended in an infinite stagnant fluid. The smaller the sphere, the larger the flux. Small algal cells profit from that. In fact, the total flux into such a sphere is:

$$\Sigma F = -4\pi r_0^2 \cdot \frac{D_{bl}}{r_0}\Delta C = -4\pi r_0 D_{bl}\Delta C, \quad \text{if } \delta \gg r_0 \tag{19-66}$$

Yet, the important quantity is the flux per volume; it clearly favors small cells:

$$\frac{\Sigma F}{V} = \frac{3}{4\pi r_0^3}\Sigma F = -\frac{3}{r_0^2}D_{bl}\Delta C, \quad \text{if } \delta \gg r_0 \tag{19-67}$$

3. As an intermediate case, we consider $r_0 = \delta$. Then $\delta_{eff} = \frac{r_0}{2}$.

The following thought experiment may help our understanding of the spherical boundary layer problem. Assume that a flat boundary layer with thickness δ is limiting transport of a chemical with diffusivity D_{bl}. Then the corresponding exchange velocity is $v_{flat} = D_{bl}/\delta$, hence identical with the exchange velocity at the surface of a sphere with infinite radius. Now imagine that the interface begins the bend into a spherical surface with decreasing radius r_0 while δ remains constant. With Eq. 19-65 we can calculate how the exchange velocity behaves relative to the flat case:

$$\frac{v_{eff}(r_0)}{v_{flat}} = \frac{v_{eff}(r_0)}{v_{eff}(r \to \infty)} = 1 + \frac{\delta}{r_0} \tag{19-68}$$

This result demonstrates how the spherical shape of the boundary enhances the flux. In the extreme case, $r_0 \to 0$, the enhancement becomes infinitely large.

Sorption Kinetics for Porous Particles Surrounded by Water

Next we analyze the sorption kinetics of a sorbate with constant aqueous concentration, C_w^0, sorbing into a porous spherical aggregate with radius r_0. More precisely, the "macroparticle" is a homogeneous aggregate of "microparticles" which are separated by micropores filled with water (Fig. 19.17). The sorbate diffuses in these pores and sorbs to the microparticles. It is not relevant whether sorption occurs at the surface or in the interior of the microparticles as long as we can assume that sorption equilibrium between the solute concentration and the microparticles at each position within the aggregate is attained instantaneously.

Our aim is to derive a model that describes the temporal evolution of the ratio of the average concentration per unit particle mass measured on the dried macroparticle, $C_s(t)$, and the dissolved concentration outside the macroparticles, C_w^0:

$$C_s(t) = K_d(t) \cdot C_w^0 \tag{19-69}$$

where $K_d(t)$ is the time-dependent macroscopic solid–water distribution ratio. In

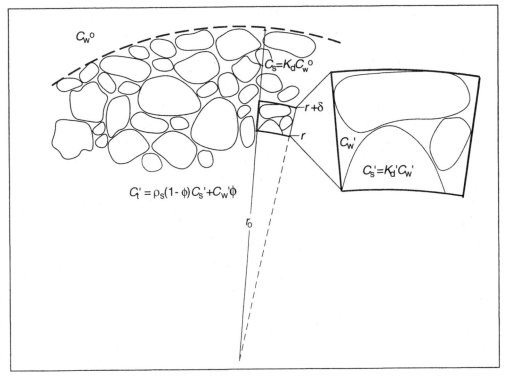

Figure 19.17 Spherical macroparticle with radius r_o consisting of an aggregate of microparticles separated by micropores filled with water. A chemical with constant concentration C_w^o diffuses into the pore volume of the macroparticle. The local dissolved pore concentration C_w' is at instantaneous equilibrium with the local sorbed phase C_s' (K_d' is microscopic equilibrium coefficient). Note that the macroscopic distribution coefficient K_d is time dependent (see Eq. 19-78.)

contrast, the solute concentration inside the aggregate pores, C_w', is always at equilibrium with the local sorbed phase, C_s':

$$C_s' = K_d' C_w' \qquad (19\text{-}70)$$

where K_d' is the (microscopic) distribution coefficient that applies within the aggregate. The primed variables refer to microscopic quantities which depend on t and r. The total local concentration per bulk volume, C_t', is the weighted sum of molecules in dissolved and sorbed phases:

$$C_t' = \rho_s(1-\phi)C_s' + \phi C_w' = \left[\rho_s(1-\phi)K_d' + \phi\right]C_w' \qquad (19\text{-}71)$$

where ρ_s is the density of dry solid, ϕ is the porosity of the particle aggregate, and we have made use of Eq. 19-70.

Assuming that the sorbed phase does not diffuse, the radial flux of the dissolved sorbate inside the particle aggregate is:

$$F = -\phi D_{pm} \frac{\partial C_w'}{\partial r} \qquad (19\text{-}72)$$

where D_{pm} is the diffusivity in the porous matter which due to tortuosity (see Eq. 18-57) or the Renkin effect is usually smaller than the open-water diffusivity D_w. Combined with the spherical boundary conditions, the flux leads to the same concentration distribution and integrated flux ratio as shown in Fig. 18.7 and given by Eqs. 18-33 and 18-35, except for the fact that diffusivity in the dissolved phase is slowed down by the ongoing equilibration with the sorbed phase. In Box 18.5

(Eq. 11) we found that the open-water diffusivity is reduced by the dissolved fraction f_w. Adapted to diffusion in the particle aggregate, we get the *effective diffusivity*:

$$D_{eff} = \frac{\phi D_{pm}}{K'_d(1-\phi)\rho_s + \phi} = \frac{D_{pm}}{K'_d r_{sw} + 1} \tag{19-73}$$

where $r_{sw} = \rho_s \dfrac{1-\phi}{\phi}$ is the solid-to-water-phase ratio of the particle aggregate.

Note that in the notation of Fig. 18.6, C_A becomes the aqueous concentration in the pores of the particle aggregate, C'_w, while C_B^o becomes the time-constant fluid concentration C_w^o. For simplicity we consider only the case where the initial concentration in the sphere is zero ($C_A^o = 0$). The dissolved concentration in the sphere, $C'_w(r,t)$, follows from Eq. 18-33 by putting $C_A^o = 0$ and replacing C_B^o by C_w^o. The *half-saturation time* $t_{1/2}$ is given by Eq. 18-37 (D replaced by the effective diffusivity D_{eff}):

$$t_{1/2} = 0.03 \frac{r_o^2}{D_{eff}} = 0.03 \frac{r_o^2}{D_{pm}} \left(1 + K'_d r_{sw}\right) \tag{19-74}$$

To calculate the (macroscopic) distribution coefficient $K_d(t)$ from Eq. 19-69 we have to remember that $C_s(t)$ is defined as the total mass of the chemical in the dried particle divided by the particle mass. This includes the dissolved fraction of the chemical in the pore water when the particles are dried. Thus from Eq. 19-71:

$$C_s(t) = \frac{\overline{C'_t(t)}}{(1-\phi)\rho_s} = \left[K'_d + \frac{\phi}{(1-\phi)\rho_s} \right] \overline{C'_w(t)} \tag{19-75}$$

where the overbars indicate the spatial average over the particle aggregate. From Eq. 19-70:

$$K_d(t) = \frac{C_s(t)}{C_w^o} = \left[K'_d + \frac{\phi}{(1-\phi)\rho_s} \right] \frac{\overline{C'_w(t)}}{C_w^o} \tag{19-76}$$

For $t \to \infty$, the pore water concentration is everywhere equal to the external solute concentration. Thus, $\overline{C'_w(t)} = C_w^o$ and:

$$K_d(t \to \infty) \equiv K_d^\infty = \left[K'_d + \frac{\phi}{(1-\phi)\rho_s} \right] = K'_d + \frac{1}{r_{sw}} \tag{19-77}$$

which, for compounds with $K'_d \gg 10$, is almost equal to K'_d. The concentration ratio, $\overline{C'_w(t)} / C_w^o$, is equal to the expression on the right-hand side of Eq. 18-35. Thus putting everything together yields:

$$K_d(t) = K_d^\infty \left[1 - \frac{6}{\pi^2} \sum_{k=1}^{\infty} \frac{1}{k^2} \exp\left(-k^2 \pi^2 t^*\right) \right] \tag{19-78}$$

where K_d^∞ is defined in Eq. 19-77 and the nondimensional time is:

$$t^* = \frac{D_{eff} t}{r_o^2} = \frac{D_{pm} t}{(1 + K'_d r_{sw}) r_o^2} \tag{19-79}$$

Note that the ratio $K_d(t)/K_d^\infty$ corresponds exactly to the curve $\mathcal{M}(t)/\mathcal{M}^\infty$ of the radial diffusion model defined in Eq. 18-35 and shown in Fig. 18.7b.

The radial diffusion model can be approximated by a linear uptake model of the form:

$$\frac{dK_d}{dt} = k_{lin}(K_d^\infty - K_d) \tag{19-80}$$

where the specific linear sorption rate, k_{lin} (dimension T^{-1}), is chosen such that the half-saturation time, $t_{1/2}$, is equal for both models. According to Eq. 6 of Box 12.1, the solution of Eq. 19-80 is:

$$K_d(t) = K_d^\infty \left(1 - e^{-k_{lin}t}\right) \tag{19-81}$$

with $t_{1/2} = \ln 2 / k_{lin}$ where $t_{1/2}$ must be the same as $t_{1/2}$ of Eq. 19-74. Thus:

$$k_{lin} = \frac{\ln 2}{t_{1/2}} = \frac{\ln 2}{0.03}\frac{D_{eff}}{r_o^2} \cong 23\frac{D_{pm}}{r_o^2(1 + K_d'r_{sw})} \tag{19-82}$$

The radial diffusion model and the linear sorption model are compared in Fig. 18.7b. Since according to Eq. 19-76 the total mass of the chemical associated with the particle aggregate, $\mathcal{M}(t)$, and the macroscopic solid–water distribution ratio, $K_d(t)$, are linearly related:

$$\mathcal{M}(t) = \frac{4\pi}{3}r_o^3\rho_s(1 - \phi)\,C_s(t) = \frac{4\pi}{3}r_o^3\rho_s(1 - \phi)\,C_w^o K_d(t), \tag{19-83}$$

the mass ratio shown in Fig. 18.7b and the sorption ratio, $K_d(t)/K_d^\infty$, are identical. Note that the linear model underpredicts sorption at short times and overpredicts it at long times. Since in Fig. 18.7b time is shown on a logarithmic scale, at first sight it is not obvious that the mass flux across the aggregate surface has its maximum for $t = 0$ and then steadily decreases until saturation is approached for $t \to \infty$. In fact, the total flux, ΣF, is given by:

$$\Sigma F = -\frac{d\mathcal{M}(t)}{dt} = -\frac{4\pi}{3}r_o^3\rho_s(1 - \phi)\,C_w^o\frac{dK_d(t)}{dt}, \tag{19-84}$$

where we have used the sign convention introduced in Eq. 19-63 according to which ΣF is positive if the flux is directed in the positive r-direction (i.e., from the sphere to the fluid). To show how ΣF changes with time we insert the result of the linear approximation (Eqs. 19-80 and 19-81) and then use Eq. 19-77:

$$\Sigma F_{lin} = -\frac{4\pi}{3}r_o^3\rho_s(1 - \phi)C_w^o k_{lin}[K_d^\infty - K_d(t)]$$

$$= -\frac{4\pi}{3}r_o^3\rho_s(1 - \phi)C_w^o k_{lin}K_d^\infty e^{-k_{lin}t} = -\frac{4\pi}{3}r_o^3 C_w^o k_{lin}\phi(1 + K_d'r_{sw})\,e^{-k_{lin}t} \tag{19-85}$$

The absolute value of ΣF_{lin} has its maximum at $t = 0$. Note from Fig. 18.7b that ΣF of the radial diffusion model is even larger than ΣF_{lin}.

Given the possibility that the particle aggregate is surrounded by a diffusive boundary layer, the question arises whether during the initial phase of the sorption process transport across the bottleneck may limit the flux and, if so, at what time the flux would change from a boundary layer-controlled process to a process determined by diffusion in the particle aggregate. Remember that in Section 19.3 we dealt with a similar problem, that is, with the combination of a wall boundary and a boundary layer at a *flat surface* (see Eqs. 19-40 to 19-45). The mathematics for the spherical boundary layer is derived in Box 19.3. As it turns out the aquatic boundary layer may slow down the kinetics of sorption during the initial phase, that is, for times smaller than $t_{1/2}$. Neglecting this effect creates an error which is of similar size as replacing the radial diffusion model by the linear model (see Fig. 18.7b). Fortunately, these errors have opposite signs and thus compensate each other at least partially. The situation is further discussed in Illustrative Example 19.5 (see below).

Box 19.3   Spherical Wall Boundary with Boundary Layer

We consider a spherical particle aggregate with radius r_o surrounded by a concentric boundary layer of thickness δ (Fig. 19.16). Transport into the aggregate is described by the linear approximation of the radial diffusion model. Thus, the total flux from the particle to the fluid is given by Eq. 19-85:

$$\Sigma F_{\text{lin}} = -\frac{4\pi}{3} r_o^3 C_w^{\text{eq}} k_{\text{lin}} \phi (1 + K_d' r_{\text{sw}}) \, e^{-k_{\text{lin}} t} \tag{1}$$

where we have replaced C_w^{o} by the water-side concentration at the interface between boundary layer and particle, C_w^{eq}. According to Eq. 19-63 the flux through the boundary layer is:

$$\Sigma F_{\text{bl}} = -4\pi D_{\text{bl}} (C_w^{\text{o}} - C_w^{\text{eq}}) \delta^* \tag{2}$$

where:

$$\delta^* = \left(\frac{1}{r_o} - \frac{1}{r_o + \delta} \right)^{-1} = \frac{r_o (r_o + \delta)}{\delta} \tag{3}$$

At quasi-steady state, the two fluxes, Eqs. 1 and 2, have to be equal. This allows us to calculate the boundary layer concentration C_w^{eq}:

$$C_w^{\text{eq}} = \frac{D_{\text{bl}} \delta^*}{D_{\text{bl}} \delta^* + \frac{1}{3} r_o^3 k_{\text{lin}} \phi (1 + K_d' r_{\text{sw}}) e^{-k_{\text{lin}} t}} C_w^{\text{o}} \tag{4}$$

If the concentration difference across the boundary, $C_w^{\text{o}} - C_w^{\text{eq}}$, is small compared to C_w^{o}, that is, if C_w^{eq} and C_w^{o} are about equal, then the water-side boundary layer has no significant influence on the flux across the particle surface. According to Eq. 4 this is the case if the following condition holds:

$$D_{\text{bl}} \delta^* \gg \frac{1}{3} r_o^3 k_{\text{lin}} \phi (K_d' r_{\text{sw}} + 1) e^{-k_{\text{lin}} t} \tag{5}$$

Since the right-hand side of this expression decreases with elapsed time t, the sorption process always reaches a point beyond which Eq. 5 is valid, that is, beyond which the boundary layer has no influence on the sorption kinetics. We call the time when both sides of Eq. 5 are equal, the critical time t_{crit}. Then:

$$e^{k_{lin}t} = \frac{r_o^3 k_{lin}\phi(1+K_d'r_{sw}')}{3D_{bl}\delta^*} \quad \rightarrow \quad t = t_{crit} \equiv \frac{1}{k_{lin}} \ln\left(\frac{r_o^3 k_{lin}\phi(1+K_d'r_{sw}')}{3D_{bl}\delta^*}\right) \tag{6}$$

The influence of the boundary layer on sorption kinetics disappears for times $t \gg t_{crit}$. Note that Eq. 6 makes sense only if the expression in the logarithm is greater than 1. If this is not the case, then the right-hand side of Eq. 5 is already smaller than the left-hand side when sorption begins. In order to see when this is the case, we rewrite the logarithmic expression by replacing k_{lin} by Eq. 19-82 and δ^* by Eq. 3 from above. After some algebraic manipulations we get:

$$\frac{r_o^3 k_{lin}\phi(1+K_d'r_{sw}')}{3D_{bl}\delta^*} = \frac{23}{3}\phi\frac{\delta}{\delta+r_0}\frac{D_{pm}}{D_{bl}} \approx 8\phi\frac{\delta}{\delta+r_0}\frac{1}{\tau} \tag{7}$$

where τ is tortuosity of the porous particle aggregate. Note that we have used Eq. 18-57 with $D_{bl} = D$. With ϕ typically lying between 0.2 and 0.5, τ between 1.5 and 6, it turns out that the expression in Eq. 7 is often smaller than 1. Yet, this does not mean that the aquatic boundary layer has absolutely no influence on sorption. In fact, as shown in Fig. 18.7b, during the initial phase of sorption the radial diffusion model would yield a much larger diffusive flux than the linear model. This "real" flux may be significantly reduced by the aqueous boundary layer, even in these cases when the linear model indicates little or no influence. In other words: replacing the radial flux model by the linear model may create a similar error as neglecting the influence of the aquatic boundary layer. Fortunately, these errors have opposite signs and thus tend to compensate each other.

Finite Bath Sorption

Until now we have tacitly assumed that the initial concentration in the fluid outside the sphere, C_w^o, remains constant during the process of sorptive equilibration between fluid and particle. This situation is called the *infinite bath* case (Wu and Gschwend, 1988), since it requires an infinite fluid volume. In the case where the fluid volume is finite and thus a significant portion of the dissolved load is transferred to the particle, things become more complicated. The finite bath situation can be characterized by the nondimensional number:

$$\gamma = K_d' S_t \tag{19-86}$$

where S_t is the solid concentration per total volume (dimension ML^{-3}). γ corresponds to the sorbed fraction of the chemical in the total system (particle aggregate and surrounding fluid) when complete sorption equilibrium has been reached. Note that for the infinite bath case the water volume is assumed to be infinitely large, thus $\gamma = 0$.

Figure 19.18 shows how for different values of γ sorption proceeds as a function of the nondimensional time t^* defined by Eq. 19-79. As γ increases from the infinite bath case, we see that the time required to reach equilibrium decreases. This can be understood by recognizing that while the chemical is diffusing into the sphere, the concentration in the surrounding fluid drops. Hence, the total mass exchange needed to attain equilibrium between the fluid and the sphere is smaller than in the infinite bath case in which the external concentration remains constant.

Examining the curves for the finite bath cases allows us to specify k_{lin} of Eq. 19-82 when the fluid concentration, C_w^o, is not constant. For example, we may be interested

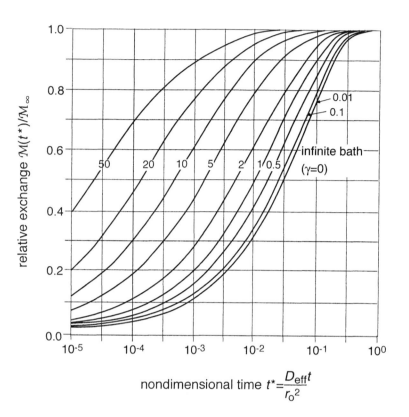

Figure 19.18 Dynamics of diffusive uptake of a chemical by a spherical particle suspended in a fluid of finite volume. The numbers on the curves correspond to γ defined by Eq. 19-86, that is, to the fraction of the chemical taken up by the sphere when equilibrium is reached. From Wu and Gschwend (1988).

in a case where γ is 5. From Fig. 19.18 we learn that the nondimensional time, t^*, required to reach 50% equilibration, is about 1.5×10^{-3}. Hence:

$$t_{1/2}(\gamma = 5) = 1.5 \times 10^{-3} \frac{r_o^2}{D_{\mathrm{eff}}} = 1.5 \times 10^{-3} \frac{(1 + K_d' r_{\mathrm{sw}}) r_o^2}{D_{\mathrm{pm}}} \qquad (19\text{-}87)$$

The approximate first-order rate constant should be:

$$k_{\mathrm{lin}}(\gamma = 5) = \frac{\ln 2}{t_{1/2}(\gamma = 5)} = \frac{\ln 2}{0.0015} \frac{D_{\mathrm{eff}}}{r_o^2} = 460 \frac{D_{\mathrm{pm}}}{r_o^2} \left(1 + K_d' r_{\mathrm{sw}}\right) \qquad (19\text{-}88)$$

which is 20 times larger than k_{lin} of the infinite bath case (Eq. 19-82). The two expressions, Eqs. 19-82 and 19-88, can be written in the general form:

$$k_{\mathrm{lin}}(\gamma) = \alpha_\gamma \frac{D_{\mathrm{eff}}}{r_o^2} \qquad (19\text{-}89)$$

where the factor α_γ increases from $\alpha_\gamma = 23$ at $\gamma = 0$ to $\alpha_\gamma = 460$ at $\gamma = 5$. The increase of α_γ with γ indicates that the equilibration process between the sphere and the water becomes faster when the bath volume surrounding the particle gets smaller. This is confirmed by Fig. 19.19, where the radial diffusion model and the first-order sorption model are compared for different γ. The curves move to smaller times when γ gets larger. Based on the investigation by Wu and Gschwend (1988), $k_{\mathrm{lin}}(\gamma)$ can be approximated by the second-order polynomial:

$$k_{\mathrm{lin}}(\gamma) = (11\gamma^2 + 34\gamma + 23) \frac{D_{\mathrm{eff}}}{r_o^2} \qquad (19\text{-}90)$$

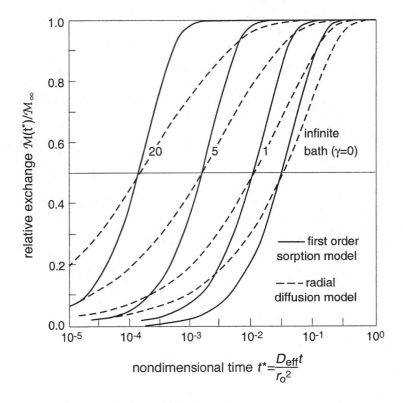

Figure 19.19 Comparison of the solution of the linear sorption model with the radial diffusion model. Numbers on curves show γ defined in Eq. 19-86. γ is the fraction of the chemical taken up by the sphere when equilibrium is reached. After Wu and Gschwend (1988).

Figure 19.19 clearly demonstrates that the linear approximation is well suited for describing the midregion of the exchange process, but does poorly for early and late times.

Illustrative Example 19.5

Desorption Kinetics of an Organic Chemical from Contaminated Sediments

During dredging operations in an estuary, polluted sediment material spills and then settles through the water column. It is feared that polychlorinated biphenyls (PCBs) might desorb while the sedimentary particles remain in suspension and thus pollute the water column. Due to their larger solid–water distribution ratio the heavy PCB congeners accumulate more strongly on sediments than the less chlorinated congeners. Therefore, the heavy congeners are the more likely candidates for redissolution. Their typical physicochemical properties are listed below. The mean depth of the estuary is 20 m. The PCB concentration in the estuarian water before dredging is below detection limit.

Problem

(a) As a first step, use the half-saturation time, $t_{1/2}$, from the linear sorption model to decide whether desorption of the PCBs from the sediment particles is relevant.

(b) How would the result from (a) be altered if the radial diffusion model were used instead?

(c)   Would the existence of an aquatic boundary layer around the particle aggre gate with $\delta = 50$ μm have a significant influence on these results?

Typical properties of heavy PCB congeners

Aqueous diffusivity,
$D_w = 6 \times 10^{-6}$ cm^2 s^{-1}

Octanol–water partition constant,
$K_{ow} = 10^7$

Properties of sedimentary particle aggregates:

Radius, $r_o = 50$ μm

Porosity, $\phi = 0.2$

Tortuosity, $\tau = 6$

Organic carbon content,
$f_{oc} = 0.04$

Density of solid material,
$\rho_s = 2.5$ g cm^{-3}

Estimated settling velocity,
$v_s = 0.5$ m d^{-1}

Answer (a)

The (microscopic) distribution coefficient of the PCB congeners, K_d', can be esti mated from K_{ow} with the relations developed in Chapter 9 (Eqs. 9-29a and 9-22):

$$\log K_{oc} = 0.74 \log K_{ow} + 0.15 = 0.74 \times 7 + 0.15 = 5.33$$

$$\rightarrow \quad K_{oc} = 2.1 \times 10^5 \text{ L kg}^{-1} = 2.1 \times 10^2 \text{ m}^3\text{kg}^{-1}$$

$$\rightarrow \quad K_d' = f_{oc} K_{oc} = 0.04 \times 2.1 \times 10^2 \text{ m}^3\text{kg}^{-1} = 8.4 \text{ m}^3\text{kg}^{-1}$$

The solid-to-water-phase ratio is:

$$r_{sw} = \frac{\rho_s(1-\phi)}{\phi} = \frac{2500 \text{ kg m}^{-3} \times 0.8}{0.2} = 1 \times 10^4 \text{ kg m}^{-3}$$

According to Eqs. 18-57 and 19-73, the effective diffusivity is:

$$D_{eff} = \frac{D_w}{\tau(K_d' r_{sw} + 1)} \approx \frac{6 \times 10^{-6} \text{ cm}^2\text{s}^{-1}}{6 \times 8.4 \times 10^4} = 12 \times 10^{-12} \text{ cm}^2\text{s}^{-1}$$

Since the water mass getting into contact with the suspended sediment particles is probably very large compared to the mass of solids that will spill and settle through the water body, the infinite bath approximation can be used ($\gamma = 0$). Thus from Eq. 19-74:

$$t_{1/2} = 0.03 \frac{(5 \times 10^{-3} \text{ cm})^2}{12 \times 10^{-12} \text{ cm}^2\text{s}^{-1}} = 6.3 \times 10^4 \text{s} = 0.7 \text{ day}$$

The average time a particle remains in the water column is 20 m/0.5 m day^{-1} = 40 day. Thus the above result indicates that a significant portion of the PCBs desorb while the particles are suspended.

Answer (b)

According to Eq. 19-79 the nondimensional time characterizing the radial diffusion model is:

$$t^* = \frac{12 \times 10^{-12} \text{ cm}^2\text{s}^{-1} \times 40 \text{ d} \times 8.64 \times 10^4 \text{s d}^{-1}}{(5 \times 10^{-3} \text{ cm})^2} = 1.7$$

According to Fig. 18.7b, about 98% of the PCBs can desorb while the particles are suspended.

Answer (c)

In order to find out whether an aquatic boundary layer around the suspended particle may significantly slow down desorption, we evaluate Eq. 7 of Box 19.3:

$$8\phi \frac{\delta}{\delta + r_0} \frac{1}{\tau} = 8 \times 0.2 \times \frac{50}{50 + 50} \times \frac{1}{6} = 0.13 < 1$$

Thus, when desorption begins ($t = 0$), the term which accounts for the influence of the boundary layer around the particles (second term in the denominator of Eq. 4 of Box 19.3) is only 13% of the term describing diffusion into the particle aggregate and decreases when time grows. We conclude that the existence of an aquatic boundary layer does not significantly alter the above result, according to which most of the PCB desorbs into the water while the sediment particles are suspended in the water column.

Note: The contribution of particle resuspension followed by desorption to the total sediment–water exchange will be further discussed in Chapter 23 (Box 23.2 and Table 23.6).

19.6   Questions and Problems

Q 19.1

Explain the difference between a boundary and an interface. Give examples for both.

Q 19.2

We have classified boundaries of environmental systems in just three different types. What are they and how are they characterized?

Q 19.3

The flux across a bottleneck boundary can be expressed either in terms of Fick's first law or by a transfer velocity. Explain how the two views are related.

Q 19.4

In the expression for the total exchange velocity across a two-layer bottleneck boundary between different media (Eq. 19-20) the transfer velocity v_B is multiplied by the extra factor $K_{A/B}$. What is $K_{A/B}$? Can you imagine a scheme in which v_A carries an extra factor instead? How are the two schemes related?

Q 19.5

Assume that the concentrations on either side of a boundary, C_A and C_B, are constant. You calculate the flux across this boundary by treating it (*a*) as a bottleneck boundary and (*b*) as a wall boundary, respectively. How does the flux evolve as a function of time in these two models?

Q 19.6

Explain the concept of the *half-concentration penetration depth* of the wall boundary model.

Q 19.7

In the wall boundary model with additional boundary layer we have defined a critical time, t_{crit}. Explain the meaning of this time.

Q 19.8

Water temperature in a lake fluctuates annually around a mean value. How far into the sediment column do these temperature changes penetrate? Use the tools developed for the wall boundary model to make a first estimate and assume that the sediment is mostly pure water.

Q 19.9

What are the advantages for an algal cell to be small?

Q 19.10

Explain in words the difference between the curves shown in Fig. 19.18.

Q 19.11

Explain the difference between the two types of curves shown in Fig. 19.19, which are labeled *radial diffusion model* and *sorption model*, respectively.

Problems

P 19.1   The Thermocline of a Lake as a Two-Layer Bottleneck Boundary

In Illustrative Example 19.1, we calculated the vertical exchange of water across the thermocline in a lake by assuming that transport from the epilimnion into the hypolimnion is controlled by a bottleneck layer with thickness $\delta = 4m$. From experimental data the vertical diffusivity was estimated to lie between 0.01 and 0.04 cm^{2}s^{-1}. Closer inspection of the temperature profiles (see figure in Illustrative Example 19.1) suggests that it would be more adequate to subdivide the bottleneck boundary in two or more sublayers, each with its own diffusivity.

You are interested in the question whether such a refinement of the model would have a significant influence on the total exchange velocity v_{ex}^{tot}. Therefore you decide to calculate v_{ex}^{tot} in two different ways. First, you assume *one* bottleneck boundary of 4 m with an average diffusivity of 0.025 cm^{2}s^{-1}. Second, you divide the bottleneck into two layers which are 2 m thick each. In the upper layer where the temperature gradient is steeper (see figure in Illustrative Example 19.1) you assume a diffusivity of 0.01 cm^{2}s^{-1}, in the lower a diffusivity of 0.04 cm^{2}s^{-1}. Compare the overall exchange velocity with the value from the first method. (Note that the mean diffusivity in the thermocline is equal in both cases.)

P 19.2   *How Fast Does a Patch of Hexachlorobenzene (HCB) Disappear from the Surface of the Ocean?*

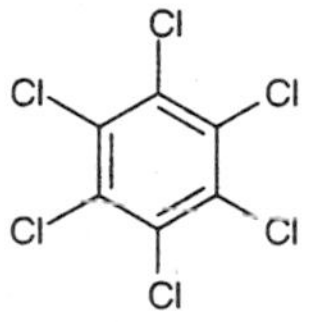

hexachlorobenzene

As a result of an accident a cloud of hexachlorobenzene (HCB) of approximatively circular shape floats in the surface-mixed layer of the ocean. At time t_0 when the patch is first detected, it has approximately the shape of a two-dimensional normal distribution:

$$C(r, t_0) = C_0 \exp\left(-\frac{r^2}{2R_0^2}\right)$$

where r is distance from the center of the cloud

$R_o = 5$ km is a characteristic length scale for the size of the cloud at time t_o

$C_o = 25$ nmol L^{-1} is the concentration at the center of the cloud at time t_o

Estimate the maximum concentration, $C_{max}(t)$, 24 hours and one week later.

Hints and Help: Assume that as a first approximation horizontal turbulent diffusivity can be considered to be isotropic, that is, $E_x = E_y = 2 \times 10^4 cm^2 s^{-1}$. Disregard the loss of HCB across the thermocline and to the atmosphere. In Chapter 22.3 we will see that because of horizontal water currents horizontal diffusion is, in fact, not isotropic. Nonetheless the above approximation yields reasonable results. Finally, note that in order to keep the total mass of HCB constant, $C_o(t)$ must decrease as $R(t)$ grows such that $C_o R^2 = $ constant.

P 19.3   *Three-Layer Bottleneck Model*

(a) Derive a general expression for the total exchange velocity, v_{tot}, across an interface which is composed of three bottleneck boundaries. As an example, consider the air–water exchange for the case that the water surface is covered by a thin organic surface layer. Define the water phase as system A and use it as the reference system. Then the organic layer is system B, the air is system C. The transfer velocities in these systems are v_A, v_B, and v_C. The chemical equilibrium between the three phases is defined by $K_{B/A}$ and $K_{C/A}$, respectively.

(b) Apply the model developed in (a) to assess the influence of an organic surface layer (thickness $\delta_{osl} = 0.1$ cm) on the air–water exchange of a chemical. Use the following information (the origin of some of the figures will become clearer in Chapter 20.3 which deals in greater detail with air–water exchange models): $v_{iw} = 1 \times 10^{-4} cm\ s^{-1}$, $v_{ia} = 0.5$ cm s^{-1}. The molecular diffusion coefficient in the organic layer is $D_{i\ osl} = 0.5 \times 10^{-5} cm^2 s^{-1}$. The subscripts stand for w = water, a = air, osl = organic surface layer. Consider first "nonvolatile" compounds ($K_{ia/w} < 10^{-4}$) and determine the range of $K_{iosl/w}$ for which the organic surface layer has an influence on the overall transfer velocity v_{tot}. Second, make the same analysis for "volatile" compounds ($K_{ia/w} > 10^{-4}$). Is it likely to find such compounds and would they belong to the "nonvolatile" or the "volatile" category or to both?

Chapter 20

AIR–WATER EXCHANGE

20.1 Introduction
Illustrated Example 20.1: *Evaluating the Direction of Air–Water Exchange*

20.2 Measurement of Air–Water Transfer Velocities
Transfer Velocities in Air Deduced from Evaporation of Water
Illustrative Example 20.2: *Estimating Evaporation Rates of Pure Organic Liquids*
Transfer Velocities in the Water Phase Deduced from Compounds with Large Henry's Law Constants
Box 20.1: *Influence of Wind Speed Variability on the Mean Air–Water Exchange Velocity of Volatile Compounds*

20.3 Air–Water Exchange Models
Film Model
Surface Renewal Model
Boundary Layer Model
Box 20.2: *Temperature Dependence of Air–Water Exchange Velocity v_{iw} of Volatile Compounds Calculated with Different Models*
Overall Air–Water Exchange Velocities
Illustrative Example 20.3: *Estimating the Overall Air–Water Transfer Velocity from Wind Speed for Different Water Temperatures*

20.4 Air–Water Exchange in Flowing Waters
Box 20.3: *Eddy Models for Air–Water Exchange in Rivers*
Small Roughness and Small-Eddy Model
Rough River Flow and Large-Eddy Model
Enhanced Air–Water Exchange Through Bubbles
Illustrative Example 20.4: *Air–Water Exchange of Benzene in Rivers*

20.5 **Influence of Surface Films and Chemical Reactions on Air–Water Exchange (*Advanced Topic*)**
Surface Films
Influence of Chemical Reactions on Air–Water Exchange Rates
Fast Reaction
Reaction and Diffusion of Similar Magnitude
Illustrative Example 20.5: *Air–Water Exchange Enhancement for Formaldehyde and Acetaldehyde*

20.6 **Questions and Problems**

20.1   Introduction

Air and water are the two most important fluids on earth. The atmosphere and the hydrosphere are complementary in their role of transporting and transforming chemicals. The atmosphere is the fastest and most efficient global conveyor belt. Yet, certain chemicals prefer the aquatic milieu which is of global dimension, as well. In the hydrosphere typical transport velocities are significantly smaller than in the atmosphere. Therefore, residence times of chemicals in the water are usually much larger than in the atmosphere.

The size of the interface between atmosphere and hydrosphere is immense (see Appendix E): 71% of the earth's surface (361×10^6 km^2) is covered by water. In addition, the atmosphere contains about 13×10^{15} kg of water vapor. Expressed as liquid volume, this amounts to 13×10^{12} m^3 or 2.5 cm per m^2 of earth surface. This is a small volume compared to the total ocean volume of 1.37×10^{18} m^3, but it is important in terms of the additional interfacial area between water and air. Although most of the water in the atmosphere is present as water vapor, roughly 50% of the earth's surface is covered by clouds which contain between 0.1 and 1 g of liquid water per cubic meter of air. The water is present in droplets with a typical diameter of 20 μm. Thus, clouds supply an air–water interface area of the order of 0.1 m^2 per cubic meter of air (Seinfeld, 1986). For a cloud cover 500 m thick this would yield an air–water contact zone of 50 m^2 per m^2 of earth surface.

Often we are concerned with the transfer of chemicals between gaseous and liquid phases which are not at equilibrium. On the one hand, this is due to the large sizes of the involved systems, which do not quickly transfer materials from their bulk interior to adjacent phases. On the other hand, in these systems natural biogeochemical reactions as well as man-made processes are continuously driving the global systems away from their equilibrium. And finally, the environmental chemist is often faced with extraordinary (catastrophic) situations in which chemicals are spilled into the environment and transported across different compartments.

All the necessary tools to develop kinetic models for air–water exchange have been derived already in Chapters 18 and 19. However, we don't yet understand in detail the physical processes which act at the water surface and which are relevant for the exchange of chemicals between air and water. In fact, we are not even able to clearly classify the air–water interface either as a bottleneck boundary, a diffusive boundary, or even something else. Therefore, for a quantitative description of mass fluxes at this interface, we have to make use of a mixture of theoretical concepts and empirical knowledge. Fortunately, the former provide us with information which is independent of the exact nature of the exchange process. As it turned out, the different flux equations which we have derived so far (see Eqs. 19-3, 19-12, 19-57) are all of the form:

$$F_{i\,a/w} = v_{i\,a/w}(C_{iw} - C_{iw}^{eq}) \qquad [\mathrm{ML^{-2}T^{-1}}] \qquad (20\text{-}1)$$

where

$$C_{iw}^{eq} = \frac{C_{ia}}{K_{i\,a/w}} \qquad (20\text{-}2)$$

is the aqueous concentration in equilibrium with the atmospheric concentration, C_{ia}, and the indices a,w stand for air and water, respectively. In this chapter the distinction between different compounds will be important, thus all compound-specific quantities are marked by the extra subscript i. Note that in accordance with the notation introduced in Chapters 6, 18, and 19, the equilibrium air–water partition constant (the nondimensional Henry's law constant) is designed as $K_{ia/w}$ It is related to the Henry's law constant, K_{iH}, by:

$$K_{ia/w} = \frac{K_{iH}}{RT} \tag{6-15}$$

In Eq. 20-1 we have chosen the sign of $F_{ia/w}$ such that a positive value indicates a net flux from the water into the atmosphere. As demonstrated in Illustrative Example 20.1, subtle changes of the environmental conditions, such as water temperature, may lead to a reversal of the flux.

Illustrative Example 20.1

i = 1,1,1-trichloroethane
(methyl chloroform, MCF)

M_i = 133.4 g mol^{-1}

K_{iH} (25°C) = 19.5 bar L mol^{-1}

$K_{iH}^{\text{air/seawater}}$(0°C) = 6.5 bar L mol^{-1}

$K_{iH}^{\text{air/seawater}}$(25°C) = 23.8 bar L mol^{-1}

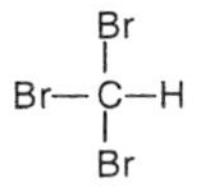

i = tribromomethane (BF)

M_i = 252.8 g mol^{-1}

K_{iH} (25°C) = 0.60 bar L mol^{-1}

$K_{iH}^{\text{air/seawater}}$(0°C) = 0.20 bar L mol^{-1}

$K_{iH}^{\text{air/seawater}}$(25°C) = 0.86 bar L mol^{-1}

Evaluating the Direction of Air–Water Exchange

C_1 and C_2 halocarbons of natural and anthropogenic origin are omnipresent in the atmosphere and the ocean. For example, in the eighties, typical concentrations in the northern hemisphere air and in Arctic seawater of 1,1,1-trichloroethane (also called methyl chloroform, MCF) and tribromomethane (bromoform, BF) were measured by Fogelqvist (1985):

	MCF	BF
Concentration in air (ng L^{-1})	0.93	0.05
Concentration in Arctic Ocean (ng L^{-1})		
surface (0-10 m)	2.5	9.8
at 200 m depth	1.6	3.0

The air–seawater Henry's law coefficients, $K_{iH}^{\text{air/seawater}}$, and the usual Henry's law constants are given in the margin.

Problem

Using the concentrations of MCF and BF given above, evaluate whether there are net fluxes of these compounds between the air and the surface waters of the Arctic Ocean assuming water temperatures of (a) 0°C and (b) 10°C. In each case, if there is a net flux, indicate its direction (i.e., sea-to-air or air-to-sea).

Answer

Independent of the model that is used to describe air–water exchange at the sea surface, the flux F_i is proportional to (Eqs. 20-1, 20-2, where subscript w is replaced by sw for seawater):

$$F_i = \text{constant} \left(C_{iw} - \frac{C_{ia}}{K_{ia/sw}} \right) , \quad \text{where} \quad K_{ia/sw} = \frac{K_{iH}^{\text{air/seawater}}}{RT}$$

A positive sign of this expression indicates a net flux from the water into the air.

(a) $T = 0°C$.

The resulting $K_{ia/sw}$ values are:

$$K_{ia/sw} \ (\text{MCF, } 0°C) \quad = 0.29$$

$$K_{ia/sw} \ (\text{BF, } 0°C) \quad = 0.0089$$

Determine the sign of the flux by using these values together with the measured concentrations given above:

Compound	C_{iw} (ng L^{-1})	$C_{ia}/K_{ia/sw}$ (ng L^{-1})	$C_{iw} - C_a/K_{ia/sw}$ (ng L^{-1})
MCF	2.5	3.2	-0.7
BF	9.8	5.6	$+4.2$

Thus, at $0°C$, there is a net flux of MCF from the atmosphere to the sea, while for BF a net transfer occurs from the water to the atmosphere.

(b) $T = 10°C$

Estimate first $K_{iH}^{\text{air/seawater}}$ at $10°C$ from the $K_{iH}^{\text{air/seawater}}$ values given for $0°C$ and $25°C$ using a temperature dependence of the form (see Eq. 6-8):

$$\ln K_{iH}^{\text{air/seawater}} (T) = -\frac{B_i}{T} + A_i$$

where T is the temperature in Kelvin and A_i, B_i are constant parameters. Since:

$$\ln \left(\frac{K_{iH}^{\text{air/seawater}} (T_1)}{K_{iH}^{\text{air/seawater}} (T_2)} \right) = \ln K_{iH}^{\text{air/seawater}} (T_1) - K_{iH}^{\text{air/seawater}} (T_2)$$

$$= B_i \left(\frac{1}{T_2} - \frac{1}{T_1} \right)$$

the parameter B_i can be calculated from:

$$B_i = \ln \frac{K_{iH}^{\text{air/seawater}} (298.2 \text{ K})}{K_{iH}^{\text{air/seawater}} (273.2 \text{ K})} \cdot \left(\frac{1}{273.2 \text{ K}} - \frac{1}{298.2 \text{ K}} \right)^{-1}$$

Inserting the $K_{iH}^{\text{air/seawater}}$ values given above yields:

$$B_{\text{MCF}} = 4230 \text{ K}$$

$$B_{\text{BF}} \quad = 4750 \text{ K}$$

Use these B_i values with, for example, the $K_{iH}^{air/seawater}$ at 25°C to calculate $K_{iH}^{air/seawater}$ values of the two compounds at 10°C:

$$\ln K_{MCF\ H}^{air/seawater}(10°C) = \ln(23.8) + \left(\frac{1}{298.2} - \frac{1}{283.2}\right) = 2.42$$

Hence:

$$K_{MCF\ H}^{air/seawater}(10°C) = 11.2 \text{ bar L mol}^{-1}$$

and therefore:

$$K_{MCF\ a/sw}(10°C) = 0.48$$

Similarly:

$$\ln K_{BF\ H}^{air/seawater}(10°C) = \ln(0.86) + 4750\left(\frac{1}{298.2} - \frac{1}{283.2}\right) = -0.99$$

or:

$$K_{BF\ H}^{air/seawater}(10°C) = 0.37 \text{ bar L mol}^{-1}$$

and:

$$K_{BF\ a/sw}(10°C) = 0.016$$

Inserting $K_{ia/sw}$ values in the flux equations yields:

Compound	C_{iw} (ng L^{-1})	$C_a/K_{ia/sw}$ (ng L^{-1})	$C_{iw}-C_a/K_{ia/sw}$ (ng L^{-1})
MCF	2.5	1.9	+ 0.6
BF	9.8	3.1	+ 6.7

Comparison with the results obtained above for 0°C shows that, with the same air and seawater concentrations, the net flux of MCF is now directed from the water to the atmosphere.

Note: The consistent tendency of BF to show seawater-to-atmosphere fluxes has been interpreted as evidence for a natural source of bromoform in the sea.

All the physics is hidden in the coefficient $v_{ia/w}$ which, because it has the dimension of a velocity (LT^{-1}), is called the (overall) *air–water exchange velocity*. Air–water exchange occurs due to random motion of molecules. Equation 20-1 is a particular version of Eq. 18-4 in which the air–water exchange velocity adopts the role of the mass transfer velocity, $v_{A/B}$.

Generally, the total exchange velocity, $v_{ia/w}$, can be interpreted as resulting from a two-component (air, water) interface with phase change. Independently of the chosen model (bottleneck or wall boundary), if we choose water as the reference phase, $v_{ia/w}$ is always of the form (see Eqs. 19-13 and 19-58):

$$\frac{1}{v_{ia/w}} = \frac{1}{v_{iw}} + \frac{1}{v_{ia} K_{ia/w}} \tag{20-3}$$

The main part of this chapter deals with models for describing $v_{ia/w}$ as a function of different environmental factors such as wind speed, water temperature, flow velocity, and others. None of these models is able to totally depict the complexity of the processes acting at the surface of a natural water body. Therefore, theoretical predictions of the exchange velocity always meet severe limitations. Nonetheless, two properties stick out from Eqs. 20-1 to 20-3:

1. The concentration difference $(C_{iw} - C_{iw}^{eq})$ determines the size and the sign of the flux. Thus, even without detailed knowledge of the processes at the water surface, it is usually possible to identify a given water body either as a source or a sink of a specific chemical.

2. The structure of Eq. 20-3 allows us to identify ranges of $K_{ia/w}$ for which the transfer velocity, $v_{ia/w}$, depends on just one of the two single-phase exchange velocities.

To separate the two ranges of $K_{ia/w}$ where $v_{ia/w}$ can be approximated by one of the two single-phase velocities alone, we define the following compound-independent critical Henry's law constant:

$$K_{a/w}^{critical} = \frac{K_H^{critical}}{RT} = \frac{v_w^{typical}}{v_a^{typical}} \tag{20-4}$$

where $v_a^{typical}$ and $v_w^{typical}$ are typical single-phase exchange velocities. As it turns out, the compound-specific exchange velocities, v_{ia} and v_{iw}, vary by less than one order of magnitude between different compounds. These values are approximately related to the inverse of the densities of the two phases, air and water. Since the latter is about 1000 times larger than the former, we deduce that:

$$K_{a/w}^{critical} \sim 10^{-3} \quad \text{and} \quad K_H^{critical} \sim 0.025 \text{ L bar mol}^{-1}$$

In the next section we will show that:

$$v_a^{typical} \sim 1 \text{ cm s}^{-1} \quad ; \quad v_w^{typical} \sim 10^{-3} \text{ cm s}^{-1}$$

Note that the two terms on the right-hand side of Eq. 20-3 are of the same magnitude for a compound whose air–water partition coefficient is equal to $K_{a/w}^{critical}$ and whose single-phase exchange velocities have the typical size as given above. Therefore, single-phase controlled substances must have $K_{ia/w}$ values which are either much smaller or much larger than $K_{a/w}^{critical}$:

(a) Air-phase-controlled regime:

$$K_{ia/w} \ll K_{a/w}^{critical} = 10^{-3} ; \text{ thus } v_{ia/w} \approx K_{ia/w} v_{ia} \tag{20-5a}$$

(b) Water-phase-controlled regime:

$$K_{ia/w} \gg K_{a/w}^{critical} = 10^{-3} ; \text{ thus } v_{ia/w} \approx v_{iw} \tag{20-5b}$$

At first sight, there seems to be a basic difference between the two regimes with respect to the influence of $K_{ia/w}$. In the water-phase-controlled regime, the overall exchange velocity, $v_{ia/w}$, is independent of $K_{ia/w}$, whereas in the air–phase controlled regime $v_{ia/w}$ is linearly related to $K_{ia/w}$. Yet, this asymmetry is just a consequence of our decision to relate all concentrations to the water phase. In fact, for substances with small $K_{ia/w}$ values, the aqueous phase is not the ideal reference system to describe air–water exchange. This can be best demonstrated for the case of exchange of water itself ($K_{ia/w} = 2.3 \times 10^{-5}$ at 25°C), that is, for the evaporation of water.

Let us rewrite Eq. 20-1:

$$F_{ia/w} = v_{ia/w}\left(C_{iw} - \frac{C_{ia}}{K_{ia/w}}\right) = \frac{v_{ia/w}}{K_{ia/w}}\left(K_{ia/w}\,C_{iw} - C_{ia}\right)$$

$$= v^*_{ia/w}\left(C_{ia}^{eq} - C_{ia}\right) \tag{20-6}$$

where:

$$v^*_{ia/w} = \frac{v_{ia/w}}{K_{ia/w}} \tag{20-7}$$

is the exchange velocity if the concentrations refer to the air phase, and:

$$C_{ia}^{eq} = K_{ia/w}\,C_{iw} \tag{20-8}$$

is the air phase concentration in equilibrium with the aqueous concentration C_{iw}. The two regimes defined in Eq. 20-5 are now characterized by:

(a*) Air-phase-controlled regime:

$$K_{ia/w} \ll K_{a/w}^{critical} = 10^{-3}, \quad \text{then } v^*_{ia/w} \approx v_{ia} \tag{20-9a}$$

(b*) Water-phase-controlled regime:

$$K_{ia/w} \gg K_{a/w}^{critical} = 10^{-3}, \quad \text{then } v^*_{ia/w} \approx \frac{v_{iw}}{K_{ia/w}} \tag{20-9b}$$

Note that compared to Eq. 20-5, the role of $K_{ia/w}$ is now reversed. Now it is the air-phase-controlled regime where $v^*_{ia/w}$ becomes independent of $K_{ia/w}$.

The total transfer velocities, $v_{ia/w}$ and $v^*_{ia/w}$, are plotted in Fig. 20.1 as a function of $K_{ia/w}$. Note that these curves give approximate values which are valid if the real exchange velocities are close to the chosen typical values, $v_a^{typical}$ and $v_w^{typical}$. The figure shows nicely that $v_{ia/w}$ and $v^*_{ia/w}$, respectively, become independent of $K_{ia/w}$ if that phase is chosen as the reference system which dominates the overall exchange velocity. A more refined picture of the overall transfer velocities will be shown in Fig. 20.7.

Before looking closer at the experimental information on single-phase exchange velocities, let us briefly examine the assumption that we made regarding the exist-

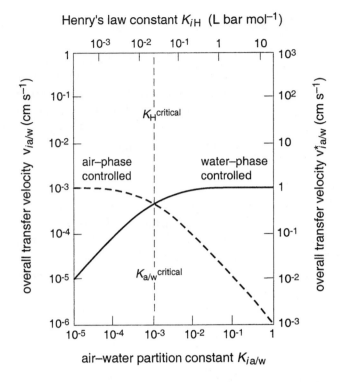

Henry's law constant K_{iH} (L bar mol^{-1})

air–water partition constant $K_{ia/w}$

Figure 20.1 Schematic view of the overall air–water exchange velocity, $v_{ia/w}$, as a function of the air–water partition coefficient, $K_{ia/w}$, calculated from Eq. 20-3 with typical single-phase transfer velocities $v_{ia} = 1$ cm s^{-1}, $v_{iw} = 10^{-3}$ cm s^{-1}. The broken line shows the exchange velocity $v^*_{ia/w}$ (air chosen as the reference system). The upper scale gives the Henry's Law coefficient at 25°C, $K_{iH} = 24.7$ (L bar mol^{-1}) × $K_{ia/w}$.

ence of an instantaneous chemical equilibrium across the air–water interface. From the molecular theory of gases, we can estimate the speed at which molecules cross the interface. We consider the mean molecular velocity along a fixed coordinate axis perpendicular to the interface of those molecules which move *toward* the interface. According to Eq. 18-38, this velocity is of order:

$$u(\text{toward interface}) = \frac{1}{2}\left(\frac{2RT}{\pi M_i}\right)^{1/2} \tag{20-10}$$

For $T = 298$ K (25°C) and $M_i = 78$ g mol^{-1} (benzene), this velocity is 7×10^3 cm s^{-1}. Even if only a small fraction of the molecules, say one out of every thousand, actually penetrates into the other phase and stays there, the molecular exchange velocity is still of the order of 10 cm s^{-1}. This is much larger than the largest transfer velocities $v_{ia/w}$ shown in Fig. 20.1. Hence, the molecular transfer right at the interface is not the limiting step. At the interface equilibrium conditions can be assumed, indeed.

20.2   Measurement of Air–Water Transfer Velocities

Let us now analyze the information on air–water exchange velocities which has been gained from observations both in the field and in the laboratory. We are especially interested in those extreme situations which are either solely water-phase or solely air-phase controlled, since they allow us to separate the influences of the two phases. We start with the latter case, the air-phase-controlled exchange.

Transfer Velocities in Air Deduced from Evaporation of Water

Traditionally, water is used as the test substance for determining v_{ia}. Its air–water partition constant at 25°C is $K_{ia/w} = 2.3 \times 10^{-5}$, which is much smaller than $K_{a/w}^{critical}$ of Eq. 20-4. Thus, the exchange of water vapor at the air–water interface is solely controlled by physical phenomena in the air above the water surface. The flux of water into air (evaporation) is given by (see Eqs. 20-6, 20-7, 20-9a):

$$F_{evap} = v_{water\,a/w}^{*}\left(C_{water\,a}^{eq} - C_{water\,a}\right) = v_{water\,a}\left(C_{water\,a}^{eq} - C_{water\,a}\right)$$

$$= v_{water\,a}\, C_{water\,a}^{eq}\, (1 - RH) \qquad (20\text{-}11)$$

where $C_{water\,a}^{eq}$ is the (temperature dependent) water vapor concentration in air (g m^{-3}) at equilibrium with liquid water. The relative humidity, RH, is defined as:

$$RH = \frac{C_{water\,a}}{C_{water\,a}^{eq}} \qquad (20\text{-}12)$$

$v_{water\,a}$ values can be determined from the corresponding observed evaporative fluxes, F_{evap}:

$$v_{water\,a} = \frac{F_{evap}}{C_{water\,a}^{eq}\,(1 - RH)} \quad (cm\ s^{-1}) \qquad (20\text{-}13)$$

where the evaporative flux, F_{evap}, is expressed in units of g cm^{-2}s^{-1}.

Hydrologists have long recognized the relationship between $v_{water\,a}$ and environmental conditions such as wind speed (Dalton, 1802; Fitzgerald, 1886; Rohwer, 1931; Sverdrup et al., 1942). In laboratory experiments (Liss, 1973; Penman, 1948; Münnich et al., 1978; Mackay and Yeun, 1983), increasing wind velocities u distinctly enhance $v_{water\,a}$ approximately linearly (Fig. 20.2), although as winds exceed 10 m s^{-1} the exchange velocity may increase more quickly, possibly due to aerosol ejection and wave breaking (see data by Mackay and Yeun, 1983, in Fig. 20.2 and Table 20.1).

In the experiments used to draw Fig. 20.2, wind speeds were measured at different heights above the water surface. Since wind speed generally decreases when approaching the water surface, these experiments can be compared only if we find a means to transform the wind speeds to a standard height (usually 10 m). Mackay and Yeun (1983) use the standard boundary layer theory with a roughness height of 0.03 cm and a wind stress coefficient of 1.5×10^{-3} to get:

Figure 20.2 Impact of wind speed u on the air-phase mass transfer velocity $v_{water\,a}$, as observed by the evaporation of water in laboratory experiments and as predicted using various empirical expressions (see Table 20.1). Note that the wind speeds refer to different heights z above the water surface (subscript indicates height in meters). Since generally wind speed decreases when the water surface is approached, it explains the higher $v_{water\,a}$ values for wind measured at smaller height z. The solid lines give $v_{water\,a}$ values which are adjusted to wind speeds at the standard height $z = 10$ m using the logarithmic boundary layer theory (Eq. 20-14).

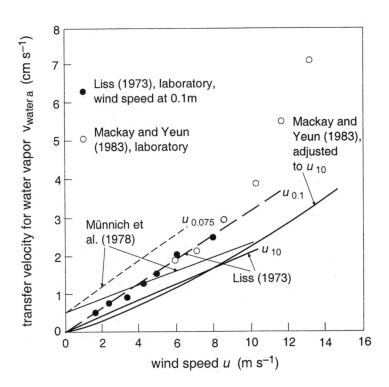

$$u_{10} = \left(\frac{10.4}{\ln z + 8.1} \right) u_z \tag{20-14}$$

where u_z and u_{10} are wind speeds at heights z and 10 m, respectively, and z is height measured in meters. This expression is used in Fig. 20.2 and Table 20.1 to transform the measured values to the standard height of 10 m.

The interpretation of these data in terms of the discussed air–water exchange models will be given in the following section. At this point we conclude that no matter what the exact physical mechanism of transport is, the lumped information from oceanographers, hydrologists, and engineers gives clear evidence that $v_{water\,a}$ is positively correlated with wind speed. Perhaps a suitable approximation deduced from all these investigations is:

$$v_{water\,a} \text{ (cm s}^{-1}) \sim 0.2\, u_{10} \text{ (m s}^{-1}) + 0.3 \tag{20-15}$$

where u_{10} is the wind speed measured 10 m above the water surface and the intercept value of 0.3 cm s^{-1} reflects the minimum value of $v_{water\,a}$. In Section 20.3 we will see that $v_{water\,a}$ depends only weakly on air temperature.

To demonstrate the feasibility of extrapolating such laboratory results to the field, we use Eq. 20-15 to calculate the order of magnitude of the annual evaporation rates from surface waters. Let us assume a typical relative humidity of 80% (RH = 0.8), wind speeds between 0 and 15 m s^{-1}, and a water temperature of 15°C. Water vapor saturation of air at 15°C is $C_{water\,a}^{eq} = 12.8 \times 10^{-6}$ g cm^{-3} (Appendix B, Table B.3). Thus, from Eq. 20-11:

$$F_{evap} = v_{water\,a}\, C^{eq}_{water\,a}\,(1-RH)$$

$$= (0.3 \text{ to } 3 \text{ cm s}^{-1})\,(12.8 \times 10^{-6} \text{ g cm}^{-3})\,(0.2)$$

$$\sim (0.8 \text{ to } 8)\,10^{-6} \text{ g cm}^{-2}\text{s}^{-1}$$

This corresponds to an annual evaporation rate of 0.25 to 2.5 m yr^{-1}, which is in accordance with observed evaporative loss rates of water from lakes and oceans (Sverdrup et al., 1942; Miller, 1977).

As shown in Illustrative Example 20.2, another prominent example of a gas-phase-controlled exchange flux is the evaporation of pure organic liquids.

Table 20.1 Various Empirical Relationships Between Wind Velocity Measured at Heigth z Above the Water, u_z, and Air-Phase Transfer Velocity of Water[a], v_a, Deduced from Observations of Water Evaporation Rates [b]

Reference and *Type of Data*	Original Data	v_a Related to u_{10} [c]
(1) Sverdrup et al. (1942) *Field*	$v_a = 0.16\, u_6$	$v_a = 0.15\, u_{10}$
(2) Penman (1948) *Evaporation pans*	$v_a = 0.59 + 0.27\, u_2$	$v_a = 0.59 + 0.23\, u_{10}$
(3) Rohwer (1931) *Evaporation pans*		$v_a = 0.35 + 0.12\, u_{10}$
(4) Liss (1973) *Wind-water tunnel*	$v_a = 0.005 + 0.32\, u_{0.1}$	$v_a = 0.005 + 0.21\, u_{10}$
(5) Münnich et al. (1978) *Circular wind-water tank*	$v_a = 0.5 + 0.35\, u_{0.075}$	$v_a = 0.5 + 0.185\, u_{10}$
(6) Mackay and Yeun (1983) *Wind-wave tank*		$v_a = 0.065\,(6.1 + 0.63\, u_{10})^{0.5}\, u_{10}$

[a] To simplify the equations, $v_{water\,a}$ of Eq. 20-13 is written as v_a. [b] As shown in Section 20.3, v_a depends only weakly on air temperature. [c] u_z is transformed to u_{10} with Eq. 20-14.

Illustrative Example 20.2   Estimating Evaporation Rates of Pure Organic Liquids

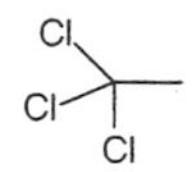

i = 1,1,1-trichloroethane (methyl chloroform, MCF)

M_i = 133.4 g mol^{-1}
T_m = – 30.4 °C
T_b = 74.1 °C
ρ_{iL} = 1.34 g cm^{-3}

Another prominent example of a gas-phase-controlled exchange flux is the evaporation of pure organic liquids. To illustrate this case, we again use the example of methylchloroform (MCF). In the following the subscript i stands for MCF.

Problem

Consider a spill of $\mathcal{M}_i = 2$ kg of pure liquid MCF forming a puddle on the ground of about 0.3 m^2 surface area in a closed hall of volume $V = 200$ m^3. Estimate how long it takes for the liquid MCF to completely evaporate at ambient temperatures of 0°C and 25°C. Assume that initially the MCF concentration in the air is zero. There is no observable motion of air in the hall.

The *Handbook of Physics and Chemistry* (Lide, 1995) gives the following saturation vapor pressures of liquid MCF, p^*_{iL}:

$$p^*_{iL}(0°C) = 4.8 \text{ kPa} \quad ; \quad p^*_{iL}(25°C) = 16.5 \text{ kPa}$$

Note that these values could also be estimated with Eq. 4-33 using the boiling temperature of MCF and $K_F = 1$ (nonpolar compound).

Answer

As a first rough estimate, you can make several simplifications. First, use Eq. 20-15 with $u_{10} = 0$ to calculate v_{ia} of water vapor: $v_{ia} = 0.3 \text{ cm s}^{-1}$. Second, assume that the evaporation velocity of MCF is approximately the same. Third, assume that the surface area of the puddle remains constant until all the MCF is evaporated.

The concentration of MCF in the air right at the interface above the liquid MCF, C^{eq}_{ia}, is determined by its saturation vapor pressure, p^*_{iL}:

$$C^{eq}_{ia} = \frac{p^*_{iL}}{RT}$$

With $R = 8.31 \text{ J mol}^{-1}\text{K}^{-1} = 8.31 \text{ Pa m}^3\text{mol}^{-1}\text{K}^{-1}$ you get:

$$0°C: \quad C^{eq}_{ia} = \frac{4.8 \times 10^3 \text{ Pa}}{8.31 \times 273 \text{ Pa m}^3\text{mol}^{-1}} = 2.1 \text{ mol m}^{-3}$$

$$25°C: \quad C^{eq}_{ia} = \frac{16.5 \times 10^3 \text{ Pa}}{8.31 \times 298 \text{ Pa m}^3\text{mol}^{-1}} = 6.7 \text{ mol m}^{-3}$$

Obviously, the transfer between the pure organic liquid and the air must be air-phase controlled. Since the MCF concentrations at the interface are expressed in terms of the concentrations in the air, apply Eq. 20-9a, where the pure MCF takes over the role of the water:

$$v^*_{ia/MCF} \approx v_{ia} \approx v_{\text{water a}}$$

Since initially the concentration of MCF in air, C_{ia}, is zero, the evaporation flux calculated from Eq. 20-6 is ($A = 0.3 \text{ m}^2$, surface area of puddle):

$$F_{\text{tot}} = A \, v_{\text{water a}} (C^{eq}_{ia} - C_{ia}) = A \, v_{ia} \, C^{eq}_{ia}$$

$$0°C: \quad F_{\text{tot}} = 0.3 \text{ m}^2 \times 3 \times 10^{-3} \text{ m s}^{-1} \times 2.1 \text{ mol m}^{-3} = 1.9 \times 10^{-3} \text{ mol s}^{-1}$$

$$25°C: \quad F_{\text{tot}} = 0.3 \text{ m}^2 \times 3 \times 10^{-3} \text{ m s}^{-1} \times 6.7 \text{ mol m}^{-3} = 6.0 \times 10^{-3} \text{ mol s}^{-1}$$

There are two reasons why F_{tot} might decrease during the process of evaporation: (1) the surface area of the puddle will decrease, and (2) the concentration of MCF in the ambient air, C_{ia}, will increase. The influence of the latter effect can be estimated from the maximum concentration to be reached in the closed room when all MCF has evaporated:

$$C_{ia}^{max} = \frac{\text{total initial mass}}{\text{room volume}} = \frac{\mathcal{M}_i}{V} = \frac{2 \times 10^3\,g}{200\,m^3}$$

$$= 10\,g\,m^{-3} = 0.075\,mol\,m^{-3}$$

Since this value is much smaller than C_{ia}^{eq}, the second term in the parentheses of Eq. 20-6 can always be neglected.

If F_{tot} remained constant, the times needed for total evaporation are:

$$t_{evap} = \frac{\mathcal{M}_i}{F_{tot}} = \begin{cases} 7.9 \times 10^3\,s = 2.2\,h & (0°C) \\ 2.5 \times 10^3\,s = 0.7\,h & (25°C) \end{cases}$$

The decrease of the puddle area A is difficult to estimate without additional information on the physical properties of the surface from which the MCF evaporates, but it would hardly increase t_{evap} by more than 50%.

A final remark: In Section 20.3 we will see that due to the difference in air diffusivities the evaporation velocity of MCF is smaller than v_{ia} of water vapor. A rough estimate will be given by the ratio of the molecular masses (combine Eqs. 20-27 and 18-46):

$$\frac{v_{MCF\,a}}{v_{water\,a}} \sim \left(\frac{M_{CFM}}{M_{water}} \right)^{-1/3} \sim 0.5$$

Thus, the real evaporation velocity of MCF may be reduced by a factor of 2.

Transfer Velocities in the Water Phase Deduced from Compounds with Large Henry's Law Constants

As a next step we analyze experimental information on air–water exchange for substances for which only the water-side exchange velocity is relevant. According to Eq. 20-5b, we find these reference substances among compounds with large Henry's law constants. In many publications such substances are called "volatile," although this is not very precise. If volatile means having a high vapor pressure, then water would be volatile as well. What we really mean is a compound with a high vapor pressure *and* a low solubility, that is, according to Eq. 6-16 a substance with a large Henry's law constant. The reader has to remember this subtlety when the term "volatile" is used.

Early workers on air–water exchange pursued laboratory experiments for gases such as O_2 or CO_2 (e.g., Kanwisher, 1963; Liss, 1973). Later, geochemical investigations expanded the database to include natural radioactive tracers such as ^{3}He and ^{222}Rn (Emerson et al., 1973; Broecker and Peng, 1974; Peng et al., 1979; Torgersen et al., 1982; Jähne et al., 1987a) and artificial chemical tracers such as SF_6 (Wanninkhof et al., 1987). Additionally, many laboratory experiments on mixed or wind-blown waters containing organic chemicals have been reported (e.g., Dilling et al., 1975; Dilling 1977; Smith et al., 1980; Mackay and Yeun, 1983). Some of these data are shown in Fig. 20.3 and the resulting empirical relationships are listed in Table 20.2. Note that the wind velocities are all transformed to u_{10} values using Eq. 20-14.

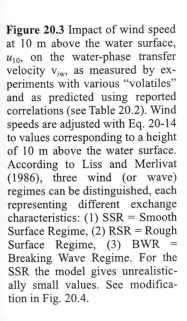

Figure 20.3 Impact of wind speed at 10 m above the water surface, u_{10}, on the water-phase transfer velocity v_{iw}, as measured by experiments with various "volatiles" and as predicted using reported correlations (see Table 20.2). Wind speeds are adjusted with Eq. 20-14 to values corresponding to a height of 10 m above the water surface. According to Liss and Merlivat (1986), three wind (or wave) regimes can be distinguished, each representing different exchange characteristics: (1) SSR = Smooth Surface Regime, (2) RSR = Rough Surface Regime, (3) BWR = Breaking Wave Regime. For the SSR the model gives unrealistically small values. See modification in Fig. 20.4.

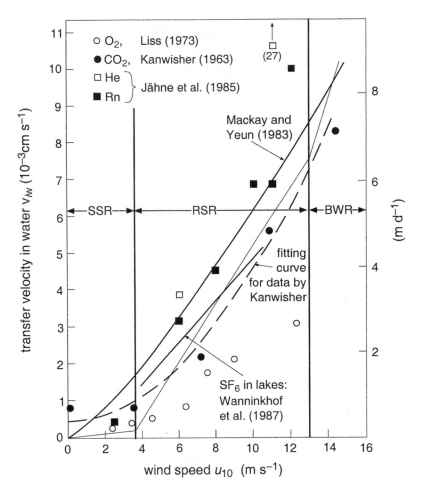

What makes these data difficult to interpret is that they originate from studies in which different substances, mainly CO_2 and O_2, but also radon and sulfur hexafluoride (SF_6), have been used. If the exact nature of the exchange process is not known, it is not immediately evident how data from different gases should be compared. A thorough discussion has to be postponed to Section 20.3 where models of air–water exchange are presented. Then we will also tackle the question of how water temperature affects the transfer velocity.

When we attempt to interpret Fig. 20.3, we must realize that, as opposed to the case of the air-phase mass transfer velocity, wind speed is not the only important parameter that controls the magnitude of v_{iw}. As demonstrated by Jähne et al. (1984, 1987a), v_{iw} is also affected by the wave field, which itself depends in a complicated way on wind stress, wind fetch, surface contamination (affecting the surface tension of the water), and water currents. Thus, not surprisingly field and laboratory data give fairly different results. Generally, laboratory experiments tend to overestimate air–water exchange rates of volatile compounds occurring under natural conditions.

The influence of the waves becomes particularly evident at wind speeds above about 10 m s^{-1}, that is, above the onset of wave breaking and formation of air bubbles (Blanchard and Woodcock, 1957; Monahan, 1971; Kolovayev, 1976; Johnson and

Table 20.2 Empirical Relationships Between Wind Velocity u_{10} and Water-Phase Air–Water Transfer Velocity v_{iw} [a]

Source and *Type of Data*	v_{iw} Related to u_{10} [b]	
(1) Kanwisher (1963) *CO₂, wind–water tunnel*	$v_{iw} = (4.1 + 0.41\ u_{10}^2) \cdot 10^{-4}$	
(2) Liss (1973) *Laboratory tank, O₂*	See Fig. 20.3	
(3) Mackay and Yeun (1983) *Lab, organic solutes*	$v_{iw} = 1.75 \cdot 10^{-4}\ (6.1 + 0.63\ u_{10})^{0.5}\ u_{10}$ (for O_2)	
(4) Wanninkhof et al. (1987) *SF₆ in lakes*	$v_{iw} = (-0.89 + 5.8\ u_{10}) \cdot 10^{-4}$	
(5) Liss and Merlivat (1986) [c] *Valid for $Sc_{iw} \sim 600$ (e.g., CO₂ at 20 °C)* [d]	$v_{iw} = 0.047 \times 10^{-3}\ u_{10}$	for $u_{10} < 3.6$ m s⁻¹ (SSR)
	$v_{iw} = (0.79\ u_{10} - 2.68) \times 10^{-3}$	for 3.6 m s⁻¹ $< u_{10} \leq$ 13 m s⁻¹ (RSR)
	$v_{iw} = (1.64\ u_{10} - 13.69) \times 10^{-3}$	for $u_{10} > 13$ m s⁻¹ (BWR)
(6) Modification of Liss and Merlivat model		
Adapted from Livingstone and Imboden (1993), combined with Liss and Merlivat (1986) $Sc_{iw} = 600$	$v_{iw} = 0.65 \times 10^{-3}$ for $u_{10} \leq 4.2$ m·s⁻¹ (SSR) $v_{iw} = (0.79\ u_{10} - 2.68) \times 10^{-3}$ for $4.2 < u_{10} \leq 13$ m·s⁻¹ (RSR) $v_{iw} = (1.64\ u_{10} - 13.69) \times 10^{-3}$ for $u_{10} > 13$ m·s⁻¹ (BWR)	

[a] u_{10} = wind speed (m s⁻¹) measured 10 m above the water surface, v_w = transfer velocity in cm s⁻¹.
[b] Some experiments report wind speed at a different height. Then u_z is transformed to u_{10} with Eq. 20-14. [c] Three wind/wave regimes: SSR = Smooth Surface Regime, RSR = Rough Surface Regime, BWR = Breaking Wave Regime. [d] Sc_{iw} is the water-phase Schmidt Number of substance i defined in Eq. 20-23.

Cooke, 1979; Wu, 1981). In this range transfer velocities determined from natural systems are possibly distorted by an additional effect called *wind pumping*. In this situation, bubbles are injected deep below the water surface and experience pressures in excess of atmospheric. As a result, larger quantities of the chemicals contained in the bubbles are dissolved in the water than are required for equilibrium at the water surface. This leads to supersaturation of O_2, N_2, and CO_2 of up to 15% (Smith and Jones, 1985). Note that this process not only influences the deduced sizes of v_{iw} and $v_{ia/w}$, but it may also invalidate the general form of Eq. 20-1 according to which the sign of the net air–water flux is determined by the sign of the concentration difference $(C_{iw} - C_{iw}^{eq})$. In order to produce supersaturation, $F_{ia/w}$ must be directed into the water $(F_{ia/w} < 0)$ even if $C_{iw}^{eq} > C_{iw}$. We will come back to this phenomenon in Section 20.5.

Liss and Merlivat (1986) distinguish between three regimes, each representing a different structure of the water surface. These regimes are:

SSR: Smooth Surface Regime $u_{10} \leq 3.6$ m s^{-1}

RSR: Rough Surface Regime 3.6 m s$^{-1} < u_{10} \leq 13$ m s^{-1}

BWR: Breaking Wave Regime $u_{10} > 13$ m s^{-1}

Since the structure of the wave field also depends on the wind history and on the size and exposure of the water body, the wind speeds which separate the regimes vary between different water bodies. The above limits reflect average conditions for the ocean.

According to the model of Liss and Merlivat, each wind regime is characterized by its own linear relationship between u_{10} and v_{iw} (see Fig. 20.3 and Table 20.2). Yet, transfer velocities in the SSR ($u_{10} \leq 3.6$ m s^{-1}) are extremely small (v_{iw} less than 0.17×10^{-3} cm s^{-1}) and contradict the few reported experimental data at low wind speeds which show that v_{iw} is finite even if u_{10} is zero (Fig. 20.4). Since typical wind speeds over land (and thus over many lakes) are less than 5 m s^{-1}, the model of Liss and Merlivat leads to a significant underestimation of air–water exchange of volatile compounds in many lakes. Livingstone and Imboden (1993) used the Weibull distribution to describe wind-speed probabilities in combination with the Liss and Merlivat model for v_{iw} (Box 20.1). They constructed plots relating average wind speeds, $\bar{u}_{10}$, with average exchange velocities, $\bar{v}_{iw}$, and concluded that, although the nonlinearity of the $\bar{v}_{iw}/\bar{u}_{10}$ relationship explains part of the low wind speed problem, clear evidence remains from various long-term field studies that shows that at $u_{10} < 2$ m s^{-1}, v_{iw} is 5 times or more larger than in the Liss and Merlivat model.

Box 20.1 Influence of Wind Speed Variability on the Mean Air–Water Exchange Velocity of Volatile Compounds

For compounds with $K_{ia/w}$ larger than about 10^{-2} the overall air–water transfer velocity is approximately equal to the water-phase exchange velocity v_{iw} The latter is related to wind speed u_{10} by a nonlinear relation (Table 20.2, Eq. 20-16). The annual mean of v_{iw} calculated from Eq. 20-16 with the annual mean wind speed u_{10} would underestimate the real mean air–water exchange velocity. Thus, we need information not only on the average wind speed, but also on the wind-speed probability distribution.

The two-dimensional Weibull distribution (Weibull, 1951) is often used to describe the cumulative frequency distribution of wind speeds (see Livingstone and Imboden (1993) for a review):

$$F(u_{10}) = \exp\left[-\left(\frac{u_{10}}{u_o} \right)^{\xi} \right] \tag{1}$$

$F(u_{10})$ is the probability of a measured wind speed exceeding a given value u_{10}, u_o is a scaling factor, and the exponent ξ describes the form of the distribution curve.

We restrict the following discussion to the typical case $\xi = 1$. Then, the probability density function $f(u_{10})$ is:

$$f(u_{10}) = -\frac{dF(u_{10})}{du_{10}} = \frac{1}{u_o} \exp\left[-\frac{u_{10}}{u_o} \right] \tag{2}$$

$f(u_{10})du_{10}$ gives the probability that the wind speed lies between u_{10} and $(u_{10} + du_{10})$. The distribution is characterized by:

$$\overline{u_{10}} = u_o \qquad \text{mean wind speed}$$

$$\sigma_u^2 = u_o^2 \qquad \text{variance}$$

$$\eta = \sigma_u / \overline{u_{10}} = 1 \quad \text{coefficient of variation}$$

To facilitate the mathematical discussion we use the power-law by Kanwisher (1963) to describe the functional relationship between v_{iw} and u_{10} (Table 20.2), that is, not the trilinear expression of Eq. 20-16. The relationships are fairly similar (see Fig. 20.3). Thus:

$$v_{iw}(u_{10}) = A + B u_{10}^2 \tag{3}$$

where A and B are the fitting parameters of Kanwisher's power law. The mean exchange velocity $\overline{v}_{iw}$ is:

$$\overline{v}_{iw} = \int_0^\infty v_{iw}(u_{10}) f(u_{10}) du_{10} = \int_0^\infty (A + Bu_{10}^2) \cdot \frac{1}{u_o} \exp\left[-\frac{u_{10}}{u_o}\right] du_{10}$$

Evaluation of the integral yields:

$$\overline{v}_{iw} = A + 2Bu_o^2 = A + 2B\overline{u}_{10}^2 \tag{4}$$

In contrast, if the average wind speed, $\overline{u}_{10} = u_o$, is directly inserted into Eq. 3 we get:

$$\overline{v}_{iw}^* = A + B\overline{u}_{10}^2 \tag{5}$$

The difference between Eqs. 4 and 5 becomes significant for large $\overline{u}_{10}$ values. Similar considerations for other Weibull-shape parameters ξ are given in Livingstone and Imboden (1993).

Numerical Example

From Table 20.2: $A = 4.1 \times 10^{-4}$ cm s^{-1}, $B = 0.41 \times 10^{-4}$ (m s^{-1})$^{-2}$ cm s^{-1}:

Mean Wind Speed $\overline{u}_{10}$ (m s^{-1})	Effective Mean (Eq. 4) $\overline{v}_{iw}$ (10^{-3} cm s^{-1})	Mean from Eq. 5 $\overline{v}_{iw}^*$ (10^{-3} cm s^{-1})
0.5	0.47	0.44
1	0.49	0.45
5	0.59	0.50
10	0.67	0.54
20	0.78	0.59

Figure 20.4 Modified water-phase transfer velocity v_{iw} for CO_2 at 20°C (Schmidt Number $Sc_{iw} = 600$, see Eq. 20-23 and explanations below) for small wind speeds (SSR = Smooth Surface Regime) according to Livingstone and Imboden (1993). For wind speeds $u_{10} \leq 4.2$ m s^{-1} the Liss and Merlivat model (Fig. 20.3) is replaced by the constant value, $v_{iw} = 0.65 \times 10^{-3}$ cm s^{-1}. The SSR is slightly extended to $u_{10} = 4.2$ m s^{-1}. For larger u_{10} values, v_{iw} corresponds to the Liss and Merlivat model. Solid circles show laboratory data by Liss et al. (1981) at low wind speeds converted to $Sc_{iw} = 600$ and wind measured at 10 m (u_{10}).

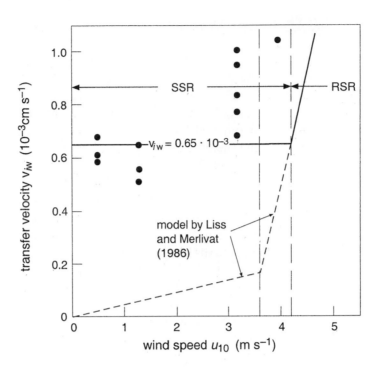

An alternative model is presented in Fig. 20.4. Based on the results by Livingstone and Imboden, v_{iw} is replaced by a constant value of $v_{iw} = 0.65 \times 10^{-3}$ cm s^{-1} for $u_{10} < 4.2$ m s^{-1} (i.e., for the SSR). This yields the following model:

$$v_{iw} = 0.65 \times 10^{-3} \text{ cm s}^{-1} \qquad \text{for } u_{10} \leq 4.2 \text{ m s}^{-1} \qquad \text{(SSR)}$$

$$v_{iw} = (0.79\, u_{10} - 2.68) \times 10^{-3} \quad \text{for } 4.2 \text{ m s}^{-1} < u_{10} \leq 13 \text{ m s}^{-1} \quad \text{(RSR)} \quad \text{(20-16)}$$

$$v_{iw} = (1.64\, u_{10} - 13.69) \times 10^{-3} \quad \text{for } u_{10} > 13 \text{ m s}^{-1} \qquad \text{(BWR)}$$

Note that Eq. 20-16 is valid for CO_2 at 20°C. In the next section we discuss how these data can be applied to other water temperatures and other chemicals.

The physical reason behind the modification at low wind speed as suggested by field data remains unclear. Yet, we should not forget that at low wind speed, the instantaneous wind is not the only significant source of motion at the water surface. Water motions caused by wind do not stop as soon as the wind ceases. Furthermore, thermal processes lead to density instabilities and convective motion, even if there is absolutely no wind. In fact, natural surface water bodies are hardly ever at rest.

To conclude this section, in analogy to the empirical equation for v_{ia}, we offer a simple tool to estimate v_{iw} provided that no detailed analysis is needed. The following relationship is based on the data of Kanwisher (see Table 20.2):

$$v_{iw} = 4 \times 10^{-4} + 4 \times 10^{-5} u_{10}^2 \qquad \text{(cm s}^{-1}) \qquad \text{(20-17)}$$

where u_{10} is given in (m s^{-1}). This equation is typical for CO_2 at 20°C.

20.3   Air–Water Exchange Models

In the preceding discussion, we presented experimental information on the "single-phase" air–water exchange velocities. Water vapor served as the test substance for the air-phase velocity v_{ia}, while O_2, CO_2 or other compounds yielded information on v_{iw}. Now, we need to develop a model with which these data can be extrapolated to other chemicals which either belong also to the single-phase group or are intermediate cases in which both v_{ia} and v_{iw} affect the overall exchange velocity $v_{ia/w}$ (Eq. 20-3).

To this end we use the boundary models derived in Chapter 19. Since each model has its own characteristic dependence on substance-specific properties (primarily molecular diffusivity in air or water), the experimental data from different compounds help us recognize the strengths and limitations of the various theoretical concepts.

Air–water exchange models have a long history. The first attempts to understand and describe the process have their roots in chemical engineering where the design of chemical production lines required a basic understanding of the physicochemical parameters controlling air–water exchange (Liss and Merlivat, 1986). It was recognized that the transfer at gas–liquid interfaces is governed by a complex combination of molecular diffusion and turbulent transport.

The first model, the *film model* by Whitman (1923), depicted the interface as a (single- or two-layer) bottleneck boundary. Although many aspects of this model are outdated in light of our improved knowledge of the physical processes occurring at the interface, its mathematical simplicity keeps the model popular.

An alternative approach, developed by chemical engineers as well, is the *surface renewal model* by Higbie (1935) and Danckwerts (1951). It applies to highly turbulent conditions in which new surfaces are continuously formed by breaking waves, by air bubbles entrapped in the water, and by water droplets ejected into the air. Here the interface is described as a diffusive boundary.

In the seventies, the growing interest in global geochemical cycles and in the fate of man-made pollutants in the environment triggered numerous studies of air–water exchange in natural systems, especially between the ocean and the atmosphere. In micrometeorology the study of heat and momentum transfer at water surfaces led to the development of detailed models of the structure of turbulence and momentum transfer close to the interface. The best-known outcome of these efforts, Deacon's (1977) *boundary layer model*, is similar to Whitman's film model. Yet, Deacon replaced the step-like drop in diffusivity (see Fig. 19.8a) by a continuous profile as shown in Fig. 19.8b. As a result the transfer velocity loses the simple form of Eq. 19-4. Since the turbulence structure close to the interface also depends on the viscosity of the fluid, the model becomes more complex but also more powerful (see below).

Figure 20.5 gives an overview of the basic ideas behind these three models. The upper picture shows the situation as depicted in the film model and the boundary

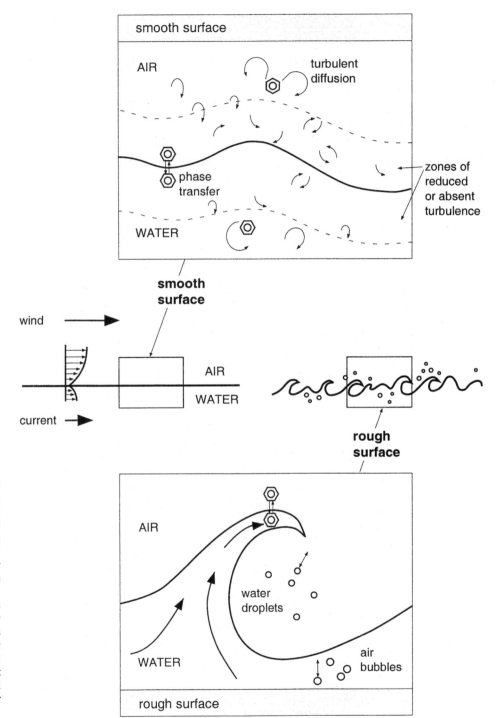

Figure 20.5 Physical processes at the air–water interface. For calm (smooth) surfaces the horizontal velocities on both sides of the interface decrease toward the boundary. The turbulent eddies become smaller and disappear completely at the interface (boundary layer characteristics). For rough conditions new surfaces are continuously formed by breaking waves, by air bubbles entrapped in the water, and by water droplets ejected into the air. Generally, these surfaces do not last long enough to reach chemical equilibrium between air and water phase.

layer model. In the former diffusivity drops discontinuously to molecular values, while in the latter the change is smooth from the fully turbulent zone through the boundary layer to the very interface. The lower picture depicts the continuous formation of new interfacial areas as described by the surface renewal model. Let us now analyze these models more closely.

Film Model (Whitman, 1923)

In the film model the air–water interface is described as a one- or two-layer *bottleneck boundary* of thicknesses δ_a and δ_w, respectively. Thus, according to Eq. 19-9:

$$v_{ia} = \frac{D_{ia}}{\delta_a} \quad ; \quad v_{iw} = \frac{D_{iw}}{\delta_w} \tag{20-18}$$

Using the approximate value for water vapor, $D_{water\,a} \sim 0.3 \text{ cm}^2 \text{ s}^{-1}$, and for CO_2, $D_{CO2\,w} \sim 2 \times 10^{-5} \text{ cm}^2 \text{ s}^{-1}$, as well as the approximate size of the exchange velocities discussed in Section 20.1 ($v_{water\,a} \sim 1 \text{ cm s}^{-1}$, $v_{CO_2\,w} \sim 10^{-3} \text{ cm s}^{-1}$), the implied film thicknesses are typically:

$$\delta_a = \frac{D_{water\,a}}{v_{water\,a}} \sim 0.3 \text{ cm} \quad ; \quad \delta_w = \frac{D_{CO_2w}}{v_{CO_2w}} \sim 0.02 \text{ cm} \tag{20-18a}$$

One essential assumption is that all substances experience the same film thickness. Therefore, the model predicts that for given conditions the exchange velocities of different compounds, *i* and *j*, should be linearly related to their molecular diffusivities:

$$\frac{v_{i\alpha}}{v_{j\alpha}} = \frac{D_{i\alpha}}{D_{j\alpha}} \quad ; \quad \alpha = a, w \tag{20-19}$$

Because the diffusivity ratio, $D_{i\alpha}/D_{j\alpha}$, is not exactly identical for air and water and since Eq. 20-3 also contains $K_{ia/w}$, Eq. 20-19 does not hold for the composite (overall) exchange velocity $v_{ia/w}$. It can be applied to classes of substances which are either solely water-phase- or air-phase-controlled.

In the last 20 years, considerable efforts have been made to measure air–water exchange rates either in the laboratory or in the field. One central goal of these investigations was to check the validity of Eq. 20-19 or of alternative expressions. Thus let us see how corresponding forms of Eq. 20-19 look for other models.

Surface Renewal Model (Higbie, 1935; Danckwerts, 1951)

In this model the interface is described as a diffusive boundary. From Eq. 19-59 we get:

$$v_{ia} = \left(\frac{D_{ia}}{\pi t_{exp}^a} \right)^{1/2} \quad ; \quad v_{iw} = \left(\frac{D_{iw}}{\pi t_{exp}^w} \right)^{1/2} \tag{20-20}$$

where the exposure times, t_{exp}^a and t_{exp}^w, now adopt the role of the free parameters which in the film model were the film thicknesses, δ_a and δ_w. Using the same exchange velocities as before, we get for the air side:

$$t_{exp}^a = \frac{D_{water\,a}}{\pi\, v_{water\,a}^2} \sim \frac{0.3 \text{ cm}^2\text{s}^{-1}}{3\,(1 \text{ cm s}^{-1})^2} = 0.1 \text{ s} \tag{20-21a}$$

and for the water-side:

$$t_{exp}^w = \frac{D_{CO_2w}}{\pi\, v_{CO_2w}^2} \sim \frac{2 \times 10^{-5} \text{ cm}^2\text{s}^{-1}}{3\,(10^{-3} \text{ cm s}^{-1})^2} \sim 7 \text{ s} \tag{20-21b}$$

The above results suggest that the air is replaced more often than the water. On the one hand, given the different densities and viscosities of these fluids, this result looks reasonable. On the other hand, if the exposure times are different, the picture which we have developed for the diffusive boundary model does not strictly apply. When the replacement of fluid does not occur simultaneously on either side of the boundary, the concentration profile across the interface does not exhibit the symmetric shape shown in Fig. 19.15. For instance, if a new air parcel is brought to the interface while the adjacent water is not replaced, the air meets with water that already carries a certain depletion structure. Obviously, this influences the exchange across the interface. The mathematical description of such a situation would be complicated and lead well beyond the intention of this discussion. Yet, whatever the outcome of a more refined model, the net flux would always depend on the square root of the diffusivities in the two media, although the numerical factors in Eq. 19-59 may change. Thus, if one of the two fluids, air or water, controls the overall exchange, the surface renewal model suggests a relationship between the exchange velocities of two compounds, i and j, of the form:

$$\frac{v_{i\alpha}}{v_{j\alpha}} = \left(\frac{D_{i\alpha}}{D_{j\alpha}}\right)^{1/2} \quad ; \quad \alpha = a, w \tag{20-22}$$

with an exponent of 1/2 in contrast to 1 in the film model (Eq. 20-19).

Before we discuss the experimental data which were collected to distinguish between the two models, we discuss a third model which, as we will see, lies in between.

Boundary Layer Model (Deacon, 1977)

To understand the principal idea of Deacon's model we have to remember the key assumption of the film model according to which a bottleneck boundary is described by an abrupt drop of diffusivity, for instance, from turbulent to molecular conditions (see Fig. 19.3a). Yet, theories on turbulence at a boundary derived from fluid dynamics show that this drop is gradual and that the thickness of the transition zone from fully turbulent to molecular conditions depends on the viscosity of the fluid. In Whitman's film model this effect is incorporated in the film thicknesses, δ_a and δ_w (Eq. 20-17). In addition, the film thickness depends on the intensity of turbulent kinetic energy production at the interface as, for instance, demonstrated by the relationship between wind velocity and exchange velocity (Figs. 20.2 and 20.3).

Deacon's intention was to separate the viscosity effect from the wind effect, so that the new model would be able to describe the change of $v_{i\alpha}$ due to a change of water or air temperature (i.e., of viscosity) at constant wind speed. Deacon concluded that mass transfer at the interface must be controlled by the simultaneous influence of two related processes, that is, by the transport of chemicals (described by molecular diffusivity $D_{i\alpha}$), and by the transport of turbulence (described by the coefficient of kinematic viscosity v_α). Note that v_α has the same dimension as $D_{i\alpha}$. Thus, the ratio between the two quantities is nondimensional. It is called the *Schmidt Number*, $Sc_{i\alpha}$:

$$Sc_{i\alpha} = \frac{v_\alpha}{D_{i\alpha}} \quad ; \quad \alpha = a, w \tag{20-23}$$

Deacon derived his model for volatile compounds whose air–water transfer velocities solely depend on the conditions in the water phase. In its original form, which is valid for a smooth and rigid water surface and for Schmidt Numbers larger than 100, it has the form:

$$v_{iw} = \text{constant } (Sc_{iw})^{-2/3} \quad \text{for } Sc_{iw} > 100 \qquad (20\text{-}24)$$

According to this expression, v_{iw} increases for increasing D_{iw} and decreasing v_w. It would remain unchanged if both viscosity and diffusivity would increase or decrease by the same relative amount.

The following picture may help us to understand this result at least qualitatively. Imagine a border between two states which—for whatever reason—can only be crossed on foot. People use taxis to get to the border, yet when approaching the border the streets become increasingly narrow and the cars get stuck. The passengers in the taxis (they must all be trained mathematicians!) know exactly the optimal time to jump out of the cars, in order to walk or run the remaining distance and to cross the border after the shortest possible time. Obviously, the distance from the border where people leave the taxis is not the same for all persons and all road conditions. People who are fast runners (that is, have "large diffusivities") leave the cars earlier than people who can walk only with difficulty ("small diffusivities"). The latter will remain in their taxis as long as possible, even if the cars move only very slowly through the congested streets, but they have to get out of their vehicles at some time as well. In turn, one and the same person does not always leave the taxi at the same distance from the border. In some areas the roads leading to the border are narrower and thus more strongly congested ("large viscosity" damping the motion of the cars, that is, of the eddies); in others they are broader (small viscosity).

To summarize, the time needed to cross the border, that is, the *border transfer velocity*, depends on both the individual mobility on foot (diffusivity) and the quality of the roads (viscosity). Or stated differently: the distance from the border where the passengers leave the taxi since the speed of the cars (water movement) drops below the speed of the individual pedestrian (molecular transport), depends on the relative size of pedestrian mobility and car mobility. Transfer velocities are large for fast runners and permeable road systems and small for physically handicapped passengers and narrow streets.

This picture makes the role of the Schmidt Number, Sc_{iw}, at least plausible, although it obviously does not explain the size of the exponent in Eq. 20-24. In fact, this exponent depends on the existence of a rigid wall (a motionless water surface). Although the model remains valid as long as the water surface is not too much distorted by waves, at larger wind speeds the exponent in Eq. 20-24 changes to $-1/2$ (Jähne et al., 1987a). This transition takes place at wind speeds of about 5 m s^{-1} (Liss and Merlivat, 1986). Hence, we rewrite Deacon's model in a more general form:

$$v_{iw} = \text{constant } (Sc_{iw})^{-a_{Sc}} \text{ with } a_{Sc} = \begin{cases} 2/3 & \text{for } u_{10} \leq 5\,\text{m s}^{-1} \\ \\ 1/2 & \text{for } u_{10} > 5\,\text{m s}^{-1} \end{cases} \qquad (20\text{-}24a)$$

Table 20.3 Molecular Diffusivities, Kinematic Viscosities, and Schmidt Numbers (Sc$_{iw}$) in Water for Selected Chemicals

T (°C)	Viscosity[a] ν_w (cm²s⁻¹)	Oxygen (O₂)[b] D_{iw} (cm²s⁻¹)	Sc$_{iw}$ (–)	Carbon Dioxide (CO₂)[c] D_{iw} (cm²s⁻¹)	Sc$_{iw}$ (–)	Methane (CH₄)[c] D_{iw} (cm²s⁻¹)	Sc$_{iw}$ (–)	Helium (He)[c] D_{iw} (cm²s⁻¹)	Sc$_{iw}$ (–)	Benzene[d] D_{iw} (cm²s⁻¹)	Sc$_{iw}$ (–)	Decane[d] D_{iw} (cm²s⁻¹)	Sc$_{iw}$ (–)	Acetone[d] D_{iw} (cm²s⁻¹)	Sc$_{iw}$ (–)
0	1.787×10⁻²	1.11×10⁻⁵	1600	0.933×10⁻⁵	1910	0.940×10⁻⁵	1890	4.74×10⁻⁵	380	0.39×10⁻⁵	4600	0.29×10⁻⁵	6100	0.81×10⁻⁵	2200
5	1.518	1.3	1170	1.09	1390	1.09	1390	5.20	290						
10	1.307	1.52	860	1.26	1040	1.25	1050	5.68	230						
15	1.139	1.77	640	1.46	780	1.43	800	6.19	180						
20	1.002	2.05	490	1.68	600	1.63	610	6.73	150	1.06×10⁻⁵	840	0.63×10⁻⁵	1400	1.30×10⁻⁵	690
25	0.890	2.36	380	1.92	470	1.85	480	7.30	120						
30	0.797	2.70	300	2.18	370	2.09	380	7.89	100						

[a] From Appendix B, Table B.3. [b] Diffusivities from Himmelblau (1964). [c] Diffusivities from Jähne et al. (1987b). [d] Diffusivities from Oelkers (1991).

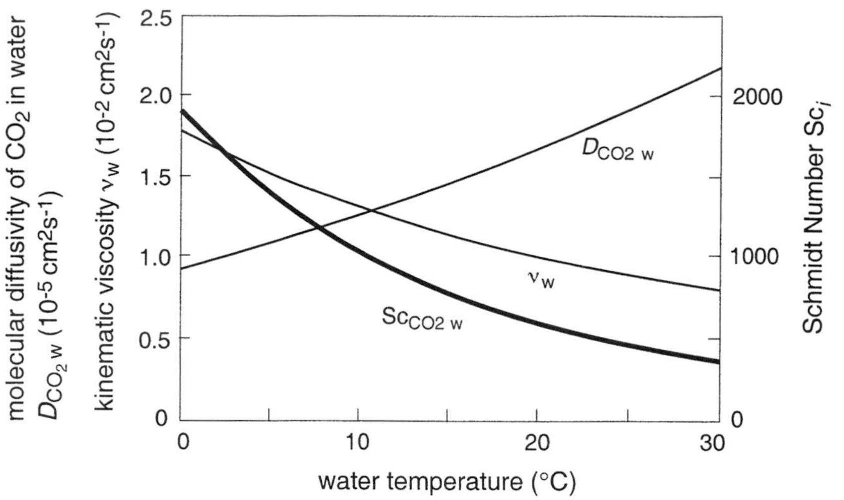

Figure 20.6 Variation with water temperature of kinematic viscosity ν_w, aqueous diffusivity D_{CO2w} of CO_2, and Schmidt Number, Sc_{CO2w} $= \nu_w/D_{CO2w}$.

Although the scientific literature is full of particular data sets which yield exponents, a_{Sc}, between 0 and 1.2 [see the review by Frost and Upstill-Goddard (1999) which contains an extended literature survey], we consider the modified Deacon model to be the most convenient one. It is more general than the other models since it allows us to evaluate the temperature dependence of ν_{iw} and even to extrapolate the model to other liquids. If it is applied to different chemicals in the same liquid and under the same hydrodynamic condition, the influence of viscosity ν is eliminated and we can directly compare Deacon's result with Eqs. 20-19 and 20-22, which were obtained earlier. The following equation combines all these models into one expression:

$$\frac{\nu_{iw}}{\nu_{jw}} = \left(\frac{D_{iw}}{D_{jw}}\right)^{a_D} \tag{20-25}$$

where the exponent a_D is given by the different models as:

Film model	$a_D =$	1
Surface renewal model	$a_D =$	1/2
Boundary layer model	$a_D =$	2/3 for $u_{10} \leq 5$ m s^{-1}
	$a_D =$	1/2 for $u_{10} > 5$ m s^{-1}

As pointed out by Livingstone and Imboden (1993), the change in the exponent a_{Sc} (Eq. 20-24a), when moving from the SSR to the RSR, has the "blemish" that the abrupt transition between different Sc-dependences must result in a discontinuity in ν_{iw} for all Schmidt Numbers different from the reference number of 600. In view of the other uncertainties involved, this should, however, be of little practical significance.

As we can learn from Table 20.3, the variation of Sc_{iw} between different chemicals is by more than one order of magnitude (compare helium and decane). Apparently, the lower limit of Sc_{iw} set in Eq. 20-24 does not even exclude helium with its large diffusivity as long as the water temperature does not exceed 30°C. The influence of water temperature on v_{iw} is fairly strong since the temperature dependencies of viscosity of water, v_w, and molecular diffusivity of an arbitrary compound i, D_{iw}, have opposite signs (Fig. 20.6). The exponent a_{Sc} of Eq. 20-24a determines how strongly this effect is transmitted to the air–water exchange velocity, v_{iw}.

With Eq. 20-24a the temperature dependence of v_{iw} can be written as:

$$\frac{d}{dT}(\ln v_{iw}) = -a_{Sc}\frac{d}{dT}(\ln Sc_{iw}) \quad,$$

that is:

$$\frac{1}{v_{iw}}\frac{dv_{iw}}{dT} = -a_{Sc}\left(\frac{1}{Sc_{iw}}\frac{dSc_{iw}}{dT}\right) \tag{20-26}$$

Box 20.2 summarizes the numerical evaluation of this expression with experimental data obtained for trichlorofluoromethane (CFC-11) by Zheng et al. (1998) (see also Table 20.3). The behavior of CFC-11 is typical for many volatile organic substances. The relative effect of temperature on v_{iw} decreases from 4 percent per Kelvin at 5°C and low wind speed to 2.4 percent per Kelvin at 25°C and high wind speed. The total increase of v_{iw} between 5°C and 25°C amounts to the factors 2 (low wind speed) and 1.7 (high wind speed), respectively.

Box 20.2 Temperature Dependence of Air–Water Exchange Velocity v_{iw} of Volatile Compounds Calculated with Different Models (T is in Kelvin if not stated otherwise)

"Volatile" compounds are characterized here by (see Eq. 20-9a): $K_{ia/w} \gg K_{a/w}^{critical} = 10^{-3}$. For these compounds, the air–water exchange is controlled by the water phase: $\Rightarrow v_{ia/w} \sim v_{iw}$.

Relative temperature variation of Schmidt Number $Sc_{iw} \equiv \dfrac{v_w}{D_{iw}}$ is:

$$\frac{1}{Sc_{iw}}\frac{dSc_{iw}}{dT} = \frac{1}{v_w}\frac{dv_w}{dT} - \frac{1}{D_{iw}}\frac{dD_{iw}}{dT} \tag{1}$$

A. Relative temperature dependence of kinematic viscosity v_w (see Fig. 20.6 and Appendix B, Table B.3)

Since the variation of water density with temperature is extremely small, the relative temperature variation of kinematic viscosity and dynamic viscosity η_w are approximately equal.

T (°C)	5	15	25
$\dfrac{1}{v_w}\dfrac{dv_w}{dT}$ (K^{-1})	-3.12×10^{-2}	-2.64×10^{-2}	-2.26×10^{-2}

B. Relative temperature dependence of aqueous diffusivity D_{iw}

Several methods to evaluate the temperature variation of D_{iw} are given in Box 18.4. As an example we use the values for *trichlorofluoromethane* (CFC-11) calculated from the activation theory model (see Box 18.4, Eq. 2):

T (°C)	5	15	25
$\dfrac{1}{D_{iw}}\dfrac{dD_{iw}}{dT}$ (K^{-1})	2.82×10^{-2}	2.63×10^{-2}	2.45×10^{-2}

C. Relative temperature dependence of Sc_{iw} and v_{ia} calculated for trichlorofluoromethane (CFC-11) from Eqs. 1 and the boundary layer model for moderate and high wind speed (Eq. 20-16):

	T (°C)	5	15	25
$\dfrac{1}{Sc_{iw}}\dfrac{dSc_{iw}}{dT}$ (K^{-1})		-5.94×10^{-2}	-5.27×10^{-2}	-4.71×10^{-2}
$\dfrac{1}{v_{ia}}\dfrac{dv_{ia}}{dT}$ (K^{-1})	low wind speed (a_{Sc}=2/3)	3.96×10^{-2}	3.51×10^{-2}	3.14×10^{-2}
	high wind speed (a_{Sc}=1/2)	2.97×10^{-2}	2.64×10^{-2}	2.36×10^{-2}

Deacon's model has also been applied to the air-phase exchange velocity, but the physical basis for such an extension is weak since typical Schmidt Numbers in air are about 1 ($Sc_{ia} \sim 0.57$ for water vapor at 20°C). Furthermore, the temperature dependence of Sc_{ia} is small since both v_a and D_{ia} increase with air temperature. In fact, for most substances Sc_{ia} varies by less than 10% for temperatures between 0°C and 25°C.

Therefore, instead of Sc_{ia} we use the diffusivity ratios to compare v_{ia} of different substances. According to the empirical observations of Mackay and Yeun (1983), the appropriate exponent is 2/3. That is, it lies between the film model and the surface replacement model:

$$\frac{v_{ia}}{v_{ja}} = \left(\frac{D_{ia}}{D_{ja}}\right)^{0.67} \tag{20-27}$$

To summarize, the theoretical understanding of the physical processes which control air–water exchange has made significant progress during the last 20 years. However, this insight also explains why the hope of finding simple relationships between wind speed, Schmidt Number, and air–water exchange velocity must ultimately fail. As shown by numerous laboratory investigations (e.g., Jähne et al., 1987a), at higher wind speed the mean square wave slope controls the size of v_{iw}. Yet, the wave field not only depends on the instantaneous local wind speed, but also on the wind history of the whole water body over the time period during which wave motion and turbulence are stored. Such time scales extend from the order of one hour in small lakes to several days in the ocean.

Let us now combine these results to estimate the overall air–water exchange velocity $v_{ia/w}$.

Table 20.4 Air–Water Exchange Velocities: Summary of Wind Speed and Compound-Specific Dependence

u_{10} are in $[\text{m s}^{-1}]$, v_{ia} and v_{iw} in $[\text{cm s}^{-1}]$

Air-phase

Reference substance: water vapor at air temperature between 0°C and 25°C

$$v_{\text{water a}}[\text{cm s}^{-1}] \quad = 0.2\,u_{10} + 0.3 \qquad\qquad\qquad\qquad\qquad\qquad \text{Eq. 20-15}$$

$$v_{ia} = \left(\frac{D_{ia}}{D_{\text{H}_2\text{O a}}}\right)^{0.67} v_{\text{water a}} \qquad\qquad \text{Mackay and Yeun (1983)} \qquad \text{Eq. 20-27}$$

Water-phase

Reference substance: CO_2 at 20°C ($Sc_{iw} = 600$)

$$v_{\text{CO}_2\text{w}}[\text{cm s}^{-1}] = \begin{cases} 0.65\times10^{-3} & \text{for } u_{10} \le 4.2\,\text{m s}^{-1} & \text{(SSR)} \\ (0.79\,u_{10} - 2.68)\times10^{-3} & \text{for } 4.2 < u_{10} \le 13\,\text{m s}^{-1} & \text{(RSR)} \\ (1.64\,u_{10} - 13.69)\times10^{-3} & \text{for } u_{10} > 13\,\text{m s}^{-1} & \text{(BWR)} \end{cases} \qquad \text{Eq. 20-16}$$

$$v_{iw} = \left(\frac{Sc_{iw}}{600}\right)^{-a_{Sc}} v_{\text{CO}_2\text{w}} \quad \text{with } a_{Sc} = \begin{cases} 0.67 & \text{for SSR} \\ 0.50 & \text{for RSR and BWR} \end{cases} \qquad \text{Eq. 20-24a}$$

Overall Air–Water Exchange Velocities

The foregoing discussion has demonstrated that the physics of air–water exchange is extremely complex. The combination of different system-specific and substance-specific influences on $v_{ia/w}$ and the mismatch of time scales of the external forcing (wind) and of the system's response explain why observations in the laboratory and field data have never given the unique picture which early investigators hoped to achieve. Although it is important to realize that experimental data often are ambiguous, we should not be discouraged from developing useful recipes for estimating overall air–water transfer velocities for different environmental conditions and different compounds. A summary of such recipes is given in Table 20.4. All these equations have been extensively discussed before.

Let us now examine some specific examples using a set of compounds with air–water partition constants, $K_{ia/w}$, between 30 and 1×10^{-5} (Table 20.5a) and by choosing two extreme wind velocities, $u_{10} = 1$ m s^{-1} and 20 m s^{-1}. From Table 20.4 we get the following reference transfer velocities:

For $u_{10} = 1$ m s^{-1}: $v_{\text{water a}} = 0.5$ cm s^{-1}, v_{iw} ($Sc_{iw} = 600$) $= 0.65 \times 10^{-3}$ cm s^{-1}

For $u_{10} = 20$ m s^{-1}: $v_{\text{water a}} = 4.3$ cm s^{-1}, v_{iw} ($Sc_{iw} = 600$) $= 19.1 \times 10^{-3}$ cm s^{-1}

Table 20.5a Calculation of Overall Air–Water Transfer Velocities for Different Organic Compounds at 25°C: Substance-Specific Properties

Substance	M_i (g mol^{-1})	$K_{ia/w}$ (25°C)	D_{ia} [a] (cm^2 s^{-1})	D_{iw} [b] (cm^2 s^{-1})	Sc_{iw} [c]
1 Methane	16.0	27	0.28	1.85×10^{-5} [d]	480
2 Trichlorofluoromethane	137.4	5.3	0.94	1.0×10^{-5} [e]	890
3 Octadecane	254.4	1.1	0.69	0.80×10^{-5}	1120
4 Tetrachloroethene	165.8	0.73	0.86	0.99×10^{-5}	900
5 Benzene	78.1	0.23	0.12	1.06×10^{-5} [f]	840
6 1,2,4-Trichlorobenzene	181.5	0.11	0.082	0.95×10^{-5}	940
7 Naphthalene	128.2	0.018	0.097	0.97×10^{-5} [f]	920
8 2,2',4,4',5,5'-Hexachlorobiphenyl	360.9	3×10^{-3}	0.058	0.67×10^{-5}	1330
9 1-Hexanol	102.2	5×10^{-4}	0.11	1.26×10^{-5}	710
10 Benzo(a)pyrene	252.3	5×10^{-5}	0.069	0.80×10^{-5}	1110
11 Phenol	94.1	1.7×10^{-5}	0.11	1.31×10^{-5}	680

[a] From Eq. 18-45 and $D_{water\,a} = 0.26$ cm s^{-1} (25°C). [b] From Eq. 18-55 and $D_{CO_2w} = 1.92 \times 10^{-5}$ cm s^{-1} (25°C), if not otherwise stated. [c] Schmidt Number of water phase $Sc_{iw} = v_i / D_{iw}$, v_i = kinematic viscosity of water, 0.893×10^{-2} cm^2 s^{-1} at 25°C. [d] From Jähne et al. (1987b) see also Table 20.3. [e] From Zheng et al. (1998). [f] From Oelkers (1991).

The contribution from the air-side boundary is given by:

$$v'_{ia} \equiv K_{ia/w}\, v_{ia} = K_{ia/w} \left(\frac{D_{ia}}{D_{water\,a}} \right)^{0.67} v_{water\,a} \tag{20-28}$$

where $D_{water\,a} = 0.26$ cm^2 s^{-1} at 25°C.

To calculate the water-side transfer velocity, v_{iw}, we must know the Schmidt Number of the selected compound i at 25°C. If the diffusivity needed to calculate Sc_{iw} is not known, it is estimated from Eq. 18-55 and the diffusivity of CO_2 at 25°C: D_{CO2w} (25°C) $= 1.92 \times 10^{-5}$ cm s^{-1}. Then:

$$v_{iw}(25°C) = \left(\frac{Sc_{iw}(25°C)}{600} \right)^{-a_{Sc}} v_{iw}\,(Sc = 600) \tag{20-29}$$

where $a_{Sc} = 0.67$ for $u_{10} = 1$ m s^{-1} and $a_{Sc} = 0.5$ for $u_{10} = 20$ m s^{-1}. For the two wind speeds, the resulting overall transfer velocities, $v_{ia/w}$, calculated from Eq. 20-3 are listed in Table 20.5b and shown in Fig. 20.7. Note that the deviations of the individual transfer velocities from the curves, calculated for average compound properties and a continuously varying Henry's Law coefficient, are relatively small. The magnitude of the air–water transfer velocity is determined mainly by wind speed and Henry's law coefficient of the individual compounds.

Inverse transfer velocities can be interpreted as layer resistances. Thus, the resistance ratio:

$$R_{ia/w} = \frac{R_{ia}}{R_{iw}} = \frac{1/v'_{ia}}{1/v_{iw}} = \frac{v_{iw}}{v'_{ia}} \tag{20-30}$$

Table 20.5b Calculation of Overall Air–Water Transfer Velocities at Two Wind Speeds for Different Organic Compounds at 25°C: Single-Phase and Overall Velocities (See also Table 20.5a)[a]

	$u_{10} = 1$ m s^{-1} [b]				$u_{10} = 20$ m s^{-1} [c]			
	v'_{ia} (cm s^{-1})	v_{iw} (cm s^{-1})	$v_{ia/w}$ (cm s^{-1})	$R_{ia/w}$	v'_{ia} (cm s^{-1})	v_{iw} (cm s^{-1})	$v_{ia/w}$ (cm s^{-1})	$R_{ia/w}$
1 Methane	14	7.5×10^{-4}	7.5×10^{-4}	5×10^{-5}	1.2×10^{-2}	2.1×10^{-2}	2.1×10^{-2}	2×10^{-4}
2 Trichlorofluoromethane	1.3	5.0×10^{-4}	5.0×10^{-4}	4×10^{-4}	1.2×10^{1}	1.6×10^{-2}	1.6×10^{-2}	1×10^{-3}
3 Octadecane	0.23	4.3×10^{-4}	4.2×10^{-4}	2×10^{-3}	1.9	1.4×10^{-2}	1.4×10^{-2}	7×10^{-3}
4 Tetrachloroethene	0.17	5.0×10^{-4}	4.9×10^{-4}	3×10^{-3}	1.5	1.6×10^{-2}	1.5×10^{-2}	1×10^{-2}
5 Benzene	6.9×10^{-2}	5.2×10^{-4}	5.2×10^{-4}	8×10^{-3}	5.9×10^{-1}	1.6×10^{-2}	1.6×10^{-2}	3×10^{-2}
6 1,2,4-Trichlorobenzene	2.5×10^{-2}	4.8×10^{-4}	4.7×10^{-4}	1.9×10^{-2}	2.2×10^{-1}	1.5×10^{-2}	1.4×10^{-2}	7×10^{-2}
7 Naphthalene	4.6×10^{-2}	4.9×10^{-4}	4.4×10^{-4}	0.11	4.0×10^{-2}	1.5×10^{-2}	1.4×10^{-2}	0.4
8 2,2',4,4',5,5'-Hexachlorobiphenyl	5.5×10^{-4}	3.8×10^{-4}	2.3×10^{-4}	0.69	4.7×10^{-3}	1.3×10^{-2}	3.5×10^{-3}	2.7
9 1-Hexanol	1.4×10^{-4}	5.8×10^{-4}	1.1×10^{-4}	4.1	1.2×10^{-3}	1.8×10^{-2}	1.1×10^{-3}	15
10 Benzo(a)pyrene	1.0×10^{-5}	4.3×10^{-4}	1.0×10^{-5}	42	8.8×10^{-5}	1.4×10^{-2}	8.8×10^{-5}	160
11 Phenol	4.8×10^{-6}	6.0×10^{-4}	4.7×10^{-6}	125	4.1×10^{-5}	1.8×10^{-2}	4.1×10^{-5}	440

[a] v'_{ia} from Eq. 20-28, v_{iw} from Eq. 20-29, $v_{ia/w}$ from Eq. 20.3, $R_{ia/w}$ (resistance ratio) from Eq. 20-30. [b] Based on $v_{water\ a} = 4.3$ cm s^{-1}, v_{iw} ($Sc_{iw} = 600$) = 19.1×10^{-3} cm s^{-1}. [c] Based on $v_{water\ a} = 0.5$ cm s^{-1}, v_{iw} ($Sc_{iw} = 600$) = 0.65×10^{-3} cm s^{-1}.

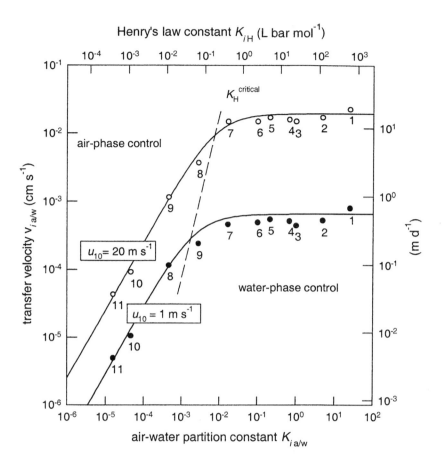

Figure 20.7 Overall air–water transfer velocity $v_{ia/w}$ as a function of Henry's Law coefficient for two very different wind conditions, u_{10} = 1 m s^{-1} (calm overland condition) and u_{10} = 20 m s^{-1} (rough ocean conditions). The solid lines are calculated for average compound properties: D_{ia} = 0.1 cm^2 s^{-1} and Sc$_{iw}$ = 600. The dashed line indicates the boundary between air-phase- and water-phase-controlled transfer velocities. See Table 20.5 for definitions of parameters and substances.

measures the relative importance of the air side-resistance compared to the water-side resistance. According to Table 20.5b, only large (e.g., benzo(a)pyrene) or polar (e.g., phenol) chemicals are air-phase controlled ($R_{ia/w}$ is large). Hence we expect that compounds like gasoline hydrocarbons and nonpolar solvents can be modeled with liquid-phase velocities only. In contrast, fluxes of polar compounds like acetone or fairly nonvolatile and nonpolar compounds like PCBs or PAHs are likely to be governed by air-phase dynamics, or a combination of both resistances. Typically, the $K_{ia/w}$ value below which air-phase control becomes dominant is 10^{-3}, but at larger wind speeds it shifts to slightly larger values ($\sim 5 \times 10^{-3}$ at 20 m s^{-1}) owing to the faster increase of the water transfer velocity v_{iw} with increasing wind speed.

Illustrative Example 20.3

Estimating the Overall Air–Water Transfer Velocity from Wind Speed for Different Water Temperatures

Problem

Calculate the total air–water transfer velocity, $v_{ia/w}$, of 1,1,1-trichloroethane (methyl chloroform, MCF) and tribromomethane (bromoform, BF) at the surface of the ocean for a wind speed of 15 m s^{-1} measured 3 m above the water surface at seawater temperatures of 25°C and 0°C, respectively.

The necessary information (molar mass, Henry's law constants) is given in Illustrative Example 20.1.

Answer

The overall transfer velocity is

$$\frac{1}{v_{ia/w}} = \frac{1}{v_{iw}} + \frac{1}{v'_{ia}} \tag{20-3}$$

where $v'_{ia} = K_{ia/sw} \, v_{ia}$ and $K_{ia/sw}$ is the air–seawater distribution coefficient. Note that changing water temperatures affect v_{iw} and v'_{ia} (through the Schmidt Number Sc_{iw} and $K_{ia/sw}$) but leave v_{ia} approximately unchanged. Problem 20.3 deals with the question of why it is water temperature, not air temperature, that determines the temperature at the very air–water interface and thus the temperature for which $K_{ia/sw}$ has to be evaluated.

(a) *Calculating* $K_{ia/sw}(T)$

From the information given in Illustrative Example 20.1:

	MCF	BF
$K_{ia/sw}$ (0°C)	0.29	8.9×10^{-3}
$K_{ia/sw}$ (25°C)	0.97	0.035

(b) *Calculating* v_{ia}

Convert the wind speed measured at 3 m, u_3, to the standard height u_{10} using Eq. 20-14:

$$u_{10} = \frac{10.4}{\ln 3 + 8.1} u_3 = \frac{10.4}{1.10 + 8.1} \, 15 \text{ m s}^{-1} = 17 \text{ m s}^{-1}$$

From Eq. 20.15 you get $v_{water\,a}$ for water vapor:

$$v_{water\,a} = (0.2 \times 17 + 0.3) \text{ cm s}^{-1} = 3.7 \text{ cm s}^{-1}$$

To calculate v_{ia} from $v_{water\,a}$ we use Eq. 20-27 and the simple molar mass relationship (Eq. 18-45) to estimate molecular diffusivity in air, D_{ia}:

$$v_{ia} = v_{water\,a} \left(\frac{D_{ia}}{D_{water\,a}} \right)^{0.67} = v_{water\,a} \left(\frac{M_i}{M_{water}} \right)^{-0.5 \times 0.67}$$

$$= 3.7 \text{ cm s}^{-1} \left(\frac{M_i}{18 \text{ g mol}^{-1}} \right)^{-0.355}$$

Thus:

$$v_{ia} = 1.89 \text{ cm s}^{-1} \text{ for MCB } (M_i = 133.4 \text{ g mol}^{-1})$$

$$v_{ia} = 1.53 \text{ cm s}^{-1} \text{ for BF } (M_i = 252.8 \text{ g mol}^{-1})$$

(c) *Calculating* v_{iw}

From Eq. 20-16 we get for $Sc_{iw} = 600$ (CO_2 at 20°C):

$$v_{iw}(600) = [164 \times 17 - 13.69] \times 10^{-3} \text{ cm s}^{-1} = 14.2 \times 10^{-3} \text{ cm s}^{-1}$$

To transform this value to other compounds and other temperatures we use Eq. 20-24a and calculate the Schmidt Number Sc_{iw} for both compounds at both temperatures. We do this in three steps.

(1) Calculate D_{iw} (25°C) and then Sc_{iw} (25°C)

We estimate D_{iw} (25°C) from molar mass and the regression line shown in Fig. 18.10*b*:

$$\text{MCF: } D_{iw}(25°C) = \frac{2.7 \times 10^{-4}}{(133.4)^{0.71}} \text{ (cm}^2\text{s}^{-1}) = 8.4 \times 10^{-6} \text{ cm}^2\text{s}^{-1}$$

$$\text{BF: } D_{iw}(25°C) = \frac{2.7 \times 10^{-4}}{(252.8)^{0.71}} \text{ (cm}^2\text{s}^{-1}) = 5.3 \times 10^{-6} \text{ cm}^2\text{s}^{-1}$$

The Schmidt Numbers Sc_{iw} are (use v_w (25°C) = 0.893×10^{-2} cm^{2}s^{-1}):

$$\text{MCF: } Sc_{iw}(25°C) = \frac{0.893 \times 10^{-2} \text{ cm}^2\text{s}^{-1}}{8.4 \times 10^{-6} \text{ cm}^2\text{s}^{-1}} = 1060$$

$$\text{BF: } Sc_{iw}(25°C) = 1680$$

(2) Calculate Sc_{iw}(0°C) from the Stokes-Einstein relation (Eq. 18-52):

$$Sc_{iw}(T) = \frac{v_w(T)}{D_{iw}(T)} = \frac{6\pi \rho_w r_i v_w^2}{kT}$$

Thus, if the temperature dependence of r_i is neglected:

$$Sc_{iw}(T) = Sc_{iw}(T_1) \times \left(\frac{v_w(T_2)}{v_w(T_1)} \right)^2 \times \frac{T_1}{T_2}$$

For $T_1 = 298$ K (25°C), $T_2 = 273$ K (0°C):

$$Sc_{iw}(0°C) = Sc_{iw}(25°C) \times \left(\frac{1.79 \times 10^{-2}}{0.893 \times 10^{-2}} \right)^2 \times \frac{298}{273}$$

$$\text{MCF: } Sc_{iw}(0°C) = 4630 \qquad \text{BF: } Sc_{iw}(0°C) = 7340$$

(3) Remember that v_{iw} at $Sc = 600$ plays the same role as a reference like $v_{water\,a}$ for the air-phase transfer velocity. Thus from Eq. 20-24a:

$$v_{iw}(Sc_{iw}) = v_{iw}(600) \times \left(\frac{Sc_{iw}}{600} \right)^{-0.5} = 14.2 \times 10^{-3} \text{ cm s}^{-1} \times \left(\frac{Sc_{iw}}{600} \right)^{-0.5}$$

To summarize:

	$Sc_{iw}(0°C)$	$v_{iw}(0°C)$ (cm s^{-1})	$Sc_{iw}(25°C)$	$v_{iw}(25°C)$ (cm s^{-1})
MCF	4630	5.1×10^{-3}	1060	10.7×10^{-3}
BF	7340	4.1×10^{-3}	1680	8.5×10^{-3}

(d) *Calculating the overall exchange velocity $v_{ia/w}$ from Eq. 20-3*

Note: $R_{ia/w}$ measures the size of the air-side resistance relative to the water-side resistance (Eq. 20-30).

		$K_{ia/sw}$	v_{ia} (cm s^{-1})	v'_{ia} (cm s^{-1})	v_{iw} (cm s^{-1})	$v_{ia/w}$ (cm s^{-1})	$R_{ia/w}$
MCF	25°C	0.97	1.89	1.83	10.7×10^{-3}	10.6×10^{-3}	5.8×10^{-3}
	0°C	0.29	1.89	0.55	5.1×10^{-3}	5.1×10^{-3}	9.3×10^{-3}
BF	25°C	0.035	1.53	0.054	8.5×10^{-3}	7.3×10^{-3}	0.16
	0°C	8.9×10^{-3}	1.53	0.0136	4.1×10^{-3}	3.2×10^{-3}	0.30

Note that the air–water exchange velocity decreases by about a factor of 2 when the temperature decreases from 25°C to 0°C. For MCF, $v_{ia/w}$ is completely water-side controlled, while for BF the relative resistance between air and water increases from 16% to 30% when the temperature decreases from 25°C to 0°C. This change of $R_{ia/w}$ is due to the fact that $K_{ia/sw}$ and thus v'_{ia} drops more strongly with T than v_{iw}.

20.4   Air–Water Exchange in Flowing Waters

Now we turn our attention to flowing waters. Here the physics of the boundary is influenced by two kinds of motion, the motions induced by the wind and the water currents, respectively. The latter will be extensively discussed in Chapter 24. At this point it is sufficient to introduce the most important concept in fluid dynamics to quantify the intensity of turbulent motion and to assess the relative importance of several simultaneous processes of turbulent kinetic energy production.

Recall our short discussion in Section 18.5 where we learned that turbulence is kind of an analytical trick introduced into the theory of fluid flow to separate the large-scale motion called advection from the small-scale fluctuations called turbulence. Since the turbulent velocities are deviations from the mean, their average size is zero, but not their kinetic energy. The kinetic energy is proportional to the mean value of the squared turbulent velocities, $\overline{u_{turb}^2}$, that is, of the variance of the turbulent velocity (see Box 18.2). The square root of this quantity (the standard deviation of the turbulent velocities) has the dimension of a velocity. Thus, we can express the turbulent kinetic energy content of a fluid by a quantity with the dimension of a velocity. In the boundary layer theory, which is used to describe wind-induced turbulence, this quantity is called *friction velocity* and denoted by u^*. In contrast, in river hydraulics turbulence is mainly caused by the friction at the bottom of a river. The corresponding u^* is called *shear velocity* (see Chapter 24, Eq. 24-5). In both cases u^* is proportional to the standard deviation of the resulting turbulent velocities.

It has been a central goal of a great number of experiments both in the laboratory and in the field to find relations between the water-phase transfer velocity v_{iw} and the shear velocity u^*. Moog and Jirka (1999a,b) have summarized these results and combined them with their own experiments (see Box 20.3).

First we note that there exists a continuum of situations from nearly stagnant water bodies (e.g., a slowly flowing river) to cases of great turbulence (e.g., an extremely rough mountain stream). We cannot expect that simple mathematical expressions could be developed which relate the air–water exchange velocity, $v_{ia/w}$, to just a few stream parameters, such as mean flow velocity, $\bar{u}$, or river depth, h, and which are valid for all kinds of rivers and streams from the Mississippi to an Alpine creek. However, if we restrict ourselves to flowing water bodies with a well-defined water surface (though this surface may be deformed by waves and turbulent eddies), the water-phase exchange velocity indeed can be written as a function f of just a few parameters:

$$v_{iw} = f(Sc_{iw}, u_{10}, \bar{u}, u^*, h, S_o) \tag{20-31}$$

where $\quad Sc_{iw} = v_i / D_{iw}$ is the water-phase Schmidt Number (Eq. 20-23)

$\quad\quad u_{10}\quad$ is wind speed 10 m above the water surface

$\quad\quad \bar{u}\quad$ is mean flow velocity in the river

$\quad\quad h\quad$ is mean river depth

$\quad\quad S_o\quad$ is slope of the river bed (i.e., the ratio of horizontal distance and change of altitude, see Table 24.1)

The above expression is very general and includes both the case of stagnant waters ($\bar{u} = 0$, e.g., Eq. 20-24) as well as situations in which the water flow-induced turbulence dominates the exchange velocity relative to the influence of the wind. Obviously, as wind speed changes, for a given river the situation may switch between current-dominated and wind-dominated regimes. Another factor which influences the shape of the empirical function f of Eq. 20-31 is the typical size of the turbulent structures (the eddies) relative to the water depth. This leads to two different models, the *small-eddy* and the *large-eddy model,* respectively (Fig. 20.8 and Box 20.3).

Before we discuss these models, we note that, in contrast to v_{iw}, the air-phase exchange velocity, v_{ia}, is not strongly affected by the flow. Thus, the following considerations are not relevant for compounds with very small Henry's law coefficients. This is no longer true when the air–water interface is broken up by bubbles and droplets. Some models attempt to incorporate the effect of air bubbles into the exchange velocity v_{iw} (see Eq. 20-38 below), yet air bubbles also lead to a modification of Eq. 20-3 describing the overall exchange velocity, $v_{ia/w}$. In the context of river flow, this situation will be treated in Section 24.4.

Box 20.3   Eddy Models for Air–Water Exchange in Rivers

In rivers the water-side air–water exchange velocity v_{iw} is influenced by both turbulence produced by the wind and turbulence produced by the flow due to friction at the river bed. In this box we summarize the important concepts for flow-induced friction (see Moog and Jirka (1999a, b) for details).

The predominant model is based on the surface renewal concept by Higbie (1935). Thus, according to Eqs. 20-20 and 20-24a, v_{iw} is proportional to $(Sc_{iw})^{-1/2}$. Furthermore, v_{iw} is scaled by the friction velocity u^*:

$$v_{iw} = \text{constant} \left[u^*(Sc_{iw})^{-1/2} R_*^n \right] \tag{1}$$

where the nondimensional *shear Reynolds Number* R_* is defined as:

$$R_* = \frac{u^* h}{v_w} \tag{2}$$

h is mean depth of the river and v_w is kinematic viscosity of water. The size of the exponent n of Eq. 1 is controlled by the mechanism which produces the turbulent eddies that are responsible for the surface renewal. Two cases are distinguished, the *small-eddy situation* and the *large-eddy situation*.

Small-Eddy Model by Lamont and Scott (1970) (Fig. 20.8.a)

The small-eddy model applies, if the nondimensional roughness parameter d^* fulfills:

$$d^* = \frac{d_s u^*}{v_w} < 136, \tag{20-36}$$

where d_s is the so-called *equivalent sand-grain diameter*, that is, the diameter of particles (from boulders to clay) which would form a sediment surface with the same roughness as in the particular river bed. Typical diameters are:

Name of Particle	Typical Diameter d_s [m]
Clay	$< 4 \times 10^{-6}$
Silt	4×10^{-6} to 1×10^{-4}
Sand	1×10^{-4} to 2×10^{-3}
Pebble	2×10^{-3} to 0.1
Cobble	0.1 to 0.3
Boulder	> 0.3

From their experiments, Moog and Jirka (1999a) deduced the following air–water transfer velocity for the small-eddy regime:

$$v_{iw} = 0.161 (Sc_{iw})^{-1/2} \left(\frac{v_w u^{*3}}{h} \right)^{1/4} = 0.161 (Sc_{iw})^{-1/2} u^* \left(\frac{u^* h}{v_w} \right)^{-1/4} = 0.161 (Sc_{iw})^{-1/2} u^* R_*^{-1/4} \tag{3}$$

Thus, this equation has indeed the form of Eq. 1 with the exponent $n = 1/4$. Note that the shear velocity for river flow can be calculated either from the slope and geometry of the river bed (Eq. 20-33a) or from the mean flow velocity $\bar{u}$ (Eq. 20-33b).

Large-Eddy Model by O'Connor and Dobbins (1958) (Fig. 20.8*b*)

If the roughness parameter d^* increases beyond 136, the turbulent eddies become larger and begin to feel the limited vertical extension of the water column. As a result, the exponent n in Eq. 1 steadily increases until it reaches a new constant value at 1/2. At this point the eddies have reached the full depth of the river and are thus able to transport water fast between surface and bottom. This is the situation described by the large-eddy model of O'Connor and Dobbins (1958). We formulate it in the general form of Eq. 1:

$$v_{iw} = \text{constant} \times u^* (\text{Sc}_{iw})^{-1/2} \left(\frac{u^* h}{v_w} \right)^{-1/2} = \text{constant} \times u^* \left(\frac{D_{iw}}{v_w} \cdot \frac{v_w}{u^* h} \right)^{1/2}$$

$$= \text{constant} \times \left(\frac{D_{iw} u^*}{h} \right)^{1/2} \tag{4}$$

Since according to Eq. 20-33b, u^* is proportional to the mean river velocity $\bar{u}$, Eq. 4 can also be written as:

$$v_{iw} = \text{constant} \times \left(\frac{D_{iw} \bar{u}}{h} \right)^{1/2} \tag{5}$$

The constant in Eq. (5) turns out to be of order 1. This yields the large-eddy model by O'Connor and Dobbins (1958):

$$v_{iw} \approx \left(\frac{D_{iw} \bar{u}}{h} \right)^{1/2} \tag{6}$$

Small Roughness and Small-Eddy Model

Qualitatively, a river with *small roughness* can be visualized as a water body in which the turbulent eddies produced by the water currents flowing over the uneven bottom of the river bed made up by particles of different size (e.g., pebble, sand, silt etc.) are much smaller than the depth of the river (Fig. 20.8*a*). When these eddies travel upward from the bottom, they transport turbulent energy to the water surface and thus influence the air–water exchange. According to Moog and Jirka (1999a), the resulting air–water transfer velocity can be best described by the model of Lamont and Scott (1970). The conditions for the validity of the small-eddy model and further details are given in Box 20.3. The model leads to the following air–water transfer velocity:

$$v_{iw} = 0.161 (\text{Sc}_{iw})^{-1/2} \left(\frac{v_w u^{*3}}{h} \right)^{1/4} \qquad [\text{LT}^{-1}] \tag{20-32}$$

where the friction velocity u^* can either be estimated from the river slope S_o (see Eq. 24-5):

$$u^* = (g S_o R_h)^{1/2} \quad , \qquad [\text{LT}^{-1}] \tag{20-33a}$$

or from the mean flow velocity $\bar{u}$ (Eq. 24-6):

$$u^* = \frac{\bar{u}}{\alpha} \quad , \qquad [\text{LT}^{-1}] \tag{20-33b}$$

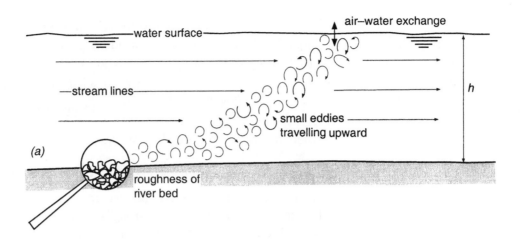

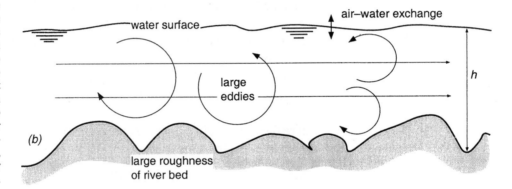

Figure 20.8 Depending on the roughness of the river bed, the production of turbulence leads either to (*a*) eddies which are much smaller than the river depth *h*, or to (*b*) large eddies which are able to transport dissolved chemicals fast from and to the water surface. Both situations can be described by two different models for air–water exchange; (*a*) the small-eddy model by Lamont and Scott (1970), and (*b*) the large-eddy model by O'Connor and Dobbins (1958). See Box 20.3 for details.

Here R_h is the *hydraulic radius* of the river bed (see Table 24.1) which for wide rivers (that is, for rivers that are much wider than they are deep) is about equal to mean depth *h*. α is a nondimensional factor which typically lies between 10 (rough river bed) and 20 (smooth river bed). Note that Eq. 20-32 is dimensionally correct. Hence the factor 0.161 is independent of a specific choice of units for the involved parameters.

Interpretation of Eq. 20-32 yields the following points:

(1) v_{iw} increases with turbulence intensity (quantified by u^*). The influence of a given turbulence intensity on v_{iw} increases with decreasing water depth *h*.

(2) Through the factor $(Sc_{iw})^{-1/2}$ the exchange velocity v_{iw} is proportional to $(D_{iw})^{1/2}$. This indicates that air–water exchange results from a process of surface renewal (see Eq. 20-20).

The second point allows us to compare the influence on air–water exchange of the flow and the wind, respectively. If we insert Eq. 20-33b into Eq. 20.32, and choose $\alpha = 20$ and v_w at 20°C (1×10^{-6} m^{2}s^{-1}), we can calculate the theoretical exposure time, t_{exp}, produced by the flowing water and compare it with the exposure time produced by wind-induced turbulence. According to Eq. 20-21b the latter is of order 10 s to 0.1 s for weak and strong winds, respectively. It turns out that current-induced turbulence surpasses the influence of the wind on the air–water exchange velocity, v_{iw}, if the mean flow velocity in the river is larger than a critical velocity, $\bar{u}_{crit}$, defined by:

$$\bar{u}_{crit}\,(\text{m s}^{-1}) \sim \begin{cases} 0.2\ h^{1/3} & \text{weak winds (up to 4 m s}^{-1}\text{)} \\ 3\ h^{1/3} & \text{strong winds (order 10 m s}^{-1}\text{)} \end{cases} \qquad (20\text{-}34)$$

where h is mean river depth in meters.

Rough River Flow and Large-Eddy Model

As the roughness of the river bed increases, the eddies induced by the irregularities become larger until the *macro-roughness elements* (see Box 20.3) extend over the whole water body (Fig. 20.8b). These macro-eddies allow water to be transported from the bottom to the water surface within short times, thus enhancing the exchange between air and water. According to Moog and Jirka (1999b), then the size of the exponent $n = 1/4$ in Eq. 1 of Box 20.3 gradually increases to 1/2. Eventually, the air–water exchange velocity v_{iw} is transformed into the large-eddy model of O'Connor and Dobbins (1958) :

$$v_{iw} = \text{constant} \times \left(\frac{D_{iw}\bar{u}}{h} \right)^{1/2} \approx \left(\frac{D_{iw}\bar{u}}{h} \right)^{1/2} \qquad (20\text{-}35)$$

where the constant in the above expression turns out to be about 1 (Box 20.3, Eqs. 4 to 6). Note that this expression can also be written in the form:

$$v_{iw} = \left(\frac{D_{iw}}{t_{river}} \right)^{1/2}, \qquad \text{with } t_{river} = h/\bar{u} \qquad (20\text{-}35a)$$

where t_{river} is the transport time over distance h at velocity $\bar{u}$. In the light of the surface renewal model (Eq. 20-20), t_{river} adopts the role of the replacement time (except for the factor π). Thus, according to the large eddy model air–water exchange is controlled by eddies extending over the whole water depth which circle at a speed scaled by the mean current velocity, $\bar{u}$.

The transition between the two regimes is controlled by the nondimensional roughness parameter d^*:

$$d^* \equiv \frac{d_s u^*}{v_w} \begin{cases} <136 & \text{small eddy} \\ \gg 136 & \text{large eddy} \end{cases} \qquad (20\text{-}36)$$

where d_s [L] is the *equivalent sand-grain diameter*, a measure for the size of the particles (grains, pebbles, stones, etc.) covering the river bed. Typical particle diameters are given in Box 20.3.

Enhanced Air–Water Exchange Through Bubbles

If the roughness of the river bed increases even further, the water surface loses its intact form. Air bubbles are entrained into the water, water droplets are ejected into the air, and foam and spray are formed. The parameter which determines the onset of this kind of enhanced air–water exchange is the nondimensional *element Froude*

Number F_E (see Moog and Jirka, 1999b):

$$F_E = \frac{h \cdot \bar{u}}{[g(h-h_E)^3]^{1/2}} \qquad h > h_E \qquad (20\text{-}37)$$

where h_E is the typical height of "roughness elements", a length scale which gradually turns into the equivalent grain-size diameter d_s if h_E becomes smaller. Note that Eq. 20-37 makes sense only if $h_E < h$, that is, if all "roughness elements" are submerged. In fact, for an alpine stream in which the typical size of the boulders forming the stream bed is larger than the water depth, the above relation is not valid.

In laboratory experiments Moog and Jirka (1996b) found that enhanced air–water exchange sets in for $F_E > 1.4$ and is then described by:

$$v_{iw} = Sc_{iw}^{-1/2} u^* (0.0071 + 0.023 F_E), \qquad F_E > 1.4 \qquad (20\text{-}38)$$

In Illustrative Example 20.4, air–water exchange velocities are calculated for benzene in a river for different roughness conditions. A more radical influence of air bubbles on air–water exchange will be discussed in Chapter 24.4.

Illustrative Example 20.4

Air–Water Exchange of Benzene in Rivers

You analyze air–water exchange of benzene along a river which consists of three stretches of different bed roughness. The characteristics of the river sections are given in the following table.

River characteristics

Stretch	$\bar{u}$ (m s^{-1})	α (Eq. 20-33b)	h (m)	d_s or h_E (m)
A	1	20	1	10^{-3}
B	1	10	1	0.1
C	1	5	1	0.8

Problem

Calculate the air–water exchange velocity of benzene at 25°C due to the flow-induced turbulence for the stretches A, B, C. Compare the result with the wind-induced exchange velocity v_{iw} produced by a wind speed $u_{10} = 3$ m s^{-1}.

i = benzene

$M_i = 78.1$ g mol^{-1}
$K_{ia/w}$ (25°) = 0.23
$D_{iw} = 1.06 \times 10^{-5}$ cm^{2}s^{-1}

Answer

According to Table 20.5a the Schmidt Number of benzene at 25°C is $Sc_{iw} = 840$ (note v_w (25°C) = 0.893×10^{-2} cm^2 s^{-1}).

In order to decide which model is relevant for the different river sections, we first calculate the roughness parameters d^* (Eq. 20-36) and the element Froude Numbers F_E (Eq. 20-37).

Stretch A: $u^* = \bar{u}/a = 1 \text{ m s}^{-1} / 20 = 0.05 \text{ m s}^{-1}$

$$d^* = \frac{10^{-3} \text{m} \times 0.05 \text{ m s}^{-1}}{0.893 \times 10^{-6} \text{ m}^2\text{s}^{-1}} = 56 \quad \text{(note the appropriate units of } v_w\text{)}$$

$$F_E = \frac{1 \text{ m} \times 1 \text{ m s}^{-1}}{[9.81 \text{ m s}^{-2} \cdot (1 \text{ m})^3]^{1/2}} \sim 0.3$$

Thus, the appropriate description for stretch A is the small-eddy model.

Stretch B: $u^* = 1 \text{ m s}^{-1} / 10 = 0.1 \text{ m s}^{-1}$

$$d^* = \frac{0.1 \text{ m} \times 0.1 \text{ m s}^{-1}}{0.893 \times 10^{-6} \text{ m}^2\text{s}^{-1}} \sim 10^4$$

$$F_E = \frac{1 \text{ m} \times 1 \text{ m s}^{-1}}{[9.81 \text{ ms}^{-2} \times (0.9 \text{ m})^3]^{1/2}} \sim 0.4$$

This river stretch would be treated with the large-eddy model without bubble enhancement.

Stretch C: $u^* = 1 \text{ m s}^{-1} / 5 = 0.2 \text{ m s}^{-1}$

$$d^* = \frac{0.8 \text{ m} \times 0.2 \text{ m s}^{-1}}{0.893 \times 10^{-6} \text{ m}^2\text{s}^{-1}} = 1.8 \times 10^5$$

$$F_E = \frac{1 \text{ m} \times 1 \text{ m s}^{-1}}{[9.81 \text{ m s}^{-2} \times (0.2 \text{ m})^3]^{1/2}} = 3.5 > 1.4$$

Now we have bubble enhancement.

Next we calculate the current-induced exchange velocities and compare it with the wind-induced value. The latter is (see Table 20.4, and Eq. 20-24a):

$$v_{iw}(\text{wind}) = \left(\frac{840}{600}\right)^{-2/3} 0.65 \times 10^{-3} \text{cm s}^{-1} = 0.52 \times 10^{-3} \text{cm s}^{-1}$$

Stretch A: From Eq. 20-32:

$$v_{iw} = 0.161 \times (840)^{-1/2} \left(\frac{0.893 \times 10^{-6} \text{m}^2\text{s}^{-1} \times (0.05 \text{ m s}^{-1})^3}{1\text{m}}\right)^{1/4} = 1.8 \times 10^{-5} \text{m s}^{-1}$$

$$= 1.8 \times 10^{-3} \text{cm s}^{-1}$$

Stretch B: From Eq. 20-35:

$$v_{iw} \sim \left(\frac{1.06 \times 10^{-9} \text{m}^2\text{s}^{-1} \times 1 \text{m s}^{-1}}{1\text{m}}\right)^{1/2} = 3.3 \times 10^{-5} \text{m s}^{-1} = 3.3 \times 10^{-3} \text{cm s}^{-1}$$

Stretch C: From Eq. 20-382

$$v_{iw} = (840)^{-1/2} \times 0.2\,\mathrm{m\ s}^{-1} \times (0.0071 + 0.023 \times 3.5)$$
$$= 6.0 \times 10^{-4}\,\mathrm{m\ s}^{-1} = 6.0 \times 10^{-2}\,\mathrm{cm\ s}^{-1}$$

To summarize, the air–water exchange velocity increases by about a factor 2 from stretch A to B, that is, when the small eddies are replaced by large ones. A more spectacular increase (factor 20) takes place at the transition to enhanced exchange due to bubble formation (stretch C). In all cases the flow-induced air–water exchange is more important than the influence from the wind. This is in accordance with Eq. 20-34. Note that for $u_{10} = 10$ m s^{-1}, the wind-induced turbulence would dominate in stretches A and B, but not in C.

Advanced Topic　　**20.5** ## Influence of Surface Films and Chemical Reactions on Air–Water Exchange

Surface Films

Surface active substances (*surfactants*) are chemicals which accumulate at the water surface and reduce the air–water interfacial tension. The influence of such films on air–water exchange is twofold: (1) they create an additional transport barrier, and (2) they change the hydrodynamics at the water surface such that the transport of solutes by eddies approaching the water surface is reduced (hydrodynamic damping).

The first effect can be described by an additional term in Eq. 20-3 (see also Problem 19.3):

$$\frac{1}{v_{ia/w}} = \frac{1}{v_{iw}} + \frac{1}{v_{ia}K_{ia/w}} + \frac{1}{v_{if}K_{if/w}} \tag{20-39}$$

where v_{if} is the transfer velocity in the surfactant layer which in the film model can be described as (see Eq. 20-17):

$$v_{if} = \frac{D_{if}}{\delta_f} \tag{20-40}$$

D_{if} is molecular diffusivity of compound i in the film, δ_f is surface film thickness, and $K_{if/w}$ is the nondimensional equilibrium distribution coefficient of substance i in the film relative to the water.

Since δ_f is often extremely thin (sometimes a monomolecular layer), usually the third term on the right-hand side of Eq. 20-39 does not significantly influence $v_{ia/w}$. However, if the water surface is polluted by a thick layer of oil, air–water exchange may be strongly suppressed. Fig. 20.9 shows laboratory experiments by Downing and Truesdale (1955) for oil films of different thickness. The authors measured the overall transfer velocity of molecular oxygen. The measurements demonstrate that $v_{ia/w}$ is significantly reduced when δ_f reaches a value of about 100 μm (10^{-4} m). According to Eq. 20-18a, this is of the same order as the typical thickness of the water-side stagnant film, δ_w. Since the air–water exchange of O_2 is water-film

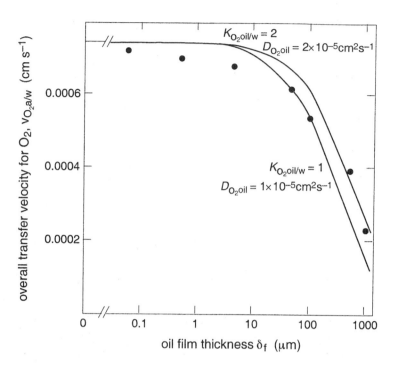

Figure 20.9 Decreasing air–water exchange rate of O_2 for a water surface covered with an increasingly thick film of oil. Experimental data by Downing and Truesdale (1955) (dots) are compared with curves calculated from Eqs. 20-39 and 20-40. $K_{O_2\ oil/w}$ is the oil–water distribution coefficient of O_2; $D_{O_2\ oil}$ is molecular diffusivity of O_2 in oil. Note that the term containing the air-side exchange velocity, v_{ia}, is neglected.

controlled, the second term on the right-hand side of Eq. 20-39 can be neglected. The curves in Fig. 20.9 show the calculated overall exchange velocity for a water surface covered with an oil film of thickness δ_f. The distribution coefficient of O_2 in oil, $K_{O_2\ oil/w}$, is about 2; diffusivity of O_2 in the oil is 2×10^{-5} cm²s⁻¹. As it turns out, the simple model (Eqs. 20-39 and 20-40) corresponds well to the experimental data.

Regarding hydrodynamic damping, the second effect of surfactant films on air–water exchange, it has long been realized that the addition of minute amounts of oils to the sea surface calms the waves and yields a "glassy" slick (Plinius and Benjamin Franklin have been quoted in this regard!). Such action appears to require a continuous monolayer of substances like oleyl alcohol or oleic acid, corresponding to films of approximately nanometer thickness and coverages of about 100 ng cm⁻² of organic chemical (Garrett, 1967; Jarvis et al., 1967). By extracting organic matter from seawater, Jarvis et al. (1967) have shown that natural organic materials can also coat a water–air interface and damp capillary waves. If such surface-calming reflects diminished mixing in the water just below the surface, we would anticipate that water–air exchange will become more and more limited by the ability of chemicals to diffuse rather than be carried by microscopic eddies; thus exchange velocities should decrease. This is exactly the result seen by Broecker et al. (1978) for exchange of CO_2 as a function of wind speed (Fig. 20.10). At very low winds (≤ 2 m s⁻¹), the importance of a monolayer of oleyl alcohol is negligible, since such conditions are insufficient to ruffle a water in any case. However, above the wind velocity at which capillary waves begin to form ($2 - 3$ m s⁻¹), the oleyl alcohol was seen both to "prohibit" surface roughening (up to speeds of 12 m s⁻¹) and concomitantly to reduce CO_2 gas exchange relative to the "clean surface" case by as much as a factor of 4 at winds of 10 m s⁻¹.

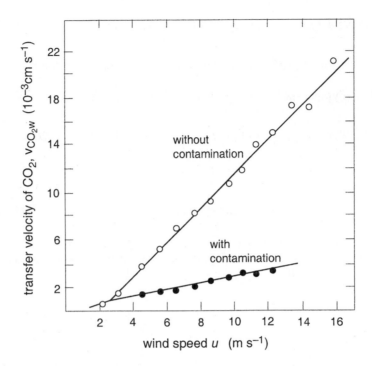

Figure 20.10 Effect of wind speed on the air–water exchange velocity of CO_2 with and without contamination of the water with a monolayer of oleyl alcohol measured in a wind–wave tank. Lines indicate linear approximation of experimental data. From Broecker et al. (1978).

As reported by Romano (1996), surfactant films may be more common than previously assumed. In the Indian Ocean he found such films in 30% of coastal and 11% of open sea water. At wind speeds $u_{10} > 6$ m s^{-1} these films seem to be destroyed by turbulence, but they are able to form again on a time scale of a few hours. Frost and Upstill-Goddard (1999) give an overview of available information on the composition of surfactant films.

Influence of Chemical Reactions on Air–Water Exchange Rates

In certain situations, a chemical of interest may be involved in a rapid reversible transformation in the water phase. Such a reaction would affect the concentration in the boundary zone and thus would alter the transfer rate. The reaction time t_r (defined by the inverse of the first-order reaction rate constant, $t_r = k_r^{-1}$) determines whether air–water exchange is influenced by the reaction. Three cases can be distinguished.

(1) The reaction is slow relative to the time t_w needed to transport the chemical across the water-phase boundary ($t_r \gg t_w$). In this case we can reasonably assume that such a transformation has no significant impact on the molecules during the time they spend diffusing in the boundary zone. Thus, the equations derived above remain valid.

(2) The reaction is fast compared to the transport time ($t_r \ll t_w$). In this case we should include the newly formed species in our thinking, but we can do it in a simplified way by assuming immediate equilibrium between the species throughout the boundary zone.

(3) Reaction and transport times are of the same order of magnitude ($t_r \sim t_w$). This situation requires a more detailed analysis of both the fluxes and the concentration

Table 20.6 Typical Transport Times for Organic Molecules Traversing the Water-Phase Boundary Zone

Typical exchange velocity $v_{iw} = 5 \times 10^{-4}$ cm s^{-1} to 5×10^{-3} cm s^{-1} (Table 20.5b)

Film Model   Typical molecular diffusivity $D_{iw} = 1 \times 10^{-5}$ cm s^{-1} (Table 20.5a)

Typical transport time $t_w = \dfrac{\delta_w^2}{2D_{iw}} = \dfrac{D_{iw}}{2v_{iw}^2} = 0.2$ s to 20 s   (from Eqs. 18-8 and 20-18)

Surface Renewal Model

Typical exposure time $t_{rep} = t_w = 0.2$ to 20 s $\dfrac{D_{iw}}{\pi v_{iw}^2} = 0.1$ s to 10 s   (Eq. 20-21b)

Note: With respect to transport time, the film and the surface renewal model are consistent except for the slightly different numerical factor in the denominator.

Boundary Layer Model

Similar as for film model [a]

---

[a] The boundary layer model is a modified version of the film model in which diffusivity varies more smoothly than in the film or bottleneck model. Thus, similar typical transport times occur. The main difference between the film model and the boundary model is the way the transfer velocities for different compounds are related (diffusivity vs. Schmidt number, size of exponent).

profiles within the boundary zone. We will see that this case may extend to situations with $t_r \ll t_w$ (Eq. 20-55).

Before we discuss the second and third situation in greater depth, we should have approximate values for the typical transport times t_w with which the reaction time t_r has to be compared. Table 20.6 summarizes the situation for the water-phase boundary zone for the three models which we have used to assess the transfer processes. Note that for two reasons we do not consider reactions in the air-phase boundary zones. First, transport times in this zone are about one order of magnitude smaller than transport times in the water-phase. Second and more importantly, there are generally no transformations in the air which are fast enough to compete with the speed of the transfer (see Section 16.3, Fig. 16.7).

Fast Reaction (Case 2)

From Table 20.6 we conclude that independent of which model we use, typical transfer times are between a few tenths of a second and a minute. Proton exchange reactions of the form (see Section 8.2):

$$HA \iff A^- + H^+ \qquad (8\text{-}6)$$

have reaction times in water that are much smaller than these transport times. Such acid-base reactions can be characterized by equilibrium constants of the form:

$$K_a = \frac{[H^+][A^-]}{[HA]} \qquad (20\text{-}41)$$

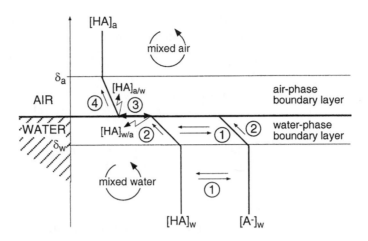

Figure 20.11 Air–water exchange of an organic compound HA undergoing a proton exchange reaction. The conjugate base A⁻ cannot leave the water, but it contributes to the diffusive transport across the water-phase boundary layer. 1 = fast acid/base equilibrium (Eq. 8-6), 2 = diffusive transport of HA and A⁻ across water-phase boundary layer, 3 = Henry's law equilibrium of HA between water and air, 4 = diffusive transport of HA across air-phase boundary layer.

where [HA], [H⁺], [A⁻] are the corresponding concentrations and the activity coefficients γ_i are assumed to be 1.

Since the conjugate base A⁻ is charged, it has virtually no ability to leave the polar solvent and enter the air. Thus, only neutral molecules diffuse into the air and exhibit a distribution equilibrium with the molecules in the water just at the interface:

$$K_{HA\,a/w} = \frac{[HA]_{a/w}}{[HA]_{w/a}} \tag{20-42}$$

The proton exchange reaction can be assumed to be at equilibrium everywhere in the water. Thus the ratio of the total and neutral compound concentration is given by:

$$K_{HAtot/HA} = \frac{[HA]+[A^-]}{[HA]} = 1+\frac{K_a}{[H^+]} \tag{20-43}$$

Note that the inverse of $K_{HAtot/HA}$ is identical with α_a which was introduced in Eq. 8-21. Here we choose the K-notation to indicate that the ratio is like a partition coefficient which appears in the flux (Eq. 20-1) if different phases or different chemical species are involved (see section 19.2 and Eq.19-20). In order to show how the combination of both partitioning relationships, one between air and water (Eq. 20-42), the other between neutral and total concentration (Eq. 20-43), affect the air–water exchange of [HA], we choose the simplest air–water transfer model, the film or bottleneck model. Figure 20.11 helps to understand the following derivation.

First, we note that in the air only the neutral species participates in the flux:

$$F_a = \frac{D_{HA\,a}([HA]_{a/w} - [HA]_a)}{\delta_a} \tag{20-44}$$

while in the water the flux consists of the sum of the fluxes of the protonated and deprotonated species:

$$F_w = \frac{D_{HA\,w}([HA]_w - [HA]_{w/a})}{\delta_w} + \frac{D_{A^-\,w}([A^-]_w - [A^-]_{w/a})}{\delta_w} \tag{20-45}$$

Typically it is reasonable to assume that the diffusivities of HA and A⁻ are the same since these species differ little in size. Recalling the acid–base equilibrium relationship and using the usual definition of the transfer velocity, $v_{HAw} = D_{HAw}/\delta_w$, yields:

$$F_w = v_{HA\,w}\,([HA]_w - [HA]_{w/a}) + v_{HA\,w}\left(\frac{K_a[HA]_w}{[H^+]} - \frac{K_a[HA]_{w/a}}{[H^+]}\right)$$

$$= v_{HA\,w}\left(1 + \frac{K_a}{[H^+]}\right)([HA]_w - [HA]_{w/a}) \tag{20-46}$$

$$= v_{HA\,w}\,K_{HAtot/HA}\,([HA]_w - [HA]_{w/a})$$

Here we have assumed that the solution pH and temperature are the same throughout the boundary region. Effectively, we have obtained the same flux expression as Eq. 19-17, except for the fact that the water-side exchange velocity is modified by the factor containing the acid–base equilibrium constant and pH. Thus, to combine Eqs. 20-44 and 20-46 into a single flux expression we can use the rules given in Eq. 19-21. However, in the situation discussed here the reference concentration is $[HA]_w$ and this is neither the diffusing species in the air-phase (which is $[HA]_a$) nor the one in the water which is $[A_{tot}] = [HA] + [A^-]$). Following the substitution law given in Eq. 19-21, we simple have to multiply the single-phase transfer velocities, v_{ia} and v_{iw}, by the corresponding partitioning coefficient, i.e.:

$$v_{ia} \rightarrow v_{ia}K_{HAa/w} \qquad v_{iw} \rightarrow v_{iw}K_{HAtot/HA} \tag{20-47}$$

Combining these substitutions yields the following flux equation for a fast reacting species:

$$F = v_{HAtot\,a/w}\left([HA]_w - \frac{[HA]_a}{K_{HA\,a/w}}\right) \tag{20-48}$$

with:

$$\frac{1}{v_{HAtot\,a/w}} = \frac{1}{v_{HAa}\,K_{HA\,a/w}} + \frac{1}{v_{HAw}\,K_{HAtot/HA}} \tag{20-49}$$

Two important lessons can be gained from these results (see also Fig. 20.11): First, the conductance through the water film of the reactive species is enhanced by the factor $K_{HAtot/HA}$, since the diffusive transport is accomplished not only by the neutral but also by the charged species. The multiplication factor may be very big if much of the compound is ionized ($K_a/[H^+]$ large). Second, it is the concentration difference of the neutral species only, $[HA]$, which drives air–water exchange.

It can be shown that if the flux is derived from the surface renewal model (Chapter 19.4), the result is identical with Eq. 20-49. Again, the corresponding transfer velocity v_{iw} (this time given by Eq. 19-59 or 20-20) is enhanced by the factor $K_{HAtot/HA}$.

In spite of the above result, which supports our intuition (fast deprotonation helps to "draw" the molecule into the water), there is hardly an example where the modified transfer Eq. 20-49 is of practical importance. Why? Not because the enhancement of

v_{iw} is small (in fact, it can easily exceed a factor 100), but because polar substances usually have extremely small Henry's Law coefficients. That is, their air–water transfer is air-phase controlled. Therefore, the enhancement of v_{iw}—large as it may be—does not really affect $v_{i\,a/w}$.

Reaction and Diffusion of Similar Magnitude (Case 3)

This conclusion does not hold in the following example in which t_r and t_w are of similar size (case 3 in the above list). In Chapter 12 we discussed the hydration/ dehydration of formaldehyde as a pseudo-first-order two-way reaction (Eqs. 12-15 to 12-24). The reaction time of hydration is of the order of 10 s and thus similar to the air–water transfer time.

Let us again exemplify the theory using the film model. For simplicity, we consider one (water-phase) film (see Fig. 20-12). We adopt the notation introduced in Eq. 12-16, where [A] stands for the aldehyde and [D] for the diol concentration (the hydrated aldehyde), and assume that only the aldehyde diffuses into the atmosphere, and that in the aqueous mixed layer the two species are in equilibrium:

$$\frac{[D]}{[A]} = K_r \qquad (12\text{-}17)$$

As a first step we have to formulate the diffusion/reaction equation of A and D in the water-film. For this purpose we combine Fick's second law (Eq. 18-14) with the forward/backward reaction of the aldehyde (Eq. 12-16):

$$\frac{\partial[A]}{\partial t} = D_{Aw}\frac{\partial^2[A]}{\partial z^2} - k_1[A] + k_2[D]$$

$$\frac{\partial[D]}{\partial t} = D_{Dw}\frac{\partial^2[D]}{\partial z^2} + k_1[A] - k_2[D] \qquad (20\text{-}50)$$

where D_{Aw}, D_{Dw} are molecular diffusion coefficients in water and k_1, k_2 are forward and backward first-order reaction rate constants.

If the boundary conditions (i.e., the aldehyde concentration in the atmosphere, $[A]_a$, and in the interior of the water body, $[A]_w$) are given and held constant, steady-state conditions are quickly established in the film: $\partial[A]/\partial t = \partial[D]/\partial t = 0$. Since we assume that the diol cannot escape into the atmosphere, the slope of the [D]-profile must be zero at the water surface. Note that any spatial gradient at $z = 0$ would mean transport by molecular diffusion from or to the boundary.

Though in principle the steady-state solution of Eq. 20-50 together with the mentioned boundary conditions can be derived by well-known techniques (see Chapter 22), we will spare the reader the derivation. Instead, we prefer to discuss the qualitative aspects of the concentration of species A and D across the stagnant film. In order to make it easier to read Fig. 20.12, we draw the concentrations of A and D as if the equilibrium constant K_r of Eq. 12-17 and the Henry's law constant of compound A were 1. Thus, [A] and [D] at equilibrium are equal, and [A] does not show a concentration jump at the air–water interface. Note that the following

Figure 20.12 Air–water exchange of an aldehyde A converting to a diol D by a hydration/dehydration reaction. Since the diol D cannot leave the water, the slope of its concentration at the air/water interface is zero. For simplicity, the scales of A and D are chosen such that the equilibrium constant of hydration, K_r, and the Henry's law constant of the aldehyde, $K_{Aa/w}$, are 1. The dashed straight line marked [A]$_{nonreactive}$ helps to picture the modification due to the reactivity of A.

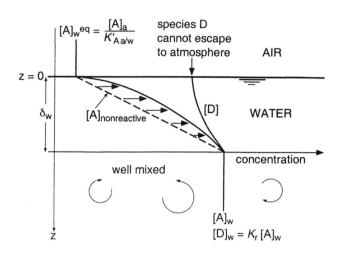

equations are formulated for the general case in which K_r and $K_{Aa/w}$ are different from 1.

As discussed in Section 19.2 (see Fig. 19.4), the concentration profile of a nonreactive species across a boundary layer with constant diffusivity is a straight line which connects the corresponding boundary concentrations, $[A]_w^{aq}$ and $[A]_w$. This is the dashed line in Fig. 20.12 marked as [A]$_{nonreactive}$. However, if A is produced from another species D, the A-profile is no longer a straight line. Intuitively we can understand this with the following consideration. Since species D cannot cross the air–water interface, it is continuously transformed into species A while approaching the interface. At the interface the D-flux becomes zero (thus the zero gradient of $[D]_w$ at the interface). Since at steady-state everywhere in the boundary layer the total flux (A and D species) must be constant, the A species must take care of an increasing fraction of the flux the closer it gets to the interface. Since the diffusion coefficient across the boundary layer is constant, the only way to increase the flux of species A is by increasing its spatial gradient, that is, by making the concentration curve deform upward relative to the straight line of the nonreactive species. As a consequence of Fick's first law the flux of species at the very interface must therefore be larger than the flux of a hypothetical nonreactive species $A_{nonreactive}$. Commonly, this is expressed in terms of the flux enhancement coefficient, ψ, defined as the ratio between the flux of a reactive and nonreactive species, respectively.

For the situation described by Eq. 20-47, the flux enhancement is given by the expression:

$$\psi = \frac{\text{Flux reactive species}}{\text{Flux nonreactive species}} = \frac{K_r + 1}{1 + (K_r / q) \cdot \tanh q} \qquad (20\text{-}51)$$

where K_r is defined in Eq. 12-17 and the nondimensional parameter q (*reaction/diffusion parameter*) is given by:

$$q = \left(\frac{\delta_w^2 / D_{Aw}}{t_r} \right)^{1/2} = \left(\frac{D_{Aw} / v_{Aw}^2}{t_r} \right)^{1/2} = \left(\frac{2t_w}{t_r} \right)^{1/2} \qquad (20\text{-}52)$$

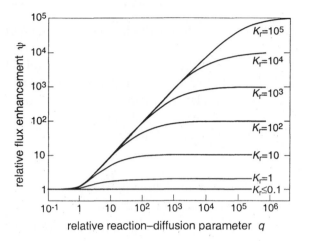

Figure 20.13 Air–water flux enhancement ψ for reactive species as a function of the reaction/diffusion parameter $q = (2\, t_w / t_r)^{1/2}$ for different equilibrium constants K_r. See Eqs. 12-17, 20-51, 20-52, and 20-54.

Note that, up to a factor of 2, q^2 is the ratio between diffusion time across the film (Table 20.6) and time of reaction, t_r. The latter is given by (see Eq. 12-21):

$$t_r = \frac{1}{k_1 + k_2} \qquad (20\text{-}53)$$

For the derivation of Eq. 20-52, we have made use of the fact that in the absence of other reactive species, diffusion of both aldehyde and diol is coupled, thus $D_{Aw} = D_{Dw}$.

In Fig. 20.13 flux enhancement ψ is shown as a function of the reaction/diffusion parameter q for different equilibrium constants K_r. Remember that q^2 is basically the ratio of reaction time k_r and diffusion time k_w (Eq. 20-52). Thus, $q \ll 1$ corresponds to case (1) mentioned at the beginning of this section; flux enhancement should not occur ($\psi = 1$). The other extreme ($\psi \gg 1$, that is $t_r \ll t_w$) was discussed with the example of proton exchange reactions (Eq. 8-6). We found from Eq. 20-49 that for this case the water-side exchange velocity v_{iw} is enhanced by the factor $(1 + K_a / [H^+])$. By comparing Eqs. 8-6 and 12-17 we see that for the case of proton exchange $K_a/[H^+]$ plays the role of the equilibrium constant K_r between the two species. Thus, flux enhancement is:

$$\psi = 1 + K_a / [H^+] = 1 + K_r \qquad \text{if} \quad q \to \infty \qquad (20\text{-}54)$$

The $\psi(q)$ curves in Fig. 20.13 indeed reach constant values for $q \to \infty$ which are (approximately) equal to K_r. This result also follows directly from evaluating Eq. 20-51 for $q \to \infty$. Since $\tanh(q \to \infty) = 1$, the second term in the denominator approaches zero for $q \to \infty$. Finally, Fig. 20.13 nicely demonstrates how $\psi(q)$ increases from 1 at $q = 0$ to its maximum value $(1 + K_r)$ for $q \to \infty$. Note that for large K_r values, this transition extends from $q \sim 1$ to fairly large q values. Thus, for these cases flux enhancement remains q-dependent in situations where t_r is much smaller than t_w. Thus, our first attempt to characterize the intermediate case with $t_r \sim t_w$ should rather be replaced by:

$$1 < (t_w / t_r) < K_r \qquad (20\text{-}55)$$

In conclusion, Eqs. 20-51 and 20-52 include the quantitative description of all three cases. A numerical example for the flux enhancement of formaldehyde is given in Illustrative Example 20.5.

Illustrative Example 20.5   Air–Water Exchange Enhancement for Formaldehyde and Acetaldehyde

Consider two aldehydes at neutral pH, formaldehyde and acetaldehyde. The hydration/dehydration (pseudo-) first-order rate constants and the nondimensional Henry's law constants are summarized below. Since in the following discussion you are interested in orders of magnitude only, you assume that aqueous molecular diffusivities of all involved species are the same as the value for CO_2, $(D_{iw} = 2 \times 10^{-5}$ cm^{2}s^{-1}) and that the corresponding values in air are the same as the value for water vapor $(D_{water\ a} = 0.26$ cm^{2}s^{-1}). This allows us (as a rough estimate) to calculate v_{iw} and v_{ia} directly from Eqs. 20-15 and 20-16, respectively.

formaldehyde

acetaldehyde

	Hydration k_1	Dehydration k_2	t_r (Eq. 20-53)	$K_r = \dfrac{k_1}{k_2}$	$K_{ia/w}$
Formaldehyde	10 s^{-1}	5×10^{-3} s^{-1}	0.1 s	2000	0.016
Acetaldehyde	10 s^{-1}	10 s^{-1}	0.05 s	1	8.4×10^{-3}

Problem

For both aldehydes you are interested in the flux enhancement of the water-phase exchange velocity v_{iw} as well as of the overall exchange velocity $v_{ia/w}$. Evaluate these numbers for two wind velocities, $u_{10} = 1$ m s^{-1} and 10 m s^{-1}.

Answer

The single-phase exchange velocities are calculated from Eqs. 20-15 and 20-16 (Table 20.4), the diffusion time is $t_w = (1/2)D_{iw} / v_{iw}^2$ (see Eq. 20-52).

	$u_{10} = 1$ m s^{-1}	$u_{10} = 10$ m s^{-1}
v_{ia}	0.5 cm s^{-1}	2.3 cm s^{-1}
v_{iw}	0.65×10^{-3} cm s^{-1}	5.3×10^{-3} cm s^{-1}
t_w	24 s	0.36 s

The following table summarizes the results. ψ is the *water-phase* flux enhancement (Eq. 20-52) and ψ^* is the *overall* flux enhancement, that is, the ratio of the enhanced $v_{ia/w}$ and the normal $v_{ia/w}$.

	q (Eq. 20-52)	ψ (Eq. 20-51)	ψv_{iw} (cm s^{-1})	$v_{ia} K_{ia/w}$ (cm s^{-1})	$v_{ia/w}^{enhanced}$ (cm s^{-1})	$v_{ia/w}$ (cm s^{-1})	ψ^*
Formaldehyde							
$u_{10} = 1$ m s^{-1}	22	21.5	0.014	8.0×10^{-3}	5.1×10^{-3}	6.0×10^{-4}	8.5
$u_{10} = 10$ m s^{-1}	2.7	2.7	0.014	3.7×10^{-2}	1.0×10^{-2}	4.6×10^{-3}	2.2
Acetaldehyde							
$u_{10} = 1$ m s^{-1}	31	1.94	1.3×10^{-3}	4.2×10^{-3}	1.0×10^{-3}	5.6×10^{-4}	1.7
$u_{10} = 10$ m s^{-1}	3.8	1.58	8.4×10^{-3}	1.9×10^{-2}	5.9×10^{-3}	4.1×10^{-3}	1.4

There are a few interesting things which we can learn from this simple example:

(1) Although both aldehydes have fairly large reaction/diffusion parameters q, the enhancement ψ of acetaldehyde remains small due to the small equilibrium constant K_r of acetaldehyde.

(2) The increase of the wind velocity to $u_{10} = 10 \text{ m s}^{-1}$ reduces ψ of formaldehyde by about 8. This results from the significantly reduced boundary layer thickness which makes diffusion time t_w small. In contrast, wind velocity does not affect ψ of acetaldehyde very much since it is already close to 1.

(3) Since the nondimensional Henry's law constants $K_{i\,a/w}$ of both aldehydes are fairly small (order 10^{-2}), the influence of the air-side boundary layer on $v_{ia/w}$ is not completely negligible. The considerable flux enhancement of formaldehyde reverses the role of the boundary layer; the resistance in the air-phase becomes dominant. Therefore, the enhancement of v_{iw} ($\psi = 21.5$) is reduced to $\psi^* = 8.5$ when the overall transfer velocity $v_{ia/w}$ is considered.

To summarize, the effective size of flux enhancement is controlled by a fairly complex interplay of different compound properties and environmental factors.

20.6 Questions and Problems

Questions

Q 20.1

Whatever the detailed physicochemical model of the interface, most models of the air–water exchange flux are written as the product of two factors, one describing the physics, the other the chemistry. What are these factors?

Q 20.2

Usually evaporation of water is formulated as a function of relative humidity. Explain how this formalism is linked to the usual two-phase air–water model in which the nondimensional Henry's law constant (i.e., the air–water partition constant) of the exchanged chemical appears.

Q 20.3

Why is wind speed important to describe the air–water exchange velocity of a chemical? Explain why a simple relationship between wind speed and air–water exchange cannot be expected when wind speed is highly variable.

Q 20.4

Why do we have to specify the height above the water surface at which wind speed is measured when we formulate an expression which relates the air–water exchange velocity to wind speed?

Q 20.5

In Fig. 20.3 several empirical relations between wind speed and air–water exchange velocities of volatile chemicals are plotted. How would these curves shift if on the horizontal axis we were to plot u_1 instead of u_{10} without changing the scale of the axis?

Q 20.6

Somebody wants to calculate monthly means of air–water exchange velocities from monthly mean wind speed data. What is the problem?

Q 20.7

In the film model of Whitman the water-phase exchange velocity, v_{iw}, is a function of the molecular diffusion coefficient of the chemical, while in Deacon's boundary layer model v_{iw} depends on the Schmidt Number Sc_{iw}. Explain the reason for this difference.

Q 20.8

The temperature dependence of the water-phase exchange velocity, v_{iw}, is stronger for the boundary layer model than for the film model. Why?

Q 20.9

Air–water exchange in rivers depends on both the river flow and the wind speed. In Eq. 20-34 we have defined the critical wind speed, $\bar{u}_{crit}$, at which the influences of both forces are equal. Explain why $\bar{u}_{crit}$ increases with mean water depth h.

Q 20.10

Explain how a change in water temperature could reverse the direction of the net air–water exchange flux even if all other parameters remain unchanged.

Problems

P 20.1   *What Is the Source of Benzene in the Water of a Pond?*

benzene

Part of your job as a consultant to the State Water Authority is to survey the water quality of several ponds located in a recreation area. Among the volatile organic compounds, your laboratory monitors the concentration of benzene in the water of various ponds. When inspecting the results, you realize that on certain weekends during the summer, the benzene concentration in the surface water of these ponds is up to ten times higher than in the same ponds in the middle of the week or during the winter. For example, in one pond you measure a peak concentration of 1 μg L^{-1}. You wonder whether this elevated benzene concentration is due to air pollution by the heavy car traffic during the summer weekends, or whether the input occurs primarily by leakage of gasoline and oil from the numerous boats cruising on the ponds. You realize that for assessing this question you need to know something about the benzene concentration in the air. Since you have no measurements from the area, you search the literature and find in a review by Field et al. (1992) the following typical benzene concentrations in air reported for different areas:

Location	Mean benzene concentration in air (ppbv) [a]
Remote areas	
Brasil	0.5
Pacific	0.5
Urban areas	
Hamburg	3.2 (peak: 20) [b]
London	8.8

[a] Parts per billion on a volume base; use ideal gas law to convert the numbers to molar concentrations. [b] Bruckmann et al. (1988).

Answer the following questions:

1. Is the atmosphere a likely source for the elevated benzene concentrations during the summer weekends? Assume a water temperature of 25°C.

2. Estimate the direction and rate of air–water exchange of benzene for a wind speed $u_{10} = 2$ m s^{-1}. Use an air concentration of 10 ppbv, a concentration in the water of 1 μg L^{-1}, and a water temperature of 25°C.

3. Assess the situation for a pond located in the center of a big city for a typical winter day. There are no motor boats on the pond. Use the following conditions: air concentration of benzene 10 ppbv, water concentration 0.1 μg L^{-1}, water temperature 5°C, wind speed $u_{10} = 2$ m s^{-1}.

P 20.2 *A Lindane Accident in a Drinking-Water Reservoir*

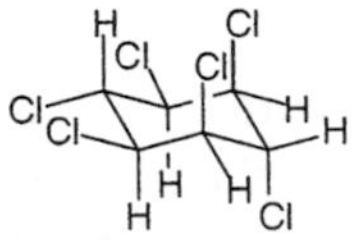

lindane

Due to an accident, an unknown amount of the insecticide lindane (γ-HCH) is introduced into a well-mixed pond that is used as the drinking-water reservoir for a small town. The water inflow and the intake by the water works are immediately stopped and the water is analyzed for lindane. In water samples taken at various locations and depths, an average concentration of (5.0 ± 0.2) μg L^{-1} is determined. As a resident of the area, you ask the person in charge of the water works what they intend to do about this problem, since 5.0 μg L^{-1} is far above the drinking-water limit. "Oh, no problem!" the person tells you, "Within a few days, all lindane will escape into the atmosphere." Being well-trained in environmental organic chemistry you are very suspicious about this answer. So you decide to make a calculation. The people from the water works provide you with the following additional information: Pond surface area $A = 5 \times 10^5$ m^2, volume $V = 2.5 \times 10^6$ m^3, water temperature $T = 20$°C, evaporation rate of water at prevailing wind conditions and humidity of 70% is 5 liter per m^2 and per day. (Note that with this information and the absolute water vapor pressures listed in Appendix B you can calculate $K_{ia/w}$ of water and thus the corresponding air-phase exchange velocity $v_{water\,a}$.) Relevant information on lindane can be found in Appendix C.

1. Calculate the half-life of lindane in the pond (days), assuming that the reservoir volume remains constant. The one-box model introduced in Chapter 12 may help.

2. Compare the relative loss of lindane from the reservoir to the relative loss by evaporation. If the evaporated water were not to be replaced (i.e., no inlets and outlets), would the lindane concentration increase or decrease as a result of the simultaneous water–air fluxes of water and lindane?

P 20.3   *The Influence of Water Temperature on Air–Water Exchange*

In Illustrative Example 20.3 it is shown how water temperature influences the air–water exchange velocity, $v_{ia/w}$. An additional temperature effect of the air–water flux results from the temperature dependence of the air–water partition constant, $K_{ia/w}$. If water and air temperatures are different, the question arises whether the equilibrium between the air and water phase is determined by water temperature, by air temperature, or even by a mixture of both. Explain why $K_{ia/w}$ should be evaluated for the temperature of the water, not the air.

Hint: The process of heat exchange across an interface can be treated in the same way as the exchange of a chemical at the interface. To do so, we must express the molecular thermal heat conductivity by a molecular diffusivity of heat in water and air, $D_{th\,w}$ and $D_{th\,a}$, respectively. At 20°C, we have (see Appendix B): $D_{th\,w} = 1.43 \times 10^{-3}\,\text{cm}^2\,\text{s}^{-1}$, $D_{th\,a} = 0.216\,\text{cm}^2\,\text{s}^{-1}$. Use the film model for air–water exchange with the typical film thicknesses of Eq. 20-18a.

tetrachloroethene
(PCE)

1,4-dichlorobenzene
(1,4-DCB)

P 20.4   *Experimental Determination of the Total Air-Water Exchange Rate for Two Chlorinated Hydrocarbons in a River*

In a field study in the River Glatt in Switzerland, the concentrations of tetrachloroethene (PCE) and 1,4-dichlorobenzene (1,4-DCB) were measured at four locations along a river section of 2.4 km length. Any input of water and chemicals can be excluded in this river section. The average depth ($h = 0.4$ m) and the mean flowing velocity ($\bar{u} = 0.67$ m s^{-1}) were fairly constant in the section. The measurements were made in a way that a specific water parcel was followed downstream and sampled at the appropriate distances. Calculate the mean total air–water exchange velocity, $v_{ia/w}$, of the two substances by assuming that exchange to the atmosphere is the only elimination process from the water, and that the atmospheric concentration of PCE and 1,4 DCB are very small.

Distance (m)	[PCE] (ng L^{-1})	[1,4-DCB] (ng L^{-1})
0	690 ± 40	234 ± 5
600	585 ± 30	201 ± 5
1200	505 ± 6	180 ± 8
2400	365 ± 10	130 ± 5

Data from Schwarzenbach (1983)

P 20.5   *An Inadvertant Air–Water Exchange Experiments*

You have worked hard to study the internal dynamics of tetrachloroethene (PCE) and to calculate vertical turbulent diffusion coefficients in lakes. A friend of yours is more interested in the process of air–water exchange. One day, she sees some of your PCE data lying on your desk. She is very happy with the table below and immediately calculates the average $v_{ia/w}$ value for the period between June 10 and July 10. You ask her how she gets the result, but she just tells you that she needs to assume that input and output of PCE can be neglected. In addition, for simplicity she assumes that the cross-section of the lake, $A(z)$, is independent of depth z, that is, that the lake has the form of a swimming pool with vertical walls. What air–water exchange velocity $v_{ia/w}$ does she get?

Depth z	PCE-concentration (10^{-9} mol L^{-1})	
(m)	June 10	July 10
0	1.0	0.5
2	5.0	2.0
4	9.5	3.5
6	5.5	5.5
8	3.9	4.7
10	3.1	3.6
12	2.6	3.0
14	2.3	2.7
16	2.1	2.4
18	2.0	2.2
20 (bottom)	1.9	2.1

P 20.6 *Air–Water Exchange of Nitrophenols*

Phenols are organic acids. Their dissociation in water is very fast, that is, for most processes we can consider the acid–base equilibrium as an instantaneous equilibrium. The question arises whether the dissociation influences the air–water exchange velocity, and if yes, under which conditions. Before going into much detail you try to get a first approximate answer based on the information which you have at hand for the two compounds, 2-nitrophenols and 4-nitrophenols. Specifically, you want to answer the question whether for these two substances air–water exchange is affected by dissociation at pH $=7$. Use $u_{10} = 1$m s^{-1} and the usual approximations for the molecular diffusivities. Would your answer be modified for pH$=9.5$?

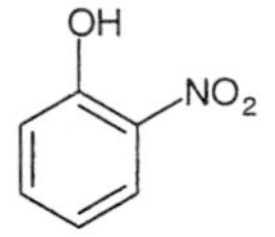

i = 2-nitrophenol

$M_i = 139.1$ g mol^{-1}
p$K_a = 7.15$
$K_{ia/w} = 7.9 \times 10^{-4}$

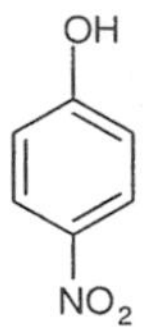

i = 4-nitrophenol

$M_i = 139.1$ g mol^{-1}
p$K_a = 7.06$
$K_{ia/w} = 3.2 \times 10^{-8}$

Chapter 21

BOX MODELS

21.1 Principles of Modeling
Models in Environmental Sciences
Prediction, Inference, Uncertainty
Environmental Systems
Box 21.1: *Terminology for Dynamic Models of Environmental Systems*
Space and Time, Transport and Transformation

21.2 One-Box Models
Linear One-Box Model with One Variable
Box 21.2: *One-Box Model for Tetrachloroethene (PCE) in Greifensee (Switzerland)*
Illustrative Example 21.1: *Assessing the Behavior of Nitrilotriacetic Acid (NTA) in a Lake*
Time Variability and Response
Box 21.3: *Time-Dependent External Forcing of Linear One-Box-Model*
Box 21.4: *Temporal Variability of PCE-Input and Response of Concentration in Greifensee*
Illustrative Example 21.2: *How Certain Is the Degradation Rate of Nitrilotriacetic Acid (NTA) in Greifensee? (Advanced Topic)*
Nonlinear One-Box Models *(Advanced Topic)*
Box 21.5: *One-Box Model with Water Throughflow and Second-Order Reaction (Advanced Topic)*
Illustrative Example 21.3: *Higher-Order Reaction of Nitrilotriacetic Acid (NTA) in Greifensee (Advanced Topic)*
One-Box Model with Two Variables
Box 21.6: *Solution of Two Coupled First-Order Linear Inhomogeneous Differential Equations (Coupled FOLIDEs)*
Illustrative Example 21.4: *Fate of a Pesticide and its Decomposition Product in a Small Lake*

21.3   Two-Box Models
Linear Two-Box Model with One Variable
Linear Two-Box Model of a Stratified Lake
Box 21.7: *Linear Two-Box Model for Stratified Lake*
Illustrative Example 21.5: *Tetrachloroethene (PCE) in Greifensee:*
    From the One-Box to the Two-Box Model
Linear Two-Box Models with Two and More Variables
Nonlinear Two-Box Models

21.4   Dynamic Properties of Linear Multidimensional Models
(Advanced Topic)
Linear *n*-Dimensional Systems and Their Eigenvalues
Box 21.8: *Eigenvalues and Eigenfunctions of Linear Systems*
Illustrative Example 21.6: *Dynamic Behavior of Tetrachloroethene*
    (PCE) in Greifensee
From Box Models to Continuous Models

21.5   Questions and Problems

This is the first of several chapters which deal with the construction of models of environmental systems. Rather than focusing on the physical and chemical processes themselves, we will show how these processes can be *combined*. The importance of modeling has been repeatedly mentioned before, for instance, in Chapter 1 and in the introduction to Part IV. The rationale of modeling in environmental sciences will be discussed in more detail in Section 21.1. Section 21.2 deals with both linear and nonlinear one-box models. They will be further developed into two-box models in Section 21.3. A systematic discussion of the properties and the behavior of linear multibox models will be given in Section 21.4. This section leads to Chapter 22, in which variation in space is described by continuous functions rather than by a series of homogeneous boxes. In a sense the continuous models can be envisioned as box models with an infinite number of boxes.

21.1   Principles of Modeling

Models in Environmental Sciences

Scientific knowledge is acquired by an intelligent combination of empirical information with models. When we talk about models, we have the following simple definition in mind:

A model is an imitation of reality that stresses those aspects that are assumed to be important and omits all properties considered to be nonessential.

tetrachloroethene
(PCE)

The construction of models is like formulating hypotheses, which, in turn, must be tested by new observations. As an example we consider the following experiment. Different small amounts of tetrachloroethene (PCE) are added to identical glass flasks, each containing the same water-to-air volume ratio. The flasks are sealed and shaken to attain an equilibrium between aqueous and gaseous phases. Then the PCE concentration in both phases of each flask is determined. Whether the results are listed in a table or represented by a two-dimensional plot, it will not be very difficult to find that for all flasks the ratios of the concentrations in the two equilibrated phases are approximately the same. So we may formulate a linear expression relating the concentration in the water and the partial pressure in the air and give it the name Henry's law (Chapter 6). Based on our knowledge of thermodynamics, we may then be able to propose a relation between the newly found Henry's law constant, the activity coefficient of PCE in water, and the vapor pressure of pure liquid PCE (Eq. 6-16). This relation can then be tested against data.

A next step may involve testing the law with other substances, or we may explore the law's temperature or pressure dependence, and so forth. In doing so we follow the classical tradition of controlled experiments. The laboratory serves as a synthetic world in which the external influences can be controlled such that the outcome of an experiment can be explained by a few mathematical equations.

As we have seen in Parts II and III, environmental chemistry is based on laboratory experiments as well. For instance, we can determine the relevant physicochemical properties of a compound, such as its aqueous solubility, its vapor pressure, or its molecular diffusion coefficient in water. However, additional information is needed for understanding the chemical's fate in the environment that does not allow us to retreat entirely to the laboratory. Even with the best physicochemical knowledge, it would not have been possible to predict the rapid distribution of DDT throughout the global environment (Chapter 1). There was only one global DDT experiment, complex and unplanned in its setup and not easy to reconstruct. If we want to learn something from this experiment (at least *a posteriori*), the field observations must be combined with "model experiments," that is, with scenarios which we can invoke as alternative explanations for the one existing real-time experiment. Thus, in environmental sciences modeling serves as a substitute for the controlled experiment which cannot be conducted in a natural system. For instance, a typical question for a global DDT model could be to calculate the hypothetical DDT concentrations in different environmental systems provided that the use of DDT were still completely unrestricted.

Running a model under different external conditions helps to identify the essential processes, the key parameters, and the most important empirical information needed to understand a given situation. Therefore, models also serve as a guideline for setting up environmental monitoring programs. Models that are developed for the sole purpose of reproducing a given set of field data are, however, of limited scientific use, since real observations are always superior to the output of a fancy computer program.

Prediction, Inference, Uncertainty

In essence, models are used in two intertwined ways: (1) to calculate the evolution of a system from known external conditions, and (2) to infer the internal structure of a system by comparing various external conditions with the corresponding behavior of the system. The first is usually called *prediction*, the second *diagnosis* or *inference* (Kleindorfer et al., 1993).

Although prediction is often considered to be the ultimate goal of modeling, it is neither the only nor the most crucial one. In fact, the above example of Henry's law is a highly idealistic one. For instance, it precludes the existence of contradictory information. We know that real life is different for two major reasons. First, observations bear uncertainties which are linked to various factors, such as the limited precision of our analytical tools. Quantum mechanics yields an insurmountable theoretical reason for why we cannot make an "absolutely precise" observation. But we don't even have to invoke the uncertainty principle. We can just argue that data are never absolutely exact.

Second, there is no unique scheme of data interpretation. The process of inference always remains arbitrary to some extent. In fact, all the existing DDT data combined still allow for an infinite number of models that could reproduce these data, even if we were to disregard the measurement uncertainties and take the data as "absolute" numbers. Although this may sound strange, it is less so if we think in terms of degrees of freedom. Let us assume that there are one million measurements of DDT concentration in the environment. Then a model which contains one million adjustable parameters can, in principle, exactly (that is, without residual error) reproduce these data. If we included models with more adjustable parameters than observations, then an infinite number of models would reproduce the measurements. Obviously, these are not the kind of models we have in mind. To the contrary, a good model contains significantly fewer adjustable parameters than data to be explained. Newton's law of gravitation is made from one single constant, the constant of gravity, and it seems to work in billions and trillions of situations.

A typical worldwide DDT model might only contain 100 parameters. Again, there must be a great number of very different models to explain existing measurements, but none of these will be able to exactly reproduce them. Thus, on one hand the parameters of a given DDT model will be known only within certain limits of uncertainty, and on the other hand there will be a great number of alternative DDT models. Now imagine that several of the many possible models are used to predict the future development of global DDT concentrations. It is easy to anticipate the consequences: different models will produce different forecasts, although they may have been equivalent in explaining the past, that is, the data from which they were deduced. Some of the forecasts may even contradict each other in fundamental ways; one may predict a further increase, another a drop of DDT levels. Then the nice picture of a logically structured science seems to crumble. This is exactly what occurs with such complex issues as the prediction of global climate due to man-made greenhouse gas emissions. The fact that different models predict different outcomes is gloatingly acknowledged by all those who resent the possible consequences of cutting back emissions. For their defense we may assume that they don't understand the conduct of science under uncertainties, but often the scientists themselves are not careful enough to point out the potential *and* limits of their modeling efforts.

Although a more detailed treatment of data and model uncertainty would go far beyond the goals of this book, those who engage in the task of modeling should be reminded of this point. In fact, during the last decades, the science of interpretation of uncertain data has developed into an elaborate field (e.g., Chatfield, 1995).

Environmental Systems

The construction of a model starts out with the proper choice of a system. Thus, we have to define what the term "system" shall mean in the context of environmental models. Some disciplines use it in a fairly abstract way. The Oxford English Dictionary characterizes a system as "a set or assemblage of things connected, associated, or interdependent, so as to form a complex unity." We prefer a more concrete definition (see Fig. 21.1*a*):

An environmental system is a subunit of the environment separated from the rest of the world by a boundary. The system is characterized by a specific choice of state variables (such as temperature, pressure, concentration of compound i, etc.), by the relations among these variables, and by the action of the outside world on these variables.

Let us discuss the meaning of some of the expressions which appear in the above definition (see Box 21.1). A *subunit of the environment* can be any spatial compartment of the world, from the whole planet to a single algal cell floating in the ocean. The term *state variable* refers to those properties which are used to characterize the

---

Box 21.1   Terminology for Dynamic Models of Environmental Systems

State or model variables	Quantities described by the model as a function of time and (for some modes) as a function of space *Example*: Concentration of chemical i in a lake
External forces	Processes (constant or time-variable) which influence the state variables without being influenced by them *Example*: Light intensity at the surface of a pond which determines the rate of photolysis in the water
Input or input variable	A special kind of external force, usually a mass input of a chemical *Example*: Addition of chemical i into a lake per unit time
Output	Effect of state variable on the outside world, often a mass flux out of the system *Example*: Flux of chemical i from a lake to the atmosphere
Internal process	Transport or transformation process which affects one or several state variables. An output flux is a special kind of internal process *Example*: Biodegradation of chemical i in a lake
Model parameter	Parameter used to describe internal processes *Example*: Rate constant of biodegradation of chemical i
Box	Spatial subunit to approximate the continuous spatial variation of state variables. A box can be characterized by the spatially averaged concentration of one or several state variables *Example*: Well-mixed reactor
One-box/ two-box/ multibox models	Model consisting of one or several boxes. Each box is characterized by one or several state variables *Example*: Two-box model of a lake consisting of the boxes *epilimnion* and *hypolimnion*

---

selected subunit of the world. As an example, we choose the (hypothetical) Lake Y as our subunit. Obviously, there exists a nearly infinite number of quantities that, in principle, could be used to define the state of Lake Y, such as water temperature in every cubic meter of the lake, the same for the concentrations of dissolved oxygen, nitrate, tetrachloroethene or any other chemical we can imagine, the same for any biological parameter such as algal species, bacteria, and so on. Remember, however, that simplicity was mentioned as one of the key elements of modeling. Hence, the question is, which of all these parameters shall we choose?

The answer depends on the purpose of the model. To illustrate this point, we consider a model capable of assessing the fate of phenanthrene in Lake Y (Fig. 21.1*b*).

Figure 21.1 (*a*) An environmental system is a subunit ("IN") of the world separated from the rest of the world ("OUT") by a boundary (bold line). The dynamics of the system is determined by internal processes and by external forces (see Box 21.1 for definitions). There is an output of the system to the environment, whose effect on the external forces is neglected in the model (no feedback from "IN" to "OUT".

(*b*) The concentration of phenanthrene in a lake serves as an example. It is influenced by (1) sorption–desorption on suspended solids, (2) photolysis, (3) biodegradation, (4) input by inlets and precipitation, (5) loss through outlet, (6) exchange at air–water interface, (7) exchange at sediment–water interface, (8) mixing. For the case of phenanthrene, neglecting the feedback from the sediments to the lake may not be reasonable. By shifting the model boundary into the sediment (bold intermittent line), the interaction between sediment and water turns into an internal process and the concentration of phenanthrene in the sediments becomes an additional state variable.

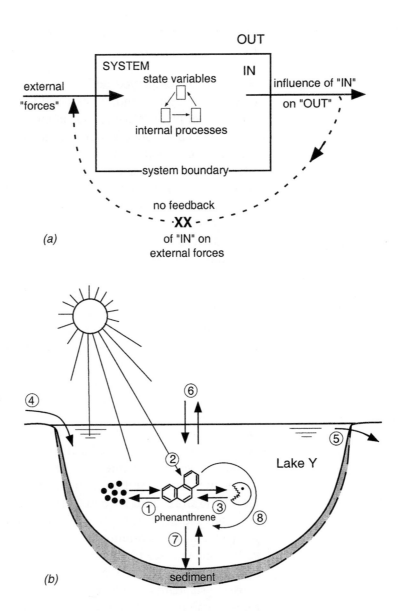

From the physicochemical properties of this compound we expect that the following internal processes are relevant: (1) sorption–desorption between the aqueous solution and solids (suspended particles or sediments), (2) direct and indirect photolysis, and (3) biodegradation. Relevant transport mechanisms across the system boundaries may include: (4) input of phenanthrene by inlets and precipitation onto the water surface, (5) loss through the outlet, (6) exchange at the air–water interface, and (7) exchange at the sediment–water interface. Factors that may be chosen as state variables are concentration of phenanthrene in the lake (either a whole-lake mean value or values for different subvolumes), concentration of suspended particles, and water temperature (due to its influence on reaction kinetics). In contrast, "action from the outside world" (called *external forces*) may include solar radiation (due to its control of photolysis), water throughflow, and wind velocity over the lake (for describing air–water exchange). The so-called *internal processes*, that is, the relations among the state variables and the influence of the external parameters on them, are described based on our understanding of the relevant environmental chemistry, biology, and physics as developed in the preceding chapters of this book.

Note that the employed definition of a system contains an important asymmetry between the system and the external world. The system's description includes the influence of the external forces on the system, but not the reverse, the influence of the system on the outside world. From a mathematical viewpoint the equations that describe the system's behavior contain the external forces, but the latter have to be taken from information outside the model. As will been shown, the specific hierarchy between system and outside world can often (but not always) be justified based on the respective strength of the interactions. Take the system of the earth. Solar radiation is a very strong driving force for the earth, but the back-radiation from the earth to the sun is so tiny that nobody would want to include it as a feedback mechanism in a radiation model of the sun.

Construction of an environmental model, like the model of phenanthrene in Lake Y, proceeds in several steps. In the first step, the system is characterized by choosing a specific boundary between "in" and "out". For the phenanthrene model (Fig. 21.1b) an obvious candidate is the interface between the water body of the lake and its surroundings (lake sediments, atmosphere, inlets). The resulting model needs information on the inputs and outputs of phenanthrene by rivers, on the exchange with the atmosphere and with the sediments of the lake. Obviously, the quality of the model also depends on the quality of the input data. To assess the sedimentary input, information is needed regarding the phenanthrene content of the sediments. Since the boundary between water and sediment is not as distinct as in the sun–earth system (in fact, the sediment content of phenanthrene strongly depends on the history of this compound's concentration in the lake), we may decide to include the sediments in the system and to shift the boundary of the model to below the sediment–water interface into the sediments. In a further step we may conclude that predicting the phenanthrene concentration in the lake requires a description of processes occurring in the drainage area of the lake. Thus, the system boundary may be extended again. Eventually, we may even include into the system the political mechanisms by which emission standards for phenanthrene are fixed. So, instead of building a lake model we may end up constructing a socioeconomic–geochemical model of the drainage basin of Lake Y. The point at which the process of increasing the system complexity is stopped depends on the purpose of the model and the complexity of the available data. Often, it is useless to build a highly complex model for which only rather coarse data are available.

The second step in the model construction involves choosing the complexity of both the *internal processes* and the *external forces*. According to the simplicity postulate, the most complex model is not necessarily the best one. To the contrary, a good model is like a caricature in which the cartoonist enhances the characteristics of a person's face that, in a given context, are most relevant. If we intend to model the mean concentration of phenanthrene in Lake Y over several decades, the adequate model would certainly be different from a model designed to describe the daily spatial concentration variations of phenanthrene. Choosing the model structure is the cartoonist's task, and he or she solves it based on what the message of the cartoon should be. In other words, before a model can be designed or selected we should know its purpose.

As a third step, the relations between the various model components have to be specified in terms of mathematical expressions, once the model structure is fixed. In contrast to the common chemical reaction models which describe the reaction kinetics under laboratory conditions (e.g., in a test tube), environmental models usually contain two kinds of processes: (1) the familiar reaction processes discussed in Parts II and III of this book, and (2) the transport processes. These processes are linked by the concept of *mass balance*.

Space and Time, Transport and Transformation

The concentration of a compound at a given location depends on (1) the rate of *transformation* of the compound (positive for production and negative for consumption), and (2) the rate of *transport* to or from the location. In Part III we discussed different kinds of transformation processes. Internal transport rates were introduced in Chapter 18. Remember that we have divided them into just two categories, the directed transport called advection and the random transport called diffusion or dispersion. The second Fickian law (Eq. 18-14) describes the local rate of change due to diffusion. The corresponding law for advective processes will be introduced in Chapter 22. In Chapter 19 we discussed transport processes across boundaries.

In this chapter we will keep the description of transport simpler than Fick's law, which would eventually lead to partial differential equations and thus to rather complex models. Instead of allowing the concentration of a chemical to change continuously in space, we assume that the concentration distribution exhibits some coarse structure. As an extreme, but often sufficient, approximation we go back to the example of phenanthrene in a lake and ask whether it would be adequate to describe the mass balance of phenanthrene by using just the average concentration in the lake, a value calculated by dividing the total phenanthrene mass in the lake by the lake volume. If the measured concentration in the lake at any location or depth would not deviate too much from the mean (say, less than 20%), then it may be reasonable to replace the complex three-dimensional concentration distribution of phenanthrene (which would never be adequately known anyway) by just one value, the average lake concentration. In other words, in this approach we would describe the lake as a *well-mixed reactor* and could then use the fairly simple mathematical equations which we have introduced in Section 12.4 (see Fig 12.7). The model which results from such an approach is called a *one-box model*.

In case the actual concentration of phenanthrene would exhibit significant spatial variations, the one-box model would not be the ideal description. Instead, it may be adequate to subdivide the lake into two or more boxes in such a way that within the defined subvolumes, phenanthrene concentration would be fairly homogeneous. So we would end up with a two- or multi-box model. In certain situations this box model approach may still not be sufficient. We may need a model which allows for a continuous concentration variation in time and in space. Such models will be discussed in Chapter 22.

Figure 21.2 The interplay of transport and reaction, exemplified by the hypothetical vertical concentration profile of phenanthrene in a lake. (*a*) The rate of photolysis decreases with depth due to the diminishing light intensity with water depth. (*b*) Two possible vertical profiles of phenanthrene concentration: if vertical mixing in the water column is strong, the profile is constant (profile 1). If vertical mixing is slow, a distinct vertical concentration gradient develops with small values at the water surface where photolysis is strongest (profile 2).

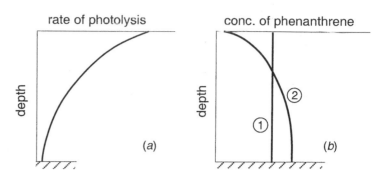

Interestingly enough, we cannot decide once and forever whether the description of phenanthrene in a lake needs a one-box, two-box, *n*-box, or continuous model. The answer depends on certain conditions which change, not only from lake to lake, but also in a given lake over time. As an example, we continue the discussion of phenanthrene in Lake Y and consider two hypothetical situations. We assume that all relevant processes that influence the spatial distribution of phenanthrene (see Fig. 21.1*b*) are constant except for the intensity of turbulent mixing. We can imagine that in one situation the water column of Lake Y is exposed to strong winds while at another time there is no wind. Obviously, the different internal transformation processes are spatially not constant in the lake. Fig. 21.2*a* shows the vertical variation of the rate of photolysis resulting from the decreasing light intensity in the water column. If mixing is strong, the consequences of the spatially inhomogeneous transformation rates are continuously erased; hence the vertical concentration profile of phenanthrene is constant and determined by the mean value of all the transformation rates (Fig. 21.2*b*, profile 1). If vertical mixing is slow, a distinct vertical concentration gradient develops with small values at the water surface where photolysis is strongest (profile 2).

In this and the following chapter we develop a series of tools to describe such situations. Of course, we could imagine building just one complex model which can deal with all situations at once. Indeed such models exist for different environmental systems (lakes, rivers, aquifers, atmosphere etc.), but they are complex, need information on many parameters, and require a lot of a priori knowledge about the system. Here we pursue the approach of tailor-made models, that is, models that for every situation are able to handle just those features which we want to know and which are resolved by the available data, not more and not less. In this sense, situation 2 of Fig. 21.2 needs a more complex model than situation 1. In fact, for the latter the one-box model that we discussed earlier (Chapter 12.4) may be adequate; for the former we need something more, yet still not the most sophisticated lake model we can imagine. For instance, having measured just one vertical profile of phenanthrene it would not be reasonable to use a model which, besides vertical mixing, includes horizontal mixing as well, since we have no data on horizontal concentration gradients. However, if we were interested in the question whether horizontal gradients of phenanthrene may be important and whether it may be worthwhile to organize a field campaign to search for them, then a three-dimensional lake model may help, provided that the simple estimates that will be discussed in Chapter 22 do not already give an unambiguous answer.

21.2   One-Box Models

The simplest and often most suitable modeling tool is the *one-box model*. One-box models describe the system as a single spatially homogeneous entity. Homogeneous means that no further spatial variation is considered. However, one-box models can have one or several state variables, for instance, the mean concentration of one or several compounds i which are influenced both by external "forces" (or inputs) and by internal processes (removal or transformation). A particular example, the model of the *well-mixed reactor* with one state variable, has been discussed in Section 12.4 (see Fig. 12.7). The mathematical solution of the model has been given for the special case that the model equation is linear (Box 12.1). It will be the starting point for our discussion on box models.

Linear One-Box Model with One Variable

Imagine a well-mixed system characterized by the concentration of one or several chemicals i, C_i (Fig. 21.3). The concentrations are influenced by inputs I_i, by outputs O_i, and by internal transformation processes which occur either between state variables (for instance, between C_i and C_j) or between other chemicals X, Y,... which are not part of the set of state variables.

If the system has the constant volume V, the mass balance of compound i can be written as (see Eq. 12-43):

$$\frac{dC_i}{dt} = \frac{1}{V}\left(I_i - O_i - \Sigma R_i + \Sigma P_i \right) \qquad [\text{ML}^{-3}\text{T}^{-1}] \qquad (21\text{-}1)$$

where ΣR_i, and ΣP_i stand for the sum of all internal consumption and production rates of processes which involve chemical i.

Let us consider the case where we model just one chemical i, for instance, phenanthrene. Other chemicals may determine the size of the right-hand terms of Eq. 21-1 (I_i, O_i, ΣR_i, ΣP_i), but we describe their influence as external force. Thus:

$$I_i,\ O_i,\ \Sigma R_i,\ \Sigma P_i = \{\text{function of external forces and of } C_i\} = f_p(C_i) \qquad (21\text{-}2)$$

where the subscript p points to the different processes on the left-hand side of Eq. 21-2. Note that some of these functions $f_p(C_i)$, for instance, the one describing the external input I_i, are usually independent of C_i, that is, depend only on the external forces.

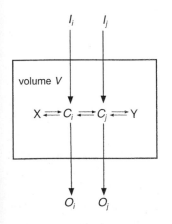

Figure 21.3 One-box model or well-mixed reactor model. State variables are the concentrations, $C_i, C_j,...$, of chemicals $i, j,...$ They are influenced by inputs ($I_i, I_j, ...$), outputs ($O_i, O_j, ...$), and internal transformations between the state variables or between other chemicals X, Y,... which do not appear as state variables in the model.

If we insert all these functions $f_p(C_i)$ into Eq. 21-1, we may end up with a rather complicated expression. Yet, in this section we are interested only in models in which all functions $f_p(C_i)$ are linear, that is, have the following form:

$$f_p(C_i) = a_p(t) + b_p(t)C_i \qquad (21\text{-}3)$$

where $a_p(t)$ and $b_p(t)$ are arbitrary functions of time t, but *not* of C_i. Some of these functions may be simpler than Eq. 21-3, that is, $a_p(t)$ or $b_p(t)$ may be independent of time t or even be zero. In any case, putting all these terms into Eq. 21-1 results in an expression of the form:

$$\frac{dC_i}{dt} = \sum_p a_p(t) + \left\{ \sum_p b_p(t) \right\} C_i = J_i(t) - k_i(t)\, C_i \qquad (21\text{-}4)$$

where $J_i(t)$ sums up all $a_p(t)$, and $k_i(t)$ is the *negative* sum of all $b_p(t)$. As we will see below, the minus sign in the definition of $k_i(t)$ is introduced for convenience since the b_p terms are usually negative. Equation 21-4 is the most general form of a *first-order linear inhomogeneous differential equation* (FOLIDE). Its solutions are given in Box 12.1.

Essentially, what we have just discussed is a more general and thus more abstract version of the one-box model derived in Section 12.4. The advantage of the little detour is that Eqs. 21-3 and 21-4 provide a universal tool to assess the linearity (or nonlinearity) of a model. Or stated differently, one single function $f_p(C_i)$ which cannot be brought into the form of Eq. 21-3 makes the box model (Eq. 21-4) nonlinear.

Linear models play a central role in the theory of differential equations. They provide the basis for the discussion of the more complicated nonlinear equations and always have analytical solutions composed of exponential functions. In contrast, the solutions of nonlinear models embrace a much larger variety and thus significantly more possibilities (and surprises).

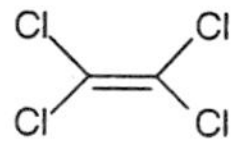

tetrachloroethene
(PCE)

In order to demonstrate how Eq. 21-4 evolves from a specific environmental system, we look at the example of tetrachloroethene (or perchloroethylene, PCE) in Greifensee, a small Swiss lake southeast of Zurich. Since we will use this lake in other examples as well, some characteristic data are summarized in Table 21.1.

From repeated measurements of PCE concentrations at different depths we can calculate total mass and mean concentration in Greifensee. (Note: This calculation includes some a priori knowledge regarding the spatial distribution of PCE in Greifensee; horizontal concentration gradients in the lake must be small so that total mass can be calculated as a volume-weighted average of the concentrations measured along a vertical profile at the deepest location of the lake. As will be discussed in Chapter 23, in certain cases the validity of this assumption can be taken for granted even without actual measurements of horizontal concentration profiles.) Furthermore, we must also know the input of PCE through the inlets and its loss at the outlet. With this information we would like to answer the following question:

How long would it take for the mean PCE concentration to drop from its initial value, $C_{io} = 0.55\ \mu\text{mol m}^{-3}$, to $0.1\ \mu\text{mol m}^{-3}$, provided that the total input of PCE into Greifensee could be stopped at once?

From the physicochemical properties of PCE it appears that the only significant removal mechanism (other than loss at the outlet) is air–water exchange. *In situ* reactions and removal to the sediments after sorption on particles are not important. Thus, the mass balance of PCE consists of just three terms which have the following form (here the subscript i refers to PCE):

Table 21.1 Characteristic Data of Greifensee (Switzerland)[a]

Volume:	Total	V	$150 \times 10^6 \text{ m}^3$
	Epilimnion	V_E	$50 \times 10^6 \text{ m}^3$
	Hypolimnion	V_H	$100 \times 10^6 \text{ m}^3$
Area:	Surface	A_o	$8.6 \times 10^6 \text{ m}^2$
	At the thermocline	A_{th}	$7.5 \times 10^6 \text{ m}^2$
Mean depth:	Total lake	$h = V/A_o$	17.4 m
	Epilimnion	$h_E = V_E/A_o$	5.8 m
	Hypolimnion	$h_H = V_H/A_{th}$	13.3 m
Throughflow of water		Q	$0.34 \times 10^6 \text{ m}^3\text{d}^{-1}$
Water exchange rate		$k_w = Q/V$	$2.3 \times 10^{-3}\text{d}^{-1}$
Mean water residence time		$\tau_w = k_w^{-1}$	440 d
Flushing velocity		$q = Q/A_o$	0.040 m d^{-1}
Turbulent diffusivity in thermocline		E_{th}	$0.2 \text{ m}^2\text{d}^{-1}$
Thermocline thickness		h_{th}	4 m
Turbulent exchange velocity between epilimnion and hypolimnion		$v_{th} = E_{th}/h_{th}$	0.05 m d^{-1}
Typical wind speed 10 m above lake surface		u_{10}	1 m s^{-1}
Air–water exchange velocity at typical wind speed (from Table 20.5b)		$v_{a/w}$	$4.9 \times 10^{-6} \text{ m s}^{-1}$

[a] Numbers for characteristic lake data (volume, surface, area, etc.) and input data are rounded off to facilitate quantitative considerations.

1. Input at inlets: I_{iw} (subscript w: water throughflow)

2. Loss at outlet (see Eq. 12-44; $C_{i\text{ surface}}$ is PCE–concentration at lake surface):

$$O_{iw} = Q C_{i\text{ surface}} \tag{21-5a}$$

3. Exchange at air-water interface (see Eq. 20-6; C_{iw} is replaced by the water surface concentration $C_{i\text{ surface}}$; C_{ia} is atmospheric PCE concentration):

$$-A_o F_{ia/w} = -A_o v_{ia/w}\left(C_{i\text{ surface}} - \frac{C_{ia}}{K_{ia/w}}\right) = -O_{ia/w} + I_{ia/w} \tag{21-5b}$$

The above equation demonstrates that air–water exchange as well as other boundary exchange processes can be interpreted as a combination of an input flux ($I_{ia/w} = A_o v_{ia/w} C_{ia}/K_{ia/w}$) and an output flux ($O_{ia/w} = A_o v_{ia/w} C_{i\text{ surface}}$). It is a matter of personal taste whether one prefers to keep both terms separated or add them to get the familiar form of a net exchange flux.

Combining the three parts of Eq. 21-5 into a mass balance of the average lake concentration, C_i, does not yield the linear one-box equation (Eq. 21-4) which we would like to get. Equations 21-5a and 21-5b still contain an "alien variable", the surface concentration, $C_{i\,surface}$. So we need an expression which relates average and surface concentration. Here we assume that the lake is well mixed, that is, that $C_{i\,surface}$ and C_i are equal. Then the mass balance equation becomes:

$$\frac{dC_i}{dt} = \frac{1}{V}\left(I_{iw} - QC_i - A_o v_{ia/w} C_i + A_o v_{ia/w} \frac{C_{ia}}{K_{ia/w}} \right) \tag{21-6}$$

Comparison with Eq. 21-4 allows one to identify J_i and k_i:

$$J_i = \frac{I_{iw}}{V} + \frac{A_o v_{ia/w}}{V} \frac{C_{ia}}{K_{ia/w}} = k_w C_{iin} + k_{ia/w} \frac{C_{ia}}{K_{ia/w}} \tag{21-7a}$$

$$k_i = \frac{Q}{V} + \frac{A_o v_{ia/w}}{V} = k_w + k_{ia/w} \tag{21-7b}$$

where we have used the following definitions:

$$\frac{I_{iw}}{V} = \frac{QC_{iin}}{V} = k_w C_{iin} \quad \text{and} \quad C_{iin} = \frac{I_{iw}}{Q} \tag{21-8a}$$

$$k_w = \frac{Q}{V} \tag{21-8b}$$

$$k_{ia/w} = \frac{A_o v_{ia/w}}{V} = \frac{v_{ia/w}}{h} \tag{21-8c}$$

$h = V / A_o$ is mean depth of the lake, and $C_{i\,in}$ is the average input concentration.

If all external forces (I_{iw}, $v_{ia/w}$, C_{ia}, Q) and model parameters (V, A_o, $K_{ia/w}$) are constant over time, the solution of Eq. 21-6 is (see Box 12.1, case b):

$$C_i(t) = C_{io}\, e^{-(k_w + k_{ia/w})t} + C_{i\infty}(1 - e^{-(k_w + k_{ia/w})t}) \tag{21-9}$$

with the steady-state PCE concentration:

$$C_{i\infty} = \frac{k_w C_{iin} + k_{ia/w} \dfrac{C_{ia}}{K_{ia/w}}}{k_w + k_{ia/w}} \tag{21-10}$$

To calculate the response of the PCE concentration in the lake to a sudden stop of the PCE input, we have to find values for the different coefficients in Eqs. 21-9 and 21-10. The procedure is summarized in Box 21.2. As it turns out, the time needed to lower the PCE concentration from its initial value of 0.55 µmol m⁻³ to 0.1 µmol m⁻³ is 160 days, that is, about 5 months.

With this example we have shown how the mathematical formulations of the processes which are part of the model can be combined into one single linear differential equation. Illustrative Example 21.1 deals with another example, the dynamics of nitrilotriacetic acid (NTA) in Greifensee.

Box 21.2   One-Box Model for Tetrachloroethene (PCE) in Greifensee (Switzerland)

In this box we demonstrate the construction and application of a simple one-box model to a small lake like Greifensee to analyze the dynamic behavior of a chemical such as PCE. Characteristic data of Greifensee are given in Table 21.1. Measurements of PCE in the water column of the lake yield the following information:

Total mass of PCE in the lake: $M_i = 83$ mol

Mean concentration of PCE $C_i = M_i / V = 0.55 \times 10^{-6}$ mol m^{-3}

River input $I_{iw} = 0.90$ mol d^{-1}

River output $O_{iw} = Q\overline{C_i} = 0.19$ mol d^{-1}

Nondimensional Henry's law constant $K_{ia/w} = 0.73$ (see Table 20.5a)

Mean concentration in air $C_{ia} < 10^{-8}$ mol m^{-3}

Since repeated measurements show no significant temporal change of M_i with time, we evaluate Eq. 21-6 for steady-state ($dC_i / dt = 0$). Furthermore, since $C_{ia} / K_{ia/w} < 1.4 \times 10^{-8}$ mol m^{-3}, we neglect the last term on the right-hand side of Eq. 21-6. Solving for the atmospheric loss term yields:

$$\frac{A_o v_{ia/w}}{V} C_i = \frac{1}{V}\left(I_{iw} - O_{iw}\right) = \frac{(0.90 - 0.19)\text{mol d}^{-1}}{150 \times 10^6 \text{ m}^3} = 4.7 \times 10^{-9} \text{ mol m}^{-3}\text{d}^{-1}$$

Thus, the specific loss rate to the atmosphere is:

$$k_{ia/w} = \frac{A_o v_{ia/w}}{V} = \frac{4.7 \times 10^{-9} \text{ mol m}^{-3}\text{d}^{-1}}{0.55 \times 10^{-6} \text{ mol m}^{-3}} = 8.6 \times 10^{-3} \text{d}^{-1}$$

With $h = V / A_o = 17.4$ m (Table 21.1), the air–water exchange velocity $v_{ia/w} = h\, k_{ia/w} = 0.15$ m d^{-1} = 1.7×10^{-4} cm s^{-1}, which is significantly smaller than expected from Table 20.5b. We come back to this point later.

Furthermore, the specific rate of loss at the outlet

$$k_w = \frac{Q}{V} = 2.3 \times 10^{-3} \text{d}^{-1}$$

is about four times smaller than the specific loss rate to the atmosphere. The (overall) rate constant is $k = k_w + k_{ia/w} = 10.9 \times 10^{-3}$d^{-1}.

As a check the reader can now calculate $C_{i\infty}$ from Eq. 21-10. Since we have started the calculation by assuming that the measured mean lake concentration represents a steady-state, the value should obviously yield the measured mean concentration, $C_{i\infty} = C_i = 0.55 \times 10^{-6}$ mol m^{-3}.

If the input by the river were stopped, the new steady-state value $C_{i\infty}$ would become zero. (Remember that the atmospheric input term was neglected. If we want to be absolutely correct, this approximation would not be justified any more if the atmosphere were the only PCE input to the lake. Yet, for the following calculation this does not matter.)

With $C_{i\infty}^o = 0$, Eq. 21-9 becomes:

$$C_i(t) = C_{io}\, e^{-\left(k_{iw} + k_{i\,a/w}\right)t}$$

We are looking for the time t_i^* at which $C_i(t)$ takes the value 0.1×10^{-6} mol m^{-3}:

$$t_i^* = \frac{-\ln\left(C_i(t^*)/C_{io}\right)}{k_w + k_{ia/w}} = \frac{-\ln(0.1/0.55)}{(2.3 + 8.6) \times 10^{-3} \text{d}^{-1}} = 160 \text{ d}$$

From Eq. 12-50 we can also calculate the 5% response time:

$$t_{i5\%} = \frac{3}{k_w + k_{ia/w}} = 280 \text{ d}$$

Since $C_i = 0.1$ μmol m^{-3} is still larger than 5% of the original lake concentration (0.55 μmol m^{-3}), t_i^* is smaller than $t_{i5\%}$.

Illustrative Example 21.1   **Assessing the Behavior of Nitrilotriacetic Acid (NTA) in a Lake**

Around 1983, typical concentrations of nitrilotriacetic acid (NTA) in Greifensee (see Table 21.1 for characteristic data of the lake) were found to be 3.7×10^{-9} M (3.7×10^{-6} mol m^{-3}). NTA is a cation complexation agent which at that time was mainly used in detergents. Based on measurements made in 1982 and 1983 on the major rivers and sewage inlets of Greifensee, total NTA loading was estimated as $I_{NTA} = 13$ mol d^{-1}. From laboratory experiments in which NTA was added to samples taken from natural waters (estuaries, rivers), first-order biodegradation rate constants, k_{NTA}, between 0.02 and 1 d^{-1} were determined for NTA concentrations between 2×10^{-8} and 5×10^{-6} M (Larson and Davidson, 1982; Bartholomew and Pfaender, 1983). Rate constants varied with temperature ($Q_{10} = 2$, i.e., k_{NTA} increases by 2 for a temperature increase of 10°C) and were reduced for oxygen concentrations below 10^{-5} M (Larson et al., 1981). No data are available for NTA degradation at concentrations below 2×10^{-8} M.

Use the information on NTA concentrations in Greifensee to estimate the first-order degradation rate of NTA in this lake at concentrations below 2×10^{-8} M.

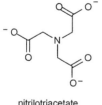

nitrilotriacetate
(NTA^{3-})

Answer

Assume that (1) the measured concentration of 3.7×10^{-9} M is typical for the whole lake, (2) this value corresponds to the steady-state concentration, C_∞, caused by a NTA input of 13 mol d^{-1}, and (3) flushing and biodegradation are the only relevant removal mechanisms. Solve Eq. 12-49b for the unknown reaction rate constant $k_{tot} = k_{NTA}$:

$$k_{NTA} = \frac{I_{NTA}}{V C_{NTA}} - k_w = \frac{13 \text{ mol d}^{-1}}{150 \times 10^6 \text{ m}^3 \times 3.7 \times 10^{-6} \text{ mol m}^{-3}} - 2.3 \times 10^{-3} \text{d}^{-1} \qquad (1)$$

$$= (23.4 \times 10^{-3} - 2.3 \times 10^{-3}) \text{d}^{-1} = 2.1 \times 10^{-2} \text{d}^{-1}$$

This value corresponds to the lower limit of the rates measured under laboratory conditions (albeit at larger concentrations).

Given the various assumptions which had to be made to get the above result, check how robust the conclusion is regarding NTA degradation in the lake. Therefore calculate the hypothetical steady-state concentration of NTA in Greifensee provided that NTA was conservative ($k_{NTA} = 0$). From Eq. 12-49b:

$$C_\infty = \frac{I_{NTA}}{V k_w} = \frac{I_{NTA}}{Q} = \frac{13 \text{ mol d}^{-1}}{0.34 \times 10^{-6} \text{ m}^3 \text{d}^{-1}} = 38 \times 10^{-9} \text{M} \qquad (2)$$

This number is 10 times larger than the measured concentration. Thus, NTA must be removed by some process in the lake. We will continue the discussion on NTA in Greifensee in Illustrative Example 21.2.

Time Variability and Response

External and internal conditions are seldom constant in natural systems. For instance, in the previous case of PCE in Greifensee, external input (I_{iw}) and air–water exchange velocity ($v_{ia/w}$) are probably the most important source of time variability. The latter depends on wind speed, u_{10}, that is, on the variability of the weather, the former on the consumer's behavior regarding the use of products containing PCE. Hence, although we are in principle able to solve Eq. 21-6 for time-variable coefficients (Box 12.1, Eq. 9), often this capability is not of great value. If the model is employed for diagnostic purposes, often the needed input data either do not exist or do not have the necessary temporal resolution. In the case of a predictive use of the model the situation is even worse, as human behavior and weather are difficult fields for prognostication.

Fortunately, not every wiggle affects a system in the same way. Linear systems have particular ranges of temporal variability which they feel strongly while others are averaged out. Here we want to derive a simple tool which allows the modeler to distinguish between relevant and irrelevant time variation.

To do so we observe that typical temporal variations of quantities like air temperature, wind speed, flushing rate, but also of man-made parameters like the input of some anthropogenic organic compound into the environment, can be described by just two distinct types of variability: (1) by long-term trends, and (2) by period fluctuations. In the following the subscript i is omitted for brevity.

To describe the long-term trend we use an exponential growth or decay curve and characterize it by the rate constant α [T^{-1}]:

$$J(t) = J_o \, e^{\alpha t} \qquad (21\text{-}11)$$

where α [T^{-1}] is positive for growth and negative for decay. For the fluctuations we employ the example of a sinusoidal function with time period τ_p:

$$J(t) = \bar{J}(1 + A_J \sin \omega \, t) \qquad (21\text{-}12)$$

where $\bar{J}$ is the mean, A_J the relative amplitude, and ω the angular frequency which is related to the period τ_p by:

$$\omega = \frac{2\pi}{\tau_p} \quad [T^{-1}] \qquad (21\text{-}13)$$

If $J(t)$ from Eq. 21-11 or 21-12 is inserted into Eq. 21-4, we get a linear differential equation with a time variable inhomogeneous term but constant rate k. The corresponding solution is given in Box 12.1, Eq. 8. Application of the general solution to the above case is described in Box 21.3. The reader who is not interested in the mathematics can skip the details but should take a moment to digest the message which summarizes our analytical exercise.

Box 21.3　　**Time-Dependent External Forcing of Linear One-Box Model**

We discuss the solutions of the linear differential equation:

$$\frac{dC}{dt} = J(t) - k\,C_i \tag{1}$$

with constant k and either of the following input functions:

(a)　$J(t) = J_0\,e^{\alpha t}$ (2)

(b)　$J(t) = \bar{J}(1 + A_J \sin \omega t)$ (3)

The general solution of Eq.1 is given in Box 12.1 (Eq. 8).

Case a.

$$C(t) = C_0\,e^{-kt} + J_0 \int_0^t e^{-k(t-t')} e^{\alpha t'}\,dt'$$

$$= C_0\,e^{-kt} + J_0\,\frac{e^{\alpha t} - e^{-kt}}{k + \alpha} = \left(C_0 - \frac{J_0}{k+\alpha}\right) e^{-kt} + \frac{J_0 e^{\alpha t}}{k+\alpha} \tag{4}$$

For further discussion we consider only the case $k > 0$; the case $k < 0$ is not meaningful if Eq. 1 is a mass balance model. Then for large times ($kt \gg 1$), we can neglect the term containing e^{-kt}, hence:

$$C(t) = \frac{J_0\,e^{\alpha t}}{k + \alpha} = \frac{J(t)}{k + \alpha}\,, \qquad kt \gg 1 \tag{5}$$

Case b.

$$C(t) = C_0\,e^{-kt} + \bar{J} \int_0^t e^{-k(t-t')}\,dt' + \bar{J}A_J \int_0^t e^{-k(t-t')} \sin \omega t'\,dt'$$

$$= C_0\,e^{-kt} + \frac{\bar{J}}{k}\left(1 - e^{-kt}\right) + \frac{\bar{J}A_J}{\left(k^2 + \omega^2\right)^{1/2}} \sin(\omega t - \varphi) + \frac{\bar{J}A_J\,\omega}{k^2 + \omega^2}\,e^{-kt} \tag{6}$$

where $\varphi = \tan^{-1}(\omega/k)$. Again we are interested in large times and neglect all terms containing e^{-kt}:

$$C(t) = \frac{\bar{J}}{k} + \frac{\bar{J}A_J}{\left(k^2 + \omega^2\right)^{1/2}} \sin(\omega t - \varphi)\,, \qquad kt \gg 1 \tag{7}$$

Exponential change. For large times ($kt \gg 1$), the solution of Eq. 21-4 with exponential external forcing, Eq. 21-11, is (Box 21.3, Eq. 5):

$$C(t) = \frac{J_o\,e^{\alpha t}}{k+\alpha} = \frac{J(t)}{k+\alpha} \quad (kt \gg 1) \tag{21-14}$$

This looks almost like the time-dependent steady-state solution of Eq. 21-4 which one gets by putting $dC/dt = 0$ and solving for $C = C_\infty$:

$$C_\infty = \frac{J}{k} \tag{21-15}$$

If $J(t)$ is variable, we can still define such a hypothetical steady-state (even if it is never reached) and design it as:

$$C_\infty(t) = \frac{J(t)}{k} = \frac{J_o\,e^{\alpha t}}{k} \tag{21-16}$$

The time argument behind C_∞ points to the fact that it is related to a variable input, $J(t)$. $C_\infty(t)$, would be attained only if J remained constant long enough. (In fact, "long enough" would mean that $kt \gg 1$.) Now we can compare the *actual* concentration of the system (Eq. 21-14) with the *hypothetical steady-state* (Eq. 21-16):

$$\frac{C(t)}{C_\infty(t)} = \frac{k}{k+\alpha} \tag{21-17}$$

For $\alpha \ll k$, the right-hand side is close to 1. In this case, the growth of $J(t)$ is sufficiently slow so that the actual concentration, $C(t)$, and the corresponding steady-state, $C_\infty(t)$, are approximately equal.

In contrast, if α is of the same order of magnitude as k or larger, then the actual state of the system, $C(t)$, lags behind the hypothetical steady-state, $C_\infty(t)$, which corresponds to the actual input. If α is positive, that is, if $J(t)$ grows exponentially, the lag causes $C(t)$ to remain smaller than the corresponding $C_\infty(t)$. If α is negative, that is, if $J(t)$ is decreasing, $C(t)$ is larger than $C_\infty(t)$. The time lag grows with the ratio α/k. Figure 21.4 shows $C(t)$ and $C_\infty(t)$ for two different positive α/k ratios. If $|\alpha| \ll k$, actual and steady-state are approximately equal.

Figure 21.4 Response of linear system to external exponential perturbation (Eq. 21-11). Solid lines show hypothetical steady-state (Eq. 21-16); dashed lines show the system response (Box 21.3, Eq. 4). The rate constant $k = 4$ yr^{-1} is the same for both calculations; it corresponds to the one-box model of PCE in Greifensee (Box 21.2). The exponential growth rate of the external forcing, α, is 0.1 and 1 yr^{-1}, respectively. The former corresponds to a slow external perturbation ($\alpha \ll k$) which leaves the system close to its steady-state.

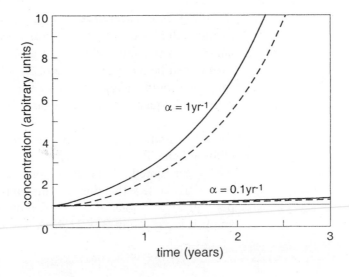

Periodic perturbation. For large times ($kt \gg 1$), the solution of Eq. 21-4 under the influence of Eq. 21-12 is (Box 21.3, Eq.7):

$$C(t) = \frac{\bar{J}}{k} + \frac{\bar{J} A_J}{\left(k^2 + \omega^2\right)^{1/2}} \sin\left(\omega t - \varphi\right) , \quad kt \gg 1 \qquad (21\text{-}18)$$

where $\varphi = \tan^{-1}(\omega/k)$. Since $\tan^{-1}$(inverse tangent function) converges to zero for $\omega/k \to 0$, a limiting case of Eq. 21-18 is:

$$C(t) \approx \frac{\bar{J}}{k}\left(1 + A_J \sin\omega t\right) = \frac{J(t)}{k} = C_\infty(t) , \quad \omega \ll k \qquad (21\text{-}19)$$

where we have used Eqs. 21-12 and 21-15 . Hence, if the frequency of the disturbance ω is much smaller than the rate constant k, the actual state $C(t)$ closely follows the steady-state corresponding to the actual external force, $J(t)$. This is another example of a slow disturbance. An alternative way to characterize a slow perturbation is $2\pi/\omega = \tau_p \gg \tau_{5\%} = 3/k$ (see Box 12.1, Eq. 4). This means that if the period of the perturbation, τ_p, is much larger than the response time of the system, $\tau_{5\%}$, then actual state and steady-state are approximately the same.

The other extreme case is given by $\omega \gg k$ (or $\tau_p \ll \tau_{5\%}$). Then, $\tan^{-1}(\omega/k) = \tan^{-1}(\sim \infty) = \pi/2$, and Eq. 21-18 becomes:

$$\begin{aligned} C(t) &\sim \frac{\bar{J}}{k} + \frac{\bar{J}A_J}{\omega}\sin\left(\omega t - \frac{\pi}{2}\right) \\ &= \frac{\bar{J}}{k}\left[1 + \frac{k}{\omega}A_J \sin\left(\omega t - \frac{\pi}{2}\right)\right] , \quad \omega \gg k \end{aligned} \qquad (21\text{-}20)$$

The particular form of the last expression helps to interpret the result. The influence of the oscillatory input term is reduced by the factor (k/ω) and goes to zero when ω grows. The system does not respond to fast input variations but remains at the steady-state of the mean input, $\bar{J}$.

Arbitrary input variation. We have now developed recipes with which we can estimate the influence of temporal variations of the external force in general. One can always separate the trend of a function $J(t)$ from its oscillatory variation. The former can be analyzed with Eq. 21-14, the latter with 21-18. Since the system is linear we could make a frequency analysis of $J(t)$ and then treat the influences of the different frequencies on $C(t)$ separately.

To demonstrate the procedure, we continue the discussion on the behavior of tetrachloroethene (PCE) in Greifensee (see Box 21.2). Based on repeated measurements we have assumed that the PCE concentration in the lake is at steady-state. Now we want to make sure that this conclusion is not biased either by a long-term trend or by periodic fluctuations of the PCE input into the lake. The numerical details of the analysis are given in Box 21.4.

Box 21.4   Temporal Variability of PCE Input and Response of Concentration in Greifensee

The (overall) rate constant for PCE in Greifensee is (Box 21.2): $k = 10.9 \times 10^{-3} d^{-1} = 4.0$ yr^{-1}.

Trend Variation.

We consider two different rates of exponential increases, α:
(a) $\alpha = 0.1$ yr^{-1} (10% per year)
(b) $\alpha = 1$ yr^{-1} (270% per year)

The ratio between actual concentration, $C(t)$, and hypothetical steady-state concentration, $C_\infty(t)$, is calculated from Eq. 21-17:

(a) $\dfrac{C(t)}{C_\infty(t)} = \dfrac{4.0}{4.0+0.1} = 0.976 \approx 1$ Valid for $kt \gg 1$, i.e. $t \gg \dfrac{1}{4\ \text{yr}^{-1}} = 0.25$ yr = 3 months

(b) $\dfrac{C(t)}{C_\infty(t)} = \dfrac{4.0}{4.0+1.0} = 0.80$ Valid for $t \gg 3$ months

Periodic Variations.

Diurnal: $\omega = \dfrac{2\pi}{1\,d} = 6.3\,d^{-1} \Rightarrow \dfrac{\omega}{k} = 580 \Rightarrow \varphi = \tan^{-1}(580) = \dfrac{\pi}{2}$

If the relative input variation is $\pm 100\%$ ($A_J = 1$), the relative variation of the lake concentration $C(t)$ is $\pm (100\%/580) = \pm 0.17\%$ with a time lag relative to the input signal of 6 hours (one-fourth of a period).
Note: The problem with diurnal input variation is underestimated by the one-box approach in which spatial homogeneity is assumed. In fact, the limiting factor is spatial mixing.

Annual: $\omega = \dfrac{2\pi}{1\,\text{yr}} = 0.017\,d^{-1} \Rightarrow \dfrac{\omega}{k} = 1.6 \Rightarrow \varphi = \tan^{-1}(1.6) = 1.00$

Hence, the time lag is $1.00/2\pi = 0.16$ (16%) of the annual period (i.e., about 2 months).
A relative input variation of $\pm 100\%$ results in a relative variation of the lake concentration of

$$100\% \left(\frac{4}{\left[4^2 + (2\pi)^2 \right]^{1/2}} \right) = 54\% .$$

As it turns out, the influence of an exponential input rise of 10% per year keeps $C(t)$ close to the steady-state concentration (Fig. 21.4). In contrast, an exponential increase of the input of 270% per year leads to a concentration reduction of 80% relative to the hypothetical steady-state. To find such an enormous rate as a long-term trend is hardly realistic. Thus we conclude that a possible trend in the input does not really put the steady-state assumption into question for this particular case.

Regarding periodic input variations, there are two obvious periods to look at: the diurnal period, which is typical for inputs from sewage effluents, and the annual

Figure 21.5 Response of linear system to external periodic perturbation (Eq. 21-12). Full lines show hypothetical steady-state (Eq. 21-19); dashed lines give system response (Eq. 21-18). The system rate constant $k = 4.0 \text{ yr}^{-1}$ corresponds to the behavior of PCE in Greifensee (Box 21.2). Curve A corresponds to an annual variation with relative amplitude $A_J = 0.5$, curve B to a variation with period of 4 years and $A_J = 1$.

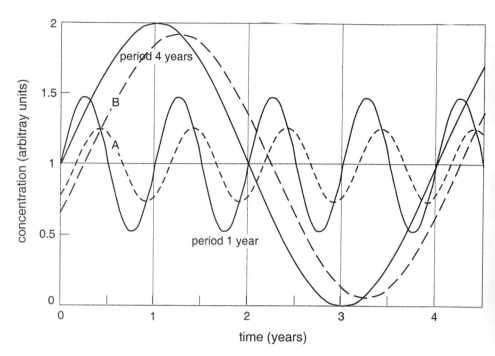

period, which has both natural and anthropogenic causes. Specifically, we want to know whether the measurement of the PCE content of Greifensee would depend on the sampling time (time during the day or selected month during the year), provided that the input varies periodically. As shown in Box 21.4, diurnal variations do not affect the PCE concentration in the lake. (Note: The one-box model is overoptimistic here since complete spatial mixing is assumed. In fact, the limitations of the model due to the final mixing rate are more important.) In turn, 54% of the annual input variation is felt in the lake with a time lag of two months. Figure 21.5 shows the effect of a (hypothetical) annual input variation with a relative amplitude $A_J = 0.5$ on the PCE concentration in Greifensee (curve A). For a periodicity of 4 years, the response of PCE would be nearly at steady-state (curve B, $A_J = 1$).

In this section we have only discussed the effect of the temporal variability of the inhomogeneous term (the external force) on the solution of Eq. 21-4. The solution of the FOLIDE given in Box 21.1 includes the effect of a variable rate constant, $k(t)$. A situation in which both terms (J and k) vary with time is described in Illustrative Example 21.2.

(*Text continues on page 970*)

Illustrative Example 21.2

Advanced Topic

How Certain Is the Degradation Rate of Nitrilotriacetic Acid (NTA) in Greifensee?

Problem

In order to derive a first-order degradation rate constant, k_{NTA}, for NTA in Greifensee (Illustrative Example 21.1), you had to make several restrictive assumptions regarding the characteristics of the lake and the behavior of NTA. Identify possible sources of uncertainty and formulate the necessary conditions: (1) to keep the uncertainty of k_{NTA} below 10%, and (2) to prove that NTA is not conservative (nonreactive) in Greifensee. Note: Here we do not consider the analytical uncertainty of the measurement of NTA in water.

Answer

Some important sources of uncertainty for the indirect determination of the degradation rate, k_{NTA}, from the linear one-box model include:

(a)   The input of NTA into the lake, I_{NTA}, may be variable.

(b)   The rate constant k may be variable, either because the flushing rate k_w or the reaction rate k_{NTA} or both are variable.

(c)   The degradation rate of NTA may not be a first-order (linear) reaction.

(d)   Spatial variation of the NTA concentration in the lake may render the calculation of the average NTA concentration (or the total mass of NTA in the lake) difficult.

The effect of a nonlinear degradation rate (point c) will be described in Illustrative Example 21.3. The discussion on spatial inhomogencity (point d) follows in Section 21.3 and in Chapters 22 and 23.

(a) Variable input

From Table 21.1 and Illustrative Example 21.1 calculate the (overall) linear rate constant of NTA in Greifensee:

$$k = k_w + k_{NTA} = 0.23 \times 10^{-2} d^{-1} + 2.1 \times 10^{-2} d^{-1}$$
$$\cong 0.023 d^{-1} = 8.5 yr^{-1} \qquad \text{(reactive NTA)} \tag{1}$$

One could argue that, in fact, NTA is nonreactive and that the calculated rate constant, k_{NTA}, originates from a wrong interpretation of the data. This would mean that the overall rate constant was:

$$k = k_w = 0.0023 d^{-1} = 0.84 yr^{-1} \qquad \text{(nonreactive NTA)} \tag{2}$$

In order to check the "nonreactivity hypothesis" calculate the sensitivity of the average NTA concentration in Greifensee due to a possible variability in the NTA input function, I_{NTA}.

Exponential increase of input

(1) From Eq. 21-17 you can conclude that in order to keep the difference between $C(t)$ and $C_\infty(t)$ below 10%, the absolute value of the exponential growth (or decay) rate α must obey the condition:

$$|\alpha| < 0.1 k = 0.85 \text{ yr}^{-1} \qquad \text{(reactive NTA)} \tag{3}$$

(2) For the "nonreactivity hypothesis" to be compatible with the measured concentrations, the difference between the measured NTA concentration in the lake (3.7×10^{-9} M) and the expected concentration of a conservative chemical with the observed input of 13 mol d^{-1} (38×10^{-9} M, see Illustrative Example 21.1, Eq. 2) had to be interpreted as the lag effect from an exponentially increasing input (see Fig. 1 below). Thus from Eq. 21-17:

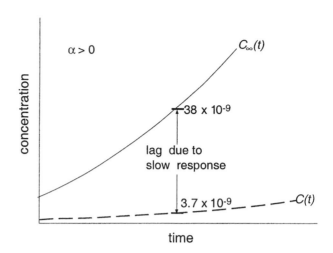

Figure 1 If NTA in Greifensee were nonreactive, then the difference between the measured concentration (3.7×10^{-9} M) and the predicted steady-state concentration (38×10^{-9} M) had to be interpreted as the retarded response to an exponential increase of NTA input into the lake with the unrealistically large growth rate $\alpha = 7.5$ yr^{-1}.

$$\frac{C(t)}{C_\infty(t)} = \frac{3.7 \times 10^{-9}\,\text{M}}{38 \times 10^{-9}\,\text{M}} \cong 0.1 = \frac{k}{k+\alpha} \tag{4}$$

where $k = 0.84$ yr^{-1} is the overall rate provided that NTA is nonreactive. Solving for α yields:

$$\alpha = 9k = 9 \times 0.84\ \text{yr}^{-1} = 7.5\ \text{yr}^{-1} \tag{5}$$

Hence, the growth rate of the NTA input would have to be unrealistically large to make the "nonreactivity hypothesis" compatible with the measured NTA concentrations. It would cause an enormous concentration increase in the lake, which would not go undetected for very long.

Periodic input variation

Similar considerations can be made for the case of periodic input variations. The relevant concentration ratio follows from Eqs. 21-18 and 21-19:

$$\frac{C(t)}{C_\infty(t)} = \frac{1 + \dfrac{kA_{\text{J}}}{\left(k^2 + \omega^2\right)^{1/2}} \sin\left(\omega t - \varphi\right)}{1 + A_{\text{J}}\sin\left(\omega t\right)}, \qquad \varphi = \tan^{-1}\left(\frac{\omega}{k}\right) \tag{6}$$

Although it is formally possible to find values for A_{J} and ω which for a certain time t would make the concentration ratio as small as 0.1, such solutions are highly unlikely.

(b) Variable rate constant k

The solution of Eq. 21-4 with variable rate constant, $k(t)$, is given in Box 12.1 (Eq. 9). Since the steady-state concentration C_∞ is inversely related to k, the concentration on which $C(t)$ converges rises and falls when k is changing. For the case of a conservative compound, the only cause for a changing rate k is the variation of the flushing rate of the lake. To demonstrate this effect, consider three different $k(t)$ evolutions and the corresponding temporal development of the NTA concentration in Greifensee during a period of 60 days. Initial concentration ($C_{\text{NTAo}} = 3.7 \times 10^{-9}$ M), NTA input ($I_{\text{NTA}} = 13$ mol d^{-1}), and average k-value shall be identical for all cases.

The situations differ only in their temporal variation of k.

1. $k = 0.023$ d^{-1} during the whole period of 60 days
2. $k = 0.04$ d^{-1} for the first 30 days and 0.006 d^{-1} for the second 30 days
3. $k = 0.006$ d^{-1} for the first 30 days and 0.04 d^{-1} for the second 30 days

To get the solution of Eq. 21-4 for a time-dependent rate $k(t)$ you could use Eq. 9 of Box 12.1. Since in the above examples, $k(t)$ is constant except for an abrupt change at $t = 30$ d, you can also use the solution for time-independent coefficients J and k (Box 12.1, Eq. 6) for the first 30 days and then insert the final concentration as the initial value for the second 30 days. The three concentration curves are shown in Fig. 2. Their behavior can be understood qualitatively with the following considerations:

Case 1. This is the reference case which you calculated in Illustrative Example 21.1 by assuming that the concentration in the lake is at steady-state with the measured input. Hence C_{NTA} remains constant: $C_{NTAo} = C_{NTA}(60) = 3.7 \times 10^{-9}$ M.

Case 2. Since the steady-state concentration is inversely proportional to k, $C(t)$ first drops toward the lower steady-state $C_{A\infty}$, which belongs to the rate constant $k = 0.04$ d^{-1}. From Eq. 12-49b:

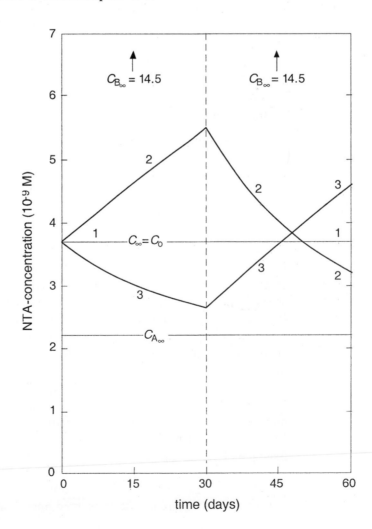

Figure 2 Computed hypothetical NTA concentration for three different histories of k variation (see text). NTA input is equal for all cases. C_∞, $C_{A\infty}$, $C_{B\infty}$ are steady-state concentration for $k = 0.023$ d^{-1}, 0.04 d^{-1} and 0.006 d^{-1}, respectively.

$$C_{A\infty} = \frac{I_{NTA}/V}{k} = \frac{13 \text{ mol d}^{-1}}{150 \times 10^6 \text{m}^3 \times 0.04 \text{d}^{-1}} = 2.2 \times 10^{-6} \text{mol m}^{-3} = 2.2 \times 10^{-9} \text{M}$$

The concentration between $t = 0$ and $t = 30$ d is described by (Box 12.1, Eq. 6):

$$C(t) = C_{A\infty} + \left(C_{NTA0} - C_{A\infty}\right) e^{-kt} = \left[2.2 + \left(3.7 - 2.2\right) e^{-0.04t}\right] \times 10^{-9} \text{M}$$

which for $t = 30$ d yields $C(30) = 2.65 \times 10^{-9}$ M.

During the second months, the concentration increases toward the steady-state $C_{B\infty}$, which belongs to $k = 0.006$ d^{-1} and is thus much larger than $C_{A\infty}$:

$$C_{B\infty} = \frac{13 \text{ mol d}^{-1}}{150 \times 10^6 \text{m}^3 \times 0.006 \text{d}^{-1}} = 14.5 \times 10^{-6} \text{mol m}^{-3} = 14.5 \times 10^{-9} \text{M}$$

We leave it to the reader to calculate the corresponding solution. At $t = 60$ d one gets $C(60) = 4.6 \times 10^{-9}$ M.

Case 3. The result of the calculation is shown in Fig. 2. The final value is $C(60) = 3.2 \times 10^{-9}$ M.

The above example demonstrates how sensitively the concentration responds to changes of I or k. Hence, repeated measurements of C and I are needed to confirm the *exact size* of the degradation rate of NTA in Greifensee, although based on the available measurements the existence of some NTA degradation in Greifensee can hardly by denied.

Advanced Topic

Nonlinear One-Box Model

The apparent rates of transformation processes in natural systems are not necessarily zeroth or first-order. If the application of a simple model (such as the one-box model) to a real environmental system suggests that an internal transformation rate is of second-order, this does not necessarily mean that the underlying process is really of that order. For instance, in Illustrative Example 21.1 we compared degradation rates of nitrilotriacetic acid (NTA) measured in the laboratory at concentrations between 2×10^{-8} and 5×10^{-6} M with rates derived from lake data (typical concentration 4×10^{-9} M). The result suggested that the (apparent) first-order reaction rate coefficient, k_{NTA}, increases with concentration. This seems to point to a reaction of higher than first order, although from a biochemical point of view we would expect that the degradation of NTA by microorganisms is rather of Michaelis-Menten type (see Box 12.2), that is, first order at low concentration and zeroth order at large concentrations. As will be indicated in Illustrative Example 21.3, for the case of NTA in Greifensee an apparent second-order degradation may result from a combination of the mass balance of the microorganisms degrading the NTA with the mass balance of the NTA itself. If only the latter is explicitly considered in the model (as is the case in Illustrative Example 21.3), the apparent NTA degradation indeed may be of second order.

As long as the range of the concentration variation of a chemical is small, the non-linearity of reactions is often not recognized; and even if it is, the reaction can sometimes be approximated by a pseudo-first-order reaction. For instance, if the actual concentration can be written as small deviations from the mean, $C_i = \overline{C}_i + \delta C_i$, where $\delta C_i \ll \overline{C}_i$, the second-order reaction:

$$\left(\frac{dC_i}{dt}\right)_{\text{reaction}} = -k_{i2}\, C_i^2 \tag{21-21}$$

can be approximated by

$$\left(\frac{dC_i}{dt}\right)_{\text{reaction}} \cong -k_{i2}\, \overline{C}_i\left(\overline{C}_i + \delta C_i\right) = -\hat{k}_{i1}\, C_i \tag{21-22}$$

where

$$\hat{k}_{i1} = k_{i2}\, \overline{C}_i \tag{21-23}$$

is the pseudo-first-order rate constant (see also Chapter 12.3). Yet, this approximation fails if a model that was calibrated for a narrow concentration range is extrapolated to another situation in which the concentrations are very different from the values used for calibration.

To illustrate this point, we consider a chemical in a completely mixed reactor (or lake) with water exchange rate Q. The chemical is degraded by a second-order reaction (Eq. 21-21). Compared to Eq. 12-53, the mass balance equation is slightly modified:

$$\frac{dC_i}{dt} = J_i - k_w\, C_i - k_{i2}\, C_i^2 = k_w\, C_{i\text{in}} - k_w\, C_i - k_{i2}\, C_i^2 \tag{21-24}$$

where $J_i = I_i / V$ is the input of the chemical per volume and time, and $C_{i\text{in}} = I_i / Q$ is the average input concentration.

For constant parameters k_w, $C_{i\text{in}}$, k_{i2}, the solution of the nonlinear differential Equation 21-24 is given in Box 21.5. The steady-state, $C_{i\infty}$, results from the solution of a quadratic equation. It is controlled by the characteristic concentration

$$C_i^* = \frac{k_w}{k_{i2}} \tag{21-25}$$

which separates two ranges of different behavior. For $C_i \ll C_i^*$, the in-situ reaction is negligible compared to the loss at the outlet. In contrast, for $C_i \gg C_i^*$ the reaction dominates the removal of the compound from the system. For $C_i = C_i^*$, second-order reaction and loss at the outlet are equal.

(*Text continues on page 974*)

Box 21.5   One-Box Model with Water Throughflow and Second-Order Reaction (Advanced Topic)

The dynamic equation (index i omitted for brevity)

$$\frac{dC}{dt} = k_w C_{\text{in}} - k_w C - k_2 C^2 \tag{21-24}$$

has the steady-state solution:

$$C_\infty = \frac{1}{2k_2}\left[-k_w \pm \left(k_w^2 + 4k_w k_2 C_{\text{in}}\right)^{1/2}\right] = \frac{C^*}{2}\left[-1 \pm \left(1 + 4\frac{C_{\text{in}}}{C^*}\right)^{1/2}\right] \tag{1}$$

with the characteristic concentration $C^* = k_w / k_2$. Note that the minus sign in front of the square root of Eq. 1 would make the concentration C_∞ negative and thus meaningless.

The relative size of the input concentration, C_{in}, and the characteristic concentration, C^*, determines how C_∞ depends on C_{in}. The two extreme situations are:

$$C_\infty \approx \begin{cases} C_{in} & \text{for } C_{in} \ll C^* \\ \left[C_{in} C^* \right]^{1/2} = \left[C_{in} k_w / k_2 \right]^{1/2} & \text{for } C_{in} \gg C^* \end{cases} \tag{2}$$

where we have used the approximation:

$$\left\{ 1 + 4 \frac{C_{in}}{C^*} \right\}^{1/2} \approx 1 + 2 \frac{C_{in}}{C^*} \ , \ \text{for } C_{in} \ll C^* \tag{3}$$

Thus, for small input the compound behaves like a conservative substance ($C_\infty \approx C_{in}$), since the (second-order) reaction rate is negligible compared to flushing. In contrast, a large input is balanced by the second-order reaction alone, thus C_∞ increases as $(C_{in})^{1/2}$.

The rate constant k_2 and the corresponding (pseudo-)first-order rate constant k_1 are related by:

$$k_2 C^2 = k_1 C \quad \Rightarrow \quad k_1 = k_2 C \tag{4}$$

For $C = C^*$ the pseudo-first-order rate constant k_1 is:

$$k_1 = k_2 C^* = k_w \tag{5}$$

Hence, C^* is that concentration at which removal by flushing and by *in situ* reaction are equal. For $C > C^*$, reaction is the dominant removal process.

If the actual concentration deviates only little from steady-state, that is, if:

$$\left| \delta C \right| = \left| C - C_\infty \right| \ll C_\infty \tag{6}$$

the change of δC can be approximated by the linear expression:

$$\frac{d(\delta C)}{dt} \approx - \left(k_w + 2 k_2 C_\infty \right) \delta C \tag{7}$$

Thus, according to Eq. 4 of Box 12.1, time to steady-state

$$\tau_{5\%} = \frac{3}{k_w + 2 k_2 C_\infty} \tag{8}$$

depends on C_∞ and thus on the input, J or C_{in}.

Illustrative Example 21.3
Advanced Topic

Higher-Order reaction of Nitrilotriacetic Acid (NTA) in Greifensee

Problem

In Illustrative Example 21.2 we identified the possibility of a nonlinear (apparent) degradation reaction of NTA in Greifensee as one of several sources of uncertainty regarding the interpretation of the fate of NTA in the lake. In Illustrative Example 21.1, we found that at the concentration $C_{NTA} = 3.7 \times 10^{-9}$ M the (pseudo-)first-order rate constant is $k_{NTA1} = 2.1 \times 10^{-2}$ d^{-1}. From Eq. 21-23 the corresponding second-order rate constant would be $k_{NTA2} = 5.7 \times 10^{6}$ M^{-1} d^{-1}. Compare the steady-state concentration $C_{NTA\infty}$ of NTA in Greifensee for an NTA input, I_{NTA}, between 10^{-2} and 10^{5} mol d^{-1} for the following three different assumptions: (a) NTA is conservative, (b) NTA is degraded by a first-order reaction with rate $k_{NTA1} = 2.1 \times 10^{-2}$ d^{-1}, (c) NTA is degraded by an apparent second-order reaction with rate $k_{NTA2} = 5.7 \times 10^{6}$ M^{-1} d^{-1}. (d) Give a simple argument that the real degradation rate can be first-order while the *apparent* degradation in the lake looks like a second-order reaction.

Answer

(a) The steady-state concentration for a conservative substance is (Eq. 12-51 with $k_{tot} = 0$):

$$C^{(a)}_{NTA\infty} = C_{NTA\,in} = I_{NTA} / Q = \left(3.4 \times 10^{5} \, m^{3} d^{-1}\right)^{-1} I_{NTA} \tag{1}$$

$$I_{NTA} \text{ in } [mol\,d^{-1}], \quad C_{NTA} \text{ in } [mol\,m^{-3}]$$

(b) The steady-state concentration for degradation by first-order reaction is (Eq. 12-51 with $k_{tot} = k_{NTA1}$):

$$C^{(b)}_{NTA\infty} = C_{NTA\,in} \frac{k_w}{k_w + k_{NTA1}} = C_{NTA\,in} \frac{2.3 \times 10^{-3}}{2.3 \times 10^{-2}} = 0.1\,C_{NTA\,in} = 0.1\,C^{(a)}_{NTA\infty} \tag{2}$$

Note that $C^{(b)}_{NTA\infty}$ is 10% of $C^{(a)}_{NTA\infty}$ and both concentrations are proportional to I_{NTA}.

(c) Degradation by second-order reaction (see Box 21.5): the characteristic concentration is:

$$C^{*} = \frac{k_w}{k_{NTA\,2}} = \frac{2.3 \times 10^{-3} d^{-1}}{5.7 \times 10^{6} M^{-1} d^{-1}} = 4.0 \times 10^{-10} M$$

and

$$C^{(c)}_{NTA\infty} = \frac{4.0 \times 10^{-10} M}{2} \left[-1 + \left(1 + \frac{4\,C_{NTA\,in}}{4.0 \times 10^{-10} M}\right)^{1/2} \right] \tag{3}$$

The three models are compared in the figure below (note the double logarithmic scale). As predicted by Eq. 3 of Box 21.5, if the mean input concentration, $C_{NTA\,in}$, is much smaller than $C^{*} = 4.0 \times 10^{-10}$ M, then $C^{(c)}_{NTA\infty} \approx C_{NTA\,in}$; that is, NTA behaves like a conservative compound. In contrast, for $C_{NTA\,in} \gg C^{*}$, $C^{(c)}_{NTA\infty}$ increases as the square root of the input. Note that for the three models an interpolation from the measured input rate (13 mol d^{-1}) to other input rates predict very different $C_{NTA\infty}$ values.

(d) As long as the NTA concentrations in the lake are fairly constant, the biomass concentration of the microorganisms that degrade NTA would probably reach a steady-state, B_∞, determined by a growth rate which is proportional to the available NTA concentration, C, and a loss rate (flushing through the outlet, sedimentation) which is proportional to B_∞. As a result, B_∞ turns out to be proportional to the NTA concentration, C. In turn, the NTA degradation rate is first order with respect to C and proportional to the concentration of the degrading microorganisms, B_∞, thus:

$$\left(\frac{dC}{dt}\right)_{\text{degradation}} = -\text{constant } C B_\infty = -\text{constant } C^2$$

since B_∞ is proportional to C.

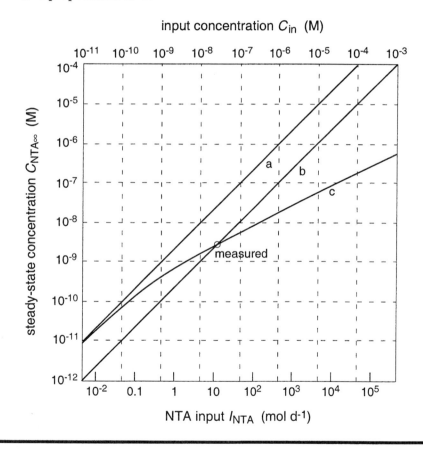

There are still other causes of nonlinearities than (apparent or real) higher-order transformation kinetics. In Section 12.3 we discussed catalyzed reactions, especially the enzyme kinetics of the Michaelis-Menten type (see Box 12.2). We may also be interested in the modeling of chemicals which are produced by a nonlinear autocatalytic reaction, that is, by a production rate function, $p(C_i)$, which depends on the product concentration, C_i. Such a production rate can be combined with an elimination rate function, $r(C_i)$, which may be linear or nonlinear and include different processes such as flushing and chemical transformations. Then the model equation has the general form:

$$\frac{dC_i}{dt} = p(C_i) - r(C_i) \tag{21-26}$$

The functions $p(C_i)$ and $r(C_i)$ can be plotted in the same graph (Fig. 21.6). Wherever the two curves cross, that is, for concentrations obeying

$$p(C_i) = r(C_i) \tag{21-27}$$

the system has a steady-state. Thus, we designate these concentrations by $C_{i\infty}^{(\alpha)}$, where the superscript α serves to number the different steady-states in case there are more than one (see Figs. 21.6b and c). To exemplify the case shown in Fig. 21.6c, we can choose the following analytical expressions:

$$p(C_i) = J \frac{C_i}{C_i + J/k_p} \quad \text{and} \quad r(C_i) = k_r C_i \tag{21-28}$$

All parameters (J, k_p, k_r) shall be positive. $p(C_i)$ describes an autocatalytic production function which is first-order at small C_i and zeroth order at large C_i. Formally, it looks like the Michaelis-Menten expression (Eq. 12-32), although it does not have the same mechanistic basis.

Inserting into Eq. 21-27 and solving for C_i yields the following quadratic equation for $C_{i\infty}^{(\alpha)}$:

$$k_r \left(C_{i\infty}^{(\alpha)}\right)^2 + J\left(k_r / k_p - 1\right) C_{i\infty}^{(\alpha)} = 0$$

with the two solutions:

$$C_{i\infty}^{(1)} = 0 \; ; \quad C_{i\infty}^{(2)} = J\left(\frac{1}{k_r} - \frac{1}{k_p}\right) \tag{21-29}$$

The trivial steady-state, $C_{i\infty}^{(1)}$, simply means that both $p(C_i)$ and $r(C_i)$ are zero if the concentration C_i is zero. The second solution, $C_{i\infty}^{(2)}$, makes sense only if $k_p > k_r$; otherwise $C_{i\infty}^{(2)}$ would be negative, that is, the production function would never exceed the decay function and the concentration would always fall back to zero.

Steady-states can be stable, unstable, or indifferent. A stable steady-state holds the system fast; that is, if the concentration C_i slightly deviates from the steady-state, the system reacts such as to bring the concentration back to the steady-state. At an unstable steady-state, small excursions are self-perpetuating such that the system moves away from its original state. At indifferent steady-states the system does not respond to small excursions away from the steady-state.

To demonstrate the difference between stable and unstable steady-states we compare the nontrivial steady states, $C_{i\infty}^{(2)}$, of the cases shown in Fig. 21.6b and c, respectively. Imagine that the actual concentration is slightly smaller than $C_{i\infty}^{(2)}$ (i.e., in Fig. 21.6 the system is to the left of $C_{i\infty}^{(2)}$). Then for case b production $p(C_i)$ is smaller than decay $r(C_i)$. According to Eq. 21-26, C_i decreases and moves even further away from its original position until it finally comes to a halt at the other steady-state, $C_{i\infty}^{(1)} = 0$. In contrast, for case c a slight deviation to the left causes the concentration to grow ($p > r$)

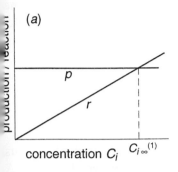

(a)

production / reaction

p

r

concentration C_i $C_{i\infty}^{(1)}$

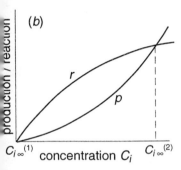

(b)

production / reaction

r

p

$C_{i\infty}^{(1)}$ concentration C_i $C_{i\infty}^{(2)}$

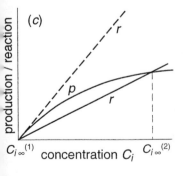

(c)

production / reaction

r

p

r

$C_{i\infty}^{(1)}$ concentration C_i $C_{i\infty}^{(2)}$

Figure 21.6 One-box models described by different production and reaction functions, $p(C_i)$ and $r(C_i)$, respectively (Eq. 21-26). The steady-state(s) ($C_{i\infty}^{(1)}$, $C_{i\infty}^{(2)}$, ...) are given by the crossing points of the two functions: (a) linear system (Eq. 21-4) with one steady-state; (b) nonlinear system with two steady-states ($C_{i\infty}^{(2)}$ unstable); (c) nonlinear system (Eq. 21-28) with two steady-states ($C_{i\infty}^{(2)}$ stable). The dashed r line shows the situation when there is no positive $C_{i\infty}^{(2)}$.

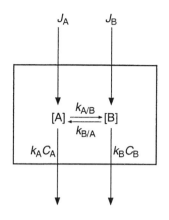

Figure 21.7 Linear one-box model for two chemicals with concentrations C_A and C_B. J_A and J_B are external inputs, k_A and k_B are the first-order output rate constants of A and B, respectively, and $k_{A/B}$, $k_{B/A}$ the first-order forward and backward reaction rate constants.

until it reaches $C_{i\infty}^{(2)}$ again. Thus, in case *b* we find that $C_{i\infty}^{(2)}$ is an unstable steady-state, while in case *c* it is stable. We leave it to the reader to show that deviations from $C_{i\infty}^{(2)}$ to the right show the same stability behavior (see Problem 21.4).

It is true that the above considerations and examples given in Fig. 21.6 must look somewhat far-fetched in the context of the dynamics of organic chemicals in the environment. This is no longer the case for models with more than one variable. Such models are often nonlinear and have multiple steady-states. In this respect, the purpose of Fig. 21.6 is primarily to open the door to a world of complexity which itself leads far beyond the scope of this book. The interested reader is referred to the corresponding literature (e.g., Arrowsmith and Place, 1992).

One-Box Model with Two Variables

Until now we have dealt with one-box models with only one state variable. Let us now look at a model with two variables, say the concentration of two chemicals, A and B, where A is transformed into B by a chemical process and vice versa. We have already encountered such a model in Section 12.3, where we discussed the hydration of aldehydes. In the following, we restrict the discussion to *linear models* with two variables; they exhibit all the important properties of linear multivariable systems. Nonlinear models and models with three or more variables quickly become too complicated for an analytical discussion. Such models have to be solved numerically.

The system is shown in Fig. 21.7. It is described by two concentrations (state variables), C_A and C_B, by two zero-order input functions, J_A and J_B (input per volume and time), by two first-order output functions, $k_A C_A$ and $k_B C_B$ (output per volume and time), and by the first-order transformations from A to B and vice versa. The inputs and outputs can be the sum of two or more processes, for instance, the sum of the input through different inlets and from the atmosphere (as in Eq. 21-7a), or the sum of the output at the outlet and by exchange to the atmosphere (as in Eq. 21-7b.).

If there were no transformation between A and B, we could describe each chemical separately by a linear one-box equation (Eq. 21-4). However, the chemical reaction links the two equations:

$$\frac{dC_A}{dt} = J_A - k_A C_A - k_{A/B} C_A + k_{B/A} C_B \tag{21-30a}$$

$$\frac{dC_B}{dt} = J_B - k_B C_B - k_{B/A} C_B + k_{A/B} C_A \tag{21-30b}$$

These are two *coupled first-order linear inhomogeneous differential equations* (coupled FOLIDEs). In their most general form they can be condensed to a form in which the right-hand side of each equation consists of not more than three terms, that is, an inhomogeneous term, a linear term depending on the first variable and one depending on the second variable. This general form and the solution of the system for constant coefficients are given in Box 21.6. As it turns out, generally the solution is still composed of exponential functions (Eq. 1, Box 21.6).

Box 21.6   Solution of Two Coupled First-Order Linear Inhomogeneous Differential Equations (Coupled FOLIDEs)

The system of two first-order linear differential equations has the general form:

$$\frac{dy_1}{dt} = J_1 - k_{11}y_1 + k_{12}y_2$$

$$\frac{dy_2}{dt} = J_2 + k_{21}y_1 - k_{22}y_2 \tag{1}$$

Its solutions are determined by two constant terms, $y_{1\infty}$ and $y_{2\infty}$, and by two exponential functions with rates k_1 and k_2:

$$y_1(t) = y_{1\infty} + A_{11}\,e^{-k_1 t} + A_{12}\,e^{-k_2 t}$$

$$y_2(t) = y_{2\infty} + A_{21}\,e^{-k_1 t} + A_{22}\,e^{-k_2 t} \tag{2}$$

The rate constants, k_1 and k_2, are the *negative eigenvalue* of the linear system (see Box 21.8):

$$k_1 = \frac{1}{2}\left[k_{11} + k_{22} - q\right], \qquad k_2 = \frac{1}{2}\left[k_{11} + k_{22} + q\right]$$

$$q = \left[(k_{11} - k_{22})^2 + 4k_{12}k_{21}\right]^{1/2} \tag{3}$$

If k_1 and k_2 are positive, the constant terms correspond to the steady-state values of the system:

$$y_{1\infty} = \frac{k_{22}J_1 + k_{12}J_2}{k_{11}k_{22} - k_{12}k_{21}} \qquad y_{2\infty} = \frac{k_{21}J_1 + k_{11}J_2}{k_{11}k_{22} - k_{12}k_{21}} \tag{4}$$

The integration constants are (y_{1o} and y_{2o} are the initial values at $t = 0$):

$$A_{11} = 1/q\left[-(k_{11} - k_2)(y_{1o} - y_{1\infty}) + k_{12}(y_{2o} - y_{2\infty})\right]$$

$$A_{12} = 1/q\left[\ (k_{11} - k_1)(y_{1o} - y_{1\infty}) - k_{12}(y_{2o} - y_{2\infty})\right]$$

$$A_{21} = 1/q\left[-(k_{22} - k_2)(y_{2o} - y_{2\infty}) + k_{21}(y_{1o} - y_{1\infty})\right]$$

$$A_{22} = 1/q\left[\ (k_{22} - k_1)(y_{2o} - y_{2\infty}) - k_{21}(y_{1o} - y_{1\infty})\right] \tag{5}$$

It can be shown that all the coefficients k_{ij} of Eq. 1 are positive or zero, if the model equations result from a mass balance scheme. Then $k_i \geq 0$, where k_1 is the smaller of the two eigenvalues. By analogy to Eq. 4 of Box 12.1, the time to steady-state can be defined by:

$$t_{5\%} = \frac{3}{\min\{k_i\}} = \frac{6}{k_{11} + k_{22} - q} \tag{6}$$

To bring Eq. 21-30 into the form of Box 21.6, the following substitutions have to be made:

$$y_1 \to C_A , \qquad y_2 \to C_B$$
$$J_1 \to J_A , \qquad J_2 \to J_B \tag{21-31}$$
$$k_{11} \to (k_A + k_{A/B}), \quad k_{12} \to k_{B/A}, \quad k_{21} \to k_{A/B}, \quad k_{22} \to (k_B + k_{B/A})$$

A full discussion of these models will follow in Section 21.3. At this point we will deal only with a special version of Eq. 21-30 in which there is no back-reaction from B to A and no external input of chemical B. As an example imagine the case of the pollution of a pond by a pesticide A that in the water is transformed into another compound B that is more toxic than the original compound. For instance, thiophosphoric acid ester is transformed to the corresponding phosphoric acid ester (see Chapters 2 and 13). Then it would be important to predict the maximum concentrations of A and B in the pond and to estimate the necessary time until the pollutant concentration has fallen below a certain threshold. Such a case will be discussed in Illustrative Example 21.4 below.

First, let us derive the necessary mathematical formalism. Obviously, we could apply the equations given in Box 21.6 and the substitution rules of Eq. 21-31 for the special case $J_B = 0$, $k_{B/A} = 0$. Yet, a simpler approach makes use of the fact that now Eq. 21-30a no longer depends on the second variable C_B, hence it can be solved independently of the solution of Eq. 21-30b. The result, $C_A(t)$, is inserted into Eq. 21-30b. Systems which are coupled in such a way are called hierarchical (B depends on A, but A does not depend on B).

If $k_{B/A} = 0$, the solution of Eq. 21-30a follows from Eq. 6 of Box 12.1:

$$C_A(t) = C_{A\infty} + \left(C_{A0} - C_{A\infty}\right)e^{-(k_A + k_{A/B})t}$$
$$C_{A\infty} = \frac{J_A}{k_A + k_{A/B}} \tag{21-32}$$

where C_{A0} is the initial concentration of chemical A. Inserting this result into Eq. 21-30b yields (remember that we assume $J_B = 0$ and $k_{B/A} = 0$):

$$\frac{dC_B}{dt} = k_{A/B} C_{A\infty} + k_{A/B}\left(C_{A0} - C_{A\infty}\right)e^{-(k_A + k_{A/B})t} - k_B C_B \tag{21-33}$$

Thus, we have reduced Eq. 21-30b to the already-familiar case of a FOLIDE with variable inhomogeneous term. Its solution is given by Eq. 8 of Box 12.1. If the initial concentration of B is assumed to be zero ($C_{B0} = 0$), we get after some algebraic manipulation:

$$C_B(t) = C_{B\infty}(1 - e^{-k_B t}) + \frac{k_{A/B}}{k_A - k_B + k_{A/B}}\left(C_{A0} - C_{A\infty}\right)\left(e^{-k_B t} - e^{-(k_A + k_{A/B})t}\right)$$
$$\text{where} \quad C_{B\infty} = \frac{k_{A/B}}{k_B} C_{A\infty} \tag{21-34}$$

The above model is applied in Illustrative Example 21.4.

Illustrative Example 21.4

cis-1,3-dichloro-1-propene (DCP)

$M_i = 111.0$ g mol^{-1}

K_{iH} (25°C) = 2.4 bar mol^{-1}L^{-1}

$i = DCP = A$

cis-3-chloro-2-propene-1-ol (CPO)

$M_i = 92.5$ g mol^{-1}

K_{iH} (25°C) = 7.6×10^{-3} bar mol^{-1}L^{-1}

$i = CPO = B$

First-order rate constant of abiotic hydrolysis of DCP at 25°C: $k_{A/B} = 0.06$ d^{-1} (independent of pH)

Atmospheric concentrations of DCP and CPO are negligible.

Pond

Volume $V = 5 \times 10^4$ m^3

Surface area $A = 1 \times 10^4$ m^2

Average throughflow
$Q = 5 \times 10^3$ m^{3}d^{-1}

Water temperature $T = 25°$

Mean wind velocity over pond
$u_{10} = 0.5$ m s^{-1}

Fate of a Pesticide and its Decomposition Product in a Small Lake

Problem

Due to an accident, $M_A = 30$ kg of the pesticide *cis-1,3-dichlor-1-propene* (*DCP*) finds its way into a small, well-mixed pond. (1) You are asked to estimate how long it will take until the DCP concentration has fallen below 1 µg L^{-1}. You also worry about the buildup of *cis-3-chloro-2-propene-1-ol* (*CPO*), the product of abiotic hydrolysis of DCP. (2) What maximum concentration of CPO do you expect in the pond and (3) what is its concentration at the time when the concentration of the original pollutant (DCP) has fallen below 1 µg L^{-1}?

Answer

Since there is no back reaction from B to A, the resulting differential equations are hierarchical. Furthermore, since after the accident the input of A and B is zero, the inhomogeneous terms J_A and J_B (see Eq. 21-30) are zero. The initial concentrations are:

$$C_{Ao} = \frac{M_A}{V} = \frac{30 \times 10^3 \text{g}}{5 \times 10^4 \text{m}^3} = 0.6 \text{g m}^{-3} = 600 \text{µg L}^{-1}$$

$$C_{Bo} = 0$$

Thus, Eqs. 21-32 and 21-34 can be simplified to:

$$C_A(t) = C_{Ao} \, e^{-\left(k_A + k_{A/B}\right)t} \quad (1)$$

$$C_B(t) = \frac{k_{A/B}}{k_A - k_B + k_{A/B}} \, C_{Ao} \left(e^{-k_B t} - e^{-\left(k_A + k_{A/B}\right)t}\right) \quad (2)$$

where $k_{A/B} = 0.06$ d^{-1}, $k_A = k_w + k_{A \, a/w}$, $k_B = k_w + k_{B \, a/w}$

and $k_w = \dfrac{Q}{V} = \dfrac{5 \times 10^3 \text{m}^3\text{d}^{-1}}{5 \times 10^4 \text{m}^3} = 0.1$ d^{-1}

The only parameters to be determined are the air–water exchange rates, $k_{A \, a/w}$ and $k_{B \, a/w}$.

For the wind velocity $u_{10} = 0.5$ m s^{-1}, we get from Table 20.4:

$$v_{\text{water a}} = (0.2u_{10} + 0.3)\text{cm s}^{-1} = 0.4 \text{ cm s}^{-1}$$

$$v_w (\text{Sc} = 600) = 0.65 \times 10^{-3} \text{cm s}^{-1}$$

We approximate the relative diffusivities in air and water by Eqs. 18-45 and 18-55, respectively:

$$v_{ia} = v_{\text{H}_2\text{O a}}\left(\frac{D_{ia}}{D_{\text{H}_2\text{O a}}}\right)^{0.67} = 0.4 \text{ cm s}^{-1}\left(\frac{M_{\text{H}_2\text{O}}}{M_i}\right)^{0.335} = 0.4 \text{ cm s}^{-1}\left(\frac{18}{M_i}\right)^{0.335}$$

$$v_{iw} = v_w(600)\left(\frac{D_{iw}}{D_{CO_2}}\right)^{0.67} = 0.65\times10^{-3}\,\text{cm s}^{-1}\left(\frac{M_{CO_2}}{M_i}\right)^{0.335} = 0.65\times10^{-3}\,\text{cm s}^{-1}\left(\frac{44}{M_i}\right)$$

Applied to DCP and CPO:

	$v_{ia}(\text{cm s}^{-1})$	$K_{ia/w}$ (at 25°C)	$v'_{ia} = v_{ia}K_{ia/w}$ (cm s^{-1})	v_{iw} (cm s^{-1})	$v_{ia/w}$ (cm s^{-1})	(Eq. 20-3) (m d^{-1})
DCP (A)	0.22	0.1	0.022	0.48×10^{-3}	0.47×10^{-3}	0.41
CPO (B)	0.23	3.1×10^{-4}	7.1×10^{-5}	0.51×10^{-3}	6.3×10^{-5}	0.054

By analogy to Eq. 21-8c, we get:

$$k_{A\,a/w} = \frac{0.41\,\text{md}^{-1}}{A_o/V} = \frac{0.41\,\text{md}^{-1}}{5\,\text{m}} = 0.082\,\text{d}^{-1}$$

$$k_{B\,a/w} = \frac{0.054\,\text{md}^{-1}}{5\,\text{m}} = 0.011\,\text{d}^{-1}$$

Hence, the rates appearing in Eqs. 1 and 2 are:

$$k_1 \equiv k_A + k_{A/B} = k_w + k_{A\,a/w} + k_{A/B} = (0.1+0.082+0.06)\,\text{d}^{-1} = 0.24\,\text{d}^{-1}$$

$$k_2 \equiv k_B = k_w + k_{B\,a/w} = (0.1+0.011)\,\text{d}^{-1} = 0.11\,\text{d}^{-1}$$

(1) How long does it take for C_A to drop below 1 mg L^{-1}?
From Eq. 1:

$$t = -\frac{1}{k_1}\ln\left(\frac{C_A(t)}{C_{Ao}}\right) = -\frac{1}{0.24}\ln\left(\frac{1\,\mu\text{g L}^{-1}}{600\,\mu\text{g L}^{-1}}\right) = 27\,\text{d} \equiv t_A$$

(2) At time t_A, $C_B(t)$ has reached the value (see Eq. 2):

$$C_B(t_A) = \frac{k_{A/B}}{k_1 - k_2}C_{Ao}\left(e^{-k_2 t_A} - e^{-k_1 t_A}\right)$$

$$= \frac{0.06}{0.13}\,600\,\mu\text{g L}^{-1}[\exp(-0.11\times27) - \exp(-0.24\times27)]$$

$$= 277\,\mu\text{g L}^{-1}(0.0513 - 0.0015) = 13.8\,\mu\text{g L}^{-1}$$

(3) The maximum value of $C_B(t)$ is reached when:

$$\frac{dC_B}{dt} = \frac{k_{A/B}}{k_1 - k_2}C_{Ao}\left[-k_2 e^{-k_2 t} + k_1 e^{-k_1 t}\right] = 0$$

This expression is zero if:

$$k_2 e^{-k_2 t} = k_1 e^{-k_1 t}$$

Solving for t and calling this time t_B yields:

$$t_B = \frac{1}{k_1 - k_2} \ln\left(\frac{k_1}{k_2}\right) = \frac{1}{0.13\,\mathrm{d}^{-1}} \ln\left(\frac{0.24}{0.11}\right) = 6.0\,\mathrm{d}$$

Inserting time t_B into Eq. 2 yields the maximum value of C_B:

$$C_B(t_B = 6.0\,\mathrm{d}) = \frac{0.06}{0.13}\ 600\ \mu\mathrm{g\ L^{-1}}\left[\exp(-0.22 \times 6) - \exp(-0.24 \times 6)\right] = 77.5\ \mu\mathrm{g\ L^{-1}}$$

The temporal development of C_A and C_B is shown in the figure below.

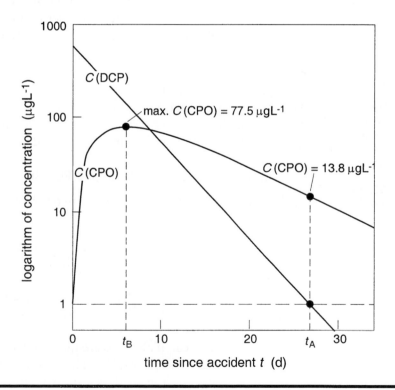

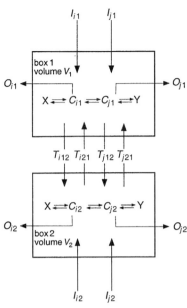

Figure 21.8 Two-box model of two completely mixed environmental compartments. Definition of subscripts: first subscript (i or j) designs the compound, the second (1 or 2) the box. Transfer fluxes T carry three subscripts. For instance, T_{i12} describes the interbox flux of variable i from box 1 to box 2. X and Y denote other chemicals which are not state variables.

21.3   Two-Box Models

The derivation of the two-box model follows naturally from the one-box model. It is useful for describing systems consisting of two spatial subsystems which are connected by one or several transport processes. The mass balance equations for the individual boxes look like Eq. 21-1 with the addition of terms describing mass fluxes between the boxes. Each box can be characterized by one or several state variables. Thus, the dimension of the system of coupled differential equations is the product of the number of boxes and the number of variables per box.

Figure 21.8 exemplifies the adopted nomenclature by using the case of a two-box model with two variables. Unlike Figure 21.3, the external flux terms carry an extra subscript designating the box number. It is placed after the index i which refers to the chemical. In addition, the model includes the interbox fluxes $F_{i\alpha\beta}$ where i designates the chemical and the pair $\alpha\beta$ indicates that the flux occurs from box α to box β. Depending on the physical nature of the boxes, the fluxes can depend either on the flow rate of a fluid (water, air, etc.) or on an interface exchange velocity, discussed in Chapter 19. As an example of the former we mention a chain of two lakes connected by a river. The vertical separation of a lake into an epilimnion and a hypolimnion (see Fig. 19.1a) is an example of the latter.

From a mathematical point of view, the distinction between boxes and chemicals is not relevant. In fact, a one-box model with two chemicals and a two-box model with one chemical lead to the same system of differential equations. Therefore, Box 21.4 also helps in solving a linear two-box model with one variable.

Linear Two-Box Model with One Variable

The mass balance equations of the two boxes 1 and 2 are an extension of Eq. 21-1

$$\frac{dC_{i1}}{dt} = \frac{1}{V_1}\left(I_{i1} - O_{i1} - \sum R_{i1} + \sum P_{i1} - T_{i12} + T_{i21}\right) \tag{21-35a}$$

$$\frac{dC_{i2}}{dt} = \frac{1}{V_2}\left(I_{i2} - O_{i2} - \sum R_{i2} + \sum P_{i2} - T_{i21} + T_{i12}\right) \tag{21-35b}$$

The model is linear, if all terms are linear functions of the concentrations C_{i1} and C_{i2}. For the transfer mass fluxes, T_{i12} and T_{i21}, this implies that the fluxes are proportional to the concentration in the box from which the flux originates. These flux terms couple the two differential equations.

As a first example we analyze a system consisting of a completely mixed water volume in contact with a finite and completely mixed air space (Fig 21.9). The two compartments can have a throughflow (of air or water, respectively). This particular setup could be anything from a large drinking-water cavern to a glass flask with a head space filled with air to measure the Henry's law constant (Section 6.4). We analyze the concentration of a volatile compound, such as tetrachloroethene or benzene, in both water and air, C_{iw} and C_{ia}.

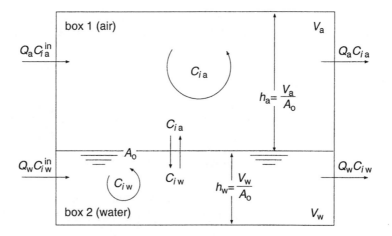

Figure 21.9 Linear two-box model for a volatile compound in a coupled air–water system. Box 1 is the air volume (1 $\equiv$ a), box 2 the water volume (2 $\equiv$ w). Both volumes (V_a and V_b) are flushed by the volumetric rates, Q_a and Q_b, respectively. The system is described by Eq. 21-38. See text for further explanations.

In this example the subdivision of the system is obvious since the air–water interface clearly defines the boundary between the two boxes. In Chapter 20 we learned that independently of the specific air–water exchange model used, the exchange across the interface is always described mathematically by Eq. 20-1. This expression can be separated into two unidirectional fluxes with the form:

$$T_{i12} \equiv T_{iaw} = v_{ia/w} \frac{C_{ia}}{K_{ia/w}} A_o \quad \left[M\,T^{-1} \right] \tag{21-36a}$$

$$T_{i21} = T_{iwa} = v_{ia/w}\, C_{iw}\, A_o \quad \left[M\,T^{-1} \right] \tag{21-36b}$$

where, according to Figure 21.9, box 1 corresponds to the air space (1 $\equiv$ a) and box 2 to the water volume (2 $\equiv$ w). $K_{ia/w}$ is the nondimensional Henry's law constant, and A_o is the interface area. We assume that the substance is conservative (non-reactive) in both boxes. There can be a through flow in both systems (air and water exchange rates are Q_w and Q_a) as well as an external input described by the respective input concentrations, C_{ia}^{in} and C_{iw}^{in}.

Mass balance of compound i in the coupled air–water system is expressed by:

$$\frac{dC_{ia}}{dt} = \frac{1}{V_a} \left(Q_a\, C_{ia}^{in} - Q_a\, C_{ia} - v_{ia/w}\, A_o \frac{C_{ia}}{K_{ia/w}} + v_{ia/w}\, A_o\, C_{iw} \right) \tag{21-37a}$$

$$\frac{dC_{iw}}{dt} = \frac{1}{V_w} \left(Q_w\, C_{iw}^{in} - Q_w\, C_{iw} + v_{ia/w}\, A_o \frac{C_{ia}}{K_{ia/w}} - v_{ia/w}\, A_o\, C_{iw} \right) \tag{21-37b}$$

We introduce the following rate constants:

$$k_{qa} = \frac{Q_a}{V_a} \qquad\qquad \text{specific exchange rate of air volume}$$

$$k_{qw} = \frac{Q_w}{V_w} \qquad\qquad \text{specific exchange rate of water volume}$$

$$k_{ia/w}^{air} = \frac{v_{ia/w}}{h_a} \qquad\qquad \text{specific rate of air–water exchange for air volume}$$

$$k_{ia/w}^{water} = \frac{v_{ia/w}}{h_w} \qquad\qquad \text{specific rate of air–water exchange for water}$$

$h_a = V_a / A_o$ and $h_w = V_w / A_o$ are mean depth of the air and water box, respectively. Inserting into Eq. 21-37 and rearranging the terms yields:

$$\frac{dC_{ia}}{dt} = k_{qa} C_{ia}^{in} - \left(k_{qa} + \frac{k_{ia/w}^{air}}{K_{ia/w}} \right) C_{ia} + k_{ia/w}^{air} C_{iw} \qquad (21\text{-}38a)$$

$$\frac{dC_{iw}}{dt} = k_{qw} C_{iw}^{in} + \frac{k_{ia/w}^{water}}{K_{ia/w}} C_{ia} - \left(k_{qw} + k_{ia/w}^{water} \right) C_{iw} \qquad (21\text{-}38b)$$

The equations are linear and of the same form as Eq. 1 of Box 21.6.

As a special case we consider the situation where the chemical enters the system through the air ($C_{ia}^{in} \neq 0$, $C_{iw}^{in} = 0$). Then the steady-state solution of Eq. 21-38 is:

$$C_{ia}^{\infty} = \frac{k_{qa} \left(k_{qw} + k_{ia/w}^{water} \right) K_{ia/w}}{k_{qa} \left(k_{qw} + k_{ia/w}^{water} \right) K_{ia/w} + k_{ia/w}^{air} k_{qw}} C_{ia}^{in} \qquad (21\text{-}39a)$$

$$C_{iw}^{\infty} = \frac{k_{ia/w}^{water}}{k_{qw} + k_{ia/w}^{water}} \frac{C_{ia}^{\infty}}{K_{ia/w}} \qquad (21\text{-}39b)$$

Note that if the flushing rate in the water compartment is not zero ($k_{qw} > 0$), C_{ia}^{∞} is smaller than C_{ia}^{in}. Furthermore, the concentration ratio at steady-state, $C_{ia}^{\infty} / C_{iw}^{\infty}$, is larger than the chemical equilibrium (which is $K_{ia/w}$). An application of that model is given in Problem 21.6.

Linear Two-Box Model of a Stratified Lake

As a second example of a two-box model we discuss the case of a stratified lake which is divided into the surface layer (epilimnion E, box 1) and the deep-water layer (hypolimnion H, box 2). The model and its parameters are shown in Fig. 21.10. It includes the following processes (numbers as in the figure):

1. Input of compound by inlets and other sources: The parameter η ($0 \leq \eta \leq 1$) defines the fraction of the input going to the hypolimnion

2. Air–water exchange at the water surface

3. Loss of chemical at the outlet of the lake

4. Chemical reaction in the lake; the rate constants in the epilimnion and hypolimnion (k_{irE} and k_{irH}) may be different

5. Loss of chemical to the sediments (sedimentation)

6. Two-way mass exchange across the thermocline (i.e., across the interface between epilimnion and hypolimnion)

The mass balance equations for the epilimnion and hypolimnion look like Eq. 21-38, except for the air–water exchange fluxes which are replaced by the vertical fluxes across the thermocline, T_{iEH} and T_{iHE}. According to the general form of mass transfer models (Eq.18-4), we can express these fluxes as:

$$T_{iEH} = v_{ex} A_{th} C_{iE} ; \qquad T_{iHE} = v_{ex} A_{th} C_{iH} \qquad \left[M\, T^{-1} \right] \qquad (21\text{-}40)$$

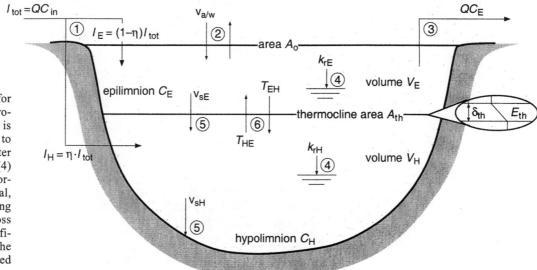

Figure 21.10 Two-box model for stratified lake. The numbered processes are: (1) input by inlets (η is relative fraction of input going to the hypolimnion), (2) air–water exchange, (3) loss at the outlet, (4) loss by *in situ* chemical transformation (chemical, photochemical, biological), (5) flux on settling solid matter, (6) exchange across the thermocline. See text for definition of parameters. Note that the substance subscript i is omitted for brevity.

where v_{ex} is the exchange velocity and A_{th} the thermocline area. Unlike the case of air–water exchange, v_{ex} does not usually depend on the chemical i, since it results from the (turbulent) exchange of fluid elements between the two boxes. As shown in Illustrative Example 19.1, we can either interpret v_{ex} as a volumetric exchange rate of water per unit time, Q_{ex}:

$$v_{ex} = \frac{Q_{ex}}{A_{th}} \qquad \left[L\,T^{-1}\right] \qquad (21\text{-}41)$$

or visualize the thermocline as a bottleneck boundary with thickness δ_{th} and (turbulent) diffusivity E_{th}, which, in analogy to Eq. 19-4, leads to

$$v_{ex} = \frac{E_{th}}{\delta_{th}} \qquad \left[L\,T^{-1}\right] \qquad (21\text{-}42)$$

If the size of v_{ex} cannot be determined by fitting the two-box model to observed concentration profiles, then Eq. 21-42 is the appropriate expression to estimate this model parameter (see Illustrative Example 19.1).

The vertical mass fluxes caused by the settling of suspended matter (process 5 in Fig. 21.10) are relevant whenever the chemical i is either incorporated in the suspended particles or sorbed to their surface. A thorough discussion on particle settling will follow in Chapter 23. At this point we just note that the total flux per unit time leaving the epilimnion can be written as:

$$S_{iE}^{out} = v_{sE}\,A_o\left(1 - f_{iwE}\right)C_{iE} \qquad \left[M\,T^{-1}\right] \qquad (21\text{-}43)$$

where v_{sE} $[L\,T^{-1}]$ is particle-settling velocity in the epilimnion

         A_o $[L^2]$ is surface area of the lake

         f_{iwE} $[-]$ is the dissolved fraction of chemical i in the epilimnion (Eq. 9-12)

Due to the funnel-like shape of the lake basin, only the fraction (A_{th}/A_o) of the flux is received by the hypolimnion while the rest of the material is added to the sediment column which is directly exposed to the epilimnion. Thus, the corresponding input into the hypolimnion is:

$$S_{iH}^{in} = v_{sE}\, A_{th}\, (1 - f_{iwE})\, C_{iE} \tag{21-44}$$

where A_{th} is the lake-cross-section at the thermocline. Finally, the flux from the hypolimnion to the hypolimnetic sediments is

$$S_{iH}^{out} = v_{sH}\, A_{th}\, (1 - f_{iwH})\, C_{iH} \tag{21-45}$$

where the definitions are analogous to Eq. 21-43. All these mass fluxes now can be combined into the mass balance equations of C_{iE} and C_{iH}. The resulting expressions are summarized in Box 21.7.

The equations given in Box 21.7 are general. Often, one or several of the described processes are absent, that is, some of the terms disappear. In this respect Box 21.7 serves as the starting point from which simplifications can be easily made. The procedure is demonstrated in Illustrative Example 21.5, in which we analyze the fate of tetrachloroethene in Greifensee. In addition, in Problem 21.7, we take another look at the distribution of NTA in the same lake.

(*Text continues on page 990*)

Box 21.7   Linear Two-Box Model for Stratified Lake

The mass balance equations for the epilimnion and hypolimnion are (see Eq. 21-35 and Fig. 21.10)

$$\frac{dC_{iE}}{dt} = (1-\eta)\frac{I_{itot}}{V_E} + k_{ia/w}\frac{C_{ia}}{K_{ia/w}} - \left(k_{wE} + k_{ia/w} + k_{isE} + k_{irE} + k_{exE}\right)C_{iE} + k_{exE}\,C_{iH} \tag{1}$$

$$\qquad\qquad\quad (1)\qquad\qquad (2)\qquad\quad (3)\quad (2)\quad\ (5)\ \ (4)\ \ (6)\qquad\qquad (6)$$

$$\frac{dC_{iH}}{dt} = \eta\,\frac{I_{itot}}{V_H} - \left(k_{isH} + k_{irH} + k_{exH}\right)C_{iH} + \left(k_{isH}^* + k_{exH}\right)C_{iE} \tag{2}$$

$$\qquad\quad (1)\qquad\ (5)\quad (4)\quad (6)\qquad\ (5)\quad (6)$$

The numbers below the different terms refer to the processes identified by the same numbers in Fig. 21.10. The various rate constants (all with dimension T^{-1}) are defined by:

$$k_{wE} = \frac{Q}{V_E} \qquad\qquad\qquad\qquad \text{flushing rate of epilimnion}$$

$$k_{ia/w} = f_{iwE}\frac{v_{ia/w}}{h_E} = f_{iwE}\frac{v_{ia/w}\,A_o}{V_E} \qquad \text{air–water exchange; } h_E = V_E/A_o \text{ mean depth of epilimnion}$$

$$k_{isE} = \frac{A_o}{V_E}\,v_{sE}\left(1 - f_{iwE}\right) \qquad\qquad \text{removal with particles from the epilimnion (see Section 23.2)}$$

$$k_{isH}^* = \frac{A_{th}}{V_H} v_{sE}\left(1 - f_{iwE}\right)$$

loss with particles falling into the hypolimnion

$$k_{isH} = \frac{A_{th}}{V_H} v_{sH}\left(1 - f_{iwH}\right)$$

removal with particles from the hypolimnion

$$f_{iwE} k_{irE} \, , f_{iwH} k_{irH}$$

chemical, photochemical, and biological transformations in the epilimnion and hypolimnion, respectively

$$k_{exE} = \frac{v_{ex} A_{th}}{V_E}$$

loss from epilimnion by mixing across thermocline

$$k_{exH} = \frac{v_{ex} A_{th}}{V_H}$$

loss from hypolimnion by mixing across thermocline

Note that in the above definitions we assume that air–water exchange is due to the dissolved fraction of the chemical i, sedimentation is due to the particulate fraction ($[1-f_{iwE}]$, $[1-f_{iwH}]$), and the chemical transformation affects the dissolved fraction only. The latter can be easily extended to include transformations of the sorbed fraction.

The solution of the coupled differential equations follows from Box 21.6 with the following definitions (1 = epilimnion, 2 = hypolimnion):

$$J_1 = \left(1 - \eta\right)\frac{I_{itot}}{V_E} + k_{ia/w}\frac{C_{ia}}{K_{ia/w}}$$

$$J_2 = \eta \frac{I_{itot}}{V_H}$$

$$k_{11} = k_{wE} + k_{ia/w} + k_{isE} + k_{irE} + k_{exE}$$

$$k_{12} = k_{exE}$$

$$k_{21} = k_{isH}^* + k_{exH}$$

$$k_{22} = k_{isH} + k_{irH} + k_{exH}$$

Illustrative Example 21.5   **Tetrachloroethene (PCE) in Greifensee: From the One-Box to the Two-Box Model**

Problem

In Box 21.2 we analyzed the dynamics of PCE in Greifensee and found that the air–water exchange velocity, $v_{ia/w}$, which explains the observed concentration, turned out to be unrealistically small. One reason could be that the lake is stratified during the summer and thus the surface concentration of PCE, which determines the size of both air–water exchange and loss at the outlet, is overestimated. Use the physical characteristics of Greifensee listed in Table 21.1. Total input of PCE is $I_{itot} = 0.90$ mol d^{-1}. The PCE concentration in air is smaller than 10^{-8} mol m^{-3}. Note that the air–water exchange velocity of PCE at $u_{10} = 1$ m s^{-1} is $v_{ia/w} = 4.9 \times 10^{-6}$ m s^{-1} (Table 21.1).

tetrachloroethene
(PCE)

M_i = 165.8 g mol^{-1}

$K_{i\,a/w}$ = 0.73 at 25°C

Calculate the steady-state concentrations of PCE during stratification, C_{iE}^∞ and C_{iH}^∞, for the following two cases: (1) all the input of PCE is in the epilimnion; (2) all the input is in the hypolimnion. Neglect the (small) discharge of water that brings the PCE to the hypolimnion.

Answer

The two cases differ only in the size of the parameter η (case 1: $\eta = 0$; case 2: $\eta = 1$). Most of the terms in the differential equations 1 and 2 of Box 21.7 are zero. The remaining ones are:

$$(1-\eta)\frac{I_{itot}}{V_E} = (1-\eta)\,\frac{0.90\,\text{mol d}^{-1}}{50\times10^6\,\text{m}^3} = (1-\eta)\times1.80\times10^{-8}\,\text{mol m}^{-3}\text{d}^{-1}$$

$$\eta\frac{I_{itot}}{V_E} = \eta\,\frac{0.90\,\text{mol d}^{-1}}{100\times10^6\,\text{m}^3} = \eta\times0.90\times10^{-8}\,\text{mol m}^{-3}\text{d}^{-1}$$

$$k_{wE} = \frac{Q}{V_E} = \frac{0.34\times10^6\,\text{m}^3\text{d}^{-1}}{50\times10^6\,\text{m}^3} = 6.80\times10^{-3}\,\text{d}^{-1}$$

$$k_{i\,a/w} = \frac{v_{i\,a/w}A_o}{V_E} = \frac{4.9\times10^{-6}\,\text{m s}^{-1}\,8.6\times10^6\,\text{m}^2}{50\times10^6\,\text{m}^3} = 8.43\times10^{-7}\,\text{s}^{-1} = 7.28\times10^{-2}\,\text{d}^{-1}$$

$$k_{exE} = \frac{v_{ex}A_{th}}{V_E} = \frac{0.05\,\text{m d}^{-1}\,7.5\times10^6\,\text{m}^2}{50\times10^6\,\text{m}^3} = 7.50\times10^{-3}\,\text{d}^{-1}$$

$$k_{exH} = \frac{v_{ex}A_{th}}{V_E} = \frac{0.05\,\text{m d}^{-1}\,7.5\times10^6\,\text{m}^2}{100\times10^6\,\text{m}^3} = 3.75\times10^{-3}\,\text{d}^{-1}$$

Note that the atmospheric PCE concentration is so small that the input from the atmosphere can be neglected.

The coefficients of the coupled linear differential equations of Box 21.6 become:

$$J_1 = (1-\eta)\times1.8\times10^{-8}\,\text{mol m}^{-3}\text{d}^{-1}$$

$$J_2 = \eta\times0.9\times10^{-8}\,\text{mol m}^{-3}\text{d}^{-1}$$

$$k_{11} = k_{wE} + k_{i\,a/w} + k_{exE} = (6.8+72.8+7.5)\times10^{-3}\,\text{d}^{-1} = 87.1\times10^{-3}\,\text{d}^{-1}$$

$$k_{12} = k_{exE} = 7.5\times10^{-3}\,\text{d}^{-1}$$

$$k_{21} = k_{exH} = 3.75\times10^{-3}\,\text{d}^{-1}$$

$$k_{22} = k_{exH} = 3.75\times10^{-3}\,\text{d}^{-1}$$

Useful quantity: $k_{11}k_{22} - k_{12}k_{21} = 2.99 \times 10^{-4} \, \text{d}^{-2}$

Case a: Input into the epilimnion ($\eta = 0$)

From Box 21.6 with $J_2 = 0$:

$$C_{iE}^{\infty} = \frac{k_{22} J_1}{k_{11} k_{22} - k_{12} k_{21}} = \frac{3.75 \times 10^{-3} \text{d}^{-1} \; 1.8 \times 10^{-8} \text{mol m}^{-3} \text{d}^{-1}}{2.99 \times 10^{-4} \text{d}^{-2}} = 0.23 \times 10^{-6} \text{mol m}^{-3}$$

$$C_{iH}^{\infty} = \frac{k_{21} J_1}{k_{11} k_{22} - k_{12} k_{21}} = C_{iE}^{\infty} \quad (\text{since } k_{22} = k_{21})$$

At first sight, the result is puzzling, and it becomes even more so if we calculate the corresponding steady-state concentration of the one-box model using the same lake parameters. From Eq. 7 of Box 12.1:

$$C_i^{\infty} = \frac{I_{itot} / V}{k_w^* + k_{ia/w}^*} = \frac{0.6 \times 10^{-8} \text{mol m}^{-3} \text{d}^{-1}}{2.3 \times 10^{-3} \text{d}^{-1} + 24.3 \times 10^{-3} \text{d}^{-1}} = 0.23 \times 10^{-6} \text{mol m}^{-3}$$

Note that k_w^* and $k_{ia/w}^*$ are one-third of k_{wE} and $k_{ia/w}$, respectively, since they refer to the total volume instead of the epilimnion volume.

Hence, the one-box and two-box models yield the same result. There is a simple reason for that. Since the only removal processes of PCE act at the lake surface, at steady-state the surface concentration in both models (C_i^{∞} for the one-box model, C_{iE}^{∞} for the two-box model) must attain the same value to compensate for the input $I_{i \text{ tot}}$. Furthermore, since the hypolimnion has neither source nor sink, the net exchange flux across the thermocline must be zero, and this requires $C_{iE} = C_{iH}$.

As discussed below (Illustrative Example 21.6), the real advantage of the two-box model is the description of the transient behavior of the concentration. That is, such models allow us to examine the dynamic change from the initial value to the steady-state, rather than the computation of the steady-state itself.

Case b: Input into the hypolimnion ($\eta = 1$)

From Box 21.6 with $J_1 = 0$:

$$C_{iE}^{\infty} = \frac{k_{12} J_2}{k_{11} k_{22} - k_{12} k_{21}} = \frac{7.5 \times 10^{-3} \text{d}^{-1} \; 0.9 \times 10^{-8} \text{mol m}^{-3} \text{d}^{-1}}{2.99 \times 10^{-4} \text{d}^{-2}} = 0.23 \times 10^{-6} \text{mol m}^{-3}$$

$$C_{iH}^{\infty} = \frac{k_{11} J_2}{k_{11} k_{22} - k_{12} k_{21}} = \frac{87.1 \times 10^{-3} \text{d}^{-1} \; 0.9 \times 10^{-8} \text{mol m}^{-3} \text{d}^{-1}}{2.99 \times 10^{-4} \text{d}^{-2}} = 2.6 \times 10^{-6} \text{mol m}^{-3}$$

The average concentration in the lake at steady-state is:

$$\overline{C_i} = \frac{1}{V}\left(V_E C_{iE}^\infty + V_H C_{iH}^\infty\right) = 1.8 \times 10^6 \, \text{mol m}^{-3}$$

The following table summarizes these results:

(values in 10^{-6} mol m^{-3})	C_{iE}^∞	C_{iH}^∞	$\overline{C_i}$
Measured [a]	–	–	0.55 [a]
Calculated 1-box model [b]	–	–	0.23
2-box model [b]			
input in E	0.23	0.23	0.23
input in H	0.23	2.6	1.8

[a] From this value the air–water exchange velocity was calculated as $v_{ia/w} = 0.15$ m d^{-1}. [b] Calculated with realistic air–water exchange velocity of $v_{ia/w} = 0.42$ m d^{-1}.

One could try to determine η such that the mean concentration $\overline{C_i}$, calculated with the realistic air–water exchange velocity, corresponds to the measured value. In fact, PCE in Greifensee is never at steady-state. Thus, rather than trying to optimize η, in Illustrative Example 21.6 we will analyze the dynamic behavior of the PCE concentration in the lake under the influence of changing stratification regimes.

Linear Two-Box Models with Two and More Variables

From a conceptual point of view, nothing new is needed to further extend the above approach. For instance, the one-box model with two variables shown in Fig. 21.7 can be combined with the two-box (epilimnion/ hypolimnion) model. This results in four coupled differential equations. Even if the equations are linear, it is fairly complicated to solve them analytically. Computers can deal more efficiently with such problems, thus we refrain from adding another example. But we should always remember that independently from how many equations we couple, the solutions of linear models always consist of the sum of a number of exponential terms which have exactly one steady-state, although it may be at infinity. In Section 21.4 we will discuss the general structure of linear differential equations.

Nonlinear Two-Box Models

The world of linear models is mathematically simple and transparent. There is no room for surprises. We could even say (without questioning their analytical power) that linear models are simple-minded and a little bit boring. Natural systems are not linear, although under certain conditions and within a limited range of variation their behavior can be approximated by linear equations. If we include nonlinear elements, the evolving properties of models may become extremely complex and puzzling. They may have multiple steady-states to which the system may be attracted or around which it may circle in endless loops. The direction in which a model moves may strongly depend on its initial state, that is, on the initial size of the variables. As demonstrated by the simple examples of Fig. 21.6, some of the steady-states may be stable and "attract" the system; others are unstable. In systems with three and more dimensions, new types of behavior and steady-state points evolve, such as *strange attractors*. As nicely demonstrated by Lorenz (1963) for the case of modeling the flow in the atmosphere, a set of just three nonlinear differential equations is sufficient to produce *chaotic behavior*. This means that it may be impossible to predict the behavior of the system over longer time periods, since tiny differences in the initial values may lead to very different system states after some time. The interested reader is referred to the vast literature on nonlinear systems (e.g., Arrowsmith and Place, 1992).

Advanced Topic 21.4   Dynamic Properties of Linear Multidimensional Models

In this final section we no longer distinguish between boxes and variables. We just consider the total number of system variables. For instance, a system with four variables (that is, a four-dimensional system) can either describe four chemical species in one box, two chemical species in two boxes, or one species in four boxes. Only linear systems are discussed; multi-dimensional nonlinear systems can be extremely complex and do not allow for a short and concise systematic discussion.

Linear *n*-Dimensional Systems and Their Eigenvalues

In Illustrative Example 21.5 we discussed the behavior of tetrachloroethene (PCE) in a stratified lake. As mentioned before, our conclusions suffer from the assumption that the concentrations of PCE in the lake reach a steady-state. Since in the moderate climate zones (most of Europe and North America) a lake usually oscillates between a state of stratification in the summer and of mixing in the winter, we must now address the question whether the system has enough time to reach a steady-state in either condition (mixed or stratified lake). To find an answer we need a tool like the recipe for one-dimensional models (Eq. 4, Box 12.1) to estimate the time to steady-state for multidimensional systems.

We restrict our discussion to those systems of n linear differential equations that evolve from the construction of mass balance models for one or several chemicals in one or several environmental compartments (boxes). Such systems are always of the form:

$$\frac{dy_1}{dt} = J_1 - k_{11}\, y_1 + k_{12}\, y_2 + \ldots\ldots + k_{1n}\, y_n$$

$$\frac{dy_2}{dt} = J_2 + k_{21}\, y_1 - k_{22}\, y_2 + \ldots\ldots + k_{2n}\, y_n \qquad (21\text{-}46)$$

$$\ldots\ldots$$

$$\frac{dy_n}{dt} = J_n + k_{n1}\, y_1 + k_{n2}\, y_2 + \ldots\ldots - k_{nn}\, y_n$$

All the coefficients $k_{\alpha\beta}$ (where α, β run from 1 to n) are positive or zero. Note that in analogy to the two-dimensional system introduced in Eq. 1 of Box 21.6, we define these equations such that the diagonal coefficients (the $k_{\alpha\beta}$ with $\alpha = \beta$) have a negative sign while all *off-diagonal* coefficients ($\alpha \neq \beta$) have a positive sign. The terms "diagonal" and "off-diagonal" are used since the k-values can be written as a square matrix with size ($n \times n$) in which the $k_{\alpha\alpha}$'s are positioned in the diagonal of the matrix:

$$\mathbf{K} = \begin{pmatrix} -k_{11} & k_{12} & ...... & k_{1n} \\ k_{21} & -k_{22} & ...... & k_{2n} \\ \vdots & & & \\ k_{n1} & k_{n2} & ...... & -k_{nn} \end{pmatrix}, \quad k_{\alpha\beta} \geq 0 \ , \ \alpha, \beta = 1, ...n \qquad (21\text{-}47)$$

In textbooks on linear algebra and dynamical modeling, it is shown that the *eigenvalues* λ of such matrixes are negative or zero (see, e. g., Arrowsmith and Place, 1992). Eigenvalues and eigenfunctions of quadratic matrixes are defined in Box 21.8.

The solutions of Eq. 21-45 are always of the form:

$$y_\alpha(t) = y_\alpha^* + \sum_{\beta=1}^{n} A_{\alpha\beta} \, e^{-k_\beta t}, \quad \alpha = 1......n \qquad (21\text{-}48)$$

Here y_α^* and $A_{\alpha\beta}$ are constant coefficients which depend on the matrix $\mathbf{K}$, on the inhomogeneous terms J_α, and, for the case of the $A_{\alpha\beta}$'s, on the initial values of the variables, y_α°. The k_α are the *negative eigenvalues* of the matrix $\mathbf{K}$. As mentioned above, these eigenvalues are negative or zero, so that the k_α are positive or zero:

$$k_\alpha \geq 0, \quad \alpha = 1..... n \qquad (21\text{-}49)$$

In the context of our discussion, the value of Eq. 21-48 is not to solve Eq. 21-46 explicitly, but to discuss some general properties of the solution. We start with a model for which all J_α are zero. In Box 12.3 we called this model homogeneous.

(*Text continues on page 995*)

Box 21.8   Eigenvalues and Eigenfunctions of Linear Systems

The system of n linear coupled first-order differential equations (Eq. 21-46) can be written as a matrix equation:

$$\frac{d\mathbf{y}}{dt} = \mathbf{J} + \mathbf{K} \cdot \mathbf{y} \qquad (1)$$

where

$$\mathbf{y} = \begin{pmatrix} y_1 \\ y_2 \\ \vdots \\ y_n \end{pmatrix}, \quad \frac{d\mathbf{y}}{dt} = \begin{pmatrix} \dfrac{dy_1}{dt} \\ \dfrac{dy_2}{dt} \\ \vdots \\ \dfrac{dy_n}{dt} \end{pmatrix}, \quad \mathbf{J} = \begin{pmatrix} J_1 \\ J_2 \\ \vdots \\ J_n \end{pmatrix} \qquad (2)$$

The matrix $\mathbf{K}$ is defined by Eq. 21-47. Its *eigenvalues* $\lambda_1, \lambda_2, ... \lambda_n$ are the solution of the determinant equation:

$$\det \begin{pmatrix} (\lambda + k_1) & -k_{12} & \cdots & -k_{1n} \\ -k_{21} & (\lambda + k_{22}) & \cdots & -k_{2n} \\ \vdots & \vdots & \vdots & \vdots \\ -k_{n1} & -k_{n2} & \cdots & (\lambda + k_{nn}) \end{pmatrix} = 0 \tag{3}$$

where det(...) means determinant of (...). Eq. 3 is a polynomial of order n. If $k_{\alpha\beta} \geq 0$, it has the n solutions $\lambda_1, \lambda_2, \ldots \lambda_n$. For simplicity we consider only the case where all λ_α are different (no multiple eigenvalues).

The central idea for solving Eq. 1 can be explained without the details of the mathematical formalism. With an appropriate linear combination of the variables y_α we can define new variables z_α (called *eigenfunctions*) such that the resulting differential equation is:

$$\frac{d\mathbf{z}}{dt} = \mathbf{J}^* + \mathbf{\Lambda} \cdot \mathbf{z} \tag{4}$$

The matrix $\mathbf{\Lambda}$ is diagonal with the eigenvalues λ_α along the diagonal:

$$\mathbf{\Lambda} = \begin{pmatrix} \lambda_1 & 0 & \cdots & 0 \\ 0 & \lambda_2 & \cdots & 0 \\ \vdots & \cdots & \ddots & 0 \\ 0 & 0 & \cdots & \lambda_n \end{pmatrix} \tag{5}$$

$\mathbf{J}^*$ is the vector of the transformed inhomogeneous terms. With the special form of $\mathbf{\Lambda}$, Eq. 4 decays into n uncoupled differential equations:

$$\frac{dz_\alpha}{dt} = J_\alpha^* + \lambda_\alpha z_\alpha \quad , \quad \alpha = 1, \ldots n \tag{6}$$

which can easily be solved with the equations given in Box 12.1:

$$z_\alpha(t) = z_\alpha^o e^{\lambda_\alpha t} - \frac{J_\alpha^*}{\lambda_\alpha}\left(1 - e^{\lambda_\alpha t}\right) \tag{7}$$

For $\lambda_\alpha < 0$, $z_\alpha(t \to \infty)$ converges toward $(-J_\alpha^* / \lambda_\alpha)$. By applying the reverse transformation one gets the original variables $y_\alpha(t)$ from the functions $z_\alpha(t)$.

To demonstrate the power of this seemingly abstract procedure, we apply it to the homogeneous forward/ backward reaction we discussed in Chapter 12 (see Eq. 12-16). By replacing [A] and [D] by y_1 and y_2, respectively, we get:

$$\frac{dy_1}{dt} = -k_1 y_1 + k_2 y_2$$
$$\frac{dy_2}{dt} = k_1 y_1 - k_2 y_2 \tag{8}$$

Thus:

$$\mathbf{K} = \begin{pmatrix} -k_1 & k_2 \\ k_1 & -k_2 \end{pmatrix} \tag{9}$$

Equation 3 becomes:

$$\begin{vmatrix} \lambda + k_1 & -k_2 \\ -k_1 & \lambda + k_2 \end{vmatrix} = (\lambda + k_1)(\lambda + k_2) - k_1 k_2 = \lambda^2 + \lambda(k_1 + k_2) = 0$$

with the solution (eigenvalues) $\lambda_1 = 0$, $\lambda_2 = -(k_1 + k_2)$.

Next, we have to find the new functions z_α whose dynamic equations have the simple form of Eq. 6. There are algebraic recipes for how these eigenfunctions can be derived (see any textbook on linear algebra), but we skip this tedious calculation. Let us assume that we just found out by intuition that the following functions do the job:

$$\begin{aligned} z_1 &= y_1 + y_2 \\ z_2 &= k_1 y_1 - k_2 y_2 \end{aligned} \tag{10}$$

The dynamic equations of z_α can be calculated from Eq. 8:

$$\frac{dz_1}{dt} = \frac{dy_1}{dt} + \frac{dy_2}{dt} = 0 \tag{11a}$$

$$\begin{aligned} \frac{dz_2}{dt} &= k_1 \frac{dy_1}{dt} - k_2 \frac{dy_2}{dt} = k_1(-k_1 y_1 + k_2 y_2) - k_2(-k_1 y_1 - k_2 y_2) \\ &= -(k_1 + k_2)(k_1 y_1 - k_2 y_2) = -(k_1 + k_2) z_2 \end{aligned} \tag{11b}$$

Since $\lambda_1 = 0$, Eq. 11a can also be written as:

$$\frac{dz_1}{dt} = -\lambda_1 z_1 = 0 \tag{12a}$$

and the second equation is:

$$\frac{dz_2}{dt} = -\lambda_2 z_2 \tag{12b}$$

The solution of these equations is:

$$\begin{aligned} z_1(t) &= z_1^o = y_1^o + y_2^o \\ z_2(t) &= z_2^o e^{-\lambda_2 t} = (k_1 y_1^o - k_2 y_2^o) e^{-\lambda_2 t} \end{aligned} \tag{13}$$

The inverse transformation of Eq. 10 is:

$$y_1 = \frac{k_2 z_1 + z_2}{k_1 + k_2}, \qquad y_2 = \frac{k_1 z_1 + z_2}{k_1 + k_2} \tag{14}$$

Inserting the solution for the variables $z_i(t)$ from Eq. 13 yields:

$$y_1(t) = \frac{k_2(y_1^o + y_2^o) + (k_1 y_1^o - k_2 y_2^o)\, e^{-\lambda_2 t}}{k_1 + k_2} \tag{15}$$

which can also be written as:

$$y_1(t) = y_1^o\, e^{-\lambda_2 t} + \frac{k_2}{k_1 + k_2}\, (y_1^o + y_2^o)\left(1 - e^{-\lambda_2 t}\right), \qquad \lambda_2 = k_1 + k_2 \tag{16a}$$

Correspondingly,

$$y_2(t) = y_2^o\, e^{-\lambda_2 t} + \frac{k_1}{k_1 + k_2}\, (y_1^o + y_2^o)\left(1 - e^{-\lambda_2 t}\right) \tag{16b}$$

The steady-state values are

$$y_1^\infty = \frac{k_2}{k_1 + k_2}\left(y_1^o + y_2^o\right), \quad y_2^\infty = \frac{k_1}{k_1 + k_2}\left(y_1^o + y_2^o\right). \tag{17}$$

They are equivalent to the result derived in Chapter 12 (Eqs. 12-21 and 12-22).

Homogeneous system *(all J_α are zero).* An example of a two-dimensional homogeneous system is the first-order reversible reaction between two chemical species discussed in Chapter 12 using the example of the hydration of an aldehyde (see Eq. 12-16). Again, matrix theory provides us with a very useful rule which states that for such systems the resulting matrix is singular (that is, its determinant is zero, see Box 21.8) and thus at least one eigenvalue must be zero. Furthermore, in Eq. 21-48 all y_α^* are zero.

Let us assume that we can number the eigenvalues such that $\lambda_1 = -k_1 = 0$ and all other eigenvalues are negative. Then for large times t, Eq. 21-48 approaches the steady-state

$$y_\alpha(t \to \infty) \equiv y_\alpha^\infty = A_{\alpha 1} \tag{21-50}$$

since all the other terms contain an exponential function of the form $e^{-k_\alpha t}$ which for positive k_α goes to zero if $t \to \infty$.

We have to remember that one does not usually calculate the steady-state values by hand by solving Eq. 21-46 explicitly. However, our general discussion helps to decide whether such steady-states exist and how long it takes to reach them. Since it is the exponential function with the smallest k_α value which decreases most slowly we conclude that, in analogy to Eq. 4 of Box 12.1, the overall time to steady-state of the system is determined by

$$t_{5\%} = \frac{3}{\min\{k_\alpha\}} \tag{21-51}$$

where $\min\{k_\alpha\}$ means the smallest of all k_α values except $k_1 = 0$.

Furthermore, let us assume that the coefficients are numbered such that k_2 is the smallest one. Now it could be that for some variable y_α, the coefficient $A_{\alpha 2}$ which according to Eq. 21-48 links the solution $y_\alpha(t)$ to the particular exponential function, $\exp(-k_2 t)$, is zero or very small. Then for that state variable, y_α, the slow disappearance of exp($-k_2t$) is not relevant for its time to reach steady-state. In other words, while Eq. 21-51 gives an overall time to steady-state, individual variables may react faster. Mathematicians have developed tools to analyze in detail the approach of the system toward its steady-state. Their key instruments are the *eigenfunctions*. Generally, every eigenvalue has its proper eigenfunction. The interested reader will find more in Box 21.8.

Inhomogeneous systems. If Eq. 21-46 is an inhomogeneous system, that is, if at least one J_α is different from zero, then usually all eigenvalues are different from zero and negative, at least if the equations are built from mass balance considerations. Again, the eigenvalue with the smallest absolute size determines time to steady-state for the overall system, but some of the variables may reach steady-state earlier. In Illustrative Example 21.6 we continue the discussion on the behavior of tetrachloroethene (PCE) in a stratified lake (see also Illustrative Example 21.5). Problem 21.8 deals with a three-box model for which time to steady-state is different for each box.

Illustrative Example 21.6

Dynamic Behavior of Tetrachloroethene (PCE) in Greifensee

In Illustrative Example 21.5 we calculated the steady-state concentrations of tetrachloroethene (PCE) in the epilimnion and hypolimnion of Greifensee for two different input situations. In *case a*, all the PCE is put into the surface water (epilimnion) whereas in *case b* the PCE is added only to the hypolimnion. In reality, Greifensee is not stratified during the whole year. Periods of stratification during the warm season are separated by periods of complete mixing (winter). Thus, the lake switches between two distinctly different stages. It seems that the steady-state considerations made in Illustrative Example 21.5 do not adequately reflect the real behavior of Greifensee.

Problem

Calculate the dynamic response of PCE concentrations in Greifensee provided that PCE input occurs only to the hypolimnion (*case b* of Illustrative Example 21.5). Assume that all parameters (flushing rate, air–water exchange velocity, PCE input) are held constant except for vertical mixing. Assume that the latter switches between a period of stagnation (duration 9 months, turbulent velocity $v_{ex} = 0.05$ m d^{-1} as in Table 21.1) and complete mixing (duration 3 months, $v_{ex} \to \infty$). Start the calculation at the beginning of stagnation and assume $C_E^o = C_H^o = 0$. Extend the calculation over two stagnation periods which are interrupted by a mixing period. Does PCE reach an approximate cyclic steady-state? Assess the typical response time during both periods.

Answer

The *negative* eigenvalues of the two-dimensional matrix are given by Eq. 3 of Box 21.6:

$$k_\alpha = \frac{1}{2}\left[(k_{11} + k_{22}) \mp q\right] \quad \text{where} \quad q = \left[(k_{11} - k_{22})^2 + 4k_{12}k_{21}\right]^{\frac{1}{2}}, \quad \alpha = 1,2$$

The box numbers are 1 for the epilimnion and 2 for the hypolimnion. From Illustrative Example 21.5 we get the following numbers (units $10^{-3}d^{-1}$):

$$k_{11} = 87.1, \quad k_{12} = 7.5, \quad k_{21} = 3.75, \quad k_{22} = 3.75$$

Thus, $k_\alpha = \dfrac{1}{2}[90.85 \mp 84.0] \times 10^{-3}d^{-1}$

$$\Rightarrow \quad k_1 = 3.41 \times 10^{-3}d^{-1}; \quad k_2 = 87.4 \times 10^{-3}d^{-1}$$

In order to understand the physical meaning of the eigenvalues we compare them with the rates characterizing the two-box model (see Illustrative Example 21.5).

$k_1 \approx k_{exH}$:   Corresponds to the water exchange rate in the hypolimnion due to vertical exchange with the epilimnion

$k_2 \approx k_{11} = k_{wE} + k_{a/w} + k_{exE}$:   Corresponds to the total removal rate of PCE in the epilimnion due to the combined action of flushing, air–water exchange, and vertical mixing

At the end of stagnation (time $t_{st} = 9$ months $= 270$ d), the exponential functions in Eq. 2 of Box 21.6 are:

$$e^{-k_1 t_{st}} = \exp\left(-3.41 \times 10^{-3} \times 270\right) = e^{-0.92} = 0.40$$

$$e^{-k_2 t_{st}} = \exp\left(-87.4 \times 10^{-3} \times 270\right) = e^{-23.6} = 5.6 \times 10^{-11}$$

Thus, at the end of the stagnation period, the second eigenvalue has completely lost its influence on the behavior of the system. Therefore, it is not necessary to calculate the coefficients A_{12} and A_{22} of Eq. 2 (Box 21.6).

For *case b* of Illustrative Example 21.5 we have:

$$J_1 = 0, \quad J_2 = 9.0 \times 10^{-9} \, \text{mol m}^{-3}\text{d}^{-1}$$

From Eq. 5 (Box 21.6), with $y_1^o = y_2^o = 0$ and with the steady-state values calculated in Illustrative Example 21.5 (*case b*):

$$y_1^\infty = C_E^\infty = 0.226 \, \mu\text{mol m}^{-3}$$
$$y_2^\infty = C_H^\infty = 2.63 \, \mu\text{mol m}^{-3}$$

we get (k_{ij}, q in $[10^{-3}d^{-1}]$, C in $[\mu\text{mol m}^{-3}]$):

$$A_{11} = \frac{1}{q}\left[+\left(k_{11} - k_2\right)C_E^\infty - k_{12} \, C_H^\infty\right] = \frac{1}{84}\left[(87.1 - 87.4) \times 0.226 - 7.5 \times 2.63\right]$$

$$= -0.236 \; (\mu\text{mol m}^{-3})$$

$$A_{21} = \frac{1}{q}\left[-k_{21}\,C_E^{\infty} + \left(k_{22} - k_2\right)C_H^{\infty}\right] = \frac{1}{84}\left[-3.75 \times 0.226 + \left(3.75 - 87.4\right) \times 2.63\right]$$

$$= -2.63 \quad (\mu\text{mol m}^{-3})$$

Thus, at time $t_{st} = 9$ months (end of stagnation) the concentrations are:

$$C_E(t_{st}) = C_E^{\infty} + A_{11}\,e^{-k_1 t_{st}} = (0.226 - 0.236 \times 0.4) \times 10^{-6}\,\text{mol m}^{-3}$$

$$= 0.132\,\mu\text{mol m}^{-3}$$

$$C_H(t_{st}) = C_H^{\infty} + A_{21}\,e^{-k_1 t_{st}} = (2.63 - 2.63 \times 0.40) \times 10^{-6}\,\text{mol m}^{-3}$$

$$= 1.58\,\mu\text{mol m}^{-3}$$

Since it is assumed that mixing sets in at once, the initial concentration in the mixed lake is the weighted average of $C_E(t_{st})$ and $C_H(t_{st})$. From Table 21.1 the relative volume fractions are $V_E/V = 1/3$, $V_H/V = 2/3$, thus:

$$C_{mix}(t_{st}) = \frac{1}{3}C_E(t_{st}) + \frac{2}{3}C_H(t_{st}) = \left(\frac{1}{3}0.132 + \frac{2}{3}1.58\right) \times 10^{-6}\,\text{mol m}^{-3} = 1.10 \times 10^{-6}\,\text{mo}$$

During the circulation period (duration 3 months), the mean concentration changes according to the one-box equation (Eqs. 21-6 to 21-10 and Box 21.2). With $t_c = 3$ months the duration of the circulation period and $t_1 = 1$ yr, we get:

$$C_{mix}(t_1) = C_{mix}(t_{st})\,e^{-kt_c} + (1 - e^{-kt_c})\,C_{mix}^{\infty}$$

The steady-state of the circulation period, C_{mix}^{∞}, is (Eq. 21-10):

$$C_{mix}^{\infty} = \frac{I/Q}{k} = \frac{I/Q}{k_w + k_{a/w}}$$

with $k_w = 2.3 \times 10^{-3}\,\text{d}^{-1}$ (Table 21.1) and

$$k_{a/w} = \frac{v_{a/w}}{h} = \frac{4.9 \times 10^{-6}\,\text{m s}^{-1}}{17.4\,\text{m}} = 2.82 \times 10^{-7}\,\text{s}^{-1} = 24.3 \times 10^{-3}\,\text{d}^{-1}$$

where we have used $v_{a/w} = 4.9 \times 10^{-6}$ m s^{-1} from Table 21.1. Thus:

$$k = k_w + k_{a/w} = 26.6 \times 10^{-3}\,\text{d}^{-1}\,,\quad e^{-kt_C} = e^{-0.0266 \times 90} = e^{-2.39} = 0.091$$

$$C_{mix}^{\infty} = \frac{0.90\,\text{mol d}^{-1}}{150 \times 10^6\,\text{m}^3 \times 0.0266\,\text{d}^{-1}} = 0.226\,\mu\text{mol m}^{-3}$$

$$C(t_1) = \left[1.10 \times 0.091 + (1 - 0.091) \times 0.226\right]\mu\text{mol m}^{-3} = 0.31\,\mu\text{mol m}^{-3}$$

This value can be taken as the initial concentrations for both C_E and C_H for the second stagnation period. We leave it to the reader to check that at the end of the second stagnation period the concentrations are:

$$C_E\left(t_1 + t_{st}\right) = 0.144 \ \mu mol \ m^{-3}$$

$$C_H\left(t_1 + t_{st}\right) = 1.71 \ \mu mol \ m^{-3}$$

$$C_{mix}\left(t_1 + t_{st}\right) = 1.19 \ \mu mol \ m^{-3}$$

Since the concentration at the beginning of the second mixing period is not very different from the concentration one year before, the system has almost reached a cyclic steady-state.

The behavior of the model is graphically summarized in the figure below. At first sight it seems to be paradoxical that during the second stagnation period C_E drops below its steady-state C_E^∞. In fact, this happens because C_E is coupled to C_H which, in turn, remains below C_H^∞ during the whole stagnation period.

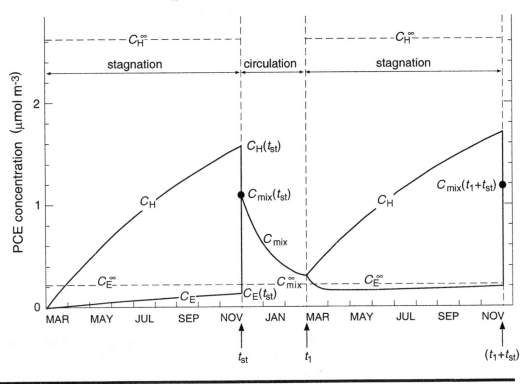

From Box Models to Continuous Models

In the epilimnion/hypolimnion two-box model the vertical concentration profile of a chemical adopts the shape of two zones with constant values separated by a thin zone with an abrupt concentration gradient. Often vertical profiles in lakes and oceans exhibit a smoother and more complex structure (see, e.g., Figs. 19.1a and 19.2). Obviously, the two-box model can be refined by separating the water body into three or more horizontal layers which are connected by vertical exchange rates.

Compound concentrations within the layers would be characterized by the mean layer concentration $C_{i\alpha}$ ($\alpha = 1.....n$ counts the layers), but as the layers get thinner the concentration differences between adjacent layers would become smaller until they blend into a smooth vertical concentration profile.

This approach of subdividing space into an increasing number of discrete pieces provides the basis for many numerical computer models (e.g., the so-called finite difference models); an example will be discussed in Chapter 23. Although these models are extremely powerful and convenient for the analysis of field data, they often conceal the basic principles which are responsible for a given result. Therefore, in the next chapter we will discuss models which are not only continuous in time, but also continuous along one or several space axes. In this context "continuous in space" means that the concentrations are given not only as steadily varying functions in time [$C_i(t)$], but also as functions in space [$C_i(t,x)$ or $C_i(t,x,y,z)$]. Such models lead to partial differential equations. A prominent example is Fick's second law (Eq. 18-14).

21.5 Questions and Problems

Q 21.1

Why does environmental science need models? What can we do with models?

Q 21.2

Give reasons why different models of the same system often predict different "futures", even if these models were constructed based on the same data.

Q 21.3

Why is it important to distinguish between *external* forces and *internal* processes?

Q 21.4

What type of model would you choose to describe the concentration of phenanthrene in a small pond (maximum depth 2m, area 1000 m^2)?

Q 21.5

What type of model would you choose to describe the concentration of phenanthrene in a very large lake (depth more than 100 m, area more than 500 km^2), if the only field data you have are a few concentrations from the surface waters taken during the last 10 years?

Q 21.6

How do you recognize that your one-box model is linear?

Q 21.7

Give some typical properties of linear models.

Q 21.8

You are proud of having built a linear one-box model for the fate of dichloro-difluoromethane (CCl_2F_2, CFC-12) in a lake. Now, your collaborator confronts you with a new data set of atmospheric CFC-12 measurements which show that during the last 10 years the concentration decreased as $a + (b / t)$, where t is time since 1980. Do you have to give up your linear model?

Q 21.9

You have constructed a linear two-box model for tetrachloroethene (PCE) in a lake in which the only input of PCE is from the outlet of a sewage treatment plant. The atmospheric PCE concentration is assumed to be zero in your model. How will the steady-state of the model be altered if the PCE input from sewage is reduced by 27%?

Q 21.10

Why do we have to introduce a convention (such as $t_{5\%}$) to define the response time of a linear model? Are there no "natural" or unique definitions?

Q 21.11

What is a hierarchical n-box model? How can you define it in words, and how mathematically?

Problems

P 21.1 *Search for the Adequate Model to Assess the Possible Influence of PCBs in the Outlet of a Polluted Lake*

During many decades a factory has spilled polychlorinated biphenyls (PCBs) into a small creek which leads to a small lake. Meanwhile the pollution has stopped, yet the local authorities are afraid that PCB concentrations in the outlet of the lake may still be dangerous for the drinking-water supply operating on the aquifer further down-stream. You are asked to make a first guess whether this fear may be substantiated. You decide to use a simple box model. Where would you draw the boundaries of the model and which subcompartments (if any) would you choose?

P 21.2 *Steady-State Concentration of PCE in an Evaporative Lake*

Calculate the steady-state concentration of tetrachloroethene (PCE) in an evaporative lake with constant volume (lake without outlet in which the inflowing water is balanced by evaporation). Assume that the lake is kept completely mixed and use the following information:

PCE input into lake	J	$= 0.2$ mol d^{-1}
Air–water exchange velocity of PCE	$v_{ia/w}$	$= 0.6$ m d^{-1}
Surface area of lake	A_o	$= 10^6$ m^2
Lake volume	V	$= 8 \times 10^6$ m^3
Water inflow	Q	$= 4 \times 10^3$ m^{3}d^{-1}

Do we really need all this information?

P 21.3 *Influence of Input Variation on a Linear System*

Consider a linear one-box model for concentration C. Estimate the maximum tolerable relative input variation A_J of a sinusoidal external force (Eq. 21-12) which keeps the relative deviation of $C(t)$ from the mean value $\overline{C} = C_\infty$ smaller than 10%. The total rate constant of the linear system is $k = 0.3$ yr^{-1}. Assume that the period of the external force is (a) one year, (b) one week.
Hint: Illustrative Example 21.2 may give some help.

P 21.4 *Stability Analysis of Steady-States of Nonlinear Systems*

Analyze the stability of the steady-state $C_1^\infty = 0$ of Figs. 21.6a and b.

P 21.5   *Behavior of Chemical Produced by a Nonlinear Reaction*

Consider a chemical in a completely mixed reactor with a constant water exchange rate Q and volume V. The chemical is produced by a Michaelis-Menten type of auto-catalytic reaction (see Eq. 12-26):

$$\left(\frac{dC}{dt}\right)_{production} = J_r\,\frac{C}{C+C_M}$$

The concentration of the chemical in the inflowing water is zero. Calculate the steady-state(s) of the system and discuss its/their stability.

Numbers: Q = 40 L h^{-1}
           V = 100 L
           J_r = 40 mmol L^{-1}
           C_M = 50 mmol L^{-1} (Michaelis half-saturation concentration)

What happens to the system if the flushing rate is increased to $Q = 100$ L h^{-1}?

Hint: It may help to first draw a qualitative graph which shows the different processes as a function of the concentration in the reactor, C.

i = benzene

P 21.6 *Is There a Danger that the Drinking-Water Supply System of a City is Affected by Air Contaminants?*

An underground cavern of the city drinking-water supply system consists of a water reservoir (area $A_o = 2000$ m^2, depth $h_w = 4$ m) and an air space for maintenance above it (mean height $h_a = 2.5$ m). The flow rate of the water is $Q_w = 1600$ m^3 h^{-1}. The air space is exchanged in 2 hours.

The question arose whether contaminants in the fairly dirty city air could pollute the drinking water by air–water exchange. You remember the two-box model shown in Fig. 21.9 and decide to make a first assessment by using the steady-state solution of this model. As an example you use the case of benzene, which can reach a partial pressure in air of up to $p_i = 10$ ppbv in polluted areas. You use a water temperature of 10°C and the corresponding Henry's law constant $K_{iH} = 3.1$ L bar mol^{-1}. The air–water exchange velocity of benzene under these conditions is estimated as $v_{i\,a/w} = 5 \times 10^{-4}$ cm s^{-1}.

P 21.7 *Two-Box Model for NTA in Greifensee*

In Illustrative Examples 21.1 and 21.2 we have studied the fate of nitrilotriacetic acid (NTA) in Greifensee, especially its *in situ* degradation rate. Now we want to refine our analysis using the two-box model developed in Illustrative Example 21.5.

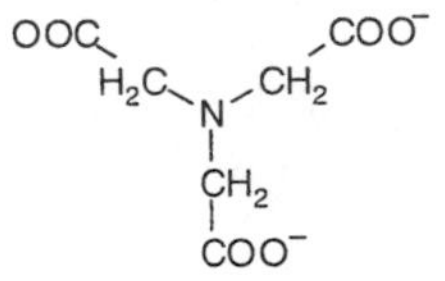

i = nitrilotriacetate
(NTA³⁻)

During the stagnation period, the following NTA concentrations are found in Greifensee:

$$C_E = 5.2 \times 10^{-6} \text{ mol m}^{-3}$$
$$C_H = 2.1 \times 10^{-6} \text{ mol m}^{-3}$$

Assume that the input ($I_{NTA} = 13$ mol d^{-1}) is to the epilimnion only. Assume steady-state conditions and calculate the first-order degradation rates in the epilimnion (k_{rE}) and hypolimnion (k_{rH}) which would explain the measured NTA-concentrations. Use the data from Table 21.1 and Illustrative Example 21.5.

Advanced Topic

P 21.8 *A Conservative Chemical in a Chain of Three Lakes*

Consider a conservative chemical (no air–water exchange, no *in situ* reaction, no removal to the sediments) entering a chain of three lakes. A constant rate of water is flowing into lake 1 and then through lake 2 and 3 into the outlet. The mean residence times in the three lakes are 10 days, 1000 days, and 5 days, respectively. At time $t = 0$, the concentration of the chemical in the inlet to lake 1 suddenly rises from 0 to $C_{in} = 100$ µg L^{-1} and then remains constant. There are no other direct inputs into lakes 2 and 3. Calculate the time to steady-state for each lake.

Hint: Analyze the characteristic rate constant (eigenvalues) for each lake and consider the size of the factor in front of the corresponding exponential functions.

Chapter 22

MODELS IN SPACE AND TIME

22.1 One-Dimensional Diffusion/Advection/Reaction Models
Transport Processes and Gauss' Theorem
One-Dimensional Diffusion/Advection/Reaction Equation
Box 22.1: *One-Dimensional Diffusion/Advection/Reaction Equation at Steady-State*
Peclet Number and Damköhler Number
Box 22.2: *Alternative Definitions of the Damköhler Number*
Illustrative Example 22.1: *Vertical Distribution of Dichlorodifluoromethane (CFC-12) in a Small Lake*

22.2 Turbulent Diffusion
Turbulent Exchange Model
Reynolds' Splitting Model
Vertical Turbulent Diffusion
Illustrative Example 22.2: *Vertical Turbulent Diffusion Coefficient in a Lake*

22.3 Horizontal Diffusion: Two–Dimensional Mixing
Turbulence Theory and the "4/3 Law"
Box 22.3: *Turbulent Diffusion and Spatial Scale*
The Shear Diffusion Model
Box 22.4: *The Shear Diffusion Model*
Illustrative Example 22.3: *A Patch of the Pesticide Atrazine Below the Surface of a Lake*

22.4 Dispersion (*Advanced Topic*)
Dispersion and the Fickian Laws
The Dispersion Coefficient
Illustrative Example 22.4: *Transport of a Volatile Compound From the Groundwater Through the Unsaturated Zone Into the Atmosphere— Illustrative Example 19.2 Reconsidered*
Concluding Remarks

22.5 Questions and Problems

In this chapter models are discussed in which the model variables (e.g., the concentrations C_i) depend *on time and on space*. Solving these differential equations analytically is difficult and outdated in the age of powerful computers. Thus, the following discussion on the analytical solution of these models is not meant as a substitute for computer models. To the contrary, every scientist should be able to handle computer models, although he or she may not know all the mathematical and numerical details which have been put into the construction of these models. However, the interpretation of the results of such models becomes easier if one has some understanding of the properties of the applied concepts. The main mechanisms are *transformation*, *directed transport* (advection, flow, settling of particles), and *random transport* (diffusion, dispersion). In this chapter we develop simple tools for assessing the relative importance of these processes. Such knowledge helps to design and make optimal use of numerical models.

During the last several decades, the field of three-dimensional space-time modeling has undergone a rapid development. On the one hand, the dramatic increase of cheap and fast computing facilities has created new possibilities which scientists could have only dreamed of in the sixties and before. On the other hand, the increasing interest in the fluid systems of the earth, atmosphere, and ocean, triggered by man's concern regarding global climate change, led to the construction of sophisticated coupled atmosphere-ocean models. In this chapter we discuss a few basic elements from which these models are built and look at the characteristics of their solutions. Most of the discussion will be restricted to *one* spatial dimension. In fact, in most cases the emerging properties of one-dimensional models are sufficient to demonstrate the relevant features of more complex situations. Again we will omit the substance-specific subscript i wherever the context is clear.

22.1   One-Dimensional Diffusion/Advection/Reaction Models

In Section 21.1 we discussed the simultaneous influence of transport and transformation processes on the spatial distribution of a chemical in an environmental system. As an example we used the case of phenanthrene in the surface water of a lake. In Fig. 21.2*b* two situations were distinguished which differed by the relative importance of the rate of vertical mixing versus the rate of photolysis. Yet, neither was a quantitative method given to calculate the resulting vertical concentration profile (profiles 1 and 2 in Fig. 21.2*b*), nor did we explain how the rates of such diverse mechanisms as diffusion, advection, and photolysis should be compared in order to calculate their relative importance. In this section we will develop the mathematical tools which are needed for dealing with such situations.

Transport Processes and Gauss' Theorem

Recall the distinction between advective and diffusive transport, which we made in Section 18.1 while traveling in the dining car through the Swiss Alps. We then introduced Fick's first law to describe the mass flux per unit area and time by diffusion or by any other random process (Eq. 18-6). Rewritten in terms of partial derivatives, the diffusive flux along the *x*-axis is:

$$F_x^{\text{diff}} = -D\frac{\partial C}{\partial x} \quad \left[\text{M L}^{-2}\text{T}^{-1}\right] \tag{22-1}$$

where D is diffusivity. The superscript *diff* indicates that the flux is by diffusion.

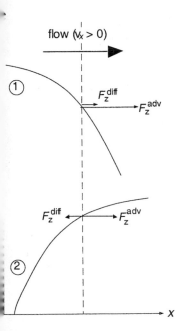

Now we need the corresponding expression for advective transport. Note that the advective velocity along the x-axis, v_x, can be interpreted as a volume flux (of water, air, or any other fluid) per unit area and time. Thus, to calculate the flux of a dissolved chemical we must multiply the fluid volume flux with the concentration of the chemical, C:

$$F_x^{\text{adv}} = v_x C \quad \left[\text{M L}^{-2}\text{T}^{-1}\right] \tag{22-2}$$

Note the difference between Eqs. 22-1 and 22-2 regarding the sign (that is, the direction) of the flux. Since C is always positive or zero, the sign of F_x^{adv} is determined by the sign of the velocity v_x. All chemicals in the fluid are transported in the same direction, which is given by the direction of the fluid flow. In contrast, the sign of F_x^{diff} is determined by the sign of the concentration gradient since D is positive (see Fig. 18.3). Thus the signs of the diffusive fluxes of two compounds in the same system can be different if their concentration profiles are different. Figure 22.1 gives two examples.

Figure 22.1 Relation between concentration profile and flux by diffusion and advection, respectively. For profile 1 both fluxes are positive; for profile 2 they have opposite signs.

In Chapter 18 we derived the *Gauss' theorem* (Eq. 18-12), which allows us to relate a general mass flux, F_x, to the temporal change of the local concentration, $\partial C/\partial t$ (see Fig. 18.4). Applying this law to the total flux,

$$F_x^{\text{tot}} = F_x^{\text{adv}} + F_x^{\text{diff}} = v_x C - D\frac{\partial C}{\partial x} \tag{22-3}$$

yields (with $v_x, D = \text{const.}$):

$$\left(\frac{\partial C}{\partial t}\right)_{\text{transport}} = -\frac{\partial F_x^{\text{tot}}}{\partial x} = -v_x\frac{\partial C}{\partial x} + D\frac{\partial^2 C}{\partial x^2} \quad \left[\text{M L}^{-3}\text{T}^{-1}\right] \tag{22-4}$$

In Fig. 22.2, the profiles of Fig. 22.1 are analyzed in terms of the sign of the two terms on the far right-hand side of Eq. 22-4. Note that for profile 1 the two terms have opposite signs while for profile 2 they are both negative.

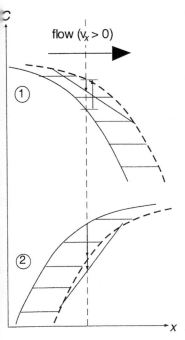

Figure 22.2 Relation between concentration profile and temporal concentration change, $\partial C/\partial t$, for the same profiles as shown in Fig. 22.1. As indicated by the straight lines, diffusion always shifts a concentration profile to its concave side.

The dependency of the pairs, *advection/diffusion* and *flux/concentration change*, on the concentration profile $C(x)$ can be represented in a two-dimensional scheme which helps to remember these relationships (Table 22.1). Note that in this scheme every move to the right (from flux to concentration change) or downward (from advection to diffusion) involves a sign change as well as an additional differentiation of C with respect to x.

One-Dimensional / Diffusion / Advection / Reaction Equation

To extend the transport equation (22-4) to other processes affecting the spatial distribution of a chemical we introduce a zero-order production rate, J [M L^{-3}T^{-1}], and a first-order decay rate (specific reaction rate k_r [T^{-1}]):

Table 22.1 Scheme to Describe Flux and Temporal Concentration Change Due to Advection and Diffusion (or Dispersion)

	Flux F $[ML^{-2}T^{-1}]$	Concentration Change $\partial C / \partial t$ $[ML^{-3}T^{-1}]$
Advection	vC	$-v\dfrac{\partial C}{\partial x}$
Diffusion	$-D\dfrac{\partial C}{\partial x}$	$+D\dfrac{\partial^2 C}{\partial x^2}$

v: Advection velocity $[LT^{-1}]$
D: Diffusivity $[L^2T^{-1}]$
C: Concentration $[ML^{-3}]$

$$\left(\frac{\partial C}{\partial t}\right)_{\text{reaction}} = J - k_{\text{r}} C \tag{22-5}$$

Combining Eqs. 22-4 and 22-5 yields the transport/reaction equation:

$$\frac{\partial C}{\partial t} = \left(\frac{\partial C}{\partial t}\right)_{\text{transport}} + \left(\frac{\partial C}{\partial t}\right)_{\text{reaction}} = D\frac{\partial^2 C}{\partial x^2} - v_x\frac{\partial C}{\partial x} - k_{\text{r}}C + J \tag{22-6}$$

This is a *second-order linear partial differential equation*. Note that the transport terms (Eq. 22-4) are linear *per se*, while the reaction term (Eq. 22-5) has been intentionally restricted to a linear expression. For simplicity, nonlinear reaction kinetics (see Section 21.2) will not be discussed here. For the same reason we will not deal with the time-dependent solution of Eq. 22-6; the interested reader is referred to the standard textbooks (e.g. Carslaw and Jaeger, 1959; Crank, 1975).

We will now discuss the steady-state solutions of Eq. 22-6. Remember that steady-state does not mean that all individual processes (diffusion, advection, reaction) are zero, but that their combined effect is such that at every location along the *x*-axis the concentration *C* remains constant. Thus, the left-hand side of Eq. 22-6 is zero. Since at steady-state time no longer matters, we can simplify $C(x,t)$ to $C(x)$ and replace the partial derivatives on the right-hand side of Eq. 22-6 by ordinary ones:

$$D\frac{d^2 C}{dx^2} - v_x\frac{dC}{dx} - k_{\text{r}}C + J = 0 \tag{22-7}$$

In contrast to all the differential equations which we have discussed in Chapter 21, here the state variable *C* is not treated as a function of time *t* but of space *x*. For the solution of Eq. 22-7, this is irrelevant, except for the type of boundary conditions that can

Table 22.2 Solution of Diffusion/Advection/Reaction Equation at Steady-State: A_1 and A_2 of Eq. 22-9 for Different Types of Boundary Conditions

Boundaries at $x = 0$ and $x = L$. Boundary conditions are given either as values or as spatial derivatives (see Eqs. 22-8a, b). See Box 22.1 for definition of variables.

Given Boundary Conditions[a]	A_1	A_2
(1) C_o, C_L	$\dfrac{\left(C_L - \dfrac{J}{k_r}\right) - \left(C_o - \dfrac{J}{k_r}\right)e^{\lambda_2 L}}{e^{\lambda_1 L} - e^{\lambda_2 L}}$	$\dfrac{-\left(C_L - \dfrac{J}{k_r}\right) + \left(C_o - \dfrac{J}{k_r}\right)e^{\lambda_1 L}}{e^{\lambda_1 L} - e^{\lambda_2 L}}$
(2) C_o', C_L'	$\dfrac{C_L' - C_o' e^{\lambda_2 L}}{\lambda_1(e^{\lambda_1 L} - e^{\lambda_2 L})}$	$\dfrac{-C_L' + C_o' e^{\lambda_1 L}}{\lambda_2(e^{\lambda_1 L} - e^{\lambda_2 L})}$
(3) C_o, C_o'	$\dfrac{C_o' - \lambda_2\left(C_o - \dfrac{J}{k_r}\right)}{\lambda_1 - \lambda_2}$	$\dfrac{-C_o' + \lambda_1\left(C_o - \dfrac{J}{k_r}\right)}{\lambda_1 - \lambda_2}$
(4) C_o, C_L'	$\dfrac{C_L' - \left(C_o - \dfrac{J}{k_r}\right)\lambda_2 e^{\lambda_2 L}}{\lambda_1 e^{\lambda_1 L} - \lambda_2 e^{\lambda_2 L}}$	$\dfrac{\pm C_L' + \left(C_o - \dfrac{J}{k_r}\right)\lambda_1 e^{\lambda_1 L}}{\lambda_1 e^{\lambda_1 L} - \lambda_2 e^{\lambda_2 L}}$

If the system is unbounded on one side (e.g., $L \to \infty$), only one boundary condition is needed, C_o or C_o':

(5) C_o	0	$C_o - J/k_r$
(6) C_o'	0	C_o'/λ_2

[a] The two omitted combinations (C_L, C_L' and C_o', C_L) can be derived by redefining the coordinate x such that $x = 0$ becomes $x = L$, and vice versa.

occur. The solution of Eq. 22-7 requires two boundary conditions, that is, two explicit values of either a concentration, C, or a concentration derivative, dC/dx. If C were considered as a function of time, then the only meaningful choice for the required boundary conditions would be to give the *initial values* at the initial time t_o, $C(t_o=0)$ and $(dC/dt)_{t=t_o}$. Usually, the initial time is defined as $t_o = 0$.

In contrast, if C is considered as a function of x, our choice is greater. Imagine that we are looking for the solution of Eq. 22-7 within a given spatial range along the x-axis, say between $x = 0$ and $x = L$. Then we have six choices to specify the boundary conditions: we can either prescribe the two boundary values,

$$C_o \equiv C(x = 0); \quad C_L \equiv C(x_L) \tag{22-8a}$$

or the two boundary gradients,

$$C'_o \equiv \frac{dC}{dx}\bigg|_{x=0} \quad ; \quad C'_L \equiv \frac{dC}{dx}\bigg|_{x=x_L} \tag{22-8b}$$

or we give a mixture of the two (one value, one gradient). These possibilities influence the solution of Eq. 22-7 and reflect the different real situations for which Eq. 22-7 has to be analyzed. Now we also understand why differential equations in time are simpler. In the physical world of causality it does not make sense to use boundary conditions at a later time to describe the behavior of a system at an earlier time. The future cannot influence the past; teleological concepts have been banned from modern science.

But let us go back to Eq. 22-7. As shown in Box 22.1, its general solution can be written in the form:

$$C(x) = \frac{J}{k_r} + A_1 e^{\lambda_1 x} + A_2 e^{\lambda_2 x} \tag{22-9}$$

where A_1 and A_2 are determined by the model parameters (D, v_x, k_r, J) and by the boundary conditions (Table 22.2). λ_1, λ_2 are the *eigenvalues* of Eq. 22-7. They are determined by D, v_x, and k_r alone and depend neither on the inhomogeneous term J nor on the boundary conditions.

Box 22.1   One-Dimensional Diffusion/Advection/Reaction Equation at Steady-State

The one-dimensional diffusion/advection/reaction equation at steady-state is (22-7):

$$D\frac{d^2C}{dx^2} - v_x\frac{dC}{dx} - k_r C + J = 0 \tag{22-7}$$

We assume D, $k_r \geq 0$; v_x can have either sign. Substituting C by a new variable, $C^* = C - J/k_r$, the equation becomes homogeneous:

$$D\frac{d^2C^*}{dx^2} - v_x\frac{dC^*}{dx} - k_r C^* = 0 \tag{1}$$

The general solution of Eq. 1 can be written as a linear combination of two basic functions of the form $e^{\lambda_\alpha x}$ ($\alpha = 1,2$):

$$C^*(x) = A_1 e^{\lambda_1 x} + A_2 e^{\lambda_2 x} \tag{2}$$

where λ_α [L^{-1}] are the *eigenvalues* of Eq. 1. Inserting the basic functions, $e^{\lambda_\alpha x}$, into Eq. 1 yields a quadratic equation for λ_α,

$$D\lambda_\alpha^2 - v_x \lambda_\alpha - k_r = 0 \tag{3}$$

which has the two solutions:

$$\lambda_\alpha = \frac{1}{2D}\left[v_x \pm \{v_x^2 + 4Dk_r\}^{1/2}\right] \quad , \quad \alpha = 1,2 \tag{4}$$

If $v_x \neq 0$, Eq. 4 can be written in the form ($\mathrm{sgn}(v_x) = \pm 1$ is the sign of v_x):

$$\lambda_\alpha = \frac{|v_x|}{2D}\left[\mathrm{sgn}(v_x) \pm \{1 + 4\cdot\mathrm{Da}\}^{1/2}\right] \tag{5}$$

where Da is the nondimensional Damköhler Number:

$$\text{Da} \equiv \frac{D k_r}{v_x^2} \geq 0 \tag{6}$$

Since Da ≥ 0, λ_1 is always positive while λ_2 is negative.

Depending on the size of Da, the eigenvalues can be approximated by:

(a) For Da $\gg 1 \ \leftrightarrow \ v_x^2 \ll D k_r$:

$$\lambda_j = \pm \left(\frac{k_r}{D} \right)^{1/2} \tag{7}$$

(b) For Da $\ll 1 \ \leftrightarrow \ v_x^2 \gg D k_r$, the sign of v_x leads to two different situations:

$$v_x > 0: \qquad \lambda_1 = \frac{v_x}{D}, \qquad \lambda_2 = -\frac{k_r}{v_x} \tag{8a}$$

$$v_x < 0: \qquad \lambda_1 = -\frac{k_r}{v_x}, \qquad \lambda_2 = \frac{v_x}{D} \tag{8b}$$

Convince yourself that λ_1 is always positive and λ_2 is always negative.

For case (a) the eigenvalues have the same absolute size, while for case (b) the role of λ_1 and λ_2 change according to the sign of v_x:

$$\text{If Da} \ll 1: \qquad v_x > 0: \qquad \left| \frac{\lambda_1}{\lambda_2} \right| = \frac{v_x^2}{D k_r} = \frac{1}{\text{Da}} \gg 1 \tag{9a}$$

$$v_x < 0: \qquad \left| \frac{\lambda_1}{\lambda_2} \right| = \frac{D k_r}{v_x^2} = \text{Da} \ll 1 \tag{9b}$$

The coefficients A_1 and A_2 of Eq. 2 are determined by the boundary conditions (see Table 22.2). Thus:

$$C(x) = \frac{J}{k_r} + C^*(x) = \frac{J}{k_r} + A_1 e^{\lambda_1 x} + A_2 e^{\lambda_2 x} \tag{22-9}$$

If $C(x)$ is confined to the finite interval $x = \{0, L\}$, we can rewrite Eq. 22-9 with the nondimensional coordinate $\xi = x / L$ (ξ varies between 0 and 1):

$$C(\xi) = \frac{J}{k_r} + A_1 e^{\lambda_1^* \xi} + A_2 e^{\lambda_2^* \xi} \tag{22-10}$$

The nondimensional eigenvalues λ_α^* are:

$$\lambda_\alpha^* = \lambda_\alpha L = \frac{\text{Pe}}{2} \left[\text{sgn}(v_x) \pm \{1 + 4\text{Da}\}^{1/2} \right] , \quad \alpha = 1, 2 \tag{10}$$

where Pe is the *Peclet Number*: $\qquad\qquad\qquad\qquad\qquad\qquad\qquad\qquad\qquad$ (11)

$$\text{Pe} = \frac{|v_x| L}{D}$$

Peclet Number and Damköhler Number

The size of the eigenvalues λ_α ($\alpha = 1,2$) determines the shape of the profile $C(x)$. To illustrate this point we assume that the profile is bound to the interval $x = \{0, L\}$. As shown in Box 22.1, Eq. 22-9 can then be written as a function of the nondimensional coordinate $\xi = x / L$ which varies between 0 and 1:

$$C(\xi) = \frac{J}{k_r} + A_1 e^{\lambda_1^* \xi} + A_2 e^{\lambda_2^* \xi}, \qquad 0 \leq \xi \leq 1 \tag{22-10}$$

where the nondimensional eigenvalues λ_α^* (Eq. 10 of Box 22.1) can be expressed in terms of two nondimensional numbers:

$$\text{Pe} = \frac{|v_x| L}{D} \geq 0 \qquad \text{Peclet Number} \tag{22-11a}$$

$$\text{Da} = \frac{D\, k_r}{v_x^2} \geq 0 \qquad \text{Damköhler Number} \tag{22-11b}$$

The Peclet Number can be interpreted as the ratio between the transport time over the distance L by diffusion and by advection, respectively. Transport time by diffusion is expressed by the relation of Einstein and Smoluchowski (Eq. 18-8):

$$t_{\text{diff}} \approx \frac{L^2}{D}, \tag{22-12a}$$

where the factor 2 in the denominator of Eq. 18-8 is conventionally omitted. Transport time by advection is simply:

$$t_{\text{adv}} = \frac{L}{|v_x|}, \tag{22-12b}$$

Thus:

$$\frac{t_{\text{diff}}}{t_{\text{adv}}} = \frac{L^2 / D}{L / |v_x|} = \frac{L |v_x|}{D} = \text{Pe} \tag{22-13}$$

In other words: The Peclet Number Pe measures the relative pace of transport by advection and diffusion, respectively. If Pe $\ll$ 1, diffusion is much faster than advection; if Pe $\gg$ 1, advection outruns diffusion. Note that Pe depends on the distance L over which transport takes place. If L is large enough, advection always becomes dominant since the time of advection t_{adv} grows linearly with L whereas t_{diff} grows as L^2. In turn, for extremely small distances diffusion is always dominant.

The situation can also be viewed another way: given specific values for D and v_x, we can calculate the critical distance, L_{crit}, at which advection is as important as diffusion, that is, where Pe = 1. Inserting Pe = 1 into Eq. 22-11a and solving for $L = L_{\text{crit}}$ yields:

$$L_{\text{crit}} = \frac{D}{|v_x|} \tag{22-14}$$

Typical values of L_{crit} for various situations are given in Table 22.3. They clearly

Table 22.3 Critical Distance L_{crit} at Which the Influence of Diffusion and Advection Is Equal (Eq. 22-14) For $L \gg L_{crit}$, transport by diffusion is negligible compared to advective transport.

				Critical Length, L_{crit} [m]				
Diffusivity [a] D (m²s⁻¹)			Advection $\lvert v_x \rvert$ (m s⁻¹)	10^{-4}	10^{-2}	1	10^2	10^3
Molecular	in water	10^{-9}		10^{-5}	10^{-7}	10^{-9}	10^{-11}	10^{-12}
	in air	10^{-5}		10^{-1}	10^{-3}	10^{-5}	10^{-7}	10^{-8}
Turbulent	in water							
	vertical	10^{-6}		10^{-2}	10^{-4}	10^{-6}	10^{-8}	10^{-9}
	horizontal	1		10^4	10^2	1	10^{-2}	10^{-3}
Turbulent in atmosphere		10^2		(10^6)[b]	(10^4)	10^2	1	0.1

[a] Typical sizes of (molecular and turbulent) diffusivities are taken from Table 18.4. [b] Such small advection velocities do not occur in a turbulent atmosphere.

show that molecular diffusion plays a role only if the advection velocity is extremely small or zero.

In a similar way the Damköhler Number (Eq. 22-11b) measures the relative pace of diffusion versus advection in the time period $t_r = k_r^{-1}$, which corresponds to the mean life of the reactive chemical. In fact:

$$\text{Da} = \frac{D k_r}{v_x^2} = \frac{D/k_r}{\left(v_x/k_r\right)^2} = \left(\frac{x_{diff}}{x_{adv}}\right)^2, \qquad (22\text{-}15)$$

where

$$x_{diff} = \left(D/k_r\right)^{1/2} = \left(D t_r\right)^{1/2} \qquad (22\text{-}16)$$

is the diffusion distance in time t_r, and

$$x_{adv} = \frac{v_x}{k_r} = v_x t_r \qquad (22\text{-}17)$$

is advection distance in time t_r. Note that in the scientific literature there are several other definitions of the Damköhler Number (see Box 22.2).

Situations in which either Da or Pe are much larger or much smaller than 1 indicate that in the diffusion-advection-reaction equation some of these processes are dominant while others can be disregarded. Figure 22.3 gives an overview of such cases. A first distinction is made according to the size of Da:

(a) Da $\gg 1$, i.e., $v_x^2 \ll D k_r$: The nondimensional eigenvalues (Eq. 10, Box 22.1) become

$$\lambda_\alpha^* \cong \pm \text{Pe Da}^{1/2} = \pm L \left(\frac{k_r}{D}\right)^{1/2} \qquad (22\text{-}18)$$

Box 22.2   Alternative Definitions of the Damköhler Number

Several authors (e.g., Domenico and Schwartz, 1998) distinguish between two (or more) Damköhler Numbers:

$$\text{Da}_{\text{adv}} = \frac{k_r L}{|v_x|} = \frac{L/|v_x|}{t_r} \tag{1}$$

is the ratio of the time needed for advection over distance L vs. the mean life of the chemical due to reaction, $t_r = k_r^{-1}$. In contrast,

$$\text{Da}_{\text{diff}} = \frac{k_r L^2}{D} = \frac{L^2/D}{t_r} \tag{2}$$

is the ratio of diffusion time vs. reaction time. Note that:

$$\text{Pe} = \frac{\text{Da}_{\text{diff}}}{\text{Da}_{\text{adv}}} \tag{3}$$

The Damköhler Number used here (Eq. 22-11b) is related to Da_{adv} and Da_{diff} by:

$$\text{Da} = \frac{\text{Da}_{\text{adv}}^2}{\text{Da}_{\text{diff}}}$$

Note that the advection velocity, v_x, no longer appears in this expression. Therefore, this situation is called the *diffusion-reaction regime*.

Remember that in Eq. 22-10 the variation of ξ is restricted to the range $\{0, 1\}$. Thus, the largest absolute value which the argument in the exponential functions of Eq. 22-10 can assume is $|\lambda_\alpha^*| = L (k_r / D)^{1/2}$. If the latter is much smaller than 1, that is, if:

$$L(k_r/D)^{1/2} \ll 1 \quad \leftrightarrow \quad k_r \ll \frac{D}{L^2} \tag{22-19}$$

then $e^{\lambda_\alpha^* \xi} = e^{\lambda_\alpha x}$ can be approximated by $(1 + \lambda_\alpha x)$. Equation 22-9 becomes:

$$
\begin{aligned}
C(x) &\approx \frac{J}{k_r} + A_1 \left(1 + \lambda_1 x\right) + A_2 \left(1 + \lambda_2 x\right) \\
&= \frac{J}{k_r} + A_1 \left(1 + \lambda x\right) + A_2 \left(1 - \lambda x\right) \\
&= \frac{J}{k_r} + A_1 + A_2 + \lambda \left(A_1 - A_2\right) x
\end{aligned}
\tag{22-20}
$$

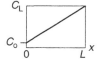

where $\lambda = (k_r/D)^{1/2}$. Note that Eq. 22-20 is a linear function. The concentration profile is a straight line which connects the two boundary values at $x = 0$ and $x = L$

Figure 22.3 One-dimensional concentration profiles at steady-state calculated from the diffusion/advection/reaction equation (Eq. 22-7) for different parameter values D (diffusivity), v_x (advection velocity), and k_r (first-order reaction rate constant). Boundary conditions at $x = 0$ and $x = L$ are C_o and C_L, respectively. $Pe = L|v_x|/D$ is the Peclet Number, $Da = Dk_r/v_x^2$ is the Damköhler Number. See text for further explanations.

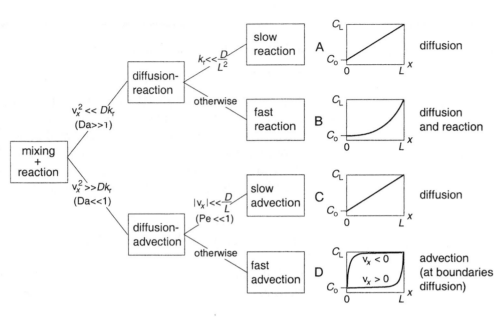

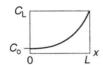

(see Fig. 22.3, case A). The profile is purely diffusive; neither advection nor reaction plays a significant role. If Eq. 22-19 does *not* hold, but Da is still large, the profile adopts an exponential slope (Fig. 22.3, case B); this is the diffusion/reaction case.

(b) Da $\ll 1$, i.e., $v_x^2 \gg Dk_r$: As shown in Box 22.1, now the absolute size of the two eigenvalues is very different. Which one of the two, λ_1 or λ_2, has a larger absolute size depends on the sign of v_x (see Box 22.1, Eq. 8), but in both cases the larger eigenvalue has the value v_x/D.

As before we consider a finite range along the x-axis and ask for the condition to keep the absolute values of the exponents in Eq. 22-19 significantly below 1. This is the case if:

$$\left| L \frac{v_x}{D} \right| \ll 1 \quad \leftrightarrow \quad |v_x| \ll \frac{D}{L} \tag{22-21}$$

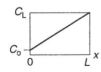

Since according to Eq. 9 of Box 22.1 the absolute value of the other eigenvalue is even smaller, we can replace both exponential functions by their linear approximations. Thus, Eq. 22-9 turns into Eq. 22-20 indicating a linear concentration profile. We call this the case of slow advection (Fig. 22.3, case C).

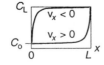

In contrast, if Eq. 22-21 does not hold, we have the situation of fast advection (Fig. 22.3, case D). Note that in both cases (C and D) the reaction rate k_r has no significant influence on the slope of the profile.

A first application of Eq. 22-7 is given in Illustrative Example 22.1.

(*Text continues on page 1019*)

Illustrative Example 22.1   **Vertical Distribution of Dichlorodifluoromethane (CFC-12) in a Small Lake**

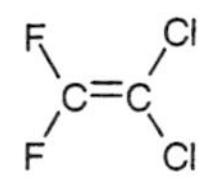

Problem

Dichlorodifluoromethane (CCl_2F_2, CFC-12) enters a small lake (surface area $A_o = 2 \times 10^4$ m^2, maximum depth $z_m = 10$ m) from the atmosphere by air–water exchange. The top 2 m of the lake are well mixed. Vertical turbulent diffusivity between 2 and 10 m is estimated to be $E_z = 1 \times 10^{-5}$ m^2 s^{-1} (more about turbulent diffusion in Section 22.2). Groundwater infiltrates at the bottom of the lake adding fresh water at the rate of $Q_{gw} = 100$ L s^{-1}. The only outlet of the lake is at the surface.

The CFC-12 concentration in the mixed surface water is $C_o = 10 \times 10^{-12}$ mol L^{-1} and below detection limit in the infiltrating groundwater.

(a) Calculate the vertical profile of CFC-12 between 2 and 10 m depth at steady-state provided that all relevant processes (turbulent mixing, discharge rate of groundwater, mixed-layer concentration) remain constant.

(b) How long would it take for the profile to reach steady-state?

(c) Calculate the vertical flux of CFC-12 at 2 and 6 m depth.

(d) Somebody claims that CFC-12 might not be stable in the water column. To check this possibility you compare your model with a vertical CFC-12 profile measured in the lake. How big would a hypothetical first-order reaction rate constant, k_r, have to be in order to be detected by your model? Assume that the absolute accuracy of your CFC-12 analysis is $\pm$ 10%.

Note: Disregard the variable bathymetry of the lake. Assume that the cross-sectional area of the lake is A_o at all depths. See Problem 22.4 for variable cross-sectional area.

Answer (a)

To calculate the vertical profile of CFC-12, define a vertical coordinate z which is oriented downward such that $z = 0$ at 2 m depth (lower boundary of mixed layer) and $z = L = 8$ m at the bottom of the lake. At steady-state, the profile can be calculated from Eq. 22-7 with k_r, $J = 0$ and D replaced by E_z:

$$E_z \frac{d^2C}{dz^2} - v_z \frac{dC}{dz} = 0$$

The vertical advection (upwelling) velocity is:

$$v_z = -\frac{Q_{gw}}{A_o} = -\frac{100 \text{ L s}^{-1}}{2 \times 10^4 \text{ m}^2} = -\frac{0.1 \text{ m}^3 \text{s}^{-1}}{2 \times 10^4 \text{ m}^2} = -0.5 \times 10^{-5} \text{ m s}^{-1}$$

Since the Damköhler Number Da = 0 (Eq. 22-11b), the eigenvalues are (Eq. 4, Box 22.1):

$$\lambda_1 = \frac{v_z}{E_z} = -\frac{0.5 \times 10^{-5} \text{ m s}^{-1}}{1 \times 10^{-5} \text{ m}^2 \text{s}^{-1}} = -0.5 \text{ m}^{-1}; \qquad \lambda_2 = 0$$

With $L = 8$ m, the Peclet Number is (Eq. 22-11a):

$$\text{Pe} = \frac{L|v_z|}{E_z} = \frac{8 \text{ m} \cdot 0.5 \times 10^{-5} \text{m s}^{-1}}{1 \times 10^{-5} \text{m}^2 \text{s}^{-1}} = 4$$

indicating an intermediate advection regime (Fig. 22.3).

To calculate the profile you need two boundary conditions. The first one is given by $C(z = 0) = C_o = 10 \times 10^{-12}$ mol L^{-1}. The second one is less obvious. You observe that no CFC-12 is entering the lake at the bottom (nor at another depth). If CFC-12 is nonreactive, the sum of the vertical advective and diffusive fluxes anywhere in the profile must be zero. Thus, from Eq. 22-3:

$$F_z^{\text{tot}} = v_z C - E_z \frac{\partial C}{\partial z} = 0$$

This equation also must be valid at $z = 0$:

$$\left.\frac{\partial C}{\partial z}\right|_{z=0} \equiv C_o' = \frac{v_z}{E_z} C_o = \lambda_1 C_o$$

This is the second boundary condition. The two conditions correspond to case 3 of Table 22.2. Since $J = 0$, $\lambda_2 = 0$, you get:

$$A_1 = \frac{C_o'}{\lambda_1} = C_o ; \qquad A_2 = \frac{1}{\lambda_1}\left(-C_o' + \lambda_1 C_o\right) = 0$$

Thus, from Eq. 22-9 the profile has a simple exponential shape:

$$C(z) = C_o e^{\lambda_1 z} = (10 \times 10^{-12} \text{mol L}^{-1}) \exp(-0.5\, z)$$

Note that at the bottom ($z = L = 8$ m), the concentration has dropped to $C(8 \text{ m}) = (10 \times 10^{-12} \text{ mol L}^{-1}) e^{-4} = (10 \times 10^{-12} \text{ mol L}^{-1})\, 0.018 = 0.18 \times 10^{-12} \text{ mol L}^{-1}$.

Answer (b)

Since Pe > 1, advection is more effective to distribute the chemical in the water column (see Eq. 22-13). Diffusion would then just be needed for the local adjustment of the profile. Thus:

$$t_{\text{adv}} = \frac{L}{|v_z|} = \frac{8 \text{ m}}{0.5 \times 10^{-5} \text{m s}^{-1}} = 1.6 \times 10^6 \text{s} \quad \text{or about } 19 \text{ days}$$

Answer (c)

As mentioned before, the vertical net flux is zero throughout the profile.

Answer (d)

It is helpful to make some approximate considerations before going too much into the details of this question. First note that even without degradation of CFC-12, the concentration at the lake bottom has dropped to less than 2% of the value in the mixed layer. Obviously, any decomposition of CFC-12 would reduce the vertical penetration of CFC-12 even more. Thus, you can treat the problem as if the lower boundary did not exist. Then you are left with case 5 of Table 22.2 with $v_z < 0$. The "reactive" profile, C_r, has the form:

$$C_r(z) = C_o e^{\lambda_r z}$$

where λ_r is the negative eigenvalue. From Eq. 4 of Box 22.1:

$$\lambda_r = -\frac{v_z}{2E_z}\left(1 + \{1 + 4Da\}^{1/2}\right) = \frac{\lambda_1}{2}(1 + q)$$

where $\lambda_1 = -0.5$ m^{-1} is the eigenvalue of the nonreactive case and $q = \{1+Da\}^{1/2}$.

Both profiles, $C(z)$ and $C_r(z)$, are pure exponentials. The relative difference between both curves,

$$\Delta \equiv \frac{C(z) - C_r(z)}{C(z)} = 1 - e^{(\lambda_r - \lambda_1)z}$$

becomes largest at the bottom of the lake ($z = 8$ m). Taking $\Delta = 0.1$ as the criterion that both profiles are different, you get

$$\lambda_r - \lambda_1 = \frac{1}{z}\ln 0.9 = -0.013 \text{ m}^{-1}$$

In turn,

$$\lambda_r - \lambda_1 = \frac{\lambda_1}{2}(q - 1) = -0.25 \text{ m}^{-1}\left(\{1 + Da\}^{1/2} - 1\right)$$

Thus, $\{1 + Da\}^{1/2} = 1.052$. Finally

$$Da = \frac{E_z k_r}{v_z^2} \cong 0.1$$

The critical reaction rate constant that CFC-12 would have to exceed in order to make the profile different from the conservative curve is:

$$k_r \geq 0.1 \frac{v_z^2}{E_z} = 0.02 \text{ d}^{-1}$$

22.2 Turbulent Diffusion

In Section 18.5 we discussed the distinction between laminar and turbulent flow. The Reynolds Number Re (Eq. 18-69) serves to separate the two flow regimes. Large Reynolds Numbers indicate turbulent flow.

It is difficult to define turbulence. Intuitively, we associate it with the fine-structure of the fluid motion, as opposed to the flow pattern of the large-scale currents. Although it is not possible to describe exactly the distribution in space and time of this small-scale motion, we can characterize it in terms of certain statistical parameters such as the variance of the current velocity at some fixed location. A similar approach has been adopted to describe the motion at the molecular level. It is not possible to describe the movement of some "individual" molecule, but groups of molecules obey certain characteristic laws. In this way the individual behavior of many molecules sums to yield the average motion in response to macroscopic forces.

In light of this analogy, we anticipate that the effect of turbulence may be dealt with in a similar manner like the random motion of molecules for which the gradient-flux law of diffusion (Eq. 18-6) has been developed. In addition, the mass transfer model (Eq. 18-4) may provide an alternative tool for describing the effect of turbulence on transport.

Turbulent Exchange Model

Let us start with the exchange model and assume that the concentration of a compound along the x-axis, $C(x)$, is influenced by turbulent velocity fluctuations. For simplicity the mean velocity is assumed to be zero. The effect of the fluctuations can be pictured by the occasional exchange of water parcels across a virtual plane located at x_0 (Fig. 22.4). Owing to the continuity of water flow, any water transport in the positive x-direction has to be compensated for by a corresponding flow in the negative x-direction. In a simplified way we can thus visualize the effect of turbulence as occasional exchange events of water volumes q_{ex} over distance L_x across the interface x_0. A single exchange event causes a net compound mass transport of the magnitude $(C_1 - C_2) q_{ex}$. The concentration difference $(C_1 - C_2)$ can be approximated by the linear gradient of the concentration curve, $(-L_x \, \partial C / \partial x)$. Summation over all exchange events across some area Δa within a given time interval Δt yields the mean turbulent flux of the compound per unit area and time:

$$F_{\text{turb}, x} = -\frac{\Sigma L_x q_{ex}}{\Delta a \, \Delta t} \frac{\partial C}{\partial x} = -E_x \frac{\partial C}{\partial x} \quad \left[\text{M L}^2\text{T}^{-1} \right] \quad (22\text{-}22)$$

The coefficient E_x is called the turbulent (or eddy) diffusion coefficient; it has the same dimension as the molecular diffusion coefficient $[\text{L}^2\text{T}^{-1}]$. The index x indicates the coordinate axis along which the transport occurs. Note that the turbulent diffusion coefficient can be interpreted as the product of a mean transport distance $\overline{L_x}$ times a mean velocity $\overline{v} = (\Delta a \, \Delta t)^{-1} \Sigma q_{ex}$, as found in the random walk model, Eq. 18-7.

Reynolds' Splitting Model

Though Eq. 22-22 may be adequate as a qualitative model for turbulent diffusion, it is certainly not suitable for quantifying E_x or relating it to measurable quantities such

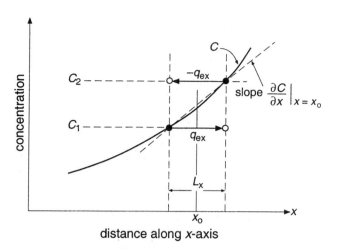

Figure 22.4 Exchange model for mass transport by turbulent diffusion. C is mean concentration along the x-axis. Turbulence causes "exchange events" of water (q_{ex}) over distance L_x. The resulting net mass transport is proportional to the mean slope, $\partial C/\partial x$ (see Eq. 22-1).

as current velocities. A different model going back to Reynolds (1894) and Schmidt (1917) casts more light on the concept of turbulent diffusion. The central idea is to distinguish turbulence from mean motion by isolating the temporal (or spatial) fine structure of any field variable f, such as current velocity, v_x, or concentration, C. This is done by separating f into a mean value $\bar{f}$ and a residual fluctuation f':

$$f(t) = \bar{f} + f'(t) \tag{22-23}$$

Here f' is the time average of the variable measured at some fixed point over the time interval s:

$$\bar{f} = \frac{1}{s} \int_{t-s/2}^{t+s/2} f(t')dt' \tag{22-24}$$

The residual $f'(t)$ varies with time since sometimes $f(t)$ is bigger than $\bar{f}$ and sometimes it is smaller. Note that the average of all deviations of the instantaneous value $f(t)$ from its mean $\bar{f}$ is zero:

$$\overline{f'} = \overline{(f - \bar{f})} = \bar{f} - \bar{f} = 0 \tag{22-25}$$

The fluctuation model, Eq. 22-23, can be employed to describe advective transport, Eq. 22-2. For simplicity, we restrict ourselves to the x-component:

$$F_x^{adv} = v_x C = \left(\overline{v_x} + v_x'\right)\left(\overline{C} + C'\right) = \overline{v_x}\,\overline{C} + v_x'C' + \overline{v_x}C' + v_x'\overline{C} \tag{22-26}$$

If the time average is taken on both sides of Eq. 22-26, the last two terms of the right-hand side become zero because $\overline{C'}$ and $\overline{v_x'}$ are zero (see Eq. 22-25). Thus,

$$\overline{F_x^{adv}} = \overline{v_x}\,\overline{C} + \overline{v_x'C'} \tag{22-27}$$

and similarly for the other two flux components ($\overline{F_y^{adv}}$, $\overline{F_z^{adv}}$). Thus, the average advective mass flux $\overline{F_x^{adv}}$ consists of two contributions: (1) the product of the mean concentration $\overline{C}$ and the mean velocity $\overline{v_x}$ and (2) the "turbulent flux":

$$F_x^{turb} = \overline{v_x'C'} \tag{22-28}$$

This flux is the mean value of the product of two fluctuating quantities, the concentration and the velocity. Unlike the last two terms of the far right-hand side of Eq. 22-26, which contain only one fluctuation quantity, the product of two fluctuations does not generally disappear when the temporal mean is taken. This is because there may exist some correlation between the fluctuating part of the two quantities. As an example, consider the concentration profile shown in Figure 22.4. It is likely that if the velocity, v_x, is a little more than average to the right $(v_x' > 0)$, then the corresponding water parcel carries with it a concentration which is below average $(C' < 0)$ since in our example concentration is generally smaller at small x-values. Similarly, we expect situations with $v_x' < 0$ to be associated with $C' > 0$. Thus the two fluctuations "covary." That is why mathematicians call the right-hand side of Eq. 22-28 "covariance." In the above case covariance would be negative; positive flux fluctuations coincide with negative concentration fluctuations.

Consequently, F_x^{turb} would only be zero either in the absence of turbulence $(v_x' = 0)$ or if the concentration distribution were completely homogeneous $(\partial C / \partial x = 0$, thus $C' = 0)$. In turn, we expect C' (and thus F_x^{turb}) to increase with increasing mean gradient $(\partial \overline{C} / \partial x)$. We may even assume a linear relationship between F_x^{turb} and the mean gradient: $F_x^{turb} = \text{constant} \, (\partial \overline{C} / \partial x)$. This brings us back to the expression which we derived for the exchange model (Fig. 22.4). In fact, we can interpret the concentration gradient in Eq. 22-22 as a mean value; thus we get similar expressions for F_x^{turb} in both models:

$$F_x^{turb} = \overline{v_x' C'} = -E_x \frac{\partial \overline{C}}{\partial x} \qquad (22\text{-}29)$$

The empirical coefficient E_x, formally defined by

$$E_x = -\frac{\overline{v_x' C'}}{\partial \overline{C} / \partial x} \qquad (22\text{-}30)$$

depends only on the fluid motion (the turbulence structure of the fluid, to be more precise), and not on the substance described by the concentration C. This is an important difference from the case of molecular diffusion which is specific to the physicochemical properties of the substance and the medium. Since the intensity of turbulence must strongly depend on the forces driving the currents (wind, solar radiation, river flow, etc.), the coefficients of turbulent diffusion constantly vary in space and time. Of course, we cannot tabulate them in chemical or physical handbooks as we do their molecular counterparts.

The decomposition of turbulent motion into mean and random fluctuations resulting in the separation of the flux, Eq. 22-27, leaves us a serious problem of ambiguity. It concerns the question of how to choose the averaging interval s introduced in Eq. 22-24. In a schematic manner we can visualize turbulence to consist of eddies of different sizes. Their velocities overlap to yield the turbulent velocity field. When these eddies are passing a fixed point, they cause fluctuation in the local velocity. We expect that some relationship should exist between the spatial dimension of those eddies and the typical frequencies of velocity fluctuations produced by them. Small eddies would be connected to high frequencies and large eddies to low frequencies.

Consequently, the choice of the averaging time s determines which eddies appear in the mean advective transport term and which ones appear in the fluctuating part (and thus are interpreted as turbulence). The scale dependence of turbulent diffusivity is relevant mainly in the case of horizontal diffusion where eddies come in very different sizes, basically from the millimeter scale to the size of the ring structures related to ocean currents like the Gulf Stream, which exceed the hundred-kilometer scale. Horizontal diffusion will be further discussed in Section 22.3; here we first discuss vertical diffusivity where the scale problem is less relevant.

Vertical Turbulent Diffusivity

While molecular diffusivity is commonly independent of direction (isotropic, to use the correct expression), turbulent diffusivity in the horizontal direction is usually much larger than vertical diffusion. One reason is the involved spatial scales. In the troposphere (the lower part of the atmosphere) and in surface waters, the vertical distances that are available for the development of turbulent structures, that is, of eddies, are generally smaller than the horizontal distances. Thus, for pure geometrical reasons the eddies are like flat pancakes. Needless to say, they are more effective in turbulent mixing along their larger axes than along their smaller vertical extension.

Yet, there is another and often more important factor which distinguishes vertical from horizontal diffusion, that is, the influence of vertical density gradients. Water bodies are usually vertically stratified, meaning that water density is increasing with depth. The strength of the stratification can be quantified by the vertical density gradient of the fluid, $(d\rho/dz)$. Yet, the rather peculiar units which such a quantity would have $(kg\ m^{-4})$ do not really help to understand its relationship to the classical concept of stability used in mechanics. According to this concept, a stable system can be characterized by a restoring force, which brings the system back to its original state every time a perturbation drives it away from the stability point. Examples are the pendulum or a mass hanging on a spring. The "restoration capacity" (that is, the stability) can be described by the time needed to move the system back: *small* restoring times would then indicate *large* stabilities. Conversely, we get a direct correlation if we use the inverse of the restoring time (the restoring rate or frequency): *large* restoring frequencies mean *large* stabilities.

Therefore, the so-called stability frequency is often used to describe the stability of a fluid system. In the case of a vertically stratified water column, the appropriate quantity is called the Brunt-Väisälä frequency, N. It is defined by

$$N = \left(\frac{g}{\rho} \frac{d\rho}{dz} \right)^{1/2} \qquad \left[T^{-1} \right] \qquad (22\text{-}31)$$

where $d\rho/dz$ is the vertical gradient of the water density ρ, and g is the acceleration of gravity $(9.81\ m\ s^{-2})$. In this expression the vertical coordinate z is increasing downward. Thus, a stable stratification of the water column means that density, ρ, increases with depth, z.

If the density gradient depends only on temperature and not on dissolved chemicals, N can be written as:

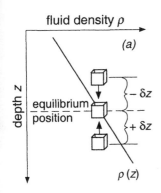

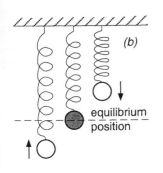

Figure 22.5 Analogy between stability oscillation in (*a*) a stratified water column and (*b*) the motion of a body hanging on a spring (linear oscillator). For small vertical displacements, ± δ*z*, the restoring forces are proportional to δ*z* and to the specific restoring force constant (density gradient or spring constant, respectively). This triggers an oscillation with a characteristic frequency.

$$N = \left(-g\alpha \frac{dT}{dz} \right)^{1/2} \qquad (22\text{-}31a)$$

where $\alpha = -(1/\rho)(d\rho/dT)$ is the thermal expansivity of water (see Appendix B). The meaning of N can be understood by looking at a small water parcel which moves vertically within a stratified water column without exchanging heat or solutes with its environment, that is, without changing its density (Fig. 22.5*a*). If the water parcel is displaced upward, its own density is larger than the density of the surrounding fluid. Therefore, the parcel experiences a "restoring" force downward which is proportional to the vertical excursion δ*z* and the vertical density gradient dρ/d*z*. The reverse happens if the parcel moves below its equilibrium depth. Thus, the so-called buoyancy forces act on the water parcel in an analogous way as gravity acts on a sphere hanging on a spring (the so-called linear oscillator); both situations result in an oscillation around the equilibrium point. The stronger the spring (or the density gradient), the larger will be the frequency of oscillations about the equilibrium point. Therefore, large stability frequencies indicate strong stratification.

In a stratified fluid, the turbulent motion is concentrated along the planes of constant density (which are practically horizontal surfaces), while across these planes (in the vertical direction) the turbulent motion is suppressed. Given a fixed mechanical energy input (e.g., by wind), a large density gradient makes the typical vertical size of the eddies, L_z, small and thus, according to Eq. 22-22, reduces the vertical diffusivity E_z, while a small gradient allows for large eddy sizes.

Based on theoretical considerations on the nature of turbulence (Welander, 1968) quantified this relationship by an equation of the form:

$$E_z = a\left(N^2\right)^{-q} \qquad (22\text{-}32)$$

where the parameter a depends on the overall level of kinetic energy input and the exponent q depends on the mechanism that transforms this energy into turbulent motion. Welander distinguished between two extreme cases: (1) $q = 0.5$, for shear-generated turbulence (i.e., turbulence produced by the friction between waters flowing at different velocity), and (2) $q = 1$, for turbulence generated by energy cascading from the large-scale motion (such as tidal motion) down to the small-scale turbulent motion.

Before we analyze the validity of Welander's equation, we have to discuss how turbulent diffusion coefficients are actually measured. Although the relation between E_z and turbulent fluctuations (Eq. 22-30) would, in principle, provide a basis how to determine E_z from measurements of current velocity and temperature, such experiments are not trivial. Furthermore, they only yield information on the instantaneous and often highly variable size of E_z. In contrast, the dispersion of organic compounds in natural systems is usually a long-term process influenced by the action of turbulence integrated over some space and time interval (days, months, or years). Thus, we are interested in a method which integrates or averages the influence of turbulent diffusion rather than giving an instantaneous but not representative value.

In closed basins (lakes, estuaries), the long-term dynamics of the vertical temperature distribution provides such a tool (Fig. 22.6). Let us assume that in a lake the

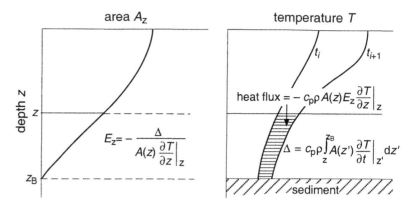

Fig. 22.6 Comparison of vertical temperature profiles measured at consecutive times t_i and t_{i+1} can be used to determine the vertical turbulent diffusivity E_z. From Imboden et al. (1979).

temperature varies only in the vertical, not in the horizontal. This means that at a fixed depth, for instance at 10 m, the water temperature is constant across the lake. (In fact, this assumption is not true for an instantaneous view of the lake since the surfaces of constant temperature move up and down like waves at the water surface; yet for the time-averaged temperatures the picture of a horizontally homogeneous temperature field is mostly reasonable.) Consequently, if we obtain a vertical temperature profile taken at time t_i we can calculate the heat content of the whole lake or part of it, for instance, of the zone between depth z and the lake bottom at z_B:

$$\text{heat content} = c_p \rho \int_z^{z_B} A(z')T(z')\,dz' \quad [\text{Joule}] \tag{22-33}$$

where the integration over the depth variable z' goes from depth z to the lake bottom at z_B. The vertical temperature profile, $T(z')$, is weighted by the local cross-sectional area $A(z')$. It is assumed that the specific heat, c_p, and density, ρ, of water are constant. Note that $c_p\rho$ is the specific heat per water volume.

Owing to vertical (turbulent) diffusion, heat is transported from regions of warm water to adjacent colder layers. Mathematically this appears as a heat flux against the vertical temperature gradient (remember Fick's first law, Eq. 18-6). Thus, at a later time, t_{i+1}, we expect to find warmer water between z and z_B. The change of the heat content with time Δ is:

$$\Delta \equiv \frac{\partial}{\partial t}\left(\begin{array}{c}\text{heat}\\\text{content}\end{array}\right) = \frac{\partial}{\partial t}\left\{c_p\rho\int_z^{z_B} A(z')\,T(z')\,dz'\right\}$$

$$= c_p\rho\int_z^{z_B} A(z')\frac{\partial T(z')}{\partial t}\,dz' \quad [\text{Watt}] \tag{22-34}$$

(Note that $A(z')$ does not change with time. Furthermore, differentiation with respect to time and integration over space can be interchanged.)

If other sources and sinks of thermal energy can be excluded, the changing heat content, Δ, must be related to the total turbulent heat flux through the area $A(z)$. According to Eq. 18-6 this flux is:

$$\text{total heat flux per unit time} = -A(z)\, E_z \frac{\partial}{\partial z}(c_p \rho T)\bigg|_z \qquad (22\text{-}35)$$

$$= -c_p \rho\, A(z) E_z \frac{\partial T}{\partial z}\bigg|_z$$

where $(c_p \rho T)$ is heat energy per unit volume. Note that D from Eq. 18-6 was replaced by the vertical eddy diffusivity, E_z.

Since the left-hand sides of Eqs. 22-34 and 2-35 must be equal, we can solve the equation for E_z:

$$E_z = -\frac{\Delta}{c_p \rho\, A(z) \dfrac{\partial T}{\partial t}\bigg|_z} = -\frac{\displaystyle\int_z^{z_B} A(z')\frac{\partial T}{\partial t}\bigg|_{z'} dz'}{A(z)\dfrac{\partial T}{\partial t}\bigg|_z} \qquad (22\text{-}36)$$

The E_z values calculated by this method represent the average turbulent diffusivity at depth z for the time interval between t_i and t_{i+1}.

Application of Eq. 22-36 is demonstrated in Illustrative Example 22.2. Note that the same method could also be applied to the vertical profile of any chemical substance provided that in situ reactions are absent or can be quantified. An example is given in Problem 22.3.

(*Text continues on page 1028*)

Illustrative Example 22.2   **Vertical Turbulent Diffusion Coefficient in a Lake**

Problem

You are responsible for the water quality monitoring in Lake X (surface area $A_o = 16$ km^2, maximum depth $z_m = 40$ m). Among the various physical and chemical parameters which you measure regularly is the vertical distribution of water temperature T. The following table gives two profiles measured during the same summer. The table also gives some information on lake topography, i.e., on the variation of lake cross-section with depth, $A(z)$.

(a) Calculate the vertical turbulent diffusion coefficient, E_z, at the following depths: 7.5, 12.5, 17.5, 25 and 35 meters.

(b) Since you are interested in understanding the physical processes in the lake, you compare the calculated E_z values with the vertical stability frequency, N, by applying Eq. 22-32. What can you learn from the size of the exponent q?

Note: Assume that the density gradient in the water column is not influenced by dissolved solids.

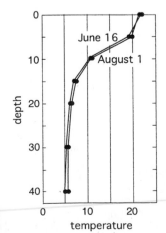

Lake X

Depth z	Area A	Water Temperature T (C°)	
(m)	(km²)	June 16	August 1
0	16	22.0	21.7
5	14	19.0	19.8
10	12	10.5	10.7
15	10	7.0	7.4
20	8	6.1	6.3
30	4	5.3	5.6
40	0	4.9	5.5

Answer (a)

To apply Eq. 22-36, you need a method for evaluating the integral in the numerator from the temperature data and lake areas which are available at discrete depths only. You know that in the age of computers nobody would really execute such a computation by hand anymore. Yet, even if you use a computer program you make a certain choice as to how you are going to approximate the integral, although in many cases you are not aware of it. Thus it may be instructive to learn from a simple example, step by step, how the calculation proceeds.

First, you note that $A(z)$ decreases linearly with depth z according to:

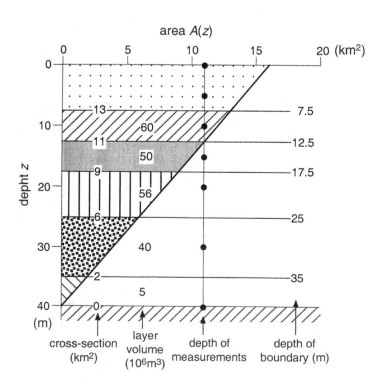

Figure 1 The subdivision of the lake volume into sublayers centered around the depths of measurements (black dots) serves to evaluate Eq. 22-36 by numerical approximation. The black line shows the cross-sectional area as a function of depth. The numbers in the figure arranged in three columns give (from left to right) the cross-sectional area at the interfaces, the volume of the sublayers, and the depth of the sublayer boundaries.

$$A(z) = A_o\left(1 - \frac{z}{z_m}\right) = 16 \text{ km}^2\left(1 - \frac{z}{40 \text{ m}}\right)$$

Next, you divide the lake into zones which are bounded by planes located halfway between the depths for which temperature data exist (see Fig. 1). Since $A(z)$ is linear, the volume for each zone is simply the arithmetic mean of the cross-sections above and below the layer multiplied by the thickness of the layer. These volumes are multiplied by the corresponding temporal temperature changes. The results are summarized in the following table. The last column, the cumulative sum of the weighted temporal temperature gradient summed from below, is an approximation of the numerator of Eq. 22-36.

Zone α (m)	Volume V_α ($10^6\,\text{m}^3$)	$\partial T_\alpha/\partial t$ (K d^{-1})	$V_\alpha(\partial T_\alpha/\partial t)$ ($10^6\,\text{K m}^3\text{d}^{-1}$)	$\Sigma V_\alpha(\partial T_\alpha/\partial t)$ ($10^6\,\text{K m}^3\text{d}^{-1}$)
35 - 40	5	0.0133	0.067	0.067
25 -.35	40	0.0067	0.268	0.335
17.5 - 25	56	0.0044	0.246	0.581
12.5 - 17.5	50	0.0089	0.445	1.026
7.5 - 12.5	60	0.0044	0.264	1.290

Note: The absolute temperature changes within 45 days at a given depth are small. Thus, it would be convenient to have an instrument which allows us to measure T with an accuracy of at least $\pm\,0.01$ K. Furthermore, an absolute calibration stability of $\pm\,0.01$ K is a prerequisite to calculate E_z from T.

The following table summarizes the remaining steps to evaluate Eq. 22-36.

depth (m)	area $A(z)$ ($10^6\,\text{m}^2$)	$\partial T_\alpha/\partial z$ [a] (K m^{-1})	$\Sigma V_\alpha(\partial T_\alpha/\partial t)$ [b] ($10^6\,\text{K m}^3\text{d}^{-1}$)	E_z [c] (m^{2}d^{-1})	(cm^{2}s^{-1})
7.5	13	-1.66	1.290	0.060	0.0069
12.5	11	-0.68	1.026	0.14	0.016
17.5	9	-0.20	0.581	0.32	0.037
25	6	-0.08	0.335	0.70	0.081
35	2	-0.025	0.067	1.3	0.16

[a] Arithmetic mean of gradient of the two temperature profiles. [b] From preceding table.
[c] From Eq. 22-36 with 1 m^{2}d^{-1} = 0.116 cm^{2}s^{-1}

Answer (b)

Since it is assumed that the density gradient is not influenced by dissolved chemicals, the stability frequency, N, can be calculated from the temperature profiles alone:

Depth (m)	T^a (°C)	α^b $(10^{-6}\,K^{-1})$	$\partial T / \partial z$ $(K\,m^{-1})$	N^c $(10^{-3}\,s^{-1})$	$E_z{}^d$ $(cm^2\,s^{-1})$
7.5	15	151	−1.66	49.6	0.0069
12.5	9	74	−0.68	22.2	0.016
17.5	6.7	39	−0.20	8.7	0.037
25	5.8	26	−0.08	4.5	0.081
35	5.3	19	−0.025	2.2	0.16

^{a} Mean value of the two profiles interpolated. ^{b} From Appendix B. ^{c} From Eq. 22-31a. ^{d} From preceding table.

Equation 22-32 can be written in the form

$$\ln E_z = \ln a - 2q \ln N$$

Figure 2 shows a nearly perfect correlation with $q = 0.5$ and $a = \exp(-8.01) = 3.3 \times 10^{-4}$, which is indicative for turbulence produced by local shear.

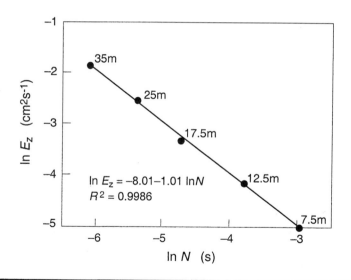

Figure 2 The correlation between stability frequency, N, and vertical turbulent diffusivity, E_z, according to Eq. 22-32 yields $q = 0.5$.

Let us now come back to the relationship between vertical diffusivity, E_z, and density stratification quantified by the Brunt-Väisälä frequency, N (Eq. 22-31). Numerous examples from lakes and oceans show that the exponent q of Eq. 22-32 commonly lies between 0.5 and 0.7. Two different situations are shown in Fig. 22.7. Both data sets originate from two fairly deep Swiss lakes (Imboden and Wüest, 1995). Figure 22.7a shows the long-term average E_z values from a basin of Lake Lucerne with $q = 0.5$ indicating that turbulence is mainly shear-produced. In contrast, the data from Lake Zug (Fig. 22.7b) were obtained during an extreme storm event. Note that beside the different q value, the turbulent diffusivity, E_z, is two to three orders of magnitude larger than for the case of Lake Lucerne. This indicates the enormous influence which single events can have as well as the difficulty of using models for the prediction of mixing.

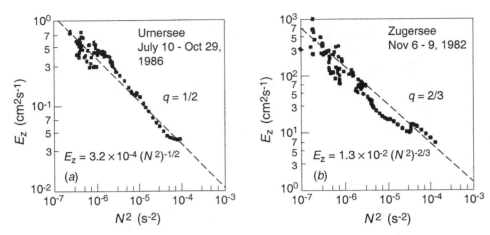

Figure 22.7 Vertical turbulent diffusivity E_z versus square of stability frequency N^2 in two Swiss lakes (see Eq. 22-32). (*a*) For Urnersee (maximum depth 196 m), a basin of Lake Lucerne, the data refer to 10–100 m depth and indicate shear-produced turbulence. (*b*) For Zugersee (maximum depth 198 m) the values are calculated for an extreme storm of about two days duration. The data refer to the depth interval between 10 and 70 m; they show a mixture between turbulence production by local shear and large-scale motion. (From Imboden and Wüest., 1995.)

Radioactive or stable isotopes of noble gases are also used to determine vertical turbulent diffusion in natural water bodies. For instance, the decay of tritium (^{3}H)—either produced by cosmic rays in the atmosphere or introduced into the hydrosphere by anthropogenic sources—causes the natural stable isotope ratio of helium, ^{3}He/^{4}He, to increase. Only if water contacts the atmosphere can the helium ratio be set back to its atmospheric equilibrium value. Thus the combined measurement of the ^{3}H-concentration and the ^{3}He/^{4}He ratio yields information on the so-called water age, that is, the time since the analyzed water was last exposed to the atmosphere (Aeschbach-Hertig et al., 1996). The vertical distribution of water age in lakes and oceans allows us to quantify vertical mixing.

Another procedure is based on the measurement of the radioactive isotope radon-222 (half-life 3.8 days), the decay product of natural radium-226. At the bottom of lakes and oceans, radon diffuses from the sediment to the overlying water where it is transported upward by turbulence. Broecker (1965) was among the first to use the vertical profile of ^{222}Rn in the deep sea to determine vertical turbulent diffusivity in the ocean.

The one-dimensional diffusion-decay equation of the excess radon activity, C_{Rn}^{exc} (i.e., the radon activity exceeding the activity of its parent isotope radium-226) is given by:

$$\frac{\partial C_{Rn}^{exc}}{\partial t} = E_z \frac{\partial^2 C_{Rn}^{exc}}{\partial h^2} - \lambda_{Rn} C_{Rn}^{exc} \tag{22-37}$$

where $\lambda_{Rn} = 0.181$ d^{-1} is the decay constant of radon-222. In Eq. 22-37 horizontal transport is disregarded, and the eddy diffusivity E_z is assumed to be independent of height above the sediment surface, h.

Because of the relatively short half-life of radon-222, the vertical radon distribution approaches steady-state within about a week, once the turbulent conditions remain constant. With $\partial C_{Rn}^{exc} / \partial t = 0$, the solution of Eq. 22-37 is:

$$C_{Rn}^{exc}(h) = C_{Rn}^{exc}(h=0) \exp\left[-\left(\frac{\lambda_{Rn}}{E_z}\right)^{1/2} h\right] \tag{22-38}$$

Thus, a plot of $\ln\left\{C_{Rn}^{exc}(h)\right\}$ versus h should yield a straight line with slope $-(\lambda_{Rn} / E_z)^{1/2}$. Since λ_{Rn} is known, E_z can be determined from the slope.

Figure 22.8 Vertical profile of dissolved excess radon-222 activity (i.e., the radon-222 activity exceeding the activity of its parent nucleus radium-226) in the bottom waters of Greifensee (Switzerland) serves to compute vertical turbulent diffusivity E_z. Activity units are "decay per minute per liter" (dpm L^{-1}). Data from Imboden and Emerson (1978).

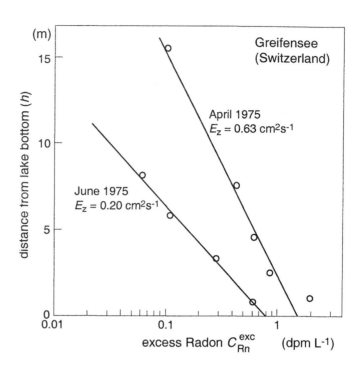

Figure 22.8 gives an example from Greifensee, Switzerland (see Table 21.1 for characteristic data of the lake). Though there are limits to the one-dimensional interpretation of the data brought about by lateral transport from the sides (Imboden and Emerson, 1978), under certain conditions the method still can be useful to yield insight into the vertical mixing regime at the bottom of lakes and oceans. We see that in this small lake E_z was between 0.1 and 1 cm^{2}s^{-1}, thereby indicating vertical Rn-222 transport far greater than that explainable by molecular diffusion.

22.3   Horizontal Diffusion: Two-Dimensional Mixing

Turbulence Theory and the "4/3 Law"

As mentioned earlier, turbulent motion is usually more intensive along the horizontal than the vertical axis. Turbulent structures (eddies) can be horizontally very large. For instance, the eddies or gyres produced by the Gulf Stream are more than 100 km wide. Thus, for horizontal transport the separation between random and directed motion plays a more crucial role than for the case of vertical diffusion.

Let us make this point clearer by the following hypothetical experiment. At some initial time t_o a droplet of dye is put on the surface of a turbulent fluid (Fig. 22.9). At some later time t_1 the large-scale fluid motion has moved the dye patch to a new location which can be characterized by the position of the center of mass of the patch. In addition, the patch has grown in size because of the small (turbulent) eddies, more precisely, those eddies with sizes similar to or smaller than the patch size.

If horizontal diffusivity were isotropic ($E_x = E_y$) and constant with time, then according

Figure 22.9 Horizontal growth and movement of a tracer patch under the influence of turbulent currents. While the mean currents move the patch as a whole (represented by the center of mass; black dots), the turbulent components increase the size of the patch. Usually, the spreading is faster in the direction of the mean current. Therefore, the patch develops approximately into an ellipse with major and minor principle axes, σ_{ma} and σ_{mi}. From Peeters et al. (1996).

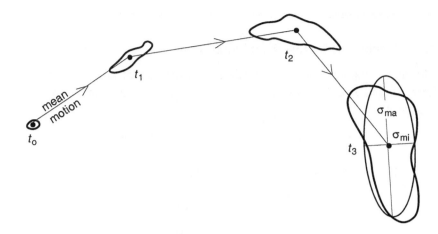

to Box 18.3, an initial small dye patch would develop into a two-dimensional circular normal distribution with variance $\sigma = 2\sigma_x^2 = 2\sigma_y^2$. According to Eq. 18-17 the patch would grow with time as:

$$\frac{d\sigma_x^2}{dt} = \frac{d\sigma_y^2}{dt} = 2E_x = 2E_y \qquad (22\text{-}39)$$

Thus, observing the growth of a tracer patch would allow us to calculate the horizontal diffusion coefficient. The above expression would still hold if the growing effect of random motion with patch size were considered as inferred by the picture of a spectrum of turbulent structures of different size. The turbulent diffusivities, E_x and E_y, would then depend on σ_x or σ_y.

In fact, the scale-dependence of the horizontal diffusivity E_h with length scale, L, as expressed in a well-known figure by Okubo (1971), is still an extremely popular model for describing mixing conditions in the ocean. Okubo's interpretation is based on the so-called *inertial subrange* theory of turbulence (Kolmogorov, 1941), which leads to the famous "4/3 law." According to this law, horizontal diffusivity depends on the length scale as:

$$E_h = (0.2 \text{ to } 1) \, L^{4/3} \qquad (22\text{-}40)$$

where E_h is horizontal diffusivity in cm^2s^{-1}
            L is length scale in m

This empirical relation is based on data extending from $L = 10$ m to $L = 1000$ km, yielding horizontal diffusivities between $E_h = 10$ cm^2s^{-1} and 10^8 cm^2s^{-1}, respectively.

Yet, Okubo's law and the physical model on which it is based disregard two important properties of measured tracer distributions. The first one concerns the shape of tracer clouds. As indicated in Fig. 22.9, clouds usually develop into elongated structures which can be approximated by ellipses with major and minor principal axes, σ_{ma} and σ_{mi}. The major axis points in the direction of the mean flow. When Okubo

---

Box 22.3   Turbulent Diffusion and Spatial Scale

The following considerations are based on an overview of horizontal diffusion theories by Peeters et al. (1996). Turbulence theory suggests a power law with unknown exponent m to describe the growth of cloud size σ^2 with elapsed time t:

$$\sigma^2(t) = \text{const. } t^m \qquad (1)$$

Due to Eq. 18-17, horizontal turbulent diffusivity, E_h, and σ^2 are related by:

$$\frac{d\sigma^2}{dt} = \text{const. } E_h \qquad (2)$$

Combining Eq. 1 and 2 yields:

$$\frac{d\sigma^2}{dt} = \text{const. } t^{m-1} = \text{const. } \left(\sigma^2\right)^{\frac{m-1}{m}} = E_h \qquad (3)$$

If σ is used as a measure for the horizontal scale, L:

$$E_h = \text{const. } L^{2\left(1-\frac{1}{m}\right)} \qquad (4)$$

According to Kolmogorov's (1941) inertial subrange theory of turbulence, the exponent in Eq. 1 is $m = 3$. Inserting into Eq. 4:

$$E_h = \text{const. } L^{4/3} \qquad (m = 3) \qquad (5)$$

This is Okubo's "4/3 law" (Eq. 22-40).

Experiments show that real tracer clouds can be approximated by ellipses with major and minor principal axes, σ_{ma} and σ_{mi}. Their growth with time yields:

$$\text{Major axis: } \quad m = 1.5 \pm 0.2 \quad \rightarrow \quad E_{ma} = \text{const. } L^{(0.67 \pm 0.2)} \qquad (6a)$$

$$\text{Minor axis: } \quad m = 1.0 \pm 0.1 \quad \rightarrow \quad E_{mi} = \text{const. } L^{(0 \pm 0.2)} \qquad (6b)$$

The growth of σ_{mi} seems to obey the normal Fickian law with constant diffusivity E_{mi}.

The composite size, $\sigma^2 = 2\sigma_{ma}\,\sigma_{mi}$ (Eq. 22-41), grows with $m = \frac{1}{2}(1.5 + 1.0) = 1.25 = 5/4$, thus from Eq. 4:

$$E_\sigma = \text{const. } L^{2/5} \qquad (7)$$

---

derived his "4/3 law," he circumvented the problem posed by noncircular patch shape by defining a composite patch size of the form:

$$\sigma^2 = 2\sigma_{ma}\,\sigma_{mi} \qquad (22\text{-}41)$$

For two-dimensional normal distributions (see Appendix A, Eq. A-3), the lines of equal concentrations are ellipses. The ellipse that corresponds to e^{-1} (37%) of the maximum concentration encloses 63% of the total mass (Appendix A, Table A.1) and has an area $A = 2\pi\,\sigma_{ma}\sigma_{mi} = \pi\sigma^2$.

Although the definition of Eq. 22-41 seems to be a successful recipe for making even elliptical tracer clouds suitable for the "4/3 law," there is a more fundamental problem with Eq. 22-40. As shown in Box 22.3, this expression in combination with the corresponding law which relates diffusivity and growth of patch size (see Eq. 22-39) would be equivalent to the following power law for σ^2 as a function of elapsed time t:

$$\sigma^2(t) = \text{const. } t^3 \qquad (22\text{-}42)$$

The growth rate of the patch size $\sigma^2(t)$ determines the dilution of a chemical into the environment and thus the drop of the maximum concentration of the chemical. Therefore, the power law relating $\sigma^2(t)$ and t is of great practical importance, for instance, to predict the behavior of a pollutant cloud in the environment. As it turns out, the specific power law which follows from Okubo's theory, Eq. 22-42, greatly exaggerates the effect of dilution compared to observations made in the field. The shear diffusion model which will be discussed next gives a more realistic picture.

The Shear Diffusion Model

Peeters et al. (1996) have made a series of experiments with artificial tracers added to the upper hypolimnion of several Swiss lakes or basins of lakes (basin size between 5 and 220 km²). They always found elongated cloud shapes which they approximated by ellipses. The principal axes grew with time as:

$$\sigma_{ma}^2 = \text{const. } t^{(1.5\pm0.2)} \qquad (22\text{-}43a)$$

$$\sigma_{mi}^2 = \text{const. } t^{(1.0\pm0.1)} \qquad (22\text{-}43b)$$

Consequently, the ratio, $\sigma_{ma}^2 : \sigma_{mi}^2$, increases with time.

The different exponents found in Eq. 22-43 indicate that the spreading of a tracer cloud is caused by two processes: perpendicular to the flow direction, that is, along the minor principal axis, the spreading is compatible with normal Fickian diffusion with scale-independent horizontal diffusivity E_h (see Box 22.3, Eq. 6b). An additional effect is important along the axis of flow, the process of longitudinal *dispersion*.

Dispersion will be discussed in the subsequent section. At this point we only mention that dispersion always occurs in fluids with a distinct direction of advective flow. It originates from the velocity difference between adjacent streamlines. This effect is called *velocity shear*.

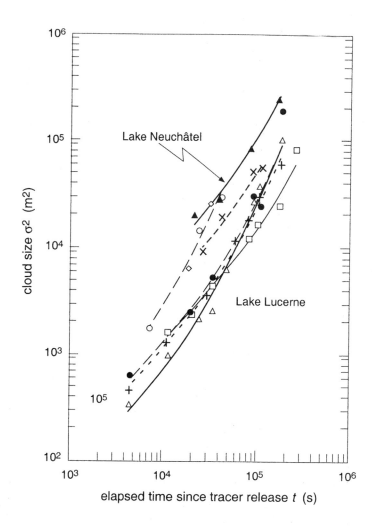

Figure 22.10 Fit of measured horizontal cloud size, σ^2, to the shear diffusion model by Carter and Okuto (1965), Eq. 22-44, of several tracer experiments in different basins of Lake Lucerne (Switzerland) and in Lake Neuchâtel (Switzerland) (different symbols). From Peeters et al. (1996).

As pointed out by Peeters et al. (1996), based on their own experiments and on the reinterpretation of published field data, the adequate model to describe horizontal diffusion in lakes and oceans is the shear diffusion model by Carter and Okubo (1965). The model is described in Box 22.4. The most important consequence of this model is that the "4/3 law" and the equivalent t^3-power law for $\sigma^2(t)$ expressed by Eq. 22-42 are replaced by an equation which corresponds to a continuous increase of the exponent m from 1 to 2 (Box 22.4, Eq.1):

$$\sigma^2(t) = 2\left[4\,A_1^2 t^2 + \frac{1}{3}\,A_1 A_2 t^4\right]^{1/2} \tag{22-44}$$

The meaning and typical sizes of the coefficients A_1 and A_2 are discussed in Box 22.4. From Eq. 22-44 we note that for small times t, $\sigma^2(t)$ grows as t, whereas for large times it grows as t^2. The critical time, t_{crit}, defined in Eq. 3 of Box 22.4 separates the two regimes. Figure 22.10 shows $\sigma^2(t)$ curves from different experiments conducted in Swiss lakes. In Illustrative Example 22.3 the shear diffusion model is applied to the case of an accident in which a pollutant is added to the thermocline of a lake.

Box 22.4   The Shear Diffusion Model

As shown by Peeters et al. (1996), horizontal diffusion experiments in lakes and oceans can be best described with the shear-diffusion model of Carter and Okubo (1965). The model yields the following relation between cloud size, σ^2, and time, t:

$$\sigma^2(t) = 2\left[A_1^2 t^2 + \frac{1}{3}A_1 A_2 t^4\right]^{1/2} \tag{1}$$

The spreading results from a combination of two processes, (1) Fickian horizontal diffusion with scale-independent diffusivity E_h, and (2) dispersion by velocity shear in the direction of the mean flow. The process of dispersion is discussed in Section 22.4. It is related to the flow velocity difference (called "velocity shear") between adjacent streamlines. Since water parcels traveling on different streamlines (e.g., at different depth) have different velocities, a tracer cloud is elongated along the direction of the mean flow (see figure below).

Equation (1) contains two parameters, A_1 and A_2. In the framework of the shear diffusion model they can be identified with real physical quantities:

$$A_1 = E_h \qquad \left[L^2 T^{-1}\right] : \quad \text{horizontal diffusivity} \tag{2a}$$

$$A_2 = E_h\left(\frac{\partial v_x}{\partial y}\right)^2 + 4E_z\left(\frac{\partial v_x}{\partial z}\right)^2 \qquad \left[L^2 T^{-3}\right] \tag{2b}$$

where v_x is velocity of flow along the x-direction

$\dfrac{\partial v_x}{\partial y}$ is horizontal shear perpendicular to the flow

$\dfrac{\partial v_x}{\partial z}$ is vertical shear

E_h is horizontal diffusivity

E_z is vertical diffusivity

Note that according to Eq. 1, the exponent m of the power law $\sigma^2(t) = \text{const.}\ t^m$ (see Box 22.3) increases from $m = 1$ to $m = 2$. The critical time, t_{crit}, when both terms in the square root of Eq. 1 are equal, is:

$$t_{crit} = \left(\frac{12A_1}{A_2}\right)^{1/2} \tag{3}$$

Typical sizes of A_1, A_2, and t_{crit} are given in the table below. The values are taken from Peeters et al. (1996); these authors successfully reanalyzed several published tracer experiments in terms of the shear-diffusion model. An example is shown in Fig. 22.10.

Typical Values for Coefficients of Eq. 1

		A_1 (m²s⁻¹)	A_2 (10^{-10}m²s⁻³)	t_{crit} (Eq. 3) (s)	(hours)
Lakes,	calm	0.04	1	7.7×10^4	22
	windy	0.2	100	1.5×10^4	4
Ocean		0.1	500	0.5×10^4	1.4

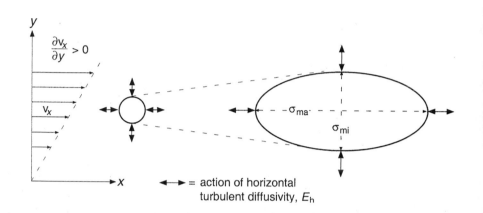

$\frac{\partial v_x}{\partial y} > 0$

v_x

σ_{ma}

σ_{mi}

$\longleftrightarrow$ = action of horizontal turbulent diffusivity, E_h

---

Illustrative Example 22.3   **A Patch of the Pesticide Atrazine Below the Surface of a Lake**

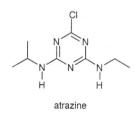

atrazine

$M_i = 215.7 \ \text{g mol}^{-1}$

Problem

A total amount of 5 kg of the herbicide atrazine is spilled from a cornfield into a creek only a small distance upstream of the point where it discharges into Lake G. An employee from the local water authority happens to be on a boat taking water samples. With her turbidity meter she detects the turbid water from the accident at 12 m depth. Based on a few profiles she estimates the pollutant cloud to be 1 m thick, 40 m long, and 10 m wide. The cloud is drifting at a speed of about 5 cm s⁻¹ toward the intake of a local drinking-water supply, which is about 5 km away from the river mouth. Is it necessary to stop the pumps of the water plant if she wants to prevent atrazine concentration larger than the tolerable drinking-water concentration (0.1 μg L⁻¹) from entering the plant?

Answer

Obviously, there is not much time to organize an elaborate measurement campaign and to run a sophisticated lake model. Thus a worst-case scenario must be sufficient for a quick decision. Based on the estimated cloud size the initial concentration of atrazine in the lake is about:

$$C_o = \frac{5 \ \text{kg}}{1 \ \text{m} \times 40 \ \text{m} \times 10 \ \text{m}} = \frac{5 \times 10^3 \text{g} / 215.7 \text{g mol}^{-1}}{4 \times 10^5 \text{L}} \cong 0.5 \times 10^{-4} \ \text{mol L}^{-1}$$

Note: The aqueous solubility of atrazine at 25°C is 1.5×10^{-4} mol L⁻¹.

If the water currents transport the pollutant cloud directly to the intake of the water plant, the transport time would be about:

$$t_{transport} = \frac{5 \times 10^3 \, m}{0.05 \, m \, s^{-1}} = 1 \times 10^5 \, s = 28 \text{ hours}$$

During this time the cloud will grow in size, and the atrazine will be diluted. The following estimates are lower limits and thus on the safe side:

(a) Vertical mixing: $E_z \approx 10^{-2} \, cm^2 \, s^{-1}$

$$\rightarrow \sigma_z = \left(2 \, E_z \, t_{transport} \right)^{1/2} = \left(2 \times 10^{-2} \, cm^2 s^{-1} \times 10^5 \, s \right)^{1/2} = 0.45 \, m$$

Since the cloud is already 1 m thick, the additional thickening can be disregarded.

(b) Horizontal mixing:
Use Eq. 22-44 with $A_1 = 0.02 \, m^2 s^{-1}$, $A_2 = 0.1 \times 10^{-10} \, m^2 s^{-3}$. According to the table in Box 22.4, these values are small and thus safe.

Critical time (Eq. 3, Box 22.4):

$$t_{crit} = \left(\frac{12 \times 0.02 \, m^2 s^{-1}}{1 \times 10^{-11} m^2 s^{-3}} \right)^{1/2} = 1.5 \times 10^5 \, s = 43 \text{ hours}$$

Thus, while the cloud is traveling toward the intake, its size, σ^2, mainly grows like t.

To calculate the initial cloud size from Eq. 22-41, we use $\sigma_{ma} = 20$ m, $\sigma_{mi} = 5$ m (half of observed length and width):

$$\sigma_o^2 = 2 \times 20 \, m \times 5 \, m = 200 \, m^2$$

Note that the initial time t_o which corresponds to the initial cloud with size σ_o is not known. There are two possibilities:

(1) Assume $t_o = 0$ and apply Eq. 22-44 for $t = t_{transport} = 10^5 \, s$. Then:

$$\sigma^2 (t_{transport})$$

$$= 2 \left[4 \times (0.02 \, m^2 s^{-1})^2 \times (10^5 s)^2 + \frac{1}{3} \times 0.02 \, m^2 s^{-1} \times 10^{-11} m^2 s^{-3} \times (10^5 s)^4 \right]^{1/2}$$

$$= 2 \left[1.6 \times 10^7 \, m^4 + 0.67 \times 10^7 \, m^4 \right]^{1/2} = 4.8 \times 10^3 \, m^2$$

(2)   Formally, one could use the known initial cloud size σ_o^2 and solve Eq. 22-44 for $t = t_o$. The result is: $t_o = 2.5 \times 10^3 \, s = 0.7$ h. This would add very little to $t_{transport}$, which was used above to calculate σ^2.

Since the vertical spreading of the pollutant cloud can be neglected, the maximum concentration of atrazine at $t = 10^5$ s would be:

$$C(t_{\text{transport}}) \approx \frac{\sigma_o^2}{\sigma^2(t_{\text{transport}})} C_o = \frac{400 \text{ m}^2}{4.8 \times 10^3 \text{ m}^2} \times 0.5 \times 10^{-4} \text{ mol L}^{-1}$$

$$\approx 5 \times 10^{-6} \text{ mol L}^{-1}$$

Since this value is still clearly larger than the tolerable drinking-water concentration ($C_{\text{tot}} = 0.1$ mg L$^{-1} \approx 0.5 \times 10^{-6}$ mol L^{-1}), it would be wise to stop the operation of the pumps and survey the movement of the cloud.

Note: Obviously, the cloud is still fairly small after one day and thus the chance is small that it would exactly drift to the intake. Furthermore, if the intake were not at the same depth as the drifting cloud, it might never affect the quality of the drinking water. Yet, this is no argument against precautionary measures.

Advanced Topic   **22.4**   # Dispersion

Although dispersion can be described by the same law as diffusion, its nature is different. Dispersion is the result of the velocity shear, that is, of the velocity difference between adjacent streamlines in an advective flow. Due to turbulent exchange perpendicular to the direction of flow, water parcels continuously change the streamline along which they move. Since these streamlines move at different speeds, each water parcel has its own individual "history of speed" and thus its individual mean velocity.

To render this point less abstract, imagine the various streamlines to be railway tracks running in parallel from A to B. On some tracks run the fast trains, on others trains of intermediate speed, and on still others the slow freight trains. People traveling on these trains randomly jump between these trains, thus changing the speed of their journey. At time $t = 0$, several people start their trip at A on different tracks. Since each person spends different amounts of time on the different tracks, they will arrive at B at different times (Fig. 22.11). Imagine that several observers register the time of arrival at B and also at an intermediate location X. From their data they calculate the average arrival time as well as the variance σ^2 of the individual times relative to the mean (see Fig. 22.11). It then will turn out that the variance grows with increasing time, that is, it will be larger at B than at X. If the observers compare their data more closely, they will notice that σ^2 increases approximately linearly with the mean traveling time $\bar{t}$: $\sigma_t^2 = \text{const.} \, \bar{t}$.

Does this sound familiar? Yes, it should remind us of the Einstein-Smoluchowski law, Eq. 18-8. In fact, the analogy suggests that dispersion can be described by the same mathematical formalism as diffusion, that is, by the first and second Fickian laws (Eqs. 18-6 ad 18-14). We just have to replace diffusivity D by the dispersion

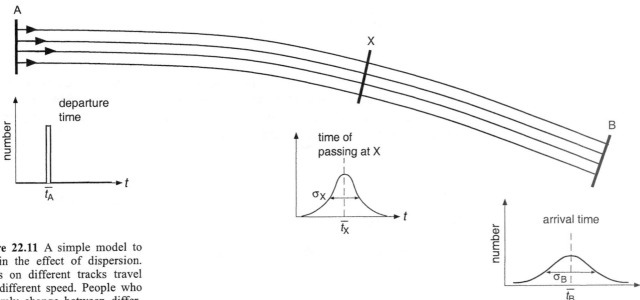

Figure 22.11 A simple model to explain the effect of dispersion. Trains on different tracks travel with different speed. People who randomly change between different trains have different travel times from A to B. The variance of their arrival time increases from location X (σ_X) to location B (σ_B).

coefficient, E_{dis}. Once we have qualitatively understood the nature of dispersion, we now can take a short look at its mathematical description.

Dispersion and the Fickian Laws

We analyze the transport of a chemical (concentration C) along a fluid flow with mean velocity $\overline{v}_x$ (x designates the direction of the flow). Examples are transport in a river (Chapter 24) or in an aquifer (Chapter 25). Imagine that at time $t = 0$, per cross-sectional area the total amount $\mathcal{M}^*$ (dimension ML^{-2}) of a chemical is added to the flow at location $x = 0$. Then, the one-dimensional concentration distribution, $C(x, t = 0)$ can be described by the δ-function (see Eq. 18-15):

$$C(x,\ t = 0) = \mathcal{M}^* \delta(x) = \begin{cases} 0 & \text{for } x \neq 0 \\ \\ \infty & \text{for } x = 0 \end{cases} \quad \text{with } \int_{-\infty}^{\infty} \delta(x)dx = 1 \quad (22\text{-}45)$$

The action of dispersion upon $C(x,t)$ can be characterized by the following points:

1. The center of mass, x_{CM}, travels with the velocity $\overline{v}_x$:

$$x_{CM}(t) = \overline{v}_x t \quad (22\text{-}46)$$

2. Relative to the center of mass, the concentration is spread according to the law:

$$\frac{\partial C}{\partial t} = E_{dis} \frac{\partial^2 C}{\partial x^2} \quad (22\text{-}47)$$

3. By analogy to Eq. 18-6, the solution of Eq. 22-47 with initial condition Eq. 22-45 has the form:

$$C(x,t) = \frac{\mathcal{M}^*}{2\left(\pi E_{dis}\, t\right)^{1/2}} \exp\left\{-\frac{\left(x - \overline{v}_x\, t\right)^2}{4 E_{dis}\, t}\right\} \qquad (22\text{-}48)$$

Note the difference to Eq. 18-16: The maximum of the normal distribution, characterized by that x which makes the argument of the exponential function zero, moves with time according to Eq. 22-46. At a fixed time t, the concentration distribution along the x-axis is a normal distribution. Its variance increases as:

$$\sigma_t^2 = 2 E_{dis}\, t \qquad (22\text{-}49)$$

where the dispersion coefficient E_{dis} has the same dimension as the diffusion coefficient (L^2T^{-1}). The analogy between Eqs. 18-8 and 22-49 is obvious.

4. The description of dispersion also has an analogy to Fick's first law. Relative to a water parcel moving at the mean velocity $\overline{v}_x$ there is a mass flux due to dispersion which takes the form:

$$F_{dis} = -E_{dis} \frac{\partial C}{\partial x} \qquad (22\text{-}50)$$

It is this flux which dilutes concentration patches moving along the streamlines. At this point, the connection to the shear diffusion model of horizontal mixing becomes clearer. An example of how dispersion is causing the dilution of a pollutant patch with its environment has been discussed in Illustrative Example 22.3; others will follow in Chapters 24 and 25.

The Dispersion Coefficient

Once the mathematical description of dispersion has been clarified, we are left with the task of quantifying the dispersion coefficient, E_{dis}. Obviously, E_{dis} depends on the characteristics of the flow field, particularly on the velocity shear, $\partial v_x / \partial y$ and $\partial v_x / \partial z$. As it turns out, the shear is directly related to the mean flow velocity $\overline{v}_x$. In addition, the probability that the water parcels change between different streamlines must also influence dispersion. This probability must be related to the turbulent diffusivity perpendicular to the flow, that is, to vertical and lateral diffusion. At this point it is essential to know whether the lateral and vertical extension of the system is finite or whether the flow is virtually unlimited. For the former (a situation typical for river flow), the dispersion coefficient is proportional to $(\overline{v}_x)^2$:

$$E_{dis} = \beta(\overline{v}_x)^2 \qquad \text{(flow limited)} \qquad (22\text{-}51)$$

where the parameter β (dimension T) is inversely related to the lateral turbulent diffusivity, E_y. In other words, if water parcels change frequently between streamlines (E_y large), dispersion is small since all water parcels seem to travel at the average velocity $\overline{v}_x$. Typical orders of magnitude of E_{dis} in river flow will be given in Chapter 24. In Illustrative Example 22.4 a case is discussed in which the behavior of a chemical in a river is influenced by dispersion.

If the flow is virtually unlimited, as in the case of currents in the ocean or in aquifers, E_{dis} is directly proportional to $\overline{v}_x$:

$$E_{dis} = \alpha_L \, \bar{v}_x \quad \text{(flow unlimited)} \tag{22-52}$$

For the case of groundwater flow, α_L (dimension L) is called *dispersivity*. A further discussion of this type of dispersion follows in Chapter 25.

In Illustrative Example 22.4 we reanalyze the case of the flux of trichloroethene (TCE) from a contaminated aquifer through the unsaturated zone into the atmosphere. As pointed out in Illustrative Example 19.2, measurements of TCE in the soil indicate that vertical advection of air may influence the TCE profile and the flux.

Illustrative Example 22.4
Advanced Topic

Transport of a Volatile Compound From the Groundwater Through the Unsaturated Zone Into the Atmosphere—Illustrative Example 19.2 Reconsidered

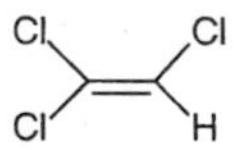

Problem

In Illustrative Example 19.2 we discussed the flux of trichloroethene (TCE) from a contaminated aquifer through the unsaturated zone into the atmosphere. The example was based on a real case of a polluted aquifer in New Jersey (Smith et al., 1996). These authors compared the diffusive fluxes, calculated from measured TCE vapor concentration gradients, with total fluxes measured with a vertical flux chamber. They found that the measured fluxes were often several orders of magnitude larger than the fluxes calculated from Fick's first law. In these situations the vapor profiles across the unsaturated zone were not always linear. The authors attributed this to the influence of advective transport through the unsaturated zone. In order to test this hypothesis you are asked to make the following checks:

trichloroethene
(TCE)

$M_i = 131.4$ g mol^{-1}

(a) Estimate the relative flux enhancement due to advective flow compared to the pure diffusive flux.

(b) Give an upper limit for the effective advection velocity, v_{eff}, below which the flux would be indeed purely diffusive.

Relevant information from Illustrative Example 19.2

Unsaturated zone of aquifer

Depth $L = 4$ m

Porosity $\phi = 0.3$

Volumetric gas content $\theta_g = 0.15$

Diffusivity of TCE in air of unsaturated zone: $D_{uz} = 6.7 \times 10^{-3}$ m^{2}d^{-1}

Fraction of TCE present in gas phase $f_a = 0.05$

Thus: Effective diffusivity $D_{eff} = D_{uz} f_a = 0.05 \times 6.7 \times 10^{-3}$ m^{2}d^{-1} = 3.4×10^{-4} m^{2}d^{-1}

TCE concentrations:

In equilibrium with groundwater $C_o = 25$ μg L^{-1}

In open air $C_a = 0$

Answer (a)

Choose a coordinate system with the z-axis pointing upward, $z = 0$ at the upper boundary of the groundwater table, $z = L = 4$ m at the soil surface. If transport is by diffusion alone, the steady-state concentration profile is linear:

$$C(z) = C_o\left(1 - \frac{z}{L}\right) \tag{1}$$

The flux *per gas-filled pore cross-sectional area* is:

$$F_{diff}^* = D_{eff}\frac{C_o}{L} \tag{2}$$

If transport is by diffusion and advection (effective velocity), the transport equation is

$$\frac{\partial C}{\partial t} = E_{dis}\frac{\partial^2 C}{\partial z} - v_{eff}\frac{\partial C}{\partial z} \tag{3}$$

where v_{eff} is effective velocity (taking into account the effect of tortuosity and sorption of TCE on solids), and E_{dis} combines both the effect of diffusion and of dispersion. For the latter use Eq. 22-52, thus:

$$E_{dis} = D_{eff} + \alpha_L v_{eff} \tag{4}$$

where α_L is the dispersivity of the unsaturated zone.

The solution of Eq. 2 at steady-state leads to the same eigenvalues as in Illustrative Example 22.1:

$$\lambda_1 = \frac{v_{eff}}{E_{dis}}, \qquad \lambda_2 = 0 \tag{5}$$

The boundary conditions correspond to case 1 of Table 22.2 ($C_o = 25$ µg L^{-1}, $C_L = 0$). Thus, the profile for the diffusive-advective case has the form:

$$C^*(z) = \frac{C_o}{1 - e^{-\lambda_1 L}}\left[1 - e^{-\lambda_1(L-z)}\right] \tag{6}$$

The corresponding net flux (diffusion and advection) per gas-filled area is:

$$\begin{aligned}
F_{net}^*(z) &= -E_{dis}\frac{dC^*}{dz} + v_{eff}C^* \\
&= E_{dis}\frac{\lambda_1 C_o}{1 - e^{-\lambda_1 L}}e^{-\lambda_1(L-z)} + \frac{v_{eff}C_o}{1 - e^{-\lambda_1 L}}\left[1 - e^{-\lambda_1(L-z)}\right]
\end{aligned} \tag{7}$$

In fact, at steady-state the net flux, $F_{net}^*(z)$, must be independent of z. Replacing v_{eff} by Eq. 5 yields:

$$F_{net}^* = E_{dis}\frac{\lambda_1 C_o}{1 - e^{-\lambda_1 L}} = \text{constant} \tag{8}$$

The flux ratio between the net (advective + diffusive) and the pure diffusive flux is:

$$\psi \equiv \frac{F_{net}^*}{F_{diff}^*} = \frac{E_{dis}}{D_{eff}} \frac{\lambda_1 L}{1-e^{-\lambda_1 L}} \quad , \qquad \lambda_1 = \frac{v_{eff}}{E_{dis}} \tag{9}$$

Consider the following extreme cases:

(1) $\lambda_1 L \gg 1 \quad \leftrightarrow \quad \dfrac{v_{eff}L}{E_{dis}} = Pe \gg 1$

According to the typology of Fig. 22.3, this corresponds to the case of fast advection ($Da = 0$, Pe not much smaller than 1) with $v_{eff} > 0$.

Then:

$$\psi \approx \frac{E_{dis}}{D_{eff}} \lambda_1 L = \frac{E_{dis}}{D_{eff}} \frac{v_{eff}L}{E_{dis}} = \frac{v_{eff}L}{D_{eff}} \quad (\textit{fast advection}) \tag{10}$$

Note that dispersion no longer appears in this expression.

(2) $\lambda_1 L \ll 1 \quad \leftrightarrow \quad \dfrac{v_{eff}L}{E_{dis}} = Pe \ll 1$

(slow advection, that is, diffusion controlled)

Then use $e^{-\lambda_1 L} \approx 1 - \lambda_1 L$ to evaluate Eq. 9:

$$\psi \approx \frac{E_{dis}}{D_{eff}} \frac{\lambda_1 L}{\lambda_1 L} = \frac{E_{dis}}{D_{eff}} = \frac{D_{eff} + \alpha_L v_{eff}}{D_{eff}} \quad (\textit{slow advection}) \tag{11}$$

Answer (b)

Strong flux enhancement (case 1, Eq. 10) occurs if

$$v_{eff} \gg \frac{E_{dis}}{L} = \frac{D_{eff}}{L} + \frac{\alpha_L}{L} v_{eff}$$

As will be discussed in Chapter 25, α_L/L is typically of the order 10^{-1}. Thus, you can approximately write:

$$v_{eff}\left(1 - \frac{\alpha_L}{L}\right) \approx v_{eff} \gg \frac{D_{eff}}{L} = \frac{3.4 \times 10^{-4} \ m^2 d^{-1}}{4 \ m} = 8.5 \times 10^{-5} \ m \ d^{-1}$$

Then it can be concluded that an extremely small advection velocity is sufficient to produce a significant flux enhancement and to deform the TCE profile correspondingly. Note that case (2) (Pe $\ll$ 1) only occurs under extremely quiet conditions.

Concluding Remarks

This concludes Part IV of the book in which we discussed a series of modeling tools for describing mixing and transport in environmental systems. We found, on the one hand, that random motion is an important agent for transporting chemicals within environmental systems and across boundaries which separates them. On the other hand, directed flow patterns, either in the atmosphere or the hydrosphere, were identified as the main mechanism for the large-scale distribution of chemicals in the environment.

Some of the mathematical tools, such as the linear one-box model, are both fairly simple and nonetheless sufficient for handling a great variety of situations. More complex systems require the use of multibox models. In some cases, continuous time-space models are needed. The mathematics of the latter involve partial differential equations and quickly lead beyond the scope of this book. In this chapter, a few important concepts were discussed which allow the reader both to make some approximative calculations and to critically analyze the results from computer models in which time-space processes are employed.

22.5 Questions and Problems

Questions

Q 22.1

Is it possible that the advective transport of two solutes can simultaneously occur in opposite directions? What about diffusive fluxes?

Q 22.2

Why does the incorporation of advection and diffusion into a model lead to partial differential equations? For what conditions do these equations become normal differential equations again?

Q 22.3

How can diffusive transport be incorporated into box models? Are the resulting expressions partial differential equations?

Q 22.4

Explain Gauss' theorem in words.

Q 22.5

Explain qualitatively the significance of the Damköhler Number and of the Peclet Number.

Q 22.6

In two of the four cases shown in Fig. 22.3, diffusion is mentioned as the dominant process. Explain the difference between these two cases.

Q 22.7

In none of the four cases shown in Fig. 22.3 can diffusion be completely ignored. Why?

Q 22.8

Where does molecular diffusion (in air or water) actually play a role in the environment? Give reasons why it is not relevant in systems which you exclude in your answer.

Q 22.9

Give two frequently used models for turbulent transport in the environment and explain how they relate to the turbulent diffusion coefficient.

Q 22.10

Why is turbulent diffusion scale-dependent?

Q 22.11

Which assumptions are necessary for determining vertical turbulent diffusion coefficients from repeated vertical temperature measurements made at a single location in the middle of a lake?

Q 22.12

Explain the relationship between vertical turbulent diffusivity in surface waters and vertical stratification of the water column.

Q 22.13

Explain the physical meaning of the Brunt-Väisälä frequency.

Q 22.14

What is the problem whith Okubo's diffusion diagram?

Q 22.15

Explain the difference between the dispersion coefficient, E_{dis}, in a river and in the atmosphere. How is E_{dis} related to the mean advection velocity and to lateral turbulent diffusivity in each case?

Problems

P 22.1 *Diffusive Fluxes and Concentration Changes*

Consider the concentration profile $C(x) = C_o e^{-\alpha x}$ along the positive x-axis ($0 \leq x < \infty$), where C_o and α are constant positive parameters. (a) Calculate size and direction of the diffusive flux as a function of x produced by the constant diffusivity D. (b) Calculate the corresponding *in situ* concentration change due to diffusion, $\partial C / \partial t$.

Numbers: $C_o = 1 \text{ mg L}^{-1}$
$\alpha = 0.02 \text{ m}^{-1}$
$D = 10^{-9} \text{ m}^2\text{s}^{-1}$

Evaluate flux, F, and concentration change, $\partial C / \partial t$, at $x = 0$, 10 m, 100 m, 1 km.

P 22.2 *Net Diffusive and Advective Fluxes*

Consider the same profile as in P 22.1. In addition to diffusion, an advective velocity v acts on the profile. (a) Calculate the corresponding additional contribution to the flux and to $\partial C / \partial t$. (b) Determine the relation between v and the other parameters (D, α, C_o) such that the profile, given in P22.1 between $x = 0$ and $x = \infty$ corresponds to a steady-state. Is such a steady-state possible if v > 0?

P 22.3 *Determine Vertical Turbulent Diffusivity in a Lake from Measurements of Tetrachloroethene (PCE)*

tetrachloroethene
(PCE)

In a lake (maximum depth: 20 m) two vertical profiles of tetrachloroethene were measured at a time interval of one month (see table below). Calculate the vertical turbulent diffusivity, E_z, at 8, 12, and 16 m depth. For simplicity assume that the cross-section of the lake, $A(z)$, is independent of depth z. (Note that the same data were used in Problem 20.5 to calculate the air–water exchange rate of PCE.)

Depth z	PCE concentration (10^{-9}mol L^{-1})	
(m)	June 10	July 10
0	1.0	0.5
2	5.0	2.0
4	9.5	3.5
6	5.5	5.5
8	3.9	4.7
10	3.1	3.6
12	2.6	3.0
14	2.3	2.7
16	2.1	2.4
18	2.0	2.2
20	1.9	2.1

Advanced Topic

P 22.4 *Vertical Steady-State Benzene Profile in a Lake with Depth-Dependent Cross Section*

benzene

In Illustrative Example 19.4, the dissolution of a non-aqueous-phase liquid (NAPL) into groundwater was discussed. Here we consider a similar (although somewhat hypothetical) case. Assume that a mixture of chlorinated solvents totally covers the flat bottom of a small pond (maximum depth $z_{max} = 4$ m, surface area $A_{surface} = 10^4 \text{ m}^2$) forming a dense non-aqueous-phase liquid (DNAPL). The DNAPL is contaminated by benzene which dissolves into the water column and is vertically transported by turbulent diffusion. The pond is horizontally well mixed. The vertical turbulent diffusion coefficient is $E_z = 0.1 \text{ cm}^2\text{s}^{-1}$ and approximately constant over the whole water column.

Calculate the vertical steady-state profile of benzene as well as the total vertical flux of benzene, $\Sigma F_{benzene}$ (expressed as mass per unit time), if the following conditions hold:

Lake bottom: $z = 0$ m $A_{bottom} = 2 \times 10^3$ m^2

Lake surface: $z = z_{max} = 4$ m $A_{surface} = 10^4$ m^2

Lake cross section increases linearly from A_{bottom} to $A_{surface}$.

Aqueous benzene concentration in contact with DNAPL at $z = 0$:   0.4 mg/L

in contact with air at $z = 4$ m: 0 mg/L

Hint: Write the lake cross section in the form $A(z) = A_{bottom} + a_1 z$, and determine a_1 from the information given above. Use Fick's first law to express $\Sigma F_{benzene}$. This yields an algebraic expression for $dC_{benzene} / dz$. Then, $C(z)$ and $\Sigma F_{benzene} =$ constant can be determined from the boundary conditions.

Additional Question:
You may have realized that in the above problem the assumption that $C_{surface} = 0$ is not realistic. In fact, even if $C_{air} = 0$, at steady-state the aqueous concentration at the boundary must be larger than 0 in order to induce the air–water flux which compensates for $\Sigma F_{benzene}$ in the water. Use Table 20.5b and the wind speed $u_{10} = 1$ m s^{-1} to determine a realistic value for $C_{surface}$.

Advanced Topic

P 22.5 *Shear Diffusion Model and Apparent Diffusivity*

The shear diffusion model (Eq. 22-44) relates the size of a pollutant patch to the elapsed time since the patch was formed. Apply the relation

$$E_{app} = \frac{1}{4} \frac{d\sigma^2}{dt}$$

for the apparent diffusivity and calculate E_{app} after one day and one week, respectively. Analyze two different conditions (see Box 22.4):

	A_1 (m s^{-1})	A_2 (10^{-10} m^{2}s^{-3})
calm	0.04	1
windy	0.2	100

Calculate also the relative patch dilution between day 1 and day 7.

Distance from Bottom (m)	Activity of Excess Radon (dpm/100 kg)[a]
4	41.6 ± 1.1
11	35.1 ± 3.8
16	30.2 ± 2.0
21	24.7 ± 1.2
27	17.5 ± 1.1
34	12.3 ± 1.8

[a] dpm/100 kg = decay per minute per 100 kg sea water

P 22.6 *Radon Profiles and Vertical Turbulent Diffusivity at the Bottom of the Ocean*

in the benthic boundary layer in the Hatteras Abyssal Plain (North Atlantic). The excess radon activity, that is, the activity of radon-222 in excess of the activity of radium-226, from which radon-222 originates, allows us to determine the vertical turbulent diffusivity, E_z, in the benthic boundary layer. The following table gives one of the profiles, taken on August 6, 1977, at 28°35'N, 70°34'W. Calculate the average E_z value for the lowest 40 m of the water column.

PART V

ENVIRONMENTAL SYSTEMS AND CASE STUDIES

The last part of the book is devoted to the development of simple integrative models of organic chemicals in real environmental systems. The compound-specific tools derived in Parts II and III will be combined with the modeling tools of Part IV. As before, the aim of the discussion is not to compete with the many rather sophisticated models of environmental systems which can be found in the literature, but to concentrate on the simple models which can be analyzed by hand or with desk calculators.

The variety of environmental systems is immense, thus a deliberate choice was necessary. The remaining three chapters of the book are devoted to "typical" cases, that is to standing surface waters (mainly ponds and lakes, although applications to oceans are possible) running waters and subsurface aquifers (mainly the saturated groundwater zone).

To simplify the notation, the compound-specific subscript i is omitted wherever the context is clear.

Chapter 23

PONDS, LAKES, AND OCEANS

23.1 Linear One-Box Models of Lakes, Ponds, and Oceans
Box 23.1: *Linear One-Box Model for Well-Mixed Volumes of Ponds, Lakes, and Oceans*
Illustrative Example 23.1: A *Vinyl Acetate Spill Into a Pond*

23.2 The Role of Particles and the Sediment–Water Interface
Partitioning Between Dissolved and Particulate Phases
Particle Settling
PCBs in Lake Superior (Part 1)
Exchange at the Sediment–Water Interface
Box 23.2: *Model of Sediment–Water Exchange*
PCBs in Lake Superior (Part 2)

23.3 Two-Box Models of Lakes
Two-Box Model for Lake/Sediment System
Box 23.3: *Solution of Linear Water–Sediment Model*
PCBs in Lake Superior (Part 3)
Other Two-Box Models

23.4 One-Dimensional Continuous Lake Models (*Advanced Topic*)
Internal Transport versus Reaction and Boundary Fluxes
One-Dimensional Vertical (1DV) Lake Model
Numerical Models
Box 23.4: *Numerical Approximation of Partial Differential Equations of One-Dimensional Vertical Lake Model*
Application of the 1DV Lake Model

23.5 Questions and Problems

The world oceans cover 361×10^6 km^2 or 71% of the surface of this planet. They have a total volume of 1370×10^6 km^3 and a mean depth of 3.8 km (Table 23.1).

There are about 5 million lakes with a surface area greater than 0.1 km^2 (Table 23.2) and countless smaller lakes and ponds. Their total volume exceeds 180×10^3 km^3 (0.013% of the ocean volume), and their total surface area is about 1.6×10^6 km^2 (0.44% of the ocean surface or 0.31% of the earth surface). Yet, nearly all of the total lake volume and close to 90% of the surface area is concentrated in the roughly 250 lakes with surface areas greater than 500 km^2.

The largest lake, the Caspian Sea, has a volume of 78,200 km^3 (43% of total lake volume). Among the remaining (mostly freshwater) lakes, Lake Baikal has the largest volume (about 22% of the total volume of all freshwater lakes) and the greatest depth (1,630 m).

This chapter deals with the integrative modeling of organic pollutants in standing surface waters. Our approach is similar to the description of engineered systems like reactors. From a didactic point of view, ponds and lakes are well suited to demonstrate how the interplay of transport and reaction processes determines the spatial and temporal distribution of organic compounds in natural environments. For such systems, simple back-of-the-envelope calculations using one- or two-box models may already help one answer important questions, including a quick assessment of the most important

Table 23.1 Physical, Chemical, and Biological Characteristics of the World Oceans [a]

Surface area	361×10^6 km^2
Volume	1370×10^6 km^3
Mean depth	3.8 km
Thermocline depth (typical values)	50 – 200 m
Typical residence time of deep ocean water	1000 yr
Salinity mean	34.7 ‰
$\quad\quad\quad$ 90% of ocean water between 34 and 35‰	
pH of seawater	8.0
Concentration of suspended solids (solid-to-water phase ratio, r_{sw})	
$\quad$ surface water	$10^{-4} - 5 \times 10^{-3}$ kg m^{-3}
$\quad$ deep water	$(1-5) \times 10^{-5}$ kg m^{-3}
Fraction of organic matter of suspended solids (expressed as organic carbon), f_{oc}	$0.02 - 0.2$ kg$_{oc}$kg$_s^{-1}$
Typical decadal attenuation coefficient, $\alpha(\lambda)$, in seawater at $\lambda = 500$ nm (see Section 15.2)	
$\quad$ coastal water	0.2 m^{-1}
$\quad$ open water	0.05 m^{-1}

[a] The figures originate from different sources; most of them are from Neumann and Pierson (1966) and from Horne (1969)

Table 23.2 Orographic Characteristics of Lakes

Lakes with area > 0.1 km² [a]					
	Number		$\approx 5 \times 10^6$		
	Total surface area		$1.6 \times 10^6\,\mathrm{km}^2$		
	Total volume		$> 180 \times 10^3\,\mathrm{km}^3$		
Large lakes (area > 500 km²) [b]					
	Number		253		
	Total surface area		$1.4 \times 10^6\,\mathrm{km}^2$		
	Total volume		$179 \times 10^3\,\mathrm{km}^3$		

		Volume (km³)	Area (km²)	Max. Depth (m)	
1	Caspian	78,200	374,000	1025	brackish
2	Baikal	23,000	31,500	1630	
3	Tanganyika	18,900	32,900	1470	
4	Superior	12,230	82,100	407	
5	Nyasa	6,140	22,490	706	
6	Michigan	4,920	57,750	282	
7	Huron	3,537	59,500	229	
8	Victoria	2,700	68,460	92	
9	Great Bear	2,381	31,326	452	
10	Great Slave	2,088	28,568	625	
11	Issykkul	1,730	6,240	702	brackish
12	Ontario	1,637	19,000	245	

[a] Meybeck (1995). [b] Herdendorf (1990)

processes. To this end, it is useful to express each process by a first-order or a pseudo-first-order rate law exhibiting a characteristic rate constant (see Chapter 21). By doing so, the relative importance of each of the processes may be determined immediately by a direct comparison of the characteristic (pseudo-)first-order rate constants.

In Section 23.1, this procedure will be applied to just one completely mixed water body. This "control volume" may represent the lake as a whole or some part of it (e.g., the mixed surface layer). Section 23.2 deals with the dynamics of particles in lakes and their influence on the behavior of organic chemicals. Particles to which chemicals are sorbed may be suspended in the water column and eventually settle to the lake bottom. In addition, particles already lying at the sediment–water interface may act as source or sink for the dissolved chemical. In Section 23.3, two-box models of lakes are discussed, particularly a model consisting of the water body as one box and the sediment bed as the other. Finally, in Section 23.4, one-dimensional vertical models of lakes and oceans are discussed.

The reader will see that this chapter does not really introduce new mathematical concepts. In fact, all tools were derived in Chapters 18 to 22. To facilitate the reading, the necessary equations will be summarized at the beginning of each section.

23.1   Linear One-Box Models of Ponds, Lakes, and Oceans

In Part IV we repeatedly used box models for describing the dynamics of chemicals in lakes. In this chapter we will summarize this information. As a first step, Fig. 23.1 illustrates the one-box model approach for the average total concentration of a chemical, C_t, in a well-mixed water body such as a pond, a shallow lake, a subcompartment of a deep lake or ocean (e.g., the mixed surface layer), or even an engineered system like a completely stirred reactor.

The mathematical formulation of the model and the definition of the variables and parameters are given in Box 23.1. Note that the model is based on the following simplifying assumptions:

1. The total concentration is given by the sum of the dissolved and particle-bound concentrations, $C_t = C_d + C_p$. All concentrations are expressed as mass per bulk volume. It is assumed that the dissolved and particulate phases are always equilibrated (see Chapter 9).

2. All internal transformations (chemical, photochemical, biological) in which the chemical is consumed, are described by first-order processes. The *in situ* production of the compound is described as a zero-order process, j. This includes the case in which the compound may be produced by (e.g., first-order) transformation of another compound, provided that the concentration of the latter is approximately constant.

3. The external processes (boundary fluxes) can be combined into four pairs of generalized exchange fluxes; that is: (a) input/output by streams, rivers, or ground-water, (b) air–water exchange, (c) sediment–water exchange, (d) exchange with adjacent water compartments. If the box represents a pond or lake as a whole, flux (d) does not exist. The fluxes into the system are controlled by external parameters such as the concentration in the inlets, the atmospheric and the sedimentary concentrations. These concentrations can be constant or variable with time.

4. The fluxes out of the system and the internal transformations are all formulated as functions of the total concentration C_t, although certain processes are controlled by C_d or C_p alone. This is taken into account by multiplying the rate constants by the appropriate equilibrium fraction, f_w or $(1 - f_w)$, respectively (see Eq. 9-12).

Most processes listed in Box 23.1 have been discussed before. However, the term describing the settling of suspended particles, $k_s^* C_t$, and the exchange between the water and the sediments, $k_{sedex}[C_{sed}^{eq} - C_t]$, need further explanation. A detailed discussion follows in Section 23.2. First, we discuss the fate of vinyl acetate in a pond (Illustrative Example 23.1) for which the solute–particle interaction is not important.

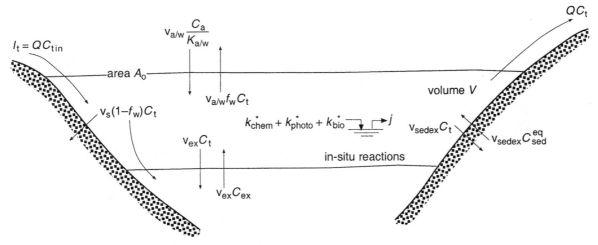

Figure 23.1 General view of a linear one-box model of a well-mixed pond, a lake, or part of a lake or ocean. See Box 23.1 for definitions. If the box represents the complete water body of a lake, the terms with v_{ex} (water exchange with adjacent boxes) do not exist. Similarly, if the box is not in contact with the atmosphere, the air–water exchange flux ($v_{a/w}$) is absent.

Box 23.1 Linear One-Box Model for Well-Mixed Volumes of Ponds, Lakes, and Oceans

Note: To simplify the notation, the subscript i indicating a specific chemical is omitted.

The equation for the linear one-box model of a mixed water body (Fig. 23.1) is given by:

$$\frac{dC_t}{dt} = J - k^* C_t \tag{1}$$

where:

$$J = j + k_w C_{tin} + k_{a/w} \frac{C_a}{K_{a/w}} + k_{sedex} C_{sed}^{eq} + k_{ex} C_{ex} \tag{2}$$

$$k^* = k_w + k_{a/w}^* + k_s^* + k_{sedex} + k_{ex} + k_{chem}^* + k_{photo}^* + k_{bio}^* \tag{3}$$

The asterisk indicates that the specific rate constants refer to the total concentration C_t. Some processes involve only the dissolved phase (like air–water exchange $k_{a/w}^*$), others the particle phase (like sedimentation k_s^*), and still others may be a mixture of both phases, e.g.,

$$k_{bio}^* = f_w k_{bio,d} + (1 - f_w) k_{bio,p} \tag{4}$$

Definitions:

V	$[L^3]$	volume of water body (constant)
A_o	$[L^2]$	surface area of water body
$h = V/A_o$	$[L]$	mean depth
Q	$[L^3 T^{-1}]$	discharge rate

$$C_t = C_d + C_p$$

[ML^{-3}]   total mean concentration (subscript t) of chemical in the system (box) expressed as the sum of dissolved (d) and sorbed (p) concentration

J

[ML^{-3}T^{-1}]   total external input of chemical per unit volume and time

j

[ML^{-3}T^{-1}]   in-situ production of chemical per unit volume and time

k^*

[T^{-1}]   total specific elimination rate

$k_w = Q / V$

[T^{-1}]   specific river input/output rate (Eq. 12-50)

$k_{a/w}^* = f_w k_{a/w} = f_w \dfrac{v_{a/w}}{h}$

[T^{-1}]   specific rate constant of air–water exchange (Eq. 21-8c)

$k_s^* = (1 - f_w) k_s = (1 - f_w) \dfrac{v_s}{h}$

[T^{-1}]   specific removal rate by particle settling (Eq 23-19)

k_{sedex}

[T^{-1}]   specific exchange rate between water and sediment column (diffusion and sediment resuspension, see Box 23.2, Eq. 6)

$k_{ex} = \dfrac{v_{ex}}{h}$

[T^{-1}]   specific removal rate by water exchange with adjacent box or boxes (Eq. 21-40 or 21-41)

k_{chem}^*

[T^{-1}]   specific chemical transformation rate (hydrolysis etc.)

k_{photo}^*

[T^{-1}]   specific photochemical transformation rate

k_{bio}^*

[T^{-1}]   specific biological transformation rate

$C_{tin} = I_t / Q$

[ML^{-3}]   mean total concentration in inflow (Eq. 21-8a)

C_a

[ML^{-3}]   atmospheric concentration

C_{sed}^{eq}

[ML^{-3}]   apparent sediment concentration for input (Box 23.2, Eq. 7)

C_{ex}

[ML^{-3}]   total concentration in adjacent box

$f_w = \dfrac{1}{1 + r_{sw} K_d}$

[–]   dissolved fraction of chemical (Box 18.5, Eq. 3)

K_d

[M^{-1}L^3]   solid–water distribution coefficient (Eq. 9-7)

r_{sw}

[ML^{-3}]   solid-to-water phase ratio

Solution of Eq. 1 for constant parameters (Box 12.1, Eq. 6)

$$C_t(t) = C_{to}\,e^{-k^* t} + C_{t\infty}\left(1 - e^{-k^* t}\right) \tag{4}$$

C_{to} concentration in system at time $t = 0$

$$C_{t\infty} = \frac{J}{k^*} \quad \text{steady-state concentration} \tag{5}$$

Illustrative Example 23.1   A Vinyl Acetate Spill Into a Pond

Problem

Due to an accident during the summer holidays, an unknown amount of vinyl acetate is introduced into a small, well-mixed pond located in the center of a city. Working in a consulting firm, you are asked (1) to determine the concentration of vinyl acetate in the pond, (2) to estimate how much vinyl acetate has entered the pond, and, most importantly, (3) to say something about the half-life of this compound in the pond.

Because your laboratory technician is out of town, it takes you 10 days until you are ready to make the first measurement. At this time (i.e., 10 days after the input), the measured vinyl acetate concentration in the pond water is 50 µg L^{-1}. Ultimately, the people from the town want to know from you how long it will take until the concentration drops below 1 µg L^{-1}. Try to answer this question by making the (worst-case) assumption. Also calculate how much vinyl acetate has at least been introduced into the pond. Base your answers on the data given below .

Characteristic Data of Pond

Volume	V	$= 10^4\,\text{m}^3$
Surface area	A_o	$= 2 \times 10^3\,\text{m}^2$
Water throughflow	Q	$= 10^2\,\text{m}^3\text{d}^{-1}$
Average wind speed above the pond	u_{10}	$= 1.5\ \text{m s}^{-1}$
Water temperature	T_w	$= 25°\text{C}$
	pH	$= 7.0$
Geographical latitude		$47.5°\text{N}$
Rate constants of hydrolysis	$k_A(25°\text{C})$	$= 1.4 \times 10^{-4}\text{M}^{-1}\text{s}^{-1}$
	$k_N(25°\text{C})$	$= 1.1 \times 10^{-7}\text{s}^{-1}$
	$k_B(25°\text{C})$	$= 10\,\text{M}^{-1}\text{s}^{-1}$
Vapor pressure	$p_{iL}^{\cdot}(25°\text{C})$	$= 0.14\ \text{bar}$
Melting temperature	T_m	$= -92.8°\text{C}$
Boiling temperature	T_b	$= 72.5°\text{C}$
Solubility	C_{iw}^{sat}	$= 0.23\ \text{mol L}^{-1}$

Answer

The relevant processes are: (a) flushing, (b) air–water exchange, (c) hydrolysis.

(a) Flushing: $k_w = Q/V = 10^2\text{m}^3\text{d}^{-1}/10^4\text{m}^3 = 0.01\ \text{d}^{-1}$

(b) Air–water exchange: The Henry's law coefficient of vinyl acetate can be estimated from p_L^* and C_w^{sat}:

$$K_H \approx p_L^* / C_w^{\text{sat}} = 0.14\ \text{bar} / 0.23\ \text{mol L}^{-1} = 0.61\ \text{L bar mol}^{-1}$$

According to Fig. 20.7, $v_{a/w}$ is water-phase controlled. From Eq. 20-16:

$$v_{CO_2 w}(u_{10} = 1.5\ \text{m s}^{-1}) = 0.65 \times 10^{-3}\text{cm s}^{-1}$$

Thus, for vinyl acetate (see Eqs. 20-24a, 20-25, 18-55):

$$v_w = \left(\frac{D_w}{D_{CO_2 w}}\right)^{2/3} v_{CO_2 w} = \left(\frac{M_i}{M_{CO_2}}\right)^{-1/3} v_{CO_2 w}$$

$$= 0.65 \times 10^{-3}\text{cm s}^{-1}\left(\frac{86.1}{44}\right)^{-1/3} = 0.52 \times 10^{-3}\text{cm s}^{-1}$$

Finally:

$$k_{a/w} = \frac{v_w}{h} = \frac{v_w A_o}{V} = \frac{0.52 \times 10^{-5} \text{m s}^{-1} \times 2 \times 10^3 \text{m}^2}{10^4 \text{m}^3}$$

$$= 1.04 \times 10^{-6} \text{s}^{-1} = 0.09 \text{ d}^{-1}$$

(c) Hydrolysis:

$$k_{chem} = k_A[\text{H}^+] + k_N + k_B[\text{OH}^-] \cong (1.1+10)\ 10^{-7}\text{s}^{-1} = 0.096 \text{ d}^{-1}$$

Since after the spill the input of vinyl acetate in the pond is zero, the one-box model has the form (note $C_t = C_w$, thus $k^* = k$):

$$\frac{dC_w}{dt} = -k\,C_w, \qquad k = k_w + k_{a/w} + k_{chem} \approx 0.20 \text{ d}^{-1}$$

$$\Rightarrow C_w(t) = C_{wo}\,e^{-kt} \qquad \text{with } t = 10 \text{ d}$$

$$C_{wo} = C_w(t = 10 \text{ d})\,e^{kt} = 50 \text{ µg L}^{-1} \times 7.4 = 370 \text{ µg L}^{-1}$$

Thus, the input, I, of vinyl acetate was at least?

$$I = VC_{wo} = 370 \text{ mg m}^{-3} \times 10^4 \text{m}^3 \cong 3.7 \text{ kg}$$

The time, t_1, needed for the concentration to drop to 1 µg L^{-1} can be calculated from the relation:

$$1 \text{ µg L}^{-1} = 370 \text{ µg L}^{-1}\,e^{-kt_1}$$

$$t_1 = -\frac{1}{k}\ln\left(\frac{1}{370}\right) = \frac{1}{0.2 \text{ d}^{-1}} \times 5.9 = 30 \text{ d}$$

$$t_1: \text{time from accident until } C_w = 1 \text{ µg L}^{-1}$$

Problem

Since you are interested in checking whether your predictions were right, you measure the vinyl acetate concentration in the pond again 20 days after the accident. The value that you obtain is now 4 µg L^{-1}. Compare this result with your predictions. Try to explain any discrepancies, and if necessary, revise your answers given 10 days after the accident.

Answer

From the two measurements (at $t = 10$ and 20 d) the real total elimination rate k_{real} can be calculated:

$$4 \text{ µg L}^{-1} = 50 \text{ µg L}^{-1}\,e^{k_{real}\,10d}$$

$$\Rightarrow k_{real} = -\frac{1}{10d}\ln\left(\frac{4}{50}\right) = 0.25 \text{ d}^{-1}$$

The difference between k and k_{real} is most probably due to biodegradation. When using the experimentally determined total rate constant, the estimated input is calculated as:

$$I = 50 \text{ mg m}^{-3}\,e^{2.5} \times 10^4 \text{m}^3 = 6.1 \times 10^6 \text{mg} = 6.1 \text{ kg}$$

The concentration 1 $\mu g\ L^{-1}$ is reached at time t_{1real}:

$$t_{\text{1real}} = -\frac{1}{0.25\ \text{d}^{-1}} \ln\left(\frac{1}{610}\right) \cong 26\ \text{d}$$

(time after the accident).

The Role of Particles and the Sediment–Water Interface

Partitioning Between Dissolved and Particulate Phases

In Chapter 21 on box models no distinction was made between a compound being present as a dissolved species or sorbed to solid surfaces (e.g., suspended particles, sediment–water interface). In Boxes 18.5 and 19.1, and also in Illustrative Example 19.6, we learned that several of the transport and transformation processes may selectively act on either the dissolved or the sorbed form of a constituent. For instance, a molecule sitting on the surface of a sedimentary particle at the lake bottom does not feel the effect of turbulent flow in the lake water, while the dissolved chemical species is passively moved around by the currents. In contrast, a molecule sorbed to a suspended particle (e.g., an algal cell) can sink through the water column because of gravity, unlike its dissolved counterpart.

In this section the distinction between dissolved and sorbed species is introduced into the box model concept in the simplest possible manner, that is, by assuming a reversible linear equilibrium relationship between the dissolved concentration, $C_w(\text{mol m}_w^{-3})$, and the species sorbed on solids, $C_s(\text{mol kg}_s^{-1})$. (The units m_w^3 and kg_s refer to water volume and solid mass, respectively.) The sorption/desorption process shall be fast compared to other processes which affect the chemical (e.g., mixing, chemical transformation). As discussed in Chapter 9 (Eq. 9-7), the (observed) solid–water distribution ratio K_d is defined by:

$$K_d = \frac{C_s}{C_w} \qquad (m_w^3 kg_s^{-1}) \tag{23-1}$$

Note that in this chapter the volume is expressed in cubic meters and not, as in Chapter 9 and in most other chapters, in liters (L). Thus, K_d values are numerically smaller by 10^{-3}; densities ρ and the solid-to-solution phase ratio r_{sw} are larger by 10^3.

From Chapter 9 we remember that K_d may represent a composite of several kinds of sorption processes (Eq. 9-16). For most neutral apolar or weakly monopolar chemicals, however, only sorption to natural organic matter may be relevant. In that case, we can use Eq. 9-22 to express K_d:

$$K_d = f_{oc} K_{oc} \qquad (m_w^3 kg_{oc}^{-1}) \tag{9-22}$$

where f_{oc} is the fraction of the solid which is natural organic carbon, and K_{oc} is the sorption coefficient reflecting partitioning between particulate organic carbon and solution phases.

In box models the concentration variables have to be related to the same reference volume. We use total (bulk) volume (indicated by m^3, without index). Thus, the dissolved and sorbed (particulate) concentrations per total volume, C_d and C_p are

$$C_d = \frac{V_w}{V_t} C_w = \phi\, C_w \qquad (\text{mol m}^{-3}) \qquad (23\text{-}2)$$

$$C_p = C_s (M_s / V_t) \qquad (\text{mol m}^{-3}) \qquad (23\text{-}3)$$

where $\phi = V_w/V_t$ is the volumetric fraction occupied by water (in environments with many particles, such as sediments, ϕ is called porosity), and (M_s/V_t) is the mass of solids per total volume. We can express the "single phase concentrations" C_d and C_p in terms of the total concentration, $C_t = C_d + C_p$, and the fraction of the chemical in solution, f_w, introduced in Eq. 9-12:

$$f_w = \frac{1}{1 + r_{sw} K_d} \qquad (9\text{-}12)$$

or for many neutral apolar and weakly monopolar chemicals:

$$f_w = \frac{1}{1 + f_{oc} r_{sw} K_{oc}} \qquad (23\text{-}4)$$

where r_{sw} is the solid-to-solution phase ratio (now in kilograms per cubic meter). Thus,

$$C_d = f_w C_t \quad \text{and} \quad C_p = (1 - f_w) C_t \qquad (23\text{-}5)$$

In open water bodies the volumetric fraction occupied by particles is so tiny that the "porosity" ϕ is equal to 1. Also, since $V_t \sim V_w$, the mass of solids per total volume (M_s/V_t) is about equal to $r_{sw} = M_s/V_w$. Thus, we can approximate Eqs. 23-2 and 23-3 by:

$$C_d \sim C_w \quad \text{and} \quad C_p \sim r_{sw} C_s \qquad (23\text{-}6)$$

The degree of partitioning between the dissolved and particulate phases depends on both system-specific properties (solid-to-solution phase ratio, composition of solids) and compound-specific properties (e.g., K_{oc}). In the open water column, r_{sw} ranges typically between 10^{-5} kg_sm^{-3} (deep sea) and 5×10^{-3} kg_sm^{-3} (surface waters, lakes) with organic carbon weight fractions, f_{oc}, between 0.02 and 0.2 (Table 23.1). Assuming a maximum particulate organic matter–to-solution ratio, $f_{oc}r_{sw}$, of 10^{-3} $\text{kg}_{oc}\text{m}^{-3}$, compounds with K_{oc} values below 1 $\text{m}^3\text{kg}_{oc}^{-1}$ have solid fractions, $(1 - f_w)$, less than 10^{-3} and can thus, from the point of view of transport processes, be considered to be totally dissolved. However, for transformation processes the sorbed fraction, in spite of its small size, may be relevant for compounds with important sorbed-phase reaction.

Note that in sediments, soils, and aquifers, $f_{oc}r_{sw}$ can reach values up to 100 $\text{kg}_{oc}\text{m}^{-3}$. This means that, even for "nonsorptive" compounds with fairly small K_{oc} values of 0.1 $\text{m}^3\text{kg}_{oc}^{-1}$, about 90% of the total concentration can be bound to particles ($f_w \sim 0.1$).

Particle Settling

In order to understand how the association of compounds with particles may influence their transport, we need to discuss how particles move in the water column. Besides the passive motion with the water in which they are suspended, particles feel the force of gravity. For laminar conditions, the settling velocity v_s of a particle with density ρ_s through a fluid with density $\rho_w < \rho_s$ can be described by Stokes law (Lerman, 1979):

$$v_s = \alpha B r^2 \quad (\text{m s}^{-1}) \tag{23-7}$$

where r is a linear dimension characterizing the size of a particle, α is a nondimensional form factor, and B is a factor that depends on the nature of the fluid and the particle density:

$$B = \frac{2g(\rho_s - \rho_w)}{9\eta} \quad (\text{m}^{-1}\text{s}^{-1}) \tag{23-8}$$

where $g = 9.81$ m s^{-2} is the acceleration due to gravity, and η is the dynamic viscosity of the fluid (kg m^{-1}s^{-1}). For a sphere, the form factor α is 1 and r is the sphere radius. Using quartz spheres ($\rho_s = 2650$ kg m^{-3}) and water at 20°C ($\eta = 10^{-3}$ kg m^{-1}s^{-1}), the typical size of B is 3.6×10^6 m^{-1}s^{-1}. Thus for spheres with $r = 1$ μm, v_s is 3.6×10^{-6} m s^{-1} (0.3 m d^{-1}).

The particle-specific Reynolds number (see also Eq. 18-69),

$$\text{Re} = \frac{2r v_s}{\eta / \rho_w} \tag{23-9}$$

determines whether the flow past the particle is laminar or turbulent. If Re is smaller than the critical value 0.1, the flow is laminar and Eq. 23-7 applies. For turbulent conditions (Re > 0.1), the resisting force of the water on a settling particle is different from the Stokes drag used to derive Eq. 23-7 (see e.g., Håkanson and Jansson, 1983). Combining Eqs. 23-7 and 23-9, replacing Re by the critical value (0.1), and solving for the particle radius r yields the following condition for laminar flow past the suspended particle:

$$r < \left\{ \frac{0.1(\eta / \rho_w)}{2\alpha B} \right\}^{1/3} = \left(\frac{0.225}{\alpha g} \frac{\eta^2}{\rho_w(\rho_s - \rho_w)} \right)^{1/3} \quad (\text{laminar settling}) \tag{23-10}$$

For quartz spheres, the maximum particle radius compatible with Eq. 23-10 is only 24 μm. In contrast, biogenic particles have a much smaller excess density, $\rho_s - \rho_w$, and complicated shapes (α small). Typical settling velocities for these particles are 0.1 m d^{-1} for 1-μm particles and 100 m d^{-1} for 100-μm particles (Lerman, 1979). The 100-μm particles just meet the laminar flow limit (Re $\approx$ 0.1).

In Box 21.7 and again in Box 23.1, we assumed that the removal of suspended solids from the water column (and thus the removal of the sorbed concentration of a chemical, C_s) is a linear process. It is to be expected that the corresponding compound-specific removal rate to the sediments, k_s^*, is related to the suspended particle settling velocity, v_s, of Eq. 23-7. We will derive this relation by calculating the particle removal rate from a water body due to settling. First, consider a rectangular tank in which particles of equal size and density (i.e., of equal sinking velocity) are initially homogeneously suspended (Fig 23.2a). The initial suspended matter concentration is r_{sw}. In the absence of any currents in the tank, after some time Δt, the top layer of depth

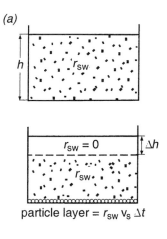

Figure 23.2 Removal of suspended particles (described by the solid-to-water phase ratio r_{sw}) with uniform sinking velocity v_s. (a) No mixing: constant particle flux $F_s = r_{sw} v_s$ until upper "horizon" reaches the bottom after time $t = h / v_s$; (b) homogeneously mixed system: exponential decrease of r_{sw}; (c) change of mean particle flux across level z_o for the case of heterogeneous distribution of particles and a spatially variable vertical velocity component.

$\Delta h = v_s \Delta t$ will be devoid of particles, and a particle layer of areal mass density $r_{sw} \Delta h = r_{sw} v_s \Delta t$ ($\mathrm{kg_s m^{-2}}$) covers the bottom. Thus, the initial relative particle removal rate k_s is given by:

$$k_s = \frac{\text{particles removed per area and time } \Delta t}{\text{particles in system per area}}$$

$$= \frac{r_{sw} v_s}{r_{sw} h} = \frac{v_s}{h} \quad (\mathrm{s^{-1}}) \tag{23-11}$$

The corresponding particle mass flux per unit area through any horizontal cross section is:

$$F_s = r_{sw} v_s \quad (\mathrm{kg_s m^{-2} s^{-1}}) \tag{23-12}$$

Now, let us assume that the water body is kept well mixed by turbulent fluid motion in the tank (Fig. 23.2b). The vertical motion of a particle is now given by the algebraic sum of the local vertical current velocity v_z and the settling velocity v_s: $v_{z,tot} = v_s + v_z$. The total particle flux, ΣF_s, across an interface at an arbitrary depth z_o is given by the integral of Eq. 23-12 over the area A_o, whereby v_s is replaced by the total vertical particle velocity, $v_{z,tot}$:

$$\sum F_s = \int_{A_o} r_{sw}(x, y, z_o)\left[v_s + v_z(x, y, z_o)\right] dxdy \qquad (23\text{-}13)$$

Note that the settling velocity v_s does not depend on the space coordinates (x, y, z) since we have assumed particles of equal size and mass. If the concentration of the suspended particles is spatially constant, $r_{sw}(x, y, z) = r_{sw}$, Eq. 23-13 becomes:

$$\sum F_s = r_{sw} v_s A_o + r_{sw} \int_{A_o} v_z(x, y, z_o) dxdy \qquad (23\text{-}14)$$

In the absence of any subsurface source or sink of water, the law of mass conservation applied to the water itself requires the remaining integral in Eq. 23-14 to be zero. Thus, in spite of the water currents, the mean flux per unit area at z_o, $\overline{F}_s = \sum F_s / A_o$, looks like the flux in a quiescent water body (Eq. 23-12).

Let us assume that turbulence in the tank keeps the suspended particle concentration homogeneous, but that at the bottom of the tank the particles can sink through some screen below which no water currents exist (Fig. 23.2b). In the absence of any external particle fluxes or *in situ* production/removal of particles, the mass balance equation for suspended particle mass is given by equating the rate of change of particle mass in the water volume V with time with the rate of loss due to settling:

$$V\frac{dr_{sw}}{dt} = -A_o v_s r_{sw} \qquad (23\text{-}15)$$

or after division by V:

$$\frac{dr_{sw}}{dt} = -k_s r_{sw} \qquad \text{with } k_s = \frac{v_s}{h} \qquad (23\text{-}16)$$

where $h = V/A_o$ is the mean depth and k_s [T^{-1}] is a first-order particle removal rate. Note the similarity between Eqs. 23-11 and 23-16. For the nonmixed system the sedimentation flux remains constant until all particles have disappeared from the system (this occurs after time h/v_s), whereas the removal for the mixed system is first order, leading to an exponential decrease of r_{sw}. There is a strong resemblance between the mixed-particle removal process and the process of air–water transfer. Therefore, the corresponding rates, k_s and $k_{a/w}$ (see Eq. 21-8c), have the same form.

The dynamic equation of r_{sw} (Eq. 23-16) now can be used to describe the effect of particle settling on a chemical which is sorbed to the suspended particles:

$$\left(\frac{dC_p}{dt}\right)_{settling} = -k_s C_p \qquad (23\text{-}17)$$

If the sorbed and aqueous phase are always at equilibrium, we can use Eq. 23-5 to express the effect of particle settling on the total concentration C_t:

$$\left(\frac{dC_t}{dt}\right)_{settling} = -k_s\left(1 - f_w\right)C_t = -k_s^* C_t \qquad (23\text{-}18)$$

where

$$k_s^* = \left(1 - f_w\right) k_s = \left(1 - f_w\right)\frac{v_s}{h} \qquad (23\text{-}19)$$

is the substance-specific removal rate by particle settling.

We have now developed the necessary equations to formulate box models in which a distinction is made between the dissolved and sorbed phase of a given compound. It seems that knowledge of the settling velocity, v_s, together with the mean depth of the box, is all that is needed to calculate particle fluxes and thus fluxes of the compound sorbed to the particles. The settling velocities, in turn, can be estimated by Stokes law (Eq. 23-7). Unfortunately, nature is not as simple. Settling velocities, which have been indirectly determined by looking at the fate of sorbed species in natural water bodies (so-called *apparent settling velocities*), not only vary a great deal with time, but also are generally much smaller than the velocities calculated from Stokes law (Bloesch and Sturm, 1986). There are several reasons for this discrepancy. One has to do with the omnipresence of horizontal water currents. What we have discussed so far as a vertical displacement is, in fact, mainly a horizontal journey of particles with a tiny vertical net component. A typical sinking velocity of biogenic particles of 1 m d^{-1} or about 10^{-3} cm s^{-1} coexists in a water body with typical horizontal velocities of the order of 1 cm s^{-1}. Thus, the process of particle sinking looks like snow flakes during a wind storm! Snow does not accumulate homogeneously during a storm if the wind velocities are 1000 times larger than the sinking velocity of the flakes.

To be more specific, let us assume that under the influence of a specific current pattern the particle concentration and the vertical current velocities, v_z, in a given horizontal plane are not constant. We assume above-average r_{sw} values in those areas where the currents are predominantly upward, and below-average r_{sw} values at downwelling areas (Fig 23.2c). As a result, the mean particle flux (calculated from Eq. 23-13) is smaller than the value calculated from Eq. 23-12. As first shown by Stommel (1949), such current structures supporting inhomogeneous particle distributions do indeed exist in lakes and oceans. Langmuir circulation is one example of a circular current system in which small particles can be trapped and exhibit apparent settling velocities of zero (Ledbetter, 1979).

We should also note that particles in natural water bodies do not come in a single size class; thus sinking velocities are not the same for all suspended solids. Colloidal particles do not sink at all, while fecal pellets can sink several hundred meters per day. Therefore, organisms have a strong influence on v_s, since they constantly change the size and nature of the suspended solids. In addition, particle coagulation (O'Melia, 1985) and breakup constantly change the particle size spectrum. Finally, particles can be taken up by organisms or are organisms themselves which can influence their sinking velocity either by active swiming or by adjusting their density.

In conclusion, the first-order particle removal model, Eq. 23-16, is a reasonable approximation to describe the influence of suspended particles on sorbed chemical species. However, the sinking velocity is not necessarily identical with the Stokes law velocity, but rather is an empirical parameter which may strongly depend on lake currents and biological processes and may vary over short time periods.

Table 23.3 Characteristic Data of Lake Superior (North America)[a]

Volume	V	12,230 km^3
Surface area	A_o	82,100 km^2
Mean depth	$h = V/A_o$	149 m
Maximum depth	h_{max}	406 m
Throughflow of water (including precipitation)	Q	71 km^{3}yr^{-1}
Flushing rate	$k_w = Q/V$	5.8×10^{-3} yr^{-1}
Mean water residence time	$\tau_\omega = k_w^{-1}$	172 yr
Flushing velocity	$q = Q/A_o$	0.86 m yr^{-1}
Solid-to-water phase ratio[b]	r_{sw}	0.4×10^{-3} kg$_s$m^{-3}
Organic carbon content of suspended solids	f_{oc}	0.2 kg$_{oc}$kg$_s^{-1}$
Organic carbon content of settled solids	f_{ocs}	0.03 kg$_{oc}$kg$_s^{-1}$
Porosity of surficial sediments	ϕ	0.9
Sediment accumulation		
on 50% of lake area		0.2 kg$_s$m^{-2}yr^{-1}
mean for total lake	F_s	0.1 kg$_s$m^{-2}yr^{-1}
Mean particle-settling velocity (Eq. 23-12)	$v_s = F_s/r_{sw}$	250 m yr^{-1}
Particle removal rate (Eq. 23-16)	$k_s = v_s/h$	1.7 yr^{-1}
Mean wind speed	u_{10}	5 m s^{-1}
Air–water transfer velocity for		
water vapor (Eq. 20-15)	v_a (H$_2$O)	1100 m d^{-1}
Air–water transfer velocity for CO$_2$ (Eq. 20-17)	v_w (CO$_2$)	1.2 m d^{-1}

[a] From Eisenreich et al. (1989). [b] For surface waters, r_{sw} is approximately equal to the suspended solid concentration.

PCBs in Lake Superior (Part 1)

Let us demonstrate the power of the one-box lake model by analyzing the fate of two different polychlorinated biphenyl congeners (PCBs) in Lake Superior (North America). Characteristic data of the lake are given in Table 23.3.

The PCBs usually come as mixtures of more than 35 important congeners, each having different physicochemical properties (Henry's law constants, aqueous solubilities, etc.). One does not learn much by applying the formalism of Box 23.1 to this mixture. Fortunately, in the Laurentian Great Lakes individual PCB congeners have been studied in the necessary detail to evaluate the mass balance equation for single congeners (e.g., Baker and Eisenreich, 1990). Two of them are chosen as examples, 2',3,4-trichlorobiphenyl (PCB33) and 2,2',3,4,5,5',6-heptachlorobiphenyl (PCB185), the latter being much more hydrophobic than the former.

2',3,4-trichlorobiphenyl (PCB33)

2,2',3,4,5,5',6 heptachlorobiphenyl (PCB185)

Their physicochemical properties and the individual factors that we need to evaluate the equations of Box 23.1 are listed in Table 23.4. Except for the measurements that are specific for Lake Superior (input rates, concentrations of PCBs, composition of the particles determining K_d, etc.), all the data were derived from information given in this book either in tables (e.g., Henry's Law constants) or indirectly by approximative relationships (e.g., $K_d = f_{oc} K_{oc}$). More details are given in the footnotes to Table 23.4.

Table 23.4 Physicochemical Properties and Model Parameters for Two Selected PCB Congeners

IUPAC No.			2',3,4-Trichlorobiphenyl PCB33	2,2',3,4,5,5',6-Heptachlorobiphenyl PCB185
Molar mass	M_i	(g mol^{-1})	257.5	395.4
Air–water partition constant at 15°C[a]	$K_{a/w} = K_H / RT$	(–)	0.003	0.007
Air–water transfer velocity				
Air[b]	v_a	(m d^{-1})	450	390
Water[b]	v_w	(m d^{-1})	0.67	0.58
Total, Eq. 20-3	$v_{a/w}$	(m d^{-1})	0.45	0.48
Air–water exchange rate	$k_{a/w} = v_{a/w} / h$	(yr^{-1})	1.10	1.17
Atmospheric concentration[a]	C_a	(mol m^{-3})	1.8×10^{-13}	1.2×10^{-14}
Octanol–water partition constant	K_{ow}	(–)	6.0×10^5	2.0×10^7
Natural organic matter–water partition coefficient	K_{oc}	(m^3 kg$_{oc}^{-1}$)	27	360
Distribution coefficient				
Suspended solids[c]	K_d	(m^3 kg$_s^{-1}$)	5	70
Settled solids[c]	K_{ds}	(m^3 kg$_s^{-1}$)	0.8	10
Fraction dissolved in the water column, Eq. 9-12	f_w	(–)	0.998	0.973
Total input rate of PCB congener[d]	I_t	(mol yr^{-1})	100	30
Terms of Box 23.1[e]				
Input per unit time and volume				
River, precipitation (wet+dry)	$I_t / V = k_w C_{tin}$	(mol m^{-3}yr^{-1})	8.2×10^{-12}	2.5×10^{-12}
From the atmosphere[f]	$k_{a/w} C_a / K_{a/w}$	(mol m^{-3}yr^{-1})	66×10^{-12}	2.0×10^{-12}
Removal rates				
Flushing	k_w	(yr^{-1})	5.8×10^{-3}	5.8×10^{-3}
To the atmosphere	$k_{a/w}^* = f_w k_{a/w}$	(yr^{-1})	1.10	0.83
Sedimentation	$k_s^* = (1 - f_w) k_s$	(yr^{-1})	3.4×10^{-3}	0.046
Steady-state concentration (calculated)				
Total, Eq. 23-20	$C_{t\infty}$	(mol m^{-3})	6.7×10^{-11}	0.38×10^{-11}
Dissolved	$C_{d\infty} = f_w C_{t\infty}$	(mol m^{-3})	6.7×10^{-11}	0.37×10^{-11}
On particles, Eq. 23-6	$C_{s\infty} = \dfrac{1 - f_w}{r_{sw}} C_{t\infty}$	(mol kg$_s^{-1}$)	0.33×10^{-9}	0.26×10^{-9}
Measured concentrations				
Dissolved				
Surface waters[a]	C_w	(mol m^{-3})	$(10 \pm 9) \times 10^{-11}$	$(2.5 \pm 1.9) \times 10^{-11}$
Sorbed				
Epilimnetic particles[g]	C_s	(mol kg$_s^{-1}$)	$(5.4 \pm 2.3) \times 10^{-9}$	$(1.6 \pm 0.8) \times 10^{-9}$
Sediment surface	C_{ss}	(mol kg$_s^{-1}$)	$(0.19 \pm 0.04) \times 10^{-9}$	$(0.48 \pm 0.13) \times 10^{-9}$
Apparent distribution coefficient	$K_d = C_s / C_w$	(kg$_s$ m^{-3})	50 (20...200)	60 (20...200)

[a] From Baker and Eisenreich (1990). [b] Calculated for wind speed $u_{10} = 5$ m s^{-1} (Table 23.3) using Eqs. 20-25 and 20-27 with exponents 0.67. Molecular diffusivities are approximated from molar mass (see Eqs. 18-45 and 18-55). [c] K_d calculated from K_{ow} (Appendix C) and f_{oc} values (Table 23.3) using Eqs. 9-26a and 9-22. [d] From Eisenreich et al. (1989). [e] No *in-situ* degradation (k_{chem}, k_{photo}, $k_{bio} = 0$). [f] Note that for both congeners, given the measured lake concentration, the net input (input minus removal) is negative, that is, directed from the lake into the atmosphere. [g] From Baker and Eisenreich (1989)

Assuming that no significant in-situ degradation of PCBs occurs ($k^*_{\text{chem}} = k^*_{\text{photo}} = k^*_{\text{bio}} = 0$), three elimination pathways remain which, if described in terms of first-order reaction rates, can be directly compared with respect to their relative importance for the elimination of each PCB congener from the water column. As shown by the removal rates listed in Table 23.4, for both compounds the flux to the atmosphere is by far the most important process. Because of its larger K_d value, removal of the heptachlorobiphenyl to the sediments is predicted to be also of some importance. By the way, from this simple model we would expect to find the heptachlorobiphenyl relatively enriched in the sediments compared to the trichlorobiphenyl. We shall see later whether this is true.

Given the PCBs' inputs, I_t, and the atmospheric concentrations, C_a, we can now calculate the total steady-state concentration in the lake for the two congeners from Eq. 5 of Box 23.1. Note that among the input processes (nominator of Eq. 5) only the input from the rivers and the atmosphere are different from zero, whereas among the removal processes (denominator of Eq. 5) flushing, air/water exchange, and sedimentation are relevant. Thus, we can formulate the steady-state explicitly for the case of the two PCB congeners:

$$C_{t\infty} = \frac{k_w C_{\text{tin}} + k_{a/w} C_a / K_{a/w}}{k_w + f_w k_{a/w} + (1 - f_w) k_s}$$ (23-20)

From $C_{t\infty}$ we can also calculate the dissolved steady-state concentrations, $C_{d\infty} = f_w C_{t\infty}$ and compare them to concentrations measured in the surface waters of Lake Superior (Table 23.4). Average measured values of the trichlorobiphenyl congener are about 30% larger than the calculated steady-state value, a remarkable consistency if we consider the many simplifying assumptions and estimates made to derive Eq. 23-20. A more severe discrepancy is found for the heptachlorobiphenyl congener; the measured values are about 7 times larger than the calculated steady-state concentration.

There are numerous reasons why the model calculation could be wrong. The following discussion demonstrates how modelers should proceed from simple to more refined schemes by comparing their calculations to field data in order to decide which processes to include in their model. Let us discuss some of the factors affecting the results yielded by use of Eq. 23-20:

1. The input I_t may be wrong. Since the input estimation is based on PCB mixtures and typical relative congener compositions, an error of 30% is not unlikely and could thus explain the discrepancy found for the trichlorobiphenyl. For the heavier congener, the discrepancy seems to be too large to be solely explained by an input error.

2. The concentration measured in the surface waters may not represent the mean lake concentration. This hypothesis is supported by concentrations of sorbed PCBs which are very different for solids collected from different depths (Baker and Eisenreich, 1989). This point is discussed below.

3. The air–water exchange rate, $k_{a/w}$, the dominant removal rate for both congeners, may be overestimated by taking a mean wind velocity of 5 m s⁻¹. Since for PCB33

air–water exchange is dominant for both the input terms (nominator of Eq. 23-20) and the loss terms (denominator of Eq. 23-20), a decrease of $k_{a/w}$ by, for example, 10% would affect both numerator and denominator of Eq. 23-20 about equally and thus leave the size of $C_{t\infty}$ practically unchanged. In contrast, for PCB185 the atmospheric input is less than 50% of the total input, whereas removal to the atmosphere is still more than 95% (see values in Table 23.4). Thus, a reduction of $k_{a/w}$ by 10% would reduce the nominator of Eq. 23-20 by about 5%, and the denominator by about 10%, thus leading to an increase of $C_{t\infty}$, of about 5%. Although the trend may be right, the air–water exchange flux alone can hardly explain the sevenfold discrepancy between model and measurements.

4. The influence of water temperature on K_{oc} (see Eqs. 9-29 and 9-30) and thus on K_d was neglected. In fact, at 5°C K_d of the PCB33 and PCB 185 is, respectively, 1.8 and 2.5 times larger than at 25°C.

5. The presence of organic colloids may give rise to a third fraction of the biphenyl molecules, the fraction sorbed to nonsettling microparticles and macromolecules. This fraction neither contributes to the air–water exchange equilibrium nor participates in the process of particle settling. By generalizing the notation introduced in Eq. 9-23 one can write:

$$C_t = C_d + C_p + C_{DOC} \tag{23-21}$$

and define:

$$f_w = \frac{1}{1 + r_{sw} K_d + K_{DOC}[DOC]} = \frac{C_d}{C_t} \tag{23-22a}$$

$$f_{DOC} = \frac{K_{DOC}[DOC]}{1 + r_{sw} K_d + K_{DOC}[DOC]} = \frac{C_{DOC}}{C_t} \tag{23-22b}$$

$$f_s = \frac{r_{sw} K_d}{1 + r_{sw} K_d + K_{DOC}[DOC]} = \frac{C_p}{C_t} \tag{23-22c}$$

where C_{DOC} is the amount of chemical sorbed to colloidal organic matter per unit bulk volume, [DOC] is the concentration of organic colloids in the water, and K_{DOC} is the colloid–fluid distribution coefficient.

In Table 23.5 the characteristic parameters of the modified three-phase model for the two selected PCB congeners are listed. The sorption to colloids slightly reduces the dissolved fraction of the PCB33, f_w, makes the air–water exchange, $f_w k_{a/w}$, a little bit less effective, and thus slightly increases the steady-state concentration $C_{t\infty}$. The changes for PCB185 are more spectacular. About 35% of the congener is now sorbed to the colloids and "feels" the drive to participate neither in air–water exchange nor in sedimentation. Realizing that the "dissolved" fraction reported in the literature includes the colloidal fraction as well, the measured and calculated "dissolved" concentrations come closer, although there is still a factor of 4 between them.

Table 23.5 Three-Phase Model for Two Selected PCB Congeners in Lake Superior[a]

IUPAC No.		Unit	2',3,4- Trichlorobiphenyl PCB33	2,2',3,4,5,5',6- Heptachlorobiphenyl PCB185
Concentration of colloidal organic carbon	$[DOC]^b$	(kg m^{-3})	1.6×10^{-3}	1.6×10^{-3}
Distribution coefficient	$K_{DOC}{}^c$	(m^3 kg^{-1})	27	360
	$K_{DOC}[DOC]$	(–)	0.043	0.576
	$r_{sw} K_d{}^d$	(–)	0.002	0.028
(Equation 23-22a)	f_w	(–)	0.957	0.623
(Equation 23-22b)	f_{DOC}	(–)	0.041	0.359
	$f_s = 1 - f_w - f_{DOC}$	(–)	0.002	0.018
Modified terms of Eq. 23-20 and Table 23.4				
Removal rates				
Flushing	k_w	(yr^{-1})	5.8×10^{-3}	5.8×10^{-3}
To the atmosphere	$k_{a/w}^* = f_w k_{a/w}$	(yr^{-1})	1.05	0.729
Sedimentation	$k_s^* = f_s k_s$	(yr^{-1})	3.4×10^{-3}	0.031
Steady-state concentration (calculated)				
Total	$C_{t\infty}$	(mol m^{-3})	7.0×10^{-11}	0.59×10^{-11}
"Dissolved" (incl. colloids)	$(f_w + f_{DOC}) C_{t\infty}$	(mol m^{-3})	7.0×10^{-11}	0.58×10^{-11}
On non-colloidal ("large") particles	$C_{s\infty} = (f_s / r_{sw}) C_{t\infty}$	(mol kg$_s^{-1}$)	0.35×10^{-9}	0.26×10^{-9}
Apparent distribution, calculated[e]	K_d^{app}	(kg m^{-3})	5	45

[a] If not stated otherwise, all parameter values are as in Tables 23.3 and 23.4. [b] From Baker and Eisenreich (1989). [c] Assuming $K_{DOC} = K_{oc}$. [d] See Tables 23.3 and 23.4. [e] See Eq. 23-23.

Based on the three-phase model we can define an apparent K_d value:

$$K_d^{app} = \frac{C_s}{C_w + C_{DOC}} = \frac{C_s}{C_w} \frac{1}{1 + (f_{DOC} / f_w)} \qquad (23\text{-}23)$$

As shown in the last line of Table 23.5, for PCB33 the three-phase model has little effect on the calculated K_d value. Both values are much smaller than the observed one, although the uncertainty of the latter is large. In contrast, for PCB185 the new calculated apparent distribution coefficient is reduced as compared to the "large particle" value of Table 23.4 (from 70 to 45 m^{3}kg^{-1}).

6. There is still another point to be discussed, which may limit the calculations presented in Tables 23.4 and 23.5. In 1986, when the concentrations were measured, the lake may not have been at steady-state. In fact, the PCB input, which mainly occurred through the atmosphere, dropped by about a factor of 5 between 1965 and 1980. However, the response time of Lake Superior (time to steady-state, calculated according to Eq. 4 of Box 12.1 from the inverse sum of all removal rates listed in Table 23.4) for both congeners would be less than 3 years. This is quite short, especially if we use the model developed for an exponentially changing input (Chapter 21.2, Eq. 21-17) with a specific rate of change $\alpha = -0.1$ yr^{-1} (that is, the rate which

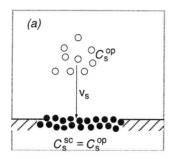
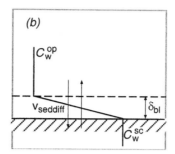
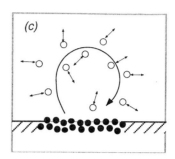

Figure 23.3 The processes which contribute to the exchange flux between open water (op) and sediment column (sc): (a) settling of suspended particles; (b) exchange flux of dissolved phase across a stagnant bottom boundary layer, (c) particle resuspension followed by equilibration between particle and open water.

reduces the input by a factor of 5 every 15 years) to formally analyze the time delay between input and lake concentration.

However, there are indications that the lake as a whole may possess a much longer memory than the one that we calculated from taking account of the water column alone. The obvious candidate for the additional memory is the sediment. Remember that there is a significant flux to the sediment for at least the higher molecular mass congener. The lake water "feels" the sediment memory, which means that we have to consider a process that mediates the exchange of sedimentary constituents back into the free-water column. Let us thus discuss how to expand our one-box model to account for this additional process

Exchange at the Sediment–Water Interface

In this section we treat the exchange at the sediment–water interface in the same manner as the air-water exchange. That is, we assume that the concentration in the sediments is a given quantity (an "external force," to use the terminology of Box 21.1). In Section 23.3 we will discuss the lake/sediment system as a two-box model in which both the concentration in the water and in the sediments are model variables.

The flux between open water (superscript op) and sediment column (sc) results from three different processes:

1. The settling of particles (Eqs. 23-18 and 23-19) is a directed flux (like the advective flux of Eq. 22-2). It is always directed from the water into the sediments (Fig. 23.3a).

2. Molecular diffusion of the dissolved phase of the chemical between the open water and the pore water accompanied by sorption/desorption with the "local" particles can be described by an exchange flux (see Fig. 23.3b and Box 23.2):

$$F_{\text{sed diff}} = -v_{\text{sed diff}}\left(C_w^{\text{op}} - C_w^{\text{sc}}\right) \quad \left[\text{M L}^{-2}\text{T}^{-1}\right] \quad (23\text{-}24)$$

where

$\quad v_{\text{sed diff}}$ is the sediment–water exchange velocity

$\quad C_w^{\text{op}}$ is the aqueous concentration in the open water

$\quad C_w^{\text{sc}}$ is the aqueous concentration in the sediment pore water in equilibrium with the concentration of the sediment particles, C_s^{sc}

Box 23.2 Model of Sediment–Water Exchange

The diffusive flux of the aqueous species of a chemical between the open water (op) and the pore water of the sediment column (sc) can be described by the linear expression:

$$F_{\text{sed diff}} = -v_{\text{sed diff}}\left(C_w^{op} - C_w^{sc}\right) \quad \left[ML^{-2}T^{-1}\right] \tag{23-24}$$

where a positive flux corresponds a sediment-to-water-column transfer. For equilibrium sorption/desorption, C_w^{op} is related to the total concentration per bulk volume, C_t^{op}, by (see Box 18.5):

$$C_w^{op} = \frac{C_d^{op}}{\phi^{op}} = \frac{f_w^{op}}{\phi^{op}} C_t^{op} \approx f_w^{op} C_t^{op} \tag{1}$$

with $\phi_{op} \approx 1$. The aqueous concentration in the pore water of the sediment bed, C_w^{sc}, is related to the sorbed concentration on the sediment particles by:

$$C_w^{sc} = \frac{C_s^{sc}}{K_d^{sc}} \tag{2}$$

Inserting Eqs. 1 and 2 into Eq. 23-24 yields:

$$F_{\text{sed diff}} = -v_{\text{sed diff}}\left(f_w^{op} C_t^{op} - \frac{C_s^{sc}}{K_d^{sc}}\right) \tag{3}$$

The flux resulting from sediment resuspension is assumed to be proportional to the difference between the concentration on the resuspended sediment particles, C_s^{sc}, and the sorbed concentration, C_s^{op}, which the particles would have in equilibrium with the aqueous concentration in the water column, C_w^{op}:

$$\begin{aligned}
F_{\text{resuspension}} &= \mu_{\text{res}}\left(C_s^{sc} - C_s^{op}\right) = \mu_{\text{res}}\left(C_s^{sc} - K_d^{sc} C_w^{op}\right) \\
&\approx \mu_{\text{res}}\left(C_s^{sc} - K_d^{sc} f_w^{op} C_t^{op}\right)
\end{aligned} \tag{4}$$

The parameter μ_{res} (units: particle mass per unit area and time) depends on the sedimentary mass which is resuspended per unit area and time. If the sediment particles were to completely equilibrate with the open water during the equilibration process, μ_{res} would be exactly the resuspended mass. If equilibration is only partial, μ_{res} is the resuspended mass multiplied with the relative degree of equilibration.

Combining the two fluxes yields:

$$\begin{aligned}
F_{\text{sedex}} &= F_{\text{sed diff}} + F_{\text{resuspension}} \\
&= -v_{\text{sed diff}}\left(f_w^{op} C_t^{op} - \frac{C_s^{sc}}{K_d^{sc}}\right) - \mu_{\text{res}}\left(K_d^{sc} f_w^{op} C_t^{op} - C_s^{sc}\right) \\
&= -\left(v_{\text{sed diff}} + K_d^{sc} \mu_{\text{res}}\right) f_w^{op} C_t^{op} + \left(\frac{v_{\text{sed diff}}}{K_d^{sc}} + \mu_{\text{res}}\right) C_s^{sc}
\end{aligned}$$

$$\Rightarrow \quad F_{\text{sedex}} = -\left(v_{\text{sed diff}} + K_d^{\text{sc}} \mu_{\text{res}} \right) f_w^{\text{op}} \left(C_t^{\text{op}} - \frac{C_s^{\text{sc}}}{f_w^{\text{op}} K_d^{\text{sc}}} \right) \tag{5}$$

$$= -v_{\text{sedex}} \left(C_t^{\text{op}} - C_{\text{sed}}^{\text{eq}} \right)$$

where

$$v_{\text{sedex}} = f_w^{\text{op}} \left(v_{\text{sed diff}} + K_d^{\text{sc}} \mu_{\text{res}} \right) \tag{6}$$

$$C_{\text{sed}}^{\text{eq}} = \frac{C_s^{\text{sc}}}{f_w^{\text{op}} K_d^{\text{sc}}} \tag{7}$$

Note that if the sediment surface were to consist of freshly sedimented particles with concentration $C_s^{\text{sc}} = C_s^{\text{op}}$, then the pore water in equilibrium with these particles would have the aqueous concentration $C_w^{\text{sc}} = C_w^{\text{op}}$, and thus according to Eq. 23-24 the diffusive exchange flux $F_{\text{sed diff}}$ would be zero. However, in most cases the sediment surface is not in equilibrium with the water column, because diagenetic processes change the physicochemical properties of the sediments and thus its solid–water distribution ratio, K_d^{sc}, relative to K_d^{op}. Furthermore, the sediment surface usually reflects a longer history of exposure to the chemical under consideration than the water column. Therefore, water and sediments would approach equilibrium only if the external loading to the lake has changed very slowly in the past. For man-made chemicals this is usually not the case.

In Section 19.3 we treated the sediment–water interface as a wall boundary. Unfortunately, in this scheme the sediment–water exchange flux, and thus the exchange velocity $v_{\text{sed diff}}$, depend on time (see Eq. 19-33). As a result, the specific elimination rate k^* of the linear box-model (Box 23.1, Eq. 1) would be time dependent. Although explicit solutions can still be formulated, they would be complicated. Yet, for a strongly sorbing chemical we can (at least for some time) visualize the exchange as a bottleneck flux (Fig. 19.10) and use Eq. 19-42 for the case $t \ll t_{\text{crit}}$. Then $v_{\text{sed diff}} \approx v_{\text{bl}} = D_{\text{bl}} / \delta_{\text{bl}}$ could be interpreted as resulting from molecular diffusion (diffusivity D_{bl}) through a bottom boundary layer with thickness δ_{bl} (Fig. 23.3b). However, the following discussion does not depend on the specific scheme which we apply to the boundary flux.

3. The third exchange process is related to the possible resuspension and resettling of sedimentary particles (Fig. 23.3c). As shown in Box 23.2, processes 2 and 3 can be combined into a single exchange flux with one single specific exchange rate, v_{sedex}, which combines both mechanisms (Box 23.2, Eqs. 5 to 7):

$$F_{\text{sedex}} = -v_{\text{sedex}} \left(C_t^{\text{op}} - C_{\text{sed}}^{\text{eq}} \right) \tag{23-25}$$

where $C_{\text{sed}}^{\text{eq}}$ is the total concentration per bulk volume which is in equilibrium with the sorbed concentration on the sediment particles, C_s^{sc} (Box 23.2, Eq. 7). Although v_{sedex} can be interpreted as resulting from the combined effect of boundary diffusion and sediment resuspension, we can also simply treat it as an empirical parameter describing the coupling between water and sediment.

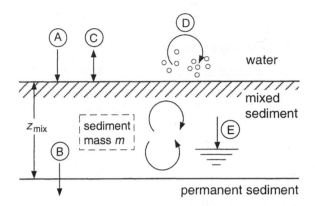

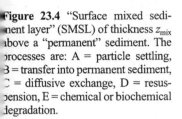

Figure 23.4 "Surface mixed sediment layer" (SMSL) of thickness z_{mix} above a "permanent" sediment. The processes are: A = particle settling, B = transfer into permanent sediment, C = diffusive exchange, D = resuspension, E = chemical or biochemical degradation.

PCBs in Lake Superior (Part 2)

We have now developed the necessary tool to continue our analysis of the PCBs in Lake Superior. Remember that we have asked the question how fast the PCB concentration in the lake would respond to a change of the PCB input. We found that the water column alone has a typical response time of about 3 years and then speculated whether the PCBs in the sediments would significantly alter this time.

In order to get an idea of the approximate size of the quantities which characterize the water-sediment exchange (Box 23.2), we introduce the concept of a "surface mixed sediment layer" (SMSL), which is supported by observed profiles of ^{210}Pb, ^{137}Cs and other radioactive isotopes. According to Robbins (1986), the mixing power can originate either from water currents or animals (bioturbation) or by a combination of both. The SMSL is defined by a mixing depth, z_{mix}, or by the solid mass per unit area m (Fig. 23.4). Both are related by:

$$m = z_{mix}(1 - \phi^{sc})\rho_s^{sc} \qquad [ML^{-2}] \qquad (23\text{-}26)$$

where ρ_s^{sc} is dry density of the sediment column, and ϕ^{sc} is porosity. It is assumed that below the SMSL lies the "permanent" sediment, that is, the zone from which no feedback into the water column is possible. Thus, the sedimentary memory of the chemical is assumed to be completely confined to the SMSL.

Typical values for the sediment model are given in Table 23.6. The sizes of z_{mix} and μ_{res} should be understood as rough estimates chosen with the aim of analyzing the possible influence of the sediment memory on the PCB concentrations in the open water of Lake Superior. We are interested in the size of k_{sedex} relative to the other removal rates (Box 23.1, Eq. 3), and in the relative contributions of diffusive exchange and particulate resuspension to this exchange coefficient.

According to Table 23.6, for both congeners the sediment–water interaction contributes 21% to the total removal rate. Most of this (about 98%) is attributed to diffusive exchange.

At this stage we have treated the SMSL concentration as a given external parameter, that is, as a quasi-infinite PCB reservoir, which, in turn, is not significantly affected

Table 23.6 Characteristic Quantities of the Sediment–Water Exchange and SMSL Model (Fig. 23.4)

Sediment Mixed Surface Layer (SMSL)

Aqueous boundary layer thickness	δ_{bl} [a]	(m)	5×10^{-4}	
Mixing depth	z_{mix}	(m)	0.04	
Density of solids	ρ_s	(kg$_s$m^{-3})	2.5×10^3	
Porosity	ϕ^{sc}	(–)	0.85	
Mixed layer mass (Eq. 23-26)	m	(kg$_s$m^{-2})	15	
Resuspension rate [b]	μ_{res}	(kg$_s$m^{-2}yr^{-1})	1.5	
Solid-to-water phase ratio [c]	r_{sw}^{sc}	(kg$_s$m^{-3})	440	
Preservation factor of organic carbon in SMSL	β	(–)	0.1	

The PCB Congeners in the SMSL

			PCB33	PCB185
Distribution coefficient in SMSL [d]	K_d^{sc}	(m$_3$kg$_s^{-1}$)	0.8	10
Fraction dissolved in open water [e]	f_w^{op}	(–)	0.957	0.623
in pore water	f_w^{sc}	(–)	4.5×10^{-4}	3.2×10^{-5}
Molecular diffusivity in water [f]	D_w	(m^{2}s^{-1})	7×10^{-10}	6×10^{-10}
Diffusive exchange velocity	$v_{sed\,diff} = D_w / \delta_{bl}$	(m yr^{-1})	44	38
Exchange velocity due to resuspension	$K_d^{sc}\mu_{res}$	(m yr^{-1})	1.2	15
Total exchange velocity (Box 23.2, Eq.6)	v_{sedex}	(m yr^{-1})	43	33
Diffusive penetration into SMSL in one year ($t = 3.15 \times 10^7$s)	$z_{diff} \approx (f_w^{sc} D_w t)^{1/2}$ (m)		0.01	3×10^{-3}

Dynamic Model of PCB Congeners in SMSL

			PCB33	PCB185
Specific sediment–water exchange rate [g]	$k_{sedex} = v_{sedex} / h$	(yr^{-1})	0.29	0.22
Relative contribution of k_{sedex} to k^* (Box 23.1) [h]			0.21	0.22
Relative contribution of resuspension to k_{sedex} [i]			0.027	0.31

Mass Fluxes Through SMSL

		PCB33	PCB185
Flux by particle settling [j]	(mol m^{-2}yr^{-1})	0.5×10^{-9}	0.2×10^{-9}
Exchange flux water→ sediment [k]	(mol m^{-2}yr^{-1})	4×10^{-9}	0.8×10^{-9}
PCB in SMSL [l]	(mol m^{-2})	3×10^{-9}	7×10^{-9}

[a] In Boston Harbor (Illustrative Example 19.3), a boundary layer thickness of 1.5×10^{-4} m was assumed. This explains the larger exchange fluxes of PCBs. Note that turbulence in Boston Harbor must be much larger than at the bottom of Lake Superior. [b] Mixed layer depth z_{mix} and resuspension rate μ_{res} must be physically related. Therefore, μ_{res} is taken as some fraction of m (here 10% per year). [c] Calculated from Eq. 4 of Box 18.5. [d] See Table 23.4. [e] From Table 23.5, excluding the colloid fraction. [f] Approximated from molecular mass (Eqs. 18-45 and 18-55). [g] See Box 23.1. [h] For the PCBs this is given by $k_{sedex}/(k_w+k_{a/w}+k_s+k_{sedex})$. Values in Table 23.5. [i] Calculated as $K_d^{sc}\mu_{res}/(v_{sed\,diff} + K_d^{sc}\mu_{res})$. [j] Calculated from data in Tables 26.3 and 26.4 as $F_s C_s$, where C_s is measured. [k] Calculated as $v_{sedex} C_t^{op}$, C_t^{op} measured. [l] Approximated as $m C_s^{sc}$, C_s^{sc} measured (Table 23.4).

by the flux across the sediment–water interface. In order to estimate over what time period this assumption may be reasonable, we need to look at the size of the fluxes in and out of the SMSL and compare these values with the amount of PCBs stored in the SMSL. The results are summarized in the last part of Table 23.6. As it turns out, for PCB33 the exchange flux per year is of the same order of magnitude as the amount of PCB33 in the SMSL, whereas for PCB185 the flux per year is about 10% of the PCB content of the SMSL. Such large turnover rates are in contradiction with the above infinite-reservoir-assumption. Thus, it is necessary to model both systems simultaneously, that is, the water column and the sediment mixed surface layer. The resulting two-box model will be discussed in the next section and summarized in Table 23.7. The modified model is able to explain why often field investigations identify the sediment as a source, not a sink of a pollutant.

23.3   Two-Box Models of Lakes

In Chapter 21 the model of a stratified lake served as a prototype of a linear two-box model (Fig. 21.10). The necessary mathematics were developed in Boxes 21.6 and 21.7. In Illustrative Example 21.5 the fate of tetrachloroethene (PCE) in Greifensee was used to demonstrate that for the case of a two-box model it is still possible to carry out back-of-the-envelope calculations. Further examples are given in Problems 23.2 and 23.3, where the behavior of anthracene in a mixed as well as in a stratified lake is assessed.

Two-Box Model for Lake/Sediment System

Now we want to apply the box model approach to a two-box system which consists of a completely mixed water body in contact with a sediment box. Although the sediment column can hardly be visualized as being completely mixed, the concept of a surface mixed sediment layer (SMSL) introduced in the previous section is an approximate view of the sediments as mixed box. In fact, for strongly sorbing chemicals the diffusive penetration into the sediment column is so slow and the storage capacity of the top 1 to 2 cm so large, that the deeper parts of the sediments can be treated as sort of a permanent sink from which no feedback to the SMSL and to the open water column is possible.

Let us formulate the dynamic mass balance equation of the chemical in the SMSL. Fig.23.5 summarizes all processes. At this point we have to select the variable which shall characterize the SMSL. (Remember for the open water box we have chosen the total concentration C_t^{op}.) Due to the large solid-to-water ratio of sediments r_{sw}^{sc}, chemicals with moderate to large distribution coefficients ($K_d > 0.1 \, \text{m}^3\text{kg}_s^{-1}$) are predominantly sorbed to the solid phase. Therefore, C_s^{sc} offers itself as the natural choice for the second state variable.

To do the mass balance for the SMSL, we have to consider the total compound per unit bulk volume, C_t^{sc}. That is, we sum the fraction sorbed on the particles and the fraction dissolved in the pore water:

$$C_t^{sc} = \phi^{sc} C_w^{sc} + (1 - \phi^{sc}) \rho_s \, C_s^{sc} \qquad (\text{mol m}^{-3}) \qquad (23\text{-}27)$$

If the aqueous and sorbed phase are always at equilibrium, we can use Eq. 23-1 to replace C_w^{sc}:

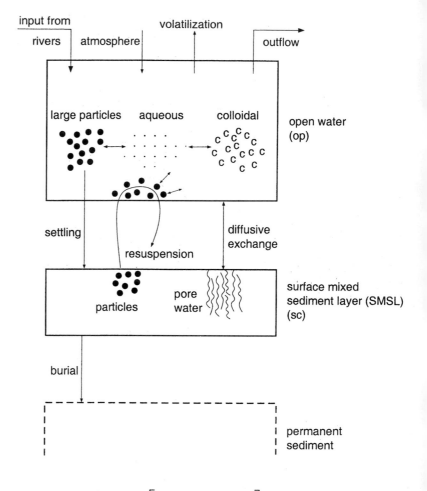

Figure 23.5 Processes considered for the combined sediment–water two-box model to describe the fate of PCB congeners in lakes.

$$C_t^{sc} = C_s^{sc} \left[\frac{\phi^{sc}}{K_d^{sc}} + (1 - \phi^{sc}) \rho_s \right]$$ (23-28a)

Let us compare the relative size of the two terms on the right-hand side of Eq. 23-28a for a chemical with $K_d^{sc} = 0.1 \ \mathrm{m^3 kg_s^{-1}}$. (Note that for still smaller K_d values not much of the compound is being removed to the sediments, thus the sediment model is not relevant anyway.) As an example we choose $\phi^{sc} = 0.9$, $\rho_s = 2.5 \times 10^3 \ \mathrm{kg_s m^{-3}}$ and get $\phi^{sc} / K_d^{sc} = 9 \ \mathrm{kg_s m^{-3}}$ and $(1-\phi^{sc}) \rho_s = 250 \ \mathrm{kg_s m^{-3}}$. Thus the sorbed phase contains more than 96% of the chemical. For $K_d^{sc} > 0.1 \ \mathrm{m^3 kg_s^{-1}}$, the fraction is even larger. It is therefore justified to approximate Eq. 23-28a by (see Eq. 23-26):

$$C_t^{sc} \approx (1 - \phi^{sc}) \, C_s^{sc} = \frac{m}{z_{mix}} \, C_s^{sc}$$ (23-28b)

The following processes contribute to the mass balance of the chemical in the SMSL (see Fig. 23.4):

A. Input as sorbed species on settling particles, $v_s \, r_{sw}^{op} \, C_s^{op} = v_s \, f_s^{op} \, C_t^{op}$.

B. Transfer into the permanent sediment layer, $\beta v_s \, r_{sw}^{sc} \, C_s^{sc}$, where β is a preservation factor indicating the fraction of settled particulate matter eventually reaching the permanent sediment layer (i.e., $\beta \le 1$). The size of β depends on the sorptive properties of the chemical. Imagine that sorption depends on the organic carbon content of the sediment material, f_{oc}^{sc}, as expressed by Eq. 23-4. If due to biodegradation f_{oc}^{sc} of the material that is transferred to the permanent sediment layer is smaller than the average f_{oc}^{sc} of the SMSL, then the transfer of the sorbed chemical is reduced by the corresponding factor, that is, β is smaller than 1. In contrast, the flux of total solids to the permanent sediment must be approximately equal to the flux of fresh sediments from the lake. Thus, a chemical which sorbs unspecifically to the solid fraction—not typical for organic compounds—would have a preservation factor $\beta \approx 1$.

C. Diffusive exchange between the dissolved phase in the lake water and the pore water (Box 23.2, Eq. 3).

D. Resuspension and resettling of sedimentary particles (Box 23.2, Eq. 4). Remember that the parameter μ_{res} is the resuspended sediment mass per unit area and time multiplied with the relative degree of equilibration between suspended particle and surrounding water.

As explained before, processes C and D can be combined into a flux described by a single exchange velocity, v_{sedex} (Box 23.2, Eq. 6).

E. Chemical or biochemical degradation in the SMSL, $k_r^{sc} m C_s^{sc}$. Since the dissolved phase concentration is proportional to C_s^{sc}, the reaction rate, k_r^{sc}, if multiplied with the corresponding fraction, f_w^{sc}, could also be used to account for degradation occurring in the dissolved fraction.

Putting all these processes into the mass balance equation for the chemical in the SMSL yields (units: mol m^{-2}s^{-1}):

$$\frac{d}{dt}\left(m\, C_s^{sc}\right) = m\,\frac{d\,C_s^{sc}}{dt}$$

$$= v_s\, f_s^{op}\, C_t^{op} - \beta\, v_s\, r_{sw}^{op}\, C_s^{sc} + v_{sedex}\left(C_t^{op} - \frac{C_s^{sc}}{f_w^{op}\, K_d^{sc}}\right) - k_r^{sc}\, m\, C_s^{sc} \tag{23-29}$$

After division by m and sorting the terms yields:

$$\frac{dC_s^{sc}}{dt} = \frac{1}{m}\left(v_s f_s^{op} + v_{sedex}\right)C_t^{op} - \left(\beta\,\frac{v_s}{m}\,r_{sw}^{op} + \frac{v_{sedex}}{m f_w^{op} K_d^{sc}} + k_r^{sc}\right)C_s^{sc} \tag{23-30}$$

Note that the only remaining variables are C_t^{op} and C_s^{sc}, that is, the chosen state variables of the two-box model.

The mass balance equation for C_t^{op} was derived in Box 23.1. Although Eq. 1 represents a one-box model for C_t^{op} alone, it can also be interpreted as part of a two-box model. The link to C_s^{sc} is hidden in the inhomogeneous term J (Eq. 2, Box 23.1). As shown in Box 23.2 (Eq. 7), the term $k_{sedex}\, C_{sed}^{eq}$ is linked to C_s^{sc} by

$$k_{sedex}\, C_{sed}^{eq} = \frac{k_{sedex}}{f_w^{op}\, K_d^{sc}}\, C_s^{sc} \tag{23-31}$$

We substitute this expression into Eq. 2 of Box 23.1. Furthermore, we set the in-situ production term j and k_{ex} equal to zero and define $k_r^{op} \equiv k_{chem}^* + k_{photo}^* + k_{bio}^*$.

Sorting the terms in the usual manner yields the dynamic equation of the total lake concentration, C_t^{op}. Note that the equation is now coupled with the sediment concentration C_s^{sc}.

$$\frac{dC_t^{op}}{dt} = k_w C_{tin} + k_{a/w} \frac{C_a}{K_{a/w}}$$
$$- \left(k_w + k_{a/w}^* + k_s^* + k_{sedex} + k_r^{op} \right) C_t^{op} + \frac{k_{sedex}}{f_w^{op} K_d^{sc}} C_s^{sc} \tag{23-32a}$$

where from Eq. 6 of Box 23.2 we have:

$$k_{sedex} = \frac{v_{sedex}}{h} = \frac{f_w^{op}}{h} \left(v_{seddiff} + K_d^{sc} \mu_{res} \right) \tag{23-32b}$$

The two coupled differential equations, Eqs. 23-30 and 23-32, are summarized in Box 23.3 and solved according to the recipe given in Box 21.6.

Box 23.3 Solution of Linear Water–Sediment Model

State variables: C_t^{op} Total concentration in lake (mol m^{-3})

C_s^{sc} Concentration on solids in surface mixed sediment layer (SMSL) (mol kg$_s^{-1}$)

The two coupled linear differential equations are (see Eqs. 23-30 and 23-32):

$$\frac{dC_t^{op}}{dt} = J_t - k_{11} C_t^{op} + k_{12} C_s^{sc} \qquad \text{(mol m}^{-3}\text{yr}^{-1}) \tag{1a}$$

$$\frac{dC_s^{sc}}{dt} = k_{21} C_t^{op} - k_{22} C_s^{sc} \qquad \text{(mol kg}_s^{-1}\text{yr}^{-1}) \tag{1b}$$

where

$$J_t = k_w C_{t\,in} + k_{a/w} \frac{C_a}{K_{a/w}} \qquad \text{(mol m}^{-3}\text{yr}^{-1}) \tag{2a}$$

$$\begin{aligned} k_{11} &= k_w + k_{a/w}^* + k_s^* + k_{sedex} + k_r^{op} \\ (k_r^{op} &\equiv k_{chem}^* + k_{photo}^* + k_{bio}^*) \end{aligned} \qquad \text{(yr}^{-1}) \tag{2b}$$

$$k_{12} = \frac{k_{sedex}}{f_w^{op} K_d^{sc}} \qquad \text{(kg}_s\text{m}^{-3}\text{yr}^{-1}) \tag{2c}$$

$$k_{21} = \frac{1}{m} \left(v_s f_s^{op} + v_{sedex} \right) \qquad \text{(kg}_s^{-1}\text{m}^3\text{yr}^{-1}) \tag{2d}$$

$$k_{22} = k_r^{sc} + \frac{1}{m}\left(\beta v_s r_{sw}^{op} + \frac{v_{sedex}}{f_w^{op} K_d^{sc}}\right) \qquad (yr^{-1}) \qquad (2e)$$

Note that j and k_{ex} of Box 23.1 are zero.

Comparison with Box 21.6 yields the following correspondence:

Box 21.6	A_{ij}	J_1	J_2
Box 23.3	k_{ij}	J_t	0

The model is applied to the case of PCBs in Lake Superior (see Table 23.7).

PCBs in Lake Superior (Part 3)

This is the third and final part of the story about PCBs in Lake Superior.

Application of the dynamic water/sediment model to the fate of the PCBs in Lake Michigan is summarized in Table 23.7. As it turns out, for both congeners the steady-state concentrations are virtually unchanged relative to the values calculated for the three-phase one-box model of Table 23.5. We also note that in the model the sorbed concentrations are still significantly smaller than the measured values. The same is

Table 23.7 PCBs in Lake Superior Described by a Coupled Sediment–Water Model

Application of steady-state solution of linear water–sediment model (Box 23.3) to two PCB congeners in Lake Superior. The steady-state is calculated from Box 21.6.

		Notation Box 21.6	PCB33	PCB185
Parameters from Box 23.3				
J_t	(mol m^{-3}yr^{-1})	J_1	74×10^{-12}	4.5×10^{-12}
k_{11}	(yr^{-1})	A_{11}	1.35	0.986
k_{22}	(yr^{-1})	A_{22}	3.7	0.353
k_{12}	(kg$_s$ m^{-3}yr^{-1})	A_{12}	0.379	0.0353
k_{21}	(kg$_s^{-1}$m^{3}yr^{-1})	A_{21}	2.9	2.5
Steady-state Solution				
$C_{t\infty}^{op}$	(mol m^{-3})	$y_{1\infty}$	7.0×10^{-11}	0.61×10^{-11}
$C_{s\infty}^{op} = f_w^{op} C_{t\infty}^{op} K_d^{op}$	(mol kg$_s^{-1}$)		0.34×10^{-9}	0.27×10^{-9}
$C_{s\infty}^{sc}$	(mol kg$_s^{-1}$)	$y_{2\infty}$	0.055×10^{-9}	0.043×10^{-9}
Eigenvalues and Time to Steady-State				
k_1	(yr^{-1})	k_1	0.95	0.235
k_2	(yr^{-2})	k_2	4.1	1.10
$t_{ss} = \dfrac{3}{\min\{k_1, k_2\}}$	(yr)		3	13

true for the dissolved concentration of PCB185; the model cannot explain the much larger measurements. Yet, the important new property of the combined water column/sediment model is a more realistic estimation of the response time, t_{ss}, of the two congeners in the lake/sediment system. Not surprisingly, the heavier congener is found to have a longer memory (13 years) than the lighter PCB33 (3 years). Given the loading history of the PCBs in that area, it may be that at least the heavier congeners have never been completely at steady-state with their actual loading.

Let us summarize the whole PCB story by going through the different steps we have undertaken so far (Table 23.8): In Part 1 we modeled the two selected PCB congeners with a one-box model in which two phases (C_w and C_s) were distinguished, and these were assumed to be at equilibrium (Table 23.4). We found that for the lighter congener (PCB33) the modeled dissolved concentration falls within the rather broad range of observed values. The calculated sorbed concentration lies between the large concentration on epilimnetic particles and the about 30 times smaller concentration on particles at the sediment surface. Since in the one-box model the chemical composition of the suspended particles is the same at all depths and these particles also form the fresh sediments, the model cannot reproduce the observed vertical gradient of the sorbed PCB concentration. Things were different for the heavy congener (PCB185). The calculated dissolved concentration turned out to be six times smaller than the observed values, and the sorbed concentration to be smaller than the concentration measured on epilimnetic as well as on sedimentary particles.

Table 23.8 Two PCB Congeners in Lake Superior: Summary of Models and Measurements

		PCB33	PCB185
A. Measured concentrations (from Table 23.4)			
Dissolved	(mol m^{-3})	$(10 \pm 9) \times 10^{-11}$	$(2.5 \pm 1.9) \times 10^{-11}$
Sorbed	(mol kg$_s^{-1}$)		
Epilimnetic particles		$(5.4 \pm 2.3) \times 10^{-9}$	$(1.6 \pm 0.8) \times 10^{-9}$
Particles at sediment surface		$(0.19 \pm 0.04) \times 10^{-9}$	$(0.48 \pm 0.13) \times 10^{-9}$
B. Two-phase one-box lake model (from Table 23.4)			
Dissolved	(mol m^{-3})	6.7×10^{-11}	0.37×10^{-11}
On suspended solids	(mol kg$_s^{-1}$)	0.33×10^{-9}	0.26×10^{-9}
C. Three-phase one-box lake model (from Table 23.5)			
"Dissolved" (aqueous and colloidal)	(mol m^{-3})	7.0×10^{-11}	0.59×10^{-11}
On noncolloidal ("large") particles	(mol kg$_s^{-1}$)	0.35×10^{-9}	0.26×10^{-9}
D. Two-box sediment/water model combined with three-phase equilibrium (Table 23.7)			
"Dissolved" in water column	(mol m^{-3})	7.0×10^{-11}	0.61×10^{-11}
Sorbed on noncolloidal ("large") particles	(mol kg$_s^{-1}$)		
Lake		0.34×10^{-9}	0.27×10^{-9}
Surface sediments		0.055×10^{-9}	0.043×10^{-9}

In order to explain the discrepancy between measured and modeled PCB concentrations, we then introduced colloids as a third phase to which the PCBs can also sorb. Because of the rather small distribution coefficient, K_d, of PCB33, predictions for this congener are not expected to respond strongly to such a model change. In fact, it turns out that none of the model refinements were relevant for weakly sorbing compounds such as PCB33 (see Table 23.8). In contrast, for PCB185 the changes were more significant and point in the right direction. The colloidal phase represents about 35% of the total PCB185 in the water column and thus doubles the measured "dissolved" (aqueous and colloidal) concentration. Yet, the "real" dissolved concentration, C_d, remained virtually unchanged and so did the sorbed concentration on "large" particles.

In Part 2 of the PCB story, we introduced the exchange between the water column and the surface sediments in exactly the same way as we describe air/water exchange. That is, we used an exchange velocity, v_{sedex}, or the corresponding exchange rate, k_{sedex} (Table 23.6). Since at this stage the sediment concentration was treated as an external parameter (like the concentration in the air, C_a), this model refinement is not meant to produce new concentrations. Rather we wanted to find out how much the sediment–water interaction would contribute to the total elimination rate of the PCBs from the lake and how it would affect the time to steady-state of the system. As shown in Table 23.6, the contribution of k_{sedex} to the total rate is about 20% for both congeners. Furthermore, it turned out that diffusion between the lake and the sediment pore water was much more important than sediment resuspension and reequilibration, at least for the specific assumptions made to describe the physics and sorption equilibria at the sediment surface.

In the last step (Part 3), the sedimentary compartment (the "surface mixed sediment layer", SMSL) was treated as an independent box (Table 23.7). The steady-state solution of the combined sediment/water system explained another characteristic of the observed concentrations, which, as mentioned above, could not be resolved by the one-box model. As shown in Table 23.8, for both congeners the concentration measured on particles suspended in the lake is larger than on sediment particles. The two-box model explained this difference in terms of the different relative organic carbon content of epilimnetic and sedimentary particles. This model also gave a more realistic value for the response time of the combined lake/sediment system with respect to changes in external loading of PCBs. However, major differences between modeled and observed concentrations remained unexplained.

The possible depth of the analysis has by no means reached its limit. If one wants to move beyond the coupled two-box model shown in Fig. 23.5, one has to rely on computer models. This lies outside the aim of this book.

At this point we should remember that the main purpose of these considerations is not to find the "true model" for the PCBs in Lake Superior. Given the very crude assumptions which had to be made at different stages of the modeling exercise, such an aim would be unrealistic. Yet, the example helps to develop a feeling for the importance of different assumptions and processes. For instance, the effect of introducing the colloidal phase was more important for PCB185, which sorbs more strongly to particles. The effect would be still greater for compounds with K_d values of 10^3 kg m^{-3} or larger.

Other Two-Box Models

There are numerous other examples of two-box models. For instance, a two-box epilimnion/hypolimnion model was discussed in Chapter 21, and additional examples are given as problems at the end of this chapter. We must remember that as long as these models are linear, their solutions can be constructed with the help of Box 21.6. They always consist of the sum of not more than two exponential functions and are thus fairly simple. This situation changes drastically if we allow the differential equations to become nonlinear. A system of two or more nonlinear differential equations rarely can be solved analytically, yet the available computer tools (such as MATLAB) make their solution easy.

Advanced Topic   23.4 **One-Dimensional Continuous Lake Models**

In this last section we discuss lake models which allow us to calculate continuous vertical concentration profiles as a function of time. The necessary mathematical instruments were given in Chapter 22. Since vertical mixing in lakes and oceans is usually much slower than horizontal mixing, as for water temperature the concentration variations of most chemical substances (inorganic or organic) are usually much larger in the vertical than in the horizontal direction. This distinguishes lakes from rivers. Rivers are usually well mixed vertically and laterally, but may show large concentration variations along the direction of the flow (see Chapter 24).

Internal Transport versus Reaction and Boundary Fluxes

In order to decide whether a specific lake has to be described with a continuous model or whether a box model would be adequate, we can either analyze the spatial and temporal variations of field data or—if the necessary data do not exist—rely on the concepts introduced in Chapter 22.

First, recall that the nondimensional Damköhler number, Da (Eq. 22-11b), allows us to decide whether advection is relevant relative to the influence of diffusion and reaction. As summarized in Fig. 22.3, if Da $\gg$ 1, advection can be neglected (in vertical models this is often the case). Second, if advection is not relevant, we can decide whether mixing by diffusion is fast enough to eliminate all spatial concentration differences that may result from various reaction processes in the system (see the case of photolysis of phenanthrene in a lake sketched in Fig. 21.2). To this end, the relevant expression is $L(k_r / E_z)^{1/2}$, where L is the vertical extension of the system, E_z the vertical turbulent diffusivity, and k_r the first-order reaction rate constant (Eq. 22-13). If this number is much smaller than 1, that is, if

$$k_r \ll E_z / L^2 \tag{22-19}$$

then mixing is fast relative to reaction, and spatial concentration variations due to reaction are not relevant. In the absence of large boundary fluxes, the concentration in the system is fairly constant; thus the box model approach is appropriate.

Boundary fluxes are another potential source of spatial inhomogeneity. For instance, many volatile chemicals, such as tetrachloroethene (PCE), enter a lake mainly through its inlets, whereas the main losses are due to air–water exchange at the lake surface. If we were interested in the PCE concentration at midlake compared with the concentration close to the shore, we must therefore compare the time needed to transport PCE from the shore to the center of the lake with the typical residence time of PCE in the surface waters. If the horizontal transport time is short compared to the residence time, the concentration at the lake surface should be fairly constant from the shore to the center. From Section 22.3 we know that horizontal mixing results from a combination of advection and dispersion by shear. The transport away from the inlet occurs by advection with typical velocities between 1 and 10 cm s^{-1}. The local mixing of the advected cloud occurs by shear diffusion (Box 22.3). According to the experiments by Peeters et al. (1996), even in calm conditions the cloud size grows by about a factor of 100 within one day; that is, the input concentration is strongly reduced by lateral mixing. Usually, these crude estimates suffice to calculate the time (or the distance from the input location) where mixing has eliminated significantly horizontal concentration gradients.

If the lake is stratified, vertical transport is commonly the time-limiting step for complete mixing. This was the reason for applying the two-box model to the case of PCE in Greifensee (Illustrative Example 21.5). Now we go one step further. We consider a vertical water column of mean depth h with a constant vertical eddy diffusion coefficient E_z. The flux $F_{a/w}$ of PCE escaping to the atmosphere is given by Eq. 20-1a:

$$F_{a/w} = v_{a/w}(C_w - C_w^{eq}) = v_{a/w} C_w^{excess} , \qquad C_w^{excess} = C_w - C_w^{eq} \qquad (23\text{-}33)$$

where C_w^{excess} is the concentration at the water surface in excess of the equilibrium with the PCE concentration in the atmosphere. Within the water column PCE is transported upward to compensate for the loss at the surface. This flux is described by Fick's first law (Eq. 18-6):

$$F_{diff} = -E_z \frac{dC_w}{dz} \qquad (23\text{-}34)$$

Let us disregard the PCE sources for the moment and calculate the vertical concentration gradient needed to compensate for the loss at the water surface. Combining Eqs. 23-33 and 23-34 and solving for dC_w/dz yields:

$$\frac{dC_w}{dz} = -\frac{v_{a/w}}{E_z} C_w^{excess} \qquad (23\text{-}35)$$

The vertical coordinate is chosen positive upward. The negative sign indicates that C_w is decreasing toward the surface if PCE in the water is oversaturated with respect to the PCE concentration in the air ($C_w^{excess} > 0$).

Next, we can define homogeneity by requiring that the total concentration difference between lake bottom and lake surface, ΔC, may not be larger than, for instance, 10% of C_w^{excess}. Thus,

$$\Delta C = h \left| \frac{dC_w}{dz} \right| \le 0.1\, C_w^{excess} \qquad (23\text{-}36)$$

Inserting Eq. 23-35 into 23-36 and rearranging the terms yields:

$$\frac{h\,v_{a/w}}{E_z} \leq 0.1 \qquad (23\text{-}37)$$

Remembering the definition and meaning of the Peclet number (Eq. 22-11a), this result is not really a big surprise. In fact, the left-hand side of Eq. 23-37 is the Peclet number for an advective ($v_{a/w}$)/diffusive (E_z) flux over distance h.

We apply Eq. 23-37 to the case of PCE in Greifensee. From Table 21.1 we have $h = 17.4$ m, $v_{a/w} = 4.9 \times 10^{-6}$ m s^{-1} = 0.42 m d^{-1}. Turbulent diffusivity in the thermocline is 0.2 m^{2}d^{-1}, but from Fig. 22.8 we learn that in the hypolimnion it may be as large as 0.6 cm^{2}s$^{-1} \approx 5$ m^{2}d^{-1}. Inserting these numbers into Eq. 23-37 makes the left-hand side 37 and 1.5, respectively. Thus, even the larger E_z value is too small to keep PCE homogeneously mixed in the whole water column of Greifensee. In fact, Eq. 23-37 would need a vertical diffusivity E_z exceeding 7 m^{2}d$^{-1} \approx 0.8$ cm^{2}s^{-1}. Such values are often exceeded during the winter when the density stratification of the water column is weak.

From these semiquantitative considerations we conclude that competition between, on one hand, processes which eliminate concentration gradients like vertical mixing, and on the other hand, processes which produce concentration gradients like in-situ reactions and boundary fluxes, can be highly variable in time. Sometimes vertical mixing is intensive enough to warrant the box-model approach, other times we would need a model that allows us to describe spatially continuous concentration profiles. Sometimes we need a tool which could handle both situations. Such a model will be discussed next.

One-Dimensional Vertical (1DV) Lake Model

The one-dimensional vertical (1DV) lake model shown in Fig. 23.6 (Imboden and Schwarzenbach, 1985) is adequate for lakes with surface areas up to 100 km^2 in which horizontal concentration variations can be disregarded. Sometimes, this also applies to much larger lakes and even to parts of the ocean away from major inlets and other local sources. Note that in spite of the one-dimensional structure of the model, the three-dimensional bottom topography of the lake is not ignored. In a rectangular lake (mean depth and maximum depth are identical), only the deepest water layer would be in direct contact with the sediment surface. In contrast, in the model shown in Fig. 23.6 every layer has its (albeit reduced) contact area with the sediments where the chemicals can be directly exchanged. This is essential for describing the fate of a strongly sorbing chemical in a lake.

The goal of the 1DV-model is to calculate the time-dependent continuous vertical concentration profile of a compound, $C_w(z,t)$, where the depth coordinate z is the height above the deepest point of the lake (thus the vertical coordinate z is chosen as positive upward). Let us consider a horizontal layer of thickness Δz confined by the cross sections at depth z and $z + \Delta z$, $A(z)$ and $A(z + \Delta z)$, respectively (Fig. 23.6). The volume of the layer, ΔV, can be approximated by $A(z)\Delta z$, and the sediment contact area ΔA by $[A(z + \Delta z) - A(z)]$. Note that bottom slopes of lakes are commonly so small that ΔA, the horizontal projection of the inclined sediment surface, is usually a good approximation for the real contact area between water and sediment surface. In drawings of lake basins (such as Fig 23.6), the vertical dimensions are often strongly exaggerated so that the slopes look much steeper than they really are.

We define the specific flux rate of a chemical between sediment and water, $F_{s/w}$ (mol m^{-2}s^{-1}), as positive if the flux is directed from the sediment into the water. Then the mean concentration change due to sediment–water exchange in the layer of thickness Δz is:

$$\left(\frac{dC}{dt} \right)_{\text{sed/water exchange}} = \frac{\Delta A}{\Delta V} F_{s/w} = \frac{A(z+\Delta z)-A(z)}{A(z)\Delta z} F_{s/w}$$

If we decrease the thickness of the box, Δz, toward zero, we get:

$$\left(\frac{dC}{dt} \right)_{\text{sed/water exchange}} = a(z) F_{s/w} \qquad (23\text{-}38)$$

where the characteristic topographic function,

$$a(z) = \frac{1}{A} \left(\frac{A(z+\Delta z)-A(z)}{\Delta z} \right)_{\Delta z \to 0} = \frac{1}{A} \frac{dA}{dz} = \frac{dA}{dV} \qquad [\text{L}^{-1}] \qquad (23\text{-}39)$$

is the change of lake cross section per lake volume. It is an important parameter to describe the effect of lake topography on the spatial distribution of chemical compounds. Frequently, the depth-dependent lake cross section $A(z)$ can be approximated by (Imboden, 1973):

$$A(z) = A_0 \left(\frac{z}{z_0} \right)^{\eta} \qquad (23\text{-}40)$$

where z_0 is the maximum depth and the exponent η lies typically between 0.5 (flat littoral zone, "hole" at deepest part) and 2 (steep littoral zone, flat bottom). From Eqs. 23-39 and 23-40 follows:

$$a(z) = \frac{\eta}{z} \qquad (23\text{-}41)$$

Thus, the characteristic topographic function increases continuously from the surface ($z = z_0$) to the lake bottom ($z = 0$) where it becomes infinitely large. In fact, at the lake bottom a tiny lake volume stays in contact with a finite sediment area. This explains the great spatial and temporal gradients often found close to the bottom of lakes for compounds which are exchanged at the sediment–water interface (oxygen, phosphorus, methane, etc.).

The model (Fig. 23.6) consists of three compartments, (a) the surface mixed water layer (SMWL) or epilimnion, (b) the remaining open water column (OP), and (c) the surface mixed sediment layer (SMSL). SMWL and OP are assumed to be completely mixed; their mass balance equations correspond to the expressions derived in Box 23.1, although the different terms are not necessarily linear. The open water column is modeled as a spatially continuous system described by a diffusion/advection/ reaction

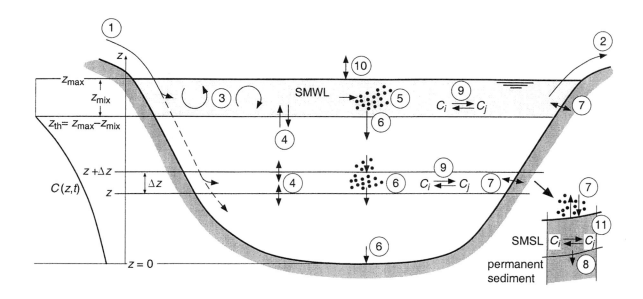

Figure 23.6 One-dimensional vertical (1DV) lake model with depth-dependent cross section. Processes are labeled with the same numbers as used in Table 23.9. The simulation tools MASAS (Ulrich et al., 1995) and AQUASIM (Reichert, 1994) are based on the same modeling concept.

equation as given by Eq. 22-6. We want to describe the dynamics of a chemical compound being present in different phases (aqueous, sorbed on "large" particles, sorbed in colloids, etc.) which are always at equilibrium. In the water compartments SMWL and OP, the total concentrations, $C_t^{mix}(t)$, $C_t^{op}(z,t)$, are chosen as the model functions. For the sediment box we use the sorbed concentration, $C_s^{sc}(z,t)$. Note that in contrast to the water/sediment model of Box 23.3, every depth zone has its own surface mixed sediment layer (SMSL); therefore $C_s^{sc}(z,t)$ depends also on depth z. In addition, for the SMWL and the water column we also need a mass balance equation for the suspended particles. Their concentrations are given by the solid-to-water phase ratio, r_{sw}^{mix} and $r_{sw}^{op}(z)$.

Let us first develop the dynamic equation of the total concentration in the SMWL, $C_t^{mix}(t)$. Not much additional effort is needed. We can just copy the structure of Eqs. 1 to 3 of Box 23.1, although the boundary flux terms have a slightly different form. The numbers shown below the different terms in Eq. 23-42 correspond to the notation introduced in Table 23.9 and also used in Fig. 23.6.

$$\frac{dC_t^{mix}}{dt} = \underset{(1)}{\frac{I^{mix}}{V^{mix}}} - \underset{(2)}{QC_t^{mix}} - \underset{(4)}{\frac{A_{th}}{V_{mix}}\frac{\partial C_t^{op}}{\partial z}\bigg|_{z_{th}}} - \underset{(6)}{\frac{A_o}{V_{mix}}v_s^{mix}f_s^{mix}C_t^{mix}}$$

$$\underset{(7)}{-\frac{(A_o - A_{th})}{V_{mix}}v_{sedex}\left(C_t^{mix} - \frac{C_s^{sc,mix}}{f_w^{mix}K_d^{sc,mix}}\right)} - \underset{(9)}{R^{mix}} \tag{23-42}$$

$$\underset{(10)}{-\frac{A_o}{V_{mix}}v_{a/w}\left(f_w C_t^{mix} - \frac{C_a}{K_{a/w}}\right)} \qquad \left[\text{mol m}^{-3}\text{s}^{-1}\right]$$

To facilitate the understanding of this equation we give a few references to other sections of the book. Term 4 results from Fick's first law (Eq. 18-6) formulated at the

Table 23.9 Processes of One-Dimensional Vertical (1DV) Lake Model (see Fig. 23.6)

#	Variable	Unit	Description [a]
Lake-Specific Processes			
1	I^{mix}	mol s^{-1}	Input of chemical by rivers and effluents into SMWL
	$I^{op}(z)$	mol m^{-1}s^{-1}	Input of chemical (per unit lake depth) by rivers and effluents into OP at depth z
2	QC_t^{mix}	mol s^{-1}	Loss from SMWL through outlet
3[b]	–	–	Complete mixing of SMWL
4	$E_z(z,t)$	m^{2}s^{-1}	Vertical turbulent diffusivity in OP
5	$p^{mix}, p(z)$	kg$_s$m^{-3}s^{-1}	In-situ production of particulate matter, mainly in the SMWL (e.g., phytoplankton growth)
6	$v_s^{mix}, v_s(z)$	m s^{-1}	Settling velocity of suspended particles
7	$v_{sedex}^{mix}, v_{sedex}(z)$	m s^{-1}	Exchange velocity between water and SMSL by diffusion and resuspension
8	$\left.\begin{array}{l}\beta^{mix}\, v_s^{mix}\, r_{sw}^{mix} \\ \beta(z)\, v_s(z) r_{sw}^{op}(z)\end{array}\right\}$	kg$_s$m^{-2}s^{-1}	Transfer of sediment mass into "permanent" (noninteracting) sediment column
Compound-Specific Processes			
9	$R^{mix}, R(z)$	mol m^{-3}s^{-1}	In-situ transformation of chemical (hydrolysis, photolysis, biotransformation etc.)
10	$v_{a/w}$	m s^{-1}	air–water transfer velocity at lake surface
11	R_{SMSL}	mol kg$_s^{-1}$s^{-1}	Transformation of chemical in SMSL

[a] SMWL = surface mixed water layer, SMSL = surface mixed sediment layer, OP = open water (deep water). [b] Variable does not explicitly appear in Eq. 23-42 (see text).

upper boundary of the open water column (thermocline depth $z_{th} = z_{max} - z_{mix}$). Term 5 (settling particles) was derived in Eqs. 23-11 to 23-19. Term 7 (sediment–water exchange) follows from Eqs. 5 to 7 of Box 23.2. The nondimensional equilibrium partition functions (f_w, f_s) are constructed according to Eq. 23-22. Remember that the vertical coordinate z is positive upward. We should also mention that vertical mixing in the SMWL (term 3) does not explicitly appear in Eq. 23-42. It is hidden in the assumption that the concentration within the SMWL is spatially constant.

The corresponding balance equation for the concentration of suspended solids in the SMWL (expressed as solid-to-water phase ratio, r_{sw}^{mix}) is:

$$\frac{d\, r_{sw}^{mix}}{dt} = \underset{(1)}{\frac{I_s^{mix}}{V_{mix}}} - \underset{(2)}{Q r_{sw}^{mix}} - \underset{(4)}{\frac{A_{th}}{V_{mix}}\frac{\partial r_{sw}^{op}}{\partial z}\bigg|_{z_{th}}}$$

$$+\underset{(5)}{p^{mix}} - \underset{(6)}{\frac{A_o}{V_{mix}}\, v_s^{mix}\, r_{sw}^{mix}} \quad \left[kg_s m^{-3} s^{-1}\right] \tag{23-43}$$

I_s^{mix} is the input of suspended particles into the SMWL (units: $\mathrm{kg_s s^{-1}}$), p^{mix} the in-situ production of particulate matter ($\mathrm{kg_s m^{-3} s^{-1}}$) such as from phytoplankton growth.

The corresponding dynamic equations of the open water column are constructed from Eq. 22-6. They are completed by the sediment-water boundary flux derived in Eq. 23-38. We assume that the *net* vertical advection of water is zero. Thus, the vertical water movement is incorporated in the turbulent diffusivity, E_z. The assumption implies that if chemicals are directly introduced at depth z (term 1), they would not be accompanied by significant quantities of water. Typically, such inputs are due to sewage outlets (treated or untreated) into the lake. We get:

$$\underset{(1)}{\frac{\partial C_t^{\mathrm{op}}(z)}{\partial t}} = \underset{(1)}{\frac{I^{\mathrm{op}}(z)}{A(z)}} + \underset{(4)}{\frac{1}{A(z)}\frac{\partial}{\partial z}\left[A(z)E_z(z,t)\frac{\partial C_t^{\mathrm{op}}(z)}{\partial z}\right]}$$

$$+\underset{(6)}{\frac{\partial}{\partial z}\left[v_s(z)f_s(z)C_t^{\mathrm{op}}(z)\right]} - \underset{(7)}{a(z)v_{\mathrm{sedex}}(z)\left[C_t^{\mathrm{op}} - \frac{C_s^{\mathrm{sc}}(z)}{f_w(z)K_d^{\mathrm{sc}}(z)}\right]} \qquad (23\text{-}44)$$

$$\underset{(9)}{-R(z)} \qquad \left[\mathrm{mol\ m^{-3} s^{-1}}\right]$$

Note that the particular choice of the vertical coordinate pointing upward makes the sign of the advective term 6 positive since the settling velocity v_s was chosen as a positive number in spite the fact that it is directed in the negative z-direction.

The corresponding equation for the suspended particles is:

$$\underset{(1)}{\frac{\partial r_{\mathrm{sw}}^{\mathrm{op}}(z)}{\partial t}} = \underset{(1)}{\frac{I_s(z)}{A(z)}} + \underset{(4)}{\frac{1}{A(z)}\frac{\partial}{\partial z}\left[A(z)E(z,t)\frac{\partial r_{\mathrm{sw}}^{\mathrm{op}}(z)}{\partial z}\right]}$$

$$\underset{(5)}{+p(z)+} \underset{(6)}{\frac{\partial}{\partial z}\left[v_s(z)r_{\mathrm{sw}}^{\mathrm{op}}(z)\right]} \qquad \left[\mathrm{kg_s m^{-3} s^{-1}}\right] \qquad (23\text{-}45)$$

Finally, the balance equation for the SMSL looks like Eq. 23-30, except for the reaction term which is not necessarily linear now. Note that C_s^{sc} depends also on z (the depth in the lake, *not* the depth in the sediment column). Every depth zone has its own SMSL, but it is assumed that these layers do not interact with each other. In fact, the distance between them is much too large for lateral molecular diffusion in the sediments to play any role. There is no equation for the particles in the SMSL. Their balance is indirectly included in the preservation factor $\beta(z)$. Remember that if total solids are used to describe the solid phase, β is about 1, whereas if particulate organic carbon (POC) is used, β is smaller than one because part of the POC is degraded in the SMSL. Finally, we get:

$$\frac{dC_s^{sc}(z)}{dt} = \underbrace{\frac{v_s(z)}{m(z)} f_s(z) C_t^{op}(z)}_{(6)} + \underbrace{\frac{v_{sedex}(z)}{m(z)} \left[C_t^{op}(z) - \frac{C_s^{sc}(z)}{f_w(z) K_d^{sc}(z)} \right]}_{(7)}$$

$$\underbrace{- \beta(z) \frac{v_s(z)}{m(z)} r_{sw}^{op}(z) C_s^{sc}(z)}_{(8)} - \underbrace{R_{SMSL}}_{(11)} \qquad \left[\text{mol kg}_s^{-1} \text{s}^{-1} \right] \qquad (23\text{-}46)$$

where

$m(z)$ [kg$_s$m^{-2}] is the particle mass in the SMSL

$\beta(z)$ is the particle preservation factor

$K_d^{sc}(z)$ is the solid–water distribution ratio in the SMSL at lake depth z

All other variables were defined before.

Numerical Models

With these five equations (Eqs. 23-42 to 23-46), two of them partial differential equations, the limits of the analytical approach and the goals of this book are clearly exceeded. However, at this point we take the occasion to look at how such equations are solved numerically. User-friendly computer programs, such as MASAS (Modeling of Anthropogenic Substances in Aquatic Systems, Ulrich et al., 1995) or AQUASIM (Reichert, 1994), or just a general mathematical tool like MATLAB and MATHE-MATICA, can be used to solve these equations for arbitrary constant or variable parameters and boundary conditions.

Most of these tools use the finite difference method (at least for one-dimensional models) in which the continuous space coordinate is divided into a number of boxes. So we are back to the box-model technique. To demonstrate the procedure, in Box 23.4 we show how the partial differential Eqs. 23-44 and 23-45 are transformed into discrete (box) equations.

Box 23.4 Numerical Approximation of Partial Differential Equations of One-Dimensional Vertical Lake Model

Basis: Dynamic equations for total compound concentration, $C_t^{op}(z)$, and suspended particle concentration, $r_{sw}^{op}(z)$, in the water column between $z = 0$ (lake bottom) and z_{th} (depth of thermocline): Eqs. 23-44 and 23-45.

Procedure: Subdivide the vertical axis in n horizontal layers each with depth $\Delta z = z_{th}/n$. Layer 1 is at lake bottom, layer is just below the surface mixed water layer (SMWL).

Simplified notation:
C_j average C_t^{op} in box j $(j = 1....n)$
r_j average r_{sw}^{op} in box j $(j = 1....n)$
C_{mix}, r_{mix} corresponding values in SMWL (depth z_{mix})

Approximation of spatial derivatives:

The boundary between layer $(j - 1)$ and j is denoted as $(j - 1/2)$, the boundary between layer j and $(j + 1)$ as $(j + 1/2)$. Term 6 (Eqs. 23-44 and 23-45) is approximated as:

$$\frac{1}{\Delta z}\left(v_{s\,j+1/2}\,f_{s\,j+1/2}\,C_{j+1/2} - v_{s\,j-1/2}\,f_{s\,j-1/2}\,C_{j-1/2}\right)$$

$$\approx \frac{1}{\Delta z}v_{sj}f_{sj}\left[\frac{1}{2}\left(C_{j+1}+C_j\right) - \frac{1}{2}\left(C_j + C_{j-1}\right)\right]$$

$$= \frac{1}{2\Delta z}v_{sj}f_{sj}\left(C_{j+1}+C_{j-1}\right)$$

Term 4 (Eqs. 23-44 and 23-45) is approximated as follows:

$$A(z)E_z(z,t)\frac{\partial C}{\partial z}\ \text{ at boundary } (j-1/2) \ \rightarrow\ A_{j-1/2}\,E_{j-1/2}\left.\frac{\partial C}{\partial z}\right|_{j-1/2} \approx A_j\,E_{j-1/2}\frac{C_j - C_{j-1}}{\Delta z}$$

and similarly for the boundary $(j + 1/2)$.

With $V_j \approx A_j\,\Delta z$, term 4 becomes:

$$\text{term 4}\ \rightarrow\ \frac{1}{\Delta z^2}\left[E_{j+1/2}\left(C_{j+1}-C_j\right) - E_{j-1/2}\left(C_j - C_{j-1}\right)\right]$$

and correspondingly for r_j.

The topographic function $a(z)$ (Eq. 23-39) is approximated by

$$a_j \approx \frac{A_{j+1/2} - A_{j-1/2}}{A_j\,\Delta z}$$

With these substitutions Eq. 23-44 transforms into a set of n ordinary differential equations in time:

$$\underset{(1)}{\frac{dC_j}{dt}} = \underset{\,}{\frac{I_j}{A_j}} + \underset{(4)}{\frac{1}{\Delta z^2}\left[E_{j+1/2}\left(C_{j+1}-C_j\right) - E_{j-1/2}\left(C_j - C_{j-1}\right)\right]}$$

$$+ \underset{(6)}{\frac{v_{sj}f_{sj}}{2\Delta z}\left(C_{j+1}-C_{j-1}\right)} - \underset{(7)}{a_j\,v_{\text{sedex }j}\left[C_j - \frac{C_{sj}}{f_{wj}\,K^{sc}_{sj}}\right]} - \underset{(9)}{R_j}$$

Correspondingly for the suspended particle concentration r_j (Eq. 23-45):

$$\frac{dr_j}{dt} = \frac{I_{sj}}{A_j} + \frac{1}{\Delta z^2}\left[E_{j+1/2}\left(r_{j+1} - r_j\right) - E_{j-1/2}\left(r_j - r_{j-1}\right)\right] + p_i + \frac{v_{sj}}{2\Delta z}\left(r_{j+1} - r_{j-1}\right)$$

$$(1) \qquad\qquad (4) \qquad\qquad (5) \qquad\qquad (6)$$

For box $j = n$ just below the SMWL, the same equations can be used by setting the variables with subscript $(n + 1)$ equal to the corresponding SMWL values.

Note: In order to avoid a phenomenon called *numerical diffusion*, the time step Δt for the iteration of the above equations has to be much smaller than either $(\Delta z / v_{sj})$ or $(\Delta z^2 / E_{j\pm1/2})$.

Application of the 1DV Lake Model

The application of the continuous lake model is illustrated by continuing the story of tetrachloroethylene (PCE) in Greifensee. Remember that PCE is a compound which is quasi-conservative in the water and is not significantly sorbed by particles. Besides flushing, exchange at the air–water interface is the only relevant process to be considered (see Box 21.2 and Illustrative Examples 21.5, 21.6)

We discuss the traces of an "input event" detected in Greifensee in 1985 (Ulrich et al., 1994). Until May 1985, the PCE content of Greifensee was relatively small (between about 40 and 80 moles, see Fig. 23.7). On June 3, 1985, a PCE content of more than 200 moles was found in the lake. The PCE was mainly detected in the surface mixed layer (about 4 m deep) and in the thermocline. On July 1, 1985, the PCE content was still around 200 moles, but the concentration maximum had moved from the surface to the thermocline. This concentration peak remained visible throughout the summer and fall until the PCE content had returned to its "normal" level.

The calculations were done with the modeling software MASAS (Ulrich et al., 1995). The model consisted of up to 32 layers, each one meter deep. Depending on the varying mixing depth, z_{mix}, a variable number of boxes were incorporated in the surface mixed water layer (SMWL). Air–water exchange velocities were calculated as described earlier. The spatial and temporal variations of vertical diffusivity, $E_z(z,t)$, were determined from regular temperature measurements according to the method described in Section 22.2. The model was used to determine the unknown mean PCE input for the periods between two sampling dates.

As shown in Figure 23.7, the continuous lake model nicely describes the concentration maximum, which slowly moved to greater depth due to the deepening of the surface mixed layer. From the model calculation we can conclude that the processes involved in producing this maximum were the combination of riverine PCE input into the surface mixed layer and loss to the atmosphere by gas transfer. The extra input of PCE into the lake between May 6 and July 1, 1985 had to be about 360 moles. The model calculations suggest that the input had dropped to virtually zero after July 1. Part of the compound was quickly and continuously lost to the atmosphere so that the PCE content of the lake never increased much beyond 200 moles.

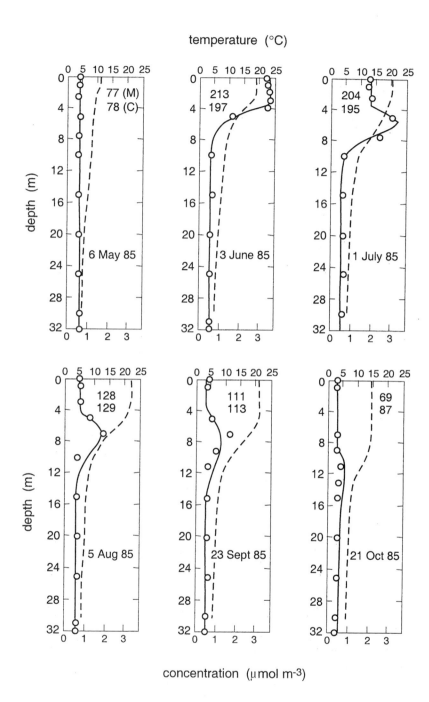

Figure 23.7 Vertical profiles of water temperature (dotted line) and of measured (circles) and calculated (solid line) PCE concentration in Greifensee (Switzerland) for the period May to October 1985. Numbers give PCE inventory in moles (M = measured, C = calculated). From the model calculation it can be concluded that between May 6 and July 1, 1985, about 360 moles of PCE entered the lake, thus leading to a significant increase of the concentration in the lake during several months. After July 1, the input was virtually zero.

As mentioned before, the 1DV lake model, although still relatively simple compared to the three-dimensional nature of real transport and reaction processes, predicts concentrations and inventories which in most cases are not matched by available field data in terms of chemical, spatial, and temporal resolution. In fact, in a time when powerful computers are ubiquitously available, it is not unusual to find publications in which highly sophisticated model outputs are compared to poor data sets for which much simpler models would have been adequate. However, this is not an argument against the development of good models, but a plea for their wise use. To compare measured and calculated data is not the only task for which models are good. One reason for using models is, for instance, to evaluate the relative importance of various processes for the dynamics of chemical compounds in aquatic systems. As an example, based on such a model one could assess the influence of changing particle concentrations on a strongly sorbing compound, such as the PCB185 congener, in the water column of a lake. The change of the trophic state of a water body, that is, the frequency and intensity of algal blooms in the water, greatly affect the concentration of sorbing chemical species. Thus, a model can help one find connections between different components of an aquatic ecosystem and to evaluate possible inadvertent effects of human interventions in complex systems. In addition, models that are more advanced than the present techniques of analytical chemistry can be used to find optimal strategies for the protection of aquatic systems against pollution by xenobiotic compounds. This can be done by identifying the mechanisms to which the system is most vulnerable, as well as by evaluating alternative restoration procedures in case the compound has already entered the system.

23.5   Questions and Problems

Questions

Q 23.1

List the processes that govern the mass balance of a sorbing chemical in a lake or pond.

Q 23.2

What assumptions have to be made to describe the behavior of a sorbing chemical in a lake by just *one* state variable (e.g., by total concentration)?

Q 23.3

Give reasons why particle-settling velocities calculated from Stokes law (Eq. 23-7) do not necessarily correspond to the apparent settling velocities determined from the particle removal rates in lakes.

Q 23.4

Why should we not use Stokes law to calculate the settling velocity of a fecal pellet with diameter of 1 mm and density of $1.2 \ g \ cm^{-3}$?

Q 23.5

In what respect can we compare the settling of particles in a lake with the settling of snowflakes in a snowstorm?

Q 23.6

Why can sorption on colloids cause a significant discrepancy between calculated and measured "apparent dissolved concentrations" (concentrations determined after filtration) for strongly sorbing chemicals?

Q 23.7

In Box 23.2 it is shown that the combined exchange of a chemical at the sediment–water interface due to sediment resuspension and to diffusion, respectively, can be expressed by a single exchange velocity. Explain why this result is a direct consequence of the assumption that the lake model can be formulated in terms of linear differential equations.

Q 23.8

Why do we need at least a two-box model to explain the long-term memory of the PCBs in Lake Superior?

Q 23.9

In what sense is the 1DV lake model more than one-dimensional, in spite of the fact that the resulting differential equation is one-dimensional? Why is this important for describing sediment–water exchange?

Q 23.10

Explain the meaning of the topographic function, $a(z)$. Why does $a(z)$ generally increase toward the lake bottom? Imagine a chemical with a constant sediment boundary flux and give a qualitative picture of the vertical concentration profiles in the water column which evolve as a result of such a flux. Assume that the chemical does not react in the water column.

Problems

P 23.1 *How Fast Is Benzene Biodegraded in This Pond?*

In the same well-mixed pond in which you already had to deal with a vinyl acetate spill (Illustrative Example 23.1), somebody monitors the benzene concentration during the summer. Interestingly, the concentration of benzene in the pond water does not vary much over time, and is always in the order of $0.05 \ \mu g \ L^{-1}$. A colleague of yours who is responsible for air pollution measurements in the area tells you that the average benzene concentration in the air is $0.04 \ \mu g \ L^{-1}$. He claims that input of benzene by gas exchange from the air is the predominant source of this compound in the pond. You remember vaguely that you already dealt with such a problem in Chapter 20. Assuming he is right, and assuming that biodegradation is the only transformation process for benzene in the pond, calculate the characteristic biodegradation rate constant (yr^{-1}) for benzene by using the pond characteristics and conditions given in Illustrative Example 23.1.

P 23.2   *What Steady-State Concentration of Anthracene Is Established in the Epilimnion of This Lake?*

With a leachate from a coal tar site, each day 2 kg of anthracene are introduced continuously into the epilimnion of a eutrophic lake. The lake is stratified between April and November. As an employee of the state water authority you are asked to monitor the anthracene concentration in the epilimnion of this lake. In order to get an idea of how sensitive your analytical technique has to be, you wonder what anthracene concentration (order of magnitude) you have to expect. To this end, answer the following questions using the average epilimnion characteristics and conditions given below. Neglect any water exchange between the epilimnion and the hypolimnion.

(a) What would be the anthracene concentration at steady-state if anthracene exhibited conservative behavior in the epilimnion of the lake?

(b) What anthracene steady-state concentration can actually be expected assuming that biodegradation can be neglected?

(c) What is the time required to reach this steady-state?

You can find all relevant physical-chemical properties of anthracene in Appendix C. Note that $\Delta_{aw}H$ is about 50 kJ mol^{-1} (Table 6.3). Furthermore, when inspecting Fig.15.3 and Table 15.7 you realize that direct photolysis could be an important process. Before going through tedious calculations, you remember that a friend of yours who works at the EPA laboratory in Athens (Georgia) told you that the near-surface photolysis half-life (averaged over 24 h) of anthracene under clear skies in the summer at 40°N latitude (by coincidence, exactly the same latitude as your lake!) is about 5 days. You also remember from a course in environmental photochemistry that the average light intensity between April and October is about 50% of that observed on a clear midsummer day. This estimate includes both the seasonal variation in light intensity as well as the effect of clouds.

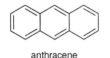

anthracene

Average Epilimnion Characteristics and Conditions

Volume	V	$= 5 \times 10^7$ m^3
Surface area	A	$= 1 \times 10^7$ m^2
Mean water residence time	τ_w	$= 100$ days
Average particle concentration	r_{sw}	$= 5 \times 10^{-3}$ kg$_s$m^{-3}
Organic carbon content of suspended particles	f_{oc}	$= 0.4$ kg$_{oc}$kg$_s^{-1}$
Average particle flux	F_s	$= 5 \times 10^{-3}$kg$_s$m^{-2}d^{-1}
Average wind speed 10 m above the lake surface	u_{10}	$= 1.5$ m s^{-1}
Average water temperature	T	$= 20°$C
	pH	$= 8.2$
Concentration of colloidal organic carbon	[DOC]	$= 4$ mgC L^{-1}

Decadal beam attenuation coefficients $\alpha(\lambda)$ of the water at various wavelengths (λ)

λ (nm)	300	325	350	375	400	450
$\alpha(\lambda)$ (m^{-1})	4.0	2.5	1.5	1.0	0.6	0.3

P 23.3 *Extending the Anthracene Case*

In Problem 23.2 you were asked to calculate the steady-state concentration of anthracene in the epilimnion of a lake under the assumption that 2 kg of this compound were introduced into the epilimnion every day. For your calculation you neglected any water exchange between the epilimnion and the hypolimnion. Extend your calculation now to the whole lake, and consider various input scenarios. In particular, answer the following questions:

(a) What are the anthracene steady-state concentrations in the epilimnion and in the hypolimnion, respectively, if (i) all input (i.e., 2 kg d^{-1}) goes into the epilimnion, (ii) all input goes into the hypolimnion, and (iii) half of the input goes into the epilimnion and half into the hypolimnion? Assume that during stratification, all water input goes into the epilimnion.

(b) Considering that the lake is stratified between the middle of April and the end of November, is there enough time to reach steady-state? Discuss each of the three cases.

(c) Take case (i) and assume that steady-state is reached at the end of November. Furthermore, assume that at this time the whole lake mixes instantaneously. What anthracene concentration do you expect at the end of March? How close is this concentration to the steady-state concentration that would establish under typical winter conditions?

Additional lake data (see also table in P 23.2)

Total lake volume	V_t	$= 15 \times 10^7 \ m^3$
Lake area at thermocline	A_{th}	$= 0.8 \times 10^7 \ m^2$

Summer conditions

Turbulent exchange velocity between epilimnion and hypolimnion	v_{th}	$= 0.05 \ m \ d^{-1}$
Average particle concentration		
in the epilimnion	r_{sw}^E	$= 5 \times 10^{-3} kg_s m^{-3}$
in the hypolimnion	r_{sw}^H	$= 2 \times 10^{-3} kg_s m^{-3}$
Organic carbon content of suspended particles	f_{oc}^E, f_{oc}^H	$= 0.4 \ kg_{oc} kg_s^{-1}$
Average settling velocity of particles	v_{sE}, v_{sH}	$= 1 \ m \ d^{-1}$
Average wind speed 10 m above the lake surface	u_{10}	$= 1.5 \ m \ s^{-1}$
Average water temperature	T_E	$= 20°C$
	T_H	$= 5°C$
Average pH	pH_E	$= 8.2$
	pH_H	$= 7.5$
Dissolved organic carbon	$[DOC]_E$	$= 4 \ mg_{oc} \ L^{-1}$
	$[DOC]_H$	$= 2 \ mg_{oc} \ L^{-1}$
Decadal beam attenuation coefficients	$\alpha(\lambda)_E$	see Problem 23.2

Winter conditions (whole lake)

Average particle concentration	r_{sw}	$= 2 \times 10^{-3} kg_s m^{-3}$
Organic carbon content of suspended particles	f_{oc}	$= 0.2 \ kg_{oc} kg_s^{-1}$
Average settling velocity of particles	v_s	$= 1 \ m \ d^{-1}$
Average wind speed 10 m above the lake surface	u_{10}	$= 2 \ m \ s^{-1}$
Average water temperature	T	$= 5°C$
Average pH	pH	$= 7.5$
Concentration of colloidal organic carbon	$[DOC]$	$= 2 \ mg_{oc} \ L^{-1}$

Hints and Help

Note that some of the characteristic rate constants calculated in P 23.2 are also valid for this model. For estimating the rate of direct photolysis in the winter, assume that $\alpha(\lambda)$ is proportional to [DOC] and that the average light intensity is about 3–4 times lower as compared to the summer (compare 24-hour averaged light intensity values in Table 15.3 with those in Table 15.4). For the mathematics of the two-box model it may be helpful to look at Table 21.7.

P 23.4 *Determining the Fate of Hexachlorobenzene in a Pond*

It remains unnoticed for several years that the pesticide hexachlorobenzene (HCB) has continuously entered the small well-mixed pond described in Illustrative Example 23.1. One day, your colleague determines the HCB concentration in the surface sediments of the pond and finds a C_s^{sc} value of 1.3 μmol kg_s^{-1}. You are alarmed by this value, because you fear that the HCB concentration in the water column could be dangerously high. Since you cannot get a water sample at once, you first try to calculate the concentration in the water column, C_t^{op}, by assuming a steady-state situation. From earlier investigations the following data on the sediments are available.

Sediment characteristics

Temperature	T	$= 5°C$
Depth of SMSL	z_{mix}	$= 2$ cm
Particle density	ρ_S	$= 2.5$ $g_s cm^{-3}$
Porosity in SMSL	ϕ	$= 0.92$
Preservation factor	β	$= 0.8$
Organic carbon content of particles in SMSL	f_{oc}^{sc}	$= 0.05$ $kg_{oc} kg_s^{-1}$
Sediment/water exchange rate (see Eq. 23-32b)	k_{sedex}	$= 0.011$ d^{-1}

(a) Calculate the steady-state concentration in the lake, $C_{t\infty}$.

(b) Estimate the ratio between the total HCB mass in the water column and in the SMSL, respectively.

(c) What fraction of the HCB entering the pond at any given time is eventually leaving the pond by gas exchange and through the outlet? What happens to the rest?

(d) Estimate the input of HCB per unit time into the pond. After a thorough investigation you locate the HCB source and stop it.

(e) Estimate the time needed to eliminate 95% of the HCB from the combined water column/SMSL system.

hexachlorobenzene
(HCB)
$M_i = 284.8$ g·mol⁻¹

P 23.5 *A Lake's Long-Term Memory Former Pollution by a Heptachlorobiphenyl*

Over a long time period, a lake has been exposed to pollution by different polychlorinated biphenyl (PCB) congeners. As discussed in Chapter 23.3, a fraction of the PCBs introduced into a lake is stored in the sediments.

You are responsible for the PCB monitoring program in the lake. Despite the fact that all external PCB inputs have been stopped, you still find significant PCB concentrations in the water column. A detailed survey of the sediments shows that in the top 10 cm of the sediment layer the mean concentration, C_s^{sc}, of 2,2',3,4,5,5',6-heptachlorobiphenyl (PCB185) is 4.0 nmol kg_s^{-1}.

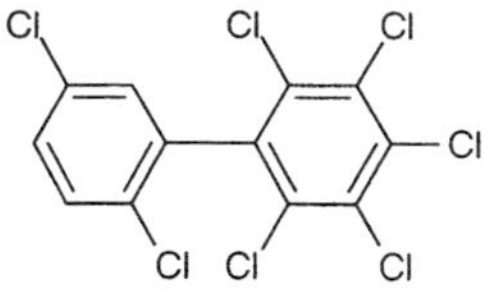

2,2',3,4,5,5',6
heptachlorobiphenyl
(PCB185)

(a) Calculate the mean total concentration in the water column, C_t^{op}, which is in equilibrium with C_s^{sc}.

(b) How long does it take until C_s^{sc} and C_t^{op} have dropped to 95% of their present values?

The characteristic data describing the lake are summarized in the following table. Note that the physicochemical properties of PCB185 can be found in Table 23.4.

Lake

Total volume	V	$= 7.5 \times 10^8 \text{ m}^3$
Surface area	A	$= 30 \times 10^6 \text{ m}^2$
Throughflow of water	Q	$= 7.5 \times 10^6 \text{ m}^3\text{d}^{-1}$
Average water temperature at lake bottom	T	$= 5°C$
Settling velocity of particles	v_s	$= 0.7 \text{ m d}^{-1}$
Suspended particle concentration	r_{sw}	$= 1 \times 10^{-3}\text{kg m}^3$
Organic carbon content of suspended particle	f_{oc}	$= 0.2 \text{ kg}_{oc}\text{kg}_s^{-1}$
Average total air/water exchange velocity of HCBP	$v_{a/w}$	$= 0.4 \text{ m d}^{-1}$

Sediment

Depth of surface mixed sediment layer (SMSL)	z_{mix}	$= 0.1 \text{ m}$
Particle density	ρ_S	$= 2500 \text{ kg}_s \text{ m}^{-3}$
Porosity of SMSL	ϕ	$= 0.9$
Preservation factor	β	$= 0.8$
Organic carbon content of particles in SMSL	f_{oc}^{sc}	$= 0.05$
Sediment/water exchange rate for PCB185 (Eq. 23-32b)	k_{sedex}	$= 2.6 \times 10^{-5} \text{ d}^{-1}$

Hints:
When calculating the characteristic coefficient, k_{11}, k_{12}, k_{21}, k_{22}, of Box 23.3 you will find that the response velocity of the sediment reservoir is much smaller than that of the open water column. In order to predict the decrease of both concentrations, C_s^{sc} and C_t^{op}, you can assume a quasi-steady state between the two concentrations in which the system is controlled by the decrease of the slowly reacting sediment reservoir.

After you have done your model calculation, you realize that the measured C_t^{op} value is nearly twice as big as the value you have calculated. You remember that colloidal organic matter can significantly change the behavior of the PCBs in the water column (see Table 23.5). Your measurements give an average concentration of colloidal organic matter of $[DOC] = 3.2 \times 10^{-3} \text{ kg m}^{-3}$. Modify the above calculations and check whether the colloids can explain the discrepancy between observation and model calculation.

P 23.6 *Another Aspect of the PCB Story in Lake Michigan*

For many years now, we have seen the concentrations of banned compounds like the polychlorinated biphenyls (PCBs) decrease (or at least remain constant) in Lake

Michigan. Given that direct inputs in effluents have ceased, we are interested in identifying the phenomena that maintain the current levels in the lake. One possibility is that exchanges with polluted air account for the lake water levels. Observations over and in Lake Michigan indicate that (*a*) during the summer, PCBs are coming into the water from the air; (*b*) during the winter, the water is a source of PCBs to the air. Moreover, the absolute values of the summertime fluxes are three times greater than the absolute values of the wintertime fluxes. You wonder how to explain these observations. Wisely, you choose to focus on a specific congener, 2,2',4,4',5,5'-hexachlorobiphenyl (PCB153). Its properties are shown below.

You want to answer the following questions:

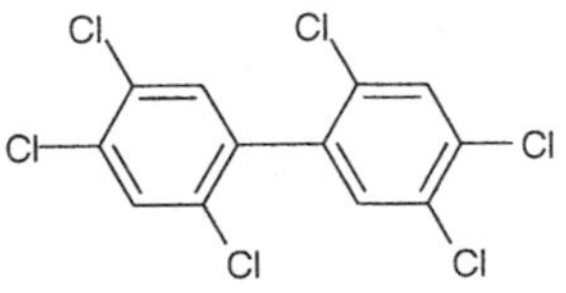

(1) If the measured dissolved PCB153 concentrations in lake water samples are always 1×10^{-8} g m^{-3}, but the air concentrations over the lake are 2×10^{-11} g m^{-3} in the summer and 1×10^{-11} g m^{-3} in the winter, does the change in air concentrations explain the seasonal reversal of fluxes?

(2) What wintertime flux (mol m^{-2}s^{-1}) do you expect for PCB153 and what is the direction of the flux?

(3) Would your estimate of the summertime flux change if you knew that suspended solids at 2 mg L^{-1} exhibiting 20% organic carbon content occurred in Lake Michigan surface water? Give flux directions both with and without considering solids.

2,2',4,4',5,5',-
hexachlorobiphenyl
(PCB153)
M_i = 360.9 g·mol^{-1}

Properties of Lake Michigan	Summer	Winter
Air temperature (°C)	T_a = 25	2
Water temperature (°C)	T_w = 25	2
Wind speed (m s^{-1})	u_{10} = 3	7

Most properties of PCB153 can be found in Appendix C. Some additional data for PCB153 (for 25°C):

$$D_{ia} = 0.058 \text{ cm}^2\text{s}^{-1}$$
$$D_{iw} = 0.67 \times 10^{-5} \text{ cm}^2\text{s}^{-1}$$
$$Sc_{iw} = 1330$$

1100   Notes

Chapter 24

RIVERS

24.1 Transport and Reaction in Rivers
Characterization of River Flow
Transport by the Mean Flow
Illustrative Example 24.1: *Mean Flow Velocity of River G for Different Discharge Rates*
Box 24.1: *Stationary Solution for the Advection/Reaction Equation in a River*
Mean Flow and Air–Water Exchange
Illustrative Example 24.2: *Air–Water Exchange in River G*
From Reaction Times to Reaction Distances
Sediment–Water Interaction
Illustrative Example 24.3: *A Spill of Atrazine in River G*
Box 24.2: *Sediment–Water Exchange in Sandy Sediments*

24.2 Turbulent Mixing and Dispersion in Rivers
Vertical Mixing by Turbulence
Lateral Mixing by Turbulence
Longitudinal Mixing by Dispersion
Illustrative Example 24.4: *Turbulent Diffusion and Longitudinal Dispersion in River G*
Dilution of a Concentration Patch by Longitudinal Dispersion
Illustrative Example 24.5: *A Second Look at the Atrazine Spill in River G: The Effect of Dispersion*

24.3 A Linear Transport/Reaction Model for Rivers
All Things Considered: The Reaction of Total Load and Maximum Concentration
First Case Study: Chloroform in the Mississippi River
Second Case Study: Chemical Pollution of the River Rhine Due to a Fire in a Storehouse

24.4 Questions and Problems

Compared to lakes, the contribution of rivers to the water inventory of the earth is not very impressive. Their estimated water volume is 1.2×10^3 km^3, that is, only about 1% of the water stored in freshwater lakes. Rivers and streams cover an area of about 10^5 km^2, near 10% of the total lake area. Yet, rivers are more equally distributed on the globe than lakes.

In this chapter tools are presented for the construction of simple river models. The flows of rivers that often extend over thousands of kilometers, are extremely important for the transfer of water into areas with little precipitation. Furthermore, humans have always used the transport capacity of rivers to get rid of their waste. Due to industrialization, this waste has changed its composition from purely natural inorganic and organic compounds to the complex cocktail of chemicals that characterizes our present world. Thus rivers, together with the atmosphere and the ocean, became important conveyor belts for the distribution of xenobiotic chemicals on the globe. As many examples have demonstrated in the past, rivers may quickly spread pollutants from a local source to areas far away, thus strongly affecting the health of humans and animals and restricting the use of the water for drinking and even for irrigation.

24.1   Transport and Reaction in Rivers

Characterization of River Flow

Unlike lakes which, as a first approximation, can be described as completely mixed boxes (Chapter 23.1), transport processes in rivers and streams are dominated by the unidirectional flow of the water which forces all chemicals to spread downstream. The goal of this section is to describe the dynamic behavior of organic pollutants, either dissolved or sorbed on suspended particles, as they are transported along the river and undergo all kinds of transformations and exchange processes between the water and the adjacent environmental compartments, the atmosphere and the river bed.

The description of river flow and mixing can be extremely complicated. The following discussion will be restricted to the case of stationary flow, that is, to a situation in which the water discharge rate Q, at every cross section of the river, $A(x)$, is constant with time (Fig. 24.1). In reality, $Q(x)$ along the longitudinal axis of the river, x, is not necessarily constant because of merging rivers and water exchange with the underlying aquifer. Nevertheless, a river can always be looked at as consisting of "river reaches" in between river junctions for which the above assumption holds.

Stationary flow does not necessarily imply that the concentration of a chemical along the river is stationary as well. Often one has to assess the fate of chemicals that are accidentally spilled into a river and are transported downstream as a concentration cloud. It may be important to predict the temporal change of the concentration of the chemicals at a given location downstream of the spill, especially the time when the concentration starts to increase and when the maximum concentration of the spill passes by that location.

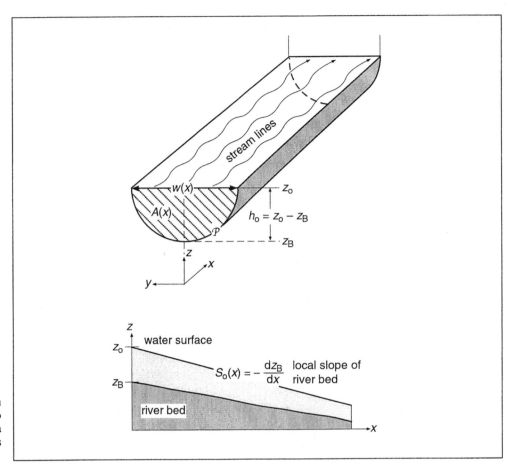

Figure 24.1 Coordinate system and geometric parameters used to characterize mixing processes in a river. See Table 24.1 for definitions of symbols.

The following explanations cannot replace the large literature on hydraulics and mixing processes in rivers (see, e.g., Fischer et al., 1979; Rutherford, 1994). As a first step toward "full-scale" river modeling, we shall explain and describe the following processes:

(a) Advective transport along the river caused by the mean flow of the water

(b) Chemical reactions in the water ("volumetric transformations")

(c) Exchange processes at the boundary of the water body, for instance, air–water exchange ("surface reaction")

(d) Vertical mixing by turbulence

(e) Lateral mixing by turbulence

(f) Longitudinal mixing by dispersion

(g) Sedimentation and sediment–water interaction

Note that this list does not include longitudinal mixing by turbulence. In fact, its influence is masked by longitudinal dispersion. Dispersion, an inevitable byproduct of advective transport, is commonly much larger than turbulent diffusion along the direction of flow.

Table 24.1 Definition of Characteristic River Parameters (Fig. 24.1)

Choice of local coordinates

	x-axis	Main direction of flow (horizontal)
	y-axis	Lateral direction (horizontal)
	z-axis	Vertical, positive upward

Geometry and Discharge (all variables are subjected to changes if Q changes with time)

$A(x)$	m^2	Cross section of river filled by water
$w(x)$	m	Width at water surface
$z_B(x)$	m	Height of deepest point of river bed at cross section $A(x)$, relative to arbitrary base
$z_0(x)$	m	Height of water surface at cross section $A(x)$
$h_0 = z_0 - z_B$	m	Maximum depth
$h = A/w$	m	Mean depth
$S_0 = -dz_B/dx$	–	Local slope of riverbed
$\mathcal{P}$	m	Wetted perimeter (wet area per unit length)
$R_h = A/\mathcal{P}$	m	Hydraulic radius
Q	m^{3}s^{-1}	Discharge of water
$\bar{u} = Q/A$	ms^{-1}	Mean flow velocity along the x-axis
f^*	–	Friction factor

Mixing Parameters

E_y, E_z	m^{2}s^{-1}	Turbulent diffusion coefficients
E_{dis}	m^{2}s^{-1}	Coefficient of longitudinal dispersion

Often the flow of water in the river bed is accompanied by a subsurface flow through the aquifer below and beside the surface flow. Although this flow is much slower than the surface flow, its cross section can be very large and its discharge rate can exceed the discharge of the surface flow, especially in dry areas. The interaction of water and chemicals between surface flow and aquifer is called *hyporheic exchange*. Here we will not deal with the influence of the hyporheic flow on the surface flow. The opposite, the impact of the river water on the chemical characteristics of the groundwater will be discussed in Chapter 25.

Figure 24.1 and Table 24.1 give all the necessary definitions. The coordinate system at any location is chosen such that x always points in the direction of the main flow, y across the river, and z vertically upward.

Transport by the Mean Flow

The mathematical description of the time-dependent, three-dimensional concentration distribution of a chemical in a river, $C(x,y,z,t)$, is not trivial. However, depending on the actual situation and the questions to be answered, significant simplifications can be made. In the following discussion, complexity is added step by step in order to allow the reader to build up to what is otherwise a confusing picture.

Let us consider a water volume defined by two river cross sections at x and $x + dx$, respectively. As a starting point, we assume that the mean total (dissolved plus sorbed) concentration of a chemical i in the water of this "slice" of the river, $C_t(x)$, is controlled by only two types of processes: (1) in-situ production or consumption of the chemical, R; and (2) fluxes at the water surface or at the sediment–water interface, $\mathcal{F}$:

$$\left(\frac{\partial C_t}{\partial t} \right)_{\text{Lagrange}} = R(C_t) - \mathcal{F}(C_t) \qquad \left[\text{ML}^{-3}\text{T}^{-1} \right] \qquad (24\text{-}1)$$

In order to be consistent with other chapters, $R(C_t)$ is defined as a positive number if the chemical is produced in the river and $\mathcal{F}(C_t)$ is positive if the net flux is directed from the river into the atmosphere or sediment. Note that $\mathcal{F}(C_t)$ is a flux per unit volume; its relation to the usual flux per area as defined, for instance, in Chapter 20, is given below (Eq. 24-15). Again we suppress the compound subscript i wherever the context is clear. The subscript "Lagrange" refers to what fluid dynamicists call the *Lagrangian representation* of the flow in which the observer travels with a selected water volume (the "river slice") and watches the concentration changes in the volume while moving downstream. Later the notion of an isolated water volume will be modified when mixing due to diffusion and dispersion across the boundaries of the volume is taken into account.

Although it seems natural to formulate the dynamic equations of a chemical in a river in terms of the Langrangian picture, the field data are usually made in the *Eulerian reference system*. In this system we consider the changes at a *fixed point* in space, for instance, at a fixed river cross section located at x_o. In Eq. 22-6 we adopted the Eulerian system and found that this representation combines the influence from in-situ reactions (the Langrangian picture) with the influence from transport. The latter appears in the additional advective transport term $-\bar{u}\,\partial C_t / \partial x$, where the mean flow velocity:

$$\bar{u} = \frac{Q}{A} \qquad (24\text{-}2)$$

replaces v_x of Eq. 22-6. Thus:

$$\left(\frac{\partial C_t}{\partial t} \right)_{\text{Eulerian}} = \left(\frac{\partial C_t}{\partial x} \right)_{\text{Lagrangian}} - \bar{u}\,\frac{\partial C_t}{\partial x}$$

$$= R(C_t) - \mathcal{F}(C_t) - \bar{u}\,\frac{\partial C_t}{\partial x} \qquad (24\text{-}3)$$

Prediction or measurement of $\bar{u}$ is of paramount importance for assessing transport of chemicals in rivers. The following discussion will be restricted to the special case of stationary uniform flow. This means that *at a fixed location* the discharge Q is constant, the cross section A does not change in size or shape, and the surface slope remains constant. With these assumptions, an equilibrium between the gravitational force pushing the water downhill and the frictional force, which increases in proportion to $\bar{u}^2$, can be formulated. The first force is represented by the product of the slope of the river bed, S_o, the water volume per unit length (equal to the cross-sectional area A), and the acceleration due to gravity, g. The second force is expressed by the product of the nondimensional friction factor, f^*, the square of the mean velocity, and the contact area per unit length of the water with the river bed, $\mathcal{P}$ (the wetted perimeter). This leads to the *Darcy-Weisbach equation for stationary uniform flow* (see e.g., Chow, 1959):

$$\bar{u} = \left(\frac{8gA}{f^*\mathcal{P}} S_o\right)^{1/2} = \left(\frac{8g}{f^*} R_h S_o\right)^{1/2} \tag{24-4}$$

where $R_h = A/\mathcal{P}$ is called the *hydraulic radius* (see Fig. 24.1 and Table 24.1). If the river is wide (i.e., if its width, w, is much larger than its mean depth, $h = A/w$), $\mathcal{P}$ is approximately equal to w, and Eq 24-4 becomes:

$$\bar{u} = \left(\frac{8g}{f^*} h S_o\right)^{1/2} \qquad \text{wide river} \tag{24-4a}$$

The friction factor typically lies between $f^* = 0.02$ (smooth river bed, like a man-made channel) and $f^* = 0.1$ (rough river bed, like a mountain stream, or a small river with large dunes or sand bars). For a given river bed, f^* decreases with increasing depth, h (i.e., with increasing discharge Q).

Because of friction, the flow velocity of a river remains fairly small compared to the speed of a freely falling object under the influence of the acceleration due to gravity. At the end the potential energy of the water is completely transformed into heat by internal friction (molecular viscosity) and friction at the river bed. Before the kinetic energy reaches the small-scale motion at which the molecular forces become relevant, the energy passes through the whole spatial scale of turbulent motion (see Chapter 22.2). Thus, friction is always related to turbulent mixing. As will be shown in Section 24.2, the stronger the friction, the larger the accompanying turbulence. The latter can be scaled by the so-called *shear velocity u^**, which is defined by (e.g., Fischer et al., 1979):

$$u^* = \left(g S_o R_h\right)^{1/2} \tag{24-5}$$

For wide rivers, the approximation $R_h \sim h$ can be used again, and thus:

$$u^* = \left(g S_o h\right)^{1/2} \qquad \text{wide river} \tag{24-5a}$$

The ratio between mean flow velocity, $\bar{u}$, and friction velocity, u^*, is called α^*. According to Eqs. 24-4 and 24-5, α^* is related to the friction factor f^*:

$$\alpha^* = \frac{\bar{u}}{u^*} = \left(\frac{8}{f^*}\right)^{1/2} \tag{24-6}$$

Using the typical range of $f^* = 0.02$ (smooth river bed) to $f^* = 0.1$ (rough river bed), the following range for α^* is obtained:

$$\alpha^* = 20 \text{ (smooth)} \dots 10 \text{ (rough)} \tag{24-7}$$

In Illustrative Example 24.1, using the case of a hypothetical small river (*River G*), we address the question how the river hydraulics change if the amount of water to be discharged through a certain (natural or artificial) channel changes. In fact, a river can adapt to a changing discharge in two ways, that is, (1) by changing its average flow velocity, $\bar{u}$, and (2) by changing its cross section (depth). The Darcy-Weisbach Equation tells us how a given river actually behaves. As it turns out, for the case of River G, which is supposed to flow in a rather narrow channel so that the river width does not change strongly with water depth, $\bar{u}$ is roughly proportional to $Q^{1/3}$ (Q is discharge rate). In flood plains, rivers do not flow much faster if Q is large, but they take much more land (or cross-sectional area). In this chapter, we will use two different flow regimes of River G (*Regime I and II*) to exemplify various aspects of the behavior of chemicals in rivers.

Illustrative Example 24.1

Mean Flow Velocity of River G for Different Discharge Rates

Problem

Consider the River G, which has a slope of 0.4 m per kilometer, a width of $w = 10$ m, and is bounded by rather steep slopes (see figure). The normal discharge is $Q = 0.8 \text{ m}^3\text{s}^{-1}$ (Regime I). It reaches maximum values of $10 \text{ m}^3\text{s}^{-1}$ during heavy rain (Regime II). The friction factor is assumed to be $f^* = 0.04$ for Regime I and $f^* = 0.026$ for Regime II. Calculate the mean flow velocity $\bar{u}$ and the mean depth h for both regimes.

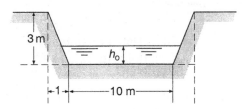

Answer

As a first approximation the river bed is described as a "wide" rectangular channel. Using this result, it can then be verified in a second step whether the original assumption was justified. If not, the calculation can be continued iteratively with more realistic river cross sections, that is, with $A = hw$ and $Q = \bar{u} A = \bar{u} hw$. The latter relationship is used to replace h by $Q/(\bar{u} w)$ in the shallow river Darcy-Weisbach Eq. 24-4a:

$$\bar{u} = \left(\frac{8g}{f^*} h S_0 \right)^{1/2} = \left(\frac{8g}{f^*} \frac{Q}{\bar{u} w} S_0 \right)^{1/2}$$

Solving for $\bar{u}$ yields:

$$\bar{u} = \left(\frac{8g Q S_0}{f^* w} \right)^{1/3} \tag{1}$$

With $S_0 = 0.4 \text{ m km}^{-1} = 4 \times 10^{-4}$ and $h = Q/\bar{u} w$ one obtains:

Regime I $Q = 0.8\ \text{m}^3\text{s}^{-1}$ $\bar{u} = 0.40\ \text{m s}^{-1}$ $h = 0.20\ \text{m}$

Regime II $Q = 10\ \text{m}^3\text{s}^{-1}$ $\bar{u} = 1.06\ \text{m s}^{-1}$ $h = 0.94\ \text{m}$

Regime I does not pose any problems with respect to the rectangular channel assumption. Note that for regime II, due to the slope of 3:1 on both sides of the river, the surface width w increases to about $[10 + 0.94 \cdot (2/3)]\ \text{m} = 10.6\ \text{m}$. Thus the cross section is slightly larger than $10\ [\text{m}] \times h$. Yet, this effect is more than compensated for by the fact that the wetted perimeter P is greater than $w = 10\ \text{m}$ ($P = 12.0\ \text{m}$). From Eq. 24-4 it can be concluded that the true $\bar{u}$ is slightly smaller and thus h larger than the values derived from the rectangular wide river approximation. In fact, the corresponding correction of $\bar{u}$ is -6% ($\bar{u} = 1.00\ \text{m s}^{-1}$) not much in comparison with the uncertainty caused by assuming an appropriate friction factor f^*. The following rounded values will be used when the example is continued below:

Regime I $Q = 0.8\ \text{m}^3\text{s}^{-1}$ $\bar{u} = 0.40\ \text{m s}^{-1}$ $h = 0.20\ \text{m}$ $w = 10\ \text{m}$

Regime II $Q = 10\ \text{m}^3\text{s}^{-1}$ $\bar{u} = 1.0\ \text{m s}^{-1}$ $h = 1.0\ \text{m}$ $w = 10\ \text{m}$

Note that Eq. 1 tells us that $\bar{u}$ changes roughly as $Q^{1/3}$ if width w and friction factor f^* do not change much with water depth h. This is typical for rivers flowing through narrow (often man-made) channels. In contrast, rivers flowing across flood plains increase their cross-sectional area by flooding the surrounding land while their average flow velocity remains the same or may even decrease.

As a next step we combine the Lagrangian Equation (24-1) of an in-situ production/consumption process, $R(C_t)$, with the Eulerian view (Eq. 24-4) of transport and reaction. There are two ways to analyze the resulting situation, a more intuitive method and a formal approach. We elaborate on the former and explain the latter in Box 24.1.

From the Lagrangian equation we calculate the temporal evolution of the concentration in a fixed water parcel, C_t. Given the production/consumption function $R(C_t)$, C_t can be calculated from the (analytical or numerical) solution of the Lagrangian equation, $\partial C_t / \partial t = R(C_t)$, with initial concentration $C_t(t_0)$, where t_0 marks the time when the water volume has passed by a given cross section at x_0 (see Fig. 24.2). For instance, if $R(C_t)$ is a linear function:

$$R(C_t) = J - k_r C_t , \qquad J,\ k_r = \text{constants} \tag{24-8}$$

then according to Box 12.1 we have:

$$C_t(t) = C_t(t_0)\,e^{-k_r(t-t_0)} + \frac{J}{k_r}\left(1 - e^{-k_r(t-t_0)}\right) \tag{24-9}$$

We could also use a nonlinear $R(C_t)$ function, for instance, the second-order function discussed in Box 21.5. Next, the resulting time-dependent concentration, $C_t(t)$, is

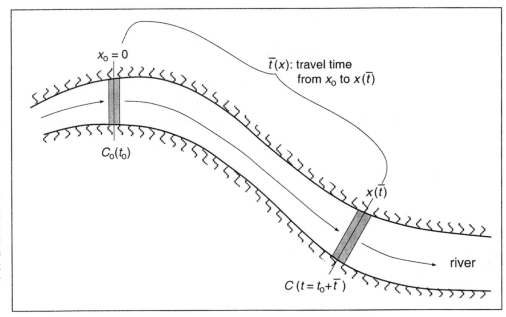

Figure 24.2 Calculation of the time-dependent concentration along the river axis x of a reactive compound. The effect of the reaction is calculated in the Lagrangian framework; the effect of advection is accounted for by the relation between flow time $\bar{t}$ and distance x, $x(\bar{t})$.

transformed into the space-function $C_t(x)$ by using the relation between the mean flow time, $\bar{t} = t - t_o$, and distance x:

$$\bar{t}(x) = \int_0^x \frac{1}{\bar{u}(x')}\,dx' \qquad (24\text{-}10)$$

Remember that $\bar{u}(x)$ is not necessarily constant along the river axis x, even if Q is constant, since the cross section $A(x)$ may change (see Eq. 24-2). Then:

$$C_t(x) = C_t\left[t_o + \bar{t}(x)\right] \qquad (24\text{-}11)$$

As an example, we use the linear rate function Eq. 24-8. From Eq. 24-9 with $t = t_o + \bar{t}(x)$:

$$C_t(x) = C_t(t_o)e^{-k_r\bar{t}(x)} + \frac{J}{k_r}\left(1 - e^{-k_r\bar{t}(x)}\right) \qquad (24\text{-}12)$$

Note that due to Eq. 24-10 the mean flow velocity $\bar{u}(x)$ is hidden in the function $\bar{t}(x)$. To make this more explicit, we assume $\bar{u}(x) = \text{constant} = \bar{u}$. Then Eq. 24-10 becomes:

$$\bar{t}(x) = \frac{x}{\bar{u}}, \qquad \text{for } \bar{u} = \text{constant} \qquad (24\text{-}13)$$

Furthermore, if the initial concentration at $x = 0$, $C_t(t_o)$, and the parameters J, k_r are independent of time, the concentration along the river is stationary; that is, $C_t(x)$ at a fixed location x does not change with time:

$$C_t(x) = C_t^o e^{-\varepsilon_r x} + \frac{J}{k_r}\left(1 - e^{-\varepsilon_r x}\right), \qquad \varepsilon_r = \frac{k_r}{\bar{u}}, \qquad C_t^o = C_t(t_o) \qquad (24\text{-}14)$$

Note that incomplete mixing of the chemical in the river does not alter the above result (Eq. 24-14) provided that the reactions are linear. Then $C_t(x)$ means the average cross-sectional concentration. In contrast, for nonlinear reactions the degree of homogeneity influences the average reaction rate.

In Box 24.1, Eq. 24-12 is derived from the Eulerian Equation 24-3.

Box 24.1   Stationary Solution for the Advection/Reaction Equation in a River

The advection/reaction equation (see Eq. 24-3, $\mathcal{F}(C_t) = 0$):

$$\frac{\partial C_t}{\partial t} = -\bar{u}\frac{\partial C_t}{\partial x} + R(C_t) \tag{1}$$

is solved for stationary conditions, $\partial C_t/\partial t = 0$, $\bar{u}(x,t) = \bar{u}(x)$.

The resulting ordinary differential equation:

$$\frac{dC_t}{dx} = \frac{R(C_t)}{\bar{u}(x)} \tag{2}$$

can be solved by variable separation:

$$\frac{dC_t}{R(C_t)} = \frac{dx}{\bar{u}(x)} \tag{3}$$

Integration from $x = 0$ to x, that is, from C_t^o to $C_t(x)$, yields:

$$\int_{C_t^o}^{C_t(x)} \frac{dC_t'}{R(C_t')} = \int_0^x \frac{dx'}{\bar{u}(x')} = \bar{t}(x) \tag{4}$$

where $\bar{t}(x)$ is the flow time from $x = 0$ to x. The integral on the left-hand side of Eq. (4) can be solved for an explicit reaction function, $R(C_t)$. For instance, for the linear function:

$$R(C_t) = J - k_r C_t, \qquad J, \ k_r = \text{constants} \tag{5}$$

we have:

$$\int_{C_t^o}^{C_t(x)} \frac{dC_t'}{J - k_r C_t'} = -\frac{1}{k_r} \ln\left(\frac{J - k_r C_t(x)}{J - k_r C_t^o}\right) \tag{6}$$

Inserting into Eq. (4) and taking the exponential function on both sides:

$$e^{-k_r \bar{t}(x)} = \frac{J - k_r C_t(x)}{J - k_r C_t^o} \tag{7}$$

After some algebraic manipulations this becomes:

$$C_t(x) = C_t^o e^{-k_r \bar{t}(x)} + \frac{J}{k_r}\left(1 - e^{-k_r \bar{t}(x)}\right) \tag{8}$$

Eq. 4 corresponds to Eq. 24-12 for constant initial concentration, $C_t(t_o) = C_t^o$.

Mean Flow and Air–Water Exchange

In Eq. 24-1 a distinction has been made between the in-situ production/consumption rate function, $R(C_t)$, and the rate of change due to boundary fluxes, $\mathcal{F}(C_t)$. According to the discussion in Chapter 19, the different types of boundary fluxes can always be written as the product of an exchange velocity and a concentration difference. The latter describes the disequilibrium between the two phases on either side of the boundary, for instance, between water and air or between water and sediments. The latter will be discussed in Section 24.3.

The rate of change due to air–water exchange in a well-mixed water body, $\mathcal{F}(C_t)$, is equal to the flux divided by the mean depth h (see Eqs. 20-1 and 21-8c). Note that for air–water exchange, $\mathcal{F}$ depends on the aqueous concentration C_w:

$$\mathcal{F}(C_t) = \mathcal{F}(C_w) = \frac{F_{a/w}}{h} = \frac{v_{a/w}}{h}\left(C_w - C_{eq}\right) = k_{a/w}\left(C_w - C_{eq}\right) \quad (24\text{-}15)$$

where $C_{eq} = C_a / K_{a/w}$

C_a: concentration in the atmosphere

$K_{a/w}$: nondimensional Henry's law constant

$k_{a/w} = v_{a/w} / h$: air–water exchange rate

$h = A / w$: mean depth of river (Table 24.1)

At first sight, Eqs. 24-8 and 24-15 look equivalent ($J \to k_{a/w} C_{eq}$, $k_r \to k_{a/w}$). Yet, since $k_{a/w}$ depends on depth h, which, in turn, depends on river width w and cross section A, the coefficients of Eq. 24-15 change with x. However, the variation of $k_{a/w}$ is linked to the river parameters in a very peculiar manner which makes things simpler again.

To show this, we insert Eq. 24-15 into Eq. 24-3. Let us, at this point, consider a nonsorbing chemical ($C_t = C_w$). Thus:

$$\frac{\partial C_w}{\partial t} = -\bar{u}\frac{\partial C_w}{\partial x} - \frac{v_{a/w}}{h}\left(C_w - C_{eq}\right) \quad (24\text{-}16)$$

For steady-state ($\partial C_w / \partial t = 0$), the partial differential equation becomes an ordinary differential equation:

$$\frac{dC_w}{dx} = -\frac{v_{a/w}}{\bar{u}\,h}\left(C_w - C_{eq}\right) = -\frac{v_{a/w}}{Q}w\left(C_w - C_{eq}\right) \quad (24\text{-}17)$$

with $\bar{u} = Q / A$ and $h = A / w$.

As it turns out, the appropriate coordinate to describe the concentration change along the river due to a boundary exchange flux is the *integrated surface area s(x)*, measured from an arbitrary cross section, where $s(0) = 0$, to the coordinate x (Fig. 24.3). The infinitesimal increment ds is related to dx by:

$$ds = w\,dx, \quad \text{thus:} \quad \frac{dx}{ds} = \frac{1}{w} \quad (24\text{-}18)$$

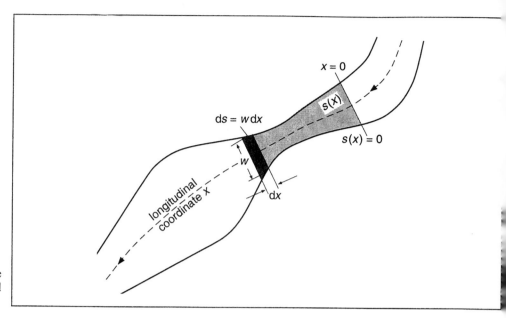

Figure 24.3 The surface area of the river can serve as a longitudinal coordinate along the river.

The change of C_w along the integrated surface area, $s(x)$, is:

$$\frac{dC_w}{ds} = \frac{dC_w}{dx}\frac{dx}{ds} = -\frac{1}{w}\frac{v_{a/w}}{Q}w\left(C_w - C_{eq}\right)$$

$$= -\frac{v_{a/w}}{Q}\left(C_w - C_{eq}\right) \tag{24-19}$$

Thus, even if depth h, width w, and cross section A are changing along the river, the rate ($v_{a/w}/Q$) may still be fairly constant for sections with constant discharge Q. (Note: According to Section 20.4 on air–water exchange velocities in rivers, $v_{a/w}$ depends on characteristic river data such as $\bar{u}$, h, and so on (see Eq. 20-31), but often this indirect effect is negligible compared to the changing geometry of the river bed.)

For constant ($v_{a/w}/Q$), the solution of Eq. 24-17 is:

$$C_w[s(x)] = C_{eq} + \left(C_w^o - C_{eq}\right)e^{-(v_{a/w}/Q)s(x)} \tag{24-20}$$

Here the integrated surface area, $s(x)$, adopts the role of a coordinate along the river (Fig. 24.3). For constant width w, we have $s(x) = wx$, thus:

$$C_w(x) = C_{eq} + \left(C_w^o - C_{eq}\right)e^{-\varepsilon_{a/w}x} \tag{24-21}$$

where:

$$\varepsilon_{a/w} = \frac{v_{a/w}\,w}{Q} \qquad \left[L^{-1}\right] \tag{24-22}$$

is the inverse of a characteristic distance for air–water equilibration.

Note that incomplete *lateral* mixing of C_w does not alter the average surface concentration provided that *vertical* mixing is complete. Thus, Eq. 24-21 remains valid as long as vertical homogeneity is attained. In Illustrative Example 24.2, the linear air–water exchange model is applied to two chemicals in River G.

Illustrative Example 24.2

Air–Water Exchange in River G

Problem

Wastewater is introduced into the River G (see Illustrative Example 24.1) at location $x = 0$. It causes a continuous input of $J = 0.25$ mol h^{-1} of benzyl chloride (BC) and $J = 0.08$ mol h^{-1} of tetrachloroethene (PCE).

(a) Calculate for the two flow regimes (I and II) which were introduced in Illustrative Example 24.1 the concentrations of BC and PCE in the River G, 20 km downstream of the wastewater input, provided that air–water exchange is the only elimination mechanism. Assume that the water temperature is 25°C and the concentration in air of both substances negligible.

(b) If for flood control the width of the river were enlarged to $w = 20$ m, how would the above result be altered? Just analyze the case of PCE for Regime I.

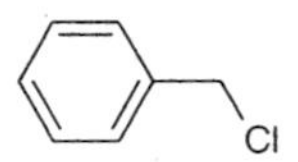

Answer (a)

Provided that mixing of the chemicals in the river is fast, the initial concentration C_o can be calculated from $C_o = J / Q$. Note: In Chapter 24.2 the mixing hypothesis will be substantiated by giving explicit estimates for the relevant mixing parameters.

The following initial concentrations are obtained:

benzylchloride
(BC)

$M_i = 126.6$ g mol^{-1}
$K_{a/w}$ (25°C) $= 0.015$

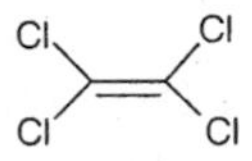

tetrachloroethene
(PCE)

$M_i = 165.8$ g mol^{-1}
$K_{a/w}$ (25°C) $= 0.73$

Equivalent sand-grain
diameter of River G (Eq. 20-36):
$d_s = 0.01$ m

	Input J (mol s^{-1})	Initial Concentration C_o (mol L^{-1})	
		I $(Q = 0.8$ m^{3}s$^{-1})$	II $(Q = 10$ m^{3}s$^{-1})$
BC	6.9×10^{-5}	87×10^{-9}	6.9×10^{-9}
PCE	2.2×10^{-5}	28×10^{-9}	2.2×10^{-9}

Air–water exchange of PCE is liquid-film controlled (Table 20.5b). For BC, the size of the Henry's law coefficient ($K_{a/w} = 0.015$) suggests a slight influence of the air-film which you may disregard. To relate $v_{a/w} \sim v_w$ of PCE and BC, combine Eqs. 18-55 and 20-29 into the simple expression:

$$v_{BC\,a/w} = v_{PCE\,a/w} \left(\frac{M_{PCE}}{M_{BC}} \right)^{0.5 \times 0.67} = 1.09 \; v_{PCE\,a/w}$$

To calculate $v_{PCE\,a/w}$ we note that for both discharge regimes the nondimensional roughness parameter d^* (Eq. 20-36) is larger than 136. Thus, we use the large-eddy model (Eq. 20-35):

$$v_{PCE\,a/w} \approx v_{PCE\,w} = \left(\frac{D_{PCE\,w} \, \bar{u}}{h} \right)^{1/2}$$

where $D_{PCE\,w} = 0.99 \times 10^{-5}$ cm^{2}s^{-1} (Table 20.5a).

Air–water exchange velocities for PCE in the two flow regimes are:

	$f^{*\,a}$	$\bar{u}\,(\text{m s}^{-1})^{\,a}$	$h\,(\text{m})^{\,a}$	$\alpha^{*\,b}$	$u^{*}\,(\text{m s}^{-1})^{\,b}$	$d^{*\,c}$	$v_{\text{PCE w}}\,(\text{m s}^{-1})$
I	0.04	0.4	0.2	14	0.028	310	4.4×10^{-5}
II	0.026	1.0	1.0	18	0.056	630	3.1×10^{-5}

[a] From Illustrative Example 24.1. [b] From Eq. 24-6. [c] From Eq. 20-36 with $v_w(25°C) = 0.893 \times 10^{-6}\,\text{m s}^-$ $d_s = 0.01$ m. [d] From Eq. 20-35.

To calculate C_w at $x = 10$ km we use Eq. 24-14, the simplified version of Eq. 24-1: valid for w = constant. Note that beyond $x = 0$ the input term $J = 0$ is zero, since the atmospheric concentrations of both PCE and BC are negligible.

		$v_{\text{a/w}}$	$\varepsilon_{\text{a/w}} = \dfrac{v_{a/w}\,w}{Q}$	C_o	$C\,(x = 10\text{ km})^{\,a}$
		(m s^{-1})	(m^{-1})	$(10^{-9}\text{mol L}^{-1})$	$(10^{-9}\text{mol L}^{-1})$
PCE	I	4.4×10^{-5}	5.5×10^{-4}	28	0.11
	II	3.1×10^{-5}	3.1×10^{-5}	2.2	1.6
BC	I	4.8×10^{-5}	6.0×10^{-4}	87	$(0.22)^{\,b}$
	II	3.4×10^{-5}	3.4×10^{-5}	6.9	$(4.9)^{\,b}$

[a] $C(10\text{km}) = C_o \exp(-\varepsilon_{a/w} \times 10^4\,\text{m})$; [b] Air–water exchange is not the only elimination process for BC (see Table 24.2 and also Illustrative Example 25.7).

The loss to the atmosphere is significantly reduced during periods of large discharge. In spite of the greater dilution of the mass input during the peak discharge, the concentration 10 km downstream of the input is larger than for the small discharge. Three factors are involved in this effect: (1) the decrease of $v_{i\,a/w}$ with increasing Q, (2) the decreasing surface-to-volume ratio (i.e., the increase of depth) with increasing Q, (3) the decrease of flowing time from $x = 0$ to $x = 10$ km. Since $h\,\bar{u} = Q/w$, for fixed discharge rate Q and width w, the flowing time and the mean depth are inversely related and thus act in opposite directions. This explains why neither h nor $\bar{u}$ explicitly appear in Eq. 24-22.

Answer (b)

If the width of the River G were to be increased to $w = 20$ m, but the mean flow velocity $\bar{u}$ remain unchanged, then the mean depth h would drop by 2. According to the large-eddy model (Eq. 20-35), the air–water exchange velocity would then increase by $2^{1/2} = 1.4$. Simultaneously, the surface area of the river between $x = 0$ and $x = 10$ km would double. Thus, according to Eq. 24-22 we expect $\varepsilon_{\text{a/w}}$ to grow by $2 \times 2^{1/2} = 2^{3/2} = 2.8$. For discharge regime I and PCE, $\varepsilon_{\text{a/w}} = 1.6 \times 10^{-3}\,\text{m}^{-1}$. Thus, at $x = 10^4$ m, C_o would drop by the factor $e^{-16} = 1.1 \times 10^{-7}$ or virtually to zero.

In fact, according to Eq. 1 of Illustrative Example 24.1, $\bar{u}$ would slightly decrease when w increases (from 0.40 m s^{-1} to 0.32 m s^{-1} for discharge regime I). Thus, the increase of $\varepsilon_{\text{a/w}}$ would be somewhat smaller, but $\varepsilon_{\text{a/w}}$ would be still so large that the solutes would be completely lost to the atmosphere at $x = 10$ km.

From Reaction Times to Reaction Distances

Remember that the solutions of linear box models usually comprise exponential functions in time of the form e^{-kt}, where the specific rate constant, k, has the dimension of T^{-1}. As Eq. 24-21 demonstrates, in the case of concentration profiles along a river, the relevant functions are of the form $e^{-\varepsilon x}$, where x is length along the river (see Eq. 24-14). The parameter ε, which describes one or several processes in the moving water volume, has dimension L^{-1}. Then $x_p = \varepsilon_p^{-1}$ is the characteristic distance of the particular process p or of the sum of all processes. The relative size of different ε_p values describes the relative importance of different processes occurring along the river.

As an example, we analyze the fate of benzyl chloride (BC) along the River G. In Illustrative Example 24.2 we calculated the characteristic distance for the air–water exchange of BC. Now we include the hydrolysis of BC, which can be described by a first-order rate constant, $k_r = 0.05 \ h^{-1}$ at 25°C (Table 13.6). From Eq. 24-14 we learn that the *characteristic distance of reaction*, x_r, is given by:

$$x_r = \varepsilon_r^{-1} = \frac{\bar{u}}{k_r} \tag{24-23}$$

The differential equation of the combined action of air–water exchange and reaction at steady-state is:

$$\frac{dC_w}{dx} = -\left(\varepsilon_r + \varepsilon_{a/w}\right)C_w + \varepsilon_{a/w}C_{eq} \tag{24-24}$$

where ε_r and $\varepsilon_{a/w}$ are given by Eqs. 24-22 and 24-23.

As shown in Table 24.2, both $\varepsilon_{a/w}$ and ε_r decrease with discharge rate Q, and thus the characteristic distance of benzyl chloride increases with Q. The relative contribution of the two processes changes as well. Although air–water exchange is always the most important elimination process, the relative importance of hydrolysis increases with increasing Q. Note that, in spite of dilution into an increasingly large water body, at $x = 10$ km the concentration is greater at large discharge rates as compared to small rates.

Table 24.2 Reaction Distances of Benzyl Chloride (BC) in River G

Flow Regime	I ($Q = 0.8$ m³s⁻¹)	II ($Q = 10$ m³s⁻¹)
Initial concentration C_o (mol L⁻¹) [a]	87×10^{-9}	6.9×10^{-9}
Effect of air–water exchange $\varepsilon_{a/w}$ (m⁻¹) [a]	6.0×10^{-4}	3.4×10^{-5}
Effect of hydrolysis ε_r (m⁻¹) [b]	3.5×10^{-5}	1.4×10^{-5}
Total characteristic distance $x_t = (\varepsilon_{a/w} + \varepsilon_r)^{-1}$ (km)	1.6	21
Conc. at $x = 10$ km; $C(10$ km) (mol L⁻¹) [c]	2.5×10^{-12}	4.3×10^{-9}
Relative weight of air–water flux $\dfrac{\varepsilon_{a/w}}{\varepsilon_{a/w} + \varepsilon_r} \times 100\%$	94.5%	70.8%

[a] From Illustrative Example 24.2. [b] From Eq. 24-23 with $k_r = 0.05 \ h^{-1}$, $\bar{u}$ from Illustrative Example 24.1.
[c] $C(10$ km$) = C_0 \exp[-(\varepsilon_r + \varepsilon_{a/w}) \times 10^4 \ m]$.

Sediment–Water Interaction

Compared to the situation in lakes, the sediment–water interactions in rivers are more complex. Because the flow velocity is constantly changing, particles may either settle at the bottom or be resuspended and deposited again further downstream. In order to adequately describe the effect of these processes on the concentration of a chemical in the river, we would need a coupled water–sediment model with which the profile of the chemical along the river of both the aqueous concentration in the river and the concentration in the sediment bed are described. This is a task to be left to numerical modeling. We choose a simpler approach by approximating the net deposition of the particles and the chemicals sorbed to them as a linear process (see Eqs. 23-16 and 23-17):

$$\left(\frac{dC_t}{dt}\right)_{settling} = -k_s(1-f_w)C_t = -k_s^* C_t \tag{24-25}$$

where $k_s^* = k_s(1-f_w)$, and f_w is the dissolved fraction of the chemical (Eq. 9-12).

Yet, in view of the often strong currents in rivers, it would be difficult to interpret k_s as a vertical settling velocity divided by the mean depth of the river bed as done in Eq. 23-16. In fact, k_s should be understood as an empirical coefficient describing the specific first-order removal rate of suspended particles from the river.

For stationary conditions, the temporal concentration change due to removal on settling particles can be transformed into a variation along the river. By analogy to Eq. 24-23 we get:

$$\left(\frac{dC_t}{dx}\right)_{settling} = -\varepsilon_{settling} C_t \; ; \qquad \varepsilon_{settling} = \frac{k_s}{u}(1-f_w) \tag{24-26}$$

Besides interacting with suspended particles, a chemical also undergoes direct exchange at the sediment surface by diffusion and advection into the hyporheic zone. Furthermore, resuspension followed by exchange between water and particles also adds to the sediment–water interaction. These processes have been extensively discussed in Chapter 23, especially in Box 23.2. There we concluded that the effect from the different mechanisms can be combined into a flux of the form (see Eq. 23-25):

$$F_{sedex} = -v_{sedex}\left(C_t - C_{sed}^{eq}\right) \tag{24-27}$$

where v_{sedex} is the overall sediment–water exchange velocity (Box 23.2, Eq. 6) and C_{sed}^{eq} is proportional to the concentration of the chemical on the sediment particles (Box 23.2, Eq. 7). (Note that we have replaced C_t^{op} of Eq. 23-25 by the simpler notation C_t.)

Since in most cases detailed information on the characteristics of the sediments along the river is missing, it is hardly justified to attempt to calculate v_{sedex} from a mechanistic model of the various processes involved. However, there are situations in which we should at least remember that v_{sedex} may depend on the exposure history of the river sediments. In Section 24.3 we will discuss such a case: the pollution of the River Rhine by a pesticide after a fire in a storehouse. In this and similar cases, when a pollutant cloud of finite length (and thus of finite duration Δt) drifts along a river, Δt determines how much of the dissolved pollutant is taken up per unit time by the sediments.

To understand this point, we must recall that the sediment surface acts as a wall boundary with respect to diffusion (Chapter 19). According to Eq. 19-33, the exchange flux, F_{sed}, and thus the corresponding apparent exchange velocity, v_{sedex}, decreases with elapsed time t. From Eq. 19-33, with C_w^{op} replaced by $f_w C_t$, we get:

$$v_{sedex}(t) = \frac{F_{sed}(t)}{C_t} = \left(\frac{1}{\pi}\right)^{1/2} \left(\frac{D_w^{sed}}{f_w^{sed} t}\right)^{1/2} \phi f_w \qquad (24\text{-}28)$$

where D_w^{sed} is the aqueous molecular diffusivity of the chemical in the pore water of the sediments, ϕ is the sediment porosity, and f_w^{sed} is the dissolved fraction of the chemical in the pore water. For simplicity we assumed that the initial concentration in the pore water is zero. The corresponding change of the concentration in the river, averaged over the time Δt during which the pollution cloud passes by, can be written as a first-order process (h is mean depth):

$$\left(\frac{dC_t}{dt}\right)_{sedex} = -\frac{\overline{F}_{sed}(t)}{h} = -\overline{k}_{sedex} C_t \qquad (24\text{-}29a)$$

where:

$$\overline{k}_{sedex} = 2\left(\frac{1}{\pi}\right)^{1/2} \frac{\phi}{h} \left(\frac{D_w^{sed}}{f_w^{sed} \Delta t}\right)^{1/2} f_w \qquad (24\text{-}29b)$$

The additional factor of 2 is due to the averaging of Eq. 24-28 over the time interval Δt. For stationary conditions we get:

$$\left(\frac{dC}{dt}\right)_{sedex} = -\varepsilon_{sedex} C_t; \qquad \varepsilon_{sedex} = \frac{\overline{k}_{sedex}}{\overline{u}} \qquad (24\text{-}30)$$

To summarize, $\overline{k}_{sedex}$ can be either derived from an explicit model of sediment–water exchange (for instance, Eq. 6 of Box 23.2, or Eq. 24-29b) or treated as an empirical parameter to be determined from field data. Note that in any case, $\overline{k}_{sedex}$ is proportional to f_w of the chemical. Strongly sorbing chemicals are mainly removed on settling particles, while for weakly sorbing chemicals diffusion at the sediment surface is more important. The formalism of Eqs. 24-28 and 24-29 is applied in Illustrative Example 24.3.

Eylers (1994) has described a further mechanism of sediment–water exchange which is especially important for sandy river beds. The bottom of such rivers is often shaped by ripples and dunes which lead to horizontal pressure gradients. As a result, river water is forced through the pore space of the sediments where chemicals are exchanged. The mathematics of this process is summarized in Box 24.2.

(Text continues on page 1120)

Illustrative Example 24.3   **A Spill of Atrazine in River G**

Problem

From an adjacent cornfield a total amount of 200 kg of the herbicide *atrazine* is accidentally spilled into the River G over a time span of 30 minutes. Atrazine is fairly conservative in the water. Calculate what fraction of the atrazine is held back by the sediments on the first 10 km downstream of the spill. Consider both flow regimes (I and II) introduced in Illustrative Example 23.1.

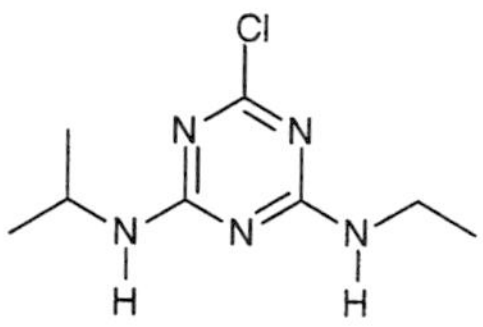

atrazine

$M_i = 215.7$ g mol^{-1}

Organic carbon-water partition coefficient $K_{oc} = 340$ L kg$_{oc}^{-1}$

River G

$r_{sw} = 50$ mg$_s$L^{-1}:	Suspended solids in River G
$f_{oc}^{sed} = 0.2$	Relative organic carbon content of suspended particles
$k_s = 1$ m d^{-1}	Sedimentation rate (estimated)
$\phi = 0.8$	Porosity of sediments
$r_s = 2.5$ kg$_s$L^{-1}:	Density of sediment particles
$f_{oc}^{sed} = 0.005$ kg$_{oc}$ kg$_s^{-1}$	Relative organic carbon content of sediment particles

Note that the process of dissolution of atrazine in the water takes some time. To make the situation not too complicated, it is assumed that the particles containing undissolved atrazine are small enough to be kept in suspension until an equilibrium between the particulate and dissolved phase is reached.

Answer

Calculate first the fraction of atrazine in particulate form (Eq. 23-4):

$$1 - f_w = \frac{f_{oc}^{sus} r_{sw} K_{oc}}{1 + f_{oc}^{sus} r_{sw} K_{oc}} = 3.4 \times 10^{-3}$$

From Eq. 24-25 the loss rate due to particle settling is then:

$$k_s^* = (1 - f_w) k_s = 3.4 \times 10^{-3} \text{d}^{-1} = 3.9 \times 10^{-8} \text{s}^{-1}$$

The loss of atrazine due to diffusion into the sediment is estimated from Eq. 24-29. Note that in spite of the fairly small K_{oc} value of atrazine, the chemical sorbs well in the sediments due to the large r_{sw}^{sed} value:

From Eq. 9-15:

$$r_{sw}^{sed} = \rho_s \frac{1 - \phi}{\phi} = 2.5 \text{ kg}_s\text{L}^{-1} \frac{0.2}{0.8} = 0.63 \text{ kg}_s\text{L}^{-1}$$

Thus:

$$f_w^{sed} = \frac{1}{1 + r_{sw}^{sed} f_{oc}^{sed} K_{oc}} = 0.48$$

D_w^{sed} is estimated from D_w of CO_2 (1.9×10^{-5} cm s^{-1} at 25°C) and the molecular mass ratio of the two substances (Eq. 18-55) to be about 8×10^{-6} cm^{2}s^{-1}. Inserting these values into Eq. 24-29 with $\Delta t = 30$ min $= 1800$ s:

$$\bar{k}_{sedex} = 1.13 \frac{0.8}{h} \left(\frac{8 \times 10^{-10}\, m^2 s^{-1}}{0.48 \times 1800s} \right)^{1/2} f_w = \frac{8.7 \times 10^{-7}\, m\,s^{-1}}{h}$$

After division by the mean flow velocity $\bar{u}$ we get the corresponding ε-values. The results are summarized in the following table.

	Flow regime	
	I	II
h (m)	0.2	1
$\bar{u}$ (m s^{-1})	0.4	1
$\varepsilon_s = k_s^* / \bar{u}$ (m^{-1})	1.0×10^{-7}	4×10^{-8}
Reduction at $x = 10$ km: $e^{-\varepsilon_s x}$	0.999	1.0
$\bar{k}_{sedex}$ (s^{-1})	4.3×10^{-6}	8.7×10^{-7}
$\varepsilon_{sedex} = \bar{k}_{sedex} / \bar{u}$ (m^{-1})	1.1×10^{-5}	8.7×10^{-7}
Reduction at $x = 10$ km: $e^{-\varepsilon_{sedex} x}$	0.900	0.991

Note: All these elimination rates are fairly small. If River G had a sandy bed with ripples, then according to Box 24.2 the sediment-water exchange by pore water flow could become the largest process.

The discussion is continued in Illustrative Example 24.5.

Box 24.2   Sediment–Water Exchange in Sandy Sediments

According to Eylers (1994), ripples and dunes lead to horizontal pressure gradients along the riverbed (high pressure in front of dunes, low pressure behind the dunes). These gradients force river water to flow through the pore space of the sediments where chemicals are sorbed and desorbed. The process can be quantified by the following exchange velocity:

$$v_{sedflow} = \frac{0.42\, K_q\, \bar{u}^2}{g} \frac{H}{\lambda} \left(\frac{H}{h} \right)^{3/8} \tag{1}$$

where K_q is hydraulic conductivity [LT^{-1}] (see Eq. 25-5)

   λ is wavelength (horizontal distance) of ripples or dunes [L]

   H is the height of the dunes [L]

   h is the mean depth of the river bed [L]

   $g = 9.81$ m s^{-2} is acceleration due to gravity

The removal rate for C_t in the river also depends on the fraction of the substance that is retained on the sediment particles. The latter is given by $(1 - f_w^{sed})$, where f_w^{sed} is the fraction of the compound in the pore water that is dissolved. Thus:

$$k_{sedflow} = \frac{v_{sedflow}}{h} (1 - f_w^{sed}) \tag{2}$$

For stationary conditions:

$$\varepsilon_{sedflow} = \frac{k_{sedflow}}{\bar{u}} = \frac{v_{sedflow}}{h\bar{u}}\left(1 - f_w^{sed}\right)$$

(3)

Example (corresponds roughly to River G at flow regime II and with a sandy bed)

K_q $= 10^{-3}$ m s^{-1}

h $= 1$ m

$\bar{u}$ $= 1$ m s^{-1}

λ $= 3\,h = 3$ m

H $= h/10 = 0.1$ m

$$v_{sedflow} = \frac{0.42}{9.81 \text{ m s}^{-2}} \frac{10^{-3}\text{m s}^{-1}(1 \text{ m s}^{-1})^2}{3\text{m}}\left(\frac{0.1\text{m}}{1\text{m}}\right)^{3/8} = 6.0\times10^{-6}\text{m s}^{-1}$$

$$\varepsilon_{sedflow} = \frac{v_{sedflow}}{h\bar{u}} \frac{6.0\times10^{-6}\text{m s}^{-1}}{1\text{m} \times 1\text{m s}^{-1}} = 6.0\times10^{-6}\text{m}^{-1}$$

Note: Compared with $\varepsilon_{sedflow}$ calculated for atrazine in River G (Illustrative Example 24.3), $\varepsilon_{sedflow}$ is fairly large. Thus, for sandy river beds this process could not be neglected.

24.2   Turbulent Mixing and Dispersion in Rivers

Vertical Mixing by Turbulence

Currents in rivers and streams are turbulent. Turbulent mixing can be described by the Fickian laws (Eqs.18-6 and 18-14) and by empirical turbulent diffusion coefficients E_α, where α stands for x, y, z (Chapter 22). The main source of turbulence is the friction between the water and the river bed. It can be expected that increasing roughness of the river leads to increasing turbulence, much in the same way as a large roughness causes the mean flow $\bar{u}$ to become slow (see the effect on Eq. 24-4 if the friction coefficient f^* increases). In fact, turbulence in rivers can be scaled by the shear velocity, u^*, defined in Eq. 24-5.

The coefficient of vertical turbulent diffusivity, E_z, is depth dependent. It vanishes at the two boundaries, $h = 0$ and $h = h_o$ (Fig. 24.4), and reaches its maximum value at mid-depth,

$$E_z(h) = \kappa u^* h\left(1 - \frac{h}{h_o}\right)$$

(24-31)

where $\kappa = 0.41$ is the *von Kàrmàn Constant*.

The mean value of E_z averaged over the total water depth is:

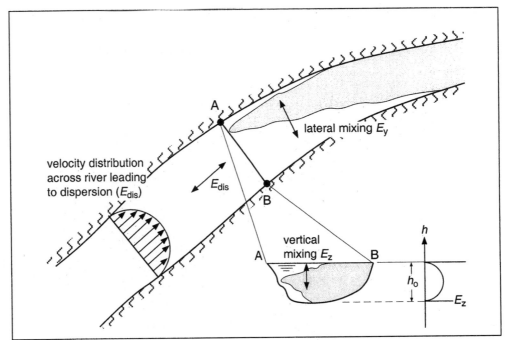

$$\overline{E}_z = \frac{\kappa}{6} u^* h_o = 0.07 \, u^* h_o = \frac{0.07}{\alpha^*} \overline{u} \, h_o \tag{24-32}$$

$$= (0.0035 \, [\text{smooth}] ... \, 0.007 \, [\text{rough}]) \overline{u} \, h_o$$

where we have used α^* defined in Eq. 24-6 to replace u^* by $\overline{u}$. According to Eq. 24-7, α^* lies typically between 10 (rough riverbed) and 20 (smooth riverbed). The expression $(u^* h_o)$ is another example of a case where diffusion is expressed as the product of the relevant velocity times the related diffusion distance (see Eq. 18-7).

The time needed to completely mix the water column over the total depth h_o is of the order (Eq. 18-8):

$$t_z^{\text{diff}} = \frac{h_o^2}{2 E_z} = \frac{\alpha^*}{0.14} \frac{h_o}{\overline{u}} = (140 \, [\text{smooth}] ... \, 70 \, [\text{rough}]) \frac{h_o}{\overline{u}} \tag{24-33}$$

which can also be expressed as a "vertical mixing distance" along the river:

$$x_z^{\text{diff}} = t_z^{\text{diff}} \overline{u} = (140 \, [\text{smooth}] ... \, 70 \, [\text{rough}]) h_o \tag{24-34}$$

Lateral Mixing by Turbulence

Like E_z, the lateral turbulent diffusion coefficient E_y is also characterized by u^*:

$$E_y = \theta u^* h_o \tag{24-35}$$

where θ is an empirical parameter. In straight channels with constant cross section, θ is between 0.1 and 0.2. In natural rivers and streams it lies typically between 0.4 and 0.8. Fischer et al. (1979) suggest using:

$$\theta = 0.6 \tag{24-36}$$

as the best estimate. In meandering rivers with variable cross section, θ can increase to 1 to 2.

From Eqs. 24-7 and 24-35 with $\theta = 0.6$ one obtains:

$$E_y = \frac{0.6}{\alpha^*}\, \bar{u}\, h_o = (0.03\ [\text{smooth}]... 0.06\ [\text{rough}])\bar{u}\, h_o \tag{24-37}$$

By analogy to vertical mixing, the time and distance characteristic of lateral mixing can be expressed as:

$$t_y^{\text{diff}} = \frac{w^2}{2E_y} = \frac{\alpha^* w^2}{1.2\ \bar{u} h_o} = (17\ [\text{smooth}]... 8\ [\text{rough}])\frac{w^2}{\bar{u} h_o} \tag{24-38}$$

$$x_y^{\text{diff}} = \bar{u}\, t_y^{\text{diff}} = \frac{\alpha^* w^2}{1.2\ h_o} = (17\ [\text{smooth}]... 8\ [\text{rough}])\frac{w^2}{h_o} \tag{24-39}$$

Longitudinal Mixing by Dispersion

Longitudinal dispersion occurs whenever fluids are transported along a predominant direction of flow (see Section 22.4). Dispersion results from the different current velocities associated with the various streamlines passing through a given river cross section. Since friction at the bottom of the river acts as the main "brake" that balances the gravitational forces, currents are strongly depth dependent, and the vertical velocity gradient, the so-called vertical shear, is large. Nevertheless, the vertical shear usually has a smaller influence on longitudinal dispersion than the lateral shear, since mixing between streamlines at different depths usually takes much less time than mixing from one side of the river bed to the other (see above). This is especially true at bends and for cross sections with asymmetric depth contours (Fig. 24.5).

The technique developed in Section 22.2 (Reynolds' splitting) to describe transport by turbulent diffusion can also be applied to dispersion. By analogy to Eq. 22-28, the

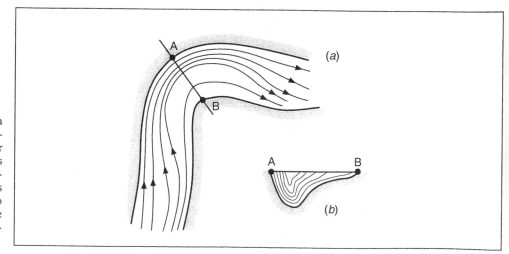

Figure 24.5 (*a*) Streamlines at a bend are asymmetrically distributed across the river. (*b*) Contour lines of equal velocity in a cross section indicate that flow velocities strongly vary laterally as well as from the water surface to the bottom. Such variations are responsible for longitudinal dispersion.

mass flux relative to a cross-sectional disc of water which moves at mean velocity $\bar{u}$ along the river is:

$$F_{dis} = \overline{u'C'} \tag{24-40}$$

where u' and C' are the spatial deviations from the mean values, $\bar{u}$ and $\bar{C}$ respectively, in a given river cross section. According to Gauss' theorem (Eq. 18-12), the mass balance equation for a water parcel which is bounded by two moving cross-sections is given by:

$$\left(\frac{\partial \bar{C}}{dt} \right)_{Lagrange} = -\frac{\partial F_{dis}}{\partial x} = -\frac{\partial}{\partial x}(\overline{u'C'}) \tag{24-41}$$

As for turbulent diffusion (Eq. 22-29), the longitudinal dispersion coefficient E_{dis} (dimension L^2T^{-1}) is defined by:

$$F_{dis} = \overline{u'C'} = -E_{dis}\frac{\partial \bar{C}}{\partial x} \tag{24-42}$$

Thus:

$$\left(\frac{\partial \bar{C}}{\partial t} \right)_{Lagrange} = \frac{\partial}{\partial x}\left(E_{dis}\frac{\partial \bar{C}}{\partial x} \right) \tag{24-43}$$

and for E_{dis} = constant:

$$\left(\frac{\partial \bar{C}}{\partial t} \right)_{Lagrange} = E_{dis}\frac{\partial^2 \bar{C}}{\partial x^2} \tag{24-44}$$

This relationship already has been used in Section 22.4 (see Eq- 22-47).

It is not possible to calculate E_{dis} for a natural river from first principles alone. However, starting from experiments with uniform channel flow, Fischer et al. (1979) developed concepts to relate E_{dis} to other characteristic parameters of the river flow such as u^*, E_z, E_y, which were introduced to describe turbulent mixing in the river. There are two qualitative arguments for the way E_{dis} should depend on other river parameters:

First, E_{dis} should increase with the maximum velocity difference across the river. Since the velocity of the streamlines that are close to the boundaries (i.e., near the river bank or the river bed) is practically zero and the maximum velocity is on the order of the mean velocity $\bar{u}$, one would expect to find E_{dis} growing with $\bar{u}$. As it turns out, E_{dis} grows in proportion to $\bar{u}^2$ (see also Eq. 22-51).

Second, remembering the picture of the railway tracks we used in Section 22.4 to describe dispersion, we concluded that if the number of streamlines (tracks) is limited, then the dispersion coefficient should be inversely related to lateral diffusivity. Thus, from the concept of lateral mixing time (Eq. 24-38) one expects E_{dis} to vary as w^2/E_y.

These considerations can be combined into an empirical equation:

$$E_{dis} = \text{constant } \theta(\bar{u})^2 \frac{w^2}{E_y} \tag{24-45}$$

Substitution of E_y from Eq. 24-35 yields:

$$E_{dis} = \text{constant } (\bar{u})^2 \frac{w^2}{h_o u^*} \tag{24-46}$$

According to Fischer et al. (1979), the best choice for the numerical factor in Eqs. 24-45 and 24-46 is 0.011:

$$E_{dis} = 0.011 \, (\bar{u})^2 \frac{w^2}{h_o u^*} \tag{24-47}$$

In Illustrative Example 24.4 we calculate the longitudinal dispersion coefficient E_{dis} for the flow regimes I and II of River G.

Illustrative Example 24.4   **Turbulent Diffusion and Longitudinal Dispersion in River G**

Problem

For the two flow regimes of River G discussed in Illustrative Example 24.1, calculate (a) the characteristic time and length scale for vertical mixing; (b) the characteristic time and length scale for transversal mixing; and (c) the dispersion coefficient.

Answer (a)

The coefficient of vertical diffusivity is calculated from Eq. 24-32 and from α^* evaluated in Illustrative Example 24.2. The characteristic time and length scales for vertical mixing are given by Eqs. 24-33 and 24-34. The following table summarizes the results:

	α^*	$h_o \approx h$	$\bar{u}$	$\overline{E_z}$	t_z^{diff}	x_z^{diff}
	$(-)$	(m)	$(m\,s^{-1})$	$(m^2 s^{-1})$	(s)	(m)
I	14	0.20	0.40	4.0×10^{-4}	50	20
II	18	1.0	1.0	3.9×10^{-3}	130	130

Note: The shape of the riverbed (see Illustrative Example 24.1) suggests that it is reasonable to approximate the maximum depth h_o by mean depth h.

Answer (b)

Use Eqs. 24-37 to 24-39 and $w \approx 10$ m for both regimes and get the following result for lateral mixing:

	E_y (m^2s^{-1})	t_y^{diff} (s)	x_y^{diff} (m)
I	3.4×10^{-3}	14×10^3 s (4 h)	5.8×10^3
II	3.3×10^{-2}	1.5×10^3 s (0.4 h)	1.5×10^3

Note: Time and distance of lateral mixing are significantly larger than those of vertical mixing, although E_y is larger than E_z, because the width w of a river is commonly much larger than its depth h_o. Yet, the extremely slow lateral mixing (especially in regime I) is not real if there are cascades or sharp bends along the river. The above values apply only for the hypothetical case of a long, uniform riverbed.

Answer (c)

The longitudinal dispersion coefficient is calculated from Eq. 24-47:

Regime I: $u^* = 0.029$ m s^{-1} $E_{dis} = 31$ m^{2}s^{-1}

Regime II: $u^* = 0.056$ m s^{-1} $E_{dis} = 20$ m^{2}s^{-1}

Dilution of a Concentration Patch by Longitudinal Dispersion

Longitudinal concentration gradients of a chemical that is introduced into a river at a constant rate are small (except for a chemical with a very large in-situ reaction rate). Thus, according to Eq. 24-44 the effect of dispersion on $C(x,t)$ is small. In contrast, the concentration profile resulting from a *nonstationary* input as caused, for instance, by an accidental spill is strongly affected by dispersion. In fact, often dispersion is the most important mechanism to reduce the maximum concentration of a concentration patch that moves along the river.

To quantify the effect of dispersion, we assume that at some location and time (for which x and t are arbitrarily set to 0), a total amount $\mathcal{M}$ of some chemical is spilled into the river and mixed uniformly in the constant cross section A. The discharge Q along x shall be constant. The flow is assumed to be stationary (i.e., not changing with time).

To solve Eq. 24-44 in a coordinate system moving at speed $\bar{u}$, we use Eq. 18-16:

$$\overline{C}(\xi,t) = \frac{\mathcal{M}/A}{2(\pi E_{dis}t)^{1/2}} \exp\left(-\frac{\xi^2}{4E_{dis}t}\right) \qquad (24\text{-}48)$$

where ξ is the distance from the center of mass of the concentration cloud, x_m, which moves as:

$$x_m(t) = \bar{u}t \qquad (24\text{-}49)$$

In order to transform Eq. 24-49 into absolute coordinates along the river, ξ is written as:

$$\xi = x - x_m = x - \bar{u}t \qquad (24\text{-}50)$$

Thus:

$$\overline{C}(x,t) = \frac{\mathcal{M}/A}{2(\pi E_{\text{dis}}\,t)^{1/2}}\,\exp\left(-\frac{(x-\overline{u}t)^2}{4E_{\text{dis}}\,t}\right) \qquad (24\text{-}51)$$

Imagine that it were possible to measure simultaneously the concentration profile along the river at a fixed time t, $\overline{C}(x,t)$. Plotted along the river coordinate x, the concentration profile would be a normal distribution with center at $x_m(t) = \overline{u}t$ and standard deviation $\sigma(t_o) = (2E_{\text{dis}}\,t)^{1/2}$ (see Eq. 18-17). The maximum concentration, located at $x_m(t)$, would be:

$$\overline{C_{\max}}(t) = \frac{\mathcal{M}/A}{2(\pi E_{\text{dis}}\,t)^{1/2}} = \frac{(\mathcal{M}/Q)\,\overline{u}}{2(\pi E_{\text{dis}}\,t)^{1/2}} \qquad (24\text{-}52)$$

This concentration maximum decreases with the elapsed time as $(t)^{-1/2}$. Furthermore, the total amount of the chemical along the river remains constant (Eq. 18-18).

Note that the normal distribution for a fixed time (Eq. 24-51) does not cause a normal *concentration variation with time at a fixed location* x_o. Imagine the concentration cloud passing at x_o. Because of the finite longitudinal extension of the cloud, the passage of the cloud (or at least of most of it, since mathematically speaking the cloud has infinite size) takes some time during which the width of the cloud, σ, is growing. Thus, when the concentration at x_o is rising, σ is smaller (and the temporal increase of $\overline{C}(x_o,t)$ is larger) than when the concentration is decreasing after the center of the cloud has passed x_o.

Yet, this is not the only, and usually not even the most important reason why the concentration measured at a fixed location is asymmetric in time. In many cases chemicals enter the river from outfalls (see Fig. 24.4). Remember that vertical mixing usually occurs over a short distance, whereas lateral mixing may need more time (or distance). As discussed before, it is mostly the lateral mixing (or rather its slowness!) which allows longitudinal dispersion. As long as not all streamlines are "occupied", the dispersion coefficient is small and the concentration front steep.

Observations in rivers show that there exists an initial phase during which the cloud is skewed. The concentration at a fixed location as a function of time has a rising slope and a long falling tail (Fig. 24.6). During the initial phase, the standard deviation σ grows more slowly than $t^{1/2}$. The time needed for the transition from the initial state of dispersion, the so-called advective period, to the Gaussian dispersion can be related to the transverse mixing time, t_y^{diff} (Eq. 24-38). According to Fischer et al. (1979), the concentration is skewed for times:

$$t < t_{\text{skewed}} = 0.8 t_y^{\text{diff}} = 0.4\frac{w^2}{E_y} \qquad (24\text{-}53)$$

The standard deviation σ grows as $t^{1/2}$ for times:

$$t > t_o = 0.4 t_y^{\text{diff}} = 0.2\frac{w^2}{E_y} \qquad (24\text{-}54)$$

A first step toward the full two-dimensional (longitudinal/lateral) modeling of

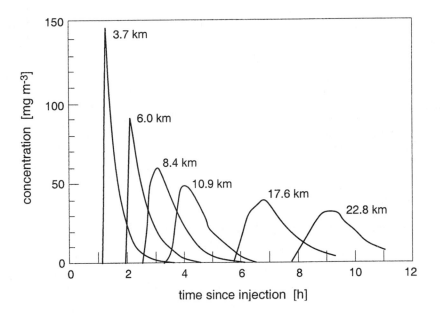

Figure 24.6 Temporal evolution of a concentration cloud along a river. The curves show cross-sectional averaged dye concentration measured at six sites in the Waikato River (New Zealand) below an instantaneous transverse line source. From Rutherford (1994).

dispersion in rivers is given, for instance, by the "enhanced one-dimensional model" developed by Reichert and Wanner (1991).

Until now it has been assumed that the chemical enters the river during a very short time period. This leads to a narrow and sharp initial concentration peak along the river (e.g., to the δ-function of Eq. 18-15) which is then gradually transformed into a normal distribution with growing standard deviation σ. Often the compound is added continuously or—even if it originates from an accidental spill—during a finite time Δt. We look at two extreme scenarios. In the first scenario, the input is instantaneously "switched on" to a constant input rate, J (mass per unit time), and "turned off" after time Δt (Fig. 24.7a). Even if the advective period of dispersion can be disregarded, dispersion would first smooth out the sharp concentration edges only at the onset and end of the input event, while the concentration in the middle part of the event, $C_{in} = J / Q$, would not be altered. The "initial time" until dispersion from both sides would meet at the center of the cloud, t_{ini}, can be estimated with Eq. 18-17:

$$t_{ini} = \frac{(\Delta x / 2)^2}{2 E_{dis}} = \frac{(\Delta t / \overline{u})^2}{8 E_{dis}} \tag{24-55}$$

where $\Delta x = \Delta t \, \overline{u}$ is the original spatial extension of the concentration cloud along x. For $t > t_{ini}$, the cloud can be approximated by a Gaussian distribution with maximum concentration $C_{in} = J / Q$, which for increasing times spreads according to the expression given by Eq. 24-51. In fact, by inserting $t = t_{ini}$ and the total mass input of the event, $M = C_{in} Q \Delta t = C_{in} A \overline{u} \Delta t$, into the expression that describes the maximum concentration of a Gaussian distribution, Eq. 24-52, one gets:

$$\overline{C}_{max}(t) = \frac{C_{in}}{(\pi / 2)^{1/2}} = 0.8 \, C_{in} \tag{24-56}$$

Thus, Eq. 24-51 is an adequate approximation for times $t > t_{ini}$, where $t = 0$ is the time

Figure 24.7 Two extreme input scenarios for chemical being spilled into a river. (a) Constant input rate J (mass per unit time) during time Δt leading to a "rectangular" concentration profile, $C_{in} = J/Q$. The dashed line shows how dispersion acts on the edges and leaves the concentration in the middle of the cloud unchanged. (b) Gaussian input scenario. The time integral between $t = -2\sigma_o$ and $t = 2\sigma_o$ comprises 95% of the total input $\mathcal{M}$ (see Box 18.2). Dispersion causes the variance to increase according to Eq. 24-60. The maximum concentration is given by Eq. 24-62.

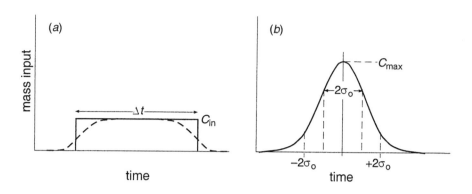

when half of the input has reached the river, and $\mathcal{M}/A$ is replaced by $C_{in}\,\bar{u}\,\Delta t$. For the situation shown in Fig. 24.7a:

$$\overline{C}(x,t) = \frac{C_{in}\,\bar{u}\,\Delta t}{2(\pi E_{dis}\,t)^{1/2}}\exp\left(-\frac{(x-\bar{u}\,t)^2}{4\,E_{dis}\,t}\right), \qquad \text{for } t > t_{ini} \qquad (24\text{-}57)$$

The second input scenario, simpler from a mathematical viewpoint but less probable to occur, is a generalization of the instantaneous (δ-)input. It is assumed that the temporal variation of the input is Gaussian and leads to an initial longitudinal variation of the concentration cloud with standard deviation:

$$\sigma_o = \frac{\Delta t\,\bar{u}}{4} \qquad (24\text{-}58)$$

Δt is the time period during which 95% of the substance is added to the river (see Fig. 24.7b). Eq. 22-39 describes how the variance σ^2 grows with time:

$$\frac{d\sigma^2}{dt} = 2E_{dis} \qquad (24\text{-}59)$$

For constant E_{dis}, the solution is:

$$\sigma^2(t) = \sigma_o^2 + 2\,E_{dis}\,t \qquad (24\text{-}60)$$

Then the generalized version of Eq. 24-51 for the scenario of a Gaussian (or normal) input event is:

$$\overline{C}(x,t) = \frac{\mathcal{M}/A}{(2\pi)^{1/2}(\sigma_o^2 + 2\,E_{dis}\,t)^{1/2}}\exp\left(-\frac{(x-\bar{u}\,t)^2}{2\sigma_o^2 + 4\,E_{dis}\,t}\right) \qquad (24\text{-}61)$$

which is again a normal distribution with time-dependent variance. The maximum concentration in the river as a function of time is:

$$\overline{C_{max}}(t) = \frac{\mathcal{M}/A}{(2\pi)^{1/2}(\sigma_o^2 + 2\,E_{dis}\,t)^{1/2}} \qquad (24\text{-}62)$$

In Illustrative Example 24.5 we look again at the atrazine spill in River G of Illustrative Example 24.3 and ask how strongly dispersion would reduce the maximum atrazine concentration while the pollution cloud is moving downstream.

Illustrative Example 24.5   ## A Second Look at the Atrazine Spill in River G: The Effect of Dispersion

In Illustrative Example 24.3 we considered an accidental spill of atrazine into the River G. We analyzed the possible removal of atrazine by the solid–water interaction and found that only a very small fraction of atrazine would be removed to the sediments.

Problem

Calculate for both flow regimes I and II (Illustrative Example 24.1) the initial maximum atrazine concentration and its reduction at $x = 10$ km due to dispersion. Use the dispersion coefficients calculated in Illustrative Example 24.4. Remember that 200 kg of atrazine enters the river within 30 minutes.

Answer

The average input rate of atrazine is $J = 200$ kg $/1800$ s $= 0.11$ kg s^{-1} = 0.52 mol s^{-1}. Thus, the (initial) input concentration, $C_{in} = J/Q$, is 6.4×10^{-4} mol L^{-1} and 5.2×10^{-5} mol L^{-1} for $Q_I = 0.8$ m^{3}s^{-1} and $Q_{II} = 10$ m^{3}s^{-1}, respectively.

The relevant mixing characteristics calculated in Illustrative Example 24.3 are summarized below:

	C_{in} (mol m^{-3})	$\bar{u}$ (m s^{-1})	E_{dis} [a] (m^{2}s^{-1})	t_y^{diff} [a] (s)	t_{ini} [b] (s)	$x_{ini} = \bar{u}\, t_{ini}$ (m)
I	0.64 [c]	0.40	31	14×10^3	2.1×10^3	840
II	0.052	1.0	20	1.5×10^3	2.0×10^4	20×10^3

[a] From Illustrative Example 24.4. [b] From Eq. 24-55 with $\Delta t = 30$ minutes. [c] Note that C_o is larger than the aqueous solubility of atrazine (0.15 mol m^{-3} at 25°C, see Appendix C). Thus, dispersion has to "pull the cloud apart" before all atrazine is dissolved.

For regime I, dispersion begins to affect the initial atrazine concentration less than 1 km downstream of the spill. In contrast, for flow regime II, x_{ini} is 20 km. Thus, at $x = 10$ km the atrazine cloud still looks like Fig 24.7a; dispersion does not yet lower $\overline{C}_{max}$.

For regime I, the time t_{diff} during which dispersion is effective is the total flowing time from $x = 0$ to $x = 10$ km reduced by $t_{ini} = 2.1 \times 10^3$: $t_{diff} = (2.5 \times 10^4 - 2.1 \times 10^3)$ s $\approx 2.2 \times 10^4$ s. Thus, from Eq. 24-57, the maximum atrazine concentration when the cloud passes at $x = 10$ km is:

$$\overline{C}_{max} = \frac{C_{in}\, \bar{u}\, \Delta t}{2\left(\pi E_{dis}\, t_{diff}\right)^{1/2}} = \frac{0.64 \text{ mol m}^{-3} \times 0.40 \text{ m s}^{-1} \times 1800 \text{ s}}{2\left(3.141 \times 20 \text{ m s}^{-1} \times 2.2 \times 10^4 \text{ s}\right)^{1/2}} = 0.20 \text{ mol m}^{-3}$$

To summarize, in flow regime I the maximum atrazine concentration at $x = 10$ km is reduced to (0.20/0.63) 100% = 32% of its original value. This reduction is significantly larger than that from any other possible mechanism to reduce the concentration. In contrast, in flow regime II dispersion is not yet felt at $x = 10$ km.

A Linear Transport/Reaction Model for Rivers

All Things Considered: The Reduction of Total Load and Maximum Concentration

As shown above, the concentration of a compound that is transported in a river is affected by various mixing and elimination processes. Their relative importance for reducing the riverborne mass flow and the maximum concentration depends on the characteristics of the river as well as on the compound under consideration. This section gives a summary of the relevant rate constants by emphasizing the simplest descriptions. In order to compare their relative importance, all processes will be approximated by first-order rate constants, either in time (k-rates, dimension T^{-1}) or in space along the river (ε-rates, dimension L^{-1}).

A distinction has to be made between processes that (a) affect both the total mass and the concentration in the river, and (b) affect only the concentration but leave the mass in the river unchanged. Processes which move the compound from the open river water to another environmental compartment (atmosphere, sediment) are listed among (a) as well. The following processes belong to the first category:

1. Chemical and biological reactions, such as hydrolysis, biodegradation, photolysis, and other chemical transformations. The relevant expressions are given by Eqs. 24-8 and 24-23.

2. Air–water exchange, see Eq. 24-22.

3. Removal to the sediments by particle settling, see Eq. 24-25.

4. Diffusion into the pore space of the sediments, see Eq. 24-30.

The following processes belong to the second category (no reduction of total mass in the river water):

5. Longitudinal dispersion

6. Dilution by merging rivers and by infiltrating groundwater

Vertical and lateral mixing are not considered here. It is assumed that a given mass flux of a compound is mixed instantaneously into the corresponding volume flux of water. Of course, close to the input, vertical and lateral mixing are the most important mechanisms to reduce the concentration in the river. Also note that processes 3, 4, and 5 are only effective in reducing the concentration of the compound for the case of an episodic input. If a compound is continuously added to the river, the sediments reach an equilibrium with the river water and thus do not act as a sink anymore. Similarly, dispersion does not reduce the concentration of a compound that is permanently added to the river.

In order to compare the effects of dispersion and dilution with the other processes, they will both be described by pseudolinear models. In the following expressions, the overbar used to describe the mean concentration along the river will be omitted. Instead, the subscript t reminds the reader that all equations are expressed in terms of the total concentration.

The rate constant of longitudinal dispersion is defined as the relative change with time of the maximum concentration of a concentration cloud:

$$\left(\frac{\partial C_{t\,max}}{\partial t}\right)_{dis} = -k_{dis}C_{t\,max} \tag{24-63}$$

For the case of a δ-input, k_{dis} can be calculated from Eq. 24-52 as:

$$k_{dis}(t) = \frac{1}{2t} \tag{24-64}$$

Since according to Eq. 24-57 the concentration cloud of a chemical that is introduced into a river at a constant rate during time Δt assumes normal shape for times $t > t_{ini}$ (see Eq. 24-55 for definition), Eq. 24-64 also holds for this case. For the case of a normal input with initial variance σ_o^2, the maximum concentration can be derived from Eq. 24-57 with $x = \bar{u}t$:

$$C_{t\,max}(t) = C_t(x = \bar{u}t, t) = \frac{\mathcal{M}/A}{(2\pi)^{1/2}(\sigma_o^2 + 2E_{dis}t)^{1/2}} \tag{24-65}$$

Differentiation with respect to time t and using Eq. 24-63 yields:

$$k_{dis}(t) = \frac{E_{dis}}{\sigma_o^2 + 2E_{dis}t} \tag{24-66}$$

Note that k_{dis} varies with time. This means that over a long time period dispersion does not behave as a real first-order-process.

The rate constant of dilution is calculated from the definition of concentration, C_t, as mass flux per unit time, m, divided by volume flux of water, Q:

$$C_t = \frac{m}{Q} \tag{24-67}$$

The derivative of this equation with respect to time, where m is constant, yields:

$$\left(\frac{\partial C_t}{\partial t}\right)_{dil} = -\frac{m}{Q^2}\frac{dQ}{dt} = -\frac{C_t}{Q}\frac{dQ}{dt} = -k_{dil}C_t \tag{24-68}$$

with:

$$k_{dil} = \frac{1}{Q}\frac{dQ}{dt} \tag{24-69}$$

Again, k_{dil} may vary with time.

In this section, we have put together a toolbox consisting of six transport and transformation processes which allow us to make a first and rather simple assessment of the fate of an organic compound in a river. Since all processes were described (or at least approximated) by linear equations, the resulting first-order rate constants k_α (or the corresponding first order ε_α values) allow us to compare the relative importance of the different processes and to concentrate a further analysis on the most important ones. In order to demonstrate the strength of this method, we will now discuss two case studies in some detail.

CI—C—H structure

trichloromethane
(chloroform)

M_i = 119.4 g mol^{-1}

T_m = -63.5°C

T_b = 61.7°C

C_w^{sat} (25°C) = 6.5 × 10^{-2} mol L^{-1}

K_H (25°C) = 3.6 L bar mol^{-1}

K_{ow} (25°C) = 85

ρ_L (20°C) = 1.489 g cm^{-3}

Hydrological characteristics of the Mississippi River at the time of the accident:

Q = 7.5 × 10^3 m^3 s^{-1} discharge

$\bar{u}$ = 0.57 m s^{-1} mean flow velocity

w = 1200 m width

h = 11 m mean depth

h_o = 20 m maximum depth (estimated)

S_o = 2 × 10^{-5} slope of river bed

First Case Study: Chloroform in the Mississippi River

On August 19, 1973, a barge carrying three tanks of trichloromethane (chloroform) was damaged on the Mississippi River at Baton Rouge, Louisiana (Neely et al. 1976). Two tanks containing a total of 7.8×10^5 kg of chloroform were damaged and their contents spilled to the river.

Further information about the accident is:

1. The first tank ($\mathcal{M} = 3.9 \times 10^5$ kg) ruptured at 2:40 P.M. (Note: This time is chosen as reference time, $t = 0$.) All the chloroform was lost within a very short time.

2. The second tank (3.9×10^5 kg) began to leak at 10:00 P.M. ($t = 7.3$ h); its contents were lost over a 45-minute period.

3. Water samples were taken at 26 km (station A) and 195 km (station B, New Orleans) from the point of the accident (see Fig. 24.8).

4. It was found that at the time when chloroform first was detected at station A, the chloroform was evenly distributed across the river.

In the first step, we want to answer the following questions:

Which processes determine the fate of the chloroform on its journey along the Mississippi River? What kind of information (about the chloroform or the river) is needed in order to quantify these processes? Assess separately the change (1) of the total mass and (2) of the maximum concentration of chloroform along the river.

Air–water exchange deserves closer inspection. On one hand, regarding the depth and size of the river the small-eddy model (Eq. 20-32) may be appropriate to calculate $v_{a/w}$ of chloroform. Since the Mississippi River is wide and its waters flow rather slowly, we can, on the other hand estimate air–water exchange from wind speed while noting that the Henry's law coefficient of chloroform indicates a water side-controlled process. The reader is invited to compare the two approaches.

As a third possibility, we rely on the value $v_{a/w} = 2.2 \times 10^{-3}$ cm s^{-1} which Neely et al. derived from their observations. From Eq. 24-22:

$$\varepsilon_{a/w} = \frac{v_{a/w}w}{Q} = \frac{2.2 \times 10^{-5} \text{m s}^{-1} \times 1200 \text{m}}{7.5 \times 10^3 \text{m}^3 \text{ s}^{-1}} = 3.5 \times 10^{-6} \text{m}^{-1}$$

This yields the following mass reductions due to air–water exchange:

At station A (x = 26 km): $\exp(-3.5 \times 10^{-6} \times 2.6 \times 10^4) = 0.91$

At station B (x = 195 km): $\exp(-3.5 \times 10^{-6} \times 1.95 \times 10^5) = 0.51$

Based on the physicochemical properties and the persistence of chloroform with respect to transformation reactions, one can conclude that there are no other significant mechanisms of mass reduction. Furthermore, dilution cannot be relevant for a river the size of the Mississippi. Thus, besides air–water exchange, the decrease of the maximum concentration (Fig. 24.8) must also be influenced by dispersion.

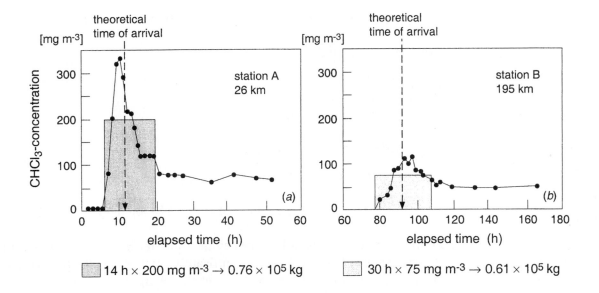

Figure 24.8 Concentration of chloroform as a function of time measured at (*a*) Station A (26 km downstream of spill) and (*b*) Station B (195 km downstream of spill at New Orleans). The rectangular areas serve to estimate the total mass flux originating from the rupture of the first tank. Data from Neely et al. (1976).

To estimate the dispersion coefficient E_{dis} we need the lateral turbulent diffusivity E_y (see Eq. 24-45) which in turn is calculated from the friction velocity, u^*. Problem 24.4 deals with the calculation of E_{dis}. As it turns out, a realistic value which agrees with the measured concentrations should be $E_{dis} = 800 \ m^2 s^{-1}$.

As a first attempt we consider the rupture of the first tank as an individual event during which all of the chloroform was spilled into the river during a very short period (say within less than 5 minutes). We assume that the chloroform is immediately dissolved in the water (is this a reasonable assumption?) and want to calculate the maximum concentrations at stations A and B.

The times of arrival of the chloroform peak at stations A and B are:

$$t_A = \frac{x_A}{\overline{u}} = \frac{26 \times 10^3 \, m}{0.57 \ m \ s^{-1}} = 4.56 \times 10^4 \, s = 12.7 \ h$$

$$t_B = \frac{x_B}{\overline{u}} = \frac{195 \times 10^3 \, m}{0.57 \ m \ s^{-1}} = 3.4 \times 10^5 \, s = 95 \ h$$

Figure 24.8 suggests that the chloroform peak arrived somwhat earlier at Station A. Apparently, the chloroform was first flowing in the middle of the river where the flow velocity is greater than the mean, $\overline{u}$. For the following discussion, we use $t_A^{obs} = 10 \ h$.

The effect of dispersion on the maximum concentration at station A is:

$$\overline{C_{max}}(A) = \frac{\mathcal{M}/A}{2\left(\pi E_{dis} t_A^{obs}\right)^{1/2}} = \frac{3.9 \times 10^{11} mg / (1200 \ m \times 11 \ m)}{2 \, (3.141 \times 800 \ m^2 s^{-1} \times 3.6 \times 10^4 \, s)^{1/2}} = 1.6 \times 10^3 mg \ m^{-3}$$

Combined with the effect of air–water exchange:

$$C_{eff}(A) = 0.91 \times 1.6 \times 10^3 mg \ m^{-3} = 1.4 \times 10^3 mg \ m^{-3}$$

In contrast, the observed maximum concentration at A is 320 mg m^{-3} (Fig. 24.8a):

Correspondingly at Station B:

Due to dispersion:

$$\overline{C_{max}}(B) = 500 \text{ mg m}^{-3}$$

Combined with air–water exchange:

$$C_{eff}(B) = 0.51 \times 500 \text{ mg m}^{-3} = 260 \text{ mg m}^{-3}$$

The observed maximum concentration at B is 110 mg m^{-3} (Fig. 24.8b).

From these calculations it becomes evident that the peak concentration of chloroform measured at both stations is significantly smaller than the predicted values. In order to explain this discrepancy, Neely et al. (1976) assume that there was a pool of pure chloroform (DNAPL) laying on the bottom of the river at the place of the accident from which chloroform slowly dissolved into the water and drifted downstream. The essence of the Neely model is the following elements:

• 20% of the first tank (20% of 3.9×10^5 kg, i.e., about 0.8×10^5 kg) is immediately released to the river water. Evidence for this number is given in Fig. 24.8. The cloud drifts downstream and is broadened by dispersion. Air–water exchange leads to some reduction of the chloroform concentration.

• The remaining portion of the first tank (about 3.1×10^5 kg) rests on the bottom of the river from which it is introduced by a first-order reaction into the flowing water (rate constant $k_b = 3 \times 10^{-3}$ h^{-1}). Dispersion of this "cloud" can be disregarded, but not air–water exchange.

• All of the second tank (3.9×10^5 kg) is added to the bottom reservoir from which it leaks out at the same rate starting at time $t_2 = 7.6$ h.

Since all the relevant reactions are linear, each can be calculated separately and then added up. As it turns out, the consideration of these processes leads to concentrations which are fairly consistent with the observed values. We leave it to the reader to confirm this conclusion.

To summarize, the chloroform concentration in the Mississippi River from the accident at Baton Rouge was controlled by the following processes (ranged according to their importance):

(a) Retarded seepage from the tank and from the bulk chloroform resting on the bottom of the river

(b) Dispersion of initial concentration peak

(c) Air–water exchange

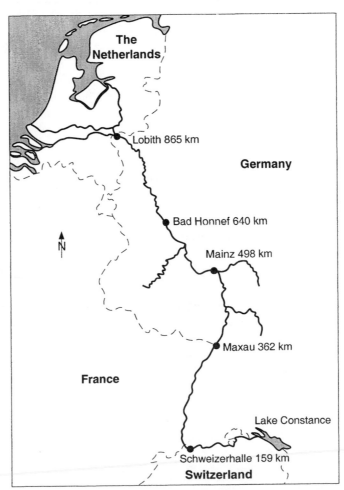

disulfoton

Second Case Study: Chemical Pollution of the River Rhine Due to a Fire in a Storehouse

The second example includes the influence of sorption and sediment–water interaction, processes which were not relevant for the case of chloroform. We choose the real case of a chemical pollution of the River Rhine. On November 1, 1986, a fire destroyed a storehouse at Schweizerhalle near Basel (Switzerland). During the fighting of the fire, several tons of various pesticides and other chemicals were flushed into the River Rhine (Wanner et al., 1989). One of the major constituents discharged into the river was disulfoton, an insecticide. An estimated quantity of 3.3 metric tons reached the river within a time period of about 12 hours leading to a massive killing of fish and other aquatic organisms.

On their trip of about 8 days to Lobith, 700 km downstream of Schweizerhalle at the border between Germany and the Netherlands (Fig. 24.9), the spilled chemicals underwent various transformations, were transferred to other environmental compartments (atmosphere, sediments), and were diluted by dispersion and merging rivers. In the following we will calculate how much these processes contribute to the reduction of disulfoton from Schweizerhalle to Lobith (a) in terms of the maximum concentration in the river, and (b) in terms of the total load. All relevant information is given in Table 24.3.

Fig. 24.9 Map of River Rhine showing the site of the accident (Schweizerhalle) and the sampling stations at Maxau, Mainz, Bad Honnef, and Lobith. From Wanner et al. (1989).

Table 24.3 Relevant Information on the Behavior of Disulfoton in the River Rhine

River Rhine

Distance from Schweizerhalle to Lobith	$x_o = 700$ km
Mean flow velocity	$\bar{u} = 1$ m s^{-1}
Mean depth	$h = 5$ m
Discharge of Rhine at Schweizerhalle	$Q_o = 750$ m^{3}s^{-1}
Discharge of Rhine at Lobith	$Q_1 = 2300$ m^{3}s^{-1}
Concentration of suspended solids	$r_{sw} = 0.4$ kg$_s$m^{-3}
Relative organic carbon content of suspended solid	$f_{oc}^{sus} = 0.005$ kg$_{oc}$kg$_s^{-1}$
Settling velocity of particles	$v_s = 2$ m d^{-1}
Coefficient of longitudinal dispersion	$E_{dis} = 2.8 \times 10^3$ m^{2}s^{-1}
Air–water transfer velocity in air for H$_2$O	$v_a(H_2O) = 5 \times 10^{-3}$ m s^{-1}
in water for O$_2$	$v_w(O_2) = 2 \times 10^{-5}$ m s^{-1}
Water temperature (approximative)	$T_w = 10°C$
pH	7.9

Disulfoton

Molar mass	$M_i = 274.4$ g mol^{-1}
Henry's Law constant at 10°C	$K_H = 1 \times 10^{-3}$ bar L mol^{-1}
Air–water transfer velocity in air (v_a) and in water (v_w)	reduced by a factor of 5 relative to H$_2$O and O$_2$, respectively
Organic carbon–water partition coefficient	$K_{oc} = 1.5$ m^{3}kg$_{oc}^{-1}$
Molecular diffusion coefficient in water at 10°C	$D_w = 3.8 \times 10^{-10}$ m^{2}s^{-1}
Rate constants for abiotic hydrolysis at pH 7.9 [a] (see Problem 13.9)	$k_{hyd} = 4.1 \times 10^{-3}$ d^{-1}
Photochemical transformation: indirect photolysis (reaction with O$_2$)	$k_{photo} = 6.9 \times 10^{-4}$ d^{-1}
Biological transformation (at 5µg L^{-1})[b]	$k_{biol} = 0.14$ d^{-1}

Accident

Duration of spill (approximative)	$\Delta t_o = 12$ h
Total initial input	$\mathcal{M} = 3.3 \times 10^3$ kg

[a] For details see Wanner et al. (1989); [b] Measured disulfoton concentrations were between 0 and 30 µg L^{-1}, except in the first kilometers downstream of the spill.

First, we calculate the fraction in particulate form, $(1-f_w)$, at sorption equilibrium by assuming that hydrophobic partitioning is the dominant sorption mechanism. From Eq. 23-4 and using the values given in Table 24.3 we get:

$$(1-f_w) = \frac{f_{oc}^{sus} r_{sw} K_{oc}}{1 + f_{oc}^{sus} r_{sw} K_{oc}} = \frac{0.005 \times 0.4 \times 1.5}{1 + 0.005 \times 0.4 \times 1.5} = 3 \times 10^{-3} \qquad (24\text{-}70)$$

Hence, $f_w \approx 1$. Therefore, for the processes affecting the dissolved species, the dissolved concentration can be approximated by the total concentration.

Since not enough information is available on the exact input mode of the polluted water into the River Rhine, the initial mixing phase will not be considered. In fact, a hydroelectric power station located a few kilometers downstream of the spill significantly accelerated vertical and lateral mixing. The total mass of disulfoton, which had entered the river, was indirectly calculated from water samples taken at Village Neuf located about 10 km downstream of the power station. In these samples the concentration of disulfoton was about 100 μg L^{-1}.

The following evaluation is based on the data from Table 24.3 and the list of transformation and mixing processes given in the previous paragraph:

1. *Chemical and biological transformations.* By far the largest reduction of the total load is due to biological transformation; hydrolysis and photochemical reactions combined contribute less than 4% to the total reaction rate constant:

$$k_r = k_{hyd} + k_{photo} + k_{bio} \approx k_{bio} = 0.14 \ \text{d}^{-1} \qquad (24\text{-}71)$$

Note that the biological transformation is assumed to be first order. If river sections with higher concentrations were considered, a Michaelis-Menten approach (Box 12.2) would have to be taken (see Wanner et al., 1989).

2. *Air–water exchange.* From the exchange velocities adjusted for disulfoton, $v_a = 1 \times 10^{-3}$ m s^{-1} and $v_w = 4 \times 10^{-6}$ m s^{-1}, and the nondimensional Henry's Law constant at 10°C, $K_{a/w} = 4.3 \times 10^{-5}$, it turns out that the exchange is air-side controlled, thus $v_{a/w} \cong v_a K_{a/w} = 4.3 \times 10^{-8}$ m s^{-1} and:

$$k_{a/w}^* = k_{a/w} f_w = \frac{v_{a/w}}{h} f_w = 8.6 \times 10^{-9} \text{s}^{-1} = 7.4 \times 10^{-4} \text{d}^{-1} \qquad (24\text{-}72)$$

3. *Particle sedimentation.* The removal rate of particles, $k_s = v_s / h = 2$ m d^{-1} / 5 m = 0.4 d^{-1}, is fairly large, but only a minor fraction of disulfoton is sorbed to the particles:

$$k_s^* = k_s(1 - f_w) = 0.4 \ \text{d}^{-1} \times 3 \times 10^{-3} = 1.2 \times 10^{-3} \text{d}^{-1} \qquad (24\text{-}73)$$

4. *Sediment–water exchange.* We do not have enough information on the sediments of the River Rhine in order to evaluate Eq. 24-29 in detail. To estimate the order of magnitude of k_{sedex}, we use the same parameters as in Illustrative Example 24.3 (River G), that is: porosity $\phi = 0.8$, density of sediment particles $\rho_s = 2500$ kg$_s$m^{-3}, relative organic carbon content of sediment particles $f_{oc}^{sed} = 0.005$. According to Eq. 9-15:

$$r_{sw}^{sed} = \rho_s \frac{1-\phi}{\phi} = 2500 \text{ kg}_s\text{m}^{-3} \frac{0.2}{0.8} = 625 \text{ kg}_s\text{m}^{-3}$$

and:

$$f_w^{sed} = \frac{1}{1 + f_{oc}^{sed} r_{sw}^{sed} K_{oc}} = \frac{1}{1 + 0.005 \times 625 \times 1.5} = 0.176$$

We also need information on the length of time until the spill has passed by a fixed location in the river. Initially, Δt_o is 0.5 days (Table 24.3). The time until longitudinal dispersion becomes effective is calculated from Eq. 24-55:

$$t_{ini} = \frac{(\Delta t_o / \overline{u})^2}{8 E_{dis}} = \frac{(12 \times 3600 \text{ s} / 1 \text{ m s}^{-1})^2}{8 \times 2.8 \times 10^3 \text{ m}^2\text{s}^{-1}} = 8.3 \times 10^4 \text{s} \quad (23 \text{ hours})$$

For $t > t_{ini}$, the size of the cloud grows as Eq. 24-60 with $\sigma_o = 0$. The total flowing time from Schweizerhalle to Lobith is $x_o / \overline{u} = 7 \times 10^5 \text{m} / 1 \text{ m s}^{-1} = 7 \times 10^5 \text{ s} \cong 8$ days. Thus:

$$\Delta t(\text{Lobith}) = \frac{\sigma(\text{Lobith})}{\overline{u}} = \frac{(2 E_{dis} \times 7.5 \times 10^5 \text{s})^{1/2}}{1 \text{ m s}^{-1}} = 6.5 \times 10^4 \text{s} \quad (0.7 \text{ d})$$

Thus, the temporal duration of the cloud increases only slightly from 0.5 days at Schweizerhalle to about 0.7 days at Lobith. We use an average $\overline{\Delta t} = 0.6 \text{ d} = 5.2 \times 10^4 \text{s}$. Finally, we approximate D_w^{sed} by $D_w = 3.8 \times 10^{-10} \text{ m}^2\text{s}^{-1}$. Inserting these value into Eq. 24-29b yields ($f_w \approx 1$):

$$\overline{k}_{sedex} = 2 \left(\frac{1}{\pi}\right)^{1/2} \frac{\phi}{h} \left(\frac{D_w^{sed}}{f_w^{sed} \overline{\Delta t}}\right)^{1/2} f_w$$

$$\text{(24-74)}$$

$$= 1.13 \frac{0.8}{5 \text{ m}} \left(\frac{3.8 \times 10^{-10} \text{m}^2\text{s}^{-1}}{0.176 \times 5.2 \times 10^4 \text{s}}\right)^{1/2} = 3.7 \times 10^{-8} \text{s}^{-1} = 3.2 \times 10^{-3} \text{d}^{-1}$$

Although not considered in the original analysis by Wanner et al. (1989), it is instructive to compare the size of $\overline{k}_{sedex}$ to the possible effect of the flow of water through the sediments calculated with the model by Eylers (1994) given in Box 24.2. Due to the lack of adequate data, the following parameters are to be considered as rough estimates: $K_q = 10^{-3} \text{m s}^{-1}$, $\lambda \approx 3$ $h = 15 \text{ m}$, $H \approx h / 10 = 0.5 \text{ m}$. Equation 1 of Box 24.2 yields $v_{sedflow} = 1.2 \times 10^{-6} \text{m s}^{-1}$. Thus, from Eq. 2 of Box 24.2 with $f_w^{sed} = 0.176$:

$$k_{sedflow} = \frac{v_{sedflow}}{h} \left(1 - f_w^{sed}\right) = 2.0 \times 10^{-7} \text{s} = 1.7 \times 10^{-2} \text{d}^{-1} \quad \text{(24-75)}$$

which is five times larger than $\overline{k}_{sedex}$. Due to the speculative character of this number, it will not be included in the summary table below, although the result suggests the possible importance of this still-not-well-understood process.

The following processes reduce only the maximum concentration, not the mass flux of disulfoton.

5. *Longitudinal dispersion.* According to Eq. 24-64 the rate k_{dis} at which the maximum concentration is reduced decreases with flow time t. For $t = 4$ d, that is, for half the travel time from Schweizerhalle to Lobith, we get:

$$\bar{k}_{dis} \approx k_{dis}(t = 4 \text{ d}) = 0.125 \text{ d}^{-1}$$

Since $k_{dis}(t)$ is not a linear function of the flow time t, a better way to evaluate $\bar{k}_{dis}$ is to use the ratio of the maximum concentrations for $t = 8$ d and 0 d, respectively. The latter is simply ($\Delta t_o = 12$ h = initial duration of spill):

$$C_{t \max}(t = 0) = \frac{\mathcal{M}}{Q_o \Delta t_o}$$

$$= \frac{3.3 \times 10^6 \text{g}}{750 \text{ m}^3\text{s}^{-1} \times 4.32 \times 10^4 \text{s}} = 0.10 \text{ g m}^{-3} = 100 \text{ µg L}^{-1}$$

The former follows from Eq. 24-52 with $t = 8$ days $= 6.9 \times 10^5$ s:

$$C_{t \max}(t = 8 \text{ days}) = \frac{(\mathcal{M} / Q_o)\, \bar{u}}{2(\pi E_{dis}\, t)^{1/2}}$$

$$= \frac{(3.3 \times 10^6 \text{g} / 750 \text{ m}^3\text{s}^{-1}) \times 1 \text{ m s}^{-1}}{2(3.141 \times 2.8 \times 10^3 \text{m}^2\text{s}^{-1} \times 6.9 \times 10^5 \text{s})^{1/2}} = 2.8 \times 10^{-2} \text{g m}^{-3} = 28 \text{ µg L}^{-1}$$

Note that since in our analysis we separate the influences from dispersion and dilution, respectively, we have inserted the discharge rate at Schweizerhalle, $Q_o = 750$ m³s⁻¹. The effect of dilution follows below. Thus:

$$\frac{C_{t \max}(t = 8 \text{ d})}{C_{t \max}(t = 0)} = 0.28$$

Interpreted in terms of the linear model (Eq. 24-63), $C_{t \max}(t) = C_{t \max}(0) \exp\{-\bar{k}_{dis}t\}$, we get:

$$\bar{k}_{dis} = -\frac{1}{8 \text{ d}} \ln(0.28) = 0.16 \text{ d}^{-1} \qquad (24\text{-}76)$$

This value will be used for the final result given in Table 24.4.

6. *Dilution by merging rivers.* The rate constant is given by Eq. 24-69. Of course, the discharge $Q(t)$ increases stepwise along the Rhine, not continuously. However, we still can calculate a mean dilution rate, $\bar{k}_{dil}$, by solving Eq. 24-68 for constant k_{dil}:

$$Q(t) = Q(0)\, e^{k_{dil}t}$$

The information in Table 24.3 yields: $Q(t)/Q(0) = 2300/750 = 3.1$. Thus:

$$\bar{k}_{dil} = \frac{1}{t} \ln[Q(t)/Q(0)] = \frac{1}{8 \text{ d}} \ln(3.1) = 0.14 \text{ d}^{-1} \qquad (24\text{-}77)$$

The different rate constants (Eqs. 24-71 to 24-77) are summarized in Table 24.4. Roughly two-thirds of the disulfoton disappeared from the water between

Table 24.4 Transport, Dilution, and Transformation of Disulfoton Spill Between Schweizerhalle and Lobith (River Rhine) [k_p is the (pseudo-)first-order rate constant of (transformation or transport) process p.]

	Eq.	Rate k_p (d^{-1})	% of Load Reduction[a]	% of Reduction of Max.Concentration[b]
Load Reduction (Transfer and Transformation)				
Hydrolysis, k_{hyd}	24-71	4.1×10^{-3}	2.7	0.91
Photochemical transformation, k_{photo}	24-71	6.9×10^{-4}	0.5	0.15
Biological transformation, k_{biol}	24-71	0.14	93.4	31.1
Air–water exchange, $k_{a/w}^*$	24-72	7.4×10^{-4}	0.5	0.16
Particle sedimentation, k_s^*	24-73	1.2×10^{-3}	0.8	0.27
Sediment–water exchange, $\bar{k}_{sedex}$	24-74	3.2×10^{-3}	2.1	0.71
Water flow through sediments, $k_{sedflow}$	24-75	(1.7×10^{-2}) [c]	(–)	(–)
Total load reduction,[c] k_{load}		**0.150**	**100**	**33.3**
Dispersion and Dilution				
Dispersion, $\bar{k}_{dis}$	24-76	0.16	–	35.6
Dilution, $\bar{k}_{dil}$	24-77	0.14	–	31.1
Total of all processes,[c] k_{tot}		**0.45**	–	**100**
Remaining total load at Lobith[d]		$\mathcal{M} = 1.0 \times 10^3$ kg		(30%)
Remaining maximal concentration at Lobith[e]		$C_{tmax} = 2.7\ \mu g\ L^{-1}$		(2.7%)

[a] $(k_p/k_{load}) \times 100\%$; [b] $(k_p/k_{tot}) \times 100\%$; [c] Water flow through sediments not included in k_{load} and k_{tot}
[d] $\mathcal{M}(Lobith) = \mathcal{M}_o\ e^{-k_{load}t_o}$ with $\mathcal{M}_o = 3.3 \times 10^3$ kg, $t_o = 8$ days; [e] $C_{tmax}(Lobith) = C_{tmax}\ (t = 0)\ e^{-k_{tot}t_o}$ with $C_{tmax}(t = 0) = 100\ \mu g\ L^{-1}$

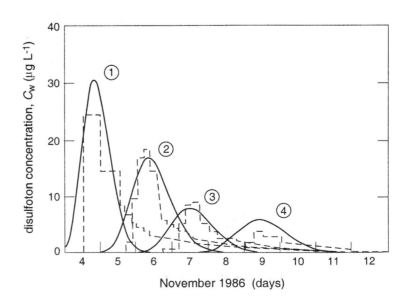

Figure 24.10 Comparison of model calculations (solid lines) by Wanner et al. (1989) with measured dissolved concentrations C_w (dashed lines) of disulfoton in the River Rhine at (1) Maxau (km 362), (2) Mainz (km 498), (3) Bad Honne (km 640), and (4) Lobith (km 865). Schweizerhalle is located at km 159. Note that C_w is approximatively equal to the total concentration, C_t. From Wanner et al. (1989).

Schweizerhalle and Lobith; 93.5% of this loss was due to biological transformation. According to the model, the maximum concentration decreased from its original value of 100 $\mu g\,L^{-1}$ to 2.7 $\mu g\,L^{-1}$ (2.7%). The drop of the maximum concentration is mainly due to dispersion ($\approx$ 36%), dilution ($\approx$ 31%), and biodegradation (31%).

Measurements taken at various locations between Schweizerhalle and Lobith are in good agreement with these calculations. In Fig. 24.10, the results of a more refined hydraulic model are compared to measured disulfoton concentrations. It is remarkable that the fairly simple linear model summarized in Table 24.4 is capable of duplicating the major results of the advanced model.

24.4 Questions and Problems

Questions

Q 24.1

What is the meaning of "stationary flow"? Can it be that under the condition of stationary flow the mean velocity $\bar{u}$ changes along the river? Is $\partial C/\partial t \neq 0$ compatible with stationary flow?

Q 24.2

Explain the difference between the Lagrangian and the Eulerian reference systems.

Q 24.3

Give a qualitative explanation for the physical meaning of the hydraulic radius R_h and why it is relevant for the mean flow velocity of a river, $\bar{u}$.

Q 24.4

Why do rivers with small α^* (Eq. 24-6) have large vertical and lateral diffusivities?

Q 24.5

Explain the physical meaning of the integrated surface area, $s(x)$, defined in Eq. 24-18.

Q 24.6

Explain the difference between longitudinal diffusion and longitudinal dispersion. Which one is more important in rivers?

Q 24.7

Equation 24-47 gives an empirical relation for the longitudinal dispersion coefficient. Explain qualitatively the role of the different parameters in this expression and why they appear in the nominator or denominator, respectively.

Q 24.8

What is the physical meaning of the time t_{skewed} of Eq. 24-53? In what respect does a pollution cloud behave differently for times $t < t_{skewed}$ than for $t > t_{skewed}$?

Q 24.9

List all processes which contribute to the decrease of the maximum concentration c
a pollutant in a river.

Q 24.10

List all processes that contribute to the decrease of the total mass of a pollutant in
river.

Q 24.11

Explain the difference between diffusive sediment–water exchange and the
sediment–water interaction caused by ripples on a sandy river bottom.

Problems

P 24.1 *Continuous Pollution of a Small River*

dichloromethylacetate
DCMA
$M_i = 143.0 \; \text{g·mol}^{-1}$

The outflow of a sewage pipe continuously adds dissolved dichloroacetic acid
methyl ester (dichloromethylacetate, DCMA) to a small river. The mean
concentration in the effluent is 400 µg L^{-1}. In the river, DCMA undergoes a base-
catalyzed reaction (see Table 13.8). At a water temperature of 15°C, k_h is 0.2 h^{-1} for
pH = 7 and 2.0 h^{-1} for pH = 8.

The air–water partition coefficient,
$K_{i \; a/w}$, of DCMA is not exactly
known, but it should be smaller
than 0.1.

River data

pH = 8
Q_{eff} = 0.05 m^{3}s^{-1} (effluent)
Q = 0.2 m^{3}s^{-1} (river)
w = 5 m
h = 0.2 m
h_o = 0.4 m
S_o = 2 × 10^{-4}

(a) Estimate the distance, x_{mix}, until the sewage effluent is completely mixed in the
river cross section.

(b) Calculate the concentration of DCMA at x_{mix}. Is the chemical degradation
relevant?

(c) Which other processes may be relevant to reduce the DCMA concentration in the
river? Illustrative Example 24.2 may help to estimate the size of one process which
may play some role.

(d) How would the above answers be changed if the pH in the river dropped to 7?

P 24.2 *Gas Exchange Experiments in a River with Different Halogenated Compounds*

In order to study the air–water exchange at river cascades, Cirpka et al. (1993)
performed an experiment in the River Glatt near Zurich (Switzerland). On January
28, 1992, five halogenated compounds (Table 24.5) with very different Henry's law
constants were injected into the river at a constant rate during 2.5 hours. To
guarantee vertical and transverse mixing, the injection site was chosen about 1 km
upstream of the first of four cascades. The concentration of each compound was
measured just upstream and downstream of each cascade.

For the following considerations one of the river sections in between the cascades,
Section # 2, is chosen. The relevant data for this section are summarized in Table
24.6. The concentrations in air, C_{ia}, of all compounds are small and can be
disregarded.

Table 24.5 Characteristic Data of Chemical Tracers Used for the Gas Exchange Experiment: Nondimensional Henry's Law Constant, $K_{i\,a/w}$, Molecular Diffusion Coefficient in Air, D_{ia}, and Water, D_{iw}, Octanol–Water Partition Coefficient, $K_{i\,ow}$, All valid for 4°C (Data from Cirpka et al., 1993)

Compound	$K_{i\,a/w}$ (–)	D_{iw} (cm^{2}s^{-1})	D_{ia} (cm^{2}s^1)	$K_{i\,ow}$ (–)
Sulfur hexafluoride (SF$_6$)	82.4	6.23×10^{-6}	0.187	< 50
1,1,2-Trichlorotrifluoroethane (R 113)	8.58	4.41×10^{-6}	0.127	100
Trichloroethene (C$_2$HCl$_3$)	0.137	5.26×10^{-6}	0.160	240
Trichloromethane (CHCl$_3$)	0.0564	5.51×10^{-6}	0.178	93.3
Tribromomethane (CHBr$_3$)	0.00717	5.04×10^{-6}	0.165	251

Table 24.6 Geometric, Hydraulic, and Chemical Data for Section #2 of the River Glatt (Data from Cirpka et al., 1993.)

River discharge	Q	33.3 m^{3}s^{-1}
Water depth	$h_o \approx h$	0.48 m
Mean flow velocity	$\bar{u}$	0.45 m s^{-1}
Distance from injection site		
Beginning of Section #2	x_o	1440 m
End of Section #2	x_1	2661 m
Difference in altitude	Δz_o	8.56 m
Width	w	16.2 m
Water temperature	T_w	4°C
Air temperature	T_a	4°C

Measured Concentration at Boundaries of Section #2

	$C_{i\,up}$ (upstream)	$C_{i\,down}$ (downstream)
SF$_6$ (ng L^{-1})	13.8 ± 0.6	7.25 ± 0.40
R 113 (µg L^{-1})	1.03 ± 0.016	0.57 ± 0.015
C$_2$HCl$_3$ (µg L^{-1})	2.36 ± 0.010	1.68 ± 0.025
CHCl$_3$ (µg L^{-1})	2.92 ± 0.033	2.26 ± 0.060
CHBr$_3$ (µg L^{-1})	4.18 ± 0.046	3.59 ± 0.073

Problem (a)

Check whether the assumption of vertical and lateral homogeneity from the point of tracer injection to the first cascade (1 km downstream) is justified. At the start of the injection, a short pulse of uranin (sodium fluorescein, a fluorescent dye) was added to the river in order to measure the travel time of the water. At each station, the samples for analyzing the halogenated compounds were taken 1.5 hours after the uranin peak had passed by. Based on this information, justify why longitudinal dispersion can be disregarded in the evaluation of the experiment.

Hints and Help: The friction factor, f^*, calculated from Eq. 24-4 may turn out to be extremely large, which would indicate a very rough riverbed. In fact, in the River Glatt in between the cascades there are small drops every 50 m. Use this fact to explain the rather extreme value of f^*.

Problem (b)

Calculate the "excess ratio," $R_i = C_{i\text{up}} / C_{i\text{down}}$, for all five compounds: (1) from the concentration measured above and below the river section (R_i is then called R_i^{meas}), and (2) from the linear air–water exchange model, Eqs. 24-21 and 24-22 (R_i is then called R_i^{model}). Compare R_i^{meas} with R_i^{model}.

Hints and Help: For the evaluation of Eq. 24-22 you need to know the total transfer velocity $v_{i\,\text{a/w}}$ for all compounds. Use Eq. 20-35 for the waterside transfer velocity, v_{iw}, and Table 20.4 with a wind speed of $u_{10} \approx 1 \text{ m s}^{-1}$ for the air-side velocity, v_{ia}, and combine these parameters with the compound-specific information given in Table 24.5.

Problem (c)

As it will turn out, all the R_i^{meas} values are systematically larger than the corresponding R_i^{model} values. This means that the mechanism of air–water exchange is significantly underestimated by the theory represented by Eq. 24-21. Consider two effects that could explain the observed discrepancy.

1. The drops in the river bed as well as surface waves significantly increase the effective area of gas exchange per unit river length relative to the geometric value, wx.

2. There exists an additional mechanism of air–water exchange which is based on the injection of air bubbles into the water, and subsequent gas exchange between bubbles and water. As shown by Cirpka et al. (1993), this effect can be quantified by the excess ratio:

$$R_i^{\text{bubbles}} = 1 + A\, K_{i\,\text{a/w}} \left(1 - e^{-B/K_{i\,\text{a/w}}}\right) \qquad (24\text{-}78)$$

where A and B are two parameters to be determined by observation. R_i^{bubbles} increases monotonically from $R_i^{\text{bubbles}} = 1$ (if $K_{i\,\text{a/w}} = 0$) to $R_i^{\text{bubbles}} = 1 + AB$ (if $K_{i\,\text{a/w}} \to \infty$).

Assume that for $CHBr_3$, which has the smallest $K_{i\,\text{a/w}}$, the discrepancy between R_i^{meas} and R_i^{mod} is completely due to the first effect. Calculate the modified exchange area $(wx)_{\text{eff}}$ and the modified R_i^{eff} for all other compounds. Calculate the remaining difference between R_i^{meas} and R_i^{eff} that may still exist for the compounds (except for $CHBr_3$), and express it as:

$$R_i^{\text{bubbles}} = \frac{R_i^{\text{meas}}}{R_i^{\text{eff}}}$$

Convince yourself that R_i^{bubbles} increases with $K_{i\,\text{a/w}}$ and estimate the fitting parameters A and B of Eq. 24-78.

Hints and Help: As shown by Cirpka et al. (1993) in greater detail, the process of bubble injection is important for rivers with steps and cascades. Note that in this process the influence of the Henry's Law constant on the overall air–water exchange flux is extended to much greater $K_{i\,a/w}$ values than is the case for the surface exchange two-film model. As shown by Fig. 20.7, for the latter $v_{i\,a/w}$ becomes independent of K_{iH}, if $K_{iH} > \approx 1$ L bar mol^{-1} (corresponding to $K_{i\,a/w} > 0.04$ for $T = 4°C$). In contrast, evaluation of the River Glatt case study along the guidelines described above shows that for the process of bubble-induced gas exchange the influence of $K_{i\,a/w}$ is extended to much larger values. However, for small $K_{i\,a/w}$ values the bubble process is not effective. Develop a qualitative argument which supports this finding.

For further information on air–water exchange at river cascades read Cirpka et al. (1993).

P 24.3 *Using a Volatile Organic Compound as Tracer for Quantifying Oxygen Demand in a Polluted River*

Problem (a)

A colleague who has to survey the water quality in River A tells you that he needs to quantify the oxygen consumption in the river downstream of a sewage discharge. He shows you the following measurements of dissolved oxygen as well as concentration data for tetrachloroethene (PCE) that is introduced into the river with a constant input by the sewage plant.

tetrachloroethene
(PCE)

River A: (rectangular cross section)

River discharge	Q	20 m³ s⁻¹
Width	w	15 m
Slope	S_o	0.002
Friction factor	f^*	0.1
Water temperature	T_w	10°C

Assumption: No other major inlets along the river section considered.

Distance from the Sewage Outlet (km)	O_2 Concentration (mg L⁻¹)	PCE Concentration (ng L⁻¹)
2.0	8.9	810
7.0	9.1	690
14.0	9.0	570
19.0	9.1	490

He recognizes that along the whole river section considered, O_2 seems to be roughly constant, but about 2.2 to 2.4 mg L⁻¹ below the atmospheric equilibrium concentration of 11.3 mg L⁻¹ (the temperature of the river water is 10°C). He also tells you that the PCE concentration in air is virtually zero. However, your colleague is unable to quantify the oxygen consumption, J (mgO₂ L⁻¹d⁻¹), in the well-mixed river. Realizing that in the river section considered there are no cascades present that could complicate calculations of gas exchange rates (see P 24.2, above), and using the information given below for river A, you sit down and, an hour later, you give your friend the answer. How large is J?

Problem (b)

Your colleague also tells you about a flood prevention project in which it is planned to double the width of river A to 30 m, and to lower the friction factor to $f^* = 0.025$. He wants to know how the concentration of O_2 and PCE would be affected in the river when assuming that everything else (including the PCE input and the oxygen consumption rate) would remain the same. Try to answer this question. Semiquantitative arguments are sufficient!

P 24.4 *Mixing and Dispersion in the Mississippi River*

The following characteristic data on the Mississippi river between Baton Rouge and New Orleans are available: Discharge $Q = 7.5 \times 10^3$ m^3 s^{-1}, mean flow velocity $\bar{u} = 0.57$ ms^{-1}, width $w = 1200$ m, mean depth $h = 11$ m, maximum depth $h_0 = 20$ m, slope of river bed $S_0 = 2 \times 10^{-5}$ m.

(a) Estimate the vertical mixing distances, x_z^{diff}. *Hint:* The given information allows to calculate the friction factor f^* and the friction velocity u^*.

(b) Estimate the lateral mixing distance, x_y^{diff}. *Hint:* Calculate first the lateral tubulent diffusivity, E_y, and then assure that mixing proceeds from the center of the river to the banks, thus use $w/2 = 600$ m in Eq. 24-39.

(c) The lateral mixing distance, x_y^{diff}, calculated above will be extremely large. However, observations confirm that x_y^{diff} is not larger than 26 km. Use $x_y^{\text{diff}} = 26$ km to recalculate the lateral diffusivity, E_y^{obs}. Calculate the longitudinal dispersion coefficient, E_{dis}, from E_y^{obs}.

Chapter 25

GROUNDWATER

25.1 Groundwater Hydraulics
Physical Characteristics
Groundwater System S
Transport by the Mean Flow and Darcy's Law
Longitudinal Dispersion
Advection versus Dispersion
Illustrative Example 25.1: *Darcy's Law in Groundwater Systems S*
Illustrative Example 25.2: *Dispersion and Advection in Groundwater System S*

25.2 Time-Dependent Input into an Aquifer (*Advanced Topic*)
The Pulse Input
The Step Input
The Sinusoidal Input with Period τ
Illustrative Example 25.3: *Infiltration of Polluted River Water into Groundwater System S*
Illustrative Example 25.4: *A Sudden Rise of the 2,4-Dinitrophenol Concentration in River R*
Illustrative Example 25.5: *Continuing the Case of 2,4-Dinitrophenol in River R*

25.3 Sorption and Transformations
Effect of Sorption (*Advanced Topic*)
Effect of Transformation Processes
The Role of Colloids in Pollutant Transport
Models and Reality
Illustrative Example 25.6: *Transport of Tetrachloroethene in Groundwater System S*
Illustrative Example 25.7: *Infiltration of Benzylchloride into Groundwater System S*

25.4 Questions and Problems

25.1 Groundwater Hydraulics

Physical Characteristics

Among the three major aqueous environments that are discussed in this book (lakes, rivers, porous media), the porous media represent the biggest challenge to those who are looking for simple principles for assessing the behavior of organic contaminants. To fall back on the simplifications that were successfully employed for lakes and rivers seems to be rather naive. Regarding the complex physical structure of soils and aquifers, it is difficult to imagine that concepts like the completely mixed box model, the one-phase (pure liquid) model, or the one-dimensional diffusion/advection approach could lead to meaningful results. Every porous system is an individual case for which general principles seem to have little value.

And yet, without the courage for the "terrible simplification" it is difficult to extract, from the information gained for one particular system, the general principles that can be used for other situations. The aim of this chapter is to demonstrate that simplifications are necessary not only because of the limits of our understanding, but also because they help to shift these limits and to gain new insights into the basic mechanisms that determine the fate of organic compounds in porous media.

Deliberately, this chapter remains simple—too simple, as some of the readers who are familiar with the subject may think. However, the advantage of the simple approach will be to demonstrate, step by step, how the various processes contribute to the "final" story and also to reveal the similarities to other environmental systems treated in this book. The chapter does not try to compete with the vast specialized literature on the physics and chemistry of porous media (e.g., Domenico and Schwartz, 1998; Hillel, 1998).

Simplicity requires choice; to choose wisely, the options should be known among which the choice is possible. The following characteristics make the porous media special:

1. There are always at least two phases involved, that is, *water* and *solids*, that are of comparable importance for the fate of organic compounds in porous media. A third phase (*air*) becomes important in the so-called unsaturated zone; recent studies show that a fourth phase, *colloids,* are important for the transport of chemicals in porous media. Furthermore, if extreme situations are analyzed, such as chemical dump-sites, phases like nonaqueous phase liquids (NAPLs) also have to be considered.

2. Compared to rivers and lakes, *transport* in porous media is generally *slow, three-dimensional,* and *spatially variable* due to heterogeneities in the medium. The velocity of transport differs by orders of magnitude among the phases of air, water, colloids, and solids. Due to the small size of the pores, transport is seldom turbulent. Molecular diffusion and dispersion along the flow are the main producers of "randomness" in the mass flux of chemical compounds.

3. Heterogeneities exist on all spatial scales, from the micropores, the larger pores and particles to the macrostructure of the aquifer (Fig. 25.1). Therefore, the description of

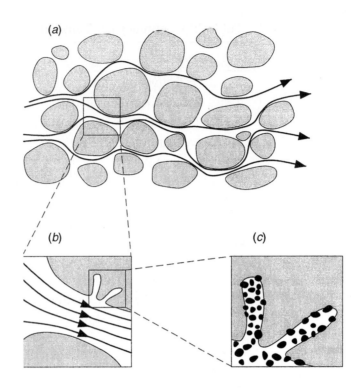

Figure 25.1 Heterogeneity is one of the main properties of porous media; it not only characterizes the scales shown in the figure, but also occurs on larger scales up to the size of the whole porous system. Three important mechanisms of transport and mixing in porous media are (*a*) interpore dispersion caused by mixing of pore channels; (*b*) intrapore dispersion caused by nonuniform velocity distribution and mixing in individual channels; (*c*) dispersion and retardation of solute transport caused by molecular diffusion between open and dead-end pores as well as between the water and the surface of the solids.

transport in porous media strongly depends on the scale of interest. For instance, in natural aquifers dispersion is much larger than in artificial porous media that are frequently used in the laboratory to study the behavior of chemicals. The reason is that in natural systems there are large-scale heterogeneities that cannot be included in laboratory systems since they are usually too small. Therefore, quantification of dispersion as deduced from measurements in laboratory systems cannot be easily extrapolated to field situations.

4. Field information on a specific porous medium remains scarce compared to its complexity. Obviously, it is much simpler to take water samples in a river or in a lake than drilling holes that are spaced closely enough in order to depict the spatial variation of the groundwater system. Ironically, in many cases it is easier to invent complex models than to collect the field data that would be needed to validate the model for a given system. Therefore, mathematical models of aquifers are important additional tools to explore the potential behavior of aquifers under different conditions.

Porous media share certain properties with lakes, others with rivers. They are three-dimensional like lakes, but nonturbulent. They have much larger solid-to-solution phase ratios, r_{sw}, than lakes; typical values lie between 1 and 10 kg L^{-1}, compared to about 1×10^{-6} kg L^{-1} for lakes. Porous media usually have a predominant direction of flow like rivers, but the flow is slower and often not confined by well-defined vertical and lateral boundaries.

Most of the sections will be restricted to the two major phases of the *saturated zone*, water and solids. The structure of this chapter is similar to that of Chapter 24 on rivers. All the equations will be written for *one dimension* only, that is, for the *x*-axis,

Table 25.1 Definition of Characteristic Parameters for Porous Media

			Typical values in GWS[a]
ϕ	(Effective) porosity ($\phi < 1$)	–	0.31
ρ_s	Density of solid material	kg m^{-3}	2500
χ	Tortuosity ($\chi > 1$)	–	1.5
K_q	Hydraulic conductivity	m s^{-1}	1×10^{-3}
k	Permeability	m^2	1×10^{-10}
h_{wt}	Height of water table (relative to some fixed depth, e.g., above sea level)	m	
x	Distance along the main flow	m	
$S_{wt} =$	$-dh_{wt}/dx$, slope of the water table	–	0.004
q	Flow of water per unit bulk area (specific discharge)	m s^{-1}	
$\bar{u}$	$= q / \phi$ effective mean flow velocity in pores along the x-axis	m s^{-1}	
ν	Kinematic viscosity of water (Appendix B) $\quad \nu(5°C) = 1.52 \times 10^{-6}$ $\quad \nu(15°C) = 1.14 \times 10^{-6}$	m^{2}s^{-1}	
Mixing parameters			
E_{dis}	Coefficient of longitudinal dispersion	m^{2}s^{-1}	$(4-40) \times 10^{-5}$
α_L	Dispersivity [b]	m	3

[a] GWS = Groundwater System S; [b] α_L depends strongly on the length scale L over which the transport of chemicals is analyzed.

which is chosen in the direction of the mean flow. This explains and (hopefully) justifies the choice of a rather special groundwater system, *Groundwater System S (GWS)*, as the main playground to exemplify the processes that will be discussed in this and the following section:

(a) Transport by mean flow

(b) Longitudinal dispersion and diffusion

(c) Sorption/desorption between water and solids

(d) Chemical and biological transformation

(e) Transport on colloids

The relevant parameters are summarized in Table 25.1.

Groundwater System S

In this chapter, all phenomena and processes that are introduced step by step will be exemplified for a specific aquifer, *Groundwater System S (GWS)*. It is assumed that the flow in GWS is one-dimensional, consisting of parallel streamlines crossing River R at a right angle (Fig. 25.2a). Water infiltrates from River R into the aquifer. Groundwater systems that are fed by a river are often used for drinking-water supply by wells located in the vicinity of the river. If the height of the water table is locally

Figure 25.2 (*a*) The streamlines in *Groundwater System S* are crossing River R at a right angle. The slope of the water table in the GWS is $S_{wt} = 0.004$. The aquifer is assumed to be homogeneous with porosity $\phi = 0.31$ and permeability $k = 1 \times 10^{-10}$ m^2. Water is infiltrating from the river into the aquifer. (*b*) Pumping of water from a single well located at distance $x_w = 30$ m from the river causes a local distortion of the streamlines in the groundwater flow. (*c*) Pumping from a number of wells, located along the river at distance $x_w = 30$ m, each $d = 3$ m apart from its neighbors, keeps the flow field quasi-linear.

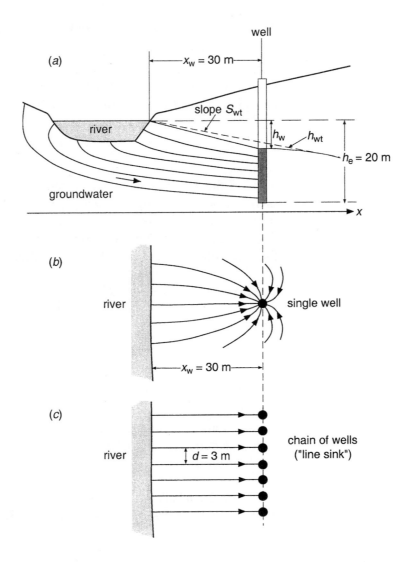

reduced by pumping water from a single well, the groundwater flow is distorted. Only part of the pumped water originates from the river. Due to the local perturbation of the hydraulic gradient some water flows from the opposite side into the well (see Fig. 25.2*b*). Finally, the well is also fed by water that crosses the riverbed at greater depth and is not directly influenced by the river water.

An adequate quantitative description of such a situation requires a two- or even three-dimensional approach. Today, a great variety of numerical models are available that allow us to solve such models almost routinely. However, from a didactic point of view numerical models are less suitable as illustrative examples than equations that can still be solved analytically. Therefore, an alternative approach is chosen. In order to keep the flow field quasi-one-dimensional, the single well is replaced by a dense array of wells located along the river at a fixed distance x_w (Fig. 25.2*c*). Ultimately, the set of wells can be looked at as a *line sink*. This is certainly not the usual method to exploit aquifers! Nonetheless, from a qualitative point of view a single well has properties very similar to the line sink.

Transport by the Mean Flow and Darcy's Law

In order to describe the flow of water through a porous medium, some basic geometric characteristics have to be discussed. In the saturated zone a given bulk volume V_{tot} (e.g., 1 m^3) consists of the volume V_w filled with water and the volume $V_s = V_{\text{tot}} - V_w$ occupied by solids. The porosity ϕ is defined by (see Chapter 9)

$$\phi = \frac{V_w}{V_{\text{tot}}} = \frac{V_w}{V_w + V_s} \qquad (9\text{-}13)$$

Most of the pore volume is connected so that water can flow through. However, there are pores that are completely cut off from the main pore volume or that are dead ends that are barely participating in the flow of the water (Fig. 25.1c). In the following, the parameter ϕ always refers to the so-called *effective porosity*, that is, to the pore volume that is available for the bulk flow.

The flow rate of water through the porous medium per unit total (bulk) area perpendicular to the direction of flow, the so-called *specific discharge q*, is related to the *effective* mean flow velocity in the pores along the x-axis, $\bar{u}$, by

$$q = \phi \bar{u} \qquad (\text{m}^3\text{s}^{-1}\text{m}^{-2} = \text{m s}^{-1}) \qquad (25\text{-}1)$$

Note that q has the units of a velocity and is always smaller than the effective velocity in the pores, $\bar{u}$, since $\phi < 1$. It is assumed that the x-axis of the (local) coordinate system is always pointing in the direction of the mean flow. Since the mean flow usually does not follow a straight line, the x-axis does not necessarily point in a fixed direction either.

The flux per unit bulk area and time of a chemical compound along the x-axis due to the mean flow follows from a modified version of Eq. 22-2 (see Table 25.1 for definitions):

$$F_{\text{flow}} = \phi \bar{u} C_w = q C_w \qquad (25\text{-}2)$$

where C_w is the dissolved mass of the chemical per unit pore water volume. Here we assume that the pore area available for the flow in the x-direction is equal to the (effective) porosity ϕ (isotropic pore structure). It is easy to imagine situations for which this is not the case, for instance, a porous medium with pores which are all aligned along one direction. Obviously, in spite of a porosity $\phi > 0$, this medium would have zero transport perpendicular to the pores.

Using Gauss' theorem (Eq. 18-12) for the concentration per unit bulk volume, ϕC_w, yields:

$$\frac{\partial}{\partial t}(\phi C_w) = \phi \frac{\partial C_w}{\partial t} = -\frac{\partial F_{\text{flow}}}{\partial x} = -\phi \bar{u} \frac{\partial C_w}{\partial x} \qquad (25\text{-}3)$$

provided that ϕ and $\bar{u}$ are constant. Thus:

$$\left(\frac{\partial C_w}{\partial t} \right)_{\text{flow}} = -\bar{u} \frac{\partial C_w}{\partial x} \qquad (25\text{-}4)$$

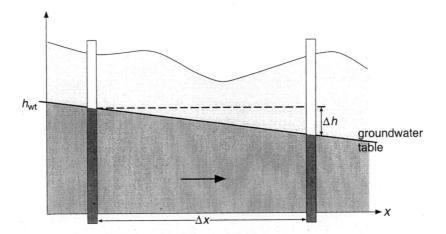

Figure 25.3 The hydraulic gradient, defined as the slope of the (unconfined) groundwater table, $S_{wt} = \Delta h_{wt}/\Delta x$, is a measure for the horizontal pressure gradient that drives the flow through a porous medium from high to low pressure.

This expression describes the fastest and most important mode of transport in groundwater. In fact, an important task of the hydrologist is to develop models to predict the effective velocity $\bar{u}$ (or the specific flow rate q). Like the Darcy-Weisbach equation for rivers (Eq. 24-4), for this purpose there is an important equation for groundwater flow, *Darcy's Law*. In its original version, formulated by Darcy in 1856, the equation describes the one-dimensional flow through a vertical filter column. The characteristic properties of the column (i.e., of the aquifer) are described by the so-called hydraulic conductivity, K_q (units: m s^{-1}). Based on Darcy's Law, Dupuit derived an approximate equation for quasi-horizontal flow:

$$q = -K_q \frac{dh_{wt}}{dx} \quad (\text{m s}^{-1}) \quad (25\text{-}5)$$

where dh_{wt}/dx is the hydraulic gradient, that is, the slope of the height of the (unconfined) water table along the x-axis (Fig. 25.3). In fact, from a physical point of view it is the horizontal *pressure* gradient which drives the flow, yet in hydrology this gradient is commonly expressed by the slope of the water table, since it can be easily determined by comparing the water levels in adjacent wells located along the direction of the mean flow.

In the literature the hydraulic conductivity is commonly designed by K; here K_q is used to distinguish it from the various partition coefficients, K_d, K_{om}, etc.

K_q depends not only on the structure of the porous medium (i.e., porosity, tortuosity, and grain size distribution), but also on the viscosity of the liquid that flows through the pores, and finally on the acceleration of gravity, g. In contrast, the *permeability*, k,

$$k = \frac{\nu}{g} K_q \quad (\text{m}^2) \quad (25\text{-}6)$$

is independent of gravity and of the viscosity of the liquid and thus yields a better characterization of the physical structure of the aquifer than K_q. Water at $T = 20°C$ has a kinematic viscosity $\nu = 10^{-6}$ m^{2}s^{-1} (Appendix B, Table B.3). With $g \approx 10$ m s^{-2}, k and K_q are numerically related by $k(\text{m}^2) \approx 10^{-7}(\text{m s}) \times K_q (\text{m s}^{-1})$.

grain radius r

grain radius $2r$

Figure 25.4 A simplified groundwater matrix consisting of spherical grains of equal radius r serves to demonstrate why the hydraulic conductivity (i.e., the flow per unit area for constant $\Delta h_{wt}/\Delta x$) is proportional to r^2. Note the following proportionalities:

(1) Pore size $\sim r^2$

(2) Flow per pore $Q(r) \sim r^4$
    (Hagen-Poiseuille)

(3) Number of pores per unit area
    $N(r) \sim r^{-2}$

(4) Total flow per unit area
    $q = Q(r)\, N(r) \sim r^2$

As mentioned above, k depends only on the geometry and structure of the porous medium. Several investigations demonstrate that for the case of perfect spheres of uniform size, k increases as the square of the particle radius, r:

$$k = a\, r^2 \qquad (25\text{-}7)$$

where a is an empirical constant. Figure 25.4 illustrates how the r^2 dependence of k can be understood based on the law of *Hagen-Poiseuille,* stating that the (laminar) flow through a tube with radius r is proportional to r^4 provided that all other parameters (pressure, viscosity of liquid, length of tube) are kept constant.

If r is defined as the mean particle radius, Eq. 25-7 also can be used for aquifers that consist of particles of different size. Then the parameter a depends on the particle size distribution. It increases with porosity ϕ, which in turn is linked to the so-called sorting coefficient, So (see Fig. 25.5). Small So values indicate greater uniformity of the particles; large So values indicate a greater variance of the particle size, that is, a denser packing (small particles fill the space in between the larger ones). Therefore,

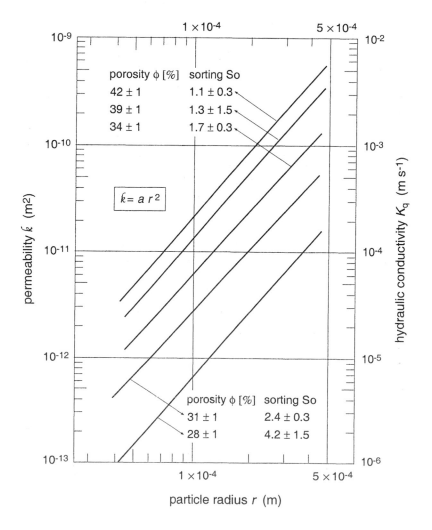

Figure 25.5 Permeability k as a function of (mean) particle radius r for different aquifer porosity ϕ, which, in turn, depends on the sorting coefficient So $= (r_{75}/r_{25})^{1/2}$, where r_{75} and r_{25} characterize the particle radii larger than, respectively, 75% and 25% of the radii of all the aquifer particles. The hydraulic conductivity K_q (right scale) refers to water at 20°C. Redrawn from Lerman (1979).

porosity ϕ and permeability k decrease with increasing sorting coefficient, So. More sophisticated theories include additional characteristics of the particle size distribution (see Freeze and Cherry, 1979). A first application of Darcy's Law is given in Illustrative Example 25.1.

Longitudinal Dispersion

In porous media the flow of water and the transport of solutes is complex and three-dimensional on all scales (Fig. 25.1). A one-dimensional description needs an empirical correction that takes account of the three-dimensional structure of the flow. Due to the different length and irregular shape of the individual pore channels, the flow time between two (macroscopically separated) locations varies from one channel to another. As discussed for rivers (Section 24.2), this causes dispersion, the so-called *interpore dispersion*. In addition, the nonuniform velocity distribution within individual channels is responsible for *intrapore dispersion*. Finally, molecular diffusion along the direction of the main flow also contributes to the longitudinal dispersion/ diffusion process. For simplicity, transversal diffusion (as discussed for rivers) is not considered here. The discussion is limited to the one-dimensional linear case for which simple calculations without sophisticated computer programs are possible.

By analogy to the description of dispersion in rivers, the dispersive flux relative to the mean flow, F_{dis}, can be described by an equation of the First Fickian Law type (see Eq. 24-42):

$$F_{dis} = -\phi\, E_{dis}\, \frac{\partial C_w}{\partial x} \qquad (25\text{-}8)$$

where E_{dis} is the coefficient of longitudinal dispersion and C_w is the (mean) concentration in the pore water. This flux leads to an additional term in Eq. 25-4:

$$\left(\frac{\partial C_w}{\partial t} \right)_{trans} = \left(\frac{\partial C_w}{\partial t} \right)_{flow} + \left(\frac{\partial C_w}{\partial t} \right)_{dis} = -\bar{u}\, \frac{\partial C_w}{\partial x} + \frac{\partial}{\partial x}\left(E_{dis}\, \frac{\partial C_w}{\partial x} \right) \qquad (25\text{-}9)$$

which for constant E_{dis} becomes

$$\left(\frac{\partial C_w}{\partial t} \right)_{trans} = -\bar{u}\, \frac{\partial C_w}{\partial x} + E_{dis}\, \frac{\partial^2 C_w}{\partial x^2} \qquad (25\text{-}10)$$

E_{dis} combines the effects of the different processes causing dispersion. For the case of inter/intrapore dispersion it is usually written as (Freeze and Cherry, 1979):

$$E_{dis} = \frac{D_{iw}}{\tau} + \alpha_L\, \bar{u} \qquad (25\text{-}11)$$

D_{iw} is the molecular diffusion coefficient of the chemical in water, τ is tortuosity, and α_L is the (longitudinal) dispersivity (dimension: L). The first term describes molecular diffusion in a porous medium (Eq. 18-57), the second the effect of dispersion (Eq. 22-52). Typical values of the dispersivity α_L for field systems with flow distances of up to about 100 m lie between 1 and 100 m. Since α_L depends strongly on the scale of the aquifer, test columns used in the laboratory have much smaller α_L values, which are between 0.001 and 0.01 m. Such values should not be extrapolated to the field. Brusseau (1993) summarizes the information regarding the dependence of α_L on the characteristic structure of the aquifer.

It appears that in most natural groundwater systems the influence of molecular diffusivity on dispersion is negligible, unless flow in the aquifer is extremely small or even absent. Thus, Eq. 25-11 is often used in the simplified form:

$$E_{dis} = \alpha_L \, \bar{u} \qquad (25\text{-}12)$$

An application of Eqs. 25-11 and 25-12 is given in Illustrative Example 25.2.

Advection versus Dispersion

The relative importance of advective versus dispersive transport for a given distance L can be expressed by the nondimensional *Peclet Number* (Eq. 22-11a):

$$\text{Pe} = \frac{L\,\bar{u}}{E_{dis}} \qquad (25\text{-}13)$$

With Eq. 25-12 this reduces to:

$$\text{Pe} = \frac{L}{\alpha_L} \qquad (25\text{-}14)$$

The ratio between transport distance and dispersivity determines whether transport is dominated by dispersion ($L \ll \alpha_L$, Pe $\ll$ 1) or by advection ($L \gg \alpha_L$, Pe $\gg$ 1). Since the scale of the heterogeneities grows with the size of the system, α_L increases with L, as well. Thus, Pe does not necessarily increase linearly with distance L. In Illustrative Example 25.2 we calculate the dispersion coefficient E_{dis} for Groundwater System S and compare transport by dispersion and by advection, respectively. As indicated by Eq. 25-14, the relative importance of dispersion vs. advection (that is, the Peclet Number) is independent of the flow regime, provided that molecular diffusion can be neglected in Eq. 25-11. In other words: If the flow q is accelerated by increasing the pumping rate, then dispersion increases roughly proportionally and Pe remains constant.

(Text continues on page 1160)

Illustrative Example 25.1   **Darcy's Law in Groundwater System S**

(a) Problem

Consider the River R, from which water (at $T = 20°C$) is infiltrating through a saturated zone into Groundwater System S (GWS) (Fig. 25.2). The flow in the GWS crosses the river at a right angle. The slope of the (undisturbed) water table is $S_{wt} = -dh_{wt}/dx = 0.004$ (Fig. 25.2a). The aquifer is assumed to be homogeneous with porosity $\phi = 0.31$. Its permeability k is 1×10^{-10} m^2. Calculate the specific discharge of the groundwater system.

Answer (a)

Inserting the permeability $k = 1 \times 10^{-10}$ m^2 and $g/\nu \approx 10^7$ m^{-1}s^{-1} into Eq. 25-6 yields the hydraulic conductivity $K_q = (g/\nu)\, k \approx 1 \times 10^{-3}$ m s^{-1}. Thus, the specific discharge calculated from Eq. 25-5 is:

$$q = 1 \times 10^{-3}\, \mathrm{m\,s^{-1}}\,(-0.004) = 4 \times 10^{-6}\, \mathrm{m\,s^{-1}}$$

According to the orientation of the x-axis (see Fig. 25.2a), the gradient dh_{wt}/dx is negative; thus, q is positive, indicating a flow along the positive x-axis. In the following discussion this situation is called the *Natural Flow Regime*.

Note: A real aquifer is usually not homogeneous. Often the permeability at the transition from the river into the aquifer is reduced due to a process called *colmatation*. As a result, the slope of the water table is significantly larger at the point of infiltration than farther away from the river. For simplification, such effects are disregarded.

(b) Problem

At the distance $x_w = 30$ m from the River R, parallel to the riverbank, there is an array of wells regularly spaced with distance $d = 3$ m in between the wells (Fig. 25.2c). The water level in the wells lies 60 cm below the water surface of the river ($h_w = 0.6$ m). The lower end of the wells is located 20 m below the river water surface ($h_e = 20$ m). Estimate how much water is pumped from each well.

Hints and Help

The geometry of the flow modified by the wells is a superposition of linear and radial flow and thus not ideal for an analytical discussion. As explained above, a one-dimensional approximation is justified, nonetheless, provided that the distance in between the wells is small compared to x_w. As a first approximation it can then be assumed that each well captures the water from the groundwater that flows through a section with width equal to the spacing of the wells, d. Furthermore, the mean slope of the water table that is drawing water into the well can be approximated by h_w/x_w.

Answer (b)

The area from which water is captured by each well is $a_x = (h_e - h_w)\, d = 19.4\ \mathrm{m} \times 3\ \mathrm{m} = 58$ m^2. According to the approximate one-dimensional model described above, the flow per unit area to the wells is (Fig. 25.2b):

$$q = -1 \times 10^{-3}\,\mathrm{ms^{-1}}\left(-\frac{0.6\ \mathrm{m}}{30\ \mathrm{m}}\right) = 2 \times 10^{-5}\,\mathrm{ms^{-1}}$$

Thus, the pump rate in each well, Q_{well}, is:

$$Q_{\mathrm{well}} = q\,a_x = 2 \times 10^{-5}\,\mathrm{m\,s^{-1}} \times 58\ \mathrm{m}^2$$
$$= 1.16 \times 10^{-3}\ \mathrm{m^3s^{-1}} \approx 1.2\ \mathrm{L\,s^{-1}}$$

In the following discussion this situation is called the *Pump Regime I*.

(c) Problem

The pump rate of the individual pumps is increased to $Q_{well} = 2.5$ L s^{-1}. How big is h_w, once a new steady-state is reached?

Answer (c)

Intuitively, it is clear that the water level in the well drops until the slope of the surface of the water table, h_w/x_w, becomes large enough in order to produce a flow into the well that equals Q_{well}. This flow can be expressed by:

$$Q_{well} = K_q \frac{h_w}{x_w} d(h_e - h_w) = \gamma h_w (h_e - h_w)$$

with $\gamma = K_q d/x_w = 1 \times 10^{-4}$ m s^{-1}. Solving the quadratic equation for h_w yields:

$$h_w = \frac{1}{2}\left(h_e \pm \left[h_e^2 - 4 Q_{well}/\gamma \right]^{1/2} \right)$$

Inserting $h_e = 20$ m, $Q_{well} = 0.0025$ m^{3}s^{-1}, yields the two solutions $h_w = 1.3$ m and $h_w = 18.7$ m. The first value is the one that the system will attain if the pump rate is continuously increased to 2.5 L s^{-1}. This situation will be called *Pump Regime II*. The second solution, although formally correct, is physically unrealistic; it yields the same Q_{well}, in spite of the much steeper slope of the water table, since the thickness of the layer reaching the well is smaller. However, the physical conditions underlying the validity of Eq. 25-5 are no longer fulfilled since at such large hydraulic gradients the flow breaks off and becomes unsaturated. To summarize, the regimes are characterized as follows:

Pump Regime I $Q_{well} = 1.2$ L s^{-1}, $h_w = 0.6$ m

Pump Regime II $Q_{well} = 2.5$ L s^{-1}, $h_w = 1.3$ m

(d) Problem

A cloud of a nonreactive (conservative) pollutant is transported along the river. How much time does it take for the pollutant to reach the wells? Discuss all three regimes (i.e., the Natural Regime as well as the Pump Regimes I and II).

Answer (d)

According to Eq. 25-1, the effective flow velocity along the x-axis is:

$$\bar{u} = \frac{q}{\phi} = K_q \frac{h_w}{x_w} \frac{1}{\phi}$$

Since the travel time of a pollutant to the wells is $t_w = x_w/\bar{u}$, the following results are obtained:

Specific Discharge q, Effective Mean Flow Velocity $\bar{u}$, and Travel Time t_w for Different Flow Regimes in Groundwater System S

Regime	q (m s^{-1}) ^{a}	$\bar{u}$ (m s^{-1}) ^{b}	$\bar{u}$ (m d^{-1})	t_w (days)
Natural	4×10^{-6}	1.3×10^{-5}	1.1	27
I	2×10^{-5}	6.5×10^{-5}	5.6	5.4
II	4.5×10^{-6}	14.5×10^{-5}	12.5	2.4

^{a} With $K_q = 1 \times 10^{-3}$ m s^{-1}. ^{b} With $\phi = 0.31$.

Note: The travel time calculated by this simple model is not quite correct. More realistic results are obtained by including longitudinal dispersion as discussed in the next section.

Illustrative Example 25.2

Dispersion and Advection in Groundwater System S

(a) Problem

Calculate the dispersion coefficient, E_{dis}, for 2,4-dinitrophenol (2,4-DNP) at 15°C in Groundwater System S (GWS, Fig. 25.2) for the three flow regimes given in Illustrative Example 25.1. Dispersivity is $\alpha_L = 3$m, tortuosity is $\tau = 1.5$. The pH of the groundwater is 7.5.

Answer (a)

The pK_a of 2,4-DNP is 3.94 (Schwarzenbach et al., 1988). Thus, at pH = 7.5, more than 99.99% of the chemical is present in its anionic form, and sorption can be neglected. When further assuming that 2,4-DNP does not undergo any transformation, the compound can be assumed to exhibit a conservative behavior in the aquifer.

2,4-dinitrophenol
(2,4-DNP)
$M_i = 184.1$ g·mol^{-1}

The molecular diffusion coefficient in water of 2,4-DNP is estimated from $D_{iw}(O_2)$ using Eq. 18-55. At 25°C, $D_{iw}(O_2) = 2.1 \times 10^{-5}$ cm^{2}s^{-1}. At 15°C, $D_{iw}(O_2)$ is slightly smaller: 2.0×10^{-5} cm^{2}s^{-1}. (Note: Box 18.4 helps in understanding the temperature dependence of molecular diffusivity.) Thus,

$$D_{iw}(2,4\text{-DNP, }15°C) = \left(\frac{M_i(O_2)}{M_i(2,4\text{-DNP})} \right)^{1/2} D_{iw}(O_2, 15°C)$$

$$= \left(\frac{32}{184} \right)^{1/2} \times 2.0 \times 10^{-5} \text{cm}^2\text{s}^{-1} = 8.3 \times 10^{-6} \text{cm}^2\text{s}^{-1} = 8.3 \times 10^{-10} \text{m}^2\text{s}^{-1}$$

From Eq. 25-11 for the Natural Regime ($\bar{u} = 1.1$ m d^{-1} = 1.3×10^{-5} m s^{-1}) one obtains:

$$E_{dis}(2,4\text{-DNP, }15°C) = \frac{8.3 \times 10^{-10} \text{m s}^{-1}}{1.5} + 3 \text{ m} \times 1.3 \times 10^{-5} \text{m s}^{-1}$$

$$= 5.5 \times 10^{-10} \text{m}^2\text{s}^{-1} + 3.9 \times 10^{-5} \text{m}^2\text{s}^{-1} \approx 3.9 \times 10^{-5} \text{m}^2\text{s}^{-1}$$

It appears that molecular diffusivity is negligible. The following table summarizes the values for all regimes.

Regime	$\bar{u}\ (\text{m s}^{-1})$	$E_{\text{dis}}\ (\text{m}^2\text{s}^{-1})$
Natural Regime	1.3×10^{-5}	3.9×10^{-5}
Pump Regime I	6.5×10^{-5}	2.0×10^{-4}
Pump Regime II	14.5×10^{-5}	4.4×10^{-4}

(b) Problem

Assess the relative importance of dispersion versus advection in the GWS for 2,4-dinitrophenol infiltrating from the river and detected in one of the wells. Discuss all three regimes.

Answer (b)

Note that if E_{dis} is approximated by Eq. 25-12, the Peclet Number, Pe, does not depend on the flow regime. From Eq. 25-14 with $x_o = x_w = 30$ m:

$$\text{Pe} = \frac{30\ \text{m}}{3\ \text{m}} = 10 \gg 1$$

Thus, for the transport from the river to the well, advection is more important than dispersion.

Advanced Topic   25.2   ## Time-Dependent Input into an Aquifer

Aquifers which are closely connected to a river (as for the case of Groundwater System S) may be influenced by abrupt input variations that are driven by the corresponding concentration changes in the river. In order to analyze the resulting concentrations in the aquifer, the solution of the transport equation (Eq. 25-10) will be discussed for different time-dependent input concentrations at $x = 0$, $C_{\text{in}}(t)$. In the river typical time scales of change are of the order of minutes (in case of an accidental spill) to days, but seasonal variations also exist for some chemicals. In contrast, transport within the aquifer is slower than most riverine concentration variations. The question arises how the river dynamics are transmitted to the groundwater. Three different cases are discussed:

1. The pulse input

2. The step input

3. The fluctuating (sinusoidal) input with period T

Although the partial differential equation Eq. 25-10 is linear and looks rather simple, explicit analytical solutions can be derived only for special cases. They are characterized by the size of certain nondimensional numbers that completely determine the shape of the solutions in space and time. A reference distance x_o and a reference time t_o are chosen that are linked by:

$$t_o = \frac{x_o}{\bar{u}} \tag{25-15}$$

t_o is the time needed for the water in the aquifer to flow with the effective mean velocity $\bar{u}$ from the point of infiltration ($x = 0$) to x_o. The following nondimensional coordinates are introduced:

$$\xi = \frac{x}{x_o}, \qquad \theta = \frac{t}{t_o} = \frac{\bar{u}}{x_o} t \qquad\qquad (25\text{-}16)$$

By using the transformation rules:

$$\frac{\partial}{\partial t} = \frac{\partial}{\partial \theta} \frac{d\theta}{dt} = \frac{1}{t_o} \frac{\partial}{\partial \theta} \qquad\qquad (25\text{-}17a)$$

$$\frac{\partial}{\partial x} = \frac{\partial}{\partial \xi} \frac{d\xi}{dx} = \frac{1}{x_o} \frac{\partial}{\partial \xi} \qquad\qquad (25\text{-}17b)$$

Eq. 25-10 can be transformed into:

$$\left(\frac{\partial C_w}{\partial \theta} \right)_{trans} = -\frac{\partial C_w}{\partial \xi} + \frac{1}{Pe} \frac{\partial^2 C_w}{\partial \xi^2} \qquad\qquad (25\text{-}18)$$

where we have replaced L by x_o in the definition of the Peclet Number, Pe (Eq. 25-13). Given the geometry of the system, the solutions of Eq. 25-10, if expressed in terms of the relative distance, ξ, and the relative time, θ, are thus completely determined by the Peclet Number.

The Pulse Input

Consider the case of a pollution cloud in the river passing by the infiltration location during time Δt, which shall be very short compared to the time t_w needed for the ground-water to travel from the river to the wells. During the event, the concentration in the river is C_{in}; before and after the event, the riverine concentration is approximately zero.

Note: As discussed in Chapter 24, a pollution cloud caused by an accidental spill and traveling along a river often has the shape of a normal distribution (see Fig. 24.7*b*). In order to keep the following considerations simple, it is assumed that the variance of the cloud in the river is still small at the time when infiltration takes place. Otherwise, the following considerations would have to be modified in a similar way as explained in Eqs. 24-55 and 24-56.

The total mass input by infiltration into the aquifer per unit area perpendicular to the flow is (see Eq. 25-1):

$$\mathcal{M} = \Delta t\, q\, C_{in} = \Delta t\, \bar{u}\, \phi\, C_{in} \quad (\text{mol m}^{-2}) \qquad\qquad (25\text{-}19)$$

The fate of the pollutant moving in the aquifer along the streamlines is determined by the advection-dispersion equation, Eq. 25-10 or 25-18. For Pe $\gg$ 1, that is, for locations $x \gg E_{dis} / \bar{u}$, the concentration cloud can be envisioned to originate from an infinitely short input at $x = 0$ of total mass $\mathcal{M}$ (a so-called δ input) that by dispersion is turned into a normal distribution function along the x-axis with growing standard deviation. Since the arrival of the main pollution cloud at some distance x is determined by the effective flow velocity, $\bar{u}$, the symmetry point (or maximum value) of the normal distribution moves as $\bar{u}\,t$. Thus, the solution is identical to the one derived for dispersion in rivers (Eq. 24-51 and Fig. 24.6):

$$C_w(x,t) = \frac{M/\phi}{2(\pi E_{dis}\, t)^{1/2}} \exp\left(-\frac{(x-\bar{u}t)^2}{4\,E_{dis}\,t}\right) \qquad (25\text{-}20)$$

The integral over x of this expression is constant (mass conservation), and its time-dependent standard deviation is (see Eq. 22-49):

$$\sigma(t) = (2\,E_{dis}\,t)^{1/2} \qquad (25\text{-}21)$$

Its maximum moves along x at the effective mean flow velocity, $\bar{u}$ (Eq. 24-49):

$$x_m = \bar{u}\,t \qquad (25\text{-}22)$$

Obviously, everything that has been said in Chapter 24 on dispersion in rivers also applies to the one-dimensional flow in aquifers.

Remember that from the nondimensional version of the advection-dispersion equation, Eq. 25-18, the Peclet Number Pe was identified as the only parameter that determines the shape of the concentration distribution in the aquifer. By introducing relative coordinates for space (ξ) and time (θ) as defined in Eq. 25-16, Eq. 25-20 takes the form:

$$C_w(x,t) = C_w(\xi,\theta) = \frac{\hat{C}_w}{(2\pi)^{1/2}\, s} \exp\left(-\frac{(\xi-\theta)^2}{2s^2}\right) \qquad (25\text{-}23)$$

where:

$$s = \left(\frac{2\theta}{\text{Pe}}\right)^{1/2} \qquad \text{(nondimensional standard deviation)} \qquad (25\text{-}24)$$

$$\hat{C}_w = \frac{M}{x_o\,\phi} = \frac{\Delta t\,\bar{u}}{x_o}\,C_{in} \qquad (25\text{-}25)$$

Note that Eq. 25-23 describes a normal distribution with maximum at $\xi = \theta$:

$$C_{max} = \frac{\hat{C}_w}{(2\pi)^{1/2}\, s} = \frac{1}{2}\left(\frac{\text{Pe}}{\pi\theta}\right)^{1/2}\hat{C}_w \qquad (25\text{-}26)$$

or:

$$\frac{C_{max}}{C_{in}} = \frac{1}{2}\left(\frac{\Delta t\,\bar{u}\,\phi}{x_o}\right)\left(\frac{\text{Pe}}{\pi\theta}\right)^{1/2} \qquad (25\text{-}27)$$

A first application is discussed in Illustrative Example 25.3 (*see page 1166*).

The Step Input

Next, we considered the case in which at time $t = 0$ the concentration of a chemical in the infiltrating river water suddenly changes from C_o to C_1 and then remains at C_1 (Fig. 25.6a). Since in this context only concentration *changes* are relevant, it can be assumed that the initial concentration C_o is 0. This situation is called a step input.

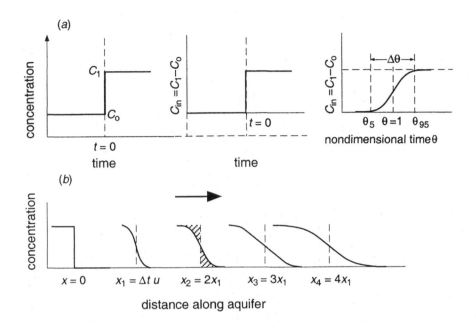

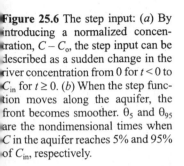

Figure 25.6 The step input: (*a*) By introducing a normalized concentration, $C - C_o$, the step input can be described as a sudden change in the river concentration from 0 for $t < 0$ to C_{in} for $t \geq 0$. (*b*) When the step function moves along the aquifer, the front becomes smoother. θ_5 and θ_{95} are the nondimensional times when C in the aquifer reaches 5% and 95% of C_{in}, respectively.

As for the pulse input, the evolving concentration, $C_w(x,t)$, can be envisioned as the result of two simultaneous processes: (1) advective transport along x at effective flow velocity $\bar{u}$ and (2) smoothing by dispersion relative to the moving front, $x = \bar{u}\,t$. The front is a diffusive boundary (Section 19.4). The concentration relative to its mean position, $x = \bar{u}\,t$, can be described by Eq. 19-52. Note that in Eq. 19-52 x is the distance relative to the moving front whereas in Eq. 25-28 we replace it by $x = \bar{u}\,t$ where now x is a fixed coordinate along the flow:

$$C_w(x,t) = \frac{C_{in}}{2}\,\mathrm{erfc}\!\left(\frac{x - \bar{u}\,t}{2(E_{dis}t)^{1/2}}\right) \tag{25-28}$$

where erfc is the complement error function (Eq. 18-21 and Table A.2 of Appendix A).

In order to discuss the properties of this solution, again it is helpful to transform Eq. 25-28 into a nondimensional form by choosing a reference distance x_o and applying Eq. 25-16:

$$\frac{C_w(x,t)}{C_{in}} = \frac{1}{2}\,\mathrm{erfc}\!\left(\frac{\xi - \theta}{2(\theta / \mathrm{Pe})^{1/2}}\right) \tag{25-29}$$

The temporal change of $C_w(x,t)$ at a fixed location, say at $x = x_o$ (i.e., $\xi = 1$), can be qualitatively described by distinguishing three phases:

1. The actual time t is still much smaller than t_o, the time needed for the front to reach x_o by traveling at the effective flow velocity $\bar{u}$; the argument of the error function,

$$y = \frac{x - \bar{u}\,t}{2(E_{dis}\,t)^{1/2}} = \frac{\xi - \theta}{2(\theta / \mathrm{Pe})^{1/2}} \tag{25-30}$$

is still much larger than 1, i.e., $C_w(x,t) \approx 0$.

2. The central part of the front passes by x_o (i.e., $y \approx 0$ and $\mathrm{erfc}(y) \approx 1$, i.e., $C_w(x,t) \approx C_{in}/2$)

3. The actual time t is much larger than t_o (i.e., the main portion of the front has passed x_o). Then $y \rightarrow -\infty$ (i.e., $\mathrm{erfc}(y) \approx 2$ and $C_w \approx C_{in}$).

To quantify the time interval during which the front is moving past x_o, we choose the two times t_1 and t_2 when $C_w(x,t)$ reaches 5% and 95% of C_{in}, respectively (Fig. 25.6a). Thus, we are looking for the critical values y_5 and y_{95} for which:

$$\mathrm{erfc}(y_5) = 0.1, \qquad \mathrm{erfc}(y_{95}) = 1.9 \qquad (25\text{-}31)$$

Note that $y_5 = -y_{95}$. From Table A.2 of Appendix A we find by interpolation $y_5 = 1.16$ and $y_{95} = -1.16$. Solving Eq. 25-30 for θ with $y = \pm 1.16$ and $\xi = 1$ yields the corresponding (nondimensional) times, θ_5 and θ_{95}, which mark the passing of the concentration front. The duration of the front passage, $\Delta\theta = \theta_{95} - \theta_5$, decreases with the Peclet Number, Pe. Some values are given in Table 25.2. The model is applied in Illustrative Example 25.4 (*see page 1168*).

The Sinusoidal Input with Period τ

Often the concentration of a chemical substance in a river shows periodic fluctuations with typical periods of a day (diurnal fluctuations), a week, or even a year (annual fluctuations). In Chapter 21 we have discussed the response of a completely mixed box to a periodic input (see Box 21.3). Here the situation is more complex since the fluctuations propagate into the aquifer by diffusion as well as by advection. In addition to the Peclet Number, the period τ determines how far the fluctuations are felt within the aquifer and how much the phase shifts relative to the phase of the "driving" river concentration. Remember that the angular frequency ω is related to the period τ by $\omega = 2\pi/\tau$. If the concentration in the river varies as:

$$C_{in}(t) = C_o + C_1 \sin(\omega t) \,, \qquad (25\text{-}32)$$

then the solution of Eq. 25-10 can be written in the form:

$$C(x,t) = C_o + e^{-\alpha x} C_1 \sin(\omega t - \beta x) \qquad (25\text{-}33)$$

The coefficients α and β, which depend on the size of $\bar{u}$, E_{dis}, and ω, describe the attenuation of the fluctuations in the aquifer and the corresponding phase shift, respectively. Note the similarity with Eq. 7 of Box 21.3, which describes the periodic disturbance of a box. In Eq. 25-33 the phase shift (βx) and the attenuation of the concentration amplitude ($e^{-\alpha x}$) both depend on the distance from the input, x. The nondimensional version of Eq. 25-33 is:

$$\frac{C(\xi, \theta)}{C_o} = 1 + \frac{C_1}{C_o} e^{-\hat{\alpha}\xi} \sin(\Omega\theta - \hat{\beta}\xi) \qquad (25\text{-}34)$$

with $\hat{\alpha} = \alpha x_o$, $\hat{\beta} = \beta x_o$. The nondimensional space and time coordinates, ξ and θ, are defined in Eq. 25-16, and the nondimensional angular frequency is:

$$\Omega = \frac{\omega x_o}{\bar{u}} \qquad (25\text{-}35)$$

Table 25.2 Nondimensional Time Interval of the Passage of a Step Front at $\xi = 1$ for Different Peclet Numbers, Pe

θ_5 and θ_{95} are the nondimensional times when the concentration reaches 5% and 95% of C_{in}, respectively:

$$\theta_{5/95} = \frac{1}{2}\left[2+\frac{5.382}{\text{Pe}}\pm\left\{\left(2+\frac{5.382}{\text{Pe}}\right)^2-4\right\}^{1/2}\right]$$

Pe	θ_5	θ_{95}	$\Delta\theta = \theta_{95} - \theta_5$
1 [a]	0.14	7.2	7
3	0.28	3.5	3.2
10	0.49	2.1	1.6
30	0.657	1.522	0.87
100	0.793	1.261	0.47
300	0.874	1.143	0.269
10'00	0.929	1.076	0.147
100'00	0.977	1.024	0.047

[a] Note that for the case of fixed boundary conditions at $x = 0$, Eq. 25-29 is only an approximate solution of the differential equation (25-10). The approximation is not appropriate if Pe ≤ 1.

The nondimensional attenuation coefficient, $\hat{\alpha}$, and the phase shift coefficient, $\hat{\beta}$, are (Roberts and Valocchi, 1981):

$$\hat{\alpha} = \frac{\text{Pe}}{2}\left\{\left[\frac{a+1}{2}\right]^{1/2}-1\right\} \tag{25-36a}$$

$$\hat{\beta} = \frac{\text{Pe}}{2}\left[\frac{a-1}{2}\right]^{1/2} \tag{25-36b}$$

with:

$$a = \left\{1+\left(\frac{4\Omega}{\text{Pe}}\right)^2\right\}^{1/2} \tag{25-36c}$$

Figure 25.7 shows how $\hat{\alpha}$ and $\hat{\beta}$ vary with Pe and Ω. If Pe and Ω are extremely different, then Eq. 25-36 can be approximated by:

$$\Omega \ll \text{Pe:} \quad \hat{\alpha} = \frac{\Omega^2}{\text{Pe}}, \quad \hat{\beta} = \Omega \tag{25-37a}$$

$$\Omega \gg \text{Pe:} \quad \hat{\alpha} = \hat{\beta} = \left(\frac{\text{Pe }\Omega}{2}\right)^{1/2} \tag{25-37b}$$

Since the sinus has a period of $2\pi = 6.28$, for a given (nondimensional) distance, ξ, the time lag of oscillation can be expressed as fraction of the period τ,

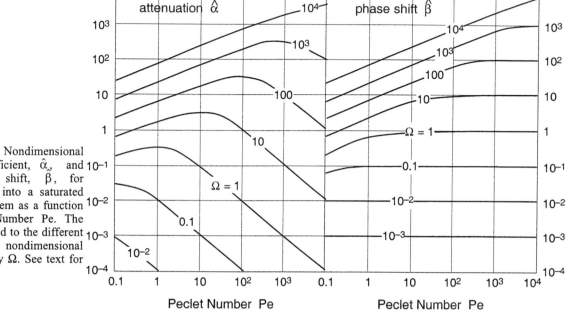

Figure 25.7 Nondimensional attenuation coefficient, $\hat{\alpha}_z$, and specific phase shift, $\hat{\beta}$, for sinusoidal input into a saturated groundwater system as a function of the Peclet Number Pe. The numbers attributed to the different lines give the nondimensional angular frequency Ω. See text for definitions.

or as absolute time:

$$\hat{\Delta}_\tau = -\frac{\hat{\beta}\xi}{2\pi} \qquad (25\text{-}38a)$$

$$\Delta_\tau = \hat{\Delta}_\tau \, \tau \qquad (25\text{-}38b)$$

In Illustrative Example 25.5 we continue our discussion on the infiltration of 2,4-dinitrophenol into an aquifer.

(Text continues on page 1170)

Illustrative Example 25.3

Infiltration of Polluted River Water into Groundwater System S

(a) Problem

In River R a pollution cloud of 2,4-dinitrophenol (see Illustrative Example 25.2) of duration $\Delta t = 1$ h and concentration $C_{in} = 50$ ng L^{-1} is passing by Groundwater System S. Calculate the maximum concentration reached at the wells for the three regimes. Compare these values to the maximum concentrations reached 3 m away from the river.

Answer

It is convenient to choose $x_o = x_w = 30$ m as the point of reference, although the following calculation does not depend on this choice. Thus Eq. 25-23 has to be solved for $\xi = 1$, where $x = x_o$. The maximum concentration is approximately reached when $\theta = \xi = 1$. (Note: It is true that $C_w(x,t)$ as a function of x for *a fixed time* t has its maximum at $x_m = \bar{u} \, t$. In contrast, the maximum of $C_w(x,t)$ as a function of t for a *fixed location* x is only approximately reached at time $t_m = x / \bar{u}$, but for Pe ≥ 3 the error involved is small and will be disregarded.) Since the Peclet Number Pe = 10 is independent of the pump regime, C_{max} depends on the pump regime only via the dependence of $\hat{C}$ on $\bar{u}$ (Eq.25-25). For instance, for the Natural Regime with $\bar{u} = 1.3 \times 10^{-5}$ m s^{-1}:

$$\hat{C} = \frac{3600 \text{ s} \times 1.3 \times 10^{-5} \text{ m s}^{-1}}{30 \text{ m}} \times 50 \text{ ng L}^{-1} = 7.8 \times 10^{-2} \text{ng L}^{-1}$$

The nondimensional standard deviation s follows from Eq. 25-24:

$$s_{30} = \left(\frac{2}{10}\right)^{1/2} = 0.45 \qquad (\text{at } x_w = 30 \text{ m})$$

At $x = 3$ m, which is 10% along the stretch from the river to the wells, the nondimensional coordinates which characterize the transition of the maximum concentration are $\xi = \theta = 0.1$. Thus:

$$s_3 = \left(\frac{2 \times 0.1}{10}\right)^{1/2} = 0.14 \qquad (\text{at } x = 3 \text{ m})$$

The maximum concentrations are calculated from Eq. 25-26:

Regime	$\hat{C}$ (ng L^{-1})	C_{max} (ng L^{-1}) [a] x = 30 m	C_{max} (ng L^{-1}) [a] x = 3 m
Natural	0.078	0.069	0.22
I	0.39	0.35	1.10
II	0.90	0.80	2.5

[a] With $C_{in} = 50$ ng L^{-1}

C_{max} increases with increasing pumping intensity. At $x = 3$ m, C_{max} reaches 2.5 ng L^{-1} (Regime II), but this is still only 5% of the concentration in the river ($C_{in} = 50$ ng L^{-1}). Since dilution of the infiltrating river water with uncontaminated groundwater is disregarded, the real concentrations are even smaller.

Remember that according to the approximation made for the dispersion coefficient, E_{dis} (Eq. 25-12), the Peclet Number does not depend on $\bar{u}$. Therefore, the influence of the pump regime on the concentration at the well is *not* caused by a change of the transport conditions in the aquifer (e.g., flow time t_w), but simply by the fact that more dinitrophenol enters the groundwater during the passage of the pollution cloud if $\bar{u}$ is large.

Note: The fact that $x_w = 30$ m was chosen as the reference scale to transform Eq. 25-20 into a nondimensional form (Eq. 25-23) may cause some confusion. The reader may have been tempted to change the reference point in order to calculate the concentration at $x = 3$ m. Yet, once x_o is chosen, the concentration can be calculated for any distance by using the relative scales, ξ and θ. In fact, any other choice would have been appropriate. Particularly, one could also use $x_o = 3$ m. Then, the Peclet Number would decrease by a factor of 10 (Pe = 1, see Eq. 25-14). Also, the relative time for the passing of the concentration maximum at $x = 3$ m would now be $\theta = 1$. Combining these changes into Eq. 25-24 increases s by a factor of 10 (θ increases by 10, Pe decreases by 10, and the square root leaves a factor of 10 in the nominator). This change is compensated for by a corresponding tenfold increase of $\hat{C}$ (Eq. 25-25) caused by the change of x_o from 30 m to 3 m. Thus, $C(x,t)$ of Eq. 25-23 and C_{max} (Eq. 25-26) remain unchanged, as should be the case.

Illustrative Example 25.4

A Sudden Rise of the 2,4-Dinitrophenol Concentration in River R

Problem

On April 5, at noon, the 2,4-dinitrophenol concentration in River R at Groundwater System S (GWS, see Illustrative Example 25.1 and 25.2) suddenly increases from 0 to 50 ng L^{-1} and then remains constant. At what time does the concentration in the wells of the GWS reach 25 ng L^{-1} and 47.5 ng L^{-1}, respectively? Calculate the time for all three flow regimes. When are these concentrations reached 3 m from the river if no water is pumped from the wells (Natural Regime)?

Answer

Again, $x_w = 30$ m is chosen as the point of reference, x_o. The concentration 25 ng L^{-1} (this is half of C_{in}) is reached at $\theta = 1$ (i.e., at time $t_{50} = t_w = x_w / \bar{u}$). The corresponding times for all regimes are summarized in the table at the end of Illustrative Example 25.1.

Since Pe = 10 is constant for all flow regimes, θ_{95} can be directly taken from Table 25.2: $\theta_{95} = 2.1$; thus $\tau_{95} = 2.1\, t_w = 2.1\, x_w / \bar{u}$. The results are summarized below.

Regime	Time Since April 5		Absolute Time	
	t_{50} (days)	t_{95} (days)	t_{50}	t_{95}
Natural	27	57	May 2	June 1
I	5.4	11.3	April 10, at night	April 16
II	2.4	5.0	April 7, at night	April 10

There are two ways to calculate θ_{50} and θ_{95} for the location $x = 3$ m. One way is to solve Eq. 25-30 for $\xi = 3$ m / 30 m = 0.1 and $y = \pm 1.16$. An alternative possibility is to choose $x_o = 3$ m as the new reference point. Then Eq. 25-14 yields Pe = 3m / 3m = 1 and Table 25.2 yields $\theta_{95} = 7.2$. Since the mean flow time $t_o = 3$m/1.1 m d^{-1} = 2.7 d, it follows:

$$t_{95} = 7.2 \times 2.7\ \text{d} = 20\ \text{d} \quad \text{(April 25)}$$

Illustrative Example 25.5

Continuing the Case of 2,4-Dinitrophenol in River R

Problem

Continuous measurements of 2,4-dinitrophenol in River R (see Illustrative Examples 25.1 to 25.4) passing by the GWS show a superposition of sinusoidal diurnal and annual concentration variations (Eq. 25-33). Calculate how much of this variation is observed at one of the wells and determine the phase shift (in days) between the oscillations in the river and in the well. Include into the analysis also a monthly oscillation ($\tau = 30$ days). Make all calculations for the three flow regimes.

Answer

Remember that the Peclet Number does not depend on the flow regime (Pe = 10). In contrast, the nondimensional angular frequency, Ω, is proportional to $(\bar{u})^{-1}$ (Eq. 25-35).

The point of reference is chosen as $x_o = 30$ m. The angular frequencies for the diurnal, monthly, and annual variations are, respectively:

$$\omega_1 = \frac{2\pi}{1d} = 6.28 \text{ d}^{-1}, \quad \omega_{30} = 0.209 \text{ d}^{-1}, \quad \omega_{365} = 0.0172 \text{ d}^{-1}.$$

As an example, Ω is calculated from Eq. 25-35 for the Natural Regime and the diurnal input variation:

$$\Omega = \frac{30 \text{ m} \times 6.28 \text{ d}^{-1}}{1.1 \text{ m d}^{-1}} = 171$$

The following table summarizes the characteristic numbers for all cases. The mean flow velocities, $\bar{u}$, are taken from Illustrative Example 25.1.

Regime	Pe	Nondimensional Angular Frequency, Ω		
		Diurnal	Monthly	Annual
Natural	10	171	5.7	0.47
Pump Regime I	10	34	1.1	0.092
Pump Regime II	10	15	0.50	0.041

Next, $\hat{\alpha}$ and $\hat{\beta}$ are calculated from Eq. 25-36. For instance, the daily variation in the Pump Regime II yields the following coefficients (use the table above: Pe = 10, $\Omega = 15$, i.e., $(4\Omega/\text{Pe})^2 = 36$, thus $a = \sqrt{37} = 6.1$):

$$\hat{\alpha} = \frac{10}{2} \left\{ \left[\frac{6.1+1}{2} \right]^{1/2} - 1 \right\} = 4.4$$

$$\hat{\beta} = \frac{10}{2} \left\{ \left[\frac{6.1+1}{2} \right]^{1/2} \right\} = 8.0$$

Since $x_o = x_w = 30$ m was chosen as the reference point, Eq. 25-34 has to be evaluated at $\xi = 1$. Thus, the sinusoidal variation at the well is attenuated by the factor:

$$\exp(-\hat{\alpha}\,\xi) = \exp(-4.4 \times 1) = 0.012, \quad \tau = 1 \text{ day}, \quad \text{Regime II}$$

that is, to 1.2% of the full amplitude C_1 of Eq. 25-32. The phase shift is (Eq. 25-38):

$$\hat{\Delta}_\tau = -\frac{\hat{\beta}\,\xi}{2\pi} = -\frac{8.0 \times 1}{6.28} = -1.27, \quad \Delta_\tau = -1.27 \text{ d} \quad \text{for} \quad \tau = 1 \text{ day}, \quad \text{Regime II}$$

Note that Δ_τ is less than the mean advective traveling time from the river to the well, $t_w = 2.4$ d (for Pump Regime II). The table below summarizes the relevant parameters for all periods and pump regimes.

	Natural Regime	Pump Regime I	Pump Regime II
Attenuation, $e^{-\hat{\alpha}\xi}$, (at $\xi = 1$)			
Period [a] day	2.4×10^{-11}	2.0×10^{-4}	1.2×10^{-2}
month	0.20	0.89	0.98
year	0.98	1.00	1.00
Time lag of oscillation, Δ_τ (days), Eq. 25-38b			
Period [a] day	-4.6	-2.0	-1.27
month	-21	-5.1	-2.4
year	-27	-5.3	-2.4
For comparison			
advective flow time, t_w (days)	27	5.4	2.4

[a] Corresponds to $\omega_1 = 6.28$ d^{-1}, $\omega_{30} = 0.209$ d^{-1}, $\omega_{365} = 0.0172$ d^{-1}.

The following features can be extracted from these results:

1. The attenuation strongly decreases with increasing length of the characteristic period τ. Particularly, for periods τ that are smaller than the mean advective flow time t_w, the signal of the concentration variation is practically absent at x_w.

2. The attenuation decreases from the Natural Regime to Pump Regime II. This is consistent with (1), since the mean advective flow time t_w is smallest for Pump Regime II. A short advective flow time increases the change of the oscillations to reach the well.

3. The time lag of the oscillation increases with increasing period τ and approaches (but does not surpass) the advective flow time, t_w.

25.3   Sorption and Transformations

Effect of Sorption *(Advanced Topic)*

Porous media have much larger solid-to-solution phase ratios (r_{sw}) than surface waters (lakes and rivers). Therefore, even the transport of a chemical with moderate to small solid–water distribution ratios (K_d) may be influenced by sorption processes. The basic mathematical tools which are needed to quantify the effect of sorption on transport are described in Section 18.4 and summarized in Box 18.5.

Although the case of finite sorption/desorption dynamics has earned much research interest (see, e.g., Brusseau, 1992), the following discussion will be restricted to the case of a linear and instantaneous local equilibrium between the compound sorbed on solids, C_s (mol kg^{-1}), and the dissolved compound concentration, C_w (mol m^{-3}):

$$K_d = \frac{C_s}{C_w} \qquad (\text{m}^3\text{kg}^{-1}) \qquad (9\text{-}7)$$

For porous media it is convenient to choose C_w as the key variable since it is this concentration that is determined in filtered water samples taken at a well. The dynamic equation for C_w is derived from Eq. 25-10 according to the same procedure as in Box 18.5. Note that now both transport by advection and by diffusion are reduced by the relative fraction in dissolved form, f_w:

$$\frac{\partial C_w}{\partial t} = -f_w \bar{u} \, \frac{\partial C_w}{\partial x} + f_w E_{dis} \, \frac{\partial^2 C_w}{\partial x^2} \qquad (25\text{-}39)$$

where from Eqs. 9-12 and 9-15 we have:

$$f_w = \left(1 + K_d \, \rho_s \, \frac{1-\phi}{\phi} \right)^{-1} \qquad (25\text{-}40)$$

For a very weakly sorbing chemical, f_w is equal to 1 ($K_d \approx 0$) and smaller than 1 otherwise.

If f_w does not change along x, that is, if the aquifer matrix is homogeneous, then the transformation

$$E_{dis}^* = f_w E_{dis}, \qquad \bar{u}^* = f_w \bar{u} \qquad (25\text{-}41)$$

brings Eq. 25-39 back into the familiar form of the advection-diffusion equation of a nonsorbing compound (Eq. 25-10). Note that the Peclet Number Pe (Eq. 25-13) is not altered by Eq. 25-41; a sorbing species has the same Pe value as water. Thus, the solutions of Eq. 25-10 derived in Section 25.2 for the three different input scenarios (peak, step, sinusoidal) are also valid for sorbing chemicals, provided that E_{dis} and $\bar{u}$ are replaced by E_{dis}^* and $\bar{u}^*$, respectively. The consequences of this substitution for the behavior of a chemical in groundwater are discussed separately for the three cases.

1. *Pulse input*: The nondimensional solution, Eq. 25-23, remains valid if the following modified definition is used:

$$t_o \rightarrow t_o^* = \frac{x_o}{\bar{u}f_w} = \frac{t_o}{f_w} \qquad (25\text{-}42)$$

Note that the relative time, $\theta = t/t_o$, for the peak concentration to reach a given location, is the same for all chemicals, although t and t_o depend on f_w. Thus the growth of the nondimensional standard deviation s with the nondimensional time θ (Eq. 25-24) is independent of sorption. All the compounds have the same nondimensional standard deviation, s, when they pass by some fixed location x. Of course, the less sorbing substances arrive sooner at x than the more sorbing ones. In contrast, an observer measuring C_w as a function of time at a *fixed location* x, for instance, at the well, has a different view: since the concentration clouds are passing by the point of observation with their individual velocity, $f_w \bar{u}$, the $C_w(t)$ curve appears as a broader signal for the slow (sorbing) compound than for the fast (nonsorbing) one.

Liquid chromatography can be used to picture the transport of sorbing chemicals in an aquifer. If at a given time the concentration distributions of various compounds along the column of an ideal system could instantaneously be measured, then their variance, σ^2, would be linearly related to the distance that they have traveled. When the concentration cloud reaches the detector at the end of the column, the concentration is determined as a function of time. Thus, the early peaks are narrower than the later ones.

Sorption also affects the absolute value of C_w, particularly C_{max}. Given the total mass input, $\mathcal{M}$ (Eq. 25-25), only the fraction $f_w \mathcal{M}$ is dissolved; thus, all concentrations in the dissolved phase are reduced by the factor f_w.

2. *Step input*: All the conclusions drawn for the pulse input can directly be transferred to the step input. The concentrations, if expressed for the nondimensional coordinates ξ and θ, are not affected by sorption; the shape of the concentration curve along x for a fixed time is independent of sorption. Yet, when the front passes by a fixed location x, the time needed for the concentration to increase from, say, 5% to 95% of the maximum concentration, grows as $(f_w)^{-1}$. This can be directly deduced from Table 25.2, where the duration of the passage of the front is quantified by the nondimensional time interval $\Delta\theta = \theta_{95} - \theta_5$. This value does not depend on sorption, but after transformation back into real time it does (see Eq. 25-42):

$$\Delta t^* = t_o^* \Delta\theta = t_o \Delta\theta / f_w = \Delta t / f_w \qquad (25\text{-}43)$$

Of course the absolute time for the front to reach a given location increases as $(f_w)^{-1}$.

(3) *Sinusoidal input*: All properties of the sinusoidal input scenario can be expressed by the two nondimensional numbers, Pe and Ω. Whereas Pe is not affected by sorption, Ω is modified (Eq. 25-35):

$$\Omega \to \Omega^* = \frac{\omega \, x_o}{\bar{u} \, f_w} = \Omega / f_w \qquad (25\text{-}44)$$

Thus, sorption affects $C(\xi,\theta)$ in the same way as if the angular frequency ω were increased by the factor $(f_w)^{-1}$ or the period τ were reduced by f_w. As shown in Fig. 25.7, such a change causes both the attenuation $\hat{\alpha}$ and the specific phase shift $\hat{\beta}$ to grow. In Illustrative Example 25.6 we look at the influence of sorption on the transport of tetrachloroethene (PCE) in an aquifer.

Effect of Transformation Processes

As a last step, a first-order (linear) reaction is added to the advective-diffusive equation of a sorbing substance, Eq. 25-39:

$$\frac{\partial C_w}{\partial t} = \bar{u} f_w \frac{\partial C_w}{\partial x} + E_{dis} f_w \frac{\partial^2 C_w}{\partial x^2} - k_{r,w} C_w - \frac{1 - f_w}{f_w} k_{r,s} C_w \qquad (25\text{-}45)$$

where $k_{r,w}$ and $k_{r,s}$ are the first-order reaction rates for the dissolved and sorbed phase, respectively. This equation can also be written as:

$$\frac{\partial C_w}{\partial t} = \bar{u}^* \frac{\partial C_w}{\partial x} + E_{dis}^* \frac{\partial^2 C_w}{\partial x^2} - k_r^* C_w \qquad (25\text{-}46)$$

where $\bar{u}^*, E_{dis}^*$ were defined in Eq. 25-41, and with the apparent first-order rate constant:

$$k_r^* = k_{r,w} + \frac{1 - f_w}{f_w} k_{r,s} \qquad (25\text{-}47)$$

Steady-state solutions of Eq. 25-46 are discussed in Box 22.1. The solution depends on two nondimensional numbers, on the Peclet Number Pe (Eq. 22-11a) and on the Damköhler Number (Eq. 22-11b):

$$Da = \frac{E_{dis}^* k_r^*}{(\bar{u}^*)^2} \qquad (25\text{-}48)$$

In nondimensional form, the steady-state version of Eq. 25-46 is:

$$\frac{d^2 C_w}{d\xi^2} - Pe \frac{dC_w}{d\xi} - Pe\, Da^2\, C_w = 0 \qquad (25\text{-}49)$$

It has the solution (see Eq. 22-10 and Box 22.1 with $J = 0$):

$$C_w(\xi) = A_1 e^{\lambda_1^* \xi} + A_1 e^{\lambda_2^* \xi}; \qquad \xi = \frac{x}{x_o} \qquad (25\text{-}50)$$

The coefficients A_1 and A_2 are fixed by two boundary conditions. The nondimensional eigenvalues, λ_α^*, are (Box 22.1, Eq. 10):

$$\lambda_\alpha^* = \frac{Pe}{2} \left[sgn(\bar{u}^*) \pm \{1 + 4Da\}^{1/2} \right] , \quad \alpha = 1,2 \qquad (25\text{-}51)$$

Figure 22.3 shows schematically the shape of the solution $C_w(\xi)$ for different values of Pe and Da. In Illustrative Example 25.7 we discuss the infiltration of benzylchloride (BC) into an aquifer; BC is hydrolyzed in the aqueous phase.

Solutions of Eq. 25-46 for *nonsteady-state conditions* are difficult to obtain analytically, yet numerical procedures are straightforward. For the case of slow reactions, more precisely for Da << 1, the solution can be approximated by multiplying the time-dependent solution for a conservative substance with the exponential factor $exp(-k_r^* t)$. For instance, for the pulse input Eq. 25-20 is modified into:

$$C(x,t) \approx \frac{(M/\phi) e^{-k_r^* t}}{2(\pi E_{dis}^* t)^{1/2}} \exp\left(-\frac{(x - \bar{u}^* t)^2}{4 E_{dis}^* t} \right), \qquad Da \ll 1 \qquad (25\text{-}52)$$

and correspondingly for Eq. 25-28 (step input) and Eq. 25-33 (sinusoidal input).

(see Illustrative Example 25.7 page 1177)

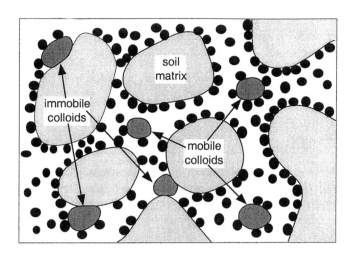

Figure 25.8 Conceptual view of a saturated porous medium with colloids. The black dots represent the contaminant that is dissolved, sorbed on soil matrix particles, or sorbed on colloids. The colloids themselves can sorb on the soil matrix. Adapted from Corapcioglu and Jiang (1993).

The Role of Colloids in Pollutant Transport

The influence of colloids on the transport of chemicals in groundwater has recently become an important issue in groundwater research. Colloids can be important for the transport of contaminants; they may also influence remediation efforts of contaminated aquifers. A review of the topic is given by Corapcioglu and Jiang (1993).

It would lie far beyond the aim of this chapter to introduce the state-of-the art concepts that have been developed to quantify the influence of colloids on transport and reaction of chemicals in an aquifer. Instead, a few effects will be discussed on a purely qualitative level. In general, the presence of colloidal particles, like dissolved organic matter (DOM), enhances the transport of chemicals in groundwater. Figure 25.8 gives a conceptual view of the relevant interaction mechanisms of colloids in saturated porous media. A simple model consists of just three phases, the dissolved (aqueous) phase, the colloid (carrier) phase, and the solid matrix (stationary) phase. The distribution of a chemical between the phases can be, as first step, described by an equilibrium relation as introduced in Section 23.2 to discuss the effect of colloids on the fate of polychlorinated biphenyls (PCBs) in Lake Superior (see Table 23.5).

For the case of the PCBs it has been shown that sorption on colloids enhances the "dissolved" concentration which is usually determined on filtered samples in which most colloids are still present. Since the colloidal fraction contributes neither to air/water exchange nor to sedimentation (only the large particles, not the colloids, sink to the bottom), the effect of the colloids can result in an increase of the residence time of the chemical in the lake.

In contrast, transport of chemicals through groundwater systems is generally enhanced by colloids. Though colloids may become immobile due to attachment to the solid matrix, part of them are moving with the water. Thus, chemicals that are sorbed on colloids and not on the solid matrix are not (or not as strongly) retarded as one would calculate by using the relative dissolved fraction, f_w.

An additional process that enhances transport is due to the typical velocity distribution within the pores of the aquifer. Since colloids mainly move in the central parts of the larger pores, their effective flow velocity may be larger than $\bar{u}$. Remember that $\bar{u}$ represents an average of all the velocities in the aquifer. Colloids have the tendency to select the faster streamlines. This phenomenon is called hydrodynamic chromatography (Small, 1974; Stoistis et al. 1976).

Even a simple mathematical model for transport on colloids in an aquifer must include dynamic equations for the dissolved phase and for the colloids. The latter equation describes the migration, immobilization, and detachment of the colloids. More sophisticated models include dynamic equations for sorption and desorption of the chemical onto colloids and the stationary solid phase.

Models and Reality

The aim of this chapter was to present some of the basic mechanisms that determine the fate of organic pollutants in porous media. In order to achieve a general understanding, simple models were presented that have explicit mathematical solutions.

The price that is paid for this kind of simplicity is obvious to the specialist. There are numerous reasons why real systems do not behave as described above. However, one can understand the complex case only once the simple one has been analyzed.

Brusseau (1994) gives an overview on the real behavior of reactive contaminants in heterogeneous porous media. The reader can also find a large updated list of references on transport in porous media. Just a few phenomena are shortly mentioned here:

1. The presence of macropores as well as of dead end zones causes a real breakthrough curve to look quite differently from that given by Eq. 25-28. Real breakthrough curves may rise earlier (early breakthrough) and smear out later. It may take longer to reach the full concentration since the exchange of water and chemicals is slow between the macropores and the pores which are less well connected to the main flow.

2. Sorption kinetics: Sorption and desorption may not always be fast compared to other processes, such as advection and dispersion. In addition, the sorption equilibrium does not necessarily follow a linear relationship.

3. Transport in unsaturated aquifers, especially in their gaseous phase, may lead to rather complicated situations.

4. Kinetics of microbial growth: In the case of a sudden contamination of an aquifer by a chemical compound, the buildup of the appropriate population of microorganisms able to decompose the compound may be a slow and complicated process. Such situations have to be simultaneously analyzed by field observations and models, but the latter may become rather complicated.

5. The heterogeneity of porous media with respect to their hydraulic permeability poses one of the most difficult problems. This is especially true for aquifers formed by glacial and fluvial deposits. Prediction of breakthrough curves may become impossible if a few long macropores or highly conducting layers are present in which water moves at a speed 10 or 100 times faster than the effective mean velocity. Such situations are still full of surprises, even to the specialist.

Illustrative Example 25.6   **Transport of Tetrachloroethene in Groundwater System S**

(a) Problem

Calculate the retardation factor, f_w, for tetrachloroethene (PCE, see Box 21.2 and Illustrative Example 21.5) in Groundwater System S. The fraction of organic carbon in the solid aquifer material is 0.8‰; the density of the solids is $\rho_s = 2.5$ g cm^{-3}.

Cl Cl
Cl Cl
tetrachloroethene
PCE
M_i = 165.8 g·mol^{-1}

Answer

The K_d value of PCE is determined by the fraction of organic carbon in the solid aquifer material, $f_{oc} = 0.0008$, and the corresponding distribution coefficient, K_{oc} (Eq. 9-22):

$$K_d = f_{oc} K_{oc}$$

Estimate K_{oc} from the octanol–water partition constant, K_{ow}, using the linear free energy relationship (Eq. 9-26f), with log $K_{ow} = 2.88$ (Appendix C):

$$\log K_{oc} = 0.96 \log K_{ow} - 0.23 = 2.53$$

thus, $K_{oc} = 340$ L kg$_{oc}^{-1}$ and $K_d = 0.27$ L kg^{-1}. Since the empirical relations given for K_d and K_{oc} in Chapter 9 refer to values expressed in units of L kg^{-1}, the solid density ρ_s has to be expressed in units of kg L^{-1} = g cm^{-3}. From Eq. 25-40, the dissolved fraction of PCE is ($\phi = 0.31$):

$$f_w = \left(1 + 0.27 \text{ L kg}^{-1} \times 2.5 \text{ kg L}^{-1} \times \frac{0.69}{0.31}\right)^{-1} = 0.40$$

(b) Problem

Calculate the *annual* sinusoidal variation of the PCE concentration in the wells of Groundwater System S relative to the variation in River R. Compare this number with the relative variation of a nonsorbing chemical such as 2,4-dinitrophenol (see Illustrative Example 25.5). Determine the time lag of oscillation in the well relative to the variation in the river. Use all three flow regimes of Illustrative Example 25.1.

Answer

With the dissolved fraction of PCE, $f_w = 0.40$, the Attenuation Number Ω^* for the annual variations becomes (see Eq. 25-44 and table in Illustrative Example 25.5): $\Omega_{NR}^* = 1.2$, $\Omega_I^* = 0.23$, $\Omega_{II}^* = 0.10$, respectively. Attenuation and retardation are calculated from Eq. 25-36 as explained for the case of the nonsorbing solute. The result is summarized in the table below. It demonstrates that the attenuation of PCE is not affected by the rather weak sorption of PCE. In contrast, the time lag strongly increases and in all cases reaches nearly the advective flow time.

Measurements in an aquifer at the River Aare in Switzerland by Schwarzenbach et al. (1983) show a time lag of about 4 months between the concentration in the river and in a well situated about 13 m from the river. The corresponding amplitude ratio is about 0.8. In the same aquifer, the annual temperature variation shows a shift of 1 month from the river to the observation well. This demonstrates the more conservative (or less "sorbing") nature of water temperature versus PCE.

Attenuation and Retardation of an *Annual* Input Signal of PCE with $f_w = 0.40$ at the Wells of the Groundwater System

	Natural Regime	Pump Regime I	Pump Regime II
Attenuation, $e^{-\hat{\alpha}\xi}$ (at $\xi = 1$)			
nonsorbing[a]	0.98	1.00	1.00
PCE	0.87	0.99	1.00
Time lag of oscillation, Δ_τ (days), Eq. 25-38b			
nonsorbing[a]	−27	−5.4	−2.4
PCE	−68	−13	−5.8
Advective flow time (days)			
nonsorbing[b]	27	5.4	2.4
sorbing	68	14	6.0

[a] From Illustrative Example 25.5. [b] From Illustrative Example 25.1.

Illustrative Example 25.7

Infiltration of Benzylchloride into Groundwater System S

Problem

The concentration of benzylchloride in River R has a constant value of 3 µg L^{-1}. When the infiltrating river water is flowing through the GWS, benzyl chloride is hydrolyzed and also sorbs to the particles of the aquifer. The rate of hydrolysis is assumed to be the same for the dissolved and the sorbed phase. Possible biodegradation of benzylchloride is disregarded. Calculate for the Natural Regime the concentration of benzylchloride at $x = 3$ m and at the wells ($x = 30$ m, see Fig. 25.2). The water temperature is 10°C. Assume that the rate of hydrolysis at 10°C is four times smaller than at 25°C. The octanol–water partition coefficient of benzylchloride is log $K_{ow} = 2.30$ (Hansch and Leo, 1979). Use $\phi = 0.31$, $\rho_s = 2.5$ g cm^{-3}, and $f_{oc} = 0.0008$ kg$_{oc}$kg$_s^{-1}$ to characterize the solid phase of the aquifer (Table 25.1).

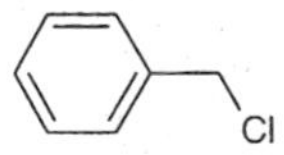

benzylchloride
(BC)
$M_i = 126.6$ g·mol^{-1}

Hints and Help: One of the boundary conditions can be chosen at $x = 0$: $C_w = C_{River}$. Due to hydrolysis and biodegradation, benzylchloride eventually disappears completely from the groundwater. Therefore, it can be assumed that $C_w = 0$ for $x \to \infty$. This serves to formulate the second boundary condition.

Answer

The half-life of benzylchloride with respect to hydrolysis at 25°C is 15 hours (see Table 13.6). Thus, at 10°C the first-order rate constant is:

$$k_r = k_{r,w} = k_{r,s} = \frac{\ln 2}{4 \times 15 \text{ h}} = \frac{0.693}{60 \text{ h}} = 0.012 \text{ h}^{-1} = 3.2 \times 10^{-6} \text{s}^{-1}.$$

Use Eq. 9-26a to calculate K_{oc}:

$$\log K_{oc} = 0.74 \times 2.30 + 0.15 = 1.85 \ (\text{L kg}^{-1})$$

thus $K_{oc} = 70 \ \text{L kg}^{-1}$ and $K_d = 0.0008 \times 70 \ \text{L kg}^{-1} = 0.056 \ \text{L kg}^{-1}$. The dissolved fraction of benzylchloride is calculated from Eq. 25-40:

$$f_w = \left(1 + 0.056 \ \text{L kg}^{-1} \times 2.5 \ \text{kg L}^{-1} \times \frac{0.69}{0.31}\right)^{-1} = 0.76 \ .$$

The relevant coefficients of Eq. 25-46 for the Natural Regime are (see Illustrative Example 25.2):

$$\bar{u}^* = 0.76 \times 1.3 \times 10^{-5} \ \text{m s}^{-1} = 0.99 \times 10^{-5} \ \text{m s}^{-1}$$

$$E_{dis}^* = 0.76 \times 3.9 \times 10^{-5} \ \text{m}^2 \text{s}^{-1} \approx 3.0 \times 10^{-5} \ \text{m}^2 \text{s}^{-1}$$

$$k_r^* = 4.2 \times 10^{-6} \ \text{s}^{-1} \ .$$

Note that for the special case $k_{r,w} = k_{r,s}$, k_r^* is $k_{r,w}/f_w$. The extra factor $(f_w)^{-1}$ compensates for the fact that in Eq. 25-46 k_r^* is multiplied only with the dissolved BC concentration, whereas here we assume that the hydrolysis affects the total (dissolved and sorbed) BC concentration. The Peclet Number (Eq. 25-13, with $x_o = 30$ m) is affected neither by sorption nor by reaction: Pe = 10. The Damköhler Number is (Eq. 25-48):

$$\text{Da} = \frac{3.0 \times 10^{-5} \times 4.2 \times 10^{-6}}{(0.99 \times 10^{-5})^2} = 1.3$$

The eigenvalues, λ_α^*, are (Eq. 25-51):

$$\lambda_\alpha^* = \frac{10}{2} \left[1 \pm \{1 + 4 \times 1.3\}^{1/2}\right] \ ,$$

$$\lambda_1^* = 17.5 \quad ; \quad \lambda_2^* = -7.4$$

Since $C_w(x \to \infty) = 0$, the coefficient A_1 of Eq. 25-50 that belongs to the positive eigenvalue must be zero. Then, A_2 is just equal to the concentration in the infiltrating water: $A_2 = 3 \ \mu\text{g L}^{-1}$. Thus:

$$C_w(\xi) = 3 \ \mu\text{g L}^{-1} \ \exp(-7.4\xi)$$

Evaluated at $\xi = 1$ ($x_o = 30$ m) and $\xi = 0.1$ ($x = 3$ m):

$$C_w(x = 30 \ \text{m}) = 3 \ \mu\text{g L}^{-1} \times \exp(-7.4) = 0.002 \ \mu\text{g L}^{-1}$$

$$C_w(x = 3 \ \text{m}) = 3 \ \mu\text{g L}^{-1} \times \exp(-0.74) = 1.4 \ \mu\text{g L}^{-1}$$

This result demonstrates that it may be worthwhile to drill wells not too close to the river even at the expense of a reduced yield of water.

25.4   Questions and Problems

Questions

Q 25.1

Which characteristic properties make porous media different compared to other aqueous systems?

Q 25.2

Compare the typical transport processes in porous media with transport processes in lakes and in rivers.

Q 25.3

Explain the difference between specific discharge and effective mean flow velocity. What is effective porosity?

Q 25.4

What is the physical reason behind Darcy's Law? Why can the specific discharge be related to the slope of the groundwater table?

Q 25.5

Explain the relation between hydraulic conductivity and permeability. Why can the latter be related to the square of the mean particle radius of the aquifer?

Q 25.6

Which mechanisms contribute to longitudinal dispersion in an aquifer? Why does the dispersivity depend on the scale over which transport in the aquifer is analyzed?

Q 25.7

The transport of a conservative tracer in an aquifer can be characterized by just one nondimensional number. By which one? If the chemical is consumed by a first-order reaction, a second nondimensional number becomes relevant as well. Which one?

Q 25.8

If a periodic concentration variation (e.g., in a river) penetrates into an aquifer, the amplitude is reducd and the signal retarded. Explain qualitatively which factors favor the amplitude reduction and which ones the retardation.

Q 25.9

The time of retardation of a periodic concentration variation at a fixed distance from the boundary (e.g., from the site of infiltration) has an upper limit. Which one?

Q 25.10

Explain the effect of sorption on the transport of a chemical in an aquifer.

Q 25.11

How can colloids affect the transport of a chemical in an aquifer?

Q 25.12

Which properties of the aquifer or of the chemicals may limit the validity of the linear transport equation (e.g., Eq. 25-45)?

Q 25.13

What can we learn from a breakthrough curve of a chemical in an aquifer?

Problems

P 25.1 *Assessing the Effect of River Water Pollution on the Quality of the Bank Filtrate Used for Drinking-Water Supply*

You are responsible for the safe operation of a drinking-water supply system that gets its raw water from a well located close to a river. From tracer experiments you know that the effective mean flow velocity is $\bar{u} = 3$ m d^{-1} and that the distance along the streamline from the point of infiltration to the well is $x_w = 18$ m. The dispersivity of the aquifer for this distance of flow is $\alpha_L = 5$ m. In order to be prepared for a possible pollution event in the river you are interested in the following questions:

Problem (a)

In order to prevent polluted river water from reaching your drinking-water system, you want to know how much time you have to turn off the pumps once a pollution cloud in the river has reached the location adjacent to the well. In your considerations you assume that the concentration of the pollutant suddenly increases from 0 to a value 10 times above the maximum tolerable drinking-water concentration and then remains at this level. (1) Take the worst-case scenario and calculate how much time you have to turn off the pumps. (2) How much does this time change if you assume that the concentration in the river reaches 1000 times the maximum tolerable drinking-water concentration?

Hints and Help: Use the equations derived for the step input to model the pollution scenario. Think about the appropriate choice of f_w in order to cover the worst case.

Problem (b)

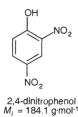

OH
NO$_2$
NO$_2$

2,4-dinitrophenol
$M_i = 184.1$ g·mol^{-1}

One day, measurements in the river suddenly show rather high concentrations of 2,4-dinitrophenol. Further observations show that the values vary on a weekly basis and reach maximum concentrations $C_{max} = 3.5$ µg L^{-1} during the week and minimum concentrations $C_{min} = 1.3$ µg L^{-1} over the weekend. Calculate the maximum concentration of 2,4-dinitrophenol to be expected at the drinking-water well. The pH in the infiltrating water is 7.8.

Hints and Help: Assume first that the concentration in the river can be described by a sinusoidal function with a period of one week. Then remember that any deviation from the sinus mode can be accounted for by superposition of harmonic oscillations of higher frequencies (Fourier series). Use Fig. 25.7 to prove that the attenuation of these faster modes is larger than the attenuation of the weekly mode and that therefore they do not affect the result calculated for the weekly sinusoidal input variation.

Problem (c)

You wonder how much the weekly variation of trichloroethene, another pollutant, would be reduced from the river to the well. Use the following information on the aquifer: porosity $\phi = 0.37$, solid phase density $\rho_s = 2.5$ g cm^{-3}, organic matter content of solid phase $f_{oc} = 0.01$ kg$_{oc}$ kg$_s^{-1}$. Compare this attenuation with the case of a concentration variation at the well that is caused by a short pollution pulse of trichloroethene in the river. To what fraction is the maximum concentration reduced from the river to the pumping well? Assume $\Delta t = 1$ hour for pulse duration.

trichloroethene
(TCE)
$M_i = 131.4$ g·mol^{-1}

tetrachloroethene
(PCE)
$M_i = 165.8$ g·mol^{-1}

P 25.2 *Determining the Hydraulic Properties of an Aquifer from Observed Concentration Time Series in Two Adjacent Wells*

You are interested in the hydraulic and geochemical properties of an aquifer, particularly in its dispersivity and average organic matter content. Fortunately, at your disposition you have long-term time series of water temperature and tetrachloroethene (PCE) measured in two adjacent wells, 15 m apart from each other along the main flow direction in the aquifer. From earlier tracer experiments you are pretty sure that the two wells are located on the same streamline, that is, they "see" the same water passing by, although not at the same time.

Both water temperature and PCE concentration show a clear annual variation. The maximum (minimum) water temperature is registered in the first well about 18 days earlier than in the second one, and the temperature amplitudes are approximately equal in both wells. In contrast, the time lag observed for the concentration variation of PCE between the two wells is 6 months. The concentration amplitude of PCE in the downstream well, that is, the difference between maximum and mean concentration, is only about 65% of the amplitude in the upstream well, but the annual mean concentrations are equal in both wells.

Estimate the dispersivity of the aquifer, α_L, and the relative dissolved fraction of PCE, f_w. Calculate the organic carbon content, f_{oc}, of the aquifer material by using a rough estimate for the solid phase density, $\rho_s = 2.5$ g cm^{-1}, and for the effective porosity $\phi = 0.35$.

Hints and Help: Assume, as a first approximation, that water temperature is a conservative, nonretarded parameter. You can then interpret the f_w value calculated for PCE as a retardation factor *relative* to water temperature. Then you may try to estimate the retardation of water temperature. To do so it may be helpful to assume, as a rough estimate, an average value for the specific heat of the solids in the aquifer, c_p, of 1 J g^{-1}K^{-1}.

P 25.3 *Determining the Characteristic Transport Properties from Measurements of Breakthrough Curves in Laboratory Columns*

You work in a research laboratory, and part of your duty is to determine the sorption behavior of a series of polycyclic aromatic hydrocarbons in laboratory columns. In one series of experiments you use a column that is 1 m long and has a diameter of

Table 25.3 Breakthrough Experiment in the Laboratory

Data from Chloride Experiment		Data from Naphthalene Experiment	
Time Since Beginning (hours)	Chloride Concentration (mg L⁻¹)	Time Since Beginning (hours)	Naphthalene Concentration (nmol L⁻¹)
12	0	100	0
16	1	200	30
20	4	250	100
24	10	300	200
28	18	350	380
32	28	400	490
36	39	425	560
40	51	450	620
44	60	475	660
48	67	500	710
52	75	550	790
56	80	600	850
64	89		
80	96		
90	100		

5 cm. The flow of water through the column is regulated by a pump at a constant discharge rate of 20 mL per hour. The column material is sterile to avoid biological transformation processes. Prior to every experiment the column is carefully flushed and then kept saturated with pure water.

Problem (a)

With the first experiment, conducted with chloride, you want to determine the hydraulic parameters of the column. At time $t = 0$, the chloride concentration at the input is set to $C_{in} = 100$ mg L⁻¹ and then kept constant. Time series of chloride concentrations are measured at the outlet of the column. The results are given in Table 25.3. Determine the porosity ϕ and the dispersion coefficient E_{dis} of the column.

Problem (b)

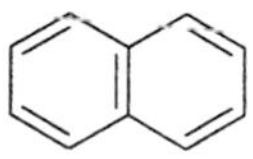

The second experiment is conducted with naphthalene. At $t = 0$, the concentration of naphthalene is set to a constant value of $C_{in} = 1$ µmol L⁻¹ and the naphthalene concentration measured in the outlet as a function of time (Table 25.3). Calculate the relative dissolved fraction f_w and the relative organic content f_{oc} of the column material. Use $\rho_s = 2.5$ g cm⁻³ and the parameters determined for chloride.

naphthalene
$M_i = 128.2$ g·mol⁻¹

P 23.4 *Interpreting Real Breakthrough Curves*

A field experiment was conducted at the Canadian Air Forces Base Borden, Ontario, to study the behavior of organic pollutants in a sand aquifer under natural conditions (Mackay et al., 1986). Figure 25.9 shows the results of two experiments, the first one for tetrachloroethene, the second one for chloride. Both substances were added as short pulses to the aquifer. The curves marked as "ideal" were computed according to Eqs. 25-20 or 25-23. The measured data clearly deviate from the ideal curve. The "nonideal" curves were constructed by Brusseau (1994) with a mathematical model that includes various factors causing nonideal behavior.

Give qualitative reasons to explain the experimental data.

Hints and Help: The concentration curves in Fig. 25.9 are plotted versus pore volumes, V_p, instead of time. V_p sums up the amount of water (per unit area) flowing through the aquifer expressed in units of pore space (per unit area) contained in the aquifer between input and output (location of measurement). Convince yourself that this variable is equivalent to time and that for ideal flow of a nonsorbing substance the maxima of the input and output curves, respectively, should be shifted by $\Delta V_p = 1$.

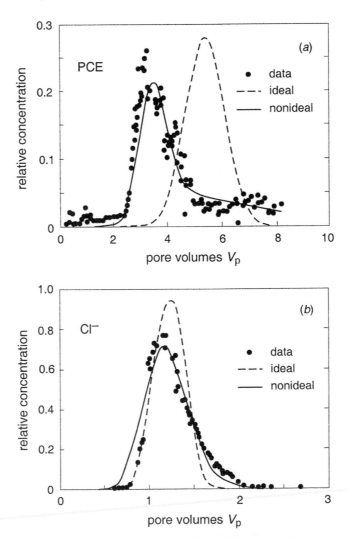

Figure 25.9 Measured breakthrough curves from the Borden field experiment for tetrachloroethene (PCE) and chloride (Cl⁻). From Brusseau (1994).

P 25.5 *Migration of Chemical Pollutants from Dump Site Through Clay Liner into the Groundwater*

Clay liner

Mainly kaolinite; assume that 1% is present as K^+-kaolinite, the rest is mainly Na^+- and Ca^{2+}-kaolinite (see Ilustrative Example 11.4 and Table 11.2).

$\rho_s = 2.65$ g cm^{-3}

tortuosity $\tau = 1.5$

$f_{oc} = 0.0005$ kg$_{oc}$kg$_s^{-1}$

$\phi = 0.4$

pH = 7.2

$T = 10°C$

A clay liner of 1 m thickness protects the underlying groundwater from chemically polluted soil. The characteristics of the clay liner are given in the margin. The hydrogeologists assure you that no water can flow from the polluted area into the groundwater. However, you are also concerned with transport by diffusion and decide to estimate the minimum time needed for the four compounds listed below to reach the groundwater. Assume that negatively charged species do not sorb significantly to the clay.

Hints and Help: Estimate the transport time by assuming that the chemicals diffuse from a reservoir of constant concentration into the liner that at time $t = 0$ is assumed to be uncontaminated. Equation 25-41 may give you an idea how to calculate the effective diffusion coefficient through the liner. Select the most critical compound among the four. What is the criterion?

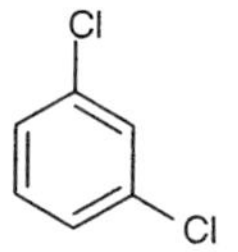

1,3-dichlorobenzene
log K_{ow} = 3.47
(Appendix C)

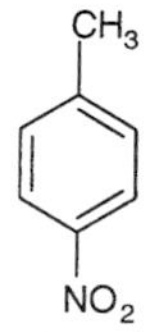

4-nitrotoluene
log K_{ow} = 2.38
(Appendix C)

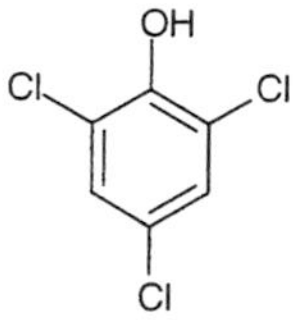

2,4,6-trichlorophenol
log K_{ow} = 3.72
pK_a = 6.13
(Appendix C)

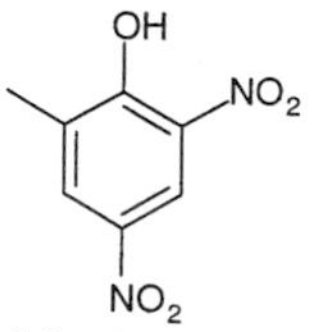

2,4-dinitro-6-methylphenol
(DNOC)
log K_{ow} = 2.12
pK_a = 4.31
(Appendix C)

Appendix

APPENDIX A: MATHEMATICS

APPENDIX B: PHYISCAL CONSTANTS AND UNITS

APPENDIX C: PHYSICOCHEMICAL PROPERTIES OF ORGANIC COMPOUNDS

APPENDIX D: TEMPERATURE DEPENDENCE OF EQUILIBRIUM AND KINETIC CONSTANTS

APPENDIX E: PROPERTIES OF THE EARTH

Appendix A

MATHEMATICS

A.1   **The Normal Distribution**

A.2   **The Error Function and its Complement**

A.1 The Normal Distribution

One-dimensional Normal (or Gaussian) Distribution:

$$p_\sigma^{(1)}(x) = \frac{1}{(2\pi)^{1/2}\,\sigma}\exp\left(-\frac{x^2}{2\sigma^2}\right) \tag{A-1}$$

Probability to find a point inside the interval $\{-x, +x\}$:

$$I_1(x) = \int_{-x}^{x} p_\sigma^{(1)}(x')\,dx' = \mathrm{erf}\left(\frac{x}{\sqrt{2}\sigma}\right) \tag{A-2}$$

The error function $\mathrm{erf}(x)$ is defined in Section A.2. Table A.1 gives specific numerical values of the normal distribution and its integral $I_1(x)$.

Two-dimensional Isotropic Normal Distribution:

Use $r^2 = x^2 + y^2$
Isotropy means $\sigma_x = \sigma_y \equiv \sigma$

$$p_\sigma^{(2)}(r) = \frac{1}{2\pi\,\sigma^2}\exp\left(-\frac{r^2}{2\sigma^2}\right) \tag{A-3}$$

Probability to find a point inside circle with radius r:

$$I_2(r) \equiv \int_0^r p_\sigma^{(2)}(r')\,2\pi r'\,dr' = \frac{1}{\sigma^2}\int_0^r \exp\left(-\frac{r'^2}{2\sigma^2}\right)r'\,dr'$$

$$= 1 - \exp\left(-\frac{r^2}{2\sigma^2}\right) \tag{A-4}$$

Integrals are listed in Table A.1.

Three-dimensional Isotropic Normal Distribution:

$r^2 = x^2 + y^2 + z^2, \qquad \sigma_x = \sigma_y = \sigma_z \equiv \sigma$

$$p_\sigma^{(3)}(r) = \frac{1}{(2\pi)^{3/2}\,\sigma^3}\exp\left(-\frac{r^2}{2\sigma^2}\right) \tag{A-5}$$

Probability to find a point inside a sphere with radius r:

$$I_3(r) \equiv \int_0^r p_\sigma^{(3)}(r')\,4\pi r'^2\,dr' = \left(\frac{2}{\pi}\right)^{1/2}\frac{1}{\sigma^3}\int_0^r \exp\left(-\frac{r'^2}{2\sigma^2}\right)r'^2\,dr'$$

$$= \mathrm{erf}\left(\frac{r}{\sqrt{2}\sigma}\right) - \left(\frac{2}{\pi}\right)^{1/2}\frac{r}{\sigma}\exp\left(-\frac{r^2}{2\sigma^2}\right) \tag{A-6}$$

Integrals are listed in Table A.1

A.2   **The Error Function and its Complement**

Error Function erf(x):

$$\text{erf}(x) = \frac{2}{\sqrt{\pi}} \int_0^x \exp(-y^2)\,dy \tag{A-7}$$

Complement of the Error Function erfc(x)

$$\text{erfc}(x) = 1 - \text{erf}(x) = \frac{2}{\sqrt{\pi}} \int_x^\infty \exp(-y^2)\,dy \tag{A-8}$$

Special values: $\text{erf}(0)$ $= \text{erfc}(\infty) = 0$

   $\text{erf}(\infty)$ $= \text{erfc}(0) = 1$

   $\text{erf}(-x)$ $= -\text{erf}(x)$

   $\text{erfc}(-x)$ $= 2 - \text{erfc}(x)$

Numerical values of erf(x) and erfc(x) are given in Table A.2.

For $x = \varepsilon \ll 1$, the error function can be approximated by:

$$\text{erf}(\varepsilon) \approx \frac{2}{\sqrt{\pi}}\varepsilon \tag{A-9}$$

Table A.1 The Normal Distribution $p_\sigma(\xi)$ and its Integral. $\xi = x/\sigma$ is a Normalized Coordinate. For $\sigma = 1, x = \xi$

$\xi \equiv x/\sigma$ or $\xi \equiv r/\sigma$	$p_\sigma^{(1)}(\xi) = p_\sigma^{(1)}(-\xi)$	$I_1(\xi) = \int_{-\xi}^{\xi} p_\sigma^{(1)}(\xi')\,d\xi'$	$I_2(\xi)$ (Eq. A-4)	$I_3(\xi)$ (Eq. A-6)
0.00	0.3989	0.0000	0.0000	0.00000
0.05	0.3984	0.0398	0.0012	0.00003
0.10	0.3970	0.0796	0.0050	0.00027
0.15	0.3945	0.1192	0.0112	0.00089
0.20	0.3910	0.1586	0.0198	0.00210
0.25	0.3867	0.1974	0.0308	0.00408
0.30	0.3814	0.2358	0.0440	0.00699
0.35	0.3752	0.2736	0.0594	0.01099
0.40	0.3683	0.3108	0.0769	0.01623
0.45	0.3605	0.3472	0.0963	0.02282
0.50	0.3503	0.3830	0.1175	0.03086
0.55	0.3429	0.4176	0.1404	0.04044
0.60	0.3332	0.4514	0.1647	0.05162
0.65	0.3230	0.4844	0.1904	0.06444
0.70	0.3123	0.5160	0.2173	0.07892
0.75	0.3011	0.5468	0.2452	0.09504
0.80	0.2897	0.5762	0.2739	0.11278
0.85	0.2780	0.6046	0.3032	0.13210
0.90	0.2661	0.6318	0.3330	0.15293
0.95	0.2541	0.6578	0.3632	0.17518
1.00	0.2420	0.6826	0.3935	0.19875
1.10	0.2179	0.7286	0.4539	0.24939
1.20	0.1942	0.7698	0.5133	0.30381
1.30	0.1714	0.8064	0.5704	0.36084
1.40	0.1497	0.8384	0.6247	0.41925
1.50	0.1295	0.8664	0.6754	0.47783
1.60	0.1109	0.8904	0.7220	0.53545
1.70	0.0941	0.9108	0.7643	0.59110
1.80	0.0790	0.9282	0.8021	0.64392
1.90	0.0656	0.9426	0.8355	0.69323
2.00	0.0540	0.9544	0.8647	0.73854
2.10	0.0440	0.9642	0.8898	0.77954
2.20	0.0355	0.9739	0.9111	0.81610
2.30	0.0283	0.9784	0.9290	0.84825
2.40	0.0224	0.9836	0.9439	0.87611
2.50	0.0175	0.9874	0.9561	0.89994
2.60	0.0136	0.9906	0.9660	0.92005
2.70	0.0104	0.9931	0.9739	0.93679
2.80	0.0079	0.9948	0.9802	0.95056
2.90	0.0060	0.9962	0.9851	0.96174
3.00	0.0044	0.9972	0.9889	0.97071
4.00	0.0001	0.9998	0.9997	0.99887

Table A.2 The Error Function and its Complement

x	$\text{erf}(x)$	$\text{erfc}(x)$
0	0	1.0
0.05	0.056372	0.943628
0.1	0.112463	0.887537
0.15	0.167996	0.832004
0.2	0.222703	0.777297
0.25	0.276326	0.723674
0.3	0.328627	0.671373
0.35	0.379382	0.620618
0.4	0.428392	0.571608
0.45	0.475482	0.524518
0.5	0.520500	0.479500
0.55	0.563323	0.436677
0.6	0.603856	0.396144
0.65	0.642029	0.357971
0.7	0.677801	0.322199
0.75	0.711156	0.288844
0.8	0.742101	0.257899
0.85	0.770668	0.229332
0.9	0.796908	0.203092
0.95	0.820891	0.179109
1.0	0.842701	0.157299
1.1	0.880205	0.119795
1.2	0.910314	0.089686
1.3	0.934008	0.065992
1.4	0.952285	0.047715
1.5	0.966105	0.033895
1.6	0.976348	0.023652
1.7	0.983790	0.016210
1.8	0.989091	0.010909
1.9	0.992790	0.007210
2.0	0.995322	0.004678
2.1	0.997021	0.002979
2.2	0.998137	0.001863
2.3	0.998857	0.001143
2.4	0.999311	0.000689
2.5	0.999593	0.000407
2.6	0.999764	0.000236
2.7	0.999866	0.000134
2.8	0.999925	0.000075
2.9	0.999959	0.000041
3.0	0.999978	0.000022

Appendix B

PHYSICAL CONSTANTS AND UNITS

B.1   **Some Useful Constants (IUPAC, 1988)**

B.2   **Dimensions and Units of Physical Quantities**

B.3   **Specific Properties of Water at Different Temperatures**

B.4   **Water Phase Schmidt Numbers of Solutes**

B.5   **Specific Properties of Dry Air**

B.1   Some Useful Constants (IUPAC, 1988)

Atomic mass	m_u	$\cong 1.6605402 \times 10^{-27}$ kg
Avogadro's number	N_A	$\cong 6.0221367 \times 10^{23}$ mol^{-1}
Boltzmann's constant	k	$\cong 1.380658 \times 10^{-23}$ J K^{-1}
Elementary charge	e	$\cong 1.60217733 \times 10^{-19}$ C
Faraday's constant	F	$\cong 9.6485309 \times 10^4$ C mol^{-1}
Gas (molar) constant	R	$= k \times N = 8.314510$ J mol^{-1}K^{-1}
		$= 0.083145$ L bar mol^{-1}K^{-1}
Gravitational acceleration	g	$= 9.80665$ m s^{-2}
Molar volume of an ideal gas at 1 bar and 25°C	$\overline{V}_{ideal\ gas}$	$= 24.465$ L mol^{-1}
Permittivity of vacuum	ε_o	$\cong 8.854187 \times 10^{-12}$ C V^{-1}m^{-1}
Planck's constant	h	$\cong 6.6260755 \times 10^{-34}$ J s
Zero of Celsius scale	0°C	$= 273.15$ K

B.2   Dimensions and Units of Physical Quantities

Dimensions are: M = mass; L = length; T = time; I = current.

Physical Quantity	Name of Unit	Symbol	Dimensions	SI Units
Amount of photons	einstein	einstein		mol photons
Concentration	molar	M	M L^{-3}	10^3 mol m^{-3}
Dipole moment	debye	D	I T L	$\sim 3.34 \times 10^{30}$ C m
Electric charge	coulomb	C	I T	A s
Electrical potential	volt	V	M L^{2}T^{-3}I^{-1}	J C^{-1} = kg m^{2}s^{-3}A^{-1}
Energy	joule	J	M L^{2}T^{-2}	N m = kg m^{2}s^{-2}
	volt-coulomb	VC		J
	watt-second	Ws		J
	erg	erg		10^{-7} J
	liter-atmosphere	L-atm		101.325 J
	calorie	cal		4.184 J
	electron-volt	eV		$\sim 1.60 \times 10^{-19}$ J
Force	newton	N	M L T^{-2}	kg m s^{-2}
	dyne	dyn		10^{-5} N
Frequency	hertz	Hz	T^{-1}	s^{-1}
Length	ångström	Å	L	10^{-10} m
	nanometer	nm		10^{-9} m
	micrometer	μm		10^{-6} m
	millimeter	mm		10^{-3} m
	centimeter	cm		10^{-2} m
	kilometer	km		10^3 m

Physical Quantity	Name of Unit	Symbol	Dimensions	SI Units
Mass	ton (metric)	t	M	10^3 kg
Pressure	pascal	Pa	$M L^{-1}T^{-2}$	$N m^{-2} = kg\ m^{-1}s^{-2}$
	bar	bar		10^5 Pa
	atmosphere	atm		101,325 Pa
	torr	torr		133.32 Pa
	millimeter mercury	mm Hg		~ 133.32 Pa
	pounds per square inch	psi		~ 6.89×10^3 Pa
Time	minute	min	T	60s
	hour	h		3 600 s
	day	d		86 400 s
	year (365.25 d)	yr		31 557 600 s
Dynamic viscosity	centipoise	cp	$M L^{-1}T^{-1}$	10^{-3} kg $m^{-1}s^{-1}$
Volume	liter	L	L^3	10^{-3} m^3
	milliliter	mL		10^{-6} m^3
	microliter	μL		10^{-9} m^3

B.3   Specific Properties of Water Depending on Temperature

ρ_w $(kg\ m^{-3})$ density

$\alpha = -\dfrac{1}{\rho}\left(\dfrac{\partial \rho}{\partial T}\right)_P$ (K^{-1}) thermal expansivity

η_w $(kg\ m^{-1}s^{-1})$ dynamic viscosity

$\nu_w = h_w / r_w$ (m^2s^{-1}) kinematic viscosity

c_{pw} $(J\ kg^{-1}K^{-1})$ specific heat (at constant pressure)

$c_{pw} = 4.18 \times 10^3 J\ kg^{-1}K^{-1}$ for $T = 20°C$ (Variation less than 1% from 0 to 30°C)

λ_w $(W\ m^{-1}K^{-1})$ specific conductivity of heat

$D_{th\ w} = \dfrac{\lambda_w}{c_{pw}\rho_w}$ (m^2s^{-1}) thermal diffusivity

$C^{eq}_{water\ a}$ $(g\ m^{-3})$ equilibrium concentration of water vapor in air in contact with liquid water

Some of these properties are listed in Table B.3 for temperatures between 0 and 30°C.

Table B.3 Specific properties of water as a function of temperature T for $p = 1.013$ bar

T (°C)	α (10^{-6} K^{-1})	η (10^{-3} kg m^{-1}s^{-1})	ν (10^{-6} m^{2}s^{-1})	D_{th} (10^{-6} m^{2}s^{-1})	$C_{water\ a}^{eq}$ (g m^{-3})
0	−68.05	1.787	1.787	0.134	4.84
1	−50.09	1.728	1.728		5.2
2	−32.74	1.671	1.672		5.6
3	−15.97	1.618	1.618		6.0
4	0.27	1.567	1.567		6.4
5	16.00	1.518	1.519		6.8
6	31.24	1.472	1.472		7.3
7	46.04	1.428	1.428		7.8
8	60.41	1.386	1.386		8.3
9	74.38	1.346	1.346		8.8
10	87.97	1.307	1.308	0.138	9.4
11	101.20	1.270	1.271		10.0
12	114.09	1.235	1.236		10.7
13	126.65	1.202	1.202		11.4
14	138.90	1.169	1.170		12.1
15	150.87	1.139	1.140		12.8
16	162.56	1.109	1.110		13.6
17	173.98	1.081	1.082		14.5
18	185.15	1.053	1.055		15.4
19	196.08	1.027	1.029		16.3
20	206.78	1.002	1.004	0.142	17.3
21	217.27	0.978	0.980		18.3
22	227.54	0.955	0.957		19.4
23	237.62	0.932	0.934		20.6
24	247.50	0.911	0.913		21.8
25	257.21	0.890	0.893		23.0
26	266.73	0.870	0.873		24.4
27	276.10	0.851	0.854		25.8
28	285.30	0.832	0.835		27.2
29	294.35	0.814	0.818		28.7
30	303.25	0.797	0.800	0.146	30.3

B.4   Water Phase Schmidt Numbers of Solutes

Water phase Schmidt number of solute i:

$$Sc_{iw} = \frac{\nu_w}{D_{iw}} \tag{B-1}$$

Table B.4 Schmidt Numbers $Sc_{iw} = \nu_w / D_{iw}$ of Gases Dissolved in Water

T (°C)	Carbone dioxide (CO_2) Sc_{iw} (−)	Methane (CH_4) Sc_{iw} (−)	Oxygen (O_2) Sc_{iw} (−)
0	1915	1902	1611
1	1795	1785	1507
2	1683	1677	1412
3	1580	1577	1323
4	1484	1484	1241
5	1395	1397	1165
6	1312	1317	1094
7	1235	1242	1028
8	1163	1172	968
9	1097	1107	911
10	1035	1046	858
11	977	989	809
12	923	936	763
13	872	886	721
14	825	840	681
15	781	796	643
16	740	755	609
17	701	717	576
18	665	681	546
19	631	647	517
20	599	615	490
21	569	585	465
22	540	557	441
23	514	531	419
24	489	506	398
25	466	482	379
26	443	460	360
27	422	439	343
28	403	419	327
29	384	400	311
30	367	383	297

B.5 Specific Properties of Dry Air

p_o = 1.0133 × 10⁵ Pa (mean pressure at sea level). See Section B.3 for definitions.

ρ_a	=	1.293 kg m⁻³	$T = 0°C$
		1.270	$T = 5°C$
		1.247	$T = 10°C$
		1.226	$T = 15°C$
		1.205	$T = 20°C$
		1.184	$T = 25°C$

| ν_a | = | 0.13 cm²s⁻¹ | $T = 0°C$ |
| | | 0.15 | $T = 20°C$ |

$c_{p\,a}$ = 1.005 × 10³ J kg⁻¹K⁻¹ $T = 20°C$

λ_a = 2.56 × 10² W m⁻¹K⁻¹ $T = 20°C$

$D_{th\,a}$ = 2.11 × 10⁻⁵ m²s⁻¹ $T = 20°C$

Appendix C

PHYSICOCHEMICAL PROPERTIES OF ORGANIC COMPOUNDS

Appendix C contains the names, molecular formula, molar mass (M_i), density (ρ_i), melting point (T_m), boiling point (T_b), vapor pressure (p_i^*), aqueous solubility (C_{iw}^sat), air–water partition constant (K_{iaw}), octanol–water partition constant (K_{iow}), and acidity constant (K_{ia}, where appropriate) of some environmentally relevant organic chemicals. Except for density (20°C), all data are given for 25°C. The data have been collected from various data compilations (and references cited therin) including Abraham et al. (1994a and b), Hansch et al. (1995), Lide (1998), Mackay et al. (1992-97), Mitchell and Jurs (1998), Montgomery (1997), and Ruelle and Kesselring (1997a and b).

Compound Name	Molecular Formula	M_i (g·mol⁻¹)	ρ_i (g·cm⁻³)	T_m (°C)	T_b (°C)	log p_i^*/Pa	$-\log C_{iw}^{sat}$	$-\log K_{iaw}$ calculated (experimental)	log K_{iow}	pK_{ia}
n-Alkanes										
Methane	CH$_4$	16.0		−182.5	−164.0	7.45	2.82 (1bar)	−1.43	1.09	
Ethane	C$_2$H$_6$	30.1		−183.3	−88.6	6.61	2.69 (1bar)	−1.30	1.81	
Propane	C$_3$H$_8$	44.1		−189.7	−42.1	5.98	2.85 (1bar)	−1.46	2.36	
n-Butane	C$_4$H$_{10}$	58.1	0.58 (L)	−138.4	−0.5	5.40	2.98 (1bar)	−1.58	2.89	
n-Pentane	C$_5$H$_{12}$	72.2	0.63	−129.7	36.1	4.83	3.25	−1.69	3.39	
n-Hexane	C$_6$H$_{14}$	86.2	0.66	−95.0	69.0	4.30	3.83	−1.74	4.00	
n-Heptane	C$_7$H$_{16}$	100.2	0.68	−90.6	98.4	3.79	4.53	−1.93	4.66	
n-Octane	C$_8$H$_{18}$	114.2	0.70	−56.8	125.7	3.26	5.20	−2.07	5.15	
n-Nonane	C$_9$H$_{20}$	128.3	0.72	−51.0	150.8	2.76	5.77	−2.14	5.65	
n-Decane	C$_{10}$H$_{22}$	142.3	0.73	−29.7	174.1	2.24	6.42	−2.27	6.25	
n-Undecane	C$_{11}$H$_{24}$	156.3	0.74	−25.6	195.9	1.72				
n-Dodecane	C$_{12}$H$_{26}$	170.3	0.75	−9.6	216.3	1.19	7.52	−2.32		
n-Hexadecane	C$_{16}$H$_{34}$	226.4	0.77	18.2	287.0	−0.73	7.80	−0.68		
n-Octadecane	C$_{18}$H$_{38}$	254.4	0.78	28.2	316.1	−1.78	8.08	−0.09		
Branched Alkanes										
2-Methylpropane (isobutane)	C$_4$H$_{10}$	58.1	0.56 (L)	−159.4	−11.7	5.56	3.07 (1bar)	−1.68	2.82	
2-Methylbutane (isopentane)	C$_5$H$_{12}$	72.2	0.62	−159.9	27.9	4.96	3.18	−1.75		
2,2-Dimethylpropane (neopentane)	C$_5$H$_{12}$	72.2	0.59 (L)	−16.6	9.5	5.24	3.34 (1bar)	−1.95	3.11	
2-Methylpentane (isohexane)	C$_6$H$_{14}$	86.2	0.64	−153.8	60.1	4.45	3.80	−1.86	3.60	
2,2-Dimethylbutane (neohexane)	C$_6$H$_{14}$	86.2	0.65	−99.8	49.7	4.63	3.65	−1.89	3.42	
2,2,4-Trimethylpentane (isooctane)	C$_8$H$_{18}$	114.2	0.69	−107.4	99.2	3.81	4.67	−2.09		
Unsaturated and Alicyclic Hydrocarbons										
1,3-Butadiene	C$_4$H$_6$	54.1	0.62 (L)	−108.9	−4.4	5.45	1.86	−0.47	1.99	
2-Methyl-1,3-butadiene (isoprene)	C$_5$H$_8$	68.1	0.68	−146.0	34.0	4.86	2.04	−0.51	2.05	
1,4-Pentadiene	C$_5$H$_8$	68.1	0.66	−148.3	26.0	5.00	2.08	−0.69		
Cyclopentene	C$_5$H$_8$	68.1	0.77	−135.1	44.2	4.70	2.09	−0.40		
Cyclopentane	C$_5$H$_{10}$	70.1	0.74	−92.9	49.2	4.63	2.64	−0.88	3.00	
1,4-Cyclohexadiene	C$_6$H$_8$	80.1	0.85	−49.2	81.5	3.95	2.03	0.41	2.30	
Cyclohexene	C$_6$H$_{10}$	82.1	0.81	−103.5	83.0	4.07	2.60	−0.28	2.86	
Cyclohexane	C$_6$H$_{12}$	84.2	0.78	6.5	80.7	4.10	3.17	−0.89	3.44	
1-Hexene	C$_6$H$_{12}$	84.2	0.67	−139.8	63.4	4.40	3.22	−1.22	3.40	
Methylcyclohexane	C$_7$H$_{14}$	98.2	0.77	−126.6	100.6	3.79	3.81	−1.21	3.88	

Alkylated Benzenes

Benzene	C₆H₆	78.1	0.88	5.5	80.1	4.10	1.65	0.65 (0.65 exp.)	2.17
Methylbenzene (toluene)	C₇H₈	92.2	0.87	−95.0	110.6	3.57	2.22	0.60 (0.62 exp.)	2.69
Ethylbenzene	C₈H₁₀	106.2	0.86	−95.0	136.2	3.09	2.80	0.50 (0.47 exp.)	3.20
Vinylbenzene (styrene)	C₈H₈	104.2	0.91	−31.0	145.1	2.94	2.53	0.93 (0.92 exp.)	2.95
1,2-Dimethylbenzene (*o*-xylene)	C₈H₁₀	106.2	0.88	−25.2	144.4	2.95	2.75	0.69 (0.68 exp.)	3.16
1,3-Dimethylbenzene (*m*-xylene)	C₈H₁₀	106.2	0.86	−47.8	139.1	3.04	2.82	0.53 (0.52 exp.)	3.30
1,4-Dimethylbenzene (*p*-xylene)	C₈H₁₀	106.2	0.86	13.3	138.1	3.07	2.77	0.55 (0.54 exp.)	3.27
n-Propylbenzene	C₉H₁₂	120.2	0.86	−99.6	159.2	2.65	3.34	0.40	3.69
Isopropylbenzene	C₉H₁₂	120.2	0.86	−96.6	154.2	2.79	3.33	0.27	3.66
1,2,3-Trimethylbenzene	C₉H₁₂	120.2	0.89	−25.4	176.1	2.30	3.23	0.86	3.60
1,2,4-Trimethylbenzene	C₉H₁₂	120.2	0.88	−43.8	169.4	2.42	3.33	0.65	3.65
1,3,5-Trimethylbenzene	C₉H₁₂	120.2	0.88	−44.7	164.7	2.52	3.38	0.50	3.42
1-Ethyl-2-methylbenzene	C₉H₁₂	120.2	0.88	−83.8	165.2	2.52	3.20	0.67	3.53
1-Ethyl-4-methylbenzene	C₉H₁₂	120.2	0.86	−62.4	162.0	2.60	3.10	0.69	3.63
n-Butylbenzene	C₁₀H₁₄	134.2	0.86	−88.0	183.0	2.15	3.95	0.29	4.38
s-Butylbenzene	C₁₀H₁₄	134.2	0.86	−75.0	174.0	2.40	3.73	0.26	4.44
t-Butylbenzene	C₁₀H₁₄	134.2	0.87	−58.0	169.0	2.46	3.60	0.33	4.11
1,2,3,4-Tetramethylbenzene	C₁₀H₁₄	134.2	0.91	−6.3	205.0	1.65			4.00
1,2,4,5-Tetramethylbenzene	C₁₀H₁₄	134.2	0.84	79.5	195.9	1.11	4.58	0.70	4.10
n-Pentylbenzene	C₁₁H₁₆	148.3	0.86	−78.3	202.2	1.65	4.59	0.15	4.90
n-Hexylbenzene	C₁₂H₁₈	162.3	0.86	−61.0	226.0	1.13	5.20	0.06	5.52
Hexamethylbenzene	C₁₂H₁₈	162.3	0.86	166.7	265.0	−0.80	5.84	1.35	4.75

Polycyclic Aromatic Hydrocarbons and Related Compounds

Indane	C₉H₁₀	118.2	0.96	−51.4	178.0	2.30	3.03	1.06	3.33
Naphthalene	C₁₀H₈	128.2	1.16	80.2	218.0	1.05	3.60	1.74 (1.72 exp.)	3.33
1-Methylnaphthalene	C₁₁H₁₀	142.2	1.02	−22.0	244.0	0.92	3.71	1.76	3.87
2-Methylnaphthalene	C₁₁H₁₀	142.2	1.01	35.0	241.0	0.95	3.75	1.69	3.99
Acenaphthene	C₁₂H₁₀	154.2	1.05	96.2	278.0	−0.51	4.61	2.29 (2.28 exp.)	4.20
Acenaphthylene	C₁₂H₈	152.2	0.90	92.5	270.0	−0.05	4.59	1.85	4.00
Fluorene	C₁₃H₁₀	166.2	1.20	116.0	295.0	−1.02	4.94	2.47 (2.39 exp.)	4.32

Compound Name	Molecular Formula	M_i (g·mol⁻¹)	ρ_i (g·cm⁻³)	T_m (°C)	T_b (°C)	log p_i^*/Pa	$-\log C_{iw}^{sat}$	$-\log K_{iaw}$ calculated (experimental)	log K_{iow}	pK_{ia}
Phenanthrene	$C_{14}H_{10}$	178.2	0.98	101.0	339.0	−1.66	5.20	2.85 (2.93 exp.)	4.57	
Anthracene	$C_{14}H_{10}$	178.2	1.25	217.5	341.0	−3.01	6.60	2.80 (2.77 exp.)	4.68	
Fluoranthene	$C_{16}H_{10}$	202.3	1.25	110.5	384.0	−2.91	5.96	3.34 (3.35 exp.)	5.23	
Pyrene	$C_{16}H_{10}$	202.3	1.27	156.0	403.0	−3.09	6.16	3.32 (3.36 exp.)	5.13	
Benzo(a)anthracene	$C_{18}H_{12}$	228.3	1.25	160.6	437.5	−4.60	7.32	3.68	5.91	
Chrysene	$C_{18}H_{12}$	228.3	1.28	255.0	448.0	−6.22	8.05	4.56	5.81	
Benzo(a)pyrene	$C_{20}H_{12}$	252.3		176.5	496.0	−6.15	8.14	4.79 (4.86 exp.)	6.13	
Perylene	$C_{20}H_{12}$	252.3		277.0	503.0	−7.85	8.80	5.44	6.25	
Chlorinated C_1- to C_4-Compounds										
Chloromethane	CH_3Cl	50.5	0.92 (L)	−97.7	−24.2	5.76	0.98 (1bar)	0.42	0.91	
Dichloromethane	CH_2Cl_2	84.9	1.33	−95.1	40.1	4.76	0.70	0.93 (0.97 exp.)	1.31	
Trichloromethane	$CHCl_3$	119.4	1.48	−63.3	61.4	4.40	1.15	0.84 (0.97 exp.)	1.95	
Tetrachloromethane	CCl_4	153.8	1.59	−23.0	76.7	4.16	2.27	−0.04 (0.82 exp.)	2.77	
1,1-Dichloroethane	$C_2H_4Cl_2$	99.0	1.18	−97.3	57.3	4.49	1.29	0.61 (0.05 exp.)	1.79	
1,2-Dichloroethane	$C_2H_4Cl_2$	99.0	1.25	−35.5	83.6	4.05	1.07	1.27 (0.62 exp.)	1.46	
1,1,1-Trichloroethane	$C_2H_3Cl_3$	133.4	1.34	−31.4	73.9	4.22	2.01	−0.16 (1.30 exp.)	2.49	
1,1,2-Trichloroethane	$C_2H_3Cl_3$	133.4	1.44	−37.0	113.6	3.60	1.47	1.32 (0.15 exp.)	2.34	
1,1,2,2-Tetrachloroethane	$C_2H_2Cl_4$	167.9	1.60	−40.8	146.3	2.90	1.71	1.78 (1.40 exp.)	2.39	
Pentachloroethane	C_2HCl_5	202.3	1.68	−29.0	160.8	2.79	2.61	0.99 (1.85 exp.)	3.22	
Hexachloroethane	C_2Cl_6	236.7	2.09	187.5	187.5	1.70	3.68	1.01	3.93	
Chloroethene	C_2H_3Cl	62.5	0.91 (L)	−153.8	−13.7	5.55	1.35 (1bar)	−0.04 (−0.05 exp.)	1.27	
1,1-Dichloroethene	$C_2H_2Cl_2$	96.9	1.22	−122.0	31.7	4.90	1.59	−0.10 (−0.03 exp.)	1.48	
cis-1,2-Dichloroethene	$C_2H_2Cl_2$	96.9	1.27	−81.0	60.0	4.45	1.28	0.66 (0.79 exp.)	1.86	
trans-1,2-Dichloroethene	$C_2H_2Cl_2$	96.9	1.27	−50.0	48.0	4.61	1.19	0.59 (0.46 exp.)	2.09	

Compound	Formula								
Trichloroethene	C_2HCl_3	131.4	1.46	−73.0	67.0	4.00	2.00	(0.40 exp.)	2.12
Tetrachloroethene	C_2Cl_4	165.8	1.62	−22.4	121.1	3.40	3.07	−0.08 (0.12 exp.)	2.88
3-Chloro-1-propene	C_3H_5Cl	76.5	0.94	−135.0	44.9	4.70	1.33	0.36 (0.42 exp.)	1.45
Brominated and Iodated C_1- and C_2-Compound									
Bromomethane	CH_3Br	94.9	1.68 (L)	−93.6	3.6	5.34	0.79 (1bar)	0.35	1.19
Dibromomethane	CH_2Br_2	173.9	2.50	−52.6	96.7	3.81	1.18	1.40 (1.20 exp.)	1.88
Tribromomethane	$CHBr_3$	252.8	2.89	8.3	149.6	2.86	1.91	1.62 (1.72 exp.)	2.67
Bromoethane	C_2H_5Br	109.0	1.45	−119.0	38.4	4.80	1.09	1.50	1.61
1,2-Dibromoethane	$C_2H_4Br_2$	187.9	2.18	9.8	131.5	3.21	1.63	1.55	1.96
Bromoethene	C_2H_3Br	107.0	1.49 (L)	−139.5	15.8	5.15			1.57
Iodomethane	CH_3I	141.9	2.28	−66.5	42.5	4.73	1.01	0.66 (0.65 exp.)	1.51
Iodoethane	C_2H_5I	156.0	1.94	−111.0	72.4	4.26	1.60	0.53	2.00
Mixed Halogenated C_1- and C_2-Compounds									
Bromochloromethane	CH_2BrCl	129.4	1.93	−88.0	68.1	4.29	0.94	1.16	1.41
Bromodichloromethane	$CHBrCl_2$	163.8	1.97	−57.1	90.0	3.82	1.55	1.01	2.10
Dichlorodifluoromethane (Freon 12)	CCl_2F_2	120.9	1.33 (L)	−158.0	−29.8	5.75	2.64 (1bar)	−1.25	2.16
Trichlorofluoromethane (Freon 11)	CCl_3F	137.4	1.18	−111.0	23.8	5.03	2.10 (1bar)	−0.71 (−1.18 exp.)	2.53
1,1,2-Trichloro-1,2,2-trifluoroethane (Freon 113)	$C_2Cl_3F_3$	187.4	1.57	−35.0	47.6	4.68	3.04	−1.29 (−0.61 exp.)	3.16
1,1,2,2-Tetrachloro-1,2-difluoroethane (Freon 112)	$C_2Cl_4F_2$	203.8	1.64	−25.5	92.8	3.87	3.23	−0.71 (−1.11 exp.)	3.73
Chlorobenzenes									
Chlorobenzene	C_6H_5Cl	112.6	1.11	−45.2	132.0	3.20	2.39	0.80 (0.82 exp.)	2.78
1,2-Dichlorobenzene	$C_6H_4Cl_2$	147.0	1.30	−17.0	180.0	2.30	3.05	1.04 (1.00 exp.)	3.40
1,3-Dichlorobenzene	$C_6H_4Cl_2$	147.0	1.29	−24.7	173.0	2.45	3.08	0.86 (0.82 exp.)	3.47
1,4-Dichlorobenzene	$C_6H_4Cl_2$	147.0	1.25	53.1	174.0	2.05	3.30	1.04 (1.10 exp.)	3.45
1,2,3-Trichlorobenzene	$C_6H_3Cl_3$	181.5	1.69	52.0	218.5	1.45	3.94	1.01	4.14
1,2,4-Trichlorobenzene	$C_6H_3Cl_3$	181.5	1.45	17.0	213.5	1.58	3.78	1.03	4.06
1,3,5-Trichlorobenzene	$C_6H_3Cl_3$	181.5	1.39	63.5	208.0	1.50	4.53	0.36	4.19

Compound Name	Molecular Formula	M_i (g·mol⁻¹)	ρ_i (g·cm⁻³)	T_m (°C)	T_b (°C)	log p_i^*/Pa	-log C_{iw}^{sat}	-log K_{iaw} calculated (experimental)	log K_{iow}	pK_{ia}
1,2,3,4-Tetrachlorobenzene	$C_6H_2Cl_4$	215.9		47.5	254.0	0.75	4.69	0.95	4.64	
1,2,3,5-Tetrachlorobenzene	$C_6H_2Cl_4$	215.9	1.86	54.5	246.0	1.00	4.79	0.60	4.66	
1,2,4,5-Tetrachlorobenzene	$C_6H_2Cl_4$	215.9		140.0	243.0	-0.14	5.23	1.30	4.72	
Pentachlorobenzene	C_6HCl_5	250.3	1.83	86.0	277.0	-0.66	5.58	1.47	5.18	
Hexachlorobenzene	C_6Cl_6	284.8	2.08	230.0	322.0	-2.60	7.55	1.44 (1.54 exp.)	5.80	
Polychlorinated Biphenyls (PCBs), Selected Congeners										
Biphenyl	$C_{12}H_{10}$	154.2	0.87	71.0	255.9	0.11	4.34	1.93	4.01	
2-CBP (PCB1)	$C_{12}H_9Cl$	188.6	0.98	34.0	274.0	-0.30	4.54	1.55	4.53	
4-CBP (PCB3)	$C_{12}H_9Cl$	188.6	0.98	77.7	291.0	-0.57	5.19	1.77	4.61	
2,2'-CBP (PCB4)	$C_{12}H_8Cl_2$	223.1	1.05	61.0		-0.58	5.35	1.62	4.97	
2,4-CBP (PCB7)	$C_{12}H_8Cl_2$	223.1	1.05	24.4		-0.60	5.25	1.74	5.30	
2,4'-CBP (PCB8)	$C_{12}H_8Cl_2$	223.1	1.05	45.0		-1.10	5.35	2.14	5.10	
4,4'-CBP (PCB15)	$C_{12}H_8Cl_2$	223.1	1.05	148.5		-2.30	6.57	2.12	5.33	
2,2',5-CBP (PCB18)	$C_{12}H_7Cl_3$	257.5	1.15	44.0		-0.84	5.80	1.43	5.60	
2,4,4'-CBP (PCB28)	$C_{12}H_7Cl_3$	257.5	1.15	57.5		-1.70	6.20	1.89	5.62	
2,4,5-CBP (PCB29)	$C_{12}H_7Cl_3$	257.5	1.15	78.5		-1.95	6.26	2.08	5.90	
2,2',4,4'-CBP (PCB47)	$C_{12}H_6Cl_4$	292.0	1.20	41.5		-2.00	6.51	1.88	6.29	
2,2',5,5'-CBP (PCB52)	$C_{12}H_6Cl_4$	292.0	1.20	86.5		-2.30	6.99	1.70	6.09	
3,3',4,4'-CBP (PCB77)	$C_{12}H_6Cl_4$	292.0	1.20	180.0			7.47		6.50	
2,2',3,4,5'-CBP (PCB87)	$C_{12}H_5Cl_5$	326.4	1.28	112.0		-3.52	7.91	2.00	6.37	
2,2',4,5,5'-CBP (PCB101)	$C_{12}H_5Cl_5$	326.4	1.28	77.0		-2.96	7.51	1.83	6.36	
2,2',4,4',5,5'-CBP (PCB153)	$C_{12}H_4Cl_6$	360.9		103.5		-3.92	8.55	1.76	7.15	
2,2',3,4,4',5,5'-CBP (PCB180)	$C_{12}H_3Cl_7$	395.4		109.5				2.36	7.36	
Miscellaneous Polychlorinated Compounds										
α-hexachlorocyclohexane (α-HCH)	$C_6H_6Cl_6$	290.8		158.0		-2.52	5.28	3.63	3.81	
β-hexachlorocyclohexane (β-HCH)	$C_6H_6Cl_6$	290.8		309.5		-4.40	6.46	4.33	3.80	
γ-hexachlorocyclohexane (lindane, γ-HCH)	$C_6H_6Cl_6$	290.8		112.0		-2.15	4.60	3.94	3.78	
p,p'-DDT	$C_{14}H_9Cl_5$	354.5	1.55	109.0		-4.70	7.80	3.30	6.36	
p,p'-DDE	$C_{14}H_8Cl_4$	318.0		89.0		-3.20	6.90	2.69	5.70	
p,p'-DDD	$C_{14}H_{10}Cl_4$	320.0		109.5		-3.90	6.80	3.49	5.50	

Aliphatic Ethers

Dimethyl ether	C_2H_6O	46.1	0.67	−141.5	−24.8	5.77		(1.40 exp.)	0.10
Diethyl ether	$C_4H_{10}O$	74.1	0.71	−116.3	34.5	4.85	0.05	1.49	0.89
Methyl-t-butyl-ether (MBTE)	$C_5H_{12}O$	88.2	0.74	−109.0	55.2	4.51	0.34	1.54	0.94
Di-n-propyl ether	$C_6H_{14}O$	102.2	0.75	−123.2	90.1	3.92	1.50	0.97	2.03
Di-isopropyl ether	$C_6H_{14}O$	102.2	0.73	−85.5	68.5	4.30	1.10	0.98	1.52
n-Butyl-ethyl ether	$C_6H_{14}O$	102.2	0.75	−103.0	92.2	3.90	1.20	1.29	2.03
D-n-butyl ether	$C_8H_{18}O$	130.2	0.77	−95.2	140.3	2.95	2.75	0.69	3.21

Miscellaneous Ethers Including Epoxides

Ethylene oxide (epoxyethane)	C_2H_4O	44.1	0.87	−111.0	10.7	5.16	−0.93 (1bar)		−0.30
Propyleneoxide (epoxypropane)	C_3H_6O	58.1	0.83	−112.1	34.2	4.85	−0.91		0.03
Tetrahydrofuran	C_4H_8O	72.1	0.89	−108.5	66.0	4.33	miscible	(2.55 exp.)	0.46
1,4-Dioxane	$C_4H_8O_2$	88.1	1.03	11.0	101.1	3.70	miscible	(3.71 exp.)	−0.27
1-Chloro-2,3-epoxypropane (epichlorohydrine)	C_3H_5ClO	92.5	1.18	−57.2	116.2	3.36	0.15	2.88	0.30
Di-2-chloroethyl ether	$C_4H_8Cl_2O$	143.0	1.22	−46.8	178.0	2.31	1.15	2.93	1.29
Methoxybenzene (anisole)	C_7H_8O	108.2	0.99	−37.5	153.6	2.67	1.80	1.92	2.11
Ethoxybenzene (phenetole)	$C_8H_{10}O$	122.2	1.07	−31.0	169.5	2.31	2.18	1.90	2.51
Styreneoxide	C_8H_8O	120.2	1.05	−35.6	194.1	1.60	1.63	3.16	1.55

Polychlorinated Dibenzo-p-Dioxins (PCDDs), Selected Congeners

Dibenzo-p-dioxin (DD)	$C_{12}H_8O_2$	184.0		123.0	283.5	−1.26	5.33	2.32	4.30
1-CDD	$C_{12}H_7ClO_2$	218.5		105.5	315.5	−1.92	5.72	2.59	4.90
2,7-DCDD	$C_{12}H_6Cl_2O_2$	253.0		210.0	373.5	−3.92	7.83	2.48	5.70
1,2,3,4-TCDD	$C_{12}H_4Cl_4O_2$	322.0		190.0	419.0	−5.20	8.77	2.82	6.60
2,3,7,8-TCDD	$C_{12}H_4Cl_4O_2$	322.0		305.0	446.5	−6.70	10.22	2.87	6.80
1,2,3,4,7-PCDD	$C_{12}H_3Cl_5O_2$	356.4		196.0	464.7	−7.05	9.48	3.96	7.40
1,2,3,4,7,8-HCDD	$C_{12}H_2Cl_6O_2$	391.0		273.0	487.7	−8.29	10.94	3.74	7.80
Octachloro-DD	$C_{12}Cl_8O_2$	460.0		322.0	510.0	−10.00	12.79	3.60	8.20

Polychlorinated Dibenzofurans (PCDFs), Selected Congeners

Dibenzofuran (DF)	$C_{12}H_8O$	168.2		86.5	287.0	−0.52	4.55	2.36	4.31
2,8-DCDF	$C_{12}H_6Cl_2O$	237.1		184.0	375.0	−3.41	7.21	2.59	5.44
2,3,7,8-TCDF	$C_{12}H_4Cl_4O$	306.0		227.0	438.0	−5.70	8.86	3.23	6.10
2,3,4,7,8-PCDF	$C_{12}H_3Cl_5O$	340.4		196.0	464.7	−6.46	9.16	3.87	6.50
1,2,3,4,7,8-HCDF	$C_{12}H_2Cl_6O$	374.9		225.5	487.7	−7.50	10.66	3.23	7.00
Octachloro-DF	$C_{12}Cl_8O_2$	443.8		258.0	537.0	−9.30	11.58	4.11	8.00

Compound Name	Molecular Formula	M_i (g·mol⁻¹)	ρ_i (g·cm⁻³)	T_m (°C)	T_b (°C)	$\log p_i^*$/Pa	$-\log C_{iw}^{sat}$	$-\log K_{iaw}$ calculated (experimental)	$\log K_{iow}$	pK_{ia}
Alkylated Phenols										
Phenol	C_6H_6O	94.1	1.05	40.9	181.8	1.79	0.005	4.59 (4.79 exp.)	1.44	9.95
2-Methylphenol (*o*-cresol)	C_7H_8O	108.1	1.05	30.7	191.0	1.60	0.61	4.18 (4.30 exp.)	2.07	10.28
3-Methylphenol (*m*-cresol)	C_7H_8O	108.1	1.03	11.9	202.1	1.30	0.67	4.42	1.98	10.05
4-Methylphenol (*p*-cresol)	C_7H_8O	108.1	1.03	35.2	201.9	1.20	0.75	4.44 (4.49 exp.)	1.93	10.25
4-Ethylphenol	$C_8H_{10}O$	122.2	1.03	47.0	219.0	0.80	1.18	4.40	2.50	10.00
2,6-Dimethylphenol	$C_8H_{10}O$	122.2	1.13	49.0	203.0	1.28	1.29	3.70 (3.56 exp.)	2.36	10.63
3,4-Dimethylphenol	$C_8H_{10}O$	122.2	1.14	67.0		−0.07	1.40	5.06	2.23	10.34
2,4,6-Trimethylphenol	$C_9H_{12}O$	136.2		72.0		0.82	2.10	3.47	2.73	10.90
4-*n*-Butylphenol	$C_{10}H_{14}O$	150.2	0.98	22.0	248.0	0.08	2.31	4.20	3.64	
4-*t*-Butylphenol	$C_{10}H_{14}O$	150.2		99.0	238.0		2.11		3.14	9.90
4-*n*-Octylphenol	$C_{14}H_{22}O$	206.3		41.5		−1.14	4.18	3.35		
4-*n*-Nonylphenol	$C_{15}H_{24}O$	220.4	1.51	96.0	295.0	−1.15	4.64	2.89	5.76	
Chlorinated Phenols										
2-Chlorophenol	C_6H_5ClO	128.6	1.26	9.8	175.2	2.50	0.65	3.24	2.19	8.44
3-Chlorophenol	C_6H_5ClO	128.6	1.25	32.6	214.0	1.54	0.69	4.16	2.48	8.98
4-Chlorophenol	C_6H_5ClO	128.6	1.31	42.7	219.0	1.27	0.68	4.43	2.42	9.29
2,4-Dichlorophenol	$C_6H_4Cl_2O$	163.0	1.38	43.7	213.0	1.20	1.57	3.61	3.09	7.85
2,4,5-Trichlorophenol	$C_6H_3Cl_3O$	197.5	1.50	62.5		0.62	2.22	3.55	3.90	6.91
2,4,6-Trichlorophenol	$C_6H_3Cl_3O$	197.5		68.8	243.5	0.37	2.37	3.65	3.67	6.19
2,3,4,5-Tetrachlorophenol	$C_6H_2Cl_4O$	231.9		117.0		−1.00	3.15	4.24	4.87	6.35
2,3,4,6-Tetrachlorophenol	$C_6H_2Cl_4O$	231.9		69.5		−0.55	3.10	3.84	4.45	5.40
Pentachlorophenol	C_6HCl_5O	266.3		189.3		−2.04	4.15	4.28	5.24	4.83
Nitrophenols										
2-Nitrophenol	$C_6H_5NO_3$	139.1	1.55	44.7	215.0	1.26	2.03	3.10 (3.40 exp.)	1.78	7.15
3-Nitrophenol	$C_6H_5NO_3$	139.1		96.5			1.03		2.00	8.36
4-Nitrophenol	$C_6H_5NO_3$	139.1	1.48	114.0		−2.26	0.98	7.66	1.96	7.06
2,4-Dinitrophenol	$C_6H_4N_2O_5$	184.1		114.1			2.74		1.66	4.01
2,4-Dinitro-6-methylphenol (dinitro-*o*-cresol; DNOC)	$C_7H_6N_2O_5$	198.1		86.5		−1.14	3.00	4.53	2.12	4.31

Miscellaneous Phenolic Compounds

Compound	Formula	MW		mp (°C)	bp (°C)					pKa
1,2-Dihydroxybenzene (catechol)	$C_6H_6O_2$	110.1	1.15	104.0	245.0	-0.65	0.39	(6.65)	0.88	9.34 / 12.60
1,3-Dihydroxybenzene (resorcinol)	$C_6H_6O_2$	110.1	1.27	110.0	277.0	-2.93	0.00	(9.32)	0.80	9.32 / 11.10
1,4-Dihydroxybenzene (hydroquinone)	$C_6H_6O_2$	110.1	1.33	172.0	287.0	-2.59	0.20	(8.78)	0.59	9.85 / 11.40
2-Methoxyphenol (guaiacol)	$C_7H_8O_2$	124.1		32.0	205.0	1.32	0.70	4.38 (4.28 exp.)	1.32	
4,5-Dichloro-2-methoxyphenol (4,5-dichloroguaiacol)	$C_7H_6Cl_2O_2$	193.0		74.0		-0.24	2.52	4.11	3.26	8.52
3,4,5-trichloro-2-methoxyphenol (3,4,5-Trichloro-guaiacol)	$C_7H_5Cl_3O_2$	227.5		85.5		-0.79	2.86	4.32	3.77	7.56
4,5,6-Trichloro-2-methoxyphenol (4,5,6-trichloro-guaiacol)	$C_7H_5Cl_3O_2$	227.5		113.5		-1.49	3.62	4.26	3.74	7.07
Tetrachloro-2-methoxyphenol (tetrachloroguaiacol)	$C_7H_4Cl_4O_2$	261.9		121.5		-1.80	4.00	4.19	4.45	6.26
Aldehydes										
Methanal (formaldehyde)	CH_2O	30.0		-92.0	-21.00	5.72		(4.90 exp.)	0.45	
Ethanal (acetaldehyde)	C_2H_4O	44.1	0.78	-123.0	20.8	5.08		(2.52 exp.)	0.59	
Propanal	C_3H_6O	58.1	0.87	-80.0	48.0	4.63		(2.50 exp.)	0.88	
n-Butanal (butyraldehyde)	C_4H_8O	72.1	0.80	-96.4	74.8	4.19	0.01	2.20		
iso-Butanal (isobutyraldehyde)	C_4H_8O	72.1	0.81	-65.0	64.1	4.36		(2.33 exp.)		
n-Pentanal (valeraldehyde)	$C_5H_{10}O$	86.1	0.81	-93.5	103.0	3.66		(2.22 exp.)		
Hexanal	$C_6H_{12}O$	100.2	0.85	-56.0	131.0	3.17	1.30	1.92	1.78	
2-Propenal (acrolein)	C_3H_4O	56.1	0.84	-87.3	52.6	4.56		(2.06 exp.)	-0.01	
Benzaldehyde	C_7H_6O	106.1	1.04	-26.0	178.8	2.24	1.55	2.60 (2.95 exp.)	1.48	
Ketones										
Propanone (acetone)	C_3H_6O	58.1	0.79	-94.7	56.1	4.50	miscible	(2.80 exp.)	-0.24	
Butanone	C_4H_8O	72.1	0.81	-87.0	79.6	4.09		(2.60 exp.)	0.29	
2-Pentanone	$C_5H_{10}O$	86.1	0.81	-76.9	102.3	3.70	0.16	2.53 (2.58 exp.)	0.90	
2-Hexanone	$C_6H_{12}O$	100.2	0.81	-55.8	128.6	3.20	0.76	2.43	1.38	
Cyclohexanone	$C_6H_{10}O$	98.1	0.95	-32.1	155.6	2.78	0.63	2.98	0.71	
Methyl-phenyl-ketone (acetophenone)	C_8H_8O	120.2	1.03	19.6	202.0	1.67	1.35	3.37	1.63	
Diphenylketone (benzophenone)	$C_{13}H_{10}O$	182.2		48.0	305.4	-1.05	2.82	4.62	3.18	

Compound Name	Molecular Formula	M_i (g·mol⁻¹)	ρ_i (g·cm⁻³)	T_m (°C)	T_b (°C)	$\log p_i^*$/Pa	$-\log C_{iw}^{sat}$	$-\log K_{iaw}$ calculated (experimental)	$\log K_{iow}$	pK_{ia}
Carboxylic Acids										
Acetic acid	$C_2H_4O_2$	60.1		16.7	117.9	3.32	miscible	(4.95 exp.)	−0.25	4.75
Propanoic acid	$C_3H_6O_2$	74.1	0.99	−20.7	141.1	2.70	miscible	(4.74 exp.)	0.33	4.87
Butanoic acid (butyric acid)	$C_4H_8O_2$	88.1	0.96	−5.7	163.7	2.18	0.19	4.02	0.79	4.85
Hexanoic acid	$C_6H_{12}O_2$	116.1	0.93	−3.5	205.0	0.70		(4.66 exp.)	1.92	4.87
Benzoic acid	$C_7H_6O_2$	122.1	1.27	122.4	249.2	−0.96	1.55	5.80	1.89	4.19
Phenylacetic acid	$C_8H_8O_2$	136.2		77.0	265.0	−0.08	0.92	5.55	1.41	4.31
Salicylic acid (2-hydroxy benzoic acid)	$C_7H_6O_3$	138.1	1.44	159.0	211.0	−1.70	1.78	5.31	2.24	2.97 13.40
o-Phthalic acid	$C_8H_6O_4$	166.1		210.0			1.38		0.73	2.89 5.51
Carboxylic Acid Esters										
Methyl acetate	$C_3H_6O_2$	74.1	0.93	−98.1	56.9	4.46	−0.52	2.45 (2.04 exp.)	0.20	
Ethyl acetate	$C_4H_8O_2$	88.1	0.90	−41.5	77.1	4.10	−0.04	2.33 (2.16 exp.)	0.69	
Propyl acetate	$C_5H_{10}O_2$	102.1	0.89	−95.0	101.5	3.65	0.67	2.07 (2.05 exp.)	1.24	
Butyl acetate	$C_6H_{12}O_2$	116.2	0.88	−73.5	126.1	3.20	1.28	1.91 (1.94 exp.)	1.80	
Hexyl acetate	$C_8H_{16}O_2$	144.2	0.88	−80.9	171.5	2.27	2.46	1.66	2.83	
Vinyl acetate	$C_4H_6O_2$	86.1	0.93	−92.8	72.5	4.15	0.64	1.60	0.73	
Methyl benzoate	$C_8H_8O_2$	136.2	1.09	−12.1	199.5	1.72	1.81	2.86	2.20	
Ethyl benzoate	$C_9H_{10}O_2$	150.2	1.05	−34.0	212.4	1.38	2.63	2.38	2.64	
Phthalates										
Dimethylphthalate	$C_{10}H_{10}O_4$	194.2	1.19	5.5	283.7	0.38	1.66	4.35	1.53	
Diethylphthalate	$C_{12}H_{14}O_4$	222.2	1.23	−40.5	296.0	−0.66	2.44	4.61	2.39	
Di-*n*-propyl-phthalate	$C_{14}H_{18}O_4$	250.3			304.5		3.36		3.27	
Di-*n*-butyl-phthalate	$C_{16}H_{22}O_4$	278.3	1.05	−35.0	340.0	−2.28	4.36	4.31	4.61	
Benzyl-*n*-butyl-phthalate	$C_{19}H_{20}O_4$	312.4	1.12	<−35.0	370.0	−2.94	5.08	4.25	4.91	
Di-(2-ethylhexyl)-phthalate	$C_{24}H_{38}O_4$	390.6	0.98	−50.0	385.0	−2.72	7.13	1.98	7.48	
Aromatic Amines										
Aminobenzene (aniline)	C_6H_7N	93.1	1.01	−6.3	184.4	1.92	0.44	4.03	0.95	4.63
2-Methylaniline (o-toluidine)	C_7H_9N	107.2	1.00	−16.3	200.3	1.55	0.82	4.02	1.32	4.44

Name	Formula	MW		mp (°C)	bp (°C)					
3-Methylaniline (m-toluidine)	C7H9N	107.4								
4-Methylaniline (p-toluidine)	C7H9N	107.2	1.04	43.7	200.4	1.60	1.15	3.64	1.39	5.17
2,6-Dimethylaniline	C8H11N	121.2	0.98	11.2	214.0	1.30	1.41	3.68	1.84	3.95
3,4-Dimethylaniline	C8H11N	121.2		51.0	228.0		1.65		1.84	5.28
N,N-Dimethylaniline	C8H11N	121.2		2.5	194.0	1.95	2.04	2.40	2.31	5.12
2-Chloroaniline	C6H6ClN	127.6	1.21	-14.0	208.8	1.54	1.53	3.32	1.88	2.65
3-Chloroaniline	C6H6ClN	127.6	1.22	-10.3	229.5	0.98	1.37	4.04	1.99	3.52
4-Chloroaniline	C6H6ClN	127.6	1.43	71.0	231.3	0.40	1.64	4.35	1.83	4.00
3,4-Dichloroaniline	C6H5Cl2N	162.0		71.0	272.0	0.36	3.24	2.79	2.70	2.97
N-Phenylaniline (diphenylamine)	C12H11N	169.2	1.16	53.0	302.0	-1.22	3.53	4.08	3.50	0.90
4,4'-Diaminobiphenyl (benzidine)	C12H12N2	184.2		120.0	401.0		2.66		1.34	3.57 / 4.66
1-Naphthylamine	C10H9N	143.2	1.13	49.5	300.9		1.92	5.34	2.25	3.92
2-Naphthylamine	C10H9N	143.2	1.06	111.5	306.0			5.48	2.30	4.15
Aliphatic Amines										
Methylamine	CH5N	31.1		-92.5	-6.5	5.55		(3.34 exp.)	-0.57	
Dimethylamine	C2H7N	45.1		-96.0	7.4	5.31		(3.15 exp.)	-0.38	
Trimethylamine	C3H9N	59.1		-117.2	2.9	5.34		(2.56 exp.)	0.27	
Ethylamine	C2H7N	45.1		-81.0	16.6	5.15		(3.30 exp.)	-0.13	
Diethylamine	C4H11N	73.1	0.71	-49.5	56.1	4.50		(2.98 exp.)	0.50	
n-Propylamine	C3H9N	59.1	0.72	-83.0	48.7	4.62		(3.22 exp.)	0.32	
n-Butylamine	C4H11N	73.1	0.74	-50.0	77.9	4.11		(3.11 exp.)	0.98	
n-Hexylamine	C6H15N	101.2	0.77	-22.9	132.8	3.07	0.25	3.07 (2.90 exp.)	2.06	
Heterocyclic N-Compounds										
Pyridine	C5H5N	79.1	0.98	-41.6	115.3	3.44		(3.44 exp.)	0.65	
2-Methylpyridine (2-picoline)	C6H7N	93.1	0.94	-66.7	129.4	3.18		(3.40 exp.)	1.11	
3-Methylpyridine (3-picoline)	C6H7N	93.1	0.95	-18.1	144.1	3.12		(3.50 exp.)	1.22	
4-Methylpyridine (4-picoline)	C6H7N	93.1	0.95	3.7	145.4	2.88		(3.62 exp.)	1.22	
2,3-Dimethylpyridine (2,3-lutidine)	C7H9N	107.2	0.94	-15.5	163.5	2.62		(3.54 exp.)	1.65	
2,6-Dimethylpyridine (2,6-lutidine)	C7H9N	107.2	0.92	-6.1	144.1	2.87		(3.37 exp.)	1.68	
Quinoline	C9H7N	129.2	1.10	-14.8	237.1	0.08	1.33	4.98 (4.20 exp.)	2.06	
Isoquinoline	C9H7N	129.2	1.10	26.5	243.2			5.40		
Nitrobenzenes										
Nitrobenzene	C6H5NO2	123.1	1.20	5.7	210.8	1.48	1.79	3.12	1.85	
2-Methylnitrobenzene (2-nitrotoluene)	C7H7NO2	137.1	1.16	-9.9	222.0	1.43	2.35	2.61	2.30	

Compound Name	Molecular Formula	M_i (g·mol⁻¹)	ρ_i (g·cm⁻³)	T_m (°C)	T_b (°C)	log p_i^*/Pa	-log C_{iw}^{sat}	-log K_{iaw} calculated (experimental)	log K_{iow}	pK_{ia}
3-Methylnitrobenzene (3-nitrotoluene)	$C_7H_7NO_2$	137.1		15.9	232.0	1.28	2.43	2.68	2.43	
4-Methylnitrobenzene (4-nitrotoluene)	$C_7H_7NO_2$	137.1	1.39	51.6	238.3	1.20	2.43	2.76	2.38	
2-Chloronitrobenzene	$C_6H_4ClNO_2$	157.6		33.0	245.3	0.70	2.69	3.00	2.45	
3-Chloronitrobenzene	$C_6H_4ClNO_2$	157.6		45.0	235.6	0.27	2.73	3.39 (2.74 exp.)	2.48	
4-Chloronitrobenzene	$C_6H_4ClNO_2$	157.6		83.3	240.5	0.44	2.92	3.03 (2.66 exp.)	2.50	
1,3-Dinitrobenzene	$C_6H_4N_2O_4$	168.1		89.9	301.9	-1.92	2.54	5.77	1.49	
1-Methyl-2,4-dinitrobenzene (2,4-dinitrotoluene)	$C_7H_6N_2O_4$	182.1	1.52	70.3		-1.54	2.86	5.25	2.00	
1,3-Dinitro-2-methylbenzene (2,6-dinitrotoluene)	$C_7H_6N_2O_4$	182.1	1.28	65.5		-1.12	3.00	4.51 (4.80 exp.)	2.03	
1,3,5-Trinitrobenzene	$C_6H_3N_3O_6$	213.1		122.9		-1.82	2.81	5.40	1.18	
2-Methyl-1,3,5-trinitrobenzene (2,4,6-trinitrotoluene "TNT")	$C_7H_5N_3O_6$	227.1		80.8		-3.07	3.24	6.22 (5.98 exp.)	1.98	

Triazine-, Carbamide-, Carbamate-, and Urea Pesticides

Compound Name	Molecular Formula	M_i (g·mol⁻¹)	ρ_i (g·cm⁻³)	T_m (°C)	T_b (°C)	log p_i^*/Pa	-log C_{iw}^{sat}	-log K_{iaw} calculated (experimental)	log K_{iow}	pK_{ia}
Simazine	$C_7H_{12}ClN_5$	201.7	1.30	226.0		-5.07	4.55	6.91	2.18	
Atrazine	$C_8H_{10}ClN_5$	215.7	1.19	175.0		-4.40	3.86	6.93	2.65	
Cyanazine	$C_9H_{13}ClN_6$	240.7		167.0		-6.67	3.15	9.90	2.22	
Propachlor	$C_{11}H_{14}ClNO$	211.7	1.13	71.0		-1.70	2.55	5.54	2.20	
Alachlor	$C_{14}H_{20}ClNO$	269.8	1.13	41.0		-2.70	3.05	6.04	2.95	
Metolachlor	$C_{15}H_{22}ClNO_2$	283.8		<25		-2.40	2.75	6.04	3.13	
Carbaryl	$C_{12}H_{11}NO_2$	201.2	1.23	142.0		-4.57	3.22	7.74	2.36	
Carbofuran	$C_{12}H_{15}NO_3$	221.3	1.18	152.5		-4.12	2.81	7.70	1.52	
Fenuron	$C_9H_{12}N_2O$	164.2	1.08	135.5			1.63		0.98	
Diuron	$C_9H_{10}Cl_2N_2O$	233.1	1.48	158.5		<-4.0	3.76		2.60	
Isoproturon	$C_{12}H_{18}N_2O$	206.3	1.16	155.5		<-4.0	3.57		2.50	

Phosphoric- and Thiophosphoric Acid (Thio)Esters

Compound Name	Molecular Formula	M_i (g·mol⁻¹)	ρ_i (g·cm⁻³)	T_m (°C)	T_b (°C)	log p_i^*/Pa	-log C_{iw}^{sat}	-log K_{iaw} calculated (experimental)	log K_{iow}	pK_{ia}
Tributylphosphate	$C_{12}H_{27}O_4P$	263.3	0.98	<0		-0.10	2.65	3.84	2.50	
Tri-o-cresylphosphat	$C_{21}H_{21}O_4P$	368.4	1.21	50		-3.00	5.31	4.08	5.11	
Parathion	$C_{10}H_{14}NO_5PS$	291.3	1.27	6.1		-3.22	4.30	5.31	3.81	
Methylparathion	$C_8H_{10}NO_5PS$	263.5	1.36	37.5		-2.70	4.00	5.10	3.00	
Fenthion	$C_{10}H_{15}O_3PS_2$	278.3	1.25	7.5		-2.70	3.74	5.35	4.10	
Disulfoton	$C_8H_{19}O_2PS_3$	274.4	1.14	<25		-1.80	4.20	4.00	4.02	

Appendix D

Temperature Dependence of Equilibrium Constants and Rate Constants

D.1   Temperature Dependence of Equilibrium Constants and Rate Constants as a Function of the Corresponding Enthalpy Changes and Activation Energies, Respectively

D.2   Temperature Dependence of the Ion Product of Water

Table D.1 Temperature Dependence of Equilibrium Constants (K_{i12}, K_r) or Rate Constants (k) as a Function of the Corresponding Enthalpy Changes [$\Delta_{12}H_i$ (Eq. 3-51), $\Delta_r H^0$ (Eq. 8-20)] or Activation Energies [E_a (Eq. 12-30)], Respectively. Values Given as Percent of the Value at 25°C ($T = 298$ K)

$\Delta_{12}H_i$, $\Delta_r H^0$, or E_a (kJ·mol⁻¹)	0°C	5°C	10°C	15°C	20°C	25°C	30°C	35°C	40°C	Average factor for a change in temperature of 10°C
10	69.1	74.8	80.7	86.9	93.3	100.0	106.9	114.0	121.3	1.2
15	57.4	64.7	72.6	81.0	90.2	100.0	110.5	121.7	133.7	1.2
20	47.7	55.9	65.2	75.6	87.1	100.0	114.2	130.0	147.2	1.3
25	39.7	48.4	58.6	70.4	84.2	100.0	118.1	138.8	162.2	1.4
30	33.0	41.8	52.6	65.7	81.3	100.0	122.1	148.2	178.6	1.5
35	27.4	36.2	47.3	61.2	78.6	100.0	126.3	158.2	196.8	1.6
40	22.8	31.3	42.5	57.1	75.9	100.0	130.5	168.9	216.8	1.8
45	19.0	27.1	38.2	53.2	73.3	100.0	134.9	180.3	238.8	1.9
50	15.8	23.4	34.3	49.6	70.9	100.0	139.5	192.6	263.0	2.0
55	13.1	20.3	30.8	46.3	68.5	100.0	144.2	205.6	289.7	2.2
60	10.9	17.5	27.7	43.1	66.2	100.0	149.1	219.5	319.2	2.3
65	9.1	15.1	24.9	40.2	63.9	100.0	154.2	234.4	351.6	2.5
70	7.5	13.1	22.4	37.5	61.7	100.0	159.4	250.2	387.3	2.7
75	6.3	11.3	20.1	35.0	59.7	100.0	164.8	267.2	426.6	2.9
80	5.2	9.8	18.1	32.6	57.6	100.0	170.4	285.3	469.9	3.1
85	4.3	8.5	16.2	30.4	55.7	100.0	176.1	304.6	517.6	3.3
90	3.6	7.3	14.6	28.3	53.8	100.0	182.1	325.2	570.2	3.6
95	3.0	6.3	13.1	26.4	52.0	100.0	188.3	347.2	628.1	3.8
100	2.5	5.5	11.8	24.6	50.2	100.0	194.6	370.8	691.8	4.1
105	2.1	4.7	10.6	23.0	48.5	100.0	201.2	395.9	762.1	4.4
110	1.7	4.1	9.5	21.4	46.9	100.0	208.1	422.7	839.5	4.7
115	1.4	3.5	8.5	20.0	45.3	100.0	215.1	451.3	924.7	5.1
120	1.2	3.1	7.7	18.6	43.8	100.0	222.4	481.9	1018.6	5.4
125	1.0	2.7	6.9	17.3	42.3	100.0	229.9	514.5	1122.0	5.8
130	0.8	2.3	6.2	16.2	40.8	100.0	237.7	549.3	1236.0	6.3
135	0.7	2.0	5.6	15.1	39.5	100.0	245.7	586.5	1361.5	6.7
140	0.6	1.7	5.0	14.1	38.1	100.0	254.1	626.2	1499.7	7.2
145	0.5	1.5	4.5	13.1	36.8	100.0	262.7	668.6	1652.0	7.8
150	0.4	1.3	4.0	12.2	35.6	100.0	271.6	713.9	1819.8	8.3

Table D.2 Temperature Dependence of the Ion Product of Water [a] (from Stumm and Morgan, 1996)

$T/°C$	K_w	pK_w	$T/°C$	K_w	pK_w
0	0.12×10^{-14}	14.93	20	0.68×10^{-14}	14.17
5	0.18×10^{-14}	14.73	25	1.01×10^{-14}	14.00
10	0.29×10^{-14}	14.53	30	1.47×10^{-14}	13.83
15	0.45×10^{-14}	14.35	35	5.48×10^{-14}	13.26

[a] $\log K_w = -4471 / T + 6.0875 - 0.01706 \, T$ (T in K).

Appendix E

Properties of the Earth

E.1   Properties of the Earth

E.2   Stocks and Flows of Water on Earth

Table E.1 Masses, Volumes, and Areas of the Earth

Mass of Earth	**5.98×10^{24} kg**
Atmosphere	5.14×10^{18} kg
Stratosphere	0.5×10^{18} kg
Oceans	1.4×10^{21} kg
Water in atmosphere	1.3×10^{16} kg
Surface fresh water	1.26×10^{17} kg
Area total	**5.10×10^{14} m^2**
Total land	**1.48×10^{14} m^2**
Eurasia	0.536×10^{14} m^2
Africa	0.298×10^{14} m^2
America	0.417×10^{14} m^2
Antarctica	0.149×10^{14} m^2
Oceania	0.089×10^{14} m^2
Ice-free land	1.33×10^{14} m^2
Total ocean	**3.61×10^{14} m^2**
Ice-free Pacific Ocean	1.66×10^{14} m^2
Ice-free Atlantic Ocean	0.83×10^{14} m^2
Indian Ocean	0.65×10^{14} m^2
Ice-free Arctic ocean	0.14×10^{14} m^2
Sea ice (average)	0.33×10^{14} m^2
Volume of oceans	**1.37×10^{18} m^3**
Volume of mixed ocean layer	2.7×10^{16} m^3
Mean ocean depth	3740 m
Mean depth of oceanic mixed surface layer	75 m

Table E.2 Stocks and Flows of Water on Earth

Volume of oceans	1370×10^{15} m^3
Ice	29×10^{15} m^3
Groundwater[a]	8.3×10^{15} m^3
Freshwater lakes	0.125×10^{15} m^3
Saline lakes and inland seas	0.104×10^{15} m^3
Soil water	0.067×10^{15} m^3
Atmosphere	0.013×10^{15} m^3
Water in living biomass	0.003×10^{15} m^3
Average amount in stream channels	0.001×10^{15} m^3

Global flow rates of water	total flux $(10^{12}$ m^{3}yr$^{-1})$	flux per area[b] (m yr^{-1})
Precipitation on the sea	401	1.11
Evaporation from the sea	456	1.26
Precipitation on land	108	0.73
Evapotranspiration from the land	62	0.42
Runoff	46	0.31

[a] About one half of the stock lies within a depth of 1 km. [b] per land or ocean area, respectively

BIBLIOGRAPHY

Aber, J. D., and M. Martin, *ACCP Leaf Chemistry Data.*, http://www.eosdis.ornl.gov.1999.

Abraham, M. H., G. S. Whiting, R. M. Doherty, W. J. Shuely, "XVI. A new solute solvation parameter, π_2^H, from gas chromatographic data", *J. of Chromatography,* **587**, 213-228 (1991).

Abraham, H. M., G. S. Whiting, P. W. Carr, and H. Ouyang, "Hydrogen bonding. Part 45. The solubility of gases and vapours in methanol at 298 K: an LFER anylysis", *J. Chem. Soc. Perkin Trans.*, **2**, 1283-1530 (1998).

Abraham, M. H., "Scales of solute hydrogen bonding: Their construction and application to physicochemical and biochemical processes", *Chem. Soc. Rev.*, **22**, 73-83 (1993).

Abraham, M. H., J. Andonian-Haftan, G. S. Whiting, A. Leo, and R. S. Taft, "Hydrogen bonding, Part 34. The factors that influence the solubility of gases and vapors in water at 298 K, and a new method for its determination", *J. Chem. Soc. Perkin Trans.*, **2**, 1777-1791 (1994a).

Abraham, M. H., H. C. Chadha, R. A. E. Leitao, R. C. Mitchell, W. J. Lambert, R. Kaliszan, A. Nasal, and P. Haber, "Determination of solute lipophilicity, as log P (octanol) and log P (alkane) using poly(styrene-divinylbenzene) and immobilized artificial membrane stationary phases in reversed-phase high performance liquid chromatography", *J. Chromatogr. A*, **766**, 35-47 (1997).

Abraham, M. H., S. C. Chadha, G. S. Whiting, and R. C. Mitchel, "Hydrogen bonding. 32. An analysis of water–octanol and water–alkane partitioning and the Δlog P parameter of Seiler", *J. Pharm. Sci.*, **83**, 1085-1100 (1994b).

Abraham, M. H., P. L. Grellier, and R. A. McGill, "Determination of olive oil–gas and hexadecane–gas partition coefficients, and calculation of the corresponding olive oil-water and hexadecane-water partition coefficients." *J. Chem. Soc. Perkin Trans.*, **2**, 797-803 (1987).

Abraham, M. H., J. Le, W. E. Acree, P. W. Carr, and A. J. Dallas, "The solubility of gases and vapors in dry octan-1-ol at 298 K." *Chemosphere*, **44**, 855-863 (2001).

Abraham, M. H., and J. C. McGowan, "The use of characteristic volumes to measure cavity terms in reversed phase liquid chromatography", *Chromatographia*, **25**, 243-246 (1987).

Abraham, M. H., J. A. Platts, A. Hersey, A. J. Leo, and R. W. Taft, "Correlation and estimation of gas–chloroform and water–chloroform partition coefficients by a linear energy relationship method", *J. Pharmaceutical Sci.*, **88**, 670-679 (1999).

Abraham, M. H., G. S. Whiting, R. Fuchs, and E. J. Chambers, "Thermodynamics of Solute Transfer from Water to Hexadecane", *J. Chem. Soc. Perkin Trans.*, **2**, 291-300 (1990).

Accardi-Dey, A., and P. M. Gschwend, "Assessing the combined roles of natural organic matter and black carbon as sorbents in sediments", *Environ. Sci. Technol.*, **36**, 21-29 (2002).

Achtnich, C., H. Lenke, U. Klaus, M. Spiteller, and H.-J. Knackmuss, "Stability of immobilized TNT derivatives in soil as a function of nitro group reduction", *Environ. Sci. Technol.*, **34**, 3698-3704 (2000).

Aeschbach-Hertig, W., R. Kipfer, M. Hofer, D. M. Imboden, and H. Baur, "Density-driven exchange between the basins of Lake Lucerne (Switzerland) traced with 3H-3He method", *Limnol. Oceanogr.*, **41**, 4, 707-721 (1996).

Agrawal, A., and P. G. Tratnyek, "Reduction of nitro aromatic compounds by zero-valent iron metal", *Environ. Sci. Technol.*, **30**, 153-160 (1996).

Ahel, M., C. Schaffner, and W. Giger, "Behaviour of alkylphenol polyethoxylate surfactants in the aquatic environment – III. Occurrence and elimination of their persistant metabolites during infiltration of river water to groundwater", *Water Res.*, **30**, 37-46 (1996).

Ahti, T., and T. L. Hepburn, *Preliminary Studies on Woodland Caribou Range, Especially on LichenSstands in Ontario*, Res. Rep. (Wildlife) No. 74, Ontario Dept. Lands and Forests, Research Branch, Toronto, ON, 1967.

Ainsworth, S. J., "Soaps and detergents", *Chem. Eng. News*, **84**, 32-54 (1996).

Alaee, M., R. M. Whittal, and W. M. J. Strachan, "The effect of water temperature and composition on Henry's law constant for various PAHs", *Chemosphere*, **32**, 1153-1164 (1996).

Alcock, R. E., and K. C. Jones, "Dioxins in the environment: A review of trend data", *Environ. Sci. Technol.*, **30**, 3133-3143 (1996).

Alexander, M., "Biodegradation of chemicals of environmental concern", *Science*, **211**, 132-138 (1981).

Ali, M. A., and D. A. Dzombak, "Competitive sorption of simple organic acids and sulfate on goethite", *Environ. Sci. Technol.*, **30**, 1061-1071 (1996a).

Ali, M. A., and D. A. Dzombak, "Effects of simple organic acids on sorption of Cu^{2+} and Ca^{2+}", *Geochim. Cosmochim. Acta*, **60**, 291-304 (1996b).

Allen, J. M., S. Lucas, and S. K. Allen, "Formation of hydroxyl radical in illuminated surface waters contaminated with acidic mine drainage", *Environ. Tox. Chem.*, **15**, 107-113 (1996).

Almeida, M. B., A. M. Alvarez, E. M. deMiguel, and E. S. delHoyo, "Setschenow coefficients for naphthols by distribution method", *Can. J. Chem.*, **61**, 244-248 (1983).

Altenbach, and W. Giger, "Determination of benzenesulfonates and naphthalenesulfonates in waste-water by solid-phase extraction with graphitized carbon-black and ion-pair liquid-chromatography with UV-detection", *Anal. Chem.*, **67**, 2325-2333 (1995).

Aluwihare, L. I., and D. J. Repeta, "A comparison of the chemical characteristics of oceanic DOM and extracellular DOM produced by marine algae", *Marine Ecology-Progress Series*, **186**, 105-117 (1999).

Aluwihare, L. I., D. J. Repeta, and R. F. Chen, "A major biopolymeric component to dissolved organic carbon in surface sea water", *Nature*, **387**, 166-169 (1997).

Alvarez-Cohen, L., and G. E. J. Speitel, "Kintics of aerobic cometabolism of chlorinated solvents", *Biodegradation*, **12**, 105-126 (2001).

Aminabhavi, T. M., and H. G. Naik, "Chemical compatibility study of geomembranes-sorption/desorption, diffusion and swelling phenomena", *J. Hazardous Materials*, **60**, 175-203 (1998).

Amonette, J. E., D. J. Workman, D. W. Kennedy, J. Fruchter, and Y. A. Gorbi, "Dechlorination of carbon tetrachloride by Fe(II) associated with goethite", *Environ. Sci. Technol.*, **34**, 4606-4613 (2000).

Anderson, P. N., and R. A. Hites, "System to measure relative rate constants of semivolatile organic compounds with hydroxyl-radicals", *Environ. Sci. Technol.*, **30**, 301-306 (1996).

Annweiler, E., A. Materna, M. Safinowski, A. Kappler, H. H. Richnow, W. Michaelis, and R. U. Meckenstock, "Anaerobic degradation of 2-methylnaphthalene by a sulfate-reducing enrichment culture", *Appl. Environ. Microbiol.*, **66**, 5329-5333 (2000).

Aochi, Y. O., and W. J. Farmer, "Role of microstructural properties in the time-dependent sorption/desorption behavior of 1,2-dichloroethane on humic substances", *Environ. Sci. Technol.*, **31**, 2520-2526 (1997).

Appelo, C. A. J., and D. Postma, *Geochemistry, Groundwater and Pollution*, A. A. Balkema, Rotterdam, 1993.

Appleton, H., T., S. Banerjee, and H. C. Sikka, "Fate of 3,3'-dichlorobenzidine in the aquatic environment". In *Dynamics, Exposure, and Hazard Assessment of Toxic Chemicals*, H. Rizwanul, Ed., Ann Arbor Science, Ann Arbor, MI, 1980, pp. 251-272.

Arcand, Y., J. Hawari, and S. R. Guiot, "Solubility of pentachlorophenol in aqueous solutions: The pH effect", *Wat. Res.*, **29**, 131-136 (1995).

Arnold, C. G., A. Ciani, S. R. Müller, A. Amirbahman, and R. P. Schwarzenbach, "Association of triorganotin compounds with dissolved humic acids", *Environ. Sci. Technol.*, **32**, 2976-2983 (1998).

Arnold, C. G., A. Weidenhaupt, M. M. David, S. R. Muller, S. B. Haderlein, and R. P. Schwarzenbach, "Aqueous speciation and 1-octanol–water partitioning of tributyl- and triphenyltin: Effect of pH and ion composition", *Environ. Sci. Technol.*, **31**, 2596-2602 (1997).

Arnold, W. A., and A. L. Roberts, "Pathways of chlorinated ethylene and chlorinated acetylene reaction with Zn(0)", *Environ. Sci. Technol.*, **32**, 3017-3025 (1998).

Arnold, W. A., and A. L. Roberts, "Pathways and kinetics of chlorinated ethylene and chlorinated acetylene reaction with Fe(0) particles", *Environ. Sci. Technol.*, **34**, 1794-1805 (2000).

Arrowsmith, D. K., and C. M. Place, *Dynamical Systems – Differential Equations, Maps and Chaotic Behaviour*, Chapman and Hall, London, 1992.

Atkins, P. W., *Physical Chemistry*, Oxford University Press, Oxford, 1998.

Atkinson, D., and G. Curthoys, "The determination of heats of adsorption by gas-solid chromatography", *J. Chem. Educ.*, **55**, 564-566 (1978).

Atkinson, R., "Kinetics and mechanisms of the gas-phase reactions of the hydroxyl radical with organic compounds under atmospheric conditions", *Chem. Rev.*, **86**, 69-201 (1986).

Atkinson, R., "Kinetics and mechanisms of the gas-phase reactitions of the hydroxyl radical with organic compounds", *J. Phys. Chem. Ref. Data, Monograph No. 1*, (1989).

Atkinson, R., "Gas-phase tropospheric chemistry of organic compounds", *J. Phys. Chem. Ref. Data, Monograph No. 2*, (1994).

Atkinson, R., "Atmospheric Chemistry of VOCs and NOx", *Atmos. Environ.*, **34**, 2063-2101 (2000).

Atkinson, R., R. Guichevit, R. A. Hites, W.-U. Palm, J. N. Seiber, and P. de Voogt, "Transformations of pesticides in the atmosphere: A state of the art", *Water, Air, and Soil Pollution*, **115**, 219-243 (1999).

Atkinson, R., E. C. Tuazon, and S. M. Aschmann, "Products of the gas phase reactions of O3 with alkenes", *Environ. Sci. Technol.*, **29**, 1860-1866 (1995).

Bachmann, A., P. Walet, P. Wijnen, W. De Bruin, L. M. Huntjens, W. Roelossen, and A. J. B. Zehnder, "Biodegradation of α- and β-hexachlorocyclohexane in a soil slurry under different redox conditions", *Appl. Environ. Microbiol.*, **54**, 143-149 (1988).

Backhus, D. A., *Colloids in Groundwater: Laboratory and Field Studies of Their Influence on Hydrophobic Organic Contaminants*, Ph.D. thesis, Massachusetts Institute of Technology, Cambridge, MA, 1990.

Backhus, D. A., and P. M. Gschwend, "Fluorescent polycyclic aromatic hydrocarbons as probes for studying the impact of colloids on pollutant transport in groundwater", *Environ. Sci. Technol.*, **24**, 1214-1223 (1990).

Bahadur, N. P., W.-Y. Shiu, D. G. B. Boocock, and D. Mackay, "Tricaprylin-water partition coefficients and their temperature dependence for selected chlorobenzenes", *J. Chem. Eng. Data*, **44**, 40-43 (1999).

Bailey, G. W., J. L. White, and T. Rothberg, "Adsorption of organic herbicides by montmorillonite: Role of pH and chemical character of adsorbate", *Soil Sci. Soc. Am. Proc.*, **32**, 222-234 (1968).

Baker, J. E., and S. J. Eisenreich, "PCBs and PAHs as tracers of particulate dynamics in large lakes", *J. Great Lakes Res.*, **15**, 84-103 (1989).

Baker, J. E., and S. J. Eisenreich, "Concentrations and fluxes of polycyclic aromatic hydrocarbons and polychlorinated biphenyls across the air–water interface of Lake Superior", *Environ. Sci. Technol.*, **24**, 342-352 (1990).

Baker, K. S., and R. C. Smith, "Bio-optical classification and model of natural waters", *Limnol. Oceanogr.*, **27**, 500-509 (1982).

Balistrieri, L. S., and J. W. Murray, "The influence of the major ions of seawater on the adsorption of simple organic acids by goethite", *Geochim. Cosmochim. Acta*, **51**, 1151-1160 (1987).

Ballapragada, B. S., H. D. Stensel, J. A. Puhakka, and J. F. Ferguson, "Effect of hydrogen on reductive dechlorination of chlorinated ethenes", *Environ. Sci. Technol.*, **31**, 1728-1734 (1997).

Ballschmiter, K., "Transport and fate of organic compounds in the global environment", *Angew. Chem. Int. Ed. Engl.*, **31**, 487-515 (1992).

Ballschmiter, K., R. Bacher, A. Mennel, R. Fischer, U. Riehle, and M. Swerev, "The determination of chlorinated biphenyls, chlorinated dibenzodioxins, and chlorinated dibenzofurans by GC-MS", *J. High Res. Chromatogr.*, **15**, 260-270 (1992).

Balmer, M., K.-U. Goss, and R. P. Schwarzenbach, "Photolytic transformation of organic pollutants on soil surfaces – An experimental approach", *Environ. Sci. Technol.*, **34**, 1240-1245 (2000).

Bamford, H. A., D. L. Poster, and J. E. Baker, "Temperature dependence of Henry's law constants of thirteen polycyclic aromatic hydrocarbons between 4°C and 31°C", *Environ. Toxicol. Chem.*, **18**, 1905-1912 (1999).

Banerjee, P., M. D. Piwoni, and K. Ebeid, "Sorption of organic contaminants to a low carbon subsurface core", *Chemosphere*, **8**, 1057-1067 (1985).

Bank, S., and R. J. Tyrell, "Copper(II)-promoted aqueous decomposition of aldicarb", *J. Org. Chem.*, **50**, 4938-4943 (1985).

Barlin, G. B., and D. D. Perrin, "Prediction of the strengths of organic acids", *Quart. Rev. Chem. Soc.*, **20**, 75-101 (1966).

Barnard, P. W. C., C. A. Burton, D. R. Llewellyn, C. A. Vernon, and V. A. Welch, "The reactions of organic phosphates. Part V. The hydrolysis of triphenyl and trimethyl phosphate", *J. Chem. Soc.*, **Part III**, 2670-2676 (1961).

Barrer, R. M., and R. R. Fergusson, "Diffusion of benzene in rubber and polyethylene", *Trans. Faraday Soc.*, **54**, 989-1000 (1958).

Barron, M. G., J. A. Hansen, and J. Lipton, "Association between contaminant tissue residues and effects in aquatic organisms", *Rev. Environ. Contam.*, **173**, 1-37 (2002).

Bartels, I., H.-J. Knackmuss, and W. Reinecke, "Suicide inactivation of catechol 2,3-dioxygenase from *Pseudomonas putida* mt-2 by 3-halocatechols", *Appl. Environ. Microbiol.*, **47**, 500-505 (1984).

Bartha, R., "Fate of herbicide-derived chloroanilines in soil", *J. Agric. Food. Chem.*, **19**, 385-387 (1971).

Bartholomew, G. W., and F. K. Pfaender, "Influence of spatial and temporal variations on organic pollutant biodegradation rates in an estuarine environment", *Appl. Environ. Microbiol.*, **45**, 102-109 (1983).

Bartnicki, E. W., N. O. Belser, and C. E. Castro, "Oxidation of heme proteins by alkyl halides: A probe for axial inner sphere redox capacity in solution and in whole cells", *Biochemistry*, **17**, 5582-5586 (1978).

Barton, A. F. M., *Handbook of Solubility Parameters and Other Cohesion Parameters*, CRC, Boca Raton, 1991.

Beattie, B. M., "Life-cycle and changes in carbohydrates, proteins and lipids of *Pentapedilum uncinatum Goet. (Diptera; Chironomidae)*", *Freshwater Biology*, **8**, 109-113 (1978).

Bell, R. P., and A. O. McDougall, "Hydration equilibrium of some aldehydes and ketones", *Trans. Faraday Soc.*, **56**, 1281-1285 (1960).

Beller, H., and A. M. Spormann, "Benzylsuccinate formation as a means of anaerobic toluene activation by sulfate-reducing strain PRTOL1." *Appl. Environ. Microbiol.*, **63**, 3729-3731 (1997).

Bender, M. L., and R. B. Homer, "The mechanism of the alkaline hydrolysis of *p*-nitrophenol *N*-methylcarbamate", *J. Org. Chem.*, **30**, 3975-3978 (1965).

Berezin, I. V., A. V. Levashov, and K. Martinek, "On the modes of interaction between competitive inhibitors and the alpha-chymotrypsin active centre", *FEBS Lett.*, **7**, 20-22 (1970).

Berg, M., C. G. Arnold, S. R. Müller, M. Schärer, and R. P. Schwarzenbach, "Sorption/disorption behavior of organotin compounds in sediment-pore water systems", *Environ. Sci. Technol.*, **35**, 3151-3157 (2001).

Berg, M., S. R. Müller, J. Mühlemann, A. Wiedmer, and R. P. Schwarzenbach, "Concentrations and mass fluxes of chloroacetic acids and trifluoroacetic acid in rain and natural waters in Switzerland", *Environ. Sci. Technol.*, **34**, 2675-2683 (2000).

Berglund, O., P. Larsson, G. Ewald, and L. Ohla, "Bioaccumulation and differential partitioning of polychlorinated biphenyls in freshwater, planktonic food webs", *Can. J. Fish. Aquat. Sci.*, **57**, 1160-1168 (2000).

Bisaillon, J.-G., F. Lepine, R. Beaudet, and M. Sylvestre, "Carboxylation of *o*-cresol by an anaerobic consortium under methanogenic conditions", *Appl. Environ. Microbiol.*, **57**, 2131-2134 (1991).

Blanchard, D. C., and A. H. Woodcock, "Bubble formation and modification in the sea and its meteorological significance", *Tellus*, **9**, 145-158 (1957).

Blehert, D. S., B. G. Fox, and G. H. Chambliss, "Cloning and sequence analysis of two Pseudomonas flavoprotein xenobiotic reductases", *J. Bacteriol.*, **181**, 6254-6263 (1999).

Bloesch, J., and M. Sturm, "Settling flux and sinking velocities of particulate phosphorus (PP) and particulate organic carbon (POC) in Lake Zug, Switzerland". In *Sediments and Water Interactions*, P. G. Sly, Ed., Springer, New York, 1986, pp. 481-490.

Blokzijl, W., and J. B. F. N. Engberts, "Hydrophobic effects – Opinions and facts", *Angew. Chem. Int. Ed.*, **32**, 1545-1579 (1993).

Blough, N. V., and R. G. Zepp, "Reactive oxygen species in natural waters". In *Active Oxygen in Chemistry*, C. Foote and J. Valentine, Eds., Blackie Academic and Professional, London, 1995, pp. 280-333.

Blumer, M., "Organic compounds in nature: Limits of our knowledge", *Angewandte Chemie*, **14**, 507-514 (1975).

Boehncke, A., K. Martin, M. G. Müller, and H. K. Cammenga, "The vapor pressure of lindane (g-1,2,3,4,5,6-hexachlorocyclohexane) – A comparison of Knudsen Effusion Measurements with data from other techniques", *J. Chem. Eng. Data*, **41**, 543-545 (1996).

Boethling, R. S., P. H. Howard, W. Meyland, W. Stiteler, J. Beauman, and N. Tirado, "Group contribution method for predicting probability and rate of aerobic biodegradation", *Environ. Sci. Technol.*, **28**, 459-465 (1994).

Boethling, R. S., and D. Mackay, Eds., *Handbook of Property Estimation Methods for Chemicals, Environmental and Health Sciences*, Lewis Publishers, CRC Press, Boca Raton, FL, 2000.

Böhme, F., K. Welsch-Pausch, and M. S. McLachlan, "Uptake of airborne semivolatile organic compounds in agricultural plants: Field measurements of interspecies variability", *Environ. Sci. Technol.*, **33**, 1805-1813 (1999).

Bosshardt, H., "Entwicklungstendenzen in der Bekämpfung von Schadorganismen in der Landwirtschaft", *Vierteljahresschr. Naturforsch. Ges. Zurich*, **133**, 225-240 (1988).

Bouchard, D. C., "Organic cosolvent effects on the sorption and transport of neutral organic chemicals", *Chemosphere*, **36**, 1883-1892 (1998).

Boucher, F. R., and G. F. Lee, "Adsorption of lindane and dieldrin pesticides on unconsolidated aquifer sands", *Environ. Sci. Technol.*, **6**, 538-543 (1972).

Boule, P., Ed. *The Handbook of Environmental Chemistry*, Springer, Berlin, 1999.

Bouwer, E. J., and P. L. McCarty, "Transformations of 1- and 2-carbon halogenated aliphatic organic compounds under methanogenic conditions", *Applied Environ. Microbiol.*, **45**, 1286-1294 (1983a).

Bouwer, E. J., and P. L. McCarty, "Transformations of halogenated organic compounds under denitrification conditions", *Appl. Environ. Mibrobiol.*, **45**, 1295-1299 (1983b).

Boyd, S. A., M. M. Mortland, and C. T. Chiou, "Sorption characteristics of organic compounds on hexadecyltrimethyl ammonium–smectite", *Soils Sci. Soc. Am. J.*, **52**, 652-657 (1988).

Brennan, R. A., N. Nirmalakhandan, and R. E. Speece, "Comparison of predictive methods for Henry's law coefficients of organic chemicals", *Wat. Res.*, **32**, 1901-1911 (1998).

Brett, J. R., and T. D. D. Groves, "Physiological energetics". In *Fish Physiology*, Vol. III, Academic Press, New York, 1979, pp. 279-352.

Brezonik, P. L., *Chemical Kinetics and Process Dynamics in Aquatic Systems*, Lewis, Boca Raton, 1994.

Brezonik, P. L., and J. Fulkerson-Brekken, "Nitrate-induced photolysis in natural waters: Controls on concentrations of hydroxyl radical photo-intermediates by natural scavenging agents", *Environ. Sci. Technol.*, **32**, 3004-3010 (1998).

Broecker, H. C., J. Petermann, and W. Seims, "The influence of wind on CO_2-exchange in a wind-wave tunnel, including the effects of monolayers", *J. Mar. Res.*, **36**, 595-610 (1978).

Broecker, W. S., "An application of natural radon to problems in ocean circulation". In *Diffusion in Oceans and Fresh Waters*, T. Ichiye, Ed., Lamont Doherty Geol. Obs., Palisades, 1965, pp. 116-145.

Broecker, W. S., and T.-H. Peng, "Gas exchange rates between air and sea", *Tellus*, **26**, 21-35 (1974).

Broholm, K., and S. Feenstra, "Laboratory measurements of the aqueous solubility of mixtures of chlorinated solvents", *Environ. Toxicol. Chem.*, **14**, 9-15 (1995).

Brønsted, J. N., and K. Pedersen, "Die katalytische Zersetzung des Nitramids und ihre physikalisch-chemische Bedeutung", *Z. Phys. Chem.*, **108**, 185-235 (1924).

Brownawell, B. J., *The Role of Colloid Organic Matter in the Marine Geochemistry of PCBs*, Ph.D. thesis, Woods Hole–MIT Joint Program, Cambridge, MA, 1986.

Brownawell, B. J., H. Chen, J. M. Collier, and J. C. Westall, "Adsorption of organic cations to natural materials", *Environ. Sci. Technol.*, **24**, 1234-1241 (1990).

Brownlow, A. H., *Geochemistry*, Prentice-Hall, Englewood Cliffs, NJ., 1979.

Brubacker, W. W. Jr., and R. A. Hites, "Polychlorinated dibenzo-*p*-dioxins and dibenzofurans: Gas phase hydroxyl radical reactions and related atmospheric removal", *Environ. Sci. Technol.*, **31**, 1805-1810 (1997).

Bruckmann, P., W. Kersten, W. Funcke, E. Balfanz, J. König, J. Theisen, M. Ball, and O. Päpke, "The occurrence of chlorinated and other organic trace compounds in urban air", *Chemosphere*, **17**, 2363-2380 (1988).

Bruice, T. C., T. H. Fife, J. J. Bruno, and N. E. Brandon, "Hydroxyl group catalysis. II. The reactivity of the hydroxyl group to serine. The nucleophilicity of alcohols and the ease of hydrolysis of their acetyl esters nucleophilicity as related to their pK's", *Biochemistry*, **1**, 7-12 (1962).

Brunel, F., and J. Davison, "Cloning and sequencing of Pseudomanoas genes", *J. Bacteriol.*, **170**, 4924-4930 (1988).

Brunner, W., D. Staub, and T. Leisinger, "Bacterial degradation of dichloromethane", *Appl. Environ. Microbiol.*, **40**, 950-958 (1980).

Brusseau, M. L., "Transport of rate-limited sorbing solutes in heterogeneous porous media: Application of a one-dimensional multifactor nonideality model to field data", *Water Resour. Res.*, **29**, 2485-2487 (1992).

1218   Bibliography

Brusseau, M. L., "The influence of solute size, pore water velocity, and intraparticle porosity on solute dispersion and transport in soil", *Water Resour. Res.*, **29**, 1071-1080 (1993).

Brusseau, M. L., "Transport of reactive contaminants in heterogeneous porous media", *Rev. Geophys.*, **32**, 285-313 (1994).

Brzuzy, L. P., and R. A. Hites, "Global mass balance for polychlorinated dibenzo-*p*-dioxins and dibenzofurans", *Environ. Sci. Technol.*, **30**, 1797-1804 (1996).

Bucheli, T. D., and Ö. Gustafsson, "Quantification of the soot–water distribution coefficient of PAHs provides mechanistic basis for enhanced sorption observations", *Environ. Sci. Technol.*, **34**, 5144-5151 (2000).

Bucheli, T. D., S. R. Müller, S. Heberle, and R. P. Schwarzenbach, "Occurrence and behavior of pesticides in rainwater roof runoff, and artificial stormwater infiltration", *Environ. Sci. Technol.*, **32**, 3457-3464 (1998a).

Bucheli, T. D., S. R. Müller, A. Voegelin, and R. P. Schwarzenbach, "Bituminous roof sealing membranes as major source of the herbicide (R,S)-mecoprop in roof runoff waters: Potential contamination of groundwater and surface waters", *Environ. Sci. Technol.*, **32**, 3465-3471 (1998b).

Bucheli-Witschel, M., and T. Egli, "Environmental fate and microbial degradation of aminopolycarboxylic acids", *FEMS Microbiol. Reviews*, **25**, 69-106 (2001).

Buckingham, D. A., and C. R. Clark, "Metal-hydroxyde-promoted hydrolysis of activated esters. Hydrolysis of 2,4-dinitrophenyl acetate and 4-nitrophenyl acetate", *Aust. J. Chem.*, **35**, 431-436 (1982).

Bugg, T., *An Introduction to Enzyme and Coenzyme Chemistry*, Blackwell Science, Oxford, UK, 1997.

Bugg, T., J. M. Focht, M. A. Pickard, and M. R. Gray, "Uptake and active efflux of polycyclic aromatic hydrocarbons by *Pseudomonas fluorescens* LP6a", *Appl. Environ. Microbiol.*, **66**, 5387-5392 (2000).

Bumpus, J. A., M. Tier, D. Wrigth, and S. D. Aust, "Oxidation of persistent environmental pollutants by a white rot fungus", *Science*, **228**, 1434-1436 (1985).

Burdick, G. E., E. J. Harris, H. J. Dean, T. M. Walker, J. Skea, and D. Colby, "The accumulation of DDT in lake trout and the effect on reproduction", *Trans. Amer. Fisheries Society*, **93**, 127-136 (1964).

Burgos, W. E., J. T. Novak, and D. F. Berry, "Reversible sorption and irreversible binding of naphthalene and α-naphthol to soil: Elucidation of processes", *Environ. Sci. Technol.*, **30**, 1205-1211 (1996).

Burlinson, N. E., L. A. Lee, and D. H. Rosenblatt, "Kinetics and products of hydrolysis of 1,2-dibromo-3-chloropropane", *Environ. Sci. Technol.*, **16**, 627-632 (1982).

Burrow, P. D., K. Aflatooni, and G. A. Gallup, "Dechlorination rate constants on iron and the correlation with electron attachment energies", *Environ. Sci. Technol.*, **34**, 3368-3371 (2000).

Burrows, K. J., A. Cornish, D. S. Scott, and I. J. Higgins, "Substrate specificities of the soluble and particulate methane monooxygenases of *Methylosinus trichosporium* OB3b." *Mol. Microbiol.*, **6**, 335-342 (1984).

Buschmann, J., W. Angst, and R. P. Schwarzenbach, "Iron porphyrin and cysteine mediated reduction of ten polyhalogenated methanes in homogeneous aqueous solution: Product analyses and mechanistic considerations", *Environ. Sci. Technol.*, **33**, 1015-1020 (1999).

Bustamante, P., S. Romero, A. Pena, B. Escalera, and A. Reillo, "Enthalpy-entropy compensation for the solubility of drugs in solvent mixtures: Paracetamol, acetanilide, and nalidixic acid in dioxane–water", *J. Pharm. Sci.*, **87**, 1590-1596 (1998).

Butler, E. C., and K. F. Hayes, "Kinetics of the transformation of trichloroethylene and tetrachloroethylene by iron sulfide", *Environ. Sci. Technol.*, **33**, 2021-2027 (1999).

Butler, E. C., and K. F. Hayes, "Kinetics of the transformation of halogenated aliphatic compounds by iron sulfide", *Environ. Sci. Technol.*, **34**, 422-429 (2000).

Button, D. K., "Nutrient uptake by microorganisms according to kinetic parameters from theory as related to cytoarchitecture", *Microbiol. Molecular Biol. Rev.*, **62**, 636-645 (1998).

Calamari, D., P. Tremolada, A. DiGuardo, and M. Vighi, "Chlorinated hydrocarbons in pine needles in Europe – Fingerprint for the past and recent use", *Environ. Sci. Technol.*, **28**, 429-434 (1994).

Calamari, D., P. Tremolada, and V. Notarianni, "Relationships between chlorinated hydrocarbons in vegetation and socioeconomic indexes on a global scale", *Environ. Sci. Technol.*, **29**, 2267-2272 (1995).

Calvert, J. G., and J. N. Pitts, Jr., *Photochemistry*, Wiley, New York, 1967.

Canonica, S., and M. Freiburghaus, "Electron-rich phenols for probing the photochemical reactivity of freshwaters", *Environ. Sci. Technol.*, **35**, 690-695 (2001).

Canonica, S., B. Hellrung, and J. Wirz, "Oxidation of phenols by triplet aromatic ketones in aqueous solution", *J. Phys. Chem. A.*, **104**, 1226-1232 (2000).

Canonica, S., U. Jans, K. Stemmler, and J. Hoigne, "Transformation kinetics of phenols in water: Photosensitization by dissolved organic material and aromatic ketones", *Environ. Sci. Technol.*, **29**, 1822-1831 (1995).

Canonica, S., and P. G. Tratnyek, "Quantitative structure reactivity relationships (QSARs) for oxidation reactions of organic chemicals in water", *Environ. Tox. Chem.*, **21**, in press (2003).

Capel, P. D., W. Giger, P. Reichert, and O. Wanner, "Accidental input of pesticides into the Rhine River", *Environ. Sci. Technol.*, **22**, 992-996 (1988).

Carlsson, H., U. Nilsson, G. Becker, and C. Östman, "Organophosphate ester flame retardants and plasticizers in the indoor environment: Analytical methodology and occurrence", *Environ. Sci. Technol.*, **31**, 2931-2936 (1997).

Carmo, A. M., L. S. Hundal, and M. L. Thompson, "Sorption of hydrophobic organic compounds by soil materials: Application of unit equivalent Freundlich coefficients", *Environ. Sci. Technol.*, **34**, 4363-4369 (2000).

Carslaw, H. S., and J. C. Jaeger, *Conduction of Heat in Solids, 2nd ed.*, Oxford University Press, Oxford, UK, 1959.

Carter, C. W., and I. H. Suffet, "Binding of DDT to dissolved humic materials", *Environ. Sci. Technol.*, **16**, 735-740 (1982).

Carter, H. H., and A. Okubo, "A study of the physical processes of the movement and dispersion in the Cape Kennedy area", Final report under the U.S. Atomic Energy Commission, Rep. NYO-2973-1, Chesapeake Bay Inst., Johns Hopkins Univ., Baltimore, 1965, p. 164.

Castro, C. E., and E. W. Bartnicki, "Conformational isomerism and effective redox geometry in the oxidation of heme proteins by alkyl halides, cytochrome C, and cytochrome oxidase", *Biochem.*, **14**, 498-503 (1975).

Castro, C. E., R. S. Wade, and N. O. Belser, "Biodehalogenation: Reactions of cytochrome P-450 with polyhalomethanes", *Biochem.*, **24**, 204-210 (1985).

Cavaletto, J. F., and W. S. Gardner, "Seasonal dynamics of lipids in freshwater benthic invertebrates". In *Lipids in Freshwater Ecosystems,* M. T. Arts and B. C. Wainman, Eds., Springer, New York, 1999, pp. 109-131.

Chandar, P., P. Somasundaran, and N. J. Turro, "Fluorescent probe studies on the structure of the absorbed layer of dodecyl sulfate at the alumina–water interface", *J. Colloid Int. Sci.*, **117**, 31-46 (1987).

Chandar, S., D. W. Fuerstenau, and D. Stigler, "On hemimicelle formation at oxide/water interfaces". In *Adsorption from Solution,* R. H. Ottewill, Ed., Academic Press, London, 1983, pp. 197-210.

Chang, H.-L., and L. Alvarez-Cohen, "Model for the co-metabolic biodegradation of chlorinated organics", *Environ. Sci. Technol.*, **29**, 2357-2367 (1995).

Chang, H.-L., and L. Alvarez-Cohen, "Biodegradation of individual and multiple chlorinated aliphatic hydrocarbons of methane-oxidizing cultures", *Appl. Environ. Microbiol.*, **62**, 3371-3377 (1996).

Chang, W.-K., and C. S. Criddle, "Experimental evaluation of a model for cometabolism: Prediction of simultaneous degradation of trichloroethylene and methane by a methanotrophic mixed culture", *Biotech. Bioeng.*, **56**, 492-501 (1997).

Chapman, S., and T. G. Cowling, *The Mathematical Theory of Non-Uniform Gases*, Cambridge University Press, Cambridge, 1970.

Chatfield, C., *Problem Solving – A Statistician's Guide*, Chapman and Hall, London, 1995.

Chaudhry, G. R., and A. N. Ali, "Bacterial metabolism of carbofuran", *Appl. Environ. Microbiol.*, **54**, 1414-1419 (1988).

Chin, J., "Developing artificial hydrolytic metalloenzymes by a unified mechanistic approach", *Acc. Chem. Res.*, **24**, 145-152 (1991).

Chin, Y.-P., G. Aiken, and E. O'Loughlin, "Molecular weight, polydispersity, and spectroscopic properties of aquatic humic substances", *Environ. Sci. Technol.*, **28**, 1853-1858 (1994).

Chin, Y. P., G. R. Aiken, and K. M. Danielsen, "Binding of pyrene to aquatic and commercial humic substances: The role of molecular weight and aromaticity", *Environ. Sci. Technol.,* **31**, 1630-1635 (1997).

Chiou, C. T., "Partition coefficients of organic compounds in lipid-water systems and correlations with fish bioconcentration factors", *Environ. Sci. Technol.*, **19**, 57-62 (1985).

Chiou, C. T., and D. E. Kile, "Deviations from sorption linearity on soils of polar and nonpolar organic compounds at low relative concentrations", *Environ. Sci. Technol.*, **32**, 338-343 (1998).

Chiou, C. T., D. E. Kile, D. W. Rutherford, G. Shung, and S. A. Boyd, "Sorption of selected organic compounds from water to a peat soil and its humic-acid and humin fractions: Potential sources of the sorption nonlinearity", *Environ. Sci. Technol.*, **34**, 1254-1258 (2000).

Chiou, C. T., J.-F. Lee, and S. A. Boyd, "The surface area of soil organic matter", *Environ. Sci. Technol.*, **24**, 1164-1166 (1990).

Chiou, C. T., R. L. Malcolm, T. I. Brinton, and D. E. Kile, "Water solubility enhancement of some organic pollutants and pesticides by dissolved humic and fulvic acids", *Environ. Sci. Technol.*, **20**, 502-508 (1986).

Chiou, C. T., S. E. McGroddy, and D. E. Kile, "Partition characteristics of polycyclic aromatic hydrocarbons on soils and sediments", *Envion. Sci. Technol.*, **32**, 264-269 (1998).

Chiou, C. T., G. Sheng, and M. Manes, "A partition-limited model for the plant uptake of organic contaminants from soil and water", *Environ. Sci. Technol.*, **35**, 1437-1444 (2001).

Chiou, C. T., and T. D. Shoup, "Soil sorption of organic vapors and effects of humidity on sorptive mechanism and capacity", *Environ. Sci. Technol.*, **19**, 1196-1200 (1985).

Chisaka, H., and P. C. Kearney, "Metabolism of propanil in soils", *J. Agric. Food. Chem.*, **18**, 854-858 (1970).

Chou, J. T., and P. C. Jurs, "Computer assisted computation of partition coefficients from molecular structures using fragment constants", *J. Chem. Inf. Computer Sci.*, **19**, 172-178 (1979).

Chow, V. T., *Open-Channel Hydraulics*, McGraw-Hill, New York, 1959.

Christensen, T. H., P. L. Bjerg, S. A. Banwart, R. Jakobsen, G. Heron, and H.-J. Albrechtsen, "Characterization of redox conditions in groundwater contaminant plumes", *J. Contam. Hydrol.*, **45**, 165-241 (2000).

Christensen, T. H., P. Kjeldsen, P. L. Bjerg, D. L. Jensen, J. B. Christensen, A. Baun, H.-J. Albrechtsen, and G. Heron, "Biogeochemistry of landfill leachate plumes", *Appl. Geochem.*, **16**, 659-718 (2001).

Christenson, I., "Alkaline hydrolysis of some carbonic acid esters", *Acta Chem. Scand.*, **4**, 904-922 (1964).

Church, C. D., J. F. Pankow, and P. G. Tratnyek, "Hydrolysis of *tert*-butyl formate: Kinetics, products, and implications for the environmental impact of methyl *tert*-butyl ether." *Environ. Sci. Technol.*, **18**, 2789-2796 (1999).

Cirpka, O., P. Reichert, O. Wanner, S. R. Müller, and R. P. Schwarzenbach, "Gas exchange at river cascades: Field experiments and model calculations", *Environ. Sci. Technol.*, **27**, 2086-2097 (1993).

Clark, J., and D. D. Perrin, "Prediction of the strengths of organic bases", *Quart. Revs. Chem. Soc.*, **18**, 295-320 (1964).

Clarke, J. U., and V. A. McFarland, "Uncertainty analysis for an equilibrium partitioning-based estimator of polynuclear aromatic hydrocarbon bioaccumulation potential in sediments", *Environ. Tox. Chem.*, **19**, 360-367 (2000).

Cline, P., J. J. Delfino, and P. S. C. Rao, "Partitioning of aromatic constituents into water from gasoline and other complex solvent mixtures", *Environ. Sci. Technol.*, **25**, 914-920 (1991).

Colby, J., D. I. Stirling, and H. Dalton, "The soluble methane monooxygenase from *Methylococcus copsulatus* bath – Its ability to oxygenate n-alkanes, ethers and alicyclic, aromatic, and heterocyclic compounds", *Biochem. J.*, **165**, 395-402 (1977).

Conrad, H. E., K. Lieb, and I. C. Gunsalus, "Mixed function oxidation. III. An electron transport complex in camphor ketolactonization", *J. Biol. Chem.*, **240**, 4029-4037 (1965).

Corapcioglu, M. Y., and S. Jiang, "Colloid-facilitated groundwater contaminant transport", *Water Resour. Res.*, **29**, 2215-2226 (1993).

Cornelissen, G., P. C. M. van Noort, G. Nachtegaal, and A. P. M. Kentgens, "A solid-state fluorine-NMR study on hexafluorobenzene sorbed by sediments, polymers, and active carbon", *Environ. Sci. Technol.*, **34**, 645-649 (2000).

Cowan, C. T., and D. White, "The mechanism of exchange reactions occurring between sodium montmorillonite and various *n*-primary aliphatic amine salts", *Trans. Faraday Soc.*, **54**, 691-697 (1958).

Coyle, G. T., T. C. Harmon, and I. H. Suffet, "Aqueous solubility depressions for hydrophobic organic chemicals in the presence of partially miscible organic solvents", *Environ. Sci. Technol.*, **31**, 384-389 (1997).

Crank, J., *The Mathematics of Diffusion*, Clarendon Press, Oxford, 1975.

Criddle, C. S., P. L. McCarty, M. C. Elliot, and J. F. Barker, "Reduction of hexachloroethane to tetrachloroethylene in groundwater", *J. Contam. Hydrol.*, **1**, 133-142 (1986).

Csanady, G. T., *Turbulent Diffusion in the Environment*, Reidel, Dordrecht, Holland, 1973.

Currie, J. A., "Movement of gases in soil respiration", *SCI-Monograph Soc. Chem. Ind.*, **37**, 152-171 (1970).

Cussler, E. L., *Diffusion – Mass Transfer in Fluid Systems*, Cambridge University Press, Cambridge, 1984.

Dallas, A. J., and P. W. Carr, "A thermodynamic and solvatochromic investigation of the effect of water on phase–transfer properties of octan-1-ol", *J. Chem. Soc. Perkin Trans.*, **2**, 2155-2161 (1992).

Dalton, J., "Experimental essays on the constitution of mixed gases; on the force of steam or vapor from waters and other liquids, both in a Torricellian vacuum and in air; on evaporation; and on the expansion of gases by heart", *Mem. Proc. Manchester Lit. Phil. Soc.*, **5**, 535-602 (1802).

Damborsky, J., K. Manova, and M. Kuty, "A mechanistic approach to deriving quantitative structure biodegradability relationships. A case study: Dehalogenation of haloaliphatic compounds". In *Biodegradability Prediction*, W. J. G. M. Peijnenburg and J. Damborsky, Eds., Kluwer Academic, Dordrecht, 1996.

Danckwerts, P. V., "Significance of liquid–film coefficients in gas absorption", *Ind. Eng. Chem.*, **43**, 1460-1467 (1951).

Dannenberg, A., and S. O. Pekkonen, "Investigation of the heterogeneously catalyzed hydrolysis of organophosphorous pesticides", *J. Agric. Food Chem.*, **46**, 325-334 (1998).

Dannenfelser, R. M., N. Surendran, and S. H. Yalkowski, "Molecular symmetry and related properties", *SAR QSAR Environ. Res.*, **1**, 273-292 (1993).

Dannenfelser, R. M., and S. H. Yalkowsky, "Estimation of entropy of melting from molecular structure: A non-group contribution method", *Ind. Eng. Chem. Res.*, **35**, 1483-1486 (1996).

Daubert, T. E., *Physical and Thermodynamic Properties of Pure Chemicals*, National Standard Reference Data System, New York, 1997.

Davies, G., and E. Ghabbour, Eds., *Humic Substances; Structures, Properties and Uses*, Royal Society of Chemistry, Cambridge, UK, 1998.

Davis, J. A., "Adsorption of natural dissolved organic matter at the oxide/water interface", *Geochim. Cosmochim. Acta*, **46**, 2381-2393 (1982).

Davis, J. W., "Physico-chemical factors influencing ethyleneamine sorption to soil", *Environ. Toxicol. Chem.*, **12**, 27-35 (1993).

de Boer, J., K. de Boer, and J. P. Boon, "Polybrominated biphenyls and diphenylethers". In *The Handbook of Environmental Chemistry,* Vol. 3, Part K, J. Paasivirta, Ed., Springer, Berlin, 2000, pp. 61-95.

Deacon, E. L., "Gas transfer to and across an air/water interface", *Tellus*, **29**, 363-374 (1977).

Dean, J. A., Ed. *Lange's Handbook of Chemistry*, McGraw-Hill, New York, 1985.

DeBruyn, W. J., and E. S. Saltzman, "The solubility of methyl bromide in pure water, 35‰ sodium chloride, and seawater", *Mar. Chem*, **56**, 51-57 (1997).

Dec, J., and J.-M. Bollag, "Dehalogenation of chlorinated phenols during oxidative coupling", *Environ. Sci. Technol.*, **28**, 484-490 (1994).

Dec, J., and J.-M. Bollag, "Effect of various factors on dehalogenation of chlorinated phenols and anilines during oxidative coupling", *Environ. Sci. Technol.*, **29**, 657-663 (1995).

Delle Site, A., "The vapor pressure of environmentally significant organic chemicals: A review of methods and data at ambient temperature", *J. Phys. Chem. Ref. Data*, **26**, 157-193 (1997).

Demianov, P., D. DeStefano, A. Gianguzza, and S. Sammartano, "Equilibrium studies in natural waters: Speciation of phenolic compounds in synthetic seawater at different salinities", *Environ. Toxicol. Chem*, **14**, 767-773 (1995).

Demond, A. H., and A. S. Lindner, "Estimation of interfacial tension between organic liquids and water", *Environ. Sci. Technol.*, **27**, 2318-2331 (1993).

Devlin, J. F., and D. Müller, "Field and laboratorys studies of carbon tetrachloride transformation in a sandy aquifer under sulfate reducing conditions", *Environ. Sci. Technol.*, **33**, 1021-1027 (1999).

Dewulf, J., D. Drijvers, and H. vanLangenhove, "Measurement of Henry's law constant as function of temperature and salinity for the low temperature range", *Atmos. Environ.*, **29**, 323-331 (1995).

Dickhut, R. M., A. W. Andren, and D. E. Armstrong, "Naphthalene solubility in selected organic solvent/water mixture", *J. Chem. Eng. Data*, **34**, 438-443 (1989).

Dilling, W. L., N. B. Tefertiller, and G. J. Kallos, "Evaporation rates and reactivities of methylene chloride, chloroform, 1,1,1-trichloroethane, trichloroethylene, tetrachloroethylene, and other chlorinated compounds in dilute aqueous solutions", *Environ. Sci. Technol.*, **9**, 833-838 (1975).

Dilling, W. L., "Interphase Transfer Processes. II. Evaporation rates of chloromethanes, ethanes, ethylenes, propanes, and propylenes from dilute aqueous solutions. Comparisons with theoretical predictions", *Environ. Sci. Technol.,* **11**, 405-409 (1977).

Dittert, L. W., and T. Higuchi, "Rates of hydrolysis of carbonate and carbonate esters in alkaline solution", *J. Pharm. Sci.*, **52**, 852-857 (1963).

Dohnal, V., and D. Fenclová, "Air-water partitioning and aqueous solubility of phenols", *J. Chem. Eng. Data*, **40**, 478-483 (1995).

Dolfing, J., and J. M. Tiedje, "Hydrogen cycling in a three-tiered food web growing on the methanogenic conversion of 3-chlorobenzoate", *FEMS Microbiol. Ecol.*, **38**, 293-298 (1986).

Doll, T. E., F. H. Frimmel, M. U. Kumke, and G. Ohlenbusch, "Interaction between natural organic matter (NOM) and polycyclic aromatic compounds (PAC) – comparison of fluorescence quenching and solid phase micto extraction (SPME)", *Fresenius J. Anal. Chem.*, **364**, 313-319 (1999).

Domenico, P. A., and F. W. Schwartz, *Physical and Chemical Hydrology*, 2nd ed., Wiley, New York, 1998.

Doong, S. J., and W. S. W. Ho, "Diffusion of hydrocarbons in polyethylene", *Ind. Eng. Chem. Res.*, **31**, 1050-1060 (1992).

Downing, A. L., and G. A. Truesdale, "Some factors affecting the rate of solution of oxygen in water", *J. Appl. Chem.*, **5**, 570-581 (1955).

Drever, J. I., *The Geochemistry of Natural Waters*, Prentice Hall, Englewood Cliffs, NJ, 1988.

Drossman, H., H. Johnson, and T. Mill, "Structure activity relationships for environmental processes 1: Hydrolysis of esters and carbamates", *Chemosphere*, **17**, 1509-1530 (1988).

Drost-Hansen, W., "Structure of water near solid interfaces", *Ind. Eng. Chem.*, **61**, 10-47 (1969).

Dugas, H., *A Chemical Approach to Enzyme Action*, Springer, New York, 1996.

Dulin, D., and T. Mill, "Development and evaluation of sunlight actinometers", *Environ. Sci. Technol.*, **16**, 815-820 (1982).

Dunnivant, F. M., D. Macalady, and R. P. Schwarzenbach, "Reduction of substituted nitrobenzenes in aqueous solutions containing natural organic matter", *Environ. Sci. Technol.*, **26**, 2133-2141 (1992).

Dzombak, D. A., and F. M. M. Morel, *Surface Complexation Modeling*, Wiley-Interscience, New York, 1990.

Eberson, L., *Electron Transfer Reactions in Organic Chemistry*, Springer, Berlin, 1987.

Eganhouse, R. P., and J. A. Calder, "The solubility of medium molecular weight aromatic hydrocarbons and the effects of hydrocarbon co-solutes and salinity", *Geochim. Cosmochim. Acta*, **40**, 555-561 (1976).

Egli, C., R. Scholtz, A. M. Cook, and T. Leisinger, "Anaerobic dechlorination of tetrachloromethane and 1,2-dichloroethane to degradable products by pure cultures of *Desulfobacterium* sp. and *Methanobacterium* sp." *FEMS Microbiol. Lett.*, **43**, 257-261 (1987).

Egli, T., "Biodegradation of metal-complexing aminopolycarboxylic acids", *J. Biosci. and Bioeng.*, **92**, 89-97 (2001).

Einstein, A., "Ueber die von der molekularkinetischen Theorie der Wärme geforderte Bewegung von in ruhenden Flüssigkeiten suspendierten Teilchen", *Annalen der Physik*, **17**, 549-560 (1905).

Eisenreich, S. J., W. A. Willford, and W. M. J. Strachan, "The role of atmospheric deposition in organic contaminant cycling in the Great Lakes". In *Intermedia Pollutant Transport: Modelling and Field Measurements*, D. Allen, Ed., Plenum Press, New York, 1989, pp. 19-40.

Eklund, H., and M. Fontecave, "A conservative structural basis for radicals", *Structure Fold. Des.*, **7**, 257-262 (1999).

El-Amamy, M. M., and T. Hill, "Hydrolysis kinetics of organic chemicals on montmorillonite and kaolinite surfaces as related to moisture content", *Clays and Minerals*, **32**, 67-73 (1984).

Elovitz, M. S., and W. Fish, "Redox interactions of Cr(VI) and substituted phenols: Kinetic investigation", *Environ. Sci. Technol.*, **28**, 2161-2169 (1994).

Elovitz, M. S., and W. Fish, "Redox interactions of Cr(VI) and substituted phenols: Products and mechanisms", *Environ. Sci. Technol.*, **29**, 1993-1943 (1995).

Elovitz, M. S., and W. J. Weber, "Sediment-mediated reduction of 2,4,6-trinitrotoluene and fate of the resulting aromatic (poly)amines", *Environ. Sci. Technol.*, **33**, 2617-2625 (1999).

Emerson, S., W. S. Broecker, and D. W. Schindler, "Gas exchange rate in a small lake as determined by the radon method", *J. Fish. Res. Bd. Can.*, **30**, 1475-1484 (1973).

Englehardt, G., P. R. Wallnofer, and R. Plapp, "Purification and properties of an aryl acylamidase of *Bacillus sphaericus*, catalyzing the hydrolysis of various phenylamide herbicides and fungicides", *Appl. Microbiol.*, **26**, 709-718 (1973).

Epand, R. M., and I. B. Wilson, "Evidence for the formation of hippuryl chymotrypsin during the hydrolysis of hippuric acid esters", *J. Biol. Chem.*, **238**, 1718-1723 (1963).

Escher, B. I., R. W. Hunziker, and R. P. Schwarzenbach, "Interaction of phenolic uncouplers in binary mixtures: Concentration-additive and synergistic effects", *Environ. Sci. Technol.*, **35**, 3905-3914 (2001).

Escher, B. I., and R. P. Schwarzenbach, "Partitioning of substituted phenols in liposome-water, biomembrane-water, and octanol-water systems", *Environ. Sci. Technol.*, **30**, 260-270 (1996).

Escher, B. I., R. P. Schwarzenbach, and J. C. Westall, "Evaluation of liposome-water partitioning of organic acid and bases. 1. Development of a sorption model", *Environ. Sci. Technol.*, **34**, 3954-3961 (2000).

Estes, T. J., R. V. Shah, and V. L. Vilker, "Adsorption of low molecular weight halocarbons by montmorillonite", *Environ. Sci. Technol.*, **22**, 377-381 (1988).

Eto, M., *Organophosphorous Pesticides: Organic and Biological Chemistry*, CRC Press, Boca Raton, 1979.

Evanko, C. R., and D. A. Dzombak, "Influence of structural features on sorption of NOM-analogue organic acids to goethite", *Environ. Sci. Technol.*, **32**, 2846-2855 (1998).

Evanko, C. R., and D. A. Dzombak, "Surface complexation modeling of organic acid sorption to goethite", *J. Coll. Int. Sci.*, **214**, 189-206 (1999).

Ewald, G., *Role of Lipids in the Fate of Organochlorine Compounds in Aquatic Ecosystems*, Ph.D. thesis, Lund University, Lund Sweden, 1996.

Exner, O., *Correlation Analysis of Chemical Data*, Plenum Press, New York and London, 1988.

Eylers, H., *Transport of Adsorbing Metal Ions Between Stream and Sediment Bed in a Metal Laboratory Flume*, Ph.D. thesis, California Institute of Technology, Pasadena, CA, 1994.

Eyring, H., "Plasticity and diffusion as example of absolute reaction rates", *J. Chem. Phys*, **4**, 283-291 (1936).

Fabrega, J. R., C. Jafvert, H. Li, and L. S. Lee, "Modeling short-term soil-water distribution of aromatic amines", *Environ. Sci. Technol.*, **32**, 2788-2794 (1998).

Fabrega-Duque, J. R., C. T. Jafvert, H. H. Li, and L. S. Lee, "Modeling abiotic processes of aniline in water-saturated soils", *Environ. Sci. Technol.*, **34**, 1687-1693 (2000).

Fan, C., and C. T. Jafvert, "Margules equations applied to PAH solubilities in alcohol-water mixtures", *Environ. Sci. Technol.*, **31**, 3516-3522 (1997).

Farrell, J., D. Grassian, and M. Jones, "Investigation of mechanisms contributing to slow desorption of hydrophobic organic compounds from mineral solids", *Environ. Sci. Technol.*, **33**, 1237-1243 (1999).

Faust, B. C., "Aquatic photochemical reactions in atmospheric surface, and marine waters: Influences on oxidant formation and pollutant degradation". In *The Handbook of Environmental Chemistry,* Vol. 2, Part L, P. Boule, Ed., Springer, Berlin, 1999, pp. 101-122.

Faust, B. C., and J. Hoigné, "Sensitized photo-oxidation of alkylphenols by fulvic acid and natural water", *Environ. Sci. Technol.*, **21**, 957-964 (1987).

Faust, S. D., and H. M. Gomaa, "Chemical hydrolysis of some organic phosphorus and carbamate pesticides in aquatic environments", *Environ. Lett.*, **3**, 171-201 (1972).

Fay, A. A., B. J. Brownawell, A. A. Elskus, and A. E. McElroy, "Critical body residues in the marine amphipod *Ampelisca abdita*: Sediment exposures with nonionic organic contaminants", *Environ. Toxicol. Chem.*, **19**, 1028-1035 (2000).

Feldmann, R. S., and J. E. Titus, "Polychlorinated biphenyl accumulation differs among pumpkinseed sunfish during experimental field exposure: The role of invertebrate prey", *Aquat. Toxicol.*, **51**, 389-404 (2001).

Fennelly, J. P., and A. L. Roberts, "Reaction of 1,1,1-trichloroethane with zero-valent metals and bimetallic reductants", *Environ. Sci. Technol.*, **32**, 1980-1988 (1998).

Fent, K., "Ecotoxicology of organotin compounds", *Crit. Rev. Toxicol.*, **26**, 3-117 (1996).

Fersht, A., *Enzyme Structure and Mechanism*, Freeman, New York, 1985.

Fesch, C., W. Simon, S. B. Haderlein, P. Reichert, and R. P. Schwarzenbach, "Nonlinear sorption and nonequilibrium solute transport in aggregated porous media: Experiments, process identification and modeling", *J. Contam. Hydrol.*, **31**, 373-407 (1998).

Fetzner, S., "Bacterial dehalogenation", *Appl. Microbiol. Biotechnol.*, **50**, 633-657 (1998).

Fiedler, H., and W. Mücke, "Nitro derivatives of polycyclic aromatic hydrocarbons". In *The Handbook of Environmental Chemistry,* Vol. 3, Part G, O. Hutzinger, Ed., Springer, Berlin, 1991, pp. 97-137.

Field, J. A., and E. M. Thurman, "Glutathione conjugation and contaminant transformation", *Environ. Sci. Technol.*, **30**, 1413-1418 (1996).

Field, R. A., M. E. Goldstone, J. N. Lester, and R. Perry, "The sources and behaviour of tropospheric anthropogenic volatile hydrocarbons", *Atmos. Environ.*, **26A**, 2983-2996 (1992).

Finlayson-Pitts, B. J., and J. N. Pitts, Jr., *Atmospheric Chemistry: Fundamental and Experimental Techniques*, Wiley-Interscience, New York, 1986.

Fischer, H. B., J. Imberger, E. J. List, R. C. Y. Koh, and N. H. Brooks, *Mixing in Inland and Coastal Waters*, Academic Press, New York, 1979.

Fischer, M., and P. Warneck, "Photodecomposition of nitrite and undissociated nitrous acid in aqueous solution", *J. Phys. Chem.*, **100**, 18749-18756 (1996).

Fishbein, L., "Aromatic amines". In *The Handbook of Environmental Chemistry,* Vol. 3, Part C, O. Hutzinger, Ed., Springer, Berlin, 1984, pp. 1-40.

Fishtine, S. H., "Reliable latent heats of vaporization", *Ind. Eng. Chem.*, **55**, 47-56 (1963).

Fitzgerald, D., "Evaporation", *Trans. ASCE*, **15**, 581-646 (1986).

Fleck, G. M., *Equilibria in Solution*, Holt, Rinehart, and Winston, New York, 1966.

Flint, D. H., and R. M. Allen, "Iron-sulfur proteins with nonredox functions", *Chem. Rev.*, **96**, 2315-2334 (1996).

Flynn, J. H., "A collection of kinetic data for the diffusion of organic compounds in polyolefins", *Polymer*, **23**, 1325-1344 (1982).

Focardi, S., C. Fassi, C. Leonzio, S. Corsolini, and O. Parra, "Persistent organochlorine residues in fish and water birds from the Biobio River, Chile", *Environmental Monitoring and Assessment*, **43**, 73-92 (1996).

Fogel, M. N., A. R. Taddeo, and S. Fogel, "Biodegradation of chlorinated ethenes by a methane-utilizing mixed culture", *Appl. Environ. Microbiol.*, **54**, 720-724 (1986).

Fogelqvist, E., "Carbon tetrachloride, tetrachloroethylene, 1,1,1-trichloroethane and bromoform in arctic seawater", *J. Geophys. Res.*, **90**, 9181-9183 (1985).

Fowkes, F. M., "Attractive forces at interfaces", *Ind. Eng. Chem.*, **56**, 40-52 (1964).

Fowkes, F. M., "Quantitative characterization of the acid-base properties of solvents, polymers, and inorganic surfaces". In *Acid-Base Interactions – Relevance to Adhesion Science and Technology*, K. L. Mittal and H. R. Anderson, Eds., VSP, Utrecht, 1991, pp. 93-115.

Fowkes, F. M., F. L. Riddle Jr., W. E. Pastore, and A. E. Weber, "Interfacial interactions between self-associated polar liquids and squalane used to test equations for solid–liquid interactions", *Colloids and Surfaces*, **43**, 367-387 (1990).

Frank, H. S., and M. W. Evans, "Free volume and entropy in condensed systems. III. Entropy in binary liquid mixtures; partial molal entropy in dilute solution; structure and thermodynamics in aqueous electrolytes", *J. Chem. Phys.*, **13**, 507-532 (1945).

Frank, R., and H. Rau, "Photochemical transformation in aqueous solution and possible environmental fate of ethylenediaminetetraacetic acid (EDTA)", *Ecotoxicology Environ. Safety*, **19**, 55-63 (1990).

Freeze, R. A., and J. A. Cherry, *Groundwater*, Prentice-Hall, Englewood Cliffs, NJ, 1979.

Friesen, J. J., and G. R. B. Webster, "Temperature dependence of the aqueous solubilities of highly chlorinated dibenzo-*p*-dioxins", *Environ. Sci. Technol.*, **24**, 97-101 (1990).

Froese, K. L., A. Wolanski, and S. E. Hruday, "Factors governing odorous aldehyde formation as disinfection by-products in drinking water", *Wat. Res.*, **33**, 1355-1364 (1999).

Frost, A. A., and R. G. Pearson, *Kinetics and Mechanism*, Wiley-Interscience, New York, 1961.

Frost, T., and R. C. Upstill-Goddard, "Air-sea gas exchange into the millennium: Progress and uncertainties", *Oceangr. Marine Biol.: Annual Rev.*, **37**, 1-45 (1999).

Fu, J.-K., and R. G. Luthy, "Aromatic compound solubility in solvent/water mixtures", *J. Environ. Eng.*, ASCE, **112**, 346-346 (1986).

Fuerstenau, D. W., "Streaming potential studies on quartz in solutions of ammonium acetates in relation to the formation of hemimicelles at the quartz-solution interface", *J. Phys. Chem.*, **60**, 981-985 (1956).

Fuerstenau, D. W., and T. Wakamatsu, "Effect of pH on the adsorption of sodium dodecanesulphonate at the alumina/water interface", *Faraday Disc. Chem. Soc.*, **59**, 157-168 (1975).

Fuller, E. N., P. D. Schettler, and J. C. Giddings, "A new method for prediction of binary gas-phase diffusion coefficient", *Ind. Eng. Chem.*, **58**, 19-27 (1966).

Gaffney, J. S., N. A. Marley, and M. M. Cunningham, "Measurement of the adsorption constants for nitrate in water between 270 and 335 nm", *Environ. Sci. Technol.*, **26**, 207-209 (1992).

Gamberger, D., S. Sekusak, Z. Medven, and A. Sabljic, "Application of artificial intelligence in biodegradation modelling". In *Biodegradability Prediction*, W. J. G. M. Peijnenburg and J. Damborsky, Eds., Kluwer Academic, Dordrecht, 1996, pp. 41-50.

Gan, J., S. R. Yates, M. A. Anderson, W. F. Spencer, F. F. Ernst, and M. V. Yates, "Effect of soil properties on degradation and sorption of methyl bromide in soil", *Chemosphere*, **29**, 2685-2700 (1994).

Gantzer, C. J., and L. P. Wackett, "Reductive dechlorination catalyzed by bacterial transition-metal coenzymes", *Environ. Sci. Technol.*, **25**, 715-722 (1991).

Garbarini, D. R., and L. W. Lion, "Evaluation of sorptive partitioning of nonionic pollutants in closed systems by headspace analysis", *Environ. Sci. Technol.*, **19**, 1122-1138 (1985).

Garbarini, D. R., and L. W. Lion, "The influence of the nature of soil organics on the sorption of toluene and trichloroethylene", *Environ Sci. Technol.*, **20**, 1263-1269 (1986).

Garg, S., and W. G. Rixey, "The dissolution of benzene, toluene, m-xylene and naphthalene from a residually trapped non-aqueous phase liquid under mass transfer limited conditions", *J. Contam. Hydrol.*, **36**, 313-331 (1999).

Garrett, W. D., "Damping of capillary waves at the air–sea interface by organic surface active material", *J. Mar. Res.*, **25**, 279-291 (1967).

Gauthier, T. D., W. R. Seitz, and C. L. Grant, "Effect of structural and compositional variations of dissolved humic materials on pyrene K_{oc} values", *Environ. Sci. Technol.*, **21**, 243-248 (1987).

Gauthier, T. D., E. C. Shane, W. F. Guerin, W. R. Seltz, and C. L. Grant, "Fluorescence quenching method for determining equilibrium constants for polycyclic aromatic hydrocarbons binding to dissolved humic materials", *Environ Sci. Technol.*, **20**, 1162-1166 (1986).

Gawlik, B. M., N. Sotiriou, E. A. Feicht, S. Schulte-Hostede, and A. Kettrup, "Alternatives for the determination of the soil adsorption coefficient, K_{oc}, of nonionic organic compounds – A review", *Chemosphere*, **34**, 2525-2551 (1997).

Georgi, A., *Sorption of Hydrophobic Organic Compounds by Dissolved Humic Materials*, Umweltforschungszentrum Leipzig-Halle, Leipzig, Germany, 1998 (in German).

Gerecke, A. C., S. Canonica, S. R. Müller, M. Schärer, and R. P. Schwarzenbach, "Quantification of dissolved organic matter (DOM) mediated phototransformation of phenylurea herbicides in lakes", *Environ. Sci. Technol.*, **35**, 3915-3928 (2001).

Gerstl, Z., "Estimation of organic chemical sorption by soils", *J. Contam. Hydrol*, **6**, 357-375 (1990).

Gevao, B., T. Harner, and K. Jones, "Sedimentary record of polychlorinated naphthalene concentrations and deposition fluxes in a dated Lake Core", *Environ. Sci. Technol.*, **34**, 1, 33-38 (2000).

Gianessi, L. P., and J. E. Anderson, *Pesticide Use in U.S. Crop Production*, National Center for Food and Agricultural Policy, Washington, DC, 1995.

Gibbs, J. W., "XI. Graphical methods in the thermodynamics of fluids", *Trans. Conn. Acad.*, **2**, 309-342 (1873).

Gibbs, J. W., "V. On the equilibrium of heterogeneous substances", *Trans. Conn. Acad.*, **3**, 108-248 (1876).

Gibson, D. T., M. Hensley, H. Yoshioka, and T. J. Mabrys, "Formation of (+)-*cis*-2,3-dihydroxy-1-methyl cyclohexa-4,6-diene from toluene by *Pseudomonas putida*", *Biochem.*, **9**, 1626-1630 (1970).

Giger, W., P. H. Brunner, and C. Schaffner, "4-Nonylphenol in sewage sludge: Accumulation of toxic metabolites from nonionic surfactants", *Science*, **225**, 623-625 (1984).

Gilliland, E. R., "Diffusion coefficients in gaseous systems", *Ind. Eng. Chem.*, **26**, 681-685 (1934).

Glod, G., W. Angst, C. Holliger, and R. P. Schwarzenbach, "Corrinoid-mediated reduction of tetrachloroethene, trichloroethene, and trichlorofluoroethene in homogeneous aqueous solution: Reaction kinetics and reaction mechanisms", *Environ. Sci. Technol.*, **31**, 253-260 (1997a).

Glod, G., U. Brodmann, W. Angst, C. Holliger, and R. P. Schwarzenbach, "Cobalamin-mediated reduction of *cis*- and *trans*-dichlorethene, 1,1-dichloroethene, and vinyl chloride in homogeneous aqueous solution: Reaction kinetics and mechanistic considerations", *Environ. Sci. Technol.*, **31**, 3154-3160 (1997b).

Gmehling, J., J. Menke, and M. Schiller, *Activity Coefficients at Infinite Dilution*, DECHEMA, Frankfurt, 1994.

Goar, B. G., "Today's gas-treating processes", *Oil Gas. J.*, **69**, 75-78 (1971).

Gobas, F. A. P. C., J. M. Lahittete, G. Garofalo, W.-Y. Shiu, and D. Mackay, "A novel method for measuring membrane–water partition coefficients of hydrophobic organic chemicals: Comparison with 1-octanol–water partitioning", *J. Pharm. Sci.*, **77**, 265-272 (1988).

Gobas, F. A. P. C., and H. A. Morrison, "Bioconcentration and biomagnification in the aquatic environment". In *Handbook of Property Estimation Methods for Chemicals – Environmental and Health Sciences,* R. S. Boethling and D. Mackay, Eds., Lewis Publishers, Boca Raton, FL, 2000, pp. 198-231.

Gobas, F. A. P. C., J. P. Wilcockson, R. W. Russel, and G. D. Haffner, "Mechanism of biomagnification in fish under laboratory and field conditions", *Environ. Sci. Technol.*, **33**, 133-141 (1999).

Goldberg, E. D., *Black Carbon in the Environment*, Wiley, New York, 1985.

Goldman, P., "The enzymatic cleavage of the carbon-fluorine bond in fluoroacetate", *J. Biol. Chem.*, **240**, 3434-3438 (1965).

Good, R. J., and M. K. Chaudhury, "Theory of adhesive forces across interfaces". In *Fundamentals of Adhesion,* L. H. Lee, Ed., Plenum Press, New York, 1991, pp. 137-172.

Gordon, J. E., and R. L. Thorne, "Salt effects on the activity coefficient of naphthalene in mixed aqueous electrolyte solutions. I. Mixtures of two salts", *J. Phys. Chem.*, **71**, 4390-4399 (1967a).

Gordon, J. E., and R. L. Thorne, "Salt effects on non-electrolyte activity coefficients in mixed aqueous electrolyte solutions. II. Artificial and natural sea waters", *Geochim. Cosmochim. Acta*, **31**, 2433-2443 (1967b).

Goss, K.-U., "Predicting the enrichment of organic compounds in fog caused by adsorption on the water surface", *Atmospheric Environ.*, **28**, 3513-3517 (1994).

Goss, K.-U., "Considerations about the adsorption of organic molecules from the gas phase to surfaces: Implications for inverse gas chromatography and the prediction of adsorption coefficients", *J. Colloid Interface Sci.*, **190**, 241-249 (1997a).

Goss, K.-U., "Conceptual model for the adsorption of organic compounds from the gas phase to liquid and solid surfaces", *Environ. Sci. Technol.*, **31**, 3600-3605 (1997b).

Goss, K.-U., and R. P. Schwarzenbach, "Gas/solid and gas/liquid partitioning of organic compounds: Critical evaluation of the interpretation of equilibrium constants", *Environ. Sci. Technol.*, **32**, 2025-2032 (1998).

Goss, K.-U., and R. P. Schwarzenbach, "Empirical prediction of heats of vaporization and heats of adsorption of organic compounds", *Environ. Sci. Technol.*, **33**, 3390-3393 (1999a).

Goss, K.-U., and R. P. Schwarzenbach, "Quantification of the effect of humidity on the gas–mineral oxide and gas–salt adsorption of organic compounds", *Environ. Sci. Technol.*, **33**, 4073-4078 (1999b).

Goss, K.-U., and R. P. Schwarzenbach, "Linear free energy relationships used to evaluate equilibrium partitioning of organic compounds", *Environ. Sci. Technol.*, **35**, 1-9 (2001).

Goss, K.-U., and R. P. Schwarzenbach, "Adsorption of a diverse set of organic vapors on Quartz, $CaCO_3$, and α-Al_2O_3 at different relative humidities", *J. Colloid Int. Sci. (in press)*, (2002).

Gottschalk, G., *Bacterial Metabolism*, Springer, New York, 1986.

Grain, C. F., "Vapor Pressure". In *Handbook of Chemical Property Estimation Methods; Environmental Behavior of Organic Compounds,* W. J. Lyman, W. F. Reehl and D. H. Rosenblatt, Eds., McGraw-Hill, New York, 1982a, pp. 14-1 – 14-20.

Grain, C. F., "Activity Coefficient". In *Handbook of Chemical Property Estimation Methods; Environmental Behavior of Organic Compounds,* W. J. Lyman, W. F. Reehl and D. H. Rosenblatt, Eds., McGraw-Hill, New York, NY, 1982b, pp. 11-1 – 11-53.

Gray, P. H. H., and H. G. Thornton, "Soil bacteria that decompose certain aromatic compounds", *Zentralblatt Bakteriologie Parasitenkunde Infektionskrankheiten*, **3**, 74-96 (1928).

Greizerstein, H. B., C. Stinson, P. Mendola, G. M. Buck, P. J. Kostyniak, and J. E. Vena, "Comparison of PCB congeners and pesticide levels between serum and milk from lactating women", *Environ. Res. A.*, **80**, 280-286 (1999).

Griebler, C., and D. Slezak, "Microbial activity in aquatic environments measured by dimethyl sulfoxide reduction and intercomparison with commonly used methods", *Appl. Environ. Microbiol.*, **67**, 100-109 (2001).

Grim, R. E., *Clay Mineralogy*, McGraw-Hill, New York, 1968.

Groenewegen, P. E. J., J. M. Driessen, W. N. Konings, and J. A. M. de Bont, "Energy-dependent uptake of 4-chlorobenzoate in the coryne-form bacterium NTB-1", *J. Bacteriol.*, **172**, 419-423 (1990).

Grosjean, E., J. Bittercourt de Andrade, and D. Grosjean, "Carbonyl products of the gas-phase reaction of ozone with simple alkanes", *Environ. Sci. Technol.*, **30**, 975-983 (1996a).

Grosjean, E., D. Grosjean, and J. H. Seinsfeld, "Atmospheric chemistry of 1-octene, 1-decene, and cyclohexene: Gas phase carbonyl and peroxyacyl nitrate products", *Environ. Sci. Technol.*, **30**, 1038-1047 (1996b).

Groves, F. R. J., "Effect of cosolvents on the solubility of hydrocarbons in water", *Environ. Sci. Technol.*, **22**, 282-286 (1988).

Groves, J. T., and Y.-Z. Han, "Models and mechanisms of cytochrome P450 action". In *Cytochrome P450, Structure, Mechanism, and Biochemistry,* P. R. Ortiz de Montellano, Ed., Plenum Press, New York, 1995, pp. 3-48.

Gruber, D., D. Langenheim, J. Gmehling, and W. Moollan, "Measurement of activity coefficients at infinite dilution using gas–liquid chromatography. 6. Results for systems exhibiting gas–liquid interface adsorption with 1-octanol", *J. Chem. Eng. Data*, **42**, 882-885 (1997).

Gschwend, P. M., and R. A. Hites, "Fluxes of polycyclic aromatic hydrocarbons to marine and lacustrine sediments in the Northeastern United States", *Geochim. Cosmochim. Acta*, **45**, 2359-2367 (1981).

Gschwend, P. M., and S.-C. Wu, "On the constancy of sediment–water partition coefficients of hydrophobic organic pollutants", *Environ. sci. Technol.*, **19**, 90-96 (1985).

Gu, B. H., O. R. West, and R. L. Siegrist, "Using C-14-labeled radiochemicals can cause experimental error in studies of the behavior of volatile organic compounds", *Environ. Sci. Technol.*, **29**, 1210-1214 (1995).

Guengerich, F. P., and T. L. MacDonald, "Chemical mechanisms of catalysis by cytochromes P-450: A unified view", *Acc. Chem. Res.*, **17**, 9-16 (1984).

Guha, S., C. A. Peters, and P. R. Jaffe, "Multisubstrate biodegradation kintics of naphthalene, phenanthrene, and pyrene mixtures", *Biotech. Bioeng.*, **65**, 491-499 (1999).

Gunatilleka, A. D., and C. F. Poole, "Models for estimating the non-specific aquatic toxicity of organic compounds", *Anal. Commun.*, **36**, 235-242 (1999).

Gundersen, J. L., W. G. Macintyre, and R. C. Hale, "pH-dependent sorption of chlorinated guaiacols on estuarine sediments: The effects of humic acids and TOC", *Environ. Sci. Technol.*, **31**, 188-193 (1997).

Gustafson, C. G., "PCBs-prevalent and persistent", *Environ. Sci. Technol.*, **4**, 814-819 (1970).

Gustafsson, O., and P. M. Gschwend, "The flux of black carbon to surface sediments on the New England continental shelf", *Geochim. Cosmochim. Acta*, **62**, 465-472 (1998).

Gustafsson, O., and P. M. Gschwend, "Hydrophobic organic compound partitioning from bulk water to the water/air interface", *Atmos. Environ.*, **33**, 163-167 (1999).

Gustafsson, O., F. Haghseta, C. Chan, J. MacFarlane, and P. M. Gschwend, "Quantification of the dilute sedimentary soot phase: Implications for PAH speciation and bioavailability", *Environ Sci. Technol.*, **31**, 203-209 (1997).

Haag, W. R., and J. Hoigné, "Singlet oxygen in surface waters. 3. Photochemical formation and steady-state concentrations in various types of waters", *Environ. Sci. Technol.*, **20**, 341-348 (1986).

Haag, W. R., J. Hoigné, E. Gassmann, and A. M. Braun, "Singlet oxygen in surface waters – Part I: Furfuryl alcohol as a trapping agent", *Chemosphere*, **13**, 631-640 (1984a).

Haag, W. R., J. Hoigné, E. Gassmann, and A. M. Braun, "Singlet oxygen in surface waters – Part II: Quantum yields of its production by some natural humic materials as a function of wavelength", *Chemosphere*, **13**, 641-650 (1984b).

Haag, W. R., and T. Mill, "Direct and indirect photolysis of water-soluble azodyes: Kinetic measurements and structure-activity relationship", *Environ. Toxicol. Chem.*, **6**, 359-369 (1987).

Haag, W. R., and T. Mill, "Some reactions of naturally occurring nucleophiles with haloalkanes in water", *Environ. Toxicol. Chem.*, **7**, 917-924 (1988a).

Haag, W. R., and C. C. D. Yao, "Rate constants for reaction of hydroxyl radicals with several drinking water contaminants", *Environ. Sci. Technol.*, **26**, 1005-1013 (1992).

Haderlein, S. B., T. B. Hofstetter, and R. P. Schwarzenbach, "Subsurface chemistry of nitroaromatic compounds". In *Biodegradation of Nitroaromatic Compounds and Explosives,* J. C. Spain, J. B. Hughes and H.-J. Knackmuss, Eds., Lewis Publishers, Boca Raton, FL, 2000, pp. 312-356.

Haderlein, S. B., and R. P. Schwarzenbach, "Adsorption of substituted nitrobenzenes and nitrophenols to mineral surfaces", *Environ. Sci. Technol.*, **27**, 316-326 (1993).

Haderlein, S. B., and R. P. Schwarzenbach, "Environmental processes influencing the rate of abiotic reduction of nitroaromatic compounds in the subsurface". In *Biodegradation of Nitroaromatic Compounds,* J. C. Spain, Ed., Plenum Press, New York, 1995, pp. 199-225.

Haderlein, S. B., K. W. Weissmahr, and R. P. Schwarzenbach, "Specific adsorption of nitroaromatic explosives and pesticides to clay minerals", *Envion. Sci. Technol.*, **30**, 612-622 (1996).

Haitzer, M., G. Abbt-Braun, W. Traunspurger, and C. E. W. Steinberg, "Effects of humic substances on the bioconcentration of polycyclic aromatic hydrocarbons: Correlations with spectroscopic and chemical properties of humic substances", *Environ. Toxicol. Chem.*, **18**, 2782-2788 (1999).

Håkanson, L., and M. Jansson, *Principles of Lake Sedimentology*, Springer Verlag, Heidelberg, 1983.

Hall, J. C., J. S. Wickenden, and K. F. Yau, "Biochemical conjugation of pesticides in plants and microorganisms: An overview of similarities and divergences". In *Pesticide Biotransformation in Plants and Microorganisms, Similarities and Divergences,* J. C. Hall, R. E. Hoagland and R. M. Zablotowicz, Eds., American Chemical Society, Washington DC, 2001, pp. 89-118.

Halling-Sørensen, B., S. N. Nielsen, P. F. Lanzky, F. Ingerslev, H. C. Holten Lützhøff, and S. E. Jørgensen, "Occurrence, fate and effects of pharmaceutical substances in the environment – A review", *Chemosphere*, **36**, 357-393 (1998).

Hammel, K. E., B. Kalyanaraman, and T. K. Kirk, "Oxidation of polycyclic aromatic hydrocarbons and dibenzo(p) dioxins by *Phanerochaete chrysosporium* ligninase", *J. Biol. Chem.*, **261**, 6948-6952 (1986).

Hammett, L. P., *Physical Organic Chemistry*, McGraw-Hill, New York, NY, 1940.

Hansch, C., and A. Leo, *Substituent Constants for Correlation Analysis in Chemistry and Biology*, Wiley-Interscience, New York, 1979.

Hansch, C., A. Leo, and S. R. Heller, *Fundamentals and Applications in Chemistry and Biology*, ACS, Washington, DC, 1995.

Hansch, C., A. Leo, and D. Hoekman, *Exploring QSAR, Hydrophobic, Electronic and Steric Constants*, ACS, Washington DC, 1995.

Hansch, C., A. Leo, and R. W. Taft, "A survey of Hammett substituent constants and resonance and field parameters", *Chem. Rev.*, **91**, 165-195 (1991).

Hanselmann, K. W., "Microbial energetics applied to waste repositories", *Experientia*, **47**, 645-687 (1991).

Hansen, H. K., P. Rasmussen, A. Fredenslund, M. Schiller, and J. Gmehling, "Vapor-liquid equilibria by UNIFAC group contribution. 5. Revision and extension", *Ind. Eng. Chem. Res.*, **30**, 2352-2355 (1991).

Hanzlik, R. P., "Reactivity and toxicity among halogenated methanes and related compounds. A physicochemical correlate with predictive value", *Biochem. Pharmacol.*, **30**, 3027-3030 (1981).

Harding, G. C., R. J. LeBlanc, W. P. Vass, R. F. Addison, B. T. Hargrave, S. Pearre, A. Dupuis, and P. F. Brodie, "Bioaccumulation of polychlorinated biphenyls (PCBs) in the marine pelagic food web, based on a seasonal study in the southern Gulf of St. Lawrence, 1976-1977", *Mar. Chem.*, **56**, 145-179 (1997).

Harner, T., and T. F. Bidleman, "Measurement of octanol–air partition coefficients for polycyclic aromatic hydrocarbons and polychlorinated naphthalenes", *J. Chem. Eng. Data*, **43**, 40-46 (1998a).

Harner, T., and T. F. Bidleman, "Octanol–air partition coefficient for describing particle/gas partitioning of aromatic compounds in urban air", *Envion. Sci. Technol.*, **32**, 194-1502 (1998b).

Harner, T., and D. Mackay, "Measurement of octanol–air partition coefficients for chlorobenzenes, PCBs, and DDT", *Environ. Sci. Technol.*, **29**, 1599-1606 (1995).

Harris, J. C., and M. J. Hayes, "Acid dissociation constant". In *Handbook of Chemical Property Estimation Methods*, W. J. Lyman, W. F. Reehl and D. H. Rosenblatt, Eds., McGraw-Hill, New York, 1982, pp. 6-1 – 6-28.

Harris, J. M., and C. C. Wamser, *Fundamentals of Organic Reaction Mechanisms*, Wiley-Interscience, New York, 1978.

Hartmans, S., and J. A. M. de Bont, "Aerobic vinyl chloride metabolism in *Mycobacterium aurum* L1", *Appl. Environ. Microbiol.*, **58**, 1220-1226 (1992).

Harwood, C. S., and J. Gibson, "Shedding light on anaerobic benzene ring degradation: a process unique to prokaryotes", *J. Bacteriol.*, **179**, 301-309 (1997).

Hashimoto, Y., K. Tokura, H. Kishi, and W. M. J. Strachan, "Prediction of seawater solubility of aromatic compounds", *Chemosphere*, **13**, 881-888 (1984).

Hassall, K. A., *The Biochemistry and Uses of Pesticides*, VCH, Weinheim, 1990.

Hassett, J. J., J. C. Means, W. L. Banwart, S. G. Wood, S. Ali, and A. Khan, "Sorption of dibenzothiophene by soils and sediments", *J. Environ. Qual.*, **9**, 184-186 (1980).

Hassett, J. P., and E. Milicic, "Determination of equilibrium and rate constants for binding of a polychlorinated biphenyl congener by dissolved humic substances", *Envion. Sci. Technol*, **19**, 638-643 (1985).

Haston, Z. C., and P. L. McCarty, "Chlorinated ethene half-velocity coefficient (K_s) for reductive dechlorination", *Environ. Sci. Technol.*, **33**, 223-226 (1999).

Hatcher, P. G., J. M. Bortiatynski, R. D. Minard, J. Dec, and J.-M. Bollag, "Use of high-resolution 13C-NMR to the enzymatic covalent binding of 13C-labelled 2,4-dichlorophenol to humic substances", *Environ. Sci. Technol.*, **27**, 2098-2103 (1993).

Hay, R. W., and P. J. Morris, "Metal ion-promoted hydrolysis of amino acid esters and peptides". In *Metal Ions in Biological Systems*, Vol. 5, H. Siegel, Ed., Marcel Dekker, New York, 1976, pp. 173-243.

Hayduk, W., and H. Laudie, "Prediction of diffusion coefficients for non-electrolytes in dilute aqueous solutions", *AIChE J.*, **20**, 611-615 (1974).

Hayes, M. H. B., "Progress towards more realistic concepts of structures". In *Humic Substances; Structures, Properties and Uses*, G. Davies and E. A. Ghabbour, Eds., Royal Society of Chemistry, Cambridge, UK, 1998, pp. 1-28.

Hayes, M. H. B., and W. S. Wilson, Eds., *Humic Substances in Soils, Peats and Waters*, Royal Society of Chemistry, Cambridge, GB, 1997.

Hedegaard, J., and I. C. Gunsalus, "Mixed function oxidation. IV. An induced methylene hydroxylase in camphor oxidation", *J. Biol. Chem.*, **240**, 4038-4043 (1965).

Heermann, S. E., and S. E. Powers, "Modeling the partitioning of BTEX in water-reformulated gasoline systems containing ethanol." *J. Contam. Hydrol.*, **34**, 315-341 (1998).

Heider, J., and G. Fuchs, "Microbial anaerobic aromatic metabolism", *Anaerobe*, **3**, 1-22 (1997).

Heijman, C. G., E. Grieder, C. Holliger, and R. P. Schwarzenbach, "Reduction of nitroaromatic compounds coupled to microbial iron reduction in laboratory aquifer columns", *Environ. Sci. Technol.*, **29**, 775-783 (1995).

Helmer, F., K. Kiehs, and C. Hansch, "The linear free-energy relationship between partition coefficients and the binding and conformational perturbation of macromolecules by small organic compounds", *Biochemistry*, **7**, 2858-2863 (1965).

Hemptinne, J.-C., H. Delepine, C. Jose, and J. Jose, "Aqueous solubility of hydrocarbon mixtures", *Revue de l'institut français du pétrole*, **53**, 309-419 (1998).

Henderson, R. J., and D. R. Tocher, "The lipid composition and biochemistry of freshwater fish", *Prog. Lip. Res.*, **26**, 281-347 (1987).

Herbert, B. J., and J. G. Dorsey, "*n*-Octanol–water partition coefficient estimation by micellar electrokinetic capillary chromatography", *Anal. Chem.*, **67**, 744-749 (1995).

Herbert, V. R., and G. C. Miller, "Depth dependence of direct and indirect photolysis on soil surfaces", *J. Agric. Food Chem.*, **38**, 913-918 (1990).

Herdendorf, C. E., *Distribution of the World's Large Lakes*, Springer Verlag, Berlin, Heidelberg, 1990.

Higbie, R., "The rate of adsorption of a pure gas into a still liquid during short periods of exposure", *Trans. Am. Ins. Chem. Eng.*, **35**, 365-389 (1935).

Hill, D. W., and P. L. McCarty, "Anaerobic degradation of selected chlorinated hydrocarbon pesticides", *J. Water Pollution Control Fed.*, **39**, 1259-1277 (1967).

Hill, W. E., J. Szechi, C. Hofstee, and J. H. Dane, "Fate of a highly strained hydrocarbon in aqueous soil environment", *Environ. Sci. Technol.*, **31**, 651-655 (1997).

Hillel, D., *Environmental Soil Physics*, Academic Press, San Diego, 1998.

Himmelblau, D. M., "Diffusion of dissolved gases in liquids", *Chem. Rev.*, **64**, 527-550 (1964).

Hinckley, D. A., T. F. Bidleman, W. T. Foreman, and J. R. Tuschall, "Determination of vapor pressure for nonpolar and semipolar organic compounds from gas chromatographic retention data", *J. Chem. Eng. Data*, **35**, 232-237 (1990).

Hine, J., *Physical Organic Chemistry*, McGraw-Hill, New York, 1962.

Hine, J., and P. K. Mookerjee, "The intrinsic hydrophilic character of organic compounds. Correlations in terms of structural contributions", *J. Org. Chem.*, **40**, 292-298 (1975).

Hinedi, Z. R., A. C. Chang, and D. B. Borchardt, "Probing the association of fluorobenzene with dissolved organic matter using NMR spectroscopy", *Wat. Res.*, **31**, 877-883 (1997).

Hoffman, M., "Catalysis in aquatic environments". In *Aquatic Chemical Kinetics,* W. Stumm, Ed., Wiley-Interscience, New York, 1990, pp. 71-111.

Hofstetter, T. B., C. G. Heijman, C. Holliger, and R. P. Schwarzenbach, "Complete reduction of TNT and other (poly)nitroaromatic compounds under iron-reducing subsurface conditions", *Environ. Sci. Technol.*, **33**, 1479-1487 (1999).

Hoigne, J., "Inter-calibration of OH radical sources and water quality parameters", *Wat. Sci. Technol.*, **35**, 1-8 (1997).

Hoigné, J., B. C. Faust, W. R. Haar, F. E. Scully, Jr., and R. G. Zepp, "Aquatic humic substances as sources and sinks of photochemically produced transient reactants". In *Aquatic Humic Substances: Influence on Fate and Treatment of Pollutants,* I. H. Suffet and P. MacCarthy, Eds., Advances in Chemistry Series 219, American Chemical Society, Washington, DC, 1989, pp. 363-381.

Holliger, C., G. Schraa, E. Stupperich, A. J. M. Stams, and A. J. B. Zehnder, "Evidence for the involvement of corrinoids and factor F430 in the reductive dechlorination of 1,2-dichloroethane by *Methonasarcina barkeri*", *J. Bacteriol.*, **174**, 4427-4434 (1992).

Holmes, D. C., J. H. Simmons, and J. O. G. Tatton, "Chlorinated hydrocarbons in British wildlife", *Nature*, **216**, 227-229 (1967).

Hong, F., and S. O. Pekkonen, "Hydrolysis of phorate using simulated environmental conditions: Rates, mechanisms and product analysis", *J. Agric. Food Chem.*, **46**, 1192-1199 (1998).

Hornsby, A. G., R. D. Wauchope, and A. E. Herner, *Pesticide Properties in the Environment*, Springer, New York, 1996.

Horvath, R. S., "Microbial cometabolism and the degradation of organics in nature", *Bacteriol. Rev.*, **36**, 146-155 (1972).

Horvath, A. L., *Halogenated Hydrocarbons. Solubility-Miscibility with Water*, Marcel Dekker, New York, 1982.

Hsu, T. S., and R. Bartha, "Interaction of pesticide-derived chloroaniline residues with soil organic matter", *Soil Sci.*, **116**, 444-452 (1974).

Hsu, T. S., and R. Bartha, "Hydrolyzable and nonhydrolyzable 3,4-dichloroaniline-humus complexes and their respective rates of biodegradation", *J. Agric. Food Chem.*, **24**, 118-122 (1976).

Huang, C. H., *Hydrolysis of Amide, Carbamate, Hydrazide, and Sulfonylurea Agrochemicals*, Ph.D. thesis, Johns Hopkins University, Baltimore, MD, 1997.

Huang, C. H., and A. T. Stone, "Hydrolysis of naptalan and structurally related amides: Inhibition by dissolved metal ions and metal (hydr)oxide surfaces", *J. Agric. Food Sci.*, **47**, 4425-4434 (1999).

Huang, J., Z. He, and J. Wiegel, "Cloring, characterization, and expression of a novel gene encoding a reversible 4-hydroxybenzoate decarboxylase from *Clostridium hydroxybenzoicum*", *J. Bacteriol.*, **181**, 5119-5122 (1999).

Huang, J., and S. A. Mabury, "The role of carbonate radical in limiting the persistence of sulfur-containing chemicals in sunlit natural waters", *Chemosphere*, **41**, 1775-1782 (2000).

Huang, P. M., N. Senesi, and J. Buffle, *Structure and Surface Reactions of Soil Particles*, Wiley, Chichester, UK, 1998.

Huang, W., T. M. Young, M. A. Schlautman, H. Yu, and W. J. Weber Jr., "A distributed reactivity model for sorption by soils and sediments. 9. General isotherm nonlinearity and applicability of the dual reactive domain model", *Environ. Sci. Technol.*, **31**, 1703-1710 (1997).

Hughes, E. A. M., *The Chemical Statics and Kinetics of Solutions*, Academic, London, 1971.

Hui, Y. M., "Edible oil and fat products: general applications". In *Bailey's Industrial Oil and Fat Products,* Y. M. Hui, Ed., Vol. 1, Wiley, New York, 1996, pp. 399-400.

Huibers, P. D. T., and A. R. Katritzky, "Correlation of the aqueous solubility of hydrocarbons and halogenated hydrocarbons with molecular structure", *J. Chem. Inf. Comput. Sci.*, **38**, 283-292 (1998).

Hunchak-Kariouk, K., L. Schweitzer, and I. H. Suffet, "Partitioning of 2,2',4,4'-tetrachlorobiphenyl by the dissolved organic matter in oxic and anoxic porewaters", *Envion. Sci. Technol.*, **31**, 639-645 (1997).

Hundal, L. S., M. L. Thompson, D. A. Laird, and A. M. Carmo, "Sorption of phenanthrene by reference smectites", *Environ. Sci. Technol.*, **35**, 3456-3461 (2001).

Hung, H., C. J. Halsall, P. Blanchard, H. H. Li, P. Fellin, G. Stern, and B. Rosenberg, "Are PCBs in the Canadian Arctic atmosphere declining? Evidence from 5 years of monitoring", *Environ. Sci. Technol.*, **35**, 1303-1311 (2001).

Hunkeler, D., P. Höhener, A. Häner, T. Bregnard, and J. Zeyer, "Quantification of hydrocarbon mineralization in a diesel fuel contaminated aquifer treated by *in situ* biorestoration". In *Groundwater Quality: Remediation and Protection,* K. Kovar and J. Krasny, Eds., IAHS Press, Wallingford, UK, 1995, pp. 411-420.

Hunt, J. M., *Petroleum Geochemistry and Geology*, W.H. Freeman and Company, San Francisco, 1979.

Hunt, J. R., N. Sitar, and K. S. Udell, "Nonaqueous phase liquid transport and cleanup. 1. Analysis of mechanisms", *Water Resourc. Res.*, **24**, 1247-1258 (1988).

Huston, J. E., Grasses, Forbs, and Browse, "http://www.inform.umd.edu/EdRes/Topic/AgrEnv/ndd/goat/ GRASSES_FORBS_AND_BROWSE.html", (2001).

Hutchinson, T. C., J. A. Hellebust, D. Tam, D. Mackay, R. A. Mascarenhas, and W.-Y. Shiu, "The correlation of the toxicity to algae of hydrocarbons and halogenated hydrocarbons with their physical-chemical properties". In *Hydrocarbons and Halogenated Hydrocarbons in the Aquatic Environment,* B. K. Afghan and D. Mackay, Eds.,Plenum Press, New York, 1978, pp. 577-586.

Ilan, Y. A., G. Czapski, and D. Meisel, "The one-electron transfer redox potential of free radicals: 1. The oxygen/superoxide system", *Biochim. Biophys. Acta*, **430**, 209-244 (1987).

Imboden, D. M., "Limnologische Transport- und Nährstoffmodelle", *Schweiz. Z. Hydrol.*, **35**, 29-68 (1973).

Imboden, D. M., B. S. F. Eid, T. Joller, M. Schuster, and J. Wetzel, "MELIMEX, an experimental heavy-metal pollution study. 2. Vertical mixing in a large limnocorral", *Schweiz. Z. Hydrol.*, **47**, 177-189 (1979).

Imboden, D. M., and S. Emerson, "Natural radon and phosphorus as limnological tracers: Horizontal and vertical eddy diffusion in Greifensee", *Limnol. Oceanogr.*, **23**, 77-90 (1978).

Imboden, D. M., and R. P. Schwarzenbach, "Spatial and temporal distribution of chemical substances in lakes: Modeling concepts". In *Chemical Processes in Lakes,* W. Stumm, Ed.,Wiley-Interscience, New York, 1985, pp. 1-30.

Imboden, D. M., and A. Wüest, "Mixing mechanisms in lakes". In *Physics and Chemistry of Lakes,* A. Lerman, D. M. Imboden and J. Gat, Eds., Springer, Heidelberg, 1995, pp. 83-138.

Israelachvili, J. N., *Intermolecular and Surface Forces*, Academic Press, London, 1992.

Jackson, L. J., S. R. Carpenter, J. Manchester-Neesvig, and C. A. Stow, "PCB congeners in Lake Michigan coho and chinook salmon", *Environ. Sci. Technol.*, **35**, 856-862 (2001).

Jafvert, C. T., "Sorption of organic acid compounds to sediments: Initial model development", *Environ. Toxicol. Chem.*, **9**, 1259-1268 (1990).

Jafvert, C. T., and J. K. Heath, "Sediment- and saturated-soil-associated reactions involving an anionic surfactant (dodecylsulfate). 1. Precipitation and micelle formation", *Environ. Sci. Technol.*, **25**, 1031-1039 (1991).

Jafvert, C. T., J. C. Westall, E. Grieder, and R. P. Schwarzenbach, "Distribution of hydrophobic ionogenic organic compounds between octanol and water: Organic acids", *Environ. Sci. Technol.*, **24**, 1795-1803 (1990).

Jafvert, C. T., and N. L. Wolfe, "Degradation of selected halogenated ethanes in anoxic sediment–water systems", *Environ. Toxicol. and Chem.*, **6**, 827-837 (1987).

Jähne, B., K. H. Fisher, J. Ilmberger, P. Libner, W. Weiss, D. Imboden, U. Lemin, and J. M. Jaquet, "Parameterization of air/lake exchange". In *Gas Transfer at Water Surfaces,* W. Brutsaert and G. H. Jirka, Eds., D. Reidel, Boston, 1984, pp. 459-466.

Jähne, B., K. O. Münnich, R. Bösinger, A. Dutzi, W. Huber, and P. Libner, "On the parameters influencing air-water gas exchange", *J. Geophys. Res.*, **92**, 1937-1949 (1987a).

Jähne, B., G. Heinz, and W. Dietrich, "Measurement of the diffusion coefficient of sparingly soluble gases in water", *J. Geophys. Res.*, **92**, 10767-10776 (1987b).

Jankowski, J. J., D. J. Kieber, and K. Mopper, "Nitrate and nitrite ultraviolet actinometers", *Photochem. Photobiol.*, **70**, 319-328 (1999).

Jannasch, H. W., "Enrichments of aquatic bacteria in continuous colture", *Arch. Mikrobiol.*, **59**, 165-173 (1967).

Jannasch, H. W., "Competitive elimination of *Enterobacteriaceae* from seawater", *Appl. Microbiol.*, **16**, 1616-1618 (1968).

Janssen, D. B., J. E. Oppentocht, and G. J. Poelarend, "Microbial dehalogenation", *Curr. Opinion Biotechn.*, **12**, 254-258 (2001).

Jarvis, N. L., W. D. Garret, M. A. Scheiman, and C. O. Timmons, "Surface chemical characterization of surface-active material in seawater", *Limnol. Oceanogr.*, **12**, 88-96 (1967).

Jayasinghe, D. S., B. J. Brownawell, H. Chen, and J. C. Westall, "Determination of Henry's constants of organic compounds of low volatility: Methylanilines in methanol-water", *Environ. Sci. Technol.*, **26**, 2275-2281 (1992).

Jeans, J., *Dynamic Theory of Gases*, Cambridge University Press, Cambridge, 1921.

Jeffers, P. M., C. Brenner, and N. L. Wolfe, "Hydrolysis of carbon tetrachloride", *Environ. Toxicol. Chem.*, **15**, 1064-1065 (1996).

Jeffers, P. M., L. M. Ward, L. M. Woytowitch, and N. L. Wolfe, "Homogeneous hydrolysis rate constants for selected chlorinated methanes, ethanes, ethenes, and propanes", *Environ. Sci. Technol.*, **23**, 965-969 (1989).

Jeffers, P. M., and N. L. Wolfe, "Homogeneous hydrolysis rate constants – Part II: Additions, corrections and halogen effects", *Environ. Toxicol. Chem.*, **15**, 1066-1070 (1996).

Jenal-Wanner, U., and P. L. McCarty, "Development and evaluation of semicontinuous slurry microcosms to simulate in situ biodegradation of trichloroethylene in contaminated aquifers", *Environ. Sci. Technol.*, **31**, 2915-2922 (1997).

Jensen, J., "Fate and effects of linear alkylbenzene sulfonates (LAS) in the terrestrial environment", *Sci. Total Environ.*, **226**, 93-111 (1999).

Johnson, B. D., and R. C. Cooke, "Bubble populations and spectra in coastal waters: A photographic approach", *J. Geophys. Res.*, **84**, 3761-3766 (1979).

Johnson, C. A., and J. C. Westall, "Effect of pH and KCl concentration on the octanol–water distribution of methyl anilines", *Environ. Sci. Technol.*, **24**, 1869-1875 (1990).

Jones, A. S., "Metabolism of aldicarb by five soil fungi", *J. Agric. Food Chem.*, **24**, 115-117 (1976).

Kallenborn, R., M. Oehme, D. D. Wynn-Williams, M. Schlabach, and J. Harvis, "Ambient air levels and atmospheric long-range transport of persistent organochlorines to Signy Island, Antarctica", *Sci. Total Environ.*, **220**, 167-180 (1998).

Kamlet, M. J., J.-L. M. Abboud, M. H. Abraham, and R. W. Taft, "Linear solvation energy relationships. 23. A comprehensive collection of the solvatochromic parameters, π^*, α and β, and some methods for simplifying the generalized solvatochromic equation." *J. Org. Chem.*, **48**, 2877-2887 (1983).

Kan, A. T., and M. B. Tomson, "UNIFAC prediction of aqueous and nonaqueous solubilities of chemicals with enviromental interest", *Environ. Sci. Technol.*, **30**, 1367-1376 (1996).

Kanwisher, J., "Effect of wind on CO_2 exchange across the sea surface", *J. Geophys. Res.*, **68**, 3921-3927 (1963).

Karapanagioti, H. K., S. Kleineidam, D. A. Sabatini, P. Grathwohl, and B. Ligouis, "Impacts of heterogeneous organic matter on phenanthrene sorption: Equilibrium and kinetic studies with aquifer material", *Environ. Sci. Technol.*, **34**, 406-414 (2000).

Kari, F. G., and W. Giger, "Modeling the photochemical degradation of ethylenediamine tetraacetate in the River Glatt", *Environ. Sci. Technol.*, **29**, 2814-2827 (1995).

Kari, F. G., S. Hilger, and S. Canonica, "Determination of the reaction quantum yield for the photochemical degradation of Fe(III)-EDTA: Implications for the environmental fate of EDTA in surface waters", *Environ. Sci. Technol.*, **29**, 1008-1017 (1995).

Karickhoff, S. W., "Organic pollutant sorption in aquatic system", *J. Hydraulic Eng.*, **110**, 707-735 (1984).

Kaschabek, S. R., T. Kasberg, M. D., A. E. Mars, D. B. Janssen, and W. Reinecke, "Degradation of chloroaromatics: Purification and characterization of a novel type of chlorocatechol 2,3-dioxygenase of *Pseudomonas putida* GJ31", *J. Bacteriol.*, **180**, 296-302 (1998).

Keith, L., *Environmental Endocrine Disruptors – A Handboook of Property Data*, Wiley, New York, 1997.

Kelly, B. C., and F. A. P. C. Gobas, "Bioaccumulation of persistent organic pollutants in lichens-caribou-wolf food chains of Canada's Central and Western Arctic", *Environ. Sci. Technol.*, **35**, 325-334 (2001).

Keuning, S., D. B. Janssen, and B. Witholt, "Purification and characterization of hydrolytic haloalkane dehalogenase from *Xanthobacter autophicus* GJ10", *J. Bacteriol.*, **163**, 635-639 (1985).

Key, B. D., R. D. Howell, and C. R. Criddle, "Fluorinated organics in the biosphere", *Environ. Sci. Technol.*, **31**, 2445-2454 (1997).

Khan, S. U., *Pesticides in the Soil Environment*, Elsevier, Amsterdam, 1980.

Khossravi, D., and K. A. Connors, "Solvent effects on chemical processes. I: Solubility of aromatic and heterocyclic compounds in binary aqueous-organic solvents", *J. Pharm. Sci.*, **81**, 371-379 (1992).

Kiehs, K., C. Hansch, and L. Moore, "The role of hydrophobic bonding in the binding of organic compounds by bovine hemoglobin", *Biochemistry*, **5**, 2602-2605 (1966).

Kile, D. E., C. T. Chiou, H. Zhou, H. Li, and O. Xu, "Partition of nonpolar organic pollutants from water to soil and sediment organic matters", *Environ. Sci. Technol.*, **29**, 1401-1406 (1995).

Kim, H.-Y., H. F. Hemond, L. R. Krumholz, and B. A. Cohen, "In-situ biodegradation of toluene in a contaminated stream: Part 1. Field studies", *Environ. Sci. Technol.*, **29**, 108-116 (1995).

Kim, J. Y., J. K. Park, and T. B. Edil, "Sorption of organic compounds in the aqueous phase onto tire rubber", *J. Environ. Eng. ASCE*, **123**, 827-835 (1997).

Kim, M. H., and O. J. Hao, "Cometabolic degradation of chlorophenols by *Acinetobacter* species", *Water Res.*, **33**, 562-574 (1999).

Kimble, K. D., and Y.-P. Chin, "The sorption of polycyclic aromatic hydrocarbons by soils in low-methanol/water mixtures", *J. Contam. Hydrol.*, **17**, 129-143 (1994).

Kirby, A. J., "Hydrolysis and formation of esters of organic acids". In *Comprehensive Chemical Kinetics,* Vol. 10, C. H. Brandford and C. F. H. Tipper, Eds., Elsevier, Amsterdam, 1972, pp. 57-207.

Kirby, A. J., and S. G. Warren, *The Organic Chemistry of Phosphorus*, Elsevier, New York, 1967.

Kirk-Othmer, *Encyclopedia of Chemical Technology*, Wiley, New York, 1992.

Kirsch, J. F., W. Clevell, and A. Simon, "Multiple structure-reactivity correlations. The alkaline hydrolysis of acyl- and aryl-substituted phenyl benzoates", *J. Org. Chem.*, **33**, 127-132 (1968).

Kishino, T., and K. Kobayashi, "Relation between the chemical structures of chlorophenols and their dissociation constants and partition coefficients in several solvent water systems", *Wat. Res.*, **28**, 1547-1552 (1994).

Kistiakowsky, W., "Über Verdampfungswärme und einige Gleichungen, welche die Eigenschaften der unassoziierten Flüssigkeiten bestimmen", *Z. Phys. Chem.*, **107**, 65-73 (1923).

Klausen, J., S. B. Haderlein, and R. P. Schwarzenbach, "Oxidation of substituted anilines by aqueous MnO_2: Effect of co-solutes on initial and quasi-steady-state kinetics", *Environ. Sci. Technol.*, **31**, 2642-2649 (1997).

Klausen, J., M. A. Meier, and R. P. Schwarzenbach, "Assessing the fate of organic contaminants in aquatic environments: Mechanism and kinetics of hydrolysis of a carboxylic ester", *J. Chem. Educ.*, **74**, 1440-1444 (1997).

Klausen, J., S. P. Tröber, S. B. Haderlein, and R. P. Schwarzenbach, "Reduction of substituted nitrobenzenes by Fe(II) in aqueous mineral suspensions", *Environ. Sci. Technol.*, **29**, 2396-2404 (1995).

Kleindorfer, P. R., H. C. Kunreuther, and P. J. H. Dchoemaker, *Decision Sciences*, Cambridge University Press, Cambridge, 1993.

Kleineidam, S., H. Rügner, B. Ligouis, and P. Grathwohl, "Organic matter facies and equilibrium sorption of phenanthrene", *Environ. Sci. Technol.*, **33**, 1637-1644 (1999).

Klopman, G., and M. Tu, "Structure-biodegradability study and computer-automated prediction of aerobic biodegradation of chemicals", *Environ. Toxicol. Chem.*, **16**, 1829-1835 (1997).

Knackmuss, H.-J., "Degradation of halogenated and sulfonated hydrocarbons". In *Microbial Degradation of Xenobiotics and Recalcitrant Compounds,* Academic, T. Leisinger, R. Hütter, A. M. Cook and J. Nüesch, Eds., New York, 1981, pp. 189-212.

Kobori, T., H. Sasaki, W. C. Lee, S. Zenno, K. Saigo, M. E. P. Murphy, and M. Tanokura, "Structure and site-directed mutagenesis of a flavoprotein from *Escherichia coli* that reduces nitrocompounds", *J. Biol. Chem.*, **276**, 2816-2823 (2001).

Koeman, J. H., M. C. T. DeBrauw, and R. H. DeVos, "Chlorinated biphenyls in fish, mussels and birds from the River Rhine and the Netherlands coastal area", *Nature*, **221**, 1126-1128 (1969).

Koh, S.-C., J. P. Bowman, and G. S. Sayler, "Soluble methane monooxygenase production and trichloroethylene degradation by a type I methanotroph, *Methylomonas methanica* 68-1", *Appl. Environ. Microbiol.*, **59**, 960-967 (1993).

Kohler, H.-P. E., D. Kohler-Staub, and D. D. Focht, "Cometabolism of polychlorinated biphenyls, enhanced transformation of Arochlor 1254 by growing cells", *Appl. Environ. Microbiol.*, **54**, 1940-1945 (1988).

Kohler-Staub, D., and T. Leisinger, "Dichloromethane dehalogenase of *Hyphomicrobium* sp. strain DM2", *J. Bacteriol.*, **162**, 676-681 (1985).

Kolmogorov, N. A., "Dissipation of energy under locally isotropic turbulence", *Dokl. Akad. Nauk SSSR*, **30**, 16-18 (1941).

Kolovayev, D. A., "Investigation of the concentration and statistical size distribution of wind-produced bubbles in the near-surface ocean", *Oceanology*, **15**, 659-661 (1976).

Kolpin, D. W., and S. J. Kolkhoff, "Atrazine degradation in a small stream in Iowa", *Environ. Sci. Technol.*, **27**, 134-139 (1993).

Kolpin, D. W., E. T. Furlong, M. T. Meyer, E. M. Thurman, S. D. Zaugg, L. B. Barber, and H. T. Buxton, "Pharmaceuticals, hormones and other organic waste water contaminants in US streams, 1999-2000: A national reconnaissance", *Environ. Sci. Technol.*, **36**, 1202-1211 (2002).

Kömp, P., and M. S. McLachlan, "Octanol/air partitioning of polychlorinated biphenyls", *Environ. Tox. Chem.*, **16**, 2433-2437 (1997a).

Kömp, P., and M. S. McLachlan, "Interspecies variability of the plant/air partitioning of polychlorinated biphenyls", *Environ. Sci. Technol.*, **31**, 2944-2948 (1997b).

Konings, W. N., K. J. Hellingwerf, and G. T. Robillard, "Transport across bacterial membranes". In *Membrane Transport*, S. L. Bonting and J. J. H. H. M. d. Pont, Eds., Elsevier, New York, 1981, pp. 257-283.

Kopinke, F.-D., J. Pörschmann, and A. Georgi, "Application of SPME to study sorption phenomena on dissolved humic organic matter". In *Applications of Solid Phase Microextraction,* J. Pawliszyn, Ed., Royal Society of Chemistry, Cambridge. UK, 1999, pp. 111-128.

Kortüm, G., W. Vogel, and K. Andrussow, *Dissociation Constants of Organic Acids in Aqueous Solution*, Butterworths, London, 1961.

Kramer, J. B., S. Canonica, J. Hoigne, and J. Kaschig, "Degradation of fluorescent whitening agents in sunlit natural waters", *Environ. Sci. Technol.*, **30**, 2227-2234 (1996).

Krauss, M., W. Wilcke, and W. Zech, "Availability of polycyclic aromatic hydrocarbons (PAHs) and polychlorinated biphenyls (PCBs) to earthworms in urban soils", *Environ. Sci. Technol.*, **34**, 4335-4340 (2000).

Krieger, C. J., W. Roseboom, S. P. J. Albracht, and A. M. Spormann, "A stable organic free radical in anaerobic benzylsuccinate synthase of *Azoracus* sp. strain T", *J. Biol. Chem.*, **276**, 12924-12927 (2001).

Kriegman-King, M. R., and M. Reinhard, "Transformation of carbon tetrachloride in the presence of sulfide, biotite, and vermiculite", *Environ. Sci. Technol.*, **26**, 2198-2206 (1992).

Kriegman-King, M. R., and M. Reinhard, "Transformation of carbon tetrachloride by pyrite in aqueous solution", *Environ. Sci. Technol.*, **28**, 692-700 (1994).

Krone, K. E., R. K. Thauer, and H. P. C. Hogenkamp, "Reductive dehalogenation of chlorinated C1-hydrocarbons mediated by corrinoids", *Biochemistry*, **28**, 4908-4914 (1989a).

Krone, U. E., K. Laufer, and R. K. Thauer, "Coenzyme F430 as a possible catalyst for the reductive dehalogenation of chlorinated C1-hydrocarbons in methanogenic bacteria", *Biochemistry*, **28**, 10061-10065 (1989b).

Krop, H. B., L. C. M. Commandeur, and H. A. J. Govers, "Prediction of redox potentials and isomer distribution yields for reductive dechlorination of chlorobenzenes", *SAR QSAR Environ. Res.*, **2**, 271-287 (1994).

Kropp, K. G., I. A. Davidova, and J. M. Suflita, "Anaerobic oxidation of *n*-dodecane by an addition reaction in a sulfate-reducing bacterial enrichment culture", *Appl. Environ. Microbiol.*, **66**, 5393-5398 (2000).

Kucklick, J. R., and J. E. Baker, "Organochlorines in Lake Superior's food web", *Environ. Sci. Technol.*, **32**, 1192-1998 (1998).

Kumke, M. U., H.-G. Löhmannsröben, and T. Roch, "Fluorescence quenching of polycyclic aromatic compounds by humic acid", *Analyst*, **119**, 997-1001 (1994).

Kung, K.-H., and M. B. McBride, "Adsorption of *para*-substituted benzoates on iron oxides", *Soil Sci. Soc. Am. J.*, **53**, 1673-1678 (1989).

Kwok, E. S. C., J. Arey, and R. Atkinson, "Gas-phase atmospheric chemistry of dibenzo-p-dioxin and dibenzofuran", *Environ. Sci. Technol.*, **28**, 528-533 (1994).

Kwok, E. S. C., and R. Atkinson, "Estimation of hydroxyl radical reaction rate constants for gas phase organic compounds using a structure-reactivity relationship: An update", *Atmos. Environ.*, **29**, 1685-1695 (1995).

Laha, S., and R. G. Luthy, "Oxidation of aniline and other primary aromatic amines by manganese dioxide", *Environ. Sci. Technol.*, **24**, 363-373 (1990).

Laidler, J. J., *Chemical Kinetics*, McGraw-Hill, New York, 1965.

Lambert, W. J., "Modeling oil–water partitioning and membrane permeation using reversed phase chromatography", *J. Chromatogr. A*, **656**, 469-484 (1993).

Lamont, J. C., and D. S. Scott, "An eddy cell model of mass transfer into the surface of a turbulent liquid", *AIChE J.*, **16**, 4, 513-519 (1970).

Lane, W. F., and R. C. Loehr, "Estimating the equilibrium aqueous concentrations of polynuclear aromatic hydrocarbons in complex mixtures", *Environ. Sci. Technol.*, **26**, 983-990 (1992).

Laor, Y., W. J. Farmer, Y. Aochi, and P. F. Strom, "Phenanthrene binding and sorption to dissolved and to mineral-associated humic acid", *Wat. Res.*, **32**, 1923-1931 (1998).

Larson, R., and D. Davidson, "Acclimation to and biodegradation of NTA at trace concentrations in natural waters", *Water Res.*, **16**, 1597-1604 (1982).

Larson, R. A., and E. J. Weber, *Reaction Mechanisms in Environmental Organic Chemistry*, Lewis, Boca Raton, FL, 1994.

Larson, R. A., and R. Zepp, "Reactivity of the carbonate radical with aniline derivatives", *Environ. Toxicol. Chem.*, **7**, 265-274 (1988).

Larson, R. J., G. G. Clinckemaillie, and L. VanBelle, "Effect of temperature and dissolved oxygen on biodegradation and nitrilotriacetate", *Water Res.*, **15**, 615-620 (1981).

Leberman, R., and A. K. Soper, "Effect of high salt concentrations on water structure", *Nature*, **378**, 364-366 (1995).

Leblanc, G. A., "Trophic-level differences in the bioconcentration of chemicals: Implications in assessing biomagnification", *Environ. Sci. Technol.*, **29**, 154-160 (1995).

Leboeuf, E. J., and W. J. Weber Jr., "A distributed reactivity model for sorption by soils and sediments. 8. Sorbent organic domains: Discovery of a humic acid glass transition and an argument for a polmer-based model", *Environ. Sci. Technol.*, **31**, 1697-1702 (1997).

Ledbetter, M., "Langmuir circulations and plankton patchiness", *Ecol. Modeling*, **7**, 289-310 (1979).

Lee, L. S., C. A. Bellin, R. Pinal, and P. S. C. Rao, "Cosolvent effects on sorption of organic acids by soils from mixed solvents", *Environ. Sci. Technol.*, **27**, 165-171 (1993).

Lee, L. S., M. Hagwall, J. J. Delfino, and P. S. C. Rao, "Partitioning of polycyclic aromatic hydrocarbons from diesel fuel into water", *Environ. Sci. Technol.*, **26**, 2104-2110 (1992a).

Lee, L. S., P. S. C. Rao, and I. Okuda, "Equilibrium partitioning of polycyclic aromatic hydrocarbons from coal tar into water", *Environ. Sci. Technol.*, **26**, 2110-2115 (1992b).

Lee, L. S., A. K. Nyman, H. H. Li, M. C. Nyman, and C. T. Jafvert, "Initial sorption of aromatic amines to surface soils", *Environ. Toxicol. Chem.*, **16**, 1575-1582 (1997).

Lee, L. S., and P. S. C. Rao, "Impact of several water-miscible organic solvents on sorption of benzoic acid by soil", *Environ. Sci. Technol.*, **30**, 1533-1539 (1996).

Lee, L. S., P. S. C. Rao, P. Nkedi-Kizza, and J. J. Delfino, "Influence of solvent and sorbent characteristics on distribution of pentachlorophenol in octanol–water and soil–water systems", *Environ. Sci. Technol.*, **24**, 654-661 (1990).

Lee, S. S., A. K. Nyman, H. Li, M. C. Nyman, and C. Jafvert, "Initial sorption of aromatic amines to surface soils", *Environ. Toxicol. Chem.*, **16**, 1575-1582 (1997).

Legierse, K. C. H. M., W. H. J. Vaes, T. Sinnige, and J. L. M. Hermens, *Distribution of chlorobenzenes and the organophosphorous pesticide chlorthion in the pond snail (Lymnaea stagnalis): A multi-compartment model based on partitioning to polar lipids, neutral lipids and pond snail protein*, Ph.D. thesis, University of Utrecht, Utrecht, 1998.

Leifer, A., *The Kinetics of Environmental Aquatic Photochemistry*, American Chemical Society, Washington, DC, 1988.

Lemaire, J., J. A. Guth, O. Klais, J. Leahey, W. Merz, J. Philp, R. Wilmes, and C. J. M. Wolff, "Ring test of a method for assessing the phototransformation of chemicals in water", *Chemosphere*, **14**, 53-77 (1985).

Lenke, H., C. Achtnich, and H.-J. Knackmuss, "Perspectives of bioelimination of polynitroaromatic compounds". In *Biodegradation of Nitroaromatic Compounds and Explosives,* J. C. Spain, J. B. Hughes and H.-J. Knackmuss, Eds., Lewis Publishers, Boca Raton, FL, 2000, pp. 92-126.

Lerman, A., *Geochemical Processes. Water and Sediment Environments*, Wiley, New York, 1979.

Leveau, J. H. J., A. J. B. Zehnder, and J. R. van der Meer, "The *tfdK* gene product facilitates the uptake of 2,4-dichlorophenoxyacetate by *Ralstonia eutropha* JMP134(pJP4)", *J. Bacteriol.*, **180**, 2237-2243 (1998).

Leversee, G. J., P. F. Landrum, J. P. Giesy, and T. Fannin, "Humic acids reduce bioaccumulation of some polycyclic aromatic hydrocarbons", *Can. J. Fish. Aquat. Sci.*, **40**, Suppl. 2, 63-69 (1983).

Lewis, G. N., "The law of physico-chemical change", *Proc. Am. Acad.*, **37**, 49-69 (1901).

Lewis, T. A., M. J. Morra, and P. D. Brown, "Comparative product analysis of carbon tetrachloride dehalogenation catalyzed by cobalt corrins in the presence of thiol or titanium(III) reducing agents", *Environ. Sci. Technol.*, **30**, 292-300 (1996).

Li, A., and A. W. Andren, "Solubility of polychlorinated biphenyls in water/alcohol mixtures. 1. Experimental data", *Environ. Sci. Technol.*, **28**, 47-52 (1994).

Li, A., and A. W. Andren, "Solubility of polychlorinated biphenyls in water/alcohol mixtures. 2. Predictive methods", *Envion. Sci. Technol.*, **29**, 3001-3006 (1995).

Li, A., A. W. Andren, and S. H. Yalkowsky, "Choosing a cosolvent: Solubilization of naphthalene and cosolvent property", *Environ. Toxicol. Chem.*, **15**, 2233-2239 (1996).

Li, H., and L. S. Lee, "Sorption and abiotic transformation of aniline and α-naphthylamine by surface soils", *Environ. Sci. Technol.*, **33**, 1864-1870 (1999).

Li, H., L. S. Lee, C. T. Jafvert, and J. G. Gravel, "Effect of substitution on irreversible binding and transformation of aromatic amines with soils in aqueous systems", *Environ. Sci. Technol.*, **34**, 3674-3680 (2000).

Li, Y.-F., Y. Hata, T. Fujii, T. Hisano, M. Nishihara, T. Kurihara, and N. Esaki, "Crystal structures of reaction intermediates of L-2-haloacid dehalogenase and implications for the reaction mechanisms", *J. Biol. Chem.*, **273**, 15035-15044 (1998).

Liang, C., and D. A. Gallagher, "QSPR prediction of vapor pressure from solely theoretically-derived descriptors", *J. Chem. Inf. Comput. Sci.*, **38**, 321-324 (1998).

Liang, C., J. F. Kankow, J. R. Odum, and J. H. Seinfeld, "Gas/particle partitioning of semivolatile organic compounds to model inorganic, organic, and ambient smog aerosols", *Environ. Sci. Technol.*, **31**, 3086-3092 (1997).

Liang, C., and J. F. Pankow, "Gas/particle partitioning of organic compounds to environmental tabacco smoke: Partition coefficient measurements by desorption and comparison to urban particulate matter", *Environ. Sci. Technol.*, **30**, 2800-2805 (1996).

Lide, D. R., Ed. *CRC Handbook of Chemistry and Physics*, CRC Press, Boca Raton, FL, 1995/96.

Lipnick, R. L., "Structure-activity relationships". In *Fundamentals of Aquatic Toxicology,* 2nd Ed., G. M. Rand, Ed., Taylor and Francis Publishers, Washington, DC, 1995, pp. 609-655.

Lippa, K. A., *Reactions of Chloro-s-triazine and Chloroacetanilide Agrochemicals with Reduced Sulfur Species*, Ph.D. Thesis, The Johns Hopkins University, Baltimore, MD, 2002.

Liss, P. S., "Processes of gas exchange across an air–water interface", *Deep Sea Res.*, **20**, 221-238 (1973).

Liss, P. S., P. W. Balls, F. N. Martinelli, and M. Coantic, "The effect of evaporation and condensation on gas transfer across an air–water interface", *Oceanologica Acta*, **4**, 129-138 (1981).

Liss, P. S., and L. Merlivat, Eds., *Air-Sea Gas Exchange Rates: Introduction and Sythesis, The Role of Air-sea Exchange in Geochemical Cycling*, D. Reidel Publishing, Dordrecht, The Netherlands, 1986.

Little, C. D., A. V. Palumbo, S. E. Herbes, M. E. Lidstrom, R. L. Tyndall, and P. J. Gilmer, "Trichloroethylene biodegradation by a methane-oxidizing bacterium", *Microbiol.*, **54**, 951-956 (1988).

Liu, Z., E. A. Betterton, and R. G. Arnold, "Electrolytic reduction of low molecular weight chlorinated aliphatic compounds: Structural and thermodynamic effects on process kinetics", *Environ. Sci. Technol.*, **34**, 804-811 (2000).

Livingstone, D. M., and D. M. Imboden, "The non-linear influence of wind-speed variability on gas transfer in lakes", *Tellus*, **45B**, 275-295 (1993).

Locher, H. H., B. Poolmann, A. M. Cook, and W. N. Konings, "Uptake of 4-toluene sulfonate by *Comamonas testosteroni* T-2", *J. Bacteriol.*, **175**, 1075-1080 (1993).

Lohmann, R., and K. C. Jones, "Dioxins and furans in air and deposition: A review of levels, behaviour and processes", *Sci. Total Environ.*, **219**, 53-81 (1998).

Loonen, H., F. Lindgren, B. Hansen, W. Karcher, J. Niemela, K. Hiromatsu, M. Takatsuki, W. J. G. M. Peijnenburg, E. Rorije, and J. Struijs, "Prediction of biodegradability from chemical structure: Modeling of ready biodegradation test data", *Environ. Toxicol. Chem.*, **18**, 1763-1768 (1999).

Lorenz, E., "Deterministic non-periodic flows", *J. Atmos. Sci.*, **20**, 130-141 (1963).

Lövgren, L., "Complexation reactions of phthalic acid and aluminium(III) with the surface of goethite", *Geochim. Cosmochim. Acta*, **55**, 3639-3645 (1991).

Lowry, T. H., and K. Schueller-Richardson, *Mechansisms and Theory in Organic Chemistry*, Harper and Row, New York, 1981.

Lüers, F., and T. E. M. tenHulscher, "Temperature effect on the partitioning of polycyclic aromatic hydrocarbons between natural organic carbon and water", *Chemosphere*, **33**, 643-657 (1996).

Luther, S. M., M. J. Dudas, and P. M. Fedorak, "Sorption of sulfolane and diisopropanolamine by soils, clays and aquifer materials", *J. Contam. Hydrol.*, **32**, 159-176 (1998).

Luthy, R. G., D. A. Dzombak, M. J. R. Shannon, R. Unterman, and J. R. Smith, "Dissolution of PCB congeners from an aroclor and an aroclor/hydraulic oil mixture", *Wat. Res.*, **31**, 561-573 (1997a).

Luthy, R. G., G. R. Aiken, M. L. Brusseau, S. D. Cunningham, P. M. Gschwend, J. J. Pignatello, M. Reinhard, S. J. Traina, W. J. Weber Jr., and J. C. Westall, "Sequestration of hydrophobic organic contaminants by geosorbents", *Environ. Sci. Technol.*, **31**, 3341-3447 (1997b).

Ma, W.-C., A. vanKleunen, J. Immerzeel, and P. G.-J. deMaagd, "Bioaccumulation of polycyclic aromatic hydrocarbons by earthworms: Assessment of equilibrium partitioning in in situ studies and water experiments", *Environ. Toxicol. Chem.*, **17**, 1730-1737 (1998).

Mabey, W., and T. Mill, "Critical review of hydrolysis of organic compounds in water under environmental conditions", *J. Phys. Ref. Data*, **7**, 2, 383-415 (1978).

Mabey, W. R., D. Tse, A. Baraze, and T. Mill, "Photolysis of nitroaromatics in aquatic systems. I. 2,4,6-Trinitrotoluene", *Chemosphere*, **12**, 3-16 (1983).

Macalady, D. L., P. G. P. G. Tratnyek, and T. J. Grundl, "Abiotic reduction reactions of anthropogenic organic chemicals in anaerobic systems: A critical review", *J. Contam. Hydrol.*, **1**, 1-28 (1986).

Mackay, A. A., Y.-P. Chin, J. K. MacFarlane, and P. M. Gschwend, "Laboratory assessment of BTEX soil flushing", *Environ. Sci. Technol.*, **30**, 3223-3231 (1996).

MacKay, A. A., and P. M. Gschwend, "Sorption of monoaromatic hydrocarbons to wood", *Environ. Sci. Technol.*, **34**, 839-845 (2000).

Mackay, D., "Finding fugacity feasible", *Environ. Sci. Technol.*, **13**, 1218-1223 (1979).

Mackay, D., A. Bobra, D. W. Chan, and W. Y. Shiu, "Vapor pressure correlations for low-volatility environmental chemicals", *Environ. Sci. Technol.*, **16**, 645-649 (1982).

Mackay, D., and W. Y. Shiu, "A critical review of Henry's Law constants for chemicals of environmental interest", *J. Phys. Chem. Ref. Data*, **10**, 1175-1199 (1981).

Mackay, D., W. Y. Shiu, and K. C. Ma, Eds., *Illustrated Handbook of Physical-Chemical Properties and Environmental Fate for Organic Chemicals, Vol. I: Monoaromatic Hydrocarbons, Chlorobenzenes, and PCBs*, CRC Press, Boca Raton, FL, 1992a.

Mackay, D., W. Y. Shiu, and K. C. Ma, Eds., *Illustrated Handbook of Physical-Chemical Properties and Environmental Fate for Organic Chemicals, Vol. II: Polynuclear Aromatic Hydrocarbons, Polychlorinated Dioxins, Dibenzofurans*, CRC Press, Boca Raton, FL, 1992b.

Mackay, D., W. Y. Shiu, and K. C. Ma, Eds., *Illustrated Handbook of Physical-Chemical Properties and Environmental Fate for Organic Chemicals, Vol. III: Volatile Organic Compounds*, CRC Press, Boca Raton, FL, 1993.

Mackay, D., W. Y. Shiu, and K. C. Ma, Eds., *Illustrated Handbook of Physical-Chemical Properties and Environmental Fate for Organic Chemicals, Vol. IV: Oxygen, Nitrogen, and Sulfur Containing Compounds*, CRC Press, Boca Raton, FL, 1995.

Mackay, D., W. Y. Shiu, and K. C. Ma, Eds., *Illustrated Handbook of Physical-Chemical Properties and Environmental Fate for Organic Chemicals, Vol. V: Pesticide Chemicals*, CRC Press, Boca Raton, FL, 1997.

Mackay, D., and A. T. K. Yeun, "Mass transfer coefficients correlations for volatilization of organic solutes from water", *Environ. Sci. Technol.*, **17**, 211-233 (1983).

Mackay, D. M., and J. A. Cherry, "Groundwater contamination: Pump-and-treat remediation", *Environ. Sci. Technol.*, **23**, 630-636 (1989).

Mackay, D. M., D. L. Freyberg, P. V. Roberts, and J. A. Cherry, "A natural gradient experiment on solute transport in a sand aquifer, 1. Approach and overview of plume movement", *Water Resour. Res.*, **22**, 2017-2029 (1986).

Mackay, D. W., D. Y. Shiu, and R. P. Sutherland, "Determination of air–water Henry's Law constants for hydrophobic pollutants", *Environ. Sci. Technol.*, **13**, 333-337 (1979).

MacPhee, J. A., A. Panaye, and J.-E. Dubois, "A critical examination of the Taft steric parameter-E_s. Definition of a revised, broader and homogeneous scale. Extension to highly congested alkyl groups", *Tetrahedron*, **34**, 3553-3562 (1978).

Mader, B. T., K.-U. Goss, and S. J. Eisenreich, "Sorption of nonionic, hydrophobic organic chemicals to mineral surfaces", *Environ. Sci. Technol.*, **31**, 1079-1086 (1997).

Madigan, M. T., J. M. Martinko, and J. Parker, *Brock Biology of Microorganisms*, Upper Saddle River, NJ, 2000.

Magnuson, J. K., M. F. Romine, D. R. Burris, and M. T. Kingsley, "Trichloroethene reductive dehalogenase from *Dehalococcoides ethenogenes*: Sequence of tceA and substrate range characterization", *Appl. Environ. Microbiol.*, **66**, 5141-5147 (2000).

Magnuson, J. K., R. V. Stern, J. M. Gossett, S. H. Zinder, and D. R. Burris, "Reductive dechlorination of tetrachloroethene to ethene by a two-component enzyme pathway", *Appl. Environ. Microbiol.*, **64**, 1270-1275 (1998).

Maher, P. J., and B. D. Smith, "Vapor-liquid equilibrium data for binary systems of chlorobenzene with acetone, acetonitrile, ethyl acetate, ethylbenzene, methanol and 1-pentene", *J. Chem. Eng. Data*, **24**, 363-377 (1979).

Maloney, S. E., A. Maule, and A. R. W. Smith, "Microbial transformation of pyrethroid insecticides: Permethrin, deltamethrin, fastac, fenvalerate, and fluvalinate", *Appl. Environ. Microbiol*, **54**, 2874-2876 (1988).

Mannervik, B., "The isoenzymes of glutathione transfers", *Adv. Enzymol.*, **57**, 357-417 (1985).

March, J., *Advanced Organic Chemistry, 4th ed.*, Wiley, New York, 1992.

Marks, T. S., J. D. Allpress, and A. Maule, "Dehalogenation of lindane by a variety of porphyrins and corrins", *Appl. Environ. Microbiol.*, **55**, 1258-1261 (1989).

Marrero, T. R., and E. A. Mason, "Gaseous diffusion coefficients", *J. Chem. Phys. Ref. Dat.*, **1**, 3-118 (1972).

Mars, A. E., T. Kasberg, S. R. Kaschabek, M. H. van Agteren, D. B. Janssen, and W. Reinecke, "Microbial degradation of chloroaromatics: Use of the *meta*-cleavage pathway for Mineralization of chlorobenzene", *J. Bacteriol.*, **179**, 4530-4537 (1997).

Martell, A. E., and R. M. Smith, *Critical Stability Constants*, Plenum, New York, 1977.

Masters, G. M., *Introduction to Environmental Engineering*, Prentice Hall, Upper Saddle River, 1998.

Matthis, E., M. S. Holt, A. Kiewiet, and G. B. J. Rijs, "Environmental monitoring for linear alkylbenzene sulfonate, alcohol ethoxylate, alcohol ethoxy sulfate, alcohol sulfate, and soap", *Environ Toxicol. Chem.*, **18**, 2634-2644 (1999).

McCarthy, J. F., and B. D. Jimenez, "Interactions between polycyclic aromatic hydrocarbons and dissolved humic material: Binding and dissociation", *Environ. Sci. Technol.*, **19**, 1072-1076 (1985).

McCarty, P. L., "Breathing with chlorinated solvents", *Science*, **276**, 1521-1522 (1997).

McCarty, P. L., M. N. Goltz, M. E. Hopkins, M. E. Dolan, and I. P. Allan, B. T. Kawakami, and T. J. Carrothers, "Full-scale evaluation of in situ cometabolic degradation of trichlorethylene in groundwater through toluene injection", *Environ. Sci. Technol.*, **32**, 88-100 (1998).

McCarty, P. L., D. Mackay, A. D. Smith, G. W. Ozburn, and D. G. Dixon, "Residue-based interpretation of toxicity and bioconcentration QSARs from aquatic bioassays. Neutral narcotic organics", *Environ. Toxicol. Chem.*, **11**, 917-930 (1992).

McDevit, W. F., and F. A. Long, "The activity coefficient of benzene in aqueous salt solutions", *J. Am. Chem. Soc.*, **74**, 1772-1777 (1952).

McGovern, E. W., "Chlorohydrocarbons solvents", *Ind. Eng. Chem.*, **35**, 1230-1239 (1943).

McGowan, J. C., and A. Mellors, *Molecular Volumes in Chemistry and Biology – Applications Including Partitioning and Toxicity*, Ellis Horwood, Chichester, UK, 1986.

McGroddy, S. E., "*Sediment–pore water partitioning of PAHs and PCBs in Boston Harbor, MA*", Ph.D. thesis, Univ. of Massachusetts, Boston, 1993, 255 p.

McGroddy, S. E., J. W. Farrington, and P. M. Gschwend, "Comparison of *in situ* and desorption sediment–water partitioning of polycyclic aromatic hydrocarbons and polychlorinated biphenyls", *Environ. Sci. Technol.*, **30**, 172-177 (1996).

McLachlan, M. S., "Bioaccumulation of hydrophobic chemicals in agricultural food chains", *Environ. Sci. Technol.*, **30**, 252-259 (1996).

McLachlan, M. S., "Framework for the interpretation of measurements of SOCs in plants", *Environ. Sci. Technol.*, **33**, 1799-1804 (1999).

Meallier, P., "Phototransformation of pesticides in aqueous solution." In *The Handbook of Environmental Chemistry*, Vol. 2, Part L, P. Boule, Ed., Springer, Berlin, 1999, pp. 241-262.

Means, J. C., S. G. Wood, J. J. Hassett, and W. L. Banwart, "Sorption of polynuclear aromatic hydrocarbons by sediments and soils", *Environ Sci. Technol.*, **14**, 1524-1528 (1980).

Meckenstock, R. U., R. Krieger, S. Ensign, P. M. H. Kroneck, and B. Schink, "Acetylene hydratase of *Pelobacter acetylenicus*", *Eur. J. Biochem.*, **264**, 176-182 (1999).

Meng, E. C., and P. A. Kollman, "Molecular dynamics studies of the properties of water around simple organic solutes", *J. Phys. Chem.*, **100**, 11460-11470 (1996).

Mesuere, K., and W. Fish, "Chromate and oxalate adsorption on goethite. 1. Calibration of surface complexation models", *Environ. Sci. Technol.*, **26**, 2357-2364 (1992).

Meybeck, M., "Global distribution of lakes". In A. Lerman, D. M. Imboden and J. Gat, Eds., *Physics and Chemistry of Lakes*, Springer Verlag, Heidelberg, 1995, pp. 1-35.

Meyer, H., "Zur Theorie der Alkoholnarkose. I. Welche Eigenschaft der Anesthetica bedingt ihre narkotische Wirkung", *Arch. Exp. Pathol. Pharmakol.*, **42**, 109-118 (1899).

Meylan, W. M., and P. H. Howard, "Bond contribution method for estimating Henry's law constants", *Environ. Toxicol. Chem.*, **10**, 1283-1293 (1991).

Meylan, W. M., and P. H. Howard, "Atom/fragment contribution method for estimating octanol–water partition coefficients", *J. Pharm. Sci.*, **84**, 83-92 (1995).

Mihelcic, J. R., and R. G. Luthy, "Degradation of polycyclic aromatic hydrocarbon compounds under various redox conditions in soil–water systems." *Appl. Environ. Microbiol.*, **54**, 1182-1887 (1988).

Mikhail, R. S., S. Brunauer, and E. E. Broder, "Investigations of a complete pore structure analysis. I. Analysis of macropores", *J. Colloid Int. Sci.*, **26**, 45-53 (1968a).

Mikhail, R. S., S. Brunauer, and E. E. Broder, "Investigations of a complete pore structure analysis. II. Analysis of four silica gels", *J. Colloid Int. Sci.*, **26**, 54-61 (1968b).

Mill, T., "Structure–activity relationships for photooxidation processes in the environment". *Environ. Toxicol. Chem.*, **8**, 31-43 (1989).

Mill, T., and W. Mabey, "Photochemical transformations". In W. B. Neely and G. E. Blau, Eds., *Environmental Exposure from Chemicals*, Vol. 1, CRC Press, Boca Raton, FL, 1985, pp. 175-216.

Mill, T., W. M. Mabey, B. K. Lan, and A. Baraze, "Photolysis of polycyclic aromatic hydrocarbons in water", *Chemosphere*, **10**, 1281-1290 (1981).

Miller, D. H., *Water at the Surface of the Earth. An Introduction to Ecosystem Hydrodynamics*, Academic, New York, 1977.

Miller, G. C., and R. G. Zepp, "Photoreactivity of aquatic pollutants sorbed on suspended sediments", *Environ. Sci. Technol*, **13**, 860-863 (1979a).

Miller, G. C., and R. G. Zepp, "Effects of suspended sediments on photolysis rates of dissolved pollutants", *Water Res.*, **13**, 453-459 (1979b).

Miller, P. L., D. Vasudevan, P. M. Gschwend, and A. L. Roberts, "Transformation of hexachloroethane in a sulfidic natural water", *Environ. Sci. Technol.*, **32**, 1269-1257 (1998).

Mills, A. C., and J. W. Biggar, "Solubility–temperature effect on the adsorption of gamma- and beta-BHC from aqueous and hexane solutions by soil materials", *Soil Sci. Soc. Am. Proc.*, **33**, 210-216 (1969a).

Mills, A. C., and J. W. Biggar, "Adsorption of 1,2,3,4,5,6-hexachlorocyclohexane from solution: The differential heat of adsorption applied to adsorption from dilute solutions on organic and inorganic surfaces", *J. Coll. Int. Sci.*, **29**, 720-731 (1969b).

Mitchell, B. E., and P. C. Jurs, "Prediction of infinite dilution activity coefficients of organic compounds in aqueous solution from molecular structure", *J. Chem. Inf. Comput. Sci.*, **38**, 200-209 (1998).

Mohn, W. W., and J. M. Tiedje, "Microbial reductive dehalogenation", *Microbiol. Reviews*, **56**, 482-507 (1992).

Monahan, E. C., "Oceanic whitecaps", *J. Phys. Oceanogr.*, **1**, 139-144 (1971).

Monfort, J.-P., and J.-L. Pellegatta, "Diffusion coefficients of the halocarbons CCl_2F_2 and $C_2Cl_2F_4$ with simple gases", *J. Chem. Eng. Data*, **36**, 136-137 (1991).

Monod, J., "The growth of bacterial cultures", *Ann. Rev. Microbiol.*, **3**, 371-394 (1949).

Montgomery, J. H., *Agrochemicals Desk Reference*, Lewis Publishers, Boca Raton, FL, 1997.

Moog, D. B., and G. H. Jirka, "Air-water gas transfer in uniform channel flow", *J. Hydraulic Eng.*, **125**, 3-10 (1999a).

Moog, D. B., and G. H. Jirka, "Stream reaeration in nonuniform flow: Macroroughness enhancement", *J. Hydraulic Eng.*, **125**, 11-16 (1999b).

Morel, F. M. M., *Principles of Aquatic Chemistry*, Wiley-Interscience, New York, 1983.

Morel, F. M. M., and J. G. Hering, *Principles and Applications of Aquatic Chemistry*, Wiley-Interscience, New York, 1993.

Morris, K. R., R. Abramowitz, R. Pinal, P. Davis, and S. H. Yalkowsky, "Solubility of aromatic pollutants in mixed solvents", *Chemosphere*, **17**, 285-298 (1988).

Morrison, H. A., F. A. P. C. Gobas, R. Lazar, and G. D. Haffner, "Development and verification of a bioaccumulation model for organic contaminants in benthic invertebrates", *Environ. Sci. Technol.*, **30**, 3377-3384 (1996).

Mueller, E. J., P. J. Loida, and S. G. Sligar, "Twenty-five years of P450cam research". In *Cytochrome P450, Structure, Mechanism, and Biochemistry,* P. R. Ortiz de Montellano, Ed., Plenum Press, New York, NY, 1995.

Muhlmann, R., and A. Schrader, "Hydrolyse der insektiziden Phosphorsäureester", *Z. Naturforsch.*, **12b**, 196-208 (1957).

Müller, J. A., A. S. Galushko, A. Kappler, and B. Schink, "Anaerobic degradation of *m*-cresol by *Desulfobacterium cetonicum* is initiated by formation of 3-hydroxybenzylsuccinate", *Arch. Microbiol.*, **172**, 287-294 (1999).

Müller, J. A., A. S. Galushko, A. Kappler, and B. Schink, "Initiation of anaerobic degradation of *p*-cresol by formation of 4-hydroxybenzylsuccinate in *desulfobacterium cetonicum*", *J. Bacteriol.*, **183**, 752-757 (2001).

Munnecke, D. M., "Enzymatic hydrolysis of organophosphate insecticides, a possible pesticide disposal method", *Appl. Environ. Microbiol.*, **32**, 7-13 (1976).

Münnich, K. O., W. B. Clarke, K. H. Fisscher, D. Flothmann, B. Kromer, W. Roether, U. Siegenthaler, Z. Top, and W. Weiss, "Gas exchange and evaporation studies in a circular wind tunnel, continuous radon-222 measurements at sea, and tritium/helium-3 measurements in a lake". In *Turbulent Fluxes Through the Sea Surface, Wave Dynamics and Predictions,* H. Favre and K. Hasselmann, Eds., Plenum, New York, 1978, pp. 151-165.

Munz, C. H., and P. V. Roberts, "The effects of solute concentration and cosolvents on the aqueous activity coefficients of low molecular weight halogenated hydrocarbons", *Environ. Sci. Technol.*, **20**, 830-836 (1986).

Myrdal, P., G. H. Ward, P. Simamora, and S. H. Yalkowsky, "AQUAFAC: Aqueous functional group activity coefficients", *SAR QSAR Environ. Res.*, **1**, 53-61 (1993).

Myrdal, P. B., J. F. Krzyzaniak, and S. H. Yalkowsky, "Modified Trouton's rule for predicting the entropy of boiling", *Ind. Eng. Chem. Res.*, **35**, 1788-1792 (1996).

Myrdal, P. B., and S. H. Yalkowski, "Estimating pure component vapor pressures of complex organic molecules", *Ind. Eng. Chem. Res.*, **36**, 2494-2499 (1997).

Naes, K., J. Axelman, C. Näf, and D. Broman, "Role of soot carbon and other carbon matrices in the distribution of PAHs among particles, DOC, and dissolved phase in the effluent and recipient waters of an aluminum reduction plant", *Environ. Sci. Technol.*, **32**, 1786-1792 (1998).

Napela, T. F., J. F. Cavaletto, M. Ford, W. M. Gordon, and M. Wimmer, "Seasonal and annual variation in weight and biochemical content of zebra mussel, *Dreissena polymorpha* in Lake St. Clair", *J. Great Lakes Res.*, **19**, 541-552 (1993).

Neely, W. B., G. E. Blau, and J. Alfrey Jr., "Mathematical models predict concentration-time profiles resulting from chemical spill in a river", *Environ. Sci. Technol.*, **10**, 72-76 (1976).

Neidhardt, F. G., J. L. Ingraham, and M. Schaechter, Eds., *Physiology of the Bacterial Cell. A Molecular Approach*, Sinauer Associates, Sunderland, MA, 1990.

Nelson, D. L., and M. M. Cox, *Lehninger Principles of Biochemistry*, Worth Publishers, New York, NY, 2000.

Neta, P., and D. Meisel, "Substituent effects of nitroaromatic radical anions in aqueous solution", *J. Phys. Chem.*, **80**, 519-524 (1976).

Neumann, A., G. Wohlfarth, and G. Diekert, "Purification and characterization of tetrachloroethene reductive dehalogenase from *Dehalospirillum multivorans*", *J. Biol. Chem.*, **271**, 16515-16519 (1996).

Neumann, G., and W. J. J. Pierson, *Principles of Physical Oceanography*, Prentice-Hall, Englewood Cliffs, NJ, 1966.

Nielson, T., K. Siigu, C. Helweg, O. Jørgensen, P. E. Hansen, and U. Kirso, "Sorption of polycyclic aromatic compounds to humic acid as studied by high-performance liquid chromatography", *Environ. Sci. Technol.*, **31**, 1102-1108 (1997).

Niessen, W. R., Ed. *Handbook of Solid Waste Management*, Van Nostrand Reinhold, New York, 1977.

Nikaido, H., "Nonspecific transport through the outer membrane". In *Bacterial Outer Membranes, Biogenesis and Functions,* M. Inouye, Ed., Wiley-Interscience, New York, 1979, pp. 361-407.

Nivinskas, H., R. L. Koder, Z. Anusevicius, J. Sarlauskas, A.-F. Miller, and N. Cenas, "Quantitative structure activity relationships in two-electron reduction of nitroaromatic compounds by *Entrobacter cloacae* NAD(P)H:nitroreductase", *Arch. Biochem. Biophys.*, **385**, 170-178 (2001).

Nkedi-Kizza, P., P. S. C. Rao, and A. G. Hornsby, "Influence of organic cosolvents on sorption of hydrophobic organic chemicals by soils", *Environ. Sci. Technol.*, **19**, 975-979 (1985).

Nowack, B., "The behavior of phosphonates in wastewater treatment plants of Switzerland", *Wat. Res.*, **32**, 1271-1279 (1998).

Nowack, B., and A. T. Stone, "Degradation of nitrilotris(methylene-phosphoric acid) and related amino(phosphonate) chelating agents in the presence of manganese and molecular oxygen", *Environ. Sci. Technol.*, **34**, 4759-4765 (2000).

Nzengung, V. A., E. A. Voudrias, and P. Nkedi-Kizza, "Organic cosolvent effects on sorption equilibrium of hydrophobic organic chemicals by organoclays", *Environ. Sci. Technol.*, **30**, 89-96 (1996).

O'Connor, D. J., and W. E. Dobbins, "Mechanisms of reaeration in natural streams", *Trans. Am. Soc. Civ. Eng.*, **123**, 641-684 (1958).

O'Melia, C. R., "The influence of coagulation and sedimentation on the fate of particles, associated pollutants, and nutrients in lakes". In *Chemical Processes in Lakes,* W. Stumm, Ed., Wiley-Interscience, New York, 1985, pp. 207-224.

Ocampo, J., and J. Klinger, "Modification of the structure of ice during ageing", *J. Phys. Chem.*, **87**, 4167-4170 (1983).

Oelkers, E. H., "Calculation of diffusion coefficients for aqueous organic species at temperatures from 0 to 350°C", *Geochimica et Cosmochimica Acta*, **55**, 3515-3529 (1991).

Ogram, A. V., R. E. Jessup, L. T. Ou, and P. S. C. Rao, "Effects of sorption on biological degradation rates of (2,4-dichlorophenoxy) acetic acid in soils", *Appl Environ. Microbiol.*, **49**, 582-587 (1985).

Okubo, A., "Oceanic diffusion diagrams", *Deep Sea Res.*, **18**, 789-802 (1971).

Oliver, B. G., and A. J. Nimi, "Trophodynamic analysis of polychlorinated biphenyl congeners and other chlorinated hydrocarbons in the Lake Ontario ecosystem", *Environ. Sci. Technol.*, **22**, 388-397 (1988).

Ong, J. H., and C. E. Castro, "Oxidation of iron(II) porphyrins and hemoproteins by nitro aromatics", *J. Am. Chem. Soc.*, **99**, 6740-6745 (1977).

Opperhuizen, A., E. W. van der Velde, F. A. P. C. Gobas, D. A. K. Liem, and J. M. D. van der Steen, "Relationship between bioconcentration in fish and steric factors of hydrophobic chemicals", *Chemosphere*, **14**, 1871-1896 (1985).

Oser, B. L., Ed. *Hawk's Physiological Chemistry*, McGraw-Hill, New York, 1965.

Othmer, D. F., and M. S. Thakar, "Correlating diffusion coefficients in liquids", *Ind. Eng. Chem.*, **45**, 589-593 (1953).

Overton, E., "Über die allgemeinen osmotischen Eigenschaften der Zelle, ihre vermutlichen Ursachen und ihre Bedeutung für die Physiologie", *Vierteljahresschr. Naturforsch. Ges. Zürich*, **44**, 88-135 (1899).

Owen, K., Ed. *Critical Report on Applied Chemistry, Gasoline and Diesel Fuel Additives*, Wiley, Chichester, 1989.

Pagni, R. M., and M. E. Sigman, "The photochemistry of PAHs and PCBs in water and on solids". In *The Handbook of Environmental Chemistry,* Vol. 2, Part L., P. Boule, Ed., Springer, Berlin, 1999, pp. 139-179.

Pak, J. W., K. L. Knoke, D. R. Nogeuera, B. G. Fox, and G. H. Chambliss, "Transformation of 2,4,6-trinitrotoluene by purified xenobiotic reductase B from *Pseudomonas fluorescens* I-C", *Appl. Environ. Microbiol.*, **66**, 4742-4750 (2000).

Paris, D. F., D. L. Lewis, and N. L. Wolfe, "Rates of degradation of malathion by bacteria isolated from aquatic systems", *Environ. Sci. Technol.*, **9**, 135-138 (1975).

Paris, D. F., W. C. Steen, G. L. Baughman, and J. T. J. Barnett, "Second-order model to predict microbial degradation of organic compounds in natural waters", *Appl. Environ. Microbiol*, **41**, 603-609 (1981).

Park, J. H., A. Hussam, P. Couasnon, D. Fritz, and P. W. Carr, "Experimental reexamination of selected partition coefficients from Rohrschneider's data set", *Anal. Chem.*, **59**, 1970-1976 (1987).

Parks, G. A., "The isoelectric points of solid oxides, solid hydroxides, and aqueous hydroxo complex systems", *Chem. Rev.*, **65**, 177-198 (1965).

Parlar, H., "Photochemistry at surfaces and interfaces". In *The Handbook of Environmental Chemistry,* Vol. 2, Part A, O. Hutzinger, Ed., Springer, Berlin, 1980.

Parr, J. F., and S. Smith, "Degradation of toxaphene in selected anaerobic soil environments", *Soil Sci.*, **121**, 52-57 (1976).

Parsons, T. R., and M. Takahashi, *Biological Oceanographic Processes*, Pergamon Press, Oxford, 1973.

Paterson, M. J., D. C. G. Muir, B. Rosenberg, E. J. Fee, C. Anema, and W. Franzin, "Does lake size affect concentrations of atmospherically derived polychlorinated biphenyls in water, sediment, zooplankton, and fish?" *Can. J. Fish. Aquat. Sci.*, **55**, 544-553 (1998).

Pauling, L., *The Nature of the Chemical Bond*, Cornell University Press, New York, 1960.

Pavelich, W. H., and R. W. Taft, Jr., "The evaluation of inductive and steric effects on reactivity. The methoxide ion-catalyzed rates of methanolysis of *l*-menthyl esters in methanol", *J. Am. Chem. Soc.*, **79**, 4935-4940 (1957).

Pearson, C. R., "C$_1$- and C$_2$-Halocarbons". In *The Handbook of Environmental Chemistry,* Vol. 3, Part B., O. Hutzinger, Ed., Springer, Berlin, 1982a, pp. 69-88.

Pearson, C. R., "Halogenated aromatics". In *The Handbook of Environmental Chemistry,* Vol. 3, Part B, O. Hutzinger, Ed., Springer, Berlin, 1982b, pp. 89-116.

Pearson, R. G., "Hard and soft acids and bases", *J. Am. Chem. Soc.,* **85**, 3533-3539 (1963).

Pearson, R. G., H. Sobel, and J. Songstad, "Nucleophilic reactivity constants toward methyl iodide and trans[Pt(py)$_2$Cl$_2$]", *J. Am. Chem. Soc.,* **90**, 319-326 (1968).

Pecher, K., S. B. Haderlein, and R. P. Schwarzenbach, "Reduction of polyhalogenated methanes by surface bound Fe(II) in aqueous suspensions of iron oxides", *Environ. Sci. Technol.,* **36**, 1734-1741 (2002).

Peeters, F., A. Wüest, G. Piepke, and D. M. Imboden, "Horizontal mixing in lakes", *J. Geophys. Res.,* **101**, C8, 18361-18375 (1996).

Peng, J., and A. Wang, "Effect of ionic strength on Henry's constants of volatile organic compounds", *Chemosphere,* **36**, 2731-2740 (1998).

Peng, T.-H., W. S. Broecker, G. G. Mathieu, Y.-H. Li, and A. E. Bainbridge, "Radon evasion rates in the Atlantic and Pacific Oceans as determined during the GEOSECS program", *J. Geophys. Res.,* **84**, 2471-2586 (1979).

Penman, H. L., "Natural evaporation from open water, bare soil, and grass", *Proc. Roy. Soc. London, Ser. A,* **193**, 120-146 (1948).

Perlinger, J. A., W. Angst, and R. P. Schwarzenbach, "Kinetics of the reduction of hexachloroethane by juglone in solutions containing hydrogen sulfide", *Environ. Sci. Technol.,* **30**, 3408-3417 (1996).

Perlinger, J. A., J. Buschmann, W. Angst, and R. P. Schwarzenbach, "Iron porphyrin and mercaptojuglane mediated reduction of polyhalogenated methanes and ethanes in homogeneous aqueous solution", *Environ. Sci. Technol.,* **32**, 2431-2437 (1998).

Perlinger, J. A., R. Venkatapathy, and J. F. Harrison, "Linear free energy relationships for polyhalogenated alkane transformation by electron-transfer mediators in model aqueous systems", *J. Phys. Chem. A.,* **104**, 2752-2763 (2000).

Perona, M. J., "The solubility of hydrophobic compounds in aqueous droplets", *Atmospheric Environ.,* **26**, 2549-2553 (1992).

Perrin, D. D., *Dissociation Constants of Organic Bases in Aqueous Solution: Supplement 1972,* Butterworths, London, 1972.

Perrin, D. D., "Prediction of pK_a values". In *Physical Chemical Properties of Drugs,* S. H. Yalkowsky, A. A. Sinkula and S. C. Valvani, Eds., Dekker, New York, 1980.

Peters, C. A., C. D. Knightes, and D. G. Brown, "Long-term composition dynamics of PAH-containing NAPLs and implications for risk assessment", *Environ. Sci. Technol.,* **33**, 4499-4507 (1999a).

Peters, C. A., S. Mukherji, and W. J. Weber Jr., "UNIFAC modeling of multicomponent nonaqueous phase liquids containing polycyclic aromatic hydrocarbons", *Environ. Toxicol. Chem.,* **18**, 426-429 (1999b).

Peterson, J. A., and S. E. Graham-Lorence, "Bacterial P450s. Structural similarities and functional differences". In *Cytochrome P450, Structure, Mechanism, and Biochemistry,* P. R. Ortiz de Montellano, Ed., Plenum Press, New York, 1995.

Pfennig, B. W., and R. L. Frock, "The use of molecular modeling and VSEPR theory in the undergraduate curriculum to predict the three-dimensional structure of molecules", *J. Chem. Educ.,* **76**, 1018-1022 (1999).

Piccolo, A., and P. Conte, "Molecular size of humic substances. Supramolecular associations versus macromolecular polymers", *Advances in Environ. Res.,* **3**, 508-521 (2000).

Picel, K. C., V. C. Stamoudis, and M. S. Simmons, "Distribution coefficients for chemical components of a coal-oil/water system", *Water Res.,* **22**, 1189-1199 (1988).

Pignatello, J. J., and B. Xing, "Mechanisms of slow sorption of organic chemicals to natural particles", *Environ. Sci. Technol.,* **30**, 1-11 (1996).

Pinal, R., L. S. Lee, and P. S. C. Rao, "Prediction of the solubility of hydrophobic compounds in nonideal solvent mixtures", *Chemosphere,* **22**, 939-951 (1991).

Pinal, R., P. S. C. Rao, L. S. Lee, P. V. Cline, and S. H. Yalkowsky, "Cosolvency of partially miscible organic solvents on the solubility of hydrophobic organic chemicals", *Environ. Sci. Technol.,* **24**, 639-647 (1990).

Pinkart, H. C., J. W. Wolfram, R. Rogers, and D. C. White, "Cell envelope changes in solvent-tolerant and solvent-sensitive *Pseudomonas putida* strains following exposure to *o*-xylene", *Appl. Environ. Microbiol.,* **62**, 1129-1132 (1996).

Piwoni, M. D., and P. Banerjee, "Sorption of volatile organic solvents from aqueous solution onto subsurface solids", *J. Contam. Hydrol.,* **4**, 163-179 (1989).

Plastourgou, M., and M. R. Hoffmann, "Transformation and fate of organic esters in layered-flow systems: The role of trace metal catalysis", *Environ. Sci. Technol.*, **18**, 756-764 (1984).

Poelarends, G. J., J. E. T. van Hylckama, J. R. Marchesi, L. M. Freitas dos Santos, and D. B. Janssen, "Degradation of 1,2-dibromoethane by *Mycobacterium sp. strain GP1*", *J. Bacteriol.*, **181**, 2050-2058 (1999).

Poole, S. K., and C. F. Poole, "Chromatographic models for sorption of neutral organic compounds by soil from water and air", *J. Chromatogr. A*, **845**, 381-400 (1999).

Pörschmann, J., Z. Zhang, F.-D. Kopinke, and J. Pawliszyn, "Solid–phase microextraction for determining the distribution of chemicals in aqueous matrices", *Anal. Chem.*, **69**, 597-608 (1997).

Poulsen, M., L. Lemon, and J. F. Barker, "Dissolution of mono-aromatic hydrocarbons into groundwater from gasoline–oxygenate mixtures", *Environ. Sci. Technol.*, **26**, 2483-2489 (1992).

Prausnitz, J. M., *Thermodynamics of Fluid-Phase Equilibria*, Prentice-Hall, Englewood Cliffs, NJ, 1969.

Pretsch, E., P. Bühlmann, and C. Affolter, *Structure Determination of Organic Compounds*, Springer, Berlin, 2000.

Pretsch, E., J. T. Clerc, J. Seibl, and W. Simon, *Spectral Data for Structure Determination of Organic Compounds*, Springer, Berlin, 1983.

Price, J. F., M. O'Neil Baringer, R. G. Lueck, G. C. Johnson, I. Ambar, G. Parrilla, A. Cantos, M. A. Kennelly, and T. B. Sanford, "Mediterranean outflow mixing and dynamics", *Science*, **259**, 1277-1282 (1993).

Quinn, J. A., J. L. Anderson, W. S. Ho, and W. J. Petzny, "Model pores of molecular dimension – Preparation and characterization of track-etched membranes", *Biophysical Journal*, **12**, 990-1007 (1972).

Rabus, R., H. Wilkes, A. Behrends, A. Armswtroff, T. Fischer, A. J. Pierik, and F. Widdel, "Anaerobic initial reaction of *n*-alkanes in a denitrifying bacterium: Evidence for (1-methylpentyl)succinate as initial product and for the involvement of an organic radical in *n*-hexane metabolism", *J. Bacteriol.*, **183**, 1707-1715 (2001).

Ragle, C. S., R. R. Engebredson, and R. von Wandruska, "Sequestration of hydrophobic micropollutants by dissolved humic acids", *Soil. Sci.*, **162**, 106-114 (1997).

Ramanand, K., M. Sharmila, and N. Sethunathan, "Mineralization of carbofuran by a soil bacterium", *Appl. Environ. Microbiol.*, **54**, 2129-2133 (1988).

Ramos, J. L., E. Duque, J.-J. Rodriguez-Herva, P. Godoy, A. Haidour, F. Reyes, and A. Fernandez-Barrero, "Mechanisms for solvent tolerance in bacteria", *J. Biol. Chem.*, **272**, 3887-3890 (1997).

Ramos, J. L., E. Duque, P. Godoy, and A. Segura, "Efflux pumps involved in toluene tolerance in *Pseudomonas putida* DOT-TIE", *J. Bacteriol.*, **180**, 3323-3329 (1998).

Rao, P. S., L. S. Lee, and R. Pinal, "Cosolvency and sorption of hydrophobic organic chemicals", *Environ. Sci. Technol.*, **24**, 647-654 (1990).

Rao, P. S. C., A. G. Hornsby, D. P. Kilcrease, and P. Nkedi-Kizza, "Sorption and transport of hydrophobic organic chemicals in aqueous and mixed solvent systems: Model development and preliminary evaluation", *J. Environ. Qual.*, **14**, 376-383 (1985).

Ratledge, C., "Microbial conversions of alkanes and fatty acids", *A. Am. Oil Chem. Soc.*, **61**, 447-453 (1984).

Reichardt, C., *Solvents and Solvent Effects in Organic Chemistry, 2nd ed.*, VCH, Weinheim, Germany, 1988.

Reichardt, P. B., B. L. Chadwick, M. A. Cole, B. R. Robertson, and D. K. Button, "Kinetic study of the biodegradation of biphenyl and its monochlorinated analogues by a mixed marine microbial community", *Environ. Sci. Technol.*, **15**, 75-79 (1981).

Reichert, P., "AQUASIM – A tool for simulation and data analysis of aquatic systems", *Wat. Sci. Tech.*, **30**, 2, 21-30 (1994).

Reichert, P., and O. Wanner, "Enhanced one-dimensional modeling of transport in rivers", *J. Hydr. Div. ASCE*, **117**, 9, 1165-1183 (1991).

Reid, R. C., J. M. Prausnitz, and T. K. Sherwood, *The Properties of Gases and Liquids,* 3rd ed., McGraw-Hill, New York, 1977.

Rekker, R. F., *The Hydrophobic Fragment Constant*, Elsevier, New York, 1977.

Renkin, E. M., "Filtration, diffusion and molecular sieving through porous cellulose membranes", *J. Gen. Physiol.*, **38**, 225-243 (1954).

Reynolds, O., "On the dynamical theory of incompressible viscous fluids and the determination of the criterion", *Phil. Trans. Roy. Soc.*, **186**, 123 (1894).

Richard, C., and G. Grabner, "Mechanism of phototransformation of phenol and its derivatives in aqueous solution". In *The Handbook of Environmental Chemistry,* Vol. 2, Part L, P. Boule, Ed., Springer, Berlin, 1999, pp. 217-240.

Riefler, R. G., and B. F. Smets, "Enzymatic reduction of 2,4,6-dinitrotoluene and related nitroarenes: Kinetics linked to one-electron redox potentials", *Environ. Sci. Technol.*, **34**, 3900-3906 (2000).

Riegert, U., G. Heiss, P. Fischer, and A. Stolz, "Distal cleavage of 3-chlorocatechol by an extradiol dioxygenase to 3-chloro-2-hydroxymuconic semialdehyde", *J. Bacteriol.*, **180**, 2849-2853 (1998).

Rippen, G., Ed. *Handbook of Environmental Chemicals (in German)*, Ecomed, Landsberg, Germany, 1989-97.

Risebrough, R. W., P. Rieche, D. B. Peakall, S. G. Herman, and M. N. Kirven, "Polychlorinated biphenyls in the global ecosystem", *Nature*, **220**, 1098-1102 (1968).

Rittmann, B. E., and P. L. McCarty, *Environmental Biotechnology: Principles and Applications*, McGraw-Hill, New York, NY, 2001.

Robbins, J. A., "A model for particle-selective transport of tracers in sediments with conveyor belt deposit feeders", *J. Geophys. Res.*, **91**, 8542-8558 (1986).

Roberts, A. L., *Workbook for Environmental Organic Chemistry*, Johns Hopkins University, Baltimore, MD, 1995.

Roberts, A. L., P. N. Sanborn, and P. M. Gschwend, "Nucleophilic substitution reactions of dihalomethanes with hydrogen sulfide species", *Environ. Sci. Technol.*, **26**, 2263-2274 (1992).

Roberts, A. L., P. M. Jeffers, N. L. Wolfe, and P. M. Gschwend, "Structure-reactivity relationships in dehydrohalogenation reactions of polychlorinated and polybrominated alkanes", *Crit. Rev. Environ. Sci. Technol.*, **23**, 1-39 (1993).

Roberts, A. L., L. A. Totten, W. A. Arnold, D. R. Burris, and T. J. Campbell, "Reductive elimination of chlorinated ethylenes by zero-valent metals", *Environ. Sci. Technol.*, **30**, 2654-2659 (1996).

Roberts, P. V., and A. J. Valocchi, "Principles of organic contaminant behavior during artificial recharge". In *Quality of Groundwater*, W. v. Dujvenbooden and P. Glasbergen, Eds., Elsevier, Amsterdam, 1981, pp. 439-450.

Robertson, R. E., "Solvolysis in water", *Prog. Phys. Org. Chem.*, **4**, 213-280 (1967).

Rogers, C. E., V. Stannett, and M. Szwarc, "The sorption, diffusion and permeation of organic vapors in polyethylene", *J. Polymer Sci.*, **45**, 61-82 (1960).

Rohwer, E., "Evaporation from free water surfaces", *Technical Bulletin*, **271**, US Department of Agriculture (1931).

Romano, J. C., "Sea-surface slick occurrence in the open sea (Mediterranean, Red Sea, Indian Ocean) in relation to wind-speed", *Deep-Sea Research Part 1 – Oceanographic Research Papers*, **43**, 4, 411-423 (1996).

Rombauer, I. S., M. R. Becker, and F. Becker, *Joy of Cooking*, Scribner, New York, NY, 1997.

Roof, A. A. M., "Basic principles of environmental photochemistry". In *The Handbook of Environmental Chemistry,* Vol. 2, Part B, O. Hutzinger, Ed., Springer, Berlin, 1982, pp. 1-17.

Rorije, E., W. J. G. M. Peijnenburg, and G. Klopman, "Structural requirements for anaerobic biodegradation of organic chemicals: A fragment model analysis", *Environ. Toxicol. Chem.*, **17**, 1943-1950 (1998).

Rorije, E., H. Loonen, M. Muller, G. Klopman, and W. J. G. M. Peijnenburg, "Evaluation and application of models for the prediction of ready biodegradability in the MITI-I test", *Chemosphere*, **38**, 1409-1417 (1999).

Rosenberg, A., and M. Alexander, "Microbial cleavage of various organophosphorus insecticides", *Appl. Environ. Microbiol.*, **37**, 886-891 (1979).

Roth, C., K.-U. Goss, and R. P. Schwarzenbach, "Adsorption of a diverse set of organic vapors on the bulk water surface", *J. Colloid Int. Sci.* (in press), (2002).

Rowe, E. S., F. Zhang, T. W. Leung, J. S. Parr, and P. T. Guy, "Thermodynamics of membrane partitioning for a series of *n*-alcohols determined by titration calorimetry: Role of hydrophobic effects", *Biochemistry*, **37**, 2430-2440 (1998).

Rudolphi, A., A. Tschech, and G. Fuchs, "Anaerobic degradation of cresols by denitrifying bacteria", *Arch. Microbiol.*, **155**, 238-248 (1991).

Ruelle, P., and V. W. Kesselring, "Aqueous solubility prediction of environmentally important chemicals from the mobile order thermodynamics", *Chemosphere*, **34**, 275-298 (1997a).

Ruelle, P., and V. W. Kesselring, "The hydrophobic propensity of water towards amphiprotic solutes: Prediction of molecular origin of the aqueous solubility of aliphatic alcohols", *J. Pharm. Sci.*, **86**, 179-186 (1997b).

Rügge, K., T. B. Hofstetter, S. B. Haderlein, P. L. Berg, S. Knudsen, C. Zraunig, H. Mosbaek, and T. H. Christensen, "Characterization of predominant reductants in an anaerobic leachate-contaminated aquifer by nitroaromatic probe compounds", *Environ. Sci. Technol.*, **32**, 23-31 (1998).

Russel, R. W., R. Lazar, and G. D. Haffner, "Biomagnification of organochlorines in Lake Erie white bass", *Environ. Toxicol. Chem.*, **14**, 719-724 (1995).

Rutherford, J. C., *River Mixing*, Wiley, Chichester, 1994.

Ruus, A., K. I. Ugland, O. Espeland, and J. V. Skaare, "Organochlorine contaminants in a local marine food chain from Jarfjord, Northern Norway", *Marine Environ. Res.*, **48**, 131-146 (1999).

Saarikoski, J., and M. Viluksela, "Relation between physicochemical properties of phenolics and their toxicites and accumulation in fish", *Ecotoxicol. Environ Safety*, **6**, 501-512 (1982).

Sabljic, A., H. Güsten, H. Verhaaar, and J. Hermens, "QSAR modelling of soil sorption. Improvements and systematics of log K_{oc} vs. log K_{ow} correlations", *Chemosphere*, **31**, 4489-4514 (1995).

Sada, E., S. Kito, and Y. Ito, "Solubility of toluene in aqueous salt solutions", *J. Chem. Eng. Data*, **20**, 373-375 (1975).

Sanemasa, I., S. Arakawa, M. Araki, and T. Deguchi, "The effects of salts on the solubilities of benzene, toluene, ethylbenzene, and propylbenzene in water", *Bull. Chem. Soc. Jpn*, **57**, 1539-1544 (1984).

Sarmiento, J. L., and P. E. Biscaye, "Radon 222 in the benthic boundary layer", *J. Geophysic. Res.*, **91**, 833-844 (1986).

Sasaki, J., S. M. Aschmann, E. S. C. Kwok, R. Atkinson, and J. Arey, "Products of the gas-phase OH and NO_3 radical-initiated reactions of naphthalene", *Environ. Sci. Technol.*, **31**, 3173-3179 (1997).

Schellenberg, K., C. Leuenberger, and R. P. Schwarzenbach, "Sorption of chlorinated phenols by natural sediments and aquifer materials", *Environ. Sci. Technol.*, **18**, 652-657 (1984).

Scherer, M. M., B. A. Balko, D. A. Gallagher, and P. G. Tratnyek, "Correlation analysis of rate constants for dechlorination by zero-valent iron", *Environ. Sci. Technol.*, **32**, 3026-3033 (1998).

Scherer, M. M., K. M. Johnson, J. C. Westall, and P. G. Tratnyek, "Mass transport effects on the kinetics of nitrobenzene reduction by iron metal", *Environ. Sci. Technol.*, **35**, 2804-2811 (2001).

Scherer, M. M., S. Richter, R. L. Valentine, and P. J. J. Alvarez, "Chemistry and microbiology of permeable reactive barriers for *in situ* groundwater clean up", *Critical Reviews in Environmental Sci. Technol.*, **30**, 363-411 (2000).

Schindler, P. W., and W. Stumm, "The surface chemistry of oxides, hydroxides, and oxide minerals". In *Aquatic Surface Chemistry*, W. Stumm, Ed., Wiley-Interscience, New York, 1987, pp. 83-110.

Schink, B., "Degradation of unsaturated hydrocarbons by methanogenic enrichment cultures", *FEMS Microbiol. Ecol.*, **31**, 69-77 (1985a).

Schink, B., "Fermentation of acetylene by an obligate anaerobe, *Pelobacter acetylenicus* sp.nov." *Arch. Microbiol.*, **142**, 295-301 (1985b).

Schink, B., "Energetics of syntrophic cooperation in methanogenic degradation", *Microbiol. Mol. Biol. Rev.*, **61**, 226-280 (1997).

Schlautman, M. A., and J. J. Morgan, "Effects of aqueous chemistry on the binding of polycyclic aromatic hydrocarbons by dissolved humic materials", *Environ. Sci. Technol*, **27**, 961-969 (1993).

Schlautman, M. A., and J. J. Morgan, "Sorption of perylene on a nonporous inorganic silica surface – Effects of aqueous chemistry on sorption rates", *Environ. Sci. Technol.*, **28**, 2184-2190 (1994).

Schlichting, J., J. Berendzen, K. Chu, A. M. Stock, S. A. Maves, D. E. Benson, R. M. Sweet, D. Ringe, G. A. Petsko, and S. G. Sligar, "The catalytic pathway of cytochrome P450cam at atomic resolution", *Science*, **287**, 1615-1622 (2000).

Schluep, M., D. M. Imboden, R. Gälli, and J. Zeyer, "Mechanisms affecting the dissolution of nonaqueous phase liquids into the aqueous phase in slow-stirring batch systems", *Env. Tox. Chem.*, **20**, 459-466 (2001).

Schmidt, W., "Wirkungen der ungeordneten Bewegungen im Wasser der Meere und Seen", *Ann. Hydrogr. Marit. Meteorol.*, **45**, 367-381 (1917).

Schmidt, T. C., P. Kleinert, C. Stengel, K.-U. Goss, and S. B. Haderlein, "Polar fuel constituents – Compound identification and equilibrium partitioning between non-aqueous phase liquids and water", *Environ. Sci. Technol.*, in press (2002).

Schnitzer, M., and S. U. Khan, *Humic Substances in the Environment*, Dekker, New York, 1972.

Scholtz, R., T. Leisinger, F. Suter, and A. M. Cook, "Characterization of 1-chlorohexane halidohydrolase, a dehalogenase of wide substrate range from an *Arthrobacter* sp." *J. Bacteriol.*, **169**, 5016-5021 (1987a).

Scholtz, R., A. Schmuckle, A. M. Cook, and T. Leisinger, "Degradation of eighteen 1-mono-haloalkanes by *Arthobacter* sp. strain HA1", *J. Gen. Microbiol.*, **133**, 267-274 (1987b).

Schulten, H.-R., and M. Schnitzer, "Chemical model structures for soil organic matter and soils", *Soil Sci.*, **162**, 115-130 (1997).

Schumacher, W., C. Holliger, A. J. B. Zehnder, and W. R. Hagen, "Redox chemistry of cobalamin and iron-sulfur cofactors in the tetrachloroethene reductase of *Dehalobacter restrictus*", *FEBS Lett.*, **409**, 421-425 (1997).

Schüth, C., and M. Reinhard, "Hydrodechlorination and hydrogenation of aromatic compounds over palladium on alumina in hydrogen-saturated water", *Appl. Catal. B: Environ.*, **18**, 215-221 (1998).

Schwarzenbach, R. P., W. Angst, C. Holliger, S. J. Hug, and J. Klausen, "Reductive transformations of anthropogenic chemicals in natural and technical systems", *Chimia*, **51**, 908-914 (1997).

Schwarzenbach, R. P., W. Giger, E. Hoehn, and J. K. Schneider, "Behavior of organic compounds during infiltration of river water to ground-water. Field studies", *Environ. Sci. Technol.*, **17**, 472-479 (1983).

Schwarzenbach, R. P., W. Giger, C. Schaffner, and O. Wanner, "Groundwater contamination by volatile halogenated alkanes: Abiotic formation of volatile sulfur compounds under anaerobic conditions", *Environ. Sci. Technol.*, **19**, 322-327 (1985).

Schwarzenbach, R. P., R. Stierli, B. R. Folsom, and J. Zeyer, "Compound properties relevant for assessing the environmental partitioning of nitrophenols", *Environ. Sci. Technol.*, **22**, 83-92 (1988).

Schwarzenbach, R. P., R. Stierli, K. Lanz, and J. Zeyer, "Quinone and iron porphyrin mediated reduction of nitroaromatic compounds in homogeneous aqueous solution", *Environ. Sci. Technol.*, **24**, 1566-1574 (1990).

Schwarzenbach, R. P., and J. Westall, "Transport of nonpolar organic compounds from surface water to groundwater: Laboratory sorption studies", *Environ. Sci. Technol.*, **15**, 1360-1367 (1981).

Scully, F. E. J., and J. Hoigné, "Rate constants for reactions of singlet oxygen and phenols and other compounds in water", *Chemosphere*, **16**, 681-694 (1987).

Sedlak, D. L., J. L. Gray, and K. E. Pinkston, "Understanding microcontaminants in recycled water", *Environ. Sci. Technol.*, **34**, 509A-515A (2000).

Seinfeld, J. H., *Atmospheric Chemistry and Physics of Air Pollution*, Wiley, New York, 1986.

Semprini, L., "Strategies for the aerobic co-metabolism of chlorinated solvents", *Curr. Opinion Biotechn.*, **8**, 296-308 (1997).

Senior, E., A. T. Bull, and J. H. Slater, "Enzyme evolution in a microbial community growing on the herbicede Dalapon", *Nature*, **263**, 476-479 (1976).

Sergides, C. A., J. A. Jassim, A. R. Chughtal, and D. M. Smith, "The structure of hexane soot. Part III: Ozonation studies", *Appl. Spectrosc.* **41**, 482-492 (1987).

Serjeant, E. P., and B. Dempsey, *Ionization Constants of Organic Acids in Aqueous Solution*, Pergamon, New York, 1979.

Setschenow, J., "Über die Konstitution der Salzlösungen auf Grund ihres Verhaltens zu Kohlensäure", *Z. Phys. Chem., Vierter Band*, **1**, 117-125 (1889).

Severtson, S. J., and S. Banerjee, "Sorption of chlorophenols to wood pulp", *Environ. Sci. Technol.*, **30**, 1961-1969 (1996).

Shao, C., W. J. Cooper, and D. R. S. Lean, "Singlet oxygen formation in lake waters from mid-latitudes". In *Aquatic and Surface Photochemistry,* G. R. Helz, R. G. Zepp and D. G. Crosby, Eds., CRC Press, Boca Raton, 1994, pp. 215-221.

Sherman, S. R., D. B. Trampe, D. M. Bush, M. Schiller, C. A. Eckert, A. J. Dallas, J. Li, and P. W. Carr, "Compilation and correlation of limiting activity coefficients of nonelectrolytes in water", *Ind. Eng. Chem. Res.*, **35**, 1044-1058 (1996).

Shey, J., and W. A. van der Donk, "Mechanistcic studies on the vitamine B_{12}-catalyzed dechlorination of chlorinated alkenes", *J. Amer. Chem. Soc.*, **122**, 12403-12404 (2000).

Shifrin, N. S., and S. W. Chisholm, "Phytoplankton lipids: Interspecific differences and effects of nitrate, silicate, and light-dark cycles", *J. Phycol.*, **17**, 374-384 (1981).

Shinoda, K., "Iceberg formation and solubility", *J. Phys. Chem.*, **81**, 1300-1302 (1977).

Shiu, W.-Y., F. Wania, H. Hung, and D. Mackay, "Temperature dependence of aqueous solubility of selected chlorobenzenes, polychlorinated biphenyls, and dibenzofuran", *J. Chem. Eng. Data*, **42**, 293-297 (1997).

Shorter, J., "Values of σm and σp based on the ionization of substituted benzoic acids in water at 25°C", *Pure Appl. Chem.*, **66**, 2451-2468 (1994).

Shorter, J., "Compilation and critical evaluation of structure-reactivity parameters and equations: Part 2", *Pure Appl. Chem.*, **69**, 2497-2510 (1997).

Sijm, D. T. H. M., and J. L. M. Hermens, "Internal effect concentration: Link between bioaccumulation and ecotoxicity for organic chemicals". In *The Handbook of Environmental Chemistry,* Vol. 2, Part J, B. Beek, Ed., Springer, Berlin, 2000, pp. 167-199.

Sijm, D. T. H. M., M. Schipper, and A. Opperhuizen, "Toxicokinetics of halogenated benzenes in fish: Lethal body burden as toxicological end point", *Environ. Tox. Chem.*, **12**, 1117-1127 (1993).

Sikkema, J., J. A. M. deBont, and B. Poolman, "Mechanisms of membrane toxicity of hydrocarbons", *Microbiol. Reviews*, **59**, 201-222 (1995).

Silverstein, R. M., G. C. Bassler, and T. C. Morill, *Spectrometric Identification of Organic Compounds*, 5th ed., Wiley, New York, 1991.

Simmons, M. S., and R. G. Zepp, "The influence of humic substances on the photolysis of nitroaromatic compounds in aqueous systems", *Water Res.*, **20**, 899-904 (1986).

Simonich, S. L., and R. A. Hites, "Organic pollutant accumulation in vegetation", *Environ. Sci. Technol.*, **29**, 2905-2914 (1995).

Simonich, S. L., and R. A. Hites, "Relationships between socioeconomic indicators and concentrations of organochlorine pesticides in tree bark", *Environ. Sci. Technol.*, **31**, 999-1003 (1997).

Small, H., "Hydrodynamic chromatography – A technique for size analysis of colloidal particles", *J. Colloid Interface Sci.*, **48**, 147-161 (1974).

Smatlak, C. R., J. M. Gossett, and S. H. Zinder, "Comparative kinetics of hydrogen utilization for reductive dechlorination of tetrachloroethene and methanogenesis in an anaerobic enrichment culture", *Environ. Sci. Technol.*, **30**, 2850-2858 (1996).

Smith, J. A., A. K. Trisdale, and H. J. Cho, "Quantification of natural vapor fluxes of trichlorethene in the unsaturated zone and Picatinny Arsenal, New Jersey", *Environ. Sci. Technol.*, **30**, 2243-2250 (1996).

Smith, J. H., D. C. Bomberger, Jr., and D. L. Haynes, "Prediction of the volatilization rates of high-volatility chemicals from natural water bodies", *Environ. Sci. Technol.*, **14**, 1332-1337 (1980).

Smith, J. H., W. R. Mabey, N. Bohonos, B. R. Hoh, S. S. Lee, T.-W. Chou, D. C. Bomberger, and T. Mill, *Environmental pathways of selected chemicals in freshwater systems. Part II. Laboratory Studies*, EPA-600/7-78-074, Washington, DC, 1978.

Smith, J. H., and F. M. Menger, "Concerning back-bonding and polar effects in aminoalkylbenzene derivatives", *J. Org. Chem.*, **34**, 77-80 (1969).

Smith, R. C., and J. E. Tyler, "Transmission of solar radiation into natural waters", *Photochem. Photobiol. Rev.*, **1**, 117-155 (1976).

Smith, S. D., and E. P. Jones, "Evidence for wind pumping of air–sea gas exchange based on direct measurements of CO_2-fluxes", *J. Geophys. Res.*, **90**, 869-875 (1985).

Smolen, J. M., and A. T. Stone, "Divalent metal ion-catalyzed hydrolysis of phosphorothionate ester pesticides and their corresponding oxonates", *Environ. Sci. Technol.*, **31**, 1664-1673 (1997a).

Smolen, J. M., and A. T. Stone, "Metal(hydr)oxide surface catalyzed hydrolysis of chlorpyrifos-methyl, chlorpyrifos-methyl oxon, and paraoxon", *Soil Sci. Am. J.*, **62**, 636-643 (1997b).

Smolen, J. M., and A. T. Stone, "Metal(hydr)oxide surface catalyzed hydrolysis of chlorpyrifos-methyl, chlorpyrifos-methyl oxon, and paraoxon", *Soil Sci. Soc. Amer. J.*, **62**, 636-643 (1998).

So, C. M., and L. Y. Young, "Initial reactions in anaerobic alkane degradation by a sulfate reducer, strain AK-10", *Appl. Environ. Microbiol.*, **65**, 5532-5540 (1999).

Somasundaran, P., and G. E. Agar, "The zero point of charge of calcite", *J. Colloid Interface Sci.*, **24**, 433-440 (1967).

Somasundaran, P., T. W. Healy, and D. W. Fuerstenau, "Surfactant adsorption at the solid–liquid interface – Dependence of mechanism on chain length", *J. Phys. Chem.*, **68**, 3562-3566 (1964).

Somasundaran, P., R. Middleton, and K. V. Viswanathan, "Relationship between surfactant structure and adsorption". In *Structure/Performance Relationship in Surfactants*, M. J. Rosen, Ed., American Chemical Society, Washington, DC, 1984, pp. 269-290.

Somerville, C. C., S. F. Nishino, and J. C. Spain, "Purification and characterization of nitrobenzene nitroreductase from *Pseudomonas pseudoalcaligenes* JS45", *J. Bacteriol.*, **177**, 3837-3842 (1995).

Spain, J. C., J. B. Hughes, and H.-J. Knackmuss, Eds., *Biodegradation of Nitroaromatic Compounds and Explosives*, Lewis Publishers, CRC Press, Boca Raton, FL, 2000.

Spormann, A. M., and F. Widdel, "Metabolism of alkylbenzenes, alkanes, and other hydrocarbons in anaerobic bacteria", *Biodegradation*, **11**, 85-105 (2000).

Sposito, G., L. Martin-Neto, and A. Yang, "Atrazine complexation by soil humic acids", *J. Environ. Qual.*, **25**, 1203-1209 (1996).

Spurlock, F. C., "Estimation of humic-based sorption enthalpies from nonlinear isotherm temperature dependence: Theoretical development and applications to substituted phenylureas", *J. Environ. Qual.*, **24**, 42-49 (1995).

Spurlock, F. C., and J. W. Biggar, "Thermodynamics of organic chemical partition in soils. 2. Nonlinear partition of substituted phenylureas from aqueous solution", *Environ. Sci. Technol.*, **28**, 996-1002 (1994a).

Spurlock, F. C., and J. W. Biggar, "Thermodynamics of organic chemical partition in soils. 3. Nonlinear partition from water-miscible cosolvent solutions", *Environ. Sci. Technol.*, **28**, 1002-1009 (1994b).

Squillace, P. J., J. F. Pankow, N. E. Korte, and J. S. Zogorsky, "Review of the environmental behavior and fate of methyl *t*-butyl ether", *Environ. Toxicol. Chem.*, **16**, 1836-1844 (1997).

Srivastava, R., G. Natarajan, and B. D. Smith, "Total pressure vapor-liquid equilibrium data for binary systems of diethyl ether with acetone, acetonitrile, and methanol." *J. Chem. Eng. Data*, **31**, 89-93 (1986).

Stamper, D. M., S. J. Traina, and O. H. Tuovinen, "Anaerobic transformation of alachlor, propachlor, and metalachlor with sulfide", *J. Environ. Qual.*, **26**, 488-494 (1997).

Stange, K., and D. L. Swackhamer, "Factors affecting phytoplankton species-specific differences in accumulation of 40 polychlorinateed biphenyls (PCBs)", *Environ. Toxicol. Chem.*, **13**, 1849-1860 (1994).

Stapleton, M. G., D. L. Sparks, and S. K. Deutel, "Sorption of pentachlorophenol to HDTMA-clay as a function of ionic strength and pH", *Environ. Sci. Technol.*, **28**, 2330-2335 (1994).

Staudinger, J., and P. V. Roberts, "A critical review of Henry's law constants for environmental application", *Crit. Rev. Environ. Sci. Technol.*, **26**, 205-297 (1996).

Steinberg, S. M., and F. Lena, "Hydrolysis of several substituted methyl benzoates in the aqueous solution", *Wat. Res.*, **29**, 965-969 (1995).

Stevenson, F. J., "Organic matter reactions involving pesticides in soil". In *Bound and Conjugated Pesticide Residues,* D. D. Kaufman, G. G. Still, G. D. Paulson and S. K. Bandal, Eds., American Chemical Society, Washington, DC, 1976, pp. 180-207.

Stewart, P. S., "A review of experimental measurements of effective diffusive permeabilities and effective diffusion coefficients in biofilms", *Biotechn. Bioeng.*, **59**, 261-272 (1998).

Stoistis, R. F., G. W. Poehlein, and J. W. Venderhoft, "Mathematical modeling of hydrodynamic chromatography", *J. Colloid Interface Sci.*, **57**, 337-344 (1976).

Stoll, J.-M. A., and W. Giger, "Mass balance for detergent-derived fluorescent whitening agents in surface waters of Switzerland", *Wat. Res.*, **32**, 2041-2050 (1998).

Stommel, H., "Trajectories of small bodies sinking slowly through convection cells", *J. Mar. Res.*, **8**, 24-29 (1949).

Stone, A. T., "Reductive dissolution of mangenese(III/IV) oxides by substituted phenols", *Environ. Sci. Technol*, **21**, 979-988 (1987).

Stone, A. T., "Enhanced rates of monophenyl terephthalate hydrolysis in aluminium oxide suspensions", *J. Colloid. Interface Sci.*, **127**, 429-441 (1989).

Stone, A. T., A. Torrents, J. Smolen, D. Vasudevan, and J. Hadley, "Adsorption of organic compounds possessing ligand donor groups at the oxide/water interface", *Environ. Sci. Technol.*, **27**, 895-909 (1993).

Storer, T. I., "DDT and wildlife", *Wildlife Management*, **10**, 181-183 (1946).

Storey, J. M. E., W. Luo, L. M. Isabelle, and J. F. Pankow, "Gas/solid partitioning of semivolatile organic compounds to model atmospheric solid surfaces as a function of relative humidity. 1. Clean quartz", *Environ. Sci. Technol.*, **29**, 2420-2428 (1995).

Strathmann, T. J., and C. T. Jafvert, "Ion-pair association of substituted phenolates with K$^+$ in octanol", *Environ. Toxicol. Chem.*, **17**, 369-376 (1998).

Stucki, G., R. Gälli, H.-R. Ebersold, and T. Leisinger, "Dehalogenation of dichloromethane by cell extracts of *Hyphomicorbium* DM2", *Arch. Microbiol.*, **130**, 366-376 (1981).

Stumm, W., R. Kummert, and L. Sigg, "A ligand exchange model for adsorption of inorganic and organic ligands at hydrous oxide interfaces", *Croat. Chem. Acta*, **53**, 291-312 (1980).

Stumm, W., and J. J. Morgan, *Aquatic Chemistry, 3rd ed.*, Wiley-Interscience, New York, 1996.

Stumm, W., R. P. Schwarzenbach, and L. Sigg, "From environmental analytical chemistry to ecotoxicology – A plea for more concepts and less monitoring and testing", *Angewandte Chemie*, **22**, 380-389 (1983).

Suatoni, J. C., R. E. Snyder, and R. O. Clark, "Voltametric studies of phenol and aniline using ring substitution", *Anal. Chem.*, **33**, 1894-1897 (1961).

Suflita, J. M., J. A. Robinson, and J. M. Tiedja, "Kinetics of microbial dehalogenation of haloaromatic substrates in methanogenic environments", *Appl. Environ. Microbiol.*, **45**, 1466-1473 (1983).

Suh, J., "Model studies of metalloenzymes involving metal ions as Lewis acid catalysts", *Acc. Chem. Res.*, **25**, 273-279 (1992).

Sulzberger, B., S. Canonica, T. Egli, W. Giger, J. Klausen, and U. von Gunten, "Oxidative transformations of contaminants in natural and technical systems", *Chimia*, **51**, 900-907 (1997).

Surber, E. W., "Effects of DDT on fish", *J. Wildlife Management*, **10**, 183-191 (1946).

Suske, W. A., M. Held, A. Schmidt, T. Fleischmann, M. G. Wubbolts, and H.-P. E. Kohler, "Purification and characterization of 2-hydroxybiphenyl 3-monooxygenase, a novel NADH-dependent, FAD-containing aromatic hydroxylase from *Pseudomonas azelaica* HBP1", *J. Biol. Chem.*, **272**, 24257-24265 (1997).

Suske, W. A., W. J. H. van Berkel, and H.-P. E. Kohler, "Catalytic mechanism of 2-hydroxybiphenyl 3-monooxygenase, a flavoprotein from *Pseudomonas azelaica* HBP1", *J. Biol. Chem.*, **247**, 33355-33365 (1999).

Suter, M. J.-F., S. Riediker, and W. Giger, "Selective determination of aromatic sulfonates in landfill leachates and groundwater using microbore liquid chromatography coupled with mass spectrometry", *Anal. Chem.*, **71**, 897-904 (1999).

Sutton, P. A., and D. A. Buckingham, "Cobalt(III)-promoted hydrolysis of amino acid esters and peptides and the synthesis of small peptides", *Acc. Chem. Res.*, **20**, 357-364 (1987).

Suzuki, K. T. Gomi, and E. Itagaki, "Intermediate and mechanism of hydroxylation of *ortho*-iodophenol by salicylate hydroxylase", *J. Biochem. (Tokyo)*, **109**, 791-797 (1991).

Sverdrup, H. U., W. Johnson, and R. H. Fleming, *The Oceans: Their Physics, Chemistry and General Biology*, Prentice Hall, Englewood Cliffs, NJ, 1942.

Swain, C. G., and C. B. Scott, "Quantitative correlation of relative rates. Comparison of hydroxide ion with other nucleophilic reagents toward alkyl halides, esters, epoxides, and acyl halides", *J. Am. Chem. Soc.*, **75**, 141-147 (1953).

Swisher, R. D., *Surfactant Biogradation*, Dekker, New York, 1987.

Szecsody, J. E., and R. C. Bales, "Sorption kinetics of low-molecular-weight hydrophobic organic compounds on surface-modified silica", *J. Contam. Hydrol.*, **4**, 181-203 (1989).

Tables of Physical and Chemical Constants, Longman, London, 1973.

Taft, R. W., Jr., *Steric Effects in Organic Chemistry*, Wiley-Interscience, New York, 1956.

Takatsuki, M., Y. Takayanagi, and M. Kitano, "An attempt to SAR of biodegradation". In *Proceedings of the Workshop, Quantitative Structure Activity Relationships for Biodegradation,* W. J. G. M. Peijnenburg and W. Karcher, Eds., National Institute of Public Health and Environmental Protection (RIVM), Bilthoven, The Netherlands, 1995, pp. 67-103.

Talbot, R. J. E., "The hydrolysis of carboxylic acid derivates". In *Comprehensive Chemical Kinetics,* C. H. Branford and C. F. H. Tipper, Eds., Elsevier, Amsterdam, 1972, pp. 209-293.

Tam, D. D., W.-Y. Shiu, K. Qiang, and D. Mackay, "Uptake of chlorobenzenes by tissues of the soybean plant: Equilibria and kinetics", *Environ. Toxicol. Chem.*, **15**, 489-494 (1996).

Tenhulscher, T. E. M., L. E. Van der Velde, and W. A. Bruggeman, "Temperature-dependence of Henry law constants for selected chlorobenzenes, polychlorinated biphenyls and polycyclic aromatic hydrocarbons", *Environ. Toxicol. Chem.*, **11**, 1595-1603 (1992).

Ternes, T., and R.-D. Wilken, Eds., *Drugs and Hormones as Pollutants of the Aquatic Environment: Determination and Ecotoxicological Impacts*, Elsevier, Lausanne, Switzerland, 1999.

Thauer, R. K., K. Jungermann, and K. Decker, "Energy conservation in chemotrophic anaerobic bacteria", *Bacteriol. Rev.*, **41**, 100-180 (1977).

Thomann, R. V., and J. P. Connolly, "Model of PCB in the Lake-Michigan lake trout food-chain", *Environ. Sci. Technol.*, **18**, 65-71 (1984).

Thomas, G. O., A. J. Sweetman, and K. C. Jones, "Input-output balance of polychlorinated biphenyls in a long-term study of lactating diary cows", *Environ. Sci. Technol.*, **33**, 104-112 (1999).

Thomas, G. O., A. J. Sweetman, R. Lohmann, and K. C. Jones, "Derivation and field testing of air-milk and feed–milk transfer factors for PCBs", *Environ. Sci. Technol.*, **32**, 3522-3528 (1998a).

Thomas, G. O., A. J. Sweetman, W. A. Ockenden, D. Mackay, and K. C. Jones, "Air–pasture transfer of PCBs", *Environ. Sci. Technol.*, **32**, 936-942 (1998b).

Thompson, N. S., Ed. *Encyclopedia of Chemical Technology*, Wiley, New York, 1996.

Thompson, P. L., L. A. Ramer, and J. L. Schnoor, "Uptake and transformation of TNT by hybrid polar trees", *Environ. Sci. Technol.*, **32**, 975-980 (1998).

Thomson, N. S., Ed. *Encyclopedia of Chemical Technology*, Vols. 1-14, Wiley, New York, 1996.

Thorn, K. A., P. J. Pettigrew, W. S. Goldenberg, and E. J. Weber, "Covalent binding of aniline to humic substances. 2. 15N-NMR studies of nucleophilic addition reactions", *Environ. Sci. Technol.*, **30**, 2764-2775 (1996).

Thurman, E. M., *Organic Geochemistry of Natural Waters*, Martinus Nijhoff, Boston, 1985.

Tiegs, D., J. Gmehling, A. Medina, M. Soares, J. Bastos, P. Alessi, and I. Kikic, *Activity Coefficients at Infinite Dilution,* DECHEMA, Frankfurt, 1986.

Tinsley, I. J., *Chemical Concepts in Pollutant Behavior*, Wiley-Interscience, New York, 1979.

Tipping, E., "The adsorption of aquatic humic substances by iron oxides", *Geochim. Cosmochim Acta*, **45**, 191-199 (1981).

Tokunaga, T. K., L. J. Waldron, and J. Nemson, "A closed tube method for measuring gas diffusion coefficients", *Soil Sci. Soc. Am. J.*, **52**, 17-23 (1988).

Tolls, J., and M. S. McLachlan, "Partitioning of semivolatile organic compounds between air and *Lolium multiflorum* (Welsch Ray Grass)", *Environ. Sci. Technol.*, **28**, 159-166 (1994).

Tomlin, C., Ed. *The Pesticide Manual*, The British Crop Protection Council and the Royal Society of Chemistry, Cambridge, UK, 1994.

Torgersen, T., G. Mathieu, R. H. Hesslein, and W. S. Broecker, "Gas-exchange dependency on diffusion-coefficient-direct Rn-222 and He-3 comparisons in a small lake", *J. Geophys. Res.*, **87**, C1, 546-556 (1982).

Torrents, A., and A. T. Stone, "Hydrolysis of phenyl picolinate at the mineral/water interface", *Environ. Sci. Technol.*, **25**, 143-149 (1991).

Torrents, A., and A. T. Stone, "Catalysis of picolinate ester hydrolysis at the oxide/water interface: Inhibition by coadsorbed species", *Environ. Sci. Technol.*, **27**, 1060-1067 (1993a).

Torrents, A., and A. T. Stone, "Catalysis of picolinate ester hydrolysis at the oxide/water interface: Inhibition by adsorbed natural organic matter", *Environ. Sci. Technol.*, **27**, 2381-2386 (1993b).

Torrents, A., and A. T. Stone, "Oxide surface-catalyzed hydrolysis of carboxylate esters and phosphorothionate esters", *Soil Sci. Soc. Am. J.*, **58**, 738-745 (1994).

Totten, L. A., and A. L. Roberts, "Calculated one- and two-electron reduction potentials and related molecular descriptors for reduction of alkyl and vinyl halides in water", *Crit. Rev. Environ. Sci. Technol.*, **31**, 175-221 (2001).

Tracey, G. A., and D. J. Hansen, "Use of biota-sediment accumulation factors to assess similarity of nonionic organic chemical exposure to benthically coupled organisms of differing trophic mode", *Environ. Toxicol. Chem.*, **30**, 467-476 (1996).

Traina, S. J., D. C. McAvoy, and D. J. Versteeg, "Association of linear alkylbenzenesulfonates with dissolved humic substances and its effect on bioavailability", *Environ. Sci. Technol.*, **30**, 1300-1309 (1996).

Trapp, S., K. S. B. Miglioranza, and H. Mosbaek, "Sorption of lipophilic organic compounds to wood and implications for their environmental fate", *Environ. Sci. Technol.*, **35**, 1561-1566 (2001).

Tratnyek, P. G., "Putting corrosion to use: Remediating contaminated groundwater with zero-valent metals", *Chem. Ind.*, **1 July**, 499-503 (1996).

Tratnyek, P. G., and J. Hoigné, "Oxidation of phenols in the environment: A QSAR analysis of rate constants for reaction with singlet oxygen", *Environ. Sci. Technol.*, **25**, 1596-1604 (1991).

Tratnyek, P. G., J. Hoigne, J. Zeyer, and R. P. Schwarzenbach, "QSAR analysis of oxidation and reduction rates of environmental organic pollutants in model systems", *The Science of the Total Environment*, **109/110**, 327-341 (1991).

Tremolada, P., V. Burnett, D. Calamari, and K. C. Jones, "Spatial distribution of PAHs in the UK atmosphere using pine needles", *Environ. Sci. Technol.*, **30**, 3570-3577 (1996).

Tremp, J., P. Mattrel, S. Fingler, and W. Giger, "Phenols and nitrophenols as tropospheric pollutants: Emissions from automobile exhausts and phase transfer in the atmosphere", *Water Air Soil Pollut.*, **68**, 113-123 (1993).

Tros, M., T. N. P. Bosma, G. Schraa, and A. J. B. Zehnder, "Measurement of minimum substrate concentration (S_{min}) in a recycling fermentor and its prediction from the kinetic parameters of *Pseudomonas* sp. strain B13 from batch and chemostat cultures", *Appl. Environ. Microbiol.*, **62**, 3655-3661 (1996).

Trouton, F., "IV. On molecular latent heat", *Phil. Mag.*, **18**, 54-57 (1884).

Tschech, A., and G. Fuchs, "Anaerobic degradation of phenol via carboxylation to 4-hydroxybenzoate: *In vitro* study of isotope exchange between $^{14}CO_2$ and 4-hydroxybenzoate", *Arch. Microbiol.*, **152**, 594-599 (1989).

Tunkel, J., P. H. Howard, R. S. Boethling, W. Stiteler, and H. Loonen, "Predicting ready biodegradability in the Japanese Ministry of International Trade and Industry test", *Environ. Toxicol. Chem.*, **19**, 2478-2485 (2000).

Turro, N. J., *Modern Molecular Photochemistry*, Benjamin/Cummings, Menlo Park, CA, 1991.

Ullrich, V., "Enzymatic hydroxylations with molecular oxygen", *Ange. Chem. Int. Ed.*, **11**, 701-712 (1972).

Ulrich, H.-J., and A. T. Stone, "Oxidation of chlorophenols adsorbed to manganese oxide surfaces", *Environ. Sci. Technol.*, **23**, 421-428 (1989).

Ulrich, M. M., S. R. Müller, H. P. Singer, D. M. Imboden, and R. P. Schwarzenbach, "Input and dynamic behavior of the organic pollutants tetrachloroethene, atrazine, and NTA in a lake: A study combining mathematical modeling and field measurements", *Environ. Sci. Technol.*, **28**, 1674-1685 (1994).

Ulrich, M. M., D. M. Imboden, and R. P. Schwarzenbach, "MASAS – A user-friendly simulation tool for modeling the fate of anthropogenic substances in lakes", *Environ. Software*, **10**, 177-198 (1995).

Uraizee, F. A., A. D. Venosa, and M. T. Suidan, "A model for diffusion controlled bioavailability of crude oil components", *Biodegradation*, **8**, 287-296 (1998).

Vaes, W. H. J., E. U. Ramos, C. Hamwijk, I. von Holsteijn, B. J. Balaauboer, W. Seinen, H. J. M. Verhaar, and J. L. M. Hermens, "Solid phase microextraction as a tool to determine membrane/water partition coefficients and bioavailable concentrations in *in vitro* system", *Chem. Res. Toxicol.*, **10**, 1067-1072 (1997).

Vaes, W. H. J., E. U. Ramos, H. J. M. Verhaar, and J. L. M. Hermens, "Acute toxicity of nonpolar versus polar narcosis: Is there a difference", *Environ. Toxicol. Chem.*, **16**, 1380-1384 (1998).

Vaes, W. H. J., E. U. Ramos, H. J. M. Verhaar, W. Seinen, and J. L. M. Hermens, "Measurement of the free concentration using solid-phase microextraction: Binding to protein", *Anal. Chem.*, **68**, 4463-4467 (1996).

Vallack, H. W., D. J. Bakker, I. Brandt, E. Broström-Lundén, A. Brouwer, K. R. Bull, C. Gough, R. Guardan, I. Holoubek, B. Jansson, R. Koch, J. Kuylenstierna, A. Lecloux, D. Mackay, P. McCutcheon, P. Mocarelli, and R. D. F. Taalman, "Controlling persistent organic pollutants – What next?" *Environ. Toxicol. Pharmacol.*, **6**, 143-175 (1998).

Valsaraj, K. T., G. J. Thoma, D. D. Reible, and L. J. Thibodeaux, "On the enrichment of hydrophobic organic compounds in fog droplets", *Atmospheric Environ.*, **27**, 203-210 (1993).

van Bladel, R., and A. Moreale, "Adsorption of fenuron and monuron (substituted ureas) by two montmorillonite clays", *Soil Sci. Soc. Am. Proc.*, **38**, 244-249 (1974).

Van Hooidonk, C., and L. Ginjaar, "On the reactivity of organophosphorous compounds. Part III", *Rec. Trav. Chim.*, **86**, 449-457 (1967).

van Wezel, A. P., G. Cornelissen, J. K. van Miltenburg, and A. Opperhuizen, "Membrane burdens of chlorinated benzenes lower the main phase transition temperature in dipalmitoyl-phosphatidylcholine vesicles: Implications for toxicity by narcotic chemicals", *Environ. Toxicol. Chem.*, **15**, 203-212 (1996a).

van Wezel, A. P., D. A. M. deVries, D. T. H. M. Sijm, and A. Opperhuizen, "Use of lethal body burden in the evaluation of mixture toxicity", *Ecotoxicol. Environ. Safety*, **35**, 236-241 (1996b).

van Wezel, A. P., and A. Opperhuizen, "Narcosis due to environmental pollutants in aquatic organisms: Residue-based toxicity, mechanisms, and membrane burdens", *Critical Reviews in Toxicology*, **25**, 255-279 (1995).

van Wezel, A. P., S. S. Punte, and A. Opperhuizen, "Lethal body burdens of polar narcotics: Chlorophenols", *Environ. Tox. Chem.*, **14**, 1579-1585 (1995).

Vasudevan, D., and A. T. Stone, "Adsorption of catechols, 2-aminophenols, and 1,2-phenylenediamines at the metal (hydr)oxide-water interface: Effects of ring substituents on the adsorption onto TiO_2", *Environ. Sci. Technol.*, **30**, 1604-1613 (1996).

Vaugham, P. P., and N. V. Blough, "Photochemical formation of hydroxyl radical by constituents of natural waters", *Environ. Sci. Technol.*, **32**, 2947-2953 (1998).

Verhaar, H. J. M., C. J. vonLeeuwen, and J. L. M. Hermens, "Classifying environmental pollutants. 1: Structure-activity relationships for prediction of aquatic toxicity", *Chemosphere*, **25**, 471-491 (1992).

Vitha, M. F., and P. W. Carr, "The chemical meaning of the standard free energy of transfer: Use of van der Waals' equation of state to unravel the interplay between free volume, volume entropy and the role of standard states", *J. Phys. Chem. B*, **104**, 5343-5349 (2000).

Vogel, T. M., C. S. Criddle, and P. L. McCarty, "Transformation of halogenated aliphatic compounds", *Environ. Sci. Technol.*, **21**, 722-736 (1987).

Vontor, T., J. Socha, and M. Vecera, "Kinetics and mechanisms of hydrolysis of 1-naphthyl-N-methyl- and N,N-dimethylcarbamates", *Coll. Czechoslow. Chem. Commun.*, **37**, 2183-2196 (1972).

Vorbeck, C., H. Lenke, P. Fischer, J. C. Spain, and H.-J. Knackmuss, "Initial reactions in aerobic microbial metabolism of 2,4,6-trinitrotoluene", *Appl. Environ. Microbiol.*, **64**, 246-252 (1998).

Vuilleumier, S., "Bacterial glutathione S-transferases: What are they good for?" *J. Bacteriol.*, **179**, 1431-1441 (1997).

Wackett, L. P., "Co-metabolism: is the emperor wearing any clothes?" *Curr. Opinion Biotechn.*, **7**, 321-325 (1996).

Wade, R. S., and C. E. Castro, "Oxidation of iron(II) prophyrins by alkyl halides", *J. Am. Chem. Soc.*, **95**, 226-230 (1973a).

Wade, R. S., and C. E. Castro, "Oxidation of heme proteins by alkyl halides", *J. Am. Chem. Soc.*, **95**, 231-234 (1973b).

Wallington, T. J., W. F. Schneider, D. R. Worsnop, O. J. Nielsen, J. Sehested, W. J. Debruyn, and J. A. Shorter, "The environmental impact of CFC replacements – HFCs and HCFCs", *Environ. Sci. Technol.*, **28**, 320A-326A (1994).

Wang, Q., J. Gan, S. K. Papiernik, and S. R. Yates, "Transformation and detoxification of halogenated fumigants by ammonium thiosulfate", *Environ. Sci. Technol.*, **34**, 3717-3721 (2000).

Wania, F., and M. S. McLachlan, "Estimating the influence of forests on the overall fate of semivolatile organic compounds using a multimedia fate model", *Environ. Sci. Technol.*, **35**, 582-590 (2001).

Wania, F., W.-Y. Shiu, and D. Mackay, "Measurements of the vapor pressure of several low-volatility organochlorine chemicals at low temperatures with a gas saturation method", *J. Chem. Eng. Data*, **39**, 572-577 (1994).

Wanner, O., T. Egli, T. Fleischmann, K. Lanz, P. Reichert, and R. P. Schwarzenbach, "The behavior of the insecticides disulfoton and thiometon in the Rhine River – A chemodynamic study", *Environ. Sci. Technol*, **23**, 1232-1242 (1989).

Wanninkhof, R., J. P. Leswell, and W. S. Broecker, "Gas exchange on Mono Lake and Crowley Lake, California", *J. Geophys. Res.*, **92**, 14567-14580 (1987).

Warner, M. J., and R. F. Weiss, "Solubilities of chlorofluorocarbons 11 and 12 in water and seawater", *Deep-Sea Res.*, **32**, 1485-1497 (1985).

Weber, E. J., D. Colon, and G. L. Baughman, "Sediment-associated reactions of aromatic amines. 1. Elucidation of sorption mechanisms", *Environ. Sci. Technol.*, **35**, 2470-2475 (2001).

Weber, E. J., D. L. Spidle, and K. A. Thorn, "Covalent binding of aniline to humic substances. 1. Kinetic studies", *Environ. Sci. Technol.*, **30**, 2755-2763 (1996).

Weber, E. J., and N. L. Wolfe, "Kinetic studies of the reduction of aromatic azo compounds in anaerobic sediment/water systems", *Environ. Toxicol. Chem.*, **6**, 911-919 (1987).

Weber, J. B., and H. D. Coble, "Microbial decomposition of diquat adsorbed on montmorillonite and kaolinite clays", *J. Agric. Food Chem.*, **16**, 475-478 (1968).

Weber, K., "Degradation of parathion in seawater", *Water Res.*, **10**, 237-241 (1976).

Weber, W. J. Jr., and F. A. DiGiano, *Process Dynamics in Environmental Systems*, Wiley, New York, 1996.

Weber, W. J. Jr., P. M. McGinley, and L. E. Katz, "A distributed reactivity model for sorption by soils and sediments. 1. Conceptual basis and equilibrium assessments", *Envion. Sci. Technol.*, **26**, 1955-1962 (1992).

Weibull, W., "A statistical distribution function of wide applicability", *J. Appl. Mech.*, **18**, 293-297 (1951).

Weidenhaupt, A., C. G. Arnold, S. R. Müller, S. B. Haderlein, and R. P. Schwarzenbach, "Sorption of organotin biocides to mineral surfaces", *Environ. Sci. Technol*, **31**, 2603-2609 (1997).

Weinbach, E. C., "Biochemical basis for the toxicity of pentachlorophenol", *Biochemistry*, **43**, 393-397 (1957).

Weiner, J. H., D. P. MacIsaac, R. E. Bishop, and P. T. Bilons, "Purification and properties of *Escherichia coli* dimethyl sulfoxide reductase, an iron–sulfur molybdoenzyme with broad substrate specificity", *J. Bacteriol.*, **170**, 1505-1510 (1988).

Weissmahr, K. W., S. B. Haderlein, R. P. Schwarzenbach, R. Hany, and R. Nüsch, "*In situ* spectroscopic investigations of adsorption mechanisms of nitroaromatic compounds at clay minerals", *Environ. Sci. Technol.*, **31**, 240-247 (1997).

Weissmahr, K. W., M. Hildenbrand, R. P. Schwarzenbach, and S. B. Haderlein, "Laboratory and field scale evaluation of geochemical controls on groundwater transport of nitroaromatic ammunition residues", *Environ. Sci. Technol.*, **33**, 2593-2600 (1999).

Welander, P., "Theoretical forms for the vertical exchange coefficients in a startified fluid with application to lakes and seas", *Geophysica Gothoburgensia*, **1**, 1-27 supplement (1968).

Welke, B., K. Ettlinger, and M. Riederer, "Sorption of volatile organic chemicals in plant surfaces", *Environ. Sci. Technol.*, **32**, 1099-1104 (1998).

Westall, J. C., W. Zhang, and B. J. Brownawell, "Sorption of linear alkylbenzenesulfonates on sediment materials", *Environ. Sci. Technol.*, **33**, 3110-3118 (1999).

Wester, P. G., J. deBoer, and U. A. T. Brinkman, "Determination of polychlorinated terphenyls in aquatic biota and sediment with gas chromatography/mass spectrometry using negative chemical ionization", *Environ. Sci. Technol.*, **30**, 473-480 (1996).

Weston, D. P., D. L. Penry, and L. K. Gulman, "The role of ingestion as a route of contaminant bioaccumulation in a deposit-feeding polycheate", *Arch. Environ. Contam. Toxicol*, **38**, 446-454 (2000).

White, R. E., and M. J. Coon, "Oxygen activation by cytochrome P-450", *Ann. Rev. Biochem.*, **49**, 315-356 (1980).

Whitehouse, B. G., "The effects of temperature and salinity on the aqueous solubility of polynuclear aromatic hydrocarbons", *Mar. Chem.*, **14**, 319-332 (1984).

Whitman, W. G., "The two-film theory of gas absorption", *Chem. Met. Eng.*, **29**, 146-148 (1923).

Willett, K. L., E. M. Ulrich, and R. A. Hites, "Differential toxicity and environmental fates of hexachlorocylohexane isomers", *Environ. Sci. Technol.*, **32**, 2197-2207 (1998).

Williams, A., "Alkaline hydrolysis of substituted phenylcarbamates – Structure-reactivity relationships consistent with an E1B mechanism", *J. Chem. Soc. Perkins*, **II**, 808-812 (1972).

Williams, A., "Participation of an elimination mechanism in alkaline hydrolysis of alkyl N-phenylcarbamates", *J. Chem. Soc. Perkins*, **II**, 1244-1247 (1973).

Williams, A., "Free-energy correlations and reaction mechanisms". In *The Chemistry of Enzyme Action,* M. I. Page, Ed., Elsevier, Amsterdam, 1984, pp. 127-201.

Williams, D. H., and I. Fleming, *Spectroscopic Methods in Organic Chemistry*, McGraw-Hill, New York, 1980.

Williams, D. J. A., and K. P. Williams, "Electrophoresis and zeta potential of kaolinite", *J. Colloid Int. Sci.*, **65**, 79-87 (1978).

Williams, J., and S. A. Elder, *Fluid Physics for Oceanographers and Physicists – An Introduction to Incompressible Flow,* Pergamon Press, Oxford, 1989.

Wilson, J. T., and B. H. Wilson, "Biotransormation of trichloroethylene in soil", *Appl. Environ. Microbiol*, **49**, 242-243 (1985).

Winterle, J. S., D. Tse, and W. R. Mabey, "Measurement of attenuation coefficients in natural water columns", *Environ. Toxicol. Chem.*, **6**, 663-672 (1987).

Wohlfarth, G., and G. Diekert, "Reductive dehalogenases". In *Chemistry and Biochemistry of B_{12},* R. Banerjee, Ed., Wiley-Interscience, New York, 1999, pp. 871-893.

Wolfe, N. L., "Organophosphate and organophosphothionate esters: Application of LFERs to estimate hydrolysis rate constants for use in environmental fate assessment", *Chemosphere*, **9**, 571-579 (1980).

Wolfe, N. L., D. F. Paris, W. C. Steen, and G. L. Baughman, "Correlation of microbial degradation rates with chemical structure", *Environ. Sci. Technol.*, **14**, 1143-1144 (1980).

Wolfe, N. L., R. G. Zepp, D. F. Paris, G. L. Baughman, and R. C. Hollis, "Methoxychlor and DDT degradation in water: Rates and products", *Environ. Sci. Technol.*, **11**, 1077-1081 (1977).

Wood, A. L., D. C. Bouchard, M. L. Brusseau, and P. S. C. Rao, "Cosolvent effects on sorption and mobility of organic chemicals in soils", *Chemosphere*, **21**, 575-587 (1990).

Wood, J. M., F. S. Kennedy, and R. S. Wolfe, "The reaction of multihalogenated hydrocarbons with free and bound reduced vitamin B12", *Biochemistry*, **7**, 1707-1713 (1968).

Wu, J., "Bubble populations and spectra in near-surface ocean: Summary and review of field measurements", *J. Geophys. Res.*, **86**, 457-463 (1981).

Wu, S.-C., and P. M. Gschwend, "Numerical modeling of sorption kinetics of organic compounds to soil and sediment particles", *Water Resources Res.*, **24**, 1373-1383 (1988).

Wüest, A., "Interaktionen in Seen: Die Biologie als Quelle dominanter physikalischer Kräfte", *Limnologica*, **24**, 93-104 (1994).

Xia, G., and W. P. Ball, "Adsorption–partitioning uptake of nine low-polarity organic chemicals on a natural sorbent", *Environ. Sci. Technol.*, **33**, 262-269 (1999).

Xia, G., and W. P. Ball, "Polany-based models for the competitive sorption of low-polarity organic contaminants on a natural sorbent", *Environ. Sci. Technol.*, **34**, 1246-1253 (2000).

Xiao, S., C. Moresoli, J. Bovenkamp, and D. DeKee, "Sorption and permeation of organic environmental contaminants through PVC geomembranes", *J. Applied Polymer Sci.*, **63**, 1189-1197 (1997).

Xie, W.-H., W.-Y. Shiu, and D. Mackay, "A review of the effect of salts on the solubility of organic compounds in seawater", *Marine Environ. Res.*, **44**, 429-444 (1997).

Xing, B., W. B. McGill, and J. J. Dudas, "Sorption of α-naphthol onto organic sorbents varying in polarity and aromaticity", *Chemosphere*, **28**, 145-153 (1994).

Xing, B., and J. J. Pignatello, "Dual-mode sorption of low-polarity compounds in glassy polyvinylchloride and soil organic matter", *Environ. Sci. Technol.*, **31**, 792-799 (1997).

Xing, B., J. J. Pignatello, and B. Gigliotti, "Competitive sorption between atrazine and other organic compounds in soils and model sorbents", *Environ. Sci. Technol.*, **30**, 2432-2440 (1996).

Yalkowsky, S. H., "Estimation of entropies of fusion of organic compounds", *Ind. Eng. Chem. Fundam.*, **18**, 108-111 (1979).

Yalkowsky, S. H., and S. Banerjee, *Aqueous Solubility. Methods of Estimation for Organic Compounds*, Marcel Dekker, New York, 1992.

Yang, W., L. Dostal, and J. P. N. Rosazza, "Stereospecificity of microbial hydrations of oleic acid to 10-hydroxystearic acid", *Appl. Environ. Microbiol.*, **59**, 281-284 (1993).

Yaron, B., A. R. Swoboda, and G. W. Thomas, "Aldrin adsorption by soils and clays", *J. Agric. Food Chem.*, **15**, 671-675 (1967).

Yeh, W. K., D. T. Gibson, and E. Liug, "Toluene dioxygenase: A multicomponent enzyme system", *Biochem. Biophys. Res. Comm.*, **78**, 401-410 (1977).

Yost, E. C., C. I. Tejedor-Tejedor, and M. A. Anderson, "*In situ* CIR–FTIR characterization of salicylate complexes at the goethite/aqueous solution interface", *Environ. Sci. Technol.*, **24**, 822-828 (1990).

Yu, J., K. E. Taylor, H. Zou, N. Biswas, and J. K. Bewtra, "Phenol conversion and dimeric intermediates in horseradish peroxidase-catalyzed phenol removal from water", *Environ. Sci. Technol.*, **28**, 2154-2160 (1994).

Yuan, H. D., R. Ranatung, P. W. Carr, and J. Pawliszyn, "Determination of equilibrium constant of alkylbenzenes binding to bovine serum albumin by solid phase microextraction", *Analyst*, **124**, 1443-1448 (1999).

Zachara, J. M., C. C. Ainsworth, L. J. Felice, and C. T. Resch, "Quinoline sorption to subsurface materials: Role of pH and retention of the organic cation", *Environ. Sci. Technol.*, **20**, 620-627 (1986).

Zafiriou, O. C., "Reactions of methyl halides with seawater and marine aerosols", *J. Marine Res.*, **33**, 75-81 (1975).

Zafiriou, O. C., "Natural water photochemistry". In *Chemical Oceanography,* Vol. 8, Academic Press, London, 1983, pp. 339-379.

Zafiriou, O. C., J. Joussot-Dubien, R. G. Zepp, and R. G. Zika, "Photochemistry of natural waters", *Environ. Sci. Technol.*, **18**, 358A-371A (1984).

Zepp, R. G., "Quantum yields for reaction of pollutants in dilute aqueous solution", *Environ. Sci. Technol.*, **12**, 327-329 (1978).

Zepp, R. G., "Assessing the photochemistry of organic pollutants in aquatic environments". In *Dynamics, Exposure and Hazard Assessment of Toxic Chemicals,* R. Haque, Ed., Ann Arbor Science, Ann Arbor, MI, 1980, pp. 69-109.

Zepp, R. G., "Experimental approaches to environmental photochemistry". In *The Handbook of Environmental Chemistry,* Vol. 2, Part B, SO. Hutzinger, Ed., pringer, Berlin, 1982, pp. 19-41.

Zepp, R. G., and G. L. Baughman, "Prediction of photochemical transformation of pollutants in the aquatic environment". In *Aquatic Pollutants, Transformations and Biological Effects,* O. Hutzinger, I. M. v. Lelyveld and B. C. J. Zoeteman, Eds., Pergamon, New York, 1978, pp. 237-263.

Zepp, R. G., and D. M. Cline, "Rates of direct photolysis in aquatic environment", *Environ. Sci. Technol.*, **11**, 359-366 (1977).

Zepp, R. G., and L. F. Ritmiller, "Photoreactions providing sinks and sources of halocarbons in aquatic environments". In *Aquatic Chemistry – Inerfacial and Interspecies Processes,* C. P. Huang, C. R. O Melia and J. J. Morgan, Eds., American Chemical Society, Washington, DC, 1995, pp. 253-278.

Zepp, R. G., and P. F. Schlotzhauer, "Photoreactivity of selected aromatic hydrocarbons in water". In *Polynuclear Aromatic Hydrocarbons,* W. Jones and T. Leber, Eds., Ann Arbor Science, Ann Arbor, MI, 1979, pp. 141-158.

Zepp, R. G., P. F. Schlotzhauer, and R. M. Sink, "Photosensitized transformations involving electronic energy transfer in natural waters: Role of humic substances", *Environ. Sci. Technol.*, **19**, 74-81 (1985).

Zepp, R. G., N. L. Wolfe, G. L. Baughman, and R. C. Hollis, "Singlet oxygen in natural waters", *Nature*, **267**, 421-423 (1977).

Zerner, B., R. P. M. Band, and M. L. Bender, "Kinetic evidence for the formation of acylenzyme intermediates in the a-chymotrypsin-catalyzed hydrolysis of specific substrates", *J. Am. Chem. Soc.*, **86**, 3674-3679 (1964).

Zeyer, J., and H. P. Kocher, "Purification and characterization of bacterial nitrophenol oxygenase which convert *ortho*-nitrophenol to catechol and nitrite", *J. Bacteriol.*, **170**, 1789-1794 (1988).

Zhang, X., and L. Y. Young, "Carboxylation as an initial reaction in the anaerobic metabolism of naphthalene and phenanthrene by sulfidogenic consortia", *Appl. Environ. Microbiol.*, **63**, 4759-4764 (1997).

Zheng, M., W. J. De Bruyn, and E. S. Saltzman, "Measurements of the diffusion coefficients of CFC11 and CFC12 in pure water and seawater", *J. Geophys. Res.*, **103**, 1375-1379 (1998).

Zhou, J. L., S. Rowland, and F. C. Montoura, "Partition of synthetic pyrethroid insecticides between dissolved and particular phases", *Wat. Res.*, **29**, 1023-1031 (1995).

Zhou, X., and K. Mopper, "Apparent partition coefficients of 15 carbonyl compounds between air and seawater and between air and freshwater; Implications for air-sea exchange", *Environ. Sci. Technol.*, **24**, 1864-1869 (1990).

Zierath, D. L., J. J. Hassett, W. L. Banwart, S. G. Wood, and J. C. Means, "Sorption of benzidine by sediments and soils", *Soil Sci.*, **129**, 277-281 (1980).

Zinder, S. H., and T. D. Brock, "Dimethyl sulphoxide reduction by microorganisms", *J. Gen. Microbiol.*, **105**, 335-342 (1978).

Zipper, C., M. Bunk, A. J. B. Zehnder, and H.-P. E. Kohler, "Enantioselective uptake and degradation of the chiral herbicide dichlorprop [(*RS*)-2-(2,4-dichlorophenoxy)propanoic acid] by *Sphingomonas herbicidovorans* MH", *J. Bacteriol.*, **180**, 3368-3374 (1998).

Zoro, J. A., J. M. Hunter, G. Englinton, and G. C. Ware, "Degradation of *p,p'*-DDT in reducing environments", *Nature*, **247**, 235-237 (1974).

Zullig, J. J., and J. W. Morse, "Interaction of organic acids with carbonate mineral surfaces in seawater and related solutions. I. Fatty acid adsorption", *Geochim. Cosmochim. Acta*, **52**, 1667-1678 (1988).

SUBJECT INDEX

A

Abiotic transformation of compound
definition; 459, 556

Absorbance
definition; 615
determination of; 616

Absorption
definition; 277

Absorption of compound from gas phase
definition; 59
into liquid phase; 67

Absorption of radiation; *See Light absorption*

Absorption spectrum of compound; *See Electronic absorption spectrum*

Acid
definition; 246
air–water exchange, scheme of processes, Fig.; 933
calculation of air–water distribution, Ill. Ex.,; 269
effect of deprotonation on light absorption; 622
fractions of neutral and charged forms; 255
Lewis; *See Lewis acid*
multiprotic, calculation of fractions, Ill. Ex; 254
reaction with water; 246
sorption to natural organic matter; 321
sorption to NOM as function of pH, Fig.; 322
strong and weak, definition; 249
uptake of by liposomes and baseline toxicity; 377

Acid–base reaction
characteristics of; 246
organic acid with water as base; 246
of oxide surfaces; 419
proton and water, definition as standard; 247

Acid-to-base concentrations ratio; *See Fraction of amount of acid*
and pH of solution; 249

Acid catalyzed hydrolysis reaction
effects of compound's substituents on rate; 522
mechanism and scheme of carboxylic acid esters; 521
of (thio-)phosphoric acid esters; 539

Acid catalyzed hydrolysis reaction rate
function of pH, for carboxylic acid esters, Fig.; 514

Acid derivative
hydrolysis reaction, general mechanism; 513
hydrolysis reaction products; 513

Acidity constant of compound
definition; 248
acid-to-base concentration ratio and pH; 249
as criterion for ease of base as leaving group; 491
calculation from Hammett correlations, Ill. Ex.; 266
of the conjugate acid of a base; 249
determination of; 248
determination of and literature to data; 261
effect of substituent; 261
effect of substituent and effect of type of acid; 263
effect of substituent at aromatic acids, Fig.; 262
estimation of value; *See Hammett correlation method*
interpretation of; 249
microscopic and macroscopic in zwitterions; 255
temperature dependence; 252
thermodynamic and operational; 248
values at various temperatures, Tab.; 252
values of selected organic acids, Tab.; 250
values of selected organic bases, Tab.; 251
values of water at various temperatures, Tab.; 252

Acidity constant of oxide surface; 420
as function of pH; 420

Activated complex
definition; 479
bimolecular elimination reaction, Fig.; 510
first-order nucleophil substitution, Fig.; 496
second-order nucleophil substitution, Fig.; 495
and universal reaction rate constant; 479

Activation energy of reaction
definition in Arrhenius equation; 478
calculation from laboratory data, Ill. Ex.; 515
change of by enzyme catalysis; 696
nonenzymatic and enzymatically catalyzed, Fig.; 696
in one-electron reduction reaction, Fig.; 584
value range of hydrolysis of carbamates; 531
value range of hydrolysis of carboxylic acid amide; 527
value range of hydrolysis of carboxylic acid ester; 520
values, substitution and elimination reaction, Tab; 508

Active uptake by microorganism
by system of exterior membrane; 691
examples and energetic mechanism; 738

Activity coefficient of compound
and cosolvent type, Fig.; 166, 167
calculation for binary mixture, Ill. Ex.; 170
and cavity formation in solvent; 81
in different solvents, values, Tab. 3.2; 82
and excess free energy; 82
as function of cosolvent fraction, Fig.; 169, 167
as function of solutes molar volume, Fig.; 167
in ideal solution or ideal mixture; 183
in ideally diluted solution; 183
influence of phase mixing; 186
in liquid solution, as function of cosolvent; 311
as pure liquid and as solute; 80
value and reference state; 80
values for ideal solution in various solvents, Tab; 188
values for selected cosolvents, Tab.; 168

Activity of compound
definition; 80
calculation for water, sediment, organism, Ill. Ex; 359
calculation from actual and saturation concentr.; 357
change in biomagnification along food chain; 368
and concentration in equilibrium constant; 66
and concentration in Freundlich constant; 281
as a measure of disequilibrium; 356
and reference state; 80
at saturation pressure of pure liquid; 356
at solubility concentration in liquid solution; 356
temperature dependence; 357

Acute toxicity of compound
and baseline toxicity; 377
relation to baseline toxicity, Fig.; 379
toxic ratio, definition; 378

Acyclic
definition; 30

Addition reaction, microbial
carbon dioxide addition (carboxylation); 731
fumarate addition; 730
hydration with water of substrate double bond; 734
initial step in degradation reactions; 702
reaction sequence of benzylsuccinate synthase Fig; 732
reaction sequence of carboxylation; 733

Adsorption
definition; 59, 277

Adsorption of compound
from air on surface; *See Air–surface adsorption*

Advection; 779

Advection/diffusion equation; *See Diffusion/ advection equation*

Advection/reaction equation in river
stationary solution of; 1110

Advective transport of compound
concentration profile and flux, Fig.; 1007
distance of; 1013
time of; 1012
time, relative to diffusion time; 1012
and turbulent diffusion, patch development,
Fig.; 1031

Aerobic respiration
definition; 571
role of oxygen in biologically catalyzed
reactions; 571
step in organic compound oxidation, Fig.; 570

Aerosol
salt surface as sorbent in marine environment; 400
as sorption surface; *See Air–surface adsorption*

Air filter
as sorption surface; *See Air–surface adsorption*

Air pollution
contaminating laboratory samples, Ill. Ex.; 197
contaminating olive oil, Ill. Ex.; 196

**Air–apolar surface partition constant of
compound**
as function of London dispersive energy,
Fig.; 72

Air–aqueous phase; *See Air–water*

**Air–hexadecane partition constant of com-
pound**
as function of London dispersive energy,
Fig.; 71
LFER parameters for saturation pressure,
Fig.; 117

Air–organic solvent partition constant
definition; 185
calculated versus experimental values, Fig.; 193
comparison of different solvents; 186
derived from two connected partition constants; 215
dimension of; 184
estimation from molecular data, LFER; 192
in dry versus in wet octanol, Fig.; 187
LFER for two apolar solvents, Fig.; 190, 191
LFER for two polar solvents, Fig.; 191
LFER for two solvents; 189
multiparameter LFER, values for solvents,
Tab.; 194
for organic solvent/water mixture; 201
and saturation pressure; 189
schematic picture of relations, Fig.; 186
temperature dependence; 185, 195
values for various compounds and solvents,
Tab.; 188

**Air–organic solvent partitioning of com-
pound; 185**
applications in nature and in laboratory; 195
olive oil contamination by air pollution, Ill.
Ex.; 196

sample contamination in the laboratory, Ill.
Ex.; 197
schematic illustration of processes, Fig.; 186
solvent effect; 185
temperature effect; 185

Air–organic solvent transfer of compound
examples of; 182
standard enthalpy change; 185

Air-phase transfer velocity of compound
impact of wind speed on water vapor, Fig.; 897
of water vapor and wind speed, relationships,
Tab.; 898
pure liquid, estimation of evaporation, Ill.
Ex.; 898
summary for reference and optional compound,
Tab.; 915
values for two wind speeds, various compounds,
Tab; 917
of water vapor; 896

Air–plant partition coefficient of compound
definition; 362
for calculating activity of compound in
plant; 363
deviation from equilibrium; 371
effect of seasonal temperature; 363
prediction from air–octanol partitioning,
Tab.; 362

**Air–polar surface partition constant of
compound**
as function of London dispersive energy,
Fig.; 72

**Air–pure liquid partition constant of com-
pound**
definition; 68
as function of London dispersive energy,
Fig.; 69
and saturation pressure; 68

Air–solid partitioning of compound
ab- or adsorption to smoke particles? Ill.
Ex.; 406

**Air–solid surface partitioning of com-
pound; 71**

Air–surface adsorption of compound
competition of water and organic sorbate; 392
notation of process; 394
schematic illustration, Fig.; 390
on soil at various relative humidities, Fig.; 395
survey of important sorption phenomena; 391

Air–surface partition coefficient of compound
definition; 394
calculation from molecular and surface
data; 397
calculation of sorption on vessel walls, Ill.
Ex.; 402
LFER for any surface and monopolar
sorbates; 401
LFER for apolar surfaces or for apolar
sorbates; 401
LFER using saturation pressure of pure
liquid; 396, 400
multiparameter LFER from sorbate and surface
data; 396

from sum of individual sorption interactions; 395
temperature dependence; 397

Air–water distribution ratio of acid or base
definition; 269
and air–water partition constant of compound; 269

Air–water exchange of compound
air-phase controlled; 893, 894
and motion of molecules; 892
boundary layer model; 906, 909, 912
boundary layer model, typical transport time,
Tab.; 932
direction of flux evaluation, Ill. Ex.; 890
effect of chemical reaction, Ill. Ex.; 938
effect of chemical reaction, scheme of,
Fig.; 936
effect of oil film thickness, Fig.; 930
film model; 906, 908, 912
film model, typical transport time, Tab.; 932
flux enhancement by reactive species; 936
flux equations; 889
impact of wind speed; 901
influence of chemical reactions; 931
influence of surface film; 929
influence of waves; 901
low wind speed, Fig.; 905
in one-box modelling of a pond, Ill. Ex.; 1056
physical processes at interface; 907
surface renewal model; 869, 906, 908, 912
surface renewal model, typical transport time,
Tab; 932
temperature dependence; 913
water-phase controlled; 893, 894, 900
wave regimes and wind speed; 903
wind pumping; 902
wind speed probability distribution; 903

Air–water exchange of compound in river; 921
bubble enhancement; 926
calculation for two flow regimes, Ill. Ex.; 1113
calculation for various roughnesses, Ill.
Ex.; 927
eddy models; 922, 923
flux of compound; 1111
large eddy model, definition; 924
of pesticide in the river Rhine, case study; 1137
small eddy model, definition; 923

Air–water exchange velocity of compound; 892
as function of Henry's constant, Fig.; 895
in air, reference compound and wind speed,
Tab.; 914
in water, reference compound and wind speed,
Tab.; 914
and shear velocity in rivers; 922
total; 915
total, for two wind speeds, various compounds,
Tab; 917
total, function of the Henry's law constant,
Fig.; 918
total, given wind speed and temperature, Ill.
Ex.; 918
total, influence of surfactant film; 929
total, properties of various compounds,
Tab.; 916
total, water as reference phase; 893

Air–water interface
   characterized by the Henry's law equilibrium; 837
   clouds in atmosphere; 889

Air–water partition constant; *See also The Henry's Law constant*
   derived from two connected partition constants; 215

Air–water partition constant of compound
   of acid or base; *See Air–water distribution ratio of acid or base*
   bond contribution data for estimation, Tab.; 206
   critical evaluation of data; 203
   data compilations, references to; 203
   definition with Henry's Law constant; 197
   determination by purge-and-trap method; 204
   estimation from bond contributions, Ill. Ex.; 207
   estimation from bond contributions LFER; 205
   estimation from molecular data by LFER; 205
   estimation from solubility and vapor pressure; 198
   experimental determination methods; 203
   as function of London dispersive energy, Fig.; 71
   for saltwater; 199
   schematic illustration of partitioning, Fig.; 186

Air–water partitioning of compound; 197
   aqueous concentration as function of time; 204
   effect of methanol concentration, Ill. Ex.; 202
   effect of solution composition; 199
   effect of solution composition, Ill. Ex.; 202
   of organic acids and bases; 269
   salt concentration effect, Ill. Ex.; 202
   schematic picture of processes, Fig.; 186
   temperature effect; 199
   and the Henry's Law constant; 197

Air–water transfer of compound
   in aqueous activity coefficient formulation; 148
   examples of; 182
   standard enthalpy change; 199

Alicyclic
   definition; 32

Aliphatic
   definition; 32

Alkane
   definition; 32

Alkene
   definition; 32

Alkyl group
   definition; 32

Amagat's Law; 85

Amount concentration of compound
   definition; 85

Amphiphilic compound
   definition; 417
   micelle and hemimicelle formation at surface, Fig.; 440

Anabolism
   definition; 697

Anion exchange capacity (AEC)
   definition; 423

Anoxic condition
   definition; 571, 694
   functional groups favoring biodegradation; 706
   microbial DDT degradation products, Ill. Ex.; 729
   selection criterion in biodegradation pathway; 694, 703

Anthropogenic chemical
   definition; 4

Antibonding electron orbital
   definition and symbolism in drawing structure; 613

Antoine equation; 105

Apolar compound
   definition; 62
   adsorbed on mineral surface, illustration, Fig.; 393
   adsorption on water surface of gasoline; 72
   lipid–water vs. octanol–water partitioning, Fig.; 342
   partition constant as funct. of vdW parameter, Fig; 69, 71, 72
   partitioning in organic solvent–water systems; 216
   protein–water vs. octanol–water partitioning, Fig; 341
   as surrogate for apolar living medium; 339
   in water, reordering of water molecules; 144

Apolar surface
   intermolecular forces with sorbent; 72

AQUASIM
   computer program for lake modeling, reference to; 1089

Aqueous activity coefficient at saturation
   calculation from solubility; 137, Ill. Ex.; 140, 156

Aqueous activity coefficient of compound
   and Setschenow constant; 160
   calculated versus experimental data, Fig.; 151
   calculation for seawater, Ill. Ex.; 164
   calculation from molecular data; 149, 150, Ill. Ex; 153
   concentration dependence; 141
   concentration dependence, determination of; 139
   concentration independency; 142
   in dilute salt solution, determination of; 160
   estimation from chemical structure data; 174
   experimental determination methods; 173
   LFER from molecular size; 146
   optimizing of calculation; 151
   in organic mixture–water partitioning; 236
   problems when calculation from simple LFER; 152
   of pure gas at saturation; 139
   of pure liquid at saturation; 137
   of pure liquid in own phase saturated with water; 136
   of pure solid at saturation; 138
   from saturation pressure and the Henry constant; 148
   values at infinite dilution and saturation, Tab.; 141

Aqueous activity coefficient of solute
   in acid–base reaction; 248
   approximate value; 247
   in saline water and in seawater; 248
   in transformation reaction; 463

Aqueous solubility of compound
   definition; 135
   definition for acids and bases; 268
   definition of total acid and total base; 268
   calculation for seawater, Ill. Ex.; 164
   calculation from molecular parameters; 150
   data compilations, reference to; 173
   experimental determination methods; 172
   LFER from molecular size; 146
   mole fraction expression; 136
   and Setschenow constant; 160

Aqueous solubility of pure gaseous compound
   and activity coefficient in water; 139
   calculation from excess enthalpy, Ill. Ex.; 156
   temperature dependence; 156

Aqueous solubility of pure liquid compound
   and activity coefficient in water; 137
   calculation from excess enthalpy, Ill. Ex.; 156
   calculation of, Ill. Ex.; 140
   as function of molar volume, Fig.; 147
   and molar volume, parameters of correlation, Tab.; 148
   schematic picture of relations, Fig.; 186
   temperature dependence; 155

Aqueous solubility of pure solid compound
   and activity coefficient in water; 138
   calculation from excess enthalpy, Ill. Ex; 156
   temperature dependence; 156

Aqueous solubility of subcooled liquid
   temperature dependence; 155

Aqueous solubility of superheated liquid
   and saturation pressure; 139
   temperature dependence, solubility minimum; 155

Aqueous solution of compound
   reference state and standard state, definition; 247

Aquifer; *See Groundwater*

Aromatic compound
   definition; 32
   examples, Fig.; 30

Aromatic system
   definition; 29
   delocalization of electrons; 29
   effect of size on light absorption spectrum; 621

Aromaticity
   definition; 30
   characterizing quality of dissolved organic matter; 316

Arrhenius equation; 478
   and transition state theory of elementary reaction; 480

Atmosphere
   compartments of, illustration, Fig.; 835
   water content; 889

Atmospheric lifetime
   parameters affecting it; 672

Atom/fragment contribution method
calculation for new compound, Ill. Ex.; 232
calculation for related compound, Ill. Ex.; 233
calculation of octanol–water partition constant; 229, 231
description of; 228
values of atom/fragment coefficients, Tab.; 230
values of atom/fragment correction factors, Tab.; 231

Atomic mass
and isotope; 15
unit, definition and value; 15

Atomic volume
values of characteristic volumes of elements; 149

Attenuation coefficient
nondimensional, as function of Peclet number, Fig.; 1166
nondimensional in groundwater; 1165
of light; *See Beam attenuation coefficient*

Avogadro's number
value; App. B; 15

Axial substituent in ring
definition; 28

B

Bacteria
average mass and carbon content of a cell; 745
building up a chemocline in lake; 837
phototrophic sulfide oxidizing; 836

Base
definition; 246
calculation of air–water distribution, Ill. Ex.; 269
effect of protonation on light absorption; 622
as leaving group in nucleophilic reaction; 491
Lewis; *See Lewis acid and base*
reaction with water; 249
sorption to humic acid as function of pH, Fig.; 324, 325
sorption to natural organic matter; 323

Base catalyzed hydrolysis rate constant
derivation from experimental data, Ill. Ex.; 515
effect of solution composition, Tab.; 482

Base catalyzed hydrolysis reaction
of acid derivative, general reaction mechanism; 513
of carbamates; 528
mechanism and scheme of carboxylic acid esters; 523
of (thio-)phosphoric acid esters; 538
reaction scheme of carbamates, elimination, Fig.; 530
reaction scheme of carbamates, Fig.; 531

Base catalyzed hydrolysis reaction rate
calculation from Hammett correlation, Ill. Ex.; 534
function of acidity constant, Brønsted plot, Fig.; 536
pH dependence of carboxylic acid esters, Fig; 514

Baseline toxicity of compound
correlation with accumulation in cell membrane; 375
from octanol–water partition constant LFER; 375

Basicity constant of compound; *See also Acidity constant of compound*
definition; 249

Bathochromic shift
definition; 620

Beam attenuation coefficient
definition; 616
as function of wavelength in natural waters, Fig.; 637
in seawater, Tab.; 1052
at various wavelengths in Greifensee, Tab.; 638

Beer-Lambert law; 615

Bernoulli coefficients
and normal distribution, Fig.; 783

Beta-elimination; *See Elimination reaction*

Bimolecular elimination reaction; *See Elimination reaction, second-order*

Bimolecular nucleophilic substitution; *See Nucleophilic substitution, second-order*

Bioaccumulation factor of compound
definition and illustration, Fig.; 345
calculated and measured values, Tab.; 355
calculation from concentration in water; 354
disequilibrium quantity in dynamic process; 349
from uptake and depuration mechanisms; 351

Bioaccumulation of compound in aquatic system
description by one-box model; 350
observed versus equilibrium partitioning; 349
parameters used for description; 345
a dynamic process, illustration, Fig.; 350

Bioaccumulation of compound in terrestrial system
and air–plant partitioning; 361
in atmospheric pollution monitoring; 361
temperature effect versus seasons; 361

Bioavailability of compound
definition; 735
effect of dissolved organic matter on; 314
fraction of in biotransformation rate law; 739
fractions of compound bioavailable and dissolved; 735

Bioconcentration factor of compound; *See also Bioaccumulation factor*
calculation for various biological matter, Ill. Ex; 320

Biodegradability of compound
fragment coefficients for estimation, Tab.; 705
prediction from simple rules; 706
prediction from structural groups; 702
probability of ease of biodegradation function; 704
readily degradable compound, definition; 702
structure-condition-mechanism, overview, Tab.; 703

Biofilm
definition; 736
diffusive transport of compound limited; 690
kinetics of compound uptake, literature to; 736

Biological material
components and their elemental composition, Tab.; 296

Biomagnification of compound
definition; 366
factor of, definition and illustration, Fig.; 345
in aquatic food chain and human fish consumption; 370
in aquatic food chain, values and scheme, Fig.; 367
change of activity and fugacity in food chain; 368
concentrations along terrestrial food chain, Tab.; 371
and hydrophobicity of compound; 370
temperature effect in terrestrial food chain; 371, 373
in terrestrial food chain; 370
values for various aquatic food chains, Fig.; 369

Biota–sediment bioaccumulation factor of compound
definition and illustration, Fig.; 345
for benthic organism, illustration, Fig.; 352
calculated and measured values, Tab.; 355
calculation from sediment organic carbon concentr.; 353
disequilibrium for organisms at interfaces; 351

Biota–soil bioaccumulation of compound; 365

Biotransformation processes
sequence of events, schematic illustration, Fig.; 690
thermodynamics vs. kinetics of reactions; 689

Biotransformation rate of compound
bioavailability as limiting process; 735
degradation in co-metabolism of methane, Ill. Ex.; 762
degradation time for substrate in pond, Ill. Ex.; 749
desorption of compound as rate-limiting process; 735
estimation for compound in groundwater, Ill. Ex.; 765
estimation of for Michaelis-Menten kinetics; 761
Monod kinetics and steady state in CSTR, Ill. Ex.; 747
possible rate limiting processes; 734
rate coefficient, definition; 739
rate law for fully available substrate; 738
uptake as rate-limiting process; 735

Biotransformation reaction
definition; 689
additions, 703, 730
carboxylation; 703, 731
hydration; 703, 734
hydrogen abstraction; 703
hydrolysis; 703, 706
by isolated cells; 690
limiting processes; 691

oxidation; 703, 715
reduction; 703, 721
structure-condition-mechanism, overview, Tab.; 703
thermodynamics of, calculation, Ill. Ex.; 692

Biotransformation reaction of compound
of pesticide in the river Rhine, case study; 1137

Bipolar compound
definition; 62
adsorbed on mineral surface, illustration, Fig.; 393
lipid–water vs. octanol–water partitioning, Fig.; 342
partition constant as funct. of vdW parameter, Fig; 69, 71
partitioning in organic solvent–water systems; 218
protein–water vs. octanol–water partitioning,Fig; 341
quantification with H-bond descriptors; 115
as surrogate for bipolar living medium; 339

Bipolar group of compound
effect on vapor pressure and boiling point; 114

Black carbon
fraction of in sediment; 304
schematic structure, Fig.; 297
as solid phase organic matter; 298

Black carbon–water partition coefficient
in dual-mode model of sorption; 304
estimation from LFER; 305

Boat conformation
picture, Fig.; 28

Boiling point of compound
definition of normal and standard; 102

Boiling point temperature of compound
enthalpy and entropy changes; 104, 120
estimation of, literature to; 120
illustration in phase diagram, Fig.; 101
and total external pressure; 102

Bond
covalent; *See Covalent bond*
double; *See Double bond*
polar; *See Polar bond*
single; *See Single bond*
triple; *See Triple bond*

Bond angle
examples of in molecules, Fig.; 25
rules for determination; 25

Bond contribution method
for estimation of air–water partition constant; 205
values for air–water partition constant, Tab.; 206

Bond enthalpy
definition; 19
effect on reaction rates of substitution; 504
energy of molecular motion; 22
for estimation of reaction enthalpies; 22
values and wavelength of corresponding light, Tab.; 615
values of single and multiple bonds, Tab.; 20

Bond length
definition; 19
values of single and multiple bonds, Tab.; 20

Bonding electron pair
electronic ground state and excited state; 613
symbolism in drawing structure, Fig.; 18

Boston Harbor
aqueous boundary thickness of SMSL, Tab.; 1074
sediments, polluted, Ill. Ex.; 859

Bottleneck boundary
definition; 837
around a spherical structure; 872
between different phases; 844
diffusivity profile across boundary, Fig.; 837
multilayer; 842
permeability of cell membrane; 691
simple; 839
simple noninterface; 841
simple, scheme of concentration profile, Fig.; 840
thickness; 840
two-layer; 842
two-layer, concentration profile, Fig.; 845
two-layer, scheme of concentration profile, Fig.; 843

Boundary; 835
definition; 837
bottleneck; 837
diffusive; 837
wall; 837

Bovine serum albumin–water partitioning; *See Protein-water partitioning*

Box model
one-box; *See One-box model*
two-box; *See Two-box model*

Breakthrough time
definition; 819
calculation for aquifer pollution case, Ill. Ex.; 847

Brønsted relationship; 535

Brunt-Väisälä frequency; *See Stability frequency*

Bulk density
of a solid–water system, definition; 287

C

Capacity factor of solid; *See Freundlich constant*

Carbohydrate
mass fraction in living organisms, Tab.; 337

Carboxylation reaction, microbial
of aromatic alcohols; 732
at anoxic conditions, with bicarbonate; 731
reaction sequence, Fig.; 733

Case study
accidental input of compound in Mississipi river; 1132
chemical pollution of the river Rhine; 1135

Catabolism
definition; 697

Catalyst
definition; 475

Catalyzed reaction; *See also Michaelis-Menten enzyme kinetics*
general mathematical formulation; 475
reaction rates depending on model, Fig.; 476

Cation exchange capacity (CEC)
definition; 423
of aluminosilicates' edges and faces; 425

Cavity formation
in absorption from gas phase; 67
and activity coefficient of solute; 81
in air–solid surface partitioning; 71
in apolar and bipolar solvents; 70

Cavity formation in water
for apolar solute; 144
for polar solute; 145

Cell envelope of bacteria
active transport of substrate; 738
boundary of transport of compound to enzyme system; 736
efflux pump for active transport out of cell; 738
gram-negative, schematic illustration, Fig.; 737
gram-positive, schematic illustration, Fig.; 737
porins for passive entrance of hydrophilic species; 738

Chair conformation
picture of, Fig.; 28

Chapman-Enskog theory
of molecular diffusivity; 801

Charged mineral surface
acid–base behavior of oxides; 419
adsorption of organic ion, equations; 428
anion and cation exchange capacities of clays; 423
calculation of charge concentration of oxides; 422
charge density; 419
schematic picture of water-side solution, Fig.; 419

Charged mineral–water partitoning of organic ion; 417
reaction formulation; 426
schematic illustration, Fig.; 390

Chemical actinometer
binary, for light intensity determination; 647
effect of vessel geometry; 648
light intensity measurements; 646

Chemical potential of aqueous solute
at infinite dilution reference state; 246, 463
species of acid–base reaction; 247

Chemical potential of compound
definition; 73
absolute value and change of; 77
entropy and excess free energy terms; 82
illustration of, Fig.; 74
and partial molar enthalpy and entropy; 74
relative and fugacity of compound; 75
at standard conditions; *See Standard chemical potential*
in two phases, at equilibrium; 85

Chemical potential of gaseous compound
and fugacity; 76
over pure liquid; 104
and partial pressure; 75

Chemical potential of pure liquid compound; 104
in organic phase, saturated with water; 136
and standard chemical potential of liquid; 80
in water phase, saturated with organic liquid; 136

Chemical reaction
definition; 459
effect on air–water exchange rate of compound; 931
from excited state of molecule, Fig.; 624
of nucleophile at polar bond; 491
with nucleophilic species; *See Nucleophilic substitution reaction*

Chemical structure properties for LFER of
acidity constant; *See Hammett correlation method*
air–water partition constant; *See Bond contribution method*
biodegradability of compound; *See Biodegradability of compound*
calculation of entropy of vaporization; 113
calculation of free energy change of transfer; 92
estimation of equilibrium partition constant; 91
hydroxyl radical reaction rate in atmosphere; *See Group rate constant; Group substituent factor*
octanol–water partition constant; *See Atom/ fragment contribution method*

Chemisorption
definition; 389, *441*
on charged mineral surface; 425

Chemocline
noninterface boundary, graph, Fig.; 836

Chemostat; 748

Chirality
definition; 26

Chromatography
for determination of partition constants; 226

Chromophore; *See also Unknown chromophore*
effect of double bonds on light absorption; 620

Cis-/trans- isomerism
definition; 27
at double bond, examples, Fig.; 27
in rings, examples, Fig.; 28

Clausius-Clapeyron equation; 105, 120
pure solid–vapor equilibrium; 123

Cloud
air-to-water volume ratio; 209
in atmosphere, air–water interface; 889
calculation of acid and base distribution, Ill. Ex; 269

Co-ion of ionic organic sorbate
in near surface water of charged mineral, Fig.; 419

Co-metabolism
definition; 697
of benzo[f]quinoline/quinoline; 752
biodegradation of TCE by methanotrophes, Ill. Ex.; 762
example, reaction pathway, Fig.; 698
limiting the overall substrate removal rate; 750
second-order rate law of removal of co-substrate; 753

Co-substrate
influence on overall growth rate; 743
specific growth rate; 742

Coagulation of particles
and particle settling velocity; 1064

Coenzyme
in microbially mediated oxidation reaction; 715
reagent with enzyme; 702

Collision rate model
of elementary bimolecular reaction; 478

Colloid; 1064
in porous media, illustration, Fig.; 1174
role in compound transport in porous media; 1174
sorption on; 1068
transport in groundwater; 1150

Colloid, organic; *See Dissolved organic matter*

Combustion-derived material
main components and elemental composition, Tab; 296

Competing ion of ionic organic sorbate
in near surface water of charged mineral, Fig.; 419

Competition quotient
definition; 589
function of one-electron reduction potential, Fig.; 587

Competitive inhibitor
definition and example; 697
in compound mixtures as oil spills; 698

Completely mixed reactor; 953

Completely water-miscible organic solvent; *See Cosolvent*

Compound; *See Organic compound*

Concentration change, spatial
along axis, diffusive flux direction Fig.; 787
diffusion at flat boundary, Fig.; 792

Concentration of compound
and activity, in equilibrium constant; 66
and activity, in sorption equilibrium; 281
effective; *See Effective concentration of compound*
estimation time course from field data, Ill. Ex.; 484
as function of time in first order reaction; 470
lethal; *See Lethal concentration*
at steady state in one-box model, constant input; 484
temporal change of; *See Temporal change of concentration of compound*

Concentration of compound in living organism
definition; 344
lipid normalized, definition; 345
lipid normalized, calculation, Ill. Ex.; 380

Concentration of compound in medium
definition; 344

Concentration of compound in organic phase
definition; 344

Concentration of gaseous compound
calculation from mixing ratio of volume; 673
calculation from partial pressure, Ill. Ex.; 109

Concentration of solute in aqueous solution; 247

Concentration of surface charge of oxide
definition; 421
calculation; 422
calculation example for environmental conditions; 421
typical values for environmental conditions; 423

Concentration of surface reactive sites of mineral
definition; *443*

Concentration patch of compound
dilution by dispersion in river; 1125
horizontal advection and diffusive growth, Fig.; 1031
horizontal mixing in a lake, Ill. Ex.; 1036
temporal evolution of concentration in river, Fig.; 1127

Concurrent flow technique
air–water partitioning; 204

Condensation; *See Evaporation*

Configuration of compound
definition; 28
of six-membered rings; 28

Conformation of compound
definition; 28
boat, chair, twist, pictures, Fig.; 28
standard entropy change of; 125

Conjugate acid/base pair
definitions; 246
base as leaving group and dissociation of acid; 491

Conjugated electrons; *See Delocalization of electrons*

Conjugated pi-bond
definition; 28
and delocalization; *See Delocalization of electrons*
effect on light absorption of compound; 620

Connectivity
of atoms in a structure, definition; 15

Consortium of microorganisms
definition and example in biodegradation; 695

Constitution of compound
definition; 15

Constitutive enzyme system
definition; 691

Contact area of sorbate with surface
definition; 395

Contamination
of olive oil by air pollution, Ill. Ex.; 196
of samples in the refrigerator, Ill. Ex.; 197

Continuous model; 999
application for compound in Greifensee; 1091
one-dimensional of lake; *See One-dimensional contiuous lake model*

Continuously stirred tank reactor(CSTR)
purifying waste water by co-metabolism, Ill. Ex.; 762
substrate concentration at steady state, Ill. Ex.; 747

Cosolvency power
definition; 170
effect on sorption partition coefficient; 311

Cosolvent in aqueous solution
in analytical chemistry; 165
calculation of mole fraction from volume fraction; 170
completely water-miscible, examples of; 166
data compilations, references to; 166
effect of molar volume of compound, Fig.; 167
effect of type of cosolvent, Fig.; 167
effect of volume fraction of cocolvent, Fig.; 167
effect on activity coefficient of compound; 166
effect on solubility of organic compound; 165
effect on sorption; 311
effect on water from molecular properties; 169
effects of, compared to dissolved organic matter; 314
and excess free energy of compound; 169
in organic mixture–water partitioning; 236
partially miscible organic solvent (PMOS); 170
quantitative models; 169
typical occurrence of; 165
values of solubility for selected cosolvents, Tab.; 168
in remediation techniques; 165

Counterion
definition; 419
of charged mineral in double layer, Fig.; 419

Coupled FOLIDE; *See First-order linear inhomogeneous differential eq.*

Covalent bond
definition; 17
polarity and reactivity; 20
symbolism in drawing structure, Fig.; 18

Critical body burden of compound
definition; 378

Critical micelle concentration
definition; 439

Critical point of compound
definition; 103

D

Damköhler number
definition; 1011
alternative definitions; 1014
and vertical mixing in lake; 1082
interpretation of; 1013

Darcy-Weisbach equation of stationary uniform flow; 1106

Darcy's law; 1153
in groundwater system, Ill. Ex.; 1156

Dead-end metabolite
definition and example; 700

Debye energy; *See Dipole:induced dipole force*

Decadic absorption coefficient of medium
definition; 615

Decadic molar absorption coefficient of compound
definition; 615
calculation from absorbance measurement, Ill. Ex.; 616

Dehydrohalogenation reaction; 556

Delocalization of electrons
aromatic systems; 29
effects of substituent on acidity constant; 258
effect on absorption spectrum of compound; 620
effect on bathochromic shift; 620
symbolism in drawing structure; 29

Delocalized chemical bond
definition; 28
relative energy of; 28

Denitrification
calculation of standard potential, Ill. Ex.; 565
reduction reaction equation, Ill. Ex.; 565
step in organic compound oxidation, Fig.; 570
at suboxic conditions; 571

Dense non aqueous phase liquid (DNAPL); *See Organic mixture–water partitioning of Component*

Derepression of enzymatic activity
definition; 691

Derepression of enzyme activity
and lag period of compound degradation; 701

Desorption kinetics
from contaminated sediment, Ill. Ex.; 881

Desorption of compound
rate law of transfer into a bioavailable species; 735

Diagenesis
of natural organic matter; 294

Diastereoisomerism
definition; 26
cis- and trans- isomers; 27
of double bonds and rings; 27
multiple chirality centers; 26

Diagenic material
components and their elemental composition, Tab.; 296

Diffuse attenuation coefficient of medium
definition; 628
calculation from beam attenuation coefficient; 629

Diffuse double layer; *See Near surface water layer*

Diffusion; 779

Diffusion volume, molar of compound
values of elemental volume contributions, Tab.; 804

Diffusive boundary; 866
definition; 837
between different phases; 869
diffusivity profile across boundary, Fig.; 837
exchange velocity; 871
exposure time; 871
mass flux; 870

Diffusive transport of compound
concentration profile and flux, Fig.; 1007
diffusion distance, Einstein-Smoluchowsky law; 788
distance and time, Einstein-Smoluchowsky law, Tab.; 826
distance of; 1013
at flat boundaries; 791
in gas-filled porous media; 817
horizontal; *See Horizontal diffusion*
in biofilms and flocs; 690
into and from a spherical particle; 795
into sphere, concentration progress, Fig.; 795
Knudsen effect; 817
length scales; 827
in liquid-filled porous media; 817
mass exchange and nondimensional time, Fig.; 797
penetration distance; 793, 795
in porous media; 815, 818
radial model; 877
random walk model, first Fick's law; 787
reference system; 798
Renkin effect; 817
of sorbing compound; 818
time of; 1012
time, relative to advection time; 1012
turbulent; *See Turbulent diffusion*
in unsaturated zone; 818
unsaturated zone, estimation of flux, Ill. Ex.; 847

Diffusive transport regime
evaluation and flow chart, Fig.; 1015

Diffusive transport/reaction regime
evaluation and flow chart, Fig.; 1015

Diffusive/advective transport /reaction equation; 1007
critical distance, definition; 1012
critical distance, values, Tab.; 1013
solution for steady state concentration; 1009, 1010

Diffusive/advective transport equation
in porous media; 1171

Diffusive/advective transport regime
evaluation and flow chart, Fig.; 1015

Diffusivity; *See Molecular diffusivity*
effective; *See Effective diffusivity*

Dihalo-elimination reaction; 557

Dimensionless Henry constant; *See Air–water partition constant*

Dipolarity; *See Polarizability of compound*

Dipole
definition; 21

Dipole moment
of a molecule, definition; 21

Dipole:dipole force
definition; 61
effect of solvent on air–solvent partition
const.; 187

Dipole:induced dipole force
definition; 60
effect of solvent on air–solvent partition
const.; 187

Direct photolysis
definition; 613
effect of particles and surfaces; 649
reaction pathways of selected compounds; 625
solute and solution composition effects; 624
temperature effect; 626

Direct photolysis first-order rate constant
definition; 641
calculation, near-surface and bulk water, Ill.
Ex.; 643
determination of; 646

Dispersion; 827

Dispersion coefficient
calculation for acid in GWS, Ill.; 1159
in layer of groundwater system, Tab.; 1150
limited flow conditions; 1040
longitudinal in groundwater; 1155
unlimited flow conditions; 1041

Dispersive interactions
effect of solvent on air–solvent partition
const.; 187

Dispersive intermolecular force; *See London*
dispersive force

Dispersive transport; 1038
at edge of pollutant front; 867
and the Fickian laws; 1039
in aquifer; 1149
in porous media, schematic illustration,
Fig.; 1149
longitudinal in groundwater; 1155
longitudinal, in layer of groundwater system,
Tab.; 1150
origin of; 1033
in rivers; 867
schematic illustration to simple model,
Fig.; 1039

Dispersive transport in river; 1122
calculation of coefficient in river G, Ill.
Ex.; 1124
dilution of concentration of compound
patch; 1125
flow velocities, lateral and vertical, Fig.; 1122
initial state; 1126
of pesticide in the river Rhine, case study; 1139

Dissociation constant of water
definition; 252
values at various temperatures, Tab.; App.
D2; 252

Dissociation enthalpy
of a bond, definition; 19

Dissociative electron transfer reaction
definition; 581

Dissolved organic carbon–water distribution
coeff
definition; 315
calculated versus measured values, Fig.; 317
determination methods; 315
effect of inorganic salts; 317
effect of pH; 318
influence of DOM structure properties; 316
LFER for compound set and various humic
acids, Fig; 319
LFER for various compounds and one humic
acid, Fig; 319
LFER using octanol–water partition constant; 318
and solid–water distribution coefficient; 300
temperature dependence; 318

Dissolved organic carbon–water distribution
coeff.
estimation in bioaccumulation problem, Ill.
Ex.; 346

Dissolved organic matter (DOM); *See also*
Natural organic matter (NOM)
definition; 314
characterization; 314
characterization by sorption affinity of
pyrene; 316
characterization criteria; 316
effect of, on dissolved compound; 314
effect on beam attenuation coefficient, Fig.; 637
effect on bioavailability of compound; 314
effect on reduction rate of compound; 586
effect on reduction rate of compound,
Fig; 581, 583
effect on singlet oxygen concentration,
Fig.; 667
effect on solid–water distribution coefficient; 299
excited triplet DOM (3DOM*); *See Photosensi-*
tized reaction
quencher in indirect photolysis; 660
role in indirect photolysis; *See Unknown*
chromophore
size range of; 314

Distribution function of light paths
definition; 629
near-surface conditions; *See Near-surface*
distribution function of light paths
value, average for nonturbid water; 639
values for various media and wavelengths; 629

Distribution ratio
definition of the term; 90

Double bond; *See also Pi-bond*
definition; 18
and dia-stereoisomerism; 27
lengths and enthalpies, Tab.; 20
picture of atomic orbital, Fig.; 27
symbolism in formula; 19
type of the bond; 27
value of bathochromic shift; 620

Dual-mode model of sorption isotherm; 304

Dye
light absorption wavelength and color; 621

Dynamic method
for measuring partition constants; 204

Dynamic variable in FOLIDE
definition; 471
as function of time, graph; 472

Dynamic viscosity of water
temperature dependence; App. B; 913

E

Eddy diffusion; *See Turbulent diffusion*

Effective concentration of compound
in baseline toxicity, definition; 375

Effective diffusivity; 818
estimation in gas phase, unsaturated zone, Ill.
Ex; 847

Effective mean flow
in groundwater; 1152
in layer of groundwater system, Tab.; 1150

Effective porosity
in groundwater; 1152

Efflux pump
definition; 738

Eigenfunctions; 992

Eigenvalue; 992
in advection/diffusion/reaction equation; 1010
approximation of; 1011
linear n-dimensional systems; 991
nondimensional, Dahmköhler and Peclet
numbers; 1013
size and shape of concentration profile; 1012

Einstein
amount of photons, definition of unit; 614

Einstein-Smoluchowsky law
diffusion distance of molecules; 788
transport time by diffusion; 1012

Electrochemical potential difference of reaction
definition; 561
for reduction half reaction against SHE; *See*
Reduction potential of half-reaction
sign convention; 561

Electromagnetic radiation; *See Light*

Electron acceptor/Electron donor; *See also H-*
acceptor/H-donor
definition; 62

Electron donor–acceptor (EDA) force; *See also*
H-donor–acceptor (HDA) force
definition; 62

Electron donor-acceptor (EDA) force
enthalpy of adsorption on mineral surface; 412
specific adsorption on mineral surface; 412

Electron in aqueous solution
standard chemical potential of, definition; 562

Electron pair
and covalent bond; 17
shared; *See Bonding electron pair*
unshared; *See Nonbonding electron pair*

Electron sharing
in covalent bond; 17

Electron transfer mediator
as catalyst in environmental redox processes; 559
DOM in electron transfer of reduction; 586
schematic illustration, Fig; 557

Electron transfer reaction; *See Dissociative electron transfer; One-electron transfer reaction; Redox reaction*

Electronegativity
definition; 19
Pauling's, definition; 20
of atoms in a bond and bond polarity; 491
and bond polarity; 19
and oxidation state; 23
values for elements, Tab.; 21

Electronic absorption spectrum of compound
from absorbance measurement, Fig.; 616
effects of aqueous or organic solvents; 617
examples for aromatic compounds, Fig.; 622
examples of aromatic compounds, Fig.; 618, 619, 620

Electrophilic species
definition; 491
concentration relative to nucleophilic species; 491
hard and soft Lewis acids; 500
photooxidant reactions with compound; 668
reaction at polar bond; 491

Electrostatic attraction
in covalent bond; 17

Element
as thermodynamic reference state; 78
characteristic natural isotope mixture; 15
rules of binding; 15

Elemental composition
of a compound, definition; 14

Elimination reaction of compound
of acid ester to acid and olefin; 526
of acid ester to alcohol and ketene intermediate; 526
catalyzed by metal species in aqueous solution; 543
in hydrolysis of carbamates, reaction scheme, Fig.; 531
in hydrolysis of (thio-)phosphoric acid ester; 540
kinetic data to nonreductive elimination, Tab.; 508
role of leaving group of compound; 510

Elimination reaction, first-order (E1)
definition; 511
competing reactions; 511
effect of compound structure on reaction rate; 511

Elimination reaction, second-order (E2)
competing reactions; 511
effect of compound structure on reaction rate; 510
kinetic data for selected compounds, Tab.; 508
standard free energy pathway, Fig.; 510

Enantiomorphism; 26
in microbially mediated reactions; 697

Endothermic reaction
definition; 23

Energy
of a bond; *See Bond enthalpy*
of light, single photon and per mole of photons; 614

Energy state of molecule
electronic ground state and electron orbitals; 613
electronically excited; *See Excited energy state*
temperature effect on different states; 613

Enol/keto-forms of functional group; 715
in microbial carboxylation reactions; 732

Enthalpy
of a bond; *See Bond enthalpy*
contribution to excess free energy; 82
and intermolecular energies; 56

Enthalpy change; *See Standard enthalpy change*

Entropy
contribution to excess free energy; 82
and freedom of molecular motion; 56

Environmental behavior of organic compound
scheme of, Fig.; 8

Enzyme
definition; 696
effectiveness in lowering reaction rate; 696
nonspecific in biodegradation reactions; 698
specific; 697

Enzyme-catalyzed reaction; *See Michaelis-Menten enzyme kinetics*

Epilimnion
definition; 836

Equatorial substituent
in ring, definition; 28

Equilibrium concentration of compound
from kinetic reaction formulation; 473

Equilibrium constant of reaction
definition from a kinetic approach; 473
definition from thermodynamics; 465
of acid–base reaction in aqueous solution; *See Acidity constant*
calculation for a given reaction system, Ill. Ex.; 467
and standard free energy change of reaction; 248, 465
temperature dependence; App. D1; 252, 465

Equilibrium partition constant of compound
definition of the term; 90
definition, thermodynamic; 66
on amount concentration basis; 86
concentration and activity; 66
estimation from LFER; 89
estimation from structural group contributions; 91
LFER for compound in different two-phase systems; 90
LFER for different two-phase systems, Tab.; 91
and molar volumes of phases; 86
on mole fraction basis; 85
sediment–water; 853
and standard free energy change; 66

temperature dependence; 87
temperature dependence; App. D1, Tab.; 88
values from excess free energy, Tab.; 87

Equilibrium partitioning of compound
at phase boundary, selection of reference phase; 844
at phase boundary; 844
pressure effect; 86
temperature effect; 86

Equivalent sand-grain diameter; 926

Eulerian reference system of water flow
definition; 1105

Evaporation of compound
and condensation at equilibrium; 99
value of entropy change of position; 125

Evaporation rate of compound
pure liquid, calculation for quiet air, Ill. Ex.; 898
water as test compound of transfer velocity in air; 896

Excess enthalpy of compound
and intermolecular forces; 83
as gas and in solvents, values, Tab.; 83

Excess enthalpy of compound in organic solvent
and compound properties; 216
organic solvent to air transfer of compound; 185
water to organic solvent transfer of compound; 216

Excess enthalpy of compound in water
for apolar and polar solute; 144
calculation from solubility data, Ill. Ex.; 156
determination for diluted and saturated solution; 142
magnitude and compound properties; 216
and intermolecular forces; 142
in water to air transfer of compound; 199
and sorption to dissolved organic matter; 318
and sorption to solid phase organic matter; 310
values for diluted and saturated solution, Tab.; 143

Excess enthalpy of compound sorbed
to dissolved organic matter; 318
to solid phase organic matter; 310

Excess entropy of compound
and intermolecular forces; 83
as gas and in solvents, values, Tab.; 83

Excess entropy of compound in water
for apolar and polar solute; 145
determination for diluted and saturated solution; 142
values for diluted and saturated solution, Tab.; 143

Excess free energy of compound
definition; 82
and activity coefficient; 82
and chemical potential in solution; 82
excess enthalpy and excess entropy; 82
as gas, and pure liquid state as reference; 104
as gas, and standard free energy of vaporization; 104
as solid and free energy of fusion; 107
values as gas and in solvents, Tab.; 83

Excess free energy of compound in water; *See also Aqueous activity coefficient*
calculation at saturation, Ill. Ex.; 140
calculation from molecular data, Ill. Ex.; 153
in cosolvent mixture; 169
dilute and saturated condition, difference; 142
enthalpic and entropic parts; 84, 142
function of compound characteristics; 146
and mole fraction at saturation; 137
values at infinite dilution and saturation, Tab.; 141, 143

Exchange; *See Transfer*

Excited energy state of molecule
bonding, (non-) and antibonding electron orbitals; 613
chemical reactions to product, Fig.; 624
physical processes to ground state, Fig.; 624

Exothermic reaction
definition; 23

Exponential growth of cell population
as function of time, graph, Fig.; 740
degradation time for substrate in pond, Ill. Ex.; 749
rate law in growth limited case; 740
time of population doubling; 740

External force
definition; 950

External variable
definition; 483

Extrathermodynamic functions; *See Linear free energy relationship (LFER)*

F

Faraday constant
definition and value; 561

Fecal pellet; 1064

Fick's law
first law; 787
second law; 788
second law in three dimensions; 791

First-order linear homogeneous differential eq.
definition and solution; 471

First-order linear inhomogeneous differential eq.
arbitrary input variation; 964
coupled (coupled FOLIDE); 976
coupled, solution of, in sediment–water system; 1078
coupled, solution of two coupled FOLIDEs; 977
definition; 471
exponential perturbation; 963
general form; 956
periodic perturbation; 964
periodic perturbations, response to, Fig.; 966
temporal change of concentration in one-box model; 483
time-variable coefficients; 961

First-order rate constant of reaction
definition; 470
direct photolysis reaction; 641
indirect photolysis of phenyl ureas, Tab.; 670

First-order rate law
of particle removal, model of; 1063, 1064
removal of compound by particle settling; 1063

First-order rate law of reaction
definition; 469
bioaccumulation of compound; 350
of biodegradation of fully available substrate; 738
desorption of bio-unavailabe species; 735
including back reaction; 473
in Michaelis-Menten kinetics case; 751
in microbial population growth limited case; 740
unimolecular nucleophilic substitution reaction; 497
in uptake of bioavailable species; 735

First-order reaction
concentration of reactant vs. time, plot, Fig.; 470
elimination of substituent; *See Elimination reaction, first order*
Michaelis-Menten enzyme kinetics; 476
nucleophilic substitution; *See Nucleophilic substitution reaction, first-order*

Fishtine equation and factor; 113

Floc
definition; 736
diffusive transport of compound limited; 690

Flushing rate
in one-box modeling of a pond, Ill. Ex.; 1056

Flushing rate of box
definition; 483

Flux of compound
by advective transport, one-dimensional; 1007
by advective transport, three-dimensional; 1020
by air–water exchange; 889, 893
by diffusive transport, one-dimensional; 1007
by dispersive transport in porous media; 1155
relative enhancement due to advection, Ill. Ex.; 1041
by sediment–water exchange in river; 1116
by sediment–water exchange, models; 1071
total transport, one-dimensional; 1007
vertical mixing, continuous model; 1083

Flux of particle mass
definition; 1062

Fog
air-to-water volume ratio; 274
water droplets as sorption surface; *See Air–surface adsorption of compound*

FOLIDE; *See First-order linear inhomogeneous differential eq.*

Fraction in acid form
definition; 253
calculation for cloud water, Ill. Ex.; 269
calculation for different pHs, Ill. Ex.; 254
as function of pH, Fig.; 253
of multiprotic acid, function of pH; 256
and total aqueous concentration of compound; 268
values at pH 7, Tab.; 250

Fraction in base form
calculation for cloud water, Ill. Ex.; 269
calculation for different pHs, Ill. Ex.; 254
of multiprotic base, function of pH; 256
in neutral base form, values at pH 7, Tab.; 251

Fraction of black carbon
definition; 304
in sediment; 304

Fraction of compound
calculation for two- and multi-phase systems; 93

Fraction of compound in air
calculation of exchange from soil, Ill. Ex.; 404

Fraction of compound in water
calculation for various organic matter, Ill. Ex.; 320
in groundwater; 288
in lakes; 287
from solid–water distribution and phase ratio; 287
in two phase solid–water system; 286

Fraction of compound on solid
in two phase solid–water system; 287

Fraction of light absorbed by compound
definition; 630

Fraction of organic carbon on solid
definition; 292
relation to fraction of organic matter; 292

Fraction of organic matter on solid
relation to fraction of organic carbon; 292

Fraction of organic phase
definition; 343

Fraction of radiation absorbed
in Lambert's law; 615

Free energy change; *See Standard free energy change*

Free energy change of reaction
definition; 464
calculation for redox reaction, Ill. Ex.; 577
calculation from reactant activities, Ill. Ex.; 465
and electrochemical potential difference; 561
at equilibrium; 248, 465
and Nernst equation; 562
proton transfer in aqueous solution; 248
redox reaction in aqueous solution, pH 7; 561

Free energy of a system
definition; 73

Freezing; *See Fusion*

Frequency factor; *See Preexponential factor of Arrhenius equation*

Frequency of electromagnetic radiation
definition; 614

Freundlich constant of compound
definition; 281
deduction from experimental data; 281
dimension of; 281
and solid–water distribution coefficient; 283

Freundlich exponent
definition; 281
deduction from experimental data; 281
thermodynamic interpretation; 281

reundlich isotherm of compound
definition; 281
in dual-mode model of sorption; 304
and Freundlich exponent, schematic plot,
Fig.; 281

riction factor in river
in Darcy-Weisbach equation; 1106

riction velocity
in rivers, definition; 921
in rivers, estimation of; 924

ricton factor in river
typical value range; 1106

roude number
definition; 926

ugacity of compound
definition; 75
in air, calculation of, Ill. Ex; 363
calculation for plant, Ill. Ex; 363
calculation for water, sediment, organism, Ill.
Ex; 359
calculation from actual and saturation fugacities; 357
calculation of air–plant disequilibrium, Ill.
Ex.; 363
change in biomagnification along food
chain; 368
coefficient, definition; 76
coefficient at atmospheric pressure; 76
in ideal and nonideal phases, illustration,
Fig.; 79
in ideal and real solution, definitions; 183
inside/outside a cell and compound uptake
rate; 735
in liquid mixture; 78
as measure of disequilibrium; 356
temperature change effect; 358
values for air, plant, cow and human milk,
Fig.; 372
values for arctic plant, caribou and wolf ,
Fig.; 373
values for water, sediment, organism, Fig.; 358

Fugacity of gaseous compound
and chemical potential; 76
and partial pressure; 77
and pressure; 76

Fugacity of pure condensed compound
and saturation pressure; 78

Fulvic acid
definition; 295
components of and their elemental composition,
Tab; 296
effect of pH on distribution coefficient; 318
effect on aqueous fraction of compound, Ill.
Ex.; 320
light absorption and color of natural waters; 621
sorption of compounds, from LFER, Fig.; 319

Functional group of compound
definition; 31

G

Gas; *See also Air*
as a phase; *See Air*
ideal; *See Ideal gas*
nonideal; *See Real gas*

Gas chromatography
determination of vapor pressure of pure
compound; 118

Gas stripping technique
air–water partitioning; 204

Gaseous compound; *See Pure gaseous
compound*

Gasoline
composition and other data; 243

Gauss' theorem
in groundwater system; 1152
mathematical relation of space and time; 789
relation of flux and temporal concentration
change; 1007

Generator column method
in determination of octanol–water partitioning; 226

Geometrical isomerism; *See Cis-/trans-
isomerism*

Gibbs energy change; *See Standard free energy
change*

Gibbs–Helmholtz equation; 88

Gradient flux law
definition; 785
physical processes, equations and units,
Tab.; 786

Gram-negative bacteria
definition; 737
cell envelope, schematic illustration, Fig.; 737

Gram-positive bacteria
definition; 737
cell envelope, schematic illustration, Fig.; 737

Greifensee
application of continuous model MASAS; 1091
characteristic physical data; 957
estimation of degradation of NTA, Ill. Ex.; 960
one-box model, analysis of dynamics of
PCE; 959
second-order degradation of NTA, Ill. Ex.; 973
spectral light data of epilimnion; 638
stratified, two-box model, steady state of PCE,
Il; 987
stratified/mixed cycling response times, Ill.
Ex.; 996
time-dependent input of PCE and response; 965
verification of degradation of NTA, Ill. Ex.; 966

Groundwater
effect of transformation processes; 1172
hydraulics; 1148
nondimensional coordinates; 1161
pulse input of compound; 1161, 1171
reference distance; 1160
reference time; 1160
saturated zone; 1149
sinusoidal input of compound; 1164, 1172
step input of compound; 1162, 1172

step input of compound, illustration, Fig.; 1163
time dependent input; 1160
transport by mean flow; 1152
transport characteristics; 1148
values of nondimensional time, step input,
Tab.; 1165

Groundwater system S (GWS); 1150
Darcy's law, Ill. Ex.; 1156
hydraulics, schematic illustration, Fig.; 1151
infiltration from river, Ill. Ex.; 1166, 1168
infiltration of benzylchloride from river, Ill.
Ex; 1177
Peclet number, Ill. Ex.; 1159
pump regimes, Ill. Ex.; 1156, 1168
sinusoidal input, Ill. Ex.; 1168
transport of tetrachloroethene, Ill. Ex.; 1176

Group rate constant
H abstraction, differently bonded H, Tab.; 676
HO° addition, different structural units,
Tab.; 678
HO° interaction, heteroatom containing groups,
Tab; 679

Group substituent factor
electrophilic aromatic substituents, Tab.; 679
H abstraction, various substituent groups,
Tab.; 676
HO° addition, various substituent groups,
Tab.; 678

H

H-acceptor/H-donor; *See also Electron
acceptor/Electron donor*
definition; 62

H-acceptor/H-donor descriptor of compound
definition; 115
in air–surface partition coefficient LFER; 396
values for selected compounds, Tab.; 116

H-acceptor/H-donor descriptor of surface
in air–surface partition coefficient LFER; 396
as function of relative humidity of air, Fig.; 399
values for selected liquids and solids, Tab; 398

H-acceptor/H-donor interaction
in air–surface partition coefficient; 395
in aqueous activity coefficient description; 149
effect of solvent on air–solvent partition
const.; 187
of mineral surface; 392
of mineral surface, Fig.; 393
in organic solvent–water partition constant; 218

H-donor–acceptor (HDA) force; *See also
Electron donor–acceptor (EDA) force*
definition; 62

Hagen-Poiseuille law; 1154

Half-life of compound
definition; 18
in atmosphere, indirect photolysis; 674
estimation of in atmosphere, Ill. Ex.; 680
first-order reaction; 470
hydrolysis reaction; *See Hydrolysis half-life*
indirect photolysis with hydroxyl radical, Ill.
Ex; 665
indirect photolysis with singlet oxygen; 669

Half-reaction of redox process
definition of notation; 560

Hammett constants; *See Sigma constant of substituent; Susceptibility factor of acid*

Hammett correlation method
and acidity constant of a substituted acid; 263
direct resonance of para substituents; 264
direct resonance of para substituents, Fig.; 259
hydrolysis constant of benzoic acid ethyl ester; 532
photosensitized rate constants of phenyl ureas; 671
sigma constant of substituent, definition; 261
for substituted aromatic acids, plot, Fig.; 264
susceptibility factor of acid, definition; 263
thermodynamic functions of; 261
values of sigma constants of substituents, Tab.; 263, 266, 670
values of susceptibility factors of aromatic acids; 266

Hammett equation
acidity constant; 263
reaction rate constant; 532

Hayduk and Laudie relation
of molecular diffusivity in water; 812

Hemimicelle formation
at mineral surface, mechanism, Fig.; 440

Henry's law; 183
original and derived formulations; 184

Henry's law constant
between air and seawater, Ill. Ex.; 890
critical value of single phase control; 893
temperature dependence, Ill. Ex.; 890

Henry's law constant of compound in a phase
estimation from partition constants; 357
and fugacity of compound in phase; 357
temperature effect; 357
water as solvent; *See The Henry's Law constant*

Heteroaromatic ring
examples, Fig.; 30
resonance with nonbonding electrons; 30

Heteroatom
definition; 14, 31

Hexadecane–water partition constant
from LFER of octane–water partition constant; 219

Horizontal diffusive transport of compound
and horizontal advection, patch developement, Fig.; 1031
and shear diffusion model; 1033
turbulence theory; 1030

Horizontal diffusivity
scale dependence; 1031

Humic acid
definition; 295
components of and their elemental composition, Tab; 296
effect on aqueous fraction of compound, Ill. Ex.; 320
light absorption and color of natural waters; 621
schematic structure, Fig.; 297

sorption of compounds on, from LFER, Fig.; 319

Humic substance
definition; 294

Humin
definition; 295
elemental composition, Tab.; 296

Humus
properties of its surface, Tab.; 424

Hydration reaction, microbial
addition of water at double bond; 734

Hydraulic conductivity
of groundwater system; 1153
of groundwater system, Tab.; 1150
solid matrix dependence, illustration, Fig.; 1154

Hydraulic gradient
definition; 1153

Hydraulic radius
definition; 925, 1106

Hydrogen bond
definition; 21
intramolecular; 22
intramolecular, effect on acidity constant; 260
intramolecular, illustration, Fig.; 260
symbolism in drawing structure formula; 22

Hydrolysis half-life of compound
definition; 513
calculation for given pH and temperature, Ill. Ex.; 518
data for (thio-)phosphoric acid esters, Tab.; 537
data for carbamates, Tab.; 529
data for carboxylic acid amides, Tab.; 527
data for carboxylic acid esters, Tab.; 520
as function of pH, carboxylic acid esters, Fig.; 514
as function of pH, thiophosporic acid esters, Fig.; 540

Hydrolysis rate constant, pseudo-first-order
calculation from laboratory data, Ill. Ex.; 515
dependence of pH, schematic diagram, Fig.; 515
sum of neutral, acid- and base catalyzed rates; 514

Hydrolysis reaction
definition; 491
acid catalyzed; *See Acid catalyzed hydrolysis*
base catalyzed; *See Base catalyzed hydrolysis*
calculation of thermodynamics, Ill. Ex.; 493
of carbamates; 528
of carbamates, Tab.; 529
of carboxylic acid amides; 527
of carboxylic acid amides, Tab.; 527
of carboxylic acid esters, Tab.; 520
competing reactions; 511
effect of aqueous metal species; 540
of halogenated compounds, Tab; 505
of (thio-)phosphoric acid derivatives; 539
of (thio-)phosphoric acid ester, Tab; 537
reaction product characteristics; 493
thermodynamics at environmental conditions; 493
water as a nucleophilic reactant; 491

Hydrolysis reaction, microbial
alkyl halide hydrolysis with glutathione; 708
and abiotic hydrolysis reaction of compound; 708
degradation of alkyl halides in cell free extract; 759
enzyme characteristics; 706
ester hydrolysis and analogy to fatty acid esters; 710
ester hydrolysis, effect of pH; 711
ester hydrolysis, schematic reaction sequence, Fig; 712, 713
initial step in degradation of compound; 706
initial step in degradation reactions; 702
mechanism of nucleophilic attack by water; 708
nucleophilic attack by glutathione, mechanism; 709
products from an alkyl halide, Ill. Ex.; 710
rates of hydrolysis of families of esters, Fig.; 760
selected substrates and hydrolysis products, Tab.; 707
typical nucleophilic species in initial attack; 708

Hydrolysis reaction of compound
in one-box modeling of a pond, Ill. Ex.; 1056

Hydrophobic compound
definition; 33
correct use of term; 72

Hydrosphere; 889

Hydroxyl radical
characteristics as reactant; 665
in atmosphere, peak concentration and variations; 673
in atmosphere, tropospheric source; 673
in water, calculation of production rate, Ill. Ex.; 665
in water, reaction mechanisms with compound; 664
in water, steady state concentration Ill. Ex.; 662, 665

Hydroxyl radical rate constant of reaction
correlation of aqueous and atmospheric; 675
estimation for various compounds in air, Ill. Ex.; 680
estimation from structure-reactivity relations; 675
temperature dependence in atmosphere; 675
values for compounds in atmosphere, Tab.; 674
values for compounds in water, Tab.; 664

Hypolimnion
definition; 836

Hypsochromic shift
definition; 623

I

Ice surface; *See Air–surface adsorption of compound*

Ideal gas; *See also Air*
fugacity of, illustration, Fig.; 79
molecular theory and diffusivity; 798
molecular theory and mean velocity of molecules; 895

and pressure of; 75
and saturation pressure of compound; 99

Ideal mixture
definition; 183
and Raoult's Law; 183

Ideal solution
definition; 183
fugacity of compound, illustration, Fig.; 79
and Henry's Law; 183

Indirect photolysis
definition; 656
in atmosphere, with hydroxyl radical; 672
kinetics in water; 660
reaction schemes, illustration, Fig.; 659
wavelength dependence in water; 658

Inducible enzyme; 701

Induction of enzymatic system
definition; 691

Inductive effect of substituent on acidity const.
description; 257
substituents with positive or negative effect; 257

Inductive effect of substituent on reaction rate
stabilizing/destabilizing activated complex; 532

Inertial subrange of turbulence; 1031

Infinite dilution activity coefficient; *See Activity coefficient*

Infinite dilution state of compound
as reference state; 78, 246
as reference state in transformation reactions; 463

Inhomogeneous term in FOLIDE; *See Input term in FOLIDE*

Inner-sphere complex
definition; 389

Inner-sphere mechanism
of transition state in one-electron transfer; 581

Inorganic salts in aqueous solution
effect on solubility of compound; *See Setschenow constant of compound*

Inorganic surface; *See Mineral surface*

Input term in FOLIDE
definition; 471
zero, constant or time dependent; 471

Input variable
definition; 950

Integrated surface area of river
definition; 1111
as a rivers' longitudinal coordinate, Fig.; 1112
change of compound concentration along river; 1112

Interface
definition and examples; 835
air–water interface and the Henry's law; 837

Intermolecular force; 60
and activity coefficient of solute; 81
aqueous activity coefficient from LFER; 149
and classification of compounds; 62
Debye energy; 60
determining free energy of vaporization; 120
dipole:dipole force, definition; 61

dipole:induced dipole force, definition; 60
and enthalpy of vaporization; 110
enthalpy and entropy; 56
excess enthalpy and excess entropy; 83
and excess enthalpy in water; 142
illustrations of; 61
London dispersive force, definition; 60
polar force, definition; 62
in pure liquid; 68
and standard free energy change of phase transfer; 67
van der Waals (vdW) force, definition; 60
and vapor pressure of pure liquid; 110
and vapor pressure of pure compound; 114

Internal conversion of compound
definition; 623

Internal energy
definition; 73

Internal process
definition; 950

Interpore dispersion
in porous media; 1155

Intrapore dispersion
in porous media; 1155

Intrinsic acidity constant of oxide surface
definition; 420

Ionic character
of a bond and polarity; 20

Ionic organic species
adsorption to charged mineral, equations; 428
behavior at wet mineral surface; 426
chemically bound to mineral surface, Fig.; 419
electrostatically fixed to charged surface, Fig.; 419

Ionization constant of water
definition; 252
values at various temperatures; App. D2, Tab.; 252

Irreversible reaction
definition; 473
hydrolysis reactions at environmental conditions; 493

Iso-compound (iso-)
definition; 32

Isoelectric pH
definition; 256

Isomerism
definition; 15
cis-/trans-; *See Enantiomorphism*
effect on properties and reactivity of compound; 18
optical; *See Enantiomorphism*
stereoisomerism; *See Stereoisomerism*

Isotope
definition; 15
mixture in natural elements; 15

K

Keesom energy; *See Dipole:dipole force*

Kernel
definition; 17

Kerogen
definition; 295
elemental composition, Tab.; 296

Kinematic viscosity of water
temperature dependence, graph, Fig.; 912
temperature dependence; App. B; 913
values at various temperatures, Tab.; 911

Kinetically controlled reaction
and thermodynamically controlled reaction; 494
definition; 494

Kistiakowsky equation; 113

Knudsen effect
diffusion in gas filled porous media; 817

Knudsen number; 817

L

Labile substrate; 739

Lag period of compound degradation
enzyme induction; 701
possible reasons for; 701
time-scale of; 701

Lagrangian reference system of water flow; 1105

Lake
chemocline, Fig.; 836
compartments in one-dimensional vertical model; 1085
epilimnion; 836
Greifensee; *See Greifensee*
hypolimnion; 836
number, surface area and volume of total; 1052
numerical models, computer programs for; 1089
one-box models; 1054
orographic characteristics, Tab.; 1053
thermocline; 836
two-box models; 1075
vertical distribution of compound, Ill. Ex.; 1016

Lake Superior
characteristic data, Tab.; 1065
one-box model for fate of PCBs; 1065
parameter and concentration of modeling PCBs, Tab.; 1079
sediment–water exchange model of PCBs; 1073
summary of applied models for PCBs in system; 1080
summary of models and measurements of PCBs, Tab.; 1080
three-phase model for PCB congeners, Tab.; 1069
two-box sediment–water model for fate of PCBs; 1079

Langmuir constant of compound
definition; 282
derivation from experimental data; 282

Langmuir isotherm of compound
definition; 281

Lateral mixing by turbulence in river; 1121
calculation of time and length scale, Ill. Ex.; 1124
coefficient, definition; 1121
schematic illustration, Fig.; 1121

Lateral mixing of compound
in river; 1112

Layer resistance ratio of compound
definition; 916
values for two wind speeds, various compounds, Tab; 917

Leaving group
definition; 491
of acid derivative hydrolysis reaction; 513
characteristics of; 491
effect on acid catalyzed hydrolysis rate; 520
effect on hydrolysis of acid ester, Tab.; 525
effect on rate constants of substitutions, Fig.; 498
effect on sensitivity for nucleophilic reactions; 499
examples from reaction with nucleophiles, Tab.; 492
in hydrolysis of acid amides; 527
from hydrolysis of carbamates; 528
mechanism of hydrolysis of phosphoric acid esters; 538
role in elimination reactions; 510
strength of bond and reaction rate; 504

Lethal body burden of compound
definition; 378
values for fish, Ill. Ex.; 380

Lethal concentration of compound
in baseline toxicity, definition; 375
LFER for apolar and polar narcotics, Fig.; 376

Lewis acid and Lewis base
definition of hard and soft, examples; 500
rules of hard and soft acid, base; 500

LFER; *See Linear free energy relationship*

Light
behavior as wave and as particle; 614
speed of in vacuum, approximate value; 614
wavelength and frequency of, definitions; 614

Light absorbtion of a solution
as function of wavelength, Fig.; 616

Light absorption of compound
absorbed wavelength and color; 621
delocalized system of pi-electrons; 620
transition to excited state; 617

Light energy
and wavelength of corresponding bond enthalpy, Tab; 615
and wavelength of photons; 614

Light intensity at given wavelength
incident, at water surface; *See Spectral photon fluence rate*

Light intensity at given wavelength and depth
definition; 628
determination method; 646

Light intensity at wavelength in solution
definition in Beer-Lambert law; 615

Light non aqueous phase liquid (LNAPL); *See Organic mixture–water partitioning of Component*

Light-screening factor
definition; 639
calculation for a water body, Ill. Ex.; 640

Limiting substrate; 739, 741, 742

Linear free energy relationship (LFER)
definition; 89
acidity constant estimation; *See Hammett correlation*
air–organic solvent partition constant; 189, 192
air–surface partitioning from saturation pressure; 396
air–water partition constant, bond contributions; 205
air–water partition constant, from molecular data; 205
of aqueous activity coefficient; 149
aqueous solubility from molecular size; 146
atmospheric photolysis rate with hydroxyl radical; 676
for baseline toxicity of compound; 375
calculation of octanol–water partition, Ill. Ex.; 232, 233
calculation of solvent–water partition, Ill. Ex.; 222
as check for data consistency; 91
criteria for good and bad correlations; 90
determination of slope and constant term; 91
DOC- and octanol–water partition constants; 318
enthalpy of vaporization and vapor pressure; 119
for estimation of reaction rates, principles; 481
free energy of transfer in two-phase system; 89
half-wave potential and reaction rate; 601
hexadecane- and octanol–water partitioning, Fig; 219
humic acid– and octanol–water partition constant; 319
hydrolysis rate constants of benzoic acid esters; 531
information about badly characterized phase; 91
limitations of prediction from one-parameter LFER; 219
lipid–water partition, one-parameter; 342
multiparameter, air–surface partition coefficient; 397
octanol–water partitioning and aqueous solubility; 224
one-electron reduction potential and reaction rate; 585, 587
one-electron transfer reaction; 568
one-parameter of two-phase systems, Tab.; 91
org carbon– and octanol–water partition constant; 301
organic solvent–water partitioning, multiparam.; 220
partition constant of compound in two-phase system; 90
photosensitized rate constant and sigma value, Fig; 671
prediction of partition constant; 89
protein–water partition, one-parameter; 341
various solvent–water coefficients, one compound; 218
Swain-Scott relation of relative nucleophilities; 497
values of air–organic solvent partitioning, Tab.; 194

values of aqueous solubility and molar volume, Tab; 148
values of octanol–water partition parameters, Tab; 225
values of solvent parameters for partitioning, Tab; 221

Linear reaction; *See First-order reaction*

Linear sorption isotherm
description; 280

Lipid
mass fraction in living organisms, Tab.; 337

Lipid–water partition coefficient of compound
calculation from octanol–water partition, Ill. Ex; 346
calculation from one-parameter LFER; 342
values for selected compounds, Tab.; 339
versus octanol–water partition coefficient, Fig.; 342

Liposomes
reference phase for acute toxicity of acids, Fig.; 377
as surrogate for membranes in baseline toxicity; 376
as surrogate for polar lipids and membranes; 342
uptake of neutral and charged forms of acid/base; 377

Liquid compound; *See Pure liquid compound*

Liquid solution of compound
definition; 166

Liquid surface
values of H-acceptor/donor, vdW parameters, Tab.; 398

Liquid–vapor equilibrium; *See Saturation pressure of pure liquid*

Living organic media
composition; 335
macromolecules in living structures; 335
polymers in living structures; 335

Living organism
chemical composition of dry weight, Tab.; 337

London dispersive force
definition; 60
and air–apolar surface partition constant, Fig.; 72
and air–hexadecane partition constant, Fig.; 71
and air–polar surface partition constant, Fig.; 72
and air–pure liquid partition constant, Fig.; 69
and air–water partition constant, Fig.; 71
description on a molecular level; 63
standard free energy change of; 64

Longitudinal mixing by dispersion in river; *See Dispersion*

Lorentz-Lorenz equation; 64

Luminescent processes
fluorescence and phosphorescence; 623

A

Macropore
effect of compound transport in porous
    media; 1175

MASAS
application for compound in Greifensee; 1091
computer program for lake modeling, reference
    to; 1089

Mass flux
mass balance over volume, schematic, Fig.; 789

Mass transfer across boundary
diffusion model; 837
transfer model; 785

Maximal velocity of reaction of substrate
definition; 751
first step is rate limiting; 757
second step is rate limiting; 757
values for degradation of various substrates,
    Tab.; 758

Maximum growth rate of cell population
definition; 741
values for various growth limiting
    substrates; 746

Maximum surface concentration of compound
calculation from experimental data, Ill. Ex.; 283
derivation from experimental data; 282
and total number of surface sites of sorbent; 281

Mean flow time of river water
over distance, definition; 1109

Mean flow velocity in river
definition; 1105
calculation for two flow regimes, Ill. Ex.; 1107
in Darcy-Weisbach equation; 1106
for wide river; 1106

Mean free path
of molecules, schematic illustration, Fig.; 800

Melting point of compound
enthalpy and entropy changes of fusion; 107
estimation of, literature to; 120
illustration in phase diagram, Fig.; 101
normal and standard, definition; 100

Membrane
and walls of gram positive, gram negative
    bacteria; 737
boundary of cell for compound uptake; 691
building up and function in bacteria; 737

Meta- (m-)
position of substituents, definition; 32

Metabolic pathway
by nonspecific enzyme; 699
detailed, web adress to; 692
of dead-end metabolite; 700
of substituted benzene by nonspecific en-
    zyme; 699
secondary, definition; 705
of suicide metabolite formation; 700

Metal oxide surface
catalysis of hydrolysis; 544
effect of surface types on hydrolysis rate,
    Fig.; 545

effect on reduction rate, ferrogenic condit.,
    Fig.; 589
effect on transformation mechanism and
    rate; 544
illustation of compound coordination, Fig.; 544
reaction scheme of bound iron as reductant,
    Fig.; 590
survey of mechanisms affecting transforma-
    tions; 545

Metal species in aqueous solution
effect on hydrolysis reactions; 540, 542
leading to elimination reaction mechanism; 543
light absorption of organic complexes; 623
metal ligand complex, acting as a
    nucleophile; 541

Methanogenesis
in organic compound oxidation, Fig.; 570

Methanotroph cell population
degradation of TCE in co-metabolism, Ill.
    Ex.; 762

Micelle formation
in surface vicinal water, mechanism, Fig.; 440

Michaelis-Menten enzyme kinetics; 476
definition; 739
co-substrate removal rate; 753
concept of reaction steps; 751
estimation of biotransformation rate, Ill.
    Ex.; 765
estimation of biotransformation rates; 761
example for hydrolases; 754
first step is rate limiting; 754
hydrolysis of alkyl halides in cell free ex-
    tract; 759
parameters for degradation of substrates,
    Tab.; 758
second step is rate limiting; 755
simplifying assumptions, summary of expres-
    sions; 756

Michaelis-Menten half-saturation constant
definition; 751
first step is rate limiting; 756
second step is rate limiting; 756
values for degradation of various substrates,
    Tab.; 758

Microbial activity
affecting redox conditions; 569

Microbial population dynamics
and rate of compound biotransformation; 691
in Michaelis-Menten kinetics biodegradation,
    Fig.; 753
and lag period of compound degradation; 701
population growth as funct. of enzyme
    reaction; *See Michaelis-Menten enzyme
    kinetics*
population growth as funct. of limiting
    substrate; *See Monod population growth
    kinetics*

Microbially mediated reaction
carboxylation of compound; *See Carboxylation
    reaction, microbial*
hydration of compound; *See Hydration
    reaction, microbial*

hydrolysis of compound; *See Hydrolysis
    reaction, microbial*
oxidation of compound; *See Oxidation reaction,
    microbial*
reduction of compound; *See Reduction reaction,
    microbial*

Microorganism
optimation mechanisms for environmental
    adaption; 696
species mixture and environmental condi-
    tions; 694
types of and presence in environment; 694

Mineral surface
anion and cation exchange capacities of
    clays; 423
characterization; 392
charged; *See Charged mineral surface*
effects on compound transformation; *See Metal
    oxide surface*
properties of various solids, Tab.; 424
and sorbates of various polarities, Fig.; 393
vicinal water at; *See Surface vicinal water*
water layer thickness and relative humidity; 392

Mineral surface–water sorption of compound
competition with natural organic mat-
    ter; 409, 412
EDA interaction with surface, illustration,
    Fig.; 390
EDA interactions, calculation of, Ill. Ex.; 415
EDA interactions, definition; 413
EDA interactions, effects of surface and
    sorbate; 413
for EDA interactions, values, Tab.; 414
EDA interactions with surface, Fig.; 413
environmental conditions for; 408
as function of aqueous activity coefficient,
    Fig.; 411
possible mechanisms; 410

Mineralization of compound
definition; 459, 689
structure–biodegradability relationship; 702

Mississippi River
accidental chloroform input, case study; 1132

Mixing ratio of gases
definition and various units; 673
calculation as gas concentration; 673

Mixture; *See also* **Ideal mixture**
mixture and solution, definitions; 183

Model
definition; 947
for inference; 948
for prediction; 948
linear; 956
parameter, definition; 950

Molar concentration of compound
definition; 85

Molar mass of compound
definition; 15
and molecular diffusion coefficient in air,
    Fig.; 802
and molecular diffusion coefficient in water,
    Fig.; 809

Molar volume of compound
diffusion volume; *See Diffusion volume*
elemental contributions to, Tab.; 804
estimation from various methods, Ill. Ex.; 805
from elemental contributions; 803
from liquid density; 803
and molecular diffusion coefficient in air, Fig.; 802
and molecular diffusion coefficient in water, Fig.; 809
values from estimation methods, Tab.; 804
values of elemental volume contributions, Tab.; 804

Molar volume of liquid mixture
definition; 236

Molar volume of solute
calculation from atomic volumes; 149
depending on calculation method; 150
LFER parameter of aqueous solubility, Tab.; 148
parameter of aqueous solubility estimation, Fig.; 147
phase mixing influence; 186

Molar volume of solvent
approximation for aqueous solution; 85
calculation for liquid mixture; 85

Mole
definition; 15

Mole fraction concentration of compound
conversion to amount concentration; 85
calculation from volume fraction for two solvents; 170

Molecular diffusion coefficient in air
estimation from various methods, Ill. Ex.; 806
from molar mass; 803
temperature dependence; 805
values as function of mass, of volume, Fig; 802

Molecular diffusion coefficient in water
Hayduk and Laudie relation; 812
Stokes-Einstein relation; 810, 812
temperature dependence; 811, 913
temperature dependence, graph, Fig.; 912
values as function of mass, of volume, Fig.; 809
values for selected compounds, Tab.; 911

Molecular diffusivity
in air; 799
Chapman-Enskog theory; 801
in porous media; 816
and molecular theory of gas; 798
in water; 808

Molecular formula of compound
definition; 15
drawing representations, Fig.; 18

Molecular mass of compound
definition; 15

Molecular size of compound
effect on the Henry's law constant; 198

Monod constant of growth of cell population
definition; 741
values for various growth limiting substrates; 746

Monod population growth kinetics
first order rate constant; *See Specific growth rate of microbial population*
first order rate law, growth limited case; 740
graphical representation, Fig.; 741
kinetic data for various limiting substrates; 746
one, two and n limiting substrates; 742
substrate dependence; 740

Monopolar compound
definition; 62
adsorbed on mineral surface, illustration, Fig.; 393
partition constant as funct. of vdW parameter, Fig; 69, 71, 72
partitioning in organic solvent–water systems; 218
quantification with H-bond descriptors; 115
as surrogate for monopolar living medium; 339

Multibox model
exchange velocity at box interfaces, Fig.; 841

Multidimensional model
linear; 991

Mutation of microorganism
and lag period of compound degradation; 701

N

Normal compound (n-)
definition; 32

Narcosis; *See Baseline toxicity*

Natural attenuation
definition; 689

Natural organic carbon–water distribution ratio
definition; 322
of acid, as function of pH, Fig.; 322
of base, as function of pH, Fig.; 325, 324
and sorption of cationic species of base; 324
and sorption of neutral species of acid; 323
and sorption of neutral species of base; 324

Natural organic matter (NOM); *See also Dissolved organic matter (DOM); Solid phase organic matter (POM)*
characteristics of sorption of charged species; 325
charge of and inorganic cation in solution; 323
charge of and pH; 323
charge of at ambient pH; 321
components of and their elemental composition, Tab; 296
relative nucleophilicity, range of values; 498

Natural organic phase
surrogate for; 182

Near-surface specific light absorption rate
definition; 631
calculated from field and compound data, Tab.; 635
calculation example; 631
from Z-function and absorption coefficient; 632
wavelength dependence, Fig.; 634

Near-surface total specific light absorption rate
definition; 634

calculated from wavelength specific values, Tab.; 635
as function of season and latitude, Fig.; 636
integral over effective wavelength range, Fig.; 634
per day relative to per second; 635

Near-surface water
depth of for direct photolysis; 631
distribution function of light paths, definition; 631

Neo compound (neo-)
definition; 32

Nernst equation; 562

Neutral hydrolysis reaction
mechanism, (thio-)phosphoric acid esters; 538
mechanism and scheme, carboxylic acid esters; 524
mechanisms and rates of alkyl halides, Tab.; 505
scheme, carboxylic acid esters, Fig.; 524

Neutral hydrolysis reaction rate
definition; 515
function of pH, carboxylic acid esters, Fig.; 514
rate constant from experimental data, Ill. Ex.; 515

Non-aqueous phase liquid (NAPL)
exchange of compounds with river water; 869
exchange with aqueous phase; 860
exchange with aqueous phase, Ill. Ex.; 821

Nonbonding electron pair
attacked in microbial oxidation reaction; 715
electronic ground and excited states; 613
in resonance with aromatic system; 30
influence on bond angles; 26
in nucleophilic species; 491
symbolism in drawing structure, Fig.; 18

Nondimensional coordinate
in groundwater flow; 1161
time of passage, step input into groundwater, Tab.; 1165

Nonideal solution; *See Real solution*

Noninterface boundary
examples in the environment, Fig.; 835, 836

Nonlinear sorption isotherm; 304
calculation of dual-mode sorption type, Ill. Ex.; 305

Nonpolar compound
definition; 62
effect on Setschenow constant; 161

Normal distribution; 782
definition; 784
and Bernoulli coefficients; 782
solution of Fick's second law; 790
variance; 784

Nucleophilic species
definition; 491
calculation of first-order rate constant, Ill. Ex.; 502
concentration of equal reaction rate, Tab.; 501
hard and soft Lewis bases; 500
hydride, formed by NAD(P)H in bioreductions; 722

important ones in aqueous environment; 491
metal-ligand complexes; 541
qualitative criteria for nucleophilicity; 498
reactant with relevant compound group,
  Tab.; 492
reaction at polar bond; 491
reactions of; *See Nucleophilic substitution
  reaction*
reference nucleophiles in Swain-Scott relationship; 497
relative effectiveness in displacing leaving
  group; 497
valence shell characteristics; 491

Nucleophilic substitution reaction
effect of compound structure on mechanism and
  rate; 504
initial step in microbial degradation of
  compound; 702
kinetic data for selected compounds, Tab.; 508
reaction mechanisms; 495

Nucleophilic substitution reaction, first-order
effect of compound structure on mechanism and
  rate; 505
environmental conditions and competing
  mechanisms; 511
favorised products; 497
mechanism, description of; 496
rate law; 497
standard free energy pathway, Fig.; 496

**Nucleophilic substitution reaction, second-
  order**
calculation of pseudo-first-order const., Ill.
  Ex.; 502, 503
effect of compound structure on mechanism and
  rate; 505
environmental conditions and competing
  mechanisms; 511
mechanism, description of; 495
mechanisms at (thio-)phosphoric acid esters; 538
rate law; 496
under reducing conditions; 501
in seawater; 501
standard free energy pathway, Fig.; 495

Nucleophilicity of nucleophilic species
qualitative criteria; 498
quantitative measure in Swain-Scott relation; 497
relative, of relevant nucleophilic species,
  Tab.; 498, 499

Numerical models of lakes; 1089

O

Ocean
characteristics of, Tab.; 1052
one-box models; 1054

**Octanol–water distribution ratio of acid or
  base**
definition; 271
as function of pH, Fig.; 271

Octanol–water partition constant of compound
activity coefficient in octanol; 224

as function of aqueous activity coefficient,
  Fig.; 224
and atom/fragment contribution estimation; 228
and baseline toxicity of compound; 375
calculation from fragment contributions,
  examples; 229, 231, 232, 233
data compilations, reference to; 226
determination methods; 226
estimation methods; 226; *See also Atom/
  fragment contribution method*
LFER with aqueous activity coefficient; 224
values of atom/fragment coefficients, Tab.; 230
values of atom/fragment corrections, Tab.; 231
values of LFER parameters of activity coeff.,
  Tab.; 225

Octanol–water partitioning of compound
characteristics of octanol; 223
surrogate for organic phases–water partitioning; 223

Octet rule
in covalent bonding; 17

Olefin; *See Alkene*

One-box model; 953, 955
definition; 482, 950
analysis of dynamics of compound in lake; 959
for bioaccumulation in living organism; 350
estimation of degradation of compound, Ill.
  Ex.; 960
with input and output, schematic illustration,
  Fig; 955
linear, with one variable; 955
linear with two variables; 976
mass balance; 482
nonlinear; 970
nonlinear, stable and unstable steady state,
  Fig.; 975
reactant degradation, product formation, Ill.
  Ex..; 979
schematic illustration, Fig.; 483
second-order degradation of compound, Ill.
  Ex.; 973
second-order reactions of compound; 971
time-dependent external forcing; 962
time-dependent input of compound numerical
  example; 965
verification of degradation of compound, Ill.
  Ex.; 966

One-box model of lakes and oceans; 1054; *See
  also Greifensee*
boundaries of fluxes; 1054
flushing, air–water exchange, hydrolysis, Ill.
  Ex; 1056
modeling assumptions; 1054
schematic picture of boxes, fluxes, reactions,
  Fig; 1055
summary of modeling equations and parameters; 1055

One-dimensional continuous lake model; 1082

**One-dimensional vertical (1DV) lake
  model; 1084**
applicability and characteristics of model; 1084
concentration profiles, calculated, measured,
  Fig.; 1092

goal of model; 1084
lake topography; 1085
numerical model, application information; 1089
schematic illustration, Fig.; 1086
use of continuous model for compound in
  Greifensee; 1091

One-electron oxidation potential
definition; 601
and reaction rate constant of oxidation,
  Fig.; 601

One-electron reduction potential
definition; 568
and reaction rate constant of reduction; 585
and reaction rate constant of reduction,
  Fig.; 587
value for molecular oxygen; 598
values for photooxidants, Tab.; 656
values for substituted nitrobenzenes, Tab.; 585

One-electron transfer reaction
free energy of first step and of overall reaction; 568
microbial reduction of nitroaromatic compounds; 725
multi-electron redox reactions; 569
and rate limiting step of redox reaction; 568
reaction scheme; 580
reaction scheme and free energy profile,
  Fig.; 584
reduction of molecular oxygen; 598
reduction of nitroaromatic compounds,
  Fig.; 584
stability of formed radical; 568

Open water column (OP)
in one-dimensional vertical lake model; 1085
rate law of compound total concentration; 1088
rate law of particle concentration; 1088

Optical isomerism; *See Enantiomorphism*

Organic acid and base; *See Acid, see Base*

Organic compound
apolar; *See Apolar compound*
bipolar; *See Bipolar compound*
as cosolvent in aqueous solution; *See Cosolvent*
environmental beahvior of, scheme, Fig.; 8
fate of in a lake, overview, Fig.; 7
general environmental risks; 6
global production rate; 5
ionic; *See Ionic organic species*
main elements in; 14
monopolar; *See Monopolar compound*
nonpolar; *See Nonpolar compound*
number of in daily use; 5
polar; *See Polar compound*
processes in the environment, overview; 6
pure, definition; 99
reactivity in general; 22
size of; *See Size of compound*
as substrate in biotransformation processes; *See
  Substrate*

Organic matter
on solid phase; *See Solid phase organic
  carbon; Solid phase organic matter*

Organic mixture–water partitioning of component
activity coefficients in organic phase; 237
aqueous solubility of mixture components; 236
aqueous solubility of PCBs, Ill. Ex.; 238
cosolvent effect in aqueos phase; 236
description of common mixtures; 235
equilibrium coefficient of; 237

Organic phase
as surrogate for living media; 339

Organic phase–water partition coefficient
values for various phases and compounds, Tab.; 339

Organic solvent–water distribution ratio
of organic acid or base; 271

Organic solvent–water partition constant
definition; 215
activity coefficient expression; 215
air–solvent and air–water partition constants; 215
calculation from LFER parameters, Ill. Ex.; 222
influence of aqueous solubility of compound; 215
LFER for other organic solvents, same compound; 218
molar concentration expression; 215
mole fraction expression; 215
from molecular data, multiparameter LFER; 220
and salt concentration; 216
schematic picture of relations, Fig.; 186
temperature dependence; 215
values for various solvents and compounds, Tab.; 217
values of molecular LFER parameters, Tab.; 221

Organic solvent–water partitioning of compound
definition; 214
acid or base distribution; 270
comparison of different organic solvents; 216
comparison of salt concentrations; 216
general applicability of; 214
model of multiparameter LFER description; 220
schematic illustration of processes, Fig.; 186
temperature effect; 215

Organism–environmental medium partitioning
definition; 343
definition from singular partition coefficients; 344
calculation from lipid–water partitioning, Ill. E; 346
calculation from org. phases partitioning, Ill. Ex; 348
calculation from organic phases partition, Ill. Ex; 349
dynamic process and equilibrium; 343
estimated and measured, interpretation, Ill. Ex.; 346, 348
lipid normalized, definition; 345
and observed bioaccumulation factor; 349
schematic illustration of interrelations, Fig.; 345
units; 344

Ortho- (o-)
position of substituents, definition; 32

Outer-sphere mechanism
of transition state in one-electron transfer; 581

Output
definition; 950

Overall first-order rate constant in FOLIDE
definition; 471
constant or time dependent; 471

Overall quantum efficiency photooxidant production; 660

Overall rate of transformation of compound
for multiple order and multiple mechanisms; 512

Oxic condition
definition; 571, 694
degradation rate compared to anoxic condition, Fig; 694
forming electrophilic form of oxygen; 715
functional groups favorising biodegradation; 705
functional groups inhibiting biodegradation; 705
mastervariable in selection of compound biodegrada; 694
selection criterion in biodegradation pathway; 703
vinyl chloride biodegradation products, Ill. Ex.; 720

Oxidant
definition; 559
alcohol dehydrogenase in biooxidation reactions; 722
generated from light absorption; *See Photooxidant*
sequence of in organic compound oxidation, Fig.; 570

Oxidation reaction
definition; 556
notation; 23
one-electron, anilins and coupling products, Fig; 600
one-electron, phenols and coupling products, Fig; 599

Oxidation reaction, microbial
electrophilic attack at double bond; 715
electrophilic attack at nonbonding electron pair; 715
electrophilic attack at sigma bond; 715
example of cytochrome P450 monooxygenase; 718
initial step in degradation; 700
initial step in degradation reactions; 702
involving electrophilic oxygen-bioreactants; 715
reaction sequence of a monooxygenase, Fig.; 719
selected reactants and oxidation products, Tab.; 717
specialized organisms executing oxidations; 719
transfer of one or two oxygen atoms; 715

Oxidation state

definition; 23
assigning in molecules, Ill. Ex.; 24
rules of assigning; 23

Oxidative coupling
anilines, mechanisms and products of, Fig.; 600
phenols, mechanisms and products of, Fig.; 599

Oxygen
atom state, forming hydroxyl radicals; 673
co-substrate in microbial oxidation reaction; *715*
transition and reaction cycle, scheme, Fig.; 659

Oxygen, singlet state
concentration determination method; 667
generation from triplet and reaction scheme, Fig.; 659
in biochemical conversion of triplet oxygen; 696
near-surface and DOM concentrations, Fig.; 667
potential reaction partners in indirect photolysis; 668
rate constants as function of pH, Fig.; 669
wavelength range of production; 660

Oxygen, triplet state
excitation to singlet state, reaction scheme, Fig.; 659
promotion energy to singlet state; 658
as quencher of excited chromophores; 658

P

Para- (p-)
position of substituents, definition; 32

Paraffin; *See Alkane*

Partial differential equation
first-order linear; *See* First-order linear
second-order linear; *See* Second-order linear

Partial molar enthalpy of compound
and chemical potential of compound; 74

Partial molar entropy of compound
and chemical potential of compound; 74

Partial molar excess enthalpy of compound; *See Excess enthalpy of compound*

Partial molar excess entropy of compound; *See Excess entropy of compound*

Partial molar excess free energy of compound; *See Excess free energy of compound*

Partial molar free energy of compound; *See chemical potential of compound*

Partial pressure
of water in air and relative humidity; 392

Partial pressure of gaseous compound
definition; 76
and fugacity; 77
mixing ratio of volume expression; 673

Partially miscible organic solvent (PMOS); *See Cosolvent*

Particle; *See Suspended solids*
coagulation of; 1064
colloidal; 1064

concentration, ocean surface and deep water, Tab.; *1052*
first-order removal model; 1064
first-order removal rate constant; 1063
rate law in open water column; 1088
rate law in surface mixed water layer; 1087
removal from water column; 1061
removal from water, schematic illustration, Fig.; 1062
resuspension from sediment; 1070
resuspension from sediment, flux of compound; 1072
in surface mixed sediment layer; 1088
variation in origin, size, settling velocity; 1064

Particle aggregate; *See Porous particle*

Particle settling
compound input into surface mixed sediment layer; 1076
laminar; 1061
mass flux per unit area, definition; 1062
on lake sediment, schematic illustration, Fig.; 1070
transport mechanism of compound; 1059
vertical mass flux of compound; 985

Particle settling in river
in river Rhine, case study; 1137
rate law for length axis; 1116
rate law for time axis; 1116

Particle settling velocity
apparent; 1064
Stokes law; 1061, 1064
typical values; 1061
typical values and horizontal water currents; 1064

Particulate organic carbon; *See Solid phase organic carbon*

Particulate organic matter (POM); *See Solid phase organic matter*

Partition coefficient of compound
definition of the term coefficient; 90

Partition constant, general; *See Equilibrium partition constant*

Partition constant of compound
from dry and wet solvent; 186
indirect determination of; 186
and transfer rate across interface; 55

Partitioning of compound
between phases, definition; 59
between phases, in open water column; 1060
and equilibrium assumption; 55
equilibrium notation; 66
excess of intermolecular forces; 66
reaction formulation; 59

Passive uptake of compound by microorgansims
limiting overall transformation rate; 691
of hydrated molecules and charged species; 738

Path length of light in solution
definition; 615

Peclet number
definition; 1011
and nondimensional attenuation coefficient,
Fig.; 1166
and phase shift coefficient, Fig.; 1166
in groundwater transport; 1156, 1161
in groundwater transport, Ill. Ex.; 1159
interpretation of; 1012

Permeability
as function of porosity and particle size, Fig.; 1154
in layer of groundwater system, Tab.; 1150
of groundwater system; 1153

Permeation of compound into bacterium cell
time scale for nonpolar compounds; 737

Petroleum
global production; 33

pH
definition; 249
and acid-to-base concentration ratio; 249
buffering in natural waters; 253
change of by organic acid contamination; 253
reaction rate sensitivity and reaction mechanism; 511
of seawater, Tab.; 1052
stability of in natural waters; 253

pH of intersection of equal reaction rates
definition; 515
acid/neutral, neutral/base, acid/base points, Fig.; 515
calculation from laboratory data, Ill. Ex.; 515
data for (thio-)phosphoric acid esters, Tab.; 537
data for carboxylic acid amides, Tab.; 527
data for carboxylic acid esters, Tab.; 520
temperature dependence; 515

pH of zero point of charge
definition; 421
calculation from intrinsic acidity constants; 421

Phase diagram of pure compound
illustration of the four phases, Fig.; 101
description; 99

Phase rule, Gibbs; 99

Phase shift coefficient
in groundwater and Peclet number, Fig.; 1166
sinusoidal input into groundwater; 1165

Phase transfer process
and equilibrium assumption; 55

Photolysis
direct; *See Direct photolysis*
in a lake and vertical mixing, Fig.; 954

Photon
energy of at given wavelength; 614
interaction with electromagnetic field of molecule; 614

Photooxidant
calculation of production rate, Ill. Ex.; 665
calculation of steady state concentration, Ill. Ex; 662, 665
concentration range in environment, Fig.; 657
determination of concentration in natural water; 661
hydroxyl radical; *See Hydroxyl radical*
in atmosphere, types and reaction characteristics; 672
inorganic ions' role in indirect photolysis; 658

oxygen, singlet and triplet state; *See Oxygen*
production rate law; 660
rate law of total consummation; 660
reactants in atmosphere; 656
reactants in water treatment plants; 656
steady state concentration determination; 657
steady state concentration, near surface; 661
values of one-electron reduction potentials, Tab.; 656

Photosensitization
definition; 623

Photosensitized reaction
definition; 670
first-order rate constants of some herbicides, Tab; 670
first-order rate constants vs. sigma constant, Fig; 671

Photosensitizer
definition; 623

Photosynthesis
redox conditions in aquatic environment; 569

Physisorption
definition; 389

Pi-bond; *See also Double bond*
conjugated, definition; 28
in electronic ground and excited states; 613
picture of, in multiple bonds, Fig.; 27

Pi-electron
delocalized, picture of orbitals, Fig.; 29
in conjugated Pi-bonds; 28
transition from bonding to antibonding orbital; 620

Planck's constant
definition and value; 614

Plant–air partition coefficient of compound
calculated and measured values, Ill. Ex.; 349

Plasmid
definition; 695
exchange of and lag period of compound degradation; 701
mechanism of DNA exchange and consequences; 695
occurrence in bacteria and role in biodegradation; 695

Polar bond
definition; 19
and electronegativity difference of atoms; 19, 491
and enthalpy change; 23
and reactivity of bond; 21, 491
reactivity with electrophile and nucleophile; 491
symbolism in drawing structure; 19

Polar compound
definition; 62
in water, reordering of water molecules; 145
partitioning into polar and nonpolar lipids; 343

Polar intermolecular force; *See also Electron donor-acceptor force*
definition; 62

Polar surface
intermolecular forces with adsorbents; 72

Polarizability of compound
in aqueous activity coefficient calculation; 150
and London dispersive energy; 60
of a molecule, definition; 63
values of parameter of selected compounds, Tab.; 152

Pollution cloud
dilution along one dimension; 868

Polycyclic aromatic compound
definition; 30
examples, Fig.; 30

Pore water concentration of compound
calculation, linear and nonlin. sorption, Ill. Ex.; 308

Porin
definition; 737
passive uptake of compound and selectivity; 738

Porosity of porous medium
definition; 815
unit; 816

Porosity of solid–water system
definition; 287
and solid–water phase ratio; 287
typical value range in groundwater; 287

Porous media; *See also Groundwater*
characteristic parameters; 1150
diffusivity of compound; 816
dispersion types, schematic illustration, Fig.; 1149
effect of sorption; 1170
gas-filled, diffusion of compound; 817
heterogeneity of; 1175
heterogeneity, schematic illustration, Fig.; 1149
liquid-filled, diffusion of compound; 817
porosity, definition; 815
tortuosity; 816

Porous particle
half-saturation time; 876
sorption kinetics; 874

Potential; *See Chemical potential; Electrochemical potential*

Preexponential factor of Arrhenius equation
definition; 478

Pressure of gas
and fugacity; 76

Primary carbon atom
definition; 32

Production rate
global, of organic chemicals; 5

Protein
mass fraction in living organisms, Tab.; 337

Protein–water partition coefficient of compound
calculation from one-parameter LFER; 341
values for selected compounds, Tab.; 339
versus octanol–water partition coefficient, Fig.; 341

Proton in aqueous solution
reactant or product of standard hydrogen electrode; 560
reaction with water as base; 247

standard chemical potential of, per definition; 247

Proton transfer reaction; *See Acid–base reaction*

Proximity effect of substituent on acidity const.
intramolecular and steric, description of; 259
intramolecular and steric, Fig.; 260

Pseudo-first-order rate constant of reaction
definition; 470
calculation at given temperature, Ill. Ex.; 484
calculation for multiple SN-2 reactions, Ill. Ex.; 502, 503
observed, for multiple order and mechanisms; 512
photooxidant consumption; 660

Pseudo-first-order rate law of reaction
definition; 470
Michaelis-Menten enzyme kinetics; *See First-order rate law of reaction*
Monod population growth kinetics; *See First-order rate law of reaction*

Pump regime in groundwater
in groundwater system S, Ill. Ex.; 1156, 1168
time of compound reaching wells, Ill. Ex.; 1156

Pure compound
definition; 99
notation of; 68

Pure compound–water partitioning; 135

Pure gaseous compound
aqueous solubility and activity coefficient; 139
aqueous solubility as function of temperature, Fig; 155
calculation of aqueous solubility, Ill. Ex.; 140
chemical potential of; 104
phase diagram, Fig.; 101

Pure liquid compound
aqueous activity coefficient, calculation, Ill. Ex; 153
aqueous solubility; 136
aqueous solubility and activity coefficient; 135
aqueous solubility as function of temperature, Fig; 155
calculation of aqueous solubility, Ill. Ex.; 140
chemical potential of; 80, 104
enthalpy and entropy terms in; 99
fugacity of, illustration, Fig.; 79
phase diagram, Fig.; 101
in pure liquid–water mixture; 136
as reference for compound in liquid solution; 78
as reference state of compound; 77
standard chemical potential of; 80
symbol of; 68
vapor pressure versus temperature, Fig.; 106

Pure liquid compound–water equilibrium
definition; 135
activity coefficient of liquid compound; 136
mole fraction values of compound in own phase, Tab; 136
schematic picture of processes, Fig.; 186
temperature dependence; 155

Pure liquid–vapor boundary of compound
saturation pressure and temperature, Fig.; 101

Pure solid compound
aqueous activity coefficient, calculation, Ill. Ex; 153
aqueous solubility and activity coefficient; 138
aqueous solubility as function of temperature, Fig; 155
aqueous solubility with cosolvent; 166
calculation of aqueous solubility, Ill. Ex.; 140
phase diagram, Fig.; 101
vapor pressure versus temperature, Fig.; 106

Pure solid–vapor boundary of compound
saturation pressure and temperature, Fig.; 101

Purge-and-trap method
in air–water partitioning; 204

Q

Quantitative structure activity relationship, QSAR; *See Linear free energy relationship*

Quantum efficiency
determination in natural water; 647

Quantum yield of photolysis of compound; *See also Reaction quantum yield of photolysis*
definition; 626

Quenching
definition; 658
physical and chemical in water; 660
quencher, definition; 623
water for singlet oxygen; 666

R

R-, as substituent
definition; 32

Radical formation
from light irradiation; *See Photooxidant*
in one-electron transfer reaction; 568
in oxidation of anilines and phenols; 600

Random motion
of object through discrete boxes, Fig.; 781

Raoult's law; 183
activity coefficient in ideal solution; 79
for components in organic mixture; 237

Rate constant of light absorption of compound; *See Specific rate of light absorption*

Rate constant of reaction; *See also First-order rate constant; Pseudo-first-order rate constant; Second-order rate constant*
definition; 469
effect of solution composition; 481
temperature dependence; App. D1; 478
total of various transformation processes; *See Total reaction rate constant of compound*
in transition state theory of elementary reaction; 480

Rate law of transformation reaction; *See also First-order rate law; Pseudo-first-order rate law; Second-order rate law*
definition; 468
direct photolysis reaction; 641
equation for disappearance of compound; 469

first-order; 469
first-order including back reaction; 473
and mechanism of reaction; 469
Michaelis-Menten enzyme kinetics; 477
orders of reaction, definition; 469
photooxidant consumption; 660
photooxidant formation in indirect photolysis; 660
pseudo-first-order; 470
second-order; 470
temperature dependence; App. D1; 478
in transition state theory; 479

Rate limiting step of transformation
definition; 478

Rate of light absorption by aqueous medium
per unit surface area and per unit volume; 629

Rate of light absorption by compound
definition; 630

Rate of transformation of compound
accelerated by metal species in solution; 542
controlling factors in collision model; 478
factors of influence in redox reactions; 581
by indirect photolysis at near surface; 661
inhibited by metal species in solution; 543
overall of multiple mechanisms; *See Overall rate of transformation of compound*
relative, in kinetically controlled reactions; 494
and transition state theory; 479

Reaction
definition; 459
chemical, definition; 459

Reaction constant; *See Equilibrium constant of reaction*

Reaction distance of compound in river
definition; 1115
values for two flow regimes of river G, Tab.; 1115

Reaction enthalpy
estimation from bond enthalpies; 22
sign convention; 23

Reaction order
definition; 469

Reaction process
total of first order, in one-box model; 483
total, in one-box model; 482

Reaction product formation
one-box model, coupled FOLIDEs approach, Ill. Ex.; 979

Reaction quantum yield in aqueous solution
definition; 626
determination methods; 645
values for various compounds, Tab.; 642
wavelength dependence; 627

Reaction quotient
definition; 465
calculation for given conditions, Ill. Ex.; 467
influence on free energy change of reaction; 465
in Nernst equation of redox reactions; 562

Reactive oxygen species (ROS); 656

Reactivity
of a bond and its polarity; 21
of organic compounds in general; 22

Real gas
fugacity and pressure; 76

Real solution
fugacity of compound, illustration, Fig.; 79

Redox conditions in aquatic environment
sequence of redox processes; 569, 574
sequence of redox processes, Fig.; 570

Redox couples
inorganic, standard potential values, Tab.; 563, 575
mediators, standard potential values, Tab.; 575
organic, standard potential values, Tab.; 564, 575

Redox reaction
definition; 23, 556
calculation of amount of oxidants needed, Ill. Ex.; 573
calculation of redox capacities, Ill. Ex.; 571
change of bond polarity; 23
combination of two reduction half-reactions; 569
e-transfer and potential in galvanic cell, Fig.; 560
kinetics; 580
problems in quantification; 559
sequence of oxidants in aquatic environment; 570
thermodynamics; 559

Reductant
definition; 559
alcohol dehydrogenase in bioreduction reactions; 722
NAD(P)H, direct or indirect in bioreductions; 724

Reduction potential of half-reaction
definition; 562
calculation at given concentration and pH, Ill. Ex; 576
and Nernst equation; 562
one-electron transfer; *See One-electron reduction potential*
at standard conditions; *See Standard reduction potential*

Reduction potential of natural aquatic system
measuring for oxic and for suboxic/anoxic state; 574
problems from nonequilibrium state; 574

Reduction rate constant of compound
function of half-wave potential, Fig.; 601
function of one-electron reduction potential, Fig.; 587
and half-wave oxidation potential; 601
and one-electron reduction potential; 585

Reduction rate of compound
dechlorination, as function of bond energy, Fig.; 597
dechlorination, as function of potential, Fig.; 597
dechlorination at metal electrodes, Fig.; 597

dehalogenation of halogenated methanes, Fig.; 596
in presence of dissolved organic matter, Fig.; 583
in presence of surface, iron(II) as mediator, Fig; 589
and reaction mechanism; 580
relative rates for various reductants, Fig.; 581

Reduction reaction
abiotic, schematic illustration, Fig.; 557
biological, schematic illustration, Fig.; 557
calculation for DOM-mediated e-transfer, Ill. Ex.; 590
definition and notation; 23
reaction rate; *See Reaction rate of reduction*
various pathways to final product, Fig.; 582

Reduction reaction, microbial
bioreduction by alcohol dehydrogenase; 722
initial step in degradation reactions; 702
involving nucleophilic bioreactants; 721
nitroaromatic compound reduction by flavoprotein; 725
point of reductive attack of bioreactant; 721
reaction sequence of a dehydrogenase, Fig.; 723
reaction sequence of metal containing enzyme, Fig.; 728
reduced metal in enzymes; 725
selected reactants and oxidation products, Tab.; 721

Reductive dehalogenation reaction
definition; 556
complexity of kinetics; 595
and dehydrohalogenation reaction, same compound; 556
environmental conditions for; 592
microbial, membrane permeability limited; 691
possible mechanisms for chlorinated ethenes; 598
relative reaction rates in different systems, Fig.; 596, 597
variability of mechanisms and products, Fig.; 593, 594

Reference state of compound
definition; 77
and activity of compound; 80
in environmental chemistry; 77
infinite dilution state, definition; 78, 246
pure liquid state; 77
and reference state of the elements; 78
and standard state; 77
value of activity coefficient; 80

Refractive index of compound
as measure for polarizability; 64
and polarizability of molecules; 64
values for selected liquids, Tab.; 65

Relative humidity of air
definition; 392
effect on H-donor and vdW forces of surface, Fig.; 399

Relative retention time in chromatography
definition; 227
parameter in partition constant estimation; 227

Renkin effect; 875
diffusion in liquid filled porous media; 817

Resonance
symbolism in drawing structure; 29

Resonance effect of substituent on acidity const.
description; 258
direct resonance, definition; 264
direct resonance, Fig.; 259
and position at aromatic compound, Fig; 259
substituents with positive or negative effect; 257

Resonance energy
definition; 29
value of benzene; 29

Retardation of compound in solid–water system
definition; 288
calculation for mineral–water adsorption, Ill. Ex; 415
effect of cosolvent, Ill. Ex.; 312
illustration for groundwater situation, Fig.; 288, 1149

Reynolds number
definition; 826
and diffusion type; 1019
particle specific, definition; 1061

Reynolds' splitting; 1019

Ring structure of compound
and dia-stereoisomerism; 27

River
advection/reaction equation, stationary solution; 1110
air–water exchange of compound; 1111
change of maximum concentration of compound; 1130
change of total load of compound; 1130
characteristic parameters, scheme, Fig.; 1103
characteristic parameters, Tab.; 1104
dispersion, lateral and vertical, picture, Fig.; 1122
distance of reaction, definition; 1115
flow and reaction of compound; 1108
flow and reaction of compound, illustration, Fig.; 1109
friction factor, typical value range; 1106
groundwater crossing, schematic illustration, Fig.; 1151
infiltration into groundwater system S, Ill. Ex.; 1166, 1168
input of chemical, different input scenarios, Fig.; 1128
integrated surface area as coordinate, Fig.; 1112
lateral mixing of compound; 1112
lateral turbulent mixing; 1121
longitudinal mixing by dispersion; 1122
mixing processes, lateral and vertical, Fig.; 1121
processes in; 1103
reference systems of flow; 1105
schematic illustration of geometry, Fig.; 1103
sediment–water exchange of compound; 1116

stationary flow; 1102
temporal evolution of compound patch, graph, Fig.; 1127
total water volume; 1102
transport of compound by the mean flow; 1104
transport/reaction model; 1130
vertical mixing of compound; 1112
vertical turbulent mixing; 1120

River G
air–water exchange for two flow regimes, Ill. Ex.; 1113
effect of dispersion on atrazine spill, Ill. Ex.; 1129
mean flow velocity for two flow regimes, Ill. Ex.; 1107
reaction distance for two flow regimes, Tab.; 1115
sediment–water exchange of atrazine, Ill. Ex.; 1118
turbulent diffusion and dispersion, Ill. Ex.; 1124

River R
exfiltration of compound into groundwater, Ill. Ex; 1177

River Rhine
accidential pesticide input, case study; 1135
longitudinal dispersion of a pesticide, case study; 1139
pesticide concentrations, measured and modeled, Fi; 1140
physico-chemical data of river and pesticide; 1136
summary of processes and data of pesticide, Tab.; 1140

Rotation of compound
freedom of and molecular symmetry; 126
standard entropy change of; 126

Rotational symmetry number; 123

Roughness of river bed
rough river, definition; 926
small and large, illustration of eddies, Fig.; 925
small, definition; 924

Roughness parameter of river
definition; 926

S

Salinity
of seawater, Tab.; 1052

Salting constant; *See Setschenow constant*

Salting-in
definition; 161

Salting-out
definition; 159

Saturated calomel electrode
standard reduction potential value; 574

Saturated carbon
definition; 26

Saturated compound
definition; 32

Saturation concentration of compound in water; *See Aqueous solubility of compound*

Saturation pressure of pure compound
and air–solvent partition constant; 189
and air–pure liquid partition constant; 68
in aqueous activity coefficient formulation; 148
data compilations, reference to; 119
description of equilibrium with condensed state; 98
experimental methods for determination; 118
as function of temperature, Fig.; 106
at given temperature, calculation of, Ill. Ex.; 108
information about compound properties; 98
and intermolecular forces; 110
of normal alkanes, Fig.; 100
partial and total pressures of pure compound; 99
and total pressure in closed and open vessels; 101

Saturation pressure of pure liquid compound
and activity of compound in gas phase; 356
calculated values for selected compounds; 122
calculated versus experimental values, Fig.; 117
calculation at another temperature; 119
calculation from air–hexadecane partitioning, Fig; 117
calculation from molecular data; 118
calculation from vdW force and surface area; 396
describing van der Waals force of compound; 396
estimation from melting and boiling points; 121
extrapolation to subcooled liquid; 105
and free energy change of vaporization; 104
and its fugacity; 78
and fugacity of compound in liquid solution; 183
influence of compound's properties on; 114
LFER for air–surface partition coefficient; 396
at liquid–vapor equilibrium; 104
of water, and relative humidity of air; 392
as subcooled liquid; 103
temperature dependence; App. D1; 105
thermodynamic description; 103

Saturation pressure of pure solid compound
calculated values for selected compounds, Tab.; 122
calculation from subcooled liquid; 124
and its fugacity; 78
temperature dependence; 105, 107, 123
thermodynamic description; 105

Saturation pressure of subcooled liquid
calculation from pure solid and melting point; 124

Saturation pressure of superheated liquid
and aqueous solubility of compound; 139

Schmidt number
definition; 909
factor in water-phase exchange velocity in river; 925
temperature dependence; 914
in water, values for selected compounds, Tab.; 911
in water, temperature dependence, graph, Fig.; 912

Seawater
  artificial, salt composition of; 161
  calculation of solubility of solid in, Ill. Ex.; 164
  inorganic salt composition, Tab.; 160
  nucleophilic substitution reactions in; 501

Second-order linear partial differential equation
  transport and chemical reaction of compound; 1008

Second-order rate constant of reaction
  definition; 470
  hydroxyl radical in atmosphere, Fig.; 674
  hydroxyl radical in water, Fig.; 664
  metal-ligand complex as competing nucleophile; 542
  in Swain-Scott relation; 497
  for various nucleophiles and leaving groups, Fig.; 498

Second-order rate law of reaction
  definition; 470
  bimolecular nucleophilic substitution reaction; 496
  biodegradation rate of co-substrate; 753
  conditions for reduction to pseudo-first-order; 475
  elimination of substituent; *See Elimination reaction, second-order*
  nucleophilic substitution; *See Nucleophilic substitution reaction, second-order*

Secondary carbon atom (sec-)
  definition; 32

Sediment
  desorption kinetics of contaminant, Ill. Ex.; 881
  of lakes, main organic components of; 294
  permanent, compound transfer from SMSL; 1076
  permanent, in SMSL model; 1073
  porosity; 851
  preservation factor of settling particles; 1076
  stratigraphic profile interpretation, Ill. Ex.; 823

Sediment–water exchange of compound
  in one-dimensional vertical lake model; 1085

Sediment–water exchange of compound in river; 1116
  flux of compound; 1116
  from particle settling; 1116
  of pesticide in the river Rhine, case study; 1137
  in sandy sediments; 1119
  spill of atrazine into river G; 1118

Sediment–water interface; 851
  exchange flux, schematic illustration, Fig.; 1070
  exchange models; 1071
  exchange of compound; 1070
  exchange of sorbing compound, example, Tab.; 852
  surface mixed layer; *See Surface mixed sediment layer (SMSL)*
  as wall boundary; 1072

Sediment–water system; *See also Surface mixed sediment layer (SMSL)*
  parameters, concentration of PCBs in Lake Superior; 1079
  particle resuspension; 1072

solution of linear two-box model; 1078
  two-box model, schematic illustration, Fig.; 1076

Sensitivity of compound for nuleophilic reaction
  compound structure and reactivity; 499
  in Swain-Scott relation; 497
  reference compound and standard value; 499

Sensitized photolysis; *See Indirect photolysis*

Sequestration
  definition; 735

Setschenow constant of compound
  definition; 160
  and aqueous activity coefficient; 160
  calculation for sodium chloride, Ill. Ex.; 164
  determination; 160
  effect of type of inorganic ion; 160
  effect on sorption; 310
  in salt mixture, from individual constants; 162
  in seawater, calculation for solid compound; 164
  in seawater, calculation from salt mixture in seaw; 162
  in seawater, values for selected compounds, Tab.; 163
  temperature dependence; 162
  values for different salts, Tab.; 160

Setschenow equation; 159

Shake flask method
  in determination of octanol–water partitioning; 226

Shared electrons; *See Bonding electron pair*

Shear diffusion model
  and horizontal diffusion; 1033
  size of compound cloud and transport time, Fig.; 1034
  summary of equations; 1035

Shear velocity
  definition; 922
  in rivers, definition; 1106

Sigma constant of substituent
  effect on biodegradation rate; 716

Sigma-bond; *See Single bond*
  in conjugated Pi-bonds; 28
  in electronic ground and excited state; 613

Sigma-Hammett constant of substituent
  definition; 261
  effect on hydrolysis rate of an acid ester, Fig.; 532
  in hydrolysis rate expression; 532
  values for phenyl urea herbicides, Tab.; 670
  values for substituents and positions, Tab.; 263, 266

Silver–silver chloride electrode
  standard reduction potential value; 574

Single bond
  lengths and enthalpies, Tab.; 20
  rotation of substituents, Fig.; 26

Size of compound
  estimation from molar volume; 149
  for estimation of aqueous solubilities; 146
  parameters for estimation of; 146

Snow
  as sorption surface; *See Air–surface adsorption of compound*

Soil
  effect on direct photolysis on sorbed compound; 649
  frequency diagram for various compounds, Fig.; 294

Solar radiation
  intensity as function of season and latitude, Fig.; 636
  intensity as function of wavelength, Fig.; 634
  intensity per day relative to per second; 635
  wavelength range of ultraviolet and visible light; 613

Solid compound; *See Pure solid compound*

Solid phase organic carbon–water partition coeff.
  definition; 292
  definition by activity coefficients in both phases; 311
  calculation for cosolvent–water system, Ill. Ex.; 312
  cosolvent effect; 311
  data compilation, references to; 300
  determination methods and problems of; 299
  estimation methods; 300
  as function of compound, Fig.; 292
  from octanol–water partition constant, LFER; 301
  and octanol–water partition constant, plot, Fig.; 301, 303
  and origin of organic matter, Fig.; 293, 294
  parameters of LFER of octanol–water, Tab.; 302
  pH dependence; 310
  salt concentration dependence; 310
  temperature dependence; 310
  unit of in modeling concepts; 1059

Solid phase organic matter (POM); *See also Natural organic matter (NOM)*
  components of and their elemental composition, Tab; 296
  difference between aquatic and terrestrial POM; 293
  main components of; 294
  method of determining organic material; 291
  molecular composition, size and structure; 295
  relation to organic carbon mass; 292
  relative nucleophilicity, range of values; 498
  role as sorbent in aquatic environment; 291

Solid surface; *See also Charged mineral; Mineral surface*
  charged particulate organic matter; 425
  chemical and morphological characteristics; 397
  properties for various minerals and humus, Tab.; 424
  values of H-acceptor/donor, vdW parameters, Tab.; 398

Solid-to-water phase ratio
  definition; 286
  and porosity of solid–water system; 287
  typical values in lakes; 287

in ocean, surface and deep water, Tab.; 1052
of sediment relative to open water; 851

Solid–vapor equilibrium; *See Saturation pressure of pure solid*

Solid–water distribution coefficient of compound
definition; 282
apparent, and dissolved organic matter; 300
apparent and true; 300
apparent, colloids as sorbent; 1069
apparent, for weakly sorbing compound; 300
apparent, in experimental determination; 299
calculation for water and cosolvent–water, Ill Ex.; 312
calculation from experimental data, Ill. Ex.; 283
calculation of anion sorption on oxide, Ill. Ex.; 446
of EDA force adsorbed sorbate to mineral surface; 414
and Freundlich constant of compound; 283
as function of compound concentration; 283
as function of organic matter of solid, Fig.; 292
and quality of organic matter; 293
relative change to change of concentration; 283
as sum of various individual sorption mechanisms; 289, 389
unit of, used in modeling; 1059

Solid–water distribution of compound
calculation of concentration on solid, Ill. Ex.; 283
evaluation of experimental data, Ill. Ex.; 282, 284
ion exchange–water exchange; 291
mineral surface–water exchange; 290
most important mechanisms of exchange; 290
qualitative descripton; 280
solid phase organic matter–water exchange; 290
surface reaction–water exchange; 290

Solubility of compound; *See also Activity coefficient of compound*

Solubility of compound in liquid solution
and activity of compound; 356
calculation for binary mixture, Ill. Ex.; 170
as function of cosolvent fraction; 169
values for selected cosolvents, Tab.; 168

Solubility of compound in water; *See Aqueous solubility*

Solubility of water in organic liquid
definition; 136

Solution; *See also Ideal solution; Real solution*
solution and mixture, definitions; 183

Solvatochromic parameters, definition; 174

Sorbate
definition; 277

Sorbed-to-dissolved distribution ratio of ion
calculation of anion sorption on oxide, Ill. Ex.; 446
effect of hydrophobicity of ion on constant, Fig.; 434
of organic ion on charged mineral; 426

Sorbent
definition; 277

Sorption
definition; 277
and bioavailability of compound; 690
finite bath; 879
linear model, Fig.; 881
sorbed versus dissolved species, illustration, Fig; 278
sorbent–sorbate interactions, illustrations, Fig.; 279
survey of mechanisms; 277

Sorption coefficient of acid
charged species and natural organic matter; 322
neutral species and natural organic matter; 322

Sorption isotherm of compound
definition; 280
of amphiphilic sorbate on charged surface, Fig.; 440
distributed reactivity model; 282
dual-mode model; 282, 304
dual-mode model, estimated vs. measured values; 307
evaluation of nonlinear sorption, Ill. Ex.; 305
Freundlich; *See Freundlich isotherm*
Langmuir; *See Langmuir isotherm*
nonlinear and concentration dependence; 304
nonlinear for charged mineral surface, Fig.; 418
of organic ion sorption on charged mineral; 428
schematic illustration of various types, Fig.; 280
on soil at various relative humidities, Fig.; 395
of water adsorption on mineral surface, Fig.; 394
water–mineral surface, linear and nonlinear, Fig.; 409

Sorption of compound to inorganic surface
schematic illustration of sorption phenomena, Fig.; 390
survey of important sorption phenomena; 389

Sorting coefficient
of porosity of aquifer; 1154

Specific discharge
in groundwater; 1152

Specific growth rate of cell population
definition; 740
as funct. of limiting substrate concentration, Fig; 741
as funct. of yield; 743
calculation for steady state in CSTR, Ill. Ex.; 747
many limiting substrates; 742
one limiting substrate; 742
two limiting substrates; 742

Specific intermolecular force; *See also Polar intermolecular force*
definition; 61

Specific rate of light absorption of compound
definition and unit; 630
chromophores forming photooxidants; 660
from light screening factor and near-surface rate; 639
near-surface conditions; *See Near-surface specific rate of light absorption*
near-surface total light conditions; *See Near-surface total specific light absorption rate*

total light absorption conditions; *See Total specific rate of light absorption*

Spectral photon fluence rate
definition; 627
integration over wavelength range; 631
values at noon, Tab.; 632, 633
values for whole day , Tab.; 638

Spherical boundary
with boundary layer; 878
exchange velocity; 874
mass flux; 871

Stability frequency
definition; 1022
and turbulent diffusivity; 1023
vertical of lake and turbulent diffusivity,Ill. Ex; 1025

Standard chemical potential
definition; 78

Standard chemical potential of aqueous solute
of electron, definition; 562
at infinite dilution reference state; 463, 247
of proton, definition; 247

Standard chemical potential of gaseous compound
definition; 75

Standard chemical potential of pure liquid
and chemical potential of pure liquid; 80

Standard condition
definition; 78
standard concentration of solute in solution; 247

Standard deviation
nondimensional in groundwater transport; 1162

Standard enthalpy change of activation
and Arrhenius activation energy; 480

Standard enthalpy change of fusion
calculation from pressure/temperature, Ill. Ex.; 109
estimated and measured values for compounds, Tab.; 124
and excess enthalpy in water; 142
at melting point temperature; 107

Standard enthalpy change of phase transfer
air–water and and excess enthalpy in water; 142
organic solvent to air transfer of compound; 185
water to air transfer of compound; 199
determination of; 88
lipid to air transfer of compound; 358, 371
natural organic matter to air transfer of compound; 358
solid–water and and excess enthalpy in water; 142
surface to air transfer of compound; 397
temperature dependence; App. D1; 88
value of plant to air transfer of compound; 363
values for water to air transfer of compound, Tab.; 200
water to dissolved organic matter transfer; 318
water to mineral surface transfer and EDA force; 412
water to mineral surface transfer of compound; 411
water to solid phase organic matter transfer; 310

water–organic solvent transfer of compound; 216

Standard enthalpy change of reaction
and acidity constant; 252
temperature dependence; App. D1; 465

Standard enthalpy change of sublimation
definition; 106

Standard enthalpy change of vaporization
at boiling point temperature; 104, 120
co-variation to entropy of vaporization, Fig.; 111
correlation to saturation pressure of pure liquid; 119
estimation of at other temperatures; 121
and excess enthalpy in water; 142
in organic solvent to air transfer of compound; 185
in water to air transfer of compound; 199
and intermolecular forces; 110
temperature dependence; 105
values for selected organic compounds, Tab.; 200

Standard entropy change of fusion
definition; 123
calculation from pressure/temperature, Ill. Ex.; 109
estimated and measured values for compounds, Tab.; 124
estimation from molecular data; 123
at melting point; 107

Standard entropy change of phase transfer
positional, conformational and rotational parts; 125
values for fusion and evaporation; 125

Standard entropy change of sublimation
definition; 107

Standard entropy change of vaporization
definition; 110
at boiling point temperature; 104, 120
calculation of at boiling point temperature; 113
co-variation to enthalpy of vaporization, Fig.; 111
effect of intermolecular forces on; 113
estimation at boiling point using Trouton's rule; 111

Standard free energy change
of adsorbing compound at surface, Fig.; 67
of cavity formation in liquid, Fig.; 67
electrostatic attraction of ionic species; 426
hydrophobic attraction, funct. of chain length; 435
hydrophobic attraction of ionic species; 426
of insertion compound in liquid cavity, Fig.; 67
of London dispersive energy; 64

Standard free energy change of activation
definition; 480
for LFER estimation of reaction rates; 481
and LFERs of reaction rates; 531
nonenzymatic and enzymatically catalysed, Fig.; 696
in one-electron transfer reaction; 584

Standard free energy change of adsorption
value for water on silica surface; 392

values for sorbates of various polarities, Fig.; 393

Standard free energy change of fusion
definition; 106
calculation from pressure/temperature, Ill. Ex.; 109
estimation from molecular data and melting point; 124
and subcooled liquid state; 107

Standard free energy change of phase transfer
definition; 66
air to liquid transfer of compound, Fig.; 67
air to surface transfer of compound, Fig.; 67
air–solid surface exchange of compound; 71
calculation from structural group contributions; 92
as difference of excess free energies of compound; 85
and equilibrium partition constant; 66
and intermolecular forces; 67
pure liquid to air transfer of compound; 68
water to mineral surface transfer of ion; 426

Standard free energy change of proton transfer
definition; 248
and Hammett correlation; 261
of unsubstituted and substituted acid; 261, 263

Standard free energy change of reaction
definition; 464
activated complex in E2 reaction, Fig.; 510
activated complex in SN-1 reaction, Fig.; 496
activated complex in SN-2 reaction, Fig.; 495
and equilibrium reaction constant; 465
redox process and one-electron reduction potential; 569
and standard reduction potential; 562
standard hydrogen electrode (SHE); 561
substrate-enzyme hydrolysis reaction, Fig.; 755

Standard free energy change of reaction (W)
definition; 562
of environmentally relevant aquatic reactions; 563
of organic compounds in aqueous solution; 564

Standard free energy change of sublimation
definition; 106
and excess free energies; 106

Standard free energy change of vaporization
enthalpy and entropy contributions, Fig.; 111
and excess free energy of gas; 104
and intermolecular forces; 110, 120
polar part of, estimation from molecular data; 115
proportionality to enthalpy change of vaporization; 119
and saturation pressure of pure liquid; 104

Standard free energy of formation of compound
calculation for change of reference state, Ill. Ex; 467, 566, 567
electron in aqueous solution, definition; 562
proton in aqueous solution, definition; 247
of solute in aqueous solution; 247

Standard heat capacity change of vaporization
estimation of for pure liquid; 121

Standard hydrogen electrode (SHE)
thermodynamic definition; 561
description; 560
schematic illustration of, in redox reaction, Fig.; 560

Standard pressure
value of; 78

Standard redox potential of reaction
calculation from reduction potentials; 569

Standard reduction potential in natural water (W)
definition; 562
calculation for denitrification reaction, Ill. Ex.; 565, 566, 567
environmentally relevant aquatic reactions; 563
and one-electron reduction potentials; 569
organic compounds in aqueous solution; 564
organic, inorganic, mediators redox couples, Tab.; 575

Standard reduction potential of half-reaction
definition; 562
calculation for denitrification reaction, Ill. Ex.; 565, 566, 567
environmentally relevant aquatic reactions; 563
and one-electron reduction potentials; 569
organic compounds in aqueous solution; 564

Standard state of compound
definition; 78
as aqueous solute; 247
and reference state; 77

State variable
definition; 949, 950

Static equilibrium method
for measuring partition constants; 203

Steady state
indifferent, stable, unstable; *975*

Steady state concentration of compound
advection, diffusion and reaction parts; 1008
and bioaccumulation in living organism; 350
calculation for hydroxyl radical, Ill. Ex.; 662, 665
in first-order rate law; 473
Monod-type degradation in CSTR, Ill. Ex.; 747
photooxidant in indirect photolysis reaction; 661

Steady state of dynamic system
definition; 471
time to reach; *See Time to steady state*

Stereoisomerism
definition; 15, 26
dia-stereoisomerism, definition; 26

Stereoselectivity
and dia-stereoisomerism; 26

Steric effect
of substituents on acidity constant; 260
of substituents on acidity constant, Fig.; 260

Stokes law; 1061

Stokes-Einstein relation
of molecular diffusivity in water; 810, 812

Stratification
analogy to mechanics, schematic illustration, Fig.; 1023

strength of and mechanical stability; 1022

Stratopause
a noninterface boundary, graph, Fig.; 835

Structural isomerism; 17

Structure of compound
definition; 15

Structure-biodegradability relationship; *See Biodegradability of compound*

Subcooled state of liquid compound
definition; 103
aqueous solubility, function of temperature, Fig.; 155
illustration in phase diagram, Fig.; 101
saturation pressure of; 103; *See also Saturation pressure of pure liquid*
as solute of a solid pure compound; 103
and standard free energy change of fusion; 107
vapor pressure versus temperature, Fig.; 106

Suboxic condition
definition; 571

Substituent
definition; 25
effect on hydrolysis rate of an acid ester, Fig.; 532
values of Hammett constants, Tab.; 263, 266

Substrate disappearance rate
as function of yield; *743*
in Michaelis-Menten enzyme kinetics; 751
Michaelis-Menten-type kinetics, Fig.; 753
in Monod-type kinetics; 741

Substrate specificity
and co-metabolism; 753
of microbially mediated reactions; 697

Suicide metabolite
definition and example; 700

Supercritical fluid
definition; 103
illustration in phase diagram, Fig.; 101

Superheated state of liquid compound
definition; 102
aqueous solubility and activity coefficient; 139
aqueous solubility, function of temperature, Fig.; 155
illustration in phase diagram, Fig.; 101
saturation pressure of; *See Saturation pressure of pure liquid*
vapor pressure versus temperature, Fig.; 106

Surface area of river
integrated area; *See Integrated surface area of river*

Surface film on water interface
effect on air–water exchange rate of compound; 929

Surface mixed sediment layer (SMSL)
definition; 1073
characteristic quantities, Tab.; 1074
dynamic model for compound, Tab.; 1074
in one-box model of PCBs in Lake Superior; 1073
mass balance equation of compound; 1075
mass fluxes through layer, Tab.; 1074
in one-dimensional vertical lake model; 1085

processes in, schematic illustration, Fig.; 1073
rate law of compound total concentration; 1089

Surface mixed water layer (SMWL)
in one-dimensional vertical lake model; 1085
rate law of compound total concentration; 1086
rate law of particle concentration; 1087

Surface reaction of sorbate
calculation of anion sorption on oxide, Ill. Ex.; 446
equilibrium constant, definition; *443*
with mineral surface; 441
with mineral surface, illustration, Fig.; 390
with natural organic matter (NOM); 441
reaction formulation; *443*
selected sorbates reacting with minerals, Tab.; *442*

Surface reactive sites of mineral
concentration on surface; *443*

Surface sites, total of sorbent
in Langmuir isotherm; 281
and maximum surface concentration of compound; 281

Surfactant
effect on air–water exchange of compound; 929

Susceptibility factor
in Hammett equation of acidity constant; 263
in Hammett equation of hydrolysis rate constant; 532
values for some aromatic acids, Tab.; 266

Suspended solids
concentration, ocean surface and deep water, Tab.; 1052
effect on direct photolysis of compound; 649
in lakes, main organic components of; 294
in oceans, main organic components of; 294
physical behavior of; *See Particle*
settling, vertical mass flux of compound; 985
typical concentrations in lakes; 287

Swain-Scott relationship; 497
temperature effect; 499

Syntrophy
definition; 695

System
definition; 949
definition and schematic illustration, Fig.; 951
boundary; 952

T

Taft correlation
literature references to; 266

Tautomerism; 715

Temperature
effect on different states of molecules; 613
and energy of molecular motion; 22
reaction rate sensitivity and reaction mechanism; 511
reference value at standard conditions; 78
saturation pressure and states of pure compound; 102

Temporal change of concentration
in one-box model; 483

Tertiary carbon atom (tert-)
definition; 32

The Henry's law
characterizing an interface boundary; 837

The Henry's law constant
definition; 184, 197
approximaton of; 198
between air and seawater, Ill. Ex.; 890
critical value of single phase control; 893
effects of compound structure on; 198
temperature dependence, Ill. Ex.; 890

Theoretical bioaccumulation potential (TBP)
definition; 344
lipid normalized, calculation, Ill. Ex.; 380
lipid normalized, definition; 345

Thermal expansivity
of water; 1023

Thermal motion
of atoms and molecules; 780

Thermocline
definition; 836
noninterface boundary, graph, Fig.; 835

Thermocline depth
in ocean, Tab.; 1052

Time to steady state
dependence on reaction rate constants; 474
diffusive flux in unsaturated zone, Ill. Ex.; 847
effect of flushing rate and reaction rates; 484
in one-box model at constant input; 484
theoretical and practical definition; 471

Topographic function
definition; 1085

Tortuosity
in layer of groundwater system, Tab.; 1150
in porous media diffusion; 816
in porous media dispersion; 1155

Total aqueous solubility of acid or base
definition; 268
as function of pH, Fig.; 269

Total rate of light absorption by compound
definition; 636

Total reaction rate constant of compound
definition; 483
calculation from field data, Ill. Ex.; 484

Total specific rate of light absorption
definition; 636
calculated from field and compound data, Tab.; 638

Total specific rate of total light absorption
definition; 637
calculation for a water body, Ill. Ex.; 640

Total surface area of compound
and contact area of sorbate with surface; 395
estimation of; 115

Toxic ratio
definition; 378

Toxicity of compound; *See also Acute toxicity; Baseline toxicity*
specific and nonspecific toxicity; 374

Trans- isomer; *See Cis-/trans- isomerism*

Transfer resistance
  definition; 844
  total; 844
  total, at equilibrium partitioning of compound; 846
Transfer velocity
  at bottleneck boundary, definition; 840
  at bottleneck boundary, calculation, Ill.Ex.; 841
  at interface; 840
  at two-phase boundary, equilibrium partitioning; 846
Transformation reaction of compound
  indirect photolysis in atmosphere; 672
  indirect photolysis in surface water; 658
  calculation of direction, Ill. Ex.; 467
  catalyzed; 475
  direct photolytic process; *See Direct photolysis*
  environmental quality of products; 459
  formulation of reversible reaction equation; 463
  indirect photolytic process; *See Indirect photolysis*
  kinetics of; 468
  reference and standard state in aqueous solution; 463
  survey of environmental processes; 459
  thermodynamics of; 463
Transient photooxidant; *See Photooxidant*
Transition state theory
  of elementary reaction; 479
Transport
  advective; 953
  by random motion; 953
  by directed motion, Tab.; 1008
  by directed motion; 779
  Gauss' theorem; 789; *See also Gauss' theorem*
  mean and fluctuating parts; 1022
  by random motion; 779
  by random motion, Tab.; 1008
  turbulent; 1019
Triple bond
  definition; 18
  lengths and enthalpies; 20
  picture of electron orbitals, Fig.; 27
  symbolism in drawing structure; 19
  type of the third bond; 27
Triple point temperature of compound
  definition; 100
  illustration in phase diagram, Fig.; 101
Tropopause
  definition; 836
  a noninterface boundary, graph, Fig.; 835
Trouton's rule; 111
Turbulence
  internal friction generated; 1023
  shear generated; 1023
Turbulent diffusion coefficient of compound; *See Turbulent diffusivity*
  definition; 1021, 826
Turbulent diffusive transport of compound; 825
  in atmosphere, in lakes and in ocean; 827

  description of turbulence; 1019
  exchange model; 1019
  exchange model, Fig.; 1020
  and Reynolds number; 1019
Turbulent diffusivity
  and density gradients; 1022
  horizontal and vertical; 1022
  and spatial scale, summary of equations; 1032
  turbulent structure; 1022
  vertical; *See Vertical turbulent diffusivity*
Turbulent flux of compound
  definition; 1019, 1021
  part of advective flux; 1020
Turbulent mixing; 954
Twist conformation
  picture, Fig.; 28
Two-box model; 953, 982
  definition; 950
  in a coupled air–water system, Fig.; 983
  of lake; 1075; *See also Lake Superior*
  of lake, general remarks to various box evaluation; 1082
  linear with one variable; 982
  linear with two or more variables; 990
  nonlinear; 990
  of sediment–water system; 1075
  sediment–water system, scheme, Fig.; 1076
  stratified lake, equations of processes; 986
  stratified lake, relevant processes; 984
  stratified lake, schematic illustration, Fig.; 985

U

Unified atomic mass unit
  definition and value; 15
Unimolecular elimination reaction; *See Elimination reaction, first-order*
Unimolecular nucleophilic substitution; *See Nucleophilic substitution, first-order*
Unknown chromophore (UC)
  concentration estimation in Greifensee; 658
  excited state (UC*), effect on triplet oxygen; 658
  excitement and reaction cycle, scheme, Fig.; 659
  excitement cycles per day in a eutrophic lake; 658
Unsaturated compound
  definition; 32
Unsaturated zone
  diffusion of compound; 818
  transport of compound; 1175
Unshared electron; *See Nonbonding electron pair*
Uptake rate of compound by microorganism
  compound structure and cell permeation; 691
  limiting overall transformation rate; 691
  meanings of uptake rate constant; 736
  rate law for uptake as limiting process; 735

V

Valence electron
  definition; 17
  symbolization in formula drawing, Fig.; 18
Valence shell
  definition; 17
  and octet rule; 17
Valency of an atom
  and number of covalent bonds; 17
Van der Waals (vdW) force of compound
  definition; 60
  in air–surface partition coefficient; 395
  role in adsorption of compound on surface; 392
  and saturation pressure of pure liquid; 396
Van der Waals (vdW) force of surface
  in air–surface partition coefficient LFER; 396
  as function of relative humidity of air, Fig.; 399
  values for selected liquids and solids, Tab; 398
Van der Waals (vdW) parameters controlling
  air–apolar surface partition constant, Fig.; 72
  air–hexadecane partition constant, Fig.; 71
  air–polar surface partition constant, Fig.; 72, 393
  air–pure liquid partition constant, Fig.; 69
  air–water partition constant, Fig.; 71
  organic solvent–water partition constant; 216
Van't Hoff equation; 88
  temperature dependence of saturation pressure; 105
Vapor pressure of compound; *See Saturation pressure*
Velocity of reaction of substrate
  definition; 751
  as function of substrate concentration; 751
  as function of substrate concentration, Fig.; 752
  in Michaelis-Menten enzyme kinetics; 751
Velocity shear
  origin of; 1033
Vertical mixing by turbulence in river; 1120
  calculation of time and length scale, Ill. Ex.; 1124
  coefficient, definition; 1120
  schematic illustration, Fig.; 1121
Vertical mixing of compound
  and boundary fluxes; 1083
  in lakes from temperature, Ill. Ex..; 841
  in river; 1112
Vertical turbulent diffusivity
  and excess radon-222 activity; 1030
  in a lake, calculation, Ill. Ex.; 1025
  measurement from noble gas isotopes; 1029
  from temperature profile; 1024
  from temperature profile, calculation, Ill. Ex.; 1025
  and vertical stability frequency; 1023
  and vertical stability frequency in a lake, Fig.; 1028, 1029
Vicinal water
  definition; 410, 419
  amount of at inorganic surfaces; 410
  of charged mineral surface, schematic picture, Fig; 419

and near surface water layer; 419
sorption into, schematic illustration, Fig; 390
thickness in natural waters; 419

Vicinal water–water partitioning of comp.
formulation; 410

Viscosity of water
at various temperatures; App. B; *See also
Dynamic viscosity; Kinematic viscosity*

Volatile compound
enrichment from water; 204

Volume fraction concentration of compound
calculation from mole fraction for two
solvents; 170

Von Kàrmàn constant
in vertical turbulent mixing in river; 1120

W

Wall boundary; 848
definition; 837
between different phases; 850
between identical phases; 849
critical time, definition; 856
critical time, gas and stagnant liquid phases,
Tab; 858
critical time, water and porous media phases,
Tab.; 858
diffusivity and concentration profiles, Fig.; 849
diffusivity profile across boundary, Fig.; 837
half-concentration penetration depth; 854
mass flux at; 849
time-variable boundary concentration; 864
turbulence at; 849
with boundary layer; 854
with phase change, concentration profile,
Fig.; 850

Water
in acid–base reaction, standard definition; 247
adsorbed on mineral surface, illustration,
Fig.; 393
competing with organic sorbate on mineral
surface; 399
concentration change as reaction partner; 470
droplets, as sorption surface; *See Air–surface
adsorption of compound*
in microbial hydration reaction of double
bond; 734
in microbially mediated hydrolysis; *See
Microbially mediated hydrolysis reaction*
ionisation constant, definition; 252
ionisation constant, various temperatures; App.
D2; 252
layer on mineral surface and humidity, Fig.; 394
molecular order in bulk and around cavity; 144
as nucleophilic reactant; *See Hydrolysis
reaction*
quencher of photooxidants; 660, 666
in reaction with an organic acid; 246
and relative humidity of air; 392
sorbed on mineral surface; 392
as test compound of transfer velocity in air; 896
thermal expansivity; 1023
transfer velocity in air and wind speed,
Fig.; 897

Water solubility of compound; *See* **Aqueous
solubility of compound**

Water surface
adsorption of apolar compound; 72
values of H-acceptor/donor, vdW parameters,
Tab.; 398

Water-phase transfer velocity of compound
impact of wind speed, Fig.; 901
for low wind speed, Fig.; 905
reference compound and wind speed; 914
summary for reference and optional compound,
Tab.; 915
temperature dependence; 913
values for two wind speeds, various compounds,
Tab; 917
of references and wind speed, relationships,
Tab.; 902

**Water-phase transfer velocity of compound in
river**
parameters of function; 922

Wavelength of electromagnetic radiation
definition; 614
solar radiation, range of chemically effective; 613

Weibull distribution; 903

Weight
molar; *See Molar mass*
molecular; *See Molecular mass*

Well-mixed reactor; *See One-box model*

Wind speed
air-phase transfer velocity of water, Tab.; 898
effect on air–water exchange velocity, Ill.
Ex.; 918
impact on air-phase transfer of water, Fig.; 897
impact on evaporation of water, Fig.; 897
impact on exchange rate of contaminated water,
Fig; 931
impact on water-phase transfer of compound,
Fig.; 901
Weibull distribution; 903

Y

Yield of biomass enhancement
common value; 743
definition; 743
values for various growth-limiting
substrates; 746

Z

Z-function, near-surface photolysis
values at noon and per day, Tab.; 632, 633

**Z-function, near-surface photolysis rate
constant**
definition; 631
integration over wavelength range; 631

Zero-order reaction
step in Michaelis-Menten enzyme kinetics; 477

Zwitterion
fraction of as function of pH; 256
microscopic and macroscopic acidity constants; 255

COMPOUND INDEX

---

used name in text

abbreviation

2,2',3,4,4',5,5'-Heptachloro biphenyl   PCB180 **E10.1** —— structure information

2,3,4,5,2',4',5'-Heptachloro-biphenyl —— IUPAC name

air-lipid partition coefficient   *T10.5* —— data in table 10.5

bioconcentration factor in   *E10.1* —— data in illustrative example 10.1
phytoplankton

concentration in air, pasture and   *T10.5* —— data in table 10.5
milk

fugacities in air and terrestrial food   *F10.11* —— data in figure 10.11
chain

Legend:

B	box
E	illustrative example
F	figure
P	problem
Q	question
T	table

Bold type entries (e.g. **E20.5, 1065**) mean **structure** information
Italic type entries (e.g. *T14.2, 176*) mean *data* information
plain text entries (e.g. Q3.4, 72) mean general information

Examples:

F10.11 means a general information in figure no 11 of chapter 10
1065 means structure information on page 1065
P13.8 means data are provided in problem no 8 of chapter 13

A

Acenaphthene
Acenaphthene

H-donor (α) and   *T4.3*
H-acceptor property (β)

polarizability *T5.5*

Acetaldehyde   **F2.17**
Acetaldehyde

air-water partition constant *E20.5*

chemical intermediate 41

formation by enzymatic hydration of 734
ethyne

hydration rate constant *E20.5*

microbial reduction to ethanol 721

overall air-water exchange velocity *E20.5*
at different wind speeds,
calculation of

rate constant with hydroxyl radical *F16.3*

rate constant with hydroxyl radical *F16.7*
in troposphere

Acetamide   **T13.10**
Acetamide

rate constant with hydroxyl radical *F16.3*

Acetic acid   **F2.17**
Acetic acid

acetate as nucleophile *T13.3*
    T13.1

acidity constant *T8.1*

acidity constant at different *T8.3*
temperatures

in methanogenic degradation, *E12.2*
calculations

microbial degradation to methane 695

molecular diffusivity in water *F18.10*

presence in the environment by   42
direct input and by hydrolysis of
esters

product in ethyl acetate hydrolysis *T13.2*
reaction

rate constant with hydroxyl radical *F16.7*
in troposphere

refractive index *T3.1*

standard Gibbs energy of acetate in *E12.2,*
aqueous solution   *E13.1*

value of μ_{max} and K_{iM} *T17.6*

Acetic acid 2,4-dinitrophenyl   **F13.8**
ester
Acetic acid 2,4-dinitro-phenyl ester

activation energy of hydrolysis, *E13.4*
experimental data and calculations

half life time of hydrolysis at given *E13.5*
pH and temperature, calculations

hydrolysis rate constant and pH *F13.8*

hydrolysis rate constants, *E13.4*
experimental data and calculations

mechanism of metal catalyzed 541 hydrolysis

rate constants of hydrolysis reaction *T13.8, T13.9*

Acetic acid amide   T13.10
Acetamide

rate constants of hydrolysis reaction *T13.10*

Acetic acid *tert*-butyl ester   F13.8
Acetic acid *tert*-butyl ester

hydrolysis rate constant and pH *F13.8*

rate constants of hydrolysis reaction *T13.8*

Acetic acid ethenyl ester   T13.8
Acetic acid vinyl ester

rate constants of hydrolysis reaction *T13.8*

Acetic acid ethyl ester   F2.17
Acetic acid ethyl ester

extent of hydrolysis reaction in *E13.1* groundwater, calculations

hydrolysis rate constant and pH *F13.8*

hydrolysis reaction *T13.2*

molar volume, calculation *T18.3*

molecular diffusivity in air *F18.9*

rate constant with hydroxyl radical *F16.3*

rate constant with hydroxyl radical *P16.3* in troposphere

rate constants of hydrolysis reaction *T13.8, T13.9*

solvent *F2.17*

standard Gibbs energy in aqueous *E13.1* solution

water saturated, mole fraction *T5.1*

Acetic acid *N,N*-dimethyl   T13.10 amide
N,N-Dimethyl-acetamide

rate constants of hydrolysis reaction *T13.10*

Acetic acid *N*-methyl amide   T13.10
N-Methyl-acetamide

rate constants of hydrolysis reaction *T13.10*

Acetic acid phenyl ester   F13.8
Acetic acid phenyl ester

hydrolysis rate constant and pH *F13.8*

rate constants of hydrolysis reaction *T13.8, T13.9*

Acetone   F2.2
Propan-2-one

aqueous solution data *T5.2*

as cosolvent *T5.8*

bond angles *F2.3*

molecular diffusivity at different *T20.3* temperatures

molecular diffusivity in water *F18.10*

rate constant with hydroxyl radical *F16.3*

rate constant with hydroxyl radical *F16.7* in troposphere

refractive index *T3.1*

Schmidt number in water at *T20.3* different temperatures

solvent and chemical intermediate 41

solvent-water partition constants for *T7.1* five solvents

Acetonitrile   T5.8
Acetonitrile

as cosolvent *T5.8*

solvent fitting coefficients for   *T6.2* air-acetonitrile partition constant of organic compounds

17α-Acetoxy-progesterone   F2.25
Acetic acid 17-acetyl-10,13-dimethyl-3-oxo-2,3, 6,7,8,9,10,11,12,13,14,15,16,17-tetradeca hydro-1*H*-cyclopenta[a]phenanthren-17-yl ester

hormone F2.25

Acetylchloride   F2.3
Acetyl-chloride

bond angles *F2.3*

***N*-Acetyl-phenylalanine   F17.19**
2-Acetylamino-3-phenyl-propionic acid

formation by enzymatic hydrolysis *F17.19* of *N*-Acetyl-phenylalanine esters

***N*-Acetyl-phenylalanine   F17.19 4-nitro-phenyl ester**
2-Acetylamino-3-phenyl-propionic acid 4-nitro-phenyl ester

rate of enzymatic hydrolysis *F17.19*

***N*-Acetyl-phenylalanine ethyl   F17.19 ester**
2-Acetylamino-3-phenyl-propionic acid ethyl ester

rate of enzymatic hydrolysis *F17.19*

***N*-Acetyl-phenylalanine   F17.19 methyl ester**
2-Acetylamino-3-phenyl-propionic acid methyl ester

rate of enzymatic hydrolysis *F17.19*

Acetylene
see Ethyne

2-Acetylnitrobenzene 2-Ac-NB
1-(2-Nitro-phenyl)-ethanone

one-electron reduction potential *T14.4*

relative reduction rates in different *F14.10* reduction media

3-Acetylnitrobenzene 3-Ac-NB
1-(3-Nitro-phenyl)-ethanone

one-electron reduction potential *T14.4*

relative reduction rates in different *F14.10* reduction media

4-Acetylnitrobenzene 4-Ac-NB
1-(4-Nitro-phenyl)-ethanone

one-electron reduction potential *T14.4*

relative reduction rates in different *F14.10* reduction media

relative reduction rates in different *F14.5* systems

Acrolein   F2.10
Propenal

delocalization of π-electrons F2.10 29

Air

typical transfer velocity of a *893* substance in air

water content and air-water 889 interface in atmosphere

Alachlor   F2.21
2-Chloro-*N*-(2,6-diethyl-phenyl)-*N*-methoxymethyl-acetamide

precursor of sulfonic acid F2.21 metabolite

reaction with nucleophiles 501

Aldicarb   F2.24
1-Methyl-3-(2-methyl-2-methylsulfanyl-propylidene)-urea

influence of metals on   543 decomposition mechanism

insecticide and nematicide F2.24

microbial oxidation to sulfoxide and 717 sulfone

4-Amino-2,6-dinitrotoluene   F14.6
4-Methyl-3,5-dinitro-phenylamine

exchange constant   *T11.2* water-aluminosilicate surface ($K_{NAC,EDA}$)

formation by microbial reduction of 721 *N*-(4-Methyl-3,5-dinitro- phenyl)-hydroxylamine

intermediate in TNT reduction F14.6

one-electron reduction potential *T14.4*

2-Amino-4,6-dinitrotoluene   F14.6
2-Methyl-3,5-dinitro-phenylamine

exchange constant   *T11.2* water-aluminosilicate surface ($K_{NAC,EDA}$)

intermediate in TNT reduction F14.6

one-electron reduction potential *T14.4*

2-Amino-naphthalene-4,8-disulfonic acid

azo dye component F2.20

3-Aminonitrobenzene 3-NH$_2$-NB
3-Nitro-phenylamine

one-electron reduction potential *T14.4*

4-Aminonitrobenzene 4-NH$_2$-NB
4-Nitro-phenylamine

one-electron reduction potential *T14.4*

2-Aminopropylbenzene   Q7.4
1-Methyl-2-phenyl-ethylamine

solvent-water partition constants for *P7.3* five solvents

3-Aminophenyl   **P13.6**
N-methyl-N-phenyl
carbamate
Methyl-phenyl-carbamic acid 3-amino-phenyl
ester

   rate constant of base catalyzed *P13.6*
   hydrolysis

4-Aminopyridine
Pyridin-4-ylamine

   acidity constant of corresponding *T8.3*
   acid at different temperatures

4-Aminotoluene
p-Tolylamine

   salting constant, seawater *T5.7*

Ammonia   **F2.3**
Ammonia

   bond angles *F2.3*

Aniline An   **F2.18**
Phenylamine

   acidity constant of corresponding *T8.2*
   acid   *T8.6*
       F8.3

   aqueous solution data *T5.2,*
       T5.3

   biological organic phases-water *T10.2*
   partition coefficients

   chemical intermediate 43

   decadic molar absorption   *F15.5*
   coefficient of neutral and cationic
   species

   Fishtine and structural parameters *T4.4*
   for estimating vapor pressure

   H-donor (*α*) and   *T4.3*
   H-acceptor property (*β*)

   Hammett constant for ortho *T8.7*
   substitution

   Hammett susceptibility factor of *T8.6*
   corresponding acid

   molecular diffusivity in air *F18.9*

   molecular diffusivity in water *F18.10*

   octanol-water partition constant *T10.2*

   radicals of oxidation and products of F14.19
   coupling

   rate constant with hydroxyl radical *F16.3*

   rate constant with hydroxyl radical *F16.7*
   in troposphere

   relative oxidation rate by *F14.20*
   manganese oxide

   resonance of nonbonded electrons 30

   solvent-water partition constants for *T7.1*
   five solvents

   sorption isotherm of cation on *F11.9*
   natural solid

   standard enthalpies of vaporization *T6.3*
   and of water-air transfer

   standard enthalpy and standard *T4.2*
   entropy of evaporation

   standard enthalpy of water-air *P8.2*
   transfer

   standard oxidation potential *T14.3,*
       F14.4

   water saturated, mole fraction *T5.1*

Anthracene   **F2.13**
Anthracene

   aqueous solution data *T5.2,*
       T5.3

   decadic molar absorption *F15.3*
   coefficient

   H-donor (*α*) and   *T4.3*
   H-acceptor property (*β*)

   parameters for estimating vapor *T4.4*
   pressure

   rate constant with hydroxyl radical *F16.7*
   in troposphere

   reaction quantum yield of direct *T15.7*
   photolysis

   salting constant, seawater *T5.7*

   standard enthalpies of vaporization *T6.3*
   and of water-air transfer

   standard enthalpy and standard *T4.2*
   entropy of evaporation

Argon
Argon

   molecular diffusivity in water *F18.10*

Atrazine   **F2.18**
6-Chloro-*N*-ethyl-*N'*-isopropyl-[1,3,5]triazine-2,4-
diamine

   aqueous solubility *E22.3*

   herbicide F2.18

   horizontal transport time in lake *E22.3*
   from spill cloud, calculations

   indirect photolysis half-life, *E16.2*
   calculation of

   initial concentration in river after *E24.5*
   peak input

   rate constant of indirect photolysis *E16.2*
   with hydroxyl radical

   rate constant with hydroxyl radical *F16.3*

   reduction of concentration by   *E24.5*
   dispersion in river after peak input

   sediment–water exchange along *E24.3*
   river flow, peak input

   variability of natural organic *F9.9*
   carbon-water distribution
   coefficient of

Azobenzene   **F2.18**
Diphenyl-diazene

   chemical intermediate F2.18

   decadic molar absorption *F15.2*
   coefficient

   n to *π** electron transition and dyes 618

   product of oxidative coupling of F14.19
   aniline oxidation

   standard reduction potential *T14.3,*
       F14.4

B

Benzaldehyde   **F2.17**
Benzaldehyde

   aqueous solution data *T5.3*

   H-donor (*α*) and   *T4.3*
   H-acceptor property (*β*)

   polarizability *T5.5*

   salting constant, seawater *T5.7*

   standard enthalpies of vaporization *T6.3*
   and of water-air transfer

Benzene   **F2.13**
Benzene

   activity coefficients in organic *T3.2*
   solvents and in water

   air-quartz partition constant and *F3.7*
   dispersive vdW parameter

   air-water exchange at 5 and 25 °C, *E6.2*
   calculations

   air-water partition constant *E20.4*
       T20.5

   air-water partition constant E6.4
   estimation, calculations

   aqueous solution data *T5.2,*
       T5.3

   as air pollutant in olive oil, E6.1
   calculations

   concentration in diesel fuel *T19.3*

   decadic molar absorption *F15.2*
   coefficient

   degradation under methanogenic *E17.1*
   conditions

   delocalization of electrons 34

   density *T19.3*

   dissolution from diesel fuel into *E19.4*
   aqueous phase, calculations

   energies of air-hexadecane,   *T3.4*
   air-water and hexadecane-water
   transfers

   excess functions in gas phase, *T3.3*
   hexadecane and water

   flux from gasoline through porous *E18.4*
   media to atmosphere, calculation

   H-donor (*α*) and   *T4.3*
   H-acceptor property (*β*)

   microbial oxidation to catechol 699

   molar volume, calculation *T18.3*

   molecular diffusivity *E20.4*

   molecular diffusivity at different *T20.3*
   temperatures

   molecular diffusivity in air *T20.5*

   molecular diffusivity in air, *F18.9,*
   calculation   *803*

   molecular diffusivity in water *F18.10*
       T20.5

molecular diffusivity in water and in *T19.3* diesel fuel

parameters for estimating vapor *T4.4* pressure

polarizability *T5.5*

rate constant with hydroxyl radical *F16.3*

rate constant with hydroxyl radical *F16.7* in troposphere

refractive index *T3.1*

salting constant, seawater *T5.7*

salting constants, single salts *T5.6*

Schmidt number in water *T20.5*

Schmidt number in water at *T20.3* different temperatures

solvent fitting coefficients for *T6.2* air-benzene partition constant of organic compounds

standard enthalpies of vaporization *T6.3* and of water-air transfer

standard enthalpy and entropy of *T4.5* fusion, experimental and predicted data

standard enthalpy and standard *T4.2* entropy of evaporation

transfer velocities in air, in water *T20.5*, and overall at two wind speeds *F20.7*

typical concentrations in air of *P20.1* remote and urban areas

value of K_{fMM} and V_{max} *T17.7*

vapor pressure as function of *F4.4* temperature

water phase transfer velocity with *E20.4* flow turbulence, calculation of

water saturated, mole fraction *T5.1*

Benzene sulfonic acid
Benzene sulfonic acid

surfactant 48

Benzidine   **F14.19**
Biphenyl-4,4'-diamine

product of oxidative coupling of F14.19 aniline oxidation

sorption isotherm data for a *P11.10* sediment

Benzimidazole   **T8.2**
1*H*-Benzimidazole

acidity constant of corresponding *T8.2* acid

Benz[*a*]anthracene
Benzo[a]anthracene

decadic molar absorption *F15.3* coefficient

H-donor (α) and *T4.3* H-acceptor property (β)

salting constant, seawater *T5.7*

vapor pressure as function of *F4.4* temperature

reaction quantum yield of direct *T15.7* photolysis

Benzo[*a*]fluorene
11*H*-Benzo[a]fluorene

H-donor (α) and *T4.3* H-acceptor property (β)

Benzo[*a*]pyrene BP   **F2.13**
Benzo[def]chrysene

air-water partition constant *T20.5*

aqueous solution data *T5.2*, *T5.3*

concentrations in earthworm and *P10.2* soil

enthalpy of evaporation *P11.1*

fraction in water at varying *E9.5* dissolved organic carbon content, calculations

H-donor (α) and *T4.3* H-acceptor property (β)

molecular diffusivity in air *T20.5*

molecular diffusivity in water *T20.5*

reaction quantum yield of direct *T15.7* photolysis

salting constant, seawater *T5.7*

Schmidt number in water *T20.5*

transfer velocities in air, in water *T20.5*, and overall at two wind speeds *F20.7*

Benzo[*f*]quinoline BQ   **753**
Benzo[f]quinoline

co-metabolic degradation with F17.17 quinoline

Benzo[*ghi*]perylene
Benzo[ghi]perylene

H-donor (α) and *T4.3* H-acceptor property (β)

Benzoic acid   **F2.17**
Benzoic acid

acidity constant *T8.1* *T8.6* *F8.6*

correlation of biodegradability with 716 $\Sigma\sigma_j$

goethite surface-water distribution *E11.7* coefficient, calculations

Hammett susceptibility factor *T8.6*

microbial degradation to acetic acid 695

presence in the environment by 42 direct input and by hydrolysis of esters

reaction of anion with mineral T11.4 surfaces

standard enthalpy and entropy of *T4.5* fusion, experimental and predicted data

Benzoic acid CoA ester   **731**
microbial pathway intermediate 731

Benzoic acid ethyl ester   **481**
Benzoic acid ethyl ester

rate constants in water and water/ *T12.1* cosolvent mixtures

Benzonitrile
Benzonitrile

polarizability *T5.5*

1,4-Benzoquinone BQ   **561**
[1,4]-Benzoquinone

standard reduction potential *T14.3*

Benzotriazole   **T8.2**
1*H*-Benzotriazole

acidity constant of corresponding *T8.2* acid

N-Benzoyl-glycine   **F17.19**
Benzoylamino-acetic acid

formation by enzymatic hdrolysis of *F17.19* N-Benzoyl-glycine esters

N-Benzoyl-glycine ethyl ester   **F17.19**
Benzoylamino-acetic acid ethyl ester

rate of enzymatic hydrolysis *F17.19*

N-Benzoyl-glycine isobutyl ester   **F17.19**
Benzoylamino-acetic acid isobutyl ester

rate of enzymatic hydrolysis *F17.19*

N-Benzoyl-glycine isopropyl ester   **F17.19**
Benzoylamino-acetic acid isopropyl ester

rate of enzymatic hydrolysis *F17.19*

N-Benzoyl-glycine methyl ester   **F17.19**
Benzoylamino-acetic acid methyl ester

rate of enzymatic hydrolysis *F17.19*

N-Benzoyl-glycine pyridin-4-ylmethyl ester   **F17.19**
Benzoylamino-acetic acid pyridin-4-ylmethyl ester

rate of enzymatic hydrolysis *F17.19*

Benzylalcohol
Phenyl-methanol

aqueous solution data *T5.3*

Fishtine and structural parameters *T4.4* for estimating vapor pressure

standard enthalpy and standard *T4.2* entropy of evaporation

Benzyl chloride   **F12.1**
Chloromethyl-benzene

activation energy of hydrolysis *E12.3*

characteristic distances of air–water *T24.2* exchange and hydrolysis in river

characteristic times of air–water *E24.2* exchange in river, two flow regimes

concentration change in *E25.7* groundwater, effects of hydrolysis, sorption and flow distance

concentration change in river due to *E24.2* flow time and air exchange

hydrolysis, concentration decreases *F2.2* at different pHs

one box model calculations for *E12.3*
removal mechanisms in pond

reaction rate constant of hydrolysis *E12.3*

2-Benzyl-succinic acid   **704**
2-Benzyl-succinic acid

formation by addition of toluene to 730
fumarate

formation by enzymatic addition of 704
toluene to fumarate

Biphenyl   **F2.13**
Biphenyl

molecular diffusivity in air *F18.9*

value of K_{iMM} and V_{max} *T17.7*

Biphenyl-2,2'-diol   **F14.18**
Biphenyl-2,2'-diol

product of oxidative coupling of F14.18
phenol oxidation

Biphenyl-2,4'-diol   **F14.18**
Biphenyl-2,4'-diol

product of oxidative coupling of F14.18
phenol oxidation

Biphenyl-4,4'-diol   **F14.18**
Biphenyl-4,4'-diol

product of oxidative coupling of F14.18
phenol oxidation

4,4'-Bis (2-sulfostyryl) DSBP   **F2.20**
biphenyl
2,2'-[(1,1'-Biphenyl)-4,4'-diyldivinylene]bis-
(benzenesulfonic) acid

fluorescent whitening agent *F2.20*

2-Bromo-2-chloropropane   **P13.5**
2-Bromo-2-chloro-propane

kinetic data of neutral hydrolysis *P13.5*

Bromobenzene   **F16.3**
Bromo-benzene

rate constant with hydroxyl radical *F16.3*

4-Bromobenzene sulfonic acid   **P13.10**
methyl ester
4-Bromo-benzenesulfonic acid methyl ester

rate constant of neutral hydrolysis *P13.10*

3-Bromo-benzoic acid   **716**
3-Bromo-benzoic acid

correlation of biodegradability with 716
$\Sigma\sigma_j$

4-Bromo-benzoic acid   **716**
4-Bromo-benzoic acid

correlation of biodegradability with 716
$\Sigma\sigma_j$

Bromobutane
Bromo-butane

value of V_{max} *T17.8*

Bromodichloromethane   **F14.16**
Bromo-dichloro-methane

relative rates of reduction in *F14.16*
different media

Bromoethane
Bromo-ethane

value of V_{max} *T17.8*

2-Bromoethanol   **F16.3**
2-Bromo-ethanol

formation by hydrolysis of E17.2
1,2-dibromo-ethanol

rate constant with hydroxyl radical *F16.3*

Bromomethane   **F4.4**
Bromo-methane

3-Bromophenol   **P13.9**
3-Bromo-phenol

acidity constant *P13.9*

4-Bromophenol   **P13.9**
4-Bromo-phenol

acidity constant *P13.9*

1-Bromopropane   **P13.5**
1-Bromo-propane

kinetic data of neutral hydrolysis *P13.5*

value of K_{iMM} and V_{max} *T17.7*

value of V_{max} *T17.8*

2-Bromopropane   **P13.5**
2-Bromo-propane

kinetic data of neutral hydrolysis *P13.5*

Bromotrichloromethane   **F14.16**
Bromo-trichloro-methane

relative rates of reduction in *F14.16*
different media

1,3-Butadiene   **F2.12**
Buta-1,3-diene

rate constant with hydroxyl radical *F16.7*
in troposphere

Butanal
Butyraldehyde

salting constant, seawater *T5.7*

n-Butane
Butane

critical temperature and critical *T4.1*
pressure

1-Butanol   **F18.9**
Butan-1-ol

aqueous solution data *T5.2*

molecular diffusivity in air *F18.9*

water saturated, mole fraction *T5.1*

2-Butanone   **F2.17**

Butan-2-one

activity coefficients in six organic *T6.1*
solvents

air-solvent partition constants for six *T6.1*
organic solvents

aqueous solution data *T5.3*

salting constant, seawater *T5.7*

solvent 41

standard enthalpies of vaporization *T6.3*
and of water-air transfer

water saturated, mole fraction *T5.1*

Butylacetate
Acetic acid butyl ester

standard enthalpies of vaporization *T6.3*
and of water-air transfer

water saturated, mole fraction *T5.1*

n-Butylchloride   **F2.1**
1-Chloro-butane

four isomers 18

rate constant with hydroxyl radical *F16.3*

value of V_{max} *T17.8*

2-sec-Butyl-4,6-dinitrophenol Dinoseb **Q11.7**
2-sec-Butyl-4,6-dinitro-phenol

distribution ratio of anion into polar *F10.14*
lipids and into octanol   377

exchange constant   *T11.2*
water-aluminosilicate surface
($K_{NAC,EDA}$)

lethal concentration vs.   *F10.15*
liposome-water distribution ratio

n-Butylbenzene
Butyl-benzene

standard enthalpy and entropy of   *T4.5*
fusion, experimental and predicted
data

standard enthalpy and standard *T4.2*
entropy of evaporation

2,6-Di-tert-butyl-p-cresol DBPC **F2.15**
2,6-Di-tert-butyl-4-methyl-phenol

as antioxidant 38

2-sec-Butyl-phenol
2-sec-Butyl-phenol

value of K_{iMM} and V_{max} *T17.7*

1-Butylamine   **F11.13**
Butylamine

free energy change of cation of ion *F11.14*
exchange on montmorillonite

sorption isotherm of cation on *F11.13*
montmorillonite

2-Butyne   **F2.3**
But-2-yne

bond angles *F2.3*

C

Camphor   **717**
1,7,7-Trimethyl-bicyclo[2.2.1]heptan-2-one

microbial oxidation to   717
5-exo-hydroxy-bornan-2-one

Carbaryl   **P15.4**
Methyl-carbamic acid naphthalen-1-yl ester

decadic molar absorption *P15.4*
coefficients

hydrolysis rate constants *P15.4*

reaction quantum yield of direct *P15.4*
photolysis

Carbofuran   T13.2
Methyl-carbamic acid
2,2-dimethyl-2,3-dihydro-benzofuran-7-yl-ester

enzymatic hydrolysis 707

hydrolysis reaction

Carbon dioxide
Carbon dioxide

molecular diffusivity at different T20.3
temperatures

molecular diffusivity in water F20.3,
F20.4

Schmidt number in water at T20.3
different temperatures

transfer velocity in water and wind F20.3,
speed   F20.4
905

transfer velocity in water and wind F20.10
speed for contaminated water

water phase transfer velocity and T20.4
wind speed

Catechol   699
Benzene-1,2-diol

formation by biodegradation of E17.9
2-nitrophenol

formation by enzymatic oxidation of 699
benzene

formation by enzymatic oxidation of 716
phenol with monooxygenase

formation by enzymatic oxidation of 699
salicylic acid

reaction of anion with mineral T11.4
surfaces

CFC-11
see Trichloro-fluoro-methane

CFC-12
see Dichloro-difluoro-methane

3-Chloronitrobenzene 3-Cl-NB
1-Chloro-3-nitro-benzene

one-electron reduction potential T14.4

relative reduction rates in different F14.10
reduction media

4-Chloronitrobenzene 4-Cl-NB
1-Chloro-4-nitro-benzene

one-electron reduction potential T14.4

reduction rate in iron(II) / magnetite F14.12
solution

reduction to 4-chloroaniline in F14.11
ferrogenic columns

relative reduction rates in different F14.10
reduction media

relative reduction rates in different F14.5
systems

1-Chloro-1,1-difluoroethane   T13.7
1-Chloro-1,1-difluoro-ethane

kinetic data for substitution and T13.7
elimination reactions

1-Chloro-2,3-epoxypropane   F2.15
2-Chloromethyl-oxirane

rate constant with hydroxyl radical F16.3

rate constant with hydroxyl radical P16.3
in troposphere

5-Chloro-2-hydroxymuconic   700
acid semialdehyde
5-Chloro-2-hydroxy-6-oxo-hexa-2,4-dienoic acid

formation by meta cleavage of 700
4-chloro-catechol

2-Chloro-2-methylpropane   P13.5
2-Chloro-2-methyl-propane

kinetic data of neutral hydrolysis P13.5

cis-3-Chloro-2-propene-1-ol CPO   E21.4
(Z)-3-Chloro-prop-2-ene-1-ol

concentration change by in situ E21.4
formation and flushing,
calculations

formation rate constant from E21.4
hydrolysis of precursor

Henry constant E21.4

4-Chloro-α-(4-chlorophenyl) DDA   271
benzene acetic acid
Bis-(4-chloro-phenyl)-acetic acid

acidity constant F8.9

n-octanol-water distribution ratio F8.9
and pH

Chloro-acetic acid
Chloro-acetic acid

value of K_{iMM} T17.7

3-Chloroaniline   F14.20
3-Chloro-phenylamine

relative oxidation rate by F14.20
manganese oxide

4-Chloroaniline 4-Cl-An   T8.2
4-Chloro-phenylamine

acidity constant of corresponding T8.2
acid

air-water distribution at different pH E8.3
and T, calculations

air-water partition constant E8.3

product of 4-chloronitrobenzene F14.11
reduction

relative oxidation rate by F14.20
manganese oxide

standard enthalpy of air-water E8.3
transfer

Chlorobenzene CB   P9.4
Chloro-benzene

air-quartz partition constant and F3.7
dispersive vdW parameter

air-water exchange for different E6.3
water compositions, calculations

aqueous solution data T5.2

biological organic phases-water T10.2
partition coefficients

H-donor (α) and   T4.3
H-acceptor property (β)

molecular diffusivity in air F18.9

octanol-water partition constant T10.2

polarizability T5.5

rate constant with hydroxyl radical F16.3

refractive index T3.1

salting constant, seawater T5.7

solvent-water partition constants for T7.1
five solvents

standard enthalpies of vaporization T6.3
and of water-air transfer

standard enthalpy and standard T4.2
entropy of evaporation

standard free energy of formation P14.1

standard reduction potential T14.3

value of K_{iMM} and V'_{max} T17.7

water saturated, mole fraction T5.1

3-Chloro-benzoic acid   F8.6
3-Chloro-benzoic acid

acidity constant F8.6

correlation of biodegradability with 716
$\Sigma \sigma_j$

microbial degradation to benzoic 695
acid

value of K_{iMM} and V'_{max} T17.7

value of μ_{max} and K_{iM} T17.6

4-Chloro-benzoic acid   F8.6
4-Chloro-benzoic acid

acidity constant F8.6

correlation of biodegradability with 716
$\Sigma \sigma_j$

uptake by organisms 738

4-Chlorobiphenyl   P16.3
4-Chloro-biphenyl

rate constant with hydroxyl radical P16.3
in troposphere

1-Chloro-dibenzo[1,4]dioxin   P16.7
1-Chloro-dibenzo[1,4]dioxin

rate constant with hydroxyl radical F16.7
in troposphere

2-Chloro-ethanol   T17.3
2-Chloro-ethanol

formation by microbial hydrolysis of 707
1,2-dichloro-ethane

2-Chloronitrobenzene 2-Cl-NB
1-Chloro-2-nitro-benzene

one-electron reduction potential T14.4

relative reduction rates in different F14.10
reduction media

Chloro-oxirane   E 17.4
2-Chloro-oxirane

formation by microbial degradation E17.4
of vinyl chloride

3-Chlorophenyl *N*-phenyl   E13.6
carbamate
Phenyl-carbamic acid 3-chloro-phenyl ester

rate constant of base catalyzed E13.6
hydrolysis

4-Chlorophenyl *N*-phenyl   **E13.6**
carbamate
Phenyl-carbamic acid 4-chloro-phenyl ester

rate constant of base catalyzed *E13.6*
hydrolysis

3-Chloro phenylacetic acid   **F8.6**
(3-Chloro-phenyl)-acetic acid

acidity constant *F8.6*

4-Chloro phenylacetic acid   **F8.6**
(4-Chloro-phenyl)-acetic acid

acidity constant *F8.6*

2-Chloropropane   **P13.5**
2-Chloro-propane

kinetic data of neutral hydrolysis *P13.5*

4-Chloropyridine   **T8.2**
4-Chloro-pyridine

acidity constant of corresponding *T8.2*
acid

Chloroacetic acid amide   **T13.10**
2-Chloro-acetamide

rate constants of hydrolysis reaction *T13.10*

Chloroacetic acid methyl   **F13.8**
ester
Chloro-acetic acid methyl ester

hydrolysis rate constant and pH *F13.8*

rate constants of hydrolysis reaction *T13.8,*
T13.9

4-Chlorocatechol   **700**
4-Chloro-benzene-1,2-diol

meta cleavage to   700
5-chloro-2-hydroxymuconic acid
semialdehyde

Chloroethene VC   **E5.1**
Chloro-ethene

higly toxic reaction product in 459
landfills

microbial degradation to E17.4
chloro-oxirane

molecular diffusivity in water *F18.10*

pathways of reduction by zinc *F14.15*

rate constant with hydroxyl radical *F16.3*

rate constant with hydroxyl radical *F16.7*
in troposphere

value of K_{iMM} and V_{max} *T17.7*

value of μ_{max} and K_{iM} *T17.6*

Chloroethyne   **F14.15**
Chloro-ethyne

intermediate in reduction of *F14.15*
chlorinated ethenes

Chlorohexane   **T17.3**
Chloro-hexane

hydrolysis to hexan-1-ol 707

value of K_{iMM} and V_{max} *T17.7*

2-Chlorophenol   **P16.2**
2-Chloro-phenol

acidity constant *P16.2*

polarizability *T5.5*

rate constants of indirect photolysis *P16.2*
with singlet oxygen

3-Chlorophenol   **F8.6**
3-Chloro-phenol

acidity constant *F8.6*

standard oxidation   *P14.7*

4-Chlorophenol   **F16.3**
4-Chloro-phenol

acidity constant *F8.6*

biological organic phases-water *T10.2*
partition coefficients

octanol-water partition constant *T10.2*

polarizability *T5.5*

rate constant with hydroxyl radical *F16.3*

1-Chloropropane
Chloro-propane

value of K_{iMM} and V_{max} *T17.7*

Chlorothion   **E10.2**
Thiophosphoric acid O-(3-chloro-4-nitro-phenyl)
ester O',O''-dimethyl ester

biological organic phases-water *T10.2*
partition coefficients

equilibrium bioaccumulation factor, *E10.1*
K_{ibio}, calculations

octanol-water partition constant *T10.2*

Chlorthiamide   **F2.20**
2,6-Dichloro-thiobenzamide

pesticide 48

Chrysene
Chrysene

H-donor (α) and   *T4.3*
H-acceptor property (β)

salting constant, seawater *T5.7*

Clarithromycin   **F2.25**
6-(4-Dimethylamino-3-hydroxy-6-methyl-tetra
hydro-pyran-2-yloxy)- 14-ethyl-12,13-dihydroxy-
4-(5-hydroxy-4-methoxy-4,6-dimethyl-tetrahydro-
pyran-2-yloxy)-7-methoxy-3,5,7,9,11,13-
hexamethyl-oxa-cyclotetradecane-2,10-dione

antibiotic F2.25

Cobalamin
Cobalamin

reduction potential of 728

standard reduction and oxidation *F14.4*
potentials

Cyclohexane   **F2.12**
Cyclohexane

contaminated with tetrachloroethene E6.1
from air at 5 °C, calculations

standard enthalpies of vaporization *T6.3*
and of water-air transfer

Cyclohexanol
Cyclohexanol

standard enthalpies of vaporization *T6.3*
and of water-air transfer

Cypermethrin   **F2.24**
3-(2,2-Dichloro-vinyl)-2,2-dimethyl-cyclopropane
carboxylic acid
cyano-(3-phenoxy-phenyl)-methyl ester

insecticide F2.24

Cyprofloxacin   **F2.25**
1-Cyclopropyl-6-fluoro-4-oxo-7-piperazin-1-yl-
1,4-dihydro-quinoline-3-carboxylic acid

antibiotic F2.25

Cysteine
2-Amino-3-mercapto-propionic acid

standard oxidation potential *T14.3,*
F14.4

Cystine
2-Amino-3-(2-amino-2-carboxy-ethyldisulfanyl)-p
ropionic acid

standard reduction potential *T14.3,*
F14.4

D

1,1-DCE
see 1,1-Dichloroethene

***cis*-1,2-DCE**
see *cis*-1,2-Dichloroethene

***trans*-1,2-DCE**
see *trans*-1,2-Dichloroethene

DDD
*see p,p'-*Dichlorodiphenyl- dichloroethane

DDE
*see p,p'-*Dichlorodiphenyl- dichloroethene

DDT
*see p,p'-*Dichlorodiphenyl- trichloroethane

Decanal
Decanal

salting constant, seawater *T5.7*

***n*-Decane**   **P5.1**
Decane

critical temperature and critical *T4.1*
pressure

molecular diffusivity at different *T20.3*
temperatures

rate constant with hydroxyl radical *F16.7*
in troposphere

Schmidt number in water at *T20.3*
different temperatures

standard enthalpy and entropy of   *T4.5*
fusion, experimental and predicted
data

standard enthalpy and standard *T4.2*
entropy of evaporation

water saturated, mole fraction *T5.1*

1-Decylamine   **F11.13**
Decylamine

free energy change of cation of ion *F11.14*
exchange on montmorillonite

sorption isotherm of cation on *F11.13*
montmorillonite

2,6-Diamino-4-nitrotoluene   F14.6
2-Methyl-5-nitro-benzene-1,3-diamine

exchange constant   T11.2
water-aluminosilicate surface
($K_{NAC.EDA}$)

intermediate in TNT reduction F14.6

one-electron reduction potential T14.4

2,4-Diamino-6-nitrotoluene   F14.6
4-Methyl-5-nitro-benzene-1,3-diamine

intermediate in TNT reduction F14.6

one-electron reduction potential T14.4

1,2-Diaminopropane   F8.2
Propane-1,2-diamine

acidity constants and species F8.2
distribution

Diazinon   T13.12
Thiophosphoric acid O,O'-diethyl ester
O''-(2-isopropyl-6-methyl-pyrimidin-4-yl) ester

rate constants of hydrolysis reaction T13.12

Diazoxon   T13.12
Phosphoric acid diethyl ester
2-isopropyl-6-methyl-pyrimidin-4-yl ester

rate constants of hydrolysis reaction T13.12

Dibenzo[1,4]dioxin   E16.3
Dibenzo[1,4]dioxin

indirect photolysis half-life by   E16.3
hydroxyl radical in troposphere,
calculation of

1,2-Dibromo-3-chloro DBCP   T13.7
propane
1,2-Dibromo-3-chloro-propane

different leaving group in 510
elimination reactions

kinetic data for substitution and T13.7
elimination reactions

rate constant with hydroxyl radical P16.3
in troposphere

Dibromochloromethane   F14.16
Dibromo-chloro-methane

relative rates of reduction in F14.16
different media

Dibromodichloromethane   F14.16
Dibromo-dichloro-methane

relative rates of reduction in F14.16
different media

2,2-Dibromopropane   P13.5
2,2-Dibromo-propane

kinetic data of neutral hydrolysis P13.5

1,2-Dibromoethane EDB   T13.7
1,2-Dibromo-ethane

biodegradation to 2-bromo-ethanol E17.2

kinetic data for substitution and T13.7
elimination reactions

neutral hydrolyis constant and E13.3
activation energy

nucleophilic substitution reaction E13.3
with polysulfides

rate constant with hydroxyl radical F16.3

rate constant with hydroxyl radical F16.7
in troposphere

Dibromomethane   F16.3
Dibromo-methane

rate constant with hydroxyl radical F16.3

value of K_{iMM} and V_{max} T17.7

Dibutyl ether
1-Butoxy-butane

air-quartz partition constant and F3.7
dispersive vdW parameter

Di-n-butyl phthalate   E5.1
Phthalic acid dibutyl ester

vapor pressure as function of F4.4
temperature

1,1-Dichloro-1-fluoroethane   T13.7
1,1-Dichloro-1-fluoro-ethane

kinetic data for substitution and T13.7
elimination reactions

major use T2.4

cis-1,3-Dichloro-1-propene DCP   E21.4
(Z)-1,3-Dichloro-propene

concentration change by reaction E21.4
and flushing, calculations

Henry constant E21.4

1,1-Dichloro-2,2,2-trifluoro   511
ethane
2,2-Dichloro-1,1,1-trifluoro-ethane

rate constant with hydroxyl radical F16.7
in troposphere

Dichloroacetic acid methyl   F13.8
ester
2,2-Dichloro-acetic acid methyl ester

hydrolysis rate constant and pH F13.8

rate constants of hydrolysis reaction T13.8,
T13.9

Dichloroacetic acid phenyl   T13.8
ester
2,2-Dichloro-acetic acid phenyl ester

rate constants of hydrolysis reaction T13.8,
T13.9

3,4-Dichloroaniline   T17.3
3,4-Dichloro-phenylamine

formation by enzymatic degradation 714
of linuron

formation by enzymatic hydrolysis 707
of diuron

formation by enzymatic hydrolysis
of propanil

reactant of a higly toxic reaction 459
product in soils

reaction quantum yield of direct T15.7
photolysis

3,5-Dichloroaniline
3,5-Dichloro-phenylamine

reaction quantum yield of direct T15.7
photolysis

1,2-Dichlorobenzene 1,2-DCB   P4.5
1,2-Dichloro-benzene

aqueous solution data T5.2

bioaccumulation factor from water P10.1
into soybeans leaves and roots

H-donor (α) and   T4.3
H-acceptor property (β)

organic carbon-water distribution F9.8
coefficients for various solids

polarizability T5.5

standard enthalpy and standard T4.2
entropy of evaporation

standard free energy of formation P14.1

vapor pressure as function of F4.4
temperature

1,3-Dichlorobenzene 1,3-DCB
1,3-Dichloro-benzene

air-soil sorption isotherm at various F11.4
relative humidities

H-donor (α) and   T4.3
H-acceptor property (β)

1,4-Dichlorobenzene 1,4-DCB   P4.4
1,4-Dichloro-benzene

concentrations in a river at different P20.4
flow distances

polarizability T5.5

salting constant, seawater T5.7

standard enthalpy and entropy of   T4.5
fusion, experimental and predicted
data

standard enthalpy and standard T4.2
entropy of evaporation

vapor pressures at six temperatures P4.4

3,4-Dichloro-benzoic acid   716
3,4-Dichloro-benzoic acid

correlation of biodegradability with 716
$\Sigma\sigma_j$

3,5-Dichloro-benzoic acid   697
3,5-Dichloro-benzoic acid

correlation of biodegradability with 716
$\Sigma\sigma_j$

microbial degradation to 698
3-chloro-benzoic acid

value of K_{iMM} and V_{max} T17.7

2,5-Dichloro-biphenyl
2,5-Dichloro-biphenyl

standard enthalpies of vaporization T6.3
and of water-air transfer

4,4'-Dichloro-biphenyl   F16.7
4,4'-Dichloro-biphenyl

rate constant with hydroxyl radical F16.7
in troposphere

Dichloro-difluoro-methane CFC-12   P4.2
Dichloro-difluoro-methane

establishing a concentration profile E22.1
in lake from atmospheric input,
calculations

molar volume, calculations E18.1

molecular diffusivity in air, E18.2
calculations

molecular diffusivity in water, *E18.3*
calculations

salting constant, seawater *T5.7*

standard enthalpies of vaporization *T6.3*
and of water-air transfer

vapor pressures at five temperatures *P4.2*

vertical flux in lake, calculations *E22.1*

world production rate, major use T2.4

p,p'-Dichlorodiphenyl DDD   556
dichloroethane
1,1-Dichloro-2,2-bis(4-chloro-phenyl)-ethane

formation by microbial reduction of E17.5
p,p'-DDT

product from reductive   556
dechlorination reaction of DDT

p,p'-Dichlorodiphenyl DDE   556
dichloroethene
1,1-Dichloro-2,2-bis(4-chloro-phenyl)-ethene

product from dehydrochlorination 556
reaction of DDT

p,p'-Dichlorodiphenyl DDT   F2.14
trichloroethane
1,1,1-Trichloro-2,2-bis-(4-chloro-phenyl)-ethane

accumulation rates in sediments of *F1.4*
lake Ontario

as compound for illustration of 948
principal modelling problems

concentrations in aquatic organisms *F10.10*

dehydrochlorination reaction 556

early findings in environment 4

hydrolysis rate constants *P13.3*

kinetic data for substitution and *T13.7*
elimination reactions

microbial degradation to *p,p'*-DDD E17.5

production rates in the United States *F1.4*

reductive dechlorination reaction 556

standard enthalpy and entropy of   *T4.5*
fusion, experimental and predicted
data

1,1-Dichloroethane
1,1-Dichloro-ethane

salting constant, seawater *T5.7*

1,2-Dichloroethane   T13.7
1,2-Dichloro-ethane

kinetic data for substitution and *T13.7*
elimination reactions

microbial hydrolysis to 707
2-chloro-ethanol

microbial reduction to ethene 721

salting constant, seawater *T5.7*

1,1-Dichloroethene   F14.15
1,1-Dichloro-ethene

intermediate in reduction of F14.15
chlorinated ethenes

relative rates of reduction by three *F14.17*
metals

1,2-Dichloroethene   27
1,2-Dichloro-ethene

properties of cis- and trans- 27
structures

***cis*-1,2-Dichloroethene   F14.15**
(Z)-1,2-Dichloro-ethene

intermediate in reduction of F14.15
chlorinated ethenes

relative rates of reduction by three *F14.17*
metals

standard free energy of formation *P14.1*

value of K_{iMM} and V_{max} *T17.7*

***trans*-1,2-Dichloroethene DCE   F14.15**
(E)-1,2-Dichloro-ethene

intermediate in reduction of F14.15
chlorinated ethenes

rate constant with hydroxyl radical *F16.3*

rate constant with hydroxyl radical *F16.7*
in troposphere

relative rates of reduction by three *F14.17*
metals

Dichloroethyne   F14.15
1,2-Dichloro-ethyne

intermediate in reduction of F14.15
tetrachloroethene

Dichloro-fluoromethane   F16.7
Dichloro-fluoro-methane

rate constant with hydroxyl radical *F16.7*
in troposphere

Dichloromethane   F2.3
Dichloro-methane

activity coefficients of five solutes *T6.1*
in dichloromethane

air-dichloromethane partition *T6.1*
constants for five solutes

bond angles *F2.3*

H-donor (α) and   *T4.3*
H-acceptor property (β)

hydrolysis to formaldehyde with 709
glutathione

kinetic data for substitution and *T13.7*
elimination reactions

major use T2.4

microbial hydrolysis to 707
formaldehyde

rate constant with hydroxyl radical *F16.3*

rate constant with hydroxyl radical *F16.7*
in troposphere

standard enthalpies of vaporization *T6.3*
and of water-air transfer

standard oxidation potential *T14.3,*
F14.4

value of K_{iMM} and V_{max} *T17.7*

value of μ_{max} and K_{iM} *T17.6*

2,4-Dichlorophenol   F16.5
2,4-Dichloro-phenol

biological organic phases-water *T10.2*
partition coefficients

octanol-water partition constant *T10.2*

rate constant with hydroxyl radical *F16.7*
in troposphere

rate constant with singlet oxygen *F16.5*

2,4-Dichlorophenoxy butyric   P11.8
acid
4-(2,4-Dichloro-phenoxy)-butyric acid

acidity constant *P11.8*

octanol-water partition constant *P11.8*

4-(2,4-Dichlorophenoxy)   F11.9
butyric acid
4-(2,4-Dichloro-phenoxy)-butyric acid

sorption isotherm of anion on natural *F11.9*
solid

1,2-Dichloropropane   P9.3
1,2-Dichloro-propane

rate constant with hydroxyl radical *F16.3*

2,2-Dichloropropane   P13.4
2,2-Dichloro-propane

kinetic data of neutral hydrolysis *P13.4*

***cis*-1,3-Dichloro-1-propene DCP   E21.4**
(Z)-1,3-Dichloro-propene

hydrolysis rate constant *E21.4*

indirect photolysis half-life by   *E16.3*
hydroxyl radical in troposphere,
calculation of

2-(2,2-Dichloro-vinyl)-cyclo   T17.3
propanecarboxylic acid
2-(2,2-Dichloro-vinyl)-cyclopropanecarboxylic
acid

formation by enzymatic hydrolysis 707
of permethrin

Dichlorprop
2-(2,4-Dichloro-phenoxy)-propionic acid

uptake by organisms 738

Diethyl ether   F2.15
Ethoxy-ethane

activity coefficients in organic *T3.2*
solvents and in water

air-teflon and air-quartz partition *F3.7*
constants and dispersive vdW
parameter

air-water partition constant E6.4
estimation, calculations

aqueous solution data *T5.3*

diethyl ether-water partition   *T7.1*
constants for nine organic solutes

energies of air-hexadecane,   *T3.4*
air-water and hexadecane-water
transfers

excess functions in gas phase, *T3.3*
hexadecane and water

rate constant with hydroxyl radical *F16.7*
in troposphere

solvent fitting coefficients for   *T7.2*
diethylether-water partition
constant of organic compounds

standard enthalpies of vaporization *T6.3*
and of water-air transfer

standard excess enthalpy in water *216*

water saturated, mole fraction *T5.1*

Diethyl sulfide
Ethylsulfanyl-ethane

aqueous solution data *T5.3*

rate constant with singlet oxygen *F16.5*

standard enthalpies of vaporization *T6.3*
and of water-air transfer

N,N-Diethylaniline   **F8.5**
Diethyl-phenyl-amine

acidity constant of corresponding *F8.5*
acid and steric interactions

Diethylamine   **499**
Diethyl-amine

nucleophilicity in water *499*

Diethylphosphate   **T17.3**
Phosphoric acid diethyl ester

formation by enzymatic hydrolysis *707*
of paraoxon

O,O'-Diethylthiophosphoric   **T13.2**
acid
Thiophosphoric acid O,O'-diethyl ester

product in parathion hydrolysis *T13.2*
reaction

2,3-Dihydro-3,3-dimethyl-7-   **T13.2**
benzofuranol
2,2-Dimethyl-2,3-dihydro-benzofuran-7-ol

product in carbofuran hydrolysis *T13.2*
reaction

***cis*-1,2-Dihydroxy-3-methyl-**   **717**
cyclohexa-3,5-diene
cis-3-Methyl-cyclohexa-3,5-diene-1,2-diol

formation by microbial oxidation of *717*
toluene

3,4-Dihydroxy-benzoic acid   **P8.1**
3,4-Dihydroxy-benzoic acid

acidity constants *P8.1*

Di-isopropanol amine DIPA   **E11.5**
1-(2-Hydroxy-propylamino)-propan-2-ol

acidity constant of corresponding
acid

cation exchange on mineral *E11.5*
surfaces in water, calculations

Diisopropyl ether
2-Isopropoxy-propane

air-quartz partition constant and *F3.7*
dispersive vdW parameter

3,5-Dimethoxyphenol
3,5-Dimethoxy-phenol

standard oxidation *P14.9*

Dimethyl ether   **F2.3**
Methoxy-methane

bond angles *F2.3*

rate constant with hydroxyl radical *F16.3*

rate constant with hydroxyl radical *F16.7*
in troposphere

Dimethyl phosphate   **T13.2**
Phosphoric acid dimethyl ester

product in trimethylphosphate *T13.2*
substitution reaction

Dimethyl sulfide   **F2.3**
Methylsulfanyl-methane

bond angles *F2.3*

molar volume, calculation *T18.3*

rate constant with hydroxyl radical *F16.7*
in troposphere

salting constant, seawater *T5.7*

solvent *48*

standard enthalpies of vaporization *T6.3*
and of water-air transfer

standard oxidation potential *T14.3,*
F14.4

2,2-Dimethyl-2,3-dihydro-   **T17.3**
benzofuran-6-ol
2,2-Dimethyl-2,3-dihydro-benzofuran-6-ol

formation by enzymatic hydrolysis *707*
of carbofuran

2,3-Dimethyl-2-butene   **F16.5**
2,3-Dimethyl-but-2-ene

rate constant with singlet oxygen *F16.5*

3,4-Dimethylaniline 3,4-DMA 290
3,4-Dimethyl-phenylamine

acidity constant, calculation *E8.2*

acidity constant of corresponding *E8.1*
acid

parameters for indirect photolysis *P16.2*
with carbonate radical

polymechanistic sorption behavior *279*
for neutral and charged species

polymechanistic sorption depending *290*
on surface characteristics

species fractions at different pHs, *E8.1*
calculations

N,N-Dimethylaniline   **T8.2**
Dimethyl-phenyl-amine

acidity constant of corresponding *T8.2*
acid

acidity constant of corresponding *F8.5*
acid and steric interactions

1,2-Dimethylbenzene
o-Xylene

biological organic phases-water *T10.2*
partition coefficients

octanol-water partition constant *T10.2*

polarizability *T5.5*

salting constant, seawater *T5.7*

standard enthalpies of vaporization *T6.3*
and of water-air transfer

1,3-Dimethylbenzene   **F 17.11**
m-Xylene

addition to fumarate *F17.11*

salting constant, seawater *T5.7*

water saturated, mole fraction *T5.1*

1,4-Dimethylbenzene 1,4-DMB
p-Xylene

aqueous solution data *T5.3*

polarizability *T5.5*

salting constant, seawater *T5.7*

solid-water distribution in lakes and *287*
in groundwater, calculations

standard enthalpies of vaporization *T6.3*
and of water-air transfer

3,4-Dimethylbenzene sulfonic   **P13.10**
acid methyl ester
3,4-Dimethyl-benzenesulfonic acid methyl ester

rate constant of neutral hydrolysis *P13.10*

Dimethyl disulfide   **F2.20**
Methyldisulfanylmethane

from anthropogenic and biogenic *48*
sources

rate constant with hydroxyl radical *F16.7*
in troposphere

Dimethyl-formamide DMF   **T5.8**
N,N-Dimethyl-formamide

as cosolvent *T5.8*

2,5-Dimethylfuran 2,5-DMF 667
2,5-Dimethyl-furan

rate constant with singlet oxygen *F16.5*

trapping agent in singlet oxygen *667*
measurement

N,O-Dimethyl-hydroxylamine   **E 17.3**
Methoxy-methyl-amine

formation by decarboxylation of *E17.3*
methoxy-methyl-carbamic acid

N,N-Dimethyl-N'-(3,4-   **P9.2**
dichlorophenyl) urea
3-(3,4-Dichloro-phenyl)-1,1-dimethyl-urea

octanol-water and organic *P9.2*
carbon-water partition constants

N,N-Dimethyl-N'-(3,5-   **P9.2**
dimethylphenyl) urea
3-(3,5-Dimethyl-phenyl)-1,1-dimethyl-urea

octanol-water and organic *P9.2*
carbon-water partition constants

N,N-Dimethyl-N'-(3-fluoro   **P9.2**
phenyl) urea
3-(3-Fluoro-phenyl)-1,1-dimethyl-urea

octanol-water and organic *P9.2*
carbon-water partition constants

N,N-Dimethyl-N'-(4-chloro   **P9.2**
phenyl) urea
3-(4-Chloro-phenyl)-1,1-dimethyl-urea

octanol-water and organic *P9.2*
carbon-water partition constants

N,N-Dimethyl-*N'*-(4-methox   P9.2
y phenyl) urea
3-(4-Methoxy-phenyl)-1,1-dimethyl-urea

octanol-water and organic   *P9.2*
carbon-water partition constants

N,N-Dimethyl-*N'*-(4-methyl   P9.2
phenyl) urea
1,1-Dimethyl-3-*p*-tolyl-urea

octanol-water and organic   *P9.2*
carbon-water partition constants

2,4-Dimethylphenol   **16.7**
2,4-Dimethyl-phenol

rate constant with hydroxyl radical *F16.7*
in troposphere

2,5-Dimethylphenol   **P16.3**
2,5-Dimethyl-phenol

rate constant with hydroxyl radical *P16.3*
in troposphere

Dimethyl-phosphate   **T17.3**
Phosphoric acid dimethyl ester

formation by enzymatic hydrolysis 707
of malathion

Dimethyl sulfide   **F2.3**
Methylsulfanyl-methane

formation by microbial reduction of 721
dimethylsulfoxide

N,N'-Dimethylurea
1,3-Dimethyl-urea

rate constant with hydroxyl radical *F16.3*

Dimethylamine   **T18.3**
Dimethyl-amine

formation by enzymatic hydrolysis 707
of diuron

molar volume, calculation *T18.3*

Dimethylsulfone DMSF   **T14.3**
Methanesulfonylmethane

standard reduction potential *T14.3*

Dimethylsulfoxide DMSO   **F2.20**
Methanesulfinylmethane

as cosolvent *T5.8*

microbial reduction to dimethyl 721
sulfide

standard oxidation potential *T14.3*

standard reduction potential *T14.3,*
F14.4

valence shell expansion in double 46
bond

L-α-Dimyristoylphosphatidyl
choline
Phosphoric acid-((*S*)-2,3-bis-myristoyloxy-propyl
ester)-2-trimethylammonio-ethyl ester)-betaine

L-α-Dimyristoylphosphatidylcholine *F10.4*
-water vs. octanol-water
partitioning of organic compounds

1,2-Dinitrobenzene 1,2-DNB **Q11.7**
1,2-Dinitro-benzene

exchange constant   *T11.2*
water-aluminosilicate surface
($K_{NAC,EDA}$)

one-electron reduction potential *T14.4*

1,3-Dinitrobenzene 1,3-DNB
1,3-Dinitro-benzene

one-electron reduction potential *T14.4*

1,4-Dinitrobenzene 1,4-DNB **E9.1**
1,4-Dinitro-benzene

clay mineral-water distribution, *E9.1*
measured concentrations

exchange constant   *T11.2*
water-aluminosilicate surface
($K_{NAC,EDA}$)

one-electron reduction potential *T14.4*

solid-water distribution coefficient, *E9.1*
calculation

2,6-Dinitrobenzoic acid   **T8.1**
2,6-Dinitro-benzoic acid

acidity constant *T8.1*

2,4-Dinitro-*o*-cresol DNOC **F2.18**
2-Methyl-4,6-dinitro-phenol

acidity constant *E8.3*

acidity constant *F8.9*

air-water distribution at different pH *E8.3*
and *T*, calculations

air-water partition constant *E8.3*

exchange constant   *T11.2*
water-aluminosilicate surface
($K_{NAC,EDA}$)

herbicide F2.18

n-octanol-water distribution ratio *F8.9*
and pH

nonlinear isotherm with mineral 412
surfaces

standard enthalpy of air-water *E8.3*
transfer

2,4-Dinitrophenol   **E25.2**
2,4-Dinitrophenol

comparison of dispersive and *E25.2*
advective transport in
Groundwater

dispersion coefficient in   *E25.2*
Groundwater, calculation for
natural and pump regimes

maximum concentration at   *E25.3*
groundwater well from river
infiltration

time to reach groundwater well *E25.4*
from river step infiltration

time variation of concentration at *E25.5*
groundwater well from river
time-variable infiltration

2,4-Dinitrotoluene 2,4-DNT **Q11.7**
1-Methyl-2,4-dinitro-benzene

exchange constant   *T11.2*
water-aluminosilicate surface
($K_{NAC,EDA}$)

one-electron reduction potential *T14.4*

reaction quantum yield of direct *T15.7*
photolysis

2,6-Dinitrotoluene 2,6-DNT
1-Methyl-2,6-dinitro-benzene

exchange constant   *T11.2*
water-aluminosilicate surface
($K_{NAC,EDA}$)

one-electron reduction potential *T14.4*

organic carbon-water partition *P11.6*
constant

Dinoseb
see 2-*sec*-Butyl-4,6-dinitrophenol

1,4-Dioxane   **F2.15**
[1,4]-Dioxane

activity coefficients in six organic *T6.1*
solvents

air-solvent partition constants for six *T6.1*
organic solvents

as cosolvent *T5.8*

Disperse Blue 79   **F2.18**
Acetic acid
2-{(2-acetoxy-ethyl)-[5-acetylamino-4-
(2-bromo-4,6-dinitro- phenylazo)-2-methoxy-
phenyl]-amino}-ethyl ester

azo dye 45

Disulfoton   **F2.22**
Dithiophosphoric acid *O,O'*-diethyl ester
S-(2-ethylsulfanyl-ethyl) ester

air–water exchange rates in River *1137*
Rhine

biological and abiotic transformation *1137*
rates in River Rhine

characteristic physico-chemical *T24.3*
data

fate of along 700 km flow in River *1135*
Rhine after accidental spill, case
study

insecticide and acaricide F2.22

longitudinal dispersion for four days *1137*
flow in River Rhine

measured and modeled change of *1139*
peak concentration along River
Rhine

processes and effects of load   *T24.4*
reduction in River Rhine, case
study

rate constants of hydrolysis at five *P13.8*
temperatures and three pH's

rate constants of hydrolysis reaction *T13.12*

sediment–water exchange rates in *1137*
River Rhine

Diuron   **308**
3-(3,4-Dichloro-phenyl)-1,1-dimethyl-urea

enzymatic hydrolysis 707

standard excess enthalpy in *310*
particulate organic matter

1,3-DMB
see 1,3-Dimethylbenzene

1,4-DMB
see 1,4-Dimethylbenzene

DMSO
see Dimethylsulfoxide

DNOC
see 2,4-Dinitro-o-cresol

2,4-DNT
see 2,4-Dinitrotoluene

2,6-DNT
see 2,6-Dinitrotoluene

Dodecyl sulfonate   E11.6
Dodecane-1-sulfonic acid anion

alumina surface-water distribution E11.6
coefficient, calculations

N-Dodecylpyridinium ion   F11.15
1-Dodecyl-pyridinium

sorption isotherm of cation on a soil F11.15

4-(3-Dodecyl)-   717
benzenesulfonic acid
4-(1-Ethyl-decyl)-benzenesulfonic acid

microbial oxidation to   717
4-(1-Ethyl-10- hydroxy-decyl)-
benzenesulfonic acid

E

n-Eicosane
Eicosane

as model compound: pressure   100
behavior with increasing
temperature

critical temperature and critical   T4.1
pressure

standard enthalpy and entropy of   T4.5
fusion, experimental and predicted
data

Ethane   F14.15
Ethane

product in reduction of chlorinated F14.15
ethenes, pathway

rate constant with hydroxyl radical F16.3

rate constant with hydroxyl radical F16.7
in troposphere

Ethanethiol   T8.1
Ethanethiol

acidity constant T8.1

Ethanol   F2.15
Ethanol

activity coefficients in organic T3.2
solvents and in water

activity coefficients in six organic T6.1
solvents

activity coefficients of four organic T3.2
solutes in ethanol

air-solvent partition constants for six T6.1
organic solvents

air-water partition constant E6.4
estimation, calculations

aqueous solution data T5.2

as cosolvent T5.8

energies of air-hexadecane,   T3.4
air-water and hexadecane-water
transfers

excess functions in gas phase,   T3.3
hexadecane and water

formation by microbial reduction of 721
acetaldehyde

in methanogenic degradation, E12.2
calculations

molecular diffusivity in air F18.9

product in ethyl acetate hydrolysis T13.2
reaction

rate constant with hydroxyl radical F16.3

rate constant with hydroxyl radical F16.7
in troposphere

refractive index T3.1

standard enthalpy and standard T4.2
entropy of evaporation

standard Gibbs energy E12.2

Ethene   F2.3
Ethene

bond angles F2.3

cloud of π electrons, double bond 27

formation by microbial reduction of 721
1,2-dichloro-ethane

intermediate in reduction of F14.15
chlorinated ethenes

rate constant with hydroxyl radical F16.7
in troposphere

Ethyl acetate
see Acetic acid ethyl ester

4-(1-Ethyl-10-hydroxy-decyl)   717
-benzenesulfonic acid
4-(1-Ethyl-10-hydroxy-decyl)-benzenesulfonic
acid

formation by microbial oxidation of 717
4-(3-Dodecyl)-benzenesulfonic
acid

4-Ethyl-2,6-Dimethylpyridine EDMP   P8.3
4-Ethyl-2,6-dimethyl-pyridine

acidity constant P8.3

trichloromethane-water partition P8.3
constant

Ethyl-benzene   F2.13
Ethyl-benzene

H-donor (α) and   T4.3
H-acceptor property (β)

molecular diffusivity in water F18.10

salting constant, seawater T5.7

standard enthalpies of vaporization T6.3
and of water-air transfer

standard enthalpy and standard T4.2
entropy of evaporation

Ethyl-cyclopentane   P16.3
Ethyl-cyclopentane

rate constant with hydroxyl radical P16.3
in troposphere

Ethyl N,N-dimethyl   T13.11
carbamate
Dimethyl-carbamic acid ethyl ester

rate constants of hydrolysis reaction T13.11

Ethyl N-methyl carbamate   T13.11
Methyl-carbamic acid ethyl ester

rate constants of hydrolysis reaction T13.11

Ethyl N-methyl-N-phenyl   T13.11
carbamate
Methyl-phenyl-carbamic acid ethyl ester

rate constants of hydrolysis reaction T13.11

Ethyl N-phenyl carbamate   T13.11
Phenyl-carbamic acid ethyl ester

rate constants of hydrolysis reaction T13.11

4-Ethylphenol   P13.9
4-Ethyl-phenol

acidity constant P13.9

Ethylamine EA   429
Ethylamine

free energy change of cation of ion F11.14
exchange on montmorillonite

sorption isotherm of cation on F11.12,
montmorillonite   F11.13

Ethylenediamine-tetraacetic EDTA   738
acid
[[2-(Bis-carboxymethyl-amino)-ethyl]-2-hydroxy-
acetyl)-amino]-acetic acid

uptake by organisms 738

Ethyleneglycol   F2.15
Ethane-1,2-diol

activity coefficients of five solutes T6.1
in ethylene glycol

air-ethylene glycol partition T6.1
constants for five solutes

as cosolvent T5.8

refractive index T3.1

Ethylmercaptan   F2.20
Ethanethiol

odorant and chemical intermediate F2.20

Ethyne   F2.6
Ethyne

cloud of π electrons, triple bond F2.6

hydration to acetaldehyde 734

intermediate in reduction of F14.15
chlorinated ethenes

17-Ethynyl-estradiol   F2.25
17-Ethynyl-13-methyl-7,8,9,11,12,13,14,15,16,
17-decahydro-6H-cyclopenta[a]phenanthrene-
3,17-diol

birth control pill F2.25

F

FAD   724

reduction potential of flavoenzymes 725

FFA
see Furfuryl alcohol

Fluoranthene
Fluoranthene

H-donor (α) and   T4.3
H-acceptor property (β)

salting constant, seawater T5.7

standard enthalpy and entropy of   *T4.5*
fusion, experimental and predicted
data

Fluorene
9*H*-Fluorene

H-donor (α) and   *T4.3*
H-acceptor property (β)

salting constant, seawater *T5.7*

Fluoro-acetic acid
Fluoro-acetic acid

value of K_{iMM} *T17.7*

Fluorobenzene
Fluoro-benzene

standard enthalpy and standard *T4.2*
entropy of evaporation

Fluorotribromomethane   **F14.16**
Tribromo-fluoro-methane

relative rates of reduction in *F14.16*
different media

Formaldehyde   **F2.17**
Formaldehyde

air-water partition constant *E20.5*

dehydration rate constant *E20.5*

formation by enzymatic cleavage 699
reaction of vanillic acid

formation by enzymatic hydrolysis 707
of dichloromethane

formation by hydrolysis of   709
dichloromethane with glutathione

hydration rate constant *E20.5*

overall air-water exchange velocity *E20.5*
at different wind speeds,
calculation of

rate constant with hydroxyl radical *F16.7*
in troposphere

Fumaric acid   **F2.7**
(2*E*)-But-2-enedioic acid

enzyme catalized addition of 730
organic compounds to

property of 27

Furan   **F2.2**
Furan

conjugated π-electron system 30

Furfuryl alcohol FFA   **667**
Furan-2-yl-methanol

rate constant with singlet oxygen *F16.5*

reaction and products with singlet 667
oxygen

trapping agent in singlet oxygen 667
measurement

G

Glucose
6-Hydroxymethyl-tetrahydro-pyran-2,3,4,5-tetraol

molecular diffusivity in water *F18.10*

standard free energy change in *569*
photosynthesis,calculations

standard oxidation potential *T14.2*

value of μ_{max} and K_{iM} *T17.6*

Glutathione GSH   **708**
2-Amino-4-[1-(carboxymethyl-carbamoyl)-2-
mercapto- ethylcarbamoyl]-butyric acid

hydrolysis of dichloromethane to 709
formaldehyde

Glycerol   **T5.8**
Propane-1,2,3-triol

as cosolvent *T5.8*

degradation in a well mixed tank *E17.6*

value of μ_{max} and K_{iM} *T17.6*

H

Haloxyfop   **F2.24**
2-[4-(3-Chloro-5-trifluoromethyl-pyridin-
2-yloxy)-phenoxy]-propionic acid

herbicide *F2.24*

HCA
see Hexachloroethane

HCB
see Hexachlorobenzene

HCFC-141B
see 1,1-Dichloro-1-fluoroethane

HCH
see 1,2,3,4,5,6-Hexachlorocyclohexane

Helium
Helium

molecular diffusivity at different *T20.3*
temperatures

Schmidt number in water at *T20.3*
different temperatures

transfer velocity in water and wind *F20.3*
speed

2,2',3,3',4,4',5-Heptachloro PCB170 **E19.3**
biphenyl
2,3,4,5,2',3',4'-Heptachloro-biphenyl

molar volume and molecular *E19.3*
diffusivity in water

partition constants *E19.3*

sediment to open water diffusion, *E19.3*
calculations

2,2',3,4,4',5,5'-Heptachloro PCB180 **E10.1**
biphenyl
2,3,4,5,2',4',5'-Heptachloro-biphenyl

air-lipid partition coefficient *T10.5*

air-octanol partition constant *E10.5*

bioconcentration factor in *E10.1*
phytoplankton

biomagnification factors from *T10.6*
pasture to animal fat

concentrations in air and in pasture *E10.5*

concentrations in air, pasture and *T10.5*
milk

equilibrium bioaccumulation factor, *E10.1*
K_{ibio}, calculations

fugacities in air and pasture,   *E10.5*
equilibrium concentration in
pasture, calculations

fugacities in air and terrestrial food *F10.11*
chain

fugacities in arctic terrestrial food *F10.12*
chain

standard enthalpy of air-lipid *T10.5*
transfer

standard enthalpy of air-pasture *E10.5*
transfer

2,2',3,4,5,5',6-Heptachloro PCB185   **1065**
biphenyl
2,3,4,5,6,2',5'-Heptachloro-biphenyl

Summary of models and   *T23.8*
measurements in Lake Superior

two-box coupled water/sediment *T23.7*
system in Lake Superior

concentrations, inputs and outputs of *T23.4*
Lake Superior

one-box three phase model in Lake 1068
Superior

one-box two phase model in Lake *T23.4*
Superior   1065

physico-chemical properties *T23.4*

surface mixed sediment layer model 1065
in Lake Superior

two-box water/sediment model in 1079
Lake Superior

Heptanal
Heptanal

salting constant, seawater *T5.7*

n-**Heptane**
Heptane

air-teflon partition constant and *F3.7*
dispersive vdW parameter

standard enthalpies of vaporization *T6.3*
and of water-air transfer

water saturated, mole fraction *T5.1*

1-Heptene   **P5.4**
Hept-1-ene

indirect photolysis half-life by   *E16.3*
hydroxyl radical in troposphere,
calculation of

1-Heptylamine
Heptylamine

free energy change of cation of ion *F11.14*
exchange on montmorillonite

Hexachlorobenzene HCB   **F2.14**
1,2,3,4,5,6-Hexachloro-benzene

aqueous solution data *T5.2,*
T5.3

concentrations in air, soil, grass and *P10.4*
milk

concentrations in aquatic organisms *F10.10*

standard enthalpies of vaporization *T6.3*
and of water-air transfer

standard enthalpy of evaporation *P10.4*

standard reduction potential *T14.3*

2,2',4,4',5,5'-Hexachloro PCB153 biphenyl
2,4,5,2',4',5'-Hexachloro-biphenyl

air-lipid partition coefficient *T10.5*

air-octanol partition constant *E10.5*

air-water partition constant *T20.5*

biomagnification factors from *T10.6* pasture to animal fat

concentrations in air and in pasture *E10.5*

concentrations in air, pasture and *T10.5* milk

concentrations in aquatic organisms *F10.10*

concentrations in lake water and *T10.3* sediment

concentrations in water, sediment *E10.4* and mussel

fugacites in water, sediment and *F10.8* biolipid

fugacities and activities in water, *E10.4* sediment and mussel, calculations

fugacities in air and pasture, *E10.5* equilibrium concentration in pasture, calculations

fugacities in air and terrestrial food *F10.11* chain

fugacities in arctic terrestrial food *F10.12* chain

molecular diffusivity in air *T20.5*

molecular diffusivity in water *T20.5*

partition constants for octanol, air, *E10.4* organic C and lipids with water

partition properties and *T10.3* accumulation factors in aquatic organisms

saturation concentration in water *E10.4*

Schmidt number in water *T20.5*

standard enthalpy of air-lipid *T10.5* transfer

standard enthalpy of air-pasture *E10.5* transfer

standard enthalpy of air-plant *363* transfer

transfer velocities in air, in water *T20.5*, and overall at two wind speeds *F20.7*

1,2,3,4,5,6-Hexachloro HCH   27 cyclohexane
1,2,3,4,5,6-Hexachloro-cyclohexane

α-, β-, γ- isomers and environmental *27* activity

aliases for the γ-isomer *31*

kinetic data for substitution and *T13.7* elimination reactions

rate constant with hydroxyl radical *F16.3*

rate constant with hydroxyl radical *F16.7* in troposphere

standard enthalpies of vaporization *T6.3* and of water-air transfer

γ-Hexachloro-cyclohexane

see 1,2,3,4,5,6-Hexachloro-cyclohexane

Hexachloroethane HCA   P12.1
1,1,1,2,2,2-Hexachloro-ethane

formation by microbial reduction of *726* tetrachloromethane

Henry constant *P12.1*

microbial reduction to   *726* tetrachloroethene by heme proteins

reductive dihalo-elimination *557* reaction

standard Gibbs energy *P12.1*

standard reduction potential *T14.3*, *F14.4*

n-Hexadecane   F2.12
Hexadecane

activity coefficients of five solutes *T6.1* in hexadecane

activity coefficients of four organic *T3.2* solutes in hexadecane

air-hexadecane partition constants *T6.1* for five solutes

air-quartz and air-teflon surface *E4.2* partition constants, calculations

Fishtine and structural parameters *T4.4* for estimating vapor pressure

fitting coefficients for   *T6.2* air-hexadecane partition constant

molar volume *187*

solvent fitting coefficients for   *T7.2* hexadecane-water partition constant of organic compounds

standard enthalpy and standard *T4.2* entropy of evaporation

vapor pressure of liquid at given *E4.2* temperature, calculation

water saturated, mole fraction *T5.1*

Hexadecanoic acid   698
Hexadecanoic acid

formation by microbial degradation *698* of octadecanoic acid

Hexanal
Hexanal

salting constant, seawater *T5.7*

n-Hexane   F2.12
Hexane

activity coefficients in organic *T3.2* solvents and in water

air-water partition constant *E6.4* estimation, calculations

aqueous solution data *T5.3*

energies of air-hexadecane,   *T3.4* air-water and hexadecane-water transfers

excess functions in gas phase, *T3.3* hexadecane and water

Fishtine and structural parameters *T4.4* for estimating vapor pressure

hexane-water partition constants for *T7.1* nine organic solutes

molecular diffusivity in polyethylene *P18.8*

molecular diffusivity in water *P18.8*

rate constant with hydroxyl radical *F16.3*

rate constant with hydroxyl radical *F16.7* in troposphere

refractive index *T3.1*

salting constant, seawater *T5.7*

standard enthalpies of vaporization *T6.3* and of water-air transfer

standard enthalpy and standard *T4.2* entropy of evaporation

water saturated, mole fraction *T5.1*

Hexanoic acid   T8.1
Hexanoic acid

acidity constant *T8.1*

solvent-water partition constants for *T7.1* five solvents

1-Hexanol   P5.3
Hexan-1-ol

standard enthalpies of vaporization *T6.3* and of water-air transfer

air-water partition constant *T20.5*

aqueous solution data *T5.2*

molecular diffusivity in air *T20.5*

molecular diffusivity in water *T20.5*

rate constant with hydroxyl radical *F16.7* in troposphere

Schmidt number in water *T20.5*

solvent-water partition constants for *T7.1* five solvents

standard enthalpies of vaporization *T6.3* and of water-air transfer

transfer velocities in air, in water *T20.5*, and overall at two wind speeds *F20.7*

water saturated, mole fraction *T5.1*

n-Hexanol
see 1-Hexanol

2-Hexanone
Hexan-2-one

standard enthalpies of vaporization *T6.3* and of water-air transfer

water saturated, mole fraction *T5.1*

n-Hexylamine   T8.2
Hexylamine

acidity constant of corresponding *T8.2* acid

free energy change of cation of ion *F11.14* exchange on montmorillonite

sorption isotherm of cation on *F11.13*
montmorillonite

HFC-134a
see 1,1,1,2-Tetrafluoroethane

Hydrazobenzene   **F14.19**
N,N'-Diphenyl-hydrazine

product of oxidative coupling of F14.19
aniline oxidation

Hydroquinone HQ   **561**
Benzene-1,4-diol

standard oxidation potential *T14.3*

7-Hydroxy isoquinoline   **F8.2**
Isoquinolin-7-ol

acidity constants and species *F8.2*
distribution

2-Hydroxy-benzoic acid   **F8.5**
2-Hydroxy-benzoic acid

acidity constants and proximity *F8.5*
effects

4-Hydroxy-benzoic acid   **F8.2**
4-Hydroxy-benzoic acid

acidity constants and species *F8.2*
distribution

2-Hydroxy-biphenyl
Biphenyl-2-ol

value of K_{iMM} and V_{max} *T17.7*

5-*exo*-Hydroxy-bornan-2-one   **717**
5-*exo*-Hydroxy-1,7,7-trimethyl-bicyclo
[2.2.1]heptan-2-one

formation by microbial oxidation of 717
camphor

2-Hydroxymuconic acid   **700**
semialdehyde
2-Hydroxy-6-oxo-hexa-2,4-dienoic acid

degradation to 2-oxo-pent-4-enoic 700
acid

1-Hydroxyethyl-1,1- HEDP   **F2.22**
diphosphonic acid
(1-Hydroxy-1-phosphono-ethyl)-phosphonic acid

complexing agent F2.22

I

Imidazole   **T8.2**
1*H*-Imidazole

acidity constant of corresponding *T8.2*
acid

acidity constant of corresponding *T8.3*
acid at different temperatures

Indane
Indan

H-donor (α) and   *T4.3*
H-acceptor property (β)

Indole   **F2.11**
1-*H*-Indole

conjugated π-electron system 30

Iodo-propane
value of V_{max} *T17.8*

Iodobutane
Iodo-butane

value of V_{max} *T17.8*

Iodoethane
Iodo-ethane

value of V_{max} *T17.8*

Iodopropane
Iodo-propane

value of K_{iMM} and V_{max} *T17.7*

Ironporphyrin

standard reduction and oxidation *F14.4*
potentials

Isobutyraldehyde   **F2.17**
2-Methyl-propionaldehyde

formation of, usage 40

Isobutyric acid amide   **T13.10**
Isobutyramide

rate constants of hydrolysis reaction *T13.10*

Isooctane   **P5.4**
2,2,4-Trimethyl-pentane

indirect photolysis half-life by   *E16.3*
hydroxyl radical in troposphere,
calculation of

Isoquinoline   **T8.2**
Isoquinoline

acidity constant of corresponding *T8.2*
acid

J

Juglone   **F14.4**
5-Hydroxy-[1,4]naphthoquinone

standard reduction potential *F14.4*

L

Lactate
2-Hydroxy-propionic acid

standard oxidation potential *T14.2*

Levothyroxine   **F2.25**
2-Amino-3-[4-(4-hydroxy-3,5-diiodo-phenoxy)-3,5
-diiodo-phenyl]-propionic acid

hormone F2.25

Linuron   **E17.3**
N'-(3,4-Dichloro-phenyl)-*N*-methoxy-
N-methyl-urea

value of K_{iMM} and V_{max} *T17.7*

M

Malathion   **T17.3**
2-(Dimethoxy-thiophosphorylsulfanyl)-succinic
acid diethyl ester

enzymatic hydrolysis 707

value of μ_{max} and K_{iM} *T17.6*

Maleic acid   **F2.7**
(2*Z*)-But-2-enedioic acid

property of 27

Mecoprop   **F2.5**
2-(4-Chloro-2-methyl-phenoxy)-propionic acid

biological activity of enantiomers 26

roof runoff 42

uptake by organisms 738

***p*-Menth-2-ene**   **F16.7**
3-Isopropenyl-6-methyl-cyclohexene

rate constant with hydroxyl radical *F16.7*
in troposphere

2-Mercapto-succinic acid   **T17.3**
diethyl ester
2-Mercapto-succinic acid diethyl ester

formation by enzymatic hydrolysis 707
of malathion

Methane   **F2.3**
Methane

air-water partition constant *T20.5*

bond angles *F2.3*

critical temperature and critical *T4.1*
pressure

formation by degradation of *E17.1*
benzene

in methanogenic degradation *E12.2*

methanogenesis in anoxic F14.3
groundwater

molecular diffusivity at different *T20.3*
temperatures

molecular diffusivity in air *F18.9*
T20.5

molecular diffusivity in water *T20.5*

rate constant with hydroxyl radical *F16.3*

rate constant with hydroxyl radical *F16.7*
in troposphere

Schmidt number in water *T20.5*

Schmidt number in water at *T20.3*
different temperatures

standard oxidation potential *T14.2*

transfer velocities in air, in water *T20.5*,
and overall at two wind speeds *F20.7*

value of μ_{max} and K_{iM} *T17.6*

Methane thiol   **T13.2**
Methane thiol

product in methylchloride *T13.2*
substitution reaction

Methanol   **F2.3**
Methanol

acidity constant *F8.3*

activity coefficients of five solutes *T6.1*
in methanol

air-methanol partition constants for *T6.1*
five solutes

aqueous solution data *T5.2*

as cosolvent *T5.8*

bond angles *F2.3*

industrial chemical 38

molar volume *187*

product in methylbromide *T13.2*
substitution reaction

product in trimethylphosphate *T13.2*
substitution reaction

rate constant with hydroxyl radical *F16.3*

refractive index *T3.1*

solvent fitting coefficients for   *T6.2*
air-methanol partition constant of

solvent for relative nucleophilicities *T13.4*
against methyl iodide   499

standard enthalpies of vaporization *T6.3*
and of water-air transfer

standard Gibbs energy in aqueous *E13.1*
solution

Methidathion   F2.24
Dithiophosphoric acid
S-(5-methoxy-2-oxo-[1,3,4]thiadiazol-3-ylmethyl)
ester O,O'-dimethyl ester

insecticide and acaricide F2.24

2-Methoxyaniline   F14.20
2-Methoxy-phenylamine

relative oxidation rate by *F14.20*
manganese oxide

3-Methoxyaniline   F14.20
3-Methoxy-phenylamine

relative oxidation rate by *F14.20*
manganese oxide

4-Methoxyaniline   F14.20
4-Methoxy-phenylamine

relative oxidation rate by *F14.20*
manganese oxide

Methoxybenzene   F16.3
Methoxy-benzene

rate constant with hydroxyl radical *F16.3*

rate constant with hydroxyl radical *F16.7*
in troposphere

water saturated, mole fraction *T5.1*

4-Methoxybenzene sulfonic   P13.10
acid methyl ester
4-Methoxy-benzenesulfonic acid methyl ester

rate constant of neutral hydrolysis *P13.10*

Methoxy-methyl-carbamic   E17.3
acid
Methoxy-methyl-carbamic acid

formation by enzymatic hydrolysis 714
of linuron

3-Methoxyphenol   P13.9
3-Methoxy-phenol

acidity constant *P13.9*

4-Methoxyphenol   P13.9
4-Methoxy-phenol

acidity constant *P13.9*
     P16.2

rate constants of indirect photolysis *P16.2*
with singlet oxygen

4-Methoxyphenyl N-phenyl   E13.6
carbamate
Phenyl-carbamic acid 4-methoxy-phenyl ester

rate constant of base catalyzed *E13.6*
hydrolysis

Methoxychlor   P13.3
1,1'-(2,2,2-Trichloroethane-1,1-diyl)-bis-(4-
methoxybenzene)

hydrolysis rate constants *P13.3*

Methyl bromide   F4.4
Bromo-methane

as reference for relative   *T13.3*
nucleophilicity, n, of nucleophile in 498
water

equilibrium with methylchloride *E12.1*
under natural conditions,
calculations

extent of hydrolysis reaction in *E13.1*
groundwater, calculations

half life in water with nucleophiles, *E13.1*
calculations

kinetic data of neutral hydrolysis *P13.4*

nucleophile substitution reaction *T13.2*

rate constant with hydroxyl radical *F16.7*
in troposphere

salting constant, seawater *T5.7*

standard Gibbs energy *E12.1*

standard Gibbs energy in aqueous *E13.1*
solution

vapor pressure as function of *F4.4*
temperature

Methyl chloride   E12.1
Chloro-methane

equilibrium with methylbromide *E12.1*
under natural conditions,
calculations

kinetic data of neutral hydrolysis *P13.4*

molecular diffusivity in water *F18.10*

nucleophile substitution reaction *T13.2*

rate constant with hydroxyl radical *F16.7*
in troposphere

standard Gibbs energy *E12.1*

Methyl iodide   499
Iodo-methane

as reference for relative   *T13.4*
nucleophilicity, n, of nucleophile in 498
methanol

transformation in seawater 501

value of V'_{max} *T17.8*

Methyl sulfide   F2.3
Methanethiol

bond angles *F2.3*

2-Methyl-1,5-hexadiene   P16.3
2-Methyl-hexa-1,5-diene

rate constant with hydroxyl radical *P16.3*
in troposphere

N-(4-Methyl-3,5-dinitro-   721
phenyl)-hydroxylamine
N-(4-Methyl-3,5-dinitro-phenyl)-hydroxylamine

formation by microbial reduction of 721
2,6-dinitro-4-nitroso-toluene

microbial reduction to   721
4-amino-2,6-dinitro-toluene

3-Methyl-4-nitrophenyl   P13.6
N-methyl-N-phenyl
carbamate
Methyl-phenyl-carbamic acid
3-methyl-4-nitro-phenyl ester

rate constant of base catalyzed *P13.6*
hydrolysis

N-Methyl-acetamide   F2.3
N-Methyl-acetamide

bond angles *F2.3*

2-Methylaniline   F14.20
o-Tolylamine

relative oxidation rate by *F14.20*
manganese oxide

3-Methylaniline   F14.20
m-Tolylamine

relative oxidation rate by *F14.20*
manganese oxide

4-Methylaniline   F14.20
p-Tolylamine

relative oxidation rate by *F14.20*
manganese oxide

4-Methylbenzene sulfonic   P13.10
acid methyl ester
Toluene-4-sulfonic acid methyl ester

rate constant of neutral hydrolysis *P13.10*

3-Methyl-benzoic acid   716
3-Methyl-benzoic acid

correlation of biodegradability with 716
$\Sigma\sigma_j$

4-Methyl-benzoic acid   F8.6
4-Methyl-benzoic acid

acidity constant *F8.6*

correlation of biodegradability with 716
$\Sigma\sigma_j$

2-(3-Methyl-benzyl)-succinic   F17.11
acid
2-(3-Methyl-benzyl)-succinic acid

enzymatic formation from fumarate F 17.11
and 1,3-dimethyl- benzene

Methyl *tert*-butyl ether MTBE   F2.15
2-Methoxy-2-methyl-propane

high volume pollutant 39

rate constant with hydroxyl radical *F16.7*
in troposphere

standard enthalpies of vaporization *T6.3*
and of water-air transfer

N-Methyl-N'-phenyl urea,
substituted

rate constants in sensitised photolysis *T16.2*
and Hammett constants

1-Methylnaphthalene 1-MeNa E5.2
1-Methyl-naphthalene

reaction quantum yield of direct *T15.7*
photolysis

2-Methylnaphthalene
2-Methyl-naphthalene

reaction quantum yield of direct *T15.7*
photolysis

2-Methyl-nitrobenzene 2-CH₃-NB

1-Methyl-2-nitro-benzene

one-electron reduction potential *T14.4*

relative reduction rates in different *F14.5*
systems   *F14.10*

3-Methyl-nitrobenzene 3-CH₃-NB

1-Methyl-3-nitro-benzene

one-electron reduction potential *T14.4*

relative reduction rates in different *F14.10*
reduction media

4-Methyl-nitrobenzene 4-CH₃-NB

1-Methyl-4-nitro-benzene

one-electron reduction potential *T14.4*

relative reduction rates in different *F14.10*
reduction media

2-Methylphenol

2-Methyl-phenol

standard enthalpies of vaporization *T6.3*
and of water-air transfer

3-Methylphenol

3-Methyl-phenol

acidity constant *P13.9*

aqueous solution data *T5.2*

value of K_{iMM} and V_{max} *T17.7*

4-Methylphenol   **P13.9**

4-Methyl-phenol

acidity constant *F8.6*

aqueous solution data *T5.3*

biodegradation of spilled compound *E17.7*
in a pond

determination of μ_{max} 744

formation of biomass from 740

rate constant with singlet oxygen *F16.5*

standard enthalpies of vaporization *T6.3*
and of water-air transfer

value of μ_{max} and K_{iM} *T17.6*

Methyl-phenyl-ketone   **F2.17**

1-Phenyl-ethanone

solvent and chemical intermediate *F2.17*

4-Methylphenyl N-methyl   **T13.11**
carbamate

Methyl-carbamic acid *p*-tolyl-ester

rate constants of hydrolysis reaction *T13.11*

4-Methylphenyl N-phenyl   **E13.6**
carbamate

Phenyl-carbamic acid *p*-tolyl ester

rate constant of base catalyzed *E13.6*
hydrolysis

4-Methyl phenylacetic acid   **F8.6**

p-Tolyl-acetic acid

acidity constant *F8.6*

Methylamine   **F2.3**

Methylamine

acidity constant of corresponding *F8.3*
acid and resonance structure

bond angles *F2.3*

formation by enzymatic hydrolysis 707
of carbofuran

product in carbofuran hydrolysis *T13.2*
reaction

Methylbenzene

see Toluene

Methylchloropyriphos   **542**

Thiophosphoric acid *O,O'*-dimethyl ester
O''-(3,5,6-trichloro-pyridin-2-yl) ester

influence of metals on hydrolysis 542
rate and mechanism

Methylchloropyriphos oxon   **542**

Phosphoric acid dimethyl ester
3,5,6-trichloro-pyridin-2-yl ester

influence of metals on hydrolysis 542
rate and mechanism

Methylparathion   **T13.12**

Thiophosphoric acid *O,O'*-dimethyl ester
O''-(4-nitro-phenyl) ester

rate constants of hydrolysis reaction *T13.12*

value of μ_{max} and K_{iM} *T17.6*

Metobenzuron   **F2.24**

N-Methoxy-*N'*-(4-[(2-methoxy-2,4,4-trimethyl-3,4-
dihydro-2*H*- chromen-7-yl)oxy]phenyl)-
N-methylurea

herbicide *F2.24*

MTBE

see Methyl tert-butyl ether

N

Naphthacene   **F15.3**

Naphthacene

decadic molar absorption *F15.3*
coefficient

Naphthalene   **F2.11**

Naphthalene

activity coefficient in water/ethanol 236
mixture

air-water partition constant *T20.5*

aqueous solubilities with cosolvents *T5.8*

aqueous solution data *T5.2,*
                                       T5.3

aromatism

biological organic phases-water *T10.2*
partition coefficients

carboxylation to 2-naphthoic acid 733

concentration in diesel fuel *T19.3*

decadic molar absorption *F15.3*
coefficient

degradation by microorganisms *F17.2*

density *T19.3*

dissolution from diesel fuel into *E19.4*
aqueous phase, calculations

Fishtine and structural parameters *T4.4*
for estimating vapor pressure

H-donor (α) and   *T4.3*
H-acceptor property (β)

molecular diffusivity in air *T20.5*

molecular diffusivity in water *T20.5*

molecular diffusivity in water and in *T19.3*
diesel fuel

octanol-water partition constant *T10.2*

partition between fat, water and air *E3.1*
in a soup bowl, calculations

polarizability *T5.5*

rate constant with hydroxyl radical *F16.3*

rate constant with hydroxyl radical *F16.7*
in troposphere

reaction quantum yield of direct *T15.7*
photolysis

refractive index *T3.1*

salting constant, seawater *T5.7*

salting constants, single salts *T5.6*

Schmidt number in water *T20.5*

standard enthalpies of vaporization *T6.3*
and of water-air transfer

standard enthalpy and entropy of   *T4.5*
fusion, experimental and predicted
data

standard enthalpy and standard *T4.2*
entropy of evaporation

transfer velocities in air, in water *T20.5,*
and overall at two wind speeds *F20.7*

value of K_{iMM} and V_{max} *T17.7*

value of μ_{max} and K_{iM} *T17.6*

vapor pressure as function of *F4.4*
temperature

1-Naphthalene-sulfonic acid   **T8.1**

Naphthalene-1-sulfonic acid

surfactant 48

acidity constant *T8.1*

1-Naphthol

Naphthalen-1-ol

salting constants, single salts *T5.6*

salting constant, seawater *T5.7*

solvent-water partition constants for *P7.3*
five solvents

2-Naphthol   **T8.1**

Naphthalen-2-ol

acidity constant *T8.1*

1,2-Naphthoquinone

[1,2]Naphthoquinone

decadic molar absorption *F15.4*
coefficient

1,4-Naphthoquinone

[1,4]Naphthoquinone

decadic molar absorption *F15.4*
coefficient

1-Naphthyl-*N*-methyl   **T13.11**
carbamate

Methyl-carbamic acid naphthalen-1-yl ester

rate constants of hydrolysis reaction *T13.11*

1-Naphthylamine   *T8.2*
Naphthalen-1-ylamine

acidity constant of corresponding *T8.2*
acid

Hammett susceptibility factor *Q8.5*

Naptalam   **544**
N-Naphthalen-1-yl-phthalamic acid

influence of metals on hydrolysis 544
rate and mechanism

Nitrilotriacetic acid NTA   **E21.1**
(Bis-carboxymethyl-amino)-acetic acid

characteristic biodegradation rate *E21.1*
constants

characteristic concentration in a *E21.1*
Swiss lake

reaction rate constant with variable *E21.2*
input, calculation of

reaction rate for second order   *E21.3*
reaction rate constant, calculation
of

reaction rate for variable reaction *E21.2*
rate constant, calculation of

total reaction rate constant in a lake, *E21.1*
calculation of

para-Nitroacetophenone PNAP   **631**
1-(4-Nitro-phenyl)-ethanone

24-h-averaged total specific light *T15.6*
absorption rate, calculations

decadic molar absorption *T15.4*
coefficients

reactant in binary actinometer 648

specific light absorption rate at   *T15.4,*
wavelength and total, calculations *F15.9*

specific light absorption rate, 631
calculation example

specific light absorption rate,   *F15.10*
seasonal and latitudinal variations

total specific light absorption rate, *E15.2*
calculation

3-Nitroaniline
3-Nitro-phenylamine

biological organic phases-water *T10.2*
partition coefficients

octanol-water partition constant *T10.2*

4-Nitroaniline   **T8.2**
4-Nitro-phenylamine

acidity constant of corresponding *T8.2*
acid

parameters for indirect photolysis *P16.2*
with carbonate radical

para-Nitroanisole PNA   **648**
1-Methoxy-4-nitro-benzene

reactant in binary actinometer 648

Nitro-benzene NB   **T14.3**
Nitro-benzene

absorbance spectrum of light *E15.1*

biological organic phases-water *T10.2*
partition coefficients

decadic molar absorption *F15.1,*
coefficient   *P15.3*

decadic molar extinction *E15.1*
coefficients, calculated

electronic absorption spectrum *F15.1*

exchange constant   *T11.2*
water-aluminosilicate surface
($K_{NAC,EDA}$)

molecular diffusivity in air *F18.9*

octanol-water partition constant *T10.2*

one-electron reduction potential *T14.4*

polarizability *T5.5*

rate constant with hydroxyl radical *F16.3*

rate constant with hydroxyl radical *F16.7*
in troposphere

reaction quantum yield of direct *T15.7*
photolysis

reduction potentials of single *F14.9*
electron steps

reduction rates with sulfide and *F14.7*
DOM

refractive index *T3.1*

relative reduction rates in different *F14.10*
reduction media

relative reduction rates in different
systems

standard enthalpy and standard *T4.2*
entropy of evaporation

standard reduction potential *T14.3,*
F14.4

value of K_{iMM} and V_{max} *T17.7*

water saturated, mole fraction *T5.1*

3-Nitrobenzene sulfonic acid   **P13.10**
methyl ester
3-Nitro-benzenesulfonic acid methyl ester

rate constant of neutral hydrolysis *P13.10*

4-Nitrobenzene sulfonic acid   **P13.10**
methyl ester
4-Nitro-benzenesulfonic acid methyl ester

rate constant of neutral hydrolysis *P13.10*

3-Nitro-benzoic acid   **F8.6**
3-Nitro-benzoic acid

acidity constant *F8.6*

4-Nitrobenzoic acid   **T8.1**
4-Nitro-benzoic acid

acidity constant *T8.1*

acidity constant at different *T8.3*
temperatures

2-Nitrophenol   **T8.1**
2-Nitro-phenol

acidity constant *T8.1*
P20.6

acidity constant at different *T8.3*
temperatures

air-water partition constant *P20.6*

biodegradation to catechol *E17.9*

polarizability *T5.5*

salting constant, seawater *T5.7*

value of K_{iMM} and V_{max} *T17.7*

3-Nitrophenol
3-Nitro-phenol

acidity constant and resonance *F8.4*
structures   *F8.6*

salting constant, seawater *T5.7*

4-Nitrophenol   **T13.2**
4-Nitro-phenol

acidity constant and resonance *F8.4*
structures   *P20.6*

acidity constant, calculation *E8.2*

air-water partition constant *P20.6*

decadic molar absorption *F15.5*
coefficient

decadic molar absorption *F15.5*
coefficient of anion

direct photolysis rate at near surface *E15.3*
and in water body, calculation

direct photolysis rate of anion at   *E15.3*
near surface and in water body,
calculation

formation by enzymatic hydrolysis 707
of paraoxon

polarizability *T5.5*

product in parathion hydrolysis *T13.2*
reaction

rate constant with hydroxyl radical *F16.7*
in troposphere

rate constant with singlet oxygen *F16.5*

reaction quantum yield of direct *T15.7*
photolysis

reaction quantum yield of direct *T15.7*
photolysis of anion

salting constant, seawater *T5.7*

4-Nitrophenyl *N*-phenyl   **528**
carbamate
Phenyl-carbamic acid 4-nitro-phenyl ester

hydrolysis reaction pathway 528

rate constants of hydrolysis reaction *T13.11*

4-Nitrophenyl *N*-methyl-   **528**
***N*-phenyl carbamate**
Methyl-phenyl-carbamic acid 4-nitro-phenyl
ester

hydrolysis reaction pathway 528

rate constant of base catalyzed *P13.6*
hydrolysis

rate constants of hydrolysis reaction *T13.11*

4-Nitrophenyl *N,N*-dimethyl   **T13.11**
carbamate
Dimethyl-carbamic acid 4-nitro-phenyl ester

rate constants of hydrolysis reaction *T13.11*

3-Nitrophenyl *N*-phenyl   E13.6
carbamate
Phenyl-carbamic acid 3-nitro-phenyl ester
  rate constant of base catalyzed   *E13.6*
  hydrolysis

3-Nitro-phenylacetic acid   F8.6
(3-Nitro-phenyl)-acetic acid
  acidity constant *F8.6*

Nitropropane   F16.3
1-Nitro-propane
  rate constant with hydroxyl radical *F16.3*

1-Nitropyrene   F2.18
1-Nitro-pyrene
  fuel combustion product *F2.18*

4-Nitro-pyridine   T8.2
4-Nitro-pyridine
  acidity constant of corresponding *T8.2*
  acid

2-Nitrotoluene   Q11.7
1-Methyl-2-nitro-benzene
  exchange constant   *T11.2*
  water-aluminosilicate surface
  ($K_{NAC,EDA}$)

3-Nitrotoluene   E16.3
1-Methyl-3-nitro-benzene
  exchange constant   *T11.2*
  water-aluminosilicate surface
  ($K_{NAC,EDA}$)
  indirect photolysis half-life by   *E16.3*
  hydroxyl radical in troposphere,
  calculation of

4-Nitrotoluene
1-Methyl-4-nitro-benzene
  exchange constant   *T11.2*
  water-aluminosilicate surface
  ($K_{NAC,EDA}$)
  reaction quantum yield of direct *T15.7*
  photolysis
  salting constant, seawater *T5.7*

4-Nitrophenyl *N*-phenyl   528
carbamate
Phenyl-carbamic acid 4-nitro-phenyl ester
  rate constants of hydrolysis reaction *T13.11*

Nonanal
Nonanal
  salting constant, seawater *T5.7*

***n*-Nonane**
Nonane
  air-quartz partition constant and *F3.7*
  dispersive vdW parameter

4-Nonyl-phenol   F2.15
4-Nonyl-phenol
  a biodegradation product 39

4-Nonylphenol-   F2.16
diethyleneglycol ether
2-[2-(4-Nonyl-phenoxy)-ethoxy]-ethanol
  a biodegradation product 39

4-Nonylphenol-ethyleneglycol   F2.16
ether
2-(4-Nonyl-phenoxy)-ethanol
  a biodegradation product 39

4-Nonylphenol-
polyethyleneglycol ether
  biodegradadation of 39

NTA
see Nitrilotriacetic acid

O

***n*-Octadecane**
Octadecane
  air-water partition constant *T20.5*
  molecular diffusivity in air *T20.5*
  molecular diffusivity in water *T20.5*
  Schmidt number in water *T20.5*
  transfer velocities in air, in water *T20.5*,
  and overall at two wind speeds *F20.7*

Octanal
Octanal
  salting constant, seawater *T5.7*

***n*-Octane**   E5.2
Octane
  activity coefficients in six organic *T6.1*
  solvents
  air-quartz partition constant and *F3.7*
  dispersive vdW parameter
  air-solvent partition constants for six *T6.1*
  organic solvents
  aqueous solution data *T5.3*
  solvent-water partition constants for *T7.1*
  five solvents
  standard enthalpies of vaporization *T6.3*
  and of water-air transfer
  water saturated, mole fraction *T5.1*

***n*-Octanol**   182
Octan-1-ol
  activity coefficients of five solutes *T6.1*
  in octanol
  air-octanol partition constants for *T6.1*
  five solutes
  as surrogate for natural organic 182
  phases
  character as solvent 223
  cosolute in water, effect on activity *187*
  coefficient
  molar volumes of dry and wet *186, 215*
  solvent
  octanol-water partition constants for *T7.1*
  nine organic solutes
  partition constants for dry and wet *F6.2*
  solvent
  rate constant with hydroxyl radical *F16.3*
  refractive index *T3.1*

  solvent fitting coefficients for *T6.2*
  air-octanol partition constant
  solvent fitting coefficients for   *T7.2*
  octanol-water partition constant of
  aqueous solution data *T5.3*
  standard enthalpies of vaporization *T6.3*
  and of water-air transfer
  water saturated, mole fraction *T5.1*

1-Octylamine   F11.13
Octylamine
  free energy change of cation of ion *F11.14*
  exchange on montmorillonite

Olive oil
  contaminated with benzene from E6.1
  air, calculations
  solvent fitting coefficients for   *T6.2*
  air-olive oil partition constant *i*

Oxadiazon   F2.24
5-*tert*-Butyl-3-(2,4-dichloro-5-isopropoxy-phenyl)-
3*H*-[1,3,4]oxadiazol-2-one
  herbicide F2.24

Oxalic acid   T11.4
Oxalic acid
  reaction of anion with mineral T11.4
  surfaces

Oxirane   F16.3
Oxirane
  rate constant with hydroxyl radical *F16.3*

2-Oxo-pent-4-enoic acid   700
2-Oxo-pent-4-enoic acid
  formation by enzymatic degradation 700
  of 2-hydroxymuconic acid
  semialdehyde

Oxygen
Oxygen
  air-water overall transfer velocity *F20.9*
  and oil film thickness
  molecular diffusivity at different *T20.3*
  temperatures
  molecular diffusivity in oil *930*
  oil-water distribution coefficient *930*
  Schmidt number in water at *T20.3*
  different temperatures
  transfer velocity in water and wind *F20.3*
  speed

P

Palmitic acid
see Hexadecanoic acid

Paraoxon   T13.12
Phosphoric acid diethyl ester 4-nitro-phenyl
ester
  enzymatic hydrolysis 707
  rate constants of hydrolysis reaction *T13.12*

Parathion   F2.22
Thiophosphoric acid *O,O*'-diethyl ester
O''-(4-nitro-phenyl) ester
  insecticide and acaricide F2.22

mechanism of base catalyzed 538
hydrolysis

rate constants of hydrolysis reaction *T13.12*

PCB
see Biphenyl, polychlorinated

PCB33
see 2',3,4-Trichlorobiphenyl

PCB47
see 2,2',4,4'-Tetrachlorobiphenyl

PCB52
see 2,2',5,5'-Tetrachlorobiphenyl

PCB101
see 2,2',4,5,5'-Pentachloro- biphenyl

PCB153
see 2,2',4,4',5,5'-Hexachloro- biphenyl

PCB170
see 2,2',3,3',4,4',5-Heptachloro- biphenyl

PCB180
see 2,2',3,4,4',5,5'-Heptachloro- biphenyl

PCB185
see 2,2',3,4,5,5',6-Heptachloro- biphenyl

PCDD
see Dibenzo-*p*-dioxine, polychlorinated

PCDF
see Dibenzo-furan, polychlorinated

PCE
see Tetrachloroethene

PCP
see Pentachlorophenol

Penicillin V   F2.25
3,3-Dimethyl-7-oxo-6-(2-phenoxy-acetylamino)-4 - thia-1-aza-bicyclo[3.2.0]heptane-2-carboxylic acid

antibiotic F2.25

2,2',4,4',5-Pentachloro biphenyl
2,4,5,2',4'-Pentachloro-biphenyl

vapor pressure as function of *F4.4* temperature

2,2',4,5,5'-Pentachloro PCB101   E9.3 biphenyl
2,4,5,2',5'-Pentachloro-biphenyl

concentration in pore water from *E9.3* sediment concentration, calculation

concentrations in lake water and *T10.3* sediment

molar volume and molecular *E19.3* diffusivity in water

organic carbon-water distribution *E9.3* ratio, calculations

partition constants *E19.3*

partition properties and   *T10.3* accumulation factors in aquatic organisms

sediment to open water diffusion, *E19.3* calculations

standard enthalpy and entropy of   *T4.5* fusion, experimental and predicted data

2,3,4,5,6-Pentachloro-   726 biphenyl
2,3,4,5,6-Pentachloro-biphenyl

microbial reduction catalyzed by 726 metalloenzymes

Pentachlorobenzene
Pentachloro-benzene

biological organic phases-water *T10.2* partition coefficients

equilibrium bioaccumulation factor, *E10.1* K_{ibio}, calculations

octanol-water partition constant *T10.2*

Pentachloroethane   T13.7
1,1,1,2,2-Pentachloro-ethane

β-elimination reaction *507*

kinetic data for substitution and *T13.7* elimination reactions

Pentachlorophenol PCP   39
2,3,4,5,6-Pentachloro-phenol

absorbance spectrum of light *P15.5*

absorbance spectrum of light of *P15.5* anion

acidic biocide 39

acidity constant *T8.1, F8.9*

acidity constant, calculation *E8.2*

distribution ratio of anion into polar *F10.14* lipids and into octanol   377

lethal concentration vs.   *F10.15* liposome-water distribution ratio

n-octanol-water distribution ratio *F8.9* and pH

organic carbon-water distribution *F9.17* ratio and pH

phenolate anion as biocide 39

reaction quantum yield of direct *P15.5* photolysis

reaction quantum yield of direct *T15.7* photolysis of anion

species fractions at different pHs, *E8.1* calculations

Pentanal
Pentanal

salting constant, seawater *T5.7*

***n*-Pentane**
Pentane

air-teflon partition constant and *F3.7* dispersive vdW parameter

aqueous solution data *T5.3*

salting constant, seawater *T5.7*

water saturated, mole fraction *T5.1*

1-Pentanol
Pentan-1-ol

aqueous solution data *T5.3*

rate constant with hydroxyl radical *F16.7* in troposphere

water saturated, mole fraction *T5.1*

molecular diffusivity in air *F18.9*

2-Pentanone
Pentan-2-one

aqueous solution data *T5.3*

water saturated, mole fraction *T5.1*

1-Pentylamine
Pentylamine

free energy change of cation of ion *F11.14* exchange on montmorillonite

Permethrin   T17.3
3-(2,2-dichloro-vinyl)-2,2-dimethyl-cyclopropane carboxylic acid 3-phenoxy-benzyl ester

enzymatic hydrolysis 707

Perylene   F2.13
Perylene

biological organic phases-water *T10.2* partition coefficients

H-donor (α) and   *T4.3* H-acceptor property (β)

octanol-water partition constant *T10.2*

Phenanthrene   F2.11
Phenanthrene

air-quartz and air-teflon surface *E4.2* partition constants, calculations

aqueous solution data *T5.2, T5.3*

aromatism

as model compound in aquatic 950 environment, qualitative description

carboxylation to phenanthrene 733 carboxylic acid

concentrations in earthworm and *P10.2* soil

concentrations in ryegrass, yarrow *P10.3* and air

decadic molar absorption *F15.3* coefficient

H-donor (α) and   *T4.3* H-acceptor property (β)

measured and predicted   *E9.2* concentrations on soils and sediments

organic carbon-water partition *E9.2* constant

rate constant with hydroxyl radical *F16.7* in troposphere

reaction quantum yield of direct *T15.7* photolysis

retardation factor in groundwater, *E9.4* calculations

retardation factor in water with *E9.4* cosolvent, calculations

salting constant, seawater *T5.7*

soil-water and sediment-water *E9.2* distributions, calculations

sorption on teflon and on quartz *E11.1* surfaces from air, calculations

standard enthalpies of vaporization *T6.3* and of water-air transfer

standard enthalpy and entropy of   *T4.5* fusion, experimental and predicted data

standard enthalpy and standard *T4.2* entropy of evaporation

standard enthalpy of air-plant *P10.3* transfer

value of μ_{max} and K_{iM} *T17.6*

vapor pressure of subcooled liquid *E4.2* at given temperature, calculation

Phenol   **F2.15**
Phenol

   acidity constant *T8.1*
                    T8.6
                    F8.6

   acidity constant and resonance *F8.3* structur

   air-water partition constant *198*
                                T20.5

   aqueous solution data *T5.2,*
                         T5.3

   H-donor (α) and   *T4.3* H-acceptor property (β)

   Hammett constant for ortho *T8.7* substitution

   Hammett susceptibility factor *T8.6,* *F8.7*

   molecular diffusivity in air *T20.5*

   molecular diffusivity in water *T20.5*

   nucleophilicity of phenolate in water *499*

   polarizability *T5.5*

   radicals of oxidation and products of *F14.18* coupling

   rate constant with hydroxyl radical *F16.3*

   rate constant with hydroxyl radical *F16.7* in troposphere

   salting constant, seawater *T5.7*

   Schmidt number in water *T20.5*

   solvent-water partition constants for *T7.1* five solvents

   standard enthalpies of vaporization *T6.3* and of water-air transfer

   standard enthalpy and standard *T4.2* entropy of evaporation

transfer velocities in air, in water *T20.5,* and overall at two wind speeds *F20.7*

value of μ_{max} and K_{iM} *T17.6*

Phenoxyacetic acid   **T8.6**
Phenoxy-acetic acid

   acidity constant *T8.6*

   Hammett susceptibility factor *T8.6*

3-Phenoxy-benzylalcohol   **T17.3**
(3-Phenoxy-phenyl)-methanol

   formation by enzymatic hydrolysis *707* of permethrin

2-Phenoxy-phenol   **F14.18**
2-Phenoxy-phenol

   product of oxidative coupling of F14.18 phenol oxidation

4-Phenoxy-phenol   **F14.18**
4-Phenoxy-phenol

   product of oxidative coupling of F14.18 phenol oxidation

N-Phenyl-[1,4]benzoquinone-   **F14.19**
imine
(4-Imino-cyclohexa-2,5-dienylidene)-phenyl-amine

   product of oxidative coupling of F14.19 aniline oxidation

Phenyl sulfide
Benzenethiol

   nucleophilicity in water *T13.3*

N-Phenyl-1,4-   **F14.19**
phenylenediamine
N-Phenyl-benzene-1,4-diamine

   product of oxidative coupling of F14.19 aniline oxidation

Phenylacetic acid   **T8.6**
Phenyl-acetic acid

   acidity constant *T8.6*
                    F8.6

   Hammett susceptibility factor *T8.6,* *F8.7*

   rate constant with hydroxyl radical *F16.3*

Phenyl N-methyl-N-phenyl   **P13.6**
carbamate
Methyl-phenyl-carbamic acid phenyl ester

   rate constant of base catalyzed *P13.6* hydrolysis

5-Phenyl-pentanoic acid   **F17.4**
5-Phenyl-pentanoic acid

   β-oxidation and degradation to *F17.4* 3-phenyl-propionic acid

3-Phenyl-propionic acid   **T8.6**
3-Phenyl-propionic acid

   acidity constant *T8.6*

   formation by degradation of *F17.4* 5-Phenyl-pentanoic acid

   Hammett susceptibility factor *T8.6,* *F8.7*

Phenylamine
see Aniline

nucleophilicity in water *499*

Phorate   **P13.11**
Dithiophosphoric acid O,O'-diethyl ester S-ethylsulfanylmethyl ester

   rate constants of hydrolysis *P13.11*

Phosphoric acid diethyl   **P13.9**
3-bromophenyl ester
Phosphoric acid 3-bromo-phenyl ester diethyl ester

   rate constant of base catalyzed *P13.9* hydrolysis

Phosphoric acid diethyl   **P13.9**
3-chlorophenyl ester
Phosphoric acid 3-chloro-phenyl ester diethyl ester

   rate constant of base catalyzed *P13.9* hydrolysis

Phosphoric acid diethyl   **P13.9**
3-methoxyphenyl ester
Phosphoric acid diethyl ester 3-methoxy-phenyl ester

   rate constant of base catalyzed *P13.9* hydrolysis

Phosphoric acid diethyl   **P13.9**
3-methylyphenyl ester
Phosphoric acid diethyl ester 3-methyl-phenyl ester

   rate constant of base catalysed *P13.9* hydrolysis

Phosphoric acid diethyl   **P13.9**
3-nitrophenyl ester
Phosphoric acid diethyl ester 3-nitro-phenyl ester

   rate constant of base catalyzed *P13.9* hydrolysis

Phosphoric acid diethyl   **P13.9**
4-acetylphenyl ester
Phosphoric acid 4-acetyl-phenyl ester diethyl ester

   rate constant of base catalyzed *P13.9* hydrolysis

Phosphoric acid diethyl   **P13.9**
4-bromophenyl ester
Phosphoric acid 4-bromo-phenyl ester diethyl ester

   rate constant of base catalyzed *P13.9* hydrolysis

Phosphoric acid diethyl   **P13.9**
4-chlorophenyl ester
Phosphoric acid 4-chloro-phenyl ester diethyl ester

   rate constant of base catalyzed *P13.9* hydrolysis

Phosphoric acid diethyl   **P13.9**
4-cyanophenyl ester
Phosphoric acid 4-cyano-phenyl ester diethyl ester

   rate constant of base catalyzed *P13.9* hydrolysis

Phosphoric acid diethyl   **P13.9**
4-ethylyphenyl ester
Phosphoric acid diethyl ester 4-ethyl-phenyl ester

   rate constant of base catalyzed *P13.9* hydrolysis

Phosphoric acid diethyl P13.9
4-methoxyphenyl ester
Phosphoric acid diethyl ester 4-methoxy-phenyl
ester

   rate constant of base catalyzed *P13.9*
   hydrolysis

Phosphoric acid diethyl P13.9
4-nitrophenyl ester
Phosphoric acid diethyl ester 4-nitro-phenyl
ester

   rate constant of base catalyzed *P13.9*
   hydrolysis

Phosphoric acid diethyl phenyl P13.9
ester
Phosphoric acid diethyl ester phenyl ester

   rate constant of base catalyzed *P13.9*
   hydrolysis

Phthalates F2.17

   production rate, usage as plasticizer 42

o-**Phthalic acid** *o*-PA E8.1
Phthalic acid

   reaction of anion with mineral T11.4
   surfaces

   sorption on goethite *P11.12*

   acidity constants *E8.1*

   species fractions at different pHs, *E8.1*
   calculations

1,2-Phthalic acid
see o-Phthalic acid

Picolinate PHP F13.21
Pyridine-2-carboxylic acid phenyl ester

   oxide surface catalyzed hydrolysis *F13.21,*
                                    F13.22

Piperidine T8.2
Piperidine

   acidity constant of corresponding *T8.2*
   acid

   acidity constant of corresponding *T8.3*
   acid at different temperatures

PNA
see para-Nitroanisol

PNAP
see para-Nitroacetophenone

Polychlorinated PCDD F2.15
dibenzo-[1,4]-dioxins

   environmental source and fate 41

Polychlorinated PCDF F2.15
dibenzo-furans

   environmental source and fate 41

Preventol F2.17

   pesticide 42

Propachlor 501
2-Chloro-*N*-isopropyl-*N*-phenyl-acetamide

   reaction with nucleophiles 501

Propane F2.3
Propane

   bond angles *F2.3*

   value of μ_{max} and K_{iM} *T17.6*

Propanil T17.3
N-(3,4-Dichloro-phenyl)-propionamide

   enzymatic hydrolysis 707

1-Propanol T5.8
Propan-1-ol

   as cosolvent *T5.8*

2-Propanol F18.9
Propan-2-ol

   molecular diffusivity in air *F18.9*

Propionic acid T17.3
Propionic acid

   formation by enzymatic hydrolysis 707
   of propanil

n-**Propylbenzene**
Propyl-benzene

   aqueous solution data *T5.3*

   salting constant, seawater *T5.7*

   standard enthalpy and standard *T4.2*
   entropy of evaporation

   water saturated, mole fraction *T5.1*

Purine F16.3
7*H*-Purine

   rate constant with hydroxyl radical *F16.3*

Pyrene F2.13
Pyrene

   black carbon-water distribution *E9.3*
   coefficient

   concentration in pore water from *E9.3*
   sediment concentration,
   calculation

   concentrations in ryegrass, yarrow *P10.3*
   and air

   dissolved organic carbon-water *F9.14*
   distribution coefficient for
   different organic matters

   Fishtine and structural parameters *T4.4*
   for estimating vapor pressure

   H-donor (α) and *T4.3*
   H-acceptor property (β)

   organic carbon- and black *304*
   carbon-water distribution
   coefficients

   organic carbon-water distribution *E9.3*
   ratio, calculations

   reaction quantum yield of direct *T15.7*
   photolysis

   refractive index *T3.1*

   salting constant, seawater *T5.7*

   sorption isotherm to kaolinite in *F11.6*
   aqueous solution

   standard enthalpies of vaporization *T6.3*
   and of water-air transfer

   standard enthalpy and entropy of *T4.5*
   fusion, experimental and predicted
   data

   standard enthalpy of air-plant *P10.3*
   transfer

   value of μ_{max} and K_{iM} *T17.6*

Pyridine F2.11
Pyridine

   acidity constant of corresponding *T8.2*
   acid *T8.6*

   conjugated π-electron system 30

   H-donor (α) and *T4.3*
   H-acceptor property (β)

   Hammett susceptibility factor of *T8.6*
   corresponding acid

   nucleophilicity in water *499*

   rate constant with hydroxyl radical *F16.3*

   rate constant with hydroxyl radical *F16.7*
   in troposphere

   reactant in binary actinometer 648

   solvent-water partition constants for *T7.1*
   five solvents

Pyrimidine F16.3
Pyrimidine

   rate constant with hydroxyl radical *F16.3*

Pyrrole F16.7
1*H*-Pyrrole

   rate constant with hydroxyl radical *F16.7*
   in troposphere

Pyruvate
2-Oxo-propionic acid

   standard redox potential *T14.2*

Q

Quadricyclane P13.11
Tetracyclo[3.2.0.0^{2,7}.0^{4,6}]heptane

   rate constants in water at various *P13.11*
   pHs

Quartz

   as bipolar surface sorbent for 71
   sorbates of different polarity

Quaternary ammonium salts F2.18

   cationic surfactants F2.18

Quinoline Q8.5
Quinoline

   acidity constant of corresponding *Q8.5*
   acid

   distribution to humic acid and pH *F9.18*

   Hammett susceptibility factor *Q8.5*

   sorption isotherm of cation on *F11.9*
   natural solid

   value of μ_{max} and K_{iM} *T17.6*

R

Radon
Radon

   profile of lake for vertical turbulent *F22.8*
   diffusivity calculations 1029

   profile of the bottom of the north *P22.6*
   atlantic

transfer velocity in water and wind *F20.3*
speed

Ronnel   542
Thiophosphoric acid *O,O'*-dimethyl ester
O''-(2,4,5-trichloro-phenyl) ester

influence of metals on hydrolysis *542*
rate and mechanism

S

Salicylic acid   **T11.4**
2-Hydroxy-benzoic acid

microbial oxidation to catechol 699

reaction of anion with mineral *T11.4*
surfaces

Sarin   **F2.22**
Isopropyl-methylphosphono-fluoridoate

nerve poison *F2.22*

Semiquinone SQ   **568**
Semiquinone

intermediate in redox reaction 568

Stearic acid
see Octadecanoic acid

***trans*-Stilbene**
1,1'-(*E*)-Ethene-1,2-diyl-dibenzene

decadic molar absorption *F15.2*
coefficient

Styrene   **F2.13**
Vinyl-benzene

decadic molar absorption *F15.2*
coefficient

rate constant with hydroxyl radical *F16.7*
in troposphere

Sucrose
2-(3,4-Dihydroxy-2,5-bis-hydroxymethyl-
tetrahydro-furan-2-yloxy)-6-hydroxymethyl-
tetrahydro-pyran-3,4,5-triol

molecular diffusivity in water *F18.10*

Sulcotrion   **Q8.7**
2-(2-Chloro-4-methanesulfonyl-benzoyl)-
cyclohexane-1,3-dione

acidity constant *Q8.7*

Sulfadiazine   **F2.20**
4-Amino-*N*-(4,6-dimethyl-pyrimidin-2-yl)-
benzenesulfonamide

therapeutical drug *F2.20*

Sulfomethuron   **F2.20**
2-[(4,6-Dimethyl-pyrimidin-2-ylcarbamoyl)-
sulfamoyl]-benzoic acid

herbicide *F2.20*

Sulfur hexafluoride
Sulfur hexafluoride

transfer velocity in water and wind *F20.3*
speed

T

TBT
see Tributyltin

1,2,3,4-TCDD
see 1,2,3,4-Tetrachloro- dibenzo[1,4]dioxin

1,2,3,7-TCDD
see 1,2,3,7-Tetrachloro-dibenzo[1,4]dioxin

2,3,7,8-TCDD
see 2,3,7,8-Tetrachloro-dibenzo[1,4]dioxin

TCE
see Trichloro-ethene

2,4,5-TCP
see 2,4,5-Trichlorophenol

2,4,6-TCP
see 2,4,6-Trichlorophenol

Teflon
Poly-tetrafluoro-ethene

as nonpolar surface sorbent for F 3.7
sorbates of different polarity

α-Terpinene   **P16.3**
p-Cymene

rate constant with hydroxyl radical *P16.3*
in troposphere

Tetrabromo-methane   **T14.3**
Tetrabromo-methane

standard reduction potential *T14.3,*
F14.4

1,1,2,2-Tetrachloro-   **T13.7**
1-fluoroethane
1,1,2,2-Tetrachloro-1-fluoro-ethane

kinetic data for substitution and *T13.7*
elimination reactions

3,3',4,4'-Tetrachloro-   **459**
azobenzene
Bis-(3,4-dichloro-phenyl)-diazene

highly toxic reaction product in soils 459

1,2,3,4-Tetrachlorobenzene   **P10.1**
1,2,3,4-Tetrachloro-benzene

bioaccumulation factor of soybean *P10.1*
leaves and roots to water

biological organic phases-water *T10.2*
partition coefficients

octanol-water partition constant *T10.2*

polarizability *T5.5*

1,2,4,5-Tetrachlorobenzene   **P9.5**
1,2,4,5-Tetrachloro-benzene

Freundlich isotherm parameters *P9.5*

organic carbon-water partition *P9.5*
constant

polarizability *T5.5*

2,2',5,5'-Tetrachlorobiphenyl PCB52 **E10.3**
2,5,2',5'-Tetrachloro-biphenyl

air-lipid partition coefficient *T10.5*

air-octanol partition constant *E10.5*

air-plant equilibrium coefficient *E10.3*

aqueous solution data *T5.2*

bioconcentration factor in *E10.1*
phytoplankton

biomagnification factors from *T10.6*
pasture to animal fat

concentrations in air and in pasture *E10.5*

concentrations in air, pasture and *T10.5*
milk

concentrations in lake water and *T10.3*
sediment

concentrations in water, sediment *E10.4*
and mussel

equilibrium bioaccumulation factor, *E10.1*
K_{ibio}, calculations   *E10.2*

food chain concentrations and *F10.9*
biomagnification in trout

fugacites in water, sediment and *F10.8*
biolipid

fugacities and activities in water, *E10.4*
sediment and mussel, calculations

fugacities in air and pasture, *E10.5*
equilibrium concentration in
pasture, calculations

fugacities in air and terrestrial food *F10.11*
chain

fugacities in arctic terrestrial food *F10.12*
chain

partition constants for octanol, air, *E10.4*
org.C and lipids with water

partition properties and   *T10.3*
accumulation factors in aquatic
organisms

saturation concentration in water *E10.4*

standard enthalpies of vaporization *T6.3*
and of water-air transfer

standard enthalpy of air-lipid *T10.5*
transfer

standard enthalpy of air-pasture *E10.5*
transfer

standard enthalpy of air-plant 363
transfer

1,2,3,4-Tetrachloro-   **F16.7**
dibenzo[1,4]dioxin
1,2,3,4-Tetrachloro-dibenzo[1,4]dioxin

rate constant with hydroxyl radical *F16.7*
in troposphere   *P16.3*

1,2,3,7-Tetrachloro   **P5.5**
-dibenzo[1,4]dioxin
1,2,3,7-Tetrachloro-dibenzo[1,4]dioxin

2,3,7,8-Tetrachloro TCDD   **41**
dibenzo[1,4]dioxin
2,3,7,8-Tetrachloro-dibenzo[1,4]dioxin

concentrations in air, soil, grass and *P10.4*
milk

rate constant with hydroxyl radical *P16.3*
in troposphere

standard enthalpy of evaporation *P10.4*

1,1,1,2-Tetrachloroethane   **T13.7**
1,1,1,2-Tetrachloro-ethane

H-donor (α) and   *T4.3*
H-acceptor property (β)

kinetic data for substitution and *T13.7*
elimination reactions

1,1,2,2-Tetrachloroethane   **T13.2**
1,1,2,2-Tetrachloro-ethane

β-elimination reaction *T13.2*

β-elimination reaction 510

kinetic data for substitution and *T13.7* elimination reactions

polarizability *T5.5*

Tetrachloroethene PCE   F2.2
1,1,2,2-Tetrachloro-ethene

air-water partition constant *T20.5*

aqueous solution data *T5.2, T5.3*

as air pollutant in cyclohexane at E6.1 5 °C, calculations

as model compound in different 947 one-box model calculations

as terminal electron acceptor 729

bond angles *F2.3*

characteristic amount, input and *B21.2* output data of a Swiss lake

concentration in air over a Swiss *B21.2* lake

concentrations in a river at different *P20.4* flow distances

Fishtine and structural parameters *T4.4* for estimating vapor pressure

formation by microbial reduction of 726 hexaxhloroethane by heme proteins

H-donor (α) and   *T4.3* H-acceptor property (β)

Henry constant *P12.1*

major use T2.4

microbial reduction catalyzed by F17.10 cobalamin

microbial reduction to 721 trichloroethene

molecular diffusivity in air *T20.5*

molecular diffusivity in water *T20.5*

one-box lake model, calculations *B21.2*

one-box lake model with variable *B21.2* input, calculations

overall air-water exchange velocity *P21.2*

pathways of reduction by zinc F14.15

polarizability *T5.5*

product of reductive   557 dihalo-elimination reaction of HCE

rate constant with hydroxyl radical *F16.3*

rate constant with hydroxyl radical *F16.7* in troposphere

reactant of a higly toxic reaction 459 product in landfills

reduction to trichloroethene by 704 metallo-enzymes

relative rates of reduction by three *F14.17* metals

salting constant, seawater *T5.7*

Schmidt number in water *T20.5*

soil surface-air equilibrium, *E11.2* calculations

standard enthalpies of vaporization T6.3 and of water-air transfer

standard Gibbs energy *P12.1*

standard reduction potential *T14.3, F14.4*

transfer velocities in air, in water *T20.5,* and overall at two wind speeds *F20.7*

two-box lake model (stratification), *E21.5* calculations

two-box lake model with   *E21.6* stratification and mixing periods, calculations

value of μ_{max} and K_{iM} *T17.6*

vapor pressure as function of *F4.4* temperature

vapor pressures at four *P4.1* temperatures

water saturated, mole fraction *T5.1*

Tetrachloromethane CT   292
Tetrachloro-methane

aqueous solution data *T5.3*

H-donor (α) and   *T4.3* H-acceptor property (β)

kinetic data for substitution and *T13.7* elimination reactions

microbial reduction to chloroform 721

organic carbon-water distribution *F9.7* and quality of solid

pathways of reduction to various *F14.14* products

polarizability *T5.5*

reduction by heme proteins to the 726 trichloro-methyl radical

relative rates of reduction in *F14.16* different media

salting constant, seawater *T5.7*

solid-water distribution and organic *F9.7* carbon content of solid

standard reduction potential *T14.3, F14.4*

water saturated, mole fraction *T5.1*

2,3,4,6-Tetrachlorophenol
2,3,4,6-Tetrachloro-phenol

standard enthalpy of air-water *P8.2* transfer

1,1,2,3-Tetrachloro-propene   767
1,1,2,3-Tetrachloro-propene

Tetrachloroethene PCE   F2.2
1,1,2,2-Tetrachloro-ethene

change of a time-variable input in *E25.6* well due to sorption in groundwater system

characteristic times of air–water *E24.2* exchange in river, two flow regimes

concentration change in river due to *E24.2* flow time and air exchange

one-dimensional vertical lake *F23.7* model, calculated vs. measured values for Greifensee

one-dimensional vertical lake   1091 model, numerical calculations with MASAS

retardation by sorption in *E25.6* groundwater system

1,1,1,2-Tetrafluoroethane   P4.3
1,1,1,2-Tetrafluoro-ethane

vapor pressures at six temperatures *P4.3*

world production rate, major use T2.4

Tetrahydrofuran   E16.3
Tetrahydro-furan

indirect photolysis half-life by *E16.3* hydroxyl radical in troposphere, calculation of

1,2,4,5-Tetramethylbenzene TeMB
1,2,4,5-Tetramethyl-benzene

Fishtine and structural parameters *T4.4* for estimating vapor pressure

standard enthalpy and standard *E4.1* entropy of fusion, calculations

vapor pressure at given *E4.1* temperatures, calculation

Thioacetic acid   T8.1
Thioacetic acid

acidity constant *T8.1*

Thiometon   T13.12
Dithiophosphoric acid *S*-(2-ethylsulfanyl-ethyl) ester *O,O'*-dimethyl ester

rate constants of hydrolysis reaction *T13.12*

Thiophenol   F2.20
Benzenethiol

acidity constant *T8.1*

chemical intermediate F2.20

TNT
see 2,4,6-Trinitrotoluene

Toluene   F2.13
Toluene

activity coefficients in six organic *T6.1* solvents

activity coefficients of five solutes *T6.1* in toluene

addition to fumarate 730

air-solvent partition constants for six *T6.1* organic solvents

air-teflon and air-quartz partition *F3.7* constants and dispersive vdW parameter

air-toluene partition constants for *T6.1* five solutes

air-water partition constant *198*

aqueous solution data *T5.3*

biological organic phases-water *T10.2* partition coefficients

enzymatic addition to fumarate 704

H-donor (α) and H-acceptor property (β) *T4.3*

methylbenzene-water partition constants for nine organic solutes *T7.1*

microbial oxidation to 717 *cis*-1,2-dihydroxy-3-methyl-cyclohexa-3,5-diene

molecular diffusivity in air *F18.9*

molecular diffusivity in water *F18.10*

octanol-water partition constant *T10.2*

polarizability *T5.5*

rate constant with hydroxyl radical *F16.7* in troposphere

salting constant, seawater *T5.7*

solvent-water partition constants for five solvents *T7.1*

standard enthalpies of vaporization and of water-air transfer *T6.3*

standard enthalpy and standard entropy of evaporation *T4.2*

value of K_{iMM} and V'_{max} *T17.7*

value of μ_{max} and K_{iM} *T17.6*

water saturated, mole fraction *T5.1*

Toluene-4-sulfonic acid **T8.1**
Toluene-4-sulfonic acid

acidity constant *T8.1*

anion as hardener and stabilizer F 2.20

uptake by organisms 738

TPT
see Triphenyltin

1,2,3-TrCB
see 1,2,3-Trichlorobenzene

1,2,4-TrCB
see 1,2,4-Trichlorobenzene

1,3,5-TrCB
see 1,3,5-Trichlorobenzene

2,4,6-Triaminotoluene TAT **F14.6**
2-Methyl-benzene-1,3,5-triamine

endproduct in TNT reduction F14.6

1,1,2-Tribromoethane **T13.7**
1,1,2-Tribromo-ethane

kinetic data for substitution and elimination reactions *T13.7*

Tribromo-methane **T13.7**
Tribromo-methane

concentration in air *E20.1*

concentration in arctic ocean *E20.1*

Henry constant *E20.1*

Henry constant in seawater *E20.1*

Henry constant in seawater, temperature dependence *E20.1*

kinetic data for substitution and elimination reactions *T13.7*

molecular diffusivity, calculation of *E20.3*

overall transfer velocity at wind speed, calculation of *E20.3*

polarizability *T5.5*

refractive index *T3.1*

relative rates of reduction in different media *F14.16*

Schmidt number in water at different temperatures, calculation of *E20.3*

seawater-air flux, calculation of *E20.1*

standard reduction potential *T14.3*, *F14.4*

Tributyltin TBT **325**

humic acid-water distribution ratio and pH *F9.19*

Tricapryline
Octanoic acid 2-octanoyloxy-1-octanoyloxymethyl-ethyl ester

Tricapryline-water vs. octanol-water partition *F10.4*

Trichlorfon **F2.22**
(2,2,2-Trichloro-1-hydroxy-ethyl)-phosphonic acid dimethyl ester

insecticide F2.22

Trichloroacetic acid **F2.17**
Trichloro-acetic acid

presence in the environment by direct input and by hydrolysis of esters 42

rate constant with hydroxyl radical *F16.3*

1,2,3-Trichlorobenzene
1,2,3-Trichloro-benzene

lethal body burden (*LBB*) for fish *E10.6*

polarizability *T5.5*

theoretical bioaccumulation potential (*TBP*) in fish, calculations *E10.6*

volume fraction in fish membrane lipids, calculations *E10.6*

1,2,4-Trichlorobenzene **F16.7**
1,2,4-Trichloro-benzene

air-water partition constant *T20.5*

molecular diffusivity in air *T20.5*

molecular diffusivity in water *T20.5*

rate constant with hydroxyl radical in troposphere *F16.7*

Schmidt number in water *T20.5*

transfer velocities in air, in water, and overall at two wind speeds *T20.5*, *F20.7*

1,3,5-Trichlorobenzene **P9.4**
1,3,5-Trichloro-benzene

biological organic phases-water partition coefficients *T10.2*

octanol-water partition constant *T10.2*

polarizability *T5.5*

value of K_{iMM} and V'_{max} *T17.7*

2',3,4-Trichlorobiphenyl PCB33 **1065**
3,4,2'-Trichloro-biphenyl

Summary of models and measurements in Lake Superior *T23.8*

two-box coupled water/sediment system in Lake Superior *T23.7*

concentrations, inputs and outputs in Lake Superior *T23.4*

one-box three phase model in Lake Superior 1068

one-box two phase model in Lake Superior 1065

physico-chemical properties *T23.4*

surface mixed sediment layer model in Lake Superior 1065

two-box water/sediment model in Lake Superior 1079

2,4,4'-Trichlorobiphenyl
2,4,4'-Trichloro-biphenyl

aqueous solution data *T5.2*

standard enthalpies of vaporization and of water-air transfer *T6.3*

1,1,1-Trichloroethane TrCE **T13.7**
1,1,1-Trichloro-ethane

concentration in air *E20.1*

concentration in arctic ocean *E20.1*

density *E20.2*

evaporation time, calculation of *E20.2*

H-donor (α) and H-acceptor property (β) *T4.3*

Henry constant *E20.1*

Henry constant in seawater *E20.1*

Henry constant in seawater, temperature dependence *E20.1*

kinetic data for substitution and elimination reactions *T13.7*

major use T2.4

molecular diffusivity, calculation of *E20.3*

overall transfer velocity at wind speed, calculation of *E20.3*

rate constant with hydroxyl radical *F16.3*

rate constant with hydroxyl radical in troposphere *F16.7*

salting constant, seawater *T5.7*

saturation pressure of pure liquid *E20.2*

Schmidt number in water at different temperatures, calculation *E20.3*

seawater-air flux, calculation of *E20.1*

typical concentration in air *P6.3*

typical concentration in Arctic surface waters *P6.3*

1,1,2-Trichloroethane T13.7
1,1,2-Trichloro-ethane

indirect photolysis half-life by *E16.3*
hydroxyl radical in troposphere,
calculation of

kinetic data for substitution and *T13.7*
elimination reactions

rate constant with hydroxyl radical *F16.3*

Trichloroethene TCE **F14.15**
1,1,2-Trichloro-ethene

aqueous solution data *T5.3*

co-metabolic degradation with *E17.8*
methane

diffusion from groundwater to *E19.2*
atmosphere, calculations

diffusion plus advection from *E22.4*
groundwater to atmosphere,
calculations

formation by microbial reduction of 721
tetrachloroethene

H-donor (α) and *T4.3*
H-acceptor property (β)

microbial oxidation to 717
trichloro-oxirane

molar volume, calculations *804*
T18.3

molecular diffusivity in water, *813*
calculations

pathways of reduction by zinc F14.15

product in tetrachloroethane *T13.2*
β-elimination reaction

rate constant with hydroxyl radical *F16.3*

rate constant with hydroxyl radical *F16.7*
in troposphere

relative rates of reduction by three *F14.17*
metals

salting constant, seawater *T5.7*

standard enthalpies of vaporization *T6.3*
and of water-air transfer

standard free energy of formation *P14.1*

standard oxidation potential *T14.3,*
F14.4

value of K_{iMM} and V'_{max} *T17.7*

water saturated, mole fraction *T5.1*

world production rate, major use T2.4

Trichloro-fluoromethane CFC-11 **T2.4**
Trichloro-fluoro-methane

air-water partition constant *T20.5*

major use T2.4

molecular diffusivity in air *T20.5*

molecular diffusivity in water *T20.5*

salting constant, seawater *T5.7*

Schmidt number in water *T20.5*

standard enthalpies of vaporization *T6.3*
and of water-air transfer

temperature dependence of *B18.4*
molecular diffusivity in water

transfer velocities in air, in water *T20.5,*
and overall at two wind speeds *F20.7*

Trichloromethane **E5.3**
Trichloro-methane

activity coefficients of four organic *T3.2*
solutes in trichloromethane

air-teflon partition constant and *F3.7*
dispersive vdW parameter

aqueous solution data *T5.2,*
T5.3

formation by microbial reduction of 721
tetrachloromethane

H-donor (α) and *T4.3*
H-acceptor property (β)

kinetic data for substitution and *T13.7*
elimination reactions

molecular diffusivity in water *P18.4*

polarizability *T5.5*

rate constant with hydroxyl radical *F16.3*

rate constant with hydroxyl radical *F16.7*
in troposphere

refractive index *T3.1*

salting constant, seawater *T5.7*

solvent fitting coefficients for *T6.2*
air-trichloromethane partition
constant *i*

solvent fitting coefficients for *T7.2*
trichloromethane-water partition
constant of organic compounds

standard enthalpies of vaporization *T6.3*
and of water-air transfer

standard reduction potential *T14.3,*
F14.4

trichloromethane-water partition *T7.1*
constants for nine organic solutes

water saturated, mole fraction *T5.1*

Trichloro-oxirane 717
2,2,3-Trichloro-oxirane

formation by microbial oxidation of 717
trichloroethene

2,4,5-Trichlorophenol 323
2,4,5-Trichloro-phenol

organic carbon-water distribution *F9.17*
ratio and pH

2,4,6-Trichlorophenol **Q8.6**
2,4,6-Trichloro-phenol

acidity constant *Q8.6,*
P8.3

butyl acetate-water partition *P8.3*
constant

3,4,5-Trichlorophenol **Q8.6**
3,4,5-Trichloro-phenol

acidity constant *Q8.6*

acidity constant, calculation *E8.2*

2,4,5-Trichlorophenoxy acetic 2,4,5-T 271
acid
(2,4,5-Trichloro-phenoxy)-acetic acid

acidity constant *F8.9*

acidity constant, calculation *E8.2*

n-octanol-water distribution ratio *F8.9,*
and pH and K^+-concentration *F8.10*

3,4,5-Trichlorophenyl **E13.6**
N-phenyl carbamate
Phenyl-carbamic acid 3,4,5-trichloro-phenyl
ester

rate constant of hydrolysis, *E13.6*
calculations

Trichloromethane **E5.3**
Trichloro-methane

fate of along Mississippi River after *1132*
two peak spills, case study

maximum concentration change *1132*
along flow in Mississippi River,
case study

total mass change along flow in *1132*
Mississippi River, case study

Trichloromethyl radical 726
Trichloromethyl

formation by microbial reduction of 726
tetrachloromethane with heme
proteins

Triethylphosphate **T13.12**
Phosphoric acid triethyl ester

mechanism of base catalyzed 538
hydrolysis

rate constant with hydroxyl radical *F16.3*

rate constants of hydrolysis reaction *T13.12*

Triethylamine **F2.18**
Triethyl-amine

solvent, wetting agent F2.18

Trifluoroacetic acid **T8.1**
Trifluoro-acetic acid

acidity constant *T8.1*

Trifluralin **F15.7**
(2,6-Dinitro-4-trifluoromethyl-phenyl)-
dipropyl-amide

reaction pathway of direct *F15.7*
photolysis

Trimethyl-[3-(methyl-phenyl- **P13.6**
carbamoyloxy)-phenyl]-
ammonium
N,N,N-Trimethyl-3-(([methyl-(phenyl)-amino]-
carbonyl)-oxy)benzenaminium

rate constant of base catalyzed *P13.6*
hydrolysis

Trimethylphosphate **T13.2**
Phosphoric acid trimethyl ester

mechanism of base catalyzed 538
hydrolysis

nucleophile substitution reaction *T13.2*

rate constant with hydroxyl radical *F16.3*

rate constants of hydrolysis reaction *T13.12*

1,2,3-Trimethylbenzene
1,2,3-Trimethyl-benzene
polarizability *T5.5*

1,3,5-Trimethylbenzene   **F16.7**
Mesitylene
aqueous solution data *T5.2,*
                    T5.3
polarizability *T5.5*

rate constant with hydroxyl radical *F16.7*
in troposphere

water saturated, mole fraction *T5.1*

2,3,5-Trimethylhexane   **P16.3**
2,3,5-Trimethyl-hexane
rate constant with hydroxyl radical *P16.3*
in troposphere

2,4,6-Trimethylphenol   **T8.1**
2,4,6-Trimethyl-phenol
acidity constant *T8.1*

Trimethylamine   **F2.3**
Trimethyl-amine
acidity constant of corresponding *T8.2*
acid

bond angles *F2.3*

molar volume, calculation *T18.3*

rate constant with hydroxyl radical *F16.3*

1,3,5-Trinitrobenzene   **F11.6**
1,3,5-Trinitro-benzene
sorbed to clay mineral, schematic F11.8
drawing

sorption isotherm to kaolinite in *F11.6*
aqueous solution

2,4,6-Trinitrophenol   **T8.1**
Picric acid
acidity constant *T8.1*

2,4,6-Trinitrotoluene TNT   **F2.18**
2-Methyl-1,3,5-trinitro-benzene
exchange constant   *T11.2*
water-aluminosilicate surface
($K_{\text{NAC,EDA}}$)

explosive 45

microbial reduction to   721
2,6-Dinitro-4-nitroso-toluene

nonlinear isotherm with mineral 409
surfaces

one-electron reduction potential *T14.4*

organic carbon-water partition *P11.5*
constant

pathways of reduction to F14.6
triaminotoluene

reaction quantum yield of direct *T15.7*
photolysis

sorption on illite from water, *E11.4*
calculations

Trinitroglycerol   **F2.18**
1,2,3-Tris-nitrooxy-propane
explosive F2.18

Triolein
Octadec-9-enoic acid
2-octadec-9-enoyloxy-1-octadec-9-
enoyloxymethyl-ethyl ester
Triolein-water vs. octanol-water *F10.4*
partition

Triphenylphosphate   **F2.22**
Phosphoric acid triphenyl ester
plasticizer and fire retardant F2.22

rate constants of hydrolysis reaction *T13.12*

Triphenyltin TPT   **325**
humic acid-water distribution ratio *F9.19*
and pH

U

Urea   **F18.10**
Urea
molecular diffusivity in water *F18.10*

V

Vanillic acid   **699**
4-Hydroxy-3-methoxy-benzoic acid
oxidative cleavage to 699
protocatechuic acid

Vinyl acetate
see Acetic acid ethenyl ester

Vinyl chloride VC
see Chloroethene

W

Water   **F2.3**
Water
acidity constant at different *T8.3*
temperatures

activity coefficients of four organic *T3.2*
solutes in water

air phase transfer velocity and wind *T20.4*
speed

amount adsorbed on mineral oxide *F11.3*
surface and relative humidity

amount of in a cloud *P6.2*

bond angles *F2.3*

characteristics of lakes *T23.1*

characteristics of oceans *T23.1*

content in atmosphere 889

film renewal time in substance *908*
exchange models

film thickness in substance *908*
exchange models

flux of evaporation, annual *898*
evaporation rate

H-bonding structures in pure liquid *F3.5*

H-donor (α) and   *T4.3*
H-acceptor property (β)

kinematic viscosity at different *T20.3,*
temperatures   *F20.6*

layer thickness on mineral surface 394
and affinity to organic sorbates

molecular diffusivity in air *F18.9*

molecular order in bulk and cavity 10

mono- and multilayer film on 392
mineral surfaces

sorbed to clay mineral, schematic F11.8
drawing

standard Gibbs energy of pure *E13.1*
water

surface of droplets in atmospheric 889
clouds

transfer velocity at evaporation 896

transfer velocity of vapor in air and *F20.2*
wind speed

typical transfer velocity of a *893*
substance in water

X

***m*-Xylene**   **F17.11**
m-Xylene
addition to fumarate 731

***m/p*-Xylene**   **F2.13**
m/p-Xylene
air-quartz partition constant and *F3.7*
dispersive vdW parameter

concentration in diesel fuel *T19.3*

density *T19.3*

dissolution from diesel fuel into *E19.4*
aqueous phase, calculations

molecular diffusivity in water and in *T19.3*
diesel fuel

LIST OF ILLUSTRATIVE EXAMPLES

Chapter 2   2.1   Determining the Oxidation States of the Carbon Atoms Present in Organic Molecules

Chapter 3   3.1   The "Soup Bowl" Problem

Chapter 4   4.1   Basic Vapor Pressure Calculations

Chapter 5   5.1   Deriving Liquid Aqueous Solubilities, Aqueous Activity Coefficients, and Excess Free Energies in Aqueous Solution from Experimental Solubility Data

          5.2   Evaluating the Factors that Govern the Aqueous Activity Coefficient of a Given Compound

          5.3   Evaluating the Effect of Temperature on Aqueous Solubilities and Aqueous Activity Coefficients

          5.4   Quantifying the Effect of Inorganic Salts on Aqueous Solubility and Aqueous Activity Coefficients

          5.5   Estimating the Solubilities and Activity Coefficients of Organic Pollutants in Organic Solvent–Water Mixtures

Chapter 6   6.1   Assessing the Contamination of Organic Liquids by Air Pollutants

          6.2   Evaluating the Direction of Air–Water Gas Exchange at Different Temperatures

          6.3   Assessing the Effect of Solution Composition on Air–Aqueous Phase Partitioning

          6.4   Estimating Air–Water Partition Constants by the Bond Contribution Method

Chapter 7   7.1   Evaluating the Factors that Govern the Organic Solvent–Water Partitioning of a Compound

          7.2   Estimating Octanol–Water Partition Constants from Structure Using the Atom/Fragment Contribution Method

          7.3   Estimating Octanol–Water Partition Constants Based on Experimental K_{iow}'s of Structurally Related Compounds

          7.4   Estimating the Concentrations of Individual PCB Congeners in Water that Is in Equilibrium with an Aroclor and an Aroclor/Hydraulic Oil Mixture

Chapter 8   8.1   Assessing the Speciation of Organic Acids and Bases in Natural Waters

          8.2   Estimating Acidity Constants of Aromatic Acids and Bases Using the Hammett Equation

          8.3   Assessing the Air–Water Distribution of Organic Acids and Bases in a Cloud

Chapter 9   9.1   Determining K_{id} Values from Experimental Data

9.2   Evaluating the Concentration Dependence of Sorption of Phenanthrene to Soil and Sediment POM

9.3   Estimating Pore Water Concentrations in a Polluted Sediment

9.4   How Much Does the Presence of 20% Methanol in the "Aqueous" Phase Affect the Retardation of Phenanthrene in an Aquifer?

9.5   Evaluating the Effect of DOM on the Bioavailability of Benzo(a)pyrene

Chapter 10   10.1   Evaluating Bioaccumulation from a Colloid-Containing Aqueous Solution

10.2   Estimating Equilibrium Bioaccumulation Factors from Water

10.3   Estimating Equilibrium Bioaccumulation Factors from Air

10.4   Calculating Fugacities or Chemical Activities to Evaluate Bioaccumulation

10.5   Evaluating Air–Pasture Partitioning of PCBs

10.6   Evaluating Lethal Body Burdens of Chlorinated Benzenes in Fish

Chapter 11   11.1   Estimating the Fraction of Phenanthrene Present in the Gas Phase and Sorbed to the Walls of a Vessel

11.2   Sorption of Tetrachloroethene from Air to Moist and Dry Soil

11.3   Gas–Particle Partitioning of Organic Compounds to Environmental Tobacco Smoke Adsorption or Absorption?

11.4   Estimating the Retardation of Trinitrotoluene Transport in Groundwater

11.5   Transport of Di-Isopropanol-Amine (DIPA) in Groundwater from a Sour Gas Processing Plant

11.6   Estimating Dodecyl Sulfonate Sorption to Alumina at Different pHs

11.7   Estimating the Adsorption of Benzoic Acid to Goethite

Chapter 12   12.1   Energetics of Syntrophic Cooperation in Methanogenic Degradation

12.2   Transformation of Methyl Bromide to Methyl Chloride and Vice Versa

12.3   A Benzyl Chloride Spill Into a Pond

Chapter 13   13.1   Evaluating the Thermodynamics of Hydrolysis Reactions

13.2   Some More Reactions Involving Methyl Bromide

13.3   1,2-Dibromoethane in the Hypolimnion of the Lower Mystic Lake, Massachusetts

13.4   Deriving Kinetic Parameters for Hydrolysis Reactions from Experimental Data

13.5   Calculating Hydrolysis Reaction Times as a Function of Temperature and pH

13.6   Estimating Hydrolysis Rate Constants Using the Hammett Relationship

Chapter 14   14.1   Calculating Standard Reduction Potentials from Free Energies of Formation

14.2   Establishing Mass Balances for Oxygen and Nitrate in a Given System

14.3   Calculating the Reduction Potential of an Aqueous Hydrogen Sulfide (H_2S) Solution as a Function of pH and Total H_2S Concentration

14.4   Calculating Free Energies of Reaction from Half Reaction Reduction Potentials

	14.5	Estimating Rates of Reduction of Nitroaromatic Compounds by DOM Components in the Presence of Hydrogen Sulfide
Chapter 15	15.1	Determining Decadic Molar Extinction Coefficients of Organic Pollutants
	15.2	Using the Screening Factor $S(\lambda_m)$ to Estimate the Total Specific Light Absorption Rate of PNAP in the Epilimnion of a Lake
	15.3	Estimating the Photolysis Half-Life of a Weak Organic Acid in the Well-Mixed Epilimnion of a Lake
Chapter 16	16.1	Estimating Near-Surface Hydroxyl Radical Steady-State Concentrations in Sunlit Natural Waters
	16.2	Estimating the Indirect Photolysis Half-Life of Atrazine in a Shallow Pond
	16.3	Estimating Tropospheric Half-Lives of Organic Pollutants
Chapter 17	17.1	Is a Proposed Microbial Transformation Thermodynamically Feasible?
	17.2	What Products Do You Expect from the Microbial Degradation of EDB?
	17.3	What Products Do You Expect from the Microbial Degradation of Linuron?
	17.4	What Products Do You Expect from the Microbial Degradation of Vinyl Chloride in an Oxic Environment?
	17.5	What Products Do You Expect from the Microbial Degradation of DDT in a Reducing Environment?
	17.6	Evaluating the Biodegradation of Glycerol by Microorganisms Growing on that Substrate in a Well-Mixed Tank
	17.7	Estimating the Time to Degrade a Spilled Chemical
	17.8	Evaluating the Co-Metabolic Biodegradation of Trichloroethene by Microorganisms Growing on Methane in a Well-Mixed Tank
	17.9	Estimating Biotransformation Rates of an Organic Pollutant in a Natural System
Chapter 18	18.1	Estimating Molar Volumes
	18.2	Estimating Molecular Diffusivity in Air Diffusivities in Water
	18.3	Estimating Molecular Diffusivity in Water
	18.4	Evaluating the Steady-State Flux of Benzene from Spilled Gasoline Through Soil to the Atmosphere
	18.5	Interpreting Stratigraphic Profiles of Polychlorinated Naphthalenes in Lake Sediments
Chapter 19	19.1	Vertical Exchange of Water in a Lake
	19.2	Diffusion of a Volatile Compound from the Groundwater Through the Unsaturated Zone into the Atmosphere
	19.3	Release of PCBs from the Historically Polluted Sediments of Boston Harbor
	19.4	Dissolution of a Non-Aqueous-Phase Liquid (NAPL) into the Aqueous Phase
	19.5	Desorption Kinetics of an Organic Chemical from Contaminated Sediments
Chapter 20	20.1	Evaluating the Direction of Air–Water Exchange
	20.2	Estimating Evaporation Rates of Pure Organic Liquids

	20.3	Estimating the Overall Air–Water Transfer Velocity from Wind Speed for Different Water Temperatures
	20.4	Air–Water Exchange of Benzene in Rivers
	20.5	Air–Water Exchange Enhancement for Formaldehyde and Acetaldehyde
Chapter 21	21.1	Assessing the Behavior of Nitrilotriacetic Acid (NTA) in a Lake
	21.2	How Certain Is the Degradation Rate of Nitrilotriacetic Acid (NTA) in Greifensee? (Advanced Topic)
	21.3	Higher-Order Reaction of Nitrilotriacetic Acid (NTA) in Greifensee (Advanced Topic)
	21.4	Fate of a Pesticide and its Decomposition Product in a Small Lake
	21.5	Tetrachloroethene (PCE) in Greifensee: From the One-Box to the Two-Box Model
	21.6	Dynamic Behavior of Tetrachloroethene (PCE) in Greifensee
Chapter 22	22.1	Vertical Distribution of Dichlorodifluoromethane (CFC-12) in a Small Lake
	22.2	Vertical Turbulent Diffusion Coefficient in a Lake
	22.3	A Patch of the Pesticide Atrazine Below the Surface of a Lake
	22.4	Transport of a Volatile Compound From the Groundwater Through the Unsaturated Zone Into the Atmosphere—Illustrative Example 19.2 Reconsidered
Chapter 23	23.1	A Vinyl Acetate Spill Into a Pond
Chapter 24	24.1	Mean Flow Velocity of River G for Different Discharge Rates
	24.2	Air–Water Exchange in River G
	24.3	A Spill of Atrazine in River G
	24.4	Turbulent Diffusion and Longitudinal Dispersion in River G
	24.5	A Second Look at the Atrazine Spill in River G The Effect of Dispersion
Chapter 25	25.1	Darcy's Law in Groundwater Systems S
	25.2	Dispersion and Advection in Groundwater System S
	25.3	Infiltration of Polluted River Water into Groundwater System S
	25.4	A Sudden Rise of the 2,4-Dinitrophenol Concentration in River R
	25.5	Continuing the Case of 2,4-Dinitrophenol in River R
	25.6	Transport of Tetrachloroethene in Groundwater System S
	25.7	Infiltration of Benzylchloride into Groundwater System S